VECTOR IDENTITIES

If $\mathbf{u} = u_1\mathbf{i} + u_2\mathbf{j} + u_3\mathbf{k}$, $\mathbf{v} = v_1\mathbf{i} + v_2\mathbf{j} + v_3\mathbf{k}$, $\mathbf{w} = w_1\mathbf{i} + w_2\mathbf{j} + w_3\mathbf{k}$ then

(dot product) $\mathbf{u} \bullet \mathbf{v} = u_1v_1 + u_2v_2 + u_3v_3$

(cross product) $$\mathbf{u} \times \mathbf{v} = \begin{vmatrix} \mathbf{i} & \mathbf{j} & \mathbf{k} \\ u_1 & u_2 & u_3 \\ v_1 & v_2 & v_3 \end{vmatrix} = (u_2v_3 - u_3v_2)\mathbf{i} + (u_3v_1 - u_1v_3)\mathbf{j} + (u_1v_2 - u_2v_1)\mathbf{k}$$

length of $\mathbf{u} = |\mathbf{u}| = \sqrt{\mathbf{u} \bullet \mathbf{u}} = \sqrt{u_1^2 + u_2^2 + u_3^2}$

angle between $\mathbf{u}$ and $\mathbf{v}$ $= \cos^{-1}\left(\dfrac{\mathbf{u} \bullet \mathbf{v}}{|\mathbf{u}||\mathbf{v}|}\right)$

triple product identities $\quad \mathbf{u} \bullet (\mathbf{v} \times \mathbf{w}) = \mathbf{v} \bullet (\mathbf{w} \times \mathbf{u}) = \mathbf{w} \bullet (\mathbf{u} \times \mathbf{v}) \qquad \mathbf{u} \times (\mathbf{v} \times \mathbf{w}) = (\mathbf{u} \bullet \mathbf{w})\mathbf{v} - (\mathbf{u} \bullet \mathbf{v})\mathbf{w}$

IDENTITIES INVOLVING GRADIENT, DIVERGENCE, CURL, AND LAPLACIAN

$\nabla = \mathbf{i}\dfrac{\partial}{\partial x} + \mathbf{j}\dfrac{\partial}{\partial y} + \mathbf{k}\dfrac{\partial}{\partial z}$ ("del" or "nabla" operator)

$\nabla\phi(x, y, z) = \mathbf{grad}\,\phi(x, y, z) = \dfrac{\partial\phi}{\partial x}\mathbf{i} + \dfrac{\partial\phi}{\partial y}\mathbf{j} + \dfrac{\partial\phi}{\partial z}\mathbf{k}$

$\mathbf{F}(x, y, z) = F_1(x, y, z)\,\mathbf{i} + F_2(x, y, z)\,\mathbf{j} + F_3(x, y, z)\,\mathbf{k}$

$\nabla \bullet \mathbf{F}(x, y, z) = \mathbf{div}\,\mathbf{F}(x, y, z) = \dfrac{\partial F_1}{\partial x} + \dfrac{\partial F_2}{\partial y} + \dfrac{\partial F_3}{\partial z}$

$$\nabla \times \mathbf{F}(x, y, z) = \mathbf{curl}\,\mathbf{F}(x, y, z) = \begin{vmatrix} \mathbf{i} & \mathbf{j} & \mathbf{k} \\ \dfrac{\partial}{\partial x} & \dfrac{\partial}{\partial y} & \dfrac{\partial}{\partial z} \\ F_1 & F_2 & F_3 \end{vmatrix}$$
$$= \left(\frac{\partial F_3}{\partial y} - \frac{\partial F_2}{\partial z}\right)\mathbf{i} + \left(\frac{\partial F_1}{\partial z} - \frac{\partial F_3}{\partial x}\right)\mathbf{j} + \left(\frac{\partial F_2}{\partial x} - \frac{\partial F_1}{\partial y}\right)\mathbf{k}$$

$\nabla(\phi\psi) = \phi\nabla\psi + \psi\nabla\phi$

$\nabla \bullet (\phi\mathbf{F}) = (\nabla\phi) \bullet \mathbf{F} + \phi(\nabla \bullet \mathbf{F})$

$\nabla \times (\phi\mathbf{F}) = (\nabla\phi) \times \mathbf{F} + \phi(\nabla \times \mathbf{F})$

$\nabla \times (\nabla\phi) = \mathbf{0}$ $\quad$ (**curl grad** $= \mathbf{0}$)

$\nabla^2\phi(x, y, z) = \nabla \bullet \nabla\phi(x, y, z) = \mathbf{div}\,\mathbf{grad}\,\phi = \dfrac{\partial^2\phi}{\partial x^2} + \dfrac{\partial^2\phi}{\partial y^2} + \dfrac{\partial^2\phi}{\partial z^2}$

$\nabla \bullet (\mathbf{F} \times \mathbf{G}) = (\nabla \times \mathbf{F}) \bullet \mathbf{G} - \mathbf{F} \bullet (\nabla \times \mathbf{G})$

$\nabla \times (\mathbf{F} \times \mathbf{G}) = \mathbf{F}(\nabla \bullet \mathbf{G}) - \mathbf{G}(\nabla \bullet \mathbf{F}) - (\mathbf{F} \bullet \nabla)\mathbf{G} + (\mathbf{G} \bullet \nabla)\mathbf{F}$

$\nabla(\mathbf{F} \bullet \mathbf{G}) = \mathbf{F} \times (\nabla \times \mathbf{G}) + \mathbf{G} \times (\nabla \times \mathbf{F}) + (\mathbf{F} \bullet \nabla)\mathbf{G} + (\mathbf{G} \bullet \nabla)\mathbf{F}$

$\nabla \bullet (\nabla \times \mathbf{F}) = 0$ $\quad$ (**div curl** $= 0$)

$\nabla \times (\nabla \times \mathbf{F}) = \nabla(\nabla \bullet \mathbf{F}) - \nabla^2\mathbf{F}$ $\quad$ (**curl curl** $=$ **grad div** $-$ laplacian)

VERSIONS OF THE FUNDAMENTAL THEOREM OF CALCULUS

$\displaystyle\int_a^b f'(t)\,dt = f(b) - f(a)$ $\quad$ (the one-dimensional **Fundamental Theorem**)

$\displaystyle\int_C \mathbf{grad}\,\phi \bullet d\mathbf{r} = \phi\big(\mathbf{r}(b)\big) - \phi\big(\mathbf{r}(a)\big)$ if C is the curve $\mathbf{r} = \mathbf{r}(t)$, $(a \le t \le b)$.

$\displaystyle\iint_R \left(\frac{\partial F_2}{\partial x} - \frac{\partial F_1}{\partial y}\right) dA = \oint_C \mathbf{F} \bullet d\mathbf{r} = \oint_C F_1(x, y)\,dx + F_2(x, y)\,dy$ where C is the positively oriented boundary of R $\quad$ **(Green's Theorem)**

$\displaystyle\iint_S \mathbf{curl}\,\mathbf{F} \bullet \hat{\mathbf{N}}\,dS = \oint_C \mathbf{F} \bullet d\mathbf{r} = \oint_C F_1(x, y, z)\,dx + F_2(x, y, z)\,dy + F_3(x, y, z)\,dz$ where C is the oriented boundary of S. $\quad$ **(Stokes's Theorem)**

Three-dimensional versions: S is the closed boundary of D, with outward normal $\hat{\mathbf{N}}$

$\displaystyle\iiint_D \mathbf{div}\,\mathbf{F}\,dV = \oiint_S \mathbf{F} \bullet \hat{\mathbf{N}}\,dS$ $\quad$ **Divergence Theorem**

$\displaystyle\iiint_D \mathbf{curl}\,\mathbf{F}\,dV = -\oiint_S \mathbf{F} \times \hat{\mathbf{N}}\,dS$

$\displaystyle\iiint_D \mathbf{grad}\,\phi\,dV = \oiint_S \phi\hat{\mathbf{N}}\,dS$

FORMULAS RELATING TO CURVES IN 3-SPACE

Curve: $\mathbf{r} = \mathbf{r}(t) = x(t)\mathbf{i} + y(t)\mathbf{j} + z(t)\mathbf{k}$

Velocity: $\mathbf{v} = \dfrac{d\mathbf{r}}{dt} = v\hat{\mathbf{T}}$

Speed: $v = |\mathbf{v}| = \dfrac{ds}{dt}$

Arc length: $s = \displaystyle\int_{t_0}^{t} v\,dt$

Acceleration: $\mathbf{a} = \dfrac{d\mathbf{v}}{dt} = \dfrac{d^2\mathbf{r}}{dt^2}$

Tangential and normal components: $\mathbf{a} = \dfrac{dv}{dt}\hat{\mathbf{T}} + v^2\kappa\hat{\mathbf{N}}$

Unit tangent: $\hat{\mathbf{T}} = \dfrac{\mathbf{v}}{v}$

Binormal: $\hat{\mathbf{B}} = \dfrac{\mathbf{v} \times \mathbf{a}}{|\mathbf{v} \times \mathbf{a}|}$

Normal: $\hat{\mathbf{N}} = \hat{\mathbf{B}} \times \hat{\mathbf{T}} = \dfrac{d\hat{\mathbf{T}}/dt}{|d\hat{\mathbf{T}}/dt|}$

Curvature: $\kappa = \dfrac{|\mathbf{v} \times \mathbf{a}|}{v^3}$

Radius of curvature: $\rho = \dfrac{1}{\kappa}$

Torsion: $\tau = \dfrac{(\mathbf{v} \times \mathbf{a}) \bullet (d\mathbf{a}/dt)}{|\mathbf{v} \times \mathbf{a}|^2}$

The Frenet-Serret formulas: $\quad \dfrac{d\hat{\mathbf{T}}}{ds} = \kappa\hat{\mathbf{N}}, \qquad \dfrac{d\hat{\mathbf{N}}}{ds} = -\kappa\hat{\mathbf{T}} + \tau\hat{\mathbf{B}}, \qquad \dfrac{d\hat{\mathbf{B}}}{ds} = -\tau\hat{\mathbf{N}}$

ORTHOGONAL CURVILINEAR COORDINATES

transformation: $x = x(u, v, w), \quad y = y(u, v, w), \quad z = z(u, v, w)$

position vector: $\mathbf{r} = x(u, v, w)\mathbf{i} + y(u, v, w)\mathbf{j} + z(u, v, w)\mathbf{k}$

scale factors: $h_u = \left|\frac{\partial \mathbf{r}}{\partial u}\right|, \quad h_v = \left|\frac{\partial \mathbf{r}}{\partial v}\right|, \quad h_w = \left|\frac{\partial \mathbf{r}}{\partial w}\right|$

local basis: $\hat{\mathbf{u}} = \frac{1}{h_u}\frac{\partial \mathbf{r}}{\partial u}, \quad \hat{\mathbf{v}} = \frac{1}{h_v}\frac{\partial \mathbf{r}}{\partial v}, \quad \hat{\mathbf{w}} = \frac{1}{h_w}\frac{\partial \mathbf{r}}{\partial w}$

volume element: $dV = h_u h_v h_w \, du \, dv \, dw$

scalar field: $f(u, v, w)$

vector field: $\mathbf{F}(u, v, w) = F_u(u, v, w)\hat{\mathbf{u}} + F_v(u, v, w)\hat{\mathbf{v}} + F_w(u, v, w)\hat{\mathbf{w}}$

gradient: $\nabla f = \frac{1}{h_u}\frac{\partial f}{\partial u}\hat{\mathbf{u}} + \frac{1}{h_v}\frac{\partial f}{\partial v}\hat{\mathbf{v}} + \frac{1}{h_w}\frac{\partial f}{\partial w}\hat{\mathbf{w}}$

divergence: $\nabla \bullet \mathbf{F} = \frac{1}{h_u h_v h_w}\left[\frac{\partial}{\partial u}\Big(h_v h_w F_u\Big) + \frac{\partial}{\partial v}\Big(h_u h_w F_v\Big) + \frac{\partial}{\partial w}\Big(h_u h_v F_w\Big)\right]$

$$\nabla^2 f = \frac{1}{h_u h_v h_w}\left[\frac{\partial}{\partial u}\left(\frac{h_v h_w}{h_u}\frac{\partial f}{\partial u}\right) + \frac{\partial}{\partial v}\left(\frac{h_u h_w}{h_v}\frac{\partial f}{\partial v}\right) + \frac{\partial}{\partial w}\left(\frac{h_u h_v}{h_w}\frac{\partial f}{\partial w}\right)\right]$$

curl: $\nabla \times \mathbf{F} = \frac{1}{h_u h_v h_w}\begin{vmatrix} h_u\hat{\mathbf{u}} & h_v\hat{\mathbf{v}} & h_w\hat{\mathbf{w}} \\ \frac{\partial}{\partial u} & \frac{\partial}{\partial v} & \frac{\partial}{\partial w} \\ F_u h_u & F_v h_v & F_w h_w \end{vmatrix}$

PLANE POLAR COORDINATES

transformation: $x = r\cos\theta, \quad y = r\sin\theta$

position vector: $\mathbf{r} = r\cos\theta\,\mathbf{i} + r\sin\theta\,\mathbf{j}$

scale factors: $h_r = \left|\frac{\partial \mathbf{r}}{\partial r}\right| = 1, \quad h_\theta = \left|\frac{\partial \mathbf{r}}{\partial \theta}\right| = r$

local basis: $\hat{\mathbf{r}} = \cos\theta\mathbf{i} + \sin\theta\mathbf{j}, \quad \hat{\boldsymbol{\theta}} = -\sin\theta\mathbf{i} + \cos\theta\mathbf{j}$

area element: $dA = r\,dr\,d\theta$

scalar field: $f(r, \theta)$

vector field: $\mathbf{F}(r, \theta) = F_r(r, \theta)\hat{\mathbf{r}} + F_\theta(r, \theta)\hat{\boldsymbol{\theta}}$

gradient: $\nabla f = \frac{\partial f}{\partial r}\hat{\mathbf{r}} + \frac{1}{r}\frac{\partial f}{\partial \theta}\hat{\boldsymbol{\theta}}$

divergence: $\nabla \bullet \mathbf{F} = \frac{\partial F_r}{\partial r} + \frac{1}{r}F_r + \frac{1}{r}\frac{\partial F_\theta}{\partial \theta}$

laplacian: $\nabla^2 f = \frac{\partial^2 f}{\partial r^2} + \frac{1}{r}\frac{\partial f}{\partial r} + \frac{1}{r^2}\frac{\partial^2 f}{\partial \theta^2}$

curl: $\nabla \times \mathbf{F} = \left[\frac{\partial F_\theta}{\partial r} + \frac{F_\theta}{r} - \frac{1}{r}\frac{\partial F_r}{\partial \theta}\right]\mathbf{k}$

CYLINDRICAL COORDINATES

transformation: $x = r\cos\theta, \quad y = r\sin\theta, \quad z = z$

position vector: $\mathbf{r} = r\cos\theta\,\mathbf{i} + r\sin\theta\,\mathbf{j} + z\mathbf{k}$

scale factors: $h_r = \left|\frac{\partial \mathbf{r}}{\partial r}\right| = 1, \quad h_\theta = \left|\frac{\partial \mathbf{r}}{\partial \theta}\right| = r, \quad h_z = \left|\frac{\partial \mathbf{r}}{\partial z}\right| = 1$

local basis: $\hat{\mathbf{r}} = \cos\theta\mathbf{i} + \sin\theta\mathbf{j}, \quad \hat{\boldsymbol{\theta}} = -\sin\theta\mathbf{i} + \cos\theta\mathbf{j}, \quad \hat{\mathbf{z}} = \mathbf{k}$

volume element: $dV = r\,dr\,d\theta\,dz$

surface area element (on $r = a$): $dS = a\,d\theta\,dz$

scalar field: $f(r, \theta, z)$

vector field: $\mathbf{F}(r, \theta, z) = F_r(r, \theta, z)\hat{\mathbf{r}} + F_\theta(r, \theta, z)\hat{\boldsymbol{\theta}} + F_z(r, \theta, z)\mathbf{k}$

gradient: $\nabla f = \frac{\partial f}{\partial r}\hat{\mathbf{r}} + \frac{1}{r}\frac{\partial f}{\partial \theta}\hat{\boldsymbol{\theta}} + \frac{\partial f}{\partial z}\mathbf{k}$

divergence: $\nabla \bullet \mathbf{F} = \frac{\partial F_r}{\partial r} + \frac{1}{r}F_r + \frac{1}{r}\frac{\partial F_\theta}{\partial \theta} + \frac{\partial F_z}{\partial z}$

laplacian: $\nabla^2 f = \frac{\partial^2 f}{\partial r^2} + \frac{1}{r}\frac{\partial f}{\partial r} + \frac{1}{r^2}\frac{\partial^2 f}{\partial \theta^2} + \frac{\partial^2 f}{\partial z^2}$

curl: $\nabla \times \mathbf{F} = \frac{1}{r}\begin{vmatrix} \hat{\mathbf{r}} & r\hat{\boldsymbol{\theta}} & \mathbf{k} \\ \frac{\partial}{\partial r} & \frac{\partial}{\partial \theta} & \frac{\partial}{\partial z} \\ F_r & rF_\theta & F_z \end{vmatrix}$

SPHERICAL COORDINATES

transformation: $x = \rho\sin\phi\cos\theta, \quad y = \rho\sin\phi\sin\theta, \quad z = \rho\cos\phi$

position vector: $\mathbf{r} = \rho\sin\phi\cos\theta\,\mathbf{i} + \rho\sin\phi\sin\theta\,\mathbf{j} + \rho\cos\phi\mathbf{k}$

scale factors: $h_\rho = \left|\frac{\partial \mathbf{r}}{\partial \rho}\right| = 1, \quad h_\phi = \left|\frac{\partial \mathbf{r}}{\partial \phi}\right| = \rho, \quad h_\theta = \left|\frac{\partial \mathbf{r}}{\partial \theta}\right| = \rho\sin\phi$

local basis: $\hat{\boldsymbol{\rho}} = \sin\phi\cos\theta\,\mathbf{i} + \sin\phi\sin\theta\,\mathbf{j} + \cos\phi\,\mathbf{k}, \quad \hat{\boldsymbol{\phi}} = \cos\phi\cos\theta\,\mathbf{i} + \cos\phi\sin\theta\,\mathbf{j} - \sin\phi\,\mathbf{k}, \quad \hat{\boldsymbol{\theta}} = -\sin\theta\,\mathbf{i} + \cos\theta\,\mathbf{j}$

volume element: $dV = \rho^2\sin\phi\,d\rho\,d\phi\,d\theta$

surface area element (on $\rho = a$): $dS = a^2\sin\phi\,d\theta\,d\phi$

scalar field: $f(\rho, \phi, \theta)$

vector field: $\mathbf{F}(\rho, \phi, \theta) = F_\rho(\rho, \phi, \theta)\hat{\boldsymbol{\rho}} + F_\phi(\rho, \phi, \theta)\hat{\boldsymbol{\phi}} + F_\theta(\rho, \phi, \theta)\hat{\boldsymbol{\theta}}$

gradient: $\nabla f = \frac{\partial f}{\partial \rho}\hat{\boldsymbol{\rho}} + \frac{1}{\rho}\frac{\partial f}{\partial \phi}\hat{\boldsymbol{\phi}} + \frac{1}{\rho\sin\phi}\frac{\partial f}{\partial \theta}\hat{\boldsymbol{\theta}}$

divergence: $\nabla \bullet \mathbf{F} = \frac{\partial F_\rho}{\partial \rho} + \frac{2}{\rho}F_\rho + \frac{1}{\rho}\frac{\partial F_\phi}{\partial \phi} + \frac{\cot\phi}{\rho}F_\phi + \frac{1}{\rho\sin\phi}\frac{\partial F_\theta}{\partial \theta}$

laplacian: $\nabla^2 f = \frac{\partial^2 f}{\partial \rho^2} + \frac{2}{\rho}\frac{\partial f}{\partial \rho} + \frac{1}{\rho^2}\frac{\partial^2 f}{\partial \phi^2} + \frac{\cot\phi}{\rho^2}\frac{\partial f}{\partial \phi} + \frac{1}{\rho^2\sin^2\phi}\frac{\partial^2 f}{\partial \theta^2}$

curl: $\nabla \times \mathbf{F} = \frac{1}{\rho^2\sin\phi}\begin{vmatrix} \hat{\boldsymbol{\rho}} & \rho\hat{\boldsymbol{\phi}} & \rho\sin\phi\hat{\boldsymbol{\theta}} \\ \frac{\partial}{\partial \rho} & \frac{\partial}{\partial \phi} & \frac{\partial}{\partial \theta} \\ F_\rho & \rho F_\phi & \rho\sin\phi F_\theta \end{vmatrix}$

A COMPLETE COURSE

Calculus

A COMPLETE COURSE

Calculus

Robert A. Adams
UNIVERSITY OF BRITISH COLUMBIA
Christopher Essex
UNIVERSITY OF WESTERN ONTARIO

Eighth Edition

PEARSON
Toronto

Vice-President, Editorial Director: Gary Bennett
Editor-in-Chief: Nicole Lukach
Acquisitions Editor: Cathleen Sullivan
Marketing Manager: Michelle Bish
Developmental Editor: Karen Townsend
Project Manager: Marissa Lok
Production Editor: Leanne Rancourt
Copy Editor: Valerie Adams
Proofreader: Leanne Rancourt
Compositor: Robert Adams
Art Director: Julia Hall
Cover and Interior Designer: Miguel Acevedo
Cover Image: Gloria H. Chomica/Masterfile

10 9 8 V011

Library and Archives Canada Cataloguing in Publication

Adams, Robert A. (Robert Alexander), 1940-
Calculus : a complete course / Robert A. Adams, Christopher Essex. — 8th ed.

Includes index.
ISBN 978-0-321-78107-9

1. Calculus—Textbooks. I. Essex, Christopher II. Title.

QA303.2.A33 2013 515 C2012-904554-3

ISBN 978-0-32-178107-9

To Noreen and Sheran

Contents

Preface

The word "tears" is found in mathematical titles surprisingly often. One reads of "mathematics without tears," "geometry without tears," "topology without tears," "statistics without tears," and, of course, "calculus without tears," among others. Compare these juxtapositions of tears and mathematics with what the late Fields Medalist and member of Bourbaki, Laurent Schwartz, once famously wrote,"Il n'y a pas de mathematiques sans larmes . . . " [There is no mathematics without tears. . . .] It seems that there has been much weeping over mathematics, as well as disagreement over whether the weeping is necessary or not. Perhaps people have carried the tear metaphor too far, but the underlying sentiment behind tearlessness has also been long expressed in other ways too. Over the twentieth century we have had all manner of mathematics titles using words like, "outline," "nutshell," "simple," or "dummies."

For example, S. P. Thompson published *Calculus Made Easy* in 1910. It retains enough of a following today, more than a century later, that you can still buy fresh printings of the 1914 second edition on Amazon. Not even a more recent edition rewritten by Martin Gardener, no less, has been able to push the second edition into oblivion. Mathematics texts are like that. At the beginning of the twentieth century, some students were still taught from Euclid's *Elements*, in the original ancient Greek. In mathematics the basics last. This makes the modern mathematics instructor a bit nihilistic about choosing which textbook to use in teaching calculus. "Does it really matter which text I use?" they ask. At some level it indeed does not, so they just employ whatever text was used last year. "More of the same" becomes the standard operating procedure. But this cannot always hold true or we might be using Thompson today instead of any number of modern texts.

One reason more-of-the-same doesn't always hold is that the audience changes even when the basics don't. We explained in the seventh edition that modern students, with a lifelong exposure to modern graphical computer interfaces, cannot help but look at mathematics differently than previous generations. It is all too easy for them to have the impression that mathematics is an application of computer science, existing as a mere icon on desktops. Why learn mathematics, they reason, when it seems we can have it all at a "click?" Learning that computer science, not mathematics, is actually the application is a good first step. Learning that computers are not something to be believed, except in a conditional manner, is the definitive lesson for the computer age. That alone is reason enough to learn mathematics. The seventh edition dealt with this through a new thematic topic called "Numerical Monsters" (marked by ⚠ in this new edition). These represent a form of *anti-numerical analysis*—"anti" because the topic aims to make computer errors as large as possible instead of minimizing them. They provide natural, self-contained mathematical applications that play off the finite representation of numbers within all computers. What is learned is fundamental, and qualitatively independent of code or platform. All of this was inconceivable to Euclid, Newton, or S. P. Thompson.

There is another reason why more-of-the-same is problematic. Much of the basic mathematics used in mathematically based fields was set about a century and half ago. At the time of Thompson, one hundred years ago, there was a gap between the application fields and the calculus exposition of that time. More-of-the-same has preserved that gap over all the subsequent years. When students move from calculus to the mathematics of one of these fields, they enter a strange world with mathematical customs that may even contradict what they have been taught in their calculus courses. For example, what are physics, chemistry, or engineering students supposed to make of the famous equation, $dE = \delta W + \delta Q$? It depicts a differential equalling the sum of two things that are not actually differentials. Puzzling to say the least. But what are these things on the right side? Students are told not to be alarmed in their field-specific texts because E is not actually a function of W or Q. Well, since δW and δQ are not actually differentials anyway, maybe that is okay then, or is it? Sometimes the δ's are replaced by d's with little bars through them to emphasize that some kind of unique-to-thermodynamics "mathematics" is in play. This representation is an anachronism dating from the nineteenth century, recalling dubious attempts to depict everything, in addition to functions, in terms of differentials alone. Calculus texts have simply not ventured to show how to proceed without the nineteenth century awkwardness still in play today.

This is not the only example of this phenomenon by any means. Such things have generated more confusion and "tears" than any mathematics course ever has. A good introductory applied calculus textbook ought to lead the reader to where the calculus properly connects to actual fields that calculus students may actually encounter in their subsequent training. This not only helps to stave off unnecessary "tears" but ultimately can lead to a more lucid standard of exposition for the fields in question. We described these connections with the thematic title "Gateway Applications" in the seventh edition. We have marked them by the symbol 🏠 in the eighth edition. They should not be confused with "applications" that appear as tamed examples and staged problems typical in all textbooks. Instead, they take the reader from a calculus topic at hand directly to a mathematical tool often overlooked in calculus texts but crucial to an actual field, or they take the reader to an insight on how calculus sets the structure of an entire field, without actually pursuing the field. Now that's application!

In the seventh edition we introduced a number of these gateway applications, from Liapunov functions to thermodynamics and Legendre transformations. We also sketched out why these things are important and how they are actually used. In the eighth edition we have added a calculus-based explanation of entropy as a gateway application, showing how it naturally arises from simple calculus properties and how it fits both into statistical mechanics in physics and information theory. Gateway applications are not meant to replace those traditional "applications." They are meant to enhance the possibilities within a calculus course, either as source material for independent projects, or as enrichment for a course that an instructor may choose to explore to make a point to a class. Moreover, they are value-added when viewing the book as a future reference work. When students encounter the gaps between their calculus training and their subsequent courses, chances are they will find answers on how to bridge the gaps not available any place else. The eighth edition is no crib sheet to be discarded after the course is done.

There is yet one more reason why more-of-the-same is problematic. Over long enough timescales our best understanding of mathematics and how it is used does actually change. Unanswered questions linger even in conventional calculus that are not answered by more-of-the-same. These questions bother students, impeding their learning, and may even have bothered instructors when they were students. A relatively straightforward example is how we know that a minimization in a Lagrange multiplier problem really provides a minimum. Does the famous maximum entropy principle really provide a maximum? How does one know? An answer to this cannot be found in other mainstream calculus textbooks, oddly enough, but you can find it in the eighth edition. A subtler question on the minds of students is what is the difference between a differential of a function and the differential appearing in a multiple integral (i.e., think of $dy = f'(x)\,dx$ versus, say, $dV = dx\,dy\,dz$)? If you read Thompson, or nearly any other textbook, you may be forgiven for concluding that they are really just the same. But they are not the same. In fact, the differences turn out to have great significance, revealing structures that simplify and unify advanced calculus while having stunning implications for the fundamental differential equations governing many fields of science. All of this is found within the subject of *Exterior Calculus*.

Despite the power of exterior calculus it is not part of the normal more-of-the-same approach, at least in part because some of its development was after, or contemporary with, works like Thompson. But that still makes its current form nearly a hundred years old—just like yesterday from the perspective of mathematics pedagogy. Until now, a basic textbook-style treatment of the subject has been unavailable because it has been frozen out by more-of-the-same, causing it to be thought of as strictly an advanced topic. Technical monographs only reinforce this idea, but the structures while not low level, are not that high level in principle either. Vectors, linear algebra, basic calculus, some abstract imagination, and some clarifications about terminological customs is pretty much all that is needed. Maybe at some future date, the advanced calculus curriculum will be structured differently, but for now we offer an answer to the student's question about differentials in the form of the new Chapter 17 of the eighth edition. This is consistent with being innovative and living up to the subtitle: *A Complete Course*, while maintaining continuity and respecting tradition as much as is practical.

But, as with other value-added features of this volume, an instructor can simply ignore the material and teach a conventional program. There are few treatments of that traditional material more straightforward or succinct. But if the instructor wants to go a little further or a lot further; if the instructor wants to make a simple pedagogical point or identify a project for students; if the instructor just wants to point to a place where a student's questions can be answered, the eighth edition can help accomplish all of these things. Chapter 17 has been tried out in an advanced calculus course at the University of British Columbia, with positive results. But what is really intriguing to us is how many colleagues expressed interest in reading a textbook-style treatment of something that they always wanted to learn about themselves.

There is nothing wrong with books like *Calculus Made Easy*. We encourage readers to look at some, not only because alternative treatments can be helpful at times, but also to disabuse readers of any impression that they contain any sort of special educational magic beyond the usual protocol of diagrams, explanations, statements, derivations, definitions, worked examples, and exercises. There is also nothing wrong with modern interactive computer treatments of mathematics either. There are many advantages, such as hauling around software rather than paper, but even here the false allure of a mythical magic road to learning mathematics is tempting. Programmed learning still boils down to the same protocol as in a textbook: diagrams, explanations, etc. Add an instructor and answers in the back of the book, then you span the same pedagogical space. No matter how it is delivered, learning mathematics that is new to a student takes work and maybe enduring some hopefully temporary tears. There is no magic road. Therefore, instructors should call for fearlessness from students rather than worrying about their tearlessness. Challenge marks the path of greatest gain, particularly in a textbook where the path is fully known and clearly marked. Fearless rather than tearless students learn quickly what S. P. Thompson meant when he began his text, "What one fool can do, another can."

To the Student

You are holding what has become known as a "high-end" calculus text in the book trade. You are lucky. Think of it as having a high-end touring car instead of a compact economy car. But it is not high end in the material sense. It does not have scratch-and-sniff pages, sparkling radioactive ink, or anything else like that. It's the contents that set it apart. Unlike the car business, "high-end" book content is not priced any higher than any other books. It is one of the few consumer items where anyone can afford to buy into the high end. But there is a catch. Unlike cars, you have to do the work to achieve the promise of the book. So in that sense "high end" is more like a form of "secret" martial arts for your mind that the economy version cannot deliver. If you practise, your mind will become stronger. You will become more confident and disciplined. Secrets of the ages will become open to you. You will become fearless, as your mind longs to tackle any new mathematical challenge.

But hard work is the watchword. Practise, practise, practise. Think of how bees work busily to get their honey. There is a sort of "honey" in calculus. It is sweet when you finally get a new idea that you did not understand before. There are few experiences as great as figuring things out. That is one of the reasons why there has always been a booming world puzzle industry. In a high-end book there is more honey to be had than in recreational puzzles or lesser calculus texts. Doing exercises and checking against solutions in the back of the book are how you practise mathematics with a text. You can do essentially the same thing on a computer interface: you still do the problems and check the answers. However you do it, more exercises mean more practice and better performance.

There are numerous exercises in this text—too many for you to try them all perhaps, but be ambitious. Some are "drill" exercises to help you develop your skills in calculation. More important, however, are the problems that develop reasoning skills and your ability to apply the techniques you have learned to concrete situations. In some cases you will have to plan your way through a problem that requires several different "steps" before you can get to the answer. Other exercises are designed to extend the theory developed in the text and therefore enhance your understanding of the concepts of calculus. Think of the problems as a tool to help you correctly wire your mind. You may have a lot of great components in your head, but if you don't wire the components together properly, your "home theatre" won't work.

The exercises vary greatly in difficulty. Usually, the more difficult ones occur toward the end of exercise sets, but these sets are not strictly graded in this way because exercises on a specific topic tend to be grouped together. Also, "difficulty" can be subjective. For some students, exercises designated difficult may seem easy, while exercises designated easy may seem difficult. Nonetheless, some exercises in the regular sets are marked with the symbols **!** , which indicates that the exercise is somewhat more difficult than most, or **?** , which indicates a more theoretical exercise. The theoretical ones need not be difficult; sometimes they are quite easy. Most of the problems in the *Challenging Problems* section forming part of the *Chapter Review* at the end of most chapters are also on the difficult side.

It is not a bad idea to review the background material in Chapter P (Preliminaries), even if your instructor does not refer to it in class.

If you find some of the concepts in the book difficult to understand, *re-read* the material slowly, if necessary several times; *think about it*; formulate questions to ask fellow students, your TA, or your instructor. Don't delay. It is important to resolve your problems as soon as possible. If you don't understand today's topic, you may not understand how it applies to tomorrow's either. Mathematics builds from one idea to the next. Testing your understanding of the later topics also tests your understanding of the earlier ones. Do not be discouraged if you can't do *all* the exercises. Some are very difficult indeed. The range of exercises ensures that nearly all students can find a comfortable level to practise at, while allowing for greater challenges as skill grows.

Answers for most of the odd-numbered exercises are provided at the back of the book. Exceptions are exercises that don't have short answers: for example, "Prove that ..." or "Show that ..." problems where the answer is the whole solution. A *Student Solutions Manual* that contains detailed solutions to even-numbered exercises is available.

Besides **!** and **?** used to mark more difficult and theoretical problems, the following symbols are used to mark exercises of special types:

- Exercises pertaining to differential equations and initial-value problems. (It is not used in sections that are wholly concerned with DEs.)
- Problems requiring the use of a calculator. Often a scientific calculator is needed. Some such problems may require a programmable calculator.
- Problems requiring the use of either a graphing calculator or mathematical graphing software on a personal computer.
- Problems requiring the use of a computer. Typically, these will require either computer algebra software (e.g., Maple, Mathematica) or a spreadsheet program such as Microsoft Excel.

To the Instructor

This book covers the material usually encountered in a three- or four-semester real-variable calculus program, involving real-valued functions of a single real variable (differential calculus in Chapters 1–4 and integral calculus in Chapters 5–8), as well as vector-valued functions of a single real variable (covered in Chapter 11), real-valued functions of several real variables (in Chapters 12–14), and vector-valued functions of several real variables (in Chapters 15 and 16). Chapter 9 concerns sequences and series, and its position is rather arbitrary.

Chapter 10 contains necessary background on vectors and geometry in 3-space as well as a bit of linear algebra that is useful, although not absolutely essential, for the understanding of subsequent multivariable material. Most of the material requires only a reasonable background in high school algebra and analytic geometry. (See Chapter P—Preliminaries for a review of this material.) However, some optional material is more subtle and/or theoretical and is intended for stronger students, special topics, and reference purposes. It also allows instructors considerable flexibility in making points, answering questions, and selective enrichment of a course. Chapter 18, for example, is a compact treatment of linear ordinary differential equations which may provide supplementary material or become a major topic in a multi-topic course.

Changes in the eighth edition include numerous improvements and clarifications throughout, including notational adjustments and corrections. Major additions include Taylor's formula in terms of functions of n variables (Section 12.9); the classification of extrema for functions with constraints (Section 13.4); the gateway application, entropy in statistical mechanics, and information theory (Section 13.9); and Chapter 17, "Differential Forms and Exterior Calculus."

There is a wealth of material here—too much to include in any course. It was never intended to be otherwise. You must select what material to include and what to omit, taking into account the background and needs of your students. At the University of British Columbia, where one author taught for 34 years, and at the University of Western Ontario, where the other author continues to teach, calculus is divided into four semesters, the first two covering single-variable calculus, the third covering functions of several variables, and the fourth covering vector calculus. In none of these courses was there enough time to cover all the material in the appropriate chapters; some sections are always omitted. The text is designed to allow students and instructors to conveniently find their own level while enhancing any course from general calculus to courses focused on science and engineering students.

Several supplements are available for use with *Calculus: A Complete Course, 8th Edition.* Available to students is the **Student Solutions Manual** (ISBN: 9780321862938): This manual contains detailed solutions to all the even-numbered exercises, prepared by the authors. There are also such Manuals for the split volumes, *Single Variable Calculus* (ISBN: 9780321877468), and *Calculus of Several Variables* (ISBN: 9780321877475).

The following supplements are available for download from a password-protected section of Pearson Education Canada's online catalogue (catalogue.pearsoned.ca):

- **Instructor's Solutions Manual**,
- **Text Solutions in online-publishable form**,
- **Pearson TestGen**. TestGen is testing software that enables instructors to view and edit the existing questions, (over 1,500 test questions are provided), and add questions, generate tests, and distribute the tests in a variety of formats.
- **Image Library**, which contains all of the figures in the text provided as individual enlarged .pdf files suitable for printing to transparencies.

Navigate to this book's catalogue page to view a list of those supplements that are available. See your local sales representative for details and access.

Also available to qualified instructors are **MyMathLab**® and **MathXL**® Online Courses for which access codes are required.

MyMathLab helps improve individual students' performance. It has a consistently positive impact on the quality of learning in higher-education math instruction. MyMathLab's comprehensive online gradebook automatically tracks your students' results on tests, quizzes, homework, and in the study plan. MyMathLab provides engaging experiences that personalize, stimulate, and measure learning for each student. The homework and practice exercises in MyMathLab are correlated to the exercises in the textbook. The software offers immediate, helpful feedback when students enter incorrect answers. Exercises include guided solutions, sample problems, animations, and eText clips for extra help. MyMathLab comes from an experienced partner with educational expertise and an eye on the future. Knowing that you are using a Pearson product means knowing that you are using quality content. That means that our eTexts are accurate and our assessment tools work. To learn more about how MyMathLab combines proven learning applications with powerful assessment, visit www.mymathlab.com or contact your Pearson representative.

MathXL is the homework and assessment engine that runs MyMathLab. (MyMathLab is MathXL plus a learning management system.) MathXL is available to qualified adopters. For more information, visit our website at www.mathxl.com, or contact your Pearson representative.

Acknowledgments

We are grateful to many colleagues and students, at the University of British Columbia, the University of Western Ontario, and at many other institutions where these books have been used, for their encouragement and useful comments and suggestions.

In preparing this edition, we have had guidance from several dedicated reviewers who provided new insight and direction, namely:

Angelina Chin Yan Mui	University of Malaya
Jimmy Chi-Hung Fung	Hong Kong University of Science and Technology
Elena Devdariani	Carleton University
Yousry Elsabrouty	University of Calgary
Sean Graves	University of Alberta
Alexandre Karassev	Nipissing University
Leung Pui Fai	National University of Singapore
Mariya Svishchuk	Mount Royal University

We are also greatly appreciative of the comments and suggestions made by Professor Brian Marcus and his students in Math 227 (Honours Vector Calculus) at the University of British Columbia, in March 2012, as well as the following reviewers who made valuable comments on drafts of the new Chapter 17:

Hichem Ben-El-Mechaiekh	Brock University
Michael Haslam	York University
Robert Israel	University of British Columbia
Martin Lgar	University of Alberta
Peter Lawrence	Ryerson University
Mitja Mastnak	Saint Mary's University
Cristian Rios	University of Calgary
Robert Steacy	University of Victoria

Finally, we wish to thank the sales and marketing staff of all Addison-Wesley (now Pearson Canada) divisions around the world for making the previous editions so successful, and the editorial and production staff in Toronto, in particular, Acquisitions Editor Cathleen Sullivan, Developmental Editor Karen Townsend, Project Manager Marissa Lok, Production Editor Leanne Rancourt, and Copy Editor Valerie Adams, for their assistance and encouragement.

This volume was typeset by Robert Adams using TEX on an iMac running OSX version 10.6. Most of the figures were generated using the mathematical graphics software package **MG** developed by Robert Israel and Robert Adams. Some were produced with Maple 10. Miguel Acevedo provided the cover design.

The expunging of errors and obscurities in a text is an ongoing and asymptotic process; hopefully each edition is better than the previous one. Nevertheless, some such imperfections always remain, and we will be grateful to any readers who call them to our attention, or give us other suggestions for future improvements.

May 2012

R.A.A.	*C.E.*
Vancouver, Canada	London, Canada
`adms@math.ubc.ca`	`essex@uwo.ca`

What Is Calculus?

Early in the seventeenth century, the German mathematician Johannes Kepler analyzed a vast number of astronomical observations made by Danish astronomer Tycho Brahe and concluded that the planets must move around the sun in elliptical orbits. He didn't know why. Fifty years later, the English mathematician and physicist Isaac Newton answered that question.

Why do the planets move in elliptical orbits around the sun? Why do hurricane winds spiral counterclockwise in the northern hemisphere? How can one predict the effects of interest rate changes on economies and stock markets? When will radioactive material be sufficiently decayed to enable safe handling? How do warm ocean currents in the equatorial Pacific affect the climate of eastern North America? How long will the concentration of a drug in the bloodstream remain at effective levels? How do radio waves propagate through space? Why does an epidemic spread faster and faster and then slow down? How can I be sure the bridge I designed won't be destroyed in a windstorm?

These and many other questions of interest and importance in our world relate directly to our ability to analyze motion and how quantities change with respect to time or each other. Algebra and geometry are useful tools for describing relationships between *static* quantities, but they do not involve concepts appropriate for describing how a quantity *changes*. For this we need new mathematical operations that go beyond the algebraic operations of addition, subtraction, multiplication, division, and the taking of powers and roots. We require operations that measure the way related quantities change.

Calculus provides the tools for describing motion quantitatively. It introduces two new operations called *differentiation* and *integration*, which, like addition and subtraction, are opposites of one another; what differentiation does, integration undoes.

For example, consider the motion of a falling rock. The height (in metres) of the rock t seconds after it is dropped from a height of h_0 m is a function $h(t)$ given by

$$h(t) = h_0 - 4.9t^2.$$

The graph of $y = h(t)$ is shown in the figure below:

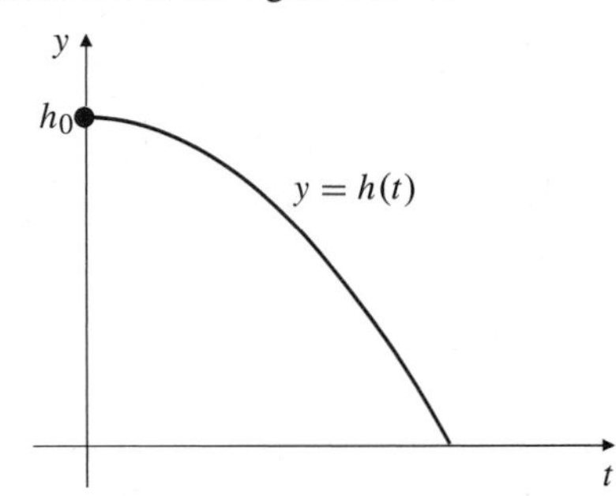

The process of differentiation enables us to find a new function, which we denote $h'(t)$ and call the *derivative* of h with respect to t, that represents the *rate of change* of the height of the rock, that is, its *velocity* in metres/second:

$$h'(t) = -9.8t.$$

Conversely, if we know the velocity of the falling rock as a function of time, integration enables us to find the height function $h(t)$.

Calculus was invented independently and in somewhat different ways by two seventeenth-century mathematicians: Isaac Newton and Gottfried Wilhelm Leibniz. Newton's motivation was a desire to analyze the motion of moving objects. Using his calculus, he was able to formulate his laws of motion and gravitation and to *conclude from them* that the planets must move around the sun in elliptical orbits.

Many of the most fundamental and important "laws of nature" are conveniently expressed as equations involving rates of change of quantities. Such equations are called *differential equations*, and techniques for their study and solution are at the heart of calculus. In the falling rock example, the appropriate law is Newton's Second Law of Motion:

$$\text{force} = \text{mass} \times \text{acceleration}.$$

The *acceleration*, -9.8 m/s^2, is the rate of change (the *derivative*) of the velocity, which is in turn the rate of change (the *derivative*) of the height function.

Much of mathematics is related indirectly to the study of motion. We regard *lines*, or *curves*, as geometric objects, but the ancient Greeks thought of them as paths traced out by moving points. Nevertheless, the study of curves also involves geometric concepts such as tangency and area. The process of differentiation is closely tied to the geometric problem of finding tangent lines; similarly, integration is related to the geometric problem of finding areas of regions with curved boundaries.

Both differentiation and integration are defined in terms of a new mathematical operation called a **limit**. The concept of the limit of a function will be developed in Chapter 1. That will be the real beginning of our study of calculus. In the chapter called "Preliminaries" we will review some of the background from algebra and geometry needed for the development of calculus.

CHAPTER P

Preliminaries

> 'Reeling and Writhing, of course, to begin with,' the Mock Turtle replied, 'and the different branches of Arithmetic — Ambition, Distraction, Uglification, and Derision.'

Lewis Carroll (Charles Lutwidge Dodgson) 1832–1898
from *Alice's Adventures in Wonderland*

Introduction

This preliminary chapter reviews the most important things you should know before beginning calculus. Topics include the real number system, Cartesian coordinates in the plane, equations representing straight lines, circles, and parabolas, functions and their graphs, and, in particular, polynomials and trigonometric functions.

Depending on your precalculus background, you may or may not be familiar with these topics. If you are, you may want to skim over this material to refresh your understanding of the terms used; if not, you should study this chapter in detail.

P.1 Real Numbers and the Real Line

Calculus depends on properties of the real number system. **Real numbers** are numbers that can be expressed as decimals, for example,

$$\begin{aligned} 5 &= 5.00000\ldots \\ -\tfrac{3}{4} &= -0.750000\ldots \\ \tfrac{1}{3} &= 0.3333\ldots \\ \sqrt{2} &= 1.4142\ldots \\ \pi &= 3.14159\ldots \end{aligned}$$

In each case the three dots (...) indicate that the sequence of decimal digits goes on forever. For the first three numbers above, the patterns of the digits are obvious; we know what all the subsequent digits are. For $\sqrt{2}$ and π there are no obvious patterns.

The real numbers can be represented geometrically as points on a number line, which we call the **real line**, shown in Figure P.1. The symbol $\mathbb{R}$ is used to denote either the real number system or, equivalently, the real line.

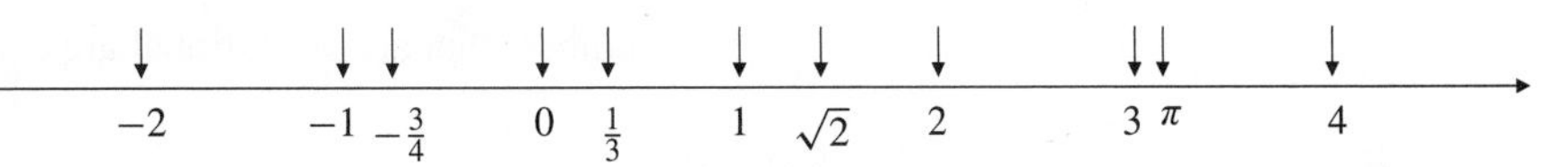

Figure P.1 The real line

The properties of the real number system fall into three categories: algebraic properties, order properties, and completeness. You are already familiar with the *algebraic properties*; roughly speaking, they assert that real numbers can be added, subtracted, multiplied, and divided (except by zero) to produce more real numbers and that the usual rules of arithmetic are valid.

The *order properties* of the real numbers refer to the order in which the numbers appear on the real line. If x lies to the left of y, then we say that "x is less than y" or "y is greater than x." These statements are written symbolically as $x < y$ and $y > x$, respectively. The inequality $x \leq y$ means that either $x < y$ or $x = y$. The order properties of the real numbers are summarized in the following *rules for inequalities*:

The symbol $\Longrightarrow$ means "implies."

Rules for inequalities

If a, b, and c are real numbers, then:

1. $a < b \quad \Longrightarrow \quad a + c < b + c$
2. $a < b \quad \Longrightarrow \quad a - c < b - c$
3. $a < b$ and $c > 0 \quad \Longrightarrow \quad ac < bc$
4. $a < b$ and $c < 0 \quad \Longrightarrow \quad ac > bc$; in particular, $-a > -b$
5. $a > 0 \quad \Longrightarrow \quad \dfrac{1}{a} > 0$
6. $0 < a < b \quad \Longrightarrow \quad \dfrac{1}{b} < \dfrac{1}{a}$

Rules 1–4 and 6 (for $a > 0$) also hold if $<$ and $>$ are replaced by $\leq$ and $\geq$.

Note especially the rules for multiplying (or dividing) an inequality by a number. If the number is positive, the inequality is preserved; if the number is negative, the inequality is reversed.

The *completeness* property of the real number system is more subtle and difficult to understand. One way to state it is as follows: if A is any set of real numbers having at least one number in it, and if there exists a real number y with the property that $x \leq y$ for every x in A (such a number y is called an **upper bound** for A), then there exists a *smallest* such number, called the **least upper bound** or **supremum** of A, and denoted $\sup(A)$. Roughly speaking, this says that there can be no holes or gaps on the real line—every point corresponds to a real number. We will not need to deal much with completeness in our study of calculus. It is typically used to prove certain important results—in particular, Theorems 8 and 9 in Chapter 1. (These proofs are given in Appendix III but are not usually included in elementary calculus courses; they are studied in more advanced courses in mathematical analysis.) However, when we study infinite sequences and series in Chapter 9, we will make direct use of completeness.

The set of real numbers has some important special subsets:

(i) the **natural numbers** or **positive integers**, namely, the numbers $1, 2, 3, 4, \ldots$

(ii) the **integers**, namely, the numbers $0, \pm 1, \pm 2, \pm 3, \ldots$

(iii) the **rational numbers**, that is, numbers that can be expressed in the form of a fraction m/n, where m and n are integers, and $n \neq 0$.

The rational numbers are precisely those real numbers with decimal expansions that are either:

(a) terminating, that is, ending with an infinite string of zeros, for example, $3/4 = 0.750000\ldots$, or

(b) repeating, that is, ending with a string of digits that repeats over and over, for example, $23/11 = 2.090909\ldots = 2.\overline{09}$. (The bar indicates the pattern of repeating digits.)

Real numbers that are not rational are called *irrational numbers*.

EXAMPLE 1 Show that each of the numbers (a) $1.323232\cdots = 1.\overline{32}$ and (b) $0.3405405405\ldots = 0.3\overline{405}$ is a rational number by expressing it as a quotient of two integers.

Solution

(a) Let $x = 1.323232\ldots$ Then $x - 1 = 0.323232\ldots$ and

$$100x = 132.323232\ldots = 132 + 0.323232\ldots = 132 + x - 1.$$

Therefore, $99x = 131$ and $x = 131/99$.

(b) Let $y = 0.3405405405\ldots$ Then $10y = 3.405405405\ldots$ and $10y - 3 = 0.405405405\ldots$ Also,

$$10,000y = 3,405.405405405\ldots = 3,405 + 10y - 3.$$

Therefore, , $9990y = 3,402$ and $y = 3,402/9,990 = 63/185$.

The set of rational numbers possesses all the algebraic and order properties of the real numbers but not the completeness property. There is, for example, no rational number whose square is 2. Hence, there is a "hole" on the "rational line" where $\sqrt{2}$ should be.[1] Because the real line has no such "holes," it is the appropriate setting for studying limits and therefore calculus.

Intervals

A subset of the real line is called an **interval** if it contains at least two numbers and also contains all real numbers between any two of its elements. For example, the set of real numbers x such that $x > 6$ is an interval, but the set of real numbers y such that $y \neq 0$ is not an interval. (Why?) It consists of two intervals.

If a and b are real numbers and $a < b$, we often refer to

(i) the **open interval** from a to b, denoted by (a, b), consisting of all real numbers x satisfying $a < x < b$.

(ii) the **closed interval** from a to b, denoted by $[a, b]$, consisting of all real numbers x satisfying $a \leq x \leq b$.

(iii) the **half-open interval** $[a, b)$, consisting of all real numbers x satisfying the inequalities $a \leq x < b$.

(iv) the **half-open interval** $(a, b]$, consisting of all real numbers x satisfying the inequalities $a < x \leq b$.

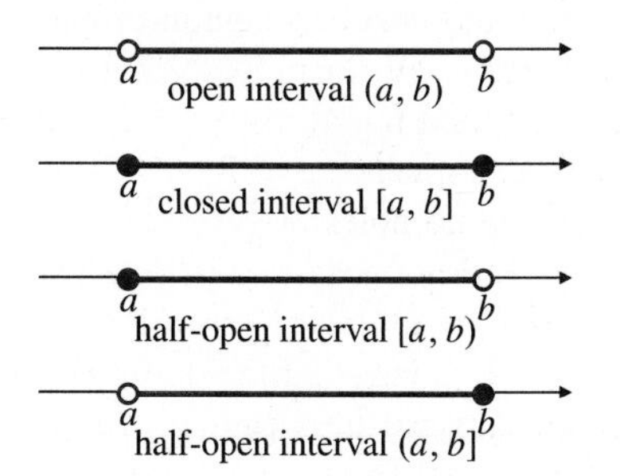

Figure P.2 Finite intervals

These are illustrated in Figure P.2. Note the use of hollow dots to indicate endpoints of intervals that are not included in the intervals, and solid dots to indicate endpoints that are included. The endpoints of an interval are also called **boundary points.**

The intervals in Figure P.2 are **finite intervals**; each of them has finite length $b - a$. Intervals can also have infinite length, in which case they are called **infinite intervals.** Figure P.3 shows some examples of infinite intervals. Note that the whole real line $\mathbb{R}$ is an interval, denoted by $(-\infty, \infty)$. The symbol ∞ ("infinity") does *not* denote a real number, so we never allow ∞ to belong to an interval.

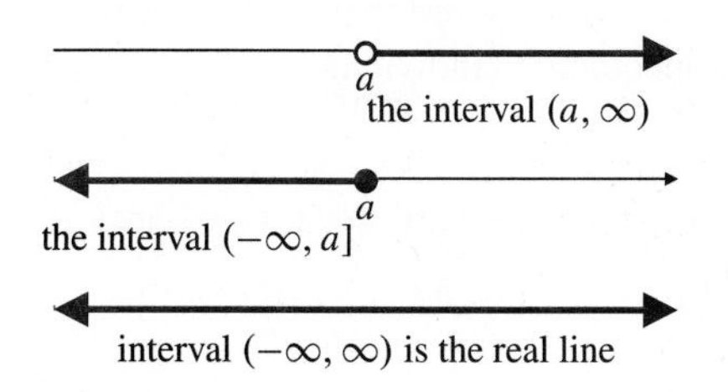

Figure P.3 Infinite intervals

[1] How do we know that $\sqrt{2}$ is an irrational number? Suppose, to the contrary, that $\sqrt{2}$ is rational. Then $\sqrt{2} = m/n$, where m and n are integers and $n \neq 0$. We can assume that the fraction m/n has been "reduced to lowest terms"; any common factors have been cancelled out. Now $m^2/n^2 = 2$, so $m^2 = 2n^2$, which is an even integer. Hence m must also be even. (The square of an odd integer is always odd.) Since m is even, we can write $m = 2k$, where k is an integer. Thus $4k^2 = 2n^2$ and $n^2 = 2k^2$, which is even. Thus n is also even. This contradicts the assumption that $\sqrt{2}$ could be written as a fraction m/n in lowest terms; m and n cannot both be even. Accordingly, there can be no rational number whose square is 2.

EXAMPLE 2 Solve the following inequalities. Express the solution sets in terms of intervals and graph them.

(a) $2x - 1 > x + 3$ (b) $-\dfrac{x}{3} \ge 2x - 1$ (c) $\dfrac{2}{x-1} \ge 5$

Solution

(a)

$2x - 1 > x + 3$	Add 1 to both sides.
$2x > x + 4$	Subtract x from both sides.
$x > 4$	The solution set is the interval $(4, \infty)$.

(b)

$-\dfrac{x}{3} \ge 2x - 1$	Multiply both sides by -3.
$x \le -6x + 3$	Add $6x$ to both sides.
$7x \le 3$	Divide both sides by 7.
$x \le \dfrac{3}{7}$	The solution set is the interval $(-\infty, 3/7]$.

(c) We transpose the 5 to the left side and simplify to rewrite the given inequality in an equivalent form:

$$\frac{2}{x-1} - 5 \ge 0 \iff \frac{2 - 5(x-1)}{x-1} \ge 0 \iff \frac{7-5x}{x-1} \ge 0.$$

The fraction $\dfrac{7-5x}{x-1}$ is undefined at $x = 1$ and is 0 at $x = 7/5$. Between these numbers it is positive if the numerator and denominator have the same sign, and negative if they have opposite sign. It is easiest to organize this sign information in a chart:

x		1		7/5	→
$7-5x$	+	+	+	0	−
$x-1$	−	0	+	+	+
$(7-5x)/(x-1)$	−	undef	+	0	−

Thus the solution set of the given inequality is the interval $(1, 7/5]$. See Figure P.4 for graphs of the solutions.

The symbol $\iff$ means "if and only if" or "is equivalent to." If A and B are two statements, then $A \iff B$ means that the truth of either statement implies the truth of the other, so either both must be true or both must be false.

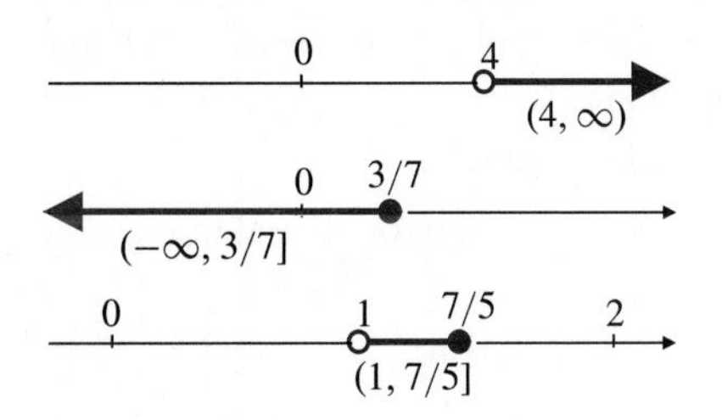

Figure P.4 The intervals for Example 2

Sometimes we will need to solve systems of two or more inequalities that must be satisfied simultaneously. We still solve the inequalities individually and look for numbers in the intersection of the solution sets.

EXAMPLE 3 Solve the systems of inequalities:
(a) $3 \le 2x + 1 \le 5$ (b) $3x - 1 < 5x + 3 \le 2x + 15$.

Solution

(a) Using the technique of Example 2, we can solve the inequality $3 \le 2x + 1$ to get $2 \le 2x$, so $x \ge 1$. Similarly, the inequality $2x + 1 \le 5$ leads to $2x \le 4$, so $x \le 2$. The solution set of system (a) is therefore the closed interval $[1, 2]$.

(b) We solve both inequalities as follows:

$$\left.\begin{aligned} 3x - 1 &< 5x + 3 \\ -1 - 3 &< 5x - 3x \\ -4 &< 2x \\ -2 &< x \end{aligned}\right\} \quad \text{and} \quad \left\{\begin{aligned} 5x + 3 &\le 2x + 15 \\ 5x - 2x &\le 15 - 3 \\ 3x &\le 12 \\ x &\le 4 \end{aligned}\right.$$

The solution set is the interval $(-2, 4]$.

Solving quadratic inequalities depends on solving the corresponding quadratic equations.

EXAMPLE 4 **Quadratic inequalities**
Solve: (a) $x^2 - 5x + 6 < 0$ (b) $2x^2 + 1 > 4x$.

Solution

(a) The trinomial $x^2 - 5x + 6$ factors into the product $(x-2)(x-3)$, which is negative if and only if exactly one of the factors is negative. Since $x - 3 < x - 2$, this happens when $x - 3 < 0$ and $x - 2 > 0$. Thus we need $x < 3$ and $x > 2$; the solution set is the open interval $(2, 3)$.

(b) The inequality $2x^2 + 1 > 4x$ is equivalent to $2x^2 - 4x + 1 > 0$. The corresponding quadratic equation $2x^2 - 4x + 1 = 0$, which is of the form $Ax^2 + Bx + C = 0$, can be solved by the quadratic formula (see Section P.6):

$$x = \frac{-B \pm \sqrt{B^2 - 4AC}}{2A} = \frac{4 \pm \sqrt{16 - 8}}{4} = 1 \pm \frac{\sqrt{2}}{2},$$

so the given inequality can be expressed in the form

$$\left(x - 1 + \tfrac{1}{2}\sqrt{2}\right)\left(x - 1 - \tfrac{1}{2}\sqrt{2}\right) > 0.$$

This is satisfied if both factors on the left side are positive or if both are negative. Therefore, we require that either $x < 1 - \frac{1}{2}\sqrt{2}$ or $x > 1 + \frac{1}{2}\sqrt{2}$. The solution set is the *union* of intervals $\left(-\infty, 1 - \frac{1}{2}\sqrt{2}\right) \cup \left(1 + \frac{1}{2}\sqrt{2}, \infty\right)$.

Note the use of the symbol $\cup$ to denote the **union** of intervals. A real number is in the union of intervals if it is in at least one of the intervals. We will also need to consider the **intersection** of intervals from time to time. A real number belongs to the intersection of intervals if it belongs to *every one* of the intervals. We will use $\cap$ to denote intersection. For example,

$$[1, 3) \cap [2, 4] = [2, 3) \quad \text{while} \quad [1, 3) \cup [2, 4] = [1, 4].$$

EXAMPLE 5 Solve the inequality $\dfrac{3}{x-1} < -\dfrac{2}{x}$ and graph the solution set.

Solution We would like to multiply by $x(x-1)$ to clear the inequality of fractions, but this would require considering three cases separately. (What are they?) Instead, we will transpose and combine the two fractions into a single one:

$$\frac{3}{x-1} < -\frac{2}{x} \iff \frac{3}{x-1} + \frac{2}{x} < 0 \iff \frac{5x-2}{x(x-1)} < 0.$$

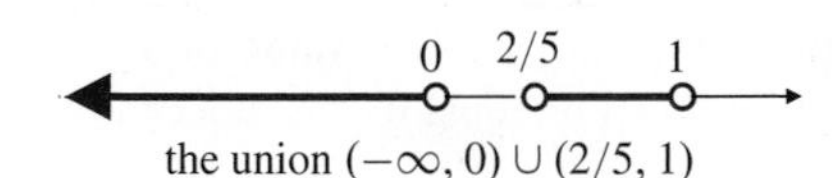

Figure P.5 The solution set for Example 5

We examine the signs of the three factors in the left fraction to determine where that fraction is negative:

x		0		2/5		1	
$5x - 2$	−	−	−	0	+	+	+
x	−	0	+	+	+	+	+
$x - 1$	−	−	−	−	−	0	+
$\dfrac{5x-2}{x(x-1)}$	−	undef	+	0	−	undef	+

The solution set of the given inequality is the union of these two intervals, namely, $(-\infty, 0) \cup (2/5, 1)$. See Figure P.5.

The Absolute Value

The **absolute value**, or **magnitude**, of a number x, denoted $|x|$ (read "the absolute value of x"), is defined by the formula

$$|x| = \begin{cases} x & \text{if } x \geq 0 \\ -x & \text{if } x < 0 \end{cases}$$

The vertical lines in the symbol $|x|$ are called **absolute value bars**.

EXAMPLE 6 $|3| = 3, \quad |0| = 0, \quad |-5| = 5.$

Note that $|x| \geq 0$ for every real number x, and $|x| = 0$ only if $x = 0$. People sometimes find it confusing to say that $|x| = -x$ when x is negative, but this is correct since $-x$ is positive in that case. The symbol $\sqrt{a}$ always denotes the *nonnegative* square root of a, so an alternative definition of $|x|$ is $|x| = \sqrt{x^2}$.

It is important to remember that $\sqrt{a^2} = |a|$. Do not write $\sqrt{a^2} = a$ unless you already know that $a \geq 0$.

Geometrically, $|x|$ represents the (nonnegative) distance from x to 0 on the real line. More generally, $|x - y|$ represents the (nonnegative) distance between the points x and y on the real line, since this distance is the same as that from the point $x - y$ to 0 (see Figure P.6):

$$|x - y| = \begin{cases} x - y, & \text{if } x \geq y \\ y - x, & \text{if } x < y. \end{cases}$$

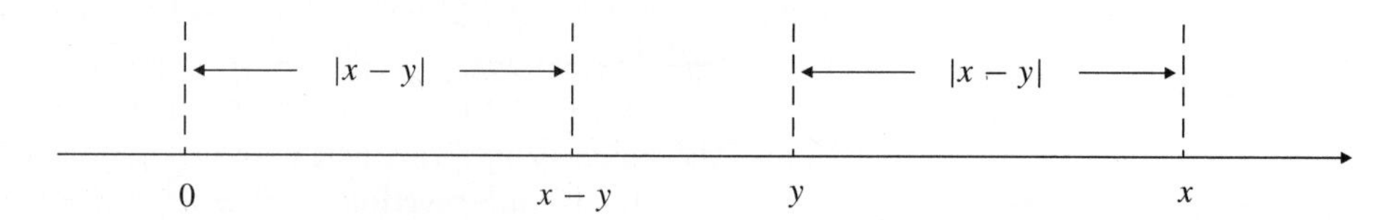

Figure P.6
$|x - y|$ = distance from x to y

The absolute value function has the following properties:

Properties of absolute values

1. $|-a| = |a|$. A number and its negative have the same absolute value.
2. $|ab| = |a||b|$ and $\left|\frac{a}{b}\right| = \frac{|a|}{|b|}$. The absolute value of a product (or quotient) of two numbers is the product (or quotient) of their absolute values.
3. $|a \pm b| \leq |a| + |b|$ (the **triangle inequality**). The absolute value of a sum of or difference between numbers is less than or equal to the sum of their absolute values.

The first two of these properties can be checked by considering the cases where either of a or b is either positive or negative. The third property follows from the first two because $\pm 2ab \leq |2ab| = 2|a||b|$. Therefore, we have

$$\begin{aligned} |a \pm b|^2 = (a \pm b)^2 &= a^2 \pm 2ab + b^2 \\ &\leq |a|^2 + 2|a||b| + |b|^2 = (|a| + |b|)^2, \end{aligned}$$

and taking the (positive) square roots of both sides, we obtain $|a \pm b| \leq |a| + |b|$. This result is called the "triangle inequality" because it follows from the geometric fact that the length of any side of a triangle cannot exceed the sum of the lengths of the other two sides. For instance, if we regard the points 0, a, and b on the number line as the vertices of a degenerate "triangle," then the sides of the triangle have lengths $|a|$, $|b|$, and $|a - b|$. The triangle is degenerate since all three of its vertices lie on a straight line.

Equations and Inequalities Involving Absolute Values

The equation $|x| = D$ (where $D > 0$) has two solutions, $x = D$ and $x = -D$: the two points on the real line that lie at distance D from the origin. Equations and inequalities involving absolute values can be solved algebraically by breaking them into cases according to the definition of absolute value, but often they can also be solved geometrically by interpreting absolute values as distances. For example, the inequality $|x - a| < D$ says that the distance from x to a is less than D, so x must lie between $a - D$ and $a + D$. (Or, equivalently, a must lie between $x - D$ and $x + D$.) If D is a positive number, then

$$\begin{aligned}
|x| = D &\iff \text{either } x = -D \text{ or } x = D\\
|x| < D &\iff -D < x < D\\
|x| \le D &\iff -D \le x \le D\\
|x| > D &\iff \text{either } x < -D \text{ or } x > D
\end{aligned}$$

More generally,

$$\begin{aligned}
|x - a| = D &\iff \text{either } x = a - D \text{ or } x = a + D\\
|x - a| < D &\iff a - D < x < a + D\\
|x - a| \le D &\iff a - D \le x \le a + D\\
|x - a| > D &\iff \text{either } x < a - D \text{ or } x > a + D
\end{aligned}$$

EXAMPLE 7 Solve: (a) $|2x + 5| = 3$ (b) $|3x - 2| \le 1$.

Solution

(a) $|2x + 5| = 3 \iff 2x + 5 = \pm 3$. Thus, either $2x = -3 - 5 = -8$ or $2x = 3 - 5 = -2$. The solutions are $x = -4$ and $x = -1$.

(b) $|3x - 2| \le 1 \iff -1 \le 3x - 2 \le 1$. We solve this pair of inequalities:

$$\left\{\begin{array}{c} -1 \le 3x - 2\\ -1 + 2 \le 3x\\ 1/3 \le x \end{array}\right\} \quad \text{and} \quad \left\{\begin{array}{r} 3x - 2 \le 1\\ 3x \le 1 + 2\\ x \le 1 \end{array}\right\}.$$

Thus the solutions lie in the interval $[1/3, 1]$.

Remark Here is how part (b) of Example 7 could have been solved geometrically, by interpreting the absolute value as a distance:

$$|3x - 2| = \left|3\left(x - \frac{2}{3}\right)\right| = 3\left|x - \frac{2}{3}\right|.$$

Thus, the given inequality says that

$$3\left|x - \frac{2}{3}\right| \le 1 \quad \text{or} \quad \left|x - \frac{2}{3}\right| \le \frac{1}{3}.$$

This says that the distance from x to $2/3$ does not exceed $1/3$. The solutions for x therefore lie between $1/3$ and 1, including both of these endpoints. (See Figure P.7.)

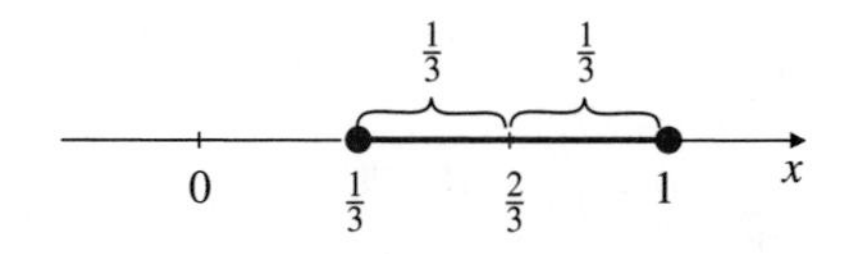

Figure P.7 The solution set for Example 7(b)

EXAMPLE 8 Solve the equation $|x+1| = |x-3|$.

Solution The equation says that x is equidistant from -1 and 3. Therefore, x is the point halfway between -1 and 3; $x = (-1+3)/2 = 1$. Alternatively, the given equation says that either $x+1 = x-3$ or $x+1 = -(x-3)$. The first of these equations has no solutions; the second has the solution $x = 1$.

EXAMPLE 9 What values of x satisfy the inequality $\left|5 - \frac{2}{x}\right| < 3$?

Solution We have

$$\begin{aligned} \left|5 - \frac{2}{x}\right| < 3 \quad \iff \quad & -3 < 5 - \frac{2}{x} < 3 && \text{Subtract 5 from each member.} \\ & -8 < -\frac{2}{x} < -2 && \text{Divide each member by } -2. \\ & 4 > \frac{1}{x} > 1 && \text{Take reciprocals.} \\ & \frac{1}{4} < x < 1. \end{aligned}$$

In this calculation we manipulated a system of two inequalities simultaneously, rather than split it up into separate inequalities as we have done in previous examples. Note how the various rules for inequalities were used here. Multiplying an inequality by a negative number reverses the inequality. So does taking reciprocals of an inequality in which both sides are positive. The given inequality holds for all x in the open interval $(1/4, 1)$.

EXERCISES P.1

In Exercises 1–2, express the given rational number as a repeating decimal. Use a bar to indicate the repeating digits.

1. $\frac{2}{9}$ **2.** $\frac{1}{11}$

In Exercises 3–4, express the given repeating decimal as a quotient of integers in lowest terms.

3. $0.\overline{12}$ **4.** $3.2\overline{7}$

5. Express the rational numbers $1/7$, $2/7$, $3/7$, and $4/7$ as repeating decimals. (Use a calculator to give as many decimal digits as possible.) Do you see a pattern? Guess the decimal expansions of $5/7$ and $6/7$ and check your guesses.

6. Can two different decimals represent the same number? What number is represented by $0.999\ldots = 0.\overline{9}$?

In Exercises 7–12, express the set of all real numbers x satisfying the given conditions as an interval or a union of intervals.

7. $x \ge 0$ and $x \le 5$ **8.** $x < 2$ and $x \ge -3$

9. $x > -5$ or $x < -6$ **10.** $x \le -1$

11. $x > -2$ **12.** $x < 4$ or $x \ge 2$

In Exercises 13–26, solve the given inequality, giving the solution set as an interval or union of intervals.

13. $-2x > 4$ **14.** $3x + 5 \le 8$

15. $5x - 3 \le 7 - 3x$ **16.** $\frac{6-x}{4} \ge \frac{3x-4}{2}$

17. $3(2-x) < 2(3+x)$ **18.** $x^2 < 9$

19. $\frac{1}{2-x} < 3$ **20.** $\frac{x+1}{x} \ge 2$

21. $x^2 - 2x \le 0$ **22.** $6x^2 - 5x \le -1$

23. $x^3 > 4x$ **24.** $x^2 - x \le 2$

25. $\frac{x}{2} \ge 1 + \frac{4}{x}$ **26.** $\frac{3}{x-1} < \frac{2}{x+1}$

Solve the equations in Exercises 27–32.

27. $|x| = 3$ **28.** $|x-3| = 7$

29. $|2t+5| = 4$ **30.** $|1-t| = 1$

31. $|8-3s| = 9$ **32.** $\left|\frac{s}{2} - 1\right| = 1$

In Exercises 33–40, write the interval defined by the given inequality.

33. $|x| < 2$ **34.** $|x| \le 2$

35. $|s-1| \le 2$

36. $|t+2| < 1$

37. $|3x-7| < 2$

38. $|2x+5| < 1$

39. $\left|\frac{x}{2} - 1\right| \le 1$

40. $\left|2 - \frac{x}{2}\right| < \frac{1}{2}$

In Exercises 41–42, solve the given inequality by interpreting it as a statement about distances on the real line.

41. $|x+1| > |x-3|$

42. $|x-3| < 2|x|$

43. Do not fall into the trap $|-a| = a$. For what real numbers a is this equation true? For what numbers is it false?

44. Solve the equation $|x-1| = 1-x$.

45. Show that the inequality

$$|a-b| \ge \big||a| - |b|\big|$$

holds for all real numbers a and b.

P.2 Cartesian Coordinates in the Plane

The positions of all points in a plane can be measured with respect to two perpendicular real lines in the plane intersecting at the 0-point of each. These lines are called **coordinate axes** in the plane. Usually (but not always) we call one of these axes the x-axis and draw it horizontally with numbers x on it increasing to the right; then we call the other the y-axis, and draw it vertically with numbers y on it increasing upward. The point of intersection of the coordinate axes (the point where x and y are both zero) is called the **origin** and is often denoted by the letter O.

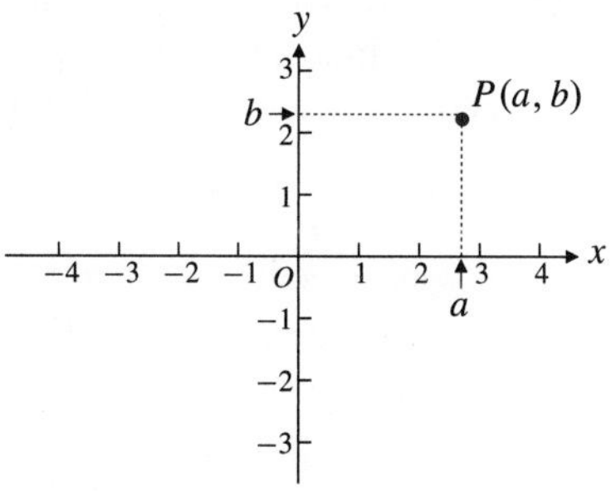

If P is any point in the plane, we can draw a line through P perpendicular to the x-axis. If a is the value of x where that line intersects the x-axis, we call a the **x-coordinate** of P. Similarly, the **y-coordinate** of P is the value of y where a line through P perpendicular to the y-axis meets the y-axis. The **ordered pair** (a, b) is called the **coordinate pair**, or the **Cartesian coordinates**, of the point P. We refer to the point as $P(a, b)$ to indicate both the name P of the point and its coordinates (a, b). (See Figure P.8.) Note that the x-coordinate appears first in a coordinate pair. Coordinate pairs are in one-to-one correspondence with points in the plane; each point has a unique coordinate pair, and each coordinate pair determines a unique point. We call such a set of coordinate axes and the coordinate pairs they determine a **Cartesian coordinate system** in the plane, after the seventeenth-century philosopher René Descartes, who created analytic (coordinate) geometry. When equipped with such a coordinate system, a plane is called a **Cartesian plane**. Note that we are using the same notation (a, b) for the Cartesian coordinates of a point in the plane as we use for an open interval on the real line. However, this should not cause any confusion because the intended meaning will be clear from the context.

Figure P.8 The coordinate axes and the point P with coordinates (a, b)

Figure P.9 shows the coordinates of some points in the plane. Note that all points on the x-axis have y-coordinate 0. We usually just write the x-coordinates to label such points. Similarly, points on the y-axis have $x = 0$, and we can label such points using their y-coordinates only.

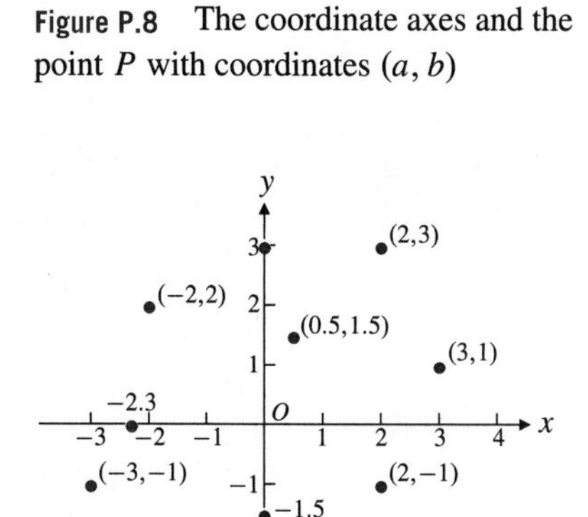

Figure P.9 Some points with their coordinates

The coordinate axes divide the plane into four regions called **quadrants**. These quadrants are numbered I to IV, as shown in Figure P.10. The **first quadrant** is the upper right one; both coordinates of any point in that quadrant are positive numbers. Both coordinates are negative in quadrant III; only y is positive in quadrant II; only x is positive in quadrant IV.

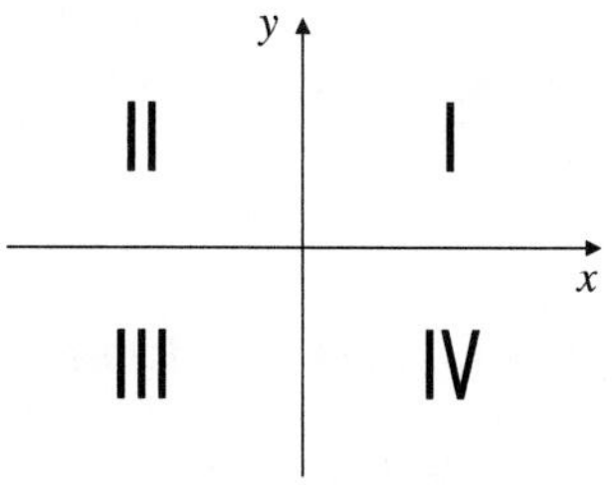

Figure P.10 The four quadrants

Axis Scales

When we plot data in the coordinate plane or graph formulas whose variables have different units of measure, we do not need to use the same scale on the two axes. If, for example, we plot height versus time for a falling rock, there is no reason to place the mark that shows 1 m on the height axis the same distance from the origin as the mark that shows 1 s on the time axis.

When we graph functions whose variables do not represent physical measurements and when we draw figures in the coordinate plane to study their geometry or trigonometry, we usually make the scales identical. A vertical unit of distance then looks the same as a horizontal unit. As on a surveyor's map or a scale drawing, line segments that are supposed to have the same length will look as if they do, and angles that are supposed to be equal will look equal. Some of the geometric results we obtain later, such as the relationship between the slopes of perpendicular lines, are valid only if equal scales are used on the two axes.

Computer and calculator displays are another matter. The vertical and horizontal scales on machine-generated graphs usually differ, with resulting distortions in distances, slopes, and angles. Circles may appear elliptical, and squares may appear rectangular or even as parallelograms. Right angles may appear as acute or obtuse. Circumstances like these require us to take extra care in interpreting what we see. High-quality computer software for drawing Cartesian graphs usually allows the user to compensate for such scale problems by adjusting the *aspect ratio* (the ratio of vertical to horizontal scale). Some computer screens also allow adjustment within a narrow range. When using graphing software, try to adjust your particular software/hardware configuration so that the horizontal and vertical diameters of a drawn circle appear to be equal.

Increments and Distances

When a particle moves from one point to another, the net changes in its coordinates are called increments. They are calculated by subtracting the coordinates of the starting point from the coordinates of the ending point. An **increment** in a variable is the net change in the value of the variable. If x changes from x_1 to x_2, then the increment in x is $\Delta x = x_2 - x_1$.

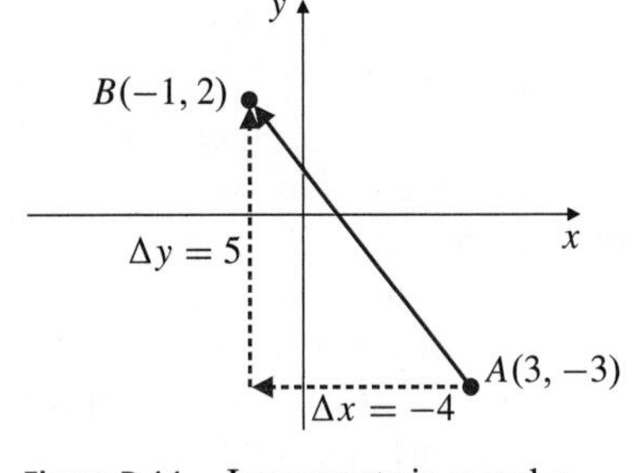

Figure P.11 Increments in x and y

EXAMPLE 1 Find the increments in the coordinates of a particle that moves from $A(3, -3)$ to $B(-1, 2)$.

Solution The increments (see Figure P.11) are:

$$\Delta x = -1 - 3 = -4 \qquad \text{and} \qquad \Delta y = 2 - (-3) = 5.$$

If $P(x_1, y_1)$ and $Q(x_2, y_2)$ are two points in the plane, the straight line segment PQ is the hypotenuse of a right triangle PCQ, as shown in Figure P.12. The sides PC and CQ of the triangle have lengths

$$|\Delta x| = |x_2 - x_1| \qquad \text{and} \qquad |\Delta y| = |y_2 - y_1|.$$

These are the *horizontal distance* and *vertical distance* between P and Q. By the Pythagorean Theorem, the length of PQ is the square root of the sum of the squares of these lengths.

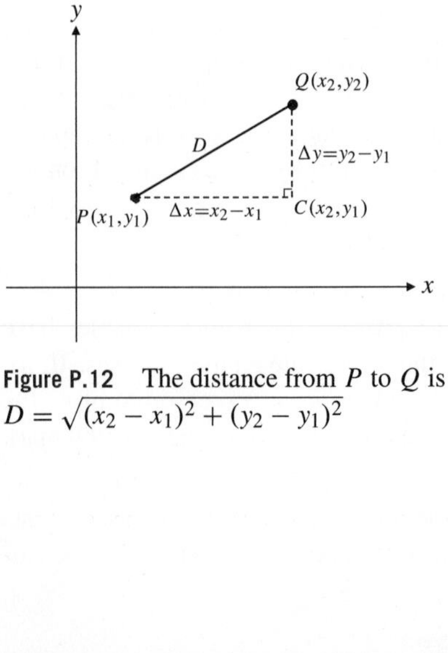

Figure P.12 The distance from P to Q is $D = \sqrt{(x_2 - x_1)^2 + (y_2 - y_1)^2}$

Distance formula for points in the plane

The distance D between $P(x_1, y_1)$ and $Q(x_2, y_2)$ is

$$D = \sqrt{(\Delta x)^2 + (\Delta y)^2} = \sqrt{(x_2 - x_1)^2 + (y_2 - y_1)^2}.$$

EXAMPLE 2 The distance between $A(3, -3)$ and $B(-1, 2)$ in Figure P.11 is

$$\sqrt{(-1 - 3)^2 + (2 - (-3))^2} = \sqrt{(-4)^2 + 5^2} = \sqrt{41} \text{ units.}$$

EXAMPLE 3 The distance from the origin $O(0, 0)$ to a point $P(x, y)$ is

$$\sqrt{(x-0)^2 + (y-0)^2} = \sqrt{x^2 + y^2}.$$

Graphs

The **graph** of an equation (or inequality) involving the variables x and y is the set of all points $P(x, y)$ whose coordinates satisfy the equation (or inequality).

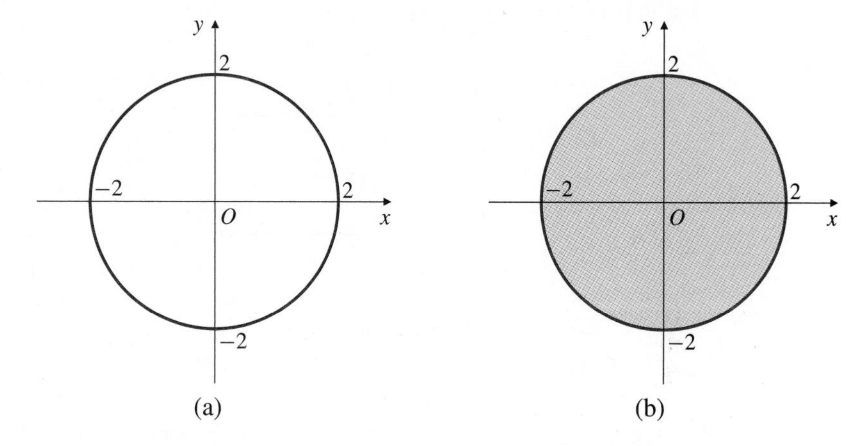

Figure P.13
(a) The circle $x^2 + y^2 = 4$
(b) The disk $x^2 + y^2 \le 4$

EXAMPLE 4 The equation $x^2 + y^2 = 4$ represents all points $P(x, y)$ whose distance from the origin is $\sqrt{x^2 + y^2} = \sqrt{4} = 2$. These points lie on the **circle** of radius 2 centred at the origin. This circle is the graph of the equation $x^2 + y^2 = 4$. (See Figure P.13(a).)

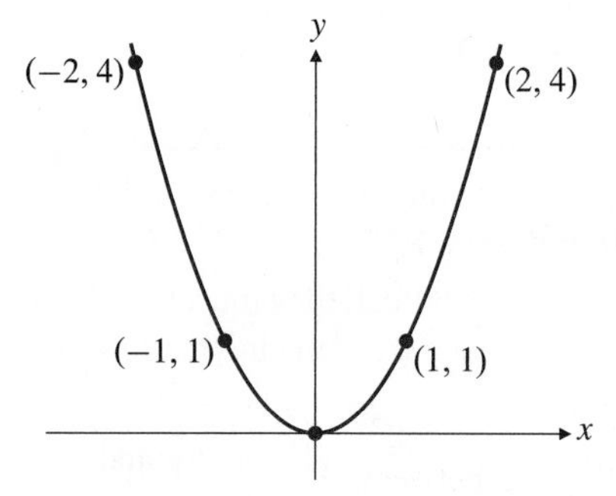

Figure P.14 The parabola $y = x^2$

EXAMPLE 5 Points (x, y) whose coordinates satisfy the inequality $x^2 + y^2 \le 4$ all have distance ≤ 2 from the origin. The graph of the inequality is therefore the disk of radius 2 centred at the origin. (See Figure P.13(b).)

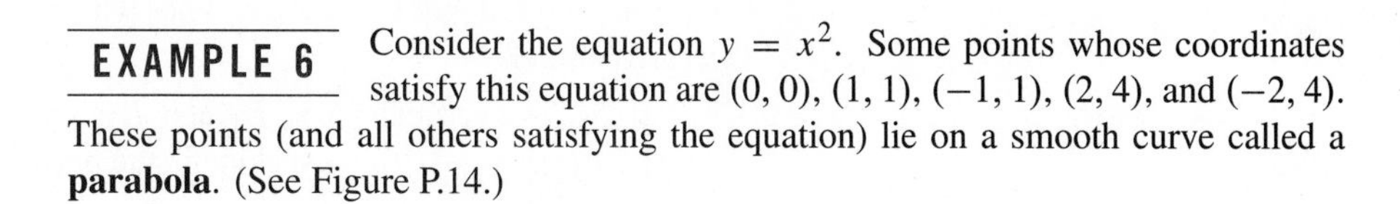

EXAMPLE 6 Consider the equation $y = x^2$. Some points whose coordinates satisfy this equation are $(0, 0)$, $(1, 1)$, $(-1, 1)$, $(2, 4)$, and $(-2, 4)$. These points (and all others satisfying the equation) lie on a smooth curve called a **parabola**. (See Figure P.14.)

Straight Lines

Given two points $P_1(x_1, y_1)$ and $P_2(x_2, y_2)$ in the plane, we call the increments $\Delta x = x_2 - x_1$ and $\Delta y = y_2 - y_1$, respectively, the **run** and the **rise** between P_1 and P_2. Two such points always determine a unique **straight line** (usually called simply a **line**) passing through them both. We call the line P_1P_2.

Any nonvertical line in the plane has the property that the ratio

$$m = \frac{\text{rise}}{\text{run}} = \frac{\Delta y}{\Delta x} = \frac{y_2 - y_1}{x_2 - x_1}$$

has the *same value* for every choice of two distinct points $P_1(x_1, y_1)$ and $P_2(x_2, y_2)$ on the line. (See Figure P.15.) The constant $m = \Delta y/\Delta x$ is called the **slope** of the nonvertical line.

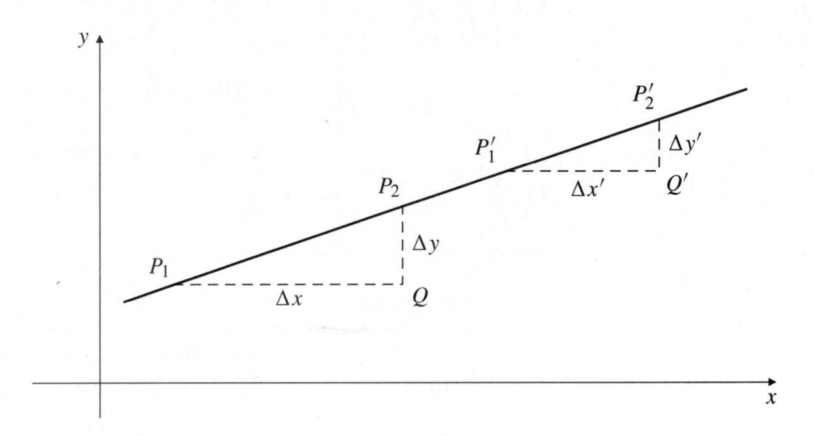

Figure P.15 $\Delta y/\Delta x = \Delta y'/\Delta x'$ because triangles P_1QP_2 and $P_1'Q'P_2'$ are similar

EXAMPLE 7 The slope of the line joining $A\,(3,-3)$ and $B\,(-1,2)$ is

$$m = \frac{\Delta y}{\Delta x} = \frac{2-(-3)}{-1-3} = \frac{5}{-4} = -\frac{5}{4}.$$

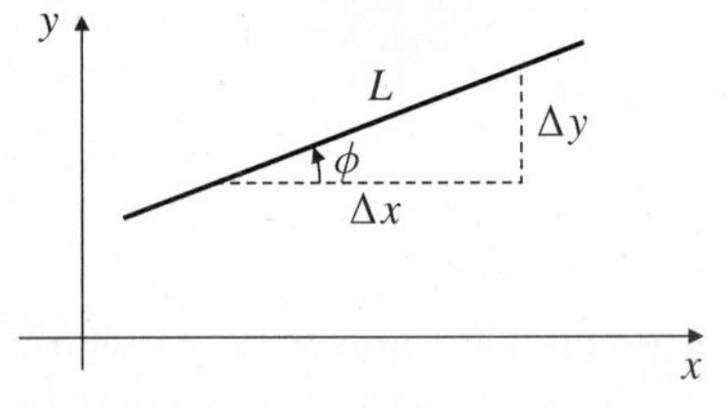

Figure P.16 Line L has inclination ϕ

The slope tells us the direction and steepness of a line. A line with positive slope rises uphill to the right; one with negative slope falls downhill to the right. The greater the absolute value of the slope, the steeper the rise or fall. Since the run Δx is zero for a vertical line, we cannot form the ratio m; the slope of a vertical line is *undefined.*

The direction of a line can also be measured by an angle. The **inclination** of a line is the smallest counterclockwise angle from the positive direction of the x-axis to the line. In Figure P.16 the angle ϕ (the Greek letter "phi") is the inclination of the line L. The inclination ϕ of any line satisfies $0° \le \phi < 180°$. The inclination of a horizontal line is 0° and that of a vertical line is 90°.

Provided equal scales are used on the coordinate axes, the relationship between the slope m of a nonvertical line and its inclination ϕ is shown in Figure P.16:

$$m = \frac{\Delta y}{\Delta x} = \tan\phi.$$

(The trigonometric function tan is defined in Section P.7.)

Parallel lines have the same inclination. If they are not vertical, they must therefore have the same slope. Conversely, lines with equal slopes have the same inclination and so are parallel.

If two nonvertical lines, L_1 and L_2, are perpendicular, their slopes m_1 and m_2 satisfy $m_1m_2 = -1$, so each slope is the *negative reciprocal* of the other:

$$m_1 = -\frac{1}{m_2} \quad \text{and} \quad m_2 = -\frac{1}{m_1}.$$

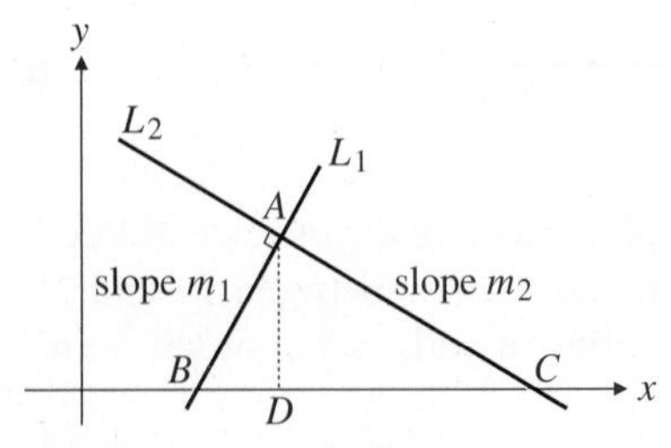

Figure P.17 $\triangle ABD$ is similar to $\triangle CAD$

(This result also assumes equal scales on the two coordinate axes.) To see this, observe in Figure P.17 that

$$m_1 = \frac{AD}{BD} \quad \text{and} \quad m_2 = -\frac{AD}{DC}.$$

Since $\triangle ABD$ is similar to $\triangle CAD$, we have $\dfrac{AD}{BD} = \dfrac{DC}{AD}$, and so

$$m_1m_2 = \left(\frac{DC}{AD}\right)\left(-\frac{AD}{DC}\right) = -1.$$

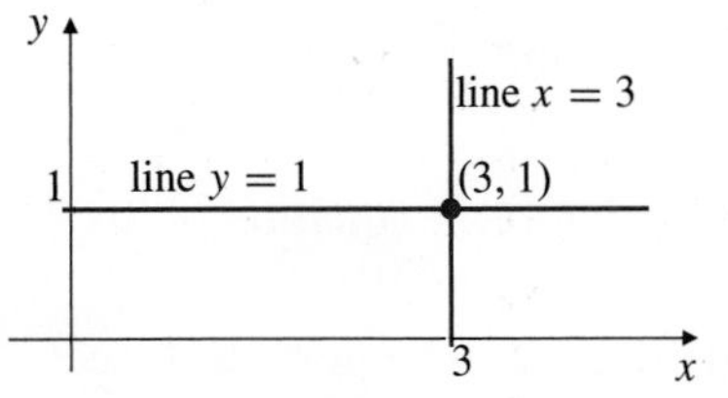

Figure P.18 The lines $y = 1$ and $x = 3$

Equations of Lines

Straight lines are particularly simple graphs, and their corresponding equations are also simple. All points on the vertical line through the point a on the x-axis have their x-coordinates equal to a. Thus $x = a$ is the equation of the line. Similarly, $y = b$ is the equation of the horizontal line meeting the y-axis at b.

EXAMPLE 8 The horizontal and vertical lines passing through the point $(3, 1)$ (Figure P.18) have equations $y = 1$ and $x = 3$, respectively.

To write an equation for a nonvertical straight line L, it is enough to know its slope m and the coordinates of one point $P_1(x_1, y_1)$ on it. If $P(x, y)$ is any other point on L, then

$$\frac{y - y_1}{x - x_1} = m,$$

so that

$$y - y_1 = m(x - x_1) \qquad \text{or} \qquad y = m(x - x_1) + y_1.$$

The equation

$$y = m(x - x_1) + y_1$$

is the **point-slope equation** of the line that passes through the point (x_1, y_1) and has slope m.

EXAMPLE 9 Find an equation of the line that has slope -2 and passes through the point $(1, 4)$.

Solution We substitute $x_1 = 1$, $y_1 = 4$, and $m = -2$ into the point-slope form of the equation and obtain

$$y = -2(x - 1) + 4 \qquad \text{or} \qquad y = -2x + 6.$$

EXAMPLE 10 Find an equation of the line through the points $(1, -1)$ and $(3, 5)$.

Solution The slope of the line is $m = \dfrac{5 - (-1)}{3 - 1} = 3$. We can use this slope with either of the two points to write an equation of the line. If we use $(1, -1)$ we get

$$y = 3(x - 1) - 1, \qquad \text{which simplifies to} \quad y = 3x - 4.$$

If we use $(3, 5)$ we get

$$y = 3(x - 3) + 5, \qquad \text{which also simplifies to} \quad y = 3x - 4.$$

Either way, $y = 3x - 4$ is an equation of the line.

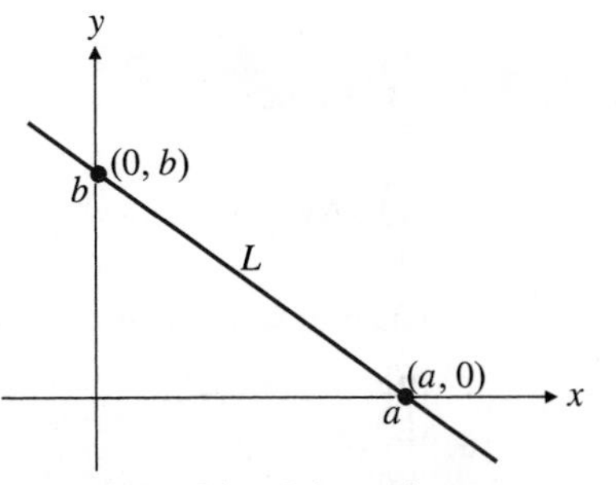

Figure P.19 Line L has x-intercept a and y-intercept b

The y-coordinate of the point where a nonvertical line intersects the y-axis is called the **y-intercept** of the line. (See Figure P.19.) Similarly, the **x-intercept** of a nonhorizontal line is the x-coordinate of the point where it crosses the x-axis. A line with slope m and y-intercept b passes through the point $(0, b)$, so its equation is

$$y = m(x - 0) + b \qquad \text{or, more simply,} \quad y = mx + b.$$

A line with slope m and x-intercept a passes through $(a, 0)$, and so its equation is

$$y = m(x - a).$$

The equation $y = mx + b$ is called the **slope–y-intercept equation** of the line with slope m and y-intercept b.

The equation $y = m(x - a)$ is called the **slope–x-intercept equation** of the line with slope m and x-intercept a.

EXAMPLE 11 Find the slope and the two intercepts of the line with equation $8x + 5y = 20$.

Solution Solving the equation for y we get

$$y = \frac{20 - 8x}{5} = -\frac{8}{5}x + 4.$$

Comparing this with the general form $y = mx + b$ of the slope–y-intercept equation, we see that the slope of the line is $m = -8/5$, and the y-intercept is $b = 4$. To find the x-intercept, put $y = 0$ and solve for x, obtaining $8x = 20$, or $x = 5/2$. The x-intercept is $a = 5/2$.

The equation $Ax + By = C$ (where A and B are not both zero) is called the **general linear equation** in x and y because its graph always represents a straight line, and every line has an equation in this form.

Many important quantities are related by linear equations. Once we know that a relationship between two variables is linear, we can find it from any two pairs of corresponding values, just as we find the equation of a line from the coordinates of two points.

EXAMPLE 12 The relationship between Fahrenheit temperature (F) and Celsius temperature (C) is given by a linear equation of the form $F = mC + b$. The freezing point of water is $F = 32°$ or $C = 0°$, while the boiling point is $F = 212°$ or $C = 100°$. Thus

$$32 = 0m + b \qquad \text{and} \qquad 212 = 100m + b,$$

so $b = 32$ and $m = (212 - 32)/100 = 9/5$. The relationship is given by the linear equation

$$F = \frac{9}{5}C + 32 \qquad \text{or} \qquad C = \frac{5}{9}(F - 32).$$

EXERCISES P.2

In Exercises 1–4, a particle moves from A to B. Find the net increments Δx and Δy in the particle's coordinates. Also find the distance from A to B.

1. $A(0, 3)$, $B(4, 0)$

2. $A(-1, 2)$, $B(4, -10)$

3. $A(3, 2)$, $B(-1, -2)$

4. $A(0.5, 3)$, $B(2, 3)$

5. A particle starts at $A(-2, 3)$ and its coordinates change by $\Delta x = 4$ and $\Delta y = -7$. Find its new position.

6. A particle arrives at the point $(-2, -2)$ after its coordinates experience increments $\Delta x = -5$ and $\Delta y = 1$. From where did it start?

Describe the graphs of the equations and inequalities in Exercises 7–12.

7. $x^2 + y^2 = 1$

8. $x^2 + y^2 = 2$

9. $x^2 + y^2 \le 1$

10. $x^2 + y^2 = 0$

11. $y \geq x^2$ **12.** $y < x^2$

In Exercises 13–14, find an equation for (a) the vertical line and (b) the horizontal line through the given point.

13. $(-2, 5/3)$ **14.** $(\sqrt{2}, -1.3)$

In Exercises 15–18, write an equation for the line through P with slope m.

15. $P(-1, 1), \quad m = 1$ **16.** $P(-2, 2), \quad m = 1/2$

17. $P(0, b), \quad m = 2$ **18.** $P(a, 0), \quad m = -2$

In Exercises 19–20, does the given point P lie on, above, or below the given line?

19. $P(2, 1), \quad 2x + 3y = 6$ **20.** $P(3, -1), \quad x - 4y = 7$

In Exercises 21–24, write an equation for the line through the two points.

21. $(0, 0), \quad (2, 3)$ **22.** $(-2, 1), \quad (2, -2)$

23. $(4, 1), \quad (-2, 3)$ **24.** $(-2, 0), \quad (0, 2)$

In Exercises 25–26, write an equation for the line with slope m and y-intercept b.

25. $m = -2, \quad b = \sqrt{2}$ **26.** $m = -1/2, \quad b = -3$

In Exercises 27–30, determine the x- and y-intercepts and the slope of the given lines, and sketch their graphs.

27. $3x + 4y = 12$ **28.** $x + 2y = -4$

29. $\sqrt{2}x - \sqrt{3}y = 2$ **30.** $1.5x - 2y = -3$

In Exercises 31–32, find equations for the lines through P that are (a) parallel to and (b) perpendicular to the given line.

31. $P(2, 1), \quad y = x + 2$ **32.** $P(-2, 2), \quad 2x + y = 4$

33. Find the point of intersection of the lines $3x + 4y = -6$ and $2x - 3y = 13$.

34. Find the point of intersection of the lines $2x + y = 8$ and $5x - 7y = 1$.

35. **(Two-intercept equations)** If a line is neither horizontal nor vertical and does not pass through the origin, show that its equation can be written in the form $\dfrac{x}{a} + \dfrac{y}{b} = 1$, where a is its x-intercept and b is its y-intercept.

36. Determine the intercepts and sketch the graph of the line $\dfrac{x}{2} - \dfrac{y}{3} = 1$.

37. Find the y-intercept of the line through the points $(2, 1)$ and $(3, -1)$.

38. A line passes through $(-2, 5)$ and $(k, 1)$ and has x-intercept 3. Find k.

39. The cost of printing x copies of a pamphlet is $\$C$, where $C = Ax + B$ for certain constants A and B. If it costs \$5,000 to print 10,000 copies and \$6,000 to print 15,000 copies, how much will it cost to print 100,000 copies?

40. **(Fahrenheit versus Celsius)** In the FC-plane, sketch the graph of the equation $C = \dfrac{5}{9}(F - 32)$ linking Fahrenheit and Celsius temperatures found in Example 12. On the same graph sketch the line with equation $C = F$. Is there a temperature at which a Celsius thermometer gives the same numerical reading as a Fahrenheit thermometer? If so, find that temperature.

Geometry

41. By calculating the lengths of its three sides, show that the triangle with vertices at the points $A(2, 1)$, $B(6, 4)$, and $C(5, -3)$ is isosceles.

42. Show that the triangle with vertices $A(0, 0)$, $B(1, \sqrt{3})$, and $C(2, 0)$ is equilateral.

43. Show that the points $A(2, -1)$, $B(1, 3)$, and $C(-3, 2)$ are three vertices of a square and find the fourth vertex.

44. Find the coordinates of the midpoint on the line segment P_1P_2 joining the points $P_1(x_1, y_1)$ and $P_2(x_2, y_2)$.

45. Find the coordinates of the point of the line segment joining the points $P_1(x_1, y_1)$ and $P_2(x_2, y_2)$ that is two-thirds of the way from P_1 to P_2.

46. The point P lies on the x-axis and the point Q lies on the line $y = -2x$. The point $(2, 1)$ is the midpoint of PQ. Find the coordinates of P.

In Exercises 47–48, interpret the equation as a statement about distances, and hence determine the graph of the equation.

47. $\sqrt{(x - 2)^2 + y^2} = 4$

48. $\sqrt{(x - 2)^2 + y^2} = \sqrt{x^2 + (y - 2)^2}$

49. For what value of k is the line $2x + ky = 3$ perpendicular to the line $4x + y = 1$? For what value of k are the lines parallel?

50. Find the line that passes through the point $(1, 2)$ and through the point of intersection of the two lines $x + 2y = 3$ and $2x - 3y = -1$.

P.3 Graphs of Quadratic Equations

This section reviews circles, parabolas, ellipses, and hyperbolas, the graphs that are represented by quadratic equations in two variables.

Circles and Disks

The **circle** having **centre** C and **radius** a is the set of all points in the plane that are at distance a from the point C.

The distance from $P(x, y)$ to the point $C(h, k)$ is $\sqrt{(x - h)^2 + (y - k)^2}$, so that the equation of the circle of radius $a > 0$ with centre at $C(h, k)$ is

$$\sqrt{(x-h)^2 + (y-k)^2} = a.$$

A simpler form of this equation is obtained by squaring both sides.

Standard equation of a circle

The circle with centre (h, k) and radius $a \ge 0$ has equation

$$(x-h)^2 + (y-k)^2 = a^2.$$

In particular, the circle with centre at the origin $(0, 0)$ and radius a has equation

$$x^2 + y^2 = a^2.$$

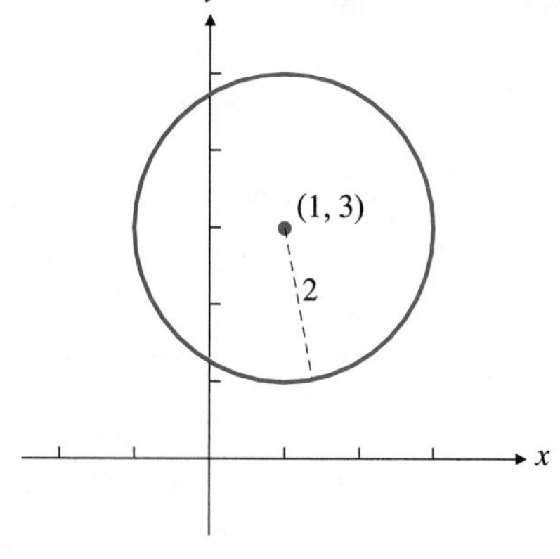

Figure P.20 Circle $(x-1)^2 + (y-3)^2 = 4$

EXAMPLE 1 The circle with radius 2 and centre $(1, 3)$ (Figure P.20) has equation $(x-1)^2 + (y-3)^2 = 4$.

EXAMPLE 2 The circle having equation $(x+2)^2 + (y-1)^2 = 7$ has centre at the point $(-2, 1)$ and radius $\sqrt{7}$. (See Figure P.21.)

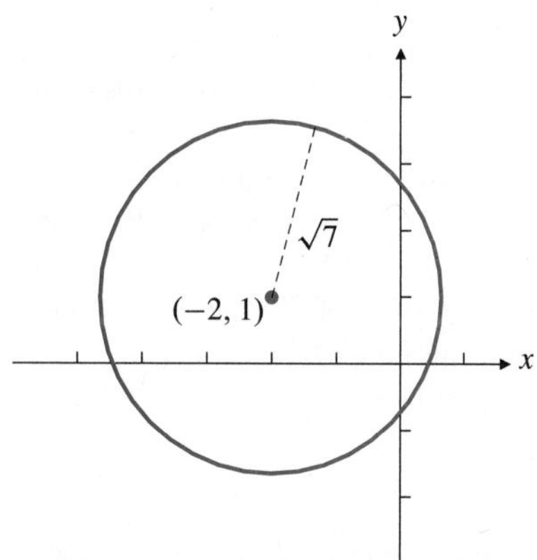

Figure P.21 Circle $(x+2)^2 + (y-1)^2 = 7$

If the squares in the standard equation $(x-h)^2 + (y-k)^2 = a^2$ are multiplied out, and all constant terms collected on the right-hand side, the equation becomes

$$x^2 - 2hx + y^2 - 2ky = a^2 - h^2 - k^2.$$

A quadratic equation of the form

$$x^2 + y^2 + 2ax + 2by = c$$

must represent a circle, which can be a single point if the radius is 0, or no points at all. To identify the graph, we complete the squares on the left side of the equation. Since $x^2 + 2ax$ are the first two terms of the square $(x+a)^2 = x^2 + 2ax + a^2$, we add a^2 to both sides to complete the square of the x terms. (Note that a^2 is *the square of half the coefficient of* x.) Similarly, add b^2 to both sides to complete the square of the y terms. The equation then becomes

$$(x+a)^2 + (y+b)^2 = c + a^2 + b^2.$$

If $c + a^2 + b^2 > 0$, the graph is a circle with centre $(-a, -b)$ and radius $\sqrt{c + a^2 + b^2}$. If $c + a^2 + b^2 = 0$, the graph consists of the single point $(-a, -b)$. If $c + a^2 + b^2 < 0$, no points lie on the graph.

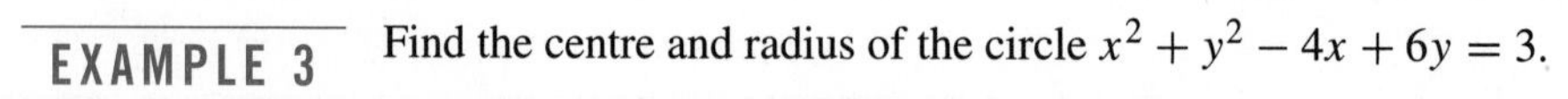

EXAMPLE 3 Find the centre and radius of the circle $x^2 + y^2 - 4x + 6y = 3$.

Solution Observe that $x^2 - 4x$ are the first two terms of the binomial square $(x-2)^2 = x^2 - 4x + 4$, and $y^2 + 6y$ are the first two terms of the square $(y+3)^2 = y^2 + 6y + 9$. Hence we add $4 + 9$ to both sides of the given equation and obtain

$$x^2 - 4x + 4 + y^2 + 6y + 9 = 3 + 4 + 9 \quad \text{or} \quad (x-2)^2 + (y+3)^2 = 16.$$

This is the equation of a circle with centre $(2, -3)$ and radius 4.

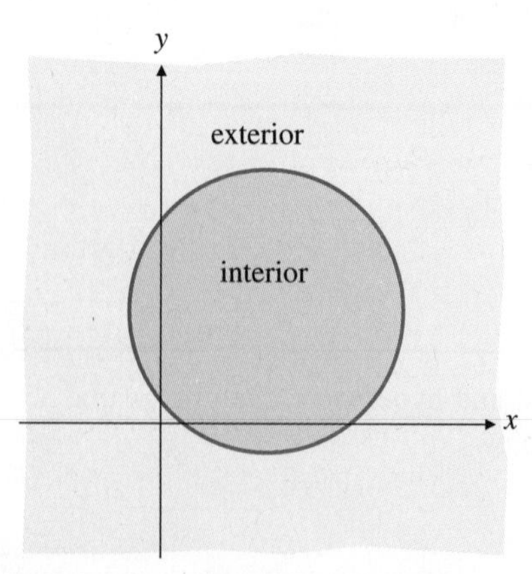

Figure P.22 The interior of a circle (darkly shaded) and the exterior (lightly shaded)

The set of all points *inside* a circle is called the **interior** of the circle; it is also called an **open disk**. The set of all points *outside* the circle is called the **exterior** of the circle. (See Figure P.22.) The interior of a circle together with the circle itself is called a **closed disk**, or simply a **disk**. The inequality

$$(x-h)^2 + (y-k)^2 \le a^2$$

represents the disk of radius $|a|$ centred at (h, k).

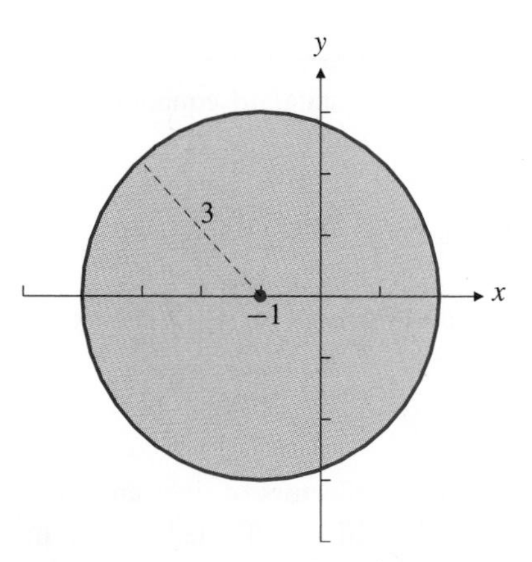

Figure P.23 The disk $x^2 + y^2 + 2x \le 8$

EXAMPLE 4 Identify the graphs of:

(a) $x^2 + 2x + y^2 \le 8$ (b) $x^2 + 2x + y^2 < 8$ (c) $x^2 + 2x + y^2 > 8$.

Solution We can complete the square in the equation $x^2 + y^2 + 2x = 8$ as follows:

$$x^2 + 2x + 1 + y^2 = 8 + 1$$
$$(x + 1)^2 + y^2 = 9.$$

Thus the equation represents the circle of radius 3 with centre at $(-1, 0)$. Inequality (a) represents the (closed) disk with the same radius and centre. (See Figure P.23.) Inequality (b) represents the interior of the circle (or the open disk). Inequality (c) represents the exterior of the circle.

Equations of Parabolas

A **parabola** is a plane curve whose points are equidistant from a fixed point F and a fixed straight line L that does not pass through F. The point F is the **focus** of the parabola; the line L is the parabola's **directrix**. The line through F perpendicular to L is the parabola's **axis**. The point V where the axis meets the parabola is the parabola's **vertex**.

Observe that the vertex V of a parabola is halfway between the focus F and the point on the directrix L that is closest to F. If the directrix is either horizontal or vertical, and the vertex is at the origin, then the parabola will have a particularly simple equation.

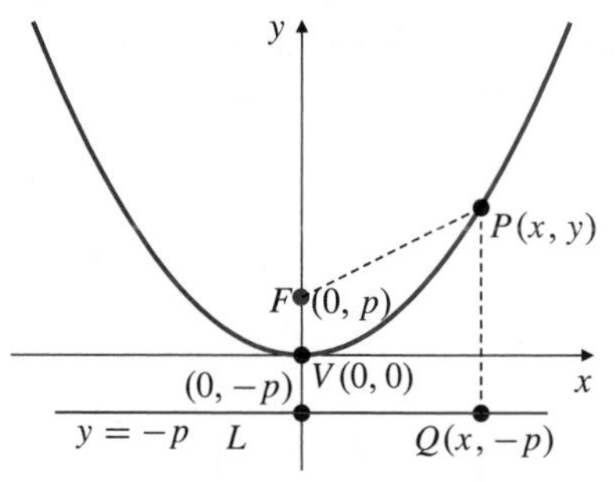

Figure P.24 The parabola $4py = x^2$ with focus $F(0, p)$ and directrix $y = -p$

EXAMPLE 5 Find an equation of the parabola having the point $F(0, p)$ as focus and the line L with equation $y = -p$ as directrix.

Solution If $P(x, y)$ is any point on the parabola, then (see Figure P.24) the distances from P to F and to (the closest point Q on) the line L are given by

$$PF = \sqrt{(x-0)^2 + (y-p)^2} = \sqrt{x^2 + y^2 - 2py + p^2}$$
$$PQ = \sqrt{(x-x)^2 + (y-(-p))^2} = \sqrt{y^2 + 2py + p^2}.$$

Since P is on the parabola, $PF = PQ$ and so the squares of these distances are also equal:

$$x^2 + y^2 - 2py + p^2 = y^2 + 2py + p^2,$$

or, after simplifying,

$$x^2 = 4py \qquad \text{or} \qquad y = \frac{x^2}{4p} \qquad \text{(called \textbf{standard forms})}.$$

Figure P.24 shows the situation for $p > 0$; the parabola opens upward and is symmetric about its axis, the y-axis. If $p < 0$, the focus $(0, p)$ will lie below the origin and the directrix $y = -p$ will lie above the origin. In this case the parabola will open downward instead of upward.

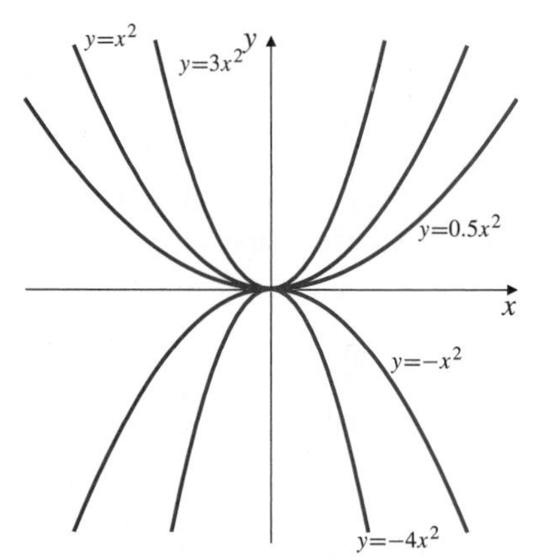

Figure P.25 Some parabolas $y = ax^2$

Figure P.25 shows several parabolas with equations of the form $y = ax^2$ for positive and negative values of a.

EXAMPLE 6 An equation for the parabola with focus $(0, 1)$ and directrix $y = -1$ is $y = x^2/4$, or $x^2 = 4y$. (We took $p = 1$ in the standard equation.)

EXAMPLE 7 Find the focus and directrix of the parabola $y = -x^2$.

Solution The given equation matches the standard form $y = x^2/(4p)$ provided $4p = -1$. Thus $p = -1/4$. The focus is $(0, -1/4)$, and the directrix is the line $y = 1/4$.

Interchanging the roles of x and y in the derivation of the standard equation above shows that the equation

$$y^2 = 4px \qquad \text{or} \qquad x = \frac{y^2}{4p} \qquad \text{(standard equation)}$$

represents a parabola with focus at $(p, 0)$ and vertical directrix $x = -p$. The axis is the x-axis.

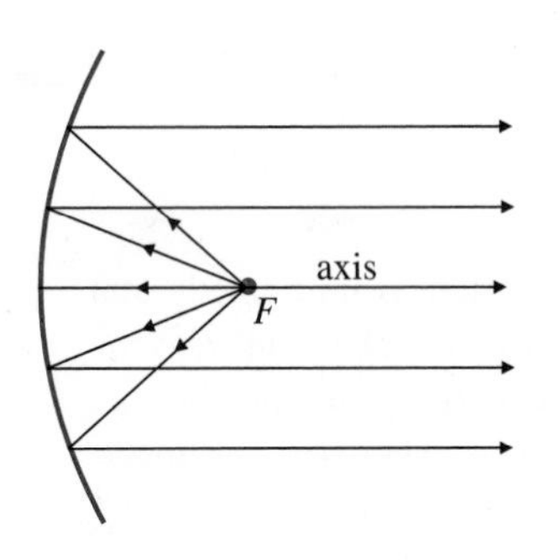

Figure P.26 Reflection by a parabola

Reflective Properties of Parabolas

One of the chief applications of parabolas is their use as reflectors of light and radio waves. Rays originating from the focus of a parabola will be reflected in a beam parallel to the axis, as shown in Figure P.26. Similarly, all the rays in a beam striking a parabola parallel to its axis will reflect through the focus. This property is the reason why telescopes and spotlights use parabolic mirrors and radio telescopes and microwave antennas are parabolic in shape. We will examine this property of parabolas more carefully in Section 8.1.

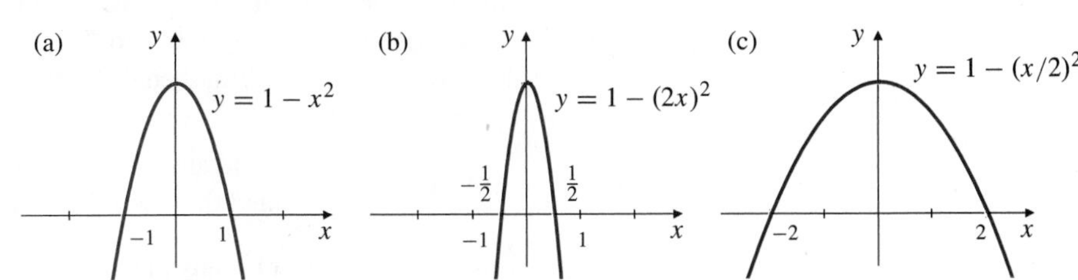

Figure P.27 Horizontal scaling:
(a) the graph $y = 1 - x^2$
(b) graph of (a) compressed horizontally
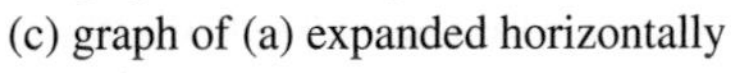
(c) graph of (a) expanded horizontally

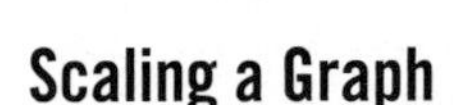

Scaling a Graph

The graph of an equation can be compressed or expanded horizontally by replacing x with a multiple of x. If a is a positive number, replacing x with ax in an equation multiplies horizontal distances in the graph of the equation by a factor $1/a$. (See Figure P.27.) Replacing y with ay will multiply vertical distances in a similar way.

You may find it surprising that, like circles, all parabolas are *similar* geometric figures; they may have different sizes, but they all have the same shape. We can change the *size* while preserving the shape of a curve represented by an equation in x and y by scaling both the coordinates by the same amount. If we scale the equation $4py = x^2$ by replacing x and y with $4px$ and $4py$, respectively, we get $4p(4py) = (4px)^2$, or $y = x^2$. Thus the general parabola $4py = x^2$ has the same shape as the specific parabola $y = x^2$, as shown in Figure P.28.

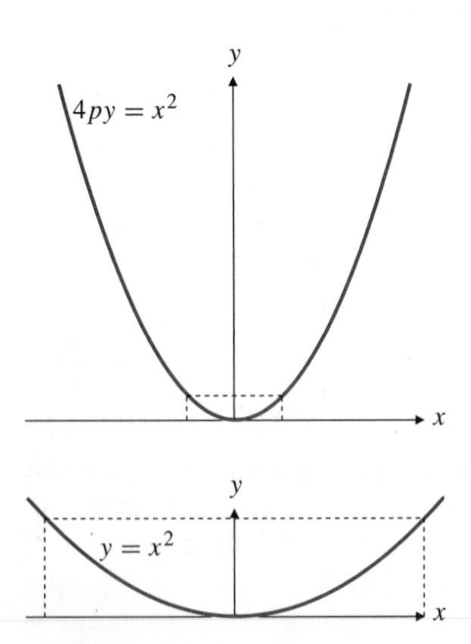

Figure P.28 The two parabolas are similar. Compare the parts inside the rectangles

Shifting a Graph

The graph of an equation (or inequality) can be shifted c units horizontally by replacing x with $x - c$ or vertically by replacing y with $y - c$.

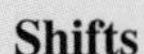

Shifts

To shift a graph c units to the right, replace x in its equation or inequality with $x - c$. (If $c < 0$, the shift will be to the left.)

To shift a graph c units upward, replace y in its equation or inequality with $y - c$. (If $c < 0$, the shift will be downward.)

EXAMPLE 8 The graph of $y = (x-3)^2$ is the parabola $y = x^2$ shifted 3 units to the right. The graph of $y = (x+1)^2$ is the parabola $y = x^2$ shifted 1 unit to the left. (See Figure P.29(a).)

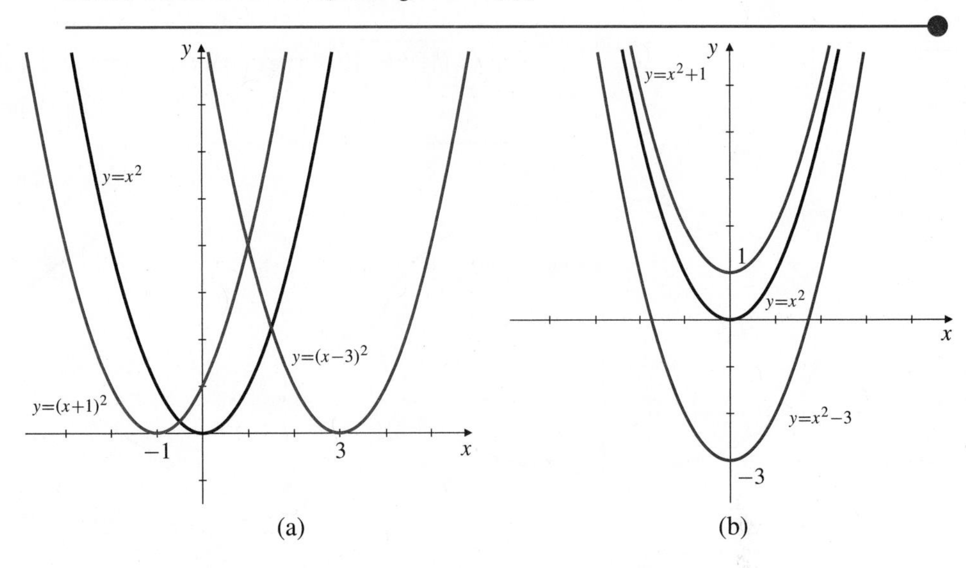

Figure P.29
(a) Horizontal shifts of $y = x^2$
(b) Vertical shifts of $y = x^2$

EXAMPLE 9 The graph of $y = x^2 + 1$ (or $y - 1 = x^2$) is the parabola $y = x^2$ shifted upward 1 unit. The graph of $y = x^2 - 3$ (or $y - (-3) = x^2$), is the parabola $y = x^2$ shifted downward 3 units. (See Figure P.29(b).)

EXAMPLE 10 The circle with equation $(x-h)^2 + (y-k)^2 = a^2$ having centre (h, k) and radius a can be obtained by shifting the circle $x^2 + y^2 = a^2$ of radius a centred at the origin h units to the right and k units upward. These shifts correspond to replacing x with $x - h$ and y with $y - k$.

The graph of $y = ax^2 + bx + c$ is a parabola whose axis is parallel to the y-axis. The parabola opens upward if $a > 0$ and downward if $a < 0$. We can complete the square and write the equation in the form $y = a(x-h)^2 + k$ to find the vertex (h, k).

EXAMPLE 11 Describe the graph of $y = x^2 - 4x + 3$.

Solution The equation $y = x^2 - 4x + 3$ represents a parabola, opening upward. To find its vertex and axis we can complete the square:

$$y = x^2 - 4x + 4 - 1 = (x-2)^2 - 1, \qquad \text{so} \quad y - (-1) = (x-2)^2.$$

This curve is the parabola $y = x^2$ shifted to the right 2 units and down 1 unit. Therefore, its vertex is $(2, -1)$, and its axis is the line $x = 2$. Since $y = x^2$ has focus $(0, 1/4)$, the focus of this parabola is $(0+2, (1/4)-1)$, or $(2, -3/4)$. (See Figure P.30.)

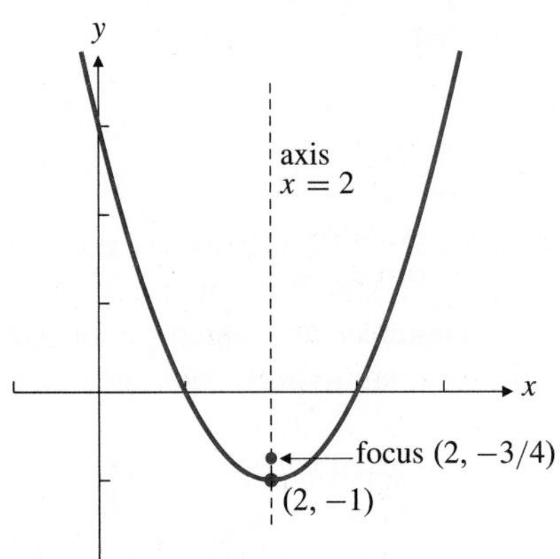

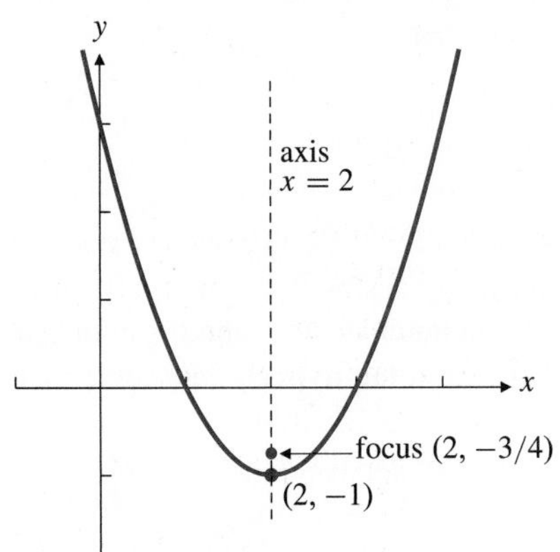

Figure P.30 The parabola $y = x^2 - 4x + 3$

Ellipses and Hyperbolas

If a and b are positive numbers, the equation

$$\frac{x^2}{a^2} + \frac{y^2}{b^2} = 1$$

represents a curve called an **ellipse** that lies wholly within the rectangle $-a \le x \le a$, $-b \le y \le b$. (Why?) If $a = b$, the ellipse is just the circle of radius a centred at the origin. If $a \ne b$, the ellipse is a circle that has been squashed by scaling it by different amounts in the two coordinate directions.

The ellipse has centre at the origin, and it passes through the four points $(a, 0)$, $(0, b)$, $(-a, 0)$, and $(0, -b)$. (See Figure P.31.) The line segments from $(-a, 0)$ to $(a, 0)$ and from $(0, -b)$ to $(0, b)$ are called the **principal axes** of the ellipse; the longer of the two is the **major axis**, and the shorter is the **minor axis**.

EXAMPLE 12 The equation $\frac{x^2}{9} + \frac{y^2}{4} = 1$ represents an ellipse with major axis from $(-3, 0)$ to $(3, 0)$ and minor axis from $(0, -2)$ to $(0, 2)$.

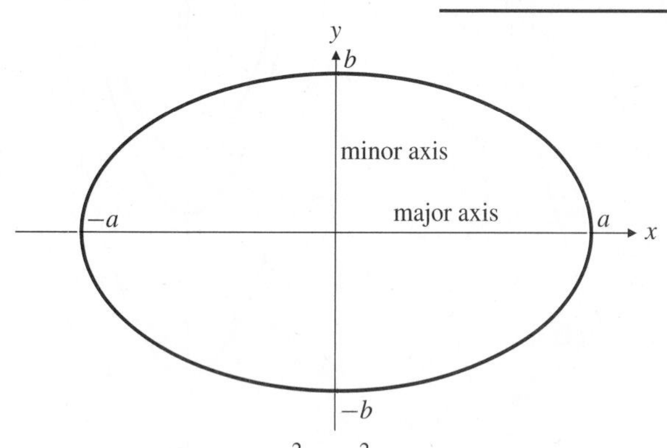

Figure P.31 The ellipse $\frac{x^2}{a^2} + \frac{y^2}{b^2} = 1$

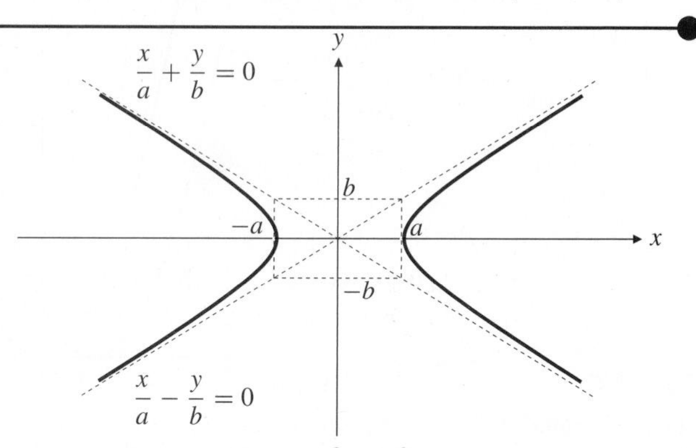

Figure P.32 The hyperbola $\frac{x^2}{a^2} - \frac{y^2}{b^2} = 1$ and its asymptotes

The equation

$$\frac{x^2}{a^2} - \frac{y^2}{b^2} = 1$$

represents a curve called a **hyperbola** that has centre at the origin and passes through the points $(-a, 0)$ and $(a, 0)$. (See Figure P.32.) The curve is in two parts (called **branches**). Each branch approaches two straight lines (called **asymptotes**) as it recedes far away from the origin. The asymptotes have equations

$$\frac{x}{a} - \frac{y}{b} = 0 \quad \text{and} \quad \frac{x}{a} + \frac{y}{b} = 0.$$

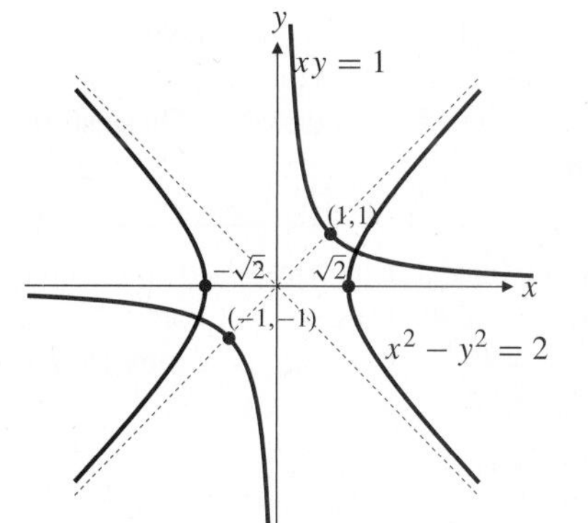

Figure P.33 Two rectangular hyperbolas

The equation $xy = 1$ also represents a hyperbola. This one passes through the points $(-1, -1)$ and $(1, 1)$ and has the coordinate axes as its asymptotes. It is, in fact, the hyperbola $x^2 - y^2 = 2$ rotated 45° counterclockwise about the origin. (See Figure P.33.) These hyperbolas are called **rectangular hyperbolas**, since their asymptotes intersect at right angles.

We will study ellipses and hyperbolas in more detail in Chapter 8.

EXERCISES P.3

In Exercises 1–4, write an equation for the circle with centre C and radius r.

1. $C(0, 0), \quad r = 4$

2. $C(0, 2), \quad r = 2$

3. $C(-2, 0), \quad r = 3$

4. $C(3, -4), \quad r = 5$

In Exercises 5–8, find the centre and radius of the circle having the given equation.

5. $x^2 + y^2 - 2x = 3$

6. $x^2 + y^2 + 4y = 0$

7. $x^2 + y^2 - 2x + 4y = 4$

8. $x^2 + y^2 - 2x - y + 1 = 0$

Describe the regions defined by the inequalities and pairs of inequalities in Exercises 9–16.

9. $x^2 + y^2 > 1$

10. $x^2 + y^2 < 4$

11. $(x + 1)^2 + y^2 \le 4$

12. $x^2 + (y - 2)^2 \le 4$

13. $x^2 + y^2 > 1, \quad x^2 + y^2 < 4$

14. $x^2 + y^2 \le 4, \quad (x + 2)^2 + y^2 \le 4$

15. $x^2 + y^2 < 2x, \quad x^2 + y^2 < 2y$

16. $x^2 + y^2 - 4x + 2y > 4, \quad x + y > 1$

17. Write an inequality that describes the interior of the circle with centre $(-1, 2)$ and radius $\sqrt{6}$.

18. Write an inequality that describes the exterior of the circle with centre $(2, -3)$ and radius 4.

19. Write a pair of inequalities that describe that part of the interior of the circle with centre $(0, 0)$ and radius $\sqrt{2}$ lying on or to the right of the vertical line through $(1, 0)$.

20. Write a pair of inequalities that describe the points that lie outside the circle with centre $(0, 0)$ and radius 2, and inside the circle with centre $(1, 3)$ that passes through the origin.

In Exercises 21–24, write an equation of the parabola having the given focus and directrix.

21. Focus: $(0, 4)$ Directrix: $y = -4$

22. Focus: $(0, -1/2)$ Directrix: $y = 1/2$

23. Focus: $(2, 0)$ Directrix: $x = -2$

24. Focus: $(-1, 0)$ Directrix: $x = 1$

In Exercises 25–28, find the parabola's focus and directrix, and make a sketch showing the parabola, focus, and directrix.

25. $y = x^2/2$ **26.** $y = -x^2$

27. $x = -y^2/4$ **28.** $x = y^2/16$

29. Figure P.34 shows the graph $y = x^2$ and four shifted versions of it. Write equations for the shifted versions.

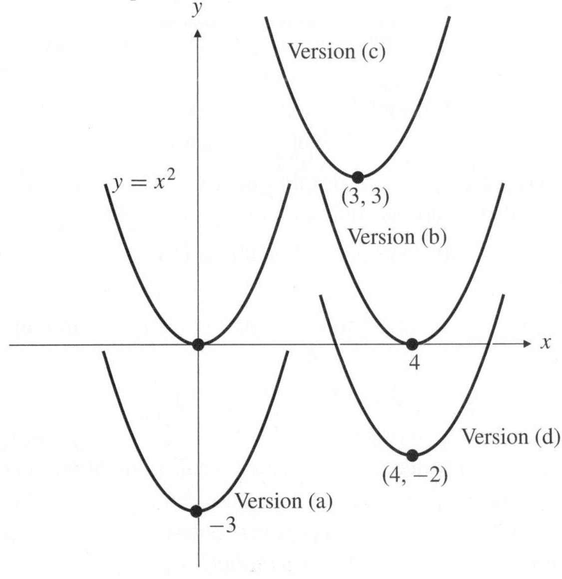

Figure P.34

30. What equations result from shifting the line $y = mx$
(a) horizontally to make it pass through the point (a, b)
(b) vertically to make it pass through (a, b)?

In Exercises 31–34, the graph of $y = \sqrt{x+1}$ is to be scaled in the indicated way. Give the equation of the graph that results from the scaling.

31. horizontal distances multiplied by 3

32. vertical distances divided by 4

33. horizontal distances multiplied by 2/3

34. horizontal distances divided by 4 and vertical distances multiplied by 2

In Exercises 35–38, write an equation for the graph obtained by shifting the graph of the given equation as indicated.

35. $y = 1 - x^2$ down 1, left 1

36. $x^2 + y^2 = 5$ up 2, left 4

37. $y = (x-1)^2 - 1$ down 1, right 1

38. $y = \sqrt{x}$ down 2, left 4

Find the points of intersection of the pairs of curves in Exercises 39–42.

39. $y = x^2 + 3, \quad y = 3x + 1$

40. $y = x^2 - 6, \quad y = 4x - x^2$

41. $x^2 + y^2 = 25, \quad 3x + 4y = 0$

42. $2x^2 + 2y^2 = 5, \quad xy = 1$

In Exercises 43–50, identify and sketch the curve represented by the given equation.

43. $\dfrac{x^2}{4} + y^2 = 1$ **44.** $9x^2 + 16y^2 = 144$

45. $\dfrac{(x-3)^2}{9} + \dfrac{(y+2)^2}{4} = 1$ **46.** $(x-1)^2 + \dfrac{(y+1)^2}{4} = 4$

47. $\dfrac{x^2}{4} - y^2 = 1$ **48.** $x^2 - y^2 = -1$

49. $xy = -4$ **50.** $(x-1)(y+2) = 1$

51. What is the effect on the graph of an equation in x and y of
(a) replacing x with $-x$?
(b) replacing y with $-y$?

52. What is the effect on the graph of an equation in x and y of replacing x with $-x$ and y with $-y$ simultaneously?

53. Sketch the graph of $|x| + |y| = 1$.

P.4 Functions and Their Graphs

The area of a circle depends on its radius. The temperature at which water boils depends on the altitude above sea level. The interest paid on a cash investment depends on the length of time for which the investment is made.

Whenever one quantity depends on another quantity, we say that the former quantity is a function of the latter. For instance, the area A of a circle depends on the radius r according to the formula

$$A = \pi r^2,$$

so we say that the area is a function of the radius. The formula is a *rule* that tells us how to calculate a *unique* (single) output value of the area A for each possible input value of the radius r.

The set of all possible input values for the radius is called the **domain** of the function. The set of all output values of the area is the **range** of the function. Since circles cannot have negative radii or areas, the domain and range of the circular area function are both the interval $[0, \infty)$ consisting of all nonnegative real numbers.

The domain and range of a mathematical function can be any sets of objects; they do not have to consist of numbers. Throughout much of this book, however, the domains and ranges of functions we consider will be sets of real numbers.

In calculus we often want to refer to a generic function without having any particular formula in mind. To denote that y is a function of x we write

$$y = f(x),$$

which we read as "y equals f of x." In this notation, due to the eighteenth-century mathematician Leonhard Euler, the function is represented by the symbol f. Also, x, called the **independent variable**, represents an input value from the domain of f, and y, the **dependent variable**, represents the corresponding output value $f(x)$ in the range of f.

DEFINITION 1

A **function** f on a set D into a set S is a rule that assigns a *unique* element $f(x)$ in S to each element x in D.

In this definition $D = \mathcal{D}(f)$ (read "D of f") is the domain of the function f. The range $\mathcal{R}(f)$ of f is the subset of S consisting of all *values* $f(x)$ of the function. Think of a function f as a kind of machine (Figure P.35) that produces an output value $f(x)$ in its range whenever we feed it an input value x from its domain.

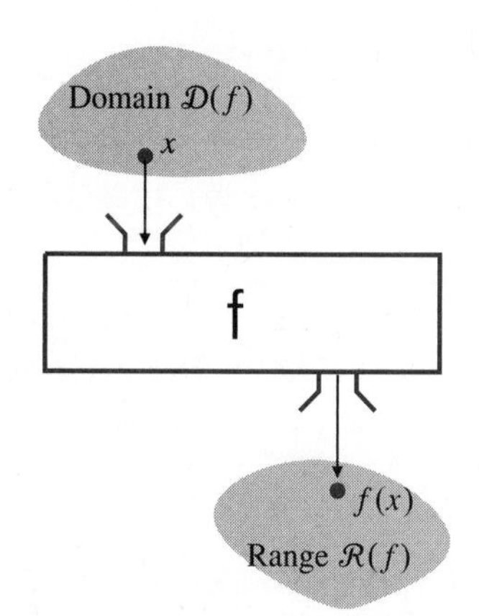

Figure P.35 A function machine

There are several ways to represent a function symbolically. The squaring function that converts any input real number x into its square x^2 can be denoted:

(a) by a formula such as $y = x^2$, which uses a dependent variable y to denote the value of the function;

(b) by a formula such as $f(x) = x^2$, which defines a function symbol f to name the function; or

(c) by a mapping rule such as $x \longrightarrow x^2$. (Read this as "x goes to x^2.")

In this book we will usually use either (a) or (b) to define functions. Strictly speaking, we should call a function f and not $f(x)$, since the latter denotes the value of the function at the point x. However, as is common usage, we will often refer to the function as $f(x)$ in order to name the variable on which f depends. Sometimes it is convenient to use the same letter to denote both a dependent variable and a function symbol; the circular area function can be written $A = f(r) = \pi r^2$ or as $A = A(r) = \pi r^2$. In the latter case we are using A to denote both the dependent variable and the name of the function.

EXAMPLE 1 The volume of a ball of radius r is given by the function

$$V(r) = \frac{4}{3}\pi r^3$$

for $r \ge 0$. Thus the volume of a ball of radius 3 ft is

$$V(3) = \frac{4}{3}\pi(3)^3 = 36\pi \text{ ft}^3.$$

Note how the variable r is replaced by the special value 3 in the formula defining the function to obtain the value of the function at $r = 3$.

EXAMPLE 2 A function F is defined for all real numbers t by

$$F(t) = 2t + 3.$$

Find the output values of F that correspond to the input values 0, 2, $x + 2$, and $F(2)$.

Solution In each case we substitute the given input for t in the definition of F:

$$\begin{aligned} F(0) &= 2(0) + 3 = 0 + 3 = 3 \\ F(2) &= 2(2) + 3 = 4 + 3 = 7 \\ F(x+2) &= 2(x+2) + 3 = 2x + 7 \\ F(F(2)) &= F(7) = 2(7) + 3 = 17. \end{aligned}$$

The Domain Convention

A function is not properly defined until its domain is specified. For instance, the function $f(x) = x^2$ defined for all real numbers $x \geq 0$ is different from the function $g(x) = x^2$ defined for all real x because they have different domains, even though they have the same values at every point where both are defined. In Chapters 1–9 we will be dealing with real functions (functions whose input and output values are real numbers). When the domain of such a function is not specified explicitly, we will assume that the domain is the largest set of real numbers to which the function assigns real values. Thus, if we talk about the function x^2 without specifying a domain, we mean the function $g(x)$ above.

The domain convention

When a function f is defined without specifying its domain, we assume that the domain consists of all real numbers x for which the value $f(x)$ of the function is a real number.

In practice, it is often easy to determine the domain of a function $f(x)$ given by an explicit formula. We just have to exclude those values of x that would result in dividing by 0 or taking even roots of negative numbers.

EXAMPLE 3 **The square root function.** The domain of $f(x) = \sqrt{x}$ is the interval $[0, \infty)$, since negative numbers do not have real square roots. We have $f(0) = 0$, $f(4) = 2$, $f(10) \approx 3.16228$. Note that, although there are *two* numbers whose square is 4, namely, -2 and 2, only *one* of these numbers, 2, is the square root of 4. (Remember that a function assigns a *unique* value to each element in its domain; it cannot assign two different values to the same input.) The **square root function** $\sqrt{x}$ always denotes the *nonnegative* square root of x. The two solutions of the equation $x^2 = 4$ are $x = \sqrt{4} = 2$ and $x = -\sqrt{4} = -2$.

EXAMPLE 4 The domain of the function $h(x) = \dfrac{x}{x^2 - 4}$ consists of all real numbers except $x = -2$ and $x = 2$. Expressed in terms of intervals,

$$\mathcal{D}(h) = (-\infty, -2) \cup (-2, 2) \cup (2, \infty).$$

Most of the functions we encounter will have domains that are either intervals or unions of intervals.

EXAMPLE 5 The domain of $S(t) = \sqrt{1 - t^2}$ consists of all real numbers t for which $1 - t^2 \geq 0$. Thus we require that $t^2 \leq 1$, or $-1 \leq t \leq 1$. The domain is the closed interval $[-1, 1]$.

Graphs of Functions

An old maxim states that "a picture is worth a thousand words." This is certainly true in mathematics; the behaviour of a function is best described by drawing its graph.

The **graph of a function** f is just the graph of the *equation* $y = f(x)$. It consists of those points in the Cartesian plane whose coordinates (x, y) are pairs of input–output values for f. Thus (x, y) lies on the graph of f provided x is in the domain of f and $y = f(x)$.

Drawing the graph of a function f sometimes involves making a table of coordinate pairs $(x, f(x))$ for various values of x in the domain of f, then plotting these points and connecting them with a "smooth curve."

EXAMPLE 6 Graph the function $f(x) = x^2$.

Table 1.

x	$y = f(x)$
-2	4
-1	1
0	0
1	1
2	4

Solution Make a table of (x, y) pairs that satisfy $y = x^2$. (See Table 1.) Now plot the points and join them with a smooth curve. (See Figure P.36(a).)

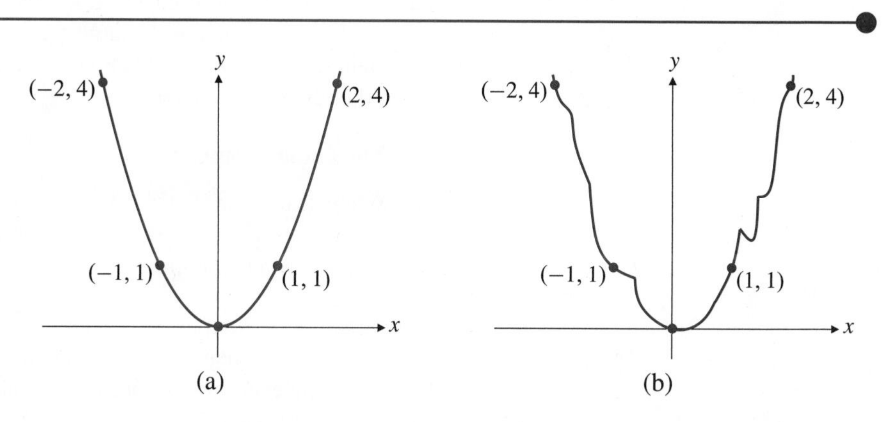

Figure P.36
(a) Correct graph of $f(x) = x^2$
(b) Incorrect graph of $f(x) = x^2$

How do we know the graph is smooth and doesn't do weird things between the points we have calculated, for example, as shown in Figure P.36(b)? We could, of course, plot more points, spaced more closely together, but how do we know how the graph behaves between the points we have plotted? In Chapter 4, calculus will provide useful tools for answering these questions.

Some functions occur often enough in applications that you should be familiar with their graphs. Some of these are shown in Figures P.37–P.46. Study them for a while; they are worth remembering. Note, in particular, the graph of the **absolute value function**, $f(x) = |x|$, shown in Figure P.46. It is made up of the two half-lines $y = -x$ for $x < 0$ and $y = x$ for $x \geq 0$.

If you know the effects of vertical and horizontal shifts on the equations representing graphs (see Section P.3), you can easily sketch some graphs that are shifted versions of the ones in Figures P.37–P.46.

EXAMPLE 7 Sketch the graph of $y = 1 + \sqrt{x - 4}$.

Solution This is just the graph of $y = \sqrt{x}$ in Figure P.40 shifted to the right 4 units (because x is replaced by $x - 4$) and up 1 unit. See Figure P.47.

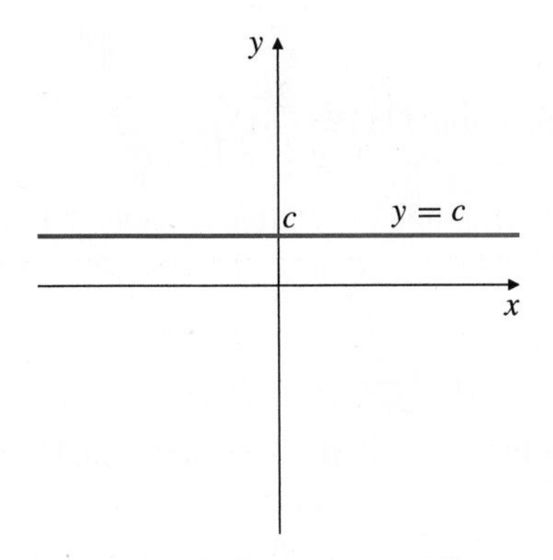

Figure P.37 The graph of a constant function $f(x) = c$

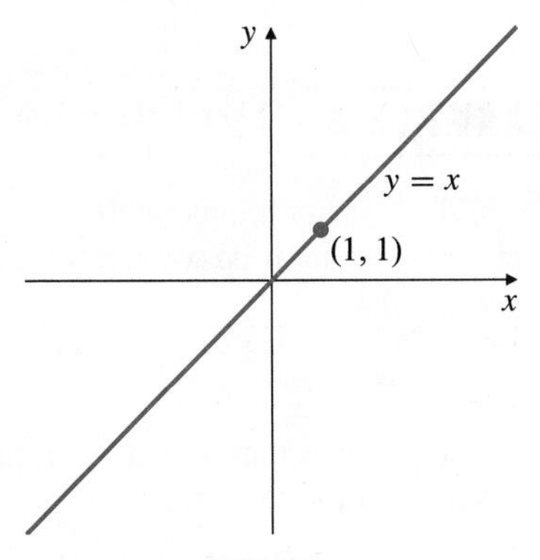

Figure P.38 The graph of $f(x) = x$

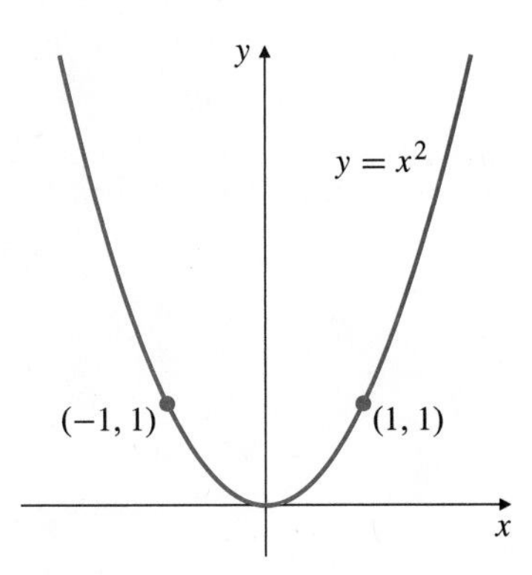

Figure P.39 The graph of $f(x) = x^2$

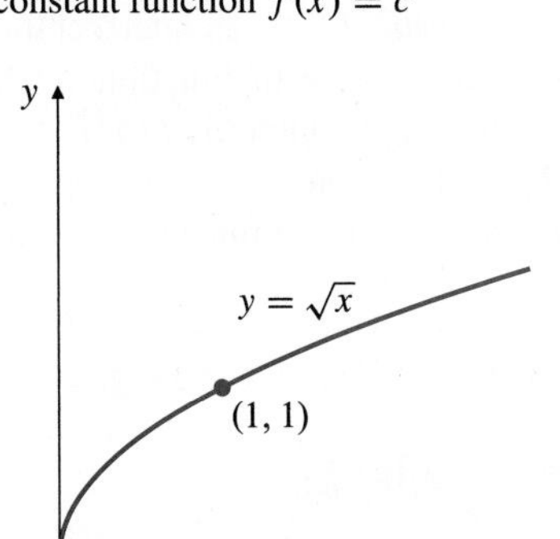

Figure P.40 The graph of $f(x) = \sqrt{x}$

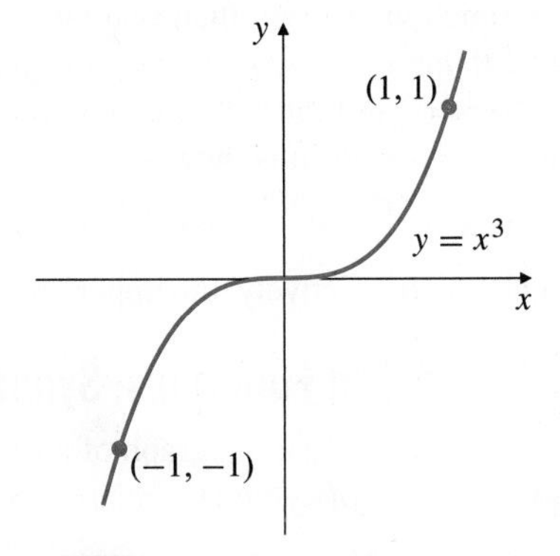

Figure P.41 The graph of $f(x) = x^3$

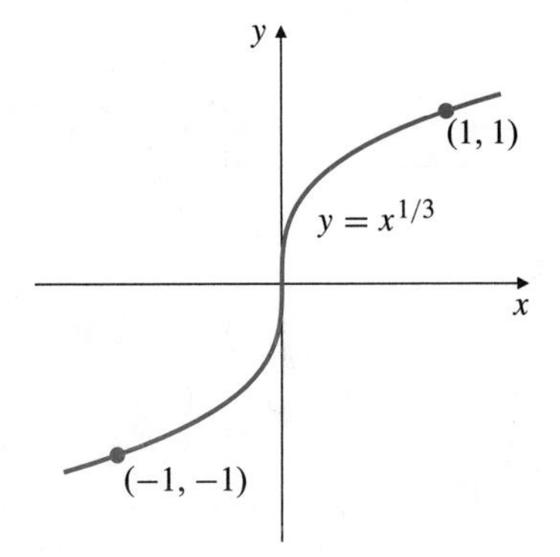

Figure P.42 The graph of $f(x) = x^{1/3}$

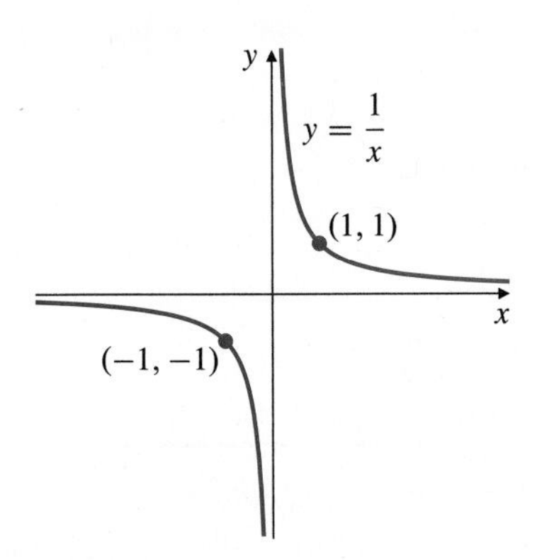

Figure P.43 The graph of $f(x) = 1/x$

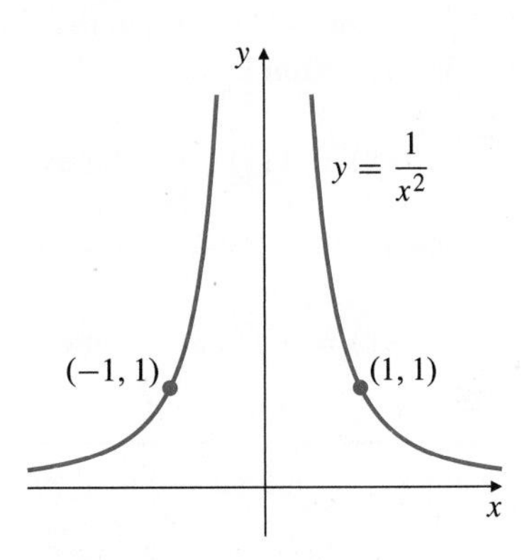

Figure P.44 The graph of $f(x) = 1/x^2$

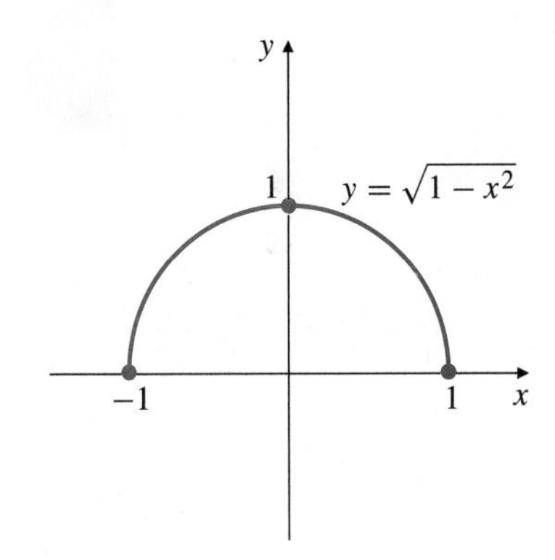

Figure P.45 The graph of $f(x) = \sqrt{1 - x^2}$

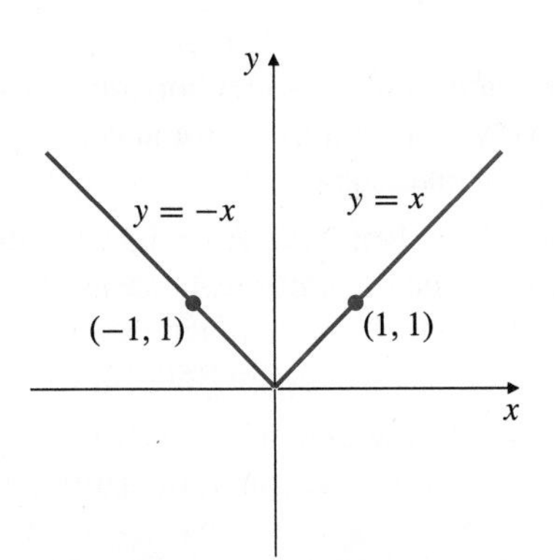

Figure P.46 The graph of $f(x) = |x|$

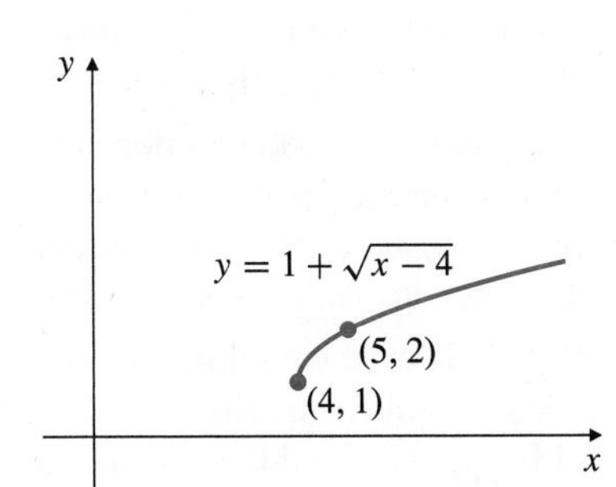

Figure P.47 The graph of $y = \sqrt{x}$ shifted right 4 units and up 1 unit

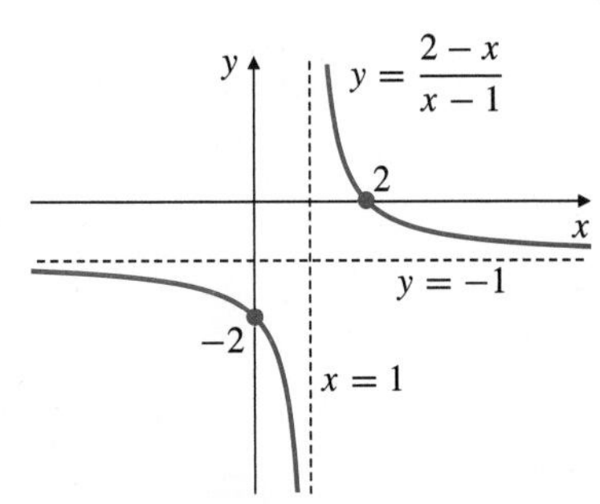

Figure P.48 The graph of $\dfrac{2 - x}{x - 1}$

EXAMPLE 8 Sketch the graph of the function $f(x) = \dfrac{2-x}{x-1}$.

Solution It is not immediately obvious that this graph is a shifted version of a known graph. To see that it is, we can divide $x - 1$ into $2 - x$ to get a quotient of -1 and a remainder of 1:

$$\frac{2-x}{x-1} = \frac{-x+1+1}{x-1} = \frac{-(x-1)+1}{x-1} = -1 + \frac{1}{x-1}.$$

Thus, the graph is that of $1/x$ from Figure P.43 shifted to the right 1 unit and down 1 unit. See Figure P.48.

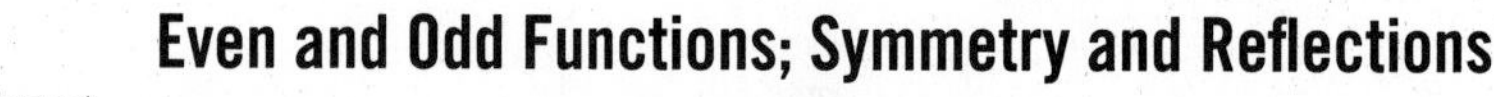

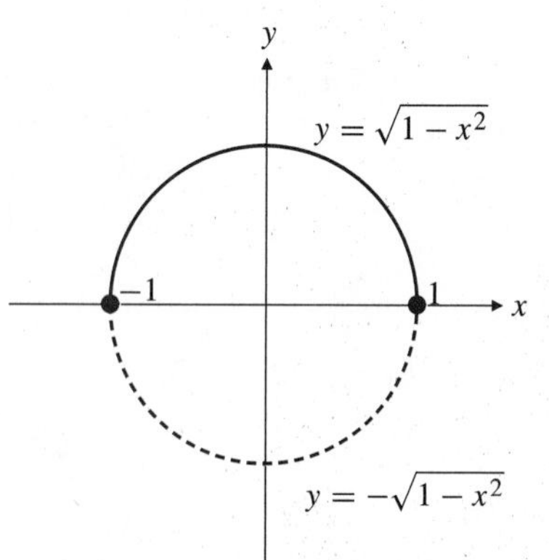

Figure P.49 The circle $x^2 + y^2 = 1$ is not the graph of a function

Not every curve you can draw is the graph of a function. A function f can have only one value $f(x)$ for each x in its domain, so no *vertical line* can intersect the graph of a function at more than one point. If a is in the domain of function f, then the vertical line $x = a$ will intersect the graph of f at the single point $(a, f(a))$. The circle $x^2 + y^2 = 1$ in Figure P.49 cannot be the graph of a function since some vertical lines intersect it twice. It is, however, the union of the graphs of two functions, namely,

$$y = \sqrt{1-x^2} \qquad \text{and} \qquad y = -\sqrt{1-x^2},$$

which are, respectively, the upper and lower halves (semicircles) of the given circle.

Even and Odd Functions; Symmetry and Reflections

It often happens that the graph of a function will have certain kinds of symmetry. The simplest kinds of symmetry relate the values of a function at x and $-x$.

DEFINITION 2

Even and odd functions

Suppose that $-x$ belongs to the domain of f whenever x does. We say that f is an **even function** if

$$f(-x) = f(x) \qquad \text{for every } x \text{ in the domain of } f.$$

We say that f is an **odd function** if

$$f(-x) = -f(x) \qquad \text{for every } x \text{ in the domain of } f.$$

The names *even* and *odd* come from the fact that even powers such as $x^0 = 1, x^2, x^4, \ldots, x^{-2}, x^{-4}, \ldots$ are even functions, and odd powers such as $x^1 = x, x^3, \ldots, x^{-1}, x^{-3}, \ldots$ are odd functions. Observe, for example, that $(-x)^4 = x^4$ and $(-x)^{-3} = -x^{-3}$.

Since $(-x)^2 = x^2$, any function that depends only on x^2 is even. For instance, the absolute value function $y = |x| = \sqrt{x^2}$ is even.

The graph of an even function is *symmetric about the y-axis.* A horizontal straight line drawn from a point on the graph to the y-axis will, if continued an equal distance on the other side of the y-axis, come to another point on the graph. (See Figure P.50(a).)

The graph of an odd function is *symmetric about the origin.* A straight line drawn from a point on the graph to the origin will, if continued an equal distance on the other side of the origin, come to another point on the graph. If an odd function f is defined at $x = 0$, then its value must be zero there: $f(0) = 0$. (See Figure P.50(b).)

If $f(x)$ is even (or odd), then so is any constant multiple of $f(x)$ such as $2f(x)$ or $-5f(x)$. Sums (and differences) of even functions are even; sums (and differences) of odd functions are odd. For example, $f(x) = 3x^4 - 5x^2 - 1$ is even, since it is the sum of three even functions: $3x^4$, $-5x^2$, and $-1 = -x^0$. Similarly, $4x^3 - (2/x)$ is an odd function. The function $g(x) = x^2 - 2x$ is the sum of an even function and an odd function and is itself neither even nor odd.

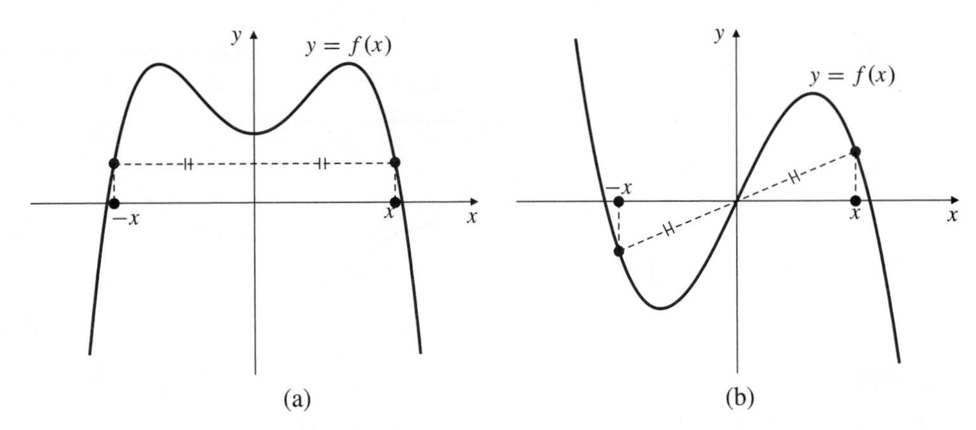

Figure P.50

(a) The graph of an even function is symmetric about the y-axis

(b) The graph of an odd function is symmetric about the origin

Other kinds of symmetry are also possible. For example, the function $g(x) = x^2 - 2x$ can be written in the form $g(x) = (x-1)^2 - 1$. This shows that the values of $g(1 \pm u)$ are equal, so the graph (Figure P.51(a)) is symmetric about the vertical line $x = 1$; it is the parabola $y = x^2$ shifted 1 unit to the right and 1 unit down. Similarly, the graph of $h(x) = x^3 + 1$ is symmetric about the point $(0, 1)$ (Figure P.51(b)).

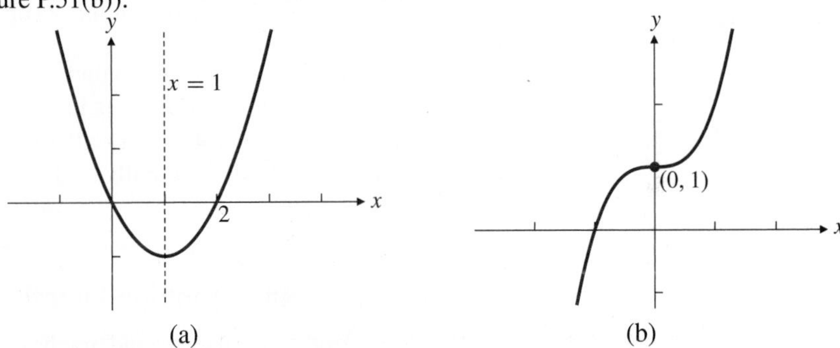

Figure P.51

(a) The graph of $g(x) = x^2 - 2x$ is symmetric about $x = 1$

(b) The graph of $y = h(x) = x^3 + 1$ is symmetric about $(0, 1)$

Reflections in Straight Lines

The image of an object reflected in a plane mirror appears to be as far behind the mirror as the object is in front of it. Thus, the mirror bisects at right angles the line from a point in the object to the corresponding point in the image. Given a line L and a point P not on L, we call a point Q the **reflection**, or the **mirror image**, of P in L if L is the right bisector of the line segment PQ. The reflection of any graph G in L is the graph consisting of the reflections of all the points of G.

Certain reflections of graphs are easily described in terms of the equations of the graphs:

Reflections in special lines

1. Substituting $-x$ in place of x in an equation in x and y corresponds to reflecting the graph of the equation in the y-axis.
2. Substituting $-y$ in place of y in an equation in x and y corresponds to reflecting the graph of the equation in the x-axis.
3. Substituting $a - x$ in place of x in an equation in x and y corresponds to reflecting the graph of the equation in the line $x = a/2$.
4. Substituting $b - y$ in place of y in an equation in x and y corresponds to reflecting the graph of the equation in the line $y = b/2$.
5. Interchanging x and y in an equation in x and y corresponds to reflecting the graph of the equation in the line $y = x$.

EXAMPLE 9 Describe and sketch the graph of $y = \sqrt{2-x} - 3$.

Solution The graph of $y = \sqrt{2-x}$ is the reflection of the graph of $y = \sqrt{x}$

(Figure P.40) in the vertical line $x = 1$. The graph of $y = \sqrt{2-x} - 3$ is the result of lowering this reflection by 3 units. See Figure P.52(a).

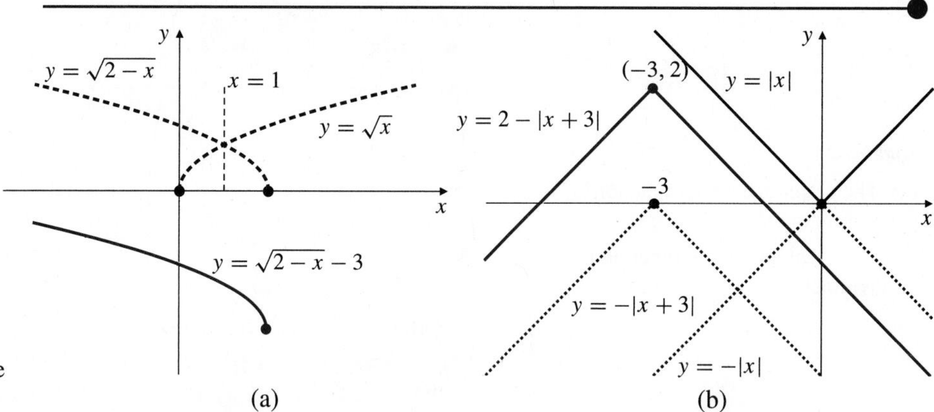

Figure P.52
(a) Constructing the graph of $y = \sqrt{2-x} - 3$
(b) Transforming $y = |x|$ to produce the coloured graph

EXAMPLE 10 Express the equation of the coloured graph in Figure P.52(b) in terms of the absolute value function $|x|$.

Solution We can get the coloured graph by first reflecting the graph of $|x|$ (Figure P.46) in the x-axis and then shifting the reflection left 3 units and up 2 units. The reflection of $y = |x|$ in the x-axis has equation $-y = |x|$, or $y = -|x|$. Shifting this left 3 units gives $y = -|x+3|$. Finally, shifting up 2 units gives $y = 2 - |x+3|$, which is the desired equation.

Defining and Graphing Functions with Maple

Many of the calculations and graphs encountered in studying calculus can be produced using a computer algebra system such as Maple or Mathematica. Here and there, throughout this book, we will include examples illustrating how to get Maple to perform such tasks. (The examples were done with Maple 10, but most of them will work with earlier or later versions of Maple as well.)

We begin with an example showing how to define a function in Maple and then plot its graph. We show in colour the input you type into Maple and in black Maple's response. Let us define the function $f(x) = x^3 - 2x^2 - 12x + 1$.

```
> f := x -> x^3-2*x^2-12*x+1; <enter>
```

$$f := x \longrightarrow x^3 - 2x^2 - 12x + 1$$

Note the use of := to indicate the symbol to the left is being defined and the use of -> to indicate the rule for the construction of $f(x)$ from x. Also note that Maple uses the asterisk * to indicate multiplication and the caret ^ to indicate an exponent. A Maple instruction should end with a semicolon ; (or a colon : if no output is desired) before the Enter key is pressed. Hereafter we will not show the `<enter>` in our input.

We can now use f as an ordinary function:

```
> f(t)+f(1);
```

$$t^3 - 2t^2 - 12t - 11$$

The following command results in a plot of the graph of f on the interval $[-4, 5]$ shown in Figure P.53.

```
> plot(f(x), x=-4..5);
```

We could have specified the expression `x^3-2*x^2-12*x+1` directly in the plot command instead of first defining the function $f(x)$. Note the use of two dots `..` to separate the left and right endpoints of the plot interval. Other options can be included in the plot command; all such options are separated with commas. You can specify the range of values of y in addition to that for x (which is required), and you can specify `scaling=CONSTRAINED` if you want equal unit distances on both axes. (This would be a bad idea for the graph of our $f(x)$. Why?)

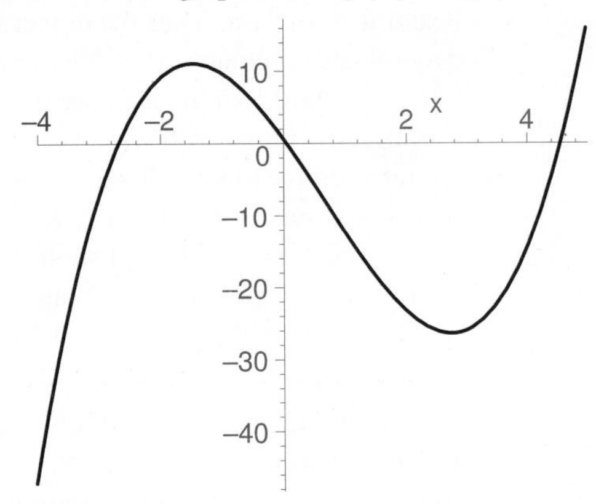

Figure P.53 A Maple plot

When using a graphing calculator or computer graphing software things can go horribly wrong in some circumstances. The following example illustrates the catastrophic effects that **round-off error** can have.

EXAMPLE 11 Consider the function $g(x) = \dfrac{|1+x|-1}{x}$.

If $x > -1$, then $|1+x| = 1+x$, so the formula for $g(x)$ simplifies to $g(x) = \dfrac{(1+x)-1}{x} = \dfrac{x}{x} = 1$, at least provided $x \neq 0$. Thus the graph of g on an interval lying to the right of $x = -1$ should be the horizontal line $y = 1$, possibly with a hole in it at $x = 0$. The Maple commands

```
> g := x -> (abs(1+x)-1)/x:  plot(g(x), x=-0.5..0.5);
```

lead, as expected, to the graph in Figure P.54. But plotting the same function on a very tiny interval near $x = 0$ leads to quite a different graph. The command

```
> plot([g(x),1],x=-7*10^(-16)..5*10^(-16),
     style=[point,line],numpoints=4000);
```

produces the graph in Figure P.55.

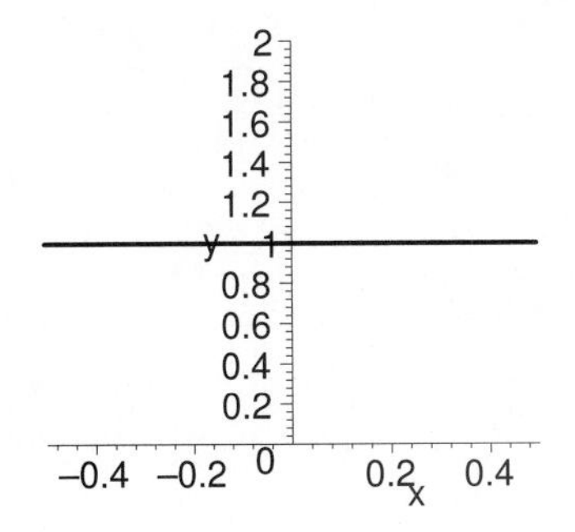

Figure P.54 The graph of $y = g(x)$ on the interval $[-0.5, 0.5]$

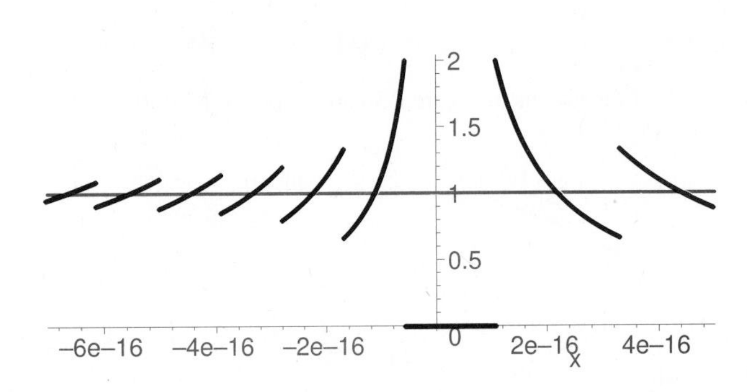

Figure P.55 The graphs of $y = g(x)$ (colour) and $y = 1$ (black) on the interval $[-7 \times 10^{-16}, 5 \times 10^{-16}]$

The coloured arcs and short line through the origin are the graph of $y = g(x)$ plotted as 4,000 individual points over the interval from -7×10^{-16} to 5×10^{-16}. For comparison sake, the black horizontal line $y = 1$ is also plotted. What makes the graph of g so strange on this interval is the fact that Maple can only represent finitely many real numbers in its finite memory. If the number x is too close to zero, Maple cannot tell the difference between $1 + x$ and 1, so it calculates $1 - 1 = 0$ for the numerator, and uses $g(x) = 0$ in the plot. This seems to happen between about -0.5×10^{-16}

and 0.8×10^{-16} (the coloured horizontal line). As we move further away from the origin, Maple can tell the difference between $1 + x$ and 1, but loses most of the significant figures in the representation of x when it adds 1, and these remain lost when it subtracts 1 again. Thus the numerator remains constant over short intervals while the denominator increases as x moves away from 0. In those intervals the fraction behaves like $constant/x$ so the arcs are hyperbolas, sloping downward away from the origin. The effect diminishes the farther x moves away from 0, as more of its significant figures are retained by Maple. It should be noted that the reason we used the absolute value of $1 + x$ instead of just $1 + x$ is that this forced Maple to add the x to the 1 before subtracting the second 1. (If we had used $(1+x) - 1$ as the numerator for $g(x)$, Maple would have simplified it algebraically and obtained $g(x) = 1$ before using any values of x for plotting.)

In later chapters we will encounter more such strange behaviour (which we call **numerical monsters** and denote by the symbox ⚠) in the context of calculator and computer calculations with floating point (i.e., real) numbers. They are a necessary consequence of the limitations of such hardware and software, and are not restricted to Maple, though they may show up somewhat differently with other software. It is necessary to be aware of how calculators and computers do arithmetic in order to be able to use them effectively without falling into errors that you do not recognize as such.

One final comment about Figure P.55: the graph of $y = g(x)$ was plotted as individual points, rather than a line as was $y = 1$, in order to make the jumps between consecutive arcs more obvious. Had we omitted the `style=[point,line]` option in the plot command, the default line style would have been used for both graphs and the arcs in the graph of g would have been connected with vertical line segments. Note how the command called for the plotting of two different functions by listing them within square brackets, and how the corresponding styles were correspondingly listed.

EXERCISES P.4

In Exercises 1–6, find the domain and range of each function.

1. $f(x) = 1 + x^2$

2. $f(x) = 1 - \sqrt{x}$

3. $G(x) = \sqrt{8 - 2x}$

4. $F(x) = 1/(x - 1)$

5. $h(t) = \dfrac{t}{\sqrt{2 - t}}$

6. $g(x) = \dfrac{1}{1 - \sqrt{x - 2}}$

7. Which of the graphs in Figure P.56 are graphs of functions $y = f(x)$? Why?

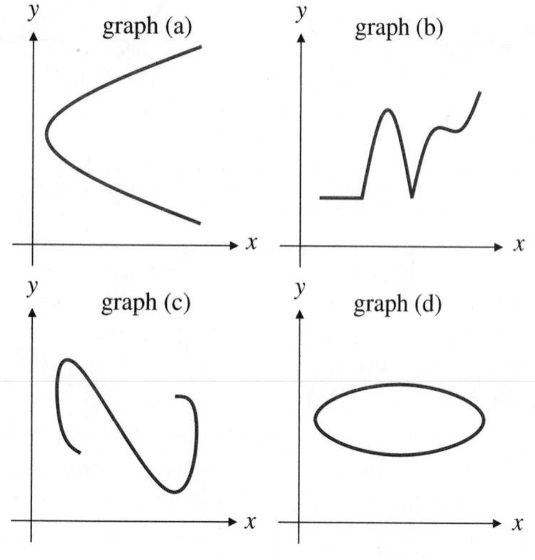

Figure P.56

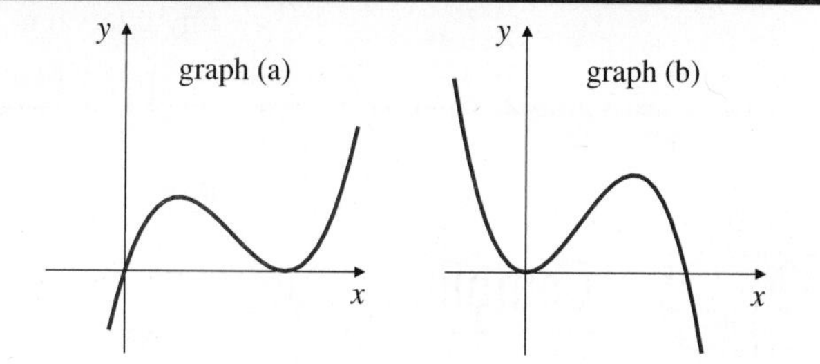

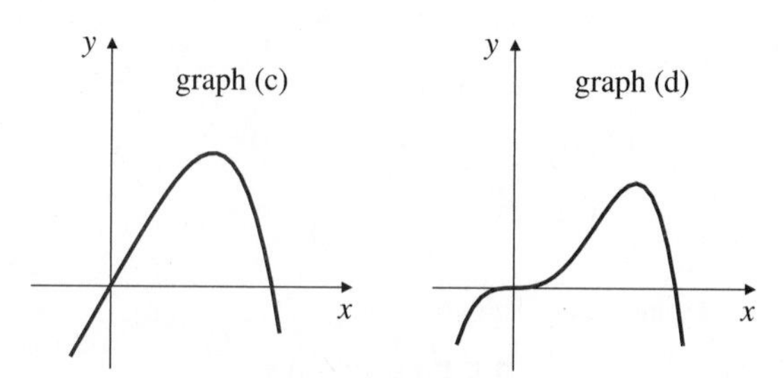

Figure P.57

8. Figure P.57 shows the graphs of the functions: (i) $x - x^4$, (ii) $x^3 - x^4$, (iii) $x(1 - x)^2$, (iv) $x^2 - x^3$. Which graph corresponds to which function?

In Exercises 9–10, sketch the graph of the function f by first making a table of values of $f(x)$ at $x = 0$, $x = \pm 1/2$, $x = \pm 1$, $x = \pm 3/2$, and $x = \pm 2$.

9. $f(x) = x^4$

10. $f(x) = x^{2/3}$

In Exercises 11–22, what (if any) symmetry does the graph of f possess? In particular, is f even or odd?

11. $f(x) = x^2 + 1$ **12.** $f(x) = x^3 + x$

13. $f(x) = \dfrac{x}{x^2 - 1}$ **14.** $f(x) = \dfrac{1}{x^2 - 1}$

15. $f(x) = \dfrac{1}{x - 2}$ **16.** $f(x) = \dfrac{1}{x + 4}$

17. $f(x) = x^2 - 6x$ **18.** $f(x) = x^3 - 2$

19. $f(x) = |x^3|$ **20.** $f(x) = |x + 1|$

21. $f(x) = \sqrt{2x}$ **22.** $f(x) = \sqrt{(x - 1)^2}$

Sketch the graphs of the functions in Exercises 23–38.

23. $f(x) = -x^2$ **24.** $f(x) = 1 - x^2$

25. $f(x) = (x - 1)^2$ **26.** $f(x) = (x - 1)^2 + 1$

27. $f(x) = 1 - x^3$ **28.** $f(x) = (x + 2)^3$

29. $f(x) = \sqrt{x} + 1$ **30.** $f(x) = \sqrt{x + 1}$

31. $f(x) = -|x|$ **32.** $f(x) = |x| - 1$

33. $f(x) = |x - 2|$ **34.** $f(x) = 1 + |x - 2|$

35. $f(x) = \dfrac{2}{x + 2}$ **36.** $f(x) = \dfrac{1}{2 - x}$

37. $f(x) = \dfrac{x}{x + 1}$ **38.** $f(x) = \dfrac{x}{1 - x}$

In Exercises 39–46, f refers to the function with domain $[0, 2]$ and range $[0, 1]$, whose graph is shown in Figure P.58. Sketch the graphs of the indicated functions and specify their domains and ranges.

39. $f(x) + 2$ **40.** $f(x) - 1$

41. $f(x + 2)$ **42.** $f(x - 1)$

43. $-f(x)$ **44.** $f(-x)$

45. $f(4 - x)$ **46.** $1 - f(1 - x)$

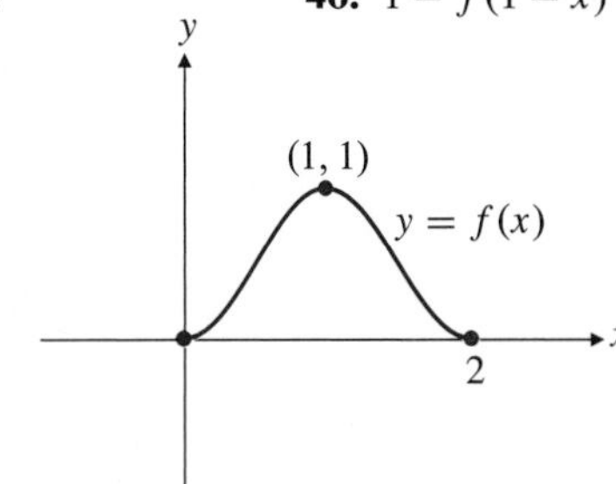

Figure P.58

It is often quite difficult to determine the range of a function exactly. In Exercises 47–48, use a graphing utility (calculator or computer) to graph the function f, and by zooming in on the graph determine the range of f with accuracy of 2 decimal places.

47. $f(x) = \dfrac{x + 2}{x^2 + 2x + 3}$ **48.** $f(x) = \dfrac{x - 1}{x^2 + x}$

In Exercises 49–52, use a graphing utility to plot the graph of the given function. Examine the graph (zooming in or out as necessary) for symmetries. About what lines and/or points are the graphs symmetric? Try to verify your conclusions algebraically.

49. $f(x) = x^4 - 6x^3 + 9x^2 - 1$

50. $f(x) = \dfrac{3 - 2x + x^2}{2 - 2x + x^2}$

51. $f(x) = \dfrac{x - 1}{x - 2}$ **52.** $f(x) = \dfrac{2x^2 + 3x}{x^2 + 4x + 5}$

53. What function $f(x)$, defined on the real line $\mathbb{R}$, is both even and odd?

P.5 Combining Functions to Make New Functions

Functions can be combined in a variety of ways to produce new functions. We begin by examining algebraic means of combining functions, that is, addition, subtraction, multiplication, and division.

Sums, Differences, Products, Quotients, and Multiples

Like numbers, functions can be added, subtracted, multiplied, and divided (except where the denominator is zero) to produce new functions.

DEFINITION 3

If f and g are functions, then for every x that belongs to the domains of both f and g we define functions $f + g$, $f - g$, fg, and f/g by the formulas:

$$(f + g)(x) = f(x) + g(x)$$
$$(f - g)(x) = f(x) - g(x)$$
$$(fg)(x) = f(x)g(x)$$
$$\left(\frac{f}{g}\right)(x) = \frac{f(x)}{g(x)}, \qquad \text{where } g(x) \neq 0.$$

A special case of the rule for multiplying functions shows how functions can be multiplied by constants. If c is a real number, then the function cf is defined for all x in the domain of f by

$$(cf)(x) = c \cdot f(x).$$

EXAMPLE 1 Figure P.59(a) shows the graphs of $f(x) = x^2$, $g(x) = x - 1$, and their sum $(f + g)(x) = x^2 + x - 1$. Observe that the height of the graph of $f + g$ at any point x is the sum of the heights of the graphs of f and g at that point.

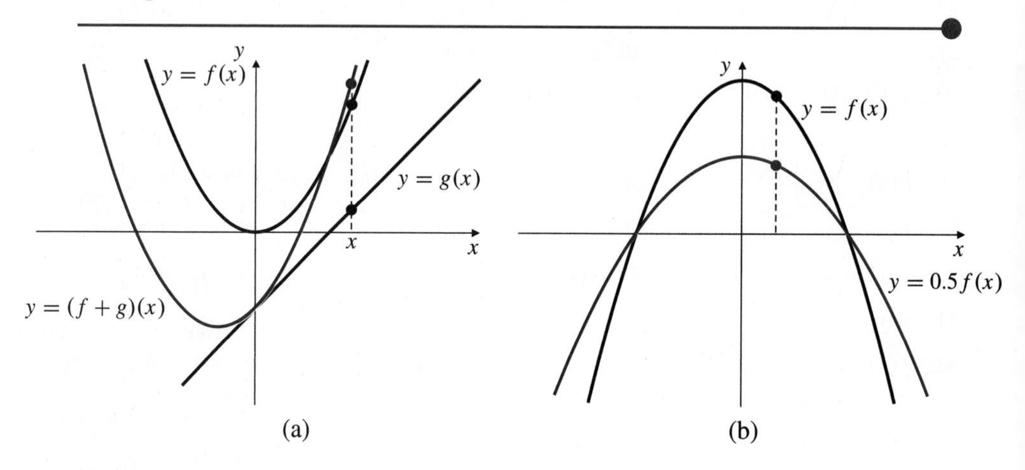

Figure P.59
(a) $(f + g)(x) = f(x) + g(x)$
(b) $g(x) = (0.5)f(x)$

EXAMPLE 2 Figure P.59(b) shows the graphs of $f(x) = 2 - x^2$ and the multiple $g(x) = (0.5)f(x)$. Note how the height of the graph of g at any point x is half the height of the graph of f there.

EXAMPLE 3 The functions f and g are defined by the formulas

$$f(x) = \sqrt{x} \qquad \text{and} \qquad g(x) = \sqrt{1 - x}.$$

Find formulas for the values of $3f$, $f + g$, $f - g$, fg, f/g, and g/f at x, and specify the domains of each of these functions.

Solution The information is collected in Table 2:

Table 2. Combinations of f and g and their domains

Function	Formula	Domain
f	$f(x) = \sqrt{x}$	$[0, \infty)$
g	$g(x) = \sqrt{1 - x}$	$(-\infty, 1]$
$3f$	$(3f)(x) = 3\sqrt{x}$	$[0, \infty)$
$f + g$	$(f + g)(x) = f(x) + g(x) = \sqrt{x} + \sqrt{1 - x}$	$[0, 1]$
$f - g$	$(f - g)(x) = f(x) - g(x) = \sqrt{x} - \sqrt{1 - x}$	$[0, 1]$
fg	$(fg)(x) = f(x)g(x) = \sqrt{x(1 - x)}$	$[0, 1]$
f/g	$\frac{f}{g}(x) = \frac{f(x)}{g(x)} = \sqrt{\frac{x}{1 - x}}$	$[0, 1)$
g/f	$\frac{g}{f}(x) = \frac{g(x)}{f(x)} = \sqrt{\frac{1 - x}{x}}$	$(0, 1]$

Note that most of the combinations of f and g have domains

$$[0, \infty) \cap (-\infty, 1] = [0, 1],$$

the intersection of the domains of f and g. However, the domains of the two quotients f/g and g/f had to be restricted further to remove points where the denominator was zero.

Composite Functions

There is another method, called **composition**, by which two functions can be combined to form a new function.

DEFINITION

Composite functions

If f and g are two functions, the **composite** function $f \circ g$ is defined by

$$f \circ g(x) = f(g(x)).$$

The domain of $f \circ g$ consists of those numbers x in the domain of g for which $g(x)$ is in the domain of f. In particular, if the range of g is contained in the domain of f, then the domain of $f \circ g$ is just the domain of g.

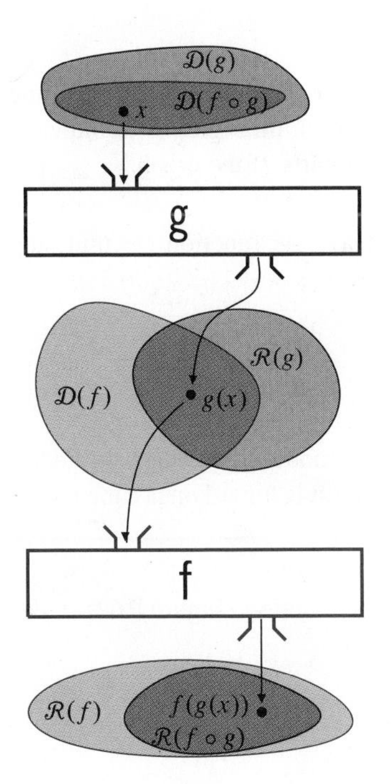

Figure P.60 $f \circ g(x) = f(g(x))$

As shown in Figure P.60, forming $f \circ g$ is equivalent to arranging "function machines" g and f in an "assembly line" so that the output of g becomes the input of f.

In calculating $f \circ g(x) = f(g(x))$, we first calculate $g(x)$ and then calculate f of the result. We call g the *inner* function and f the *outer* function of the composition. We can, of course, also calculate the composition $g \circ f(x) = g(f(x))$, where f is the inner function, the one that gets calculated first, and g is the outer function, which gets calculated last. The functions $f \circ g$ and $g \circ f$ are usually quite different, as the following example shows.

EXAMPLE 4 Given $f(x) = \sqrt{x}$ and $g(x) = x + 1$, calculate the four composite functions $f \circ g(x)$, $g \circ f(x)$, $f \circ f(x)$, and $g \circ g(x)$, and specify the domain of each.

Solution Again, we collect the results in a table. (See Table 3.)

Table 3. Composites of f and g and their domains

Function	Formula	Domain
f	$f(x) = \sqrt{x}$	$[0, \infty)$
g	$g(x) = x + 1$	$\mathbb{R}$
$f \circ g$	$f \circ g(x) = f(g(x)) = f(x+1) = \sqrt{x+1}$	$[-1, \infty)$
$g \circ f$	$g \circ f(x) = g(f(x)) = g(\sqrt{x}) = \sqrt{x} + 1$	$[0, \infty)$
$f \circ f$	$f \circ f(x) = f(f(x)) = f(\sqrt{x}) = \sqrt{\sqrt{x}} = x^{1/4}$	$[0, \infty)$
$g \circ g$	$g \circ g(x) = g(g(x)) = g(x+1) = (x+1) + 1 = x + 2$	$\mathbb{R}$

To see why, for example, the domain of $f \circ g$ is $[-1, \infty)$, observe that $g(x) = x + 1$ is defined for all real x but belongs to the domain of f only if $x + 1 \geq 0$, that is, if $x \geq -1$.

EXAMPLE 5 If $G(x) = \dfrac{1-x}{1+x}$, calculate $G \circ G(x)$ and specify its domain.

Solution We calculate

$$G \circ G(x) = G(G(x)) = G\left(\frac{1-x}{1+x}\right) = \frac{1 - \dfrac{1-x}{1+x}}{1 + \dfrac{1-x}{1+x}} = \frac{1 + x - 1 + x}{1 + x + 1 - x} = x.$$

Because the resulting function, x, is defined for all real x, we might be tempted to say that the domain of $G \circ G$ is $\mathbb{R}$. This is wrong! To belong to the domain of $G \circ G$, x must satisfy two conditions:

(i) x must belong to the domain of G, and

(ii) $G(x)$ must belong to the domain of G.

The domain of G consists of all real numbers *except* $x = -1$. If we exclude $x = -1$ from the domain of $G \circ G$, condition (i) will be satisfied. Now observe that the equation $G(x) = -1$ has no solution x, since it is equivalent to $1 - x = -(1 + x)$ or $1 = -1$. Therefore, all numbers $G(x)$ belong to the domain of G, and condition (ii) is satisfied with no further restrictions on x. The domain of $G \circ G$ is $(-\infty, -1) \cup (-1, \infty)$, that is, all real numbers except -1.

Piecewise Defined Functions

Sometimes it is necessary to define a function by using different formulas on different parts of its domain. One example is the absolute value function

$$|x| = \begin{cases} x & \text{if } x \geq 0 \\ -x & \text{if } x < 0. \end{cases}$$

Another would be the tax rates applied to various levels of income. Here are some other examples. (Note how we use solid and hollow dots in their graphs to indicate, respectively, which endpoints do or do not lie on various parts of the graph.)

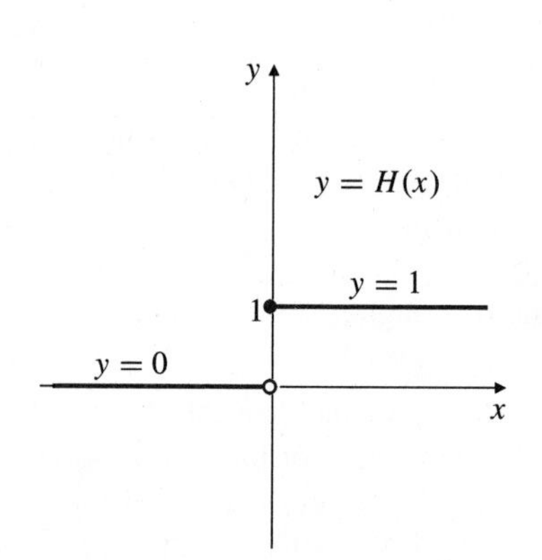

Figure P.61 The Heaviside function

EXAMPLE 6 **The Heaviside function.** The Heaviside function (or unit step function) (Figure P.61) is defined by

$$H(x) = \begin{cases} 1 & \text{if } x \geq 0 \\ 0 & \text{if } x < 0. \end{cases}$$

For instance, if t represents time, the function $6H(t)$ can model the voltage applied to an electric circuit by a 6-volt battery if a switch in the circuit is turned on at time $t = 0$.

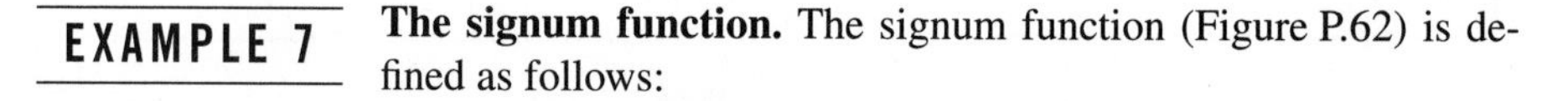

EXAMPLE 7 **The signum function.** The signum function (Figure P.62) is defined as follows:

$$\operatorname{sgn}(x) = \frac{x}{|x|} = \begin{cases} 1 & \text{if } x > 0, \\ -1 & \text{if } x < 0, \\ \text{undefined} & \text{if } x = 0. \end{cases}$$

Figure P.62 The signum function

The name *signum* is the Latin word meaning "sign." The value of the sgn(x) tells whether x is positive or negative. Since 0 is neither positive nor negative, sgn (0) is not defined. The signum function is an odd function.

EXAMPLE 8 The function

$$f(x) = \begin{cases} (x+1)^2 & \text{if } x < -1, \\ -x & \text{if } -1 \leq x < 1, \\ \sqrt{x-1} & \text{if } x \geq 1, \end{cases}$$

is defined on the whole real line but has values given by three different formulas depending on the position of x. Its graph is shown in Figure P.63(a).

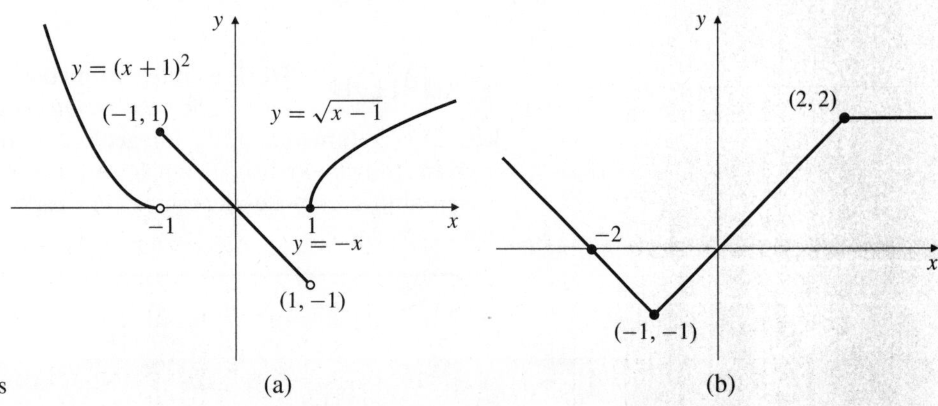

Figure P.63 Piecewise defined functions

EXAMPLE 9 Find a formula for function $g(x)$ graphed in Figure P.63(b).

Solution The graph consists of parts of three lines. For the part $x < -1$, the line has slope -1 and x-intercept -2, so its equation is $y = -(x+2)$. The middle section is the line $y = x$ for $-1 \le x \le 2$. The right section is $y = 2$ for $x > 2$. Combining these formulas, we write

$$g(x) = \begin{cases} -(x+2) & \text{if } x < -1 \\ x & \text{if } -1 \le x \le 2 \\ 2 & \text{if } x > 2. \end{cases}$$

Unlike the previous example, it does not matter here which of the two possible formulas we use to define $g(-1)$, since both give the same value. The same is true for $g(2)$.

The following two functions could be defined by different formulas on every interval between consecutive integers, but we will use an easier way to define them.

EXAMPLE 10 **The greatest integer function.** The function whose value at any number x is the *greatest integer less than or equal to* x is called the **greatest integer function**, or the **integer floor function**. It is denoted $\lfloor x \rfloor$, or, in some books, $[x]$ or $[[x]]$. The graph of $y = \lfloor x \rfloor$ is given in Figure P.64(a). Observe that

$$\lfloor 2.4 \rfloor = 2, \quad \lfloor 1.9 \rfloor = 1, \quad \lfloor 0 \rfloor = 0, \quad \lfloor -1.2 \rfloor = -2,$$
$$\lfloor 2 \rfloor = 2, \quad \lfloor 0.2 \rfloor = 0, \quad \lfloor -0.3 \rfloor = -1, \quad \lfloor -2 \rfloor = -2.$$

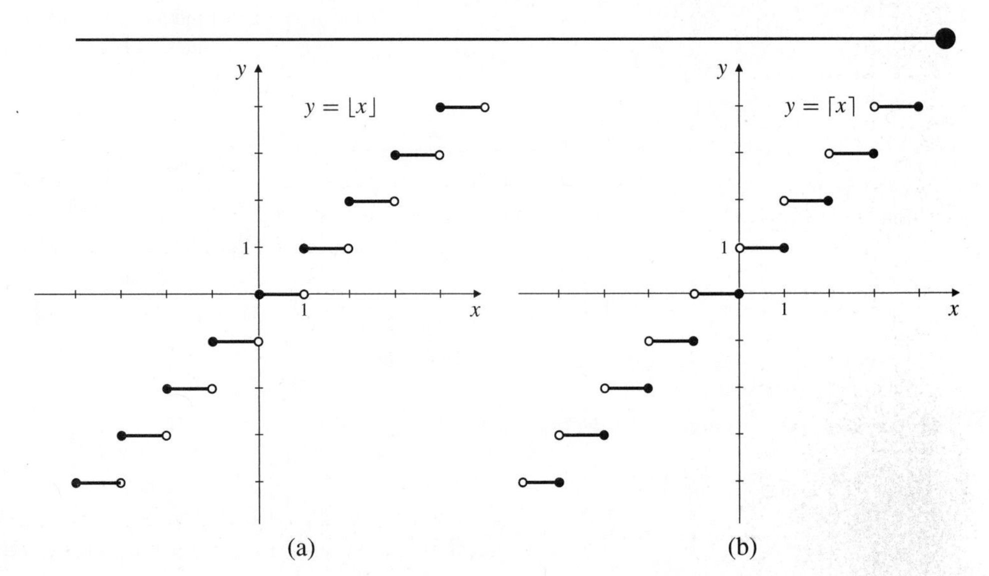

Figure P.64
(a) The greatest integer function $\lfloor x \rfloor$
(b) The least integer function $\lceil x \rceil$

EXAMPLE 11 **The least integer function.** The function whose value at any number x is the *smallest integer greater than or equal to* x is called the **least integer function**, or the **integer ceiling function**. It is denoted $\lceil x \rceil$. Its graph is given in Figure P.64(b). For positive values of x, this function might represent, for example, the cost of parking x hours in a parking lot that charges \$1 for each hour or part of an hour.

EXERCISES P.5

In Exercises 1–2, find the domains of the functions $f+g$, $f-g$, fg, f/g, and g/f, and give formulas for their values.

1. $f(x)=x, \quad g(x)=\sqrt{x-1}$

2. $f(x)=\sqrt{1-x}, \quad g(x)=\sqrt{1+x}$

Sketch the graphs of the functions in Exercises 3–6 by combining the graphs of simpler functions from which they are built up.

3. $x-x^2$ **4.** x^3-x

5. $x+|x|$ **6.** $|x|+|x-2|$

7. If $f(x)=x+5$ and $g(x)=x^2-3$, find the following:

(a) $f\circ g(0)$ (b) $g(f(0))$
(c) $f(g(x))$ (d) $g\circ f(x)$
(e) $f\circ f(-5)$ (f) $g(g(2))$
(g) $f(f(x))$ (h) $g\circ g(x)$

In Exercises 8–10, construct the following composite functions and specify the domain of each.

(a) $f\circ f(x)$ (b) $f\circ g(x)$
(c) $g\circ f(x)$ (d) $g\circ g(x)$

8. $f(x)=2/x, \quad g(x)=x/(1-x)$

9. $f(x)=1/(1-x), \quad g(x)=\sqrt{x-1}$

10. $f(x)=(x+1)/(x-1), \quad g(x)=\operatorname{sgn}(x)$

Find the missing entries in Table 4 (Exercises 11–16).

Table 4.

	$f(x)$	$g(x)$	$f\circ g(x)$
11.	x^2	$x+1$	
12.		$x+4$	x
13.	$\sqrt{x}$		$\lvert x\rvert$
14.		$x^{1/3}$	$2x+3$
15.	$(x+1)/x$		x
16.		$x-1$	$1/x^2$

17. Use a graphing utility to examine in order the graphs of the functions

$$y=\sqrt{x}, \qquad y=2+\sqrt{x},$$
$$y=2+\sqrt{3+x}, \qquad y=1/(2+\sqrt{3+x}).$$

Describe the effect on the graph of the change made in the function at each stage.

18. Repeat the previous exercise for the functions

$$y=2x, \qquad y=2x-1, \qquad y=1-2x,$$
$$y=\sqrt{1-2x}, \qquad y=\frac{1}{\sqrt{1-2x}}, \qquad y=\frac{1}{\sqrt{1-2x}}-1.$$

In Exercises 19–24, f refers to the function with domain $[0, 2]$ and range $[0, 1]$, whose graph is shown in Figure P.65. Sketch the graphs of the indicated functions, and specify their domains and ranges.

19. $2f(x)$ **20.** $-(1/2)f(x)$

21. $f(2x)$ **22.** $f(x/3)$

23. $1+f(-x/2)$ **24.** $2f((x-1)/2)$

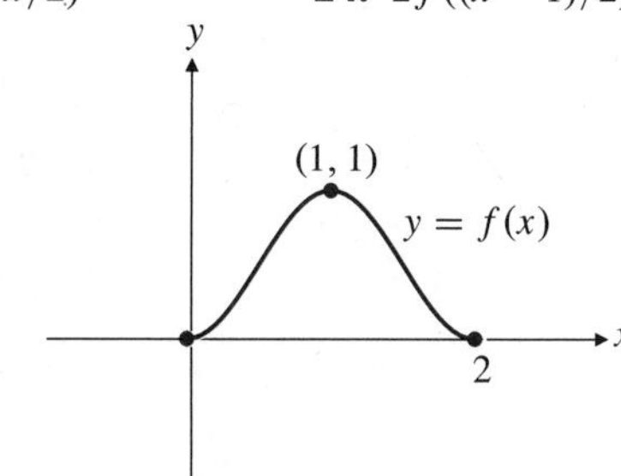

Figure P.65

In Exercises 25–26, sketch the graphs of the given functions.

25. $f(x)=\begin{cases} x & \text{if } 0\le x\le 1 \\ 2-x & \text{if } 1<x\le 2 \end{cases}$

26. $g(x)=\begin{cases} \sqrt{x} & \text{if } 0\le x\le 1 \\ 2-x & \text{if } 1<x\le 2 \end{cases}$

27. Find all real values of the constants A and B for which the function $F(x)=Ax+B$ satisfies:

(a) $F\circ F(x)=F(x)$ for all x.

(b) $F\circ F(x)=x$ for all x.

Greatest and least integer functions

28. For what values of x is (a) $\lfloor x\rfloor=0$? (b) $\lceil x\rceil=0$?

29. What real numbers x satisfy the equation $\lfloor x\rfloor=\lceil x\rceil$?

30. True or false: $\lceil -x\rceil=-\lfloor x\rfloor$ for all real x?

31. Sketch the graph of $y=x-\lfloor x\rfloor$.

32. Sketch the graph of the function

$$f(x)=\begin{cases} \lfloor x\rfloor & \text{if } x\ge 0 \\ \lceil x\rceil & \text{if } x<0. \end{cases}$$

Why is $f(x)$ called *the integer part of* x?

Even and odd functions

33. Assume that f is an even function, g is an odd function, and both f and g are defined on the whole real line $\mathbb{R}$. Is each of the following functions even, odd, or neither?

$$f+g, \quad fg, \quad f/g, \quad g/f, \quad f^2 = ff, \quad g^2 = gg$$

$$f \circ g, \quad g \circ f, \quad f \circ f, \quad g \circ g$$

34. If f is both an even and an odd function, show that $f(x) = 0$ at every point of its domain.

35. Let f be a function whose domain is symmetric about the origin, that is, $-x$ belongs to the domain whenever x does.

(a) Show that f is the sum of an even function and an odd function:

$$f(x) = E(x) + O(x),$$

where E is an even function and O is an odd function. *Hint:* Let $E(x) = (f(x) + f(-x))/2$. Show that $E(-x) = E(x)$, so that E is even. Then show that $O(x) = f(x) - E(x)$ is odd.

(b) Show that there is only one way to write f as the sum of an even and an odd function. *Hint:* One way is given in part (a). If also $f(x) = E_1(x) + O_1(x)$, where E_1 is even and O_1 is odd, show that $E - E_1 = O_1 - O$ and then use Exercise 34 to show that $E = E_1$ and $O = O_1$.

P.6 Polynomials and Rational Functions

Among the easiest functions to deal with in calculus are polynomials. These are sums of terms each of which is a constant multiple of a nonnegative integer power of the variable of the function.

DEFINITION 5

A **polynomial** is a function P whose value at x is

$$P(x) = a_n x^n + a_{n-1} x^{n-1} + \cdots + a_2 x^2 + a_1 x + a_0,$$

where $a_n, a_{n-1}, \ldots, a_2, a_1$, and a_0, called the **coefficients** of the polynomial, are constants and, if $n > 0$, then $a_n \neq 0$. The number n, the degree of the highest power of x in the polynomial, is called the **degree** of the polynomial. (The degree of the zero polynomial is not defined.)

For example,

3	is a polynomial of degree 0.
$2 - x$	is a polynomial of degree 1.
$2x^3 - 17x + 1$	is a polynomial of degree 3.

Generally, we assume that the polynomials we deal with are **real polynomials**; that is, their coefficients are real numbers rather than more general complex numbers. Often the coefficients will be integers or rational numbers. Polynomials play a role in the study of functions somewhat analogous to the role played by integers in the study of numbers. For instance, just as we always get an integer result if we add, subtract, or multiply two integers, we always get a polynomial result if we add, subtract, or multiply two polynomials. Adding or subtracting polynomials produces a polynomial whose degree does not exceed the larger of the two degrees of the polynomials being combined. Multiplying two polynomials of degrees m and n produces a product polynomial of degree $m + n$. For instance, for the product

$$(x^2 + 1)(x^3 - x - 2) = x^5 - 2x^2 - x - 2,$$

the two factors have degrees 2 and 3, so the result has degree 5.

The following definition is analogous to the definition of a rational number as the quotient of two integers.

DEFINITION 6

If $P(x)$ and $Q(x)$ are two polynomials and $Q(x)$ is not the zero polynomial, then the function

$$R(x) = \frac{P(x)}{Q(x)}$$

is called a **rational function**. By the domain convention, the domain of $R(x)$ consists of all real numbers x except those for which $Q(x) = 0$.

Two examples of rational functions and their domains are

$$R(x) = \frac{2x^3 - 3x^2 + 3x + 4}{x^2 + 1} \quad \text{with domain } \mathbb{R}\text{, all real numbers.}$$

$$S(x) = \frac{1}{x^2 - 4} \quad \text{with domain all real numbers except } \pm 2.$$

Remark If the numerator and denominator of a rational function have a common factor, that factor can be cancelled out just as with integers. However, the resulting simpler rational function may not have the same domain as the original one, so it should be regarded as a different rational function even though it is equal to the original one at all points of the original domain. For instance,

$$\frac{x^2 - x}{x^2 - 1} = \frac{x(x-1)}{(x+1)(x-1)} = \frac{x}{x+1} \quad \text{only if } x \neq \pm 1,$$

even though $x = 1$ is in the domain of $x/(x+1)$.

When we divide a positive integer a by a smaller positive integer b, we can obtain an integer quotient q and an integer remainder r satisfying $0 \le r < b$ and hence write the fraction a/b (in a unique way) as the sum of the integer q and another fraction whose numerator (the remainder r) is smaller than its denominator b. For instance,

$$\frac{7}{3} = 2 + \frac{1}{3}; \qquad \text{the quotient is 2, the remainder is 1.}$$

Similarly, if A_m and B_n are polynomials having degrees m and n, respectively, and if $m > n$, then we can express the rational function A_m/B_n (in a unique way) as the sum of a quotient polynomial Q_{m-n} of degree $m - n$ and another rational function R_k/B_n where the numerator polynomial R_k (the remainder in the division) is either zero or has degree $k < n$:

$$\frac{A_m(x)}{B_n(x)} = Q_{m-n}(x) + \frac{R_k(x)}{B_n(x)}. \qquad \textbf{(The Division Algorithm)}$$

We calculate the quotient and remainder polynomials by using long division or an equivalent method.

EXAMPLE 1 Write the division algorithm for $\dfrac{2x^3 - 3x^2 + 3x + 4}{x^2 + 1}$.

Solution **METHOD I.** Use long division:

$$\begin{array}{r|l}
 & 2x \quad - 3 \\ \hline
x^2 + 1 & 2x^3 - 3x^2 + 3x + 4 \\
 & 2x^3 \qquad\quad + 2x \\ \hline
 & \quad -3x^2 + x + 4 \\
 & \quad -3x^2 \qquad - 3 \\ \hline
 & \qquad\qquad x + 7
\end{array}$$

Thus,

$$\frac{2x^3 - 3x^2 + 3x + 4}{x^2 + 1} = 2x - 3 + \frac{x + 7}{x^2 + 1}.$$

The quotient is $2x - 3$, and the remainder is $x + 7$.

METHOD II. Use short division; add appropriate lower-degree terms to the terms of the numerator that have degrees not less than the degree of the denominator to enable factoring out the denominator, and then subtract those terms off again.

$$\begin{aligned} & 2x^3 - 3x^2 + 3x + 4 \\ = \; & 2x^3 + 2x - 3x^2 - 3 + 3x + 4 - 2x + 3 \\ = \; & 2x(x^2 + 1) - 3(x^2 + 1) + x + 7, \end{aligned}$$

from which it follows at once that

$$\frac{2x^3 - 3x^2 + 3x + 4}{x^2 + 1} = 2x - 3 + \frac{x + 7}{x^2 + 1}.$$

Roots, Zeros, and Factors

A number r is called a **root** or **zero** of the polynomial P if $P(r) = 0$. For example, $P(x) = x^3 - 4x$ has three roots: 0, 2, and -2; substituting any of these numbers for x makes $P(x) = 0$. In this context the terms "root" and "zero" are often used interchangeably. It is technically more correct to call a number r satisfying $P(r) = 0$ a *zero* of the polynomial *function* P and a *root* of the *equation* $P(x) = 0$, and later in this book we will follow this convention more closely. But for now, to avoid confusion with the *number* zero, we will prefer to use "root" rather than "zero" even when referring to the polynomial P rather than the equation $P(x) = 0$.

The **Fundamental Theorem of Algebra** (see Appendix II) states that every polynomial of degree at least 1 has a root (although the root might be a complex number). For example, the linear (degree 1) polynomial $ax + b$ has the root $-b/a$ since $a(-b/a) + b = 0$. A constant polynomial (one of degree zero) cannot have any roots unless it is the zero polynomial, in which case every number is a root.

Real polynomials need not always have real roots; the polynomial $x^2 + 4$ is never zero for any real number x, but it is zero if x is either of the two complex numbers $2i$ and $-2i$, where i is the so-called imaginary unit satisfying $i^2 = -1$. (See Appendix I for a discussion of complex numbers.) The numbers $2i$ and $-2i$ are *complex conjugates of each other*. Any complex roots of a real polynomial must occur in conjugate pairs. (See Appendix II for a proof of this fact.)

In our study of calculus we will often find it useful to factor polynomials into products of polynomials of lower degree, especially degree 1 or 2 (linear or quadratic polynomials). The following theorem shows the connection between linear factors and roots.

THEOREM 1 **The Factor Theorem**

The number r is a root of the polynomial P of degree not less than 1 if and only if $x - r$ is a factor of $P(x)$.

PROOF By the division algorithm there exists a quotient polynomial Q having degree one less than that of P and a remainder polynomial of degree 0 (i.e., a constant c) such that

$$\frac{P(x)}{x - r} = Q(x) + \frac{c}{x - r}.$$

Thus $P(x) = (x - r)Q(x) + c$, and $P(r) = 0$ if and only if $c = 0$, in which case $P(x) = (x - r)Q(x)$ and $x - r$ is a factor of $P(x)$.

It follows from Theorem 1 and the Fundamental Theorem of Algebra that every polynomial of degree $n \geq 1$ has n roots. (If P has degree $n \geq 2$, then P has a zero r and $P(x) = (x - r)Q(x)$, where Q is a polynomial of degree $n - 1 \geq 1$, which in turn has a root, etc.) Of course, the roots of a polynomial need not all be different. The 4th degree polynomial $P(x) = x^4 - 3x^3 + 3x^2 - x = x(x - 1)^3$ has four roots; one is 0 and the other three are each equal to 1. We say that the root 1 has **multiplicity** 3 because we can divide $P(x)$ by $(x - 1)^3$ and still get zero remainder.

If P is a real polynomial having a complex root $r_1 = u + iv$, where u and v are real and $v \neq 0$, then, as asserted above, the complex conjugate of r_1, namely, $r_2 = u - iv$, will also be a root of P. (Moreover, r_1 and r_2 will have the same multiplicity.) Thus, both $x - u - iv$ and $x - u + iv$ are factors of $P(x)$, and so, therefore, is their product

$$(x - u - iv)(x - u + iv) = (x - u)^2 + v^2 = x^2 - 2ux + u^2 + v^2,$$

which is a quadratic polynomial having no real roots. It follows that every real polynomial can be factored into a product of real (possibly repeated) linear factors and real (also possibly repeated) quadratic factors having no real zeros.

EXAMPLE 2 What is the degree of $P(x) = x^3(x^2 + 2x + 5)^2$? What are the roots of P and, what is the multiplicity of each root?

Solution If P is expanded, the highest power of x present in the expansion is $x^3(x^2)^2 = x^7$, so P has degree 7. The factor $x^3 = (x - 0)^3$ indicates that 0 is a root of P having multiplicity 3. The remaining four roots will be the two roots of $x^2 + 2x + 5$, each having multiplicity 2. Now

$$\begin{aligned}\left[x^2 + 2x + 5\right]^2 &= \left[(x + 1)^2 + 4\right]^2 \\ &= \left[(x + 1 + 2i)(x + 1 - 2i)\right]^2.\end{aligned}$$

Hence the seven roots of P are:

$$\begin{cases} 0,\ 0,\ 0 & 0 \text{ has multiplicity } 3, \\ -1 - 2i,\ -1 - 2i & -1 - 2i \text{ has multiplicity } 2, \\ -1 + 2i,\ -1 + 2i & -1 + 2i \text{ has multiplicity } 2. \end{cases}$$

Roots and Factors of Quadratic Polynomials

There is a well-known formula for finding the roots of a quadratic polynomial.

The Quadratic Formula

The two solutions of the quadratic equation

$$Ax^2 + Bx + C = 0,$$

where A, B, and C are constants and $A \neq 0$, are given by

$$x = \frac{-B \pm \sqrt{B^2 - 4AC}}{2A}.$$

To see this, just divide the equation by A and complete the square for the terms in x:

$$\begin{aligned} x^2 + \frac{B}{A}x + \frac{C}{A} &= 0 \\ x^2 + \frac{2B}{2A}x + \frac{B^2}{4A^2} &= \frac{B^2}{4A^2} - \frac{C}{A} \\ \left(x + \frac{B}{2A}\right)^2 &= \frac{B^2 - 4AC}{4A^2} \\ x + \frac{B}{2A} &= \pm\frac{\sqrt{B^2 - 4AC}}{2A}. \end{aligned}$$

The quantity $D = B^2 - 4AC$ that appears under the square root in the quadratic formula is called the **discriminant** of the quadratic equation or polynomial. The nature of the roots of the quadratic depends on the sign of this discriminant.

(a) If $D > 0$, then $D = k^2$ for some real constant k, and the quadratic has two distinct roots, $(-B + k)/(2A)$ and $(-B - k)/(2A)$.

(b) If $D = 0$, then the quadratic has only the root $-B/(2A)$, and this root has multiplicity 2. (It is called a *double root.*)

(c) If $D < 0$, then $D = -k^2$ for some real constant k, and the quadratic has two complex conjugate roots, $(-B + ki)/(2A)$ and $(-B - ki)/(2A)$.

EXAMPLE 3 Find the roots of these quadratic polynomials and thereby factor the polynomials into linear factors:

(a) $x^2 + x - 1$ (b) $9x^2 - 6x + 1$ (c) $2x^2 + x + 1$.

Solution We use the quadratic formula to solve the corresponding quadratic equations to find the roots of the three polynomials.

(a) $A = 1, \quad B = 1, \quad C = -1$

$$x = \frac{-1 \pm \sqrt{1+4}}{2} = -\frac{1}{2} \pm \frac{\sqrt{5}}{2}$$

$$x^2 + x - 1 = \left(x + \frac{1}{2} - \frac{\sqrt{5}}{2}\right)\left(x + \frac{1}{2} + \frac{\sqrt{5}}{2}\right).$$

(b) $A = 9, \quad B = -6, \quad C = 1$

$$x = \frac{6 \pm \sqrt{36 - 36}}{18} = \frac{1}{3} \quad \text{(double root)}$$

$$9x^2 - 6x + 1 = 9\left(x - \frac{1}{3}\right)^2 = (3x - 1)^2.$$

(c) $A = 2, \quad B = 1, \quad C = 1$

$$x = \frac{-1 \pm \sqrt{1 - 8}}{4} = -\frac{1}{4} \pm \frac{\sqrt{7}}{4}i$$

$$2x^2 + x + 1 = 2\left(x + \frac{1}{4} - \frac{\sqrt{7}}{4}i\right)\left(x + \frac{1}{4} + \frac{\sqrt{7}}{4}i\right).$$

Remark There exist formulas for calculating exact roots of cubic (degree 3) and quartic (degree 4) polynomials, but, unlike the quadratic formula above, they are very complicated and almost never used. Instead, calculus will provide us with very powerful and easily used tools for approximating roots of polynomials (and solutions of much more general equations) to any desired degree of accuracy.

Miscellaneous Factorings

Some quadratic and higher-degree polynomials can be (at least partially) factored by inspection. Some simple examples include:

(a) Common Factor: $ax^2 + bx = x(ax + b)$.

(b) Difference of Squares: $x^2 - a^2 = (x - a)(x + a)$.

(c) Difference of Cubes: $x^3 - a^3 = (x - a)(x^2 + ax + a^2)$.

(d) More generally, a difference of nth powers for any positive integer n:

$$x^n - a^n = (x - a)(x^{n-1} + ax^{n-2} + a^2x^{n-3} + \cdots + a^{n-2}x + a^{n-1}).$$

Note that $x - a$ is a factor of $x^n - a^n$ for any positive integer n.

(e) It is also true that if n is an *odd positive integer*, then $x + a$ is a factor of $x^n + a^n$. For example,

$$x^3 + a^3 = (x + a)(x^2 - ax + a^2)$$
$$x^5 + a^5 = (x + a)(x^4 - ax^3 + a^2x^2 - a^3x + a^4).$$

Finally, we mention a trial-and-error method of factoring quadratic polynomials sometimes called *trinomial factoring.* Since

$$(x + p)(x + q) = x^2 + (p + q)x + pq,$$
$$(x - p)(x - q) = x^2 - (p + q)x + pq, \qquad \text{and}$$
$$(x + p)(x - q) = x^2 + (p - q)x - pq,$$

we can sometimes spot the factors of $x^2 + Bx + C$ by looking for factors of $|C|$ for which the sum or difference is B. More generally, we can sometimes factor

$$Ax^2 + Bx + C = (ax + b)(cx + d)$$

by looking for factors a and c of A and factors b and d of C for which $ad + bc = B$. Of course, if this fails you can always resort to the quadratic formula to find the roots and, therefore, the factors, of the quadratic polynomial.

EXAMPLE 4

$$x^2 - 5x + 6 = (x - 3)(x - 2) \qquad p = 3, q = 2, pq = 6, p + q = 5$$
$$x^2 + 7x + 6 = (x + 6)(x + 1) \qquad p = 6, q = 1, pq = 6, p + q = 7$$
$$x^2 + x - 6 = (x + 3)(x - 2) \qquad p = 3, q = -2, pq = -6, p + q = 1$$
$$2x^2 + x - 10 = (2x + 5)(x - 2) \qquad a = 2, b = 5, c = 1, d = -2$$
$$ac = 2, bd = -10, ad + bc = 1.$$

EXAMPLE 5

Find the roots of the following polynomials:

(a) $x^3 - x^2 - 4x + 4$, (b) $x^4 + 3x^2 - 4$, (c) $x^5 - x^4 - x^2 + x$.

Solution (a) There is an obvious common factor:

$$x^3 - x^2 - 4x + 4 = (x - 1)(x^2 - 4) = (x - 1)(x - 2)(x + 2).$$

The roots are 1, 2, and -2.

(b) This is a trinomial in x^2 for which there is an easy factoring:

$$x^4 + 3x^2 - 4 = (x^2 + 4)(x^2 - 1) = (x + 2i)(x - 2i)(x + 1)(x - 1).$$

The roots are 1, -1, $2i$, and $-2i$.

(c) We start with some obvious factorings:

$$\begin{aligned} x^5 - x^4 - x^2 + x &= x(x^4 - x^3 - x + 1) = x(x - 1)(x^3 - 1) \\ &= x(x - 1)^2(x^2 + x + 1). \end{aligned}$$

Thus 0 is a root, and 1 is a double root. The remaining two roots must come from the quadratic factor $x^2 + x + 1$, which cannot be factored easily by inspection so we use the formula:

$$x = \frac{-1 \pm \sqrt{1 - 4}}{2} = -\frac{1}{2} \pm \frac{\sqrt{3}}{2} i.$$

EXAMPLE 6 For what values of the real constant b will the product of the real polynomials $x^2 - bx + a^2$ and $x^2 + bx + a^2$ be equal to $x^4 + a^4$? Use your answer to express $x^4 + 1$ as a product of two real quadratic polynomials each having no real roots.

Solution We have

$$\begin{aligned} (x^2 - bx + a^2)(x^2 + bx + a^2) &= \left(x^2 + a^2\right)^2 - b^2x^2 \\ &= x^4 + 2a^2x^2 + a^4 - b^2x^2 = x^4 + a^4 \end{aligned}$$

provided that $b^2 = 2a^2$, that is, $b = \pm\sqrt{2}a$.

If $a = 1$, then $b = \pm\sqrt{2}$ and we have

$$x^4 + 1 = (x^2 - \sqrt{2}x + 1)(x^2 + \sqrt{2}x + 1).$$

EXERCISES P.6

Find the roots of the polynomials in Exercises 1–12. If a root is repeated, give its multiplicity. Also, write each polynomial as a product of linear factors.

1. $x^2 + 7x + 10$

2. $x^2 - 3x - 10$

3. $x^2 + 2x + 2$

4. $x^2 - 6x + 13$

5. $16x^4 - 8x^2 + 1$

6. $x^4 + 6x^3 + 9x^2$

7. $x^3 + 1$

8. $x^4 - 1$

9. $x^6 - 3x^4 + 3x^2 - 1$

10. $x^5 - x^4 - 16x + 16$

11. $x^5 + x^3 + 8x^2 + 8$

12. $x^9 - 4x^7 - x^6 + 4x^4$

In Exercises 13–16, determine the domains of the given rational functions.

13. $\dfrac{3x + 2}{x^2 + 2x + 2}$

14. $\dfrac{x^2 - 9}{x^3 - x}$

15. $\dfrac{4}{x^3 + x^2}$

16. $\dfrac{x^3 + 3x^2 + 6}{x^2 + x - 1}$

In Exercises 17–20, express the given rational function as the sum of a polynomial and another rational function whose numerator is either zero or has smaller degree than the denominator.

17. $\dfrac{x^3 - 1}{x^2 - 2}$

18. $\dfrac{x^2}{x^2 + 5x + 3}$

19. $\dfrac{x^3}{x^2 + 2x + 3}$

20. $\dfrac{x^4 + x^2}{x^3 + x^2 + 1}$

In Exercises 21–22 express the given polynomial as a product of real quadratic polynomials with no real roots.

21. $P(x) = x^4 + 4$

22. $P(x) = x^4 + x^2 + 1$

23. Show that $x - 1$ is a factor of a polynomial P of positive degree if and only if the sum of the coefficients of P is zero.

24. What condition should the coefficients of a polynomial satsify to ensure that $x + 1$ is a factor of that polynomial?

25. The complex conjugate of a complex number $z = u + iv$ (where u and v are real numbers) is the complex number $\bar{z} = u - iv$. It is shown in Appendix I that the complex conjugate of a sum (or product) of complex numbers is the sum (or product) of the complex conjugates of those numbers. Use this fact to verify that if $z = u + iv$ is a complex root of a polynomial P having real coefficients, then its conjugate $\bar{z}$ is also a root of P.

26. Continuing the previous exercise, show that if $z = u + iv$ (where u and v are real numbers) is a complex root of a polynomial P with real coefficients, then P must have the real quadratic factor $x^2 - 2ux + u^2 + v^2$.

27. Use the result of Exercise 26 to show that if $z = u + iv$ (where u and v are real numbers) is a complex root of a polynomial P with real coefficients, then z and $\bar{z}$ are roots of P having the same multiplicity.

P.7 The Trigonometric Functions

Most people first encounter the quantities $\cos t$ and $\sin t$ as ratios of sides in a right-angled triangle having t as one of the acute angles. If the sides of the triangle are labelled "hyp" for hypotenuse, "adj" for the side adjacent to angle t, and "opp" for the side opposite angle t (see Figure P.66), then

$$\cos t = \frac{\text{adj}}{\text{hyp}} \qquad \text{and} \qquad \sin t = \frac{\text{opp}}{\text{hyp}}. \qquad (*)$$

hyp
opp
t
adj

Figure P.66 $\cos t = \text{adj}/\text{hyp}$
$\sin t = \text{opp}/\text{hyp}$

These ratios depend only on the angle t, not on the particular triangle, since all right-angled triangles having an acute angle t are similar.

In calculus we need more general definitions of $\cos t$ and $\sin t$ as functions defined for *all real numbers* t, not just acute angles. Such definitions are phrased in terms of a circle rather than a triangle.

Let C be the circle with centre at the origin O and radius 1; its equation is $x^2 + y^2 = 1$. Let A be the point $(1, 0)$ on C. For any real number t, let P_t be the point on C at distance $|t|$ from A, measured along C in the counterclockwise direction if $t > 0$, and the clockwise direction if $t < 0$. For example, since C has circumference 2π, the point $P_{\pi/2}$ is one-quarter of the way counterclockwise around C from A; it is the point $(0, 1)$.

We will use the arc length t as a measure of the size of the angle AOP_t. See Figure P.67.

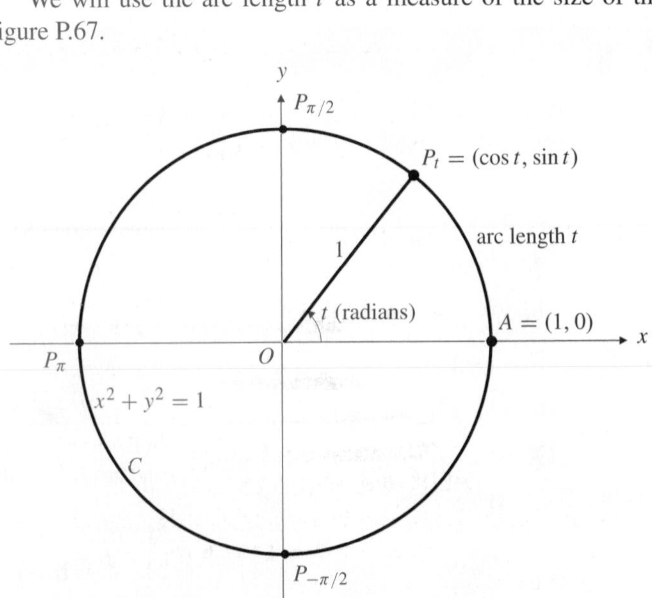

Figure P.67 If the length of arc AP_t is t units, then angle $AOP_t = t$ radians

DEFINITION 7

The **radian measure** of angle AOP_t is t radians:

$$\angle AOP_t = t \text{ radians.}$$

We are more used to measuring angles in **degrees**. Since P_π is the point $(-1, 0)$, halfway (π units of distance) around C from A, we have

$$\pi \text{ radians} = 180°.$$

To convert degrees to radians, multiply by $\pi/180$; to convert radians to degrees, multiply by $180/\pi$.

Angle convention

In calculus it is assumed that all angles are measured in radians unless degrees or other units are stated explicitly. When we talk about the angle $\pi/3$, we mean $\pi/3$ radians (which is 60°), not $\pi/3$ degrees.

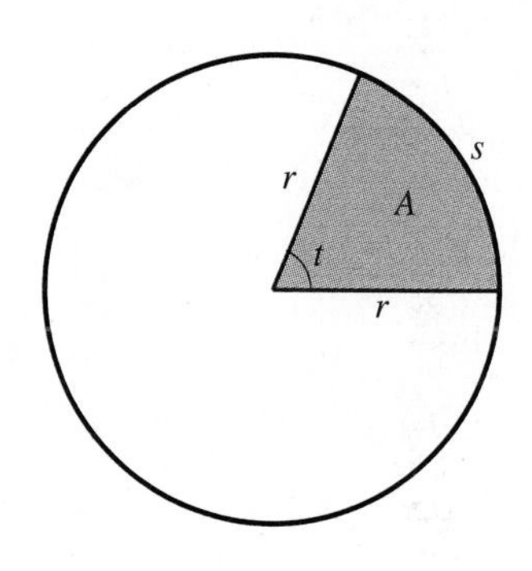

Figure P.68 Arc length $s = rt$
Sector area $A = r^2t/2$

EXAMPLE 1 **Arc length and sector area.** An arc of a circle of radius r subtends an angle t at the centre of the circle. Find the length s of the arc and the area A of the sector lying between the arc and the centre of the circle.

Solution The length s of the arc is the same fraction of the circumference $2\pi r$ of the circle that the angle t is of a complete revolution 2π radians (or 360°). Thus,

$$s = \frac{t}{2\pi}(2\pi r) = rt \text{ units.}$$

Similarly, the area A of the circular sector (Figure P.68) is the same fraction of the area πr^2 of the whole circle:

$$A = \frac{t}{2\pi}(\pi r^2) = \frac{r^2 t}{2} \text{ units}^2.$$

(We will show that the area of a circle of radius r is πr^2 in Section 1.1.)

Using the procedure described above, we can find the point P_t corresponding to any real number t, positive or negative. We define $\cos t$ and $\sin t$ to be the coordinates of P_t. (See Figure P.69.)

DEFINITION 8

Cosine and sine

For any real t, the **cosine** of t (abbreviated $\cos t$) and the **sine** of t (abbreviated $\sin t$) are the x- and y-coordinates of the point P_t.

$$\cos t = \text{the } x\text{-coordinate of } P_t$$
$$\sin t = \text{the } y\text{-coordinate of } P_t$$

Because they are defined this way, cosine and sine are often called the **circular functions**. Note that these definitions agree with the ones given earlier for an acute angle. (See formulas (∗) at the beginning of this section.) The triangle involved is $P_t O Q_t$ in Figure P.69.

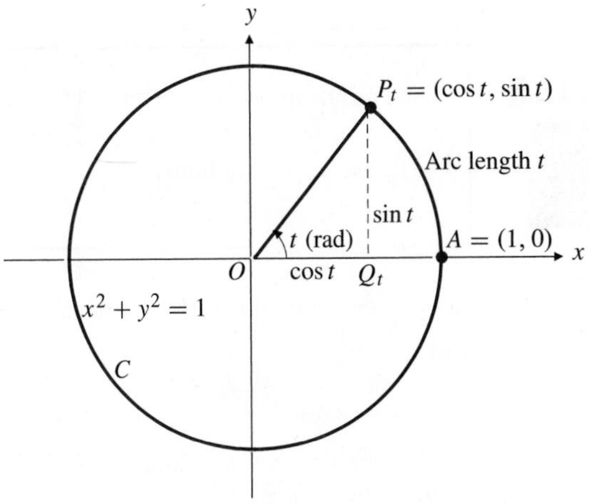

Figure P.69 The coordinates of P_t are $(\cos t, \sin t)$

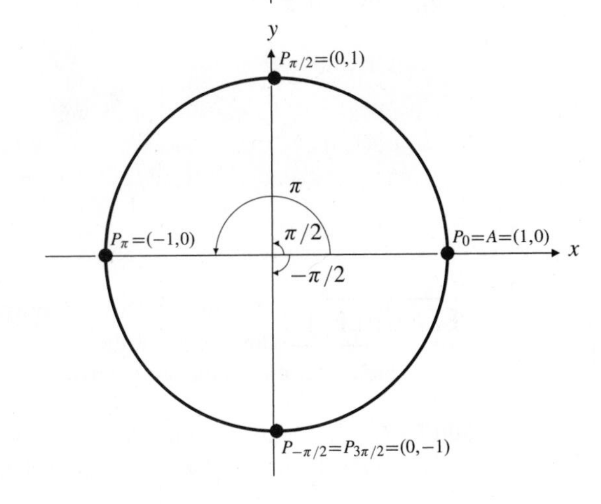

Figure P.70 Some special angles

EXAMPLE 2 Examining the coordinates of $P_0 = A$, $P_{\pi/2}$, P_π, and $P_{-\pi/2} = P_{3\pi/2}$ in Figure P.70, we obtain the following values:

$$\cos 0 = 1 \quad \cos\frac{\pi}{2} = 0 \quad \cos\pi = -1 \quad \cos\left(-\frac{\pi}{2}\right) = \cos\frac{3\pi}{2} = 0$$

$$\sin 0 = 0 \quad \sin\frac{\pi}{2} = 1 \quad \sin\pi = 0 \quad \sin\left(-\frac{\pi}{2}\right) = \sin\frac{3\pi}{2} = -1$$

Some Useful Identities

Many important properties of $\cos t$ and $\sin t$ follow from the fact that they are coordinates of the point P_t on the circle C with equation $x^2 + y^2 = 1$.

The range of cosine and sine. For every real number t,

$$-1 \le \cos t \le 1 \qquad \text{and} \qquad -1 \le \sin t \le 1.$$

The Pythagorean identity. The coordinates $x = \cos t$ and $y = \sin t$ of P_t must satisfy the equation of the circle. Therefore, for every real number t,

$$\cos^2 t + \sin^2 t = 1.$$

(Note that $\cos^2 t$ means $(\cos t)^2$, not $\cos(\cos t)$. This is an unfortunate notation, but it is used everywhere in technical literature, so you have to get used to it!)

Periodicity. Since C has circumference 2π, adding 2π to t causes the point P_t to go one extra complete revolution around C and end up in the same place: $P_{t+2\pi} = P_t$. Thus, for every t,

$$\cos(t + 2\pi) = \cos t \qquad \text{and} \qquad \sin(t + 2\pi) = \sin t.$$

This says that cosine and sine are **periodic** with period 2π.

Cosine is an even function. Sine is an odd function. Since the circle $x^2 + y^2 = 1$ is symmetric about the x-axis, the points P_{-t} and P_t have the same x-coordinates and opposite y-coordinates (Figure P.71).

$$\cos(-t) = \cos t \qquad \text{and} \qquad \sin(-t) = -\sin t.$$

Complementary angle identities. Two angles are complementary if their sum is $\pi/2$ (or 90°). The points $P_{(\pi/2)-t}$ and P_t are reflections of each other in the line $y = x$ (Figure P.72), so the x-coordinate of one is the y-coordinate of the other and vice versa. Thus,

$$\cos\left(\frac{\pi}{2} - t\right) = \sin t \qquad \text{and} \qquad \sin\left(\frac{\pi}{2} - t\right) = \cos t.$$

Supplementary angle identities. Two angles are supplementary if their sum is π (or 180°). Since the circle is symmetric about the y-axis, $P_{\pi-t}$ and P_t have the same y-coordinates and opposite x-coordinates. (See Figure P.73.) Thus,

$$\cos(\pi - t) = -\cos t \qquad \text{and} \qquad \sin(\pi - t) = \sin t.$$

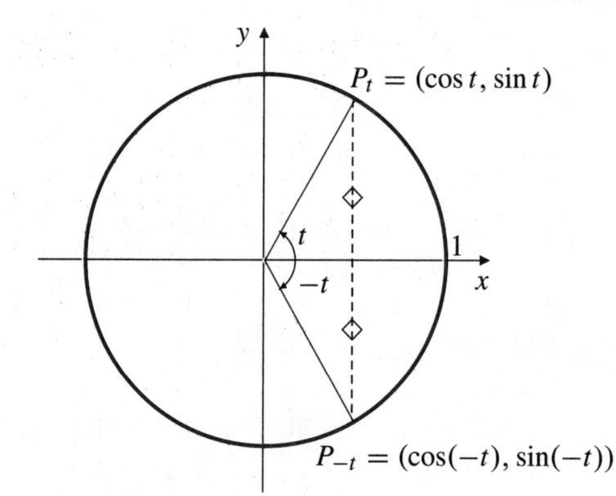

Figure P.71 $\cos(-t) = \cos t$
$\sin(-t) = -\sin t$

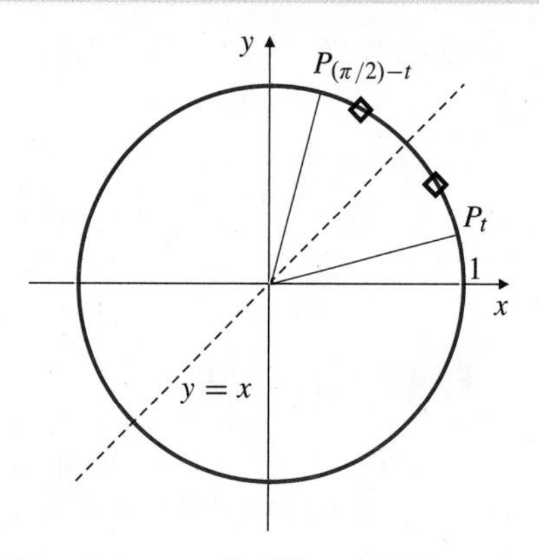

Figure P.72 $\cos((\pi/2) - t) = \sin t$
$\sin((\pi/2) - t) = \cos t$

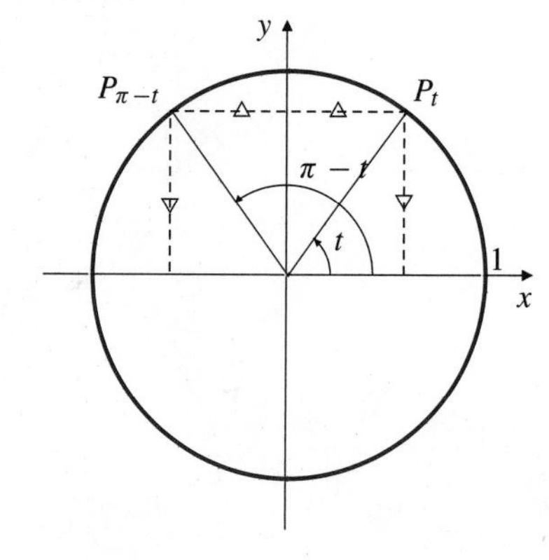

Figure P.73 $\cos(\pi - t) = -\cos t$
$\sin(\pi - t) = \sin t$

Some Special Angles

EXAMPLE 3 Find the sine and cosine of $\pi/4$ (i.e., 45°).

Solution The point $P_{\pi/4}$ lies in the first quadrant on the line $x = y$. To find its coordinates, substitute $y = x$ into the equation $x^2 + y^2 = 1$ of the circle, obtaining $2x^2 = 1$. Thus $x = y = 1/\sqrt{2}$ (see Figure P.74), and

$$\cos(45°) = \cos\frac{\pi}{4} = \frac{1}{\sqrt{2}}, \qquad \sin(45°) = \sin\frac{\pi}{4} = \frac{1}{\sqrt{2}}.$$

Figure P.74 $\sin\frac{\pi}{4} = \cos\frac{\pi}{4} = \frac{1}{\sqrt{2}}$

EXAMPLE 4 Find the values of sine and cosine of the angles $\pi/3$ (or 60°) and $\pi/6$ (or 30°).

Solution The point $P_{\pi/3}$ and the points $O(0,0)$ and $A(1,0)$ are the vertices of an equilateral triangle with edge length 1 (see Figure P.75). Thus $P_{\pi/3}$ has x-coordinate 1/2 and y-coordinate $\sqrt{1-(1/2)^2} = \sqrt{3}/2$, and

$$\cos(60°) = \cos\frac{\pi}{3} = \frac{1}{2}, \qquad \sin(60°) = \sin\frac{\pi}{3} = \frac{\sqrt{3}}{2}.$$

Since $\frac{\pi}{6} = \frac{\pi}{2} - \frac{\pi}{3}$, the complementary angle identities now tell us that

$$\cos(30°) = \cos\frac{\pi}{6} = \sin\frac{\pi}{3} = \frac{\sqrt{3}}{2}, \qquad \sin(30°) = \sin\frac{\pi}{6} = \cos\frac{\pi}{3} = \frac{1}{2}.$$

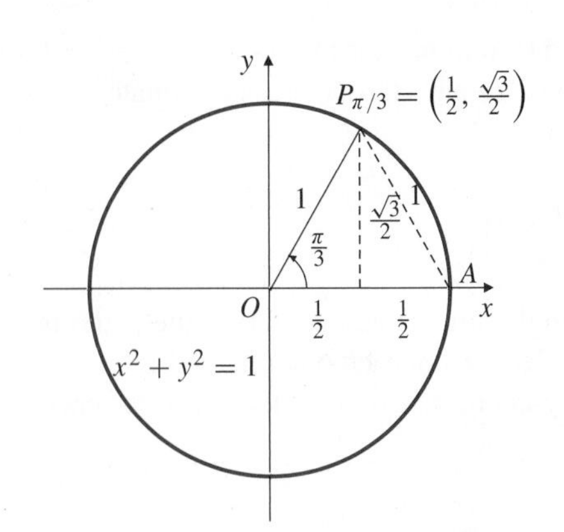

Figure P.75 $\cos\pi/3 = 1/2$
$\sin\pi/3 = \sqrt{3}/2$

Table 5 summarizes the values of cosine and sine at multiples of 30° and 45° between 0° and 180°. The values for 120°, 135°, and 150° were determined by using the supplementary angle identities; for example,

$$\cos(120°) = \cos\left(\frac{2\pi}{3}\right) = \cos\left(\pi - \frac{\pi}{3}\right) = -\cos\left(\frac{\pi}{3}\right) = -\cos(60°) = -\frac{1}{2}.$$

Table 5. Cosines and sines of special angles

Degrees	0°	30°	45°	60°	90°	120°	135°	150°	180°
Radians	0	$\frac{\pi}{6}$	$\frac{\pi}{4}$	$\frac{\pi}{3}$	$\frac{\pi}{2}$	$\frac{2\pi}{3}$	$\frac{3\pi}{4}$	$\frac{5\pi}{6}$	π
Cosine	1	$\frac{\sqrt{3}}{2}$	$\frac{1}{\sqrt{2}}$	$\frac{1}{2}$	0	$-\frac{1}{2}$	$-\frac{1}{\sqrt{2}}$	$-\frac{\sqrt{3}}{2}$	-1
Sine	0	$\frac{1}{2}$	$\frac{1}{\sqrt{2}}$	$\frac{\sqrt{3}}{2}$	1	$\frac{\sqrt{3}}{2}$	$\frac{1}{\sqrt{2}}$	$\frac{1}{2}$	0

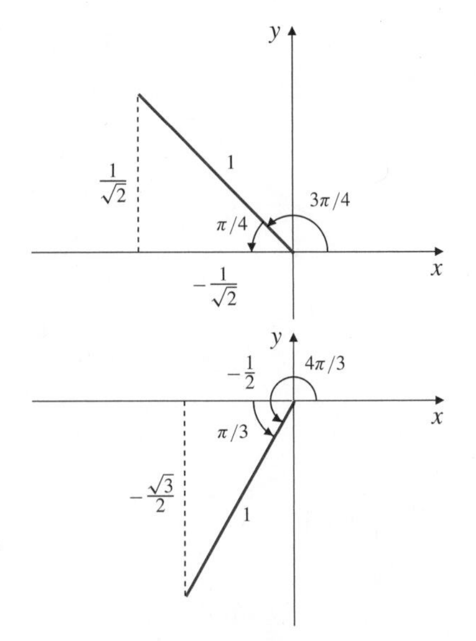

Figure P.76 Using suitably placed triangles to find trigonometric functions of special angles

EXAMPLE 5 Find: (a) $\sin(3\pi/4)$ and (b) $\cos(4\pi/3)$.

Solution We can draw appropriate triangles in the quadrants where the angles lie to determine the required values. See Figure P.76.

(a) $\sin(3\pi/4) = \sin(\pi - (\pi/4)) = 1/\sqrt{2}$.

(b) $\cos(4\pi/3) = \cos(\pi + (\pi/3)) = -\frac{1}{2}$.

While decimal approximations to the values of sine and cosine can be found using a scientific calculator or mathematical tables, it is useful to remember the exact values in the table for angles 0, $\pi/6$, $\pi/4$, $\pi/3$, and $\pi/2$. They occur frequently in applications.

When we treat sine and cosine as functions, we can call the variable they depend on anything we want (e.g., x, as we do with other functions), rather than t. The graphs of $\cos x$ and $\sin x$ are shown in Figures P.77 and P.78. In both graphs the pattern between $x = 0$ and $x = 2\pi$ repeats over and over to the left and right. Observe that the graph of $\sin x$ is the graph of $\cos x$ shifted to the right a distance $\pi/2$.

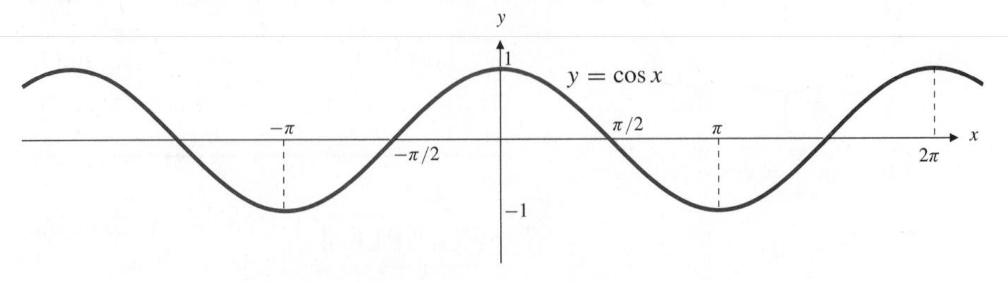

Figure P.77 The graph of $\cos x$

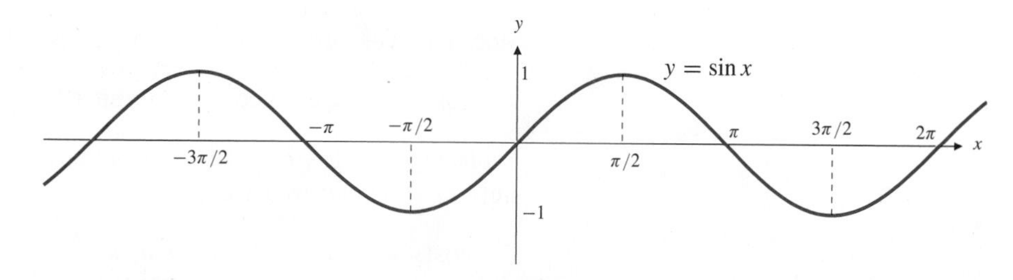

Figure P.78 The graph of $\sin x$

Remember this!

When using a scientific calculator to calculate any trigonometric functions, be sure you have selected the proper angular mode: degrees or radians.

The Addition Formulas

The following formulas enable us to determine the cosine and sine of a sum or difference of two angles in terms of the cosines and sines of those angles.

THEOREM 2 **Addition Formulas for Cosine and Sine**

$$\cos(s+t) = \cos s \cos t - \sin s \sin t$$
$$\sin(s+t) = \sin s \cos t + \cos s \sin t$$
$$\cos(s-t) = \cos s \cos t + \sin s \sin t$$
$$\sin(s-t) = \sin s \cos t - \cos s \sin t$$

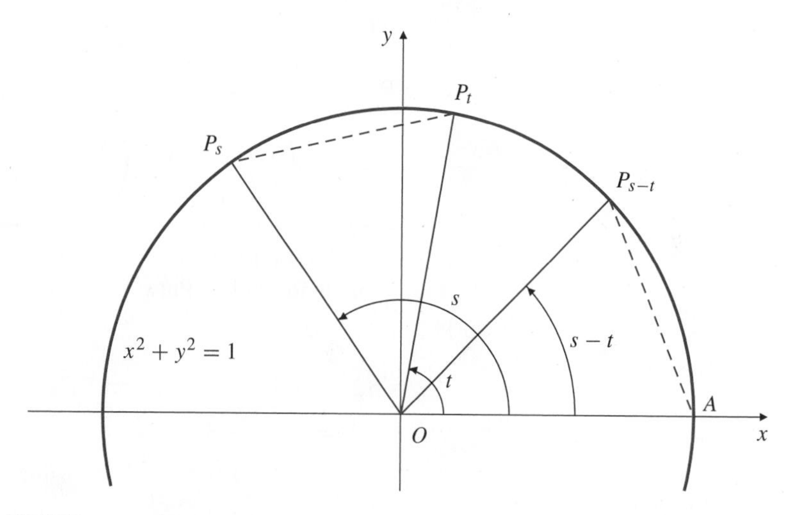

Figure P.79 $P_sP_t = P_{s-t}A$

PROOF We prove the third of these formulas as follows: Let s and t be real numbers and consider the points

$$P_t = (\cos t, \sin t) \qquad P_{s-t} = (\cos(s-t), \sin(s-t))$$
$$P_s = (\cos s, \sin s) \qquad A = (1, 0),$$

as shown in Figure P.79.

The angle $P_tOP_s = s - t$ radians = angle AOP_{s-t}, so the distance P_sP_t is equal to the distance $P_{s-t}A$. Therefore, $(P_sP_t)^2 = (P_{s-t}A)^2$. We express these squared distances in terms of coordinates and expand the resulting squares of binomials:

$$(\cos s - \cos t)^2 + (\sin s - \sin t)^2 = (\cos(s-t) - 1)^2 + \sin^2(s-t),$$

$$\cos^2 s - 2\cos s \cos t + \cos^2 t + \sin^2 s - 2\sin s \sin t + \sin^2 t$$
$$= \cos^2(s-t) - 2\cos(s-t) + 1 + \sin^2(s-t).$$

Since $\cos^2 x + \sin^2 x = 1$ for every x, this reduces to

$$\cos(s - t) = \cos s \cos t + \sin s \sin t.$$

Replacing t with $-t$ in the formula above, and recalling that $\cos(-t) = \cos t$ and $\sin(-t) = -\sin t$, we have

$$\cos(s + t) = \cos s \cos t - \sin s \sin t.$$

The complementary angle formulas can be used to obtain either of the addition formulas for sine:

$$\begin{aligned}
\sin(s + t) &= \cos\left(\frac{\pi}{2} - (s + t)\right) \\
&= \cos\left(\left(\frac{\pi}{2} - s\right) - t\right) \\
&= \cos\left(\frac{\pi}{2} - s\right)\cos t + \sin\left(\frac{\pi}{2} - s\right)\sin t \\
&= \sin s \cos t + \cos s \sin t,
\end{aligned}$$

and the other formula again follows if we replace t with $-t$.

EXAMPLE 6 Find the value of $\cos(\pi/12) = \cos 15°$.

Solution

$$\begin{aligned}
\cos\frac{\pi}{12} &= \cos\left(\frac{\pi}{3} - \frac{\pi}{4}\right) = \cos\frac{\pi}{3}\cos\frac{\pi}{4} + \sin\frac{\pi}{3}\sin\frac{\pi}{4} \\
&= \left(\frac{1}{2}\right)\left(\frac{1}{\sqrt{2}}\right) + \left(\frac{\sqrt{3}}{2}\right)\left(\frac{1}{\sqrt{2}}\right) = \frac{1 + \sqrt{3}}{2\sqrt{2}}
\end{aligned}$$

From the addition formulas, we obtain as special cases certain useful formulas called **double-angle formulas**. Put $s = t$ in the addition formulas for $\sin(s + t)$ and $\cos(s + t)$ to get

$$\begin{aligned}
\sin 2t &= 2 \sin t \cos t \qquad \text{and} \\
\cos 2t &= \cos^2 t - \sin^2 t \\
&= 2\cos^2 t - 1 \qquad \text{(using } \sin^2 t + \cos^2 t = 1\text{)} \\
&= 1 - 2\sin^2 t
\end{aligned}$$

Solving the last two formulas for $\cos^2 t$ and $\sin^2 t$, we obtain

$$\cos^2 t = \frac{1 + \cos 2t}{2} \qquad \text{and} \qquad \sin^2 t = \frac{1 - \cos 2t}{2},$$

which are sometimes called **half-angle formulas** because they are used to express trigonometric functions of half of the angle $2t$. Later we will find these formulas useful when we have to integrate powers of $\cos x$ and $\sin x$.

Other Trigonometric Functions

There are four other trigonometric functions—tangent (tan), cotangent (cot), secant (sec), and cosecant (csc)—each defined in terms of cosine and sine. Their graphs are shown in Figures P.80–P.83.

DEFINITION 9

Tangent, cotangent, secant, and cosecant

$$\tan t = \frac{\sin t}{\cos t} \qquad \sec t = \frac{1}{\cos t}$$

$$\cot t = \frac{\cos t}{\sin t} = \frac{1}{\tan t} \qquad \csc t = \frac{1}{\sin t}$$

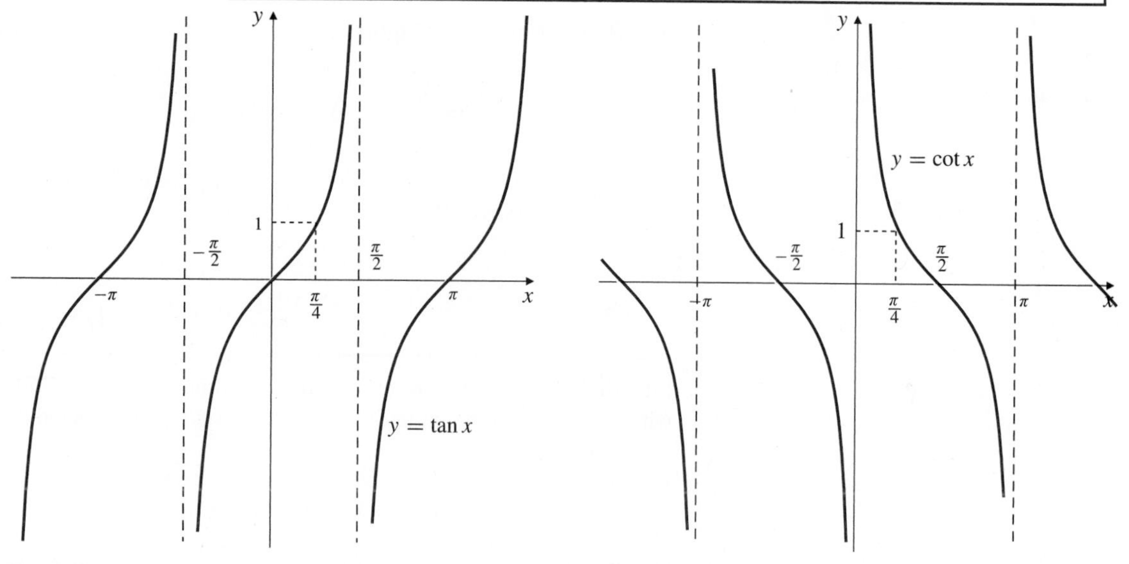

Figure P.80 The graph of $\tan x$

Figure P.81 The graph of $\cot x$

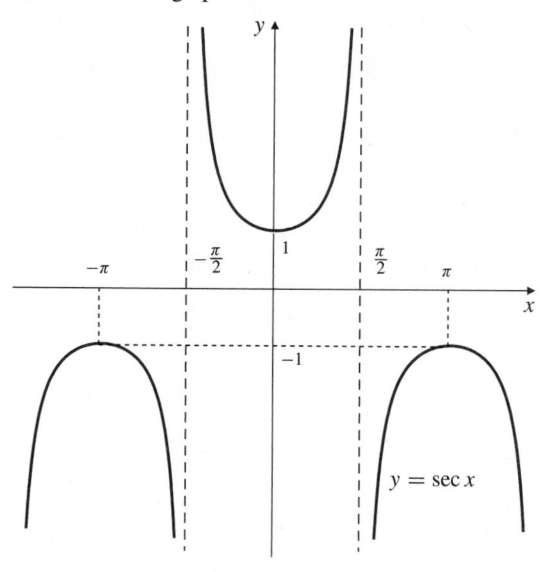

Figure P.82 The graph of $\sec x$

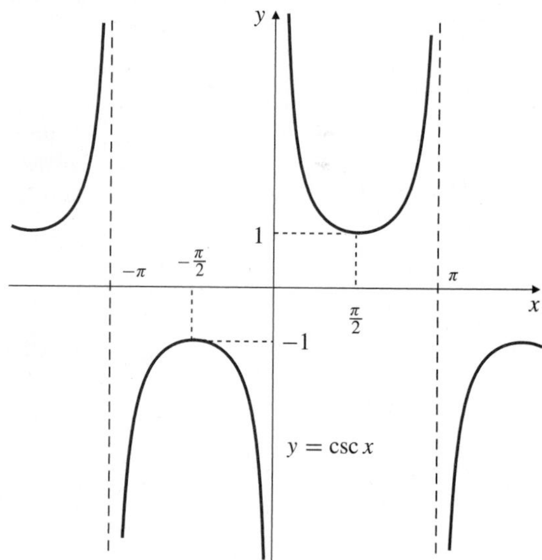

Figure P.83 The graph of $\csc x$

Observe that each of these functions is undefined (and its graph approaches vertical asymptotes) at points where the function in the denominator of its defining fraction has value 0. Observe also that tangent, cotangent, and cosecant are odd functions and secant is an even function. Since $|\sin x| \leq 1$ and $|\cos x| \leq 1$ for all x, $|\csc x| \geq 1$ and $|\sec x| \geq 1$ for all x where they are defined.

The three functions sine, cosine, and tangent are called the **primary trigonometric functions**, while their reciprocals cosecant, secant, and cotangent are called the **secondary trigonometric functions**. Scientific calculators usually just implement the primary functions; you can use the reciprocal key to find values of the corresponding secondary functions. Figure P.84 shows a useful pattern called the "CAST rule" to help you remember where the primary functions are positive. All three are positive in the first quadrant, marked A. Of the three, only sine is positive in the second quadrant S, only tangent in the third quadrant T, and only cosine in the fourth quadrant C.

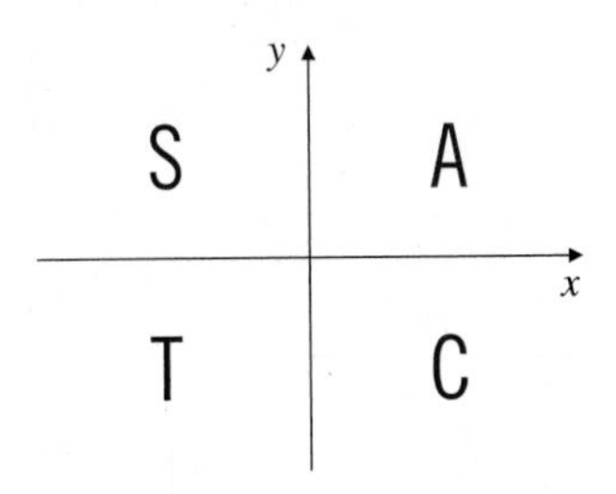

Figure P.84 The CAST rule

EXAMPLE 7 Find the sine and tangent of the angle θ in $\left[\pi, \frac{3\pi}{2}\right]$ for which we have $\cos\theta = -\frac{1}{3}$.

Solution From the Pythagorean identity $\sin^2\theta + \cos^2\theta = 1$, we get

$$\sin^2\theta = 1 - \frac{1}{9} = \frac{8}{9}, \qquad \text{so} \quad \sin\theta = \pm\sqrt{\frac{8}{9}} = \pm\frac{2\sqrt{2}}{3}.$$

The requirement that θ should lie in $[\pi, 3\pi/2]$ makes θ a third quadrant angle. Its sine is therefore negative. We have

$$\sin\theta = -\frac{2\sqrt{2}}{3} \qquad \text{and} \qquad \tan\theta = \frac{\sin\theta}{\cos\theta} = \frac{-2\sqrt{2}/3}{-1/3} = 2\sqrt{2}.$$

Like their reciprocals cosine and sine, the functions secant and cosecant are periodic with period 2π. Tangent and cotangent, however, have period π because

$$\tan(x+\pi) = \frac{\sin(x+\pi)}{\cos(x+\pi)} = \frac{\sin x\cos\pi + \cos x\sin\pi}{\cos x\cos\pi - \sin x\sin\pi} = \frac{-\sin x}{-\cos x} = \tan x.$$

Dividing the Pythagorean identity $\sin^2 x + \cos^2 x = 1$ by $\cos^2 x$ and $\sin^2 x$, respectively, leads to two useful alternative versions of that identity:

$$1 + \tan^2 x = \sec^2 x \qquad \text{and} \qquad 1 + \cot^2 x = \csc^2 x.$$

Addition formulas for tangent and cotangent can be obtained from those for sine and cosine. For example,

$$\tan(s+t) = \frac{\sin(s+t)}{\cos(s+t)} = \frac{\sin s\cos t + \cos s\sin t}{\cos s\cos t - \sin s\sin t}.$$

Now divide the numerator and denominator of the fraction on the right by $\cos s\cos t$ to get

$$\tan(s+t) = \frac{\tan s + \tan t}{1 - \tan s\tan t}.$$

Replacing t by $-t$ leads to

$$\tan(s-t) = \frac{\tan s - \tan t}{1 + \tan s\tan t}.$$

Maple Calculations

Maple knows all six trigonometric functions and can calculate their values and manipulate them in other ways. It assumes the arguments of the trigonometric functions are in radians.

```
> evalf(sin(30)); evalf(sin(Pi/6));
```

$$-.9880316241$$

$$.5000000000$$

Note that the constant Pi (with an uppercase P) is known to Maple. The `evalf()` function converts its argument to a number expressed as a floating point decimal with 10 significant digits. (This precision can be changed by defining a new value for the variable `Digits`.) Without it, the sine of 30 radians would have been left unexpanded because it is not an integer.

```
> Digits := 20; evalf(100*Pi); sin(30);
```

$$Digits := 20$$

$$314.15926535897932385$$

$$sin(30)$$

It is often useful to expand trigonometric functions of multiple angles to powers of sine and cosine, and vice versa.

```
> expand(sin(5*x));
```

$$16\sin(x)\cos(x)^4 - 12\sin(x)\cos(x)^2 + \sin(x)$$

```
> combine((cos(x))^5, trig);
```

$$\frac{1}{16}\cos(5x) + \frac{5}{16}\cos(3x) + \frac{5}{8}\cos(x)$$

Other trigonometric functions can be converted to expressions involving sine and cosine.

```
> convert(tan(4*x)*(sec(4*x))^2, sincos); combine(%,trig);
```

$$\frac{\sin(4x)}{\cos(4x)^3}$$

$$4\frac{\sin(4x)}{\cos(12x) + 3\cos(4x)}$$

The % in the last command refers to the result of the previous calculation.

Trigonometry Review

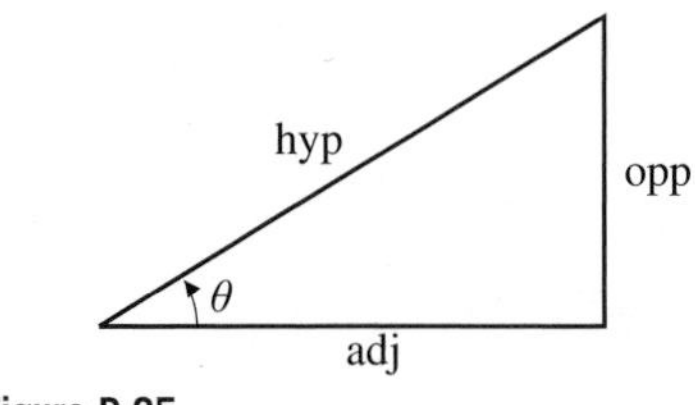

Figure P.85

The trigonometric functions are so called because they are often used to express the relationships between the sides and angles of a triangle. As we observed at the beginning of this section, if θ is one of the acute angles in a right-angled triangle, we can refer to the three sides of the triangle as adj (side adjacent θ), opp (side opposite θ), and hyp (hypotenuse). (See Figure P.85.) The trigonometric functions of θ can then be expressed as ratios of these sides, in particular:

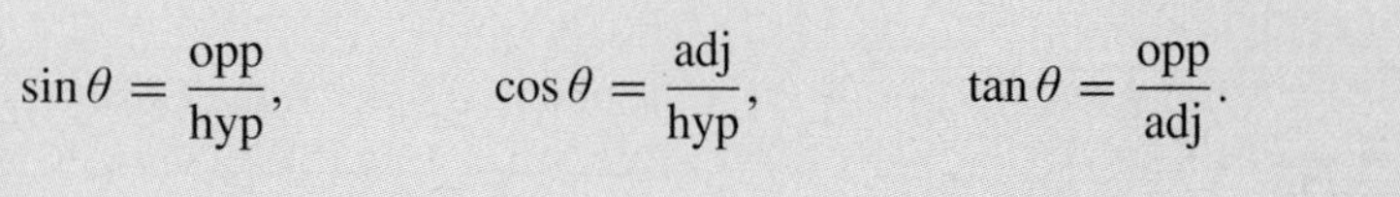

$$\sin\theta = \frac{\text{opp}}{\text{hyp}}, \qquad \cos\theta = \frac{\text{adj}}{\text{hyp}}, \qquad \tan\theta = \frac{\text{opp}}{\text{adj}}.$$

EXAMPLE 8 Find the unknown sides x and y of the triangle in Figure P.86.

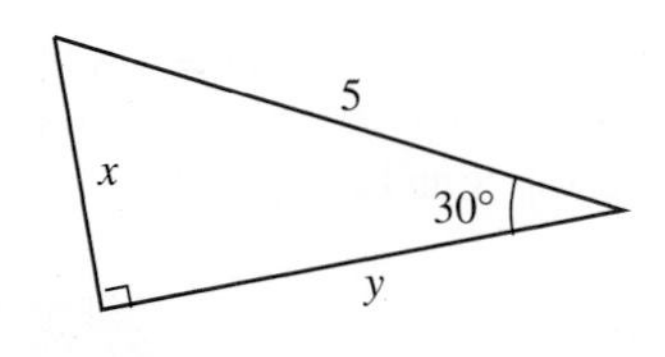

Figure P.86

Solution Here, x is the side opposite and y is the side adjacent the 30° angle. The hypotenuse of the triangle is 5 units. Thus,

$$\frac{x}{5} = \sin 30^\circ = \frac{1}{2} \quad \text{and} \quad \frac{y}{5} = \cos 30^\circ = \frac{\sqrt{3}}{2},$$

so $x = \frac{5}{2}$ units and $y = \frac{5\sqrt{3}}{2}$ units.

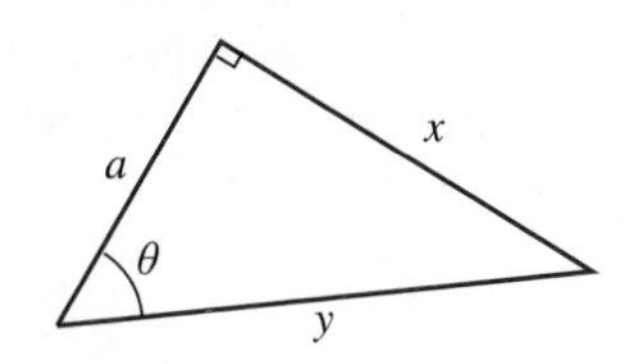

Figure P.87

EXAMPLE 9 For the triangle in Figure P.87, express sides x and y in terms of side a and angle θ.

Solution The side x is opposite the angle θ, and y is the hypotenuse. The side adjacent θ is a. Thus,

$$\frac{x}{a} = \tan\theta \quad \text{and} \quad \frac{a}{y} = \cos\theta.$$

Hence, $x = a\,\tan\theta$ and $y = \dfrac{a}{\cos\theta} = a\,\sec\theta.$

When dealing with general (not necessarily right-angled) triangles, it is often convenient to label the vertices with capital letters, which also denote the angles at those vertices, and refer to the sides opposite those vertices by the corresponding lowercase letters. See Figure P.88. Relationships between the sides a, b, and c and opposite angles A, B, and C of an arbitrary triangle ABC are given by the following formulas, called the **Sine Law** and the **Cosine Law**.

THEOREM 3

Sine Law: $$\frac{\sin A}{a} = \frac{\sin B}{b} = \frac{\sin C}{c}$$

Cosine Law: $$a^2 = b^2 + c^2 - 2bc\cos A$$
$$b^2 = a^2 + c^2 - 2ac\cos B$$
$$c^2 = a^2 + b^2 - 2ab\cos C$$

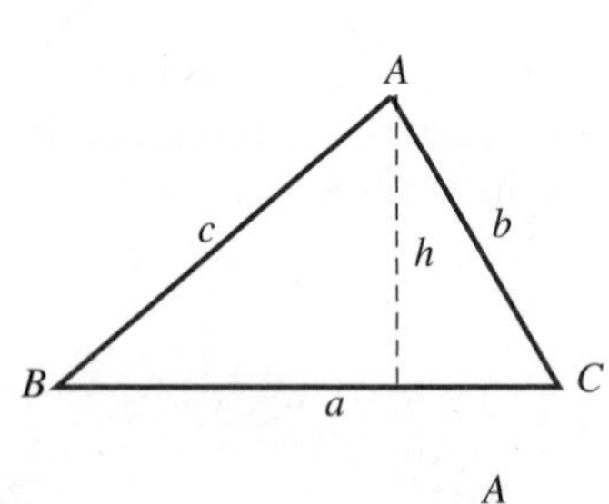

Figure P.88 In this triangle the sides are named to correspond to the opposite angles

PROOF See Figure P.89. Let h be the length of the perpendicular from A to the side BC. From right-angled triangles (and using $\sin(\pi - t) = \sin t$ if required), we get $c\sin B = h = b\sin C$. Thus $(\sin B)/b = (\sin C)/c$. By the symmetry of the formulas (or by dropping a perpendicular to another side), both fractions must be equal to $(\sin A)/a$, so the Sine Law is proved. For the Cosine Law, observe that

$$c^2 = \begin{cases} h^2 + (a - b\cos C)^2 & \text{if } C \le \dfrac{\pi}{2} \\[2ex] h^2 + (a + b\cos(\pi - C))^2 & \text{if } C > \dfrac{\pi}{2} \end{cases}$$
$$= h^2 + (a - b\cos C)^2 \qquad (\text{since } \cos(\pi - C) = -\cos C)$$
$$= b^2\sin^2 C + a^2 - 2ab\cos C + b^2\cos^2 C$$
$$= a^2 + b^2 - 2ab\cos C.$$

The other versions of the Cosine Law can be proved in a similar way.

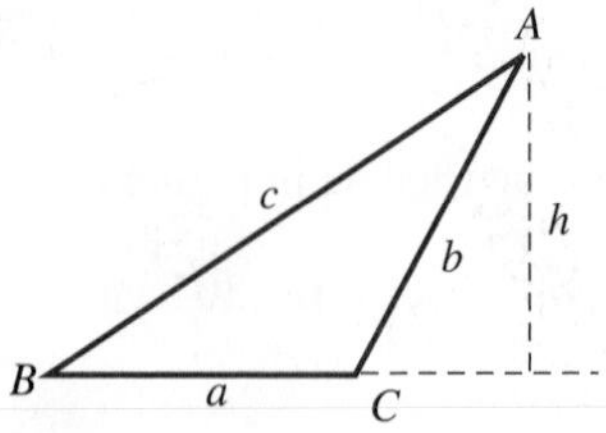

Figure P.89

EXAMPLE 10 A triangle has sides $a = 2$ and $b = 3$ and angle $C = 40°$. Find side c and the sine of angle B.

Solution From the third version of the Cosine Law:

$$c^2 = a^2 + b^2 - 2ab\cos C = 4 + 9 - 12\cos 40° \approx 13 - 12 \times 0.766 = 3.808.$$

Side c is about $\sqrt{3.808} = 1.951$ units in length. Now using Sine Law we get

$$\sin B = b\,\frac{\sin C}{c} \approx 3 \times \frac{\sin 40°}{1.951} \approx \frac{3 \times 0.6428}{1.951} \approx 0.988.$$

A triangle is uniquely determined by any one of the following sets of data (which correspond to the known cases of congruency of triangles in classical geometry):

1. two sides and the angle contained between them (e.g., Example 10);
2. three sides, no one of which exceeds the sum of the other two in length;
3. two angles and one side; or
4. the hypotenuse and one other side of a right-angled triangle.

In such cases you can always find the unknown sides and angles by using the Pythagorean Theorem or the Sine and Cosine Laws, and the fact that the sum of the three angles of a triangle is 180° (or π radians).

A triangle is not determined uniquely by two sides and a noncontained angle; there may exist no triangle, one right-angled triangle, or two triangles having such data.

EXAMPLE 11 In triangle ABC, angle $B = 30°$, $b = 2$, and $c = 3$. Find a.

Solution This is one of the ambiguous cases. By the Cosine Law,

$$\begin{aligned} b^2 &= a^2 + c^2 - 2ac\cos B \\ 4 &= a^2 + 9 - 6a(\sqrt{3}/2). \end{aligned}$$

Therefore, a must satisfy the equation $a^2 - 3\sqrt{3}a + 5 = 0$. Solving this equation using the quadratic formula, we obtain

$$\begin{aligned} a &= \frac{3\sqrt{3} \pm \sqrt{27 - 20}}{2} \\ &\approx 1.275 \quad \text{or} \quad 3.921 \end{aligned}$$

There are two triangles with the given data, as shown in Figure P.90.

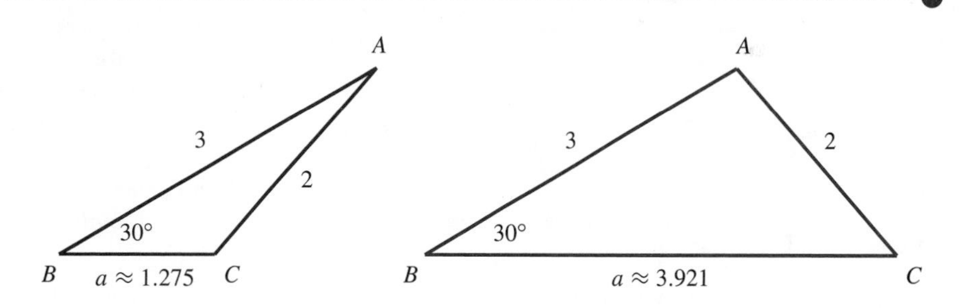

Figure P.90 Two triangles with $b = 2$, $c = 3$, $B = 30°$

EXERCISES P.7

Find the values of the quantities in Exercises 1–6 using various formulas presented in this section. Do not use tables or a calculator.

1. $\cos\dfrac{3\pi}{4}$ **2.** $\tan -\dfrac{3\pi}{4}$ **3.** $\sin\dfrac{2\pi}{3}$

4. $\sin\dfrac{7\pi}{12}$ **5.** $\cos\dfrac{5\pi}{12}$ **6.** $\sin\dfrac{11\pi}{12}$

In Exercises 7–12, express the given quantity in terms of $\sin x$ and $\cos x$.

7. $\cos(\pi + x)$ **8.** $\sin(2\pi - x)$ **9.** $\sin\left(\dfrac{3\pi}{2} - x\right)$

10. $\cos\left(\dfrac{3\pi}{2} + x\right)$ **11.** $\tan x + \cot x$ **12.** $\dfrac{\tan x - \cot x}{\tan x + \cot x}$

In Exercises 13–16, prove the given identities.

13. $\cos^4 x - \sin^4 x = \cos(2x)$

14. $\dfrac{1 - \cos x}{\sin x} = \dfrac{\sin x}{1 + \cos x} = \tan\dfrac{x}{2}$

15. $\dfrac{1 - \cos x}{1 + \cos x} = \tan^2\dfrac{x}{2}$

16. $\dfrac{\cos x - \sin x}{\cos x + \sin x} = \sec 2x - \tan 2x$

17. Express $\sin 3x$ in terms of $\sin x$ and $\cos x$.

18. Express $\cos 3x$ in terms of $\sin x$ and $\cos x$.

In Exercises 19–22, sketch the graph of the given function. What is the period of the function?

19. $f(x) = \cos 2x$

20. $f(x) = \sin \dfrac{x}{2}$

21. $f(x) = \sin \pi x$

22. $f(x) = \cos \dfrac{\pi x}{2}$

23. Sketch the graph of $y = 2\cos\left(x - \dfrac{\pi}{3}\right)$.

24. Sketch the graph of $y = 1 + \sin\left(x + \dfrac{\pi}{4}\right)$.

In Exercises 25–30, one of $\sin\theta$, $\cos\theta$, and $\tan\theta$ is given. Find the other two if θ lies in the specified interval.

25. $\sin\theta = \dfrac{3}{5}$, θ in $\left[\dfrac{\pi}{2}, \pi\right]$

26. $\tan\theta = 2$, θ in $\left[0, \dfrac{\pi}{2}\right]$

27. $\cos\theta = \dfrac{1}{3}$, θ in $\left[-\dfrac{\pi}{2}, 0\right]$

28. $\cos\theta = -\dfrac{5}{13}$, θ in $\left[\dfrac{\pi}{2}, \pi\right]$

29. $\sin\theta = \dfrac{-1}{2}$, θ in $\left[\pi, \dfrac{3\pi}{2}\right]$

30. $\tan\theta = \dfrac{1}{2}$, θ in $\left[\pi, \dfrac{3\pi}{2}\right]$

Trigonometry Review

In Exercises 31–42, ABC is a triangle with a right angle at C. The sides opposite angles A, B, and C are a, b, and c, respectively. (See Figure P.91.)

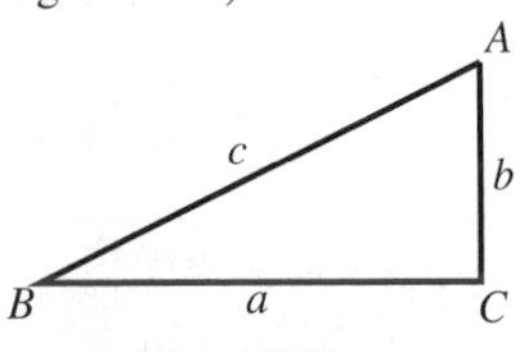

Figure P.91

31. Find a and b if $c = 2$, $B = \dfrac{\pi}{3}$.

32. Find a and c if $b = 2$, $B = \dfrac{\pi}{3}$.

33. Find b and c if $a = 5$, $B = \dfrac{\pi}{6}$.

34. Express a in terms of A and c.

35. Express a in terms of A and b.

36. Express a in terms of B and c.

37. Express a in terms of B and b.

38. Express c in terms of A and a.

39. Express c in terms of A and b.

40. Express $\sin A$ in terms of a and c.

41. Express $\sin A$ in terms of b and c.

42. Express $\sin A$ in terms of a and b.

In Exercises 43–50, ABC is an arbitrary triangle with sides a, b, and c, opposite to angles A, B, and C, respectively. (See Figure P.92.) Find the indicated quantities. Use tables or a scientific calculator if necessary.

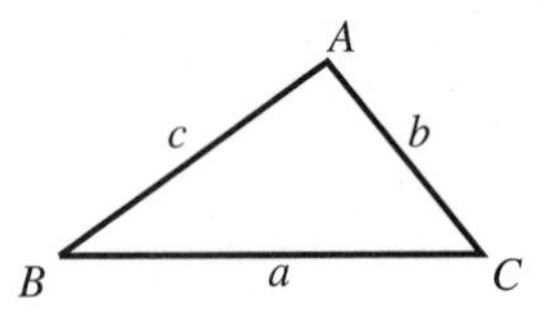

Figure P.92

43. Find $\sin B$ if $a = 4$, $b = 3$, $A = \dfrac{\pi}{4}$.

44. Find $\cos A$ if $a = 2$, $b = 2$, $c = 3$.

45. Find $\sin B$ if $a = 2$, $b = 3$, $c = 4$.

46. Find c if $a = 2$, $b = 3$, $C = \dfrac{\pi}{4}$.

47. Find a if $c = 3$, $A = \dfrac{\pi}{4}$, $B = \dfrac{\pi}{3}$.

48. Find c if $a = 2$, $b = 3$, $C = 35°$.

49. Find b if $a = 4$, $B = 40°$, $C = 70°$.

50. Find c if $a = 1$, $b = \sqrt{2}$, $A = 30°$. (There are two possible answers.)

51. Two guy wires stretch from the top T of a vertical pole to points B and C on the ground, where C is 10 m closer to the base of the pole than is B. If wire BT makes an angle of 35° with the horizontal, and wire CT makes an angle of 50° with the horizontal, how high is the pole?

52. Observers at positions A and B 2 km apart simultaneously measure the angle of elevation of a weather balloon to be 40° and 70°, respectively. If the balloon is directly above a point on the line segment between A and B, find the height of the balloon.

53. Show that the area of triangle ABC is given by $(1/2)ab\sin C = (1/2)bc\sin A = (1/2)ca\sin B$.

! 54. Show that the area of triangle ABC is given by $\sqrt{s(s-a)(s-b)(s-c)}$, where $s = (a+b+c)/2$ is the semi-perimeter of the triangle.

! This symbol is used throughout the book to indicate an exercise that is somewhat more difficult than most exercises.

? This symbol is used throughout the book to indicate an exercise that is somewhat theoretical in nature. It does not imply difficulty.

CHAPTER 1

Limits and Continuity

> Every body continues in its state of rest, or of uniform motion in a right line, unless it is compelled to change that state by forces impressed upon it.

Isaac Newton 1642–1727
from *Principia Mathematica, 1687*

> It was not until Leibniz and Newton, by the discovery of the differential calculus, had dispelled the ancient darkness which enveloped the conception of the infinite, and had clearly established the conception of the continuous and continuous change, that a full productive application of the newly found mechanical conceptions made any progress.

Hermann von Helmholtz 1821–1894

Introduction

Calculus was created to describe how quantities change. It has two basic procedures that are opposites of one another:

- *differentiation,* for finding the rate of change of a given function, and
- *integration,* for finding a function having a given rate of change.

Both of these procedures are based on the fundamental concept of the *limit* of a function. It is this idea of limit that distinguishes calculus from algebra, geometry, and trigonometry, which are useful for describing static situations.

In this chapter we will introduce the limit concept and develop some of its properties. We begin by considering how limits arise in some basic problems.

1.1 Examples of Velocity, Growth Rate, and Area

In this section we consider some examples of phenomena where limits arise in a natural way.

Average Velocity and Instantaneous Velocity

The position of a moving object is a function of time. The average velocity of the object over a time interval is found by dividing the change in the object's position by the length of the time interval.

EXAMPLE 1 **(The average velocity of a falling rock)** Physical experiments show that if a rock is dropped from rest near the surface of the earth, in the first t s it will fall a distance

$$y = 4.9t^2 \text{ m}.$$

(a) What is the average velocity of the falling rock during the first 2 s?

(b) What is its average velocity from $t = 1$ to $t = 2$?

Solution The *average velocity* of the falling rock over any time interval $[t_1, t_2]$ is the change Δy in the distance fallen divided by the length Δt of the time interval:

$$\text{average velocity over } [t_1, t_2] = \frac{\Delta y}{\Delta t} = \frac{4.9t_2^2 - 4.9t_1^2}{t_2 - t_1}.$$

(a) In the first 2 s (time interval $[0, 2]$), the average velocity is

$$\frac{\Delta y}{\Delta t} = \frac{4.9(2^2) - 4.9(0^2)}{2 - 0} = 9.8 \text{ m/s}.$$

(b) In the time interval $[1, 2]$, the average velocity is

$$\frac{\Delta y}{\Delta t} = \frac{4.9(2^2) - 4.9(1^2)}{2 - 1} = 14.7 \text{ m/s}.$$

EXAMPLE 2 How fast is the rock in Example 1 falling (a) at time $t = 1$? (b) at time $t = 2$?

Solution We can calculate the average velocity over any time interval, but this question asks for the *instantaneous velocity* at a given time. If the falling rock had a speedometer, what would it show at time $t = 1$? To answer this, we first write the average velocity over the time interval $[1, 1 + h]$ starting at $t = 1$ and having length h:

$$\text{Average velocity over } [1, 1 + h] = \frac{\Delta y}{\Delta t} = \frac{4.9(1 + h)^2 - 4.9(1^2)}{h}.$$

We can't calculate the instantaneous velocity at $t = 1$ by substituting $h = 0$ in this expression, because we can't divide by zero. But we can calculate the average velocities over shorter and shorter time intervals and see whether they seem to get close to a particular number. Table 1 shows the values of $\Delta y/\Delta t$ for some values of h approaching zero. Indeed, it appears that these average velocities get closer and closer to 9.8 m/s as the length of the time interval gets closer and closer to zero. This suggests that the rock is falling at a rate of 9.8 m/s one second after it is dropped.

Similarly, Table 2 shows values of the average velocities over shorter and shorter time intervals $[2, 2 + h]$ starting at $t = 2$. The values suggest that the rock is falling at 19.6 m/s two seconds after it is dropped.

Table 1. Average velocity over $[1, 1 + h]$

h	$\Delta y/\Delta t$
1	14.7000
0.1	10.2900
0.01	9.8490
0.001	9.8049
0.0001	9.8005

Table 2. Average velocity over $[2, 2 + h]$

h	$\Delta y/\Delta t$
1	24.5000
0.1	20.0900
0.01	19.6490
0.001	19.6049
0.0001	19.6005

In Example 2 the average velocity of the falling rock over the time interval $[t, t + h]$ is

$$\frac{\Delta y}{\Delta t} = \frac{4.9(t + h)^2 - 4.9t^2}{h}.$$

To find the instantaneous velocity (usually just called *the velocity*) at the instants $t = 1$ and $t = 2$, we examined the values of this average velocity for time intervals whose lengths h became smaller and smaller. We were, in fact, finding the *limit of the average velocity as h approaches zero.* This is expressed symbolically in the form

$$\text{velocity at time } t = \lim_{h \to 0} \frac{\Delta y}{\Delta t} = \lim_{h \to 0} \frac{4.9(t + h)^2 - 4.9t^2}{h}.$$

Read "$\lim_{h \to 0} \dots$" as "the limit as h approaches zero of ..." We can't find the limit of the fraction by just substituting $h = 0$ because that would involve dividing by zero. However, we can calculate the limit by first performing some algebraic simplifications on the expression for the average velocity.

EXAMPLE 3 Simplify the expression for the average velocity of the rock over $[t, t+h]$ by first expanding $(t+h)^2$. Hence, find the velocity $v(t)$ of the falling rock at time t directly, without making a table of values.

Solution The average velocity of the rock over time interval $[t, t+h]$ is

$$\begin{aligned}\frac{4.9(t+h)^2 - 4.9t^2}{h} &= \frac{4.9(t^2 + 2th + h^2 - t^2)}{h}\\ &= \frac{4.9(2th + h^2)}{h}\\ &= 9.8t + 4.9h.\end{aligned}$$

The final form of the expression no longer involves division by h. It approaches $9.8t + 4.9(0) = 9.8t$ as h approaches 0. Thus, t s after the rock is dropped, its velocity is $v(t) = 9.8t$ m/s. In particular, at $t = 1$ and $t = 2$ the velocities are $v(1) = 9.8$ m/s and $v(2) = 19.6$ m/s, respectively.

The Growth of an Algal Culture

In a laboratory experiment, the biomass of an algal culture was measured over a 74-day period by measuring the area in square millimetres occupied by the culture on a microscope slide. These measurements m were plotted against the time t in days and the points joined by a smooth curve $m = f(t)$, as shown in Figure 1.1.

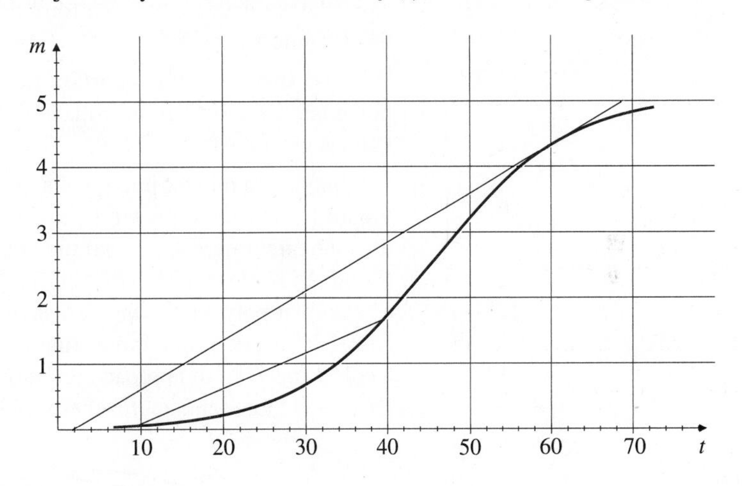

Figure 1.1 The biomass m of an algal culture after t days

Observe that the biomass was about 0.1 mm^2 on day 10 and had grown to about 1.7 mm^2 on day 40, an increase of $1.7 - 0.1 = 1.6$ mm^2 in a time interval of $40 - 10 = 30$ days. The average rate of growth over the time interval from day 10 to day 40 was therefore

$$\frac{1.7 - 0.1}{40 - 10} = \frac{1.6}{30} \approx 0.053 \text{ mm}^2\text{/d}.$$

This average rate is just the slope of the line joining the points on the graph of $m = f(t)$ corresponding to $t = 10$ and $t = 40$. Similarly, the average rate of growth of the algal biomass over any time interval can be determined by measuring the slope of the line joining the points on the curve corresponding to that time interval. Such lines are called **secant lines** to the curve.

EXAMPLE 4 How fast is the biomass growing on day 60?

Solution To answer this question, we could measure the average rates of change over shorter and shorter times around day 60. The corresponding secant lines become shorter and shorter, but their slopes approach a *limit*, namely, the slope of the **tangent line** to the graph of $m = f(t)$ at the point where $t = 60$. This tangent line is sketched in Figure 1.1; it seems to go through the points (2, 0) and (69, 5), so that its slope is

$$\frac{5-0}{69-2} \approx 0.0746 \text{ mm}^2/\text{d}.$$

This is the rate at which the biomass was growing on day 60.

The Area of a Circle

All circles are similar geometric figures; they all have the same shape and differ only in size. The ratio of the circumference C to the diameter $2r$ (twice the radius) has the same value for all circles. The number π is defined to be this common ratio:

$$\frac{C}{2r} = \pi \qquad \text{or} \qquad C = 2\pi r.$$

In school we are taught that the area A of a circle is this same number π times the square of the radius:

$$A = \pi r^2.$$

How can we deduce this area formula from the formula for the circumference that is the definition of π?

The answer to this question lies in regarding the circle as a "limit" of regular polygons, which are in turn made up of triangles, figures about whose geometry we know a great deal.

Suppose a regular polygon having n sides is inscribed in a circle of radius r. (See Figure 1.2.) The perimeter P_n and the area A_n of the polygon are, respectively, less than the circumference C and the area A of the circle, but if n is large, P_n is *close to* C and A_n is *close to* A. (In fact, the "circle" in Figure 1.2 was drawn by a computer as a regular polygon having 180 sides, each subtending a 2° angle at the centre of the circle. It is very difficult to distinguish this 180-sided polygon from a real circle.) We would expect P_n to approach the limit C and A_n to approach the limit A as n grows larger and larger and approaches infinity.

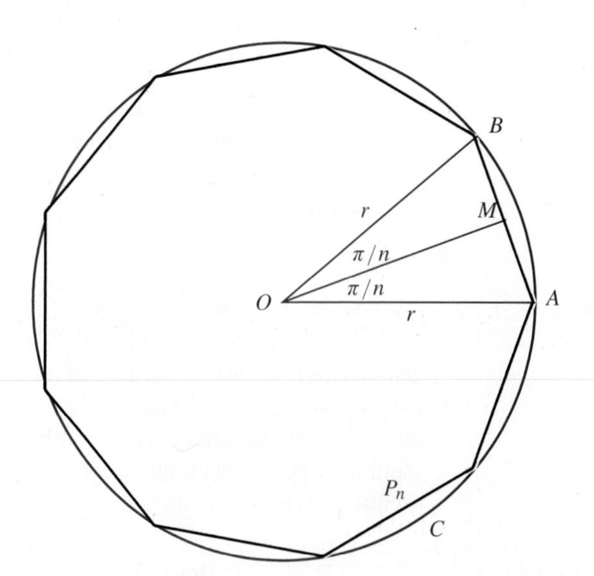

Figure 1.2 A regular polygon of n sides inscribed in a circle

A regular polygon of n sides is the union of n nonoverlapping, congruent, isosceles triangles having a common vertex at O, the centre of the polygon. One of these triangles, $\triangle OAB$, is shown in Figure 1.2. Since the total angle around the point O is 2π radians (we are assuming that a circle of radius 1 has circumference 2π), the angle AOB is $2\pi/n$ radians. If M is the midpoint of AB, then OM bisects angle AOB. Using elementary trigonometry, we can write the length of AB and the area of triangle OAB in terms of the radius r of the circle:

$$\begin{aligned}|AB| &= 2|AM| = 2r\sin\frac{\pi}{n}\\ \text{area } OAB &= \frac{1}{2}|AB||OM| = \frac{1}{2}\left(2r\sin\frac{\pi}{n}\right)\left(r\cos\frac{\pi}{n}\right)\\ &= r^2\sin\frac{\pi}{n}\,\cos\frac{\pi}{n}.\end{aligned}$$

The perimeter P_n and area A_n of the polygon are n times these expressions:

$$\begin{aligned}P_n &= 2rn\sin\frac{\pi}{n}\\ A_n &= r^2 n\sin\frac{\pi}{n}\,\cos\frac{\pi}{n}.\end{aligned}$$

Solving the first equation for $rn\sin(\pi/n) = P_n/2$ and substituting into the second equation, we get

$$A_n = \left(\frac{P_n}{2}\right) r\cos\frac{\pi}{n}.$$

Now the angle $AOM = \pi/n$ approaches 0 as n grows large, so its cosine, $\cos(\pi/n) = |OM|/|OA|$, approaches 1. Since P_n approaches $C = 2\pi r$ as n grows large, the expression for A_n approaches $(2\pi r/2)r(1) = \pi r^2$, which must therefore be the area of the circle.

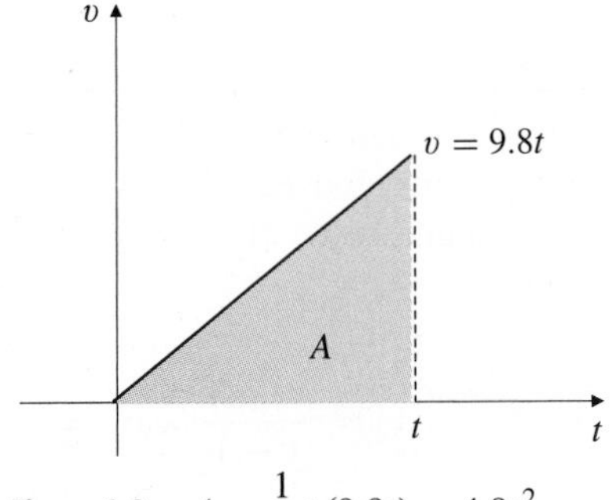

Figure 1.3 $A = \frac{1}{2}t\,(9.8t) = 4.9t^2$

Remark There is a fundamental relationship between the problem of finding the area under the graph of a function f and the problem of finding another function g whose rate of change is f. It will be explored fully beginning in Chapter 5. As an example, for the falling rock of Example 1–Example 3, the area A under the graph of the velocity function $v = 9.8t$ m/s and above the interval $[0, t]$ on the t-axis is the area of a triangle of base length t s and height 9.8 m/s, and so is (see Figure 1.3)

$$A = \frac{1}{2}(t)(9.8t) = 4.9t^2 \text{ m},$$

which is exactly the distance y that the rock falls during the first t seconds. The rate of change of the area function $A(t)$ (that is, of the distance function y) is the velocity function $v(t)$.

EXERCISES 1.1

Exercises 1–4 refer to an object moving along the x-axis in such a way that at time t s its position is $x = t^2$ m to the right of the origin.

1. Find the average velocity of the object over the time interval $[t, t+h]$.

2. Make a table giving the average velocities of the object over time intervals $[2, 2+h]$, for $h = 1, 0.1, 0.01, 0.001$, and 0.0001 s.

3. Use the results from Exercise 2 to guess the instantaneous velocity of the object at $t = 2$ s.

4. Confirm your guess in Exercise 3 by calculating the limit of the average velocity over $[2, 2+h]$ as h approaches zero, using the method of Example 3.

Exercises 5–8 refer to the motion of a particle moving along the x-axis so that at time t s it is at position $x = 3t^2 - 12t + 1$ m.

5. Find the average velocity of the particle over the time intervals $[1, 2]$, $[2, 3]$, and $[1, 3]$.

6. Use the method of Example 3 to find the velocity of the particle at $t = 1$, $t = 2$, and $t = 3$.

7. In what direction is the particle moving at $t = 1$? $t = 2$? $t = 3$?

8. Show that for any positive number k, the average velocity of the particle over the time interval $[t - k, t + k]$ is equal to its velocity at time t.

In Exercises 9–11, a weight that is suspended by a spring bobs up and down so that its height above the floor at time t s is y ft, where

$$y = 2 + \frac{1}{\pi}\sin(\pi t).$$

9. Sketch the graph of y as a function of t. How high is the weight at $t = 1$ s? In what direction is it moving at that time?

10. What is the average velocity of the weight over the time intervals $[1, 2]$, $[1, 1.1]$, $[1, 1.01]$, and $[1, 1.001]$?

11. Using the results of Exercise 10, estimate the velocity of the weight at time $t = 1$. What is the significance of the sign of your answer?

Exercises 12–13 refer to the algal biomass graphed in Figure 1.1.

12. Approximately how fast is the biomass growing on day 20?

13. On about what day is the biomass growing fastest?

14. The profits of a small company for each of the first five years of its operation are given in Table 3.

Table 3.

Year	Profit ($1,000s)
2008	6
2009	27
2010	62
2011	111
2012	174

(a) Plot points representing the profit as a function of year on graph paper, and join them by a smooth curve.

(b) What is the average rate of increase in the profits between 2010 and 2012?

(c) Use your graph to estimate the rate of increase in the profits in 2010.

1.2 Limits of Functions

In order to speak meaningfully about rates of change, tangent lines, and areas bounded by curves, we have to investigate the process of finding limits. Indeed, the concept of *limit* is the cornerstone on which the development of calculus rests. Before we try to give a definition of a limit, let us look at more examples.

EXAMPLE 1 Describe the behaviour of the function $f(x) = \dfrac{x^2 - 1}{x - 1}$ near $x = 1$.

Solution Note that $f(x)$ is defined for all real numbers x except $x = 1$. (We can't divide by zero.) For any $x \neq 1$ we can simplify the expression for $f(x)$ by factoring the numerator and cancelling common factors:

$$f(x) = \frac{(x-1)(x+1)}{x-1} = x + 1 \qquad \text{for} \quad x \neq 1.$$

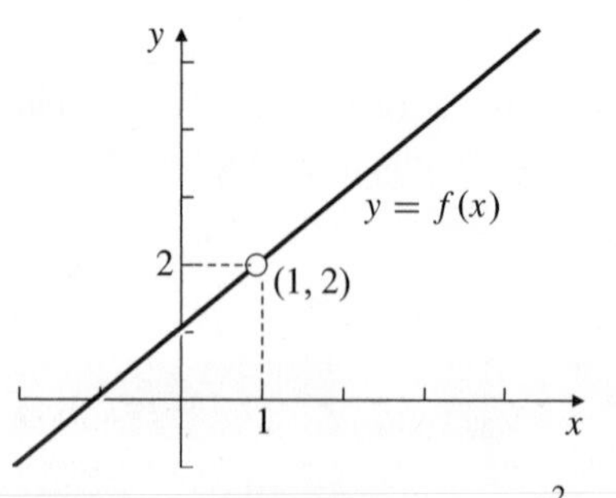

Figure 1.4 The graph of $f(x) = \dfrac{x^2 - 1}{x - 1}$

The graph of f is the line $y = x + 1$ with one point removed, namely, the point $(1, 2)$. This removed point is shown as a "hole" in the graph in Figure 1.4. Even though $f(1)$ is not defined, it is clear that we can make the value of $f(x)$ *as close as we want* to 2 by choosing x *close enough* to 1. Therefore, we say that $f(x)$ approaches arbitrarily close to 2 as x approaches 1, or, more simply, $f(x)$ approaches *the limit* 2 as x approaches 1. We write this as

$$\lim_{x\to 1} f(x) = 2 \qquad \text{or} \qquad \lim_{x\to 1} \frac{x^2 - 1}{x - 1} = 2.$$

EXAMPLE 2 What happens to the function $g(x) = (1+x^2)^{1/x^2}$ as x approaches zero?

Table 4.

x	$g(x)$
± 1.0	2.0000 00000
± 0.1	2.7048 13829
± 0.01	2.7181 45927
± 0.001	2.7182 80469
± 0.0001	2.7182 81815
± 0.00001	1.0000 00000

Solution Note that $g(x)$ is not defined at $x = 0$. In fact, for the moment it does not appear to be defined for any x whose square x^2 is not a rational number. (Recall that if $r = m/n$, where m and n are integers and $n > 0$, then x^r means the nth root of x^m.) Let us ignore for now the problem of deciding what $g(x)$ means if x^2 is irrational and consider only rational values of x. There is no obvious way to simplify the expression for $g(x)$ as we did in Example 1. However, we can use a scientific calculator to obtain approximate values of $g(x)$ for some rational values of x approaching 0. (The values in Table 4 were obtained with such a calculator.)

Except for the last value in the table, the values of $g(x)$ seem to be approaching a certain number, $2.71828\ldots$, as x gets closer and closer to 0. We will show in Section 3.4 that

$$\lim_{x\to 0} g(x) = \lim_{x\to 0}(1+x^2)^{1/x^2} = e = 2.7\,1828\,1828\,45\,90\,45\ldots.$$

The number e turns out to be very important in mathematics.

Observe that the last entry in the table appears to be wrong. This is important. It is because the calculator can only represent a finite number of numbers. The calculator was unable to distinguish $1 + (0.00001)^2 = 1.0000000001$ from 1, and it therefore calculated $1^{10{,}000{,}000{,}000} = 1$. While for many calculations on computers this reality can be minimized, it cannot be eliminated. The wrong value warns us of something called round-off error. We can explore with computer graphics what this means for g near 0. As was the case for the *numerical monster* encountered in Section P.4, the computer can produce rich and beautiful behaviour in its failed attempt to represent g, which is very different from what g actually does. While it is possible to get computer algebra software like Maple to evaluate limits correctly (as we will see in the next section), we cannot use computer graphics or floating-point arithmetic to study many mathematical notions such as limits. In fact, we will need mathematics to understand what the computer actually does so that we can be the master of our tools.

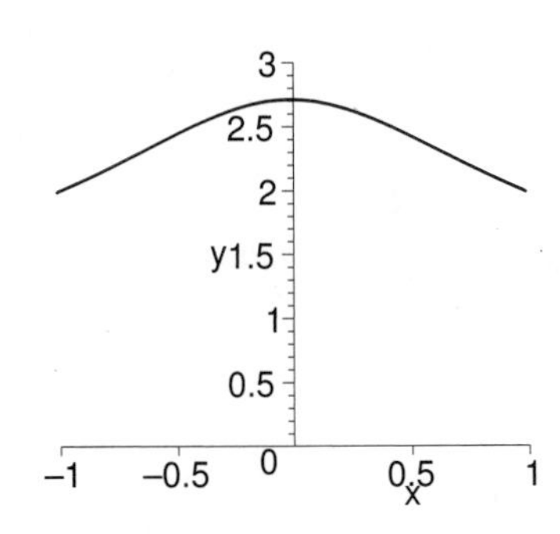

Figure 1.5 The graph of $y = g(x)$ on the interval $[-1, 1]$

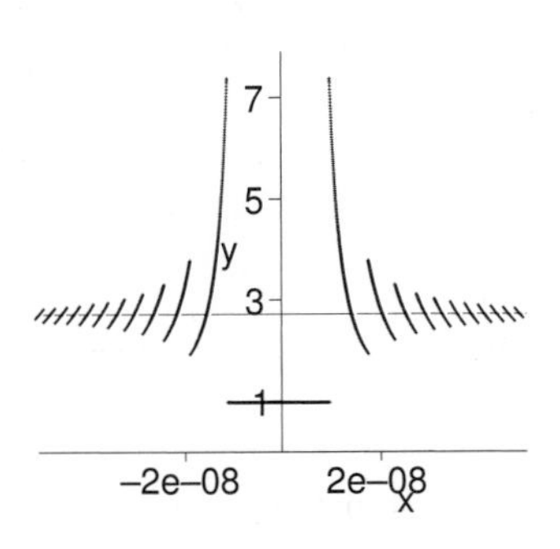

Figure 1.6 The graphs of $y = g(x)$ (colour) and $y = e \approx 2.718$ (black) on the interval $[-5 \times 10^{-8}, 5 \times 10^{-8}]$

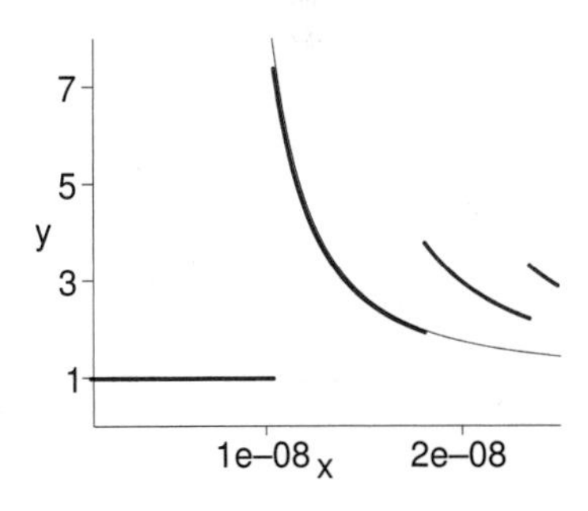

Figure 1.7 The graphs of $y = g(x)$ (colour) and $y = (1 + 2 \times 10^{-16})^{1/x^2}$ (black) on the interval $[10^{-9}, 2.5 \times 10^{-8}]$

Figures 1.5–1.7 illustrate this fascinating behaviour of g with three plots made with Maple using its default 10-significant-figure precision in representing floating-point (i.e., real) numbers. Figure 1.5 is a plot of the graph of g on the interval $[-1, 1]$. The graph starts out at height 2 at either endpoint $x = \pm 1$ and rises to height approximately $2.718\cdots$ as x decreases in absolute value, as we would expect from Table 4. Figure 1.6 shows the graph of g restricted to the tiny interval $[-5 \times 10^{-8}, 5 \times 10^{-8}]$. It consists of many short arcs decreasing in height as $|x|$ increases, and clustering around the line $y = 2.718\cdots$, and a horizontal part at height 1 between approximately -10^{-8} and 10^{-8}. Figure 1.7 zooms in on the part of the graph to the right of the origin up to $x = 2.5 \times 10^{-8}$. Note how the arc closest to 0 coincides with the graph of $y = \left(1 + 2 \times 10^{-16}\right)^{1/x^2}$ (shown in black), indicating that $1 + 2 \times 10^{-16}$ may be the smallest number greater than 1 that Maple can distinguish from 1. Both figures show that the breakdown in the graph of g is not sudden, but becomes more and more pronounced as $|x|$ decreases until the breakdown is complete near $\pm 10^{-8}$.

The examples above and those in Section 1.1 suggest the following *informal* definition of limit.

DEFINITION 1

An informal definition of limit

If $f(x)$ is defined for all x near a, except possibly at a itself, and if we can ensure that $f(x)$ is as close as we want to L by taking x close enough to a, but not equal to a, we say that the function f approaches the **limit** L as x approaches a, and we write

$$\lim_{x \to a} f(x) = L.$$

This definition is *informal* because phrases such as *close as we want* and *close enough* are imprecise; their meaning depends on the context. To a machinist manufacturing a piston, *close enough* may mean *within a few thousandths of an inch*. To an astronomer studying distant galaxies, *close enough* may mean *within a few thousand light-years*. The definition should be clear enough, however, to enable us to recognize and evaluate limits of specific functions. A more precise "formal" definition, given in Section 1.5, is needed if we want to *prove* theorems about limits like Theorems 2–4, stated later in this section.

EXAMPLE 3 Find (a) $\lim_{x \to a} x$ and (b) $\lim_{x \to a} c$ (where c is a constant).

Solution In words, part (a) asks: "What does x approach as x approaches a?" The answer is surely a.

$$\lim_{x \to a} x = a.$$

Similarly, part (b) asks: "What does c approach as x approaches a?" The answer here is that c approaches c; you can't get any closer to c than by *being* c.

$$\lim_{x \to a} c = c.$$

Example 3 shows that $\lim_{x \to a} f(x)$ can *sometimes* be evaluated by just calculating $f(a)$. This will be the case if $f(x)$ is defined in an open interval containing $x = a$ and the graph of f passes unbroken through the point $(a, f(a))$. The next example shows various ways algebraic manipulations can be used to evaluate $\lim_{x \to a} f(x)$ in situations where $f(a)$ is undefined. This usually happens when $f(x)$ is a fraction with denominator equal to 0 at $x = a$.

EXAMPLE 4 Evaluate:

(a) $\displaystyle\lim_{x \to -2} \frac{x^2 + x - 2}{x^2 + 5x + 6}$, (b) $\displaystyle\lim_{x \to a} \frac{\frac{1}{x} - \frac{1}{a}}{x - a}$, and (c) $\displaystyle\lim_{x \to 4} \frac{\sqrt{x} - 2}{x^2 - 16}$.

Solution Each of these limits involves a fraction whose numerator and denominator are both 0 at the point where the limit is taken.

(a) $\displaystyle \lim_{x\to-2} \frac{x^2+x-2}{x^2+5x+6}$ — fraction undefined at $x=-2$. Factor numerator and denominator. (See Section P.6.)

$\displaystyle = \lim_{x\to-2} \frac{(x+2)(x-1)}{(x+2)(x+3)}$ — Cancel common factors.

$\displaystyle = \lim_{x\to-2} \frac{x-1}{x+3}$ — Evaluate this limit by substituting $x=-2$.

$$= \frac{-2-1}{-2+3} = -3.$$

(b) $\displaystyle \lim_{x\to a} \frac{\frac{1}{x}-\frac{1}{a}}{x-a}$ — fraction undefined at $x=a$. Simplify the numerator.

$$= \lim_{x\to a} \frac{\frac{a-x}{ax}}{x-a}$$

$\displaystyle = \lim_{x\to a} \frac{-(x-a)}{ax(x-a)}$ — Cancel the common factor.

$$= \lim_{x\to a} \frac{-1}{ax} = -\frac{1}{a^2}.$$

(c) $\displaystyle \lim_{x\to 4} \frac{\sqrt{x}-2}{x^2-16}$ — fraction undefined at $x=4$. Multiply numerator and denominator by the conjugate of the expression in the numerator.

$$= \lim_{x\to 4} \frac{(\sqrt{x}-2)(\sqrt{x}+2)}{(x^2-16)(\sqrt{x}+2)}$$

$$= \lim_{x\to 4} \frac{x-4}{(x-4)(x+4)(\sqrt{x}+2)}$$

$$= \lim_{x\to 4} \frac{1}{(x+4)(\sqrt{x}+2)} = \frac{1}{(4+4)(2+2)} = \frac{1}{32}$$

(a)

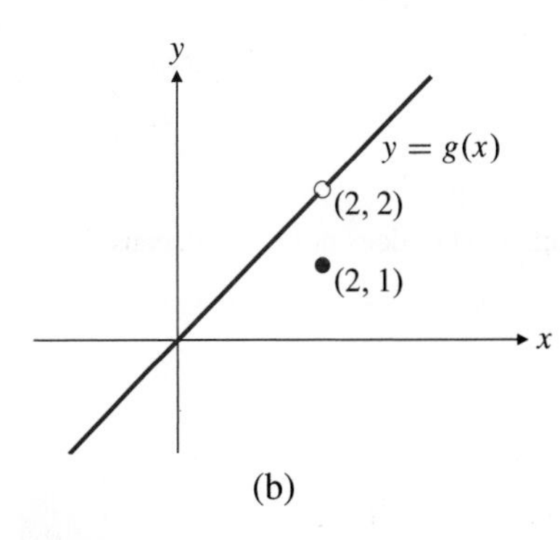

(b)

Figure 1.8

(a) $\displaystyle \lim_{x\to 0} \frac{1}{x}$ does not exist

(b) $\displaystyle \lim_{x\to 2} g(x) = 2$, but $g(2)=1$

A function f may be defined on both sides of $x=a$ but still not have a limit at $x=a$. For example, the function $f(x)=1/x$ has no limit as x approaches 0. As can be seen in Figure 1.8(a), the values $1/x$ grow ever larger in absolute value as x approaches 0; there is no single number L that they approach.

The following example shows that even if $f(x)$ is defined at $x=a$, the limit of $f(x)$ as x approaches a may not be equal to $f(a)$.

BEWARE! Always be aware that the existence of $\lim_{x\to a} f(x)$ does not require that $f(a)$ exist and does not depend on $f(a)$ even if $f(a)$ does exist. It depends only on the values of $f(x)$ for x *near but not equal to a*.

EXAMPLE 5 Let $g(x) = \begin{cases} x & \text{if } x \ne 2 \\ 1 & \text{if } x = 2. \end{cases}$ (See Figure 1.8(b).) Then

$$\lim_{x\to 2} g(x) = \lim_{x\to 2} x = 2, \quad \text{although} \quad g(2)=1.$$

One-Sided Limits

Limits are *unique*; if $\lim_{x\to a} f(x) = L$ and $\lim_{x\to a} f(x) = M$, then $L = M$. (See Exercise 31 in Section 1.5.) Although a function f can only have one limit at any particular point, it is, nevertheless, useful to be able to describe the behaviour of functions that approach different numbers as x approaches a from one side or the other. (See Figure 1.9.)

DEFINITION 2

Informal definition of left and right limits

If $f(x)$ is defined on some interval (b, a) extending to the left of $x = a$, and if we can ensure that $f(x)$ is as close as we want to L by taking x to the left of a and close enough to a, then we say $f(x)$ has **left limit** L at $x = a$, and we write

$$\lim_{x\to a-} f(x) = L.$$

If $f(x)$ is defined on some interval (a, b) extending to the right of $x = a$, and if we can ensure that $f(x)$ is as close as we want to L by taking x to the right of a and close enough to a, then we say $f(x)$ has **right limit** L at $x = a$, and we write

$$\lim_{x\to a+} f(x) = L.$$

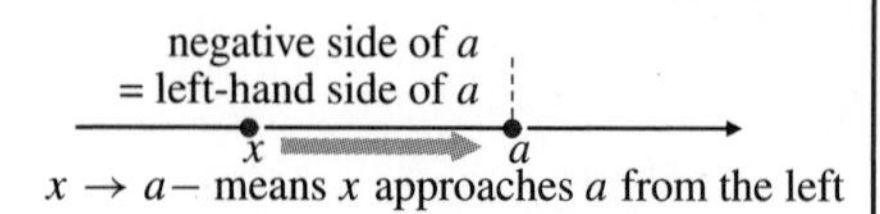

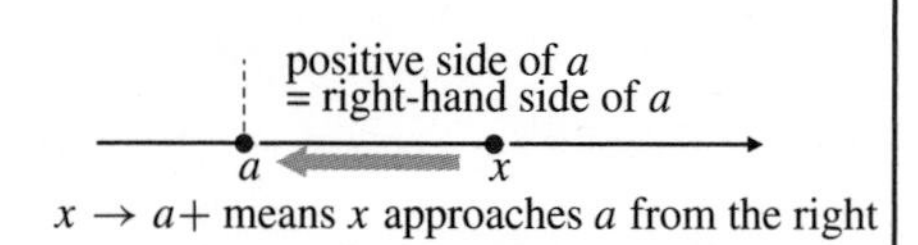

Figure 1.9 One-sided approach

Note the use of the suffix $+$ to denote approach from the right (the *positive* side) and the suffix $-$ to denote approach from the left (the *negative* side).

EXAMPLE 6 The signum function $\operatorname{sgn}(x) = x/|x|$ (see Figure 1.10) has left limit -1 and right limit 1 at $x = 0$:

$$\lim_{x\to 0-} \operatorname{sgn}(x) = -1 \quad \text{and} \quad \lim_{x\to 0+} \operatorname{sgn}(x) = 1$$

because the values of $\operatorname{sgn}(x)$ approach -1 (they *are* -1) if x is negative and approaches 0, and they approach 1 if x is positive and approaches 0. Since these left and right limits are not equal, $\lim_{x\to 0} \operatorname{sgn}(x)$ *does not exist.*

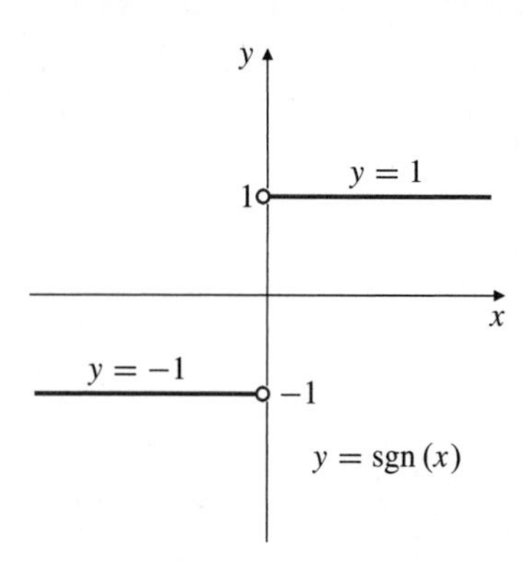

Figure 1.10
$\lim_{x\to 0} \operatorname{sgn}(x)$ does not exist, because $\lim_{x\to 0-} \operatorname{sgn}(x) = -1$, $\lim_{x\to 0+} \operatorname{sgn}(x) = 1$

As suggested in Example 6, the relationship between ordinary (two-sided) limits and one-sided limits can be stated as follows:

THEOREM 1

Relationship between one-sided and two-sided limits

A function $f(x)$ has limit L at $x = a$ if and only if it has both left and right limits there and these one-sided limits are both equal to L:

$$\lim_{x\to a} f(x) = L \iff \lim_{x\to a-} f(x) = \lim_{x\to a+} f(x) = L.$$

EXAMPLE 7 If $f(x) = \dfrac{|x-2|}{x^2+x-6}$, find: $\lim_{x\to 2+} f(x)$, $\lim_{x\to 2-} f(x)$, and $\lim_{x\to 2} f(x)$.

Solution Observe that $|x-2| = x-2$ if $x > 2$, and $|x-2| = -(x-2)$ if $x < 2$.

Therefore,

$$\lim_{x\to 2+} f(x) = \lim_{x\to 2+} \frac{x-2}{x^2+x-6} = \lim_{x\to 2+} \frac{x-2}{(x-2)(x+3)} = \lim_{x\to 2+} \frac{1}{x+3} = \frac{1}{5},$$

$$\lim_{x\to 2-} f(x) = \lim_{x\to 2-} \frac{-(x-2)}{x^2+x-6} = \lim_{x\to 2-} \frac{-(x-2)}{(x-2)(x+3)} = \lim_{x\to 2-} \frac{-1}{x+3} = -\frac{1}{5}.$$

Since $\lim_{x\to 2-} f(x) \neq \lim_{x\to 2+} f(x)$, the limit $\lim_{x\to 2} f(x)$ does not exist.

EXAMPLE 8 What one-sided limits does $g(x) = \sqrt{1-x^2}$ have at $x = -1$ and $x = 1$?

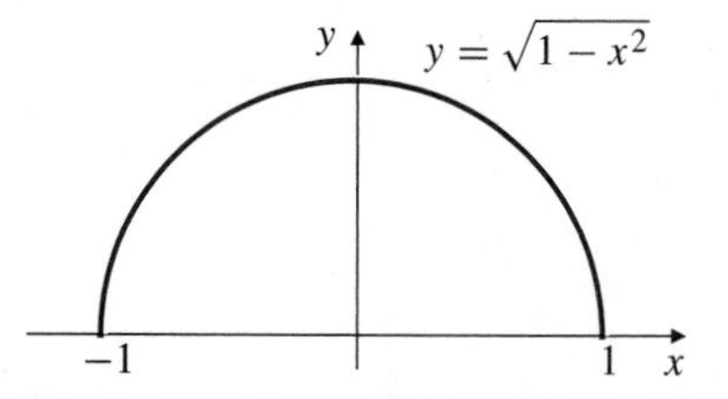

Figure 1.11 $\sqrt{1-x^2}$ has right limit 0 at -1 and left limit 0 at 1

Solution The domain of g is $[-1, 1]$, so $g(x)$ is defined only to the right of $x = -1$ and only to the left of $x = 1$. As can be seen in Figure 1.11,

$$\lim_{x\to -1+} g(x) = 0 \qquad \text{and} \qquad \lim_{x\to 1-} g(x) = 0.$$

$g(x)$ has no left limit or limit at $x = -1$ and no right limit or limit at $x = 1$.

Rules for Calculating Limits

The following theorems make it easy to calculate limits and one-sided limits of many kinds of functions when we know some elementary limits. We will not prove the theorems here. (See Section 1.5.)

THEOREM 2 **Limit Rules**

If $\lim_{x\to a} f(x) = L$, $\lim_{x\to a} g(x) = M$, and k is a constant, then

1. **Limit of a sum:** $\lim_{x\to a} [f(x) + g(x)] = L + M$
2. **Limit of a difference:** $\lim_{x\to a} [f(x) - g(x)] = L - M$
3. **Limit of a product:** $\lim_{x\to a} f(x)g(x) = LM$
4. **Limit of a multiple:** $\lim_{x\to a} kf(x) = kL$
5. **Limit of a quotient:** $\lim_{x\to a} \frac{f(x)}{g(x)} = \frac{L}{M}$, if $M \neq 0$.

If m is an integer and n is a positive integer, then

6. **Limit of a power:** $\lim_{x\to a} \left[f(x)\right]^{m/n} = L^{m/n}$, provided $L > 0$ if n is even, and $L \neq 0$ if $m < 0$.

If $f(x) \leq g(x)$ on an interval containing a in its interior, then

7. **Order is preserved:** $L \leq M$

Rules 1–6 are also valid for right limits and left limits. So is Rule 7, under the assumption that $f(x) \leq g(x)$ on an open interval extending in the appropriate direction from a.

In words, rule 1 of Theorem 2 says that the limit of a sum of functions is the sum of their limits. Similarly, rule 5 says that the limit of a quotient of two functions is the quotient of their limits, provided that the limit of the denominator is not zero. Try to state the other rules in words.

We can make use of the limits (a) $\lim_{x\to a} c = c$ (where c is a constant) and (b) $\lim_{x\to a} x = a$, from Example 3, together with parts of Theorem 2 to calculate limits of many combinations of functions.

EXAMPLE 9 Find: (a) $\lim_{x \to a} \dfrac{x^2 + x + 4}{x^3 - 2x^2 + 7}$ and (b) $\lim_{x \to 2} \sqrt{2x + 1}$.

Solution

(a) The expression $\dfrac{x^2 + x + 4}{x^3 - 2x^2 + 7}$ is formed by combining the basic functions x and c (constant) using addition, subtraction, multiplication, and division. Theorem 2 assures us that the limit of such a combination is the same combination of the limits a and c of the basic functions, provided the denominator does not have limit zero. Thus,

$$\lim_{x \to a} \frac{x^2 + x + 4}{x^3 - 2x^2 + 7} = \frac{a^2 + a + 4}{a^3 - 2a^2 + 7} \qquad \text{provided } a^3 - 2a^2 + 7 \neq 0.$$

(b) The same argument as in (a) shows that $\lim_{x \to 2} (2x + 1) = 2(2) + 1 = 5$. Then the Power Rule (rule 6 of Theorem 2) assures us that

$$\lim_{x \to 2} \sqrt{2x + 1} = \sqrt{5}.$$

The following result is an immediate corollary of Theorem 2. (See Section P.6 for a discussion of polynomials and rational functions.)

THEOREM 3

Limits of Polynomials and Rational Functions

1. If $P(x)$ is a polynomial and a is any real number, then

$$\lim_{x \to a} P(x) = P(a).$$

2. If $P(x)$ and $Q(x)$ are polynomials and $Q(a) \neq 0$, then

$$\lim_{x \to a} \frac{P(x)}{Q(x)} = \frac{P(a)}{Q(a)}.$$

The Squeeze Theorem

The following theorem will enable us to calculate some very important limits in subsequent chapters. It is called the *Squeeze Theorem* because it refers to a function g whose values are squeezed between the values of two other functions f and h that have the same limit L at a point a. Being trapped between the values of two functions that approach L, the values of g must also approach L. (See Figure 1.12.)

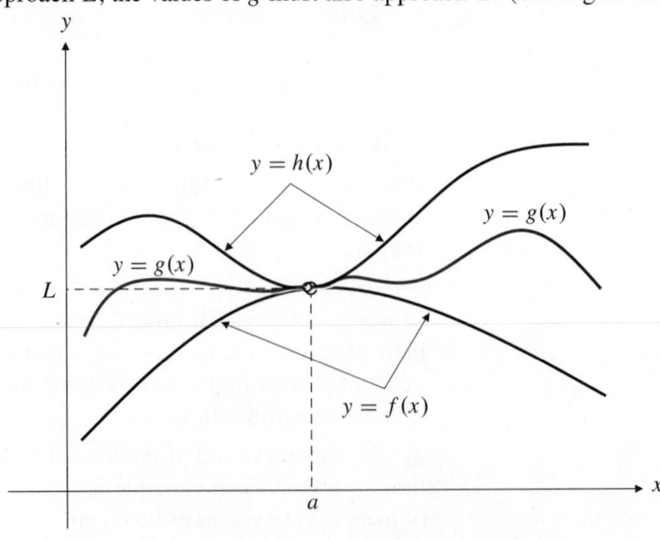

Figure 1.12 The graph of g is squeezed between those of f and h

THEOREM 4

The Squeeze Theorem

Suppose that $f(x) \le g(x) \le h(x)$ holds for all x in some open interval containing a, except possibly at $x = a$ itself. Suppose also that

$$\lim_{x\to a} f(x) = \lim_{x\to a} h(x) = L.$$

Then $\lim_{x\to a} g(x) = L$ also. Similar statements hold for left and right limits.

EXAMPLE 10 Given that $3 - x^2 \le u(x) \le 3 + x^2$ for all $x \ne 0$, find $\lim_{x\to 0} u(x)$.

Solution Since $\lim_{x\to 0}(3 - x^2) = 3$ and $\lim_{x\to 0}(3 + x^2) = 3$, the Squeeze Theorem implies that $\lim_{x\to 0} u(x) = 3$.

EXAMPLE 11 Show that if $\lim_{x\to a} |f(x)| = 0$, then $\lim_{x\to a} f(x) = 0$.

Solution Since $-|f(x)| \le f(x) \le |f(x)|$, and $-|f(x)|$ and $|f(x)|$ both have limit 0 as x approaches a, so does $f(x)$ by the Squeeze Theorem.

EXERCISES 1.2

1. Find: (a) $\lim_{x\to -1} f(x)$, (b) $\lim_{x\to 0} f(x)$, and (c) $\lim_{x\to 1} f(x)$, for the function f whose graph is shown in Figure 1.13.

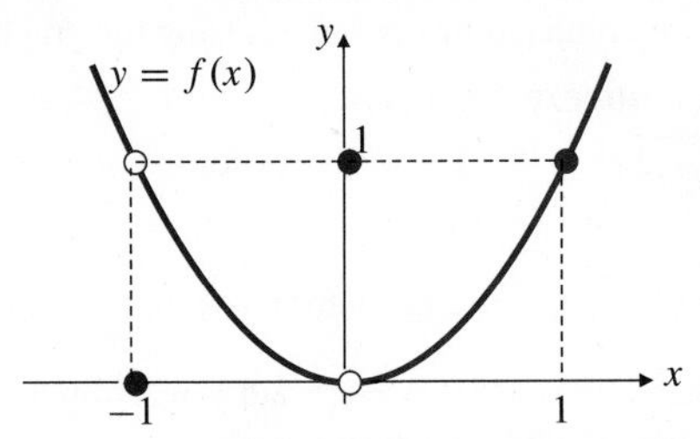

Figure 1.13

2. For the function $y = g(x)$ graphed in Figure 1.14, find each of the following limits or explain why it does not exist.
(a) $\lim_{x\to 1} g(x)$, (b) $\lim_{x\to 2} g(x)$, (c) $\lim_{x\to 3} g(x)$

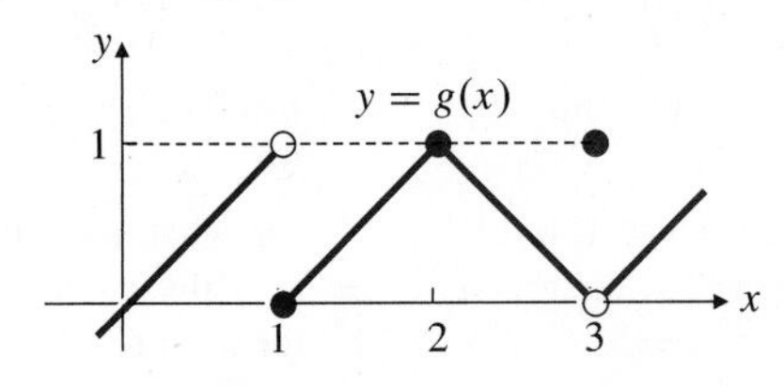

Figure 1.14

In Exercises 3–6, find the indicated one-sided limit of the function g whose graph is given in Figure 1.14.

3. $\lim_{x\to 1-} g(x)$

4. $\lim_{x\to 1+} g(x)$

5. $\lim_{x\to 3+} g(x)$

6. $\lim_{x\to 3-} g(x)$

In Exercises 7–36, evaluate the limit or explain why it does not exist.

7. $\lim_{x\to 4}(x^2 - 4x + 1)$

8. $\lim_{x\to 2} 3(1 - x)(2 - x)$

9. $\lim_{x\to 3} \dfrac{x+3}{x+6}$

10. $\lim_{t\to -4} \dfrac{t^2}{4-t}$

11. $\lim_{x\to 1} \dfrac{x^2-1}{x+1}$

12. $\lim_{x\to -1} \dfrac{x^2-1}{x+1}$

13. $\lim_{x\to 3} \dfrac{x^2-6x+9}{x^2-9}$

14. $\lim_{x\to -2} \dfrac{x^2+2x}{x^2-4}$

15. $\lim_{h\to 2} \dfrac{1}{4-h^2}$

16. $\lim_{h\to 0} \dfrac{3h+4h^2}{h^2-h^3}$

17. $\lim_{x\to 9} \dfrac{\sqrt{x}-3}{x-9}$

18. $\lim_{h\to 0} \dfrac{\sqrt{4+h}-2}{h}$

19. $\lim_{x\to \pi} \dfrac{(x-\pi)^2}{\pi x}$

20. $\lim_{x\to -2} |x - 2|$

21. $\lim_{x\to 0} \dfrac{|x-2|}{x-2}$

22. $\lim_{x\to 2} \dfrac{|x-2|}{x-2}$

23. $\lim_{t\to 1} \dfrac{t^2-1}{t^2-2t+1}$

24. $\lim_{x\to 2} \dfrac{\sqrt{4-4x+x^2}}{x-2}$

25. $\lim_{t\to 0} \dfrac{t}{\sqrt{4+t}-\sqrt{4-t}}$

26. $\lim_{x\to 1} \dfrac{x^2-1}{\sqrt{x+3}-2}$

27. $\lim_{t\to 0} \dfrac{t^2+3t}{(t+2)^2-(t-2)^2}$

28. $\lim_{s\to 0} \dfrac{(s+1)^2-(s-1)^2}{s}$

29. $\lim_{y\to 1} \dfrac{y-4\sqrt{y}+3}{y^2-1}$

30. $\lim_{x\to -1} \dfrac{x^3+1}{x+1}$

31. $\lim_{x\to 2} \dfrac{x^4-16}{x^3-8}$

32. $\lim_{x\to 8} \dfrac{x^{2/3}-4}{x^{1/3}-2}$

33. $\lim_{x\to 2}\left(\frac{1}{x-2}-\frac{4}{x^2-4}\right)$ **34.** $\lim_{x\to 2}\left(\frac{1}{x-2}-\frac{1}{x^2-4}\right)$

35. $\lim_{x\to 0}\frac{\sqrt{2+x^2}-\sqrt{2-x^2}}{x^2}$ **36.** $\lim_{x\to 0}\frac{|3x-1|-|3x+1|}{x}$

The limit $\lim_{h\to 0}\frac{f(x+h)-f(x)}{h}$ occurs frequently in the study of calculus. (Can you guess why?) Evaluate this limit for the functions f in Exercises 37–42.

37. $f(x)=x^2$ **38.** $f(x)=x^3$

39. $f(x)=\frac{1}{x}$ **40.** $f(x)=\frac{1}{x^2}$

41. $f(x)=\sqrt{x}$ **42.** $f(x)=1/\sqrt{x}$

Examine the graphs of $\sin x$ and $\cos x$ in Section P.7 to determine the limits in Exercises 43–46.

43. $\lim_{x\to\pi/2}\sin x$ **44.** $\lim_{x\to\pi/4}\cos x$

45. $\lim_{x\to\pi/3}\cos x$ **46.** $\lim_{x\to 2\pi/3}\sin x$

47. Make a table of values of $f(x)=(\sin x)/x$ for a sequence of values of x approaching 0, say ± 1.0, ± 0.1, ± 0.01, ± 0.001, ± 0.0001, and ± 0.00001. Make sure your calculator is set in *radian mode* rather than degree mode. Guess the value of $\lim_{x\to 0} f(x)$.

48. Repeat Exercise 47 for $f(x)=\frac{1-\cos x}{x^2}$.

In Exercises 49–60, find the indicated one-sided limit or explain why it does not exist.

49. $\lim_{x\to 2-}\sqrt{2-x}$ **50.** $\lim_{x\to 2+}\sqrt{2-x}$

51. $\lim_{x\to -2-}\sqrt{2-x}$ **52.** $\lim_{x\to -2+}\sqrt{2-x}$

53. $\lim_{x\to 0}\sqrt{x^3-x}$ **54.** $\lim_{x\to 0-}\sqrt{x^3-x}$

55. $\lim_{x\to 0+}\sqrt{x^3-x}$ **56.** $\lim_{x\to 0+}\sqrt{x^2-x^4}$

57. $\lim_{x\to a-}\frac{|x-a|}{x^2-a^2}$ **58.** $\lim_{x\to a+}\frac{|x-a|}{x^2-a^2}$

59. $\lim_{x\to 2-}\frac{x^2-4}{|x+2|}$ **60.** $\lim_{x\to 2+}\frac{x^2-4}{|x+2|}$

Exercises 61–64 refer to the function

$$f(x)=\begin{cases} x-1 & \text{if } x\le -1\\ x^2+1 & \text{if } -1<x\le 0\\ (x+\pi)^2 & \text{if } x>0.\end{cases}$$

Find the indicated limits.

61. $\lim_{x\to -1-} f(x)$ **62.** $\lim_{x\to -1+} f(x)$

63. $\lim_{x\to 0+} f(x)$ **64.** $\lim_{x\to 0-} f(x)$

65. Suppose $\lim_{x\to 4} f(x)=2$ and $\lim_{x\to 4} g(x)=-3$. Find:

(a) $\lim_{x\to 4}\big(g(x)+3\big)$ (b) $\lim_{x\to 4} xf(x)$

(c) $\lim_{x\to 4}\big(g(x)\big)^2$ (d) $\lim_{x\to 4}\frac{g(x)}{f(x)-1}$.

66. Suppose $\lim_{x\to a} f(x)=4$ and $\lim_{x\to a} g(x)=-2$. Find:

(a) $\lim_{x\to a}\big(f(x)+g(x)\big)$ (b) $\lim_{x\to a} f(x)\cdot g(x)$

(c) $\lim_{x\to a} 4g(x)$ (d) $\lim_{x\to a} f(x)/g(x)$.

67. If $\lim_{x\to 2}\frac{f(x)-5}{x-2}=3$, find $\lim_{x\to 2} f(x)$.

68. If $\lim_{x\to 0}\frac{f(x)}{x^2}=-2$, find $\lim_{x\to 0} f(x)$ and $\lim_{x\to 0}\frac{f(x)}{x}$.

Using Graphing Utilities to Find Limits

Graphing calculators or computer software can be used to evaluate limits at least approximately. Simply "zoom" the plot window to show smaller and smaller parts of the graph near the point where the limit is to be found. Find the following limits by graphical techniques. Where you think it justified, give an exact answer. Otherwise, give the answer correct to 4 decimal places. Remember to ensure that your calculator or software is set for radian mode when using trigonometric functions.

69. $\lim_{x\to 0}\frac{\sin x}{x}$ **70.** $\lim_{x\to 0}\frac{\sin(2\pi x)}{\sin(3\pi x)}$

71. $\lim_{x\to 1-}\frac{\sin\sqrt{1-x}}{\sqrt{1-x^2}}$ **72.** $\lim_{x\to 0+}\frac{x-\sqrt{x}}{\sqrt{\sin x}}$

73. On the same graph plot the three functions $y=x\sin(1/x)$, $y=x$, and $y=-x$ for $-0.2\le x\le 0.2$, $-0.2\le y\le 0.2$. Describe the behaviour of $f(x)=x\sin(1/x)$ near $x=0$. Does $\lim_{x\to 0} f(x)$ exist, and if so, what is its value? Could you have predicted this before drawing the graph? Why?

Using the Squeeze Theorem

74. If $\sqrt{5-2x^2}\le f(x)\le\sqrt{5-x^2}$ for $-1\le x\le 1$, find $\lim_{x\to 0} f(x)$.

75. If $2-x^2\le g(x)\le 2\cos x$ for all x, find $\lim_{x\to 0} g(x)$.

76. (a) Sketch the curves $y=x^2$ and $y=x^4$ on the same graph. Where do they intersect?

(b) The function $f(x)$ satisfies:

$$\begin{cases} x^2\le f(x)\le x^4 & \text{if } x<-1 \text{ or } x>1\\ x^4\le f(x)\le x^2 & \text{if } -1\le x\le 1\end{cases}$$

Find (i) $\lim_{x\to -1} f(x)$, (ii) $\lim_{x\to 0} f(x)$, (iii) $\lim_{x\to 1} f(x)$.

77. On what intervals is $x^{1/3}<x^3$? On what intervals is $x^{1/3}>x^3$? If the graph of $y=h(x)$ always lies between the graphs of $y=x^{1/3}$ and $y=x^3$, for what real numbers a can you determine the value of $\lim_{x\to a} h(x)$? Find the limit for each of these values of a.

78. What is the domain of $x\sin\frac{1}{x}$? Evaluate $\lim_{x\to 0} x\sin\frac{1}{x}$.

79. Suppose $|f(x)|\le g(x)$ for all x. What can you conclude about $\lim_{x\to a} f(x)$ if $\lim_{x\to a} g(x)=0$? What if $\lim_{x\to a} g(x)=3$?

1.3 Limits at Infinity and Infinite Limits

In this section we will extend the concept of limit to allow for two situations not covered by the definitions of limit and one-sided limit in the previous section:

(i) limits at infinity, where x becomes arbitrarily large, positive or negative;

(ii) infinite limits, which are not really limits at all but provide useful symbolism for describing the behaviour of functions whose values become arbitrarily large, positive or negative.

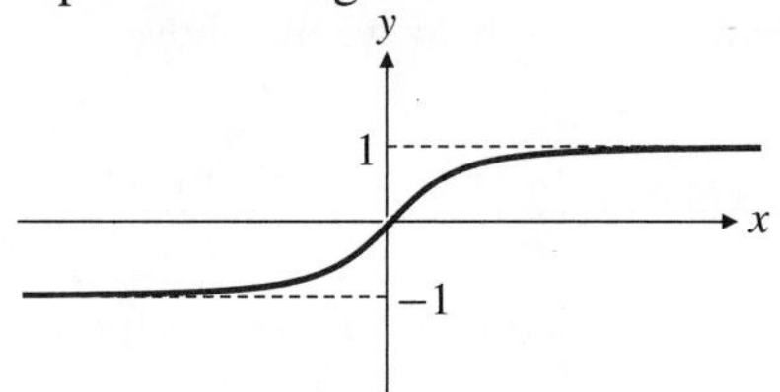

Figure 1.15 The graph of $x/\sqrt{x^2+1}$

Limits at Infinity

Table 5.

x	$f(x) = x/\sqrt{x^2+1}$
$-1{,}000$	-0.9999995
-100	-0.9999500
-10	-0.9950372
-1	-0.7071068
0	0.0000000
1	0.7071068
10	0.9950372
100	0.9999500
$1{,}000$	0.9999995

Consider the function

$$f(x) = \frac{x}{\sqrt{x^2+1}}$$

whose graph is shown in Figure 1.15 and for which some values (rounded to 7 decimal places) are given in Table 5. The values of $f(x)$ seem to approach 1 as x takes on larger and larger positive values, and -1 as x takes on negative values that get larger and larger in absolute value. (See Example 2 below for confirmation.) We express this behaviour by writing

$$\lim_{x\to\infty} f(x) = 1 \qquad \text{“}f(x)\text{ approaches 1 as }x\text{ approaches infinity.”}$$

$$\lim_{x\to-\infty} f(x) = -1 \qquad \text{“}f(x)\text{ approaches }-1\text{ as }x\text{ approaches negative infinity.”}$$

The graph of f conveys this limiting behaviour by approaching the horizontal lines $y = 1$ as x moves far to the right and $y = -1$ as x moves far to the left. These lines are called **horizontal asymptotes** of the graph. In general, if a curve approaches a straight line as it recedes very far away from the origin, that line is called an **asymptote** of the curve.

DEFINITION 3

Limits at infinity and negative infinity (informal definition)

If the function f is defined on an interval (a, ∞) and if we can ensure that $f(x)$ is as close as we want to the number L by taking x large enough, then we say that $f(x)$ **approaches the limit L as x approaches infinity**, and we write

$$\lim_{x\to\infty} f(x) = L.$$

If f is defined on an interval $(-\infty, b)$ and if we can ensure that $f(x)$ is as close as we want to the number M by taking x negative and large enough in absolute value, then we say that $f(x)$ **approaches the limit M as x approaches negative infinity**, and we write

$$\lim_{x\to-\infty} f(x) = M.$$

Recall that the symbol ∞, called **infinity**, does *not* represent a real number. We cannot use ∞ in arithmetic in the usual way, but we can use the phrase "approaches ∞" to mean "becomes arbitrarily large positive" and the phrase "approaches $-\infty$" to mean "becomes arbitrarily large negative."

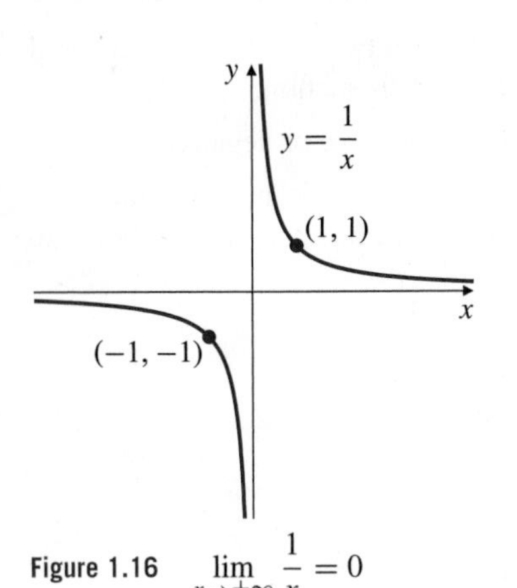

Figure 1.16 $\lim_{x\to\pm\infty} \frac{1}{x} = 0$

EXAMPLE 1 In Figure 1.16, we can see that $\lim_{x\to\infty} 1/x = \lim_{x\to-\infty} 1/x = 0$. The x-axis is a horizontal asymptote of the graph $y = 1/x$.

The theorems of Section 1.2 have suitable counterparts for limits at infinity or negative infinity. In particular, it follows from the example above and from the Product Rule for limits that $\lim_{x\to\pm\infty} 1/x^n = 0$ for any positive integer n. We will use this fact in the following examples. Example 2 shows how to obtain the limits at $\pm\infty$ for the function $x/\sqrt{x^2+1}$ by algebraic means, without resorting to making a table of values or drawing a graph, as we did above.

EXAMPLE 2 Evaluate $\lim_{x\to\infty} f(x)$ and $\lim_{x\to-\infty} f(x)$ for $f(x) = \dfrac{x}{\sqrt{x^2+1}}$.

Solution Rewrite the expression for $f(x)$ as follows:

$$f(x) = \frac{x}{\sqrt{x^2\left(1+\dfrac{1}{x^2}\right)}} = \frac{x}{\sqrt{x^2}\sqrt{1+\dfrac{1}{x^2}}} \qquad \text{Remember } \sqrt{x^2} = |x|.$$

$$= \frac{x}{|x|\sqrt{1+\dfrac{1}{x^2}}}$$

$$= \frac{\operatorname{sgn} x}{\sqrt{1+\dfrac{1}{x^2}}}, \qquad \text{where } \operatorname{sgn} x = \frac{x}{|x|} = \begin{cases} 1 & \text{if } x > 0 \\ -1 & \text{if } x < 0. \end{cases}$$

The factor $\sqrt{1+(1/x^2)}$ approaches 1 as x approaches ∞ or $-\infty$, so $f(x)$ must have the same limits as $x \to \pm\infty$ as does $\operatorname{sgn}(x)$. Therefore (see Figure 1.15),

$$\lim_{x\to\infty} f(x) = 1 \qquad \text{and} \qquad \lim_{x\to-\infty} f(x) = -1.$$

Limits at Infinity for Rational Functions

The only polynomials that have limits at $\pm\infty$ are constant ones, $P(x) = c$. The situation is more interesting for rational functions. Recall that a rational function is a quotient of two polynomials. The following examples show how to render such a function in a form where its limits at infinity and negative infinity (if they exist) are apparent. The way to do this is to *divide the numerator and denominator by the highest power of x appearing in the denominator.* The limits of a rational function at infinity and negative infinity either both fail to exist or both exist and are equal.

EXAMPLE 3 **(Numerator and denominator of the same degree)** Evaluate $\lim_{x\to\pm\infty} \dfrac{2x^2-x+3}{3x^2+5}$.

Solution Divide the numerator and the denominator by x^2, the highest power of x appearing in the denominator:

$$\lim_{x\to\pm\infty} \frac{2x^2-x+3}{3x^2+5} = \lim_{x\to\pm\infty} \frac{2-(1/x)+(3/x^2)}{3+(5/x^2)} = \frac{2-0+0}{3+0} = \frac{2}{3}.$$

EXAMPLE 4 **(Degree of numerator less than degree of denominator)** Evaluate $\lim_{x\to\pm\infty} \dfrac{5x+2}{2x^3-1}$.

Solution Divide the numerator and the denominator by the largest power of x in the denominator, namely, x^3.

$$\lim_{x\to\pm\infty} \frac{5x+2}{2x^3-1} = \lim_{x\to\pm\infty} \frac{(5/x^2)+(2/x^3)}{2-(1/x^3)} = \frac{0+0}{2-0} = 0.$$

The limiting behaviour of rational functions at infinity and negative infinity is summarized at the left.

Summary of limits at $\pm\infty$ for rational functions

Let $P_m(x) = a_m x^m + \cdots + a_0$ and $Q_n(x) = b_n x^n + \cdots + b_0$ be polynomials of degree m and n, respectively, so that $a_m \neq 0$ and $b_n \neq 0$. Then

$$\lim_{x\to\pm\infty} \frac{P_m(x)}{Q_n(x)}$$

(a) equals zero if $m < n$,

(b) equals $\dfrac{a_m}{b_n}$ if $m = n$,

(c) does not exist if $m > n$.

The technique used in the previous examples can also be applied to more general kinds of functions. The function in the following example is not rational, and the limit seems to produce a meaningless $\infty - \infty$ until we resolve matters by rationalizing the numerator.

EXAMPLE 5 Find $\lim_{x\to\infty}\left(\sqrt{x^2+x}-x\right)$.

Solution We are trying to find the limit of the difference of two functions, each of which becomes arbitrarily large as x increases to infinity. We rationalize the expression by multiplying the numerator and the denominator (which is 1) by the conjugate expression $\sqrt{x^2+x}+x$:

$$\begin{aligned}
\lim_{x\to\infty}\left(\sqrt{x^2+x}-x\right) &= \lim_{x\to\infty} \frac{\left(\sqrt{x^2+x}-x\right)\left(\sqrt{x^2+x}+x\right)}{\sqrt{x^2+x}+x} \\
&= \lim_{x\to\infty} \frac{x^2+x-x^2}{\sqrt{x^2\left(1+\dfrac{1}{x}\right)}+x} \\
&= \lim_{x\to\infty} \frac{x}{x\sqrt{1+\dfrac{1}{x}}+x} = \lim_{x\to\infty} \frac{1}{\sqrt{1+\dfrac{1}{x}}+1} = \frac{1}{2}.
\end{aligned}$$

(Here, $\sqrt{x^2} = x$ because $x > 0$ as $x \to \infty$.)

Remark The limit $\lim_{x\to-\infty}(\sqrt{x^2+x}-x)$ is not nearly so subtle. Since $-x > 0$ as $x \to -\infty$, we have $\sqrt{x^2+x}-x > \sqrt{x^2+x}$, which grows arbitrarily large as $x \to -\infty$. The limit does not exist.

Infinite Limits

A function whose values grow arbitrarily large can sometimes be said to have an infinite limit. Since infinity is not a number, infinite limits are not really limits at all, but they provide a way of describing the behaviour of functions that grow arbitrarily large positive or negative. A few examples will make the terminology clear.

EXAMPLE 6 **(A two-sided infinite limit)** Describe the behaviour of the function $f(x) = 1/x^2$ near $x = 0$.

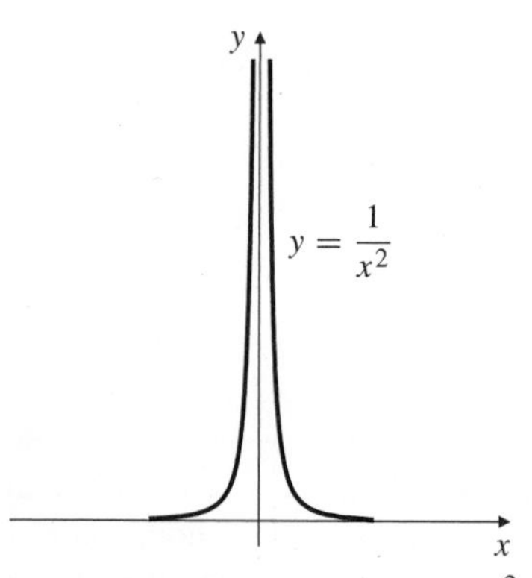

Figure 1.17 The graph of $y = 1/x^2$ (not to scale)

Solution As x approaches 0 from either side, the values of $f(x)$ are positive and grow larger and larger (see Figure 1.17), so the limit of $f(x)$ as x approaches 0 *does not exist.* It is nevertheless convenient to describe the behaviour of f near 0 by saying that $f(x)$ *approaches* ∞ as x approaches zero. We write

$$\lim_{x\to0} f(x) = \lim_{x\to0} \frac{1}{x^2} = \infty.$$

Note that in writing this we are *not* saying that $\lim_{x\to0} 1/x^2$ *exists*. Rather, we are saying that that limit *does not exist because* $1/x^2$ *becomes arbitrarily large near* $x = 0$. Observe how the graph of f approaches the y-axis as x approaches 0. The y-axis is a **vertical asymptote** of the graph.

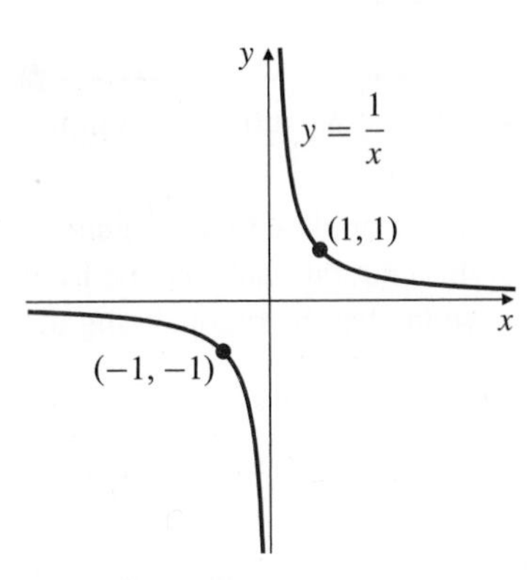

Figure 1.18 $\lim_{x\to 0-} 1/x = -\infty$, $\lim_{x\to 0+} 1/x = \infty$

EXAMPLE 7 **(One-sided infinite limits)** Describe the behaviour of the function $f(x) = 1/x$ near $x = 0$. (See Figure 1.18.)

Solution As x approaches 0 from the right, the values of $f(x)$ become larger and larger positive numbers, and we say that f has right-hand limit infinity at $x = 0$:

$$\lim_{x\to 0+} f(x) = \infty.$$

Similarly, the values of $f(x)$ become larger and larger negative numbers as x approaches 0 from the left, so f has left-hand limit $-\infty$ at $x = 0$:

$$\lim_{x\to 0-} f(x) = -\infty.$$

These statements do not say that the one-sided limits *exist*; they do not exist because ∞ and $-\infty$ are not numbers. Since the one-sided limits are not equal even as infinite symbols, all we can say about the two-sided $\lim_{x\to 0} f(x)$ is that it does not exist.

EXAMPLE 8 **(Polynomial behaviour at infinity)**

(a) $\lim_{x\to\infty} (3x^3 - x^2 + 2) = \infty$ (b) $\lim_{x\to -\infty} (3x^3 - x^2 + 2) = -\infty$

(c) $\lim_{x\to\infty} (x^4 - 5x^3 - x) = \infty$ (d) $\lim_{x\to -\infty} (x^4 - 5x^3 - x) = \infty$

The highest-degree term of a polynomial dominates the other terms as $|x|$ grows large, so the limits of this term at ∞ and $-\infty$ determine the limits of the whole polynomial. For the polynomial in parts (a) and (b) we have

$$3x^3 - x^2 + 2 = 3x^3\left(1 - \frac{1}{3x} + \frac{2}{3x^3}\right).$$

The factor in the large parentheses approaches 1 as x approaches $\pm\infty$, so the behaviour of the polynomial is just that of its highest-degree term $3x^3$.

We can now say a bit more about the limits at infinity and negative infinity of a rational function whose numerator has higher degree than the denominator. Earlier in this section we said that such a limit *does not exist.* This is true, but we can assign ∞ or $-\infty$ to such limits, as the following example shows.

EXAMPLE 9 **(Rational functions with numerator of higher degree)** Evaluate $\displaystyle\lim_{x\to\infty} \frac{x^3 + 1}{x^2 + 1}$.

Solution Divide the numerator and the denominator by x^2, the largest power of x in the denominator:

$$\lim_{x\to\infty} \frac{x^3 + 1}{x^2 + 1} = \lim_{x\to\infty} \frac{x + \frac{1}{x^2}}{1 + \frac{1}{x^2}} = \frac{\lim_{x\to\infty}\left(x + \frac{1}{x^2}\right)}{1} = \infty.$$

A polynomial $Q(x)$ of degree $n > 0$ can have at most n *zeros*; that is, there are at most n different real numbers r for which $Q(r) = 0$. If $Q(x)$ is the denominator of a rational function $R(x) = P(x)/Q(x)$, that function will be defined for all x except those finitely many zeros of Q. At each of those zeros, $R(x)$ may have limits, infinite limits, or one-sided infinite limits. Here are some examples.

EXAMPLE 10

(a) $\displaystyle \lim_{x\to 2} \frac{(x-2)^2}{x^2-4} = \lim_{x\to 2} \frac{(x-2)^2}{(x-2)(x+2)} = \lim_{x\to 2} \frac{x-2}{x+2} = 0.$

(b) $\displaystyle \lim_{x\to 2} \frac{x-2}{x^2-4} = \lim_{x\to 2} \frac{x-2}{(x-2)(x+2)} = \lim_{x\to 2} \frac{1}{x+2} = \frac{1}{4}.$

(c) $\displaystyle \lim_{x\to 2+} \frac{x-3}{x^2-4} = \lim_{x\to 2+} \frac{x-3}{(x-2)(x+2)} = -\infty.$ (The values are negative for $x > 2$, x near 2.)

(d) $\displaystyle \lim_{x\to 2-} \frac{x-3}{x^2-4} = \lim_{x\to 2-} \frac{x-3}{(x-2)(x+2)} = \infty.$ (The values are positive for $x < 2$, x near 2.)

(e) $\displaystyle \lim_{x\to 2} \frac{x-3}{x^2-4} = \lim_{x\to 2} \frac{x-3}{(x-2)(x+2)}$ does not exist.

(f) $\displaystyle \lim_{x\to 2} \frac{2-x}{(x-2)^3} = \lim_{x\to 2} \frac{-(x-2)}{(x-2)^3} = \lim_{x\to 2} \frac{-1}{(x-2)^2} = -\infty.$

In parts (a) and (b) the effect of the zero in the denominator at $x = 2$ is cancelled because the numerator is zero there also. Thus a finite limit exists. This is not true in part (f) because the numerator only vanishes once at $x = 2$, while the denominator vanishes three times there.

Using Maple to Calculate Limits

Maple's `limit` procedure can be easily used to calculate limits, one-sided limits, limits at infinity, and infinite limits. Here is the syntax for calculating

$$\lim_{x\to 2} \frac{x^2-4}{x^2-5x+6}, \quad \lim_{x\to 0} \frac{x \sin x}{1-\cos x}, \quad \lim_{x\to -\infty} \frac{x}{\sqrt{x^2+1}}, \quad \lim_{x\to \infty} \frac{x}{\sqrt{x^2+1}},$$

$$\lim_{x\to 0} \frac{1}{x}, \quad \lim_{x\to 0-} \frac{1}{x}, \quad \lim_{x\to a-} \frac{x^2-a^2}{|x-a|}, \quad \text{and} \quad \lim_{x\to a+} \frac{x^2-a^2}{|x-a|}.$$

```
> limit((x^2-4)/(x^2-5*x+6),x=2);
```

$$-4$$

```
> limit(x*sin(x)/(1-cos(x)),x=0);
```

$$2$$

```
> limit(x/sqrt(x^2+1),x=-infinity);
```

$$-1$$

```
> limit(x/sqrt(x^2+1),x=infinity);
```

$$1$$

```
> limit(1/x,x=0); limit(1/x,x=0,left);
```

undefined

$$-\infty$$

```
> limit((x^2-a^2)/(abs(x-a)),x=a,left);
```

$$-2a$$

```
> limit((x^2-a^2)/(abs(x-a)),x=a,right);
```

$$2a$$

Finally we use Maple to confirm the limit discussed in Example 2 in Section 1.2

```
> limit((1+x^2)^(1/x^2), x=0); evalf(%);
```

$$e$$

$$2.718281828$$

We will learn a great deal about this very important number in Chapter 3.

EXERCISES 1.3

Find the limits in Exercises 1–10.

1. $\lim_{x\to\infty} \frac{x}{2x-3}$

2. $\lim_{x\to\infty} \frac{x}{x^2-4}$

3. $\lim_{x\to\infty} \frac{3x^3-5x^2+7}{8+2x-5x^3}$

4. $\lim_{x\to-\infty} \frac{x^2-2}{x-x^2}$

5. $\lim_{x\to-\infty} \frac{x^2+3}{x^3+2}$

6. $\lim_{x\to\infty} \frac{x^2+\sin x}{x^2+\cos x}$

7. $\lim_{x\to\infty} \frac{3x+2\sqrt{x}}{1-x}$

8. $\lim_{x\to\infty} \frac{2x-1}{\sqrt{3x^2+x+1}}$

9. $\lim_{x\to-\infty} \frac{2x-1}{\sqrt{3x^2+x+1}}$

10. $\lim_{x\to-\infty} \frac{2x-5}{|3x+2|}$

In Exercises 11–34 evaluate the indicated limit. If it does not exist, is the limit ∞, $-\infty$, or neither?

11. $\lim_{x\to 3} \frac{1}{3-x}$

12. $\lim_{x\to 3} \frac{1}{(3-x)^2}$

13. $\lim_{x\to 3-} \frac{1}{3-x}$

14. $\lim_{x\to 3+} \frac{1}{3-x}$

15. $\lim_{x\to -5/2} \frac{2x+5}{5x+2}$

16. $\lim_{x\to -2/5} \frac{2x+5}{5x+2}$

17. $\lim_{x\to -(2/5)-} \frac{2x+5}{5x+2}$

18. $\lim_{x\to -(2/5)+} \frac{2x+5}{5x+2}$

19. $\lim_{x\to 2+} \frac{x}{(2-x)^3}$

20. $\lim_{x\to 1-} \frac{x}{\sqrt{1-x^2}}$

21. $\lim_{x\to 1+} \frac{1}{|x-1|}$

22. $\lim_{x\to 1-} \frac{1}{|x-1|}$

23. $\lim_{x\to 2} \frac{x-3}{x^2-4x+4}$

24. $\lim_{x\to 1+} \frac{\sqrt{x^2-x}}{x-x^2}$

25. $\lim_{x\to\infty} \frac{x+x^3+x^5}{1+x^2+x^3}$

26. $\lim_{x\to\infty} \frac{x^3+3}{x^2+2}$

27. $\lim_{x\to\infty} \frac{x\sqrt{x+1}\left(1-\sqrt{2x+3}\right)}{7-6x+4x^2}$

28. $\lim_{x\to\infty} \left(\frac{x^2}{x+1}-\frac{x^2}{x-1}\right)$

29. $\lim_{x\to-\infty} \left(\sqrt{x^2+2x}-\sqrt{x^2-2x}\right)$

30. $\lim_{x\to\infty} (\sqrt{x^2+2x}-\sqrt{x^2-2x})$

31. $\lim_{x\to\infty} \frac{1}{\sqrt{x^2-2x}-x}$

32. $\lim_{x\to-\infty} \frac{1}{\sqrt{x^2+2x}-x}$

33. What are the horizontal asymptotes of $y=\frac{1}{\sqrt{x^2-2x}-x}$? What are its vertical asymptotes?

34. What are the horizontal and vertical asymptotes of $y=\frac{2x-5}{|3x+2|}$?

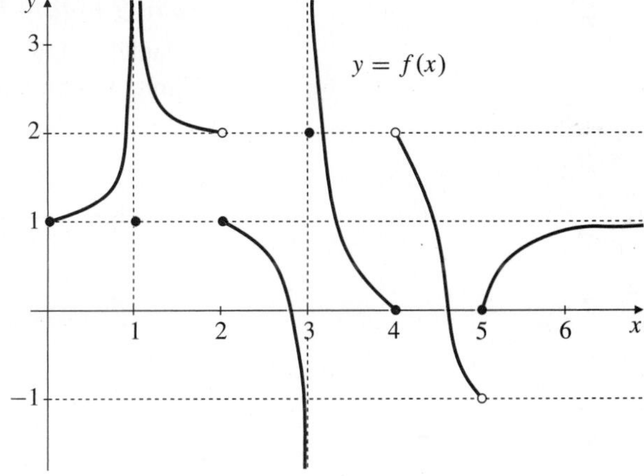

Figure 1.19

The function f whose graph is shown in Figure 1.19 has domain $[0,\infty)$. Find the limits of f indicated in Exercises 35–45.

35. $\lim_{x\to 0+} f(x)$

36. $\lim_{x\to 1} f(x)$

37. $\lim_{x\to 2+} f(x)$

38. $\lim_{x\to 2-} f(x)$

39. $\lim_{x\to 3-} f(x)$

40. $\lim_{x\to 3+} f(x)$

41. $\lim_{x\to 4+} f(x)$

42. $\lim_{x\to 4-} f(x)$

43. $\lim_{x\to 5-} f(x)$

44. $\lim_{x\to 5+} f(x)$

45. $\lim_{x\to\infty} f(x)$

46. What asymptotes does the graph in Figure 1.19 have?

Exercises 47–52 refer to the **greatest integer function** $\lfloor x\rfloor$ graphed in Figure 1.20. Find the indicated limit or explain why it does not exist.

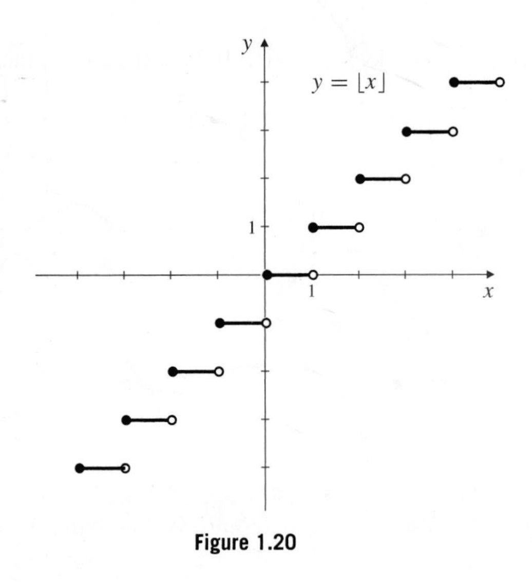

Figure 1.20

47. $\lim_{x\to 3+} \lfloor x \rfloor$

48. $\lim_{x\to 3-} \lfloor x \rfloor$

49. $\lim_{x\to 3} \lfloor x \rfloor$

50. $\lim_{x\to 2.5} \lfloor x \rfloor$

51. $\lim_{x\to 0+} \lfloor 2 - x \rfloor$

52. $\lim_{x\to -3-} \lfloor x \rfloor$

53. Parking in a certain parking lot costs \$1.50 for each hour or part of an hour. Sketch the graph of the function $C(t)$ representing the cost of parking for t hours. At what values of t does $C(t)$ have a limit? Evaluate $\lim_{t\to t_0-} C(t)$ and $\lim_{t\to t_0+} C(t)$ for an arbitrary number $t_0 > 0$.

54. If $\lim_{x\to 0+} f(x) = L$, find $\lim_{x\to 0-} f(x)$ if (a) f is even, (b) f is odd.

55. If $\lim_{x\to 0+} f(x) = A$ and $\lim_{x\to 0-} f(x) = B$, find

(a) $\lim_{x\to 0+} f(x^3 - x)$ (b) $\lim_{x\to 0-} f(x^3 - x)$

(c) $\lim_{x\to 0-} f(x^2 - x^4)$ (d) $\lim_{x\to 0+} f(x^2 - x^4)$.

1.4 Continuity

When a car is driven along a highway, its distance from its starting point depends on time in a *continuous* way, changing by small amounts over short intervals of time. But not all quantities change in this way. When the car is parked in a parking lot where the rate is quoted as "\$2.00 per hour or portion," the parking charges remain at \$2.00 for the first hour and then suddenly jump to \$4.00 as soon as the first hour has passed. The function relating parking charges to parking time will be called *discontinuous* at each hour. In this section we will define continuity and show how to tell whether a function is continuous. We will also examine some important properties possessed by continuous functions.

Continuity at a Point

Most functions that we encounter have domains that are intervals, or unions of separate intervals. A point P in the domain of such a function is called an **interior point** of the domain if it belongs to some *open* interval contained in the domain. If it is not an interior point, then P is called an **endpoint** of the domain. For example, the domain of the function $f(x) = \sqrt{4 - x^2}$ is the closed interval $[-2, 2]$, which consists of interior points in the interval $(-2, 2)$, a left endpoint -2, and a right endpoint 2. The domain of the function $g(x) = 1/x$ is the union of open intervals $(-\infty, 0) \cup (0, \infty)$ and consists entirely of interior points. Note that although 0 is an endpoint of each of those intervals, it does not belong to the domain of g and so is not an endpoint of that domain.

DEFINITION 4

Continuity at an interior point

We say that a function f is **continuous** at an interior point c of its domain if

$$\lim_{x\to c} f(x) = f(c).$$

If either $\lim_{x\to c} f(x)$ fails to exist or it exists but is not equal to $f(c)$, then we will say that f is **discontinuous** at c.

In graphical terms, f is continuous at an interior point c of its domain if its graph has no break in it at the point $(c, f(c))$; in other words, if you can draw the graph through that point without lifting your pen from the paper. Consider Figure 1.21. In (a), f is continuous at c. In (b), f is discontinuous at c because $\lim_{x\to c} f(x) \neq f(c)$. In (c), f

is discontinuous at c because $\lim_{x\to c} f(x)$ does not exist. In both (b) and (c) the graph of f has a break at $x = c$.

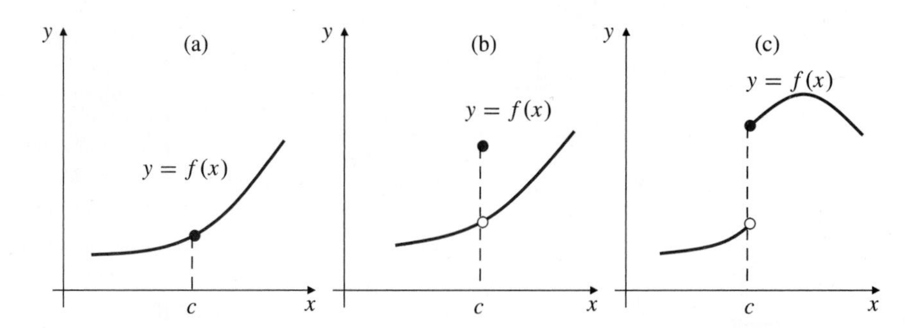

Figure 1.21
(a) f is continuous at c
(b) $\lim_{x\to c} f(x) \neq f(c)$
(c) $\lim_{x\to c} f(x)$ does not exist

Although a function cannot have a limit at an endpoint of its domain, it can still have a one-sided limit there. We extend the definition of continuity to provide for such situations.

DEFINITION 5

Right and left continuity

We say that f is **right continuous** at c if $\lim_{x\to c+} f(x) = f(c)$.

We say that f is **left continuous** at c if $\lim_{x\to c-} f(x) = f(c)$.

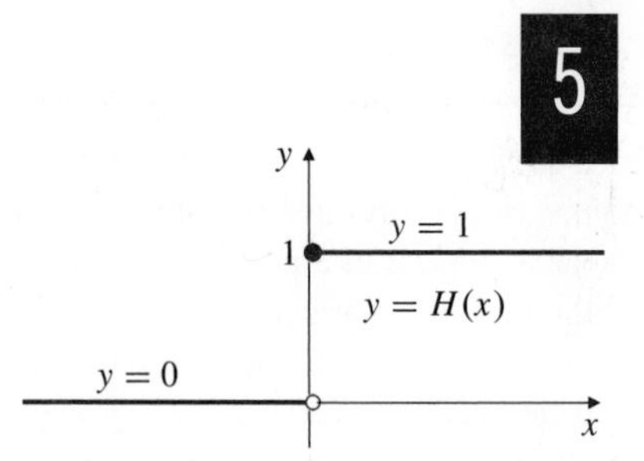

Figure 1.22 The Heaviside function

EXAMPLE 1 The Heaviside function $H(x)$, whose graph is shown in Figure 1.22, is continuous at every number x except 0. It is right continuous at 0 but is not left continuous or continuous there.

The relationship between continuity and one-sided continuity is summarized in the following theorem.

THEOREM 5

Function f is continuous at c if and only if it is both right continuous and left continuous at c.

DEFINITION 6

Continuity at an endpoint

We say that f is continuous at a left endpoint c of its domain if it is right continuous there.

We say that f is continuous at a right endpoint c of its domain if it is left continuous there.

EXAMPLE 2 The function $f(x) = \sqrt{4 - x^2}$ has domain $[-2, 2]$. It is continuous at the right endpoint 2 because it is left continuous there, that is, because $\lim_{x\to 2-} f(x) = 0 = f(2)$. It is continuous at the left endpoint -2 because it is right continuous there: $\lim_{x\to -2+} f(x) = 0 = f(-2)$. Of course, f is also continuous at every interior point of its domain. If $-2 < c < 2$, then $\lim_{x\to c} f(x) = \sqrt{4 - c^2} = f(c)$. (See Figure 1.23.)

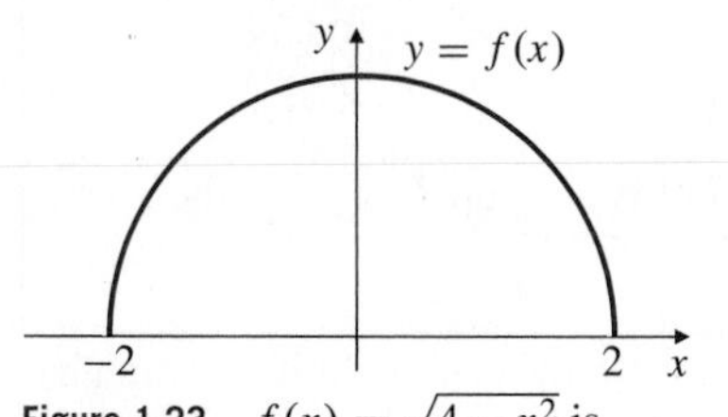

Figure 1.23 $f(x) = \sqrt{4 - x^2}$ is continuous at every point of its domain

Continuity on an Interval

We have defined the concept of continuity at a point. Of greater importance is the concept of continuity on an interval.

DEFINITION 7

Continuity on an interval

We say that function f is **continuous on the interval** I if it is continuous at each point of I. In particular, we will say that f is a **continuous function** if f is continuous at every point of its domain.

EXAMPLE 3 The function $f(x) = \sqrt{x}$ is a continuous function. Its domain is $[0, \infty)$. It is continuous at the left endpoint 0 because it is right continuous there. Also, f is continuous at every number $c > 0$ since $\lim_{x\to c} \sqrt{x} = \sqrt{c}$.

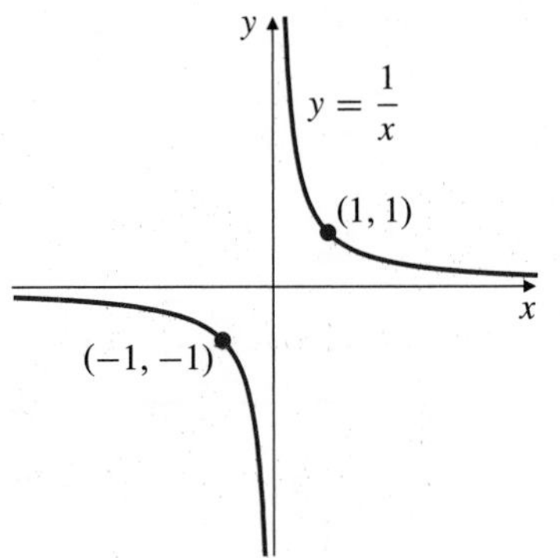

Figure 1.24 $1/x$ is continuous on its domain

EXAMPLE 4 The function $g(x) = 1/x$ is also a continuous function. This may seem wrong to you at first glance because its graph is broken at $x = 0$. (See Figure 1.24.) However, the number 0 is not in the domain of g, so we will prefer to say that g is undefined rather than discontinuous there. (Some authors would say that g is discontinuous at $x = 0$.) If we were to define $g(0)$ to be some number, say 0, then we would say that $g(x)$ is discontinuous at 0. There is no way of defining $g(0)$ so that g becomes continuous at 0.

EXAMPLE 5 The greatest integer function $\lfloor x \rfloor$ (see Figure 1.20) is continuous on every interval $[n, n + 1)$, where n is an integer. It is right continuous at each integer n but is not left continuous there, so it is discontinuous at the integers.

$$\lim_{x\to n+} \lfloor x \rfloor = n = \lfloor n \rfloor, \qquad \lim_{x\to n-} \lfloor x \rfloor = n - 1 \neq n = \lfloor n \rfloor.$$

There Are Lots of Continuous Functions

The following functions are continuous wherever they are defined:

(a) all polynomials;

(b) all rational functions;

(c) all rational powers $x^{m/n} = \sqrt[n]{x^m}$;

(d) the sine, cosine, tangent, secant, cosecant, and cotangent functions defined in Section P.7; and

(e) the absolute value function $|x|$.

Theorem 3 of Section 1.2 assures us that every polynomial is continuous everywhere on the real line, and every rational function is continuous everywhere on its domain (which consists of all real numbers except the finitely many where its denominator is zero). If m and n are integers and $n \neq 0$, the rational power function $x^{m/n}$ is defined for all positive numbers x, and also for all negative numbers x if n is odd. The domain includes 0 if and only if $m/n \geq 0$.

The following theorems show that if we combine continuous functions in various ways, the results will be continuous.

THEOREM 6

Combining continuous functions

If the functions f and g are both defined on an interval containing c and both are continuous at c, then the following functions are also continuous at c:

1. the sum $f + g$ and the difference $f - g$;
2. the product fg;
3. the constant multiple kf, where k is any number;
4. the quotient f/g (provided $g(c) \neq 0$); and

5. the nth root $(f(x))^{1/n}$, provided $f(c) > 0$ if n is even.

The proof involves using the various limit rules in Theorem 2 of Section 1.2. For example,

$$\lim_{x\to c}\big(f(x)+g(x)\big) = \lim_{x\to c} f(x) + \lim_{x\to c} g(x) = f(c) + g(c),$$

so $f + g$ is continuous.

THEOREM 7

Composites of continuous functions are continuous

If $f(g(x))$ is defined on an interval containing c, and if f is continuous at L and $\lim_{x\to c} g(x) = L$, then

$$\lim_{x\to c} f(g(x)) = f(L) = f\left(\lim_{x\to c} g(x)\right).$$

In particular, if g is continuous at c (so $L = g(c)$), then the composition $f \circ g$ is continuous at c:

$$\lim_{x\to c} f(g(x)) = f(g(c)).$$

(See Exercise 37 in Section 1.5.)

EXAMPLE 6 The following functions are continuous everywhere on their respective domains:

(a) $3x^2 - 2x$ (b) $\dfrac{x-2}{x^2-4}$ (c) $|x^2-1|$

(d) $\sqrt{x}$ (e) $\sqrt{x^2-2x-5}$ (f) $\dfrac{|x|}{\sqrt{|x+2|}}$.

Continuous Extensions and Removable Discontinuities

As we have seen in Section 1.2, a rational function may have a limit even at a point where its denominator is zero. If $f(c)$ is not defined, but $\lim_{x\to c} f(x) = L$ exists, we can define a new function $F(x)$ by

$$F(x) = \begin{cases} f(x) & \text{if } x \text{ is in the domain of } f \\ L & \text{if } x = c. \end{cases}$$

$F(x)$ is continuous at $x = c$. It is called the **continuous extension** of $f(x)$ to $x = c$. For rational functions f, continuous extensions are usually found by cancelling common factors.

EXAMPLE 7 Show that $f(x) = \dfrac{x^2-x}{x^2-1}$ has a continuous extension to $x = 1$, and find that extension.

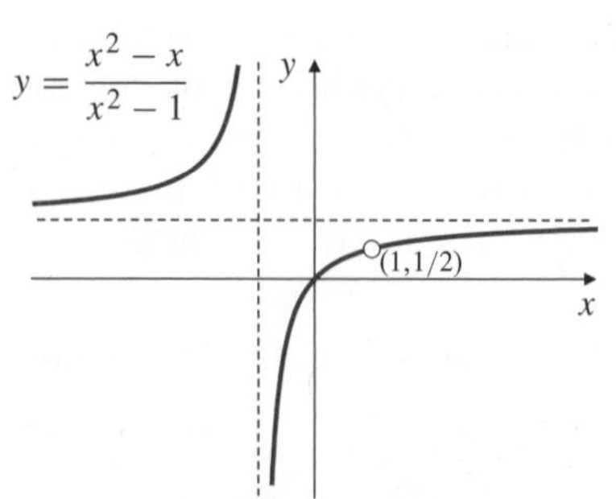

Figure 1.25 This function has a continuous extension to $x = 1$

Solution Although $f(1)$ is not defined, if $x \neq 1$ we have

$$f(x) = \frac{x^2-x}{x^2-1} = \frac{x(x-1)}{(x+1)(x-1)} = \frac{x}{x+1}.$$

The function

$$F(x) = \frac{x}{x+1}$$

is equal to $f(x)$ for $x \neq 1$ but is also continuous at $x = 1$, having there the value $1/2$. The graph of f is shown in Figure 1.25. The continuous extension of $f(x)$ to $x = 1$ is $F(x)$. It has the same graph as $f(x)$ except with no hole at $(1, 1/2)$.

If a function f is undefined or discontinuous at a point a but can be (re)defined at that *single point* so that it becomes continuous there, then we say that f has a **removable discontinuity** at a. The function f in the above example has a removable discontinuity at $x = 1$. To remove it, define $f(1) = 1/2$.

EXAMPLE 8 The function $g(x) = \begin{cases} x & \text{if } x \neq 2 \\ 1 & \text{if } x = 2 \end{cases}$ has a removable discontinuity at $x = 2$. To remove it, redefine $g(2) = 2$. (See Figure 1.26.)

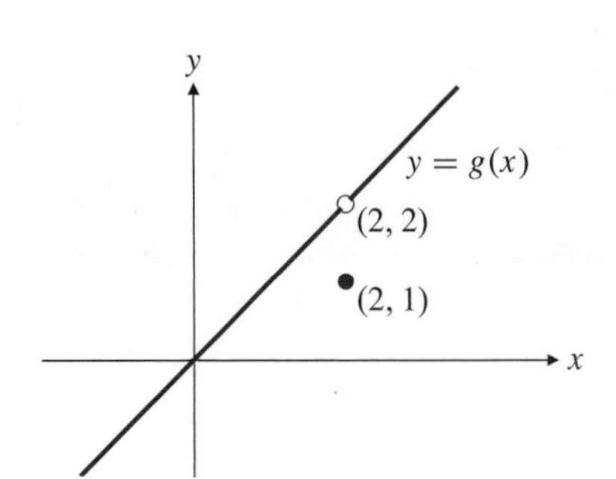

Figure 1.26 g has a removable discontinuity at 2

Continuous Functions on Closed, Finite Intervals

Continuous functions that are defined on *closed, finite intervals* have special properties that make them particularly useful in mathematics and its applications. We will discuss two of these properties here. Although they may appear obvious, these properties are much more subtle than the results about limits stated earlier in this chapter; their proofs (see Appendix III) require a careful study of the implications of the completeness property of the real numbers.

The first of the properties states that a function $f(x)$ that is continuous on a closed, finite interval $[a, b]$ must have an **absolute maximum value** and an **absolute minimum value**. This means that the values of $f(x)$ at all points of the interval lie between the values of $f(x)$ at two particular points in the interval; the graph of f has a highest point and a lowest point.

THEOREM 8

The Max-Min Theorem

If $f(x)$ is continuous on the closed, finite interval $[a, b]$, then there exist numbers p and q in $[a, b]$ such that for all x in $[a, b]$,

$$f(p) \leq f(x) \leq f(q).$$

Thus f has the absolute minimum value $m = f(p)$, taken on at the point p, and the absolute maximum value $M = f(q)$, taken on at the point q.

Many important problems in mathematics and its applications come down to having to find maximum and minimum values of functions. Calculus provides some very useful tools for solving such problems. Observe, however, that the theorem above merely asserts that minimum and maximum values *exist;* it doesn't tell us how to find them. In Chapter 4 we will develop techniques for calculating maximum and minimum values of functions. For now, we can solve some simple maximum and minimum value problems involving quadratic functions by completing the square without using any calculus.

EXAMPLE 9 What is the largest possible area of a rectangular field that can be enclosed by 200 m of fencing?

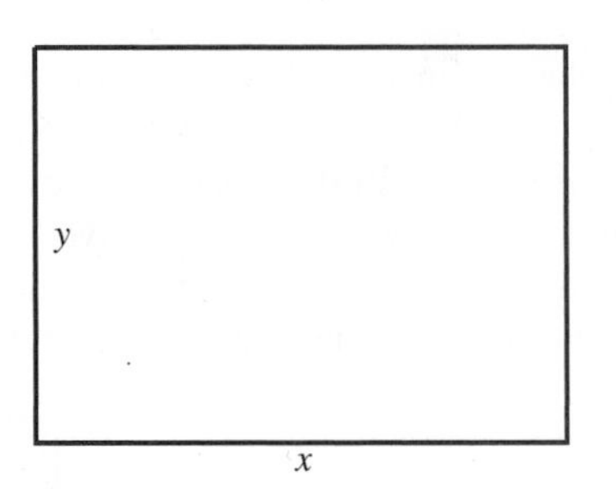

Figure 1.27 Rectangular field: perimeter $= 2x + 2y$, area $= xy$

Solution If the sides of the field are x m and y m (Figure 1.27), then its perimeter is $P = 2x + 2y$ m, and its area is $A = xy$ m^2. We are given that $P = 200$, so $x + y = 100$, and $y = 100 - x$. Neither side can be negative, so x must belong to the closed interval $[0, 100]$. The area of the field can be expressed as a function of x by substituting $100 - x$ for y:

$$A = x(100 - x) = 100x - x^2.$$

We want to find the maximum value of the quadratic function $A(x) = 100x - x^2$ on the interval $[0, 100]$. Theorem 8 assures us that such a maximum exists.

To find the maximum, we complete the square of the function $A(x)$. Note that $x^2 - 100x$ are the first two terms of the square $(x - 50)^2 = x^2 - 100x + 2{,}500$. Thus,

$$A(x) = 2{,}500 - (x - 50)^2.$$

Observe that $A(50) = 2,500$ and $A(x) < 2{,}500$ if $x \neq 50$, because we are subtracting a positive number $(x - 50)^2$ from 2,500 in this case. Therefore, the maximum value of $A(x)$ is 2,500. The largest field has area 2,500 m^2 and is actually a square with dimensions $x = y = 50$ m.

Theorem 8 implies that a function that is continuous on a closed, finite interval is **bounded**. This means that it cannot take on arbitrarily large positive or negative values; there must exist a number K such that

$$|f(x)| \le K; \quad \text{that is,} \quad -K \le f(x) \le K.$$

In fact, for K we can use the larger of the numbers $|f(p)|$ and $|f(q)|$ in the theorem.

The conclusions of Theorem 8 may fail if the function f is not continuous or if the interval is not closed. See Figures 1.28–1.31 for examples of how such failure can occur.

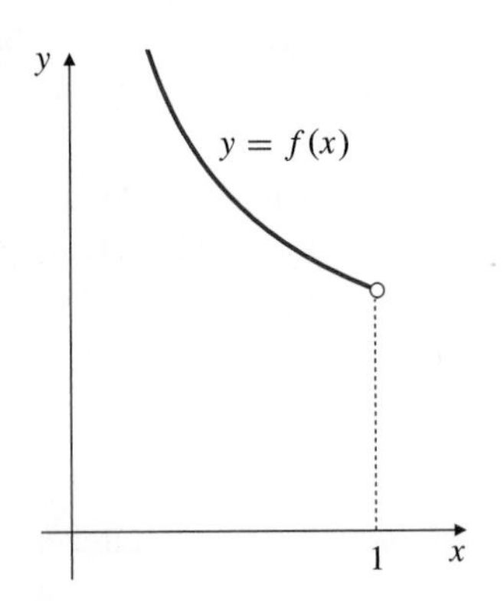

Figure 1.28 $f(x) = 1/x$ is continuous on the open interval $(0, 1)$. It is not bounded and has neither a maximum nor a minimum value

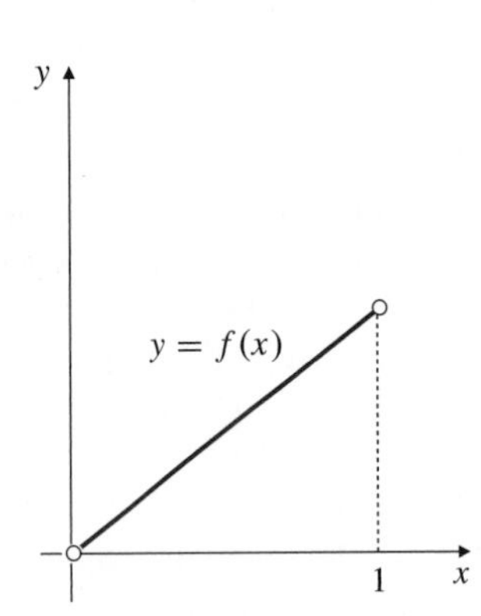

Figure 1.29 $f(x) = x$ is continuous on the open interval $(0, 1)$. It is bounded but has neither a maximum nor a minimum value

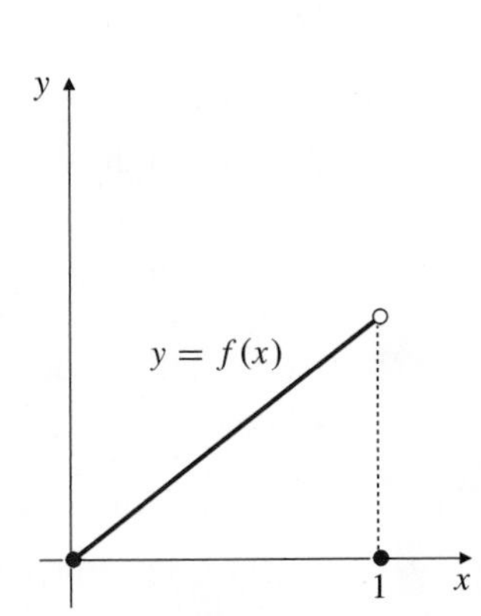

Figure 1.30 This function is defined on the closed interval $[0, 1]$ but is discontinuous at the endpoint $x = 1$. It has a minimum value but no maximum value

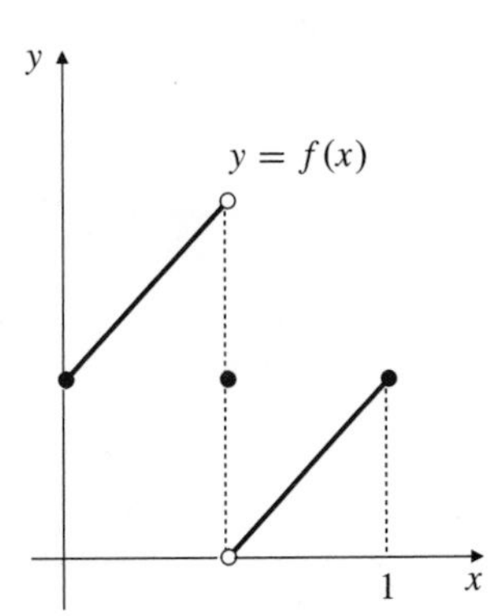

Figure 1.31 This function is discontinuous at an interior point of its domain, the closed interval $[0, 1]$. It is bounded but has neither maximum nor minimum values

Finding Maxima and Minima Graphically

Remark Graphing utilities can be used to find maximum and minimum values of functions on intervals where they are continuous. In particular, the "zoom box" and "trace" facilities of graphing calculators are helpful. Figure 1.32(a) shows the graph of the function

$$y = f(x) = \frac{x+1}{x^2+1}$$

on the window $-5 \le x \le 5$, $-2 \le y \le 2$. Observe that f appears to have a maximum value near $x = 0.5$ and a minimum value near $x = -2.5$. Figure 1.32(b) shows the result of expanding the part of the graph in (a) enclosed in the small rectangle (zoom box) to fill the whole screen. Tracing the curve to its highest point gives a more accurate estimate of the maximum value, showing that $f(x)$ has maximum value 1.2071 at $x = 0.4149$, each to 4 significant figures. Further zooming enables us to get even greater accuracy.

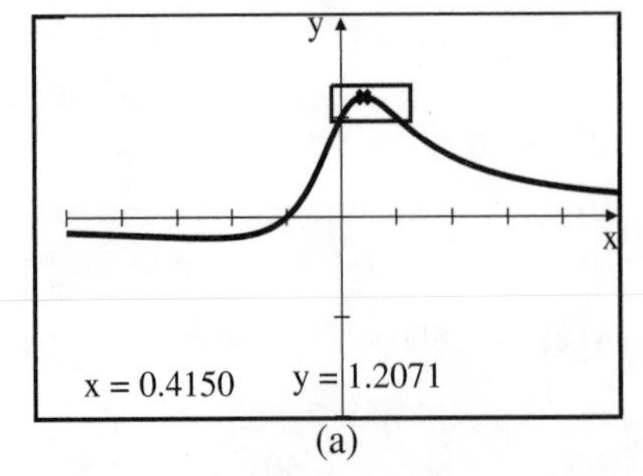

(a)

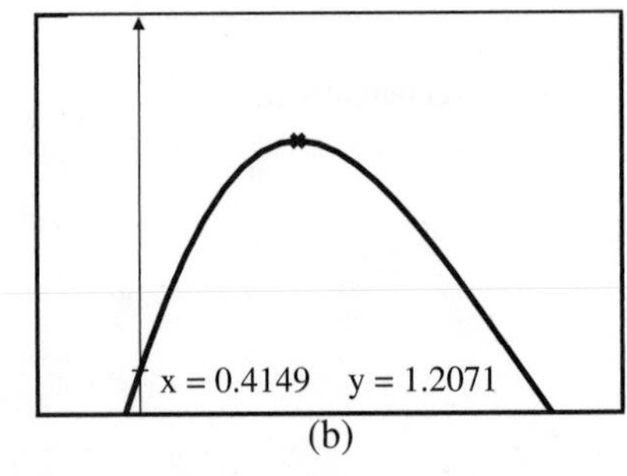

(b)

Figure 1.32 Using a "zoom box" to zoom part of a curve (a) near a maximum value to fill the screen (b) without allowing the curve to become flattened

The second property of a continuous function defined on a closed, finite interval is that the function takes on all real values between any two of its values. This property is called the **intermediate-value property**.

THEOREM 9

The Intermediate-Value Theorem

If $f(x)$ is continuous on the interval $[a, b]$ and if s is a number between $f(a)$ and $f(b)$, then there exists a number c in $[a, b]$ such that $f(c) = s$.

In particular, a continuous function defined on a closed interval takes on all values between its minimum value m and its maximum value M, so its range is also a closed interval, $[m, M]$.

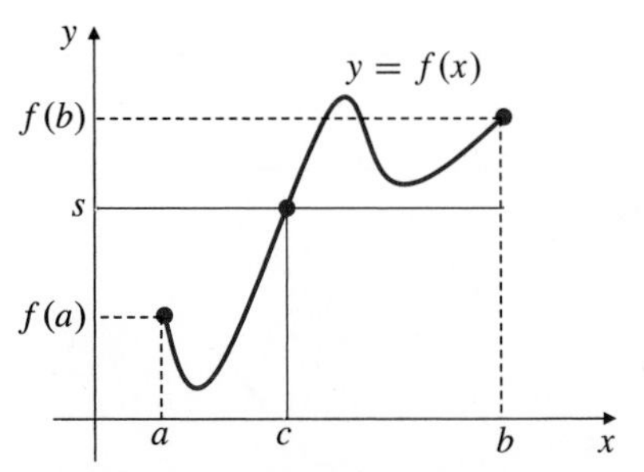

Figure 1.33 The continuous function f takes on the value s at some point c between a and b

Figure 1.33 shows a typical situation. The points $(a, f(a))$ and $(b, f(b))$ are on opposite sides of the horizontal line $y = s$. Being unbroken, the graph $y = f(x)$ must cross this line in order to go from one point to the other. In the figure, it crosses the line only once, at $x = c$. If the line $y = s$ were somewhat higher, there might have been three crossings and three possible values for c.

Theorem 9 is the reason why the graph of a function that is continuous on an interval I cannot have any breaks. It must be **connected**, a single, unbroken curve with no jumps.

EXAMPLE 10 Determine the intervals on which $f(x) = x^3 - 4x$ is positive and negative.

Solution Since $f(x) = x(x^2 - 4) = x(x - 2)(x + 2)$, $f(x) = 0$ only at $x = 0$, 2, and -2. Because f is continuous on the whole real line, it must have constant sign on each of the intervals $(-\infty, -2)$, $(-2, 0)$, $(0, 2)$, and $(2, \infty)$. (If there were points a and b in one of those intervals, say in $(0, 2)$, such that $f(a) < 0$ and $f(b) > 0$, then by the Intermediate-Value Theorem there would exist c between a and b, and therefore between 0 and 2, such that $f(c) = 0$. But we know f has no such zero in $(0, 2)$.)

To find whether $f(x)$ is positive or negative throughout each interval, pick a point in the interval and evaluate f at that point.

Since $f(-3) = -15 < 0$, $f(x)$ is negative on $(-\infty, -2)$.
Since $f(-1) = 3 > 0$, $f(x)$ is positive on $(-2, 0)$.
Since $f(1) = -3 < 0$, $f(x)$ is negative on $(0, 2)$.
Since $f(3) = 15 > 0$, $f(x)$ is positive on $(2, \infty)$.

Finding Roots of Equations

Among the many useful tools that calculus will provide are ones that enable us to calculate solutions to equations of the form $f(x) = 0$ to any desired degree of accuracy. Such a solution is called a **root** of the equation, or a **zero** of the function f. Using these tools usually requires previous knowledge that the equation has a solution in some interval. The Intermediate-Value Theorem can provide this information.

EXAMPLE 11 Show that the equation $x^3 - x - 1 = 0$ has a solution in the interval $[1, 2]$.

Solution The function $f(x) = x^3 - x - 1$ is a polynomial and is therefore continuous everywhere. Now $f(1) = -1$ and $f(2) = 5$. Since 0 lies between -1 and 5, the Intermediate-Value Theorem assures us that there must be a number c in $[1, 2]$ such that $f(c) = 0$.

One method for finding a zero of a function that is continuous and changes sign on an interval involves bisecting the interval many times, each time determining which half of the previous interval must contain the root, because the function has opposite signs at the two ends of that half. This method is slow. For example, if the original interval has length 1, it will take 11 bisections to cut down to an interval of length less than 0.0005 (because $2^{11} > 2{,}000 = 1/(0.0005)$), and thus to ensure that we have found the root correct to 3 decimal places. But this method requires no graphics hardware and is easily implemented with a calculator, preferably one into which the formula for the function can be programmed.

EXAMPLE 12 **(The Bisection Method)** Solve the equation $x^3 - x - 1 = 0$ of Example 11 correct to 3 decimal places by successive bisections.

Solution We start out knowing that there is a root in [1, 2]. Table 6 shows the results of the bisections.

Table 6. The Bisection Method for $f(x) = x^3 - x - 1 = 0$

Bisection Number	x	$f(x)$	Root in Interval	Midpoint
	1	−1		
	2	5	[1, 2]	1.5
1	1.5	0.8750	[1, 1.5]	1.25
2	1.25	−0.2969	[1.25, 1.5]	1.375
3	1.375	0.2246	[1.25, 1.375]	1.3125
4	1.3125	−0.0515	[1.3125, 1.375]	1.3438
5	1.3438	0.0826	[1.3125, 1.3438]	1.3282
6	1.3282	0.0147	[1.3125, 1.3282]	1.3204
7	1.3204	−0.0186	[1.3204, 1.3282]	1.3243
8	1.3243	−0.0018	[1.3243, 1.3282]	1.3263
9	1.3263	0.0065	[1.3243, 1.3263]	1.3253
10	1.3253	0.0025	[1.3243, 1.3253]	1.3248
11	1.3248	0.0003	[1.3243, 1.3248]	1.3246
12	1.3246	−0.0007	[1.3246, 1.3248]	

The root is 1.325, rounded to 3 decimal places. In Section 4.2, calculus will provide us with much faster methods of solving equations such as the one above.

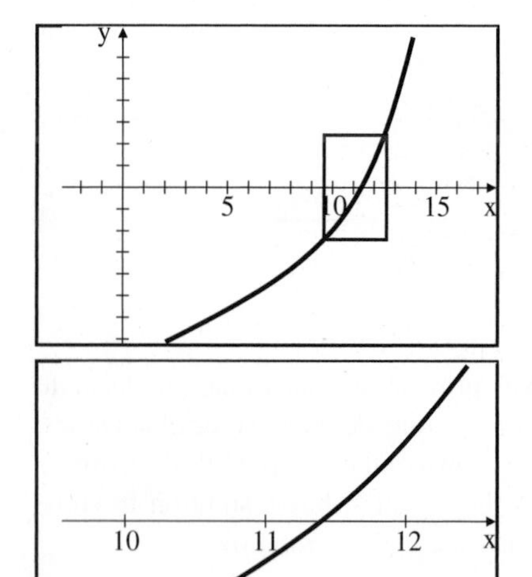

Figure 1.34 The function graphed in the upper window has a root between 11 and 12. The small rectangle (zoom box) is zoomed to fill the screen in the lower window, enabling us to estimate that the root is about 11.4. Successive zooms can provide greater precision

You can use a graphing utility to solve an equation $f(x) = 0$. Just graph the function $f(x)$ over a large enough interval so that you can see roughly where its zeros are. Then select one zero at a time, and zoom in on it by successively expanding the part of the viewing window near the zero to fill the whole viewing window. (See Figure 1.34.) Keep zooming until you can estimate the zero to as many decimal places as you want (or as the calculator or computer will allow).

Many programmable calculators and computer algebra software packages have built-in routines for solving equations. For example, Maple's `fsolve` routine can be used to find the real solution of $x^3 - x - 1 = 0$ in [1, 2]. (See Example 11.)

```
> fsolve(x^3-x-1=0,x=1..2);
```

$$1.324717957$$

Remark The Max-Min Theorem and the Intermediate-Value Theorem are examples of what mathematicians call **existence theorems**. Such theorems assert that something exists without telling you how to find it. Students sometimes complain that mathematicians worry too much about proving that a problem has a solution and not enough about how to find that solution. They argue: "If I can calculate a solution to a problem, then surely I do not need to worry about whether a solution exists." This is, however, false logic. Suppose we pose the problem: "Find the largest positive integer." Of course, this problem has no solution; there is no largest positive integer because we can add 1 to any integer and get a larger integer. Suppose, however, that we forget this and try to calculate a solution. We could proceed as follows:

Let N be the largest positive integer.
Since 1 is a positive integer, we must have $N \ge 1$.
Since N^2 is a positive integer, it cannot exceed the largest positive integer.

Therefore, $N^2 \leq N$ and so $N^2 - N \leq 0$.
Thus, $N(N-1) \leq 0$ and we must have $N - 1 \leq 0$.
Therefore, $N \leq 1$. Since also $N \geq 1$, we have $N = 1$.
Therefore, 1 is the largest positive integer.

The only error we have made here is in the assumption (in the first line) that the problem has a solution. It is partly to avoid logical pitfalls like this that mathematicians prove existence theorems.

EXERCISES 1.4

Exercises 1–3 refer to the function g defined on $[-2, 2]$, whose graph is shown in Figure 1.35.

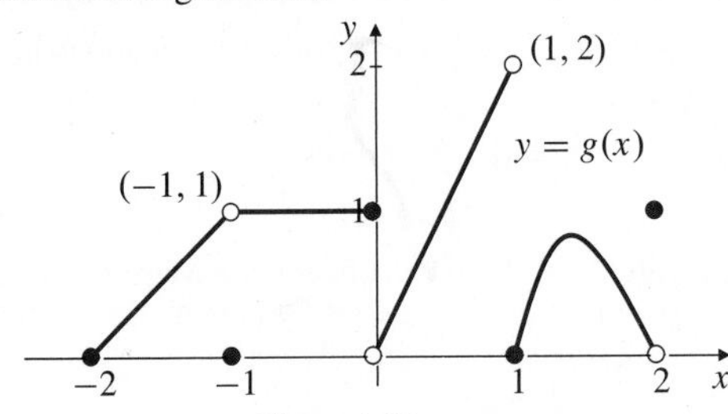

Figure 1.35

1. State whether g is (a) continuous, (b) left continuous, (c) right continuous, and (d) discontinuous at each of the points -2, -1, 0, 1, and 2.
2. At what points in its domain does g have a removable discontinuity, and how should g be redefined at each of those points so as to be continuous there?
3. Does g have an absolute maximum value on $[-2, 2]$? an absolute minimum value?

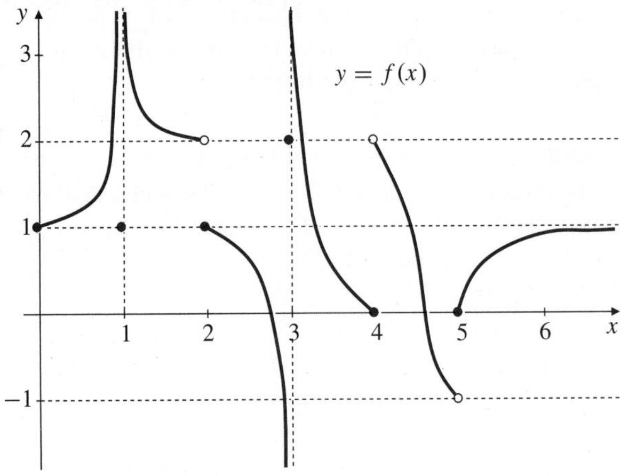

Figure 1.36

4. At what points is the function f, whose graph is shown in Figure 1.36, discontinuous? At which of those points is it left continuous? right continuous?
5. Can the function f graphed in Figure 1.36 be redefined at the single point $x = 1$ so that it becomes continuous there?
6. The function $\operatorname{sgn}(x) = x/|x|$ is neither continuous nor discontinuous at $x = 0$. How is this possible?

In Exercises 7–12, state where in its domain the given function is continuous, where it is left or right continuous, and where it is just discontinuous.

7. $f(x) = \begin{cases} x & \text{if } x < 0 \\ x^2 & \text{if } x \geq 0 \end{cases}$

8. $f(x) = \begin{cases} x & \text{if } x < -1 \\ x^2 & \text{if } x \geq -1 \end{cases}$

9. $f(x) = \begin{cases} 1/x^2 & \text{if } x \neq 0 \\ 0 & \text{if } x = 0 \end{cases}$

10. $f(x) = \begin{cases} x^2 & \text{if } x \leq 1 \\ 0.987 & \text{if } x > 1 \end{cases}$

11. The least integer function $\lceil x \rceil$ of Example 11 in Section P.5.
12. The cost function $C(t)$ of Exercise 53 in Section 1.3.

In Exercises 13–16, how should the given function be defined at the given point to be continuous there? Give a formula for the continuous extension to that point.

13. $\dfrac{x^2 - 4}{x - 2}$ at $x = 2$

14. $\dfrac{1 + t^3}{1 - t^2}$ at $t = -1$

15. $\dfrac{t^2 - 5t + 6}{t^2 - t - 6}$ at 3

16. $\dfrac{x^2 - 2}{x^4 - 4}$ at $\sqrt{2}$

17. Find k so that $f(x) = \begin{cases} x^2 & \text{if } x \leq 2 \\ k - x^2 & \text{if } x > 2 \end{cases}$ is a continuous function.
18. Find m so that $g(x) = \begin{cases} x - m & \text{if } x < 3 \\ 1 - mx & \text{if } x \geq 3 \end{cases}$ is continuous for all x.
19. Does the function x^2 have a maximum value on the open interval $-1 < x < 1$? a minimum value? Explain.
20. The Heaviside function of Example 1 has both absolute maximum and minimum values on the interval $[-1, 1]$, but it is not continuous on that interval. Does this violate the Max-Min Theorem? Why?

Exercises 21–24 ask for maximum and minimum values of functions. They can all be done by the method of Example 9.

21. The sum of two nonnegative numbers is 8. What is the largest possible value of their product?
22. The sum of two nonnegative numbers is 8. What is (a) the smallest and (b) the largest possible value for the sum of their squares?
23. A software company estimates that if it assigns x programmers to work on the project, it can develop a new product in T days, where
$$T = 100 - 30x + 3x^2.$$
How many programmers should the company assign in order to complete the development as quickly as possible?
24. It costs a desk manufacturer $\$(245x - 30x^2 + x^3)$ to send a shipment of x desks to its warehouse. How many desks should it include in each shipment to minimize the average shipping cost per desk?

Find the intervals on which the functions $f(x)$ in Exercises 25–28 are positive and negative.

25. $f(x) = \dfrac{x^2 - 1}{x}$

26. $f(x) = x^2 + 4x + 3$

27. $f(x) = \dfrac{x^2 - 1}{x^2 - 4}$

28. $f(x) = \dfrac{x^2 + x - 2}{x^3}$

29. Show that $f(x) = x^3 + x - 1$ has a zero between $x = 0$ and $x = 1$.

30. Show that the equation $x^3 - 15x + 1 = 0$ has three solutions in the interval $[-4, 4]$.

31. Show that the function $F(x) = (x - a)^2(x - b)^2 + x$ has the value $(a + b)/2$ at some point x.

32. **(A fixed-point theorem)** Suppose that f is continuous on the closed interval $[0, 1]$ and that $0 \le f(x) \le 1$ for every x in $[0, 1]$. Show that there must exist a number c in $[0, 1]$ such that $f(c) = c$. (c is called a fixed point of the function f.) *Hint:* If $f(0) = 0$ or $f(1) = 1$, you are done. If not, apply the Intermediate-Value Theorem to $g(x) = f(x) - x$.

33. If an even function f is right continuous at $x = 0$, show that it is continuous at $x = 0$.

34. If an odd function f is right continuous at $x = 0$, show that it is continuous at $x = 0$ and that it satisfies $f(0) = 0$.

Use a graphing utility to find maximum and minimum values of the functions in Exercises 35–38 and the points x where they occur. Obtain 3 decimal place accuracy for all answers.

35. $f(x) = \dfrac{x^2 - 2x}{x^4 + 1}$ on $[-5, 5]$

36. $f(x) = \dfrac{\sin x}{6 + x}$ on $[-\pi, \pi]$

37. $f(x) = x^2 + \dfrac{4}{x}$ on $[1, 3]$

38. $f(x) = \sin(\pi x) + x(\cos(\pi x) + 1)$ on $[0, 1]$

Use a graphing utility or a programmable calculator and the Bisection Method to solve the equations in Exercises 39–40 to 3 decimal places. As a first step, try to guess a small interval that you can be sure contains a root.

39. $x^3 + x - 1 = 0$

40. $\cos x - x = 0$

Use Maple's `fsolve` routine to solve the equations in Exercises 41–42.

41. $\sin x + 1 - x^2 = 0$ (two roots)

42. $x^4 - x - 1 = 0$ (two roots)

43. Investigate the difference between the Maple routines `fsolve(f,x)`, `solve(f,x)`, and `evalf(solve(f,x))`, where `f := x^3-x-1=0`. Note that no interval is specified for x here.

1.5 The Formal Definition of Limit

The material in this section is optional.

The *informal* definition of limit given in Section 1.2 is not precise enough to enable us to prove results about limits such as those given in Theorems 2–4 of Section 1.2. A more precise *formal* definition is based on the idea of controlling the input x of a function f so that the output $f(x)$ will lie in a specific interval.

EXAMPLE 1 The area of a circular disk of radius r cm is $A = \pi r^2$ cm^2. A machinist is required to manufacture a circular metal disk having area 400π cm^2 within an error tolerance of ± 5 cm^2. How close to 20 cm must the machinist control the radius of the disk to achieve this?

Solution The machinist wants $|\pi r^2 - 400\pi| < 5$, that is,

$$400\pi - 5 < \pi r^2 < 400\pi + 5,$$

or, equivalently,

$$\sqrt{400 - (5/\pi)} < r < \sqrt{400 + (5/\pi)}$$
$$19.96017 < r < 20.03975.$$

Thus, the machinist needs $|r - 20| < 0.03975$; she must ensure that the radius of the disk differs from 20 cm by less than 0.4 mm so that the area of the disk will lie within the required error tolerance.

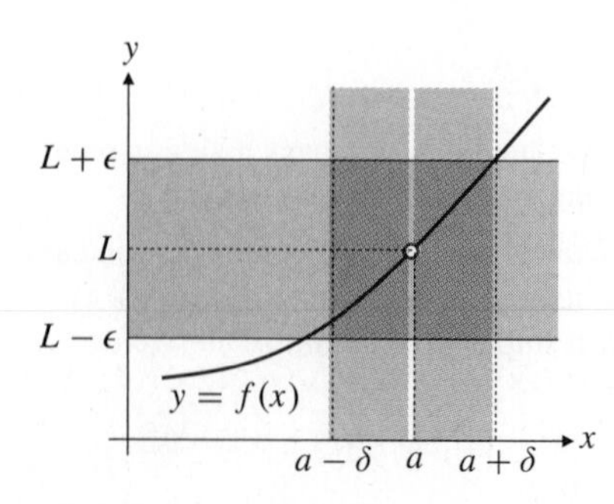

Figure 1.37 If $x \ne a$ and $|x - a| < \delta$, then $|f(x) - L| < \epsilon$

When we say that $f(x)$ has limit L as x approaches a, we are really saying that we can ensure that the *error* $|f(x) - L|$ will be less than *any* allowed tolerance, no matter how small, by taking x *close enough* to a (but not equal to a). It is traditional to use ϵ, the Greek letter "epsilon," for the size of the allowable *error* and δ, the Greek letter "delta," for the *difference* $x - a$ that measures how close x must be to a to ensure that the error is within that tolerance. These are the letters that Cauchy and Weierstrass used in their pioneering work on limits and continuity in the nineteenth century.

If ϵ is any positive number, *no matter how small*, we must be able to ensure that $|f(x) - L| < \epsilon$ by restricting x to be *close enough to* (but not equal to) a. How close is close enough? It is sufficient that the distance $|x - a|$ from x to a be less than a positive number δ that depends on ϵ. (See Figure 1.37.) If we can find such a δ for any positive ϵ, we are entitled to conclude that $\lim_{x\to a} f(x) = L$.

DEFINITION 8

A formal definition of limit

We say that $f(x)$ **approaches the limit** L as x **approaches** a, and we write

$$\lim_{x\to a} f(x) = L,$$

if the following condition is satisfied:
for every number $\epsilon > 0$ there exists a number $\delta > 0$, possibly depending on ϵ, such that if $0 < |x - a| < \delta$, then x belongs to the domain of f and

$$|f(x) - L| < \epsilon.$$

The formal definition of limit does not tell you how to find the limit of a function, but it does enable you to verify that a suspected limit is correct. The following examples show how it can be used to verify limit statements for specific functions. The first of these gives a formal verification of the two limits found in Example 3 of Section 1.2.

EXAMPLE 2 **(Two important limits)** Verify that:
(a) $\lim_{x\to a} x = a$ and (b) $\lim_{x\to a} k = k$ (k = constant).

Solution

(a) Let $\epsilon > 0$ be given. We must find $\delta > 0$ so that

$$0 < |x - a| < \delta \quad \text{implies} \quad |x - a| < \epsilon.$$

Clearly, we can take $\delta = \epsilon$ and the implication above will be true. This proves that $\lim_{x\to a} x = a$.

(b) Let $\epsilon > 0$ be given. We must find $\delta > 0$ so that

$$0 < |x - a| < \delta \quad \text{implies} \quad |k - k| < \epsilon.$$

Since $k - k = 0$, we can use any positive number for δ and the implication above will be true. This proves that $\lim_{x\to a} k = k$.

EXAMPLE 3 Verify that $\lim_{x\to 2} x^2 = 4$.

Solution Here $a = 2$ and $L = 4$. Let ϵ be a given positive number. We want to find $\delta > 0$ so that if $0 < |x - 2| < \delta$, then $|f(x) - 4| < \epsilon$. Now

$$|f(x) - 4| = |x^2 - 4| = |(x + 2)(x - 2)| = |x + 2||x - 2|.$$

We want the expression above to be less than ϵ. We can make the factor $|x - 2|$ as small as we wish by choosing δ properly, but we need to control the factor $|x + 2|$ so that it does not become too large. If we first assume $\delta \le 1$ and require that $|x - 2| < \delta$, then we have

$$\begin{aligned} |x - 2| < 1 \quad &\Rightarrow \quad 1 < x < 3 \quad \Rightarrow \quad 3 < x + 2 < 5 \\ &\Rightarrow \quad |x + 2| < 5. \end{aligned}$$

Hence,

$$|f(x) - 4| < 5|x - 2| \quad \text{if} \quad |x - 2| < \delta \le 1.$$

But $5|x-2| < \epsilon$ if $|x-2| < \epsilon/5$. Therefore, if we take $\delta = \min\{1, \epsilon/5\}$, the *minimum* (the smaller) of the two numbers 1 and $\epsilon/5$, then

$$|f(x) - 4| < 5|x - 2| < 5 \times \frac{\epsilon}{5} = \epsilon \quad \text{if} \quad |x - 2| < \delta.$$

This proves that $\lim_{x\to 2} f(x) = 4$.

Using the Definition of Limit to Prove Theorems

We do not usually rely on the formal definition of limit to verify specific limits such as those in the two examples above. Rather, we appeal to general theorems about limits, in particular Theorems 2–4 of Section 1.2. The definition is used to prove these theorems. As an example, we prove part 1 of Theorem 2, the *Sum Rule*.

EXAMPLE 4 **(Proving the rule for the limit of a sum)** If $\lim_{x\to a} f(x) = L$ and $\lim_{x\to a} g(x) = M$, prove that $\lim_{x\to a} \big(f(x) + g(x)\big) = L + M$.

Solution Let $\epsilon > 0$ be given. We want to find a positive number δ such that

$$0 < |x - a| < \delta \quad \Rightarrow \quad \big|\big(f(x) + g(x)\big) - (L + M)\big| < \epsilon.$$

Observe that

$$\begin{aligned} &\big|\big(f(x)+g(x)\big) - (L + M)\big| && \text{Regroup terms.} \\ &\quad = \big|\big(f(x) - L\big) + \big(g(x) - M\big)\big| && \text{(Use the triangle inequality:} \\ & && |a + b| \le |a| + |b|\text{).} \\ &\quad \le |f(x) - L| + |g(x) - M|. \end{aligned}$$

Since $\lim_{x\to a} f(x) = L$ and $\epsilon/2$ is a positive number, there exists a number $\delta_1 > 0$ such that

$$0 < |x - a| < \delta_1 \quad \Rightarrow \quad |f(x) - L| < \epsilon/2.$$

Similarly, since $\lim_{x\to a} g(x) = M$, there exists a number $\delta_2 > 0$ such that

$$0 < |x - a| < \delta_2 \quad \Rightarrow \quad |g(x) - M| < \epsilon/2.$$

Let $\delta = \min\{\delta_1, \delta_2\}$, the smaller of δ_1 and δ_2. If $0 < |x - a| < \delta$, then $|x - a| < \delta_1$, so $|f(x) - L| < \epsilon/2$, and $|x - a| < \delta_2$, so $|g(x) - M| < \epsilon/2$. Therefore,

$$\big|\big(f(x) + g(x)\big) - (L + M)\big| < \frac{\epsilon}{2} + \frac{\epsilon}{2} = \epsilon.$$

This shows that $\lim_{x\to a} \big(f(x) + g(x)\big) = L + M$.

Other Kinds of Limits

The formal definition of limit can be modified to give precise definitions of one-sided limits, limits at infinity, and infinite limits. We give some of the definitions here and leave you to supply the others.

DEFINITION 9

Right limits

We say that $f(x)$ has **right limit** L at a, and we write

$$\lim_{x\to a+} f(x) = L,$$

if the following condition is satisfied:
for every number $\epsilon > 0$ there exists a number $\delta > 0$, possibly depending on ϵ, such that if $a < x < a + \delta$, then x belongs to the domain of f and

$$|f(x) - L| < \epsilon.$$

Notice how the condition $0 < |x - a| < \delta$ in the definition of limit becomes $a < x < a + \delta$ in the right limit case (Figure 1.38). The definition for a left limit is formulated in a similar way.

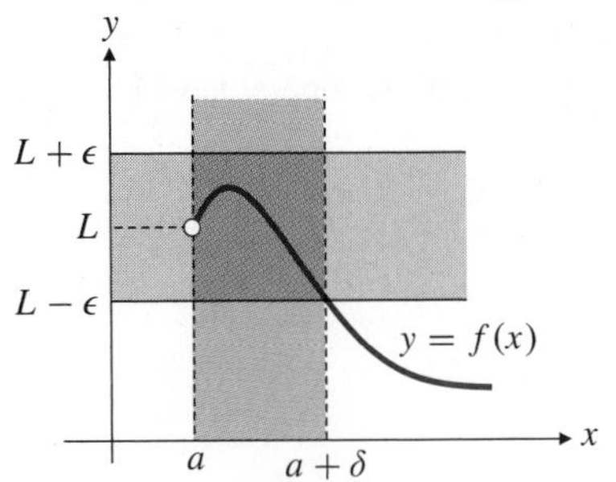

Figure 1.38 If $a < x < a + \delta$, then $|f(x) - L| < \epsilon$

EXAMPLE 5 Show that $\lim_{x\to 0+} \sqrt{x} = 0$.

Solution Let $\epsilon > 0$ be given. If $x > 0$, then $|\sqrt{x} - 0| = \sqrt{x}$. We can ensure that $\sqrt{x} < \epsilon$ by requiring $x < \epsilon^2$. Thus we can take $\delta = \epsilon^2$ and the condition of the definition will be satisfied:

$$0 < x < \delta = \epsilon^2 \quad \text{implies} \quad |\sqrt{x} - 0| < \epsilon.$$

Therefore, $\lim_{x\to 0+} \sqrt{x} = 0$.

To claim that a function f has a limit L at infinity, we must be able to ensure that the error $|f(x) - L|$ is less than any given positive number ϵ by restricting x to be *sufficiently large*, that is, by requiring $x > R$ for some positive number R depending on ϵ.

DEFINITION 10

Limit at infinity

We say that $f(x)$ **approaches the limit** L **as** x **approaches infinity**, and we write

$$\lim_{x\to\infty} f(x) = L,$$

if the following condition is satisfied:
for every number $\epsilon > 0$ there exists a number R, possibly depending on ϵ, such that if $x > R$, then x belongs to the domain of f and

$$|f(x) - L| < \epsilon.$$

You are invited to formulate a version of the definition of a limit at negative infinity.

EXAMPLE 6 Show that $\lim_{x\to\infty} \dfrac{1}{x} = 0$.

Solution Let ϵ be a given positive number. For $x > 0$ we have

$$\left|\frac{1}{x} - 0\right| = \frac{1}{|x|} = \frac{1}{x} < \epsilon \quad \text{provided} \quad x > \frac{1}{\epsilon}.$$

Therefore, the condition of the definition is satisfied with $R = 1/\epsilon$. We have shown that $\lim_{x\to\infty} 1/x = 0$.

To show that $f(x)$ has an infinite limit at a, we must ensure that $f(x)$ is larger than any given positive number (say B) by restricting x to a sufficiently small interval centred at a, and requiring that $x \neq a$.

DEFINITION 11

Infinite limits

We say that $f(x)$ approaches infinity as x approaches a and write

$$\lim_{x \to a} f(x) = \infty,$$

if for every positive number B we can find a positive number δ, possibly depending on B, such that if $0 < |x - a| < \delta$, then x belongs to the domain of f and $f(x) > B$.

Try to formulate the corresponding definition for the concept $\lim_{x \to a} f(x) = -\infty$. Then try to modify both definitions to cover the case of infinite one-sided limits and infinite limits at infinity.

EXAMPLE 7 Verify that $\lim_{x \to 0} \frac{1}{x^2} = \infty$.

Solution Let B be any positive number. We have

$$\frac{1}{x^2} > B \qquad \text{provided that} \quad x^2 < \frac{1}{B}.$$

If $\delta = 1/\sqrt{B}$, then

$$0 < |x| < \delta \quad \Rightarrow \quad x^2 < \delta^2 = \frac{1}{B} \quad \Rightarrow \quad \frac{1}{x^2} > B.$$

Therefore $\lim_{x \to 0} 1/x^2 = \infty$.

EXERCISES 1.5

1. The length L of a metal rod is given in terms of the temperature T (°C) by $L = 39.6 + 0.025T$ cm. Within what range of temperature must the rod be kept if its length must be maintained within ± 1 mm of 40 cm?

2. What is the largest tolerable error in the 20 cm edge length of a cubical cardboard box if the volume of the box must be within $\pm 1.2\%$ of 8,000 cm^3?

In Exercises 3–6, in what interval must x be confined if $f(x)$ must be within the given distance ϵ of the number L?

3. $f(x) = 2x - 1$, $L = 3$, $\epsilon = 0.02$
4. $f(x) = x^2$, $L = 4$, $\epsilon = 0.1$
5. $f(x) = \sqrt{x}$, $L = 1$, $\epsilon = 0.1$
6. $f(x) = 1/x$, $L = -2$, $\epsilon = 0.01$

In Exercises 7–10, find a number $\delta > 0$ such that if $|x - a| < \delta$, then $|f(x) - L|$ will be less than the given number ϵ.

7. $f(x) = 3x + 1$, $a = 2$, $L = 7$, $\epsilon = 0.03$
8. $f(x) = \sqrt{2x + 3}$, $a = 3$, $L = 3$, $\epsilon = 0.01$
9. $f(x) = x^3$, $a = 2$, $L = 8$, $\epsilon = 0.2$
10. $f(x) = 1/(x + 1)$, $a = 0$, $L = 1$, $\epsilon = 0.05$

In Exercises 11–20, use the formal definition of limit to verify the indicated limit.

11. $\lim_{x \to 1} (3x + 1) = 4$
12. $\lim_{x \to 2} (5 - 2x) = 1$
13. $\lim_{x \to 0} x^2 = 0$
14. $\lim_{x \to 2} \frac{x - 2}{1 + x^2} = 0$
15. $\lim_{x \to 1/2} \frac{1 - 4x^2}{1 - 2x} = 2$
16. $\lim_{x \to -2} \frac{x^2 + 2x}{x + 2} = -2$
17. $\lim_{x \to 1} \frac{1}{x + 1} = \frac{1}{2}$
18. $\lim_{x \to -1} \frac{x + 1}{x^2 - 1} = -\frac{1}{2}$
19. $\lim_{x \to 1} \sqrt{x} = 1$
20. $\lim_{x \to 2} x^3 = 8$

Give formal definitions of the limit statements in Exercises 21–26.

21. $\lim_{x \to a-} f(x) = L$
22. $\lim_{x \to -\infty} f(x) = L$
23. $\lim_{x \to a} f(x) = -\infty$
24. $\lim_{x \to \infty} f(x) = \infty$
25. $\lim_{x \to a+} f(x) = -\infty$
26. $\lim_{x \to a-} f(x) = \infty$

Use formal definitions of the various kinds of limits to prove the statements in Exercises 27–30.

27. $\lim_{x\to 1+} \frac{1}{x-1} = \infty$

28. $\lim_{x\to 1-} \frac{1}{x-1} = -\infty$

29. $\lim_{x\to\infty} \frac{1}{\sqrt{x^2+1}} = 0$

30. $\lim_{x\to\infty} \sqrt{x} = \infty$

Proving Theorems with the Definition of Limit

31. Prove that limits are unique; that is, if $\lim_{x\to a} f(x) = L$ and $\lim_{x\to a} f(x) = M$, prove that $L = M$. *Hint:* Suppose $L \neq M$ and let $\epsilon = |L - M|/3$.

32. If $\lim_{x\to a} g(x) = M$, show that there exists a number $\delta > 0$ such that

$$0 < |x-a| < \delta \quad \Rightarrow \quad |g(x)| < 1 + |M|.$$

(*Hint:* Take $\epsilon = 1$ in the definition of limit.) This says that the values of $g(x)$ are **bounded** near a point where g has a limit.

33. If $\lim_{x\to a} f(x) = L$ and $\lim_{x\to a} g(x) = M$, prove that $\lim_{x\to a} f(x)g(x) = LM$ (the Product Rule part of Theorem 2). *Hint:* Reread Example 4. Let $\epsilon > 0$ and write

$$\begin{aligned} |f(x)g(x) - LM| &= |f(x)g(x) - Lg(x) + Lg(x) - LM| \\ &= |(f(x) - L)g(x) + L(g(x) - M)| \\ &\le |(f(x) - L)g(x)| + |L(g(x) - M)| \\ &= |g(x)||f(x) - L| + |L||g(x) - M| \end{aligned}$$

Now try to make each term in the last line less than $\epsilon/2$ by taking x close enough to a. You will need the result of Exercise 32.

34. If $\lim_{x\to a} g(x) = M$, where $M \neq 0$, show that there exists a number $\delta > 0$ such that

$$0 < |x-a| < \delta \quad \Rightarrow \quad |g(x)| > |M|/2.$$

35. If $\lim_{x\to a} g(x) = M$, where $M \neq 0$, show that

$$\lim_{x\to a} \frac{1}{g(x)} = \frac{1}{M}.$$

Hint: You will need the result of Exercise 34.

36. Use the facts proved in Exercises 33 and 35 to prove the Quotient Rule (part 5 of Theorem 2): if $\lim_{x\to a} f(x) = L$ and $\lim_{x\to a} g(x) = M$, where $M \neq 0$, then

$$\lim_{x\to a} \frac{f(x)}{g(x)} = \frac{L}{M}.$$

37. Use the definition of limit twice to prove Theorem 7 of Section 1.4; that is, if f is continuous at L and if $\lim_{x\to c} g(x) = L$, then

$$\lim_{x\to c} f(g(x)) = f(L) = f\left(\lim_{x\to c} g(x)\right).$$

38. Prove the Squeeze Theorem (Theorem 4 in Section 1.2). *Hint:* If $f(x) \le g(x) \le h(x)$, then

$$\begin{aligned} |g(x) - L| &= |g(x) - f(x) + f(x) - L| \\ &\le |g(x) - f(x)| + |f(x) - L| \\ &\le |h(x) - f(x)| + |f(x) - L| \\ &= |h(x) - L - (f(x) - L)| + |f(x) - L| \\ &\le |h(x) - L| + |f(x) - L| + |f(x) - L| \end{aligned}$$

Now you can make each term in the last expression less than $\epsilon/3$ and so complete the proof.

CHAPTER REVIEW

Key Ideas

- **What do the following statements and phrases mean?**
  - ◇ the average rate of change of $f(x)$ on $[a, b]$
  - ◇ the instantaneous rate of change of $f(x)$ at $x = a$
  - ◇ $\lim_{x\to a} f(x) = L$
  - ◇ $\lim_{x\to a+} f(x) = L, \quad \lim_{x\to a-} f(x) = L$
  - ◇ $\lim_{x\to\infty} f(x) = L, \quad \lim_{x\to-\infty} f(x) = L$
  - ◇ $\lim_{x\to a} f(x) = \infty, \quad \lim_{x\to a+} f(x) = -\infty$
  - ◇ f is continuous at c.
  - ◇ f is left (or right) continuous at c.
  - ◇ f has a continuous extension to c.
  - ◇ f is a continuous function.
  - ◇ f takes on maximum and minimum values on interval I.
  - ◇ f is bounded on interval I.
  - ◇ f has the intermediate-value property on interval I.
- **State as many "laws of limits" as you can.**
- **What properties must a function have if it is continuous and its domain is a closed, finite interval?**
- **How can you find zeros (roots) of a continuous function?**

Review Exercises

1. Find the average rate of change of x^3 over $[1, 3]$.

2. Find the average rate of change of $1/x$ over $[-2, -1]$.

3. Find the rate of change of x^3 at $x = 2$.

4. Find the rate of change of $1/x$ at $x = -3/2$.

Evaluate the limits in Exercises 5–30 or explain why they do not exist.

5. $\lim_{x\to 1} (x^2 - 4x + 7)$

6. $\lim_{x\to 2} \frac{x^2}{1 - x^2}$

7. $\lim_{x\to 1} \frac{x^2}{1-x^2}$

8. $\lim_{x\to 2} \frac{x^2-4}{x^2-5x+6}$

9. $\lim_{x\to 2} \frac{x^2-4}{x^2-4x+4}$

10. $\lim_{x\to 2-} \frac{x^2-4}{x^2-4x+4}$

11. $\lim_{x\to -2+} \frac{x^2-4}{x^2+4x+4}$

12. $\lim_{x\to 4} \frac{2-\sqrt{x}}{x-4}$

13. $\lim_{x\to 3} \frac{x^2-9}{\sqrt{x}-\sqrt{3}}$

14. $\lim_{h\to 0} \frac{h}{\sqrt{x+3h}-\sqrt{x}}$

15. $\lim_{x\to 0+} \sqrt{x-x^2}$

16. $\lim_{x\to 0} \sqrt{x-x^2}$

17. $\lim_{x\to 1} \sqrt{x-x^2}$

18. $\lim_{x\to 1-} \sqrt{x-x^2}$

19. $\lim_{x\to\infty} \frac{1-x^2}{3x^2-x-1}$

20. $\lim_{x\to -\infty} \frac{2x+100}{x^2+3}$

21. $\lim_{x\to -\infty} \frac{x^3-1}{x^2+4}$

22. $\lim_{x\to\infty} \frac{x^4}{x^2-4}$

23. $\lim_{x\to 0+} \frac{1}{\sqrt{x-x^2}}$

24. $\lim_{x\to 1/2} \frac{1}{\sqrt{x-x^2}}$

25. $\lim_{x\to\infty} \sin x$

26. $\lim_{x\to\infty} \frac{\cos x}{x}$

27. $\lim_{x\to 0} x \sin \frac{1}{x}$

28. $\lim_{x\to 0} \sin \frac{1}{x^2}$

29. $\lim_{x\to -\infty} [x + \sqrt{x^2-4x+1}]$

30. $\lim_{x\to\infty} [x + \sqrt{x^2-4x+1}]$

At what, if any, points in its domain is the function f in Exercises 31–38 discontinuous? Is f left or right continuous at these points? In Exercises 35 and 36, H refers to the Heaviside function: $H(x) = 1$ if $x \ge 0$ and $H(x) = 0$ if $x < 0$.

31. $f(x) = x^3 - 4x^2 + 1$

32. $f(x) = \frac{x}{x+1}$

33. $f(x) = \begin{cases} x^2 & \text{if } x > 2 \\ x & \text{if } x \le 2 \end{cases}$

34. $f(x) = \begin{cases} x^2 & \text{if } x > 1 \\ x & \text{if } x \le 1 \end{cases}$

35. $f(x) = H(x-1)$

36. $f(x) = H(9-x^2)$

37. $f(x) = |x| + |x+1|$

38. $f(x) = \begin{cases} |x|/|x+1| & \text{if } x \ne -1 \\ 1 & \text{if } x = -1 \end{cases}$

Challenging Problems

1. Show that the average rate of change of the function x^3 over the interval $[a, b]$, where $0 < a < b$, is equal to the instantaneous rate of change of x^3 at $x = \sqrt{(a^2+ab+b^2)/3}$. Is this point to the left or to the right of the midpoint $(a+b)/2$ of the interval $[a, b]$?

2. Evaluate $\lim_{x\to 0} \frac{x}{|x-1|-|x+1|}$.

3. Evaluate $\lim_{x\to 3} \frac{|5-2x|-|x-2|}{|x-5|-|3x-7|}$.

4. Evaluate $\lim_{x\to 64} \frac{x^{1/3}-4}{x^{1/2}-8}$.

5. Evaluate $\lim_{x\to 1} \frac{\sqrt{3+x}-2}{\sqrt[3]{7+x}-2}$.

6. The equation $ax^2 + 2x - 1 = 0$, where a is a constant, has two roots if $a > -1$ and $a \ne 0$:

$$r_+(a) = \frac{-1+\sqrt{1+a}}{a} \quad \text{and} \quad r_-(a) = \frac{-1-\sqrt{1+a}}{a}.$$

(a) What happens to the root $r_-(a)$ when $a \to 0$?

(b) Investigate numerically what happens to the root $r_+(a)$ when $a \to 0$ by trying the values $a = 1, \pm 0.1, \pm 0.01, \ldots$. For values such as $a = 10^{-8}$, the limited precision of your calculator may produce some interesting results. What happens, and why?

(c) Evaluate $\lim_{a\to 0} r_+(a)$ mathematically by using the identity

$$\sqrt{A} - \sqrt{B} = \frac{A-B}{\sqrt{A}+\sqrt{B}}.$$

7. TRUE or FALSE? If TRUE, give reasons; if FALSE, give a counterexample.

(a) If $\lim_{x\to a} f(x)$ exists but $\lim_{x\to a} g(x)$ does not exist, then $\lim_{x\to a} (f(x) + g(x))$ does not exist.

(b) If neither $\lim_{x\to a} f(x)$ nor $\lim_{x\to a} g(x)$ exists, then $\lim_{x\to a} (f(x) + g(x))$ does not exist.

(c) If f is continuous at a, then so is $|f|$.

(d) If $|f|$ is continuous at a, then so is f.

(e) If $f(x) < g(x)$ for all x in an interval around a, and if $\lim_{x\to a} f(x)$ and $\lim_{x\to a} g(x)$ both exist, then $\lim_{x\to a} f(x) < \lim_{x\to a} g(x)$.

8. (a) If f is a continuous function defined on a closed interval $[a, b]$, show that $R(f)$ is a closed interval.

(b) What are the possibilities for $R(f)$ if $D(f)$ is an open interval (a, b)?

9. Consider the function $f(x) = \frac{x^2-1}{|x^2-1|}$. Find all points where f is not continuous. Does f have one-sided limits at those points, and if so, what are they?

10. Find the minimum value of $f(x) = 1/(x-x^2)$ on the interval $(0, 1)$. Explain how you know such a minimum value must exist.

11. (a) Suppose f is a continuous function on the interval $[0, 1]$, and $f(0) = f(1)$. Show that $f(a) = f\left(a + \frac{1}{2}\right)$ for some $a \in \left[0, \frac{1}{2}\right]$.

Hint: Let $g(x) = f\left(x + \frac{1}{2}\right) - f(x)$, and use the Intermediate-Value Theorem.

(b) If n is an integer larger than 2, show that $f(a) = f\left(a + \frac{1}{n}\right)$ for some $a \in \left[0, 1 - \frac{1}{n}\right]$.

CHAPTER 2

Differentiation

"'All right,' said Deep Thought. 'The Answer to the Great Question ...'
'Yes ...!'
'Of Life, the Universe and Everything ...' said Deep Thought.
'Yes ...!'
'Is...' said Deep Thought, and paused.
'Yes ...!...?'
'Forty-two,' said Deep Thought, with infinite majesty and calm.
...
'Forty-two!' yelled Loonquawl. 'Is that all you've got to show for seven and a half million years' work?'
'I checked it very thoroughly,' said the computer, 'and that quite definitely is the answer. I think the problem, to be quite honest with you, is that you've never actually known what the question is.'"

Douglas Adams 1952–2001
from *The Hitchhiker's Guide to the Galaxy*

Introduction

Two fundamental problems are considered in calculus. The **problem of slopes** is concerned with finding the slope of (the tangent line to) a given curve at a given point on the curve. The **problem of areas** is concerned with finding the area of a plane region bounded by curves and straight lines. The solution of the problem of slopes is the subject of **differential calculus**. As we will see, it has many applications in mathematics and other disciplines. The problem of areas is the subject of **integral calculus**, which we begin in Chapter 5.

2.1 Tangent Lines and Their Slopes

This section deals with the problem of finding a straight line L that is tangent to a curve C at a point P. As is often the case in mathematics, the most important step in the solution of such a fundamental problem is making a suitable definition.

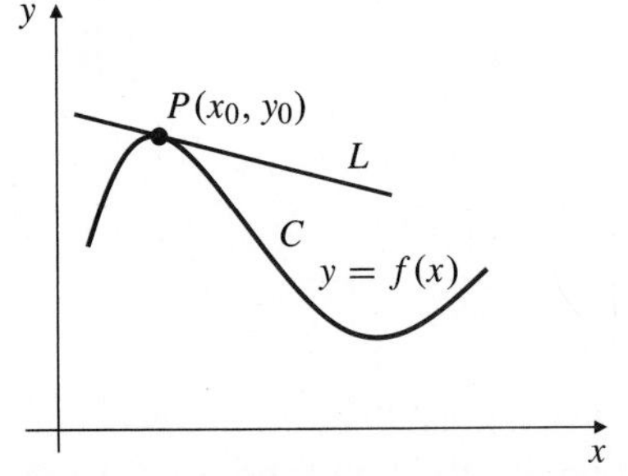

Figure 2.1 L is tangent to C at P

For simplicity, and to avoid certain problems best postponed until later, we will not deal with the most general kinds of curves now, but only with those that are the *graphs of continuous functions*. Let C be the graph of $y = f(x)$ and let P be the point (x_0, y_0) on C, so that $y_0 = f(x_0)$. We assume that P is not an endpoint of C. Therefore, C extends some distance on both sides of P. (See Figure 2.1.)

What do we mean when we say that the line L is tangent to C at P? Past experience with tangent lines to circles does not help us to define tangency for more general curves. A tangent line to a circle has the following properties (see Figure 2.2):

(i) It meets the circle at only one point.

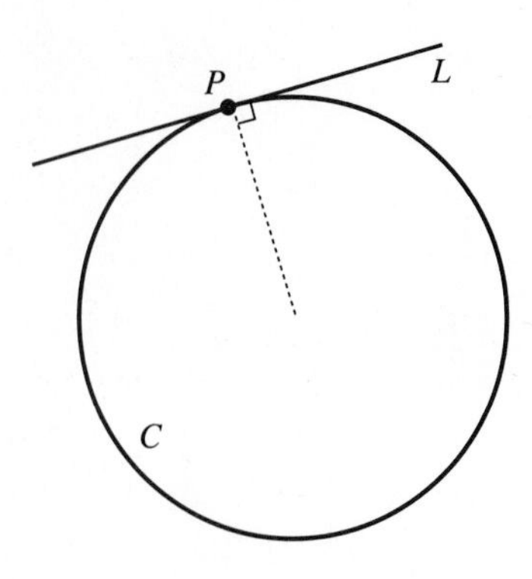

Figure 2.2 L is tangent to C at P

(ii) The circle lies on only one side of the line.

(iii) The tangent is perpendicular to the line joining the centre of the circle to the point of contact.

Most curves do not have obvious *centres*, so (iii) is useless for characterizing tangents to them. The curves in Figure 2.3 show that (i) and (ii) cannot be used to define tangency either. In particular, Figure 2.3(d) is not "smooth" at P so that curve should not have any tangent line there. A tangent line should have the "same direction" as the curve does at the point of tangency.

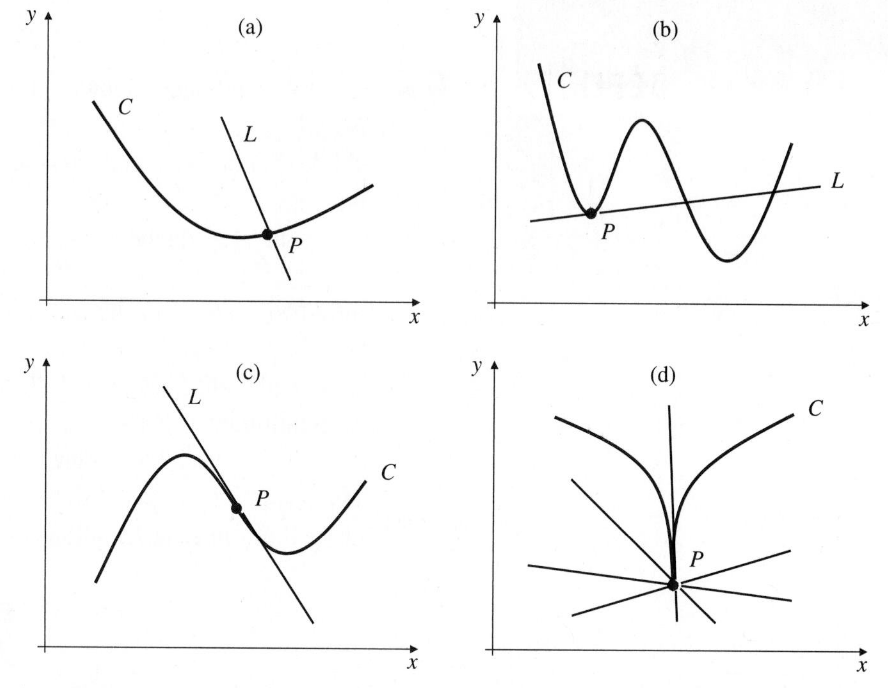

Figure 2.3

(a) L meets C only at P but is not tangent to C

(b) L meets C at several points but is tangent to C at P

(c) L is tangent to C at P but crosses C at P

(d) Many lines meet C only at P but none of them is tangent to C at P

A reasonable definition of tangency can be stated in terms of limits. If Q is a point on C different from P, then the line through P and Q is called a **secant line** to the curve. This line rotates around P as Q moves along the curve. If L is a line through P whose slope is the limit of the slopes of these secant lines PQ as Q approaches P along C (Figure 2.4), then we will say that L is tangent to C at P.

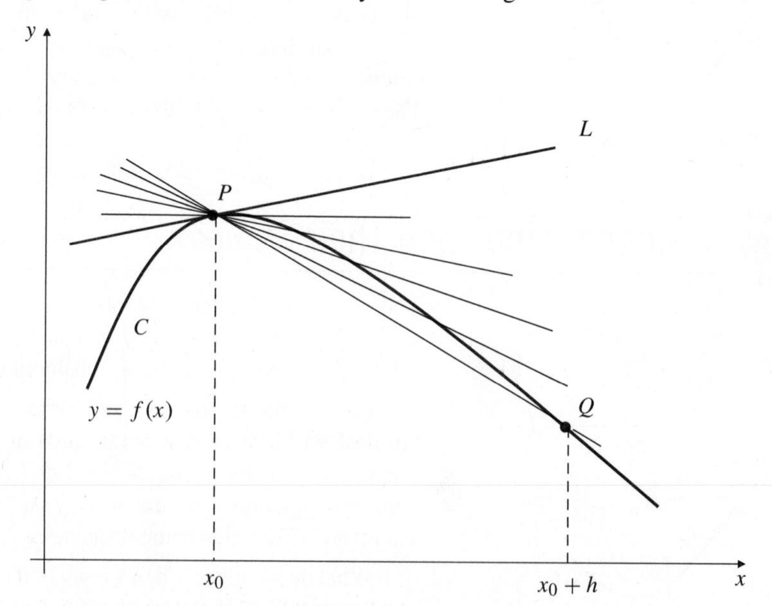

Figure 2.4 Secant lines PQ approach tangent line L as Q approaches P along the curve C

Since C is the graph of the *function* $y = f(x)$, then vertical lines can meet C only once. Since $P = (x_0, f(x_0))$, a different point Q on the graph must have a different

x-coordinate, say $x_0 + h$, where $h \neq 0$. Thus $Q = (x_0 + h, f(x_0 + h))$, and the slope of the line PQ is

$$\frac{f(x_0 + h) - f(x_0)}{h}.$$

This expression is called the **Newton quotient** or **difference quotient** for f at x_0. Note that h can be positive or negative, depending on whether Q is to the right or left of P.

DEFINITION 1

Nonvertical tangent lines

Suppose that the function f is continuous at $x = x_0$ and that

$$\lim_{h \to 0} \frac{f(x_0 + h) - f(x_0)}{h} = m$$

exists. Then the straight line having slope m and passing through the point $P = (x_0, f(x_0))$ is called the **tangent line** (or simply the **tangent**) to the graph of $y = f(x)$ at P. An equation of this tangent is

$$y = m(x - x_0) + y_0.$$

EXAMPLE 1 Find an equation of the tangent line to the curve $y = x^2$ at the point $(1, 1)$.

Solution Here $f(x) = x^2$, $x_0 = 1$, and $y_0 = f(1) = 1$. The slope of the required tangent is:

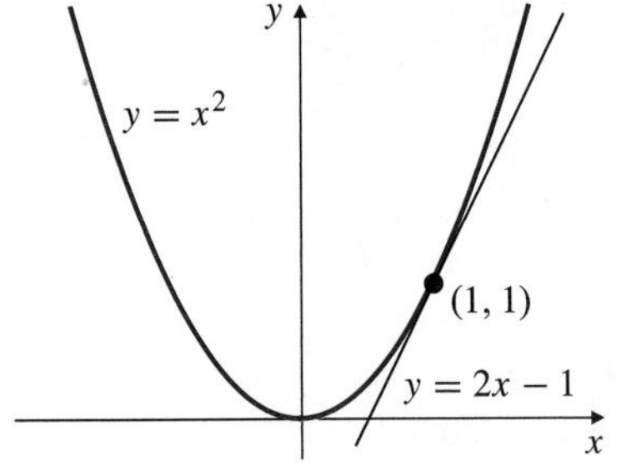

Figure 2.5 The tangent to $y = x^2$ at $(1, 1)$

$$\begin{aligned} m = \lim_{h \to 0} \frac{f(1 + h) - f(1)}{h} &= \lim_{h \to 0} \frac{(1 + h)^2 - 1}{h} \\ &= \lim_{h \to 0} \frac{1 + 2h + h^2 - 1}{h} \\ &= \lim_{h \to 0} \frac{2h + h^2}{h} = \lim_{h \to 0} (2 + h) = 2. \end{aligned}$$

Accordingly, the equation of the tangent line at $(1, 1)$ is $y = 2(x-1)+1$, or $y = 2x-1$. See Figure 2.5.

Definition 1 deals only with tangents that have finite slopes and are, therefore, not vertical. It is also possible for the graph of a continuous function to have a *vertical* tangent line.

EXAMPLE 2 Consider the graph of the function $f(x) = \sqrt[3]{x} = x^{1/3}$, which is shown in Figure 2.6. The graph is a smooth curve, and it seems evident that the y-axis is tangent to this curve at the origin. Let us try to calculate the limit of the Newton quotient for f at $x = 0$:

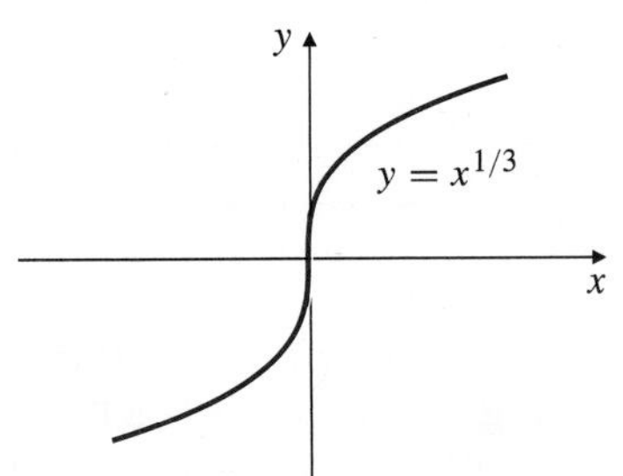

Figure 2.6 The y-axis is tangent to $y = x^{1/3}$ at the origin

$$\lim_{h \to 0} \frac{f(0 + h) - f(0)}{h} = \lim_{h \to 0} \frac{h^{1/3}}{h} = \lim_{h \to 0} \frac{1}{h^{2/3}} = \infty.$$

Although the limit does not exist, the slope of the secant line joining the origin to another point Q on the curve approaches infinity as Q approaches the origin from either side.

EXAMPLE 3 On the other hand, the function $f(x) = x^{2/3}$, whose graph is shown in Figure 2.7, does not have a tangent line at the origin because it is not "smooth" there. In this case the Newton quotient is

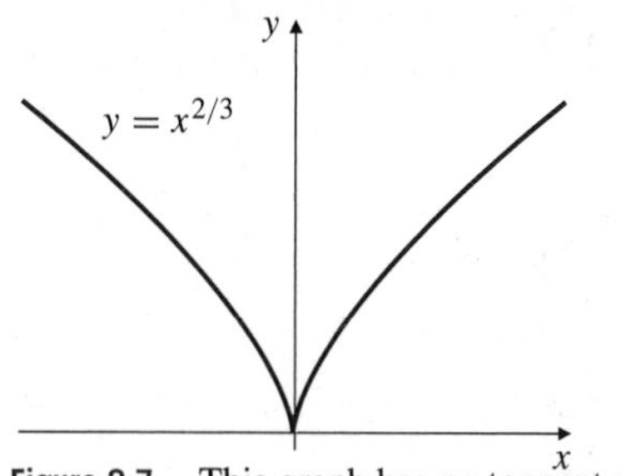

Figure 2.7 This graph has no tangent at the origin

$$\frac{f(0+h) - f(0)}{h} = \frac{h^{2/3}}{h} = \frac{1}{h^{1/3}},$$

which has no limit as h approaches zero. (The right limit is ∞; the left limit is $-\infty$.) We say this curve has a **cusp** at the origin. A cusp is an infinitely sharp point; if you were travelling along the curve, you would have to stop and turn 180° at the origin.

In the light of the two preceding examples, we extend the definition of tangent line to allow for vertical tangents as follows:

DEFINITION 2

Vertical tangents

If f is continuous at $P = (x_0, y_0)$, where $y_0 = f(x_0)$, and if either

$$\lim_{h\to 0} \frac{f(x_0+h) - f(x_0)}{h} = \infty \quad \text{or} \quad \lim_{h\to 0} \frac{f(x_0+h) - f(x_0)}{h} = -\infty,$$

then the vertical line $x = x_0$ is tangent to the graph $y = f(x)$ at P. If the limit of the Newton quotient fails to exist in any other way than by being ∞ or $-\infty$, the graph $y = f(x)$ has no tangent line at P.

EXAMPLE 4 Does the graph of $y = |x|$ have a tangent line at $x = 0$?

Solution The Newton quotient here is

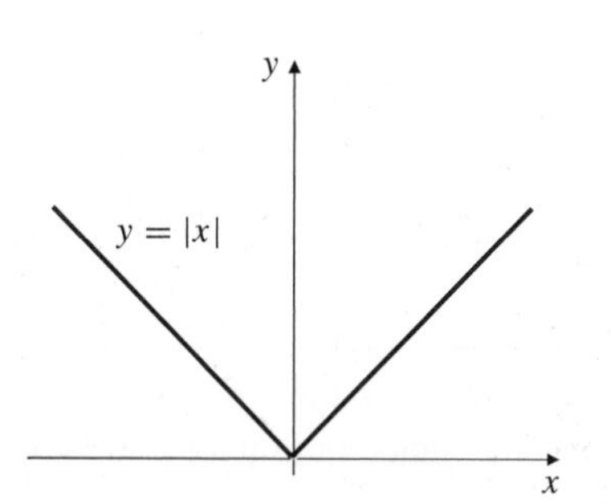

Figure 2.8 $y = |x|$ has no tangent at the origin

$$\frac{|0+h| - |0|}{h} = \frac{|h|}{h} = \operatorname{sgn} h = \begin{cases} 1, & \text{if } h > 0 \\ -1, & \text{if } h < 0. \end{cases}$$

Since $\operatorname{sgn} h$ has different right and left limits at 0 (namely, 1 and -1), the Newton quotient has no limit as $h \to 0$, so $y = |x|$ has no tangent line at $(0, 0)$. (See Figure 2.8.) The graph does not have a cusp at the origin, but it is kinked at that point; *it suddenly changes direction and is not smooth.* Curves have tangents only at points where they are smooth. The graphs of $y = x^{2/3}$ and $y = |x|$ have tangent lines everywhere except at the origin, where they are not smooth.

DEFINITION 3

The slope of a curve

The **slope** of a curve C at a point P is the slope of the tangent line to C at P if such a tangent line exists. In particular, the slope of the graph of $y = f(x)$ at the point x_0 is

$$\lim_{h\to 0} \frac{f(x_0+h) - f(x_0)}{h}.$$

EXAMPLE 5 Find the slope of the curve $y = x/(3x + 2)$ at the point $x = -2$.

Solution If $x = -2$, then $y = 1/2$, so the required slope is

$$\begin{aligned}
m &= \lim_{h\to 0} \frac{\dfrac{-2+h}{3(-2+h)+2} - \dfrac{1}{2}}{h} \\
&= \lim_{h\to 0} \frac{-4+2h-(-6+3h+2)}{2(-6+3h+2)h} \\
&= \lim_{h\to 0} \frac{-h}{2h(-4+3h)} = \lim_{h\to 0} \frac{-1}{2(-4+3h)} = \frac{1}{8}.
\end{aligned}$$

Normals

If a curve C has a tangent line L at point P, then the straight line N through P perpendicular to L is called the **normal** to C at P. If L is horizontal, then N is vertical; if L is vertical, then N is horizontal. If L is neither horizontal nor vertical, then, as shown in Section P.2, the slope of N is the negative reciprocal of the slope of L; that is,

$$\text{slope of the normal} = \frac{-1}{\text{slope of the tangent}}.$$

EXAMPLE 6 Find an equation of the normal to $y = x^2$ at $(1, 1)$.

Solution By Example 1, the tangent to $y = x^2$ at $(1, 1)$ has slope 2. Hence, the normal has slope $-1/2$, and its equation is

$$y = -\frac{1}{2}(x-1)+1 \qquad \text{or} \qquad y = -\frac{x}{2}+\frac{3}{2}.$$

EXAMPLE 7 Find equations of the straight lines that are tangent and normal to the curve $y = \sqrt{x}$ at the point $(4, 2)$.

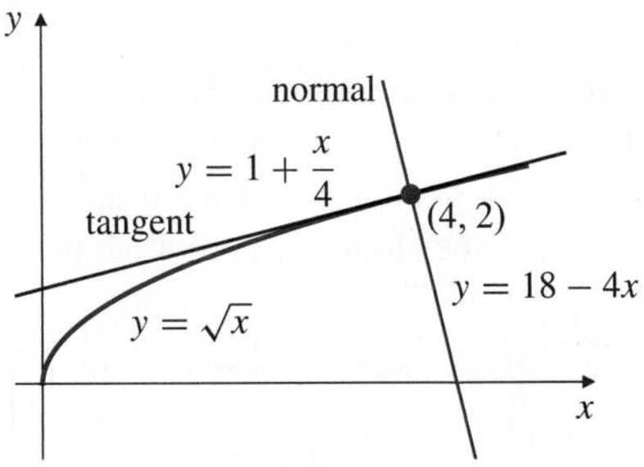

Solution The slope of the tangent at $(4, 2)$ (Figure 2.9) is

$$\begin{aligned}
m &= \lim_{h\to 0} \frac{\sqrt{4+h}-2}{h} = \lim_{h\to 0} \frac{(\sqrt{4+h}-2)(\sqrt{4+h}+2)}{h(\sqrt{4+h}+2)} \\
&= \lim_{h\to 0} \frac{4+h-4}{h(\sqrt{4+h}+2)} \\
&= \lim_{h\to 0} \frac{1}{\sqrt{4+h}+2} = \frac{1}{4}.
\end{aligned}$$

Figure 2.9 The tangent and normal to $y = \sqrt{x}$ at $(4, 2)$

The tangent line has equation

$$y = \frac{1}{4}(x-4)+2 \qquad \text{or} \qquad x - 4y + 4 = 0,$$

and the normal has slope -4 and, therefore, equation

$$y = -4(x-4)+2 \qquad \text{or} \qquad y = -4x + 18.$$

EXERCISES 2.1

In Exercises 1–12, find an equation of the straight line tangent to the given curve at the point indicated.

1. $y = 3x - 1$ at $(1, 2)$

2. $y = x/2$ at $(a, a/2)$

3. $y = 2x^2 - 5$ at $(2, 3)$

4. $y = 6 - x - x^2$ at $x = -2$

5. $y = x^3 + 8$ at $x = -2$

6. $y = \dfrac{1}{x^2 + 1}$ at $(0, 1)$

7. $y = \sqrt{x + 1}$ at $x = 3$

8. $y = \dfrac{1}{\sqrt{x}}$ at $x = 9$

9. $y = \dfrac{2x}{x + 2}$ at $x = 2$

10. $y = \sqrt{5 - x^2}$ at $x = 1$

11. $y = x^2$ at $x = x_0$

12. $y = \dfrac{1}{x}$ at $\left(a, \dfrac{1}{a}\right)$

Do the graphs of the functions f in Exercises 13–17 have tangent lines at the given points? If yes, what is the tangent line?

13. $f(x) = \sqrt{|x|}$ at $x = 0$

14. $f(x) = (x - 1)^{4/3}$ at $x = 1$

15. $f(x) = (x + 2)^{3/5}$ at $x = -2$

16. $f(x) = |x^2 - 1|$ at $x = 1$

17. $f(x) = \begin{cases} \sqrt{x} & \text{if } x \ge 0 \\ -\sqrt{-x} & \text{if } x < 0 \end{cases}$ at $x = 0$

18. Find the slope of the curve $y = x^2 - 1$ at the point $x = x_0$. What is the equation of the tangent line to $y = x^2 - 1$ that has slope -3?

19. (a) Find the slope of $y = x^3$ at the point $x = a$.

(b) Find the equations of the straight lines having slope 3 that are tangent to $y = x^3$.

20. Find all points on the curve $y = x^3 - 3x$ where the tangent line is parallel to the x-axis.

21. Find all points on the curve $y = x^3 - x + 1$ where the tangent line is parallel to the line $y = 2x + 5$.

22. Find all points on the curve $y = 1/x$ where the tangent line is perpendicular to the line $y = 4x - 3$.

23. For what value of the constant k is the line $x + y = k$ normal to the curve $y = x^2$?

24. For what value of the constant k do the curves $y = kx^2$ and $y = k(x - 2)^2$ intersect at right angles? *Hint:* Where do the curves intersect? What are their slopes there?

Use a graphics utility to plot the following curves. Where does the curve have a horizontal tangent? Does the curve fail to have a tangent line anywhere?

25. $y = x^3(5 - x)^2$

26. $y = 2x^3 - 3x^2 - 12x + 1$

27. $y = |x^2 - 1| - x$

28. $y = |x + 1| - |x - 1|$

29. $y = (x^2 - 1)^{1/3}$

30. $y = ((x^2 - 1)^2)^{1/3}$

31. If line L is tangent to curve C at point P, then the smaller angle between L and the secant line PQ joining P to another point Q on C approaches 0 as Q approaches P along C. Is the converse true: if the angle between PQ and line L (which passes through P) approaches 0, must L be tangent to C?

32. Let $P(x)$ be a polynomial. If a is a real number, then $P(x)$ can be expressed in the form

$$P(x) = a_0 + a_1(x - a) + a_2(x - a)^2 + \cdots + a_n(x - a)^n$$

for some $n \ge 0$. If $\ell(x) = m(x - a) + b$, show that the straight line $y = \ell(x)$ is tangent to the graph of $y = P(x)$ at $x = a$ provided $P(x) - \ell(x) = (x - a)^2 Q(x)$, where $Q(x)$ is a polynomial.

2.2 The Derivative

A straight line has the property that its slope is the same at all points. For any other graph, however, the slope may vary from point to point. Thus the slope of the graph of $y = f(x)$ at the point x is itself a function of x. At any point x where the graph has a finite slope, we say that f is differentiable, and we call the slope the derivative of f. The derivative is therefore the limit of the Newton quotient.

DEFINITION 4

The **derivative** of a function f is another function f' defined by

$$f'(x) = \lim_{h \to 0} \frac{f(x + h) - f(x)}{h}$$

at all points x for which the limit exists (i.e., is a finite real number). If $f'(x)$ exists, we say that f is **differentiable** at x.

The domain of the derivative f' (read "f prime") is the set of numbers x in the domain of f where the graph of f has a *nonvertical* tangent line, and the value $f'(x_0)$ of f' at such a point x_0 is the slope of the tangent line to $y = f(x)$ there. Thus, the equation of the tangent line to $y = f(x)$ at $(x_0, f(x_0))$ is

$$y = f(x_0) + f'(x_0)(x - x_0).$$

The domain $\mathcal{D}(f')$ of f' may be smaller than the domain $\mathcal{D}(f)$ of f because it contains only those points in $\mathcal{D}(f)$ at which f is differentiable. Values of x in $\mathcal{D}(f)$ where f is not differentiable and that are not endpoints of $\mathcal{D}(f)$ are **singular points** of f.

Remark The value of the derivative of f at a particular point x_0 can be expressed as a limit in either of two ways:

$$f'(x_0) = \lim_{h\to 0} \frac{f(x_0 + h) - f(x_0)}{h} = \lim_{x\to x_0} \frac{f(x) - f(x_0)}{x - x_0}.$$

In the second limit $x_0 + h$ is replaced by x, so that $h = x - x_0$ and $h \to 0$ is equivalent to $x \to x_0$.

The process of calculating the derivative f' of a given function f is called **differentiation**. The graph of f' can often be sketched directly from that of f by visualizing slopes, a procedure called **graphical differentiation**. In Figure 2.10 the graphs of f' and g' were obtained by measuring the slopes at the corresponding points in the graphs of f and g lying above them. The height of the graph $y = f'(x)$ at x is the slope of the graph of $y = f(x)$ at x. Note that -1 and 1 are singular points of f. Although $f(-1)$ and $f(1)$ are defined, $f'(-1)$ and $f'(1)$ are not defined; the graph of f has no tangent at -1 or at 1.

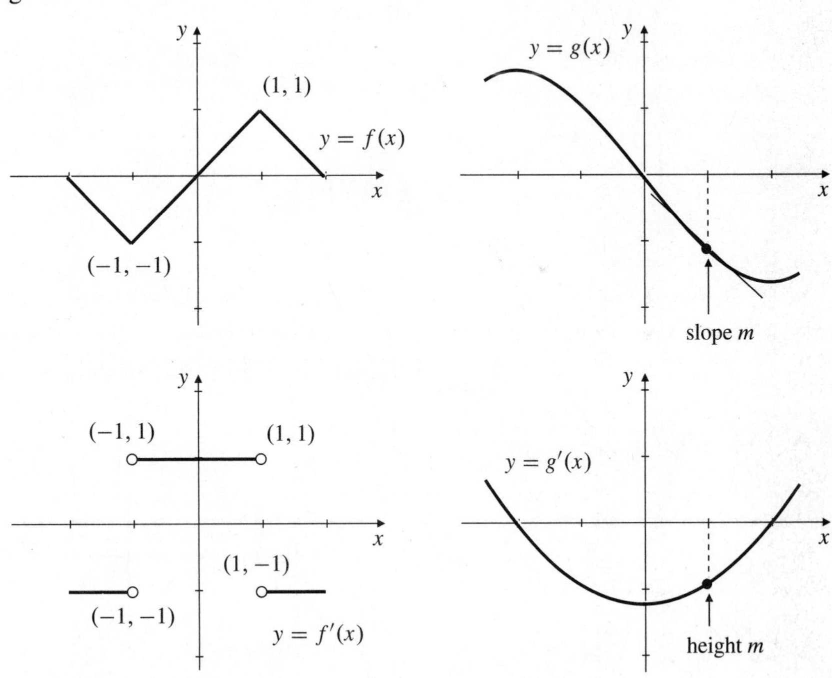

Figure 2.10 Graphical differentiation

A function is differentiable on a set S if it is differentiable at every point x in S. Typically, the functions we encounter are defined on intervals or unions of intervals. If f is defined on a closed interval $[a, b]$, Definition 4 does not allow for the existence of a derivative at the endpoints $x = a$ or $x = b$. (Why?) As we did for continuity in Section 1.4, we extend the definition to allow for a **right derivative** at $x = a$ and a **left derivative** at $x = b$:

$$f'_+(a) = \lim_{h\to 0+} \frac{f(a + h) - f(a)}{h}, \qquad f'_-(b) = \lim_{h\to 0-} \frac{f(b + h) - f(b)}{h}.$$

We now say that f is **differentiable** on $[a, b]$ if $f'(x)$ exists for all x in (a, b) and $f'_+(a)$ and $f'_-(b)$ both exist.

Some Important Derivatives

We now give several examples of the calculation of derivatives algebraically from the definition of derivative. Some of these are the basic building blocks from which more complicated derivatives can be calculated later. They are collected in Table 1 later in this section and should be memorized.

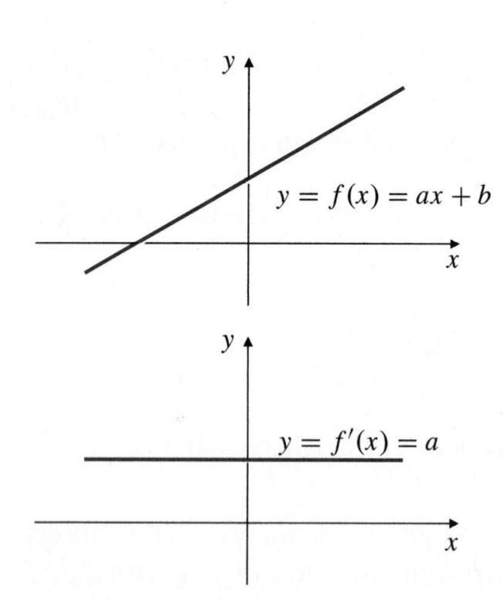

Figure 2.11 The derivative of the linear function $f(x) = ax + b$ is the constant function $f'(x) = a$

EXAMPLE 1 **(The derivative of a linear function)** Show that if $f(x) = ax+b$, then $f'(x) = a$.

Solution The result is apparent from the graph of f (Figure 2.11), but we will do the calculation using the definition:

$$\begin{aligned} f'(x) &= \lim_{h\to 0} \frac{f(x+h) - f(x)}{h} \\ &= \lim_{h\to 0} \frac{a(x+h) + b - (ax+b)}{h} \\ &= \lim_{h\to 0} \frac{ah}{h} = a. \end{aligned}$$

An important special case of Example 1 says that the derivative of a constant function is the zero function:

> If $g(x) = c$ (constant), then $g'(x) = 0$.

EXAMPLE 2 Use the definition of the derivative to calculate the derivatives of the functions:

$$\text{(a) } f(x) = x^2, \quad \text{(b) } g(x) = \frac{1}{x}, \quad \text{and} \quad \text{(c) } k(x) = \sqrt{x}.$$

Solution Figures 2.12–2.14 show the graphs of these functions and their derivatives.

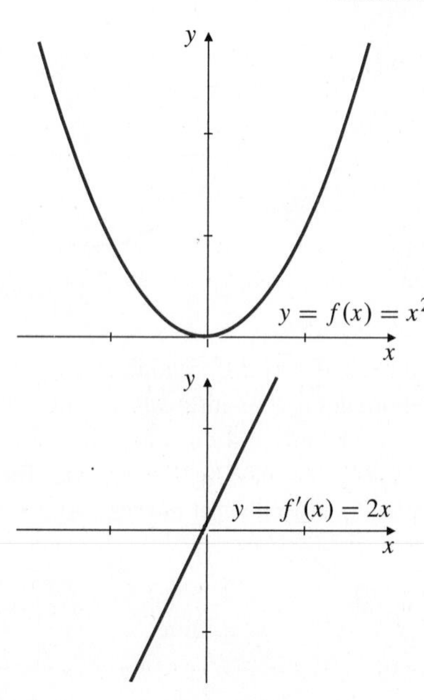

Figure 2.12 The derivative of $f(x) = x^2$ is $f'(x) = 2x$

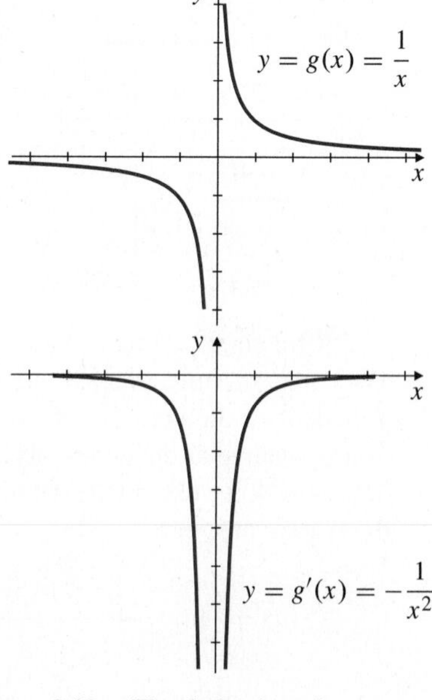

Figure 2.13 The derivative of $g(x) = 1/x$ is $g'(x) = -1/x^2$

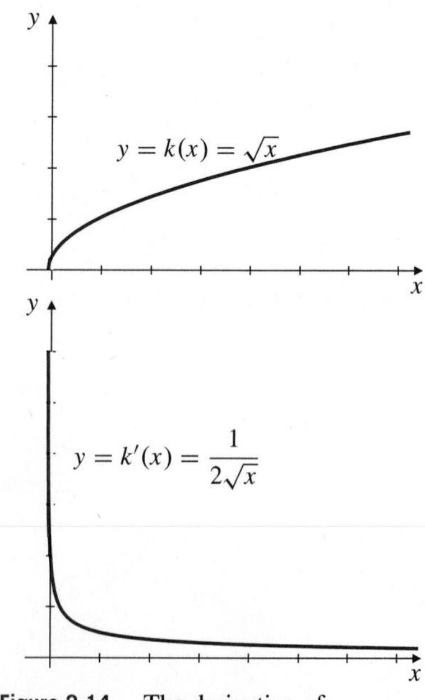

Figure 2.14 The derivative of $k(x) = \sqrt{x}$ is $k'(x) = 1/(2\sqrt{x})$

$$\text{(a)}\quad f'(x) = \lim_{h\to 0} \frac{f(x+h) - f(x)}{h} = \lim_{h\to 0} \frac{(x+h)^2 - x^2}{h} = \lim_{h\to 0} \frac{2hx + h^2}{h} = \lim_{h\to 0} (2x + h) = 2x.$$

$$\text{(b)}\quad g'(x) = \lim_{h\to 0} \frac{g(x+h) - g(x)}{h} = \lim_{h\to 0} \frac{\dfrac{1}{x+h} - \dfrac{1}{x}}{h} = \lim_{h\to 0} \frac{x - (x+h)}{h(x+h)x} = \lim_{h\to 0} -\frac{1}{(x+h)x} = -\frac{1}{x^2}.$$

$$\text{(c)}\quad k'(x) = \lim_{h\to 0} \frac{k(x+h) - k(x)}{h} = \lim_{h\to 0} \frac{\sqrt{x+h} - \sqrt{x}}{h} = \lim_{h\to 0} \frac{\sqrt{x+h} - \sqrt{x}}{h} \times \frac{\sqrt{x+h} + \sqrt{x}}{\sqrt{x+h} + \sqrt{x}} = \lim_{h\to 0} \frac{x+h-x}{h(\sqrt{x+h} + \sqrt{x})} = \lim_{h\to 0} \frac{1}{\sqrt{x+h} + \sqrt{x}} = \frac{1}{2\sqrt{x}}.$$

Note that k is not differentiable at the endpoint $x = 0$.

The three derivative formulas calculated in Example 2 are special cases of the following **General Power Rule**:

> If $f(x) = x^r$, then $f'(x) = r\,x^{r-1}$.

This formula, which we will verify in Section 3.3, is valid for *all values of r and x for which x^{r-1} makes sense as a real number.*

EXAMPLE 3 **(Differentiating powers)**

If $f(x) = x^{5/3}$, then $f'(x) = \frac{5}{3}x^{(5/3)-1} = \frac{5}{3}x^{2/3}$ for all real x.

If $g(t) = \frac{1}{\sqrt{t}} = t^{-1/2}$, then $g'(t) = -\frac{1}{2}t^{-(1/2)-1} = -\frac{1}{2}t^{-3/2}$ for $t > 0$.

Eventually, we will prove all appropriate cases of the General Power Rule. For the time being, here is a proof of the case $r = n$, a positive integer, based on the *factoring of a difference of nth powers*:

$$a^n - b^n = (a-b)(a^{n-1} + a^{n-2}b + a^{n-3}b^2 + \cdots + ab^{n-2} + b^{n-1}).$$

(Check that this formula is correct by multiplying the two factors on the right-hand side.) If $f(x) = x^n$, $a = x + h$, and $b = x$, then $a - b = h$ and

$$f'(x) = \lim_{h\to 0} \frac{(x+h)^n - x^n}{h} = \lim_{h\to 0} \frac{h\overbrace{[(x+h)^{n-1} + (x+h)^{n-2}x + (x+h)^{n-3}x^2 + \cdots + x^{n-1}]}^{n \text{ terms}}}{h} = nx^{n-1}.$$

An alternative proof based on the product rule and mathematical induction will be given in Section 2.3. The factorization method used above can also be used to demonstrate the General Power Rule for negative integers, $r = -n$, and reciprocals of integers, $r = 1/n$. (See Exercises 52 and 54 at the end of this section.)

EXAMPLE 4 **(Differentiating the absolute value function)** Verify that:

$$\text{If } f(x) = |x|, \quad \text{then} \quad f'(x) = \frac{x}{|x|} = \operatorname{sgn} x.$$

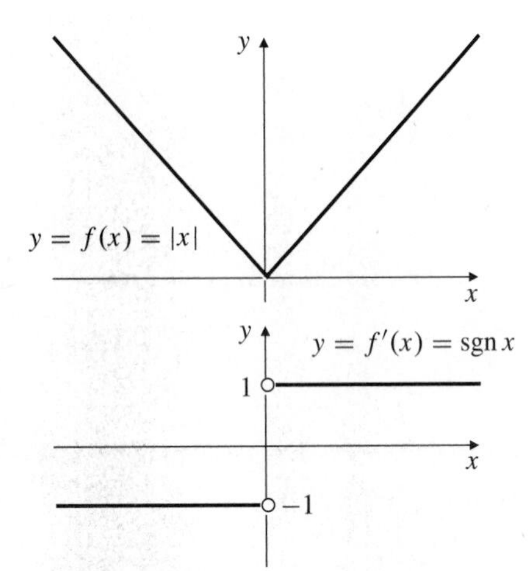

Figure 2.15 The derivative of $|x|$ is $\operatorname{sgn} x = x/|x|$

Solution We have

$$f(x) = \begin{cases} x, & \text{if } x \ge 0 \\ -x, & \text{if } x < 0 \end{cases}.$$

Thus, from Example 1 above, $f'(x) = 1$ if $x > 0$ and $f'(x) = -1$ if $x < 0$. Also, Example 4 of Section 2.1 shows that f is not differentiable at $x = 0$, which is a singular point of f. Therefore (see Figure 2.15),

$$f'(x) = \begin{cases} 1, & \text{if } x > 0 \\ -1, & \text{if } x < 0 \end{cases} = \frac{x}{|x|} = \operatorname{sgn} x.$$

Table 1 lists the elementary derivatives calculated above. Beginning in Section 2.3 we will develop general rules for calculating the derivatives of functions obtained by combining simpler functions. Thereafter, we will seldom have to revert to the definition of the derivative and to the calculation of limits to evaluate derivatives. It is important, therefore, to remember the derivatives of some elementary functions. Memorize those in Table 1.

Table 1. Some elementary functions and their derivatives

$f(x)$	$f'(x)$
c (constant)	0
x	1
x^2	$2x$
$\dfrac{1}{x}$	$-\dfrac{1}{x^2} \quad (x \ne 0)$
$\sqrt{x}$	$\dfrac{1}{2\sqrt{x}} \quad (x > 0)$
x^r	$r\,x^{r-1} \quad (x^{r-1} \text{ real})$
$\lvert x\rvert$	$\dfrac{x}{\lvert x\rvert} = \operatorname{sgn} x$

Leibniz Notation

Because functions can be written in different ways, it is useful to have more than one notation for derivatives. If $y = f(x)$, we can use the dependent variable y to represent the function, and we can denote the derivative of the function with respect to x in any of the following ways:

$$D_x y = y' = \frac{dy}{dx} = \frac{d}{dx} f(x) = f'(x) = D_x f(x) = Df(x).$$

(In the forms using "D_x," we can omit the subscript x if the variable of differentiation is obvious.) Often the most convenient way of referring to the derivative of a function given explicitly as an expression in the variable x is to write $\frac{d}{dx}$ in front of that expression. The symbol $\frac{d}{dx}$ is a *differential operator* and should be read "the derivative with respect to x of ..." For example,

$$\frac{d}{dx}x^2 = 2x \quad \text{(the derivative with respect to } x \text{ of } x^2 \text{ is } 2x\text{)}$$

$$\frac{d}{dx}\sqrt{x} = \frac{1}{2\sqrt{x}}$$

$$\frac{d}{dt}t^{100} = 100\,t^{99}$$

$$\text{if } y = u^3, \text{ then } \frac{dy}{du} = 3u^2.$$

Do not confuse the expressions

$$\frac{d}{dx}f(x) \text{ and } \frac{d}{dx}f(x)\bigg|_{x=x_0}.$$

The first expression represents a *function*, $f'(x)$. The second represents a *number*, $f'(x_0)$.

The value of the derivative of a function at a particular number x_0 in its domain can also be expressed in several ways:

$$D_x y\bigg|_{x=x_0} = y'\bigg|_{x=x_0} = \frac{dy}{dx}\bigg|_{x=x_0} = \frac{d}{dx}f(x)\bigg|_{x=x_0} = f'(x_0) = D_x f(x_0).$$

The symbol $\bigg|_{x=x_0}$ is called an **evaluation symbol**. It signifies that the expression preceding it should be evaluated at $x = x_0$. Thus,

$$\frac{d}{dx}x^4\bigg|_{x=-1} = 4x^3\bigg|_{x=-1} = 4(-1)^3 = -4.$$

Here is another example in which a derivative is computed from the definition, this time for a somewhat more complicated function.

EXAMPLE 5 Use the definition of derivative to calculate $\frac{d}{dx}\left(\frac{x}{x^2+1}\right)\bigg|_{x=2}$.

Solution We could calculate $\frac{d}{dx}\left(\frac{x}{x^2+1}\right)$ and then substitute $x = 2$, but it is easier to put $x = 2$ in the expression for the Newton quotient before taking the limit:

$$\begin{aligned}
\frac{d}{dx}\left(\frac{x}{x^2+1}\right)\bigg|_{x=2} &= \lim_{h\to 0}\frac{\dfrac{2+h}{(2+h)^2+1} - \dfrac{2}{2^2+1}}{h} \\
&= \lim_{h\to 0}\frac{\dfrac{2+h}{5+4h+h^2} - \dfrac{2}{5}}{h} \\
&= \lim_{h\to 0}\frac{5(2+h) - 2(5+4h+h^2)}{5(5+4h+h^2)h} \\
&= \lim_{h\to 0}\frac{-3h-2h^2}{5(5+4h+h^2)h} \\
&= \lim_{h\to 0}\frac{-3-2h}{5(5+4h+h^2)} = -\frac{3}{25}.
\end{aligned}$$

The notations dy/dx and $\frac{d}{dx}f(x)$ are called **Leibniz notations** for the derivative, after Gottfried Wilhelm Leibniz (1646–1716), one of the creators of calculus, who used such notations. The main ideas of calculus were developed independently by Leibniz and Isaac Newton (1642–1727); Newton used notations similar to the prime (y') notations we use here.

The Leibniz notation is suggested by the definition of derivative. The Newton quotient $[f(x+h)-f(x)]/h$, whose limit we take to find the derivative dy/dx, can be written in the form $\Delta y/\Delta x$, where $\Delta y = f(x+h)-f(x)$ is the increment in y, and $\Delta x = (x+h)-x = h$ is the corresponding increment in x as we pass from the point $(x, f(x))$ to the point $(x+h, f(x+h))$ on the graph of f. (See Figure 2.16.) Δ is the uppercase Greek letter Delta. Using symbols:

$$\frac{dy}{dx} = \lim_{\Delta x \to 0} \frac{\Delta y}{\Delta x}.$$

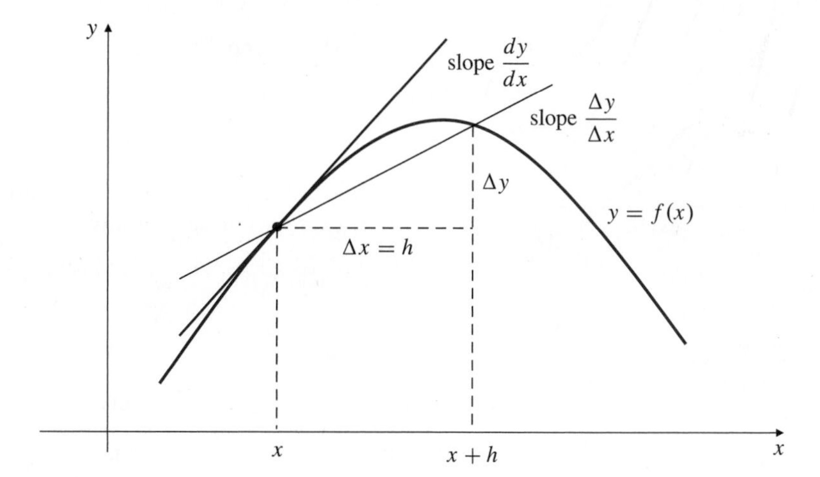

Figure 2.16 $\frac{dy}{dx} = \lim_{\Delta x \to 0} \frac{\Delta y}{\Delta x}$

Differentials

The Newton quotient $\Delta y/\Delta x$ is actually the quotient of two quantities, Δy and Δx. It is not at all clear, however, that the derivative dy/dx, the limit of $\Delta y/\Delta x$ as Δx approaches zero, can be regarded as a quotient. If y is a continuous function of x, then Δy approaches zero when Δx approaches zero, so dy/dx appears to be the meaningless quantity $0/0$. Nevertheless, it is sometimes useful to be able to refer to quantities dy and dx in such a way that their quotient is the derivative dy/dx. We can justify this by regarding dx as a new *independent* variable (called **the differential of x**) and defining a new *dependent* variable dy (**the differential of y**) as a function of x and dx by

$$dy = \frac{dy}{dx}\,dx = f'(x)\,dx.$$

For example, if $y = x^2$, we can write $dy = 2x\,dx$ to mean the same thing as $dy/dx = 2x$. Similarly, if $f(x) = 1/x$, we can write $df(x) = -(1/x^2)\,dx$ as the equivalent differential form of the assertion that $(d/dx)f(x) = f'(x) = -1/x^2$. This *differential notation* is useful in applications (see Sections 2.7 and 12.6), and especially for the interpretation and manipulation of integrals beginning in Chapter 5.

Note that, defined as above, differentials are merely variables that may or may not be small in absolute value. The differentials dy and dx were originally regarded (by Leibniz and his successors) as "infinitesimals" (infinitely small but nonzero) quantities whose quotient dy/dx gave the slope of the tangent line (a secant line meeting the graph of $y = f(x)$ at two points infinitely close together). It can be shown that such "infinitesimal" quantities cannot exist (as real numbers). It is possible to extend the number system to contain infinitesimals and use these to develop calculus, but we will not consider this approach here.

Derivatives Have the Intermediate-Value Property

Is a function f defined on an interval I necessarily the derivative of some other function defined on I? The answer is no; some functions are derivatives and some are not. Although a derivative need not be a continuous function (see Exercise 28 in Section 2.8), it must, like a continuous function, have the intermediate-value property: on an interval $[a, b]$, a derivative $f'(x)$ takes on every value between $f'(a)$ and $f'(b)$. (See Exercise 29 in Section 2.8 for a proof of this fact.) An everywhere-defined step function such as the Heaviside function $H(x)$ considered in Example 1 in Section 1.4 (see Figure 2.17) does not have this property on, say, the interval $[-1, 1]$, so cannot be the derivative of a function on that interval. This argument does not apply to the signum function, which is the derivative of the absolute value function on any interval where it is defined. (See Example 4.) Such an interval cannot contain the origin as $\operatorname{sgn}(x)$ is not defined at $x = 0$.

If $g(x)$ is continuous on an interval I, then $g(x) = f'(x)$ for some function f that is differentiable on I. We will discuss this fact further in Chapter 5 and prove it in Appendix IV.

Figure 2.17 This function is not a derivative on $[-1, 1]$; it does not have the intermediate-value property.

EXERCISES 2.2

Make rough sketches of the graphs of the derivatives of the functions in Exercises 1–4.

1. The function f graphed in Figure 2.18(a).
2. The function g graphed in Figure 2.18(b).
3. The function h graphed in Figure 2.18(c).
4. The function k graphed in Figure 2.18(d).
5. Where is the function f graphed in Figure 2.18(a) differentiable?
6. Where is the function g graphed in Figure 2.18(b) differentiable?

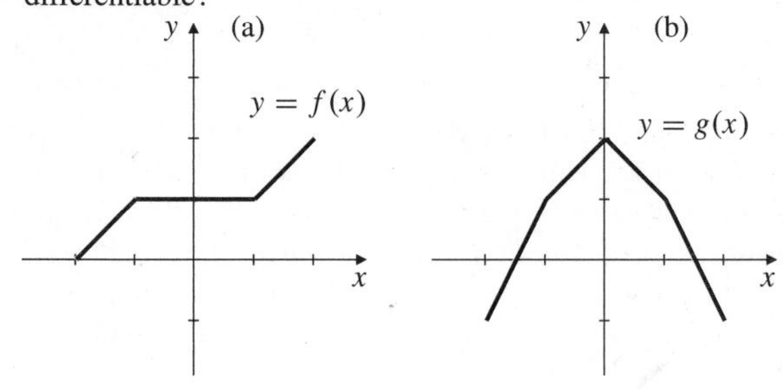

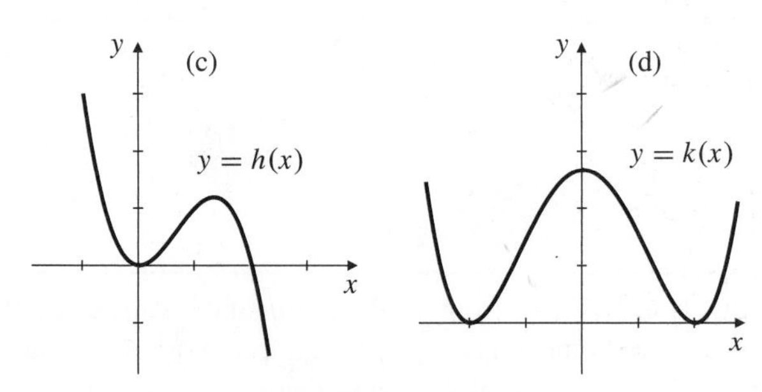

Figure 2.18

Use a graphics utility with differentiation capabilities to plot the graphs of the following functions and their derivatives. Observe the relationships between the graph of y and that of y' in each case. What features of the graph of y can you infer from the graph of y'?

7. $y = 3x - x^2 - 1$
8. $y = x^3 - 3x^2 + 2x + 1$
9. $y = |x^3 - x|$
10. $y = |x^2 - 1| - |x^2 - 4|$

In Exercises 11–24, (a) calculate the derivative of the given function directly from the definition of derivative, and (b) express the result of (a) using differentials.

11. $y = x^2 - 3x$
12. $f(x) = 1 + 4x - 5x^2$
13. $f(x) = x^3$
14. $s = \dfrac{1}{3 + 4t}$
15. $g(x) = \dfrac{2 - x}{2 + x}$
16. $y = \dfrac{1}{3}x^3 - x$
17. $F(t) = \sqrt{2t + 1}$
18. $f(x) = \dfrac{3}{4}\sqrt{2 - x}$
19. $y = x + \dfrac{1}{x}$
20. $z = \dfrac{s}{1 + s}$
21. $F(x) = \dfrac{1}{\sqrt{1 + x^2}}$
22. $y = \dfrac{1}{x^2}$
23. $y = \dfrac{1}{\sqrt{1 + x}}$
24. $f(t) = \dfrac{t^2 - 3}{t^2 + 3}$
25. How should the function $f(x) = x \operatorname{sgn} x$ be defined at $x = 0$ so that it is continuous there? Is it then differentiable there?
26. How should the function $g(x) = x^2 \operatorname{sgn} x$ be defined at $x = 0$ so that it is continuous there? Is it then differentiable there?
27. Where does $h(x) = |x^2 + 3x + 2|$ fail to be differentiable?
28. Using a calculator, find the slope of the secant line to $y = x^3 - 2x$ passing through the points corresponding to $x = 1$ and $x = 1 + \Delta x$, for several values of Δx of decreasing size, say $\Delta x = \pm 0.1, \pm 0.01, \pm 0.001, \pm 0.0001$. (Make a table.) Also, calculate $\dfrac{d}{dx}\left(x^3 - 2x\right)\Big|_{x=1}$ using the definition of derivative.
29. Repeat Exercise 28 for the function $f(x) = \dfrac{1}{x}$ and the points $x = 2$ and $x = 2 + \Delta x$.

Using the definition of derivative, find equations for the tangent lines to the curves in Exercises 30–33 at the points indicated.

30. $y = 5 + 4x - x^2$ at the point where $x = 2$
31. $y = \sqrt{x + 6}$ at the point $(3, 3)$
32. $y = \dfrac{t}{t^2 - 2}$ at the point where $t = -2$

33. $y = \dfrac{2}{t^2 + t}$ at the point where $t = a$

Calculate the derivatives of the functions in Exercises 34–39 using the General Power Rule. Where is each derivative valid?

34. $f(x) = x^{-17}$

35. $g(t) = t^{22}$

36. $y = x^{1/3}$

37. $y = x^{-1/3}$

38. $t^{-2.25}$

39. $s^{119/4}$

In Exercises 40–50, you may use the formulas for derivatives established in this section.

40. Calculate $\left.\dfrac{d}{ds}\sqrt{s}\right|_{s=9}$.

41. Find $F'(\frac{1}{4})$ if $F(x) = \dfrac{1}{x}$.

42. Find $f'(8)$ if $f(x) = x^{-2/3}$.

43. Find $\left.dy/dt\right|_{t=4}$ if $y = t^{1/4}$.

44. Find an equation of the straight line tangent to the curve $y = \sqrt{x}$ at $x = x_0$.

45. Find an equation of the straight line normal to the curve $y = 1/x$ at the point where $x = a$.

46. Show that the curve $y = x^2$ and the straight line $x + 4y = 18$ intersect at right angles at one of their two intersection points. *Hint:* Find the product of their slopes at their intersection points.

47. There are two distinct straight lines that pass through the point $(1, -3)$ and are tangent to the curve $y = x^2$. Find their equations. *Hint:* Draw a sketch. The points of tangency are not given; let them be denoted (a, a^2).

48. Find equations of two straight lines that have slope -2 and are tangent to the graph of $y = 1/x$.

49. Find the slope of a straight line that passes through the point $(-2, 0)$ and is tangent to the curve $y = \sqrt{x}$.

50. Show that there are two distinct tangent lines to the curve $y = x^2$ passing through the point (a, b) provided $b < a^2$. How many tangent lines to $y = x^2$ pass through (a, b) if $b = a^2$? if $b > a^2$?

51. Show that the derivative of an odd differentiable function is even and that the derivative of an even differentiable function is odd.

52. Prove the case $r = -n$ (n is a positive integer) of the General Power Rule; that is, prove that

$$\frac{d}{dx}x^{-n} = -n\,x^{-n-1}.$$

Use the factorization of a difference of nth powers given in this section.

53. Use the factoring of a difference of cubes:

$$a^3 - b^3 = (a - b)(a^2 + ab + b^2),$$

to help you calculate the derivative of $f(x) = x^{1/3}$ directly from the definition of derivative.

54. Prove the General Power Rule for $\frac{d}{dx}x^r$, where $r = 1/n$, n being a positive integer. (*Hint:*

$$\begin{aligned}\frac{d}{dx}x^{1/n} &= \lim_{h\to 0}\frac{(x+h)^{1/n} - x^{1/n}}{h}\\ &= \lim_{h\to 0}\frac{(x+h)^{1/n} - x^{1/n}}{((x+h)^{1/n})^n - (x^{1/n})^n}.\end{aligned}$$

Apply the factorization of the difference of nth powers to the denominator of the latter quotient.)

55. Give a proof of the power rule $\frac{d}{dx}x^n = nx^{n-1}$ for positive integers n using the Binomial Theorem:

$$\begin{aligned}(x+h)^n &= x^n + \frac{n}{1}x^{n-1}h + \frac{n(n-1)}{1\times 2}x^{n-2}h^2\\ &\quad + \frac{n(n-1)(n-2)}{1\times 2\times 3}x^{n-3}h^3 + \cdots + h^n.\end{aligned}$$

56. Use right and left derivatives, $f'_+(a)$ and $f'_-(a)$, to define the concept of a half-line starting at $(a, f(a))$ being a right or left tangent to the graph of f at $x = a$. Show that the graph has a tangent line at $x = a$ if and only if it has right and left tangents that are opposite halves of the same straight line. What are the left and right tangents to the graphs of $y = x^{1/3}$, $y = x^{2/3}$, and $y = |x|$ at $x = 0$?

2.3 Differentiation Rules

If every derivative had to be calculated directly from the definition of derivative as in the examples of Section 2.2, calculus would indeed be a painful subject. Fortunately, there is an easier way. We will develop several general *differentiation rules* that enable us to calculate the derivatives of complicated combinations of functions easily if we already know the derivatives of the elementary functions from which they are constructed. For instance, we will be able to find the derivative of $\dfrac{x^2}{\sqrt{x^2+1}}$ if we know the derivatives of x^2 and $\sqrt{x}$. The rules we develop in this section tell us how to differentiate sums, constant multiples, products, and quotients of functions whose derivatives we already know. In Section 2.4 we will learn how to differentiate composite functions.

Before developing these differentiation rules we need to establish one obvious but very important theorem which states, roughly, that the graph of a function cannot possibly have a break at a point where it is smooth.

THEOREM 1

Differentiability implies continuity

If f is differentiable at x, then f is continuous at x.

PROOF Since f is differentiable at x, we know that

$$\lim_{h\to 0} \frac{f(x+h)-f(x)}{h} = f'(x)$$

exists. Using the limit rules (Theorem 2 of Section 1.2), we have

$$\lim_{h\to 0}\big(f(x+h)-f(x)\big) = \lim_{h\to 0}\left(\frac{f(x+h)-f(x)}{h}\right)(h) = \big(f'(x)\big)(0) = 0.$$

This is equivalent to $\lim_{h\to 0} f(x+h) = f(x)$, which says that f is continuous at x. ∎

Sums and Constant Multiples

The derivative of a sum (or difference) of functions is the sum (or difference) of the derivatives of those functions. The derivative of a constant multiple of a function is the same constant multiple of the derivative of the function.

THEOREM 2

Differentiation rules for sums, differences, and constant multiples

If functions f and g are differentiable at x, and if C is a constant, then the functions $f+g$, $f-g$, and Cf are all differentiable at x and

$$(f+g)'(x) = f'(x) + g'(x),$$
$$(f-g)'(x) = f'(x) - g'(x),$$
$$(Cf)'(x) = Cf'(x).$$

PROOF The proofs of all three assertions are straightforward, using the corresponding limit rules from Theorem 2 of Section 1.2. For the sum, we have

$$\begin{aligned}(f+g)'(x) &= \lim_{h\to 0}\frac{(f+g)(x+h)-(f+g)(x)}{h}\\ &= \lim_{h\to 0}\frac{\big(f(x+h)+g(x+h)\big)-\big(f(x)+g(x)\big)}{h}\\ &= \lim_{h\to 0}\left(\frac{f(x+h)-f(x)}{h}+\frac{g(x+h)-g(x)}{h}\right)\\ &= f'(x)+g'(x),\end{aligned}$$

because the limit of a sum is the sum of the limits. The proof for the difference $f-g$ is similar. For the constant multiple, we have

$$\begin{aligned}(Cf)'(x) &= \lim_{h\to 0}\frac{Cf(x+h)-Cf(x)}{h}\\ &= C\lim_{h\to 0}\frac{f(x+h)-f(x)}{h} = Cf'(x).\end{aligned}$$

∎

The rule for differentiating sums extends to sums of any finite number of terms:

$$(f_1+f_2+\cdots+f_n)' = f_1'+f_2'+\cdots+f_n'. \qquad (*)$$

To see this we can use a technique called **mathematical induction**. (See the note in the margin.) Theorem 2 shows that the case $n = 2$ is true; this is STEP 1. For STEP 2, we must show that *if* the formula (∗) holds for some integer $n = k \geq 2$, *then* it must also hold for $n = k + 1$. Therefore, *assume* that

$$(f_1 + f_2 + \cdots + f_k)' = f_1' + f_2' + \cdots + f_k'.$$

Then we have

$$\begin{aligned}(f_1 + f_2 + \cdots + f_k + f_{k+1})'\\ &= \big(\underbrace{(f_1 + f_2 + \cdots + f_k)}_{\text{Let this function be } f} + f_{k+1}\big)'\\ &= (f + f_{k+1})' \qquad \text{(Now use the known case } n = 2.)\\ &= f' + f_{k+1}'\\ &= f_1' + f_2' + \cdots + f_k' + f_{k+1}'.\end{aligned}$$

With both steps verified, we can claim that (∗) holds for any $n \geq 2$ *by induction*. In particular, therefore, the derivative of any polynomial is the sum of the derivatives of its terms.

Mathematical Induction

Mathematical induction is a technique for proving that a statement about an integer n is true for every integer n greater than or equal to some starting integer n_0. The proof requires us to carry out two steps:

STEP 1. Prove that the statement is true for $n = n_0$.

STEP 2. Prove that if the statement is true for some integer $n = k$, where $k \geq n_0$, then it is also true for the next larger integer, $n = k + 1$.

Step 2 prevents there from being a smallest integer greater than n_0 for which the statement is false. Being true for n_0, the statement must therefore be true for all larger integers.

EXAMPLE 1 Calculate the derivatives of the functions:

(a) $2x^3 - 5x^2 + 4x + 7$, (b) $f(x) = 5\sqrt{x} + \dfrac{3}{x} - 18$, (c) $y = \dfrac{1}{7}t^4 - 3t^{7/3}$.

Solution Each of these functions is a sum of constant multiples of functions that we already know how to differentiate.

(a) $\dfrac{d}{dx}(2x^3 - 5x^2 + 4x + 7) = 2(3x^2) - 5(2x) + 4(1) + 0 = 6x^2 - 10x + 4.$

(b) $f'(x) = 5\left(\dfrac{1}{2\sqrt{x}}\right) + 3\left(-\dfrac{1}{x^2}\right) - 0 = \dfrac{5}{2\sqrt{x}} - \dfrac{3}{x^2}.$

(c) $\dfrac{dy}{dt} = \dfrac{1}{7}(4t^3) - 3\left(\dfrac{7}{3}t^{4/3}\right) = \dfrac{4}{7}t^3 - 7t^{4/3}.$

EXAMPLE 2 Find an equation of the tangent to the curve $y = \dfrac{3x^3 - 4}{x}$ at the point on the curve where $x = -2$.

Solution If $x = -2$, then $y = 14$. The slope of the curve at $(-2, 14)$ is

$$\left.\frac{dy}{dx}\right|_{x=-2} = \left.\frac{d}{dx}\left(3x^2 - \frac{4}{x}\right)\right|_{x=-2} = \left.\left(6x + \frac{4}{x^2}\right)\right|_{x=-2} = -11.$$

An equation of the tangent line is $y = 14 - 11(x + 2)$, or $y = -11x - 8$.

The Product Rule

The rule for differentiating a product of functions is a little more complicated than that for sums. It is *not* true that the derivative of a product is the product of the derivatives.

THEOREM 3 **The Product Rule**

If functions f and g are differentiable at x, then their product fg is also differentiable at x, and

$$(fg)'(x) = f'(x)g(x) + f(x)g'(x).$$

PROOF We set up the Newton quotient for fg and then add 0 to the numerator in a way that enables us to involve the Newton quotients for f and g separately:

$$\begin{aligned}(fg)'(x) &= \lim_{h\to 0}\frac{f(x+h)g(x+h)-f(x)g(x)}{h}\\ &= \lim_{h\to 0}\frac{f(x+h)g(x+h)-f(x)g(x+h)+f(x)g(x+h)-f(x)g(x)}{h}\\ &= \lim_{h\to 0}\left(\frac{f(x+h)-f(x)}{h}g(x+h)+f(x)\frac{g(x+h)-g(x)}{h}\right)\\ &= f'(x)g(x)+f(x)g'(x).\end{aligned}$$

To get the last line, we have used the fact that f and g are differentiable and the fact that g is therefore continuous (Theorem 1), as well as limit rules from Theorem 2 of Section 1.2. A graphical proof of the Product Rule is suggested by Figure 2.19.

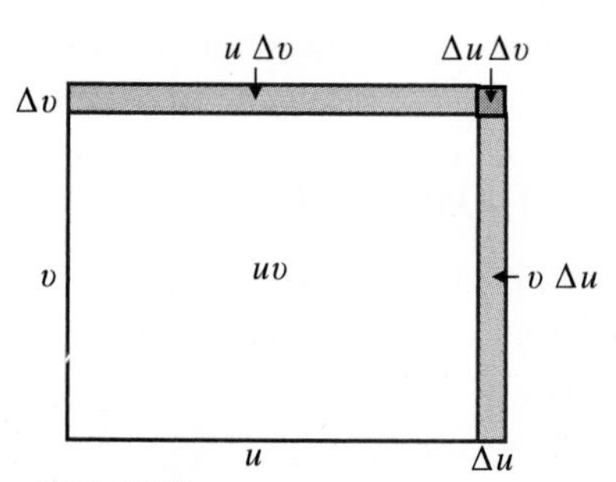

Figure 2.19

A graphical proof of the Product Rule

Here $u = f(x)$ and $v = g(x)$, so that the rectangular area uv represents $f(x)g(x)$. If x changes by an amount Δx, the corresponding increments in u and v are Δu and Δv. The change in the area of the rectangle is

$$\begin{aligned}&\Delta(uv)\\ &= (u+\Delta u)(v+\Delta v)-uv\\ &= (\Delta u)v+u(\Delta v)+(\Delta u)(\Delta v),\end{aligned}$$

the sum of the three shaded areas. Dividing by Δx and taking the limit as $\Delta x \to 0$, we get

$$\frac{d}{dx}(uv)=\left(\frac{du}{dx}\right)v+u\left(\frac{dv}{dx}\right),$$

since

$$\lim_{\Delta x\to 0}\frac{\Delta u}{\Delta x}\Delta v=\frac{du}{dx}\times 0=0.$$

EXAMPLE 3 Find the derivative of $(x^2+1)(x^3+4)$ using and without using the Product Rule.

Solution Using the Product Rule with $f(x) = x^2+1$ and $g(x) = x^3+4$, we calculate

$$\frac{d}{dx}\big((x^2+1)(x^3+4)\big)=2x(x^3+4)+(x^2+1)(3x^2)=5x^4+3x^2+8x.$$

On the other hand, we can calculate the derivative by first multiplying the two binomials and then differentiating the resulting polynomial:

$$\frac{d}{dx}\big((x^2+1)(x^3+4)\big)=\frac{d}{dx}(x^5+x^3+4x^2+4)=5x^4+3x^2+8x.$$

EXAMPLE 4 Find $\dfrac{dy}{dx}$ if $y=\left(2\sqrt{x}+\dfrac{3}{x}\right)\left(3\sqrt{x}-\dfrac{2}{x}\right)$.

Solution Applying the Product Rule with f and g being the two functions enclosed in the large parentheses, we obtain

$$\begin{aligned}\frac{dy}{dx}&=\left(\frac{1}{\sqrt{x}}-\frac{3}{x^2}\right)\left(3\sqrt{x}-\frac{2}{x}\right)+\left(2\sqrt{x}+\frac{3}{x}\right)\left(\frac{3}{2\sqrt{x}}+\frac{2}{x^2}\right)\\ &=6-\frac{5}{2x^{3/2}}+\frac{12}{x^3}.\end{aligned}$$

EXAMPLE 5 Let $y = uv$ be the product of the functions u and v. Find $y'(2)$ if $u(2) = 2$, $u'(2) = -5$, $v(2) = 1$, and $v'(2) = 3$.

Solution From the Product Rule we have

$$y'=(uv)'=u'v+uv'.$$

Therefore,

$$y'(2)=u'(2)v(2)+u(2)v'(2)=(-5)(1)+(2)(3)=-5+6=1.$$

EXAMPLE 6 Use mathematical induction to verify the formula $\dfrac{d}{dx}x^n=n\,x^{n-1}$ for all positive integers n.

Solution For $n = 1$ the formula says that $\frac{d}{dx}x^1 = 1 = 1x^0$, so the formula is true in this case. We must show that if the formula is true for $n = k \geq 1$, then it is also true for $n = k + 1$. Therefore, assume that

$$\frac{d}{dx}x^k = kx^{k-1}.$$

Using the Product Rule we calculate

$$\frac{d}{dx}x^{k+1} = \frac{d}{dx}(x^k x) = (kx^{k-1})(x) + (x^k)(1) = (k+1)x^k = (k+1)x^{(k+1)-1}.$$

Thus, the formula is true for $n = k + 1$ also. The formula is true for all integers $n \geq 1$ *by induction.*

The Product Rule can be extended to products of any number of factors; for instance,

$$\begin{aligned}(fgh)'(x) &= f'(x)(gh)(x) + f(x)(gh)'(x)\\ &= f'(x)g(x)h(x) + f(x)g'(x)h(x) + f(x)g(x)h'(x).\end{aligned}$$

In general, the derivative of a product of n functions will have n terms; each term will be the same product but with one of the factors replaced by its derivative:

$$(f_1 f_2 f_3 \cdots f_n)' = f_1' f_2 f_3 \cdots f_n + f_1 f_2' f_3 \cdots f_n + \cdots + f_1 f_2 f_3 \cdots f_n'.$$

This can be proved by mathematical induction. See Exercise 54 at the end of this section.

The Reciprocal Rule

THEOREM 4

The Reciprocal Rule

If f is differentiable at x and $f(x) \neq 0$, then $1/f$ is differentiable at x, and

$$\left(\frac{1}{f}\right)'(x) = \frac{-f'(x)}{(f(x))^2}.$$

PROOF Using the definition of the derivative, we calculate

$$\begin{aligned}\frac{d}{dx}\frac{1}{f(x)} &= \lim_{h\to 0}\frac{\dfrac{1}{f(x+h)} - \dfrac{1}{f(x)}}{h}\\ &= \lim_{h\to 0}\frac{f(x) - f(x+h)}{hf(x+h)f(x)}\\ &= \lim_{h\to 0}\left(\frac{-1}{f(x+h)f(x)}\right)\frac{f(x+h) - f(x)}{h}\\ &= \frac{-1}{(f(x))^2}f'(x).\end{aligned}$$

Again we have to use the continuity of f (from Theorem 1) and the limit rules from Section 1.2.

EXAMPLE 7 Differentiate the functions

(a) $\dfrac{1}{x^2+1}$ and (b) $f(t) = \dfrac{1}{t + \dfrac{1}{t}}$.

Solution Using the Reciprocal Rule:

(a) $$\frac{d}{dx}\left(\frac{1}{x^2+1}\right) = \frac{-2x}{(x^2+1)^2}.$$

(b) $$f'(t) = \frac{-1}{\left(t+\frac{1}{t}\right)^2}\left(1-\frac{1}{t^2}\right) = \frac{-t^2}{(t^2+1)^2}\,\frac{t^2-1}{t^2} = \frac{1-t^2}{(t^2+1)^2}.$$

We can use the Reciprocal Rule to confirm the General Power Rule for negative integers:

$$\frac{d}{dx}x^{-n} = -n\,x^{-n-1},$$

since we have already proved the rule for positive integers. We have

$$\frac{d}{dx}x^{-n} = \frac{d}{dx}\frac{1}{x^n} = \frac{-n\,x^{n-1}}{(x^n)^2} = -n\,x^{-n-1}.$$

EXAMPLE 8 **(Differentiating sums of reciprocals)**

$$\begin{aligned}\frac{d}{dx}\left(\frac{x^2+x+1}{x^3}\right) &= \frac{d}{dx}\left(\frac{1}{x}+\frac{1}{x^2}+\frac{1}{x^3}\right)\\ &= \frac{d}{dx}(x^{-1}+x^{-2}+x^{-3})\\ &= -x^{-2}-2x^{-3}-3x^{-4} = -\frac{1}{x^2}-\frac{2}{x^3}-\frac{3}{x^4}.\end{aligned}$$

The Quotient Rule

The Product Rule and the Reciprocal Rule can be combined to provide a rule for differentiating a quotient of two functions. Observe that

$$\begin{aligned}\frac{d}{dx}\left(\frac{f(x)}{g(x)}\right) = \frac{d}{dx}\left(f(x)\frac{1}{g(x)}\right) &= f'(x)\frac{1}{g(x)} + f(x)\left(-\frac{g'(x)}{(g(x))^2}\right)\\ &= \frac{g(x)f'(x)-f(x)g'(x)}{(g(x))^2}.\end{aligned}$$

Thus we have proved the following Quotient Rule.

THEOREM 5

The Quotient Rule

If f and g are differentiable at x, and if $g(x) \neq 0$, then the quotient f/g is differentiable at x and

$$\left(\frac{f}{g}\right)'(x) = \frac{g(x)f'(x)-f(x)g'(x)}{(g(x))^2}.$$

Sometimes students have trouble remembering this rule. (Getting the order of the terms in the numerator wrong will reverse the sign.) Try to remember (and use) the Quotient Rule in the following form:

$$\begin{aligned}&(\text{quotient})'\\ &= \frac{(\text{denominator})\times(\text{numerator})' - (\text{numerator})\times(\text{denominator})'}{(\text{denominator})^2}\end{aligned}$$

EXAMPLE 9 Find the derivatives of

$$\text{(a) } y = \frac{1-x^2}{1+x^2}, \quad \text{(b) } \frac{\sqrt{t}}{3-5t}, \quad \text{and} \quad \text{(c) } f(\theta) = \frac{a+b\theta}{m+n\theta}.$$

Solution We use the Quotient Rule in each case.

$$\text{(a) } \frac{dy}{dx} = \frac{(1+x^2)(-2x)-(1-x^2)(2x)}{(1+x^2)^2} = -\frac{4x}{(1+x^2)^2}.$$

$$\text{(b) } \frac{d}{dt}\left(\frac{\sqrt{t}}{3-5t}\right) = \frac{(3-5t)\dfrac{1}{2\sqrt{t}} - \sqrt{t}(-5)}{(3-5t)^2} = \frac{3+5t}{2\sqrt{t}(3-5t)^2}.$$

$$\text{(c) } f'(\theta) = \frac{(m+n\theta)(b)-(a+b\theta)(n)}{(m+n\theta)^2} = \frac{mb-na}{(m+n\theta)^2}.$$

In all three parts of Example 9, the Quotient Rule yielded fractions with numerators that were complicated but could be simplified algebraically. It is advisable to attempt such simplifications when calculating derivatives; the usefulness of derivatives in applications of calculus often depends on such simplifications.

EXAMPLE 10 Find equations of any lines that pass through the point $(-1, 0)$ and are tangent to the curve $y = (x-1)/(x+1)$.

Solution The point $(-1, 0)$ does not lie on the curve, so it is not the point of tangency. Suppose a line is tangent to the curve at $x = a$, so the point of tangency is $(a, (a-1)/(a+1))$. Note that a cannot be -1. The slope of the line must be

$$\left.\frac{dy}{dx}\right|_{x=a} = \left.\frac{(x+1)(1)-(x-1)(1)}{(x+1)^2}\right|_{x=a} = \frac{2}{(a+1)^2}.$$

If the line also passes through $(-1, 0)$, its slope must also be given by

$$\frac{\dfrac{a-1}{a+1} - 0}{a-(-1)} = \frac{a-1}{(a+1)^2}.$$

Equating these two expressions for the slope, we get an equation to solve for a:

$$\frac{a-1}{(a+1)^2} = \frac{2}{(a+1)^2} \quad \Longrightarrow \quad a-1 = 2.$$

Thus $a = 3$, and the slope of the line is $2/4^2 = 1/8$. There is only one line through $(-1, 0)$ tangent to the given curve, and its equation is

$$y = 0 + \frac{1}{8}(x+1) \quad \text{or} \quad x - 8y + 1 = 0.$$

Remark Derivatives of quotients of functions where the denominator is a monomial, such as in Example 8, are usually easier to do by breaking the quotient into a sum of several fractions (as was done in that example) rather than by using the Quotient Rule.

EXERCISES 2.3

In Exercises 1–32, calculate the derivatives of the given functions. Simplify your answers whenever possible.

1. $y = 3x^2 - 5x - 7$

2. $y = 4x^{1/2} - \dfrac{5}{x}$

3. $f(x) = Ax^2 + Bx + C$

4. $f(x) = \dfrac{6}{x^3} + \dfrac{2}{x^2} - 2$

5. $z = \dfrac{s^5 - s^3}{15}$

6. $y = x^{45} - x^{-45}$

7. $g(t) = t^{1/3} + 2t^{1/4} + 3t^{1/5}$

8. $y = 3\sqrt[3]{t^2} - \dfrac{2}{\sqrt{t^3}}$

9. $u = \dfrac{3}{5}x^{5/3} - \dfrac{5}{3}x^{-3/5}$

10. $F(x) = (3x - 2)(1 - 5x)$

11. $y = \sqrt{x}\left(5 - x - \dfrac{x^2}{3}\right)$

12. $g(t) = \dfrac{1}{2t - 3}$

13. $y = \dfrac{1}{x^2 + 5x}$

14. $y = \dfrac{4}{3 - x}$

15. $f(t) = \dfrac{\pi}{2 - \pi t}$

16. $g(y) = \dfrac{2}{1 - y^2}$

17. $f(x) = \dfrac{1 - 4x^2}{x^3}$

18. $g(u) = \dfrac{u\sqrt{u} - 3}{u^2}$

19. $y = \dfrac{2 + t + t^2}{\sqrt{t}}$

20. $z = \dfrac{x - 1}{x^{2/3}}$

21. $f(x) = \dfrac{3 - 4x}{3 + 4x}$

22. $z = \dfrac{t^2 + 2t}{t^2 - 1}$

23. $s = \dfrac{1 + \sqrt{t}}{1 - \sqrt{t}}$

24. $f(x) = \dfrac{x^3 - 4}{x + 1}$

25. $f(x) = \dfrac{ax + b}{cx + d}$

26. $F(t) = \dfrac{t^2 + 7t - 8}{t^2 - t + 1}$

27. $f(x) = (1 + x)(1 + 2x)(1 + 3x)(1 + 4x)$

28. $f(r) = (r^{-2} + r^{-3} - 4)(r^2 + r^3 + 1)$

29. $y = (x^2 + 4)(\sqrt{x} + 1)(5x^{2/3} - 2)$

30. $y = \dfrac{(x^2 + 1)(x^3 + 2)}{(x^2 + 2)(x^3 + 1)}$

31. $y = \dfrac{x}{2x + \dfrac{1}{3x + 1}}$

32. $f(x) = \dfrac{(\sqrt{x} - 1)(2 - x)(1 - x^2)}{\sqrt{x}(3 + 2x)}$

Calculate the derivatives in Exercises 33–36, given that $f(2) = 2$ and $f'(2) = 3$.

33. $\dfrac{d}{dx}\left(\dfrac{x^2}{f(x)}\right)\bigg|_{x=2}$

34. $\dfrac{d}{dx}\left(\dfrac{f(x)}{x^2}\right)\bigg|_{x=2}$

35. $\dfrac{d}{dx}\left(x^2 f(x)\right)\bigg|_{x=2}$

36. $\dfrac{d}{dx}\left(\dfrac{f(x)}{x^2 + f(x)}\right)\bigg|_{x=2}$

37. Find $\dfrac{d}{dx}\left(\dfrac{x^2 - 4}{x^2 + 4}\right)\bigg|_{x=-2}$.

38. Find $\dfrac{d}{dt}\left(\dfrac{t(1 + \sqrt{t})}{5 - t}\right)\bigg|_{t=4}$.

39. If $f(x) = \dfrac{\sqrt{x}}{x + 1}$, find $f'(2)$.

40. Find $\dfrac{d}{dt}\Big((1 + t)(1 + 2t)(1 + 3t)(1 + 4t)\Big)\bigg|_{t=0}$.

41. Find an equation of the tangent line to $y = \dfrac{2}{3 - 4\sqrt{x}}$ at the point $(1, -2)$.

42. Find equations of the tangent and normal to $y = \dfrac{x + 1}{x - 1}$ at $x = 2$.

43. Find the points on the curve $y = x + 1/x$ where the tangent line is horizontal.

44. Find the equations of all horizontal lines that are tangent to the curve $y = x^2(4 - x^2)$.

45. Find the coordinates of all points where the curve $y = \dfrac{1}{x^2 + x + 1}$ has a horizontal tangent line.

46. Find the coordinates of points on the curve $y = \dfrac{x + 1}{x + 2}$ where the tangent line is parallel to the line $y = 4x$.

47. Find the equation of the straight line that passes through the point $(0, b)$ and is tangent to the curve $y = 1/x$. Assume $b \neq 0$.

48. Show that the curve $y = x^2$ intersects the curve $y = 1/\sqrt{x}$ at right angles.

49. Find two straight lines that are tangent to $y = x^3$ and pass through the point $(2, 8)$.

50. Find two straight lines that are tangent to $y = x^2/(x - 1)$ and pass through the point $(2, 0)$.

51. **(A Square Root Rule)** Show that if f is differentiable at x and $f(x) > 0$, then

$$\frac{d}{dx}\sqrt{f(x)} = \frac{f'(x)}{2\sqrt{f(x)}}.$$

Use this Square Root Rule to find the derivative of $\sqrt{x^2 + 1}$.

52. Show that $f(x) = |x^3|$ is differentiable at every real number x, and find its derivative.

Mathematical Induction

53. Use mathematical induction to prove that $\dfrac{d}{dx}x^{n/2} = \dfrac{n}{2}x^{(n/2)-1}$ for every positive integer n. Then use the Reciprocal Rule to get the same result for negative integers n.

54. Use mathematical induction to prove the formula for the derivative of a product of n functions given earlier in this section.

2.4 The Chain Rule

Although we can differentiate $\sqrt{x}$ and x^2+1, we cannot yet differentiate $\sqrt{x^2+1}$. To do this, we need a rule that tells us how to differentiate *composites* of functions whose derivatives we already know. This rule is known as the Chain Rule and is the most often used of all the differentiation rules.

EXAMPLE 1 The function $\dfrac{1}{x^2-4}$ is the composite $f(g(x))$ of $f(u) = \dfrac{1}{u}$ and $g(x) = x^2 - 4$, which have derivatives

$$f'(u) = \frac{-1}{u^2} \qquad \text{and} \qquad g'(x) = 2x.$$

According to the Reciprocal Rule (which is a special case of the Chain Rule),

$$\begin{aligned}\frac{d}{dx} f(g(x)) &= \frac{d}{dx}\left(\frac{1}{x^2-4}\right) = \frac{-2x}{(x^2-4)^2} = \frac{-1}{(x^2-4)^2}(2x) \\ &= f'(g(x))g'(x).\end{aligned}$$

This example suggests that the derivative of a composite function $f(g(x))$ is the derivative of f evaluated at $g(x)$ multiplied by the derivative of g evaluated at x. This is the Chain Rule:

$$\frac{d}{dx} f(g(x)) = f'(g(x))\, g'(x).$$

THEOREM

The Chain Rule

If $f(u)$ is differentiable at $u = g(x)$, and $g(x)$ is differentiable at x, then the composite function $f \circ g(x) = f(g(x))$ is differentiable at x, and

$$(f \circ g)'(x) = f'(g(x))g'(x).$$

In terms of Leibniz notation, if $y = f(u)$ where $u = g(x)$, then $y = f(g(x))$ and:

at u, y is changing $\dfrac{dy}{du}$ times as fast as u is changing;

at x, u is changing $\dfrac{du}{dx}$ times as fast as x is changing.

Therefore, at x, $y = f(u) = f(g(x))$ is changing $\dfrac{dy}{du} \times \dfrac{du}{dx}$ times as fast as x is changing. That is,

$$\frac{dy}{dx} = \frac{dy}{du}\frac{du}{dx}, \qquad \text{where } \frac{dy}{du} \text{ is evaluated at } u = g(x).$$

It appears as though the symbol du cancels from the numerator and denominator, but this is not meaningful because dy/du was not defined as the quotient of two quantities, but rather as a single quantity, the derivative of y with respect to u.

We would like to prove Theorem 6 by writing

$$\frac{\Delta y}{\Delta x} = \frac{\Delta y}{\Delta u}\frac{\Delta u}{\Delta x}$$

and taking the limit as $\Delta x \to 0$. Such a proof is valid for most composite functions but not all. (See Exercise 46 at the end of this section.) A correct proof will be given later in this section, but first we do more examples to give a better idea of how the Chain Rule works.

EXAMPLE 2 Find the derivative of $y = \sqrt{x^2+1}$.

Solution Here $y = f(g(x))$, where $f(u) = \sqrt{u}$ and $g(x) = x^2 + 1$. Since the derivatives of f and g are

$$f'(u) = \frac{1}{2\sqrt{u}} \quad \text{and} \quad g'(x) = 2x,$$

the Chain Rule gives

$$\begin{aligned}\frac{dy}{dx} &= \frac{d}{dx} f(g(x)) = f'(g(x)) \cdot g'(x) \\ &= \frac{1}{2\sqrt{g(x)}} \cdot g'(x) = \frac{1}{2\sqrt{x^2+1}} \cdot (2x) = \frac{x}{\sqrt{x^2+1}}.\end{aligned}$$

Outside and Inside Functions

In the composite $f(g(x))$, the function f is "outside," and the function g is "inside." The Chain Rule says that the derivative of the composite is the derivative f' of the outside function evaluated at the inside function $g(x)$, multiplied by the derivative $g'(x)$ of the inside function:
$\frac{d}{dx} f(g(x)) = f'(g(x)) \times g'(x)$.

Usually, when applying the Chain Rule, we do not introduce symbols to represent the functions being composed, but rather just proceed to calculate the derivative of the "outside" function and then multiply by the derivative of whatever is "inside." You can say to yourself: "the derivative of f of something is f' of that thing, multiplied by the derivative of that thing."

EXAMPLE 3 Find derivatives of the following functions:

$$\text{(a) } (7x-3)^{10}, \quad \text{(b) } f(t) = |t^2 - 1|, \quad \text{and} \quad \text{(c) } \left(3x + \frac{1}{(2x+1)^3}\right)^{1/4}.$$

Solution

(a) Here, the outside function is the 10th power; it must be differentiated first and the result multiplied by the derivative of the expression $7x - 3$:

$$\frac{d}{dx}(7x-3)^{10} = 10(7x-3)^9(7) = 70(7x-3)^9.$$

(b) Here, we are differentiating the absolute value of something. The derivative is signum of that thing, multiplied by the derivative of that thing:

$$f'(t) = \big(\operatorname{sgn}(t^2-1)\big)(2t) = \frac{2t(t^2-1)}{|t^2-1|} = \begin{cases} 2t & \text{if } t < -1 \text{ or } t > 1 \\ -2t & \text{if } -1 < t < 1 \\ \text{undefined} & \text{if } t = \pm 1. \end{cases}$$

(c) Here, we will need to use the Chain Rule twice. We begin by differentiating the $1/4$ power of something, but the something involves the -3rd power of $2x + 1$, and the derivative of that will also require the Chain Rule:

$$\begin{aligned}\frac{d}{dx}\left(3x + \frac{1}{(2x+1)^3}\right)^{1/4} &= \frac{1}{4}\left(3x + \frac{1}{(2x+1)^3}\right)^{-3/4} \frac{d}{dx}\left(3x + \frac{1}{(2x+1)^3}\right) \\ &= \frac{1}{4}\left(3x + \frac{1}{(2x+1)^3}\right)^{-3/4}\left(3 - \frac{3}{(2x+1)^4}\frac{d}{dx}(2x+1)\right) \\ &= \frac{3}{4}\left(1 - \frac{2}{(2x+1)^4}\right)\left(3x + \frac{1}{(2x+1)^3}\right)^{-3/4}.\end{aligned}$$

When you start to feel comfortable with the Chain Rule, you may want to save a line or two by carrying out the whole differentiation in one step:

$$\begin{aligned}\frac{d}{dx}\left(3x + \frac{1}{(2x+1)^3}\right)^{1/4} &= \frac{1}{4}\left(3x + \frac{1}{(2x+1)^3}\right)^{-3/4}\left(3 - \frac{3}{(2x+1)^4}(2)\right) \\ &= \frac{3}{4}\left(1 - \frac{2}{(2x+1)^4}\right)\left(3x + \frac{1}{(2x+1)^3}\right)^{-3/4}.\end{aligned}$$

Use of the Chain Rule produces products of factors that do not usually come out in the order you would naturally write them. Often you will want to rewrite the result with the factors in a different order. This is obvious in parts (a) and (c) of the example above. In monomials (expressions that are products of factors), it is common to write the factors in order of increasing complexity from left to right, with numerical factors coming first. One time when you would *not* waste time doing this, or trying to make any other simplification, is when you are going to evaluate the derivative at a particular number. In this case, substitute the number as soon as you have calculated the derivative, before doing any simplification:

$$\frac{d}{dx}(x^2-3)^{10}\bigg|_{x=2} = 10(x^2-3)^9(2x)\bigg|_{x=2} = (10)(1^9)(4) = 40.$$

EXAMPLE 4 Suppose that f is a differentiable function on the real line. In terms of the derivative f' of f, express the derivatives of:

(a) $f(3x)$, (b) $f(x^2)$, (c) $f(\pi f(x))$, and (d) $[f(3-2f(x))]^4$.

Solution

(a) $\displaystyle \frac{d}{dx} f(3x) = \big(f'(3x)\big)(3) = 3f'(3x).$

(b) $\displaystyle \frac{d}{dx} f(x^2) = \big(f'(x^2)\big)(2x) = 2xf'(x^2).$

(c) $\displaystyle \frac{d}{dx} f(\pi f(x)) = \big(f'(\pi f(x))\big)(\pi f'(x)) = \pi f'(x) f'(\pi f(x)).$

(d)
$$\begin{aligned}\frac{d}{dx}\big[f(3-2f(x))\big]^4 &= 4\big[f(3-2f(x))\big]^3 f'(3-2f(x))\big(-2f'(x)\big)\\ &= -8f'(x)f'(3-2f(x))\big[f(3-2f(x))\big]^3.\end{aligned}$$

As a final example, we illustrate combinations of the Chain Rule with the Product and Quotient Rules.

EXAMPLE 5 Find and simplify the following derivatives:

(a) $f'(t)$ if $f(t) = \dfrac{t^2+1}{\sqrt{t^2+2}}$, and (b) $g'(-1)$ if $g(x) = \big(x^2+3x+4\big)^5\sqrt{3-2x}$.

Solution

(a)
$$\begin{aligned} f'(t) &= \frac{\sqrt{t^2+2}(2t) - (t^2+1)\dfrac{2t}{2\sqrt{t^2+2}}}{t^2+2}\\ &= \frac{2t}{\sqrt{t^2+2}} - \frac{t^3+t}{(t^2+2)^{3/2}} = \frac{t^3+3t}{(t^2+2)^{3/2}}.\end{aligned}$$

(b)
$$g'(x) = 5\big(x^2+3x+4\big)^4(2x+3)\sqrt{3-2x} + \big(x^2+3x+4\big)^5\frac{-2}{2\sqrt{3-2x}}$$

$$g'(-1) = (5)(2^4)(1)(\sqrt{5}) - \frac{2^5}{\sqrt{5}} = 80\sqrt{5} - \frac{32}{5}\sqrt{5} = \frac{368\sqrt{5}}{5}.$$

Finding Derivatives with Maple

Computer algebra systems know the derivatives of elementary functions and can calculate the derivatives of combinations of these functions symbolically, using differentiation rules. Maple's `D` operator can be used to find the derivative function `D(f)` of a function `f` of one variable. Alternatively, you can use `diff` to differentiate an expression with respect to a variable and then use the substitution routine `subs` to evaluate the result at a particular number.

```
> f := x -> sqrt(1+2*x^2);
```

$$f := x \to \sqrt{1+2x^2}$$

```
> fprime := D(f);
```

$$fprime := x \to 2\frac{x}{\sqrt{1+2x^2}}$$

```
> fprime(2);
```

$$\frac{4}{3}$$

```
> diff(t^2*sin(3*t),t);
```

$$2t\sin(3t) + 3t^2\cos(3t)$$

```
> simplify(subs(t=Pi/12, %));
```

$$\frac{1}{12}\pi\sqrt{2} + \frac{1}{96}\pi^2\sqrt{2}$$

Building the Chain Rule into Differentiation Formulas

If u is a differentiable function of x and $y = u^n$, then the Chain Rule gives

$$\frac{d}{dx}u^n = \frac{dy}{dx} = \frac{dy}{du}\frac{du}{dx} = nu^{n-1}\frac{du}{dx}.$$

The formula

$$\frac{d}{dx}u^n = nu^{n-1}\frac{du}{dx}$$

is just the formula $\frac{d}{dx}x^n = nx^{n-1}$ with an application of the Chain Rule built in, so that it applies to functions of x rather than just to x. Some other differentiation rules with built-in Chain Rule applications are:

$\frac{d}{dx}\left(\frac{1}{u}\right) = \frac{-1}{u^2}\frac{du}{dx}$	(the Reciprocal Rule)
$\frac{d}{dx}\sqrt{u} = \frac{1}{2\sqrt{u}}\frac{du}{dx}$	(the Square Root Rule)
$\frac{d}{dx}u^r = r\,u^{r-1}\frac{du}{dx}$	(the General Power Rule)
$\frac{d}{dx}\lvert u\rvert = \operatorname{sgn} u\,\frac{du}{dx} = \frac{u}{\lvert u\rvert}\frac{du}{dx}$	(the Absolute Value Rule)

Proof of the Chain Rule (Theorem 6)

Suppose that f is differentiable at the point $u = g(x)$ and that g is differentiable at x. Let the function $E(k)$ be defined by

$$E(0) = 0,$$
$$E(k) = \frac{f(u+k) - f(u)}{k} - f'(u), \qquad \text{if } k \neq 0.$$

By the definition of derivative, $\lim_{k\to 0} E(k) = f'(u) - f'(u) = 0 = E(0)$, so $E(k)$ is continuous at $k = 0$. Also, whether $k = 0$ or not, we have

$$f(u+k) - f(u) = \big(f'(u) + E(k)\big)k.$$

Now put $u = g(x)$ and $k = g(x+h) - g(x)$, so that $u + k = g(x+h)$, and obtain

$$f(g(x+h)) - f(g(x)) = \big(f'(g(x)) + E(k)\big)(g(x+h) - g(x)).$$

Since g is differentiable at x, $\lim_{h\to 0}[g(x+h) - g(x)]/h = g'(x)$. Also, g is continuous at x by Theorem 1, so $\lim_{h\to 0} k = \lim_{h\to 0}(g(x+h) - g(x)) = 0$. Since E is continuous at 0, $\lim_{h\to 0} E(k) = \lim_{k\to 0} E(k) = E(0) = 0$. Hence,

$$\begin{aligned}\frac{d}{dx} f(g(x)) &= \lim_{h\to 0} \frac{f(g(x+h)) - f(g(x))}{h} \\ &= \lim_{h\to 0} \big(f'(g(x)) + E(k)\big) \frac{g(x+h) - g(x)}{h} \\ &= \big(f'(g(x)) + 0\big) g'(x) = f'(g(x))g'(x),\end{aligned}$$

which was to be proved.

EXERCISES 2.4

Find the derivatives of the functions in Exercises 1–16.

1. $y = (2x+3)^6$

2. $y = \left(1 - \dfrac{x}{3}\right)^{99}$

3. $f(x) = (4 - x^2)^{10}$

4. $y = \sqrt{1 - 3x^2}$

5. $F(t) = \left(2 + \dfrac{3}{t}\right)^{-10}$

6. $(1 + x^{2/3})^{3/2}$

7. $\dfrac{3}{5 - 4x}$

8. $(1 - 2t^2)^{-3/2}$

9. $y = |1 - x^2|$

10. $f(t) = |2 + t^3|$

11. $y = 4x + |4x - 1|$

12. $y = (2 + |x|^3)^{1/3}$

13. $y = \dfrac{1}{2 + \sqrt{3x+4}}$

14. $f(x) = \left(1 + \sqrt{\dfrac{x-2}{3}}\right)^4$

15. $z = \left(u + \dfrac{1}{u-1}\right)^{-5/3}$

16. $y = \dfrac{x^5\sqrt{3 + x^6}}{(4 + x^2)^3}$

17. Sketch the graph of the function in Exercise 10.

18. Sketch the graph of the function in Exercise 11.

Verify that the General Power Rule holds for the functions in Exercises 19–21.

19. $x^{1/4} = \sqrt{\sqrt{x}}$

20. $x^{3/4} = \sqrt{x\sqrt{x}}$

21. $x^{3/2} = \sqrt{(x^3)}$

In Exercises 22–29, express the derivative of the given function in terms of the derivative f' of the differentiable function f.

22. $f(2t+3)$

23. $f(5x - x^2)$

24. $\left[f\left(\dfrac{2}{x}\right)\right]^3$

25. $\sqrt{3 + 2f(x)}$

26. $f\left(\sqrt{3 + 2t}\right)$

27. $f\left(3 + 2\sqrt{x}\right)$

28. $f\big(2f(3f(x))\big)$

29. $f\big(2 - 3f(4 - 5t)\big)$

30. Find $\dfrac{d}{dx}\left(\dfrac{\sqrt{x^2-1}}{x^2+1}\right)\Bigg|_{x=-2}$.

31. Find $\dfrac{d}{dt}\sqrt{3t - 7}\,\Bigg|_{t=3}$.

32. If $f(x) = \dfrac{1}{\sqrt{2x+1}}$, find $f'(4)$.

33. If $y = (x^3 + 9)^{17/2}$, find $y'\Big|_{x=-2}$.

34. Find $F'(0)$ if $F(x) = (1+x)(2+x)^2(3+x)^3(4+x)^4$.

35. Calculate y' if $y = (x + ((3x)^5 - 2)^{-1/2})^{-6}$. Try to do it all in one step.

In Exercises 36–39, find an equation of the tangent line to the given curve at the given point.

36. $y = \sqrt{1 + 2x^2}$ at $x = 2$

37. $y = (1 + x^{2/3})^{3/2}$ at $x = -1$

38. $y = (ax + b)^8$ at $x = b/a$

39. $y = 1/(x^2 - x + 3)^{3/2}$ at $x = -2$

40. Show that the derivative of $f(x) = (x-a)^m(x-b)^n$ vanishes at some point between a and b if m and n are positive integers.

Use Maple or another computer algebra system to evaluate and simplify the derivatives of the functions in Exercises 41–44.

41. $y = \sqrt{x^2+1} + \dfrac{1}{(x^2+1)^{3/2}}$

42. $y = \dfrac{(x^2-1)(x^2-4)(x^2-9)}{x^6}$

43. $\dfrac{dy}{dt}\Bigg|_{t=2}$ if $y = (t+1)(t^2+2)(t^3+3)(t^4+4)(t^5+5)$

44. $f'(1)$ if $f(x) = \dfrac{(x^2+3)^{1/2}(x^3+7)^{1/3}}{(x^4+15)^{1/4}}$

45. Does the Chain Rule enable you to calculate the derivatives of $|x|^2$ and $|x^2|$ at $x = 0$? Do these functions have derivatives at $x = 0$? Why?

46. What is wrong with the following "proof" of the Chain Rule? Let $k = g(x+h) - g(x)$. Then $\lim_{h\to 0} k = 0$. Thus,

$$\begin{aligned}\lim_{h\to 0} & \frac{f(g(x+h)) - f(g(x))}{h} \\ &= \lim_{h\to 0} \frac{f(g(x+h)) - f(g(x))}{g(x+h) - g(x)} \frac{g(x+h) - g(x)}{h} \\ &= \lim_{h\to 0} \frac{f(g(x)+k) - f(g(x))}{k} \frac{g(x+h) - g(x)}{h} \\ &= f'(g(x))\, g'(x).\end{aligned}$$

2.5 Derivatives of Trigonometric Functions

The trigonometric functions, especially sine and cosine, play a very important role in the mathematical modelling of real-world phenomena. In particular, they arise whenever quantities fluctuate in a periodic way. Elastic motions, vibrations, and waves of all kinds naturally involve the trigonometric functions, and many physical and mechanical laws are formulated as differential equations having these functions as solutions.

In this section we will calculate the derivatives of the six trigonometric functions. We only have to work hard for one of them, sine; the others then follow from known identities and the differentiation rules of Section 2.3.

Some Special Limits

First, we have to establish some trigonometric limits that we will need to calculate the derivative of sine. It is assumed throughout that the arguments of the trigonometric functions are measured in radians.

THEOREM 7

The functions $\sin\theta$ and $\cos\theta$ are continuous at every value of θ. In particular, at $\theta = 0$ we have:

$$\lim_{\theta\to 0} \sin\theta = \sin 0 = 0 \qquad \text{and} \qquad \lim_{\theta\to 0} \cos\theta = \cos 0 = 1.$$

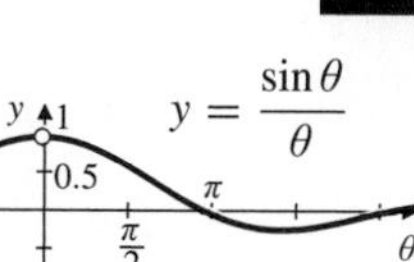

Figure 2.20 It appears that $\lim_{\theta\to 0} (\sin\theta)/\theta = 1$

This result is obvious from the graphs of sine and cosine, so we will not prove it here. A proof can be based on the Squeeze Theorem (Theorem 4 of Section 1.2). The method is suggested in Exercise 62 at the end of this section.

The graph of the function $y = (\sin\theta)/\theta$ is shown in Figure 2.20. Although it is not defined at $\theta = 0$, this function appears to have limit 1 as θ approaches 0.

THEOREM 8

An important trigonometric limit

$$\lim_{\theta\to 0} \frac{\sin\theta}{\theta} = 1 \qquad \text{(where } \theta \text{ is in radians).}$$

Figure 2.21 Area $\triangle OAP$ < Area sector OAP < Area $\triangle OAT$

PROOF Let $0 < \theta < \pi/2$, and represent θ as shown in Figure 2.21. Points $A(1, 0)$ and $P(\cos\theta, \sin\theta)$ lie on the unit circle $x^2 + y^2 = 1$. The area of the circular sector OAP lies between the areas of triangles OAP and OAT:

$$\text{Area } \triangle OAP < \text{Area sector } OAP < \text{Area } \triangle OAT.$$

As shown in Section P.7, the area of a circular sector having central angle θ (radians) and radius 1 is $\theta/2$. The area of a triangle is $(1/2) \times \text{base} \times \text{height}$, so

$$\text{Area } \triangle OAP = \frac{1}{2}(1)(\sin\theta) = \frac{\sin\theta}{2},$$

$$\text{Area } \triangle OAT = \frac{1}{2}(1)(\tan\theta) = \frac{\sin\theta}{2\cos\theta}.$$

Thus,

$$\frac{\sin\theta}{2} < \frac{\theta}{2} < \frac{\sin\theta}{2\cos\theta},$$

or, upon multiplication by the positive number $2/\sin\theta$,

$$1 < \frac{\theta}{\sin\theta} < \frac{1}{\cos\theta}.$$

Now take reciprocals, thereby reversing the inequalities:

$$1 > \frac{\sin\theta}{\theta} > \cos\theta.$$

Since $\lim_{\theta\to 0+}\cos\theta = 1$ by Theorem 7, the Squeeze Theorem gives

$$\lim_{\theta\to 0+}\frac{\sin\theta}{\theta} = 1.$$

Finally, note that $\sin\theta$ and θ are *odd functions*. Therefore, $f(\theta) = (\sin\theta)/\theta$ is an *even function*: $f(-\theta) = f(\theta)$, as shown in Figure 2.20. This symmetry implies that the left limit at 0 must have the same value as the right limit:

$$\lim_{\theta\to 0-}\frac{\sin\theta}{\theta} = 1 = \lim_{\theta\to 0+}\frac{\sin\theta}{\theta},$$

so $\lim_{\theta\to 0}(\sin\theta)/\theta = 1$ by Theorem 1 of Section 1.2.

Theorem 8 can be combined with limit rules and known trigonometric identities to yield other trigonometric limits.

EXAMPLE 1 Show that $\lim_{h\to 0}\frac{\cos h - 1}{h} = 0.$

Solution Using the half-angle formula $\cos h = 1 - 2\sin^2(h/2)$, we calculate

$$\begin{aligned}\lim_{h\to 0}\frac{\cos h - 1}{h} &= \lim_{h\to 0} -\frac{2\sin^2(h/2)}{h} \qquad \text{Let } \theta = h/2.\\ &= -\lim_{\theta\to 0}\frac{\sin\theta}{\theta}\sin\theta = -(1)(0) = 0.\end{aligned}$$

The Derivatives of Sine and Cosine

To calculate the derivative of $\sin x$, we need the addition formula for sine (see Section P.7):

$$\sin(x+h) = \sin x\cos h + \cos x\sin h.$$

THEOREM 9 **The derivative of the sine function is the cosine function.**

$$\frac{d}{dx}\sin x = \cos x$$

PROOF We use the definition of derivative, the addition formula for sine, the rules for combining limits, Theorem 8, and the result of Example 1:

$$\begin{aligned}\frac{d}{dx}\sin x &= \lim_{h\to 0}\frac{\sin(x+h)-\sin x}{h}\\ &= \lim_{h\to 0}\frac{\sin x\cos h+\cos x\sin h-\sin x}{h}\\ &= \lim_{h\to 0}\frac{\sin x(\cos h-1)+\cos x\sin h}{h}\\ &= \lim_{h\to 0}\sin x\cdot\lim_{h\to 0}\frac{\cos h-1}{h}+\lim_{h\to 0}\cos x\cdot\lim_{h\to 0}\frac{\sin h}{h}\\ &= (\sin x)\cdot(0)+(\cos x)\cdot(1)=\cos x.\end{aligned}$$

THEOREM 10

The derivative of the cosine function is the negative of the sine function.

$$\frac{d}{dx}\cos x = -\sin x$$

PROOF We could mimic the proof for sine above, using the addition rule for cosine, $\cos(x+h) = \cos x\cos h - \sin x\sin h$. An easier way is to make use of the complementary angle identities, $\sin((\pi/2)-x) = \cos x$ and $\cos((\pi/2)-x) = \sin x$, and the Chain Rule from Section 2.4:

$$\frac{d}{dx}\cos x = \frac{d}{dx}\sin\left(\frac{\pi}{2}-x\right) = (-1)\cos\left(\frac{\pi}{2}-x\right) = -\sin x.$$

Notice the minus sign in the derivative of cosine. The derivative of the sine is the cosine, but the derivative of the cosine is *minus* the sine. This is shown graphically in Figure 2.22.

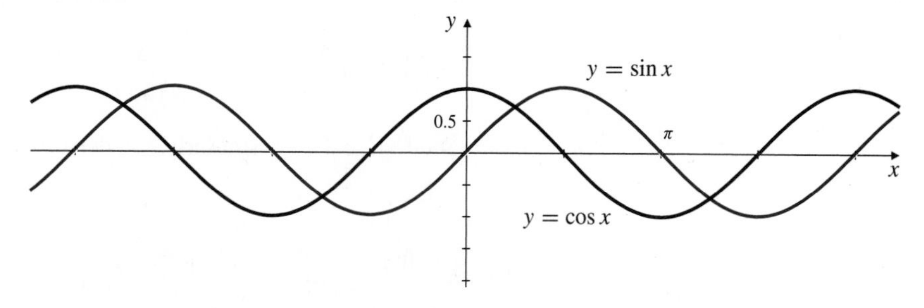

Figure 2.22 The sine and cosine plotted together. The slope of the sine curve at x is $\cos x$; the slope of the cosine curve at x is $-\sin x$

EXAMPLE 2 Evaluate the derivatives of the following functions:

(a) $\sin(\pi x)+\cos(3x)$, (b) $x^2\sin\sqrt{x}$, and (c) $\dfrac{\cos x}{1-\sin x}$.

Solution

(a) By the Sum Rule and the Chain Rule:

$$\frac{d}{dx}(\sin(\pi x)+\cos(3x)) = \cos(\pi x)(\pi)-\sin(3x)(3) = \pi\cos(\pi x)-3\sin(3x).$$

(b) By the Product and Chain Rules:

$$\frac{d}{dx}(x^2\sin\sqrt{x}) = 2x\sin\sqrt{x}+x^2\left(\cos\sqrt{x}\right)\frac{1}{2\sqrt{x}} = 2x\sin\sqrt{x}+\frac{1}{2}x^{3/2}\cos\sqrt{x}.$$

(c) By the Quotient Rule:

$$\begin{aligned}\frac{d}{dx}\left(\frac{\cos x}{1-\sin x}\right) &= \frac{(1-\sin x)(-\sin x)-(\cos x)(0-\cos x)}{(1-\sin x)^2}\\ &= \frac{-\sin x+\sin^2 x+\cos^2 x}{(1-\sin x)^2}\\ &= \frac{1-\sin x}{(1-\sin x)^2} = \frac{1}{1-\sin x}.\end{aligned}$$

We used the identity $\sin^2 x + \cos^2 x = 1$ to simplify the middle line.

Using trigonometric identities can sometimes change the way a derivative is calculated. Carrying out a differentiation in different ways can lead to different-looking answers, but they should be equal if no errors have been made.

EXAMPLE 3 Use two different methods to find the derivative of the function $f(t) = \sin t \cos t$.

Solution By the Product Rule:

$$f'(t) = (\cos t)(\cos t) + (\sin t)(-\sin t) = \cos^2 t - \sin^2 t.$$

On the other hand, since $\sin(2t) = 2\sin t \cos t$, we have

$$f'(t) = \frac{d}{dt}\left(\frac{1}{2}\sin(2t)\right) = \left(\frac{1}{2}\right)(2)\cos(2t) = \cos(2t).$$

The two answers are really the same, since $\cos(2t) = \cos^2 t - \sin^2 t$.

It is very important to remember that the formulas for the derivatives of $\sin x$ and $\cos x$ were obtained under the assumption that x is measured in *radians*. Since we know that $180° = \pi$ radians, $x° = \pi x/180$ radians. By the Chain Rule,

$$\frac{d}{dx}\sin(x°) = \frac{d}{dx}\sin\left(\frac{\pi x}{180}\right) = \frac{\pi}{180}\cos\left(\frac{\pi x}{180}\right) = \frac{\pi}{180}\cos(x°).$$

(See Figure 2.23.) Similarly, the derivative of $\cos(x°)$ is $-(\pi/180)\sin(x°)$.

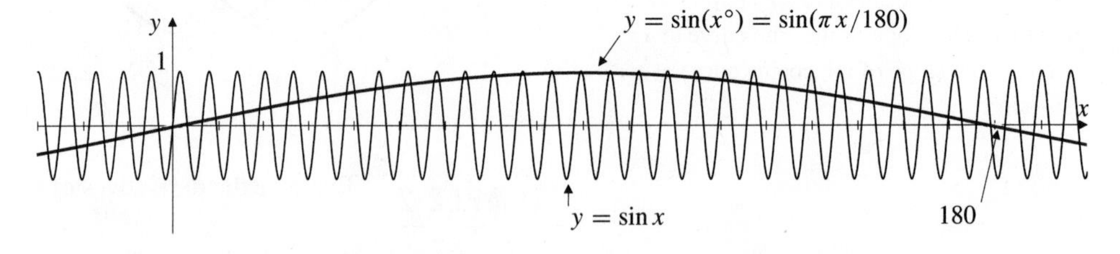

Figure 2.23 $\sin(x°)$ oscillates much more slowly than $\sin x$. Its maximum slope is $\pi/180$

Continuity

The six trigonometric functions are differentiable and, therefore, continuous (by Theorem 1) everywhere on their domains. This means that we can calculate the limits of most trigonometric functions as $x \to a$ by evaluating them at $x = a$.

The Derivatives of the Other Trigonometric Functions

Because $\sin x$ and $\cos x$ are differentiable everywhere, the functions

$$\tan x = \frac{\sin x}{\cos x} \qquad \sec x = \frac{1}{\cos x}$$

$$\cot x = \frac{\cos x}{\sin x} \qquad \csc x = \frac{1}{\sin x}$$

are differentiable at every value of x at which they are defined (i.e., where their denominators are not zero). Their derivatives can be calculated by the Quotient and Reciprocal Rules and are as follows:

The three "co-" functions (cosine, cotangent, and cosecant) have explicit minus signs in their derivatives.

$$\frac{d}{dx}\tan x = \sec^2 x \qquad \frac{d}{dx}\sec x = \sec x \tan x$$
$$\frac{d}{dx}\cot x = -\csc^2 x \qquad \frac{d}{dx}\csc x = -\csc x \cot x.$$

EXAMPLE 4 Verify the derivative formulas for $\tan x$ and $\sec x$.

Solution We use the Quotient Rule for tangent and the Reciprocal Rule for secant:

$$\begin{aligned}\frac{d}{dx}\tan x &= \frac{d}{dx}\left(\frac{\sin x}{\cos x}\right) = \frac{\cos x\frac{d}{dx}(\sin x) - \sin x\frac{d}{dx}(\cos x)}{\cos^2 x}\\ &= \frac{\cos x\cos x - \sin x(-\sin x)}{\cos^2 x} = \frac{\cos^2 x + \sin^2 x}{\cos^2 x}\\ &= \frac{1}{\cos^2 x} = \sec^2 x.\\ \frac{d}{dx}\sec x &= \frac{d}{dx}\left(\frac{1}{\cos x}\right) = \frac{-1}{\cos^2 x}\frac{d}{dx}(\cos x)\\ &= \frac{-1}{\cos^2 x}(-\sin x) = \frac{1}{\cos x}\cdot\frac{\sin x}{\cos x}\\ &= \sec x\ \tan x.\end{aligned}$$

EXAMPLE 5 (a) $\frac{d}{dx}\left[3x + \cot\left(\frac{x}{2}\right)\right] = 3 + \left[-\csc^2\left(\frac{x}{2}\right)\right]\frac{1}{2} = 3 - \frac{1}{2}\csc^2\left(\frac{x}{2}\right)$

(b) $$\begin{aligned}\frac{d}{dx}\left(\frac{3}{\sin(2x)}\right) &= \frac{d}{dx}(3\csc(2x))\\ &= 3(-\csc(2x)\cot(2x))(2) = -6\csc(2x)\cot(2x).\end{aligned}$$

EXAMPLE 6 Find the tangent and normal lines to the curve $y = \tan(\pi x/4)$ at the point $(1, 1)$.

Solution The slope of the tangent to $y = \tan(\pi x/4)$ at $(1, 1)$ is:

$$\left.\frac{dy}{dx}\right|_{x=1} = \left.\frac{\pi}{4}\sec^2(\pi x/4)\right|_{x=1} = \frac{\pi}{4}\sec^2\left(\frac{\pi}{4}\right) = \frac{\pi}{4}\left(\sqrt{2}\right)^2 = \frac{\pi}{2}.$$

The tangent is the line

$$y = 1 + \frac{\pi}{2}(x-1), \qquad \text{or} \qquad y = \frac{\pi x}{2} - \frac{\pi}{2} + 1.$$

The normal has slope $m = -2/\pi$, so its point-slope equation is

$$y = 1 - \frac{2}{\pi}(x-1), \qquad \text{or} \qquad y = -\frac{2x}{\pi} + \frac{2}{\pi} + 1.$$

EXERCISES 2.5

1. Verify the formula for the derivative of $\csc x = 1/(\sin x)$.

2. Verify the formula for the derivative of

$\cot x = (\cos x)/(\sin x)$.

Find the derivatives of the functions in Exercises 3–36. Simplify your answers whenever possible. Also be on the lookout for ways you might simplify the given expression before differentiating it.

3. $y = \cos 3x$

4. $y = \sin \dfrac{x}{5}$

5. $y = \tan \pi x$

6. $y = \sec ax$

7. $y = \cot(4 - 3x)$

8. $y = \sin((\pi - x)/3)$

9. $f(x) = \cos(s - rx)$

10. $y = \sin(Ax + B)$

11. $\sin(\pi x^2)$

12. $\cos(\sqrt{x})$

13. $y = \sqrt{1 + \cos x}$

14. $\sin(2 \cos x)$

15. $f(x) = \cos(x + \sin x)$

16. $g(\theta) = \tan(\theta \sin \theta)$

17. $u = \sin^3(\pi x/2)$

18. $y = \sec(1/x)$

19. $F(t) = \sin at \cos at$

20. $G(\theta) = \dfrac{\sin a\theta}{\cos b\theta}$

21. $\sin(2x) - \cos(2x)$

22. $\cos^2 x - \sin^2 x$

23. $\tan x + \cot x$

24. $\sec x - \csc x$

25. $\tan x - x$

26. $\tan(3x)\cot(3x)$

27. $t \cos t - \sin t$

28. $t \sin t + \cos t$

29. $\dfrac{\sin x}{1 + \cos x}$

30. $\dfrac{\cos x}{1 + \sin x}$

31. $x^2 \cos(3x)$

32. $g(t) = \sqrt{(\sin t)/t}$

33. $v = \sec(x^2)\tan(x^2)$

34. $z = \dfrac{\sin \sqrt{x}}{1 + \cos \sqrt{x}}$

35. $\sin(\cos(\tan t))$

36. $f(s) = \cos(s + \cos(s + \cos s))$

37. Given that $\sin 2x = 2 \sin x \cos x$, deduce that $\cos 2x = \cos^2 x - \sin^2 x$.

38. Given that $\cos 2x = \cos^2 x - \sin^2 x$, deduce that $\sin 2x = 2 \sin x \cos x$.

In Exercises 39–42, find equations for the lines that are tangent and normal to the curve $y = f(x)$ at the given point.

39. $y = \sin x$, $(\pi, 0)$

40. $y = \tan(2x)$, $(0, 0)$

41. $y = \sqrt{2}\cos(x/4)$, $(\pi, 1)$

42. $y = \cos^2 x$, $\left(\dfrac{\pi}{3}, \dfrac{1}{4}\right)$

43. Find an equation of the line tangent to the curve $y = \sin(x^\circ)$ at the point where $x = 45$.

44. Find an equation of the straight line normal to $y = \sec(x^\circ)$ at the point where $x = 60$.

45. Find the points on the curve $y = \tan x$, $-\pi/2 < x < \pi/2$, where the tangent is parallel to the line $y = 2x$.

46. Find the points on the curve $y = \tan(2x)$, $-\pi/4 < x < \pi/4$, where the normal is parallel to the line $y = -x/8$.

47. Show that the graphs of $y = \sin x$, $y = \cos x$, $y = \sec x$, and $y = \csc x$ have horizontal tangents.

48. Show that the graphs of $y = \tan x$ and $y = \cot x$ never have horizontal tangents.

Do the graphs of the functions in Exercises 49–52 have any horizontal tangents in the interval $0 \le x \le 2\pi$? If so, where? If not, why not?

49. $y = x + \sin x$

50. $y = 2x + \sin x$

51. $y = x + 2 \sin x$

52. $y = x + 2 \cos x$

Find the limits in Exercises 53–56.

53. $\displaystyle\lim_{x\to 0} \frac{\tan(2x)}{x}$

54. $\displaystyle\lim_{x\to \pi} \sec(1 + \cos x)$

55. $\displaystyle\lim_{x\to 0} (x^2 \csc x \cot x)$

56. $\displaystyle\lim_{x\to 0} \cos\left(\frac{\pi - \pi \cos^2 x}{x^2}\right)$

57. Use the method of Example 1 to evaluate $\displaystyle\lim_{h\to 0} \frac{1 - \cos h}{h^2}$.

58. Find values of a and b that make

$$f(x) = \begin{cases} ax + b, & x < 0 \\ 2 \sin x + 3 \cos x, & x \ge 0 \end{cases}$$

differentiable at $x = 0$.

59. How many straight lines that pass through the origin are tangent to $y = \cos x$? Find (to 6 decimal places) the slopes of the two such lines that have the largest positive slopes.

Use Maple or another computer algebra system to evaluate and simplify the derivatives of the functions in Exercises 60–61.

60. $\displaystyle\frac{d}{dx} \frac{x \cos(x \sin x)}{x + \cos(x \cos x)}\bigg|_{x=0}$

61. $\displaystyle\frac{d}{dx}\left(\sqrt{2x^2 + 3}\sin(x^2) - \frac{(2x^2 + 3)^{3/2}\cos(x^2)}{x}\right)\bigg|_{x=\sqrt{\pi}}$

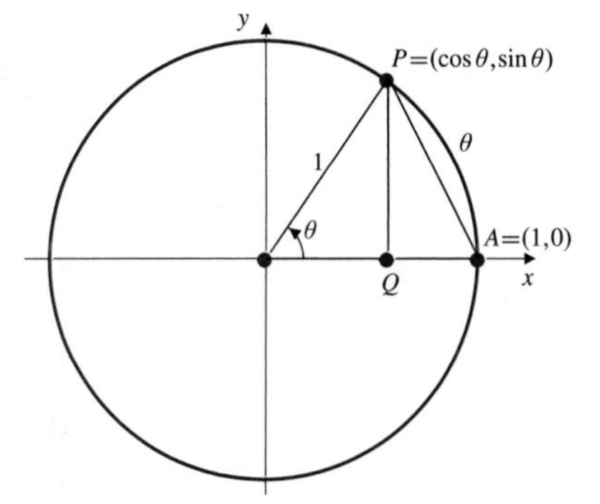

Figure 2.24

62. (The continuity of sine and cosine)

(a) Prove that

$$\lim_{\theta\to 0} \sin\theta = 0 \quad \text{and} \quad \lim_{\theta\to 0} \cos\theta = 1$$

as follows: Use the fact that the length of chord AP is less than the length of arc AP in Figure 2.24 to show that

$$\sin^2\theta + (1 - \cos\theta)^2 < \theta^2.$$

Then deduce that $0 \le |\sin\theta| < |\theta|$ and $0 \le |1 - \cos\theta| < |\theta|$. Then use the Squeeze Theorem from Section 1.2.

(b) Part (a) says that $\sin\theta$ and $\cos\theta$ are continuous at $\theta = 0$. Use the addition formulas to prove that they are therefore continuous at every θ.

2.6 Higher-Order Derivatives

If the derivative $y' = f'(x)$ of a function $y = f(x)$ is itself differentiable at x, we can calculate *its* derivative, which we call the **second derivative** of f and denote by $y'' = f''(x)$. As is the case for first derivatives, second derivatives can be denoted by various notations depending on the context. Some of the more common ones are

$$y'' = f''(x) = \frac{d^2y}{dx^2} = \frac{d}{dx}\frac{d}{dx}f(x) = \frac{d^2}{dx^2}f(x) = D_x^2 y = D_x^2 f(x).$$

Similarly, you can consider third-, fourth-, and in general nth-order derivatives. The prime notation is inconvenient for derivatives of high order, so we denote the order by a superscript in parentheses (to distinguish it from an exponent): the nth derivative of $y = f(x)$ is

$$y^{(n)} = f^{(n)}(x) = \frac{d^n y}{dx^n} = \frac{d^n}{dx^n}f(x) = D_x^n y = D_x^n f(x),$$

and it is defined to be the derivative of the $(n-1)$st derivative. For $n = 1, 2,$ and 3, primes are still normally used: $f^{(2)}(x) = f''(x)$, $f^{(3)}(x) = f'''(x)$. It is sometimes convenient to denote $f^{(0)}(x) = f(x)$, that is, to regard a function as its own zeroth-order derivative.

EXAMPLE 1 The **velocity** of a moving object is the (instantaneous) rate of change of the position of the object with respect to time; if the object moves along the x-axis and is at position $x = f(t)$ at time t, then its velocity at that time is

$$v = \frac{dx}{dt} = f'(t).$$

Similarly, the **acceleration** of the object is the rate of change of the velocity. Thus, the acceleration is the *second derivative* of the position:

$$a = \frac{dv}{dt} = \frac{d^2x}{dt^2} = f''(t).$$

We will investigate the relationships between position, velocity, and acceleration further in Section 2.11.

EXAMPLE 2 If $y = x^3$, then $y' = 3x^2$, $y'' = 6x$, $y''' = 6$, $y^{(4)} = 0$, and all higher derivatives are zero.

In general, if $f(x) = x^n$ (where n is a positive integer), then

$$\begin{aligned} f^{(k)}(x) &= n(n-1)(n-2)\cdots(n-(k-1))\,x^{n-k} \\ &= \begin{cases} \dfrac{n!}{(n-k)!}x^{n-k} & \text{if } 0 \le k \le n \\ 0 & \text{if } k > n, \end{cases} \end{aligned}$$

where $n!$ (called n **factorial**) is defined by:

$$\begin{aligned}
0! &= 1 \\
1! &= 0! \times 1 = 1 \times 1 = 1 \\
2! &= 1! \times 2 = 1 \times 2 = 2 \\
3! &= 2! \times 3 = 1 \times 2 \times 3 = 6 \\
4! &= 3! \times 4 = 1 \times 2 \times 3 \times 4 = 24 \\
&\vdots \\
n! &= (n-1)! \times n = 1 \times 2 \times 3 \times \cdots \times (n-1) \times n.
\end{aligned}$$

It follows that if P is a polynomial of degree n,

$$P(x) = a_n x^n + a_{n-1}x^{n-1} + \cdots + a_1 x + a_0,$$

where $a_n, a_{n-1}, \ldots, a_1, a_0$ are constants, then $P^{(k)}(x) = 0$ for $k > n$. For $k \le n$, $P^{(k)}$ is a polynomial of degree $n-k$; in particular, $P^{(n)}(x) = n!\, a_n$, a constant function.

EXAMPLE 3 Show that if A, B, and k are constants, then the function $y = A\cos(kt) + B\sin(kt)$ is a solution of the *second-order* **differential equation of simple harmonic motion** (see Section 3.7):

$$\frac{d^2y}{dt^2} + k^2 y = 0.$$

Solution To be a solution, the function $y(t)$ must satisfy the differential equation *identically*; that is,

$$\frac{d^2}{dt^2}\, y(t) + k^2 y(t) = 0$$

must hold for every real number t. We verify this by calculating the first two derivatives of the given function $y(t) = A\cos(kt) + B\sin(kt)$ and observing that the second derivative plus $k^2 y(t)$ is, in fact, zero everywhere:

$$\begin{aligned}
\frac{dy}{dt} &= -Ak\sin(kt) + Bk\cos(kt) \\
\frac{d^2y}{dt^2} &= -Ak^2\cos(kt) - Bk^2\sin(kt) = -k^2 y(t), \\
\frac{d^2y}{dt^2} &+ k^2 y(t) = 0.
\end{aligned}$$

EXAMPLE 4 Find the nth derivative, $y^{(n)}$, of $y = \dfrac{1}{1+x} = (1+x)^{-1}$.

Solution Begin by calculating the first few derivatives:

$$\begin{aligned}
y' &= -(1+x)^{-2} \\
y'' &= -(-2)(1+x)^{-3} = 2(1+x)^{-3} \\
y''' &= 2(-3)(1+x)^{-4} = -3!(1+x)^{-4} \\
y^{(4)} &= -3!(-4)(1+x)^{-5} = 4!(1+x)^{-5}
\end{aligned}$$

The pattern here is becoming obvious. It seems that

$$y^{(n)} = (-1)^n n!(1+x)^{-n-1}.$$

Note the use of $(-1)^n$ to denote a positive sign if n is even and a negative sign if n is odd.

We have not yet actually proved that the above formula is correct for every n, although it is clearly correct for $n = 1, 2, 3$, and 4. To complete the proof we use mathematical induction (Section 2.3). Suppose that the formula is valid for $n = k$, where k is some positive integer. Consider $y^{(k+1)}$:

$$\begin{aligned} y^{(k+1)} &= \frac{d}{dx}\,y^{(k)} = \frac{d}{dx}\left((-1)^k k!(1+x)^{-k-1}\right) \\ &= (-1)^k k!(-k-1)(1+x)^{-k-2} = (-1)^{k+1}(k+1)!(1+x)^{-(k+1)-1}. \end{aligned}$$

This is what the formula predicts for the $(k+1)$st derivative. Therefore, if the formula for $y^{(n)}$ is correct for $n = k$, then it is also correct for $n = k+1$. Since the formula is known to be true for $n = 1$, it must therefore be true for every integer $n \ge 1$ *by induction.*

EXAMPLE 5 Find a formula for $f^{(n)}(x)$, given that $f(x) = \sin(ax+b)$.

Solution Begin by calculating several derivatives:

$$\begin{aligned} f'(x) &= a\cos(ax+b) \\ f''(x) &= -a^2\sin(ax+b) = -a^2 f(x) \\ f'''(x) &= -a^3\cos(ax+b) = -a^2 f'(x) \\ f^{(4)}(x) &= a^4\sin(ax+b) = a^4 f(x) \\ f^{(5)}(x) &= a^5\cos(ax+b) = a^4 f'(x) \\ &\vdots \end{aligned}$$

The pattern is pretty obvious here. Each new derivative is $-a^2$ times the second previous one. A formula that gives all the derivatives is

$$f^{(n)}(x) = \begin{cases} (-1)^k a^n\sin(ax+b) & \text{if } n = 2k \\ (-1)^k a^n\cos(ax+b) & \text{if } n = 2k+1 \end{cases} \qquad (k = 0, 1, 2, \ldots),$$

which can also be verified by induction on k.

Our final example shows that it is not always easy to obtain a formula for the nth derivative of a function.

EXAMPLE 6 Calculate f', f'', and f''' for $f(x) = \sqrt{x^2+1}$. Can you see enough of a pattern to predict $f^{(4)}$?

Solution Since $f(x) = (x^2+1)^{1/2}$, we have

$$\begin{aligned} f'(x) &= \tfrac{1}{2}(x^2+1)^{-1/2}(2x) = x(x^2+1)^{-1/2}, \\ f''(x) &= (x^2+1)^{-1/2} + x\left(-\tfrac{1}{2}\right)(x^2+1)^{-3/2}(2x) \\ &= (x^2+1)^{-3/2}(x^2+1-x^2) = (x^2+1)^{-3/2}, \\ f'''(x) &= -\tfrac{3}{2}(x^2+1)^{-5/2}(2x) = -3x(x^2+1)^{-5/2}. \end{aligned}$$

Although the expression obtained from each differentiation simplified somewhat, the pattern of these derivatives is not (yet) obvious enough to enable us to predict the formula for $f^{(4)}(x)$ without having to calculate it. In fact,

$$f^{(4)}(x) = 3(4x^2-1)(x^2+1)^{-7/2},$$

so the pattern (if there is one) doesn't become any clearer at this stage.

Remark Computing higher-order derivatives may be useful in applications involving Taylor polynomials (see Section 4.10). As taking derivatives can be automated with a known algorithm, it makes sense to use a computer to calculate higher-order ones. However, depending on the function, the amount of memory and processor time needed may severely restrict the order of derivatives calculated in this way. Higher-order derivatives can be indicated in Maple by repeating the variable of differentiation or indicating the order by using the `$` operator:

```
> diff(x^5,x,x) + diff(sin(2*x),x$3);
```

$$20\,x^3 - 8\cos(2x)$$

The `D` operator can also be used for higher-order derivatives of a function (as distinct from an expression) by composing it explicitly or using the `@@` operator:

```
> f := x -> x^5; fpp := D(D(f)); (D@@3)(f)(a);
```

$$f := x \to x^5$$
$$fpp := x \to 20\,x^3$$
$$60\,a^2$$

EXERCISES 2.6

Find y', y'', and y''' for the functions in Exercises 1–12.

1. $y = (3 - 2x)^7$

2. $y = x^2 - \dfrac{1}{x}$

3. $y = \dfrac{6}{(x-1)^2}$

4. $y = \sqrt{ax+b}$

5. $y = x^{1/3} - x^{-1/3}$

6. $y = x^{10} + 2x^8$

7. $y = (x^2+3)\sqrt{x}$

8. $y = \dfrac{x-1}{x+1}$

9. $y = \tan x$

10. $y = \sec x$

11. $y = \cos(x^2)$

12. $y = \dfrac{\sin x}{x}$

In Exercises 13–23, calculate enough derivatives of the given function to enable you to guess the general formula for $f^{(n)}(x)$. Then verify your guess using mathematical induction.

13. $f(x) = \dfrac{1}{x}$

14. $f(x) = \dfrac{1}{x^2}$

15. $f(x) = \dfrac{1}{2-x}$

16. $f(x) = \sqrt{x}$

17. $f(x) = \dfrac{1}{a+bx}$

18. $f(x) = x^{2/3}$

19. $f(x) = \cos(ax)$

20. $f(x) = x\cos x$

21. $f(x) = x\sin(ax)$

22. $f(x) = \dfrac{1}{|x|}$

23. $f(x) = \sqrt{1-3x}$

24. If $y = \tan kx$, show that $y'' = 2k^2y(1+y^2)$.

25. If $y = \sec kx$, show that $y'' = k^2y(2y^2-1)$.

26. Use mathematical induction to prove that the nth derivative of $y = \sin(ax+b)$ is given by the formula asserted at the end of Example 5.

27. Use mathematical induction to prove that the nth derivative of $y = \tan x$ is of the form $P_{n+1}(\tan x)$, where P_{n+1} is a polynomial of degree $n+1$.

28. If f and g are twice-differentiable functions, show that $(fg)'' = f''g + 2f'g' + fg''$.

29. State and prove the results analogous to that of Exercise 28 but for $(fg)^{(3)}$ and $(fg)^{(4)}$. Can you guess the formula for $(fg)^{(n)}$?

2.7 Using Differentials and Derivatives

In this section we will look at some examples of ways in which derivatives are used to represent and interpret changes and rates of change in the world around us. It is natural to think of change in terms of dependence on time, such as the velocity of a moving object, but there is no need to be so restrictive. Change with respect to variables other than time can be treated in the same way. For example, a physician may want to know how small changes in dosage can affect the body's response to a drug. An economist may want to study how foreign investment changes with respect to variations in a country's interest rates. These questions can all be formulated in terms of rate of change of a function with respect to a variable.

Approximating Small Changes

If one quantity, say y, is a function of another quantity x, that is,

$$y = f(x),$$

we sometimes want to know how a change in the value of x by an amount Δx will affect the value of y. The exact change Δy in y is given by

$$\Delta y = f(x + \Delta x) - f(x),$$

but if the change Δx is small, then we can get a good approximation to Δy by using the fact that $\Delta y/\Delta x$ is approximately the derivative dy/dx. Thus,

$$\Delta y = \frac{\Delta y}{\Delta x}\,\Delta x \approx \frac{dy}{dx}\,\Delta x = f'(x)\,\Delta x.$$

It is often convenient to represent this approximation in terms of differentials; if we denote the change in x by dx instead of Δx, then the change Δy in y is approximated by the differential dy, that is (see Figure 2.25),

$$\Delta y \approx dy = f'(x)\,dx.$$

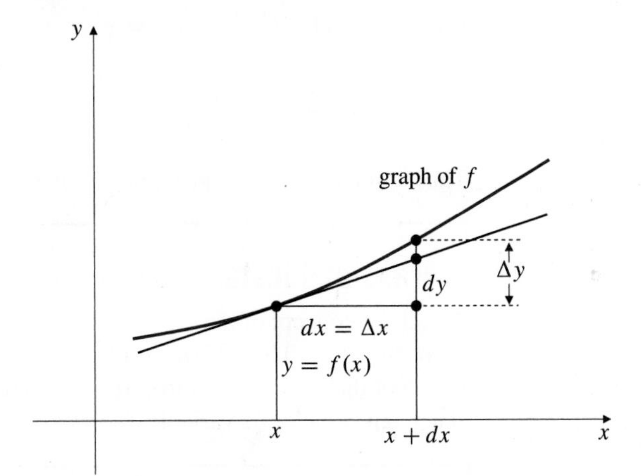

Figure 2.25 dy, the change in height to the tangent line, approximates Δy, the change in height to the graph of f

EXAMPLE 1 Without using a scientific calculator, determine by approximately how much the value of $\sin x$ increases as x increases from $\pi/3$ to $(\pi/3) + 0.006$. To 3 decimal places, what is the value of $\sin\big((\pi/3) + 0.006\big)$?

Solution If $y = \sin x$, $x = \pi/3 \approx 1.0472$, and $dx = 0.006$, then

$$dy = \cos(x)\,dx = \cos\left(\frac{\pi}{3}\right)\,dx = \frac{1}{2}(0.006) = 0.003.$$

Thus the change in the value of $\sin x$ is approximately 0.003, and

$$\sin\left(\frac{\pi}{3} + 0.006\right) \approx \sin\frac{\pi}{3} + 0.003 = \frac{\sqrt{3}}{2} + 0.003 = 0.869$$

rounded to 3 decimal places.

Whenever one makes an approximation it is wise to try and estimate how big the error might be. We will have more to say about such approximations and their error estimates in Section 4.9.

Sometimes changes in a quantity are measured with respect to the size of the quantity. The **relative change** in x is the ratio dx/x if x changes by amount dx. The **percentage change** in x is the relative change expressed as a percentage:

$$\text{relative change in } x = \frac{dx}{x}$$

$$\text{percentage change in } x = 100\,\frac{dx}{x}.$$

EXAMPLE 2 By approximately what percentage does the area of a circle increase if the radius increases by 2%?

Solution The area A of a circle is given in terms of the radius r by $A = \pi r^2$. Thus,

$$\Delta A \approx dA = \frac{dA}{dr}\,dr = 2\pi r\,dr.$$

We divide this approximation by $A = \pi r^2$ to get an approximation that links the relative changes in A and r:

$$\frac{\Delta A}{A} \approx \frac{dA}{A} = \frac{2\pi r\,dr}{\pi r^2} = 2\,\frac{dr}{r}.$$

If r increases by 2%, then $dr = \frac{2}{100}\,r$, so

$$\frac{\Delta A}{A} \approx 2 \times \frac{2}{100} = \frac{4}{100}.$$

Thus, A increases by approximately 4%.

Average and Instantaneous Rates of Change

Recall the concept of average rate of change of a function over an interval, introduced in Section 1.1. The derivative of the function is the limit of this average rate as the length of the interval goes to zero, and so represents the rate of change of the function at a given value of its variable.

DEFINITION 5

The **average rate of change** of a function $f(x)$ with respect to x over the interval from a to $a + h$ is

$$\frac{f(a+h) - f(a)}{h}.$$

The **(instantaneous) rate of change** of f with respect to x at $x = a$ is the derivative

$$f'(a) = \lim_{h \to 0} \frac{f(a+h) - f(a)}{h},$$

provided the limit exists.

It is conventional to use the word *instantaneous* even when x does not represent time, although the word is frequently omitted. When we say *rate of change*, we mean *instantaneous rate of change*.

EXAMPLE 3 How fast is area A of a circle increasing with respect to its radius when the radius is 5 m?

Solution The rate of change of the area with respect to the radius is

$$\frac{dA}{dr} = \frac{d}{dr}(\pi r^2) = 2\pi r.$$

When $r = 5$ m, the area is changing at the rate $2\pi \times 5 = 10\pi$ m^2/m. This means that a small change Δr m in the radius when the radius is 5 m would result in a change of about $10\pi\,\Delta r$ m^2 in the area of the circle.

The above example suggests that the appropriate units for the rate of change of a quantity y with respect to another quantity x are units of y per unit of x.

If $f'(x_0) = 0$, we say that f is **stationary** at x_0 and call x_0 a **critical point** of f. The corresponding point $(x_0, f(x_0))$ on the graph of f is also called a **critical point** of the graph. The graph has a horizontal tangent at a critical point, and f may or may not have a maximum or minimum value there. (See Figure 2.26.) It is still possible for f to be increasing or decreasing on an open interval containing a critical point. (See point a in Figure 2.26.) We will revisit these ideas in the next section.

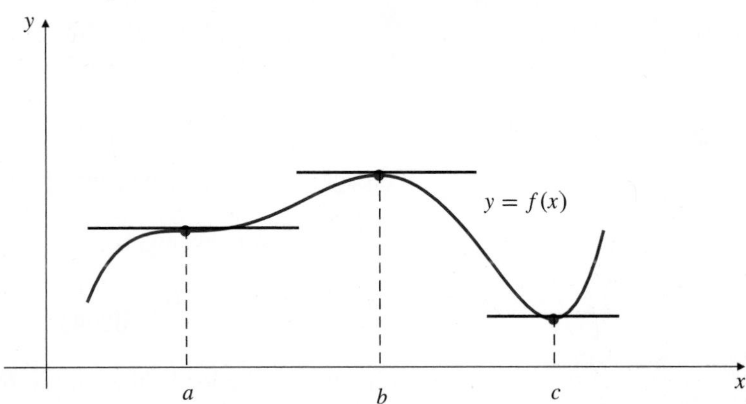

Figure 2.26 Critical points of f

EXAMPLE 4 Suppose the temperature at a certain location t hours after noon on a certain day is T °C (T degrees Celsius), where

$$T = \frac{1}{3}t^3 - 3t^2 + 8t + 10 \qquad (\text{for } 0 \le t \le 5).$$

How fast is the temperature rising or falling at 1:00 p.m.? at 3:00 p.m.? At what instants is the temperature stationary?

Solution The rate of change of the temperature is given by

$$\frac{dT}{dt} = t^2 - 6t + 8 = (t-2)(t-4).$$

If $t = 1$, then $\frac{dT}{dt} = 3$, so the temperature is *rising* at rate 3 °C/h at 1:00 p.m.

If $t = 3$, then $\frac{dT}{dt} = -1$, so the temperature is *falling* at a rate of 1 °C/h at 3:00 p.m.

The temperature is stationary when $\frac{dT}{dt} = 0$, that is, at 2:00 p.m. and 4:00 p.m.

Sensitivity to Change

When a small change in x produces a large change in the value of a function $f(x)$, we say that the function is very **sensitive** to changes in x. The derivative $f'(x)$ is a measure of the sensitivity of the dependence of f on x.

EXAMPLE 5 **(Dosage of a medicine)** A pharmacologist studying a drug that has been developed to lower blood pressure determines experimentally that the average reduction R in blood pressure resulting from a daily dosage of x mg of the drug is

$$R = 24.2\left(1 + \frac{x-13}{\sqrt{x^2-26x+529}}\right) \text{ mm Hg.}$$

(The units are millimetres of mercury (Hg).) Determine the sensitivity of R to dosage x at dosage levels of 5 mg, 15 mg, and 35 mg. At which of these dosage levels would an increase in the dosage have the greatest effect?

Solution The sensitivity of R to x is dR/dx. We have

$$\begin{aligned}\frac{dR}{dx} &= 24.2\left(\frac{\sqrt{x^2-26x+529}(1) - (x-13)\dfrac{x-13}{\sqrt{x^2-26x+529}}}{x^2-26x+529}\right)\\ &= 24.2\left(\frac{x^2-26x+529-(x^2-26x+169)}{(x^2-26x+529)^{3/2}}\right)\\ &= \frac{8{,}712}{(x^2-26x+529)^{3/2}}.\end{aligned}$$

At dosages $x = 5$ mg, 15 mg, and 35 mg, we have sensitivities of

$$\left.\frac{dR}{dx}\right|_{x=5} = 0.998 \text{ mm Hg/mg}, \qquad \left.\frac{dR}{dx}\right|_{x=15} = 1.254 \text{ mm Hg/mg},$$
$$\left.\frac{dR}{dx}\right|_{x=35} = 0.355 \text{ mm Hg/mg}.$$

Among these three levels, the greatest sensitivity is at 15 mg. Increasing the dosage from 15 to 16 mg/day could be expected to further reduce average blood pressure by about 1.25 mm Hg.

Derivatives in Economics

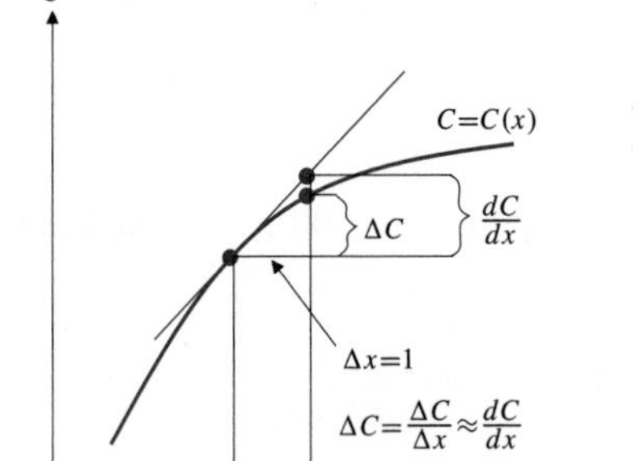

Figure 2.27 The marginal cost dC/dx is approximately the extra cost ΔC of producing $\Delta x = 1$ more unit

Just as physicists use terms such as *velocity* and *acceleration* to refer to derivatives of certain quantities, economists also have their own specialized vocabulary for derivatives. They call them marginals. In economics the term **marginal** denotes the rate of change of a quantity with respect to a variable on which it depends. For example, the **cost of production** $C(x)$ in a manufacturing operation is a function of x, the number of units of product produced. The **marginal cost of production** is the rate of change of C with respect to x, so it is dC/dx. Sometimes the marginal cost of production is loosely defined to be the extra cost of producing one more unit; that is,

$$\Delta C = C(x+1) - C(x).$$

To see why this is approximately correct, observe from Figure 2.27 that if the slope of $C = C(x)$ does not change quickly near x, then the difference quotient $\Delta C/\Delta x$ will be close to its limit, the derivative dC/dx, even if $\Delta x = 1$.

EXAMPLE 6 **(Marginal tax rates)** If your marginal income tax rate is 35% and your income increases by \$1,000, you can expect to have to pay an extra \$350 in income taxes. This does not mean that you pay 35% of your entire income in taxes. It just means that at your current income level I, the rate of increase of taxes T with respect to income is $dT/dI = 0.35$. You will pay \$0.35 out of every extra dollar you earn in taxes. Of course, if your income increases greatly, you may land in a higher tax bracket and your marginal rate will increase.

EXAMPLE 7 **(Marginal cost of production)** The cost of producing x tons of coal per day in a mine is $\$C(x)$, where

$$C(x) = 4{,}200 + 5.40x - 0.001x^2 + 0.000\,002x^3.$$

(a) What is the average cost of producing each ton if the daily production level is 1,000 tons? 2,000 tons?

(b) Find the marginal cost of production if the daily production level is 1,000 tons. 2,000 tons.

(c) If the production level increases slightly from 1,000 tons or from 2,000 tons, what will happen to the average cost per ton?

Solution

(a) The average cost per ton of coal is

$$\frac{C(x)}{x} = \frac{4,200}{x} + 5.40 - 0.001x + 0.000\,002x^2.$$

If $x = 1{,}000$, the average cost per ton is $C(1{,}000)/1{,}000 = \$10.6$ /ton. If $x = 2{,}000$, the average cost per ton is $C(2{,}000)/2{,}000 = \$13.5$ /ton.

(b) The marginal cost of production is

$$C'(x) = 5.40 - 0.002x + 0.000\,006x^2.$$

If $x = 1{,}000$, the marginal cost is $C'(1{,}000) = \$9.4$ /ton. If $x = 2{,}000$, the marginal cost is $C'(2{,}000) = \$25.4$ /ton.

(c) If the production level x is increased slightly from $x = 1{,}000$, then the average cost per ton will drop because the cost is increasing at a rate lower than the average cost. At $x = 2{,}000$ the opposite is true; an increase in production will increase the average cost per ton.

Economists sometimes prefer to measure relative rates of change that do not depend on the units used to measure the quantities involved. They use the term **elasticity** for such relative rates.

EXAMPLE 8 **(Elasticity of demand)** The demand y for a certain product (i.e., the amount that can be sold) typically depends on the price p charged for the product: $y = f(p)$. The marginal demand $dy/dp = f'(p)$ (which is typically negative) depends on the units used to measure y and p. The *elasticity of the demand* is the quantity

$$-\frac{p}{y}\frac{dy}{dp} \qquad \text{(the "$-$" sign ensures elasticity is positive),}$$

which is independent of units and provides a good measure of the sensitivity of demand to changes in price. To see this, suppose that new units of demand and price are introduced, which are multiples of the old units. In terms of the new units the demand and price are now Y and P, where

$$Y = k_1 y \qquad \text{and} \qquad P = k_2 p.$$

Thus, $Y = k_1 f(P/k_2)$ and $dY/dP = (k_1/k_2)f'(P/k_2) = (k_1/k_2)f'(p)$ by the Chain Rule. It follows that the elasticity has the same value:

$$-\frac{P}{Y}\frac{dY}{dP} = -\frac{k_2 p}{k_1 y}\frac{k_1}{k_2}f'(p) = -\frac{p}{y}\frac{dy}{dp}.$$

EXERCISES 2.7

In Exercises 1–4, use differentials to determine the approximate change in the value of the given function as its argument changes from the given value by the given amount. What is the approximate value of the function after the change?

1. $y = 1/x$, as x increases from 2 to 2.01.

2. $f(x) = \sqrt{3x+1}$, as x increases from 1 to 1.08.

3. $h(t) = \cos(\pi t/4)$, as t increases from 2 to $2 + (1/10\pi)$.

4. $u = \tan(s/4)$ as s decreases from π to $\pi - 0.04$.

In Exercises 5–10, find the approximate percentage changes in the given function $y = f(x)$ that will result from an increase of 2% in the value of x.

5. $y = x^2$

6. $y = 1/x$

7. $y = 1/x^2$

8. $y = x^3$

9. $y = \sqrt{x}$

10. $y = x^{-2/3}$

11. By approximately what percentage will the volume ($V = \frac{4}{3}\pi r^3$) of a ball of radius r increase if the radius increases by 2%?

12. By about what percentage will the edge length of an ice cube decrease if the cube loses 6% of its volume by melting?

13. Find the rate of change of the area of a square with respect to the length of its side when the side is 4 ft.

14. Find the rate of change of the side of a square with respect to the area of the square when the area is 16 m^2.

15. Find the rate of change of the diameter of a circle with respect to its area.

16. Find the rate of change of the area of a circle with respect to its diameter.

17. Find the rate of change of the volume of a sphere (given by $V = \frac{4}{3}\pi r^3$) with respect to its radius r when the radius is 2 m.

18. What is the rate of change of the area A of a square with respect to the length L of the diagonal of the square?

19. What is the rate of change of the circumference C of a circle with respect to the area A of the circle?

20. Find the rate of change of the side s of a cube with respect to the volume V of the cube.

21. The volume of water in a tank t min after it starts draining is

$$V(t) = 350(20 - t)^2 \text{ L}.$$

(a) How fast is the water draining out after 5 min? after 15 min?

(b) What is the average rate at which water is draining out during the time interval from 5 to 15 min?

22. **(Poiseuille's Law)** The flow rate F (in litres per minute) of a liquid through a pipe is proportional to the fourth power of the radius of the pipe:

$$F = kr^4.$$

Approximately what percentage increase is needed in the radius of the pipe to increase the flow rate by 10%?

23. **(Gravitational force)** The gravitational force F with which the earth attracts an object in space is given by $F = k/r^2$, where k is a constant and r is the distance from the object to the centre of the earth. If F decreases with respect to r at rate 1 *pound/mile* when $r = 4{,}000$ mi, how fast does F change with respect to r when $r = 8{,}000$ mi?

24. **(Sensitivity of revenue to price)** The sales revenue $\$R$ from a software product depends on the price $\$p$ charged by the distributor according to the formula

$$R = 4{,}000p - 10p^2.$$

(a) How sensitive is R to p when $p = \$100$? $p = \$200$? $p = \$300$?

(b) Which of these three is the most reasonable price for the distributor to charge? Why?

25. **(Marginal cost)** The cost of manufacturing x refrigerators is $\$C(x)$, where

$$C(x) = 8{,}000 + 400x - 0.5x^2.$$

(a) Find the marginal cost if 100 refrigerators are manufactured.

(b) Show that the marginal cost is approximately the difference in cost of manufacturing 101 refrigerators instead of 100.

26. **(Marginal profit)** If a plywood factory produces x sheets of plywood per day, its profit per day will be $\$P(x)$, where

$$P(x) = 8x - 0.005x^2 - 1{,}000.$$

(a) Find the marginal profit. For what values of x is the marginal profit positive? negative?

(b) How many sheets should be produced each day to generate maximum profits?

27. The cost C (in dollars) of producing n widgets per month in a widget factory is given by

$$C = \frac{80{,}000}{n} + 4n + \frac{n^2}{100}.$$

Find the marginal cost of production if the number of widgets manufactured each month is (a) 100 and (b) 300.

28. In a mining operation the cost C (in dollars) of extracting each tonne of ore is given by

$$C = 10 + \frac{20}{x} + \frac{x}{1{,}000},$$

where x is the number of tonnes extracted each day. (For small x, C decreases as x increases because of economies of scale, but for large x, C increases with x because of overloaded equipment and labour overtime.) If each tonne of ore can be sold for \$13, how many tonnes should be extracted each day to maximize the daily profit of the mine?

29. **(Average cost and marginal cost)** If it costs a manufacturer $C(x)$ dollars to produce x items, then his average cost of production is $C(x)/x$ dollars per item. Typically the average cost is a decreasing function of x for small x and an increasing function of x for large x. (Why?) Show that the value of x that minimizes the average cost makes the average cost equal to the marginal cost.

30. (Constant elasticity) Show that if demand y is related to price p by the equation $y = Cp^{-r}$, where C and r are positive constants, then the elasticity of demand (see Example 8) is the constant r.

2.8 The Mean-Value Theorem

If you set out in a car at 1:00 p.m. and arrive in a town 150 km away from your starting point at 3:00 p.m., then you have travelled at an average speed of $150/2 = 75$ km/h. Although you may not have travelled at constant speed, you must have been going 75 km/h at *at least one instant* during your journey, for if your speed was always less than 75 km/h you would have gone less than 150 km in 2 h, and if your speed was always more than 75 km/h, you would have gone more than 150 km in 2 h. In order to get from a value less than 75 km/h to a value greater than 75 km/h, your speed, which is a continuous function of time, must pass through the value 75 km/h at some intermediate time.

The conclusion that the average speed over a time interval must be equal to the instantaneous speed at some time in that interval is an instance of an important mathematical principle. In geometric terms it says that if A and B are two points on a smooth curve, then there is at least one point C on the curve between A and B where the tangent line is parallel to the chord line AB. See Figure 2.28.

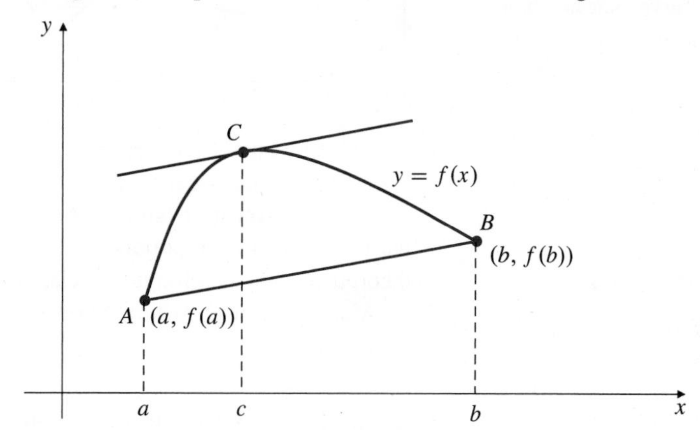

Figure 2.28 There is a point C on the curve where the tangent is parallel to the chord AB

This principle is stated more precisely in the following theorem.

THEOREM

The Mean-Value Theorem

Suppose that the function f is continuous on the closed, finite interval $[a, b]$ and that it is differentiable on the open interval (a, b). Then there exists a point c in the open interval (a, b) such that

$$\frac{f(b) - f(a)}{b - a} = f'(c).$$

This says that the slope of the chord line joining the points $(a, f(a))$ and $(b, f(b))$ is equal to the slope of the tangent line to the curve $y = f(x)$ at the point $(c, f(c))$, so the two lines are parallel.

We will prove the Mean-Value Theorem later in this section. For now we make several observations.

1. The hypotheses of the Mean-Value Theorem are all necessary for the conclusion; if f fails to be continuous at even one point of $[a, b]$ or fails to be differentiable at even one point of (a, b), then there may be no point where the tangent line is parallel to the secant line AB. (See Figure 2.29.)

2. The Mean-Value Theorem gives no indication of how many points C there may be on the curve between A and B where the tangent is parallel to AB. If the curve is itself the straight line AB, then every point on the line between A and B has the required property. In general, there may be more than one point (see Figure 2.30); the Mean-Value Theorem asserts only that there must be at least one.

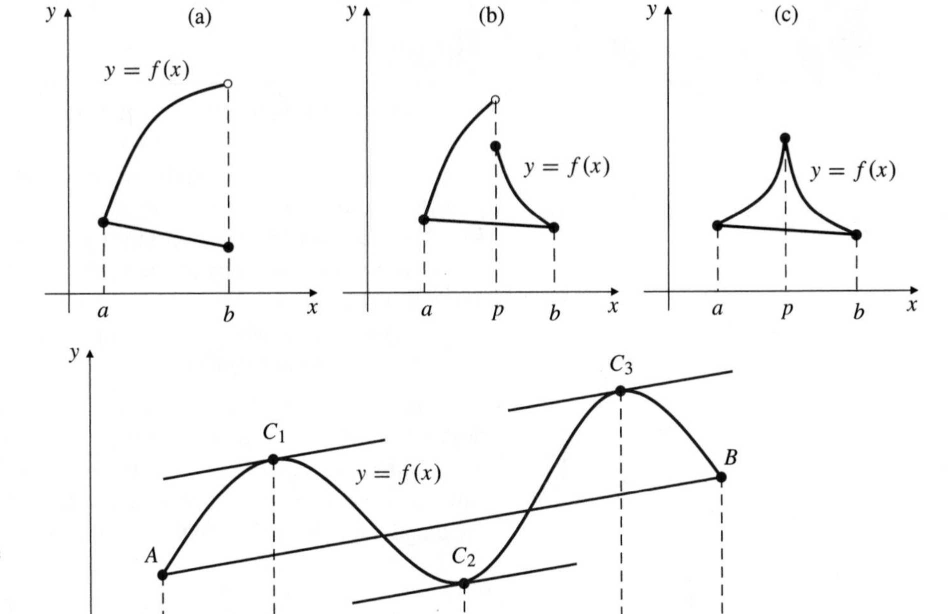

Figure 2.29 Functions that fail to satisfy the hypotheses of the Mean-Value Theorem and for which the conclusion is false:
(a) f is discontinuous at endpoint b
(b) f is discontinuous at p
(c) f is not differentiable at p

Figure 2.30 For this curve there are three points C where the tangent is parallel to the chord AB

3. The Mean-Value Theorem gives us no information on how to find the point c, which it says must exist. For some simple functions it is possible to calculate c (see the following example), but doing so is usually of no practical value. As we shall see, the importance of the Mean-Value Theorem lies in its use as a theoretical tool. It belongs to a class of theorems called *existence theorems*, as do the Max-Min Theorem and the Intermediate-Value Theorem (Theorems 8 and 9 of Section 1.4).

EXAMPLE 1 Verify the conclusion of the Mean-Value Theorem for $f(x) = \sqrt{x}$ on the interval $[a, b]$, where $0 \le a < b$.

Solution The theorem says that there must be a number c in the interval (a, b) such that

$$f'(c) = \frac{f(b) - f(a)}{b - a}$$

$$\frac{1}{2\sqrt{c}} = \frac{\sqrt{b} - \sqrt{a}}{b - a} = \frac{\sqrt{b} - \sqrt{a}}{(\sqrt{b} - \sqrt{a})(\sqrt{b} + \sqrt{a})} = \frac{1}{\sqrt{b} + \sqrt{a}}.$$

Thus, $2\sqrt{c} = \sqrt{a} + \sqrt{b}$ and $c = \left(\frac{\sqrt{b} + \sqrt{a}}{2}\right)^2$. Since $a < b$, we have

$$a = \left(\frac{\sqrt{a} + \sqrt{a}}{2}\right)^2 < c < \left(\frac{\sqrt{b} + \sqrt{b}}{2}\right)^2 = b,$$

so c lies in the interval (a, b).

The following two examples are more representative of how the Mean-Value Theorem is actually used.

EXAMPLE 2 Show that $\sin x < x$ for all $x > 0$.

Solution If $x > 2\pi$, then $\sin x \le 1 < 2\pi < x$. If $0 < x \le 2\pi$, then, by the Mean-Value Theorem, there exists c in the open interval $(0, 2\pi)$ such that

$$\frac{\sin x}{x} = \frac{\sin x - \sin 0}{x - 0} = \frac{d}{dx} \sin x \bigg|_{x=c} = \cos c < 1.$$

Thus, $\sin x < x$ in this case too.

EXAMPLE 3 Show that $\sqrt{1+x} < 1 + \frac{x}{2}$ for $x > 0$ and for $-1 \le x < 0$.

Solution If $x > 0$, apply the Mean-Value Theorem to $f(x) = \sqrt{1+x}$ on the interval $[0, x]$. There exists c in $(0, x)$ such that

$$\frac{\sqrt{1+x} - 1}{x} = \frac{f(x) - f(0)}{x - 0} = f'(c) = \frac{1}{2\sqrt{1+c}} < \frac{1}{2}.$$

The last inequality holds because $c > 0$. Multiplying by the positive number x and transposing the -1 gives $\sqrt{1+x} < 1 + \frac{x}{2}$.

If $-1 \le x < 0$, we apply the Mean-Value Theorem to $f(x) = \sqrt{1+x}$ on the interval $[x, 0]$. There exists c in $(x, 0)$ such that

$$\frac{\sqrt{1+x} - 1}{x} = \frac{1 - \sqrt{1+x}}{-x} = \frac{f(0) - f(x)}{0 - x} = f'(c) = \frac{1}{2\sqrt{1+c}} > \frac{1}{2}$$

(because $0 < 1 + c < 1$). Now we must multiply by the negative number x, which reverses the inequality, $\sqrt{1+x} - 1 < \frac{x}{2}$, and the required inequality again follows by transposing the -1.

Increasing and Decreasing Functions

Intervals on which the graph of a function f has positive or negative slope provide useful information about the behaviour of f. The Mean-Value Theorem enables us to determine such intervals by considering the sign of the derivative f'.

DEFINITION 6

Increasing and decreasing functions

Suppose that the function f is defined on an interval I and that x_1 and x_2 are two points of I.

(a) If $f(x_2) > f(x_1)$ whenever $x_2 > x_1$, we say f is **increasing** on I.
(b) If $f(x_2) < f(x_1)$ whenever $x_2 > x_1$, we say f is **decreasing** on I.
(c) If $f(x_2) \ge f(x_1)$ whenever $x_2 > x_1$, we say f is **nondecreasing** on I.
(d) If $f(x_2) \le f(x_1)$ whenever $x_2 > x_1$, we say f is **nonincreasing** on I.

Figure 2.31 illustrates these terms. Note the distinction between *increasing* and *nondecreasing*. If a function is increasing (or decreasing) on an interval, it must take different values at different points. (Such a function is called **one-to-one.**) A nondecreasing function (or a nonincreasing function) may be constant on a subinterval of its domain, and may therefore not be one-to-one. An increasing function is nondecreasing, but a nondecreasing function is not necessarily increasing.

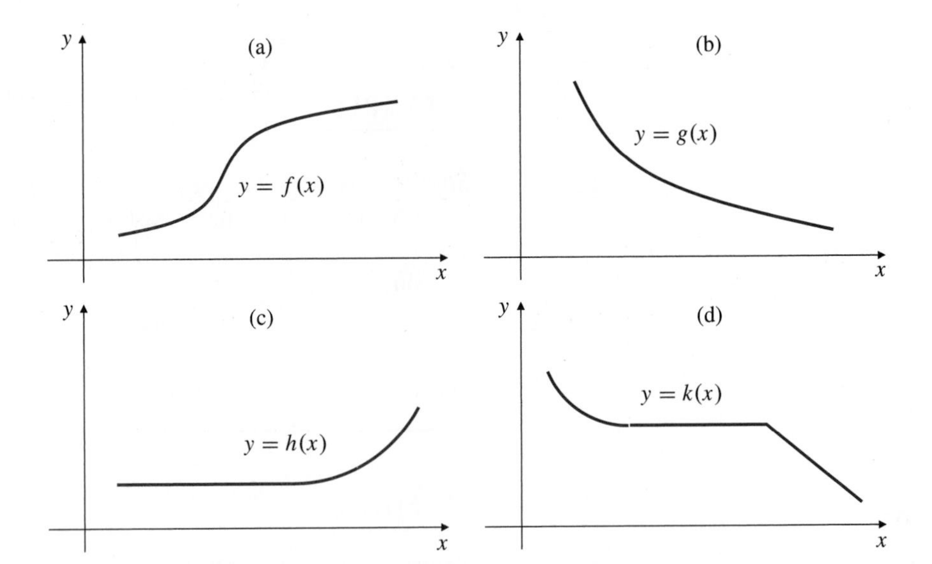

Figure 2.31
(a) Function f is increasing
(b) Function g is decreasing
(c) Function h is nondecreasing
(d) Function k is nonincreasing

THEOREM 12

Let J be an open interval, and let I be an interval consisting of all the points in J and possibly one or both of the endpoints of J. Suppose that f is continuous on I and differentiable on J.

(a) If $f'(x) > 0$ for all x in J, then f is increasing on I.
(b) If $f'(x) < 0$ for all x in J, then f is decreasing on I.
(c) If $f'(x) \geq 0$ for all x in J, then f is nondecreasing on I.
(d) If $f'(x) \leq 0$ for all x in J, then f is nonincreasing on I.

PROOF Let x_1 and x_2 be points in I with $x_2 > x_1$. By the Mean-Value Theorem there exists a point c in (x_1, x_2) (and therefore in J) such that

$$\frac{f(x_2) - f(x_1)}{x_2 - x_1} = f'(c);$$

hence, $f(x_2)-f(x_1) = (x_2-x_1)\, f'(c)$. Since $x_2-x_1 > 0$, the difference $f(x_2)-f(x_1)$ has the same sign as $f'(c)$ and may be zero if $f'(c)$ is zero. Thus, all four conclusions follow from the corresponding parts of Definition 6.

Remark Despite Theorem 12, $f'(x_0) > 0$ at a single point x_0 does *not* imply that f is increasing on *any* interval containing x_0. See Exercise 30 at the end of this section for a counterexample.

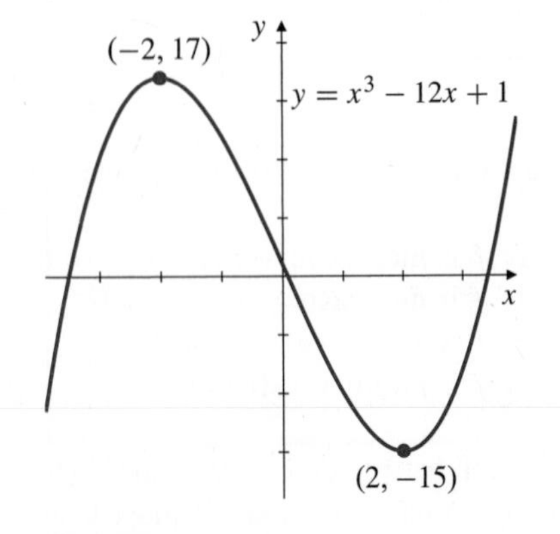

Figure 2.32

EXAMPLE 4 On what intervals is the function $f(x) = x^3 - 12x + 1$ increasing? On what intervals is it decreasing?

Solution We have $f'(x) = 3x^2 - 12 = 3(x-2)(x+2)$. Observe that $f'(x) > 0$ if $x < -2$ or $x > 2$ and $f'(x) < 0$ if $-2 < x < 2$. Therefore, f is increasing on the intervals $(-\infty, -2)$ and $(2, \infty)$ and is decreasing on the interval $(-2, 2)$. See Figure 2.32.

A function f whose derivative satisfies $f'(x) \geq 0$ on an interval can still be increasing there, rather than just nondecreasing as assured by Theorem 12(c). This will happen if $f'(x) = 0$ only at isolated points, so that f is assured to be increasing on intervals to the left and right of these points.

EXAMPLE 5 Show that $f(x) = x^3$ is increasing on any interval.

Solution Let x_1 and x_2 be any two real numbers satsifying $x_1 < x_2$. Since $f'(x) = 3x^2 > 0$ except at $x = 0$, Theorem 12(a) tells us that $f(x_1) < f(x_2)$ if either $x_1 < x_2 \le 0$ or $0 \le x_1 < x_2$. If $x_1 < 0 < x_2$, then $f(x_1) < 0 < f(x_2)$. Thus, f is increasing on every interval.

If a function is constant on an interval, then its derivative is zero on that interval. The Mean-Value Theorem provides a converse of this fact.

THEOREM 13

If f is continuous on an interval I, and $f'(x) = 0$ at every interior point of I (i.e., at every point of I that is not an endpoint of I), then $f(x) = C$, a constant, on I.

PROOF Pick a point x_0 in I and let $C = f(x_0)$. If x is any other point of I, then the Mean-Value Theorem says that there exists a point c between x_0 and x such that

$$\frac{f(x) - f(x_0)}{x - x_0} = f'(c).$$

The point c must belong to I because an interval contains all points between any two of its points, and c cannot be an endpoint of I since $c \ne x_0$ and $c \ne x$. Since $f'(c) = 0$ for all such points c, we have $f(x) - f(x_0) = 0$ for all x in I, and $f(x) = f(x_0) = C$ as claimed.

We will see how Theorem 13 can be used to establish identities for new functions encountered in later chapters. We will also use it when finding antiderivatives in Section 2.10.

Proof of the Mean-Value Theorem

The Mean-Value Theorem is one of those deep results that is based on the completeness of the real number system via the fact that a continuous function on a closed, finite interval takes on a maximum and minimum value (Theorem 8 of Section 1.4). Before giving the proof we establish two preliminary results.

THEOREM

If f is defined on an open interval (a, b) and achieves a maximum (or minimum) value at the point c in (a, b), and if $f'(c)$ exists, then $f'(c) = 0$. (Values of x where $f'(x) = 0$ are called **critical points** of the function f.)

PROOF Suppose that f has a maximum value at c. Then $f(x) - f(c) \le 0$ whenever x is in (a, b). If $c < x < b$, then

$$\frac{f(x) - f(c)}{x - c} \le 0, \qquad \text{so} \quad f'(c) = \lim_{x \to c+} \frac{f(x) - f(c)}{x - c} \le 0.$$

Similarly, if $a < x < c$, then

$$\frac{f(x) - f(c)}{x - c} \ge 0, \qquad \text{so} \quad f'(c) = \lim_{x \to c-} \frac{f(x) - f(c)}{x - c} \ge 0.$$

Thus $f'(c) = 0$. The proof for a minimum value at c is similar.

THEOREM

Rolle's Theorem

Suppose that the function g is continuous on the closed, finite interval $[a, b]$ and that it is differentiable on the open interval (a, b). If $g(a) = g(b)$, then there exists a point c in the open interval (a, b) such that $g'(c) = 0$.

PROOF If $g(x) = g(a)$ for every x in $[a, b]$, then g is a constant function, so $g'(c) = 0$ for every c in (a, b). Therefore, suppose there exists x in (a, b) such that $g(x) \ne g(a)$. Let us assume that $g(x) > g(a)$. (If $g(x) < g(a)$, the proof is similar.) By the Max-Min Theorem (Theorem 8 of Section 1.4), being continuous on $[a, b]$, g must have a maximum value at some point c in $[a, b]$. Since $g(c) \ge g(x) > g(a) = g(b)$, c cannot be either a or b. Therefore, c is in the open interval (a, b), so g is differentiable at c. By Theorem 14, c must be a critical point of g: $g'(c) = 0$.

Remark Rolle's Theorem is a special case of the Mean-Value Theorem in which the chord line has slope 0, so the corresponding parallel tangent line must also have slope 0. We can deduce the Mean-Value Theorem from this special case.

Proof of the Mean-Value Theorem Suppose f satisfies the conditions of the Mean-Value Theorem. Let

$$g(x) = f(x) - \left(f(a) + \frac{f(b) - f(a)}{b - a}(x - a) \right).$$

(For $a \le x \le b$, $g(x)$ is the vertical displacement between the curve $y = f(x)$ and the chord line

$$y = f(a) + \frac{f(b) - f(a)}{b - a}(x - a)$$

joining $(a, f(a))$ and $(b, f(b))$. See Figure 2.33.)

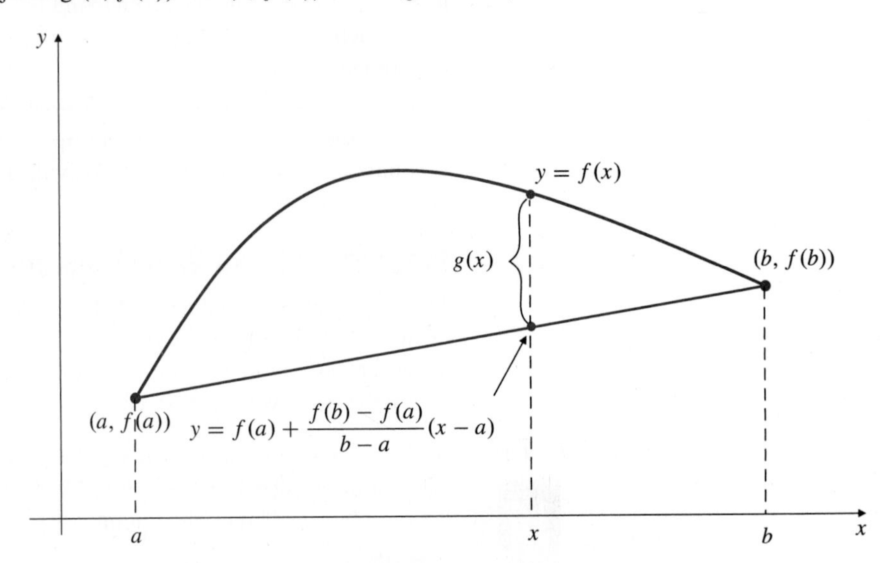

Figure 2.33 $g(x)$ is the vertical distance between the graph of f and the chord line

The function g is also continuous on $[a, b]$ and differentiable on (a, b) because f has these properties. In addition, $g(a) = g(b) = 0$. By Rolle's Theorem, there is some point c in (a, b) such that $g'(c) = 0$. Since

$$g'(x) = f'(x) - \frac{f(b) - f(a)}{b - a},$$

it follows that

$$f'(c) = \frac{f(b) - f(a)}{b - a}.$$

∎

Many of the applications we will make of the Mean-Value Theorem in later chapters will actually use the following generalized version of it.

THEOREM 16

The Generalized Mean-Value Theorem

If functions f and g are both continuous on $[a, b]$ and differentiable on (a, b), and if $g'(x) \neq 0$ for every x in (a, b), then there exists a number c in (a, b) such that

$$\frac{f(b) - f(a)}{g(b) - g(a)} = \frac{f'(c)}{g'(c)}.$$

PROOF Note that $g(b) \neq g(a)$; otherwise, there would be some number in (a, b) where $g' = 0$. Hence, neither denominator above can be zero. Apply the Mean-Value Theorem to

$$h(x) = \big(f(b) - f(a)\big)\big(g(x) - g(a)\big) - \big(g(b) - g(a)\big)\big(f(x) - f(a)\big).$$

Since $h(a) = h(b) = 0$, there exists c in (a, b) such that $h'(c) = 0$. Thus,

$$\big(f(b) - f(a)\big)g'(c) - \big(g(b) - g(a)\big)f'(c) = 0,$$

and the result follows on division by the g factors.

EXERCISES 2.8

In Exercises 1–3, illustrate the Mean-Value Theorem by finding any points in the open interval (a, b) where the tangent line to $y = f(x)$ is parallel to the chord line joining $(a, f(a))$ and $(b, f(b))$.

1. $f(x) = x^2$ on $[a, b]$

2. $f(x) = \dfrac{1}{x}$ on $[1, 2]$

3. $f(x) = x^3 - 3x + 1$ on $[-2, 2]$

4. By applying the Mean-Value Theorem to $f(x) = \cos x + \dfrac{x^2}{2}$ on the interval $[0, x]$, and using the result of Example 2, show that

$$\cos x > 1 - \frac{x^2}{2}$$

for $x > 0$. This inequality is also true for $x < 0$. Why?

5. Show that $\tan x > x$ for $0 < x < \pi/2$.

6. Let $r > 1$. If $x > 0$ or $-1 \le x < 0$, show that $(1 + x)^r > 1 + rx$.

7. Let $0 < r < 1$. If $x > 0$ or $-1 \le x < 0$, show that $(1 + x)^r < 1 + rx$.

Find the intervals of increase and decrease of the functions in Exercises 8–19.

8. $f(x) = x^3 - 12x + 1$

9. $f(x) = x^2 - 4$

10. $y = 1 - x - x^5$

11. $y = x^3 + 6x^2$

12. $f(x) = x^2 + 2x + 2$

13. $f(x) = x^3 - 4x + 1$

14. $f(x) = x^3 + 4x + 1$

15. $f(x) = (x^2 - 4)^2$

16. $f(x) = \dfrac{1}{x^2 + 1}$

17. $f(x) = x^3(5 - x)^2$

18. $f(x) = x - 2\sin x$

19. $f(x) = x + \sin x$

20. On what intervals is $f(x) = x + 2\sin x$ increasing?

21. Show that $f(x) = x^3$ is increasing on the whole real line even though $f'(x)$ is not positive at every point.

22. What is wrong with the following "proof" of the Generalized Mean-Value Theorem? By the Mean-Value Theorem, $f(b) - f(a) = (b - a)f'(c)$ for some c between a and b and, similarly, $g(b) - g(a) = (b - a)g'(c)$ for some such c. Hence, $(f(b) - f(a))/(g(b) - g(a)) = f'(c)/g'(c)$, as required.

Use a graphing utility or a computer algebra system to find the critical points of the functions in Exercises 23–26 correct to 6 decimal places.

23. $f(x) = \dfrac{x^2 - x}{x^2 - 4}$

24. $f(x) = \dfrac{2x + 1}{x^2 + x + 1}$

25. $f(x) = x - \sin\left(\dfrac{x}{x^2 + x + 1}\right)$

26. $f(x) = \dfrac{\sqrt{1 - x^2}}{\cos(x + 0.1)}$

27. If $f(x)$ is differentiable on an interval I and vanishes at $n \ge 2$ distinct points of I, prove that $f'(x)$ must vanish at at least $n - 1$ points in I.

28. Let $f(x) = x^2 \sin(1/x)$ if $x \neq 0$ and $f(0) = 0$. Show that $f'(x)$ exists at every x but f' is not continuous at $x = 0$. This proves the assertion (made at the end of Section 2.2) that a derivative, defined on an interval, need not be continuous there.

29. Prove the assertion (made at the end of Section 2.2) that a derivative, defined on an interval, must have the intermediate-value property. (*Hint:* Assume that f' exists on $[a, b]$ and $f'(a) \neq f'(b)$. If k lies between $f'(a)$ and $f'(b)$, show that the function g defined by $g(x) = f(x) - kx$ must have *either* a maximum value *or* a minimum value on $[a, b]$ occurring at an interior point c in (a, b). Deduce that $f'(c) = k$.)

30. Let $f(x) = \begin{cases} x + 2x^2 \sin(1/x) & \text{if } x \neq 0, \\ 0 & \text{if } x = 0. \end{cases}$

(a) Show that $f'(0) = 1$. (*Hint:* Use the definition of derivative.)

(b) Show that any interval containing $x = 0$ also contains points where $f'(x) < 0$, so f cannot be increasing on such an interval.

31. If $f''(x)$ exists on an interval I and if f vanishes at at least three distinct points of I, prove that f'' must vanish at some point in I.

32. Generalize Exercise 31 to a function for which $f^{(n)}$ exists on I and for which f vanishes at at least $n + 1$ distinct points in I.

33. Suppose f is twice differentiable on an interval I (i.e., f'' exists on I). Suppose that the points 0 and 2 belong to I and that $f(0) = f(1) = 0$ and $f(2) = 1$. Prove that:

(a) $f'(a) = \frac{1}{2}$ for some point a in I.

(b) $f''(b) > \frac{1}{2}$ for some point b in I.

(c) $f'(c) = \frac{1}{7}$ for some point c in I.

2.9 Implicit Differentiation

We know how to find the slope of a curve that is the graph of a function $y = f(x)$ by calculating the derivative of f. But not all curves are the graphs of such functions. To be the graph of a function $f(x)$, the curve must not intersect any vertical lines at more than one point.

Curves are generally the graphs of *equations* in two variables. Such equations can be written in the form

$$F(x, y) = 0,$$

where $F(x, y)$ denotes an expression involving the two variables x and y. For example, a circle with centre at the origin and radius 5 has equation

$$x^2 + y^2 - 25 = 0,$$

so $F(x, y) = x^2 + y^2 - 25$ for that circle.

Sometimes we can solve an equation $F(x, y) = 0$ for y and so find explicit formulas for one or more functions $y = f(x)$ defined by the equation. Usually, however, we are not able to solve the equation. However, we can still regard it as defining y as one or more functions of x *implicitly*, even it we cannot solve for these functions *explicitly*. Moreover, we still find the derivative dy/dx of these implicit solutions by a technique called **implicit differentiation**. The idea is to differentiate the given equation with respect to x, regarding y as a function of x having derivative dy/dx, or y'.

EXAMPLE 1 Find dy/dx if $y^2 = x$.

Solution The equation $y^2 = x$ defines two differentiable functions of x; in this case we know them explicitly. They are $y_1 = \sqrt{x}$ and $y_2 = -\sqrt{x}$ (See Figure 2.34), having derivatives defined for $x > 0$ by

$$\frac{dy_1}{dx} = \frac{1}{2\sqrt{x}} \quad \text{and} \quad \frac{dy_2}{dx} = -\frac{1}{2\sqrt{x}}.$$

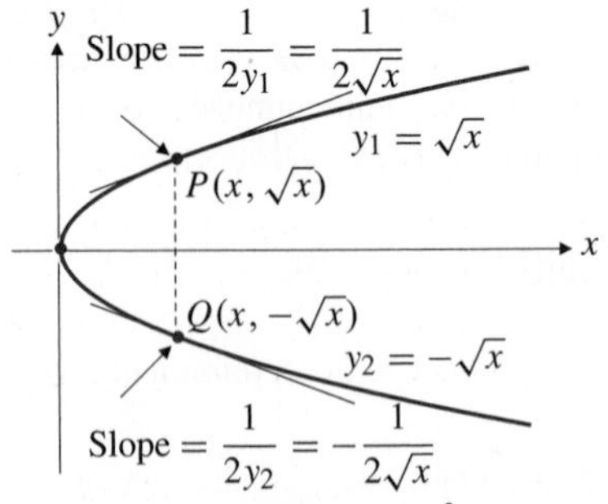

Figure 2.34 The equation $y^2 = x$ defines two differentiable functions of x on the interval $x \geq 0$

However, we can find the slope of the curve $y^2 = x$ at any point (x, y) satisfying that equation without first solving the equation for y. To find dy/dx we simply differentiate both sides of the equation $y^2 = x$ with respect to x, treating y as a differentiable function of x and using the Chain Rule to differentiate y^2:

$$\frac{d}{dx}(y^2) = \frac{d}{dx}(x) \qquad \left(\text{The Chain Rule gives } \frac{d}{dx}y^2 = 2y\frac{dy}{dx}.\right)$$

$$2y\frac{dy}{dx} = 1$$

$$\frac{dy}{dx} = \frac{1}{2y}.$$

Observe that this agrees with the derivatives we calculated above for *both* of the explicit solutions $y_1 = \sqrt{x}$ and $y_2 = -\sqrt{x}$:

$$\frac{dy_1}{dx} = \frac{1}{2y_1} = \frac{1}{2\sqrt{x}} \quad \text{and} \quad \frac{dy_2}{dx} = \frac{1}{2y_2} = \frac{1}{2(-\sqrt{x})} = -\frac{1}{2\sqrt{x}}.$$

EXAMPLE 2 Find the slope of circle $x^2 + y^2 = 25$ at the point $(3, -4)$.

Solution The circle is not the graph of a single function of x. Again, it combines the graphs of two functions, $y_1 = \sqrt{25 - x^2}$ and $y_2 = -\sqrt{25 - x^2}$ (Figure 2.35). The point $(3, -4)$ lies on the graph of y_2, so we can find the slope by calculating explicitly:

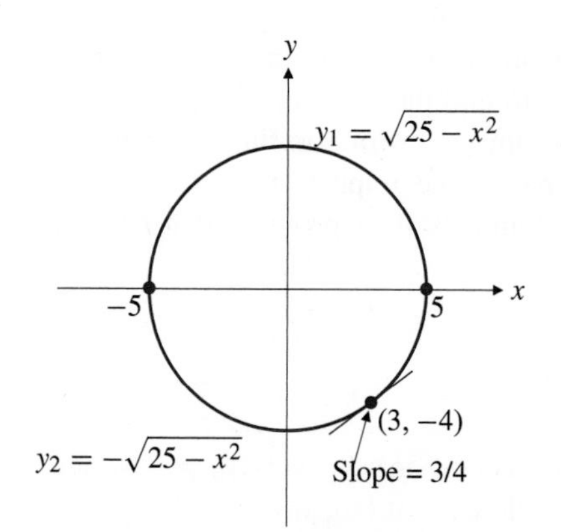

Figure 2.35 The circle combines the graphs of two functions. The graph of y_2 is the lower semicircle and passes through $(3, -4)$

$$\left.\frac{dy_2}{dx}\right|_{x=3} = -\left.\frac{-2x}{2\sqrt{25 - x^2}}\right|_{x=3} = -\frac{-6}{2\sqrt{25 - 9}} = \frac{3}{4}.$$

But we can also solve the problem more easily by differentiating the given equation of the circle implicitly with respect to x:

$$\begin{aligned}\frac{d}{dx}(x^2) + \frac{d}{dx}(y^2) &= \frac{d}{dx}(25)\\ 2x + 2y\frac{dy}{dx} &= 0\\ \frac{dy}{dx} &= -\frac{x}{y}.\end{aligned}$$

The slope at $(3, -4)$ is $-\left.\frac{x}{y}\right|_{(3,-4)} = -\frac{3}{-4} = \frac{3}{4}$.

EXAMPLE 3 Find $\frac{dy}{dx}$ if $y \sin x = x^3 + \cos y$.

Solution This time we cannot solve the equation for y as an explicit function of x, so we *must* use implicit differentiation:

$$\begin{aligned}\frac{d}{dx}(y \sin x) &= \frac{d}{dx}(x^3) + \frac{d}{dx}(\cos y) \qquad \left(\begin{array}{l}\text{Use the Product Rule}\\ \text{on the left side.}\end{array}\right)\\ (\sin x)\frac{dy}{dx} + y \cos x &= 3x^2 - (\sin y)\frac{dy}{dx}\\ (\sin x + \sin y)\frac{dy}{dx} &= 3x^2 - y \cos x\\ \frac{dy}{dx} &= \frac{3x^2 - y \cos x}{\sin x + \sin y}.\end{aligned}$$

To find dy/dx by implicit differentiation:

1. Differentiate both sides of the equation with respect to x, regarding y as a function of x and using the Chain Rule to differentiate functions of y.
2. Collect terms with dy/dx on one side of the equation and solve for dy/dx by dividing by its coefficient.

In the examples above, the derivatives dy/dx calculated by implicit differentiation depend on y, or on both y and x, rather than just on x. This is to be expected because an equation in x and y can define more than one function of x, and the implicitly calculated derivative must apply to each of the solutions. For example, in Example 2, the derivative $dy/dx = -x/y$ also gives the slope $-3/4$ at the point $(3, 4)$ on the circle. When you use implicit differentiation to find the slope of a curve at a point, you will usually have to know both coordinates of the point.

There are subtle dangers involved in calculating derivatives implicitly. When you use the Chain Rule to differentiate an equation involving y with respect to x, you are automatically assuming that the equation defines y as a differentiable function of x. This need not be the case. To see what can happen, consider the problem of finding $y' = dy/dx$ from the equation

$$x^2 + y^2 = K, \qquad (*)$$

where K is a constant. As in Example 2 (where $K = 25$), implicit differentiation gives

$$2x + 2yy' = 0 \qquad \text{or} \qquad y' = -\frac{x}{y}.$$

This formula will give the slope of the curve (∗) at any point on the curve where $y \neq 0$. For $K > 0$, (∗) represents a circle centred at the origin and having radius $\sqrt{K}$. This circle has a finite slope, except at the two points where it crosses the x-axis (where $y = 0$). If $K = 0$, the equation represents only a single point, the origin. The concept of slope of a point is meaningless. For $K < 0$, there are no real points whose coordinates satisfy equation (∗), so y' is meaningless here too. The point of this is that being able to calculate y' from a given equation by implicit differentiation does not guarantee that y' actually represents the slope of anything.

If (x_0, y_0) is a point on the graph of the equation $F(x, y) = 0$, there is a theorem that can justify our use of implicit differentiation to find the slope of the graph there. We cannot give a careful statement or proof of this **implicit function theorem** yet (see Section 12.8), but roughly speaking, it says that part of the graph of $F(x, y) = 0$ near (x_0, y_0) is the graph of a function of x that is differentiable at x_0, provided that $F(x, y)$ is a "smooth" function, and that the derivative

$$\frac{d}{dy}F(x_0, y)\bigg|_{y=y_0} \neq 0.$$

For the circle $x^2 + y^2 - K = 0$ (where $K > 0$) this condition says that $2y_0 \neq 0$, which is the condition that the derivative $y' = -x/y$ should exist at (x_0, y_0).

EXAMPLE 4 Find an equation of the tangent to $x^2 + xy + 2y^3 = 4$ at $(-2, 1)$.

Solution Note that $(-2, 1)$ does lie on the given curve. To find the slope of the tangent we differentiate the given equation implicitly with respect to x. Use the Product Rule to differentiate the xy term:

$$2x + y + xy' + 6y^2y' = 0.$$

Substitute the coordinates $x = -2$, $y = 1$, and solve the resulting equation for y':

$$-4 + 1 - 2y' + 6y' = 0 \quad \Rightarrow \quad y' = \frac{3}{4}.$$

The slope of the tangent at $(-2, 1)$ is $3/4$, and its equation is

$$y = \frac{3}{4}(x + 2) + 1 \qquad \text{or} \qquad 3x - 4y = -10.$$

A useful strategy

When you use implicit differentiation to find the value of a derivative at a particular point, it is best to substitute the coordinates of the point immediately after you carry out the differentiation and before you solve for the derivative dy/dx. It is easier to solve an equation involving numbers than one with algebraic expressions.

EXAMPLE 5 Show that for any constants a and b, the curves $x^2 - y^2 = a$ and $xy = b$ intersect at right angles, that is, at any point where they intersect their tangents are perpendicular.

Solution The slope at any point on $x^2 - y^2 = a$ is given by $2x - 2yy' = 0$, or $y' = x/y$. The slope at any point on $xy = b$ is given by $y + xy' = 0$, or $y' = -y/x$. If the two curves (they are both hyperbolas if $a \neq 0$ and $b \neq 0$) intersect at (x_0, y_0), then their slopes at that point are x_0/y_0 and $-y_0/x_0$, respectively. Clearly, these slopes are negative reciprocals, so the tangent line to one curve is the normal line to the other at that point. Hence, the curves intersect at right angles. (See Figure 2.36.)

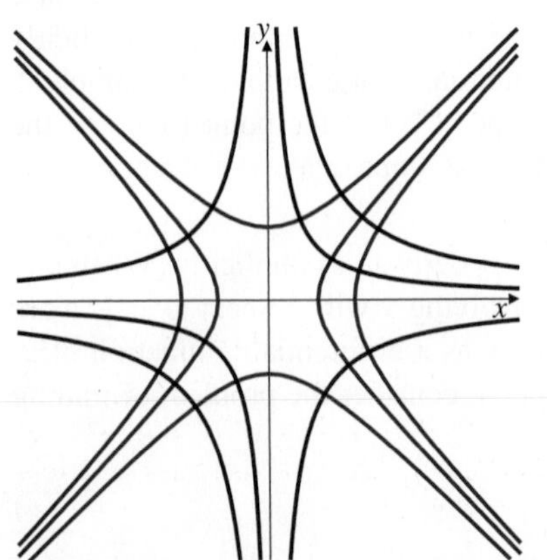

Figure 2.36 Some hyperbolas in the family $x^2 - y^2 = a$ (colour) intersecting some hyperbolas in the family $xy = b$ (black) at right angles

Higher-Order Derivatives

EXAMPLE 6 Find $y'' = \dfrac{d^2y}{dx^2}$ if $xy + y^2 = 2x$.

Solution Twice differentiate both sides of the given equation with respect to x:

$$y + xy' + 2yy' = 2$$
$$y' + y' + xy'' + 2(y')^2 + 2yy'' = 0.$$

Now solve these equations for y' and y''.

$$y' = \frac{2-y}{x+2y}$$

$$\begin{aligned} y'' &= -\frac{2y' + 2(y')^2}{x+2y} = -2\frac{2-y}{x+2y}\,\frac{1+\dfrac{2-y}{x+2y}}{x+2y} \\ &= -2\frac{(2-y)(x+y+2)}{(x+2y)^3} \\ &= -2\frac{2x - xy + 2y - y^2 + 4 - 2y}{(x+2y)^3} = -\frac{8}{(x+2y)^3}. \end{aligned}$$

(We used the given equation to simplify the numerator in the last line.)

Note that Maple uses the symbol ∂ instead of d when expressing the derivative in Leibniz form. This is because the expression it is differentiating can involve more than one variable; $(\partial/\partial x)y$ denotes the derivative of y with respect to the specific variable x rather than any other variables on which y may depend. It is called a **partial derivative**. We will study partial derivatives in Chapter 12. For the time being, just regard ∂ as a d.

Remark We can use Maple to calculate derivatives implicitly provided we show explicitly which variable depends on which. For example, we can calculate the value of y'' for the curve $xy + y^3 = 3$ at the point $(2, 1)$ as follows. First, we differentiate the equation with respect to x, writing $y(x)$ for y to indicate to Maple that it depends on x.

```
> deq := diff(x*y(x)+(y(x))^3=3, x);
```

$$deq := y(x) + x\left(\frac{\partial}{\partial x}y(x)\right) + 3y(x)^2\left(\frac{\partial}{\partial x}y(x)\right) = 0$$

Now we solve the resulting equation for y':

```
> yp := solve(deq, diff(y(x),x));
```

$$yp := -\frac{y(x)}{x + 3y(x)^2}$$

We can now differentiate yp with respect to x to get y'':

```
> ypp := diff(yp,x);
```

$$ypp := -\frac{\frac{\partial}{\partial x}y(x)}{x+3y(x)^2} + \frac{y(x)\left(1 + 6y(x)\left(\frac{\partial}{\partial x}y(x)\right)\right)}{(x+3y(x)^2)^2}$$

To get an expression depending only on x and y, we need to substitute the expression obtained for the first derivative into this result. Since the result of this substitution will involve compound fractions, let us simplify the result as well.

```
> ypp := simplify(subs(diff(y(x),x)=yp, ypp);
```

$$ypp := 2\frac{x\,y(x)}{(x+3y(x)^2)^3}$$

This is y'' expressed as a function of x and y. Now we want to substitute the coordinates $x = 2$, $y(x) = 1$ to get the value of y'' at $(2, 1)$. However, the order of the substitutions is important. *First* we must replace $y(x)$ with 1 and *then* replace x with 2. (If we replace x first, we would have to then replace $y(2)$ rather than $y(x)$ with 1.) Maple's `subs` command makes the substitutions in the order they are written.

```
> subs(y(x)=1, x=2, ypp);
```

$$\frac{4}{125}$$

The General Power Rule

Until now, we have only proven the General Power Rule

$$\frac{d}{dx}x^r = r\,x^{r-1}$$

for integer exponents r and a few special rational exponents such as $r = 1/2$. Using implicit differentiation, we can give the proof for any rational exponent $r = m/n$, where m and n are integers, and $n \neq 0$.

If $y = x^{m/n}$, then $y^n = x^m$. Differentiating implicitly with respect to x, we obtain

$$n\,y^{n-1}\frac{dy}{dx} = m\,x^{m-1}, \qquad \text{so}$$

$$\frac{dy}{dx} = \frac{m}{n}x^{m-1}\,y^{1-n} = \frac{m}{n}x^{m-1}\,x^{(m/n)(1-n)} = \frac{m}{n}x^{m-1+(m/n)-m} = \frac{m}{n}x^{(m/n)-1}.$$

EXERCISES 2.9

In Exercises 1–8, find dy/dx in terms of x and y.

1. $xy - x + 2y = 1$

2. $x^3 + y^3 = 1$

3. $x^2 + xy = y^3$

4. $x^3y + xy^5 = 2$

5. $x^2y^3 = 2x - y$

6. $x^2 + 4(y-1)^2 = 4$

7. $\dfrac{x-y}{x+y} = \dfrac{x^2}{y} + 1$

8. $x\sqrt{x+y} = 8 - xy$

In Exercises 9–16, find an equation of the tangent to the given curve at the given point.

9. $2x^2 + 3y^2 = 5$ at $(1, 1)$

10. $x^2y^3 - x^3y^2 = 12$ at $(-1, 2)$

11. $\dfrac{x}{y} + \left(\dfrac{y}{x}\right)^3 = 2$ at $(-1, -1)$

12. $x + 2y + 1 = \dfrac{y^2}{x-1}$ at $(2, -1)$

13. $2x + y - \sqrt{2}\sin(xy) = \pi/2$ at $\left(\dfrac{\pi}{4}, 1\right)$

14. $\tan(xy^2) = \dfrac{2xy}{\pi}$ at $\left(-\pi, \dfrac{1}{2}\right)$

15. $x\sin(xy - y^2) = x^2 - 1$ at $(1, 1)$

16. $\cos\left(\dfrac{\pi y}{x}\right) = \dfrac{x^2}{y} - \dfrac{17}{2}$ at $(3, 1)$

In Exercises 17–20, find y'' in terms of x and y.

17. $xy = x + y$

18. $x^2 + 4y^2 = 4$

19. $x^3 - y^2 + y^3 = x$

20. $x^3 - 3xy + y^3 = 1$

21. For $x^2 + y^2 = a^2$ show that $y'' = -\dfrac{a^2}{y^3}$.

22. For $Ax^2 + By^2 = C$ show that $y'' = -\dfrac{AC}{B^2y^3}$.

Use Maple or another computer algebra program to find the values requested in Exercises 23–26.

23. Find the slope of $x + y^2 + y\sin x = y^3 + \pi$ at $(\pi, 1)$.

24. Find the slope of $\dfrac{x+\sqrt{y}}{y+\sqrt{x}} = \dfrac{3y-9x}{x+y}$ at the point $(1, 4)$.

25. If $x + y^5 + 1 = y + x^4 + xy^2$, find d^2y/dx^2 at $(1, 1)$.

26. If $x^3y + xy^3 = 11$, find d^3y/dx^3 at $(1, 2)$.

27. Show that the ellipse $x^2 + 2y^2 = 2$ and the hyperbola $2x^2 - 2y^2 = 1$ intersect at right angles.

28. Show that the ellipse $x^2/a^2 + y^2/b^2 = 1$ and the hyperbola $x^2/A^2 - y^2/B^2 = 1$ intersect at right angles if $A^2 \le a^2$ and $a^2 - b^2 = A^2 + B^2$. (This says that the ellipse and the hyperbola have the same foci.)

29. If $z = \tan\dfrac{x}{2}$, show that

$$\frac{dx}{dz} = \frac{2}{1+z^2}, \quad \sin x = \frac{2z}{1+z^2}, \quad \text{and} \quad \cos x = \frac{1-z^2}{1+z^2}.$$

30. Use implicit differentiation to find y' if y is defined by $(x-y)/(x+y) = x/y + 1$. Now show that there are, in fact, no points on that curve, so the derivative you calculated is meaningless. This is another example that demonstrates the dangers of calculating something when you don't know whether or not it exists.

2.10 Antiderivatives and Initial-Value Problems

Throughout this chapter we have been concerned with the problem of finding the derivative f' of a given function f. The reverse problem—given the derivative f', find f—is also interesting and important. It is the problem studied in *integral calculus* and is generally more difficult to solve than the problem of finding a derivative. We will take a preliminary look at this problem in this section and will return to it in more detail in Chapter 5.

Antiderivatives

We begin by defining an antiderivative of a function f to be a function F whose derivative is f. It is appropriate to require that $F'(x) = f(x)$ on an *interval*.

DEFINITION 7

An **antiderivative** of a function f on an interval I is another function F satisfying

$$F'(x) = f(x) \quad \text{for } x \text{ in } I.$$

EXAMPLE 1

(a) $F(x) = x$ is an antiderivative of the function $f(x) = 1$ on any interval because $F'(x) = 1 = f(x)$ everywhere.

(b) $G(x) = \frac{1}{2}x^2$ is an antiderivative of the function $g(x) = x$ on any interval because $G'(x) = \frac{1}{2}(2x) = x = g(x)$ everywhere.

(c) $R(x) = -\frac{1}{3}\cos(3x)$ is an antiderivative of $r(x) = \sin(3x)$ on any interval because $R'(x) = -\frac{1}{3}(-3\sin(3x)) = \sin(3x) = r(x)$ everywhere.

(d) $F(x) = -1/x$ is an antiderivative of $f(x) = 1/x^2$ on any interval not containing $x = 0$ because $F'(x) = 1/x^2 = f(x)$ everywhere except at $x = 0$.

Antiderivatives are not unique; since a constant has derivative zero, you can always add any constant to an antiderivative F of a function f on an interval and get another antiderivative of f on that interval. More importantly, *all* antiderivatives of f on an interval can be obtained by adding constants to any particular one. If F and G are both antiderivatives of f on an interval I, then

$$\frac{d}{dx}\big(G(x) - F(x)\big) = f(x) - f(x) = 0$$

on I, so $G(x) - F(x) = C$ (a constant) on I by Theorem 13 of Section 2.8. Thus, $G(x) = F(x) + C$ on I.

Note that neither this conclusion nor Theorem 13 is valid over a set that is not an interval. For example, the derivative of

$$\operatorname{sgn} x = \begin{cases} -1 & \text{if } x < 0 \\ 1 & \text{if } x > 0 \end{cases}$$

is 0 for all $x \neq 0$, but $\operatorname{sgn} x$ is not constant for all $x \neq 0$. $\operatorname{sgn} x$ has *different* constant values on the two intervals $(-\infty, 0)$ and $(0, \infty)$ comprising its domain.

The Indefinite Integral

The *general antiderivative* of a function $f(x)$ on an interval I is $F(x) + C$, where $F(x)$ is any particular antiderivative of $f(x)$ on I and C is a constant. This general antiderivative is called the indefinite integral of $f(x)$ on I and is denoted $\int f(x)\,dx$.

DEFINITION 8

The **indefinite integral** of $f(x)$ on interval I is

$$\int f(x)\,dx = F(x) + C \qquad \text{on } I,$$

provided $F'(x) = f(x)$ for all x in I.

The symbol $\int$ is called an **integral sign**. It is shaped like an elongated "S" for reasons that will only become apparent when we study the *definite integral* in Chapter 5. Just as you regard dy/dx as a single symbol representing the derivative of y with respect to x, so you should regard $\int f(x)\,dx$ as a single symbol representing the indefinite integral (general antiderivative) of f with respect to x. The constant C is called a **constant of integration**.

EXAMPLE 2

(a) $\displaystyle\int x\,dx = \frac{1}{2}x^2 + C$ on any interval.

(b) $\displaystyle\int (x^3 - 5x^2 + 7)\,dx = \frac{1}{4}x^4 - \frac{5}{3}x^3 + 7x + C$ on any interval.

(c) $\displaystyle\int \left(\frac{1}{x^2} + \frac{2}{\sqrt{x}}\right) dx = -\frac{1}{x} + 4\sqrt{x} + C$ on any interval to the right of $x = 0$.

All three formulas above can be checked by differentiating the right-hand sides.

Finding antiderivatives is generally more difficult than finding derivatives; many functions do not have antiderivatives that can be expressed as combinations of finitely many elementary functions. However, *every formula for a derivative can be rephrased as a formula for an antiderivative*. For instance,

$$\frac{d}{dx}\sin x = \cos x; \qquad \text{therefore,} \qquad \int \cos x\,dx = \sin x + C.$$

We will develop several techniques for finding antiderivatives in later chapters. Until then, we must content ourselves with being able to write a few simple antiderivatives based on the known derivatives of elementary functions:

(a) $\displaystyle\int dx = \int 1\,dx = x + C$ (b) $\displaystyle\int x\,dx = \frac{x^2}{2} + C$

(c) $\displaystyle\int x^2\,dx = \frac{x^3}{3} + C$ (d) $\displaystyle\int \frac{1}{x^2}\,dx = \int \frac{dx}{x^2} = -\frac{1}{x} + C$

(e) $\displaystyle\int \frac{1}{\sqrt{x}}\,dx = 2\sqrt{x} + C$ (f) $\displaystyle\int x^r\,dx = \frac{x^{r+1}}{r+1} + C \quad (r \ne -1)$

(g) $\displaystyle\int \sin x\,dx = -\cos x + C$ (h) $\displaystyle\int \cos x\,dx = \sin x + C$

(i) $\displaystyle\int \sec^2 x\,dx = \tan x + C$ (j) $\displaystyle\int \csc^2 x\,dx = -\cot x + C$

(k) $\displaystyle\int \sec x \tan x\,dx = \sec x + C$ (l) $\displaystyle\int \csc x \cot x\,dx = -\csc x + C$

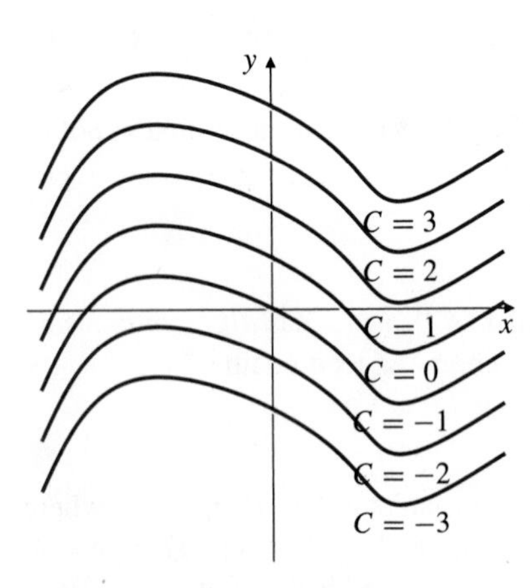

Figure 2.37 Graphs of various antiderivatives of the same function

Observe that formulas (a)–(e) are special cases of formula (f). For the moment, r must be rational in (f), but this restriction will be removed later.

The rule for differentiating sums and constant multiples of functions translates into a similar rule for antiderivatives, as reflected in parts (b) and (c) of Example 2 above.

The graphs of the different antiderivatives of the same function on the same interval are vertically displaced versions of the same curve, as shown in Figure 2.37. In general, only one of these curves will pass through any given point, so we can obtain a unique antiderivative of a given function on an interval by requiring the antiderivative to take a prescribed value at a particular point x.

EXAMPLE 3

Find the function $f(x)$ whose derivative is $f'(x) = 6x^2 - 1$ for all real x and for which $f(2) = 10$.

Solution Since $f'(x) = 6x^2 - 1$, we have

$$f(x) = \int (6x^2 - 1)\,dx = 2x^3 - x + C$$

for some constant C. Since $f(2) = 10$, we have

$$10 = f(2) = 16 - 2 + C.$$

Thus $C = -4$ and $f(x) = 2x^3 - x - 4$. (By direct calculation we can verify that $f'(x) = 6x^2 - 1$ and $f(2) = 10$.)

EXAMPLE 4 Find the function $g(t)$ whose derivative is $\dfrac{t+5}{t^{3/2}}$ and whose graph passes through the point (4, 1).

Solution We have

$$\begin{aligned} g(t) &= \int \frac{t+5}{t^{3/2}}\,dt \\ &= \int (t^{-1/2} + 5t^{-3/2})\,dt \\ &= 2t^{1/2} - 10t^{-1/2} + C \end{aligned}$$

Since the graph of $y = g(t)$ must pass through (4, 1), we require that

$$1 = g(4) = 4 - 5 + C.$$

Hence, $C = 2$ and

$$g(t) = 2t^{1/2} - 10t^{-1/2} + 2 \qquad \text{for } t > 0.$$

Differential Equations and Initial-Value Problems

A **differential equation** (DE) is an equation involving one or more derivatives of an unknown function. Any function whose derivatives satisfy the differential equation *identically on an interval* is called a **solution** of the equation on that interval. For instance, the function $y = x^3 - x$ is a solution of the differential equation

$$\frac{dy}{dx} = 3x^2 - 1$$

on the whole real line. This differential equation has more than one solution; in fact, $y = x^3 - x + C$ is a solution for any value of the constant C.

EXAMPLE 5 Show that for any constants A and B, the function $y = Ax^3 + B/x$ is a solution of the differential equation $x^2y'' - xy' - 3y = 0$ on any interval not containing 0.

Solution If $y = Ax^3 + B/x$, then for $x \neq 0$ we have

$$y' = 3Ax^2 - B/x^2 \quad \text{and} \quad y'' = 6Ax + 2B/x^3.$$

Therefore,

$$x^2y'' - xy' - 3y = 6Ax^3 + \frac{2B}{x} - 3Ax^3 + \frac{B}{x} - 3Ax^3 - \frac{3B}{x} = 0,$$

provided $x \neq 0$. This is what had to be proved.

The **order** of a differential equation is the order of the highest-order derivative appearing in the equation. The DE in Example 5 is a *second-order* DE since it involves y'' and no higher derivatives of y. Note that the solution verified in Example 5 involves two arbitrary constants, A and B. This solution is called a **general solution** to the equation, since it can be shown that every solution is of this form for some choice of the constants A and B. A **particular solution** of the equation is obtained by assigning specific values to these constants. The general solution of an nth-order differential equation typically involves n arbitrary constants.

An **initial-value problem** (IVP) is a problem that consists of:

(i) a differential equation (to be solved for an unknown function) and

(ii) prescribed values for the solution and enough of its derivatives at a particular point (the initial point) to determine values for all the arbitrary constants in the general solution of the DE and so yield a particular solution.

Remark It is common to use the same symbol, say y, to denote both the dependent variable and the function that is the solution to a DE or an IVP; that is, we call the solution function $y = y(x)$ rather than $y = f(x)$.

Remark The solution of an IVP is valid in the largest interval containing the initial point where the solution function is defined.

EXAMPLE 6 Use the result of Example 5 to solve the following initial-value problem.

$$\begin{cases} x^2y'' - xy' - 3y = 0 & (x > 0) \\ y(1) = 2 \\ y'(1) = -6 \end{cases}$$

Solution As shown in Example 5, the DE $x^2y'' - xy' - 3y = 0$ has solution $y = Ax^3 + B/x$, which has derivative $y' = 3Ax^2 - B/x^2$. At $x = 1$ we must have $y = 2$ and $y' = -6$. Therefore,

$$\begin{aligned} A + B &= 2 \\ 3A - B &= -6. \end{aligned}$$

Solving these two linear equations for A and B, we get $A = -1$ and $B = 3$. Hence, $y = -x^3 + 3/x$ for $x > 0$ is the solution of the IVP.

One of the simplest kinds of differential equation is the equation

$$\frac{dy}{dx} = f(x),$$

which is to be solved for y as a function of x. Evidently the solution is

$$y = \int f(x)\,dx.$$

Our ability to find the unknown function $y(x)$ depends on our ability to find an antiderivative of f.

EXAMPLE 7 Solve the initial-value problem

$$\begin{cases} y' = \dfrac{3 + 2x^2}{x^2} \\ y(-2) = 1. \end{cases}$$

Where is the solution valid?

Solution

$$y = \int \left(\frac{3}{x^2} + 2 \right) dx = -\frac{3}{x} + 2x + C$$

$$1 = y(-2) = \frac{3}{2} - 4 + C$$

Therefore, $C = \frac{7}{2}$ and

$$y = -\frac{3}{x} + 2x + \frac{7}{2}.$$

Although the solution function appears to be defined for all x except 0, it is only a solution of the given IVP for $x < 0$. This is because $(-\infty, 0)$ is the largest interval that contains the initial point -2 but not the point $x = 0$, where the solution y is undefined.

EXAMPLE 8 Solve the second-order IVP

$$\begin{cases} y'' = \sin x \\ y(\pi) = 2 \\ y'(\pi) = -1. \end{cases}$$

Solution Since $(y')' = y'' = \sin x$, we have

$$y'(x) = \int \sin x \, dx = -\cos x + C_1.$$

The initial condition for y' gives

$$-1 = y'(\pi) = -\cos \pi + C_1 = 1 + C_1,$$

so that $C_1 = -2$ and $y'(x) = -(\cos x + 2)$. Thus,

$$\begin{aligned} y(x) &= -\int (\cos x + 2) \, dx \\ &= -\sin x - 2x + C_2. \end{aligned}$$

The initial condition for y now gives

$$2 = y(\pi) = -\sin \pi - 2\pi + C_2 = -2\pi + C_2,$$

so that $C_2 = 2 + 2\pi$. The solution to the given IVP is

$$y = 2 + 2\pi - \sin x - 2x$$

and is valid for all x.

Differential equations and initial-value problems are of great importance in applications of calculus, especially for expressing in mathematical form certain laws of nature that involve rates of change of quantities. A large portion of the total mathematical endeavour of the last two hundred years has been devoted to their study. They are usually treated in separate courses on differential equations, but we will discuss them from time to time in this book when appropriate. Throughout this book, except in sections devoted entirely to differential equations, we will use the symbol ✱ to mark exercises about differential equations and initial-value problems.

EXERCISES 2.10

In Exercises 1–14, find the given indefinite integrals.

1. $\int 5\,dx$

2. $\int x^2\,dx$

3. $\int \sqrt{x}\,dx$

4. $\int x^{12}\,dx$

5. $\int x^3\,dx$

6. $\int (x + \cos x)\,dx$

7. $\int \tan x \cos x\,dx$

8. $\int \frac{1 + \cos^3 x}{\cos^2 x}\,dx$

9. $\int (a^2 - x^2)\,dx$

10. $\int (A + Bx + Cx^2)\,dx$

11. $\int (2x^{1/2} + 3x^{1/3})\,dx$

12. $\int \frac{6(x-1)}{x^{4/3}}\,dx$

13. $\int \left(\frac{x^3}{3} - \frac{x^2}{2} + x - 1\right)\,dx$

14. $105\int (1 + t^2 + t^4 + t^6)\,dt$

In Exercises 15–22, find the given indefinite integrals. This may require guessing the form of an antiderivative and then checking by differentiation. For instance, you might suspect that $\int \cos(5x - 2)\,dx = k \sin(5x - 2) + C$ for some k. Differentiating the answer shows that k must be $1/5$.

15. $\int \cos(2x)\,dx$

16. $\int \sin\left(\frac{x}{2}\right)\,dx$

17. $\int \frac{dx}{(1+x)^2}$

18. $\int \sec(1 - x)\tan(1 - x)\,dx$

19. $\int \sqrt{2x + 3}\,dx$

20. $\int \frac{4}{\sqrt{x+1}}\,dx$

21. $\int 2x \sin(x^2)\,dx$

22. $\int \frac{2x}{\sqrt{x^2+1}}\,dx$

Use known trigonometric identities such as $\sec^2 x = 1 + \tan^2 x$, $\cos(2x) = 2\cos^2 x - 1 = 1 - 2\sin^2 x$, and $\sin(2x) = 2\sin x \cos x$ to help you evaluate the indefinite integrals in Exercises 23–26.

23. $\int \tan^2 x\,dx$

24. $\int \sin x \cos x\,dx$

25. $\int \cos^2 x\,dx$

26. $\int \sin^2 x\,dx$

Differential equations

In Exercises 27–42, find the solution $y = y(x)$ to the given initial-value problem. On what interval is the solution valid? (Note that exercises involving differential equations are prefixed with the symbol ✱.)

27. $\begin{cases} y' = x - 2 \\ y(0) = 3 \end{cases}$

28. $\begin{cases} y' = x^{-2} - x^{-3} \\ y(-1) = 0 \end{cases}$

29. $\begin{cases} y' = 3\sqrt{x} \\ y(4) = 1 \end{cases}$

30. $\begin{cases} y' = x^{1/3} \\ y(0) = 5 \end{cases}$

31. $\begin{cases} y' = Ax^2 + Bx + C \\ y(1) = 1 \end{cases}$

32. $\begin{cases} y' = x^{-9/7} \\ y(1) = -4 \end{cases}$

33. $\begin{cases} y' = \cos x \\ y(\pi/6) = 2 \end{cases}$

34. $\begin{cases} y' = \sin(2x) \\ y(\pi/2) = 1 \end{cases}$

35. $\begin{cases} y' = \sec^2 x \\ y(0) = 1 \end{cases}$

36. $\begin{cases} y' = \sec^2 x \\ y(\pi) = 1 \end{cases}$

37. $\begin{cases} y'' = 2 \\ y'(0) = 5 \\ y(0) = -3 \end{cases}$

38. $\begin{cases} y'' = x^{-4} \\ y'(1) = 2 \\ y(1) = 1 \end{cases}$

39. $\begin{cases} y'' = x^3 - 1 \\ y'(0) = 0 \\ y(0) = 8 \end{cases}$

40. $\begin{cases} y'' = 5x^2 - 3x^{-1/2} \\ y'(1) = 2 \\ y(1) = 0 \end{cases}$

41. $\begin{cases} y'' = \cos x \\ y(0) = 0 \\ y'(0) = 1 \end{cases}$

42. $\begin{cases} y'' = x + \sin x \\ y(0) = 2 \\ y'(0) = 0 \end{cases}$

43. Show that for any constants A and B the function $y = y(x) = Ax + B/x$ satisfies the *second-order differential equation* $x^2y'' + xy' - y = 0$ for $x \neq 0$. Find a function y satisfying the initial-value problem:

$$\begin{cases} x^2y'' + xy' - y = 0 & (x > 0) \\ y(1) = 2 \\ y'(1) = 4. \end{cases}$$

44. Show that for any constants A and B the function $y = Ax^{r_1} + Bx^{r_2}$ satisfies, for $x > 0$, the differential equation $ax^2y'' + bxy' + cy = 0$, provided that r_1 and r_2 are two distinct rational roots of the quadratic equation $ar(r - 1) + br + c = 0$.

Use the result of Exercise 44 to solve the initial-value problems in Exercises 45–46 on the interval $x > 0$.

45. $\begin{cases} 4x^2y'' + 4xy' - y = 0 \\ y(4) = 2 \\ y'(4) = -2 \end{cases}$

46. $\begin{cases} x^2y'' - 6y = 0 \\ y(1) = 1 \\ y'(1) = 1 \end{cases}$

2.11 Velocity and Acceleration

Velocity and Speed

Suppose that an object is moving along a straight line (say the x-axis) so that its position x is a function of time t, say $x = x(t)$. (We are using x to represent both the dependent variable and the function.) Suppose we are measuring x in metres and t in seconds. The **average velocity** of the object over the time interval $[t, t+h]$ is the change in position divided by the change in time, that is, the Newton quotient

$$v_{\text{average}} = \frac{\Delta x}{\Delta t} = \frac{x(t+h) - x(t)}{h} \text{ m/s.}$$

The **velocity** $v(t)$ of the object at time t is the limit of this average velocity as $h \to 0$. Thus, it is the rate of change (the derivative) of position with respect to time:

$$\text{Velocity:} \quad v(t) = \frac{dx}{dt} = x'(t).$$

Besides telling us how fast the object is moving, the velocity also tells us in which direction it is moving. If $v(t) > 0$, then x is increasing, so the object is moving to the right; if $v(t) < 0$, then x is decreasing, so the object is moving to the left. At a critical point of x, that is, a time t when $v(t) = 0$, the object is instantaneously at rest—at that instant it is not moving in either direction.

We distinguish between the term *velocity* (which involves direction of motion as well as the rate) and **speed**, which only involves the rate and not the direction. The speed is the absolute value of the velocity:

$$\text{Speed:} \quad s(t) = |v(t)| = \left|\frac{dx}{dt}\right|.$$

A speedometer gives us the speed a vehicle is moving; it does not give the velocity. The speedometer does not start to show negative values if the vehicle turns around and heads in the opposite direction.

EXAMPLE 1

(a) Determine the velocity $v(t)$ at time t of an object moving along the x-axis so that at time t its position is given by

$$x = v_0 t + \frac{1}{2}at^2,$$

where v_0 and a are constants.

(b) Draw the graph of $v(t)$, and show that the area under the graph and above the t-axis, over $[t_1, t_2]$, is equal to the distance the object travels in that time interval.

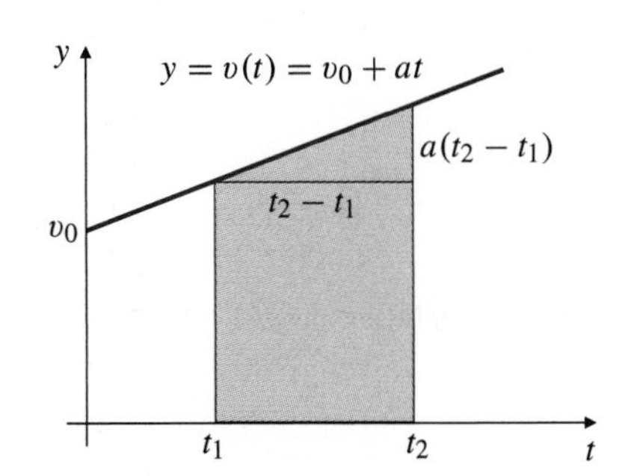

Figure 2.38 The shaded area equals the distance travelled between t_1 and t_2

Solution The velocity is given by

$$v(t) = \frac{dx}{dt} = v_0 + at.$$

Its graph is a straight line with slope a and intercept v_0 on the vertical (velocity) axis. The area under the graph (shaded in Figure 2.38) is the sum of the areas of a rectangle and a triangle. Each has base $t_2 - t_1$. The rectangle has height $v(t_1) = v_0 + at_1$, and the triangle has height $a(t_2 - t_1)$. (Why?) Thus, the shaded area is equal to

$$\begin{aligned}
\text{Area} &= (t_2 - t_1)(v_0 + at_1) + \frac{1}{2}(t_2 - t_1)[a(t_2 - t_1)] \\
&= (t_2 - t_1)\left[v_0 + at_1 + \frac{a}{2}(t_2 - t_1)\right] \\
&= (t_2 - t_1)\left[v_0 + \frac{a}{2}(t_2 + t_1)\right] \\
&= v_0(t_2 - t_1) + \frac{a}{2}(t_2^2 - t_1^2) \\
&= x(t_2) - x(t_1),
\end{aligned}$$

which is the distance travelled by the object between times t_1 and t_2.

Remark In Example 1 we differentiated the position x to get the velocity v and then used the area under the velocity graph to recover information about the position. It appears that there is a connection between finding areas and finding functions that have given derivatives (i.e., finding antiderivatives). This connection, which we will explore in Chapter 5, is perhaps the most important idea in calculus!

Acceleration

The derivative of the velocity also has a useful interpretation. The rate of change of the velocity with respect to time is the **acceleration** of the moving object. It is measured in units of distance/time^2. The value of the acceleration at time t is

$$\text{Acceleration:} \quad a(t) = v'(t) = \frac{dv}{dt} = \frac{d^2x}{dt^2}.$$

The acceleration is the *second derivative* of the position. If $a(t) > 0$, the velocity is increasing. This does not necessarily mean that the speed is increasing; if the object is moving to the left ($v(t) < 0$) and accelerating to the right ($a(t) > 0$), then it is actually slowing down. The object is speeding up only when the velocity and acceleration have the same sign. (See Table 2.)

Table 2. Velocity, acceleration, and speed

If velocity is	and acceleration is	then object is	and its speed is
positive	positive	moving right	increasing
positive	negative	moving right	decreasing
negative	positive	moving left	decreasing
negative	negative	moving left	increasing

If $a(t_0) = 0$, then the velocity and the speed are stationary at t_0. If $a(t) = 0$ during an interval of time, then the velocity is unchanging and, therefore, constant over that interval.

EXAMPLE 2 A point P moves along the x-axis in such a way that its position at time t s is given by

$$x = 2t^3 - 15t^2 + 24t \text{ ft.}$$

(a) Find the velocity and acceleration of P at time t.

(b) In which direction and how fast is P moving at $t = 2$ s? Is it speeding up or slowing down at that time?

(c) When is P instantaneously at rest? When is its speed instantaneously not changing?

(d) When is P moving to the left? to the right?

(e) When is P speeding up? slowing down?

Solution

(a) The velocity and acceleration of P at time t are

$$v = \frac{dx}{dt} = 6t^2 - 30t + 24 = 6(t-1)(t-4) \text{ ft/s} \quad \text{and}$$

$$a = \frac{dv}{dt} = 12t - 30 = 6(2t-5) \text{ ft/s}^2.$$

(b) At $t = 2$ we have $v = -12$ and $a = -6$. Thus, P is moving to the left with speed 12 ft/s, and, since the velocity and acceleration are both negative, its speed is increasing.

(c) P is at rest when $v = 0$, that is, when $t = 1$ s or $t = 4$ s. Its speed is unchanging when $a = 0$, that is, at $t = 5/2$ s.

(d) The velocity is continuous for all t so, by the Intermediate-Value Theorem, has a constant sign on the intervals between the points where it is 0. By examining the values of $v(t)$ at $t = 0, 2$, and 5 (or by analyzing the signs of the factors $(t-1)$ and $(t-4)$ in the expression for $v(t)$), we conclude that $v(t) < 0$ (and P is moving to the left) on time interval $(1, 4)$. $v(t) > 0$ (and P is moving to the right) on time intervals $(-\infty, 1)$ and $(4, \infty)$.

(e) The acceleration a is negative for $t < 5/2$ and positive for $t > 5/2$. Table 3 combines this information with information about v to show where P is speeding up and slowing down.

Table 3. Data for Example 2

Interval	$v(t)$ is	$a(t)$ is	P is
$(-\infty, 1)$	positive	negative	slowing down
$(1, 5/2)$	negative	negative	speeding up
$(5/2, 4)$	negative	positive	slowing down
$(4, \infty)$	positive	positive	speeding up

The motion of P is shown in Figure 2.39.

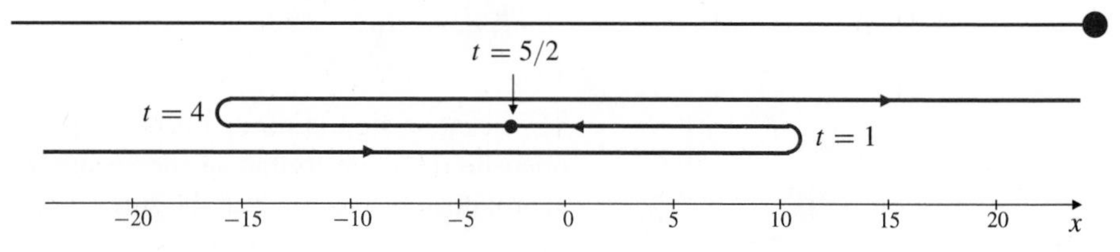

Figure 2.39 The motion of the point P in Example 2

EXAMPLE 3 An object is hurled upward from the roof of a building 10 m high. It rises and then falls back; its height above ground t s after it is thrown is

$$y = -4.9t^2 + 8t + 10 \text{ m},$$

until it strikes the ground. What is the greatest height above the ground that the object attains? With what speed does the object strike the ground?

Solution Refer to Figure 2.40. The vertical velocity at time t during flight is

$$v(t) = -2(4.9)\,t + 8 = -9.8\,t + 8 \text{ m/s}.$$

The object is rising when $v > 0$, that is, when $0 < t < 8/9.8$, and is falling for $t > 8/9.8$. Thus, the object is at its maximum height at time $t = 8/9.8 \approx 0.8163$ s, and this maximum height is

$$y_{\max} = -4.9\left(\frac{8}{9.8}\right)^2 + 8\left(\frac{8}{9.8}\right) + 10 \approx 13.27 \text{ m}.$$

The time t at which the object strikes the ground is the positive root of the quadratic equation obtained by setting $y = 0$,

$$-4.9t^2 + 8t + 10 = 0,$$

namely,

$$t = \frac{-8 - \sqrt{64 + 196}}{-9.8} \approx 2.462 \text{ s}.$$

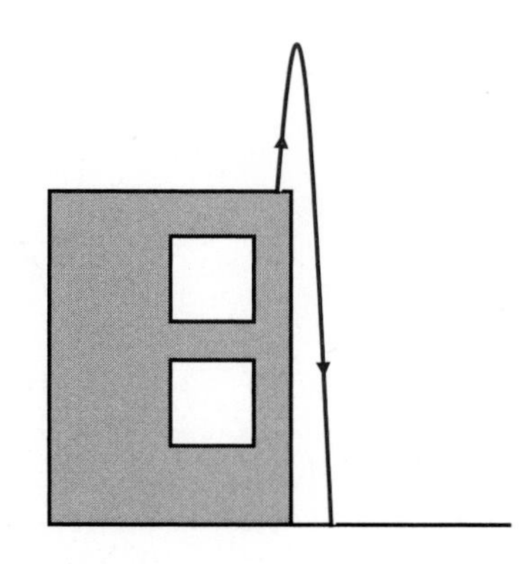

Figure 2.40

The velocity at this time is $v = -(9.8)(2.462) + 8 \approx -16.12$. Thus, the object strikes the ground with a speed of about 16.12 m/s.

Falling Under Gravity

According to Newton's Second Law of Motion, a rock of mass m acted on by an unbalanced force F will experience an acceleration a proportional to and in the same direction as F; with appropriate units of force, $F = ma$. If the rock is sitting on the ground, it is acted on by two forces: the force of gravity acting downward and the reaction of the ground acting upward. These forces balance, so there is no resulting acceleration. On the other hand, if the rock is up in the air and is unsupported, the gravitational force on it will be unbalanced and the rock will experience downward acceleration. It will fall.

According to Newton's Universal Law of Gravitation, the force by which the earth attracts the rock is proportional to the mass m of the rock and inversely proportional to the square of its distance r from the centre of the earth: $F = km/r^2$. If the relative change $\Delta r/r$ is small, as will be the case if the rock remains near the surface of the earth, then $F = mg$, where $g = k/r^2$ is approximately constant. It follows that $ma = F = mg$, and the rock experiences *constant* downward acceleration g. Since g does not depend on m, all objects experience the same acceleration when falling near the surface of the earth, provided we ignore air resistance and any other forces that may be acting on them. Newton's laws therefore imply that if the height of such an object at time t is $y(t)$, then

$$\frac{d^2y}{dt^2} = -g.$$

The negative sign is needed because the gravitational acceleration is downward, the opposite direction to that of increasing y. Physical experiments give the following approximate values for g at the surface of the earth:

$$g = 32 \text{ ft/s}^2 \quad \text{or} \quad g = 9.8 \text{ m/s}^2.$$

EXAMPLE 4 A rock falling freely near the surface of the earth is subject to a constant downward acceleration g, if the effect of air resistance is neglected. If the height and velocity of the rock are y_0 and v_0 at time $t = 0$, find the height $y(t)$ of the rock at any later time t until the rock strikes the ground.

Solution This example asks for a solution $y(t)$ to the second-order initial-value problem:

$$\begin{cases} y''(t) = -g \\ y(0) = y_0 \\ y'(0) = v_0. \end{cases}$$

We have

$$\begin{aligned} y'(t) &= -\int g\,dt = -gt + C_1 \\ v_0 &= y'(0) = 0 + C_1. \end{aligned}$$

Thus $C_1 = v_0$.

$$\begin{aligned} y'(t) &= -gt + v_0 \\ y(t) &= \int(-gt + v_0)dt = -\frac{1}{2}gt^2 + v_0t + C_2 \\ y_0 &= y(0) = 0 + 0 + C_2. \end{aligned}$$

Thus $C_2 = y_0$. Finally, therefore,

$$y(t) = -\frac{1}{2}gt^2 + v_0t + y_0.$$

EXAMPLE 5 A ball is thrown down with an initial speed of 20 ft/s from the top of a cliff, and it strikes the ground at the bottom of the cliff after 5 s. How high is the cliff?

Solution We will apply the result of Example 4. Here we have $g = 32$ ft/s^2, $v_0 = -20$ ft/s, and y_0 is the unknown height of the cliff. The height of the ball t s after it is thrown down is

$$y(t) = -16t^2 - 20t + y_0 \text{ ft.}$$

At $t = 5$ the ball reaches the ground, so $y(5) = 0$:

$$0 = -16(25) - 20(5) + y_0 \qquad \Rightarrow \qquad y_0 = 500.$$

The cliff is 500 ft high.

EXAMPLE 6 **(Stopping distance)** A car is travelling at 72 km/h. At a certain instant its brakes are applied to produce a constant deceleration of 0.8 m/s^2. How far does the car travel before coming to a stop?

Solution Let $s(t)$ be the distance the car travels in the t seconds after the brakes are applied. Then $s''(t) = -0.8$ (m/s^2), so the velocity at time t is given by

$$s'(t) = \int -0.8\,dt = -0.8t + C_1 \quad \text{m/s.}$$

Since $s'(0) = 72$ km/h $= 72 \times 1,000/3,600 = 20$ m/s, we have $C_1 = 20$. Thus,

$$s'(t) = 20 - 0.8t$$

and

$$s(t) = \int (20 - 0.8t)\,dt = 20t - 0.4t^2 + C_2.$$

Since $s(0) = 0$, we have $C_2 = 0$ and $s(t) = 20t - 0.4t^2$. When the car has stopped, its velocity will be 0. Hence, the stopping time is the solution t of the equation

$$0 = s'(t) = 20 - 0.8t,$$

that is, $t = 25$ s. The distance travelled during deceleration is $s(25) = 250$ m.

EXERCISES 2.11

In Exercises 1–4, a point moves along the x-axis so that its position x at time t is specified by the given function. In each case determine the following:

(a) the time intervals on which the point is moving to the right and (b) to the left;

(c) the time intervals on which the point is accelerating to the right and (d) to the left;

(e) the time intervals when the particle is speeding up and (f) slowing down;

(g) the acceleration at times when the velocity is zero;

(h) the average velocity over the time interval $[0, 4]$.

1. $x = t^2 - 4t + 3$ **2.** $x = 4 + 5t - t^2$

3. $x = t^3 - 4t + 1$ **4.** $x = \dfrac{t}{t^2 + 1}$

5. A ball is thrown upward from ground level with an initial speed of 9.8 m/s so that its height in metres after t s is given by $y = 9.8t - 4.9t^2$. What is the acceleration of the ball at any time t? How high does the ball go? How fast is it moving when it strikes the ground?

6. A ball is thrown downward from the top of a 100-metre-high tower with an initial speed of 2 m/s. Its height in metres above the ground t s later is $y = 100 - 2t - 4.9t^2$. How long does it take to reach the ground? What is its average velocity during the fall? At what instant is its velocity equal to its average velocity?

7. **(Takeoff distance)** The distance an aircraft travels along a runway before takeoff is given by $D = t^2$, where D is measured in metres from the starting point, and t is measured in seconds from the time the brake is released. If the aircraft will become airborne when its speed reaches 200 km/h, how long will it take to become airborne, and what distance will it travel in that time?

8. **(Projectiles on Mars)** A projectile fired upward from the surface of the earth falls back to the ground after 10 s. How long would it take to fall back to the surface if it is fired upward on Mars with the same initial velocity? $g_{\text{Mars}} = 3.72$ m/s^2.

9. A ball is thrown upward with initial velocity v_0 m/s and reaches a maximum height of h m. How high would it have gone if its initial velocity was $2v_0$? How fast must it be thrown upward to achieve a maximum height of $2h$ m?

10. How fast would the ball in the Exercise 9 have to be thrown upward on Mars in order to achieve a maximum height of $3h$ m?

11. A rock falls from the top of a cliff and hits the ground at the base of the cliff at a speed of 160 ft/s. How high is the cliff?

12. A rock is thrown down from the top of a cliff with the initial speed of 32 ft/s and hits the ground at the base of the cliff at a speed of 160 ft/s. How high is the cliff?

13. **(Distance travelled while braking)** With full brakes applied, a freight train can decelerate at a constant rate of $1/6$ m/s^2. How far will the train travel while braking to a full stop from an initial speed of 60 km/h?

14. Show that if the position x of a moving point is given by a quadratic function of t, $x = At^2 + Bt + C$, then the average velocity over any time interval $[t_1, t_2]$ is equal to the instantaneous velocity at the midpoint of that time interval.

15. **(Piecewise motion)** The position of an object moving along the s-axis is given at time t by

$$s = \begin{cases} t^2 & \text{if } 0 \le t \le 2 \\ 4t - 4 & \text{if } 2 < t < 8 \\ -68 + 20t - t^2 & \text{if } 8 \le t \le 10. \end{cases}$$

Determine the velocity and acceleration at any time t. Is the velocity continuous? Is the acceleration continuous? What is the maximum velocity and when is it attained?

(Rocket flight with limited fuel) Figure 2.41 shows the velocity v in feet per second of a small rocket that was fired from the top of a tower at time $t = 0$ (t in seconds), accelerated with constant upward acceleration until its fuel was used up, then fell back to the ground at the foot of the tower. The whole flight lasted 14 s. Exercises 16–19 refer to this rocket.

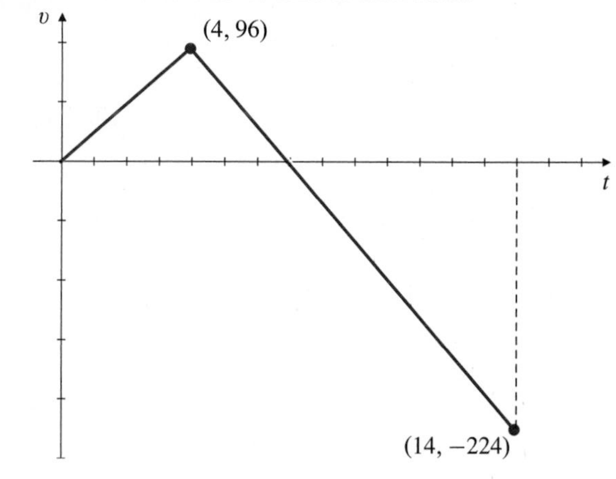

Figure 2.41

16. What was the acceleration of the rocket while its fuel lasted?

17. How long was the rocket rising?

18. What is the maximum height above ground that the rocket reached?

19. How high was the tower from which the rocket was fired?

20. Redo Example 6 using instead a nonconstant deceleration, $s''(t) = -t$ m/s^2.

CHAPTER REVIEW

Key Ideas

- **What do the following statements and phrases mean?**
  - ◇ Line L is tangent to curve C at point P.
  - ◇ the Newton quotient of $f(x)$ at $x = a$
  - ◇ the derivative $f'(x)$ of the function $f(x)$
  - ◇ f is differentiable at $x = a$.
  - ◇ the slope of the graph $y = f(x)$ at $x = a$
  - ◇ f is increasing (or decreasing) on interval I.
  - ◇ f is nondecreasing (or nonincreasing) on interval I.
  - ◇ the average rate of change of $f(x)$ on $[a, b]$
  - ◇ the rate of change of $f(x)$ at $x = a$
  - ◇ c is a critical point of $f(x)$.
  - ◇ the second derivative of $f(x)$ at $x = a$
  - ◇ an antiderivative of f on interval I
  - ◇ the indefinite integral of f on interval I
  - ◇ differential equation ◇ initial-value problem
  - ◇ velocity ◇ speed ◇ acceleration
- **State the following differentiation rules:**
  - ◇ the rule for differentiating a sum of functions
  - ◇ the rule for differentiating a constant multiple of a function
  - ◇ the Product Rule ◇ the Reciprocal Rule
  - ◇ the Quotient Rule ◇ the Chain Rule
- **State the Mean-Value Theorem.**
- **State the Generalized Mean-Value Theorem.**
- **State the derivatives of the following functions:**
  - ◇ x ◇ x^2 ◇ $1/x$ ◇ $\sqrt{x}$
  - ◇ x^n ◇ $|x|$ ◇ $\sin x$ ◇ $\cos x$
  - ◇ $\tan x$ ◇ $\cot x$ ◇ $\sec x$ ◇ $\csc x$
- **What is a proof by mathematical induction?**

Review Exercises

Use the definition of derivative to calculate the derivatives in Exercises 1–4.

1. $\dfrac{dy}{dx}$ if $y = (3x+1)^2$ **2.** $\dfrac{d}{dx}\sqrt{1-x^2}$

3. $f'(2)$ if $f(x) = \dfrac{4}{x^2}$ **4.** $g'(9)$ if $g(t) = \dfrac{t-5}{1+\sqrt{t}}$

5. Find the tangent to $y = \cos(\pi x)$ at $x = 1/6$.

6. Find the normal to $y = \tan(x/4)$ at $x = \pi$.

Calculate the derivatives of the functions in Exercises 7–12.

7. $\dfrac{1}{x - \sin x}$ **8.** $\dfrac{1 + x + x^2 + x^3}{x^4}$

9. $(4 - x^{2/5})^{-5/2}$ **10.** $\sqrt{2 + \cos^2 x}$

11. $\tan\theta - \theta\sec^2\theta$ **12.** $\dfrac{\sqrt{1+t^2}-1}{\sqrt{1+t^2}+1}$

Evaluate the limits in Exercises 13–16 by interpreting each as a derivative.

13. $\displaystyle\lim_{h\to 0}\frac{(x+h)^{20} - x^{20}}{h}$ **14.** $\displaystyle\lim_{x\to 2}\frac{\sqrt{4x+1}-3}{x-2}$

15. $\displaystyle\lim_{x\to\pi/6}\frac{\cos(2x) - (1/2)}{x - \pi/6}$ **16.** $\displaystyle\lim_{x\to -a}\frac{(1/x^2) - (1/a^2)}{x+a}$

In Exercises 17–24, express the derivatives of the given functions in terms of the derivatives f' and g' of the differentiable functions f and g.

17. $f(3 - x^2)$ **18.** $[f(\sqrt{x})]^2$

19. $f(2x)\sqrt{g(x/2)}$ **20.** $\dfrac{f(x) - g(x)}{f(x) + g(x)}$

21. $f(x + (g(x))^2)$ **22.** $f\left(\dfrac{g(x^2)}{x}\right)$

23. $f(\sin x)\, g(\cos x)$ **24.** $\sqrt{\dfrac{\cos f(x)}{\sin g(x)}}$

25. Find the tangent to the curve $x^3y + 2xy^3 = 12$ at the point $(2, 1)$.

26. Find the slope of the curve $3\sqrt{2}x\sin(\pi y) + 8y\cos(\pi x) = 2$ at the point $\left(\frac{1}{3}, \frac{1}{4}\right)$.

Find the indefinite integrals in Exercises 27–30.

27. $\displaystyle\int\frac{1+x^4}{x^2}\,dx$ **28.** $\displaystyle\int\frac{1+x}{\sqrt{x}}\,dx$

29. $\displaystyle\int\frac{2+3\sin x}{\cos^2 x}\,dx$ **30.** $\displaystyle\int(2x+1)^4\,dx$

31. Find $f(x)$ given that $f'(x) = 12x^2 + 12x^3$ and $f(1) = 0$.

32. Find $g(x)$ if $g'(x) = \sin(x/3) + \cos(x/6)$ and the graph of g passes through the point $(\pi, 2)$.

33. Differentiate $x\sin x + \cos x$ and $x\cos x - \sin x$, and use the results to find the indefinite integrals

$$I_1 = \int x\cos x\,dx \quad\text{and}\quad I_2 = \int x\sin x\,dx.$$

34. Suppose that $f'(x) = f(x)$ for every x. Let $g(x) = x\,f(x)$. Calculate the first several derivatives of g and guess a formula for the nth-order derivative $g^{(n)}(x)$. Verify your guess by induction.

35. Find an equation of the straight line that passes through the origin and is tangent to the curve $y = x^3 + 2$.

36. Find an equation of the straight lines that pass through the point $(0, 1)$ and are tangent to the curve $y = \sqrt{2 + x^2}$.

37. Show that $\dfrac{d}{dx}\Big(\sin^n x\sin(nx)\Big) = n\sin^{n-1}x\sin((n+1)x)$. At what points x in $[0, \pi]$ does the graph of $y = \sin^n x\sin(nx)$ have a horizontal tangent. Assume that $n \ge 2$.

38. Find differentiation formulas for $y = \sin^n x\cos(nx)$, $y = \cos^n x\sin(nx)$, and $y = \cos^n x\cos(nx)$ analogous to the one given for $y = \sin^n x\sin(nx)$ in Exercise 37.

39. Let Q be the point $(0, 1)$. Find all points P on the curve $y = x^2$ such that the line PQ is normal to $y = x^2$ at P. What is the shortest distance from Q to the curve $y = x^2$?

40. **(Average and marginal profit)** Figure 2.42 shows the graph of the profit $\$P(x)$ realized by a grain exporter from its sale of x tonnes of wheat. Thus, the average profit per tonne is $\$P(x)/x$. Show that the maximum average profit occurs when the average profit equals the marginal profit. What is the geometric significance of this fact in the figure?

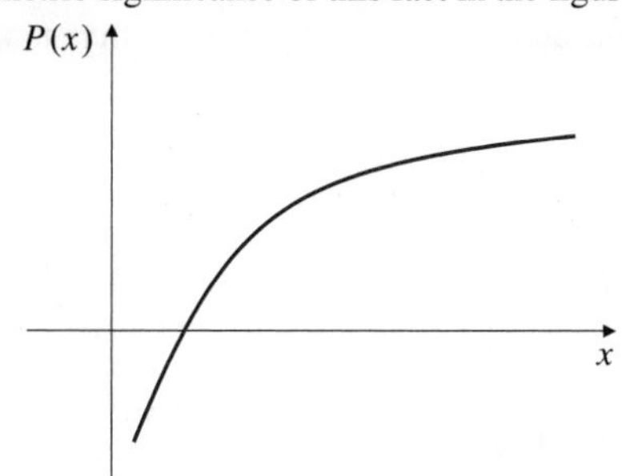

Figure 2.42

41. **(Gravitational attraction)** The gravitational attraction of the earth on a mass m at distance r from the centre of the earth is a continuous function $F(r)$ given for $r \geq 0$ by

$$F(r) = \begin{cases} \dfrac{mgR^2}{r^2} & \text{if } r \geq R \\ mkr & \text{if } 0 \leq r < R \end{cases}$$

where R is the radius of the earth, and g is the acceleration due to gravity at the surface of the earth.

(a) Find the constant k in terms of g and R.

(b) F decreases as m moves away from the surface of the earth, either upward or downward. Show that F decreases as r increases from R at twice the rate at which F decreases as r decreases from R.

42. **(Compressibility of a gas)** The isothermal compressibility of a gas is the relative rate of change of the volume V with respect to the pressure P, at a constant temperature T, that is, $(1/V)\,dV/dP$. For a sample of an ideal gas, the temperature, pressure, and volume satisfy the equation $PV = kT$, where k is a constant related to the number of molecules of gas present in the sample. Show that the isothermal compressibility of such a gas is the negative reciprocal of the pressure:

$$\frac{1}{V}\frac{dV}{dP} = -\frac{1}{P}.$$

43. A ball is thrown upward with an initial speed of 10 m/s from the top of a building. A second ball is thrown upward with an initial speed of 20 m/s from the ground. Both balls achieve the same maximum height above the ground. How tall is the building?

44. A ball is dropped from the top of a 60 m high tower at the same instant that a second ball is thrown upward from the ground at the base of the tower. The balls collide at a height of 30 m above the ground. With what initial velocity was the second ball thrown? How fast is each ball moving when they collide?

45. **(Braking distance)** A car's brakes can decelerate the car at 20 ft/s^2. How fast can the car travel if it must be able to stop in a distance of 160 ft?

46. **(Measuring variations in g)** The period P of a pendulum of length L is given by $P = 2\pi\sqrt{L/g}$, where g is the acceleration of gravity.

(a) Assuming that L remains fixed, show that a 1% increase in g results in approximately a 1/2% decrease in the period P. (Variations in the period of a pendulum can be used to detect small variations in g from place to place on the earth's surface.)

(b) For fixed g, what percentage change in L will produce a 1% increase in P?

Challenging Problems

1. René Descartes, the inventor of analytic geometry, calculated the tangent to a parabola (or a circle or other quadratic curve) at a given point (x_0, y_0) on the curve by looking for a straight line through (x_0, y_0) having only one intersection with the given curve. Illustrate his method by writing the equation of a line through (a, a^2), having arbitrary slope m, and then finding the value of m for which the line has only one intersection with the parabola $y = x^2$. Why does the method not work for more general curves?

2. Given that $f'(x) = 1/x$ and $f(2) = 9$, find:

(a) $\displaystyle\lim_{x\to 2} \frac{f(x^2+5) - f(9)}{x-2}$ (b) $\displaystyle\lim_{x\to 2} \frac{\sqrt{f(x)} - 3}{x-2}$

3. Suppose that $f'(4) = 3$, $g'(4) = 7$, $g(4) = 4$, and $g(x) \neq 4$ for $x \neq 4$. Find:

(a) $\displaystyle\lim_{x\to 4} \big(f(x) - f(4)\big)$ (b) $\displaystyle\lim_{x\to 4} \frac{f(x) - f(4)}{x^2 - 16}$

(c) $\displaystyle\lim_{x\to 4} \frac{f(x) - f(4)}{\sqrt{x} - 2}$ (d) $\displaystyle\lim_{x\to 4} \frac{f(x) - f(4)}{(1/x) - (1/4)}$

(e) $\displaystyle\lim_{x\to 4} \frac{f(x) - f(4)}{g(x) - 4}$ (f) $\displaystyle\lim_{x\to 4} \frac{f(g(x)) - f(4)}{x - 4}$

4. Let $f(x) = \begin{cases} x & \text{if } x = 1,\ 1/2,\ 1/3,\ 1/4,\ \ldots \\ x^2 & \text{otherwise.} \end{cases}$

(a) Find all points at which f is continuous. In particular, is it continuous at $x = 0$?

(b) Is the following statement true or false? Justify your answer. For any two real numbers a and b, there is some x between a and b such that $f(x) = (f(a) + f(b))/2$.

(c) Find all points at which f is differentiable. In particular, is it differentiable at $x = 0$?

5. Suppose $f(0) = 0$ and $|f(x)| > \sqrt{|x|}$ for all x. Show that $f'(0)$ does not exist.

6. Suppose that f is a function satisfying the following conditions: $f'(0) = k$, $f(0) \neq 0$, and $f(x+y) = f(x)f(y)$ for all x and y. Show that $f(0) = 1$ and that $f'(x) = k\,f(x)$ for every x. (We will study functions with these properties in Chapter 3.)

7. Suppose the function g satisfies the conditions: $g'(0) = k$, and $g(x+y) = g(x) + g(y)$ for all x and y. Show that:

(a) $g(0) = 0$, (b) $g'(x) = k$ for all x, and

(c) $g(x) = kx$ for all x. *Hint:* Let $h(x) = g(x) - g'(0)x$.

8. (a) If f is differentiable at x, show that

(i) $\displaystyle\lim_{h\to 0} \frac{f(x) - f(x-h)}{h} = f'(x)$

(ii) $\lim_{h\to 0} \frac{f(x+h) - f(x-h)}{2h} = f'(x)$

(b) Show that the existence of the limit in (i) guarantees that f is differentiable at x.

(c) Show that the existence of the limit in (ii) does *not* guarantee that f is differentiable at x. *Hint:* Consider the function $f(x) = |x|$ at $x = 0$.

9. Show that there is a line through $(a, 0)$ that is tangent to the curve $y = x^3$ at $x = 3a/2$. If $a \neq 0$, is there any other line through $(a, 0)$ that is tangent to the curve? If (x_0, y_0) is an arbitrary point, what is the maximum number of lines through (x_0, y_0) that can be tangent to $y = x^3$? the minimum number?

10. Make a sketch showing that there are two straight lines, each of which is tangent to both of the parabolas $y = x^2 + 4x + 1$ and $y = -x^2 + 4x - 1$. Find equations of the two lines.

11. Show that if $b > 1/2$, there are three straight lines through $(0, b)$, each of which is normal to the curve $y = x^2$. How many such lines are there if $b = 1/2$? if $b < 1/2$?

12. (Distance from a point to a curve) Find the point on the curve $y = x^2$ that is closest to the point $(3, 0)$. *Hint:* The line from $(3, 0)$ to the closest point Q on the parabola is normal to the parabola at Q.

13. (Envelope of a family of lines) Show that for each value of the parameter m, the line $y = mx - (m^2/4)$ is tangent to the parabola $y = x^2$. (The parabola is called the *envelope* of the family of lines $y = mx - (m^2/4)$.) Find $f(m)$ such that the family of lines $y = mx + f(m)$ has envelope the parabola $y = Ax^2 + Bx + C$.

14. (Common tangents) Consider the two parabolas with equations $y = x^2$ and $y = Ax^2 + Bx + C$. We assume that $A \neq 0$, and if $A = 1$, then either $B \neq 0$ or $C \neq 0$, so that the two equations do represent different parabolas. Show that:

(a) the two parabolas are tangent to each other if $B^2 = 4C(A - 1)$;

(b) the parabolas have two common tangent lines if and only if $A \neq 1$ and $A\big(B^2 - 4C(A - 1)\big) > 0$;

(c) the parabolas have exactly one common tangent line if either $A = 1$ and $B \neq 0$, or $A \neq 1$ and $B^2 = 4C(A - 1)$;

(d) the parabolas have no common tangent lines if either $A = 1$ and $B = 0$, or $A \neq 1$ and $A\big(B^2 - 4C(A-1)\big) < 0$.

Make sketches illustrating each of the above possibilities.

15. Let C be the graph of $y = x^3$.

(a) Show that if $a \neq 0$, then the tangent to C at $x = a$ also intersects C at a second point $x = b$.

(b) Show that the slope of C at $x = b$ is four times its slope at $x = a$.

(c) Can any line be tangent to C at more than one point?

(d) Can any line be tangent to the graph of $y = Ax^3 + Bx^2 + Cx + D$ at more than one point?

16. Let C be the graph of $y = x^4 - 2x^2$.

(a) Find all horizontal lines that are tangent to C.

(b) One of the lines found in (a) is tangent to C at two different points. Show that there are no other lines with this property.

(c) Find an equation of a straight line that is tangent to the graph of $y = x^4 - 2x^2 + x$ at two different points. Can there exist more than one such line? Why?

17. (Double tangents) A line tangent to the quartic (fourth-degree polynomial) curve C with equation $y = ax^4 + bx^3 + cx^2 + dx + e$ at $x = p$ may intersect C at zero, one, or two other points. If it meets C at only one other point $x = q$, it must be tangent to C at that point also, and it is thus a "double tangent."

(a) Find the condition that must be satisfied by the coefficients of the quartic to ensure that there does exist such a double tangent, and show that there cannot be more than one such double tangent. Illustrate this by applying your results to $y = x^4 - 2x^2 + x - 1$.

(b) If the line PQ is tangent to C at two distinct points $x = p$ and $x = q$, show that PQ is parallel to the line tangent to C at $x = (p + q)/2$.

(c) If the line PQ is tangent to C at two distinct points $x = p$ and $x = q$, show that C has two distinct inflection points R and S and that RS is parallel to PQ.

18. Verify the following formulas for every positive integer n:

(a) $\frac{d^n}{dx^n} \cos(ax) = a^n \cos\left(ax + \frac{n\pi}{2}\right)$

(b) $\frac{d^n}{dx^n} \sin(ax) = a^n \sin\left(ax + \frac{n\pi}{2}\right)$

(c) $\frac{d^n}{dx^n} \left(\cos^4 x + \sin^4 x\right) = 4^{n-1} \cos\left(4x + \frac{n\pi}{2}\right)$

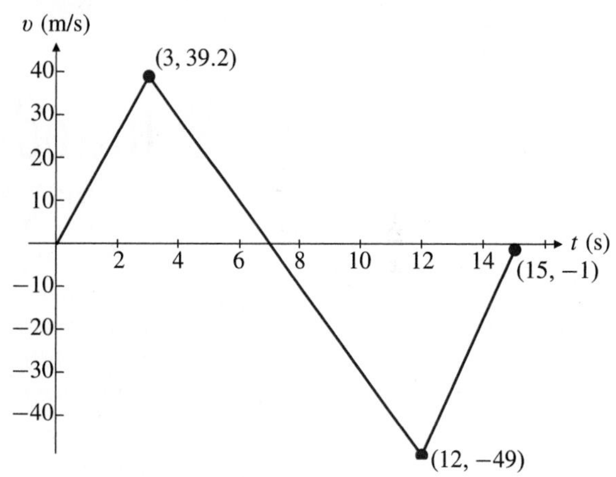

Figure 2.43

19. (Rocket with a parachute) A rocket is fired from the top of a tower at time $t = 0$. It experiences constant upward acceleration until its fuel is used up. Thereafter its acceleration is the constant downward acceleration of gravity until, during its fall, it deploys a parachute that gives it a constant upward acceleration again to slow it down. The rocket hits the ground near the base of the tower. The upward velocity v (in metres per second) is graphed against time in Figure 2.43. From information in the figure answer the following questions:

(a) How long did the fuel last?

(b) When was the rocket's height maximum?

(c) When was the parachute deployed?

(d) What was the rocket's upward acceleration while its motor was firing?

(e) What was the maximum height achieved by the rocket?

(f) How high was the tower from which the rocket was fired?

CHAPTER 3

Transcendental Functions

> It is well known that the central problem of the whole of modern mathematics is the study of the transcendental functions defined by differential equations.
>
> Felix Klein 1849–1925
> *Lectures on Mathematics (1911)*

Introduction

With the exception of the trigonometric functions, all the functions we have encountered so far have been of three main types: *polynomials, rational functions* (quotients of polynomials), and *algebraic functions* (fractional powers of rational functions). On an interval in its domain, each of these functions can be constructed from real numbers and a single real variable x by using finitely many arithmetic operations (addition, subtraction, multiplication, and division) and by taking finitely many roots (fractional powers). Functions that cannot be so constructed are called **transcendental functions**. The only examples of these that we have seen so far are the trigonometric functions.

Much of the importance of calculus and many of its most useful applications result from its ability to illuminate the behaviour of transcendental functions that arise naturally when we try to model concrete problems in mathematical terms. This chapter is devoted to developing other transcendental functions, including exponential and logarithmic functions and the inverse trigonometric functions.

Some of these functions "undo" what other ones "do" and vice versa. When a pair of functions behaves this way, we call each one the inverse of the other. We begin the chapter by studying inverse functions in general.

3.1 Inverse Functions

Consider the function $f(x) = x^3$ whose graph is shown in Figure 3.1. Like any function, $f(x)$ has only one value for each x in its domain (for x^3 this is the whole real line $\mathbb{R}$). In geometric terms, this means that any *vertical* line meets the graph of f at only one point. However, for this function f, any *horizontal* line also meets the graph at only one point. This means that different values of x always give different values $f(x)$. Such a function is said to be *one-to-one*.

DEFINITION 1

> A function f is **one-to-one** if $f(x_1) \neq f(x_2)$ whenever x_1 and x_2 belong to the domain of f and $x_1 \neq x_2$, or, equivalently, if
>
> $$f(x_1) = f(x_2) \quad \Longrightarrow \quad x_1 = x_2.$$

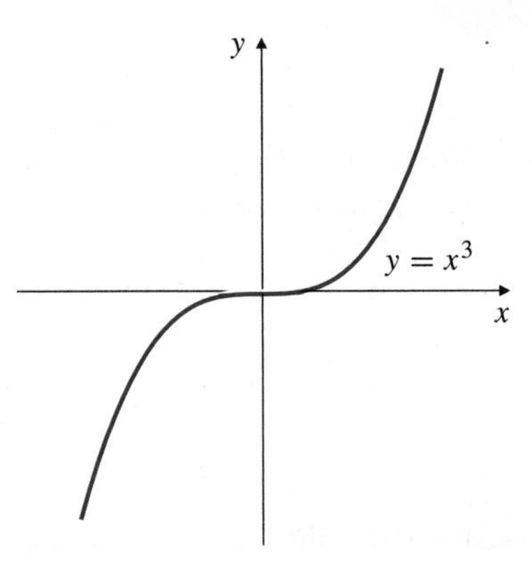

Figure 3.1 The graph of $f(x) = x^3$

A function is one-to-one if any horizontal line that intersects its graph does so at only one point. If a function defined on a single interval is increasing (or decreasing), then it is one-to-one. (See Section 2.6 for more discussion of this.)

Reconsider the one-to-one function $f(x) = x^3$ (Figure 3.1). Since the equation

$$y = x^3$$

has a unique solution x for every given value of y in the range of f, f is one-to-one. Specifically, this solution is given by

$$x = y^{1/3};$$

it defines x as a function of y. We call this new function the *inverse of* f and denote it f^{-1}. Thus,

$$f^{-1}(y) = y^{1/3}.$$

Do not confuse the -1 in f^{-1} with an exponent. The inverse f^{-1} is *not* the reciprocal $1/f$. If we want to denote the reciprocal $1/f(x)$ with an exponent we can write it as $\big(f(x)\big)^{-1}$.

In general, if a function f is one-to-one, then for any number y in its range there will always exist a single number x in its domain such that $y = f(x)$. Since x is determined uniquely by y, it is a function of y. We write $x = f^{-1}(y)$ and call f^{-1} the inverse of f. The function f whose graph is shown in Figure 3.2(a) is one-to-one and has an inverse. The function g whose graph is shown in Figure 3.2(b) is not one-to-one (some horizontal lines meet the graph twice) and so does not have an inverse.

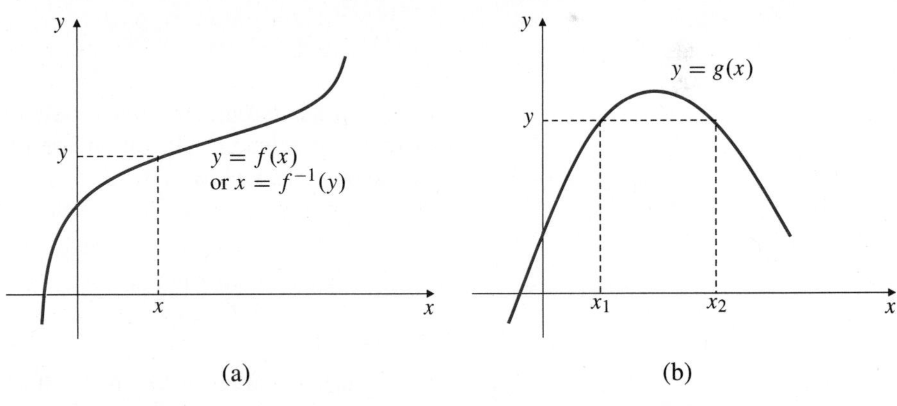

Figure 3.2
(a) f is one-to-one and has an inverse. $y = f(x)$ means the same thing as $x = f^{-1}(y)$
(b) g is not one-to-one

We usually like to write functions with the domain variable called x rather than y, so we reverse the roles of x and y and reformulate the above definition as follows.

DEFINITION 2

> If f is one-to-one, then it has an **inverse function** f^{-1}. The value of $f^{-1}(x)$ is the unique number y in the domain of f for which $f(y) = x$. Thus,
>
> $$y = f^{-1}(x) \quad \Longleftrightarrow \quad x = f(y).$$

As seen above, $y = f(x) = x^3$ is equivalent to $x = f^{-1}(y) = y^{1/3}$, or, reversing the roles of x and y, $y = f^{-1}(x) = x^{1/3}$ is equivalent to $x = f(y) = y^3$.

EXAMPLE 1 Show that $f(x) = 2x - 1$ is one-to-one, and find its inverse $f^{-1}(x)$.

Solution Since $f'(x) = 2 > 0$ on $\mathbb{R}$, f is increasing and therefore one-to-one there. Let $y = f^{-1}(x)$. Then

$$x = f(y) = 2y - 1.$$

Solving this equation for y gives $y = \dfrac{x+1}{2}$. Thus, $f^{-1}(x) = \dfrac{x+1}{2}$.

There are several things you should remember about the relationship between a function f and its inverse f^{-1}. The most important one is that the two equations

$$y = f^{-1}(x) \quad \text{and} \quad x = f(y)$$

say the same thing. They are equivalent just as, for example, $y = x + 1$ and $x = y - 1$ are equivalent. Either of the equations can be replaced by the other. This implies that the domain of f^{-1} is the range of f and vice versa.

The inverse of a one-to-one function is itself one-to-one and so also has an inverse. Not surprisingly, the inverse of f^{-1} is f:

$$y = (f^{-1})^{-1}(x) \iff x = f^{-1}(y) \iff y = f(x).$$

We can substitute either of the equations $y = f^{-1}(x)$ or $x = f(y)$ into the other and obtain the **cancellation identities**:

$$f\big(f^{-1}(x)\big) = x, \qquad f^{-1}\big(f(y)\big) = y.$$

The first of these identities holds for all x in the domain of f^{-1} and the second for all y in the domain of f. If S is any set of real numbers and I_S denotes the **identity function** on S, defined by

$$I_S(x) = x \quad \text{for all } x \text{ in } S,$$

then the cancellation identities say that if $\mathcal{D}(f)$ is the domain of f, then

$$f \circ f^{-1} = I_{\mathcal{D}(f^{-1})} \quad \text{and} \quad f^{-1} \circ f = I_{\mathcal{D}(f)},$$

where $f \circ g(x)$ denotes the composition $f\big(g(x)\big)$.

If the coordinates of a point $P = (a, b)$ are exchanged to give those of a new point $Q = (b, a)$, then each point is the reflection of the other in the line $x = y$. (To see this, note that the line PQ has slope -1, so it is perpendicular to $y = x$. Also, the midpoint of PQ is $\left(\frac{a+b}{2}, \frac{b+a}{2}\right)$, which lies on $y = x$.) It follows that the graphs of the equations $x = f(y)$ and $y = f(x)$ are reflections of each other in the line $x = y$. Since the equation $x = f(y)$ is equivalent to $y = f^{-1}(x)$, the graphs of the functions f^{-1} and f are reflections of each other in $y = x$. See Figure 3.3.

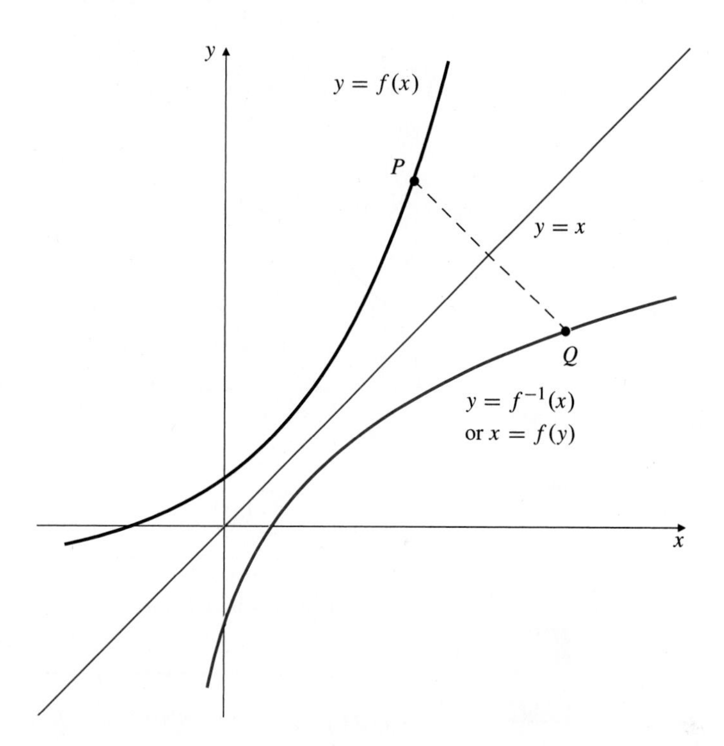

Figure 3.3 The graph of $y = f^{-1}(x)$ is the reflection of the graph of $y = f(x)$ in the line $y = x$

Here is a summary of the properties of inverse functions discussed above:

Properties of inverse functions

1. $y = f^{-1}(x) \iff x = f(y)$.
2. The domain of f^{-1} is the range of f.
3. The range of f^{-1} is the domain of f.
4. $f^{-1}\big(f(x)\big) = x$ for all x in the domain of f.
5. $f\big(f^{-1}(x)\big) = x$ for all x in the domain of f^{-1}.
6. $(f^{-1})^{-1}(x) = f(x)$ for all x in the domain of f.
7. The graph of f^{-1} is the reflection of the graph of f in the line $x = y$.

EXAMPLE 2 Show that $g(x) = \sqrt{2x+1}$ is invertible, and find its inverse.

Solution If $g(x_1) = g(x_2)$, then $\sqrt{2x_1+1} = \sqrt{2x_2+1}$. Squaring both sides we get $2x_1 + 1 = 2x_2 + 1$, which implies that $x_1 = x_2$. Thus, g is one-to-one and invertible. Let $y = g^{-1}(x)$; then

$$x = g(y) = \sqrt{2y+1}.$$

It follows that $x \ge 0$ and $x^2 = 2y + 1$. Therefore, $y = \dfrac{x^2-1}{2}$ and

$$g^{-1}(x) = \frac{x^2-1}{2} \qquad \text{for } x \ge 0.$$

(The restriction $x \ge 0$ applies since the range of g is $[0, \infty)$.) See Figure 3.4(a) for the graphs of g and g^{-1}.

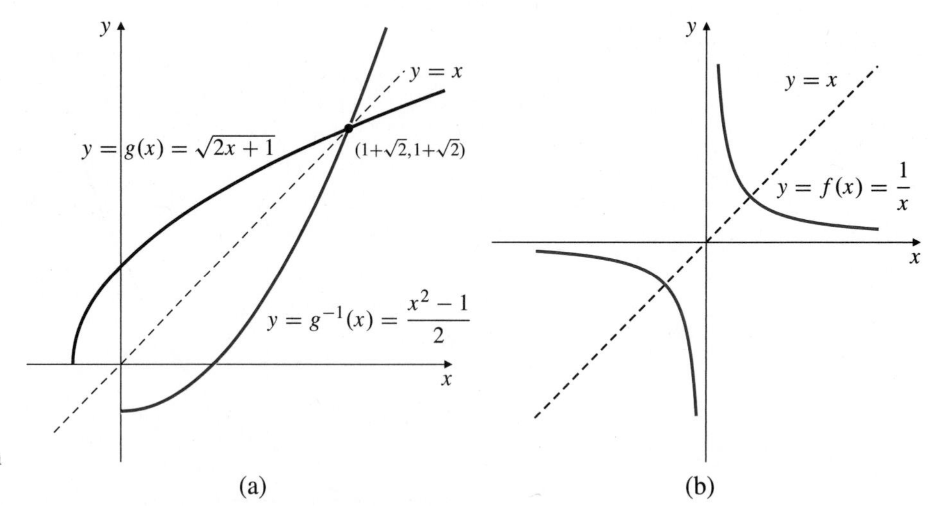

Figure 3.4

(a) The graphs of $g(x) = \sqrt{2x+1}$ and its inverse

(b) The graph of the self-inverse function $f(x) = 1/x$

DEFINITION 3

> A function f is **self-inverse** if $f^{-1} = f$, that is, if $f\big(f(x)\big) = x$ for every x in the domain of f.

EXAMPLE 3 The function $f(x) = 1/x$ is self-inverse. If $y = f^{-1}(x)$, then $x = f(y) = 1/y$. Therefore, $y = 1/x$, so $f^{-1}(x) = \dfrac{1}{x} = f(x)$. See Figure 3.4(b). The graph of any self-inverse function must be its own reflection in the line $x = y$ and must therefore be symmetric about that line.

Inverting Non–One-to-One Functions

Many important functions such as the trigonometric functions are not one-to-one on their whole domains. It is still possible to define an inverse for such a function, but we have to restrict the domain of the function artificially so that the restricted function is one-to-one.

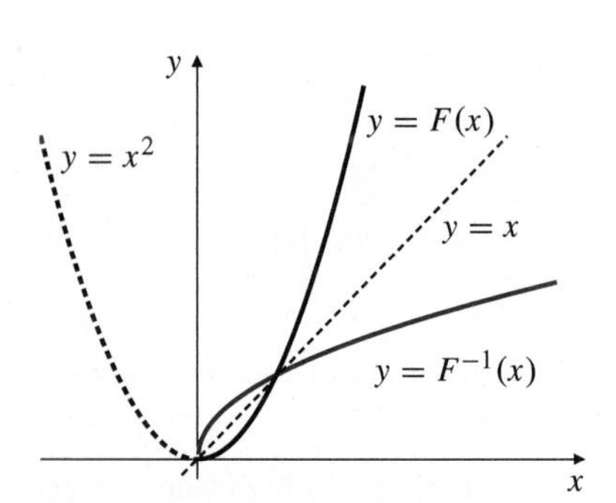

Figure 3.5 The restriction F of x^2 to $[0, \infty)$ and its inverse F^{-1}

As an example, consider the function $f(x) = x^2$. Unrestricted, its domain is the whole real line and it is not one-to-one since $f(-a) = f(a)$ for any a. Let us define a new function $F(x)$ equal to $f(x)$ but having a smaller domain, so that it is one-to-one. We can use the interval $[0, \infty)$ as the domain of F:

$$F(x) = x^2 \qquad \text{for} \quad 0 \le x < \infty.$$

The graph of F is shown in Figure 3.5; it is the right half of the parabola $y = x^2$, the graph of f. Evidently F is one-to-one, so it has an inverse F^{-1} which we calculate as follows:

Let $y = F^{-1}(x)$, then $x = F(y) = y^2$ and $y \ge 0$. Thus $y = \sqrt{x}$. Hence $F^{-1}(x) = \sqrt{x}$.

This method of restricting the domain of a non–one-to-one function to make it invertible will be used when we invert the trigonometric functions in Section 3.5.

Derivatives of Inverse Functions

Suppose that the function f is differentiable on an interval (a, b) and that either $f'(x) > 0$ for $a < x < b$, so that f is increasing on (a, b), or $f'(x) < 0$ for $a < x < b$, so that f is decreasing on (a, b). In either case f is one-to-one on (a, b) and has an inverse, f^{-1} there. Differentiating the cancellation identity

$$f\big(f^{-1}(x)\big) = x$$

with respect to x, using the Chain Rule, we obtain

$$f'\big(f^{-1}(x)\big)\,\frac{d}{dx}\,f^{-1}(x) = \frac{d}{dx}\,x = 1.$$

Thus,

$$\frac{d}{dx}f^{-1}(x) = \frac{1}{f'\left(f^{-1}(x)\right)}.$$

In Leibniz notation, if $y = f^{-1}(x)$, we have $\left.\frac{dy}{dx}\right|_{x} = \frac{1}{\left.\frac{dx}{dy}\right|_{y=f^{-1}(x)}}$.

The slope of the graph of f^{-1} at (x, y) is the reciprocal of the slope of the graph of f at (y, x). (See Figure 3.6.)

EXAMPLE 4 Show that $f(x) = x^3 + x$ is one-to-one on the whole real line, and, noting that $f(2) = 10$, find $\left(f^{-1}\right)'(10)$.

Solution Since $f'(x) = 3x^2 + 1 > 0$ for all real numbers x, f is increasing and therefore one-to-one and invertible. If $y = f^{-1}(x)$, then

$$\begin{aligned} x = f(y) = y^3 + y \quad &\Longrightarrow \quad 1 = (3y^2 + 1)y' \\ &\Longrightarrow \quad y' = \frac{1}{3y^2 + 1}. \end{aligned}$$

Now $x = f(2) = 10$ implies $y = f^{-1}(10) = 2$. Thus,

$$\left(f^{-1}\right)'(10) = \left.\frac{1}{3y^2 + 1}\right|_{y=2} = \frac{1}{13}.$$

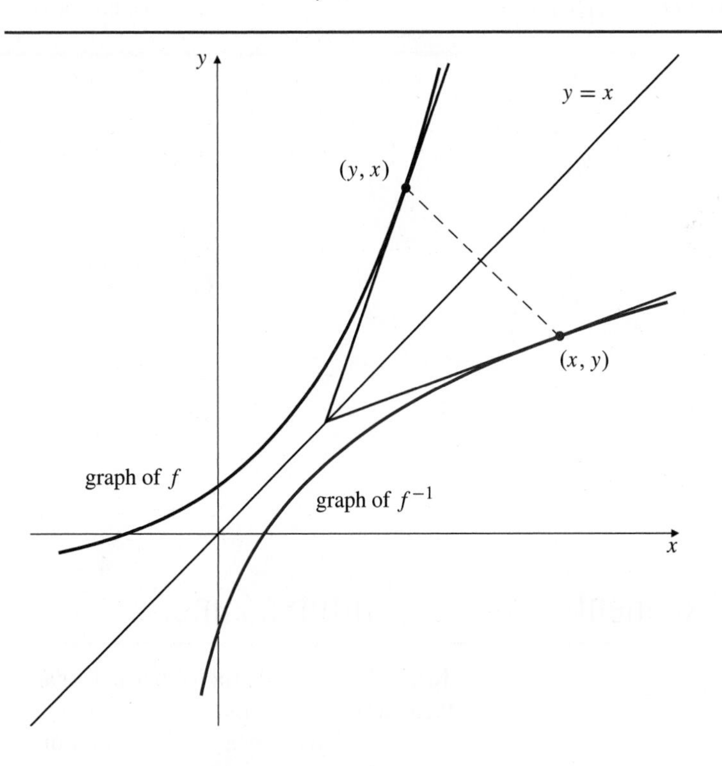

Figure 3.6 Tangents to the graphs of f and f^{-1}

EXERCISES 3.1

Show that the functions f in Exercises 1–12 are one-to-one, and calculate the inverse functions f^{-1}. Specify the domains and ranges of f and f^{-1}.

1. $f(x) = x - 1$

2. $f(x) = 2x - 1$

3. $f(x) = \sqrt{x - 1}$

4. $f(x) = -\sqrt{x - 1}$

5. $f(x) = x^3$

6. $f(x) = 1 + \sqrt[3]{x}$

7. $f(x) = x^2, \quad x \le 0$

8. $f(x) = (1 - 2x)^3$

9. $f(x) = \dfrac{1}{x + 1}$

10. $f(x) = \dfrac{x}{1 + x}$

11. $f(x) = \dfrac{1 - 2x}{1 + x}$

12. $f(x) = \dfrac{x}{\sqrt{x^2 + 1}}$

In Exercises 13–20, f is a one-to-one function with inverse f^{-1}. Calculate the inverses of the given functions in terms of f^{-1}.

13. $g(x) = f(x) - 2$

14. $h(x) = f(2x)$

15. $k(x) = -3f(x)$

16. $m(x) = f(x - 2)$

17. $p(x) = \dfrac{1}{1 + f(x)}$

18. $q(x) = \dfrac{f(x) - 3}{2}$

19. $r(x) = 1 - 2f(3 - 4x)$

20. $s(x) = \dfrac{1 + f(x)}{1 - f(x)}$

In Exercises 21–23, show that the given function is one-to-one and find its inverse.

21. $f(x) = \begin{cases} x^2 + 1 & \text{if } x \ge 0 \\ x + 1 & \text{if } x < 0 \end{cases}$

22. $g(x) = \begin{cases} x^3 & \text{if } x \ge 0 \\ x^{1/3} & \text{if } x < 0 \end{cases}$

23. $h(x) = x|x| + 1$

24. Find $f^{-1}(2)$ if $f(x) = x^3 + x$.

25. Find $g^{-1}(1)$ if $g(x) = x^3 + x - 9$.

26. Find $h^{-1}(-3)$ if $h(x) = x|x| + 1$.

27. Assume that the function $f(x)$ satisfies $f'(x) = \dfrac{1}{x}$ and that f is one-to-one. If $y = f^{-1}(x)$, show that $dy/dx = y$.

28. Find $\left(f^{-1}\right)'(x)$ if $f(x) = 1 + 2x^3$.

29. Show that $f(x) = \dfrac{4x^3}{x^2 + 1}$ has an inverse and find $\left(f^{-1}\right)'(2)$.

30. Find $\left(f^{-1}\right)'(-2)$ if $f(x) = x\sqrt{3 + x^2}$.

31. If $f(x) = x^2/(1 + \sqrt{x})$, find $f^{-1}(2)$ correct to 5 decimal places.

32. If $g(x) = 2x + \sin x$, show that g is invertible, and find $g^{-1}(2)$ and $(g^{-1})'(2)$ correct to 5 decimal places.

33. Show that $f(x) = x \sec x$ is one-to-one on $(-\pi/2, \pi/2)$. What is the domain of $f^{-1}(x)$? Find $(f^{-1})'(0)$.

34. If functions f and g have respective inverses f^{-1} and g^{-1}, show that the composite function $f \circ g$ has inverse $(f \circ g)^{-1} = g^{-1} \circ f^{-1}$.

35. For what values of the constants a, b, and c is the function $f(x) = (x - a)/(bx - c)$ self-inverse?

36. Can an even function be self-inverse? an odd function?

37. In this section it was claimed that an increasing (or decreasing) function defined on a single interval is necessarily one-to-one. Is the converse of this statement true? Explain.

38. Repeat Exercise 37 with the added assumption that f is continuous on the interval where it is defined.

3.2 Exponential and Logarithmic Functions

To begin we review exponential and logarithmic functions as you may have encountered them in your previous mathematical studies. In the following sections we will approach these functions from a different point of view and learn how to find their derivatives.

Exponentials

An **exponential function** is a function of the form $f(x) = a^x$, where the **base** a is a positive constant and the **exponent** x is the variable. Do not confuse such functions with **power** functions such as $f(x) = x^a$, where the base is variable and the exponent is constant. The exponential function a^x can be defined for integer and rational exponents x as follows:

DEFINITION 4

Exponential functions

If $a > 0$, then

$$a^0 = 1$$
$$a^n = \underbrace{a \cdot a \cdot a \cdots a}_{n \text{ factors}} \quad \text{if } n = 1, 2, 3, \ldots$$
$$a^{-n} = \frac{1}{a^n} \quad \text{if } n = 1, 2, 3, \ldots$$
$$a^{m/n} = \sqrt[n]{a^m} \quad \text{if } n = 1, 2, 3, \ldots \quad \text{and } m = \pm 1, \pm 2, \pm 3, \ldots.$$

In this definition, $\sqrt[n]{a}$ is the number $b > 0$ that satisfies $b^n = a$.

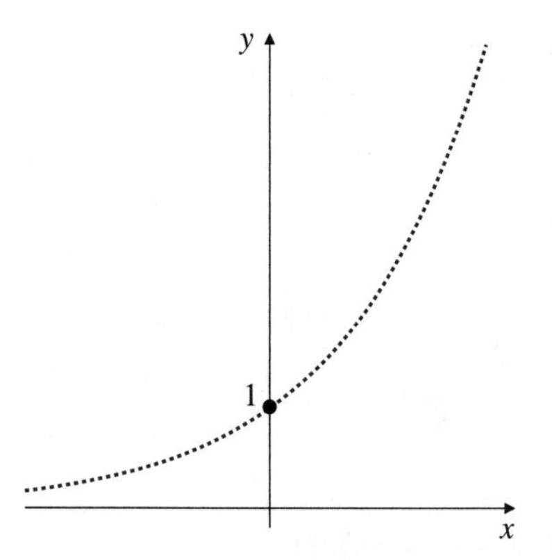

Figure 3.7 $y = 2^x$ for rational x

How should we define a^x if x is not rational? For example, what does 2^π mean? In order to calculate a derivative of a^x, we will want the function to be defined for all real numbers x, not just rational ones.

In Figure 3.7 we plot points with coordinates $(x, 2^x)$ for many closely spaced rational values of x. They appear to lie on a smooth curve. The definition of a^x can be extended to irrational x in such a way that a^x becomes a differentiable function of x on the whole real line. We will do so in the next section. For the moment, if x is irrational we can regard a^x as being the limit of values a^r for rational numbers r approaching x:

$$a^x = \lim_{\substack{r \to x \\ r \text{ rational}}} a^r.$$

EXAMPLE 1 Since the irrational number $\pi = 3.141\,592\,653\,59\ldots$ is the limit of the sequence of rational numbers

$$r_1 = 3, \quad r_2 = 3.1, \quad r_3 = 3.14, \quad r_4 = 3.141, \quad r_5 = 3.1415, \quad \ldots,$$

we can calculate 2^π as the limit of the corresponding sequence

$$2^3 = 8, \quad 2^{3.1} = 8.574\,187\,7\ldots, \quad 2^{3.14} = 8.815\,240\,9\ldots.$$

This gives $2^\pi = \lim_{n\to\infty} 2^{r_n} = 8.824\,977\,827\ldots.$

Exponential functions satisfy several identities called *laws of exponents*:

Laws of exponents

If $a > 0$ and $b > 0$, and x and y are any real numbers, then

(i) $a^0 = 1$ (ii) $a^{x+y} = a^x a^y$

(iii) $a^{-x} = \dfrac{1}{a^x}$ (iv) $a^{x-y} = \dfrac{a^x}{a^y}$

(v) $(a^x)^y = a^{xy}$ (vi) $(ab)^x = a^x b^x$

These identities can be proved for rational exponents using the definitions above. They remain true for irrational exponents, but we can't show that until the next section.

If $a = 1$, then $a^x = 1^x = 1$ for every x. If $a > 1$, then a^x is an increasing function of x; if $0 < a < 1$, then a^x is decreasing. The graphs of some typical exponential functions are shown in Figure 3.8(a). They all pass through the point (0,1) since $a^0 = 1$ for every $a > 0$. Observe that $a^x > 0$ for all $a > 0$ and all real x and that

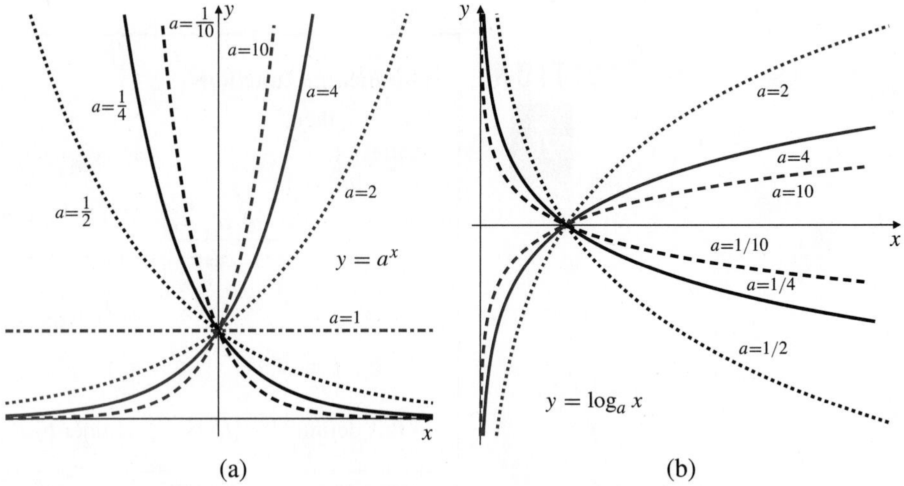

Figure 3.8
(a) Graphs of some exponential functions
(b) Graphs of some logarithmic functions

$$\text{If} \quad a > 1, \quad \text{then} \quad \lim_{x\to-\infty} a^x = 0 \quad \text{and} \quad \lim_{x\to\infty} a^x = \infty.$$

$$\text{If} \quad 0 < a < 1, \quad \text{then} \quad \lim_{x\to-\infty} a^x = \infty \quad \text{and} \quad \lim_{x\to\infty} a^x = 0.$$

The graph of $y = a^x$ has the x-axis as a horizontal asymptote if $a \neq 1$. It is asymptotic on the left (as $x \to -\infty$) if $a > 1$ and on the right (as $x \to \infty$) if $0 < a < 1$.

Logarithms

The function $f(x) = a^x$ is a one-to-one function provided that $a > 0$ and $a \neq 1$. Therefore, f has an inverse which we call a *logarithmic function.*

DEFINITION 5

If $a > 0$ and $a \neq 1$, the function $\log_a x$, called **the logarithm of x to the base a,** is the inverse of the one-to-one function a^x:

$$y = \log_a x \quad \iff \quad x = a^y, \qquad (a > 0, \quad a \neq 1).$$

Since a^x has domain $(-\infty, \infty)$, $\log_a x$ has range $(-\infty, \infty)$. Since a^x has range $(0, \infty)$, $\log_a x$ has domain $(0, \infty)$. Since a^x and $\log_a x$ are inverse functions, the following **cancellation identities** hold:

$$\log_a (a^x) = x \quad \text{for all real } x \qquad \text{and} \qquad a^{\log_a x} = x \quad \text{for all} \quad x > 0.$$

The graphs of some typical logarithmic functions are shown in Figure 3.8(b). They all pass through the point $(1, 0)$. Each graph is the reflection in the line $y = x$ of the corresponding exponential graph in Figure 3.8(a).

From the laws of exponents we can derive the following laws of logarithms:

Laws of logarithms

If $x > 0$, $y > 0$, $a > 0$, $b > 0$, $a \neq 1$, and $b \neq 1$, then

(i) $\log_a 1 = 0$

(ii) $\log_a(xy) = \log_a x + \log_a y$

(iii) $\log_a \left(\dfrac{1}{x}\right) = -\log_a x$

(iv) $\log_a \left(\dfrac{x}{y}\right) = \log_a x - \log_a y$

(v) $\log_a (x^y) = y \log_a x$

(vi) $\log_a x = \dfrac{\log_b x}{\log_b a}$

EXAMPLE 2 If $a > 0$, $x > 0$, and $y > 0$, verify that $\log_a(xy) = \log_a x + \log_a y$, using laws of exponents.

Solution Let $u = \log_a x$ and $v = \log_a y$. By the defining property of inverse functions, $x = a^u$ and $y = a^v$. Thus $xy = a^u a^v = a^{u+v}$. Inverting again, we get $\log_a(xy) = u + v = \log_a x + \log_a y$.

Logarithm law (vi) presented above shows that if you know logarithms to a particular base b, you can calculate logarithms to any other base a. Scientific calculators usually have built-in programs for calculating logarithms to base 10 and to base e, a special number that we will discover in Section 3.3. Logarithms to any base can be calculated using either of these functions. For example, computer scientists sometimes need to use logarithms to base 2. Using a scientific calculator, you can readily calculate

$$\log_2 13 = \frac{\log_{10} 13}{\log_{10} 2} = \frac{1.113\,943\,352\,31\ldots}{0.301\,029\,995\,664\ldots} = 3.700\,439\,718\,14\ldots.$$

The laws of logarithms can sometimes be used to simplify complicated expressions.

EXAMPLE 3 Simplify
(a) $\log_2 10 + \log_2 12 - \log_2 15$, (b) $\log_{a^2} a^3$, and (c) $3^{\log_9 4}$.

Solution

(a)
$$\begin{aligned} \log_2 10 + \log_2 12 - \log_2 15 &= \log_2 \frac{10 \times 12}{15} && \text{(laws (ii) and (iv))} \\ &= \log_2 8 \\ &= \log_2 2^3 = 3. && \text{(cancellation identity)} \end{aligned}$$

(b)
$$\begin{aligned} \log_{a^2} a^3 &= 3 \log_{a^2} a && \text{(law (v))} \\ &= \frac{3}{2} \log_{a^2} a^2 && \text{(law (v) again)} \\ &= \frac{3}{2}. && \text{(cancellation identity)} \end{aligned}$$

(c)
$$\begin{aligned} 3^{\log_9 4} &= 3^{(\log_3 4)/(\log_3 9)} && \text{(law (vi))} \\ &= \left(3^{\log_3 4}\right)^{1/\log_3 9} \\ &= 4^{1/\log_3 3^2} = 4^{1/2} = 2. && \text{(cancellation identity)} \end{aligned}$$

EXAMPLE 4 Solve the equation $3^{x-1} = 2^x$.

Solution We can take logarithms of both sides of the equation to any base a and get

$$\begin{aligned} (x-1)\log_a 3 &= x \log_a 2 \\ (\log_a 3 - \log_a 2)x &= \log_a 3 \\ x &= \frac{\log_a 3}{\log_a 3 - \log_a 2} = \frac{\log_a 3}{\log_a(3/2)}. \end{aligned}$$

The numerical value of x can be found using the "log" function on a scientific calculator. (This function is $\log_{10}$.) The value is $x = 2.7095\ldots$.

Corresponding to the asymptotic behaviour of the exponential functions, the logarithmic functions also exhibit asymptotic behaviour. Their graphs are all asymptotic to the y-axis as $x \to 0$ from the right:

If $a > 1$, then $\lim_{x\to 0+} \log_a x = -\infty$ and $\lim_{x\to\infty} \log_a x = \infty$.

If $0 < a < 1$, then $\lim_{x\to 0+} \log_a x = \infty$ and $\lim_{x\to\infty} \log_a x = -\infty$.

EXERCISES 3.2

Simplify the expressions in Exercises 1–18.

1. $\dfrac{3^3}{\sqrt{3^5}}$

2. $2^{1/2}8^{1/2}$

3. $\left(x^{-3}\right)^{-2}$

4. $\left(\dfrac{1}{2}\right)^x 4^{x/2}$

5. $\log_5 125$

6. $\log_4\left(\dfrac{1}{8}\right)$

7. $\log_{1/3} 3^{2x}$

8. $2^{\log_4 8}$

9. $10^{-\log_{10}(1/x)}$

10. $x^{1/(\log_a x)}$

11. $(\log_a b)(\log_b a)$

12. $\log_x\left(x(\log_y y^2)\right)$

13. $(\log_4 16)(\log_4 2)$

14. $\log_{15} 75 + \log_{15} 3$

15. $\log_6 9 + \log_6 4$

16. $2\log_3 12 - 4\log_3 6$

17. $\log_a(x^4 + 3x^2 + 2) + \log_a(x^4 + 5x^2 + 6)$
$-4\log_a \sqrt{x^2 + 2}$

18. $\log_\pi(1 - \cos x) + \log_\pi(1 + \cos x) - 2\log_\pi \sin x$

Use the base 10 exponential and logarithm functions 10^x and $\log x$ ($= \log_{10} x$) on a scientific calculator to evaluate the expressions or solve the equations in Exercises 19–24.

19. $3^{\sqrt{2}}$

20. $\log_3 5$

21. $2^{2x} = 5^{x+1}$

22. $x^{\sqrt{2}} = 3$

23. $\log_x 3 = 5$

24. $\log_3 x = 5$

Use the laws of exponents to prove the laws of logarithms in Exercises 25–28.

25. $\log_a\left(\dfrac{1}{x}\right) = -\log_a x$

26. $\log_a\left(\dfrac{x}{y}\right) = \log_a x - \log_a y$

27. $\log_a(x^y) = y\log_a x$

28. $\log_a x = (\log_b x)/(\log_b a)$

29. Solve $\log_4(x + 4) - 2\log_4(x + 1) = \dfrac{1}{2}$ for x.

30. Solve $2\log_3 x + \log_9 x = 10$ for x.

Evaluate the limits in Exercises 31–34.

31. $\lim_{x\to\infty} \log_x 2$

32. $\lim_{x\to 0+} \log_x(1/2)$

33. $\lim_{x\to 1+} \log_x 2$

34. $\lim_{x\to 1-} \log_x 2$

35. Suppose that $f(x) = a^x$ is differentiable at $x = 0$ and that $f'(0) = k$, where $k \neq 0$. Prove that f is differentiable at any real number x and that

$$f'(x) = k\,a^x = k\,f(x).$$

36. Continuing Exercise 35, prove that $f^{-1}(x) = \log_a x$ is differentiable at any $x > 0$ and that

$$(f^{-1})'(x) = \frac{1}{kx}.$$

3.3 The Natural Logarithm and Exponential

Regard this paragraph as describing a game we are going to play in this section. The result of the game will be that we will acquire two new classes of functions, logarithms, and exponentials, to which the rules of calculus will apply.

In this section we are going to define a function $\ln x$, called the *natural* logarithm of x, in a way that does not at first seem to have anything to do with the logarithms considered in Section 3.2. We will show, however, that it has the same properties as those logarithms, and in the end we will see that $\ln x = \log_e x$, the logarithm of x to a certain specific base e. We will show that $\ln x$ is a one-to-one function, defined for all positive real numbers. It must therefore have an inverse, e^x, that we will call *the* exponential function. Our final goal is to arrive at a definition of the exponential functions a^x (for any $a > 0$) that is valid for any real number x instead of just rational numbers, and that is known to be continuous and even differentiable without our having to assume those properties as we did in Section 3.2.

Table 1. Derivatives of integer powers

$f(x)$	$f'(x)$
$\vdots$	$\vdots$
x^4	$4x^3$
x^3	$3x^2$
x^2	$2x$
x^1	$1x^0 = 1$
x^0	0
x^{-1}	$-x^{-2}$
x^{-2}	$-2x^{-3}$
x^{-3}	$-3x^{-4}$
$\vdots$	$\vdots$

The Natural Logarithm

Table 1 lists the derivatives of integer powers of x. Those derivatives are multiples of integer powers of x, but one integer power, x^{-1}, is conspicuously absent from the list of derivatives; we do not yet know a function whose derivative is $x^{-1} = 1/x$. We are going to remedy this situation by defining a function $\ln x$ in such a way that it will have derivative $1/x$.

To get a hint as to how this can be done, review Example 1 of Section 2.11. In that example we showed that the area under the graph of the velocity of a moving object in a time interval is equal to the distance travelled by the object in that time interval. Since the derivative of distance is velocity, measuring the area provided a way of finding a function (the distance) that had a given derivative (the velocity). This relationship between area and derivatives is one of the most important ideas in calculus. It is called the **Fundamental Theorem of Calculus**. We will explore it fully in Chapter 5, but we will make use of the idea now to define $\ln x$, which we want to have derivative $1/x$.

DEFINITION 6

The natural logarithm

For $x > 0$, let A_x be the area of the plane region bounded by the curve $y = 1/t$, the t-axis, and the vertical lines $t = 1$ and $t = x$. The function $\ln x$ is defined by

$$\ln x = \begin{cases} A_x & \text{if } x \ge 1, \\ -A_x & \text{if } 0 < x < 1, \end{cases}$$

as shown in Figure 3.9.

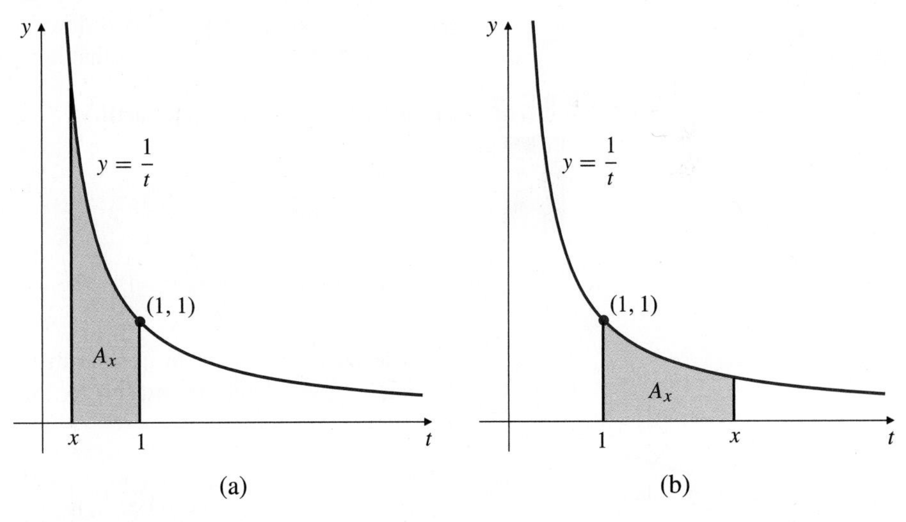

Figure 3.9

(a) $\ln x = -\text{area } A_x$ if $0 < x < 1$

(b) $\ln x = \text{area } A_x$ if $x \ge 1$

The definition implies that $\ln 1 = 0$, that $\ln x > 0$ if $x > 1$, that $\ln x < 0$ if $0 < x < 1$, and that ln is a one-to-one function. We now show that if $y = \ln x$, then $y' = 1/x$. The proof of this result is similar to the proof we will give for the Fundamental Theorem of Calculus in Section 5.5.

THEOREM 1

If $x > 0$, then

$$\frac{d}{dx} \ln x = \frac{1}{x}.$$

PROOF For $x > 0$ and $h > 0$, $\ln(x+h) - \ln x$ is the area of the plane region bounded by $y = 1/t$, $y = 0$, and the vertical lines $t = x$ and $t = x + h$; it is the shaded area in

Figure 3.10. Comparing this area with that of two rectangles, we see that

$$\frac{h}{x+h} < \text{shaded area} = \ln(x+h) - \ln x < \frac{h}{x}.$$

Hence, the Newton quotient for $\ln x$ satisfies

$$\frac{1}{x+h} < \frac{\ln(x+h) - \ln x}{h} < \frac{1}{x}.$$

Letting h approach 0 from the right, we obtain (by the Squeeze Theorem applied to one-sided limits)

$$\lim_{h \to 0+} \frac{\ln(x+h) - \ln x}{h} = \frac{1}{x}.$$

A similar argument shows that if $0 < x + h < x$, then

$$\frac{1}{x} < \frac{\ln(x+h) - \ln x}{h} < \frac{1}{x+h},$$

so that

$$\lim_{h \to 0-} \frac{\ln(x+h) - \ln x}{h} = \frac{1}{x}.$$

Figure 3.10

Combining these two one-sided limits we get the desired result:

$$\frac{d}{dx} \ln x = \lim_{h \to 0} \frac{\ln(x+h) - \ln x}{h} = \frac{1}{x}.$$

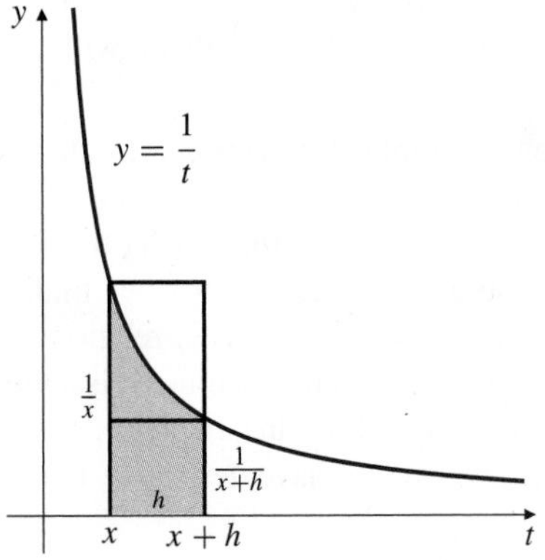

■

The two properties $(d/dx) \ln x = 1/x$ and $\ln 1 = 0$ are sufficient to determine the function $\ln x$ completely. (This follows from Theorem 13 in Section 2.8.) We can deduce from these two properties that $\ln x$ satisfies the appropriate laws of logarithms:

THEOREM 2 **Properties of the natural logarithm**

(i) $\ln(xy) = \ln x + \ln y$ (ii) $\ln\left(\frac{1}{x}\right) = -\ln x$

(iii) $\ln\left(\frac{x}{y}\right) = \ln x - \ln y$ (iv) $\ln\left(x^r\right) = r \ln x$

Because we do not want to *assume* that exponentials are continuous (as we did in Section 3.2), we should regard (iv) for the moment as only valid for exponents r that are rational numbers.

PROOF We will only prove part (i) because the other parts are proved by the same method. If $y > 0$ is a constant, then by the Chain Rule,

$$\frac{d}{dx}\big(\ln(xy) - \ln x\big) = \frac{y}{xy} - \frac{1}{x} = 0 \quad \text{for all } x > 0.$$

Theorem 13 of Section 2.8 now tells us that $\ln(xy) - \ln x = C$ (a constant) for $x > 0$. Putting $x = 1$ we get $C = \ln y$ and identity (i) follows.

■

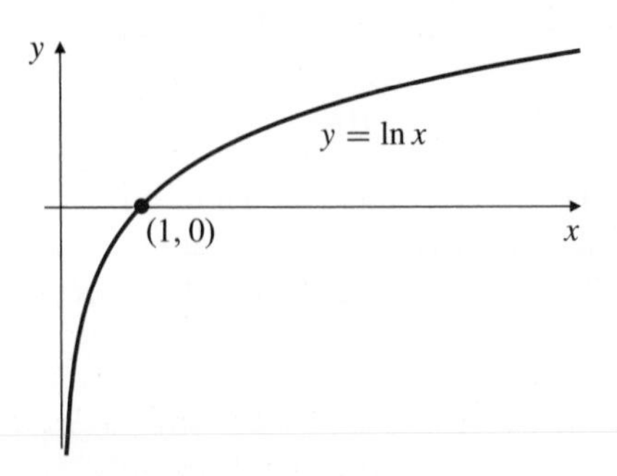

Figure 3.11 The graph of $\ln x$

Part (iv) of Theorem 2 shows that $\ln(2^n) = n \ln 2 \to \infty$ as $n \to \infty$. Therefore, we also have $\ln(1/2)^n = -n \ln 2 \to -\infty$ as $n \to \infty$. Since $(d/dx) \ln x = 1/x > 0$ for $x > 0$, it follows that $\ln x$ is increasing, so we must have (see Figure 3.11)

$$\lim_{x \to \infty} \ln x = \infty, \qquad \lim_{x \to 0+} \ln x = -\infty.$$

EXAMPLE 1 Show that $\frac{d}{dx}\ln|x| = \frac{1}{x}$ for any $x \neq 0$. Hence find $\int \frac{1}{x}\,dx$.

Solution If $x > 0$, then

$$\frac{d}{dx}\ln|x| = \frac{d}{dx}\ln x = \frac{1}{x}$$

by Theorem 1. If $x < 0$, then, using the Chain Rule,

$$\frac{d}{dx}\ln|x| = \frac{d}{dx}\ln(-x) = \frac{1}{-x}(-1) = \frac{1}{x}.$$

Therefore, $\frac{d}{dx}\ln|x| = \frac{1}{x}$, and on any interval not containing $x = 0$,

$$\int \frac{1}{x}\,dx = \ln|x| + C.$$

EXAMPLE 2 Find the derivatives of (a) $\ln|\cos x|$ and (b) $\ln\left(x + \sqrt{x^2+1}\right)$. Simplify your answers as much as possible.

Solution

(a) Using the result of Example 1 and the Chain Rule, we have

$$\frac{d}{dx}\ln|\cos x| = \frac{1}{\cos x}(-\sin x) = -\tan x.$$

(b)
$$\begin{aligned}\frac{d}{dx}\ln\left(x + \sqrt{x^2+1}\right) &= \frac{1}{x + \sqrt{x^2+1}}\left(1 + \frac{2x}{2\sqrt{x^2+1}}\right)\\ &= \frac{1}{x + \sqrt{x^2+1}}\,\frac{\sqrt{x^2+1}+x}{\sqrt{x^2+1}}\\ &= \frac{1}{\sqrt{x^2+1}}.\end{aligned}$$

The Exponential Function

The function $\ln x$ is one-to-one on its domain, the interval $(0, \infty)$, so it has an inverse there. For the moment, let us call this inverse $\exp x$. Thus,

$$y = \exp x \quad \Longleftrightarrow \quad x = \ln y \quad (y > 0).$$

Since $\ln 1 = 0$, we have $\exp 0 = 1$. The domain of exp is $(-\infty, \infty)$, the range of ln. The range of exp is $(0, \infty)$, the domain of ln. We have cancellation identities

$$\ln(\exp x) = x \quad \text{for all real } x \qquad \text{and} \qquad \exp(\ln x) = x \quad \text{for } x > 0.$$

We can deduce various properties of exp from corresponding properties of ln. Not surprisingly, they are properties we would expect an exponential function to have.

THEOREM 3 **Properties of the exponential function**

(i) $(\exp x)^r = \exp(rx)$ (ii) $\exp(x+y) = (\exp x)(\exp y)$

(iii) $\exp(-x) = \dfrac{1}{\exp(x)}$ (iv) $\exp(x-y) = \dfrac{\exp x}{\exp y}$

For the moment, identity (i) is asserted only for rational numbers r.

PROOF We prove only identity (i); the rest are done similarly. If $u = (\exp x)^r$, then, by Theorem 2(iv), $\ln u = r\ln(\exp x) = rx$. Therefore, $u = \exp(rx)$.

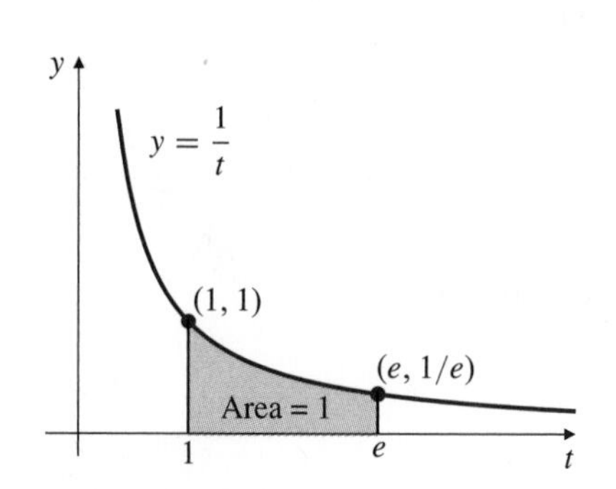

Figure 3.12 The definition of e

Now we make an important definition!

Let $e = \exp(1)$.

The number e satisfies $\ln e = 1$, so the area bounded by the curve $y = 1/t$, the t-axis, and the vertical lines $t = 1$ and $t = e$ must be equal to 1 square unit. See Figure 3.12. The number e is one of the most important numbers in mathematics. Like π, it is irrational and not a zero of any polynomial with rational coefficients. (Such numbers are called **transcendental**.) Its value is between 2 and 3 and begins

$$e = 2.7\,1828\,1828\,45\,90\,45\ldots.$$

Later on we will learn that

$$e = 1 + \frac{1}{1!} + \frac{1}{2!} + \frac{1}{3!} + \frac{1}{4!} + \cdots,$$

a formula from which the value of e can be calculated to any desired precision.

Theorem 3(i) shows that $\exp r = \exp(1r) = (\exp 1)^r = e^r$ holds for any rational number r. Now here is a crucial observation. We only know what e^r means if r is a rational number (if $r = m/n$, then $e^r = \sqrt[n]{e^m}$). But $\exp x$ is defined for all *real* x, rational or not. Since $e^r = \exp r$ when r is rational, we can use $\exp x$ as a *definition* of what e^x means for any real number x, and there will be no contradiction if x happens to be rational.

$$e^x = \exp x \qquad \text{for all real } x.$$

Theorem 3 can now be restated in terms of e^x:

(i) $(e^x)^y = e^{xy}$ (ii) $e^{x+y} = e^x e^y$

(iii) $e^{-x} = \dfrac{1}{e^x}$ (iv) $e^{x-y} = \dfrac{e^x}{e^y}$

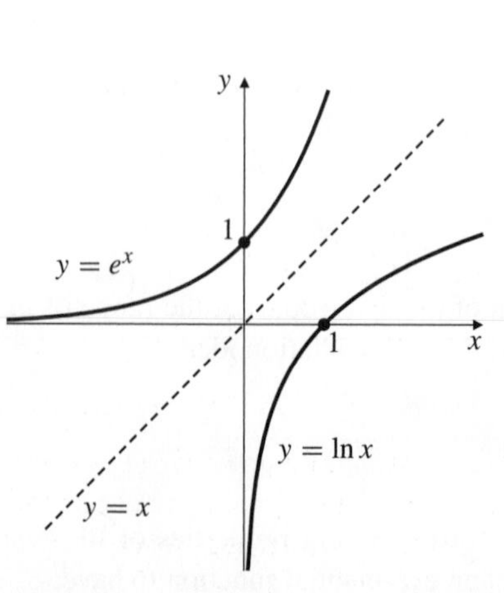

Figure 3.13 The graphs of e^x and $\ln x$

The graph of e^x is the reflection of the graph of its inverse, $\ln x$, in the line $y = x$. Both graphs are shown for comparison in Figure 3.13. Observe that the x-axis is a horizontal asymptote of the graph of $y = e^x$ as $x \to -\infty$. We have

$$\lim_{x\to-\infty} e^x = 0, \qquad \lim_{x\to\infty} e^x = \infty.$$

Since $\exp x = e^x$ actually *is* an exponential function, its inverse must actually *be* a logarithm:

$$\ln x = \log_e x.$$

The derivative of $y = e^x$ is calculated by implicit differentiation:

$$\begin{aligned} y = e^x \quad &\Longrightarrow \quad x = \ln y \\ &\Longrightarrow \quad 1 = \frac{1}{y}\frac{dy}{dx} \\ &\Longrightarrow \quad \frac{dy}{dx} = y = e^x. \end{aligned}$$

Thus, the exponential function has the remarkable property that it is its own derivative and, therefore, also its own antiderivative:

$$\frac{d}{dx}e^x = e^x, \qquad \int e^x\,dx = e^x + C.$$

EXAMPLE 3 Find the derivatives of
(a) e^{x^2-3x}, (b) $\sqrt{1+e^{2x}}$, and (c) $\dfrac{e^x - e^{-x}}{e^x + e^{-x}}$.

Solution

(a) $\dfrac{d}{dx}e^{x^2-3x} = e^{x^2-3x}(2x-3) = (2x-3)e^{x^2-3x}$.

(b) $\dfrac{d}{dx}\sqrt{1+e^{2x}} = \dfrac{1}{2\sqrt{1+e^{2x}}}\left(e^{2x}(2)\right) = \dfrac{e^{2x}}{\sqrt{1+e^{2x}}}$.

(c)
$$\begin{aligned} \frac{d}{dx}\frac{e^x - e^{-x}}{e^x + e^{-x}} &= \frac{(e^x+e^{-x})(e^x-(-e^{-x})) - (e^x-e^{-x})(e^x+(-e^{-x}))}{(e^x+e^{-x})^2} \\ &= \frac{(e^x)^2 + 2e^xe^{-x} + (e^{-x})^2 - [(e^x)^2 - 2e^xe^{-x} + (e^{-x})^2]}{(e^x+e^{-x})^2} \\ &= \frac{4e^{x-x}}{(e^x+e^{-x})^2} = \frac{4}{(e^x+e^{-x})^2}. \end{aligned}$$

EXAMPLE 4 Let $f(t) = e^{at}$. Find (a) $f^{(n)}(t)$ and (b) $\int f(t)\,dt$.

Solution (a) We have

$$\begin{aligned} f'(t) &= a\,e^{at} \\ f''(t) &= a^2\,e^{at} \\ f'''(t) &= a^3\,e^{at} \\ &\vdots \\ f^{(n)}(t) &= a^n\,e^{at}. \end{aligned}$$

(b) Also, $\displaystyle\int f(t)\,dt = \int e^{at}\,dt = \frac{1}{a}e^{at} + C$, since $\dfrac{d}{dt}\dfrac{1}{a}e^{at} = e^{at}$.

General Exponentials and Logarithms

We can use the fact that e^x is now defined for *all* real x to define the arbitrary exponential a^x (where $a > 0$) for all real x. If r is rational, then $\ln(a^r) = r\ln a$; therefore $a^r = e^{r\ln a}$. However, $e^{x\ln a}$ is defined for all real x, so we can use it as a definition of a^x with no possibility of contradiction arising if x is rational.

DEFINITION 7

The general exponential a^x

$$a^x = e^{x \ln a}, \qquad (a > 0, \quad x \text{ real}).$$

EXAMPLE 5 Evaluate 2^π, using the natural logarithm (ln) and exponential (exp or e^x) keys on a scientific calculator, but not using the y^x or ^ keys.

Solution $2^\pi = e^{\pi \ln 2} = 8.824\,977\,8\ldots$. If your calculator has a ^ key, or an x^y or y^x key, the chances are that it is implemented in terms of the exp and ln functions.

The laws of exponents for a^x as presented in Section 3.2 can now be obtained from those for e^x, as can the derivative:

$$\frac{d}{dx} a^x = \frac{d}{dx} e^{x \ln a} = e^{x \ln a} \ln a = a^x \ln a.$$

We can also verify the General Power Rule for x^a, where a is any real number, provided $x > 0$:

$$\frac{d}{dx} x^a = \frac{d}{dx} e^{a \ln x} = e^{a \ln x} \frac{a}{x} = \frac{a\, x^a}{x} = a\, x^{a-1}.$$

Do not confuse x^π, which is a power function of x, and π^x, which is an exponential function of x.

EXAMPLE 6 Show that the graph of $f(x) = x^\pi - \pi^x$ has a negative slope at $x = \pi$.

Solution

$$f'(x) = \pi\, x^{\pi - 1} - \pi^x \ln \pi$$
$$f'(\pi) = \pi\, \pi^{\pi - 1} - \pi^\pi \ln \pi = \pi^\pi (1 - \ln \pi).$$

Since $\pi > 3 > e$, we have $\ln \pi > \ln e = 1$, so $1 - \ln \pi < 0$. Since $\pi^\pi = e^{\pi \ln \pi} > 0$, we have $f'(\pi) < 0$. Thus, the graph $y = f(x)$ has negative slope at $x = \pi$.

EXAMPLE 7 Find the critical point of $y = x^x$.

Solution We can't differentiate x^x by treating it as a power (like x^a) because the exponent varies. We can't treat it as an exponential (like a^x) because the base varies. We can differentiate it if we first write it in terms of the exponential function, $x^x = e^{x \ln x}$, and then use the Chain Rule and the Product Rule:

$$\frac{dy}{dx} = \frac{d}{dx} e^{x \ln x} = e^{x \ln x} \left(\ln x + x \left(\frac{1}{x} \right) \right) = x^x (1 + \ln x).$$

Now x^x is defined only for $x > 0$, and is itself never 0. (Why?) Therefore, the critical point occurs where $1 + \ln x = 0$; that is, $\ln x = -1$, or $x = 1/e$.

Finally, observe that $(d/dx)a^x = a^x \ln a$ is negative for all x if $0 < a < 1$ and is positive for all x if $a > 1$. Thus, a^x is one-to-one and has an inverse function, $\log_a x$, provided $a > 0$ and $a \neq 1$. Its properties follow in the same way as in Section 3.2. If $y = \log_a x$, then $x = a^y$ and, differentiating implicitly with respect to x, we get

$$1 = a^y \ln a \frac{dy}{dx} = x \ln a \frac{dy}{dx}.$$

Thus, the derivative of $\log_a x$ is given by

$$\frac{d}{dx} \log_a x = \frac{1}{x \ln a}.$$

Since $\log_a x$ can be expressed in terms of logarithms to any other base, say e,

$$\log_a x = \frac{\ln x}{\ln a},$$

we normally use only natural logarithms. Exceptions are found in chemistry, acoustics, and other sciences where "logarithmic scales" are used to measure quantities for which a one-unit increase in the measure corresponds to a tenfold increase in the quantity. Logarithms to base 10 are used in defining such scales. In computer science, where powers of 2 play a central role, logarithms to base 2 are often encountered.

Logarithmic Differentiation

Suppose we want to differentiate a function of the form

$$y = (f(x))^{g(x)} \qquad (\text{for } f(x) > 0).$$

Since the variable appears in both the base and the exponent, neither the general power rule, $(d/dx)x^a = ax^{a-1}$, nor the exponential rule, $(d/dx)a^x = a^x \ln a$, can be directly applied. One method for finding the derivative of such a function is to express it in the form

$$y = e^{g(x)\ln f(x)}$$

and then differentiate, using the Product Rule to handle the exponent. This is the method used in Example 7.

The derivative in Example 7 can also be obtained by taking natural logarithms of both sides of the equation $y = x^x$ and differentiating implicitly:

$$\begin{aligned}\ln y &= x \ln x\\ \frac{1}{y}\frac{dy}{dx} &= \ln x + \frac{x}{x} = 1 + \ln x\\ \frac{dy}{dx} &= y(1+\ln x) = x^x(1+\ln x).\end{aligned}$$

This latter technique is called **logarithmic differentiation.**

EXAMPLE 8 Find dy/dt if $y = (\sin t)^{\ln t}$, where $0 < t < \pi$.

Solution We have $\ln y = \ln t \, \ln \sin t$. Thus,

$$\begin{aligned}\frac{1}{y}\frac{dy}{dt} &= \frac{1}{t}\ln\sin t + \ln t\frac{\cos t}{\sin t}\\ \frac{dy}{dt} &= y\left(\frac{\ln\sin t}{t} + \ln t\,\cot t\right) = (\sin t)^{\ln t}\left(\frac{\ln\sin t}{t} + \ln t\,\cot t\right).\end{aligned}$$

Logarithmic differentiation is also useful for finding the derivatives of functions expressed as products and quotients of many factors. Taking logarithms reduces these products and quotients to sums and differences. This usually makes the calculation easier than it would be using the Product and Quotient Rules, especially if the derivative is to be evaluated at a specific point.

EXAMPLE 9 Differentiate $y = [(x+1)(x+2)(x+3)]/(x+4)$.

Solution $\ln|y| = \ln|x+1| + \ln|x+2| + \ln|x+3| - \ln|x+4|$. Thus,

$$\frac{1}{y}y' = \frac{1}{x+1} + \frac{1}{x+2} + \frac{1}{x+3} - \frac{1}{x+4}$$
$$y' = \frac{(x+1)(x+2)(x+3)}{x+4}\left(\frac{1}{x+1} + \frac{1}{x+2} + \frac{1}{x+3} - \frac{1}{x+4}\right)$$
$$= \frac{(x+2)(x+3)}{x+4} + \frac{(x+1)(x+3)}{x+4} + \frac{(x+1)(x+2)}{x+4} - \frac{(x+1)(x+2)(x+3)}{(x+4)^2}.$$

EXAMPLE 10 Find $\left.\frac{du}{dx}\right|_{x=1}$ if $u = \sqrt{(x+1)(x^2+1)(x^3+1)}$.

Solution

$$\ln u = \frac{1}{2}\Big(\ln(x+1) + \ln(x^2+1) + \ln(x^3+1)\Big)$$
$$\frac{1}{u}\frac{du}{dx} = \frac{1}{2}\left(\frac{1}{x+1} + \frac{2x}{x^2+1} + \frac{3x^2}{x^3+1}\right).$$

At $x = 1$ we have $u = \sqrt{8} = 2\sqrt{2}$. Hence,

$$\left.\frac{du}{dx}\right|_{x=1} = \sqrt{2}\left(\frac{1}{2} + 1 + \frac{3}{2}\right) = 3\sqrt{2}.$$

EXERCISES 3.3

Simplify the expressions given in Exercises 1–10.

1. $e^3/\sqrt{e^5}$
2. $\ln\left(e^{1/2}e^{2/3}\right)$
3. $e^{5\ln x}$
4. $e^{(3\ln 9)/2}$
5. $\ln\frac{1}{e^{3x}}$
6. $e^{2\ln\cos x} + \left(\ln e^{\sin x}\right)^2$
7. $3\ln 4 - 4\ln 3$
8. $4\ln\sqrt{x} + 6\ln(x^{1/3})$
9. $2\ln x + 5\ln(x-2)$
10. $\ln(x^2 + 6x + 9)$

Solve the equations in Exercises 11–14 for x.

11. $2^{x+1} = 3^x$
12. $3^x = 9^{1-x}$
13. $\frac{1}{2^x} = \frac{5}{8^{x+3}}$
14. $2^{x^2-3} = 4^x$

Find the domains of the functions in Exercises 15–16.

15. $\ln\frac{x}{2-x}$
16. $\ln(x^2 - x - 2)$

Solve the inequalities in Exercises 17–18.

17. $\ln(2x-5) > \ln(7-2x)$
18. $\ln(x^2-2) \le \ln x$

In Exercises 19–48, differentiate the given functions. If possible, simplify your answers.

19. $y = e^{5x}$
20. $y = xe^x - x$
21. $y = \frac{x}{e^{2x}}$
22. $y = x^2e^{x/2}$
23. $y = \ln(3x-2)$
24. $y = \ln|3x-2|$
25. $y = \ln(1+e^x)$
26. $f(x) = e^{(x^2)}$
27. $y = \frac{e^x + e^{-x}}{2}$
28. $x = e^{3t}\ln t$
29. $y = e^{(e^x)}$
30. $y = \frac{e^x}{1+e^x}$
31. $y = e^x\sin x$
32. $y = e^{-x}\cos x$
33. $y = \ln\ln x$
34. $y = x\ln x - x$
35. $y = x^2\ln x - \frac{x^2}{2}$
36. $y = \ln|\sin x|$
37. $y = 5^{2x+1}$
38. $y = 2^{(x^2-3x+8)}$
39. $g(x) = t^x x^t$
40. $h(t) = t^x - x^t$
41. $f(s) = \log_a(bs+c)$
42. $g(x) = \log_x(2x+3)$
43. $y = x^{\sqrt{x}}$
44. $y = (1/x)^{\ln x}$
45. $y = \ln|\sec x + \tan x|$
46. $y = \ln|x + \sqrt{x^2-a^2}|$
47. $y = \ln\left(\sqrt{x^2+a^2} - x\right)$
48. $y = (\cos x)^x - x^{\cos x}$
49. Find the nth derivative of $f(x) = xe^{ax}$.

50. Show that the nth derivative of $(ax^2 + bx + c)e^x$ is a function of the same form but with different constants.

51. Find the first four derivatives of e^{x^2}.

52. Find the nth derivative of $\ln(2x + 1)$.

53. Differentiate (a) $f(x) = (x^x)^x$ and (b) $g(x) = x^{(x^x)}$. Which function grows more rapidly as x grows large?

54. Solve the equation $x^{x^{x^{\cdot^{\cdot^{\cdot}}}}} = a$, where $a > 0$. The exponent tower goes on forever.

Use logarithmic differentiation to find the required derivatives in Exercises 55–57.

55. $f(x) = (x - 1)(x - 2)(x - 3)(x - 4)$. Find $f'(x)$.

56. $F(x) = \dfrac{\sqrt{1+x}(1-x)^{1/3}}{(1+5x)^{4/5}}$. Find $F'(0)$.

57. $f(x) = \dfrac{(x^2-1)(x^2-2)(x^2-3)}{(x^2+1)(x^2+2)(x^2+3)}$. Find $f'(2)$. Also find $f'(1)$.

58. At what points does the graph $y = x^2 e^{-x^2}$ have a horizontal tangent line?

59. Let $f(x) = xe^{-x}$. Determine where f is increasing and where it is decreasing. Sketch the graph of f.

60. Find the equation of a straight line of slope 4 that is tangent to the graph of $y = \ln x$.

61. Find an equation of the straight line tangent to the curve $y = e^x$ and passing through the origin.

62. Find an equation of the straight line tangent to the curve $y = \ln x$ and passing through the origin.

63. Find an equation of the straight line that is tangent to $y = 2^x$ and that passes through the point $(1, 0)$.

64. For what values of $a > 0$ does the curve $y = a^x$ intersect the straight line $y = x$?

65. Find the slope of the curve $e^{xy} \ln \dfrac{x}{y} = x + \dfrac{1}{y}$ at $(e, 1/e)$.

66. Find an equation of the straight line tangent to the curve $xe^y + y - 2x = \ln 2$ at the point $(1, \ln 2)$.

67. Find the derivative of $f(x) = Ax \cos \ln x + Bx \sin \ln x$. Use the result to help you find the indefinite integrals $\int \cos \ln x \, dx$ and $\int \sin \ln x \, dx$.

68. Let $F_{A,B}(x) = Ae^x \cos x + Be^x \sin x$. Show that $(d/dx)F_{A,B}(x) = F_{A+B,B-A}(x)$.

69. Using the results of Exercise 68, find (a) $(d^2/dx^2)F_{A,B}(x)$ and (b) $(d^3/dx^3)e^x \cos x$.

70. Find $\dfrac{d}{dx}(Ae^{ax} \cos bx + Be^{ax} \sin bx)$ and use the answer to help you evaluate (a) $\int e^{ax} \cos bx \, dx$ and (b) $\int e^{ax} \sin bx \, dx$.

71. Prove identity (ii) of Theorem 2 by examining the derivative of the left side minus the right side as was done in the proof of identity (i).

72. Deduce identity (iii) of Theorem 2 from identities (i) and (ii).

73. Prove identity (iv) of Theorem 2 for rational exponents r by the same method used for Exercise 71.

74. Let $x > 0$, and let $F(x)$ be the area bounded by the curve $y = t^2$, the t-axis, and the vertical lines $t = 0$ and $t = x$. Using the method of the proof of Theorem 1, show that $F'(x) = x^2$. Hence, find an explicit formula for $F(x)$. What is the area of the region bounded by $y = t^2$, $y = 0$, $t = 0$, and $t = 2$?

75. Carry out the following steps to show that $2 < e < 3$. Let $f(t) = 1/t$ for $t > 0$.

(a) Show that the area under $y = f(t)$, above $y = 0$, and between $t = 1$ and $t = 2$ is less than 1 square unit. Deduce that $e > 2$.

(b) Show that all tangent lines to the graph of f lie below the graph. *Hint:* $f''(t) = 2/t^3 > 0$.

(c) Find the lines T_2 and T_3 that are tangent to $y = f(t)$ at $t = 2$ and $t = 3$, respectively.

(d) Find the area A_2 under T_2, above $y = 0$, and between $t = 1$ and $t = 2$. Also find the area A_3 under T_3, above $y = 0$, and between $t = 2$ and $t = 3$.

(e) Show that $A_2 + A_3 > 1$ square unit. Deduce that $e < 3$.

3.4 Growth and Decay

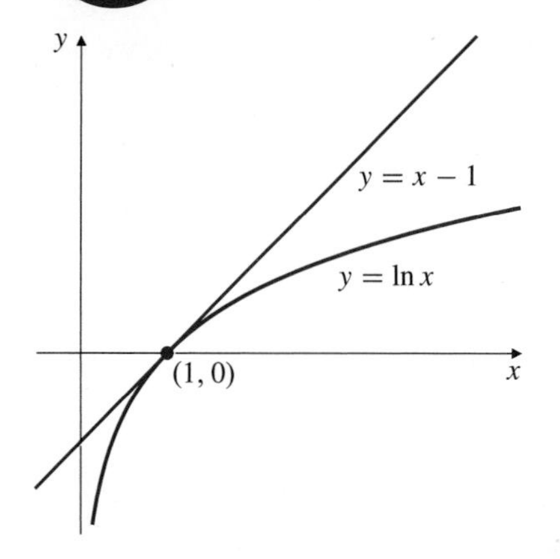

Figure 3.14 $\ln x \le x - 1$ for $x > 0$

In this section we will study the use of exponential functions to model the growth rates of quantities whose rate of growth is directly related to their size. The growth of such quantities is typically governed by differential equations whose solutions involve exponential functions. Before delving into this topic, we prepare the way by examining the growth behaviour of exponential and logarithmic functions.

The Growth of Exponentials and Logarithms

In Section 3.3 we showed that both e^x and $\ln x$ grow large (approach infinity) as x grows large. However, e^x increases very rapidly as x increases, and $\ln x$ increases very slowly. In fact, e^x increases faster than any positive power of x (no matter how large the power), while $\ln x$ increases more slowly than any positive power of x (no matter how small the power). To verify this behaviour we start with an inequality satisfied by $\ln x$. The straight line $y = x - 1$ is tangent to the curve $y = \ln x$ at the point $(1, 0)$. The following theorem asserts that the curve lies below that line. (See Figure 3.14.)

THEOREM 4

If $x > 0$, then $\ln x \leq x - 1$.

PROOF Let $g(x) = \ln x - (x - 1)$ for $x > 0$. Then $g(1) = 0$ and

$$g'(x) = \frac{1}{x} - 1 \quad \begin{cases} > 0 & \text{if } 0 < x < 1 \\ < 0 & \text{if } x > 1. \end{cases}$$

As observed in Section 2.8, these inequalities imply that g is increasing on $(0, 1)$ and decreasing on $(1, \infty)$. Thus, $g(x) \leq g(1) = 0$ for all $x > 0$ and $\ln x \leq x - 1$ for all such x.

■

THEOREM 5

The growth properties of exp and ln

If $a > 0$, then

(a) $\displaystyle\lim_{x\to\infty} \frac{x^a}{e^x} = 0,$ (b) $\displaystyle\lim_{x\to\infty} \frac{\ln x}{x^a} = 0,$

(c) $\displaystyle\lim_{x\to-\infty} |x|^a e^x = 0,$ (d) $\displaystyle\lim_{x\to 0+} x^a \ln x = 0.$

Each of these limits makes a statement about who "wins" in a contest between an exponential or logarithm and a power. For example, in part (a), the denominator e^x grows large as $x \to \infty$, so it tries to make the fraction x^a/e^x approach 0. On the other hand, if a is a large positive number, the numerator x^a also grows large and tries to make the fraction approach infinity. The assertion of (a) is that in this contest between the exponential and the power, the exponential is stronger and wins; the fraction approaches 0. The content of Theorem 5 can be paraphrased as follows:

In a struggle between a power and an exponential, the exponential wins.
In a struggle between a power and a logarithm, the power wins.

PROOF First, we prove part (b). Let $x > 1$, $a > 0$, and let $s = a/2$. Since $\ln(x^s) = s \ln x$, we have, using Theorem 4,

$$0 < s \ln x = \ln(x^s) \leq x^s - 1 < x^s.$$

Thus, $0 < \ln x < \dfrac{1}{s} x^s$ and, dividing by $x^a = x^{2s}$,

$$0 < \frac{\ln x}{x^a} < \frac{1}{s}\frac{x^s}{x^{2s}} = \frac{1}{s\,x^s}.$$

Now $1/(s\,x^s) \to 0$ as $x \to \infty$ (since $s > 0$); therefore, by the Squeeze Theorem,

$$\lim_{x\to\infty} \frac{\ln x}{x^a} = 0.$$

Next, we deduce part (d) from part (b) by substituting $x = 1/t$. As $x \to 0+$, we have $t \to \infty$, so

$$\lim_{x\to 0+} x^a \ln x = \lim_{t\to\infty} \frac{\ln(1/t)}{t^a} = \lim_{t\to\infty} \frac{-\ln t}{t^a} = -0 = 0.$$

Now we deduce (a) from (b). If $x = \ln t$, then $t \to \infty$ as $x \to \infty$, so

$$\lim_{x\to\infty} \frac{x^a}{e^x} = \lim_{t\to\infty} \frac{(\ln t)^a}{t} = \lim_{t\to\infty} \left(\frac{\ln t}{t^{1/a}}\right)^a = 0^a = 0.$$

Finally, (c) follows from (a) via the substitution $x = -t$:

$$\lim_{x\to-\infty} |x|^a e^x = \lim_{t\to\infty} |-t|^a e^{-t} = \lim_{t\to\infty} \frac{t^a}{e^t} = 0.$$

■

Exponential Growth and Decay Models

Many natural processes involve quantities that increase or decrease at a rate proportional to their size. For example, the mass of a culture of bacteria growing in a medium supplying adequate nourishment will increase at a rate proportional to that mass. The value of an investment bearing interest that is continuously compounding increases at a rate proportional to that value. The mass of undecayed radioactive material in a sample decreases at a rate proportional to that mass.

All of these phenomena, and others exhibiting similar behaviour, can be modelled mathematically in the same way. If $y = y(t)$ denotes the value of a quantity y at time t, and if y changes at a rate proportional to its size, then

$$\frac{dy}{dt} = ky,$$

where k is the constant of proportionality. The above equation is called the **differential equation of exponential growth or decay** because, for any value of the constant C, the function $y = Ce^{kt}$ satisfies the equation. In fact, if $y(t)$ is any solution of the differential equation $y' = ky$, then

$$\frac{d}{dt}\left(\frac{y(t)}{e^{kt}}\right) = \frac{e^{kt}y'(t) - ke^{kt}y(t)}{e^{2kt}} = \frac{y'(t) - ky(t)}{e^{kt}} = 0 \quad \text{for all } t.$$

Thus $y(t)/e^{kt} = C$, a constant, and $y(t) = Ce^{kt}$. Since $y(0) = Ce^0 = C$,

$$\text{The initial-value problem } \begin{cases} \dfrac{dy}{dt} = ky \\ y(0) = y_0 \end{cases} \text{ has unique solution } y = y_0 e^{kt}.$$

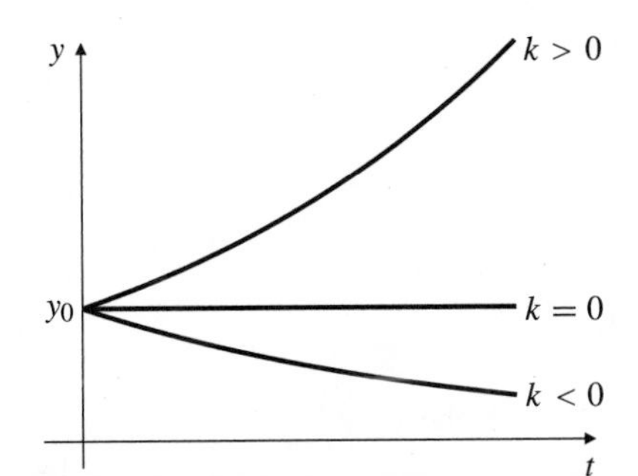

Figure 3.15 Solutions of the initial-value problem $dy/dt = ky$, $y(0) = y_0$, for $k > 0$, $k = 0$, and $k < 0$

If $y_0 > 0$, then $y(t)$ is an increasing function of t if $k > 0$ and a decreasing function of t if $k < 0$. We say that the quantity y exhibits **exponential growth** if $k > 0$ and **exponential decay** if $k < 0$. (See Figure 3.15.)

EXAMPLE 1 **(Growth of a cell culture)** A certain cell culture grows at a rate proportional to the number of cells present. If the culture contains 500 cells initially and 800 after 24 h, how many cells will there be after a further 12 h?

Solution Let $y(t)$ be the number of cells present t hours after there were 500 cells. Thus $y(0) = 500$ and $y(24) = 800$. Because $dy/dt = ky$, we have

$$y(t) = y(0)e^{kt} = 500e^{kt}.$$

Therefore, $800 = y(24) = 500e^{24k}$, so $24k = \ln \frac{800}{500} = \ln(1.6)$. It follows that $k = (1/24)\ln(1.6)$ and

$$y(t) = 500e^{(t/24)\ln(1.6)} = 500(1.6)^{t/24}.$$

We want to know y when $t = 36$: $y(36) = 500e^{(36/24)\ln(1.6)} = 500(1.6)^{3/2} \approx 1012$. The cell count grew to about 1,012 in the 12 h after it was 800.

Exponential growth is characterized by a **fixed doubling time**. If T is the time at which y has doubled from its size at $t = 0$, then $2y(0) = y(T) = y(0)e^{kT}$. Therefore, $e^{kT} = 2$. Since $y(t) = y(0)e^{kt}$, we have

$$y(t + T) = y(0)e^{k(t+T)} = e^{kT}y(0)e^{kt} = 2y(t);$$

that is, T units of time are required for y to double from any value. Similarly, exponential decay involves a fixed halving time (usually called the **half-life**). If $y(T) = \frac{1}{2}y(0)$, then $e^{kT} = \frac{1}{2}$ and

$$y(t + T) = y(0)e^{k(t+T)} = \frac{1}{2}y(t).$$

EXAMPLE 2 **(Radioactive decay)** A radioactive material has a half-life of 1,200 years. What percentage of the original radioactivity of a sample is left after 10 years? How many years are required to reduce the radioactivity by 10%?

Solution Let $p(t)$ be the percentage of the original radioactivity left after t years. Thus $p(0) = 100$ and $p(1{,}200) = 50$. Since the radioactivity decreases at a rate proportional to itself, $dp/dt = kp$ and

$$p(t) = 100e^{kt}.$$

Now $50 = p(1{,}200) = 100e^{1{,}200k}$, so

$$k = \frac{1}{1{,}200} \ln \frac{50}{100} = -\frac{\ln 2}{1{,}200}.$$

The percentage left after 10 years is

$$p(10) = 100e^{10k} = 100e^{-10(\ln 2)/1{,}200} \approx 99.424.$$

If after t years 90% of the radioactivity is left, then

$$\begin{aligned} 90 &= 100e^{kt}, \\ kt &= \ln \frac{90}{100}, \\ t &= \frac{1}{k} \ln(0.9) = -\frac{1{,}200}{\ln 2} \ln(0.9) \approx 182.4, \end{aligned}$$

so it will take a little over 182 years to reduce the radioactivity by 10%.

Sometimes an exponential growth or decay problem will involve a quantity that changes at a rate proportional to the difference between itself and a fixed value:

$$\frac{dy}{dt} = k(y - a).$$

In this case, the change of dependent variable $u(t) = y(t) - a$ should be used to convert the differential equation to the standard form. Observe that $u(t)$ changes at the same rate as $y(t)$ (i.e., $du/dt = dy/dt$), so it satisfies

$$\frac{du}{dt} = ku.$$

EXAMPLE 3 **(Newton's law of cooling)** A hot object introduced into a cooler environment will cool at a rate proportional to the excess of its temperature above that of its environment. If a cup of coffee sitting in a room maintained at a temperature of 20 °C cools from 80 °C to 50 °C in 5 minutes, how much longer will it take to cool to 40 °C?

Solution Let $y(t)$ be the temperature of the coffee t min after it was 80 °C. Thus, $y(0) = 80$ and $y(5) = 50$. Newton's law says that $dy/dt = k(y - 20)$ in this case, so let $u(t) = y(t) - 20$. Thus, $u(0) = 60$ and $u(5) = 30$. We have

$$\frac{du}{dt} = \frac{dy}{dt} = k(y - 20) = ku.$$

Thus,

$$\begin{aligned} u(t) &= 60e^{kt}, \\ 30 &= u(5) = 60e^{5k}, \\ 5k &= \ln \tfrac{1}{2} = -\ln 2. \end{aligned}$$

We want to know t such that $y(t) = 40$, that is, $u(t) = 20$:

$$\begin{aligned} 20 &= u(t) = 60e^{-(t/5)\ln 2} \\ -\frac{t}{5}\ln 2 &= \ln\frac{20}{60} = -\ln 3, \\ t &= 5\,\frac{\ln 3}{\ln 2} \approx 7.92. \end{aligned}$$

The coffee will take about $7.92 - 5 = 2.92$ min to cool from 50 °C to 40 °C.

Interest on Investments

Suppose that \$10,000 is invested at an annual rate of interest of 8%. Thus, the value of the investment at the end of 1 year will be $\$10{,}000(1.08) = \$10{,}800$. If this amount remains invested for a second year at the same rate, it will grow to $\$10{,}000(1.08)^2 = \$11{,}664$; in general, n years after the original investment was made, it will be worth $\$10{,}000(1.08)^n$.

Now suppose that the 8% rate is *compounded semiannually* so that the interest is actually paid at a rate of 4% per 6-month period. After 1 year (2 interest periods) the \$10,000 will grow to $\$10{,}000(1.04)^2 = \$10{,}816$. This is \$16 more than was obtained when the 8% was compounded only once per year. The extra \$16 is the interest paid in the second 6-month period on the \$400 interest earned in the first 6-month period. Continuing in this way, if the 8% interest is compounded *monthly* (12 periods per year and $\frac{8}{12}$% paid per period) or *daily* (365 periods per year and $\frac{8}{365}$% paid per period), then the original \$10,000 would grow in 1 year to $\$10{,}000\left(1 + \frac{8}{1{,}200}\right)^{12} = \$10{,}830$ or $\$10{,}000\left(1 + \frac{8}{36{,}500}\right)^{365} = \$10{,}832.78$, respectively.

For any given *nominal* interest rate, the investment grows more if the compounding period is shorter. In general, an original investment of $\$A$ invested at r% per annum compounded n times per year grows in one year to

$$\$A\left(1 + \frac{r}{100n}\right)^n.$$

It is natural to ask how well we can do with our investment if we let the number of periods in a year approach infinity, that is, we compound the interest *continuously*. The answer is that in 1 year the $\$A$ will grow to

$$\$A \lim_{n\to\infty}\left(1 + \frac{r}{100n}\right)^n = \$Ae^{r/100}.$$

For example, at 8% per annum compounded continuously, our \$10,000 will grow in one year to $\$10{,}000e^{0.08} \approx \$10{,}832.87$. (Note that this is just a few cents more than we get compounding daily.) To justify this result we need the following theorem.

THEOREM 6

For every real number x,

$$e^x = \lim_{n\to\infty}\left(1 + \frac{x}{n}\right)^n.$$

PROOF If $x = 0$, there is nothing to prove; both sides of the identity are 1. If $x \neq 0$, let $h = x/n$. As n tends to infinity, h approaches 0. Thus,

$$\begin{aligned}
\lim_{n\to\infty} \ln\left(1+\frac{x}{n}\right)^n &= \lim_{n\to\infty} n\ln\left(1+\frac{x}{n}\right) \\
&= \lim_{n\to\infty} x\,\frac{\ln\left(1+\frac{x}{n}\right)}{\frac{x}{n}} \\
&= x\lim_{h\to 0}\frac{\ln(1+h)}{h} && \text{(where } h = x/n\text{)} \\
&= x\lim_{h\to 0}\frac{\ln(1+h)-\ln 1}{h} && \text{(since } \ln 1 = 0\text{)} \\
&= x\left(\frac{d}{dt}\ln t\right)\Big|_{t=1} && \text{(by the definition of derivative)} \\
&= x\,\frac{1}{t}\Big|_{t=1} = x.
\end{aligned}$$

Since ln is differentiable, it is continuous. Hence, by Theorem 7 of Section 1.4,

$$\ln\left(\lim_{n\to\infty}\left(1+\frac{x}{n}\right)^n\right) = \lim_{n\to\infty}\ln\left(1+\frac{x}{n}\right)^n = x.$$

Taking exponentials of both sides gives the required formula. ∎

In the case $x = 1$ the formula given in Theorem 6 takes the following form:

$$e = \lim_{n\to\infty}\left(1+\frac{1}{n}\right)^n.$$

Table 2.

n	$\left(1+\frac{1}{n}\right)^n$
1	2
10	2.593 74 ···
100	2.704 81 ···
1,000	2.716 92 ···
10,000	2.718 15 ···
100,000	2.718 27 ···

We can use this formula to compute approximations to e, as shown in Table 2. In a sense we have cheated in obtaining the numbers in this table; they were produced using the y^x function on a scientific calculator. However, this function is actually computed as $e^{x\ln y}$. In any event, the formula in this table is not a very efficient way to calculate e to any great accuracy. Only 4 decimal places are correct for $n = 100{,}000$. A much better way is to use the series

$$e = 1 + \frac{1}{1!} + \frac{1}{2!} + \frac{1}{3!} + \frac{1}{4!} + \cdots = 1 + 1 + \frac{1}{2} + \frac{1}{6} + \frac{1}{24} + \cdots,$$

which we will establish in Section 4.8.

A final word about interest rates. Financial institutions sometimes quote *effective* rates of interest rather than *nominal* rates. The effective rate tells you what the actual effect of the interest rate will be after one year. Thus, \$10,000 invested at an effective rate of 8% will grow to \$10,800.00 in one year regardless of the compounding period. A nominal rate of 8% per annum compounded daily is equivalent to an effective rate of about 8.3278%.

Logistic Growth

Few quantities in nature can sustain exponential growth over extended periods of time; the growth is usually limited by external constraints. For example, suppose a small number of rabbits (of both sexes) is introduced to a small island where there were no rabbits previously, and where there are no predators who eat rabbits. By virtue of natural fertility, the number of rabbits might be expected to grow exponentially, but this growth will eventually be limited by the food supply available to the rabbits. Suppose the island can grow enough food to supply a population of L rabbits indefinitely. If there are $y(t)$ rabbits in the population at time t, we would expect $y(t)$ to grow at a rate proportional to $y(t)$ provided $y(t)$ is quite small (much less than L). But as the numbers increase, it will be harder for the rabbits to find enough food, and we would expect the rate of increase to approach 0 as $y(t)$ gets closer and closer to L. One possible model for such behaviour is the differential equation

$$\frac{dy}{dt} = ky\left(1 - \frac{y}{L}\right),$$

which is called the **logistic equation** since it models growth that is limited by the *supply* of necessary resources. Observe that $dy/dt > 0$ if $0 < y < L$ and that this rate is small if y is small (there are few rabbits to reproduce) or if y is close to L (there are almost as many rabbits as the available resources can feed). Observe also that $dy/dt < 0$ if $y > L$; there being more animals than the resources can feed, the rabbits die at a greater rate than they are born. Of course, the steady-state populations $y = 0$ and $y = L$ are solutions of the logistic equation; for both of these $dy/dt = 0$. We will examine techniques for solving differential equations like the logistic equation in Section 7.9. For now, we invite the reader to verify by differentiation that the solution satisfying $y(0) = y_0$ is

$$y = \frac{Ly_0}{y_0 + (L - y_0)e^{-kt}}.$$

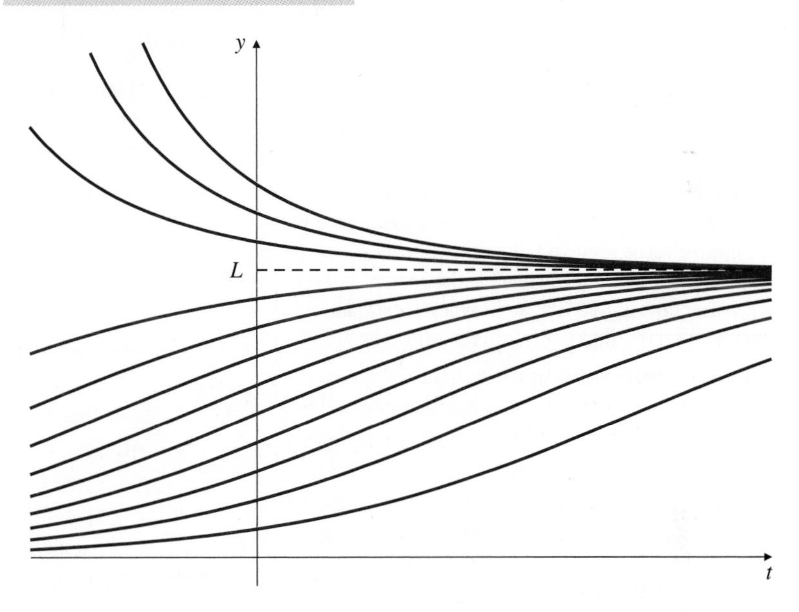

Figure 3.16 Some logistic curves

Observe that, as expected, if $0 < y_0 < L$, then

$$\lim_{t\to\infty} y(t) = L, \qquad \lim_{t\to-\infty} y(t) = 0.$$

The solution given above also holds for $y_0 > L$. However, the solution does not approach 0 as t approaches $-\infty$ in this case. It has a vertical asymptote at a certain negative value of t. (See Exercise 30 below.) The graphs of solutions of the logistic equation for various positive values of y_0 are given in Figure 3.16.

EXERCISES 3.4

Evaluate the limits in Exercises 1–8.

1. $\lim_{x\to\infty} x^3 e^{-x}$

2. $\lim_{x\to\infty} x^{-3} e^{x}$

3. $\lim_{x\to\infty} \frac{2e^x - 3}{e^x + 5}$

4. $\lim_{x\to\infty} \frac{x - 2e^{-x}}{x + 3e^{-x}}$

5. $\lim_{x\to 0+} x \ln x$

6. $\lim_{x\to 0+} \frac{\ln x}{x}$

7. $\lim_{x\to 0} x\left(\ln|x|\right)^2$

8. $\lim_{x\to\infty} \frac{(\ln x)^3}{\sqrt{x}}$

9. (Bacterial growth) Bacteria grow in a certain culture at a rate proportional to the amount present. If there are 100 bacteria present initially and the amount doubles in 1 h, how many will there be after a further $1\frac{1}{2}$ h?

10. **(Dissolving sugar)** Sugar dissolves in water at a rate proportional to the amount still undissolved. If there were 50 kg of sugar present initially, and at the end of 5 h only 20 kg are left, how much longer will it take until 90% of the sugar is disssolved?

11. **(Radioactive decay)** A radioactive substance decays at a rate proportional to the amount present. If 30% of such a substance decays in 15 years, what is the half-life of the substance?

12. **(Half-life of radium)** If the half-life of radium is 1,690 years, what percentage of the amount present now will be remaining after (a) 100 years, (b) 1,000 years?

13. Find the half-life of a radioactive substance if after 1 year 99.57% of an initial amount still remains.

14. **(Bacterial growth)** In a certain culture where the rate of growth of bacteria is proportional to the number present, the number triples in 3 days. If at the end of 7 days there are 10 million bacteria present in the culture, how many were present initially?

15. **(Weight of a newborn)** In the first few weeks after birth, babies gain weight at a rate proportional to their weight. A baby weighing 4 kg at birth weighs 4.4 kg after 2 weeks. How much did the baby weigh 5 days after birth?

16. **(Electric current)** When a simple electrical circuit containing inductance and resistance but no capacitance has the electromotive force removed, the rate of decrease of the current is proportional to the current. If the current is $I(t)$ amperes t s after cutoff, and if $I = 40$ when $t = 0$, and $I = 15$ when $t = 0.01$, find a formula for $I(t)$.

17. **(Continuously compounding interest)** How much money needs to be invested today at a nominal rate of 4% compounded continuously, in order that it should grow to $10,000 in 7 years?

18. **(Continuously compounding interest)** Money invested at compound interest (with instantaneous compounding) accumulates at a rate proportional to the amount present. If an initial investment of $1,000 grows to $1,500 in exactly 5 years, find (a) the doubling time for the investment and (b) the effective annual rate of interest being paid.

19. **(Purchasing power)** If the purchasing power of the dollar is decreasing at an effective rate of 9% annually, how long will it take for the purchasing power to be reduced to 25 cents?

20. **(Effective interest rate)** A bank claims to pay interest at an effective rate of 9.5% on an investment account. If the interest is actually being compounded monthly, what is the nominal rate of interest being paid on the account?

21. Suppose that 1,000 rabbits are introduced onto an island where they have no natural predators. During the next five years the rabbit population grows exponentially. After the first two years the population grew to 3,500 rabbits. After the first five years a rabbit virus is sprayed on the island and after that the rabbit population decays exponentially. Two years after the virus was introduced (so seven years after rabbits were introduced to the island) the rabbit population dropped to 3,000 rabbits. How many rabbits will there be on the island 10 years after they were introduced?

22. Lab rats are to be used in experiments on an isolated island. Initially R rats are brought to the island and released. Having a plentiful food supply and no natural predators on the island, the rat population grows exponentially and doubles in three months. At the end of the fifth month, and at the end of every five months thereafter, 1,000 of the rats are captured and killed. What is the minimum value of R that ensures that the scientists will never run out of rats?

Differential equations of the form $y' = a + by$

23. Suppose that $f(x)$ satisfies the differential equation

$$f'(x) = a + bf(x),$$

where a and b are constants.

(a) Solve the differential equation by substituting $u(x) = a + bf(x)$ and solving the simpler differential equation that results for $u(x)$.

(b) Solve the initial-value problem:

$$\begin{cases} \dfrac{dy}{dx} = a + by \\ y(0) = y_0 \end{cases}$$

24. **(Drug concentrations in the blood)** A drug is introduced into the bloodstream intravenously at a constant rate and breaks down and is eliminated from the body at a rate proportional to its concentration in the blood. The concentration $x(t)$ of the drug in the blood satisfies the differential equation

$$\frac{dx}{dt} = a - bx,$$

where a and b are positive constants.

(a) What is the limiting concentration $\lim_{t\to\infty} x(t)$ of the drug in the blood?

(b) Find the concentration of the drug in the blood at time t, given that the concentration was zero at $t = 0$.

(c) How long after $t = 0$ will it take for the concentration to rise to half its limiting value?

25. **(Cooling)** Use Newton's law of cooling to determine the reading on a thermometer 5 min after it is taken from an oven at 72 °C to the outdoors where the temperature is 20 °C, if the reading dropped to 48 °C after one min.

26. **(Cooling)** An object is placed in a freezer maintained at a temperature of −5 °C. If the object cools from 45 °C to 20 °C in 40 min, how many more minutes will it take to cool to 0 °C?

27. **(Warming)** If an object in a room warms up from 5 °C to 10 °C in 4 min, and if the room is being maintained at 20 °C, how much longer will the object take to warm up to 15 °C? Assume the object warms at a rate proportional to the difference between its temperature and room temperature.

The logistic equation

28. Suppose the quantity $y(t)$ exhibits logistic growth. If the values of $y(t)$ at times $t = 0$, $t = 1$, and $t = 2$ are y_0, y_1, and y_2, respectively, find an equation satisfied by the limiting value L of $y(t)$, and solve it for L. If $y_0 = 3$, $y_1 = 5$, and $y_2 = 6$, find L.

29. Show that a solution $y(t)$ of the logistic equation having $0 < y(0) < L$ is increasing most rapidly when its value is $L/2$. (*Hint:* You do not need to use the formula for the solution to see this.)

30. If $y_0 > L$, find the interval on which the given solution of the logistic equation is valid. What happens to the solution as t approaches the left endpoint of this interval?

31. If $y_0 < 0$, find the interval on which the given solution of the logistic equation is valid. What happens to the solution as t approaches the right endpoint of this interval?

32. (Modelling an epidemic) The number y of persons infected by a highly contagious virus is modelled by a logistic curve

$$y = \frac{L}{1 + Me^{-kt}},$$

where t is measured in months from the time the outbreak was discovered. At that time there were 200 infected persons, and the number grew to 1,000 after 1 month. Eventually, the number levelled out at 10,000. Find the values of the parameters L, M, and k of the model.

33. Continuing Exercise 32, how many people were infected 3 months after the outbreak was discovered, and how fast was the number growing at that time?

3.5 The Inverse Trigonometric Functions

The six trigonometric functions are periodic and, hence, not one-to-one. However, as we did with the function x^2 in Section 3.1, we can restrict their domains in such a way that the restricted functions are one-to-one and invertible.

The Inverse Sine (or Arcsine) Function

Let us define a function $\text{Sin}\, x$ (note the capital letter) to be $\sin x$, restricted so that its domain is the interval $-\frac{\pi}{2} \le x \le \frac{\pi}{2}$:

DEFINITION 8

The restricted function Sin x

$$\text{Sin}\, x = \sin x \qquad \text{if } -\frac{\pi}{2} \le x \le \frac{\pi}{2}.$$

Since its derivative $\cos x$ is positive on the interval $\left(-\frac{\pi}{2}, \frac{\pi}{2}\right)$, the function $\text{Sin}\, x$ is increasing on its domain, so it is a one-to-one function. It has domain $\left[-\frac{\pi}{2}, \frac{\pi}{2}\right]$ and range $[-1, 1]$. (See Figure 3.17.)

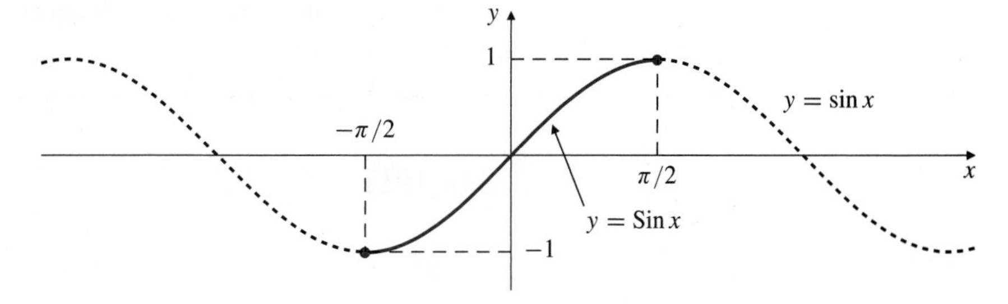

Figure 3.17 The graph of Sin x forms part of the graph of $\sin x$

Being one-to-one, Sin has an inverse function which is denoted $\sin^{-1}$ (or, in some books and computer programs, by arcsin, Arcsin, or asin) and which is called the **inverse sine** or **arcsine** function.

DEFINITION 9

The inverse sine function $\sin^{-1} x$ or $\arcsin x$

$$y = \sin^{-1} x \iff x = \text{Sin}\, y$$
$$\iff x = \sin y \quad \text{and} \quad -\frac{\pi}{2} \le y \le \frac{\pi}{2}$$

The graph of $\sin^{-1}$ is shown in Figure 3.18; it is the reflection of the graph of Sin in the line $y = x$. The domain of $\sin^{-1}$ is $[-1, 1]$ (the range of Sin), and the range of $\sin^{-1}$ is $\left[-\frac{\pi}{2}, \frac{\pi}{2}\right]$ (the domain of Sin). The **cancellation identities** for Sin and $\sin^{-1}$ are

$$\begin{aligned}
\sin^{-1}(\operatorname{Sin} x) &= \arcsin(\operatorname{Sin} x) = x && \text{for } -\frac{\pi}{2} \le x \le \frac{\pi}{2} \\
\operatorname{Sin}(\sin^{-1} x) &= \operatorname{Sin}(\arcsin x) = x && \text{for } -1 \le x \le 1
\end{aligned}$$

Since the intervals where they apply are specified, Sin can be replaced by sin in both identities above.

Remark As for the general inverse function f^{-1}, be aware that $\sin^{-1} x$ does *not* represent the *reciprocal* $1/\sin x$. (We already have a perfectly good name for the reciprocal of $\sin x$; we call it $\csc x$.) We should think of $\sin^{-1} x$ as "the angle between $-\frac{\pi}{2}$ and $\frac{\pi}{2}$ whose sine is x."

Figure 3.18 The arcsine function

EXAMPLE 1

(a) $\sin^{-1}\left(\frac{1}{2}\right) = \frac{\pi}{6}$ (because $\sin\frac{\pi}{6} = \frac{1}{2}$ and $-\frac{\pi}{2} < \frac{\pi}{6} < \frac{\pi}{2}$).

(b) $\sin^{-1}\left(-\frac{1}{\sqrt{2}}\right) = -\frac{\pi}{4}$ (because $\sin\left(-\frac{\pi}{4}\right) = -\frac{1}{\sqrt{2}}$ and $-\frac{\pi}{2} < -\frac{\pi}{4} < \frac{\pi}{2}$).

(c) $\sin^{-1}(-1) = -\frac{\pi}{2}$ (because $\sin\left(-\frac{\pi}{2}\right) = -1$).

(d) $\sin^{-1} 2$ is not defined. (2 is not in the range of sine.)

EXAMPLE 2

Find (a) $\sin\left(\sin^{-1} 0.7\right)$, (b) $\sin^{-1}(\sin 0.3)$, (c) $\sin^{-1}\left(\sin\frac{4\pi}{5}\right)$, and (d) $\cos\left(\sin^{-1} 0.6\right)$.

Solution

(a) $\sin\left(\sin^{-1} 0.7\right) = 0.7$ (cancellation identity).

(b) $\sin^{-1}(\sin 0.3) = 0.3$ (cancellation identity).

(c) The number $\frac{4\pi}{5}$ does not lie in $\left[-\frac{\pi}{2}, \frac{\pi}{2}\right]$, so we can't apply the cancellation identity directly. However, $\sin\frac{4\pi}{5} = \sin\left(\pi - \frac{\pi}{5}\right) = \sin\frac{\pi}{5}$ by the supplementary angle identity. Therefore, $\sin^{-1}\left(\sin\frac{4\pi}{5}\right) = \sin^{-1}\left(\sin\frac{\pi}{5}\right) = \frac{\pi}{5}$ (by cancellation).

(d) Let $\theta = \sin^{-1} 0.6$, as shown in the right triangle in Figure 3.19, which has hypotenuse 1 and side opposite θ equal to 0.6. By the Pythagorean Theorem, the side adjacent θ is $\sqrt{1-(0.6)^2} = 0.8$. Thus, $\cos\left(\sin^{-1} 0.6\right) = \cos\theta = 0.8$.

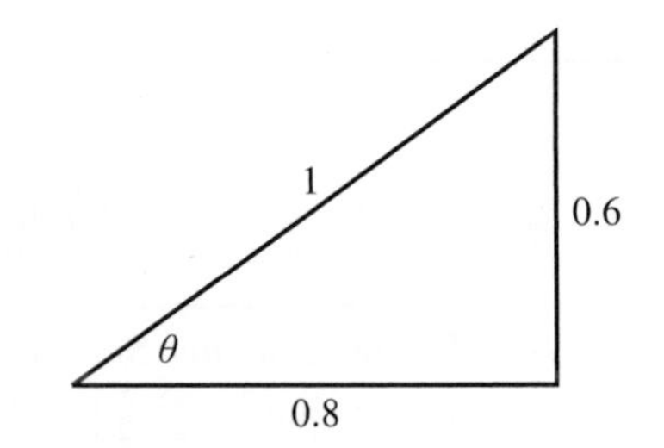

Figure 3.19

EXAMPLE 3

Simplify the expression $\tan(\sin^{-1} x)$.

Solution We want the tangent of an angle whose sine is x. Suppose first that $0 \le x < 1$. As in Example 2, we draw a right triangle (Figure 3.20) with one angle θ, and label the sides so that $\theta = \sin^{-1} x$. The side opposite θ is x, and the hypotenuse is 1. The remaining side is $\sqrt{1-x^2}$, and we have

$$\tan(\sin^{-1} x) = \tan\theta = \frac{x}{\sqrt{1-x^2}}.$$

Because both sides of the above equation are odd functions of x, the same result holds for $-1 < x < 0$.

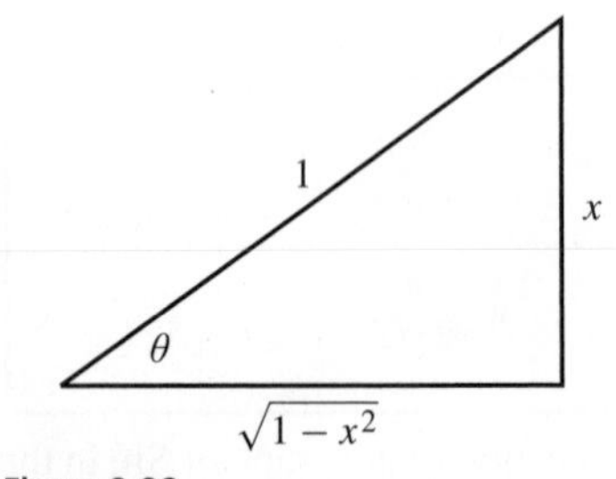

Figure 3.20

Now let us use implicit differentiation to find the derivative of the inverse sine function. If $y = \sin^{-1} x$, then $x = \sin y$ and $-\frac{\pi}{2} \le y \le \frac{\pi}{2}$. Differentiating with respect to x, we obtain

$$1 = (\cos y)\frac{dy}{dx}.$$

Since $-\frac{\pi}{2} \le y \le \frac{\pi}{2}$, we know that $\cos y \ge 0$. Therefore,

$$\cos y = \sqrt{1 - \sin^2 y} = \sqrt{1 - x^2},$$

and $dy/dx = 1/\cos y = 1/\sqrt{1 - x^2}$;

$$\frac{d}{dx} \sin^{-1} x = \frac{d}{dx} \arcsin x = \frac{1}{\sqrt{1 - x^2}}.$$

Note that the inverse sine function is differentiable only on the *open* interval $(-1, 1)$; the slope of its graph approaches infinity as $x \to -1+$ or as $x \to 1-$. (See Figure 3.18.)

EXAMPLE 4 Find the derivative of $\sin^{-1}\left(\frac{x}{a}\right)$ and hence evaluate $\int \frac{dx}{\sqrt{a^2 - x^2}}$, where $a > 0$.

Solution By the Chain Rule,

$$\frac{d}{dx} \sin^{-1} \frac{x}{a} = \frac{1}{\sqrt{1 - \dfrac{x^2}{a^2}}} \frac{1}{a} = \frac{1}{\sqrt{\dfrac{a^2 - x^2}{a^2}}} \frac{1}{a} = \frac{1}{\sqrt{a^2 - x^2}} \qquad \text{if } a > 0.$$

Hence,

$$\int \frac{1}{\sqrt{a^2 - x^2}}\, dx = \sin^{-1} \frac{x}{a} + C \qquad (a > 0).$$

EXAMPLE 5 Find the solution y of the following initial-value problem:

$$\begin{cases} y' = \dfrac{4}{\sqrt{2 - x^2}} & (-\sqrt{2} < x < \sqrt{2}) \\ y(1) = 2\pi. \end{cases}$$

Solution Using the integral from the previous example, we have

$$y = \int \frac{4}{\sqrt{2 - x^2}}\, dx = 4 \sin^{-1}\left(\frac{x}{\sqrt{2}}\right) + C$$

for some constant C. Also $2\pi = y(1) = 4 \sin^{-1}(1/\sqrt{2}) + C = 4\left(\frac{\pi}{4}\right) + C = \pi + C$. Thus, $C = \pi$ and $y = 4 \sin^{-1}(x/\sqrt{2}) + \pi$.

EXAMPLE 6 **(A sawtooth curve)** Let $f(x) = \sin^{-1}(\sin x)$ for all real numbers x.

(a) Calculate and simplify $f'(x)$.

(b) Where is f differentiable? Where is f continuous?

(c) Use your results from (a) and (b) to sketch the graph of f.

Solution (a) Using the Chain Rule and the Pythagorean identity we calculate

$$\begin{aligned} f'(x) &= \frac{1}{\sqrt{1 - (\sin x)^2}} (\cos x) \\ &= \frac{\cos x}{\sqrt{\cos^2 x}} = \frac{\cos x}{|\cos x|} = \begin{cases} 1 & \text{if } \cos x > 0 \\ -1 & \text{if } \cos x < 0. \end{cases} \end{aligned}$$

(b) f is differentiable at all points where $\cos x \neq 0$, that is, everywhere except at odd multiples of $\pi/2$, namely, $\pm\frac{\pi}{2}, \pm\frac{3\pi}{2}, \pm\frac{5\pi}{2}, \ldots$.
Since sin is continuous everywhere and has values in $[-1, 1]$, and since $\sin^{-1}$ is continuous on $[-1, 1]$, we have that f is continuous on the whole real line.

(c) Since f is continuous, its graph has no breaks. The graph consists of straight line segments of slopes alternating between 1 and -1 on intervals between consecutive odd multiples of $\pi/2$. Since $f'(x) = 1$ on the interval $\left[-\frac{\pi}{2}, \frac{\pi}{2}\right]$ (where $\cos x \geq 0$), the graph must be as shown in Figure 3.21.

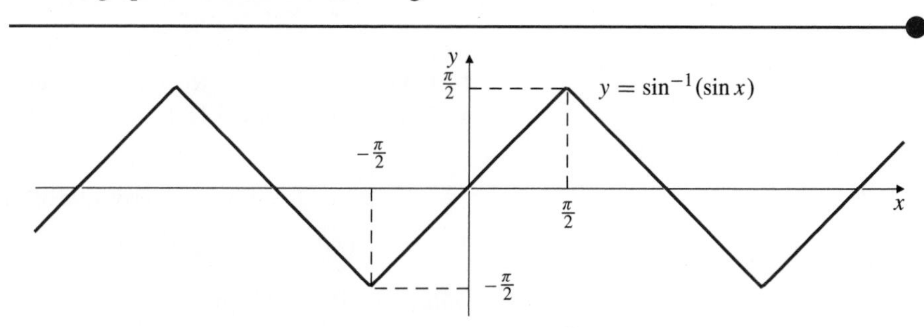

Figure 3.21 A sawtooth graph

The Inverse Tangent (or Arctangent) Function

The inverse tangent function is defined in a manner similar to the inverse sine. We begin by restricting the tangent function to an interval where it is one-to-one; in this case we use the open interval $\left(-\frac{\pi}{2}, \frac{\pi}{2}\right)$. See Figure 3.22(a).

DEFINITION 10

The restricted function Tan x

$$\text{Tan}\, x = \tan x \qquad \text{if } -\frac{\pi}{2} < x < \frac{\pi}{2}.$$

The inverse of the function Tan is called the **inverse tangent** function and is denoted $\tan^{-1}$ (or arctan, Arctan, or atan). The domain of $\tan^{-1}$ is the whole real line (the range of Tan). Its range is the open interval $\left(-\frac{\pi}{2}, \frac{\pi}{2}\right)$.

DEFINITION

The inverse tangent function $\tan^{-1} x$ or $\arctan x$

$$y = \tan^{-1} x \iff x = \text{Tan}\, y$$
$$\iff x = \tan y \quad \text{and} \quad -\frac{\pi}{2} < y < \frac{\pi}{2}$$

The graph of $\tan^{-1}$ is shown in Figure 3.22(b); it is the reflection of the graph of Tan in the line $y = x$.

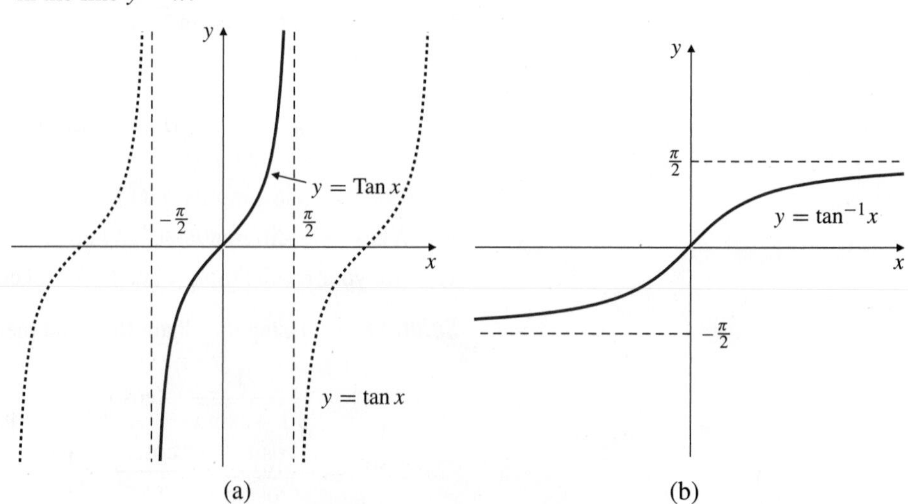

Figure 3.22
(a) The graph of Tan x
(b) The graph of $\tan^{-1} x$

The cancellation identities for Tan and $\tan^{-1}$ are

$$\tan^{-1}(\operatorname{Tan} x) = \arctan(\operatorname{Tan} x) = x \qquad \text{for } -\frac{\pi}{2} < x < \frac{\pi}{2}$$
$$\operatorname{Tan}(\tan^{-1} x) = \operatorname{Tan}(\arctan x) = x \qquad \text{for } -\infty < x < \infty$$

Again, we can replace Tan with tan above since the intervals are specified.

EXAMPLE 7 Evaluate: (a) $\tan(\tan^{-1} 3)$, (b) $\tan^{-1}\left(\tan \frac{3\pi}{4}\right)$, and (c) $\cos(\tan^{-1} 2)$.

Solution

(a) $\tan(\tan^{-1} 3) = 3$ by cancellation.

(b) $\tan^{-1}\left(\tan \frac{3\pi}{4}\right) = \tan^{-1}(-1) = -\frac{\pi}{4}$.

(c) $\cos(\tan^{-1} 2) = \cos\theta = \frac{1}{\sqrt{5}}$ via the triangle in Figure 3.23. Alternatively, we have $\tan(\tan^{-1} 2) = 2$, so $\sec^2(\tan^{-1} 2) = 1 + 2^2 = 5$. Thus $\cos^2(\tan^{-1} 2) = \frac{1}{5}$. Since cosine is positive on the range of $\tan^{-1}$, we have $\cos(\tan^{-1} 2) = \frac{1}{\sqrt{5}}$.

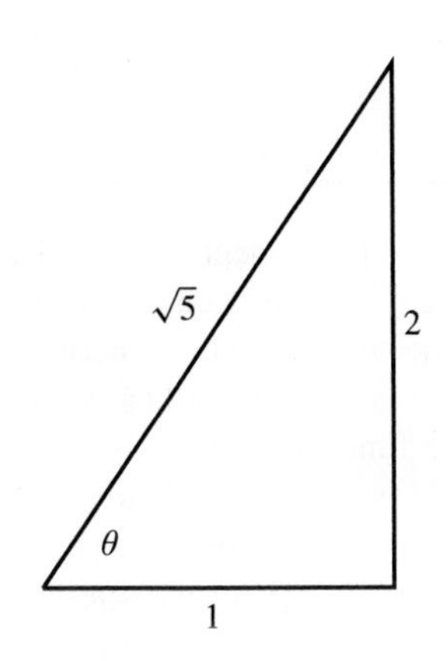

Figure 3.23

The derivative of the inverse tangent function is also found by implicit differentiation: if $y = \tan^{-1} x$, then $x = \tan y$ and

$$1 = (\sec^2 y)\frac{dy}{dx} = (1 + \tan^2 y)\frac{dy}{dx} = (1 + x^2)\frac{dy}{dx}.$$

Thus,

$$\frac{d}{dx}\tan^{-1} x = \frac{1}{1 + x^2}.$$

EXAMPLE 8 Find $\frac{d}{dx}\tan^{-1}\left(\frac{x}{a}\right)$, and hence evaluate $\int \frac{1}{x^2 + a^2}\,dx$.

Solution We have

$$\frac{d}{dx}\tan^{-1}\left(\frac{x}{a}\right) = \frac{1}{1 + \frac{x^2}{a^2}}\,\frac{1}{a} = \frac{a}{a^2 + x^2};$$

hence,

$$\int \frac{dx}{a^2 + x^2} = \frac{1}{a}\tan^{-1}\left(\frac{x}{a}\right) + C.$$

EXAMPLE 9 Prove that $\tan^{-1}\left(\frac{x-1}{x+1}\right) = \tan^{-1} x - \frac{\pi}{4}$ for $x > -1$.

Solution Let $f(x) = \tan^{-1}\left(\frac{x-1}{x+1}\right) - \tan^{-1} x$. On the interval $(-1, \infty)$ we have, by the Chain Rule and the Quotient Rule,

$$f'(x) = \frac{1}{1+\left(\frac{x-1}{x+1}\right)^2} \frac{(x+1)-(x-1)}{(x+1)^2} - \frac{1}{1+x^2}$$

$$= \frac{(x+1)^2}{(x^2+2x+1)+(x^2-2x+1)} \frac{2}{(x+1)^2} - \frac{1}{1+x^2}$$

$$= \frac{2}{2+2x^2} - \frac{1}{1+x^2} = 0.$$

Hence, $f(x) = C$ (constant) on that interval. We can find C by finding $f(0)$:

$$C = f(0) = \tan^{-1}(-1) - \tan^{-1} 0 = -\frac{\pi}{4}.$$

Hence, the given identity holds on $(-1, \infty)$.

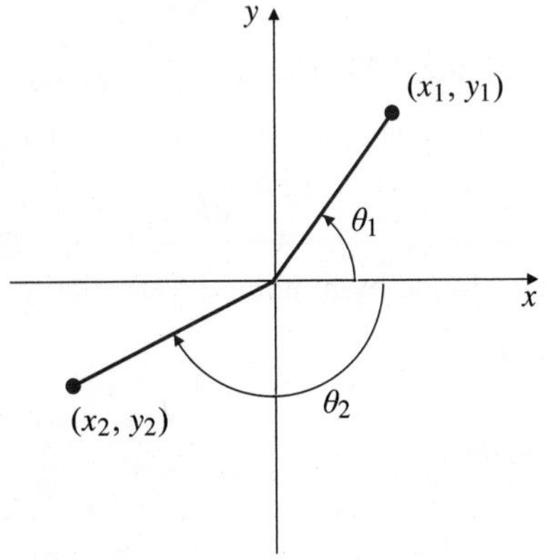

y

(x_1, y_1)

θ_1

x

(x_2, y_2)

θ_2

Figure 3.24
$\theta_1 = \tan^{-1}(y_1/x_1)$
$= \text{atan}(y_1/x_1)$
$= \text{atan2}(x_1, y_1)$
$= \arctan(y_1/x_1)$ (Maple)
$= \arctan(y_1, x_1)$ (Maple)
$\theta_2 = \text{atan2}(x_2, y_2)$
$= \arctan(y_2, x_2)$ (Maple)

Remark Some computer programs, especially spreadsheets, implement two versions of the arctangent function, usually called "atan" and "atan2." The function atan is just the function $\tan^{-1}$ that we have defined; atan(y/x) gives the angle in radians, between the line from the origin to the point (x, y) and the positive x-axis, provided (x, y) lies in quadrants I or IV of the plane. The function atan2 is a function of two variables: atan2(x, y) gives that angle for any point (x, y) not on the y-axis. See Figure 3.24. Some programs, for instance MATLAB, reverse the order of the variables x and y in their atan2 function. Maple uses `arctan(x)` and `arctan(y,x)` for the one- and two-variable versions of arctangent.

Other Inverse Trigonometric Functions

The function $\cos x$ is one-to-one on the interval $[0, \pi]$, so we could define the **inverse cosine function**, $\cos^{-1} x$ (or $\arccos x$, or $\text{Arccos}\, x$, or $\text{acos}\, x$), so that

$$y = \cos^{-1} x \iff x = \cos y \quad \text{and} \quad 0 \le y \le \pi.$$

However, $\cos y = \sin\left(\frac{\pi}{2} - y\right)$ (the complementary angle identity), and $\frac{\pi}{2} - y$ is in the interval $\left[-\frac{\pi}{2}, \frac{\pi}{2}\right]$ when $0 \le y \le \pi$. Thus, the definition above would lead to

$$y = \cos^{-1}x \iff x = \sin\left(\frac{\pi}{2} - y\right) \iff \sin^{-1}x = \frac{\pi}{2} - y = \frac{\pi}{2} - \cos^{-1}x.$$

It is easier to use this result to define $\cos^{-1}x$ directly:

DEFINITION 12

The inverse cosine function $\cos^{-1} x$ or $\arccos x$

$$\cos^{-1} x = \frac{\pi}{2} - \sin^{-1} x \quad \text{for} \quad -1 \le x \le 1.$$

The cancellation identities for $\cos^{-1}x$ are

$$\cos^{-1}(\cos x) = \arccos(\cos x) = x \qquad \text{for } 0 \le x \le \pi$$

$$\cos(\cos^{-1} x) = \cos(\arccos x) = x \qquad \text{for } -1 \le x \le 1$$

The derivative of $\cos^{-1} x$ is the negative of that of $\sin^{-1} x$ (why?):

$$\frac{d}{dx} \cos^{-1} x = -\frac{1}{\sqrt{1-x^2}}.$$

The graph of $\cos^{-1}$ is shown in Figure 3.25(a).

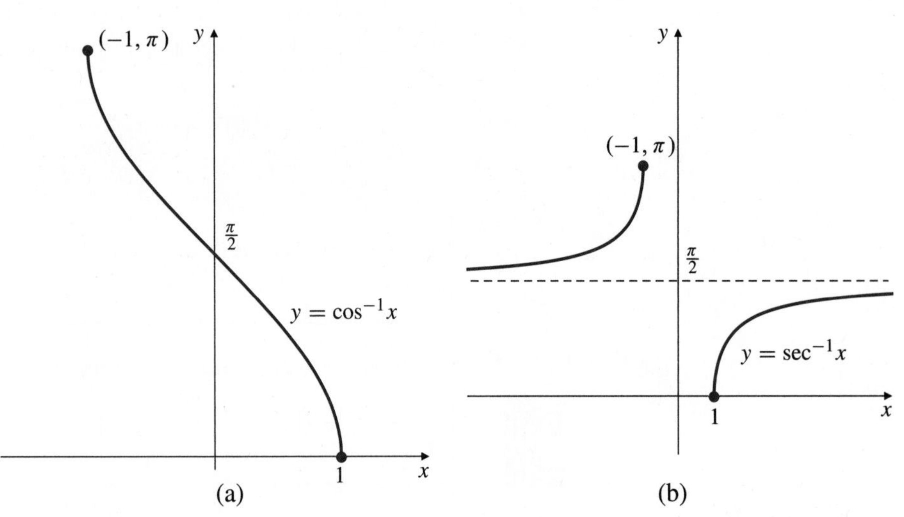

Figure 3.25 The graphs of $\cos^{-1}$ and $\sec^{-1}$

Scientific calculators usually implement only the primary trigonometric functions—sine, cosine, and tangent—and the inverses of these three. The secondary functions—secant, cosecant, and cotangent—are calculated using the reciprocal key; to calculate $\sec x$ you calculate $\cos x$ and take the reciprocal of the answer. The inverses of the secondary trigonometric functions are also easily expressed in terms of those of their reciprocal functions. For example, we define:

DEFINITION 13

The inverse secant function $\sec^{-1} x$ (or $\operatorname{arcsec} x$)

$$\sec^{-1} x = \cos^{-1}\left(\frac{1}{x}\right) \qquad \text{for} \quad |x| \ge 1.$$

The domain of $\sec^{-1}$ is the union of intervals $(-\infty, -1] \cup [1, \infty)$, and its range is $\left[0, \frac{\pi}{2}\right) \cup \left(\frac{\pi}{2}, \pi\right]$. The graph of $y = \sec^{-1}x$ is shown in Figure 3.25(b). It is the reflection in the line $y = x$ of that part of the graph of $\sec x$ for x between 0 and π. Observe that

$$\begin{aligned}
\sec(\sec^{-1} x) &= \sec\left(\cos^{-1}\left(\frac{1}{x}\right)\right) \\
&= \frac{1}{\cos\left(\cos^{-1}\left(\frac{1}{x}\right)\right)} = \frac{1}{\frac{1}{x}} = x \qquad \text{for } |x| \ge 1, \\
\sec^{-1}(\sec x) &= \cos^{-1}\left(\frac{1}{\sec x}\right) \\
&= \cos^{-1}(\cos x) = x \qquad \text{for } x \text{ in } [0, \pi],\ x \ne \frac{\pi}{2}.
\end{aligned}$$

We calculate the derivative of $\sec^{-1}$ from that of $\cos^{-1}$:

$$\begin{aligned}
\frac{d}{dx}\sec^{-1} x = \frac{d}{dx}\cos^{-1}\left(\frac{1}{x}\right) &= \frac{-1}{\sqrt{1 - \frac{1}{x^2}}}\left(-\frac{1}{x^2}\right) \\
&= \frac{1}{x^2}\sqrt{\frac{x^2}{x^2 - 1}} = \frac{1}{x^2}\frac{|x|}{\sqrt{x^2 - 1}} = \frac{1}{|x|\sqrt{x^2 - 1}}.
\end{aligned}$$

Note that we had to use $\sqrt{x^2} = |x|$ in the last line. There are negative values of x in the domain of $\sec^{-1}$. Observe in Figure 3.25(b) that the slope of $y = \sec^{-1}(x)$ is always positive.

$$\frac{d}{dx}\sec^{-1} x = \frac{1}{|x|\sqrt{x^2 - 1}}.$$

Some authors prefer to define $\sec^{-1}$ as the inverse of the restriction of $\sec x$ to the separated intervals $[0, \pi/2)$ and $[\pi, 3\pi/2)$ because this prevents the absolute value from appearing in the formula for the derivative. However, it is much harder to calculate values with that definition. Our definition makes it easy to obtain a value such as $\sec^{-1}(-3)$ from a calculator. Scientific calculators usually have just the inverses of sine, cosine, and tangent built in.

The corresponding integration formula takes different forms on intervals where $x \geq 1$ or $x \leq -1$:

$$\int \frac{1}{x\sqrt{x^2-1}}\,dx = \begin{cases} \sec^{-1}x + C & \text{on intervals where } x \geq 1 \\ -\sec^{-1}x + C & \text{on intervals where } x \leq -1 \end{cases}$$

Finally, note that $\csc^{-1}$ and $\cot^{-1}$ are defined similarly to $\sec^{-1}$. They are seldom encountered.

DEFINITION 14

The inverse cosecant and inverse cotangent functions

$$\csc^{-1}x = \sin^{-1}\left(\frac{1}{x}\right), \quad (|x| \geq 1); \qquad \cot^{-1}x = \tan^{-1}\left(\frac{1}{x}\right), \quad (x \neq 0)$$

EXERCISES 3.5

In Exercises 1–12, evaluate the given expression.

1. $\sin^{-1}\frac{\sqrt{3}}{2}$
2. $\cos^{-1}\left(\frac{-1}{2}\right)$
3. $\tan^{-1}(-1)$
4. $\sec^{-1}\sqrt{2}$
5. $\sin(\sin^{-1}0.7)$
6. $\cos(\sin^{-1}0.7)$
7. $\tan^{-1}\left(\tan\frac{2\pi}{3}\right)$
8. $\sin^{-1}(\cos 40°)$
9. $\cos^{-1}(\sin(-0.2))$
10. $\sin\left(\cos^{-1}\left(\frac{-1}{3}\right)\right)$
11. $\cos\left(\tan^{-1}\frac{1}{2}\right)$
12. $\tan(\tan^{-1}200)$

In Exercises 13–18, simplify the given expression.

13. $\sin(\cos^{-1}x)$
14. $\cos(\sin^{-1}x)$
15. $\cos(\tan^{-1}x)$
16. $\sin(\tan^{-1}x)$
17. $\tan(\cos^{-1}x)$
18. $\tan(\sec^{-1}x)$

In Exercises 19–32, differentiate the given function and simplify the answer whenever possible.

19. $y = \sin^{-1}\left(\frac{2x-1}{3}\right)$
20. $y = \tan^{-1}(ax+b)$
21. $y = \cos^{-1}\left(\frac{x-b}{a}\right)$
22. $f(x) = x\sin^{-1}x$
23. $f(t) = t\tan^{-1}t$
24. $u = z^2\sec^{-1}(1+z^2)$
25. $F(x) = (1+x^2)\tan^{-1}x$
26. $y = \sin^{-1}\frac{a}{x}$
27. $G(x) = \frac{\sin^{-1}x}{\sin^{-1}2x}$
28. $H(t) = \frac{\sin^{-1}t}{\sin t}$
29. $f(x) = (\sin^{-1}x^2)^{1/2}$
30. $y = \cos^{-1}\frac{a}{\sqrt{a^2+x^2}}$
31. $y = \sqrt{a^2-x^2} + a\sin^{-1}\frac{x}{a} \quad (a > 0)$
32. $y = a\cos^{-1}\left(1-\frac{x}{a}\right) - \sqrt{2ax-x^2} \quad (a > 0)$
33. Find the slope of the curve $\tan^{-1}\left(\frac{2x}{y}\right) = \frac{\pi x}{y^2}$ at the point $(1, 2)$.
34. Find equations of two straight lines tangent to the graph of $y = \sin^{-1}x$ and having slope 2.
35. Show that, on their respective domains, $\sin^{-1}$ and $\tan^{-1}$ are increasing functions and $\cos^{-1}$ is a decreasing function.
36. The derivative of $\sec^{-1}x$ is positive for every x in the domain of $\sec^{-1}$. Does this imply that $\sec^{-1}$ is increasing on its domain? Why?
37. Sketch the graph of $\csc^{-1}x$ and find its derivative.
38. Sketch the graph of $\cot^{-1}x$ and find its derivative.
39. Show that $\tan^{-1}x + \cot^{-1}x = \frac{\pi}{2}$ for $x > 0$. What is the sum if $x < 0$?
40. Find the derivative of $g(x) = \tan(\tan^{-1}x)$ and sketch the graph of g.

In Exercises 41–44, plot the graphs of the given functions by first calculating and simplifying the derivative of the function. Where is each function continuous? Where is it differentiable?

41. $\cos^{-1}(\cos x)$
42. $\sin^{-1}(\cos x)$
43. $\tan^{-1}(\tan x)$
44. $\tan^{-1}(\cot x)$
45. Show that $\sin^{-1}x = \tan^{-1}\left(\frac{x}{\sqrt{1-x^2}}\right)$ if $|x| < 1$.
46. Show that $\sec^{-1}x = \begin{cases} \tan^{-1}\sqrt{x^2-1} & \text{if } x \geq 1 \\ \pi - \tan^{-1}\sqrt{x^2-1} & \text{if } x \leq -1 \end{cases}$
47. Show that $\tan^{-1}x = \sin^{-1}\left(\frac{x}{\sqrt{1+x^2}}\right)$ for all x.
48. Show that $\sec^{-1}x = \begin{cases} \sin^{-1}\frac{\sqrt{x^2-1}}{x} & \text{if } x \geq 1 \\ \pi - \sin^{-1}\frac{\sqrt{x^2-1}}{x} & \text{if } x \leq -1 \end{cases}$
49. Show that the function $f(x)$ of Example 9 is also constant on the interval $(-\infty, -1)$. Find the value of the constant. *Hint:* Find $\lim_{x\to-\infty} f(x)$.
50. Find the derivative of $f(x) = x - \tan^{-1}(\tan x)$. What does your answer imply about $f(x)$? Calculate $f(0)$ and $f(\pi)$. Is there a contradiction here?
51. Find the derivative of $f(x) = x - \sin^{-1}(\sin x)$ for $-\pi \leq x \leq \pi$ and sketch the graph of f on that interval.

In Exercises 52–55, solve the initial-value problems.

52. $\begin{cases} y' = \dfrac{1}{1+x^2} \\ y(0) = 1 \end{cases}$

53. $\begin{cases} y' = \dfrac{1}{9+x^2} \\ y(3) = 2 \end{cases}$

54. $\begin{cases} y' = \dfrac{1}{\sqrt{1-x^2}} \\ y(1/2) = 1 \end{cases}$

55. $\begin{cases} y' = \dfrac{4}{\sqrt{25-x^2}} \\ y(0) = 0 \end{cases}$

3.6 Hyperbolic Functions

Any function defined on the real line can be expressed (in a unique way) as the sum of an even function and an odd function. (See Exercise 35 of Section P.5.) The **hyperbolic functions** $\cosh x$ and $\sinh x$ are, respectively, the even and odd functions whose sum is the exponential function e^x.

DEFINITION 15

The hyperbolic cosine and hyperbolic sine functions

For any real x the **hyperbolic cosine**, $\cosh x$, and the **hyperbolic sine**, $\sinh x$, are defined by

$$\cosh x = \frac{e^x + e^{-x}}{2}, \qquad \sinh x = \frac{e^x - e^{-x}}{2}.$$

(The symbol "sinh" is somewhat hard to pronounce as written. Some people say "shine," and others say "sinch.") Recall that cosine and sine are called *circular functions* because, for any t, the point $(\cos t, \sin t)$ lies on the circle with equation $x^2 + y^2 = 1$. Similarly, cosh and sinh are called *hyperbolic functions* because the point $(\cosh t, \sinh t)$ lies on the rectangular hyperbola with equation $x^2 - y^2 = 1$,

$$\cosh^2 t - \sinh^2 t = 1 \quad \text{for any real } t.$$

To see this, observe that

$$\begin{aligned} \cosh^2 t - \sinh^2 t &= \left(\frac{e^t + e^{-t}}{2}\right)^2 - \left(\frac{e^t - e^{-t}}{2}\right)^2 \\ &= \frac{1}{4}\left(e^{2t} + 2 + e^{-2t} - (e^{2t} - 2 + e^{-2t})\right) \\ &= \frac{1}{4}(2+2) = 1. \end{aligned}$$

There is no interpretation of t as an arc length or angle as there was in the circular case; however, the *area* of the *hyperbolic sector* bounded by $y = 0$, the hyperbola $x^2 - y^2 = 1$, and the ray from the origin to $(\cosh t, \sinh t)$ is $t/2$ square units (see Exercise 21 of Section 8.4), just as is the area of the circular sector bounded by $y = 0$, the circle $x^2 + y^2 = 1$, and the ray from the origin to $(\cos t, \sin t)$. (See Figure 3.26.)

Observe that, similar to the corresponding values of $\cos x$ and $\sin x$, we have

$$\cosh 0 = 1 \quad \text{and} \quad \sinh 0 = 0,$$

and $\cosh x$, like $\cos x$, is an even function, and $\sinh x$, like $\sin x$, is an odd function:

$$\cosh(-x) = \cosh x, \qquad \sinh(-x) = -\sinh x.$$

The graphs of cosh and sinh are shown in Figure 3.27. The graph $y = \cosh x$ is called a **catenary**. A chain hanging by its ends will assume the shape of a catenary.

Many other properties of the hyperbolic functions resemble those of the corresponding circular functions, sometimes with signs changed.

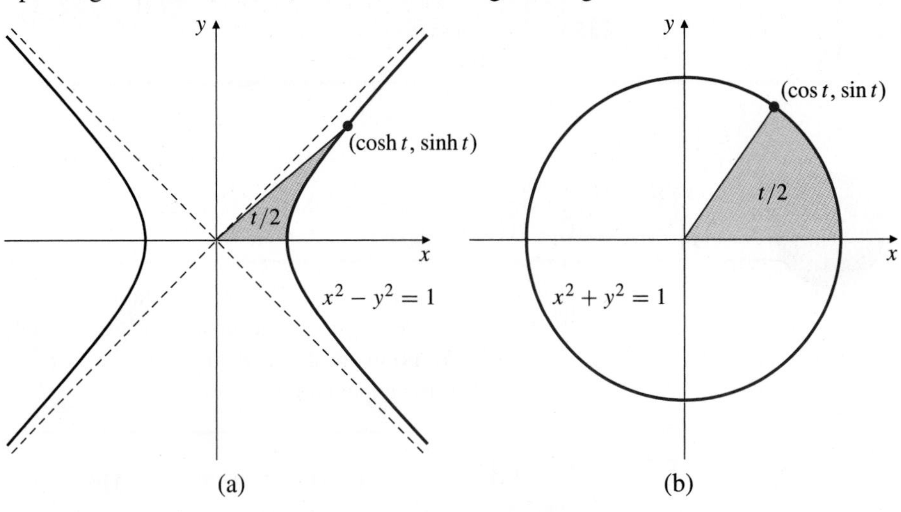

Figure 3.26 Both shaded areas are $t/2$ square units

EXAMPLE 1 Show that

$$\frac{d}{dx}\cosh x = \sinh x \quad \text{and} \quad \frac{d}{dx}\sinh x = \cosh x.$$

Solution We have

$$\frac{d}{dx}\cosh x = \frac{d}{dx}\frac{e^x + e^{-x}}{2} = \frac{e^x + e^{-x}(-1)}{2} = \sinh x$$
$$\frac{d}{dx}\sinh x = \frac{d}{dx}\frac{e^x - e^{-x}}{2} = \frac{e^x - e^{-x}(-1)}{2} = \cosh x.$$

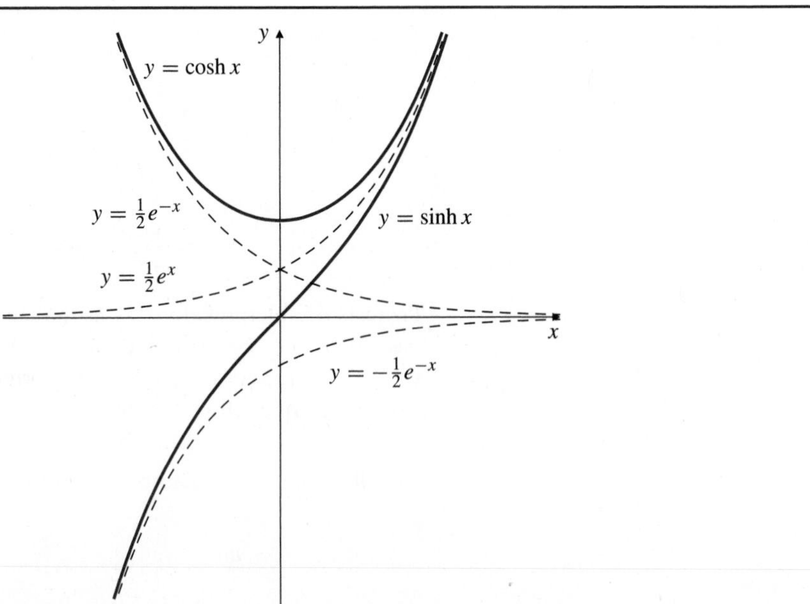

Figure 3.27 The graphs of cosh and sinh and some exponential graphs to which they are asymptotic

The following addition formulas and double-angle formulas can be checked algebraically by using the definition of cosh and sinh and the laws of exponents:

$$\cosh(x + y) = \cosh x \cosh y + \sinh x \sinh y,$$
$$\sinh(x + y) = \sinh x \cosh y + \cosh x \sinh y,$$

$$\cosh(2x) = \cosh^2 x + \sinh^2 x = 1 + 2\sinh^2 x = 2\cosh^2 x - 1,$$
$$\sinh(2x) = 2\sinh x \cosh x.$$

By analogy with the trigonometric functions, four other hyperbolic functions can be defined in terms of cosh and sinh.

DEFINITION 16

Other hyperbolic functions

$$\tanh x = \frac{\sinh x}{\cosh x} = \frac{e^x - e^{-x}}{e^x + e^{-x}} \qquad \operatorname{sech} x = \frac{1}{\cosh x} = \frac{2}{e^x + e^{-x}}$$
$$\coth x = \frac{\cosh x}{\sinh x} = \frac{e^x + e^{-x}}{e^x - e^{-x}} \qquad \operatorname{csch} x = \frac{1}{\sinh x} = \frac{2}{e^x - e^{-x}}$$

Multiplying the numerator and denominator of the fraction defining $\tanh x$ by e^{-x} and e^x, respectively, we obtain

$$\lim_{x\to\infty} \tanh x = \lim_{x\to\infty} \frac{1 - e^{-2x}}{1 + e^{-2x}} = 1 \qquad \text{and}$$
$$\lim_{x\to-\infty} \tanh x = \lim_{x\to-\infty} \frac{e^{2x} - 1}{e^{2x} + 1} = -1,$$

so that the graph of $y = \tanh x$ has two horizontal asymptotes. The graph of $\tanh x$ (Figure 3.28) resembles those of $x/\sqrt{1 + x^2}$ and $(2/\pi)\tan^{-1}x$ in shape, but, of course, they are not identical.

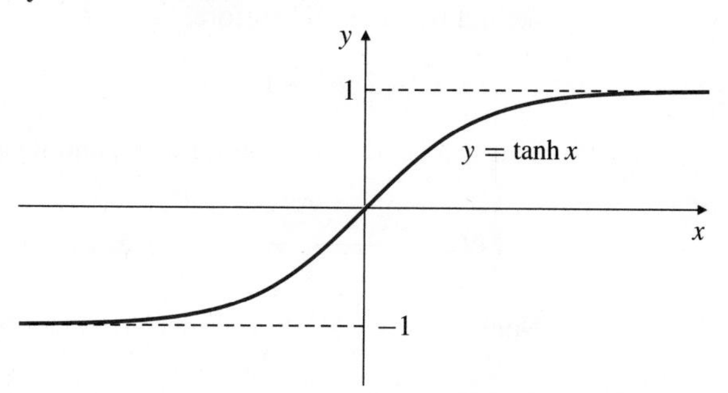

Figure 3.28 The graph of $\tanh x$

The derivatives of the remaining hyperbolic functions

$$\frac{d}{dx}\tanh x = \operatorname{sech}^2 x \qquad \frac{d}{dx}\operatorname{sech} x = -\operatorname{sech} x \tanh x$$
$$\frac{d}{dx}\coth x = -\operatorname{csch}^2 x \qquad \frac{d}{dx}\operatorname{csch} x = -\operatorname{csch} x \coth x$$

are easily calculated from those of $\cosh x$ and $\sinh x$ using the Reciprocal and Quotient Rules. For example,

$$\frac{d}{dx}\tanh x = \frac{d}{dx}\frac{\sinh x}{\cosh x} = \frac{(\cosh x)(\cosh x) - (\sinh x)(\sinh x)}{\cosh^2 x}$$
$$= \frac{1}{\cosh^2 x} = \operatorname{sech}^2 x.$$

Remark The distinction between trigonometric and hyperbolic functions largely disappears if we allow complex numbers instead of just real numbers as variables. If i is the imaginary unit (so that $i^2 = -1$), then

$$e^{ix} = \cos x + i \sin x \qquad \text{and} \qquad e^{-ix} = \cos x - i \sin x.$$

(See Appendix I.) Therefore,

$$\cosh(ix) = \frac{e^{ix} + e^{-ix}}{2} = \cos x, \qquad \cos(ix) = \cosh(-x) = \cosh x,$$
$$\sinh(ix) = \frac{e^{ix} - e^{-ix}}{2} = i \sin x, \qquad \sin(ix) = \frac{1}{i}\sinh(-x) = i \sinh x.$$

Inverse Hyperbolic Functions

The functions sinh and tanh are increasing and therefore one-to-one and invertible on the whole real line. Their inverses are denoted $\sinh^{-1}$ and $\tanh^{-1}$, respectively:

$$y = \sinh^{-1} x \iff x = \sinh y,$$
$$y = \tanh^{-1} x \iff x = \tanh y.$$

Since the hyperbolic functions are defined in terms of exponentials, it is not surprising that their inverses can be expressed in terms of logarithms.

EXAMPLE 2 Express the functions $\sinh^{-1} x$ and $\tanh^{-1} x$ in terms of natural logarithms.

Solution Let $y = \sinh^{-1} x$. Then

$$x = \sinh y = \frac{e^y - e^{-y}}{2} = \frac{(e^y)^2 - 1}{2e^y}.$$

(We multiplied the numerator and denominator of the first fraction by e^y to get the second fraction.) Therefore,

$$(e^y)^2 - 2xe^y - 1 = 0.$$

This is a quadratic equation in e^y, and it can be solved by the quadratic formula:

$$e^y = \frac{2x \pm \sqrt{4x^2 + 4}}{2} = x \pm \sqrt{x^2 + 1}.$$

Note that $\sqrt{x^2 + 1} > x$. Since e^y cannot be negative, we need to use the positive sign:

$$e^y = x + \sqrt{x^2 + 1}.$$

Hence, $y = \ln\left(x + \sqrt{x^2 + 1}\right)$, and we have

$$\sinh^{-1} x = \ln\left(x + \sqrt{x^2 + 1}\right).$$

Now let $y = \tanh^{-1} x$. Then

$$x = \tanh y = \frac{e^y - e^{-y}}{e^y + e^{-y}} = \frac{e^{2y} - 1}{e^{2y} + 1} \qquad (-1 < x < 1),$$
$$xe^{2y} + x = e^{2y} - 1,$$
$$e^{2y} = \frac{1 + x}{1 - x}, \qquad y = \frac{1}{2}\ln\left(\frac{1 + x}{1 - x}\right).$$

Thus,

$$\tanh^{-1} x = \frac{1}{2}\ln\left(\frac{1 + x}{1 - x}\right), \qquad (-1 < x < 1).$$

Since cosh is not one-to-one, its domain must be restricted before an inverse can be defined. Let us define the principal value of cosh to be

$$\operatorname{Cosh} x = \cosh x \qquad (x \geq 0).$$

The inverse, $\cosh^{-1}$, is then defined by

$$y = \cosh^{-1} x \iff x = \operatorname{Cosh} y \iff x = \cosh y \qquad (y \geq 0).$$

As we did for $\sinh^{-1}$, we can obtain the formula

$$\cosh^{-1} x = \ln\left(x + \sqrt{x^2 - 1}\right), \qquad (x \geq 1).$$

As was the case for the inverses of the reciprocal trigonometric functions, the inverses of the remaining three hyperbolic functions, coth, sech, and csch, are best defined using the inverses of their reciprocals.

$$\begin{aligned}
\coth^{-1} x = \tanh^{-1}\left(\frac{1}{x}\right) &= \frac{1}{2}\ln\left(\frac{1+\frac{1}{x}}{1-\frac{1}{x}}\right) && \text{for } \left|\frac{1}{x}\right| < 1 \\
&= \frac{1}{2}\ln\left(\frac{x+1}{x-1}\right) && \text{for } x > 1 \text{ or } x < 1 \\
\operatorname{sech}^{-1} x = \cosh^{-1}\left(\frac{1}{x}\right) &= \ln\left(\frac{1}{x} + \sqrt{\frac{1}{x^2} - 1}\right) && \text{for } \frac{1}{x} \geq 1 \\
&= \ln\left(\frac{1+\sqrt{1-x^2}}{x}\right) && \text{for } 0 < x \leq 1
\end{aligned}$$

$$\begin{aligned}
\operatorname{csch}^{-1} x = \sinh^{-1}\left(\frac{1}{x}\right) &= \ln\left(\frac{1}{x} + \sqrt{\frac{1}{x^2} + 1}\right) \\
&= \begin{cases} \ln\left(\dfrac{1+\sqrt{1+x^2}}{x}\right) & \text{if } x > 0 \\ \ln\left(\dfrac{1-\sqrt{1+x^2}}{x}\right) & \text{if } x < 0. \end{cases}
\end{aligned}$$

The derivatives of all six inverse hyperbolic functions are left as exercises for the reader. See Exercise 5 and Exercises 8–10 below.

EXERCISES 3.6

1. Verify the formulas for the derivatives of $\operatorname{sech} x$, $\operatorname{csch} x$, and $\coth x$ given in this section.

2. Verify the addition formulas

$$\cosh(x + y) = \cosh x \cosh y + \sinh x \sinh y,$$
$$\sinh(x + y) = \sinh x \cosh y + \cosh x \sinh y.$$

Proceed by expanding the right-hand side of each identity in terms of exponentials. Find similar formulas for $\cosh(x - y)$ and $\sinh(x - y)$.

3. Obtain addition formulas for $\tanh(x + y)$ and $\tanh(x - y)$ from those for sinh and cosh.

4. Sketch the graphs of $y = \coth x$, $y = \operatorname{sech} x$, and $y = \operatorname{csch} x$, showing any asymptotes.

5. Calculate the derivatives of $\sinh^{-1} x$, $\cosh^{-1} x$, and $\tanh^{-1} x$. Hence, express each of the indefinite integrals

$$\int \frac{dx}{\sqrt{x^2+1}}, \quad \int \frac{dx}{\sqrt{x^2-1}}, \quad \int \frac{dx}{1-x^2}$$

in terms of inverse hyperbolic functions.

6. Calculate the derivatives of the functions $\sinh^{-1}(x/a)$, $\cosh^{-1}(x/a)$, and $\tanh^{-1}(x/a)$ (where $a > 0$), and use your answers to provide formulas for certain indefinite integrals.

7. Simplify the following expressions: (a) $\sinh \ln x$, (b) $\cosh \ln x$, (c) $\tanh \ln x$, (d) $\dfrac{\cosh \ln x + \sinh \ln x}{\cosh \ln x - \sinh \ln x}$.

8. Find the domain, range, and derivative of $\coth^{-1} x$ and sketch the graph of $y = \coth^{-1} x$.

9. Find the domain, range, and derivative of $\text{sech}^{-1} x$ and sketch the graph of $y = \text{sech}^{-1} x$.

10. Find the domain, range, and derivative of $\text{csch}^{-1} x$, and sketch the graph of $y = \text{csch}^{-1} x$.

11. Show that the functions $f_{A,B}(x) = Ae^{kx} + Be^{-kx}$ and $g_{C,D}(x) = C \cosh kx + D \sinh kx$ are both solutions of the differential equation $y'' - k^2 y = 0$. (They are both general solutions.) Express $f_{A,B}$ in terms of $g_{C,D}$, and express $g_{C,D}$ in terms of $f_{A,B}$.

12. Show that $h_{L,M}(x) = L \cosh k(x-a) + M \sinh k(x-a)$ is also a solution of the differential equation in the previous exercise. Express $h_{L,M}$ in terms of the function $f_{A,B}$ above.

13. Solve the initial-value problem $y'' - k^2 y = 0$, $y(a) = y_0$, $y'(a) = v_0$. Express the solution in terms of the function $h_{L,M}$ of Exercise 12.

3.7 Second-Order Linear DEs with Constant Coefficients

A differential equation of the form

$$a\,y'' + b\,y' + cy = 0, \qquad (*)$$

where a, b, and c are constants and $a \neq 0$, is called a **second-order, linear, homogeneous** differential equation with constant coefficients. The *second-order* refers to the highest order derivative present; the terms *linear* and *homogeneous* refer to the fact that if $y_1(t)$ and $y_2(t)$ are two solutions of the equation, then so is $y(t) = Ay_1(t) + By_2(t)$ for any constants A and B:

$$\text{If } ay_1''(t) + by_1'(t) + cy_1(t) = 0 \text{ and } ay_2''(t) + by_2'(t) + cy_2(t) = 0,$$
$$\text{and if } y(t) = Ay_1(t) + By_2(t), \text{ then } ay''(t) + by'(t) + cy(t) = 0.$$

(See Section 18.1 for more details on this terminology.) Throughout this section we will assume that the independent variable in our functions is t rather than x, so the prime ($'$) refers to the derivative d/dt. This is because in most applications of such equations the independent variable is time.

Equations of type $(*)$ arise in many applications of mathematics. In particular, they can model mechanical vibrations such as the motion of a mass suspended from an elastic spring or the current in certain electrical circuits. In most such applications the three constants a, b, and c are positive, although sometimes we may have $b = 0$.

Recipe for Solving $ay'' + by' + cy = 0$

In Section 3.4 we observed that the first-order, constant-coefficient equation $y' = ky$ has solution $y = Ce^{kt}$. Let us try to find a solution of equation $(*)$ having the form $y = e^{rt}$. Substituting this expression into equation $(*)$, we obtain

$$ar^2e^{rt} + bre^{rt} + ce^{rt} = 0.$$

Since e^{rt} is never zero, $y = e^{rt}$ will be a solution of the differential equation $(*)$ if and only if r satisfies the quadratic **auxiliary equation**

$$ar^2 + br + c = 0, \qquad (**)$$

which has roots given by the quadratic formula:

$$r = \frac{-b \pm \sqrt{b^2 - 4ac}}{2a} = -\frac{b}{2a} \pm \frac{\sqrt{D}}{2a},$$

where $D = b^2 - 4ac$ is called the **discriminant** of the auxiliary equation $(**)$.

There are three cases to consider, depending on whether the discriminant D is positive, zero, or negative.

CASE I Suppose $D = b^2 - 4ac > 0$. Then the auxiliary equation has two different real roots, r_1 and r_2, given by

$$r_1 = \frac{-b - \sqrt{D}}{2a}, \qquad r_2 = \frac{-b + \sqrt{D}}{2a}.$$

(Sometimes these roots can be found easily by factoring the left side of the auxiliary equation.) In this case both $y = y_1(t) = e^{r_1 t}$ and $y = y_2(t) = e^{r_2 t}$ are solutions of the differential equation $(*)$, and neither is a multiple of the other. As noted above, the function

$$y = A\, e^{r_1 t} + B\, e^{r_2 t}$$

is also a solution for any choice of the constants A and B. Since the differential equation is of second order and this solution involves two arbitrary constants, we suspect it is the **general solution**, that is, that every solution of the differential equation can be written in this form. Exercise 18 at the end of this section outlines a way to prove this.

CASE II Suppose $D = b^2 - 4ac = 0$. Then the auxiliary equation has two equal roots, $r_1 = r_2 = -b/(2a) = r$, say. Certainly, $y = e^{rt}$ is a solution of $(*)$. We can find the general solution by letting $y = e^{rt}u(t)$ and calculating:

$$\begin{aligned} y' &= e^{rt}\big(u'(t) + ru(t)\big), \\ y'' &= e^{rt}\big(u''(t) + 2ru'(t) + r^2 u(t)\big). \end{aligned}$$

Substituting these expressions into $(*)$, we obtain

$$e^{rt}\big(au''(t) + (2ar + b)u'(t) + (ar^2 + br + c)u(t)\big) = 0.$$

Since $e^{rt} \neq 0$, $2ar + b = 0$ and r satisfies $(**)$, this equation reduces to $u''(t) = 0$, which has general solution $u(t) = A + Bt$ for arbitrary constants A and B. Thus, the general solution of $(*)$ in this case is

$$y = A\, e^{rt} + Bt\, e^{rt}.$$

CASE III Suppose $D = b^2 - 4ac < 0$. Then the auxiliary equation $(**)$ has complex conjugate roots given by

$$r = \frac{-b \pm \sqrt{b^2 - 4ac}}{2a} = k \pm i\omega,$$

where $k = -b/(2a)$, $\omega = \sqrt{4ac - b^2}/(2a)$, and i is the imaginary unit ($i^2 = -1$; see Appendix I). As in Case I, the functions $y_1^*(t) = e^{(k+i\omega)t}$ and $y_2^*(t) = e^{(k-i\omega)t}$ are two independent solutions of $(*)$, but they are not real-valued. However, since

$$e^{ix} = \cos x + i \sin x \qquad \text{and} \qquad e^{-ix} = \cos x - i \sin x$$

(as noted in the previous section and in Appendix II), we can find two real-valued functions that are solutions of $(*)$ by suitably combining y_1^* and y_2^*:

$$\begin{aligned} y_1(t) &= \frac{1}{2} y_1^*(t) + \frac{1}{2} y_2^*(t) = e^{kt} \cos(\omega t), \\ y_2(t) &= \frac{1}{2i} y_1^*(t) - \frac{1}{2i} y_2^*(t) = e^{kt} \sin(\omega t). \end{aligned}$$

Therefore, the general solution of $(*)$ in this case is

$$y = A\,e^{kt}\cos(\omega t) + B\,e^{kt}\sin(\omega t).$$

The following examples illustrate the recipe for solving $(*)$ in each of the three cases.

EXAMPLE 1 Find the general solution of $y'' + y' - 2y = 0$.

Solution The auxiliary equation is $r^2 + r - 2 = 0$, or $(r+2)(r-1) = 0$. The auxiliary roots are $r_1 = -2$ and $r_2 = 1$, which are real and unequal. According to Case I, the general solution of the differential equation is

$$y = A\,e^{-2t} + Be^{t}.$$

EXAMPLE 2 Find the general solution of $y'' + 6y' + 9y = 0$.

Solution The auxiliary equation is $r^2 + 6r + 9 = 0$, or $(r+3)^2 = 0$, which has equal roots $r = -3$. According to Case II, the general solution of the differential equation is

$$y = A\,e^{-3t} + Bt\,e^{-3t}.$$

EXAMPLE 3 Find the general solution of $y'' + 4y' + 13y = 0$.

Solution The auxiliary equation is $r^2 + 4r + 13 = 0$, which has solutions

$$r = \frac{-4 \pm \sqrt{16 - 52}}{2} = \frac{-4 \pm \sqrt{-36}}{2} = -2 \pm 3i.$$

Thus, $k = -2$ and $\omega = 3$. According to Case III, the general solution of the given differential equation is

$$y = A\,e^{-2t}\cos(3t) + B\,e^{-2t}\sin(3t).$$

Initial-value problems for $ay'' + by' + cy = 0$ specify values for y and y' at an initial point. These values can be used to determine the values of the constants A and B in the general solution, so the initial-value problem has a unique solution.

EXAMPLE 4 Solve the initial-value problem

$$\begin{cases} y'' + 2y' + 2y = 0 \\ y(0) = 2 \\ y'(0) = -3. \end{cases}$$

Solution The auxiliary equation is $r^2 + 2r + 2 = 0$, which has roots

$$r = \frac{-2 \pm \sqrt{4 - 8}}{2} = -1 \pm i.$$

Thus Case III applies, $k = -1$ and $\omega = 1$. Thus, the differential equation has the general solution

$$y = A\,e^{-t}\cos t + B\,e^{-t}\sin t.$$

Also,

$$\begin{aligned} y' &= e^{-t}\big(-A\cos t - B\sin t - A\sin t + B\cos t\big) \\ &= (B-A)\,e^{-t}\cos t - (A+B)\,e^{-t}\sin t. \end{aligned}$$

Applying the initial conditions $y(0) = 2$ and $y'(0) = -3$, we obtain $A = 2$ and $B - A = -3$. Hence, $B = -1$ and the initial-value problem has the solution

$$y = 2e^{-t}\cos t - e^{-t}\sin t.$$

Simple Harmonic Motion

Many natural phenomena exhibit periodic behaviour. The swinging of a clock pendulum, the vibrating of a guitar string or drum membrane, the altitude of a rider on a rotating ferris wheel, the motion of an object floating in wavy seas, and the voltage produced by an alternating current generator are but a few examples where quantities depend on time in a periodic way. Being periodic, the circular functions sine and cosine provide a useful model for such behaviour.

It often happens that a quantity displaced from an equilibrium value experiences a restoring force that tends to move it back in the direction of its equilibrium. Besides the obvious examples of elastic motions in physics, one can imagine such a model applying, say, to a biological population in equilibrium with its food supply or the price of a commodity in an elastic economy where increasing price causes decreasing demand and hence decreasing price. In the simplest models, the restoring force is proportional to the amount of displacement from equilibrium. Such a force causes the quantity to oscillate sinusoidally; we say that it executes *simple harmonic motion*.

As a specific example, suppose a mass m is suspended by an elastic spring so that it hangs unmoving in its equilibrium position with the upward spring tension force balancing the downward gravitational force on the mass. If the mass is displaced vertically by an amount y from this position, the spring tension changes; the extra force exerted by the spring is directed to restore the mass to its equilibrium position. (See Figure 3.29.) This extra force is proportional to the displacement (Hooke's Law); its magnitude is $-ky$, where k is a positive constant called the **spring constant**. Assuming the spring is weightless, this force imparts to the mass m an acceleration d^2y/dt^2 that satisfies, by Newton's Second Law, $m(d^2y/dt^2) = -ky$ (mass $\times$ acceleration = force). Dividing this equation by m, we obtain the equation

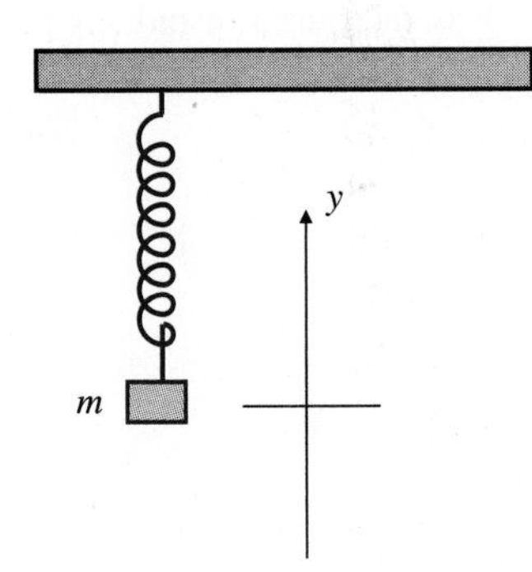

Figure 3.29

$$\frac{d^2y}{dt^2} + \omega^2 y = 0, \qquad \text{where} \quad \omega^2 = \frac{k}{m}.$$

The second-order differential equation

$$\frac{d^2y}{dt^2} + \omega^2 y = 0$$

is called the **equation of simple harmonic motion**. Its auxiliary equation, $r^2 + \omega^2 = 0$, has complex roots $r = \pm i\omega$, so it has general solution

$$y = A\cos\omega t + B\sin\omega t,$$

where A and B are arbitrary constants.

For any values of the constants R and t_0, the function

$$y = R\cos\big(\omega(t - t_0)\big)$$

is also a general solution of the differential equation of simple harmonic motion. If we expand this formula using the addition formula for cosine, we get

$$\begin{aligned} y &= R\cos\omega t_0\cos\omega t + R\sin\omega t_0\sin\omega t \\ &= A\cos\omega t + B\sin\omega t, \end{aligned}$$

where

$$\begin{aligned} A &= R\cos(\omega t_0), & B &= R\sin(\omega t_0), \\ R^2 &= A^2 + B^2, & \tan(\omega t_0) &= B/A. \end{aligned}$$

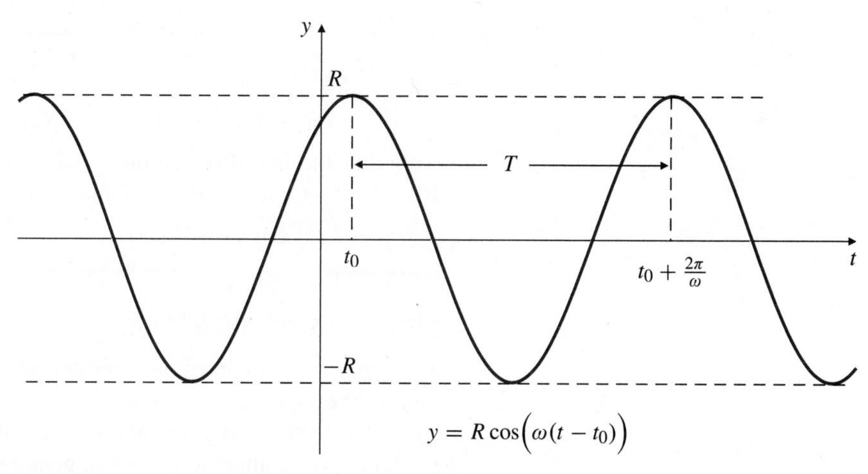

Figure 3.30 Simple harmonic motion

The constants A and B are related to the position y_0 and the velocity v_0 of the mass m at time $t = 0$:

$$y_0 = y(0) = A\cos 0 + B\sin 0 = A,$$
$$v_0 = y'(0) = -A\omega\sin 0 + B\omega\cos 0 = B\omega.$$

The constant $R = \sqrt{A^2 + B^2}$ is called the **amplitude** of the motion. Because $\cos x$ oscillates between -1 and 1, the displacement y varies between $-R$ and R. Note in Figure 3.30 that the graph of the displacement as a function of time is the curve $y = R\cos\omega t$ shifted t_0 units to the right. The number t_0 is called the **time-shift**. (The related quantity ωt_0 is called a **phase-shift**.) The **period** of this curve is $T = 2\pi/\omega$; it is the time interval between consecutive instants when the mass is at the same height moving in the same direction. The reciprocal $1/T$ of the period is called the **frequency** of the motion. It is usually measured in Hertz (Hz), that is, cycles per second. The quantity $\omega = 2\pi/T$ is called the **circular frequency**. It is measured in radians per second since 1 cycle = 1 revolution = 2π radians.

EXAMPLE 5 Solve the initial-value problem

$$\begin{cases} y'' + 16y = 0 \\ y(0) = -6 \\ y'(0) = 32. \end{cases}$$

Find the amplitude, frequency, and period of the solution.

Solution Here, $\omega^2 = 16$ so $\omega = 4$. The solution is of the form

$$y = A\cos(4t) + B\sin(4t).$$

Since $y(0) = -6$, we have $A = -6$. Also, $y'(t) = -4A\sin(4t) + 4B\cos(4t)$. Since $y'(0) = 32$, we have $4B = 32$, or $B = 8$. Thus, the solution is

$$y = -6\cos(4t) + 8\sin(4t).$$

The amplitude is $\sqrt{(-6)^2 + 8^2} = 10$, the frequency is $\omega/(2\pi) \approx 0.637$ Hz, and the period is $2\pi/\omega \approx 1.57$ s.

EXAMPLE 6 **(Spring-mass problem)** Suppose that a 100 g mass is suspended from a spring and that a force of 3×10^4 dynes (3×10^4 g-cm/s^2) is required to produce a displacement from equilibrium of 1/3 cm. At time $t = 0$ the mass is pulled down 2 cm below equilibrium and flicked upward with a velocity of 60 cm/s. Find its subsequent displacement at any time $t > 0$. Find the frequency, period, amplitude, and time-shift of the motion. Express the position of the mass at time t in terms of the amplitude and the time-shift.

Solution The spring constant k is determined from Hooke's Law, $F = -ky$. Here $F = -3 \times 10^4$ g-cm/s^2 is the force of the spring on the mass displaced 1/3 cm:

$$-3 \times 10^4 = -\frac{1}{3}k,$$

so $k = 9 \times 10^4$ g/s^2. Hence, the circular frequency is $\omega = \sqrt{k/m} = 30$ rad/s, the frequency is $\omega/2\pi = 15/\pi \approx 4.77$ Hz, and the period is $2\pi/\omega \approx 0.209$ s.

Since the displacement at time $t = 0$ is $y_0 = -2$ and the velocity at that time is $v_0 = 60$, the subsequent displacement is $y = A\cos(30t) + B\sin(30t)$, where $A = y_0 = -2$ and $B = v_0/\omega = 60/30 = 2$. Thus,

$$y = -2\cos(30t) + 2\sin(30t), \qquad (y \text{ in cm, } t \text{ in seconds}).$$

The amplitude of the motion is $R = \sqrt{(-2)^2 + 2^2} = 2\sqrt{2} \approx 2.83$ cm. The time-shift t_0 must satisfy

$$\begin{aligned} -2 = A &= R\cos(\omega t_0) = 2\sqrt{2}\cos(30t_0), \\ 2 = B &= R\sin(\omega t_0) = 2\sqrt{2}\sin(30t_0), \end{aligned}$$

so $\sin(30t_0) = 1/\sqrt{2} = -\cos(30t_0)$. Hence the phase-shift is $30t_0 = 3\pi/4$ radians, and the time-shift is $t_0 = \pi/40 \approx 0.0785$ s. The position of the mass at time $t > 0$ is also given by

$$y = 2\sqrt{2}\cos\left[30\left(t - \frac{\pi}{40}\right)\right].$$

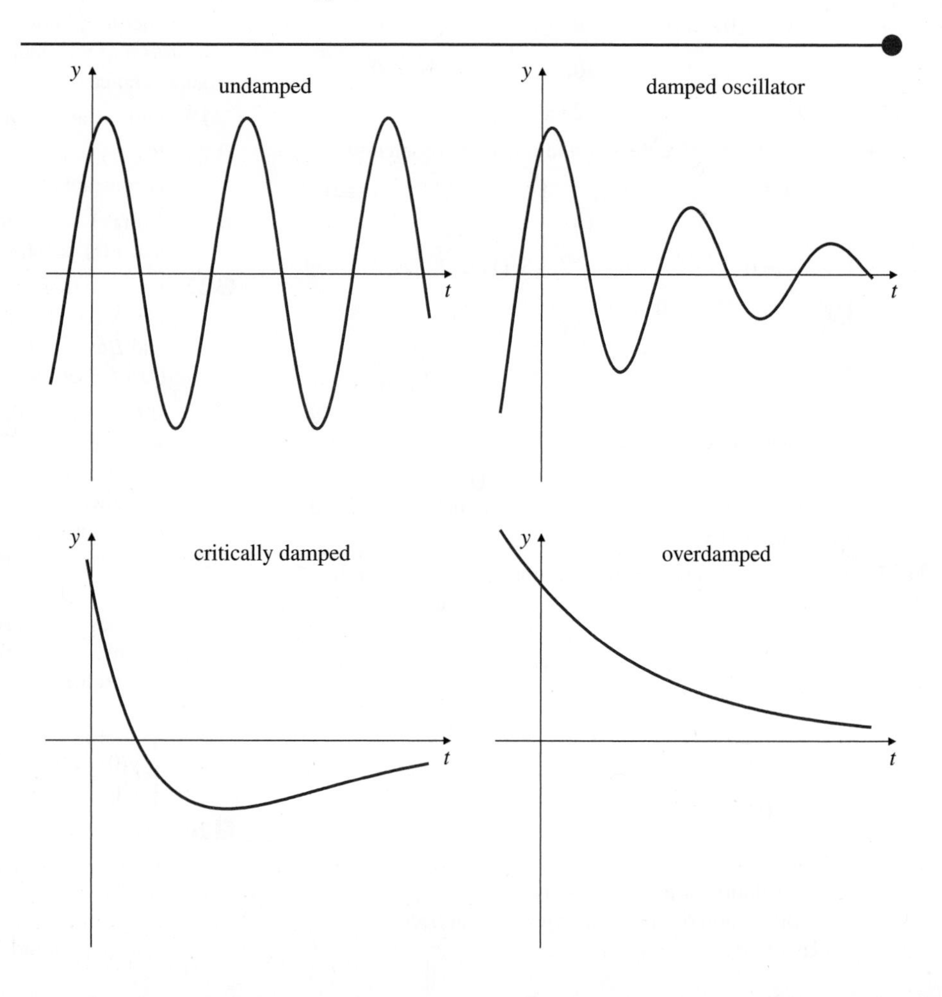

Figure 3.31
Undamped oscillator ($b = 0$)
Damped oscillator ($b > 0$, $b^2 < 4ac$)
Critically damped case ($b > 0$, $b^2 = 4ac$)
Overdamped case ($b > 0$, $b^2 > 4ac$)

Damped Harmonic Motion

If a and c are positive and $b = 0$, then equation

$$ay'' + by' + cy = 0$$

is the differential equation of simple harmonic motion and has oscillatory solutions of fixed amplitude as shown above. If $a > 0$, $b > 0$, and $c > 0$, then the roots of the auxiliary equation are either negative real numbers or, if $b^2 < 4ac$, complex numbers $k \pm i\omega$ with negative real parts $k = -b/(2a)$ (Case III). In this latter case the solutions still oscillate, but the amplitude diminishes exponentially as $t \to \infty$ because of the factor $e^{kt} = e^{-(b/2a)t}$. (See Exercise 17 below.) A system whose behaviour is modelled by such an equation is said to exhibit **damped harmonic motion**. If $b^2 = 4ac$ (Case II), the system is said to be **critically damped**, and if $b^2 > 4ac$ (Case I), it is **overdamped**. In these cases the behaviour is no longer oscillatory. (See Figure 3.31. Imagine a mass suspended by a spring in a jar of oil.)

EXERCISES 3.7

In Exercises 1–12, find the general solutions for the given equations. Since this entire section is concerned with differential equations no special symbols need be used here.

1. $y'' + 7y' + 10y = 0$

2. $y'' - 2y' - 3y = 0$

3. $y'' + 2y' = 0$

4. $4y'' - 4y' - 3y = 0$

5. $y'' + 8y' + 16y = 0$

6. $y'' - 2y' + y = 0$

7. $y'' - 6y' + 10y = 0$

8. $9y'' + 6y' + y = 0$

9. $y'' + 2y' + 5y = 0$

10. $y'' - 4y' + 5y = 0$

11. $y'' + 2y' + 3y = 0$

12. $y'' + y' + y = 0$

In Exercises 13–15, solve the given initial-value problems.

13. $\begin{cases} 2y'' + 5y' - 3y = 0 \\ y(0) = 1 \\ y'(0) = 0. \end{cases}$

14. $\begin{cases} y'' + 10y' + 25y = 0 \\ y(1) = 0 \\ y'(1) = 2. \end{cases}$

15. $\begin{cases} y'' + 4y' + 5y = 0 \\ y(0) = 2 \\ y'(0) = 2. \end{cases}$

16. Show that if $\epsilon \neq 0$, the function $y_\epsilon(t) = \dfrac{e^{(1+\epsilon)t} - e^t}{\epsilon}$ satisfies the equation $y'' - (2+\epsilon)y' + (1+\epsilon)y = 0$. Caclulate $y(t) = \lim_{\epsilon \to 0} y_\epsilon(t)$ and verify that, as expected, it is a solution of $y'' - 2y' + y = 0$.

17. If $a > 0$, $b > 0$, and $c > 0$, prove that all solutions of the differential equation $ay'' + by' + cy = 0$ satisfy $\lim_{t \to \infty} y(t) = 0$.

18. Prove that the solution given in the discussion of Case I, namely, $y = A\, e^{r_1 t} + B\, e^{r_2 t}$, is the general solution for that case as follows: First, let $y = e^{r_1 t} u$ and show that u satisfies the equation

$$u'' - (r_2 - r_1)u' = 0.$$

Then let $v = u'$, so that v must satisfy $v' = (r_2 - r_1)v$. The general solution of this equation is $v = C\, e^{(r_2 - r_1)t}$, as shown in the discussion of the equation $y' = ky$ in Section 3.4. Hence find u and y.

Simple harmonic motion

Exercises 19–22 all refer to the differential equation of simple harmonic motion:

$$\frac{d^2y}{dt^2} + \omega^2 y = 0, \qquad (\omega \neq 0). \qquad (\dagger)$$

Together they show that $y = A\cos\omega t + B\sin\omega t$ is a *general solution* of this equation, that is, every solution is of this form for some choice of the constants A and B.

19. Show that $y = A\cos\omega t + B\sin\omega t$ is a solution of (†).

20. If $f(t)$ is any solution of (†), show that $\omega^2(f(t))^2 + (f'(t))^2$ is constant.

21. If $g(t)$ is a solution of (†) satisfying $g(0) = g'(0) = 0$, show that $g(t) = 0$ for all t.

22. Suppose that $f(t)$ is any solution of the differential equation (†). Show that $f(t) = A\cos\omega t + B\sin\omega t$, where $A = f(0)$ and $B\omega = f'(0)$.
(*Hint:* Let $g(t) = f(t) - A\cos\omega t - B\sin\omega t$.)

23. If $b^2 - 4ac < 0$, show that the substitution $y = e^{kt}u(t)$, where $k = -b/(2a)$, transforms $ay'' + by' + cy = 0$ into the equation $u'' + \omega^2 u = 0$, where $\omega^2 = (4ac - b^2)/(4a^2)$. Together with the result of Exercise 22, this confirms the recipe for Case III, in case you didn't feel comfortable with the complex number argument given in the text.

In Exercises 24–25, solve the given initial-value problems. For each problem determine the circular frequency, the frequency, the period, and the amplitude of the solution.

24. $\begin{cases} y'' + 4y = 0 \\ y(0) = 2 \\ y'(0) = -5. \end{cases}$

25. $\begin{cases} y'' + 100y = 0 \\ y(0) = 0 \\ y'(0) = 3. \end{cases}$

26. Show that $y = \alpha\cos(\omega(t-c)) + \beta\sin(\omega(t-c))$ is a solution of the differential equation $y'' + \omega^2 y = 0$, and that it satisfies $y(c) = \alpha$ and $y'(c) = \beta\omega$. Express the solution in the form $y = A\cos(\omega t) + B\sin(\omega t)$ for certain values of the constants A and B depending on α, β, c, and ω.

27. Solve $\begin{cases} y'' + y = 0 \\ y(2) = 3 \\ y'(2) = -4. \end{cases}$

28. Solve $\begin{cases} y'' + \omega^2 y = 0 \\ y(a) = A \\ y'(a) = B. \end{cases}$

29. What mass should be suspended from the spring in Example 6 to provide a system whose natural frequency of oscillation is 10 Hz? Find the displacement of such a mass from its equilibrium position t s after it is pulled down 1 cm from equilibrium and flicked upward with a speed of 2 cm/s. What is the amplitude of this motion?

30. A mass of 400 g suspended from a certain elastic spring will oscillate with a frequency of 24 Hz. What would be the frequency if the 400 g mass were replaced with a 900 g mass? a 100 g mass?

31. Show that if t_0, A, and B are constants and $k = -b/(2a)$ and $\omega = \sqrt{4ac - b^2}/(2a)$, then

$$y = e^{kt}\left[A\cos\left(\omega(t - t_0)\right) + B\sin\left(\omega(t - t_0)\right)\right]$$

is an alternative to the general solution of the equation $ay'' + by' + cy = 0$ for Case III ($b^2 - 4ac < 0$). This form of the general solution is useful for solving initial-value problems where $y(t_0)$ and $y'(t_0)$ are specified.

32. Show that if t_0, A, and B are constants and $k = -b/(2a)$ and $\omega = \sqrt{b^2 - 4ac}/(2a)$, then

$$y = e^{kt}\left[A\cosh\left(\omega(t - t_0)\right) + B\sinh\left(\omega(t - t_0)\right)\right]$$

is an alternative to the general solution of the equation $ay'' + by' + cy = 0$ for Case I ($b^2 - 4ac > 0$). This form of the general solution is useful for solving initial-value problems where $y(t_0)$ and $y'(t_0)$ are specified.

Use the forms of solution provided by the previous two exercises to solve the initial-value problems in Exercises 33–34.

33. $\begin{cases} y'' + 2y' + 5y = 0 \\ y(3) = 2 \\ y'(3) = 0. \end{cases}$

34. $\begin{cases} y'' + 4y' + 3y = 0 \\ y(3) = 1 \\ y'(3) = 0. \end{cases}$

35. By using the change of dependent variable $u(x) = c - k^2 y(x)$, solve the initial-value problem

$$\begin{cases} y''(x) = c - k^2 y(x) \\ y(0) = a \\ y'(0) = b. \end{cases}$$

36. A mass is attached to a spring mounted horizontally so the mass can slide along the top of a table. With a suitable choice of units, the position $x(t)$ of the mass at time t is governed by the differential equation

$$x'' = -x + F,$$

where the $-x$ term is due to the elasticity of the spring, and the F is due to the friction of the mass with the table. The frictional force should be constant in magnitude and directed opposite to the velocity of the mass when the mass is moving. When the mass is stopped, the friction should be constant and opposed to the spring force unless the spring force has the smaller magnitude, in which case the friction force should just cancel the spring force and the mass should remain at rest thereafter. For this problem, let the magnitude of the friction force be 1/5. Accordingly,

$$F = \begin{cases} -\frac{1}{5} & \text{if } x' > 0 \text{ or if } x' = 0 \text{ and } x < -\frac{1}{5} \\ \frac{1}{5} & \text{if } x' < 0 \text{ or if } x' = 0 \text{ and } x > \frac{1}{5} \\ x & \text{if } x' = 0 \text{ and } |x| \le \frac{1}{5}. \end{cases}$$

Find the position $x(t)$ of the mass at all times $t > 0$ if $x(0) = 1$ and $x'(0) = 0$.

CHAPTER REVIEW

Key Ideas

- **State the laws of exponents.**
- **State the laws of logarithms.**
- **What is the significance of the number e?**
- **What do the following statements and phrases mean?**
  - ◇ f is one-to-one.
  - ◇ f is invertible.
  - ◇ Function f^{-1} is the inverse of function f.
  - ◇ $a^b = c$
  - ◇ $\log_a b = c$
  - ◇ the natural logarithm of x
  - ◇ logarithmic differentiation
  - ◇ the half-life of a varying quantity
  - ◇ The quantity y exhibits exponential growth.
  - ◇ The quantity y exhibits logistic growth.
  - ◇ $y = \sin^{-1}x$
  - ◇ $y = \tan^{-1}x$
  - ◇ The quantity y exhibits simple harmonic motion.
  - ◇ The quantity y exhibits damped harmonic motion.
- **Define the functions $\sinh x$, $\cosh x$, and $\tanh x$.**
- **What kinds of functions satisfy second-order differential equations with constant coefficients?**

Review Exercises

1. If $f(x) = 3x + x^3$, show that f has an inverse and find the slope of $y = f^{-1}(x)$ at $x = 0$.

2. Let $f(x) = \sec^2 x \tan x$. Show that f is increasing on the interval $(-\pi/2, \pi/2)$ and, hence, one-to-one and invertible there. What is the domain of f^{-1}? Find $(f^{-1})'(2)$. *Hint:* $f(\pi/4) = 2$.

Exercises 3–5 refer to the function $f(x) = x\,e^{-x^2}$.

3. Find $\lim_{x\to\infty} f(x)$ and $\lim_{x\to-\infty} f(x)$.

4. On what intervals is f increasing? decreasing?

5. What are the maximum and minimum values of $f(x)$?

6. Find the points on the graph of $y = e^{-x}\sin x$, $(0 \le x \le 2\pi)$, where the graph has a horizontal tangent line.

7. Suppose that a function $f(x)$ satisfies $f'(x) = x\,f(x)$ for all real x, and $f(2) = 3$. Calculate the derivative of $f(x)/e^{x^2/2}$, and use the result to help you find $f(x)$ explicitly.

8. A lump of modelling clay is being rolled out so that it maintains the shape of a circular cylinder. If the length is increasing at a rate proportional to itself, show that the radius is decreasing at a rate proportional to itself.

9. (a) What nominal interest rate, compounded continuously, will cause an investment to double in 5 years?

(b) By about how many days will the doubling time in part (a) increase if the nominal interest rate drops by 0.5%?

10. (A poor man's natural logarithm)

(a) Show that if $a > 0$, then

$$\lim_{h\to 0}\frac{a^h - 1}{h} = \ln a.$$

Hence, show that

$$\lim_{n\to\infty} n(a^{1/n} - 1) = \ln a.$$

(b) Most calculators, even nonscientific ones, have a square root key. If n is a power of 2, say $n = 2^k$, then $a^{1/n}$ can be calculated by entering a and hitting the square root key k times:

$$a^{1/2^k} = \sqrt{\sqrt{\cdots\sqrt{a}}} \qquad (k \text{ square roots}).$$

Then you can subtract 1 and multiply by n to get an approximation for $\ln a$. Use $n = 2^{10} = 1024$ and $n = 2^{11} = 2048$ to find approximations for $\ln 2$. Based on the agreement of these two approximations, quote a value of $\ln 2$ to as many decimal places as you feel justified.

11. A nonconstant function f satisfies

$$\frac{d}{dx}\Big(f(x)\Big)^2 = \Big(f'(x)\Big)^2$$

for all x. If $f(0) = 1$, find $f(x)$.

12. If $f(x) = (\ln x)/x$, show that $f'(x) > 0$ for $0 < x < e$ and $f'(x) < 0$ for $x > e$, so that $f(x)$ has a maximum value at $x = e$. Use this to show that $e^\pi > \pi^e$.

13. Find an equation of a straight line that passes through the origin and is tangent to the curve $y = x^x$.

14. (a) Find $x \ne 2$ such that $\dfrac{\ln x}{x} = \dfrac{\ln 2}{2}$.

(b) Find $b > 1$ such that there is *no* $x \ne b$ with $\dfrac{\ln x}{x} = \dfrac{\ln b}{b}$.

15. Investment account A bears simple interest at a certain rate. Investment account B bears interest at the same nominal rate but compounded instantaneously. If \$1,000 is invested in each account, B produces \$10 more in interest after one year than does A. Find the nominal rate both accounts use.

16. Express each of the functions $\cos^{-1}x$, $\cot^{-1}x$, and $\csc^{-1}x$ in terms of $\tan^{-1}$.

17. Express each of the functions $\cos^{-1}x$, $\cot^{-1}x$, and $\csc^{-1}x$ in terms of $\sin^{-1}$.

18. (A warming problem) A bottle of milk at 5 °C is removed from a refrigerator into a room maintained at 20 °C. After 12 min the temperature of the milk is 12 °C. How much longer will it take for the milk to warm up to 18 °C?

19. (A cooling problem) A kettle of hot water at 96 °C is allowed to sit in an air-conditioned room. The water cools to 60 °C in 10 min and then to 40 °C in another 10 min. What is the temperature of the room?

20. Show that $e^x > 1 + x$ if $x \ne 0$.

21. Use mathematical induction to show that

$$e^x > 1 + x + \frac{x^2}{2!} + \cdots + \frac{x^n}{n!}$$

if $x > 0$ and n is any positive integer.

Challenging Problems

1. (a) Show that the function $f(x) = x^x$ is strictly increasing on $[e^{-1}, \infty)$.

(b) If g is the inverse function to f of part (a), show that

$$\lim_{y\to\infty}\frac{g(y)\ln(\ln y)}{\ln y} = 1$$

Hint: Start with the equation $y = x^x$ and take the ln of both sides twice.

Two models for incorporating air resistance into the analysis of the motion of a falling body

2. (Air resistance proportional to speed) An object falls under gravity near the surface of the earth, and its motion is impeded by air resistance proportional to its speed. Its velocity v therefore satisfies the equation

$$\frac{dv}{dt} = -g - kv, \qquad (*)$$

where k is a positive constant depending on such factors as the shape and density of the object and the density of the air.

(a) Find the velocity of the object as a function of time t, given that it was v_0 at $t = 0$.

(b) Find the limiting velocity $\lim_{t\to\infty} v(t)$. Observe that this can be done either directly from $(*)$ or from the solution found in (a).

(c) If the object was at height y_0 at time $t = 0$, find its height $y(t)$ at any time during its fall.

3. (Air resistance proportional to the square of speed) Under certain conditions a better model for the effect of air resistance on a moving object is one where the resistance is proportional to the square of the speed. For an object falling under constant gravitational acceleration g, the equation of motion is

$$\frac{dv}{dt} = -g - kv|v|,$$

where $k > 0$. Note that $v|v|$ is used instead of v^2 to ensure that the resistance is always in the opposite direction to the velocity. For an object falling from rest at time $t = 0$, we have $v(0) = 0$ and $v(t) < 0$ for $t > 0$, so the equation of motion becomes

$$\frac{dv}{dt} = -g + kv^2.$$

We are not (yet) in a position to solve this equation. However, we can verify its solution.

(a) Verify that the velocity is given for $t \geq 0$ by

$$v(t) = \sqrt{\frac{g}{k}}\,\frac{1 - e^{2t\sqrt{gk}}}{1 + e^{2t\sqrt{gk}}}.$$

(b) What is the limiting velocity $\lim_{t\to\infty} v(t)$?

(c) Also verify that if the falling object was at height y_0 at time $t = 0$, then its height at subsequent times during its fall is given by

$$y(t) = y_0 + \sqrt{\frac{g}{k}}\,t - \frac{1}{k}\ln\left(\frac{1 + e^{2t\sqrt{gk}}}{2}\right).$$

4. **(A model for the spread of a new technology)** When a new and superior technology is introduced, the percentage p of potential clients that adopt it might be expected to increase logistically with time. However, even newer technologies are continually being introduced, so adoption of a particular one will fall off exponentially over time. The following model exhibits this behaviour:

$$\frac{dp}{dt} = kp\left(1 - \frac{p}{e^{-bt}M}\right).$$

This DE suggests that the growth in p is logistic but that the asymptotic limit is not a constant but rather $e^{-bt}M$, which decreases exponentially with time.

(a) Show that the change of variable $p = e^{-bt}y(t)$ transforms the equation above into a standard logistic equation, and hence find an explicit formula for $p(t)$ given that $p(0) = p_0$. It will be necessary to assume that $M < 100k/(b+k)$ to ensure that $p(t) < 100$.

(b) If $k = 10$, $b = 1$, $M = 90$, and $p_0 = 1$, how large will $p(t)$ become before it starts to decrease?

CHAPTER 4

More Applications of Differentiation

> "In the fall of 1972 President Nixon announced that the rate of increase of inflation was decreasing. This was the first time a sitting president used the third derivative to advance his case for reelection."

Hugo Rossi
Mathematics Is an Edifice, Not a Toolbox, Notices of the AMS, v. 43, Oct. 1996

Introduction

Differential calculus can be used to analyze many kinds of problems and situations that arise in applied disciplines. Calculus has made and will continue to make significant contributions to every field of human endeavour that uses quantitative measurement to further its aims. From economics to physics and from biology to sociology, problems can be found whose solutions can be aided by the use of some calculus.

In this chapter we will examine several kinds of problems to which the techniques we have already learned can be applied. These problems arise both outside and within mathematics. We will deal with the following kinds of problems:

1. Related rates problems, where the rates of change of related quantities are analyzed.
2. Root finding methods, where we try to find numerical solutions of equations.
3. Evaluation of limits.
4. Optimization problems, where a quantity is to be maximized or minimized.
5. Graphing problems, where derivatives are used to illuminate the behaviour of functions.
6. Approximation problems, where complicated functions are approximated by polynomials.

Do not assume that most of the problems we present here are "real-world" problems. Such problems are usually too complex to be treated in a general calculus course. However, the problems we consider, while sometimes artificial, do show how calculus can be applied in concrete situations.

4.1 Related Rates

When two or more quantities that change with time are linked by an equation, that equation can be differentiated with respect to time to produce an equation linking the rates of change of the quantities. Any one of these rates may then be determined when the others, and the values of the quantities themselves, are known. We will consider a couple of examples before formulating a list of procedures for dealing with such problems.

EXAMPLE 1 An aircraft is flying horizontally at a speed of 600 km/h. How fast is the distance between the aircraft and a radio beacon increasing 1 min after the aircraft passes 5 km directly above the beacon?

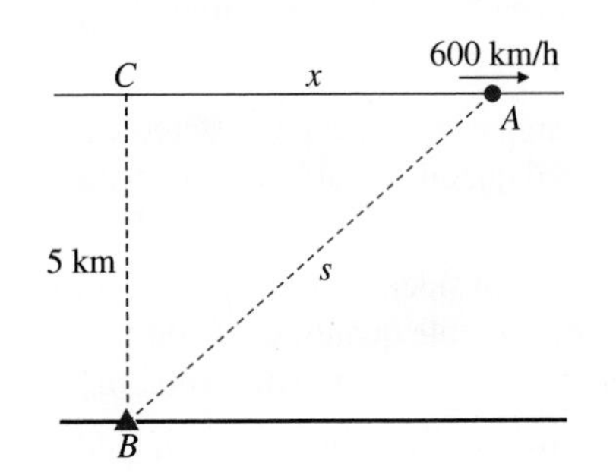

Figure 4.1

Solution A diagram is useful here; see Figure 4.1. Let C be the point on the aircraft's path directly above the beacon B. Let A be the position of the aircraft t min after it is at C, and let x and s be the distances CA and BA, respectively. From the right triangle BCA we have

$$s^2 = x^2 + 5^2.$$

We differentiate this equation implicitly with respect to t to obtain

$$2s\,\frac{ds}{dt} = 2x\,\frac{dx}{dt}.$$

We are given that $dx/dt = 600$ km/h $= 10$ km/min. Therefore, $x = 10$ km at time $t = 1$ min. At that time $s = \sqrt{10^2 + 5^2} = 5\sqrt{5}$ km and is increasing at the rate

$$\frac{ds}{dt} = \frac{x}{s}\,\frac{dx}{dt} = \frac{10}{5\sqrt{5}}(600) = \frac{1,200}{\sqrt{5}} \approx 536.7 \text{ km/h}.$$

One minute after the aircraft passes over the beacon, its distance from the beacon is increasing at about 537 km/h.

EXAMPLE 2 How fast is the area of a rectangle changing if one side is 10 cm long and is increasing at a rate of 2 cm/s and the other side is 8 cm long and is decreasing at a rate of 3 cm/s?

Solution Let the lengths of the sides of the rectangle at time t be x cm and y cm, respectively. Thus the area at time t is $A = xy$ cm^2. (See Figure 4.2.) We want to know the value of dA/dt when $x = 10$ and $y = 8$, given that $dx/dt = 2$ and $dy/dt = -3$. (Note the negative sign to indicate that y is decreasing.) Since all the quantities in the equation $A = xy$ are functions of time, we can differentiate that equation implicitly with respect to time and obtain

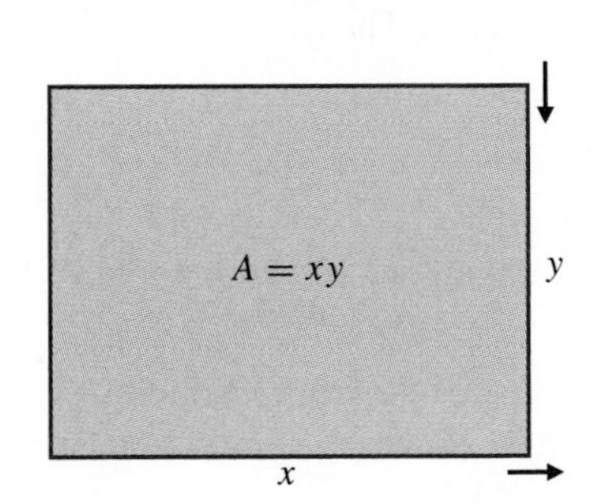

Figure 4.2 Rectangle with sides changing

$$\left.\frac{dA}{dt}\right|_{\substack{x=10\\y=8}} = \left.\left(\frac{dx}{dt}\,y + x\,\frac{dy}{dt}\right)\right|_{\substack{x=10\\y=8}} = 2(8) + 10(-3) = -14.$$

At the time in question, the area of the rectangle is decreasing at a rate of 14 cm^2/s.

Procedures for Related-Rates Problems

In view of these examples we can formulate a few general procedures for dealing with related-rates problems.

How to solve related-rates problems

1. Read the problem very carefully. Try to understand the relationships between the variable quantities. What is given? What is to be found?
2. Make a sketch if appropriate.
3. Define any symbols you want to use that are not defined in the statement of the problem. Express given and required quantities and rates in terms of these symbols.
4. From a careful reading of the problem or consideration of the sketch, identify one or more equations linking the variable quantities. (You will need as many equations as quantities or rates to be found in the problem.)
5. Differentiate the equation(s) implicitly with respect to time, regarding all variable quantities as functions of time. You can manipulate the equation(s) algebraically before the differentiation is performed (for instance, you could solve for the quantities whose rates are to be found), but it is usually easier to differentiate the equations as they are originally obtained and solve for the desired items later.
6. Substitute any given values for the quantities and their rates, then solve the resulting equation(s) for the unknown quantities and rates.
7. Make a concluding statement answering the question asked. Is your answer reasonable? If not, check back through your solution to see what went wrong.

EXAMPLE 3 A lighthouse L is located on a small island 2 km from the nearest point A on a long, straight shoreline. If the lighthouse lamp rotates at 3 revolutions per minute, how fast is the illuminated spot P on the shoreline moving along the shoreline when it is 4 km from A?

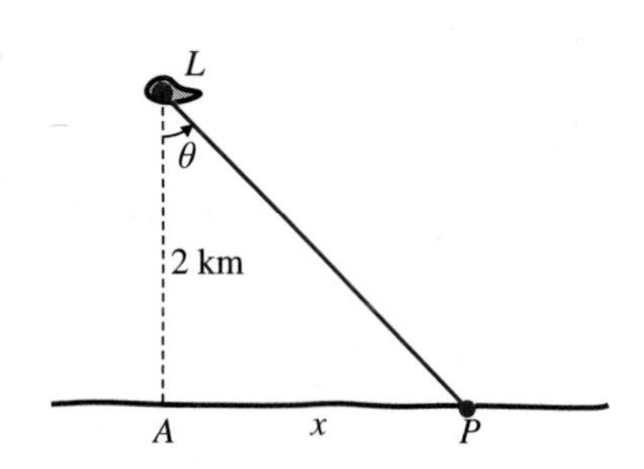

Figure 4.3

Solution Referring to Figure 4.3, let x be the distance AP, and let θ be the angle PLA. Then $x = 2\tan\theta$ and

$$\frac{dx}{dt} = 2\sec^2\theta\,\frac{d\theta}{dt}.$$

Now

$$\frac{d\theta}{dt} = (3\text{ rev/min})(2\pi\text{ radians/rev}) = 6\pi\text{ radians/min}.$$

When $x = 4$, we have $\tan\theta = 2$ and $\sec^2\theta = 1 + \tan^2\theta = 5$. Thus,

$$\frac{dx}{dt} = (2)(5)(6\pi) = 60\pi \approx 188.5.$$

The spot of light is moving along the shoreline at a rate of about 189 km/min when it is 4 km from A.

(Note that it was essential to convert the rate of change of θ from revolutions per minute to radians per minute. If θ were not measured in radians we could not assert that $(d/d\theta)\tan\theta = \sec^2\theta$.)

EXAMPLE 4 A leaky water tank is in the shape of an inverted right circular cone with depth 5 m and top radius 2 m. When the water in the tank is 4 m deep, it is leaking out at a rate of $1/12$ m^3/min. How fast is the water level in the tank dropping at that time?

Solution Let r and h denote the surface radius and depth of water in the tank at time t (both measured in metres). Thus, the volume V (in cubic metres) of water in the tank at time t is

$$V = \frac{1}{3}\pi r^2 h.$$

Using similar triangles (see Figure 4.4), we can find a relationship between r and h:

$$\frac{r}{h} = \frac{2}{5}, \qquad \text{so} \quad r = \frac{2h}{5} \quad \text{and} \quad V = \frac{1}{3}\pi\left(\frac{2h}{5}\right)^2 h = \frac{4\pi}{75}h^3.$$

Differentiating this equation with respect to t we obtain

$$\frac{dV}{dt} = \frac{4\pi}{25}h^2\frac{dh}{dt}.$$

Since $dV/dt = -1/12$ when $h = 4$, we have

$$\frac{-1}{12} = \frac{4\pi}{25}(4^2)\frac{dh}{dt}, \qquad \text{so} \quad \frac{dh}{dt} = -\frac{25}{768\pi}.$$

When the water in the tank is 4 m deep, its level is dropping at a rate of $25/(768\pi)$ m/min, or about 1.036 cm/min.

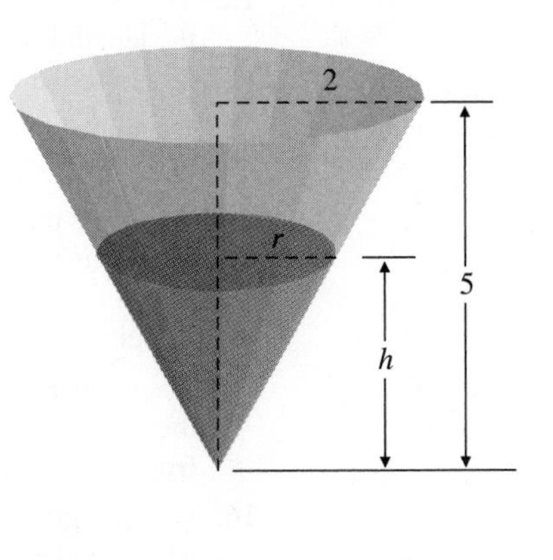

Figure 4.4 The conical tank of Example 4

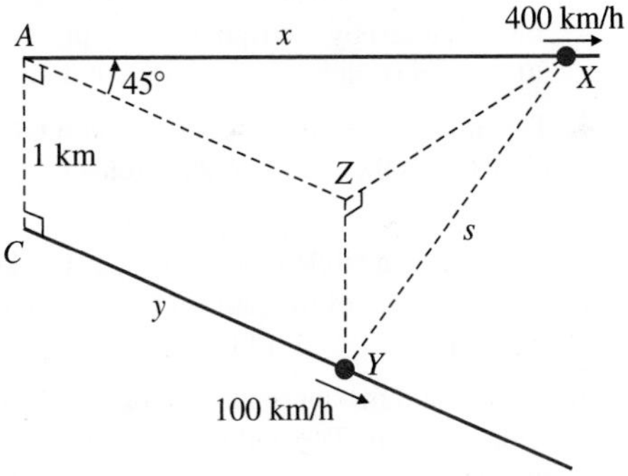

Figure 4.5 Aircraft and car paths in Example 5

EXAMPLE 5 At a certain instant an aircraft flying due east at 400 km/h passes directly over a car travelling due southeast at 100 km/h on a straight, level road. If the aircraft is flying at an altitude of 1 km, how fast is the distance between the aircraft and the car increasing 36 s after the aircraft passes directly over the car?

Solution A good diagram is essential here. See Figure 4.5. Let time t be measured in hours from the time the aircraft was at position A directly above the car at position C. Let X and Y be the positions of the aircraft and the car, respectively, at time t. Let x be the distance AX, y the distance CY, and s the distance XY, all measured in kilometres. Let Z be the point 1 km above Y. Since angle $XAZ = 45°$, the Pythagorean Theorem and Cosine Law yield

$$\begin{aligned} s^2 &= 1 + (ZX)^2 = 1 + x^2 + y^2 - 2xy\cos 45° \\ &= 1 + x^2 + y^2 - \sqrt{2}xy. \end{aligned}$$

Thus,

$$\begin{aligned} 2s\frac{ds}{dt} &= 2x\frac{dx}{dt} + 2y\frac{dy}{dt} - \sqrt{2}\frac{dx}{dt}y - \sqrt{2}x\frac{dy}{dt} \\ &= 400(2x - \sqrt{2}y) + 100(2y - \sqrt{2}x), \end{aligned}$$

since $dx/dt = 400$ and $dy/dt = 100$. When $t = 1/100$ (i.e., 36 s after $t = 0$), we have $x = 4$ and $y = 1$. Hence,

$$\begin{aligned} s^2 &= 1 + 16 + 1 - 4\sqrt{2} = 18 - 4\sqrt{2} \\ s &\approx 3.5133. \\ \frac{ds}{dt} &= \frac{1}{2s}\big(400(8 - \sqrt{2}) + 100(2 - 4\sqrt{2})\big) \approx 322.86. \end{aligned}$$

The aircraft and the car are separating at a rate of about 323 km/h after 36 s. (Note that it was necessary to convert 36 s to hours in the solution. In general, all measurements should be in compatible units.)

EXERCISES 4.1

1. Find the rate of change of the area of a square whose side is 8 cm long, if the side length is increasing at 2 cm/min.
2. The area of a square is decreasing at 2 ft^2/s. How fast is the side length changing when it is 8 ft?
3. A pebble dropped into a pond causes a circular ripple to expand outward from the point of impact. How fast is the area enclosed by the ripple increasing when the radius is 20 cm and is increasing at a rate of 4 cm/s?
4. The area of a circle is decreasing at a rate of 2 cm^2/min. How fast is the radius of the circle changing when the area is 100 cm^2?
5. The area of a circle is increasing at $1/3$ km^2/h. Express the rate of change of the radius of the circle as a function of (a) the radius r and (b) the area A of the circle.
6. At a certain instant the length of a rectangle is 16 m and the width is 12 m. The width is increasing at 3 m/s. How fast is the length changing if the area of the rectangle is not changing?
7. Air is being pumped into a spherical balloon. The volume of the balloon is increasing at a rate of 20 cm^3/s when the radius is 30 cm. How fast is the radius increasing at that time? (The volume of a ball of radius r units is $V = \frac{4}{3}\pi r^3$ cubic units.)
8. When the diameter of a ball of ice is 6 cm, it is decreasing at a rate of 0.5 cm/h due to melting of the ice. How fast is the volume of the ice ball decreasing at that time?
9. How fast is the surface area of a cube changing when the volume of the cube is 64 cm^3 and is increasing at 2 cm^3/s?
10. The volume of a right circular cylinder is 60 cm^3 and is increasing at 2 cm^3/min at a time when the radius is 5 cm and is increasing at 1 cm/min. How fast is the height of the cylinder changing at that time?
11. How fast is the volume of a rectangular box changing when the length is 6 cm, the width is 5 cm, and the depth is 4 cm, if the length and depth are both increasing at a rate of 1 cm/s and the width is decreasing at a rate of 2 cm/s?
12. The area of a rectangle is increasing at a rate of 5 m^2/s while the length is increasing at a rate of 10 m/s. If the length is 20 m and the width is 16 m, how fast is the width changing?
13. A point moves on the curve $y = x^2$. How fast is y changing when $x = -2$ and x is decreasing at a rate of 3?
14. A point is moving to the right along the first-quadrant portion of the curve $x^2y^3 = 72$. When the point has coordinates $(3, 2)$, its horizontal velocity is 2 units/s. What is its vertical velocity?
15. The point P moves so that at time t it is at the intersection of the curves $xy = t$ and $y = tx^2$. How fast is the distance of P from the origin changing at time $t = 2$?
16. **(Radar guns)** A police officer is standing near a highway using a radar gun to catch speeders. (See Figure 4.6.) He aims the gun at a car that has just passed his position and, when the gun is pointing at an angle of 45° to the direction of the highway, notes that the distance between the car and the gun is increasing at a rate of 100 km/h. How fast is the car travelling?

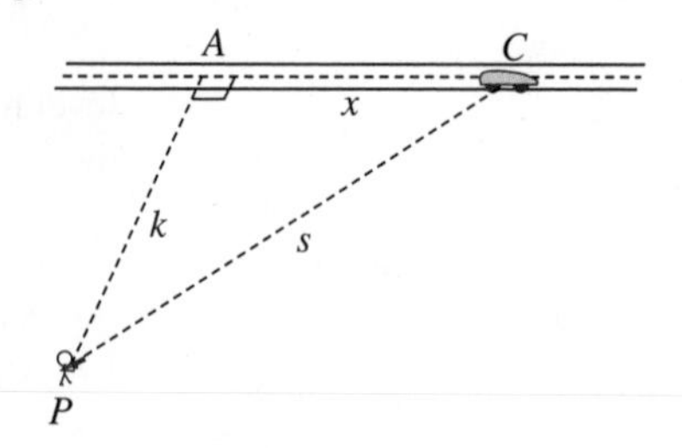

Figure 4.6

17. If the radar gun of Exercise 16 is aimed at a car travelling at 90 km/h along a straight road, what will its reading be when it is aimed making an angle of 30° with the road?

18. The top of a ladder 5 m long rests against a vertical wall. If the base of the ladder is being pulled away from the base of the wall at a rate of 1/3 m/s, how fast is the top of the ladder slipping down the wall when it is 3 m above the base of the wall?

19. A man 2 m tall walks toward a lamppost on level ground at a rate of 0.5 m/s. If the lamp is 5 m high on the post, how fast is the length of the man's shadow decreasing when he is 3 m from the post? How fast is the shadow of his head moving at that time?

20. A woman 6 ft tall is walking at 2 ft/s along a straight path on level ground. There is a lamppost 5 ft to the side of the path. A light 15 ft high on the lamppost casts the woman's shadow on the ground. How fast is the length of her shadow changing when the woman is 12 feet from the point on the path closest to the lamppost?

21. (Cost of production) It costs a coal mine owner $\$C$ each day to maintain a production of x tons of coal, where $C = 10{,}000 + 3x + x^2/8{,}000$. At what rate is the production increasing when it is 12,000 tons and the daily cost is increasing at \$600 per day?

22. (Distance between ships) At 1:00 p.m. ship A is 25 km due north of ship B. If ship A is sailing west at a rate of 16 km/h and ship B is sailing south at 20 km/h, at what rate is the distance between the two ships changing at 1:30 p.m?

23. What is the first time after 3:00 p.m. that the hands of a clock are together?

24. (Tracking a balloon) A balloon released at point A rises vertically with a constant speed of 5 m/s. Point B is level with and 100 m distant from point A. How fast is the angle of elevation of the balloon at B changing when the balloon is 200 m above A?

25. Sawdust is falling onto a pile at a rate of 1/2 m^3/min. If the pile maintains the shape of a right circular cone with height equal to half the diameter of its base, how fast is the height of the pile increasing when the pile is 3 m high?

26. (Conical tank) A water tank is in the shape of an inverted right circular cone with top radius 10 m and depth 8 m. Water is flowing in at a rate of 1/10 m^3/min. How fast is the depth of water in the tank increasing when the water is 4 m deep?

27. (Leaky tank) Repeat Exercise 26 with the added assumption that water is leaking out of the bottom of the tank at a rate of $h^3/1{,}000$ m^3/min when the depth of water in the tank is h m. How full can the tank get in this case?

28. (Another leaky tank) Water is pouring into a leaky tank at a rate of 10 m^3/h. The tank is a cone with vertex down, 9 m in depth and 6 m in diameter at the top. The surface of water in the tank is rising at a rate of 20 cm/h when the depth is 6 m. How fast is the water leaking out at that time?

29. (Kite flying) How fast must you let out line if the kite you are flying is 30 m high, 40 m horizontally away from you, and moving horizontally away from you at a rate of 10 m/min?

30. (Ferris wheel) You are on a Ferris wheel of diameter 20 m. It is rotating at 1 revolution per minute. How fast are you rising or falling when you are 6 m horizontally away from the vertical line passing through the centre of the wheel?

31. (Distance between aircraft) An aircraft is 144 km east of an airport and is travelling west at 200 km/h. At the same time, a second aircraft at the same altitude is 60 km north of the airport and travelling north at 150 km/h. How fast is the distance between the two aircraft changing?

32. (Production rate) If a truck factory employs x workers and has daily operating expenses of $\$y$, it can produce $P = (1/3)x^{0.6}y^{0.4}$ trucks per year. How fast are the daily expenses decreasing when they are \$10,000 and the number of workers is 40, if the number of workers is increasing at 1 per day and production is remaining constant?

33. A lamp is located at point $(3, 0)$ in the xy-plane. An ant is crawling in the first quadrant of the plane and the lamp casts its shadow onto the y-axis. How fast is the ant's shadow moving along the y-axis when the ant is at position $(1, 2)$ and moving so that its x-coordinate is increasing at rate 1/3 units/s and its y-coordinate is decreasing at 1/4 units/s?

34. A straight highway and a straight canal intersect at right angles, the highway crossing over the canal on a bridge 20 m above the water. A boat travelling at 20 km/h passes under the bridge just as a car travelling at 80 km/h passes over it. How fast are the boat and car separating after one minute?

35. (Filling a trough) The cross section of a water trough is an equilateral triangle with top edge horizontal. If the trough is 10 m long and 30 cm deep, and if water is flowing in at a rate of 1/4 m^3/min, how fast is the water level rising when the water is 20 cm deep at the deepest?

36. (Draining a pool) A rectangular swimming pool is 8 m wide and 20 m long. (See Figure 4.7.) Its bottom is a sloping plane, the depth increasing from 1 m at the shallow end to 3 m at the deep end. Water is draining out of the pool at a rate of 1 m^3/min. How fast is the surface of the water falling when the depth of water at the deep end is (a) 2.5 m? (b) 1 m?

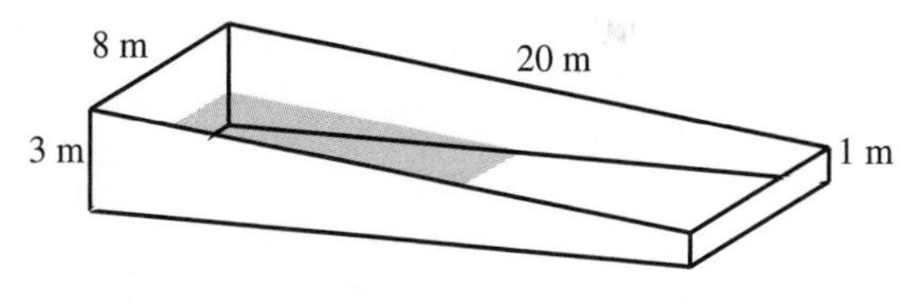

Figure 4.7

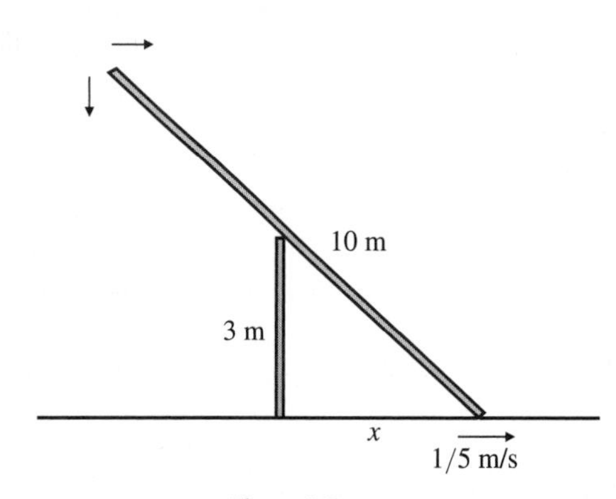

Figure 4.8

37. One end of a 10 m long ladder is on the ground and the ladder is supported partway along its length by resting on top of a 3 m high fence. (See Figure 4.8.) If the bottom of the ladder is 4 m from the base of the fence and is being dragged along the ground away from the fence at a rate of 1/5 m/s, how fast is the free top end of the ladder moving (a) vertically and (b) horizontally?

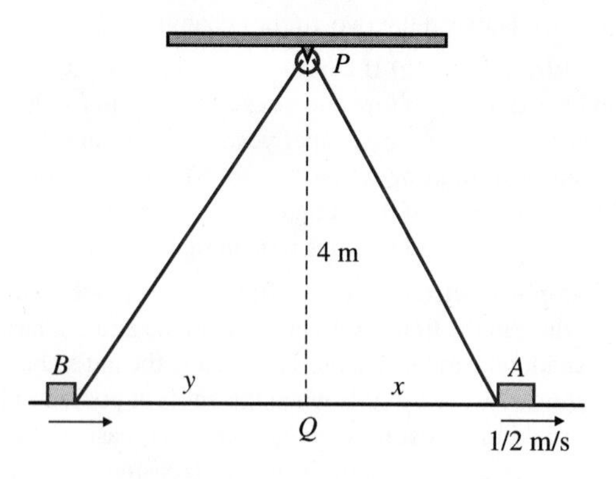

Figure 4.9

38. Two crates, A and B, are on the floor of a warehouse. The crates are joined by a rope 15 m long, each crate being hooked at floor level to an end of the rope. The rope is stretched tight and pulled over a pulley P that is attached to a rafter 4 m above a point Q on the floor directly between the two crates. (See Figure 4.9.) If crate A is 3 m from Q and is being pulled directly away from Q at a rate of 1/2 m/s, how fast is crate B moving toward Q?

39. **(Tracking a rocket)** Shortly after launch, a rocket is 100 km high and 50 km downrange. If it is travelling at 4 km/s at an angle of 30° above the horizontal, how fast is its angle of elevation, as measured at the launch site, changing?

40. **(Shadow of a falling ball)** A lamp is 20 m high on a pole. At time $t = 0$ a ball is dropped from a point level with the lamp and 10 m away from it. The ball falls under gravity (acceleration 9.8 m/s^2) until it hits the ground. How fast is the shadow of the ball moving along the ground (a) 1 s after it is dropped? (b) just as the ball hits the ground?

41. **(Tracking a rocket)** A rocket blasts off at time $t = 0$ and climbs vertically with acceleration 10 m/s^2. The progress of the rocket is monitored by a tracking station located 2 km horizontally away from the launch pad. How fast is the tracking station antenna rotating upward 10 s after launch?

4.2 Finding Roots of Equations

Finding solutions (roots) of equations is an important mathematical problem to which calculus can make significant contributions. There are only a few general classes of equations of the form $f(x) = 0$ that we can solve exactly. These include **linear equations**:

$$ax + b = 0, \quad (a \neq 0) \qquad \Rightarrow \qquad x = -\frac{b}{a}$$

and **quadratic equations**:

$$ax^2 + bx + c = 0, \quad (a \neq 0) \qquad \Rightarrow \qquad x = \frac{-b \pm \sqrt{b^2 - 4ac}}{2a}.$$

Cubic and quartic (3rd- and 4th-degree polynomial) equations can also be solved, but the formulas are very complicated. We usually solve these and most other equations approximately by using numerical methods, often with the aid of a calculator or computer.

In Section 1.4 we discussed the Bisection Method for approximating a root of an equation $f(x) = 0$. That method uses the Intermediate-Value Theorem and depends only on the continuity of f and our ability to find an interval $[x_1, x_2]$ that must contain the root because $f(x_1)$ and $f(x_2)$ have opposite signs. The method is rather slow; it requires between three and four iterations to gain one significant figure of precision in the root being approximated.

If we know that f is more than just continuous, we can devise better (i.e., faster) methods for finding roots of $f(x) = 0$. We study two such methods in this section:

(a) **Fixed-Point Iteration**, which looks for solutions of an equation of the form $x = f(x)$. Such solutions are called **fixed points** of the function f.

(b) **Newton's Method**, which looks for solutions of the equation $f(x) = 0$ as fixed points of the function $g(x) = x - \dfrac{f(x)}{f'(x)}$, i.e., points x such that $x = g(x)$. This method is usually very efficient, but it requires that f be differentiable.

Like the Bisection Method, both of these methods require that we have at the outset a rough idea of where a root can be found, and they generate sequences of approximations that get closer and closer to the root.

Discrete Maps and Fixed-Point Iteration

A **discrete map** is an equation of the form

$$x_{n+1} = f(x_n), \qquad \text{for } n = 0, 1, 2, \ldots,$$

which generates a sequence of values $x_1, x_2, x_3, \ldots$, from a given starting value x_0. In certain circumstances this sequence of numbers will converge to a limit, $r = \lim_{n\to\infty} x_n$, in which case this limit will be a fixed point of f: $r = f(r)$. (A thorough discussion of convergence of sequences can be found in Section 9.1. For our purposes here, an intuitive understanding will suffice: $\lim_{n\to\infty} x_n = r$ if $|x_n - r|$ approaches 0 as $n \to \infty$.)

For certain kinds of functions f, we can solve the equation $f(r) = r$ by starting with an initial guess x_0 and calculating subsequent values of the discrete map until sufficient accuracy is achieved. This is the **Method of Fixed-Point Iteration**. Let us begin by investigating a simple example:

EXAMPLE 1 Find a root of the equation $\cos x = 5x$.

Solution This equation is of the form $f(x) = x$, where $f(x) = \frac{1}{5}\cos x$. Since $\cos x$ is close to 1 for x near 0, we see that $\frac{1}{5}\cos x$ will be close to $\frac{1}{5}$ when $x = \frac{1}{5}$. This suggests that a reasonable first guess at the fixed point is $x_0 = \frac{1}{5} = 0.2$. The values of subsequent approximations

$$x_1 = \frac{1}{5}\cos x_0, \quad x_2 = \frac{1}{5}\cos x_1, \quad x_3 = \frac{1}{5}\cos x_2, \ldots$$

are presented in Table 1. The root is 0.196 164 28 to 8 decimal places.

Table 1.

n	x_n
0	0.2
1	0.196 013 32
2	0.196 170 16
3	0.196 164 05
4	0.196 164 29
5	0.196 164 28
6	0.196 164 28

Why did the method used in Example 1 work? Will it work for any function f? In order to answer these questions, examine the polygonal line in Figure 4.10. Starting at x_0 it goes vertically to the curve $y = f(x)$, the height there being x_1. Then it goes horizontally to the line $y = x$, meeting that line at a point whose x-coordinate must therefore also be x_1. Then the process repeats; the line goes vertically to the curve $y = f(x)$ and horizontally to $y = x$, arriving at $x = x_2$. The line continues in this way, "spiralling" closer and closer to the intersection of $y = f(x)$ and $y = x$. Each value of x_n is closer to the fixed point r than the previous value.

Now consider the function f whose graph appears in Figure 4.11(a). If we try the same method there, starting with x_0, the polygonal line spirals outward, away from the root, and the resulting values x_n will not "converge" to the root as they did in Example 1. To see why the method works for the function in Figure 4.10 but not for the function in Figure 4.11(a), observe the slopes of the two graphs $y = f(x)$, near the fixed point r. Both slopes are negative, but in Figure 4.10 the absolute value of the slope is less than 1 while the absolute value of the slope of f in Figure 4.11(a) is greater than 1. Close consideration of the graphs should convince you that it is this fact that caused the points x_n to get closer to r in Figure 4.10 and farther from r in Figure 4.11(a).

Figure 4.10 Iterations of $x_{n+1} = f(x_n)$ "spiral" toward the fixed point

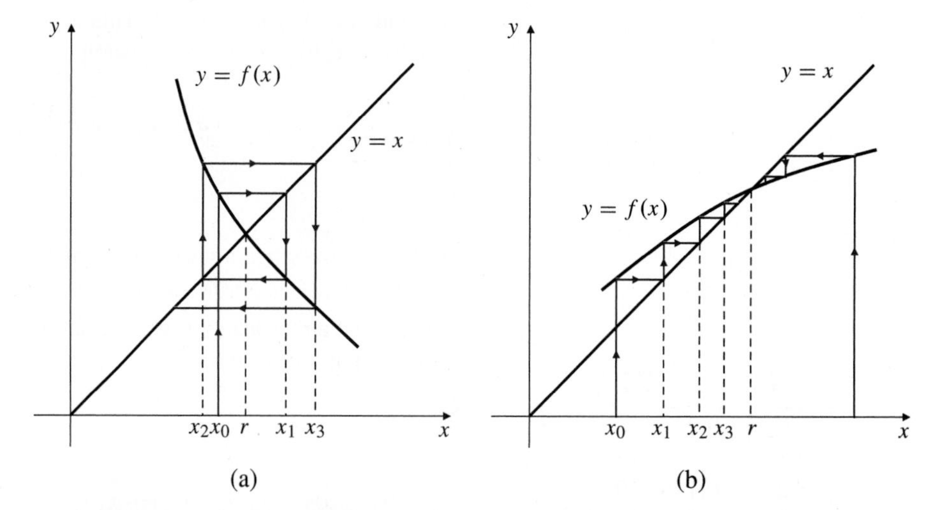

Figure 4.11

(a) A function f for which the iterations $x_{n+1} = f(x_n)$ do not converge
(b) "Staircase" convergence to the fixed point

A third example, Figure 4.11(b), shows that the method can be expected to work for functions whose graphs have positive slope near the fixed point r, provided that the slope is less than 1. In this case the polygonal line forms a "staircase" rather than a "spiral," and the successive approximations x_n increase toward the root if $x_0 < r$ and decrease toward it if $x_0 > r$.

Remark Note that if $|f'(x)| > 1$ near a fixed point r of f, you may still be able to find that fixed point by applying fixed-point iteration to $f^{-1}(x)$. Evidently $f^{-1}(r) = r$ if and only if $r = f(r)$.

The following theorem guarantees that the method of fixed-point iteration will work for a particular class of functions.

THEOREM

A fixed-point theorem

Suppose that f is defined on an interval $I = [a, b]$ and satisfies the following two conditions:

(i) $f(x)$ belongs to I whenever x belongs to I and

(ii) there exists a constant K with $0 < K < 1$ such that for every u and v in I,

$$|f(u) - f(v)| \le K|u - v|.$$

Then f has a unique fixed point r in I, that is, $f(r) = r$, and starting with any number x_0 in I, the iterates

$$x_1 = f(x_0), \quad x_2 = f(x_1), \quad \ldots$$

converge to r.

You are invited to prove this theorem by a method outlined in Exercises 26 and 27 at the end of this section.

EXAMPLE 2 Show that if $0 < k < 1$, then $f(x) = k \cos x$ satisfies the conditions of Theorem 1 on the interval $I = [0, 1]$. Observe that if $k = 1/5$, the fixed point is that calculated in Example 1 above.

Solution Since $0 < k < 1$, f maps I into I. If u and v are in I, then the Mean-Value Theorem says there exists c between u and v such that

$$|f(u) - f(v)| = |(u - v)f'(c)| = k|u - v| \sin c \le k|u - v|.$$

Thus, the conditions of Theorem 1 are satisfied and f has a fixed point r in $[0, 1]$. Of course, even if $k \ge 1$, f may still have a fixed point in I locatable by iteration, provided the slope of f near that point is less than 1.

Newton's Method

We want to find a **root** of the equation $f(x) = 0$, that is, a number r such that $f(r) = 0$. Such a number is also called a **zero** of the function f. If f is differentiable near the root, then tangent lines can be used to produce a sequence of approximations to the root that approaches the root quite quickly. The idea is as follows. (See Figure 4.12.) Make an initial guess at the root, say $x = x_0$. Draw the tangent line to $y = f(x)$ at $(x_0, f(x_0))$, and find x_1, the x-intercept of this tangent line. Under certain circumstances x_1 will be closer to the root than x_0 was. The process can be repeated over and over to get numbers $x_2, x_3, \ldots$, getting closer and closer to the root r. The number x_{n+1} is the x-intercept of the tangent line to $y = f(x)$ at $(x_n, f(x_n))$.

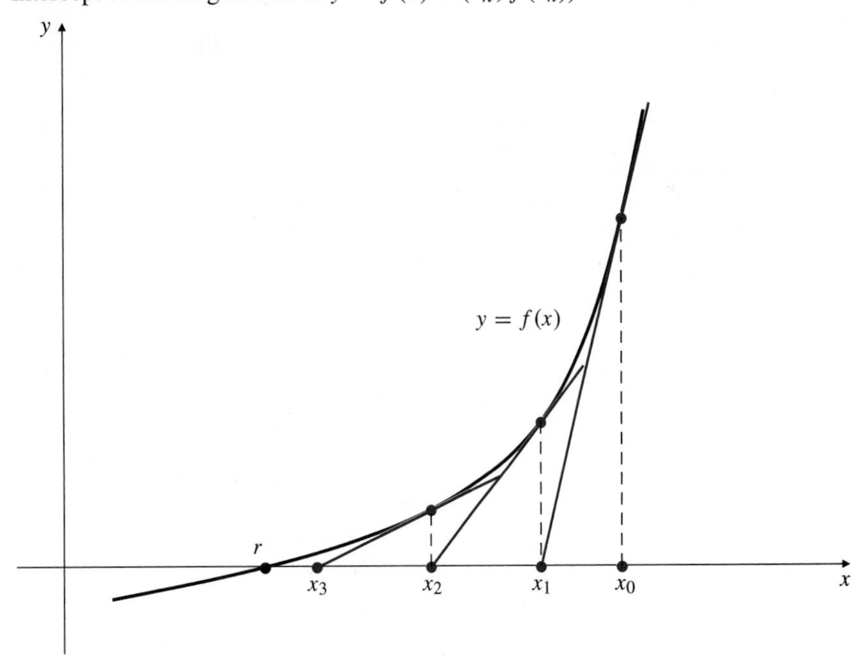

Figure 4.12

The tangent line to $y = f(x)$ at $x = x_0$ has equation

$$y = f(x_0) + f'(x_0)(x - x_0).$$

Since the point $(x_1, 0)$ lies on this line, we have $0 = f(x_0) + f'(x_0)(x_1 - x_0)$. Hence,

$$x_1 = x_0 - \frac{f(x_0)}{f'(x_0)}.$$

Similar formulas produce x_2 from x_1, then x_3 from x_2, and so on. The formula producing x_{n+1} from x_n is the discrete map $x_{n+1} = g(x_n)$, where $g(x) = x - \dfrac{f(x)}{f'(x)}$. That is,

$$x_{n+1} = x_n - \frac{f(x_n)}{f'(x_n)},$$

which is known as the **Newton's Method formula**. If r is a fixed point of g then $f(r) = 0$ and r is a zero of f. We usually use a calculator or computer to calculate the successive approximations $x_1, x_2, x_3, \ldots$, and observe whether these numbers appear to converge to a limit. Convergence will not occur if the graph of f has a horizontal or vertical tangent at any of the numbers in the sequence. However, if $\lim_{n\to\infty} x_n = r$ exists, and if f/f' is continuous near r, then r must be a zero of f. This method is known as **Newton's Method** or **The Newton-Raphson Method**. Since Newton's Method is just a special case of fixed-point iteration applied to the function $g(x)$ defined above, the general properties of fixed-point iteration apply to Newton's Method as well.

EXAMPLE 3 Use Newton's Method to find the only real root of the equation $x^3 - x - 1 = 0$ correct to 10 decimal places.

Solution We have $f(x) = x^3 - x - 1$ and $f'(x) = 3x^2 - 1$. Since f is continuous and since $f(1) = -1$ and $f(2) = 5$, the equation has a root in the interval $[1, 2]$. Figure 4.13 shows that the equation has only one root to the right of $x = 0$. Let us make the initial guess $x_0 = 1.5$. The Newton's Method formula here is

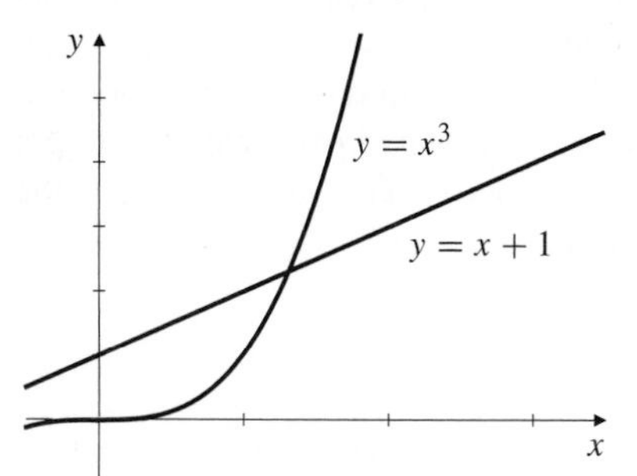

Figure 4.13 The graphs of x^3 and $x+1$ meet only once to the right of $x = 0$, and that meeting is between 1 and 2

$$x_{n+1} = x_n - \frac{x_n^3 - x_n - 1}{3x_n^2 - 1} = \frac{2x_n^3 + 1}{3x_n^2 - 1},$$

so that, for example, the approximation x_1 is given by

$$x_1 = \frac{2(1.5)^3 + 1}{3(1.5)^2 - 1} \approx 1.347\,826\ldots.$$

The values of $x_1, x_2, x_3, \ldots$ are given in Table 2.

Table 2.

n	x_n	$f(x_n)$
0	1.5	0.875 000 000 000 ···
1	1.347 826 086 96 ···	0.100 682 173 091 ···
2	1.325 200 398 95 ···	0.002 058 361 917 ···
3	1.324 718 174 00 ···	0.000 000 924 378 ···
4	1.324 717 957 24 ···	0.000 000 000 000 ···
5	1.324 717 957 24 ···	

The values in Table 2 were obtained with a scientific calculator. Evidently $r = 1.324\,717\,957\,2$ correctly rounded to 10 decimal places.

Observe the behaviour of the numbers x_n. By the third iteration, x_3, we have apparently achieved a precision of 6 decimal places, and by x_4 over 10 decimal places. It is characteristic of Newton's Method that when you begin to get close to the root the convergence can be very rapid. Compare these results with those obtained for the same equation by the Bisection Method in Example 12 of Section 1.4; there we achieved only 3 decimal place precision after 11 iterations.

EXAMPLE 4 Solve the equation $x^3 = \cos x$ to 11 decimal places.

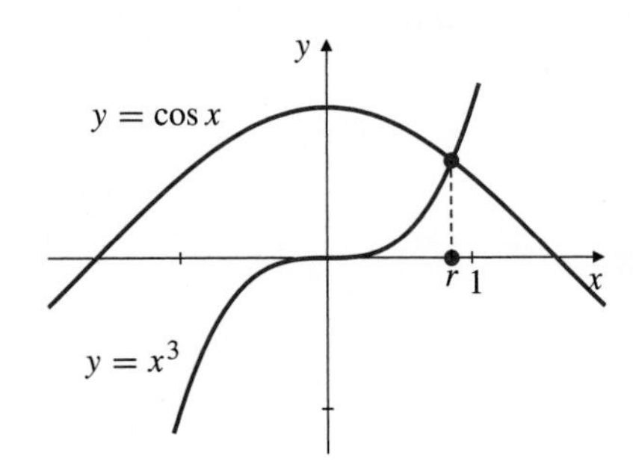

Figure 4.14 Solving $x^3 = \cos x$

Solution We are looking for the x-coordinate r of the intersection of the curves $y = x^3$ and $y = \cos x$. From Figure 4.14 it appears that the curves intersect slightly to the left of $x = 1$. Let us start with the guess $x_0 = 0.8$. If $f(x) = x^3 - \cos x$, then $f'(x) = 3x^2 + \sin x$. The Newton's Method formula for this function is

$$x_{n+1} = x_n - \frac{x_n^3 - \cos x_n}{3x_n^2 + \sin x_n} = \frac{2x_n^3 + x_n \sin x_n + \cos x_n}{3x_n^2 + \sin x_n}.$$

The approximations $x_1, x_2, \ldots$ are given in Table 3:

Table 3.

n	x_n	$f(x_n)$
0	0.8	−0.184 706 709 347 ···
1	0.870 034 801 135 ···	0.013 782 078 762 ···
2	0.865 494 102 425 ···	0.000 006 038 051 ···
3	0.865 474 033 493 ···	0.000 000 001 176 ···
4	0.865 474 033 102 ···	0.000 000 000 000 ···
5	0.865 474 033 102 ···	

The two curves intersect at $x = 0.865\,474\,033\,10$, rounded to 11 decimal places.

Remark Example 4 shows how useful a sketch can be for determining an initial guess x_0. Even a rough sketch of the graph of $y = f(x)$ can show you how many roots the equation $f(x) = 0$ has and approximately where they are. Usually, the closer the initial approximation is to the actual root, the smaller the number of iterations needed to achieve the desired precision. Similarly, for an equation of the form $g(x) = h(x)$, making a sketch of the graphs of g and h (on the same set of axes) can suggest starting approximations for any intersection points. In either case, you can then apply Newton's Method to improve the approximations.

Remark When using Newton's Method to solve an equation that is of the form $g(x) = h(x)$ (such as the one in Example 4), we must rewrite the equation in the form $f(x) = 0$ and apply Newton's Method to f. Usually we just use $f(x) = g(x) - h(x)$, although $f(x) = \big(g(x)/h(x)\big) - 1$ is also a possibility.

Remark If your calculator is programmable, you should learn how to program the Newton's Method formula for a given equation so that generating new iterations requires pressing only a few buttons. If your calculator has graphing capabilities, you can use them to locate a good initial guess.

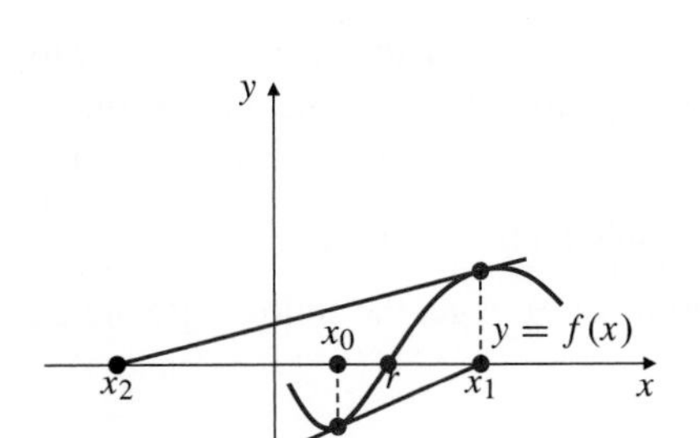

Figure 4.15 Here the Newton's Method iterations do not converge to the root

Newton's Method does not always work as well as it does in the preceding examples. If the first derivative f' is very small near the root, or if the second derivative f'' is very large near the root, a single iteration of the formula can take us from quite close to the root to quite far away. Figure 4.15 illustrates this possibility. (Also see Exercises 21 and 22 at the end of this section.)

Before you try to use Newton's Method to find a real root of a funcion f, you should make sure that a real root actually exists. If you use the method starting with a real initial guess, but the function has no real root nearby, the successive "approximations" can exhibit strange behaviour. The following example illustrates this for a very simple function.

EXAMPLE 5 Consider the function $f(x) = 1 + x^2$. Clearly f has no real roots though it does have complex roots $x = \pm i$. The Newton's Method formula for f is

$$x_{n+1} = x_n - \frac{1 + x_n^2}{2x_n} = \frac{x_n^2 - 1}{2x_n}.$$

If we start with a real guess $x_0 = 2$, iterate this formula 20,000 times, and plot the resulting points (n, x_n), we obtain Figure 4.16, which was done using a Maple procedure. It is clear from this plot that not only do the iterations not converge (as one might otherwise expect), but they do not diverge to ∞ or $-\infty$, and they are not periodic either. This phenomenon is known as **chaos**.

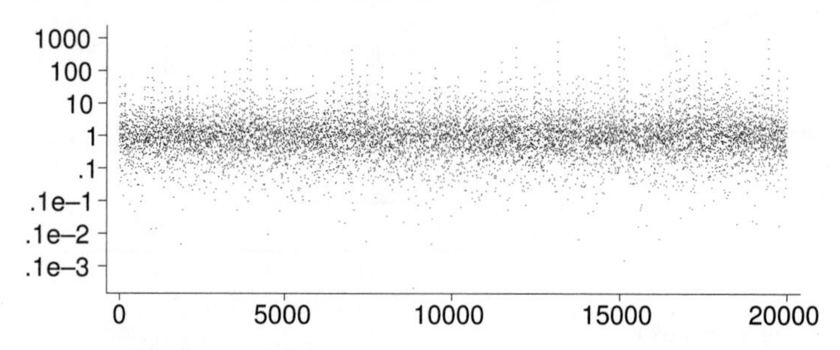

Figure 4.16 Plot of 20,000 points (n, x_n) for Example 5

A definitive characteristic of this phenomenon is sensitivity to initial conditions. To demonstrate this sensitivity in the case at hand we make a change of variables. Let

$$y_n = \frac{1}{1+x_n^2},$$

then the Newton's Method formula for f becomes

$$y_{n+1} = 4y_n(1-y_n),$$

(see Exercise 24), which is a special case of a discrete map called the **logistic map**. It represents one of the best-known and simplest examples of chaos. If, for example, $y_n = \sin^2(u_n)$, for $n = 0,\ 1,\ 2,\ \ldots,$ then it follows (see Exercise 25 below) that $u_n = 2^n u_0$. Unless u_0 is a rational multiple of π, it follows that two different choices of u_0 will lead to differences in the resulting values of u_n that grow exponentially with n. In Exercise 25 it is shown that this sensitivity is carried through to the first order in x_n.

Remark The above example does not imply that Newton's Method cannot be used to find complex roots; the formula simply cannot escape from the real line if a real initial guess is used. To accomodate a complex initial guess, $z_0 = a_0 + ib_0$, we can substitute, $z_n = a_n + ib_n$ into the complex version of Newton's Method formula $z_{n+1} = \dfrac{z_n^2 - 1}{2z_n}$ (see Appendix I for a discussion of complex arithmetic) to get the following coupled equations,

$$a_{n+1} = \frac{a_n^3 + a_n(b_n^2 - 1)}{2(a_n^2 + b_n^2)}$$

$$b_{n+1} = \frac{b_n^3 + b_n(a_n^2 + 1)}{2(a_n^2 + b_n^2)}.$$

With initial guess $z_0 = 1 + i$ the next six members of the sequence of complex numbers (in 14 figure precision) become

$$\begin{aligned}
z_1 &= 0.250\,000\,000\,000\,00 + i\,0.750\,000\,000\,000\,00\\
z_2 &= -0.075\,000\,000\,000\,00 + i\,0.975\,000\,000\,000\,00\\
z_3 &= 0.001\,715\,686\,274\,51 + i\,0.997\,303\,921\,568\,63\\
z_4 &= -0.000\,004\,641\,846\,27 + i\,1.000\,002\,160\,490\,67\\
z_5 &= -0.000\,000\,000\,010\,03 + i\,0.999\,999\,999\,991\,56\\
z_6 &= 0.000\,000\,000\,000\,00 + i\,1.000\,000\,000\,000\,00
\end{aligned}$$

converging to the root $+i$. For an initial guess, $1 - i$, the resulting sequence converges as rapidly to the root $-i$. Note that for the real initial guess $z_0 = 0 + i0$, neither a_1

nor b_1 is defined, so the process fails. This corresponds to the fact that $1 + x^2$ has a horizontal tangent $y = 1$ at $(0, 1)$, and this tangent has no finite x-intercept.

The following theorem gives sufficient conditions for the Newton approximations to converge to a root r of the equation $f(x) = 0$ if the initial guess x_0 is sufficiently close to that root.

THEOREM 2

Error bounds for Newton's Method

Suppose that f, f', and f'' are continuous on an interval I containing x_n, x_{n+1}, and a root $x = r$ of $f(x) = 0$. Suppose also that there exist constants $K > 0$ and $L > 0$ such that for all x in I we have

(i) $|f''(x)| \le K$ and

(ii) $|f'(x)| \ge L$.

Then

(a) $|x_{n+1} - r| \le \dfrac{K}{2L}|x_{n+1} - x_n|^2$ and

(b) $|x_{n+1} - r| \le \dfrac{K}{2L}|x_n - r|^2$.

Conditions (i) and (ii) assert that near r the slope of $y = f(x)$ is not too small in size and does not change too rapidly. If $K/(2L) < 1$, the theorem shows that x_n converges quickly to r once n becomes large enough that $|x_n - r| < 1$.

The proof of Theorem 2 depends on the Mean-Value Theorem. We will not give it since the theorem is of little practical use. In practice, we calculate successive approximations using Newton's formula and observe whether they seem to converge to a limit. If they do, and if the values of f at these approximations approach 0, we can be confident that we have located a root.

"Solve" Routines

Many of the more advanced models of scientific calculators and most computer-based mathematics software have built-in routines for solving general equations numerically or, in a few cases, symbolically. These "Solve" routines assume continuity of the left and right sides of the given equations and often require the user to specify an interval in which to search for the root or an initial guess at the value of the root, or both. Typically the calculator or computer software also has graphing capabilities, and you are expected to use them to get an idea of how many roots the equation has and roughly where they are located before invoking the solving routines. It may also be possible to specify a *tolerance* on the difference of the two sides of the equation. For instance, if we want a solution to the equation $f(x) = 0$, it may be more important to us to be sure that an approximate solution $\hat{x}$ satisfies $|f(\hat{x})| < 0.0001$ than it is to be sure that $\hat{x}$ is within any particular distance of the actual root.

The methods used by the solve routines vary from one calculator or software package to another and are frequently very sophisticated, making use of numerical differentiation and other techniques to find roots very quickly, even when the search interval is large. If you have an advanced scientific calculator and/or computer software with similar capabilities, it is well worth your while to read the manuals that describe how to make effective use of your hardware/software for solving equations. Applications of mathematics to solving "real-world" problems frequently require finding approximate solutions of equations that are intractable by exact methods.

EXERCISES 4.2

Use fixed-point iteration to solve the equations in Exercises 1–6. Obtain 5 decimal place precision.

1. $2x = e^{-x}$, start with $x_0 = 0.3$

2. $1 + \frac{1}{4}\sin x = x$

3. $\cos\dfrac{x}{3} = x$

4. $(x+9)^{1/3} = x$

5. $\dfrac{1}{2+x^2} = x$

6. Solve $x^3 + 10x - 10 = 0$ by rewriting it in the form $1 - \frac{1}{10}x^3 = x$.

In Exercises 7–16, use Newton's Method to solve the given equations to the precision permitted by your calculator.

7. Find $\sqrt{2}$ by solving $x^2 - 2 = 0$.

8. Find $\sqrt{3}$ by solving $x^2 - 3 = 0$.

9. Find the root of $x^3 + 2x - 1 = 0$ between 0 and 1.

10. Find the root of $x^3 + 2x^2 - 2 = 0$ between 0 and 1.

11. Find the two roots of $x^4 - 8x^2 - x + 16 = 0$ in $[1, 3]$.

12. Find the three roots of $x^3 + 3x^2 - 1 = 0$ in $[-3, 1]$.

13. Solve $\sin x = 1 - x$. A sketch can help you make a guess x_0.

14. Solve $\cos x = x^2$. How many roots are there?

15. How many roots does the equation $\tan x = x$ have? Find the one between $\pi/2$ and $3\pi/2$.

16. Solve $\dfrac{1}{1+x^2} = \sqrt{x}$ by rewriting it $(1+x^2)\sqrt{x} - 1 = 0$.

17. If your calculator has a built-in Solve routine, or if you use computer software with such a routine, use it to solve the equations in Exercises 7–16.

Find the maximum and minimum values of the functions in Exercises 18–19.

18. $\dfrac{\sin x}{1+x^2}$

19. $\dfrac{\cos x}{1+x^2}$

20. Let $f(x) = x^2$. The equation $f(x) = 0$ clearly has solution $x = 0$. Find the Newton's Method iterations x_1, x_2, and x_3, starting with $x_0 = 1$.

(a) What is x_n?

(b) How many iterations are needed to find the root with error less than 0.0001 in absolute value?

(c) How many iterations are needed to get an approximation x_n for which $|f(x_n)| < 0.0001$?

(d) Why do the Newton's Method iterations converge more slowly here than in the examples done in this section?

21. **(Oscillation)** Apply Newton's Method to

$$f(x) = \begin{cases} \sqrt{x} & \text{if } x \ge 0, \\ \sqrt{-x} & \text{if } x < 0, \end{cases}$$

starting with the initial guess $x_0 = a > 0$. Calculate x_1 and x_2. What happens? (Make a sketch.) If you ever observed this behaviour when you were using Newton's Method to find a root of an equation, what would you do next?

22. **(Divergent oscillations)** Apply Newton's Method to $f(x) = x^{1/3}$ with $x_0 = 1$. Calculate x_1, x_2, x_3, and x_4. What is happening? Find a formula for x_n.

23. **(Convergent oscillations)** Apply Newton's Method to find $f(x) = x^{2/3}$ with $x_0 = 1$. Calculate x_1, x_2, x_3, and x_4. What is happening? Find a formula for x_n.

24. Verify that the Newton's Method map for $1 + x^2$, namely $x_{n+1} = x_n - \dfrac{1+x_n^2}{2x_n}$, transforms into the logistic map $y_{n+1} = 4y_n(1 - y_n)$ under the transformation $y_n = \dfrac{1}{1+x_n^2}$.

25. Sensitivity to initial conditions is regarded as a definitive property of chaos. If the initial values of two sequences differ, and the differences between the two sequences tends to grow exponentially, the map is said to be sensitive to initial values. Growing exponentially in this sense does not require that each sequence grow exponentially on its own. In fact, for chaos the growth should only be exponential in the differential. Moreover, the growth only needs to be exponential for large n.

a) Show that the logistic map is sensitive to initial conditions by making the substitution $y_j = \sin^2 u_j$ and taking the differential, given that u_0 is not an integral multiple of π.

b) Use part (a) to show that the Newton's Method map for $1 + x^2$ is also sensitive to initial conditions. Make the reasonable assumption, based on Figure 4.16, that the iterates neither converge nor diverge.

Exercises 26–27 constitute a proof of Theorem 1.

26. Condition (ii) of Theorem 1 implies that f is continuous on $I = [a, b]$. Use condition (i) to show that f has a unique fixed point r on I. *Hint:* Apply the Intermediate-Value Theorem to $g(x) = f(x) - x$ on $[a, b]$.

27. Use condition (ii) of Theorem 1 and mathematical induction to show that $|x_n - r| \le K^n|x_0 - r|$. Since $0 < K < 1$, we know that $K^n \to 0$ as $n \to \infty$. This shows that $\lim_{n\to\infty} x_n = r$.

4.3 Indeterminate Forms

In Section 2.5 we showed that

$$\lim_{x\to 0} \frac{\sin x}{x} = 1.$$

We could not readily see this by substituting $x = 0$ into the function $(\sin x)/x$ because both $\sin x$ and x are zero at $x = 0$. We call $(\sin x)/x$ an **indeterminate form** of type $[0/0]$ at $x = 0$. The limit of such an indeterminate form can be any number. For instance, each of the quotients kx/x, x/x^3, and x^3/x^2 is an indeterminate form of type $[0/0]$ at $x = 0$, but

$$\lim_{x\to 0} \frac{kx}{x} = k, \qquad \lim_{x\to 0} \frac{x}{x^3} = \infty, \qquad \lim_{x\to 0} \frac{x^3}{x^2} = 0.$$

There are other types of indeterminate forms. Table 4 lists them together with an example of each type.

Table 4. Types of indeterminate forms

Type	Example
$[0/0]$	$\lim_{x\to 0} \frac{\sin x}{x}$
$[\infty/\infty]$	$\lim_{x\to 0} \frac{\ln(1/x^2)}{\cot(x^2)}$
$[0 \cdot \infty]$	$\lim_{x\to 0+} x \ln \frac{1}{x}$
$[\infty - \infty]$	$\lim_{x\to(\pi/2)-} \left(\tan x - \frac{1}{\pi - 2x}\right)$
$[0^0]$	$\lim_{x\to 0+} x^x$
$[\infty^0]$	$\lim_{x\to(\pi/2)-} (\tan x)^{\cos x}$
$[1^\infty]$	$\lim_{x\to\infty} \left(1 + \frac{1}{x}\right)^x$

Indeterminate forms of type $[0/0]$ are the most common. You can evaluate many indeterminate forms of type $[0/0]$ with simple algebra, typically by cancelling common factors. Examples can be found in Sections 1.2 and 1.3. We will now develop another method called **l'Hôpital's Rules**[1] for evaluating limits of indeterminate forms of the types $[0/0]$ and $[\infty/\infty]$. The other types of indeterminate forms can usually be reduced to one of these two by algebraic manipulation and the taking of logarithms. In Section 4.10 we will discover yet another method for evaluating limits of type $[0/0]$.

l'Hôpital's Rules

THEOREM 3

The first l'Hôpital Rule

Suppose the functions f and g are differentiable on the interval (a, b), and $g'(x) \neq 0$ there. Suppose also that

(i) $\lim_{x\to a+} f(x) = \lim_{x\to a+} g(x) = 0$ and

(ii) $\lim_{x\to a+} \frac{f'(x)}{g'(x)} = L$ (where L is finite or ∞ or $-\infty$).

Then

$$\lim_{x\to a+} \frac{f(x)}{g(x)} = L.$$

Similar results hold if every occurrence of $\lim_{x\to a+}$ is replaced by $\lim_{x\to b-}$ or even $\lim_{x\to c}$ where $a < c < b$. The cases $a = -\infty$ and $b = \infty$ are also allowed.

PROOF We prove the case involving $\lim_{x\to a+}$ for finite a. Define

$$F(x) = \begin{cases} f(x) & \text{if } a < x < b \\ 0 & \text{if } x = a \end{cases} \quad \text{and} \quad G(x) = \begin{cases} g(x) & \text{if } a < x < b \\ 0 & \text{if } x = a \end{cases}$$

Then F and G are continuous on the interval $[a, x]$ and differentiable on the interval (a, x) for every x in (a, b). By the Generalized Mean-Value Theorem (Theorem 16 of Section 2.8) there exists a number c in (a, x) such that

$$\frac{f(x)}{g(x)} = \frac{F(x)}{G(x)} = \frac{F(x) - F(a)}{G(x) - G(a)} = \frac{F'(c)}{G'(c)} = \frac{f'(c)}{g'(c)}.$$

[1] The Marquis de l'Hôpital (1661–1704), for whom these rules are named, published the first textbook on calculus. The circumflex (ˆ) did not come into use in the French language until after the French Revolution. The Marquis would have written his name "l'Hospital."

Since $a < c < x$, if $x \to a+$, then necessarily $c \to a+$, so we have

$$\lim_{x\to a+} \frac{f(x)}{g(x)} = \lim_{c\to a+} \frac{f'(c)}{g'(c)} = L.$$

The case involving $\lim_{x\to b-}$ for finite b is proved similarly. The cases where $a = -\infty$ or $b = \infty$ follow from the cases already considered via the change of variable $x = 1/t$:

$$\lim_{x\to\infty} \frac{f(x)}{g(x)} = \lim_{t\to 0+} \frac{f\left(\frac{1}{t}\right)}{g\left(\frac{1}{t}\right)} = \lim_{t\to 0+} \frac{f'\left(\frac{1}{t}\right)\left(\frac{-1}{t^2}\right)}{g'\left(\frac{1}{t}\right)\left(\frac{-1}{t^2}\right)} = \lim_{x\to\infty} \frac{f'(x)}{g'(x)} = L.$$

EXAMPLE 1 Evaluate $\displaystyle\lim_{x\to 1} \frac{\ln x}{x^2 - 1}$.

Solution We have

$$\lim_{x\to 1} \frac{\ln x}{x^2 - 1} \qquad \left[\frac{0}{0}\right]$$

$$= \lim_{x\to 1} \frac{1/x}{2x} = \lim_{x\to 1} \frac{1}{2x^2} = \frac{1}{2}.$$

BEWARE! Note that in applying l'Hôpital's Rule we calculate the quotient of the derivatives, *not* the derivative of the quotient.

This example illustrates how calculations based on l'Hôpital's Rule are carried out. Having identified the limit as that of a [0/0] indeterminate form, we replace it by the limit of the quotient of derivatives; the existence of this latter limit will justify the equality. It is possible that the limit of the quotient of derivatives may still be indeterminate, in which case a second application of l'Hôpital's Rule can be made. Such applications may be strung out until a limit can finally be extracted, which then justifies all the previous applications of the rule.

EXAMPLE 2 Evaluate $\displaystyle\lim_{x\to 0} \frac{2\sin x - \sin(2x)}{2e^x - 2 - 2x - x^2}$.

Solution We have (using l'Hôpital's Rule three times)

$$\lim_{x\to 0} \frac{2\sin x - \sin(2x)}{2e^x - 2 - 2x - x^2} \qquad \left[\frac{0}{0}\right]$$

$$= \lim_{x\to 0} \frac{2\cos x - 2\cos(2x)}{2e^x - 2 - 2x} \qquad \text{cancel the 2s}$$

$$= \lim_{x\to 0} \frac{\cos x - \cos(2x)}{e^x - 1 - x} \qquad \text{still } \left[\frac{0}{0}\right]$$

$$= \lim_{x\to 0} \frac{-\sin x + 2\sin(2x)}{e^x - 1} \qquad \text{still } \left[\frac{0}{0}\right]$$

$$= \lim_{x\to 0} \frac{-\cos x + 4\cos(2x)}{e^x} = \frac{-1+4}{1} = 3.$$

EXAMPLE 3 Evaluate (a) $\displaystyle\lim_{x\to(\pi/2)-} \frac{2x - \pi}{\cos^2 x}$ and (b) $\displaystyle\lim_{x\to 1+} \frac{x}{\ln x}$.

Solution

(a) $$\lim_{x\to(\pi/2)-} \frac{2x - \pi}{\cos^2 x} \qquad \left[\frac{0}{0}\right]$$

$$= \lim_{x\to(\pi/2)-} \frac{2}{-2\sin x\cos x} = -\infty$$

(b) l'Hôpital's Rule cannot be used to evaluate $\lim_{x\to 1+} x/(\ln x)$ because this is not an indeterminate form. The denominator approaches 0 as $x \to 1+$, but the numerator does not approach 0. Since $\ln x > 0$ for $x > 1$, we have, directly,

$$\lim_{x\to 1+} \frac{x}{\ln x} = \infty.$$

(Had we tried to apply l'Hôpital's Rule, we would have been led to the erroneous answer $\lim_{x\to 1+}(1/(1/x)) = 1$.)

BEWARE! Do not use l'Hôpital's Rule to evaluate a limit that is not indeterminate.

EXAMPLE 4 Evaluate $\lim_{x\to 0+} \left(\frac{1}{x} - \frac{1}{\sin x}\right)$.

Solution The indeterminate form here is of type $[\infty - \infty]$ to which l'Hôpital's Rule cannot be applied. However, it becomes $[0/0]$ after we combine the fractions into one fraction.

$$\begin{aligned}
&\lim_{x\to 0+} \left(\frac{1}{x} - \frac{1}{\sin x}\right) && [\infty - \infty] \\
&= \lim_{x\to 0+} \frac{\sin x - x}{x \sin x} && \left[\frac{0}{0}\right] \\
&= \lim_{x\to 0+} \frac{\cos x - 1}{\sin x + x \cos x} && \left[\frac{0}{0}\right] \\
&= \lim_{x\to 0+} \frac{-\sin x}{2\cos x - x \sin x} = \frac{-0}{2} = 0.
\end{aligned}$$

A version of l'Hôpital's Rule also holds for indeterminate forms of the type $[\infty/\infty]$.

THEOREM 4

The second l'Hôpital Rule

Suppose that f and g are differentiable on the interval (a, b) and that $g'(x) \neq 0$ there. Suppose also that

(i) $\lim_{x\to a+} g(x) = \pm\infty$ and

(ii) $\lim_{x\to a+} \frac{f'(x)}{g'(x)} = L$ (where L is finite, or ∞ or $-\infty$).

Then

$$\lim_{x\to a+} \frac{f(x)}{g(x)} = L.$$

Again, similar results hold for $\lim_{x\to b-}$ and for $\lim_{x\to c}$, and the cases $a = -\infty$ and $b = \infty$ are allowed.

The proof of the second l'Hôpital Rule is technically rather more difficult than that of the first Rule and we will not give it here. A sketch of the proof is outlined in Exercise 35 at the end of this section.

Remark Do *not* try to use l'Hôpital's Rules to evaluate limits that are not indeterminate of type $[0/0]$ or $[\infty/\infty]$; such attempts will almost always lead to false conclusions as observed in Example 3(b) above. (Strictly speaking, the second l'Hôpital Rule can be applied to the form $[a/\infty]$, but there is no point to doing so if a is not infinite, since the limit is obviously 0 in that case.)

Remark No conclusion about $\lim f(x)/g(x)$ can be made using either l'Hôpital Rule if $\lim f'(x)/g'(x)$ does not exist. Other techniques might still be used. For example, $\lim_{x\to 0} (x^2 \sin(1/x))/\sin(x) = 0$ by the Squeeze Theorem even though $\lim_{x\to 0} (2x \sin(1/x) - \cos(1/x))/\cos(x)$ does not exist.

EXAMPLE 5 Evaluate (a) $\lim_{x\to\infty} \frac{x^2}{e^x}$ and (b) $\lim_{x\to 0+} x^a \ln x$, where $a > 0$.

Solution Both of these limits are covered by Theorem 5 in Section 3.4. We do them here by l'Hôpital's Rule.

$$\begin{aligned}
\text{(a)}\quad &\lim_{x\to\infty} \frac{x^2}{e^x} && \left[\frac{\infty}{\infty}\right] \\
&= \lim_{x\to\infty} \frac{2x}{e^x} && \text{still } \left[\frac{\infty}{\infty}\right] \\
&= \lim_{x\to\infty} \frac{2}{e^x} = 0.
\end{aligned}$$

Similarly, one can show that $\lim_{x\to\infty} x^n/e^x = 0$ for any positive integer n by repeated applications of l'Hôpital's Rule.

$$\begin{aligned}
\text{(b)}\quad &\lim_{x\to 0+} x^a \ln x \quad (a > 0) && [0 \cdot (-\infty)] \\
&= \lim_{x\to 0+} \frac{\ln x}{x^{-a}} && \left[\frac{-\infty}{\infty}\right] \\
&= \lim_{x\to 0+} \frac{1/x}{-ax^{-a-1}} = \lim_{x\to 0+} \frac{x^a}{-a} = 0.
\end{aligned}$$

The easiest way to deal with indeterminate forms of types $[0^0]$, $[\infty^0]$, and $[1^\infty]$ is to take logarithms of the expressions involved. The next two examples illustrate the technique.

EXAMPLE 6 Evaluate $\lim_{x\to 0+} x^x$.

Solution This indeterminate form is of type $[0^0]$. Let $y = x^x$. Then

$$\lim_{x\to 0+} \ln y = \lim_{x\to 0+} x \ln x = 0,$$

by Example 5(b). Hence $\lim_{x\to 0} x^x = \lim_{x\to 0+} y = e^0 = 1$.

EXAMPLE 7 Evaluate $\lim_{x\to\infty} \left(1 + \sin\frac{3}{x}\right)^x$.

Solution This indeterminate form is of type 1^∞. Let $y = \left(1 + \sin\frac{3}{x}\right)^x$. Then, taking ln of both sides,

$$\begin{aligned}
\lim_{x\to\infty} \ln y &= \lim_{x\to\infty} x \ln\left(1 + \sin\frac{3}{x}\right) && [\infty \cdot 0] \\
&= \lim_{x\to\infty} \frac{\ln\left(1 + \sin\frac{3}{x}\right)}{\frac{1}{x}} && \left[\frac{0}{0}\right] \\
&= \lim_{x\to\infty} \frac{\frac{1}{1 + \sin\frac{3}{x}}\left(\cos\frac{3}{x}\right)\left(-\frac{3}{x^2}\right)}{-\frac{1}{x^2}} = \lim_{x\to\infty} \frac{3\cos\frac{3}{x}}{1 + \sin\frac{3}{x}} = 3.
\end{aligned}$$

Hence $\lim_{x\to\infty} \left(1 + \sin\frac{3}{x}\right)^x = e^3$.

EXERCISES 4.3

Evaluate the limits in Exercises 1–32.

1. $\lim_{x\to 0} \dfrac{3x}{\tan 4x}$

2. $\lim_{x\to 2} \dfrac{\ln(2x-3)}{x^2-4}$

3. $\lim_{x\to 0} \dfrac{\sin ax}{\sin bx}$

4. $\lim_{x\to 0} \dfrac{1-\cos ax}{1-\cos bx}$

5. $\lim_{x\to 0} \dfrac{\sin^{-1} x}{\tan^{-1} x}$

6. $\lim_{x\to 1} \dfrac{x^{1/3}-1}{x^{2/3}-1}$

7. $\lim_{x\to 0} x\cot x$

8. $\lim_{x\to 0} \dfrac{1-\cos x}{\ln(1+x^2)}$

9. $\lim_{t\to \pi} \dfrac{\sin^2 t}{t-\pi}$

10. $\lim_{x\to 0} \dfrac{10^x-e^x}{x}$

11. $\lim_{x\to \pi/2} \dfrac{\cos 3x}{\pi-2x}$

12. $\lim_{x\to 1} \dfrac{\ln(ex)-1}{\sin \pi x}$

13. $\lim_{x\to \infty} x\sin\dfrac{1}{x}$

14. $\lim_{x\to 0} \dfrac{x-\sin x}{x^3}$

15. $\lim_{x\to 0} \dfrac{x-\sin x}{x-\tan x}$

16. $\lim_{x\to 0} \dfrac{2-x^2-2\cos x}{x^4}$

17. $\lim_{x\to 0+} \dfrac{\sin^2 x}{\tan x - x}$

18. $\lim_{r\to \pi/2} \dfrac{\ln \sin r}{\cos r}$

19. $\lim_{t\to \pi/2} \dfrac{\sin t}{t}$

20. $\lim_{x\to 1-} \dfrac{\arccos x}{x-1}$

21. $\lim_{x\to \infty} x(2\tan^{-1} x - \pi)$

22. $\lim_{t\to (\pi/2)-} (\sec t - \tan t)$

23. $\lim_{t\to 0} \left(\dfrac{1}{t} - \dfrac{1}{te^{at}}\right)$

24. $\lim_{x\to 0+} x^{\sqrt{x}}$

25. $\lim_{x\to 0+} (\csc x)^{\sin^2 x}$

26. $\lim_{x\to 1+} \left(\dfrac{x}{x-1} - \dfrac{1}{\ln x}\right)$

27. $\lim_{t\to 0} \dfrac{3\sin t - \sin 3t}{3\tan t - \tan 3t}$

28. $\lim_{x\to 0} \left(\dfrac{\sin x}{x}\right)^{1/x^2}$

29. $\lim_{t\to 0} (\cos 2t)^{1/t^2}$

30. $\lim_{x\to 0+} \dfrac{\csc x}{\ln x}$

31. $\lim_{x\to 1-} \dfrac{\ln \sin \pi x}{\csc \pi x}$

32. $\lim_{x\to 0} (1+\tan x)^{1/x}$

33. **(A Newton quotient for the second derivative)** Evaluate $\lim_{h\to 0} \dfrac{f(x+h)-2f(x)+f(x-h)}{h^2}$ if f is a twice differentiable function.

34. If f has a continuous third derivative, evaluate

$$\lim_{h\to 0} \frac{f(x+3h)-3f(x+h)+3f(x-h)-f(x-3h)}{h^3}.$$

35. **(Proof of the second l'Hôpital Rule)** Fill in the details of the following outline of a proof of the second l'Hôpital Rule (Theorem 4) for the case where a and L are both finite. Let $a < x < t < b$ and show that there exists c in (x, t) such that

$$\frac{f(x)-f(t)}{g(x)-g(t)} = \frac{f'(c)}{g'(c)}.$$

Now juggle the above equation algebraically into the form

$$\frac{f(x)}{g(x)} - L = \frac{f'(c)}{g'(c)} - L + \frac{1}{g(x)}\left(f(t) - g(t)\frac{f'(c)}{g'(c)}\right).$$

It follows that

$$\left|\frac{f(x)}{g(x)} - L\right| \le \left|\frac{f'(c)}{g'(c)} - L\right| + \frac{1}{|g(x)|}\left(|f(t)| + |g(t)|\left|\frac{f'(c)}{g'(c)}\right|\right).$$

Now show that the right side of the above inequality can be made as small as you wish (say less than a positive number ϵ) by choosing first t and then x close enough to a. Remember, you are given that $\lim_{c\to a+}\left(f'(c)/g'(c)\right) = L$ and $\lim_{x\to a+} |g(x)| = \infty$.

4.4 Extreme Values

The first derivative of a function is a source of much useful information about the behaviour of the function. As we have already seen, the sign of f' tells us whether f is increasing or decreasing. In this section we use this information to find maximum and minimum values of functions. In Section 4.8 we will put the techniques developed here to use solving problems that require finding maximum and minimum values.

Maximum and Minimum Values

Recall (from Section 1.4) that a function has a maximum value at x_0 if $f(x) \le f(x_0)$ for all x in the domain of f. The maximum value is $f(x_0)$. To be more precise, we should call such a maximum value an *absolute* or *global* maximum because it is the largest value that f attains anywhere on its entire domain.

DEFINITION 1

Absolute extreme values

Function f has an **absolute maximum value** $f(x_0)$ at the point x_0 in its domain if $f(x) \leq f(x_0)$ holds for every x in the domain of f.
Similarly, f has an **absolute minimum value** $f(x_1)$ at the point x_1 in its domain if $f(x) \geq f(x_1)$ holds for every x in the domain of f.

A function can have at most one absolute maximum or minimum value, although this value can be assumed at many points. For example, $f(x) = \sin x$ has absolute maximum value 1 occurring at every point of the form $x = (\pi/2) + 2n\pi$, where n is an integer, and an absolute minimum value -1 at every point of the form $x = -(\pi/2) + 2n\pi$. A function need not have any absolute extreme values. The function $f(x) = 1/x$ becomes arbitrarily large as x approaches 0 from the right, so has no finite absolute maximum. (Remember, ∞ is not a number, and is not a value of f.) It doesn't have an absolute minimum either. Even a bounded function may not have an absolute maximum or minimum value. The function $g(x) = x$ with domain specified to be the *open* interval (0, 1) has neither; the range of g is also the interval (0, 1), and there is no largest or smallest number in this interval. Of course, if the domain of g (and therefore also its range) were extended to be the *closed* interval [0, 1], then g would have both a maximum value, 1, and a minimum value, 0.

Maximum and minimum values of a function are collectively referred to as **extreme values**. The following theorem is a restatement (and slight generalization) of Theorem 8 of Section 1.4. It will prove very useful in some circumstances when we want to find extreme values.

THEOREM 5

Existence of extreme values

If the domain of the function f is a *closed, finite interval* or a union of finitely many such intervals, and if f is *continuous* on that domain, then f must have an absolute maximum value and an absolute minimum value.

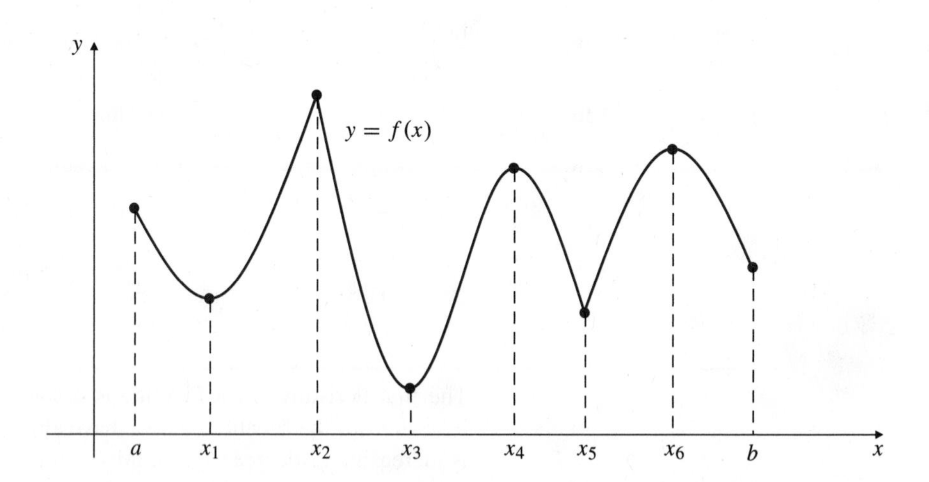

Figure 4.17 Local extreme values

Consider the graph $y = f(x)$ shown in Figure 4.17. Evidently the absolute maximum value of f is $f(x_2)$, and the absolute minimum value is $f(x_3)$. In addition to these extreme values, f has several other "local" maximum and minimum values corresponding to points on the graph that are higher or lower than neighbouring points. Observe that f has *local maximum values* at a, x_2, x_4, and x_6 and local minimum values at x_1, x_3, x_5, and b. The absolute maximum is the highest of the local maxima; the absolute minimum is the lowest of the local minima.

DEFINITION 2

Local extreme values

Function f has a **local maximum value (loc max)** $f(x_0)$ at the point x_0 in its domain provided there exists a number $h > 0$ such that $f(x) \leq f(x_0)$ whenever x is in the domain of f and $|x - x_0| < h$.
Similarly, f has a **local minimum value (loc min)** $f(x_1)$ at the point x_1 in its domain provided there exists a number $h > 0$ such that $f(x) \geq f(x_1)$ whenever x is in the domain of f and $|x - x_1| < h$.

Thus, f has a local maximum (or minimum) value at x if it has an absolute maximum (or minimum) value at x when its domain is restricted to points sufficiently near x. Geometrically, the graph of f is at least as high (or low) at x as it is at nearby points.

Critical Points, Singular Points, and Endpoints

Figure 4.17 suggests that a function $f(x)$ can have local extreme values only at points x of three special types:

(i) **critical points** of f (points x in $\mathcal{D}(f)$ where $f'(x) = 0$),

(ii) **singular points** of f (points x in $\mathcal{D}(f)$ where $f'(x)$ is not defined), and

(iii) **endpoints** of the domain of f (points in $\mathcal{D}(f)$ that do not belong to any open interval contained in $\mathcal{D}(f)$).

In Figure 4.17, x_1, x_3, x_4, and x_6 are critical points, x_2 and x_5 are singular points, and a and b are endpoints.

THEOREM 6

Locating extreme values

If the function f is defined on an interval I and has a local maximum (or local minimum) value at point $x = x_0$ in I, then x_0 must be either a critical point of f, a singular point of f, or an endpoint of I.

PROOF Suppose that f has a local maximum value at x_0 and that x_0 is neither an endpoint of the domain of f nor a singular point of f. Then for some $h > 0$, $f(x)$ is defined on the open interval $(x_0 - h, x_0 + h)$ and has an absolute maximum (for that interval) at x_0. Also, $f'(x_0)$ exists. By Theorem 14 of Section 2.8, $f'(x_0) = 0$. The proof for the case where f has a local minimum value at x_0 is similar. ∎

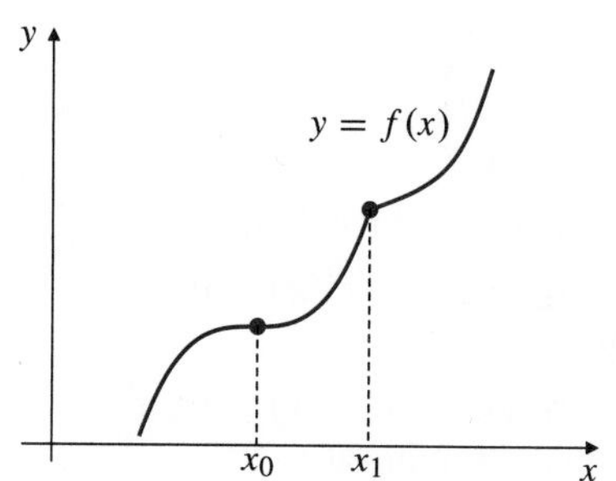

Figure 4.18 A function need not have extreme values at a critical point or a singular point

Although a function cannot have extreme values anywhere other than at endpoints, critical points, and singular points, it need not have extreme values at such points. Figure 4.18 shows the graph of a function with a critical point x_0 and a singular point x_1 at neither of which it has an extreme value. It is more difficult to draw the graph of a function whose domain has an endpoint at which the function fails to have an extreme value. See Exercise 49 at the end of this section for an example of such a function.

Finding Absolute Extreme Values

If a function f is defined on a closed interval or a union of finitely many closed intervals, Theorem 5 assures us that f must have an absolute maximum value and an absolute minimum value. Theorem 6 tells us how to find them. We need only check the values of f at any critical points, singular points, and endpoints.

EXAMPLE 1 Find the maximum and minimum values of the function $g(x) = x^3 - 3x^2 - 9x + 2$ on the interval $-2 \leq x \leq 2$.

Solution Since g is a polynomial, it can have no singular points. For critical points,

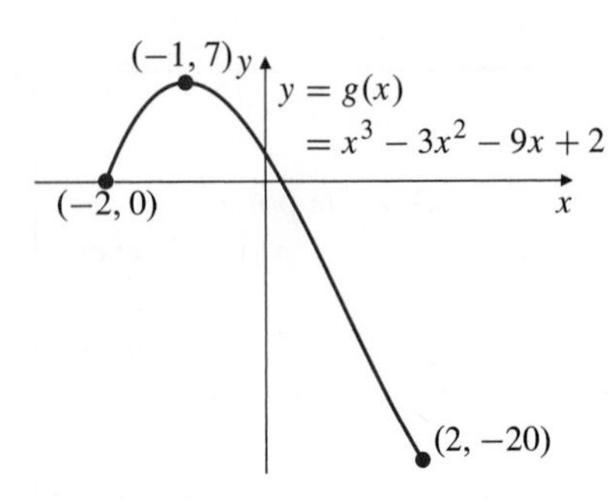

Figure 4.19 g has maximum and minimum values 7 and -20, respectively

we calculate

$$\begin{aligned} g'(x) &= 3x^2 - 6x - 9 = 3(x^2 - 2x - 3) \\ &= 3(x+1)(x-3) \\ &= 0 \quad \text{if} \quad x = -1 \text{ or } x = 3. \end{aligned}$$

However, $x = 3$ is not in the domain of g, so we can ignore it. We need to consider only the values of g at the critical point $x = -1$ and at the endpoints $x = -2$ and $x = 2$:

$$g(-2) = 0, \qquad g(-1) = 7, \qquad g(2) = -20.$$

The maximum value of $g(x)$ on $-2 \le x \le 2$ is 7, at the critical point $x = -1$, and the minimum value is -20, at the endpoint $x = 2$. See Figure 4.19.

EXAMPLE 2 Find the maximum and minimum values of $h(x) = 3x^{2/3} - 2x$ on the interval $[-1, 1]$.

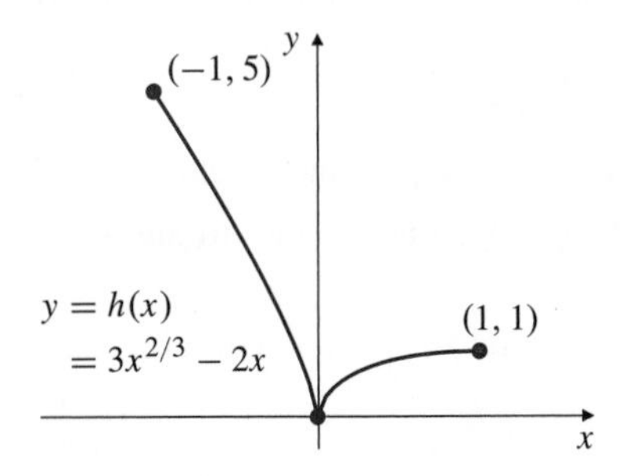

Figure 4.20 h has absolute minimum value 0 at a singular point

Solution The derivative of h is

$$h'(x) = 3\left(\frac{2}{3}\right)x^{-1/3} - 2 = 2(x^{-1/3} - 1).$$

Note that $x^{-1/3}$ is not defined at the point $x = 0$ in $\mathcal{D}(h)$, so $x = 0$ is a singular point of h. Also, h has a critical point where $x^{-1/3} = 1$, that is, at $x = 1$, which also happens to be an endpoint of the domain of h. We must therefore examine the values of h at the points $x = 0$ and $x = 1$, as well as at the other endpoint $x = -1$. We have

$$h(-1) = 5, \qquad h(0) = 0, \qquad h(1) = 1.$$

The function h has maximum value 5 at the endpoint -1 and minimum value 0 at the singular point $x = 0$. See Figure 4.20.

The First Derivative Test

Most functions you will encounter in elementary calculus have nonzero derivatives everywhere on their domains except possibly at a finite number of critical points, singular points, and endpoints of their domains. On intervals between these points the derivative exists and is not zero, so the function is either increasing or decreasing there. If f is continuous and increases to the left of x_0 and decreases to the right, then it must have a local maximum value at x_0. The following theorem collects several results of this type together.

THEOREM 7 **The First Derivative Test**

PART I. Testing interior critical points and singular points.

Suppose that f is continuous at x_0, and x_0 is not an endpoint of the domain of f.

(a) If there exists an open interval (a, b) containing x_0 such that $f'(x) > 0$ on (a, x_0) and $f'(x) < 0$ on (x_0, b), then f has a local maximum value at x_0.

(b) If there exists an open interval (a, b) containing x_0 such that $f'(x) < 0$ on (a, x_0) and $f'(x) > 0$ on (x_0, b), then f has a local minimum value at x_0.

PART II. Testing endpoints of the domain.

Suppose a is a left endpoint of the domain of f and f is right continuous at a.

(c) If $f'(x) > 0$ on some interval (a, b), then f has a local minimum value at a.

(d) If $f'(x) < 0$ on some interval (a, b), then f has a local maximum value at a.

Suppose b is a right endpoint of the domain of f and f is left continuous at b.

(e) If $f'(x) > 0$ on some interval (a, b), then f has a local maximum value at b.

(f) If $f'(x) < 0$ on some interval (a, b), then f has a local minimum value at b.

Remark If f' is positive (or negative) on *both* sides of a critical or singular point, then f has neither a maximum nor a minimum value at that point.

EXAMPLE 3 Find the local and absolute extreme values of $f(x) = x^4 - 2x^2 - 3$ on the interval $[-2, 2]$. Sketch the graph of f.

Solution We begin by calculating and factoring the derivative $f'(x)$:

$$f'(x) = 4x^3 - 4x = 4x(x^2 - 1) = 4x(x - 1)(x + 1).$$

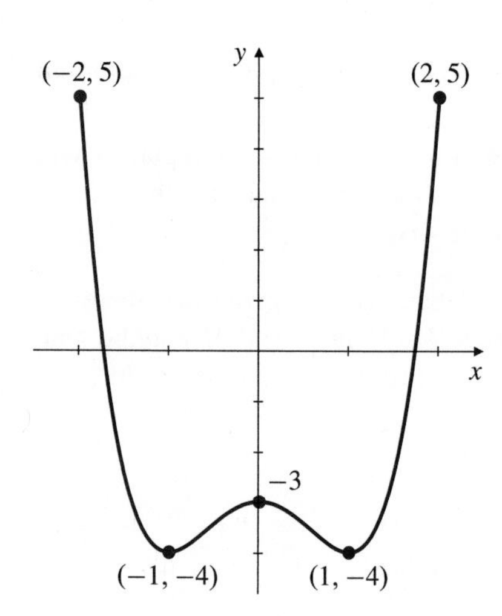

Figure 4.21 The graph $y = x^4 - 2x^2 - 3$

The critical points are 0, -1, and 1. The corresponding values are $f(0) = -3$, $f(-1) = f(1) = -4$. There are no singular points. The values of f at the endpoints -2 and 2 are $f(-2) = f(2) = 5$. The factored form of $f'(x)$ is also convenient for determining the sign of $f'(x)$ on intervals between these endpoints and critical points. Where an odd number of the factors of $f'(x)$ are negative, $f'(x)$ will itself be negative; where an even number of factors are negative, $f'(x)$ will be positive. We summarize the positive/negative properties of $f'(x)$ and the implied increasing/decreasing behaviour of $f(x)$ in chart form:

	EP		CP		CP		CP		EP
x	-2		-1		0		1		2
f'		$-$	0	$+$	0	$-$	0	$+$	
f	max	↘	min	↗	max	↘	min	↗	max

Note how the sloping arrows indicate visually the appropriate classification of the endpoints (EP) and critical points (CP) as determined by the First Derivative Test. We will make extensive use of such charts in future sections. The graph of f is shown in Figure 4.21. Since the domain is a closed, finite interval, f must have absolute maximum and minimum values. These are 5 (at ± 2) and -4 (at ± 1).

EXAMPLE 4 Find and classify the local and absolute extreme values of the function $f(x) = x - x^{2/3}$ with domain $[-1, 2]$. Sketch the graph of f.

Solution $f'(x) = 1 - \frac{2}{3}x^{-1/3} = \left(x^{1/3} - \frac{2}{3}\right)/x^{1/3}$. There is a singular point, $x = 0$, and a critical point, $x = 8/27$. The endpoints are $x = -1$ and $x = 2$. The values of f at these points are $f(-1) = -2$, $f(0) = 0$, $f(8/27) = -4/27$, and $f(2) = 2 - 2^{2/3} \approx 0.4126$ (see Figure 4.22). Another interesting point on the graph is the x-intercept at $x = 1$. Information from f' is summarized in the chart:

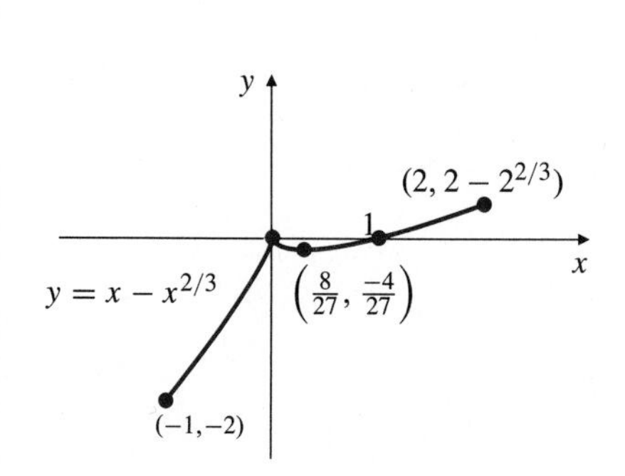

Figure 4.22 The graph for Example 4

	EP		SP		CP		EP
x	-1		0		8/27		2
f'		$+$	undef	$-$	0	$+$	
f	min	↗	max	↘	min	↗	max

There are two local minima and two local maxima. The absolute maximum of f is $2 - 2^{2/3}$ at $x = 2$; the absolute minimum is -2 at $x = -1$.

Functions Not Defined on Closed, Finite Intervals

If the function f is not defined on a closed, finite interval, then Theorem 5 cannot be used to guarantee the existence of maximum and minimum values for f. Of course, f may still have such extreme values. In many applied situations we will want to find extreme values of functions defined on infinite and/or open intervals. The following theorem adapts Theorem 5 to cover some such situations.

THEOREM 8

Existence of extreme values on open intervals

If f is continuous on the open interval (a, b), and if

$$\lim_{x \to a+} f(x) = L \qquad \text{and} \qquad \lim_{x \to b-} f(x) = M,$$

then the following conclusions hold:

(i) If $f(u) > L$ and $f(u) > M$ for some u in (a, b), then f has an absolute maximum value on (a, b).

(ii) If $f(v) < L$ and $f(v) < M$ for some v in (a, b), then f has an absolute minimum value on (a, b).

In this theorem a may be $-\infty$, in which case $\lim_{x \to a+}$ should be replaced with $\lim_{x \to -\infty}$, and b may be ∞, in which case $\lim_{x \to b-}$ should be replaced with $\lim_{x \to \infty}$. Also, either or both of L and M may be either ∞ or $-\infty$.

PROOF We prove part (i); the proof of (ii) is similar. We are given that there is a number u in (a, b) such that $f(u) > L$ and $f(u) > M$. Here, L and M may be finite numbers or $-\infty$. Since $\lim_{x \to a+} f(x) = L$, there must exist a number x_1 in (a, u) such that

$$f(x) < f(u) \qquad \text{whenever} \quad a < x < x_1.$$

Similarly, there must exist a number x_2 in (u, b) such that

$$f(x) < f(u) \qquad \text{whenever} \quad x_2 < x < b.$$

(See Figure 4.23.) Thus, $f(x) < f(u)$ at all points of (a, b) that are not in the closed, finite subinterval $[x_1, x_2]$. By Theorem 5, the function f, being continuous on $[x_1, x_2]$, must have an absolute maximum value on that interval, say at the point w. Since u belongs to $[x_1, x_2]$, we must have $f(w) \ge f(u)$, so $f(w)$ is the maximum value of $f(x)$ for all of (a, b).

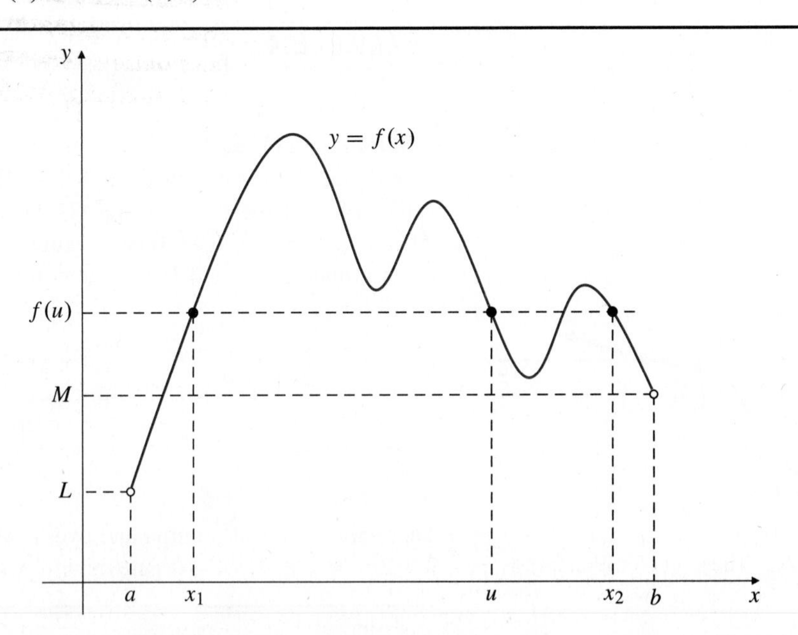

Figure 4.23

Theorem 6 still tells us where to look for extreme values. There are no endpoints to consider in an open interval, but we must still look at the values of the function at any critical points or singular points in the interval.

EXAMPLE 5 Show that $f(x) = x + (4/x)$ has an absolute minimum value on the interval $(0, \infty)$, and find that minimum value.

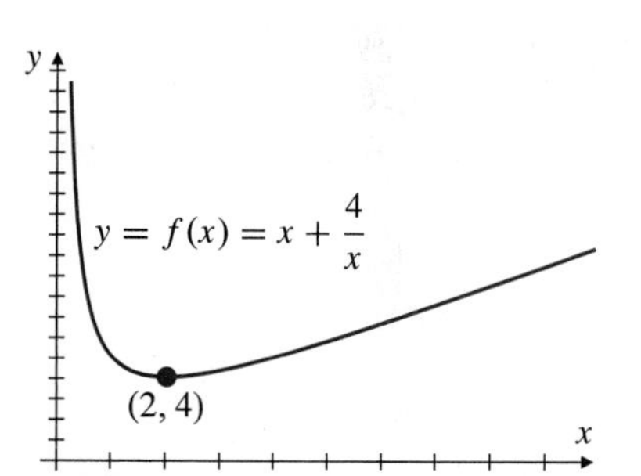

Figure 4.24 f has minimum value 4 at $x = 2$

Solution We have

$$\lim_{x \to 0+} f(x) = \infty \qquad \text{and} \qquad \lim_{x \to \infty} f(x) = \infty.$$

Since $f(1) = 5 < \infty$, Theorem 8 guarantees that f must have an absolute minimum value at some point in $(0, \infty)$. To find the minimum value we must check the values of f at any critical points or singular points in the interval. We have

$$f'(x) = 1 - \frac{4}{x^2} = \frac{x^2 - 4}{x^2} = \frac{(x-2)(x+2)}{x^2},$$

which equals 0 only at $x = 2$ and $x = -2$. Since f has domain $(0, \infty)$, it has no singular points and only one critical point, namely, $x = 2$, where f has the value $f(2) = 4$. This must be the minimum value of f on $(0, \infty)$. (See Figure 4.24.)

EXAMPLE 6 Let $f(x) = x\,e^{-x^2}$. Find and classify the critical points of f, evaluate $\lim_{x \to \pm\infty} f(x)$, and use these results to help you sketch the graph of f.

Solution $f'(x) = e^{-x^2}(1 - 2x^2) = 0$ only if $1 - 2x^2 = 0$ since the exponential is always positive. Thus, the critical points are $\pm\frac{1}{\sqrt{2}}$. We have $f\left(\pm\frac{1}{\sqrt{2}}\right) = \pm\frac{1}{\sqrt{2e}}$. f' is positive (or negative) when $1 - 2x^2$ is positive (or negative). We summarize the intervals where f is increasing and decreasing in chart form:

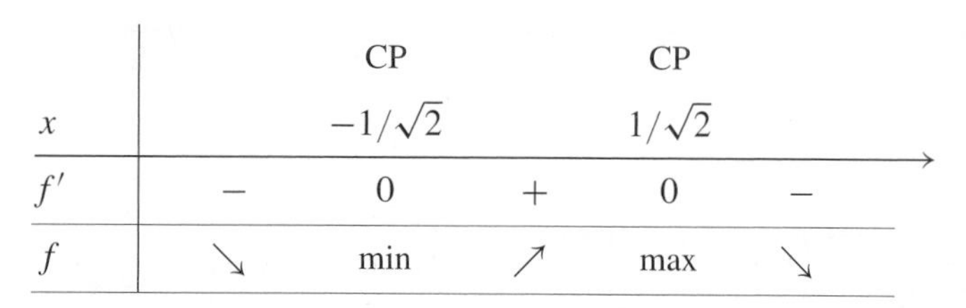

		CP		CP	
x		$-1/\sqrt{2}$		$1/\sqrt{2}$	
f'	$-$	0	$+$	0	$-$
f	↘	min	↗	max	↘

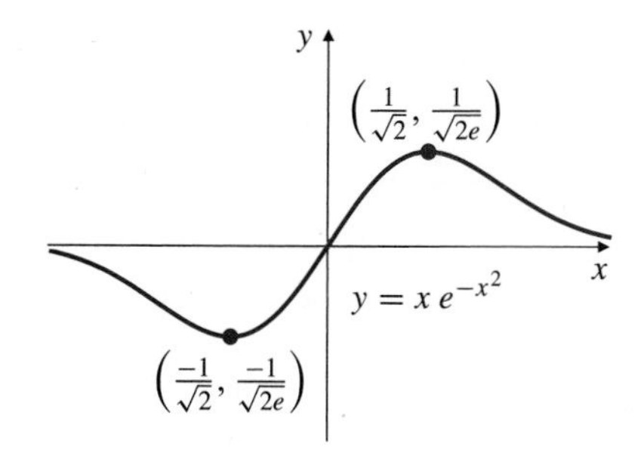

Figure 4.25 The graph for Example 6

Note that $f(0) = 0$ and that f is an odd function ($f(-x) = -f(x)$), so the graph is symmetric about the origin. Also,

$$\lim_{x \to \pm\infty} x\,e^{-x^2} = \left(\lim_{x \to \pm\infty} \frac{1}{x}\right)\left(\lim_{x \to \pm\infty} \frac{x^2}{e^{x^2}}\right) = 0 \times 0 = 0$$

because $\lim_{x \to \pm\infty} x^2 e^{-x^2} = \lim_{u \to \infty} u\,e^{-u} = 0$ by Theorem 5 of Section 3.4. Since $f(x)$ is positive at $x = 1/\sqrt{2}$ and is negative at $x = -1/\sqrt{2}$, f must have absolute maximum and minimum values by Theorem 8. These values can only be the values $\pm 1/\sqrt{2e}$ at the two critical points. The graph is shown in Figure 4.25. The x-axis is an asymptote as $x \to \pm\infty$.

EXERCISES 4.4

In Exercises 1–17, determine whether the given function has any local or absolute extreme values, and find those values if possible.

1. $f(x) = x + 2$ on $[-1, 1]$ **2.** $f(x) = x + 2$ on $(-\infty, 0]$

3. $f(x) = x + 2$ on $[-1, 1)$ **4.** $f(x) = x^2 - 1$

5. $f(x) = x^2 - 1$ on $[-2, 3]$ **6.** $f(x) = x^2 - 1$ on $(2, 3)$

7. $f(x) = x^3 + x - 4$ on $[a, b]$

8. $f(x) = x^3 + x - 4$ on (a, b)

9. $f(x) = x^5 + x^3 + 2x$ on $(a, b]$

10. $f(x) = \dfrac{1}{x - 1}$ **11.** $f(x) = \dfrac{1}{x - 1}$ on $(0, 1)$

12. $f(x) = \dfrac{1}{x - 1}$ on $[2, 3]$ **13.** $f(x) = |x - 1|$ on $[-2, 2]$

14. $|x^2 - x - 2|$ on $[-3, 3]$ **15.** $f(x) = \dfrac{1}{x^2 + 1}$

16. $f(x) = (x + 2)^{2/3}$ **17.** $f(x) = (x - 2)^{1/3}$

In Exercises 18–40, locate and classify all local extreme values of the given function. Determine whether any of these extreme values are absolute. Sketch the graph of the function.

18. $f(x) = x^2 + 2x$ **19.** $f(x) = x^3 - 3x - 2$

20. $f(x) = (x^2 - 4)^2$ **21.** $f(x) = x^3(x - 1)^2$

22. $f(x) = x^2(x - 1)^2$ **23.** $f(x) = x(x^2 - 1)^2$

24. $f(x) = \dfrac{x}{x^2 + 1}$ **25.** $f(x) = \dfrac{x^2}{x^2 + 1}$

26. $f(x) = \dfrac{x}{\sqrt{x^4 + 1}}$ **27.** $f(x) = x\sqrt{2 - x^2}$

28. $f(x) = x + \sin x$ **29.** $f(x) = x - 2\sin x$

30. $f(x) = x - 2\tan^{-1} x$ **31.** $f(x) = 2x - \sin^{-1} x$

32. $f(x) = e^{-x^2/2}$ **33.** $f(x) = x\,2^{-x}$

34. $f(x) = x^2 e^{-x^2}$ **35.** $f(x) = \dfrac{\ln x}{x}$

36. $f(x) = |x + 1|$ **37.** $f(x) = |x^2 - 1|$

38. $f(x) = \sin|x|$ **39.** $f(x) = |\sin x|$

40. $f(x) = (x - 1)^{2/3} - (x + 1)^{2/3}$

In Exercises 41–46, determine whether the given function has absolute maximum or absolute minimum values. Justify your answers. Find the extreme values if you can.

41. $\dfrac{x}{\sqrt{x^2 + 1}}$ **42.** $\dfrac{x}{\sqrt{x^4 + 1}}$

43. $x\sqrt{4 - x^2}$ **44.** $\dfrac{x^2}{\sqrt{4 - x^2}}$

45. $\dfrac{1}{x \sin x}$ on $(0, \pi)$ **46.** $\dfrac{\sin x}{x}$

47. If a function has an absolute maximum value, must it have any local maximum values? If a function has a local maximum value, must it have an absolute maximum value? Give reasons for your answers.

48. If the function f has an absolute maximum value and $g(x) = |f(x)|$, must g have an absolute maximum value? Justify your answer.

49. **(A function with no max or min at an endpoint)** Let

$$f(x) = \begin{cases} x \sin \dfrac{1}{x} & \text{if } x > 0 \\ 0 & \text{if } x = 0. \end{cases}$$

Show that f is continuous on $[0, \infty)$ and differentiable on $(0, \infty)$ but that it has neither a local maximum nor a local minimum value at the endpoint $x = 0$.

4.5 Concavity and Inflections

Like the first derivative, the second derivative of a function also provides useful information about the behaviour of the function and the shape of its graph: it determines whether the graph is *bending upward* (i.e., has increasing slope) or *bending downward* (i.e., has decreasing slope) as we move along the graph toward the right.

DEFINITION 3

We say that the function f is **concave up** on an open interval I if it is differentiable there and the derivative f' is an increasing function on I. Similarly, f is **concave down** on I if f' exists and is decreasing on I.

The terms "concave up" and "concave down" are used to describe the graph of the function as well as the function itself.

Note that concavity is defined only for differentiable functions, and even for those, only on intervals on which their derivatives are not constant. According to the above definition, a function is neither concave up nor concave down on an interval where its graph is a straight line segment. We say the function has no concavity on such an interval. We also say a function has opposite concavity on two intervals if it is concave up on one interval and concave down on the other.

The function f whose graph is shown in Figure 4.26 is concave up on the interval (a, b) and concave down on the interval (b, c).

Some geometric observations can be made about concavity:

(i) If f is concave up on an interval, then, on that interval, the graph of f lies above its tangents, and chords joining points on the graph lie above the graph.

(ii) If f is concave down on an interval, then, on that interval, the graph of f lies below its tangents, and chords to the graph lie below the graph.

(iii) If the graph of f has a tangent at a point, and if the concavity of f is opposite on opposite sides of that point, then the graph crosses its tangent at that point. (This occurs at the point $(b, f(b))$ in Figure 4.26. Such a point is called an *inflection point* of the graph of f.)

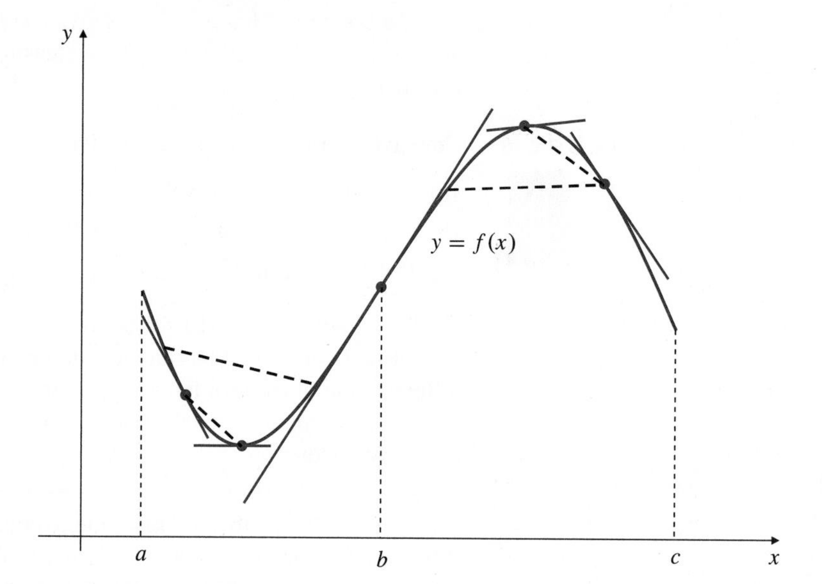

Figure 4.26 f is concave up on (a, b) and concave down on (b, c)

DEFINITION 4

Inflection points

We say that the point $(x_0, f(x_0))$ *is* an **inflection point** of the curve $y = f(x)$ (or that the function f *has* an **inflection point** at x_0) if the following two conditions are satisfied:

(a) the graph of $y = f(x)$ has a tangent line at $x = x_0$, and

(b) the concavity of f is opposite on opposite sides of x_0.

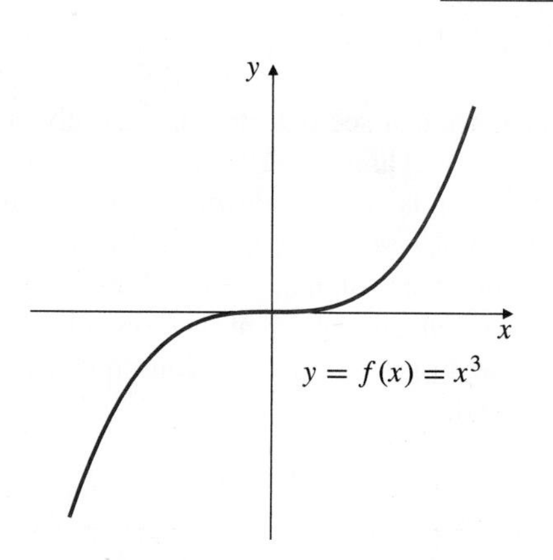

Figure 4.27 $x = 0$ is a critical point of $f(x) = x^3$, and f has an inflection point there

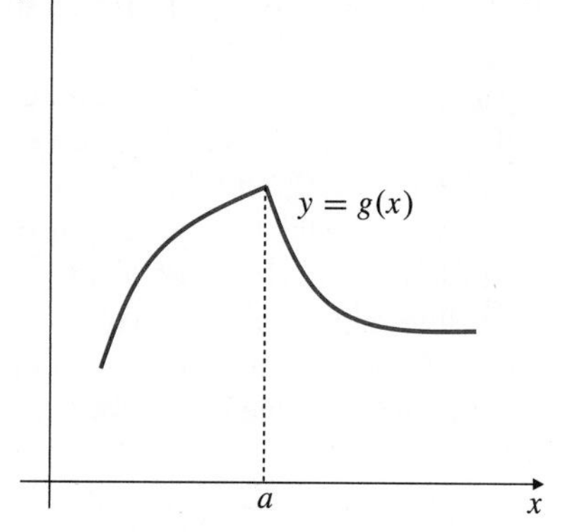

Figure 4.28 The concavity of g is opposite on opposite sides of the singular point a, but its graph has no tangent and therefore no inflection point there

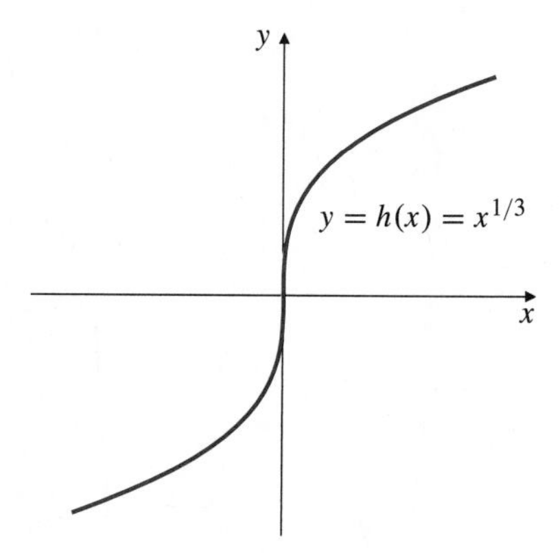

Figure 4.29 This graph of h has an inflection point at the origin even though $x = 0$ is a singular point of h

Note that (a) implies that either f is differentiable at x_0 or its graph has a vertical tangent line there, and (b) implies that the graph crosses its tangent line at x_0. An inflection point of a function f is a point on the graph of a function, rather than a point in its domain like a critical point or a singular point. A function may or may not have an inflection point at a critical point or singular point. In general, a point P is an inflection point (or simply *an inflection*) of a curve C (which is not necessarily the graph of a function) if C has a tangent at P and arcs of C extending in opposite directions from P are on opposite sides of that tangent line.

Figures 4.27–4.29 illustrate some situations involving critical and singular points and inflections.

If a function f has a second derivative f'', the sign of that second derivative tells us whether the first derivative f' is increasing or decreasing and hence determines the concavity of f.

THEOREM 9

Concavity and the second derivative

(a) If $f''(x) > 0$ on interval I, then f is concave up on I.

(b) If $f''(x) < 0$ on interval I, then f is concave down on I.

(c) If f has an inflection point at x_0 and $f''(x_0)$ exists, then $f''(x_0) = 0$.

PROOF Parts (a) and (b) follow from applying Theorem 12 of Section 2.8 to the derivative f' of f. If f has an inflection point at x_0 and $f''(x_0)$ exists, then f must be differentiable in an open interval containing x_0. Since f' is increasing on one side of x_0 and decreasing on the other side, it must have a local maximum or minimum value at x_0. By Theorem 6, $f''(x_0) = 0$. ■

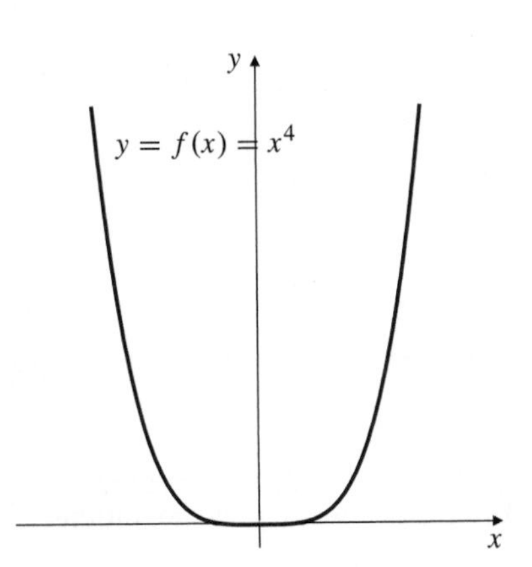

Figure 4.30 $f''(0) = 0$, but f does not have an inflection point at 0

Theorem 9 tells us that to find (the x-coordinates of) inflection points of a twice differentiable function f, we need only look at points where $f''(x) = 0$. However, not every such point has to be an inflection point. For example, $f(x) = x^4$, whose graph is shown in Figure 4.30, does not have an inflection point at $x = 0$ even though $f''(0) = 12x^2|_{x=0} = 0$. In fact, x^4 is concave up on every interval.

EXAMPLE 1 Determine the intervals of concavity of $f(x) = x^6 - 10x^4$ and the inflection points of its graph.

Solution We have

$$f'(x) = 6x^5 - 40x^3,$$
$$f''(x) = 30x^4 - 120x^2 = 30x^2(x-2)(x+2).$$

Having factored $f''(x)$ in this manner, we can see that it vanishes only at $x = -2$, $x = 0$, and $x = 2$. On the intervals $(-\infty, -2)$ and $(2, \infty)$, $f''(x) > 0$ so f is concave up. On $(-2, 0)$ and $(0, 2)$, $f''(x) < 0$ so f is concave down. $f''(x)$ changes sign as we pass through -2 and 2. Since $f(\pm 2) = -96$, the graph of f has inflection points at $(\pm 2, -96)$. However, $f''(x)$ does not change sign at $x = 0$, since $x^2 > 0$ for both positive and negative x. Thus there is no inflection point at 0. As was the case for the first derivative, information about the sign of $f''(x)$ and the consequent concavity of f can be conveniently conveyed in a chart:

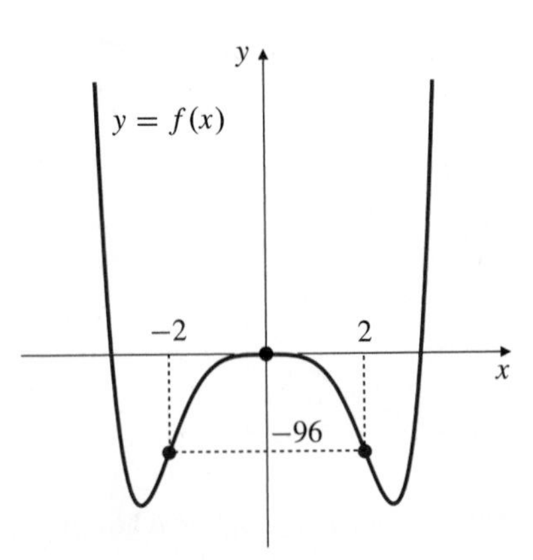

Figure 4.31 The graph of $f(x) = x^6 - 10x^4$

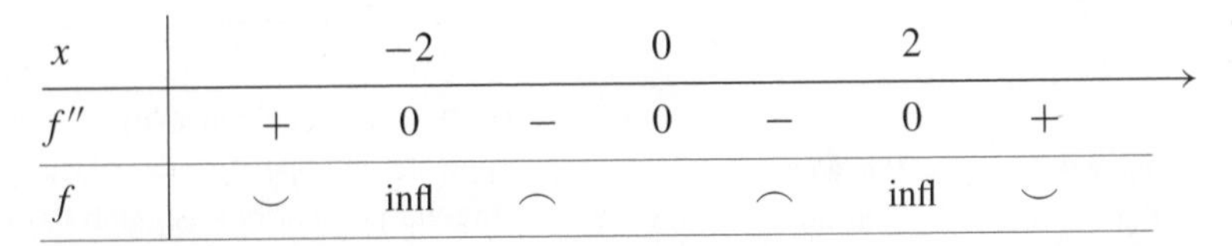

x		-2		0		2	
f''	$+$	0	$-$	0	$-$	0	$+$
f	⌣	infl	⌢		⌢	infl	⌣

The graph of f is sketched in Figure 4.31. ●

EXAMPLE 2 Determine the intervals of increase and decrease, the local extreme values, and the concavity of $f(x) = x^4 - 2x^3 + 1$. Use the information to sketch the graph of f.

Solution

$$f'(x) = 4x^3 - 6x^2 = 2x^2(2x - 3) = 0 \quad \text{at } x = 0 \text{ and } x = 3/2,$$
$$f''(x) = 12x^2 - 12x = 12x(x - 1) = 0 \quad \text{at } x = 0 \text{ and } x = 1.$$

The behaviour of f is summarized in the following chart:

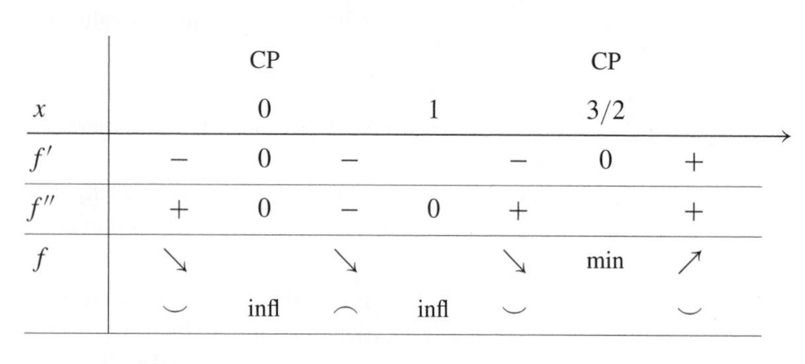

		CP				CP	
x		0		1		3/2	
f'	−	0	−		−	0	+
f''	+	0	−	0	+		+
f	↘		↘		↘	min	↗
	⌣	infl	⌢	infl	⌣		⌣

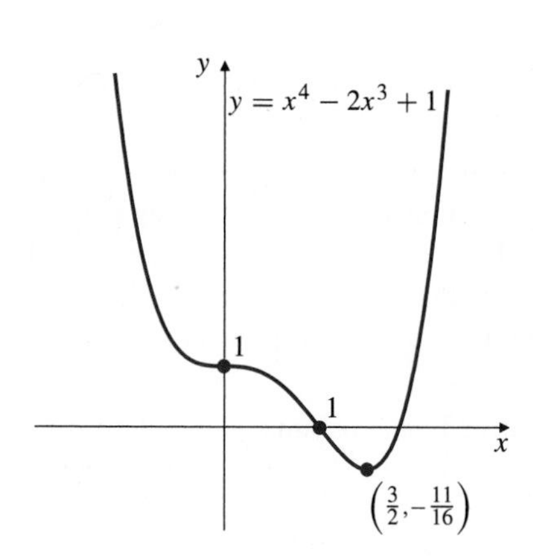

Figure 4.32 The function of Example 2

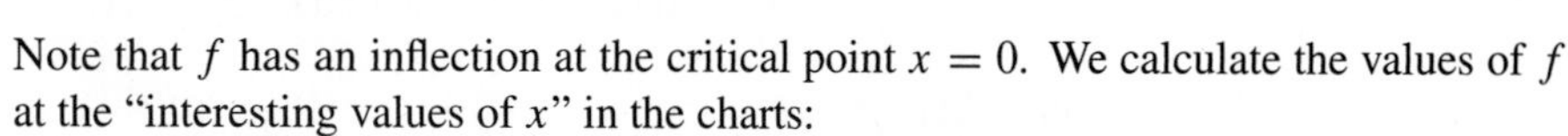

Note that f has an inflection at the critical point $x = 0$. We calculate the values of f at the "interesting values of x" in the charts:

$$f(0) = 1, \qquad f(1) = 0, \qquad f\left(\tfrac{3}{2}\right) = -\tfrac{11}{16}.$$

The graph of f is sketched in Figure 4.32.

The Second Derivative Test

A function f will have a local maximum (or minimum) value at a critical point if its graph is concave down (or up) in an interval containing that point. In fact, we can often use the value of the second derivative at the critical point to determine whether the function has a local maximum or a local minimum value there.

THEOREM

The Second Derivative Test

(a) If $f'(x_0) = 0$ and $f''(x_0) < 0$, then f has a local maximum value at x_0.
(b) If $f'(x_0) = 0$ and $f''(x_0) > 0$, then f has a local minimum value at x_0.
(c) If $f'(x_0) = 0$ and $f''(x_0) = 0$, no conclusion can be drawn; f may have a local maximum at x_0 or a local minimum, or it may have an inflection point instead.

PROOF Suppose that $f'(x_0) = 0$ and $f''(x_0) < 0$. Since

$$\lim_{h \to 0} \frac{f'(x_0 + h)}{h} = \lim_{h \to 0} \frac{f'(x_0 + h) - f'(x_0)}{h} = f''(x_0) < 0,$$

it follows that $f'(x_0 + h) < 0$ for all sufficiently small positive h, and $f'(x_0 + h) > 0$ for all sufficiently small negative h. By the first derivative test (Theorem 7), f must have a local maximum value at x_0. The proof of the local minimum case is similar.

The functions $f(x) = x^4$ (Figure 4.30), $f(x) = -x^4$, and $f(x) = x^3$ (Figure 4.27) all satisfy $f'(0) = 0$ and $f''(0) = 0$. But x^4 has a minimum value at $x = 0$, $-x^4$ has a maximum value at $x = 0$, and x^3 has neither a maximum nor a minimum value at $x = 0$ but has an inflection there. Therefore, we cannot make any conclusion about the nature of a critical point based on knowing that $f''(x) = 0$ there.

EXAMPLE 3 Find and classify the critical points of $f(x) = x^2e^{-x}$.

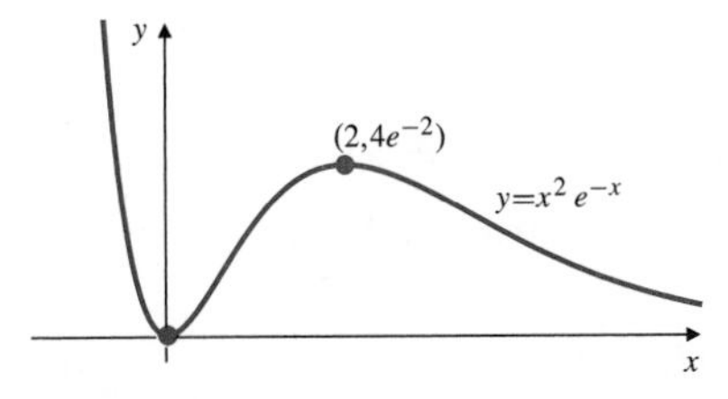

Figure 4.33 The critical points of $f(x) = x^2 e^{-x}$

Solution We begin by calculating the first two derivatives of f:

$$f'(x) = (2x - x^2)e^{-x} = x(2 - x)e^{-x} = 0 \quad \text{at } x = 0 \text{ and } x = 2,$$
$$f''(x) = (2 - 4x + x^2)e^{-x}$$
$$f''(0) = 2 > 0, \qquad f''(2) = -2e^{-2} < 0.$$

Thus, f has a local minimum value at $x = 0$ and a local maximum value at $x = 2$. See Figure 4.33.

For many functions the second derivative is more complicated to calculate than the first derivative, so the First Derivative Test is likely to be of more use in classifying critical points than is the Second Derivative Test. Also note that the First Derivative Test can classify local extreme values that occur at endpoints and singular points as well as at critical points.

It is possible to generalize the Second Derivative Test to obtain a higher derivative test to deal with some situations where the second derivative is zero at a critical point. (See Exercise 40 at the end of this section.)

EXERCISES 4.5

In Exercises 1–22, determine the intervals of constant concavity of the given function, and locate any inflection points.

1. $f(x) = \sqrt{x}$ **2.** $f(x) = 2x - x^2$

3. $f(x) = x^2 + 2x + 3$ **4.** $f(x) = x - x^3$

5. $f(x) = 10x^3 - 3x^5$ **6.** $f(x) = 10x^3 + 3x^5$

7. $f(x) = (3 - x^2)^2$ **8.** $f(x) = (2 + 2x - x^2)^2$

9. $f(x) = (x^2 - 4)^3$ **10.** $f(x) = \dfrac{x}{x^2 + 3}$

11. $f(x) = \sin x$ **12.** $f(x) = \cos 3x$

13. $f(x) = x + \sin 2x$ **14.** $f(x) = x - 2\sin x$

15. $f(x) = \tan^{-1} x$ **16.** $f(x) = x\,e^x$

17. $f(x) = e^{-x^2}$ **18.** $f(x) = \dfrac{\ln(x^2)}{x}$

19. $f(x) = \ln(1 + x^2)$ **20.** $f(x) = (\ln x)^2$

21. $f(x) = \dfrac{x^3}{3} - 4x^2 + 12x - \dfrac{25}{3}$

22. $f(x) = (x - 1)^{1/3} + (x + 1)^{1/3}$

23. Discuss the concavity of the linear function $f(x) = ax + b$. Does it have any inflections?

Classify the critical points of the functions in Exercises 24–35 using the Second Derivative Test whenever possible.

24. $f(x) = 3x^3 - 36x - 3$ **25.** $f(x) = x(x - 2)^2 + 1$

26. $f(x) = x + \dfrac{4}{x}$ **27.** $f(x) = x^3 + \dfrac{1}{x}$

28. $f(x) = \dfrac{x}{2^x}$ **29.** $f(x) = \dfrac{x}{1 + x^2}$

30. $f(x) = xe^x$ **31.** $f(x) = x \ln x$

32. $f(x) = (x^2 - 4)^2$ **33.** $f(x) = (x^2 - 4)^3$

34. $f(x) = (x^2 - 3)e^x$ **35.** $f(x) = x^2 e^{-2x^2}$

36. Let $f(x) = x^2$ if $x \ge 0$ and $f(x) = -x^2$ if $x < 0$. Is 0 a critical point of f? Does f have an inflection point there? Is $f''(0) = 0$? If a function has a nonvertical tangent line at an inflection point, does the second derivative of the function necessarily vanish at that point?

37. Verify that if f is concave up on an interval, then its graph lies above its tangent lines on that interval. *Hint:* Suppose f is concave up on an open interval containing x_0. Let $h(x) = f(x) - f(x_0) - f'(x_0)(x - x_0)$. Show that h has a local minimum value at x_0 and hence that $h(x) \ge 0$ on the interval. Show that $h(x) > 0$ if $x \ne x_0$.

38. Verify that the graph $y = f(x)$ crosses its tangent line at an inflection point. *Hint:* Consider separately the cases where the tangent line is vertical and nonvertical.

39. Let $f_n(x) = x^n$ and $g_n(x) = -x^n$, $(n = 2, 3, 4, \dots)$. Determine whether each function has a local maximum, a local minimum, or an inflection point at $x = 0$.

40. **(Higher Derivative Test)** Use your conclusions from Exercise 39 to suggest a generalization of the Second Derivative Test that applies when

$$f'(x_0) = f''(x_0) = \dots = f^{(k-1)}(x_0) = 0, \ f^{(k)}(x_0) \ne 0,$$

for some $k \ge 2$.

41. This problem shows that no test based solely on the signs of derivatives at x_0 can determine whether every function with a critical point at x_0 has a local maximum or minimum or an inflection point there. Let

$$f(x) = \begin{cases} e^{-1/x^2} & \text{if } x \ne 0 \\ 0 & \text{if } x = 0. \end{cases}$$

Prove the following:

(a) $\lim_{x \to 0} x^{-n} f(x) = 0$ for $n = 0, 1, 2, 3, \dots$.

(b) $\lim_{x \to 0} P(1/x) f(x) = 0$ for every polynomial P.

(c) For $x \neq 0$, $f^{(k)}(x) = P_k(1/x)f(x) (k = 1, 2, 3, \ldots)$, where P_k is a polynomial.

(d) $f^{(k)}(0)$ exists and equals 0 for $k = 1, 2, 3, \ldots$.

(e) f has a local minimum at $x = 0$; $-f$ has a local maximum at $x = 0$.

(f) If $g(x) = xf(x)$, then $g^{(k)}(0) = 0$ for every positive integer k and g has an inflection point at $x = 0$.

42. A function may have neither a local maximum nor a local minimum nor an inflection at a critical point. Show this by considering the following function:

$$f(x) = \begin{cases} x^2 \sin \dfrac{1}{x} & \text{if } x \neq 0 \\ 0 & \text{if } x = 0. \end{cases}$$

Show that $f'(0) = f(0) = 0$, so the x-axis is tangent to the graph of f at $x = 0$; but $f'(x)$ is not continuous at $x = 0$, so $f''(0)$ does not exist. Show that the concavity of f is not constant on any interval with endpoint 0.

4.6 Sketching the Graph of a Function

When sketching the graph $y = f(x)$ of a function f, we have three sources of useful information:

(i) **the function f itself**, from which we determine the coordinates of some points on the graph, the symmetry of the graph, and any asymptotes;

(ii) **the first derivative, f'**, from which we determine the intervals of increase and decrease and the location of any local extreme values; and

(iii) **the second derivative, f''**, from which we determine the concavity and inflection points, and sometimes extreme values.

Items (ii) and (iii) were explored in the previous two sections. In this section we consider what we can learn from the function itself about the shape of its graph, and then we illustrate the entire sketching procedure with several examples using all three sources of information.

We could sketch a graph by plotting the coordinates of many points on it and joining them by a suitably smooth curve. This is what computer software and graphics calculators do. When carried out by hand (without a computer or calculator), this simplistic approach is at best tedious and at worst can fail to reveal the most interesting aspects of the graph (singular points, extreme values, and so on). We could also compute the slope at each of the plotted points and, by drawing short line segments through these points with the appropriate slopes, ensure that the sketched graph passes through each plotted point with the correct slope. A more efficient procedure is to obtain the coordinates of only a few points and use qualitative information from the function and its first and second derivatives to determine the *shape* of the graph between these points.

Besides critical and singular points and inflections, a graph may have other "interesting" points. The **intercepts** (points at which the graph intersects the coordinate axes) are usually among these. When sketching any graph it is wise to try to find all such intercepts, that is, all points with coordinates $(x, 0)$ and $(0, y)$ that lie on the graph. Of course, not every graph will have such points, and even when they do exist it may not always be possible to compute them exactly. Whenever a graph is made up of several disconnected pieces (called **components**), the coordinates of *at least one point on each component* must be obtained. It can sometimes be useful to determine the slopes at those points too. Vertical asymptotes (discussed below) usually break the graph of a function into components.

Realizing that a given function possesses some symmetry can aid greatly in obtaining a good sketch of its graph. In Section P.4 we discussed odd and even functions and observed that odd functions have graphs that are symmetric about the origin, while even functions have graphs that are symmetric about the y-axis, as shown in Figure 4.34. These are the symmetries you are most likely to notice, but functions can have other symmetries. For example, the graph of $2 + (x - 1)^2$ will certainly be symmetric about the line $x = 1$, and the graph of $2 + (x - 3)^3$ is symmetric about the point $(3, 2)$.

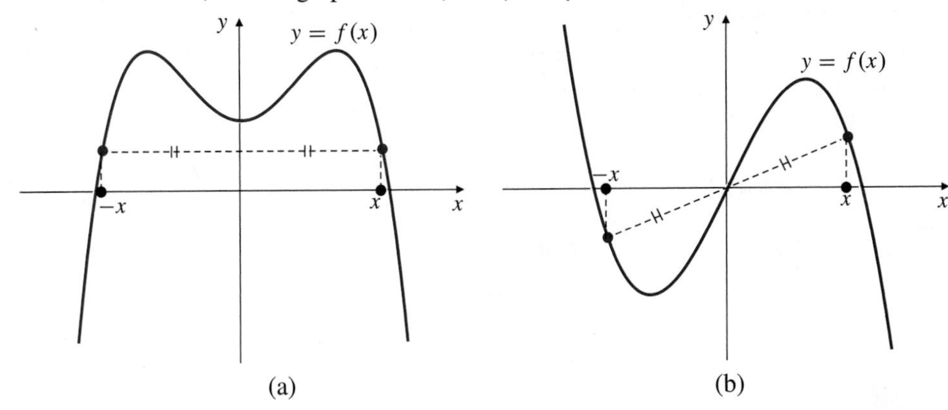

Figure 4.34

(a) The graph of an even function is symmetric about the y-axis

(b) The graph of an odd function is symmetric about the origin

Asymptotes

Some of the curves we have sketched in previous sections have had **asymptotes**, that is, straight lines to which the curve draws arbitrarily close as it recedes to infinite distance from the origin. Asymptotes are of three types: vertical, horizontal, and oblique.

DEFINITION 5

> The graph of $y = f(x)$ has a **vertical asymptote** at $x = a$ if
>
> either $\lim_{x \to a-} f(x) = \pm\infty$ or $\lim_{x \to a+} f(x) = \pm\infty$, or both.

This situation tends to arise when $f(x)$ is a quotient of two expressions and the denominator is zero at $x = a$.

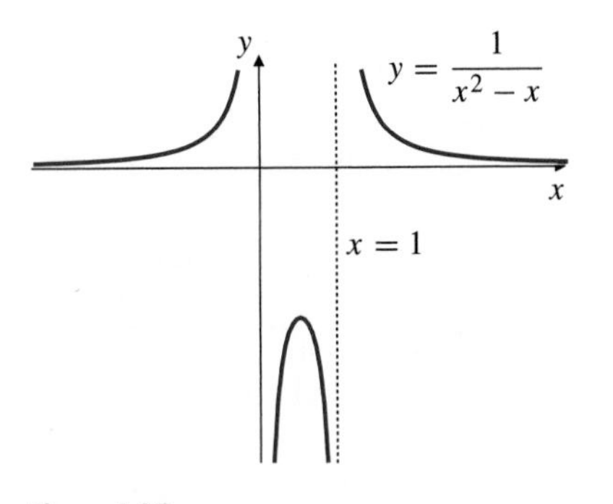

Figure 4.35

EXAMPLE 1 Find the vertical asymptotes of $f(x) = \dfrac{1}{x^2 - x}$. How does the graph approach these asymptotes?

Solution The denominator $x^2 - x = x(x - 1)$ approaches 0 as x approaches 0 or 1, so f has vertical asymptotes at $x = 0$ and $x = 1$ (Figure 4.35). Since $x(x - 1)$ is positive on $(-\infty, 0)$ and on $(1, \infty)$ and is negative on $(0, 1)$, we have

$$\lim_{x \to 0-} \frac{1}{x^2 - x} = \infty, \qquad \lim_{x \to 1-} \frac{1}{x^2 - x} = -\infty,$$

$$\lim_{x \to 0+} \frac{1}{x^2 - x} = -\infty, \qquad \lim_{x \to 1+} \frac{1}{x^2 - x} = \infty.$$

DEFINITION 6

> The graph of $y = f(x)$ has a **horizontal asymptote** $y = L$ if
>
> either $\lim_{x \to \infty} f(x) = L$ or $\lim_{x \to -\infty} f(x) = L$, or both.

EXAMPLE 2 Find the horizontal asymptotes of

(a) $f(x) = \dfrac{1}{x^2 - x}$ and (b) $g(x) = \dfrac{x^4 + x^2}{x^4 + 1}$.

Solution

(a) The function f has horizontal asymptote $y = 0$ (Figure 4.35) since

$$\lim_{x \to \pm\infty} \frac{1}{x^2 - x} = \lim_{x \to \pm\infty} \frac{1/x^2}{1 - (1/x)} = \frac{0}{1} = 0.$$

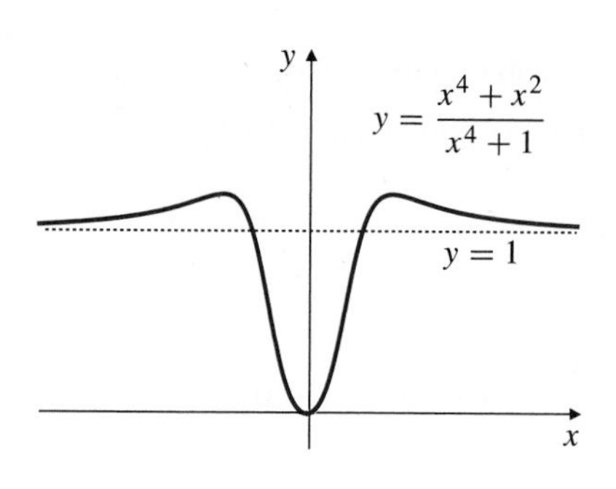

Figure 4.36

(b) The function g has horizontal asymptote $y = 1$ (Figure 4.36) since

$$\lim_{x\to\pm\infty} \frac{x^4+x^2}{x^4+1} = \lim_{x\to\pm\infty} \frac{1+(1/x^2)}{1+(1/x^4)} = \frac{1}{1} = 1.$$

Observe that the graph of g crosses its asymptote twice. (There is a popular misconception among students that curves cannot cross their asymptotes. Exercise 41 below gives an example of a curve that crosses its asymptote infinitely often.)

The horizontal asymptotes of both functions f and g in Example 2 are **two-sided**, which means that the graphs approach the asymptotes as x approaches both infinity and negative infinity. The function $\tan^{-1} x$ has two **one-sided** asymptotes, $y = \pi/2$ (as $x \to \infty$) and $y = -(\pi/2)$ (as $x \to -\infty$). See Figure 4.37.

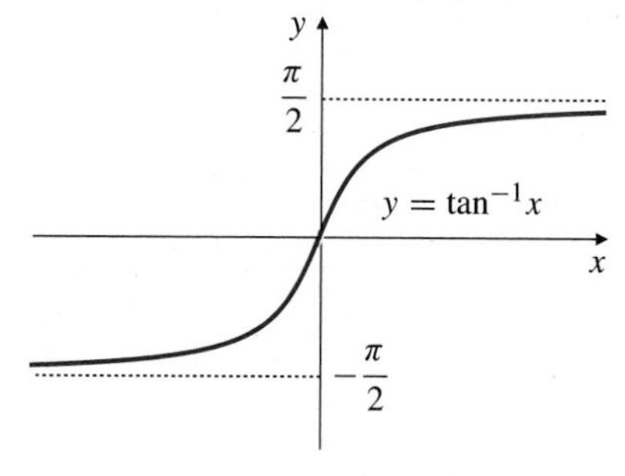

Figure 4.37 One-sided horizontal asymptotes

It can also happen that the graph of a function f approaches a nonhorizontal straight line as x approaches ∞ or $-\infty$ (or both). Such a line is called an *oblique asymptote* of the graph.

DEFINITION 7

The straight line $y = ax + b$ (where $a \neq 0$) is an **oblique asymptote** of the graph of $y = f(x)$ if

$$\text{either} \quad \lim_{x\to-\infty} \big(f(x) - (ax+b)\big) = 0 \quad \text{or} \quad \lim_{x\to\infty} \big(f(x) - (ax+b)\big) = 0,$$

or both.

EXAMPLE 3 Consider the function $f(x) = \frac{x^2+1}{x} = x + \frac{1}{x}$, whose graph is shown in Figure 4.38(a). The straight line $y = x$ is a *two-sided* oblique asymptote of the graph of f because

$$\lim_{x\to\pm\infty} \big(f(x) - x\big) = \lim_{x\to\pm\infty} \frac{1}{x} = 0.$$

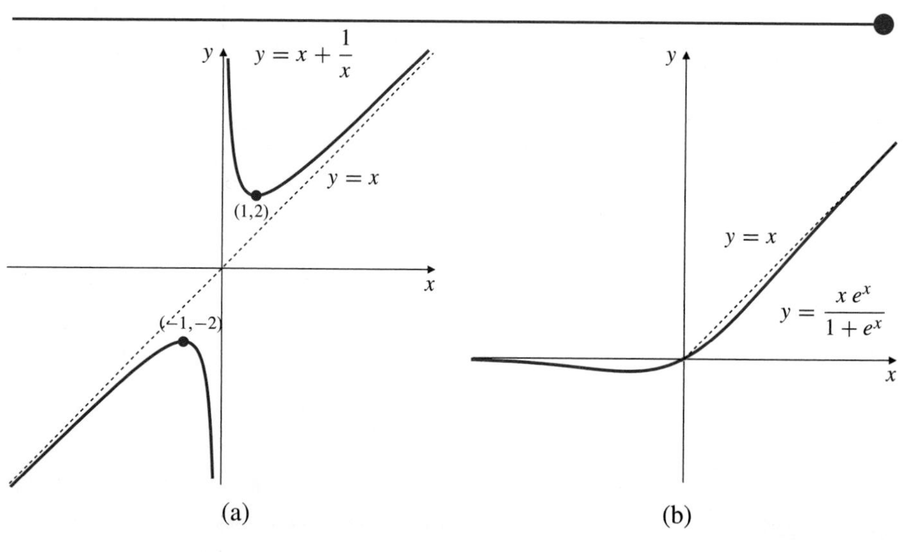

Figure 4.38

(a) The graph of $y = f(x)$ has a two-sided oblique asymptote, $y = x$

(b) This graph has a horizontal asymptote at the left and an oblique asymptote at the right

EXAMPLE 4 The graph of $y = \dfrac{x\,e^x}{1+e^x}$, shown in Figure 4.38(b), has a horizontal asymptote $y = 0$ at the left and an oblique asymptote $y = x$ at the right:

$$\lim_{x\to-\infty} \frac{x\,e^x}{1+e^x} = \frac{0}{1} = 0 \qquad \text{and}$$

$$\lim_{x\to\infty} \left(\frac{x\,e^x}{1+e^x} - x\right) = \lim_{x\to\infty} \frac{x(e^x - 1 - e^x)}{1+e^x} = \lim_{x\to\infty} \frac{-x}{1+e^x} = 0.$$

Recall that a **rational function** is a function of the form $f(x) = P(x)/Q(x)$, where P and Q are polynomials. Following observations made in Sections P.6, 1.2, and 1.3, we can be quite specific about the asymptotes of a rational function.

Asymptotes of a rational function

Suppose that $f(x) = \dfrac{P_m(x)}{Q_n(x)}$, where P_m and Q_n are polynomials of degree m and n, respectively. Suppose also that P_m and Q_n have no common linear factors. Then

(a) The graph of f has a vertical asymptote at every position x such that $Q_n(x) = 0$.

(b) The graph of f has a two-sided horizontal asymptote $y = 0$ if $m < n$.

(c) The graph of f has a two-sided horizontal asymptote $y = L$, $(L \neq 0)$ if $m = n$. L is the quotient of the coefficients of the highest degree terms in P_m and Q_n.

(d) The graph of f has a two-sided oblique asymptote if $m = n + 1$. This asymptote can be found by dividing Q_n into P_m to obtain a linear quotient, $ax + b$, and remainder, R, a polynomial of degree at most $n - 1$. That is,

$$f(x) = ax + b + \frac{R(x)}{Q_n(x)}.$$

The oblique asymptote is $y = ax + b$.

(e) The graph of f has no horizontal or oblique asymptotes if $m > n + 1$.

EXAMPLE 5 Find the oblique asymptote of $y = \dfrac{x^3}{x^2+x+1}$.

Solution We can either obtain the quotient by long division:

$$\begin{array}{r|l} & x \;-\; 1 \\ \hline x^2+x+1 & x^3 \\ & x^3 + x^2 + x \\ \hline & -x^2 - x \\ & -x^2 - x - 1 \\ \hline & \qquad\qquad 1 \end{array} \qquad \frac{x^3}{x^2+x+1} = x - 1 + \frac{1}{x^2+x+1}$$

or we can obtain the same result by short division:

$$\frac{x^3}{x^2+x+1} = \frac{x^3 + x^2 + x - x^2 - x - 1 + 1}{x^2+x+1} = x - 1 + \frac{1}{x^2+x+1}.$$

In any event, we see that the oblique asymptote has equation $y = x - 1$.

Examples of Formal Curve Sketching

Here is a checklist of things to consider when you are asked to make a careful sketch of the graph of $y = f(x)$. It will, of course, not always be possible to obtain every item of information mentioned in the list.

Checklist for curve sketching

1. Calculate $f'(x)$ and $f''(x)$, and express the results in factored form.
2. Examine $f(x)$ to determine its domain and the following items:
   (a) Any vertical asymptotes. (Look for zeros of denominators.)
   (b) Any horizontal or oblique asymptotes. (Consider $\lim_{x\to\pm\infty} f(x)$.)
   (c) Any obvious symmetry. (Is f even or odd?)
   (d) Any easily calculated intercepts (points with coordinates $(x, 0)$ or $(0, y)$) or endpoints or other "obvious" points. You will add to this list when you know any critical points, singular points, and inflection points. Eventually you should make sure you know the coordinates of at least one point on every component of the graph.
3. Examine $f'(x)$ for the following:
   (a) Any critical points.
   (b) Any points where f' is not defined. (These will include singular points, endpoints of the domain of f, and vertical asymptotes.)
   (c) Intervals on which f' is positive or negative. It's a good idea to convey this information in the form of a chart such as those used in the examples. Conclusions about where f is increasing and decreasing and classification of some critical and singular points as local maxima and minima can also be indicated on the chart.
4. Examine $f''(x)$ for the following:
   (a) Points where $f''(x) = 0$.
   (b) Points where $f''(x)$ is undefined. (These will include singular points, endpoints, vertical asymptotes, and possibly other points as well, where f' is defined but f'' isn't.)
   (c) Intervals where f'' is positive or negative and where f is therefore concave up or down. Use a chart.
   (d) Any inflection points.

When you have obtained as much of this information as possible, make a careful sketch that reflects *everything* you have learned about the function. Consider where best to place the axes and what scale to use on each so the "interesting features" of the graph show up most clearly. Be alert for seeming inconsistencies in the information—that is a strong suggestion you may have made an error somewhere. For example, if you have determined that $f(x) \to \infty$ as x approaches the vertical asymptote $x = a$ from the right, and also that f is decreasing and concave down on the interval (a, b), then you have very likely made an error. (Try to sketch such a situation to see why.)

EXAMPLE 6 Sketch the graph of $y = \dfrac{x^2 + 2x + 4}{2x}$.

Solution It is useful to rewrite the function y in the form

$$y = \frac{x}{2} + 1 + \frac{2}{x},$$

since this form not only shows clearly that $y = (x/2) + 1$ is an oblique asymptote, but also makes it easier to calculate the derivatives

$$y' = \frac{1}{2} - \frac{2}{x^2} = \frac{x^2 - 4}{2x^2}, \qquad y'' = \frac{4}{x^3}.$$

From y: Domain: all x except 0. Vertical asymptote: $x = 0$,
Oblique asymptote: $y = \frac{x}{2} + 1$, $\quad y - \left(\frac{x}{2} + 1\right) = \frac{2}{x} \to 0$ as $x \to \pm\infty$.
Symmetry: none obvious (y is neither odd nor even).
Intercepts: none. $x^2 + 2x + 4 = (x+1)^2 + 3 \geq 3$ for all x, and y is not defined at $x = 0$.

From y': Critical points: $x = \pm 2$; points $(-2, -1)$ and $(2, 3)$.
y' not defined at $x = 0$ (vertical asymptote).

From y'': $y'' = 0$ nowhere; y'' undefined at $x = 0$.

		CP		ASY		CP	
x		-2		0		2	
y'	$+$	0	$-$	undef	$-$	0	$+$
y''	$-$		$-$	undef	$+$		$+$
y	↗	max	↘	undef	↘	min	↗
	⌢		⌢		⌣		⌣

The graph is shown in Figure 4.39.

EXAMPLE 7 Sketch the graph of $f(x) = \dfrac{x^2 - 1}{x^2 - 4}$.

Solution We have

$$f'(x) = \frac{-6x}{(x^2 - 4)^2}, \qquad f''(x) = \frac{6(3x^2 + 4)}{(x^2 - 4)^3}.$$

From f: Domain: all x except ± 2. Vertical asymptotes: $x = -2$ and $x = 2$.
Horizontal asymptote: $y = 1$ (as $x \to \pm\infty$).
Symmetry: about the y-axis (y is even).
Intercepts: $(0, 1/4)$, $(-1, 0)$, and $(1, 0)$.
Other points: $(-3, 8/5)$, $(3, 8/5)$. (The two vertical asymptotes divide the graph into three components; we need points on each. The outer components require points with $|x| > 2$.)

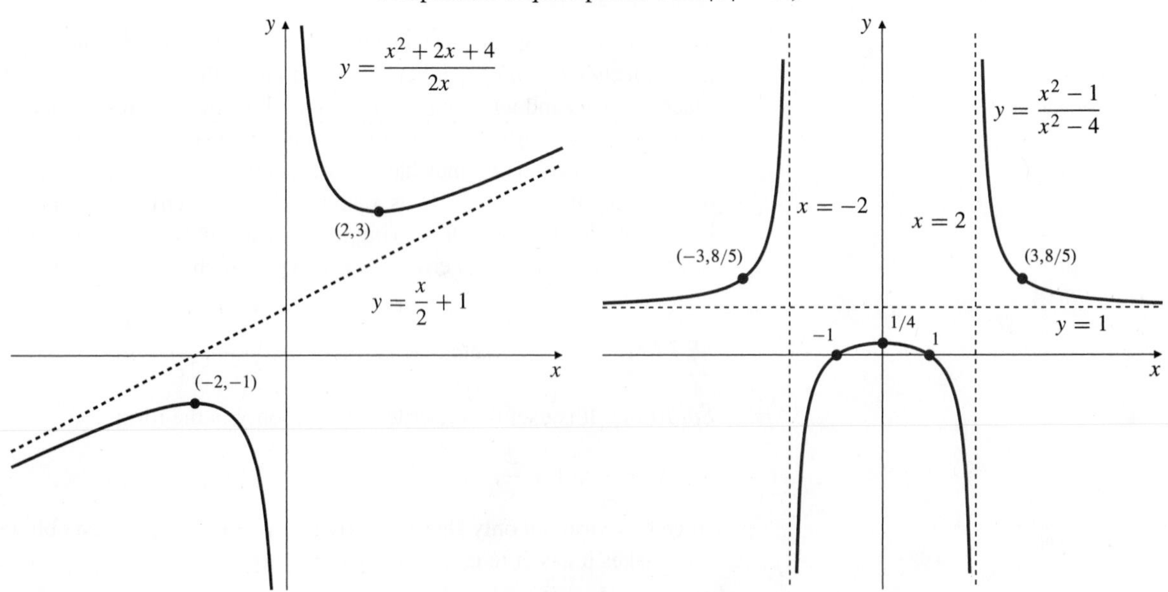

Figure 4.39

Figure 4.40

From f': Critical point: $x = 0$; f' not defined at $x = 2$ or $x = -2$.
From f'': $f''(x) = 0$ nowhere; f'' not defined at $x = 2$ or $x = -2$.

		ASY		CP		ASY	
x		-2		0		2	
f'	$+$	undef	$+$	0	$-$	undef	$-$
f''	$+$	undef	$-$		$-$	undef	$+$
f	↗	undef	↗	max	↘	undef	↘
	⌣		⌢		⌢		⌣

The graph is shown in Figure 4.40.

EXAMPLE 8 Sketch the graph of $y = xe^{-x^2/2}$.

Solution We have $y' = (1 - x^2)e^{-x^2/2}$, $\quad y'' = x(x^2 - 3)e^{-x^2/2}$.

From y: Domain: all x.
Horizontal asymptote: $y = 0$. Note that if $t = x^2/2$, then $|xe^{-x^2/2}| = \sqrt{2t}\, e^{-t} \to 0$ as $t \to \infty$ (hence as $x \to \pm\infty$).
Symmetry: about the origin (y is odd). Intercepts: $(0, 0)$.

From y': Critical points: $x = \pm 1$; points $(\pm 1, \pm 1/\sqrt{e}) \approx (\pm 1, \pm 0.61)$.

From y'': $y'' = 0$ at $x = 0$ and $x = \pm\sqrt{3}$;
points $(0, 0)$, $(\pm\sqrt{3}, \pm\sqrt{3}e^{-3/2}) \approx (\pm 1.73, \pm 0.39)$.

				CP				CP			
x		$-\sqrt{3}$		-1		0		1		$\sqrt{3}$	
y'	$-$		$-$	0	$+$		$+$	0	$-$		$-$
y''	$-$	0	$+$		$+$	0	$-$		$-$	0	$+$
y	↘		↘	min	↗		↗	max	↘		↘
	⌢	infl	⌣		⌣	infl	⌢		⌢	infl	⌣

The graph is shown in Figure 4.41.

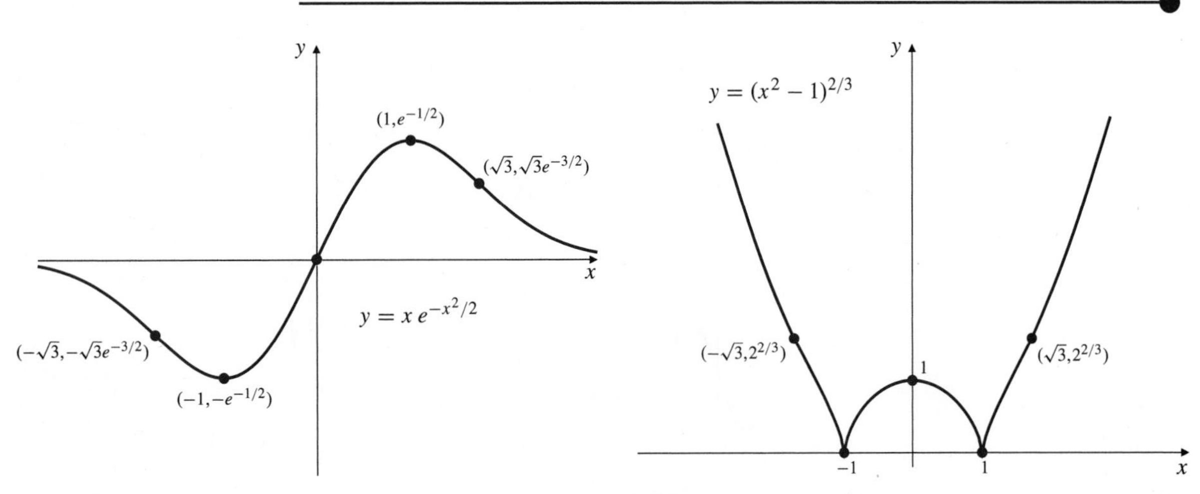

Figure 4.41

Figure 4.42

EXAMPLE 9 Sketch the graph of $f(x) = (x^2 - 1)^{2/3}$. (See Figure 4.42.)

Solution $f'(x) = \dfrac{4}{3}\dfrac{x}{(x^2-1)^{1/3}}, \qquad f''(x) = \dfrac{4}{9}\dfrac{x^2-3}{(x^2-1)^{4/3}}.$

From f: Domain: all x.
Asymptotes: none. ($f(x)$ grows like $x^{4/3}$ as $x \to \pm\infty$.)
Symmetry: about the y-axis (f is an even function).
Intercepts: $(\pm 1, 0)$, $(0, 1)$.

From f': Critical points: $x = 0$; singular points: $x = \pm 1$.

From f'': $f''(x) = 0$ at $x = \pm\sqrt{3}$; points $(\pm\sqrt{3}, 2^{2/3}) \approx (\pm 1.73, 1.59)$;
$f''(x)$ not defined at $x = \pm 1$.

				SP		CP		SP			
x		$-\sqrt{3}$		-1		0		1		$\sqrt{3}$	
f'	$-$		$-$	undef	$+$	0	$-$	undef	$+$		$+$
f''	$+$	0	$-$	undef	$-$		$-$	undef	$-$	0	$+$
f	↘		↘	min	↗	max	↘	min	↗		↗
	⌣	infl	⌢		⌢		⌢		⌢	infl	⌣

EXERCISES 4.6

1. Figure 4.43 shows the graphs of a function f, its two derivatives f' and f'', and another function g. Which graph corresponds to each function?

2. List, for each function graphed in Figure 4.43, such information that you can determine (approximately) by inspecting the graph (e.g., symmetry, asymptotes, intercepts, intervals of increase and decrease, critical and singular points, local maxima and minima, intervals of constant concavity, inflection points).

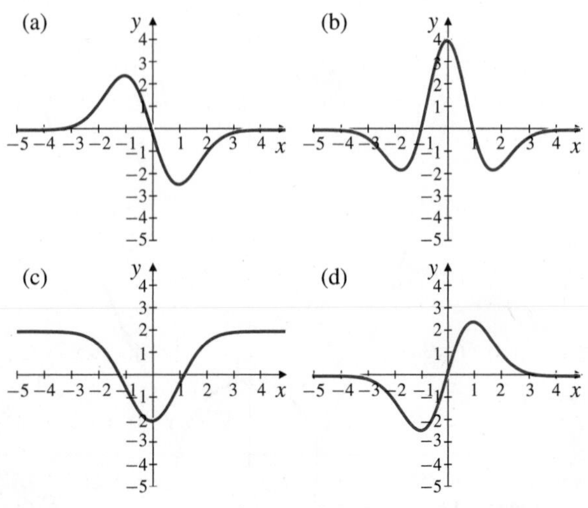

Figure 4.43

3. Figure 4.44 shows the graphs of four functions:

$$f(x) = \frac{x}{1-x^2}, \qquad g(x) = \frac{x^3}{1-x^4},$$

$$h(x) = \frac{x^3 - x}{\sqrt{x^6+1}}, \qquad k(x) = \frac{x^3}{\sqrt{|x^4-1|}}.$$

Which graph corresponds to each function?

4. Repeat Exercise 2 for the graphs in Figure 4.44.

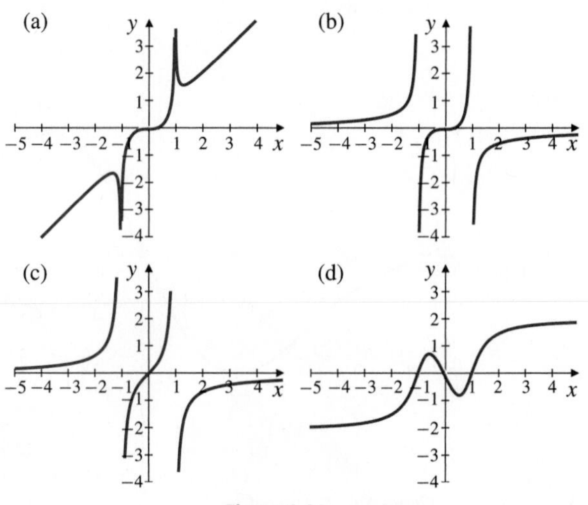

Figure 4.44

In Exercises 5–6, sketch the graph of a function that has the given properties. Identify any critical points, singular points, local maxima and minima, and inflection points. Assume that f is continuous and its derivatives exist everywhere unless the contrary is implied or explicitly stated.

5. $f(0) = 1$, $f(\pm 1) = 0$, $f(2) = 1$, $\lim_{x\to\infty} f(x) = 2$, $\lim_{x\to-\infty} f(x) = -1$, $f'(x) > 0$ on $(-\infty, 0)$ and on $(1, \infty)$, $f'(x) < 0$ on $(0, 1)$, $f''(x) > 0$ on $(-\infty, 0)$ and on $(0, 2)$, and $f''(x) < 0$ on $(2, \infty)$.

6. $f(-1) = 0$, $f(0) = 2$, $f(1) = 1$, $f(2) = 0$, $f(3) = 1$, $\lim_{x\to\pm\infty}(f(x) + 1 - x) = 0$, $f'(x) > 0$ on $(-\infty, -1)$, $(-1, 0)$ and $(2, \infty)$, $f'(x) < 0$ on $(0, 2)$, $\lim_{x\to-1} f'(x) = \infty$, $f''(x) > 0$ on $(-\infty, -1)$ and on $(1, 3)$, and $f''(x) < 0$ on $(-1, 1)$ and on $(3, \infty)$.

In Exercises 7–39, sketch the graphs of the given functions, making use of any suitable information you can obtain from the function and its first and second derivatives.

7. $y = (x^2 - 1)^3$

8. $y = x(x^2 - 1)^2$

9. $y = \dfrac{2 - x}{x}$

10. $y = \dfrac{x - 1}{x + 1}$

11. $y = \dfrac{x^3}{1 + x}$

12. $y = \dfrac{1}{4 + x^2}$

13. $y = \dfrac{1}{2 - x^2}$

14. $y = \dfrac{x}{x^2 - 1}$

15. $y = \dfrac{x^2}{x^2 - 1}$

16. $y = \dfrac{x^3}{x^2 - 1}$

17. $y = \dfrac{x^3}{x^2 + 1}$

18. $y = \dfrac{x^2}{x^2 + 1}$

19. $y = \dfrac{x^2 - 4}{x + 1}$

20. $y = \dfrac{x^2 - 2}{x^2 - 1}$

21. $y = \dfrac{x^3 - 4x}{x^2 - 1}$

22. $y = \dfrac{x^2 - 1}{x^2}$

23. $y = \dfrac{x^5}{(x^2 - 1)^2}$

24. $y = \dfrac{(2 - x)^2}{x^3}$

25. $y = \dfrac{1}{x^3 - 4x}$

26. $y = \dfrac{x}{x^2 + x - 2}$

27. $y = \dfrac{x^3 - 3x^2 + 1}{x^3}$

28. $y = x + \sin x$

29. $y = x + 2\sin x$

30. $y = e^{-x^2}$

31. $y = xe^x$

32. $y = e^{-x}\sin x$, $(x \ge 0)$

33. $y = x^2 e^{-x^2}$

34. $y = x^2 e^x$

35. $y = \dfrac{\ln x}{x}$, $(x > 0)$

36. $y = \dfrac{\ln x}{x^2}$, $(x > 0)$

37. $y = \dfrac{1}{\sqrt{4 - x^2}}$

38. $y = \dfrac{x}{\sqrt{x^2 + 1}}$

39. $y = (x^2 - 1)^{1/3}$

40. What is $\lim_{x\to 0+} x \ln x$? $\lim_{x\to 0} x \ln |x|$? If $f(x) = x \ln |x|$ for $x \ne 0$, is it possible to define $f(0)$ in such a way that f is continuous on the whole real line? Sketch the graph of f.

41. What straight line is an asymptote of the curve $y = \dfrac{\sin x}{1 + x^2}$? At what points does the curve cross this asymptote?

4.7 Graphing with Computers

The techniques for sketching, developed in the previous section, are useful for graphs of functions that are simple enough to allow you to calculate and analyze their derivatives. They are also essential for testing the validity of graphs produced by computers or calculators, which can be inaccurate or misleading for a variety of reasons, including the case of numerical monsters introduced in previous chapters. In practice, it is often easiest to first produce a graph using a computer or graphing calculator, but many times this will not turn out to be the last step. (We will use the term "computer" for both computers and calculators.) For many simple functions this can be a quick and painless activity, but sometimes functions have properties that complicate the process. Knowledge of the function, from techniques like those above, is important to guide you on what the next steps must be.

The Maple command[1] for viewing the graph of the function from Example 6 of Section 4.6, together with its oblique asymptote, is a straightforward example of plotting; we ask Maple to plot both $(x^2 + 2x + 4)/(2x)$ and $1 + (x/2)$.

```
> plot({(x^2+2*x+4)/(2*x), 1+(x/2)}, x=-6..6, y=-7..7);
```

This command sets the window $-6 \le x \le 6$ and $-7 \le y \le 7$. Why that window? To get a plot that characterizes the function, knowledge of its vertical asymptote at $x = 0$ is essential. (If $x - 10$ were substituted for x in the expression, the given window

[1] Although we focus on Maple to illustrate the issues of graphing with computers, the issues presented are general ones, pertaining to all software and computers.

would no longer produce a reasonable graph of the key features of the function. The new function would be better viewed on the interval $4 \le x \le 16$.) If the range $[-7, 7]$ were not specified, the computer would plot all of the points where it evaluates the function, including those very close to the vertical asymptote where the function is very large. The resulting plot would compress all of the features of the graph onto the x-axis. Even the asymptote would look like a horizontal line in that scaling. You might even miss the vertical asymptote which is squeezed into the y-axis.

Getting Maple to plot the curve in Example 9 of Section 4.6 is a bit trickier. Because Maple doesn't deal well with fractional powers of negative numbers, even when they have positive real values, we must actually plot $|x^2 - 1|^{2/3}$ or $((x^2 - 1)^2)^{1/3}$. Otherwise, the part of the graph between -1 and 1 will be missing. Either of the plot commands

```
> plot((abs(x^2-1))^(2/3), x=-4..4, y=-1..5);
> plot(((x^2-1)^2)^(1/3), x=-4..4, y=-1..5);
```

will produce the desired graph. In order to ensure a complete plot with all of the features of the function present, the graph of the simple expression should be viewed critically, and not taken at face value.

Numerical Monsters and Computer Graphing

The next obvious problem is that of false features and false behaviours. Functions that are mathematically well-behaved can still be computationally poorly behaved, leading to false features on graphs, as we have already seen.

EXAMPLE 1 Consider the function $f(x) = e^x \ln(1 + e^{-x})$ which has suitably simplified derivative

$$f'(x) = e^x\, g(x), \qquad \text{where} \qquad g(x) = \ln(1 + e^{-x}) - \frac{1}{e^x + 1}.$$

In turn, the derivative of $g(x)$ simplifies to

$$g'(x) = -\frac{1}{(e^x + 1)^2},$$

which is negative for all x, so g is decreasing. Since $g(0) = \ln 2 - 1/2 > 0$ and $\lim_{x\to\infty} g(x) = 0$, it follows that $g(x) > 0$ and decreasing for all x. Thus $f'(x)$ is positive, and $f(x)$ is an increasing function for all x. Furthermore, l'Hôpital's Rules show that

$$\lim_{x\to\infty} f(x) = 1 \qquad \textit{and} \qquad \lim_{x\to-\infty} f(x) = 0.$$

This gives us a pretty full picture of how the function f behaves. It grows with increasing x from 0 at $-\infty$, crosses the y-axis at $\ln 2$, and finally approaches 1 asymptotically from below as x increases toward ∞.

Now let's plot the graph of f using the Maple command

```
> plot(exp(x)*ln(1+(1/exp(x))), x=-20..45, style=point,
symbol=point, numpoints=1500);
```

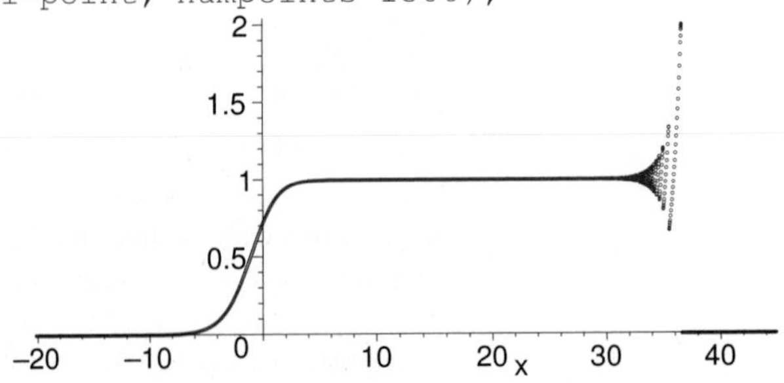

Figure 4.45 A faulty computer plot of $y = e^x \ln(1 + e^{-x})$

The result is shown in Figure 4.45. Clearly something is wrong. From $x = -20$ to about $x = 30$, the graph behaves in accordance with the mathematical analysis. However, for larger values of x, peculiarities emerge that sharply disagree with the analysis. The calculus of this chapter tells us that the function is increasing with no horizontal tangents, but the computer suggests that it decreases in some places. The calculus tells us that the function rises asymptotically to 1, but the computer suggests that the function starts to oscillate and ultimately becomes 0 at about $x = 36$.

This is another numerical monster. What a computer does can simply be wrong. In this case, it is significantly so. In practical applications an erroneous value of 0 instead of 1 could, for example, be a factor in a product, and that would change everything dramatically. If the mathematics were not known in this case, how could we even know that the computer is wrong? Another computer cannot be used to check it, as the problem is one that all computers share. Another program cannot be used because all software must use the special floating-point arithmetic that is subject to the roundoff errors responsible for the problem. Figure 4.45 is not particular to Maple. This monster, or one much like it, can be created in nearly any software package.

Floating-Point Representation of Numbers in Computers

It is necessary that you know mathematics in order to use computers correctly and effectively. It is equally necessary to understand why *all* computers fail to fully capture the mathematics. As indicated previously, the reason is that no computer can represent all numbers. Computer designers artfully attempt to minimize the effects of this by making the number of representable numbers as large as possible. But, speaking in terms of physics, a finite-sized machine can only represent a finite number of numbers. Having only a finite number of numbers leads to numbers sufficiently small, compared to 1, that the computer simply discards them in a sum. When digits are lost in this manner, the resulting error is known as **roundoff error**.

In many cases the finiteness shows up in the use of floating-point numbers and a set of corresponding arithmetic rules that approximate correct arithmetic. These approximate rules and approximate representations are not unique by any means. For example, the software package *Derive* uses so-called **slash arithmetic**, which works with a representation of numbers as continued fractions instead of decimals. This has certain advantages and disadvantages, but, in the end, finiteness forces truncation just the same.

The term "roundoff" implies that there is some kind of mitigation procedure or **rounding** done to reduce error once the smallest digits have been discarded. There are a number of different kinds of rounding practices. The various options can be quite intricate, but they all begin with the aim to slightly reduce error as a result of truncation. The truncation is the source of error, not the rounding, despite the terminology that seems to suggest otherwise. The entire process of truncation and rounding have come to be termed "roundoff," although the details of the error mitigation are immaterial for the purposes of this discussion. Rounding is beyond the scope of this section and will not be considered further.

Historically, the term "decimal" implies base ten, but the idea works the same in any base. In particular, in any base, multiplying by the base to an integral power simply shifts the position of the "decimal point." Thus, multiplying or dividing by the base is known as a **shift operation**. The term "floating-point" signifies this shifting of the point to the left or right. The general technical term for the decimal point is **radix point**. Specifically for base two, the point is sometimes called the **binary point**. However, we will use the term **decimal point** or just **decimal** for all bases, as the etymological purity is not worth having several names for one small symbol.

While computers, for the most part, work in base two, they can be and have been built in other bases. For example, there have been base-three computers, and many computers group numbers so that they work as if they were built in base eight (**octal**) or base sixteen (**hexadecimal**). (If you are feeling old, quote your age in hexadecimal. For example, $48 = 3 \times 16$ or 30 in hexadecimal. If you are feeling too young, use octal.)

In a normal binary computer, floating-point numbers approximate the mathematical *real numbers*. Several **bytes** of memory (frequently 8 bytes) are allocated for each floating-point number. Each byte consists of eight **bits**, each of which has two (physical) states and can thus store one of the two base-two digits "0" or "1," as it is the equivalent of a switch being either *off* or *on*.

Thus, an eight-byte allocation for a floating-point number can store 64 bits of data. The computer uses something similar to scientific notation, which is often used to express numbers in base ten. However, the convention is to place the decimal immediately to the left of all significant figures. For example, the computer convention would call for the base-ten number 284,070,000 to be represented as 0.28407×10^9. Here 0.28407 is called the **mantissa**, and it has 5 significant base-ten digits following the decimal point, the 2 being the most significant and the 7 the least significant digit. The 9 in the factor 10^9 is called the **exponent**, which defines the number of shift operations needed to locate the correct position of the decimal point of the actual number.

The computer only needs to represent the mantissa and the exponent, each with its appropriate sign. The base is set by the architecture and so is not stored. Neither is the decimal point nor the leading zero in the mantissa stored. These are all just implied. If the floating-point number has 64 bits, two are used for the two signs, leaving 62 bits for significant digits in the mantissa and the exponent.

As an example of base two (i.e., binary) representation, the number

$$101.011 = 1 \times 2^2 + 0 \times 2^1 + 1 \times 2^0 + 0 \times 2^{-1} + 1 \times 2^{-2} + 1 \times 2^{-3}$$

stands for the base-ten number $4 + 1 + (1/4) + (1/8) = 43/8$. On a computer the stored bits would be +101011 for the mantissa and +11 for the exponent. Thus, the base-two floating-point form is 0.101011×2^3, with mantissa 0.101011 and exponent 3. Note that we are representing the exponent in base ten (3), and not base two (11), because that is more convenient for counting shift operations.

While the base-two representation of two is 10, we will continue, for convenience, to write two as 2 when using it as the base for base-two representations. After all, any base b is represented by 10 with respect to itself as base. So, if we chose to write the number above as 0.101011×10^{11}, the numeral could as well denote a number in any base. However, for us people normally thinking in base ten, 0.101011×2^3 clearly indicates that the base is two and the decimal point is shifted 3 digits to the right of the most significant digit in the mantissa.

Now consider $x = 0.101 \times 2^{-10} = 0.0000000000101$, the base-two floating-point number whose value as a base ten fraction is $x = 5/8192$. The only significant base-two digits are the 101 in the mantissa. Now add x to 1; the result is

$$1 + x = 0.10000000000101 \times 2^1,$$

which has mantissa 0.10000000000101 and exponent 1. The mantissa now has 14 significant base-two digits; all the zeroes between the first and last 1s are significant. If your computer or calculator software only allocates, say, 12 bits for mantissas, then it would be unable to represent $1 + x$. It would have to throw away the two least significant base-two digits and save the number as

$$1 + x = 0.100000000001 \times 2^1 = \frac{2{,}049}{2{,}048},$$

thus creating a roundoff error of 1/8,192. Even worse, if only ten base-two digits were used to store mantissas, the computer would store $1 + x = 0.1000000000 \times 2^1$, (i.e., it would not be able to distinguish $1 + x$ from 1.) Of course, calculators and computer software use many more than ten or twelve base-two digits to represent mantissas of floating-point numbers, but the number of digits used is certainly finite, and so the problem of roundoff will always occur for sufficiently small floating-point numbers x.

Machine Epsilon and Its Effect on Figure 4.45

The smallest number x for which the computer recognizes that $1 + x$ is greater than 1 is called **machine epsilon** (denoted ϵ) for that computer. The computer does not return 1 when evaluating $1 + \epsilon$, but for all positive numbers x smaller than ϵ, the computer simply returns 1 when asked to evaluate $1 + x$, because the computer only keeps a finite number of (normally base 2) digits.

When using computer algebra packages like Maple, the number of digits can be increased in the software. Thus, the number of numbers that the computer can represent can be extended beyond what is native to the processor's hardware, by stringing together bits to make available larger numbers of digits for a single number. The Maple command for this is "`Digits`," which defaults to 10 (decimal digits). However, the computer remains finite in size, so there will always be an effective value for ϵ, no matter how the software is set. A hardware value for ϵ is not uniform for all devices either. Thus, for any device you may be using (calculator or computer), the value of machine epsilon may not be immediately obvious. To anticipate where a computer may be wrong, you need the value of machine epsilon, and you need to understand where the function may run afoul of it. We will outline a simple way to determine this below.

In the case of the function f in Example 1, it is clear where the computer discards digits in a sum. The factor $\ln(1 + e^{-x})$ decreases as x increases, but for sufficiently large x a computer must discard the exponential in the sum because it is too small to show up in the digits allotted for 1. When the exponential term decreases below the value of ϵ, the computer will return 1 for the argument of the natural logarithm, and the factor will be determined by the computer to be 0. Thus, f will be represented as 0 instead of nearly 1.

Of course, pathological behaviour begins to happen before the exponential e^{-x} decreases to below ϵ. When the exponential is small enough, all change with x happens in the smaller digits. The sum forces them to be discarded by the computer, so the change is discarded with it. That means for finite intervals the larger digits from the decreasing exponential term do not change until the smaller changes accrue. In the case of f, this means it behaves like an increasing exponential times a constant between corrections of the larger digits. This is confirmed in Figure 4.46, which is a close-up of the pathological region given by adjusting the interval of the plot command.

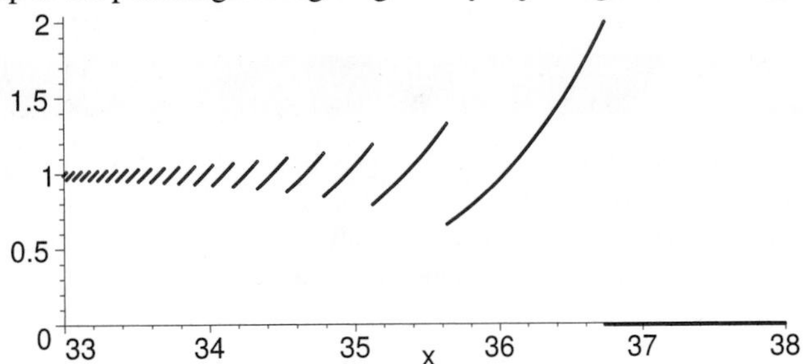

Figure 4.46 Part of the graph of f from Example 1 over the interval $[33, 38]$

Determining Machine Epsilon

A small alteration in the function f of Example 1 provides an easy way to determine the value of machine epsilon. As computers store and process data in base-two form, it is useful to use instead of f the function $h(x) = 2^x \ln(1 + 2^{-x})$. The Maple plot command

```
> plot(2^x*ln(1+1/2^x), x=50..55, style=line,
    thickness=5, xtickmarks=[50,51,52,53,54]);
```

produces the graph in Figure 4.47. The graph drops to 0 at $x = 53$. Thus, 2^{-53} is the next number below ϵ that the computer can represent. Because the first nonzero digit in a base-two number is 1, the next largest number must be up to twice as large. But because all higher digits are discarded, the effect is to have simply a change in the

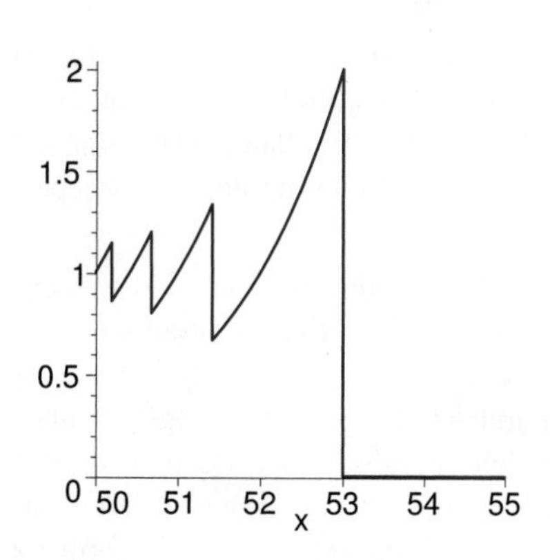

Figure 4.47 This indicates that machine epsilon is $\epsilon = 2^{-52}$

exponent of the number, a shift operation. A single shift operation larger than 2^{-53} is 2^{-52}, so $\epsilon = 2^{-52}$ in the settings for this plot.

From this we can predict when f will drop to zero in Figure 4.45 and Figure 4.46. It will be when $\epsilon/2 = 2^{-53} = e^{-x}$ or approximately $x = 36.74$. While this seems to give us a complete command of the effect for most computers, there is much more going on with computer error that depends on specific algorithms. While significant error erupts when ϵ is reached in a sum with 1, other sources of error are in play well before that for smaller values of x.

It is interesting to look at some of the complex and structured patterns of error in a close-up of what should be a single curve well before the catastrophic drop to zero. Figure 4.48 is produced by the plot instruction

```
> plot(exp(x)*ln(1+1/exp(x)), x = 29.5 ..  30,
   style = point, symbol = point, numpoints = 3000);
```

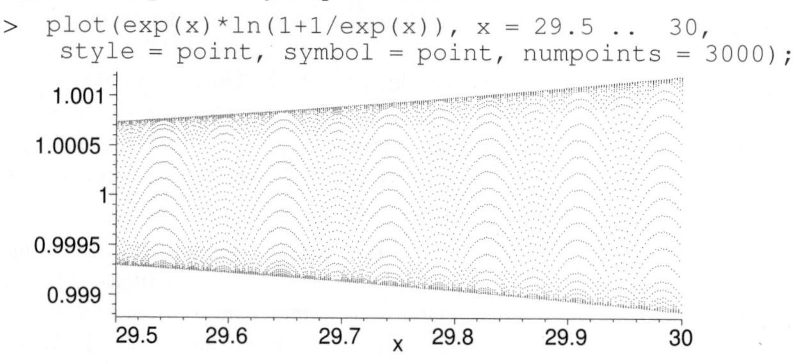

Figure 4.48 Illusions of computation

In that figure there are many fascinating and beautiful patterns created, which are completely spurious. In this region the exponential curves are collapsed together, forming what seems like a single region contained within an expanding envelope. The beautiful patterns make it easy to forget that the mathematically correct curve would appear as a single horizontal line at height 1. The patterns here are created by Maple's selection of points at which to evaluate the function and their placement in the plot. If you change the plot window, try to zoom in on them, or change the numbers of points or the interval; they will change too, or disappear. They are completely illusive and spurious features. Computers can't be trusted blindly. You can trust mathematics.

EXERCISES 4.7

1. Use Maple to get a plot instruction that plots an exponential function through one of the stripes in Figure 4.46. You can use the cursor position in the Maple display to read off the approximate coordinates of the lower left endpoint on one of the stripes.

2. Why should the expression $h(x) - \sqrt{h(x)^2}$ not be expected to be exactly zero, especially for large $h(x)$, when evaluated on a computer?

3. Consider Figure 4.49. It is the result of the plot instruction:

```
> plot([ln(2^x-sqrt(2^(2*x)-1)),
-ln(2^x+sqrt(2^(2*x)-1))], x=0..50,
y=-30..10, style=line, symbol=point,
thickness=[1,4],   color=[blue, grey],
numpoints=8000);
```

   The grey line is a plot of $f(x) = -\ln(2^x + \sqrt{2^{2x} - 1})$. The coloured line is a plot of $g(x) = \ln(2^x - \sqrt{2^{2x} - 1})$.

   (a) Show that $g(x) = f(x)$.

   (b) Why do the graphs of f and g behave differently?

   (c) Estimate a value of x beyond which the plots of f and g will behave differently. Assume machine epsilon is $\epsilon = 2^{-52}$.

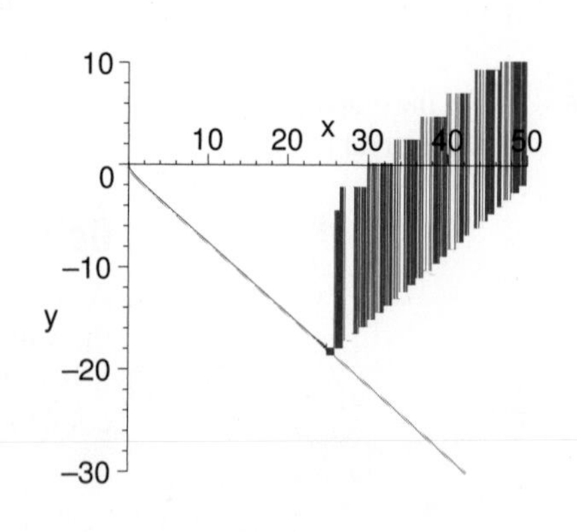

Figure 4.49

4. If you use a graphing calculator or other mathematical graphing software, try to determine machine epsilon for it.

In Exercises 5–6 assume that a computer uses 64 bits (binary digits) of memory to store a floating-point number, and that of these 64 bits 52 are used for the mantissa and one each for the signs of the mantissa and the exponent.

5. To the nearest power of 10, what is the smallest positive number that can be represented in floating-point form by the computer?

6. To the nearest power of 10, what is the largest positive number that can be represented in floating-point form by the computer?

4.8 Extreme-Value Problems

In this section we solve various word problems that, when translated into mathematical terms, require the finding of a maximum or minimum value of a function of one variable. Such problems can range from simple to very complex and difficult; they can be phrased in terminology appropriate to some other discipline, or they can be already partially translated into a more mathematical context. We have already encountered a few such problems in earlier chapters.

Let us consider a couple of examples before attempting to formulate any general principles for dealing with such problems.

EXAMPLE 1 A rectangular animal enclosure is to be constructed having one side along an existing long wall and the other three sides fenced. If 100 m of fence are available, what is the largest possible area for the enclosure?

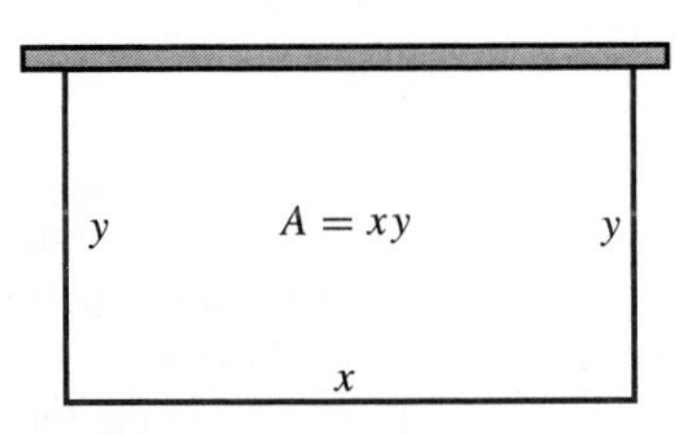

Figure 4.50

Solution This problem, like many others, is essentially a geometric one. A sketch should be made at the outset, as we have done in Figure 4.50. Let the length and width of the enclosure be x and y m, respectively, and let its area be A m^2. Thus $A = xy$. Since the total length of the fence is 100 m, we must have $x + 2y = 100$. A appears to be a function of two variables, x and y, but these variables are not independent; they are related by the *constraint* $x + 2y = 100$. This constraint equation can be solved for one variable in terms of the other, and A can therefore be written as a function of only one variable:

$$x = 100 - 2y,$$
$$A = A(y) = (100 - 2y)y = 100y - 2y^2.$$

Evidently, we require $y \geq 0$ and $y \leq 50$ (i.e., $x \geq 0$), in order that the area make sense. (It would otherwise be negative.) Thus, we must maximize the function $A(y)$ on the interval $[0, 50]$. Being continuous on this closed, finite interval, A must have a maximum value, by Theorem 5. Clearly, $A(0) = A(50) = 0$ and $A(y) > 0$ for $0 < y < 50$. Hence, the maximum cannot occur at an endpoint. Since A has no singular points, the maximum must occur at a critical point. To find any critical points, we set

$$0 = A'(y) = 100 - 4y.$$

Therefore, $y = 25$. Since A must have a maximum value and there is only one possible point where it can be, the maximum must occur at $y = 25$. The greatest possible area for the enclosure is therefore $A(25) = 1{,}250$ m^2.

EXAMPLE 2 A lighthouse L is located on a small island 5 km north of a point A on a straight east-west shoreline. A cable is to be laid from L to point B on the shoreline 10 km east of A. The cable will be laid through the water in a straight line from L to a point C on the shoreline between A and B, and from there to B along the shoreline. (See Figure 4.51.) The part of the cable lying in the water costs \$5,000/km, and the part along the shoreline costs \$3,000/km.

(a) Where should C be chosen to minimize the total cost of the cable?

(b) Where should C be chosen if B is only 3 km from A?

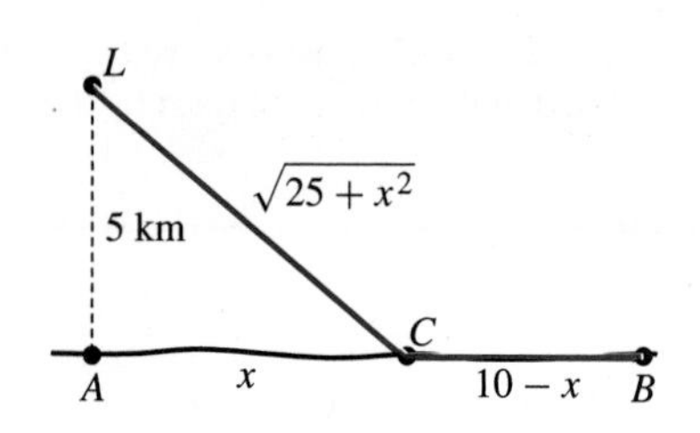

Figure 4.51

Solution

(a) Let C be x km from A toward B. Thus $0 \le x \le 10$. The length of LC is $\sqrt{25+x^2}$ km, and the length of CB is $10 - x$ km, as illustrated in Figure 4.51. Hence, the total cost of the cable is $\$T$, where

$$T = T(x) = 5{,}000\sqrt{25+x^2} + 3{,}000(10-x), \qquad (0 \le x \le 10).$$

T is continuous on the closed, finite interval $[0, 10]$, so it has a minimum value that may occur at one of the endpoints $x = 0$ or $x = 10$ or at a critical point in the interval $(0, 10)$. (T has no singular points.) To find any critical points, we set

$$0 = \frac{dT}{dx} = \frac{5{,}000x}{\sqrt{25+x^2}} - 3{,}000.$$

Thus,

$$\begin{aligned} 5{,}000x &= 3{,}000\sqrt{25+x^2} \\ 25x^2 &= 9(25+x^2) \\ 16x^2 &= 225 \\ x^2 &= \frac{225}{16} = \frac{15^2}{4^2}. \end{aligned}$$

This equation has two solutions, but only one, $x = 15/4 = 3.75$, lies in the interval $(0, 10)$. Since $T(0) = 55{,}000$, $T(15/4) = 50{,}000$, and $T(10) \approx 55{,}902$, the critical point 3.75 evidently provides the minimum value for $T(x)$. For minimal cost, C should be 3.75 km from A.

(b) If B is 3 km from A, the corresponding total cost function is

$$T(x) = 5{,}000\sqrt{25+x^2} + 3{,}000(3-x), \qquad (0 \le x \le 3),$$

which differs from the total cost function $T(x)$ of part (a) only in the added constant (9,000 rather than 30,000). It therefore has the same critical point, $x = 15/4 = 3.75$, which does not lie in the interval $(0, 3)$. Since $T(0) = 34{,}000$ and $T(3) \approx 29{,}155$, in this case we should choose $x = 3$. To minimize the total cost, the cable should go straight from L to B.

Procedure for Solving Extreme-Value Problems

Based on our experience with the examples above, we can formulate a checklist of steps involved in solving optimization problems.

Solving extreme-value problems

1. Read the problem very carefully, perhaps more than once. You must understand clearly what is given and what must be found.
2. Make a diagram if appropriate. Many problems have a geometric component, and a good diagram can often be an essential part of the solution process.
3. Define any symbols you wish to use that are not already specified in the statement of the problem.
4. Express the quantity Q to be maximized or minimized as a function of one or more variables.
5. If Q depends on n variables, where $n > 1$, find $n - 1$ equations (constraints) linking these variables. (If this cannot be done, the problem cannot be solved by single-variable techniques.)
6. Use the constraints to eliminate variables and hence express Q as a function of only one variable. Determine the interval(s) in which this variable must lie for the problem to make sense. Alternatively, regard the constraints as implicitly defining $n - 1$ of the variables, and hence Q, as functions of the remaining variable.

7. Find the required extreme value of the function Q using the techniques of Section 4.4. Remember to consider any critical points, singular points, and endpoints. Make sure to give a convincing argument that your extreme value is the one being sought; for example, if you are looking for a maximum, the value you have found should not be a minimum.
8. Make a concluding statement answering the question asked. Is your answer for the question *reasonable*? If not, check back through the solution to see what went wrong.

EXAMPLE 3 Find the length of the shortest ladder that can extend from a vertical wall, over a fence 2 m high located 1 m away from the wall, to a point on the ground outside the fence.

Solution Let θ be the angle of inclination of the ladder, as shown in Figure 4.52. Using the two right-angled triangles in the figure, we obtain the length L of the ladder as a function of θ:

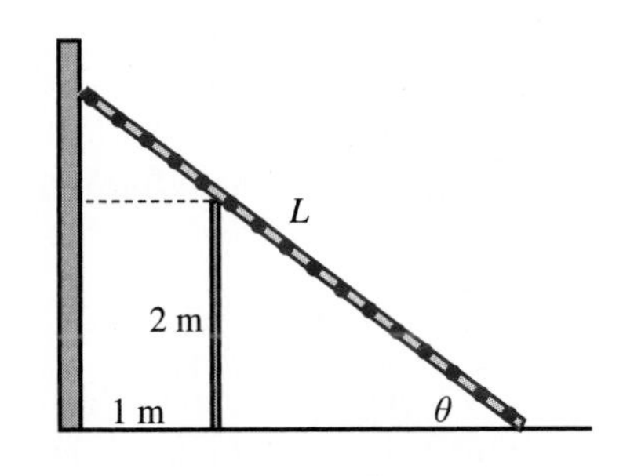

Figure 4.52

$$L = L(\theta) = \frac{1}{\cos\theta} + \frac{2}{\sin\theta},$$

where $0 < \theta < \pi/2$. Since

$$\lim_{\theta\to(\pi/2)-} L(\theta) = \infty \quad \text{and} \quad \lim_{\theta\to 0+} L(\theta) = \infty,$$

$L(\theta)$ must have a minimum value on $(0, \pi/2)$, occurring at a critical point. (L has no singular points in $(0, \pi/2)$.) To find any critical points, we set

$$0 = L'(\theta) = \frac{\sin\theta}{\cos^2\theta} - \frac{2\cos\theta}{\sin^2\theta} = \frac{\sin^3\theta - 2\cos^3\theta}{\cos^2\theta\sin^2\theta}.$$

Any critical point satisfies $\sin^3\theta = 2\cos^3\theta$, or, equivalently, $\tan^3\theta = 2$. We don't need to solve this equation for $\theta = \tan^{-1}(2^{1/3})$ since it is really the corresponding value of $L(\theta)$ that we want. Observe that

$$\sec^2\theta = 1 + \tan^2\theta = 1 + 2^{2/3}.$$

It follows that

$$\cos\theta = \frac{1}{(1 + 2^{2/3})^{1/2}} \quad \text{and} \quad \sin\theta = \tan\theta\cos\theta = \frac{2^{1/3}}{(1 + 2^{2/3})^{1/2}}.$$

Therefore, the minimal value of $L(\theta)$ is

$$\frac{1}{\cos\theta} + \frac{2}{\sin\theta} = (1 + 2^{2/3})^{1/2} + 2\,\frac{(1 + 2^{2/3})^{1/2}}{2^{1/3}} = \left(1 + 2^{2/3}\right)^{3/2} \approx 4.16.$$

The shortest ladder that can extend from the wall over the fence to the ground outside is about 4.16 m long.

EXAMPLE 4 Find the most economical shape of a cylindrical tin can.

Solution This problem is stated in a rather vague way. We must consider what is meant by "most economical" and even "shape." Without further information, we can take one of two points of view:

(i) the volume of the tin can is to be regarded as given, and we must choose the dimensions to minimize the total surface area, or

(ii) the total surface area is given (we can use just so much metal), and we must choose the dimensions to maximize the volume.

We will discuss other possible interpretations later. Since a cylinder is determined by its radius and height (Figure 4.53), its shape is determined by the ratio radius/height. Let r, h, S, and V denote, respectively, the radius, height, total surface area, and volume of the can. The volume of a cylinder is the base area times the height:

$$V = \pi r^2 h.$$

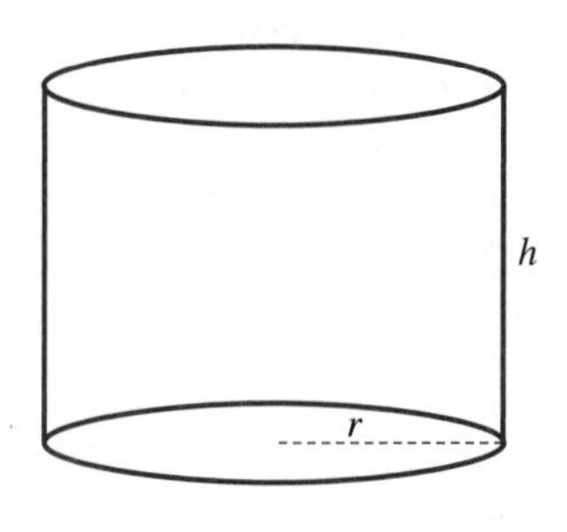

Figure 4.53

The surface of the can is made up of the cylindrical wall and circular disks for the top and bottom. The disks each have area πr^2, and the cylindrical wall is really just a rolled-up rectangle with base $2\pi r$ (the circumference of the can) and height h. Therefore, the total surface area of the can is

$$S = 2\pi rh + 2\pi r^2.$$

Let us use interpretation (i): V is a given constant, and S is to be minimized. We can use the equation for V to eliminate one of the two variables r and h on which S depends. Say we solve for $h = V/(\pi r^2)$ and substitute into the equation for S to obtain S as a function of r alone:

$$S = S(r) = 2\pi r \frac{V}{\pi r^2} + 2\pi r^2 = \frac{2V}{r} + 2\pi r^2 \qquad (0 < r < \infty).$$

Evidently, $\lim_{r\to 0+} S(r) = \infty$ and $\lim_{r\to\infty} S(r) = \infty$. Being differentiable and therefore continuous on $(0, \infty)$, $S(r)$ must have a minimum value, and it must occur at a critical point. To find any critical points,

$$0 = S'(r) = -\frac{2V}{r^2} + 4\pi r,$$

$$r^3 = \frac{2V}{4\pi} = \frac{1}{2\pi}\pi r^2 h = \frac{1}{2} r^2 h.$$

Thus, $h = 2r$ at the critical point of S. Under interpretation (i), the most economical can is shaped so that its height equals the diameter of its base. You are encouraged to show that interpretation (ii) leads to the same conclusion.

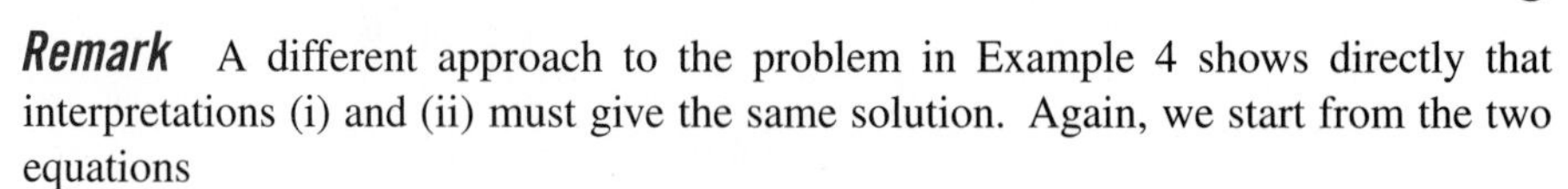

Remark A different approach to the problem in Example 4 shows directly that interpretations (i) and (ii) must give the same solution. Again, we start from the two equations

$$V = \pi r^2 h \qquad \text{and} \qquad S = 2\pi rh + 2\pi r^2.$$

If we regard h as a function of r and differentiate implicitly, we obtain

$$\frac{dV}{dr} = 2\pi rh + \pi r^2 \frac{dh}{dr},$$

$$\frac{dS}{dr} = 2\pi h + 2\pi r \frac{dh}{dr} + 4\pi r.$$

Under interpretation (i), V is constant and we want a critical point of S; under interpretation (ii), S is constant and we want a critical point of V. In *either* case, $dV/dr = 0$ and $dS/dr = 0$. Hence both interpretations yield

$$2\pi rh + \pi r^2 \frac{dh}{dr} = 0 \qquad \text{and} \qquad 2\pi h + 4\pi r + 2\pi r \frac{dh}{dr} = 0.$$

If we divide the first equation by πr^2 and the second equation by $2\pi r$ and subtract to eliminate dh/dr, we again get $h = 2r$.

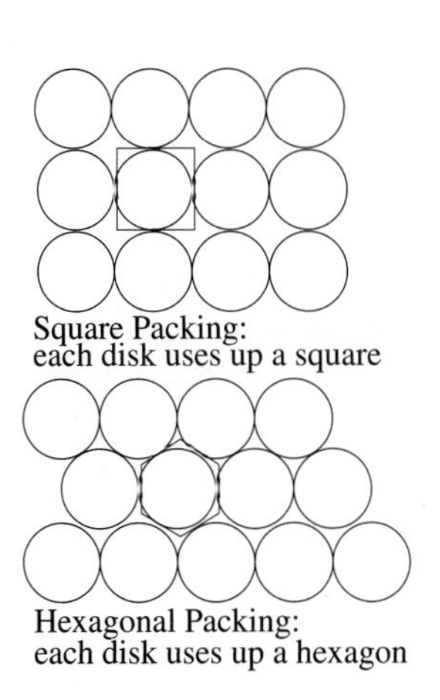

Figure 4.54 Square and hexagonal packing of disks in a plane

Remark **Modifying Example 4** Given the sparse information provided in the statement of the problem in Example 4, interpretations (i) and (ii) are the best we can do. The problem could be made more meaningful economically (from the point of view, say, of a tin can manufacturer) if more elements were brought into it. For example:

(a) Most cans use thicker material for the cylindrical wall than for the top and bottom disks. If the cylindrical wall material costs $\$A$ per unit area and the material for the top and bottom costs $\$B$ per unit area, we might prefer to minimize the total cost of materials for a can of given volume. What is the optimal shape if $A = 2B$?

(b) Large numbers of cans are to be manufactured. The material is probably being cut out of sheets of metal. The cylindrical walls are made by bending up rectangles, and rectangles can be cut from the sheet with little or no waste. There will, however, always be a proportion of material wasted when the disks are cut out. The exact proportion will depend on how the disks are arranged; two possible arrangements are shown in Figure 4.54. What is the optimal shape of the can if a square packing of disks is used? A hexagonal packing? Any such modification of the original problem will alter the optimal shape to some extent. In "real-world" problems, many factors may have to be taken into account to come up with a "best" strategy.

(c) The problem makes no provision for costs of manufacturing the can other than the cost of sheet metal. There may also be costs for joining the opposite edges of the rectangle to make the cylinder and for joining the top and bottom disks to the cylinder. These costs may be proportional to the lengths of the joins.

In most of the examples above, the maximum or minimum value being sought occurred at a critical point. Our final example is one where this is not the case.

EXAMPLE 5 A man can run twice as fast as he can swim. He is standing at point A on the edge of a circular swimming pool 40 m in diameter, and he wishes to get to the diametrically opposite point B as quickly as possible. He can run around the edge to point C, then swim directly from C to B. Where should C be chosen to minimize the total time taken to get from A to B?

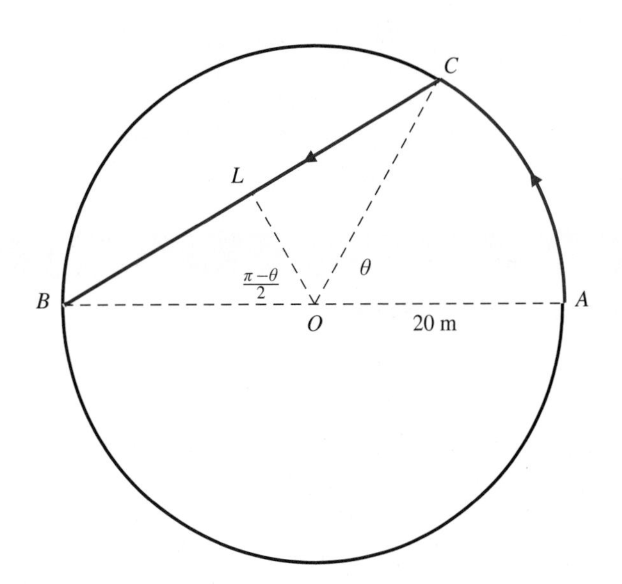

Figure 4.55 Running and swimming to get from A to B

Solution It is convenient to describe the position of C in terms of the angle AOC, where O is the centre of the pool. (See Figure 4.55.) Let θ denote this angle. Clearly $0 \le \theta \le \pi$. (If $\theta = 0$, the man swims the whole way; if $\theta = \pi$, he runs the whole way.) The radius of the pool is 20 m, so arc $AC = 20\theta$. Since angle $BOC = \pi - \theta$, we have angle $BOL = (\pi - \theta)/2$ and chord $BC = 2BL = 40 \sin\big((\pi - \theta)/2\big)$.

Suppose the man swims at a rate k m/s and therefore runs at a rate $2k$ m/s. If t is the total time he takes to get from A to B, then

$$t = t(\theta) = \text{time running} + \text{time swimming}$$
$$= \frac{20\theta}{2k} + \frac{40}{k} \sin \frac{\pi - \theta}{2}.$$

(We are assuming that no time is wasted in jumping into the water at C.) The domain of t is $[0, \pi]$ and t has no singular points. Since t is continuous on a closed, finite interval, it must have a minimum value, and that value must occur at a critical point or an endpoint. For critical points,

$$0 = t'(\theta) = \frac{10}{k} - \frac{20}{k} \cos \frac{\pi - \theta}{2}.$$

Thus,

$$\cos \frac{\pi - \theta}{2} = \frac{1}{2}, \qquad \frac{\pi - \theta}{2} = \frac{\pi}{3}, \qquad \theta = \frac{\pi}{3}.$$

This is the only critical value of θ lying in the interval $[0, \pi]$. We have

$$t\left(\frac{\pi}{3}\right) = \frac{10\pi}{3k} + \frac{40}{k} \sin \frac{\pi}{3} = \frac{10}{k}\left(\frac{\pi}{3} + \frac{4\sqrt{3}}{2}\right) \approx \frac{45.11}{k}.$$

We must also look at the endpoints $\theta = 0$ and $\theta = \pi$:

$$t(0) = \frac{40}{k}, \qquad t(\pi) = \frac{10\pi}{k} \approx \frac{31.4}{k}.$$

Evidently $t(\pi)$ is the least of these three times. To get from A to B as quickly as possible, the man should run the entire distance.

Remark This problem shows how important it is to check every candidate point to see whether it gives a maximum or minimum. Here, the critical point $\theta = \pi/3$ yielded the *worst* possible strategy: running one-third of the way around and then swimming the remainder would take the greatest time, not the least.

EXERCISES 4.8

1. Two positive numbers have sum 7. What is the largest possible value for their product?
2. Two positive numbers have product 8. What is the smallest possible value for their sum?
3. Two nonnegative numbers have sum 60. What are the numbers if the product of one of them and the square of the other is maximal?
4. Two numbers have sum 16. What are the numbers if the product of the cube of one and the fifth power of the other is as large as possible?
5. The sum of two nonnegative numbers is 10. What is the smallest value of the sum of the cube of one number and the square of the other?
6. Two nonnegative numbers have sum n. What is the smallest possible value for the sum of their squares?
7. Among all rectangles of given area, show that the square has the least perimeter.
8. Among all rectangles of given perimeter, show that the square has the greatest area.
9. Among all isosceles triangles of given perimeter, show that the equilateral triangle has the greatest area.
10. Find the largest possible area for an isosceles triangle if the length of each of its two equal sides is 10 m.
11. Find the area of the largest rectangle that can be inscribed in a semicircle of radius R if one side of the rectangle lies along the diameter of the semicircle.
12. Find the largest possible perimeter of a rectangle inscribed in a semicircle of radius R if one side of the rectangle lies along the diameter of the semicircle. (It is interesting that the rectangle with the largest perimeter has a different shape than the one with the largest area, obtained in Exercise 11.)
13. A rectangle with sides parallel to the coordinate axes is

inscribed in the ellipse

$$\frac{x^2}{a^2} + \frac{y^2}{b^2} = 1.$$

Find the largest possible area for this rectangle.

14. Let ABC be a triangle right-angled at C and having area S. Find the maximum area of a rectangle inscribed in the triangle if (a) one corner of the rectangle lies at C, or (b) one side of the rectangle lies along the hypotenuse, AB.

15. Find the maximum area of an isosceles triangle whose equal sides are 10 cm in length. Use half the length of the third side of the triangle as the variable in terms of which to express the area of the triangle.

16. Repeat Exercise 15, but use instead the angle between the equal sides of the triangle as the variable in terms of which to express the area of the triangle. Which solution is easier?

17. (Designing a billboard) A billboard is to be made with 100 m^2 of printed area and with margins of 2 m at the top and bottom and 4 m on each side. Find the outside dimensions of the billboard if its total area is to be a minimum.

18. (Designing a box) A box is to be made from a rectangular sheet of cardboard 70 cm by 150 cm by cutting equal squares out of the four corners and bending up the resulting four flaps to make the sides of the box. (The box has no top.) What is the largest possible volume of the box?

19. (Using rebates to maximize profit) An automobile manufacturer sells 2,000 cars per month, at an average profit of \$1,000 per car. Market research indicates that for each \$50 of factory rebate the manufacturer offers to buyers it can expect to sell 200 more cars each month. How much of a rebate should it offer to maximize its monthly profit?

20. (Maximizing rental profit) All 80 rooms in a motel will be rented each night if the manager charges \$40 or less per room. If he charges $\$(40 + x)$ per room, then $2x$ rooms will remain vacant. If each rented room costs the manager \$10 per day and each unrented room \$2 per day in overhead, how much should the manager charge per room to maximize his daily profit?

21. (Minimizing travel time) You are in a dune buggy in the desert 12 km due south of the nearest point A on a straight east-west road. You wish to get to point B on the road 10 km east of A. If your dune buggy can average 15 km/h travelling over the desert and 39 km/h travelling on the road, toward what point on the road should you head in order to minimize your travel time to B?

22. Repeat Exercise 21, but assume that B is only 4 km from A.

23. (Flying with least energy) At the altitude of airliners, winds can typically blow at a speed of about 100 knots (nautical miles per hour) from the west toward the east. A westward-flying passenger jet from London, England, on its way to Toronto, flies directly against this wind for 3,000 nautical miles. The energy per unit time expended by the airliner is proportional to v^3, where v is the speed of the airliner relative to the air. This reflects the power required to push aside the air exerting ram pressure proportional to v^2. What speed uses the least energy on this trip? Estimate the time it would take to fly this route at the resulting optimal speed. Is this a typical speed at which airliners travel? Explain.

24. (Energy for a round trip) In the preceding problem we found that an airliner flying against the wind at speed, v, with respect to the air consumes the least energy over a flight if it travels at $v = 3u/2$, where u is the speed of the headwind with respect to the ground. Assume the power (energy per unit time) required to push aside the air is kv^3.

(a) Write the general expression for energy consumed over a trip of distance ℓ flying with an airspeed v into a headwind of speed u. Also write the general expression for energy used on the return journey along the same path with airspeed w aided by a tailwind of speed u.

(b) Show that the energy consumed in the return journey is a strictly increasing function of w. What is the least energy consumed in the return journey if the airliner must have a minimum airspeed of s (known as "stall speed") to stay aloft?

(c) What is the least energy consumed in the round trip if $u > 2s/3$? What is the energy consumed when $u < 2s/3$?

25. A one-metre length of stiff wire is cut into two pieces. One piece is bent into a circle, the other piece into a square. Find the length of the part used for the square if the sum of the areas of the circle and the square is (a) maximum and (b) minimum.

26. Find the area of the largest rectangle that can be drawn so that each of its sides passes through a different vertex of a rectangle having sides a and b.

27. What is the length of the shortest line segment having one end on the x-axis, the other end on the y-axis, and passing through the point $(9, \sqrt{3})$?

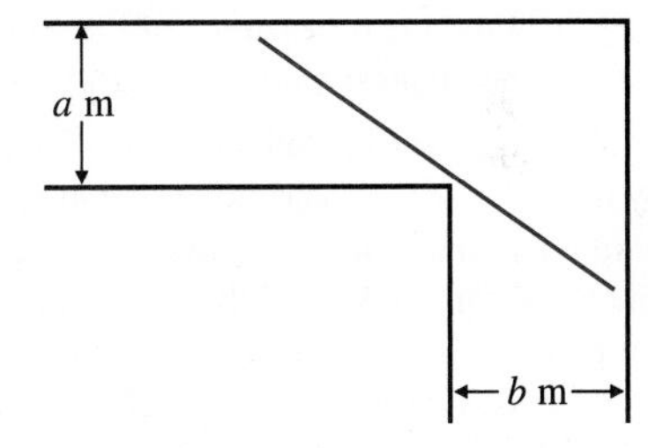

Figure 4.56

28. (Getting around a corner) Find the length of the longest beam that can be carried horizontally around the corner from a hallway of width a m to a hallway of width b m. (See Figure 4.56; assume the beam has no width.)

29. If the height of both hallways in Exercise 28 is c m, and if the beam need not be carried horizontally, how long can it be and still get around the corner? *Hint:* You can use the result of the previous exercise to do this one easily.

30. The fence in Example 3 is demolished and a new fence is built 2 m away from the wall. How high can the fence be if a 6 m ladder must be able to extend from the wall, over the fence, to the ground outside?

31. Find the shortest distance from the origin to the curve $x^2y^4 = 1$.

32. Find the shortest distance from the point $(8, 1)$ to the curve $y = 1 + x^{3/2}$.

33. Find the dimensions of the largest right-circular cylinder that can be inscribed in a sphere of radius R.

34. Find the dimensions of the circular cylinder of greatest volume that can be inscribed in a cone of base radius R and height H if the base of the cylinder lies in the base of the cone.

35. A box with square base and no top has a volume of 4 m^3. Find the dimensions of the most economical box.

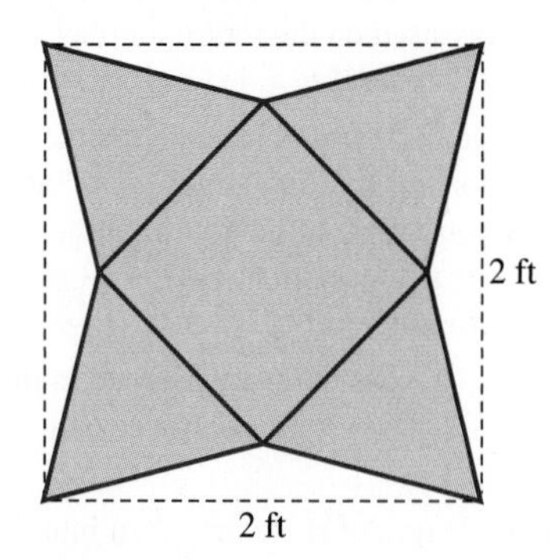

Figure 4.57

36. (Folding a pyramid) A pyramid with a square base and four faces, each in the shape of an isosceles triangle, is made by cutting away four triangles from a 2 ft square piece of cardboard (as shown in Figure 4.57) and bending up the resulting triangles to form the walls of the pyramid. What is the largest volume the pyramid can have? *Hint:* The volume of a pyramid having base area A and height h measured perpendicular to the base is $V = \frac{1}{3}Ah$.

37. (Getting the most light) A window has perimeter 10 m and is in the shape of a rectangle with the top edge replaced by a semicircle. Find the dimensions of the rectangle if the window admits the greatest amount of light.

38. (Fuel tank design) A fuel tank is made of a cylindrical part capped by hemispheres at each end. If the hemispheres are twice as expensive per unit area as the cylindrical wall, and if the volume of the tank is V, find the radius and height of the cylindrical part to minimize the total cost. The surface area of a sphere of radius r is $4\pi r^2$; its volume is $\frac{4}{3}\pi r^3$.

39. (Reflection of light) Light travels in such a way that it requires the minimum possible time to get from one point to another. A ray of light from C reflects off a plane mirror AB at X and then passes through D. (See Figure 4.58.) Show that the rays CX and XD make equal angles with the normal to AB at X. (*Remark:* You may wish to give a proof based on elementary geometry without using any calculus, or you can minimize the travel time on CXD.)

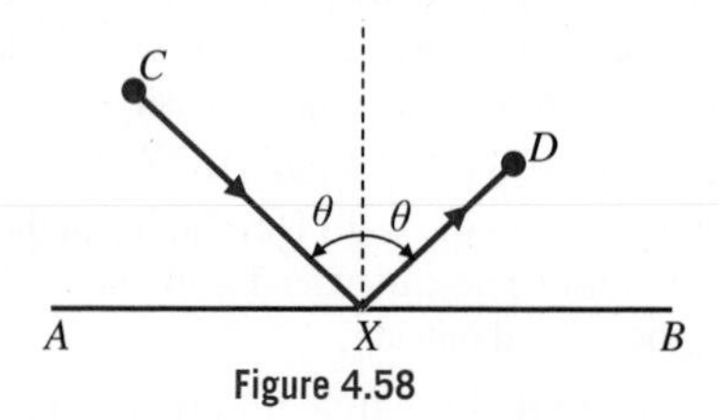

Figure 4.58

40. (Snell's Law) If light travels with speed v_1 in one medium and speed v_2 in a second medium, and if the two media are separated by a plane interface, show that a ray of light passing from point A in one medium to point B in the other is bent at the interface in such a way that

$$\frac{\sin i}{\sin r} = \frac{v_1}{v_2},$$

where i and r are the angles of incidence and refraction, as is shown in Figure 4.59. This is known as Snell's Law. Deduce it from the least-time principle stated in Exercise 39.

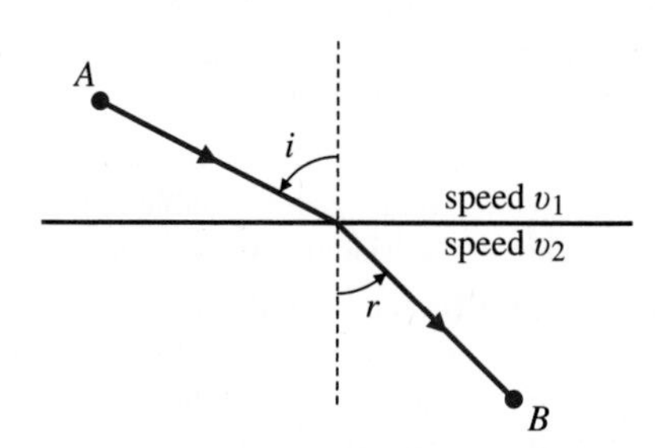

Figure 4.59

41. (Cutting the stiffest beam) The stiffness of a wooden beam of rectangular cross section is proportional to the product of the width and the cube of the depth of the cross section. Find the width and depth of the stiffest beam that can be cut out of a circular log of radius R.

42. Find the equation of the straight line of maximum slope tangent to the curve $y = 1 + 2x - x^3$.

43. A quantity Q grows according to the differential equation

$$\frac{dQ}{dt} = kQ^3(L - Q)^5,$$

where k and L are positive constants. How large is Q when it is growing most rapidly?

44. Find the smallest possible volume of a right-circular cone that can contain a sphere of radius R. (The volume of a cone of base radius r and height h is $\frac{1}{3}\pi r^2 h$.)

45. (Ferry loading) A ferry runs between the mainland and the island of Dedlos. The ferry has a maximum capacity of 1,000 cars, but loading near capacity is very time consuming. It is found that the number of cars that can be loaded in t hours is

$$f(t) = 1{,}000\,\frac{t}{e^{-t} + t}.$$

(Note that $\lim_{t\to\infty} f(t) = 1{,}000$ as expected.) Further, it is found that it takes $x/1{,}000$ hours to unload x cars. The sailing time to or from the island is 1 hour. Assume there are always more cars waiting for each sailing than can be loaded. How many cars should be loaded on the ferry for each sailing to maximize the average movement of cars back and forth to the island? (You will need to use a graphing calculator or computer software like Maple's `fsolve` routine to find the appropriate critical point.)

46. (The best view of a mural) How far back from a mural should one stand to view it best if the mural is 10 ft high and the bottom of it is 2 ft above eye level? (See Figure 4.60.)

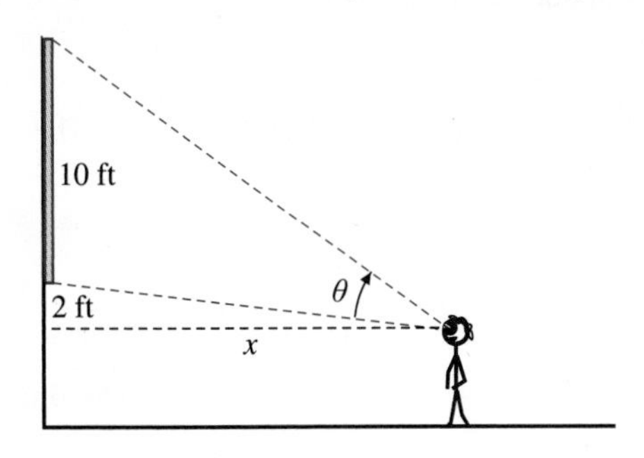

Figure 4.60

47. (Improving the enclosure of Example 1) An enclosure is to be constructed having part of its boundary along an existing straight wall. The other part of the boundary is to be fenced in the shape of an arc of a circle. If 100 m of fencing is available, what is the area of the largest possible enclosure? Into what fraction of a circle is the fence bent?

48. (Designing a Dixie cup) A sector is cut out of a circular disk of radius R, and the remaining part of the disk is bent up so that the two edges join and a cone is formed. (See Figure 4.61.) What is the largest possible volume for the cone?

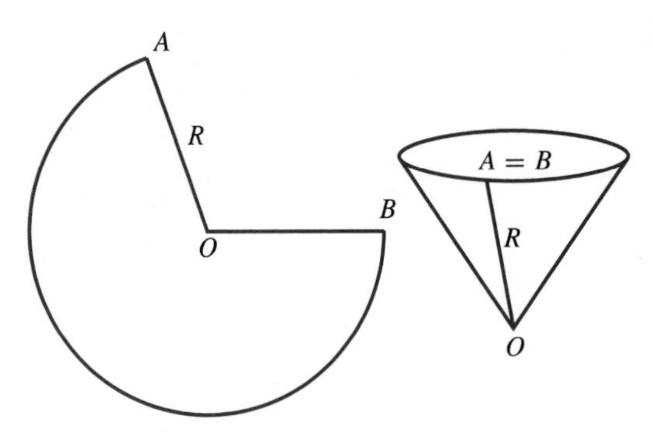

Figure 4.61

49. (Minimize the fold) One corner of a strip of paper a cm wide is folded up so that it lies along the opposite edge. (See Figure 4.62.) Find the least possible length for the fold line.

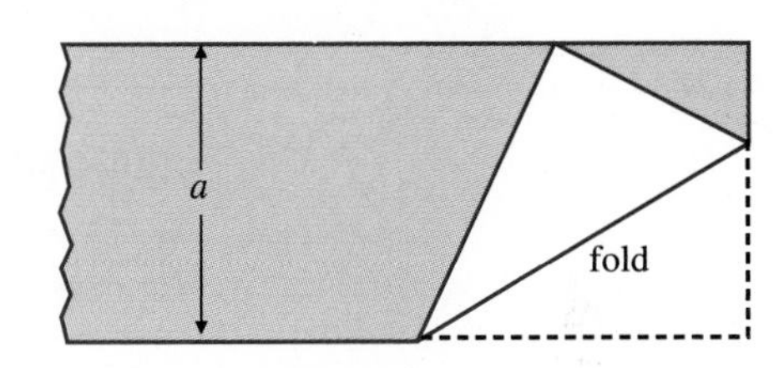

Figure 4.62

4.9 Linear Approximations

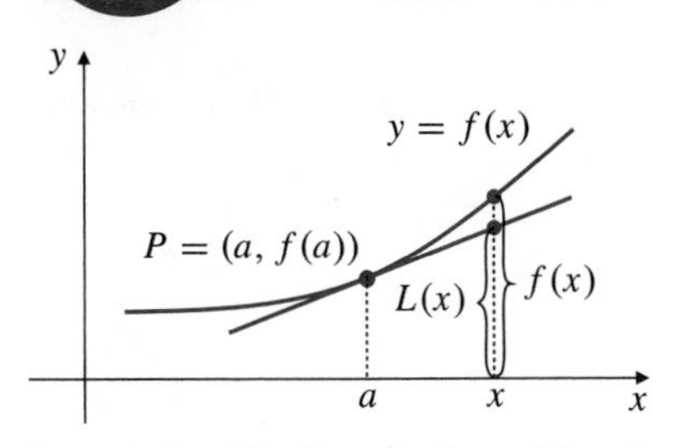

Figure 4.63 The linearization of function f about a

Many problems in applied mathematics are too difficult to be solved exactly—that is why we resort to using computers, even though in many cases they may only give approximate answers. However, not all approximation is done with machines. Linear approximation can be a very effective way to estimate values or test the plausibility of numbers given by a computer. In Section 2.7 we observed how differentials could be used to approximate (changes in) the values of functions between nearby points. In this section we reconsider such approximations in a more formal way, and obtain estimates for the size of the errors encountered when such "linear" approximations are made.

The tangent to the graph $y = f(x)$ at $x = a$ describes the behaviour of that graph near the point $P = (a, f(a))$ better than any other straight line through P, because it goes through P in the same direction as the curve $y = f(x)$. (See Figure 4.63.) We exploit this fact by using the height to the tangent line to calculate approximate values of $f(x)$ for values of x near a. The tangent line has equation $y = f(a) + f'(a)(x - a)$. We call the right side of this equation the linearization of f about a (or the linearization of $f(x)$ about $x = a$).

DEFINITION 8

The **linearization** of the function f about a is the function L defined by

$$L(x) = f(a) + f'(a)(x - a).$$

We say that $f(x) \approx L(x) = f(a) + f'(a)(x - a)$ provides **linear approximations** for values of f near a.

EXAMPLE 1 Find linearizations of (a) $f(x) = \sqrt{1+x}$ about $x = 0$ and (b) $g(t) = 1/t$ about $t = 1/2$.

Solution

(a) We have $f(0) = 1$ and, since $f'(x) = 1/(2\sqrt{1+x})$, $f'(0) = 1/2$. The linearization of f about 0 is

$$L(x) = 1 + \frac{1}{2}(x - 0) = 1 + \frac{x}{2}.$$

(b) We have $g(1/2) = 2$ and, since $g'(t) = -1/t^2$, $g'(1/2) = -4$. The linearization of $g(t)$ about $t = 1/2$ is

$$L(t) = 2 - 4\left(t - \frac{1}{2}\right) = 4 - 4t.$$

Approximating Values of Functions

We have already made use of linearization in Section 2.7, where it was disguised as the formula

$$\Delta y \approx \frac{dy}{dx}\,\Delta x$$

and used to approximate a small change $\Delta y = f(a + \Delta x) - f(a)$ in the values of function f corresponding to the small change in the argument of the function from a to $a + \Delta x$. This is just the linear approximation

$$f(a + \Delta x) \approx L(a + \Delta x) = f(a) + f'(a)\Delta x.$$

EXAMPLE 2 A ball of ice melts so that its radius decreases from 5 cm to 4.92 cm. By approximately how much does the volume of the ball decrease?

Solution The volume V of a ball of radius r is $V = \frac{4}{3}\pi r^3$, so that $dV/dr = 4\pi r^2$ and $L(r + \Delta r) = V(r) + 4\pi r^2\,\Delta r$. Thus,

$$\Delta V \approx L(r + \Delta r) = 4\pi r^2\,\Delta r.$$

For $r = 5$ and $\Delta r = -0.08$, we have

$$\Delta V \approx 4\pi(5^2)(-0.08) = -8\pi \approx -25.13.$$

The volume of the ball decreases by about 25 cm^3.

The following example illustrates the use of linearization to find an approximate value of a function near a point where the values of the function and its derivative are known.

EXAMPLE 3 Use the linearization for $\sqrt{x}$ about $x = 25$ to find an approximate value for $\sqrt{26}$.

Solution If $f(x) = \sqrt{x}$, then $f'(x) = 1/(2\sqrt{x})$. Since we know that $f(25) = 5$ and $f'(25) = 1/10$, the linearization of $f(x)$ about $x = 25$ is

$$L(x) = 5 + \frac{1}{10}(x - 25).$$

Putting $x = 26$, we get

$$\sqrt{26} = f(26) \approx L(26) = 5 + \frac{1}{10}(26 - 25) = 5.1.$$

If we use the square root function on a calculator we can obtain the "true value" of $\sqrt{26}$ (actually, just another approximation, although presumably a better one): $\sqrt{26} = 5.099\,019\,5\ldots$, but if we have such a calculator we don't need the approximation in the first place. Approximations are useful when there is no easy way to obtain the true value. However, if we don't know the true value, we would at least like to have some way of determining how good the approximation must be; that is, we want an *estimate for the error*. After all, *any number* is an approximation to $\sqrt{26}$, but the error may be unacceptably large; for instance, the size of the error in the approximation $\sqrt{26} \approx 1{,}000{,}000$ is greater than 999,994.

Error Analysis

In any approximation, the **error** is defined by

error = true value − approximate value.

If the linearization of f about a is used to approximate $f(x)$ near $x = a$, that is,

$$f(x) \approx L(x) = f(a) + f'(a)(x - a),$$

then the error $E(x)$ in this approximation is

$$E(x) = f(x) - L(x) = f(x) - f(a) - f'(a)(x - a).$$

It is the vertical distance at x between the graph of f and the tangent line to that graph at $x = a$, as shown in Figure 4.64. Observe that if x is "near" a, then $E(x)$ is small compared to the horizontal distance between x and a.

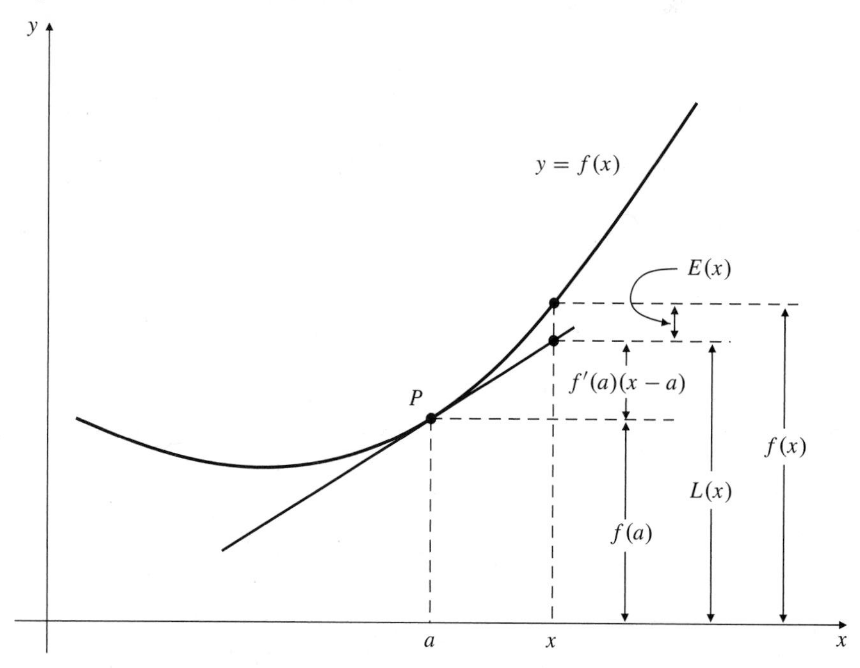

Figure 4.64 $f(x)$ and its linearization $L(x)$ about $x = a$. $E(x)$ is the error in the approximation $f(x) \approx L(x)$

The following theorem and its corollaries give us a way to estimate this error if we know bounds for the *second derivative* of f.

THEOREM 

An error formula for linearization

If $f''(t)$ exists for all t in an interval containing a and x, then there exists some point s between a and x such that the error $E(x) = f(x) - L(x)$ in the linear approximation $f(x) \approx L(x) = f(a) + f'(a)(x - a)$ satisfies

$$E(x) = \frac{f''(s)}{2}(x - a)^2.$$

PROOF Let us assume that $x > a$. (The proof for $x < a$ is similar.) Since

$$E(t) = f(t) - f(a) - f'(a)(t - a),$$

we have $E'(t) = f'(t) - f'(a)$. We apply the Generalized Mean-Value Theorem (Theorem 16 of Section 2.8) to the two functions $E(t)$ and $(t - a)^2$ on $[a, x]$. Noting that $E(a) = 0$, we obtain a number u in (a, x) such that

$$\frac{E(x)}{(x - a)^2} = \frac{E(x) - E(a)}{(x - a)^2 - (a - a)^2} = \frac{E'(u)}{2(u - a)} = \frac{f'(u) - f'(a)}{2(u - a)} = \frac{1}{2}f''(s)$$

for some s in (a, u); the latter expression is a consequence of applying the Mean-Value Theorem again, this time to f' on $[a, u]$. Thus,

$$E(x) = \frac{f''(s)}{2}(x - a)^2$$

as claimed. ■

The following three corollaries are immediate consequences of Theorem 11.

Corollary A. If $f''(t)$ has constant sign (i.e., is always positive or always negative) between a and x, then the error $E(x)$ in the linear approximation $f(x) \approx L(x)$ in the Theorem has that same sign; if $f''(t) > 0$ between a and x, then $f(x) > L(x)$; if $f''(t) < 0$ between a and x, then $f(x) < L(x)$.

Corollary B. If $|f''(t)| < K$ for all t between a and x (where K is a constant), then $|E(x)| < (K/2)(x - a)^2$.

Corollary C. If $f''(t)$ satisfies $M < f''(t) < N$ for all t between a and x (where M and N are constants), then

$$L(x) + \frac{M}{2}(x - a)^2 < f(x) < L(x) + \frac{N}{2}(x - a)^2.$$

If M and N have the same sign, a better approximation to $f(x)$ is given by the midpoint of this interval containing $f(x)$:

$$f(x) \approx L(x) + \frac{M + N}{4}(x - a)^2.$$

For this approximation the error is less than half the length of the interval:

$$|\text{Error}| < \frac{N - M}{4}(x - a)^2.$$

EXAMPLE 4 Determine the sign and estimate the size of the error in the approximation $\sqrt{26} \approx 5.1$ obtained in Example 3. Use these to give a small interval that you can be sure contains $\sqrt{26}$.

Solution For $f(t) = t^{1/2}$, we have

$$f'(t) = \frac{1}{2}t^{-1/2} \qquad \text{and} \qquad f''(t) = -\frac{1}{4}t^{-3/2}.$$

For $25 < t < 26$, we have $f''(t) < 0$, so $\sqrt{26} = f(26) < L(26) = 5.1$. Also, $t^{3/2} > 25^{3/2} = 125$, so $|f''(t)| < (1/4)(1/125) = 1/500$ and

$$|E(26)| < \frac{1}{2} \times \frac{1}{500} \times (26-25)^2 = \frac{1}{1{,}000} = 0.001.$$

Therefore, $f(26) > L(26) - 0.001 = 5.099$, and $\sqrt{26}$ is in the interval $(5.099, 5.1)$.

Remark We can use Corollary C of Theorem 11 and the fact that $\sqrt{26} < 5.1$ to find a better (i.e., smaller) interval containing $\sqrt{26}$ as follows. If $25 < t < 26$, then $125 = 25^{3/2} < t^{3/2} < 26^{3/2} < 5.1^3$. Thus,

$$M = -\frac{1}{4 \times 125} < f''(t) < -\frac{1}{4 \times 5.1^3} = N$$

$$\sqrt{26} \approx L(26) + \frac{M+N}{4} = 5.1 - \frac{1}{4}\left(\frac{1}{4 \times 125} + \frac{1}{4 \times 5.1^3}\right) \approx 5.099\,028\,8$$

$$|\text{Error}| < \frac{N-M}{4} = \frac{1}{16}\left(-\frac{1}{5.1^3} + \frac{1}{125}\right) \approx 0.000\,028\,8.$$

Thus, $\sqrt{26}$ lies in the interval $(5.099\,00, 5.099\,06)$.

EXAMPLE 5 Use a suitable linearization to find an approximate value for $\cos 36° = \cos(\pi/5)$. Is the true value greater than or less than your approximation? Estimate the size of the error, and give an interval that you can be sure contains $\cos(36°)$.

Solution Let $f(t) = \cos t$, so that $f'(t) = -\sin t$ and $f''(t) = -\cos t$. The value of a nearest to $36°$ for which we know $\cos a$ is $a = 30° = \pi/6$, so we use the linearization about that point:

$$L(x) = \cos\frac{\pi}{6} - \sin\frac{\pi}{6}\left(x - \frac{\pi}{6}\right) = \frac{\sqrt{3}}{2} - \frac{1}{2}\left(x - \frac{\pi}{6}\right).$$

Since $(\pi/5) - (\pi/6) = \pi/30$, our approximation is

$$\cos 36° = \cos\frac{\pi}{5} \approx L\left(\frac{\pi}{5}\right) = \frac{\sqrt{3}}{2} - \frac{1}{2}\left(\frac{\pi}{30}\right) \approx 0.813\,67.$$

If $(\pi/6) < t < (\pi/5)$, then $f''(t) < 0$ and $|f''(t)| < \cos(\pi/6) = \sqrt{3}/2$. Therefore, $\cos 36° < 0.813\,67$ and

$$|E(36°)| < \frac{\sqrt{3}}{4}\left(\frac{\pi}{30}\right)^2 < 0.004\,75.$$

Thus, $0.813\,67 - 0.004\,75 < \cos 36° < 0.813\,67$, so $\cos 36°$ lies in the interval $(0.808\,92, 0.813\,67)$.

Remark The error in the linearization of $f(x)$ about $x = a$ can be interpreted in terms of differentials (see Section 2.7 and the beginning of this section) as follows: If $\Delta x = dx = x - a$, then the change in $f(x)$ as we pass from $x = a$ to $x = a + \Delta x$ is $f(a + \Delta x) - f(a) = \Delta y$, and the corresponding change in the linearization $L(x)$ is $f'(a)(x - a) = f'(a)\,dx$, which is just the value at $x = a$ of the differential $dy = f'(x)\,dx$. Thus,

$$E(x) = \Delta y - dy.$$

The error $E(x)$ is small compared with Δx as Δx approaches 0, as seen in Figure 4.64. In fact,

$$\lim_{\Delta x \to 0} \frac{\Delta y - dy}{\Delta x} = \lim_{\Delta x \to 0} \left(\frac{\Delta y}{\Delta x} - \frac{dy}{dx} \right) = \frac{dy}{dx} - \frac{dy}{dx} = 0.$$

If $|f''(t)| \le K$ (constant) near $t = a$, a stronger assertion can be made:

$$\left| \frac{\Delta y - dy}{(\Delta x)^2} \right| = \left| \frac{E(x)}{(\Delta x)^2} \right| \le \frac{K}{2}, \qquad \text{so} \qquad |\Delta y - dy| \le \frac{K}{2} (\Delta x)^2.$$

EXERCISES 4.9

In Exercises 1–10, find the linearization of the given function about the given point.

1. x^2 about $x = 3$

2. x^{-3} about $x = 2$

3. $\sqrt{4 - x}$ about $x = 0$

4. $\sqrt{3 + x^2}$ about $x = 1$

5. $1/(1 + x)^2$ about $x = 2$

6. $1/\sqrt{x}$ about $x = 4$

7. $\sin x$ about $x = \pi$

8. $\cos(2x)$ about $x = \pi/3$

9. $\sin^2 x$ about $x = \pi/6$

10. $\tan x$ about $x = \pi/4$

11. By approximately how much does the area of a square increase if its side length increases from 10 cm to 10.4 cm?

12. By about how much must the edge length of a cube decrease from 20 cm to reduce the volume of the cube by 12 cm^3?

13. A spacecraft orbits the earth at a distance of 4,100 miles from the centre of the earth. By about how much will the circumference of its orbit decrease if the radius decreases by 10 miles?

14. **(Acceleration of gravity)** The acceleration a of gravity at an altitude of h miles above the surface of the earth is given by

$$a = g \left(\frac{R}{R + h} \right)^2,$$

where $g \approx 32$ ft/s^2 is the acceleration at the surface of the earth, and $R \approx 3960$ miles is the radius of the earth. By about what percentage will a decrease if h increases from 0 to 10 miles?

In Exercises 15–22, use a suitable linearization to approximate the indicated value. Determine the sign of the error and estimate its size. Use this information to specify an interval you can be sure contains the value.

15. $\sqrt{50}$

16. $\sqrt{47}$

17. $\sqrt[4]{85}$

18. $\dfrac{1}{2.003}$

19. $\cos 46^\circ$

20. $\sin \dfrac{\pi}{5}$

21. $\sin(3.14)$

22. $\sin 33^\circ$

Use Corollary C of Theorem 11 in the manner suggested in the remark following Example 4 to find better intervals and better approximations to the values in Exercises 23–26.

23. $\sqrt{50}$ as first approximated in Exercise 15.

24. $\sqrt{47}$ as first approximated in Exercise 16.

25. $\cos 36^\circ$ as first approximated in Example 5.

26. $\sin 33^\circ$ as first approximated in Exercise 22.

27. If $f(2) = 4$, $f'(2) = -1$, and $0 \le f''(x) \le 1/x$ for $x > 0$, find the smallest interval you can be sure contains $f(3)$.

28. If $f(2) = 4$, $f'(2) = -1$, and $\dfrac{1}{2x} \le f''(x) \le \dfrac{1}{x}$ for $2 \le x \le 3$, find the best approximation you can for $f(3)$.

29. If $g(2) = 1$, $g'(2) = 2$, and $|g''(x)| < 1 + (x - 2)^2$ for all $x > 0$, find the best approximation you can for $g(1.8)$. How large can the error be?

30. Show that the linearization of $\sin\theta$ at $\theta = 0$ is $L(\theta) = \theta$. How large can the percentage error in the approximation $\sin\theta \approx \theta$ be if $|\theta|$ is less than 17°?

31. A spherical balloon is inflated so that its radius increases from 20.00 cm to 20.20 cm in 1 min. By approximately how much has its volume increased in that minute?

4.10 Taylor Polynomials

The linearization of a function $f(x)$ about $x = a$, namely, the linear function

$$P_1(x) = L(x) = f(a) + f'(a)(x - a),$$

describes the behaviour of f near a better than any other polynomial of degree 1 because both P_1 and f have the same value and the same derivative at a:

$$P_1(a) = f(a) \qquad \text{and} \qquad P_1'(a) = f'(a).$$

(We are now using the symbol P_1 instead of L to stress the fact that the linearization is a polynomial of degree at most 1.)

We can obtain even better approximations to $f(x)$ by using quadratic or higher-degree polynomials and matching more derivatives at $x = a$. For example, if f is twice differentiable near a, then the polynomial

$$P_2(x) = f(a) + f'(a)(x-a) + \frac{f''(a)}{2}(x-a)^2$$

satisfies $P_2(a) = f(a)$, $P_2'(a) = f'(a)$, and $P_2''(a) = f''(a)$ and describes the behaviour of f near a better than any other polynomial of degree at most 2.

In general, if $f^{(n)}(x)$ exists in an open interval containing $x = a$, then the polynomial

$$\begin{aligned} P_n(x) = f(a) &+ \frac{f'(a)}{1!}(x-a) + \frac{f''(a)}{2!}(x-a)^2 \\ &+ \frac{f'''(a)}{3!}(x-a)^3 + \cdots + \frac{f^{(n)}(a)}{n!}(x-a)^n \end{aligned}$$

matches f and its first n derivatives at $x = a$,

$$P_n(a) = f(a), \quad P_n'(a) = f'(a), \quad \ldots, \quad P_n^{(n)}(a) = f^{(n)}(a),$$

and so describes $f(x)$ near $x = a$ better than any other polynomial of degree at most n. P_n is called the **nth-order Taylor polynomial for f about a.** (Taylor polynomials about 0 are usually called **Maclaurin** polynomials.) The 0th-order Taylor polynomial for f about a is just the constant function $P_0(x) = f(a)$. The nth-order Taylor polynomial for f about a is sometimes called the nth-*degree* Taylor polynomial, but its degree will actually be less than n if $f^{(n)}(a) = 0$.

EXAMPLE 1 Find the following Taylor polynomials:

(a) $P_2(x)$ for $f(x) = \sqrt{x}$ about $x = 25$.

(b) $P_3(x)$ for $g(x) = \ln x$ about $x = e$.

Solution (a) $f'(x) = (1/2)x^{-1/2}$, $f''(x) = -(1/4)x^{-3/2}$. Thus,

$$\begin{aligned} P_2(x) &= f(25) + f'(25)(x-25) + \frac{f''(25)}{2!}(x-25)^2 \\ &= 5 + \frac{1}{10}(x-25) - \frac{1}{1{,}000}(x-25)^2. \end{aligned}$$

(b) $g'(x) = \dfrac{1}{x}$, $g''(x) = -\dfrac{1}{x^2}$, $g'''(x) = \dfrac{2}{x^3}$. Thus,

$$\begin{aligned} P_3(x) &= g(e) + g'(e)(x-e) + \frac{g''(e)}{2!}(x-e)^2 + \frac{g'''(e)}{3!}(x-e)^3 \\ &= 1 + \frac{1}{e}(x-e) - \frac{1}{2e^2}(x-e)^2 + \frac{1}{3e^3}(x-e)^3. \end{aligned}$$

EXAMPLE 2 Find the nth-order Maclaurin polynomial $P_n(x)$ for e^x. Use $P_0(1)$, $P_1(1)$, $P_2(1)$, ... to calculate approximate values for $e = e^1$. Stop when you think you have 3 decimal places correct.

Solution Since every derivative of e^x is e^x and so is 1 at $x = 0$, the nth-order Maclaurin polynomial for e^x (i.e., Taylor polynomial at $x = 0$) is

$$P_n(x) = 1 + \frac{x}{1!} + \frac{x^2}{2!} + \frac{x^3}{3!} + \cdots + \frac{x^n}{n!}.$$

Thus, we have for $x = 1$, adding one more term at each step:

$$\begin{aligned}
P_0(1) &= 1 \\
P_1(1) &= P_0(1) + \frac{1}{1!} = 1 + 1 = 2 \\
P_2(1) &= P_1(1) + \frac{1}{2!} = 2 + \frac{1}{2} = 2.5 \\
P_3(1) &= P_2(1) + \frac{1}{3!} = 2.5 + \frac{1}{6} = 2.6666 \\
P_4(1) &= P_3(1) + \frac{1}{4!} = 2.6666 + \frac{1}{24} = 2.7083 \\
P_5(1) &= P_4(1) + \frac{1}{5!} = 2.7083 + \frac{1}{120} = 2.7166 \\
P_6(1) &= P_5(1) + \frac{1}{6!} = 2.7166 + \frac{1}{720} = 2.7180 \\
P_7(1) &= P_6(1) + \frac{1}{7!} = 2.7180 + \frac{1}{5{,}040} = 2.7182.
\end{aligned}$$

It appears that $e \approx 2.718$ to 3 decimal places. We will verify in Example 5 below that $P_7(1)$ does indeed give this much precision. The graphs of e^x and its first four Maclaurin polynomials are shown in Figure 4.65.

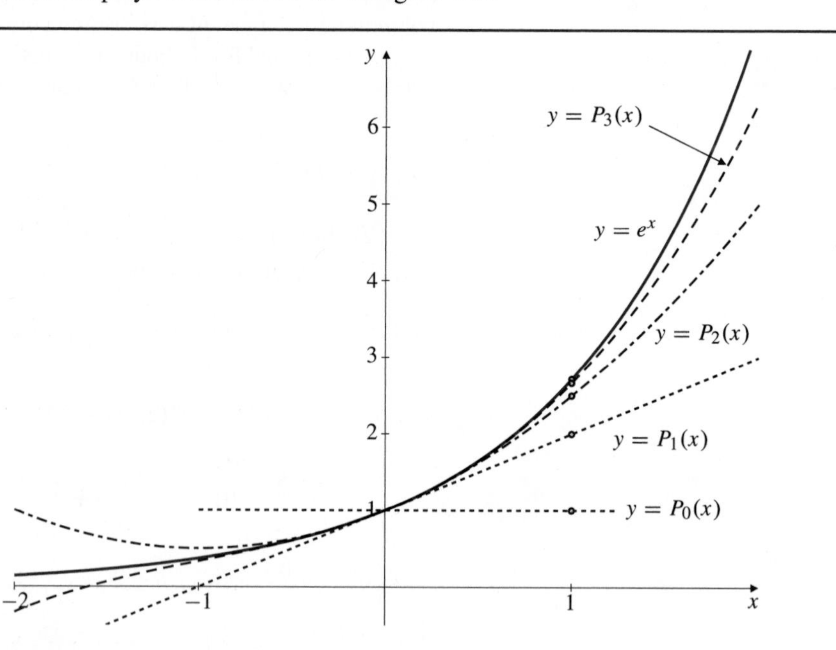

Figure 4.65 Some Maclaurin polynomials for e^x

EXAMPLE 3 Find Maclaurin polynomials $P_1(x)$, $P_2(x)$, $P_3(x)$, and $P_4(x)$ for $f(x) = \sin x$. Then write the general Maclaurin polynomials $P_{2n-1}(x)$ and $P_{2n}(x)$ for that function.

Solution We have $f'(x) = \cos x$, $f''(x) = -\sin x$, $f'''(x) = -\cos x$, and $f^{(4)}(x) = \sin x = f(x)$, so the pattern repeats for higher derivatives. Since

$$\begin{array}{llll}
f(0) = 0, & f''(0) = 0, & f^{(4)}(0) = 0, & f^{(6)}(0) = 0, \ldots \\
f'(0) = 1, & f'''(0) = -1, & f^{(5)}(0) = 1, & f^{(7)}(0) = -1, \ldots
\end{array}$$

we have

$$P_1(x) = 0 + x = x$$
$$P_2(x) = x + \frac{0}{2!}x^2 = x = P_1(x)$$
$$P_3(x) = x - \frac{1}{3!}x^3 = x - \frac{x^3}{3!}$$
$$P_4(x) = x - \frac{1}{3!}x^3 + \frac{0}{4!}x^4 = x - \frac{x^3}{3!} = P_3(x).$$

In general, $f^{(2n-1)}(0) = (-1)^{n-1}$ and $f^{(2n)}(0) = 0$, so

$$P_{2n-1}(x) = P_{2n}(x) = x - \frac{x^3}{3!} + \frac{x^5}{5!} - \cdots + (-1)^{n-1}\frac{x^{2n-1}}{(2n-1)!}.$$

Taylor's Formula

The following theorem provides a formula for the error in a Taylor approximation $f(x) \approx P_n(x)$ similar to that provided for linear approximation by Theorem 11.

THEOREM 12

Taylor's Theorem

If the $(n+1)$st-order derivative, $f^{(n+1)}(t)$, exists for all t in an interval containing a and x, and if $P_n(x)$ is the nth-order Taylor polynomial for f about a, that is,

$$P_n(x) = f(a) + f'(a)(x-a) + \frac{f''(a)}{2!}(x-a)^2 + \cdots + \frac{f^{(n)}(a)}{n!}(x-a)^n,$$

then the error $E_n(x) = f(x) - P_n(x)$ in the approximation $f(x) \approx P_n(x)$ is given by

$$E_n(x) = \frac{f^{(n+1)}(s)}{(n+1)!}(x-a)^{n+1},$$

where s is some number between a and x. The resulting formula

$$f(x) = f(a) + f'(a)(x-a) + \frac{f''(a)}{2!}(x-a)^2 + \cdots + \frac{f^{(n)}(a)}{n!}(x-a)^n + \frac{f^{(n+1)}(s)}{(n+1)!}(x-a)^{n+1}, \quad \text{for some } s \text{ between } a \text{ and } x,$$

is called **Taylor's formula with Lagrange remainder**; the Lagrange remainder term is the explicit formula given above for $E_n(x)$.

Note that the error term (Lagrange remainder) in Taylor's formula looks just like the next term in the Taylor polynomial would look if we continued the Taylor polynomial to include one more term (of degree $n+1$) EXCEPT that the derivative $f^{(n+1)}$ is not evaluated at a but rather at some (generally unknown) point s between a and x. This makes it easy to remember Taylor's formula.

PROOF Observe that the case $n = 0$ of Taylor's formula, namely,

$$f(x) = P_0(x) + E_0(x) = f(a) + \frac{f'(s)}{1!}(x-a),$$

is just the Mean-Value Theorem

$$\frac{f(x) - f(a)}{x - a} = f'(s) \quad \text{for some } s \text{ between } a \text{ and } x.$$

Also note that the case $n = 1$ is just the error formula for linearization given in Theorem 11.

We will complete the proof for higher n using mathematical induction. (See the proof of Theorem 2 in Section 2.3.) Suppose, therefore, that we have proved the case $n = k - 1$, where $k \ge 2$ is an integer. Thus, we are assuming that if f is *any* function whose kth derivative exists on an interval containing a and x, then

$$E_{k-1}(x) = \frac{f^{(k)}(s)}{k!}(x-a)^k,$$

where s is some number between a and x. Let us consider the next higher case: $n = k$. As in the proof of Theorem 11, we assume $x > a$ (the case $x < a$ is similar) and apply the Generalized Mean-Value Theorem to the functions $E_k(t)$ and $(t-a)^{k+1}$ on $[a, x]$. Since $E_k(a) = 0$, we obtain a number u in (a, x) such that

$$\frac{E_k(x)}{(x-a)^{k+1}} = \frac{E_k(x) - E_k(a)}{(x-a)^{k+1} - (a-a)^{k+1}} = \frac{E_k'(u)}{(k+1)(u-a)^k}.$$

Now

$$\begin{aligned} E_k'(u) &= \frac{d}{dt}\left(f(t) - f(a) - f'(a)(t-a) - \frac{f''(a)}{2!}(t-a)^2 \right. \\ &\qquad \left. - \cdots - \frac{f^{(k)}(a)}{k!}(t-a)^k \right)\Bigg|_{t=u} \\ &= f'(u) - f'(a) - f''(a)(u-a) - \cdots - \frac{f^{(k)}(a)}{(k-1)!}(u-a)^{k-1}. \end{aligned}$$

This last expression is just $E_{k-1}(u)$ for the function f' instead of f. By the induction assumption it is equal to

$$\frac{(f')^{(k)}(s)}{k!}(u-a)^k = \frac{f^{(k+1)}(s)}{k!}(u-a)^k,$$

for some s between a and u. Therefore,

$$E_k(x) = \frac{f^{(k+1)}(s)}{(k+1)!}(x-a)^{k+1}.$$

We have shown that the case $n = k$ of Taylor's Theorem is true if the case $n = k - 1$ is true, and the inductive proof is complete.

Remark For any value of x for which $\lim_{n\to\infty} E_n(x) = 0$, we can ensure that the Taylor approximation $f(x) \approx P_n(x)$ is as close as we want by choosing n large enough.

EXAMPLE 4 Use the 2nd-order Taylor polynomial for $\sqrt{x}$ about $x = 25$ found in Example 1(a) to approximate $\sqrt{26}$. Estimate the size of the error, and specify an interval that you can be sure contains $\sqrt{26}$.

Solution In Example 1(a) we calculated $f''(x) = -(1/4)x^{-3/2}$ and obtained the Taylor polynomial

$$P_2(x) = 5 + \frac{1}{10}(x-25) - \frac{1}{1{,}000}(x-25)^2.$$

The required approximation is

$$\sqrt{26} = f(26) \approx P_2(26) = 5 + \frac{1}{10}(26-25) - \frac{1}{1{,}000}(26-25)^2 = 5.099.$$

Now $f'''(x) = (3/8)x^{-5/2}$. For $25 < s < 26$, we have

$$|f'''(s)| \le \frac{3}{8}\frac{1}{25^{5/2}} = \frac{3}{8 \times 3{,}125} = \frac{3}{25{,}000}.$$

Thus, the error in the approximation satisfies

$$|E_2(26)| \le \frac{3}{25{,}000 \times 6}(26-25)^3 = \frac{1}{50{,}000} = 0.000\,02.$$

Therefore, $\sqrt{26}$ lies in the interval $(5.098\,98, 5.099\,02)$.

EXAMPLE 5 Use Taylor's Theorem to confirm that the Maclaurin polynomial $P_7(x)$ for e^x is sufficient to give e correct to 3 decimal places as claimed in Example 2.

Solution The error in the approximation $e^x \approx P_n(x)$ satisfies

$$E_n(x) = \frac{e^s}{(n+1)!}x^{n+1}, \quad \text{for some } s \text{ between 0 and } x.$$

If $x = 1$, then $0 < s < 1$, so $e^s < e < 3$ and $0 < E_n(1) < 3/(n+1)!$. To get an approximation for $e = e^1$ correct to 3 decimal places, we need to have $E_n(1) < 0.0005$. Since $3/(8!) = 3/40{,}320 \approx 0.000\,074$, but $3/(7!) = 3/5{,}040 \approx 0.000\,59$, we can be sure $n = 7$ will do, but we cannot be sure $n = 6$ will do:

$$e \approx 1 + 1 + \frac{1}{2!} + \frac{1}{3!} + \frac{1}{4!} + \frac{1}{5!} + \frac{1}{6!} + \frac{1}{7!} \approx 2.7183 \approx 2.718$$

to 3 decimal places.

Big-O Notation

DEFINITION 9

We write $f(x) = O\big(u(x)\big)$ as $x \to a$ (read this "$f(x)$ is big-Oh of $u(x)$ as x approaches a") provided that

$$|f(x)| \le K|u(x)|$$

holds for some constant K on some open interval containing $x = a$.

Similarly, $f(x) = g(x) + O\big(u(x)\big)$ as $x \to a$ if $f(x) - g(x) = O\big(u(x)\big)$ as $x \to a$, that is, if

$$|f(x) - g(x)| \le K|u(x)| \quad \text{near } a.$$

For example, $\sin x = O(x)$ as $x \to 0$ because $|\sin x| \le |x|$ near 0.

The following properties of big-O notation follow from the definition:

(i) If $f(x) = O\big(u(x)\big)$ as $x \to a$, then $Cf(x) = O\big(u(x)\big)$ as $x \to a$ for any value of the constant C.

(ii) If $f(x) = O\big(u(x)\big)$ as $x \to a$ and $g(x) = O\big(u(x)\big)$ as $x \to a$, then $f(x) \pm g(x) = O\big(u(x)\big)$ as $x \to a$.

(iii) If $f(x) = O\big((x-a)^k u(x)\big)$ as $x \to a$, then $f(x)/(x-a)^k = O\big(u(x)\big)$ as $x \to a$ for any constant k.

Taylor's Theorem says that if $f^{(n+1)}(t)$ exists on an interval containing a and x, and if P_n is the nth-order Taylor polynomial for f at a, then, as $x \to a$,

$$f(x) = P_n(x) + O\big((x-a)^{n+1}\big).$$

This is a statement about how rapidly the graph of the Taylor polynomial $P_n(x)$ approaches that of $f(x)$ as $x \to a$; the vertical distance between the graphs decreases as fast as $|x-a|^{n+1}$. The following theorem shows that the Taylor polynomial $P_n(x)$ is the *only* polynomial of degree at most n whose graph approximates the graph of $f(x)$ that rapidly.

THEOREM 13

If $f(x) = Q_n(x) + O\big((x-a)^{n+1}\big)$ as $x \to a$, where Q_n is a polynomial of degree at most n, then $Q_n(x) = P_n(x)$, that is, Q_n is the Taylor polynomial for $f(x)$ at $x = a$.

PROOF Let P_n be the Taylor polynomial, then properties (i) and (ii) of big-O imply that $R_n(x) = Q_n(x) - P_n(x) = O\big((x-a)^{n+1}\big)$ as $x \to a$. We want to show that $R_n(x)$ is identically zero so that $Q_n(x) = P_n(x)$ for all x. By replacing x with $a + (x - a)$ and expanding powers, we can write $R_n(x)$ in the form

$$R_n(x) = c_0 + c_1(x-a) + c_2(x-a)^2 + \cdots + c_n(x-a)^n.$$

If $R_n(x)$ is not identically zero, then there is a smallest coefficient c_k $(k \le n)$, such that $c_k \ne 0$, but $c_j = 0$ for $0 \le j \le k-1$. Thus,

$$R_n(x) = (x-a)^k\big(c_k + c_{k+1}(x-a) + \cdots + c_n(x-a)^{n-k}\big).$$

Therefore, $\lim_{x\to a} R_n(x)/(x-a)^k = c_k \ne 0$. However, by property (iii) above we have $R_n(x)/(x-a)^k = O\big((x-a)^{n+1-k}\big)$. Since $n+1-k > 0$, this says $R_n(x)/(x-a)^k \to 0$ as $x \to a$. This contradiction shows that $R_n(x)$ must be identically zero. Therefore $Q_n(x) = P_n(x)$ for all x. ∎

Table 5 lists Taylor formulas about 0 (Maclaurin formulas) for some elementary functions, with error terms expressed using big-O notation.

Table 5. Some Maclaurin Formulas with Errors in Big-O Form

	As $x \to 0$:
(a)	$e^x = 1 + x + \dfrac{x^2}{2!} + \dfrac{x^3}{3!} + \cdots + \dfrac{x^n}{n!} + O\big(x^{n+1}\big)$
(b)	$\cos x = 1 - \dfrac{x^2}{2!} + \dfrac{x^4}{4!} - \cdots + (-1)^n \dfrac{x^{2n}}{(2n)!} + O\big(x^{2n+2}\big)$
(c)	$\sin x = x - \dfrac{x^3}{3!} + \dfrac{x^5}{5!} - \cdots + (-1)^n \dfrac{x^{2n+1}}{(2n+1)!} + O\big(x^{2n+3}\big)$
(d)	$\dfrac{1}{1-x} = 1 + x + x^2 + x^3 + \cdots + x^n + O\big(x^{n+1}\big)$
(e)	$\ln(1+x) = x - \dfrac{x^2}{2} + \dfrac{x^3}{3} - \cdots + (-1)^{n-1}\dfrac{x^n}{n} + O\big(x^{n+1}\big)$
(f)	$\tan^{-1} x = x - \dfrac{x^3}{3} + \dfrac{x^5}{5} - \cdots + (-1)^n \dfrac{x^{2n+1}}{2n+1} + O\big(x^{2n+3}\big)$

It is worthwhile remembering these. The first three can be established easily by using Taylor's formula with Lagrange remainder; the other three would require much more effort to verify for general n. In Section 9.6 we will return to the subject of Taylor and Maclaurin polynomials in relation to Taylor and Maclaurin series. At that time we will have access to much more powerful machinery to establish such results. The need to calculate high-order derivatives can make the use of Taylor's formula difficult for all but the simplest functions.

The real importance of Theorem 13 is that it enables us to obtain Taylor polynomials for new functions by combining others already known; as long as the error term is of higher degree than the order of the polynomial obtained, the polynomial must be the Taylor polynomial. We illustrate this with a few examples.

EXAMPLE 6 Find the Maclaurin polynomial of order $2n$ for $\cosh x$.

Solution Write the Taylor formula for e^x at $x = 0$ (from Table 5) with n replaced by $2n + 1$, and then rewrite that with x replaced by $-x$. We get

$$e^x = 1 + x + \frac{x^2}{2!} + \frac{x^3}{3!} + \cdots + \frac{x^{2n}}{(2n)!} + \frac{x^{2n+1}}{(2n+1)!} + O\big(x^{2n+2}\big),$$

$$e^{-x} = 1 - x + \frac{x^2}{2!} - \frac{x^3}{3!} + \cdots + \frac{x^{2n}}{(2n)!} - \frac{x^{2n+1}}{(2n+1)!} + O\big(x^{2n+2}\big)$$

as $x \to 0$. Now average these two to get

$$\cosh x = \frac{e^x + e^{-x}}{2} = 1 + \frac{x^2}{2!} + \frac{x^4}{4!} + \cdots + \frac{x^{2n}}{(2n)!} + O\big(x^{2n+2}\big)$$

as $x \to 0$. By Theorem 13 the Maclaurin polynomial $P_{2n}(x)$ for $\cosh x$ is

$$P_{2n}(x) = 1 + \frac{x^2}{2!} + \frac{x^4}{4!} + \cdots + \frac{x^{2n}}{(2n)!}.$$

EXAMPLE 7 Obtain the Taylor polynomial of order 3 for e^{2x} about $x = 1$ from the corresponding Maclaurin polynomial for e^x (from Table 5).

Solution Writing $x = 1 + (x - 1)$, we have

$$\begin{aligned} e^{2x} &= e^{2+2(x-1)} = e^2 e^{2(x-1)} \\ &= e^2\left[1 + 2(x-1) + \frac{2^2(x-1)^2}{2!} + \frac{2^3(x-1)^3}{3!} + O\big((x-1)^4\big)\right] \end{aligned}$$

as $x \to 1$. By Theorem 13 the Taylor polynomial $P_3(x)$ for e^{2x} at $x = 1$ must be

$$P_3(x) = e^2 + 2e^2(x-1) + 2e^2(x-1)^2 + \frac{4e^2}{3}(x-1)^3.$$

EXAMPLE 8 Use the Taylor formula for $\ln(1 + x)$ (from Table 5) to find the Taylor polynomial $P_3(x)$ for $\ln x$ about $x = e$. (This provides an alternative to using the definition of Taylor polynomial as was done to solve the same problem in Example 1(b).)

Solution We have $x = e + (x - e) = e(1 + t)$ where $t = (x - e)/e$. As $x \to e$ we have $t \to 0$, so

$$\begin{aligned} \ln x &= \ln e + \ln(1+t) = \ln e + t - \frac{t^2}{2} + \frac{t^3}{3} + O(t^4) \\ &= 1 + \frac{x-e}{e} - \frac{1}{2}\left(\frac{x-e}{e}\right)^2 + \frac{1}{3}\left(\frac{x-e}{e}\right)^3 + O\big((x-e)^4\big). \end{aligned}$$

Therefore, by Theorem 13,

$$P_3(x) = 1 + \frac{x-e}{e} - \frac{1}{2}\left(\frac{x-e}{e}\right)^2 + \frac{1}{3}\left(\frac{x-e}{e}\right)^3.$$

Evaluating Limits of Indeterminate Forms

Taylor and Maclaurin polynomials provide us with another method for evaluating limits of indeterminate forms of type [0/0]. For some such limits this method can be considerably easier than using l'Hôpital's Rule.

EXAMPLE 9 Evaluate $\lim\limits_{x\to 0} \dfrac{2\sin x - \sin(2x)}{2e^x - 2 - 2x - x^2}$.

Solution Both the numerator and denominator approach 0 as $x \to 0$. Let us replace the trigonometric and exponential functions with their degree-3 Maclaurin polynomials plus error terms written in big-O notation:

$$\begin{aligned}
\lim_{x\to 0} \frac{2\sin x - \sin(2x)}{2e^x - 2 - 2x - x^2} &= \lim_{x\to 0} \frac{2\left(x - \dfrac{x^3}{3!} + O(x^5)\right) - \left(2x - \dfrac{2^3x^3}{3!} + O(x^5)\right)}{2\left(1 + x + \dfrac{x^2}{2!} + \dfrac{x^3}{3!} + O(x^4)\right) - 2 - 2x - x^2} \\
&= \lim_{x\to 0} \frac{-\dfrac{x^3}{3} + \dfrac{4x^3}{3} + O(x^5)}{\dfrac{x^3}{3} + O(x^4)} \\
&= \lim_{x\to 0} \frac{1 + O(x^2)}{\dfrac{1}{3} + O(x)} = \frac{1}{\dfrac{1}{3}} = 3.
\end{aligned}$$

Observe how we used the properties of big-O as listed in this section. We needed to use Maclaurin polynomials of degree at least 3 because all lower degree terms cancelled out in the numerator and the denominator.

EXAMPLE 10 Evaluate $\lim\limits_{x\to 1} \dfrac{\ln x}{x^2 - 1}$.

Solution This is also of type [0/0]. We begin by substituting $x = 1 + t$. Note that $x \to 1$ corresponds to $t \to 0$. We can use a known Maclaurin polynomial for $\ln(1+t)$. For this limit even the degree 1 polynomial $P_1(t) = t$ with error $O(t^2)$ will do.

$$\begin{aligned}
\lim_{x\to 1} \frac{\ln x}{x^2 - 1} &= \lim_{t\to 0} \frac{\ln(1+t)}{(1+t)^2 - 1} = \lim_{t\to 0} \frac{\ln(1+t)}{2t + t^2} \\
&= \lim_{t\to 0} \frac{t + O(t^2)}{2t + t^2} = \lim_{t\to 0} \frac{1 + O(t)}{2 + t} = \frac{1}{2}.
\end{aligned}$$

EXERCISES 4.10

Find the indicated Taylor polynomials for the functions in Exercises 1–8 by using the definition of Taylor polynomial.

1. for e^{-x} about $x = 0$, order 4.

2. for $\cos x$ about $x = \pi/4$, order 3.

3. for $\ln x$ about $x = 2$, order 4.

4. for $\sec x$ about $x = 0$, order 3.

5. for $\sqrt{x}$ about $x = 4$, order 3.

6. for $1/(1 - x)$ about $x = 0$, order n.

7. for $1/(2 + x)$ about $x = 1$, order n.

8. for $\sin(2x)$ about $x = \pi/2$, order $2n - 1$.

In Exercises 9–14, use second order Taylor polynomials $P_2(x)$ for the given function about the point specified to approximate the indicated value. Estimate the error, and write the smallest interval you can be sure contains the value.

9. $f(x) = x^{1/3}$ about 8; approximate $9^{1/3}$.

10. $f(x) = \sqrt{x}$ about 64; approximate $\sqrt{61}$.

11. $f(x) = \dfrac{1}{x}$ about 1; approximate $\dfrac{1}{1.02}$.

12. $f(x) = \tan^{-1} x$ about 1; approximate $\tan^{-1}(0.97)$.

13. $f(x) = e^x$ about 0; approximate $e^{-0.5}$.

14. $f(x) = \sin x$ about $\pi/4$; approximate $\sin(47°)$.

In Exercises 15–20, write the indicated case of Taylor's formula for the given function. What is the Lagrange remainder in each case?

15. $f(x) = \sin x,\ a = 0,\ n = 7$

16. $f(x) = \cos x,\ a = 0,\ n = 6$

17. $f(x) = \sin x,\ a = \pi/4,\ n = 4$

18. $f(x) = \dfrac{1}{1-x},\ a = 0,\ n = 6$

19. $f(x) = \ln x,\ a = 1, n = 6$

20. $f(x) = \tan x,\ a = 0,\ n = 3$

Find the requested Taylor polynomials in Exercises 21–26 by using known Taylor or Maclaurin polynomials and changing variables as in Examples 6–8.

21. $P_3(x)$ for e^{3x} about $x = -1$.

22. $P_8(x)$ for e^{-x^2} about $x = 0$.

23. $P_4(x)$ for $\sin^2 x$ about $x = 0$. *Hint:* $\sin^2 x = \dfrac{1 - \cos(2x)}{2}$.

24. $P_5(x)$ for $\sin x$ about $x = \pi$.

25. $P_6(x)$ for $1/(1 + 2x^2)$ about $x = 0$

26. $P_8(x)$ for $\cos(3x - \pi)$ about $x = 0$.

27. Find all Maclaurin polynomials $P_n(x)$ for $f(x) = x^3$.

28. Find all Taylor polynomials $P_n(x)$ for $f(x) = x^3$ at $x = 1$.

29. Find the Maclaurin polynomial $P_{2n+1}(x)$ for $\sinh x$ by suitably combining polynomials for e^x and e^{-x}.

30. By suitably combining Maclaurin polynomials for $\ln(1 + x)$ and $\ln(1 - x)$, find the Maclaurin polynomial of order $2n + 1$ for $\tanh^{-1}(x) = \dfrac{1}{2}\ln\left(\dfrac{1+x}{1-x}\right)$.

31. Write Taylor's formula for $f(x) = e^{-x}$ with $a = 0$, and use it to calculate $1/e$ to 5 decimal places. (You may use a calculator but not the e^x function on it.)

32. Write the general form of Taylor's formula for $f(x) = \sin x$ at $x = 0$ with Lagrange remainder. How large need n be taken to ensure that the corresponding Taylor polynomial approximation will give the sine of 1 radian correct to 5 decimal places?

33. What is the best order 2 approximation to the function $f(x) = (x - 1)^2$ at $x = 0$? What is the error in this approximation? Now answer the same questions for $g(x) = x^3 + 2x^2 + 3x + 4$. Can the constant $1/6 = 1/3!$, in the error formula for the degree 2 approximation, be improved (i.e., made smaller)?

34. By factoring $1 - x^{n+1}$ (or by long division), show that

$$\frac{1}{1-x} = 1 + x + x^2 + x^3 + \cdots + x^n + \frac{x^{n+1}}{1-x}. \qquad (*)$$

Next, show that if $|x| \le K < 1$, then

$$\left|\frac{x^{n+1}}{1-x}\right| \le \frac{1}{1-K}|x^{n+1}|.$$

This implies that $x^{n+1}/(1 - x) = O(x^{n+1})$ as $x \to 0$ and confirms formula (d) of Table 5. What does Theorem 13 then say about the nth-order Maclaurin polynomial for $1/(1 - x)$?

35. By differentiating identity (*) in Exercise 34 and then replacing n with $n + 1$, show that

$$\frac{1}{(1-x)^2} = 1 + 2x + 3x^2 + \cdots + (n+1)x^n + \frac{n + 2 - (n+1)x}{(1-x)^2}x^{n+1}.$$

Then use Theorem 13 to determine the nth-order Maclaurin polynomial for $1/(1 - x)^2$.

4.11 Roundoff Error, Truncation Error, and Computers

In Section 4.7 we introduced the idea of **roundoff error**, while in Sections 4.9 and 4.10 we discussed the result of approximating a function by its Taylor polynomials. The resulting error here is known as **truncation error.** This conventional terminology may be a bit confusing at first because rounding off is itself a kind of truncation of the digital representation of a number. However in numerical analysis "truncation" is reserved for discarding higher order terms, typically represented by big-O, often leaving a Taylor polynomial.

Truncation error is a crucial source of error in using computers to do mathematical operations. In computation with computers, many of the mathematical functions and structures being investigated are approximated by polynomials in order to make it possible for computers to manipulate them. However, the other source of error, roundoff, is ubiquitous, so it is inevitable that mathematics on computers has to involve consideration of both sources of error. These sources can sometimes be treated independently, but in other circumstances they can interact with each other in fascinating ways. In this section we look at some of these fascinating interactions in the form of Numerical Monsters using Maple. Of course, as stated previously, the issues concern all calculation on computers and not Maple in particular.

Taylor Polynomials in Maple

In much of the following discussion we will be examining the function $\sin x$. Let us begin by defining the Maple expression `s := sin(x)` to denote this function: The Maple input

```
> u := taylor(s, x=0, 5);
```

produces the Taylor polynomial of degree 4 about $x = 0$ (i.e., a Maclaurin polynomial) for $\sin(x)$ together with a big-O term of order x^5:

$$u := x - \frac{1}{6}x^3 + O(x^5)$$

The presence of the big-O term means that u is an actual representation of $\sin x$; there is no error involved. If we want to get an actual Taylor polynomial, we need to convert the expression for u to drop off the big-O term. Since the coefficient of x^4 is zero, let us call the resulting polynomial P_3:

```
> P3 := convert(u, polynom);
```

$$P3 := x - \frac{1}{6}x^3$$

Unlike u, P_3 is not an exact representation of $\sin x$; it is only an approximation. The discarded term $O(x^5) = s - P_3 = u - P_3$ is the error in this approximation. On the basis of the discussion in the previous section, this truncation error can be expected to be quite small for x close to 0, a fact that is confirmed by the Maple plot in Figure 4.66(a). The behaviour is much as expected. $\sin x$ behaves like the cubic polynomial near 0 (so the difference is nearly 0), while farther from 0 the cubic term dominates the expression.

```
> plot(s-P3, x=-1..1, style=point,
   symbol=point, numpoints=1000);
```

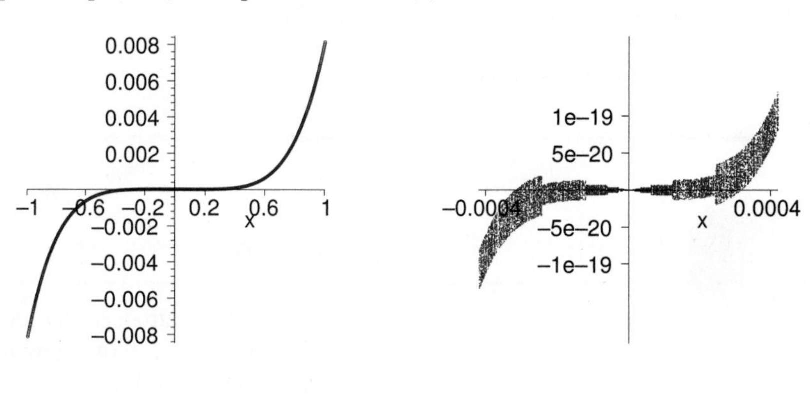

Figure 4.66 The error $\sin x - P_3(x)$ over (a) the interval $[-1, 1]$, and (b) the interval $[-4.2 \times 10^{-4}, 4.2 \times 10^{-4}]$

The limiting behaviour near 0 can be explored by changing the plot window. If the Maple plot instruction is revised to

```
> plot(s-P3, x=-0.42e-3..0.42e-3, style=point,
   symbol=point, numpoints=1000);
```

the plot in Figure 4.66(b) results. What is this structure? Clearly the distances from the x-axis are very small, and one can see the cubic-like behaviour. But why are the points not distributed along a single curve, filling out a jagged arrow-like structure instead? This is another numerical monster connected to roundoff error, as we can see if we plot $\sin(x) - P_3(x)$ together with the functions $\pm(\epsilon/2)\sin(x)$ and $\pm(\epsilon/4)\sin x$, where $\epsilon = 2^{-52}$ is machine epsilon, as calculated in Section 4.7.

```
> eps := evalf(2^(-52)):
```

```
> plot([s-P3, -eps*s/2, eps*s/2, -eps*s/4, eps*s/4],
  x=-0.1e-3,0.1e-3, colour=[blue,grey,grey,black,black],
  style=point, symbol=point, numpoints=1000);
```

The result is in Figure 4.67. The black and grey envelope curves (which appear like straight lines since the plot window is so close to the origin) link the structure of the plot to machine epsilon; the seemingly random points are not as random as they first seemed.

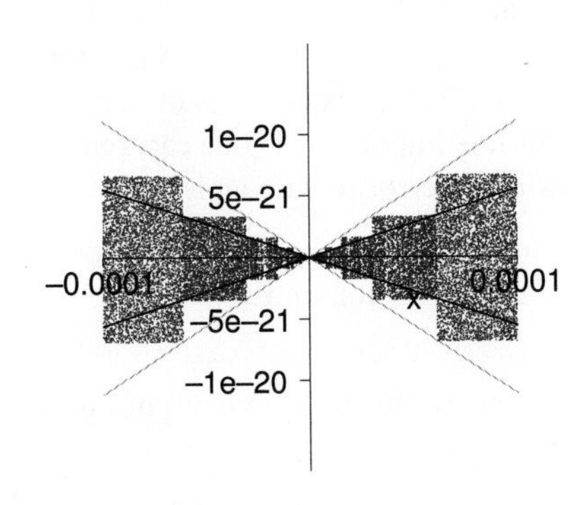

Figure 4.67 Examining the structure of the Maple plot of $\sin x - P_3(x)$ for x in $[-0.0001, 0.0001]$. Note the relationship to the envelope curves $y = \pm(\epsilon/4)\sin x$ (black), and $y = \pm(\epsilon/2)\sin x$ (grey)

Moreover, this structure is distinctive to Maple. Other software packages, such as Matlab, produce a somewhat different, but still spurious, structure for the same plotting window. Try some others. If different software produces different behaviour under the same instructions, it is certain that some type of computational error is involved. Software-dependent behaviour is one sure sign of computational error.

A distinctive aspect of this monster is that for a large plot window, the truncation error dominates, while near zero, where the truncation error approaches zero, the roundoff error dominates. This is a common relationship between truncation error and roundoff error. However, the roundoff error shows up for plot windows near zero, while the truncation error is dominant over wide ranges of plot windows. Is this always true for truncation error? No — as the next monster shows.

Persistent Roundoff Error

The trade-off between truncation error and roundoff error is distinctive, but one should not get the impression that roundoff error only matters in extreme limiting cases in certain plot windows. Consider, for example, the function $f(x) = x^2-2x+1-(x-1)^2$. It is identically 0, not just 0 in the limiting case $x = 0$. However, the computer evaluates the two mathematically equivalent parts of the function f differently, leaving different errors from rounding off the true values of the numbers inserted into the expression. The difference of the result is then not exactly 0. A plot of $f(x)$ on the interval $[-10^8, 10^8]$ is produced by the Maple command

```
> plot([eps*(x-1)^2,eps*(x-1)^2/2,-eps*(x-1)^2,
  -eps*(x-1)^2/2,(x^2-2*x+1)-(x-1)^2],
  x=-1e8..1e8,numpoints=1500,style=point,symbol=point,
  color=[black,grey,black,grey,blue],
  tickmarks=[[-1e8,-5e7,5e7,1e8],[-2,-1,1,2]]);
```

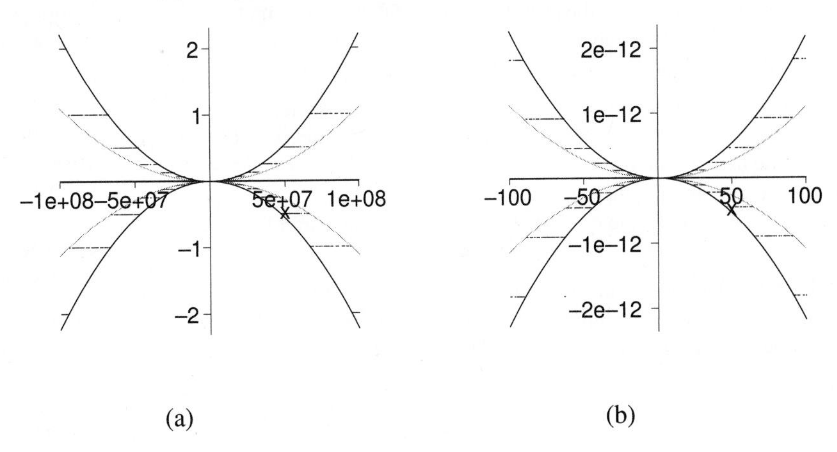

Figure 4.68 The values of $x^2 - 2x + 1 - (x - 1)^2$ (colour) lie between the parabolas $\pm\epsilon(x - 1)^2$ (black) and $\pm\epsilon(x - 1)^2/2$ (grey) for
(a) $-10^8 \le x \le 10^8$, and
(b) $-100 \le x \le 100$

It is shown in Figure 4.68(a). The spurious values of $f(x)$ seem like rungs on a ladder. Note that these false nonzero values of $f(x)$ (colour) are not small compared to 1. This is because the window is so wide. But the error is clearly due to roundoff as the grey and black envelope curves are proportional to machine epsilon. This plot is largely independent of the width of the window chosen. Figure 4.68(b) is the same plot with a window one million times narrower. Except for of a change of scale, it is virtually identical to the plot in Figure 4.68(a). This behaviour is quite different from the numerical monster involving Taylor polynomials encountered above.

Truncation, Roundoff, and Computer Algebra

One of the more modern developments in computer mathematics is the computer's ability to deal with mathematics symbolically. This important capability is known as "computer algebra." For example, Maple can generate Taylor expansions of very high order. This might appear to make the issue of error less important. If one can generate exact Taylor polynomials of very high order, how could error remain an issue?

To see how the finiteness of computers intrudes on our calculations in this case too, let us consider the Taylor (Maclaurin) polynomial of degree 99 for $\sin x$.

```
> v := taylor(s, x=0, 100):  P99 := convert(v, polynom):
```

It is good to suppress the output here; each command produces screensfull of output. Figure 4.69 shows the result of the Maple plot command

```
> plot([P99,s],x=35..39,y=-3..3,colour=[blue,black],
  style=point,symbol=point,numpoints=500,
  xtickmarks=[36,37,38,39]);
```

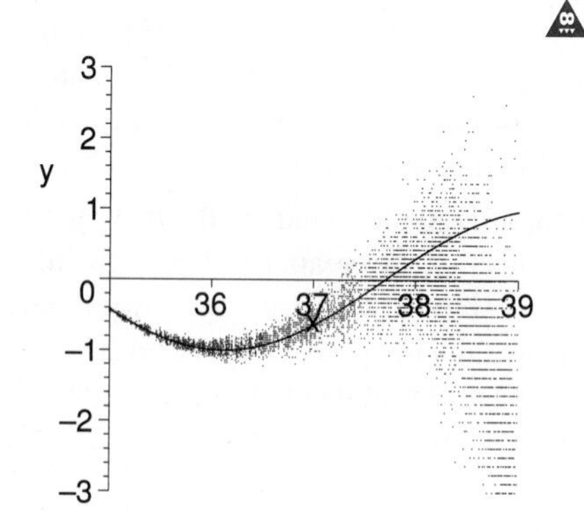

Figure 4.69 The coloured cloud results from Maple's attempt to evaluate the polynomial $P_{99}(x)$ at 500 values of x between 35 and 39

The black curve is the graph of the sine function, and the colour tornado-like cloud is the plot of $P_{99}(x)$ that Maple produces. For plotting, the polynomial must be evaluated at specific values of x. The algorithm cannot employ the large rational expressions for coefficients and high powers of input values. In order to place the result into an actual pixel on the computer screen, the value of the polynomial must be converted to a floating-point number. Then with the adding and subtracting of 100 terms involving rounded powers, roundoff error returns despite the exact polynomial that we began with.

Of course, there are often tactics to fix these types of problems, but the only way to know what the problems are that need fixing is to understand the mathematics in the first place. But this also means that careful calculations on computers constitute a full field of modern research, requiring considerable mathematical knowledge.

EXERCISES 4.11

1. Use Maple to repeat the plots of Figure 4.68, except using the mathematically equivalent function $(x-1)^2 - (x^2 - 2x + 1)$. Does the result look the same? Is the result surprising?

2. Use Maple to graph $f - P_4(x)$ where $f(x) = \cos x$ and $P_4(x)$ is the 4th degree Taylor polynomial of f about $x = 0$. Use the interval $[-10^{-3}/2, 10^{-3}/2]$ for the plot and plot 1000 points. On the same plot, graph $\pm\epsilon f/2$ and $\pm\epsilon f/4$, where ϵ is machine epsilon. How does the result differ from what is expected mathematically?

3. If a real number x is represented on a computer, it is replaced by a floating-point number $F(x)$; x is said to be "floated" by the function F. Show that the relative error in floating for a base two machine satisfies

$$|\text{error}| = |x - F(x)| \le \epsilon |x|,$$

where $\epsilon = 2^{-t}$ and t is the number of base two digits (bits) in the floating point number.

4. Consider two different, but mathematically equivalent expressions, having the value C after evaluation. On a computer, with each step in the evaluation of each of the expressions, roundoff error is introduced as digits are discarded and rounded according to various rules. In subsequent steps, resulting error is added or subtracted according to the details of the expression producing a final error that depends in detail on the expression, the particular software package, the operating system, and the machine hardware. Computer errors are not equivalent for the two expressions, even when the expressions are mathematically equivalent.

(a) If we suppose that the computer satisfactorily evaluates the expressions for many input values within an interval, all to within machine precision, why might we expect the difference of these expressions on a computer to have an error contained within an interval $[-\epsilon C, \epsilon C]$?

(b) Is it possible for exceptional values of the error to lie outside that interval in some cases? Why?

(c) Is it possible for the error to be much smaller than the interval indicates? Why?

CHAPTER REVIEW

Key Ideas

- **What do the following words, phrases, and statements mean?**
  - ◇ critical point of f
  - ◇ singular point of f
  - ◇ inflection point of f
  - ◇ f has absolute maximum value M
  - ◇ f has a local minimum value at $x = c$
  - ◇ vertical asymptote
  - ◇ horizontal asymptote
  - ◇ oblique asymptote
  - ◇ machine epsilon
  - ◇ the linearization of $f(x)$ about $x = a$
  - ◇ the Taylor polynomial of degree n of $f(x)$ about $x = a$
  - ◇ Taylor's formula with Lagrange remainder
  - ◇ $f(x) = O\big((x-a)^n\big)$ as $x \to a$
  - ◇ a root of $f(x) = 0$
  - ◇ a fixed point of $f(x)$
  - ◇ an indeterminate form
  - ◇ l'Hôpital's Rules
- **Describe how to estimate the error in a linear (tangent line) approximation to the value of a function.**
- **Describe how to find a root of an equation $f(x) = 0$ by using Newton's Method. When will this method work well?**

Review Exercises

1. If the radius r of a ball is increasing at a rate of 2 percent per minute, how fast is the volume V of the ball increasing?

2. **(Gravitational attraction)** The gravitational attraction of the earth on a mass m at distance r from the centre of the earth is a continuous function of r for $r \ge 0$, given by

$$F = \begin{cases} \dfrac{mgR^2}{r^2} & \text{if } r \ge R \\ mkr & \text{if } 0 \le r < R, \end{cases}$$

where R is the radius of the earth, and g is the acceleration due to gravity at the surface of the earth.

   (a) Find the constant k in terms of g and R.

   (b) F decreases as m moves away from the surface of the earth, either upward or downward. Show that F decreases as r increases from R at twice the rate at which F decreases as r decreases from R.

3. **(Resistors in parallel)** Two variable resistors R_1 and R_2 are connected in parallel so that their combined resistance R is given by

$$\frac{1}{R} = \frac{1}{R_1} + \frac{1}{R_2}.$$

At an instant when $R_1 = 250$ ohms and $R_2 = 1000$ ohms, R_1 is increasing at a rate of 100 ohms/min. How fast must R_2 be changing at that moment (a) to keep R constant? and (b) to enable R to increase at a rate of 10 ohms/min?

4. **(Gas law)** The volume V (in m^3), pressure P (in kilopascals, kPa), and temperature T (in kelvin, K) for a sample of a certain gas satisfy the equation $pV = 5.0T$.

   (a) How rapidly does the pressure increase if the temperature is 400 K and increasing at 4 K/min while the gas is kept confined in a volume of 2.0 m^3?

   (b) How rapidly does the pressure decrease if the volume is 2 m^3 and increases at 0.05 m^3/min while the temperature is kept constant at 400 K?

5. **(The size of a print run)** It costs a publisher \$10,000 to set up the presses for a print run of a book and \$8 to cover the material costs for each book printed. In addition, machinery servicing, labour, and warehousing add another $\$6.25 \times 10^{-7}x^2$ to the cost of each book if x copies are manufactured during the printing. How many copies should the publisher print in order to minimize the average cost per book?

6. **(Maximizing profit)** A bicycle wholesaler must pay the manufacturer \$75 for each bicycle. Market research tells the wholesaler that if she charges her customers \$$x$ per bicycle, she can expect to sell $N(x) = 4.5 \times 10^6/x^2$ of them. What price should she charge to maximize her profit, and how many bicycles should she order from the manufacturer?

7. Find the largest possible volume of a right-circular cone that can be inscribed in a sphere of radius R.

8. **(Minimizing production costs)** The cost \$$C(x)$ of production in a factory varies with the amount x of product manufactured. The cost may rise sharply with x when x is small, and more slowly for larger values of x because of economies of scale. However, if x becomes too large, the resources of the factory can be overtaxed, and the cost can begin to rise quickly again. Figure 4.70 shows the graph of a typical such cost function $C(x)$.

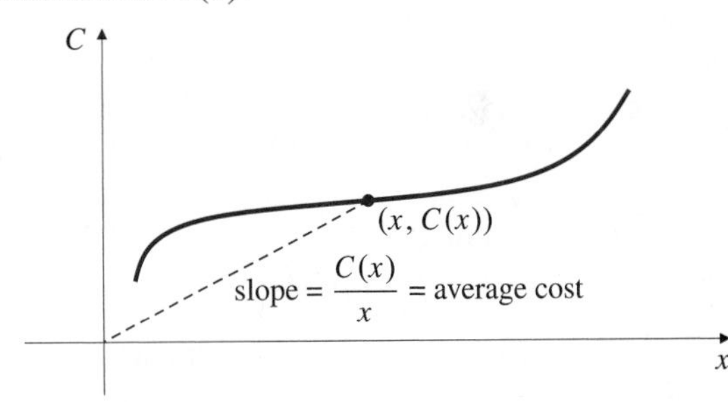

Figure 4.70

If x units are manufactured, the average cost per unit is \$$C(x)/x$, which is the slope of the line from the origin to the point $(x, C(x))$ on the graph.

   (a) If it is desired to choose x to minimize this average cost per unit (as would be the case if all units produced could be sold for the same price), show that x should be chosen to make the average cost equal to the marginal cost:

$$\frac{C(x)}{x} = C'(x).$$

   (b) Interpret the conclusion of (a) geometrically in the figure.

   (c) If the average cost equals the marginal cost for some x, does x necessarily minimize the average cost?

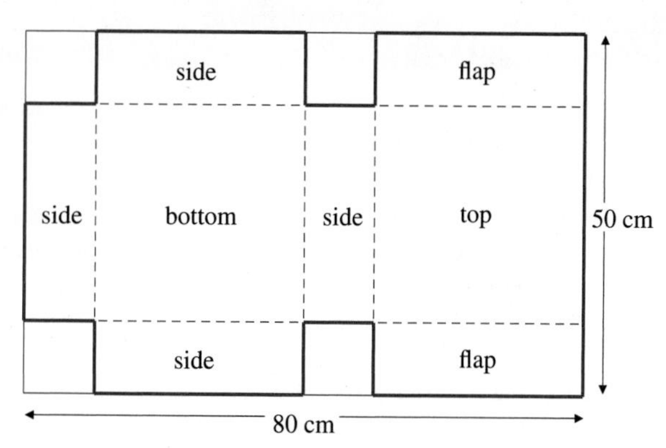

Figure 4.71

9. **(Box design)** Four squares are cut out of a rectangle of cardboard 50 cm by 80 cm, as shown in Figure 4.71, and the remaining piece is folded into a closed, rectangular box, with two extra flaps tucked in. What is the largest possible volume for such a box?

10. **(Yield from an orchard)** A certain orchard has 60 trees and produces an average of 800 apples per tree per year. If the density of trees is increased, the yield per tree drops; for each additional tree planted, the average yield per tree is reduced by 10 apples per year. How many more trees should be planted to maximize the total annual yield of apples from the orchard?

11. **(Rotation of a tracking antenna)** What is the maximum rate at which the antenna in Exercise 41 of Section 4.1 must be able to turn in order to track the rocket during its entire vertical ascent?

12. An oval table has its outer edge in the shape of the curve $x^2 + y^4 = 1/8$, where x and y are measured in metres. What is the width of the narrowest hallway in which the table can be turned horizontally through 180°?

13. A hollow iron ball whose shell is 2 cm thick weighs half as much as it would if it were solid iron throughout. What is the radius of the ball?

14. **(Range of a cannon fired from a hill)** A cannon ball is fired with a speed of 200 ft/s at an angle of 45° above the horizontal from the top of a hill whose height at a horizontal distance x ft from the top is $y = 1{,}000/(1+(x/500)^2)$ ft above sea level. How far does the cannon ball travel horizontally before striking the ground?

15. **(Linear approximation for a pendulum)** Because $\sin\theta \approx \theta$ for small values of $|\theta|$, the nonlinear equation of motion of a simple pendulum

$$\frac{d^2\theta}{dt^2} = -\frac{g}{L}\sin\theta,$$

which determines the displacement angle $\theta(t)$ away from the vertical at time t for a simple pendulum, is frequently approximated by the simpler linear equation

$$\frac{d^2\theta}{dt^2} = -\frac{g}{L}\theta,$$

when the maximum displacement of the pendulum is not large. What is the percentage error in the right side of the equation if $|\theta|$ does not exceed 20°?

16. Find the Taylor polynomial of degree 6 for $\sin^2 x$ about $x = 0$ and use it to help you evaluate

$$\lim_{x\to 0}\frac{3\sin^2 x - 3x^2 + x^4}{x^6}.$$

17. Use a second-order Taylor polynomial for $\tan^{-1} x$ about $x = 1$ to find an approximate value for $\tan^{-1}(1.1)$. Estimate the size of the error by using Taylor's formula.

18. The line $2y = 10x - 19$ is tangent to $y = f(x)$ at $x = 2$. If an initial approximation $x_0 = 2$ is made for a root of $f(x) = 0$ and Newton's Method is applied once, what will be the new approximation that results?

19. Find all solutions of the equation $\cos x = (x-1)^2$ to 10 decimal places.

20. Find the shortest distance from the point $(2, 0)$ to the curve $y = \ln x$.

21. A car is travelling at night along a level, curved road whose equation is $y = e^x$. At a certain instant its headlights illuminate a signpost located at the point $(1, 1)$. Where is the car at that instant?

Challenging Problems

1. **(Growth of a crystal)** A single cubical salt crystal is growing in a beaker of salt solution. The crystal's volume V increases at a rate proportional to its surface area and to the amount by which its volume is less than a limiting volume V_0:

$$\frac{dV}{dt} = kx^2(V_0 - V),$$

where x is the edge length of the crystal at time t.

(a) Using $V = x^3$, transform the equation above to one that gives the rate of change dx/dt of the edge length x in terms of x.

(b) Show that the growth rate of the edge of the crystal decreases with time but remains positive as long as $x < x_0 = V_0^{1/3}$.

(c) Find the volume of the crystal when its edge length is growing at half the rate it was initially.

2. **(A review of calculus!)** You are in a tank (the military variety) moving down the y-axis toward the origin. At time $t = 0$ you are 4 km from the origin, and 10 min later you are 2 km from the origin. Your speed is decreasing; it is proportional to your distance from the origin. You know that an enemy tank is waiting somewhere on the positive x-axis, but there is a high wall along the curve $xy = 1$ (all distances in kilometres) preventing you from seeing just where it is. How fast must your gun turret be capable of turning to maximize your chances of surviving the encounter?

3. **(The economics of blood testing)** Suppose that it is necessary to perform a blood test on a large number N of individuals to detect the presence of a virus. If each test costs \$$C$, then the total cost of the testing program is \$$NC$. If the proportion of people in the population who have the virus is not large, this cost can be greatly reduced by adopting the following strategy. Divide the N samples of blood into N/x groups of x samples each. Pool the blood in each group to make a single sample for that group and test it. If it tests negative, no further testing is necessary for individuals in that group. If the group sample tests positive, test all the individuals in that group.

Suppose that the fraction of individuals in the population infected with the virus is p, so the fraction uninfected is $q = 1 - p$. The probability that a given individual is unaffected is q, so the probability that all x individuals in a group are unaffected is q^x. Therefore, the probability that a pooled sample is infected is $1 - q^x$. Each group requires one test, and the infected groups require an extra x tests. Therefore the expected total number of tests to be performed is

$$T = \frac{N}{x} + \frac{N}{x}(1 - q^x)x = N\left(\frac{1}{x} + 1 - q^x\right).$$

For example, if $p = 0.01$, so that $q = 0.99$ and $x = 20$, then the expected number of tests required is $T = 0.23N$, a reduction of over 75%. But maybe we can do better by making a different choice for x.

(a) For $q = 0.99$, find the number x of samples in a group that minimizes T (i.e., solve $dT/dx = 0$). Show that the minimizing value of x satisfies

$$x = \frac{(0.99)^{-x/2}}{\sqrt{-\ln(0.99)}}.$$

(b) Use the technique of fixed-point iteration (see Section 4.2) to solve the equation in (a) for x. Start with $x = 20$, say.

4. (Measuring variations in g) The period P of a pendulum of length L is given by

$$P = 2\pi\sqrt{L/g},$$

where g is the acceleration of gravity.

(a) Assuming that L remains fixed, show that a 1% increase in g results in approximately a 0.5% decrease in the period P. (Variations in the period of a pendulum can be used to detect small variations in g from place to place on the earth's surface.)

(b) For fixed g, what percentage change in L will produce a 1% increase in P?

5. (Torricelli's Law) The rate at which a tank drains is proportional to the square root of the depth of liquid in the tank above the level of the drain: if $V(t)$ is the volume of liquid in the tank at time t, and $y(t)$ is the height of the surface of the liquid above the drain, then $dV/dt = -k\sqrt{y}$, where k is a constant depending on the size of the drain. For a cylindrical tank with constant cross-sectional area A with drain at the bottom:

(a) Verify that the depth $y(t)$ of liquid in the tank at time t satisfies $dy/dt = -(k/A)\sqrt{y}$.

(b) Verify that if the depth of liquid in the tank at $t = 0$ is y_0, then the depth at subsequent times during the draining process is $y = \left(\sqrt{y_0} - \frac{kt}{2A}\right)^2$.

(c) If the tank drains completely in time T, express the depth $y(t)$ at time t in terms of y_0 and T.

(d) In terms of T, how long does it take for half the liquid in the tank to drain out?

6. If a conical tank with top radius R and depth H drains according to Torricelli's Law and empties in time T, show that the depth of liquid in the tank at time t $(0 < t < T)$ is

$$y = y_0\left(1 - \frac{t}{T}\right)^{2/5},$$

where y_0 is the depth at $t = 0$.

7. Find the largest possible area of a right-angled triangle whose perimeter is P.

8. Find a tangent to the graph of $y = x^3 + ax^2 + bx + c$ that is not parallel to any other tangent.

9. (Branching angles for electric wires and pipes)

(a) The resistance offered by a wire to the flow of electric current through it is proportional to its length and inversely proportional to its cross-sectional area. Thus, the resistance R of a wire of length L and radius r is $R = kL/r^2$, where k is a positive constant. A long straight wire of length L and radius r_1 extends from A to B. A second straight wire of smaller radius r_2 is to be connected between a point P on AB and a point C at distance h from B such that CB is perpendicular to AB. (See Figure 4.72.) Find the value of the angle $\theta = \angle BPC$ that minimizes the total resistance of the path APC, that is, the resistance of AP plus the resistance of PC.

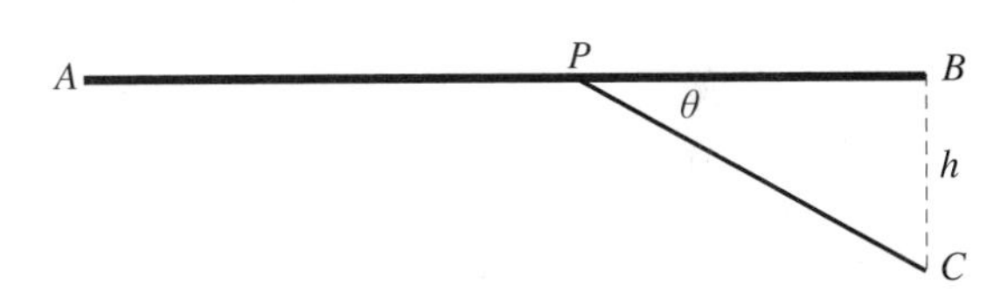

Figure 4.72

(b) The resistance of a pipe (e.g., a blood vessel) to the flow of liquid through it is, by Poiseuille's Law, proportional to its length and inversely proportional to the *fourth power* of its radius: $R = kL/r^4$. If the situation in part (a) represents pipes instead of wires, find the value of θ that minimizes the total resistance of the path APC. How does your answer relate to the answer for part (a)? Could you have predicted this relationship?

10. (The range of a spurt) A cylindrical water tank sitting on a horizontal table has a small hole located on its vertical wall at height h above the bottom of the tank. Water escapes from the tank horizontally through the hole and then curves down under the influence of gravity to strike the table at a distance R from the base of the tank, as shown in Figure 4.73. (We ignore air resistance.) Torricelli's Law implies that the speed v at which water escapes through the hole is proportional to the square root of the depth of the hole below the surface of the water: if the depth of water in the tank at time t is $y(t) > h$, then $v = k\sqrt{y - h}$, where the constant k depends on the size of the hole.

(a) Find the range R in terms of v and h.

(b) For a given depth y of water in the tank, how high should the hole be to maximize R?

(c) Suppose that the depth of water in the tank at time $t = 0$ is y_0, that the range R of the spurt is R_0 at that time, and that the water level drops to the height h of the hole in T minutes. Find, as a function of t, the range R of the water that escaped through the hole at time t.

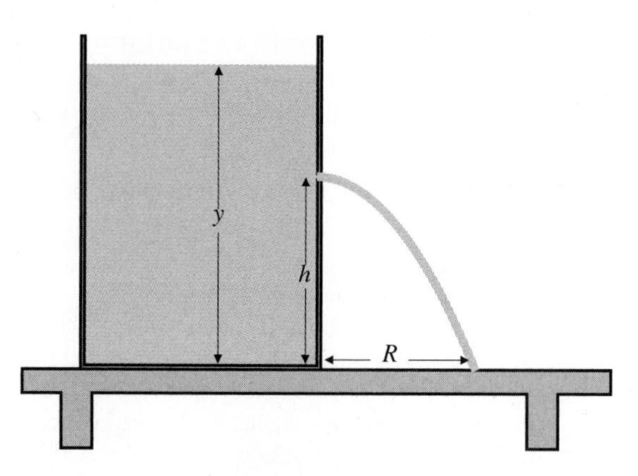

Figure 4.73

11. (Designing a dustpan) Equal squares are cut out of two adjacent corners of a square of sheet metal having sides of length 25 cm. The three resulting flaps are bent up, as shown in Figure 4.74, to form the sides of a dustpan. Find the maximum volume of a dustpan made in this way.

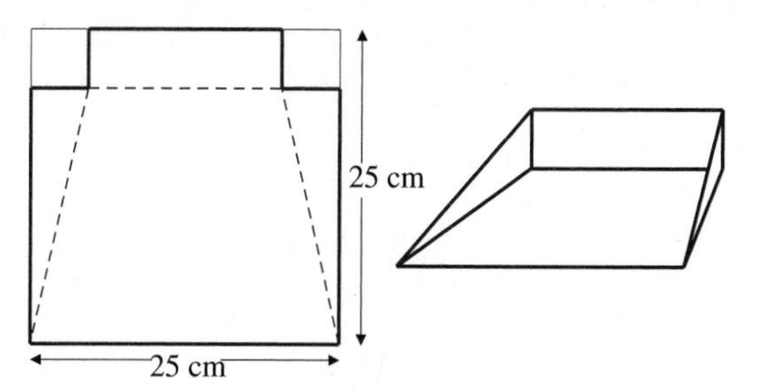

Figure 4.74

CHAPTER 5

Integration

> There are in this world optimists who feel that any symbol that starts off with an integral sign must necessarily denote something that will have every property that they should like an integral to possess. This of course is quite annoying to us rigorous mathematicians; what is even more annoying is that by doing so they often come up with the right answer.

E. J. McShane

Bulletin of the American Mathematical Society, v. 69, p. 611, 1963

Introduction

The second fundamental problem addressed by calculus is the problem of areas, that is, the problem of determining the area of a region of the plane bounded by various curves. Like the problem of tangents considered in Chapter 2, many practical problems in various disciplines require the evaluation of areas for their solution, and the solution of the problem of areas necessarily involves the notion of limits. On the surface the problem of areas appears unrelated to the problem of tangents. However, we will see that the two problems are very closely related; one is the inverse of the other. Finding an area is equivalent to finding an antiderivative or, as we prefer to say, finding an integral. The relationship between areas and antiderivatives is called the Fundamental Theorem of Calculus. When we have proved it, we will be able to find areas at will, provided only that we can integrate (i.e., antidifferentiate) the various functions we encounter.

We would like to have at our disposal a set of integration rules similar to the differentiation rules developed in Chapter 2. We can find the derivative of any differentiable function using those differentiation rules. Unfortunately, integration is generally more difficult; indeed, some fairly simple functions are not themselves derivatives of simple functions. For example, e^{x^2} is not the derivative of any finite combination of elementary functions. Nevertheless, we will expend some effort in Section 5.6 and Sections 6.1–6.4 to develop techniques for integrating as many functions as possible. Later in Chapter 6 we will examine how to approximate areas bounded by graphs of functions that we cannot antidifferentiate.

5.1 Sums and Sigma Notation

When we begin calculating areas in the next section, we will often encounter sums of values of functions. We need to have a convenient notation for representing sums of arbitrary (possibly large) numbers of terms, and we need to develop techniques for evaluating some such sums.

We use the symbol $\sum$ to represent a sum; it is an enlarged Greek capital letter S called *sigma*.

DEFINITION 1

Sigma notation

If m and n are integers with $m \leq n$, and if f is a function defined at the integers $m, m+1, m+2, \ldots, n$, the symbol $\sum_{i=m}^{n} f(i)$ represents the sum of the values of f at those integers:

$$\sum_{i=m}^{n} f(i) = f(m) + f(m+1) + f(m+2) + \cdots + f(n).$$

The explicit sum appearing on the right side of this equation is the **expansion** of the sum represented in sigma notation on the left side.

EXAMPLE 1 $\displaystyle\sum_{i=1}^{5} i^2 = 1^2 + 2^2 + 3^2 + 4^2 + 5^2 = 55.$

The i that appears in the symbol $\sum_{i=m}^{n} f(i)$ is called an **index of summation**. To evaluate $\sum_{i=m}^{n} f(i)$, replace the index i with the integers $m, m+1, \ldots, n$, successively, and sum the results. Observe that the value of the sum does not depend on what we call the index; the index does not appear on the right side of the definition. If we use another letter in place of i in the sum in Example 1, we still get the same value for the sum:

$$\sum_{k=1}^{5} k^2 = 1^2 + 2^2 + 3^2 + 4^2 + 5^2 = 55.$$

The index of summation is a *dummy variable* used to represent an arbitrary point where the function is evaluated to produce a term to be included in the sum. On the other hand, the sum $\sum_{i=m}^{n} f(i)$ does depend on the two numbers m and n, called the **limits of summation**; m is the **lower limit**, and n is the **upper limit**.

EXAMPLE 2 **(Examples of sums using sigma notation)**

$$\sum_{j=1}^{20} j = 1 + 2 + 3 + \cdots + 18 + 19 + 20$$

$$\sum_{i=0}^{n} x^i = x^0 + x^1 + x^2 + \cdots + x^{n-1} + x^n$$

$$\sum_{m=1}^{n} 1 = \underbrace{1 + 1 + 1 + \cdots + 1}_{n \text{ terms}}$$

$$\sum_{k=-2}^{3} \frac{1}{k+7} = \frac{1}{5} + \frac{1}{6} + \frac{1}{7} + \frac{1}{8} + \frac{1}{9} + \frac{1}{10}$$

Sometimes we use a subscripted variable a_i to denote the ith term of a general sum instead of using the functional notation $f(i)$:

$$\sum_{i=m}^{n} a_i = a_m + a_{m+1} + a_{m+2} + \cdots + a_n.$$

In particular, an **infinite series** is such a sum with infinitely many terms:

$$\sum_{n=1}^{\infty} a_n = a_1 + a_2 + a_3 + \cdots.$$

When no final term follows the $\cdots$, it is understood that the terms go on forever. We will study infinite series in Chapter 9.

When adding finitely many numbers, the order in which they are added is unimportant; any order will give the same sum. If all the numbers have a common factor, then that factor can be removed from each term and multiplied after the sum is evaluated: $ca + cb = c(a + b)$. These laws of arithmetic translate into the following *linearity* rule for finite sums; if A and B are constants, then

$$\sum_{i=m}^{n} \big(Af(i) + Bg(i)\big) = A\sum_{i=m}^{n} f(i) + B\sum_{i=m}^{n} g(i).$$

Both of the sums $\sum_{j=m}^{m+n} f(j)$ and $\sum_{i=0}^{n} f(i + m)$ have the same expansion, namely, $f(m) + f(m + 1) + \cdots + f(m + n)$. Therefore, the two sums are equal.

$$\sum_{j=m}^{m+n} f(j) = \sum_{i=0}^{n} f(i + m).$$

This equality can also be derived by substituting $i + m$ for j everywhere j appears on the left side, noting that $i + m = m$ reduces to $i = 0$, and $i + m = m + n$ reduces to $i = n$. It is often convenient to make such a **change of index** in a summation.

EXAMPLE 3 Express $\sum_{j=3}^{17} \sqrt{1 + j^2}$ in the form $\sum_{i=1}^{n} f(i)$.

Solution Let $j = i + 2$, Then $j = 3$ corresponds to $i = 1$ and $j = 17$ corresponds to $i = 15$. Thus,

$$\sum_{j=3}^{17} \sqrt{1 + j^2} = \sum_{i=1}^{15} \sqrt{1 + (i + 2)^2}.$$

Evaluating Sums

There is a **closed form** expression for the sum S of the first n positive integers, namely,

$$S = \sum_{i=1}^{n} i = 1 + 2 + 3 + \cdots + n = \frac{n(n + 1)}{2}.$$

To see this, write the sum forwards and backwards and add the two to get

$$\begin{array}{rcccccccccc}
S = & 1 & + & 2 & + & 3 & + \cdots + & (n-1) & + & n \\
S = & n & + & (n-1) & + & (n-2) & + \cdots + & 2 & + & 1 \\
\hline
2S = & (n+1) & + & (n+1) & + & (n+1) & + \cdots + & (n+1) & + & (n+1) = n(n+1)
\end{array}$$

The formula for S follows when we divide by 2.

It is not usually this easy to evaluate a general sum in closed form. We can only simplify $\sum_{i=m}^{n} f(i)$ for a small class of functions f. The only such formulas we will need in the next sections are collected in Theorem 1.

THEOREM 1 **Summation formulas**

(a) $\displaystyle\sum_{i=1}^{n} 1 = \underbrace{1+1+1+\cdots+1}_{n \text{ terms}} = n.$

(b) $\displaystyle\sum_{i=1}^{n} i = 1+2+3+\cdots+n = \frac{n(n+1)}{2}.$

(c) $\displaystyle\sum_{i=1}^{n} i^2 = 1^2+2^2+3^2+\cdots+n^2 = \frac{n(n+1)(2n+1)}{6}.$

(d) $\displaystyle\sum_{i=1}^{n} r^{i-1} = 1+r+r^2+r^3+\cdots+r^{n-1} = \frac{r^n-1}{r-1}$ if $r \neq 1$.

PROOF Formula (a) is trivial; the sum of n ones is n. One proof of formula (b) was given above. Three others are suggested in Exercises 34–36.

To prove (c) we write n copies of the identity

$$(k+1)^3 - k^3 = 3k^2 + 3k + 1,$$

one for each value of k from 1 to n, and add them up:

$$\begin{array}{rcrcrcrcr}
2^3 & - & 1^3 & = & 3\times 1^2 & + & 3\times 1 & + & 1 \\
3^3 & - & 2^3 & = & 3\times 2^2 & + & 3\times 2 & + & 1 \\
4^3 & - & 3^3 & = & 3\times 3^2 & + & 3\times 3 & + & 1 \\
\vdots & & \vdots & & \vdots & & \vdots & & \vdots \\
n^3 & - & (n-1)^3 & = & 3(n-1)^2 & + & 3(n-1) & + & 1 \\
(n+1)^3 & - & n^3 & = & 3n^2 & + & 3n & + & 1 \\
\hline
(n+1)^3 & - & 1^3 & = & 3\left(\sum_{i=1}^{n} i^2\right) & + & 3\left(\sum_{i=1}^{n} i\right) & + & n \\
 & & & = & 3\left(\sum_{i=1}^{n} i^2\right) & + & \dfrac{3n(n+1)}{2} & + & n.
\end{array}$$

We used formula (b) in the last line. The final equation can be solved for the desired sum to give formula (c). Note the cancellations that occurred when we added up the left sides of the n equations. The term 2^3 in the first line cancelled the -2^3 in the second line, and so on, leaving us with only two terms, the $(n+1)^3$ from the nth line and the -1^3 from the first line:

$$\sum_{k=1}^{n}\left((k+1)^3 - k^3\right) = (n+1)^3 - 1^3.$$

This is an example of what we call a **telescoping sum**. In general, a sum of the form $\sum_{i=m}^{n}\big(f(i+1) - f(i)\big)$ telescopes to the closed form $f(n+1) - f(m)$ because all but the first and last terms cancel out.

To prove formula (d), let $s = \sum_{i=1}^{n} r^{i-1}$ and subtract s from rs:

$$\begin{aligned}(r-1)s = rs - s &= (r + r^2 + r^3 + \cdots + r^n) - (1 + r + r^2 + \cdots + r^{n-1}) \\ &= r^n - 1.\end{aligned}$$

The result follows on division by $r - 1$. ■

EXAMPLE 4 Evaluate $\displaystyle\sum_{k=m+1}^{n} (6k^2 - 4k + 3)$, where $1 \le m < n$.

Solution Using the rules of summation and various summation formulas from Theorem 1, we calculate

$$\begin{aligned}\sum_{k=1}^{n}(6k^2-4k+3) &= 6\sum_{k=1}^{n}k^2-4\sum_{k=1}^{n}k+3\sum_{k=1}^{n}1\\ &= 6\,\frac{n(n+1)(2n+1)}{6}-4\,\frac{n(n+1)}{2}+3n\\ &= 2n^3+n^2+2n\end{aligned}$$

Thus,

$$\begin{aligned}\sum_{k=m+1}^{n}(6k^2-4k+3) &= \sum_{k=1}^{n}(6k^2-4k+3)-\sum_{k=1}^{m}(6k^2-4k+3)\\ &= 2n^3+n^2+2n-2m^3-m^2-2m.\end{aligned}$$

Remark Maple can find closed form expressions for some sums. For example,

```
> sum(i^4, i=1..n); factor(%);
```

$$\frac{1}{5}(n+1)^5-\frac{1}{2}(n+1)^4+\frac{1}{3}(n+1)^3-\frac{1}{30}n-\frac{1}{30}$$

$$\frac{1}{30}n(2n+1)(n+1)(3n^2+3n-1)$$

EXERCISES 5.1

Expand the sums in Exercises 1–6.

1. $\sum_{i=1}^{4} i^3$

2. $\sum_{j=1}^{100} \frac{j}{j+1}$

3. $\sum_{i=1}^{n} 3^i$

4. $\sum_{i=0}^{n-1} \frac{(-1)^i}{i+1}$

5. $\sum_{j=3}^{n} \frac{(-2)^j}{(j-2)^2}$

6. $\sum_{j=1}^{n} \frac{j^2}{n^3}$

Write the sums in Exercises 7–14 using sigma notation. (Note that the answers are not unique.)

7. $5+6+7+8+9$

8. $2+2+2+\cdots+2$ (200 terms)

9. $2^2-3^2+4^2-5^2+\cdots-99^2$

10. $1+2x+3x^2+4x^3+\cdots+100x^{99}$

11. $1+x+x^2+x^3+\cdots+x^n$

12. $1-x+x^2-x^3+\cdots+x^{2n}$

13. $1-\frac{1}{4}+\frac{1}{9}-\frac{1}{16}+\cdots+\frac{(-1)^{n-1}}{n^2}$

14. $\frac{1}{2}+\frac{2}{4}+\frac{3}{8}+\frac{4}{16}+\cdots+\frac{n}{2^n}$

Express the sums in Exercises 15–16 in the form $\sum_{i=1}^{n} f(i)$.

15. $\sum_{j=0}^{99} \sin(j)$

16. $\sum_{k=-5}^{m} \frac{1}{k^2+1}$

Find closed form values for the sums in Exercises 17–28.

17. $\sum_{i=1}^{n} \left(i^2+2i\right)$

18. $\sum_{j=1}^{1,000} (2j+3)$

19. $\sum_{k=1}^{n} (\pi^k-3)$

20. $\sum_{i=1}^{n} (2^i-i^2)$

21. $\sum_{m=1}^{n} \ln m$

22. $\sum_{i=0}^{n} e^{i/n}$

23. The sum in Exercise 8.

24. The sum in Exercise 11.

25. The sum in Exercise 12.

26. The sum in Exercise 10. *Hint:* Differentiate the sum $\sum_{i=0}^{100} x^i$.

27. The sum in Exercise 9. *Hint:* The sum is

$$\sum_{k=1}^{49}\left((2k)^2-(2k+1)^2\right)=\sum_{k=1}^{49}(-4k-1).$$

28. The sum in Exercise 14. *Hint:* apply the method of proof of Theorem 1(d) to this sum.

29. Verify the formula for the value of a telescoping sum:

$$\sum_{i=m}^{n}\left(f(i+1)-f(i)\right)=f(n+1)-f(m).$$

Why is the word "telescoping" used to describe this sum?

In Exercises 30–32, evaluate the given telescoping sums.

30. $\displaystyle\sum_{n=1}^{10}\left(n^4-(n-1)^4\right)$ **31.** $\displaystyle\sum_{j=1}^{m}(2^j-2^{j-1})$

32. $\displaystyle\sum_{i=m}^{2m}\left(\frac{1}{i}-\frac{1}{i+1}\right)$

33. Show that $\dfrac{1}{j(j+1)}=\dfrac{1}{j}-\dfrac{1}{j+1}$, and hence evaluate $\displaystyle\sum_{j=1}^{n}\frac{1}{j(j+1)}$.

34. Figure 5.1 shows a square of side n subdivided into n^2 smaller squares of side 1. How many small squares are shaded? Obtain the closed form expression for $\sum_{i=1}^{n} i$ by considering the sum of the areas of the shaded squares.

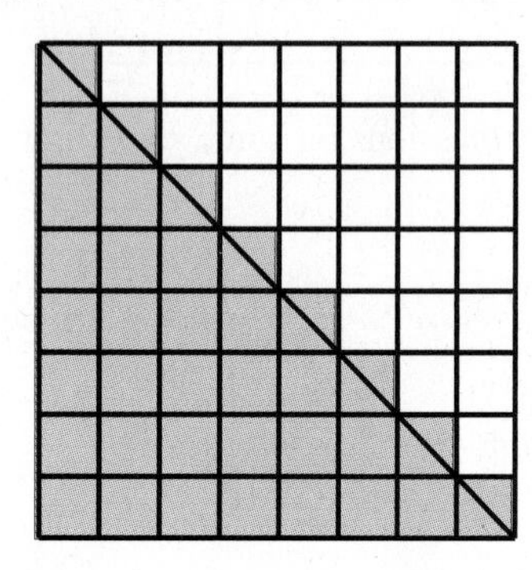

Figure 5.1

35. Write n copies of the identity $(k+1)^2-k^2=2k+1$, one for each integer k from 1 to n, and add them up to obtain the formula

$$\sum_{i=1}^{n} i=\frac{n(n+1)}{2}$$

in a manner similar to the proof of Theorem 1(c).

36. Use mathematical induction to prove Theorem 1(b).

37. Use mathematical induction to prove Theorem 1(c).

38. Use mathematical induction to prove Theorem 1(d).

39. Figure 5.2 shows a square of side $\sum_{i=1}^{n} i=n(n+1)/2$ subdivided into a small square of side 1 and $n-1$ L-shaped regions whose short edges are $2, 3, \dots, n$. Show that the area of the L-shaped region with short side i is i^3, and hence verify that

$$\sum_{i=1}^{n} i^3=\frac{n^2(n+1)^2}{4}.$$

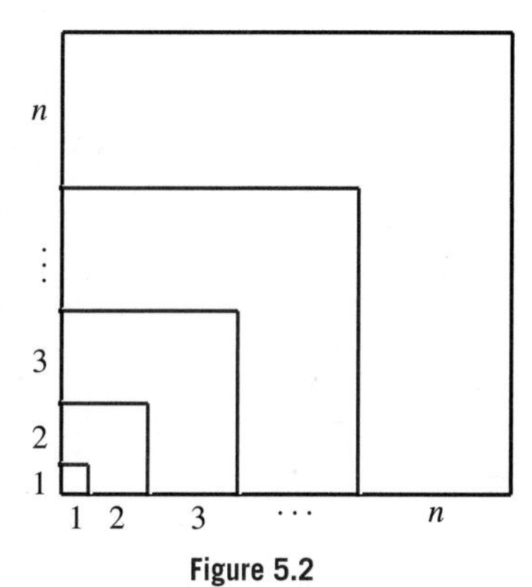

Figure 5.2

40. Write n copies of the identity

$$(k+1)^4-k^4=4k^3+6k^2+4k+1,$$

one for each integer k from 1 to n, and add them up to obtain the formula

$$\sum_{i=1}^{n} i^3=\frac{n^2(n+1)^2}{4},$$

in a manner similar to the proof of Theorem 1(c).

41. Use mathematical induction to verify the formula for the sum of cubes given in Exercise 40.

42. Extend the method of Exercise 40 to find a closed form expression for $\sum_{i=1}^{n} i^4$. You will probably want to use Maple or other computer algebra software to do all the algebra.

43. Use Maple or another computer algebra system to find $\sum_{i=1}^{n} i^k$ for $k=5, 6, 7, 8$. Observe the term involving the highest power of n in each case. Predict the highest-power term in $\sum_{i=1}^{n} i^{10}$ and verify your prediction.

5.2 Areas as Limits of Sums

We began the study of derivatives in Chapter 2 by defining what is meant by a tangent line to a curve at a particular point. We would like to begin the study of integrals by defining what is meant by the **area** of a plane region, but a definition of area is much more difficult to give than a definition of tangency. Let us assume (as we did, for example, in Section 3.3) that we know intuitively what area means and list some of its properties. (See Figure 5.3.)

(i) The area of a plane region is a nonnegative real number of *square units*.

(ii) The area of a rectangle with width w and height h is $A=wh$.

(iii) The areas of congruent plane regions are equal.

(iv) If region S is contained in region R, then the area of S is less than or equal to that of R.

(v) If region R is a union of (finitely many) nonoverlapping regions, then the area of R is the sum of the areas of those regions.

Using these five properties we can calculate the area of any **polygon** (a region bounded by straight line segments). First, we note that properties (iii) and (v) show that the area of a parallelogram is the same as that of a rectangle having the same base width and height. Any triangle can be butted against a congruent copy of itself to form a parallelogram, so a triangle has area half the base width times the height. Finally, any polygon can be subdivided into finitely many nonoverlapping triangles so its area is the sum of the areas of those triangles.

We can't go beyond polygons without taking limits. If a region has a curved boundary, its area can only be approximated by using rectangles or triangles; calculating the exact area requires the evaluation of a limit. We showed how this could be done for a circle in Section 1.1.

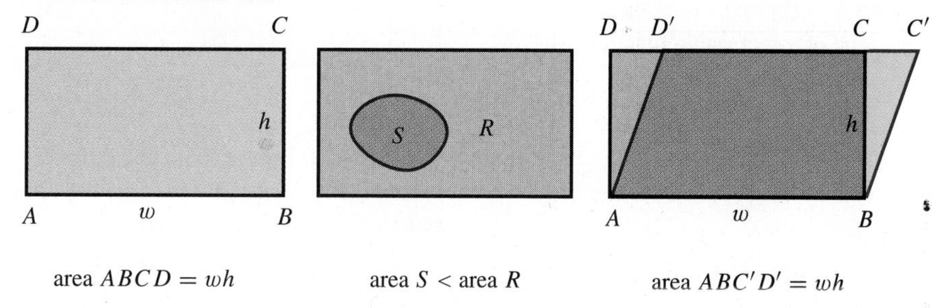

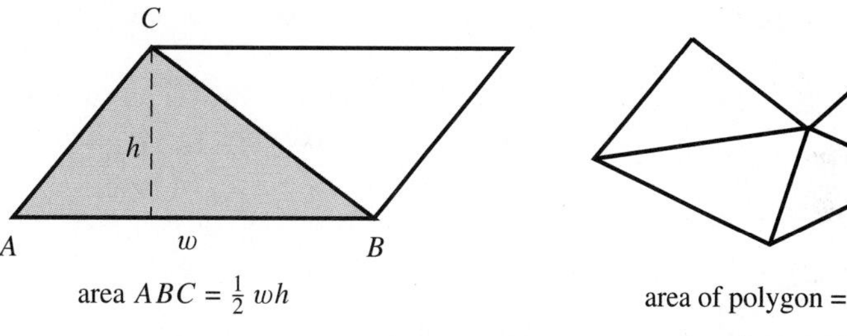

Figure 5.3 Properties of area

The Basic Area Problem

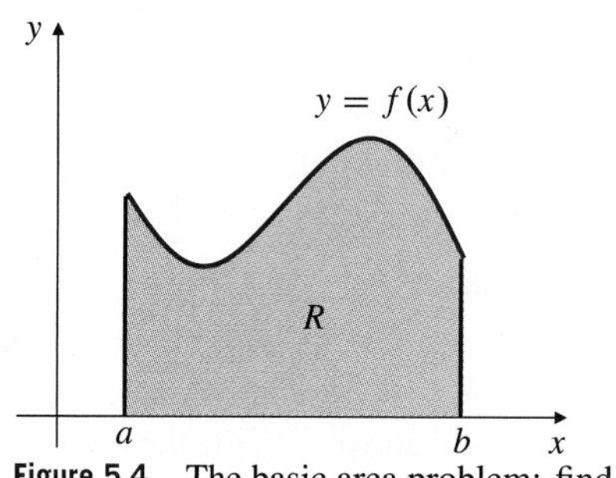

Figure 5.4 The basic area problem: find the area of region R

In this section we are going to consider how to find the area of a region R lying under the graph $y = f(x)$ of a nonnegative-valued, continuous function f, above the x-axis and between the vertical lines $x = a$ and $x = b$, where $a < b$. (See Figure 5.4.) To accomplish this we proceed as follows. Divide the interval $[a, b]$ into n subintervals by using division points:

$$a = x_0 < x_1 < x_2 < x_3 < \cdots < x_{n-1} < x_n = b.$$

Denote by Δx_i the length of the ith subinterval $[x_{i-1}, x_i]$:

$$\Delta x_i = x_i - x_{i-1}, \qquad (i = 1, 2, 3, \ldots, n).$$

Vertically above each subinterval $[x_{i-1}, x_i]$ build a rectangle whose base has length Δx_i and whose height is $f(x_i)$. The area of this rectangle is $f(x_i)\,\Delta x_i$. Form the sum of these areas:

$$S_n = f(x_1)\,\Delta x_1 + f(x_2)\,\Delta x_2 + f(x_3)\,\Delta x_3 + \cdots + f(x_n)\,\Delta x_n = \sum_{i=1}^{n} f(x_i)\,\Delta x_i.$$

The rectangles are shown shaded in Figure 5.5 for a decreasing function f. For an increasing function, the tops of the rectangles would lie above the graph of f rather than below it. Evidently, S_n is an approximation to the area of the region R, and the approximation gets better as n increases, provided we choose the points $a = x_0 < x_1 < \cdots < x_n = b$ in such a way that the width Δx_i of the widest rectangle approaches zero.

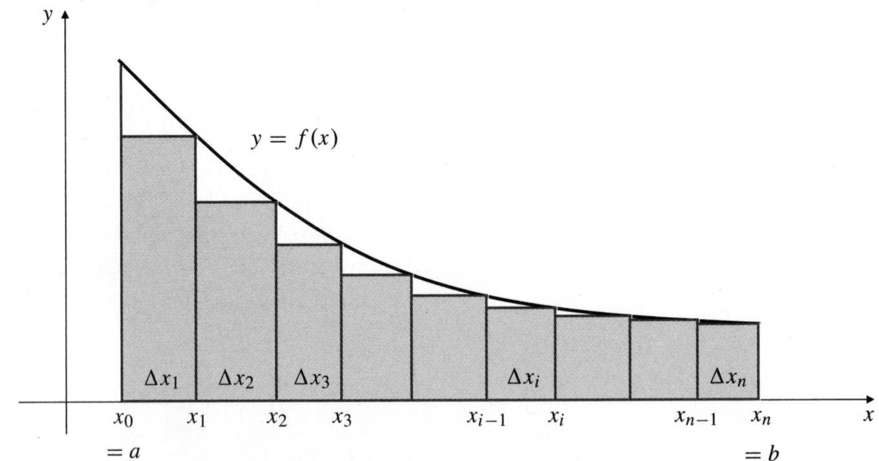

Figure 5.5 Approximating the area under the graph of a decreasing function using rectangles

Observe in Figure 5.6, for example, that subdividing a subinterval into two smaller subintervals reduces the error in the approximation by reducing that part of the area under the curve that is not contained in the rectangles. It is reasonable, therefore, to calculate the area of R by finding the limit of S_n as $n \to \infty$ with the restriction that the largest of the subinterval widths Δx_i must approach zero:

$$\text{Area of } R = \lim_{\substack{n\to\infty \\ \max \Delta x_i \to 0}} S_n.$$

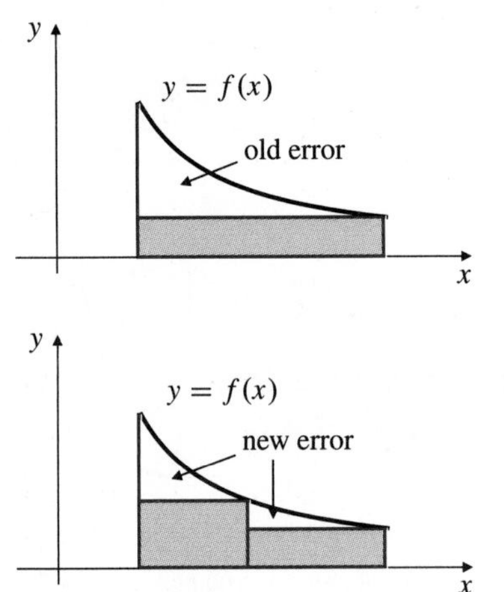

Figure 5.6 Using more rectangles makes the error smaller

Sometimes, but not always, it is useful to choose the points x_i $(0 \le i \le n)$ in $[a, b]$ in such a way that the subinterval lengths Δx_i are all equal. In this case we have

$$\Delta x_i = \Delta x = \frac{b-a}{n}, \qquad x_i = a + i\,\Delta x = a + \frac{i}{n}(b-a).$$

Some Area Calculations

We devote the rest of this section to some examples in which we apply the technique described above for finding areas under graphs of functions by approximating with rectangles. Let us begin with a region for which we already know the area so we can satisfy ourselves that the method does give the correct value.

EXAMPLE 1 Find the area A of the region lying under the straight line $y = x+1$, above the x-axis and between the lines $x = 0$ and $x = 2$.

Solution The region is shaded in Figure 5.7(a). It is a *trapezoid* (a four-sided polygon with one pair of parallel sides) and has area 4 square units. (It can be divided into a rectangle and a triangle, each of area 2 square units.) We will calculate the area as a limit of sums of areas of rectangles constructed as described above. Divide the interval $[0, 2]$ into n subintervals *of equal length* by points

$$x_0 = 0,\ x_1 = \frac{2}{n},\ x_2 = \frac{4}{n},\ x_3 = \frac{6}{n},\ \ldots\ x_n = \frac{2n}{n} = 2.$$

The value of $y = x + 1$ at $x = x_i$ is $x_i + 1 = \dfrac{2i}{n} + 1$ and the ith subinterval, $\left[\dfrac{2(i-1)}{n}, \dfrac{2i}{n}\right]$, has length $\Delta x_i = \dfrac{2}{n}$. Observe that $\Delta x_i \to 0$ as $n \to \infty$. The sum of the areas of the approximating rectangles shown in Figure 5.7(a) is

$$\begin{aligned} S_n &= \sum_{i=1}^{n} \left(\frac{2i}{n} + 1\right)\frac{2}{n} \\ &= \left(\frac{2}{n}\right)\left[\frac{2}{n}\sum_{i=1}^{n} i + \sum_{i=1}^{n} 1\right] \qquad \text{(Use parts (b) and (a) of Theorem 1.)} \\ &= \left(\frac{2}{n}\right)\left[\frac{2}{n}\,\frac{n(n+1)}{2} + n\right] \\ &= 2\,\frac{n+1}{n} + 2. \end{aligned}$$

Therefore, the required area A is given by

$$A = \lim_{n\to\infty} S_n = \lim_{n\to\infty}\left(2\,\frac{n+1}{n} + 2\right) = 2 + 2 = 4 \text{ square units.}$$

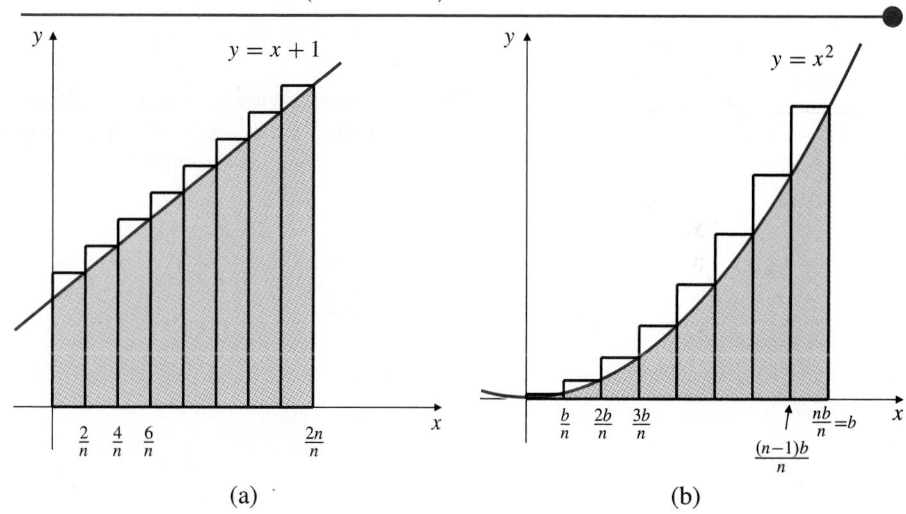

Figure 5.7
(a) The region of Example 1
(b) The region of Example 2

EXAMPLE 2 Find the area of the region bounded by the parabola $y = x^2$ and the straight lines $y = 0$, $x = 0$, and $x = b$, where $b > 0$.

Solution The area A of the region is the limit of the sum S_n of areas of the rectangles shown in Figure 5.7(b). Again we have used equal subintervals, each of length b/n. The height of the ith rectangle is $(ib/n)^2$. Thus,

$$S_n = \sum_{i=1}^{n}\left(\frac{ib}{n}\right)^2 \frac{b}{n} = \frac{b^3}{n^3}\sum_{i=1}^{n} i^2 = \frac{b^3}{n^3}\,\frac{n(n+1)(2n+1)}{6},$$

by formula (c) of Theorem 1. Hence, the required area is

$$A = \lim_{n\to\infty} S_n = \lim_{n\to\infty} b^3\,\frac{(n+1)(2n+1)}{6n^2} = \frac{b^3}{3} \text{ square units.}$$

Finding an area under the graph of $y = x^k$ over an interval I becomes more and more difficult as k increases if we continue to try to subdivide I into subintervals of equal length. (See Exercise 14 at the end of this section for the case $k = 3$.) It is, however, possible to find the area for arbitrary k if we subdivide the interval I into subintervals whose lengths increase in geometric progression. Example 3 illustrates this.

EXAMPLE 3 Let $b > a > 0$, and let k be any real number except -1. Show that the area A of the region bounded by $y = x^k$, $y = 0$, $x = a$, and $x = b$ is

$$A = \frac{b^{k+1} - a^{k+1}}{k+1} \text{ square units.}$$

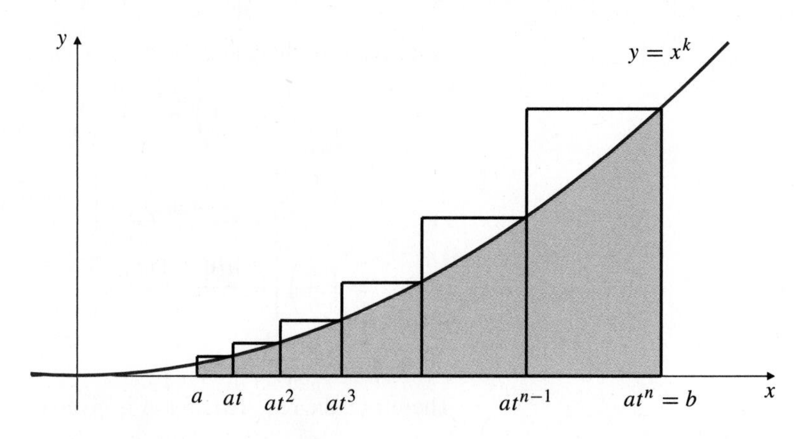

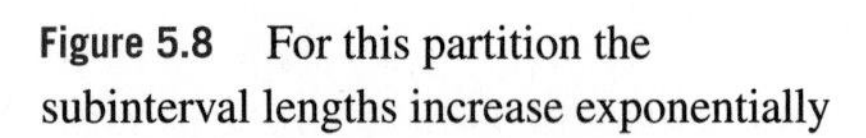
Figure 5.8 For this partition the subinterval lengths increase exponentially

BEWARE! This is a long and rather difficult example. Either skip over it or take your time and check each step carefully.

Solution Let $t = (b/a)^{1/n}$ and let

$$x_0 = a,\ x_1 = at,\ x_2 = at^2,\ x_3 = at^3,\ \ldots\ x_n = at^n = b.$$

These points subdivide the interval $[a, b]$ into n subintervals of which the ith, $[x_{i-1}, x_i]$, has length $\Delta x_i = at^{i-1}(t-1)$. If $f(x) = x^k$, then $f(x_i) = a^k t^{ki}$. The sum of the areas of the rectangles shown in Figure 5.8 is:

$$\begin{aligned}
S_n &= \sum_{i=1}^{n} f(x_i)\,\Delta x_i \\
&= \sum_{i=1}^{n} a^k\, t^{ki}\, at^{i-1}(t-1) \\
&= a^{k+1}\,(t-1)\,t^k \sum_{i=1}^{n} t^{(k+1)(i-1)} \\
&= a^{k+1}\,(t-1)\,t^k \sum_{i=1}^{n} r^{(i-1)} \qquad \text{where } r = t^{k+1} \\
&= a^{k+1}\,(t-1)\,t^k\,\frac{r^n - 1}{r - 1} \qquad \text{(by Theorem 1(d))} \\
&= a^{k+1}\,(t-1)\,t^k\,\frac{t^{(k+1)n} - 1}{t^{k+1} - 1}.
\end{aligned}$$

Now replace t with its value $(b/a)^{1/n}$ and rearrange factors to obtain

$$\begin{aligned}
S_n &= a^{k+1}\left(\left(\frac{b}{a}\right)^{1/n} - 1\right)\left(\frac{b}{a}\right)^{k/n}\frac{\left(\dfrac{b}{a}\right)^{k+1} - 1}{\left(\dfrac{b}{a}\right)^{(k+1)/n} - 1} \\
&= \left(b^{k+1} - a^{k+1}\right) c^{k/n}\,\frac{c^{1/n} - 1}{c^{(k+1)/n} - 1}, \qquad \text{where } c = \frac{b}{a}.
\end{aligned}$$

Of the three factors in the final line above, the first does not depend on n, and the second, $c^{k/n}$, approaches $c^0 = 1$ as $n \to \infty$. The third factor is an indeterminate form of type $[0/0]$, which we evaluate using l'Hôpital's Rule. First let $u = 1/n$. Then

$$\begin{aligned}
\lim_{n\to\infty} \frac{c^{1/n} - 1}{c^{(k+1)/n} - 1} &= \lim_{u\to 0+} \frac{c^u - 1}{c^{(k+1)u} - 1} \qquad \left[\frac{0}{0}\right] \\
&= \lim_{u\to 0+} \frac{c^u \ln c}{(k+1)\,c^{(k+1)u} \ln c} = \frac{1}{k+1}.
\end{aligned}$$

Therefore, the required area is

$$A = \lim_{n\to\infty} S_n = \left(b^{k+1} - a^{k+1}\right) \times 1 \times \frac{1}{k+1} = \frac{b^{k+1} - a^{k+1}}{k+1} \text{ square units.}$$

As you can see, it can be rather difficult to calculate areas bounded by curves by the methods developed above. Fortunately, there is an easier way, as we will discover in Section 5.5.

Remark For technical reasons it was necessary to assume $a > 0$ in Example 3. The result is also valid for $a = 0$ provided $k > -1$. In this case we have $\lim_{a\to 0+} a^{k+1} = 0$, so the area under $y = x^k$, above $y = 0$, between $x = 0$ and $x = b > 0$, is $A = b^{k+1}/(k+1)$ square units. For $k = 2$ this agrees with the result of Example 2.

EXAMPLE 4 Identify the limit $L = \lim_{n\to\infty} \sum_{i=1}^{n} \frac{n-i}{n^2}$ as an area, and evaluate it.

Solution We can rewrite the ith term of the sum so that it depends on i/n:

$$L = \lim_{n\to\infty} \sum_{i=1}^{n} \left(1 - \frac{i}{n}\right) \frac{1}{n}.$$

The terms now appear to be the areas of rectangles of base $1/n$ and heights $1 - x_i$, $(1 \le i \le n)$, where

$$x_1 = \frac{1}{n}, \quad x_2 = \frac{2}{n}, \quad x_3 = \frac{3}{n}, \quad \ldots, \quad x_n = \frac{n}{n}.$$

Thus, the limit L is the area under the curve $y = 1 - x$ from $x = 0$ to $x = 1$. (See Figure 5.9.) This region is a triangle having area $1/2$ square unit, so $L = 1/2$.

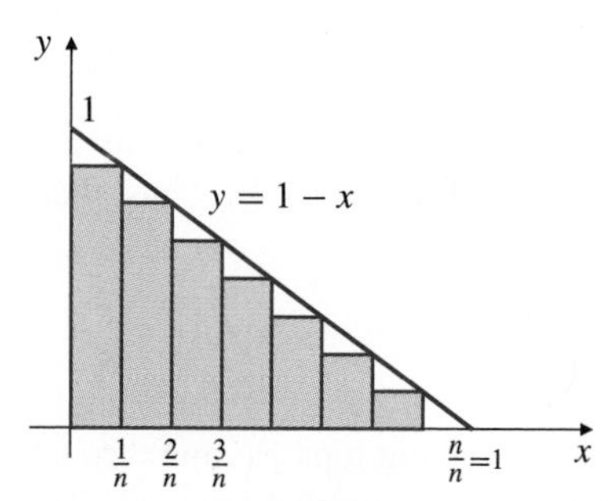

Figure 5.9 Recognizing a sum of areas

EXERCISES 5.2

Use the techniques of Examples 1 and 2 (with subintervals of equal length) to find the areas of the regions specified in Exercises 1–13.

1. Below $y = 3x$, above $y = 0$, from $x = 0$ to $x = 1$.

2. Below $y = 2x + 1$, above $y = 0$, from $x = 0$ to $x = 3$.

3. Below $y = 2x - 1$, above $y = 0$, from $x = 1$ to $x = 3$.

4. Below $y = 3x + 4$, above $y = 0$, from $x = -1$ to $x = 2$.

5. Below $y = x^2$, above $y = 0$, from $x = 1$ to $x = 3$.

6. Below $y = x^2 + 1$, above $y = 0$, from $x = 0$ to $x = a > 0$.

7. Below $y = x^2 + 2x + 3$, above $y = 0$, from $x = -1$ to $x = 2$.

8. Above $y = x^2 - 1$, below $y = 0$.

9. Above $y = 1 - x$, below $y = 0$, from $x = 2$ to $x = 4$.

10. Above $y = x^2 - 2x$, below $y = 0$.

11. Below $y = 4x - x^2 + 1$, above $y = 1$.

! 12. Below $y = e^x$, above $y = 0$, from $x = 0$ to $x = b > 0$.

! 13. Below $y = 2^x$, above $y = 0$, from $x = -1$ to $x = 1$.

14. Use the formula $\sum_{i=1}^{n} i^3 = n^2(n+1)^2/4$, from Exercises 39–41 of Section 5.1, to find the area of the region lying under $y = x^3$, above the x-axis, and between the vertical lines at $x = 0$ and $x = b > 0$.

15. Use the subdivision of $[a, b]$ given in Example 3 to find the area under $y = 1/x$, above $y = 0$ from $x = a > 0$ to $x = b > a$. Why should your answer not be surprising?

In Exercises 16–19, interpret the given sum S_n as a sum of areas of rectangles approximating the area of a certain region in the plane and hence evaluate $\lim_{n\to\infty} S_n$.

16. $S_n = \sum_{i=1}^{n} \frac{2}{n}\left(1 - \frac{i}{n}\right)$

17. $S_n = \sum_{i=1}^{n} \frac{2}{n}\left(1 - \frac{2i}{n}\right)$

18. $S_n = \sum_{i=1}^{n} \frac{2n + 3i}{n^2}$

! 19. $S_n = \sum_{j=1}^{n} \frac{1}{n}\sqrt{1 - (j/n)^2}$

5.3 The Definite Integral

In this section we generalize and make more precise the procedure used for finding areas developed in Section 5.2, and we use it to define the *definite integral* of a function f on an interval I. Let us assume, for the time being, that $f(x)$ is defined and continuous on the closed, finite interval $[a, b]$. We no longer assume that the values of f are nonnegative.

Partitions and Riemann Sums

Let P be a finite set of points arranged in order between a and b on the real line, say

$$P = \{x_0, x_1, x_2, x_3, \ldots, x_{n-1}, x_n\},$$

where $a = x_0 < x_1 < x_2 < x_3 < \cdots < x_{n-1} < x_n = b$. Such a set P is called a **partition** of $[a, b]$; it divides $[a, b]$ into n subintervals of which the ith is $[x_{i-1}, x_i]$. We call these the subintervals of the partition P. The number n depends on the particular partition, so we write $n = n(P)$. The length of the ith subinterval of P is

$$\Delta x_i = x_i - x_{i-1}, \qquad \text{(for } 1 \le i \le n)$$

and we call the greatest of these numbers Δx_i the **norm** of the partition P and denote it $\|P\|$:

$$\|P\| = \max_{1 \le i \le n} \Delta x_i.$$

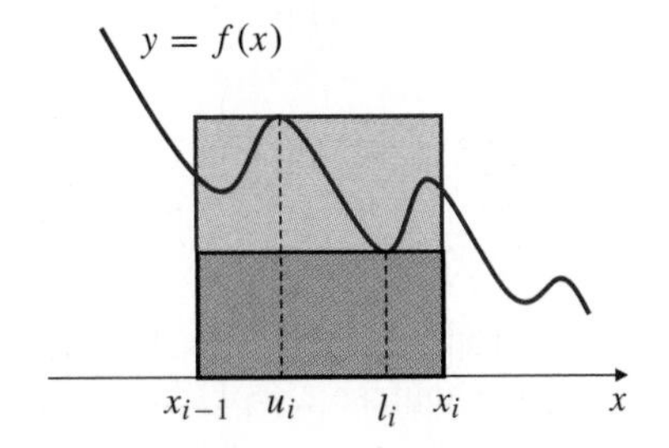

Figure 5.10

Since f is continuous on each subinterval $[x_{i-1}, x_i]$ of P, it takes on maximum and minimum values at points of that interval (by Theorem 8 of Section 1.4). Thus, there are numbers l_i and u_i in $[x_{i-1}, x_i]$ such that

$$f(l_i) \le f(x) \le f(u_i) \qquad \text{whenever } x_{i-1} \le x \le x_i.$$

If $f(x) \ge 0$ on $[a, b]$, then $f(l_i)\,\Delta x_i$ and $f(u_i)\,\Delta x_i$ represent the areas of rectangles having the interval $[x_{i-1}, x_i]$ on the x-axis as base, and having tops passing through the lowest and highest points, respectively, on the graph of f on that interval. (See Figure 5.10.) If A_i is that part of the area under $y = f(x)$ and above the x-axis that lies in the vertical strip between $x = x_{i-1}$ and $x = x_i$, then

$$f(l_i)\,\Delta x_i \le A_i \le f(u_i)\,\Delta x_i.$$

If f can have negative values, then one or both of $f(l_i)\,\Delta x_i$ and $f(u_i)\,\Delta x_i$ can be negative and will then represent the negative of the area of a rectangle lying below the x-axis. In any event, we always have $f(l_i)\,\Delta x_i \le f(u_i)\,\Delta x_i$.

DEFINITION 2

Upper and lower Riemann sums

The **lower (Riemann) sum, $L(f, P)$**, and the **upper (Riemann) sum, $U(f, P)$**, for the function f and the partition P are defined by:

$$L(f, P) = f(l_1)\,\Delta x_1 + f(l_2)\,\Delta x_2 + \cdots + f(l_n)\,\Delta x_n = \sum_{i=1}^{n} f(l_i)\,\Delta x_i,$$

$$U(f, P) = f(u_1)\,\Delta x_1 + f(u_2)\,\Delta x_2 + \cdots + f(u_n)\,\Delta x_n = \sum_{i=1}^{n} f(u_i)\Delta x_i.$$

Figure 5.11 illustrates these Riemann sums as sums of *signed* areas of rectangles; any such areas that lie below the x-axis are counted as negative.

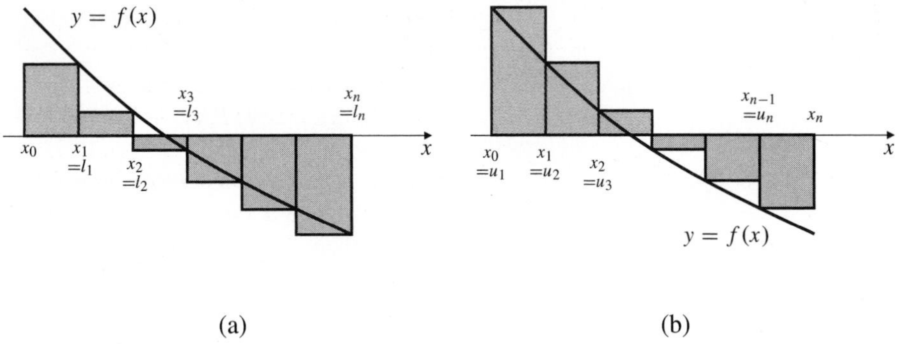

Figure 5.11 (a) A lower Riemann sum and (b) an upper Riemann sum for a decreasing function f. The areas of rectangles shaded in colour are counted as positive; those shaded in grey are counted as negative

EXAMPLE 1 Calculate lower and upper Riemann sums for the function $f(x) = 1/x$ on the interval $[1, 2]$, corresponding to the partition P of $[1, 2]$ into four subintervals of equal length.

Solution The partition P consists of the points $x_0 = 1$, $x_1 = 5/4$, $x_2 = 3/2$, $x_3 = 7/4$, and $x_4 = 2$. Since $1/x$ is decreasing on $[1, 2]$, its minimum and maximum values on the ith subinterval $[x_{i-1}, x_i]$ are $1/x_i$ and $1/x_{i-1}$, respectively. Thus, the lower and upper Riemann sums are

$$L(f, P) = \frac{1}{4}\left(\frac{4}{5} + \frac{2}{3} + \frac{4}{7} + \frac{1}{2}\right) = \frac{533}{840} \approx 0.6345,$$

$$U(f, P) = \frac{1}{4}\left(1 + \frac{4}{5} + \frac{2}{3} + \frac{4}{7}\right) = \frac{319}{420} \approx 0.7595.$$

EXAMPLE 2 Calculate the lower and upper Riemann sums for the function $f(x) = x^2$ on the interval $[0, a]$ (where $a > 0$), corresponding to the partition P_n of $[0, a]$ into n subintervals of equal length.

Solution Each subinterval of P_n has length $\Delta x = a/n$, and the division points are given by $x_i = ia/n$ for $i = 0, 1, 2, \ldots, n$. Since x^2 is increasing on $[0, a]$, its minimum and maximum values over the ith subinterval $[x_{i-1}, x_i]$ occur at $l_i = x_{i-1}$ and $u_i = x_i$, respectively. Thus, the lower Riemann sum of f for P_n is

$$\begin{aligned} L(f, P_n) &= \sum_{i=1}^{n} (x_{i-1})^2 \Delta x = \frac{a^3}{n^3} \sum_{i=1}^{n} (i-1)^2 \\ &= \frac{a^3}{n^3} \sum_{j=0}^{n-1} j^2 = \frac{a^3}{n^3} \frac{(n-1)n(2(n-1)+1)}{6} = \frac{(n-1)(2n-1)a^3}{6n^2}, \end{aligned}$$

where we have used Theorem 1(c) of Section 5.1 to evaluate the sum of squares. Similarly, the upper Riemann sum is

$$\begin{aligned} U(f, P_n) &= \sum_{i=1}^{n} (x_i)^2 \Delta x \\ &= \frac{a^3}{n^3} \sum_{i=1}^{n} i^2 = \frac{a^3}{n^3} \frac{n(n+1)(2n+1)}{6} = \frac{(n+1)(2n+1)a^3}{6n^2}. \end{aligned}$$

The Definite Integral

If we calculate $L(f, P)$ and $U(f, P)$ for partitions P having more and more points spaced closer and closer together, we expect that, in the limit, these Riemann sums will converge to a common value that will be the area bounded by $y = f(x)$, $y = 0$, $x = a$, and $x = b$ if $f(x) \geq 0$ on $[a, b]$. This is indeed the case, but we cannot fully prove it yet.

If P_1 and P_2 are two partitions of $[a, b]$ such that every point of P_1 also belongs to P_2, then we say that P_2 is a **refinement** of P_1. It is not difficult to show that in this case

$$L(f, P_1) \le L(f, P_2) \le U(f, P_2) \le U(f, P_1);$$

adding more points to a partition increases the lower sum and decreases the upper sum. (See Exercise 18 at the end of this section.) Given any two partitions, P_1 and P_2, we can form their **common refinement** P, which consists of all of the points of P_1 and P_2. Thus,

$$L(f, P_1) \le L(f, P) \le U(f, P) \le U(f, P_2).$$

Hence, every lower sum is less than or equal to every upper sum. Since the real numbers are complete, there must exist *at least one* real number I such that

$$L(f, P) \le I \le U(f, P), \qquad \text{for every partition } P.$$

If there is *only one* such number, we will call it the definite integral of f on $[a, b]$.

DEFINITION 3

The definite integral

Suppose there is exactly one number I such that for every partition P of $[a, b]$ we have

$$L(f, P) \le I \le U(f, P).$$

Then we say that the function f is **integrable** on $[a, b]$, and we call I the **definite integral** of f on $[a, b]$. The definite integral is denoted by the symbol

$$I = \int_a^b f(x)\,dx.$$

The definite integral of $f(x)$ over $[a, b]$ is a *number*; it is not a function of x. It depends on the numbers a and b and on the particular function f, but not on the variable x (which is a **dummy variable** like the variable i in the sum $\sum_{i=1}^{n} f(i)$). Replacing x with another variable does not change the value of the integral:

$$\int_a^b f(x)\,dx = \int_a^b f(t)\,dt.$$

The various parts of the symbol $\displaystyle\int_a^b f(x)\,dx$ have their own names:

(i) $\int$ is called the **integral sign**; it resembles the letter S since it represents the limit of a sum.

(ii) a and b are called the **limits of integration**; a is the **lower limit**, b is the **upper limit**.

(iii) The function f is the **integrand**; x is the **variable of integration**.

(iv) dx is the **differential** of x. It replaces Δx in the Riemann sums. If an integrand depends on more than one variable, the differential tells you which one is the variable of integration.

EXAMPLE 3 Show that $f(x) = x^2$ is integrable over the interval $[0, a]$, where $a > 0$, and evaluate $\displaystyle\int_0^a x^2\,dx$.

Solution We evaluate the limits as $n \to \infty$ of the lower and upper sums of f over $[0, a]$ obtained in Example 2 above.

$$\lim_{n\to\infty} L(f, P_n) = \lim_{n\to\infty} \frac{(n-1)(2n-1)a^3}{6n^2} = \frac{a^3}{3},$$
$$\lim_{n\to\infty} U(f, P_n) = \lim_{n\to\infty} \frac{(n+1)(2n+1)a^3}{6n^2} = \frac{a^3}{3}.$$

If $L(f, P_n) \le I \le U(f, P_n)$, we must have $I = a^3/3$. Thus, $f(x) = x^2$ is integrable over $[0, a]$, and

$$\int_0^a f(x)\,dx = \int_0^a x^2\,dx = \frac{a^3}{3}.$$

For all partitions P of $[a, b]$, we have

$$L(f, P) \le \int_a^b f(x)\,dx \le U(f, P).$$

If $f(x) \ge 0$ on $[a, b]$, then the area of the region R bounded by the graph of $y = f(x)$, the x-axis, and the lines $x = a$ and $x = b$ is A square units, where $A = \int_a^b f(x)\,dx$. If $f(x) \le 0$ on $[a, b]$, the area of R is $-\int_a^b f(x)\,dx$ square units. For general f, $\int_a^b f(x)\,dx$ is the area of that part of R lying above the x-axis minus the area of that part lying below the x-axis. (See Figure 5.12.) You can think of $\int_a^b f(x)\,dx$ as a "sum" of "areas" of infinitely many rectangles with heights $f(x)$ and "infinitesimally small widths" dx; it is a limit of the upper and lower Riemann sums.

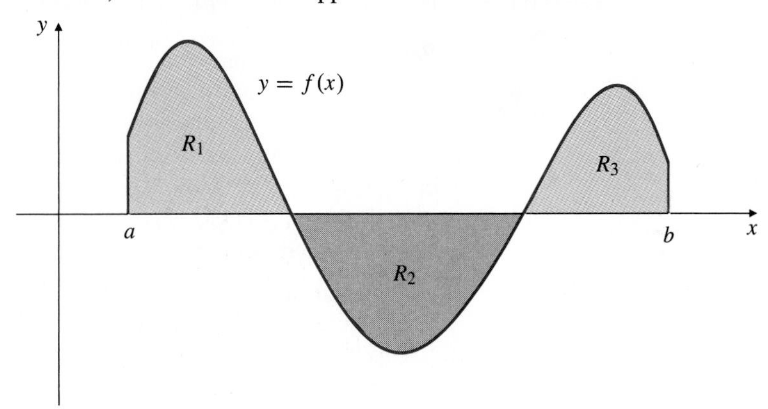

Figure 5.12 $\int_a^b f(x)\,dx$ equals area R_1 − area R_2 + area R_3

General Riemann Sums

Let $P = \{x_0, x_1, x_2, \ldots, x_n\}$, where $a = x_0 < x_1 < x_2 < \cdots < x_n = b$, be a partition of $[a, b]$ having norm $\|P\| = \max_{1\le i\le n} \Delta x_i$. In each subinterval $[x_{i-1}, x_i]$ of P pick a point c_i (called a *tag*). Let $c = (c_1, c_2, \ldots, c_n)$ denote the set of these tags. The sum

$$\begin{aligned} R(f, P, c) &= \sum_{i=1}^{n} f(c_i)\,\Delta x_i \\ &= f(c_1)\,\Delta x_1 + f(c_2)\,\Delta x_2 + f(c_3)\,\Delta x_3 + \cdots + f(c_n)\,\Delta x_n \end{aligned}$$

is called the **Riemann sum** of f on $[a, b]$ corresponding to partition P and tags c.

Note in Figure 5.13 that $R(f, P, c)$ is a sum of *signed* areas of rectangles between the x-axis and the curve $y = f(x)$. For any choice of the tags c, the Riemann sum $R(f, P, c)$ satisfies

$$L(f, P) \le R(f, P, c) \le U(f, P).$$

Therefore, if f is integrable on $[a, b]$, then its integral is the limit of such Riemann sums, where the limit is taken as the number $n(P)$ of subintervals of P increases to infinity in such a way that the lengths of all the subintervals approach zero. That is,

$$\lim_{\substack{n(P)\to\infty \\ \|P\|\to 0}} R(f, P, c) = \int_a^b f(x)\, dx.$$

As we will see in Chapter 7, many applications of integration depend on recognizing that a limit of Riemann sums is a definite integral.

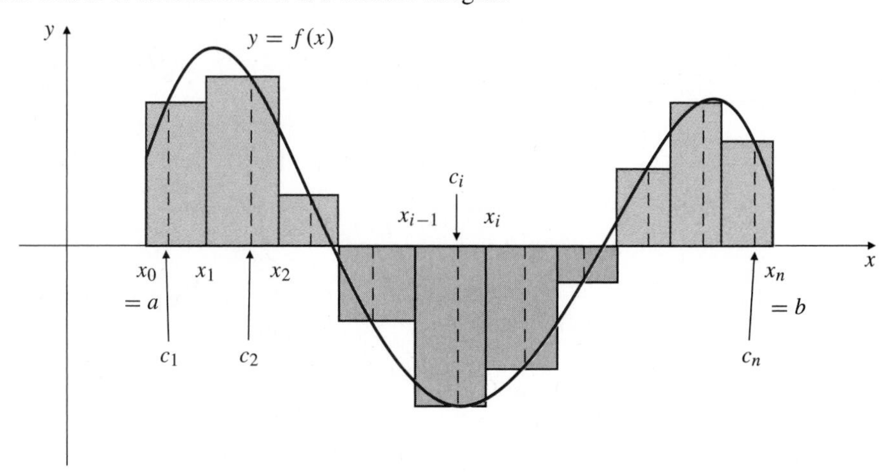

Figure 5.13 The Riemann sum $R(f, P, c)$ is the sum of areas of the rectangles shaded in colour minus the sum of the areas of the rectangles shaded in grey

THEOREM 2

If f is continuous on $[a, b]$, then f is integrable on $[a, b]$.

Remark The assumption that f is continuous in Theorem 2 may seem a bit superfluous since continuity was required throughout the above discussion leading to the definition of the definite integral. We cannot, however, prove this theorem yet. Its proof makes subtle use of the completeness property of the real numbers, and is given in Appendix IV in the context of an extended definition of definite integral that is meaningful for a larger class of functions that are not necessarily continuous. (The integral studied in Appendix IV is called the **Riemann integral**.)

We can, however, make the following observation. In order to prove that f is integrable on $[a, b]$, it is sufficient that, for any given positive number ϵ, we should be able to find a partition P of $[a, b]$ for which $U(f, P) - L(f, P) < \epsilon$. This condition prevents there being more than one number I that is both greater than every lower sum and less than every upper sum. It is not difficult to find such a partition if the function f is nondecreasing (or if it is nonincreasing) on $[a, b]$. (See Exercise 17 at the end of this section.) Therefore, nondecreasing and nonincreasing continuous functions are integrable; so, therefore, is any continuous function that is the sum of a nondecreasing and a nonincreasing function. This class of functions includes any continuous functions we are likely to encounter in concrete applications of calculus but, unfortunately, does not include all continuous functions.

Meanwhile, in Sections 5.4 and 6.5 we will extend the definition of the definite integral to certain kinds of functions that are not continuous, or where the interval of integration is not closed or not bounded.

EXAMPLE 4 Express the limit $\displaystyle\lim_{n\to\infty}\sum_{i=1}^{n}\frac{2}{n}\left(1+\frac{2i-1}{n}\right)^{1/3}$ as a definite integral.

Solution We want to interpret the sum as a Riemann sum for $f(x) = (1+x)^{1/3}$. The factor $2/n$ suggests that the interval of integration has length 2 and is partitioned into n

equal subintervals, each of length $2/n$. Let $c_i = (2i-1)/n$ for $i = 1, 2, 3, \ldots, n$. As $n \to \infty$, $c_1 = 1/n \to 0$ and $c_n = (2n-1)/n \to 2$. Thus, the interval is $[0, 2]$, and the points of the partition are $x_i = 2i/n$. Observe that $x_{i-1} = (2i-2)/n < c_i < 2i/n = x_i$ for each i, so that the sum is indeed a Riemann sum for $f(x)$ over $[0, 2]$. Since f is continuous on that interval, it is integrable there, and

$$\lim_{n\to\infty} \sum_{i=1}^{n} \frac{2}{n}\left(1 + \frac{2i-1}{n}\right)^{1/3} = \int_0^2 (1+x)^{1/3}\,dx.$$

EXERCISES 5.3

In Exercises 1–6, let P_n denote the partition of the given interval $[a, b]$ into n subintervals of equal length $\Delta x_i = (b-a)/n$. Evaluate $L(f, P_n)$ and $U(f, P_n)$ for the given functions f and the given values of n.

1. $f(x) = x$ on $[0, 2]$, with $n = 8$

2. $f(x) = x^2$ on $[0, 4]$, with $n = 4$

3. $f(x) = e^x$ on $[-2, 2]$, with $n = 4$

4. $f(x) = \ln x$ on $[1, 2]$, with $n = 5$

5. $f(x) = \sin x$ on $[0, \pi]$, with $n = 6$

6. $f(x) = \cos x$ on $[0, 2\pi]$, with $n = 4$

In Exercises 7–10, calculate $L(f, P_n)$ and $U(f, P_n)$ for the given function f over the given interval $[a, b]$, where P_n is the partition of the interval into n subintervals of equal length $\Delta x = (b-a)/n$. Show that

$$\lim_{n\to\infty} L(f, P_n) = \lim_{n\to\infty} U(f, P_n).$$

Hence, f is integrable on $[a, b]$. (Why?) What is $\int_a^b f(x)\,dx$?

7. $f(x) = x, \quad [a, b] = [0, 1]$

8. $f(x) = 1 - x, \quad [a, b] = [0, 2]$

9. $f(x) = x^3, \quad [a, b] = [0, 1]$

10. $f(x) = e^x, \quad [a, b] = [0, 3]$

In Exercises 11–16, express the given limit as a definite integral.

11. $\displaystyle\lim_{n\to\infty} \sum_{i=1}^{n} \frac{1}{n}\sqrt{\frac{i}{n}}$

12. $\displaystyle\lim_{n\to\infty} \sum_{i=1}^{n} \frac{1}{n}\sqrt{\frac{i-1}{n}}$

13. $\displaystyle\lim_{n\to\infty} \sum_{i=1}^{n} \frac{\pi}{n}\sin\left(\frac{\pi i}{n}\right)$

14. $\displaystyle\lim_{n\to\infty} \sum_{i=1}^{n} \frac{2}{n}\ln\left(1 + \frac{2i}{n}\right)$

15. $\displaystyle\lim_{n\to\infty} \sum_{i=1}^{n} \frac{1}{n}\tan^{-1}\left(\frac{2i-1}{2n}\right)$

16. $\displaystyle\lim_{n\to\infty} \sum_{i=1}^{n} \frac{n}{n^2+i^2}$

17. If f is continuous and nondecreasing on $[a, b]$, and P_n is the partition of $[a, b]$ into n subintervals of equal length ($\Delta x_i = (b-a)/n$ for $1 \le i \le n$), show that

$$U(f, P_n) - L(f, P_n) = \frac{(b-a)\big(f(b) - f(a)\big)}{n}.$$

Since we can make the right side as small as we please by choosing n large enough, f must be integrable on $[a, b]$.

18. Let $P = \{a = x_0 < x_1 < x_2 < \cdots < x_n = b\}$ be a partition of $[a, b]$, and let P' be a refinement of P having one more point, x', satisfying, say, $x_{i-1} < x' < x_i$ for some i between 1 and n. Show that

$$L(f, P) \le L(f, P') \le U(f, P') \le U(f, P)$$

for any continuous function f. (*Hint:* Consider the maximum and minimum values of f on the intervals $[x_{i-1}, x_i]$, $[x_{i-1}, x']$, and $[x', x_i]$.) Hence, deduce that

$$L(f, P) \le L(f, P'') \le U(f, P'') \le U(f, P) \text{ if } P''$$

is *any* refinement of P.

5.4 Properties of the Definite Integral

It is convenient to extend the definition of the definite integral $\int_a^b f(x)\,dx$ to allow $a = b$ and $a > b$ as well as $a < b$. The extension still involves partitions P having $x_0 = a$ and $x_n = b$ with intermediate points occurring in order between these end points, so that if $a = b$, then we must have $\Delta x_i = 0$ for every i, and hence the integral is zero. If $a > b$, we have $\Delta x_i < 0$ for each i, so the integral will be negative for positive functions f and vice versa.

Some of the most important properties of the definite integral are summarized in the following theorem.

THEOREM 3

Let f and g be integrable on an interval containing the points a, b, and c. Then

(a) An integral over an interval of zero length is zero.

$$\int_a^a f(x)\,dx = 0.$$

(b) Reversing the limits of integration changes the sign of the integral.

$$\int_b^a f(x)\,dx = -\int_a^b f(x)\,dx.$$

(c) An integral depends linearly on the integrand. If A and B are constants, then

$$\int_a^b \big(Af(x) + Bg(x)\big)\,dx = A\int_a^b f(x)\,dx + B\int_a^b g(x)\,dx.$$

(d) An integral depends additively on the interval of integration.

$$\int_a^b f(x)\,dx + \int_b^c f(x)\,dx = \int_a^c f(x)\,dx.$$

(e) If $a \le b$ and $f(x) \le g(x)$ for $a \le x \le b$, then

$$\int_a^b f(x)\,dx \le \int_a^b g(x)\,dx.$$

(f) The **triangle inequality** for sums extends to definite integrals. If $a \le b$, then

$$\left|\int_a^b f(x)\,dx\right| \le \int_a^b |f(x)|\,dx.$$

(g) The integral of an odd function over an interval symmetric about zero is zero. If f is an odd function (i.e., $f(-x) = -f(x)$), then

$$\int_{-a}^a f(x)\,dx = 0.$$

(h) The integral of an even function over an interval symmetric about zero is twice the integral over the positive half of the interval. If f is an even function (i.e., $f(-x) = f(x)$), then

$$\int_{-a}^a f(x)\,dx = 2\int_0^a f(x)\,dx.$$

The proofs of parts (a) and (b) are suggested in the first paragraph of this section. We postpone giving formal proofs of parts (c)–(h) until Appendix IV (see Exercises 5–8 in that Appendix). Nevertheless, all of these results should appear intuitively reasonable if you regard the integrals as representing (signed) areas. For instance, properties (d) and (e) are, respectively, properties (v) and (iv) of areas mentioned in the first paragraph of Section 5.2. (See Figure 5.14.)

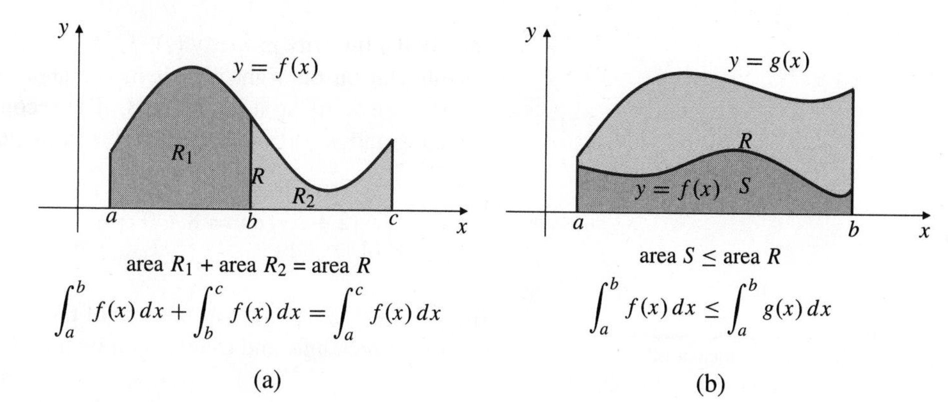

$$\int_a^b f(x)\,dx + \int_b^c f(x)\,dx = \int_a^c f(x)\,dx \qquad \int_a^b f(x)\,dx \le \int_a^b g(x)\,dx$$

Figure 5.14

(a) Property (d) of Theorem 3

(b) Property (e) of Theorem 3

Property (f) is a generalization of the triangle inequality for numbers:

$$|x+y| \le |x| + |y|, \quad \text{or more generally,} \quad \left|\sum_{i=1}^{n} x_i\right| \le \sum_{i=1}^{n} |x_i|.$$

It follows from property (e) (assuming that $|f|$ is integrable on $[a, b]$), since $-|f(x)| \le f(x) \le |f(x)|$. The symmetry properties (g) and (h), which are illustrated in Figure 5.15, are particularly useful and should always be kept in mind when evaluating definite integrals because they can save much unnecessary work.

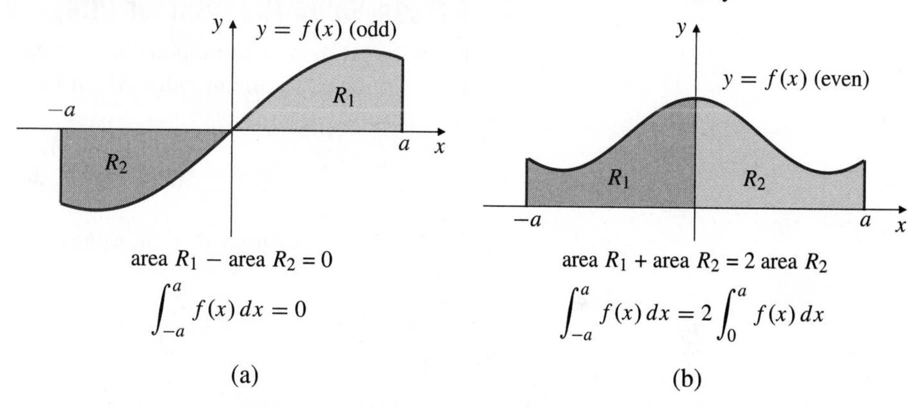

$$\int_{-a}^{a} f(x)\,dx = 0 \qquad \int_{-a}^{a} f(x)\,dx = 2\int_0^a f(x)\,dx$$

Figure 5.15

(a) Property (g) of Theorem 3

(b) Property (h) of Theorem 3

As yet we have no easy method for evaluating definite integrals. However, some such integrals can be simplified by using various properties in Theorem 3, and others can be interpreted as known areas.

EXAMPLE 1 Evaluate

$$\text{(a)} \int_{-2}^{2} (2+5x)\,dx, \quad \text{(b)} \int_0^3 (2+x)\,dx, \quad \text{and} \quad \text{(c)} \int_{-3}^{3} \sqrt{9-x^2}\,dx.$$

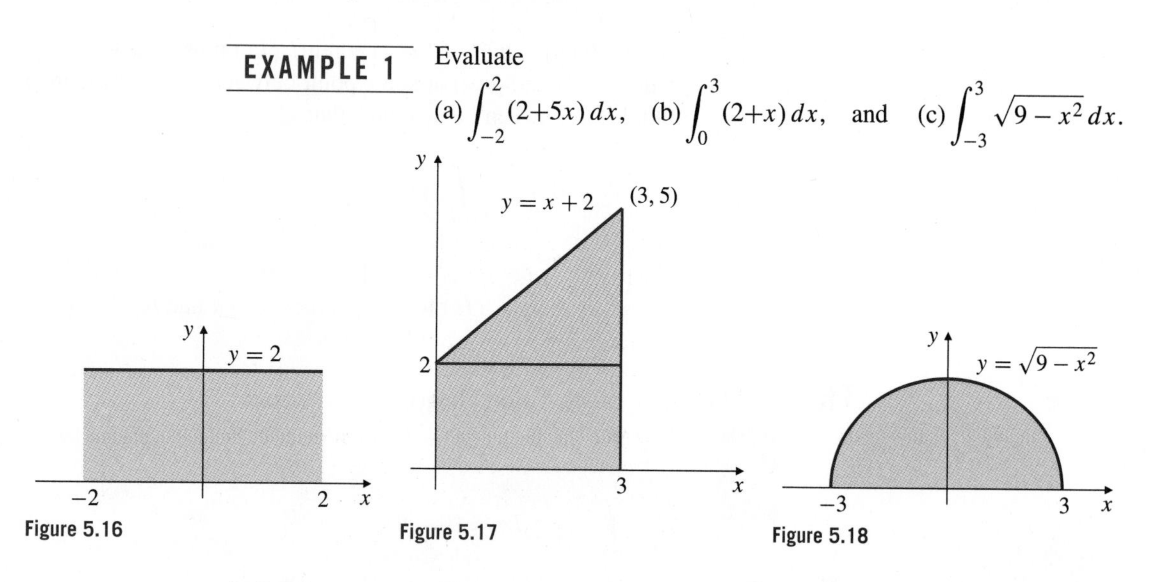

Figure 5.16 **Figure 5.17** **Figure 5.18**

Solution See Figures 5.16–5.18.

(a) By the linearity property (c), $\int_{-2}^{2}(2+5x)\,dx = \int_{-2}^{2} 2\,dx + 5\int_{-2}^{2} x\,dx$. The first integral on the right represents the area of a rectangle of width 4 and height 2 (Figure 5.16), so it has value 8. The second integral on the right is 0 because its integrand is odd and the interval is symmetric about 0. Thus,

$$\int_{-2}^{2}(2+5x)\,dx = 8+0 = 8.$$

(b) $\int_{0}^{3}(2+x)\,dx$ represents the area of the trapezoid in Figure 5.17. Adding the areas of the rectangle and triangle comprising this trapezoid, we get

$$\int_{0}^{3}(2+x)\,dx = (3\times 2) + \frac{1}{2}(3\times 3) = \frac{21}{2}.$$

While areas are measured in squared units of length, definite integrals are numbers and have no units. Even when you use an area to find an integral, do not quote units for the integral.

(c) $\int_{-3}^{3}\sqrt{9-x^2}\,dx$ represents the area of a semicircle of radius 3 (Figure 5.18), so

$$\int_{-3}^{3}\sqrt{9-x^2}\,dx = \frac{1}{2}\pi(3^2) = \frac{9\pi}{2}.$$

A Mean-Value Theorem for Integrals

Let f be a function continuous on the interval $[a,b]$. Then f assumes a minimum value m and a maximum value M on the interval, say at points $x=l$ and $x=u$, respectively:

$$m = f(l) \le f(x) \le f(u) = M \qquad \text{for all } x \text{ in } [a,b].$$

For the 2-point partition P of $[a,b]$ having $x_0 = a$ and $x_1 = b$, we have

$$m(b-a) = L(f,P) \le \int_a^b f(x)\,dx \le U(f,P) = M(b-a).$$

Therefore,

$$f(l) = m \le \frac{1}{b-a}\int_a^b f(x)\,dx \le M = f(u).$$

By the Intermediate-Value Theorem, $f(x)$ must take on every value between the two values $f(l)$ and $f(u)$ at some point between l and u (Figure 5.19). Hence, there is a number c between l and u such that

$$f(c) = \frac{1}{b-a}\int_a^b f(x)\,dx.$$

That is, $\int_a^b f(x)\,dx$ is equal to the area $(b-a)f(c)$ of a rectangle with base width $b-a$ and height $f(c)$ for some c between a and b. This is the Mean-Value Theorem for integrals.

THEOREM 4

The Mean-Value Theorem for integrals

If f is continuous on $[a,b]$, then there exists a point c in $[a,b]$ such that

$$\int_a^b f(x)\,dx = (b-a)f(c).$$

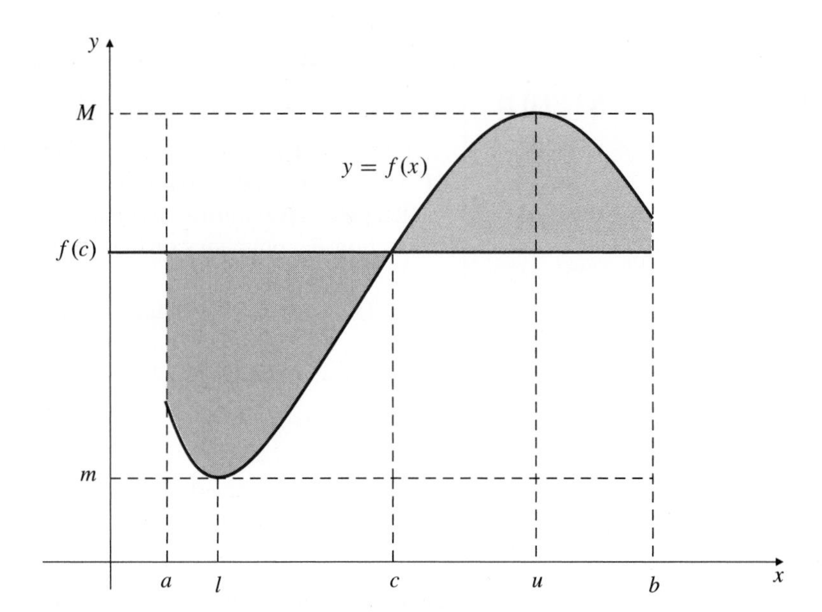

Figure 5.19 Half of the area between $y = f(x)$ and the horizontal line $y = f(c)$ lies above the line, and the other half lies below the line

Observe in Figure 5.19 that the area below the curve $y = f(x)$ and above the line $y = f(c)$ is equal to the area above $y = f(x)$ and below $y = f(c)$. In this sense, $f(c)$ is the average value of the function $f(x)$ on the interval $[a, b]$.

DEFINITION 4

Average value of a function

If f is integrable on $[a, b]$, then the **average value** or **mean value** of f on $[a, b]$, denoted by $\bar{f}$, is

$$\bar{f} = \frac{1}{b-a}\int_a^b f(x)\,dx.$$

EXAMPLE 2 Find the average value of $f(x) = 2x$ on the interval $[1, 5]$.

Solution The average value (see Figure 5.20) is

$$\bar{f} = \frac{1}{5-1}\int_1^5 2x\,dx = \frac{1}{4}\left(4 \times 2 + \frac{1}{2}(4 \times 8)\right) = 6.$$

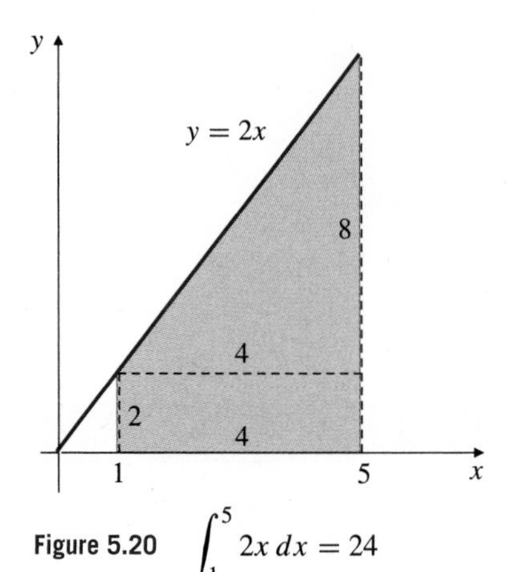

Figure 5.20 $\int_1^5 2x\,dx = 24$

Definite Integrals of Piecewise Continuous Functions

The definition of integrability and the definite integral given above can be extended to a wider class than just continuous functions. One simple but very important extension is to the class of *piecewise continuous functions*.

Consider the graph $y = f(x)$ shown in Figure 5.21(a). Although f is not continuous at all points in $[a, b]$ (it is discontinuous at c_1 and c_2), clearly the region lying under the graph and above the x-axis between $x = a$ and $x = b$ does have an area. We would like to represent this area as

$$\int_a^{c_1} f(x)\,dx + \int_{c_1}^{c_2} f(x)\,dx + \int_{c_2}^{b} f(x)\,dx.$$

This is reasonable because there are continuous functions on $[a, c_1]$, $[c_1, c_2]$, and $[c_2, b]$ equal to $f(x)$ on the corresponding open intervals, (a, c_1), (c_1, c_2), and (c_2, b).

DEFINITION 5

Piecewise continuous functions

Let $c_0 < c_1 < c_2 < \cdots < c_n$ be a finite set of points on the real line. A function f defined on $[c_0, c_n]$ except possibly at some of the points c_i, $(0 \le i \le n)$, is called **piecewise continuous** on that interval if for each i $(1 \le i \le n)$ there exists a function F_i continuous on the *closed* interval $[c_{i-1}, c_i]$ such that

$$f(x) = F_i(x) \qquad \text{on the } \textit{open} \text{ interval} \quad (c_{i-1}, c_i).$$

In this case, we define the definite integral of f from c_0 to c_n to be

$$\int_{c_0}^{c_n} f(x)\,dx = \sum_{i=1}^{n} \int_{c_{i-1}}^{c_i} F_i(x)\,dx.$$

EXAMPLE 3 Find $\int_0^3 f(x)\,dx$, where $f(x) = \begin{cases} \sqrt{1-x^2} & \text{if } 0 \le x \le 1 \\ 2 & \text{if } 1 < x \le 2 \\ x-2 & \text{if } 2 < x \le 3. \end{cases}$

Solution The value of the integral is the sum of the shaded areas in Figure 5.21(b):

$$\begin{aligned} \int_0^3 f(x)\,dx &= \int_0^1 \sqrt{1-x^2}\,dx + \int_1^2 2\,dx + \int_2^3 (x-2)\,dx \\ &= \left(\frac{1}{4} \times \pi \times 1^2\right) + (2 \times 1) + \left(\frac{1}{2} \times 1 \times 1\right) = \frac{\pi + 10}{4}. \end{aligned}$$

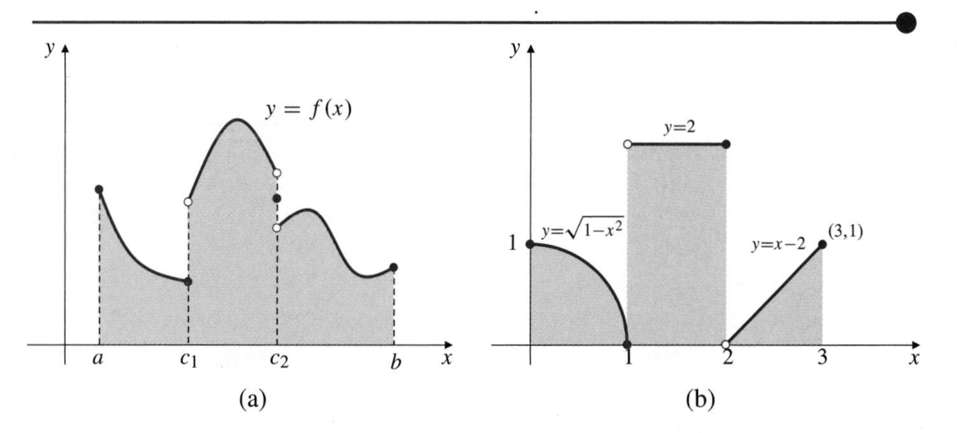

Figure 5.21 Two piecewise continuous functions

EXERCISES 5.4

1. Simplify $\int_a^b f(x)\,dx + \int_b^c f(x)\,dx + \int_c^a f(x)\,dx.$

2. Simplify $\int_0^2 3f(x)\,dx + \int_1^3 3f(x)\,dx - \int_0^3 2f(x)\,dx - \int_1^2 3f(x)\,dx.$

Evaluate the integrals in Exercises 3–16 by using the properties of the definite integral and interpreting integrals as areas.

3. $\int_{-2}^2 (x+2)\,dx$

4. $\int_0^2 (3x+1)\,dx$

5. $\int_a^b x\,dx$

6. $\int_{-1}^2 (1-2x)\,dx$

7. $\int_{-\sqrt{2}}^{\sqrt{2}} \sqrt{2-t^2}\,dt$

8. $\int_{-\sqrt{2}}^0 \sqrt{2-x^2}\,dx$

9. $\int_{-\pi}^{\pi} \sin(x^3)\,dx$

10. $\int_{-a}^a (a-|s|)\,ds$

11. $\int_{-1}^1 (u^5 - 3u^3 + \pi)\,du$

12. $\int_0^2 \sqrt{2x-x^2}\,dx$

13. $\int_{-4}^4 (e^x - e^{-x})\,dx$

14. $\int_{-3}^3 (2+t)\sqrt{9-t^2}\,dt$

15. $\int_0^1 \sqrt{4-x^2}\,dx$ **16.** $\int_1^2 \sqrt{4-x^2}\,dx$

Given that $\int_0^a x^2\,dx = \frac{a^3}{3}$, evaluate the integrals in Exercises 17–22.

17. $\int_0^2 6x^2\,dx$ **18.** $\int_2^3 (x^2-4)\,dx$

19. $\int_{-2}^2 (4-t^2)\,dt$ **20.** $\int_0^2 (v^2-v)\,dv$

21. $\int_0^1 (x^2+\sqrt{1-x^2})\,dx$ **22.** $\int_{-6}^6 x^2(2+\sin x)\,dx$

The definition of $\ln x$ as an area in Section 3.3 implies that

$$\int_1^x \frac{1}{t}\,dt = \ln x$$

for $x > 0$. Use this to evaluate the integrals in Exercises 23–26.

23. $\int_1^2 \frac{1}{x}\,dx$ **24.** $\int_2^4 \frac{1}{t}\,dt$

25. $\int_{1/3}^1 \frac{1}{t}\,dt$ **26.** $\int_{1/4}^3 \frac{1}{s}\,ds$

Find the average values of the functions in Exercises 27–32 over the given intervals.

27. $f(x) = x+2$ over $[0,4]$

28. $g(x) = x+2$ over $[a,b]$

29. $f(t) = 1+\sin t$ over $[-\pi,\pi]$

30. $k(x) = x^2$ over $[0,3]$

31. $f(x) = \sqrt{4-x^2}$ over $[0,2]$

32. $g(s) = 1/s$ over $[1/2,2]$

Piecewise continuous functions

33. Evaluate $\int_{-1}^2 \operatorname{sgn} x\,dx$. Recall that $\operatorname{sgn} x$ is 1 if $x > 0$ and -1 if $x < 0$.

34. Find $\int_{-3}^2 f(x)\,dx$, where $f(x) = \begin{cases} 1+x & \text{if } x < 0 \\ 2 & \text{if } x \ge 0. \end{cases}$

35. Find $\int_0^2 g(x)\,dx$, where $g(x) = \begin{cases} x^2 & \text{if } 0 \le x \le 1 \\ x & \text{if } 1 < x \le 2. \end{cases}$

36. Evaluate $\int_0^3 |2-x|\,dx$.

37. Evaluate $\int_0^2 \sqrt{4-x^2}\,\operatorname{sgn}(x-1)\,dx$.

38. Evaluate $\int_0^{3.5} \lfloor x \rfloor\,dx$, where $\lfloor x \rfloor$ is the greatest integer less than or equal to x. (See Example 10 of section P.5.)

Evaluate the integrals in Exercises 39–40 by inspecting the graphs of the integrands.

39. $\int_{-3}^4 \Big(|x+1| - |x-1| + |x+2|\Big)\,dx$

40. $\int_0^3 \frac{x^2-x}{|x-1|}\,dx$

41. Find the average value of the function $f(x) = |x+1|\operatorname{sgn} x$ on the interval $[-2,2]$.

42. If $a < b$ and f is continuous on $[a,b]$, show that $\int_a^b \Big(f(x) - \bar{f}\Big)\,dx = 0$.

43. Suppose that $a < b$ and f is continuous on $[a,b]$. Find the constant k that minimizes the integral $\int_a^b \Big(f(x)-k\Big)^2\,dx$.

5.5 The Fundamental Theorem of Calculus

In this section we demonstrate the relationship between the definite integral defined in Section 5.3 and the indefinite integral (or general antiderivative) introduced in Section 2.10. A consequence of this relationship is that we will be able to calculate definite integrals of functions whose antiderivatives we can find.

In Section 3.3 we wanted to find a function whose derivative was $1/x$. We solved this problem by defining the desired function ($\ln x$) in terms of the area under the graph of $y = 1/x$. This idea motivates, and is a special case of, the following theorem.

THEOREM 5

The Fundamental Theorem of Calculus

Suppose that the function f is continuous on an interval I containing the point a.

PART I. Let the function F be defined on I by

$$F(x) = \int_a^x f(t)\,dt.$$

Then F is differentiable on I, and $F'(x) = f(x)$ there. Thus, F is an antiderivative of f on I:

$$\frac{d}{dx}\int_a^x f(t)\,dt = f(x).$$

PART II. If $G(x)$ is *any* antiderivative of $f(x)$ on I, so that $G'(x) = f(x)$ on I, then for any b in I we have

$$\int_a^b f(x)\,dx = G(b) - G(a).$$

PROOF Using the definition of the derivative, we calculate

$$\begin{aligned}
F'(x) &= \lim_{h\to 0}\frac{F(x+h)-F(x)}{h}\\
&= \lim_{h\to 0}\frac{1}{h}\left(\int_a^{x+h} f(t)\,dt - \int_a^x f(t)\,dt\right)\\
&= \lim_{h\to 0}\frac{1}{h}\int_x^{x+h} f(t)\,dt \qquad \text{by Theorem 3(d)}\\
&= \lim_{h\to 0}\frac{1}{h}hf(c) \qquad \text{for some } c = c(h) \text{ (depending on } h\text{)}\\
&\qquad\qquad\qquad \text{between } x \text{ and } x+h \text{ (Theorem 4)}\\
&= \lim_{c\to x} f(c) \qquad \text{since } c \to x \text{ as } h \to 0\\
&= f(x) \qquad \text{since } f \text{ is continuous.}
\end{aligned}$$

Also, if $G'(x) = f(x)$, then $F(x) = G(x) + C$ on I for some constant C (by Theorem 13 of Section 2.8). Hence,

$$\int_a^x f(t)\,dt = F(x) = G(x) + C.$$

Let $x = a$ and obtain $0 = G(a) + C$ via Theorem 3(a), so $C = -G(a)$. Now let $x = b$ to get

$$\int_a^b f(t)\,dt = G(b) + C = G(b) - G(a).$$

Of course, we can replace t with x (or any other variable) as the variable of integration on the left-hand side.

Remark You should remember *both* conclusions of the Fundamental Theorem; they are both useful. Part I concerns the derivative of an integral; it tells you how to differentiate a definite integral with respect to its upper limit. Part II concerns the integral of a derivative; it tells you how to evaluate a definite integral if you can find an antiderivative of the integrand.

DEFINITION 6

To facilitate the evaluation of definite integrals using the Fundamental Theorem of Calculus, we define the **evaluation symbol**:

$$F(x)\Big|_a^b = F(b) - F(a).$$

Thus,

$$\int_a^b f(x)\,dx = \left(\int f(x)\,dx\right)\bigg|_a^b,$$

where $\int f(x)\,dx$ denotes the indefinite integral or general antiderivative of f. (See Section 2.10.) When evaluating a definite integral this way, we will omit the constant of integration ($+C$) from the indefinite integral because it cancels out in the subtraction:

$$(F(x)+C)\bigg|_a^b = F(b)+C-(F(a)+C) = F(b)-F(a) = F(x)\bigg|_a^b.$$

Any antiderivative of f can be used to calculate the definite integral.

EXAMPLE 1 Evaluate (a) $\displaystyle\int_0^a x^2\,dx$ and (b) $\displaystyle\int_{-1}^2 (x^2-3x+2)\,dx$.

Solution

(a) $\displaystyle\int_0^a x^2\,dx = \frac{1}{3}x^3\bigg|_0^a = \frac{1}{3}a^3 - \frac{1}{3}0^3 = \frac{a^3}{3}$ (because $\displaystyle\frac{d}{dx}\frac{x^3}{3} = x^2$).

(b) $$\int_{-1}^2 (x^2-3x+2)\,dx = \left(\frac{1}{3}x^3 - \frac{3}{2}x^2 + 2x\right)\bigg|_{-1}^2$$
$$= \frac{1}{3}(8) - \frac{3}{2}(4) + 4 - \left(\frac{1}{3}(-1) - \frac{3}{2}(1) + (-2)\right) = \frac{9}{2}.$$

BEWARE! Be careful to keep track of all the minus signs when substituting a negative lower limit.

EXAMPLE 2 Find the area A of the plane region lying above the x-axis and under the curve $y = 3x - x^2$.

Solution We need to find the points where the curve $y = 3x - x^2$ meets the x-axis. These are solutions of the equation

$$0 = 3x - x^2 = x(3-x).$$

The only roots are $x = 0$ and $x = 3$. (See Figure 5.22.) Hence, the area of the region is given by

$$A = \int_0^3 (3x - x^2)\,dx = \left(\frac{3}{2}x^2 - \frac{1}{3}x^3\right)\bigg|_0^3$$
$$= \frac{27}{2} - \frac{27}{3} - (0-0) = \frac{27}{6} = \frac{9}{2} \text{ square units.}$$

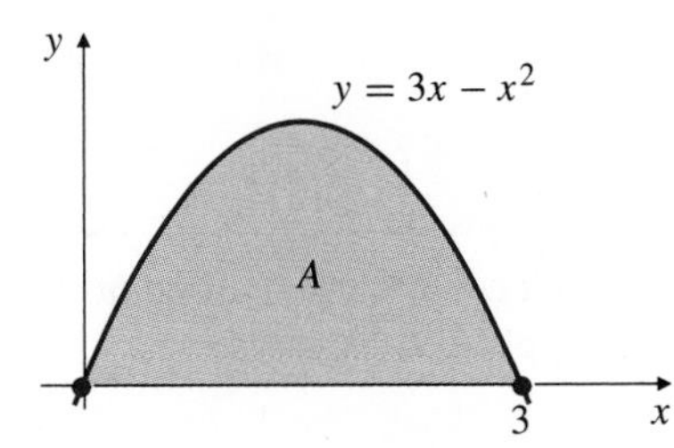

Figure 5.22

EXAMPLE 3 Find the area under the curve $y = \sin x$, above $y = 0$ from $x = 0$ to $x = \pi$.

Solution The required area, illustrated in Figure 5.23, is

$$A = \int_0^\pi \sin x\,dx = -\cos x\bigg|_0^\pi = -\big(-1-(1)\big) = 2 \text{ square units.}$$

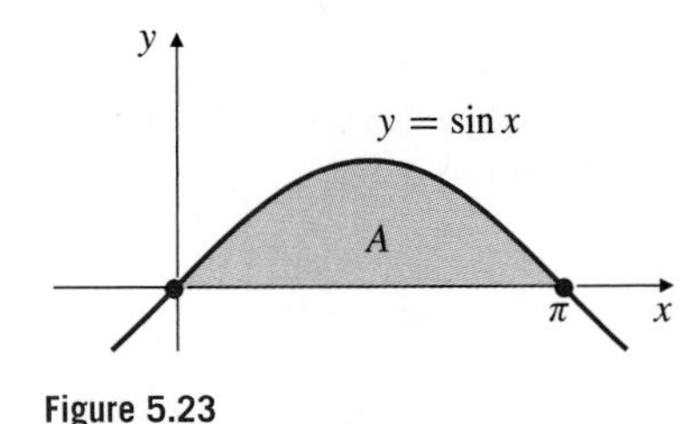

Figure 5.23

Note that while the definite integral is a pure number, an area is a geometric quantity that implicitly involves units. If the units along the x- and y-axes are, for example, metres, the area should be quoted in square metres (m^2). If units of length along the x-axis and y-axis are not specified, areas should be quoted in square units.

EXAMPLE 4 Find the area of the region R lying above the line $y = 1$ and below the curve $y = 5/(x^2 + 1)$.

Solution The region R is shaded in Figure 5.24. To find the intersections of $y = 1$ and $y = 5/(x^2 + 1)$, we must solve these equations simultaneously:

$$1 = \frac{5}{x^2 + 1},$$

so $x^2 + 1 = 5$, $x^2 = 4$, and $x = \pm 2$.

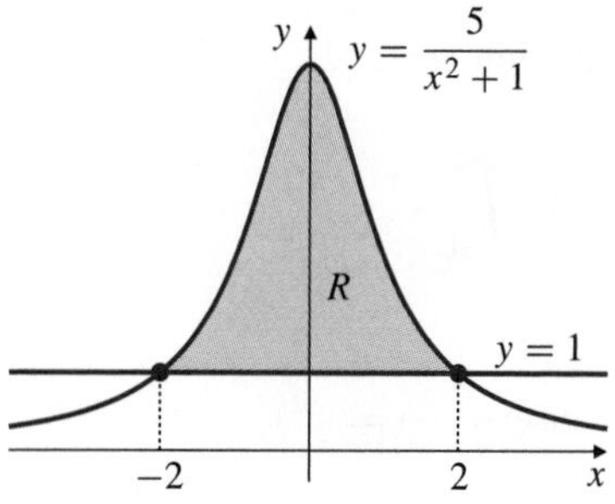

The area A of the region R is the area under the curve $y = 5/(x^2 + 1)$ and above the x-axis between $x = -2$ and $x = 2$, minus the area of a rectangle of width 4 and height 1. Since $\tan^{-1} x$ is an antiderivative of $1/(x^2 + 1)$,

$$\begin{aligned} A &= \int_{-2}^{2} \frac{5}{x^2 + 1}\,dx - 4 = 2\int_{0}^{2} \frac{5}{x^2 + 1}\,dx - 4 \\ &= 10 \tan^{-1} x \Big|_{0}^{2} - 4 = 10 \tan^{-1} 2 - 4 \text{ square units.} \end{aligned}$$

Figure 5.24

Observe the use of even symmetry (Theorem 3(h) of Section 5.4) to replace the lower limit of integration by 0. It is easier to substitute 0 into the antiderivative than -2.

EXAMPLE 5 Find the average value of $f(x) = e^{-x} + \cos x$ on the interval $[-\pi/2, 0]$.

Solution The average value is

$$\begin{aligned} \bar{f} &= \frac{1}{0 - \left(-\frac{\pi}{2}\right)} \int_{-(\pi/2)}^{0} (e^{-x} + \cos x)\,dx \\ &= \frac{2}{\pi}\left(-e^{-x} + \sin x\right)\Big|_{-(\pi/2)}^{0} \\ &= \frac{2}{\pi}\left(-1 + 0 + e^{\pi/2} - (-1)\right) = \frac{2}{\pi} e^{\pi/2}. \end{aligned}$$

Beware of integrals of the form $\int_a^b f(x)\,dx$, where f is not continuous at *all* points in the interval $[a, b]$. The Fundamental Theorem does not apply in such cases.

EXAMPLE 6 We know that $\frac{d}{dx} \ln|x| = \frac{1}{x}$ if $x \neq 0$. It is *incorrect*, however, to state that

$$\int_{-1}^{1} \frac{dx}{x} = \ln|x| \Big|_{-1}^{1} = 0 - 0 = 0,$$

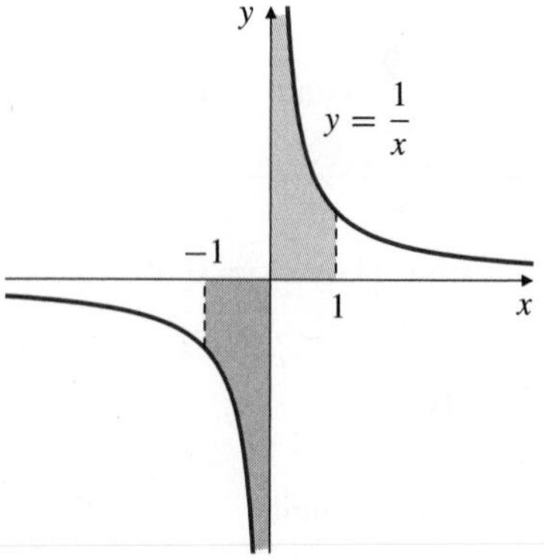

even though $1/x$ is an odd function. In fact, $1/x$ is undefined and has no limit at $x = 0$, and it is not integrable on $[-1, 0]$ or $[0, 1]$ (Figure 5.25). Observe that

$$\lim_{c \to 0+} \int_c^1 \frac{1}{x}\,dx = \lim_{c \to 0+} -\ln c = \infty,$$

Figure 5.25

so both shaded regions in Figure 5.25 have infinite area. Integrals of this type are called **improper integrals**. We deal with them in Section 6.5.

The following example illustrates, this time using definite integrals, the relationship observed in Example 1 of Section 2.11 between the area under the graph of its velocity and the distance travelled by an object over a time interval.

EXAMPLE 7 An object at rest at time $t = 0$ accelerates at a constant 10 m/s^2 during the time interval $[0, T]$. If $0 \le t_0 \le t_1 \le T$, find the distance travelled by the object in the time interval $[t_0, t_1]$.

Solution Let $v(t)$ denote the velocity of the object at time t, and let $y(t)$ denote the distance travelled by the object during the time interval $[0, t]$, where $0 \le t \le T$. Then $v(0) = 0$ and $y(0) = 0$. Also $v'(t) = 10$ and $y'(t) = v(t)$. Thus,

$$v(t) = v(t) - v(0) = \int_0^t v'(u)\,du = \int_0^t 10\,du = 10u\Big|_0^t = 10t$$

$$y(t) = y(t) - y(0) = \int_0^t y'(u)\,du = \int_0^t v(u)\,du = \int_0^t 10u\,du = 5u^2\Big|_0^t = 5t^2.$$

On the time interval $[t_0, t_1]$, the object has travelled distance

$$y(t(1)) - y(t(0)) = 5t_1^2 - 5t_0^2 = \int_0^{t_1} v(t)\,dt - \int_0^{t_0} v(t)\,dt = \int_{t_0}^{t_1} v(t)\,dt \quad \text{m.}$$

Observe that this last integral is the area under the graph of $y = v(t)$ above the interval $[t_0, t_1]$ on the t axis.

We now give some examples illustrating the first conclusion of the Fundamental Theorem.

EXAMPLE 8 Find the derivatives of the following functions:

(a) $F(x) = \int_x^3 e^{-t^2}\,dt$, (b) $G(x) = x^2 \int_{-4}^{5x} e^{-t^2}\,dt$, (c) $H(x) = \int_{x^2}^{x^3} e^{-t^2}\,dt$.

Solution The solutions involve applying the first conclusion of the Fundamental Theorem together with other differentiation rules.

(a) Observe that $F(x) = -\int_3^x e^{-t^2}\,dt$ (by Theorem 3(b)). Therefore, by the Fundamental Theorem, $F'(x) = -e^{-x^2}$.

(b) By the Product Rule and the Chain Rule,

$$\begin{aligned} G'(x) &= 2x \int_{-4}^{5x} e^{-t^2}\,dt + x^2 \frac{d}{dx} \int_{-4}^{5x} e^{-t^2}\,dt \\ &= 2x \int_{-4}^{5x} e^{-t^2}\,dt + x^2 e^{-(5x)^2}(5) \\ &= 2x \int_{-4}^{5x} e^{-t^2}\,dt + 5x^2 e^{-25x^2}. \end{aligned}$$

(c) Split the integral into a difference of two integrals in each of which the variable x appears only in the upper limit.

$$\begin{aligned} H(x) &= \int_0^{x^3} e^{-t^2}\,dt - \int_0^{x^2} e^{-t^2}\,dt \\ H'(x) &= e^{-(x^3)^2}(3x^2) - e^{-(x^2)^2}(2x) \\ &= 3x^2 e^{-x^6} - 2x\,e^{-x^4}. \end{aligned}$$

Parts (b) and (c) of Example 8 are examples of the following formulas that build the Chain Rule into the first conclusion of the Fundamental Theorem.

$$\frac{d}{dx}\int_a^{g(x)} f(t)\,dt = f\big(g(x)\big)\,g'(x)$$
$$\frac{d}{dx}\int_{h(x)}^{g(x)} f(t)\,dt = f\big(g(x)\big)\,g'(x) - f\big(h(x)\big)\,h'(x)$$

EXAMPLE 9 Solve the **integral equation** $f(x) = 2 + 3\int_4^x f(t)\,dt$.

Solution Differentiate the integral equation to get $f'(x) = 3f(x)$, the DE for exponential growth, having solution $f(x) = Ce^{3x}$. Now put $x = 4$ into the integral equation to get $f(4) = 2$. Hence $2 = Ce^{12}$, so $C = 2e^{-12}$. Therefore, the integral equation has solution $f(x) = 2e^{3x-12}$.

We conclude with an example showing how the Fundamental Theorem can be used to evaluate limits of Riemann sums.

EXAMPLE 10 Evaluate $\lim_{n\to\infty} \frac{1}{n}\sum_{j=1}^{n} \cos\left(\frac{j\pi}{2n}\right)$.

Solution The sum involves values of $\cos x$ at the right endpoints of the n subintervals of the partition

$$0,\quad \frac{\pi}{2n},\quad \frac{2\pi}{2n},\quad \frac{3\pi}{2n},\quad \ldots,\quad \frac{n\pi}{2n}$$

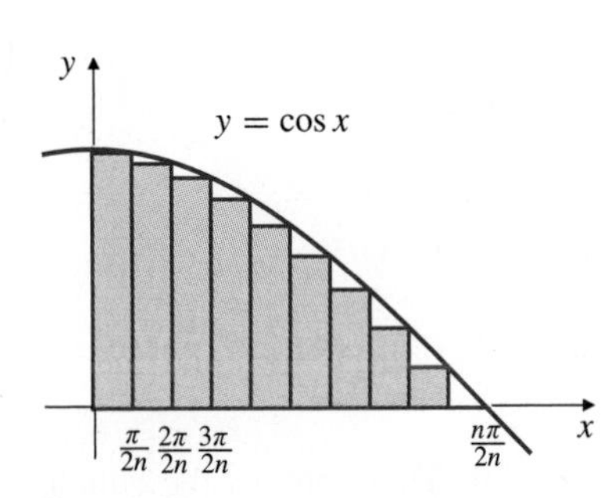

Figure 5.26

of the interval $[0, \pi/2]$. Since each of the subintervals of this partition has length $\pi/(2n)$, and since $\cos x$ is continuous on $[0, \pi/2]$, we have, expressing the limit of a Riemann sum as an integral (see Figure 5.26),

$$\lim_{n\to\infty} \frac{\pi}{2n}\sum_{j=1}^{n} \cos\left(\frac{j\pi}{2n}\right) = \int_0^{\pi/2} \cos x\,dx = \sin x\bigg|_0^{\pi/2} = 1 - 0 = 1.$$

The given sum differs from the Riemann sum above only in that the factor $\pi/2$ is missing. Thus,

$$\lim_{n\to\infty} \frac{1}{n}\sum_{j=1}^{n} \cos\left(\frac{j\pi}{2n}\right) = \frac{2}{\pi}.$$

EXERCISES 5.5

Evaluate the definite integrals in Exercises 1–20.

1. $\int_0^2 x^3\,dx$

2. $\int_0^4 \sqrt{x}\,dx$

3. $\int_{1/2}^1 \frac{1}{x^2}\,dx$

4. $\int_{-2}^{-1}\left(\frac{1}{x^2} - \frac{1}{x^3}\right)dx$

5. $\int_{-1}^2 (3x^2 - 4x + 2)\,dx$

6. $\int_1^2 \left(\frac{2}{x^3} - \frac{x^3}{2}\right)dx$

7. $\int_{-2}^2 (x^2 + 3)^2\,dx$

8. $\int_4^9 \left(\sqrt{x} - \frac{1}{\sqrt{x}}\right)dx$

9. $\int_{-\pi/4}^{-\pi/6} \cos x\,dx$

10. $\int_0^{\pi/3} \sec^2\theta\,d\theta$

11. $\int_{\pi/4}^{\pi/3} \sin\theta\,d\theta$

12. $\int_0^{2\pi} (1 + \sin u)\,du$

13. $\int_{-\pi}^{\pi} e^x\,dx$

14. $\int_{-2}^2 (e^x - e^{-x})\,dx$

15. $\int_0^e a^x\,dx \quad (a > 0)$

16. $\int_{-1}^1 2^x\,dx$

17. $\int_{-1}^{1} \frac{dx}{1+x^2}$

18. $\int_{0}^{1/2} \frac{dx}{\sqrt{1-x^2}}$

19. $\int_{-1}^{1} \frac{dx}{\sqrt{4-x^2}}$

20. $\int_{-2}^{0} \frac{dx}{4+x^2}$

Find the area of the region R specified in Exercises 21–32. It is helpful to make a sketch of the region.

21. Bounded by $y = x^4$, $y = 0$, $x = 0$, and $x = 1$

22. Bounded by $y = 1/x$, $y = 0$, $x = e$, and $x = e^2$

23. Above $y = x^2 - 4x$ and below the x-axis

24. Bounded by $y = 5 - 2x - 3x^2$, $y = 0$, $x = -1$, and $x = 1$

25. Bounded by $y = x^2 - 3x + 3$ and $y = 1$

26. Below $y = \sqrt{x}$ and above $y = \frac{x}{2}$

27. Above $y = x^2$ and to the right of $x = y^2$

28. Above $y = |x|$ and below $y = 12 - x^2$

29. Bounded by $y = x^{1/3} - x^{1/2}$, $y = 0$, $x = 0$, and $x = 1$

30. Under $y = e^{-x}$ and above $y = 0$ from $x = -a$ to $x = 0$

31. Below $y = 1 - \cos x$ and above $y = 0$ between two consecutive intersections of these graphs

32. Below $y = x^{-1/3}$ and above $y = 0$ from $x = 1$ to $x = 27$

Find the integrals of the piecewise continuous functions in Exercises 33–34.

33. $\int_{0}^{3\pi/2} |\cos x|\,dx$

34. $\int_{1}^{3} \frac{\operatorname{sgn}(x-2)}{x^2}\,dx$

In Exercises 35–38, find the average values of the given functions over the intervals specified.

35. $f(x) = 1 + x + x^2 + x^3$ over $[0, 2]$

36. $f(x) = e^{3x}$ over $[-2, 2]$

37. $f(x) = 2^x$ over $[0, 1/\ln 2]$

38. $g(t) = \begin{cases} 0 & \text{if } 0 \le t \le 1 \\ 1 & \text{if } 1 < t \le 3 \end{cases}$ over $[0, 3]$

Find the indicated derivatives in Exercises 39–46.

39. $\frac{d}{dx}\int_{2}^{x} \frac{\sin t}{t}\,dt$

40. $\frac{d}{dt}\int_{t}^{3} \frac{\sin x}{x}\,dx$

41. $\frac{d}{dx}\int_{x^2}^{0} \frac{\sin t}{t}\,dt$

42. $\frac{d}{dx}x^2\int_{0}^{x^2} \frac{\sin u}{u}\,du$

43. $\frac{d}{dt}\int_{-\pi}^{t} \frac{\cos y}{1+y^2}\,dy$

44. $\frac{d}{d\theta}\int_{\sin\theta}^{\cos\theta} \frac{1}{1-x^2}\,dx$

45. $\frac{d}{dx}F(\sqrt{x})$, if $F(t) = \int_{0}^{t} \cos(x^2)\,dx$

46. $H'(2)$, if $H(x) = 3x\int_{4}^{x^2} e^{-\sqrt{t}}\,dt$

47. Solve the integral equation $f(x) = \pi\left(1 + \int_{1}^{x} f(t)\,dt\right)$.

48. Solve the integral equation $f(x) = 1 - \int_{0}^{x} f(t)\,dt$.

49. Criticize the following erroneous calculation:

$$\int_{-1}^{1} \frac{dx}{x^2} = -\frac{1}{x}\bigg|_{-1}^{1} = -1 + \frac{1}{-1} = -2.$$

Exactly where did the error occur? Why is -2 an unreasonable value for the integral?

50. Use a definite integral to define a function $F(x)$ having derivative $\frac{\sin x}{1+x^2}$ for all x and satisfying $F(17) = 0$.

51. Does the function $F(x) = \int_{0}^{2x-x^2} \cos\left(\frac{1}{1+t^2}\right) dt$ have a maximum or a minimum value? Justify your answer.

Evaluate the limits in Exercises 52–54.

52. $\lim_{n\to\infty} \frac{1}{n}\left(\left(1+\frac{1}{n}\right)^5 + \left(1+\frac{2}{n}\right)^5 + \cdots + \left(1+\frac{n}{n}\right)^5\right)$.

53. $\lim_{n\to\infty} \frac{\pi}{n}\left(\sin\frac{\pi}{n} + \sin\frac{2\pi}{n} + \sin\frac{3\pi}{n} + \cdots + \sin\frac{n\pi}{n}\right)$.

54. $\lim_{n\to\infty} \left(\frac{n}{n^2+1} + \frac{n}{n^2+4} + \frac{n}{n^2+9} + \cdots + \frac{n}{2n^2}\right)$.

5.6 The Method of Substitution

As we have seen, the evaluation of definite integrals is most easily carried out if we can antidifferentiate the integrand. In this section and Sections 6.1–6.4 we develop some *techniques of integration*, that is, methods for finding antiderivatives of functions. Although the techniques we develop can be used for a large class of functions, they will not work for all functions we might want to integrate. If a definite integral involves an integrand whose antiderivative is either impossible or very difficult to find, we may wish, instead, to approximate the definite integral by numerical means. Techniques for doing that will be presented in Sections 6.6–6.8.

Let us begin by assembling a table of some known indefinite integrals. These results have all emerged during our development of differentiation formulas for elementary functions. You should *memorize* them.

Some elementary integrals

1. $\int 1\,dx = x + C$
2. $\int x\,dx = \frac{1}{2}x^2 + C$
3. $\int x^2\,dx = \frac{1}{3}x^3 + C$
4. $\int \frac{1}{x^2}\,dx = -\frac{1}{x} + C$
5. $\int \sqrt{x}\,dx = \frac{2}{3}x^{3/2} + C$
6. $\int \frac{1}{\sqrt{x}}\,dx = 2\sqrt{x} + C$
7. $\int x^r\,dx = \frac{1}{r+1}x^{r+1} + C \quad (r \neq -1)$
8. $\int \frac{1}{x}\,dx = \ln|x| + C$
9. $\int \sin ax\,dx = -\frac{1}{a}\cos ax + C$
10. $\int \cos ax\,dx = \frac{1}{a}\sin ax + C$
11. $\int \sec^2 ax\,dx = \frac{1}{a}\tan ax + C$
12. $\int \csc^2 ax\,dx = -\frac{1}{a}\cot ax + C$
13. $\int \sec ax\,\tan ax\,dx = \frac{1}{a}\sec ax + C$
14. $\int \csc ax\,\cot ax\,dx = -\frac{1}{a}\csc ax + C$
15. $\int \frac{1}{\sqrt{a^2 - x^2}}\,dx = \sin^{-1}\frac{x}{a} + C \quad (a > 0)$
16. $\int \frac{1}{a^2 + x^2}\,dx = \frac{1}{a}\tan^{-1}\frac{x}{a} + C$
17. $\int e^{ax}\,dx = \frac{1}{a}e^{ax} + C$
18. $\int b^{ax}\,dx = \frac{1}{a\ln b}b^{ax} + C$
19. $\int \cosh ax\,dx = \frac{1}{a}\sinh ax + C$
20. $\int \sinh ax\,dx = \frac{1}{a}\cosh ax + C$

Note that formulas 1–6 are special cases of formula 7, which holds on any interval where x^r makes sense. The linearity formula

$$\int (A\,f(x) + B\,g(x))\,dx = A\int f(x)\,dx + B\int g(x)\,dx$$

makes it possible to integrate sums and constant multiples of functions.

EXAMPLE 1 **(Combining elementary integrals)**

(a) $\displaystyle\int (x^4 - 3x^3 + 8x^2 - 6x - 7)\,dx = \frac{x^5}{5} - \frac{3x^4}{4} + \frac{8x^3}{3} - 3x^2 - 7x + C$

(b) $\displaystyle\int \left(5x^{3/5} - \frac{3}{2 + x^2}\right)dx = \frac{25}{8}x^{8/5} - \frac{3}{\sqrt{2}}\tan^{-1}\frac{x}{\sqrt{2}} + C$

(c) $\displaystyle\int (4\cos 5x - 5\sin 3x)\,dx = \frac{4}{5}\sin 5x + \frac{5}{3}\cos 3x + C$

(d) $\displaystyle\int \left(\frac{1}{\pi x} + a^{\pi x}\right)dx = \frac{1}{\pi}\ln|x| + \frac{1}{\pi\ln a}a^{\pi x} + C, \quad (a > 0).$

Sometimes it is necessary to manipulate an integrand so that the method can be applied.

EXAMPLE 2

$$\begin{aligned}\int \frac{(x+1)^3}{x}\,dx &= \int \frac{x^3 + 3x^2 + 3x + 1}{x}\,dx \\ &= \int \left(x^2 + 3x + 3 + \frac{1}{x}\right)dx \\ &= \frac{1}{3}x^3 + \frac{3}{2}x^2 + 3x + \ln|x| + C.\end{aligned}$$

When an integral cannot be evaluated by inspection, as those in Examples 1–2 can, we require one or more special techniques. The most important of these techniques is the **method of substitution**, the integral version of the Chain Rule. If we rewrite the Chain Rule, $\frac{d}{dx} f\big(g(x)\big) = f'\big(g(x)\big)\, g'(x)$, in integral form, we obtain

$$\int f'\big(g(x)\big)\, g'(x)\, dx = f\big(g(x)\big) + C.$$

Observe that the following formalism would produce this latter formula even if we did not already know it was true:

Let $u = g(x)$. Then $du/dx = g'(x)$, or in differential form, $du = g'(x)\, dx$. Thus,

$$\int f'\left(g(x)\right)\, g'(x)\, dx = \int f'(u)\, du = f(u) + C = f\left(g(x)\right) + C.$$

EXAMPLE 3 **(Examples of substitution)** Find the indefinite integrals:

(a) $\displaystyle\int \frac{x}{x^2+1}\, dx,$ (b) $\displaystyle\int \frac{\sin(3\ln x)}{x}\, dx,$ and (c) $\displaystyle\int e^x\sqrt{1+e^x}\, dx.$

Solution

(a) $\displaystyle\int \frac{x}{x^2+1}\, dx$ Let $u = x^2 + 1$.
Then $du = 2x\, dx$ and
$x\, dx = \frac{1}{2}\, du$

$$= \frac{1}{2}\int \frac{du}{u} = \frac{1}{2}\ln|u| + C = \frac{1}{2}\ln(x^2+1) + C = \ln\sqrt{x^2+1} + C.$$

(Both versions of the final answer are equally acceptable.)

(b) $\displaystyle\int \frac{\sin(3\ln x)}{x}\, dx$ Let $u = 3\ln x$.
Then $du = \dfrac{3}{x}\, dx$

$$= \frac{1}{3}\int \sin u\, du = -\frac{1}{3}\cos u + C = -\frac{1}{3}\cos(3\ln x) + C.$$

(c) $\displaystyle\int e^x\sqrt{1+e^x}\, dx$ Let $v = 1 + e^x$.
Then $dv = e^x\, dx$

$$= \int v^{1/2}\, dv = \frac{2}{3}v^{3/2} + C = \frac{2}{3}(1+e^x)^{3/2} + C.$$

Sometimes the appropriate substitutions are not as obvious as they were in Example 3, and it may be necessary to manipulate the integrand algebraically to put it into a better form for substitution.

EXAMPLE 4 Evaluate (a) $\displaystyle\int \frac{1}{x^2+4x+5}\, dx$ and (b) $\displaystyle\int \frac{dx}{\sqrt{e^{2x}-1}}$.

Solution

(a) $\displaystyle\int \frac{dx}{x^2+4x+5} = \int \frac{dx}{(x+2)^2+1}$ Let $t = x + 2$.
Then $dt = dx$.

$$= \int \frac{dt}{t^2+1}$$

$$= \tan^{-1} t + C = \tan^{-1}(x+2) + C.$$

(b) $$\int \frac{dx}{\sqrt{e^{2x}-1}} = \int \frac{dx}{e^x\sqrt{1-e^{-2x}}}$$

$$= \int \frac{e^{-x}\,dx}{\sqrt{1-(e^{-x})^2}} \qquad \text{Let } u = e^{-x}. \text{ Then } du = -e^{-x}\,dx.$$

$$= -\int \frac{du}{\sqrt{1-u^2}}$$

$$= -\sin^{-1} u + C = -\sin^{-1}\left(e^{-x}\right) + C.$$

The method of substitution cannot be *forced* to work. There is no substitution that will do much good with the integral $\int x(2+x^7)^{1/5}\,dx$, for instance. However, the integral $\int x^6(2+x^7)^{1/5}\,dx$ will yield to the substitution $u = 2+x^7$. The substitution $u = g(x)$ is more likely to work if $g'(x)$ is a factor of the integrand.

The following theorem simplifies the use of the method of substitution in definite integrals.

THEOREM 6 **Substitution in a definite integral**

Suppose that g is a differentiable function on $[a,b]$ that satisfies $g(a) = A$ and $g(b) = B$. Also suppose that f is continuous on the range of g. Then

$$\int_a^b f\big(g(x)\big)\,g'(x)\,dx = \int_A^B f(u)\,du.$$

PROOF Let F be an antiderivative of f; $F'(u) = f(u)$. Then

$$\frac{d}{dx} F\big(g(x)\big) = F'\big(g(x)\big)\,g'(x) = f\big(g(x)\big)\,g'(x).$$

Thus,

$$\int_a^b f\big(g(x)\big)\,g'(x)\,dx = F\big(g(x)\big)\Big|_a^b = F\big(g(b)\big) - F\big(g(a)\big)$$

$$= F(B) - F(A) = F(u)\Big|_A^B = \int_A^B f(u)\,du.$$

EXAMPLE 5 Evaluate the integral $I = \displaystyle\int_0^8 \frac{\cos\sqrt{x+1}}{\sqrt{x+1}}\,dx.$

Solution **METHOD I.** Let $u = \sqrt{x+1}$. Then $du = \dfrac{dx}{2\sqrt{x+1}}$. If $x = 0$, then $u = 1$; if $x = 8$, then $u = 3$. Thus

$$I = 2\int_1^3 \cos u\,du = 2\sin u\Big|_1^3 = 2\sin 3 - 2\sin 1.$$

METHOD II. We use the same substitution as in Method I, but we do not transform the limits of integration from x values to u values. Hence, we must return to the variable x before substituting in the limits:

$$I = 2\int_{x=0}^{x=8} \cos u\,du = 2\sin u\Big|_{x=0}^{x=8} = 2\sin\sqrt{x+1}\Big|_0^8 = 2\sin 3 - 2\sin 1.$$

Note that the limits *must* be written $x = 0$ and $x = 8$ at any stage where the variable is not x. It would have been *wrong* to write

$$I = 2\int_0^8 \cos u\,du$$

because this would imply that u, rather than x, goes from 0 to 8. Method I gives the shorter solution and is therefore preferable. However, in cases where the transformed limits (the u-limits) are very complicated, you might prefer to use Method II.

EXAMPLE 6 Find the area of the region bounded by $y = \left(2 + \sin \frac{x}{2}\right)^2 \cos \frac{x}{2}$, the x-axis, and the lines $x = 0$ and $x = \pi$.

Solution Because $y \geq 0$ when $0 \leq x \leq \pi$, the required area is

$$A = \int_0^\pi \left(2 + \sin \frac{x}{2}\right)^2 \cos \frac{x}{2}\, dx \qquad \text{Let } v = 2 + \sin \frac{x}{2}.$$

$$\text{Then } dv = \frac{1}{2} \cos \frac{x}{2}\, dx$$

$$= 2 \int_2^3 v^2\, dv = \frac{2}{3} v^3 \bigg|_2^3 = \frac{2}{3}(27 - 8) = \frac{38}{3} \text{ square units.}$$

Remark The condition that f be continuous on the range of the function $u = g(x)$ (for $a \leq x \leq b$) is essential in Theorem 6. Using the substitution $u = x^2$ in the integral $\int_{-1}^1 x \csc(x^2)\, dx$ leads to the erroneous conclusion

$$\int_{-1}^1 x \csc(x^2)\, dx = \frac{1}{2} \int_1^1 \csc u\, du = 0.$$

Although $x \csc(x^2)$ is an odd function, it is not continuous at 0, and it happens that the given integral represents the difference of *infinite* areas. If we assume that f is continuous on an interval containing A and B, then it suffices to know that $u = g(x)$ is one-to-one as well as differentiable. In this case the range of g will lie between A and B, so the condition of Theorem 6 will be satisfied.

Trigonometric Integrals

The method of substitution is often useful for evaluating trigonometric integrals. We begin by listing the integrals of the four trigonometric functions whose integrals we have not yet seen. They arise often in applications and should be memorized.

Integrals of tangent, cotangent, secant, and cosecant

$$\int \tan x\, dx = \ln |\sec x| + C,$$

$$\int \cot x\, dx = \ln |\sin x| + C = -\ln |\csc x| + C,$$

$$\int \sec x\, dx = \ln |\sec x + \tan x| + C,$$

$$\int \csc x\, dx = -\ln |\csc x + \cot x| + C = \ln |\csc x - \cot x| + C.$$

All of these can, of course, be checked by differentiating the right-hand sides. The first two can be evaluated directly by rewriting $\tan x$ or $\cot x$ in terms of $\sin x$ and $\cos x$ and using an appropriate substitution. For example,

$$\int \tan x\, dx = \int \frac{\sin x}{\cos x}\, dx \qquad \text{Let } u = \cos x.$$

$$\text{Then } du = -\sin x\, dx.$$

$$= -\int \frac{du}{u} = -\ln |u| + C$$

$$= -\ln |\cos x| + C = \ln \left| \frac{1}{\cos x} \right| + C = \ln |\sec x| + C.$$

The integral of $\sec x$ can be evaluated by rewriting it in the form

$$\int \sec x\, dx = \int \frac{\sec x(\sec x + \tan x)}{\sec x + \tan x}\, dx$$

and using the substitution $u = \sec x + \tan x$. The integral of $\csc x$ can be evaluated similarly. (Show that the two versions given for that integral are equivalent!)

We now consider integrals of the form

$$\int \sin^m x \cos^n x\, dx.$$

If either m or n is an odd, positive integer, the integral can be done easily by substitution. If, say, $n = 2k+1$ where k is an integer, then we can use the identity $\sin^2 x + \cos^2 x = 1$ to rewrite the integral in the form

$$\int \sin^m x\,(1 - \sin^2 x)^k \cos x\, dx,$$

which can be integrated using the substitution $u = \sin x$. Similarly, $u = \cos x$ can be used if m is an odd integer.

EXAMPLE 7 Evaluate: (a) $\displaystyle\int \sin^3 x \cos^8 x\, dx$ and (b) $\displaystyle\int \cos^5 ax\, dx$.

Solution

(a)

$$\begin{aligned}\int \sin^3 x \cos^8 x\, dx &= \int (1 - \cos^2 x)\cos^8 x \sin x\, dx \qquad \text{Let } u = \cos x,\ du = -\sin x\, dx.\\ &= -\int (1 - u^2)\, u^8\, du = \int (u^{10} - u^8)\, du\\ &= \frac{u^{11}}{11} - \frac{u^9}{9} + C = \frac{1}{11}\cos^{11} x - \frac{1}{9}\cos^9 x + C.\end{aligned}$$

(b)

$$\begin{aligned}\int \cos^5 ax\, dx &= \int (1 - \sin^2 ax)^2 \cos ax\, dx \qquad \text{Let } u = \sin ax,\ du = a\cos ax\, dx.\\ &= \frac{1}{a}\int (1 - u^2)^2\, du = \frac{1}{a}\int (1 - 2u^2 + u^4)\, du\\ &= \frac{1}{a}\left(u - \frac{2}{3}u^3 + \frac{1}{5}u^5\right) + C\\ &= \frac{1}{a}\left(\sin ax - \frac{2}{3}\sin^3 ax + \frac{1}{5}\sin^5 ax\right) + C.\end{aligned}$$

If the powers of $\sin x$ and $\cos x$ are both even, then we can make use of the *double-angle formulas* (see Section P.7):

$$\cos^2 x = \frac{1}{2}(1 + \cos 2x) \qquad \text{and} \qquad \sin^2 x = \frac{1}{2}(1 - \cos 2x).$$

EXAMPLE 8 **(Integrating even powers of sine and cosine)** Verify the integration formulas

$$\int \cos^2 x\, dx = \frac{1}{2}(x + \sin x \cos x) + C,$$

$$\int \sin^2 x\, dx = \frac{1}{2}(x - \sin x \cos x) + C.$$

These integrals are encountered frequently and are worth remembering.

Solution Each of the integrals follows from the corresponding double-angle identity. We do the first; the second is similar.

$$\begin{aligned}\int \cos^2 x\,dx &= \frac{1}{2}\int (1+\cos 2x)\,dx\\ &= \frac{x}{2}+\frac{1}{4}\sin 2x + C\\ &= \frac{1}{2}(x+\sin x\cos x) + C \quad (\text{since } \sin 2x = 2\sin x\cos x).\end{aligned}$$

EXAMPLE 9 Evaluate $\int \sin^4 x\,dx$.

Solution We will have to apply the double-angle formula twice.

$$\begin{aligned}\int \sin^4 x\,dx &= \frac{1}{4}\int (1-\cos 2x)^2\,dx\\ &= \frac{1}{4}\int (1-2\cos 2x+\cos^2 2x)\,dx\\ &= \frac{x}{4}-\frac{1}{4}\sin 2x+\frac{1}{8}\int (1+\cos 4x)\,dx\\ &= \frac{x}{4}-\frac{1}{4}\sin 2x+\frac{x}{8}+\frac{1}{32}\sin 4x + C\\ &= \frac{3}{8}x-\frac{1}{4}\sin 2x+\frac{1}{32}\sin 4x + C\end{aligned}$$

(Note that there is no point in inserting the constant of integration C until the last integral has been evaluated.)

Using the identities $\sec^2 x = 1+\tan^2 x$ and $\csc^2 x = 1+\cot^2 x$ and one of the substitutions $u=\sec x$, $u=\tan x$, $u=\csc x$, or $u=\cot x$, we can evaluate integrals of the form

$$\int \sec^m x\,\tan^n x\,dx \qquad \text{or} \qquad \int \csc^m x\,\cot^n x\,dx,$$

unless m is odd and n is even. (If this is the case, these integrals can be handled by integration by parts; see Section 6.1.)

EXAMPLE 10 **(Integrals involving secants and tangents)** Evaluate the following integrals:

(a) $\int \tan^2 x\,dx$, (b) $\int \sec^4 t\,dt$, and (c) $\int \sec^3 x\,\tan^3 x\,dx$.

Solution

(a) $\displaystyle\int \tan^2 x\,dx = \int (\sec^2 x - 1)\,dx = \tan x - x + C.$

(b) $$\begin{aligned}\int \sec^4 t\,dt &= \int (1+\tan^2 t)\sec^2 t\,dt \qquad \text{Let } u=\tan t,\ du=\sec^2 t\,dt.\\ &= \int (1+u^2)\,du = u+\frac{1}{3}u^3+C = \tan t+\frac{1}{3}\tan^3 t + C.\end{aligned}$$

(c) $$\begin{aligned}&\int \sec^3 x\,\tan^3 x\,dx\\ &= \int \sec^2 x\,(\sec^2 x-1)\sec x\tan x\,dx \qquad \text{Let } u=\sec x,\ du=\sec x\tan x\,dx.\\ &= \int (u^4-u^2)\,du = \frac{u^5}{5}-\frac{u^3}{3}+C = \frac{1}{5}\sec^5 x-\frac{1}{3}\sec^3 x + C.\end{aligned}$$

EXERCISES 5.6

Evaluate the integrals in Exercises 1–44. Remember to include a constant of integration with the indefinite integrals. Your answers may appear different from those in the Answers section but may still be correct. For example, evaluating $I = \int \sin x \cos x\, dx$ using the substitution $u = \sin x$ leads to $I = \frac{1}{2}\sin^2 x + C$; using $u = \cos x$ leads to $I = -\frac{1}{2}\cos^2 x + C$; and rewriting $I = \frac{1}{2}\int \sin(2x)\, dx$ leads to $I = -\frac{1}{4}\cos(2x) + C$. These answers are all equal except for different choices for the constant of integration C: $\frac{1}{2}\sin^2 x = -\frac{1}{2}\cos^2 + \frac{1}{2} = -\frac{1}{4}\cos(2x) + \frac{1}{4}$.

You can always check your own answer to an indefinite integral by differentiating it to get back to the integrand. This is often easier than comparing your answer with the answer in the back of the book. You may find integrals that you can't do, but you should not make mistakes in those you can do because the answer is so easily checked. (This is a good thing to remember during tests and exams.)

1. $\int e^{5-2x}\, dx$ **2.** $\int \cos(ax+b)\, dx$

3. $\int \sqrt{3x+4}\, dx$ **4.** $\int e^{2x}\sin(e^{2x})\, dx$

5. $\int \frac{x\, dx}{(4x^2+1)^5}$ **6.** $\int \frac{\sin\sqrt{x}}{\sqrt{x}}\, dx$

7. $\int x e^{x^2}\, dx$ **8.** $\int x^2 2^{x^3+1}\, dx$

9. $\int \frac{\cos x}{4+\sin^2 x}\, dx$ **10.** $\int \frac{\sec^2 x}{\sqrt{1-\tan^2 x}}\, dx$

11. $\int \frac{e^x+1}{e^x-1}\, dx$ **12.** $\int \frac{\ln t}{t}\, dt$

13. $\int \frac{ds}{\sqrt{4-5s}}$ **14.** $\int \frac{x+1}{\sqrt{x^2+2x+3}}\, dx$

15. $\int \frac{t\, dt}{\sqrt{4-t^4}}$ **16.** $\int \frac{x^2\, dx}{2+x^6}$

17. $\int \frac{dx}{e^x+1}$ **18.** $\int \frac{dx}{e^x+e^{-x}}$

19. $\int \tan x \ln \cos x\, dx$ **20.** $\int \frac{x+1}{\sqrt{1-x^2}}\, dx$

21. $\int \frac{dx}{x^2+6x+13}$ **22.** $\int \frac{dx}{\sqrt{4+2x-x^2}}$

23. $\int \sin^3 x \cos^5 x\, dx$ **24.** $\int \sin^4 t \cos^5 t\, dt$

25. $\int \sin ax \cos^2 ax\, dx$ **26.** $\int \sin^2 x \cos^2 x\, dx$

27. $\int \sin^6 x\, dx$ **28.** $\int \cos^4 x\, dx$

29. $\int \sec^5 x \tan x\, dx$ **30.** $\int \sec^6 x \tan^2 x\, dx$

31. $\int \sqrt{\tan x}\sec^4 x\, dx$ **32.** $\int \sin^{-2/3} x \cos^3 x\, dx$

33. $\int \cos x \sin^4(\sin x)\, dx$ **34.** $\int \frac{\sin^3 \ln x \cos^3 \ln x}{x}\, dx$

35. $\int \frac{\sin^2 x}{\cos^4 x}\, dx$ **36.** $\int \frac{\sin^3 x}{\cos^4 x}\, dx$

37. $\int \csc^5 x \cot^5 x\, dx$ **38.** $\int \frac{\cos^4 x}{\sin^8 x}\, dx$

39. $\int_0^4 x^3(x^2+1)^{-\frac{1}{2}}\, dx$ **40.** $\int_1^{\sqrt{e}} \frac{\sin(\pi \ln x)}{x}\, dx$

41. $\int_0^{\pi/2} \sin^4 x\, dx$ **42.** $\int_{\pi/4}^{\pi} \sin^5 x\, dx$

43. $\int_e^{e^2} \frac{dt}{t\ln t}$ **44.** $\int_{\frac{\pi^2}{16}}^{\frac{\pi^2}{9}} \frac{2^{\sin\sqrt{x}}\cos\sqrt{x}}{\sqrt{x}}\, dx$

45. Use the identities $\cos 2\theta = 2\cos^2\theta - 1 = 1 - 2\sin^2\theta$ and $\sin\theta = \cos\left(\frac{\pi}{2} - \theta\right)$ to help you evaluate the following:

$$\int_0^{\pi/2} \sqrt{1+\cos x}\, dx \quad \text{and} \quad \int_0^{\pi/2} \sqrt{1-\sin x}\, dx$$

46. Find the area of the region bounded by $y = x/(x^2+16)$, $y = 0$, $x = 0$, and $x = 2$.

47. Find the area of the region bounded by $y = x/(x^4+16)$, $y = 0$, $x = 0$, and $x = 2$.

48. Express the area bounded by the ellipse $(x^2/a^2) + (y^2/b^2) = 1$ as a definite integral. Make a substitution that converts this integral into one representing the area of a circle, and hence evaluate it.

49. Use the addition formulas for $\sin(x \pm y)$ and $\cos(x \pm y)$ from Section P.7 to establish the following identities:

$$\cos x \cos y = \frac{1}{2}\Big(\cos(x-y) + \cos(x+y)\Big),$$
$$\sin x \sin y = \frac{1}{2}\Big(\cos(x-y) - \cos(x+y)\Big),$$
$$\sin x \cos y = \frac{1}{2}\Big(\sin(x+y) + \sin(x-y)\Big).$$

50. Use the identities established in Exercise 49 to calculate the following integrals:

$\int \cos ax \cos bx\, dx$, $\int \sin ax \sin bx\, dx$,

and $\int \sin ax \cos bx\, dx$.

51. If m and n are integers, show that:

(i) $\int_{-\pi}^{\pi} \cos mx \cos nx\, dx = 0$ if $m \neq n$,

(ii) $\int_{-\pi}^{\pi} \sin mx \sin nx\, dx = 0$ if $m \neq n$,

(iii) $\int_{-\pi}^{\pi} \sin mx \cos nx\, dx = 0$.

52. (Fourier coefficients) Suppose that for some positive integer k,

$$f(x) = \frac{a_0}{2} + \sum_{n=1}^{k}(a_n \cos nx + b_n \sin nx)$$

holds for all x in $[-\pi, \pi]$. Use the result of Exercise 51 to show that the coefficients a_m $(0 \le m \le k)$ and b_m $(1 \le m \le k)$, which are called the Fourier coefficients of f on $[-\pi, \pi]$, are given by

$$a_m = \frac{1}{\pi}\int_{-\pi}^{\pi} f(x)\cos mx\,dx, \quad b_m = \frac{1}{\pi}\int_{-\pi}^{\pi} f(x)\sin mx\,dx.$$

5.7 Areas of Plane Regions

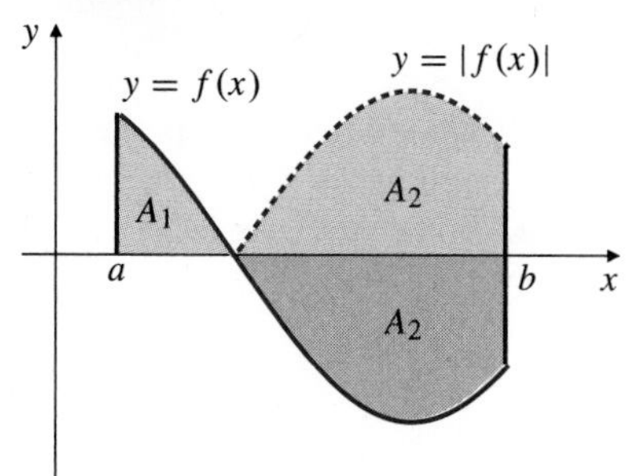
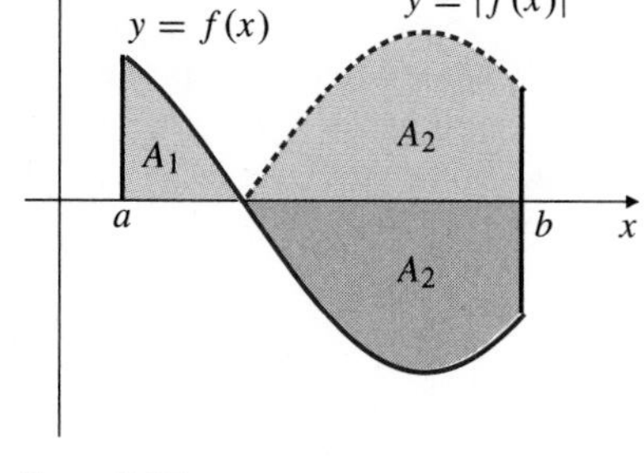

Figure 5.27

In this section we review and extend the use of definite integrals to represent plane areas. Recall that the integral $\int_a^b f(x)\,dx$ measures the area between the graph of f and the x-axis from $x = a$ to $x = b$, but treats as *negative* any part of this area that lies below the x-axis. (We are assuming that $a < b$.) In order to express the total area bounded by $y = f(x)$, $y = 0$, $x = a$, and $x = b$, counting all of the area positively, we should integrate the *absolute value* of f (see Figure 5.27):

$$\int_a^b f(x)\,dx = A_1 - A_2 \qquad \text{and} \qquad \int_a^b |f(x)|\,dx = A_1 + A_2.$$

There is no "rule" for integrating $\int_a^b |f(x)|\,dx$; one must break the integral into a sum of integrals over intervals where $f(x) > 0$ (so $|f(x)| = f(x)$), and intervals where $f(x) < 0$ (so $|f(x)| = -f(x)$).

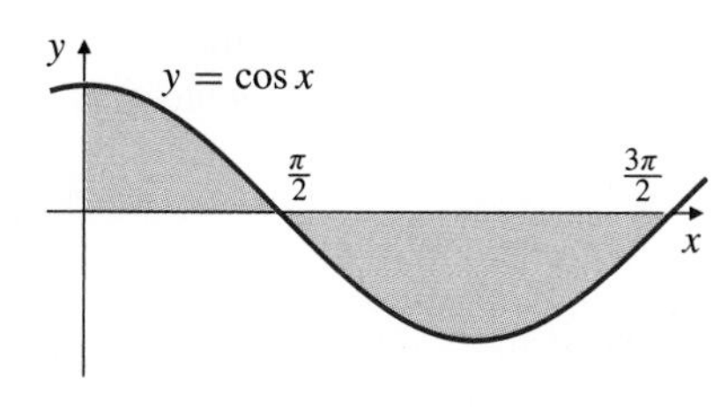

Figure 5.28

EXAMPLE 1 The area bounded by $y = \cos x$, $y = 0$, $x = 0$, and $x = 3\pi/2$ (see Figure 5.28) is

$$\begin{aligned}
A &= \int_0^{3\pi/2} |\cos x|\,dx \\
&= \int_0^{\pi/2} \cos x\,dx + \int_{\pi/2}^{3\pi/2} (-\cos x)\,dx \\
&= \sin x\Big|_0^{\pi/2} - \sin x\Big|_{\pi/2}^{3\pi/2} \\
&= (1 - 0) - (-1 - 1) = 3 \text{ square units.}
\end{aligned}$$

Areas Between Two Curves

Suppose that a plane region R is bounded by the graphs of two continuous functions, $y = f(x)$ and $y = g(x)$, and the vertical straight lines $x = a$ and $x = b$, as shown in Figure 5.29(a). Assume that $a < b$ and that $f(x) \le g(x)$ on $[a, b]$, so the graph of f lies below that of g. If $f(x) \ge 0$ on $[a, b]$, then the area A of R is the area above the x-axis and under the graph of g minus the area above the x-axis and under the graph of f:

$$A = \int_a^b g(x)\,dx - \int_a^b f(x)\,dx = \int_a^b \big(g(x) - f(x)\big)\,dx.$$

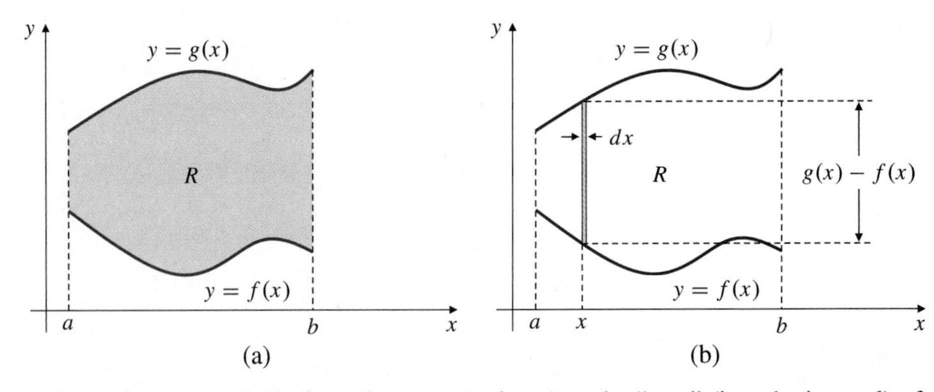

Figure 5.29
(a) The region R lying between two graphs
(b) An area element of the region R

It is useful to regard this formula as expressing A as the "sum" (i.e., the integral) of *infinitely many* **area elements**

$$dA = (g(x) - f(x))\, dx,$$

corresponding to values of x between a and b. Each such area element is the area of an infinitely thin vertical rectangle of width dx and height $g(x) - f(x)$ located at position x (see Figure 5.29(b)). Even if f and g can take on negative values on $[a, b]$, this interpretation and the resulting area formula

$$A = \int_a^b \big(g(x) - f(x)\big)\, dx$$

remain valid, provided that $f(x) \le g(x)$ on $[a, b]$ so that all the area elements dA have positive area. Using integrals to represent a quantity as a *sum* of *differential elements* (i.e., a sum of little bits of the quantity) is a very helpful approach. We will do this often in Chapter 7. Of course, what we are really doing is identifying the integral as a *limit* of a suitable Riemann sum.

More generally, if the restriction $f(x) \le g(x)$ is removed, then the vertical rectangle of width dx at position x extending between the graphs of f and g has height $|f(x) - g(x)|$ and hence area

$$dA = |f(x) - g(x)|\, dx.$$

(See Figure 5.30.) Hence the total area lying between the graphs $y = f(x)$ and $y = g(x)$ and between the vertical lines $x = a$ and $x = b > a$ is given by

$$A = \int_a^b \big|f(x) - g(x)\big|\, dx.$$

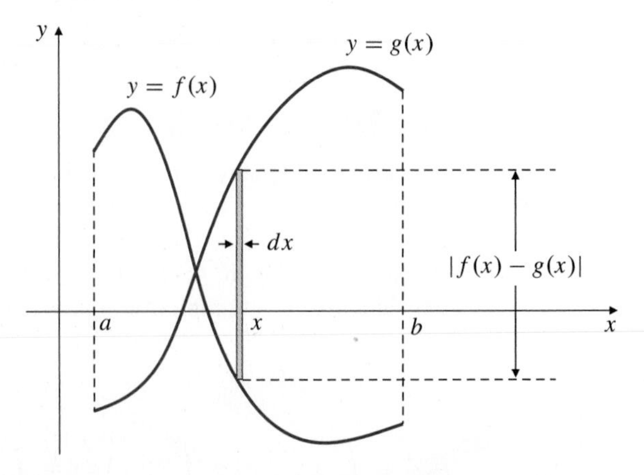

Figure 5.30 An area element for the region between $y = f(x)$ and $y = g(x)$

In order to evaluate this integral, we have to determine the intervals on which $f(x) > g(x)$ or $f(x) < g(x)$, and break the integral into a sum of integrals over each of these intervals.

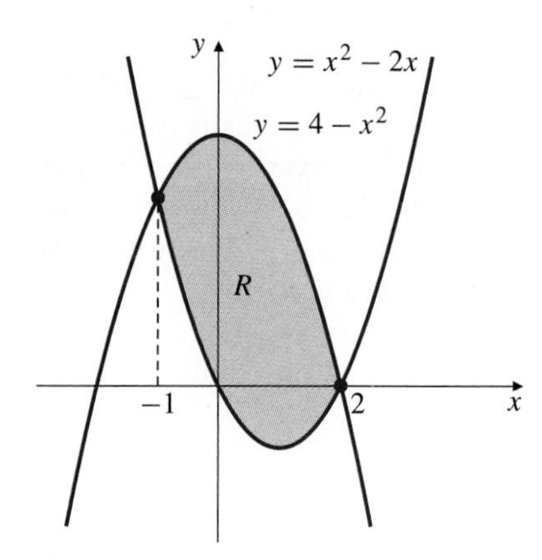

Figure 5.31

EXAMPLE 2 Find the area of the bounded, plane region R lying between the curves $y = x^2 - 2x$ and $y = 4 - x^2$.

Solution First, we must find the intersections of the curves, so we solve the equations simultaneously:

$$x^2 - 2x = y = 4 - x^2$$
$$2x^2 - 2x - 4 = 0$$
$$2(x-2)(x+1) = 0 \quad \text{so } x = 2 \text{ or } x = -1.$$

The curves are sketched in Figure 5.31, and the bounded (finite) region between them is shaded. (A sketch should always be made in problems of this sort.) Since $4 - x^2 \ge x^2 - 2x$ for $-1 \le x \le 2$, the area A of R is given by

$$\begin{aligned} A &= \int_{-1}^{2} \big((4 - x^2) - (x^2 - 2x)\big)\,dx \\ &= \int_{-1}^{2} (4 - 2x^2 + 2x)\,dx \\ &= \left(4x - \frac{2}{3}x^3 + x^2\right)\Bigg|_{-1}^{2} \\ &= 4(2) - \frac{2}{3}(8) + 4 - \left(-4 + \frac{2}{3} + 1\right) = 9 \text{ square units.} \end{aligned}$$

Note that in representing the area as an integral we *must subtract the height y to the lower curve from the height y to the upper curve* to get a positive area element dA. Subtracting the wrong way would have produced a negative value for the area.

EXAMPLE 3 Find the total area A lying between the curves $y = \sin x$ and $y = \cos x$ from $x = 0$ to $x = 2\pi$.

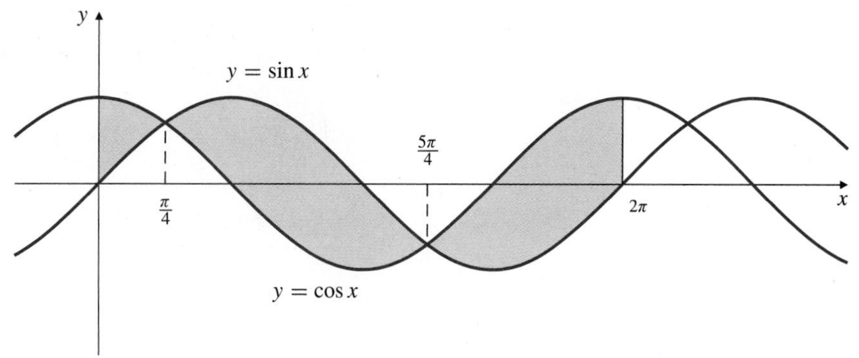

Figure 5.32

Solution The region is shaded in Figure 5.32. Between 0 and 2π the graphs of sine and cosine cross at $x = \pi/4$ and $x = 5\pi/4$. The required area is

$$\begin{aligned} A &= \int_{0}^{\pi/4} (\cos x - \sin x)\,dx + \int_{\pi/4}^{5\pi/4} (\sin x - \cos x)\,dx \\ &\quad + \int_{5\pi/4}^{2\pi} (\cos x - \sin x)\,dx \\ &= (\sin x + \cos x)\Bigg|_{0}^{\pi/4} - (\cos x + \sin x)\Bigg|_{\pi/4}^{5\pi/4} + (\sin x + \cos x)\Bigg|_{5\pi/4}^{2\pi} \\ &= (\sqrt{2} - 1) + (\sqrt{2} + \sqrt{2}) + (1 + \sqrt{2}) = 4\sqrt{2} \text{ square units.} \end{aligned}$$

It is sometimes more convenient to use horizontal area elements instead of vertical ones and integrate over an interval of the y-axis instead of the x-axis. This is usually the case if the region whose area we want to find is bounded by curves whose equations are written in terms of functions of y. In Figure 5.33(a), the region R lying to the right of $x = f(y)$ and to the left of $x = g(y)$, and between the horizontal lines $y = c$ and $y = d > c$, has area element $dA = \big(g(y) - f(y)\big)\,dy$. Its area is

$$A = \int_c^d \big(g(y) - f(y)\big)\,dy.$$

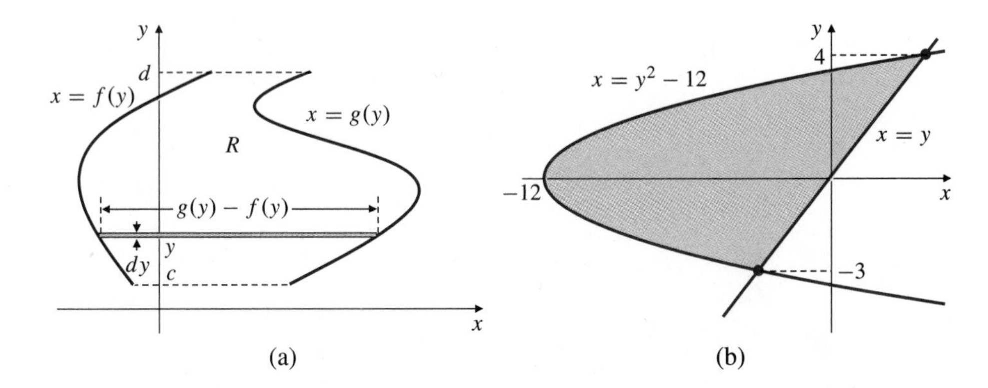

Figure 5.33
(a) A horizontal area element
(b) The finite region bounded by $x = y^2 - 12$ and $x = y$

EXAMPLE 4 Find the area of the plane region lying to the right of the parabola $x = y^2 - 12$ and to the left of the straight line $y = x$, as illustrated in Figure 5.33(b).

Solution For the intersections of the curves:

$$\begin{aligned}
&y^2 - 12 = x = y \\
&y^2 - y - 12 = 0 \\
&(y - 4)(y + 3) = 0 \quad \text{so } y = 4 \text{ or } y = -3.
\end{aligned}$$

Observe that $y^2 - 12 \le y$ for $-3 \le y \le 4$. Thus, the area is

$$A = \int_{-3}^{4} \big(y - (y^2 - 12)\big)\,dy = \left(\frac{y^2}{2} - \frac{y^3}{3} + 12y\right)\Bigg|_{-3}^{4} = \frac{343}{6} \text{ square units.}$$

Of course, the same result could have been obtained by integrating in the x direction, but the integral would have been more complicated:

$$A = \int_{-12}^{-3} \big(\sqrt{12 + x} - (-\sqrt{12 + x})\big)\,dx + \int_{-3}^{4} \big(\sqrt{12 + x} - x\big)\,dx;$$

different integrals are required over the intervals where the region is bounded below by the parabola and by the straight line.

EXERCISES 5.7

In Exercises 1–16, sketch and find the area of the plane region bounded by the given curves.

1. $y = x, \quad y = x^2$

2. $y = \sqrt{x}, \quad y = x^2$

3. $y = x^2 - 5, \quad y = 3 - x^2$

4. $y = x^2 - 2x, \quad y = 6x - x^2$

5. $2y = 4x - x^2, \quad 2y + 3x = 6$

6. $x - y = 7, \quad x = 2y^2 - y + 3$

7. $y = x^3, \quad y = x$

8. $y = x^3, \quad y = x^2$

9. $y = x^3, \quad x = y^2$

10. $x = y^2, \quad x = 2y^2 - y - 2$

11. $y = \dfrac{1}{x}, \quad 2x + 2y = 5$

12. $y = (x^2 - 1)^2, \quad y = 1 - x^2$

13. $y = \dfrac{1}{2}x^2, \quad y = \dfrac{1}{x^2 + 1}$

14. $y = \dfrac{4x}{3 + x^2}, \quad y = 1$

15. $y = \dfrac{4}{x^2}, \quad y = 5 - x^2$

16. $x = y^2 - \pi^2, \quad x = \sin y$

Find the areas of the regions described in Exercises 17–28. It is helpful to sketch the regions before writing an integral to represent the area.

17. Bounded by $y = \sin x$ and $y = \cos x$, and between two consecutive intersections of these curves

18. Bounded by $y = \sin^2 x$ and $y = 1$, and between two consecutive intersections of these curves

19. Bounded by $y = \sin x$ and $y = \sin^2 x$, between $x = 0$ and $x = \pi/2$

20. Bounded by $y = \sin^2 x$ and $y = \cos^2 x$, and between two consecutive intersections of these curves

21. Under $y = 4x/\pi$ and above $y = \tan x$, between $x = 0$ and the first intersection of the curves to the right of $x = 0$

22. Bounded by $y = x^{1/3}$ and the component of $y = \tan(\pi x/4)$ that passes through the origin

23. Bounded by $y = 2$ and the component of $y = \sec x$ that passes through the point $(0, 1)$

24. Bounded by $y = \sqrt{2}\cos(\pi x/4)$ and $y = |x|$

25. Bounded by $y = \sin(\pi x/2)$ and $y = x$

26. Bounded by $y = e^x$ and $y = x + 2$

27. Find the total area enclosed by the curve $y^2 = x^2 - x^4$.

28. Find the area of the closed loop of the curve $y^2 = x^4(2 + x)$ that lies to the left of the origin.

29. Find the area of the finite plane region that is bounded by the curve $y = e^x$, the line $x = 0$, and the tangent line to $y = e^x$ at $x = 1$.

30. Find the area of the finite plane region bounded by the curve $y = x^3$ and the tangent line to that curve at the point $(1, 1)$. *Hint:* Find the other point at which that tangent line meets the curve.

CHAPTER REVIEW

Key Ideas

- **What do the following terms and phrases mean?**
  - ◇ sigma notation
  - ◇ a partition of an interval
  - ◇ a Riemann sum
  - ◇ a definite integral
  - ◇ an indefinite integral
  - ◇ an integrable function
  - ◇ an area element
  - ◇ an evaluation symbol
  - ◇ the triangle inequality for integrals
  - ◇ a piecewise continuous function
  - ◇ the average value of function f on $[a, b]$
  - ◇ the method of substitution
- **State the Mean-Value Theorem for integrals.**
- **State the Fundamental Theorem of Calculus.**
- **List as many properties of the definite integral as you can.**
- **What is the relationship between the definite integral and the indefinite integral of a function f on an interval $[a, b]$?**
- **What is the derivative of $\int_{f(x)}^{g(x)} h(t)\,dt$ with respect to x?**
- **How can the area between the graphs of two functions be calculated?**

Review Exercises

1. Show that $\dfrac{2j+1}{j^2(j+1)^2} = \dfrac{1}{j^2} - \dfrac{1}{(j+1)^2}$; hence evaluate $\displaystyle\sum_{j=1}^{n} \frac{2j+1}{j^2(j+1)^2}$.

2. **(Stacking balls)** A display of golf balls in a sporting goods store is built in the shape of a pyramid with a rectangular base measuring 40 balls long and 30 balls wide. The next layer up is 39 balls by 29 balls, etc. How many balls are in the pyramid?

3. Let $P_n = \{x_0 = 1, x_1, x_2, \ldots, x_n = 3\}$ be a partition of $[1, 3]$ into n subintervals of equal length, and let $f(x) = x^2 - 2x + 3$. Evaluate $\int_1^3 f(x)\,dx$ by finding $\lim_{n\to\infty} \sum_{i=1}^{n} f(x_i)\,\Delta x_i$.

4. Interpret $R_n = \displaystyle\sum_{i=1}^{n} \frac{1}{n}\sqrt{1 + \frac{i}{n}}$ as a Riemann sum for a certain function f on the interval $[0, 1]$; hence evaluate $\lim_{n\to\infty} R_n$.

Evaluate the integrals in Exercises 5–8 without using the Fundamental Theorem of Calculus.

5. $\displaystyle\int_{-\pi}^{\pi} (2 - \sin x)\,dx$

6. $\displaystyle\int_0^{\sqrt{5}} \sqrt{5 - x^2}\,dx$

7. $\displaystyle\int_1^3 \left(1 - \frac{x}{2}\right) dx$

8. $\displaystyle\int_0^{\pi} \cos x\,dx$

Find the average values of the functions in Exercises 9–10 over the indicated intervals.

9. $f(x) = 2 - \sin x^3$ on $[-\pi, \pi]$

10. $h(x) = |x - 2|$ on $[0, 3]$

Find the derivatives of the functions in Exercises 11–14.

11. $f(t) = \displaystyle\int_{13}^{t} \sin(x^2)\,dx$

12. $f(x) = \displaystyle\int_{-13}^{\sin x} \sqrt{1 + t^2}\,dt$

13. $g(s) = \int_{4s}^{1} e^{\sin u}\, du$

14. $g(\theta) = \int_{e^{\sin\theta}}^{e^{\cos\theta}} \ln x\, dx$

15. Solve the integral equation $2f(x) + 1 = 3\int_x^1 f(t)\, dt$.

16. Use the substitution $x = \pi - u$ to show that

$$\int_0^{\pi} x\, f(\sin x)\, dx = \frac{\pi}{2}\int_0^{\pi} f(\sin x)\, dx$$

for any function f continuous on $[0, 1]$.

Find the areas of the finite plane regions bounded by the indicated graphs in Exercises 17–22.

17. $y = 2 + x - x^2$ and $y = 0$

18. $y = (x - 1)^2$, $y = 0$, and $x = 0$

19. $x = y - y^4$ and $x = 0$

20. $y = 4x - x^2$ and $y = 3$

21. $y = \sin x$, $y = \cos 2x$, $x = 0$, and $x = \pi/6$

22. $y = 5 - x^2$ and $y = 4/x^2$

Evaluate the integrals in Exercises 23–30.

23. $\int x^2 \cos(2x^3 + 1)\, dx$

24. $\int_1^e \frac{\ln x}{x}\, dx$

25. $\int_0^4 \sqrt{9t^2 + t^4}\, dt$

26. $\int \sin^3(\pi x)\, dx$

27. $\int_0^{\ln 2} \frac{e^u}{4 + e^{2u}}\, du$

28. $\int_1^{\sqrt[4]{e}} \frac{\tan^2 \pi \ln x}{x}\, dx$

29. $\int \frac{\sin\sqrt{2s+1}}{\sqrt{2s+1}}\, ds$

30. $\int \cos^2\frac{t}{5}\sin^2\frac{t}{5}\, dt$

31. Find the minimum value of $F(x) = \int_0^{x^2-2x} \frac{1}{1+t^2}\, dt$. Does F have a maximum value? Why?

32. Find the maximum value of $\int_a^b (4x - x^2)\, dx$ for intervals $[a, b]$, where $a < b$. How do you know such a maximum value exists?

33. An object moves along the x-axis so that its position at time t is given by the function $x(t)$. In Section 2.11 we defined the average velocity of the object over the time interval $[t_0, t_1]$ to be $v_{\text{av}} = \big(x(t_1) - x(t_0)\big)/(t_1 - t_0)$. Show that v_{av} is, in fact, the average value of the velocity function $v(t) = dx/dt$ over the interval $[t_0, t_1]$.

34. If an object falls from rest under constant gravitational acceleration, show that its average height during the time T of its fall is its height at time $T/\sqrt{3}$.

35. Find two numbers x_1 and x_2 in the interval $[0, 1]$ with $x_1 < x_2$ such that if $f(x)$ is any cubic polynomial (i.e., polynomial of degree 3), then

$$\int_0^1 f(x)\, dx = \frac{f(x_1) + f(x_2)}{2}.$$

Challenging Problems

1. Evaluate the upper and lower Riemann sums, $U(f, P_n)$ and $L(f, P_n)$, for $f(x) = 1/x$ on the interval $[1, 2]$ for the partition P_n with division points $x_i = 2^{i/n}$ for $0 \le i \le n$. Verify that $\lim_{n\to\infty} U(f, P_n) = \ln 2 = \lim_{n\to\infty} L(f, P_n)$.

2. (a) Use the addition formulas for $\cos(a + b)$ and $\cos(a - b)$ to show that

$$\cos\Big(\big(j + \tfrac{1}{2}\big)t\Big) - \cos\Big(\big(j - \tfrac{1}{2}\big)t\Big) = -2\sin\big(\tfrac{1}{2}t\big)\sin(jt),$$

and hence deduce that if $t/(2\pi)$ is not an integer, then

$$\sum_{j=1}^{n} \sin(jt) = \frac{\cos\frac{t}{2} - \cos\Big(\big(n + \frac{1}{2}\big)t\Big)}{2\sin\frac{t}{2}}$$

(b) Use the result of part (a) to evaluate $\int_0^{\pi/2} \sin x\, dx$ as a limit of a Riemann sum.

3. (a) Use the method of Problem 2 to show that if $t/(2\pi)$ is not an integer, then

$$\sum_{j=1}^{n} \cos(jt) = \frac{\sin\Big(\big(n + \frac{1}{2}\big)t\Big) - \sin\frac{t}{2}}{2\sin\frac{t}{2}}$$

(b) Use the result to part (a) to evaluate $\int_0^{\pi/3} \cos x\, dx$ as a limit of a Riemann sum.

4. Let $f(x) = 1/x^2$ and let $1 = x_0 < x_1 < x_2 < \cdots < x_n = 2$, so that $\{x_0, x_1, x_2, \ldots, x_n\}$ is a partition of $[1, 2]$ into n subintervals. Show that $c_i = \sqrt{x_{i-1}x_i}$ is in the ith subinterval $[x_{i-1}, x_i]$ of the partition, and evaluate the Riemann sum $\sum_{i=1}^{n} f(c_i)\, \Delta x_i$. What does this imply about $\int_1^2 (1/x^2)\, dx$?

5. (a) Use mathematical induction to verify that for every positive integer k, $\sum_{j=1}^{n} j^k = \frac{n^{k+1}}{k+1} + \frac{n^k}{2} + P_{k-1}(n)$, where P_{k-1} is a polynomial of degree at most $k - 1$. *Hint:* Start by iterating the identity

$$(j+1)^{k+1} - j^{k+1} = (k+1)j^k + \frac{(k+1)k}{2}j^{k-1} + \text{lower powers of } j$$

for $j = 1, 2, 3, \ldots, k$ and adding.

(b) Deduce from (a) that $\int_0^a x^k\, dx = \frac{a^{k+1}}{k+1}$.

6. Let C be the cubic curve $y = ax^3 + bx^2 + cx + d$, and let P be any point on C. The tangent to C at P meets C again at point Q. The tangent to C at Q meets C again at R. Show that the area between C and the tangent at Q is 16 times the area between C and the tangent at P.

7. Let C be the cubic curve $y = ax^3 + bx^2 + cx + d$, and let P be any point on C. The tangent to C at P meets C again at point Q. Let R be the inflection point of C. Show that R lies between P and Q on C and that QR divides the area between C and its tangent at P in the ratio 16/11.

8. **(Double tangents)** Let line PQ be tangent to the graph C of the quartic polynomial $f(x) = ax^4 + bx^3 + cx^2 + dx + e$ at two distinct points: $P = (p, f(p))$ and $Q = (q, f(q))$. Let $U = (u, f(u))$ and $V = (v, f(v))$ be the other two points where the line tangent to C at $T = ((p+q)/2, f((p+q)/2))$ meets C. If A and B are the two inflection points of C, let R and S be the other two points where AB meets C. (See Figure 5.34. Also see Challenging Problem 17 in Chapter 2 for more background.)

(a) Find the ratio of the area bounded by UV and C to the area bounded by PQ and C.

(b) Show that the area bounded by RS and C is divided at A and B into three parts in the ratio $1 : 2 : 1$.

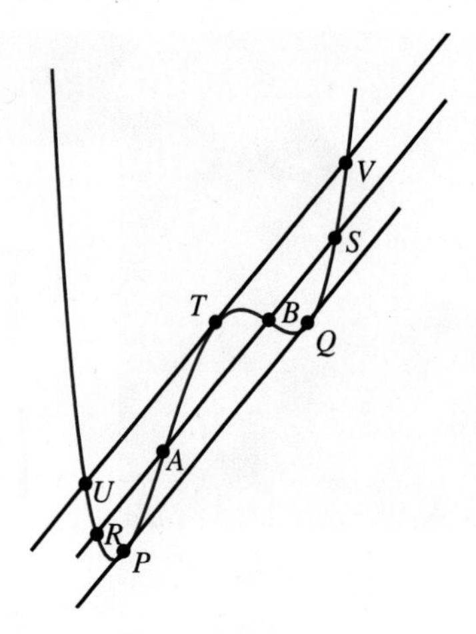

Figure 5.34

CHAPTER 6

Techniques of Integration

> I'm very good at integral and differential calculus,
> I know the scientific names of beings animalculous;
> In short, in matters vegetable, animal, and mineral,
> I am the very model of a modern Major-General.

William Schwenck Gilbert 1836–1911
from *The Pirates of Penzance*

Introduction

This chapter is completely concerned with how to evaluate integrals. The first four sections continue our search, begun in Section 5.6, for ways to find antiderivatives and, therefore, definite integrals by the Fundamental Theorem of Calculus. Section 6.5 deals with the problem of finding definite integrals of functions over infinite intervals, or over intervals where the functions are not bounded. The remaining three sections deal with techniques of *numerical integration* that can be used to find approximate values of definite integrals when an antiderivative cannot be found.

It is not necessary to cover the material of this chapter before proceeding to the various applications of integration discussed in Chapter 7, but some of the examples and exercises in that chapter do depend on techniques presented here.

6.1 Integration by Parts

Our next general method for antidifferentiation is called **integration by parts**. Just as the method of substitution can be regarded as inverse to the Chain Rule for differentiation, so the method for integration by parts is inverse to the Product Rule for differentiation.

Suppose that $U(x)$ and $V(x)$ are two differentiable functions. According to the Product Rule,

$$\frac{d}{dx}\big(U(x)V(x)\big) = U(x)\,\frac{dV}{dx} + V(x)\,\frac{dU}{dx}.$$

Integrating both sides of this equation and transposing terms, we obtain

$$\int U(x)\,\frac{dV}{dx}\,dx = U(x)V(x) - \int V(x)\,\frac{dU}{dx}\,dx$$

or, more simply,

$$\int U\,dV = UV - \int V\,dU.$$

The above formula serves as a *pattern* for carrying out integration by parts, as we will see in the examples below. In each application of the method, we break up the given integrand into a product of two pieces, U and V', where V' is readily integrated and where $\int VU'\,dx$ is usually (but not always) a *simpler* integral than $\int UV'\,dx$. The technique is called integration by parts because it replaces one integral with the sum of an integrated term and another integral that remains to be evaluated. That is, it accomplishes only *part* of the original integration.

EXAMPLE 1

$$\int xe^x\,dx \qquad \text{Let } U = x, \quad dV = e^x\,dx.$$
$$\text{Then } dU = dx, \quad V = e^x.$$
$$= xe^x - \int e^x\,dx \qquad (\text{i.e., } UV - \textstyle\int V\,dU)$$
$$= xe^x - e^x + C.$$

Note the form in which the integration by parts is carried out. We indicate at the side what choices we are making for U and dV and then calculate dU and V from these. However, we do not actually substitute U and V into the integral; instead, we use the formula $\int U\,dV = UV - \int V\,dU$ as a pattern or mnemonic device to replace the given integral by the equivalent partially integrated form on the second line.

Note also that had we included a constant of integration with V, for example, $V = e^x + K$, that constant would cancel out in the next step:

$$\int xe^x\,dx = x(e^x + K) - \int (e^x + K)\,dx$$
$$= xe^x + Kx - e^x - Kx + C = xe^x - e^x + C.$$

In general, do not include a constant of integration with V or on the right-hand side until the last integral has been evaluated.

Study the various parts of the following example carefully; they show the various ways in which integration by parts is used, and they give some insights into what choices should be made for U and dV in various situations. An improper choice can result in making an integral more difficult rather than easier. Look for a factor of the integrand that is easily integrated, and include dx with that factor to make up dV. Then U is the remaining factor of the integrand. Sometimes it is necessary to take $dV = dx$ only. When breaking up an integrand using integration by parts, choose U and dV so that, if possible, $V\,dU$ is "simpler" (easier to integrate) than $U\,dV$.

EXAMPLE 2 Use integration by parts to evaluate:

(a) $\int \ln x\,dx$, (b) $\int x^2 \sin x\,dx$, (c) $\int x \tan^{-1} x\,dx$, (d) $\int \sin^{-1} x\,dx$.

Solution

(a)

$$\int \ln x\,dx \qquad \text{Let } U = \ln x, \quad dV = dx.$$
$$\text{Then } dU = dx/x, \quad V = x.$$
$$= x\ln x - \int x\frac{1}{x}\,dx$$
$$= x\ln x - x + C.$$

(b) We have to integrate by parts twice this time:

$$\int x^2 \sin x\,dx \qquad \begin{array}{ll} \text{Let } U = x^2, & dV = \sin x\,dx. \\ \text{Then } dU = 2x\,dx, & V = -\cos x. \end{array}$$

$$= -x^2 \cos x + 2\int x \cos x\,dx \qquad \begin{array}{ll} \text{Let } U = x, & dV = \cos x\,dx. \\ \text{Then } dU = dx, & V = \sin x. \end{array}$$

$$= -x^2 \cos x + 2\left(x \sin x - \int \sin x\,dx\right)$$

$$= -x^2 \cos x + 2x \sin x + 2 \cos x + C.$$

(c)

$$\int x \tan^{-1} x\,dx \qquad \begin{array}{ll} \text{Let } U = \tan^{-1} x, & dV = x\,dx. \\ \text{Then } dU = dx/(1+x^2), & V = \tfrac{1}{2}x^2. \end{array}$$

$$= \frac{1}{2}x^2 \tan^{-1} x - \frac{1}{2}\int \frac{x^2}{1+x^2}\,dx$$

$$= \frac{1}{2}x^2 \tan^{-1} x - \frac{1}{2}\int \left(1 - \frac{1}{1+x^2}\right) dx$$

$$= \frac{1}{2}x^2 \tan^{-1} x - \frac{1}{2}x + \frac{1}{2}\tan^{-1} x + C.$$

(d)

$$\int \sin^{-1} x\,dx \qquad \begin{array}{ll} \text{Let } U = \sin^{-1} x, & dV = dx. \\ \text{Then } dU = dx/\sqrt{1-x^2}, & V = x. \end{array}$$

$$= x \sin^{-1} x - \int \frac{x}{\sqrt{1-x^2}}\,dx \qquad \begin{array}{l} \text{Let } u = 1 - x^2, \\ du = -2x\,dx \end{array}$$

$$= x \sin^{-1} x + \frac{1}{2}\int u^{-1/2}\,du$$

$$= x \sin^{-1} x + u^{1/2} + C = x \sin^{-1} x + \sqrt{1-x^2} + C.$$

The following are two useful rules of thumb for choosing U and dV:

(i) If the integrand involves a polynomial multiplied by an exponential, a sine or a cosine, or some other readily integrable function, try U equals the polynomial and dV equals the rest.

(ii) If the integrand involves a logarithm, an inverse trigonometric function, or some other function that is not readily integrable but whose derivative is readily calculated, try that function for U and let dV equal the rest.

(Of course, these "rules" come with no guarantee. They may fail to be helpful if "the rest" is not of a suitable form. There remain many functions that cannot be antidifferentiated by any standard techniques; e.g., e^{x^2}.)

The following two examples illustrate a frequently occurring and very useful phenomenon. It may happen after one or two integrations by parts, with the possible application of some known identity, that the original integral reappears on the right-hand side. Unless its coefficient there is 1, we have an equation that can be solved for that integral.

EXAMPLE 3 Evaluate $I = \int \sec^3 x\,dx$.

Solution Start by integrating by parts:

$$I = \int \sec^3 x\,dx \qquad \begin{array}{ll} \text{Let } U = \sec x, & dV = \sec^2 x\,dx. \\ \text{Then } dU = \sec x \tan x\,dx, & V = \tan x. \end{array}$$

$$= \sec x \tan x - \int \sec x \tan^2 x\,dx$$

$$= \sec x \tan x - \int \sec x (\sec^2 x - 1)\, dx$$

$$= \sec x \tan x - \int \sec^3 x\, dx + \int \sec x\, dx$$

$$= \sec x \tan x - I + \ln|\sec x + \tan x|.$$

This is an equation that can be solved for the desired integral I: Since $2I = \sec x \tan x + \ln|\sec x + \tan x|$, we have

$$\int \sec^3 x\, dx = I = \frac{1}{2} \sec x \tan x + \frac{1}{2} \ln|\sec x + \tan x| + C.$$

This integral occurs frequently in applications and is worth remembering.

EXAMPLE 4 Find $I = \int e^{ax} \cos bx\, dx$.

Solution If either $a = 0$ or $b = 0$, the integral is easy to do, so let us assume $a \neq 0$ and $b \neq 0$. We have

$$I = \int e^{ax} \cos bx\, dx \qquad \begin{array}{ll} \text{Let } U = e^{ax}, & dV = \cos bx\, dx. \\ \text{Then } dU = a\, e^{ax}\, dx, & V = (1/b) \sin bx. \end{array}$$

$$= \frac{1}{b} e^{ax} \sin bx - \frac{a}{b} \int e^{ax} \sin bx\, dx$$

$$\begin{array}{ll} \text{Let } U = e^{ax}, & dV = \sin bx\, dx. \\ \text{Then } dU = a e^{ax} dx, & V = -(\cos bx)/b. \end{array}$$

$$= \frac{1}{b} e^{ax} \sin bx - \frac{a}{b} \left(-\frac{1}{b} e^{ax} \cos bx + \frac{a}{b} \int e^{ax} \cos bx\, dx \right)$$

$$= \frac{1}{b} e^{ax} \sin bx + \frac{a}{b^2} e^{ax} \cos bx - \frac{a^2}{b^2} I.$$

Thus,

$$\left(1 + \frac{a^2}{b^2}\right) I = \frac{1}{b} e^{ax} \sin bx + \frac{a}{b^2} e^{ax} \cos bx + C_1$$

and

$$\int e^{ax} \cos bx\, dx = I = \frac{b\, e^{ax} \sin bx + a\, e^{ax} \cos bx}{b^2 + a^2} + C.$$

Observe that after the first integration by parts we had an integral that was different from, but no simpler than, the original integral. At this point we might have become discouraged and given up on this method. However, perseverance proved worthwhile; a second integration by parts returned the original integral I in an equation that could be solved for I. Having chosen to let U be the exponential in the first integration by parts (we could have let it be the cosine), we made the same choice for U in the second integration by parts. Had we switched horses in midstream and decided to let U be the trigonometric function the second time, we would have obtained

$$I = \frac{1}{b} e^{ax} \sin bx - \frac{1}{b} e^{ax} \sin bx + I;$$

we would have *undone* what we accomplished in the first step.

If we want to evaluate a definite integral by the method of integration by parts, we must remember to include the appropriate evaluation symbol with the integrated term.

EXAMPLE 5 **(A definite integral)**

$$\int_1^e x^3 (\ln x)^2\,dx \qquad \begin{array}{lll} \text{Let} & U = (\ln x)^2, & dV = x^3\,dx. \\ \text{Then} & dU = 2\ln x\,(1/x)\,dx, & V = x^4/4. \end{array}$$

$$= \frac{x^4}{4}(\ln x)^2\Big|_1^e - \frac{1}{2}\int_1^e x^3 \ln x\,dx \qquad \begin{array}{lll} \text{Let} & U = \ln x, & dV = x^3\,dx. \\ \text{Then} & dU = dx/x, & V = x^4/4. \end{array}$$

$$= \frac{e^4}{4}(1^2) - 0 - \frac{1}{2}\left(\frac{x^4}{4}\ln x\Big|_1^e - \frac{1}{4}\int_1^e x^3\,dx\right)$$

$$= \frac{e^4}{4} - \frac{e^4}{8} + \frac{1}{8}\frac{x^4}{4}\Big|_1^e = \frac{e^4}{8} + \frac{e^4}{32} - \frac{1}{32} = \frac{5}{32}e^4 - \frac{1}{32}.$$

Reduction Formulas

Consider the problem of finding $\int x^4 e^{-x}\,dx$. We can, as in Example 1, proceed by using integration by parts four times. Each time will reduce the power of x by 1. Since this is repetitive and tedious, we prefer the following approach. For $n \geq 0$, let

$$I_n = \int x^n e^{-x}\,dx.$$

We want to find I_4. If we integrate by parts, we obtain a formula for I_n in terms of I_{n-1}:

$$I_n = \int x^n e^{-x}\,dx \qquad \begin{array}{lll} \text{Let} & U = x^n, & dV = e^{-x}\,dx. \\ \text{Then} & dU = nx^{n-1}\,dx, & V = -e^{-x}. \end{array}$$

$$= -x^n e^{-x} + n\int x^{n-1} e^{-x}\,dx = -x^n e^{-x} + nI_{n-1}.$$

The formula

$$I_n = -x^n e^{-x} + nI_{n-1}$$

is called a **reduction formula** because it gives the value of the integral I_n in terms of I_{n-1}, an integral corresponding to a reduced value of the exponent n. Starting with

$$I_0 = \int x^0 e^{-x}\,dx = \int e^{-x}\,dx = -e^{-x} + C,$$

we can apply the reduction formula four times to get

$$\begin{aligned} I_1 &= -xe^{-x} + I_0 = -e^{-x}(x+1) + C_1 \\ I_2 &= -x^2 e^{-x} + 2I_1 = -e^{-x}(x^2 + 2x + 2) + C_2 \\ I_3 &= -x^3 e^{-x} + 3I_2 = -e^{-x}(x^3 + 3x^2 + 6x + 6) + C_3 \\ I_4 &= -x^4 e^{-x} + 4I_3 = -e^{-x}(x^4 + 4x^3 + 12x^2 + 24x + 24) + C_4. \end{aligned}$$

EXAMPLE 6 Obtain and use a reduction formula to evaluate

$$I_n = \int_0^{\pi/2} \cos^n x\,dx \qquad (n = 0,\ 1,\ 2,\ 3,\ \ldots).$$

Solution Observe first that

$$I_0 = \int_0^{\pi/2} dx = \frac{\pi}{2} \quad \text{and} \quad I_1 = \int_0^{\pi/2} \cos x\, dx = \sin x \Big|_0^{\pi/2} = 1.$$

Now let $n \ge 2$:

$$\begin{aligned} I_n &= \int_0^{\pi/2} \cos^n x\, dx = \int_0^{\pi/2} \cos^{n-1} x \cos x\, dx \\ &\qquad\qquad U = \cos^{n-1} x, \qquad dV = \cos x\, dx \\ &\qquad\qquad dU = -(n-1)\cos^{n-2} x \sin x\, dx, \quad V = \sin x \\ &= \sin x \cos^{n-1} x \Big|_0^{\pi/2} + (n-1)\int_0^{\pi/2} \cos^{n-2} x \sin^2 x\, dx \\ &= 0 - 0 + (n-1)\int_0^{\pi/2} \cos^{n-2} x\,(1 - \cos^2 x)\, dx \\ &= (n-1)I_{n-2} - (n-1)I_n. \end{aligned}$$

Transposing the term $-(n-1)I_n$, we obtain $nI_n = (n-1)I_{n-2}$, or

$$I_n = \frac{n-1}{n} I_{n-2},$$

which is the required reduction formula. It is valid for $n \ge 2$, which was needed to ensure that $\cos^{n-1}(\pi/2) = 0$. If $n \ge 2$ is an *even integer*, we have

$$\begin{aligned} I_n &= \frac{n-1}{n} I_{n-2} = \frac{n-1}{n} \cdot \frac{n-3}{n-2} I_{n-4} = \cdots \\ &= \frac{n-1}{n} \cdot \frac{n-3}{n-2} \cdot \frac{n-5}{n-4} \cdots \frac{5}{6} \cdot \frac{3}{4} \cdot \frac{1}{2} \cdot I_0 \\ &= \frac{n-1}{n} \cdot \frac{n-3}{n-2} \cdot \frac{n-5}{n-4} \cdots \frac{5}{6} \cdot \frac{3}{4} \cdot \frac{1}{2} \cdot \frac{\pi}{2}. \end{aligned}$$

If $n \ge 3$ is an *odd* integer, we have

$$\begin{aligned} I_n &= \frac{n-1}{n} \cdot \frac{n-3}{n-2} \cdot \frac{n-5}{n-4} \cdots \frac{6}{7} \cdot \frac{4}{5} \cdot \frac{2}{3} \cdot I_1 \\ &= \frac{n-1}{n} \cdot \frac{n-3}{n-2} \cdot \frac{n-5}{n-4} \cdots \cdot \frac{6}{7} \cdot \frac{4}{5} \cdot \frac{2}{3}. \end{aligned}$$

See Exercise 38 for an interesting consequence of these formulas.

EXERCISES 6.1

Evaluate the integrals in Exercises 1–28.

1. $\int x \cos x\, dx$

2. $\int (x+3)e^{2x}\, dx$

3. $\int x^2 \cos \pi x\, dx$

4. $\int (x^2 - 2x)e^{kx}\, dx$

5. $\int x^3 \ln x\, dx$

6. $\int x(\ln x)^3\, dx$

7. $\int \tan^{-1} x\, dx$

8. $\int x^2 \tan^{-1} x\, dx$

9. $\int x \sin^{-1} x\, dx$

10. $\int x^5 e^{-x^2}\, dx$

11. $\int_0^{\pi/4} \sec^5 x\, dx$

12. $\int \tan^2 x \sec x\, dx$

13. $\int e^{2x} \sin 3x\, dx$

14. $\int x e^{\sqrt{x}}\, dx$

15. $\int_{1/2}^{1} \frac{\sin^{-1} x}{x^2}\, dx$

16. $\int_0^1 \sqrt{x} \sin(\pi\sqrt{x})\, dx$

17. $\int x \sec^2 x\, dx$

18. $\int x \sin^2 x\, dx$

19. $\int \cos(\ln x)\, dx$

20. $\int_1^e \sin(\ln x)\, dx$

21. $\int \frac{\ln(\ln x)}{x}\, dx$

22. $\int_0^4 \sqrt{x} e^{\sqrt{x}}\, dx$

23. $\int \arccos x\, dx$

24. $\int x \sec^{-1} x\, dx$

25. $\int_1^2 \sec^{-1} x\, dx$

26. $\int (\sin^{-1} x)^2\, dx$

27. $\int x(\tan^{-1} x)^2\, dx$

28. $\int x\, e^x \cos x\, dx$

29. Find the area below $y = e^{-x} \sin x$ and above $y = 0$ from $x = 0$ to $x = \pi$.

30. Find the area of the finite plane region bounded by the curve $y = \ln x$, the line $y = 1$, and the tangent line to $y = \ln x$ at $x = 1$.

Reduction formulas

31. Obtain a reduction formula for $I_n = \int (\ln x)^n\, dx$, and use it to evaluate I_4.

32. Obtain a reduction formula for $I_n = \int_0^{\pi/2} x^n \sin x\, dx$, and use it to evaluate I_6.

33. Obtain a reduction formula for $I_n = \int \sin^n x\, dx$ (where $n \geq 2$), and use it to find I_6 and I_7.

34. Obtain a reduction formula for $I_n = \int \sec^n x\, dx$ (where $n \geq 3$), and use it to find I_6 and I_7.

35. By writing

$$\begin{aligned} I_n &= \int \frac{dx}{(x^2 + a^2)^n} \\ &= \frac{1}{a^2} \int \frac{dx}{(x^2 + a^2)^{n-1}} - \frac{1}{a^2} \int x \frac{x}{(x^2 + a^2)^n}\, dx \end{aligned}$$

and integrating the last integral by parts, using $U = x$, obtain a reduction formula for I_n. Use this formula to find I_3.

36. If f is twice differentiable on $[a, b]$ and $f(a) = f(b) = 0$, show that

$$\int_a^b (x - a)(b - x) f''(x)\, dx = -2 \int_a^b f(x)\, dx.$$

(*Hint:* Use integration by parts on the left-hand side twice.) This formula will be used in Section 6.6 to construct an error estimate for the Trapezoid Rule approximation formula.

37. If f and g are two functions having continuous second derivatives on the interval $[a, b]$, and if $f(a) = g(a) = f(b) = g(b) = 0$, show that

$$\int_a^b f(x)\, g''(x)\, dx = \int_a^b f''(x)\, g(x)\, dx.$$

What other assumptions about the values of f and g at a and b would give the same result?

38. **(The Wallis Product)** Let $I_n = \int_0^{\pi/2} \cos^n x\, dx$.

(a) Use the fact that $0 \leq \cos x \leq 1$ for $0 \leq x \leq \pi/2$ to show that $I_{2n+2} \leq I_{2n+1} \leq I_{2n}$, for $n = 0, 1, 2, \ldots$.

(b) Use the reduction formula $I_n = ((n-1)/n) I_{n-2}$ obtained in Example 6, together with the result of (a), to show that

$$\lim_{n \to \infty} \frac{I_{2n+1}}{I_{2n}} = 1.$$

(c) Combine the result of (b) with the explicit formulas obtained for I_n (for even and odd n) in Example 6 to show that

$$\lim_{n \to \infty} \frac{2}{1} \cdot \frac{2}{3} \cdot \frac{4}{3} \cdot \frac{4}{5} \cdot \frac{6}{5} \cdot \frac{6}{7} \cdots \frac{2n}{2n-1} \cdot \frac{2n}{2n+1} = \frac{\pi}{2}.$$

This interesting product formula for π is due to the seventeenth-century English mathematician John Wallis and is referred to as the Wallis Product.

6.2 Integrals of Rational Functions

In this section we are concerned with integrals of the form

$$\int \frac{P(x)}{Q(x)}\, dx,$$

where P and Q are polynomials. Recall that a **polynomial** is a function P of the form

$$P(x) = a_n x^n + a_{n-1} x^{n-1} + \cdots + a_2 x^2 + a_1 x + a_0,$$

where n is a nonnegative integer, $a_0, a_1, a_2, \ldots, a_n$ are constants, and $a_n \neq 0$. We call n the **degree** of P. A quotient $P(x)/Q(x)$ of two polynomials is called a **rational function**. (See Section P.6 for more discussion of polynomials and rational functions.) We need normally concern ourselves only with rational functions $P(x)/Q(x)$ where the degree of P is less than that of Q. If the degree of P equals or exceeds the degree of Q, then we can use division to express the fraction $P(x)/Q(x)$ as a polynomial plus another fraction $R(x)/Q(x)$, where R, the remainder in the division, has degree less than that of Q.

EXAMPLE 1 Evaluate $\displaystyle\int \frac{x^3+3x^2}{x^2+1}\,dx$.

Solution The numerator has degree 3 and the denominator has degree 2 so we need to divide. We use long division:

$$\begin{array}{r|l} & x + 3 \\ \hline x^2+1 & x^3 + 3x^2 \\ & x^3 \quad\quad + x \\ \hline & \quad 3x^2 - x \\ & \quad 3x^2 \quad + 3 \\ \hline & \quad\quad - x - 3 \end{array}$$

$$\frac{x^3+3x^2}{x^2+1} = x+3-\frac{x+3}{x^2+1}.$$

Thus,

$$\begin{aligned}\int \frac{x^3+3x^2}{x^2+1}\,dx &= \int (x+3)\,dx - \int \frac{x}{x^2+1}\,dx - 3\int \frac{dx}{x^2+1} \\ &= \frac{1}{2}x^2 + 3x - \frac{1}{2}\ln(x^2+1) - 3\tan^{-1}x + C.\end{aligned}$$

EXAMPLE 2 Evaluate $\displaystyle\int \frac{x}{2x-1}\,dx$.

Solution The numerator and denominator have the same degree, 1, so division is again required. In this case the division can be carried out by manipulation of the integrand:

$$\frac{x}{2x-1} = \frac{1}{2}\,\frac{2x}{2x-1} = \frac{1}{2}\,\frac{2x-1+1}{2x-1} = \frac{1}{2}\left(1+\frac{1}{2x-1}\right),$$

a process that we call *short division* (see Section P.6). We have

$$\int \frac{x}{2x-1}\,dx = \frac{1}{2}\int\left(1+\frac{1}{2x-1}\right)dx = \frac{x}{2}+\frac{1}{4}\ln|2x-1|+C.$$

In the discussion that follows, we always assume that any necessary division has been performed and the quotient polynomial has been integrated. The remaining basic problem with which we will deal in this section is the following:

The basic problem

Evaluate $\displaystyle\int \frac{P(x)}{Q(x)}\,dx$, where the degree of P < the degree of Q.

The complexity of this problem depends on the degree of Q.

Linear and Quadratic Denominators

Suppose that $Q(x)$ has degree 1. Thus, $Q(x) = ax+b$, where $a \neq 0$. Then $P(x)$ must have degree 0 and be a constant c. We have $P(x)/Q(x) = c/(ax+b)$. The substitution $u = ax+b$ leads to

$$\int \frac{c}{ax+b}\,dx = \frac{c}{a}\int \frac{du}{u} = \frac{c}{a}\ln|u| + C,$$

so that for $c = 1$:

The case of a linear denominator

$$\int \frac{1}{ax+b}\,dx = \frac{1}{a}\ln|ax+b| + C.$$

Now suppose that $Q(x)$ is quadratic, that is, has degree 2. For purposes of this discussion we can assume that $Q(x)$ is either of the form $x^2 + a^2$ or of the form $x^2 - a^2$, since completing the square and making the appropriate change of variable can always reduce a quadratic denominator to this form, as shown in Section 6.2. Since $P(x)$ can be at most a linear function, $P(x) = Ax + B$, we are led to consider the following four integrals:

$$\int \frac{x\,dx}{x^2+a^2}, \qquad \int \frac{x\,dx}{x^2-a^2}, \qquad \int \frac{dx}{x^2+a^2}, \qquad \text{and} \qquad \int \frac{dx}{x^2-a^2}.$$

(If $a = 0$, there are only two integrals; each is easily evaluated.) The first two integrals yield to the substitution $u = x^2 \pm a^2$; the third is a known integral. The fourth integral will be evaluated by a different method below. The values of all four integrals are given in the following box:

The case of a quadratic denominator

$$\int \frac{x\,dx}{x^2+a^2} = \frac{1}{2}\ln(x^2+a^2) + C,$$
$$\int \frac{x\,dx}{x^2-a^2} = \frac{1}{2}\ln|x^2-a^2| + C,$$
$$\int \frac{dx}{x^2+a^2} = \frac{1}{a}\tan^{-1}\frac{x}{a} + C,$$
$$\int \frac{dx}{x^2-a^2} = \frac{1}{2a}\ln\left|\frac{x-a}{x+a}\right| + C.$$

To obtain the last formula in the box, let us try to write the integrand as a sum of two fractions with linear denominators:

$$\frac{1}{x^2-a^2} = \frac{1}{(x-a)(x+a)} = \frac{A}{x-a} + \frac{B}{x+a} = \frac{Ax+Aa+Bx-Ba}{x^2-a^2},$$

where we have added the two fractions together again in the last step. If this equation is to hold identically for all x (except $x = \pm a$), then the numerators on the left and right sides must be identical as polynomials in x. The equation $(A+B)x + (Aa - Ba) = 1 = 0x + 1$ can hold for all x only if

$$\begin{aligned} A + B &= 0 && \text{(the coefficient of } x\text{)},\\ Aa - Ba &= 1 && \text{(the constant term)}. \end{aligned}$$

Solving this pair of linear equations for the unknowns A and B, we get $A = 1/(2a)$ and $B = -1/(2a)$. Therefore,

$$\begin{aligned} \int \frac{dx}{x^2-a^2} &= \frac{1}{2a}\int \frac{dx}{x-a} - \frac{1}{2a}\int \frac{dx}{x+a}\\ &= \frac{1}{2a}\ln|x-a| - \frac{1}{2a}\ln|x+a| + C\\ &= \frac{1}{2a}\ln\left|\frac{x-a}{x+a}\right| + C. \end{aligned}$$

Partial Fractions

The technique used above, involving the writing of a complicated fraction as a sum of simpler fractions, is called the **method of partial fractions**. Suppose that a polynomial $Q(x)$ is of degree n and that its highest degree term is x^n (with coefficient 1). Suppose also that Q factors into a product of n *distinct* linear (degree 1) factors, say,

$$Q(x) = (x - a_1)(x - a_2) \cdots (x - a_n),$$

where $a_i \neq a_j$ if $i \neq j$, $1 \leq i, j \leq n$. If $P(x)$ is a polynomial of degree smaller than n, then $P(x)/Q(x)$ has a **partial fraction decomposition** of the form

$$\frac{P(x)}{Q(x)} = \frac{A_1}{x - a_1} + \frac{A_2}{x - a_2} + \cdots + \frac{A_n}{x - a_n}$$

for certain values of the constants $A_1, A_2, \ldots, A_n$. We do not attempt to give any formal proof of this assertion here; such a proof belongs in an algebra course. (See Theorem 1 below for the statement of a more general result.)

Given that $P(x)/Q(x)$ has a partial fraction decomposition as claimed above, there are two methods for determining the constants $A_1, A_2, \ldots, A_n$. The first of these methods, and one that generalizes most easily to the more complicated decompositions considered below, is to add up the fractions in the decomposition, obtaining a new fraction $S(x)/Q(x)$ with numerator $S(x)$, a polynomial of degree one less than that of $Q(x)$. This new fraction will be identical to the original fraction $P(x)/Q(x)$ if S and P are identical polynomials. The constants $A_1, A_2, \ldots, A_n$ are determined by solving the n linear equations resulting from equating the coefficients of like powers of x in the two polynomials S and P.

The second method depends on the following observation: If we multiply the partial fraction decomposition by $x - a_j$, we get

$$\begin{aligned} &(x - a_j)\frac{P(x)}{Q(x)} \\ &= A_1\frac{x - a_j}{x - a_1} + \cdots + A_{j-1}\frac{x - a_j}{x - a_{j-1}} + A_j + A_{j+1}\frac{x - a_j}{x - a_{j+1}} + \cdots + A_n\frac{x - a_j}{x - a_n}. \end{aligned}$$

All terms on the right side are 0 at $x = a_j$ except the jth term, A_j. Hence,

$$\begin{aligned} A_j &= \lim_{x \to a_j} (x - a_j)\frac{P(x)}{Q(x)} \\ &= \frac{P(a_j)}{(a_j - a_1) \cdots (a_j - a_{j-1})(a_j - a_{j+1}) \cdots (a_j - a_n)}, \end{aligned}$$

for $1 \leq j \leq n$. In practice, you can use this method to find each number A_j by cancelling the factor $x - a_j$ from the denominator of $P(x)/Q(x)$ and evaluating the resulting expression at $x = a_j$.

EXAMPLE 3 Evaluate $\displaystyle\int \frac{(x + 4)}{x^2 - 5x + 6}\,dx$.

Solution The partial fraction decomposition takes the form

$$\frac{x + 4}{x^2 - 5x + 6} = \frac{x + 4}{(x - 2)(x - 3)} = \frac{A}{x - 2} + \frac{B}{x - 3}.$$

We calculate A and B by both of the methods suggested above.

METHOD I. Add the partial fractions

$$\frac{x + 4}{x^2 - 5x + 6} = \frac{Ax - 3A + Bx - 2B}{(x - 2)(x - 3)},$$

and equate the coefficient of x and the constant terms in the numerators on both sides to obtain

$$A + B = 1 \qquad \text{and} \qquad -3A - 2B = 4.$$

Solve these equations to get $A = -6$ and $B = 7$.

METHOD II. To find A, cancel $x - 2$ from the denominator of the expression $P(x)/Q(x)$ and evaluate the result at $x = 2$. Obtain B similarly.

$$A = \left.\frac{x+4}{x-3}\right|_{x=2} = -6 \qquad \text{and} \qquad B = \left.\frac{x+4}{x-2}\right|_{x=3} = 7.$$

In either case we have

$$\begin{aligned}\int \frac{(x+4)}{x^2 - 5x + 6}\,dx &= -6\int \frac{1}{x-2}\,dx + 7\int \frac{1}{x-3}\,dx \\ &= -6\ln|x-2| + 7\ln|x-3| + C.\end{aligned}$$

EXAMPLE 4 Evaluate $I = \displaystyle\int \frac{x^3+2}{x^3-x}\,dx$.

Solution Since the numerator does not have degree smaller than the denominator, we must divide:

$$I = \int \frac{x^3 - x + x + 2}{x^3 - x}\,dx = \int \left(1 + \frac{x+2}{x^3-x}\right)dx = x + \int \frac{x+2}{x^3-x}\,dx.$$

Now we can use the method of partial fractions.

$$\begin{aligned}\frac{x+2}{x^3-x} = \frac{x+2}{x(x-1)(x+1)} &= \frac{A}{x} + \frac{B}{x-1} + \frac{C}{x+1} \\ &= \frac{A(x^2-1) + B(x^2+x) + C(x^2-x)}{x(x-1)(x+1)}\end{aligned}$$

We have

$$\begin{aligned} A + B + C &= 0 \qquad &&\text{(coefficient of } x^2\text{)} \\ B - C &= 1 \qquad &&\text{(coefficient of } x\text{)} \\ -A &= 2 \qquad &&\text{(constant term).}\end{aligned}$$

It follows that $A = -2$, $B = 3/2$, and $C = 1/2$. We can also find these values using Method II of the previous example:

$$A = \left.\frac{x+2}{(x-1)(x+1)}\right|_{x=0} = -2, \qquad B = \left.\frac{x+2}{x(x+1)}\right|_{x=1} = \frac{3}{2}, \quad \text{and}$$

$$C = \left.\frac{x+2}{x(x-1)}\right|_{x=-1} = \frac{1}{2}.$$

Finally, we have

$$\begin{aligned} I &= x - 2\int \frac{1}{x}\,dx + \frac{3}{2}\int \frac{1}{x-1}\,dx + \frac{1}{2}\int \frac{1}{x+1}\,dx \\ &= x - 2\ln|x| + \frac{3}{2}\ln|x-1| + \frac{1}{2}\ln|x+1| + C.\end{aligned}$$

Next, we consider a rational function whose denominator has a quadratic factor that is equivalent to a sum of squares and cannot, therefore, be further factored into a product of real linear factors.

EXAMPLE 5 Evaluate $\displaystyle\int \frac{2+3x+x^2}{x(x^2+1)}\,dx$.

Solution Note that the numerator has degree 2 and the denominator degree 3, so no division is necessary. If we decompose the integrand as a sum of two simpler fractions, we want one with denominator x and one with denominator x^2+1. The appropriate form of the decomposition turns out to be

$$\frac{2+3x+x^2}{x(x^2+1)} = \frac{A}{x} + \frac{Bx+C}{x^2+1} = \frac{A(x^2+1)+Bx^2+Cx}{x(x^2+1)}.$$

Note that corresponding to the quadratic (degree 2) denominator we use a linear (degree 1) numerator. Equating coefficients in the two numerators, we obtain

$$\begin{aligned} A + B \quad &= 1 \qquad \text{(coefficient of } x^2\text{)} \\ C &= 3 \qquad \text{(coefficient of } x\text{)} \\ A \quad &= 2 \qquad \text{(constant term).} \end{aligned}$$

Hence $A = 2$, $B = -1$, and $C = 3$. We have, therefore,

$$\begin{aligned} \int \frac{2+3x+x^2}{x(x^2+1)}\,dx &= 2\int \frac{1}{x}\,dx - \int \frac{x}{x^2+1}\,dx + 3\int \frac{1}{x^2+1}\,dx \\ &= 2\ln|x| - \frac{1}{2}\ln(x^2+1) + 3\tan^{-1}x + C. \end{aligned}$$

We remark that addition of the fractions is the only reasonable real-variable method for determining the constants A, B, and C here. We could determine A by Method II of Example 3, but there is no simple equivalent way of finding B or C without using complex numbers.

Completing the Square

Quadratic expressions of the form Ax^2+Bx+C are often found in integrands. These can be written as sums or differences of squares using the procedure of completing the square, as was done to find the formula for the roots of quadratic equations in Section P.6. First factor out A so that the remaining expression begins with x^2+2bx, where $2b = B/A$. These are the first two terms of $(x+b)^2 = x^2+2bx+b^2$. Add the third term $b^2 = B^2/4A^2$ and then subtract it again:

$$\begin{aligned} Ax^2+Bx+C &= A\left(x^2+\frac{B}{A}x+\frac{C}{A}\right) \\ &= A\left(x^2+\frac{B}{A}x+\frac{B^2}{4A^2}+\frac{C}{A}-\frac{B^2}{4A^2}\right) \\ &= A\left(x+\frac{B}{2A}\right)^2+\frac{4AC-B^2}{4A}. \end{aligned}$$

The substitution $u = x + \dfrac{B}{2A}$ should then be made.

EXAMPLE 6 Evaluate $I = \displaystyle\int \frac{1}{x^3+1}\,dx$.

Solution Here $Q(x) = x^3+1 = (x+1)(x^2-x+1)$. The latter factor has no real roots, so it has no real linear subfactors. We have

$$\begin{aligned} \frac{1}{x^3+1} &= \frac{1}{(x+1)(x^2-x+1)} = \frac{A}{x+1} + \frac{Bx+C}{x^2-x+1} \\ &= \frac{A(x^2-x+1)+B(x^2+x)+C(x+1)}{(x+1)(x^2-x+1)} \end{aligned}$$

$$\begin{aligned} A + B \quad &= 0 \qquad \text{(coefficient of } x^2) \\ -A + B + C &= 0 \qquad \text{(coefficient of } x) \\ A \quad + C &= 1 \qquad \text{(constant term).} \end{aligned}$$

Hence, $A = 1/3$, $B = -1/3$, and $C = 2/3$. We have

$$I = \frac{1}{3}\int \frac{dx}{x+1} - \frac{1}{3}\int \frac{x-2}{x^2-x+1}\,dx.$$

The first integral is easily evaluated; in the second we complete the square in the denominator: $x^2 - x + 1 = \left(x - \frac{1}{2}\right)^2 + \frac{3}{4}$, and make a similar modification in the numerator.

$$\begin{aligned} I &= \frac{1}{3}\ln|x+1| - \frac{1}{3}\int \frac{x - \frac{1}{2} - \frac{3}{2}}{\left(x-\frac{1}{2}\right)^2 + \frac{3}{4}}\,dx \qquad \begin{array}{l}\text{Let } u = x - 1/2, \\ du = dx\end{array} \\ &= \frac{1}{3}\ln|x+1| - \frac{1}{3}\int \frac{u}{u^2+\frac{3}{4}}\,du + \frac{1}{2}\int \frac{1}{u^2+\frac{3}{4}}\,du \\ &= \frac{1}{3}\ln|x+1| - \frac{1}{6}\ln\left(u^2+\frac{3}{4}\right) + \frac{1}{2}\frac{2}{\sqrt{3}}\tan^{-1}\left(\frac{2u}{\sqrt{3}}\right) + C \\ &= \frac{1}{3}\ln|x+1| - \frac{1}{6}\ln(x^2-x+1) + \frac{1}{\sqrt{3}}\tan^{-1}\left(\frac{2x-1}{\sqrt{3}}\right) + C. \end{aligned}$$

Denominators with Repeated Factors

We require one final refinement of the method of partial fractions. If any of the linear or quadratic factors of $Q(x)$ is *repeated* (say, m times), then the partial fraction decomposition of $P(x)/Q(x)$ requires m distinct fractions corresponding to that factor. The denominators of these fractions have exponents increasing from 1 to m, and the numerators are all constants where the repeated factor is linear or linear where the repeated factor is quadratic. (See Theorem 1 below.)

EXAMPLE 7 Evaluate $\displaystyle\int \frac{1}{x(x-1)^2}\,dx$.

Solution The appropriate partial fraction decomposition here is

$$\begin{aligned} \frac{1}{x(x-1)^2} &= \frac{A}{x} + \frac{B}{x-1} + \frac{C}{(x-1)^2} \\ &= \frac{A(x^2-2x+1) + B(x^2-x) + Cx}{x(x-1)^2}. \end{aligned}$$

Equating coefficients of x^2, x, and 1 in the numerators of both sides, we get

$$\begin{aligned} A + B \quad &= 0 \qquad \text{(coefficient of } x^2) \\ -2A - B + C &= 0 \qquad \text{(coefficient of } x) \\ A \qquad &= 1 \qquad \text{(constant term).} \end{aligned}$$

Hence $A = 1$, $B = -1$, $C = 1$, and

$$\begin{aligned} \int \frac{1}{x(x-1)^2}\,dx &= \int \frac{1}{x}\,dx - \int \frac{1}{x-1}\,dx + \int \frac{1}{(x-1)^2}\,dx \\ &= \ln|x| - \ln|x-1| - \frac{1}{x-1} + C \\ &= \ln\left|\frac{x}{x-1}\right| - \frac{1}{x-1} + C. \end{aligned}$$

EXAMPLE 8 Evaluate $I = \displaystyle\int \frac{x^2+2}{4x^5+4x^3+x}\,dx.$

Solution The denominator factors to $x(2x^2+1)^2$, so the appropriate partial fraction decomposition is

$$\begin{aligned}\frac{x^2+2}{x(2x^2+1)^2} &= \frac{A}{x} + \frac{Bx+C}{2x^2+1} + \frac{Dx+E}{(2x^2+1)^2}\\ &= \frac{A(4x^4+4x^2+1)+B(2x^4+x^2)+C(2x^3+x)+Dx^2+Ex}{x(2x^2+1)^2}.\end{aligned}$$

Thus,

$$\begin{array}{lll} 4A + 2B & = 0 & \text{(coefficient of } x^4)\\ 2C & = 0 & \text{(coefficient of } x^3)\\ 4A + B + D & = 1 & \text{(coefficient of } x^2)\\ C + E & = 0 & \text{(coefficient of } x)\\ A & = 2 & \text{(constant term).}\end{array}$$

Solving these equations, we get $A = 2$, $B = -4$, $C = 0$, $D = -3$, and $E = 0$.

$$\begin{aligned} I &= 2\int\frac{dx}{x} - 4\int\frac{x\,dx}{2x^2+1} - 3\int\frac{x\,dx}{(2x^2+1)^2} \qquad \begin{array}{l}\text{Let } u = 2x^2+1,\\ du = 4x\,dx\end{array}\\ &= 2\ln|x| - \int\frac{du}{u} - \frac{3}{4}\int\frac{du}{u^2}\\ &= 2\ln|x| - \ln|u| + \frac{3}{4u} + C\\ &= \ln\left(\frac{x^2}{2x^2+1}\right) + \frac{3}{4}\frac{1}{2x^2+1} + C.\end{aligned}$$

The following theorem summarizes the various aspects of the method of partial fractions.

THEOREM 1 **Partial fraction decompositions of rational functions**

Let P and Q be polynomials with real coefficients, and suppose that the degree of P is less than the degree of Q. Then

(a) $Q(x)$ can be factored into the product of a constant K, real linear factors of the form $x - a_i$, and real quadratic factors of the form $x^2 + b_i x + c_i$ having no real roots. The linear and quadratic factors may be repeated:

$$\begin{aligned}Q(x) = K(x-a_1)^{m_1}(x-a_2)^{m_2}\cdots(x-a_j)^{m_j}(x^2+b_1x+c_1)^{n_1}\\ \cdots(x^2+b_kx+c_k)^{n_k}.\end{aligned}$$

The degree of Q is $m_1 + m_2 + \cdots + m_j + 2n_1 + 2n_2 + \cdots + 2n_k$.

(b) The rational function $P(x)/Q(x)$ can be expressed as a sum of partial fractions as follows:

(i) corresponding to each factor $(x-a)^m$ of $Q(x)$ the decomposition contains a sum of fractions of the form

$$\frac{A_1}{x-a} + \frac{A_2}{(x-a)^2} + \cdots + \frac{A_m}{(x-a)^m};$$

(ii) corresponding to each factor $(x^2+bx+c)^n$ of $Q(x)$ the decomposition contains a sum of fractions of the form

$$\frac{B_1x+C_1}{x^2+bx+c} + \frac{B_2x+C_2}{(x^2+bx+c)^2} + \cdots + \frac{B_nx+C_n}{(x^2+bx+c)^n}.$$

The constants $A_1, A_2, \ldots, A_m, B_1, B_2, \ldots, B_n, C_1, C_2, \ldots, C_n$ can be determined by adding up the fractions in the decomposition and equating the coefficients of like powers of x in the numerator of the sum with those in $P(x)$.

Part (a) of the above theorem is just a restatement of results discussed and proved in Section P.6 and Appendix II. The proof of part (b) is algebraic in nature and is beyond the scope of this text.

Note that part (a) does not tell us how to find the factors of $Q(x)$; it tells us only what form they have. We must know the factors of Q before we can make use of partial fractions to integrate the rational function $P(x)/Q(x)$. Partial fraction decompositions are also used in other mathematical situations, in particular, to solve certain problems involving differential equations.

EXERCISES 6.2

Evaluate the integrals in Exercises 1–28.

1. $\displaystyle\int \frac{2\,dx}{2x-3}$

2. $\displaystyle\int \frac{dx}{5-4x}$

3. $\displaystyle\int \frac{x\,dx}{\pi x+2}$

4. $\displaystyle\int \frac{x^2}{x-4}\,dx$

5. $\displaystyle\int \frac{1}{x^2-9}\,dx$

6. $\displaystyle\int \frac{dx}{5-x^2}$

7. $\displaystyle\int \frac{dx}{a^2-x^2}$

8. $\displaystyle\int \frac{dx}{b^2-a^2x^2}$

9. $\displaystyle\int \frac{x^2\,dx}{x^2+x-2}$

10. $\displaystyle\int \frac{x\,dx}{3x^2+8x-3}$

11. $\displaystyle\int \frac{x-2}{x^2+x}\,dx$

12. $\displaystyle\int \frac{dx}{x^3+9x}$

13. $\displaystyle\int \frac{dx}{1-6x+9x^2}$

14. $\displaystyle\int \frac{x\,dx}{2+6x+9x^2}$

15. $\displaystyle\int \frac{x^2+1}{6x-9x^2}\,dx$

16. $\displaystyle\int \frac{x^3+1}{12+7x+x^2}\,dx$

17. $\displaystyle\int \frac{dx}{x(x^2-a^2)}$

18. $\displaystyle\int \frac{dx}{x^4-a^4}$

19. $\displaystyle\int \frac{x^3\,dx}{x^3-a^3}$

20. $\displaystyle\int \frac{dx}{x^3+2x^2+2x}$

21. $\displaystyle\int \frac{dx}{x^3-4x^2+3x}$

22. $\displaystyle\int \frac{x^2+1}{x^3+8}\,dx$

23. $\displaystyle\int \frac{dx}{(x^2-1)^2}$

24. $\displaystyle\int \frac{x^2\,dx}{(x^2-1)(x^2-4)}$

25. $\displaystyle\int \frac{dx}{x^4-3x^3}$

26. $\displaystyle\int \frac{dt}{(t-1)(t^2-1)^2}$

27. $\displaystyle\int \frac{dx}{e^{2x}-4e^x+4}$

28. $\displaystyle\int \frac{d\theta}{\cos\theta(1+\sin\theta)}$

In Exercises 29–30 write the form that the partial fraction decomposition of the given rational function takes. Do not actually evaluate the constants you use in the decomposition.

29. $\displaystyle\frac{x^5+x^3+1}{(x-1)(x^2-1)(x^3-1)}$

30. $\displaystyle\frac{123-x^7}{(x^4-16)^2}$

31. Write $\displaystyle\frac{x^5}{(x^2-4)(x+2)^2}$ as the sum of a polynomial and a partial fraction decomposition (with constants left undetermined) of a rational function whose numerator has smaller degree than the denominator.

32. Show that x^4+4x^2+16 factors to $(x^2+kx+4)(x^2-kx+4)$ for a certain positive constant k. What is the value of k? Now repeat the previous exercise for the rational function $\displaystyle\frac{x^4}{x^4+4x^2+16}$.

33. Suppose that P and Q are polynomials such that the degree of P is smaller than that of Q. If

$$Q(x) = (x-a_1)(x-a_2)\cdots(x-a_n),$$

where $a_i \neq a_j$ if $i \neq j$ $(1 \le i, j \le n)$, so that $P(x)/Q(x)$ has partial fraction decomposition

$$\frac{P(x)}{Q(x)} = \frac{A_1}{x-a_1} + \frac{A_2}{x-a_2} + \cdots + \frac{A_n}{x-a_n},$$

show that

$$A_j = \frac{P(a_j)}{Q'(a_j)} \qquad (1 \le j \le n).$$

This gives yet another method for computing the constants in a partial fraction decomposition if the denominator factors completely into distinct linear factors.

6.3 Inverse Substitutions

The substitutions considered in Section 5.6 were direct substitutions in the sense that we simplified an integrand by replacing an expression appearing in it with a single variable. In this section we consider the reverse approach; we replace the variable of integration with a function of a new variable. Such substitutions, called *inverse substitutions*, would appear on the surface to make the integral more complicated. That is, substituting $x = g(u)$ in the integral

$$\int_a^b f(x)\,dx$$

leads to the more "complicated" integral

$$\int_{x=a}^{x=b} f\big(g(u)\big)\,g'(u)\,du.$$

As we will see, however, sometimes such substitutions can actually simplify an integrand, transforming the integral into one that can be evaluated by inspection or to which other techniques can readily be applied. In any event, inverse substitutions can often be used to convert integrands to rational functions, to which the methods of Section 6.2 can be applied.

The Inverse Trigonometric Substitutions

Three very useful inverse substitutions are:

$$x = a\sin\theta, \qquad x = a\tan\theta, \qquad \text{and} \qquad x = a\sec\theta.$$

These correspond to the direct substitutions:

$$\theta = \sin^{-1}\frac{x}{a}, \qquad \theta = \tan^{-1}\frac{x}{a}, \qquad \text{and} \qquad \theta = \sec^{-1}\frac{x}{a} = \cos^{-1}\frac{a}{x}.$$

The inverse sine substitution

Integrals involving $\sqrt{a^2 - x^2}$ (where $a > 0$) can frequently be reduced to a simpler form by means of the substitution

$$x = a\sin\theta \quad \text{or, equivalently,} \quad \theta = \sin^{-1}\frac{x}{a}.$$

Observe that $\sqrt{a^2 - x^2}$ makes sense only if $-a \le x \le a$, which corresponds to $-\pi/2 \le \theta \le \pi/2$. Since $\cos\theta \ge 0$ for such θ, we have

$$\sqrt{a^2 - x^2} = \sqrt{a^2(1 - \sin^2\theta)} = \sqrt{a^2\cos^2\theta} = a\cos\theta.$$

(If $\cos\theta$ were not nonnegative, we would have obtained $a|\cos\theta|$ instead.) If needed, the other trigonometric functions of θ can be recovered in terms of x by examining a right-angled triangle labelled to correspond to the substitution. (See Figure 6.1.)

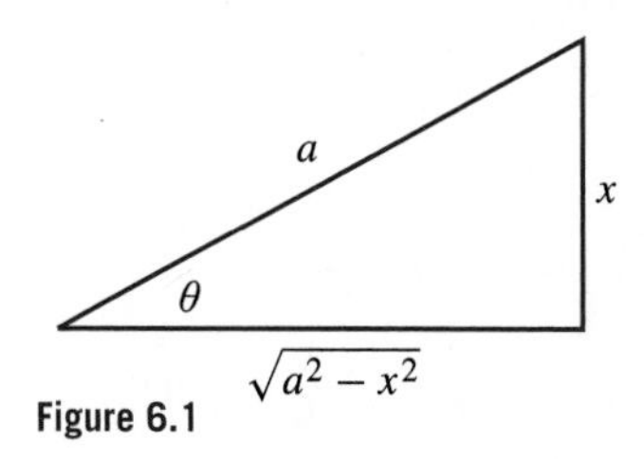

Figure 6.1

$$\cos\theta = \frac{\sqrt{a^2 - x^2}}{a} \qquad \text{and} \qquad \tan\theta = \frac{x}{\sqrt{a^2 - x^2}}.$$

EXAMPLE 1 Evaluate $\displaystyle\int \frac{1}{(5 - x^2)^{3/2}}\,dx.$

Solution Refer to Figure 6.2.

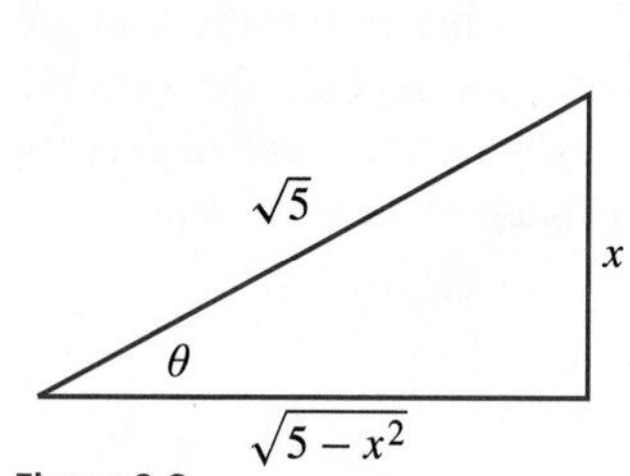

Figure 6.2

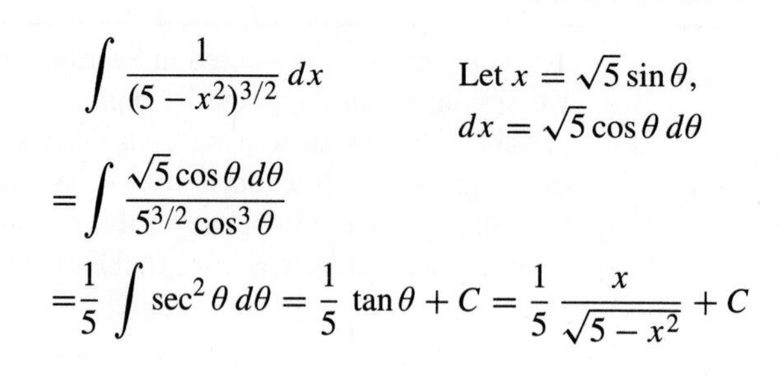

$$\int \frac{1}{(5-x^2)^{3/2}}\,dx \qquad \text{Let } x=\sqrt{5}\sin\theta,\quad dx=\sqrt{5}\cos\theta\,d\theta$$

$$=\int \frac{\sqrt{5}\cos\theta\,d\theta}{5^{3/2}\cos^3\theta}$$

$$=\frac{1}{5}\int \sec^2\theta\,d\theta=\frac{1}{5}\tan\theta+C=\frac{1}{5}\frac{x}{\sqrt{5-x^2}}+C$$

EXAMPLE 2 Find the area of the circular segment shaded in Figure 6.3.

Figure 6.3

Solution The area is

$$A=2\int_b^a \sqrt{a^2-x^2}\,dx \qquad \text{Let } x=a\sin\theta,\quad dx=a\cos\theta\,d\theta$$

$$=2\int_{x=b}^{x=a} a^2\cos^2\theta\,d\theta$$

$$=a^2(\theta+\sin\theta\cos\theta)\Big|_{x=b}^{x=a} \qquad \text{(as in Example 8 of Section 5.6)}$$

$$=a^2\left(\sin^{-1}\frac{x}{a}+\frac{x\sqrt{a^2-x^2}}{a^2}\right)\Bigg|_b^a \qquad \text{(See Figure 6.1.)}$$

$$=\frac{\pi}{2}a^2-a^2\sin^{-1}\frac{b}{a}-b\sqrt{a^2-b^2}\ \text{square units.}$$

The inverse tangent substitution

Integrals involving $\sqrt{a^2+x^2}$ or $\dfrac{1}{x^2+a^2}$ (where $a>0$) are often simplified by the substitution

$$x=a\tan\theta \quad \text{or, equivalently,} \quad \theta=\tan^{-1}\frac{x}{a}.$$

Since x can take any real value, we have $-\pi/2<\theta<\pi/2$, so $\sec\theta>0$ and

$$\sqrt{a^2+x^2}=a\sqrt{1+\tan^2\theta}=a\sec\theta.$$

Other trigonometric functions of θ can be expressed in terms of x by referring to a right-angled triangle with legs a and x and hypotenuse $\sqrt{a^2+x^2}$ (see Figure 6.4):

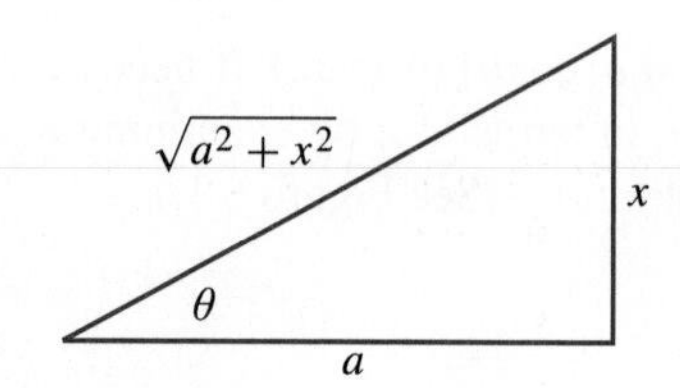

Figure 6.4

$$\sin\theta=\frac{x}{\sqrt{a^2+x^2}} \qquad \text{and} \qquad \cos\theta=\frac{a}{\sqrt{a^2+x^2}}.$$

EXAMPLE 3 Evaluate (a) $\displaystyle\int \frac{1}{\sqrt{4+x^2}}\,dx$ and (b) $\displaystyle\int \frac{1}{(1+9x^2)^2}\,dx$.

Solution Figures 6.5 and 6.6 illustrate parts (a) and (b), respectively.

(a) $\displaystyle\int \frac{1}{\sqrt{4+x^2}}\,dx$ Let $x = 2\tan\theta$, $dx = 2\sec^2\theta\,d\theta$

$$= \int \frac{2\sec^2\theta}{2\sec\theta}\,d\theta$$

$$= \int \sec\theta\,d\theta$$

$$= \ln|\sec\theta + \tan\theta| + C = \ln\left|\frac{\sqrt{4+x^2}}{2} + \frac{x}{2}\right| + C$$

$$= \ln\left(\sqrt{4+x^2} + x\right) + C_1, \qquad \text{where } C_1 = C - \ln 2.$$

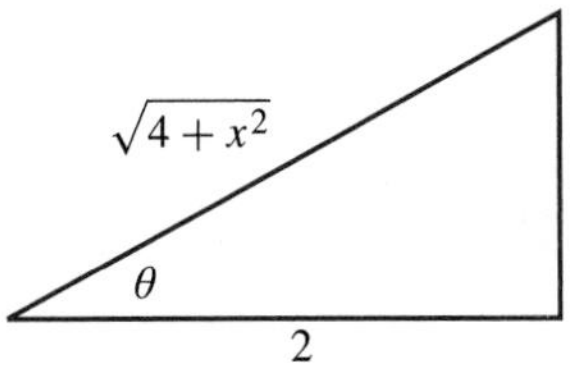

Figure 6.5

(Note that $\sqrt{4+x^2} + x > 0$ for all x, so we do not need an absolute value on it.)

(b) $\displaystyle\int \frac{1}{(1+9x^2)^2}\,dx$ Let $3x = \tan\theta$, $3dx = \sec^2\theta\,d\theta$, $1 + 9x^2 = \sec^2\theta$

$$= \frac{1}{3}\int \frac{\sec^2\theta\,d\theta}{\sec^4\theta}$$

$$= \frac{1}{3}\int \cos^2\theta\,d\theta = \frac{1}{6}\left(\theta + \sin\theta\,\cos\theta\right) + C$$

$$= \frac{1}{6}\tan^{-1}(3x) + \frac{1}{6}\frac{3x}{\sqrt{1+9x^2}}\frac{1}{\sqrt{1+9x^2}} + C$$

$$= \frac{1}{6}\tan^{-1}(3x) + \frac{1}{2}\frac{x}{1+9x^2} + C$$

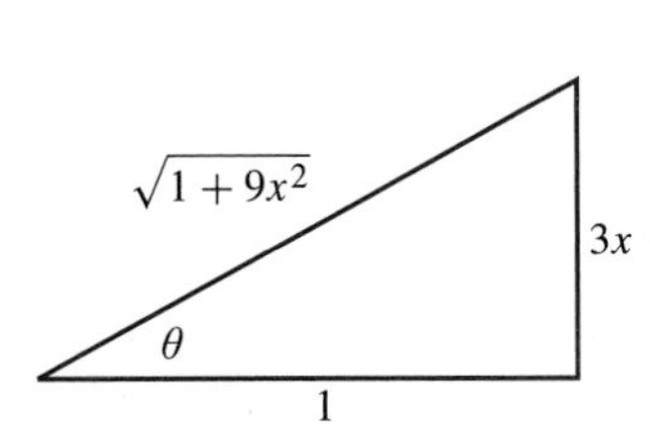

Figure 6.6

The inverse secant substitution

Integrals involving $\sqrt{x^2 - a^2}$ (where $a > 0$) can frequently be simplified by using the substitution

$$x = a\sec\theta \quad \text{or, equivalently,} \quad \theta = \sec^{-1}\frac{x}{a}.$$

We must be more careful with this substitution. Although

$$\sqrt{x^2 - a^2} = a\sqrt{\sec^2\theta - 1} = a\sqrt{\tan^2\theta} = a|\tan\theta|,$$

we cannot always drop the absolute value from the tangent. Observe that $\sqrt{x^2 - a^2}$ makes sense for $x \ge a$ and for $x \le -a$.

If $x \ge a$, then $0 \le \theta = \sec^{-1}\dfrac{x}{a} = \arccos\dfrac{a}{x} < \dfrac{\pi}{2}$, and $\tan\theta \ge 0$.

If $x \le -a$, then $\dfrac{\pi}{2} < \theta = \sec^{-1}\dfrac{x}{a} = \arccos\dfrac{a}{x} \le \pi$, and $\tan\theta \le 0$.

In the first case $\sqrt{x^2 - a^2} = a\tan\theta$; in the second case $\sqrt{x^2 - a^2} = -a\tan\theta$.

EXAMPLE 4 Find $I = \displaystyle\int \frac{dx}{\sqrt{x^2 - a^2}}$, where $a > 0$.

Solution For the moment, assume that $x \ge a$. If $x = a\sec\theta$, then $dx = a\sec\theta\tan\theta\,d\theta$ and $\sqrt{x^2 - a^2} = a\tan\theta$. (See Figure 6.7). Thus,

$$I = \int \sec\theta\,d\theta = \ln|\sec\theta + \tan\theta| + C$$

$$= \ln\left|\frac{x}{a} + \frac{\sqrt{x^2 - a^2}}{a}\right| + C = \ln|x + \sqrt{x^2 - a^2}| + C_1,$$

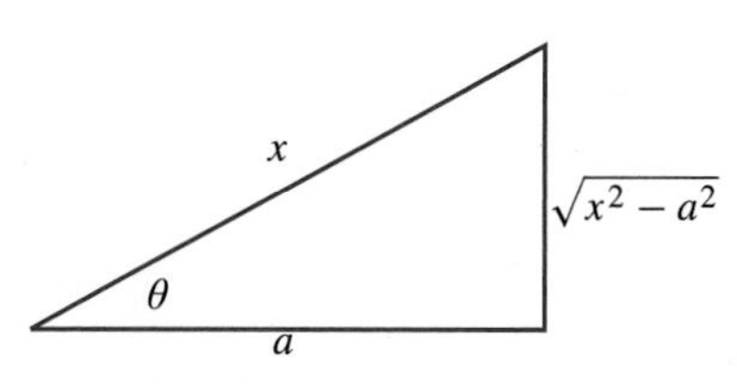

Figure 6.7

where $C_1 = C - \ln a$. If $x \le -a$, let $u = -x$ so that $u \ge a$ and $du = -dx$. We have

$$\begin{aligned} I &= -\int \frac{du}{\sqrt{u^2-a^2}} = -\ln|u+\sqrt{u^2-a^2}| + C_1 \\ &= \ln\left|\frac{1}{-x+\sqrt{x^2-a^2}}\,\frac{x+\sqrt{x^2-a^2}}{x+\sqrt{x^2-a^2}}\right| + C_1 \\ &= \ln\left|\frac{x+\sqrt{x^2-a^2}}{-a^2}\right| + C_1 = \ln|x+\sqrt{x^2-a^2}| + C_2, \end{aligned}$$

where $C_2 = C_1 - 2\ln a$. Thus, in either case, we have

$$I = \ln|x+\sqrt{x^2-a^2}| + C.$$

The following example requires the technique of completing the square as presented in Section 6.2.

EXAMPLE 5 Evaluate (a) $\displaystyle\int \frac{1}{\sqrt{2x-x^2}}\,dx$ and (b) $\displaystyle\int \frac{x}{4x^2+12x+13}\,dx$.

Solution

(a)
$$\begin{aligned} \int \frac{1}{\sqrt{2x-x^2}}\,dx &= \int \frac{dx}{\sqrt{1-(1-2x+x^2)}} \\ &= \int \frac{dx}{\sqrt{1-(x-1)^2}} \qquad \text{Let } u = x-1,\ du = dx \\ &= \int \frac{du}{\sqrt{1-u^2}} \\ &= \sin^{-1} u + C = \sin^{-1}(x-1) + C. \end{aligned}$$

(b)
$$\begin{aligned} \int \frac{x}{4x^2+12x+13}\,dx &= \int \frac{x\,dx}{4\left(x^2+3x+\frac{9}{4}+1\right)} \\ &= \frac{1}{4}\int \frac{x\,dx}{\left(x+\frac{3}{2}\right)^2+1} \qquad \text{Let } u = x+(3/2),\ du = dx,\ x = u-(3/2) \\ &= \frac{1}{4}\int \frac{u\,du}{u^2+1} - \frac{3}{8}\int \frac{du}{u^2+1} \qquad \text{In the first integral let } v = u^2+1,\ dv = 2u\,du \\ &= \frac{1}{8}\int \frac{dv}{v} - \frac{3}{8}\tan^{-1} u \\ &= \frac{1}{8}\ln|v| - \frac{3}{8}\tan^{-1} u + C \\ &= \frac{1}{8}\ln(4x^2+12x+13) - \frac{3}{8}\tan^{-1}\left(x+\frac{3}{2}\right) + C_1, \end{aligned}$$

where $C_1 = C - (\ln 4)/8$.

Inverse Hyperbolic Substitutions

As an alternative to the inverse secant substitution $x = a\sec\theta$ to simplify integrals involving $\sqrt{x^2-a^2}$ (where $x \ge a > 0$) we can use the inverse hyperbolic cosine substitution $x = a\cosh u$. Since $cosh^2 u - 1 = \sinh^2 u$, this substution produces $\sqrt{x^2-a^2} = a\sinh u$. To express u in terms of x, we need the result, noted in Section 3.6,

$$\cosh^{-1} x = \ln\left(x+\sqrt{x^2-1}\right), \qquad x \ge 1.$$

To illustrate, we redo Example 4 using the inverse hyperbolic cosine substitution.

EXAMPLE 6 Find $I = \int \frac{dx}{\sqrt{x^2 - a^2}}$, where $a > 0$.

Solution Again we assume $x \geq a$. (The case where $x \leq -a$ can be handled similarly.) Using the substitution $x = a \cosh u$, so that $dx = a \sinh u\, du$, we have

$$\begin{aligned} I &= \int \frac{a \sinh u}{a \sinh u}\, du = \int du = U + C \\ &= \cosh^{-1} \frac{x}{a} + C = \ln\left(\frac{x}{a} + \sqrt{\frac{x^2}{a^2} - 1}\right) + C \\ &= \ln\left(x + \sqrt{x^2 - a^2}\right) + C_1 \qquad \text{(where } C_1 = C - \ln a\text{)} \end{aligned}$$

Similarly, the inverse hyperbolic substitution $x = a \sinh u$ can be used instead of the inverse tangent substitution $x = a \tan\theta$ to simplify integrals involving $\sqrt{x^2 + a^2}$ or $\frac{1}{x^2 + a^2}$. In this case we have $dx = a \cosh u\, du$ and $x^2 + a^2 = a^2 \cosh^2 u$, and we may need the result

$$\sinh^{-1} x = \ln\left(x + \sqrt{x^2 + 1}\right)$$

valid for all x and proved in Section 3.6.

EXAMPLE 7 Evaluate $I = \int_0^4 \frac{dx}{(x^2 + 9)^{3/2}}$.

Solution We use the inverse substitution $x = 3 \sinh u$, so that $dx = 3 \cosh u\, du$ and $x^2 + 9 = 9 \cosh^2 u$. We have

$$\begin{aligned} I &= \int_{x=0}^{x=4} \frac{3 \cosh u}{27 \cosh^3 u}\, du = \frac{1}{9} \int_{x=0}^{x=4} \operatorname{sech}^2 u\, du = \frac{1}{9} \tanh u \Bigg|_{x=0}^{x=4} \\ &= \frac{1}{9} \frac{\sinh u}{\cosh u} \Bigg|_{x=0}^{x=4} = \frac{1}{9} \frac{x/3}{(\sqrt{x^2 + 9})/3} \Bigg|_0^4 = \frac{1}{9} \times \frac{4}{5} = \frac{4}{45}. \end{aligned}$$

Integrals involving $\sqrt{a^2 - x^2}$, where $|x| \leq a$, can be attempted with the aid of the inverse hyperbolic substitution $x = a \tanh u$, making use of the identity $1 - \tanh^2 u = \operatorname{sech}^2 u$. However, it is usually better to use the inverse sine substitution $x = a \sin\theta$ for such integrals. In general, it is better to avoid the inverse trigonometric substitutions unless you are very familiar with the identities satisfied by the hyperbolic functions as presented in Section 3.6.

Other Inverse Substitutions

Integrals involving $\sqrt{ax + b}$ can sometimes be made simpler with the substitution $ax + b = u^2$.

EXAMPLE 8

$$\int \frac{1}{1+\sqrt{2x}}\,dx \qquad \text{Let } 2x = u^2,\quad 2\,dx = 2u\,du$$

$$= \int \frac{u}{1+u}\,du$$

$$= \int \frac{1+u-1}{1+u}\,du$$

$$= \int \left(1 - \frac{1}{1+u}\right) du \qquad \text{Let } v = 1+u,\quad dv = du$$

$$= u - \int \frac{dv}{v} = u - \ln|v| + C$$

$$= \sqrt{2x} - \ln\left(1+\sqrt{2x}\right) + C$$

Sometimes integrals involving $\sqrt[n]{ax+b}$ will be much simplified by the hybrid substitution $ax+b = u^n$, $a\,dx = n\,u^{n-1}\,du$.

EXAMPLE 9

$$\int_{-1/3}^{2} \frac{x}{\sqrt[3]{3x+2}}\,dx \qquad \text{Let } 3x+2 = u^3,\quad 3\,dx = 3u^2\,du$$

$$= \int_1^2 \frac{u^3-2}{3u}\,u^2\,du$$

$$= \frac{1}{3}\int_1^2 (u^4 - 2u)\,du = \frac{1}{3}\left(\frac{u^5}{5} - u^2\right)\Bigg|_1^2 = \frac{16}{15}.$$

Note that the limits were changed in this definite integral. $u = 1$ when $x = -1/3$, and, coincidentally, $u = 2$ when $x = 2$.

If more than one fractional power is present, it may be possible to eliminate all of them at once.

EXAMPLE 10 Evaluate $\displaystyle\int \frac{1}{x^{1/2}(1+x^{1/3})}\,dx$.

Solution We can eliminate both the square root and the cube root by using the inverse substitution $x = u^6$. (The power 6 is chosen because 6 is the least common multiple of 2 and 3.)

$$\int \frac{dx}{x^{1/2}(1+x^{1/3})} \qquad \text{Let } x = u^6,\quad dx = 6u^5\,du$$

$$= 6\int \frac{u^5\,du}{u^3(1+u^2)} = 6\int \frac{u^2}{1+u^2}\,du = 6\int \left(1 - \frac{1}{1+u^2}\right) du$$

$$= 6\,(u - \tan^{-1} u) + C = 6\,(x^{1/6} - \tan^{-1} x^{1/6}) + C.$$

The tan(θ/2) Substitution

There is a certain special substitution that can transform an integral whose integrand is a rational function of $\sin\theta$ and $\cos\theta$ (i.e., a quotient of polynomials in $\sin\theta$ and $\cos\theta$) into a rational function of x. The substitution is

$$x = \tan\frac{\theta}{2} \quad \text{or, equivalently,} \quad \theta = 2\tan^{-1} x.$$

Observe that

$$\cos^2\frac{\theta}{2} = \frac{1}{\sec^2\dfrac{\theta}{2}} = \frac{1}{1+\tan^2\dfrac{\theta}{2}} = \frac{1}{1+x^2},$$

so

$$\cos\theta = 2\cos^2\frac{\theta}{2} - 1 = \frac{2}{1+x^2} - 1 = \frac{1-x^2}{1+x^2}$$

$$\sin\theta = 2\sin\frac{\theta}{2}\cos\frac{\theta}{2} = 2\tan\frac{\theta}{2}\cos^2\frac{\theta}{2} = \frac{2x}{1+x^2}.$$

Also, $dx = \frac{1}{2}\sec^2\frac{\theta}{2}\,d\theta$, so

$$d\theta = 2\cos^2\frac{\theta}{2}\,dx = \frac{2\,dx}{1+x^2}.$$

In summary:

The tan(θ/2) substitution

If $x = \tan(\theta/2)$, then

$$\cos\theta = \frac{1-x^2}{1+x^2}, \qquad \sin\theta = \frac{2x}{1+x^2}, \qquad \text{and} \qquad d\theta = \frac{2\,dx}{1+x^2}.$$

Note that $\cos\theta$, $\sin\theta$, and $d\theta$ all involve only rational functions of x. We examined general techniques for integrating rational functions of x in Section 6.2.

EXAMPLE 11 $\displaystyle\int \frac{1}{2+\cos\theta}\,d\theta$

Let $x = \tan(\theta/2)$, so

$$\cos\theta = \frac{1-x^2}{1+x^2},$$

$$d\theta = \frac{2\,dx}{1+x^2}$$

$$= \int \frac{\dfrac{2\,dx}{1+x^2}}{2+\dfrac{1-x^2}{1+x^2}} = 2\int \frac{1}{3+x^2}\,dx$$

$$= \frac{2}{\sqrt{3}}\tan^{-1}\frac{x}{\sqrt{3}} + C$$

$$= \frac{2}{\sqrt{3}}\tan^{-1}\left(\frac{1}{\sqrt{3}}\tan\frac{\theta}{2}\right) + C.$$

EXERCISES 6.3

Evaluate the integrals in Exercises 1–42.

1. $\displaystyle\int \frac{dx}{\sqrt{1-4x^2}}$

2. $\displaystyle\int \frac{x^2\,dx}{\sqrt{1-4x^2}}$

3. $\displaystyle\int \frac{x^2\,dx}{\sqrt{9-x^2}}$

4. $\displaystyle\int \frac{dx}{x\sqrt{1-4x^2}}$

5. $\displaystyle\int \frac{dx}{x^2\sqrt{9-x^2}}$

6. $\displaystyle\int \frac{dx}{x\sqrt{9-x^2}}$

7. $\displaystyle\int \frac{x+1}{\sqrt{9-x^2}}\,dx$

8. $\displaystyle\int \frac{dx}{\sqrt{9+x^2}}$

9. $\displaystyle\int \frac{x^3\,dx}{\sqrt{9+x^2}}$

10. $\displaystyle\int \frac{\sqrt{9+x^2}}{x^4}\,dx$

11. $\displaystyle\int \frac{dx}{(a^2 - x^2)^{3/2}}$

12. $\displaystyle\int \frac{dx}{(a^2 + x^2)^{3/2}}$

13. $\displaystyle\int \frac{x^2\,dx}{(a^2 - x^2)^{3/2}}$

14. $\displaystyle\int \frac{dx}{(1 + 2x^2)^{5/2}}$

15. $\displaystyle\int \frac{dx}{x\sqrt{x^2 - 4}}, \quad (x > 2)$

16. $\displaystyle\int \frac{dx}{x^2\sqrt{x^2 - a^2}} \quad (x > a > 0)$

17. $\displaystyle\int \frac{dx}{x^2 + 2x + 10}$

18. $\displaystyle\int \frac{dx}{x^2 + x + 1}$

19. $\displaystyle\int \frac{dx}{(4x^2 + 4x + 5)^2}$

20. $\displaystyle\int \frac{x\,dx}{x^2 - 2x + 3}$

21. $\displaystyle\int \frac{x\,dx}{\sqrt{2ax - x^2}}$

22. $\displaystyle\int \frac{dx}{(4x - x^2)^{3/2}}$

23. $\displaystyle\int \frac{x\,dx}{(3 - 2x - x^2)^{3/2}}$

24. $\displaystyle\int \frac{dx}{(x^2 + 2x + 2)^2}$

25. $\displaystyle\int \frac{dx}{(1 + x^2)^3}$

26. $\displaystyle\int \frac{x^2\,dx}{(1 + x^2)^2}$

■ 27. $\displaystyle\int \frac{\sqrt{1 - x^2}}{x^3}\,dx$

28. $\displaystyle\int \sqrt{9 + x^2}\,dx$

29. $\displaystyle\int \frac{dx}{2 + \sqrt{x}}$

30. $\displaystyle\int \frac{dx}{1 + x^{1/3}}$

■ 31. $\displaystyle\int \frac{1 + x^{1/2}}{1 + x^{1/3}}\,dx$

■ 32. $\displaystyle\int \frac{x\sqrt{2 - x^2}}{\sqrt{x^2 + 1}}\,dx$

33. $\displaystyle\int_{-\ln 2}^{0} e^x\sqrt{1 - e^{2x}}\,dx$

34. $\displaystyle\int_0^{\pi/2} \frac{\cos x}{\sqrt{1 + \sin^2 x}}\,dx$

35. $\displaystyle\int_{-1}^{\sqrt{3}-1} \frac{dx}{x^2 + 2x + 2}$

36. $\displaystyle\int_1^2 \frac{dx}{x^2\sqrt{9 - x^2}}$

■ 37. $\displaystyle\int \frac{t\,dt}{(t + 1)(t^2 + 1)^2}$

38. $\displaystyle\int \frac{x\,dx}{(x^2 - x + 1)^2}$

■ 39. $\displaystyle\int \frac{dx}{x(3 + x^2)\sqrt{1 - x^2}}$

■ 40. $\displaystyle\int \frac{dx}{x^2(x^2 - 1)^{3/2}}$

■ 41. $\displaystyle\int \frac{dx}{x(1 + x^2)^{3/2}}$

■ 42. $\displaystyle\int \frac{dx}{x(1 - x^2)^{3/2}}$

In Exercises 43–45, evaluate the integral using the special substitution $x = \tan(\theta/2)$ as in Example 11.

■ 43. $\displaystyle\int \frac{d\theta}{2 + \sin\theta}$

■ 44. $\displaystyle\int_0^{\pi/2} \frac{d\theta}{1 + \cos\theta + \sin\theta}$

■ 45. $\displaystyle\int \frac{d\theta}{3 + 2\cos\theta}$

46. Find the area of the region bounded by $y = (2x - x^2)^{-1/2}$, $y = 0$, $x = 1/2$, and $x = 1$.

47. Find the area of the region lying below $y = 9/(x^4 + 4x^2 + 4)$ and above $y = 1$.

48. Find the average value of the function $f(x) = (x^2 - 4x + 8)^{-3/2}$ over the interval $[0, 4]$.

49. Find the area inside the circle $x^2 + y^2 = a^2$ and above the line $y = b$, $(-a \le b \le a)$.

50. Find the area inside both of the circles $x^2 + y^2 = 1$ and $(x - 2)^2 + y^2 = 4$.

51. Find the area in the first quadrant, above the hyperbola $xy = 12$ and inside the circle $x^2 + y^2 = 25$.

52. Find the area to the left of $\dfrac{x^2}{a^2} + \dfrac{y^2}{b^2} = 1$ and to the right of the line $x = c$, where $-a \le c \le a$.

■ **53.** Find the area of the region bounded by the x-axis, the hyperbola $x^2 - y^2 = 1$, and the straight line from the origin to the point $\left(\sqrt{1 + Y^2}, Y\right)$ on that hyperbola. (Assume $Y > 0$.) In particular, show that the area is $t/2$ square units if $Y = \sinh t$.

■ **54.** Evaluate the integral $\displaystyle\int \frac{dx}{x^2\sqrt{x^2 - a^2}}$, for $x > a > 0$, using the inverse hyperbolic cosine substitution $x = a\cosh u$.

6.4 Other Methods for Evaluating Integrals

Sections 5.6 and 6.1–6.3 explore some standard methods for evaluating both definite and indefinite integrals of functions belonging to several well-defined classes. There is another such method which is often used to solve certain kinds of differential equations but which can also be helpful for evaluating integrals; after all, integrating $f(x)$ is equivalent to solving the DE $dy/dx = f(x)$. It goes by the name of the **Method of Undetermined Coefficients** or the **Method of Judicious Guessing**, and we will investigate it below.

Although anyone who uses calculus should be familiar with the basic techniques of integration, just as anyone who uses arithmetic should be familiar with the techniques of multiplication and division, technology is steadily eroding the necessity for being able to do long, complicated integrals by such methods. In fact, today there are several computer programs that can manipulate mathematical expressions symbolically (rather than just numerically) and that can carry out, with little or no assistance from us, the various algebraic steps and limit calculations that are required to calculate and simplify both derivatives and integrals. Much pain can be avoided and time saved by having the computer evaluate a complicated integral such as

$$\int \frac{1 + x + x^2}{(x^4 - 1)(x^4 - 16)^2}\, dx$$

rather than doing it by hand using partial fractions. Even without the aid of a computer, we can use tables of standard integrals such as the ones in the back endpapers of this book to help us evaluate complicated integrals. Using computers or tables can nevertheless require that we perform some simplifications beforehand and can make demands on our ability to interpret the answers we get. We also examine some such situations in this section.

The Method of Undetermined Coefficients

The method consists of guessing a family of functions that may contain the integral, then using differentiation to select the member of the family with the derivative that matches the integrand. It should be stressed that both people and machines are able to calculate derivatives with fewer complications than are involved in calculating integrals.

The method of undetermined coefficients is not so much a method as a strategy, because the family might be chosen on little more than an informed guess. But other integration methods can involve guesswork too. There can be some guesswork, for example, in deciding which integration technique will work best. What technique is best can remain unclear even after considerable effort has been expended. For undetermined coefficients, matters are clear. If the wrong family is guessed, a contradiction quickly emerges. Moreover, because of its broad nature, it provides a general alternative to other integration techniques. Often the guess is easily made. For example, if the integrand belongs to a family that remains unchanged under differentiation, then a good first guess at the form of the antiderivative is that family. A few examples will illustrate the technique.

EXAMPLE 1 Evaluate $I = \int (x^2 + x + 1)\, e^x\, dx$ using the method of undetermined coefficients.

Solution Experience tells us that the derivative of a polynomial times an exponential is a different polynomial of the same degree times the exponential. Thus, we "guess" that

$$I = (a_0 + a_1 x + a_2 x^2)\, e^x + C.$$

We differentiate I and equate the result to the integrand to determine the actual values of the coeffieients a_0, a_1, and a_2.

$$\begin{aligned}\frac{dI}{dx} &= (a_1 + 2a_2 x)\, e^x + (a_0 + a_1 x + a_2 x^2)\, e^x \\ &= \big(a_2 x^2 + (a_1 + 2a_2)x + (a_0 + a_1)\big)e^x \\ &= (x^2 + x + 1)e^x,\end{aligned}$$

provided that $a_2 = 1$, $a_1 + 2a_2 = 1$, and $a_0 + a_1 = 1$. These equations imply that $a_2 = 1$, $a_1 = -1$, and $a_0 = 2$. Thus,

$$\int (x^2 + x + 1)e^x\, dx = I = (x^2 - x + 2)e^x + C.$$

EXAMPLE 2 Evaluate $y = \int x^3 \cos(3x)\, dx$ using the method of undetermined coefficients.

Solution The derivative of a sum of products of polynomials with sine or cosine functions is a sum of products of polynomials with sine or cosine functions. Thus, we try $y = P(x)\cos(3x) + Q(x)\sin(3x) + C$, where $P(x)$ and $Q(x)$ are polynomials of degrees m and n respectively. The degrees m and n and the coefficients of the polynomials are determined by setting the derivative y' equal to the given integrand $x^3\cos(3x)$.

$$\begin{aligned} y' &= P'(x)\cos(3x) - 3P(x)\sin(3x) + Q'(x)\sin(3x) + 3Q'(x)\cos(3x) \\ &= x^3\cos 3x. \end{aligned}$$

Equating coefficients of like trigonometric functions, we find

$$P'(x) + 3Q(x) = x^3 \quad \text{and} \quad Q'(x) - 3P(x) = 0.$$

The second of these equations requires that $m = n - 1$. From the first we conclude that $n = 3$, which implies that $m = 2$. Thus, we let $P(x) = p_0 + p_1x + p_2x^2$ and $Q(x) = q_0 + q_1x + q_2x^2 + q_3x^3$ in these equations:

$$\begin{aligned} p_1 + 2p_2x + 3(q_0 + q_1x + q_2x^2 + q_3x^3) &= x^3 \\ q_1 + 2q_2x + 3q_3x^2 - 3(p_0 + p_1x + p_2x^2) &= 0. \end{aligned}$$

Comparison of coefficients with like powers yields:

$$\begin{array}{llll} p_1 + 3q_0 = 0 & 2p_2 + 3q_1 = 0 & 3q_2 = 0 & 3q_3 = 1 \\ q_1 - 3p_0 = 0 & 2q_2 - 3p_1 = 0 & 3q_3 - 3p_2 = 0, & \end{array}$$

which leads to $q_3 = 1/3$, $p_2 = 1/3$, $q_1 = -2/9$, and $p_0 = -2/27$, with $p_1 = q_0 = q_2 = 0$. Thus,

$$\int x^3\cos(3x)\,dx = y = \left(-\frac{2}{27} + \frac{x^2}{3}\right)\cos(3x) + \left(-\frac{2x}{9} + \frac{x^3}{3}\right)\sin(3x) + C.$$

EXAMPLE 3 Find the derivative of $f_{mn}(x) = x^m(\ln x)^n$ and use the result to suggest a trial formula for $I = \int x^3(\ln x)^2\,dx$. Thus evaluate this integral.

Solution We have

$$f'_{mn}(x) = mx^{m-1}(\ln x)^n + nx^m(\ln x)^{n-1}\frac{1}{x} = mx^{m-1}(\ln x)^n + nx^{m-1}(\ln x)^{n-1}.$$

This suggests that we try

$$I = \int x^3(\ln x)^2\,dx = \int f_{32}(x)\,dx = Px^4(\ln x)^2 + Qx^4\ln x + Rx^4 + C$$

for constants P, Q, R, and C. Differentiating, we get

$$\frac{dI}{dx} = 4Px^3(\ln x)^2 + 2Px^3\ln x + 4Qx^3\ln x + Qx^3 + 4Rx^3 = x^3(\ln x)^2,$$

provided $4P = 1$, $2P + 4Q = 0$, and $Q + 4R = 0$. Thus $P = 1/4$, $Q = -1/8$, and $R = 1/32$, and so

$$\int x^3(\ln x)^2\,dx = \frac{1}{4}x^4(\ln x)^2 - \frac{1}{8}x^4\ln x + \frac{1}{32}x^4 + C.$$

Remark These examples and most in the following exercises can also be done using integration by parts. Using undetermined coefficients does not replace other methods, but it does provide an alternative that gives insight into what types of functions will not work as guesses for the integral. This has implications for how computer algorithms can and cannot do antiderivatives. This issue is taken up in Exercise 20. Moreover, with access to a differentiation algorithm and a computer to manage details, this method can sometimes produce integrals more quickly and precisely than classical techniques alone.

Using Maple for Integration

Computer algebra systems are capable of evaluating both indefinite and definite integrals symbolically, as well as giving numerical approximations for those definite integrals that have numerical values. The following examples show how to use Maple to evaluate integrals.

We begin by calculating $\int 2^x\sqrt{1+4^x}\,dx$ and $\int_0^\pi 2^x\sqrt{1+4^x}\,dx$.

We use Maple's "int" command, specifying the function and the variable of integration:

```
>   int(2^x*sqrt(1+4^x),x);
```

$$\frac{e^{(x\ln(2))}\sqrt{1+(e^{(x\ln(2))})^2}}{2\ln(2)}+\frac{\operatorname{arcsinh}(e^{(x\ln(2))})}{2\ln(2)}$$

If you don't like the inverse hyperbolic sine, you can convert it to a logarithm:

```
>   convert(%,ln);
```

$$\frac{e^{(x\ln(2))}\sqrt{1+(e^{(x\ln(2))})^2}2\ln(2)+\ln\left(e^{(x\ln(2))}+\sqrt{1+(e^{(x\ln(2))})^2}\right)}{2\ln(2)}$$

The "%" there refers to the result of the previous calculation. Note how Maple prefers to use $e^{x\ln 2}$ in place of 2^x.

For the definite integral, you specify the interval of values of the variable of integration using two dots between the endpoints as follows:

```
>   int(2^x*sqrt(1+4^x),x=0..Pi);
```

$$\frac{-\sqrt{2}-\ln(1+\sqrt{2})+2^\pi\sqrt{1+4^\pi}+\ln(2^\pi+\sqrt{1+4^\pi})}{2\ln(2)}$$

If you want a decimal approximation to this exact answer, you can ask Maple to evaluate the last result as a floating point number:

```
>   evalf(%);
```

$$56.955\,421\,55$$

Remark Maple defaults to giving 10 significant digits in its floating point numbers unless you request a different precision by declaring a value for the variable "Digits":

```
>   Digits := 20; evalf(Pi);
```

$$3.141\,592\,653\,589\,793\,238\,5$$

Suppose we ask Maple to do an integral that we know we can't do ourselves:

```
>   int(exp(-x^2),x);
```

$$\frac{1}{2}\sqrt{\pi}\operatorname{erf}(x)$$

Maple expresses the answer in terms of the **error function** that is defined by

$$\operatorname{erf}(x)=\frac{2}{\sqrt{\pi}}\int_0^x e^{-t^2}\,dt.$$

But observe:

```
> Int(exp(-x^2),x=-infinity..infinity)
   = int(exp(-x^2), x=-infinity..infinity);
```

$$\int_{-\infty}^{\infty} e^{(-x^2)}\,dx = \sqrt{\pi}$$

Note the use of the *inert* Maple command "Int" on the left side to simply print the integral without any evaluation. The active command "int" performs the evaluation.

Computer algebra programs can be used to integrate symbolically many functions, but you may get some surprises when you use them, and you may have to do some of the work to get an answer useful in the context of the problem on which you are working. Such programs, and some of the more sophisticated scientific calculators, are able to evaluate definite integrals numerically to any desired degree of accuracy even if symbolic antiderivatives cannot be found. We will discuss techniques of numerical integration in Sections 6.6–6.8, but note here that Maple's `evalf(Int())` can always be used to get numerical values:

```
> evalf(Int(sin(cos(x)),x=0..1));
```

$$.738\,642\,998\,0$$

Using Integral Tables

You can get some help evaluating integrals by using an Integral Table, such as the one in the back endpapers of this book. Besides giving the values of the common elementary integrals that you likely remember while you are studying calculus, they also give many more complicated integrals, especially ones representing standard types that often arise in applications. Familiarize yourself with the main headings under which the integrals are classified. Using the tables usually means massaging your integral using simple substitutions until you get it into the form of one of the integrals in the table.

EXAMPLE 4 Use the table to evaluate $I = \displaystyle\int \frac{t^5}{\sqrt{3-2t^4}}\,dt$.

Solution This integral doesn't resemble any in the tables, but there are numerous integrals in the tables involving $\sqrt{a^2-x^2}$. We can begin to put the integral into this form with the substitution $t^2 = u$, so that $2t\,dt = du$. Thus,

$$I = \frac{1}{2}\int \frac{u^2}{\sqrt{3-2u^2}}\,du.$$

This is not quite what we want yet; let us get rid of the 2 multiplying the u^2 under the square root. One way to do this is with the change of variable $\sqrt{2}u = x$, so that $du = dx/\sqrt{2}$:

$$I = \frac{1}{4\sqrt{2}}\int \frac{x^2}{\sqrt{3-x^2}}\,dx.$$

Now the denominator is of the form $\sqrt{a^2-x^2}$ for $a = \sqrt{3}$. Looking through the part of the table (in the back endpapers) dealing with integrals involving $\sqrt{a^2-x^2}$, we find the third one, which says that

$$\int \frac{x^2}{\sqrt{a^2-x^2}}\,dx = -\frac{x}{2}\sqrt{a^2-x^2} + \frac{a^2}{2}\sin^{-1}\frac{x}{a} + C.$$

Thus,

$$\begin{aligned} I &= \frac{1}{4\sqrt{2}}\left(-\frac{x}{2}\sqrt{3-x^2} + \frac{3}{2}\sin^{-1}\frac{x}{\sqrt{3}}\right) + C_1 \\ &= -\frac{t^2}{8}\sqrt{3-2t^4} + \frac{3}{8\sqrt{2}}\sin^{-1}\frac{\sqrt{2}t^2}{\sqrt{3}} + C_1. \end{aligned}$$

Many of the integrals in the table are reduction formulas. (An integral appears on both sides of the equation.) These can be iterated to simplify integrals as in some of the examples and exercises of Section 6.1.

EXAMPLE 5 Evaluate $I = \int_0^1 \frac{1}{(x^2+1)^3}\,dx$.

Solution The fourth integral in the table of Miscellaneous Algebraic Integrals says that if $n \neq 1$, then

$$\int \frac{dx}{(a^2 \pm x^2)^n} = \frac{1}{2a^2(n-1)}\left(\frac{x}{(a^2 \pm x^2)^{n-1}} + (2n-3)\int \frac{dx}{(a^2 \pm x^2)^{n-1}}\right).$$

Using $a = 1$ and the + signs, we have

$$\begin{aligned}\int_0^1 \frac{dx}{(1+x^2)^n} &= \frac{1}{2(n-1)}\left(\frac{x}{(1+x^2)^{n-1}}\bigg|_0^1 + (2n-3)\int_0^1 \frac{dx}{(1+x^2)^{n-1}}\right)\\ &= \frac{1}{2^n(n-1)} + \frac{2n-3}{2(n-1)}\int_0^1 \frac{dx}{(1+x^2)^{n-1}}.\end{aligned}$$

Thus, we have

$$\begin{aligned}I &= \frac{1}{16} + \frac{3}{4}\int_0^1 \frac{dx}{(1+x^2)^2}\\ &= \frac{1}{16} + \frac{3}{4}\left(\frac{1}{4} + \frac{1}{2}\int_0^1 \frac{dx}{1+x^2}\right)\\ &= \frac{1}{16} + \frac{3}{16} + \frac{3}{8}\tan^{-1}x\bigg|_0^1 = \frac{1}{4} + \frac{3\pi}{32}.\end{aligned}$$

EXERCISES 6.4

In Exercises 1–4 use the method of undetermined coefficients to evaluate the given integrals.

1. $\int e^{3x}\sin(4x)\,dx$ **2.** $\int x\,e^{-x}\sin x\,dx$

3. $\int x^5 e^{-x^2}\,dx$ **4.** $\int x^2(\ln x)^4\,dx$

5. Use Maple or another computer algebra program to check any of the integrals you have done in the exercises from Sections 5.6 and 6.1–6.3, as well as any of the integrals you have been unable to do.

6. Use Maple or another computer algebra program to evaluate the integral in the opening paragraph of this section.

7. Use Maple or another computer algebra program to reevaluate the integral in Example 4.

8. Use Maple or another computer algebra program to reevaluate the integral in Example 5.

Use the integral tables to help you find the integrals in Exercises 9–18.

9. $\int \frac{x^2}{\sqrt{x^2-2}}\,dx$ **10.** $\int \sqrt{(x^2+4)^3}\,dx$

11. $\int \frac{dt}{t^2\sqrt{3t^2+5}}$ **12.** $\int \frac{dt}{t\sqrt{3t-5}}$

13. $\int x^4(\ln x)^4\,dx$ **14.** $\int x^7 e^{x^2}\,dx$

15. $\int x\sqrt{2x-x^2}\,dx$ **16.** $\int \frac{\sqrt{2x-x^2}}{x^2}\,dx$

17. $\int \frac{dx}{(\sqrt{4x-x^2})^3}$ **18.** $\int \frac{dx}{(\sqrt{4x-x^2})^4}$

19. Use Maple or another computer algebra program to evaluate the integrals in Exercises 9–18.

20. Consider the integral $I = \int e^{-x^2}\,dx$. It is known that any evaluation of the integral as a finite combination of elementary functions must take the form

$$I = \int e^{-x^2}\,dx = P(x)\,e^{-x^2} + C,$$

where $P(x)$ is a polynomial.

(a) Show that there can be no polynomial $P(x)$ for which the above formula holds.

(b) Because some elementary functions like e^{-x^2} do not possess antiderivatives that can be expressed as finite combinations of elementary functions, new nonelementary functions are defined to fulfill the need for such antiderivatives. One such function is the **error function** defined by

$$\operatorname{Erf}(x) = \frac{2}{\sqrt{\pi}} \int_0^x e^{-t^2}\, dt.$$

Express the integral $\int e^{-x^2}\, dx$ in terms of the error function.

(c) Use undetermined coefficients to evaluate $J = \int \operatorname{Erf}(x)\, dx$.

6.5 Improper Integrals

Up to this point, we have considered definite integrals of the form

$$I = \int_a^b f(x)\, dx,$$

where the integrand f is *continuous* on the *closed, finite* interval $[a, b]$. Since such a function is necessarily *bounded*, the integral I is necessarily a finite number; for positive f it corresponds to the area of a **bounded region** of the plane, a region contained inside some disk of finite radius with centre at the origin. Such integrals are also called **proper integrals**. We are now going to generalize the definite integral to allow for two possibilities excluded in the situation described above:

(i) We may have $a = -\infty$ or $b = \infty$ or both.

(ii) f may be unbounded as x approaches a or b or both.

Integrals satisfying (i) are called **improper integrals of type I**; integrals satisfying (ii) are called **improper integrals of type II**. Either type of improper integral corresponds (for positive f) to the area of a region in the plane that "extends to infinity" in some direction and therefore is *unbounded*. As we will see, such integrals may or may not have finite values. The ideas involved are best introduced by examples.

Improper Integrals of Type I

EXAMPLE 1 Find the area of the region A lying under the curve $y = 1/x^2$ and above the x-axis to the right of $x = 1$. (See Figure 6.8(a).)

Solution We would like to calculate the area with an integral

$$A = \int_1^\infty \frac{dx}{x^2},$$

which is improper of type I, since its interval of integration is infinite. It is not immediately obvious whether the area is finite; the region has an infinitely long "spike" along the x-axis, but this spike becomes infinitely thin as x approaches ∞. In order to evaluate this improper integral, we interpret it as a limit of proper integrals over intervals $[1, R]$ as $R \to \infty$. (See Figure 6.8(b).)

$$A = \int_1^\infty \frac{dx}{x^2} = \lim_{R\to\infty} \int_1^R \frac{dx}{x^2} = \lim_{R\to\infty} \left(-\frac{1}{x}\right)\Bigg|_1^R$$
$$= \lim_{R\to\infty} \left(-\frac{1}{R} + 1\right) = 1$$

Since the limit exists (is finite), we say that the improper integral *converges*. The region has finite area $A = 1$ square unit.

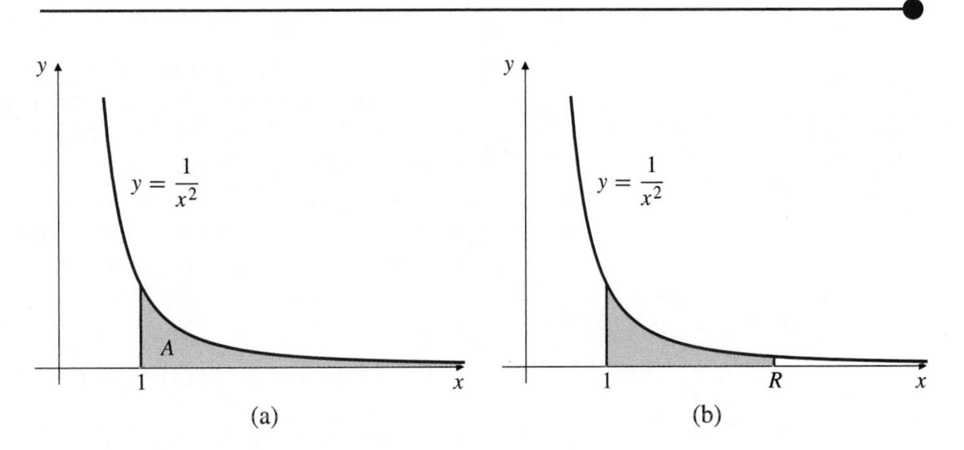

Figure 6.8

(a) $A = \int_1^\infty \frac{1}{x^2}\,dx$

(b) $A = \lim_{R\to\infty} \int_1^R \frac{1}{x^2}\,dx$

EXAMPLE 2 Find the area of the region under $y = 1/x$, above $y = 0$, and to the right of $x = 1$. (See Figure 6.9.)

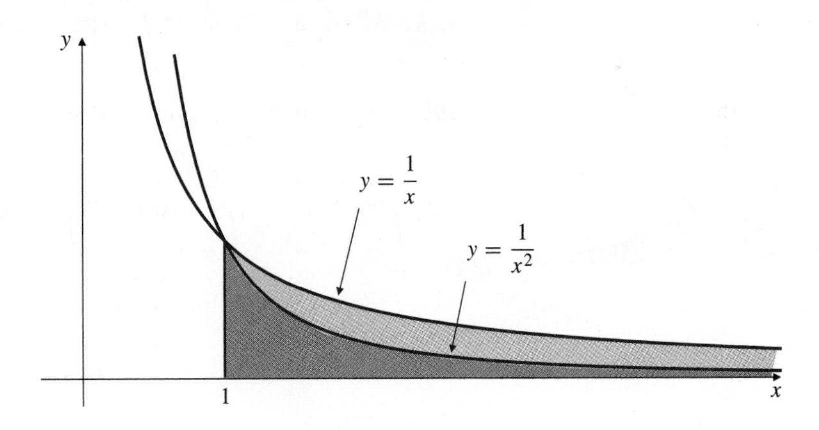

Figure 6.9 The area shaded in colour is infinite

Solution This area is given by the improper integral

$$A = \int_1^\infty \frac{dx}{x} = \lim_{R\to\infty} \int_1^R \frac{dx}{x} = \lim_{R\to\infty} \ln x\Bigg|_1^R = \lim_{R\to\infty} \ln R = \infty.$$

We say that this improper integral *diverges to infinity*. Observe that the region has a similar shape to the region under $y = 1/x^2$ considered in the above example, but its "spike" is somewhat thicker at each value of $x > 1$. Evidently, the extra thickness makes a big difference; this region has *infinite* area.

DEFINITION 1

Improper integrals of type I

If f is continuous on $[a, \infty)$, we define the improper integral of f over $[a, \infty)$ as a limit of proper integrals:

$$\int_a^\infty f(x)\,dx = \lim_{R\to\infty} \int_a^R f(x)\,dx.$$

Similarly, if f is continuous on $(-\infty, b]$, then we define

$$\int_{-\infty}^b f(x)\,dx = \lim_{R\to-\infty} \int_R^b f(x)\,dx.$$

In either case, if the limit exists (is a finite number), we say that the improper integral **converges**; if the limit does not exist, we say that the improper integral **diverges**. If the limit is ∞ (or $-\infty$), we say the improper integral **diverges to infinity** (or **diverges to negative infinity**).

The integral $\int_{-\infty}^\infty f(x)\,dx$ is, for f continuous on the real line, improper of type I at both endpoints. We break it into two separate integrals:

$$\int_{-\infty}^\infty f(x)\,dx = \int_{-\infty}^0 f(x)\,dx + \int_0^\infty f(x)\,dx.$$

The integral on the left converges if and only if *both* integrals on the right converge.

EXAMPLE 3 Evaluate $\displaystyle\int_{-\infty}^\infty \frac{1}{1+x^2}\,dx.$

Solution By the (even) symmetry of the integrand (see Figure 6.10), we have

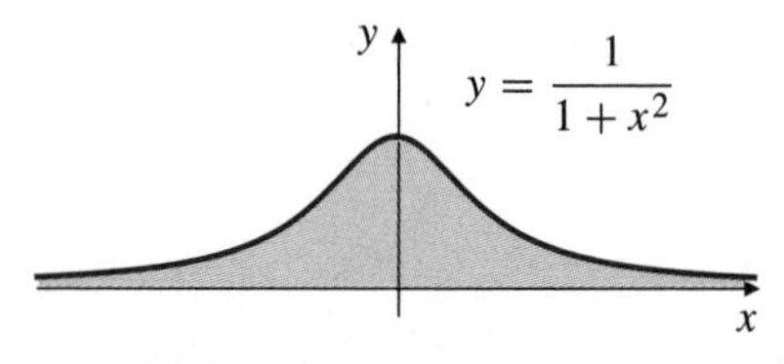

Figure 6.10

$$\begin{aligned}\int_{-\infty}^\infty \frac{dx}{1+x^2} &= \int_{-\infty}^0 \frac{dx}{1+x^2} + \int_0^\infty \frac{dx}{1+x^2} \\ &= 2\lim_{R\to\infty}\int_0^R \frac{dx}{1+x^2} \\ &= 2\lim_{R\to\infty}\tan^{-1}R = 2\left(\frac{\pi}{2}\right) = \pi.\end{aligned}$$

The use of symmetry here requires some justification. At the time we used it we did not know whether each of the half-line integrals was finite or infinite. However, since both are positive, even if they are infinite, their sum would still be twice one of them. If one had been positive and the other negative, we would not have been justified in cancelling them to get 0 until we knew that they were finite. ($\infty+\infty=\infty$, but $\infty-\infty$ is not defined.) In any event, the given integral converges to π.

EXAMPLE 4 $\displaystyle\int_0^\infty \cos x\,dx = \lim_{R\to\infty}\int_0^R \cos x\,dx = \lim_{R\to\infty}\sin R.$

This limit does not exist (and it is not ∞ or $-\infty$), so all we can say is that the given integral diverges. (See Figure 6.11.) As R increases, the integral alternately adds and subtracts the areas of the hills and valleys but does not approach any unique limit.

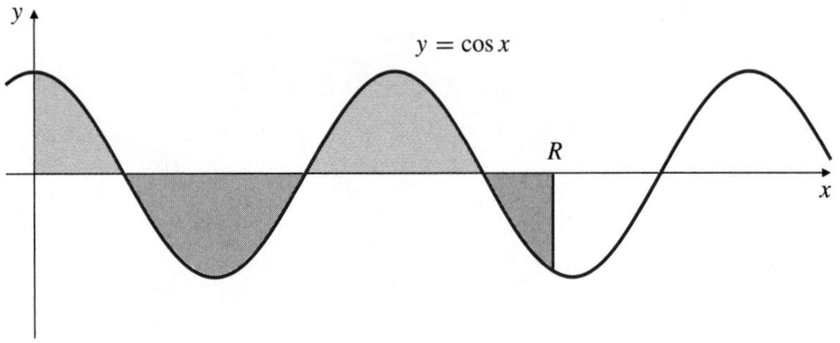

Figure 6.11 Not every divergent improper integral diverges to ∞ or $-\infty$

Improper Integrals of Type II

DEFINITION 2

Improper integrals of type II

If f is continuous on the interval $(a, b]$ and is possibly unbounded near a, we define the improper integral

$$\int_a^b f(x)\,dx = \lim_{c\to a+} \int_c^b f(x)\,dx.$$

Similarly, if f is continuous on $[a, b)$ and is possibly unbounded near b, we define

$$\int_a^b f(x)\,dx = \lim_{c\to b-} \int_a^c f(x)\,dx.$$

These improper integrals may converge, diverge, diverge to infinity, or diverge to negative infinity.

EXAMPLE 5 Find the area of the region S lying under $y = 1/\sqrt{x}$, above the x-axis, between $x = 0$ and $x = 1$.

Solution The area A is given by

$$A = \int_0^1 \frac{1}{\sqrt{x}}\,dx,$$

which is an improper integral of type II since the integrand is unbounded near $x = 0$. The region S has a "spike" extending to infinity along the y-axis, a vertical asymptote of the integrand, as shown in Figure 6.12. As we did for improper integrals of type I, we express such integrals as limits of proper integrals:

$$A = \lim_{c\to 0+} \int_c^1 x^{-1/2}\,dx = \lim_{c\to 0+} 2x^{1/2}\Big|_c^1 = \lim_{c\to 0+} (2 - 2\sqrt{c}) = 2.$$

This integral converges, and S has a finite area of 2 square units.

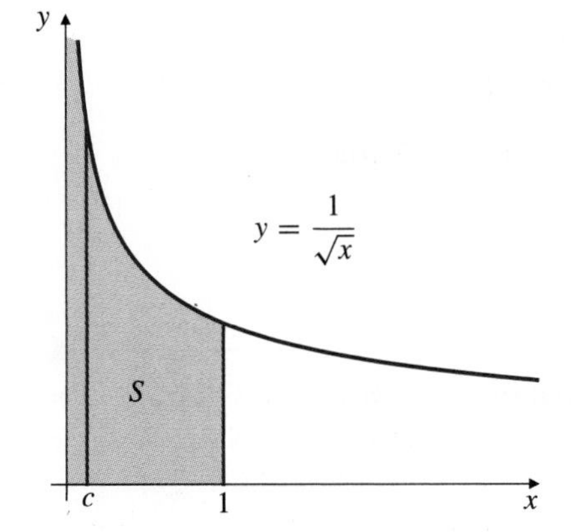

Figure 6.12 The shaded area is finite

While improper integrals of type I are always easily recognized because of the infinite limits of integration, improper integrals of type II can be somewhat harder to spot. You should be alert for singularities of integrands and especially points where they have vertical asymptotes. It may be necessary to break an improper integral into several improper integrals if it is improper at both endpoints or at points inside the interval of integration. For example,

$$\int_{-1}^1 \frac{\ln|x|\,dx}{\sqrt{1-x}} = \int_{-1}^0 \frac{\ln|x|\,dx}{\sqrt{1-x}} + \int_0^{1/2} \frac{\ln|x|\,dx}{\sqrt{1-x}} + \int_{1/2}^1 \frac{\ln|x|\,dx}{\sqrt{1-x}}.$$

Each integral on the right is improper because of a singularity at one endpoint.

EXAMPLE 6 Evaluate each of the following integrals or show that it diverges:

$$\text{(a)} \int_0^1 \frac{1}{x}\,dx, \qquad \text{(b)} \int_0^2 \frac{1}{\sqrt{2x-x^2}}\,dx, \qquad \text{and} \qquad \text{(c)} \int_0^1 \ln x\,dx.$$

Solution

(a) $\displaystyle\int_0^1 \frac{1}{x}\,dx = \lim_{c\to 0+}\int_c^1 \frac{1}{x}\,dx = \lim_{c\to 0+}(\ln 1 - \ln c) = \infty.$

This integral diverges to infinity.

(b)
$$\begin{aligned}
\int_0^2 \frac{1}{\sqrt{2x-x^2}}\,dx &= \int_0^2 \frac{1}{\sqrt{1-(x-1)^2}}\,dx && \text{Let } u = x-1,\ du = dx \\
&= \int_{-1}^1 \frac{1}{\sqrt{1-u^2}}\,du \\
&= 2\int_0^1 \frac{1}{\sqrt{1-u^2}}\,du && \text{(by symmetry)} \\
&= 2\lim_{c\to 1-}\int_0^c \frac{1}{\sqrt{1-u^2}}\,du \\
&= 2\lim_{c\to 1-}\sin^{-1}u\Big|_0^c = 2\lim_{c\to 1-}\sin^{-1}c = \pi.
\end{aligned}$$

This integral converges to π. Observe how a change of variable can be made even before an improper integral is expressed as a limit of proper integrals.

(c)
$$\begin{aligned}
\int_0^1 \ln x\,dx &= \lim_{c\to 0+}\int_c^1 \ln x\,dx && \text{(See Example 2(a) of Section 6.1 for the evaluation of the indefinite integral.)} \\
&= \lim_{c\to 0+}(x\ln x - x)\Big|_c^1 \\
&= \lim_{c\to 0+}(0 - 1 - c\ln c + c) \\
&= -1 + 0 - \lim_{c\to 0+}\frac{\ln c}{1/c} && \left[\frac{-\infty}{\infty}\right] \\
&= -1 - \lim_{c\to 0+}\frac{1/c}{-(1/c^2)} && \text{(by l'Hôpital's Rule)} \\
&= -1 - \lim_{c\to 0+}(-c) = -1 + 0 = -1.
\end{aligned}$$

The integral converges to -1.

The following theorem summarizes the behaviour of improper integrals of types I and II for powers of x.

THEOREM 2 ***p*-integrals**

If $0 < a < \infty$, then

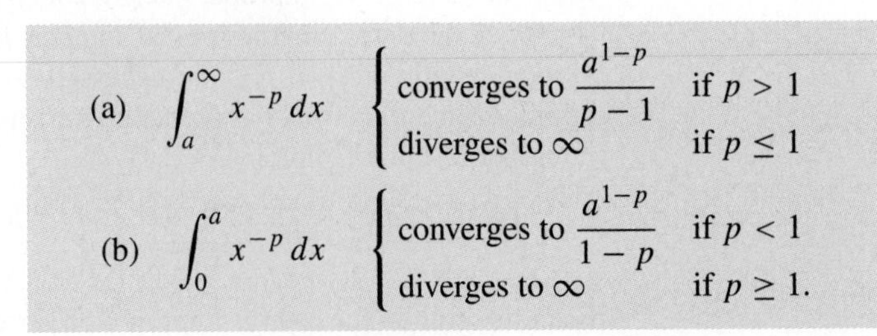

$$\text{(a)} \int_a^\infty x^{-p}\,dx \quad \begin{cases} \text{converges to } \dfrac{a^{1-p}}{p-1} & \text{if } p > 1 \\ \text{diverges to } \infty & \text{if } p \le 1 \end{cases}$$

$$\text{(b)} \int_0^a x^{-p}\,dx \quad \begin{cases} \text{converges to } \dfrac{a^{1-p}}{1-p} & \text{if } p < 1 \\ \text{diverges to } \infty & \text{if } p \ge 1. \end{cases}$$

PROOF We prove part (b) only. The proof of part (a) is similar and is left as an exercise. Also, the case $p = 1$ of part (b) is similar to Example 6(a) above, so we need consider only the cases $p < 1$ and $p > 1$. If $p < 1$, then we have

$$\begin{aligned}\int_0^a x^{-p}\,dx &= \lim_{c\to 0+}\int_c^a x^{-p}\,dx \\ &= \lim_{c\to 0+}\left.\frac{x^{-p+1}}{-p+1}\right|_c^a \\ &= \lim_{c\to 0+}\frac{a^{1-p}-c^{1-p}}{1-p} = \frac{a^{1-p}}{1-p}\end{aligned}$$

because $1 - p > 0$. If $p > 1$, then

$$\begin{aligned}\int_0^a x^{-p}\,dx &= \lim_{c\to 0+}\int_c^a x^{-p}\,dx \\ &= \lim_{c\to 0+}\left.\frac{x^{-p+1}}{-p+1}\right|_c^a \\ &= \lim_{c\to 0+}\frac{c^{-(p-1)}-a^{-(p-1)}}{p-1} = \infty.\end{aligned}$$

The integrals in Theorem 2 are called ***p*-integrals**. It is very useful to know when they converge and diverge when you have to decide whether certain other improper integrals converge or not and you can't find the appropriate antiderivatives. (See the discussion of estimating convergence below.) Note that $\int_0^\infty x^{-p}\,dx$ does not converge for any value of p.

Remark If f is continuous on the interval $[a, b]$ so that $\int_a^b f(x)\,dx$ is a proper definite integral, then treating the integral as improper will lead to the same value:

$$\lim_{c\to a+}\int_c^b f(x)\,dx = \int_a^b f(x)\,dx = \lim_{c\to b-}\int_a^c f(x)\,dx.$$

This justifies the definition of the definite integral of a piecewise continuous function that was given in Section 5.4. To integrate a function defined to be different continuous functions on different intervals, we merely add the integrals of the various component functions over their respective intervals. Any of these integrals may be proper or improper; if any are improper, all must converge or the given integral will diverge.

EXAMPLE 7 Evaluate $\displaystyle\int_0^2 f(x)\,dx$, where $f(x) = \begin{cases} 1/\sqrt{x} & \text{if } 0 < x \le 1 \\ x - 1 & \text{if } 1 < x \le 2. \end{cases}$

Solution The graph of f is shown in Figure 6.13. We have

$$\begin{aligned}\int_0^2 f(x)\,dx &= \int_0^1 \frac{dx}{\sqrt{x}} + \int_1^2 (x-1)\,dx \\ &= \lim_{c\to 0+}\int_c^1 \frac{dx}{\sqrt{x}} + \left.\left(\frac{x^2}{2} - x\right)\right|_1^2 = 2 + \left(2 - 2 - \frac{1}{2} + 1\right) = \frac{5}{2};\end{aligned}$$

the first integral on the right is improper but convergent (see Example 5 above), and the second is proper.

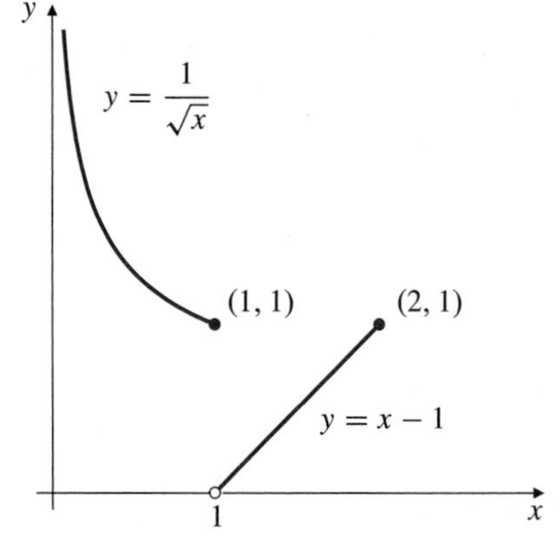

Figure 6.13 A discontinuous function

Estimating Convergence and Divergence

When an improper integral cannot be evaluated by the Fundamental Theorem of Calculus because an antiderivative can't be found, we may still be able to determine whether the integral converges by comparing it with simpler integrals. The following theorem is central to this approach.

THEOREM 3 **A comparison theorem for integrals**

Let $-\infty \le a < b \le \infty$, and suppose that functions f and g are continuous on the interval (a, b) and satisfy $0 \le f(x) \le g(x)$. If $\int_a^b g(x)\,dx$ converges, then so does $\int_a^b f(x)\,dx$, and

$$\int_a^b f(x)\,dx \le \int_a^b g(x)\,dx.$$

Equivalently, if $\int_a^b f(x)\,dx$ diverges to ∞, then so does $\int_a^b g(x)\,dx$.

PROOF Since both integrands are nonnegative, there are only two possibilities for each integral: it can either converge to a nonnegative number or diverge to ∞. Since $f(x) \le g(x)$ on (a, b), it follows by Theorem 3(e) of Section 5.4 that if $a < r < s < b$, then

$$\int_r^s f(x)\,dx \le \int_r^s g(x)\,dx.$$

This theorem now follows by taking limits as $r \to a+$ and $s \to b-$.

EXAMPLE 8 Show that $\displaystyle\int_0^\infty e^{-x^2}\,dx$ converges, and find an upper bound for its value.

Solution We can't integrate e^{-x^2}, but we can integrate e^{-x}. We would like to use the inequality $e^{-x^2} \le e^{-x}$, but this is only valid for $x \ge 1$. (See Figure 6.14.) Therefore, we break the integral into two parts.

On $[0, 1]$ we have $0 < e^{-x^2} \le 1$, so

$$0 < \int_0^1 e^{-x^2}\,dx \le \int_0^1 dx = 1.$$

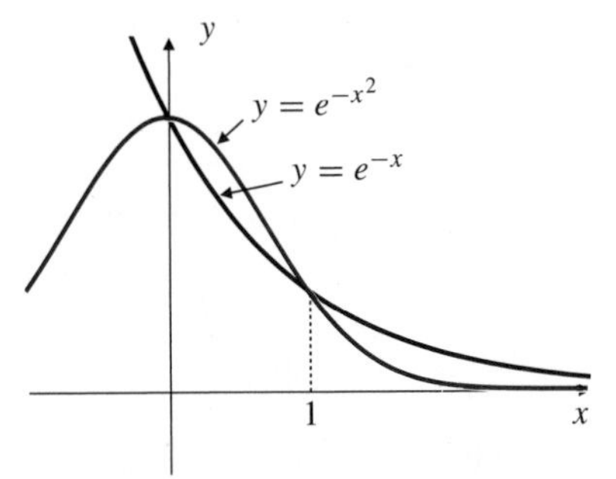

Figure 6.14 Comparing e^{-x^2} and e^{-x}

On $[1, \infty)$ we have $x^2 \ge x$, so $-x^2 \le -x$ and $0 < e^{-x^2} \le e^{-x}$. Thus,

$$\begin{aligned} 0 < \int_1^\infty e^{-x^2}\,dx \le \int_1^\infty e^{-x}\,dx &= \lim_{R\to\infty} \frac{e^{-x}}{-1}\bigg|_1^R \\ &= \lim_{R\to\infty}\left(\frac{1}{e} - \frac{1}{e^R}\right) = \frac{1}{e}. \end{aligned}$$

Hence, $\displaystyle\int_0^\infty e^{-x^2}\,dx$ converges and its value is not greater than $1 + (1/e)$.

We remark that the above integral is, in fact, equal to $\frac{1}{2}\sqrt{\pi}$, although we cannot prove this now. See Section 14.4.

For large or small values of x many integrands behave like powers of x. If so, they can be compared with p-integrals.

EXAMPLE 9 Determine whether $\displaystyle\int_0^\infty \frac{dx}{\sqrt{x + x^3}}$ converges.

Solution The integral is improper of both types, so we write

$$\int_0^\infty \frac{dx}{\sqrt{x + x^3}} = \int_0^1 \frac{dx}{\sqrt{x + x^3}} + \int_1^\infty \frac{dx}{\sqrt{x + x^3}} = I_1 + I_2.$$

On $(0, 1]$ we have $\sqrt{x + x^3} > \sqrt{x}$, so

$$I_1 < \int_0^1 \frac{dx}{\sqrt{x}} = 2 \qquad \text{(by Theorem 2).}$$

On $[1, \infty)$ we have $\sqrt{x + x^3} > \sqrt{x^3}$, so

$$I_2 < \int_1^\infty x^{-3/2}\, dx = 2 \qquad \text{(by Theorem 2).}$$

Hence, the given integral converges, and its value is less than 4.

In Section 4.10 we introduced big-O notation as a way of conveying growth-rate information in limit situations. We wrote $f(x) = O\big(g(x)\big)$ as $x \to a$ to mean the same thing as $|f(x)| \le K|g(x)|$ for some constant K on some open interval containing a. Similarly, we can say that $f(x) = O\big(g(x)\big)$ as $x \to \infty$ if for some constants a and K we have $|f(x)| \le K|g(x)|$ for all $x \ge a$.

EXAMPLE 10 $\dfrac{1 + x^2}{1 + x^4} = O\left(\dfrac{1}{x^2}\right)$ as $x \to \infty$ because, for $x \ge 1$ we have

$$\left|\frac{1 + x^2}{1 + x^4}\right| < \frac{2x^2}{x^4} = \frac{2}{x^2}.$$

EXAMPLE 11 Show that if $p > 1$ and f is continuous on $[1, \infty)$ and satisfies $f(x) = O(x^{-p})$, then $\int_1^\infty f(x)\, dx$ converges, and the error $E(R)$ in the approximation

$$\int_1^\infty f(x)\, dx \approx \int_1^R f(x)\, dx$$

satisfies $E(R) = O(R^{1-p})$ as $R \to \infty$.

Solution Since $f(x) = O(x^{-p})$ as $x \to \infty$, we have, for some $a \ge 1$ and some K, $f(x) \le K x^{-p}$ for all $x \ge a$. Thus,

$$\begin{aligned} |E(R)| &= \left|\int_R^\infty f(x)\, dx\right| \\ &\le K \int_R^\infty x^{-p}\, dx = K \frac{x^{-p+1}}{-p+1}\bigg|_R^\infty = \frac{K}{p-1} R^{1-p}, \end{aligned}$$

so $E(R) = O(R^{1-p})$ as $R \to \infty$.

EXERCISES 6.5

In Exercises 1–22, evaluate the given integral or show that it diverges.

1. $\displaystyle\int_2^\infty \frac{1}{(x-1)^3}\, dx$

2. $\displaystyle\int_3^\infty \frac{1}{(2x-1)^{2/3}}\, dx$

3. $\displaystyle\int_0^\infty e^{-2x}\, dx$

4. $\displaystyle\int_{-\infty}^{-1} \frac{dx}{x^2+1}$

5. $\displaystyle\int_{-1}^1 \frac{dx}{(x+1)^{2/3}}$

6. $\displaystyle\int_0^a \frac{dx}{a^2 - x^2}$

7. $\displaystyle\int_0^1 \frac{1}{(1-x)^{1/3}}\,dx$ **8.** $\displaystyle\int_0^1 \frac{1}{x\sqrt{1-x}}\,dx$

9. $\displaystyle\int_0^{\pi/2} \frac{\cos x\,dx}{(1-\sin x)^{2/3}}$ **10.** $\displaystyle\int_0^{\infty} x\,e^{-x}\,dx$

11. $\displaystyle\int_0^1 \frac{dx}{\sqrt{x(1-x)}}$ **12.** $\displaystyle\int_0^{\infty} \frac{x}{1+2x^2}\,dx$

13. $\displaystyle\int_0^{\infty} \frac{x\,dx}{(1+2x^2)^{3/2}}$ **14.** $\displaystyle\int_0^{\pi/2} \sec x\,dx$

15. $\displaystyle\int_0^{\pi/2} \tan x\,dx$ **16.** $\displaystyle\int_e^{\infty} \frac{dx}{x\ln x}$

17. $\displaystyle\int_1^{e} \frac{dx}{x\sqrt{\ln x}}$ **18.** $\displaystyle\int_e^{\infty} \frac{dx}{x(\ln x)^2}$

19. $\displaystyle\int_{-\infty}^{\infty} \frac{x}{1+x^2}\,dx$ **20.** $\displaystyle\int_{-\infty}^{\infty} \frac{x}{1+x^4}\,dx$

21. $\displaystyle\int_{-\infty}^{\infty} x\,e^{-x^2}\,dx$ **22.** $\displaystyle\int_{-\infty}^{\infty} e^{-|x|}\,dx$

23. Find the area below $y=0$, above $y=\ln x$, and to the right of $x=0$.

24. Find the area below $y=e^{-x}$, above $y=e^{-2x}$, and to the right of $x=0$.

25. Find the area of a region that lies above $y=0$, to the right of $x=1$, and under the curve $y=\dfrac{4}{2x+1}-\dfrac{2}{x+2}$.

26. Find the area of the plane region that lies under the graph of $y=x^{-2}e^{-1/x}$, above the x-axis, and to the right of the y-axis.

27. Prove Theorem 2(a) by directly evaluating the integrals involved.

28. Evaluate $\int_{-1}^{1}(x\,\mathrm{sgn}\,x)/(x+2)\,dx$. Recall that $\mathrm{sgn}\,x = x/|x|$.

29. Evaluate $\int_0^2 x^2\,\mathrm{sgn}\,(x-1)\,dx$.

In Exercises 30–41, state whether the given integral converges or diverges, and justify your claim.

30. $\displaystyle\int_0^{\infty} \frac{x^2}{x^5+1}\,dx$ **31.** $\displaystyle\int_0^{\infty} \frac{dx}{1+\sqrt{x}}$

32. $\displaystyle\int_2^{\infty} \frac{x\sqrt{x}\,dx}{x^2-1}$ **33.** $\displaystyle\int_0^{\infty} e^{-x^3}\,dx$

34. $\displaystyle\int_0^{\infty} \frac{dx}{\sqrt{x}+x^2}$ **35.** $\displaystyle\int_{-1}^{1} \frac{e^x}{x+1}\,dx$

36. $\displaystyle\int_0^{\pi} \frac{\sin x}{x}\,dx$ **37.** $\displaystyle\int_0^{\infty} \frac{|\sin x|}{x^2}\,dx$

38. $\displaystyle\int_0^{\pi^2} \frac{dx}{1-\cos\sqrt{x}}$ **39.** $\displaystyle\int_{-\pi/2}^{\pi/2} \csc x\,dx$

40. $\displaystyle\int_2^{\infty} \frac{dx}{\sqrt{x}\ln x}$ **41.** $\displaystyle\int_0^{\infty} \frac{dx}{xe^x}$

42. Given that $\int_0^{\infty} e^{-x^2}\,dx = \dfrac{1}{2}\sqrt{\pi}$, evaluate

(a) $\displaystyle\int_0^{\infty} x^2 e^{-x^2}\,dx$ and (b) $\displaystyle\int_0^{\infty} x^4 e^{-x^2}\,dx$.

43. Suppose f is continuous on the interval $(0,1]$ and satisfies $f(x)=O(x^p)$ as $x\to 0+$, where $p>-1$. Show that $\int_0^1 f(x)\,dx$ converges, and that if $0<\epsilon<1$, then the error $E(\epsilon)$ in the approximation

$$\int_0^1 f(x)\,dx \approx \int_\epsilon^1 f(x)\,dx$$

satisfies $E(\epsilon)=O(\epsilon^{p+1})$ as $\epsilon\to 0+$.

44. What is the largest value of k such that the error $E(\epsilon)$ in the approximation

$$\int_0^{\infty} \frac{dx}{\sqrt{x}+x^2} \approx \int_\epsilon^{1/\epsilon} \frac{dx}{\sqrt{x}+x^2},$$

where $0<\epsilon<1$, satisfies $E(\epsilon)=O(\epsilon^k)$ as $\epsilon\to 0+$.

45. If f is continuous on $[a,b]$, show that

$$\lim_{c\to a+}\int_c^b f(x)\,dx = \int_a^b f(x)\,dx.$$

Hint: A continuous function on a closed, finite interval is *bounded*: there exists a positive constant K such that $|f(x)|\le K$ for all x in $[a,b]$. Use this fact, together with parts (d) and (f) of Theorem 3 of Section 5.4, to show that

$$\lim_{c\to a+}\left(\int_a^b f(x)\,dx - \int_c^b f(x)\,dx\right) = 0.$$

Similarly, show that

$$\lim_{c\to b-}\int_a^c f(x)\,dx = \int_a^b f(x)\,dx.$$

46. (The gamma function) The gamma function $\Gamma(x)$ is defined by the improper integral

$$\Gamma(x) = \int_0^{\infty} t^{x-1}e^{-t}\,dt.$$

(Γ is the Greek capital letter gamma.)

(a) Show that the integral converges for $x>0$.

(b) Use integration by parts to show that $\Gamma(x+1)=x\Gamma(x)$ for $x>0$.

(c) Show that $\Gamma(n+1)=n!$ for $n=0,1,2,\ldots$.

(d) Given that $\int_0^{\infty} e^{-x^2}\,dx = \frac{1}{2}\sqrt{\pi}$, show that $\Gamma(\frac{1}{2})=\sqrt{\pi}$ and $\Gamma(\frac{3}{2})=\frac{1}{2}\sqrt{\pi}$.

In view of (c), $\Gamma(x+1)$ is often written $x!$ and regarded as a real-valued extension of the factorial function. Some scientific calculators (in particular, HP calculators) with the factorial function $n!$ built in actually calculate the gamma function rather than just the integral factorial. Check whether your calculator does this by asking it for 0.5!. If you get an error message, it's not using the gamma function.

6.6 The Trapezoid and Midpoint Rules

Most of the applications of integration, within and outside of mathematics, involve the definite integral

$$I = \int_a^b f(x)\,dx.$$

Thanks to the Fundamental Theorem of Calculus, we can evaluate such definite integrals by first finding an antiderivative of f. This is why we have spent considerable time developing techniques of integration. There are, however, two obstacles that can prevent our calculating I in this way:

(i) Finding an antiderivative of f in terms of familiar functions may be impossible, or at least very difficult.

(ii) We may not be given a formula for $f(x)$ as a function of x; for instance, $f(x)$ may be an unknown function whose values at certain points of the interval $[a, b]$ have been determined by experimental measurement.

In the next two sections we investigate the problem of approximating the value of the definite integral I using only the values of $f(x)$ at finitely many points of $[a, b]$. Obtaining such an approximation is called **numerical integration**. Upper and lower sums (or, indeed, any Riemann sum) can be used for this purpose, but these usually require much more calculation to yield a desired precision than the methods we will develop here. We will develop three methods for evaluating definite integrals numerically: the Trapezoid Rule, the Midpoint Rule, and Simpson's Rule (see Section 6.7). All of these methods can be easily implemented on a small computer or using a scientific calculator. The wide availability of these devices makes numerical integration a steadily more important tool for the user of mathematics. Some of the more advanced calculators have built-in routines for numerical integration.

All the techniques we consider require us to calculate the values of $f(x)$ at a set of equally spaced points in $[a, b]$. The computational "expense" involved in determining an approximate value for the integral I will be roughly proportional to the number of function values required, so that the fewer function evaluations needed to achieve a desired degree of accuracy for the integral, the better we will regard the technique. Time is money, even in the world of computers.

The Trapezoid Rule

We assume that $f(x)$ is continuous on $[a, b]$ and subdivide $[a, b]$ into n subintervals of equal length $h = (b - a)/n$ using the $n + 1$ points

$$x_0 = a, \quad x_1 = a + h, \quad x_2 = a + 2h, \quad \ldots, \quad x_n = a + nh = b.$$

We assume that the value of $f(x)$ at each of these points is known:

$$y_0 = f(x_0), \quad y_1 = f(x_1), \quad y_2 = f(x_2), \quad \ldots, \quad y_n = f(x_n).$$

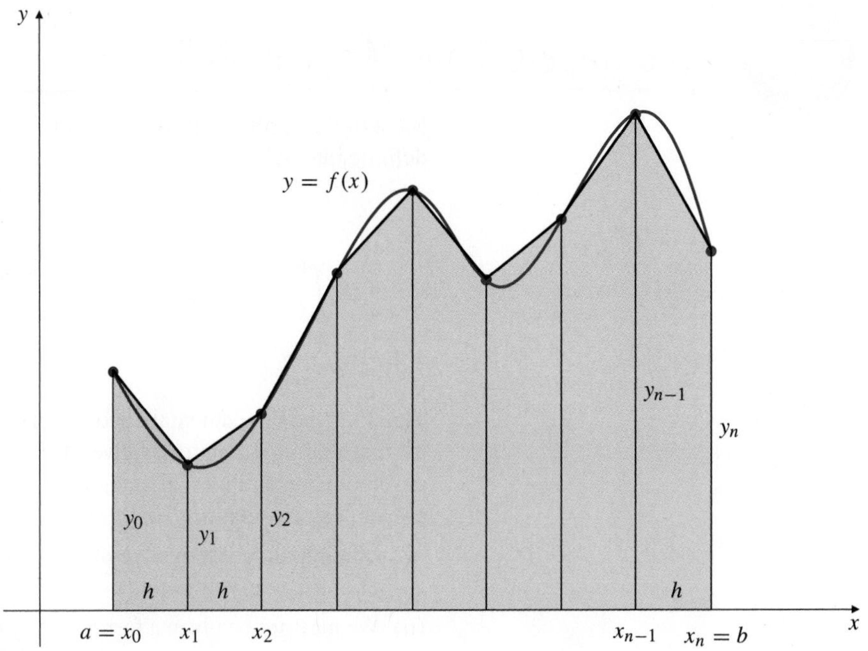

Figure 6.15 The area under $y = f(x)$ is approximated by the sum of the areas of n trapezoids

The Trapezoid Rule approximates $\int_a^b f(x)\,dx$ by using straight line segments between the points (x_{j-1}, y_{j-1}) and (x_j, y_j), $(1 \le j \le n)$, to approximate the graph of f, as shown in Figure 6.15, and summing the areas of the resulting *n trapezoids*. A **trapezoid** is a four-sided polygon with one pair of parallel sides. (For our discussion we assume f is positive so we can talk about "areas," but the resulting formulas apply to any continuous function f.)

The first trapezoid has vertices $(x_0, 0)$, (x_0, y_0), (x_1, y_1), and $(x_1, 0)$. The two parallel sides are vertical and have lengths y_0 and y_1. The perpendicular distance between them is $h = x_1 - x_0$. The area of this trapezoid is h times the average of the parallel sides:

$$h\,\frac{y_0 + y_1}{2} \text{ square units.}$$

This can be seen geometrically by considering the trapezoid as the nonoverlapping union of a rectangle and a triangle; see Figure 6.16. We use this trapezoidal area to approximate the integral of f over the first subinterval $[x_0, x_1]$:

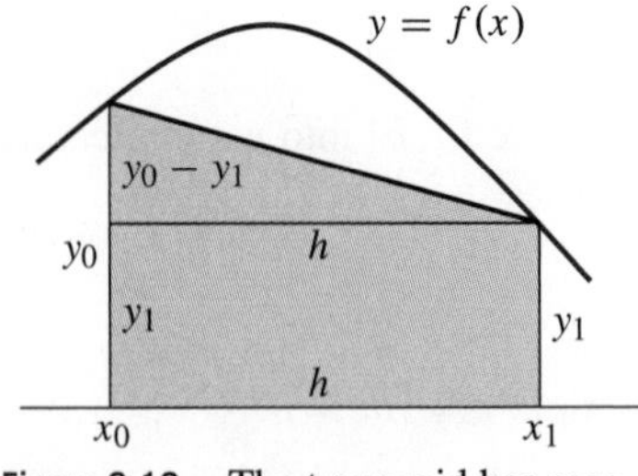

Figure 6.16 The trapezoid has area $y_1 h + \frac{1}{2}(y_0 - y_1)h = \frac{1}{2}h(y_0 + y_1)$

$$\int_{x_0}^{x_1} f(x)\,dx \approx h\,\frac{y_0 + y_1}{2}.$$

We can approximate the integral of f over any subinterval in the same way:

$$\int_{x_{j-1}}^{x_j} f(x)\,dx \approx h\,\frac{y_{j-1} + y_j}{2}, \qquad (1 \le j \le n).$$

It follows that the original integral I can be approximated by the sum of these trapezoidal areas:

$$\begin{aligned}\int_a^b f(x)\,dx &\approx h\left(\frac{y_0 + y_1}{2} + \frac{y_1 + y_2}{2} + \frac{y_2 + y_3}{2} + \cdots + \frac{y_{n-1} + y_n}{2}\right)\\ &= h\left(\frac{1}{2}\,y_0 + y_1 + y_2 + y_3 + \cdots + y_{n-1} + \frac{1}{2}\,y_n\right).\end{aligned}$$

DEFINITION 3

The Trapezoid Rule

The n-subinterval **Trapezoid Rule** approximation to $\int_a^b f(x)\,dx$, denoted T_n, is given by

$$T_n = h\left(\frac{1}{2}y_0 + y_1 + y_2 + y_3 + \cdots + y_{n-1} + \frac{1}{2}y_n\right).$$

We now illustrate the Trapezoid Rule by using it to approximate an integral whose value we already know:

$$I = \int_1^2 \frac{1}{x}\,dx = \ln 2 = 0.693\,147\,18\ldots.$$

(This value, and those of all the approximations quoted in these sections, were calculated using a scientific calculator.) We will use the same integral to illustrate other methods for approximating definite integrals later.

EXAMPLE 1 Calculate the Trapezoid Rule approximations T_4, T_8, and T_{16} for

$$I = \int_1^2 \frac{1}{x}\,dx.$$

Solution For $n = 4$ we have $h = (2-1)/4 = 1/4$; for $n = 8$ we have $h = 1/8$; for $n = 16$ we have $h = 1/16$. Therefore,

$$T_4 = \frac{1}{4}\left[\frac{1}{2}(1) + \frac{4}{5} + \frac{2}{3} + \frac{4}{7} + \frac{1}{2}\left(\frac{1}{2}\right)\right] = 0.697\,023\,81\ldots$$

$$\begin{aligned} T_8 &= \frac{1}{8}\left[\frac{1}{2}(1) + \frac{8}{9} + \frac{4}{5} + \frac{8}{11} + \frac{2}{3} + \frac{8}{13} + \frac{4}{7} + \frac{8}{15} + \frac{1}{2}\left(\frac{1}{2}\right)\right] \\ &= \frac{1}{8}\left[4\,T_4 + \frac{8}{9} + \frac{8}{11} + \frac{8}{13} + \frac{8}{15}\right] = 0.694\,121\,85\ldots \end{aligned}$$

$$\begin{aligned} T_{16} &= \frac{1}{16}\left[8\,T_8 + \frac{16}{17} + \frac{16}{19} + \frac{16}{21} + \frac{16}{23} + \frac{16}{25} + \frac{16}{27} + \frac{16}{29} + \frac{16}{31}\right] \\ &= 0.693\,391\,20\ldots. \end{aligned}$$

Note how the function values used to calculate T_4 were reused in the calculation of T_8, and similarly how those in T_8 were reused for T_{16}. When several approximations are needed, it is very useful to double the number of subintervals for each new calculation, so that previously calculated values of f can be reused.

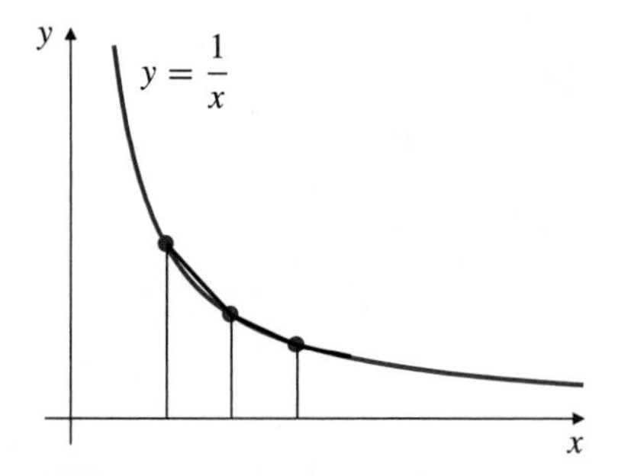

Figure 6.17 The trapezoid areas are greater than the area under the curve if the curve is concave upward

All Trapezoid Rule approximations to $I = \int_1^2 (1/x)\,dx$ are greater than the true value of I. This is because the graph of $y = 1/x$ is concave up on $[1, 2]$, and therefore the tops of the approximating trapezoids lie above the curve. (See Figure 6.17.)

We can calculate the exact errors in the three approximations since we know that $I = \ln 2 = 0.69314718\ldots$ (We always take the error in an approximation to be the true value minus the approximate value.)

$$\begin{aligned} I - T_4 &= 0.693\,147\,18\ldots - 0.697\,023\,81\ldots = -0.003\,876\,63\ldots \\ I - T_8 &= 0.693\,147\,18\ldots - 0.694\,121\,85\ldots = -0.000\,974\,67\ldots \\ I - T_{16} &= 0.693\,147\,18\ldots - 0.693\,391\,20\ldots = -0.000\,244\,02\ldots. \end{aligned}$$

Observe that the size of the error decreases to about a quarter of its previous value each time we double n. We will show below that this is to be expected for a "well-behaved" function like $1/x$.

Example 1 is somewhat artificial in the sense that we know the actual value of the integral so we really don't need an approximation. In practical applications of numerical integration we do not know the actual value. It is tempting to calculate several approximations for increasing values of n until the two most recent ones agree to within a prescribed error tolerance. For example, we might be inclined to claim that $\ln 2 \approx 0.69\ldots$ from a comparison of T_4 and T_8, and further comparison of T_{16} and T_8 suggests that the third decimal place is probably 3: $I \approx 0.693\ldots$. Although this approach cannot be justified in general, it is frequently used in practice.

The Midpoint Rule

A somewhat simpler approximation to $\int_a^b f(x)\,dx$, based on the partition of $[a, b]$ into n equal subintervals, involves forming a Riemann sum of the areas of rectangles whose heights are taken at the midpoints of the n subintervals. (See Figure 6.18.)

DEFINITION 4

The Midpoint Rule

If $h = (b-a)/n$, let $m_j = a + \left(j - \frac{1}{2}\right)h$ for $1 \le j \le n$. The **Midpoint Rule** approximation to $\int_a^b f(x)\,dx$, denoted M_n, is given by

$$M_n = h\big(f(m_1) + f(m_2) + \cdots + f(m_n)\big) = h\sum_{j=1}^{n} f(m_j).$$

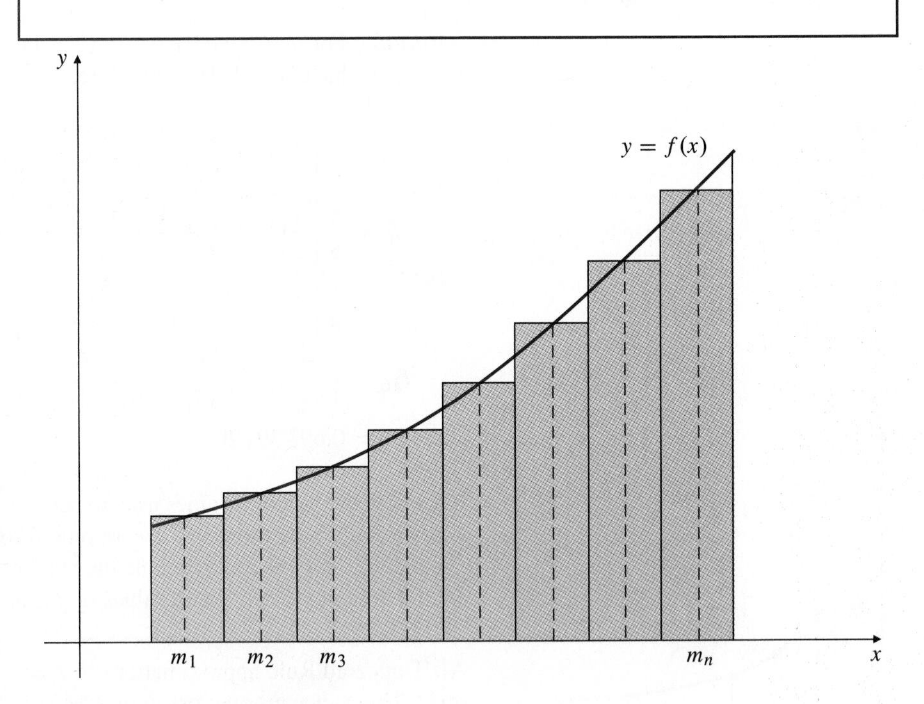

Figure 6.18 The Midpoint Rule approximation M_n to $\int_a^b f(x)\,dx$ is the Riemann sum based on the heights to the graph of f at the midpoints of the subintervals of the partition

EXAMPLE 2 Find the Midpoint Rule approximations M_4 and M_8 for the integral $I = \displaystyle\int_1^2 \frac{1}{x}\,dx$, and compare their actual errors with those obtained for the Trapezoid Rule approximations above.

Solution To find M_4, the interval $[1, 2]$ is divided into four equal subintervals,

$$\left[1, \frac{5}{4}\right], \quad \left[\frac{5}{4}, \frac{3}{2}\right], \quad \left[\frac{3}{2}, \frac{7}{4}\right], \quad \text{and} \quad \left[\frac{7}{4}, 2\right].$$

The midpoints of these intervals are $9/8$, $11/8$, $13/8$, and $15/8$, respectively. The midpoints of the subintervals for M_8 are obtained in a similar way. The required Midpoint Rule approximations are

$$M_4 = \frac{1}{4}\left[\frac{8}{9} + \frac{8}{11} + \frac{8}{13} + \frac{8}{15}\right] = 0.691\,219\,89\ldots$$

$$M_8 = \frac{1}{8}\left[\frac{16}{17} + \frac{16}{19} + \frac{16}{21} + \frac{16}{23} + \frac{16}{25} + \frac{16}{27} + \frac{16}{29} + \frac{16}{31}\right] = 0.692\,660\,55\ldots$$

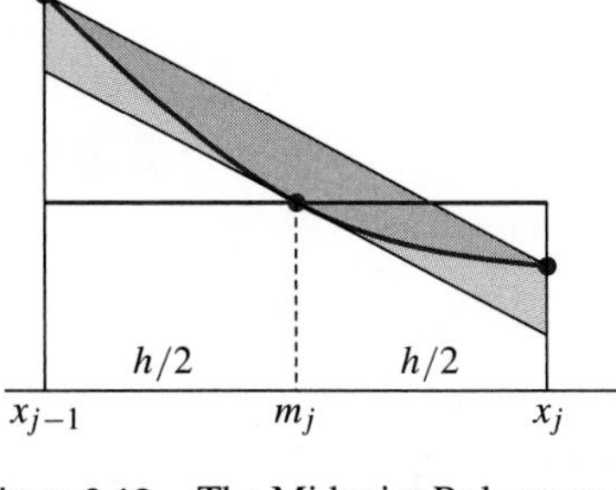
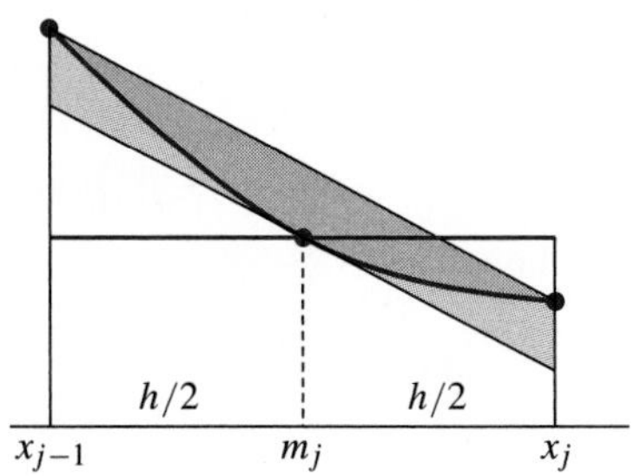

Figure 6.19 The Midpoint Rule error, the area shaded in colour, is opposite in sign and about half the size of the Trapezoid Rule error, the area shaded in grey

The errors in these approximations are

$$I - M_4 = 0.693\,147\,18\ldots - 0.691\,219\,89\ldots = 0.001\,927\,29\ldots$$

$$I - M_8 = 0.693\,147\,18\ldots - 0.692\,660\,55\ldots = 0.000\,486\,63\ldots$$

These errors are of opposite sign and about *half the size* of the corresponding Trapezoid Rule errors $I - T_4$ and $I - T_8$. Figure 6.19 suggests the reason for this. The rectangular area $hf(m_j)$ is equal to the area of the trapezoid formed by the tangent line to $y = f(x)$ at $(m_j, f(m_j))$. The shaded region above the curve is the part of the Trapezoid Rule error due to the jth subinterval. The shaded area below the curve is the corresponding Midpoint Rule error.

One drawback of the Midpoint Rule is that we cannot reuse values of f calculated for M_n when we calculate M_{2n}. However, to calculate T_{2n} we can use the data values already calculated for T_n and M_n. Specifically,

$$T_{2n} = \tfrac{1}{2}(T_n + M_n).$$

A good strategy for using these methods to obtain a value for an integral I to a desired degree of accuracy is to calculate successively:

$$T_n, \quad M_n, \quad T_{2n} = \frac{T_n + M_n}{2}, \quad M_{2n}, \quad T_{4n} = \frac{T_{2n} + M_{2n}}{2}, \quad M_{4n}, \quad \cdots$$

until two consecutive terms agree sufficiently closely. If a single quick approximation is needed, M_n is a better choice than T_n.

Error Estimates

The following theorem provides a bound for the error in the Trapezoid and Midpoint Rule approximations in terms of the second derivative of the integrand.

THEOREM 4

Error estimates for the Trapezoid and Midpoint Rules

If f has a continuous second derivative on $[a, b]$ and satisfies $|f''(x)| \le K$ there, then

$$\left|\int_a^b f(x)\,dx - T_n\right| \le \frac{K(b-a)}{12}h^2 = \frac{K(b-a)^3}{12n^2},$$

$$\left|\int_a^b f(x)\,dx - M_n\right| \le \frac{K(b-a)}{24}h^2 = \frac{K(b-a)^3}{24n^2},$$

where $h = (b - a)/n$. Note that these error bounds decrease like the square of the subinterval length as n increases.

PROOF We will prove only the Trapezoid Rule error estimate here. (The one for the Midpoint Rule is a little easier to prove; the method is suggested in Exercise 14 below.) The straight line approximating $y = f(x)$ in the first subinterval $[x_0, x_1] = [a, a + h]$ passes through the two points (x_0, y_0) and (x_1, y_1). Its equation is $y = A + B(x - x_0)$, where

$$A = y_0 \qquad \text{and} \qquad B = \frac{y_1 - y_0}{x_1 - x_0} = \frac{y_1 - y_0}{h}.$$

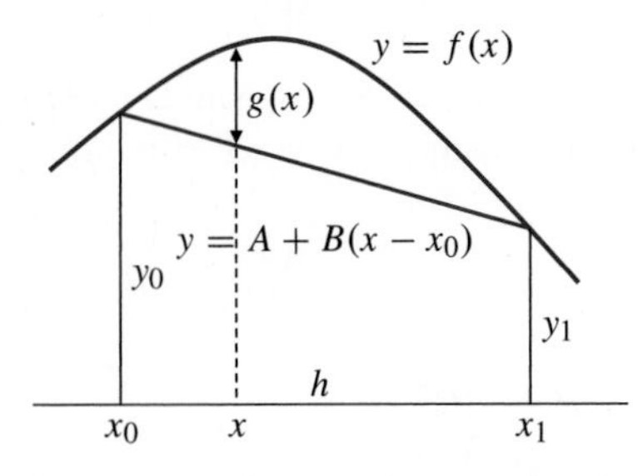

Figure 6.20 The error in approximating the area under the curve by that of the trapezoid is $\int_{x_0}^{x_1} g(x)\,dx$

Let the function $g(x)$ be the vertical distance between the graph of f and this line:

$$g(x) = f(x) - A - B(x - x_0).$$

Since the integral of $A + B(x - x_0)$ over $[x_0, x_1]$ is the area of the first trapezoid, which is $h(y_0 + y_1)/2$ (see Figure 6.20), the integral of $g(x)$ over $[x_0, x_1]$ is the error in the approximation of $\int_{x_0}^{x_1} f(x)\,dx$ by the area of the trapezoid:

$$\int_{x_0}^{x_1} f(x)\,dx - h\,\frac{y_0 + y_1}{2} = \int_{x_0}^{x_1} g(x)\,dx.$$

Now g is twice differentiable, and $g''(x) = f''(x)$. Also $g(x_0) = g(x_1) = 0$. Two integrations by parts (see Exercise 36 of Section 6.1) show that

$$\begin{aligned}\int_{x_0}^{x_1} (x - x_0)(x_1 - x)\,f''(x)\,dx &= \int_{x_0}^{x_1} (x - x_0)(x_1 - x)\,g''(x)\,dx \\ &= -2\int_{x_0}^{x_1} g(x)\,dx.\end{aligned}$$

By the triangle inequality for definite integrals (Theorem 3(f) of Section 5.4),

$$\begin{aligned}\left|\int_{x_0}^{x_1} f(x)\,dx - h\,\frac{y_0 + y_1}{2}\right| &\le \frac{1}{2}\int_{x_0}^{x_1} (x - x_0)(x_1 - x)\,|f''(x)|\,dx \\ &\le \frac{K}{2}\int_{x_0}^{x_1} \left(-x^2 + (x_0 + x_1)x - x_0x_1\right)dx \\ &= \frac{K}{12}(x_1 - x_0)^3 = \frac{K}{12}h^3.\end{aligned}$$

A similar estimate holds on each subinterval $[x_{j-1}, x_j]$ $(1 \le j \le n)$. Therefore,

$$\begin{aligned}\left|\int_a^b f(x)\,dx - T_n\right| &= \left|\sum_{j=1}^{n}\left(\int_{x_{j-1}}^{x_j} f(x)\,dx - h\,\frac{y_{j-1} + y_j}{2}\right)\right| \\ &\le \sum_{j=1}^{n}\left|\int_{x_{j-1}}^{x_j} f(x)\,dx - h\,\frac{y_{j-1} + y_j}{2}\right| \\ &= \sum_{j=1}^{n}\frac{K}{12}h^3 = \frac{K}{12}nh^3 = \frac{K(b - a)}{12}h^2,\end{aligned}$$

since $nh = b - a$. ■

We illustrate this error estimate for the approximations of Examples 1 and 2 above.

EXAMPLE 3 Obtain bounds for the errors for T_4, T_8, T_{16}, M_4, and M_8 for $I = \int_1^2 \frac{1}{x}\,dx$.

Solution If $f(x) = 1/x$, then $f'(x) = -1/x^2$ and $f''(x) = 2/x^3$. On $[1, 2]$ we have $|f''(x)| \le 2$, so we may take $K = 2$ in the estimate. Thus,

$$\begin{aligned}|I - T_4| &\le \frac{2(2-1)}{12}\left(\frac{1}{4}\right)^2 = 0.0104\ldots, \\ |I - M_4| &\le \frac{2(2-1)}{24}\left(\frac{1}{4}\right)^2 = 0.0052\ldots, \\ |I - T_8| &\le \frac{2(2-1)}{12}\left(\frac{1}{8}\right)^2 = 0.0026\ldots, \\ |I - M_8| &\le \frac{2(2-1)}{24}\left(\frac{1}{8}\right)^2 = 0.0013\ldots, \\ |I - T_{16}| &\le \frac{2(2-1)}{12}\left(\frac{1}{16}\right)^2 = 0.00065\ldots.\end{aligned}$$

The actual errors calculated earlier are considerably smaller than these bounds, because $|f''(x)|$ is rather smaller than $K = 2$ over most of the interval $[1, 2]$.

Remark Error bounds are not usually as easily obtained as they are in Example 3. In particular, if an exact formula for $f(x)$ is not known (as is usually the case if the values of f are obtained from experimental data), then we have no method of calculating $f''(x)$, so we can't determine K. Theorem 4 is of more theoretical than practical importance. It shows us that, for a "well-behaved" function f, the Midpoint Rule error is typically about half as large as the Trapezoid Rule error and that both the Trapezoid Rule and Midpoint Rule errors can be expected to decrease like $1/n^2$ as n increases; in terms of big-O notation,

$$I = T_n + O\left(\frac{1}{n^2}\right) \quad \text{and} \quad I = M_n + O\left(\frac{1}{n^2}\right) \qquad \text{as } n \to \infty.$$

Of course, actual errors are not equal to the error bounds, so they won't always be cut to exactly a quarter of their size when we double n.

EXERCISES 6.6

In Exercises 1–4, calculate the approximations T_4, M_4, T_8, M_8, and T_{16} for the given integrals. (Use a scientific calculator or computer spreadsheet program.) Also calculate the exact value of each integral, and so determine the exact error in each approximation. Compare these exact errors with the bounds for the size of the error supplied by Theorem 4.

1. $I = \int_0^2 (1 + x^2)\,dx$

2. $I = \int_0^1 e^{-x}\,dx$

3. $I = \int_0^{\pi/2} \sin x\,dx$

4. $I = \int_0^1 \frac{dx}{1 + x^2}$

5. Figure 6.21 shows the graph of a function f over the interval $[1, 9]$. Using values from the graph, find the Trapezoid Rule estimates T_4 and T_8 for $\int_1^9 f(x)\,dx$.

6. Obtain the best Midpoint Rule approximation that you can for $\int_1^9 f(x)\,dx$ from the data in Figure 6.21.

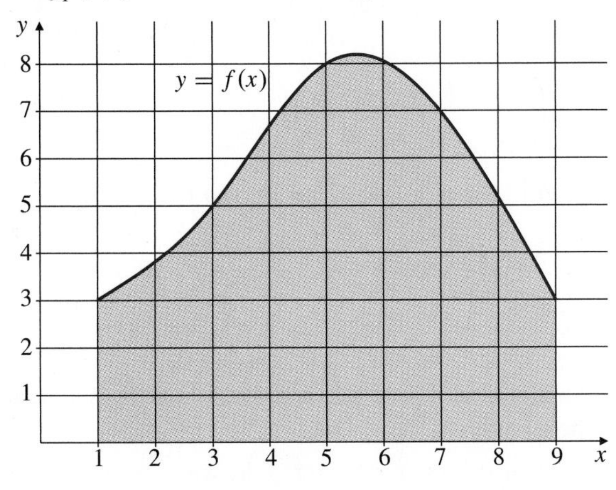

Figure 6.21

7. The map of a region is traced on the grid in Figure 6.22, where 1 unit in both the vertical and horizontal directions represents 10 km. Use the Trapezoid Rule to obtain two estimates for the area of the region.

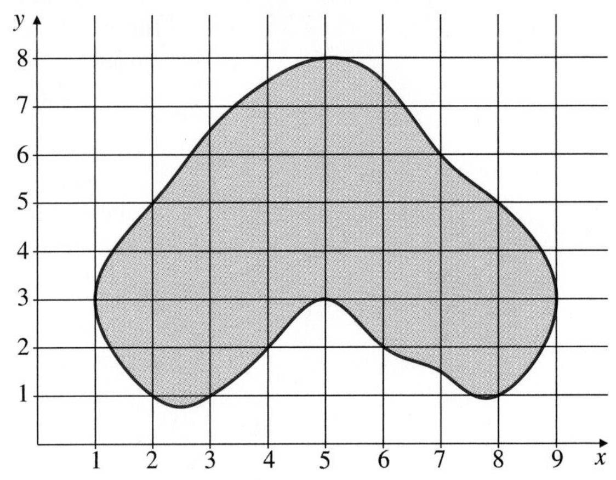

Figure 6.22

8. Find a Midpoint Rule estimate for the area of the region in Exercise 7.

9. Find T_4, M_4, T_8, M_8, and T_{16} for $\int_0^{1.6} f(x)\,dx$ for the function f whose values are given in Table 1.

Table 1.

x	$f(x)$	x	$f(x)$
0.0	1.4142	0.1	1.4124
0.2	1.4071	0.3	1.3983
0.4	1.3860	0.5	1.3702
0.6	1.3510	0.7	1.3285
0.8	1.3026	0.9	1.2734
1.0	1.2411	1.1	1.2057
1.2	1.1772	1.3	1.1258
1.4	1.0817	1.5	1.0348
1.6	0.9853		

10. Find the approximations M_8 and T_{16} for $\int_0^1 e^{-x^2}\,dx$. Quote a value for the integral to as many decimal places as you feel are justified.

11. Repeat Exercise 10 for $\int_0^{\pi/2} \frac{\sin x}{x}\,dx$.
(Assume the integrand is 1 at $x = 0$.)

12. Compute the actual error in the approximation $\int_0^1 x^2\,dx \approx T_1$ and use it to show that the constant 12 in the estimate of Theorem 4 cannot be improved. That is, show that the absolute value of the actual error is as large as allowed by that estimate.

13. Repeat Exercise 12 for M_1.

14. Prove the error estimate for the Midpoint Rule in Theorem 4 as follows: If $x_1 - x_0 = h$ and m_1 is the midpoint of $[x_0, x_1]$, use the error estimate for the tangent line approximation (Theorem 11 of Section 4.9) to show that

$$|f(x) - f(m_1) - f'(m_1)(x - m_1)| \le \frac{K}{2}(x - m_1)^2.$$

Use this inequality to show that

$$\begin{aligned} &\left| \int_{x_0}^{x_1} f(x)\,dx - f(m_1)h \right| \\ &= \left| \int_{x_0}^{x_1} \Big(f(x) - f(m_1) - f'(m_1)(x - m_1) \Big) dx \right| \\ &\le \frac{K}{24} h^3. \end{aligned}$$

Complete the proof the same way used for the Trapezoid Rule estimate in Theorem 4.

6.7 Simpson's Rule

The Trapezoid Rule approximation to $\int_a^b f(x)\,dx$ results from approximating the graph of f by straight line segments through adjacent pairs of data points on the graph. Intuitively, we would expect to do better if we approximate the graph by more general curves. Since straight lines are the graphs of linear functions, the simplest obvious generalization is to use the class of quadratic functions, that is, to approximate the graph of f by segments of parabolas. This is the basis of Simpson's Rule.

Suppose that we are given three points in the plane, one on each of three equally spaced vertical lines, spaced, say, h units apart. If we choose the middle of these lines as the y-axis, then the coordinates of the three points will be, say, $(-h, y_L)$, $(0, y_M)$, and (h, y_R), as illustrated in Figure 6.23.

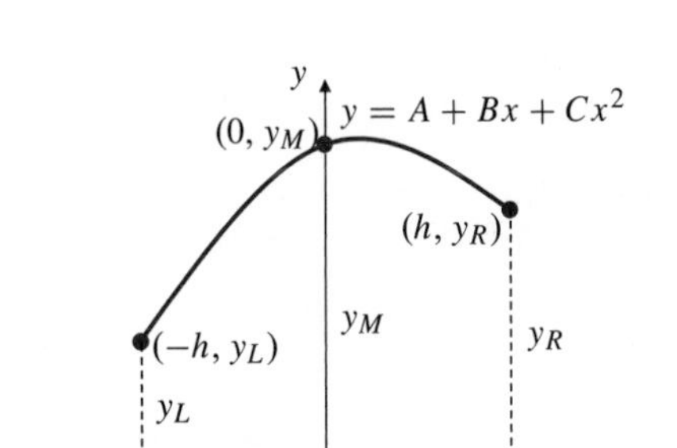

Figure 6.23 Fitting a quadratic graph through three points with equal horizontal spacing

Constants A, B, and C can be chosen so that the parabola $y = A + Bx + Cx^2$ passes through these points; substituting the coordinates of the three points into the equation of the parabola, we get

$$\left.\begin{aligned} y_L &= A - Bh + Ch^2 \\ y_M &= A \\ y_R &= A + Bh + Ch^2 \end{aligned}\right\} \quad \Rightarrow \quad A = y_M \quad \text{and} \quad 2Ch^2 = y_L - 2y_M + y_R.$$

Now we have

$$\begin{aligned} \int_{-h}^{h} (A + Bx + Cx^2)\,dx &= \left(Ax + \frac{B}{2}x^2 + \frac{C}{3}x^3 \right)\Bigg|_{-h}^{h} = 2Ah + \frac{2}{3}Ch^3 \\ &= h\left(2y_M + \frac{1}{3}(y_L - 2y_M + y_R) \right) \\ &= \frac{h}{3}(y_L + 4y_M + y_R). \end{aligned}$$

Thus, the area of the plane region bounded by the parabolic arc, the interval of length $2h$ on the x-axis, and the left and right vertical lines is equal to $(h/3)$ times the sum of the heights of the region at the left and right edges and four times the height at the middle. (It is independent of the position of the y-axis.)

Now suppose that we are given the same data for f as we were given for the Trapezoid Rule; that is, we know the values $y_j = f(x_j)$ $(0 \le j \le n)$ at $n + 1$ equally spaced points

$$x_0 = a, \quad x_1 = a + h, \quad x_2 = a + 2h, \quad \ldots, \quad x_n = a + nh = b,$$

where $h = (b - a)/n$. We can approximate the graph of f over *pairs* of the subintervals $[x_{j-1}, x_j]$ using parabolic segments, and use the integrals of the corresponding quadratic functions to approximate the integrals of f over these subintervals. Since we need to use the subintervals two at a time, we must assume that n is *even*. Using the integral computed for the parabolic segment above, we have

$$\int_{x_0}^{x_2} f(x)\,dx \approx \frac{h}{3}\left(y_0 + 4y_1 + y_2\right)$$
$$\int_{x_2}^{x_4} f(x)\,dx \approx \frac{h}{3}\left(y_2 + 4y_3 + y_4\right)$$
$$\vdots$$
$$\int_{x_{n-2}}^{x_n} f(x)\,dx \approx \frac{h}{3}\left(y_{n-2} + 4y_{n-1} + y_n\right).$$

Adding these $n/2$ individual approximations, we get the Simpson's Rule approximation to the integral $\int_a^b f(x)\,dx$.

DEFINITION 5

Simpson's Rule

The **Simpson's Rule** approximation to $\int_a^b f(x)\,dx$ based on a subdivision of $[a, b]$ into an even number n of subintervals of equal length $h = (b - a)/n$ is denoted S_n and is given by:

$$\int_a^b f(x)\,dx \approx S_n$$
$$= \frac{h}{3}\left(y_0 + 4y_1 + 2y_2 + 4y_3 + 2y_4 + \cdots + 2y_{n-2} + 4y_{n-1} + y_n\right)$$
$$= \frac{h}{3}\left(\sum y_{\text{"ends"}} + 4\sum y_{\text{"odds"}} + 2\sum y_{\text{"evens"}}\right).$$

Note that the Simpson's Rule approximation S_n requires no more data than does the Trapezoid Rule approximation T_n; both require the values of $f(x)$ at $n + 1$ equally spaced points. However, Simpson's Rule treats the data differently, weighting successive values either $1/3$, $2/3$, or $4/3$. As we will see, this can produce a much better approximation to the integral of f.

EXAMPLE 1 Calculate the approximations S_4, S_8, and S_{16} for $I = \int_1^2 \frac{1}{x}\,dx$ and compare them with the actual value $I = \ln 2 = 0.693\,147\,18\ldots$, and with the values of T_4, T_8, and T_{16} obtained in Example 1 of Section 6.6.

Solution We calculate

$$S_4 = \frac{1}{12}\left[1 + 4\left(\frac{4}{5}\right) + 2\left(\frac{2}{3}\right) + 4\left(\frac{4}{7}\right) + \frac{1}{2}\right] = 0.693\,253\,97\ldots,$$

$$S_8 = \frac{1}{24}\left[1 + \frac{1}{2} + 4\left(\frac{8}{9} + \frac{8}{11} + \frac{8}{13} + \frac{8}{15}\right) + 2\left(\frac{4}{5} + \frac{2}{3} + \frac{4}{7}\right)\right] = 0.693\,154\,53\ldots,$$

$$S_{16} = \frac{1}{48}\left[1 + \frac{1}{2} + 4\left(\frac{16}{17} + \frac{16}{19} + \frac{16}{21} + \frac{16}{23} + \frac{16}{25} + \frac{16}{27} + \frac{16}{29} + \frac{16}{31}\right) + 2\left(\frac{8}{9} + \frac{4}{5} + \frac{8}{11} + \frac{2}{3} + \frac{8}{13} + \frac{4}{7} + \frac{8}{15}\right)\right] = 0.693\,147\,65\ldots.$$

The errors are

$$\begin{aligned} I - S_4 &= 0.693\,147\,18\ldots - 0.693\,253\,97\ldots = -0.000\,106\,79, \\ I - S_8 &= 0.693\,147\,18\ldots - 0.693\,154\,53\ldots = -0.000\,007\,35, \\ I - S_{16} &= 0.693\,147\,18\ldots - 0.693\,147\,65\ldots = -0.000\,000\,47. \end{aligned}$$

These errors are evidently much smaller than the corresponding errors for the Trapezoid or Midpoint Rule approximations.

Remark Simpson's Rule S_{2n} makes use of the same $2n + 1$ data values that T_n and M_n together use. It is not difficult to verify that

$$S_{2n} = \frac{T_n + 2M_n}{3}, \qquad S_{2n} = \frac{2T_{2n} + M_n}{3}, \qquad \text{and} \qquad S_{2n} = \frac{4T_{2n} - T_n}{3}.$$

Figure 6.19 and Theorem 4 in Section 6.6 suggest why the first of these formulas ought to yield a particularly good approximation to I.

Obtaining an error estimate for Simpson's Rule is more difficult than for the Trapezoid Rule. We state the appropriate estimate in the following theorem, but we do not attempt any proof. Proofs can be found in textbooks on numerical analysis.

THEOREM 5

Error estimate for Simpson's Rule

If f has a continuous fourth derivative on the interval $[a, b]$, satisfying $|f^{(4)}(x)| \le K$ there, then

$$\left|\int_a^b f(x)\,dx - S_n\right| \le \frac{K(b-a)}{180}h^4 = \frac{K(b-a)^5}{180n^4},$$

where $h = (b - a)/n$.

Observe that, as n increases, the error decreases as the fourth power of h and, hence, as $1/n^4$. Using the big-O notation we have

$$\int_a^b f(x)\,dx = S_n + O\left(\frac{1}{n^4}\right) \qquad \text{as } n \to \infty.$$

This accounts for the fact that S_n is a much better approximation than is T_n, provided that h is small and $|f^{(4)}(x)|$ is not unduly large compared with $|f''(x)|$. Note also that for any (even) n, S_n gives the exact value of the integral of any *cubic* function $f(x) = A + Bx + Cx^2 + Dx^3$; $f^{(4)}(x) = 0$ identically for such f, so we can take $K = 0$ in the error estimate.

EXAMPLE 2 Obtain bounds for the absolute values of the errors in the approximations of Example 1.

Solution If $f(x) = 1/x$, then

$$f'(x) = -\frac{1}{x^2}, \qquad f''(x) = \frac{2}{x^3}, \qquad f^{(3)}(x) = -\frac{6}{x^4}, \qquad f^{(4)}(x) = \frac{24}{x^5}.$$

Clearly, $|f^{(4)}(x)| \le 24$ on $[1, 2]$, so we can take $K = 24$ in the estimate of Theorem 5. We have

$$\begin{aligned}
|I - S_4| &\le \frac{24(2-1)}{180}\left(\frac{1}{4}\right)^4 \approx 0.000\,520\,83, \\
|I - S_8| &\le \frac{24(2-1)}{180}\left(\frac{1}{8}\right)^4 \approx 0.000\,032\,55, \\
|I - S_{16}| &\le \frac{24(2-1)}{180}\left(\frac{1}{16}\right)^4 \approx 0.000\,002\,03.
\end{aligned}$$

Again we observe that the actual errors are well within these bounds.

EXAMPLE 3 A function f satisfies $|f^{(4)}(x)| \le 7$ on the interval $[1, 3]$, and the values $f(1.0) = 0.1860$, $f(1.5) = 0.9411$, $f(2.0) = 1.1550$, $f(2.5) = 1.4511$, and $f(3.0) = 1.2144$. Find the best possible Simpson's Rule approximation to $I = \int_1^3 f(x)\,dx$ based on these data. Give a bound for the size of the error, and specify the smallest interval you can that must contain the value of I.

Solution We take $n = 4$, so that $h = (3-1)/4 = 0.5$, and we obtain

$$\begin{aligned}
I &= \int_1^3 f(x)\,dx \\
&\approx S_4 = \frac{0.5}{3}\big(0.1860 + 4(0.9411 + 1.4511) + 2(1.1550) + 1.2144\big) \\
&= 2.2132.
\end{aligned}$$

Since $|f^{(4)}(x)| \le 7$ on $[1, 3]$, we have

$$|I - S_4| \le \frac{7(3-1)}{180}(0.5)^4 < 0.0049.$$

I must therefore satisfy

$$2.2132 - 0.0049 < I < 2.2132 + 0.0049 \quad \text{or} \quad 2.2083 < I < 2.2181.$$

EXERCISES 6.7

In Exercises 1–4, find Simpson's Rule approximations S_4 and S_8 for the given functions. Compare your results with the actual values of the integrals and with the corresponding Trapezoid Rule approximations obtained in Exercises 1–4 of Section 6.6.

1. $I = \int_0^2 (1 + x^2)\,dx$

2. $I = \int_0^1 e^{-x}\,dx$

3. $I = \int_0^{\pi/2} \sin x\,dx$

4. $I = \int_0^1 \frac{dx}{1 + x^2}$

5. Find the Simpson's Rule approximation S_8 for the integral in Exercise 5 of Section 6.6.

6. Find the best Simpson's Rule approximation that you can for the area of the region in Exercise 7 of Section 6.6.

7. Use Theorem 5 to obtain bounds for the errors in the approximations obtained in Exercises 2 and 3 above.

8. Verify that $S_{2n} = \frac{T_n + 2M_n}{3} = \frac{2T_{2n} + M_n}{3}$, where T_n and M_n refer to the appropriate Trapezoid and Midpoint Rule approximations. Deduce that $S_{2n} = \frac{4T_{2n} - T_n}{3}$.

9. Find S_4, S_8, and S_{16} for $\int_0^{1.6} f(x)\,dx$ for the function f whose values are tabulated in Exercise 9 of Section 6.6.

10. Find the Simpson's Rule approximations S_8 and S_{16} for $\int_0^1 e^{-x^2}\,dx$. Quote a value for the integral to the number of decimal places you feel is justified based on comparing the two approximations.

11. Compute the actual error in the approximation $\int_0^1 x^4\,dx \approx S_2$ and use it to show that the constant 180 in the estimate of Theorem 5 cannot be improved.

12. Since Simpson's Rule is based on quadratic approximation, it is not surprising that it should give an exact value for an integral of $A + Bx + Cx^2$. It is more surprising that it is exact for a cubic function as well. Verify by direct calculation that $\int_0^1 x^3\,dx = S_2$.

6.8 Other Aspects of Approximate Integration

The numerical methods described in Sections 6.6 and 6.7 are suitable for finding approximate values for integrals of the form

$$I = \int_a^b f(x)\,dx,$$

where $[a, b]$ is a finite interval and the integrand f is "well-behaved" on $[a, b]$. In particular, I must be a *proper* integral. There are many other methods for dealing with such integrals, some of which we mention later in this section. First, however, we consider what can be done if the function f isn't "well-behaved" on $[a, b]$. We mean by this that either the integral is improper or f doesn't have sufficiently many continuous derivatives on $[a, b]$ to justify whatever numerical methods we want to use.

The ideas of this section are best presented by means of concrete examples.

EXAMPLE 1 How can you evaluate the integral $I = \int_0^1 \sqrt{x}\, e^x\,dx$ numerically?

Solution Although I is a proper integral, with integrand $f(x) = \sqrt{x}\, e^x$ satisfying $f(x) \to 0$ as $x \to 0+$, nevertheless, the standard numerical methods can be expected to perform poorly for I because the derivatives of f are not bounded near 0. This problem is easily remedied; just make the change of variable $x = t^2$ and rewrite I in the form

$$I = 2\int_0^1 t^2 e^{t^2}\,dt,$$

whose integrand $g(t) = t^2 e^{t^2}$ has bounded derivatives near 0. The latter integral can be efficiently approximated by the methods of Sections 6.6 and 6.7.

Approximating Improper Integrals

EXAMPLE 2 Describe how to evaluate $I = \int_0^1 \frac{\cos x}{\sqrt{x}}\, dx$ numerically.

Solution The integral is improper, but convergent because, on [0, 1],

$$0 < \frac{\cos x}{\sqrt{x}} \le \frac{1}{\sqrt{x}} \qquad \text{and} \qquad \int_0^1 \frac{dx}{\sqrt{x}} = 2.$$

However, since $\lim_{x \to 0+} \dfrac{\cos x}{\sqrt{x}} = \infty$, we cannot directly apply any of the techniques developed in Sections 6.6 and 6.7. (y_0 is infinite.) The substitution $x = t^2$ removes this difficulty:

$$I = \int_0^1 \frac{\cos t^2}{t}\, 2t\, dt = 2\int_0^1 \cos t^2\, dt.$$

The latter integral is not improper and is well-behaved. Numerical techniques can be applied to evaluate it.

EXAMPLE 3 Show how to evaluate $I = \int_0^\infty \frac{dx}{\sqrt{2 + x^2 + x^4}}$ by numerical means.

Solution Here, the integral is improper of type I; the interval of integration is infinite. Although there is no singularity at $x = 0$, it is still useful to break the integral into two parts:

$$I = \int_0^1 \frac{dx}{\sqrt{2 + x^2 + x^4}} + \int_1^\infty \frac{dx}{\sqrt{2 + x^2 + x^4}} = I_1 + I_2.$$

I_1 is proper. In I_2 make the change of variable $x = 1/t$:

$$I_2 = \int_0^1 \frac{dt}{t^2\sqrt{2 + \frac{1}{t^2} + \frac{1}{t^4}}} = \int_0^1 \frac{dt}{\sqrt{2t^4 + t^2 + 1}}.$$

This is also a proper integral. If desired, I_1 and I_2 can be recombined into a single integral before numerical methods are applied:

$$I = \int_0^1 \left(\frac{1}{\sqrt{2 + x^2 + x^4}} + \frac{1}{\sqrt{2x^4 + x^2 + 1}} \right) dx.$$

Example 3 suggests that when an integral is taken over an infinite interval, a change of variable should be made to convert the integral to a finite interval.

Using Taylor's Formula

Taylor's Formula (see Section 4.10) can sometimes be useful for evaluating integrals. Here is an example.

EXAMPLE 4 Use Taylor's Formula for $f(x) = e^x$, obtained in Section 4.10, to evaluate the integral $\int_0^1 e^{x^2}\, dx$ to within an error of less than 10^{-4}.

Solution In Example 4 of Section 4.10 we showed that

$$f(x) = e^x = 1 + x + \frac{x^2}{2!} + \frac{x^3}{3!} + \cdots + \frac{x^n}{n!} + E_n(x),$$

where

$$E_n(x) = \frac{e^X}{(n+1)!}x^{n+1}$$

for some X between 0 and x. If $0 \le x \le 1$, then $0 \le X \le 1$, so $e^X \le e < 3$. Therefore,

$$|E_n(x)| \le \frac{3}{(n+1)!}x^{n+1}.$$

Now replace x by x^2 in the formula for e^x above and integrate from 0 to 1:

$$\begin{aligned}\int_0^1 e^{x^2}\,dx &= \int_0^1 \left(1 + x^2 + \frac{x^4}{2!} + \cdots + \frac{x^{2n}}{n!}\right)dx + \int_0^1 E_n(x^2)\,dx\\ &= 1 + \frac{1}{3} + \frac{1}{5 \times 2!} + \cdots + \frac{1}{(2n+1)n!} + \int_0^1 E_n(x^2)\,dx.\end{aligned}$$

We want the error to be less than 10^{-4}, so we estimate the remainder term:

$$\left|\int_0^1 E_n(x^2)\,dx\right| \le \frac{3}{(n+1)!}\int_0^1 x^{2(n+1)}\,dx = \frac{3}{(n+1)!(2n+3)} < 10^{-4},$$

provided $(2n+3)(n+1)! > 30{,}000$. Since $13 \times 6! = 9{,}360$ and $15 \times 7! = 75{,}600$, we need $n = 6$. Thus,

$$\begin{aligned}\int_0^1 e^{x^2}\,dx &\approx 1 + \frac{1}{3} + \frac{1}{5 \times 2!} + \frac{1}{7 \times 3!} + \frac{1}{9 \times 4!} + \frac{1}{11 \times 5!} + \frac{1}{13 \times 6!}\\ &\approx 1.462\,64,\end{aligned}$$

with error less than 10^{-4}.

Romberg Integration

Using Taylor's Formula, it is possible to verify that for a function f having continuous derivatives up to order $2m+2$ on $[a,b]$ the error $E_n = I - T_n$ in the Trapezoid Rule approximation T_n to $I = \int_a^b f(x)\,dx$ satisfies

$$E_n = I - T_n = \frac{C_1}{n^2} + \frac{C_2}{n^4} + \frac{C_3}{n^6} + \cdots + \frac{C_m}{n^{2m}} + O\left(\frac{1}{n^{2m+2}}\right),$$

where the constants C_j depend on the $2j$th derivative of f. It is possible to use this formula to obtain higher-order approximations to I, starting with Trapezoid Rule approximations. The technique is known as **Romberg integration** or **Richardson extrapolation.**

To begin, suppose we have constructed Trapezoid Rule approximations for values of n that are powers of 2: $n = 1, 2, 4, 8, \ldots$. Accordingly, let us define

$$T_k^0 = T_{2^k}. \qquad \text{Thus,} \quad T_0^0 = T_1, \quad T_1^0 = T_2, \quad T_2^0 = T_4, \quad \ldots.$$

Using the formula for $T_{2^k} = I - E_{2^k}$ given above, we write

$$T_k^0 = I - \frac{C_1}{4^k} - \frac{C_2}{4^{2k}} - \cdots - \frac{C_m}{4^{mk}} + O\left(\frac{1}{4^{(m+1)k}}\right) \quad \text{(as } k \to \infty).$$

Similarly, replacing k by $k+1$, we get

$$T_{k+1}^0 = I - \frac{C_1}{4^{k+1}} - \frac{C_2}{4^{2(k+1)}} - \cdots - \frac{C_m}{4^{m(k+1)}} + O\left(\frac{1}{4^{(m+1)(k+1)}}\right).$$

If we multiply the formula for T_{k+1}^0 by 4 and subtract the formula for T_k^0, the terms involving C_1 will cancel out. The first term on the right will be $4I - I = 3I$, so let us also divide by 3 and define T_{k+1}^1 to be the result. Then as $k \to \infty$, we have

$$T_{k+1}^1 = \frac{4T_{k+1}^0 - T_k^0}{3} = I - \frac{C_2^1}{4^{2k}} - \frac{C_3^1}{4^{3k}} - \cdots - \frac{C_m^1}{4^{mk}} + O\left(\frac{1}{4^{(m+1)k}}\right).$$

(The C_i^1 are new constants.) Unless these constants are much larger than the previous ones, T_{k+1}^1 ought to be a better approximation to I than T_{k+1}^0 since we have eliminated the lowest order (and therefore the largest) of the error terms, $C_1/4^{k+1}$. In fact, Exercise 8 in Section 6.7 shows that $T_{k+1}^1 = S_{2^{k+1}}$, the Simpson's Rule approximation based on 2^{k+1} subintervals.

We can continue the process of eliminating error terms begun above. Replacing $k+1$ by $k+2$ in the expression for T_{k+1}^1, we obtain

$$T_{k+2}^1 = I - \frac{C_2^1}{4^{2(k+1)}} - \frac{C_3^1}{4^{3(k+1)}} - \cdots - \frac{C_m^1}{4^{m(k+1)}} + O\left(\frac{1}{4^{(m+1)(k+1)}}\right).$$

To eliminate C_2^1 we can multiply the second formula by 16, subtract the first formula, and divide by 15. Denoting the result T_{k+2}^2, we have, as $k \to \infty$,

$$T_{k+2}^2 = \frac{16T_{k+2}^1 - T_{k+1}^1}{15} = I - \frac{C_3^2}{4^{3k}} - \cdots - \frac{C_m^2}{4^{mk}} + O\left(\frac{1}{4^{(m+1)k}}\right).$$

We can proceed in this way, eliminating one error term after another. In general, for $j < m$ and $k \geq 0$,

$$T_{k+j}^j = \frac{4^j T_{k+j}^{j-1} - T_{k+j-1}^{j-1}}{4^j - 1} = I - \frac{C_{j+1}^j}{4^{(j+1)k}} - \cdots - \frac{C_m^j}{4^{mk}} + O\left(\frac{1}{4^{(m+1)k}}\right).$$

The big-O term refers to $k \to \infty$ for fixed j. All this looks very complicated, but it is not difficult to carry out in practice, especially with the aid of a computer spreadsheet. Let $R_j = T_j^j$, called a **Romberg approximation** to I, and calculate the entries in the following scheme in order from left to right and down each column when you come to it:

Scheme for calculating Romberg approximations

$T_0^0 = T_1 = R_0 \longrightarrow$	$T_1^0 = T_2 \longrightarrow$	$T_2^0 = T_4 \longrightarrow$	$T_3^0 = T_8 \longrightarrow$	
	$\downarrow$	$\downarrow$	$\downarrow$	
	$T_1^1 = S_2 = R_1$	$T_2^1 = S_4$	$T_3^1 = S_8$	
		$\downarrow$	$\downarrow$	
		$T_2^2 = R_2$	T_3^2	
			$\downarrow$	
			$T_3^3 = R_3$	

Stop when T_j^{j-1} and R_j differ by less than the acceptable error, and quote R_j as the Romberg approximation to $\int_a^b f(x)\,dx$.

The top line in the scheme is made up of the Trapezoid Rule approximations T_1, T_2, T_4, T_8, ... Elements in subsequent rows are calculated by the formulas:

Formulas for calculating Romberg approximations

$$T_1^1 = \frac{4T_1^0 - T_0^0}{3} \qquad T_2^1 = \frac{4T_2^0 - T_1^0}{3} \qquad T_3^1 = \frac{4T_3^0 - T_2^0}{3} \quad \cdots$$

$$T_2^2 = \frac{16T_2^1 - T_1^1}{15} \qquad T_3^2 = \frac{16T_3^1 - T_2^1}{15} \quad \cdots$$

$$T_3^3 = \frac{64T_3^2 - T_2^2}{63} \quad \cdots$$

In general, if $1 \le j \le k$, then $T_k^j = \dfrac{4^j T_k^{j-1} - T_{k-1}^{j-1}}{4^j - 1}$.

Each new entry is calculated from the one above and the one to the left of that one.

EXAMPLE 5 Calculate the Romberg approximations R_0, R_1, R_2, R_3, and R_4 for the integral $I = \displaystyle\int_1^2 \frac{1}{x}\,dx$.

Solution We will carry all calculations to 8 decimal places. Since we must obtain R_4, we will need to find all the entries in the first five columns of the scheme. First we calculate the first two Trapezoid Rule approximations:

$$R_0 = T_0^0 = T_1 = \frac{1}{2} + \frac{1}{4} = 0.750\,000\,00,$$

$$T_1^0 = T_2 = \frac{1}{2}\left[\frac{1}{2}(1) + \frac{2}{3} + \frac{1}{2}\left(\frac{1}{2}\right)\right] = 0.708\,333\,33.$$

The remaining required Trapezoid Rule approximations were calculated in Example 1 of Section 6.6, so we will just record them here:

$$T_2^0 = T_4 = 0.697\,023\,81,$$
$$T_3^0 = T_8 = 0.694\,121\,85,$$
$$T_4^0 = T_{16} = 0.693\,391\,20.$$

Now we calculate down the columns from left to right. For the second column:

$$R_1 = S_2 = T_1^1 = \frac{4T_1^0 - T_0^0}{3} = 0.694\,444\,44;$$

the third column:

$$S_4 = T_2^1 = \frac{4T_2^0 - T_1^0}{3} = 0.693\,253\,97,$$

$$R_2 = T_2^2 = \frac{16T_2^1 - T_1^1}{15} = 0.693\,174\,60;$$

the fourth column:

$$S_8 = T_3^1 = \frac{4T_3^0 - T_2^0}{3} = 0.693\,154\,53,$$

$$T_3^2 = \frac{16T_3^1 - T_2^1}{15} = 0.693\,147\,90,$$

$$R_3 = T_3^3 = \frac{64T_3^2 - T_2^2}{63} = 0.693\,147\,48;$$

and the fifth column:

$$S_{16} = T_4^1 = \frac{4T_4^0 - T_3^0}{3} = 0.693\,147\,65,$$
$$T_4^2 = \frac{16T_4^1 - T_3^1}{15} = 0.693\,147\,19,$$
$$T_4^3 = \frac{64T_4^2 - T_3^2}{63} = 0.693\,147\,18,$$
$$R_4 = T_4^4 = \frac{256T_4^3 - T_3^3}{255} = 0.693\,147\,18.$$

Since T_4^3 and R_4 agree to the 8 decimal places we are calculating, we expect that

$$I = \int_1^2 \frac{dx}{x} = \ln 2 \approx 0.693\,147\,18\ldots.$$

The various approximations calculated above suggest that for any given value of $n = 2^k$, the Romberg approximation R_n should give the best value obtainable for the integral based on the $n + 1$ data values $y_0, y_1, \ldots, y_n$. This is so only if the derivatives $f^{(n)}(x)$ do not grow too rapidly as n increases.

The Importance of Higher-Order Methods

Higher-order methods, such as Romberg, remove lower-order error by manipulating series. Removing lower-order error is of enormous importance for computation. Without it, even simple computations would be impossible for all practical purposes. For example, consider again the integral $I = \int_1^2 \frac{1}{x}\,dx$.

We can use Maple to compute this integral numerically to 16 digits (classical double precision),

```
> Digits=16:
> int(1/x, x = 1 ..  2.);
```

0.6931471805599453

Comparison with ln 2

```
> ln(2.);
```

0.6931471805599453

confirms the consistency of this calculation. Furthermore, we can compute the processor time for this calculation

```
> time(int(1/x, x = 1 ..  2.));
```

0.033

which indicates that, on the system used, 16 digits of accuracy is produced in hundredths of seconds of processor time.

Now let's consider what happens without removing lower order error. If we were to estimate this integral using a simple end point Riemann sum, as we used in the original definition of a definite integral, the error is $O(h)$ or $O(1/n)$. Let the step size be 10^{-7}.

```
> 1e-7*add(1/(1+i/1e7), i = 1 ..  1e7);
```

0.6931471555599459

which has an error of 2.5×10^{-8}. The processor time used to do this sum computation is given by

```
> time(1e-7*add(1/(1+i/1e7), i = 1 ..  1e7));
```

$$175.777$$

that is, 175.577 seconds on the particular computer we used. (If you do the calculation on your machine your result will vary according to the speed of your system.) Note that we used the Maple "`add`" routine rather than "`sum`" in the calculations above. This was done to tell Maple to add the floating-point values of the terms one after another rather than to attempt a symbolic summation.

Because the computation time is proportional to the number n of rectangles used in the Riemann sum, and because the error is proportional to $1/n$, it follows that error times computation time is roughly constant. We can use this to estimate the time to compute the integral by this method to 16 digits of precision. Assuming an error of 10^{-16}, the time for the computation will be

$$175.777 \times 2.5 \times \frac{10^{-8}}{10^{-16}} \quad \text{seconds,}$$

or about 1,400 years.

Maple is not limited to 16 digits, of course. For the each additional digit of precision, the Riemann sum method corresponds to a factor-of-ten increase in time because of low-order error. The ability to compute such quantities is a powerful and important application of series expansions.

Other Methods

As developed above, the Trapezoid, Midpoint, Simpson, and Romberg methods all involved using equal subdivisions of the interval $[a, b]$. There are other methods that avoid this restriction. In particular, **Gaussian approximations** involve selecting evaluation points and weights in an optimal way so as to give the most accurate results for "well-behaved" functions. See Exercises 11–13 below. You can consult a text on numerical analysis to learn more about this method.

Finally, we note that even when you apply one of the methods of Sections 6.6 and 6.7, it may be advisable for you to break up the integral into two or more integrals over smaller intervals and then use different subinterval lengths h for each of the different integrals. You will want to evaluate the integrand at more points in an interval where its graph is changing direction erratically than in one where the graph is better behaved.

EXERCISES 6.8

Rewrite the integrals in Exercises 1–6 in a form to which numerical methods can be readily applied.

1. $\displaystyle\int_0^1 \frac{dx}{x^{1/3}(1+x)}$

2. $\displaystyle\int_0^1 \frac{e^x}{\sqrt{1-x}}\,dx$

3. $\displaystyle\int_{-1}^1 \frac{e^x}{\sqrt{1-x^2}}\,dx$

4. $\displaystyle\int_1^\infty \frac{dx}{x^2+\sqrt{x}+1}$

5. $\displaystyle\int_0^{\pi/2} \frac{dx}{\sqrt{\sin x}}$

6. $\displaystyle\int_0^\infty \frac{dx}{x^4+1}$

7. Find T_2, T_4, T_8, and T_{16} for $\int_0^1 \sqrt{x}\,dx$, and find the actual errors in these approximations. Do the errors decrease like $1/n^2$ as n increases? Why?

8. Transform the integral $I = \int_1^\infty e^{-x^2}\,dx$ using the substitution $x = 1/t$, and calculate the Simpson's Rule approximations S_2, S_4, and S_8 for the resulting integral (whose integrand has limit 0 as $t \to 0+$). Quote the value of I to the accuracy you feel is justified. Do the approximations converge as quickly as you might expect? Can you think of a reason why they might not?

9. Evaluate $I = \int_0^1 e^{-x^2}\,dx$, by the Taylor's Formula method of Example 4, to within an error of 10^{-4}.

10. Recall that $\int_0^\infty e^{-x^2}\,dx = \frac{1}{2}\sqrt{\pi}$. Combine this fact with the result of Exercise 9 to evaluate $I = \displaystyle\int_1^\infty e^{-x^2}\,dx$ to 3 decimal places.

11. (Gaussian approximation) Find constants A and u, with u between 0 and 1, such that

$$\int_{-1}^1 f(x)\,dx = Af(-u) + Af(u)$$

holds for every cubic polynomial $f(x) = ax^3 + bx^2 + cx + d$. For a general function $f(x)$ defined on $[-1, 1]$, the approximation

$$\int_{-1}^{1} f(x)\,dx \approx Af(-u) + Af(u)$$

is called a *Gaussian* approximation.

12. Use the method of Exercise 11 to approximate the integrals of (a) x^4, (b) $\cos x$, and (c) e^x, over the interval $[-1, 1]$, and find the error in each approximation.

13. (Another Gaussian approximation) Find constants A and B, and u between 0 and 1, such that

$$\int_{-1}^{1} f(x)\,dx = Af(-u) + Bf(0) + Af(u)$$

holds for every quintic polynomial $f(x) = ax^5 + bx^4 + cx^3 + dx^2 + ex + f$.

14. Use the Gaussian approximation

$$\int_{-1}^{1} f(x)\,dx \approx Af(-u) + Bf(0) + Af(u),$$

where A, B, and u are as determined in Exercise 13, to find approximations for the integrals of (a) x^6, (b) $\cos x$, and (c) e^x over the interval $[-1, 1]$, and find the error in each approximation.

15. Calculate sufficiently many Romberg approximations $R_1, R_2, R_3, \ldots$ for the integral

$$\int_0^1 e^{-x^2}\,dx$$

to be confident you have evaluated the integral correctly to 6 decimal places.

16. Use the values of $f(x)$ given in the table accompanying Exercise 9 in Section 6.6 to calculate the Romberg approximations R_1, R_2, and R_3 for the integral

$$\int_0^{1.6} f(x)\,dx$$

in that exercise.

17. The Romberg approximation R_2 for $\int_a^b f(x)\,dx$ requires five values of f, $y_0 = f(a)$, $y_1 = f(a+h)$, $\ldots$, $y_4 = f(x+4h) = f(b)$, where $h = (b-a)/4$. Write the formula for R_2 explicitly in terms of these five values.

18. Explain why the change of variable $x = 1/t$ is not suitable for transforming the integral $\displaystyle\int_\pi^\infty \frac{\sin x}{1+x^2}\,dx$ into a form to which numerical methods can be applied. Try to devise a method whereby this integral could be approximated to any desired degree of accuracy.

19. If $f(x) = \dfrac{\sin x}{x}$ for $x \neq 0$ and $f(0) = 1$, show that $f''(x)$ has a finite limit as $x \to 0$. Hence, f'' is bounded on finite intervals $[0, a]$, and Trapezoid Rule approximations T_n to $\int_0^a \dfrac{\sin x}{x}\,dx$ converge suitably quickly as n increases. Higher derivatives are also bounded (Taylor's Formula is useful for showing this) so Simpson's Rule and higher-order approximations can also be used effectively.

20. (Estimating computation time) With higher-order methods, the time to compute remains proportional to the number of intervals n, used to numerically approximate an integral. But the error is reduced. For the trapezoid rule the error goes as $O(1/n^2)$. When $n = 1 \times 10^7$, the error turns out to be 6×10^{-16}. The computation time is approximately the same as that computed for the Riemann sum approximation to $\int_1^2 (1/x)\,dx$ discussed above (175.777 seconds for our computer), because we need essentially the same number of function evaluations. How long would it take our computer to get the trapezoid approximation to have quadruple (i.e., 32 digit) precession?

21. Repeat the previous exercise, but this time using Simpson's Rule, whose error is $O(1/n^4)$. Again use the same time, 175.777 s for $n = 1 \times 10^7$, but for Simpson's Rule, the error for this calculation is 3.15×10^{-30}. How long would we expect our computer to take to achieve 32-digit accuracy (i.e., error 10^{-32})? Note, however, that Maple's integration package for the computer used took 0.134 seconds to achieve this precision. Will it have used a higher-order method than Simpson's Rule to achieve this time?

CHAPTER REVIEW

Key Ideas

- **What do the following terms and phrases mean?**
  - ◇ integration by parts
  - ◇ a reduction formula
  - ◇ an inverse substitution
  - ◇ a rational function
  - ◇ the method of partial fractions
  - ◇ a computer algebra system
  - ◇ an improper integral of type I
  - ◇ an improper integral of type II
  - ◇ a p-integral
  - ◇ the Trapezoid Rule
  - ◇ the Midpoint Rule
  - ◇ Simpson's Rule
- **Describe the inverse sine and inverse tangent substitutions.**
- **What is the significance of the comparison theorem for improper integrals?**
- **When is numerical integration necessary?**

Summary of Techniques of Integration

Students sometimes have difficulty deciding which method to use to evaluate a given integral. Often no one method will suffice to produce the whole solution, but one method may lead to a different, possibly simpler, integral that can then be dealt with on its own merits. Here are a few guidelines:

1. First, and always, be alert for simplifying substitutions. Even when these don't accomplish the whole integration, they can lead to integrals to which some other method can be applied.
2. If the integral involves a quadratic expression $Ax^2 + Bx + C$ with $A \neq 0$ and $B \neq 0$, complete the square. A simple substitution then reduces the quadratic expression to a sum or difference of squares.
3. Integrals of products of trigonometric functions can sometimes be evaluated or rendered simpler by the use of appropriate trigonometric identities such as:

$$\sin^2 x + \cos^2 x = 1$$
$$\sec^2 x = 1 + \tan^2 x$$
$$\csc^2 x = 1 + \cot^2 x$$
$$\sin x \cos x = \tfrac{1}{2} \sin 2x$$
$$\sin^2 x = \tfrac{1}{2}(1 - \cos 2x)$$
$$\cos^2 x = \tfrac{1}{2}(1 + \cos 2x).$$

4. Integrals involving $(a^2 - x^2)^{1/2}$ can be transformed using $x = a \sin\theta$. Integrals involving $(a^2 + x^2)^{1/2}$ or $1/(a^2 + x^2)$ may yield to $x = a \tan\theta$. Integrals involving $(x^2 - a^2)^{1/2}$ can be transformed using $x = a \sec\theta$ or $x = a \cosh\theta$.
5. Use integration by parts for integrals of functions such as products of polynomials and transcendental functions, and for inverse trigonometric functions and logarithms. Be alert for ways of using integration by parts to obtain formulas representing complicated integrals in terms of simpler ones.
6. Use partial fractions to integrate rational functions whose denominators can be factored into real linear and quadratic factors. Remember to divide the polynomials first, if necessary, to reduce the fraction to one whose numerator has degree smaller than that of its denominator.
7. There is a table of integrals at the back of this book. If you can't do an integral directly, try to use the methods above to convert it to the form of one of the integrals in the table.
8. If you can't find any way to evaluate a definite integral for which you need a numerical value, consider using a computer or calculator and one of the numerical methods presented in Sections 6.6–6.8.

Review Exercises on Techniques of Integration

Here is an opportunity to get more practice evaluating integrals. Unlike the exercises in Sections 5.6 and 6.1–6.3, which used only the technique of the particular section, these exercises are grouped randomly so you will have to decide which techniques to use.

1. $\displaystyle\int \frac{x\,dx}{2x^2 + 5x + 2}$
2. $\displaystyle\int \frac{x\,dx}{(x-1)^3}$
3. $\displaystyle\int \sin^3 x \cos^3 x\,dx$
4. $\displaystyle\int \frac{(1 + \sqrt{x})^{1/3}}{\sqrt{x}}\,dx$
5. $\displaystyle\int \frac{3\,dx}{4x^2 - 1}$
6. $\displaystyle\int (x^2 + x - 2) \sin 3x\,dx$
7. $\displaystyle\int \frac{\sqrt{1 - x^2}}{x^4}\,dx$
8. $\displaystyle\int x^3 \cos(x^2)\,dx$
9. $\displaystyle\int \frac{x^2\,dx}{(5x^3 - 2)^{2/3}}$
10. $\displaystyle\int \frac{dx}{x^2 + 2x - 15}$
11. $\displaystyle\int \frac{dx}{(4 + x^2)^2}$
12. $\displaystyle\int (\sin x + \cos x)^2\,dx$
13. $\displaystyle\int 2^x \sqrt{1 + 4^x}\,dx$
14. $\displaystyle\int \frac{\cos x}{1 + \sin^2 x}\,dx$
15. $\displaystyle\int \frac{\sin^3 x}{\cos^7 x}\,dx$
16. $\displaystyle\int \frac{x^2\,dx}{(3 + 5x^2)^{3/2}}$
17. $\displaystyle\int e^{-x} \sin(2x)\,dx$
18. $\displaystyle\int \frac{2x^2 + 4x - 3}{x^2 + 5x}\,dx$
19. $\displaystyle\int \cos(3 \ln x)\,dx$
20. $\displaystyle\int \frac{dx}{4x^3 + x}$
21. $\displaystyle\int \frac{x \ln(1 + x^2)}{1 + x^2}\,dx$
22. $\displaystyle\int \sin^2 x \cos^4 x\,dx$
23. $\displaystyle\int \frac{x^2}{\sqrt{2 - x^2}}\,dx$
24. $\displaystyle\int \tan^4 x \sec x\,dx$
25. $\displaystyle\int \frac{x^2\,dx}{(4x + 1)^{10}}$
26. $\displaystyle\int x \sin^{-1} \frac{x}{2}\,dx$
27. $\displaystyle\int \sin^5(4x)\,dx$
28. $\displaystyle\int \frac{dx}{x^5 - 2x^3 + x}$
29. $\displaystyle\int \frac{dx}{2 + e^x}$
30. $\displaystyle\int x^3 3^x\,dx$
31. $\displaystyle\int \frac{\sin^2 x \cos x}{2 - \sin x}\,dx$
32. $\displaystyle\int \frac{x^2 + 1}{x^2 + 2x + 2}\,dx$
33. $\displaystyle\int \frac{dx}{x^2\sqrt{1 - x^2}}$
34. $\displaystyle\int x^3 (\ln x)^2\,dx$
35. $\displaystyle\int \frac{x^3}{\sqrt{1 - 4x^2}}\,dx$
36. $\displaystyle\int \frac{e^{1/x}\,dx}{x^2}$
37. $\displaystyle\int \frac{x + 1}{\sqrt{x^2 + 1}}\,dx$
38. $\displaystyle\int e^{(x^{1/3})}\,dx$
39. $\displaystyle\int \frac{x^3 - 3}{x^3 - 9x}\,dx$
40. $\displaystyle\int \frac{10^{\sqrt{x+2}}}{\sqrt{x + 2}}\,dx$
41. $\displaystyle\int \sin^5 x \cos^9 x\,dx$
42. $\displaystyle\int \frac{x^2\,dx}{\sqrt{x^2 - 1}}$
43. $\displaystyle\int \frac{x\,dx}{x^2 + 2x - 1}$
44. $\displaystyle\int \frac{2x - 3}{\sqrt{4 - 3x + x^2}}\,dx$
45. $\displaystyle\int x^2 \sin^{-1}(2x)\,dx$
46. $\displaystyle\int \frac{\sqrt{3x^2 - 1}}{x}\,dx$
47. $\displaystyle\int \cos^4 x \sin^4 x\,dx$
48. $\displaystyle\int \sqrt{x - x^2}\,dx$
49. $\displaystyle\int \frac{dx}{(4 + x)\sqrt{x}}$
50. $\displaystyle\int x \tan^{-1} \frac{x}{3}\,dx$
51. $\displaystyle\int \frac{x^4 - 1}{x^3 + 2x^2}\,dx$
52. $\displaystyle\int \frac{dx}{x(x^2 + 4)^2}$
53. $\displaystyle\int \frac{\sin(2 \ln x)}{x}\,dx$
54. $\displaystyle\int \frac{\sin(\ln x)}{x^2}\,dx$

55. $\displaystyle\int \frac{e^{2\tan^{-1}x}}{1+x^2}\,dx$

56. $\displaystyle\int \frac{x^3+x-2}{x^2-7}\,dx$

57. $\displaystyle\int \frac{\ln(3+x^2)}{3+x^2}\,x\,dx$

58. $\displaystyle\int \cos^7 x\,dx$

59. $\displaystyle\int \frac{\sin^{-1}(x/2)}{(4-x^2)^{1/2}}\,dx$

60. $\displaystyle\int \tan^4(\pi x)\,dx$

61. $\displaystyle\int \frac{(x+1)\,dx}{\sqrt{x^2+6x+10}}$

62. $\displaystyle\int e^x(1-e^{2x})^{5/2}\,dx$

63. $\displaystyle\int \frac{x^3\,dx}{(x^2+2)^{7/2}}$

64. $\displaystyle\int \frac{x^2}{2x^2-3}\,dx$

65. $\displaystyle\int \frac{x^{1/2}}{1+x^{1/3}}\,dx$

66. $\displaystyle\int \frac{dx}{x(x^2+x+1)^{1/2}}$

67. $\displaystyle\int \frac{1+x}{1+\sqrt{x}}\,dx$

68. $\displaystyle\int \frac{x\,dx}{4x^4+4x^2+5}$

69. $\displaystyle\int \frac{x\,dx}{(x^2-4)^2}$

70. $\displaystyle\int \frac{dx}{x^3+x^2+x}$

71. $\displaystyle\int x^2\tan^{-1}x\,dx$

72. $\displaystyle\int e^x\sec(e^x)\,dx$

73. $\displaystyle\int \frac{dx}{4\sin x-3\cos x}$

74. $\displaystyle\int \frac{dx}{x^{1/3}-1}$

75. $\displaystyle\int \frac{dx}{\tan x+\sin x}$

76. $\displaystyle\int \frac{x\,dx}{\sqrt{3-4x-4x^2}}$

77. $\displaystyle\int \frac{\sqrt{x}}{1+x}\,dx$

78. $\displaystyle\int \sqrt{1+e^x}\,dx$

79. $\displaystyle\int \frac{x^4\,dx}{x^3-8}$

80. $\displaystyle\int xe^x\cos x\,dx$

Other Review Exercises

1. Evaluate $I=\int x\,e^x\cos x\,dx$ and $J=\int x\,e^x\sin x\,dx$ by differentiating $e^x\big((ax+b)\cos x+(cx+d)\sin x\big)$ and examining coefficients.

2. For which real numbers r is the following reduction formula (obtained using integration by parts) valid?

$$\int_0^\infty x^r e^{-x}\,dx = r\int_0^\infty x^{r-1}e^{-x}\,dx$$

Evaluate the integrals in Exercises 3–6, or show that they diverge.

3. $\displaystyle\int_0^{\pi/2}\csc x\,dx$

4. $\displaystyle\int_1^\infty \frac{1}{x+x^3}\,dx$

5. $\displaystyle\int_0^1 \sqrt{x}\ln x\,dx$

6. $\displaystyle\int_{-1}^1 \frac{dx}{x\sqrt{1-x^2}}$

7. Show that the integral $I=\int_0^\infty (1/(\sqrt{x}\,e^x))\,dx$ converges and that its value satisfies $I<(2e+1)/e$.

8. By measuring the areas enclosed by contours on a topographic map, a geologist determines the cross-sectional areas A (m^2) through a 60 m high hill at various heights h (m) given in Table 2.

Table 2.

h	0	10	20	30	40	50	60
A	10,200	9,200	8,000	7,100	4,500	2,400	100

If she uses the Trapezoid Rule to estimate the volume of the hill (which is $V=\int_0^{60}A(h)\,dh$), what will be her estimate, to the nearest 1,000 m^3?

9. What will be the geologist's estimate of the volume of the hill in Exercise 8 if she uses Simpson's Rule instead of the Trapezoid Rule?

10. Find the Trapezoid Rule and Midpoint Rule approximations T_4 and M_4 for the integral $I=\int_0^1\sqrt{2+\sin(\pi x)}\,dx$. Quote the results to 5 decimal places. Quote a value of I to as many decimal places as you feel are justified by these approximations.

11. Use the results of Exercise 10 to calculate the Trapezoid Rule approximation T_8 and the Simpson's Rule approximation S_8 for the integral I in that exercise. Quote a value of I to as many decimal places as you feel are justified by these approximations.

12. Devise a way to evaluate $I=\int_{1/2}^\infty x^2/(x^5+x^3+1)\,dx$ numerically, and use it to find I correct to 3 decimal places.

13. You want to approximate the integral $I=\int_0^4 f(x)\,dx$ of an unknown function $f(x)$, and you measure the following values of f:

Table 3.

x	0	1	2	3	4
$f(x)$	0.730	1.001	1.332	1.729	2.198

(a) What are the approximations T_4 and S_4 to I that you calculate with these data.

(b) You then decide to make more measurements in order to calculate T_8 and S_8. You obtain $T_8=5.5095$. What do you obtain for S_8?

(c) You have theoretical reasons to believe that $f(x)$ is, in fact, a polynomial of degree 3. Do your calculations support this theory? Why or why not?

Challenging Problems

1. (a) Some people think that $\pi=22/7$. Prove that this is not so by showing that

$$\int_0^1 \frac{x^4(1-x)^4}{x^2+1}\,dx=\frac{22}{7}-\pi.$$

(b) If $I=\int_0^1 x^4(1-x)^4\,dx$, show that

$$\frac{22}{7}-I<\pi<\frac{22}{7}-\frac{I}{2}.$$

(c) Evaluate I and hence determine an explicit small interval containing π.

2. (a) Find a reduction formula for $\int(1-x^2)^n\,dx$.

(b) Show that if n is a positive integer, then

$$\int_0^1(1-x^2)^n\,dx=\frac{2^{2n}(n!)^2}{(2n+1)!}.$$

(c) Use your reduction formula to evaluate $\int(1-x^2)^{-3/2}\,dx$.

3. (a) Show that x^4+x^2+1 factors into a product of two real quadratics, and evaluate $\int(x^2+1)/(x^4+x^2+1)\,dx$. *Hint:* $x^4+x^2+1=(x^2+1)^2-x^2$.

(b) Use the same method to find $\int (x^2+1)/(x^4+1)\,dx$.

4. Let $I_{m,n} = \int_0^1 x^m (\ln x)^n\,dx$.

(a) Show that $I_{m,n} = (-1)^n \int_0^\infty x^n e^{-(m+1)x}\,dx$.

(b) Show that $I_{m,n} = \dfrac{(-1)^n n!}{(m+1)^{n+1}}$.

5. Let $I_n = \int_0^1 x^n e^{-x}\,dx$.

(a) Show that $0 < I_n < \dfrac{1}{n+1}$ and hence that $\lim_{n\to\infty} I_n = 0$.

(b) Show that $I_n = nI_{n-1} - \dfrac{1}{e}$ for $n \ge 1$, and $I_0 = 1 - \dfrac{1}{e}$.

(c) Verify by induction that $I_n = n!\left(1 - \dfrac{1}{e}\displaystyle\sum_{j=0}^{n} \frac{1}{j!}\right)$.

(d) Deduce from (a) and (c) that $\displaystyle\lim_{n\to\infty} \sum_{j=0}^{n} \frac{1}{j!} = e$.

6. If K is very large, which of the approximations T_{100} (Trapezoidal Rule), M_{100} (Midpoint Rule), and S_{100} (Simpson's Rule) will be closest to the true value for $\int_0^1 e^{-Kx}\,dx$? Which will be farthest? Justify your answers. (*Caution:* This is trickier than it sounds!)

7. Simpson's Rule gives the exact definite integral for a cubic f. Suppose you want a numerical integration rule that gives the exact answer for a polynomial of degree 5. You might approximate the integral over the subinterval $[m-h, m+h]$ by something of the form $2h\left(af(m-h) + bf(m - \frac{h}{2}) + f(m) + bf(m + \frac{h}{2}) + af(m+h)\right)$, for some constants a, b, and c.

(a) Determine a, b, and c for which this will work. (*Hint:* Take $m = 0$ to make things simple.)

(b) Use this method to approximate $\int_0^1 e^{-x}\,dx$ using first one and then two of these intervals (thus evaluating the integrand at nine points).

8. The convergence of improper integrals can be a more delicate matter when the integrand changes sign. Here is one method that can be used to prove convergence in some cases where the comparison theorem fails.

(a) Suppose that $f(x)$ is differentiable on $[1, \infty)$, $f'(x)$ is continuous there, $f'(x) < 0$, and $\lim_{x\to\infty} f(x) = 0$. Show that $\int_1^\infty f'(x)\cos(x)\,dx$ converges. *Hint:* What is $\int_1^\infty |f'(x)|\,dx$?

(b) Under the same hypotheses, show that $\int_1^\infty f(x)\sin x\,dx$ converges. *Hint:* Integrate by parts and use (a).

(c) Show that $\int_1^\infty \dfrac{\sin x}{x}\,dx$ converges but $\int_1^\infty \dfrac{|\sin x|}{x}\,dx$ diverges. *Hint:* $|\sin x| \ge \sin^2 x = \dfrac{1-\cos(2x)}{2}$. Note that (b) would work just as well with $\sin x$ replaced by $\cos(2x)$.

CHAPTER 7

Applications of Integration

> 'It's like this,' he said. 'When you go after honey with a balloon, the great thing is not to let the bees know you're coming. Now if you have a green balloon, they might think you were only part of the tree and not notice you, and if you have a blue balloon, they might think you were only part of the sky and not notice you, and the question' [said Winnie the Pooh] 'is: Which is most likely?'

A. A. Milne 1882–1956
from *Winnie the Pooh*

Introduction

Numerous quantities in mathematics, physics, economics, biology, and indeed any quantitative science can be conveniently represented by integrals. In addition to measuring plane areas, the problem that motivated the definition of the definite integral, we can use these integrals to express volumes of solids, lengths of curves, areas of surfaces, forces, work, energy, pressure, probabilities, dollar values of a stream of payments, and a variety of other quantities that are in one sense or another equivalent to areas under graphs.

In addition, as we saw previously, many of the basic principles that govern the behaviour of our world are expressed in terms of differential equations and initial-value problems. Indefinite integration is a key tool in the solution of such problems.

In this chapter we examine some of these applications. For the most part they are independent of one another, and for that reason some of the later sections in this chapter can be regarded as optional material. The material of Sections 7.1–7.3, however, should be regarded as core because these ideas will arise again in the study of multivariable calculus.

7.1 Volumes by Slicing—Solids of Revolution

In this section we show how volumes of certain three-dimensional regions (or *solids*) can be expressed as definite integrals and thereby determined. We will not attempt to give a definition of *volume* but will rely on our intuition and experience with solid objects to provide enough insight for us to specify the volumes of certain simple solids. For example, if the base of a rectangular box is a rectangle of length l and width w (and therefore area $A = lw$), and if the box has height h, then its volume is $V = Ah = lwh$. If l, w, and h are measured in *units* (e.g., centimetres), then the volume is expressed in *cubic units* (cubic centimetres, or cm^3).

A rectangular box is a special case of a solid called a **prism** or **cylinder**. (See Figure 7.1.) Such a solid has a flat base occupying a region R in a plane, and consists of all points on parallel straight line segments having one end in R and the other end in a (necessarily congruent) region in a second plane parallel to the plane of the base. Either of these regions can be called the **base** of the prism or cylinder. If the base is bounded by straight lines, the solid is called a prism; if at least part of the boundary of the base is curved, the solid is called a cylinder. The height of the solid is the perpendicular distance between the parallel planes containing the two bases. If this height is h units and the area of a base is A square units, then the volume of the prism or cylinder is $V = Ah$ cubic units.

We use the adjective **right** to describe a prism or cylinder if the parallel line segments that constitute it are perpendicular to the base planes; otherwise, the prism or cylinder is called **oblique**. For example, a right cylinder whose bases are circular disks of radius r units and whose height is h units is called a **right circular cylinder**; its volume is $V = \pi r^2 h$ cubic units. Obliqueness has no effect on the volume $V = Ah$ of a prism or cylinder since h is always measured in a direction perpendicular to the base.

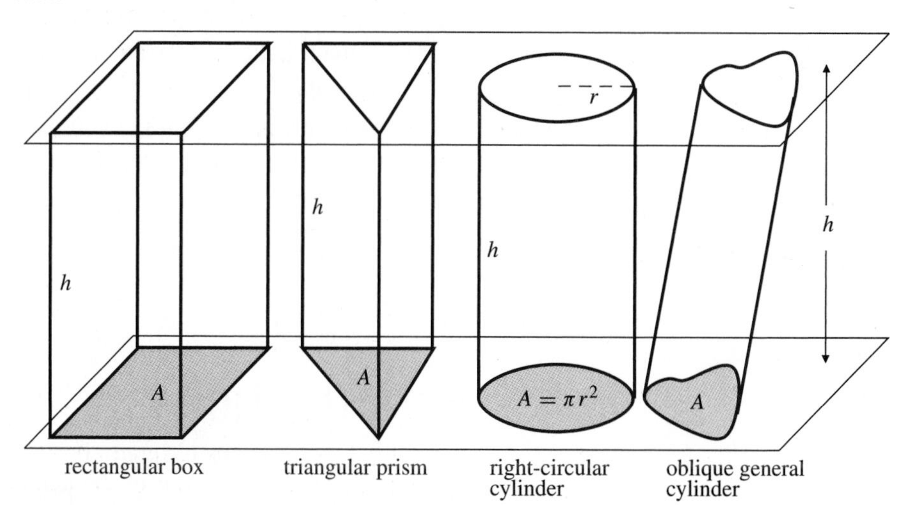

Figure 7.1 The volume of any prism or cylinder is the base area times the height (measured perpendicularly to the base): $V = Ah$

Volumes by Slicing

Knowing the volume of a cylinder enables us to determine the volumes of some more general solids. We can divide solids into thin "slices" by parallel planes. (Think of a loaf of sliced bread.) Each slice is approximately a cylinder of very small "height"; the height is the thickness of the slice. See Figure 7.2, where the height is measured horizontally in the direction of the x-axis. If we know the cross-sectional area of each slice, we can determine its volume and sum these volumes to find the volume of the solid.

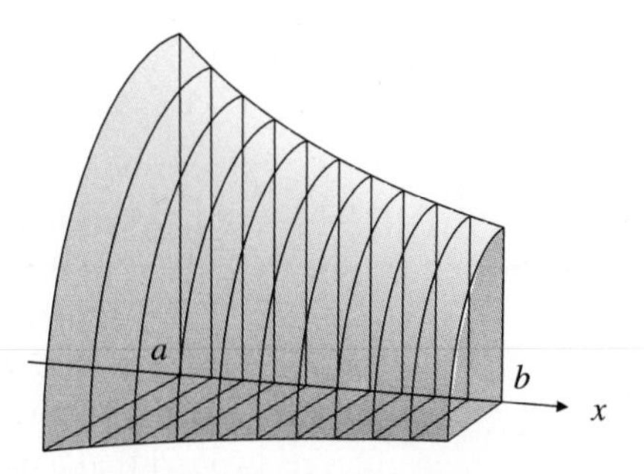

Figure 7.2 Slicing a solid perpendicularly to an axis

To be specific, suppose that the solid S lies between planes perpendicular to the x-axis at positions $x = a$ and $x = b$ and that the cross-sectional area of S in the plane perpendicular to the x-axis at x is a known function $A(x)$, for $a \le x \le b$. We assume that $A(x)$ is continuous on $[a, b]$. If $a = x_0 < x_1 < x_2 < \cdots < x_{n-1} < x_n = b$, then $P = \{x_0, x_1, x_2, \ldots, x_{n-1}, x_n\}$ is a partition of $[a, b]$ into n subintervals, and the planes perpendicular to the x-axis at $x_1, x_2, \ldots, x_{n-1}$ divide the solid into n slices of which the ith has thickness $\Delta x_i = x_i - x_{i-1}$. The volume ΔV_i of that slice lies between the maximum and minimum values of $A(x)\,\Delta x_i$ for values of x in $[x_{i-1}, x_i]$ (Figure 7.3), so

$$\Delta V_i = A(c_i)\,\Delta x_i$$

for some c_i in $[x_{i-1}, x_i]$, by the Intermediate-Value Theorem. The volume of the solid is therefore given by the Riemann sum

$$V = \sum_{i=1}^{n} \Delta V_i = \sum_{i=1}^{n} A(c_i)\,\Delta x_i.$$

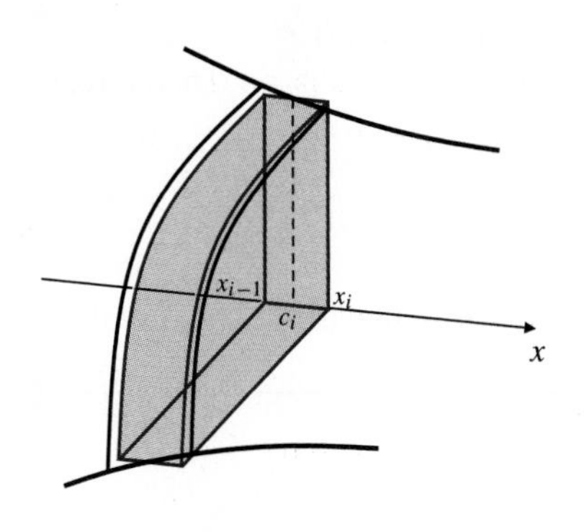

Figure 7.3 The volume of a slice

Letting n approach infinity in such a way that max Δx_i approaches 0, we obtain the definite integral of $A(x)$ over $[a, b]$ as the limit of this Riemann sum. Therefore:

> The volume V of a solid between $x = a$ and $x = b$ having cross-sectional area $A(x)$ at position x is
>
> $$V = \int_a^b A(x)\,dx.$$

There is another way to obtain this formula and others of a similar nature. Consider a slice of the solid between the planes perpendicular to the x-axis at positions x and $x + \Delta x$. Since $A(x)$ is continuous, it doesn't change much in a short interval, so if Δx is small, then the slice has volume ΔV approximately equal to the volume of a cylinder of base area $A(x)$ and height Δx:

$$\Delta V \approx A(x)\,\Delta x.$$

The error in this approximation is small compared to the size of ΔV. This suggests, correctly, that the **volume element**, that is, the volume of an infinitely thin slice of thickness dx is $dV = A(x)\,dx$, and that the volume of the solid is the "sum" (i.e., the integral) of these volume elements between the two ends of the solid, $x = a$ and $x = b$ (see Figure 7.4):

$$V = \int_{x=a}^{x=b} dV, \qquad \text{where} \qquad dV = A(x)\,dx.$$

We will use this *differential element* approach to model other applications that result in integrals rather than setting up explicit Riemann sums each time. Even though this argument does *not* constitute a proof of the formula, you are strongly encouraged to think of the formula this way; the volume is the integral of the volume elements.

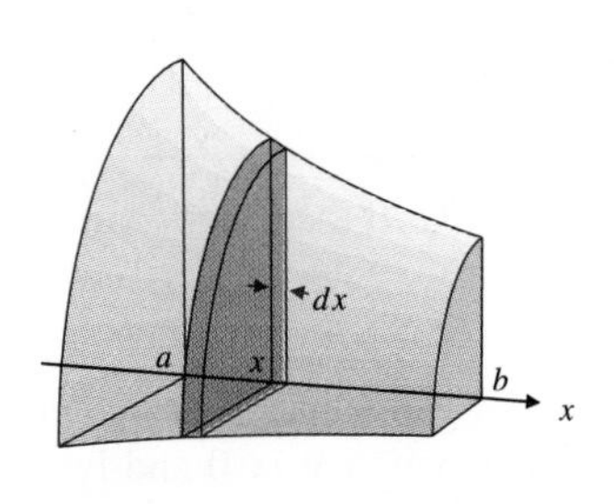

Figure 7.4 The volume element

Solids of Revolution

Many common solids have circular cross-sections in planes perpendicular to some axis. Such solids are called **solids of revolution** because they can be generated by rotating a plane region about an axis in that plane so that it sweeps out the solid. For example, a solid ball is generated by rotating a half-disk about the diameter of that half-disk (Figure 7.5(a)). Similarly, a solid right-circular cone is generated by rotating a right-angled triangle about one of its legs (Figure 7.5(b)).

If the region R bounded by $y = f(x)$, $y = 0$, $x = a$, and $x = b$ is rotated about the x-axis, then the cross-section of the solid generated in the plane perpendicular to the x-axis at x is a circular disk of radius $|f(x)|$. The area of this cross-section is $A(x) = \pi\big(f(x)\big)^2$, so the volume of the solid of revolution is

$$V = \pi \int_a^b (f(x))^2\,dx.$$

EXAMPLE 1 **(The volume of a ball)** Find the volume of a solid ball having radius a.

Solution The ball can be generated by rotating the half-disk, $0 \le y \le \sqrt{a^2 - x^2}$, $-a \le x \le a$ about the x-axis. See the cutaway view in Figure 7.5(a). Therefore, its volume is

$$\begin{aligned} V &= \pi \int_{-a}^{a} (\sqrt{a^2 - x^2})^2\, dx = 2\pi \int_0^a (a^2 - x^2)\, dx \\ &= 2\pi \left(a^2 x - \frac{x^3}{3}\right)\bigg|_0^a = 2\pi \left(a^3 - \frac{1}{3}a^3\right) = \frac{4}{3}\pi a^3 \text{ cubic units.} \end{aligned}$$

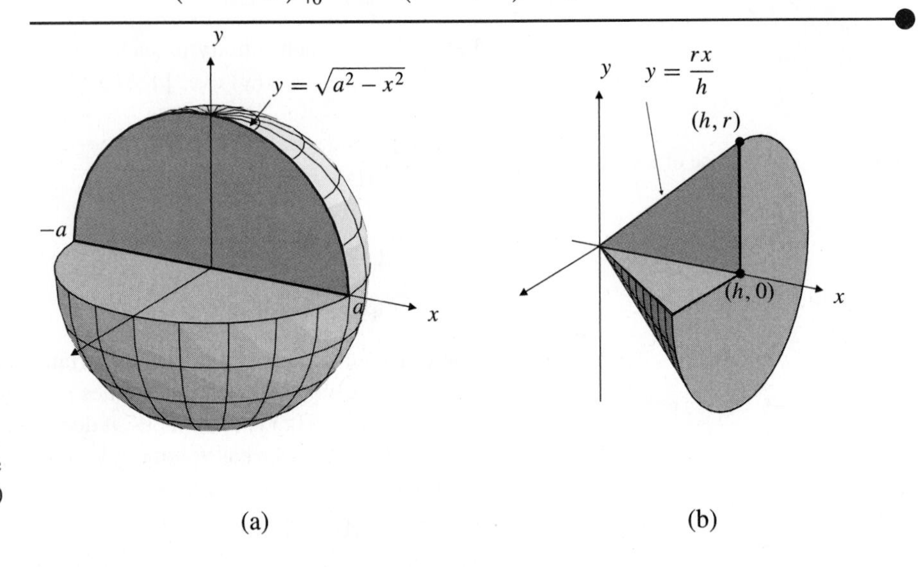

Figure 7.5

(a) The ball is generated by rotating the half-disk $0 \le y \le \sqrt{a^2 - x^2}$ (shown in colour) about the x-axis

(b) The cone of base radius r and height h is generated by rotating the triangle $0 \le x \le h$, $0 \le y \le rx/h$ (in colour) about the x-axis

EXAMPLE 2 **(The volume of a right-circular cone)** Find the volume of the right-circular cone of base radius r and height h that is generated by rotating the triangle with vertices $(0, 0)$, $(h, 0)$, and (h, r) about the x-axis.

Solution The line from $(0, 0)$ to (h, r) has equation $y = rx/h$. Thus the volume of the cone (see the cutaway view in Figure 7.5(b)) is

$$V = \pi \int_0^h \left(\frac{rx}{h}\right)^2 dx = \pi \left(\frac{r}{h}\right)^2 \frac{x^3}{3}\bigg|_0^h = \frac{1}{3}\pi r^2 h \text{ cubic units.}$$

Improper integrals can represent volumes of unbounded solids. If the improper integral converges, the unbounded solid has a finite volume.

EXAMPLE 3 Find the volume of the infinitely long horn that is generated by rotating the region bounded by $y = 1/x$ and $y = 0$ and lying to the right of $x = 1$ about the x-axis. The horn is illustrated in Figure 7.6.

Solution The volume of the horn is

$$\begin{aligned} V &= \pi \int_1^{\infty} \left(\frac{1}{x}\right)^2 dx = \pi \lim_{R\to\infty} \int_1^R \frac{1}{x^2}\, dx \\ &= -\pi \lim_{R\to\infty} \frac{1}{x}\bigg|_1^R = -\pi \lim_{R\to\infty} \left(\frac{1}{R} - 1\right) = \pi \text{ cubic units.} \end{aligned}$$

It is interesting to note that this finite volume arises from rotating a region that itself has infinite area: $\int_1^{\infty} dx/x = \infty$. We have a paradox: it takes an infinite amount of paint to paint the region but only a finite amount to fill the horn obtained by rotating the region. (How can you resolve this paradox?)

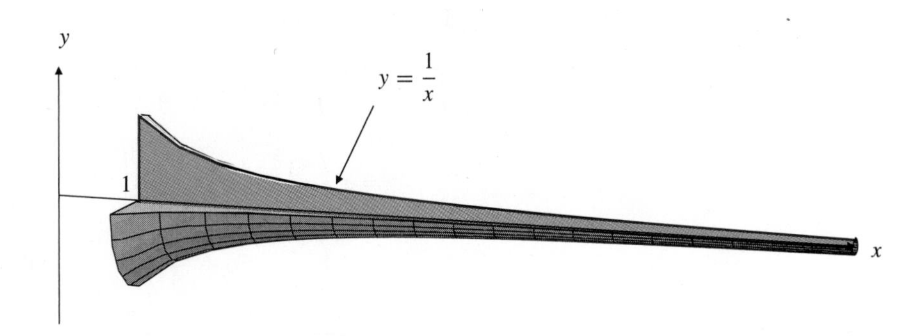

Figure 7.6 Cutaway view of an infinitely long horn

The following example shows how to deal with a problem where the axis of rotation is not the x-axis. Just rotate a suitable area element about the axis to form a volume element.

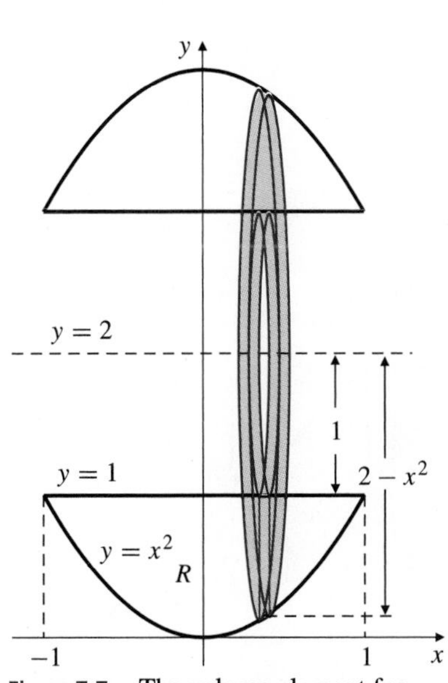

Figure 7.7 The volume element for Example 4

EXAMPLE 4 A ring-shaped solid is generated by rotating the finite plane region R bounded by the curve $y = x^2$ and the line $y = 1$ about the line $y = 2$. Find its volume.

Solution First, we solve the pair of equations $y = x^2$ and $y = 1$ to obtain the intersections at $x = -1$ and $x = 1$. The solid lies between these two values of x. The area element of R at position x is a vertical strip of width dx extending upward from $y = x^2$ to $y = 1$. When R is rotated about the line $y = 2$, this area element sweeps out a thin, washer-shaped volume element of thickness dx and radius $2 - x^2$, having a hole of radius 1 through the middle. (See Figure 7.7.) The cross-sectional area of this element is the area of a circle of radius $2 - x^2$ minus the area of the hole, a circle of radius 1. Thus,

$$dV = \left(\pi(2 - x^2)^2 - \pi(1)^2\right) dx = \pi(3 - 4x^2 + x^4)\, dx.$$

Since the solid extends from $x = -1$ to $x = 1$, its volume is

$$V = \pi \int_{-1}^{1} (3 - 4x^2 + x^4)\, dx = 2\pi \int_{0}^{1} (3 - 4x^2 + x^4)\, dx$$

$$= 2\pi \left(3x - \frac{4x^3}{3} + \frac{x^5}{5}\right)\Bigg|_0^1 = 2\pi \left(3 - \frac{4}{3} + \frac{1}{5}\right) = \frac{56\pi}{15} \text{ cubic units.}$$

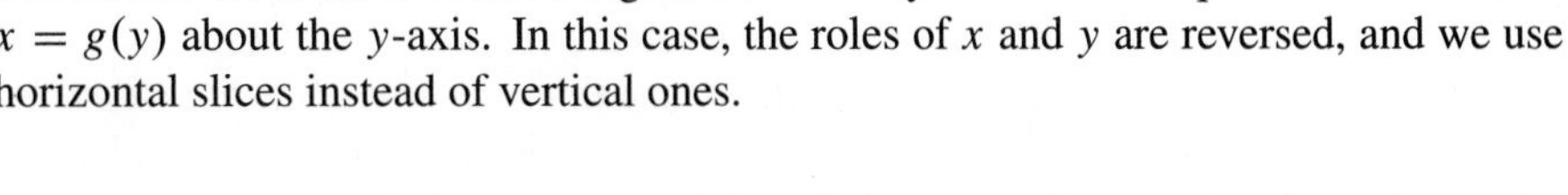

Sometimes we want to rotate a region bounded by curves with equations of the form $x = g(y)$ about the y-axis. In this case, the roles of x and y are reversed, and we use horizontal slices instead of vertical ones.

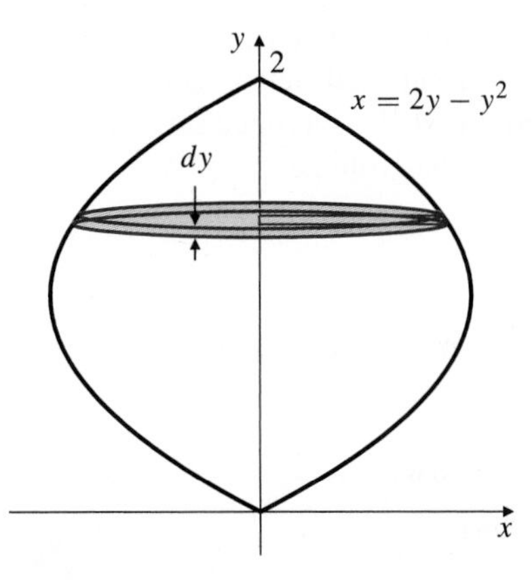

Figure 7.8 The volume element for Example 5

EXAMPLE 5 Find the volume of the solid generated by rotating the region to the right of the y-axis and to the left of the curve $x = 2y - y^2$ about the y-axis.

Solution For intersections of $x = 2y - y^2$ and $x = 0$, we have

$$2y - y^2 = 0 \quad \Longrightarrow \quad y = 0 \quad \text{or} \quad y = 2.$$

The solid lies between the horizontal planes at $y = 0$ and $y = 2$. A horizontal area element at height y and having thickness dy rotates about the y-axis to generate a thin disk-shaped volume element of radius $2y - y^2$ and thickness dy. (See Figure 7.8.) Its volume is

$$dV = \pi(2y - y^2)^2\, dy = \pi(4y^2 - 4y^3 + y^4)\, dy.$$

Thus, the volume of the solid is

$$\begin{aligned} V &= \pi \int_0^2 (4y^2 - 4y^3 + y^4)\, dy \\ &= \pi \left(\frac{4y^3}{3} - y^4 + \frac{y^5}{5} \right) \Bigg|_0^2 \\ &= \pi \left(\frac{32}{3} - 16 + \frac{32}{5} \right) = \frac{16\pi}{15} \text{ cubic units.} \end{aligned}$$

Cylindrical Shells

Suppose that the region R bounded by $y = f(x) \geq 0$, $y = 0$, $x = a \geq 0$, and $x = b > a$ is rotated about the y-axis to generate a solid of revolution. In order to find the volume of the solid using (plane) slices, we would need to know the cross-sectional area $A(y)$ in each plane of height y, and this would entail solving the equation $y = f(x)$ for one or more solutions of the form $x = g(y)$. In practice this can be inconvenient or impossible.

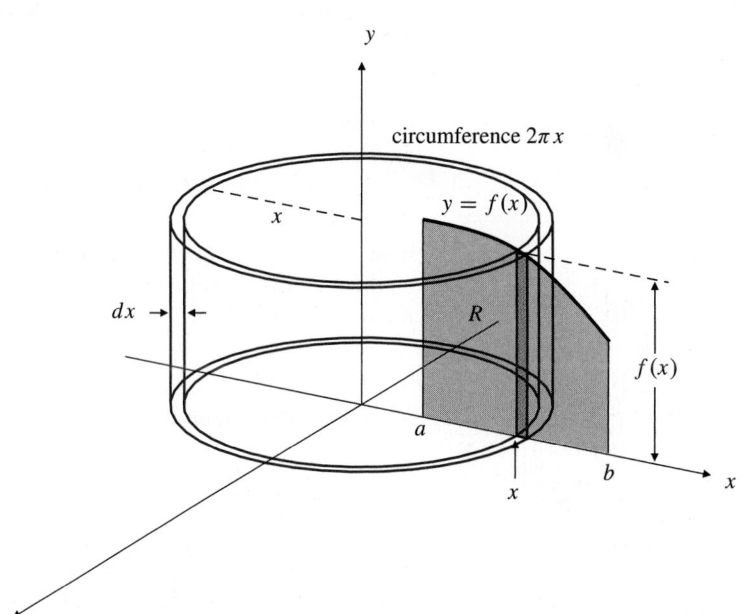

Figure 7.9 When rotated around the y-axis, the area element of width dx under $y = f(x)$ at x generates a cylindrical shell of height $f(x)$, circumference $2\pi x$, and hence volume $dV = 2\pi x\, f(x)\, dx$

The standard area element of R at position x is a vertical strip of width dx, height $f(x)$, and area $dA = f(x)\, dx$. When R is rotated about the y-axis, this strip sweeps out a volume element in the shape of a circular **cylindrical shell** having radius x, height $f(x)$, and thickness dx. (See Figure 7.9.) Regard this shell as a rolled-up rectangular slab with dimensions $2\pi x$, $f(x)$, and dx; evidently it has volume

$$dV = 2\pi x\, f(x)\, dx.$$

The volume of the solid of revolution is the sum *(integral)* of the volumes of such shells with radii ranging from a to b:

The volume of the solid obtained by rotating the plane region $0 \leq y \leq f(x)$, $0 \leq a < x < b$ about the y-axis is

$$V = 2\pi \int_a^b x\, f(x)\, dx.$$

EXAMPLE 6 **(The volume of a torus)** A disk of radius a has centre at the point $(b, 0)$, where $b > a > 0$. The disk is rotated about the y-axis to generate a **torus** (a doughnut-shaped solid), illustrated in Figure 7.10. Find its volume.

Solution The circle with centre at $(b, 0)$ and having radius a has equation $(x - b)^2 + y^2 = a^2$, so its upper semicircle is the graph of the function

$$f(x) = \sqrt{a^2 - (x - b)^2}.$$

We will double the volume of the upper half of the torus, which is generated by rotating the half-disk $0 \le y \le \sqrt{a^2 - (x - b)^2}$, $b - a \le x \le b + a$ about the y-axis. The volume of the complete torus is

$$\begin{aligned} V &= 2 \times 2\pi \int_{b-a}^{b+a} x\sqrt{a^2 - (x - b)^2}\, dx \qquad \text{Let } u = x - b, \ du = dx \\ &= 4\pi \int_{-a}^{a} (u + b)\sqrt{a^2 - u^2}\, du \\ &= 4\pi \int_{-a}^{a} u\sqrt{a^2 - u^2}\, du + 4\pi b \int_{-a}^{a} \sqrt{a^2 - u^2}\, du \\ &= 0 + 4\pi b \frac{\pi a^2}{2} = 2\pi^2 a^2 b \text{ cubic units.} \end{aligned}$$

(The first of the final two integrals is 0 because the integrand is odd and the interval is symmetric about 0; the second is the area of a semicircle of radius a.) Note that the volume of the torus is $(\pi a^2)(2\pi b)$, that is, the area of the disk being rotated times the distance travelled by the centre of that disk as it rotates about the y-axis. This result will be generalized by Pappus's Theorem in Section 7.5.

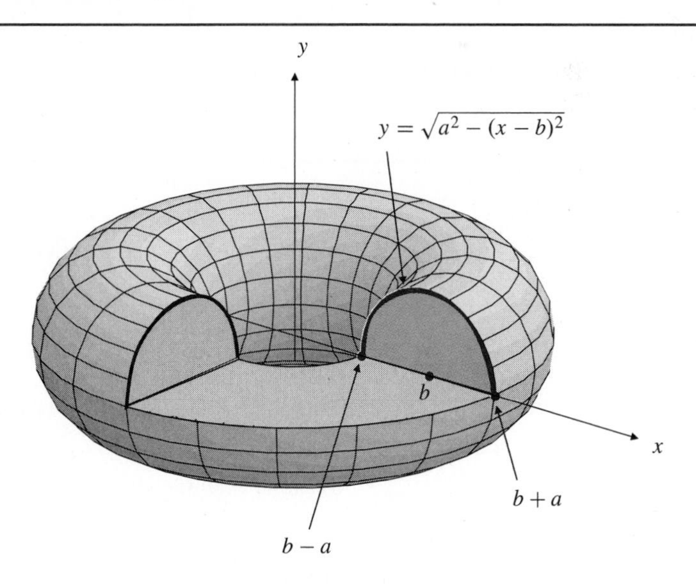

Figure 7.10 Cutaway view of a torus

EXAMPLE 7 Find the volume of a bowl obtained by revolving the parabolic arc $y = x^2$, $0 \le x \le 1$ about the y-axis.

Solution The interior of the bowl corresponds to revolving the region given by $x^2 \le y \le 1$, $0 \le x \le 1$ about the y-axis. The area element at position x has height $1 - x^2$ and generates a cylindrical shell of volume $dV = 2\pi x(1 - x^2)\, dx$. (See

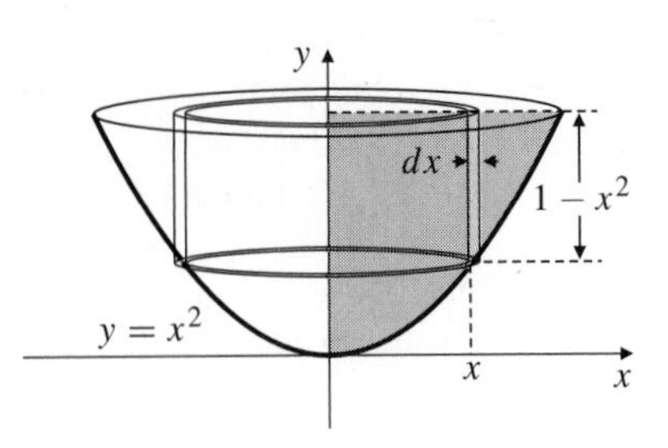

Figure 7.11 A parabolic bowl

Figure 7.11.) Thus, the volume of the bowl is

$$\begin{aligned} V &= 2\pi \int_0^1 x(1-x^2)\,dx \\ &= 2\pi \left(\frac{x^2}{2} - \frac{x^4}{4} \right) \Bigg|_0^1 = \frac{\pi}{2} \text{ cubic units.} \end{aligned}$$

We have described two methods for determining the volume of a solid of revolution, slicing and cylindrical shells. The choice of method for a particular solid is usually dictated by the form of the equations defining the region being rotated and by the axis of rotation. The volume element dV can always be determined by rotating a suitable area element dA about the axis of rotation. If the region is bounded by vertical lines and one or more graphs of the form $y = f(x)$, the appropriate area element is a vertical strip of width dx. If the rotation is about the x-axis or any other horizontal line, this strip generates a disk- or washer-shaped slice of thickness dx. If the rotation is about the y-axis or any other vertical line, the strip generates a cylindrical shell of thickness dx. On the other hand, if the region being rotated is bounded by horizontal lines and one or more graphs of the form $x = g(y)$, it is easier to use a horizontal strip of width dy as the area element, and this generates a slice if the rotation is about a vertical line and a cylindrical shell if the rotation is about a horizontal line. For very simple regions either method can be made to work easily. See the following table.

Table 1. Volumes of solids of revolution

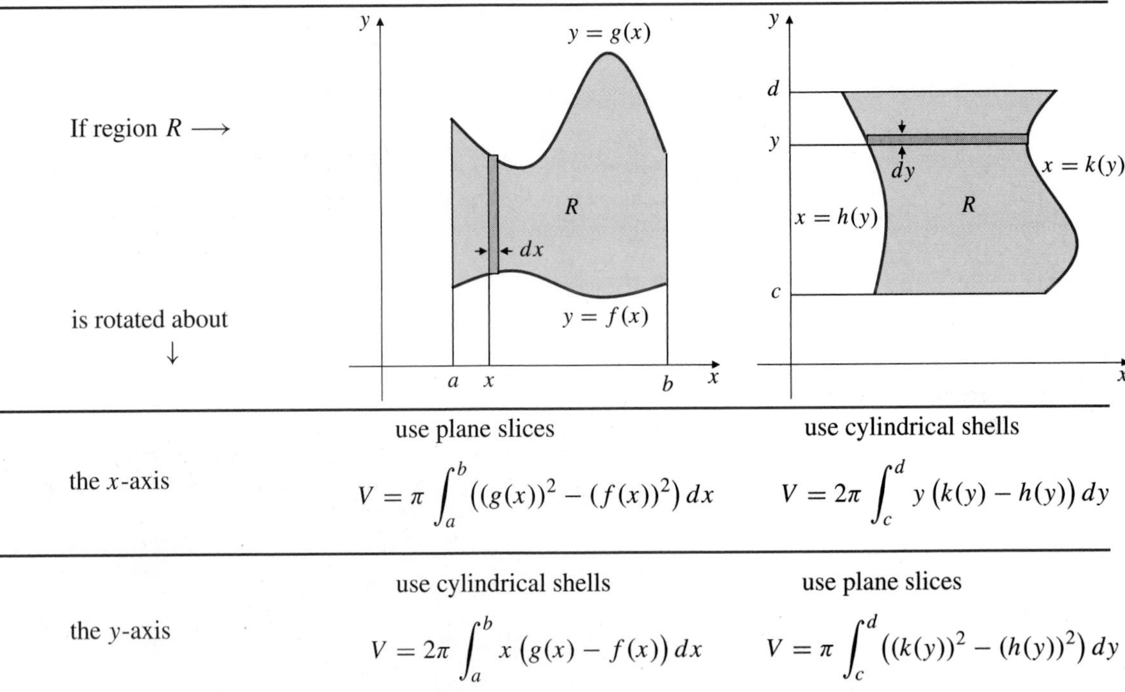

If region R $\longrightarrow$ is rotated about $\downarrow$	(region bounded by $y=f(x)$, $y=g(x)$, $x=a$, $x=b$)	(region bounded by $x=h(y)$, $x=k(y)$, $y=c$, $y=d$)
the x-axis	use plane slices $V = \pi \int_a^b \left((g(x))^2 - (f(x))^2 \right) dx$	use cylindrical shells $V = 2\pi \int_c^d y \left(k(y) - h(y) \right) dy$
the y-axis	use cylindrical shells $V = 2\pi \int_a^b x \left(g(x) - f(x) \right) dx$	use plane slices $V = \pi \int_c^d \left((k(y))^2 - (h(y))^2 \right) dy$

Our final example involves rotation about a vertical line other than the y-axis.

EXAMPLE 8 The triangular region bounded by $y = x$, $y = 0$, and $x = a > 0$ is rotated about the line $x = b > a$. (See Figure 7.12.) Find the volume of the solid so generated.

Solution Here the vertical area element at x generates a cylindrical shell of radius $b - x$, height x, and thickness dx. Its volume is $dV = 2\pi(b-x)\,x\,dx$, and the volume of the solid is

$$V = 2\pi \int_0^a (b-x)\,x\,dx = 2\pi \left(\frac{bx^2}{2} - \frac{x^3}{3} \right) \Bigg|_0^a = \pi \left(a^2 b - \frac{2a^3}{3} \right) \text{ cubic units.}$$

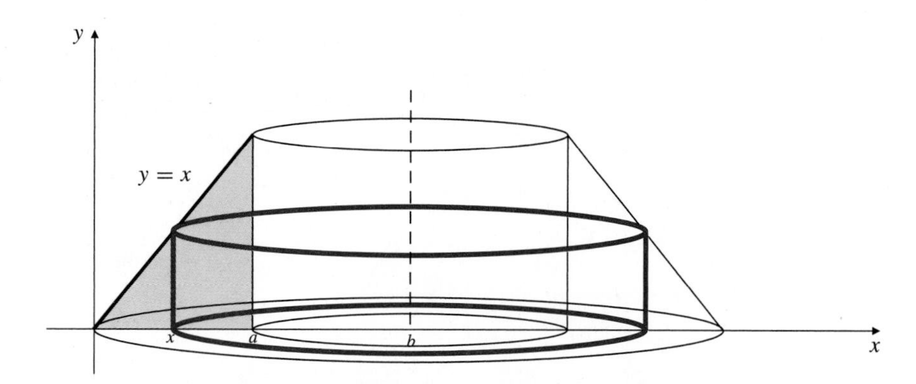

Figure 7.12 The volume element for Example 8

EXERCISES 7.1

Find the volume of each solid S in Exercises 1–4 in two ways, using the method of slicing and the method of cylindrical shells.

1. S is generated by rotating about the x-axis the region bounded by $y = x^2$, $y = 0$, and $x = 1$.

2. S is generated by rotating the region of Exercise 1 about the y-axis.

3. S is generated by rotating about the x-axis the region bounded by $y = x^2$ and $y = \sqrt{x}$ between $x = 0$ and $x = 1$.

4. S is generated by rotating the region of Exercise 3 about the y-axis.

Find the volumes of the solids obtained if the plane regions R described in Exercises 5–10 are rotated about (a) the x-axis and (b) the y-axis.

5. R is bounded by $y = x(2 - x)$ and $y = 0$ between $x = 0$ and $x = 2$.

6. R is the finite region bounded by $y = x$ and $y = x^2$.

7. R is the finite region bounded by $y = x$ and $x = 4y - y^2$.

8. R is bounded by $y = 1 + \sin x$ and $y = 1$ from $x = 0$ to $x = \pi$.

9. R is bounded by $y = 1/(1+x^2)$, $y = 2$, $x = 0$, and $x = 1$.

10. R is the finite region bounded by $y = 1/x$ and $3x + 3y = 10$.

11. The triangular region with vertices $(0, -1)$, $(1, 0)$, and $(0, 1)$ is rotated about the line $x = 2$. Find the volume of the solid so generated.

12. Find the volume of the solid generated by rotating the region $0 \le y \le 1 - x^2$ about the line $y = 1$.

13. What percentage of the volume of a ball of radius 2 is removed if a hole of radius 1 is drilled through the centre of the ball?

14. A cylindrical hole is bored through the centre of a ball of radius R. If the length of the hole is L, show that the volume of the remaining part of the ball depends only on L and not on R.

15. A cylindrical hole of radius a is bored through a solid right-circular cone of height h and base radius $b > a$. If the axis of the hole lies along that of the cone, find the volume of the remaining part of the cone.

16. Find the volume of the solid obtained by rotating a circular disk about one of its tangent lines.

17. A plane slices a ball of radius a into two pieces. If the plane passes b units away from the centre of the ball (where $b < a$), find the volume of the smaller piece.

18. Water partially fills a hemispherical bowl of radius 30 cm so that the maximum depth of the water is 20 cm. What volume of water is in the bowl?

19. Find the volume of the ellipsoid of revolution obtained by rotating the ellipse $(x^2/a^2) + (y^2/b^2) = 1$ about the x-axis.

20. Recalculate the volume of the torus of Example 6 by slicing perpendicular to the y-axis rather than using cylindrical shells.

21. The region R bounded by $y = e^{-x}$ and $y = 0$ and lying to the right of $x = 0$ is rotated (a) about the x-axis and (b) about the y-axis. Find the volume of the solid of revolution generated in each case.

22. The region R bounded by $y = x^{-k}$ and $y = 0$ and lying to the right of $x = 1$ is rotated about the x-axis. Find all real values of k for which the solid so generated has finite volume.

23. Repeat Exercise 22 with rotation about the y-axis.

24. Early editions of this text incorrectly defined a prism or cylinder as being a solid for which cross-sections parallel to the base were congruent to the base. Does this define a larger or smaller set of solids than the definition given in this section? What does the older definition say about the volume of a cylinder or prism having base area A and height h?

25. Continuing Exercise 24, consider the solid S whose cross-section in the plane perpendicular to the x-axis at x is an isosceles right-angled triangle having equal sides of length a cm with one end of the hypotenuse on the x-axis and with hypotenuse making angle x with a fixed direction. Is S a prism according to the definition given in early editions? Is it a prism according to the definition in this edition? If the height of S is b cm, what is the volume of S?

26. The region shaded in Figure 7.13 is rotated about the x-axis. Use Simpson's Rule to find the volume of the resulting solid.

27. The region shaded in Figure 7.13 is rotated about the y-axis. Use Simpson's Rule to find the volume of the resulting solid.

28. The region shaded in Figure 7.13 is rotated about the line $x = -1$. Use Simpson's Rule to find the volume of the resulting solid.

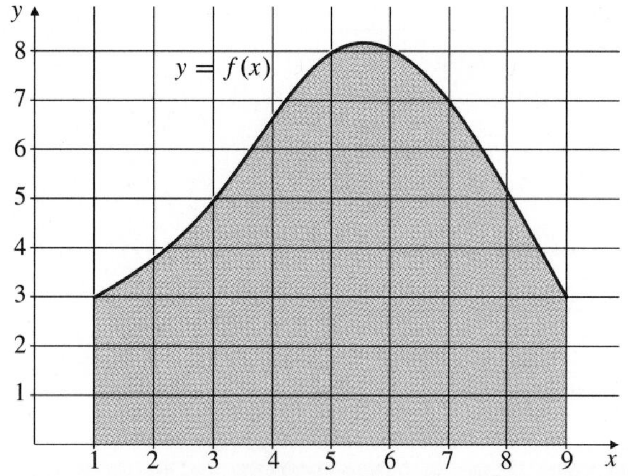

Figure 7.13

29. Find the volume of the solid generated by rotating the finite region in the first quadrant bounded by the coordinate axes and the curve $x^{2/3} + y^{2/3} = 4$ about either of the coordinate axes. (Both volumes are the same. Why?)

30. Given that the surface area of a sphere of radius r is kr^2, where k is a constant independent of r, express the volume of a ball of radius R as an integral of volume elements that are the volumes of spherical shells of thickness dr and varying radii r. Hence find k.

The following problems are *very difficult*. You will need some ingenuity and a lot of hard work to solve them by the techniques available to you now.

31. A martini glass in the shape of a right-circular cone of height h and semivertical angle α (see Figure 7.14) is filled with liquid. Slowly a ball is lowered into the glass, displacing liquid and causing it to overflow. Find the radius R of the ball that causes the greatest volume of liquid to overflow out of the glass.

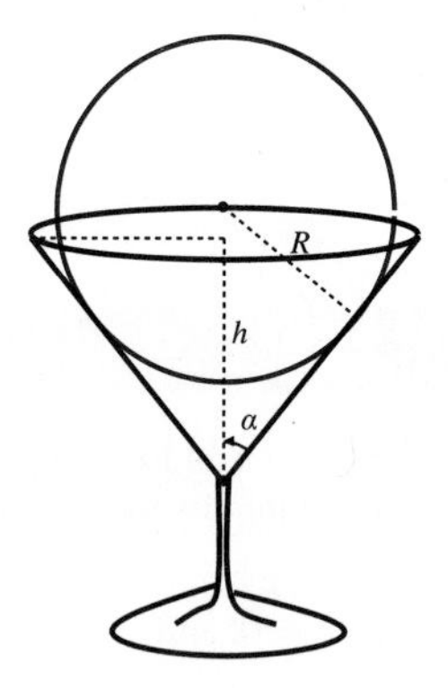

Figure 7.14

32. The finite plane region bounded by the curve $xy = 1$ and the straight line $2x + 2y = 5$ is rotated about that line to generate a solid of revolution. Find the volume of that solid.

7.2 More Volumes by Slicing

The method of slicing introduced in Section 7.1 can be used to determine volumes of solids that are not solids of revolution. All we need to know is the area of cross-section of the solid in every plane perpendicular to some fixed axis. If that axis is the x-axis, if the solid lies between the planes at $x = a$ and $x = b > a$, and if the cross-sectional area in the plane at x is the continuous (or even piecewise continuous) function $A(x)$, then the volume of the solid is

$$V = \int_a^b A(x)\,dx.$$

In this section we consider some examples that are not solids of revolution.

Pyramids and **cones** are solids consisting of all points on line segments that join a fixed point, the **vertex**, to all the points in a region lying in a plane not containing the vertex. The region is called the **base** of the pyramid or cone. Some pyramids and cones are shown in Figure 7.15. If the base is bounded by straight lines, the solid is called a pyramid; if the base has a curved boundary the solid is called a cone. All pyramids and cones have volume

$$V = \frac{1}{3}Ah,$$

where A is the area of the base region, and h is the height from the vertex to the plane of the base, measured in the direction perpendicular to that plane. We will give a very simple proof of this fact in Section 16.4. For the time being, we verify it for the case of a rectangular base.

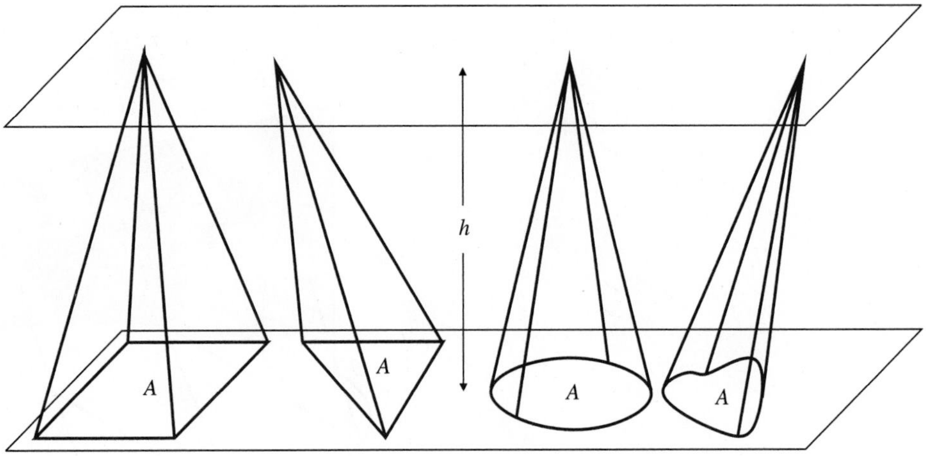

Figure 7.15 Some pyramids and cones. Each has volume $V = \frac{1}{3}Ah$, where A is the area of the base, and h is the height measured perpendicular to the base

EXAMPLE 1 Verify the formula for the volume of a pyramid with rectangular base of area A and height h.

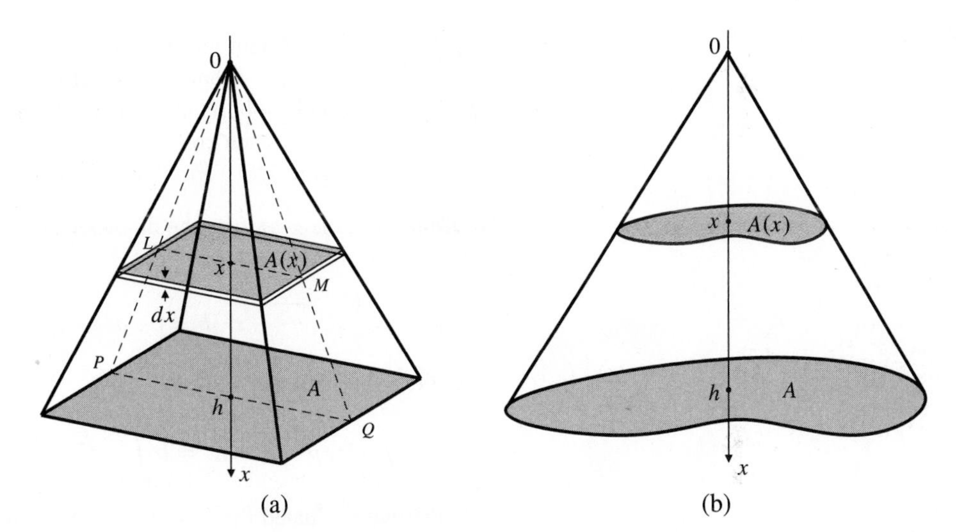

Figure 7.16
(a) A rectangular pyramid
(b) A general cone

Solution Cross-sections of the pyramid in planes parallel to the base are similar rectangles. If the origin is at the vertex of the pyramid and the x-axis is perpendicular to the base, then the cross-section at position x is a rectangle whose dimensions are x/h times the corresponding dimensions of the base. For example, in Figure 7.16(a), the length LM is x/h times the length PQ, as can be seen from the similar triangles OLM and OPQ. Thus, the area of the rectangular cross-section at x is

$$A(x) = \left(\frac{x}{h}\right)^2 A.$$

The volume of the pyramid is therefore

$$V = \int_0^h \left(\frac{x}{h}\right)^2 A\,dx = \frac{A}{h^2}\frac{x^3}{3}\bigg|_0^h = \frac{1}{3}Ah \text{ cubic units.}$$

A similar argument, resulting in the same formula for the volume, holds for a cone, that is, a pyramid with a more general (curved) shape to its base, such as that in Figure 7.16(b). Although it is not as obvious as in the case of the pyramid, the cross-section at x still has area $(x/h)^2$ times that of the base.

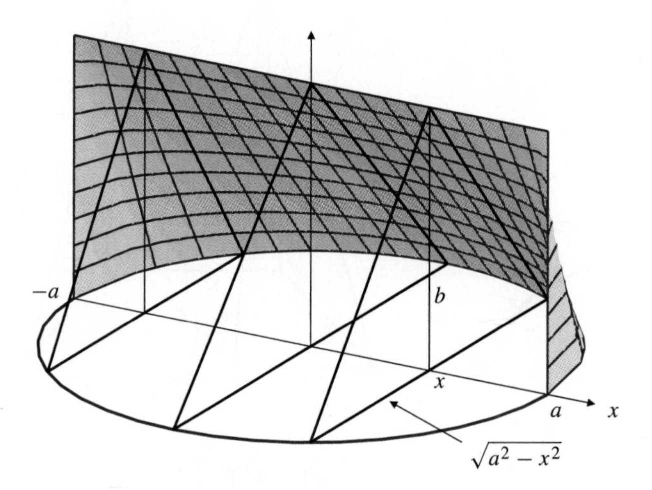

Figure 7.17 The tent of Example 2 with the front covering removed to show the shape more clearly

EXAMPLE 2 A tent has a circular base of radius a metres and is supported by a horizontal ridge bar held at height b metres above a diameter of the base by vertical supports at each end of the diameter. The material of the tent is stretched tight so that each cross-section perpendicular to the ridge bar is an isosceles triangle. (See Figure 7.17.) Find the volume of the tent.

Solution Let the x-axis be the diameter of the base under the ridge bar. The cross-section at position x has base length $2\sqrt{a^2 - x^2}$, so its area is

$$A(x) = \frac{1}{2}(2\sqrt{a^2 - x^2})b = b\sqrt{a^2 - x^2}.$$

Thus, the volume of the solid is

$$V = \int_{-a}^{a} b\sqrt{a^2 - x^2}\,dx = b\int_{-a}^{a}\sqrt{a^2 - x^2}\,dx = b\frac{\pi a^2}{2} = \frac{\pi}{2}a^2 b \text{ m}^3.$$

Note that we evaluated the last integral by inspection. It is the area of a half-disk of radius a.

EXAMPLE 3 Two circular cylinders, each having radius a, intersect so that their axes meet at right angles. Find the volume of the region lying inside both cylinders.

Solution We represent the cylinders in a three-dimensional Cartesian coordinate system where the plane containing the x- and y-axes is horizontal and the z-axis is vertical. One-eighth of the solid is represented in Figure 7.18, that part corresponding to all three coordinates being positive. The two cylinders have axes along the x- and y-axes, respectively. The cylinder with axis along the x-axis intersects the plane of the y- and z-axes in a circle of radius a.

Similarly, the other cylinder meets the plane of the x- and z-axes in a circle of radius a. It follows that if the region lying inside both cylinders (and having $x \geq 0$, $y \geq 0$, and $z \geq 0$) is sliced horizontally, then the slice at height z above the xy-plane is a square of side $\sqrt{a^2 - z^2}$ and has area $A(z) = a^2 - z^2$. The volume V of the whole region, being eight times that of the part shown, is

$$V = 8\int_0^a (a^2 - z^2)\,dz = 8\left(a^2 z - \frac{z^3}{3}\right)\Bigg|_0^a = \frac{16}{3}a^3 \text{ cubic units.}$$

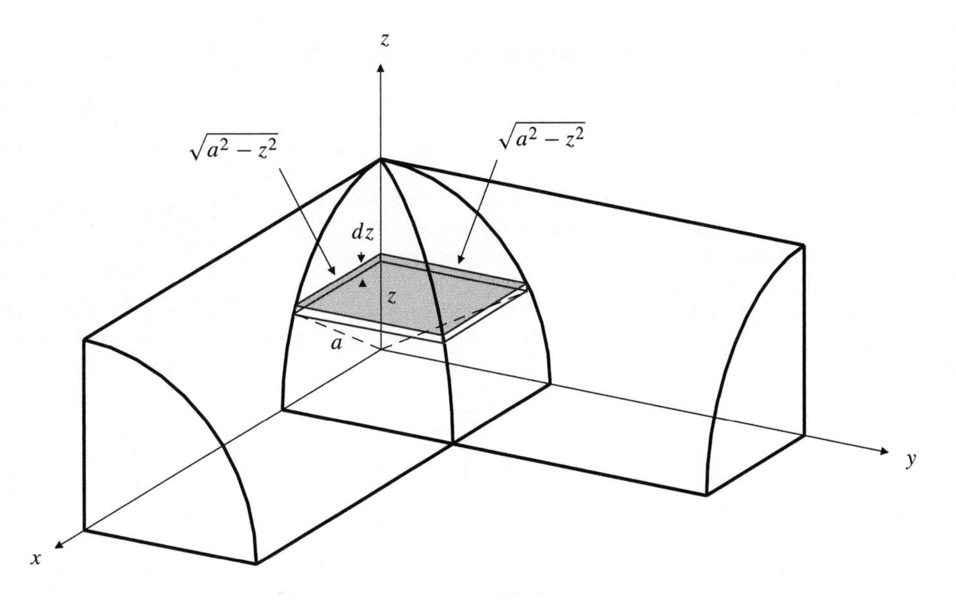

Figure 7.18 One-eighth of the solid lying inside two perpendicular cylindrical pipes. The horizontal slice shown is square

EXERCISES 7.2

1. A solid is 2 m high. The cross-section of the solid at height x above its base has area $3x$ square metres. Find the volume of the solid.
2. The cross-section at height z of a solid of height h is a rectangle with dimensions z and $h - z$. Find the volume of the solid.
3. Find the volume of a solid of height 1 whose cross-section at height z is an ellipse with semi-axes z and $\sqrt{1 - z^2}$.
4. A solid extends from $x = 1$ to $x = 3$. The cross-section of the solid in the plane perpendicular to the x-axis at x is a square of side x. Find the volume of the solid.
5. A solid is 6 ft high. Its horizontal cross-section at height z ft above its base is a rectangle with length $2 + z$ ft and width $8 - z$ ft. Find the volume of the solid.
6. A solid extends along the x-axis from $x = 1$ to $x = 4$. Its cross-section at position x is an equilateral triangle with edge length $\sqrt{x}$. Find the volume of the solid.
7. Find the volume of a solid that is h cm high if its horizontal cross-section at any height y above its base is a circular sector having radius a cm and angle $2\pi\big(1 - (y/h)\big)$ radians.
8. The opposite ends of a solid are at $x = 0$ and $x = 2$. The area of cross-section of the solid in a plane perpendicular to the x-axis at x is kx^3 square units. The volume of the solid is 4 cubic units. Find k.
9. Find the cross-sectional area of a solid in any horizontal plane at height z above its base if the volume of that part of the solid lying below any such plane is z^3 cubic units.
10. All the cross-sections of a solid in horizontal planes are squares. The volume of the part of the solid lying below any plane of height z is $4z$ cubic units, where $0 < z < h$, the height of the solid. Find the edge length of the square cross-section at height z for $0 < z < h$.
11. A solid has a circular base of radius r. All sections of the solid perpendicular to a particular diameter of the base are squares. Find the volume of the solid.
12. Repeat Exercise 11 but with sections that are equilateral triangles instead of squares.
13. The base of a solid is an isosceles right-angled triangle with equal legs measuring 12 cm. Each cross-section perpendicular to one of these legs is half of a circular disk. Find the volume of the solid.
14. **(Cavalieri's Principle)** Two solids have equal cross-sectional areas at equal heights above their bases. If both solids have the same height, show that they both have the same volume.

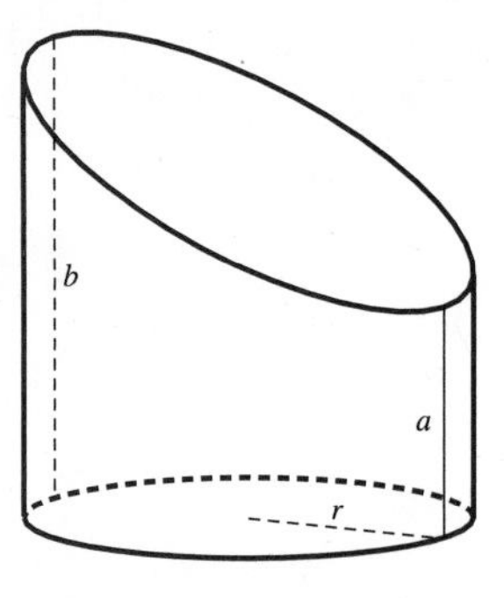

Figure 7.19

15. The top of a circular cylinder of radius r is a plane inclined at an angle to the horizontal. (See Figure 7.19.) If the lowest and highest points on the top are at heights a and b, respectively, above the base, find the volume of the cylinder. (Note that there is an easy geometric way to get the answer, but you should also try to do it by slicing. You can use either rectangular or trapezoidal slices.)

16. **(Volume of an ellipsoid)** Find the volume enclosed by the ellipsoid

$$\frac{x^2}{a^2} + \frac{y^2}{b^2} + \frac{z^2}{c^2} = 1.$$

Hint: This is not a solid of revolution. As in Example 3, the z-axis is perpendicular to the plane of the x- and y-axes. Each horizontal plane $z = k$ $(-c \le k \le c)$ intersects the ellipsoid in an ellipse $(x/a)^2 + (y/b)^2 = 1 - (k/c)^2$. Thus, $dV = dz \times$ the area of this ellipse. The area of the ellipse $(x/a)^2 + (y/b)^2 = 1$ is πab.

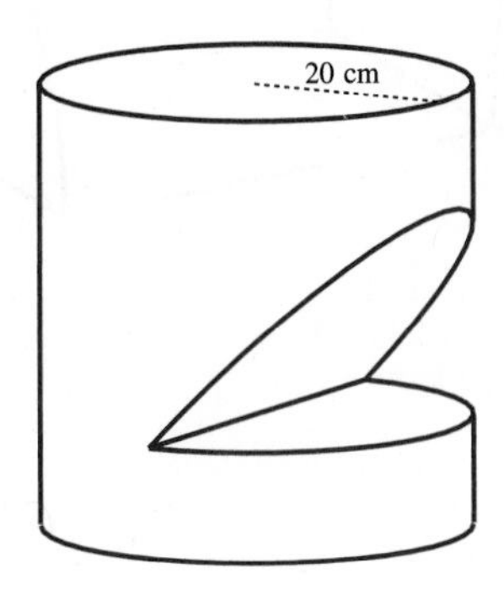

Figure 7.20

17. **(Notching a log)** A 45° notch is cut to the centre of a cylindrical log having radius 20 cm, as shown in Figure 7.20. One plane face of the notch is perpendicular to the axis of the log. What volume of wood was removed from the log by cutting the notch?

18. **(A smaller notch)** Repeat Exercise 17, but assume that the notch penetrates only one quarter way (10 cm) into the log.

19. What volume of wood is removed from a 3-in-thick board if a circular hole of radius 2 in is drilled through it with the axis of the hole tilted at an angle of 45° to board?

20. **(More intersecting cylinders)** The axes of two circular cylinders intersect at right angles. If the radii of the cylinders are a and b $(a > b > 0)$, show that the region lying inside both cylinders has volume

$$V = 8\int_0^b \sqrt{b^2 - z^2}\sqrt{a^2 - z^2}\, dz.$$

Hint: Review Example 3. Try to make a similar diagram, showing only one-eighth of the region. The integral is not easily evaluated.

21. A circular hole of radius 2 cm is drilled through the middle of a circular log of radius 4 cm, with the axis of the hole perpendicular to the axis of the log. Find the volume of wood removed from the log. *Hint:* This is very similar to Exercise 20. You will need to use numerical methods or a calculator with a numerical integration function to get the answer.

7.3 Arc Length and Surface Area

In this section we consider how integrals can be used to find the lengths of curves and the areas of the surfaces of solids of revolution.

Arc Length

If A and B are two points in the plane, let $|AB|$ denote the distance between A and B, that is, the length of the straight line segment AB.

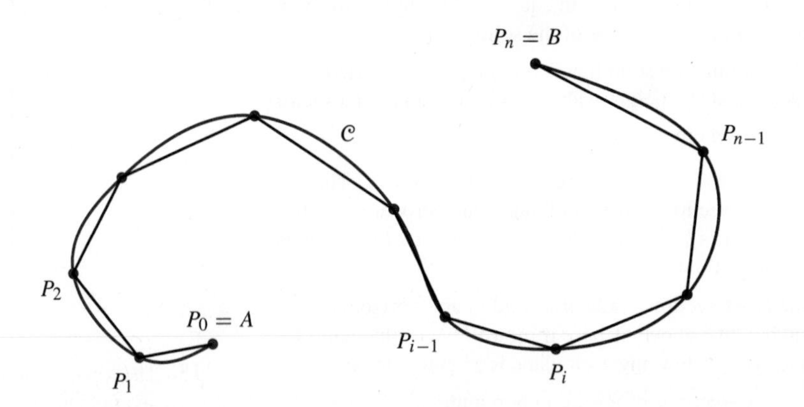

Figure 7.21 A polygonal approximation to a curve $\mathcal{C}$

Given a curve $\mathcal{C}$ joining the two points A and B, we would like to define what is meant by the *length* of the curve $\mathcal{C}$ from A to B. Suppose we choose points $A = P_0$, P_1, P_2, ..., P_{n-1}, and $P_n = B$ in order along the curve, as shown in Figure 7.21. The polygonal line $P_0P_1P_2 \ldots P_{n-1}P_n$ constructed by joining adjacent pairs of these points with straight line segments forms a *polygonal approximation* to $\mathcal{C}$, having length

$$L_n = |P_0P_1| + |P_1P_2| + \cdots + |P_{n-1}P_n| = \sum_{i=1}^{n} |P_{i-1}P_i|.$$

Intuition tells us that the shortest curve joining two points is a straight line segment, so the length L_n of any such polygonal approximation to $\mathcal{C}$ cannot exceed the length of $\mathcal{C}$. If we increase n by adding more vertices to the polygonal line between existing vertices, L_n cannot get smaller and may increase. If there exists a finite number K such that $L_n \le K$ for every polygonal approximation to $\mathcal{C}$, then there will be a smallest such number K (by the completeness of the real numbers), and we call this smallest K the arc length of $\mathcal{C}$.

DEFINITION 1

The **arc length** of the curve $\mathcal{C}$ from A to B is the smallest real number s such that the length L_n of every polygonal approximation to $\mathcal{C}$ satisfies $L_n \le s$.

A curve with a finite arc length is said to be **rectifiable**. Its arc length s is the limit of the lengths L_n of polygonal approximations as $n \to \infty$ in such a way that the maximum segment length $|P_{i-1}P_i| \to 0$.

It is possible to construct continuous curves that are bounded (they do not go off to infinity anywhere) but are not rectifiable; they have infinite length. To avoid such pathological examples, we will assume that our curves are **smooth**; they will be defined by functions having continuous derivatives.

The Arc Length of the Graph of a Function

Let f be a function defined on a closed, finite interval $[a, b]$ and having a continuous derivative f' there. If $\mathcal{C}$ is the graph of f, that is, the graph of the equation $y = f(x)$, then any partition of $[a, b]$ provides a polygonal approximation to $\mathcal{C}$. For the partition

$$\{a = x_0 < x_1 < x_2 < \cdots < x_n = b\},$$

let P_i be the point $\big(x_i, f(x_i)\big)$, $(0 \le i \le n)$. The length of the polygonal line $P_0P_1P_2 \ldots P_{n-1}P_n$ is

$$\begin{aligned} L_n &= \sum_{i=1}^{n} |P_{i-1}P_i| = \sum_{i=1}^{n} \sqrt{(x_i - x_{i-1})^2 + \big(f(x_i) - f(x_{i-1})\big)^2} \\ &= \sum_{i=1}^{n} \sqrt{1 + \left(\frac{f(x_i) - f(x_{i-1})}{x_i - x_{i-1}}\right)^2}\, \Delta x_i, \end{aligned}$$

where $\Delta x_i = x_i - x_{i-1}$. By the Mean-Value Theorem there exists a number c_i in the interval $[x_{i-1}, x_i]$ such that

$$\frac{f(x_i) - f(x_{i-1})}{x_i - x_{i-1}} = f'(c_i),$$

so we have $L_n = \displaystyle\sum_{i=1}^{n} \sqrt{1 + \big(f'(c_i)\big)^2}\, \Delta x_i$.

Thus L_n is a Riemann sum for $\int_a^b \sqrt{1 + (f'(x))^2}\, dx$. Being the limit of such Riemann sums as $n \to \infty$ in such a way that $\max(\Delta x_i) \to 0$, that integral is the length of the curve $\mathcal{C}$.

The arc length s of the curve $y = f(x)$ from $x = a$ to $x = b$ is given by

$$s = \int_a^b \sqrt{1 + \big(f'(x)\big)^2}\, dx = \int_a^b \sqrt{1 + \left(\frac{dy}{dx}\right)^2}\, dx.$$

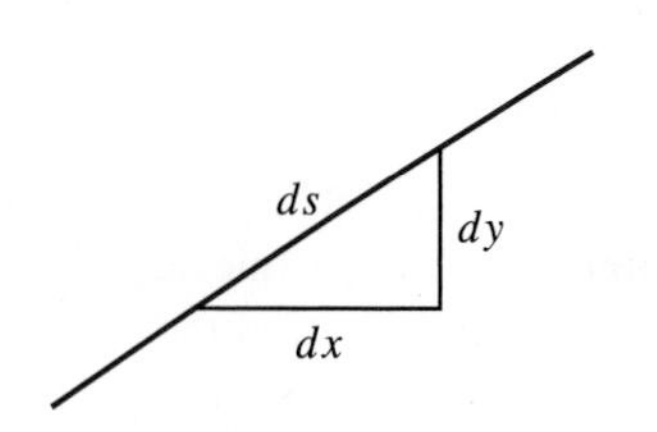

Figure 7.22 A differential triangle

You can regard the integral formula above as giving the arc length s of $\mathcal{C}$ as a "sum" of **arc length elements**

$$s = \int_{x=a}^{x=b} ds, \qquad \text{where} \qquad ds = \sqrt{1 + \big(f'(x)\big)^2}\, dx.$$

Figure 7.22 provides a convenient way to remember this; it also suggests how we can arrive at similar formulas for arc length elements of other kinds of curves. The *differential triangle* in the figure suggests that

$$(ds)^2 = (dx)^2 + (dy)^2.$$

Dividing this equation by $(dx)^2$ and taking the square root, we get

$$\left(\frac{ds}{dx}\right)^2 = 1 + \left(\frac{dy}{dx}\right)^2$$

$$\frac{ds}{dx} = \sqrt{1 + \left(\frac{dy}{dx}\right)^2}$$

$$ds = \sqrt{1 + \left(\frac{dy}{dx}\right)^2}\, dx = \sqrt{1 + \big(f'(x)\big)^2}\, dx.$$

A similar argument shows that for a curve specified by an equation of the form $x = g(y)$, $(c \le y \le d)$, the arc length element is

$$ds = \sqrt{1 + \left(\frac{dx}{dy}\right)^2}\, dy = \sqrt{1 + \big(g'(y)\big)^2}\, dy.$$

EXAMPLE 1 Find the length of the curve $y = x^{2/3}$ from $x = 1$ to $x = 8$.

Solution Since $dy/dx = \frac{2}{3}x^{-1/3}$ is continuous between $x = 1$ and $x = 8$ and $x^{1/3} > 0$ there, the length of the curve is given by

$$\begin{aligned} s &= \int_1^8 \sqrt{1 + \frac{4}{9}x^{-2/3}}\, dx = \int_1^8 \sqrt{\frac{9x^{2/3} + 4}{9x^{2/3}}}\, dx \\ &= \int_1^8 \frac{\sqrt{9x^{2/3} + 4}}{3x^{1/3}}\, dx \qquad \text{Let } u = 9x^{2/3} + 4, \\ &\qquad\qquad\qquad\qquad\quad du = 6x^{-1/3}\, dx \\ &= \frac{1}{18}\int_{13}^{40} u^{1/2}\, du = \frac{1}{27}u^{3/2}\Big|_{13}^{40} = \frac{40\sqrt{40} - 13\sqrt{13}}{27} \text{ units.} \end{aligned}$$

EXAMPLE 2 Find the length of the curve $y = x^4 + \dfrac{1}{32x^2}$ from $x = 1$ to $x = 2$.

Solution Here $\dfrac{dy}{dx} = 4x^3 - \dfrac{1}{16x^3}$ and

$$\begin{aligned} 1 + \left(\frac{dy}{dx}\right)^2 &= 1 + \left(4x^3 - \frac{1}{16x^3}\right)^2 \\ &= 1 + (4x^3)^2 - \frac{1}{2} + \left(\frac{1}{16x^3}\right)^2 \\ &= (4x^3)^2 + \frac{1}{2} + \left(\frac{1}{16x^3}\right)^2 = \left(4x^3 + \frac{1}{16x^3}\right)^2. \end{aligned}$$

The expression in the last set of parentheses is positive for $1 \le x \le 2$, so the length of the curve is

$$s = \int_1^2 \left(4x^3 + \frac{1}{16x^3}\right) dx = \left(x^4 - \frac{1}{32x^2}\right)\bigg|_1^2$$
$$= 16 - \frac{1}{128} - \left(1 - \frac{1}{32}\right) = 15 + \frac{3}{128} \text{ units.}$$

The examples above are deceptively simple; the curves were chosen so that the arc length integrals could be easily evaluated. For instance, the number 32 in the curve in Example 2 was chosen so the expression $1 + (dy/dx)^2$ would turn out to be a perfect square and its square root would cause no problems. Because of the square root in the formula, arc length problems for most curves lead to integrals that are difficult or impossible to evaluate without using numerical techniques.

EXAMPLE 3 **(Manufacturing corrugated panels)** Flat rectangular sheets of metal 2 m wide are to be formed into corrugated roofing panels 2 m wide by bending them into the sinusoidal shape shown in Figure 7.23. The period of the cross-sectional sine curve is 20 cm. Its amplitude is 5 cm, so the panel is 10 cm thick. How long should the flat sheets be cut if the resulting panels must be 5 m long?

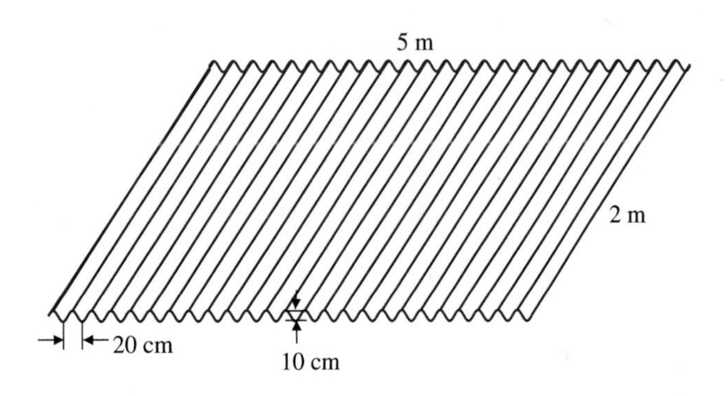

Figure 7.23 A corrugated roofing panel

Solution One period of the sinusoidal cross-section is shown in Figure 7.24. The distances are all in metres; the 5 cm amplitude is shown as 1/20 m, and the 20 cm period is shown as 2/10 m. The curve has equation

$$y = \frac{1}{20}\sin(10\pi x).$$

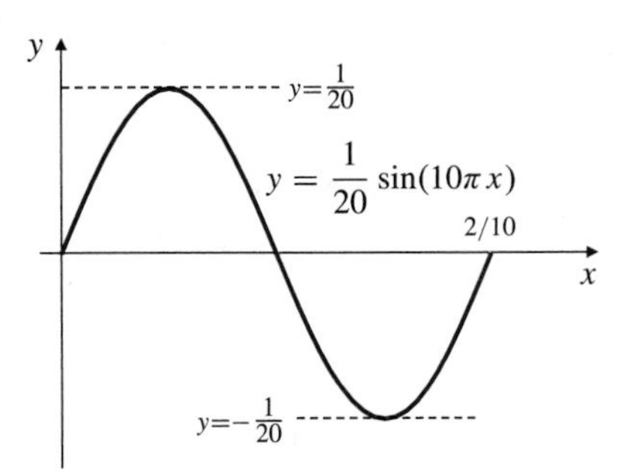

Figure 7.24 One period of the panel's cross-section

Note that 25 periods are required to produce a 5 m long panel. The length of the flat sheet required is 25 times the length of one period of the sine curve:

$$s = 25\int_0^{2/10} \sqrt{1 + \left(\frac{\pi}{2}\cos(10\pi x)\right)^2}\, dx \qquad \text{Let } t = 10\pi x, \; dt = 10\pi\, dx$$
$$= \frac{5}{2\pi}\int_0^{2\pi} \sqrt{1 + \frac{\pi^2}{4}\cos^2 t}\, dt = \frac{10}{\pi}\int_0^{\pi/2} \sqrt{1 + \frac{\pi^2}{4}\cos^2 t}\, dt.$$

The integral can be evaluated numerically using the techniques of the previous chapter or by using the definite integral function on an advanced scientific calculator. The value is $s \approx 7.32$. The flat metal sheet should be about 7.32 m long to yield a 5 m long finished panel.

If integrals needed for standard problems such as arc lengths of simple curves cannot be evaluated exactly, they are sometimes used to define new functions whose values are tabulated or built into computer programs. An example of this is the complete elliptic integral function that arises in the next example.

EXAMPLE 4 **(The circumference of an ellipse)** Find the circumference of the ellipse

$$\frac{x^2}{a^2} + \frac{y^2}{b^2} = 1,$$

where $a \geq b > 0$. See Figure 7.25.

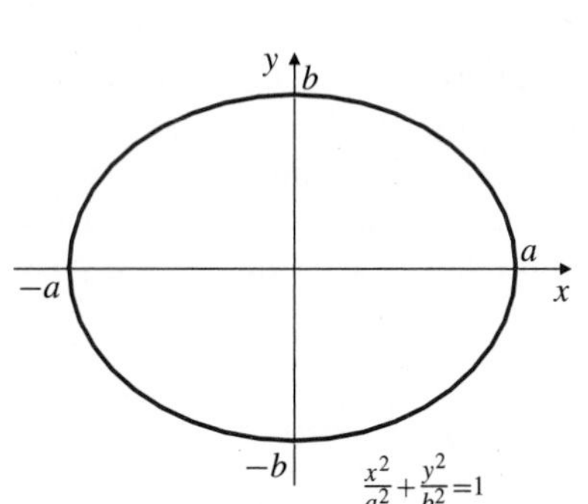

Figure 7.25 The ellipse of Example 4

Solution The upper half of the ellipse has equation $y = b\sqrt{1 - \frac{x^2}{a^2}} = \frac{b}{a}\sqrt{a^2 - x^2}$. Hence,

$$\frac{dy}{dx} = -\frac{b}{a}\frac{x}{\sqrt{a^2 - x^2}},$$

so

$$\begin{aligned} 1 + \left(\frac{dy}{dx}\right)^2 &= 1 + \frac{b^2}{a^2}\frac{x^2}{a^2 - x^2} \\ &= \frac{a^4 - (a^2 - b^2)x^2}{a^2(a^2 - x^2)}. \end{aligned}$$

The circumference of the ellipse is four times the arc length of the part lying in the first quadrant, so

$$\begin{aligned} s &= 4\int_0^a \frac{\sqrt{a^4 - (a^2 - b^2)x^2}}{a\sqrt{a^2 - x^2}}\,dx \qquad \text{Let } x = a\sin t, \; dx = a\cos t\,dt \\ &= 4\int_0^{\pi/2} \frac{\sqrt{a^4 - (a^2 - b^2)a^2\sin^2 t}}{a(a\cos t)}\,a\cos t\,dt \\ &= 4\int_0^{\pi/2} \sqrt{a^2 - (a^2 - b^2)\sin^2 t}\,dt \\ &= 4a\int_0^{\pi/2} \sqrt{1 - \frac{a^2 - b^2}{a^2}\sin^2 t}\,dt \\ &= 4a\int_0^{\pi/2} \sqrt{1 - \varepsilon^2\sin^2 t}\,dt \text{ units,} \end{aligned}$$

where $\varepsilon = (\sqrt{a^2 - b^2})/a$ is the *eccentricity* of the ellipse. (See Section 8.1 for a discussion of ellipses.) Note that $0 \leq \varepsilon < 1$. The function $E(\varepsilon)$, defined by

$$E(\varepsilon) = \int_0^{\pi/2} \sqrt{1 - \varepsilon^2\sin^2 t}\,dt,$$

is called the **complete elliptic integral of the second kind**. The integral cannot be evaluated by elementary techniques for general ε, although numerical methods can be applied to find approximate values for any given value of ε. Tables of values of $E(\varepsilon)$ for various values of ε can be found in collections of mathematical tables. As shown above, the circumference of the ellipse is given by $4aE(\varepsilon)$. Note that for $a = b$ we have $\varepsilon = 0$, and the formula returns the circumference of a circle; $s = 4a(\pi/2) = 2\pi a$ units.

Areas of Surfaces of Revolution

When a plane curve is rotated (in three dimensions) about a line in the plane of the curve, it sweeps out a **surface of revolution**. For instance, a sphere of radius a is generated by rotating a semicircle of radius a about the diameter of that semicircle.

The area of a surface of revolution can be found by integrating an area element dS constructed by rotating the arc length element ds of the curve about the given line. If the radius of rotation of the element ds is r, then it generates, on rotation, a circular band of width ds and length (circumference) $2\pi r$. The area of this band is, therefore,

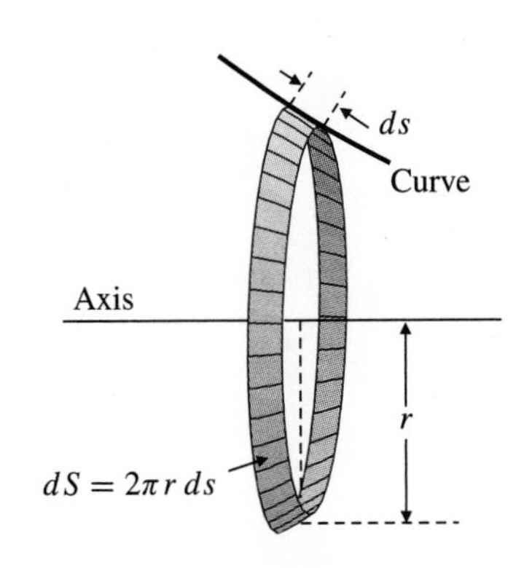

Figure 7.26 The circular band generated by rotating arc length element ds about the axis

$$dS = 2\pi r\, ds,$$

as shown in Figure 7.26. The areas of surfaces of revolution around various lines can be obtained by integrating dS with appropriate choices of r. Here are some important special cases.

Area of a surface of revolution

If $f'(x)$ is continuous on $[a, b]$ and the curve $y = f(x)$ is rotated about the x-axis, the area of the surface of revolution so generated is

$$S = 2\pi \int_{x=a}^{x=b} |y|\, ds = 2\pi \int_a^b |f(x)|\sqrt{1 + (f'(x))^2}\, dx.$$

If the rotation is about the y-axis, the surface area is

$$S = 2\pi \int_{x=a}^{x=b} |x|\, ds = 2\pi \int_a^b |x|\sqrt{1 + (f'(x))^2}\, dx.$$

If $g'(y)$ is continuous on $[c, d]$ and the curve $x = g(y)$ is rotated about the x-axis, the area of the surface of revolution so generated is

$$S = 2\pi \int_{y=c}^{y=d} |y|\, ds = 2\pi \int_c^d |y|\sqrt{1 + (g'(y))^2}\, dy.$$

If the rotation is about the y-axis, the surface area is

$$S = 2\pi \int_{y=c}^{y=d} |x|\, ds = 2\pi \int_c^d |g(y)|\sqrt{1 + (g'(y))^2}\, dy.$$

Remark Students sometimes wonder whether such complicated formulas are actually necessary. Why not just use $dS = 2\pi|y|\, dx$ for the area element when $y = f(x)$ is rotated about the x-axis instead of the more complicated area element $dS = 2\pi|y|\, ds$? After all, we are regarding dx and ds as both being infinitely small, and we certainly used dx for the width of the disk-shaped volume element when we rotated the region under $y = f(x)$ about the x-axis to generate a solid of revolution. The reason is somewhat subtle. For small thickness Δx, the volume of a slice of the solid of revolution is only approximately $\pi y^2\, \Delta x$, but the error is *small compared to the volume of this slice.* On the other hand, if we use $2\pi|y|\, \Delta x$ as an approximation to the area of a thin band of the surface of revolution corresponding to an x interval of width Δx, the error is *not small compared to the area of that band.* If, for instance, the curve $y = f(x)$ has slope 1 at x, then the width of the band is really $\Delta s = \sqrt{2}\, \Delta x$, so that the area of the band is $\Delta S = 2\pi\sqrt{2}|y|\, \Delta x$, not just $2\pi|y|\, \Delta x$. Always use the appropriate arc length element along the curve when you rotate a curve to find the area of a surface of revolution.

EXAMPLE 5 **(Surface area of a sphere)** Find the area of the surface of a sphere of radius a.

Solution Such a sphere can be generated by rotating the semicircle with equation

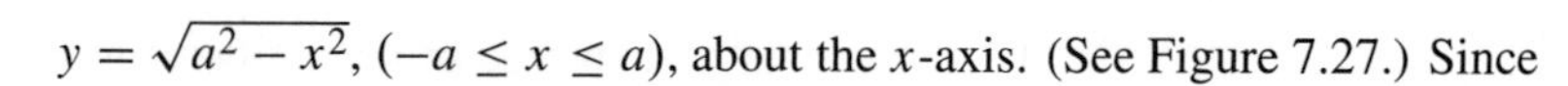

$y = \sqrt{a^2 - x^2}$, $(-a \le x \le a)$, about the x-axis. (See Figure 7.27.) Since

$$\frac{dy}{dx} = -\frac{x}{\sqrt{a^2 - x^2}} = -\frac{x}{y},$$

the area of the sphere is given by

$$\begin{aligned} S &= 2\pi \int_{-a}^{a} y\sqrt{1 + \left(\frac{x}{y}\right)^2}\, dx \\ &= 4\pi \int_0^a \sqrt{y^2 + x^2}\, dx \\ &= 4\pi \int_0^a \sqrt{a^2}\, dx = 4\pi a x \Big|_0^a = 4\pi a^2 \text{ square units.} \end{aligned}$$

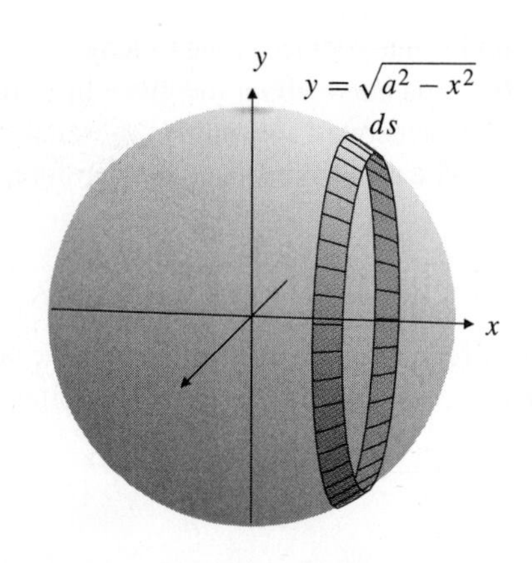

Figure 7.27 An area element on a sphere

EXAMPLE 6 **(Surface area of a parabolic dish)** Find the surface area of a parabolic reflector whose shape is obtained by rotating the parabolic arc $y = x^2$, $(0 \le x \le 1)$, about the y-axis, as illustrated in Figure 7.28.

Solution The arc length element for the parabola $y = x^2$ is $ds = \sqrt{1 + 4x^2}\, dx$, so the required surface area is

$$\begin{aligned} S &= 2\pi \int_0^1 x\sqrt{1 + 4x^2}\, dx && \text{Let } u = 1 + 4x^2, \\ & && du = 8x\, dx \\ &= \frac{\pi}{4} \int_1^5 u^{1/2}\, du \\ &= \frac{\pi}{6} u^{3/2} \Big|_1^5 = \frac{\pi}{6}(5\sqrt{5} - 1) \text{ square units.} \end{aligned}$$

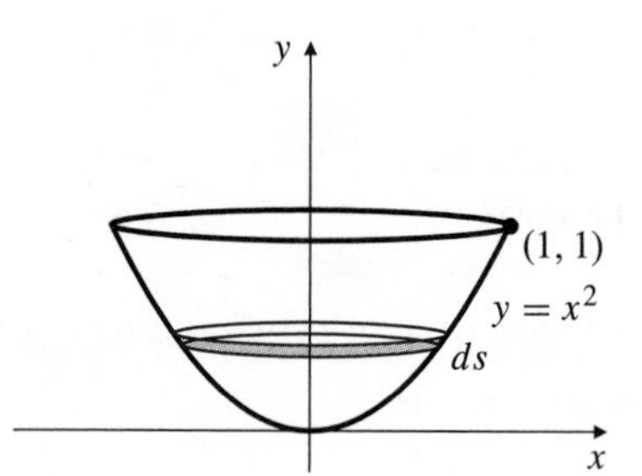

Figure 7.28 The area element is a horizontal band here

EXERCISES 7.3

In Exercises 1–14, find the lengths of the given curves.

1. $y = 2x - 1$ from $x = 1$ to $x = 3$

2. $y = ax + b$ from $x = A$ to $x = B$

3. $y = \dfrac{2}{3}x^{3/2}$ from $x = 0$ to $x = 8$

4. $y^2 = (x - 1)^3$ from $(1, 0)$ to $(2, 1)$

5. $y^3 = x^2$ from $(-1, 1)$ to $(1, 1)$

6. $2(x + 1)^3 = 3(y - 1)^2$ from $(-1, 1)$ to $(0, 1 + \sqrt{2/3})$

7. $y = \dfrac{x^3}{12} + \dfrac{1}{x}$ from $x = 1$ to $x = 4$

8. $y = \dfrac{x^3}{3} + \dfrac{1}{4x}$ from $x = 1$ to $x = 2$

9. $4y = 2\ln x - x^2$ from $x = 1$ to $x = e$

10. $y = x^2 - \dfrac{\ln x}{8}$ from $x = 1$ to $x = 2$

11. $y = \dfrac{e^x + e^{-x}}{2}$ $(= \cosh x)$ from $x = 0$ to $x = a$

12. $y = \ln\cos x$ from $x = \pi/6$ to $x = \pi/4$

13. $y = x^2$ from $x = 0$ to $x = 2$

14. $y = \ln \dfrac{e^x - 1}{e^x + 1}$ from $x = 2$ to $x = 4$

15. Find the circumference of the closed curve $x^{2/3} + y^{2/3} = a^{2/3}$. *Hint:* The curve is symmetric about both coordinate axes (why?), so one-quarter of it lies in the first quadrant.

Use numerical methods (or a calculator with an integration function) to find the lengths of the curves in Exercises 16–19 to 4 decimal places.

16. $y = x^4$ from $x = 0$ to $x = 1$

17. $y = x^{1/3}$ from $x = 1$ to $x = 2$

18. the circumference of the ellipse $3x^2 + y^2 = 3$

19. the shorter arc of the ellipse $x^2 + 2y^2 = 2$ between $(0, 1)$ and $(1, 1/\sqrt{2})$

In Exercises 20–27, find the areas of the surfaces obtained by rotating the given curve about the indicated lines.

20. $y = x^2$, $(0 \le x \le 2)$, about the y-axis

21. $y = x^3$, $(0 \le x \le 1)$, about the x-axis

22. $y = x^{3/2}$, $(0 \le x \le 1)$, about the x-axis

23. $y = x^{3/2}$, $(0 \le x \le 1)$, about the y-axis

24. $y = e^x$, $(0 \le x \le 1)$, about the x-axis

25. $y = \sin x$, $(0 \le x \le \pi)$, about the x-axis

26. $y = \dfrac{x^3}{12} + \dfrac{1}{x}$, $(1 \le x \le 4)$, about the x-axis

27. $y = \dfrac{x^3}{12} + \dfrac{1}{x}$, $(1 \le x \le 4)$, about the y-axis

28. **(Surface area of a cone)** Find the area of the curved surface of a right-circular cone of base radius r and height h by rotating the straight line segment from $(0, 0)$ to (r, h) about the y-axis.

29. **(How much icing on a doughnut?)** Find the surface area of the torus (doughnut) obtained by rotating the circle $(x - b)^2 + y^2 = a^2$ about the y-axis.

30. **(Area of a prolate spheroid)** Find the area of the surface obtained by rotating the ellipse $x^2 + 4y^2 = 4$ about the x-axis.

31. **(Area of an oblate spheroid)** Find the area of the surface obtained by rotating the ellipse $x^2 + 4y^2 = 4$ about the y-axis.

32. The ellipse of Example 4 is rotated about the line $y = c > b$ to generate a doughnut with elliptical cross-sections. Express the surface area of this doughnut in terms of the complete elliptic integral function $E(\varepsilon)$ introduced in that example.

33. Express the integral formula obtained for the length of the metal sheet in Example 3 in terms of the complete elliptic integral function $E(\epsilon)$ introduced in Example 4.

34. **(An interesting property of spheres)** If two parallel planes intersect a sphere, show that the surface area of that part of the sphere lying between the two planes depends only on the radius of the sphere and the distance between the planes, and not on the position of the planes.

35. For what real values of k does the surface generated by rotating the curve $y = x^k$, $(0 < x \le 1)$, about the y-axis have a finite surface area?

36. The curve $y = \ln x$, $(0 < x \le 1)$, is rotated about the y-axis. Find the area of the horn-shaped surface so generated.

37. A hollow container in the shape of an infinitely long horn is generated by rotating the curve $y = 1/x$, $(1 \le x < \infty)$, about the x-axis.

(a) Find the volume of the container.

(b) Show that the container has infinite surface area.

(c) How do you explain the "paradox" that the container can be filled with a finite volume of paint but requires an infinite amount of paint to cover its surface?

7.4 Mass, Moments, and Centre of Mass

Many quantities of interest in physics, mechanics, ecology, finance, and other disciplines are described in terms of densities over regions of space, the plane, or even the real line. To determine the total value of such a quantity we must add up (integrate) the contributions from the various places where the quantity is distributed.

Mass and Density

If a solid object is made of a homogeneous material, we would expect different parts of the solid that have the same volume to have the same mass as well. We express this homogeneity by saying that the object has constant density, that density being the mass divided by the volume for the whole object or for any part of it. Thus, for example, a rectangular brick with dimensions 20 cm, 10 cm, and 8 cm would have volume $V = 20 \times 10 \times 8 = 1{,}600$ cm^3, and if it was made of material having constant density $\rho = 3$ g/cm^3, it would have mass $m = \rho V = 3 \times 1{,}600 = 4{,}800$ g. (We will use the lowercase Greek letter delta (ρ) to represent density.)

By "density at a point P" of a solid object, we mean the limit $\rho(P)$ of mass/volume for the part of the solid lying in small regions containing P (for example, balls centred at P) as the dimensions of the regions approach zero. Such a density ρ is continuous at P if we can ensure that $|\rho(Q) - \rho(P)|$ is as small as we want by taking Q close enough to P.

If the density of the material constituting a solid object is not constant but varies from point to point in the object, no such simple relationship exists between mass and volume. If the density $\rho = \rho(P)$ is a *continuous* function of position P, we can subdivide the solid into many small volume elements and, by regarding ρ as approximately constant over each such element, determine the masses of all the elements and add them up to get the mass of the solid. The mass Δm of a volume element ΔV containing the point P would satisfy

$$\Delta m \approx \rho(P)\, \Delta V,$$

so the mass m of the solid can be approximated:

$$m = \sum \Delta m \approx \sum \rho(P)\, \Delta V.$$

Such approximations become exact as we pass to the limit of differential mass and volume elements, $dm = \rho(P)\,dV$, so we expect to be able to calculate masses as integrals, that is, as the limits of such sums:

$$m = \int dm = \int \rho(P)\,dV.$$

EXAMPLE 1 The density of a solid vertical cylinder of height H cm and base area A cm^2 is $\rho = \rho_0(1+h)$ g/cm^3, where h is the height in centimetres above the base and ρ_0 is a constant. Find the mass of the cylinder.

Solution See Figure 7.29(a). A slice of the solid at height h above the base and having thickness dh is a circular disk of volume $dV = A\,dh$. Since the density is constant over this disk, the mass of the volume element is

$$dm = \rho\,dV = \rho_0(1+h)\,A\,dh.$$

Therefore, the mass of the whole cylinder is

$$m = \int_0^H \rho_0 A(1+h)\,dh = \rho_0 A\left(H + \frac{H^2}{2}\right) \text{ g.}$$

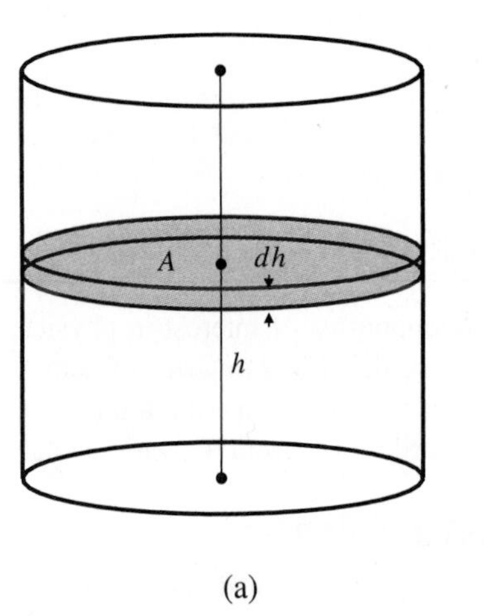

(a)

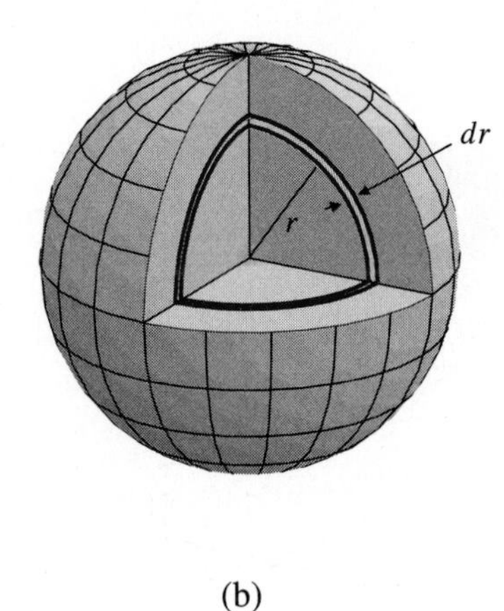

(b)

Figure 7.29

(a) A solid cylinder whose density varies with height

(b) Cutaway view of a planet whose density depends on distance from the centre

EXAMPLE 2 **(Using spherical shells)** The density of a certain spherical planet of radius R m varies with distance r from the centre according to the formula

$$\rho = \frac{\rho_0}{1+r^2} \text{ kg/m}^3.$$

Find the mass of the planet.

Solution Recall that the surface area of a sphere of radius r is $4\pi r^2$. The planet can be regarded as being composed of concentric spherical shells having radii between 0 and R. The volume of a shell of radius r and thickness dr (see Figure 7.29(b)) is equal to its surface area times its thickness, and its mass is its volume times its density:

$$dV = 4\pi r^2\,dr; \qquad dm = \rho\,dV = 4\pi\rho_0 \frac{r^2}{1+r^2}\,dr.$$

We add the masses of these shells to find the mass of the whole planet:

$$\begin{aligned} m &= 4\pi\rho_0 \int_0^R \frac{r^2}{1+r^2}\,dr = 4\pi\rho_0 \int_0^R \left(1 - \frac{1}{1+r^2}\right) dr \\ &= 4\pi\rho_0(r - \tan^{-1} r)\Big|_0^R = 4\pi\rho_0(R - \tan^{-1} R) \text{ kg.} \end{aligned}$$

Similar techniques can be applied to find masses of one- and two-dimensional objects, such as wires and thin plates, that have variable densities of the forms mass/unit length (**line density**, which we will usually denote by δ) and σ = mass/unit area (**areal density**, which we will denote by σ).

EXAMPLE 3 A wire of variable composition is stretched along the x-axis from $x = 0$ to $x = L$ cm. Find the mass of the wire if the line density at position x is $\delta(x) = kx$ g/cm, where k is a positive constant.

Solution The mass of a length element dx of the wire located at position x is given by $dm = \delta(x)\,dx = kx\,dx$. Thus, the mass of the wire is

$$m = \int_0^L kx\,dx = \left(\frac{kx^2}{2}\right)\Bigg|_0^L = \frac{kL^2}{2} \text{ g.}$$

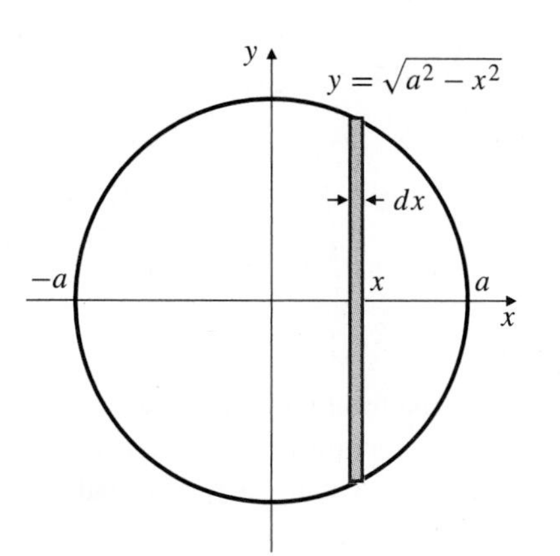

Figure 7.30 The area element of Example 4

EXAMPLE 4 Find the mass of a disk of radius a cm whose centre is at the origin in the xy-plane if the areal density at position (x, y) is $\sigma = k(2a + x)$ g/cm^2. Here k is a constant.

Solution The areal density depends only on the horizontal coordinate x, so it is constant along vertical lines on the disk. This suggests that thin vertical strips should be used as area elements. A vertical strip of thickness dx at x has area $dA = 2\sqrt{a^2 - x^2}\,dx$ (see Figure 7.30); its mass is therefore

$$dm = \sigma\,dA = 2k(2a + x)\sqrt{a^2 - x^2}\,dx.$$

Hence, the mass of the disk is

$$\begin{aligned} m &= \int_{x=-a}^{x=a} dm = 2k \int_{-a}^{a} (2a + x)\sqrt{a^2 - x^2}\,dx \\ &= 4ak \int_{-a}^{a} \sqrt{a^2 - x^2}\,dx + 2k \int_{-a}^{a} x\sqrt{a^2 - x^2}\,dx \\ &= 4ak\frac{\pi a^2}{2} + 0 = 2\pi k a^3 \text{ g.} \end{aligned}$$

We used the area of a semicircle to evaluate the first integral. The second integral is zero because the integrand is odd and the interval is symmetric about $x = 0$.

Distributions of mass along one-dimensional structures (lines or curves) necessarily lead to integrals of functions of one variable, but distributions of mass on a surface or in space can lead to integrals involving functions of more than one variable. Such integrals are studied in multivariable calculus. (See, for example, Section 14.7.) In the examples above, the given densities were functions of only one variable, so these problems, although higher dimensional in nature, led to integrals of functions of only one variable and could be solved by the methods at hand.

Moments and Centres of Mass

The **moment** about the point $x = x_0$ of a mass m located at position x on the x-axis is the product $m(x - x_0)$ of the mass and its (signed) distance from x_0. If the x-axis is a horizontal arm hinged at x_0, the moment about x_0 measures the tendency of the weight of the mass m to cause the arm to rotate. If several masses $m_1, m_2, m_3, \ldots, m_n$ are located at the points $x_1, x_2, x_3, \ldots, x_n$, respectively, then the total moment of the system of masses about the point $x = x_0$ is the sum of the individual moments (see Figure 7.31):

$$M_{x=x_0} = (x_1 - x_0)m_1 + (x_2 - x_0)m_2 + \cdots + (x_n - x_0)m_n = \sum_{j=1}^{n}(x_j - x_0)m_j.$$

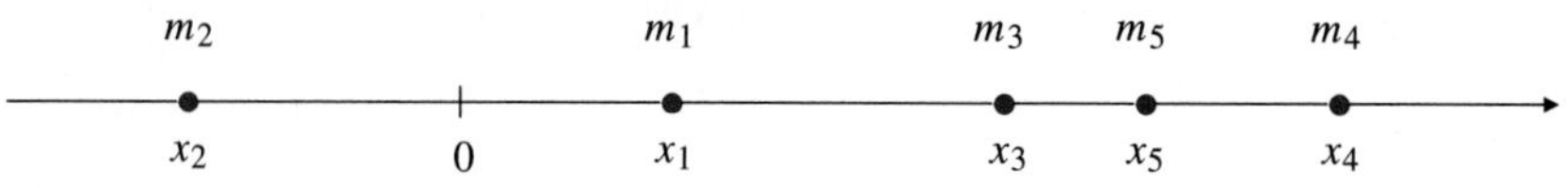

Figure 7.31 A system of discrete masses on a line

The **centre of mass** of the system of masses is the point $\bar{x}$ about which the total moment of the system is zero. Thus,

$$0 = \sum_{j=1}^{n}(x_j - \bar{x})m_j = \sum_{j=1}^{n} x_j m_j - \bar{x}\sum_{j=1}^{n} m_j.$$

The centre of mass of the system is therefore given by

$$\bar{x} = \frac{\displaystyle\sum_{j=1}^{n} x_j m_j}{\displaystyle\sum_{j=1}^{n} m_j} = \frac{M_{x=0}}{m},$$

where m is the total mass of the system and $M_{x=0}$ is the total moment about $x = 0$. If you think of the x-axis as being a weightless wire supporting the masses, then $\bar{x}$ is the point at which the wire could be supported and remain in perfect balance (equilibrium), not tipping either way. Even if the axis represents a nonweightless support, say a seesaw, supported at $x = \bar{x}$, it will remain balanced after the masses are added, provided it was balanced beforehand. For many purposes a system of masses behaves as though its total mass were concentrated at its centre of mass.

Now suppose that a one-dimensional distribution of mass with continuously variable line density $\delta(x)$ lies along the interval $[a, b]$ of the x-axis. An element of length dx at position x contains mass $dm = \delta(x)\,dx$, so its moment is $dM_{x=0} = x\,dm = x\delta(x)\,dx$ about $x = 0$. The total moment about $x = 0$ is the *sum* (integral) of these moment elements:

$$M_{x=0} = \int_a^b x\delta(x)\,dx.$$

Since the total mass is

$$m = \int_a^b \delta(x)\,dx,$$

we obtain the following formula for the centre of mass.

The centre of mass of a distribution of mass with line density $\delta(x)$ on the interval $[a, b]$ is given by

$$\bar{x} = \frac{M_{x=0}}{m} = \frac{\displaystyle\int_a^b x\delta(x)\,dx}{\displaystyle\int_a^b \delta(x)\,dx}.$$

EXAMPLE 5 At what point can the wire of Example 3 be suspended so that it will balance?

Solution In Example 3 we evaluated the mass of the wire to be $kL^2/2$ g. Its moment about $x = 0$ is

$$\begin{aligned} M_{x=0} &= \int_0^L x\delta(x)\,dx \\ &= \int_0^L kx^2\,dx = \left(\frac{kx^3}{3}\right)\bigg|_0^L = \frac{kL^3}{3}\ \text{g·cm}. \end{aligned}$$

(Note that the appropriate units for the moment are units of mass times units of distance: in this case gram-centimetres.) The centre of mass of the wire is

$$\bar{x} = \frac{kL^3/3}{kL^2/2} = \frac{2L}{3}.$$

The wire will be balanced if suspended at position $x = 2L/3$ cm.

Two- and Three-Dimensional Examples

The system of mass considered in Example 5 is one-dimensional and lies along a straight line. If mass is distributed in a plane or in space, similar considerations prevail. For a system of masses m_1 at (x_1, y_1), m_2 at (x_2, y_2), ..., m_n at (x_n, y_n), the **moment about** $x = 0$ is

$$M_{x=0} = x_1m_1 + x_2m_2 + \cdots + x_nm_n = \sum_{j=1}^{n} x_jm_j,$$

and the **moment about** $y = 0$ is

$$M_{y=0} = y_1m_1 + y_2m_2 + \cdots + y_nm_n = \sum_{j=1}^{n} y_jm_j.$$

The **centre of mass** is the point $(\bar{x}, \bar{y})$ where

$$\bar{x} = \frac{M_{x=0}}{m} = \frac{\displaystyle\sum_{j=1}^{n} x_jm_j}{\displaystyle\sum_{j=1}^{n} m_j} \quad \text{and} \quad \bar{y} = \frac{M_{y=0}}{m} = \frac{\displaystyle\sum_{j=1}^{n} y_jm_j}{\displaystyle\sum_{j=1}^{n} m_j}.$$

For continuous distributions of mass, the sums become appropriate integrals.

EXAMPLE 6 Find the centre of mass of a rectangular plate that occupies the region $0 \le x \le a$, $0 \le y \le b$, if the areal density of the material in the plate at position (x, y) is $\sigma = ky$.

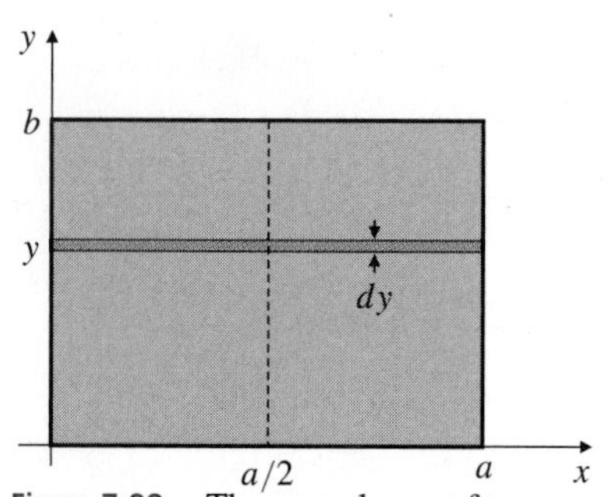

Figure 7.32 The area element for Example 6

Solution Since the areal density is independent of x and the rectangle is symmetric about the line $x = a/2$, the x-coordinate of the centre of mass must be $\bar{x} = a/2$. A thin horizontal strip of width dy at height y (see Figure 7.32) has mass $dm = aky\,dy$. The moment of this strip about $y = 0$ is $dM_{y=0} = y\,dm = kay^2\,dy$. Hence, the mass and moment about $y = 0$ of the whole plate are

$$m = ka\int_0^b y\,dy = \frac{kab^2}{2},$$

$$M_{y=0} = ka\int_0^b y^2\,dy = \frac{kab^3}{3}.$$

Therefore, $\bar{y} = M_{y=0}/m = 2b/3$, and the centre of mass of the plate is $(a/2, 2b/3)$. The plate would be balanced if supported at this point.

For distributions of mass in three-dimensional space one defines, analogously, the moments $M_{x=0}$, $M_{y=0}$, and $M_{z=0}$ of the system of mass about the planes $x = 0$, $y = 0$, and $z = 0$, respectively. The centre of mass is $(\bar{x}, \bar{y}, \bar{z})$ where

$$\bar{x} = \frac{M_{x=0}}{m}, \qquad \bar{y} = \frac{M_{y=0}}{m}, \qquad \text{and} \qquad \bar{z} = \frac{M_{z=0}}{m},$$

m being the total mass: $m = m_1 + m_2 + \cdots + m_n$. Again, the sums are replaced with integrals for continuous distributions of mass.

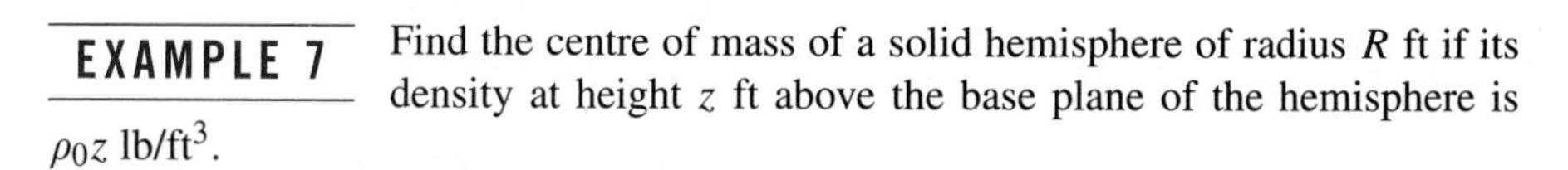

EXAMPLE 7 Find the centre of mass of a solid hemisphere of radius R ft if its density at height z ft above the base plane of the hemisphere is $\rho_0 z$ lb/ft^3.

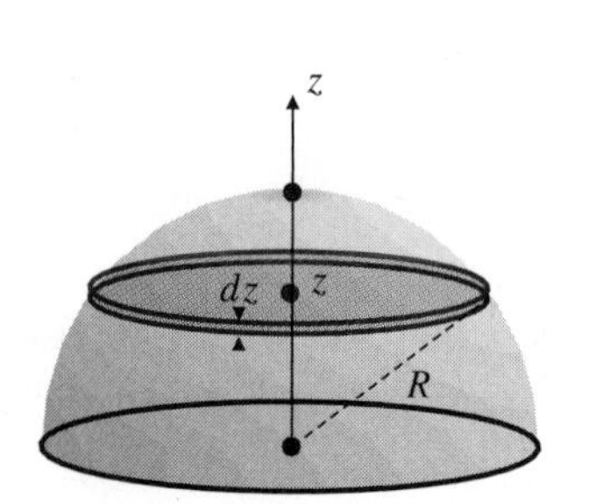

Figure 7.33 Mass element of a solid hemisphere with density depending on height

Solution The solid is symmetric about the vertical axis (let us call it the z-axis), and the density is constant in planes perpendicular to this axis. Therefore, the centre of mass must lie somewhere on this axis. A slice of the solid at height z above the base, and having thickness dz, is a disk of radius $\sqrt{R^2 - z^2}$. (See Figure 7.33.) Its volume is $dV = \pi(R^2 - z^2)\,dz$, and its mass is $dm = \rho_0 z\,dV = \rho_0\pi(R^2 z - z^3)\,dz$. Its moment about the base plane $z = 0$ is $dM_{z=0} = z\,dm = \rho_0\pi(R^2z^2 - z^4)\,dz$. The mass of the solid is

$$m = \rho_0\pi\int_0^R (R^2z - z^3)\,dz = \rho_0\pi\left(\frac{R^2z^2}{2} - \frac{z^4}{4}\right)\Bigg|_0^R = \frac{\pi}{4}\rho_0 R^4 \text{ lb}.$$

The moment of the hemisphere about the plane $z = 0$ is

$$M_{z=0} = \rho_0\pi\int_0^R (R^2z^2 - z^4)\,dz = \rho_0\pi\left(\frac{R^2z^3}{3} - \frac{z^5}{5}\right)\Bigg|_0^R = \frac{2\pi}{15}\rho_0 R^5 \text{ lb·ft}.$$

The centre of mass therefore lies along the axis of symmetry of the hemisphere at height $\bar{z} = M_{z=0}/m = 8R/15$ ft above the base of the hemisphere.

EXAMPLE 8 Find the centre of mass of a plate that occupies the region $a \le x \le b$, $0 \le y \le f(x)$, if the density at any point (x, y) is $\sigma(x)$.

Solution The appropriate area element is shown in Figure 7.34. It has area $f(x)\,dx$, mass

$$dm = \sigma(x) f(x)\,dx,$$

and moment about $x = 0$

$$dM_{x=0} = x\sigma(x)f(x)\,dx.$$

Since the density depends only on x, the mass element dm has constant density, so the y-coordinate of *its* centre of mass is at its midpoint: $\bar{y}_{dm} = \frac{1}{2}f(x)$. Therefore, the moment of the mass element dm about $y = 0$ is

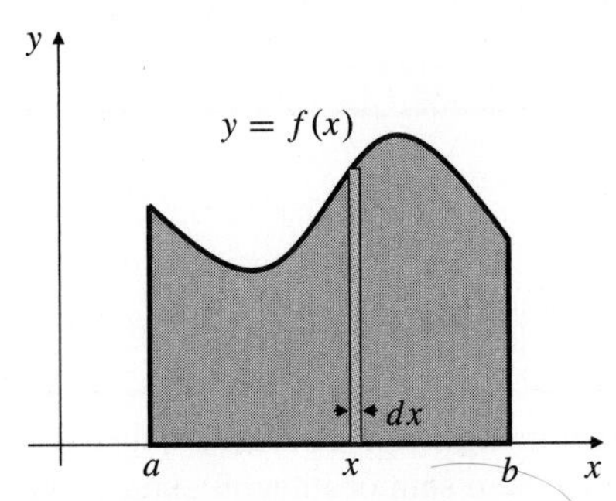

Figure 7.34 Mass element of a plate

$$dM_{y=0} = \bar{y}_{dm}\,dm = \frac{1}{2}\sigma(x)\big(f(x)\big)^2\,dx.$$

The coordinates of the centre of mass of the plate are $\bar{x} = \dfrac{M_{x=0}}{m}$ and $\bar{y} = \dfrac{M_{y=0}}{m}$, where

$$m = \int_a^b \sigma(x)f(x)\,dx,$$

$$M_{x=0} = \int_a^b x\sigma(x)f(x)\,dx,$$

$$M_{y=0} = \frac{1}{2}\int_a^b \sigma(x)\big(f(x)\big)^2\,dx.$$

Remark Similar formulas can be obtained if the density depends on y instead of x, provided that the region admits a suitable horizontal area element (e.g., the region might be specified by $c \le y \le d$, $0 \le x \le g(y)$). Finding centres of mass for plates that occupy regions specified by functions of x, but where the density depends on y, generally requires the use of "double integrals." Such problems are therefore studied in multivariable calculus. (See Section 14.7.)

EXERCISES 7.4

Find the mass and centre of mass for the systems in Exercises 1–16. Be alert for symmetries.

1. A straight wire of length L cm, where the density at distance s cm from one end is $\delta(s) = \sin \pi s/L$ g/cm

2. A straight wire along the x-axis from $x = 0$ to $x = L$ if the density is constant δ_0, but the cross-sectional radius of the wire varies so that its value at x is $a + bx$

3. A quarter-circular plate having radius a, constant areal density σ_0, and occupying the region $x^2 + y^2 \le a^2$, $x \ge 0$, $y \ge 0$

4. A quarter-circular plate of radius a occupying the region $x^2 + y^2 \le a^2$, $x \ge 0$, $y \ge 0$, having areal density $\sigma(x) = \sigma_0 x$

5. A plate occupying the region $0 \le y \le 4 - x^2$ if the areal density at (x, y) is ky

6. A right-triangular plate with legs 2 m and 3 m if the areal density at any point P is $5h$ kg/m^2, h being the distance of P from the shorter leg

7. A square plate of edge a cm if the areal density at P is kx g/cm^2, where x is the distance from P to one edge of the square

8. The plate in Exercise 7, but with areal density kr g/cm^2, where r is the distance (in centimetres) from P to one of the diagonals of the square

9. A plate of areal density $\sigma(x)$ occupying the region $a \le x \le b$, $f(x) \le y \le g(x)$

10. A rectangular brick with dimensions 20 cm, 10 cm, and 5 cm, if the density at P is kx g/cm^3, where x is the distance from P to one of the 10×5 faces

11. A solid ball of radius R m if the density at P is z kg/m^3, where z is the distance from P to a plane at distance $2R$ m from the centre of the ball

12. A right-circular cone of base radius a cm and height b cm if the density at point P is kz g/cm^3, where z is the distance of P from the base of the cone

13. The solid occupying the quarter of a ball of radius a centred at the origin having as base the region $x^2 + y^2 \le a^2$, $x \ge 0$ in the xy-plane, if the density at height z above the base is $\rho_0 z$

14. The cone of Exercise 12, but with density at P equal to kx g/cm^3, where x is the distance of P from the axis of symmetry of the cone. *Hint:* Use a cylindrical shell centred on the axis of symmetry as a volume element. This element has constant density, so its centre of mass is known, and its moment can be determined from its mass.

15. A semicircular plate occupying the region $x^2 + y^2 \le a^2$, $y \ge 0$, if the density at distance s from the origin is ks g/cm^2

16. The wire in Exercise 1 if it is bent in a semicircle

17. It is estimated that the density of matter in the neighbourhood of a gas giant star is given by $\rho(r) = Ce^{-kr^2}$, where C and k are positive constants, and r is the distance from the centre of the star. The radius of the star is indeterminate but can be taken to be infinite since $\rho(r)$ decreases very rapidly for large r. Find the approximate mass of the star in terms of C and k.

18. Find the average distance $\bar{r}$ of matter in the star of Exercise 17 from the centre of the star. $\bar{r}$ is given by $\int_0^\infty r\,dm / \int_0^\infty dm$, where dm is the mass element at distance r from the centre of the star.

7.5 Centroids

If matter is distributed uniformly in a system so that the density δ is constant, then that density cancels out of the numerator and denominator in sum or integral expressions for coordinates of the centre of mass. In such cases the centre of mass depends only on the *shape* of the object, that is, on geometric properties of the region occupied by the object, and we call it the **centroid** of the region.

Centroids are calculated using the same formulas as those used for centres of mass, except that the density (being constant) is taken to be unity, so the mass is just the length, area, or volume of the region, and the moments are referred to as **moments of the region**, rather than of any mass occupying the region. If we set $\sigma(x) = 1$ in the formulas obtained in Example 8 of Section 7.4, we obtain the following result:

The centroid of a standard plane region

The centroid of the plane region $a \le x \le b$, $0 \le y \le f(x)$, is $(\bar{x}, \bar{y})$, where
$\bar{x} = \dfrac{M_{x=0}}{A}$, $\bar{y} = \dfrac{M_{y=0}}{A}$, and

$$A = \int_a^b f(x)\,dx, \quad M_{x=0} = \int_a^b x f(x)\,dx, \quad M_{y=0} = \frac{1}{2}\int_a^b \big(f(x)\big)^2\,dx.$$

Thus, for example, $\bar{x}$ is the *average value* of the function x over the region.

The centroids of some regions are obvious by symmetry. The centroid of a circular disk or an elliptical disk is at the centre of the disk. The centroid of a rectangle is at the centre also; the centre is the point of intersection of the diagonals. The centroid of any region lies on any axes of symmetry of the region.

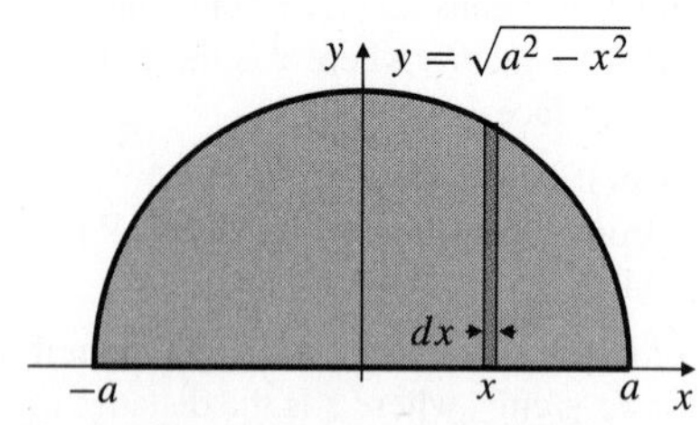

Figure 7.35 The half-disk of Example 1

EXAMPLE 1 What is the average value of y over the half-disk $-a \le x \le a$, $0 \le y \le \sqrt{a^2 - x^2}$? Find the centroid of the half-disk.

Solution By symmetry, the centroid lies on the y-axis, so its x-coordinate is $\bar{x} = 0$. (See Figure 7.35.) Since the area of the half-disk is $A = \frac{1}{2}\pi a^2$, the average value of y over the half-disk is

$$\bar{y} = \frac{M_{y=0}}{A} = \frac{2}{\pi a^2}\,\frac{1}{2}\int_{-a}^{a} (a^2 - x^2)\,dx = \frac{2}{\pi a^2}\,\frac{2a^3}{3} = \frac{4a}{3\pi}.$$

The centroid of the half-disk is $\left(0, \dfrac{4a}{3\pi}\right)$.

EXAMPLE 2 Find the centroid of the semicircle $y = \sqrt{a^2 - x^2}$.

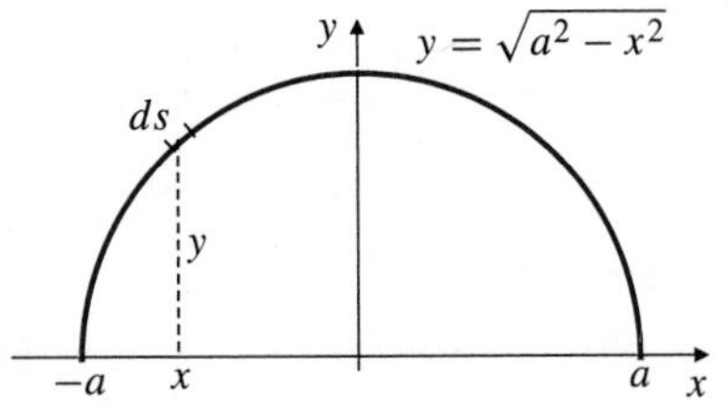

Figure 7.36 The semi-circle of Example 2

Solution Here, the "region" is a one-dimensional curve, having length rather than area. Again $\bar{x} = 0$ by symmetry. A short arc of length ds at height y on the semicircle has moment $dM_{y=0} = y\,ds$ about $y = 0$. (See Figure 7.36.) Since

$$ds = \sqrt{1 + \left(\frac{dy}{dx}\right)^2}\,dx = \sqrt{1 + \frac{x^2}{a^2 - x^2}}\,dx = \frac{a\,dx}{\sqrt{a^2 - x^2}},$$

and since $y = \sqrt{a^2 - x^2}$ on the semicircle, we have

$$M_{y=0} = \int_{-a}^{a} \sqrt{a^2 - x^2}\,\frac{a\,dx}{\sqrt{a^2 - x^2}} = a\int_{-a}^{a} dx = 2a^2.$$

Since the length of the semicircle is πa, we have $\bar{y} = \dfrac{M_{y=0}}{\pi a} = \dfrac{2a}{\pi}$, and the centroid of the semicircle is $\left(0, \dfrac{2a}{\pi}\right)$. Note that the centroid of a semicircle of radius a is not the same as that of half-disk of radius a. Note also that the centroid of the semicircle does not lie on the semicircle itself.

THEOREM 1

The centroid of a triangle

The centroid of a triangle is the point at which all three medians of the triangle intersect.

PROOF Recall that a median of a triangle is a straight line joining one vertex of the triangle to the midpoint of the opposite side. Given any median of a triangle, we will show that the centroid lies on that median. Thus, the centroid must lie on all three medians.

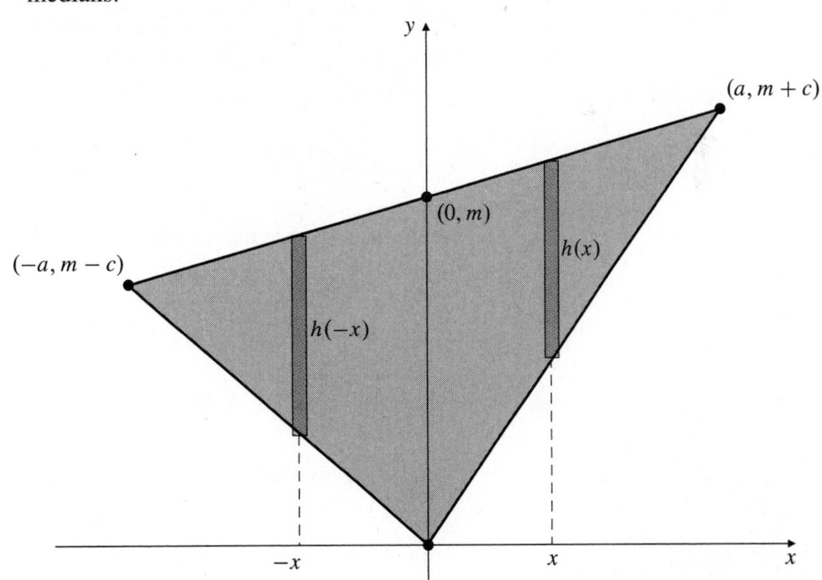

Figure 7.37 The axes of Theorem 1

Adopt a coordinate system where the median in question lies along the y-axis and such that a vertex of the triangle is at the origin. (See Figure 7.37.) Let the midpoint of the opposite side be $(0, m)$. Then the other two vertices of the triangle must have coordinates of the form $(-a, m - c)$ and $(a, m + c)$ so that $(0, m)$ will be the midpoint between them. The two vertical area elements shown in the figure are at the same distance on opposite sides of the y-axis, so they have the same heights $h(-x) = h(x)$ (by similar triangles) and the same area. The sum of the moments about $x = 0$ of these area elements is

$$dM_{x=0} = -xh(-x)\,dx + xh(x)\,dx = 0,$$

so the moment of the whole triangle about $x = 0$ is

$$M_{x=0} = \int_{x=-a}^{x=a} dM_{x=0} = 0.$$

Therefore, the centroid of the triangle lies on the y-axis.

Remark By simultaneously solving the equations of any two medians of a triangle, we can verify the following formula:

Coordinates of the centroid of a triangle

The coordinates of the centroid of a triangle are the averages of the corresponding coordinates of the three vertices of the triangle. The triangle with vertices (x_1, y_1), (x_2, y_2), and (x_3, y_3) has centroid

$$(\bar{x}, \bar{y}) = \left(\frac{x_1 + x_2 + x_3}{3}, \frac{y_1 + y_2 + y_3}{3} \right).$$

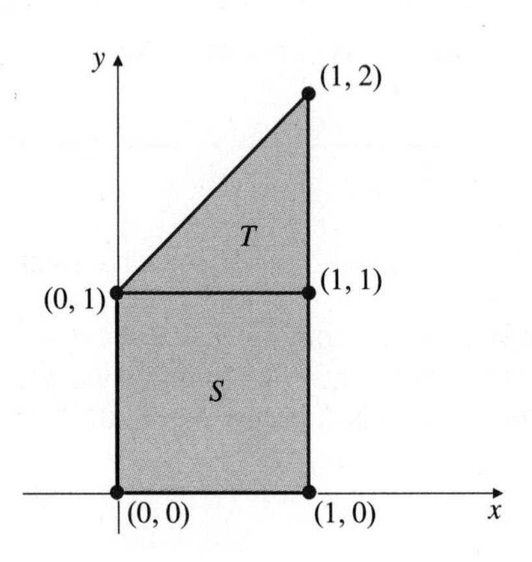

Figure 7.38 The trapezoid of Example 3

If a region is a union of nonoverlapping subregions, then any moment of the region is the sum of the corresponding moments of the subregions. This fact enables us to calculate the centroid of the region if we know the centroids and areas of all the subregions.

EXAMPLE 3 Find the centroid of the trapezoid with vertices $(0, 0)$, $(1, 0)$, $(1, 2)$, and $(0, 1)$.

Solution The trapezoid is the union of a square and a (nonoverlapping) triangle, as shown in Figure 7.38. By symmetry, the square has centroid $(\bar{x}_S, \bar{y}_S) = \left(\frac{1}{2}, \frac{1}{2}\right)$, and its area is $A_S = 1$. The triangle has area $A_T = \frac{1}{2}$, and its centroid is $(\bar{x}_T, \bar{y}_T)$, where

$$\bar{x}_T = \frac{0 + 1 + 1}{3} = \frac{2}{3} \quad \text{and} \quad \bar{y}_T = \frac{1 + 1 + 2}{3} = \frac{4}{3}.$$

Continuing to use subscripts S and T to denote the square and triangle, respectively, we calculate

$$M_{x=0} = M_{S;x=0} + M_{T;x=0} = A_S\bar{x}_S + A_T\bar{x}_T = 1 \times \frac{1}{2} + \frac{1}{2} \times \frac{2}{3} = \frac{5}{6},$$
$$M_{y=0} = M_{S;y=0} + M_{T;y=0} = A_S\bar{y}_S + A_T\bar{y}_T = 1 \times \frac{1}{2} + \frac{1}{2} \times \frac{4}{3} = \frac{7}{6}.$$

Since the area of the trapezoid is $A = A_S + A_T = \frac{3}{2}$, its centroid is

$$(\bar{x}, \bar{y}) = \left(\frac{5}{6} \Big/ \frac{3}{2}, \frac{7}{6} \Big/ \frac{3}{2} \right) = \left(\frac{5}{9}, \frac{7}{9} \right).$$

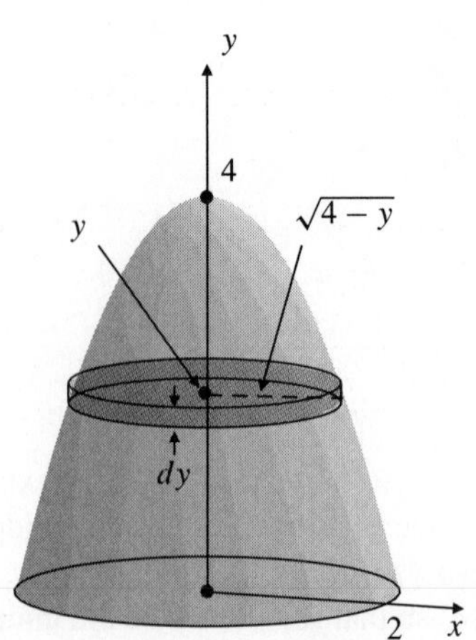

Figure 7.39 A parabolic solid

EXAMPLE 4 Find the centroid of the solid region obtained by rotating about the y-axis the first quadrant region lying between the x-axis and the parabola $y = 4 - x^2$.

Solution By symmetry, the centroid of the parabolic solid will lie on its axis of symmetry, the y-axis. A thin, disk-shaped slice of the solid at height y and having thickness dy (see Figure 7.39) has volume

$$dV = \pi x^2 \, dy = \pi(4 - y) \, dy$$

and moment about the base plane

$$dM_{y=0} = y\,dV = \pi(4y - y^2)\,dy.$$

Hence, the volume of the solid is

$$V = \pi\int_0^4 (4-y)\,dy = \pi\left(4y - \frac{y^2}{2}\right)\bigg|_0^4 = \pi(16-8) = 8\pi,$$

and its moment about $y = 0$ is

$$M_{y=0} = \pi\int_0^4 (4y - y^2)\,dy = \pi\left(2y^2 - \frac{y^3}{3}\right)\bigg|_0^4 = \pi\left(32 - \frac{64}{3}\right) = \frac{32}{3}\pi.$$

Hence, the centroid is located at $\bar{y} = \dfrac{32\pi}{3} \times \dfrac{1}{8\pi} = \dfrac{4}{3}$.

Pappus's Theorem

The following theorem relates volumes or surface areas of revolution to the centroid of the region or curve being rotated.

THEOREM

Pappus's Theorem

(a) If a plane region R lies on one side of a line L in that plane and is rotated about L to generate a solid of revolution, then the volume V of that solid is the product of the area of R and the distance travelled by the centroid of R under the rotation; that is,

$$V = 2\pi\bar{r}A,$$

where A is the area of R, and $\bar{r}$ is the perpendicular distance from the centroid of R to L.

(b) If a plane curve $\mathcal{C}$ lies on one side of a line L in that plane and is rotated about that line to generate a surface of revolution, then the area S of that surface is the length of $\mathcal{C}$ times the distance travelled by the centroid of $\mathcal{C}$:

$$S = 2\pi\bar{r}s,$$

where s is the length of the curve $\mathcal{C}$, and $\bar{r}$ is the perpendicular distance from the centroid of $\mathcal{C}$ to the line L.

PROOF We prove part (a). The proof of (b) is similar and is left as an exercise. Let us take L to be the y-axis and suppose that R lies between $x = a$ and $x = b$ where $0 \le a < b$. Thus $\bar{r} = \bar{x}$, the x-coordinate of the centroid of R. Let dA denote the area of a thin strip of R at position x and having width dx. (See Figure 7.40.) This strip generates, on rotation about L, a cylindrical shell of volume $dV = 2\pi x\,dA$, so the volume of the solid of revolution is

$$V = 2\pi\int_{x=a}^{x=b} x\,dA = 2\pi M_{x=0} = 2\pi\bar{x}A = 2\pi\bar{r}A.$$

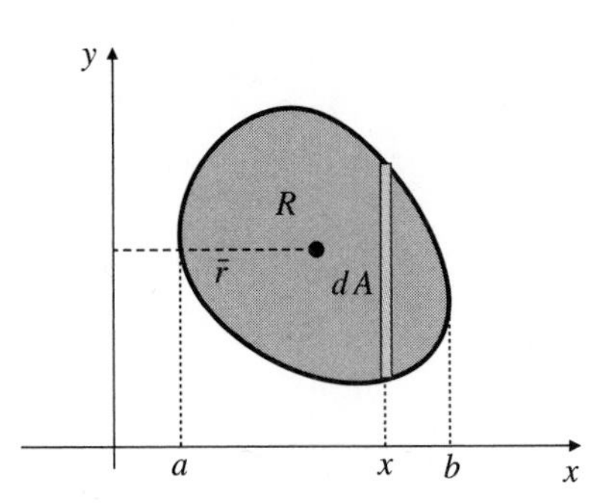

Figure 7.40 Proving Theorem 2(a)

As the following examples illustrate, Pappus's Theorem can be used in two ways: either the centroid can be determined when the appropriate volume or surface area is known, or the volume or surface area can be determined if the centroid of the rotating region or curve is known.

EXAMPLE 5 Use Pappus's Theorem to find the centroid of the semicircle $y = \sqrt{a^2 - x^2}$.

Solution The centroid of the semicircle lies on its axis of symmetry, the y-axis, so it is located at a point with coordinates $(0, \bar{y})$. Since the semicircle has length πa units and generates, on rotation about the x-axis, a sphere having area $4\pi a^2$ square units, we obtain, using part (b) of Pappus's Theorem,

$$4\pi a^2 = 2\pi(\pi a)\bar{y}.$$

Thus $\bar{y} = 2a/\pi$, as shown previously in Example 2.

EXAMPLE 6 Use Pappus's Theorem to find the volume and surface area of the torus (doughnut) obtained by rotating the disk $(x - b)^2 + y^2 \le a^2$ about the y-axis. Here $0 < a < b$. (See Figure 7.10 in Section 7.1.)

Solution The centroid of the disk is at $(b, 0)$, which is at distance $\bar{r} = b$ units from the axis of rotation. Since the disk has area πa^2 square units, the volume of the torus is

$$V = 2\pi b(\pi a^2) = 2\pi^2 a^2 b \text{ cubic units.}$$

To find the surface area S of the torus (in case you want to have icing on the doughnut), rotate the circular boundary of the disk, which has length $2\pi a$, about the y-axis and obtain

$$S = 2\pi b(2\pi a) = 4\pi^2 ab \text{ square units.}$$

EXERCISES 7.5

Find the centroids of the geometric structures in Exercises 1–21. Be alert for symmetries and opportunities to use Pappus's Theorem.

1. The quarter-disk $x^2 + y^2 \le r^2, x \ge 0, y \ge 0$
2. The region $0 \le y \le 9 - x^2$
3. The region $0 \le x \le 1, 0 \le y \le \dfrac{1}{\sqrt{1 + x^2}}$
4. The circular disk sector $x^2 + y^2 \le r^2, 0 \le y \le x$
5. The circular disk segment $0 \le y \le \sqrt{4 - x^2} - 1$
6. The semi-elliptic disk $0 \le y \le b\sqrt{1 - (x/a)^2}$
7. The quadrilateral with vertices (in clockwise order) $(0, 0)$, $(3, 1)$, $(4, 0)$, and $(2, -2)$
8. The region bounded by the semicircle $y = \sqrt{1 - (x - 1)^2}$, the y-axis, and the line $y = x - 2$.
9. A hemispherical surface of radius r
10. A solid half ball of radius r
11. A solid cone of base radius r and height h
12. A conical surface of base radius r and height h
13. The plane region $0 \le y \le \sin x,\ 0 \le x \le \pi$
14. The plane region $0 \le y \le \cos x,\ 0 \le x \le \pi/2$
15. The quarter-circle arc $x^2 + y^2 = r^2, x \ge 0,\ y \ge 0$
16. The solid obtained by rotating the region in Figure 7.41(a) about the y-axis

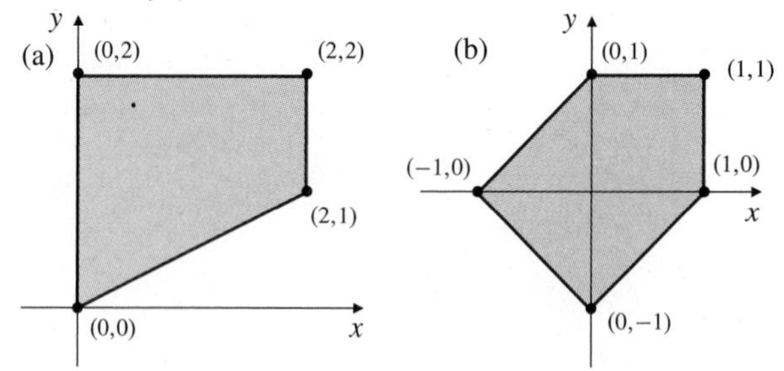

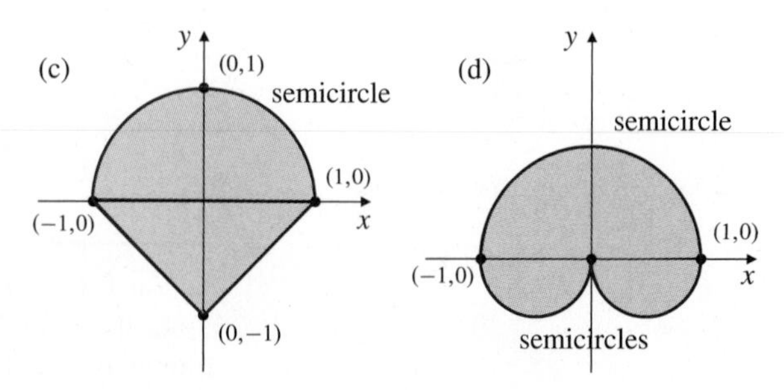

Figure 7.41

17. The region in Figure 7.41(a)
18. The region in Figure 7.41(b)
19. The region in Figure 7.41(c)
20. The region in Figure 7.41(d)
21. The solid obtained by rotating the plane region $0 \le y \le 2x - x^2$ about the line $y = -2$.
22. The line segment from $(1, 0)$ to $(0, 1)$ is rotated about the line $x = 2$ to generate part of a conical surface. Find the area of that surface.
23. The triangle with vertices $(0, 0)$, $(1, 0)$, and $(0, 1)$ is rotated about the line $x = 2$ to generate a certain solid. Find the volume of that solid.
24. An equilateral triangle of edge s cm is rotated about one of its edges to generate a solid. Find the volume and surface area of that solid.
25. Find to 5 decimal places the coordinates of the centroid of the region $0 \le x \le \pi/2$, $0 \le y \le \sqrt{x}\cos x$.
26. Find to 5 decimal places the coordinates of the centroid of the region $0 < x \le \pi/2$, $\ln(\sin x) \le y \le 0$.
27. Find the centroid of the infinitely long spike-shaped region lying between the x-axis and the curve $y = (x+1)^{-3}$ and to the right of the y-axis.
28. Show that the curve $y = e^{-x^2}$ $(-\infty < x < \infty)$ generates a surface of finite area when rotated about the x-axis. What does this imply about the location of the centroid of this infinitely long curve?
29. Obtain formulas for the coordinates of the centroid of the plane region $c \le y \le d$, $0 < f(y) \le x \le g(y)$.
30. Prove part (b) of Pappus's Theorem (Theorem 2).
31. **(Stability of a floating object)** Determining the orientation that a floating object will assume is a problem of critical importance to ship designers. Boats must be designed to float stably in an upright position; if the boat tilts somewhat from upright, the forces on it must be such as to right it again. The two forces on a floating object that need to be taken into account are its weight **W** and the balancing buoyant force **B** = −**W**. The weight **W** must be treated for mechanical purposes as being applied at the centre of mass (CM) of the object. The buoyant force, however, acts at the *centre of buoyancy* (CB), which is the centre of mass of the water displaced by the object, and is therefore the centroid of the "hole in the water" made by the object.

For example, consider a channel marker buoy consisting of a hemispherical hull surmounted by a conical tower supporting a navigation light. The buoy has a vertical axis of symmetry. If it is upright, both the CM and the CB lie on this line, as shown in Figure 7.42(left).

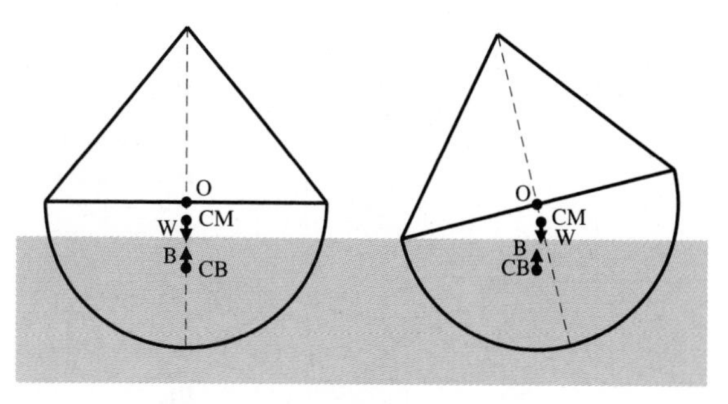

Figure 7.42

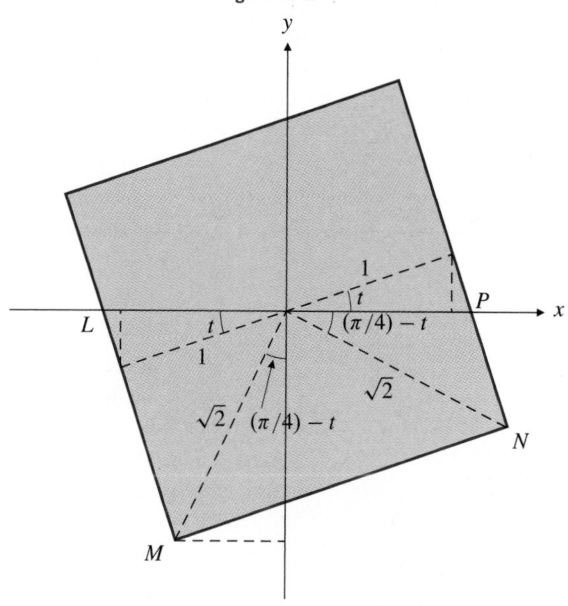

Figure 7.43

Is this upright flotation of the buoy stable? It is if the CM lies below the centre O of the hemispherical hull, as shown in the figure. To see why, imagine the buoy tilted slightly from the vertical as shown in the right figure. Observe that the CM still lies on the axis of symmetry of the buoy, but the CB lies on the vertical line through O. The forces **W** and **B** no longer act along the same line, but their torques are such as to rotate the buoy back to a vertical upright position. If CM had been above O in the left figure, the torques would have been such as to tip the buoy over once it was displaced even slightly from the vertical.

A wooden beam has a square cross-section and specific gravity 0.5, so that it will float with half of its volume submerged. (See Figure 7.43.) Assuming it will float horizontally in the water, what is the stable orientation of the square cross section with respect to the surface of the water? In particular, will the beam float with a flat face upward or an edge upward? Prove your assertions. You may find Maple or another symbolic algebra program useful.

7.6 Other Physical Applications

In this section we present some examples of the use of integration to calculate quantities arising in physics and mechanics.

Hydrostatic Pressure

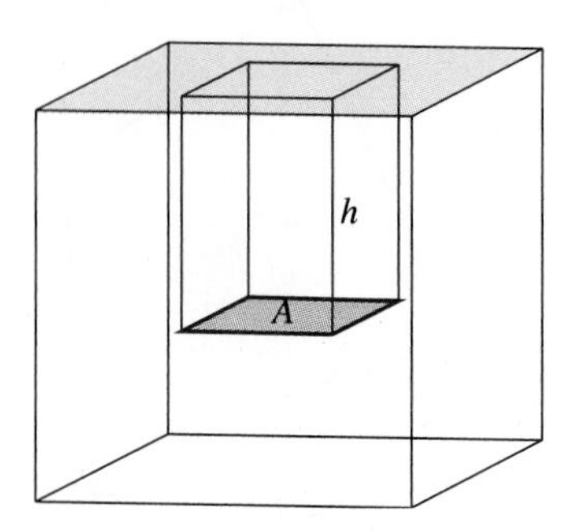

Figure 7.44 The volume of liquid above the area A is $V = Ah$. The weight of this liquid is $\rho V g = \rho g h A$, so the pressure (force per unit area) at depth h is $p = \rho g h$

The **pressure** p at depth h beneath the surface of a liquid is the *force per unit area* exerted on a horizontal plane surface at that depth due to the weight of the liquid above it. Hence p is given by

$$p = \rho g h,$$

where ρ is the density of the liquid, and g is the acceleration produced by gravity where the fluid is located. (See Figure 7.44.) For water at the surface of the earth we have, approximately, $\rho = 1{,}000$ kg/m^3 and $g = 9.8$ m/s^2, so the pressure at depth h m is

$$p = 9{,}800h \text{ N/m}^2.$$

The unit of force used here is the newton (N); 1 N = 1 kg·m/s^2, the force that imparts an acceleration of 1 m/s^2 to a mass of 1 kg.

The molecules in a liquid interact in such a way that the pressure at any depth acts equally in all directions; the pressure against a vertical surface is the same as that against a horizontal surface at the same depth. This is **Pascal's principle**.

The total force exerted by a liquid on a horizontal surface (say, the bottom of a tank holding the liquid) is found by multiplying the area of that surface by the pressure at the depth of the surface below the top of the liquid. For nonhorizontal surfaces, however, the pressure is not constant over the whole surface, and the total force cannot be determined so easily. In this case we divide the surface into area elements dA, each at some particular depth h, and we then sum (i.e., integrate) the corresponding force elements $dF = \rho g h\, dA$ to find the total force.

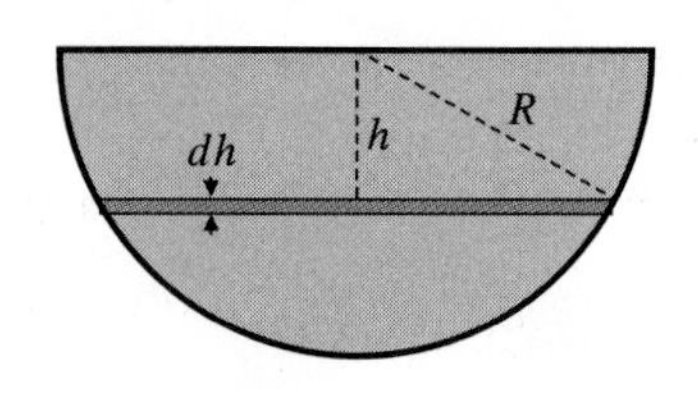

Figure 7.45 An end plate of the water trough

EXAMPLE 1 One vertical wall of a water trough is a semicircular plate of radius R m with curved edge downward. If the trough is full, so that the water comes up to the top of the plate, find the total force of the water on the plate.

Solution A horizontal strip of the surface of the plate at depth h m and having width dh m (see Figure 7.45) has length $2\sqrt{R^2 - h^2}$ m; hence, its area is $dA = 2\sqrt{R^2 - h^2}\, dh$ m^2. The force of the water on this strip is

$$dF = \rho g h\, dA = 2\rho g h \sqrt{R^2 - h^2}\, dh.$$

Thus, the total force on the plate is

$$F = \int_{h=0}^{h=R} dF = 2\rho g \int_0^R h\sqrt{R^2 - h^2}\, dh \qquad \text{Let } u = R^2 - h^2,\ du = -2h\, dh$$

$$= \rho g \int_0^{R^2} u^{1/2}\, du = \rho g \frac{2}{3} u^{3/2} \Big|_0^{R^2}$$

$$\approx \frac{2}{3} \times 9{,}800 R^3 \approx 6{,}533 R^3 \text{ N}.$$

EXAMPLE 2 **(Force on a dam)** Find the total force on a section of a dam 100 m long and having a vertical height of 10 m, if the surface holding back the water is inclined at an angle of 30° to the vertical and the water comes up to the top of the dam.

Solution The water in a horizontal layer of thickness dh m at depth h m makes contact with the dam along a slanted strip of width $dh \sec 30° = (2/\sqrt{3})\, dh$ m. (See Figure 7.46.) The area of this strip is $dA = (200/\sqrt{3})\, dh$ m^2, and the force of water against the strip is

$$dF = \rho g h\, dA = \frac{200}{\sqrt{3}} \times 1{,}000 \times 9.8h\, dh \approx 1{,}131{,}600h\, dh \text{ N}.$$

The total force on the dam section is therefore

$$F \approx 1{,}131{,}600 \int_0^{10} h\,dh = 1{,}131{,}600 \times \frac{10^2}{2} \approx 5.658 \times 10^7 \text{ N.}$$

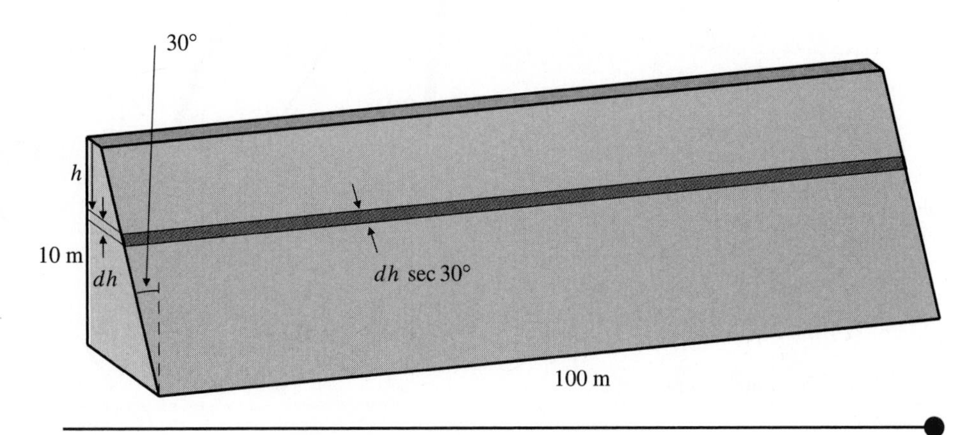

Figure 7.46 The dam of Example 2

Work

When a force acts on an object to move that object, it is said to have done **work** on the object. The amount of work done by a constant force is measured by the product of the force and the distance through which it moves the object. This assumes that the force is in the direction of the motion.

$$\text{work} = \text{force} \times \text{distance}$$

Work is always related to a particular force. If other forces acting on an object cause it to move in a direction opposite to the force F, then work is said to have been done *against* the force F.

Suppose that a force in the direction of the x-axis moves an object from $x = a$ to $x = b$ on that axis and that the force varies continuously with the position x of the object; that is, $F = F(x)$ is a continuous function. The element of work done by the force in moving the object through a very short distance from x to $x + dx$ is $dW = F(x)\,dx$, so the total work done by the force is

$$W = \int_{x=a}^{x=b} dW = \int_a^b F(x)\,dx.$$

EXAMPLE 3 **(Stretching or compressing a spring)** By **Hooke's Law**, the force $F(x)$ required to extend (or compress) an elastic spring to x units longer (or shorter) than its natural length is proportional to x:

$$F(x) = kx,$$

where k is the **spring constant** for the particular spring. If a force of 2,000 N is required to extend a certain spring to 4 cm longer than its natural length, how much work must be done to extend it that far?

Solution Since $F(x) = kx = 2{,}000$ N when $x = 4$ cm, we must have $k = 2{,}000/4 = 500$ N/cm. The work done in extending the spring 4 cm is

$$W = \int_0^4 kx\,dx = k\frac{x^2}{2}\bigg|_0^4 = 500\ \frac{\text{N}}{\text{cm}} \times \frac{4^2\ \text{cm}^2}{2} = 4{,}000\ \text{N}\cdot\text{cm} = 40\ \text{N}\cdot\text{m}.$$

Forty newton-metres (joules) of work must be done to stretch the spring 4 cm.

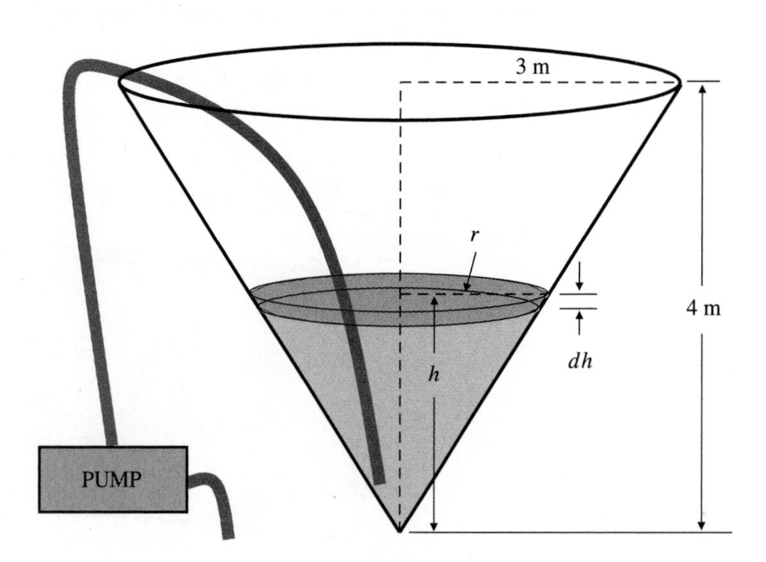

Figure 7.47 Pumping water out of a conical tank

EXAMPLE 4 **(Work done to pump out a tank)** Water fills a tank in the shape of a right-circular cone with top radius 3 m and depth 4 m. How much work must be done (against gravity) to pump all the water out of the tank over the top edge of the tank?

Solution A thin, disk-shaped slice of water at height h above the vertex of the tank has radius r (see Figure 7.47), where $r = \frac{3}{4}h$ by similar triangles. The volume of this slice is

$$dV = \pi r^2\,dh = \frac{9}{16}\pi h^2\,dh,$$

and its *weight* (the force of gravity on the mass of water in the slice) is

$$dF = \rho g\,dV = \frac{9}{16}\rho g\,\pi h^2\,dh.$$

The water in this disk must be raised (against gravity) a distance $(4 - h)$ m by the pump. The work required to do this is

$$dW = \frac{9}{16}\rho g\,\pi(4 - h)h^2\,dh.$$

The total work that must be done to empty the tank is the sum (integral) of all these elements of work for disks at depths between 0 and 4 m:

$$\begin{aligned} W &= \int_0^4 \frac{9}{16}\rho g\,\pi(4h^2 - h^3)\,dh \\ &= \frac{9}{16}\rho g\,\pi\left(\frac{4h^3}{3} - \frac{h^4}{4}\right)\Bigg|_0^4 \\ &= \frac{9\pi}{16} \times 1{,}000 \times 9.8 \times \frac{64}{3} \approx 3.69 \times 10^5\ \text{N·m}. \end{aligned}$$

EXAMPLE 5 **(Work to raise material into orbit)** The gravitational force of the earth on a mass m located at height h above the surface of the earth is given by

$$F(h) = \frac{Km}{(R + h)^2},$$

where R is the radius of the earth, and K is a constant that is independent of m and h. Determine, in terms of K and R, the work that must be done against gravity to raise an object from the surface of the earth to:

(a) a height H above the surface of the earth, and

(b) an infinite height above the surface of the earth.

Solution The work done to raise the mass m from height h to height $h + dh$ is

$$dW = \frac{Km}{(R+h)^2}\,dh.$$

(a) The total work to raise it from height $h = 0$ to height $h = H$ is

$$W = \int_0^H \frac{Km}{(R+h)^2}\,dh = \left.\frac{-Km}{R+h}\right|_0^H = Km\left(\frac{1}{R} - \frac{1}{R+H}\right).$$

If R and H are measured in metres and F is measured in newtons, then W is measured in newton-metres (N·m), or joules.

(b) The total work necessary to raise the mass m to an infinite height is

$$W = \int_0^\infty \frac{Km}{(R+h)^2}\,dh = \lim_{H\to\infty} Km\left(\frac{1}{R} - \frac{1}{R+H}\right) = \frac{Km}{R}.$$

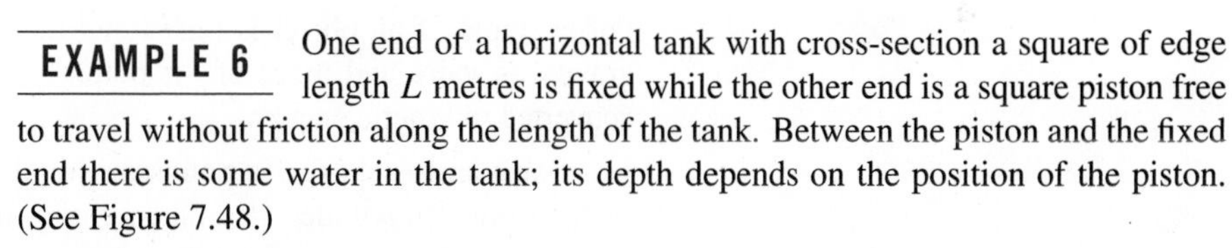

EXAMPLE 6 One end of a horizontal tank with cross-section a square of edge length L metres is fixed while the other end is a square piston free to travel without friction along the length of the tank. Between the piston and the fixed end there is some water in the tank; its depth depends on the position of the piston. (See Figure 7.48.)

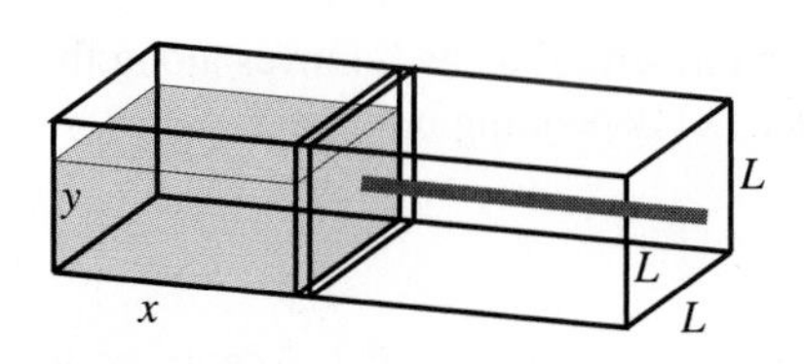

Figure 7.48 The Piston in Example 6

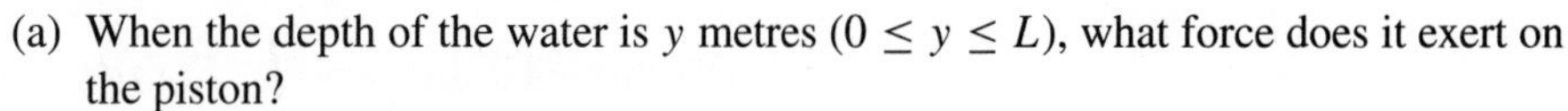

(a) When the depth of the water is y metres ($0 \le y \le L$), what force does it exert on the piston?

(b) If the piston is X metres from the fixed end of the tank when the water depth is $L/2$ metres, how much work must be done to force the piston in further to halve that distance and hence cause the water level to increase to fill the available space? Assume no water leaks out but that trapped air can escape from the top of the tank.

Solution

(a) When the depth of water in the tank is y m, a horizontal strip on the face of the piston at depth z below the surface of the water ($0 \le z \le y$) and having height dz has area $dA = L\,dz$. Since the pressure at depth z is $\rho g z = 9{,}800z$ N/m^2, the force of the water on the strip is $dF = 9{,}800\,Lz\,dz$ N. Thus, the force on the piston is

$$F = \int_0^y 9{,}800\,L\,z\,dz = 4{,}900\,L\,y^2\ \text{N}, \quad \text{where } 0 \le y \le L.$$

(b) If the distance from the fixed end of the tank to the piston is x m when the water depth is y m, then the volume of water in the tank is $V = Lxy$ m^3. But we are given that $V = L^2X/2$, so we have $u = LX/2$. Now the work done in moving the piston from x to $x - dx$ is

$$dW = 4{,}900\,Ly^2(-dx) = -4{,}900\,L\,\frac{L^2X^2}{4x^2}\,dx.$$

Thus, the work done to move the piston from position X to position $X/2$ is

$$\begin{aligned} W &= -\int_X^{X/2} 4{,}900\,\frac{L^3X^2}{4}\,\frac{dx}{x^2} \\ &= 4{,}900\,\frac{L^3X^2}{4}\left(\frac{2}{X} - \frac{1}{X}\right) = 1{,}225\ \text{N}\cdot\text{m}. \end{aligned}$$

Potential Energy and Kinetic Energy

The units of energy are the same as those of work (force $\times$ distance). Work done against a force may be regarded as storing up energy for future use or for conversion to other forms. Such stored energy is called **potential energy** (P.E.). For instance, in extending or compressing an elastic spring, we are doing work against the tension in the spring and hence storing energy in the spring. When work is done against a (variable) force $F(x)$ to move an object from $x = a$ to $x = b$, the energy stored is

$$\text{P.E.} = -\int_a^b F(x)\,dx.$$

Since the work is being done against F, the signs of $F(x)$ and $b - a$ are opposite, so the integral is negative; the explicit negative sign is included so that the calculated potential energy will be positive.

One of the forms of energy into which potential energy can be converted is **kinetic energy** (K.E.), the energy of motion. If an object of mass m is moving with velocity v, it has kinetic energy

$$\text{K.E.} = \frac{1}{2}mv^2.$$

For example, if an object is raised and then dropped, it accelerates downward under gravity as more and more of the potential energy stored in it when it was raised is converted to kinetic energy.

Consider the change in potential energy stored in a mass m as it moves along the x-axis from a to b under the influence of a force $F(x)$ depending only on x:

$$\text{P.E.}(b) - \text{P.E.}(a) = -\int_a^b F(x)\,dx.$$

(The change in P.E. is negative if m is moving in the direction of F.) According to Newton's second law of motion, the force $F(x)$ causes the mass m to accelerate, with acceleration dv/dt given by

$$F(x) = m\frac{dv}{dt} \qquad \text{(force = mass} \times \text{acceleration).}$$

By the Chain Rule we can rewrite dv/dt in the form

$$\frac{dv}{dt} = \frac{dv}{dx}\frac{dx}{dt} = v\frac{dv}{dx},$$

so $F(x) = mv\dfrac{dv}{dx}$. Hence,

$$\begin{aligned}
\text{P.E.}(b) - \text{P.E.}(a) &= -\int_a^b mv\frac{dv}{dx}\,dx \\
&= -m\int_{x=a}^{x=b} v\,dv \\
&= -\frac{1}{2}mv^2\bigg|_{x=a}^{x=b} \\
&= \text{K.E.}(a) - \text{K.E.}(b).
\end{aligned}$$

It follows that

$$\text{P.E.}(b) + \text{K.E.}(b) = \text{P.E.}(a) + \text{K.E.}(a).$$

This shows that the total energy (potential + kinetic) remains constant as the mass m moves under the influence of a force F, *depending only on position*. Such a force is said to be **conservative**, and the above result is called the **law of conservation of energy**. Conservative forces will be further discussed in Section 15.2.

EXAMPLE 7 **(Escape velocity)** Use the result of Example 5 together with the following known values,

(a) the radius R of the earth is about 6,400 km, or 6.4×10^6 m,

(b) the acceleration of gravity g at the surface of the earth is about 9.8 m/s^2,

to determine the constant K in the gravitational force formula of Example 5, and use this information to determine the escape velocity for a projectile fired vertically from the surface of the earth. The **escape velocity** is the (minimum) speed that such a projectile must have at firing to ensure that it will continue to move farther and farther away from the earth and not fall back.

Solution According to the formula of Example 5, the force of gravity on a mass m kg at the surface of the earth ($h = 0$) is

$$F = \frac{Km}{(R+0)^2} = \frac{Km}{R^2}.$$

According to Newton's second law of motion, this force is related to the acceleration of gravity (g) there by the equation $F = mg$. Thus,

$$\frac{Km}{R^2} = mg \qquad \text{and} \quad K = gR^2.$$

According to the law of conservation of energy, the projectile must have sufficient kinetic energy at firing to do the work necessary to raise the mass m to infinite height. By the result of Example 5, this required energy is Km/R. If the initial velocity of the projectile is v, we want

$$\frac{1}{2}mv^2 \geq \frac{Km}{R}.$$

Thus v must satisfy

$$v \geq \sqrt{\frac{2K}{R}} = \sqrt{2gR} \approx \sqrt{2 \times 9.8 \times 6.4 \times 10^6} \approx 1.12 \times 10^4 \text{ m/s}.$$

Thus, the escape velocity is approximately 11.2 km/s and is independent of the mass m. In this calculation we have neglected any air resistance near the surface of the earth. Such resistance depends on velocity rather than on position, so it is not a conservative force. The effect of such resistance would be to use up (convert to heat) some of the initial kinetic energy and so raise the escape velocity.

EXERCISES 7.6

1. A tank has a square base 2 m on each side and vertical sides 6 m high. If the tank is filled with water, find the total force exerted by the water (a) on the bottom of the tank and (b) on one of the four vertical walls of the tank.

2. A swimming pool 20 m long and 8 m wide has a sloping plane bottom so that the depth of the pool is 1 m at one end and 3 m at the other end. Find the total force exerted on the bottom if the pool is full of water.

3. A dam 200 m long and 24 m high presents a sloping face of 26 m slant height to the water in a reservoir behind the dam (Figure 7.49). If the surface of the water is level with the top of the dam, what is the total force of the water on the dam?

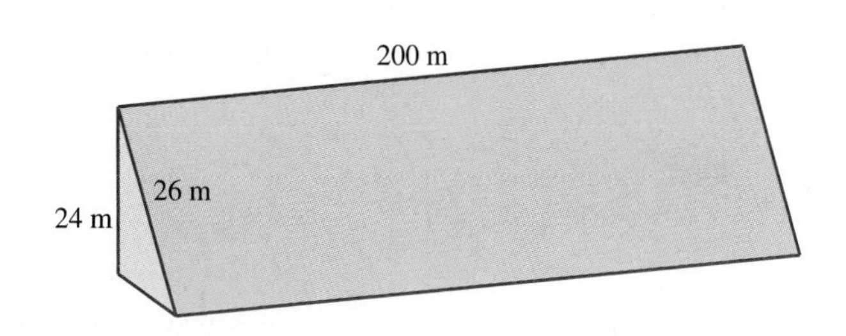

Figure 7.49

4. A pyramid with a square base, 4 m on each side and four equilateral triangular faces, sits on the level bottom of a lake at a place where the lake is 10 m deep. Find the total force of the water on each of the triangular faces.

5. A lock on a canal has a gate in the shape of a vertical rectangle 5 m wide and 20 m high. If the water on one side of the gate comes up to the top of the gate, and the water on the other side comes only 6 m up the gate, find the total force that must be exerted to hold the gate in place.

6. If 100 N·cm of work must be done to compress an elastic spring to 3 cm shorter than its natural length, how much work must be done to compress it 1 cm further?

7. Find the total work that must be done to pump all the water in the tank of Exercise 1 out over the top of the tank.

8. Find the total work that must be done to pump all the water in the swimming pool of Exercise 2 out over the top edge of the pool.

9. Find the work that must be done to pump all the water in a full hemispherical bowl of radius a m to a height h m above the top of the bowl.

10. A horizontal cylindrical tank has radius R m. One end of the tank is a fixed disk, but the other end is a circular piston of radius R m free to travel along the length of the tank. There is some water in the tank between the piston and the fixed end; its depth depends on the position of the piston. What force does the water exert on the piston when the surface of the water is y m $(-R \le y \le R)$ above the centre of the piston face? (See Figure 7.50.)

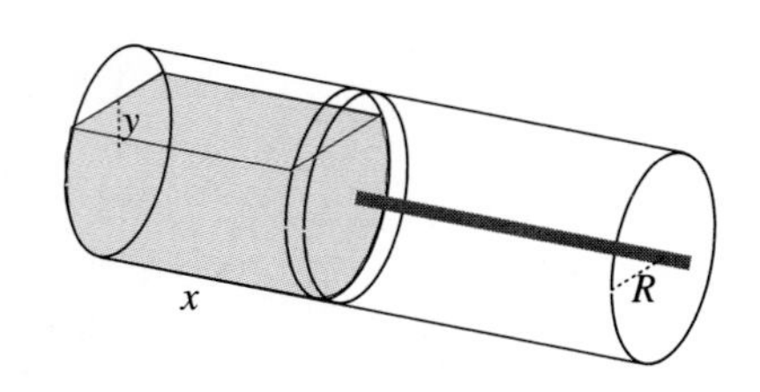

Figure 7.50

11. Continuing the previous problem, suppose that when the piston is X m from the fixed end of the tank the water level is at the centre of the piston face. How much work must be done to reduce the distance from the piston to the fixed end to $X/2$ m, and thus cause the water to fill the volume between the piston and the fixed end of the tank? As in Example 6, you can assume the piston can move without friction, and that trapped air can escape. *Hint:* The technique used to solve part (b) of Example 6 is very difficult to apply here. Instead, calculate the work done to raise the water in half of the bottom half-cylinder of length X so that it fills the top half-cylinder of length $X/2$.

12. A bucket is raised vertically from ground level at a constant speed of 2 m/min by a winch. If the bucket weighs 1 kg and contains 15 kg of water when it starts up but loses water by leakage at a rate of 1 kg/min thereafter, how much work must be done by the winch to raise the bucket to a height of 10 m?

7.7 Applications in Business, Finance, and Ecology

If the rate of change $f'(x)$ of a function $f(x)$ is known, the change in value of the function over an interval from $x = a$ to $x = b$ is just the integral of f' over $[a, b]$:

$$f(b) - f(a) = \int_a^b f'(x)\,dx.$$

For example, if the speed of a moving car at time t is $v(t)$ km/h, then the distance travelled by the car during the time interval $[0, T]$ (hours) is $\int_0^T v(t)\,dt$ km.

Similar situations arise naturally in business and economics, where the rates of change are often called marginals.

EXAMPLE 1 **(Finding total revenue from marginal revenue)** A supplier of calculators realizes a marginal revenue of $\$15 - 5e^{-x/50}$ per calculator when she has sold x calculators. What will be her total revenue from the sale of 100 calculators?

Solution The marginal revenue is the rate of change of revenue with respect to the number of calculators sold. Thus, the revenue from the sale of dx calculators after x have already been sold is

$$dR = (15 - 5e^{-x/50})\,dx \quad \text{dollars.}$$

The total revenue from the sale of the first 100 calculators is $\$R$, where

$$\begin{aligned} R = \int_{x=0}^{x=100} dR &= \int_0^{100} (15 - 5e^{-x/50})\, dx \\ &= \left(15x + 250e^{-x/50}\right)\Big|_0^{100} \\ &= 1,500 + 250e^{-2} - 250 \approx 1,283.83, \end{aligned}$$

that is, about \$1,284.

The Present Value of a Stream of Payments

Suppose that you have a business that generates income continuously at a variable rate $P(t)$ dollars per year at time t and that you expect this income to continue for the next T years. How much is the business worth today?

The answer surely depends on interest rates. One dollar to be received t years from now is worth less than one dollar received today, which could be invested at interest to yield more than one dollar t years from now. The higher the interest rate, the lower the value today of a payment that is not due until sometime in the future.

To analyze this situation, suppose that the nominal interest rate is $r\%$ per annum, but is compounded continuously. Let $\delta = r/100$. As shown in Section 3.4, an investment of \$1 today will grow to

$$\lim_{n\to\infty} \left(1 + \frac{\delta}{n}\right)^{nt} = e^{\delta t} \quad \text{dollars}$$

after t years. Therefore, a payment of \$1 after t years must be worth only $\$e^{-\delta t}$ today. This is called the *present value* of the future payment. When viewed this way, the interest rate δ is frequently called a *discount rate*; it represents the amount by which future payments are discounted.

Returning to the business income problem, in the short time interval from t to $t+dt$, the business produces income $\$P(t)\,dt$, of which the present value is $\$e^{-\delta t}P(t)\,dt$. Therefore, the present value $\$V$ of the income stream over the time interval $[0, T]$ is the "sum" of these contributions:

$$V = \int_0^T e^{-\delta t} P(t)\, dt.$$

EXAMPLE 2 What is the present value of a constant, continual stream of payments at a rate of \$10,000 per year, to continue forever, starting now? Assume an interest rate of 6% per annum, compounded continuously.

Solution The required present value is

$$V = \int_0^\infty e^{-0.06t}\, 10{,}000\, dt = 10{,}000 \lim_{R\to\infty} \frac{e^{-0.06t}}{-0.06}\Bigg|_0^R \approx \$166{,}667.$$

The Economics of Exploiting Renewable Resources

As noted in Section 3.4, the rate of increase of a biological population sometimes conforms to a logistic model[1]

$$\frac{dx}{dt} = kx\left(1 - \frac{x}{L}\right).$$

[1] This example was suggested by Professor C. W. Clark, of the University of British Columbia.

Here, $x = x(t)$ is the size (or biomass) of the population at time t, k is the natural rate at which the population would grow if its food supply were unlimited, and L is the natural limiting size of the population—the carrying capacity of its environment. Such models are thought to apply, for example, to the Antarctic blue whale and to several species of fish and trees. If the resource is harvested (say, the fish are caught) at a rate $h(t)$ units per year at time t, then the population grows at a slower rate:

$$\frac{dx}{dt} = kx\left(1 - \frac{x}{L}\right) - h(t). \qquad (*)$$

In particular, if we harvest the population at its current rate of growth,

$$h(t) = kx\left(1 - \frac{x}{L}\right),$$

then $dx/dt = 0$, and the population will maintain a constant size. Assume that each unit of harvest produces an income of $\$p$ for the fishing industry. The total annual income from harvesting the resource at its current rate of growth will be

$$T = ph(t) = pkx\left(1 - \frac{x}{L}\right).$$

Considered as a function of x, this total annual income is quadratic and has a maximum value when $x = L/2$, the value that ensures $dT/dx = 0$. The industry can maintain a stable maximum annual income by ensuring that the population level remains at half the maximal size of the population with no harvesting.

The analysis above, however, does not take into account the discounted value of future harvests. If the discount rate is δ, compounded continuously, then the present value of the income $\$ph(t)\,dt$ due between t and $t + dt$ years from now is $e^{-\delta t}ph(t)\,dt$. The total present value of all income from the fishery in future years is

$$T = \int_0^\infty e^{-\delta t}ph(t)\,dt.$$

What fishing strategy will maximize T? If we substitute for $h(t)$ from equation $(*)$ governing the growth rate of the population, we get

$$\begin{aligned} T &= \int_0^\infty pe^{-\delta t}\left[kx\left(1 - \frac{x}{L}\right) - \frac{dx}{dt}\right] dt \\ &= \int_0^\infty kpe^{-\delta t}x\left(1 - \frac{x}{L}\right) dt - \int_0^\infty pe^{-\delta t}\frac{dx}{dt}\,dt. \end{aligned}$$

Integrate by parts in the last integral above, taking $U = pe^{-\delta t}$ and $dV = \dfrac{dx}{dt}\,dt$:

$$\begin{aligned} T &= \int_0^\infty kpe^{-\delta t}x\left(1 - \frac{x}{L}\right) dt - \left[pe^{-\delta t}x\bigg|_0^\infty + \int_0^\infty p\delta e^{-\delta t}x\,dt\right] \\ &= px(0) + \int_0^\infty pe^{-\delta t}\left[kx\left(1 - \frac{x}{L}\right) - \delta x\right] dt. \end{aligned}$$

To make this expression as large as possible, we should choose the population size x to maximize the quadratic expression

$$Q(x) = kx\left(1 - \frac{x}{L}\right) - \delta x$$

at as early a time t as possible, and keep the population size constant at that level thereafter. The maximum occurs where $Q'(x) = k - (2kx/L) - \delta = 0$, that is, where

$$x = \frac{L}{2} - \frac{\delta L}{2k} = (k - \delta)\frac{L}{2k}.$$

The maximum present value of the fishery is realized if the population level x is held at this value. Note that this population level is smaller than the optimal level $L/2$ we obtained by ignoring the discount rate. The higher the discount rate δ, the smaller will be the income-maximizing population level. More unfortunately, if $\delta \geq k$, the model predicts greatest income from fishing the species to *extinction* immediately! (See Figure 7.51.)

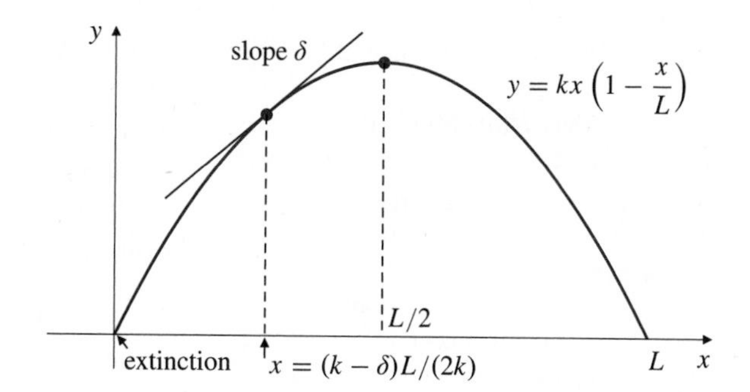

Figure 7.51 The greater the discount rate δ, the smaller the population size x that will maximize the present value of future income from harvesting. If $\delta \ge k$, the model predicts fishing the species to extinction

Of course, this model fails to take into consideration other factors that may affect the fishing strategy, such as the increased cost of harvesting when the population level is small and the effect of competition among various parts of the fishing industry. Nevertheless, it does explain the regrettable fact that, under some circumstances, an industry based on a renewable resource can find it in its best interest to destroy the resource. This is especially likely to happen when the natural growth rate k of the resource is low, as it is for the case of whales and most trees. There is good reason not to allow economics alone to dictate the management of the resource.

EXERCISES 7.7

1. **(Cost of production)** The marginal cost of production in a coal mine is $\$6 - 2 \times 10^{-3}x + 6 \times 10^{-6}x^2$ per ton after the first x tons are produced each day. In addition, there is a fixed cost of \$4,000 per day to open the mine. Find the total cost of production on a day when 1,000 tons are produced.

2. **(Total sales)** The sales of a new computer chip are modelled by $s(t) = te^{-t/10}$, where $s(t)$ is the number of thousands of chips sold per week, t weeks after the chip was introduced to the market. How many chips were sold in the first year?

3. **(Internet connection rates)** An internet service provider charges clients at a continuously decreasing marginal rate of $\$4/(1 + \sqrt{t})$ per hour when the client has already used t hours during a month. How much will be billed to a client who uses x hours in a month? (x need not be an integer.)

4. **(Total revenue from declining sales)** The price per kilogram of maple syrup in a store rises at a constant rate from \$10 at the beginning of the year to \$15 at the end of the year. As the price rises, the quantity sold decreases; the sales rate is $400/(1 + 0.1t)$ kg/year at time t years, $(0 \le t \le 1)$. What total revenue does the store obtain from sales of the syrup during the year?

(Stream of payment problems) Find the present value of a continuous stream of payments of \$1,000 per year for the periods and discount rates given in Exercises 5–10. In each case the discount rate is compounded continuously.

5. 10 years at a discount rate of 2%
6. 10 years at a discount rate of 5%
7. 10 years beginning 2 years from now at a discount rate of 8%
8. 25 years beginning 10 years from now at a discount rate of 5%
9. For all future time at a discount rate of 2%
10. Beginning in 10 years and continuing forever after at a discount rate of 5%

11. Find the present value of a continuous stream of payments over a 10-year period beginning at a rate of \$1,000 per year now and increasing steadily at \$100 per year. The discount rate is 5%.

12. Find the present value of a continuous stream of payments over a 10-year period beginning at a rate of \$1,000 per year now and increasing steadily at 10% per year. The discount rate is 5%.

13. Money flows continuously into an account at a rate of \$5,000 per year. If the account earns interest at a rate of 5% compounded continuously, how much will be in the account after 10 years?

14. Money flows continuously into an account beginning at a rate of \$5,000 per year and increasing at 10% per year. Interest causes the account to grow at a real rate of 6% (so that \$1 grows to $\$1.06^t$ in t years). How long will it take for the balance in the account to reach \$1,000,000?

15. If the discount rate δ varies with time, say $\delta = \delta(t)$, show that the present value of a payment of $\$P$ due t years from now is $\$Pe^{-\lambda(t)}$, where

$$\lambda(t) = \int_0^t \delta(\tau)\,d\tau.$$

What is the value of a stream of payments due at a rate $\$P(t)$ at time t, from $t = 0$ to $t = T$?

16. **(Discount rates and population models)** Suppose that the growth rate of a population is a function of the population size: $dx/dt = F(x)$. (For the logistic model, $F(x) = kx(1 - (x/L))$.) If the population is harvested at rate $h(t)$ at time t, then $x(t)$ satisfies

$$\frac{dx}{dt} = F(x) - h(t).$$

Show that the value of x that maximizes the present value of all future harvests satisfies $F'(x) = \delta$, where δ is the (continuously compounded) discount rate. *Hint:* Mimic the argument used above for the logistic case.

17. (Managing a fishery) The carrying capacity of a certain lake is $L = 80{,}000$ of a certain species of fish. The natural growth rate of this species is 12% per year ($k = 0.12$). Each fish is worth \$6. The discount rate is 5%. What population of fish should be maintained in the lake to maximize the present value of all future revenue from harvesting the fish? What is the annual revenue resulting from maintaining this population level?

18. (Blue whales) It is speculated that the natural growth rate of the Antarctic blue whale population is about 2% per year ($k = 0.02$) and that the carrying capacity of its habitat is about $L = 150{,}000$. One blue whale is worth, on average, \$10,000. Assuming that the blue whale population satisfies a logistic model, and using the data above, find the following:

(a) the maximum sustainable annual harvest of blue whales.

(b) the annual revenue resulting from the maximum annual sustainable harvest.

(c) the annual interest generated if the whale population (assumed to be at the level $L/2$ supporting the maximum sustainable harvest) is exterminated and the proceeds invested at 2%. (d) at 5%.

(e) the total present value of all future revenue if the population is maintained at the level $L/2$ and the discount rate is 5%.

19. The model developed above does not allow for the costs of harvesting. Try to devise a way to alter the model to take this into account. Typically, the cost of catching a fish goes up as the number of fish goes down.

7.8 Probability

Probability theory is a very important field of application of calculus. This subject cannot, of course, be developed thoroughly here—an adequate presentation requires one or more whole courses—but we can give a brief introduction that suggests some of the ways sums and integrals are used in probability theory.

In the context of probability theory the term **experiment** is used to denote a process that can result in different **outcomes**. The set of all possible outcomes is called the **sample space** for the experiment. For example, the process might be the tossing of a coin for which we could have three possible outcomes: H (the coin lands horizontal with "heads" showing on top), T (the coin lands horizontal with "tails" showing on top), or E (the coin lands and remains standing on its edge). Of course, outcome E is not very likely unless the coin is quite thick, but it can happen. So our sample space is $S = \{H, T, E\}$. Suppose we were to toss the coin a great many times, and observe that the outcomes H and T each occur on 49% of the tosses while E occurs only 2% of the time. We would say that on any one toss of the coin the outcomes H and T each have probability 0.49 and E has probability 0.02.

An **event** is any subset of the sample space. The **probability** of an event occurring is a real number between 0 and 1 that measures the proportion of times the outcome of the experiment can be expected to belong to that event if the experiment is repeated many times. If the event is the whole sample space, its occurrence is certain, and its probability is 1; if the event is the empty set $\emptyset = \{\,\}$, it cannot possibly occur, and its probability is 0. For the coin-tossing experiment, there are eight possible events; we record their probabilities as follows:

$$\begin{array}{llll} \Pr(\emptyset) = 0, & \Pr(\{T\}) = 0.49, & \Pr(\{H, T\}) = 0.98, & \Pr(\{T, E\}) = 0.51, \\ \Pr(\{H\}) = 0.49, & \Pr(\{E\}) = 0.02, & \Pr(\{H, E\}) = 0.51, & \Pr(S) = 1. \end{array}$$

Given any two events A and B (subsets of sample space S), their **intersection** $A \cap B$ consists of those outcomes belonging to both A and B; it is sometimes called the event "A and B." Two events are **disjoint** if $A \cap B = \emptyset$; no outcome can belong to two disjoint events. For instance, an event A and its **complement**, A^c, consisting of all outcomes in S that don't belong to A, are disjoint. The **union** of two events A and B (also called the event "A or B") consists of all outcomes that belong to at least one of A and B. Note that $A \cup A^c = S$.

We summarize the basic rules governing probability as follows: if S is a sample space, $\emptyset$ is the empty subset of S, and A and B are any events, then

(a) $0 \le \Pr(A) \le 1$,
(b) $\Pr(\emptyset) = 0$ and $\Pr(S) = 1$,
(c) $\Pr(A^c) = 1 - \Pr(A)$,
(d) $\Pr(A \cup B) = \Pr(A) + \Pr(B) - \Pr(A \cap B)$.

Note that just adding $\Pr(A) + \Pr(B)$ would count outcomes in $A \cap B$ twice. As an example, in our coin-tossing experiment if $A = \{H, T\}$ and $B = \{H, E\}$, then $A^c = \{E\}$, $A \cup B = \{H, T, E\} = S$, and $A \cap B = \{H\}$. We have

$$\Pr(A^c) = \Pr(\{E\}) = 0.02 = 1 - 0.98 = 1 - \Pr(\{H, T\}) = 1 - \Pr(A)$$
$$\Pr(A \cup B) = \Pr(S) = 1 = 0.51 + 0.51 - 0.02 = \Pr(A) + \Pr(B) - \Pr(A \cap B).$$

Discrete Random Variables

A **random variable** is a function defined on a sample space. We will denote random variables by using uppercase letters such as X and Y. If the sample space contains only discrete outcomes (like the sample space for the coin-tossing experiment), a random variable on it will have only discrete values and will be called a **discrete random variable**. If, on the other hand, the sample space contains all possible measurements of, say, heights of trees, then a random variable equal to that measurement can itself take on a continuum of real values and will be called a **continuous random variable**. We will study both types in this section.

Most discrete random variables have only finitely many values, but some can have infinitely many values if, say, the sample space consisted of the positive integers $\{1, 2, 3, \ldots\}$. A discrete random variable X has an associated **probability function** f defined on the range of X by $f(x) = \Pr(X = x)$ for each possible value x of X. Typically, f is represented by a bar graph; the sum of the heights of all the bars must be 1,

$$\sum_x f(x) = \sum_x \Pr(X = x) = 1,$$

since it is certain that the experiment must produce an outcome, and therefore a value of X.

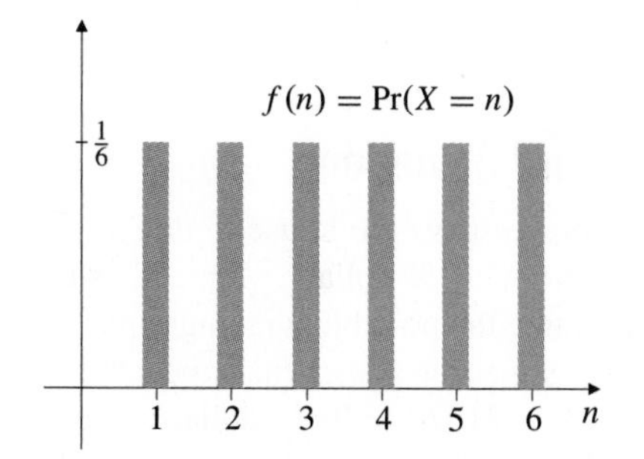

Figure 7.52 The probability function for a single rolled die

EXAMPLE 1 A single fair die is rolled so that it will show one of the numbers 1 to 6 on top when it stops. If X denotes the number showing on any roll, then X is a discrete random variable with 6 possible values. Since the die is fair, no one value of X is any more likely than any other, so the probability that the number showing is n must be 1/6 for each possible value of n. If f is the probability function of X, then

$$f(n) = \Pr(X = n) = \frac{1}{6} \qquad \text{for each } n \text{ in } \{1, 2, 3, 4, 5, 6\}.$$

The discrete random variable X is therefore said to be distributed **uniformly**. All the bars in the graph of its probability function f have the same height. (See Figure 7.52.) Note that

$$\sum_{n=1}^{6} \Pr(X = n) = 1,$$

reflecting the fact that the rolled die must certainly give one of the six possible outcomes. The probability that a roll will produce a value from 1 to 4 is

$$\Pr(1 \le X \le 4) = \sum_{n=1}^{4} \Pr(X = n) = \frac{1}{6} + \frac{1}{6} + \frac{1}{6} + \frac{1}{6} = \frac{2}{3}.$$

EXAMPLE 2 What is the sample space for the numbers showing on top when two fair dice are rolled. What is the probability that a 4 and a 2 will be showing? Find the probability function for the random variable X that gives the sum of the two numbers showing on the dice. What is the probability that that sum is less than 10?

Solution The sample space consists of all pairs of integers (m, n) satisfying $1 \leq m \leq 6$ and $1 \leq n \leq 6$. There are 36 such pairs, so the probability of any one of them is 1/36. Two of the pairs, $(4, 2)$ and $(2, 4)$, correspond to a 4 and a 2 showing, so the probability of that event is $(1/36) + (1/36) = 1/18$. The random variable X defined by $X(m, n) = m + n$ has 11 possible values, the integers from 2 to 12 inclusive. The following table lists the pairs that produce each value k of X and the probability $f(k)$ of that value, that is, the value of the probability function at k.

Table 2. Probability function for the sum of two dice

$k = m + n$	outcomes for which $X = k$	$f(k) = \Pr(X = k)$
2	$(1, 1)$	$1/36$
3	$(1, 2), (2, 1)$	$2/36 = 1/18$
4	$(1, 3), (2, 2), (3, 1)$	$3/36 = 1/12$
5	$(1, 4), (2, 3), (3, 2), (4, 1)$	$4/36 = 1/9$
6	$(1, 5), (2, 4), (3, 3), (4, 2), (5, 1)$	$5/36$
7	$(1, 6), (2, 5), (3, 4), (4, 3), (5, 2), (6, 1)$	$6/36 = 1/6$
8	$(2, 6), (3, 5), (4, 4), (5, 3), (6, 2)$	$5/36$
9	$(3, 6), (4, 5), (5, 4), (6, 3)$	$4/36 = 1/9$
10	$(4, 6), (5, 5), (6, 4)$	$3/36 = 1/12$
11	$(5, 6), (6, 5)$	$2/36 = 1/18$
12	$(6, 6)$	$1/36$

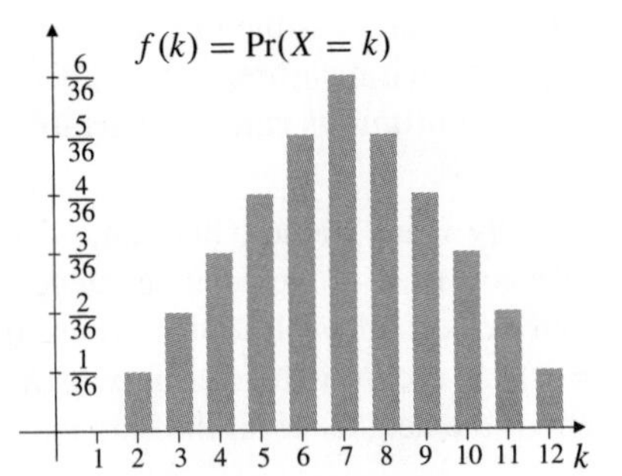

Figure 7.53 The probability function for the sum of two dice

The bar graph of the probability function f is shown in Figure 7.53. We have

$$\Pr(X < 10) = 1 - \Pr(X \geq 10) = 1 - \left(\frac{1}{12} + \frac{1}{18} + \frac{1}{36}\right) = \frac{5}{6}.$$

Expectation, Mean, Variance, and Standard Deviation

Consider a simple gambling game in which the player pays the house C dollars for the privilege of rolling a single die and in which he wins X dollars, where X is the number showing on top of the rolled die. In each game the possible winnings are 1, 2, 3, 4, 5, or 6 dollars, each with probability 1/6. In n games the player can expect to win about $n/6 + 2n/6 + 3n/6 + 4n/6 + 5n/6 + 6n/6 = 21n/6 = 7n/2$ dollars, so that his expected *average winnings per game* are 7/2 dollars, that is, \$3.50. If $C > 3.5$, the player can expect, on average, to lose money. The amount 3.5 is called the **expectation**, or **mean**, of the discrete random variable X. The mean is usually denoted by μ, the Greek letter "mu" (pronounced "mew").

DEFINITION 2

Mean or expectation

If X is a discrete random variable with range of values R and probability function f, then the **mean** (denoted μ), or **expectation** of X (denoted $E(X)$), is

$$\mu = E(X) = \sum_{x \in R} x\, f(x).$$

Also, the **expectation** of any function $g(X)$ of the random variable X is

$$E(g(X)) = \sum_{x \in R} g(x)\, f(x).$$

Note that in this usage $E(X)$ does not define a function of X but a constant (parameter) associated with the random variable X. Note also that if $f(x)$ were a mass density such as that studied in Section 7.4, then μ would be the moment of the mass about 0 and, since the total mass would be $\sum_{x \in R} f(x) = 1$, μ would in fact be the centre of mass.

Another parameter used to describe the way probability is distributed for a random variable is the variable's standard deviation.

DEFINITION 3

Variance and standard deviation

The **variance** of a random variable X with range R and probability function f is the expectation of the square of the distance of X from its mean μ. The variance is denoted σ^2 or Var(X).

$$\sigma^2 = \mathrm{Var}(X) = E\big((X - \mu)^2\big) = \sum_{x \in R} (x - \mu)^2 f(x).$$

The **standard deviation** of X is the square root of the variance and therefore is denoted σ.

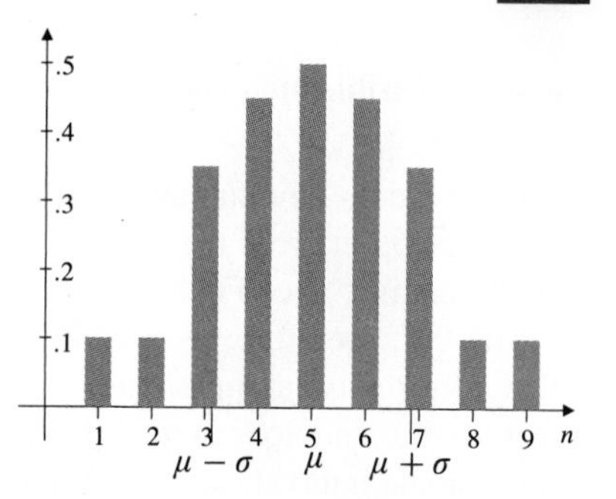

Figure 7.54 A probability function with mean $\mu = 5$ and standard deviation $\sigma = 1.86$

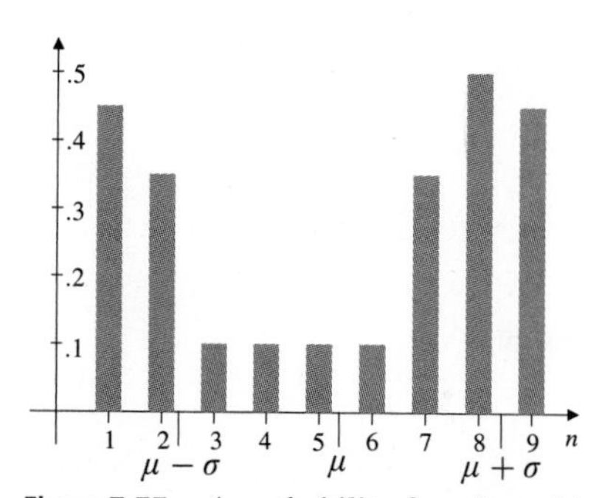

Figure 7.55 A probability function with mean $\mu = 5.38$ and standard deviation $\sigma = 3.05$

The symbol σ is the lowercase Greek letter "sigma." (The symbol Σ used for summation is an uppercase sigma.) The standard deviation gives a measure of how spread out the probability distribution of X is. The smaller the standard deviation, the more the probability is concentrated at values of X close to the mean. Figure 7.54 and Figure 7.55 illustrate the probability functions of two random variables with sample space $\{1, 2, \ldots, 9\}$, one having small σ and one with large σ. Note how a significant fraction of the total probability lies between $\mu - \sigma$ and $\mu + \sigma$ in each case. Note also the distribution of probability in Figure 7.54 is symmetric, resulting in $\mu = 5$, the midpoint of the sample space, while the distribution in Figure 7.55 is skewed a bit to the right, resulting in $\mu > 5$.

Since $\sum_{x \in R} f(x) = 1$, the expression given in the definition of variance can be rewritten as follows:

$$\begin{aligned}
\sigma^2 = \mathrm{Var}(X) &= \sum_{x \in R} (x^2 - 2\mu x + \mu^2)\, f(x) \\
&= \sum_{x \in R} x^2 f(x) - 2\mu \sum_{x \in R} x f(x) + \mu^2 \sum_{x \in R} f(x) \\
&= \sum_{x \in R} x^2 f(x) - 2\mu^2 + \mu^2 = E(X^2) - \mu^2,
\end{aligned}$$

that is,

$$\sigma^2 = \mathrm{Var}(X) = E(X^2) - \mu^2 = E(X^2) - (E(X))^2.$$

Therefore, the standard deviation of X is given by

$$\sigma = \sqrt{E(X^2) - \mu^2}.$$

EXAMPLE 3 Find the mean of the random variable X of Example 2. Also find the expectation of X^2 and the standard deviation of X.

Solution We have

$$\mu = E(X) = 2 \times \frac{1}{36} + 3 \times \frac{2}{36} + 4 \times \frac{3}{36} + 5 \times \frac{4}{36} + 6 \times \frac{5}{36} + 7 \times \frac{6}{36}$$
$$+ 8 \times \frac{5}{36} + 9 \times \frac{4}{36} + 10 \times \frac{3}{36} + 11 \times \frac{2}{36} + 12 \times \frac{1}{36} = 7,$$

a fact that is fairly obvious from the symmetry of the graph of the probability function in Figure 7.53. Also,

$$E(X^2) = 2^2 \times \frac{1}{36} + 3^2 \times \frac{2}{36} + 4^2 \times \frac{3}{36} + 5^2 \times \frac{4}{36} + 6^2 \times \frac{5}{36}$$
$$+ 7^2 \times \frac{6}{36} + 8^2 \times \frac{5}{36} + 9^2 \times \frac{4}{36} + 10^2 \times \frac{3}{36}$$
$$+ 11^2 \times \frac{2}{36} + 12^2 \times \frac{1}{36} = \frac{1{,}974}{36} \approx 54.8333.$$

The variance of X is $\sigma^2 = E(X^2) - \mu^2 \approx 54.8333 - 49 = 5.8333$, so the standard deviation of X is $\sigma \approx 2.4152$.

Continuous Random Variables

Now we consider an example with a continuous range of possible outcomes.

EXAMPLE 4 Suppose that a needle is dropped at random on a flat table with a straight line drawn on it. For each drop, let X be the number of degrees in the (acute) angle that the needle makes with the line. (See Figure 7.56(a).) Evidently X can take any real value in the interval $[0, 90]$; therefore, X is called a **continuous random variable**. The probability that X takes on any particular real value is 0. (There are infinitely many real numbers in $[0, 90]$ and none is more likely than any other.) However, the probability that X lies in some interval, say $[10, 20]$, is the same as the probability that it lies in any other interval of the same length. Since the interval has length 10 and the interval of all possible values of X has length 90, this probability is

$$\Pr(10 \le X \le 20) = \frac{10}{90} = \frac{1}{9}.$$

More generally, if $0 \le x_1 \le x_2 \le 90$, then

$$\Pr(x_1 \le X \le x_2) = \frac{1}{90}(x_2 - x_1).$$

This situation can be conveniently represented as follows: Let $f(x)$ be defined on the interval $[0, 90]$, taking at each point the constant value $1/90$:

$$f(x) = \frac{1}{90}, \qquad 0 \le x \le 90.$$

The area under the graph of f is 1, and $\Pr(x_1 \le X \le x_2)$ is equal to the area under that part of the graph lying over the interval $[x_1, x_2]$. (See Figure 7.56(b).) The function $f(x)$ is called the **probability density function** for the random variable X. Since $f(x)$ is constant on its domain, X is said to be **uniformly distributed**.

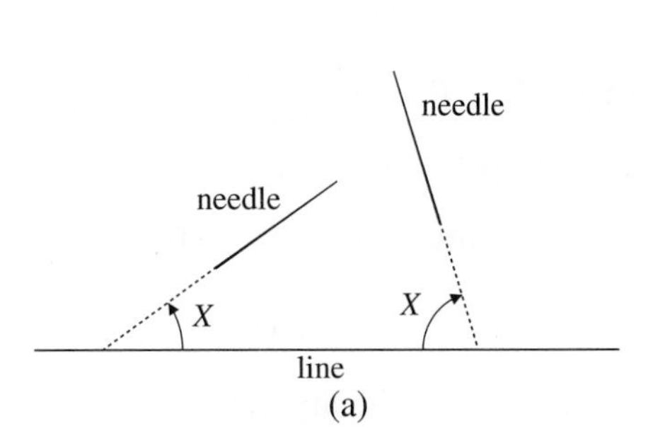

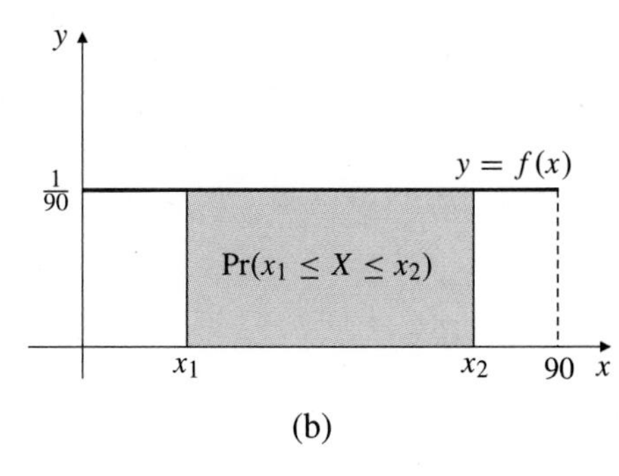

Figure 7.56

(a) X is the number of degrees in the acute angle the needle makes with the line

(b) The probability density function f of the random variable X

DEFINITION 4

Probability density functions

A function defined on an interval $[a, b]$ is a probability density function for a continuous random variable X distributed on $[a, b]$ if, whenever x_1 and x_2 satisfy $a \le x_1 \le x_2 \le b$, we have

$$\Pr(x_1 \le X \le x_2) = \int_{x_1}^{x_2} f(x)\,dx.$$

In order to be such a probability density function, f must satisfy two conditions:

(a) $f(x) \ge 0$ on $[a, b]$ (probability cannot be negative) and

(b) $\int_a^b f(x)\,dx = 1$ $(\Pr(a \le X \le b) = 1)$.

These ideas extend to random variables distributed on semi-infinite or infinite intervals, but the integrals appearing will be improper in those cases. In any event, the role played by sums in the analysis of discrete random variables is taken over by integrals for continuous random variables.

In the example of the dropping needle, the probability density function has a horizontal straight line graph, and we termed such a probability distribution uniform. The uniform probability density function on the interval $[a, b]$ is

$$f(x) = \begin{cases} \dfrac{1}{b-a} & \text{if } a \le x \le b \\ 0 & \text{otherwise.} \end{cases}$$

Many other functions are commonly encountered as density functions for continuous random variables.

EXAMPLE 5 **(The exponential distribution)** The length of time T that any particular atom in a radioactive sample survives before decaying is a random variable taking values in $[0, \infty)$. It has been observed that the proportion of atoms that survive to time t becomes small exponentially as t increases; thus,

$$\Pr(T \ge t) = Ce^{-kt}.$$

Let f be the probability density function for the random variable T. Then

$$\int_t^\infty f(x)\,dx = \Pr(T \ge t) = Ce^{-kt}.$$

Differentiating this equation with respect to t (using the Fundamental Theorem of Calculus), we obtain $-f(t) = -Cke^{-kt}$, so $f(t) = Cke^{-kt}$. C is determined by the requirement that $\int_0^\infty f(t)\,dt = 1$. We have

$$1 = Ck\int_0^\infty e^{-kt}\,dt = \lim_{R\to\infty} Ck\int_0^R e^{-kt}\,dt = -C\lim_{R\to\infty}(e^{-kR} - 1) = C.$$

Thus, $C = 1$ and $f(t) = ke^{-kt}$. Note that $\Pr(T \ge (\ln 2)/k) = e^{-k(\ln 2)/k} = 1/2$, reflecting the fact that the half-life of such a radioactive sample is $(\ln 2)/k$.

EXAMPLE 6 For what value of C is $f(x) = C(1 - x^2)$ a probability density function on $[-1, 1]$? If X is a random variable with this density, what is the probability that $X \le 1/2$?

Solution Observe that $f(x) \ge 0$ on $[-1, 1]$ if $C \ge 0$. Since

$$\int_{-1}^{1} f(x)\,dx = C\int_{-1}^{1} (1 - x^2)\,dx = 2C\left(x - \frac{x^3}{3}\right)\Bigg|_0^1 = \frac{4C}{3},$$

$f(x)$ will be a probability density function if $C = 3/4$. In this case

$$\Pr\left(X \le \frac{1}{2}\right) = \frac{3}{4}\int_{-1}^{1/2} (1 - x^2)\,dx = \frac{3}{4}\left(x - \frac{x^3}{3}\right)\Bigg|_{-1}^{1/2}$$

$$= \frac{3}{4}\left(\frac{1}{2} - \frac{1}{24} - (-1) + \frac{-1}{3}\right) = \frac{27}{32}.$$

By analogy with the discrete case, we formulate definitions for the mean (or expectation), variance, and standard deviation of a continuous random variable as follows:

DEFINITIONS 5

If X is a continuous random variable on $[a, b]$ with probability density function $f(x)$, the **mean** μ, (or **expectation** $E(X)$) of X is

$$\mu = E(X) = \int_a^b x f(x)\,dx.$$

The expectation of a function g of X is

$$E\big(g(X)\big) = \int_a^b g(x)\,f(x)\,dx.$$

Similarly, the **variance** σ^2 of X is the mean of the squared deviation of X from its mean:

$$\sigma^2 = \mathrm{Var}(X) = E((X - \mu)^2) = \int_a^b (x - \mu)^2 f(x)\,dx,$$

and the **standard deviation** is the square root of the variance.

As was the case for a discrete random variable, it is easily shown that

$$\sigma^2 = E(X^2) - \mu^2, \qquad \sigma = \sqrt{E(X^2) - \mu^2}.$$

Again the standard deviation gives a measure of how spread out the probability distribution of X is. The smaller the standard deviation, the more concentrated is the area under the density curve around the mean, and so the smaller is the probability that a value of X will be far away from the mean. (See Figure 7.57.)

EXAMPLE 7 Find the mean μ and the standard deviation σ of a random variable X distributed uniformly on the interval $[a, b]$. Find $\Pr(\mu - \sigma \le X \le \mu + \sigma)$.

Solution The probability density function is $f(x) = 1/(b-a)$ on $[a, b]$, so the mean is given by

$$\mu = E(X) = \int_a^b \frac{x}{b-a}\,dx = \frac{1}{b-a}\frac{x^2}{2}\Bigg|_a^b = \frac{1}{2}\frac{b^2 - a^2}{b-a} = \frac{b+a}{2}.$$

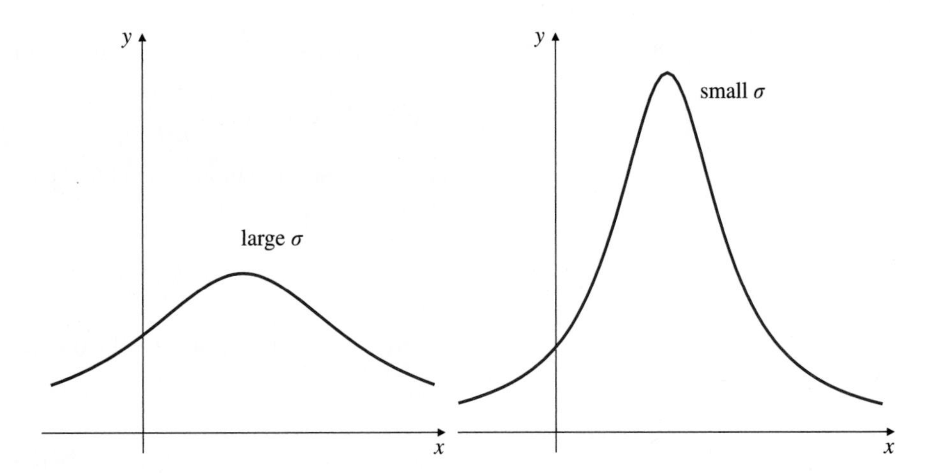

Figure 7.57 Densities with large and small standard deviations

Hence, the mean is, as might have been anticipated, the midpoint of $[a, b]$. The expectation of X^2 is given by

$$E(X^2) = \int_a^b \frac{x^2}{b-a}\,dx = \frac{1}{b-a}\frac{x^3}{3}\bigg|_a^b = \frac{1}{3}\frac{b^3-a^3}{b-a} = \frac{b^2+ab+a^2}{3}.$$

Hence, the variance is

$$\sigma^2 = E(X^2) - \mu^2 = \frac{b^2+ab+a^2}{3} - \frac{b^2+2ab+a^2}{4} = \frac{(b-a)^2}{12},$$

and the standard deviation is

$$\sigma = \frac{b-a}{2\sqrt{3}} \approx 0.29(b-a).$$

Finally,

$$\Pr(\mu - \sigma \le X \le \mu + \sigma) = \int_{\mu-\sigma}^{\mu+\sigma} \frac{dx}{b-a} = \frac{1}{b-a}\,\frac{2(b-a)}{2\sqrt{3}} = \frac{1}{\sqrt{3}} \approx 0.577.$$

EXAMPLE 8 Find the mean μ and the standard deviation σ of a random variable X distributed exponentially with density function $f(x) = ke^{-kx}$ on the interval $[0, \infty)$. Find $\Pr(\mu - \sigma \le X \le \mu + \sigma)$.

Solution We use integration by parts to find the mean:

$$\begin{aligned}
\mu = E(X) &= k\int_0^\infty xe^{-kx}\,dx \\
&= \lim_{R\to\infty} k\int_0^R xe^{-kx}\,dx \qquad \begin{array}{ll} \text{Let} & U = x, \quad dV = e^{-kx}\,dx. \\ \text{Then} & dU = dx, \quad V = -e^{-kx}/k. \end{array} \\
&= \lim_{R\to\infty}\left(-xe^{-kx}\bigg|_0^R + \int_0^R e^{-kx}\,dx\right) \\
&= \lim_{R\to\infty}\left(-Re^{-kR} - \frac{1}{k}(e^{-kR} - 1)\right) = \frac{1}{k}, \qquad \text{since } k > 0.
\end{aligned}$$

Thus, the mean of the exponential distribution is $1/k$. This fact can be quite useful in determining the value of k for an exponentially distributed random variable. A similar integration by parts enables us to evaluate

$$E(X^2) = k\int_0^\infty x^2 e^{-kx}\,dx = 2\int_0^\infty xe^{-kx}\,dx = \frac{2}{k^2},$$

so the variance of the exponential distribution is

$$\sigma^2 = E(X^2) - \mu^2 = \frac{1}{k^2},$$

and the standard deviation is equal to the mean

$$\sigma = \mu = \frac{1}{k}.$$

Now we have

$$\begin{aligned}
\Pr(\mu - \sigma \le X \le \mu + \sigma) &= \Pr(0 \le X \le 2/k) \\
&= k \int_0^{2/k} e^{-kx}\, dx \\
&= -e^{-kx} \Big|_0^{2/k} \\
&= 1 - e^{-2} \approx 0.86,
\end{aligned}$$

which is independent of the value of k. Exponential densities for small and large values of k are graphed in Figure 7.58.

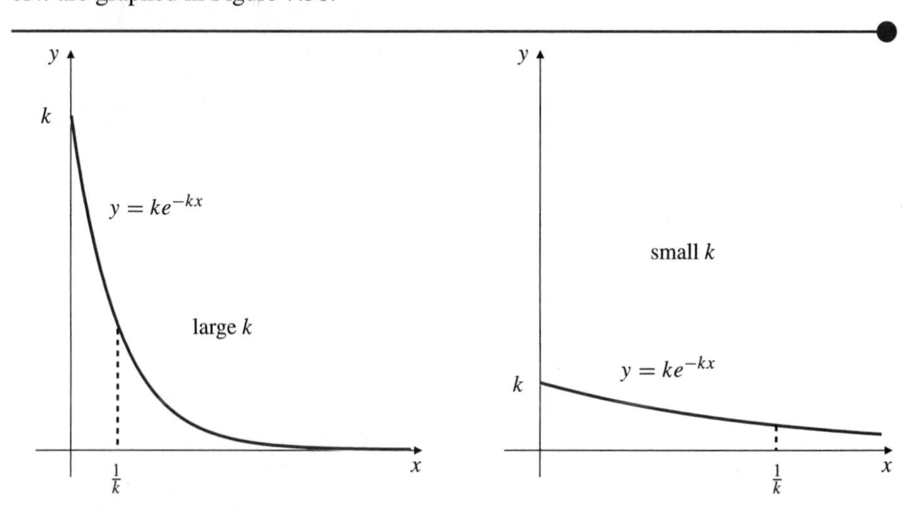

Figure 7.58 Exponential density functions

The Normal Distribution

The most important probability distributions are the so-called **normal** or **Gaussian** distributions. Such distributions govern the behaviour of many interesting random variables, in particular, those associated with random errors in measurements. There is a family of normal distributions, all related to the particular normal distribution called the **standard normal distribution**, which has the following probability density function:

DEFINITION 6

The standard normal probability density

$$f(z) = \frac{1}{\sqrt{2\pi}} e^{-z^2/2}, \qquad -\infty < z < \infty.$$

It is common to use z to denote the random variable in the standard normal distribution; the other normal distributions are obtained from this one by a change of variable. The graph of the standard normal density has a pleasant bell shape, as shown in Figure 7.59.

As we have noted previously, the function e^{-z^2} has no elementary antiderivative, so the improper integral

$$I = \int_{-\infty}^{\infty} e^{-z^2/2}\, dz$$

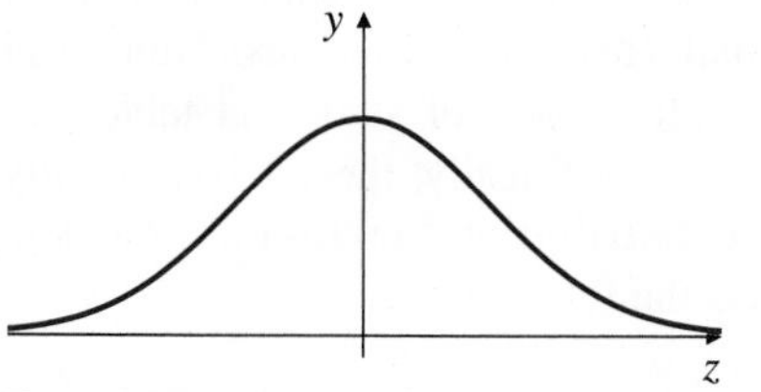

Figure 7.59 The standard normal density function $f(z) = \dfrac{1}{\sqrt{2\pi}}e^{-z^2/2}$

cannot be evaluated using the Fundamental Theorem of Calculus, although it is a convergent improper integral. The integral can be evaluated using techniques of multivariable calculus involving double integrals of functions of two variables. (We do so in Example 4 of Section 14.4.) The value is $I = \sqrt{2\pi}$, which ensures that the above-defined standard normal density $f(z)$ is indeed a probability density function:

$$\int_{-\infty}^{\infty} f(z)\,dz = \frac{1}{\sqrt{2\pi}}\int_{-\infty}^{\infty} e^{-z^2/2}\,dz = 1.$$

Since $ze^{-z^2/2}$ is an odd function of z and its integral on $(-\infty, \infty)$ converges, the mean of the standard normal distribution is 0:

$$\mu = E(Z) = \frac{1}{\sqrt{2\pi}}\int_{-\infty}^{\infty} ze^{-z^2/2}\,dz = 0.$$

We calculate the variance of the standard normal distribution using integration by parts as follows:

$$\begin{aligned}
\sigma^2 &= E(Z^2) \\
&= \frac{1}{\sqrt{2\pi}}\int_{-\infty}^{\infty} z^2 e^{-z^2/2}\,dz \\
&= \frac{1}{\sqrt{2\pi}}\lim_{R\to\infty}\int_{-R}^{R} z^2 e^{-z^2/2}\,dz \qquad \begin{array}{ll}\text{Let } U = z, & dV = ze^{-z^2/2}\,dz. \\ \text{Then } dU = dz, & V = -e^{-z^2/2}.\end{array} \\
&= \frac{1}{\sqrt{2\pi}}\lim_{R\to\infty}\left(-ze^{-z^2/2}\Big|_{-R}^{R} + \int_{-R}^{R} e^{-z^2/2}\,dz\right) \\
&= \frac{1}{\sqrt{2\pi}}\lim_{R\to\infty}(-2Re^{-R^2/2}) + \frac{1}{\sqrt{2\pi}}\int_{-\infty}^{\infty} e^{-z^2/2}\,dz \\
&= 0 + 1 = 1.
\end{aligned}$$

Hence, the standard deviation of the standard normal distribution is 1.

Other normal distributions are obtained from the standard normal distribution by a change of variable.

DEFINITION

The general normal distribution

A random variable X on $(-\infty, \infty)$ is said to be *normally distributed with mean* μ and *standard deviation* σ (where μ is any real number and $\sigma > 0$) if its probability density function $f_{\mu,\sigma}$ is given in terms of the standard normal density f by

$$f_{\mu,\sigma}(x) = \frac{1}{\sigma} f\left(\frac{x-\mu}{\sigma}\right) = \frac{1}{\sigma\sqrt{2\pi}} e^{-(x-\mu)^2/(2\sigma^2)}.$$

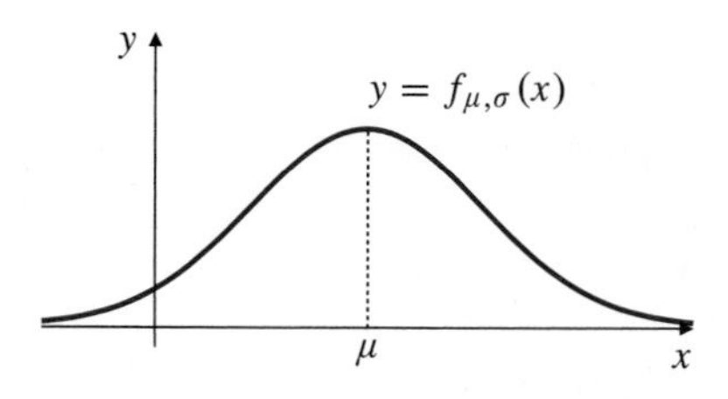

Figure 7.60 A general normal density with mean μ

(See Figure 7.60.) Using the change of variable $z = (x-\mu)/\sigma$, $dz = dx/\sigma$, we can verify that

$$\int_{-\infty}^{\infty} f_{\mu,\sigma}(x)\,dx = \int_{-\infty}^{\infty} f(z)\,dz = 1,$$

so $f_{\mu,\sigma}(x)$ is indeed a probability density function. Using the same change of variable, we can show that

$$E(X) = \mu \qquad \text{and} \qquad E((X-\mu)^2) = \sigma^2.$$

Hence, the density $f_{\mu,\sigma}$ does indeed have mean μ and standard deviation σ.

Because $e^{-z^2/2}$ cannot be easily antidifferentiated, we cannot determine normal probabilities (i.e., areas) by using the Fundamental Theorem of Calculus. Numerical integrations can be performed, or one can consult a book of statistical tables for computed areas under the standard normal curve. Specifically, these tables usually provide values for what is called the **cumulative distribution function** of a random variable with standard normal distribution. This is the function

$$F(z) = \frac{1}{\sqrt{2\pi}} \int_{-\infty}^{z} e^{-x^2/2}\,dx = \Pr(Z \le z),$$

which represents the area under the standard normal density function from $-\infty$ up to z, as shown in Figure 7.61.

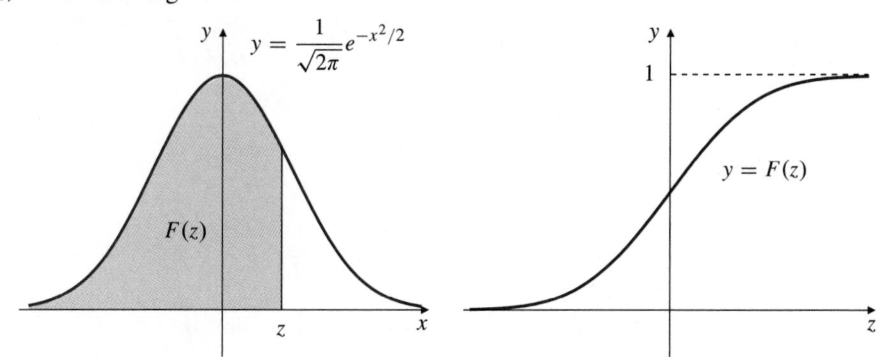

Figure 7.61 The cumulative distribution function $F(z)$ for the standard normal distribution is the area under the standard normal density function from $-\infty$ to z

For use in the following examples and exercises, we include here an abbreviated version of such a table.

Table 3. Values of the standard normal distribution function $F(z)$ (rounded to 3 decimal places)

z	**0.0**	**0.1**	**0.2**	**0.3**	**0.4**	**0.5**	**0.6**	**0.7**	**0.8**	**0.9**
−3.0	0.001	0.001	0.001	0.000	0.000	0.000	0.000	0.000	0.000	0.000
−2.0	0.023	0.018	0.014	0.011	0.008	0.006	0.005	0.003	0.003	0.002
−1.0	0.159	0.136	0.115	0.097	0.081	0.067	0.055	0.045	0.036	0.029
−0.0	0.500	0.460	0.421	0.382	0.345	0.309	0.274	0.242	0.212	0.184
0.0	0.500	0.540	0.579	0.618	0.655	0.691	0.726	0.758	0.788	0.816
1.0	0.841	0.864	0.885	0.903	0.919	0.933	0.945	0.955	0.964	0.971
2.0	0.977	0.982	0.986	0.989	0.992	0.994	0.995	0.997	0.997	0.998
3.0	0.999	0.999	0.999	1.000	1.000	1.000	1.000	1.000	1.000	1.000

EXAMPLE 9 If Z is a standard normal random variable, find (a) $\Pr(-1.2 \le Z \le 2.0)$, and (b) $\Pr(Z \ge 1.5)$.

Solution Using values from the table, we obtain

$$\begin{aligned}\Pr(-1.2 \le Z \le 2.0) &= \Pr(Z \le 2.0) - \Pr(Z < -1.2)\\ &= F(2.0) - F(-1.2) \approx 0.977 - 0.115\\ &= 0.862\end{aligned}$$

$$\begin{aligned}\Pr(Z \ge 1.5) &= 1 - \Pr(Z < 1.5)\\ &= 1 - F(1.5) \approx 1 - 0.933 = 0.067.\end{aligned}$$

EXAMPLE 10 A certain random variable X is distributed normally with mean 2 and standard deviation 0.4. Find (a) $\Pr(1.8 \le X \le 2.4)$, and (b) $\Pr(X > 2.4)$.

Solution Since X is distributed normally with mean 2 and standard deviation 0.4, $Z = (X - 2)/0.4$ is distributed according to the standard normal distribution (with mean 0 and standard deviation 1). Accordingly,

$$\begin{aligned}\Pr(1.8 \le X \le 2.4) &= \Pr(-0.5 \le Z \le 1)\\ &= F(1) - F(-0.5) \approx 0.841 - 0.309 = 0.532,\\ \Pr(X > 2.4) &= \Pr(Z > 1) = 1 - \Pr(Z \le 1)\\ &= 1 - F(1) \approx 1 - 0.841 = 0.159.\end{aligned}$$

EXERCISES 7.8

1. How much should you be willing to pay to play a game where you toss the coin discussed at the beginning of this section and win $1 if it comes up heads, $2 if it comes up tails, and $50 if it remains standing on its edge? Assume you will play the game many times and would like to at least break even.

2. A die is weighted so that if X represents the number showing on top when the die is rolled, then $\Pr(X = n) = Kn$ for $n \in \{1, 2, 3, 4, 5, 6\}$.
   (a) Find the value of the constant K.
   (b) Find the probability that $X \le 3$ on any roll of the die.

3. Find the standard deviation of your winings on a roll of the die in Exercise 1.

4. Find the mean and standard deviation of the random variable X in Exercise 2.

5. A die is weighted so that the probability of rolling each of the numbers 2, 3, 4, and 5 is still 1/6, but the probability of rolling 1 is 9/60 and the probability of rolling 6 is 11/60. What are the mean and standard deviation of the number X rolled using this die? What is the probability that $X \le 3$?

6. Two dice, each weighted like the one in Exercise 5, are thrown. Let X be the random variable giving the sum of the numbers showing on top of the two dice.
   (a) Find the probability function for X.
   (b) Determine the mean and standard deviation of X. Compare them with those found for unweighted dice in Example 3.

7. A thin but biased coin has probability 0.55 of landing heads and 0.45 of landing tails. (Standing on its edge is not possible for this coin.) The coin is tossed three times. (Determine all numerical answers to the following questions to 6 decimal places.)
   (a) What is the sample space of possible outcomes of the three tosses?
   (b) What is the probability of each of these possible outcomes?
   (c) Find the probability function for the number X of times heads comes up during the 3 tosses.
   (d) What is the probability that the number of heads is at least 1?
   (e) What is the expectation of X?

8. A sack contains 20 balls all the same size; some are red and the rest are blue. If you reach in and pull out a ball at random, the probability that it is red is 0.6.
   (a) If you reach in and pull out two balls, what is the probability they are both blue?
   (b) Suppose you reach in the bag of 20 balls and pull out three balls. Describe the sample space of possible outcomes of this experiment. What is the expectation of the number of red balls among the three balls you pulled out?

For each function $f(x)$ in Exercises 9–15, find the following:

(a) the value of C for which f is a probability density on the given interval,
(b) the mean μ, variance σ^2, and standard deviation σ of the probability density f, and
(c) $\Pr(\mu - \sigma \le X \le \mu + \sigma)$, that is, the probability that the random variable X is no further than one standard deviation away from its mean.

9. $f(x) = Cx$ on $[0, 3]$
10. $f(x) = Cx$ on $[1, 2]$
11. $f(x) = Cx^2$ on $[0, 1]$
12. $f(x) = C \sin x$ on $[0, \pi]$
13. $f(x) = C(x - x^2)$ on $[0, 1]$
14. $f(x) = C\,xe^{-kx}$ on $[0, \infty)$, $(k > 0)$
15. $f(x) = C\,e^{-x^2}$ on $[0, \infty)$. *Hint:* Use properties of the standard normal density to show that $\int_0^\infty e^{-x^2}\,dx = \sqrt{\pi}/2$.
16. Is it possible for a random variable to be uniformly distributed on the whole real line? Explain why.
17. Carry out the calculations to show that the normal density $f_{\mu,\sigma}(x)$ defined in the text is a probability density function and has mean μ and standard deviation σ.
18. Show that $f(x) = \dfrac{2}{\pi(1 + x^2)}$ is a probability density on $[0, \infty)$. Find the expectation of X for this density. If a machine generates values of a random variable X distributed with density $f(x)$, how much would you be willing to pay, per game, to play a game in which you operate the machine to produce a value of X and win X dollars? Explain.
19. Calculate $\Pr(|X - \mu| \ge 2\sigma)$ for
    (a) the uniform distribution on $[a, b]$,
    (b) the exponential distribution with density $f(x) = ke^{-kx}$ on $[0, \infty)$, and
    (c) the normal distribution with density $f_{\mu,\sigma}(x)$.

20. The length of time T (in hours) between malfunctions of a computer system is an exponentially distributed random variable. If the average length of time between successive malfunctions is 20 hours, find the probability that the system, having just had a malfunction corrected, will operate without malfunction for at least 12 hours.

21. The number X of metres of cable produced any day by a cable-making company is a normally distributed random variable with mean 5,000 and standard deviation 200. On what fraction of the days the company operates will the number of metres of cable produced exceed 5,500?

22. A spinner is made with a scale from 0 to 1. Over time it suffers from wear and tends to stick at the number 1/4. Suppose it sticks at 1/4 half of the time and the rest of the time it gives values uniformly distributed in the interval $[0, 1]$. What is the mean and standard deviation of the spinner's values? (Note: the random variable giving the spinner's value has a distribution that is partially discrete and partially continuous.)

7.9 First-Order Differential Equations

This final section on applications of integration concentrates on application of the indefinite integral rather than of the definite integral. We can use the techniques of integration developed in Chapters 5 and 6 to solve certain kinds of first-order differential equations that arise in a variety of modelling situations. We have already seen some examples of applications of differential equations to modelling growth and decay phenomena in Section 3.4.

Separable Equations

Consider the logistic equation introduced in Section 3.4 to model the growth of an animal population with a limited food supply:

$$\frac{dy}{dt} = ky\left(1 - \frac{y}{L}\right),$$

where $y(t)$ is the size of the population at time t, k is a positive constant related to the fertility of the population, and L is the steady-state population size that can be sustained by the available food supply. This equation has two particular solutions, $y = 0$ and $y = L$, that are constant functions of time.

The logistic equation is an example of a class of first-order differential equations called **separable equations** because when they are written in terms of differentials, they can be separated with only the dependent variable on one side of the equation and only the independent variable on the other. The logistic equation can be written in the form

$$\frac{L\,dy}{y(L-y)} = k\,dt,$$

and solved by integrating both sides. Expanding the left side in partial fractions and integrating, we get

$$\int \left(\frac{1}{y} + \frac{1}{L-y}\right) dy = kt + C.$$

Assuming that $0 < y < L$, we therefore obtain

$$\ln y - \ln(L-y) = kt + C,$$

$$\ln\left(\frac{y}{L-y}\right) = kt + C.$$

We can solve this equation for y by taking exponentials of both sides:

$$\frac{y}{L-y} = e^{kt+C} = C_1 e^{kt}$$

$$y = (L-y)C_1 e^{kt}$$

$$y = \frac{C_1 L e^{kt}}{1 + C_1 e^{kt}},$$

where $C_1 = e^C$.

Generally, separable equations are of the form

$$\frac{dy}{dx} = f(x)g(y).$$

We solve them by rewriting them in the form

$$\frac{dy}{g(y)} = f(x)\,dx$$

and integrating both sides. Note that the separable equation above will have a constant solution $y(x) = C$ for any constant C satisfying $g(C) = 0$.

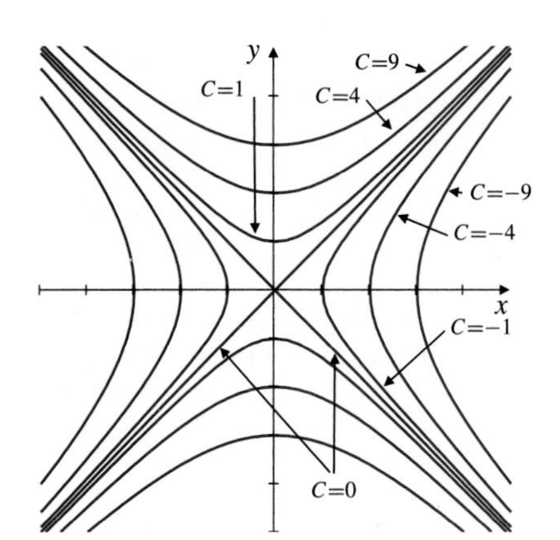

Figure 7.62 Some curves of the family $y^2 - x^2 = C$

EXAMPLE 1 Solve the equation $\dfrac{dy}{dx} = \dfrac{x}{y}$.

Solution We rewrite the equation in the form $y\,dy = x\,dx$ and integrate both sides to get

$$\frac{1}{2}y^2 = \frac{1}{2}x^2 + C_1,$$

or $y^2 - x^2 = C$, where $C = 2C_1$ is an arbitrary constant. The solution curves are rectangular hyperbolas. (See Figure 7.62.) Their asymptotes $y = x$ and $y = -x$ are also solutions corresponding to $C = 0$.

EXAMPLE 2 Solve the initial-value problem

$$\begin{cases} \dfrac{dy}{dx} = x^2y^3 \\ y(1) = 3. \end{cases}$$

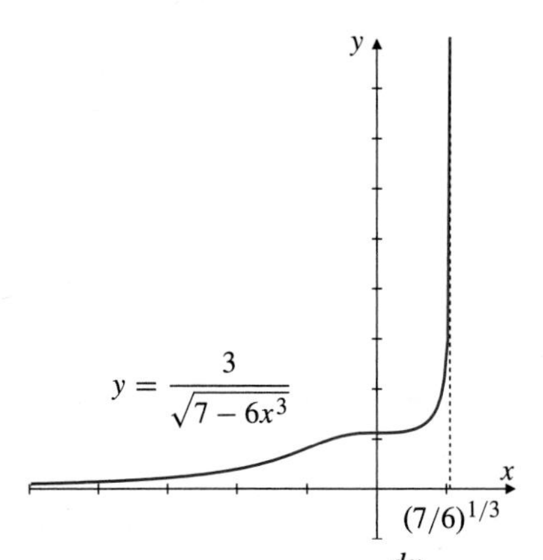

Figure 7.63 The solution of $\dfrac{dy}{dx} = x^2y^3$ satisfying $y(1) = 3$

Solution Separating the differential equation gives $\dfrac{dy}{y^3} = x^2\,dx$. Thus,

$$\int \frac{dy}{y^3} = \int x^2\,dx, \qquad \text{so} \qquad \frac{-1}{2y^2} = \frac{x^3}{3} + C.$$

Since $y = 3$ when $x = 1$, we have $-\frac{1}{18} = \frac{1}{3} + C$ and $C = -\frac{7}{18}$. Substituting this value into the above solution and solving for y, we obtain

$$y(x) = \frac{3}{\sqrt{7-6x^3}}. \qquad \text{(Only the positive square root of } y^2 \text{ satisfies } y(1) = 3.)$$

This solution is valid for $x < \left(\frac{7}{6}\right)^{1/3}$. (See Figure 7.63.)

EXAMPLE 3 Solve the **integral equation** $y(x) = 3 + 2\displaystyle\int_1^x ty(t)\,dt$.

Solution Differentiating the integral equation with respect to x gives

$$\frac{dy}{dx} = 2x\,y(x) \qquad \text{or} \qquad \frac{dy}{y} = 2x\,dx.$$

Thus $\ln|y(x)| = x^2 + C$, and solving for y, $y(x) = C_1e^{x^2}$. Putting $x = 1$ in the integral equation provides an initial value: $y(1) = 3 + 0 = 3$, so $C_1 = 3/e$ and

$$y(x) = 3e^{x^2-1}.$$

EXAMPLE 4 **(A solution concentration problem)** Initially a tank contains 1,000 L of brine with 50 kg of dissolved salt. Brine containing 10 g of salt per litre is flowing into the tank at a constant rate of 10 L/min. If the contents of the tank are kept thoroughly mixed at all times, and if the solution also flows out at 10 L/min, how much salt remains in the tank at the end of 40 min?

Solution Let $x(t)$ be the number of kilograms of salt in solution in the tank after t min. Thus $x(0) = 50$. Salt is coming into the tank at a rate of 10 g/L $\times$ 10 L/min = 100 g/min = 1/10 kg/min. At all times the tank contains 1,000 L of liquid, so the concentration of salt in the tank at time t is $x/1{,}000$ kg/L. Since the contents flow out at 10 L/min, salt is being removed at a rate of $10x/1{,}000 = x/100$ kg/min. Therefore,

$$\frac{dx}{dt} = \text{rate in} - \text{rate out} = \frac{1}{10} - \frac{x}{100} = \frac{10 - x}{100}.$$

Although $x(t) = 10$ is a constant solution of the differential equation, it does not satisfy the initial condition $x(0) = 50$, so we will find other solutions by separating variables:

$$\frac{dx}{10 - x} = \frac{dt}{100}.$$

Integrating both sides of this equation, we obtain

$$-\ln|10 - x| = \frac{t}{100} + C.$$

Observe that $x(t) \neq 10$ for any finite time t (since $\ln 0$ is not defined). Since $x(0) = 50 > 10$, it follows that $x(t) > 10$ for all $t > 0$. ($x(t)$ is necessarily continuous so it cannot take any value less than 10 without somewhere taking the value 10 by the Intermediate-Value Theorem.) Hence, we can drop the absolute value from the solution above and obtain

$$\ln(x - 10) = -\frac{t}{100} - C.$$

Since $x(0) = 50$, we have $-C = \ln 40$ and

$$x = x(t) = 10 + 40e^{-t/100}.$$

After 40 min there will be $10 + 40e^{-0.4} \approx 36.8$ kg of salt in the tank.

EXAMPLE 5 **(A rate of reaction problem)** In a chemical reaction that goes to completion in solution, one molecule of each of two reactants, A and B, combine to form each molecule of the product C. According to the law of mass action, the reaction proceeds at a rate proportional to the product of the concentrations of A and B in the solution. Thus, if there were initially present $a > 0$ molecules/cm^3 of A and $b > 0$ molecules/cm^3 of B, then the number $x(t)$ of molecules/cm^3 of C present at time t thereafter is determined by the differential equation

$$\frac{dx}{dt} = k(a - x)(b - x).$$

This equation has constant solutions $x(t) = a$ and $x(t) = b$, neither of which satisfies the initial condition $x(0) = 0$. We find other solutions for this equation by separation of variables and the technique of partial fraction decomposition under the assumption that $b \neq a$:

$$\int \frac{dx}{(a - x)(b - x)} = k \int dt = kt + C.$$

Since

$$\frac{1}{(a-x)(b-x)} = \frac{1}{b-a}\left(\frac{1}{a-x} - \frac{1}{b-x}\right),$$

and since necessarily $x \le a$ and $x \le b$, we have

$$\frac{1}{b-a}\big(-\ln(a-x) + \ln(b-x)\big) = kt + C,$$

or

$$\ln\left(\frac{b-x}{a-x}\right) = (b-a)\,kt + C_1, \quad \text{where } C_1 = (b-a)C.$$

By assumption, $x(0) = 0$, so $C_1 = \ln(b/a)$ and

$$\ln\frac{a(b-x)}{b(a-x)} = (b-a)\,kt.$$

This equation can be solved for x to yield $x = x(t) = \dfrac{ab(e^{(b-a)kt} - 1)}{be^{(b-a)kt} - a}$.

EXAMPLE 6 Find a family of curves, each of which intersects every parabola with equation of the form $y = Cx^2$ at right angles.

Solution The family of parabolas $y = Cx^2$ satisfies the differential equation

$$\frac{d}{dx}\left(\frac{y}{x^2}\right) = \frac{d}{dx}C = 0;$$

that is,

$$x^2\frac{dy}{dx} - 2xy = 0 \qquad \text{or} \qquad \frac{dy}{dx} = \frac{2y}{x}.$$

Any curve that meets the parabolas $y = Cx^2$ at right angles must, at any point (x, y) on it, have slope equal to the negative reciprocal of the slope of the particular parabola passing through that point. Thus, such a curve must satisfy

$$\frac{dy}{dx} = -\frac{x}{2y}.$$

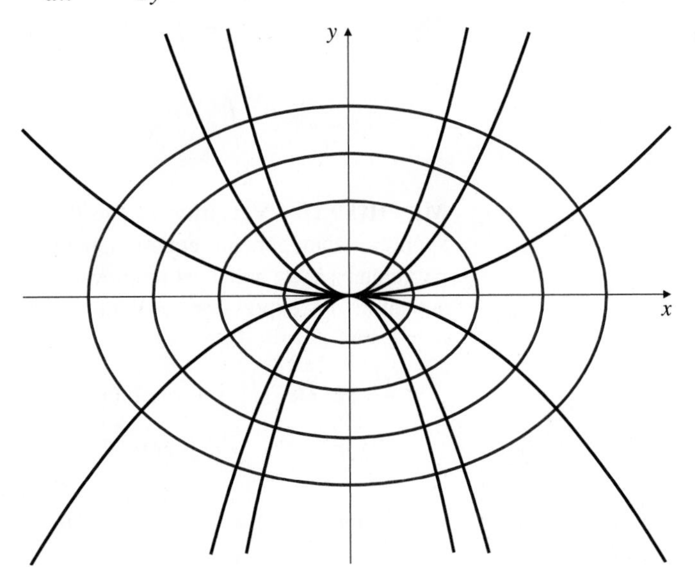

Figure 7.64 The parabolas $y = C_1x^2$ and the ellipses $x^2 + 2y^2 = C_2$ intersect at right angles

Separation of the variables leads to $2y\,dy = -x\,dx$, and integration of both sides then yields $y^2 = -\frac{1}{2}x^2 + C_1$ or $x^2 + 2y^2 = C$, where $C = 2C_1$. This equation represents a family of ellipses centred at the origin. Each ellipse meets each parabola at right angles, as shown in Figure 7.64. When the curves of one family intersect the curves of a second family at right angles, each family is called the family of **orthogonal trajectories** of the other family.

First-Order Linear Equations

A first-order **linear** differential equation is one of the type

$$\frac{dy}{dx} + p(x)y = q(x), \qquad (*)$$

where $p(x)$ and $q(x)$ are given functions, which we assume to be continuous. The equation is called **nonhomogeneous** unless $q(x)$ is identically zero. The corresponding **homogeneous** equation,

$$\frac{dy}{dx} + p(x)y = 0,$$

is separable and so is easily solved to give $y = K\,e^{-\mu(x)}$, where K is any constant and $\mu(x)$ is any antiderivative of $p(x)$:

$$\mu(x) = \int p(x)\,dx \qquad \text{and} \qquad \frac{d\mu}{dx} = p(x).$$

There are two methods for solving the nonhomogeneous equation $(*)$. Both involve the function $\mu(x)$ defined above.

METHOD I. Using an Integrating Factor. Multiply equation $(*)$ by $e^{\mu(x)}$ (which is called an **integrating factor** for the equation) and observe that the left side is just the derivative of $e^{\mu(x)}y$; by the Product Rule

$$\begin{aligned}\frac{d}{dx}\big(e^{\mu(x)}y(x)\big) &= e^{\mu(x)}\frac{dy}{dx} + e^{\mu(x)}\frac{d\mu}{dx}\,y(x)\\ &= e^{\mu(x)}\left(\frac{dy}{dx} + p(x)y\right) = e^{\mu(x)}\,q(x).\end{aligned}$$

Therefore, $e^{\mu(x)}\,y(x) = \displaystyle\int e^{\mu(x)}\,q(x)\,dx$, or

$$y(x) = e^{-\mu(x)}\int e^{\mu(x)}\,q(x)\,dx.$$

METHOD II. Variation of the Parameter. Start with the solution of the corresponding homogeneous equation, namely $y = K\,e^{-\mu(x)}$, and replace the constant (i.e., parameter) K by an as yet unknown function $k(x)$ of the independent variable. Then substitute this expression for y into the differential equation $(*)$ and simplify:

$$\begin{aligned}&\frac{d}{dx}\left(k(x)e^{-\mu(x)}\right) + p(x)k(x)e^{-\mu(x)} = q(x)\\ &k'(x)e^{-\mu(x)} - \mu'(x)k(x)e^{-\mu(x)} + p(x)k(x)e^{-\mu(x)} = q(x),\end{aligned}$$

which, since $\mu'(x) = p(x)$, reduces to

$$k'(x) = e^{\mu(x)}q(x).$$

Integrating the right side leads to the solution for $k(x)$ and thereby to the solution y for $(*)$.

EXAMPLE 7 Solve $\dfrac{dy}{dx} + \dfrac{y}{x} = 1$ for $x > 0$. Use both methods for comparison.

Solution Here, $p(x) = 1/x$, so $\mu(x) = \int p(x)\,dx = \ln x$ (for $x > 0$).
METHOD I. The ingegrating factor is $e^{\mu(x)} = x$. We calculate

$$\frac{d}{dx}(xy) = x\frac{dy}{dx} + y = x\left(\frac{dy}{dx} + \frac{y}{x}\right) = x,$$

and so

$$xy = \int x\,dx = \frac{1}{2}x^2 + C.$$

Finally,

$$y = \frac{1}{x}\left(\frac{1}{2}x^2 + C\right) = \frac{x}{2} + \frac{C}{x}.$$

This is a solution of the given equation for any value of the constant C.
METHOD II. The corresponding homogeneous equation, $\dfrac{dy}{dx} + \dfrac{y}{x} = 0$, has solution $y = Ke^{-\mu(x)} = \dfrac{K}{x}$. Replacing the constant K with the function $k(x)$ and substituting into the given differential equation we obtain

$$\frac{1}{x}k'(x) - \frac{1}{x^2}k(x) + \frac{1}{x^2}k(x) = 1,$$

so that $k'(x) = x$ and $k(x) = \dfrac{x^2}{2} + C$, where C is any constant. Therefore

$$y = \frac{k(x)}{x} = \frac{x}{2} + \frac{C}{x},$$

the same solution obtained by METHOD I.

Remark Both methods really amount to the same calculations expressed in different ways. Use whichever one you think is easiest to understand. The remaining examples in this section will be done by using integrating factors, but variation of parameters will prove useful later on (Section 18.6) to deal with nonhomogeneous linear differential equations of second or higher order.

EXAMPLE 8 Solve $\dfrac{dy}{dx} + xy = x^3$.

Solution Here, $p(x) = x$, so $\mu(x) = x^2/2$ and $e^{\mu(x)} = e^{x^2/2}$. We calculate

$$\frac{d}{dx}\left(e^{x^2/2}y\right) = e^{x^2/2}\frac{dy}{dx} + e^{x^2/2}xy = e^{x^2/2}\left(\frac{dy}{dx} + xy\right) = x^3e^{x^2/2}.$$

Thus,

$$\begin{aligned} e^{x^2/2}y &= \int x^3 e^{x^2/2}\,dx && \text{Let } U = x^2, \quad dV = x\,e^{x^2/2}\,dx. \\ & && \text{Then } dU = 2x\,dx, \quad V = e^{x^2/2}. \\ &= x^2e^{x^2/2} - 2\int x\,e^{x^2/2}\,dx \\ &= x^2e^{x^2/2} - 2e^{x^2/2} + C, \end{aligned}$$

and, finally, $y = x^2 - 2 + Ce^{-x^2/2}$.

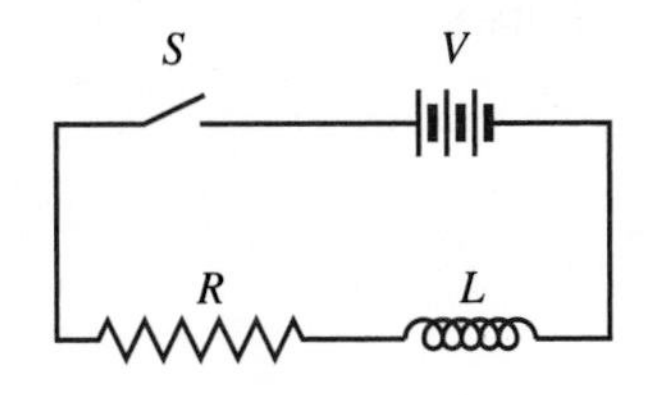

Figure 7.65 An inductance-resistance circuit

EXAMPLE 9 **(An inductance-resistance circuit)** An electric circuit (Figure 7.65) contains a constant DC voltage source of V volts, a switch, a resistor of size R ohms, and an inductor of size L henrys. The circuit has no capacitance. The switch, initially open so that no current is flowing, is closed at time $t = 0$ so that current begins to flow at that time. If the inductance L were zero, the current would suddenly jump from 0 amperes when $t < 0$ to $I = V/R$ amperes when $t > 0$. However, if $L > 0$ the current cannot change instantaneously; it will depend on time t. Let the current t seconds after the switch is closed be $I(t)$ amperes. It is known that $I(t)$ satisfies the initial-value problem

$$\begin{cases} L\dfrac{dI}{dt} + RI = V \\ \qquad I(0) = 0. \end{cases}$$

Find $I(t)$. What is $\lim_{t\to\infty} I(t)$? How long does it take after the switch is closed for the current to rise to 90% of its limiting value?

Solution The DE can be written in the form $\dfrac{dI}{dt} + \dfrac{R}{L}I = \dfrac{V}{L}$. It is linear and has integrating factor $e^{\mu(t)}$, where

$$\mu(t) = \int \frac{R}{L}\,dt = \frac{Rt}{L}.$$

Therefore,

$$\begin{aligned} \frac{d}{dt}\left(e^{Rt/L}I\right) &= e^{Rt/L}\left(\frac{dI}{dt} + \frac{R}{L}I\right) = e^{Rt/L}\frac{V}{L} \\ e^{Rt/L}I &= \frac{V}{L}\int e^{Rt/L}\,dt = \frac{V}{R}e^{Rt/L} + C \\ I(t) &= \frac{V}{R} + Ce^{-Rt/L}. \end{aligned}$$

Since $I(0) = 0$, we have $0 = (V/R) + C$, so $C = -V/R$. Thus, the current flowing at any time $t > 0$ is

$$I(t) = \frac{V}{R}\left(1 - e^{-Rt/L}\right).$$

It is clear from this solution that $\lim_{t\to\infty} I(t) = V/R$; the *steady state* current is the current that would flow if the inductance were zero.

$I(t)$ will be 90% of this limiting value when

$$\frac{V}{R}\left(1 - e^{-Rt/L}\right) = \frac{90}{100}\frac{V}{R}.$$

This equation implies that $e^{-Rt/L} = 1/10$, or $t = (L \ln 10)/R$. The current will grow to 90% of its limiting value in $(L \ln 10)/R$ seconds.

Our final example reviews a typical *stream of payments* problem of the sort considered in Section 7.7. This time we treat the problem as an initial-value problem for a differential equation.

EXAMPLE 10 A savings account is opened with a deposit of A dollars. At any time t years thereafter, money is being continually deposited into the account at a rate of $(C + Dt)$ dollars per year. If interest is also being paid into the account at a nominal rate of $100R$ percent per year, compounded continuously, find the balance $B(t)$ dollars in the account after t years. Illustrate the solution for the data $A = 5{,}000$, $C = 1{,}000$, $D = 200$, $R = 0.13$, and $t = 5$.

Solution As noted in Section 3.4, continuous compounding of interest at a nominal rate of $100R$ percent causes \$1.00 to grow to e^{Rt} dollars in t years. Without subsequent deposits, the balance in the account would grow according to the differential equation of exponential growth:

$$\frac{dB}{dt} = RB.$$

Allowing for additional growth due to the continual deposits, we observe that B must satisfy the differential equation

$$\frac{dB}{dt} = RB + (C + Dt)$$

or, equivalently, $dB/dt - RB = C + Dt$. This is a linear equation for B having $p(t) = -R$. Hence, we may take $\mu(t) = -Rt$ and $e^{\mu(t)} = e^{-Rt}$. We now calculate

$$\frac{d}{dt}\left(e^{-Rt}B(t)\right) = e^{-Rt}\frac{dB}{dt} - Re^{-Rt}B(t) = (C + Dt)\,e^{-Rt}$$

and

$$\begin{aligned} e^{-Rt}B(t) &= \int (C + Dt)e^{-Rt}\,dt \qquad \text{Let } U = C + Dt, \quad dV = e^{-Rt}\,dt. \\ &\qquad\qquad\qquad\qquad\quad \text{Then } dU = D\,dt, \qquad V = -e^{-Rt}/R. \\ &= -\frac{C + Dt}{R}e^{-Rt} + \frac{D}{R}\int e^{-Rt}\,dt \\ &= -\frac{C + Dt}{R}e^{-Rt} - \frac{D}{R^2}e^{-Rt} + K, \qquad (K = \text{constant}). \end{aligned}$$

Hence

$$B(t) = -\frac{C + Dt}{R} - \frac{D}{R^2} + Ke^{Rt}.$$

Since $A = B(0) = -\frac{C}{R} - \frac{D}{R^2} + K$, we have $K = A + \frac{C}{R} + \frac{D}{R^2}$ and

$$B(t) = \left(A + \frac{C}{R} + \frac{D}{R^2}\right)e^{Rt} - \frac{C + Dt}{R} - \frac{D}{R^2}.$$

For the illustration $A = 5{,}000$, $C = 1{,}000$, $D = 200$, $R = 0.13$, and $t = 5$, we obtain, using a calculator, $B(5) = 19{,}762.82$. The account will contain \$19,762.82, after 5 years, under these circumstances.

EXERCISES 7.9

Solve the separable equations in Exercises 1–10.

1. $\dfrac{dy}{dx} = \dfrac{y}{2x}$
2. $\dfrac{dy}{dx} = \dfrac{3y - 1}{x}$
3. $\dfrac{dy}{dx} = \dfrac{x^2}{y^2}$
4. $\dfrac{dy}{dx} = x^2y^2$
5. $\dfrac{dY}{dt} = tY$
6. $\dfrac{dx}{dt} = e^x \sin t$
7. $\dfrac{dy}{dx} = 1 - y^2$
8. $\dfrac{dy}{dx} = 1 + y^2$
9. $\dfrac{dy}{dt} = 2 + e^y$
10. $\dfrac{dy}{dx} = y^2(1 - y)$

Solve the linear equations in Exercises 11–16.

11. $\dfrac{dy}{dx} - \dfrac{2y}{x} = x^2$
12. $\dfrac{dy}{dx} + \dfrac{2y}{x} = \dfrac{1}{x^2}$
13. $\dfrac{dy}{dx} + 2y = 3$
14. $\dfrac{dy}{dx} + y = e^x$
15. $\dfrac{dy}{dx} + y = x$
16. $\dfrac{dy}{dx} + 2e^x y = e^x$

Solve the initial-value problems in Exercises 17–20.

17. $\begin{cases} \dfrac{dy}{dt} + 10y = 1 \\ y(1/10) = 2/10 \end{cases}$

18. $\begin{cases} \dfrac{dy}{dx} + 3x^2 y = x^2 \\ y(0) = 1 \end{cases}$

19. $\begin{cases} x^2 y' + y = x^2 e^{1/x} \\ y(1) = 3e \end{cases}$

20. $\begin{cases} y' + (\cos x)y = 2xe^{-\sin x} \\ y(\pi) = 0 \end{cases}$

Solve the integral equations in Exercises 21–24.

21. $y(x) = 2 + \displaystyle\int_0^x \frac{t}{y(t)}\,dt$

22. $y(x) = 1 + \displaystyle\int_0^x \frac{(y(t))^2}{1+t^2}\,dt$

23. $y(x) = 1 + \displaystyle\int_1^x \frac{y(t)\,dt}{t(t+1)}$

24. $y(x) = 3 + \displaystyle\int_0^x e^{-y(t)}\,dt$

25. If $a > b > 0$ in Example 5, find $\lim_{t\to\infty} x(t)$.

26. If $b > a > 0$ in Example 5, find $\lim_{t\to\infty} x(t)$.

27. Why is the solution given in Example 5 not valid for $a = b$? Find the solution for the case $a = b$.

28. An object of mass m falling near the surface of the earth is retarded by air resistance proportional to its velocity so that, according to Newton's Second Law of Motion,

$$m\frac{dv}{dt} = mg - kv,$$

where $v = v(t)$ is the velocity of the object at time t, and g is the acceleration of gravity near the surface of the earth. Assuming that the object falls from rest at time $t = 0$, that is, $v(0) = 0$, find the velocity $v(t)$ for any $t > 0$ (up until the object strikes the ground). Show $v(t)$ approaches a limit as $t \to \infty$. Do you need the explicit formula for $v(t)$ to determine this limiting velocity?

29. Repeat Exercise 28 except assuming that the air resistance is proportional to the square of the velocity so that the equation of motion is

$$m\frac{dv}{dt} = mg - kv^2.$$

30. Find the amount in a savings account after one year if the initial balance in the account was \$1,000, if the interest is paid continuously into the account at a nominal rate of 10% per annum, compounded continuously, and if the account is being continuously depleted (by taxes, say) at a rate of $y^2/1{,}000{,}000$ dollars per year, where $y = y(t)$ is the balance in the account after t years. How large can the account grow? How long will it take the account to grow to half this balance?

31. Find the family of curves each of which intersects all of the hyperbolas $xy = C$ at right angles.

32. Repeat the solution concentration problem in Example 4, changing the rate of inflow of brine into the tank to 12 L/min but leaving all the other data as they were in that example. Note that the volume of liquid in the tank is no longer constant as time increases.

CHAPTER REVIEW

Key Ideas

- **What do the following phrases mean?**
  - ◇ a solid of revolution
  - ◇ a volume element
  - ◇ the arc length of a curve
  - ◇ the moment of a point mass m about $x = 0$
  - ◇ the centre of mass of a distribution of mass
  - ◇ the centroid of a plane region
  - ◇ a first-order separable differential equation
  - ◇ a first-order linear differential equation
- **Let D be the plane region $0 \le y \le f(x)$, $a \le x \le b$. Use integrals to represent the following:**
  - ◇ the volume generated by revolving D about the x-axis
  - ◇ the volume generated by revolving D about the y-axis
  - ◇ the moment of D about the y-axis
  - ◇ the moment of D about the x-axis
  - ◇ the centroid of D
- **Let C be the curve $y = f(x)$, $a \le x \le b$. Use integrals to represent the following:**
  - ◇ the length of C
  - ◇ the area of the surface generated by revolving C about the x-axis
  - ◇ the area of the surface generated by revolving C about the y-axis

Review Exercises

1. Figure 7.66 shows cross-sections along the axes of two circular spools. The left spool will hold 1,000 metres of thread if wound full with no bulging. How many metres of thread of the same size will the right spool hold?

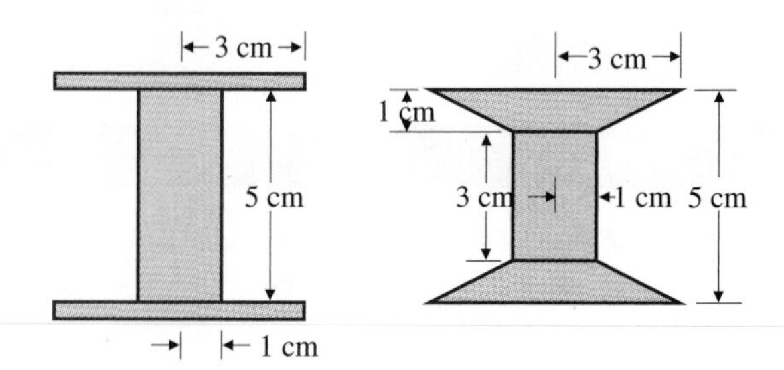

Figure 7.66

2. Water sitting in a bowl evaporates at a rate proportional to its surface area. Show that the depth of water in the bowl decreases at a constant rate, regardless of the shape of the bowl.

3. A barrel is 4 ft high and its volume is 16 cubic feet. Its top and bottom are circular disks of radius 1 ft, and its side wall is obtained by rotating the part of the parabola $x = a - by^2$ between $y = -2$ and $y = 2$ about the y-axis. Find, approximately, the values of the positive constants a and b.

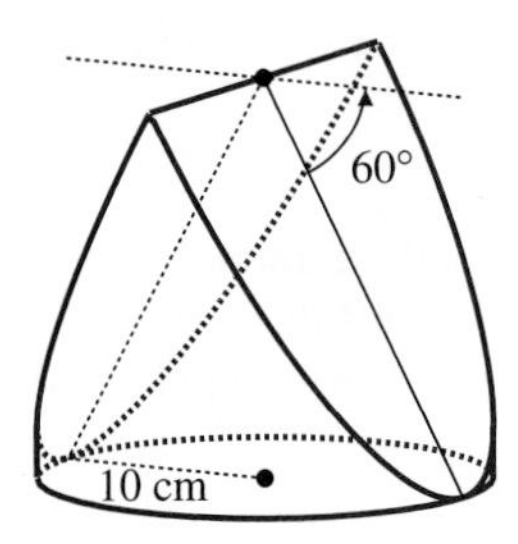

Figure 7.67

4. The solid in Figure 7.67 is cut from a vertical cylinder of radius 10 cm by two planes making angles of 60° with the horizontal. Find its volume.

5. Find to 4 decimal places the value of the positive constant a for which the curve $y = (1/a)\cosh ax$ has arc length 2 units between $x = 0$ and $x = 1$.

6. Find the area of the surface obtained by rotating the curve $y = \sqrt{x}$, $(0 \le x \le 6)$, about the x-axis.

7. Find the centroid of the plane region $x \ge 0$, $y \ge 0$, $x^2 + 4y^2 \le 4$.

8. A thin plate in the shape of a circular disk has radius 3 ft and constant areal density. A circular hole of radius 1 ft is cut out of the disk, centred 1 ft from the centre of the disk. Find the centre of mass of the remaining part of the disk.

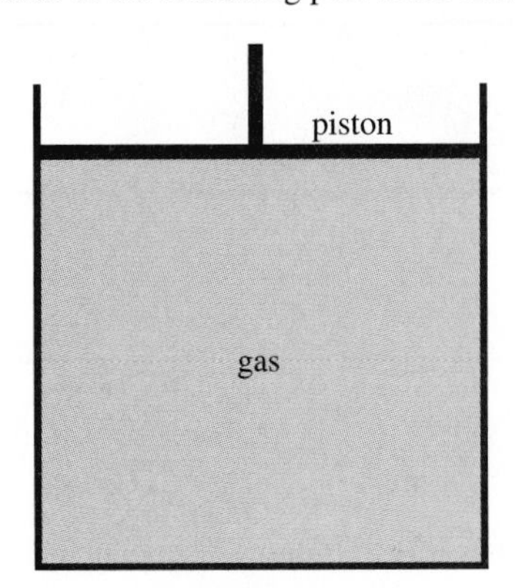

Figure 7.68

9. According to Boyle's Law, the product of the pressure and volume of a gas remains constant if the gas expands or is compressed isothermally. The cylinder in Figure 7.68 is filled with a gas that exerts a force of 1,000 N on the piston when the piston is 20 cm above the base of the cylinder. How much work is done by the piston if it compresses the gas isothermally by descending to a height of 5 cm above the base?

10. Suppose two functions f and g have the following property: for any $a > 0$, the solid produced by revolving the region of the xy-plane bounded by $y = f(x)$, $y = g(x)$, $x = 0$, and $x = a$ about the x-axis has the same volume as the solid produced by revolving the same region about the y-axis. What can you say about f and g?

11. Find the equation of a curve that passes through the point $(2, 4)$ and has slope $3y/(x-1)$ at any point (x, y) on it.

12. Find a family of curves that intersect every ellipse of the form $3x^2 + 4y^2 = C$ at right angles.

13. The income and expenses of a seasonal business result in deposits and withdrawals from its bank account that correspond to a flow rate into the account of $\$P(t)$/year at time t years, where $P(t) = 10,000\sin(2\pi t)$. If the account earns interest at an instantaneous rate of 4% per year, and has \$8,000 in it at time $t = 0$, how much is in the account two years later?

Challenging Problems

1. The curve $y = e^{-kx}\sin x$, $(x \ge 0)$, is revolved about the x-axis to generate a string of "beads" whose volumes decrease to the right if $k > 0$.
   (a) Show that the ratio of the volume of the $(n+1)$st bead to that of the nth bead depends on k, but not on n.
   (b) For what value of k is the ratio in part (a) equal to 1/2?
   (c) Find the total volume of all the beads as a function of $k > 0$.

2. **(Conservation of earth)** A landscaper wants to create on level ground a ring-shaped pool having an outside radius of 10 m and a maximum depth of 1 m surrounding a hill that will be built up using all the earth excavated from the pool. (See Figure 7.69.) She decides to use a fourth-degree polynomial to determine the cross-sectional shape of the hill and pool bottom: at distance r metres from the centre of the development the height above or below normal ground level will be

$$h(r) = a(r^2 - 100)(r^2 - k^2) \text{ metres,}$$

for some $a > 0$, where k is the inner radius of the pool. Find k and a so that the requirements given above are all satisfied. How much earth must be moved from the pool to build the hill?

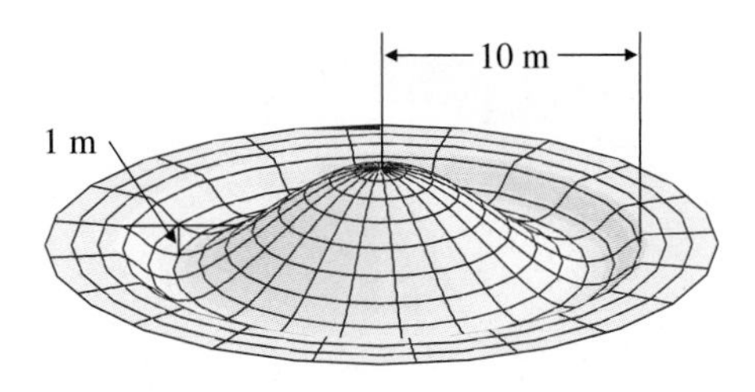

Figure 7.69

3. **(Rocket design)** The nose of a rocket is a solid of revolution of base radius r and height h that must join smoothly to the cylindrical body of the rocket. (See Figure 7.70.) Taking the origin at the tip of the nose and the x-axis along the central axis of the rocket, various nose shapes can be obtained by revolving the cubic curve

$$y = f(x) = ax + bx^2 + cx^3$$

about the x-axis. The cubic curve must have slope 0 at $x = h$, and its slope must be positive for $0 < x < h$. Find the particular cubic curve that maximizes the volume of the nose. Also show that this choice of the cubic makes the slope dy/dx at the origin as large as possible and, hence, corresponds to the bluntest nose.

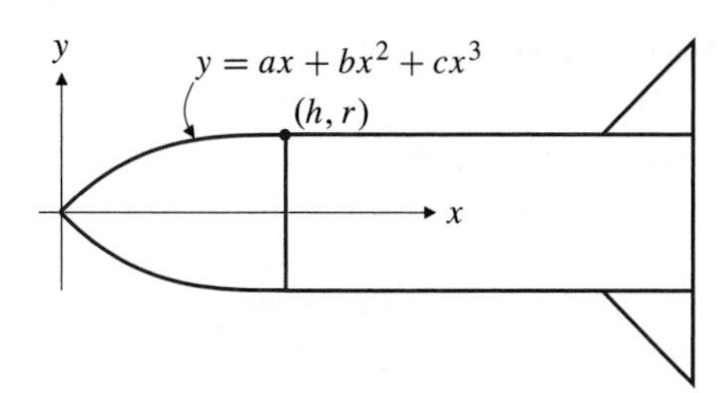

Figure 7.70

4. **(Quadratic splines)** Let $A = (x_1, y_1)$, $B = (x_2, y_2)$, and $C = (x_3, y_3)$ be three points with $x_1 < x_2 < x_3$. A function $f(x)$ whose graph passes through the three points is a *quadratic spline* if $f(x)$ is a quadratic function on $[x_1, x_2]$ and a possibly different quadratic function on $[x_2, x_3]$, and the two quadratics have the same slope at x_2. For this problem, take $A = (0, 1)$, $B = (1, 2)$, and $C = (3, 0)$.

(a) Find a one-parameter family $f(x, m)$ of quadratic splines through A, B, and C, having slope m at B.

(b) Find the value of m for which the length of the graph $y = f(x, m)$ between $x = 0$ and $x = 3$ is minimum. What is this minimum length? Compare it with the length of the polygonal line ABC.

5. A concrete wall in the shape of a circular ring must be built to have maximum height 2 m, inner radius 15 m, and width 1 m at ground level, so that its outer radius is 16 m. (See Figure 7.71.) Built on level ground, the wall will have a curved top with height at distance $15 + x$ metres from the centre of the ring given by the cubic function

$$f(x) = x(1 - x)(ax + b) \text{ m},$$

which must not vanish anywhere in the open interval $(0, 1)$. Find the values of a and b that minimize the total volume of concrete needed to build the wall.

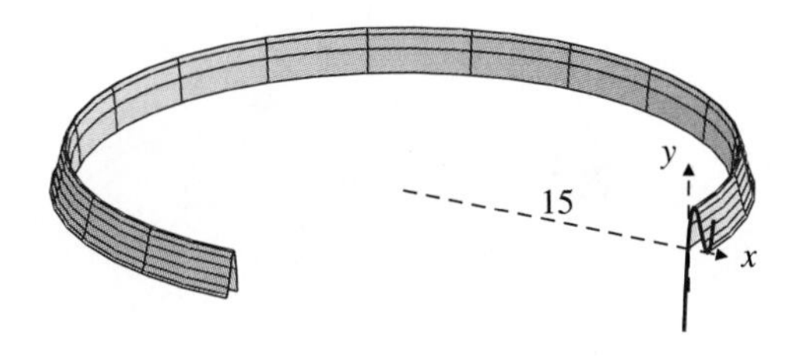

Figure 7.71

6. **(The volume of an *n*-dimensional ball)** Euclidean n-dimensional space consists of *points* $(x_1, x_2, \ldots, x_n)$ with n real coordinates. By analogy with the 3-dimensional case, we call the set of such points that satisfy the inequality $x_1^2 + x_2^2 + \cdots + x_n^2 \le r^2$ the n-dimensional *ball* centred at the origin. For example, the 1-dimensional ball is the interval $-r \le x_1 \le r$, which has *volume* (i.e., *length*) $V_1(r) = 2r$. The 2-dimensional ball is the disk $x_1^2 + x_2^2 \le r^2$, which has *volume* (i.e., *area*)

$$V_2(r) = \pi r^2 = \int_{-r}^{r} 2\sqrt{r^2 - x^2}\, dx = \int_{-r}^{r} V_1\left(\sqrt{r^2 - x^2}\right) dx.$$

The 3-dimensional ball $x_1^2 + x_2^2 + x_3^2 \le r^2$ has volume

$$V_3(r) = \frac{4}{3}\pi r^3 = \int_{-r}^{r} \pi \left(\sqrt{r^2 - x^2}\right)^2 dx = \int_{-r}^{r} V_2\left(\sqrt{r^2 - x^2}\right) dx.$$

By analogy with these formulas, the volume $V_n(r)$ of the n-dimensional ball of radius r is the integral of the volume of the $(n-1)$-dimensional ball of radius $\sqrt{r^2 - x^2}$ from $x = -r$ to $x = r$:

$$V_n(r) = \int_{-r}^{r} V_{n-1}\left(\sqrt{r^2 - x^2}\right) dx.$$

Using a computer algebra program, calculate $V_4(r)$, $V_5(r)$, $\ldots$, $V_{10}(r)$, and guess formulas for $V_{2n}(r)$ (the even-dimensional balls) and $V_{2n+1}(r)$ (the odd-dimensional balls). If your computer algebra software is sufficiently powerful, you may be able to verify your guesses by induction. Otherwise, use them to predict $V_{11}(r)$ and $V_{12}(r)$, then check your predictions by starting from $V_{10}(r)$.

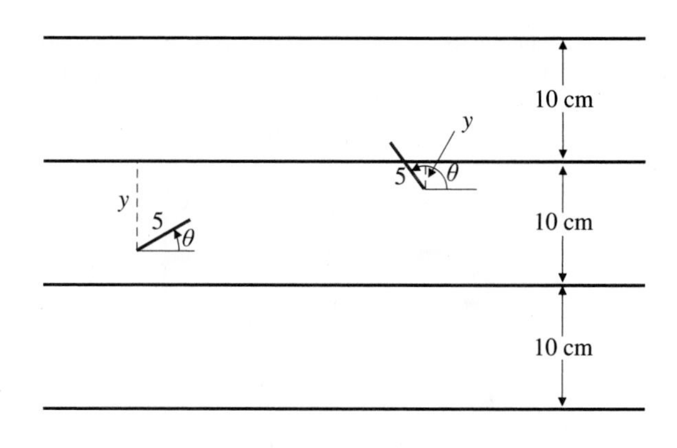

Figure 7.72

7. **(Buffon's needle problem)** A horizontal flat surface is ruled with parallel lines 10 cm apart, as shown in Figure 7.72. A needle 5 cm long is dropped at random onto the surface. Find the probability that the needle intersects one of the lines. *Hint:* Let the "lower" end of the needle (the end further down the page in the figure) be considered the reference point. (If both ends are the same height, use the left end.) Let y be the distance from the reference point to the nearest line above it, and let θ be the angle between the needle and the line extending to the right of the reference point in the figure. What are the possible values of y and θ? In a plane with Cartesian coordinates θ and y sketch the region consisting of all points (θ, y) corresponding to possible positions of the needle. Also sketch the region corresponding to those positions for which the needle crosses one of the parallel lines. The required probability is the area of the second region divided by the area of the first.

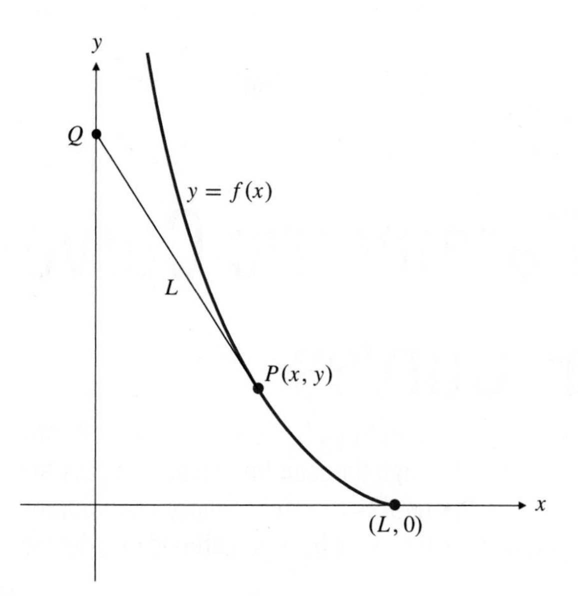

Figure 7.73

8. **(The path of a trailer)** Find the equation $y = f(x)$ of a curve in the first quadrant of the xy-plane, starting from the point $(L, 0)$, and having the property that if the tangent line to the curve at P meets the y-axis at Q, then the length of PQ is the constant L. (See Figure 7.73. This curve is called a **tractrix** after the Latin participle *tractus* meaning *dragged.* It is the path of the rear end P of a trailer of length L, originally lying along the x-axis, as the trailer is pulled (dragged) by a tractor Q moving along the y-axis away from the origin.)

9. **(Approximating the surface area of an ellipsoid)** A physical geographer studying the flow of streams around oval stones needed to calculate the surface areas of many such stones that he modelled as ellipsoids:

$$\frac{x^2}{a^2} + \frac{y^2}{b^2} + \frac{z^2}{c^2} = 1.$$

He wanted a simple formula for the surface area so that he could implement it in a spreadsheet containing the measurements a, b, and c of the stones. Unfortunately, there is no exact formula for the area of a general ellipsoid in terms of elementary functions. However, there are such formulas for ellipsoids of revolution, where two of the three semi-axes are equal. These ellipsoids are called spheroids; an *oblate spheroid* (like the earth) has its two longer semi-axes equal; a *prolate spheroid* (like an American football) has its two shorter semi-axes equal. A reasonable approximation to the area of a general ellipsoid can be obtained by linear interpolation between these two.

To be specific, assume the semi-axes are arranged in decreasing order $a \ge b \ge c$, and let the surface area be $S(a, b, c)$.

(a) Calculate $S(a, a, c)$, the area of an oblate spheroid.

(b) Calculate $S(a, c, c)$, the area of a prolate spheroid.

(c) Construct an approximation for $S(a, b, c)$ that divides the interval from $S(a, a, c)$ to $S(a, c, c)$ in the same ratio that b divides the interval from a to c.

(d) Approximate the area of the ellipsoid

$$\frac{x^2}{9} + \frac{y^2}{4} + z^2 = 1$$

using the above method.

CHAPTER 8

Conics, Parametric Curves, and Polar Curves

"Everyone knows what a curve is, until he has studied enough mathematics to become confused through the countless number of possible exceptions.... A curve is the totality of points, whose co-ordinates are functions of a parameter which may be differentiated as often as may be required."

Felix Klein 1849–1925

Introduction

Until now, most curves we have encountered have been graphs of functions, and they provided useful visual information about the behaviour of the functions. In this chapter we begin to look at plane curves as interesting objects in their own right. First, we examine conic sections, curves with quadratic equations obtained by intersecting a plane with a right-circular cone. Then we consider curves that can be described by two parametric equations that give the coordinates of points on the curve as functions of a parameter. If this parameter is time, the equations describe the path of a moving point in the plane. Finally, we consider curves described by equations in a new coordinate system called polar coordinates, in which a point is located by giving its distance and direction from the origin. In Chapter 11 we will expand our study of curves to three dimensions.

8.1 Conics

Circles, ellipses, parabolas, and hyperbolas are called **conic sections** (or, more simply, just **conics**) because they are curves in which planes intersect right-circular cones.

To be specific, suppose that a line A is fixed in space, and V is a point fixed on A. The **right-circular cone** having **axis** A, **vertex** V, and **semi-vertical angle** α is the surface consisting of all points on straight lines through V that make angle α with the line A. (See Figure 8.1.) The cone has two halves (called **nappes**) lying on opposite sides of the vertex V. Any plane P that does not pass through V will intersect the cone (one or both nappes) in a curve $\mathcal{C}$. (See Figure 8.2.) If a line normal (i.e., perpendicular) to P makes angle θ with the axis A of the cone, where $0 \le \theta \le \pi/2$, then

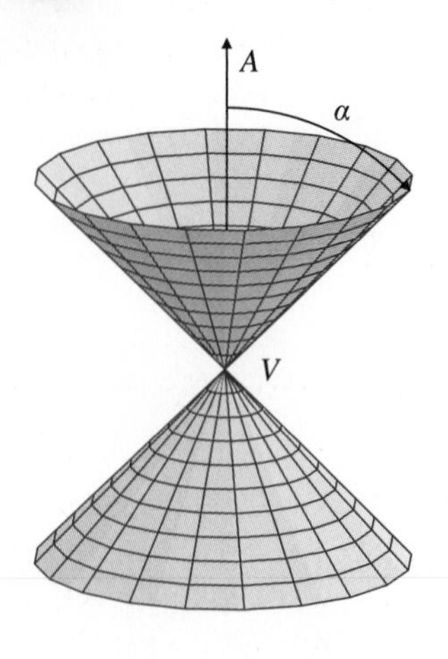

Figure 8.1 A cone with vertex V, axis A, and semi-vertical angle α

$\mathcal{C}$ is a circle if	$\theta = 0$
$\mathcal{C}$ is an ellipse if	$0 < \theta < \dfrac{\pi}{2} - \alpha$
$\mathcal{C}$ is a parabola if	$\theta = \dfrac{\pi}{2} - \alpha$
$\mathcal{C}$ is a hyperbola if	$\dfrac{\pi}{2} - \alpha < \theta \le \dfrac{\pi}{2}.$

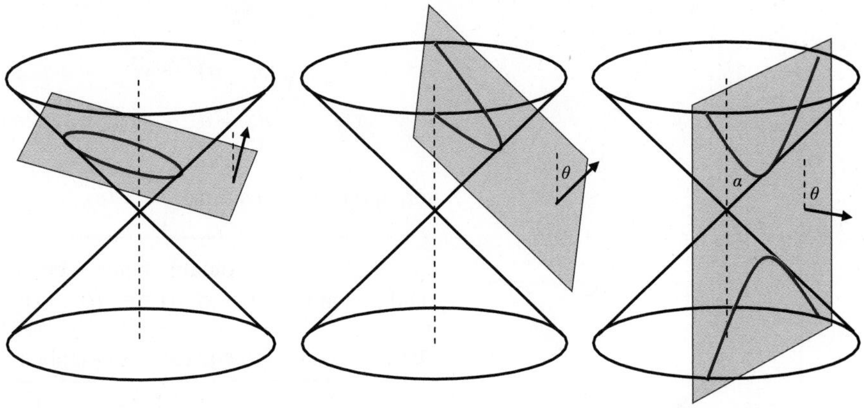

ellipse parabola hyperbola

Figure 8.2 Planes intersecting cones in an ellipse, a parabola, and a hyperbola

In Sections 10.4 and 10.5 it is shown that planes are represented by first-degree equations and cones by second-degree equations. Therefore, all conics can be represented analytically (in terms of Cartesian coordinates x and y in the plane of the conic) by a second-degree equation of the general form

$$Ax^2 + Bxy + Cy^2 + Dx + Ey + F = 0,$$

where $A, B, \ldots, F$ are constants. However, such an equation can also represent the empty set, a single point, or one or two straight lines if the left-hand side factors into linear factors:

$$(A_1x + B_1y + C_1)(A_2x + B_2y + C_2) = 0.$$

After straight lines, the conic sections are the simplest of plane curves. They have many properties that make them useful in applications of mathematics; that is why we include a discussion of them here. Much of this material is optional from the point of view of a calculus course, but familiarity with the properties of conics can be very important in some applications. Most of the properties of conics were discovered by the Greek geometer Apollonius of Perga, around 200 BC. It is remarkable that he was able to obtain these properties using only the techniques of classical Euclidean geometry; today, most of these properties are expressed more conveniently using analytic geometry and specific coordinate systems.

Parabolas

DEFINITION 1

Parabolas

A **parabola** consists of points in the plane that are equidistant from a given point (the **focus** of the parabola) and a given straight line (the **directrix** of the parabola). The line through the focus perpendicular to the directrix is called the **principal axis** (or simply **the axis**) of the parabola. The **vertex** of the parabola is the point where the parabola crosses its principal axis. It is on the axis halfway between the focus and the directrix.

EXAMPLE 1 Find an equation of the parabola whose focus is the point $F = (a, 0)$ and whose directrix is the line L with equation $x = -a$.

Solution The parabola has axis along the x-axis and vertex at the origin. (See Figure 8.3.) If $P = (x, y)$ is any point on the parabola, then the distance from P to F is equal to the distance from P to the nearest point Q on L. Thus,

$$\sqrt{(x-a)^2 + y^2} = x + a$$

$$\text{or } x^2 - 2ax + a^2 + y^2 = x^2 + 2ax + a^2,$$

or, upon simplification, $y^2 = 4ax$.

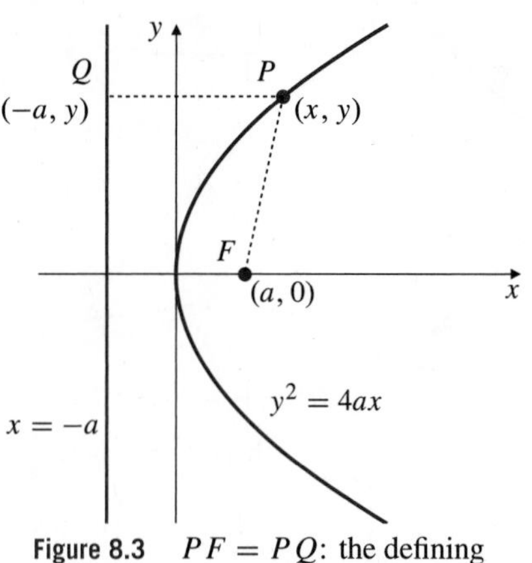

Figure 8.3 $PF = PQ$: the defining property of a parabola

Similarly, we can obtain standard equations for parabolas with vertices at the origin and foci at $(-a, 0)$, $(0, a)$, and $(0, -a)$:

Table 1. Standard equations of parabolas

Focus	Directrix	Equation
$(a, 0)$	$x = -a$	$y^2 = 4ax$
$(-a, 0)$	$x = a$	$y^2 = -4ax$
$(0, a)$	$y = -a$	$x^2 = 4ay$
$(0, -a)$	$y = a$	$x^2 = -4ay$

The Focal Property of a Parabola

All of the conic sections have interesting and useful focal properties relating to the way in which surfaces of revolution they generate reflect light if the surfaces are mirrors. For instance, a circle will clearly reflect back along the same path any ray of light incident along a line passing through its centre. The focal properties of parabolas, ellipses, and hyperbolas can be derived from the reflecting property of a straight line (i.e., a plane mirror) by elementary geometrical arguments.

Light travels in straight lines in a medium of constant optical density (one where the speed of light is constant). This is a consequence of the physical Principle of Least Action, which asserts that in travelling between two points, light takes the path requiring the minimum travel time. Given a straight line L in a plane and two points A and B in the plane on the same side of L, the point P on L for which the sum of the distances $AP + PB$ is minimum is such that AP and PB make equal angles with L, or equivalently, with the normal to L at P. (See Figure 8.4.) If B' is the point such that L is the right bisector of the line segment BB', then P is the intersection of L and AB'. Since one side of a triangle cannot exceed the sum of the other two sides,

$$AP + PB = AP + PB' = AB' \le AQ + QB' = AQ + QB.$$

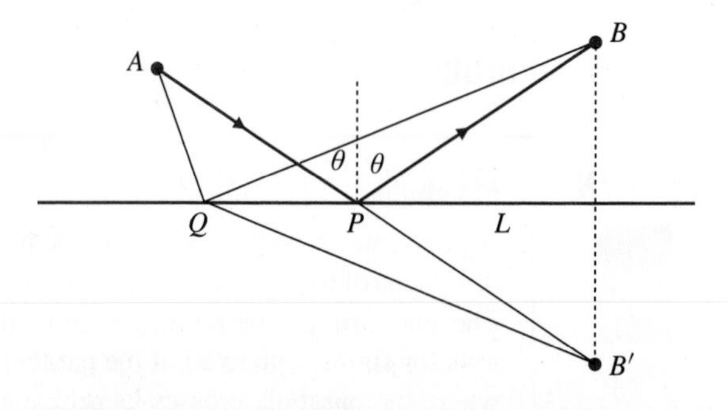

Figure 8.4 Reflection by a straight line

Reflection by a straight line

The point P on L at which a ray from A reflects so as to pass through B is the point that minimizes the sum of the distances $AP + PB$.

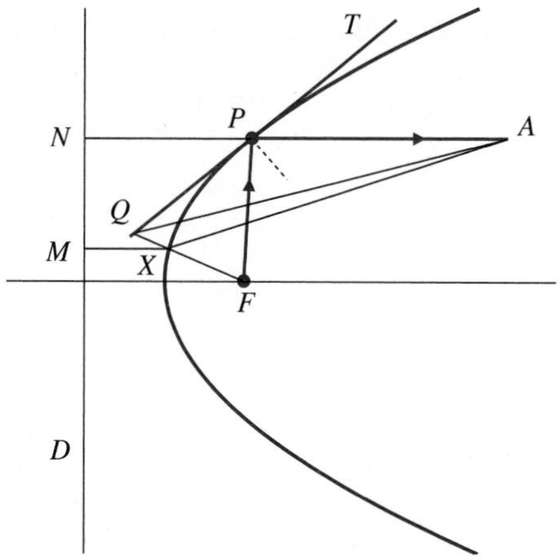

Figure 8.5 Reflection by a parabola

Now consider a parabola with focus F and directrix D. Let P be on the parabola and let T be the line tangent to the parabola at P. (See Figure 8.5.) Let Q be any point on T. Then FQ meets the parabola at a point X between F and Q. Let M and N be points on D such that MX and NP are perpendicular to D, and let A be a point on the line through N and P that lies on the same side of the parabola as F. We have

BEWARE! Consider the equalities and inequalities in this chain one at a time. Why is each one true?

$$\begin{aligned} FP + PA = NP + PA = NA \leq MX + XA = FX + XA \\ \leq FX + XQ + QA = FQ + QA. \end{aligned}$$

Thus, among all points Q on the line T, $Q = P$ is the one that minimizes the sum of distances $FQ + QA$. By the observation made for straight lines above, FP and PA make equal angles with T and so also with the normal to the parabola at P. (The parabola and the tangent line have the same normal at P.)

Reflection by a parabola

Any ray from the focus will be reflected parallel to the axis of the parabola. Equivalently, any incident ray parallel to the axis of the parabola will be reflected through the focus.

Ellipses

DEFINITION 2

Ellipses

An **ellipse** consists of all points in the plane, the sum of whose distances from two fixed points (the **foci**) is constant.

EXAMPLE 2 Find the ellipse with foci at the points $(-c, 0)$ and $(c, 0)$ if the sum of the distances from any point P on the ellipse to these two foci is $2a$ (where $a > c$).

Solution The ellipse passes through the four points $(a, 0)$, $(-a, 0)$, $(0, b)$, and $(0, -b)$, where $b^2 = a^2 - c^2$. (See Figure 8.6.) Also, if $P = (x, y)$ is on the ellipse, then

$$\sqrt{(x-c)^2 + y^2} + \sqrt{(x+c)^2 + y^2} = 2a.$$

Transposing one term from the left side to the right side and squaring, we get

$$(x-c)^2 + y^2 = 4a^2 - 4a\sqrt{(x+c)^2 + y^2} + (x+c)^2 + y^2.$$

Now we expand the squares, cancel terms, transpose, and square again:

$$a\sqrt{(x+c)^2+y^2} = a^2 + cx$$
$$a^2(x^2+2cx+c^2+y^2) = a^4 + 2a^2cx + c^2x^2$$
$$(a^2-c^2)x^2 + a^2y^2 = a^2(a^2-c^2).$$

Finally, replace $a^2 - c^2$ with b^2 and divide by a^2b^2 to get the standard equation of the ellipse:

$$\frac{x^2}{a^2} + \frac{y^2}{b^2} = 1.$$

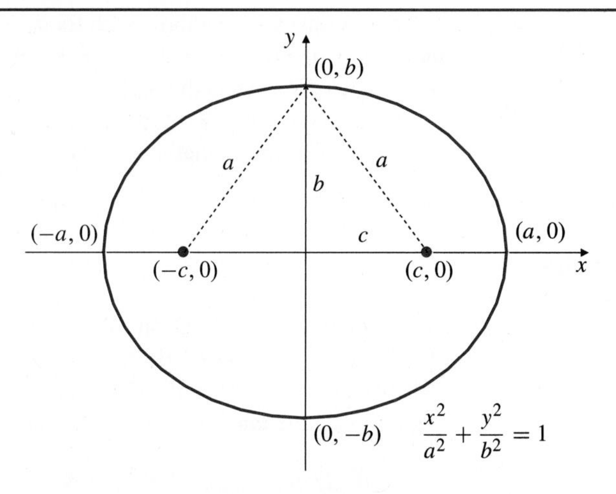

Figure 8.6 An ellipse

The following quantities describe this ellipse:

a is the **semi-major axis,**
b is the **semi-minor axis,**
$c = \sqrt{a^2 - b^2}$ is the **semi-focal separation.**

The point halfway between the foci is called the **centre** of the ellipse. In the example above it is the origin. Note that $a > b$ in this example. If $a < b$, then the ellipse has its foci at $(0, c)$ and $(0, -c)$, where $c = \sqrt{b^2 - a^2}$. The line containing the foci (the **major axis**) and the line through the centre perpendicular to that line (the **minor axis**) are called the **principal axes** of the ellipse.

The **eccentricity** of an ellipse is the ratio of the semi-focal separation to the semi-major axis. We denote the eccentricity ε. For the ellipse $\frac{x^2}{a^2} + \frac{y^2}{b^2} = 1$ with $a > b$,

$$\varepsilon = \frac{c}{a} = \frac{\sqrt{a^2 - b^2}}{a}.$$

Note that $\varepsilon < 1$ for any ellipse; the greater the value of ε, the more elongated (less circular) is the ellipse. If $\varepsilon = 0$ so that $a = b$ and $c = 0$, the two foci coincide and the ellipse is a circle.

The Focal Property of an Ellipse

Let P be any point on an ellipse having foci F_1 and F_2. The normal to the ellipse at P bisects the angle between the lines F_1P and F_2P.

Reflection by an ellipse

Any ray coming from one focus of an ellipse will be reflected through the other focus.

To see this, observe that if Q is any point on the line T tangent to the ellipse at P, then F_1Q meets the ellipse at a point X between F_1 and Q (see Figure 8.7), so

$$F_1P + PF_2 = F_1X + XF_2 \le F_1X + XQ + QF_2 = F_1Q + QF_2.$$

Among all points on T, P is the one that minimizes the sum of the distances to F_1 and F_2. This implies that the normal to the ellipse at P bisects the angle F_1PF_2.

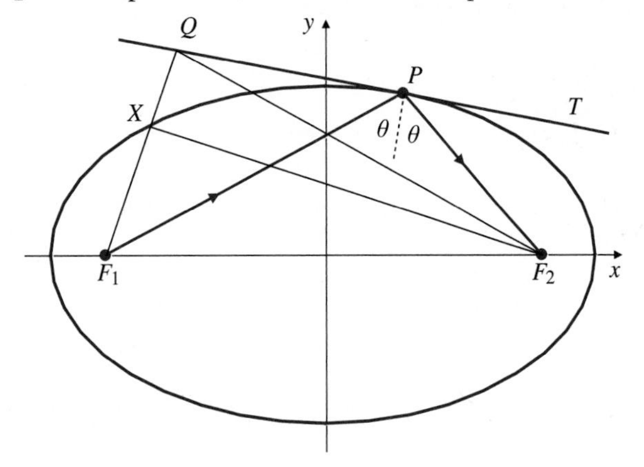

Figure 8.7 A ray from one focus of an ellipse is reflected to the other focus

The Directrices of an Ellipse

If $a > b > 0$, each of the lines $x = a/\varepsilon$ and $x = -a/\varepsilon$ is called a **directrix** of the ellipse $\dfrac{x^2}{a^2} + \dfrac{y^2}{b^2} = 1$. If P is on the ellipse, then the ratio of the distance from P to a focus to its distance from the corresponding directrix is equal to the eccentricity ε. If $P = (x, y)$, F is the focus $(c, 0)$, Q is on the corresponding directrix $x = a/\varepsilon$, and PQ is perpendicular to the directrix, then (see Figure 8.8)

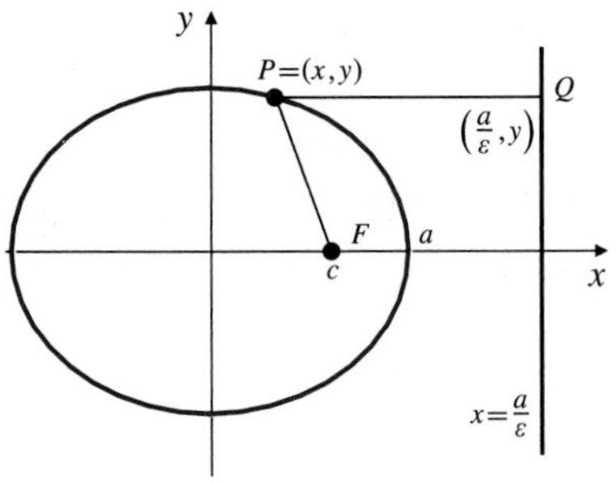

$$\begin{aligned} PF^2 &= (x - c)^2 + y^2 \\ &= x^2 - 2cx + c^2 + b^2\left(1 - \frac{x^2}{a^2}\right) \\ &= x^2\left(\frac{a^2 - b^2}{a^2}\right) - 2cx + a^2 - b^2 + b^2 \\ &= \varepsilon^2x^2 - 2\varepsilon ax + a^2 \qquad \text{(because } c = \varepsilon a\text{)} \\ &= (a - \varepsilon x)^2. \end{aligned}$$

Figure 8.8 A focus and corresponding directrix of an ellipse

Thus, $PF = a - \varepsilon x$. Also, $QP = (a/\varepsilon) - x = (a - \varepsilon x)/\varepsilon$. Therefore, $PF/QP = \varepsilon$, as asserted.

A parabola may be considered as the limiting case of an ellipse whose eccentricity has increased to 1. The distance between the foci is infinite, so the centre, one focus, and its corresponding directrix have moved off to infinity leaving only one focus and its directrix in the finite plane.

Hyperbolas

DEFINITION 3

Hyperbolas

A **hyperbola** consists of all points in the plane, the difference of whose distances from two fixed points (the **foci**) is constant.

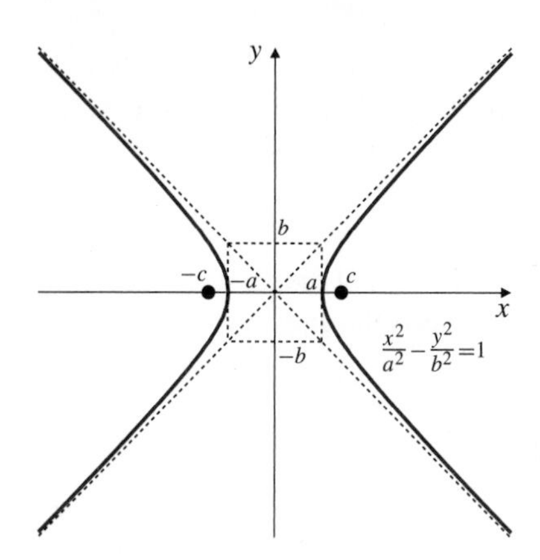

Figure 8.9 Hyperbola with foci $(\pm c, 0)$ and vertices $(\pm a, 0)$

EXAMPLE 3 If the foci of a hyperbola are $F_1 = (c, 0)$ and $F_2 = (-c, 0)$, and the difference of the distances from a point $P = (x, y)$ on the hyperbola to these foci is $2a$ (where $a < c$), then

$$PF_2 - PF_1 = \sqrt{(x+c)^2 + y^2} - \sqrt{(x-c)^2 + y^2} = \begin{cases} 2a & \text{(right branch)} \\ -2a & \text{(left branch).} \end{cases}$$

(See Figure 8.9.) Simplifying this equation by squaring and transposing as was done for the ellipse in Example 2, we obtain the standard equation for the hyperbola:

$$\frac{x^2}{a^2} - \frac{y^2}{b^2} = 1,$$

where $b^2 = c^2 - a^2$.

The points $(a, 0)$ and $(-a, 0)$ (called the **vertices**) lie on the hyperbola, one on each branch. (The two branches correspond to the intersections of the plane of the hyperbola with the two nappes of a cone.) Some parameters used to describe the hyperbola are

a	the **semi-transverse axis,**
b	the **semi-conjugate axis,**
$c = \sqrt{a^2 + b^2}$	the **semi-focal separation.**

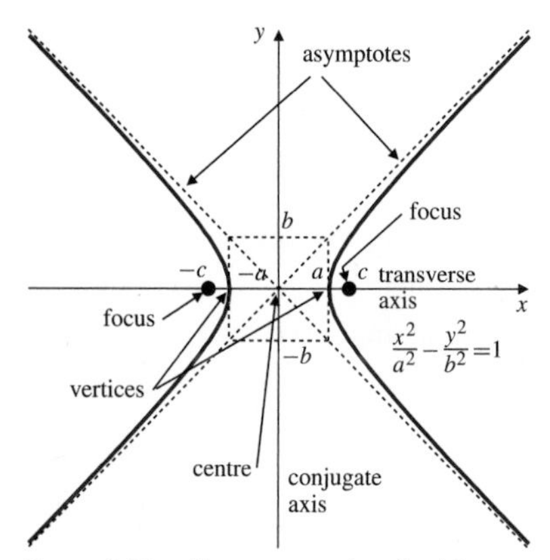

Figure 8.10 Terms associated with a hyperbola

The midpoint of the line segment F_1F_2 (in this case the origin) is called the **centre** of the hyperbola. The line through the centre, the vertices, and the foci is the **transverse axis**. The line through the centre perpendicular to the transverse axis is the **conjugate axis**. The conjugate axis does not intersect the hyperbola. If a rectangle with sides $2a$ and $2b$ is drawn centred at the centre of the hyperbola and with two sides tangent to the hyperbola at the vertices, then the two diagonal lines of the rectangle are **asymptotes** of the hyperbola. They have equations $(x/a) \pm (y/b) = 0$; that is, they are solutions of the degenerate equation

$$\frac{x^2}{a^2} - \frac{y^2}{b^2} = 0.$$

The hyperbola approaches arbitrarily close to these lines as it recedes from the origin. (See Figure 8.10.) A **rectangular** hyperbola is one whose asymptotes are perpendicular lines. (This is so if $b = a$.)

The eccentricity of the hyperbola is

$$\varepsilon = \frac{c}{a} = \frac{\sqrt{a^2 + b^2}}{a}.$$

Note that $\varepsilon > 1$. The lines $x = \pm(a/\varepsilon)$ are called the **directrices** of the hyperbola $(x^2/a^2) - (y^2/b^2) = 1$. (See Figure 8.11.) In a manner similar to that used for the ellipse, you can show that if P is on the hyperbola, then

$$\frac{\text{distance from } P \text{ to a focus}}{\text{distance from } P \text{ to the corresponding directrix}} = \varepsilon.$$

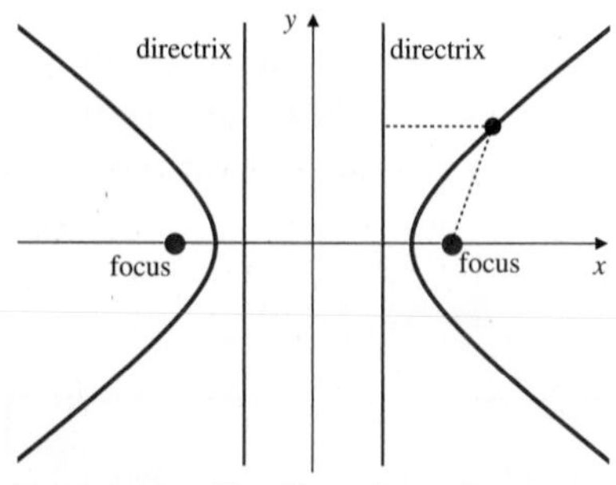

Figure 8.11 The directrices of a hyperbola

The eccentricity of a rectangular hyperbola is $\sqrt{2}$.

A hyperbola with the same asymptotes as $x^2/a^2 - y^2/b^2 = 1$, but with transverse axis along the y-axis, vertices at $(0, b)$ and $(0, -b)$, and foci at $(0, c)$ and $(0, -c)$ is represented by the equation

$$\frac{x^2}{a^2} - \frac{y^2}{b^2} = -1, \quad \text{or, equivalently,} \quad \frac{y^2}{b^2} - \frac{x^2}{a^2} = 1.$$

The two hyperbolas are said to be **conjugate** to one another. (See Figure 8.12.) The *conjugate axis* of a hyperbola is the *transverse axis* of the conjugate hyperbola. Together, the transverse and conjugate axes of a hyperbola are called its **principal axes**.

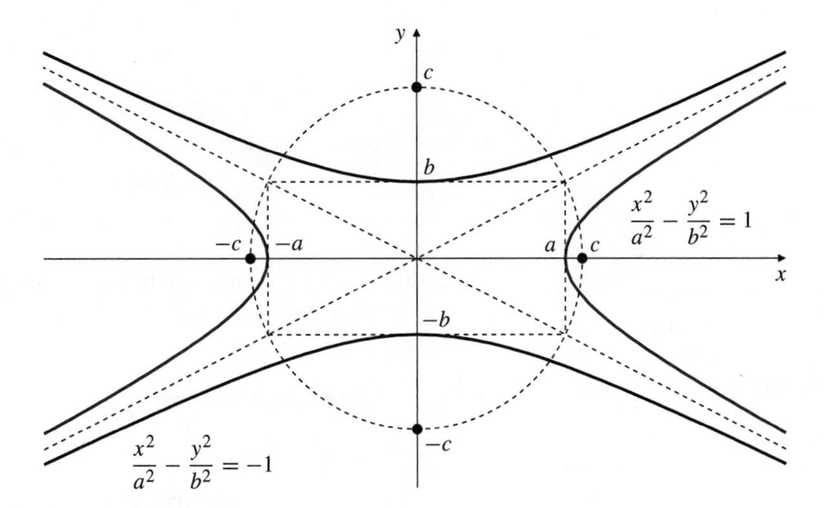

Figure 8.12 Two conjugate hyperbolas and their common asymptotes

The Focal Property of a Hyperbola

Let P be any point on a hyperbola with foci F_1 and F_2. Then the tangent line to the hyperbola at P bisects the angle between the lines F_1P and F_2P.

> **Reflection by a hyperbola**
>
> A ray from one focus of a hyperbola is reflected by the hyperbola so that it appears to have come from the other focus.

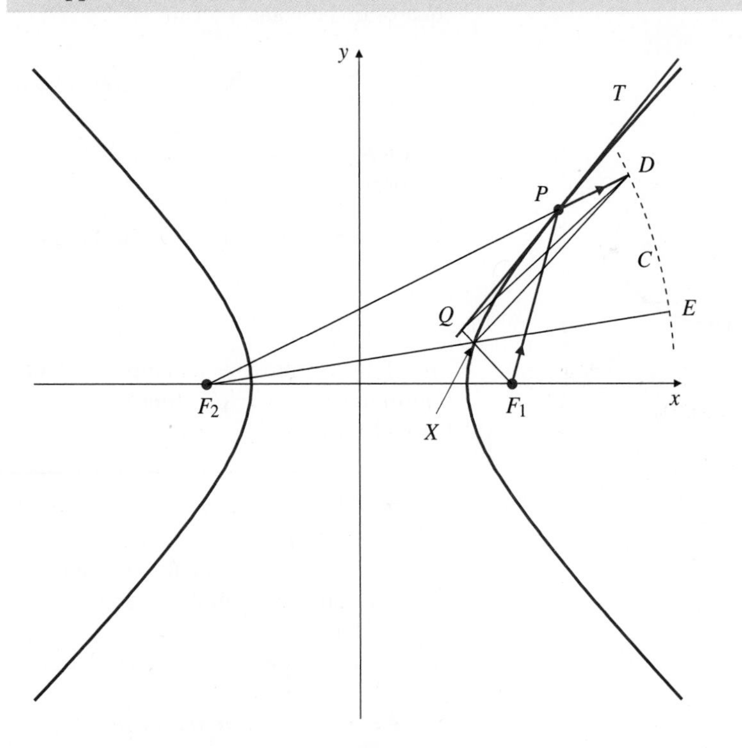

Figure 8.13 A ray from one focus is reflected along a line from the other focus

To see this, let P be on the right branch, let T be the line tangent to the hyperbola at P, and let C be a circle of large radius centred at F_2. (See Figure 8.13.) Let F_2P intersect this circle at D. Let Q be any point on T. Then QF_1 meets the hyperbola at X between Q and F_1, and F_2X meets C at E. Since X is on the radial line F_2E, it is closer to E than it is to other points on C. That is, $XE \le XD$. Thus,

$$\begin{aligned} F_1P + PD &= F_1P + F_2D - F_2P \\ &= F_2D - (F_2P - F_1P) \\ &= F_2E - (F_2X - F_1X) \end{aligned}$$

$$\begin{aligned}
&= F_1X + F_2E - F_2X \\
&= F_1X + XE \\
&\le F_1X + XD \\
&\le F_1X + XQ + QD = F_1Q + QD.
\end{aligned}$$

P is the point on T that minimizes the sum of distances to F_1 and D; therefore, the normal to the hyperbola at P bisects the angle F_1PD. Therefore, T bisects the angle F_1PF_2.

BEWARE! Check the equalities and inequalities in the above chain one at a time to make sure you understand why it is true.

Classifying General Conics

A second-degree equation in two variables,

$$Ax^2 + Bxy + Cy^2 + Dx + Ey + F = 0, \qquad (A^2 + B^2 + C^2 > 0),$$

generally represents a conic curve, but in certain degenerate cases it may represent two straight lines ($x^2 - y^2 = 0$ represents the lines $x = y$ and $x = -y$), one straight line ($x^2 = 0$ represents the line $x = 0$), a single point ($x^2 + y^2 = 0$ represents the origin), or no points at all ($x^2 + y^2 = -1$ is not satisfied by any point in the plane).

The nature of the set of points represented by a given second-degree equation can be determined by rewriting the equation in a form that can be recognized as one of the standard types. If $B = 0$, this rewriting can be accomplished by completing the squares in the x and y terms.

EXAMPLE 4 Describe the curve with equation $x^2 + 2y^2 + 6x - 4y + 7 = 0$.

Solution We complete the squares in the x and y terms, and rewrite the equation in the form

$$x^2 + 6x + 9 + 2(y^2 - 2y + 1) = 9 + 2 - 7 = 4$$

$$\frac{(x+3)^2}{4} + \frac{(y-1)^2}{2} = 1.$$

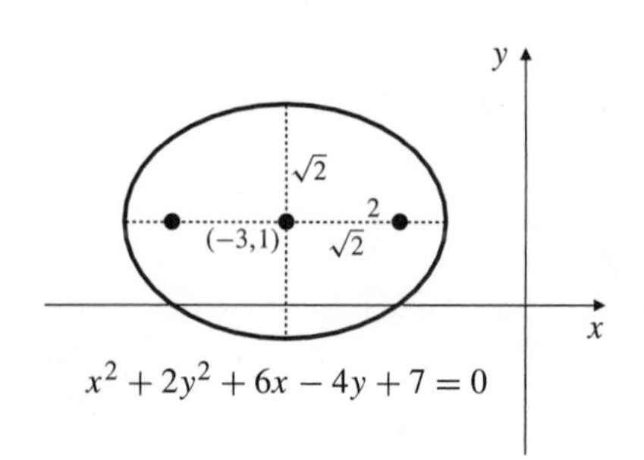

Figure 8.14 This curve is an ellipse

Therefore, it represents an ellipse with centre at $(-3, 1)$, semi-major axis $a = 2$, and semi-minor axis $b = \sqrt{2}$. Since $c = \sqrt{a^2 - b^2} = \sqrt{2}$, the foci are $(-3 \pm \sqrt{2}, 1)$. See Figure 8.14.

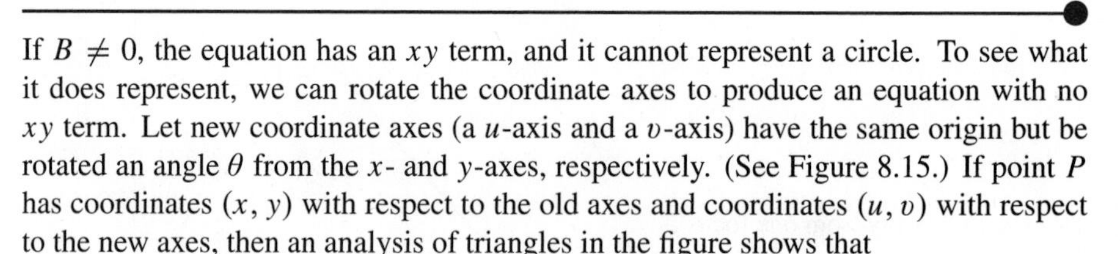

If $B \neq 0$, the equation has an xy term, and it cannot represent a circle. To see what it does represent, we can rotate the coordinate axes to produce an equation with no xy term. Let new coordinate axes (a u-axis and a v-axis) have the same origin but be rotated an angle θ from the x- and y-axes, respectively. (See Figure 8.15.) If point P has coordinates (x, y) with respect to the old axes and coordinates (u, v) with respect to the new axes, then an analysis of triangles in the figure shows that

$$\begin{aligned}
x &= OA - XA = OU\cos\theta - OV\sin\theta = u\cos\theta - v\sin\theta, \\
y &= XB + BP = OU\sin\theta + OV\cos\theta = u\sin\theta + v\cos\theta.
\end{aligned}$$

Substituting these expressions into the equation

$$Ax^2 + Bxy + Cy^2 + Dx + Ey + F = 0, \qquad (A^2 + B^2 + C^2 > 0),$$

BEWARE! A lengthy calculation is needed here. The details have been omitted.

leads to a new equation,

$$A'u^2 + B'uv + C'v^2 + D'u + E'v + F = 0,$$

where

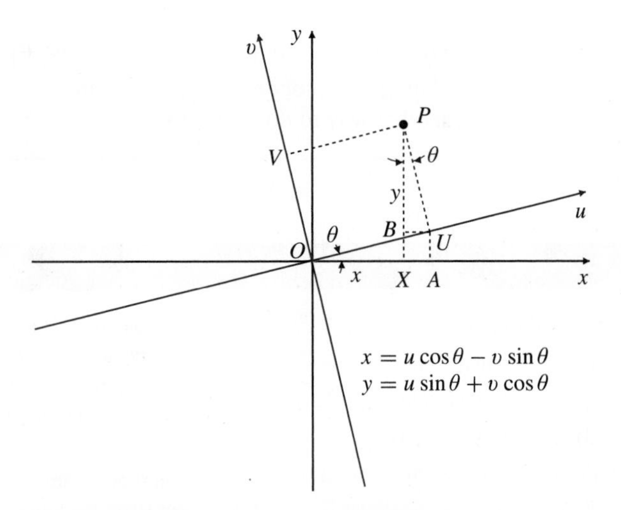

Figure 8.15 Rotation of axes

$$\begin{aligned}
A' &= \frac{1}{2}\Big(A(1+\cos 2\theta) + B\sin 2\theta + C(1-\cos 2\theta)\Big)\\
B' &= (C-A)\sin 2\theta + B\cos 2\theta\\
C' &= \frac{1}{2}\Big(A(1-\cos 2\theta) - B\sin 2\theta + C(1+\cos 2\theta)\Big)\\
D' &= D\cos\theta + E\sin\theta\\
E' &= -D\sin\theta + E\cos\theta.
\end{aligned}$$

Note that F remains unchanged. If we choose θ so that

$$\tan 2\theta = \frac{B}{A-C}, \qquad \text{or} \qquad \theta = \frac{\pi}{4} \text{ if } A = C,\ B \neq 0,$$

then $B' = 0$, and the new equation can then be analyzed as described previously.

EXAMPLE 5 Identify the curve with equation $xy = 1$.

Solution The reader is likely well aware that the given equation represents a rectangular hyperbola with the coordinate axes as asymptotes. Since the given equation involves $A = C = D = E = 0$ and $B = 1$, it is appropriate to rotate the axes through angle $\pi/4$ so that

$$x = \frac{1}{\sqrt{2}}(u - v), \qquad y = \frac{1}{\sqrt{2}}(u + v).$$

The transformed equation is $u^2 - v^2 = 2$, which is, as suspected, a rectangular hyperbola with vertices at $u = \pm\sqrt{2}$, $v = 0$, foci at $u = \pm 2$, $v = 0$, and asymptotes $u = \pm v$. Hence, $xy = 1$ represents a rectangular hyperbola with coordinate axes as asymptotes, vertices at $(1, 1)$ and $(-1, -1)$, and foci at $(\sqrt{2}, \sqrt{2})$ and $(-\sqrt{2}, -\sqrt{2})$.

EXAMPLE 6 Show that the curve $2x^2 + xy + y^2 = 2$ is an ellipse, and find the lengths of its semi-major and semi-minor axes.

Solution Here, $A = 2$, $B = C = 1$, $D = E = 0$, and $F = -2$. We rotate the axes through angle θ where $\tan 2\theta = B/(A - C) = 1$. Thus, $B' = 0$, $2\theta = \pi/4$, and $\sin 2\theta = \cos 2\theta = 1/\sqrt{2}$. We have

$$\begin{aligned}
A' &= \frac{1}{2}\left[2\left(1+\frac{1}{\sqrt{2}}\right) + \frac{1}{\sqrt{2}} + \left(1-\frac{1}{\sqrt{2}}\right)\right] = \frac{3+\sqrt{2}}{2}\\
C' &= \frac{1}{2}\left[2\left(1-\frac{1}{\sqrt{2}}\right) - \frac{1}{\sqrt{2}} + \left(1+\frac{1}{\sqrt{2}}\right)\right] = \frac{3-\sqrt{2}}{2}.
\end{aligned}$$

The transformed equation is $(3+\sqrt{2})u^2+(3-\sqrt{2})v^2=4$, which represents an ellipse with semi-major axis $2/\sqrt{3-\sqrt{2}}$ and semi-minor axis $2/\sqrt{3+\sqrt{2}}$. (We will discover another way to do a question like this in Section 13.3.)

EXERCISES 8.1

Find equations of the conics specified in Exercises 1–6.

1. ellipse with foci at $(0, \pm 2)$ and semi-major axis 3.

2. ellipse with foci at $(0, 1)$ and $(4, 1)$ and eccentricity $1/2$.

3. parabola with focus at $(2, 3)$ and vertex at $(2, 4)$.

4. parabola passing through the origin and having focus at $(0, -1)$ and axis along $y = -1$.

5. hyperbola with foci at $(0, \pm 2)$ and semi-transverse axis 1.

6. hyperbola with foci at $(\pm 5, 1)$ and asymptotes $x = \pm(y - 1)$.

In Exercises 7–15, identify and sketch the set of points in the plane satisfying the given equation. Specify the asymptotes of any hyperbolas.

7. $x^2 + y^2 + 2x = -1$

8. $x^2 + 4y^2 - 4y = 0$

9. $4x^2 + y^2 - 4y = 0$

10. $4x^2 - y^2 - 4y = 0$

11. $x^2 + 2x - y = 3$

12. $x + 2y + 2y^2 = 1$

13. $x^2 - 2y^2 + 3x + 4y = 2$

14. $9x^2 + 4y^2 - 18x + 8y = -13$

15. $9x^2 + 4y^2 - 18x + 8y = 23$

16. Identify and sketch the curve that is the graph of the equation $(x - y)^2 - (x + y)^2 = 1$.

17. Light rays in the xy-plane coming from the point $(3, 4)$ reflect in a parabola so that they form a beam parallel to the x-axis. The parabola passes through the origin. Find its equation. (There are two possible answers.)

18. Light rays in the xy-plane coming from the origin are reflected by an ellipse so that they converge at the point $(3, 0)$. Find all possible equations for the ellipse.

In Exercises 19–22, identify the conic and find its centre, principal axes, foci, and eccentricity. Specify the asymptotes of any hyperbolas.

19. $xy + x - y = 2$

20. $x^2 + 2xy + y^2 = 4x - 4y + 4$

21. $8x^2 + 12xy + 17y^2 = 20$

22. $x^2 - 4xy + 4y^2 + 2x + y = 0$

23. The *focus-directrix definition of a conic* defines a conic as a set of points P in the plane that satisfy the condition

$$\frac{\text{distance from } P \text{ to } F}{\text{distance from } P \text{ to } D} = \varepsilon,$$

where F is a fixed point, D a fixed straight line, and ε a fixed positive number. The conic is an ellipse, a parabola, or a hyperbola according to whether $\varepsilon < 1$, $\varepsilon = 1$, or $\varepsilon > 1$. Find the equation of the conic if F is the origin and D is the line $x = -p$.

Another parameter associated with conics is the **semi-latus rectum**, usually denoted ℓ. For a circle it is equal to the radius. For other conics it is half the length of the chord through a focus and perpendicular to the axis (for a parabola), the major axis (for an ellipse), or the transverse axis (for a hyperbola). That chord is called the **latus rectum** of the conic.

24. Show that the semi-latus rectum of the parabola is twice the distance from the vertex to the focus.

25. Show that the semi-latus rectum for an ellipse with semi-major axis a and semi-minor axis b is $\ell = b^2/a$.

26. Show that the formula in Exercise 25 also gives the semi-latus rectum of a hyperbola with semi-transverse axis a and semi-conjugate axis b.

27. Suppose a plane intersects a right-circular cone in an ellipse and that two spheres (one on each side of the plane) are inscribed between the cone and the plane so that each is tangent to the cone around a circle and is also tangent to the plane at a point. Show that the points where these two spheres touch the plane are the foci of the ellipse. *Hint:* All tangent lines drawn to a sphere from a given point outside the sphere are equal in length. The distance between the two circles in which the spheres intersect the cone, measured along generators of the cone (i.e., straight lines lying on the cone), is the same for all generators.

28. State and prove a result analogous to that in Exercise 27 but pertaining to a hyperbola.

29. Suppose a plane intersects a right-circular cone in a parabola with vertex at V. Suppose that a sphere is inscribed between the cone and the plane as in the previous exercises and is tangent to the plane of the parabola at point F. Show that the chord to the parabola through F which is perpendicular to FV has length equal to that of the latus rectum of the parabola. Therefore, F is the focus of the parabola.

8.2 Parametric Curves

Suppose that an object moves around in the xy-plane so that the coordinates of its position at any time t are continuous functions of the variable t:

$$x = f(t), \qquad y = g(t).$$

The path followed by the object is a curve $\mathcal{C}$ in the plane that is specified by the two equations above. We call these equations *parametric equations* of $\mathcal{C}$. A curve specified by a particular pair of parametric equations is called a *parametric curve*.

DEFINITION 4

Parametric curves

A **parametric curve** $\mathcal{C}$ in the plane consists of an ordered pair (f, g) of continuous functions each defined on the same interval I. The equations

$$x = f(t), \qquad y = g(t), \qquad \text{for } t \text{ in } I,$$

are called **parametric equations** of the curve $\mathcal{C}$. The independent variable t is called the **parameter**.

Note that the parametric curve $\mathcal{C}$ was *not* defined as a set of points in the plane, but rather as the ordered pair of functions whose range is that set of points. Different pairs of functions can give the same set of points in the plane, but we may still want to regard them as different parametric curves. Nevertheless, we will often refer to the set of points (the path traced out by (x, y) as t traverses I) as the curve $\mathcal{C}$. The axis (real line) of the parameter t is distinct from the coordinate axes of the plane of the curve. (See Figure 8.16.) We will usually denote the parameter by t; in many applications the parameter represents time, but this need not always be the case. Because f and g are assumed to be continuous, the curve $x = f(t)$, $y = g(t)$ has no breaks in it. A parametric curve has a *direction* (indicated, say, by arrowheads), namely, the direction corresponding to increasing values of the parameter t, as shown in Figure 8.16.

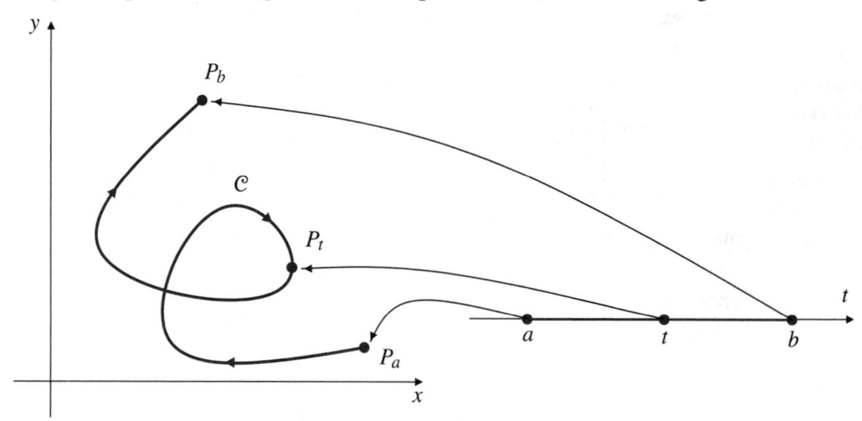

Figure 8.16 A parametric curve

EXAMPLE 1 Sketch and identify the parametric curve

$$x = t^2 - 1, \qquad y = t + 1 \qquad (-\infty < t < \infty).$$

Solution We could construct a table of values of x and y for various values of t, thus getting the coordinates of a number of points on a curve. However, for this example it is easier to *eliminate the parameter* from the pair of parametric equations, thus producing a single equation in x and y whose graph is the desired curve:

$$t = y - 1, \qquad x = t^2 - 1 = (y - 1)^2 - 1 = y^2 - 2y.$$

All points on the curve lie on the parabola $x = y^2 - 2y$. Since $y \to \pm\infty$ as $t \to \pm\infty$, the parametric curve is the whole parabola. (See Figure 8.17.)

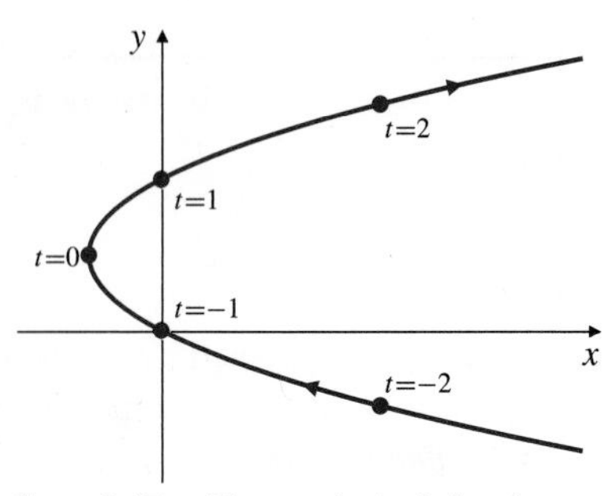

Figure 8.17 The parabola defined parametrically by $x = t^2 - 1$, $y = t + 1$, $(-\infty < t < \infty)$

Although the curve in Example 1 is more easily identified when the parameter is eliminated, there is a loss of information in going to the nonparametric form. Specifically, we lose the sense of the curve as the path of a moving point and hence also the direction of the curve. If the t in the parametric form denotes the time at which an object is at the point (x, y), the nonparametric equation $x = y^2 - 2y$ no longer tells us where the object is at any particular time t.

EXAMPLE 2 **(Parametric equations of a straight line)** The straight line passing through the two points $P_0 = (x_0, y_0)$ and $P_1 = (x_1, y_1)$ (see Figure 8.18) has parametric equations

$$\begin{cases} x = x_0 + t(x_1 - x_0) \\ y = y_0 + t(y_1 - y_0) \end{cases} \qquad (-\infty < t < \infty).$$

To see that these equations represent a straight line, note that

$$\frac{y - y_0}{x - x_0} = \frac{y_1 - y_0}{x_1 - x_0} = \text{constant} \qquad (\text{assuming } x_1 \neq x_0).$$

The point $P = (x, y)$ is at position P_0 when $t = 0$ and at P_1 when $t = 1$. If $t = 1/2$, then P is the midpoint between P_0 and P_1. Note that the line segment from P_0 to P_1 corresponds to values of t between 0 and 1.

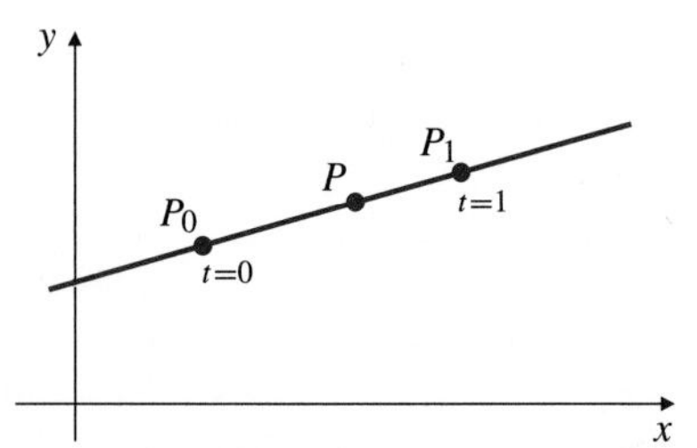

Figure 8.18 The straight line through P_0 and P_1

EXAMPLE 3 **(An arc of a circle)** Sketch and identify the curve $x = 3\cos t$, $y = 3\sin t$, $(0 \le t \le 3\pi/2)$.

Solution Since $x^2 + y^2 = 9\cos^2 t + 9\sin^2 t = 9$, all points on the curve lie on the circle $x^2 + y^2 = 9$. As t increases from 0 through $\pi/2$ and π to $3\pi/2$, the point (x, y) moves from $(3, 0)$ through $(0, 3)$ and $(-3, 0)$ to $(0, -3)$. The parametric curve is three-quarters of the circle. See Figure 8.19. The parameter t has geometric significance in this example. If P_t is the point on the curve corresponding to parameter value t, then t is the angle at the centre of the circle corresponding to the arc from the initial point to P_t.

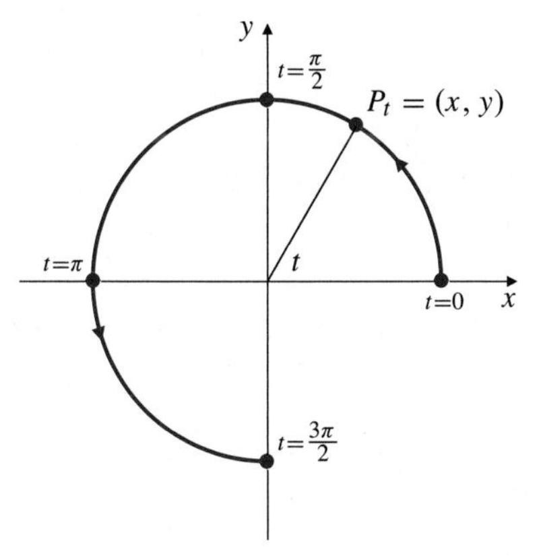

Figure 8.19 Three quarters of a circle

EXAMPLE 4 **(Parametric equations of an ellipse)** Sketch and identify the curve $x = a\cos t$, $y = b\sin t$, $(0 \le t \le 2\pi)$, where $a > b > 0$.

Solution Observe that

$$\frac{x^2}{a^2} + \frac{y^2}{b^2} = \cos^2 t + \sin^2 t = 1.$$

Therefore, the curve is all or part of an ellipse with major axis from $(-a, 0)$ to $(a, 0)$ and minor axis from $(0, -b)$ to $(0, b)$. As t increases from 0 to 2π, the point (x, y) moves counterclockwise around the ellipse starting from $(a, 0)$ and returning to the same point. Thus, the curve is the whole ellipse.

Figure 8.20(a) shows how the parameter t can be interpreted as an angle and how the points on the ellipse can be obtained using circles of radii a and b. Since the curve starts and ends at the same point, it is called a **closed curve**.

EXAMPLE 5 Sketch the parametric curve

$$x = t^3 - 3t, \qquad y = t^2, \qquad (-2 \le t \le 2).$$

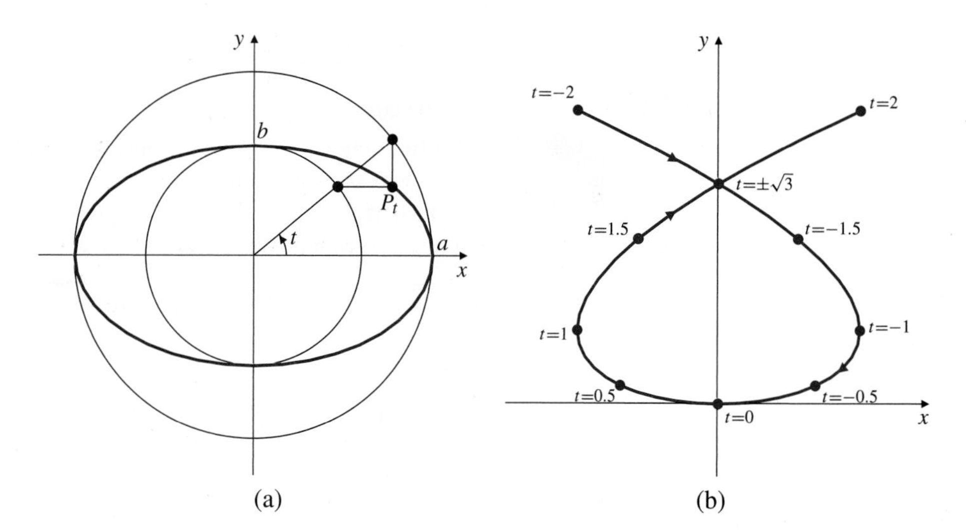

Figure 8.20

(a) An ellipse parametrized in terms of an angle and constructed with the help of two circles

(b) A self-intersecting parametric curve

Solution We could eliminate the parameter and obtain

$$x^2 = t^2(t^2 - 3)^2 = y(y - 3)^2,$$

but this doesn't help much since we do not recognize this curve from its Cartesian equation. Instead, let us calculate the coordinates of some points:

Table 2. Coordinates of some points on the curve of Example 5

t	-2	$-\frac{3}{2}$	-1	$-\frac{1}{2}$	0	$\frac{1}{2}$	1	$\frac{3}{2}$	2
x	-2	$\frac{9}{8}$	2	$\frac{11}{8}$	0	$-\frac{11}{8}$	-2	$-\frac{9}{8}$	2
y	4	$\frac{9}{4}$	1	$\frac{1}{4}$	0	$\frac{1}{4}$	1	$\frac{9}{4}$	4

Note that the curve is symmetric about the y-axis because x is an odd function of t and y is an even function of t. (At t and $-t$, x has opposite values but y has the same value.)

The curve intersects itself on the y-axis. (See Figure 8.20(b).) To find this self-intersection, set $x = 0$:

$$0 = x = t^3 - 3t = t(t - \sqrt{3})(t + \sqrt{3}).$$

For $t = 0$ the curve is at $(0, 0)$, but for $t = \pm\sqrt{3}$ the curve is at $(0, 3)$. The self-intersection occurs because the curve passes through the same point for two different values of the parameter.

Remark Here is how to get Maple to plot the parametric curve in the example above. Note the square brackets enclosing the two functions $t^3 - 3t$ and t^2, and the parameter interval, followed by the ranges of x and y for the plot.

```
> plot([t^3-3*t, t^2, t=-2..2], x=-3..3, y=-1..5);
```

General Plane Curves and Parametrizations

According to Definition 4, a parametric curve always involves a particular set of parametric equations; it is not just a set of points in the plane. When we are interested in considering a curve solely as a set of points (a *geometric object*), we need not be concerned with any particular pair of parametric equations representing that curve. In this case we call the curve simply a *plane curve*.

DEFINITION 5

Plane curves

A **plane curve** is a set of points (x, y) in the plane such that $x = f(t)$ and $y = g(t)$ for some t in an interval I, where f and g are continuous functions defined on I. Any such interval I and function pair (f, g) that generate the points of $\mathcal{C}$ is called a **parametrization** of $\mathcal{C}$.

Since a plane curve does not involve any specific parametrization, it has no specific direction.

EXAMPLE 6 The circle $x^2 + y^2 = 1$ is a plane curve. Each of the following is a possible parametrization of the circle:

(i) $x = \cos t,\ y = \sin t, \qquad (0 \le t \le 2\pi)$,
(ii) $x = \sin s^2,\ y = \cos s^2, \qquad (0 \le s \le \sqrt{2\pi})$,
(iii) $x = \cos(\pi u + 1),\ y = \sin(\pi u + 1), \qquad (-1 \le u \le 1)$,
(iv) $x = 1 - t^2,\ y = t\sqrt{2 - t^2}, \qquad (-\sqrt{2} \le t \le \sqrt{2})$.

To verify that any of these represents the circle, substitute the appropriate functions for x and y in the expression $x^2 + y^2$, and show that the result simplifies to the value 1. This shows that the parametric curve lies on the circle. Then examine the ranges of x and y as the parameter varies over its domain. For example, for (iv) we have

$$x^2 + y^2 = (1 - t^2)^2 + (t\sqrt{2 - t^2})^2 = 1 - 2t^2 + t^4 + 2t^2 - t^4 = 1,$$

and (x, y) moves from $(-1, 0)$ through $(0, -1)$ to $(1, 0)$ as t increases from $-\sqrt{2}$ through -1 to 0, and then continues on through $(0, 1)$ back to $(-1, 0)$ as t continues to increase from 0 through 1 to $\sqrt{2}$.

There are, of course, infinitely many other possible parametrizations of this curve.

EXAMPLE 7 If f is a continuous function on an interval I, then the graph of f is a plane curve. One obvious parametrization of this curve is

$$x = t, \qquad y = f(t), \qquad (t \text{ in } I).$$

Some Interesting Plane Curves

We complete this section by parametrizing two curves that arise in the physical world.

EXAMPLE 8 **(A cycloid)** If a circle rolls without slipping along a straight line, find the path followed by a point fixed on the circle. This path is called a **cycloid**.

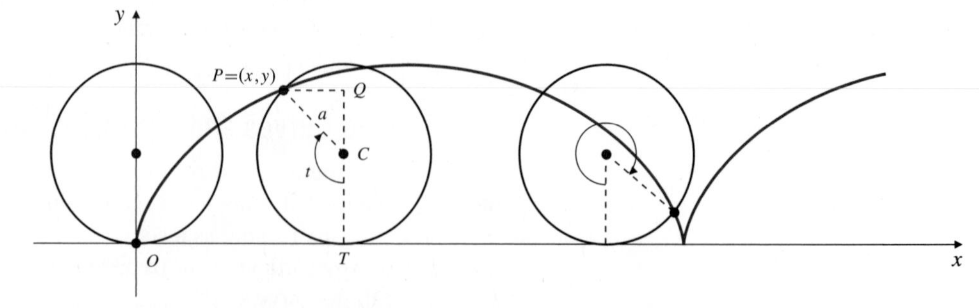

Figure 8.21 Each arch of the cycloid is traced out by P as the wheel rolls through one complete revolution

The brachistochrone and tautochrone problems
Suppose a wire is bent into a curve from point A to a lower point B and a bead can slide without friction along the wire. If the bead is released at A, it will fall toward B. What curve should be used to minimize the time it takes to fall from A to B? This problem, known as the *brachistochrone* (Greek for "shortest time") problem, has as its solution part of an upside-down arch of a cycloid. Moreover, it takes the same amount of time for the bead to slide from any point on the curve to the lowest point B, making the cycloid the solution of the *tautochrone* ("equal time") problem as well. We will examine these matters further in the Challenging Exercises at the end of Chapter 11.

Solution Suppose that the line on which the circle rolls is the x-axis, that the circle has radius a and lies above the line, and that the point whose motion we follow is originally at the origin O. See Figure 8.21. After the circle has rolled through an angle t, it is tangent to the line at T, and the point whose path we are trying to find has moved to position P, as shown in the figure. Since no slipping occurs,

$$\text{segment } OT = \text{arc } PT = at.$$

Let PQ be perpendicular to TC, as shown in the figure. If P has coordinates (x, y), then

$$x = OT - PQ = at - a\sin(\pi - t) = at - a\sin t,$$
$$y = TC + CQ = a + a\cos(\pi - t) = a - a\cos t.$$

The parametric equations of the cycloid are, therefore,

$$x = a(t - \sin t), \qquad y = a(1 - \cos t).$$

Observe that the cycloid has a cusp at the points where it returns to the x-axis, that is, at points corresponding to $t = 2n\pi$, where n is an integer. Even though the functions x and y are everywhere differentiable functions of t, the curve is not smooth everywhere. We shall consider such matters in the next section.

EXAMPLE 9 **(An involute of a circle)** A string is wound around a fixed circle. One end is unwound in such a way that the part of the string not lying on the circle is extended in a straight line. The curve followed by this free end of the string is called an **involute** of the circle. (The involute of any curve is the path traced out by the end of the curve as the curve is straightened out beginning at that end.)

Suppose the circle has equation $x^2 + y^2 = a^2$, and suppose the end of the string being unwound starts at the point $A = (a, 0)$. At some subsequent time during the unwinding let P be the position of the end of the string, and let T be the point where the string leaves the circle. The line PT must be tangent to the circle at T.

We parametrize the path of P in terms of the angle AOT, which we denote by t. Let points R on OA and S on TR be as shown in Figure 8.22. TR is perpendicular to OA and to PS. Note that

$$OR = OT\cos t = a\cos t, \qquad RT = OT\sin t = a\sin t.$$

Since angle OTP is 90°, we have angle $STP = t$. Since $PT = \text{arc } AT = at$ (because the string does not stretch or slip on the circle), we have

$$SP = TP\sin t = at\sin t, \qquad ST = TP\cos t = at\cos t.$$

If P has coordinates (x, y), then $x = OR + SP$, and $y = RT - ST$:

$$x = a\cos t + at\sin t, \qquad y = a\sin t - at\cos t, \qquad (t \ge 0).$$

These are parametric equations of the involute.

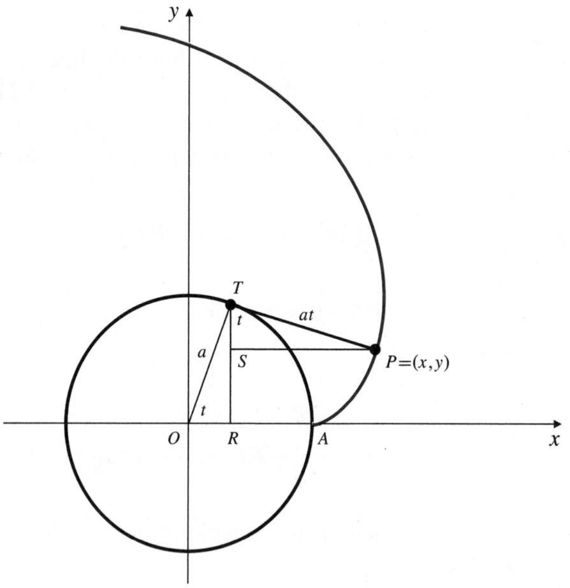

Figure 8.22 An involute of a circle

EXERCISES 8.2

In Exercises 1–10, sketch the given parametric curve, showing its direction with an arrow. Eliminate the parameter to give a Cartesian equation in x and y whose graph contains the parametric curve.

1. $x = 1 + 2t,\ y = t^2,\ (-\infty < t < \infty)$

2. $x = 2 - t,\ y = t + 1,\ (0 \le t < \infty)$

3. $x = \dfrac{1}{t},\ y = t - 1,\ (0 < t < 4)$

4. $x = \dfrac{1}{1+t^2},\ y = \dfrac{t}{1+t^2},\ (-\infty < t < \infty)$

5. $x = 3\sin 2t,\ y = 3\cos 2t,\ \left(0 \le t \le \dfrac{\pi}{3}\right)$

6. $x = a\sec t,\ y = b\tan t,\ \left(-\dfrac{\pi}{2} < t < \dfrac{\pi}{2}\right)$

7. $x = 3\sin \pi t,\ y = 4\cos \pi t,\ (-1 \le t \le 1)$

8. $x = \cos\sin s,\ y = \sin\sin s,\ (-\infty < s < \infty)$

9. $x = \cos^3 t,\ y = \sin^3 t,\ (0 \le t \le 2\pi)$

10. $x = 1 - \sqrt{4 - t^2},\ y = 2 + t,\ (-2 \le t \le 2)$

11. Describe the parametric curve $x = \cosh t$, $y = \sinh t$, and find its Cartesian equation.

12. Describe the parametric curve $x = 2 - 3\cosh t$, $y = -1 + 2\sinh t$.

13. Describe the curve $x = t\cos t$, $y = t\sin t$, $(0 \le t \le 4\pi)$.

14. Show that each of the following sets of parametric equations represents a different arc of the parabola with equation $2(x + y) = 1 + (x - y)^2$.

(a) $x = \cos^4 t,\ y = \sin^4 t$

(b) $x = \sec^4 t,\ y = \tan^4 t$

(c) $x = \tan^4 t,\ y = \sec^4 t$

15. Find a parametrization of the parabola $y = x^2$ using as parameter the slope of the tangent line at the general point.

16. Find a parametrization of the circle $x^2 + y^2 = R^2$ using as parameter the slope m of the line joining the general point to the point $(R, 0)$. Does the parametrization fail to give any point on the circle?

17. A circle of radius a is centred at the origin O. T is a point on the circle such that OT makes angle t with the positive x-axis. The tangent to the circle at T meets the x-axis at X. The point $P = (x, y)$ is at the intersection of the vertical line through X and the horizontal line through T. Find, in terms of the parameter t, parametric equations for the curve $\mathcal{C}$ traced out by P as T moves around the circle. Also, eliminate t and find an equation for $\mathcal{C}$ in x and y. Sketch $\mathcal{C}$.

18. Repeat Exercise 17 with the following modification: OT meets a second circle of radius b centred at O at the point Y. $P = (x, y)$ is at the intersection of the vertical line through X and the horizontal line through Y.

19. **(The folium of Descartes)** Eliminate the parameter from the parametric equations

$$x = \frac{3t}{1+t^3}, \qquad y = \frac{3t^2}{1+t^3}, \qquad (t \ne -1),$$

and hence find an ordinary equation in x and y for this curve. The parameter t can be interpreted as the slope of the line joining the general point (x, y) to the origin. Sketch the curve and show that the line $x + y = -1$ is an asymptote.

20. **(A prolate cycloid)** A railroad wheel has a flange extending below the level of the track on which the wheel rolls. If the radius of the wheel is a and that of the flange is $b > a$, find parametric equations of the path of a point P at the circumference of the flange as the wheel rolls along the track. (Note that for a portion of each revolution of the wheel, P is moving backward.) Try to sketch the graph of this prolate cycloid.

21. **(Hypocycloids)** If a circle of radius b rolls, without slipping, around the inside of a fixed circle of radius $a > b$, a point on the circumference of the rolling circle traces a curve called a hypocycloid. If the fixed circle is centred at the origin and the point tracing the curve starts at $(a, 0)$, show that the hypocycloid has parametric equations

$$x = (a - b)\cos t + b\cos\left(\frac{a-b}{b}t\right),$$
$$y = (a - b)\sin t - b\sin\left(\frac{a-b}{b}t\right),$$

where t is the angle between the positive x-axis and the line from the origin to the point at which the rolling circle touches the fixed circle.

If $a = 2$ and $b = 1$, show that the hypocycloid becomes a straight line segment.

If $a = 4$ and $b = 1$, show that the parametric equations of the hypocycloid simplify to $x = 4\cos^3 t$, $y = 4\sin^3 t$. This curve is called a hypocycloid of four cusps or an **astroid**. (See Figure 8.23.) It has Cartesian equation $x^{2/3} + y^{2/3} = 4^{2/3}$.

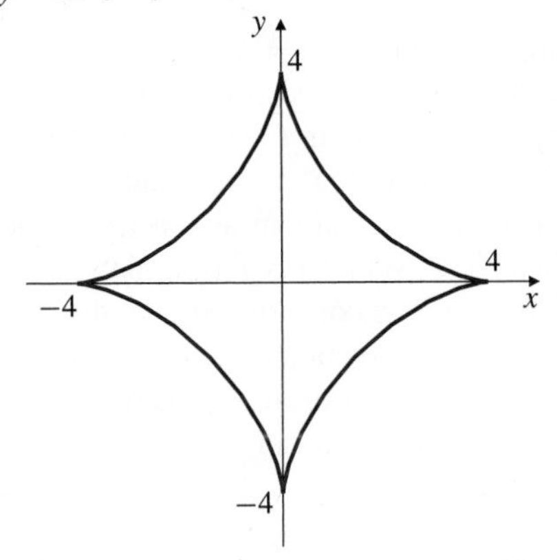

Figure 8.23 The astroid $x^{2/3} + y^{2/3} = 4^{2/3}$

Hypocycloids resemble the curves produced by a popular children's toy called Spirograph, but Spirograph curves result from following a point inside the disc of the rolling circle rather than on its circumference, and they therefore do not have sharp cusps.

22. (The witch of Agnesi)

(a) Show that the curve traced out by the point P constructed from a circle as shown in Figure 8.24 has parametric equations $x = \tan t$, $y = \cos^2 t$ in terms of the angle t shown. (*Hint:* You will need to make extensive use of similar triangles.)

(b) Use a trigonometric identity to eliminate t from the parametric equations, and hence find an ordinary Cartesian equation for the curve.

This curve is named for the Italian mathematician Maria Agnesi (1718–1799), one of the foremost women scholars of the eighteenth century and author of an important calculus text. The term *witch* is due to a mistranslation of the Italian word *versiera* ("turning curve"), which she used to describe the curve. The word is similar to *avversiera* ("wife of the devil" or "witch").

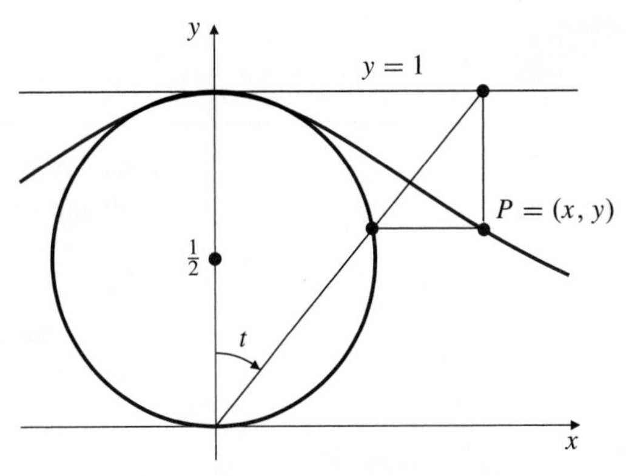

Figure 8.24 The witch of Agnesi

In Exercises 23–26, obtain a graph of the curve $x = \sin(mt)$, $y = \sin(nt)$ for the given values of m and n. Such curves are called **Lissajous figures.** They arise in the analysis of electrical signals using an oscilloscope. A signal of fixed but unknown frequency is applied to the vertical input, and a control signal is applied to the horizontal input. The horizontal frequency is varied until a stable Lissajous figure is observed. The (known) frequency of the control signal and the shape of the figure then determine the unknown frequency.

23. $m = 1, \quad n = 2$

24. $m = 1, \quad n = 3$

25. $m = 2, \quad n = 3$

26. $m = 2, \quad n = 5$

27. (Epicycloids) Use a graphing calculator or computer graphing program to investigate the behaviour of curves with equations of the form

$$x = \left(1 + \frac{1}{n}\right)\cos t - \frac{1}{n}\cos(nt)$$
$$y = \left(1 + \frac{1}{n}\right)\sin t - \frac{1}{n}\sin(nt)$$

for various integer and fractional values of $n \geq 3$. Can you formulate any principles governing the behaviour of such curves?

28. (More hypocycloids) Use a graphing calculator or computer graphing program to investigate the behaviour of curves with equations of the form

$$x = \left(1 + \frac{1}{n}\right)\cos t + \frac{1}{n}\cos((n-1)t)$$
$$y = \left(1 + \frac{1}{n}\right)\sin t - \frac{1}{n}\sin((n-1)t)$$

for various integer and fractional values of $n \geq 3$. Can you formulate any principles governing the behaviour of these curves?

8.3 Smooth Parametric Curves and Their Slopes

We say that a plane curve is *smooth* if it has a tangent line at each point P and this tangent turns in a continuous way as P moves along the curve. (That is, the angle between the tangent line at P and some fixed line, the x-axis say, is a continuous function of the position of P.)

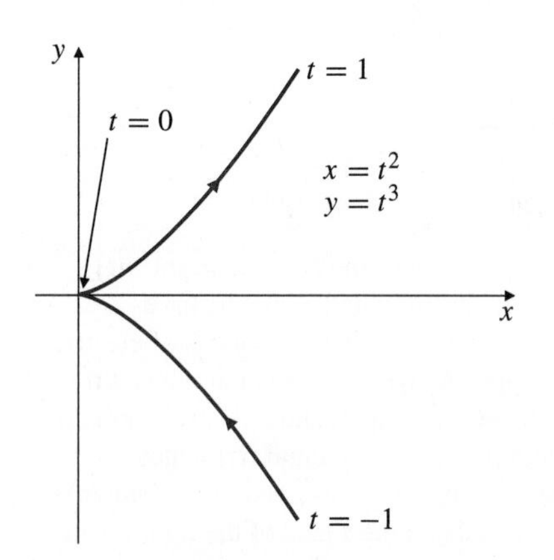

Figure 8.25 This curve is not smooth at the origin but has a cusp there

If the curve $\mathcal{C}$ is the graph of function f, then $\mathcal{C}$ is certainly smooth on any interval where the derivative $f'(x)$ exists and is a continuous function of x. It may also be smooth on intervals containing isolated singular points; for example, the curve $y = x^{1/3}$ is smooth everywhere even though dy/dx does not exist at $x = 0$.

For parametric curves $x = f(t)$, $y = g(t)$, the situation is more complicated. Even if f and g have continuous derivatives everywhere, such curves may fail to be smooth at certain points, specifically points where $f'(t) = g'(t) = 0$.

EXAMPLE 1 Consider the parametric curve $x = f(t) = t^2$, $y = g(t) = t^3$. Eliminating t leads to the Cartesian equation $y^2 = x^3$ or $x = y^{2/3}$, which is not smooth at the origin even though $f'(t) = 2t$ and $g'(t) = 3t^2$ are continuous for all t. (See Figure 8.25.) Observe that both f' and g' vanish at $t = 0$: $f'(0) = g'(0) = 0$. If we regard the parametric equations as specifying the position at time t of a moving point P, then the horizontal velocity is $f'(t)$ and the vertical velocity is $g'(t)$. Both velocities are 0 at $t = 0$, so P has come to a stop at that instant. When it starts moving again, it need not move in the direction it was going before it stopped. The cycloid of Example 8 of Section 8.2 is another example where a parametric curve is not smooth at points where dx/dt and dy/dt both vanish.

The Slope of a Parametric Curve

The following theorem confirms that a parametric curve is smooth at points where the derivatives of its coordinate functions are continuous and not both zero.

THEOREM 1

Let $\mathcal{C}$ be the parametric curve $x = f(t)$, $y = g(t)$, where $f'(t)$ and $g'(t)$ are continuous on an interval I. If $f'(t) \neq 0$ on I, then $\mathcal{C}$ is smooth and has at each t a tangent line with slope

$$\frac{dy}{dx} = \frac{g'(t)}{f'(t)}.$$

If $g'(t) \neq 0$ on I, then $\mathcal{C}$ is smooth and has at each t a normal line with slope

$$-\frac{dx}{dy} = -\frac{f'(t)}{g'(t)}.$$

Thus, $\mathcal{C}$ is smooth except possibly at points where $f'(t)$ and $g'(t)$ are both 0.

PROOF If $f'(t) \neq 0$ on I, then f is either increasing or decreasing on I and so is one-to-one and invertible. The part of $\mathcal{C}$ corresponding to values of t in I has ordinary equation $y = g\big(f^{-1}(x)\big)$ and hence slope

$$\frac{dy}{dx} = g'\big(f^{-1}(x)\big)\frac{d}{dx}f^{-1}(x) = \frac{g'\big(f^{-1}(x)\big)}{f'\big(f^{-1}(x)\big)} = \frac{g'(t)}{f'(t)}.$$

We have used here the formula

$$\frac{d}{dx}f^{-1}(x) = \frac{1}{f'\big(f^{-1}(x)\big)}$$

for the derivative of an inverse function obtained in Section 3.1. This slope is a continuous function of t, so the tangent to $\mathcal{C}$ turns continuously for t in I. The proof for $g'(t) \neq 0$ is similar. In this case the slope of the normal is a continuous function of t, so the normal turns continuously. Therefore so does the tangent.

If f' and g' are continuous, and both vanish at some point t_0, then the curve $x = f(t)$, $y = g(t)$ *may or may not* be smooth around t_0. Example 1 was an example of a curve that was not smooth at such a point.

EXAMPLE 2 The curve with parametrization $x = t^3$, $y = t^6$ is just the parabola $y = x^2$, so it is smooth everywhere, although $dx/dt = 3t^2$ and $dy/dt = 6t^5$ both vanish at $t = 0$.

Tangents and normals to parametric curves

If f' and g' are continuous and not both 0 at t_0, then the parametric equations

$$\begin{cases} x = f(t_0) + f'(t_0)(t - t_0) \\ y = g(t_0) + g'(t_0)(t - t_0) \end{cases} \qquad (-\infty < t < \infty)$$

represent the tangent line to the parametric curve $x = f(t),\ y = g(t)$ at the point $\big(f(t_0), g(t_0)\big)$. The normal line there has parametric equations

$$\begin{cases} x = f(t_0) + g'(t_0)(t - t_0) \\ y = g(t_0) - f'(t_0)(t - t_0) \end{cases} \qquad (-\infty < t < \infty).$$

Both lines pass through $\big(f(t_0), g(t_0)\big)$ when $t = t_0$.

EXAMPLE 3 Find equations of the tangent and normal lines to the parametric curve $x = t^2 - t,\ y = t^2 + t$ at the point where $t = 2$.

Solution At $t = 2$ we have $x = 2$, $y = 6$ and

$$\frac{dx}{dt} = 2t - 1 = 3, \qquad \frac{dy}{dt} = 2t + 1 = 5.$$

Hence, the tangent and the normal lines have parametric equations

$$\text{Tangent:} \begin{cases} x = 2 + 3(t - 2) = 3t - 4 \\ y = 6 + 5(t - 2) = 5t - 4. \end{cases}$$

$$\text{Normal:} \begin{cases} x = 2 + 5(t - 2) = 5t - 8 \\ y = 6 - 3(t - 2) = -3t + 12. \end{cases}$$

The concavity of a parametric curve can be determined using the second derivatives of the parametric equations. The procedure is just to calculate d^2y/dx^2 using the Chain Rule:

$$\begin{aligned} \frac{d^2y}{dx^2} = \frac{d}{dx}\frac{dy}{dx} = \frac{d}{dx}\frac{g'(t)}{f'(t)} &= \frac{d}{dt}\left(\frac{g'(t)}{f'(t)}\right)\frac{dt}{dx} \\ &= \frac{f'(t)g''(t) - g'(t)f''(t)}{(f'(t))^2}\,\frac{1}{f'(t)}. \end{aligned}$$

Concavity of a parametric curve

On an interval where $f'(t) \neq 0$, the parametric curve $x = f(t)$, $y = g(t)$ has concavity determined by

$$\frac{d^2y}{dx^2} = \frac{f'(t)g''(t) - g'(t)f''(t)}{(f'(t))^3}.$$

Sketching Parametric Curves

As in the case of graphs of functions, derivatives provide useful information about the shape of a parametric curve. At points where $dy/dt = 0$ but $dx/dt \neq 0$, the tangent is horizontal; at points where $dx/dt = 0$ but $dy/dt \neq 0$, the tangent is vertical. For points where $dx/dt = dy/dt = 0$, anything can happen; it is wise to calculate left- and right-hand limits of the slope dy/dx as the parameter t approaches one of these points. Concavity can be determined using the formula obtained above. We illustrate these ideas by reconsidering a parametric curve encountered in the previous section.

EXAMPLE 4 Use slope and concavity information to sketch the graph of the parametric curve

$$x = f(t) = t^3 - 3t, \qquad y = g(t) = t^2, \qquad (-2 \le t \le 2)$$

previously encountered in Example 5 of Section 8.2.

Solution We have

$$f'(t) = 3(t^2 - 1) = 3(t - 1)(t + 1), \qquad g'(t) = 2t.$$

The curve has a horizontal tangent at $t = 0$, that is, at $(0, 0)$, and vertical tangents at $t = \pm 1$, that is, at $(2, 1)$ and $(-2, 1)$. Directional information for the curve between these points is summarized in the following chart.

t	-2		-1		0		1		2
$f'(t)$		$+$	0	$-$	$-$	$-$	0	$+$	
$g'(t)$		$-$	$-$	$-$	0	$+$	$+$	$+$	
x		$\rightarrow$	$\cdot$	$\leftarrow$	$\leftarrow$	$\leftarrow$	$\cdot$	$\rightarrow$	
y		$\downarrow$	$\downarrow$	$\downarrow$	$\cdot$	$\uparrow$	$\uparrow$	$\uparrow$	
curve		$\searrow$	$\downarrow$	$\swarrow$	$\leftarrow$	$\nwarrow$	$\uparrow$	$\nearrow$	

For concavity we calculate the second derivative d^2y/dx^2 by the formula obtained above. Since $f''(t) = 6t$ and $g''(t) = 2$, we have

$$\begin{aligned} \frac{d^2y}{dx^2} &= \frac{f'(t)g''(t) - g'(t)f''(t)}{(f'(t))^3} \\ &= \frac{3(t^2 - 1)(2) - 2t(6t)}{[3(t^2 - 1)]^3} = -\frac{2}{9}\frac{t^2 + 1}{(t^2 - 1)^3}, \end{aligned}$$

which is never zero but which fails to be defined at $t = \pm 1$. Evidently the curve is concave upward for $-1 < t < 1$ and concave downward elsewhere. The curve is sketched in Figure 8.26.

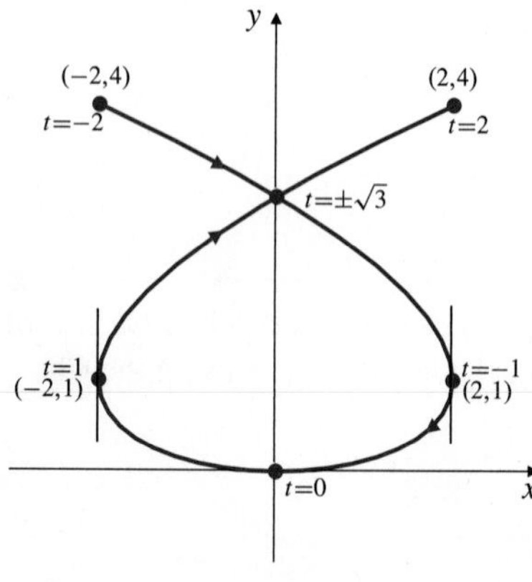

Figure 8.26 The curve $x = t^3 - 3t$, $y = t^2$, $(-2 \le t \le 2)$

EXERCISES 8.3

In Exercises 1–8, find the coordinates of the points at which the given parametric curve has (a) a horizontal tangent and (b) a vertical tangent.

1. $x = t^2 + 1,\ y = 2t - 4$ **2.** $x = t^2 - 2t,\ y = t^2 + 2t$

3. $x = t^2 - 2t,\ y = t^3 - 12t$

4. $x = t^3 - 3t,\ y = 2t^3 + 3t^2$

5. $x = te^{-t^2/2},\ y = e^{-t^2}$

6. $x = \sin t,\ y = \sin t - t\cos t$

7. $x = \sin 2t,\ y = \sin t$ **8.** $x = \dfrac{3t}{1+t^3},\ y = \dfrac{3t^2}{1+t^3}$

Find the slopes of the curves in Exercises 9–12 at the points indicated.

9. $x = t^3 + t,\ y = 1 - t^3,$ at $t = 1$

10. $x = t^4 - t^2,\ y = t^3 + 2t,$ at $t = -1$

11. $x = \cos 2t,\ y = \sin t,$ at $t = \pi/6$

12. $x = e^{2t},\ y = te^{2t},$ at $t = -2$

Find parametric equations of the tangents to the curves in Exercises 13–14 at the indicated points.

13. $x = t^3 - 2t,\ y = t + t^3,$ at $t = 1$

14. $x = t - \cos t,\ y = 1 - \sin t,$ at $t = \pi/4$

15. Show that the curve $x = t^3 - t$, $y = t^2$ has two different tangent lines at the point $(0, 1)$ and find their slopes.

16. Find the slopes of two lines that are tangent to $x = \sin t$, $y = \sin 2t$ at the origin.

Where, if anywhere, do the curves in Exercises 17–20 fail to be smooth?

17. $x = t^3,\ y = t^2$

18. $x = (t-1)^4,\ y = (t-1)^3$

19. $x = t\sin t,\ y = t^3$ **20.** $x = t^3,\ y = t - \sin t$

In Exercises 21–25, sketch the graphs of the given parametric curves, making use of information from the first two derivatives. Unless otherwise stated, the parameter interval for each curve is the whole real line.

21. $x = t^2 - 2t,\ y = t^2 - 4t$ **22.** $x = t^3,\ y = 3t^2 - 1$

23. $x = t^3 - 3t,\ y = \dfrac{2}{1+t^2}$

24. $x = t^3 - 3t - 2,\ y = t^2 - t - 2$

25. $x = \cos t + t\sin t,\ y = \sin t - t\cos t,\ (t \ge 0)$. (See Example 9 of Section 8.2.)

8.4 Arc Lengths and Areas for Parametric Curves

In this section we look at the problems of finding lengths of curves defined parametrically, areas of surfaces of revolution obtained by rotating parametric curves, and areas of plane regions bounded by parametric curves.

Arc Lengths and Surface Areas

Let $\mathcal{C}$ be a smooth parametric curve with equations

$$x = f(t), \qquad y = g(t), \qquad (a \le t \le b).$$

(We assume that $f'(t)$ and $g'(t)$ are continuous on the interval $[a, b]$ and are never both zero.) From the differential triangle with legs dx and dy and hypotenuse ds (see Figure 8.27), we obtain $(ds)^2 = (dx)^2 + (dy)^2$, so we have

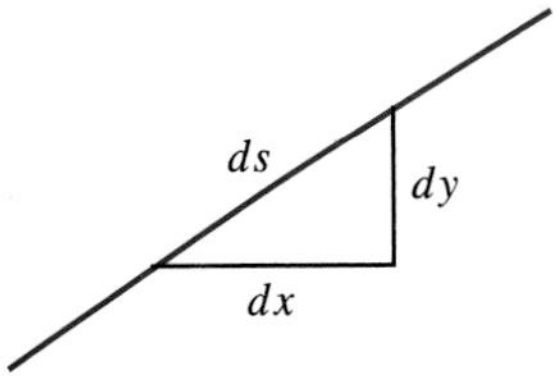

The arc length element for a parametric curve

$$ds = \frac{ds}{dt}\,dt = \sqrt{\left(\frac{ds}{dt}\right)^2}\,dt = \sqrt{\left(\frac{dx}{dt}\right)^2 + \left(\frac{dy}{dt}\right)^2}\,dt$$

Figure 8.27 A differential triangle

The length of the curve $\mathcal{C}$ is given by

$$s = \int_{t=a}^{t=b} ds = \int_a^b \sqrt{\left(\frac{dx}{dt}\right)^2 + \left(\frac{dy}{dt}\right)^2}\,dt.$$

EXAMPLE 1 Find the length of the parametric curve

$$x = e^t\cos t, \qquad y = e^t\sin t, \qquad (0 \le t \le 2).$$

Solution We have

$$\frac{dx}{dt} = e^t(\cos t - \sin t), \qquad \frac{dy}{dt} = e^t(\sin t + \cos t).$$

Squaring these formulas, adding and simplifying, we get

$$\begin{aligned}\left(\frac{ds}{dt}\right)^2 &= e^{2t}(\cos t - \sin t)^2 + e^{2t}(\sin t + \cos t)^2\\ &= e^{2t}\left(\cos^2 t - 2\cos t\sin t + \sin^2 t + \sin^2 t + 2\sin t\cos t + \cos^2 t\right)\\ &= 2e^{2t}.\end{aligned}$$

The length of the curve is, therefore,

$$s = \int_0^2 \sqrt{2e^{2t}}\,dt = \sqrt{2}\int_0^2 e^t\,dt = \sqrt{2}\,(e^2 - 1) \text{ units.}$$

Parametric curves can be rotated around various axes to generate surfaces of revolution. The areas of these surfaces can be found by the same procedure used for graphs of functions, with the appropriate version of ds. If the curve

$$x = f(t), \qquad y = g(t), \qquad (a \le t \le b)$$

is rotated about the x-axis, the area S of the surface so generated is given by

$$S = 2\pi\int_{t=a}^{t=b} |y|\,ds = 2\pi\int_a^b |g(t)|\sqrt{(f'(t))^2 + (g'(t))^2}\,dt.$$

If the rotation is about the y-axis, then the area is

$$S = 2\pi\int_{t=a}^{t=b} |x|\,ds = 2\pi\int_a^b |f(t)|\sqrt{(f'(t))^2 + (g'(t))^2}\,dt.$$

EXAMPLE 2 Find the area of the surface of revolution obtained by rotating the astroid curve $x = a\cos^3 t$, $y = a\sin^3 t$, (where $a > 0$) about the x-axis.

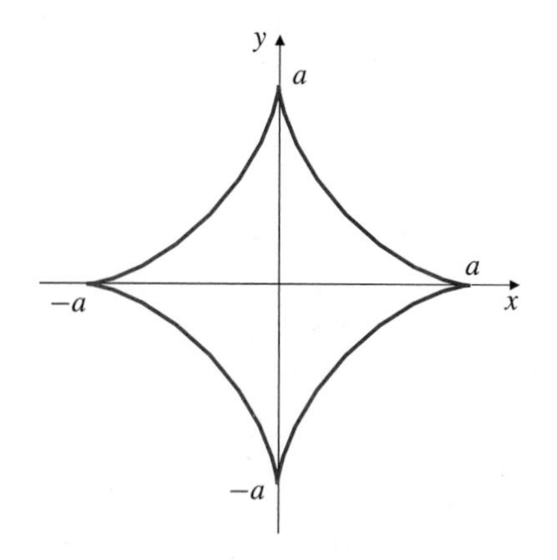

Figure 8.28 An astroid

Solution The curve is symmetric about both coordinate axes. (See Figure 8.28.) The entire surface will be generated by rotating the upper half of the curve; in fact, we need only rotate the first quadrant part and multiply by 2. The first quadrant part of the curve corresponds to $0 \le t \le \pi/2$. We have

$$\frac{dx}{dt} = -3a\cos^2 t\sin t, \qquad \frac{dy}{dt} = 3a\sin^2 t\cos t.$$

Accordingly, the arc length element is

$$\begin{aligned}ds &= \sqrt{9a^2\cos^4 t\sin^2 t + 9a^2\sin^4 t\cos^2 t}\,dt\\ &= 3a\cos t\sin t\sqrt{\cos^2 t + \sin^2 t}\,dt\\ &= 3a\cos t\sin t\,dt.\end{aligned}$$

Therefore, the required surface area is

$$\begin{aligned}S &= 2\times 2\pi\int_0^{\pi/2} a\sin^3 t\,3a\cos t\sin t\,dt\\ &= 12\pi a^2\int_0^{\pi/2}\sin^4 t\cos t\,dt \qquad \begin{array}{l}\text{Let } u = \sin t,\\ du = \cos t\,dt\end{array}\\ &= 12\pi a^2\int_0^1 u^4\,du = \frac{12\pi a^2}{5} \text{ square units.}\end{aligned}$$

Areas Bounded by Parametric Curves

Consider the parametric curve $\mathcal{C}$ with equations $x = f(t)$, $y = g(t)$, $(a \le t \le b)$, where f is differentiable and g is continuous on $[a, b]$. For the moment, let us also assume that $f'(t) \ge 0$ and $g(t) \ge 0$ on $[a, b]$, so $\mathcal{C}$ has no points below the x-axis and is traversed from left to right as t increases from a to b.

The region under $\mathcal{C}$ and above the x-axis has area element given by $dA = y\,dx = g(t)f'(t)\,dt$, so its area (see Figure 8.29) is

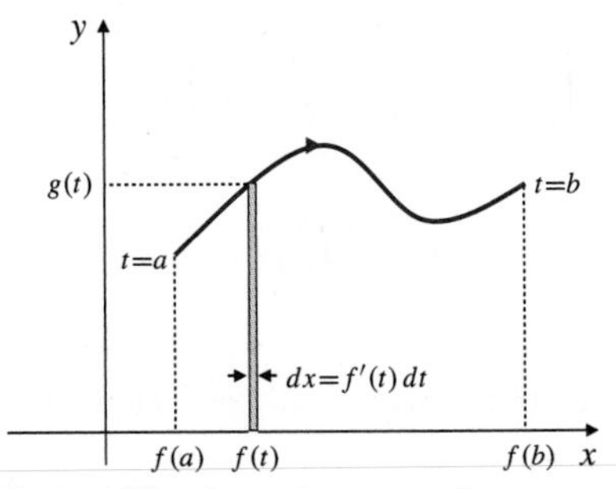

Figure 8.29 Area element under a parametric curve

$$A = \int_a^b g(t)f'(t)\,dt.$$

Similar arguments can be given for three other cases:

If $f'(t) \ge 0$ and $g(t) \le 0$ on $[a, b]$, then $A = -\int_a^b g(t)f'(t)\,dt$,

If $f'(t) \le 0$ and $g(t) \ge 0$ on $[a, b]$, then $A = -\int_a^b g(t)f'(t)\,dt$,

If $f'(t) \le 0$ and $g(t) \le 0$ on $[a, b]$, then $A = \int_a^b g(t)f'(t)\,dt$,

where A is the (positive) area bounded by $\mathcal{C}$, the x-axis, and the vertical lines $x = f(a)$ and $x = f(b)$. Combining these results we can see that

$$\int_a^b g(t)f'(t)\,dt = A_1 - A_2,$$

where A_1 is the area lying vertically between $\mathcal{C}$ and that part of the x-axis consisting of points $x = f(t)$ such that $g(t)f'(t) \ge 0$, and A_2 is a similar area corresponding to points where $g(t)f'(t) < 0$. This formula is valid for arbitrary continuous g and differentiable f. See Figure 8.30 for generic examples. In particular, if $\mathcal{C}$ is a non–self-intersecting closed curve, then the area of the region bounded by $\mathcal{C}$ is given by

$$A = \int_a^b g(t)f'(t)\,dt \qquad \text{if } \mathcal{C} \text{ is traversed clockwise as } t \text{ increases,}$$

$$A = -\int_a^b g(t)f'(t)\,dt \qquad \text{if } \mathcal{C} \text{ is traversed counterclockwise,}$$

both of which are illustrated in Figure 8.31.

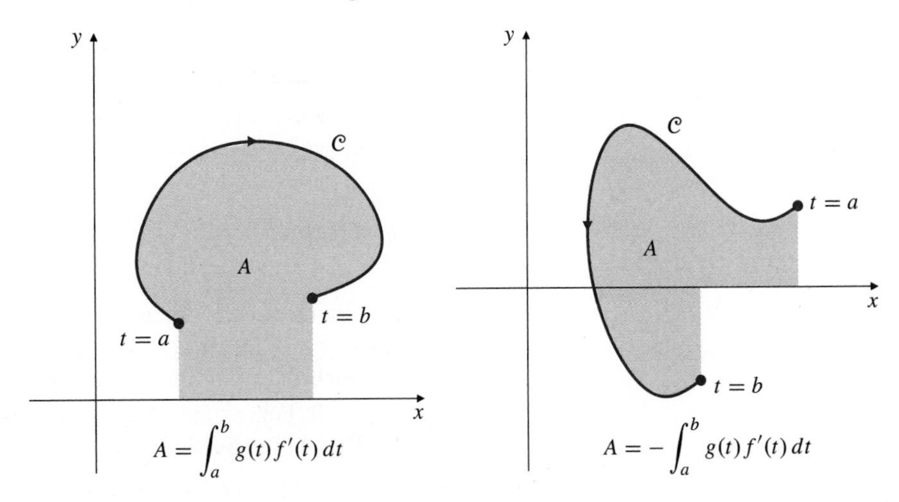

Figure 8.30 Areas defined by parametric curves

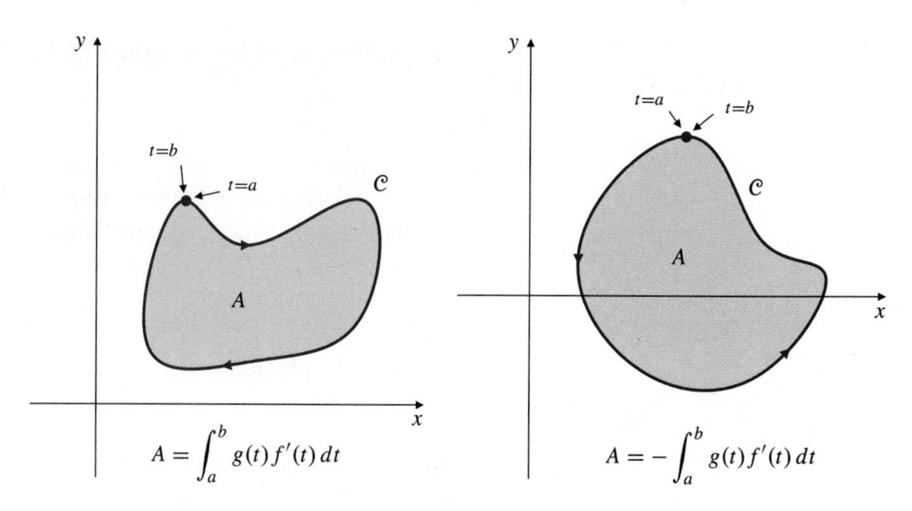

Figure 8.31 Areas bounded by closed parametric curves

EXAMPLE 3 Find the area bounded by the ellipse $x = a \cos s$, $y = b \sin s$, $(0 \le s \le 2\pi)$.

Solution This ellipse is traversed counterclockwise. (See Example 4 in Section 8.2.) The area enclosed is

$$\begin{aligned}
A &= -\int_0^{2\pi} b \sin s(-a \sin s)\, ds \\
&= \frac{ab}{2} \int_0^{2\pi} (1 - \cos 2s)\, ds \\
&= \frac{ab}{2} s \bigg|_0^{2\pi} - \frac{ab}{4} \sin 2s \bigg|_0^{2\pi} = \pi ab \text{ square units.}
\end{aligned}$$

EXAMPLE 4 Find the area above the x-axis and under one arch of the cycloid $x = at - a \sin t$, $y = a - a \cos t$.

Solution Part of the cycloid is shown in Figure 8.21 in Section 8.2. One arch corresponds to the parameter interval $0 \le t \le 2\pi$. Since $y = a(1 - \cos t) \ge 0$ and $dx/dt = a(1 - \cos t) \ge 0$, the area under one arch is

$$\begin{aligned}
A &= \int_0^{2\pi} a^2 (1 - \cos t)^2\, dt = a^2 \int_0^{2\pi} \left(1 - 2\cos t + \frac{1 + \cos 2t}{2}\right) dt \\
&= a^2 \left(t - 2 \sin t + \frac{t}{2} + \frac{\sin 2t}{4}\right)\bigg|_0^{2\pi} = 3\pi a^2 \text{ square units.}
\end{aligned}$$

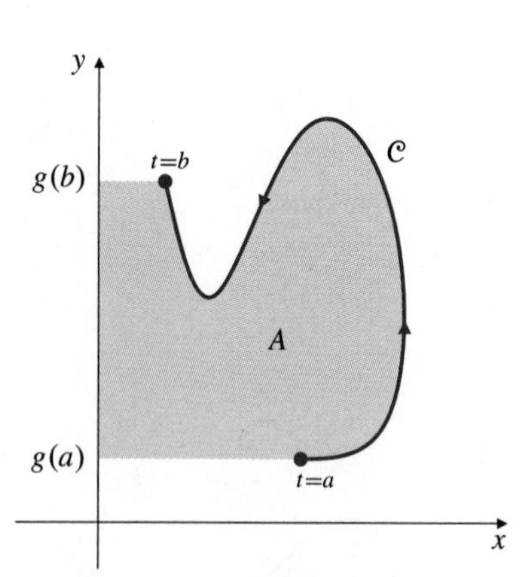

Figure 8.32 The shaded area is

$$A = \int_a^b f(t) g'(t)\, dt$$

Similar arguments to those used above show that if f is continuous and g is differentiable, then we can also interpret

$$\int_a^b f(t) g'(t)\, dt = \int_{t=a}^{t=b} x\, dy = A_1 - A_2,$$

where A_1 is the area of the region lying *horizontally* between the parametric curve $x = f(t)$, $y = g(t)$, $(a \le t \le b)$ and that part of the y-axis consisting of points $y = g(t)$ such that $f(t)g'(t) \ge 0$, and A_2 is the area of a similar region corresponding to $f(t)g'(t) < 0$. For example, the region shaded in Figure 8.32 has area $\int_a^b f(t)g'(t)\, dt$. Green's Theorem in Section 16.3 provides a more coherent approach to finding such areas.

EXERCISES 8.4

Find the lengths of the curves in Exercises 1–8.

1. $x = 3t^2,\ y = 2t^3,\ (0 \le t \le 1)$
2. $x = 1 + t^3,\ y = 1 - t^2,\ (-1 \le t \le 2)$
3. $x = a\cos^3 t,\ y = a\sin^3 t,\ (0 \le t \le 2\pi)$
4. $x = \ln(1 + t^2),\ y = 2\tan^{-1} t,\ (0 \le t \le 1)$
5. $x = t^2 \sin t,\ y = t^2 \cos t,\ (0 \le t \le 2\pi)$
6. $x = \cos t + t\sin t,\ y = \sin t - t\cos t,\ (0 \le t \le 2\pi)$
7. $x = t + \sin t,\ y = \cos t,\ (0 \le t \le \pi)$
8. $x = \sin^2 t,\ y = 2\cos t,\ (0 \le t \le \pi/2)$
9. Find the length of one arch of the cycloid $x = at - a\sin t$, $y = a - a\cos t$. (One arch corresponds to $0 \le t \le 2\pi$.)
10. Find the area of the surfaces obtained by rotating one arch of the cycloid in Exercise 9 about (a) the x-axis, (b) the y-axis.
11. Find the area of the surface generated by rotating the curve $x = e^t \cos t$, $y = e^t \sin t$, $(0 \le t \le \pi/2)$ about the x-axis.
12. Find the area of the surface generated by rotating the curve of Exercise 11 about the y-axis.
13. Find the area of the surface generated by rotating the curve $x = 3t^2$, $y = 2t^3$, $(0 \le t \le 1)$ about the y-axis.
14. Find the area of the surface generated by rotating the curve $x = 3t^2$, $y = 2t^3$, $(0 \le t \le 1)$ about the x-axis.

In Exercises 15–20, sketch and find the area of the region R described in terms of the given parametric curves.

15. R is the closed loop bounded by $x = t^3 - 4t$, $y = t^2$, $(-2 \le t \le 2)$.
16. R is bounded by the astroid $x = a\cos^3 t$, $y = a\sin^3 t$, $(0 \le t \le 2\pi)$.
17. R is bounded by the coordinate axes and the parabolic arc $x = \sin^4 t$, $y = \cos^4 t$.
18. R is bounded by $x = \cos s \sin s$, $y = \sin^2 s$, $(0 \le s \le \pi/2)$, and the y-axis.
19. R is bounded by the oval $x = (2 + \sin t)\cos t$, $y = (2 + \sin t)\sin t$.
20. R is bounded by the x-axis, the hyperbola $x = \sec t$, $y = \tan t$, and the ray joining the origin to the point $(\sec t_0, \tan t_0)$.
21. Show that the region bounded by the x-axis and the hyperbola $x = \cosh t$, $y = \sinh t$ (where $t > 0$), and the ray from the origin to the point $(\cosh t_0, \sinh t_0)$ has area $t_0/2$ square units. This proves a claim made at the beginning of Section 3.6.
22. Find the volume of the solid obtained by rotating about the x-axis the region bounded by that axis and one arch of the cycloid $x = at - a\sin t$, $y = a - a\cos t$. (See Example 8 in Section 8.2.)
23. Find the volume generated by rotating about the x-axis the region lying under the astroid $x = a\cos^3 t$, $y = a\sin^3 t$ and above the x-axis.

8.5 Polar Coordinates and Polar Curves

The **polar coordinate system** is an alternative to the rectangular (Cartesian) coordinate system for describing the location of points in a plane. Sometimes it is more important to know how far, and in what direction, a point is from the origin than it is to know its Cartesian coordinates. In the polar coordinate system there is an origin (or **pole**), O, and a **polar axis**, a ray (i.e., a half-line) extending from O horizontally to the right. The position of any point P in the plane is then determined by its polar coordinates $[r, \theta]$, where

(i) r is the distance from O to P, and

(ii) θ is the angle that the ray OP makes with the polar axis (counterclockwise angles being considered positive).

We will use square brackets for polar coordinates of a point to distinguish them from rectangular (Cartesian) coordinates. Figure 8.33 shows some points with their polar coordinates. The rectangular coordinate axes x and y are usually shown on a polar graph. The polar axis coincides with the positive x-axis.

Unlike rectangular coordinates, the polar coordinates of a point are not unique. The polar coordinates $[r, \theta_1]$ and $[r, \theta_2]$ represent the same point provided θ_1 and θ_2 differ by an integer multiple of 2π:

$$\theta_2 = \theta_1 + 2n\pi, \qquad \text{where } n = 0, \pm 1, \pm 2, \ldots.$$

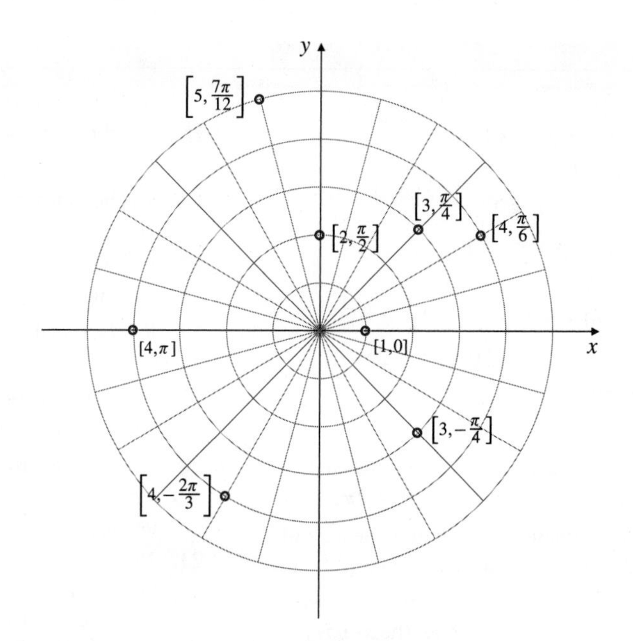

Figure 8.33 Polar coordinates of some points in the xy-plane

For instance, the polar coordinates

$$\left[3, \frac{\pi}{4}\right], \quad \left[3, \frac{9\pi}{4}\right], \quad \text{and} \quad \left[3, -\frac{7\pi}{4}\right]$$

all represent the same point with Cartesian coordinates $\left(\frac{3}{\sqrt{2}}, \frac{3}{\sqrt{2}}\right)$. Similarly, $[4, \pi]$ and $[4, -\pi]$ both represent the point with Cartesian coordinates $(-4, 0)$, and $[1, 0]$ and $[1, 2\pi]$ both represent the point with Cartesian coordinates $(1, 0)$. In addition, the origin O has polar coordinates $[0, \theta]$ for any value of θ. (If we go zero distance from O, it doesn't matter in what direction we go.)

Sometimes we need to interpret polar coordinates $[r, \theta]$, where $r < 0$. The appropriate interpretation for this "negative distance" r is that it represents a positive distance $-r$ measured in the *opposite direction* (i.e., in the direction $\theta + \pi$):

$$[r, \theta] = [-r, \theta + \pi].$$

For example, $[-1, \pi/4] = [1, 5\pi/4]$. Allowing $r < 0$ increases the number of different sets of polar coordinates that represent the same point.

If we want to consider both rectangular and polar coordinate systems in the same plane, and we choose the positive x-axis as the polar axis, then the relationships between the rectangular coordinates of a point and its polar coordinates are as shown in Figure 8.34.

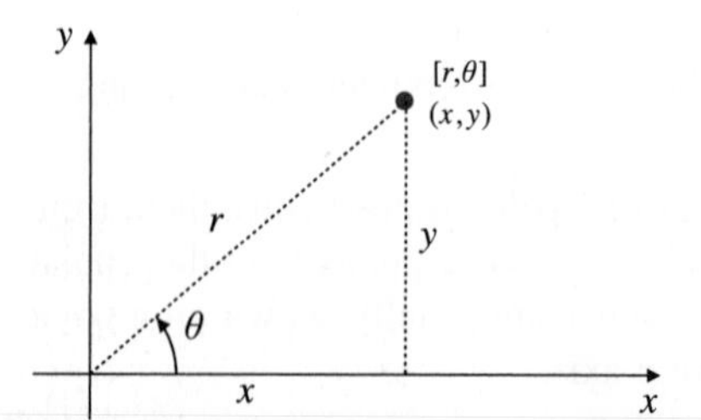

Figure 8.34 Relating Cartesian and polar coordinates of a point

Polar–rectangular conversion

$$x = r\cos\theta \qquad x^2 + y^2 = r^2$$
$$y = r\sin\theta \qquad \tan\theta = \frac{y}{x}$$

A single equation in x and y generally represents a curve in the plane with respect to the Cartesian coordinate system. Similarly, a single equation in r and θ generally represents a curve with respect to the polar coordinate system. The conversion formulas above can be used to convert one representation of a curve into the other.

EXAMPLE 1 The straight line $2x - 3y = 5$ has polar equation $r(2\cos\theta - 3\sin\theta) = 5$, or

$$r = \frac{5}{2\cos\theta - 3\sin\theta}.$$

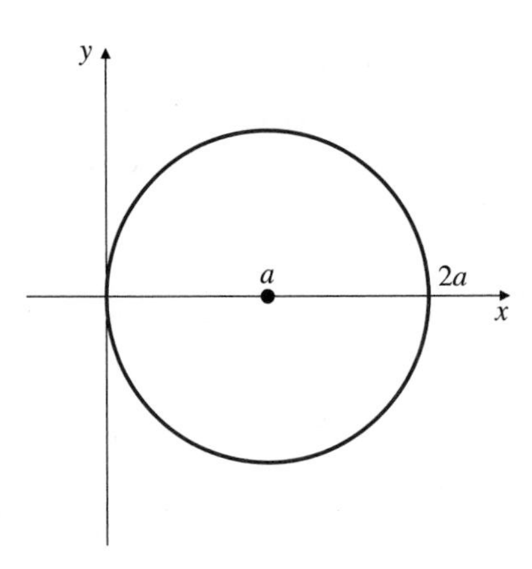

Figure 8.35 The circle $r = 2a\cos\theta$

EXAMPLE 2 Find the Cartesian equation of the curve represented by the polar equation $r = 2a\cos\theta$; hence identify the curve.

Solution The polar equation can be transformed to Cartesian coordinates if we first multiply it by r:

$$\begin{aligned} r^2 &= 2ar\cos\theta \\ x^2 + y^2 &= 2ax \\ (x-a)^2 + y^2 &= a^2 \end{aligned}$$

The given polar equation $r = 2a\cos\theta$ thus represents a circle with centre $(a, 0)$ and radius a as shown in Figure 8.35. Observe from the equation that $r \to 0$ as $\theta \to \pm\pi/2$. In the figure, this corresponds to the fact that the circle approaches the origin in the vertical direction.

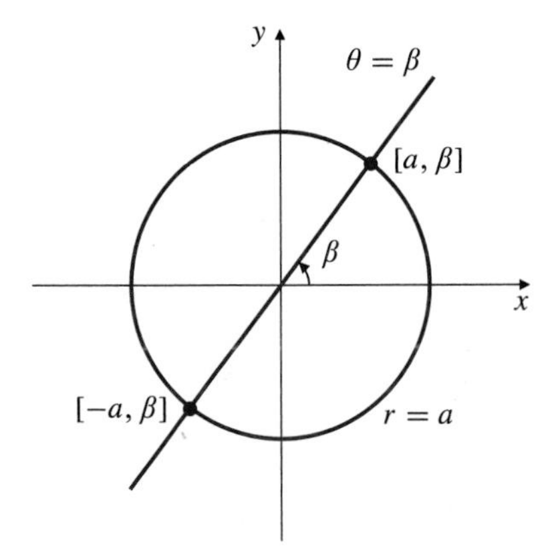

Figure 8.36 Coordinate curves for the polar coordinate system

Some Polar Curves

Figure 8.36 shows the graphs of the polar equations $r = a$ and $\theta = \beta$, where a and β (Greek "beta") are constants. These are, respectively, the circle with radius $|a|$ centred at the origin, and a line through the origin making angle β with the polar axis. Note that the line and the circle meet at two points, with polar coordinates $[a, \beta]$ and $[-a, \beta]$. The "coordinate curves" for polar coordinates, that is, the curves with equations $r =$ constant and $\theta =$ constant, are circles centred at the origin and lines through the origin, respectively. The "coordinate curves" for Cartesian coordinates, $x =$ constant and $y =$ constant, are vertical and horizontal straight lines. Cartesian graph paper is ruled with vertical and horizontal lines; polar graph paper is ruled with concentric circles and radial lines emanating from the origin, as shown in Figures 8.33 and 8.38.

The graph of an equation of the form $r = f(\theta)$ is called the **polar graph** of the function f. Some polar graphs can be recognized easily if the polar equation is transformed to rectangular form. For others, this transformation does not help; the rectangular equation may be too complicated to be recognizable. In these cases one must resort to constructing a table of values and plotting points.

EXAMPLE 3 Sketch and identify the curve $r = 2a\cos(\theta - \theta_0)$.

Solution We proceed as in Example 2.

$$\begin{aligned} r^2 &= 2ar\cos(\theta - \theta_0) = 2ar\cos\theta_0\cos\theta + 2ar\sin\theta_0\sin\theta \\ x^2 + y^2 &= 2a\cos\theta_0 x + 2a\sin\theta_0 y \\ x^2 - 2a\cos\theta_0 x &+ a^2\cos^2\theta_0 + y^2 - 2a\sin\theta_0 y + a^2\sin^2\theta_0 = a^2 \\ (x - a\cos\theta_0)^2 &+ (y - a\sin\theta_0)^2 = a^2. \end{aligned}$$

This is a circle of radius a that passes through the origin in the directions $\theta = \theta_0 \pm \frac{\pi}{2}$, which make $r = 0$. (See Figure 8.37.) Its centre has Cartesian coordinates $(a\cos\theta_0, a\sin\theta_0)$ and hence polar coordinates $[a, \theta_0]$. For $\theta_0 = \pi/2$ we have $r = 2a\sin\theta$ as the equation of a circle of radius a centred on the y-axis.

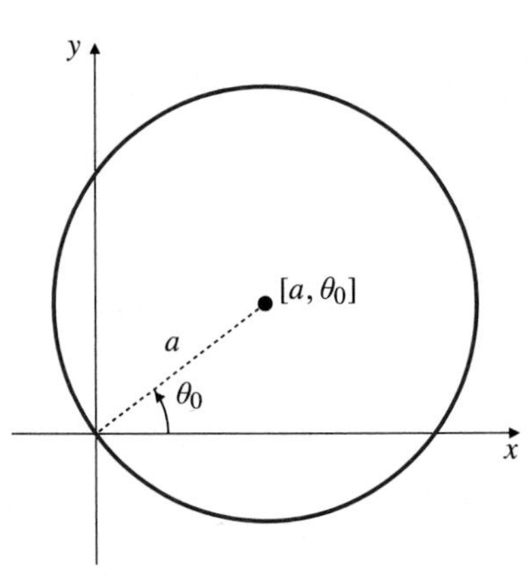

Figure 8.37 The circle $r = 2a\cos(\theta - \theta_0)$

Comparing Examples 2 and 3, we are led to formulate the following principle.

Rotating a polar graph

The polar graph with equation $r = f(\theta - \theta_0)$ is the polar graph with equation $r = f(\theta)$ rotated through angle θ_0 about the origin.

EXAMPLE 4 Sketch the polar curve $r = a(1 - \cos\theta)$, where $a > 0$.

Solution Transformation to rectangular coordinates is not much help here; the resulting equation is $(x^2+y^2+ax)^2 = a^2(x^2+y^2)$ (verify this), which we do not recognize. Therefore, we will make a table of values and plot some points.

Table 3.

θ	0	$\pm\frac{\pi}{6}$	$\pm\frac{\pi}{4}$	$\pm\frac{\pi}{3}$	$\pm\frac{\pi}{2}$	$\pm\frac{2\pi}{3}$	$\pm\frac{3\pi}{4}$	$\pm\frac{5\pi}{6}$	π
r	0	$0.13a$	$0.29a$	$0.5a$	a	$1.5a$	$1.71a$	$1.87a$	$2a$

Because it is shaped like a heart, this curve is called a **cardioid**. Observe the cusp at the origin in Figure 8.38. As in the previous example, the curve enters the origin in the directions θ that make $r = f(\theta) = 0$. In this case, the only such direction is $\theta = 0$. It is important, when sketching polar graphs, to show clearly any directions of approach to the origin.

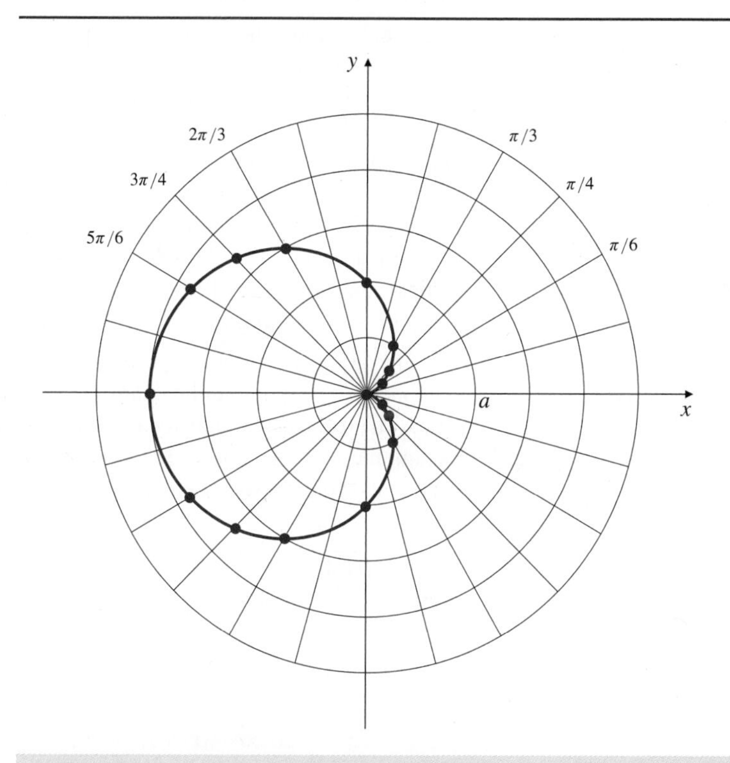

Figure 8.38 The cardioid $r = a(1 - \cos\theta)$

Direction of a polar graph at the origin

A polar graph $r = f(\theta)$ approaches the origin from the direction θ for which $f(\theta) = 0$.

The equation $r = a(1 - \cos(\theta - \theta_0))$ represents a cardioid of the same size and shape as that in Figure 8.38 but rotated through an angle θ_0 counterclockwise about the origin. Its cusp is in the direction $\theta = \theta_0$. In particular, $r = a(1 - \sin\theta)$ has a vertical cusp, as shown in Figure 8.39.

It is not usually necessary to make a detailed table of values to sketch a polar curve with a simple equation of the form $r = f(\theta)$. It is essential to determine those values of θ for which $r = 0$ and indicate them on the graph with rays. It is also useful to determine points where the curve is farthest from the origin. (Where is $f(\theta)$ maximum or minimum?) Except possibly at the origin, polar curves will be smooth wherever $f(\theta)$ is a differentiable function of θ.

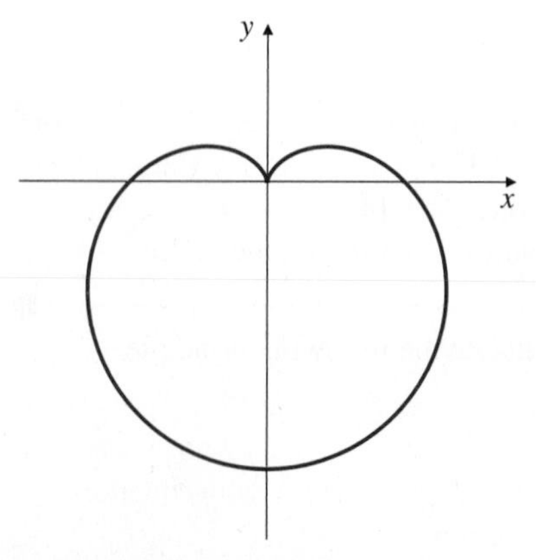

Figure 8.39 The cardioid $r = a(1 - \sin\theta)$

EXAMPLE 5 Sketch the polar graphs (a) $r = \cos(2\theta)$, (b) $r = \sin(3\theta)$, and (c) $r^2 = \cos(2\theta)$.

Solution The graphs are shown in Figures 8.40–8.42. Observe how the curves (a) and (c) approach the origin in the directions $\theta = \pm\frac{\pi}{4}$ and $\theta = \pm\frac{3\pi}{4}$, and curve (b) approaches in the directions $\theta = 0,\ \pi,\ \pm\frac{\pi}{3}$ and $\pm\frac{2\pi}{3}$. This curve is traced out twice as θ increases from $-\pi$ to π. So is curve (c) if we allow both square roots $r = \pm\sqrt{\cos(2\theta)}$. Note that there are no points on curve (c) between $\theta = \pm\frac{\pi}{4}$ and $\theta = \pm\frac{3\pi}{4}$ because r^2 cannot be negative.

Curve (c) is called a **lemniscate**. Lemniscates are curves consisting of points P such that the product of the distances from P to certain fixed points is constant. For the curve (c), these fixed points are $\left(\pm\frac{1}{\sqrt{2}}, 0\right)$.

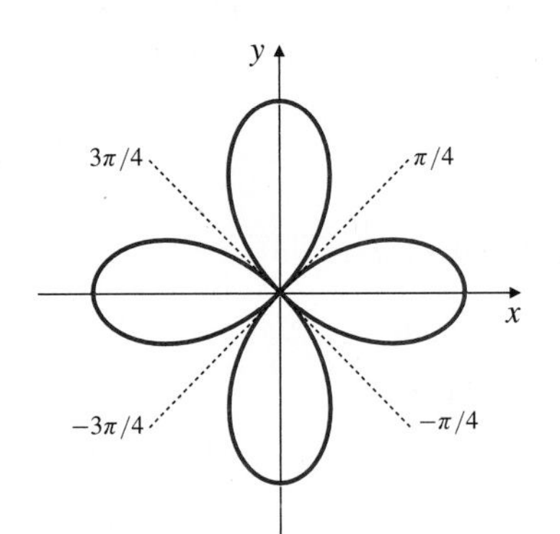

Figure 8.40 Curve (a): the polar curve $r = \cos(2\theta)$

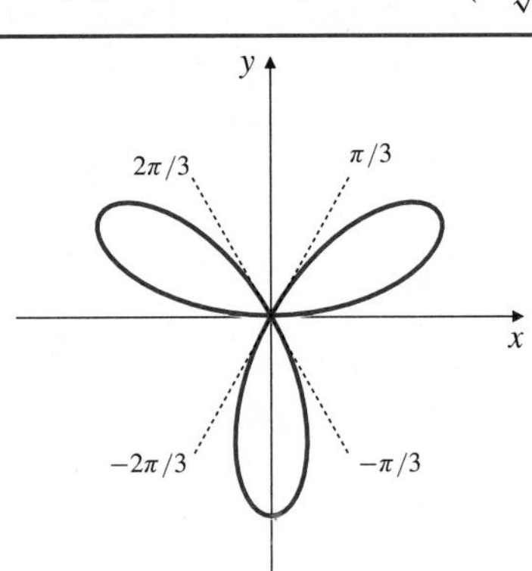

Figure 8.41 Curve (b): the polar curve $r = \sin(3\theta)$

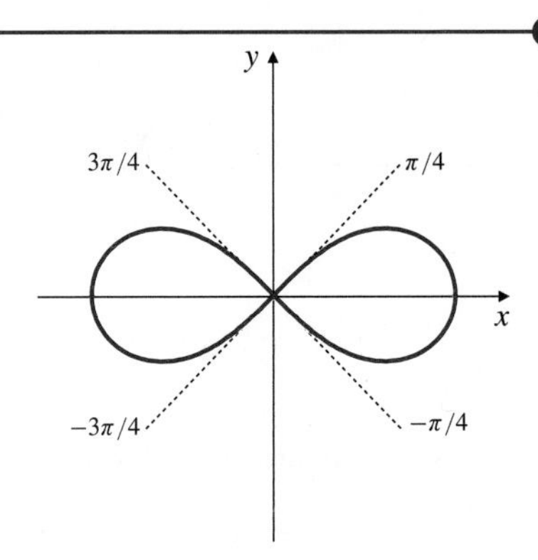

Figure 8.42 Curve (c): the lemniscate $r^2 = \cos(2\theta)$

In all of the examples above, the functions $f(\theta)$ are periodic and 2π is a period of each of them, so each line through the origin could meet the polar graph at most twice. (θ and $\theta + \pi$ determine the same line.) If $f(\theta)$ does not have period 2π, then the curve can wind around the origin many times. Two such *spirals* are shown in Figure 8.43, the **equiangular spiral** $r = \theta$ and the **exponential spiral** $r = e^{-\theta/3}$, each sketched for positive values of θ.

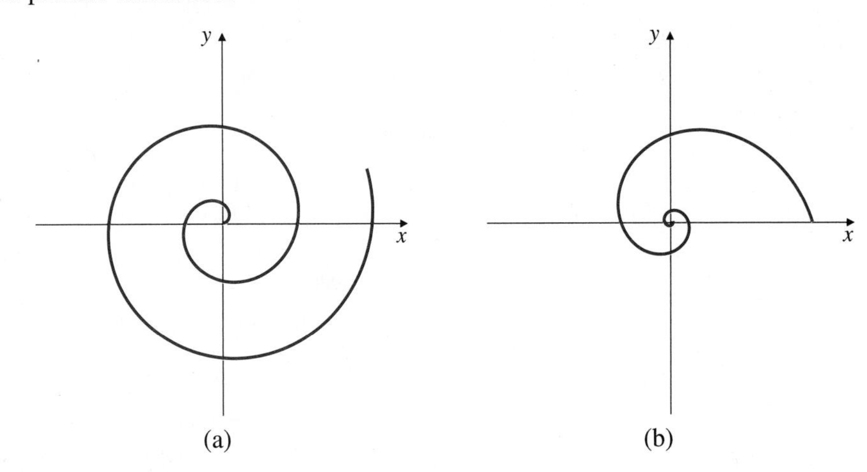

Figure 8.43
(a) The equiangular spiral $r = \theta$
(b) The exponential spiral $r = e^{-\theta/3}$

Remark Maple has a `polarplot` routine as part of its "plots" package, which must be loaded prior to the use of polarplot. Here is how to get Maple to plot on the same graph the polar curves $r = 1$ and $r = 2\sin(3\theta)$, for $0 \le \theta \le 2\pi$:

```
> with(plots):
> polarplot([1,2*sin(3*t)],t=0..2*Pi,scaling=constrained);
```

The option `scaling=constrained` is necessary with polar plots to force Maple to use the same distance unit on both axes (so a circle will appear circular).

Intersections of Polar Curves

Because the polar coordinates of points are not unique, finding the intersection points of two polar curves can be more complicated than the similar problem for Cartesian graphs. Of course, the polar curves $r = f(\theta)$ and $r = g(\theta)$ will intersect at any points $[r_0, \theta_0]$ for which

$$f(\theta_0) = g(\theta_0) \qquad \text{and} \qquad r_0 = f(\theta_0),$$

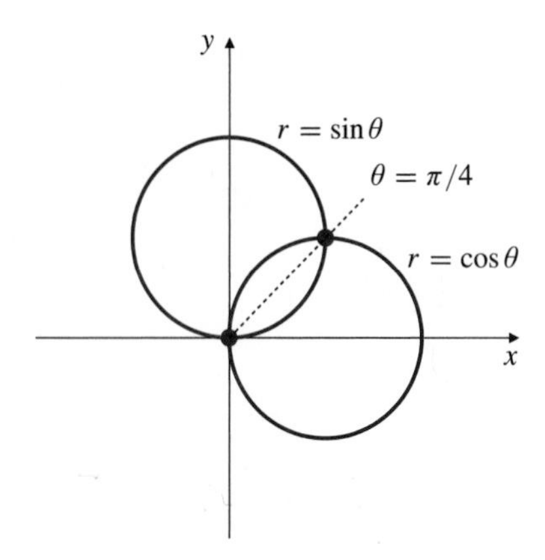

Figure 8.44 Two intersecting circles

but there may be other intersections as well. In particular, if both curves pass through the origin, then the origin will be an intersection point, even though it may not show up in solving $f(\theta) = g(\theta)$, because the curves may be at the origin for different values of θ. For example, the two circles $r = \cos\theta$ and $r = \sin\theta$ intersect at the origin and also at the point $[1/\sqrt{2}, \pi/4]$, even though only the latter point is obtained by solving the equation $\cos\theta = \sin\theta$. (See Figure 8.44.)

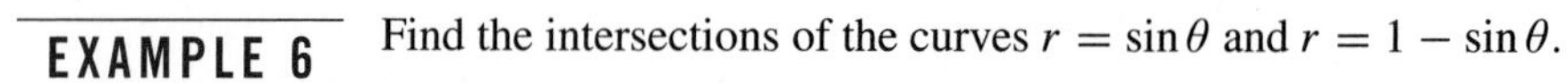

EXAMPLE 6 Find the intersections of the curves $r = \sin\theta$ and $r = 1 - \sin\theta$.

Solution Since both functions of θ are periodic with period 2π, we need only look for solutions satisfying $0 \le \theta \le 2\pi$. Solving the equation

$$\sin\theta = 1 - \sin\theta,$$

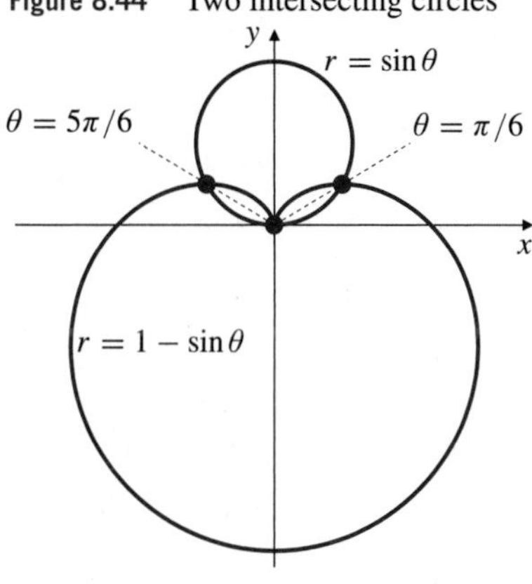

Figure 8.45 The circle and the cardioid intersect at three points

we get $\sin\theta = 1/2$, so that $\theta = \pi/6$ or $\theta = 5\pi/6$. Both curves have $r = 1/2$ at these points, so the two curves intersect at $[1/2, \pi/6]$ and $[1/2, 5\pi/6]$. Also, the origin lies on the curve $r = \sin\theta$ (for $\theta = 0$ and $\theta = 2\pi$) and on the curve $r = 1 - \sin\theta$ (for $\theta = \pi/2$). Therefore, the origin is also an intersection point of the curves. (See Figure 8.45.)

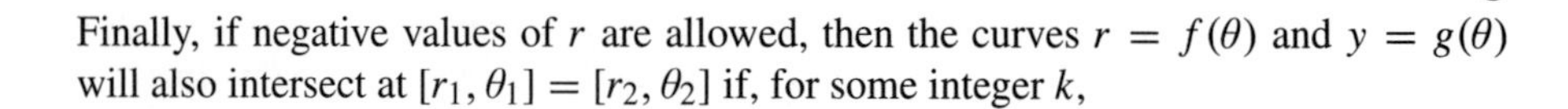

Finally, if negative values of r are allowed, then the curves $r = f(\theta)$ and $y = g(\theta)$ will also intersect at $[r_1, \theta_1] = [r_2, \theta_2]$ if, for some integer k,

$$\theta_1 = \theta_2 + (2k+1)\pi \qquad \text{and} \qquad r_1 = f(\theta_1) = -g(\theta_2) = -r_2.$$

See Exercise 28 for an example.

Polar Conics

Let D be the vertical straight line $x = -p$, and let ε be a positive real number. The set of points P in the plane that satisfy the condition

$$\frac{\text{distance of } P \text{ from the origin}}{\text{perpendicular distance from } P \text{ to } D} = \varepsilon$$

is a conic section with eccentricity ε, focus at the origin, and corresponding directrix D, as observed in Section 8.1. (It is an ellipse if $\varepsilon < 1$, a parabola if $\varepsilon = 1$, and a hyperbola if $\varepsilon > 1$.) If P has polar coordinates $[r, \theta]$, then the condition above becomes (see Figure 8.46)

$$\frac{r}{p + r\cos\theta} = \varepsilon,$$

or, solving for r,

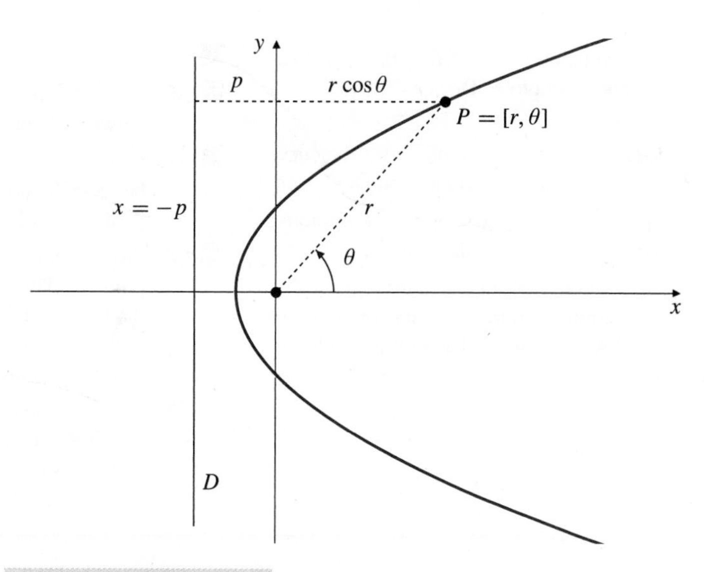

Figure 8.46 A conic curve with eccentricity ε, focus at the origin, and directrix $x = -p$

$$r = \frac{\varepsilon p}{1 - \varepsilon \cos\theta}.$$

Examples of the three possibilities (ellipse, parabola, and hyperbola) are shown in Figures 8.47–8.49. Note that for the hyperbola, the directions of the asymptotes are the angles that make the denominator $1 - \varepsilon\cos\theta = 0$. We will have more to say about polar equations of conics, especially ellipses, in Section 11.6.

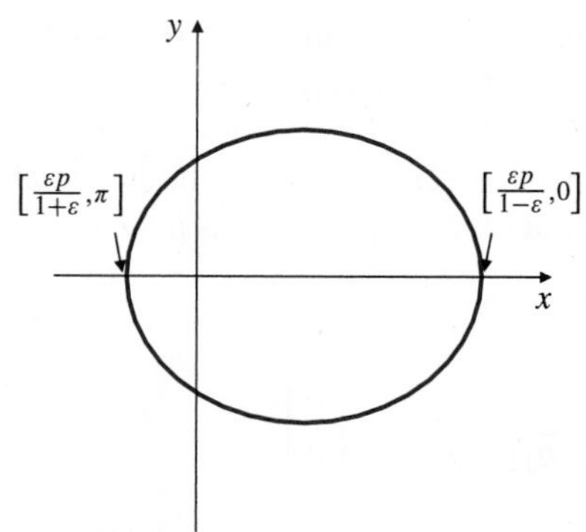

Figure 8.47 Ellipse: $\varepsilon < 1$

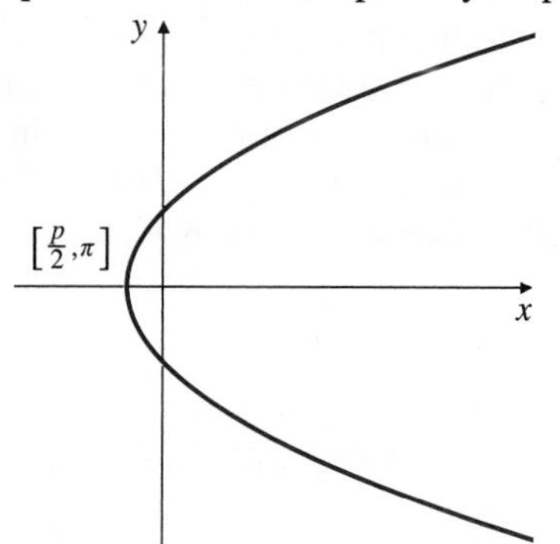

Figure 8.48 Parabola: $\varepsilon = 1$

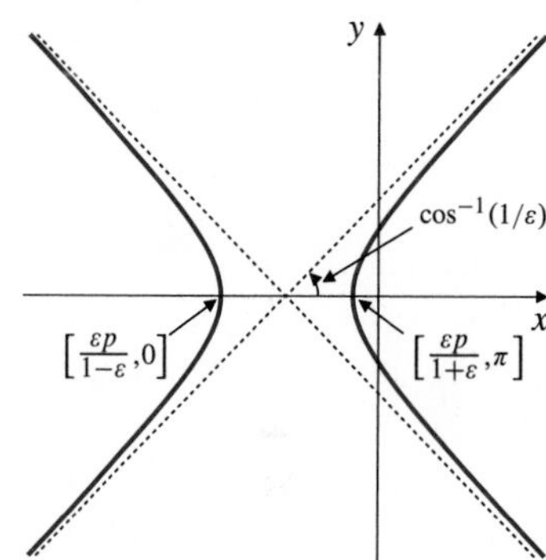

Figure 8.49 Hyperbola: $\varepsilon > 1$

EXERCISES 8.5

In Exercises 1–12, transform the given polar equation to rectangular coordinates, and identify the curve represented.

1. $r = 3\sec\theta$

2. $r = -2\csc\theta$

3. $r = \dfrac{5}{3\sin\theta - 4\cos\theta}$

4. $r = \sin\theta + \cos\theta$

5. $r^2 = \csc 2\theta$

6. $r = \sec\theta\tan\theta$

7. $r = \sec\theta(1 + \tan\theta)$

8. $r = \dfrac{2}{\sqrt{\cos^2\theta + 4\sin^2\theta}}$

9. $r = \dfrac{1}{1 - \cos\theta}$

10. $r = \dfrac{2}{2 - \cos\theta}$

11. $r = \dfrac{2}{1 - 2\sin\theta}$

12. $r = \dfrac{2}{1 + \sin\theta}$

In Exercises 13–24, sketch the polar graphs of the given equations.

13. $r = 1 + \sin\theta$

14. $r = 1 - \cos(\theta + \frac{\pi}{4})$

15. $r = 1 + 2\cos\theta$

16. $r = 1 - 2\sin\theta$

17. $r = 2 + \cos\theta$

18. $r = 2\sin 2\theta$

19. $r = \cos 3\theta$

20. $r = 2\cos 4\theta$

21. $r^2 = 4\sin 2\theta$

22. $r^2 = 4\cos 3\theta$

23. $r^2 = \sin 3\theta$

24. $r = \ln\theta$

Find all intersections of the pairs of curves in Exercises 25–28.

25. $r = \sqrt{3}\cos\theta, \quad r = \sin\theta$

26. $r^2 = 2\cos(2\theta), \quad r = 1$

27. $r = 1 + \cos\theta, \quad r = 3\cos\theta$

28. $r = \theta, \quad r = \theta + \pi$

29. Sketch the graph of the equation $r = 1/\theta, \theta > 0$. Show that this curve has a horizontal asymptote. Does $r = 1/(\theta - \alpha)$ have an asymptote?

30. How many leaves does the curve $r = \cos n\theta$ have? the curve $r^2 = \cos n\theta$? Distinguish the cases where n is odd and even.

31. Show that the polar graph $r = f(\theta)$ (where f is continuous) can be written as a parametric curve with parameter θ.

In Exercises 32–37, use computer graphing software or a graphing calculator to plot various members of the given families of polar curves, and try to observe patterns that would enable you to predict behaviour of other members of the families.

32. $r = \cos\theta\cos(m\theta), \quad m = 1, 2, 3, \ldots$

33. $r = 1 + \cos\theta\cos(m\theta), \quad m = 1, 2, 3, \ldots$

34. $r = \sin(2\theta)\sin(m\theta), \quad m = 2, 3, 4, 5, \ldots$

35. $r = 1 + \sin(2\theta)\sin(m\theta), \quad m = 2, 3, 4, 5, \ldots$

36. $r = C + \cos\theta\cos(2\theta)$ for $C = 0$, $C = 1$, values of C between 0 and 1, and values of C greater than 1

37. $r = C + \cos\theta\sin(3\theta)$ for $C = 0$, $C = 1$, values of C between 0 and 1, values of C less than 0, and values of C greater than 1

38. Plot the curve $r = \ln\theta$ for $0 < \theta \le 2\pi$. It intersects itself at point P. Thus there are two values θ_1 and θ_2 between 0 and 2π for which $[f(\theta_1), \theta_1] = [f(\theta_2), \theta_2]$. What equations must be satsified by θ_1 and θ_2? Find θ_1 and θ_2, and find the Cartesian coordinates of P correct to 6 decimal places.

39. Simultaneously plot the two curves $r = \ln\theta$ and $r = 1/\theta$, for $0 < \theta \le 2\pi$. The two curves intersect at two points. What equations must be satisfied by the θ values of these points? What are their Cartesian coordinates to 6 decimal places?

8.6 Slopes, Areas, and Arc Lengths for Polar Curves

There is a simple formula that can be used to determine the direction of the tangent line to a polar curve $r = f(\theta)$ at a point $P = [r, \theta]$ other than the origin. Let Q be a point on the curve near P corresponding to polar angle $\theta + h$. Let S be on OQ with PS perpendicular to OQ. Observe that $PS = f(\theta)\sin h$ and $SQ = OQ - OS = f(\theta+h) - f(\theta)\cos h$. If the tangent line to $r = f(\theta)$ at P makes angle ψ (Greek "psi") with the radial line OP as shown in Figure 8.50, then ψ is the limit of the angle SQP as $h \to 0$. Thus,

$$\begin{aligned}\tan\psi = \lim_{h\to 0}\frac{PS}{SQ} &= \lim_{h\to 0}\frac{f(\theta)\sin h}{f(\theta+h) - f(\theta)\cos h} && \left[\frac{0}{0}\right] \\ &= \lim_{h\to 0}\frac{f(\theta)\cos h}{f'(\theta+h) + f(\theta)\sin h} && \text{(by l'Hôpital's Rule)} \\ &= \frac{f(\theta)}{f'(\theta)} = \frac{r}{dr/d\theta}.\end{aligned}$$

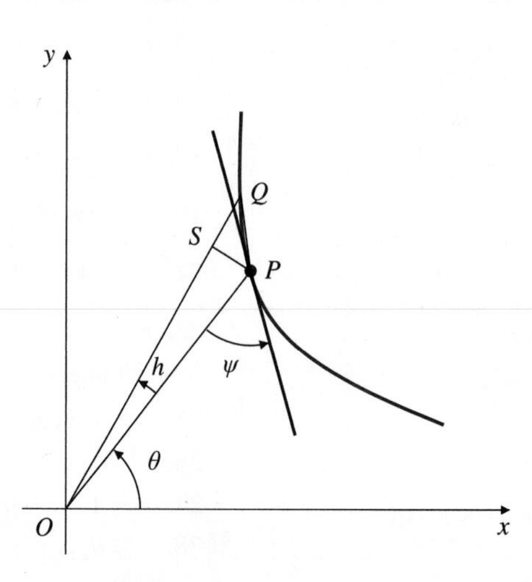

Figure 8.50 The angle ψ is the limit of angle SQP as $h \to 0$

Tangent direction for a polar curve

At any point P other than the origin on the polar curve $r = f(\theta)$, the angle ψ between the radial line from the origin to P and the tangent to the curve is given by

$$\tan\psi = \frac{f(\theta)}{f'(\theta)}.$$

In particular, $\psi = \pi/2$ if $f'(\theta) = 0$. If $f(\theta_0) = 0$ and the curve has a tangent line at θ_0, then that tangent line has equation $\theta = \theta_0$.

The formula above can be used to find points where a polar graph has horizontal or vertical tangents:

$$\psi + \theta = \pi, \qquad \text{so } \tan\psi = -\tan\theta \qquad \text{for a horizontal tangent,}$$

$$\psi + \theta = \frac{\pi}{2}, \qquad \text{so } \tan\psi = \cot\theta \qquad \text{for a vertical tangent.}$$

Remark Since for parametric curves horizontal and vertical tangents correspond to $dy/dt = 0$ and $dx/dt = 0$, respectively, it is usually easier to find the critical points of $y = f(\theta)\sin\theta$ for horizontal tangents and of $x = f(\theta)\cos\theta$ for vertical tangents.

EXAMPLE 1 Find the points on the cardioid $r = 1 + \cos\theta$, where the tangent lines are vertical or horizontal.

Solution We have $y = (1 + \cos\theta)\sin\theta$ and $x = (1 + \cos\theta)\cos\theta$. For horizontal tangents,

$$\begin{aligned} 0 = \frac{dy}{d\theta} &= -\sin^2\theta + \cos^2\theta + \cos\theta \\ &= 2\cos^2\theta + \cos\theta - 1 \\ &= (2\cos\theta - 1)(\cos\theta + 1). \end{aligned}$$

The solutions are $\cos\theta = \frac{1}{2}$ and $\cos\theta = -1$, that is, $\theta = \pm\pi/3$ and $\theta = \pi$. There are horizontal tangents at $\left[\frac{3}{2}, \pm\frac{\pi}{3}\right]$. At $\theta = \pi$, we have $r = 0$. The curve does not have a tangent line at the origin (it has a cusp). See Figure 8.51.

For vertical tangents,

$$0 = \frac{dx}{d\theta} = -\sin\theta - 2\cos\theta\sin\theta = -\sin\theta(1 + 2\cos\theta).$$

The solutions are $\sin\theta = 0$ and $\cos\theta = -\frac{1}{2}$, that is, $\theta = 0,\ \pi,\ \pm 2\pi/3$. There are vertical tangent lines at $[2, 0]$ and $\left[\frac{1}{2}, \pm\frac{2\pi}{3}\right]$.

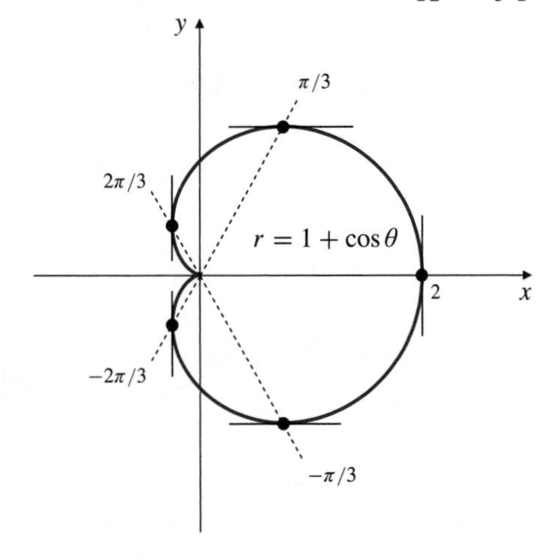

Figure 8.51 Horizontal and vertical tangents to a cardioid

Areas Bounded by Polar Curves

The basic area problem in polar coordinates is that of finding the area A of the region R bounded by the polar graph $r = f(\theta)$ and the two rays $\theta = \alpha$ and $\theta = \beta$. We assume that $\beta > \alpha$ and that f is continuous for $\alpha \le \theta \le \beta$. See Figure 8.52.

A suitable area element in this case is a sector of angular width $d\theta$, as shown in Figure 8.52. For infinitesimal $d\theta$ this is just a sector of a circle of radius $r = f(\theta)$:

$$dA = \frac{d\theta}{2\pi}\pi r^2 = \frac{1}{2}r^2\,d\theta = \frac{1}{2}\left(f(\theta)\right)^2 d\theta.$$

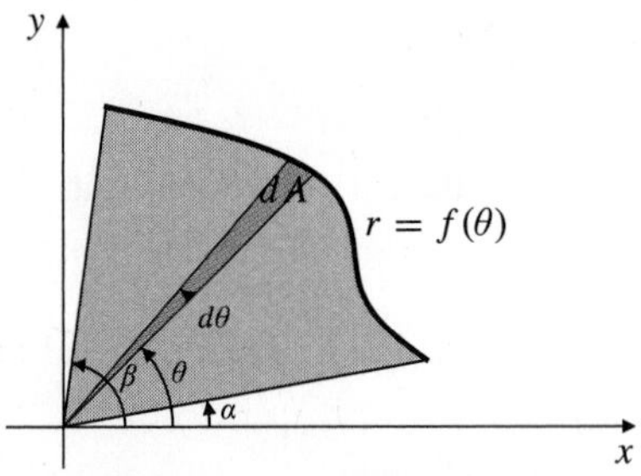

Figure 8.52 An area element in polar coordinates

Area in polar coordinates

The region bounded by $r = f(\theta)$ and the rays $\theta = \alpha$ and $\theta = \beta$, $(\alpha < \beta)$, has area

$$A = \frac{1}{2}\int_\alpha^\beta \left(f(\theta)\right)^2 d\theta.$$

EXAMPLE 2 Find the area bounded by the cardioid $r = a(1 + \cos\theta)$, as illustrated in Figure 8.53.

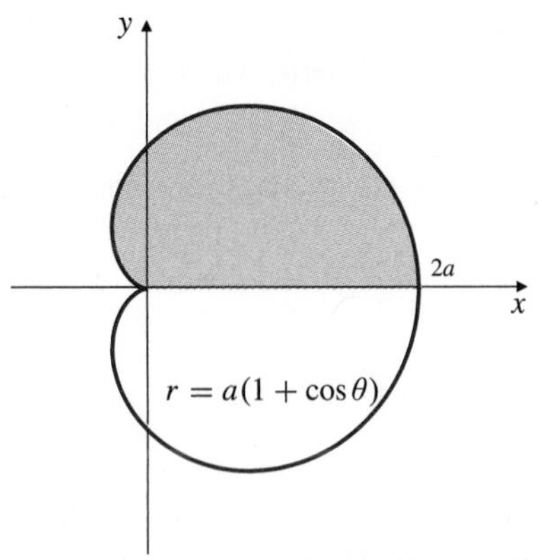

Figure 8.53 The area encosed by the cardioid is twice the shaded part

Solution By symmetry, the area is twice that of the top half:

$$\begin{aligned}
A &= 2 \times \frac{1}{2}\int_0^\pi a^2(1+\cos\theta)^2\,d\theta \\
&= a^2\int_0^\pi (1 + 2\cos\theta + \cos^2\theta)\,d\theta \\
&= a^2\int_0^\pi \left(1 + 2\cos\theta + \frac{1+\cos 2\theta}{2}\right) d\theta \\
&= a^2\left(\frac{3}{2}\theta + 2\sin\theta + \frac{1}{4}\sin 2\theta\right)\Bigg|_0^\pi = \frac{3}{2}\pi a^2 \text{ square units.}
\end{aligned}$$

EXAMPLE 3 Find the area of the region that lies inside the circle $r = \sqrt{2}\sin\theta$ and inside the lemniscate $r^2 = \sin 2\theta$.

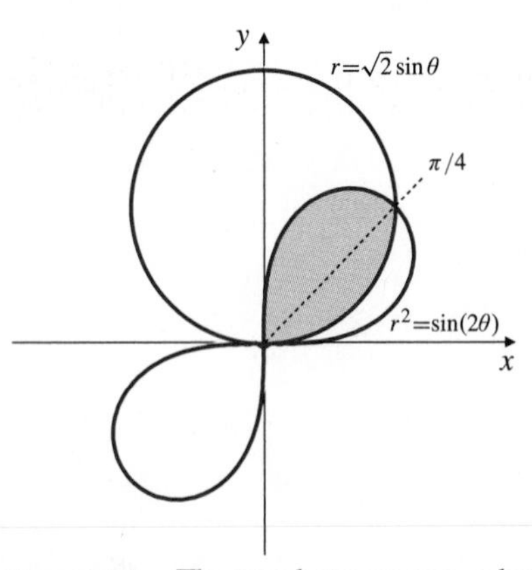

Figure 8.54 The area between two polar curves

Solution The region is shaded in Figure 8.54. Besides intersecting at the origin, the curves intersect at the first quadrant point satisfying

$$2\sin^2\theta = \sin 2\theta = 2\sin\theta\cos\theta.$$

Thus, $\sin\theta = \cos\theta$ and $\theta = \pi/4$. The required area is

$$\begin{aligned}
A &= \frac{1}{2}\int_0^{\pi/4} 2\sin^2\theta\,d\theta + \frac{1}{2}\int_{\pi/4}^{\pi/2}\sin 2\theta\,d\theta \\
&= \int_0^{\pi/4}\frac{1-\cos 2\theta}{2}\,d\theta - \frac{1}{4}\cos 2\theta\Bigg|_{\pi/4}^{\pi/2} \\
&= \frac{\pi}{8} - \frac{1}{4}\sin 2\theta\Bigg|_0^{\pi/4} + \frac{1}{4} = \frac{\pi}{8} - \frac{1}{4} + \frac{1}{4} = \frac{\pi}{8} \quad \text{square units.}
\end{aligned}$$

Arc Lengths for Polar Curves

The arc length element for the polar curve $r = f(\theta)$ can be determined from the differential triangle shown in Figure 8.55. The leg $r\,d\theta$ of the triangle is obtained as the arc length of a circular arc of radius r subtending angle $d\theta$ at the origin. We have

$$(ds)^2 = (dr)^2 + r^2(d\theta)^2 = \left[\left(\frac{dr}{d\theta}\right)^2 + r^2\right](d\theta)^2,$$

so we obtain the following formula:

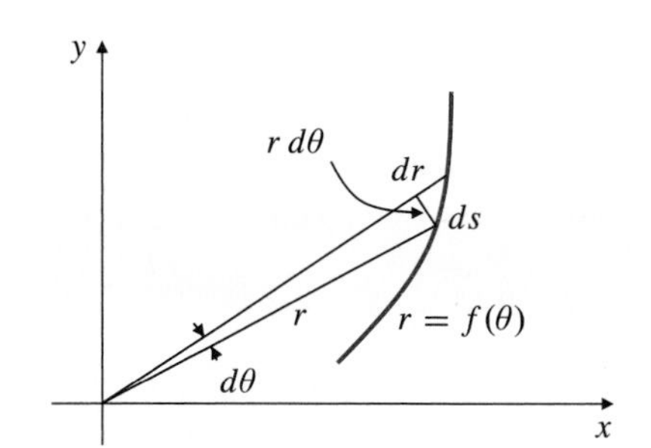

Figure 8.55 The arc length element for a polar curve

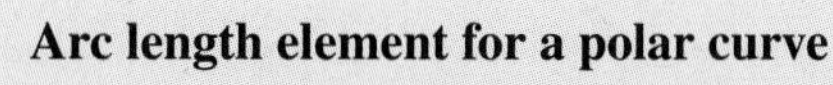

Arc length element for a polar curve

The arc length element for the polar curve $r = f(\theta)$ is

$$ds = \sqrt{\left(\frac{dr}{d\theta}\right)^2 + r^2}\,d\theta = \sqrt{\left(f'(\theta)\right)^2 + \left(f(\theta)\right)^2}\,d\theta.$$

This arc length element can also be derived from that for a parametric curve. See Exercise 26 at the end of this section.

EXAMPLE 4 Find the total length of the cardioid $r = a(1 + \cos\theta)$.

Solution The total length is twice the length from $\theta = 0$ to $\theta = \pi$. (Review Figure 8.53.) Since $dr/d\theta = -a\sin\theta$ for the cardioid, the arc length is

$$\begin{aligned}
s &= 2\int_0^{\pi} \sqrt{a^2\sin^2\theta + a^2(1+\cos\theta)^2}\,d\theta \\
&= 2\int_0^{\pi} \sqrt{2a^2 + 2a^2\cos\theta}\,d\theta \qquad (\text{but } 1+\cos\theta = 2\cos^2(\theta/2)) \\
&= 2\sqrt{2}a\int_0^{\pi} \sqrt{2\cos^2\frac{\theta}{2}}\,d\theta \\
&= 4a\int_0^{\pi} \cos\frac{\theta}{2}\,d\theta = 8a\sin\frac{\theta}{2}\bigg|_0^{\pi} = 8a \text{ units.}
\end{aligned}$$

EXERCISES 8.6

In Exercises 1–11, sketch and find the areas of the given polar regions R.

1. R lies between the origin and the spiral $r = \sqrt{\theta}$, $0 \le \theta \le 2\pi$.
2. R lies between the origin and the spiral $r = \theta$, $0 \le \theta \le 2\pi$.
3. R is bounded by the curve $r^2 = a^2\cos 2\theta$.
4. R is one leaf of the curve $r = \sin 3\theta$.
5. R is bounded by the curve $r = \cos 4\theta$.
6. R lies inside both of the circles $r = a$ and $r = 2a\cos\theta$.
7. R lies inside the cardioid $r = 1 - \cos\theta$ and outside the circle $r = 1$.
8. R lies inside the cardioid $r = a(1 - \sin\theta)$ and inside the circle $r = a$.
9. R lies inside the cardioid $r = 1 + \cos\theta$ and outside the circle $r = 3\cos\theta$.
10. R is bounded by the lemniscate $r^2 = 2\cos 2\theta$ and is outside the circle $r = 1$.
11. R is bounded by the smaller loop of the curve $r = 1 + 2\cos\theta$.

Find the lengths of the polar curves in Exercises 12–14.

12. $r = \theta^2$, $0 \le \theta \le \pi$
13. $r = e^{a\theta}$, $-\pi \le \theta \le \pi$
14. $r = a\theta$, $0 \le \theta \le 2\pi$
15. Show that the total arc length of the lemniscate $r^2 = \cos 2\theta$ is $4\int_0^{\pi/4} \sqrt{\sec 2\theta}\,d\theta$.
16. One leaf of the lemniscate $r^2 = \cos 2\theta$ is rotated (a) about the x-axis and (b) about the y-axis. Find the area of the surface generated in each case.

17. Determine the angles at which the straight line $\theta = \pi/4$ intersects the cardioid $r = 1 + \sin\theta$.

18. At what points do the curves $r^2 = 2\sin 2\theta$ and $r = 2\cos\theta$ intersect? At what angle do the curves intersect at each of these points?

19. At what points do the curves $r = 1 - \cos\theta$ and $r = 1 - \sin\theta$ intersect? At what angle do the curves intersect at each of these points?

In Exercises 20–25, find all points on the given curve where the tangent line is horizontal, vertical, or does not exist.

20. $r = \cos\theta + \sin\theta$

21. $r = 2\cos\theta$

22. $r^2 = \cos 2\theta$

23. $r = \sin 2\theta$

24. $r = e^{\theta}$

25. $r = 2(1 - \sin\theta)$

26. The polar curve $r = f(\theta)$, $(\alpha \le \theta \le \beta)$, can be parametrized:

$$x = r\cos\theta = f(\theta)\cos\theta, \qquad y = r\sin\theta = f(\theta)\sin\theta.$$

Derive the formula for the arc length element for the polar curve from that for a parametric curve.

CHAPTER REVIEW

Key Ideas

- **What do the following terms and phrases mean?**
  - a conic section
  - an ellipse
  - a parabola
  - a hyperbola
  - a parametric curve
  - a parametrization of a curve
  - a smooth curve
  - a polar curve
- **What is the focus-directrix definition of a conic?**
- **How can you find the slope of a parametric curve?**
- **How can you find the length of a parametric curve?**
- **How can you find the length of a polar curve?**
- **How can you find the area bounded by a polar curve?**

Review Exercises

In Exercises 1–4, describe the conic having the given equation. Give its foci and principal axes and, if it is a hyperbola, its asymptotes.

1. $x^2 + 2y^2 = 2$

2. $9x^2 - 4y^2 = 36$

3. $x + y^2 = 2y + 3$

4. $2x^2 + 8y^2 = 4x - 48y$

Identify the parametric curves in Exercises 5–10.

5. $x = t,\ y = 2 - t,\ (0 \le t \le 2)$

6. $x = 2\sin 3t,\ y = 2\cos 3t,\ (0 \le t \le 1/2)$

7. $x = \cosh t,\ y = \sinh^2 t,$

8. $x = e^t,\ y = e^{-2t},\ (-1 \le t \le 1)$

9. $x = \cos(t/2),\ y = 4\sin(t/2),\ (0 \le t \le \pi)$

10. $x = \cos t + \sin t,\ y = \cos t - \sin t,\ (0 \le t \le 2\pi)$

In Exercises 11–14, determine the points where the given parametric curves have horizontal and vertical tangents, and sketch the curves.

11. $x = \dfrac{4}{1+t^2},\ y = t^3 - 3t$

12. $x = t^3 - 3t,\ y = t^3 + 3t$

13. $x = t^3 - 3t,\ y = t^3$

14. $x = t^3 - 3t,\ y = t^3 - 12t$

15. Find the area bounded by the part of the curve $x = t^3 - t$, $y = |t^3|$ that forms a closed loop.

16. Find the volume of the solid generated by rotating the closed loop in Exercise 15 about the y-axis.

17. Find the length of the curve $x = e^t - t$, $y = 4e^{t/2}$ from $t = 0$ to $t = 2$.

18. Find the area of the surface obtained by rotating the arc in Exercise 17 about the x-axis.

Sketch the polar graphs of the equations in Exercises 19–24.

19. $r = \theta,\ \left(-\frac{3\pi}{2} \le \theta \le \frac{3\pi}{2}\right)$

20. $r = |\theta|,\ (-2\pi \le \theta \le 2\pi)$

21. $r = 1 + \cos 2\theta$

22. $r = 2 + \cos 2\theta$

23. $r = 1 + 2\cos 2\theta$

24. $r = 1 - \sin 3\theta$

25. Find the area of one of the two larger loops of the curve in Exercise 23.

26. Find the area of one of the two smaller loops of the curve in Exercise 23.

27. Find the area of the smaller of the two loops enclosed by the curve $r = 1 + \sqrt{2}\sin\theta$.

28. Find the area of the region inside the cardioid $r = 1 + \cos\theta$ and to the left of the line $x = 1/4$.

Challenging Problems

1. A glass in the shape of a circular cylinder of radius 4 cm is more than half filled with water. If the glass is tilted by an angle θ from the vertical, where θ is small enough that no water spills out, find the surface area of the water.

2. Show that a plane that is not parallel to the axis of a circular cylinder intersects the cylinder in an ellipse. *Hint:* You can do this by the same method used in Exercise 27 of Section 8.1.

3. Given two points F_1 and F_2 that are foci of an ellipse and a third point P on the ellipse, describe a geometric method (using a straight edge and a compass) for constructing the tangent line to the ellipse at P. *Hint:* Think about the reflection property of ellipses.

4. Let C be a parabola with vertex V, and let P be any point on the parabola. Let R be the point where the tangent to the parabola at P intersects the axis of the parabola. (Thus, the axis is the line RV.) Let Q be the point on RV such that PQ is perpendicular to RV. Show that V bisects the line segment RQ. How does this result suggest a geometric method for constructing a tangent to a parabola at a point on it, given the axis and vertex of the parabola?

5. A barrel has the shape of a solid of revolution obtained by rotating about its major axis the part of an ellipse lying between lines through its foci perpendicular to that axis. The barrel is 4 ft high and 2 ft in radius at its middle. What is its volume?

6. (a) Show that any straight line not passing through the origin can be written in polar form as

$$r = \frac{a}{\cos(\theta - \theta_0)},$$

where a and θ_0 are constants. What is the geometric significance of these constants?

(b) Let $r = g(\theta)$ be the polar equation of a straight line that does not pass through the origin. Show that

$$g^2 + 2(g')^2 - gg'' = 0.$$

(c) Let $r = f(\theta)$ be the polar equation of a curve, where f'' is continuous and $r \neq 0$ in some interval of values of θ. Let

$$F = f^2 + 2(f')^2 - ff''.$$

Show that the curve is turning toward the origin if $F > 0$ and away from the origin if $F < 0$. *Hint:* Let $r = g(\theta)$ be the polar equation of a straight line tangent to the curve, and use part (b). How do f, f', and f'' relate to g, g', and g'' at the point of tangency?

7. (Fast trip, but it might get hot) If we assume that the density of the earth is uniform throughout, then it can be shown that the acceleration of gravity at a distance $r \leq R$ from the centre of the earth is directed toward the centre of the earth and has magnitude $a(r) = rg/R$, where g is the usual acceleration of gravity at the surface ($g \approx 32$ ft/s^2), and R is the radius of the earth ($R \approx 3{,}960$ mi). Suppose that a straight tunnel AB is drilled through the earth between any two points A and B on the surface, say Atlanta and Baghdad. (See Figure 8.56.)

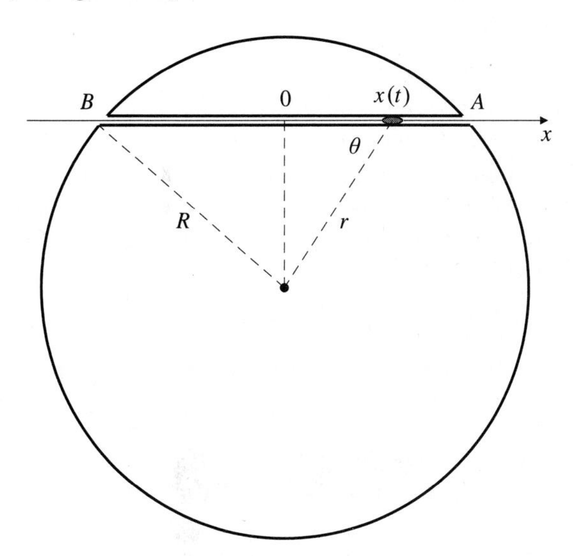

Figure 8.56

Suppose that a vehicle is constructed that can slide without friction or air resistance through this tunnel. Show that such a vehicle will, if released at one end of the tunnel, fall back and forth between A and B, executing simple harmonic motion with period $2\pi\sqrt{R/g}$. How many minutes will the round trip take? What is surprising here is that this period does not depend on where A and B are or on the distance between them. *Hint:* Let the x-axis lie along the tunnel, with origin at the point closest to the centre of the earth. When the vehicle is at position with x-coordinate $x(t)$, its acceleration along the tunnel is the component of the gravitational acceleration along the tunnel, that is, $-a(r)\cos\theta$, where θ is the angle between the line of the tunnel and the line from the vehicle to the centre of the earth.

8. (Search and Rescue) Two coast guard stations pick up a distress signal from a ship and use radio direction finders to locate it. Station O observes that the distress signal is coming from the northeast (45° east of north), while station P, which is 100 miles north of station O, observes that the signal is coming from due east. Each station's direction finder is accurate to within $\pm 3°$.

(a) How large an area of the ocean must a rescue aircraft search to ensure that it finds the foundering ship?

(b) If the accuracy of the direction finders is within $\pm\varepsilon$, how sensitive is the search area to changes in ε when $\varepsilon = 3°$? (Express your answer in square miles per degree.)

9. Figure 8.57 shows the graphs of the parametric curve $x = \sin t$, $y = \frac{1}{2}\sin(2t)$, $0 \leq t \leq 2\pi$, and the polar curve $r^2 = \cos(2\theta)$. Each has the shape of an "∞." Which curve is which? Find the area inside the outer curve and outside the inner curve.

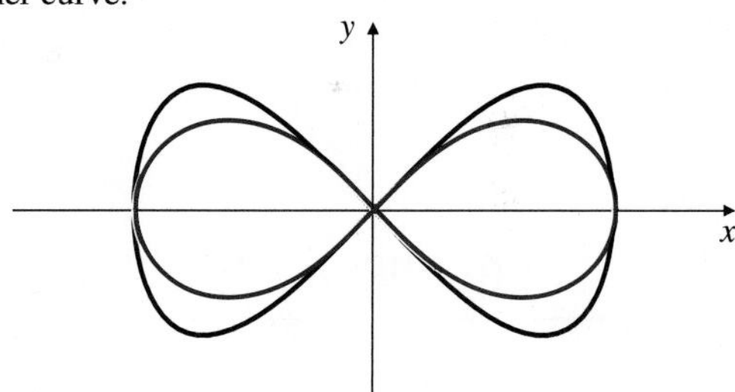

Figure 8.57

CHAPTER 9

Sequences, Series, and Power Series

> 'Then you should say what you mean,' the March Hare went on.
> 'I do,' Alice hastily replied; 'at least — at least I mean what I say — that's the same thing, you know.'
> 'Not the same thing a bit!' said the Hatter. 'Why, you might just as well say that "I see what I eat" is the same thing as "I eat what I see!"'

Lewis Carroll (Charles Lutwidge Dodgson) 1832–1898
from *Alice's Adventures in Wonderland*

Introduction

An infinite series is a sum that involves infinitely many terms. Since addition is carried out on two numbers at a time, the evaluation of the sum of an infinite series necessarily involves finding a limit. Complicated functions $f(x)$ can frequently be expressed as series of simpler functions. For example, many of the transcendental functions we have encountered can be expressed as series of powers of x so that they resemble polynomials of infinite degree. Such series can be differentiated and integrated term by term, and they play a very important role in the study of calculus.

9.1 Sequences and Convergence

By a **sequence** (or an **infinite sequence**) we mean an ordered list having a first element but no last element. For our purposes, the elements (called **terms**) of a sequence will always be real numbers, although much of our discussion could be applied to complex numbers as well. Examples of sequences are:

$\{1, 2, 3, 4, 5, \ldots\}$ the sequence of positive integers,

$\left\{-\frac{1}{2}, \frac{1}{4}, -\frac{1}{8}, \frac{1}{16}, \ldots\right\}$ the sequence of positive integer powers of $-\frac{1}{2}$.

The terms of a sequence are usually listed in braces as shown. The ellipsis points $(\ldots)$ should be read "and so on."

An infinite sequence is a special kind of function, one whose domain is a set of integers extending from some starting integer to infinity. The starting integer is usually 1, so the domain is the set of positive integers. The sequence $\{a_1, a_2, a_3, a_4, \ldots\}$ is the function f that takes the value $f(n) = a_n$ at each positive integer n. A sequence can be specified in three ways:

(i) We can list the first few terms followed by ... *if the pattern is obvious.*

(ii) We can provide a formula for the **general term** a_n as a function of n.

(iii) We can provide a formula for calculating the term a_n as a function of earlier terms $a_1, a_2, \ldots, a_{n-1}$ and specify enough of the beginning terms so the process of computing higher terms can begin.

In each case it must be possible to determine any term of the sequence, although it may be necessary to calculate all the preceding terms first.

EXAMPLE 1 **(Some examples of sequences)**

(a) $\{n\} = \{1, 2, 3, 4, 5, \ldots\}$

(b) $\left\{\left(-\frac{1}{2}\right)^n\right\} = \left\{-\frac{1}{2}, \frac{1}{4}, -\frac{1}{8}, \frac{1}{16}, \ldots\right\}$

(c) $\left\{\frac{n-1}{n}\right\} = \left\{0, \frac{1}{2}, \frac{2}{3}, \frac{3}{4}, \frac{4}{5}, \ldots\right\}$

(d) $\{(-1)^{n-1}\} = \{\cos((n-1)\pi)\} = \{1, -1, 1, -1, 1, \ldots\}$

(e) $\left\{\frac{n^2}{2^n}\right\} = \left\{\frac{1}{2}, 1, \frac{9}{8}, 1, \frac{25}{32}, \frac{36}{64}, \frac{49}{128}, \ldots\right\}$

(f) $\left\{\left(1+\frac{1}{n}\right)^n\right\} = \left\{2, \left(\frac{3}{2}\right)^2, \left(\frac{4}{3}\right)^3, \left(\frac{5}{4}\right)^4, \ldots\right\}$

(g) $\left\{\frac{\cos(n\pi/2)}{n}\right\} = \left\{0, -\frac{1}{2}, 0, \frac{1}{4}, 0, -\frac{1}{6}, 0, \frac{1}{8}, 0, \ldots\right\}$

(h) $a_1 = 1, \quad a_{n+1} = \sqrt{6 + a_n}, \quad (n = 1, 2, 3, \ldots)$
In this case $\{a_n\} = \{1, \sqrt{7}, \sqrt{6+\sqrt{7}}, \ldots\}$. Note that there is no *obvious* formula for a_n as an explicit function of n here, but we can still calculate a_n for any desired value of n provided we first calculate all the earlier values $a_2, a_3, \ldots, a_{n-1}$.

(i) $a_1 = 1, a_2 = 1, a_{n+2} = a_n + a_{n+1}, \quad (n = 1, 2, 3, \ldots)$
Here $\{a_n\} = \{1, 1, 2, 3, 5, 8, 13, 21, \ldots\}$. This is called the **Fibonacci sequence**. Each term after the second is the sum of the previous two terms.

In parts (a)–(g) of Example 1, the formulas on the left sides define the general term of each sequence $\{a_n\}$ as an explicit function of n. In parts (h) and (i) we say the sequence $\{a_n\}$ is defined **recursively** or **inductively**; each term must be calculated from previous ones rather than directly as a function of n. We now introduce terminology used to describe various properties of sequences.

DEFINITION 1

Terms for describing sequences

(a) The sequence $\{a_n\}$ is **bounded below** by L, and L is a **lower bound** for $\{a_n\}$, if $a_n \geq L$ for every $n = 1, 2, 3, \ldots$. The sequence is **bounded above** by M, and M is an **upper bound**, if $a_n \leq M$ for every such n.

The sequence $\{a_n\}$ is **bounded** if it is both bounded above and bounded below. In this case there is a constant K such that $|a_n| \leq K$ for every $n = 1, 2, 3, \ldots$. (We can take K to be the larger of $|L|$ and $|M|$.)

(b) The sequence $\{a_n\}$ is **positive** if it is bounded below by zero, that is, if $a_n \geq 0$ for every $n = 1, 2, 3, \ldots$; it is **negative** if $a_n \leq 0$ for every n.

(c) The sequence $\{a_n\}$ is **increasing** if $a_{n+1} \geq a_n$ for every $n = 1, 2, 3, \ldots$; it is **decreasing** if $a_{n+1} \leq a_n$ for every such n. The sequence is said to be **monotonic** if it is either increasing or decreasing. (The terminology here is looser than that used for functions, where we would have used *nondecreasing* and *nonincreasing* to describe this behaviour. The distinction between $a_{n+1} > a_n$ and $a_{n+1} \geq a_n$ is not as important for sequences as it is for functions defined on intervals.)

(d) The sequence $\{a_n\}$ is **alternating** if $a_n a_{n+1} < 0$ for every $n = 1, 2, \ldots$, that is, if any two consecutive terms have opposite signs. Note that this definition requires $a_n \neq 0$ for each n.

EXAMPLE 2 (Describing some sequences)

(a) The sequence $\{n\} = \{1, 2, 3, \ldots\}$ is positive, increasing, and bounded below. A lower bound for the sequence is 1 or any smaller number. The sequence is not bounded above.

(b) $\left\{\dfrac{n-1}{n}\right\} = \left\{0, \dfrac{1}{2}, \dfrac{2}{3}, \dfrac{3}{4}, \ldots\right\}$ is positive, bounded, and increasing. Here, 0 is a lower bound and 1 is an upper bound.

(c) $\left\{\left(-\dfrac{1}{2}\right)^n\right\} = \left\{-\dfrac{1}{2}, \dfrac{1}{4}, -\dfrac{1}{8}, \dfrac{1}{16}, \ldots\right\}$ is bounded and alternating. Here, $-1/2$ is a lower bound and $1/4$ is an upper bound.

(d) $\{(-1)^n n\} = \{-1, 2, -3, 4, -5, \ldots\}$ is alternating but not bounded either above or below.

When you want to show that a sequence is increasing, you can try to show that the inequality $a_{n+1} - a_n \geq 0$ holds for $n \geq 1$. Alternatively, if $a_n = f(n)$ for a differentiable function $f(x)$, you can show that f is a nondecreasing function on $[1, \infty)$ by showing that $f'(x) \geq 0$ there. Similar approaches are useful for showing that a sequence is decreasing.

EXAMPLE 3 If $a_n = \dfrac{n}{n^2+1}$, show that the sequence $\{a_n\}$ is decreasing.

Solution Since $a_n = f(n)$, where $f(x) = \dfrac{x}{x^2+1}$ and

$$f'(x) = \frac{(x^2+1)(1) - x(2x)}{(x^2+1)^2} = \frac{1-x^2}{(x^2+1)^2} \leq 0 \quad \text{for } x \geq 1,$$

the function $f(x)$ is decreasing on $[1, \infty)$; therefore, $\{a_n\}$ is a decreasing sequence.

The sequence $\left\{\dfrac{n^2}{2^n}\right\} = \left\{\dfrac{1}{2}, 1, \dfrac{9}{8}, 1, \dfrac{25}{32}, \dfrac{36}{64}, \dfrac{49}{128}, \ldots\right\}$ is positive and therefore bounded below. It seems clear that from the fourth term on, all the terms are getting smaller. However, $a_2 > a_1$ and $a_3 > a_2$. Since $a_{n+1} \leq a_n$ only if $n \geq 3$, we say that this sequence is **ultimately decreasing**. The adverb *ultimately* is used to describe any termwise property of a sequence that the terms have from some point on, but not necessarily at the beginning of the sequence. Thus, the sequence

$$\{n - 100\} = \{-99, -98, \ldots, -2, -1, 0, 1, 2, 3, \ldots\}$$

is *ultimately positive* even though the first 99 terms are negative, and the sequence

$$\left\{(-1)^n + \frac{4}{n}\right\} = \left\{3, 3, \frac{1}{3}, 2, -\frac{1}{5}, \frac{5}{3}, -\frac{3}{7}, \frac{3}{2}, \ldots\right\}$$

is *ultimately alternating* even though the first few terms do not alternate.

Convergence of Sequences

Central to the study of sequences is the notion of convergence. The concept of the limit of a sequence is a special case of the concept of the limit of a function $f(x)$ as $x \to \infty$. We say that the sequence $\{a_n\}$ **converges to the limit** L, and we write $\lim_{n\to\infty} a_n = L$, provided the distance from a_n to L on the real line approaches 0 as n increases toward ∞. We state this definition more formally as follows:

DEFINITION 2

Limit of a sequence

We say that sequence $\{a_n\}$ converges to the limit L, and we write $\lim_{n\to\infty} a_n = L$, if for every positive real number ϵ there exists an integer N (which may depend on ϵ) such that if $n \ge N$, then $|a_n - L| < \epsilon$.

This definition is illustrated in Figure 9.1.

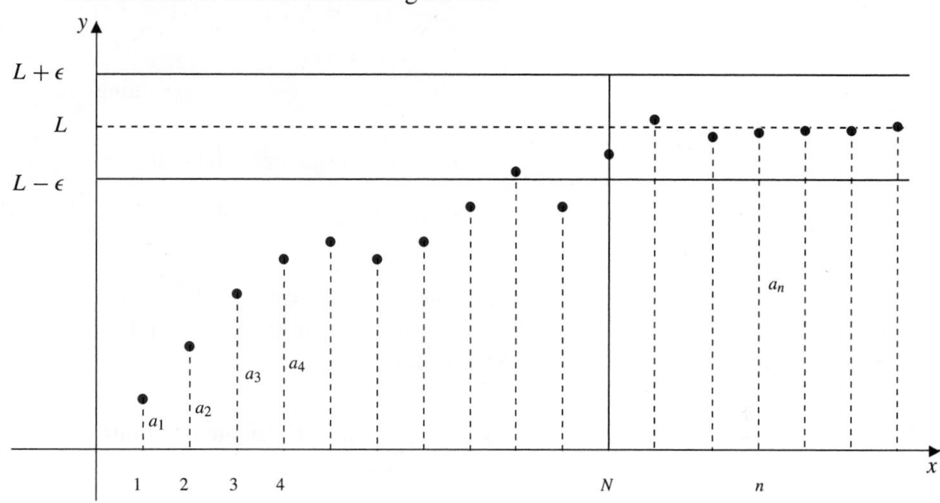

Figure 9.1 A convergent sequence

EXAMPLE 4 Show that $\lim_{n\to\infty} \dfrac{c}{n^p} = 0$ for any real number c and any $p > 0$.

Solution Let $\epsilon > 0$ be given. Then

$$\left|\frac{c}{n^p}\right| < \epsilon \quad \text{if} \quad n^p > \frac{|c|}{\epsilon},$$

that is, if $n \ge N$, the least integer greater than $(|c|/\epsilon)^{1/p}$. Therefore, by Definition 2, $\lim_{n\to\infty} \dfrac{c}{n^p} = 0$.

Every sequence $\{a_n\}$ must either **converge** to a finite limit L or **diverge**. That is, either $\lim_{n\to\infty} a_n = L$ exists (is a real number) or $\lim_{n\to\infty} a_n$ does not exist. If $\lim_{n\to\infty} a_n = \infty$, we can say that the sequence diverges to ∞; if $\lim_{n\to\infty} a_n = -\infty$, we can say that it diverges to $-\infty$. If $\lim_{n\to\infty} a_n$ simply does not exist (but is not ∞ or $-\infty$), we can only say that the sequence diverges.

EXAMPLE 5 **(Examples of convergent and divergent sequences)**

(a) $\{(n-1)/n\}$ converges to 1; $\lim_{n\to\infty}(n-1)/n = \lim_{n\to\infty}\big(1-(1/n)\big) = 1$.

(b) $\{n\} = \{1, 2, 3, 4, \ldots\}$ diverges to ∞.

(c) $\{-n\} = \{-1, -2, -3, -4, \ldots\}$ diverges to $-\infty$.

(d) $\{(-1)^n\} = \{-1, 1, -1, 1, -1, \ldots\}$ simply diverges.

(e) $\{(-1)^n n\} = \{-1, 2, -3, 4, -5, \ldots\}$ diverges (but not to ∞ or $-\infty$ even though $\lim_{n\to\infty} |a_n| = \infty$).

The limit of a sequence is equivalent to the limit of a function as its argument approaches infinity:

If $\lim\limits_{x\to\infty} f(x) = L$ and $a_n = f(n)$, then $\lim\limits_{n\to\infty} a_n = L$.

Because of this, the standard rules for limits of functions (Theorems 2 and 4 of Section 1.2) also hold for limits of sequences, with the appropriate changes of notation. Thus, if $\{a_n\}$ and $\{b_n\}$ converge, then

$$\lim_{n\to\infty}(a_n \pm b_n) = \lim_{n\to\infty} a_n \pm \lim_{n\to\infty} b_n,$$

$$\lim_{n\to\infty} ca_n = c \lim_{n\to\infty} a_n,$$

$$\lim_{n\to\infty} a_n b_n = \left(\lim_{n\to\infty} a_n\right)\left(\lim_{n\to\infty} b_n\right),$$

$$\lim_{n\to\infty} \frac{a_n}{b_n} = \frac{\lim_{n\to\infty} a_n}{\lim_{n\to\infty} b_n} \quad \text{assuming } \lim_{n\to\infty} b_n \neq 0.$$

If $a_n \le b_n$ ultimately, then $\lim_{n\to\infty} a_n \le \lim_{n\to\infty} b_n$.

If $a_n \le b_n \le c_n$ ultimately, and $\lim_{n\to\infty} a_n = L = \lim_{n\to\infty} c_n$, then $\lim_{n\to\infty} b_n = L$.

The limits of many explicitly defined sequences can be evaluated using these properties in a manner similar to the methods used for limits of the form $\lim_{x\to\infty} f(x)$ in Section 1.3.

EXAMPLE 6 Calculate the limits of the sequences

$$\text{(a) } \left\{\frac{2n^2 - n - 1}{5n^2 + n - 3}\right\}, \quad \text{(b) } \left\{\frac{\cos n}{n}\right\}, \quad \text{and} \quad \text{(c) } \{\sqrt{n^2 + 2n} - n\}.$$

Solution

(a) We divide the numerator and denominator of the expression for a_n by the highest power of n in the denominator, that is, by n^2:

$$\lim_{n\to\infty} \frac{2n^2 - n - 1}{5n^2 + n - 3} = \lim_{n\to\infty} \frac{2 - (1/n) - (1/n^2)}{5 + (1/n) - (3/n^2)} = \frac{2 - 0 - 0}{5 + 0 - 0} = \frac{2}{5},$$

since $\lim_{n\to\infty} 1/n = 0$ and $\lim_{n\to\infty} 1/n^2 = 0$. The sequence converges and its limit is 2/5.

(b) Since $|\cos n| \le 1$ for every n, we have

$$-\frac{1}{n} \le \frac{\cos n}{n} \le \frac{1}{n} \quad \text{for} \quad n \ge 1.$$

Now, $\lim_{n\to\infty} -1/n = 0$ and $\lim_{n\to\infty} 1/n = 0$. Therefore, by the sequence version of the Squeeze Theorem, $\lim_{n\to\infty} (\cos n)/n = 0$. The given sequence converges to 0.

(c) For this sequence we multiply the numerator and the denominator (which is 1) by the conjugate of the expression in the numerator:

$$\begin{aligned}\lim_{n\to\infty}(\sqrt{n^2 + 2n} - n) &= \lim_{n\to\infty} \frac{(\sqrt{n^2 + 2n} - n)(\sqrt{n^2 + 2n} + n)}{\sqrt{n^2 + 2n} + n} \\ &= \lim_{n\to\infty} \frac{2n}{\sqrt{n^2 + 2n} + n} = \lim_{n\to\infty} \frac{2}{\sqrt{1 + (2/n)} + 1} = 1.\end{aligned}$$

The sequence converges to 1.

EXAMPLE 7 Evaluate $\lim_{n\to\infty} n\tan^{-1}\left(\frac{1}{n}\right)$.

Solution For this example it is best to replace the nth term of the sequence by the corresponding function of a real variable x and take the limit as $x \to \infty$. We use l'Hôpital's Rule:

$$\begin{aligned}
\lim_{n\to\infty} n\tan^{-1}\left(\frac{1}{n}\right) &= \lim_{x\to\infty} x\tan^{-1}\left(\frac{1}{x}\right) \\
&= \lim_{x\to\infty} \frac{\tan^{-1}\left(\frac{1}{x}\right)}{\frac{1}{x}} \qquad \left[\frac{0}{0}\right] \\
&= \lim_{x\to\infty} \frac{\frac{1}{1+(1/x^2)}\left(-\frac{1}{x^2}\right)}{-\left(\frac{1}{x^2}\right)} = \lim_{x\to\infty} \frac{1}{1+\frac{1}{x^2}} = 1.
\end{aligned}$$

THEOREM 1

If $\{a_n\}$ converges, then $\{a_n\}$ is bounded.

PROOF Suppose $\lim_{n\to\infty} a_n = L$. According to Definition 2, for $\epsilon = 1$ there exists a number N such that if $n > N$, then $|a_n - L| < 1$; therefore $|a_n| < 1 + |L|$ for such n. (Why is this true?) If K denotes the largest of the numbers $|a_1|$, $|a_2|$, $\dots$, $|a_N|$, and $1 + |L|$, then $|a_n| \le K$ for every $n = 1, 2, 3, \dots$. Hence $\{a_n\}$ is bounded. ■

The converse of Theorem 1 is false; the sequence $\{(-1)^n\}$ is bounded but does not converge.

The *completeness property* of the real number system (see Section P.1) can be reformulated in terms of sequences to read as follows:

> **Bounded monotonic sequences converge**
>
> If the sequence $\{a_n\}$ is bounded above and is (ultimately) increasing, then it converges. The same conclusion holds if $\{a_n\}$ is bounded below and is (ultimately) decreasing.

Thus, a bounded, ultimately monotonic sequence is convergent. (See Figure 9.2.)

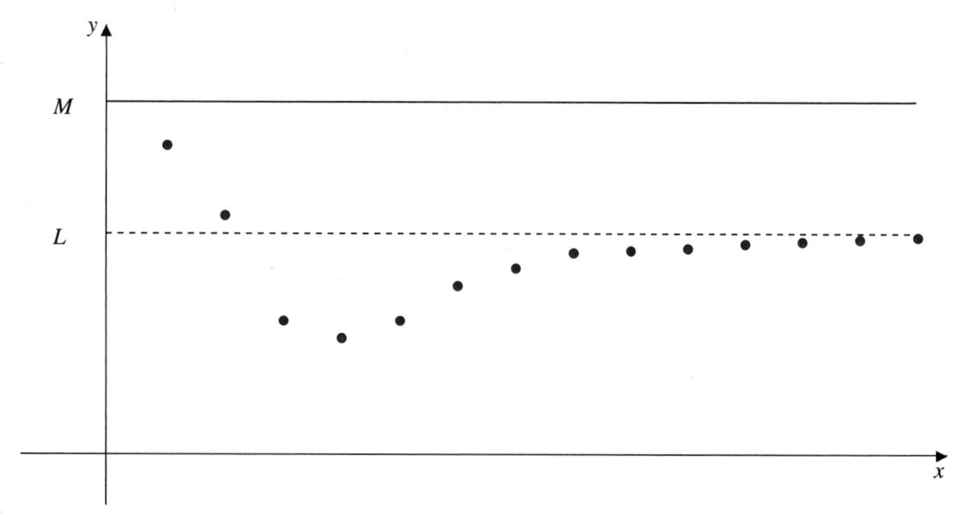

Figure 9.2 An ultimately increasing sequence that is bounded above

There is a subtle point to note in this solution. Showing that $\{a_n\}$ is increasing is pretty obvious, but how did we know to try and show that 3 (rather than some other number) was an upper bound? The answer is that we actually did the last part first and showed that *if* $\lim a_n = a$ exists, *then* $a = 3$. It then makes sense to try and show that $a_n < 3$ for all n.

Of course, we can easily show that any number greater than 3 is an upper bound.

EXAMPLE 8 Let a_n be defined recursively by

$$a_1 = 1, \qquad a_{n+1} = \sqrt{6 + a_n} \qquad (n = 1, 2, 3, \ldots).$$

Show that $\lim_{n\to\infty} a_n$ exists and find its value.

Solution Observe that $a_2 = \sqrt{6+1} = \sqrt{7} > a_1$. If $a_{k+1} > a_k$, then $a_{k+2} = \sqrt{6 + a_{k+1}} > \sqrt{6 + a_k} = a_{k+1}$, so $\{a_n\}$ is increasing, by induction. Now observe that $a_1 = 1 < 3$. If $a_k < 3$, then $a_{k+1} = \sqrt{6 + a_k} < \sqrt{6+3} = 3$, so $a_n < 3$ for every n by induction. Since $\{a_n\}$ is increasing and bounded above, $\lim_{n\to\infty} a_n = a$ exists, by completeness. Since $\sqrt{6+x}$ is a continuous function of x, we have

$$a = \lim_{n\to\infty} a_{n+1} = \lim_{n\to\infty} \sqrt{6 + a_n} = \sqrt{6 + \lim_{n\to\infty} a_n} = \sqrt{6 + a}.$$

Thus, $a^2 = 6 + a$, or $a^2 - a - 6 = 0$, or $(a-3)(a+2) = 0$. This quadratic has roots $a = 3$ and $a = -2$. Since $a_n \ge 1$ for every n, we must have $a \ge 1$. Therefore, $a = 3$ and $\lim_{n\to\infty} a_n = 3$.

EXAMPLE 9 Does $\left\{\left(1 + \dfrac{1}{n}\right)^n\right\}$ converge or diverge?

Solution We could make an effort to show that the given sequence is, in fact, increasing and bounded above. (See Exercise 32 at the end of this section.) However, we already know the answer. The sequence converges by Theorem 6 of Section 3.4:

$$\lim_{n\to\infty}\left(1 + \frac{1}{n}\right)^n = e^1 = e.$$

THEOREM 2

If $\{a_n\}$ is (ultimately) increasing, then either it is bounded above, and therefore convergent, or it is not bounded above and diverges to infinity.

The proof of this theorem is left as an exercise. A corresponding result holds for (ultimately) decreasing sequences.

The following theorem evaluates two important limits that find frequent application in the study of series.

THEOREM 3

(a) If $|x| < 1$, then $\lim_{n\to\infty} x^n = 0$.

(b) If x is any real number, then $\lim_{n\to\infty} \dfrac{x^n}{n!} = 0$.

PROOF For part (a) observe that

$$\lim_{n\to\infty} \ln|x|^n = \lim_{n\to\infty} n \ln|x| = -\infty,$$

since $\ln|x| < 0$ when $|x| < 1$. Accordingly, since e^x is continuous,

$$\lim_{n\to\infty} |x|^n = \lim_{n\to\infty} e^{\ln|x|^n} = e^{\lim_{n\to\infty} \ln|x|^n} = 0.$$

Since $-|x|^n \le x^n \le |x|^n$, we have $\lim_{n\to\infty} x^n = 0$ by the Squeeze Theorem.

For part (b), pick any x and let N be an integer such that $N > |x|$. If $n > N$ we have

$$\begin{aligned}
\left|\frac{x^n}{n!}\right| &= \frac{|x|}{1}\frac{|x|}{2}\frac{|x|}{3}\cdots\frac{|x|}{N-1}\frac{|x|}{N}\frac{|x|}{N+1}\cdots\frac{|x|}{n} \\
&< \frac{|x|^{N-1}}{(N-1)!}\frac{|x|}{N}\frac{|x|}{N}\frac{|x|}{N}\cdots\frac{|x|}{N} \\
&= \frac{|x|^{N-1}}{(N-1)!}\left(\frac{|x|}{N}\right)^{n-N+1} = K\left(\frac{|x|}{N}\right)^n,
\end{aligned}$$

where $K = \dfrac{|x|^{N-1}}{(N-1)!}\left(\dfrac{|x|}{N}\right)^{1-N}$ is a constant that is independent of n. Since $|x|/N < 1$, we have $\lim_{n\to\infty}(|x|/N)^n = 0$ by part (a). Thus $\lim_{n\to\infty}|x^n/n!| = 0$, so $\lim_{n\to\infty} x^n/n! = 0$.

EXAMPLE 10 Find $\lim_{n\to\infty} \dfrac{3^n + 4^n + 5^n}{5^n}$.

Solution $\lim_{n\to\infty} \dfrac{3^n + 4^n + 5^n}{5^n} = \lim_{n\to\infty}\left[\left(\dfrac{3}{5}\right)^n + \left(\dfrac{4}{5}\right)^n + 1\right] = 0 + 0 + 1 = 1$, by Theorem 3(a).

EXERCISES 9.1

In Exercises 1–13, determine whether the given sequence is (a) bounded (above or below), (b) positive or negative (ultimately), (c) increasing, decreasing, or alternating, and (d) convergent, divergent, divergent to ∞ or $-\infty$.

1. $\left\{\dfrac{2n^2}{n^2+1}\right\}$

2. $\left\{\dfrac{2n}{n^2+1}\right\}$

3. $\left\{4 - \dfrac{(-1)^n}{n}\right\}$

4. $\left\{\sin\dfrac{1}{n}\right\}$

5. $\left\{\dfrac{n^2-1}{n}\right\}$

6. $\left\{\dfrac{e^n}{\pi^n}\right\}$

7. $\left\{\dfrac{e^n}{\pi^{n/2}}\right\}$

8. $\left\{\dfrac{(-1)^n n}{e^n}\right\}$

9. $\left\{\dfrac{2^n}{n^n}\right\}$

10. $\left\{\dfrac{(n!)^2}{(2n)!}\right\}$

11. $\left\{n\cos\left(\dfrac{n\pi}{2}\right)\right\}$

12. $\left\{\dfrac{\sin n}{n}\right\}$

13. $\{1, 1, -2, 3, 3, -4, 5, 5, -6, \ldots\}$

In Exercises 14–29, evaluate, wherever possible, the limit of the sequence $\{a_n\}$.

14. $a_n = \dfrac{5-2n}{3n-7}$

15. $a_n = \dfrac{n^2-4}{n+5}$

16. $a_n = \dfrac{n^2}{n^3+1}$

17. $a_n = (-1)^n \dfrac{n}{n^3+1}$

18. $a_n = \dfrac{n^2 - 2\sqrt{n} + 1}{1 - n - 3n^2}$

19. $a_n = \dfrac{e^n - e^{-n}}{e^n + e^{-n}}$

20. $a_n = n\sin\dfrac{1}{n}$

21. $a_n = \left(\dfrac{n-3}{n}\right)^n$

22. $a_n = \dfrac{n}{\ln(n+1)}$

23. $a_n = \sqrt{n+1} - \sqrt{n}$

24. $a_n = n - \sqrt{n^2 - 4n}$

25. $a_n = \sqrt{n^2+n} - \sqrt{n^2-1}$

26. $a_n = \left(\dfrac{n-1}{n+1}\right)^n$

27. $a_n = \dfrac{(n!)^2}{(2n)!}$

28. $a_n = \dfrac{n^2 2^n}{n!}$

29. $a_n = \dfrac{\pi^n}{1+2^{2n}}$

30. Let $a_1 = 1$ and $a_{n+1} = \sqrt{1+2a_n}$ $(n = 1, 2, 3, \ldots)$. Show that $\{a_n\}$ is increasing and bounded above. (*Hint:* Show that 3 is an upper bound.) Hence, conclude that the sequence converges, and find its limit.

31. Repeat Exercise 30 for the sequence defined by $a_1 = 3$, $a_{n+1} = \sqrt{15 + 2a_n}$, $n = 1, 2, 3, \ldots$. This time you will have to guess an upper bound.

32. Let $a_n = \left(1 + \dfrac{1}{n}\right)^n$ so that $\ln a_n = n\ln\left(1 + \dfrac{1}{n}\right)$. Use properties of the logarithm function to show that (a) $\{a_n\}$ is increasing and (b) e is an upper bound for $\{a_n\}$.

33. Prove Theorem 2. Also, state an analogous theorem pertaining to ultimately decreasing sequences.

34. If $\{|a_n|\}$ is bounded, prove that $\{a_n\}$ is bounded.

35. If $\lim_{n\to\infty}|a_n| = 0$, prove that $\lim_{n\to\infty} a_n = 0$.

36. Which of the following statements are TRUE and which are FALSE? Justify your answers.

(a) If $\lim_{n\to\infty} a_n = \infty$ and $\lim_{n\to\infty} b_n = L > 0$, then $\lim_{n\to\infty} a_n b_n = \infty$.

(b) If $\lim_{n\to\infty} a_n = \infty$ and $\lim_{n\to\infty} b_n = -\infty$, then $\lim_{n\to\infty}(a_n + b_n) = 0$.

(c) If $\lim_{n\to\infty} a_n = \infty$ and $\lim_{n\to\infty} b_n = -\infty$, then $\lim_{n\to\infty} a_n b_n = -\infty$.

(d) If neither $\{a_n\}$ nor $\{b_n\}$ converges, then $\{a_n b_n\}$ does not converge.

(e) If $\{|a_n|\}$ converges, then $\{a_n\}$ converges.

9.2 Infinite Series

An **infinite series**, usually just called a **series**, is a formal sum of infinitely many terms; for instance, $a_1 + a_2 + a_3 + a_4 + \cdots$ is a series formed by adding the terms of the sequence $\{a_n\}$. This series is also denoted $\sum_{n=1}^{\infty} a_n$:

$$\sum_{n=1}^{\infty} a_n = a_1 + a_2 + a_3 + a_4 + \cdots.$$

For example,

$$\sum_{n=1}^{\infty} \frac{1}{n} = 1 + \frac{1}{2} + \frac{1}{3} + \frac{1}{4} + \cdots$$

$$\sum_{n=1}^{\infty} \frac{(-1)^{n-1}}{2^{n-1}} = 1 - \frac{1}{2} + \frac{1}{4} - \frac{1}{8} + \frac{1}{16} - \cdots.$$

It is sometimes necessary or useful to start the sum from some index other than 1:

$$\sum_{n=0}^{\infty} a^n = 1 + a + a^2 + a^3 + \cdots$$

$$\sum_{n=2}^{\infty} \frac{1}{\ln n} = \frac{1}{\ln 2} + \frac{1}{\ln 3} + \frac{1}{\ln 4} + \cdots.$$

Note that the latter series would make no sense if we had started the sum from $n = 1$; the first term would have been undefined.

When necessary, we can change the index of summation to start at a different value. This is accomplished by a substitution as illustrated in Example 3 of Section 5.1. For instance, using the substitution $n = m - 2$, we can rewrite $\sum_{n=1}^{\infty} a_n$ in the form $\sum_{m=3}^{\infty} a_{m-2}$. Both sums give rise to the same expansion

$$\sum_{n=1}^{\infty} a_n = a_1 + a_2 + a_3 + \cdots = \sum_{m=3}^{\infty} a_{m-2}.$$

Addition is an operation that is carried out on two numbers at a time. If we want to calculate the finite sum $a_1 + a_2 + a_3$, we could proceed by adding $a_1 + a_2$ and then adding a_3 to this sum, or else we might first add $a_2 + a_3$ and then add a_1 to the sum. Of course, the associative law for addition assures us we will get the same answer both ways. This is the reason the symbol $a_1 + a_2 + a_3$ makes sense; we would otherwise have to write $(a_1 + a_2) + a_3$ or $a_1 + (a_2 + a_3)$. This reasoning extends to any sum $a_1 + a_2 + \cdots + a_n$ of finitely many terms, but it is not obvious what should be meant by a sum with infinitely many terms:

$$a_1 + a_2 + a_3 + a_4 + \cdots.$$

We no longer have any assurance that the terms can be added up in any order to yield the same sum. In fact, we will see in Section 9.4 that in certain circumstances, changing the order of terms in a series can actually change the sum of the series. The interpretation we place on the infinite sum is that of adding from left to right, as suggested by the grouping

$$\cdots ((((a_1 + a_2) + a_3) + a_4) + a_5) + \cdots.$$

We accomplish this by defining a new sequence $\{s_n\}$, called the **sequence of partial sums** of the series $\sum_{n=1}^{\infty} a_n$, so that s_n is the sum of the first n terms of the series:

$$
\begin{aligned}
s_1 &= a_1 \\
s_2 &= s_1 + a_2 = a_1 + a_2 \\
s_3 &= s_2 + a_3 = a_1 + a_2 + a_3 \\
&\vdots \\
s_n &= s_{n-1} + a_n = a_1 + a_2 + a_3 + \cdots + a_n = \sum_{j=1}^{n} a_j \\
&\vdots
\end{aligned}
$$

We then define the sum of the infinite series to be the limit of this sequence of partial sums.

DEFINITION 3

Convergence of a series

We say that the series $\sum_{n=1}^{\infty} a_n$ **converges to the sum** s, and we write

$$\sum_{n=1}^{\infty} a_n = s,$$

if $\lim_{n\to\infty} s_n = s$, where s_n is the nth partial sum of $\sum_{n=1}^{\infty} a_n$:

$$s_n = a_1 + a_2 + a_3 + \cdots + a_n = \sum_{j=1}^{n} a_j.$$

Thus, a *series* converges if and only if the *sequence* of its partial sums converges.

Similarly, a series is said to diverge to infinity, diverge to negative infinity, or simply diverge if its sequence of partial sums does so. It must be stressed that the convergence of the series $\sum_{n=1}^{\infty} a_n$ depends on the convergence of the sequence $\{s_n\} = \{\sum_{j=1}^{n} a_j\}$, *not* the sequence $\{a_n\}$.

Geometric Series

DEFINITION 4

Geometric series

A series of the form $\sum_{n=1}^{\infty} a\, r^{n-1} = a + ar + ar^2 + ar^3 + \cdots$, whose nth term is $a_n = a\, r^{n-1}$, is called a **geometric series**. The number a is the first term. The number r is called the **common ratio** of the series, since it is the value of the ratio of the $(n+1)$st term to the nth term for any $n \ge 1$:

$$\frac{a_{n+1}}{a_n} = \frac{ar^n}{ar^{n-1}} = r, \qquad n = 1, 2, 3, \ldots.$$

The nth partial sum s_n of a geometric series is calculated as follows:

$$
\begin{aligned}
s_n &= a + ar + ar^2 + ar^3 + \cdots + ar^{n-1} \\
rs_n &= \phantom{a + {}} ar + ar^2 + ar^3 + \cdots + ar^{n-1} + ar^n.
\end{aligned}
$$

The second equation is obtained by multiplying the first by r. Subtracting these two equations (note the cancellations), we get $(1-r)s_n = a - ar^n$. If $r \ne 1$, we can divide by $1 - r$ and get a formula for s_n.

Partial sums of geometric series

If $r = 1$, then the nth partial sum of a geometric series $\sum_{n=1}^{\infty} ar^{n-1}$ is $s_n = a + a + \cdots + a = na$. If $r \neq 1$, then

$$s_n = a + ar + ar^2 + \cdots + ar^{n-1} = \frac{a(1 - r^n)}{1 - r}.$$

If $a = 0$, then $s_n = 0$ for every n, and $\lim_{n\to\infty} s_n = 0$. Now suppose $a \neq 0$. If $|r| < 1$, then $\lim_{n\to\infty} r^n = 0$, so $\lim_{n\to\infty} s_n = a/(1 - r)$. If $r > 1$, then $\lim_{n\to\infty} r^n = \infty$, and $\lim_{n\to\infty} s_n = \infty$ if $a > 0$, or $\lim_{n\to\infty} s_n = -\infty$ if $a < 0$. The same conclusion holds if $r = 1$, since $s_n = na$ in this case. If $r \leq -1$, $\lim_{n\to\infty} r^n$ does not exist and neither does $\lim_{n\to\infty} s_n$. Hence, we conclude that

$$\sum_{n=1}^{\infty} ar^{n-1} \begin{cases} \text{converges to } 0 & \text{if } a = 0 \\ \text{converges to } \dfrac{a}{1-r} & \text{if } |r| < 1 \\ \text{diverges to } \infty & \text{if } r \geq 1 \text{ and } a > 0 \\ \text{diverges to } -\infty & \text{if } r \geq 1 \text{ and } a < 0 \\ \text{diverges} & \text{if } r \leq -1 \text{ and } a \neq 0. \end{cases}$$

The representation of the function $1/(1 - x)$ as the sum of a geometric series,

$$\frac{1}{1-x} = \sum_{n=0}^{\infty} x^n = 1 + x + x^2 + x^3 + \cdots \quad \text{for } -1 < x < 1,$$

will be important in our discussion of power series later in this chapter.

EXAMPLE 1 **(Examples of geometric series and their sums)**

(a) $1 + \frac{1}{2} + \frac{1}{4} + \frac{1}{8} + \cdots = \sum_{n=1}^{\infty} \left(\frac{1}{2}\right)^{n-1} = \frac{1}{1 - \frac{1}{2}} = 2$. Here $a = 1$ and $r = \frac{1}{2}$.

Since $|r| < 1$, the series converges.

(b) $\pi - e + \frac{e^2}{\pi} - \frac{e^3}{\pi^2} + \cdots = \sum_{n=1}^{\infty} \pi \left(-\frac{e}{\pi}\right)^{n-1}$ Here $a = \pi$ and $r = -\frac{e}{\pi}$.

$$= \frac{\pi}{1 - \left(-\frac{e}{\pi}\right)} = \frac{\pi^2}{\pi + e}.$$

The series converges since $\left|-\frac{e}{\pi}\right| < 1$.

(c) $1 + 2^{1/2} + 2 + 2^{3/2} + \cdots = \sum_{n=1}^{\infty} (\sqrt{2})^{n-1}$. This series diverges to ∞ since $a = 1 > 0$ and $r = \sqrt{2} > 1$.

(d) $1 - 1 + 1 - 1 + 1 - \cdots = \sum_{n=1}^{\infty} (-1)^{n-1}$. This series diverges since $r = -1$.

(e) Let $x = 0.32\,32\,32 \cdots = 0.\overline{32}$; then

$$x = \frac{32}{100} + \frac{32}{100^2} + \frac{32}{100^3} + \cdots = \sum_{n=1}^{\infty} \frac{32}{100} \left(\frac{1}{100}\right)^{n-1} = \frac{32}{100} \frac{1}{1 - \frac{1}{100}} = \frac{32}{99}.$$

This is an alternative to the method of Example 1 of Section P.1 for representing repeating decimals as quotients of integers.

EXAMPLE 2 If money earns interest at a constant effective rate of 5% per year, how much should you pay today for an annuity that will pay you (a) \$1,000 at the end of each of the next 10 years and (b) \$1,000 at the end of every year forever?

Solution A payment of \$1,000 that is due to be received n years from now has present value $\$1{,}000 \times \left(\frac{1}{1.05}\right)^n$ (since $\$A$ would grow to $\$A(1.05)^n$ in n years). Thus, \$1,000 payments at the end of each of the next n years are worth $\$s_n$ at the present time, where

$$
\begin{aligned}
s_n &= 1{,}000\left[\frac{1}{1.05} + \left(\frac{1}{1.05}\right)^2 + \cdots + \left(\frac{1}{1.05}\right)^n\right] \\
&= \frac{1{,}000}{1.05}\left[1 + \frac{1}{1.05} + \left(\frac{1}{1.05}\right)^2 + \cdots + \left(\frac{1}{1.05}\right)^{n-1}\right] \\
&= \frac{1{,}000}{1.05}\,\frac{1 - \left(\frac{1}{1.05}\right)^n}{1 - \frac{1}{1.05}} = \frac{1{,}000}{0.05}\left[1 - \left(\frac{1}{1.05}\right)^n\right].
\end{aligned}
$$

(a) The present value of 10 future payments is $\$s_{10} = \$7{,}721.73$.

(b) The present value of future payments continuing forever is

$$\$\lim_{n\to\infty} s_n = \frac{\$1{,}000}{0.05} = \$20{,}000.$$

Telescoping Series and Harmonic Series

EXAMPLE 3 Show that the series

$$\sum_{n=1}^{\infty} \frac{1}{n(n+1)} = \frac{1}{1\times 2} + \frac{1}{2\times 3} + \frac{1}{3\times 4} + \frac{1}{4\times 5} + \cdots$$

converges and find its sum.

Solution Since $\frac{1}{n(n+1)} = \frac{1}{n} - \frac{1}{n+1}$, we can write the partial sum s_n in the form

$$
\begin{aligned}
s_n &= \frac{1}{1\times 2} + \frac{1}{2\times 3} + \frac{1}{3\times 4} + \cdots + \frac{1}{(n-1)n} + \frac{1}{n(n+1)} \\
&= \left(1 - \frac{1}{2}\right) + \left(\frac{1}{2} - \frac{1}{3}\right) + \left(\frac{1}{3} - \frac{1}{4}\right) \\
&\quad + \cdots + \left(\frac{1}{n-1} - \frac{1}{n}\right) + \left(\frac{1}{n} - \frac{1}{n+1}\right) \\
&= 1 - \frac{1}{2} + \frac{1}{2} - \frac{1}{3} + \frac{1}{3} - \cdots - \frac{1}{n} + \frac{1}{n} - \frac{1}{n+1} \\
&= 1 - \frac{1}{n+1}.
\end{aligned}
$$

Therefore, $\lim_{n\to\infty} s_n = 1$ and the series converges to 1:

$$\sum_{n=1}^{\infty} \frac{1}{n(n+1)} = 1.$$

This is an example of a **telescoping series**, so called because the partial sums *fold up* into a simple form when the terms are expanded in partial fractions. Other examples can be found in the exercises at the end of this section. As these examples show, the method of partial fractions can be a useful tool for series as well as for integrals.

EXAMPLE 4 Show that the **harmonic series**

$$\sum_{n=1}^{\infty} \frac{1}{n} = 1 + \frac{1}{2} + \frac{1}{3} + \frac{1}{4} + \cdots$$

diverges to infinity.

Solution If s_n is the nth partial sum of the harmonic series, then

$$\begin{aligned} s_n &= 1 + \frac{1}{2} + \frac{1}{3} + \cdots + \frac{1}{n} \\ &= \text{sum of areas of rectangles shaded in Figure 9.3} \\ &> \text{area under } y = \frac{1}{x} \text{ from } x = 1 \text{ to } x = n+1 \\ &= \int_1^{n+1} \frac{dx}{x} = \ln(n+1). \end{aligned}$$

Now $\lim_{n\to\infty} \ln(n+1) = \infty$. Therefore, $\lim_{n\to\infty} s_n = \infty$ and

$$\sum_{n=1}^{\infty} \frac{1}{n} = 1 + \frac{1}{2} + \frac{1}{3} + \cdots \qquad \text{diverges to infinity.}$$

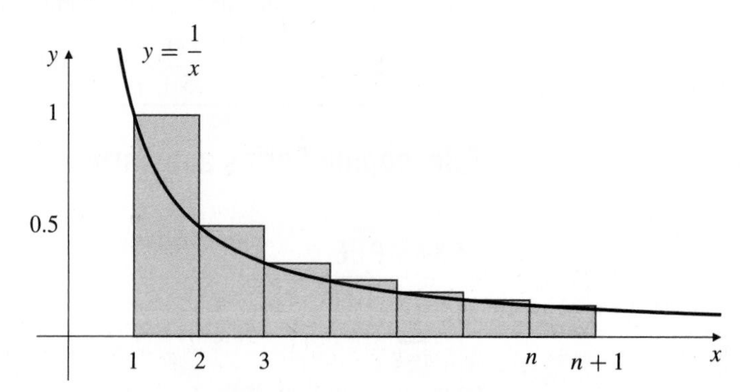

Figure 9.3 A partial sum of the harmonic series

Like geometric series, the harmonic series will often be encountered in subsequent sections.

Some Theorems About Series

THEOREM 4

If $\sum_{n=1}^{\infty} a_n$ converges, then $\lim_{n\to\infty} a_n = 0$. Therefore, if $\lim_{n\to\infty} a_n$ does not exist, or exists but is not zero, then the series $\sum_{n=1}^{\infty} a_n$ is divergent. (This amounts to an ***n*th term test for divergence of a series.**)

PROOF If $s_n = a_1 + a_2 + \cdots + a_n$, then $s_n - s_{n-1} = a_n$. If $\sum_{n=1}^{\infty} a_n$ converges, then $\lim_{n\to\infty} s_n = s$ exists, and $\lim_{n\to\infty} s_{n-1} = s$. Hence $\lim_{n\to\infty} a_n = s - s = 0$.

Remark Theorem 4 is *very important* for the understanding of infinite series. Students often err either in forgetting that *a series cannot converge if its terms do not approach zero* or in confusing this result with its *converse*, which is false. The converse would say that if $\lim_{n\to\infty} a_n = 0$, then $\sum_{n=1}^{\infty} a_n$ must converge. The harmonic series is a counterexample showing the falsehood of this assertion:

$$\lim_{n\to\infty} \frac{1}{n} = 0 \qquad \text{but} \qquad \sum_{n=1}^{\infty} \frac{1}{n} \text{ diverges to infinity.}$$

When considering whether a given series converges, the first question you should ask yourself is: "Does the nth term approach 0 as n approaches ∞?" If the answer is *no*, then the series does *not* converge. If the answer is *yes*, then the series *may or may not* converge. If the sequence of terms $\{a_n\}$ tends to a nonzero limit L, then $\sum_{n=1}^{\infty} a_n$ diverges to infinity if $L > 0$ and diverges to negative infinity if $L < 0$.

EXAMPLE 5

(a) $\displaystyle\sum_{n=1}^{\infty} \frac{n}{2n-1}$ diverges to infinity since $\lim_{n\to\infty} \dfrac{n}{2n-1} = 1/2 > 0$.

(b) $\sum_{n=1}^{\infty} (-1)^n n \sin(1/n)$ diverges since

$$\lim_{n\to\infty} \left|(-1)^n n \sin\frac{1}{n}\right| = \lim_{n\to\infty} \frac{\sin(1/n)}{1/n} = \lim_{x\to 0+} \frac{\sin x}{x} = 1 \neq 0.$$

The following theorem asserts that it is only the *ultimate* behaviour of $\{a_n\}$ that determines whether $\sum_{n=1}^{\infty} a_n$ converges. Any finite number of terms can be dropped from the beginning of a series without affecting the convergence; the convergence depends only on the *tail* of the series. Of course, the actual sum of the series depends on *all* the terms.

THEOREM 5

$\sum_{n=1}^{\infty} a_n$ converges if and only if $\sum_{n=N}^{\infty} a_n$ converges for any integer $N \geq 1$.

THEOREM 6

If $\{a_n\}$ is ultimately positive, then the series $\sum_{n=1}^{\infty} a_n$ must either converge (if its partial sums are bounded above) or diverge to infinity (if its partial sums are not bounded above).

The proofs of these two theorems are posed as exercises at the end of this section. The following theorem is just a reformulation of standard laws of limits.

THEOREM 7

If $\sum_{n=1}^{\infty} a_n$ and $\sum_{n=1}^{\infty} b_n$ converge to A and B, respectively, then

(a) $\sum_{n=1}^{\infty} ca_n$ converges to cA (where c is any constant);

(b) $\sum_{n=1}^{\infty} (a_n \pm b_n)$ converges to $A \pm B$;

(c) if $a_n \leq b_n$ for all $n = 1, 2, 3, \ldots$, then $A \leq B$.

EXAMPLE 6 Find the sum of the series $\displaystyle\sum_{n=1}^{\infty} \frac{1+2^{n+1}}{3^n}$.

Solution The given series is the sum of two geometric series,

$$\sum_{n=1}^{\infty} \frac{1}{3^n} = \sum_{n=1}^{\infty} \frac{1}{3}\left(\frac{1}{3}\right)^{n-1} = \frac{1/3}{1-(1/3)} = \frac{1}{2} \quad \text{and}$$

$$\sum_{n=1}^{\infty} \frac{2^{n+1}}{3^n} = \sum_{n=1}^{\infty} \frac{4}{3}\left(\frac{2}{3}\right)^{n-1} = \frac{4/3}{1-(2/3)} = 4.$$

Thus, its sum is $\dfrac{1}{2} + 4 = \dfrac{9}{2}$ by Theorem 7(b).

EXERCISES 9.2

In Exercises 1–18, find the sum of the given series, or show that the series diverges (possibly to infinity or negative infinity). Exercises 11–14 are telescoping series and should be done by partial fractions as suggested in Example 3 in this section.

1. $\frac{1}{3}+\frac{1}{9}+\frac{1}{27}+\cdots=\sum_{n=1}^{\infty}\frac{1}{3^n}$

2. $3-\frac{3}{4}+\frac{3}{16}-\frac{3}{64}+\cdots=\sum_{n=1}^{\infty}3\left(-\frac{1}{4}\right)^{n-1}$

3. $\sum_{n=5}^{\infty}\frac{1}{(2+\pi)^{2n}}$

4. $\sum_{n=0}^{\infty}\frac{5}{10^{3n}}$

5. $\sum_{n=2}^{\infty}\frac{(-5)^n}{8^{2n}}$

6. $\sum_{n=0}^{\infty}\frac{1}{e^n}$

7. $\sum_{k=0}^{\infty}\frac{2^{k+3}}{e^{k-3}}$

8. $\sum_{j=1}^{\infty}\pi^{j/2}\cos(j\pi)$

9. $\sum_{n=1}^{\infty}\frac{3+2^n}{2^{n+2}}$

10. $\sum_{n=0}^{\infty}\frac{3+2^n}{3^{n+2}}$

11. $\sum_{n=1}^{\infty}\frac{1}{n(n+2)}=\frac{1}{1\times 3}+\frac{1}{2\times 4}+\frac{1}{3\times 5}+\cdots$

12. $\sum_{n=1}^{\infty}\frac{1}{(2n-1)(2n+1)}=\frac{1}{1\times 3}+\frac{1}{3\times 5}+\frac{1}{5\times 7}+\cdots$

13. $\sum_{n=1}^{\infty}\frac{1}{(3n-2)(3n+1)}=\frac{1}{1\times 4}+\frac{1}{4\times 7}+\frac{1}{7\times 10}+\cdots$

14. $\sum_{n=1}^{\infty}\frac{1}{n(n+1)(n+2)}$
$=\frac{1}{1\times 2\times 3}+\frac{1}{2\times 3\times 4}+\frac{1}{3\times 4\times 5}+\cdots$

15. $\sum_{n=1}^{\infty}\frac{1}{2n-1}$

16. $\sum_{n=1}^{\infty}\frac{n}{n+2}$

17. $\sum_{n=1}^{\infty}n^{-1/2}$

18. $\sum_{n=1}^{\infty}\frac{2}{n+1}$

19. Obtain a simple expression for the partial sum s_n of the series $\sum_{n=1}^{\infty}(-1)^n$, and use it to show that the series diverges.

20. Find the sum of the series

$$\frac{1}{1}+\frac{1}{1+2}+\frac{1}{1+2+3}+\frac{1}{1+2+3+4}+\cdots.$$

21. When dropped, an elastic ball bounces back up to a height three-quarters of that from which it fell. If the ball is dropped from a height of 2 m and allowed to bounce up and down indefinitely, what is the total distance it travels before coming to rest?

22. If a bank account pays 10% simple interest into an account once a year, what is the balance in the account at the end of 8 years if $1,000 is deposited into the account at the beginning of each of the 8 years? (Assume there was no balance in the account initially.)

23. Prove Theorem 5.

24. Prove Theorem 6.

25. State a theorem analogous to Theorem 6 but for a negative sequence.

In Exercises 26–31, decide whether the given statement is TRUE or FALSE. If it is true, prove it. If it is false, give a counter-example showing the falsehood.

26. If $a_n=0$ for every n, then $\sum a_n$ converges.

27. If $\sum a_n$ converges, then $\sum(1/a_n)$ diverges to infinity.

28. If $\sum a_n$ and $\sum b_n$ both diverge, then so does $\sum(a_n+b_n)$.

29. If $a_n\ge c>0$ for every n, then $\sum a_n$ diverges to infinity.

30. If $\sum a_n$ diverges and $\{b_n\}$ is bounded, then $\sum a_n b_n$ diverges.

31. If $a_n>0$ and $\sum a_n$ converges, then $\sum(a_n)^2$ converges.

9.3 Convergence Tests for Positive Series

In the previous section we saw a few examples of convergent series (geometric and telescoping series) whose sums could be determined exactly because the partial sums s_n could be expressed in closed form as explicit functions of n whose limits as $n\to\infty$ could be evaluated. It is not usually possible to do this with a given series, and therefore it is not usually possible to determine the sum of the series exactly. However, there are many techniques for determining whether a given series converges and, if it does, for approximating the sum to any desired degree of accuracy.

In this section we deal exclusively with *positive series*, that is, series of the form

$$\sum_{n=1}^{\infty}a_n=a_1+a_2+a_3+\cdots,$$

where $a_n \ge 0$ for all $n \ge 1$. As noted in Theorem 6, such a series will converge if its partial sums are bounded above and will diverge to infinity otherwise. All our results apply equally well to *ultimately* positive series since convergence or divergence depends only on the *tail* of a series.

The Integral Test

The integral test provides a means for determining whether an ultimately positive series converges or diverges by comparing it with an improper integral that behaves similarly. Example 4 in Section 9.2 is an example of the use of this technique. We formalize the method in the following theorem.

THEOREM 8

The integral test

Suppose that $a_n = f(n)$, where f is positive, continuous, and nonincreasing on an interval $[N, \infty)$ for some positive integer N. Then

$$\sum_{n=1}^{\infty} a_n \quad \text{and} \quad \int_N^{\infty} f(t)\,dt$$

either both converge or both diverge to infinity.

PROOF Let $s_n = a_1 + a_2 + \cdots + a_n$. If $n > N$, we have

$$\begin{aligned} s_n &= s_N + a_{N+1} + a_{N+2} + \cdots + a_n \\ &= s_N + f(N+1) + f(N+2) + \cdots + f(n) \\ &= s_N + \text{ sum of areas of rectangles shaded in Figure 9.4(a)} \\ &\le s_N + \int_N^{\infty} f(t)\,dt. \end{aligned}$$

If the improper integral $\int_N^{\infty} f(t)\,dt$ converges, then the sequence $\{s_n\}$ is bounded above and $\sum_{n=1}^{\infty} a_n$ converges.

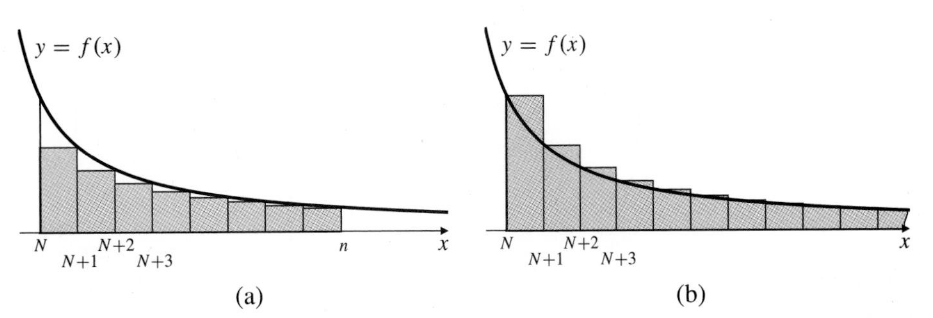

Figure 9.4 Comparing integrals and series

Conversely, suppose that $\sum_{n=1}^{\infty} a_n$ converges to the sum s. Then

$$\begin{aligned} \int_N^{\infty} f(t)\,dt &= \text{ area under } y = f(t) \text{ above } y = 0 \text{ from } t = N \text{ to } t = \infty \\ &\le \text{ sum of areas of shaded rectangles in Figure 9.4(b)} \\ &= a_N + a_{N+1} + a_{N+2} + \cdots \\ &= s - s_{N-1} < \infty, \end{aligned}$$

so the improper integral represents a finite area and is thus convergent. (We omit the remaining details showing that $\lim_{R\to\infty} \int_N^R f(t)\,dt$ exists; like the series case, the argument depends on the completeness of the real numbers.)

∎

Remark If $a_n = f(n)$, where f is positive, continuous, and nonincreasing on $[1, \infty)$, then Theorem 8 assures us that $\sum_{n=1}^{\infty} a_n$ and $\int_1^{\infty} f(x)\,dx$ both converge or both diverge to infinity. It does *not* tell us that the sum of the series is equal to the value of the integral. The two are not likely to be equal in the case of convergence. However, as we see below, integrals can help us approximate the sum of a series.

The principal use of the integral test is to establish the result of the following example concerning the series $\sum_{n=1}^{\infty} n^{-p}$, which is called a ***p*-series**. This result should be memorized; we will frequently compare the behaviour of other series with p-series later in this and subsequent sections.

EXAMPLE 1 **(*p*-series)** Show that

$$\sum_{n=1}^{\infty} n^{-p} = \sum_{n=1}^{\infty} \frac{1}{n^p} \begin{cases} \text{converges if } p > 1 \\ \text{diverges to infinity if } p \le 1. \end{cases}$$

Solution Observe that if $p > 0$, then $f(x) = x^{-p}$ is positive, continuous, and decreasing on $[1, \infty)$. By the integral test, the p-series converges for $p > 1$ and diverges for $0 < p \le 1$ by comparison with $\int_1^{\infty} x^{-p}\,dx$. (See Theorem 2(a) of Section 6.5.) If $p \le 0$, then $\lim_{n\to\infty}(1/n^p) \ne 0$, so the series cannot converge in this case. Being a positive series, it must diverge to infinity.

Remark The harmonic series $\sum_{n=1}^{\infty} n^{-1}$ (the case $p = 1$ of the p-series) is on the borderline between convergence and divergence, although it diverges. While its terms decrease toward 0 as n increases, they do not decrease *fast enough* to allow the sum of the series to be finite. If $p > 1$, the terms of $\sum_{n=1}^{\infty} n^{-p}$ decrease toward zero fast enough that their sum is finite. We can refine the distinction between convergence and divergence at $p = 1$ by using terms that decrease faster than $1/n$, but not as fast as $1/n^q$ for any $q > 1$. If $p > 0$, the terms $1/\big(n(\ln n)^p\big)$ have this property since $\ln n$ grows more slowly than any positive power of n as n increases. The question now arises whether $\sum_{n=2}^{\infty} 1/(n(\ln n)^p)$ converges. It does, provided again that $p > 1$; you can use the substitution $u = \ln x$ to check that

$$\int_2^{\infty} \frac{dx}{x(\ln x)^p} = \int_{\ln 2}^{\infty} \frac{du}{u^p},$$

which converges if $p > 1$ and diverges if $0 < p \le 1$. This process of fine-tuning Example 1 can be extended even further. (See Exercise 36 below.)

Using Integral Bounds to Estimate the Sum of a Series

Suppose that $a_k = f(k)$ for $k = n + 1, n + 2, n + 3, \ldots$, where f is a positive, continuous function, decreasing at least on the interval $[n, \infty)$. We have:

$$\begin{aligned} s - s_n &= \sum_{k=n+1}^{\infty} f(k) \\ &= \text{sum of areas of rectangles shaded in Figure 9.5(a)} \\ &\le \int_n^{\infty} f(x)\,dx. \end{aligned}$$

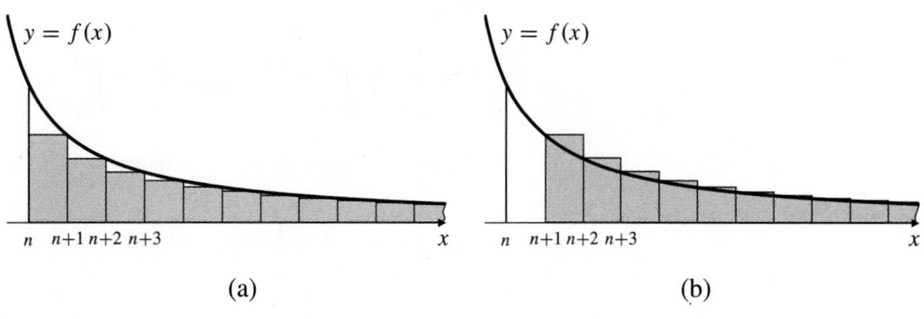

Figure 9.5 Using integrals to estimate the tail of a series

Similarly,

$$\begin{aligned} s - s_n &= \text{sum of areas of rectangles in Figure 9.5(b)} \\ &\geq \int_{n+1}^{\infty} f(x)\,dx. \end{aligned}$$

If we define

$$A_n = \int_n^{\infty} f(x)\,dx,$$

then we can combine the above inequalities to obtain

$$A_{n+1} \leq s - s_n \leq A_n,$$

or, equivalently:

$$s_n + A_{n+1} \leq s \leq s_n + A_n.$$

The error in the approximation $s \approx s_n$ satisfies $0 \leq s - s_n \leq A_n$. However, since s must lie in the interval $[s_n + A_{n+1}, s_n + A_n]$, we can do better by using the midpoint s_n^* of this interval as an approximation for s. The error is then less than half the length $A_n - A_{n+1}$ of the interval:

A better integral approximation

The error $|s - s_n^*|$ in the approximation

$$s \approx s_n^* = s_n + \frac{A_{n+1} + A_n}{2}, \qquad \text{where} \quad A_n = \int_n^{\infty} f(x)\,dx,$$

satisfies $\quad |s - s_n^*| \leq \dfrac{A_n - A_{n+1}}{2}.$

(Whenever a quantity is known to lie in a certain interval, the midpoint of that interval can be used to approximate the quantity, and the absolute value of the error in that approximation does not exceed half the length of the interval.)

EXAMPLE 2 Find the best approximation s_n^* to the sum s of the series $\sum_{n=1}^{\infty} 1/n^2$, making use of the partial sum s_n of the first n terms. How large would n have to be to ensure that the approximation $s \approx s_n^*$ has error less than 0.001 in absolute value? How large would n have to be to ensure that the approximation $s \approx s_n$ has error less than 0.001 in absolute value?

Solution Since $f(x) = 1/x^2$ is positive, continuous, and decreasing on $[1, \infty)$ for any $n = 1, 2, 3, \ldots$, we have

$$s_n + A_{n+1} \leq s \leq s_n + A_n,$$

where

$$A_n = \int_n^\infty \frac{dx}{x^2} = \lim_{R\to\infty}\left(-\frac{1}{x}\right)\Bigg|_n^R = \frac{1}{n}.$$

The best approximation to s using s_n is

$$\begin{aligned} s_n^* &= s_n + \frac{1}{2}\left(\frac{1}{n+1} + \frac{1}{n}\right) = s_n + \frac{2n+1}{2n(n+1)} \\ &= 1 + \frac{1}{4} + \frac{1}{9} + \cdots + \frac{1}{n^2} + \frac{2n+1}{2n(n+1)}. \end{aligned}$$

The error in this approximation satisfies

$$|s - s_n^*| \le \frac{1}{2}\left(\frac{1}{n} - \frac{1}{n+1}\right) = \frac{1}{2n(n+1)} \le 0.001,$$

provided $2n(n+1) \ge 1/0.001 = 1{,}000$. It is easily checked that this condition is satisfied if $n \ge 22$; the approximation

$$s \approx s_{22}^* = 1 + \frac{1}{4} + \frac{1}{9} + \cdots + \frac{1}{22^2} + \frac{45}{44 \times 23}$$

will have error with absolute value not exceeding 0.001. Had we used the approximation $s \approx s_n$ we could only have concluded that

$$0 \le s - s_n \le A_n = \frac{1}{n} < 0.001,$$

provided $n > 1{,}000$; we would need 1,000 terms of the series to get the desired accuracy.

Comparison Tests

The next test we consider for positive series is analogous to the comparison theorem for improper integrals. (See Theorem 3 of Section 6.5.) It enables us to determine the convergence or divergence of one series by comparing it with another series that is known to converge or diverge.

THEOREM 9 **A comparison test**

Let $\{a_n\}$ and $\{b_n\}$ be sequences for which there exists a positive constant K such that, ultimately, $0 \le a_n \le Kb_n$.

(a) If the series $\sum_{n=1}^\infty b_n$ converges, then so does the series $\sum_{n=1}^\infty a_n$.

(b) If the series $\sum_{n=1}^\infty a_n$ diverges to infinity, then so does the series $\sum_{n=1}^\infty b_n$.

PROOF Since a series converges if and only if its tail converges (Theorem 5), we can assume, without loss of generality, that the condition $0 \le a_n \le Kb_n$ holds for all $n \ge 1$. Let $s_n = a_1 + a_2 + \cdots + a_n$ and $S_n = b_1 + b_2 + \cdots + b_n$. Then $s_n \le KS_n$. If $\sum b_n$ converges, then $\{S_n\}$ is convergent and hence is bounded by Theorem 1. Hence $\{s_n\}$ is bounded above. By Theorem 6, $\sum a_n$ converges. Since the convergence of $\sum b_n$ guarantees that of $\sum a_n$, if the latter series diverges to infinity, then the former cannot converge either, so it must diverge to infinity too.

BEWARE! Theorem 9 does *not* say that if $\sum a_n$ converges, then $\sum b_n$ converges. It is possible that the *smaller* sum may be finite while the *larger* one is infinite. (Do not confuse a theorem with its converse.)

EXAMPLE 3 Which of the following series converge? Give reasons for your answers.

(a) $\displaystyle\sum_{n=1}^\infty \frac{1}{2^n+1}$, (b) $\displaystyle\sum_{n=1}^\infty \frac{3n+1}{n^3+1}$, (c) $\displaystyle\sum_{n=2}^\infty \frac{1}{\ln n}$.

Solution In each case we must find a suitable comparison series that we already know converges or diverges.

(a) Since $0 < \dfrac{1}{2^n + 1} < \dfrac{1}{2^n}$ for $n = 1, 2, 3, \ldots$, and since $\sum_{n=1}^{\infty} \dfrac{1}{2^n}$ is a convergent geometric series, the series $\sum_{n=1}^{\infty} \dfrac{1}{2^n + 1}$ also converges by comparison.

(b) Observe that $\dfrac{3n+1}{n^3+1}$ behaves like $\dfrac{3}{n^2}$ for large n, so we would expect to compare the series with the convergent p-series $\sum_{n=1}^{\infty} n^{-2}$. We have, for $n \geq 1$,

$$\frac{3n+1}{n^3+1} = \frac{3n}{n^3+1} + \frac{1}{n^3+1} < \frac{3n}{n^3} + \frac{1}{n^3} < \frac{3}{n^2} + \frac{1}{n^2} = \frac{4}{n^2}.$$

Thus, the given series converges by Theorem 9.

(c) For $n = 2, 3, 4, \ldots$, we have $0 < \ln n < n$. Thus $\dfrac{1}{\ln n} > \dfrac{1}{n}$. Since $\sum_{n=2}^{\infty} \dfrac{1}{n}$ diverges to infinity (it is a harmonic series), so does $\sum_{n=2}^{\infty} \dfrac{1}{\ln n}$ by comparison.

The following theorem provides a version of the comparison test that is not quite as general as Theorem 9 but is often easier to apply in specific cases.

THEOREM 10

A limit comparison test

Suppose that $\{a_n\}$ and $\{b_n\}$ are positive sequences and that

$$\lim_{n\to\infty} \frac{a_n}{b_n} = L,$$

where L is either a nonnegative finite number or $+\infty$.

(a) If $L < \infty$ and $\sum_{n=1}^{\infty} b_n$ converges, then $\sum_{n=1}^{\infty} a_n$ also converges.

(b) If $L > 0$ and $\sum_{n=1}^{\infty} b_n$ diverges to infinity, then so does $\sum_{n=1}^{\infty} a_n$.

PROOF If $L < \infty$, then for n sufficiently large, we have $b_n > 0$ and

$$0 \leq \frac{a_n}{b_n} \leq L + 1,$$

so $0 \leq a_n \leq (L+1)b_n$. Hence $\sum_{n=1}^{\infty} a_n$ converges if $\sum_{n=1}^{\infty} b_n$ converges, by Theorem 9(a).

If $L > 0$, then for n sufficiently large

$$\frac{a_n}{b_n} \geq \frac{L}{2}.$$

Therefore, $0 < b_n \leq (2/L)a_n$, and $\sum_{n=1}^{\infty} a_n$ diverges to infinity if $\sum_{n=1}^{\infty} b_n$ does, by Theorem 9(b).

EXAMPLE 4 Which of the following series converge? Give reasons for your answers.

(a) $\displaystyle\sum_{n=1}^{\infty} \frac{1}{1+\sqrt{n}}$, (b) $\displaystyle\sum_{n=1}^{\infty} \frac{n+5}{n^3 - 2n + 3}$.

Solution Again we must make appropriate choices for comparison series.

(a) The terms of this series decrease like $1/\sqrt{n}$. Observe that

$$L = \lim_{n\to\infty} \frac{\dfrac{1}{1+\sqrt{n}}}{\dfrac{1}{\sqrt{n}}} = \lim_{n\to\infty} \frac{\sqrt{n}}{1+\sqrt{n}} = \lim_{n\to\infty} \frac{1}{(1/\sqrt{n})+1} = 1.$$

Since the p-series $\sum_{n=1}^{\infty} \dfrac{1}{\sqrt{n}}$ diverges to infinity ($p = 1/2$), so does the series $\sum_{n=1}^{\infty} \dfrac{1}{1+\sqrt{n}}$, by the limit comparison test.

(b) For large n, the terms behave like n/n^3, so let us compare the series with the p-series $\sum_{n=1}^{\infty} 1/n^2$, which we know converges.

$$L = \lim_{n\to\infty} \frac{\dfrac{n+5}{n^3-2n+3}}{\dfrac{1}{n^2}} = \lim_{n\to\infty} \frac{n^3+5n^2}{n^3-2n+3} = 1.$$

Since $L < \infty$, the series $\displaystyle\sum_{n=1}^{\infty} \frac{n+5}{n^3-2n+3}$ also converges by the limit comparison test.

In order to apply the original version of the comparison test (Theorem 9) successfully, it is important to have an intuitive feeling for whether the given series converges or diverges. The form of the comparison will depend on whether you are trying to prove convergence or divergence. For instance, if you did not know intuitively that

$$\sum_{n=1}^{\infty} \frac{1}{100n+20{,}000}$$

would have to diverge to infinity, you might try to argue that

$$\frac{1}{100n+20{,}000} < \frac{1}{n} \qquad \text{for } n = 1, 2, 3, \ldots.$$

While true, this doesn't help at all. $\sum_{n=1}^{\infty} 1/n$ diverges to infinity; therefore Theorem 9 yields no information from this comparison. We could, of course, argue instead that

$$\frac{1}{100n+20{,}000} \geq \frac{1}{20{,}100n} \qquad \text{if } n \geq 1,$$

and conclude by Theorem 9 that $\sum_{n=1}^{\infty}(1/(100n+20{,}000))$ diverges to infinity by comparison with the divergent series $\sum_{n=1}^{\infty} 1/n$. An easier way is to use Theorem 10 and the fact that

$$L = \lim_{n\to\infty} \frac{\dfrac{1}{100n+20{,}000}}{\dfrac{1}{n}} = \lim_{n\to\infty} \frac{n}{100n+20{,}000} = \frac{1}{100} > 0.$$

However, the limit comparison test Theorem 10 has a disadvantage when compared to the ordinary comparison test Theorem 9. It can fail in certain cases because the limit L does not exist. In such cases it is possible that the ordinary comparison test may still work.

EXAMPLE 5 Test the series $\displaystyle\sum_{n=1}^{\infty} \frac{1+\sin n}{n^2}$ for convergence.

Solution Since

$$\lim_{n\to\infty} \frac{\dfrac{1+\sin n}{n^2}}{\dfrac{1}{n^2}} = \lim_{n\to\infty}(1+\sin n)$$

does not exist, the limit comparison test gives us no information. However, since $\sin n \le 1$, we have

$$0 \le \frac{1+\sin n}{n^2} \le \frac{2}{n^2} \qquad \text{for } n = 1,\, 2,\, 3,\, \ldots.$$

The given series does, in fact, converge by comparison with $\sum_{n=1}^{\infty} 1/n^2$, using the ordinary comparison test.

The Ratio and Root Tests

THEOREM

The ratio test

Suppose that $a_n > 0$ (ultimately) and that $\rho = \lim_{n\to\infty} \frac{a_{n+1}}{a_n}$ exists or is $+\infty$.

(a) If $0 \le \rho < 1$, then $\sum_{n=1}^{\infty} a_n$ converges.

(b) If $1 < \rho \le \infty$, then $\lim_{n\to\infty} a_n = \infty$ and $\sum_{n=1}^{\infty} a_n$ diverges to infinity.

(c) If $\rho = 1$, this test gives no information; the series may either converge or diverge to infinity.

PROOF Here ρ is the lowercase Greek letter "rho" (pronounced "roh").

(a) Suppose $\rho < 1$. Pick a number r such that $\rho < r < 1$. Since we are given that $\lim_{n\to\infty} a_{n+1}/a_n = \rho$, we have $a_{n+1}/a_n \le r$ for n sufficiently large; that is, $a_{n+1} \le r a_n$ for $n \ge N$, say. In particular,

$$\begin{aligned}
a_{N+1} &\le r a_N \\
a_{N+2} &\le r a_{N+1} \le r^2 a_N \\
a_{N+3} &\le r a_{N+2} \le r^3 a_N \\
&\;\;\vdots \\
a_{N+k} &\le r^k a_N \qquad (k = 0,\, 1,\, 2,\, 3,\, \ldots).
\end{aligned}$$

Hence, $\sum_{n=N}^{\infty} a_n$ converges by comparison with the convergent geometric series $\sum_{k=0}^{\infty} r^k$. It follows that $\sum_{n=1}^{\infty} a_n = \sum_{n=1}^{N-1} a_n + \sum_{n=N}^{\infty} a_n$ must also converge.

(b) Now suppose that $\rho > 1$. Pick a number r such that $1 < r < \rho$. Since $\lim_{n\to\infty} a_{n+1}/a_n = \rho$, we have $a_{n+1}/a_n \ge r$ for n sufficiently large, say for $n \ge N$. We assume N is chosen large enough that $a_N > 0$. It follows by an argument similar to that used in part (a) that $a_{N+k} \ge r^k a_N$ for $k = 0, 1, 2, \ldots$, and since $r > 1$, $\lim_{n\to\infty} a_n = \infty$. Therefore, $\sum_{n=1}^{\infty} a_n$ diverges to infinity.

(c) If ρ is computed for the series $\sum_{n=1}^{\infty} 1/n$ and $\sum_{n=1}^{\infty} 1/n^2$, we get $\rho = 1$ in each case. Since the first series diverges to infinity and the second converges, the ratio test cannot distinguish between convergence and divergence if $\rho = 1$.

All p-series fall into the indecisive category where $\rho = 1$, as does $\sum_{n=1}^{\infty} a_n$, where a_n is any rational function of n. The ratio test is most useful for series whose terms decrease at least exponentially fast. The presence of factorials in a term also suggests that the ratio test might be useful.

EXAMPLE 6 Test the following series for convergence:

$$\text{(a)}\ \sum_{n=1}^{\infty} \frac{99^n}{n!}, \qquad \text{(b)}\ \sum_{n=1}^{\infty} \frac{n^5}{2^n}, \qquad \text{(c)}\ \sum_{n=1}^{\infty} \frac{n!}{n^n}, \qquad \text{(d)}\ \sum_{n=1}^{\infty} \frac{(2n)!}{(n!)^2}.$$

Solution We use the ratio test for each of these series.

$$\text{(a)}\ \rho = \lim_{n\to\infty} \frac{99^{n+1}}{(n+1)!} \Bigg/ \frac{99^n}{n!} = \lim_{n\to\infty} \frac{99}{n+1} = 0 < 1.$$

Thus, $\sum_{n=1}^{\infty} (99^n/n!)$ converges.

(b) $\rho = \lim_{n\to\infty} \dfrac{(n+1)^5}{2^{n+1}} \Big/ \dfrac{n^5}{2^n} = \lim_{n\to\infty} \dfrac{1}{2}\left(\dfrac{n+1}{n}\right)^5 = \dfrac{1}{2} < 1.$

Hence, $\sum_{n=1}^{\infty}(n^5/2^n)$ converges.

(c) $$\rho = \lim_{n\to\infty} \frac{(n+1)!}{(n+1)^{n+1}} \Big/ \frac{n!}{n^n} = \lim_{n\to\infty} \frac{(n+1)!n^n}{(n+1)^{n+1}n!} = \lim_{n\to\infty}\left(\frac{n}{n+1}\right)^n$$
$$= \lim_{n\to\infty} \frac{1}{\left(1+\dfrac{1}{n}\right)^n} = \frac{1}{e} < 1.$$

Thus, $\sum_{n=1}^{\infty}(n!/n^n)$ converges.

(d) $\rho = \lim_{n\to\infty} \dfrac{(2(n+1))!}{((n+1)!)^2} \Big/ \dfrac{(2n)!}{(n!)^2} = \lim_{n\to\infty} \dfrac{(2n+2)(2n+1)}{(n+1)^2} = 4 > 1.$

Thus, $\sum_{n=1}^{\infty}(2n)!/(n!)^2$ diverges to infinity.

The following theorem is very similar to the ratio test but is less frequently used. Its proof is left as an exercise. (See Exercise 37.) For examples of series to which it can be applied, see Exercises 38 and 39.

THEOREM 12

The root test

Suppose that $a_n > 0$ (ultimately) and that $\sigma = \lim_{n\to\infty}(a_n)^{1/n}$ exists or is $+\infty$.

(a) If $0 \le \sigma < 1$, then $\sum_{n=1}^{\infty} a_n$ converges.

(b) If $1 < \sigma \le \infty$, then $\lim_{n\to\infty} a_n = \infty$ and $\sum_{n=1}^{\infty} a_n$ diverges to infinity.

(c) If $\sigma = 1$, this test gives no information; the series may either converge or diverge to infinity.

Using Geometric Bounds to Estimate the Sum of a Series

Suppose that an inequality of the form

$$0 \le a_k \le Kr^k$$

holds for $k = n+1, n+2, n+3, \ldots$, where K and r are constants and $r < 1$. We can then use a geometric series to bound the tail of $\sum_{n=1}^{\infty} a_n$.

$$\begin{aligned} 0 \le s - s_n &= \sum_{k=n+1}^{\infty} a_k \le \sum_{k=n+1}^{\infty} Kr^k \\ &= Kr^{n+1}(1 + r + r^2 + \cdots) \\ &= \frac{Kr^{n+1}}{1-r}. \end{aligned}$$

Since $r < 1$, the series converges and the error approaches 0 at an exponential rate as n increases.

EXAMPLE 7 In Section 9.6 we will show that

$$e = \frac{1}{0!} + \frac{1}{1!} + \frac{1}{2!} + \frac{1}{3!} + \cdots = \sum_{n=0}^{\infty} \frac{1}{n!}.$$

(Recall that $0! = 1$.) Estimate the error if the sum s_n of the first n terms of the series is used to approximate e. Find e to 3-decimal-place accuracy using the series.

Solution We have

$$\begin{aligned} s_n &= \frac{1}{0!} + \frac{1}{1!} + \frac{1}{2!} + \frac{1}{3!} + \cdots + \frac{1}{(n-1)!} \\ &= 1 + 1 + \frac{1}{2} + \frac{1}{6} + \frac{1}{24} + \cdots + \frac{1}{(n-1)!}. \end{aligned}$$

(Since the series starts with the term for $n = 0$, the nth term is $1/(n-1)!$.) We can estimate the error in the approximation $s \approx s_n$ as follows:

$$\begin{aligned} 0 < s - s_n &= \frac{1}{n!} + \frac{1}{(n+1)!} + \frac{1}{(n+2)!} + \frac{1}{(n+3)!} + \cdots \\ &= \frac{1}{n!}\left(1 + \frac{1}{n+1} + \frac{1}{(n+1)(n+2)} + \frac{1}{(n+1)(n+2)(n+3)} + \cdots\right) \\ &< \frac{1}{n!}\left(1 + \frac{1}{n+1} + \frac{1}{(n+1)^2} + \frac{1}{(n+1)^3} + \cdots\right) \end{aligned}$$

since $n+2 > n+1$, $n+3 > n+1$, and so on. The latter series is geometric, so

$$0 < s - s_n < \frac{1}{n!} \frac{1}{1 - \dfrac{1}{n+1}} = \frac{n+1}{n!n}.$$

If we want to evaluate e accurately to 3 decimal places, then we must ensure that the error is less than 5 in the fourth decimal place, that is, that the error is less than 0.0005. Hence, we want

$$\frac{n+1}{n} \frac{1}{n!} < 0.0005 = \frac{1}{2{,}000}.$$

Since $7! = 5{,}040$ but $6! = 720$, we can use $n = 7$ but no smaller. We have

$$\begin{aligned} e \approx s_7 &= 1 + 1 + \frac{1}{2!} + \frac{1}{3!} + \frac{1}{4!} + \frac{1}{5!} + \frac{1}{6!} \\ &= 2 + \frac{1}{2} + \frac{1}{6} + \frac{1}{24} + \frac{1}{120} + \frac{1}{720} \approx 2.718 \quad \text{to 3 decimal places.} \end{aligned}$$

It is appropriate to use geometric series to bound the tails of positive series whose convergence would be demonstrated by the ratio test. Such series converge ultimately faster than any p-series $\sum_{n=1}^{\infty} n^{-p}$, for which the limit ratio is $\rho = 1$.

EXERCISES 9.3

In Exercises 1–26, determine whether the given series converges or diverges by using any appropriate test. The p-series can be used for comparison, as can geometric series. Be alert for series whose terms do not approach 0.

1. $\displaystyle\sum_{n=1}^{\infty} \frac{1}{n^2+1}$

2. $\displaystyle\sum_{n=1}^{\infty} \frac{n}{n^4-2}$

3. $\displaystyle\sum_{n=1}^{\infty} \frac{n^2+1}{n^3+1}$

4. $\displaystyle\sum_{n=1}^{\infty} \frac{\sqrt{n}}{n^2+n+1}$

5. $\displaystyle\sum_{n=1}^{\infty} \left|\sin\frac{1}{n^2}\right|$

6. $\displaystyle\sum_{n=8}^{\infty} \frac{1}{\pi^n+5}$

7. $\displaystyle\sum_{n=2}^{\infty} \frac{1}{(\ln n)^3}$

8. $\displaystyle\sum_{n=1}^{\infty} \frac{1}{\ln(3n)}$

9. $\displaystyle\sum_{n=1}^{\infty} \frac{1}{\pi^n - n^\pi}$

10. $\displaystyle\sum_{n=0}^{\infty} \frac{1+n}{2+n}$

11. $\displaystyle\sum_{n=1}^{\infty} \frac{1+n^{4/3}}{2+n^{5/3}}$

12. $\displaystyle\sum_{n=1}^{\infty} \frac{n^2}{1+n\sqrt{n}}$

13. $\displaystyle\sum_{n=3}^{\infty} \frac{1}{n\ln n\sqrt{\ln\ln n}}$

14. $\displaystyle\sum_{n=2}^{\infty} \frac{1}{n\ln n(\ln\ln n)^2}$

15. $\displaystyle\sum_{n=1}^{\infty} \frac{1-(-1)^n}{n^4}$

16. $\displaystyle\sum_{n=1}^{\infty} \frac{1+(-1)^n}{\sqrt{n}}$

17. $\displaystyle\sum_{n=1}^{\infty} \frac{1}{2^n(n+1)}$

18. $\displaystyle\sum_{n=1}^{\infty} \frac{n^4}{n!}$

19. $\displaystyle\sum_{n=1}^{\infty} \frac{n!}{n^2e^n}$

20. $\displaystyle\sum_{n=1}^{\infty} \frac{(2n)!6^n}{(3n)!}$

21. $\displaystyle\sum_{n=2}^{\infty} \frac{\sqrt{n}}{3^n\ln n}$

22. $\displaystyle\sum_{n=0}^{\infty} \frac{n^{100}2^n}{\sqrt{n!}}$

23. $\displaystyle\sum_{n=1}^{\infty} \frac{(2n)!}{(n!)^3}$

24. $\displaystyle\sum_{n=1}^{\infty} \frac{1+n!}{(1+n)!}$

25. $\displaystyle\sum_{n=4}^{\infty} \frac{2^n}{3^n-n^3}$

26. $\displaystyle\sum_{n=1}^{\infty} \frac{n^n}{\pi^n n!}$

In Exercises 27–30, use s_n and integral bounds to find the smallest interval that you can be sure contains the sum s of the series. If the midpoint s_n^* of this interval is used to approximate s, how large should n be chosen to ensure that the error is less than 0.001?

27. $\displaystyle\sum_{k=1}^{\infty} \frac{1}{k^4}$

28. $\displaystyle\sum_{k=1}^{\infty} \frac{1}{k^3}$

29. $\displaystyle\sum_{k=1}^{\infty} \frac{1}{k^{3/2}}$

30. $\displaystyle\sum_{k=1}^{\infty} \frac{1}{k^2+4}$

For each positive series in Exercises 31–34, find the best upper bound you can for the error $s - s_n$ encountered if the partial sum s_n is used to approximate the sum s of the series. How many terms of each series do you need to be sure that the approximation has error less than 0.001?

31. $\displaystyle\sum_{k=1}^{\infty} \frac{1}{2^k k!}$

32. $\displaystyle\sum_{n=1}^{\infty} \frac{1}{(2n-1)!}$

33. $\displaystyle\sum_{n=0}^{\infty} \frac{2^n}{(2n)!}$

34. $\displaystyle\sum_{n=1}^{\infty} \frac{1}{n^n}$

35. Use the integral test to show that $\displaystyle\sum_{n=1}^{\infty} \frac{1}{1+n^2}$ converges. Show that the sum s of the series is less than $\pi/2$.

36. Show that $\sum_{n=3}^{\infty}(1/(n\ln n(\ln\ln n)^p))$ converges if and only if $p > 1$. Generalize this result to series of the form

$$\sum_{n=N}^{\infty} \frac{1}{n(\ln n)(\ln\ln n)\cdots(\ln_j n)(\ln_{j+1} n)^p},$$

where $\ln_j n = \underbrace{\ln\ln\ln\ln\cdots\ln}_{j\ \ln's} n$.

37. Prove the root test. *Hint:* Mimic the proof of the ratio test.

38. Use the root test to show that $\displaystyle\sum_{n=1}^{\infty} \frac{2^{n+1}}{n^n}$ converges.

39. Use the root test to test the following series for convergence:

$$\sum_{n=1}^{\infty} \left(\frac{n}{n+1}\right)^{n^2}.$$

40. Repeat Exercise 38, but use the ratio test instead of the root test.

41. Try to use the ratio test to determine whether $\displaystyle\sum_{n=1}^{\infty} \frac{2^{2n}(n!)^2}{(2n)!}$ converges. What happens? Now observe that

$$\frac{2^{2n}(n!)^2}{(2n)!} = \frac{[2n(2n-2)(2n-4)\cdots 6\times 4\times 2]^2}{2n(2n-1)(2n-2)\cdots 4\times 3\times 2\times 1}$$
$$= \frac{2n}{2n-1} \times \frac{2n-2}{2n-3} \times \cdots \times \frac{4}{3} \times \frac{2}{1}.$$

Does the given series converge? Why or why not?

42. Determine whether the series $\displaystyle\sum_{n=1}^{\infty} \frac{(2n)!}{2^{2n}(n!)^2}$ converges. *Hint:* Proceed as in Exercise 41. Show that $a_n \ge 1/(2n)$.

43. (a) Show that if $k > 0$ and n is a positive integer, then $n < \dfrac{1}{k}(1+k)^n$.

(b) Use the estimate in (a) with $0 < k < 1$ to obtain an upper bound for the sum of the series $\sum_{n=0}^{\infty} n/2^n$. For what value of k is this bound lowest?

(c) If we use the sum s_n of the first n terms to approximate the sum s of the series in (b), obtain an upper bound for the error $s - s_n$ using the inequality from (a). For given n, find k to minimize this upper bound.

44. (Improving the convergence of a series) We know that $\sum_{n=1}^{\infty} 1/\big(n(n+1)\big) = 1$. (See Example 3 of Section 9.2.) Since

$$\frac{1}{n^2} = \frac{1}{n(n+1)} + c_n, \quad \text{where} \quad c_n = \frac{1}{n^2(n+1)},$$

we have $\sum_{n=1}^{\infty} \frac{1}{n^2} = 1 + \sum_{n=1}^{\infty} c_n$.

The series $\sum_{n=1}^{\infty} c_n$ converges more rapidly than does $\sum_{n=1}^{\infty} 1/n^2$ because its terms decrease like $1/n^3$. Hence, fewer terms of that series will be needed to compute $\sum_{n=1}^{\infty} 1/n^2$ to any desired degree of accuracy than would be needed if we calculated with $\sum_{n=1}^{\infty} 1/n^2$ directly. Using integral upper and lower bounds, determine a value of n for which the modified partial sum s_n^* for the series $\sum_{n=1}^{\infty} c_n$ approximates the sum of that series with error less than 0.001 in absolute value. Hence, determine $\sum_{n=1}^{\infty} 1/n^2$ to within 0.001 of its true value. (The technique exhibited in this exercise is known as **improving the convergence** of a series. It can be applied to estimating the sum $\sum a_n$ if we know the sum $\sum b_n$ and if $a_n - b_n = c_n$, where $|c_n|$ decreases faster than $|a_n|$ as n tends to infinity.)

45. Consider the series $s = \sum_{n=1}^{\infty} 1/(2^n + 1)$, and the partial sum s_n of its first n terms.

(a) How large need n be taken to ensure that the error in the approximation $s \approx s_n$ is less than 0.001 in absolute value?

(b) The geometric series $\sum_{n=1}^{\infty} 1/2^n$ converges to 1. If

$$b_n = \frac{1}{2^n} - \frac{1}{2^n + 1}$$

for $n = 1, 2, 3, \ldots$, how many terms of the series $\sum_{n=1}^{\infty} b_n$ are needed to calculate its sum to within 0.001?

(c) Use the result of part (b) to calculate the $\sum_{n=1}^{\infty} 1/(2^n + 1)$ to within 0.001.

9.4 Absolute and Conditional Convergence

All of the series $\sum_{n=1}^{\infty} a_n$ considered in the previous section were ultimately positive; that is, $a_n \geq 0$ for n sufficiently large. We now drop this restriction and allow arbitrary real terms a_n. We can, however, always obtain a positive series from any given series by replacing all the terms with their absolute values.

DEFINITION 5

Absolute convergence

The series $\sum_{n=1}^{\infty} a_n$ is said to be **absolutely convergent** if $\sum_{n=1}^{\infty} |a_n|$ converges.

The series

$$s = \sum_{n=1}^{\infty} \frac{(-1)^n}{n^2} = -1 + \frac{1}{4} - \frac{1}{9} + \frac{1}{16} - \cdots$$

converges absolutely since

$$S = \sum_{n=1}^{\infty} \left| \frac{(-1)^n}{n^2} \right| = \sum_{n=1}^{\infty} \frac{1}{n^2} = 1 + \frac{1}{4} + \frac{1}{9} + \frac{1}{16} + \cdots$$

converges. It seems reasonable that the first series must converge, and its sum s should satisfy $-S \leq s \leq S$. In general, the cancellation that occurs because some terms are negative and others positive makes it *easier* for a series to converge than if all the terms are of one sign. We verify this insight in the following theorem.

THEOREM 13

If a series converges absolutely, then it converges.

PROOF Let $\sum_{n=1}^{\infty} a_n$ be absolutely convergent, and let $b_n = a_n + |a_n|$ for each n. Since $-|a_n| \leq a_n \leq |a_n|$, we have $0 \leq b_n \leq 2|a_n|$ for each n. Thus, $\sum_{n=1}^{\infty} b_n$ converges by the comparison test. Therefore, $\sum_{n=1}^{\infty} a_n = \sum_{n=1}^{\infty} b_n - \sum_{n=1}^{\infty} |a_n|$ also converges. ∎

Again you are cautioned not to confuse the statement of Theorem 13 with the converse statement, which is false. We will show later in this section that the **alternating harmonic series**

$$\sum_{n=1}^{\infty} \frac{(-1)^{n-1}}{n} = 1 - \frac{1}{2} + \frac{1}{3} - \frac{1}{4} + \frac{1}{5} - \cdots$$

converges, although it does not converge absolutely. If we replace all the terms by their absolute values, we get the divergent harmonic series

$$\sum_{n=1}^{\infty} \frac{1}{n} = 1 + \frac{1}{2} + \frac{1}{3} + \frac{1}{4} + \cdots = \infty.$$

BEWARE! Although absolute convergence implies convergence, convergence does *not* imply absolute convergence.

DEFINITION 6

Conditional convergence

If $\sum_{n=1}^{\infty} a_n$ is convergent, but not absolutely convergent, then we say that it is **conditionally convergent** or that it **converges conditionally**.

The alternating harmonic series is an example of a conditionally convergent series.

The comparison tests, the integral test, and the ratio test can each be used to test for absolute convergence. They should be applied to the series $\sum_{n=1}^{\infty} |a_n|$. For the ratio test we calculate $\rho = \lim_{n\to\infty} |a_{n+1}/a_n|$. If $\rho < 1$, then $\sum_{n=1}^{\infty} a_n$ converges absolutely. If $\rho > 1$, then $\lim_{n\to\infty} |a_n| = \infty$, so both $\sum_{n=1}^{\infty} |a_n|$ and $\sum_{n=1}^{\infty} a_n$ must diverge. If $\rho = 1$, we get no information; the series $\sum_{n=1}^{\infty} a_n$ may converge absolutely, it may converge conditionally, or it may diverge.

EXAMPLE 1 Test the following series for absolute convergence:

(a) $\displaystyle\sum_{n=1}^{\infty} \frac{(-1)^{n-1}}{2n-1}$, (b) $\displaystyle\sum_{n=1}^{\infty} \frac{n\cos(n\pi)}{2^n}$.

Solution

(a) $\displaystyle\lim_{n\to\infty} \left|\frac{(-1)^{n-1}}{2n-1}\right| \bigg/ \frac{1}{n} = \lim_{n\to\infty} \frac{n}{2n-1} = \frac{1}{2} > 0.$

Since the harmonic series $\sum_{n=1}^{\infty}(1/n)$ diverges to infinity, the comparison test assures us that $\sum_{n=1}^{\infty}((-1)^{n-1}/(2n-1))$ does not converge absolutely.

(b) $\displaystyle\rho = \lim_{n\to\infty} \left|\frac{(n+1)\cos((n+1)\pi)}{2^{n+1}} \bigg/ \frac{n\cos(n\pi)}{2^n}\right| = \lim_{n\to\infty} \frac{n+1}{2n} = \frac{1}{2} < 1.$

(Note that $\cos(n\pi)$ is just a fancy way of writing $(-1)^n$.) Therefore (ratio test) $\sum_{n=1}^{\infty}((n\cos(n\pi))/2^n)$ converges absolutely.

The Alternating Series Test

We cannot use any of the previously developed tests to show that the alternating harmonic series converges; all of those tests apply only to (ultimately) positive series, so they can test only for absolute convergence. Demonstrating convergence that is not absolute is generally more difficult to do. We present only one test that can establish such convergence; this test can only be used on a very special kind of series.

THEOREM 14

The alternating series test

Suppose $\{a_n\}$ is a sequence whose terms satisfy, for some positive integer N,

(i) $a_n a_{n+1} < 0$ for $n \ge N$,

(ii) $|a_{n+1}| \le |a_n|$ for $n \ge N$, and

(iii) $\lim_{n\to\infty} a_n = 0$,

that is, the terms are ultimately alternating in sign and decreasing in size, and the sequence has limit zero. Then the series $\sum_{n=1}^{\infty} a_n$ converges.

PROOF Without loss of generality we can assume $N = 1$ because convergence only depends on the tail of a series. We also assume $a_1 > 0$; the proof if $a_1 < 0$ is similar. If $s_n = a_1 + a_2 + \cdots + a_n$ is the nth partial sum of the series, it follows from the alternation of $\{a_n\}$ that $a_{2n+1} > 0$ and $a_{2n} < 0$ for each n. Since the terms decrease in size, $a_{2n+1} \geq -a_{2n+2}$. Therefore, $s_{2n+2} = s_{2n} + a_{2n+1} + a_{2n+2} \geq s_{2n}$ for $n = 1, 2, 3, \ldots$; the even partial sums $\{s_{2n}\}$ form an increasing sequence. Similarly, $s_{2n+1} = s_{2n-1} + a_{2n} + a_{2n+1} \leq s_{2n-1}$, so the odd partial sums $\{s_{2n-1}\}$ form a decreasing sequence. Since $s_{2n} = s_{2n-1} + a_{2n} \leq s_{2n-1}$, we can say, for any n, that

BEWARE! Read this proof slowly and think about why each statement is true.

$$s_2 \leq s_4 \leq s_6 \leq \cdots \leq s_{2n} \leq s_{2n-1} \leq s_{2n-3} \leq \cdots \leq s_5 \leq s_3 \leq s_1.$$

Hence, s_2 is a lower bound for the decreasing sequence $\{s_{2n-1}\}$, and s_1 is an upper bound for the increasing sequence $\{s_{2n}\}$. Both of these sequences therefore converge by the completeness of the real numbers:

$$\lim_{n\to\infty} s_{2n-1} = s_{\text{odd}}, \qquad \lim_{n\to\infty} s_{2n} = s_{\text{even}}.$$

Now $a_{2n} = s_{2n} - s_{2n-1}$, so $0 = \lim_{n\to\infty} a_{2n} = \lim_{n\to\infty}(s_{2n} - s_{2n-1}) = s_{\text{even}} - s_{\text{odd}}$. Therefore $s_{\text{odd}} = s_{\text{even}} = s$, say. Every partial sum s_n is either of the form s_{2n-1} or of the form s_{2n}. Thus, $\lim_{n\to\infty} s_n = s$ exists and the series $\sum(-1)^{n-1}a_n$ converges to this sum s.

Remark The proof of Theorem 14 shows that the sum s of the series always lies between any two consecutive partial sums of the series:

$$\text{either} \quad s_n < s < s_{n+1} \quad \text{or} \quad s_{n+1} < s < s_n.$$

This proves the following theorem.

THEOREM 15

Error estimate for alternating series

If the sequence $\{a_n\}$ satisfies the conditions of the alternating series test (Theorem 14), so that the series $\sum_{n=1}^{\infty} a_n$ converges to the sum s, then the error in the approximation $s \approx s_n$ (where $n \geq N$) has the same sign as the first omitted term $a_{n+1} = s_{n+1} - s_n$, and its size is no greater than the size of that term:

$$|s - s_n| \leq |s_{n+1} - s_n| = |a_{n+1}|.$$

EXAMPLE 2 How many terms of the series $\displaystyle\sum_{n=1}^{\infty} \frac{(-1)^n}{1+2^n}$ are needed to compute the sum of the series with error less than 0.001?

Solution This series satisfies the hypotheses for Theorem 15. If we use the partial sum of the first n terms of the series to approximate the sum of the series, the error will satisfy

$$|\text{error}| \leq |\text{first omitted term}| = \frac{1}{1+2^{n+1}}.$$

This error is less than 0.001 if $1 + 2^{n+1} > 1{,}000$. Since $2^{10} = 1{,}024$, $n + 1 = 10$ will do; we need 9 terms of the series to compute the sum to within 0.001 of its actual value.

When determining the convergence of a given series, it is best to consider first whether the series converges absolutely. If it does not, then there remains the possibility of conditional convergence.

EXAMPLE 3 Test the following series for absolute and conditional convergence:

$$\text{(a)}\quad \sum_{n=1}^{\infty} \frac{(-1)^{n-1}}{n}, \qquad \text{(b)}\quad \sum_{n=2}^{\infty} \frac{\cos(n\pi)}{\ln n}, \qquad \text{(c)}\quad \sum_{n=1}^{\infty} \frac{(-1)^{n-1}}{n^4}.$$

Solution The absolute values of the terms in series (a) and (b) are $1/n$ and $1/(\ln n)$, respectively. Since $1/(\ln n) > 1/n$, and $\sum_{n=1}^{\infty} 1/n$ diverges to infinity, neither series (a) nor (b) converges absolutely. However, both series satisfy the requirements of Theorem 14 and so both converge. Each of these series is conditionally convergent.

Series (c) is absolutely convergent because $|(-1)^{n-1}/n^4| = 1/n^4$, and $\sum_{n=1}^{\infty} 1/n^4$ is a convergent p-series ($p = 4 > 1$). We could establish its convergence using Theorem 14, but there is no need to do that since every absolutely convergent series is convergent (Theorem 13).

EXAMPLE 4 For what values of x does the series $\displaystyle\sum_{n=1}^{\infty} \frac{(x-5)^n}{n\,2^n}$ converge absolutely? converge conditionally? diverge?

Solution For such series whose terms involve functions of a variable x, it is usually wisest to begin testing for absolute convergence with the ratio test. We have

$$\rho = \lim_{n\to\infty} \left| \frac{(x-5)^{n+1}}{(n+1)2^{n+1}} \bigg/ \frac{(x-5)^n}{n\,2^n} \right| = \lim_{n\to\infty} \frac{n}{n+1} \left| \frac{x-5}{2} \right| = \left| \frac{x-5}{2} \right|.$$

The series converges absolutely if $|(x-5)/2| < 1$. This inequality is equivalent to $|x-5| < 2$ (the distance from x to 5 is less than 2), that is, $3 < x < 7$. If $x < 3$ or $x > 7$, then $|(x-5)/2| > 1$. The series diverges; its terms do not approach zero.

If $x = 3$, the series is $\sum_{n=1}^{\infty}((-1)^n/n)$, which converges conditionally (it is an alternating harmonic series); if $x = 7$, the series is the harmonic series $\sum_{n=1}^{\infty} 1/n$, which diverges to infinity. Hence, the given series converges absolutely on the open interval $(3, 7)$, converges conditionally at $x = 3$, and diverges everywhere else.

EXAMPLE 5 For what values of x does the series $\displaystyle\sum_{n=0}^{\infty} (n+1)^2 \left(\frac{x}{x+2}\right)^n$ converge absolutely? converge conditionally? diverge?

Solution Again we begin with the ratio test.

$$\begin{aligned}\rho &= \lim_{n\to\infty} \left| (n+2)^2 \left(\frac{x}{x+2}\right)^{n+1} \bigg/ (n+1)^2 \left(\frac{x}{x+2}\right)^n \right| \\ &= \lim_{n\to\infty} \left(\frac{n+2}{n+1}\right)^2 \left|\frac{x}{x+2}\right| = \left|\frac{x}{x+2}\right| = \frac{|x|}{|x+2|}\end{aligned}$$

The series converges absolutely if $|x|/|x+2| < 1$. This condition says that the distance from x to 0 is less than the distance from x to -2. Hence $x > -1$. The series diverges if $|x|/|x+2| > 1$, that is, if $x < -1$. If $x = -1$, the series is $\sum_{n=0}^{\infty}(-1)^n(n+1)^2$, which diverges. We conclude that the series converges absolutely for $x > -1$, converges conditionally nowhere, and diverges for $x \le -1$.

When using the alternating series test, it is important to verify (at least mentally) that *all three conditions* (i)–(iii) are satisfied.

EXAMPLE 6 Test the following series for convergence:

(a) $\displaystyle\sum_{n=1}^{\infty}(-1)^{n-1}\frac{n+1}{n}$,

(b) $\displaystyle 1-\frac{1}{4}+\frac{1}{3}-\frac{1}{16}+\frac{1}{5}-\cdots=\sum_{n=1}^{\infty}a_n$, where

$$a_n=\begin{cases}1/n & \text{if } n \text{ is odd,}\\ -1/n^2 & \text{if } n \text{ is even.}\end{cases}$$

Solution

(a) Here, the terms a_n alternate and decrease in size as n increases. However, $\lim_{n\to\infty}|a_n| = 1 \neq 0$. The alternating series test does not apply. In fact, the given series diverges because its terms do not approach 0.

(b) This series alternates and its terms have limit zero. However, the terms are not decreasing in size (even ultimately). Once again, the alternating series test cannot be applied. In fact, since

$$-\frac{1}{4}-\frac{1}{16}-\cdots-\frac{1}{(2n)^2}-\cdots \qquad \text{converges, and}$$

$$1+\frac{1}{3}+\frac{1}{5}+\cdots+\frac{1}{2n-1}+\cdots \qquad \text{diverges to infinity,}$$

it is readily seen that the given series diverges to infinity.

Rearranging the Terms in a Series

The basic difference between absolute and conditional convergence is that when a series $\sum_{n=1}^{\infty}a_n$ converges absolutely, it does so because its terms $\{a_n\}$ decrease in size fast enough that their sum can be finite even if no cancellation occurs due to terms of opposite sign. If cancellation is required to make the series converge (because the terms decrease slowly), then the series can only converge conditionally.

Consider the alternating harmonic series

$$1-\frac{1}{2}+\frac{1}{3}-\frac{1}{4}+\frac{1}{5}-\frac{1}{6}+\cdots.$$

This series converges, but only conditionally. If we take the subseries containing only the positive terms, we get the series

$$1+\frac{1}{3}+\frac{1}{5}+\frac{1}{7}+\cdots,$$

which diverges to infinity. Similarly, the subseries of negative terms

$$-\frac{1}{2}-\frac{1}{4}-\frac{1}{6}-\frac{1}{8}-\cdots$$

diverges to negative infinity.

If a series converges absolutely, the subseries consisting of positive terms and the subseries consisting of negative terms must each converge to a finite sum. If a series converges conditionally, the positive and negative subseries will both diverge, to ∞ and $-\infty$, respectively.

Using these facts we can answer a question raised at the beginning of Section 9.2. If we rearrange the terms of a convergent series so that they are added in a different order, must the rearranged series converge, and if it does will it converge to the same sum as the original series? The answer depends on whether the original series was absolutely convergent or merely conditionally convergent.

THEOREM 16 **Convergence of rearrangements of a series**

(a) If the terms of an absolutely convergent series are rearranged so that addition occurs in a different order, the rearranged series still converges to the same sum as the original series.

(b) If a series is conditionally convergent, and L is any real number, then the terms of the series can be rearranged so as to make the series converge (conditionally) to the sum L. It can also be rearranged so as to diverge to ∞ or to $-\infty$, or just to diverge.

Part (b) shows that conditional convergence is a rather suspect kind of convergence, being dependent on the order in which the terms are added. We will not present a formal proof of the theorem but will give an example suggesting what is involved. (See also Exercise 30 below.)

EXAMPLE 7 In Section 9.5 we will show that the alternating harmonic series

$$\sum_{n=1}^{\infty} \frac{(-1)^{n-1}}{n} = 1 - \frac{1}{2} + \frac{1}{3} - \frac{1}{4} + \frac{1}{5} - \frac{1}{6} + \frac{1}{7} - \cdots$$

converges (conditionally) to the sum $\ln 2$. Describe how to rearrange its terms so that it converges to 8 instead.

Solution Start adding terms of the positive subseries

$$1 + \frac{1}{3} + \frac{1}{5} + \cdots,$$

and keep going until the partial sum exceeds 8. (It will, eventually, because the positive subseries diverges to infinity.) Then add the first term $-1/2$ of the negative subseries

$$-\frac{1}{2} - \frac{1}{4} - \frac{1}{6} - \cdots.$$

This will reduce the partial sum below 8 again. Now resume adding terms of the positive subseries until the partial sum climbs above 8 once more. Then add the second term of the negative subseries and the partial sum will drop below 8. Keep repeating this procedure, alternately adding terms of the positive subseries to force the sum above 8 and then terms of the negative subseries to force it below 8. Since both subseries have infinitely many terms and diverge to ∞ and $-\infty$, respectively, eventually every term of the original series will be included, and the partial sums of the new series will oscillate back and forth around 8, converging to that number. Of course, any number other than 8 could also be used in place of 8.

EXERCISES 9.4

Determine whether the series in Exercises 1–12 converge absolutely, converge conditionally, or diverge.

1. $\displaystyle\sum_{n=1}^{\infty} \frac{(-1)^{n-1}}{\sqrt{n}}$

2. $\displaystyle\sum_{n=1}^{\infty} \frac{(-1)^{n}}{n^2 + \ln n}$

3. $\displaystyle\sum_{n=1}^{\infty} \frac{\cos(n\pi)}{(n+1)\ln(n+1)}$

4. $\displaystyle\sum_{n=1}^{\infty} \frac{(-1)^{2n}}{2^n}$

5. $\displaystyle\sum_{n=0}^{\infty} \frac{(-1)^{n}(n^2-1)}{n^2+1}$

6. $\displaystyle\sum_{n=1}^{\infty} \frac{(-2)^{n}}{n!}$

7. $\displaystyle\sum_{n=1}^{\infty} \frac{(-1)^{n}}{n\pi^n}$

8. $\displaystyle\sum_{n=0}^{\infty} \frac{-n}{n^2+1}$

9. $\displaystyle\sum_{n=1}^{\infty} (-1)^n \frac{20n^2 - n - 1}{n^3 + n^2 + 33}$

10. $\displaystyle\sum_{n=1}^{\infty} \frac{100\cos(n\pi)}{2n+3}$

11. $\displaystyle\sum_{n=1}^{\infty} \frac{n!}{(-100)^n}$ **12.** $\displaystyle\sum_{n=10}^{\infty} \frac{\sin(n+1/2)\pi}{\ln \ln n}$

For the series in Exercises 13–16, find the smallest integer n that ensures that the partial sum s_n approximates the sum s of the series with error less than 0.001 in absolute value.

13. $\displaystyle\sum_{n=1}^{\infty} (-1)^{n-1} \frac{n}{n^2+1}$ **14.** $\displaystyle\sum_{n=0}^{\infty} \frac{(-1)^n}{(2n)!}$

15. $\displaystyle\sum_{n=1}^{\infty} (-1)^{n-1} \frac{n}{2^n}$ **16.** $\displaystyle\sum_{n=0}^{\infty} (-1)^n \frac{3^n}{n!}$

Determine the values of x for which the series in Exercises 17–24 converge absolutely, converge conditionally, or diverge.

17. $\displaystyle\sum_{n=0}^{\infty} \frac{x^n}{\sqrt{n+1}}$ **18.** $\displaystyle\sum_{n=1}^{\infty} \frac{(x-2)^n}{n^2 2^{2n}}$

19. $\displaystyle\sum_{n=0}^{\infty} (-1)^n \frac{(x-1)^n}{2n+3}$ **20.** $\displaystyle\sum_{n=1}^{\infty} \frac{1}{2n-1}\left(\frac{3x+2}{-5}\right)^n$

21. $\displaystyle\sum_{n=2}^{\infty} \frac{x^n}{2^n \ln n}$ **22.** $\displaystyle\sum_{n=1}^{\infty} \frac{(4x+1)^n}{n^3}$

23. $\displaystyle\sum_{n=1}^{\infty} \frac{(2x+3)^n}{n^{1/3} 4^n}$ **24.** $\displaystyle\sum_{n=1}^{\infty} \frac{1}{n}\left(1+\frac{1}{x}\right)^n$

25. Does the alternating series test apply directly to the series $\sum_{n=1}^{\infty} (1/n) \sin(n\pi/2)$? Determine whether the series converges.

26. Show that the series $\sum_{n=1}^{\infty} a_n$ converges absolutely if $a_n = 10/n^2$ for even n and $a_n = -1/10n^3$ for odd n.

27. Which of the following statements are TRUE and which are FALSE? Justify your assertion of truth, or give a counter-example to show falsehood.

(a) If $\sum_{n=1}^{\infty} a_n$ converges, then $\sum_{n=1}^{\infty} (-1)^n a_n$ converges.

(b) If $\sum_{n=1}^{\infty} a_n$ converges and $\sum_{n=1}^{\infty} (-1)^n a_n$ converges, then $\sum_{n=1}^{\infty} a_n$ converges absolutely.

(c) If $\sum_{n=1}^{\infty} a_n$ converges absolutely, then $\sum_{n=1}^{\infty} (-1)^n a_n$ converges absolutely.

28. (a) Use a Riemann sum argument to show that

$$\ln n! \ge \int_1^n \ln t \, dt = n \ln n - n + 1.$$

(b) For what values of x does the series $\sum_{n=1}^{\infty} \frac{n! x^n}{n^n}$ converge absolutely? converge conditionally? diverge? (*Hint:* First use the ratio test. To test the cases where $\rho = 1$, you may find the inequality in part (a) useful.)

29. For what values of x does the series $\sum_{n=1}^{\infty} \frac{(2n)! x^n}{2^{2n}(n!)^2}$ converge absolutely? converge conditionally? diverge? *Hint:* See Exercise 42 of Section 9.3.

30. Devise procedures for rearranging the terms of the alternating harmonic series so that the rearranged series (a) diverges to ∞, (b) converges to -2.

9.5 Power Series

This section is concerned with a special kind of infinite series called a *power series*, which may be thought of as a polynomial of infinite degree.

DEFINITION 7

Power series

A series of the form

$$\sum_{n=0}^{\infty} a_n (x-c)^n = a_0 + a_1(x-c) + a_2(x-c)^2 + a_3(x-c)^3 + \cdots$$

is called a **power series in powers of $x - c$** or a **power series about c**. The constants $a_0, a_1, a_2, \ldots$ are called the **coefficients** of the power series.

Since the terms of a power series are functions of a variable x, the series may or may not converge for each value of x. For those values of x for which the series does converge, the sum defines a function of x. For example, if $-1 < x < 1$, then

$$1 + x + x^2 + x^3 + \cdots = \frac{1}{1-x}.$$

The geometric series on the left side is a power series *representation* of the function $1/(1-x)$ in powers of x (or about 0). Note that the representation is valid only in the open interval $(-1, 1)$ even though $1/(1-x)$ is defined for all real x except $x = 1$. For

$x = -1$ and for $|x| > 1$ the series does not converge, so it cannot represent $1/(1-x)$ at these points.

The point c is the **centre of convergence** of the power series $\sum_{n=0}^{\infty} a_n(x-c)^n$. The series certainly converges (to a_0) at $x = c$. (All the terms except possibly the first are 0.) Theorem 17 below shows that if the series converges anywhere else, then it converges on an interval (possibly infinite) centred at $x = c$, and it converges absolutely everywhere on that interval except possibly at one or both of the endpoints if the interval is finite. The geometric series

$$1 + x + x^2 + x^3 + \cdots$$

is an example of this behaviour. It has centre of convergence $c = 0$, and converges only on the interval $(-1, 1)$, centred at 0. The convergence is absolute at every point of the interval. Another example is the series

$$\sum_{n=1}^{\infty} \frac{1}{n\,2^n}(x-5)^n = \frac{x-5}{2} + \frac{(x-5)^2}{2 \times 2^2} + \frac{(x-5)^3}{3 \times 2^3} + \cdots,$$

which we discussed in Example 4 of Section 9.4. We showed that this series converges on the interval $[3, 7)$, an interval with centre $x = 5$, and that the convergence is absolute on the open interval $(3, 7)$ but is only conditional at the endpoint $x = 3$.

THEOREM

For any power series $\sum_{n=0}^{\infty} a_n\,(x-c)^n$ one of the following alternatives must hold:

(i) the series may converge only at $x = c$,

(ii) the series may converge at every real number x, or

(iii) there may exist a positive real number R such that the series converges at every x satisfying $|x - c| < R$ and diverges at every x satisfying $|x - c| > R$. In this case the series may or may not converge at either of the two *endpoints* $x = c - R$ and $x = c + R$.

In each of these cases the convergence is absolute except possibly at the endpoints $x = c - R$ and $x = c + R$ in case (iii).

PROOF We observed above that every power series converges at its centre of convergence; only the first term can be nonzero so the convergence is absolute. To prove the rest of this theorem, it suffices to show that if the series converges at any number $x_0 \neq c$, then it converges absolutely at every number x closer to c than x_0 is, that is, at every x satisfying $|x - c| < |x_0 - c|$. This means that convergence at any $x_0 \neq c$ implies absolute convergence on $(c - x_0, c + x_0)$, so the set of points x where the series converges must be an interval centred at c.

Suppose, therefore, that $\sum_{n=0}^{\infty} a_n(x_0 - c)^n$ converges. Then $\lim a_n(x_0 - c)^n = 0$, so $|a_n(x_0 - c)^n| \leq K$ for all n, where K is some constant (Theorem 1 of Section 9.1). If $r = |x - c|/|x_0 - c| < 1$, then

$$\sum_{n=0}^{\infty} |a_n(x-c)^n| = \sum_{n=0}^{\infty} |a_n(x_0 - c)^n| \left|\frac{x-c}{x_0-c}\right|^n \leq K \sum_{n=0}^{\infty} r^n = \frac{K}{1-r} < \infty.$$

Thus, $\sum_{n=0}^{\infty} a_n(x-c)^n$ converges absolutely. ■

By Theorem 17, the set of values x for which the power series $\sum_{n=0}^{\infty} a_n(x-c)^n$ converges is an interval centred at $x = c$. We call this interval the **interval of convergence** of the power series. It must have one of the following forms:

(i) the isolated point $x = c$ (a degenerate closed interval $[c, c]$),

(ii) the entire line $(-\infty, \infty)$,

(iii) a finite interval centred at c:
$[c - R, c + R]$, or $[c - R, c + R)$, or $(c - R, c + R]$, or $(c - R, c + R)$.

The number R in (iii) is called the **radius of convergence** of the power series. In case (i) we say the radius of convergence is $R = 0$; in case (ii) it is $R = \infty$.

The radius of convergence, R, can often be found by using the ratio test on the power series: if

$$\rho = \lim_{n\to\infty} \left| \frac{a_{n+1}(x-c)^{n+1}}{a_n(x-c)^n} \right| = \left(\lim_{n\to\infty} \left| \frac{a_{n+1}}{a_n} \right| \right) |x-c|$$

exists, then the series $\sum_{n=0}^{\infty} a_n(x-c)^n$ converges absolutely where $\rho < 1$, that is, where

$$|x-c| < R = 1 \Big/ \lim_{n\to\infty} \left| \frac{a_{n+1}}{a_n} \right|.$$

The series diverges if $|x-c| > R$.

Radius of convergence

Suppose that $L = \lim_{n\to\infty} \left| \dfrac{a_{n+1}}{a_n} \right|$ exists or is ∞. Then the power series $\sum_{n=0}^{\infty} a_n(x-c)^n$ has radius of convergence $R = 1/L$. (If $L = 0$, then $R = \infty$; if $L = \infty$, then $R = 0$.)

EXAMPLE 1 Determine the centre, radius, and interval of convergence of

$$\sum_{n=0}^{\infty} \frac{(2x+5)^n}{(n^2+1)3^n}.$$

Solution The series can be rewritten

$$\sum_{n=0}^{\infty} \left(\frac{2}{3}\right)^n \frac{1}{n^2+1} \left(x+\frac{5}{2}\right)^n.$$

The centre of convergence is $x = -5/2$. The radius of convergence, R, is given by

$$\frac{1}{R} = L = \lim \left| \frac{\left(\frac{2}{3}\right)^{n+1} \dfrac{1}{(n+1)^2+1}}{\left(\frac{2}{3}\right)^n \dfrac{1}{n^2+1}} \right| = \lim \frac{2}{3} \frac{n^2+1}{(n+1)^2+1} = \frac{2}{3}.$$

Thus, $R = 3/2$. The series converges absolutely on $(-5/2 - 3/2, -5/2 + 3/2) = (-4, -1)$, and it diverges on $(-\infty, -4)$ and on $(-1, \infty)$. At $x = -1$ the series is $\sum_{n=0}^{\infty} 1/(n^2+1)$; at $x = -4$ it is $\sum_{n=0}^{\infty}(-1)^n/(n^2+1)$. Both series converge (absolutely). The interval of convergence of the given power series is therefore $[-4, -1]$.

EXAMPLE 2 Determine the radii of convergence of the series

(a) $\displaystyle\sum_{n=0}^{\infty} \frac{x^n}{n!}$ and (b) $\displaystyle\sum_{n=0}^{\infty} n!\,x^n$.

Solution

(a) $L = \left| \lim \dfrac{1}{(n+1)!} \Big/ \dfrac{1}{n!} \right| = \lim \dfrac{n!}{(n+1)!} = \lim \dfrac{1}{n+1} = 0$. Thus $R = \infty$.

This series converges (absolutely) for all x. The sum is e^x, as will be shown in Example 1 in the next section.

(b) $L = \left| \lim \dfrac{(n+1)!}{n!} \right| = \lim(n+1) = \infty$. Thus $R = 0$.

This series converges only at its centre of convergence, $x = 0$.

Algebraic Operations on Power Series

To simplify the following discussion, we will consider only power series with centre of convergence 0, that is, series of the form

$$\sum_{n=0}^{\infty} a_n x^n = a_0 + a_1 x + a_2 x^2 + a_3 x^3 + \cdots.$$

Any properties we demonstrate for such series extend automatically to power series of the form $\sum_{n=0}^{\infty} a_n (y-c)^n$ via the change of variable $x = y - c$.

First, we observe that series having the same centre of convergence can be added or subtracted on whatever interval is common to their intervals of convergence. The following theorem is a simple consequence of Theorem 7 of Section 9.2 and does not require a proof.

THEOREM 18

Let $\sum_{n=0}^{\infty} a_n x^n$ and $\sum_{n=0}^{\infty} b_n x^n$ be two power series with radii of convergence R_a and R_b, respectively, and let c be a constant. Then

(i) $\sum_{n=0}^{\infty} (ca_n) x^n$ has radius of convergence R_a, and

$$\sum_{n=0}^{\infty} (ca_n) x^n = c \sum_{n=0}^{\infty} a_n x^n$$

wherever the series on the right converges.

(ii) $\sum_{n=0}^{\infty} (a_n + b_n) x^n$ has radius of convergence R at least as large as the smaller of R_a and R_b ($R \ge \min\{R_a, R_b\}$), and

$$\sum_{n=0}^{\infty} (a_n + b_n) x^n = \sum_{n=0}^{\infty} a_n x^n + \sum_{n=0}^{\infty} b_n x^n$$

wherever both series on the right converge.

The situation regarding multiplication and division of power series is more complicated. We will mention only the results and will not attempt any proofs of our assertions. A textbook in mathematical analysis will provide more details.

Long multiplication of the form

$$\begin{aligned}(a_0 + a_1 x + a_2 x^2 + \cdots)&(b_0 + b_1 x + b_2 x^2 + \cdots) \\ &= a_0 b_0 + (a_0 b_1 + a_1 b_0) x + (a_0 b_2 + a_1 b_1 + a_2 b_0) x^2 + \cdots\end{aligned}$$

leads us to conjecture the formula

$$\left(\sum_{n=0}^{\infty} a_n x^n\right)\left(\sum_{n=0}^{\infty} b_n x^n\right) = \sum_{n=0}^{\infty} c_n x^n,$$

where

$$c_n = a_0 b_n + a_1 b_{n-1} + \cdots + a_n b_0 = \sum_{j=0}^{n} a_j b_{n-j}.$$

The series $\sum_{n=0}^{\infty} c_n x^n$ is called the **Cauchy product** of the series $\sum_{n=0}^{\infty} a_n x^n$ and $\sum_{n=0}^{\infty} b_n x^n$. Like the sum, the Cauchy product also has radius of convergence at least equal to the lesser of those of the factor series.

EXAMPLE 3 Since

$$\frac{1}{1-x} = 1 + x + x^2 + x^3 + \cdots = \sum_{n=0}^{\infty} x^n$$

holds for $-1 < x < 1$, we can determine a power series representation for $1/(1-x)^2$ by taking the Cauchy product of this series with itself. Since $a_n = b_n = 1$ for $n = 0, 1, 2, \ldots$, we have

$$c_n = \sum_{j=0}^{n} 1 = n + 1 \qquad \text{and}$$

$$\frac{1}{(1-x)^2} = 1 + 2x + 3x^2 + 4x^3 + \cdots = \sum_{n=0}^{\infty} (n+1)x^n,$$

which must also hold for $-1 < x < 1$. The same series can be obtained by direct long multiplication of the series:

$$\begin{array}{rcccccccccc}
 & 1 & + & x & + & x^2 & + & x^3 & + & \cdots \\
\times & 1 & + & x & + & x^2 & + & x^3 & + & \cdots \\
\hline
 & 1 & + & x & + & x^2 & + & x^3 & + & \cdots \\
 & & & x & + & x^2 & + & x^3 & + & \cdots \\
 & & & & & x^2 & + & x^3 & + & \cdots \\
 & & & & & & & x^3 & + & \cdots \\
 & & & & & & & & & \cdots \\
\hline
 & 1 & + & 2x & + & 3x^2 & + & 4x^3 & + & \cdots
\end{array}$$

Long division can also be performed on power series, but there is no simple rule for determining the coefficients of the quotient series. The radius of convergence of the quotient series is not less than the least of the three numbers R_1, R_2, and R_3, where R_1 and R_2 are the radii of convergence of the numerator and denominator series and R_3 is the distance from the centre of convergence to the nearest *complex number* where the denominator series has sum equal to 0. To illustrate this point, observe that 1 and $1 - x$ are both power series with infinite radii of convergence:

$$\begin{aligned}
1 &= 1 + 0x + 0x^2 + 0x^3 + \cdots && \text{for all } x, \\
1 - x &= 1 - \ \ x + 0x^2 + 0x^3 + \cdots && \text{for all } x.
\end{aligned}$$

Their quotient, $1/(1-x)$, however, only has radius of convergence 1, the distance from the centre of convergence $x = 0$ to the point $x = 1$ where the denominator vanishes:

$$\frac{1}{1-x} = 1 + x + x^2 + x^3 + \cdots \qquad \text{for} \quad |x| < 1.$$

Differentiation and Integration of Power Series

If a power series has a positive radius of convergence, it can be differentiated or integrated term by term. The resulting series will converge to the appropriate derivative or integral of the sum of the original series everywhere except possibly at the endpoints of the interval of convergence of the original series. This very important fact ensures that, for purposes of calculation, power series behave just like polynomials, the easiest functions to differentiate and integrate. We formalize the differentiation and integration properties of power series in the following theorem.

THEOREM 19

Term-by-term differentiation and integration of power series

If the series $\sum_{n=0}^{\infty} a_n x^n$ converges to the sum $f(x)$ on an interval $(-R, R)$, where $R > 0$, that is,

$$f(x) = \sum_{n=0}^{\infty} a_n x^n = a_0 + a_1 x + a_2 x^2 + a_3 x^3 + \cdots, \qquad (-R < x < R),$$

then f is differentiable on $(-R, R)$ and

$$f'(x) = \sum_{n=1}^{\infty} n a_n x^{n-1} = a_1 + 2a_2 x + 3a_3 x^2 + \cdots, \qquad (-R < x < R).$$

Also, f is integrable over any closed subinterval of $(-R, R)$, and if $|x| < R$, then

$$\int_0^x f(t)\,dt = \sum_{n=0}^{\infty} \frac{a_n}{n+1} x^{n+1} = a_0 x + \frac{a_1}{2} x^2 + \frac{a_2}{3} x^3 + \cdots.$$

While understanding the statement of this theorem is very important for what follows, understanding the proof is not. Feel free to skip the proof and go on to the applications.

PROOF Let x satisfy $-R < x < R$ and choose $H > 0$ such that $|x| + H < R$. By Theorem 17 we then have[1]

$$\sum_{n=1}^{\infty} |a_n|(|x| + H)^n = K < \infty.$$

The Binomial Theorem (see Section 9.8) shows that if $n \ge 1$, then

$$(x+h)^n = x^n + nx^{n-1}h + \sum_{k=2}^{n} \binom{n}{k} x^{n-k} h^k.$$

Therefore, if $|h| \le H$ we have

$$\begin{aligned}
|(x+h)^n - x^n - nx^{n-1}h| &= \left| \sum_{k=2}^{n} \binom{n}{k} x^{n-k} h^k \right| \\
&\le \sum_{k=2}^{n} \binom{n}{k} |x|^{n-k} \frac{|h|^k}{H^k} H^k \\
&\le \frac{|h|^2}{H^2} \sum_{k=0}^{n} \binom{n}{k} |x|^{n-k} H^k \\
&= \frac{|h|^2}{H^2} (|x| + H)^n.
\end{aligned}$$

Also,

$$|nx^{n-1}| = \frac{n|x|^{n-1}H}{H} \le \frac{1}{H}(|x| + H)^n.$$

Thus,

$$\sum_{n=1}^{\infty} |na_n x^{n-1}| \le \frac{1}{H} \sum_{n=1}^{\infty} |a_n|(|x| + H)^n = \frac{K}{H} < \infty,$$

so the series $\sum_{n=1}^{\infty} na_n x^{n-1}$ converges (absolutely) to $g(x)$, say. Now

$$\begin{aligned}
\left| \frac{f(x+h) - f(x)}{h} - g(x) \right| &= \left| \sum_{n=1}^{\infty} \frac{a_n(x+h)^n - a_n x^n - na_n x^{n-1}h}{h} \right| \\
&\le \frac{1}{|h|} \sum_{n=1}^{\infty} |a_n| |(x+h)^n - x^n - nx^{n-1}h| \\
&\le \frac{|h|}{H^2} \sum_{n=1}^{\infty} |a_n| \big(|x| + H\big)^n \le \frac{K|h|}{H^2}.
\end{aligned}$$

[1] This proof is due to R. Výborný, *American Mathematical Monthly*, April 1987.

Letting h approach zero, we obtain $|f'(x) - g(x)| \le 0$, so $f'(x) = g(x)$, as required.

Now observe that since $|a_n/(n+1)| \le |a_n|$, the series

$$h(x) = \sum_{n=0}^{\infty} \frac{a_n}{n+1} x^{n+1}$$

converges (absolutely) at least on the interval $(-R, R)$. Using the differentiation result proved above, we obtain

$$h'(x) = \sum_{n=0}^{\infty} a_n x^n = f(x).$$

Since $h(0) = 0$, we have

$$\int_0^x f(t)\, dt = \int_0^x h'(t)\, dt = h(t)\Big|_0^x = h(x),$$

as required.

Together, these results imply that the termwise differentiated or integrated series have the same radius of convergence as the given series. In fact, as the following examples illustrate, the interval of convergence of the differentiated series is the same as that of the original series except for the *possible* loss of one or both endpoints if the original series converges at endpoints of its interval of convergence. Similarly, the integrated series will converge everywhere on the interval of convergence of the original series and possibly at one or both endpoints of that interval, even if the original series does not converge at the endpoints.

Being differentiable on $(-R, R)$, where R is the radius of convergence, the sum $f(x)$ of a power series is necessarily continuous on that open interval. If the series happens to converge at either or both of the endpoints $-R$ and R, then f is also continuous (on one side) up to these endpoints. This result is stated formally in the following theorem. We will not prove it here; the interested reader is referred to textbooks on mathematical analysis for a proof.

THEOREM

Abel's Theorem

The sum of a power series is a continuous function everywhere on the interval of convergence of the series. In particular, if $\sum_{n=0}^{\infty} a_n R^n$ converges for some $R > 0$, then

$$\lim_{x \to R-} \sum_{n=0}^{\infty} a_n x^n = \sum_{n=0}^{\infty} a_n R^n,$$

and if $\sum_{n=0}^{\infty} a_n (-R)^n$ converges, then

$$\lim_{x \to -R+} \sum_{n=0}^{\infty} a_n x^n = \sum_{n=0}^{\infty} a_n (-R)^n.$$

The following examples show how the above theorems are applied to obtain power series representations for functions.

EXAMPLE 4 Find power series representations for the functions

(a) $\dfrac{1}{(1-x)^2}$, (b) $\dfrac{1}{(1-x)^3}$, and (c) $\ln(1+x)$

by starting with the geometric series

$$\frac{1}{1-x} = \sum_{n=0}^{\infty} x^n = 1 + x + x^2 + x^3 + \cdots \qquad (-1 < x < 1)$$

and using differentiation, integration, and substitution. Where is each series valid?

Solution

(a) Differentiate the geometric series term by term to obtain

$$\frac{1}{(1-x)^2} = \sum_{n=1}^{\infty} nx^{n-1} = 1 + 2x + 3x^2 + 4x^3 + \cdots \qquad (-1 < x < 1).$$

This is the same result obtained by multiplication of series in Example 3 above.

(b) Differentiate again to get, for $-1 < x < 1$,

$$\frac{2}{(1-x)^3} = \sum_{n=2}^{\infty} n(n-1)\,x^{n-2} = (1 \times 2) + (2 \times 3)x + (3 \times 4)x^2 + \cdots.$$

Now divide by 2:

$$\frac{1}{(1-x)^3} = \sum_{n=2}^{\infty} \frac{n(n-1)}{2} x^{n-2} = 1 + 3x + 6x^2 + 10x^3 + \cdots \quad (-1 < x < 1).$$

(c) Substitute $-t$ in place of x in the original geometric series:

$$\frac{1}{1+t} = \sum_{n=0}^{\infty} (-1)^n t^n = 1 - t + t^2 - t^3 + t^4 - \cdots \qquad (-1 < t < 1).$$

Integrate from 0 to x, where $|x| < 1$, to get

$$\begin{aligned}\ln(1+x) &= \int_0^x \frac{dt}{1+t} = \sum_{n=0}^{\infty} (-1)^n \int_0^x t^n\,dt \\ &= \sum_{n=0}^{\infty} (-1)^n \frac{x^{n+1}}{n+1} = x - \frac{x^2}{2} + \frac{x^3}{3} - \frac{x^4}{4} + \cdots \quad (-1 < x \le 1).\end{aligned}$$

Note that the latter series converges (conditionally) at the endpoint $x = 1$ as well as on the interval $-1 < x < 1$. Since $\ln(1+x)$ is continuous at $x = 1$, Theorem 20 assures us that the series must converge to that function at $x = 1$ also. In particular, therefore, the alternating harmonic series converges to $\ln 2$:

$$\ln 2 = 1 - \frac{1}{2} + \frac{1}{3} - \frac{1}{4} + \frac{1}{5} - \cdots = \sum_{n=0}^{\infty} \frac{(-1)^n}{n+1}.$$

This would not, however, be a very useful formula for calculating the value of $\ln 2$. (Why not?)

EXAMPLE 5 Use the geometric series of the previous example to find a power series representation for $\tan^{-1} x$.

Solution Substitute $-t^2$ for x in the geometric series. Since $0 \le t^2 < 1$ whenever $-1 < t < 1$, we obtain

$$\frac{1}{1+t^2} = 1 - t^2 + t^4 - t^6 + t^8 - \cdots \qquad (-1 < t < 1).$$

Now integrate from 0 to x, where $|x| < 1$:

$$\begin{aligned}\tan^{-1} x = \int_0^x \frac{dt}{1+t^2} &= \int_0^x (1 - t^2 + t^4 - t^6 + t^8 - \cdots)\,dt \\ &= x - \frac{x^3}{3} + \frac{x^5}{5} - \frac{x^7}{7} + \frac{x^9}{9} - \cdots \\ &= \sum_{n=0}^{\infty} (-1)^n \frac{x^{2n+1}}{2n+1} \qquad (-1 < x < 1).\end{aligned}$$

However, note that the series also converges (conditionally) at $x = -1$ and 1. Since $\tan^{-1}$ is continuous at ± 1, the above series representation for $\tan^{-1} x$ also holds for these values, by Theorem 20. Letting $x = 1$ we get another interesting result:

$$\frac{\pi}{4} = 1 - \frac{1}{3} + \frac{1}{5} - \frac{1}{7} + \frac{1}{9} - \cdots.$$

Again, however, this would not be a good formula with which to calculate a numerical value of π. (Why not?)

EXAMPLE 6 Find the sum of the series $\displaystyle\sum_{n=1}^{\infty} \frac{n^2}{2^n}$ by first finding the sum of the power series

$$\sum_{n=1}^{\infty} n^2 x^n = x + 4x^2 + 9x^3 + 16x^4 + \cdots.$$

Solution Observe in Example 4(a) how the process of differentiating the geometric series produces a series with coefficients 1, 2, 3, ... Start with the series obtained for $1/(1-x)^2$ and multiply it by x to obtain

$$\sum_{n=1}^{\infty} n x^n = x + 2x^2 + 3x^3 + 4x^4 + \cdots = \frac{x}{(1-x)^2}.$$

Now differentiate again to get a series with coefficients 1^2, 2^2, 3^2, ...:

$$\sum_{n=1}^{\infty} n^2 x^{n-1} = 1 + 4x + 9x^2 + 16x^3 + \cdots = \frac{d}{dx}\frac{x}{(x-1)^2} = \frac{1+x}{(1-x)^3}.$$

Multiplication by x again gives the desired power series:

$$\sum_{n=1}^{\infty} n^2 x^n = x + 4x^2 + 9x^3 + 16x^4 + \cdots = \frac{x(1+x)}{(1-x)^3}.$$

Differentiation and multiplication by x do not change the radius of convergence, so this series converges to the indicated function for $-1 < x < 1$. Putting $x = 1/2$, we get

$$\sum_{n=1}^{\infty} \frac{n^2}{2^n} = \frac{\frac{1}{2} \times \frac{3}{2}}{\frac{1}{8}} = 6.$$

The following example illustrates how substitution can be used to obtain power series representations of functions with centres of convergence different from 0.

EXAMPLE 7 Find a series representation of $f(x) = 1/(2+x)$ in powers of $x - 1$. What is the interval of convergence of this series?

Solution Let $t = x - 1$ so that $x = t + 1$. We have

$$\begin{aligned}
\frac{1}{2+x} &= \frac{1}{3+t} = \frac{1}{3}\,\frac{1}{1+\frac{t}{3}} \\
&= \frac{1}{3}\left(1 - \frac{t}{3} + \frac{t^2}{3^2} - \frac{t^3}{3^3} + \cdots\right) && (-1 < t/3 < 1) \\
&= \sum_{n=0}^{\infty} (-1)^n \frac{t^n}{3^{n+1}} && (-3 < t < 3) \\
&= \sum_{n=0}^{\infty} (-1)^n \frac{(x-1)^n}{3^{n+1}} && (-2 < x < 4).
\end{aligned}$$

Note that the radius of convergence of this series is 3, the distance from the centre of convergence, 1, to the point -2 where the denominator is 0. We could have predicted this in advance.

Maple Calculations

Maple can find the sums of many kinds of series, including absolutely and conditionally convergent numerical series and many power series. Even when Maple can't find the formal sum of a (convergent) series, it can provide a decimal approximation to the precision indicated by the current value of its variable `Digits`, which defaults to 10. Here are some examples.

```
> sum(n^4/2^n, n=1..infinity);
```

$$150$$

```
> sum(1/n^2, n=1..infinity);
```

$$\frac{1}{6}\pi^2$$

```
> sum(exp(-n^2), n=0..infinity);
```

$$\sum_{n=0}^{\infty} e^{(-n^2)}$$

```
> evalf(%);
```

$$1.386\,318\,602$$

```
> f := x -> sum(x^(n-1)/n, n=1..infinity);
```

$$f := x \to \sum_{n=1}^{\infty}\left(\frac{x^{(n-1)}}{n}\right)$$

```
> f(1); f(-1); f(1/2);
```

$$\infty$$
$$\ln(2)$$
$$2\ln(2)$$

EXERCISES 9.5

Determine the centre, radius, and interval of convergence of each of the power series in Exercises 1–8.

1. $\displaystyle\sum_{n=0}^{\infty} \frac{x^{2n}}{\sqrt{n+1}}$

2. $\displaystyle\sum_{n=0}^{\infty} 3n\,(x+1)^n$

3. $\displaystyle\sum_{n=1}^{\infty} \frac{1}{n}\left(\frac{x+2}{2}\right)^n$

4. $\displaystyle\sum_{n=1}^{\infty} \frac{(-1)^n}{n^4 2^{2n}} x^n$

5. $\displaystyle\sum_{n=0}^{\infty} n^3(2x-3)^n$

6. $\displaystyle\sum_{n=1}^{\infty} \frac{e^n}{n^3}(4-x)^n$

7. $\displaystyle\sum_{n=0}^{\infty} \frac{(1+5^n)}{n!} x^n$

8. $\displaystyle\sum_{n=1}^{\infty} \frac{(4x-1)^n}{n^n}$

9. Use multiplication of series to find a power series representation of $1/(1-x)^3$ valid in the interval $(-1, 1)$.

10. Determine the Cauchy product of the series $1 + x + x^2 + x^3 + \cdots$ and $1 - x + x^2 - x^3 + \cdots$. On what interval and to what function does the product series converge?

11. Determine the power series expansion of $1/(1-x)^2$ by formally dividing $1 - 2x + x^2$ into 1.

Starting with the power series representation

$$\frac{1}{1-x} = 1 + x + x^2 + x^3 + \cdots, \qquad (-1 < x < 1),$$

determine power series representations for the functions indicated in Exercises 12–20. On what interval is each representation valid?

12. $\dfrac{1}{2-x}$ in powers of x

13. $\dfrac{1}{(2-x)^2}$ in powers of x

14. $\dfrac{1}{1+2x}$ in powers of x

15. $\ln(2-x)$ in powers of x

16. $\dfrac{1}{x}$ in powers of $x-1$

17. $\dfrac{1}{x^2}$ in powers of $x+2$

18. $\dfrac{1-x}{1+x}$ in powers of x

19. $\dfrac{x^3}{1-2x^2}$ in powers of x

20. $\ln x$ in powers of $x-4$

Determine the interval of convergence and the sum of each of the series in Exercises 21–26.

21. $1-4x+16x^2-64x^3+\cdots=\displaystyle\sum_{n=0}^{\infty}(-1)^n(4x)^n$

22. $3+4x+5x^2+6x^3+\cdots=\displaystyle\sum_{n=0}^{\infty}(n+3)x^n$

23. $\dfrac{1}{3}+\dfrac{x}{4}+\dfrac{x^2}{5}+\dfrac{x^3}{6}+\cdots=\displaystyle\sum_{n=0}^{\infty}\frac{x^n}{n+3}$

24. $1\times 3-2\times 4x+3\times 5x^2-4\times 6x^3+\cdots$

$$=\sum_{n=0}^{\infty}(-1)^n(n+1)(n+3)\,x^n$$

25. $2+4x^2+6x^4+8x^6+10x^8+\cdots=\displaystyle\sum_{n=0}^{\infty}2(n+1)\,x^{2n}$

26. $1-\dfrac{x^2}{2}+\dfrac{x^4}{3}-\dfrac{x^6}{4}+\dfrac{x^8}{5}-\cdots=\displaystyle\sum_{n=0}^{\infty}\frac{(-1)^n x^{2n}}{n+1}$

Use the technique (or the result) of Example 6 to find the sums of the numerical series in Exercises 27–32.

27. $\displaystyle\sum_{n=1}^{\infty}\frac{n}{3^n}$

28. $\displaystyle\sum_{n=0}^{\infty}\frac{n+1}{2^n}$

29. $\displaystyle\sum_{n=0}^{\infty}\frac{(n+1)^2}{\pi^n}$

30. $\displaystyle\sum_{n=1}^{\infty}\frac{(-1)^n n(n+1)}{2^n}$

31. $\displaystyle\sum_{n=1}^{\infty}\frac{(-1)^{n-1}}{n2^n}$

32. $\displaystyle\sum_{n=3}^{\infty}\frac{1}{n2^n}$

9.6 Taylor and Maclaurin Series

If a power series $\sum_{n=0}^{\infty} a_n(x-c)^n$ has a positive radius of convergence R, then the sum of the series defines a function $f(x)$ on the interval $(c-R, c+R)$. We say that the power series is a **representation** of $f(x)$ on that interval. What relationship exists between the function $f(x)$ and the coefficients a_0, a_1, a_2, ... of the power series? The following theorem answers this question.

THEOREM 21

Suppose the series

$$f(x)=\sum_{n=0}^{\infty}a_n(x-c)^n=a_0+a_1(x-c)+a_2(x-c)^2+a_3(x-c)^3+\cdots$$

converges to $f(x)$ for $c-R<x<c+R$, where $R>0$. Then

$$a_k=\frac{f^{(k)}(c)}{k!}\qquad\text{for } k=0,\,1,\,2,\,3,\,\ldots.$$

PROOF This proof requires that we differentiate the series for $f(x)$ term by term several times, a process justified by Theorem 19 (suitably reformulated for powers of $x-c$):

$$f'(x)=\sum_{n=1}^{\infty}na_n(x-c)^{n-1}=a_1+2a_2(x-c)+3a_3(x-c)^2+\cdots$$

$$f''(x)=\sum_{n=2}^{\infty}n(n-1)a_n(x-c)^{n-2}=2a_2+6a_3(x-c)+12a_4(x-c)^2+\cdots$$

$$\vdots$$

$$f^{(k)}(x)=\sum_{n=k}^{\infty}n(n-1)(n-2)\cdots(n-k+1)a_n(x-c)^{n-k}$$

$$=k!a_k+\frac{(k+1)!}{1!}a_{k+1}(x-c)+\frac{(k+2)!}{2!}a_{k+2}(x-c)^2+\cdots.$$

Each series converges for $c - R < x < c + R$. Setting $x = c$, we obtain $f^{(k)}(c) = k!a_k$, which proves the theorem.

Theorem 21 shows that a function $f(x)$ that has a power series representation with centre at c and positive radius of convergence must have derivatives of all orders in an interval around $x = c$, and it can have only one representation as a power series in powers of $x - c$, namely

$$f(x) = \sum_{n=0}^{\infty} \frac{f^{(n)}(c)}{n!}(x-c)^n = f(c) + f'(c)(x-c) + \frac{f''(c)}{2!}(x-c)^2 + \cdots.$$

Such a series is called a Taylor series or, if $c = 0$, a Maclaurin series.

DEFINITION 8

Taylor and Maclaurin series

If $f(x)$ has derivatives of all orders at $x = c$ (i.e., if $f^{(k)}(c)$ exists for $k = 0, 1, 2, 3, \ldots$), then the series

$$\sum_{k=0}^{\infty} \frac{f^{(k)}(c)}{k!}(x-c)^k$$
$$= f(c) + f'(c)(x-c) + \frac{f''(c)}{2!}(x-c)^2 + \frac{f^{(3)}(c)}{3!}(x-c)^3 + \cdots$$

is called the **Taylor series of f about c** (or the **Taylor series of f in powers of $x - c$**). If $c = 0$, the term **Maclaurin series** is usually used in place of Taylor series.

Note that the partial sums of such Taylor (or Maclaurin) series are just the Taylor (or Maclaurin) polynomials studied in Section 4.10.

The Taylor series is a power series as defined in the previous section. Theorem 17 implies that c must be the centre of any interval on which such a series converges, but the definition of Taylor series makes no requirement that the series should converge anywhere except at the point $x = c$ where the series is just $f(c) + 0 + 0 + \cdots$. The series exists provided all the derivatives of f exist at $x = c$; in practice this means that each derivative must exist in an open interval containing $x = c$. (Why?) However, the series may converge nowhere except at $x = c$, and if it does converge elsewhere, it may converge to something other than $f(x)$. (See Exercise 40 at the end of this section for an example where this happens.) If the Taylor series does converge to $f(x)$ in an open interval containing c, then we will say that f is analytic at c.

DEFINITION 9

Analytic functions

A function f is **analytic at c** if f has a Taylor series at c and that series converges to $f(x)$ in an open interval containing c. If f is analytic at each point of an open interval, then we say it is analytic on that interval.

Most, but not all, of the elementary functions encountered in calculus are analytic wherever they have derivatives of all orders. On the other hand, whenever a power series in powers of $x - c$ converges for all x in an open interval containing c, then its sum $f(x)$ is analytic at c, and the given series is the Taylor series of f about c.

Maclaurin Series for Some Elementary Functions

Calculating Taylor and Maclaurin series for a function f directly from Definition 8 is practical only when we can find a formula for the nth derivative of f. Examples of such functions include $(ax+b)^r$, e^{ax+b}, $\ln(ax+b)$, $\sin(ax+b)$, $\cos(ax+b)$, and sums of such functions.

EXAMPLE 1 Find the Taylor series for e^x about $x = c$. Where does the series converge to e^x? Where is e^x analytic? What is the Maclaurin series for e^x?

Solution Since all the derivatives of $f(x) = e^x$ are e^x, we have $f^{(n)}(c) = e^c$ for every integer $n \ge 0$. Thus, the Taylor series for e^x about $x = c$ is

$$\sum_{n=0}^{\infty} \frac{e^c}{n!}(x-c)^n = e^c + e^c(x-c) + \frac{e^c}{2!}(x-c)^2 + \frac{e^c}{3!}(x-c)^3 + \cdots.$$

The radius of convergence R of this series is given by

$$\frac{1}{R} = \lim_{n\to\infty} \left| \frac{e^c/(n+1)!}{e^c/n!} \right| = \lim_{n\to\infty} \frac{n!}{(n+1)!} = \lim_{n\to\infty} \frac{1}{n+1} = 0.$$

Thus, the radius of convergence is $R = \infty$ and the series converges for all x. Suppose the sum is $g(x)$:

$$g(x) = e^c + e^c(x-c) + \frac{e^c}{2!}(x-c)^2 + \frac{e^c}{3!}(x-c)^3 + \cdots.$$

By Theorem 19, we have

$$\begin{aligned} g'(x) &= 0 + e^c + \frac{e^c}{2!}2(x-c) + \frac{e^c}{3!}3(x-c)^2 + \cdots \\ &= e^c + e^c(x-c) + \frac{e^c}{2!}(x-c)^2 + \cdots = g(x). \end{aligned}$$

Also, $g(c) = e^c + 0 + 0 + \cdots = e^c$. Since $g(x)$ satisfies the differential equation $g'(x) = g(x)$ of exponential growth, we have $g(x) = Ce^x$. Substituting $x = c$ gives $e^c = g(c) = Ce^c$, so $C = 1$. Thus, the Taylor series for e^x in powers of $x - c$ converges to e^x for every real number x:

$$\begin{aligned} e^x &= \sum_{n=0}^{\infty} \frac{e^c}{n!}(x-c)^n \\ &= e^c + e^c(x-c) + \frac{e^c}{2!}(x-c)^2 + \frac{e^c}{3!}(x-c)^3 + \cdots \qquad \text{(for all } x\text{)}. \end{aligned}$$

In particular, e^x is analytic on the whole real line $\mathbb{R}$. Setting $c = 0$ we obtain the Maclaurin series for e^x:

$$e^x = \sum_{n=0}^{\infty} \frac{x^n}{n!} = 1 + x + \frac{x^2}{2!} + \frac{x^3}{3!} + \cdots \qquad \text{(for all } x\text{)}.$$

EXAMPLE 2 Find the Maclaurin series for (a) $\sin x$ and (b) $\cos x$. Where does each series converge?

Solution Let $f(x) = \sin x$. Then we have $f(0) = 0$ and

$$\begin{aligned} f'(x) &= \cos x & f'(0) &= 1 \\ f''(x) &= -\sin x & f''(0) &= 0 \\ f^{(3)}(x) &= -\cos x & f^{(3)}(0) &= -1 \\ f^{(4)}(x) &= \sin x & f^{(4)}(0) &= 0 \\ f^{(5)}(x) &= \cos x & f^{(5)}(0) &= 1 \\ &\vdots & &\vdots \end{aligned}$$

Thus, the Maclaurin series for $\sin x$ is

$$\begin{aligned} g(x) &= 0 + x + 0 - \frac{x^3}{3!} + 0 + \frac{x^5}{5!} + 0 - \cdots \\ &= x - \frac{x^3}{3!} + \frac{x^5}{5!} - \frac{x^7}{7!} + \cdots = \sum_{n=0}^{\infty} \frac{(-1)^n}{(2n+1)!} x^{2n+1}. \end{aligned}$$

We have denoted the sum by $g(x)$ since we don't yet know whether the series converges to $\sin x$. The series does converge for all x by the ratio test:

$$\begin{aligned} \lim_{n\to\infty} \left| \frac{\dfrac{(-1)^{n+1}}{(2(n+1)+1)!} x^{2(n+1)+1}}{\dfrac{(-1)^n}{(2n+1)!} x^{2n+1}} \right| &= \lim_{n\to\infty} \frac{(2n+1)!}{(2n+3)!} |x|^2 \\ &= \lim_{n\to\infty} \frac{|x|^2}{(2n+3)(2n+2)} = 0. \end{aligned}$$

Now we can differentiate the function $g(x)$ twice to get

$$\begin{aligned} g'(x) &= 1 - \frac{x^2}{2!} + \frac{x^4}{4!} - \frac{x^6}{6!} + \cdots \\ g''(x) &= -x + \frac{x^3}{3!} - \frac{x^5}{5!} + \frac{x^7}{7!} - \cdots = -g(x). \end{aligned}$$

Thus, $g(x)$ satisfies the differential equation $g''(x) + g(x) = 0$ of simple harmonic motion. The general solution of this equation, as observed in Section 3.7, is

$$g(x) = A \cos x + B \sin x.$$

Observe, from the series, that $g(0) = 0$ and $g'(0) = 1$. These values determine that $A = 0$ and $B = 1$. Thus, $g(x) = \sin x$ and $g'(x) = \cos x$ for all x.

We have therefore demonstrated that

$$\sin x = \sum_{n=0}^{\infty} \frac{(-1)^n}{(2n+1)!} x^{2n+1} = x - \frac{x^3}{3!} + \frac{x^5}{5!} - \frac{x^7}{7!} + \cdots \quad \text{(for all } x\text{)},$$

$$\cos x = \sum_{n=0}^{\infty} \frac{(-1)^n}{(2n)!} x^{2n} = 1 - \frac{x^2}{2!} + \frac{x^4}{4!} - \frac{x^6}{6!} + \cdots \quad \text{(for all } x\text{)}.$$

Theorem 21 shows that we can use any available means to find a power series converging to a given function on an interval, and the series obtained will turn out to be the Taylor series. In Section 9.5 several series were constructed by manipulating a geometric series. These include:

Some Maclaurin series

$$\frac{1}{1-x} = \sum_{n=0}^{\infty} x^n = 1 + x + x^2 + x^3 + \cdots \quad (-1 < x < 1)$$

$$\frac{1}{(1-x)^2} = \sum_{n=1}^{\infty} n x^{n-1} = 1 + 2x + 3x^2 + 4x^3 + \cdots \quad (-1 < x < 1)$$

$$\ln(1+x) = \sum_{n=1}^{\infty} \frac{(-1)^{n-1}}{n} x^n = x - \frac{x^2}{2} + \frac{x^3}{3} - \frac{x^4}{4} + \cdots \quad (-1 < x \le 1)$$

$$\tan^{-1} x = \sum_{n=0}^{\infty} \frac{(-1)^n}{2n+1} x^{2n+1} = x - \frac{x^3}{3} + \frac{x^5}{5} - \frac{x^7}{7} + \cdots \quad (-1 \le x \le 1)$$

These series, together with the intervals on which they converge, are frequently used hereafter and should be memorized.

Other Maclaurin and Taylor Series

Series can be combined in various ways to generate new series. For example, we can find the Maclaurin series for e^{-x} by replacing x with $-x$ in the series for e^x:

$$e^{-x} = \sum_{n=0}^{\infty} \frac{(-1)^n}{n!} x^n = 1 - x + \frac{x^2}{2!} - \frac{x^3}{3!} + \cdots \quad \text{(for all } x\text{)}.$$

The series for e^x and e^{-x} can then be subtracted or added and the results divided by 2 to obtain Maclaurin series for the hyperbolic functions $\sinh x$ and $\cosh x$:

$$\sinh x = \frac{e^x - e^{-x}}{2} = \sum_{n=0}^{\infty} \frac{x^{2n+1}}{(2n+1)!} = x + \frac{x^3}{3!} + \frac{x^5}{5!} + \cdots \quad \text{(for all } x\text{)}$$

$$\cosh x = \frac{e^x + e^{-x}}{2} = \sum_{n=0}^{\infty} \frac{x^{2n}}{(2n)!} = 1 + \frac{x^2}{2!} + \frac{x^4}{4!} + \cdots \quad \text{(for all } x\text{)}.$$

Remark Observe the similarity between the series for $\sin x$ and $\sinh x$ and between those for $\cos x$ and $\cosh x$. If we were to allow complex numbers (numbers of the form $z = x + iy$, where $i^2 = -1$ and x and y are real; see Appendix I) as arguments for our functions, and if we were to demonstrate that our operations on series could be extended to series of complex numbers, we would see that $\cos x = \cosh(ix)$ and $\sin x = -i \sinh(ix)$. In fact, $e^{ix} = \cos x + i \sin x$ and $e^{-ix} = \cos x - i \sin x$, so

$$\cos x = \frac{e^{ix} + e^{-ix}}{2}, \qquad \text{and} \qquad \sin x = \frac{e^{ix} - e^{-ix}}{2i}.$$

Such formulas are encountered in the study of functions of a complex variable (see Appendix II); from the complex point of view the trigonometric and exponential functions are just different manifestations of the same basic function, a complex exponential $e^z = e^{x+iy}$. We content ourselves here with having mentioned the interesting relationships above and invite the reader to verify them formally by calculating with series. (Such formal calculations do not, of course, constitute a proof, since we have not established the various rules covering series of complex numbers.)

EXAMPLE 3 Obtain Maclaurin series for the following functions:

(a) $e^{-x^2/3}$, (b) $\dfrac{\sin(x^2)}{x}$, (c) $\sin^2 x$.

Solution

(a) We substitute $-x^2/3$ for x in the Maclaurin series for e^x:

$$\begin{aligned} e^{-x^2/3} &= 1 - \frac{x^2}{3} + \frac{1}{2!}\left(\frac{x^2}{3}\right)^2 - \frac{1}{3!}\left(\frac{x^2}{3}\right)^3 + \cdots \\ &= \sum_{n=0}^{\infty} (-1)^n \frac{1}{3^n n!} x^{2n} \qquad \text{(for all real } x\text{)}. \end{aligned}$$

(b) For all $x \neq 0$ we have

$$\begin{aligned} \frac{\sin(x^2)}{x} &= \frac{1}{x}\left(x^2 - \frac{(x^2)^3}{3!} + \frac{(x^2)^5}{5!} - \cdots\right) \\ &= x - \frac{x^5}{3!} + \frac{x^9}{5!} - \cdots = \sum_{n=0}^{\infty} (-1)^n \frac{x^{4n+1}}{(2n+1)!}. \end{aligned}$$

Note that $f(x) = (\sin(x^2))/x$ is not defined at $x = 0$ but does have a limit (namely 0) as x approaches 0. If we define $f(0) = 0$ (the continuous extension of $f(x)$ to $x = 0$), then the series converges to $f(x)$ for all x.

(c) We use a trigonometric identity to express $\sin^2 x$ in terms of $\cos 2x$ and then use the Maclaurin series for $\cos x$ with x replaced by $2x$.

$$\begin{aligned}\sin^2 x &= \frac{1-\cos 2x}{2} = \frac{1}{2} - \frac{1}{2}\left(1 - \frac{(2x)^2}{2!} + \frac{(2x)^4}{4!} - \cdots\right)\\ &= \frac{1}{2}\left(\frac{(2x)^2}{2!} - \frac{(2x)^4}{4!} + \frac{(2x)^6}{6!} - \cdots\right)\\ &= \sum_{n=0}^{\infty}(-1)^n \frac{2^{2n+1}}{(2n+2)!}x^{2n+2} \qquad \text{(for all real } x).\end{aligned}$$

Taylor series about points other than 0 can often be obtained from known Maclaurin series by a change of variable.

EXAMPLE 4 Find the Taylor series for $\ln x$ in powers of $x - 2$. Where does the series converge to $\ln x$?

Solution Note that if $t = (x-2)/2$, then

$$\ln x = \ln\big(2 + (x-2)\big) = \ln\left[2\left(1 + \frac{x-2}{2}\right)\right] = \ln 2 + \ln(1+t).$$

We use the known Maclaurin series for $\ln(1+t)$:

$$\begin{aligned}\ln x &= \ln 2 + \ln(1+t)\\ &= \ln 2 + t - \frac{t^2}{2} + \frac{t^3}{3} - \frac{t^4}{4} - \cdots\\ &= \ln 2 + \frac{x-2}{2} - \frac{(x-2)^2}{2\times 2^2} + \frac{(x-2)^3}{3\times 2^3} - \frac{(x-2)^4}{4\times 2^4} + \cdots\\ &= \ln 2 + \sum_{n=1}^{\infty}\frac{(-1)^{n-1}}{n\,2^n}(x-2)^n.\end{aligned}$$

Since the series for $\ln(1+t)$ is valid for $-1 < t \le 1$, this series for $\ln x$ is valid for $-1 < (x-2)/2 \le 1$, that is, for $0 < x \le 4$.

EXAMPLE 5 Find the Taylor series for $\cos x$ about $\pi/3$. Where is the series valid?

Solution We use the addition formula for cosine:

$$\begin{aligned}\cos x &= \cos\left(x - \frac{\pi}{3} + \frac{\pi}{3}\right) = \cos\left(x - \frac{\pi}{3}\right)\cos\frac{\pi}{3} - \sin\left(x - \frac{\pi}{3}\right)\sin\frac{\pi}{3}\\ &= \frac{1}{2}\left[1 - \frac{1}{2!}\left(x-\frac{\pi}{3}\right)^2 + \frac{1}{4!}\left(x-\frac{\pi}{3}\right)^4 - \cdots\right]\\ &\quad - \frac{\sqrt{3}}{2}\left[\left(x-\frac{\pi}{3}\right) - \frac{1}{3!}\left(x-\frac{\pi}{3}\right)^3 + \cdots\right]\\ &= \frac{1}{2} - \frac{\sqrt{3}}{2}\left(x-\frac{\pi}{3}\right) - \frac{1}{2}\frac{1}{2!}\left(x-\frac{\pi}{3}\right)^2 + \frac{\sqrt{3}}{2}\frac{1}{3!}\left(x-\frac{\pi}{3}\right)^3\\ &\quad + \frac{1}{2}\frac{1}{4!}\left(x-\frac{\pi}{3}\right)^4 - \cdots.\end{aligned}$$

This series representation is valid for all x. A similar calculation would enable us to expand $\cos x$ or $\sin x$ in powers of $x - c$ for any real c; both functions are analytic at every point of the real line.

Sometimes it is quite difficult, if not impossible, to find a formula for the general term of a Maclaurin or Taylor series. In such cases it is usually possible to obtain the first few terms before the calculations get too cumbersome. Had we attempted to solve Example 3(c) by multiplying the series for $\sin x$ by itself we might have found ourselves in this bind. Other examples occur when it is necessary to substitute one series into another or to divide one by another.

EXAMPLE 6 Obtain the first three nonzero terms of the Maclaurin series for (a) $\tan x$ and (b) $\ln\cos x$.

Solution

(a) $\tan x = (\sin x)/(\cos x)$. We can obtain the first three terms of the Maclaurin series for $\tan x$ by long division of the series for $\cos x$ into that for $\sin x$:

$$
\begin{array}{r|l}
 & x + \frac{x^3}{3} + \frac{2}{15}x^5 + \cdots \\ \hline
1 - \frac{x^2}{2} + \frac{x^4}{24} & x - \frac{x^3}{6} + \frac{x^5}{120} - \cdots \\
 & x - \frac{x^3}{2} + \frac{x^5}{24} - \cdots \\ \hline
 & \qquad \frac{x^3}{3} - \frac{x^5}{30} + \cdots \\
 & \qquad \frac{x^3}{3} - \frac{x^5}{6} + \cdots \\ \hline
 & \qquad\qquad \frac{2x^5}{15} - \cdots \\
 & \qquad\qquad \frac{2x^5}{15} - \cdots \\ \hline
\end{array}
$$

Thus, $\tan x = x + \frac{1}{3}x^3 + \frac{2}{15}x^5 + \cdots$.

We cannot easily find all the terms of the series; only with considerable computational effort can we find many more terms than we have already found. This Maclaurin series for $\tan x$ converges for $|x| < \pi/2$, but we cannot demonstrate this fact by the techniques we have at our disposal now. (It is true because the complex number $z = x + iy$ closest to 0 where the "denominator" of $\tan z$, that is, $\cos z$, is zero, is, in fact, the real value $z = \pi/2$.)

(b)
$$
\begin{aligned}
\ln\cos x &= \ln\left(1 + \left(-\frac{x^2}{2!} + \frac{x^4}{4!} - \frac{x^6}{6!} + \cdots\right)\right) \\
&= \left(-\frac{x^2}{2!} + \frac{x^4}{4!} - \frac{x^6}{6!} + \cdots\right) - \frac{1}{2}\left(-\frac{x^2}{2!} + \frac{x^4}{4!} - \frac{x^6}{6!} + \cdots\right)^2 \\
&\quad + \frac{1}{3}\left(-\frac{x^2}{2!} + \frac{x^4}{4!} - \frac{x^6}{6!} + \cdots\right)^3 - \cdots \\
&= -\frac{x^2}{2} + \frac{x^4}{24} - \frac{x^6}{720} + \cdots - \frac{1}{2}\left(\frac{x^4}{4} - \frac{x^6}{24} + \cdots\right) \\
&\quad + \frac{1}{3}\left(-\frac{x^6}{8} + \cdots\right) - \cdots
\end{aligned}
$$

$$= -\frac{x^2}{2} - \frac{x^4}{12} - \frac{x^6}{45} - \cdots.$$

Note that at each stage of the calculation we kept only enough terms to ensure that we could get all the terms with powers up to x^6. Being an even function, $\ln \cos x$ has only even powers in its Maclaurin series. Again, we cannot find the general term of this series. We could try to calculate terms by using the formula $a_k = f^{(k)}(0)/k!$ but even this becomes difficult after the first few values of k.

Observe that the series for $\tan x$ could also have been derived from that of $\ln \cos x$ because we have $\tan x = -\frac{d}{dx} \ln \cos x$.

Taylor's Formula Revisited

In the examples above we have used a variety of techniques to obtain Taylor series for functions and verify that functions are analytic. As shown in Section 4.10, Taylor's Theorem provides a means for estimating the size of the error $E_n(x) = f(x) - P_n(x)$ involved when the Taylor polynomial

$$P_n(x) = \sum_{k=0}^{n} \frac{f^{(k)}(c)}{k!}(x - c)^k$$

is used to approximate the value of $f(x)$ for $x \neq c$. Since the Taylor polynomials are partial sums of the Taylor series for f at c (if the latter exists), another technique for verifying the convergence of a Taylor series is to use the formula for $E_n(x)$ provided by Taylor's Theorem to show, at least for an interval of values of x containing c, that $\lim_{n\to\infty} E_n(x) = 0$. This implies that $\lim_{n\to\infty} P_n(x) = f(x)$ so that f is indeed the sum of its Taylor series about c on that inverval, and f is analytic at c. Here is a somewhat more general version of Taylor's theroem.

THEOREM

Taylor's Theorem

If the $(n + 1)$st derivative of f exists on an interval containing c and x, and if $P_n(x)$ is the Taylor polynomial of degree n for f about the point $x = c$, then

$$f(x) = P_n(x) + E_n(x) \qquad \textbf{Taylor's Formula}$$

holds, where the error term $E_n(x)$ is given by *either* of the following formulas:

Lagrange remainder $\quad E_n(x) = \dfrac{f^{(n+1)}(s)}{(n+1)!}(x - c)^{n+1},$

for some s between c and x

Integral remainder $\quad E_n(x) = \dfrac{1}{n!}\displaystyle\int_c^x (x - t)^n f^{(n+1)}(t)\, dt.$

Taylor's Theorem with Lagrange remainder was proved in Section 4.10 (Theorem 12) by using the Mean-Value Theorem and induction on n. The Integral remainder version is also proved by induction on n. See Exercise 42 for hints on how to carry out the proof. We will not make any use of the Integral form of the remainder here.

Our final example in this section reestablishes the Maclaurin series for e^x by finding the limit of the Lagrange remainder as suggested above.

EXAMPLE 7 Use Taylor's Theorem to find the Maclaurin series for $f(x) = e^x$. Where does the series converge to $f(x)$?

Solution Since e^x is positive and increasing, $e^s \le e^{|x|}$ for any $s \le |x|$. Since $f^{(k)}(x) = e^x$ for any k we have, taking $c = 0$ in the Lagrange remainder in Taylor's Formula,

$$|E_n(x)| = \left| \frac{f^{(n+1)}(s)}{(n+1)!} x^{n+1} \right| \quad \text{for some } s \text{ between } 0 \text{ and } x$$

$$= \frac{e^s}{(n+1)!} |x|^{n+1} \le e^{|x|} \frac{|x|^{n+1}}{(n+1)!} \to 0 \text{ as } n \to \infty$$

for any real x, as shown in Theorem 3(b) of Section 9.1. Thus, $\lim_{n\to\infty} E_n(x) = 0$. Since the nth-order Maclaurin polynomial for e^x is $\sum_{k=0}^{n}(x^k/k!)$,

$$e^x = \lim_{n\to\infty}\left(\sum_{k=0}^{n} \frac{x^k}{k!} + E_n(x)\right) = \sum_{k=0}^{\infty} \frac{x^k}{k!} = 1 + x + \frac{x^2}{2!} + \frac{x^3}{3!} + \cdots,$$

and the series converges to e^x for all real numbers x.

EXERCISES 9.6

Find Maclaurin series representations for the functions in Exercises 1–14. For what values of x is each representation valid?

1. e^{3x+1}

2. $\cos(2x^3)$

3. $\sin(x - \pi/4)$

4. $\cos(2x - \pi)$

5. $x^2 \sin(x/3)$

6. $\cos^2(x/2)$

7. $\sin x \cos x$

8. $\tan^{-1}(5x^2)$

9. $\dfrac{1+x^3}{1+x^2}$

10. $\ln(2 + x^2)$

11. $\ln \dfrac{1+x}{1-x}$

12. $(e^{2x^2} - 1)/x^2$

13. $\cosh x - \cos x$

14. $\sinh x - \sin x$

Find the required Taylor series representations of the functions in Exercises 15–26. Where is each series representation valid?

15. $f(x) = e^{-2x}$ about -1

16. $f(x) = \sin x$ about $\pi/2$

17. $f(x) = \cos x$ in powers of $x - \pi$

18. $f(x) = \ln x$ in powers of $x - 3$

19. $f(x) = \ln(2 + x)$ in powers of $x - 2$

20. $f(x) = e^{2x+3}$ in powers of $x + 1$

21. $f(x) = \sin x - \cos x$ about $\dfrac{\pi}{4}$

22. $f(x) = \cos^2 x$ about $\dfrac{\pi}{8}$

23. $f(x) = 1/x^2$ in powers of $x + 2$

24. $f(x) = \dfrac{x}{1+x}$ in powers of $x - 1$

25. $f(x) = x \ln x$ in powers of $x - 1$

26. $f(x) = xe^x$ in powers of $x + 2$

Find the first three nonzero terms in the Maclaurin series for the functions in Exercises 27–30.

27. $\sec x$

28. $\sec x \tan x$

29. $\tan^{-1}(e^x - 1)$

30. $e^{\tan^{-1} x} - 1$

31. Use the fact that $(\sqrt{1+x})^2 = 1 + x$ to find the first three nonzero terms of the Maclaurin series for $\sqrt{1+x}$.

32. Does $\csc x$ have a Maclaurin series? Why? Find the first three nonzero terms of the Taylor series for $\csc x$ about the point $x = \pi/2$.

Find the sums of the series in Exercises 33–36.

33. $1 + x^2 + \dfrac{x^4}{2!} + \dfrac{x^6}{3!} + \dfrac{x^8}{4!} + \cdots$

34. $x^3 - \dfrac{x^9}{3! \times 4} + \dfrac{x^{15}}{5! \times 16} - \dfrac{x^{21}}{7! \times 64} + \dfrac{x^{27}}{9! \times 256} - \cdots$

35. $1 + \dfrac{x^2}{3!} + \dfrac{x^4}{5!} + \dfrac{x^6}{7!} + \dfrac{x^8}{9!} + \cdots$

36. $1 + \dfrac{1}{2 \times 2!} + \dfrac{1}{4 \times 3!} + \dfrac{1}{8 \times 4!} + \cdots$

37. Let $P(x) = 1 + x + x^2$. Find (a) the Maclaurin series for $P(x)$ and (b) the Taylor series for $P(x)$ about 1.

38. Verify by direct calculation that $f(x) = 1/x$ is analytic at a for every $a \ne 0$.

39. Verify by direct calculation that $\ln x$ is analytic at a for every $a > 0$.

40. Review Exercise 41 of Section 4.5. It shows that the function

$$f(x) = \begin{cases} e^{-1/x^2} & \text{if } x \ne 0 \\ 0 & \text{if } x = 0 \end{cases}$$

has derivatives of all orders at every point of the real line, and $f^{(k)}(0) = 0$ for every positive integer k. What is the Maclaurin series for $f(x)$? What is the interval of convergence of this Maclaurin series? On what interval does the series converge to $f(x)$? Is f analytic at 0?

41. By direct multiplication of the Maclaurin series for e^x and e^y show that $e^x e^y = e^{x+y}$.

42. (Taylor's Formula with integral remainder) Verify that if $f^{(n+1)}$ exists on an interval containing c and x, and if $P_n(x)$ is the nth-order Taylor polynomial for f about c, then $f(x) = P_n(x) + E_n(x)$, where

$$E_n(x) = \frac{1}{n!}\int_c^x (x-t)^n f^{(n+1)}(t)\,dt.$$

Proceed as follows:

(a) First observe that the case $n = 0$ is just the Fundamental Theorem of Calculus:

$$f(x) = f(c) + \int_c^x f'(t)\,dt.$$

Now integrate by parts in this formula, taking $U = f'(t)$ and $dV = dt$. Contrary to our usual policy of not including a constant of integration in V, here write $V = -(x-t)$ rather than just $V = t$. Observe that the result of the integration by parts is the case $n = 1$ of the formula.

(b) Use induction argument (and integration by parts again) to show that if the formula is valid for $n = k$, then it is also valid for $n = k + 1$.

43. Use Taylor's formula with integral remainder to reprove that the Maclaurin series for $\ln(1+x)$ converges to $\ln(1+x)$ for $-1 < x \le 1$.

44. (Stirling's Formula) The limit

$$\lim_{n\to\infty} \frac{n!}{\sqrt{2\pi}\,n^{n+1/2}e^{-n}} = 1$$

says that the *relative error* in the approximation

$$n! \approx \sqrt{2\pi}\,n^{n+1/2}e^{-n}$$

approaches zero as n increases. That is, $n!$ grows at a rate comparable to $\sqrt{2\pi}\,n^{n+1/2}e^{-n}$. This result, known as Stirling's Formula, is often very useful in applied mathematics and statistics. Prove it by carrying out the following steps.

(a) Use the identity $\ln(n!) = \sum_{j=1}^{n} \ln j$ and the increasing nature of ln to show that if $n \ge 1$,

$$\int_0^n \ln x\,dx < \ln(n!) < \int_1^{n+1} \ln x\,dx$$

and hence that

$$n\ln n - n < \ln(n!) < (n+1)\ln(n+1) - n.$$

(b) If $c_n = \ln(n!) - \left(n + \frac{1}{2}\right)\ln n + n$, show that

$$\begin{aligned} c_n - c_{n+1} &= \left(n + \frac{1}{2}\right)\ln\frac{n+1}{n} - 1 \\ &= \left(n + \frac{1}{2}\right)\ln\frac{1 + 1/(2n+1)}{1 - 1/(2n+1)} - 1. \end{aligned}$$

(c) Use the Maclaurin series for $\ln\dfrac{1+t}{1-t}$ (see Exercise 11) to show that

$$\begin{aligned} 0 < c_n - c_{n+1} &< \frac{1}{3}\left(\frac{1}{(2n+1)^2} + \frac{1}{(2n+1)^4} + \cdots\right) \\ &= \frac{1}{12}\left(\frac{1}{n} - \frac{1}{n+1}\right), \end{aligned}$$

and therefore that $\{c_n\}$ is decreasing and $\left\{c_n - \frac{1}{12n}\right\}$ is increasing. Hence conclude that $\lim_{n\to\infty} c_n = c$ exists, and that

$$\lim_{n\to\infty} \frac{n!}{n^{n+1/2}e^{-n}} = \lim_{n\to\infty} e^{c_n} = e^c.$$

(d) Now use the Wallis Product from Exercise 38 of Section 6.1 to show that

$$\lim_{n\to\infty} \frac{(2^n n!)^2}{(2n)!\sqrt{2n}} = \sqrt{\frac{\pi}{2}},$$

and hence deduce that $e^c = \sqrt{2\pi}$, which completes the proof.

45. (A Modified Stirling Formula) A simpler approximation to $n!$ for large n is given by

$$n! \approx n^n\,e^{-n} \quad \text{or, equivalently,} \quad \ln(n!) \approx n\ln n - n.$$

While not as accurate as Stirling's Formula, this modified version still has relative error approaching zero as $n \to \infty$ and can be useful in many applications.

(a) Prove this assertion about the relative error by using the conclusion of part (a) of the previous exercise.

(b) Compare the relative errors in the approximations for $\ln(10!)$ and $\ln(20!)$ using Stirling's Formula and the Modified Stirling Formula.

9.7 Applications of Taylor and Maclaurin Series

Approximating the Values of Functions

We saw in Section 4.10 how Taylor and Maclaurin polynomials (the partial sums of Taylor and Maclaurin series) can be used as polynomial approximations to more complicated functions. In Example 5 of that section we used the Lagrange remainder in Taylor's Formula to determine how many terms of the Maclaurin series for e^x are needed to calculate $e^1 = e$ correct to 3 decimal places. For comparison, we obtained the same result in Example 7 in Section 9.3 by using a geometric series to bound the tail of the series for e.

The following example shows how the error bound associated with the alternating series test (see Theorem 15 in Section 9.4) can also be used for such approximations. When the terms a_n of a series (i) alternate in sign, (ii) decrease steadily in size, and (iii) approach zero as $n \to \infty$, then the error involved in using a partial sum of the series as an approximation to the sum of the series has the same sign as, and is no greater in absolute value than, the first omitted term.

EXAMPLE 1 Find $\cos 43°$ with error less than $1/10{,}000$.

Solution We give two alternative solutions:

METHOD I. We can use the Maclaurin series for cosine:

$$\cos 43° = \cos\frac{43\pi}{180} = 1 - \frac{1}{2!}\left(\frac{43\pi}{180}\right)^2 + \frac{1}{4!}\left(\frac{43\pi}{180}\right)^4 - \cdots.$$

Now $43\pi/180 \approx 0.750\,49\cdots < 1$, so the series above must satisfy the conditions (i)–(iii) mentioned above. If we truncate the series after the nth term

$$(-1)^{n-1}\frac{1}{(2n-2)!}\left(\frac{43\pi}{180}\right)^{2n-2},$$

then the error E will be bounded by the size of the first omitted term:

$$|E| \le \frac{1}{(2n)!}\left(\frac{43\pi}{180}\right)^{2n} < \frac{1}{(2n)!}.$$

The error will not exceed $1/10{,}000$ if $(2n)! > 10{,}000$, so $n = 4$ will do ($8! = 40{,}320$).

$$\cos 43° \approx 1 - \frac{1}{2!}\left(\frac{43\pi}{180}\right)^2 + \frac{1}{4!}\left(\frac{43\pi}{180}\right)^4 - \frac{1}{6!}\left(\frac{43\pi}{180}\right)^6 \approx 0.731\,35\cdots$$

METHOD II. Since $43°$ is close to $45° = \pi/4$ rad, we can do a bit better by using the Taylor series about $\pi/4$ instead of the Maclaurin series:

$$\begin{aligned}\cos 43° &= \cos\left(\frac{\pi}{4} - \frac{\pi}{90}\right)\\ &= \cos\frac{\pi}{4}\cos\frac{\pi}{90} + \sin\frac{\pi}{4}\sin\frac{\pi}{90}\\ &= \frac{1}{\sqrt{2}}\left[\left(1 - \frac{1}{2!}\left(\frac{\pi}{90}\right)^2 + \frac{1}{4!}\left(\frac{\pi}{90}\right)^4 - \cdots\right)\right.\\ &\qquad \left. + \left(\frac{\pi}{90} - \frac{1}{3!}\left(\frac{\pi}{90}\right)^3 + \cdots\right)\right].\end{aligned}$$

Since

$$\frac{1}{4!}\left(\frac{\pi}{90}\right)^4 < \frac{1}{3!}\left(\frac{\pi}{90}\right)^3 < \frac{1}{20{,}000},$$

we need only the first two terms of the first series and the first term of the second series:

$$\cos 43° \approx \frac{1}{\sqrt{2}}\left(1 + \frac{\pi}{90} - \frac{1}{2}\left(\frac{\pi}{90}\right)^2\right) \approx 0.731\,358\cdots.$$

(In fact, $\cos 43° = 0.731\,353\,7\cdots$.)

When finding approximate values of functions, it is best, whenever possible, to use a power series about a point as close as possible to the point where the approximation is desired.

Functions Defined by Integrals

Many functions that can be expressed as simple combinations of elementary functions cannot be antidifferentiated by elementary techniques; their antiderivatives are not simple combinations of elementary functions. We can, however, often find the Taylor series for the antiderivatives of such functions and hence approximate their definite integrals.

EXAMPLE 2 Find the Maclaurin series for

$$E(x) = \int_0^x e^{-t^2}\,dt,$$

and use it to evaluate $E(1)$ correct to 3 decimal places.

Solution The Maclaurin series for $E(x)$ is given by

$$\begin{aligned}
E(x) &= \int_0^x \left(1 - t^2 + \frac{t^4}{2!} - \frac{t^6}{3!} + \frac{t^8}{4!} - \cdots\right) dt \\
&= \left(t - \frac{t^3}{3} + \frac{t^5}{5 \times 2!} - \frac{t^7}{7 \times 3!} + \frac{t^9}{9 \times 4!} - \cdots\right)\Bigg|_0^x \\
&= x - \frac{x^3}{3} + \frac{x^5}{5 \times 2!} - \frac{x^7}{7 \times 3!} + \frac{x^9}{9 \times 4!} - \cdots = \sum_{n=0}^{\infty} (-1)^n \frac{x^{2n+1}}{(2n+1)n!},
\end{aligned}$$

and is valid for all x because the series for e^{-t^2} is valid for all t. Therefore,

$$\begin{aligned}
E(1) &= 1 - \frac{1}{3} + \frac{1}{5 \times 2!} - \frac{1}{7 \times 3!} + \cdots \\
&\approx 1 - \frac{1}{3} + \frac{1}{5 \times 2!} - \frac{1}{7 \times 3!} + \cdots + \frac{(-1)^{n-1}}{(2n-1)(n-1)!}.
\end{aligned}$$

We stopped with the nth term. Again, the alternating series test assures us that the error in this approximation does not exceed the first omitted term, so it will be less than 0.0005, provided $(2n+1)n! > 2{,}000$. Since $13 \times 6! = 9{,}360$, $n = 6$ will do. Thus,

$$E(1) \approx 1 - \frac{1}{3} + \frac{1}{10} - \frac{1}{42} + \frac{1}{216} - \frac{1}{1{,}320} \approx 0.747,$$

rounded to 3 decimal places.

Indeterminate Forms

Examples 9 and 10 of Section 4.10 showed how Maclaurin polynomials could be used for evaluating the limits of indeterminate forms. Here are two more examples, this time using the series directly and keeping enough terms to allow cancellation of the [0/0] factors.

EXAMPLE 3 Evaluate (a) $\displaystyle\lim_{x\to 0} \frac{x - \sin x}{x^3}$ and (b) $\displaystyle\lim_{x\to 0} \frac{(e^{2x} - 1)\ln(1 + x^3)}{(1 - \cos 3x)^2}$.

Solution

(a) $\displaystyle \lim_{x\to 0}\frac{x-\sin x}{x^3}\qquad \left[\frac{0}{0}\right]$

$$= \lim_{x\to 0}\frac{x-\left(x-\dfrac{x^3}{3!}+\dfrac{x^5}{5!}-\cdots\right)}{x^3}$$

$$= \lim_{x\to 0}\frac{\dfrac{x^3}{3!}-\dfrac{x^5}{5!}+\cdots}{x^3}$$

$$= \lim_{x\to 0}\left(\frac{1}{3!}-\frac{x^2}{5!}+\cdots\right)=\frac{1}{3!}=\frac{1}{6}.$$

(b) $\displaystyle \lim_{x\to 0}\frac{(e^{2x}-1)\ln(1+x^3)}{(1-\cos 3x)^2}\qquad \left[\frac{0}{0}\right]$

$$= \lim_{x\to 0}\frac{\left(1+(2x)+\dfrac{(2x)^2}{2!}+\dfrac{(2x)^3}{3!}+\cdots-1\right)\left(x^3-\dfrac{x^6}{2}+\cdots\right)}{\left(1-\left(1-\dfrac{(3x)^2}{2!}+\dfrac{(3x)^4}{4!}-\cdots\right)\right)^2}$$

$$= \lim_{x\to 0}\frac{2x^4+2x^5+\cdots}{\left(\dfrac{9}{2}x^2-\dfrac{3^4}{4!}x^4+\cdots\right)^2}$$

$$= \lim_{x\to 0}\frac{2+2x+\cdots}{\left(\dfrac{9}{2}-\dfrac{3^4}{4!}x^2+\cdots\right)^2}=\frac{2}{\left(\dfrac{9}{2}\right)^2}=\frac{8}{81}.$$

You can check that the second of these examples is much more difficult if attempted using l'Hôpital's Rule.

EXERCISES 9.7

1. Estimate the error if the Maclaurin polynomial of degree 5 for $\sin x$ is used to approximate $\sin(0.2)$.

2. Estimate the error if the Taylor polynomial of degree 4 for $\ln x$ in powers of $x-2$ is used to approximate $\ln(1.95)$.

Use Maclaurin or Taylor series to calculate the function values indicated in Exercises 3–14, with error less than 5×10^{-5} in absolute value.

3. $e^{0.2}$
4. $1/e$
5. $e^{1.2}$
6. $\sin(0.1)$
7. $\cos 5^\circ$
8. $\ln(6/5)$
9. $\ln(0.9)$
10. $\sin 80^\circ$
11. $\cos 65^\circ$
12. $\tan^{-1} 0.2$
13. $\cosh(1)$
14. $\ln(3/2)$

Find Maclaurin series for the functions in Exercises 15–19.

15. $\displaystyle I(x)=\int_0^x \frac{\sin t}{t}\,dt$
16. $\displaystyle J(x)=\int_0^x \frac{e^t-1}{t}\,dt$
17. $\displaystyle K(x)=\int_1^{1+x}\frac{\ln t}{t-1}\,dt$
18. $\displaystyle L(x)=\int_0^x \cos(t^2)\,dt$
19. $\displaystyle M(x)=\int_0^x \frac{\tan^{-1}t^2}{t^2}\,dt$

20. Find $L(0.5)$ correct to 3 decimal places, with L defined as in Exercise 18.

21. Find $I(1)$ correct to 3 decimal places, with I defined as in Exercise 15.

Evaluate the limits in Exercises 22–27.

22. $\displaystyle \lim_{x\to 0}\frac{\sin(x^2)}{\sinh x}$
23. $\displaystyle \lim_{x\to 0}\frac{1-\cos(x^2)}{(1-\cos x)^2}$
24. $\displaystyle \lim_{x\to 0}\frac{(e^x-1-x)^2}{x^2-\ln(1+x^2)}$
25. $\displaystyle \lim_{x\to 0}\frac{2\sin 3x-3\sin 2x}{5x-\tan^{-1}5x}$
26. $\displaystyle \lim_{x\to 0}\frac{\sin(\sin x)-x}{x(\cos(\sin x)-1)}$
27. $\displaystyle \lim_{x\to 0}\frac{\sinh x-\sin x}{\cosh x-\cos x}$

9.8 The Binomial Theorem and Binomial Series

EXAMPLE 1 Use Taylor's Formula to prove the Binomial Theorem: if n is a positive integer, then

$$(a+x)^n = a^n + n\,a^{n-1}x + \frac{n(n-1)}{2!}a^{n-2}x^2 + \cdots + n\,ax^{n-1} + x^n$$
$$= \sum_{k=0}^{n} \binom{n}{k} a^{n-k}x^k,$$

where $\binom{n}{k} = \dfrac{n!}{(n-k)!k!}$.

Solution Let $f(x) = (a+x)^n$. Then

$$f'(x) = n(a+x)^{n-1} = \frac{n!}{(n-1)!}(a+x)^{n-1}$$
$$f''(x) = \frac{n!}{(n-1)!}(n-1)(a+x)^{n-2} = \frac{n!}{(n-2)!}(a+x)^{n-2}$$
$$\vdots$$
$$f^{(k)}(x) = \frac{n!}{(n-k)!}(a+x)^{n-k} \qquad (0 \le k \le n).$$

In particular, $f^{(n)}(x) = \dfrac{n!}{0!}(a+x)^{n-n} = n!$, a constant, and so

$$f^{(k)}(x) = 0 \qquad \text{for all } x, \text{ if } k > n.$$

For $0 \le k \le n$ we have $f^{(k)}(0) = \dfrac{n!}{(n-k)!}a^{n-k}$. Thus, by Taylor's Theorem with Lagrange remainder, for some s between a and x,

$$(a+x)^n = f(x) = \sum_{k=0}^{n} \frac{f^{(k)}(0)}{k!}x^k + \frac{f^{(n+1)}(s)}{(n+1)!}x^{n+1}$$
$$= \sum_{k=0}^{n} \frac{n!}{(n-k)!k!}a^{n-k}x^k + 0 = \sum_{k=0}^{n} \binom{n}{k} a^{n-k}x^k.$$

This is, in fact, the Maclaurin *series* for $(a+x)^n$, not just the Maclaurin polynomial of degree n. Since all higher-degree terms are zero, the series has only finitely many nonzero terms and so converges for all x.

Remark If $f(x) = (a+x)^r$, where $a > 0$ and r is any real number, then calculations similar to those above show that the Maclaurin polynomial of degree n for f is

$$P_n(x) = a^r + \sum_{k=1}^{n} \frac{r(r-1)(r-2)\cdots(r-k+1)}{k!}a^{r-k}x^k.$$

However, if r is not a positive integer, then there will be no positive integer n for which the remainder $E_n(x) = f(x) - P_n(x)$ vanishes identically, and the corresponding Maclaurin series will not be a polynomial.

The Binomial Series

To simplify the discussion of the function $(a+x)^r$ when r is not a positive integer, we take $a = 1$ and consider the function $(1+x)^r$. Results for the general case follow via the identity

$$(a+x)^r = a^r \left(1 + \frac{x}{a}\right)^r,$$

valid for any $a > 0$.

If r is any real number and $x > -1$, then the kth derivative of $(1+x)^r$ is

$$r(r-1)(r-2)\cdots(r-k+1)\,(1+x)^{r-k}, \qquad (k = 1, 2, \ldots).$$

Thus, the Maclaurin series for $(1+x)^r$ is

$$1 + \sum_{k=1}^{\infty} \frac{r(r-1)(r-2)\cdots(r-k+1)}{k!} x^k,$$

which is called the **binomial series**. The following theorem shows that the binomial series does, in fact, converge to $(1+x)^r$ if $|x| < 1$. We could accomplish this by writing Taylor's Formula for $(1+x)^r$ with $c = 0$ and showing that the remainder $E_n(x) \to 0$ as $n \to \infty$. (We would need to use the integral form of the remainder to prove this for all $|x| < 1$.) However, we will use an easier method, similar to the one used for the exponential and trigonometric functions in Section 9.6.

THEOREM 23 **The binomial series**

If $|x| < 1$, then

$$\begin{aligned}(1+x)^r &= 1 + rx + \frac{r(r-1)}{2!}x^2 + \frac{r(r-1)(r-2)}{3!}x^3 + \cdots \\ &= 1 + \sum_{n=1}^{\infty} \frac{r(r-1)(r-2)\cdots(r-n+1)}{n!} x^n \quad (-1 < x < 1).\end{aligned}$$

PROOF If $|x| < 1$, then the series

$$f(x) = 1 + \sum_{n=1}^{\infty} \frac{r(r-1)(r-2)\cdots(r-n+1)}{n!} x^n$$

converges by the ratio test, since

$$\begin{aligned}\rho &= \lim_{n\to\infty} \left| \frac{\dfrac{r(r-1)(r-2)\cdots(r-n+1)(r-n)}{(n+1)!} x^{n+1}}{\dfrac{r(r-1)(r-2)\cdots(r-n+1)}{n!} x^n} \right| \\ &= \lim_{n\to\infty} \left| \frac{r-n}{n+1} \right| |x| = |x| < 1.\end{aligned}$$

Note that $f(0) = 1$. We need to show that $f(x) = (1+x)^r$ for $|x| < 1$.

By Theorem 19, we can differentiate the series for $f(x)$ termwise on $|x| < 1$ to obtain

$$\begin{aligned}f'(x) &= \sum_{n=1}^{\infty} \frac{r(r-1)(r-2)\cdots(r-n+1)}{(n-1)!} x^{n-1} \\ &= \sum_{n=0}^{\infty} \frac{r(r-1)(r-2)\cdots(r-n)}{n!} x^n.\end{aligned}$$

We have replaced n with $n+1$ to get the second version of the sum from the first version. Adding the second version to x times the first version, we get

$$\begin{aligned}(1+x)f'(x) &= \sum_{n=0}^{\infty} \frac{r(r-1)(r-2)\cdots(r-n)}{n!}x^n \\ &\quad + \sum_{n=1}^{\infty} \frac{r(r-1)(r-2)\cdots(r-n+1)}{(n-1)!}x^n \\ &= r + \sum_{n=1}^{\infty} \frac{r(r-1)(r-2)\cdots(r-n+1)}{n!}x^n\big[(r-n)+n\big] \\ &= r\,f(x).\end{aligned}$$

The differential equation $(1+x)f'(x) = rf(x)$ implies that

$$\frac{d}{dx}\frac{f(x)}{(1+x)^r} = \frac{(1+x)^r f'(x) - r(1+x)^{r-1}f(x)}{(1+x)^{2r}} = 0$$

for all x satisfying $|x| < 1$. Thus, $f(x)/(1+x)^r$ is constant on that interval, and since $f(0) = 1$, the constant must be 1. Thus, $f(x) = (1+x)^r$.

Remark For some values of r the binomial series may converge at the endpoints $x = 1$ or $x = -1$. As observed above, if r is a positive integer, the series has only finitely many nonzero terms, and so converges for all x.

EXAMPLE 2 Find the Maclaurin series for $\dfrac{1}{\sqrt{1+x}}$.

Solution Here $r = -(1/2)$:

$$\begin{aligned}\frac{1}{\sqrt{1+x}} &= (1+x)^{-1/2} \\ &= 1 - \frac{1}{2}x + \frac{1}{2!}\left(-\frac{1}{2}\right)\left(-\frac{3}{2}\right)x^2 + \frac{1}{3!}\left(-\frac{1}{2}\right)\left(-\frac{3}{2}\right)\left(-\frac{5}{2}\right)x^3 + \cdots \\ &= 1 - \frac{1}{2}x + \frac{1\times 3}{2^2 2!}x^2 - \frac{1\times 3\times 5}{2^3 3!}x^3 + \cdots \\ &= 1 + \sum_{n=1}^{\infty}(-1)^n \frac{1\times 3\times 5\times\cdots\times(2n-1)}{2^n n!}x^n.\end{aligned}$$

This series converges for $-1 < x \le 1$. (Use the alternating series test to get the endpoint $x = 1$.)

EXAMPLE 3 Find the Maclaurin series for $\sin^{-1} x$.

Solution Replace x with $-t^2$ in the series obtained in the previous example to get

$$\frac{1}{\sqrt{1-t^2}} = 1 + \sum_{n=1}^{\infty}\frac{1\times 3\times 5\times\cdots\times(2n-1)}{2^n n!}t^{2n} \qquad (-1 < t < 1).$$

Now integrate t from 0 to x:

$$\begin{aligned}\sin^{-1} x &= \int_0^x \frac{dt}{\sqrt{1-t^2}} = \int_0^x \left(1 + \sum_{n=1}^{\infty}\frac{1\times 3\times 5\times\cdots\times(2n-1)}{2^n n!}t^{2n}\right)dt \\ &= x + \sum_{n=1}^{\infty}\frac{1\times 3\times 5\times\cdots\times(2n-1)}{2^n n!(2n+1)}x^{2n+1} \\ &= x + \frac{x^3}{6} + \frac{3}{40}x^5 + \cdots \qquad (-1 < x < 1).\end{aligned}$$

The Multinomial Theorem

The Binomial Theorem can be extended to provide for expansions of positive integer powers of sums of more than two quantities. Before stating this **Multinomial Theorem**, we require some new notation.

For an integer $n \geq 2$, let $m = (m_1, m_2, \ldots, m_n)$ be an n-tuple of nonnegative integers. We call m a **multiindex of order** n, and the number $|m| = m_1+m_2+\cdots+m_n$ the **degree** of the multiindex. In terms of multiindices, the Binomial Theorem can be restated in the form

$$(x_1 + x_2)^k = \sum_{|m|=k} \binom{k!}{m_1!\, m_2!} x_1^{m_1} x_2^{m_2} = \sum_{|m|=k} \frac{k!}{m_1!\, m_2!} x_1^{m_1} x_2^{m_2},$$

the sum being taken over all multiindices of order 2 having degree k. Here the binomial coefficients have been rewritten in the form

$$\binom{k}{m_1\, m_2} = \frac{k!}{m_1!\, m_2!},$$

which is correct since $m_2 = k - m_1$.

THEOREM 24

The Multinomial Theorem

If m and k are integers satisfying $n \geq 2$ and $k \geq 1$, then

$$(x_1 + x_2 + \cdots + x_n)^k = \sum_{|m|=k} \frac{k!}{m_1!\, m_2! \cdots m_n!} x_1^{m_1} x_2^{m_2} \cdots x_n^{m_n},$$

the sum being taken over all multiindices m of order n and degree k.

Evidently, the Binomial Theorem is the special case $n = 2$. The proof of the Multinomial Theorem can be carried out by induction on n. See Exercise 12 below.

The coefficients of the various products of powers of the variables x_i in the Multinomial Theorem are called **multinomial coefficients**. By analogy with the notation used for binomial coefficients, they are sometimes denoted (assuming $m_1 + \cdots + m_n = k$)

$$\binom{k}{m_1, m_2, \ldots, m_n} = \binom{m_1 + m_2 + \cdots + m_n}{m_1, m_2, \ldots, m_n} = \frac{k!}{m_1! m_2! \cdots m_n!}. \qquad (*)$$

They are useful for counting distinct arrangements of objects where not all of the objects appear to be different.

EXAMPLE 4 The number of ways that k distinct objects can be arranged in a sequence of positions $1, 2, \ldots, k$ is $k!$ because there are k choices for the object to go in position 1, then $k - 1$ choices for the object to go into position 2, etc., until there is only 1 choice for the object to go into position k. But what if the objects are not all distinct, but instead there are several objects of each of n different types, say type 1, type 2, ..., type n such that objects of the same type are indistinguishable from one another. If you just look at positions in the sequence containing objects of type j, and rearrange only those objects, you can't tell the difference. If there are m_j objects of type j, $(1 \leq j \leq n)$, then the number of distinct rearrangements of the k objects is given by the multinomial coefficient $(*)$. For example, the number of visually different arrangements of 9 balls, 2 of which are red, 3 green, and 4 blue is

$$\binom{9}{2, 3, 4} = \frac{9!}{2!3!4!} = \frac{362{,}880}{288} = 1{,}260.$$

Remark A direct proof of the Multinomial Theorem can be based on the above example. When calculating the kth power of $(x_1+x_2+\cdots+x_n)$ by long multiplication, we obtain a sum of monomials of degree k having the form $x_1^{m_1}x_2^{m_2}\cdots x_n^{m_n}$, where $m_1+m_2+\cdots+m_n=k$. The number of ways you can arrange m_1 factors x_1, m_2 factors $x_2, \dots,$ and m_n factors x_n to form that monomial is the multinomial coefficient (∗). Since $(x_1+x_2+\cdots+x_n)^k$ is the sum of all such monomials, we must have

$$(x_1+x_2+\cdots+x_n)^k = \sum_{|m|=k} \frac{k!}{m_1!\,m_2!\,\cdots\,m_n!}\,x_1^{m_1}x_2^{m_2}\cdots x_n^{m_n}.$$

EXERCISES 9.8

Find Maclaurin series representations for the functions in Exercises 1–8. Use the binomial series to calculate the answers.

1. $\sqrt{1+x}$
2. $x\sqrt{1-x}$
3. $\sqrt{4+x}$
4. $\dfrac{1}{\sqrt{4+x^2}}$
5. $(1-x)^{-2}$
6. $(1+x)^{-3}$
7. $\cos^{-1} x$
8. $\sinh^{-1} x$

9. **(Binomial coefficients)** Show that the binomial coefficients $\binom{n}{k}=\dfrac{n!}{k!\,(n-k)!}$ satisfy

(i) $\binom{n}{0}=\binom{n}{n}=1$ for every n, and

(ii) if $0\le k\le n$, then $\binom{n}{k-1}+\binom{n}{k}=\binom{n+1}{k}$.

It follows that, for fixed $n\ge 1$, the binomial coefficients

$$\binom{n}{0},\ \binom{n}{1},\ \binom{n}{2},\ \dots,\ \binom{n}{n}$$

are the elements of the nth row of **Pascal's triangle** below, where each element with value greater than 1 is the sum of the two diagonally above it.

```
                1
             1     1
          1     2     1
       1     3     3     1
    1     4     6     4     1
 1     5    10    10     5     1
```

10. **(An inductive proof of the Binomial Theorem)** Use mathematical induction and the results of Exercise 9 to prove the Binomial Theorem:

$$\begin{aligned}(a+b)^n &= \sum_{k=0}^{n}\binom{n}{k}a^{n-k}b^k\\ &= a^n+na^{n-1}b+\binom{n}{2}a^{n-2}b^2+\binom{n}{3}a^{n-3}b^3+\cdots+b^n.\end{aligned}$$

11. **(The Leibniz Rule)** Use mathematical induction, the Product Rule, and Exercise 9 to verify the Leibniz Rule for the nth derivative of a product of two functions:

$$\begin{aligned}(fg)^{(n)} &= \sum_{k=0}^{n}\binom{n}{k}f^{(n-k)}g^{(k)}\\ &= f^{(n)}g+nf^{(n-1)}g'+\binom{n}{2}f^{(n-2)}g''\\ &\quad+\binom{n}{3}f^{(n-3)}g^{(3)}+\cdots+fg^{(n)}.\end{aligned}$$

12. **(Proof of the Multinomial Theorem)** Use the Binomial Theorem and induction on n to prove Theorem 24. *Hint:* Assume the theorem holds for specific n and all k. Apply the Binomial Theorem to

$$(x_1+\cdots+x_n+x_{n+1})^k=\big((x_1+\cdots+x_n)+x_{n+1}\big)^k.$$

13. **(A Multifunction Leibniz Rule)** Use the technique of Exercise 12 to generalize the Leibniz Rule of Exercise 11 to calculate the kth derivative of a product of n functions $f_1 f_2 \cdots f_n$.

9.9 Fourier Series

As we have seen, power series representations of functions make it possible to approximate those functions as closely as we want in intervals near a particular point of interest by using partial sums of the series, that is, polynomials. However, in many important applications of mathematics, the functions involved are required to be periodic. For example, much of electrical engineering is concerned with the analysis and manipulation of *waveforms*, which are periodic functions of time. Polynomials are not periodic functions, and for this reason power series are not well suited to representing such functions.

Much more appropriate for the representations of periodic functions over extended intervals are certain infinite series of periodic functions called Fourier series.

Periodic Functions

Recall that a function f defined on the real line is **periodic** with period T if

$$f(t+T) = f(t) \quad \text{for all real } t. \tag{$*$}$$

This implies that $f(t+mT) = f(t)$ for any integer m, so that if T is a period of f, then so is any multiple mT of T. The smallest positive number T for which $(*)$ holds is called the **fundamental period**, or simply **the period** of f.

The entire graph of a function with period T can be obtained by shifting the part of the graph in any half-open interval of length T (e.g., the interval $[0, T)$) to the left or right by integer multiples of the period T. Figure 9.6 shows the graph of a function of period 2.

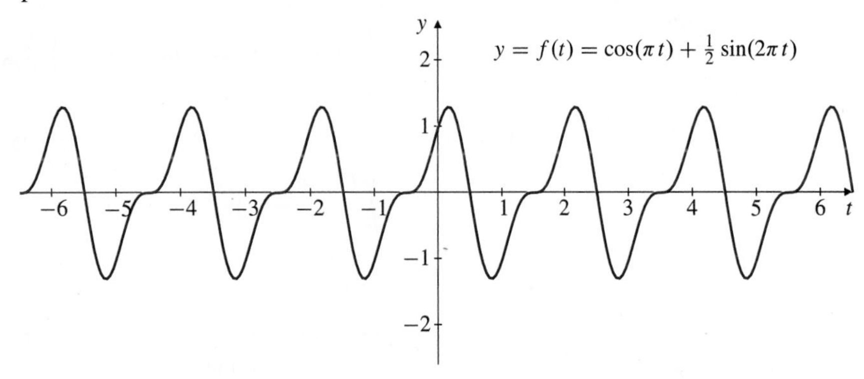

Figure 9.6 This function has period 2. Observe how the graph repeats the part in the interval $[0, 2)$ over and over to the left and right

EXAMPLE 1 The functions $g(t) = \cos(\pi t)$ and $h(t) = \sin(\pi t)$ are both periodic with period 2:

$$g(t+2) = \cos(\pi t + 2\pi) = \cos(\pi t) = g(t).$$

The function $k(t) = \sin(2\pi t)$ also has period 2, but this is not its fundamental period. The fundamental period is 1:

$$k(t+1) = \sin(2\pi t + 2\pi) = \sin(2\pi t) = k(t).$$

The sum $f(t) = g(t) + \frac{1}{2}k(t) = \cos(\pi t) + \frac{1}{2}\sin(2\pi t)$, graphed in Figure 9.6, has period 2, the least common multiple of the periods of its two terms.

EXAMPLE 2 For any positive integer n, the functions

$$f_n(t) = \cos(n\omega t) \qquad \text{and} \qquad g_n(t) = \sin(n\omega t)$$

both have fundamental period $T = 2\pi/(n\omega)$. The collection of all such functions corresponding to all positive integers n have common period $T = 2\pi/\omega$, the fundamental period of f_1 and g_1. T is an integer multiple of the fundamental periods of all the functions f_n and g_n. The subject of Fourier series is concerned with expressing general functions with period T as series whose terms are real multiples of these functions.

Fourier Series

It can be shown (but we won't do it here) that if $f(t)$ is periodic with fundamental period T, is continuous, and has a piecewise continuous derivative on the real line, then $f(t)$ is everywhere the sum of a series of the form

$$f(t) = \frac{a_0}{2} + \sum_{n=1}^{\infty}\big(a_n \cos(n\omega t) + b_n \sin(n\omega t)\big), \qquad (**)$$

called the **Fourier series** of f, where $\omega = 2\pi/T$ and the sequences $\{a_n\}_{n=0}^{\infty}$ and $\{b_n\}_{n=1}^{\infty}$ are the **Fourier coefficients** of f. Determining the values of these coefficients for a given such function f is made possible by the following identities, valid for integers m and n, which are easily proved by using the addition formulas for sine and cosine. (See Exercises 49–51 in Section 5.6.)

$$\int_0^T \cos(n\omega t)\,dt = \begin{cases} 0 & \text{if } n \neq 0 \\ T & \text{if } n = 0 \end{cases}$$
$$\int_0^T \sin(n\omega t)\,dt = 0$$
$$\int_0^T \cos(m\omega t)\cos(n\omega t)\,dt = \begin{cases} 0 & \text{if } m \neq n \\ T/2 & \text{if } m = n \end{cases}$$
$$\int_0^T \sin(m\omega t)\sin(n\omega t)\,dt = \begin{cases} 0 & \text{if } m \neq n \\ T/2 & \text{if } m = n \end{cases}$$
$$\int_0^T \cos(m\omega t)\sin(n\omega t)\,dt = 0.$$

If we multiply equation (**) by $\cos(m\omega t)$ (or by $\sin(m\omega t)$) and integrate the resulting equation over $[0, T]$ term by term, all the terms on the right except the one involving a_m (or b_m) will be 0. (The term-by-term integration requires justification, but we won't try to do that here either.) The integration results in

$$\int_0^T f(t)\cos(m\omega t)\,dt = \frac{1}{2}T a_m$$
$$\int_0^T f(t)\sin(m\omega t)\,dt = \frac{1}{2}T b_m.$$

(Note that the first of these formulas is even valid for $m = 0$ because we chose to call the constant term in the Fourier series $a_0/2$ instead of a_0.) Since the integrands are all periodic with period T, the integrals can be taken over any interval of length T; it is often convenient to use $[-T/2, T/2]$ instead of $[0, T]$. The Fourier coefficients of f are therefore given by

$$a_n = \frac{2}{T}\int_{-T/2}^{T/2} f(t)\cos(n\omega t)\,dt \quad (n = 0, 1, 2, \ldots)$$
$$b_n = \frac{2}{T}\int_{-T/2}^{T/2} f(t)\sin(n\omega t)\,dt \quad (n = 1, 2, 3, \ldots),$$

where $\omega = 2\pi/T$.

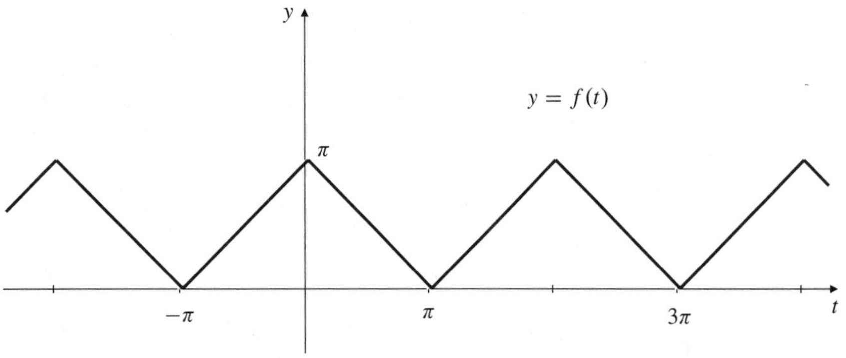

Figure 9.7 A sawtooth function of period 2π

EXAMPLE 3 Find the Fourier series of the sawtooth function $f(t)$ of period 2π whose values in the interval $[-\pi, \pi]$ are given by $f(t) = \pi - |t|$. (See Figure 9.7.)

Solution Here $T = 2\pi$ and $\omega = 2\pi/(2\pi) = 1$. Since $f(t)$ is an even function, $f(t)\sin(nt)$ is odd, so all the Fourier sine coefficients b_n are zero:

$$b_n = \frac{2}{2\pi}\int_{-\pi}^{\pi} f(t)\sin(nt)\,dt = 0.$$

Also, $f(t)\cos(nt)$ is an even function, so

$$\begin{aligned} a_n &= \frac{2}{2\pi}\int_{-\pi}^{\pi} f(t)\cos(nt)\,dt = \frac{4}{2\pi}\int_0^{\pi} f(t)\cos(nt)\,dt \\ &= \frac{2}{\pi}\int_0^{\pi} (\pi - t)\cos(nt)\,dt \\ &= \begin{cases} \pi & \text{if } n = 0 \\ 0 & \text{if } n \neq 0 \text{ and } n \text{ is even} \\ 4/(\pi n^2) & \text{if } n \text{ is odd.} \end{cases} \end{aligned}$$

Since odd positive integers n are of the form $n = 2k - 1$, where k is a positive integer, the Fourier series of f is given by

$$f(t) = \frac{\pi}{2} + \sum_{k=1}^{\infty} \frac{4}{\pi(2k-1)^2}\cos\big((2k-1)t\big).$$

Convergence of Fourier Series

The partial sums of a Fourier series are called Fourier polynomials because they can be expressed as polynomials in $\sin(\omega t)$ and $\cos(\omega t)$, although we will not actually try to write them that way. The Fourier polynomial of order m of the periodic function f having period T is

$$f_m(t) = \frac{a_0}{2} + \sum_{n=1}^{m}\big(a_n\cos(n\omega t) + b_n\sin(n\omega t)\big),$$

where $\omega = 2\pi/T$ and the coefficients a_n $(0 \le n \le m)$ and b_n $(1 \le n \le m)$ are given by the integral formulas developed earlier.

EXAMPLE 4 The Fourier polynomial of order 3 of the sawtooth function of Example 3 is

$$f_3(t) = \frac{\pi}{2} + \frac{4}{\pi}\cos t + \frac{4}{9\pi}\cos(3t).$$

The graph of this function is shown in Figure 9.8. Observe that it appears to be a reasonable approximation to the graph of f in Figure 9.7, but, being a finite sum of differentiable functions, $f_3(t)$ is itself differentiable everywhere, even at the integer multiples of π where f is not differentiable.

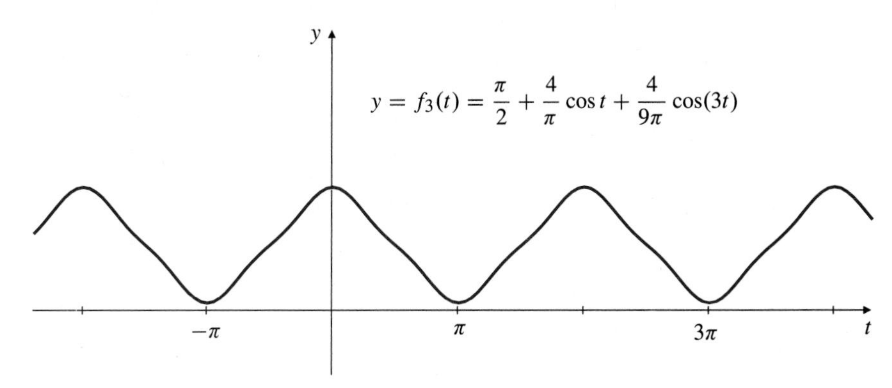

Figure 9.8 The Fourier polynomial approximation $f_3(t)$ to the sawtooth function of Example 3

As noted earlier, the Fourier series of a function $f(t)$ that is periodic, continuous, and has a piecewise continuous derivative on the real line converges to $f(t)$ at each real number t. However, the Fourier coefficients (and hence the Fourier series) can be calculated (by the formulas given above) for periodic functions with piecewise continuous derivative even if the functions are not themselves continuous, but only piecewise continuous.

Recall that $f(t)$ is piecewise continuous on the interval $[a, b]$ if there exists a partition $\{a = x_0 < x_1 < x_2 < \cdots < x_k = b\}$ of $[a, b]$ and functions $F_1, F_2, \ldots, F_k$, such that

(i) F_i is continuous on $[x_{i-1}, x_i]$, and

(ii) $f(t) = F_i(t)$ on (x_{i-1}, x_i).

The integral of such a function f is the sum of integrals of the functions F_i:

$$\int_a^b f(t)\,dt = \sum_{i=1}^{k} \int_{x_{i-1}}^{x_i} F_i(t)\,dt.$$

Since $f(t)\cos(n\omega t)$ and $f(t)\sin(n\omega t)$ are piecewise continuous if f is, the Fourier coefficients of a piecewise continuous, periodic function can be calculated by the same formulas given for a continuous periodic function. The question of where and to what the Fourier series converges in this case is answered by the following theorem, proved in textbooks on Fourier analysis.

THEOREM 25

The Fourier series of a piecewise continuous, periodic function f with piecewise continuous derivative converges to that function at every point t where f is continuous. Moreover, if f is discontinuous at $t = c$, then f has different, but finite, left and right limits at c:

$$\lim_{t \to c-} f(t) = f(c-), \qquad \text{and} \qquad \lim_{t \to c+} f(t) = f(c+).$$

The Fourier series of f converges at $t = c$ to the average of these left and right limits:

$$\frac{a_0}{2} + \sum_{n=1}^{\infty} \big(a_n \cos(n\omega c) + b_n \sin(n\omega c)\big) = \frac{f(c-) + f(c+)}{2},$$

where $\omega = 2\pi/T$.

EXAMPLE 5 Calculate the Fourier series for the periodic function f with period 2 satisfying

$$f(t) = \begin{cases} -1 & \text{if } -1 < x < 0 \\ 1 & \text{if } 0 < x < 1. \end{cases}$$

Where does f fail to be continuous? To what does the Fourier series of f converge at these points?

Solution Here $T = 2$ and $\omega = 2\pi/2 = \pi$. Since f is an odd function, its cosine coefficients are all zero:

$$a_n = \int_{-1}^{1} f(t)\cos(n\pi t)\,dt = 0. \qquad \text{(The integrand is odd.)}$$

The same symmetry implies that

$$\begin{aligned} b_n &= \int_{-1}^{1} f(t)\sin(n\pi t)\,dt \\ &= 2\int_0^1 \sin(n\pi t)\,dt = -\frac{2\cos(n\pi t)}{n\pi}\bigg|_0^1 \\ &= -\frac{2}{n\pi}\big((-1)^n - 1\big) = \begin{cases} 4/(n\pi) & \text{if } n \text{ is odd} \\ 0 & \text{if } n \text{ is even.} \end{cases} \end{aligned}$$

Odd integers n are of the form $n = 2k - 1$ for $k = 1, 2, 3, \ldots$. Therefore, the Fourier series of f is

$$\begin{aligned} &\frac{4}{\pi}\sum_{k=1}^{\infty} \frac{1}{2k-1}\sin\big((2k-1)\pi t\big) \\ &= \frac{4}{\pi}\left(\sin(\pi t) + \frac{1}{3}\sin(3\pi t) + \frac{1}{5}\sin(5\pi t) + \cdots\right). \end{aligned}$$

Note that f is continuous except at the points where t is an integer. At each of these points f jumps from -1 to 1 or from 1 to -1, so the average of the left and right limits of f at these points is 0. Observe that the sum of the Fourier series is 0 at integer values of t, in accordance with Theorem 25. See Figure 9.9.

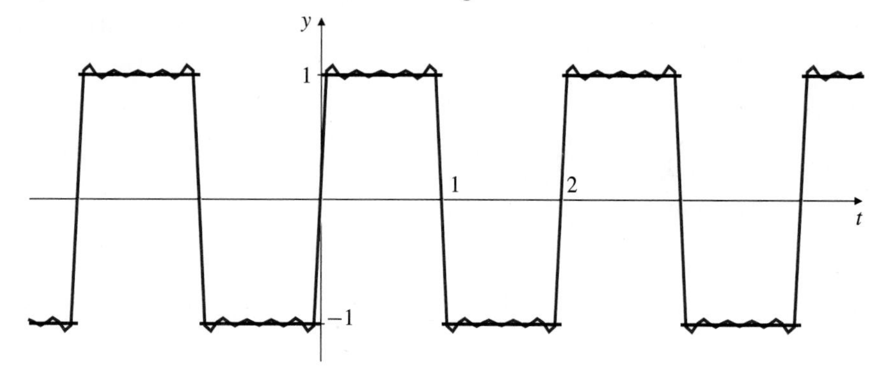

Figure 9.9 The piecewise continuous function f (black) of Example 5 and its Fourier polynomial f_{15} (colour)

$$f_{15}(t) = \sum_{k=1}^{8} \frac{4\sin\big((2k-1)\pi t\big)}{(2k-1)\pi}$$

Fourier Cosine and Sine Series

As observed in Example 3 and Example 5, even functions have no sine terms in their Fourier series, and odd functions have no cosine terms (including the constant term $a_0/2$). It is often necessary in applications to find a Fourier series representation of a given function defined on a finite interval $[0, a]$ having either no sine terms (a **Fourier cosine series**) or no cosine terms (a **Fourier sine series**). This is accomplished by extending the domain of f to $[-a, 0)$ so as to make f either even or odd on $[-a, a]$,

$$\begin{aligned} f(-t) &= f(t) \text{ if } -a \le t < 0 \text{ for the even extension} \\ f(-t) &= -f(t) \text{ if } -a \le t < 0 \text{ for the odd extension,} \end{aligned}$$

and then calculating its Fourier series considering the extended f to have period $2a$. (If we want the odd extension, we may have to redefine $f(0)$ to be 0.)

EXAMPLE 6 Find the Fourier cosine series of $g(t) = \pi - t$ defined on $[0, \pi]$.

Solution The even extension of $g(t)$ to $[-\pi, \pi]$ is the function f of Example 3. Thus, the Fourier cosine series of g is

$$\frac{\pi}{2} + \sum_{k=1}^{\infty} \frac{4}{\pi(2k-1)^2} \cos\big((2k-1)t\big).$$

EXAMPLE 7 Find the Fourier sine series of $h(t) = 1$ defined on $[0, 1]$.

Solution If we redefine $h(0) = 0$, then the odd extension of h to $[-1, 1]$ coincides with the function $f(t)$ of Example 5 except that the latter function is undefined at $t = 0$. The Fourier sine series of h is the series obtained in Example 5, namely,

$$\frac{4}{\pi} \sum_{k=1}^{\infty} \frac{1}{2k-1} \sin\big((2k-1)\pi t\big).$$

Remark Fourier cosine and sine series are treated from a different perspective in Section 13.4.

EXERCISES 9.9

In Exercises 1–4, what is the fundamental period of the given function?

1. $f(t) = \sin(3t)$
2. $g(t) = \cos(3 + \pi t)$
3. $h(t) = \cos^2 t$
4. $k(t) = \sin(2t) + \cos(3t)$

In Exercises 5–8, find the Fourier series of the given function.

5. $f(t) = t,\ -\pi < t \le \pi$, f has period 2π.
6. $f(t) = \begin{cases} 0 & \text{if } 0 \le t < 1 \\ 1 & \text{if } 1 \le t < 2, \end{cases}$ f has period 2.
7. $f(t) = \begin{cases} 0 & \text{if } -1 \le t < 0 \\ t & \text{if } 0 \le t < 1, \end{cases}$ f has period 2.
8. $f(t) = \begin{cases} t & \text{if } 0 \le t < 1 \\ 1 & \text{if } 1 \le t < 2 \\ 3-t & \text{if } 2 \le t < 3, \end{cases}$ f has period 3.
9. What is the Fourier cosine series of the function $h(t)$ of Example 7?
10. Calculate the Fourier sine series of the function $g(t)$ of Example 6.
11. Find the Fourier sine series of $f(t) = t$ on $[0, 1]$.
12. Find the Fourier cosine series of $f(t) = t$ on $[0, 1]$.
13. Use the result of Example 3 to evaluate

$$\sum_{n=1}^{\infty} \frac{1}{(2n-1)^2} = 1 + \frac{1}{3^2} + \frac{1}{5^2} + \cdots.$$

14. Verify that if f is an even function of period T, then the Fourier sine coefficients b_n of f are all zero and the Fourier cosine coefficients a_n of f are given by

$$a_n = \frac{4}{T} \int_0^{T/2} f(t)\cos(n\omega t)\, dt, \qquad n = 0, 1, 2, \ldots,$$

where $\omega = 2\pi/T$. State and verify the corresponding result for odd functions f.

CHAPTER REVIEW

Key Ideas

- **What does it mean to say that the sequence $\{a_n\}$**
  - ◇ is bounded above?
  - ◇ is ultimately positive?
  - ◇ is alternating?
  - ◇ is increasing?
  - ◇ converges?
  - ◇ diverges to infinity?
- **What does it mean to say that the series $\sum_{n=1}^{\infty} a_n$**
  - ◇ converges?
  - ◇ diverges?
  - ◇ is geometric?
  - ◇ is telescoping?
  - ◇ is a p-series?
  - ◇ is positive?
  - ◇ converges absolutely?
  - ◇ converges conditionally?

- **State the following convergence tests for series.**
  - ◇ the integral test
  - ◇ the comparison test
  - ◇ the limit comparison test
  - ◇ the ratio test
  - ◇ the alternating series test
- **How can you find bounds for the tail of a series?**
- **What is a bound for the tail of an alternating series?**
- **What do the following terms and phrases mean?**
  - ◇ a power series
  - ◇ interval of convergence
  - ◇ radius of convergence
  - ◇ centre of convergence
  - ◇ a Taylor series
  - ◇ a Maclaurin series
  - ◇ a Taylor polynomial
  - ◇ a binomial series
  - ◇ an analytic function
- **Where is the sum of a power series differentiable?**
- **Where does the integral of a power series converge?**
- **Where is the sum of a power series continuous?**
- **State Taylor's Theorem with Lagrange remainder.**
- **State Taylor's Theorem with integral remainder.**
- **What is the Binomial Theorem?**
- **What is a Fourier series?**
- **What is a Fourier cosine series? a Fourier sine series?**

Review Exercises

In Exercises 1–4, determine whether the given sequence converges, and find its limit if it does converge.

1. $\left\{\dfrac{(-1)^n e^n}{n!}\right\}$ **2.** $\left\{\dfrac{n^{100}+2^n\pi}{2^n}\right\}$

3. $\left\{\dfrac{\ln n}{\tan^{-1} n}\right\}$ **4.** $\left\{\dfrac{(-1)^n n^2}{\pi n(n-\pi)}\right\}$

5. Let $a_1 > \sqrt{2}$, and let

$$a_{n+1} = \frac{a_n}{2} + \frac{1}{a_n} \quad \text{for} \quad n = 1, 2, 3, \ldots$$

Show that $\{a_n\}$ is decreasing and that $a_n > \sqrt{2}$ for $n \geq 1$. Why must $\{a_n\}$ converge? Find $\lim_{n\to\infty} a_n$.

6. Find the limit of the sequence $\{\ln\ln(n+1) - \ln\ln n\}$.

Evaluate the sums of the series in Exercises 7–10.

7. $\displaystyle\sum_{n=1}^{\infty} 2^{-(n-5)/2}$ **8.** $\displaystyle\sum_{n=0}^{\infty} \frac{4^{n-1}}{(\pi-1)^{2n}}$

9. $\displaystyle\sum_{n=1}^{\infty} \frac{1}{n^2-\frac{1}{4}}$ **10.** $\displaystyle\sum_{n=1}^{\infty} \frac{1}{n^2-\frac{9}{4}}$

Determine whether the series in Exercises 11–16 converge or diverge. Give reasons for your answers.

11. $\displaystyle\sum_{n=1}^{\infty} \frac{n-1}{n^3}$ **12.** $\displaystyle\sum_{n=1}^{\infty} \frac{n+2^n}{1+3^n}$

13. $\displaystyle\sum_{n=1}^{\infty} \frac{n}{(1+n)(1+n\sqrt{n})}$ **14.** $\displaystyle\sum_{n=1}^{\infty} \frac{n^2}{(1+2^n)(1+n\sqrt{n})}$

15. $\displaystyle\sum_{n=1}^{\infty} \frac{3^{2n+1}}{n!}$ **16.** $\displaystyle\sum_{n=1}^{\infty} \frac{n!}{(n+2)!+1}$

Do the series in Exercises 17–20 converge absolutely, converge conditionally, or diverge?

17. $\displaystyle\sum_{n=1}^{\infty} \frac{(-1)^{n-1}}{1+n^3}$ **18.** $\displaystyle\sum_{n=1}^{\infty} \frac{(-1)^n}{2^n-n}$

19. $\displaystyle\sum_{n=10}^{\infty} \frac{(-1)^{n-1}}{\ln\ln n}$ **20.** $\displaystyle\sum_{n=1}^{\infty} \frac{n^2\cos(n\pi)}{1+n^3}$

For what values of x do the series in Exercises 21–22 converge absolutely? converge conditionally? diverge?

21. $\displaystyle\sum_{n=1}^{\infty} \frac{(x-2)^n}{3^n\sqrt{n}}$ **22.** $\displaystyle\sum_{n=1}^{\infty} \frac{(5-2x)^n}{n}$

Determine the sums of the series in Exercises 23–24 to within 0.001.

23. $\displaystyle\sum_{n=1}^{\infty} \frac{1}{n^3}$ **24.** $\displaystyle\sum_{n=1}^{\infty} \frac{1}{4+n^2}$

In Exercises 25–32, find Maclaurin series for the given functions. State where each series converges to the function.

25. $\dfrac{1}{3-x}$ **26.** $\dfrac{x}{3-x^2}$

27. $\ln(e+x^2)$ **28.** $\dfrac{1-e^{-2x}}{x}$

29. $x\cos^2 x$ **30.** $\sin(x+(\pi/3))$

31. $(8+x)^{-1/3}$ **32.** $(1+x)^{1/3}$

Find Taylor series for the functions in Exercises 33–34 about the indicated points $x = c$.

33. $1/x, \quad c = \pi$ **34.** $\sin x + \cos x, \quad c = \pi/4$

Find the Maclaurin polynomial of the indicated degree for the functions in Exercises 35–38.

35. e^{x^2+2x}, degree 3 **36.** $\sin(1+x)$, degree 3

37. $\cos(\sin x)$, degree 4 **38.** $\sqrt{1+\sin x}$, degree 4

39. What function has Maclaurin series

$$1 - \frac{x}{2!} + \frac{x^2}{4!} - \cdots = \sum_{n=0}^{\infty} \frac{(-1)^n x^n}{(2n)!}?$$

40. A function $f(x)$ has Maclaurin series

$$1 + x^2 + \frac{x^4}{2^2} + \frac{x^6}{3^2} + \cdots = 1 + \sum_{n=1}^{\infty} \frac{x^{2n}}{n^2}.$$

Find $f^{(k)}(0)$ for all positive integers k.

Find the sums of the series in Exercises 41–44.

41. $\displaystyle\sum_{n=0}^{\infty} \frac{n+1}{\pi^n}$ **42.** $\displaystyle\sum_{n=0}^{\infty} \frac{n^2}{\pi^n}$

43. $\displaystyle\sum_{n=1}^{\infty} \frac{1}{ne^n}$ **44.** $\displaystyle\sum_{n=2}^{\infty} \frac{(-1)^n \pi^{2n-4}}{(2n-1)!}$

45. If $S(x) = \displaystyle\int_0^x \sin(t^2)\,dt$, find $\displaystyle\lim_{x\to 0} \frac{x^3 - 3S(x)}{x^7}$.

46. Use series to evaluate $\displaystyle\lim_{x\to 0} \frac{(x-\tan^{-1}x)(e^{2x}-1)}{2x^2-1+\cos(2x)}$.

47. How many nonzero terms in the Maclaurin series for e^{-x^4} are needed to evaluate $\int_0^{1/2} e^{-x^4}\,dx$ correct to 5 decimal places? Evaluate the integral to that accuracy.

48. Estimate the size of the error if the Taylor polynomial of degree 4 about $x = \pi/2$ for $f(x) = \ln \sin x$ is used to approximate $\ln \sin(1.5)$.

49. Find the Fourier sine series for $f(t) = \pi - t$ on $[0, \pi]$.

50. Find the Fourier series for $f(t) = \begin{cases} 1 & \text{if } -\pi < t \le 0 \\ t & \text{if } 0 < t \le \pi. \end{cases}$

Challenging Problems

1. (A refinement of the ratio test) Suppose $a_n > 0$ and $a_{n+1}/a_n \ge n/(n+1)$ for all n. Show that $\sum_{n=1}^{\infty} a_n$ diverges. *Hint:* $a_n \ge K/n$ for some constant K.

! **2. (Summation by parts)** Let $\{u_n\}$ and $\{v_n\}$ be two sequences, and let $s_n = \sum_{k=1}^{n} v_k$.

(a) Show that $\sum_{k=1}^{n} u_k v_k = u_{n+1} s_n + \sum_{k=1}^{n} (u_k - u_{k+1}) s_n$. (*Hint:* Write $v_n = s_n - s_{n-1}$, with $s_0 = 0$, and rearrange the sum.)

(b) If $\{u_n\}$ is positive, decreasing, and convergent to 0, and if $\{v_n\}$ has bounded partial sums, $|s_n| \le K$ for all n, where K is a constant, show that $\sum_{n=1}^{\infty} u_n v_n$ converges. (*Hint:* Show that the series $\sum_{n=1}^{\infty} (u_n - u_{n+1}) s_n$ converges by comparing it to the telescoping series $\sum_{n=1}^{\infty} (u_n - u_{n+1})$.)

! **3.** Show that $\sum_{n=1}^{\infty} (1/n) \sin(nx)$ converges for every x. *Hint:* If x is an integer multiple of π, all the terms in the series are 0 so there is nothing to prove. Otherwise, $\sin(x/2) \ne 0$. In this case show that

$$\sum_{n=1}^{N} \sin(nx) = \frac{\cos(x/2) - \cos((N+1/2)x)}{2\sin(x/2)}$$

using the identity

$$\sin a \sin b = \frac{\cos(a-b) - \cos(a+b)}{2}$$

to make the sum telescope. Then apply the result of Problem 2(b) with $u_n = 1/n$ and $v_n = \sin(nx)$.

4. Let $a_1, a_2, a_3, \ldots$ be those positive integers that do not contain the digit 0 in their decimal representations. Thus $a_1 = 1$, $a_2 = 2$, ..., $a_9 = 9$, $a_{10} = 11$, ..., $a_{18} = 19$, $a_{19} = 21$, ..., $a_{90} = 99$, $a_{91} = 111$, etc. Show that the series $\sum_{n=1}^{\infty} \frac{1}{a_n}$ converges and that the sum is less than 90. (*Hint:* How many of these integers have m digits? Each term $1/a_n$, where a_n has m digits, is less than 10^{-m+1}.)

! **5. (Using an integral to improve convergence)** Recall the error formula for the Midpoint Rule, according to which

$$\int_{k-1/2}^{k+1/2} f(x)\,dx - f(k) = \frac{f''(c)}{24},$$

where $k - (1/2) \le c \le k + (1/2)$.

(a) If $f''(x)$ is a decreasing function of x, show that

$$f'(k+\tfrac{3}{2}) - f'(k+\tfrac{1}{2}) \le f''(c) \le f'(k-\tfrac{1}{2}) - f'(k-\tfrac{3}{2}).$$

(b) If (i) $f''(x)$ is a decreasing function of x, (ii) $\int_{N+1/2}^{\infty} f(x)\,dx$ converges, and (iii) $f'(x) \to 0$ as $x \to \infty$, show that

$$\frac{f'(N-\frac{1}{2})}{24} \le \sum_{n=N+1}^{\infty} f(n) - \int_{N+1/2}^{\infty} f(x)\,dx \le \frac{f'(N+\frac{3}{2})}{24}.$$

(c) Use the result of part (b) to approximate $\sum_{n=1}^{\infty} 1/n^2$ to within 0.001.

! **6. (The number e is irrational)** Start with $e = \sum_{n=0}^{\infty} 1/n!$.

(a) Use the technique of Example 7 in Section 9.3 to show that for any $n > 0$,

$$0 < e - \sum_{j=0}^{n} \frac{1}{j!} < \frac{1}{n!n}.$$

(Note that the sum here has $n+1$ terms, not n terms.)

(b) Suppose that e is a rational number, say $e = M/N$ for certain positive integers M and N. Show that $N!\left(e - \sum_{j=0}^{N}(1/j!)\right)$ is an integer.

(c) Combine parts (a) and (b) to show that there is an integer between 0 and $1/N$. Why is this not possible? Conclude that e cannot be a rational number.

7. Let

$$f(x) = \sum_{k=0}^{\infty} \frac{2^{2k} k!}{(2k+1)!} x^{2k+1}$$
$$= x + \frac{2}{3}x^3 + \frac{4}{3\times 5}x^5 + \frac{8}{3\times 5\times 7}x^7 + \cdots.$$

(a) Find the radius of convergence of this power series.

(b) Show that $f'(x) = 1 + 2xf(x)$.

(c) What is $\frac{d}{dx}\left(e^{-x^2} f(x)\right)$?

(d) Express $f(x)$ in terms of an integral.

! **8. (The number π is irrational)** Problem 6 above shows how to prove that e is irrational by assuming the contrary and deducing a contradiction. In this problem you will show that π is also irrational. The proof for π is also by contradiction but is rather more complicated, so it will be broken down into several parts.

(a) Let $f(x)$ be a polynomial, and let

$$g(x) = f(x) - f''(x) + f^{(4)}(x) - f^{(6)}(x) + \cdots$$
$$= \sum_{j=0}^{\infty} (-1)^j f^{(2j)}(x).$$

(Since f is a polynomial, all but a finite number of terms in the above sum are identically zero, so there are no convergence problems.) Verify that

$$\frac{d}{dx}\left(g'(x)\sin x - g(x)\cos x\right) = f(x)\sin x,$$

and hence that $\int_0^{\pi} f(x)\sin x\,dx = g(\pi) + g(0)$.

(b) Suppose that π is rational, say, $\pi = m/n$, where m and n are positive integers. You will show that this leads to a contradiction and thus cannot be true. Choose a positive integer k such that $(\pi m)^k/k! < 1/2$. (Why is this possible?) Consider the polynomial

$$f(x) = \frac{x^k(m - nx)^k}{k!} = \frac{1}{k!}\sum_{j=0}^{k}\binom{k}{j}m^{k-j}(-n)^j x^{j+k}.$$

Show that $0 < f(x) < 1/2$ for $0 < x < \pi$, and hence that $0 < \int_0^\pi f(x)\sin x\, dx < 1$. Thus, $0 < g(\pi) + g(0) < 1$, where $g(x)$ is defined as in part (a).

(c) Show that the ith derivative of $f(x)$ is given by

$$f^{(i)}(x) = \frac{1}{k!}\sum_{j=0}^{k}\binom{k}{j}m^{k-j}(-n)^j\frac{(j+k)!}{(j+k-i)!}x^{j+k-i}.$$

(d) Show that $f^{(i)}(0)$ is an integer for $i = 0, 1, 2, \ldots$. (*Hint:* Observe for $i < k$ that $f^{(i)}(0) = 0$, and for $i > 2k$ that $f^{(i)}(x) = 0$ for all x. For $k \le i \le 2k$, show that only one term in the sum for $f^{(i)}(0)$ is not 0, and that this term is an integer. You will need the fact that the binomial coefficients $\binom{k}{j}$ are integers.)

(e) Show that $f(\pi - x) = f(x)$ for all x, and hence that $f^{(i)}(\pi)$ is also an integer for each $i = 0, 1, 2, \ldots$. Therefore, if $g(x)$ is defined as in (a), then $g(\pi) + g(0)$ is an integer. This contradicts the conclusion of part (b) and so shows that π cannot be rational.

9. (An asymptotic series) Integrate by parts to show that

$$\int_0^x e^{-1/t}\, dt = e^{-1/x}\sum_{n=2}^{N}(-1)^n(n-1)!x^n + (-1)^{N+1}N!\int_0^x t^{N-1}e^{-1/t}\, dt.$$

Why can't you just use a Maclaurin series to approximate this integral? Using $N = 5$, find an approximate value for $\int_0^{0.1} e^{-1/t}\, dt$, and estimate the error. Estimate the error for $N = 10$ and $N = 20$.

Note that the series $\sum_{n=2}^{\infty}(-1)^n(n-1)!x^n$ **diverges** for any $x \ne 0$. This is an example of what is called an **asymptotic series**. Even though it diverges, a properly chosen partial sum gives a good approximation to our function when x is small.

CHAPTER 10

Vectors and Coordinate Geometry in 3-Space

> Lord Ronald said nothing; he flung himself from the room, flung himself upon his horse and rode madly off in all directions. ...
>
> And who is this tall young man who draws nearer to Gertrude with every revolution of the horse? ...
>
> The two were destined to meet. Nearer and nearer they came. And then still nearer. Then for one brief moment they met. As they passed Gertrude raised her head and directed towards the young nobleman two eyes so eye-like in their expression as to be absolutely circular, while Lord Ronald directed towards the occupant of the dogcart a gaze so gaze-like that nothing but a gazelle, or a gas-pipe, could have emulated its intensity.

Stephen Leacock 1869–1944
from *Gertrude the Governess: or, Simple Seventeen*

Introduction

A complete real-variable calculus program involves the study of

(i) real-valued functions of a single real variable,
(ii) vector-valued functions of a single real variable,
(iii) real-valued functions of a real vector variable,
(iv) vector-valued functions of a real vector variable.

Chapters 1–9 are concerned with item (i). The remaining chapters deal with items (ii), (iii), and (iv). Specifically, Chapter 11 deals with vector-valued functions of a single real variable. Chapters 12–14 are concerned with the differentiation and integration of real-valued functions of several real variables, that is, of a real vector variable. Chapters 15 and 16 present aspects of the calculus of functions whose domains and ranges both have dimension greater than one, that is, vector-valued functions of a vector variable. Most of the time we will limit our attention to vector functions with domains and ranges in the plane, or in 3-dimensional space.

In this chapter we will lay the foundation for multivariable and vector calculus by extending the concepts of analytic geometry to three or more dimensions and by introducing vectors as a convenient way of dealing with several variables as a single entity. We also introduce matrices, because these will prove useful for formulating some of the concepts of calculus. This chapter is not intended to be a course in linear algebra. We develop only those aspects that we will use in later chapters and omit most proofs.

10.1 Analytic Geometry in Three Dimensions

We say that the physical world in which we live is three-dimensional because through any point there can pass three, and no more, straight lines that are **mutually perpendicular**; that is, each of them is perpendicular to the other two. This is equivalent to the fact that we require three numbers to locate a point in space with respect to some reference point (the **origin**). One way to use three numbers to locate a point is by having them represent (signed) distances from the origin, measured in the directions of three mutually perpendicular lines passing through the origin. We call such a set of lines a Cartesian coordinate system, and each of the lines is called a coordinate axis. We usually call these axes the x-axis, the y-axis, and the z-axis, regarding the x- and y-axes as lying in a horizontal plane and the z-axis as vertical. Moreover, the coordinate system should have a **right-handed orientation**. This means that the thumb, forefinger, and middle finger of the right hand can be extended so as to point, respectively, in the directions of the positive x-axis, the positive y-axis, and the positive z-axis. For the more mechanically minded, a right-handed screw will advance in the positive z direction if twisted in the direction of rotation from the positive x-axis toward the positive y-axis. (See Figure 10.1(a).)

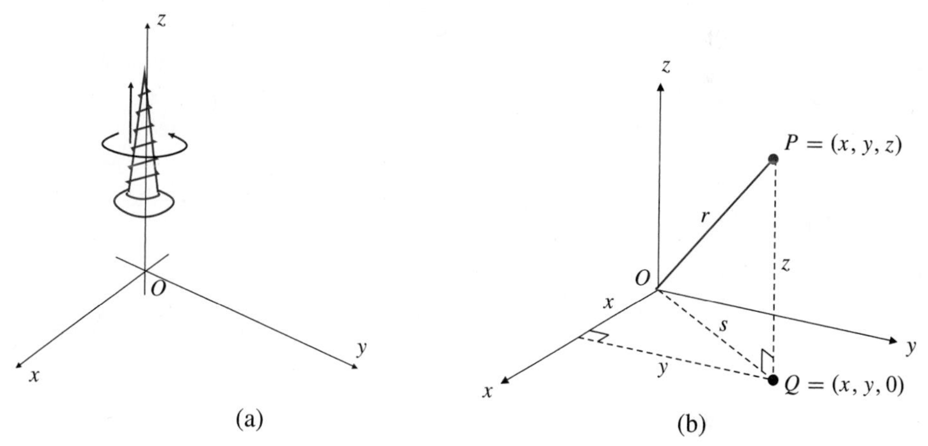

Figure 10.1

(a) The screw moves upward when twisted counterclockwise as seen from above

(b) The three coordinates of a point in 3-space

With respect to such a Cartesian coordinate system, the **coordinates** of a point P in 3-space constitute an ordered triple of real numbers, (x, y, z). The numbers x, y, and z are, respectively, the signed distances of P from the origin, measured in the directions of the x-axis, the y-axis, and the z-axis. (See Figure 10.1(b).)

Let Q be the point with coordinates $(x, y, 0)$. Then Q lies in the xy-plane (the plane containing the x- and y-axes) directly under (or over) P. We say that Q is the vertical projection of P onto the xy-plane. If r is the distance from the origin O to P and s is the distance from O to Q, then, using two right-angled triangles, we have

$$s^2 = x^2 + y^2 \qquad \text{and} \qquad r^2 = s^2 + z^2 = x^2 + y^2 + z^2.$$

Thus, the distance from P to the origin is given by

$$r = \sqrt{x^2 + y^2 + z^2}.$$

Similarly, the distance r between points $P_1 = (x_1, y_1, z_1)$ and $P_2 = (x_2, y_2, z_2)$ (see Figure 10.2) is

$$r = \sqrt{(x_2 - x_1)^2 + (y_2 - y_1)^2 + (z_2 - z_1)^2}.$$

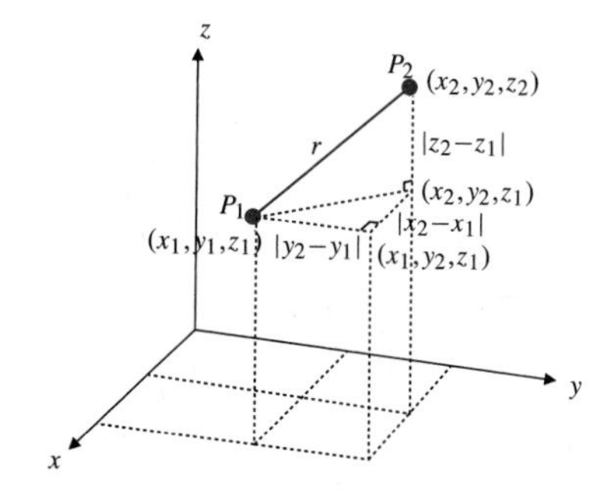

Figure 10.2 Distance between points

EXAMPLE 1 Show that the triangle with vertices $A = (1, -1, 2)$, $B = (3, 3, 8)$, and $C = (2, 0, 1)$ has a right angle.

Solution We calculate the lengths of the three sides of the triangle:

$$a = |BC| = \sqrt{(2-3)^2 + (0-3)^2 + (1-8)^2} = \sqrt{59}$$
$$b = |AC| = \sqrt{(2-1)^2 + (0+1)^2 + (1-2)^2} = \sqrt{3}$$
$$c = |AB| = \sqrt{(3-1)^2 + (3+1)^2 + (8-2)^2} = \sqrt{56}$$

By the cosine law, $a^2 = b^2 + c^2 - 2bc \cos A$. In this case $a^2 = 59 = 3 + 56 = b^2 + c^2$, so that $2bc \cos A$ must be 0. Therefore, $\cos A = 0$ and $A = 90°$.

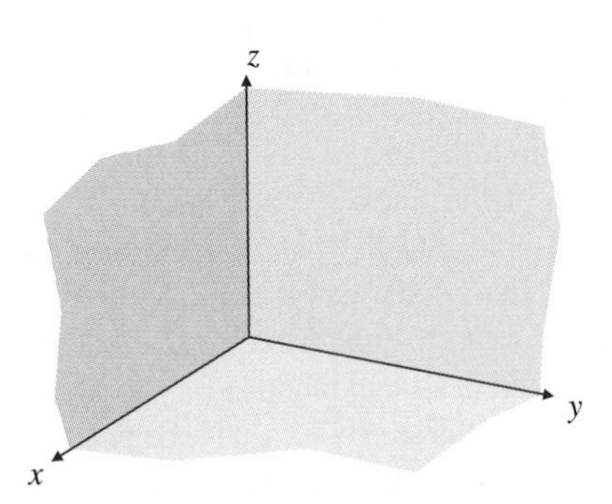

Figure 10.3 The first octant

Just as the x- and y-axes divide the xy-plane into four quadrants, so also the three **coordinate planes** in 3-space (the xy-plane, the xz-plane, and the yz-plane) divide 3-space into eight **octants**. We call the octant in which $x \geq 0$, $y \geq 0$, and $z \geq 0$ the **first octant**. When drawing graphs in 3-space it is sometimes easier to draw only the part lying in the first octant (Figure 10.3).

An equation or inequality involving the three variables x, y, and z defines a subset of points in 3-space whose coordinates satisfy the equation or inequality. A single equation usually represents a surface (a two-dimensional object) in 3-space.

EXAMPLE 2 **(Some equations and the surfaces they represent)**

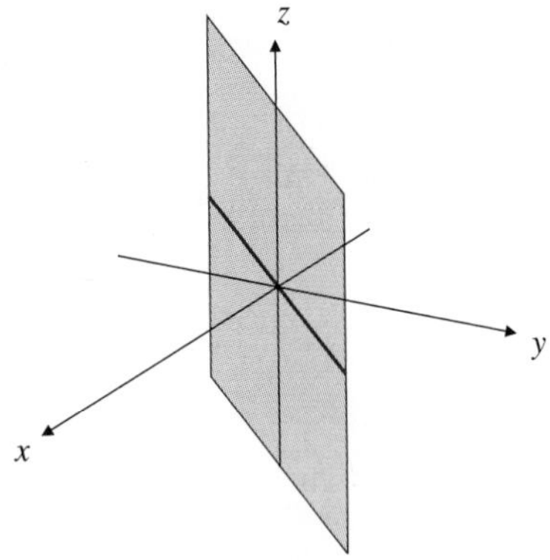

Figure 10.4 Equation $x = y$ defines a vertical plane

(a) The equation $z = 0$ represents all points with coordinates $(x, y, 0)$, that is, the xy-plane. The equation $z = -2$ represents all points with coordinates $(x, y, -2)$, that is, the horizontal plane passing through the point $(0, 0, -2)$ on the z-axis.

(b) The equation $x = y$ represents all points with coordinates (x, x, z). This is a vertical plane containing the straight line with equation $x = y$ in the xy-plane. The plane also contains the z-axis. (See Figure 10.4.)

(c) The equation $x + y + z = 1$ represents all points the sum of whose coordinates is 1. This set is a plane that passes through the three points $(1, 0, 0)$, $(0, 1, 0)$, and $(0, 0, 1)$. These points are not collinear (they do not lie on a straight line), so there is only one plane passing through all three. (See Figure 10.5.) The equation $x + y + z = 0$ represents a plane parallel to the one with equation $x + y + z = 1$ but passing through the origin.

(d) The equation $x^2 + y^2 = 4$ represents all points on the vertical circular cylinder containing the circle with equation $x^2 + y^2 = 4$ in the xy-plane. This cylinder has radius 2 and axis along the z-axis. (See Figure 10.6.)

(e) The equation $z = x^2$ represents all points with coordinates (x, y, x^2). This surface is a parabolic cylinder tangent to the xy-plane along the y-axis. (See Figure 10.7.)

(f) The equation $x^2 + y^2 + z^2 = 25$ represents all points (x, y, z) at distance 5 from the origin. This set of points is a *sphere* of radius 5 centred at the origin.

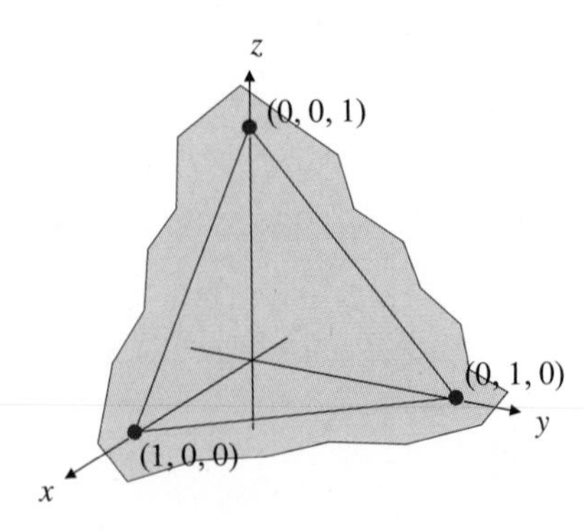

Figure 10.5 The plane with equation $x + y + z = 1$

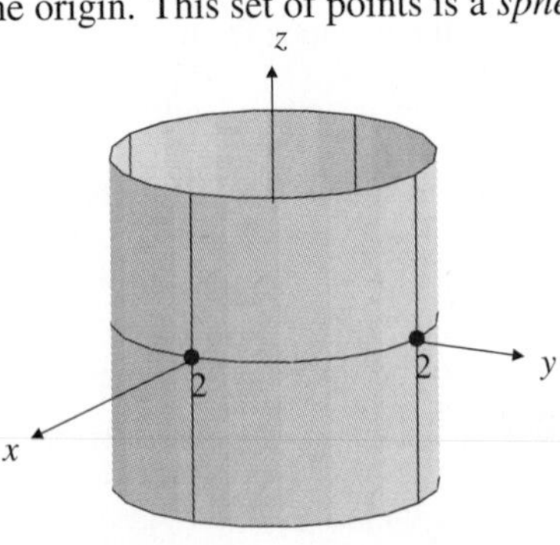

Figure 10.6 The circular cylinder with equation $x^2 + y^2 = 4$

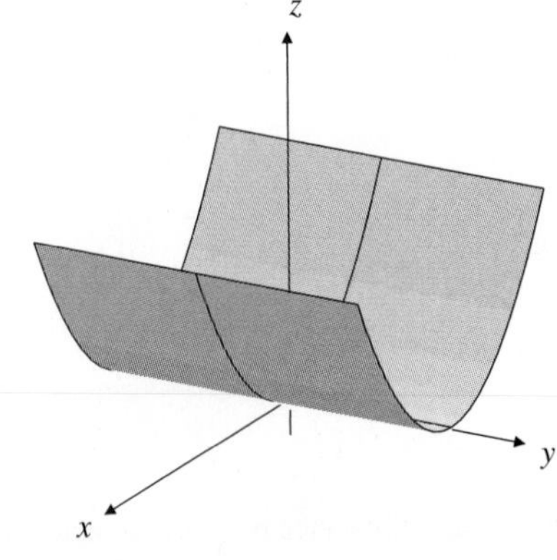

Figure 10.7 The parabolic cylinder with equation $z = x^2$

Observe that equations in x, y, and z need not involve each variable explicitly. When one of the variables is missing from the equation, the equation represents a surface *parallel to* the axis of the missing variable. Such a surface may be a plane or a cylinder. For example, if z is absent from the equation, the equation represents in 3-space a vertical (i.e., parallel to the z-axis) surface containing the curve with the same equation in the xy-plane.

Occasionally, a single equation may not represent a two-dimensional object (a surface). It can represent a one-dimensional object (a line or curve), a zero-dimensional object (one or more points), or even nothing at all.

EXAMPLE 3 Identify the graphs of: (a) $y^2+(z-1)^2=4$, (b) $y^2+(z-1)^2=0$, (c) $x^2+y^2+z^2=0$, and (d) $x^2+y^2+z^2=-1$.

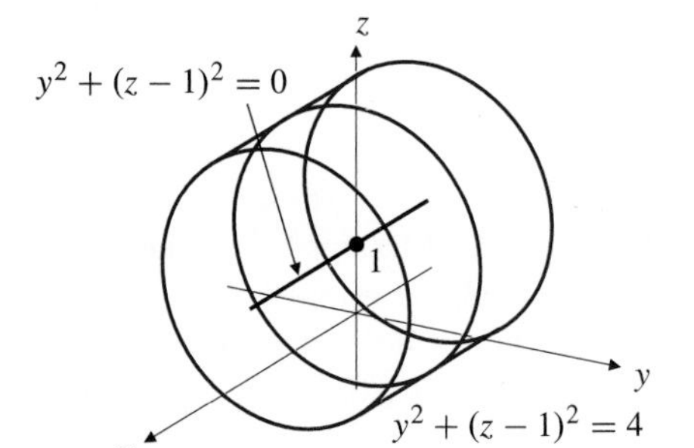

Figure 10.8 The cylinder $y^2+(z-1)^2=4$ and its axial line $y^2+(z-1)^2=0$

Solution

(a) Since x is absent, the equation $y^2+(z-1)^2=4$ represents an object parallel to the x-axis. In the yz-plane the equation represents a circle of radius 2 centred at $(y, z)=(0, 1)$. In 3-space it represents a horizontal circular cylinder, parallel to the x-axis, with axis one unit above the x-axis. (See Figure 10.8.)

(b) Since squares cannot be negative, the equation $y^2+(z-1)^2=0$ implies that $y=0$ and $z=1$, so it represents points $(x, 0, 1)$. All these points lie on the line parallel to the x-axis and one unit above it. (See Figure 10.8.)

(c) As in part (b), $x^2+y^2+z^2=0$ implies that $x=0$, $y=0$, and $z=0$. The equation represents only one point, the origin.

(d) The equation $x^2+y^2+z^2=-1$ is not satisfied by any real numbers x, y, and z, so it represents no points at all.

A single inequality in x, y, and z typically represents points lying on one side of the surface represented by the corresponding equation (together with points on the surface if the inequality is not strict).

EXAMPLE 4 (a) The inequality $z>0$ represents all points above the xy-plane.

(b) The inequality $x^2+y^2\geq 4$ says that the square of the distance from (x, y, z) to the nearest point $(0, 0, z)$ on the z-axis is at least 4. This inequality represents all points lying on or outside the cylinder of Example 2(d).

(c) The inequality $x^2+y^2+z^2\leq 25$ says that the square of the distance from (x, y, z) to the origin is no greater than 25. It represents the solid ball of radius 5 centred at the origin, which consists of all points lying inside or on the sphere of Example 2(f).

Two equations in x, y, and z normally represent a one-dimensional object, the line or curve along which the two surfaces represented by the two equations intersect. Any point whose coordinates satisfy both equations must lie on both the surfaces, so must lie on their intersection.

EXAMPLE 5 What sets of points in 3-space are represented by the following pairs of equations?

$$\text{(a)}\quad \begin{cases} x+y+z=1 \\ y-2x=0 \end{cases} \qquad \text{(b)}\quad \begin{cases} x^2+y^2+z^2=1 \\ x+y=1 \end{cases}$$

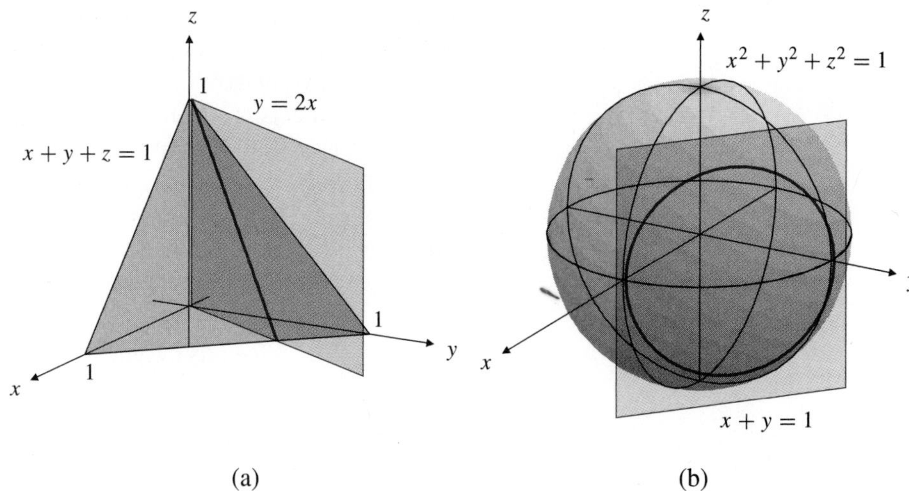

Figure 10.9

(a) The two planes intersect in a straight line

(b) The plane intersects the sphere in a circle

Solution

(a) The equation $x + y + z = 1$ represents the oblique plane of Example 2(c), and the equation $y - 2x = 0$ represents a vertical plane through the origin and the point $(1, 2, 0)$. Together these two equations represent the line of intersection of the two planes. This line passes through, for example, the points $(0, 0, 1)$ and $(\frac{1}{3}, \frac{2}{3}, 0)$. (See Figure 10.9(a).)

(b) The equation $x^2 + y^2 + z^2 = 1$ represents a sphere of radius 1 with centre at the origin, and $x + y = 1$ represents a vertical plane through the points $(1, 0, 0)$ and $(0, 1, 0)$. The two surfaces intersect in a circle, as shown in Figure 10.9(b). The line from $(1, 0, 0)$ to $(0, 1, 0)$ is a diameter of the circle, so the centre of the circle is $(\frac{1}{2}, \frac{1}{2}, 0)$, and its radius is $\sqrt{2}/2$.

In Sections 10.4 and 10.5 we will see many more examples of geometric objects in 3-space represented by simple equations.

Euclidean *n*-Space

Mathematicians and users of mathematics frequently need to consider ***n*-dimensional space**, where n is greater than 3 and may even be infinite. It is difficult to visualize a space of dimension 4 or higher geometrically. The secret to dealing with these spaces is to regard the points in n-space as *being* ordered n-tuples of real numbers; that is, $(x_1, x_2, \ldots, x_n)$ is a point in n-space instead of just being the coordinates of such a point. We stop thinking of points as existing in physical space and start thinking of them as algebraic objects. We usually denote n-space by the symbol $\mathbb{R}^n$ to show that its points are n-tuples of *real* numbers. Thus $\mathbb{R}^2$ and $\mathbb{R}^3$ denote the plane and 3-space, respectively. Note that in passing from $\mathbb{R}^3$ to $\mathbb{R}^n$ we have altered the notation a bit: in $\mathbb{R}^3$ we called the coordinates x, y, and z, while in $\mathbb{R}^n$ we called them $x_1, x_2, \ldots$ and x_n so as not to run out of letters. We could, of course, talk about coordinates (x_1, x_2, x_3) in $\mathbb{R}^3$ and (x_1, x_2) in the plane $\mathbb{R}^2$, but (x, y, z) and (x, y) are traditionally used there.

Although we think of points in $\mathbb{R}^n$ as n-tuples rather than geometric objects, we do not want to lose all sight of the underlying geometry. By analogy with the two- and three-dimensional cases, we still consider the quantity

$$\sqrt{(y_1 - x_1)^2 + (y_2 - x_2)^2 + \cdots + (y_n - x_n)^2}$$

as representing the *distance* between the points with coordinates $(x_1, x_2, \ldots, x_n)$ and $(y_1, y_2, \ldots, y_n)$. Also, we call the $(n - 1)$-dimensional set of points in $\mathbb{R}^n$ that satisfy the equation $x_n = 0$ a **hyperplane**, by analogy with the plane $z = 0$ in $\mathbb{R}^3$.

Describing Sets in the Plane, 3-Space, and *n*-Space

We conclude this section by collecting some definitions of terms used to describe sets of points in $\mathbb{R}^n$ for $n \ge 2$. These terms belong to the branch of mathematics called

topology, and they generalize the notions of open and closed intervals and endpoints used to describe sets on the real line $\mathbb{R}$. We state the definitions for $\mathbb{R}^n$, but we are most interested in the cases where $n = 2$ or $n = 3$.

A **neighbourhood** of a point P in $\mathbb{R}^n$ is a set of the form

$$B_r(P) = \{Q \in \mathbb{R}^n : \text{distance from } Q \text{ to } P < r\}$$

for some $r > 0$.

For $n = 1$, if $p \in \mathbb{R}$, then $B_r(p)$ is the **open interval** $(p - r, p + r)$ centred at p.

For $n = 2$, $B_r(P)$ is the **open disk** of radius r centred at point P.

For $n = 3$, $B_r(P)$ is the **open ball** of radius r centred at point P.

A set S is **open** in $\mathbb{R}^n$ if every point of S has a neighbourhood contained in S. Every neighbourhood is itself an open set. Other examples of open sets in $\mathbb{R}^2$ include the sets of points (x, y) such that $x > 0$, or such that $y > x^2$, or even such that $y \neq x^2$. Typically, sets defined by strict inequalities (using $>$ and $<$) are open. Examples in $\mathbb{R}^3$ include the sets of points (x, y, z) satisfying $x + y + z > 2$, or $1 < x < 3$.

The whole space $\mathbb{R}^n$ is an open set in itself. For technical reasons, the empty set (containing no points) is also considered to be open. (No point in the empty set fails to have a neighbourhood contained in the empty set.)

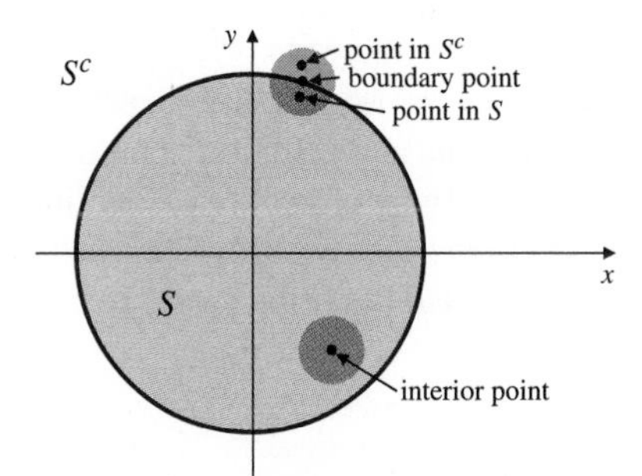

Figure 10.10 The closed disk S consisting of points $(x, y) \in \mathbb{R}^2$ that satisfy $x^2 + y^2 \leq 1$. Note the shaded neighbourhoods of the boundary point and the interior point.
bdry(S) is the circle $x^2 + y^2 = 1$
int(S) is the open disk $x^2 + y^2 < 1$
ext(S) is the open set $x^2 + y^2 > 1$

The **complement**, S^c, of a set S in $\mathbb{R}^n$ is the set of all points in $\mathbb{R}^n$ that do not belong to S. For example, the complement of the set of points (x, y) in $\mathbb{R}^2$ such that $x > 0$ is the set of points for which $x \leq 0$. A set is said to be **closed** if its complement is open. Typically, sets defined by nonstrict inequalities (using $\geq$ and $\leq$) are closed. Closed intervals are closed sets in $\mathbb{R}$. Since the whole space and the empty set are both open in $\mathbb{R}^n$ and are complements of each other, they are also both closed. They are the only sets that are both open and closed.

A point P is called a **boundary point** of a set S if every neighbourhood of P contains both points in S and points in S^c. The **boundary**, bdry(S), of a set S is the set of all boundary points of S. For example, the boundary of the closed disk $x^2 + y^2 \leq 1$ in $\mathbb{R}^2$ is the circle $x^2 + y^2 = 1$. A closed set contains all its boundary points. An open set contains none of its boundary points.

A point P is an **interior point** of a set S if it belongs to S but not to the boundary of S. P is an **exterior point** of S if it belongs to the complement of S but not to the boundary of S. The **interior**, int(S), and **exterior**, ext(S), of S consist of all the interior points and exterior points of S, respectively. Both int(S) and ext(S) are open sets. S is open if and only if int(S) $= S$. S is closed if and only if ext(S) $= S^c$. See Figure 10.10.

EXERCISES 10.1

Find the distance between the pairs of points in Exercises 1–4.

1. $(0, 0, 0)$ and $(2, -1, -2)$ **2.** $(-1, -1, -1)$ and $(1, 1, 1)$

3. $(1, 1, 0)$ and $(0, 2, -2)$ **4.** $(3, 8, -1)$ and $(-2, 3, -6)$

5. What is the shortest distance from the point (x, y, z) to (a) the xy-plane? (b) the x-axis?

6. Show that the triangle with vertices $(1, 2, 3)$, $(4, 0, 5)$, and $(3, 6, 4)$ has a right angle.

7. Find the angle A in the triangle with vertices $A = (2, -1, -1)$, $B = (0, 1, -2)$, and $C = (1, -3, 1)$.

8. Show that the triangle with vertices $(1, 2, 3)$, $(1, 3, 4)$, and $(0, 3, 3)$ is equilateral.

9. Find the area of the triangle with vertices $(1, 1, 0)$, $(1, 0, 1)$, and $(0, 1, 1)$.

10. What is the distance from the origin to the point $(1, 1, \dots, 1)$ in $\mathbb{R}^n$?

11. What is the distance from the point $(1, 1, \dots, 1)$ in n-space to the closest point on the x_1-axis?

In Exercises 12–23, describe (and sketch if possible) the set of points in $\mathbb{R}^3$ that satisfy the given equation or inequality.

12. $z = 2$ **13.** $y \geq -1$

14. $z = x$ **15.** $x + y = 1$

16. $x^2 + y^2 + z^2 = 4$

17. $(x - 1)^2 + (y + 2)^2 + (z - 3)^2 = 4$

18. $x^2 + y^2 + z^2 = 2z$

19. $y^2 + z^2 \le 4$

20. $x^2 + z^2 = 4$

21. $z = y^2$

22. $z \ge \sqrt{x^2 + y^2}$

23. $x + 2y + 3z = 6$

In Exercises 24–32, describe (and sketch if possible) the set of points in $\mathbb{R}^3$ that satisfy the given pair of equations or inequalities.

24. $\begin{cases} x = 1 \\ y = 2 \end{cases}$

25. $\begin{cases} x = 1 \\ y = z \end{cases}$

26. $\begin{cases} x^2 + y^2 + z^2 = 4 \\ z = 1 \end{cases}$

27. $\begin{cases} x^2 + y^2 + z^2 = 4 \\ x^2 + y^2 + z^2 = 4x \end{cases}$

28. $\begin{cases} x^2 + y^2 + z^2 = 4 \\ x^2 + z^2 = 1 \end{cases}$

29. $\begin{cases} x^2 + y^2 = 1 \\ z = x \end{cases}$

30. $\begin{cases} y \ge x \\ z \le y \end{cases}$

31. $\begin{cases} x^2 + y^2 \le 1 \\ z \ge y \end{cases}$

32. $\begin{cases} x^2 + y^2 + z^2 \le 1 \\ \sqrt{x^2 + y^2} \le z \end{cases}$

In Exercises 33–36, specify the boundary and the interior of the plane sets S whose points (x, y) satisfy the given conditions. Is S open, closed, or neither?

33. $0 < x^2 + y^2 < 1$

34. $x \ge 0, \quad y < 0$

35. $x + y = 1$

36. $|x| + |y| \le 1$

In Exercises 37–40, specify the boundary and the interior of the sets S in 3-space whose points (x, y, z) satisfy the given conditions. Is S open, closed, or neither?

37. $1 \le x^2 + y^2 + z^2 \le 4$

38. $x \ge 0, \quad y > 1, \quad z < 2$

39. $(x - z)^2 + (y - z)^2 = 0$

40. $x^2 + y^2 < 1, \; y + z > 2$

10.2 Vectors

A **vector** is a quantity that involves both **magnitude** (size or length) and **direction**. For instance, the *velocity* of a moving object involves its speed and direction of motion, so is a vector. Such quantities are represented geometrically by arrows (directed line segments) and are often actually identified with these arrows. For instance, the vector $\overrightarrow{AB}$ is an arrow with tail at the point A and head at the point B. In print, such a vector is usually denoted by a single letter in boldface type,

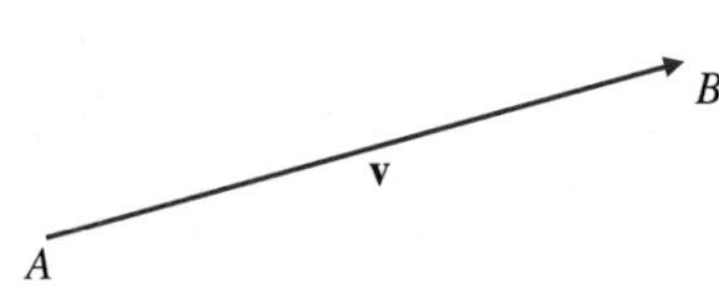

Figure 10.11 The vector $\mathbf{v} = \overrightarrow{AB}$

$$\mathbf{v} = \overrightarrow{AB}.$$

(See Figure 10.11.) In handwriting, an arrow over a letter ($\vec{v} = \overrightarrow{AB}$) can be used to denote a vector. The *magnitude* of the vector $\mathbf{v}$ is the length of the arrow and is denoted $|\mathbf{v}|$ or $|\overrightarrow{AB}|$.

While vectors have magnitude and direction, they do not generally have *position*; that is, they are not regarded as being in a particular place. Two vectors, $\mathbf{u}$ and $\mathbf{v}$, are considered *equal* if they have *the same length and the same direction*, even if their representative arrows do not coincide. The arrows must be parallel, have the same length, and point in the same direction. In Figure 10.12, for example, if $ABYX$ is a parallelogram, then $\overrightarrow{AB} = \overrightarrow{XY}$.

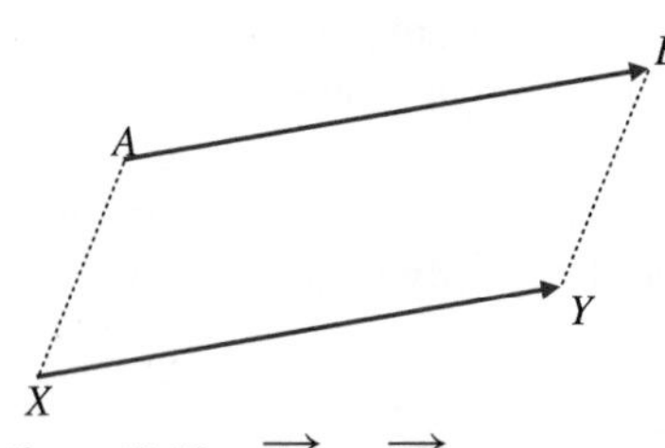

Figure 10.12 $\overrightarrow{AB} = \overrightarrow{XY}$

For the moment, we consider plane vectors, that is, vectors whose representative arrows lie in a plane. If we introduce a Cartesian coordinate system into the plane, we can talk about the x and y components of any vector. If $A = (a, b)$ and $P = (p, q)$, as shown in Figure 10.13, then the x and y components of $\overrightarrow{AP}$ are, respectively, $p - a$ and $q - b$. Note that if O is the origin and X is the point $(p - a, q - b)$, then

$$|\overrightarrow{AP}| = \sqrt{(p - a)^2 + (q - b)^2} = |\overrightarrow{OX}|$$

$$\text{slope of } \overrightarrow{AP} = \frac{q - b}{p - a} = \text{slope of } \overrightarrow{OX}.$$

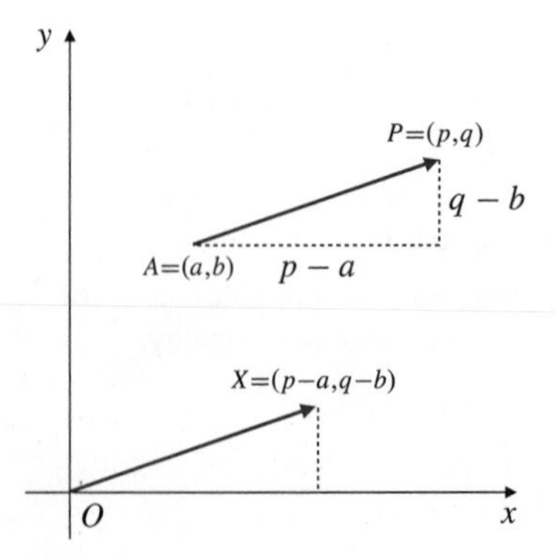

Figure 10.13 Components of a vector

Hence $\overrightarrow{AP} = \overrightarrow{OX}$. In general, two vectors are equal if and only if they have the same x components and y components.

There are two important algebraic operations defined for vectors: addition and scalar multiplication.

DEFINITION 1

Vector addition

Given two vectors **u** and **v**, their **sum u + v** is defined as follows. If an arrow representing **v** is placed with its tail at the head of an arrow representing **u**, then an arrow from the tail of **u** to the head of **v** represents **u** + **v**. Equivalently, if **u** and **v** have tails at the same point, then **u** + **v** is represented by an arrow with its tail at that point and its head at the opposite vertex of the parallelogram spanned by **u** and **v**. This is shown in Figure 10.14(a).

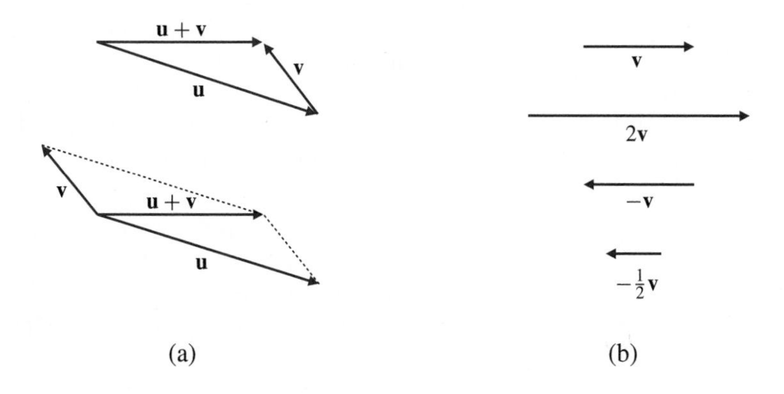

Figure 10.14
(a) Vector addition
(b) Scalar multiplication

DEFINITION 2

Scalar multiplication

If **v** is a vector and t is a real number (also called a **scalar**), then the **scalar multiple** t**v** is a vector with magnitude $|t|$ times that of **v** and direction the same as **v** if $t > 0$, or opposite to that of **v** if $t < 0$. See Figure 10.14(b). If $t = 0$, then t**v** has zero length and therefore no particular direction. It is the **zero vector**, denoted **0**.

Suppose that **u** has components a and b and that **v** has components x and y. Then the components of **u** + **v** are $a + x$ and $b + y$, and those of t**v** are tx and ty. See Figure 10.15.

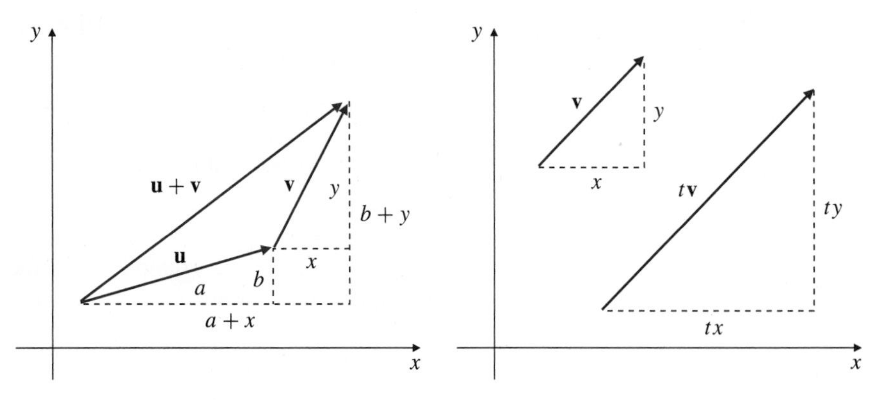

Figure 10.15 The components of a sum of vectors or a scalar multiple of a vector is the same sum or multiple of the corresponding components of the vectors

In $\mathbb{R}^2$ we single out two particular vectors for special attention. They are

(i) the vector **i** from the origin to the point (1, 0), and

(ii) the vector **j** from the origin to the point (0, 1).

Thus, **i** has components 1 and 0, and **j** has components 0 and 1. These vectors are called the **standard basis vectors** in the plane. The vector **r** from the origin to the point (x, y) has components x and y and can be expressed in the form

$$\mathbf{r} = \langle x, y \rangle = x\mathbf{i} + y\mathbf{j}.$$

In the first form we specify the vector by listing its components between angle brackets; in the second we write $\mathbf{r}$ as a **linear combination** of the standard basis vectors $\mathbf{i}$ and $\mathbf{j}$. (See Figure 10.16.) The vector $\mathbf{r}$ is called the **position vector** of the point (x, y). A position vector has its tail at the origin and its head at the point whose position it is specifying. The length of $\mathbf{r}$ is $|\mathbf{r}| = \sqrt{x^2 + y^2}$.

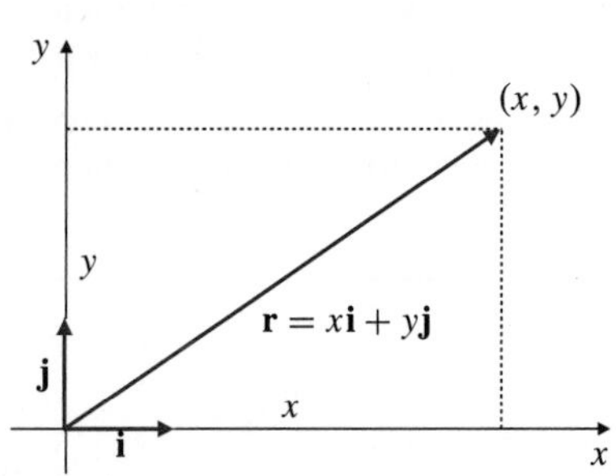

Figure 10.16 Any vector is a linear combination of the basis vectors

More generally, the vector $\overrightarrow{AP}$ from $A = (a, b)$ to $P = (p, q)$ in Figure 10.13 can also be written as a list of components or as a linear combination of the standard basis vectors:

$$\overrightarrow{AP} = \langle p - a, q - b \rangle = (p - a)\mathbf{i} + (q - b)\mathbf{j}.$$

Sums and scalar multiples of vectors are easily expressed in terms of components. If $\mathbf{u} = u_1\mathbf{i} + u_2\mathbf{j}$ and $\mathbf{v} = v_1\mathbf{i} + v_2\mathbf{j}$, and if t is a scalar (i.e., a real number), then

$$\begin{aligned} \mathbf{u} + \mathbf{v} &= (u_1 + v_1)\mathbf{i} + (u_2 + v_2)\mathbf{j}, \\ t\mathbf{u} &= (tu_1)\mathbf{i} + (tu_2)\mathbf{j}. \end{aligned}$$

The zero vector is $\mathbf{0} = 0\mathbf{i} + 0\mathbf{j}$. It has length zero and no specific direction. For any vector $\mathbf{u}$ we have $0\mathbf{u} = \mathbf{0}$. A **unit vector** is a vector of length 1. The standard basis vectors $\mathbf{i}$ and $\mathbf{j}$ are unit vectors. Given any nonzero vector $\mathbf{v}$, we can form a unit vector $\hat{\mathbf{v}}$ in the same direction as $\mathbf{v}$ by multiplying $\mathbf{v}$ by the reciprocal of its length (a scalar):

$$\hat{\mathbf{v}} = \left(\frac{1}{|\mathbf{v}|}\right)\mathbf{v}.$$

EXAMPLE 1 If $A = (2, -1)$, $B = (-1, 3)$, and $C = (0, 1)$, express each of the following vectors as a linear combination of the standard basis vectors:

(a) $\overrightarrow{AB}$ (b) $\overrightarrow{BC}$ (c) $\overrightarrow{AC}$ (d) $\overrightarrow{AB} + \overrightarrow{BC}$ (e) $2\overrightarrow{AC} - 3\overrightarrow{CB}$

(f) a unit vector in the direction of $\overrightarrow{AB}$.

Solution

(a) $\overrightarrow{AB} = (-1 - 2)\mathbf{i} + (3 - (-1))\mathbf{j} = -3\mathbf{i} + 4\mathbf{j}$

(b) $\overrightarrow{BC} = (0 - (-1))\mathbf{i} + (1 - 3)\mathbf{j} = \mathbf{i} - 2\mathbf{j}$

(c) $\overrightarrow{AC} = (0 - 2)\mathbf{i} + (1 - (-1))\mathbf{j} = -2\mathbf{i} + 2\mathbf{j}$

(d) $\overrightarrow{AB} + \overrightarrow{BC} = \overrightarrow{AC} = -2\mathbf{i} + 2\mathbf{j}$

(e) $2\overrightarrow{AC} - 3\overrightarrow{CB} = 2(-2\mathbf{i} + 2\mathbf{j}) - 3(-\mathbf{i} + 2\mathbf{j}) = -\mathbf{i} - 2\mathbf{j}$

(f) A unit vector in the direction of $\overrightarrow{AB}$ is $\dfrac{\overrightarrow{AB}}{|\overrightarrow{AB}|} = -\dfrac{3}{5}\mathbf{i} + \dfrac{4}{5}\mathbf{j}$.

Implicit in the above example is the fact that the operations of addition and scalar multiplication obey appropriate algebraic rules, such as

$$\begin{aligned} \mathbf{u} + \mathbf{v} &= \mathbf{v} + \mathbf{u}, \\ (\mathbf{u} + \mathbf{v}) + \mathbf{w} &= \mathbf{u} + (\mathbf{v} + \mathbf{w}), \\ \mathbf{u} - \mathbf{v} &= \mathbf{u} + (-1)\mathbf{v}, \\ t(\mathbf{u} + \mathbf{v}) &= t\mathbf{u} + t\mathbf{v}. \end{aligned}$$

Vectors in 3-Space

The algebra and geometry of vectors described here extends to spaces of any number of dimensions; we can still think of vectors as represented by arrows, and sums and scalar multiples are formed just as for plane vectors.

Given a Cartesian coordinate system in 3-space, we define three **standard basis vectors**, $\mathbf{i}$, $\mathbf{j}$, and $\mathbf{k}$, represented by arrows from the origin to the points $(1,0,0)$, $(0,1,0)$, and $(0,0,1)$, respectively. (See Figure 10.17.) Any vector in 3-space can be written as a *linear combination* of these basis vectors; for instance, the position vector of the point (x,y,z) is given by

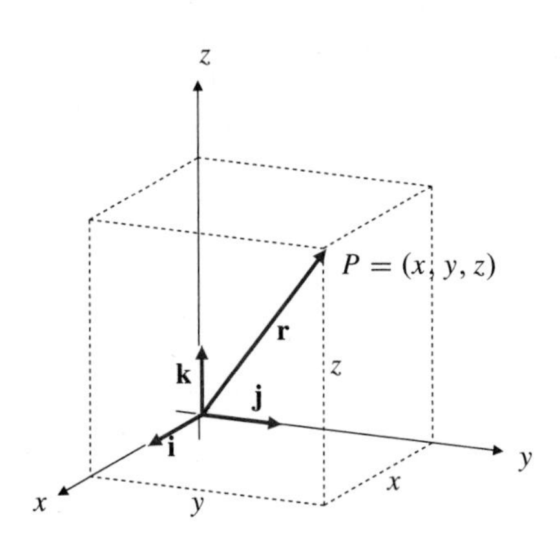

Figure 10.17 The standard basis vectors $\mathbf{i}$, $\mathbf{j}$, and $\mathbf{k}$

$$\mathbf{r} = x\mathbf{i} + y\mathbf{j} + z\mathbf{k}.$$

We say that $\mathbf{r}$ has **components** x, y, and z. The length of $\mathbf{r}$ is

$$|\mathbf{r}| = \sqrt{x^2 + y^2 + z^2}.$$

If $P_1 = (x_1, y_1, z_1)$ and $P_2 = (x_2, y_2, z_2)$ are two points in 3-space, then the vector $\mathbf{v} = \overrightarrow{P_1P_2}$ from P_1 to P_2 has components $x_2 - x_1$, $y_2 - y_1$, and $z_2 - z_1$ and is therefore represented in terms of the standard basis vectors by

$$\mathbf{v} = \overrightarrow{P_1P_2} = (x_2 - x_1)\mathbf{i} + (y_2 - y_1)\mathbf{j} + (z_2 - z_1)\mathbf{k}.$$

EXAMPLE 2 If $\mathbf{u} = 2\mathbf{i} + \mathbf{j} - 2\mathbf{k}$ and $\mathbf{v} = 3\mathbf{i} - 2\mathbf{j} - \mathbf{k}$, find $\mathbf{u}+\mathbf{v}$, $\mathbf{u}-\mathbf{v}$, $3\mathbf{u}-2\mathbf{v}$, $|\mathbf{u}|$, $|\mathbf{v}|$, and a unit vector $\hat{\mathbf{u}}$ in the direction of $\mathbf{u}$.

Solution

$$\begin{aligned}
\mathbf{u}+\mathbf{v} &= (2+3)\mathbf{i} + (1-2)\mathbf{j} + (-2-1)\mathbf{k} = 5\mathbf{i} - \mathbf{j} - 3\mathbf{k}\\
\mathbf{u}-\mathbf{v} &= (2-3)\mathbf{i} + (1+2)\mathbf{j} + (-2+1)\mathbf{k} = -\mathbf{i} + 3\mathbf{j} - \mathbf{k}\\
3\mathbf{u}-2\mathbf{v} &= (6-6)\mathbf{i} + (3+4)\mathbf{j} + (-6+2)\mathbf{k} = 7\mathbf{j} - 4\mathbf{k}\\
|\mathbf{u}| &= \sqrt{4+1+4} = 3, \qquad |\mathbf{v}| = \sqrt{9+4+1} = \sqrt{14}\\
\hat{\mathbf{u}} &= \left(\frac{1}{|\mathbf{u}|}\right)\mathbf{u} = \frac{2}{3}\mathbf{i} + \frac{1}{3}\mathbf{j} - \frac{2}{3}\mathbf{k}.
\end{aligned}$$

The following example illustrates the way vectors can be used to solve problems involving relative velocities. If A moves with velocity $\mathbf{v}_{A\,\mathrm{rel}\,B}$ relative to B, and B moves with velocity $\mathbf{v}_{B\,\mathrm{rel}\,C}$ relative to C, then A moves with velocity $\mathbf{v}_{A\,\mathrm{rel}\,C}$ relative to C, where

$$\mathbf{v}_{A\,\mathrm{rel}\,C} = \mathbf{v}_{A\,\mathrm{rel}\,B} + \mathbf{v}_{B\,\mathrm{rel}\,C}.$$

EXAMPLE 3 An aircraft cruises at a speed of 300 km/h in still air. If the wind is blowing from the east at 100 km/h, in what direction should the aircraft head in order to fly in a straight line from city P to city Q, 400 km north northeast of P? How long will the trip take?

Solution The problem is two-dimensional, so we use plane vectors. Let us choose our coordinate system so that the x- and y-axes point east and north, respectively. Figure 10.18 illustrates the three velocities that must be considered. The velocity of the air relative to the ground is

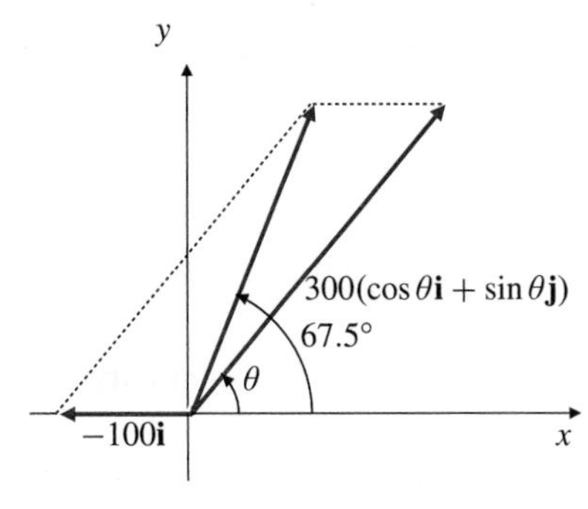

Figure 10.18 Velocity diagram for the aircraft in Example 3

$$\mathbf{v}_{\mathrm{air\ rel\ ground}} = -100\,\mathbf{i}.$$

If the aircraft heads in a direction making angle θ with the positive direction of the x-axis, then the velocity of the aircraft relative to the air is

$$\mathbf{v}_{\mathrm{aircraft\ rel\ air}} = 300\cos\theta\,\mathbf{i} + 300\sin\theta\,\mathbf{j}.$$

Thus, the velocity of the aircraft relative to the ground is

$$\begin{aligned}
\mathbf{v}_{\mathrm{aircraft\ rel\ ground}} &= \mathbf{v}_{\mathrm{aircraft\ rel\ air}} + \mathbf{v}_{\mathrm{air\ rel\ ground}}\\
&= (300\cos\theta - 100)\,\mathbf{i} + 300\sin\theta\,\mathbf{j}.
\end{aligned}$$

We want this latter velocity to be in a north-northeasterly direction, that is, in the direction making angle $3\pi/8 = 67.5°$ with the positive direction of the x-axis. Thus, we will have

$$\mathbf{v}_{\text{aircraft rel ground}} = v\left[\left(\cos 67.5°\right)\mathbf{i} + \left(\sin 67.5°\right)\mathbf{j}\right],$$

where v is the actual groundspeed of the aircraft. Comparing the two expressions for $\mathbf{v}_{\text{aircraft rel ground}}$ we obtain

$$\begin{aligned} 300\cos\theta - 100 &= v\cos 67.5° \\ 300\sin\theta &= v\sin 67.5°. \end{aligned}$$

Eliminating v between these two equations we get

$$300\cos\theta\sin 67.5° - 300\sin\theta\cos 67.5° = 100\sin 67.5°,$$

or

$$3\sin(67.5° - \theta) = \sin 67.5°.$$

Therefore, the aircraft should head in direction θ given by

$$\theta = 67.5° - \arcsin\left(\frac{1}{3}\sin 67.5°\right) \approx 49.56°,$$

that is, 49.56° north of east. The groundspeed is now seen to be

$$v = 300\sin\theta/\sin 67.5° \approx 247.15 \text{ km/h}.$$

Thus, the 400 km trip will take about $400/247.15 \approx 1.618$ hours, or about 1 hour and 37 minutes.

Hanging Cables and Chains

When it is suspended from both ends and allowed to hang under gravity, a heavy cable or chain assumes the shape of a **catenary** curve, which is the graph of the hyperbolic cosine function. We will demonstrate this now, using vectors to keep track of the various forces acting on the cable.

Suppose that the cable has line density δ (units of mass per unit length) and hangs as shown in Figure 10.19. Let us choose a coordinate system so that the lowest point L on the cable is at $(0, y_0)$; we will specify the value of y_0 later. If $P = (x, y)$ is another point on the cable, there are three forces acting on the arc LP of the cable between L and P. These are all forces that we can represent using horizontal and vertical components.

(i) The horizontal tension $\mathbf{H} = -H\mathbf{i}$ at L. This is the force that the part of the cable to the left of L exerts on the arc LP at L.

(ii) The tangential tension $\mathbf{T} = T_h\mathbf{i} + T_v\mathbf{j}$. This is the force the part of the cable to the right of P exerts on arc LP at P.

(iii) The weight $\mathbf{W} = -\delta g s\mathbf{j}$ of arc LP, where g is the acceleration of gravity and s is the length of the arc LP.

Since the cable is not moving, these three forces must balance; their vector sum must be zero:

$$\begin{aligned} &\mathbf{T} + \mathbf{H} + \mathbf{W} = \mathbf{0} \\ &(T_h - H)\mathbf{i} + (T_v - \delta g s)\mathbf{j} = \mathbf{0} \end{aligned}$$

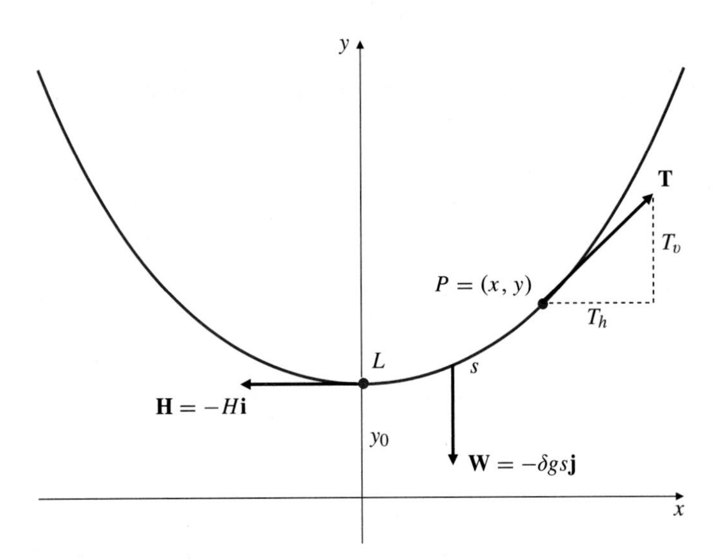

Figure 10.19 A hanging cable and the forces acting on arc LP

Thus, $T_h = H$ and $T_v = \delta gs$. Since $\mathbf{T}$ is tangent to the cable at P, the slope of the cable there is

$$\frac{dy}{dx} = \frac{T_v}{T_h} = \frac{\delta gs}{H} = as,$$

where $a = \delta g/H$ is a constant for the given cable. Differentiating with respect to x and using the fact, from our study of arc length, that

$$\frac{ds}{dx} = \sqrt{1 + \left(\frac{dy}{dx}\right)^2},$$

we obtain a second-order differential equation,

$$\frac{d^2y}{dx^2} = a\frac{ds}{dx} = a\sqrt{1 + \left(\frac{dy}{dx}\right)^2},$$

to be solved for the equation of the curve along which the hanging cable lies. The appropriate initial conditions are $y = y_0$ and $dy/dx = 0$ at $x = 0$.

Since the differential equation depends on dy/dx rather than y, we substitute $m(x) = dy/dx$ and obtain a first-order equation for m:

$$\frac{dm}{dx} = a\sqrt{1 + m^2}.$$

This equation is separable; we integrate it using the substitution $m = \sinh u$:

$$\begin{aligned}
&\int \frac{1}{\sqrt{1 + m^2}}\, dm = \int a\, dx \\
&\int du = \int \frac{\cosh u}{\sqrt{1 + \sinh^2 u}}\, du = ax + C_1 \\
&\sinh^{-1} m = u = ax + C_1 \\
&m = \sinh(ax + C_1).
\end{aligned}$$

Since $m = dy/dx = 0$ at $x = 0$, we have $0 = \sinh C_1$, so $C_1 = 0$ and

$$\frac{dy}{dx} = m = \sinh(ax).$$

This equation is easily integrated to find y. (Had we used a tangent substitution instead of the hyperbolic sine substitution for m we would have had more trouble here.)

$$y = \frac{1}{a}\cosh(ax) + C_2.$$

If we choose $y_0 = y(0) = 1/a$, then, substituting $x = 0$ we will get $C_2 = 0$. With this choice of y_0, we therefore find that the equation of the curve along which the hanging cable lies is the catenary

$$y = \frac{1}{a}\cosh(ax).$$

Remark If a hanging cable bears loads other than its own weight, it will assume a different shape. For example, a cable supporting a level suspension bridge whose weight per unit length is much greater than that of the cable will assume the shape of a parabola. See Exercise 34 below.

The Dot Product and Projections

There is another operation on vectors in any dimension by which two vectors are combined to produce a number called their *dot product*.

DEFINITION 3

The dot product of two vectors

Given two vectors, $\mathbf{u} = u_1\mathbf{i} + u_2\mathbf{j}$ and $\mathbf{v} = v_1\mathbf{i} + v_2\mathbf{j}$ in $\mathbb{R}^2$, we define their **dot product** $\mathbf{u} \bullet \mathbf{v}$ to be the sum of the products of their corresponding components:

$$\mathbf{u} \bullet \mathbf{v} = u_1v_1 + u_2v_2.$$

The terms **scalar product** and **inner product** are also used in place of dot product. Similarly, for vectors $\mathbf{u} = u_1\mathbf{i} + u_2\mathbf{j} + u_3\mathbf{k}$ and $\mathbf{v} = v_1\mathbf{i} + v_2\mathbf{j} + v_3\mathbf{k}$ in $\mathbb{R}^3$,

$$\mathbf{u} \bullet \mathbf{v} = u_1v_1 + u_2v_2 + u_3v_3.$$

The dot product has the following algebraic properties, easily checked using the definition above:

$$\begin{aligned}
&\mathbf{u} \bullet \mathbf{v} = \mathbf{v} \bullet \mathbf{u} && \text{(commutative law)},\\
&\mathbf{u} \bullet (\mathbf{v} + \mathbf{w}) = \mathbf{u} \bullet \mathbf{v} + \mathbf{u} \bullet \mathbf{w} && \text{(distributive law)},\\
&(t\mathbf{u}) \bullet \mathbf{v} = \mathbf{u} \bullet (t\mathbf{v}) = t(\mathbf{u} \bullet \mathbf{v}) && \text{(for real } t),\\
&\mathbf{u} \bullet \mathbf{u} = |\mathbf{u}|^2.
\end{aligned}$$

The real significance of the dot product is shown by the following result, which could have been used as the definition of dot product:

THEOREM 1

If θ is the angle between the directions of $\mathbf{u}$ and $\mathbf{v}$ $(0 \le \theta \le \pi)$, then

$$\mathbf{u} \bullet \mathbf{v} = |\mathbf{u}||\mathbf{v}|\cos\theta.$$

In particular, $\mathbf{u} \bullet \mathbf{v} = 0$ if and only if $\mathbf{u}$ and $\mathbf{v}$ are perpendicular. (Of course, the zero vector is perpendicular to every vector.)

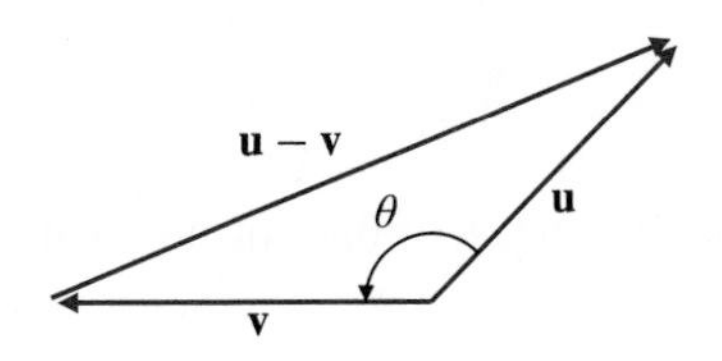

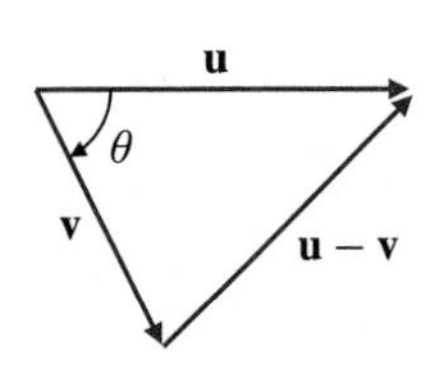

Figure 10.20 Applying the Cosine Law to a triangle reveals the relationship between dot the product and angle between vectors

PROOF Refer to Figure 10.20 and apply the Cosine Law to the triangle with the arrows **u**, **v**, and **u** − **v** as sides:

$$\begin{aligned}|\mathbf{u}|^2 + |\mathbf{v}|^2 - 2|\mathbf{u}|\,|\mathbf{v}|\cos\theta &= |\mathbf{u}-\mathbf{v}|^2 = (\mathbf{u}-\mathbf{v}) \bullet (\mathbf{u}-\mathbf{v}) \\ &= \mathbf{u} \bullet (\mathbf{u}-\mathbf{v}) - \mathbf{v} \bullet (\mathbf{u}-\mathbf{v}) \\ &= \mathbf{u} \bullet \mathbf{u} - \mathbf{u} \bullet \mathbf{v} - \mathbf{v} \bullet \mathbf{u} + \mathbf{v} \bullet \mathbf{v} \\ &= |\mathbf{u}|^2 + |\mathbf{v}|^2 - 2\mathbf{u} \bullet \mathbf{v}\end{aligned}$$

Hence $|\mathbf{u}||\mathbf{v}|\cos\theta = \mathbf{u} \bullet \mathbf{v}$, as claimed.

EXAMPLE 4 Find the angle θ between the vectors $\mathbf{u} = 2\mathbf{i} + \mathbf{j} - 2\mathbf{k}$ and $\mathbf{v} = 3\mathbf{i} - 2\mathbf{j} - \mathbf{k}$.

Solution Solving the formula $\mathbf{u} \bullet \mathbf{v} = |\mathbf{u}||\mathbf{v}|\cos\theta$ for θ, we obtain

$$\begin{aligned}\theta = \cos^{-1}\frac{\mathbf{u} \bullet \mathbf{v}}{|\mathbf{u}||\mathbf{v}|} &= \cos^{-1}\left(\frac{(2)(3) + (1)(-2) + (-2)(-1)}{3\sqrt{14}}\right) \\ &= \cos^{-1}\frac{2}{\sqrt{14}} \approx 57.69^\circ.\end{aligned}$$

It is sometimes useful to project one vector along another. We define both scalar and vector projections of **u** in the direction of **v**:

DEFINITION 4

Scalar and vector projections

The **scalar projection** s of any vector **u** in the direction of a nonzero vector **v** is the dot product of **u** with a unit vector in the direction of **v**. Thus, it is the *number*

$$s = \frac{\mathbf{u} \bullet \mathbf{v}}{|\mathbf{v}|} = |\mathbf{u}|\cos\theta,$$

where θ is the angle between **u** and **v**.

The **vector projection,** $\mathbf{u}_\mathbf{v}$, of **u** in the direction of **v** (see Figure 10.21) is the scalar multiple of a unit vector $\hat{\mathbf{v}}$ in the direction of **v**, by the scalar projection of **u** in the direction of **v**; that is,

$$\text{vector projection of } \mathbf{u} \text{ along } \mathbf{v} = \mathbf{u}_\mathbf{v} = \frac{\mathbf{u} \bullet \mathbf{v}}{|\mathbf{v}|}\hat{\mathbf{v}} = \frac{\mathbf{u} \bullet \mathbf{v}}{|\mathbf{v}|^2}\mathbf{v}.$$

Note that $|s|$ is the length of the line segment along the line of **v** obtained by dropping perpendiculars to that line from the tail and head of **u**. (See Figure 10.21.) Also, s is negative if $\theta > 90^\circ$.

It is often necessary to express a vector as a sum of two other vectors parallel and perpendicular to a given direction.

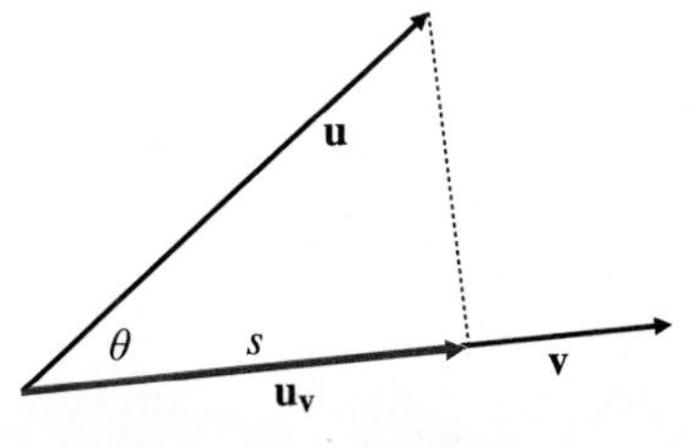

Figure 10.21 The scalar projection s and the vector projection $\mathbf{u}_\mathbf{v}$ of vector **u** along vector **v**

EXAMPLE 5 Express the vector $3\mathbf{i} + \mathbf{j}$ as a sum of vectors $\mathbf{u} + \mathbf{v}$, where **u** is parallel to the vector $\mathbf{i} + \mathbf{j}$ and **v** is perpendicular to **u**.

Solution

METHOD I (Using vector projection) Note that **u** must be the vector projection of $3\mathbf{i} + \mathbf{j}$ in the direction of $\mathbf{i} + \mathbf{j}$. Thus,

$$\begin{aligned}\mathbf{u} &= \frac{(3\mathbf{i} + \mathbf{j}) \bullet (\mathbf{i} + \mathbf{j})}{|\mathbf{i} + \mathbf{j}|^2}(\mathbf{i} + \mathbf{j}) = \frac{4}{2}(\mathbf{i} + \mathbf{j}) = 2\mathbf{i} + 2\mathbf{j} \\ \mathbf{v} &= 3\mathbf{i} + \mathbf{j} - \mathbf{u} = \mathbf{i} - \mathbf{j}.\end{aligned}$$

METHOD II (From basic principles) Since $\mathbf{u}$ is parallel to $\mathbf{i}+\mathbf{j}$ and $\mathbf{v}$ is perpendicular to $\mathbf{u}$, we have

$$\mathbf{u} = t(\mathbf{i}+\mathbf{j}) \qquad \text{and} \qquad \mathbf{v} \bullet (\mathbf{i}+\mathbf{j}) = 0,$$

for some scalar t. We want $\mathbf{u}+\mathbf{v} = 3\mathbf{i}+\mathbf{j}$. Take the dot product of this equation with $\mathbf{i}+\mathbf{j}$:

$$\mathbf{u} \bullet (\mathbf{i}+\mathbf{j}) + \mathbf{v} \bullet (\mathbf{i}+\mathbf{j}) = (3\mathbf{i}+\mathbf{j}) \bullet (\mathbf{i}+\mathbf{j})$$
$$t(\mathbf{i}+\mathbf{j}) \bullet (\mathbf{i}+\mathbf{j}) + 0 = 4.$$

Thus $2t = 4$, so $t = 2$. Therefore,

$$\mathbf{u} = 2\mathbf{i}+2\mathbf{j} \qquad \text{and} \qquad \mathbf{v} = 3\mathbf{i}+\mathbf{j}-\mathbf{u} = \mathbf{i}-\mathbf{j}.$$

Vectors in *n*-Space

All the above ideas make sense for vectors in spaces of any dimension. Vectors in $\mathbb{R}^n$ can be expressed as linear combinations of the n unit vectors

$$\begin{array}{lll} \mathbf{e_1} & \text{from the origin to the point} & (1,0,0,\ldots,0) \\ \mathbf{e_2} & \text{from the origin to the point} & (0,1,0,\ldots,0) \\ & \vdots & \\ \mathbf{e_n} & \text{from the origin to the point} & (0,0,0,\ldots,1). \end{array}$$

These vectors constitute a *standard basis* in $\mathbb{R}^n$. The n-vector $\mathbf{x}$ with components $x_1, x_2, \ldots, x_n$ is expressed in the form

$$\mathbf{x} = x_1\mathbf{e_1} + x_2\mathbf{e_2} + \cdots + x_n\mathbf{e_n}.$$

The length of $\mathbf{x}$ is $|\mathbf{x}| = \sqrt{x_1{}^2 + x_2{}^2 + \cdots + x_n{}^2}$. The angle between two vectors $\mathbf{x}$ and $\mathbf{y}$ is

$$\theta = \cos^{-1}\frac{\mathbf{x}\bullet\mathbf{y}}{|\mathbf{x}||\mathbf{y}|},$$

where

$$\mathbf{x}\bullet\mathbf{y} = x_1y_1 + x_2y_2 + \cdots + x_ny_n.$$

We will not make much use of n-vectors for $n > 3$ but you should be aware that everything said up until now for 2-vectors or 3-vectors extends to n-vectors.

EXERCISES 10.2

1. Let $A = (-1,2)$, $B = (2,0)$, $C = (1,-3)$, $D = (0,4)$. Express each of the following vectors as a linear combination of the standard basis vectors $\mathbf{i}$ and $\mathbf{j}$ in $\mathbb{R}^2$.
(a) $\overrightarrow{AB}$, (b) $\overrightarrow{BA}$, (c) $\overrightarrow{AC}$, (d) $\overrightarrow{BD}$, (e) $\overrightarrow{DA}$,
(f) $\overrightarrow{AB} - \overrightarrow{BC}$, (g) $\overrightarrow{AC} - 2\overrightarrow{AB} + 3\overrightarrow{CD}$, and
(h) $\dfrac{\overrightarrow{AB} + \overrightarrow{AC} + \overrightarrow{AD}}{3}$.

In Exercises 2–3, calculate the following for the given vectors $\mathbf{u}$ and $\mathbf{v}$:

(a) $\mathbf{u}+\mathbf{v}$, $\quad \mathbf{u}-\mathbf{v}$, $\quad 2\mathbf{u}-3\mathbf{v}$,

(b) the lengths $|\mathbf{u}|$ and $|\mathbf{v}|$,

(c) unit vectors $\hat{\mathbf{u}}$ and $\hat{\mathbf{v}}$ in the directions of $\mathbf{u}$ and $\mathbf{v}$, respectively,

(d) the dot product $\mathbf{u} \bullet \mathbf{v}$,

(e) the angle between $\mathbf{u}$ and $\mathbf{v}$,

(f) the scalar projection of $\mathbf{u}$ in the direction of $\mathbf{v}$,

(g) the vector projection of $\mathbf{v}$ along $\mathbf{u}$.

2. $\mathbf{u} = \mathbf{i} - \mathbf{j}$ and $\mathbf{v} = \mathbf{j} + 2\mathbf{k}$

3. $\mathbf{u} = 3\mathbf{i} + 4\mathbf{j} - 5\mathbf{k}$ and $\mathbf{v} = 3\mathbf{i} - 4\mathbf{j} - 5\mathbf{k}$

4. Use vectors to show that the triangle with vertices $(-1, 1)$, $(2, 5)$, and $(10, -1)$ is right-angled.

In Exercises 5–8, prove the stated geometric result using vectors.

5. The line segment joining the midpoints of two sides of a triangle is parallel to and half as long as the third side.

6. If P, Q, R, and S are midpoints of sides AB, BC, CD, and DA, respectively, of quadrilateral $ABCD$, then $PQRS$ is a parallelogram.

7. The diagonals of any parallelogram bisect each other.

8. The medians of any triangle meet in a common point. (A median is a line joining one vertex to the midpoint of the opposite side. The common point is the *centroid* of the triangle.)

9. A weather vane mounted on the top of a car moving due north at 50 km/h indicates that the wind is coming from the west. When the car doubles its speed, the weather vane indicates that the wind is coming from the northwest. From what direction is the wind coming, and what is its speed?

10. A straight river 500 m wide flows due east at a constant speed of 3 km/h. If you can row your boat at a speed of 5 km/h in still water, in what direction should you head if you wish to row from point A on the south shore to point B on the north shore directly north of A? How long will the trip take?

11. In what direction should you head to cross the river in Exercise 10 if you can only row at 2 km/h, and you wish to row from A to point C on the north shore, k km downstream from B? For what values of k is the trip not possible?

12. A certain aircraft flies with an airspeed of 750 km/h. In what direction should it head in order to make progress in a true easterly direction if the wind is from the northeast at 100 km/h? How long will it take to complete a trip to a city 1,500 km from its starting point?

13. For what value of t is the vector $2t\mathbf{i} + 4\mathbf{j} - (10 + t)\mathbf{k}$ perpendicular to the vector $\mathbf{i} + t\mathbf{j} + \mathbf{k}$?

14. Find the angle between a diagonal of a cube and one of the edges of the cube.

15. Find the angle between a diagonal of a cube and a diagonal of one of the faces of the cube. Give all possible answers.

16. (Direction cosines) If a vector $\mathbf{u}$ in $\mathbb{R}^3$ makes angles α, β, and γ with the coordinate axes, show that

$$\hat{\mathbf{u}} = \cos\alpha\,\mathbf{i} + \cos\beta\,\mathbf{j} + \cos\gamma\,\mathbf{k}$$

is a unit vector in the direction of $\mathbf{u}$, so $\cos^2\alpha + \cos^2\beta + \cos^2\gamma = 1$. The numbers $\cos\alpha$, $\cos\beta$, and $\cos\gamma$ are called the *direction cosines* of $\mathbf{u}$.

17. Find a unit vector that makes equal angles with the three coordinate axes.

18. Find the three angles of the triangle with vertices $(1, 0, 0)$, $(0, 2, 0)$, and $(0, 0, 3)$.

19. If $\mathbf{r}_1$ and $\mathbf{r}_2$ are the position vectors of two points, P_1 and P_2, and λ is a real number, show that

$$\mathbf{r} = (1 - \lambda)\mathbf{r}_1 + \lambda\mathbf{r}_2$$

is the position vector of a point P on the straight line joining P_1 and P_2. Where is P if $\lambda = 1/2$? if $\lambda = 2/3$? if $\lambda = -1$? if $\lambda = 2$?

20. Let $\mathbf{a}$ be a nonzero vector. Describe the set of all points in 3-space whose position vectors $\mathbf{r}$ satisfy $\mathbf{a} \bullet \mathbf{r} = 0$.

21. Let $\mathbf{a}$ be a nonzero vector, and let b be any real number. Describe the set of all points in 3-space whose position vectors $\mathbf{r}$ satisfy $\mathbf{a} \bullet \mathbf{r} = b$.

In Exercises 22–24, $\mathbf{u} = 2\mathbf{i} + \mathbf{j} - 2\mathbf{k}$, $\mathbf{v} = \mathbf{i} + 2\mathbf{j} - 2\mathbf{k}$, and $\mathbf{w} = 2\mathbf{i} - 2\mathbf{j} + \mathbf{k}$.

22. Find two unit vectors each of which is perpendicular to both $\mathbf{u}$ and $\mathbf{v}$.

23. Find a vector $\mathbf{x}$ satisfying the system of equations $\mathbf{x} \bullet \mathbf{u} = 9$, $\mathbf{x} \bullet \mathbf{v} = 4$, $\mathbf{x} \bullet \mathbf{w} = 6$.

24. Find two unit vectors each of which makes equal angles with $\mathbf{u}$, $\mathbf{v}$, and $\mathbf{w}$.

25. Find a unit vector that bisects the angle between any two nonzero vectors $\mathbf{u}$ and $\mathbf{v}$.

26. Given two nonparallel vectors $\mathbf{u}$ and $\mathbf{v}$, describe the set of all points whose position vectors $\mathbf{r}$ are of the form $\mathbf{r} = \lambda\mathbf{u} + \mu\mathbf{v}$, where λ and μ are arbitrary real numbers.

27. (The triangle inequality) Let $\mathbf{u}$ and $\mathbf{v}$ be two vectors.

(a) Show that $|\mathbf{u} + \mathbf{v}|^2 = |\mathbf{u}|^2 + 2\mathbf{u} \bullet \mathbf{v} + |\mathbf{v}|^2$.

(b) Show that $\mathbf{u} \bullet \mathbf{v} \le |\mathbf{u}||\mathbf{v}|$.

(c) Deduce from (a) and (b) that $|\mathbf{u} + \mathbf{v}| \le |\mathbf{u}| + |\mathbf{v}|$.

28. (a) Why is the inequality in Exercise 27(c) called a triangle inequality?

(b) What conditions on $\mathbf{u}$ and $\mathbf{v}$ imply that $|\mathbf{u} + \mathbf{v}| = |\mathbf{u}| + |\mathbf{v}|$?

29. (Orthonormal bases) Let $\mathbf{u} = \frac{3}{5}\mathbf{i} + \frac{4}{5}\mathbf{j}$, $\mathbf{v} = \frac{4}{5}\mathbf{i} - \frac{3}{5}\mathbf{j}$, and $\mathbf{w} = \mathbf{k}$.

(a) Show that $|\mathbf{u}| = |\mathbf{v}| = |\mathbf{w}| = 1$ and $\mathbf{u} \bullet \mathbf{v} = \mathbf{u} \bullet \mathbf{w} = \mathbf{v} \bullet \mathbf{w} = 0$. The vectors $\mathbf{u}$, $\mathbf{v}$, and $\mathbf{w}$ are mutually perpendicular unit vectors and as such are said to constitute an **orthonormal basis** for $\mathbb{R}^3$.

(b) If $\mathbf{r} = x\mathbf{i} + y\mathbf{j} + z\mathbf{k}$, show by direct calculation that

$$\mathbf{r} = (\mathbf{r} \bullet \mathbf{u})\mathbf{u} + (\mathbf{r} \bullet \mathbf{v})\mathbf{v} + (\mathbf{r} \bullet \mathbf{w})\mathbf{w}.$$

30. Show that if $\mathbf{u}$, $\mathbf{v}$, and $\mathbf{w}$ are any three mutually perpendicular unit vectors in $\mathbb{R}^3$ and $\mathbf{r} = a\mathbf{u} + b\mathbf{v} + c\mathbf{w}$, then $a = \mathbf{r} \bullet \mathbf{u}$, $b = \mathbf{r} \bullet \mathbf{v}$, and $c = \mathbf{r} \bullet \mathbf{w}$.

31. (Resolving a vector in perpendicular directions) If $\mathbf{a}$ is a nonzero vector and $\mathbf{w}$ is any vector, find vectors $\mathbf{u}$ and $\mathbf{v}$ such that $\mathbf{w} = \mathbf{u} + \mathbf{v}$, $\mathbf{u}$ is parallel to $\mathbf{a}$, and $\mathbf{v}$ is perpendicular to $\mathbf{a}$.

32. (Expressing a vector as a linear combination of two other vectors with which it is coplanar) Suppose that $\mathbf{u}$, $\mathbf{v}$, and $\mathbf{r}$ are position vectors of points U, V, and P, respectively, that $\mathbf{u}$ is not parallel to $\mathbf{v}$, and that P lies in the plane containing the origin, U, and V. Show that there exist numbers λ and μ such that $\mathbf{r} = \lambda\mathbf{u} + \mu\mathbf{v}$. *Hint:* Resolve both $\mathbf{v}$ and $\mathbf{r}$ as sums of vectors parallel and perpendicular to $\mathbf{u}$ as suggested in Exercise 31.

33. Given constants r, s, and t, with $r \neq 0$ and $s \neq 0$, and given a vector $\mathbf{a}$ satisfying $|\mathbf{a}|^2 > 4rst$, solve the system of equations

$$\begin{cases} r\mathbf{x} + s\mathbf{y} = \mathbf{a} \\ \mathbf{x} \bullet \mathbf{y} = t \end{cases}$$

for the unknown vectors $\mathbf{x}$ and $\mathbf{y}$.

Hanging cables

34. (A suspension bridge) If a hanging cable is supporting weight with constant horizontal line density (so that the weight supported by the arc LP in Figure 10.19 is $\delta g x$ rather than $\delta g s$, show that the cable assumes the shape of a parabola rather than a catenary. Such is likely to be the case for the cables of a suspension bridge.

35. At a point P, 10 m away horizontally from its lowest point L, a cable makes an angle 55° with the horizontal. Find the length of the cable between L and P.

36. Calculate the length s of the arc LP of the hanging cable in Figure 10.19 using the equation $y = (1/a)\cosh(ax)$ obtained for the cable. Hence, verify that the magnitude $T = |\mathbf{T}|$ of the tension in the cable at any point $P = (x, y)$ is $T = \delta g y$.

37. A cable 100 m long hangs between two towers 90 m apart so that its ends are attached at the same height on the two towers. How far below that height is the lowest point on the cable?

10.3 The Cross Product in 3-Space

There is defined, *in 3-space only*, another kind of product of two vectors called a *cross product* or *vector product*, and denoted $\mathbf{u}\times\mathbf{v}$.

DEFINITION 5

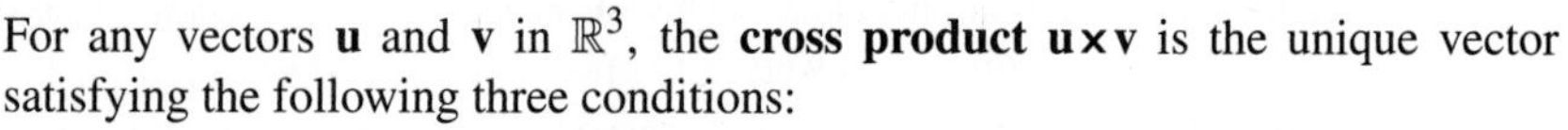

For any vectors $\mathbf{u}$ and $\mathbf{v}$ in $\mathbb{R}^3$, the **cross product** $\mathbf{u}\times\mathbf{v}$ is the unique vector satisfying the following three conditions:

(i) $(\mathbf{u} \times \mathbf{v}) \bullet \mathbf{u} = 0$ and $(\mathbf{u} \times \mathbf{v}) \bullet \mathbf{v} = 0$,

(ii) $|\mathbf{u} \times \mathbf{v}| = |\mathbf{u}||\mathbf{v}| \sin\theta$, where θ is the angle between $\mathbf{u}$ and $\mathbf{v}$, and

(iii) $\mathbf{u}$, $\mathbf{v}$, and $\mathbf{u} \times \mathbf{v}$ form a right-handed triad.

If $\mathbf{u}$ and $\mathbf{v}$ are parallel, condition (ii) says that $\mathbf{u} \times \mathbf{v} = \mathbf{0}$, the zero vector. Otherwise, through any point in $\mathbb{R}^3$ there is a unique straight line that is perpendicular to both $\mathbf{u}$ and $\mathbf{v}$. Condition (i) says that $\mathbf{u} \times \mathbf{v}$ is parallel to this line. Condition (iii) determines which of the two directions along this line is the direction of $\mathbf{u} \times \mathbf{v}$; a right-handed screw advances in the direction of $\mathbf{u} \times \mathbf{v}$ if rotated in the direction from $\mathbf{u}$ toward $\mathbf{v}$. (This is equivalent to saying that the thumb, forefinger, and middle finger of the right hand can be made to point in the directions of $\mathbf{u}$, $\mathbf{v}$, and $\mathbf{u}\times\mathbf{v}$, respectively.)

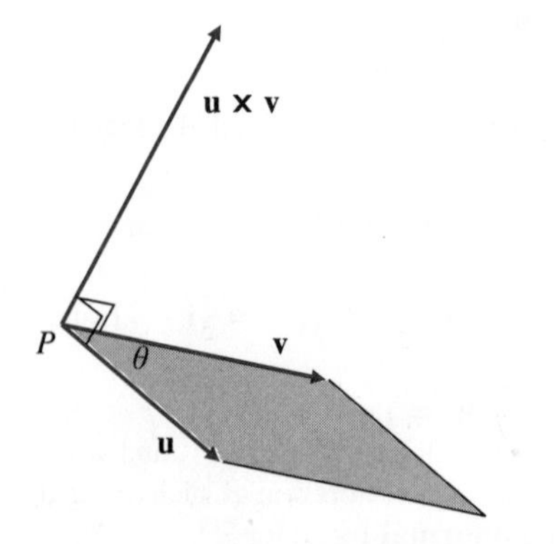

Figure 10.22 $\mathbf{u} \times \mathbf{v}$ is perpendicular to both $\mathbf{u}$ and $\mathbf{v}$ and has length equal to the area of the shaded parallelogram

If $\mathbf{u}$ and $\mathbf{v}$ have their tails at the point P, then $\mathbf{u}\times\mathbf{v}$ is normal (i.e., perpendicular) to the plane through P in which $\mathbf{u}$ and $\mathbf{v}$ lie and, by condition (ii), $\mathbf{u}\times\mathbf{v}$ has length equal to the area of the parallelogram spanned by $\mathbf{u}$ and $\mathbf{v}$. (See Figure 10.22.) These properties make the cross product very useful for the description of tangent planes and normal lines to surfaces in $\mathbb{R}^3$.

The definition of cross product given above does not involve any coordinate system and therefore does not directly show the components of the cross product with respect to the standard basis. These components are provided by the following theorem:

THEOREM 2

Components of the cross product

If $\mathbf{u} = u_1\mathbf{i} + u_2\mathbf{j} + u_3\mathbf{k}$ and $\mathbf{v} = v_1\mathbf{i} + v_2\mathbf{j} + v_3\mathbf{k}$, then

$$\mathbf{u} \times \mathbf{v} = (u_2v_3 - u_3v_2)\mathbf{i} + (u_3v_1 - u_1v_3)\mathbf{j} + (u_1v_2 - u_2v_1)\mathbf{k}.$$

PROOF First, we observe that the vector

$$\mathbf{w} = (u_2v_3 - u_3v_2)\mathbf{i} + (u_3v_1 - u_1v_3)\mathbf{j} + (u_1v_2 - u_2v_1)\mathbf{k}$$

is perpendicular to both **u** and **v** since

$$\mathbf{u} \bullet \mathbf{w} = u_1(u_2v_3 - u_3v_2) + u_2(u_3v_1 - u_1v_3) + u_3(u_1v_2 - u_2v_1) = 0,$$

and similarly $\mathbf{v} \bullet \mathbf{w} = 0$. Thus, $\mathbf{u} \times \mathbf{v}$ is parallel to **w**. Next, we show that **w** and $\mathbf{u} \times \mathbf{v}$ have the same length. In fact,

$$\begin{aligned}|\mathbf{w}|^2 &= (u_2v_3 - u_3v_2)^2 + (u_3v_1 - u_1v_3)^2 + (u_1v_2 - u_2v_1)^2 \\ &= u_2^2v_3^2 + u_3^2v_2^2 - 2u_2v_3u_3v_2 + u_3^2v_1^2 + u_1^2v_3^2 \\ &\quad - 2u_3v_1u_1v_3 + u_1^2v_2^2 + u_2^2v_1^2 - 2u_1v_2u_2v_1,\end{aligned}$$

while

$$\begin{aligned}|\mathbf{u} \times \mathbf{v}|^2 &= |\mathbf{u}|^2|\mathbf{v}|^2 \sin^2\theta \\ &= |\mathbf{u}|^2|\mathbf{v}|^2 (1 - \cos^2\theta) \\ &= |\mathbf{u}|^2|\mathbf{v}|^2 - (\mathbf{u} \bullet \mathbf{v})^2 \\ &= (u_1^2 + u_2^2 + u_3^2)(v_1^2 + v_2^2 + v_3^2) - (u_1v_1 + u_2v_2 + u_3v_3)^2 \\ &= u_1^2v_1^2 + u_1^2v_2^2 + u_1^2v_3^2 + u_2^2v_1^2 + u_2^2v_2^2 + u_2^2v_3^2 + u_3^2v_1^2 + u_3^2v_2^2 + u_3^2v_3^2 \\ &\quad - u_1^2v_1^2 - u_2^2v_2^2 - u_3^2v_3^2 - 2u_1v_1u_2v_2 - 2u_1v_1u_3v_3 - 2u_2v_2u_3v_3 \\ &= |\mathbf{w}|^2.\end{aligned}$$

Since **w** is parallel to, and has the same length as, $\mathbf{u} \times \mathbf{v}$, we must have either $\mathbf{u} \times \mathbf{v} = \mathbf{w}$ or $\mathbf{u} \times \mathbf{v} = -\mathbf{w}$. It remains to be shown that the first of these is the correct choice. To see this, suppose that the triad of vectors **u**, **v**, and **w** is rigidly rotated in 3-space so that **u** points in the direction of the positive x-axis and **v** lies in the upper half of the xy-plane. Then $\mathbf{u} = u_1\mathbf{i}$, and $\mathbf{v} = v_1\mathbf{i} + v_2\mathbf{j}$, where $u_1 > 0$ and $v_2 > 0$. By the "right-hand rule" $\mathbf{u} \times \mathbf{v}$ must point in the direction of the positive z-axis. But $\mathbf{w} = u_1v_2\mathbf{k}$ does point in that direction, so $\mathbf{u} \times \mathbf{v} = \mathbf{w}$, as asserted.

The formula for the cross product in terms of components may seem awkward and asymmetric. As we shall see, however, it can be written more easily in terms of a determinant. We introduce determinants later in this section.

EXAMPLE 1 **(Calculating cross products)**

(a)

$$\begin{aligned}&\mathbf{i} \times \mathbf{i} = \mathbf{0}, && \mathbf{i} \times \mathbf{j} = \mathbf{k}, && \mathbf{j} \times \mathbf{i} = -\mathbf{k}, \\ &\mathbf{j} \times \mathbf{j} = \mathbf{0}, && \mathbf{j} \times \mathbf{k} = \mathbf{i}, && \mathbf{k} \times \mathbf{j} = -\mathbf{i}, \\ &\mathbf{k} \times \mathbf{k} = \mathbf{0}, && \mathbf{k} \times \mathbf{i} = \mathbf{j}, && \mathbf{i} \times \mathbf{k} = -\mathbf{j}.\end{aligned}$$

(b)

$$\begin{aligned}&(2\mathbf{i} + \mathbf{j} - 3\mathbf{k}) \times (-2\mathbf{j} + 5\mathbf{k}) \\ &\quad = \big((1)(5) - (-2)(-3)\big)\mathbf{i} + \big((-3)(0) - (2)(5)\big)\mathbf{j} + \big((2)(-2) - (1)(0)\big)\mathbf{k} \\ &\quad = -\mathbf{i} - 10\mathbf{j} - 4\mathbf{k}.\end{aligned}$$

The cross product has some but not all of the properties we usually ascribe to products. We summarize its algebraic properties as follows:

Properties of the cross product

If **u**, **v**, and **w** are any vectors in $\mathbb{R}^3$, and t is a real number (a scalar), then

(i) $\mathbf{u} \times \mathbf{u} = \mathbf{0}$,

(ii) $\mathbf{u} \times \mathbf{v} = -\mathbf{v} \times \mathbf{u}$, (The cross product is **anticommutative**.)

(iii) $(\mathbf{u} + \mathbf{v}) \times \mathbf{w} = \mathbf{u} \times \mathbf{w} + \mathbf{v} \times \mathbf{w}$,

(iv) $\mathbf{u} \times (\mathbf{v} + \mathbf{w}) = \mathbf{u} \times \mathbf{v} + \mathbf{u} \times \mathbf{w}$,

(v) $(t\mathbf{u}) \times \mathbf{v} = \mathbf{u} \times (t\mathbf{v}) = t(\mathbf{u} \times \mathbf{v})$,

(vi) $\mathbf{u} \bullet (\mathbf{u} \times \mathbf{v}) = \mathbf{v} \bullet (\mathbf{u} \times \mathbf{v}) = 0$.

These identities are all easily verified using the components or the definition of the cross product or by using properties of determinants discussed below. They are left as exercises for the reader. Note the absence of an associative law. The cross product is not associative. (See Exercise 21 at the end of this section.) In general,

$$\mathbf{u} \times (\mathbf{v} \times \mathbf{w}) \neq (\mathbf{u} \times \mathbf{v}) \times \mathbf{w}.$$

Determinants

In order to simplify certain formulas such as the component representation of the cross product, we introduce 2×2 and 3×3 **determinants**. General $n \times n$ determinants are normally studied in courses on linear algebra; we will encounter them in Section 10.7. In this section we will outline enough of the properties of determinants to enable us to use them as shorthand in some otherwise complicated formulas.

A determinant is an expression that involves the elements of a square array (matrix) of numbers. The determinant of the 2×2 array of numbers

$$\begin{matrix} a & b \\ c & d \end{matrix}$$

is denoted by enclosing the array between vertical bars, and its value is the number $ad - bc$:

$$\begin{vmatrix} a & b \\ c & d \end{vmatrix} = ad - bc.$$

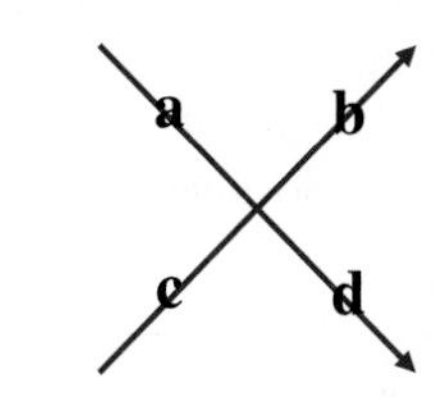

Figure 10.23 Upward and downward diagonals

This is the product of elements in the *downward diagonal* of the array minus the product of elements in the *upward diagonal*, as shown in Figure 10.23. For example,

$$\begin{vmatrix} 1 & 2 \\ 3 & 4 \end{vmatrix} = (1)(4) - (2)(3) = -2.$$

Similarly, the determinant of a 3×3 array of numbers is defined by

$$\begin{vmatrix} a & b & c \\ d & e & f \\ g & h & i \end{vmatrix} = aei + bfg + cdh - gec - hfa - idb.$$

Observe that each of the six products in the value of the determinant involves exactly one element from each row and exactly one from each column of the array. As such, each term is the product of elements in a *diagonal* of an *extended* array obtained by repeating the first two columns of the array to the right of the third column, as shown in Figure 10.24. The value of the determinant is the sum of products corresponding to the three complete *downward* diagonals minus the sum corresponding to the three *upward* diagonals. With practice you will be able to form these diagonal products without having to write the extended array.

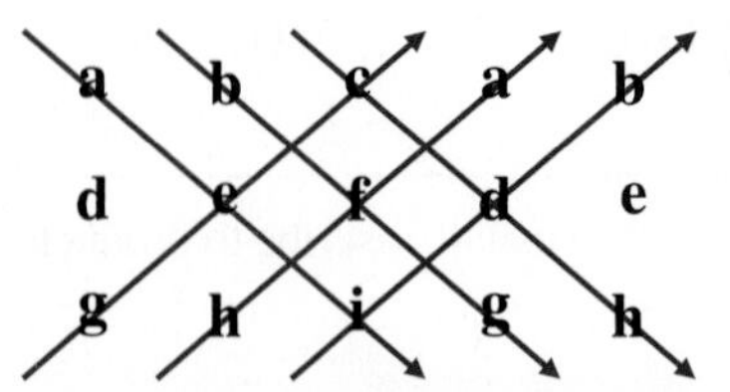

Figure 10.24 WARNING: This method does not work for 4×4 or higher-order determinants!

If we group the terms in the expansion of the determinant to factor out the elements of the first row, we obtain

$$\begin{aligned} \begin{vmatrix} a & b & c \\ d & e & f \\ g & h & i \end{vmatrix} &= a(ei - fh) - b(di - fg) + c(dh - eg) \\ &= a\begin{vmatrix} e & f \\ h & i \end{vmatrix} - b\begin{vmatrix} d & f \\ g & i \end{vmatrix} + c\begin{vmatrix} d & e \\ g & h \end{vmatrix}. \end{aligned}$$

The 2×2 determinants appearing here (called *minors* of the given 3×3 determinant) are obtained by deleting the row and column containing the corresponding element from the original 3×3 determinant. This process is called *expanding* the 3×3 determinant *in minors* about the first row.

Such expansions in minors can be carried out about any row or column. Note that if $i + j$ is an *odd* number, a minus sign appears in a term obtained by multiplying the element in the ith row and jth column and its corresponding minor obtained by deleting that row and column. For example, we can expand the above determinant in minors about the second column as follows:

The pattern of + and − signs used with the terms of an expansion in minors of a 3×3 determinant is given by

$$\begin{vmatrix} + & - & + \\ - & + & - \\ + & - & + \end{vmatrix}$$

$$\begin{aligned} \begin{vmatrix} a & b & c \\ d & e & f \\ g & h & i \end{vmatrix} &= -b\begin{vmatrix} d & f \\ g & i \end{vmatrix} + e\begin{vmatrix} a & c \\ g & i \end{vmatrix} - h\begin{vmatrix} a & c \\ d & f \end{vmatrix} \\ &= -bdi + bfg + eai - ecg - haf + hcd. \end{aligned}$$

(Of course, this is the same value as the one obtained previously.)

EXAMPLE 2

$$\begin{aligned} \begin{vmatrix} 1 & 4 & -2 \\ -3 & 1 & 0 \\ 2 & 2 & -3 \end{vmatrix} &= 3\begin{vmatrix} 4 & -2 \\ 2 & -3 \end{vmatrix} + 1\begin{vmatrix} 1 & -2 \\ 2 & -3 \end{vmatrix} \\ &= 3(-8) + 1 = -23. \end{aligned}$$

We expanded about the second row; the third column would also have been a good choice. (Why?)

Any row (or column) of a determinant may be regarded as the components of a vector. Then the determinant is a *linear function* of that vector. For example,

$$\begin{vmatrix} a & b & c \\ d & e & f \\ sx + tl & sy + tm & sz + tn \end{vmatrix} = s\begin{vmatrix} a & b & c \\ d & e & f \\ x & y & z \end{vmatrix} + t\begin{vmatrix} a & b & c \\ d & e & f \\ l & m & n \end{vmatrix}$$

because the determinant is a linear function of its third row. This and other properties of determinants follow directly from the definition. Some other properties are summarized below. These are stated for rows and for 3×3 determinants, but similar statements can be made for columns and for determinants of any order.

Properties of determinants

(i) If two rows of a determinant are interchanged, then the determinant changes sign:

$$\begin{vmatrix} d & e & f \\ a & b & c \\ g & h & i \end{vmatrix} = -\begin{vmatrix} a & b & c \\ d & e & f \\ g & h & i \end{vmatrix}.$$

(ii) If two rows of a determinant are equal, the determinant has value 0:

$$\begin{vmatrix} a & b & c \\ a & b & c \\ g & h & i \end{vmatrix} = 0.$$

(iii) If a multiple of one row of a determinant is added to another row, the value of the determinant remains unchanged:

$$\begin{vmatrix} a & b & c \\ d + ta & e + tb & f + tc \\ g & h & i \end{vmatrix} = \begin{vmatrix} a & b & c \\ d & e & f \\ g & h & i \end{vmatrix}.$$

The Cross Product as a Determinant

The elements of a determinant are usually numbers because they have to be multiplied to get the value of the determinant. However, it is possible to use vectors as the elements of *one row* (or column) of a determinant. When expanding in minors about that row (or column), the minor for each vector element is a number that determines the scalar multiple of the vector. The formula for the cross product of

$$\mathbf{u} = u_1\mathbf{i} + u_2\mathbf{j} + u_3\mathbf{k} \quad \text{and} \quad \mathbf{v} = v_1\mathbf{i} + v_2\mathbf{j} + v_3\mathbf{k}$$

presented in Theorem 2 can be expressed symbolically as a determinant with the standard basis vectors as the elements of the first row:

$$\mathbf{u} \times \mathbf{v} = \begin{vmatrix} \mathbf{i} & \mathbf{j} & \mathbf{k} \\ u_1 & u_2 & u_3 \\ v_1 & v_2 & v_3 \end{vmatrix} = \begin{vmatrix} u_2 & u_3 \\ v_2 & v_3 \end{vmatrix}\mathbf{i} - \begin{vmatrix} u_1 & u_3 \\ v_1 & v_3 \end{vmatrix}\mathbf{j} + \begin{vmatrix} u_1 & u_2 \\ v_1 & v_2 \end{vmatrix}\mathbf{k}.$$

The formula for the cross product given in that theorem is just the expansion of this determinant in minors about the first row.

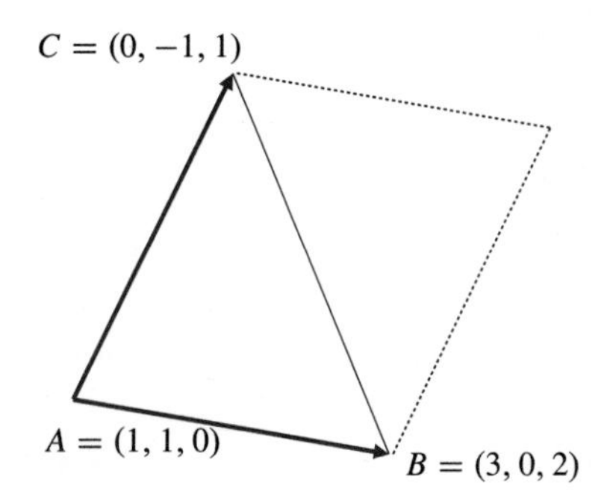

Figure 10.25

EXAMPLE 3 Find the area of the triangle with vertices at the three points $A = (1, 1, 0)$, $B = (3, 0, 2)$, and $C = (0, -1, 1)$.

Solution Two sides of the triangle (Figure 10.25) are given by the vectors:

$$\overrightarrow{AB} = 2\mathbf{i} - \mathbf{j} + 2\mathbf{k} \quad \text{and} \quad \overrightarrow{AC} = -\mathbf{i} - 2\mathbf{j} + \mathbf{k}.$$

The area of the triangle is half the area of the parallelogram spanned by $\overrightarrow{AB}$ and $\overrightarrow{AC}$. By the definition of cross product, the area of the triangle must therefore be

$$\frac{1}{2}|\overrightarrow{AB} \times \overrightarrow{AC}| = \frac{1}{2}\left|\begin{vmatrix} \mathbf{i} & \mathbf{j} & \mathbf{k} \\ 2 & -1 & 2 \\ -1 & -2 & 1 \end{vmatrix}\right|$$

$$= \frac{1}{2}|3\mathbf{i} - 4\mathbf{j} - 5\mathbf{k}| = \frac{1}{2}\sqrt{9 + 16 + 25} = \frac{5}{2}\sqrt{2} \text{ square units.}$$

A **parallelepiped** is the three-dimensional analogue of a parallelogram. It is a solid with three pairs of parallel planar faces. Each face is in the shape of a parallelogram. A rectangular brick is a special case of a parallelepiped in which nonparallel faces intersect at right angles. We say that a parallelepiped is **spanned** by three vectors coinciding with three of its edges that meet at one vertex. (See Figure 10.26.)

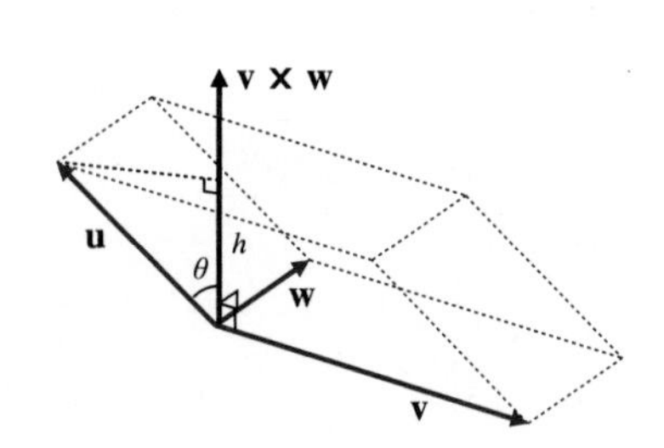

Figure 10.26

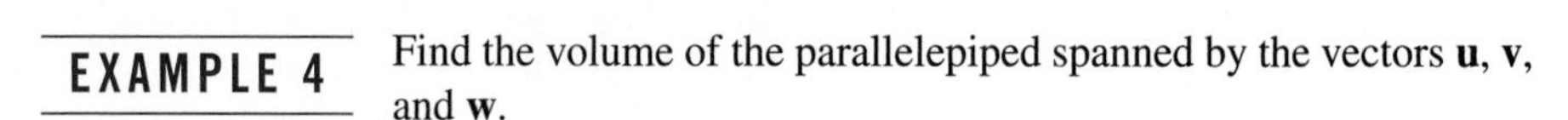

EXAMPLE 4 Find the volume of the parallelepiped spanned by the vectors **u**, **v**, and **w**.

Solution The volume of the parallelepiped is equal to the area of one of its faces, say, the face spanned by **v** and **w**, multiplied by the height of the parallelepiped measured in a direction perpendicular to that face. The area of the face is $|\mathbf{v} \times \mathbf{w}|$. Since $\mathbf{v} \times \mathbf{w}$ is perpendicular to the face, the height h of the parallelepiped will be the absolute value of the scalar projection of **u** along $\mathbf{v} \times \mathbf{w}$. If θ is the angle between **u** and $\mathbf{v} \times \mathbf{w}$, then the volume of the parallelepiped is given by

$$\text{Volume} = |\mathbf{u}||\mathbf{v} \times \mathbf{w}||\cos\theta| = |\mathbf{u} \bullet (\mathbf{v} \times \mathbf{w})| \text{ cubic units.}$$

DEFINITION 6

The quantity $\mathbf{u} \bullet (\mathbf{v} \times \mathbf{w})$ is called the **scalar triple product** of the vectors **u**, **v**, and **w**.

The scalar triple product is easily expressed in terms of a determinant. If $\mathbf{u} = u_1\mathbf{i} + u_2\mathbf{j} + u_3\mathbf{k}$, and similar representations hold for $\mathbf{v}$ and $\mathbf{w}$, then

$$\begin{aligned}\mathbf{u} \bullet (\mathbf{v} \times \mathbf{w}) &= u_1 \begin{vmatrix} v_2 & v_3 \\ w_2 & w_3 \end{vmatrix} - u_2 \begin{vmatrix} v_1 & v_3 \\ w_1 & w_3 \end{vmatrix} + u_3 \begin{vmatrix} v_1 & v_2 \\ w_1 & w_2 \end{vmatrix} \\ &= \begin{vmatrix} u_1 & u_2 & u_3 \\ v_1 & v_2 & v_3 \\ w_1 & w_2 & w_3 \end{vmatrix}.\end{aligned}$$

The volume of the parallelepiped spanned by $\mathbf{u}$, $\mathbf{v}$, and $\mathbf{w}$ is the absolute value of this determinant.

Using the properties of the determinant, it is easily verified that

$$\mathbf{u} \bullet (\mathbf{v} \times \mathbf{w}) = \mathbf{v} \bullet (\mathbf{w} \times \mathbf{u}) = \mathbf{w} \bullet (\mathbf{u} \times \mathbf{v}).$$

(See Exercise 18 below.) Note that $\mathbf{u}$, $\mathbf{v}$, and $\mathbf{w}$ remain in the same *cyclic order* in these three expressions. Reversing the order would introduce a factor -1:

$$\mathbf{u} \bullet (\mathbf{v} \times \mathbf{w}) = -\mathbf{u} \bullet (\mathbf{w} \times \mathbf{v}).$$

Three vectors in 3-space are said to be **coplanar** if the parallelepiped they span has zero volume; if their tails coincide, three such vectors must lie in the same plane.

$$\begin{aligned}\mathbf{u}, \mathbf{v}, \text{ and } \mathbf{w} \text{ are coplanar} \quad &\iff \quad \mathbf{u} \bullet (\mathbf{v} \times \mathbf{w}) = 0 \\ &\iff \quad \begin{vmatrix} u_1 & u_2 & u_3 \\ v_1 & v_2 & v_3 \\ w_1 & w_2 & w_3 \end{vmatrix} = 0.\end{aligned}$$

Three vectors are certainly coplanar if any of them is $\mathbf{0}$, or if any pair of them is parallel. If neither of these degenerate conditions apply, they are only coplanar if any one of them can be expressed as a linear combination of the other two. (See Exercise 20 below.)

Applications of Cross Products

Cross products are of considerable importance in mechanics and electromagnetic theory, as well as in the study of motion in general. For example:

(a) The linear velocity $\mathbf{v}$ of a particle located at position $\mathbf{r}$ in a body rotating with angular velocity $\mathbf{\Omega}$ about the origin is given by $\mathbf{v} = \mathbf{\Omega} \times \mathbf{r}$. (See Section 11.2 for more details.)

(b) The angular momentum of a planet of mass m moving with velocity $\mathbf{v}$ in its orbit around the sun is given by $\mathbf{h} = \mathbf{r} \times m\mathbf{v}$, where $\mathbf{r}$ is the position vector of the planet relative to the sun as origin. (See Section 11.6.)

(c) If a particle of electric charge q is travelling with velocity $\mathbf{v}$ through a magnetic field whose strength and direction are given by vector $\mathbf{B}$, then the force that the field exerts on the particle is given by $\mathbf{F} = q\mathbf{v} \times \mathbf{B}$. The electron beam in a television tube is controlled by magnetic fields using this principle.

(d) The torque $\mathbf{T}$ of a force $\mathbf{F}$ applied at the point P with position vector $\mathbf{r}$ about another point P_0 with position vector $\mathbf{r}_0$ is defined to be

$$\mathbf{T} = \overrightarrow{P_0P} \times \mathbf{F} = (\mathbf{r} - \mathbf{r}_0) \times \mathbf{F}.$$

This torque measures the effectiveness of the force $\mathbf{F}$ in causing rotation about P_0. The direction of $\mathbf{T}$ is along the axis through P_0 about which $\mathbf{F}$ acts to rotate P.

EXAMPLE 5 An automobile wheel has centre at the origin and axle along the y-axis. One of the retaining nuts holding the wheel is at position $P_0 = (0, 0, 10)$. (Distances are measured in centimetres.) A bent tire wrench with arm 25 cm long and inclined at an angle of 60° to the direction of its handle is fitted to the nut in an upright direction, as shown in Figure 10.27. If a horizontal force $\mathbf{F} = 500\mathbf{i}$ newtons (N) is applied to the handle of the wrench, what is its torque on the nut? What part (component) of this torque is effective in trying to rotate the nut about its horizontal axis? What is the effective torque trying to rotate the wheel?

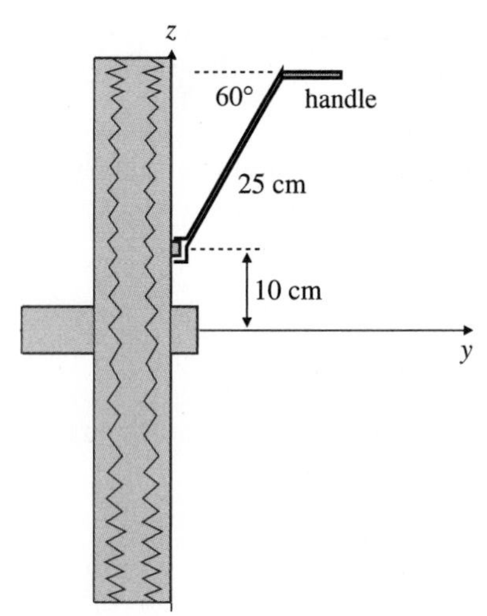

Figure 10.27 The force on the handle is 500 N in a direction directly toward you

Solution The nut is at position $\mathbf{r}_0 = 10\mathbf{k}$, and the handle of the wrench is at position

$$\mathbf{r} = 25\cos 60°\mathbf{j} + (10 + 25\sin 60°)\mathbf{k} \approx 12.5\mathbf{j} + 31.65\mathbf{k}.$$

The torque of the force $\mathbf{F}$ on the nut is

$$\begin{aligned}\mathbf{T} &= (\mathbf{r} - \mathbf{r}_0) \times \mathbf{F} \\ &\approx (12.5\mathbf{j} + 21.65\mathbf{k}) \times 500\mathbf{i} \approx 10{,}825\mathbf{j} - 6{,}250\mathbf{k},\end{aligned}$$

which is at right angles to $\mathbf{F}$ and to the arm of the wrench. Only the horizontal component of this torque is effective in turning the nut. This component is 10,825 N·cm or 108.25 N·m in magnitude. For the effective torque on the wheel itself, we have to replace $\mathbf{r}_0$ by $\mathbf{0}$, the position of the centre of the wheel. In this case the horizontal torque is

$$31.65\mathbf{k} \times 500\mathbf{i} \approx 15{,}825\mathbf{j},$$

that is, about 158.25 N·m.

EXERCISES 10.3

1. Calculate $\mathbf{u} \times \mathbf{v}$ if $\mathbf{u} = \mathbf{i} - 2\mathbf{j} + 3\mathbf{k}$ and $\mathbf{v} = 3\mathbf{i} + \mathbf{j} - 4\mathbf{k}$.
2. Calculate $\mathbf{u} \times \mathbf{v}$ if $\mathbf{u} = \mathbf{j} + 2\mathbf{k}$ and $\mathbf{v} = -\mathbf{i} - \mathbf{j} + \mathbf{k}$.
3. Find the area of the triangle with vertices $(1, 2, 0)$, $(1, 0, 2)$, and $(0, 3, 1)$.
4. Find a unit vector perpendicular to the plane containing the points $(a, 0, 0)$, $(0, b, 0)$, and $(0, 0, c)$. What is the area of the triangle with these vertices?
5. Find a unit vector perpendicular to the vectors $\mathbf{i} + \mathbf{j}$ and $\mathbf{j} + 2\mathbf{k}$.
6. Find a unit vector with positive $\mathbf{k}$ component that is perpendicular to both $2\mathbf{i} - \mathbf{j} - 2\mathbf{k}$ and $2\mathbf{i} - 3\mathbf{j} + \mathbf{k}$.

Verify the identities in Exercises 7–11, either by using the definition of cross product or the properties of determinants.

7. $\mathbf{u} \times \mathbf{u} = \mathbf{0}$
8. $\mathbf{u} \times \mathbf{v} = -\mathbf{v} \times \mathbf{u}$
9. $(\mathbf{u} + \mathbf{v}) \times \mathbf{w} = \mathbf{u} \times \mathbf{w} + \mathbf{v} \times \mathbf{w}$
10. $(t\mathbf{u}) \times \mathbf{v} = \mathbf{u} \times (t\mathbf{v}) = t(\mathbf{u} \times \mathbf{v})$
11. $\mathbf{u} \bullet (\mathbf{u} \times \mathbf{v}) = \mathbf{v} \bullet (\mathbf{u} \times \mathbf{v}) = 0$
12. Obtain the addition formula

$$\sin(\alpha - \beta) = \sin\alpha\cos\beta - \cos\alpha\sin\beta$$

by examining the cross product of the two unit vectors $\mathbf{u} = \cos\beta\mathbf{i} + \sin\beta\mathbf{j}$ and $\mathbf{v} = \cos\alpha\mathbf{i} + \sin\alpha\mathbf{j}$. Assume $0 \le \alpha - \beta \le \pi$. *Hint:* Regard $\mathbf{u}$ and $\mathbf{v}$ as position vectors. What is the area of the parallelogram they span?

13. If $\mathbf{u} + \mathbf{v} + \mathbf{w} = \mathbf{0}$, show that $\mathbf{u} \times \mathbf{v} = \mathbf{v} \times \mathbf{w} = \mathbf{w} \times \mathbf{u}$.
14. **(Volume of a tetrahedron)** A **tetrahedron** is a pyramid with a triangular base and three other triangular faces. It has four vertices and six edges. Like any pyramid or cone, its volume is equal to $\frac{1}{3}Ah$, where A is the area of the base and h is the height measured perpendicular to the base. If $\mathbf{u}$, $\mathbf{v}$, and $\mathbf{w}$ are vectors coinciding with the three edges of a tetrahedron that meet at one vertex, show that the tetrahedron has volume given by

$$\text{Volume} = \frac{1}{6}|\mathbf{u} \bullet (\mathbf{v} \times \mathbf{w})| = \frac{1}{6}\left|\begin{vmatrix} u_1 & u_2 & u_3 \\ v_1 & v_2 & v_3 \\ w_1 & w_2 & w_3 \end{vmatrix}\right|.$$

Thus, the volume of a tetrahedron spanned by three vectors is one-sixth of the volume of the parallelepiped spanned by the same vectors.

15. Find the volume of the tetrahedron with vertices $(1, 0, 0)$, $(1, 2, 0)$, $(2, 2, 2)$, and $(0, 3, 2)$.
16. Find the volume of the parallelepiped spanned by the diagonals of the three faces of a cube of side a that meet at one vertex of the cube.

17. For what value of k do the four points $(1, 1, -1)$, $(0, 3, -2)$, $(-2, 1, 0)$, and $(k, 0, 2)$ all lie in a plane?

18. (The scalar triple product) Verify the identities

$$\mathbf{u} \bullet (\mathbf{v} \times \mathbf{w}) = \mathbf{v} \bullet (\mathbf{w} \times \mathbf{u}) = \mathbf{w} \bullet (\mathbf{u} \times \mathbf{v}).$$

19. If $\mathbf{u} \bullet (\mathbf{v} \times \mathbf{w}) \neq 0$ and $\mathbf{x}$ is an arbitrary 3-vector, find the numbers λ, μ, and ν such that

$$\mathbf{x} = \lambda\mathbf{u} + \mu\mathbf{v} + \nu\mathbf{w}.$$

20. If $\mathbf{u} \bullet (\mathbf{v} \times \mathbf{w}) = 0$ but $\mathbf{v} \times \mathbf{w} \neq \mathbf{0}$, show that there are constants λ and μ such that

$$\mathbf{u} = \lambda\mathbf{v} + \mu\mathbf{w}.$$

Hint: Use the result of Exercise 19 with $\mathbf{u}$ in place of $\mathbf{x}$ and $\mathbf{v} \times \mathbf{w}$ in place of $\mathbf{u}$.

21. Calculate $\mathbf{u} \times (\mathbf{v} \times \mathbf{w})$ and $(\mathbf{u} \times \mathbf{v}) \times \mathbf{w}$, given that $\mathbf{u} = \mathbf{i} + 2\mathbf{j} + 3\mathbf{k}$, $\mathbf{v} = 2\mathbf{i} - 3\mathbf{j}$, and $\mathbf{w} = \mathbf{j} - \mathbf{k}$. Why would you not expect these to be equal?

22. Does the notation $\mathbf{u} \bullet \mathbf{v} \times \mathbf{w}$ make sense? Why? How about the notation $\mathbf{u} \times \mathbf{v} \times \mathbf{w}$?

23. (The vector triple product) The product $\mathbf{u} \times (\mathbf{v} \times \mathbf{w})$ is called a **vector triple product**. Since it is perpendicular to $\mathbf{v} \times \mathbf{w}$, it must lie in the plane of $\mathbf{v}$ and $\mathbf{w}$. Show that

$$\mathbf{u} \times (\mathbf{v} \times \mathbf{w}) = (\mathbf{u} \bullet \mathbf{w})\mathbf{v} - (\mathbf{u} \bullet \mathbf{v})\mathbf{w}.$$

Hint: This can be done by direct calculation of the components of both sides of the equation, but the job is much easier if you choose coordinate axes so that $\mathbf{v}$ lies along the x-axis and $\mathbf{w}$ lies in the xy-plane.

24. If $\mathbf{u}$, $\mathbf{v}$, and $\mathbf{w}$ are mutually perpendicular vectors, show that $\mathbf{u} \times (\mathbf{v} \times \mathbf{w}) = \mathbf{0}$. What is $\mathbf{u} \bullet (\mathbf{v} \times \mathbf{w})$ in this case?

25. Show that $\mathbf{u} \times (\mathbf{v} \times \mathbf{w}) + \mathbf{v} \times (\mathbf{w} \times \mathbf{u}) + \mathbf{w} \times (\mathbf{u} \times \mathbf{v}) = \mathbf{0}$.

26. Find all vectors $\mathbf{x}$ that satisfy the equation

$$(-\mathbf{i} + 2\mathbf{j} + 3\mathbf{k}) \times \mathbf{x} = \mathbf{i} + 5\mathbf{j} - 3\mathbf{k}.$$

27. Show that the equation

$$(-\mathbf{i} + 2\mathbf{j} + 3\mathbf{k}) \times \mathbf{x} = \mathbf{i} + 5\mathbf{j}$$

has no solutions for the unknown vector $\mathbf{x}$.

28. What condition must be satisfied by the nonzero vectors $\mathbf{a}$ and $\mathbf{b}$ to guarantee that the equation $\mathbf{a} \times \mathbf{x} = \mathbf{b}$ has a solution for $\mathbf{x}$? Is the solution unique?

10.4 Planes and Lines

A single equation in the three variables, x, y, and z, constitutes a single constraint on the freedom of the point $P = (x, y, z)$ to lie anywhere in 3-space. Such a constraint usually results in the loss of exactly one *degree of freedom* and so forces P to lie on a two-dimensional surface. For example, the equation

$$x^2 + y^2 + z^2 = 4$$

states that the point (x, y, z) is at distance 2 from the origin. All points satisfying this condition lie on a **sphere** (i.e., the surface of a ball) of radius 2 centred at the origin. The equation above therefore represents that sphere, and the sphere is the graph of the equation. In this section we will investigate the graphs of linear equations in three variables.

Planes in 3-Space

Let $P_0 = (x_0, y_0, z_0)$ be a point in $\mathbb{R}^3$ with position vector

$$\mathbf{r}_0 = x_0\mathbf{i} + y_0\mathbf{j} + z_0\mathbf{k}.$$

If $\mathbf{n} = A\mathbf{i} + B\mathbf{j} + C\mathbf{k}$ is any given *nonzero* vector, then there exists exactly one **plane** (flat surface) passing through P_0 and perpendicular to $\mathbf{n}$. We say that $\mathbf{n}$ is a **normal vector** to the plane. The plane is the set of all points P for which $\overrightarrow{P_0P}$ is perpendicular to $\mathbf{n}$. (See Figure 10.28.)

If $P = (x, y, z)$ has position vector $\mathbf{r}$, then $\overrightarrow{P_0P} = \mathbf{r} - \mathbf{r}_0$. This vector is perpendicular to $\mathbf{n}$ if and only if $\mathbf{n} \bullet (\mathbf{r} - \mathbf{r}_0) = 0$. This is the equation of the plane in vector form. We can rewrite it in terms of coordinates to obtain the corresponding scalar equation.

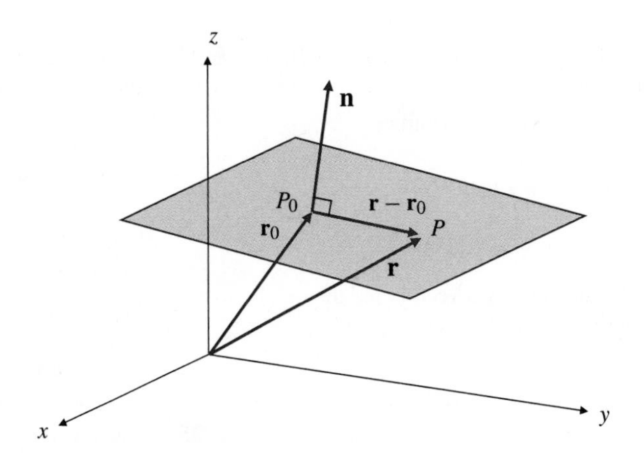

Figure 10.28 The plane through P_0 with normal **n** contains all points P for which $\overrightarrow{P_0P}$ is perpendicular to **n**

The point-normal equation of a plane

The plane having nonzero normal vector $\mathbf{n} = A\mathbf{i} + B\mathbf{j} + C\mathbf{k}$, and passing through the point $P_0 = (x_0, y_0, z_0)$ with position vector $\mathbf{r}_0$, has equation

$$\mathbf{n} \bullet (\mathbf{r} - \mathbf{r}_0) = 0$$

in vector form, or, equivalently,

$$A(x - x_0) + B(y - y_0) + C(z - z_0) = 0$$

in scalar form.

The scalar form can be written more simply in the **standard form** $Ax + By + Cz = D$, where $D = Ax_0 + By_0 + Cz_0$.

If at least one of the constants A, B, and C is not zero, then the *linear equation* $Ax + By + Cz = D$ always represents a plane in $\mathbb{R}^3$. For example, if $A \neq 0$, it represents the plane through $(D/A, 0, 0)$ with normal vector $\mathbf{n} = A\mathbf{i} + B\mathbf{j} + C\mathbf{k}$. A vector normal to a plane can always be determined from the coefficients of x, y, and z. If the constant term $D = 0$, then the plane must pass through the origin.

EXAMPLE 1 **(Recognizing and writing the equations of planes)**

(a) The equation $2x - 3y - 4z = 0$ represents a plane that passes through the origin and is normal (perpendicular) to the vector $\mathbf{n} = 2\mathbf{i} - 3\mathbf{j} - 4\mathbf{k}$.

(b) The plane that passes through the point $(2, 0, 1)$ and is perpendicular to the straight line passing through the points $(1, 1, 0)$ and $(4, -1, -2)$ has normal vector $\mathbf{n} = (4 - 1)\mathbf{i} + (-1 - 1)\mathbf{j} + (-2 - 0)\mathbf{k} = 3\mathbf{i} - 2\mathbf{j} - 2\mathbf{k}$. Therefore, its equation is $3(x - 2) - 2(y - 0) - 2(z - 1) = 0$, or, more simply, $3x - 2y - 2z = 4$.

(c) The plane with equation $2x - y = 1$ has a normal $2\mathbf{i} - \mathbf{j}$ that is perpendicular to the z-axis. The plane is therefore parallel to the z-axis. Note that the equation is independent of z. In the xy-plane, the equation $2x - y = 1$ represents a straight line; in 3-space it represents a plane containing that line and parallel to the z-axis. What does the equation $y = z$ represent in $\mathbb{R}^3$? the equation $y = -2$?

(d) The equation $2x + y + 3z = 6$ represents a plane with normal $\mathbf{n} = 2\mathbf{i} + \mathbf{j} + 3\mathbf{k}$. In this case we cannot directly read from the equation the coordinates of a particular point on the plane, but it is not difficult to discover some points. For instance, if we put $y = z = 0$ in the equation we get $x = 3$, so $(3, 0, 0)$ is a point on the plane. We say that the ***x*-intercept** of the plane is 3 since $(3, 0, 0)$ is the point where the plane intersects the x-axis. Similarly, the y-intercept is 6 and the z-intercept is 2 because the plane intersects the y- and z-axes at $(0, 6, 0)$ and $(0, 0, 2)$, respectively.

(e) In general, if a, b, and c are all nonzero, the plane with intercepts a, b, and c on the coordinate axes has equation

$$\frac{x}{a} + \frac{y}{b} + \frac{z}{c} = 1,$$

called the **intercept form** of the equation of the plane. (See Figure 10.29.)

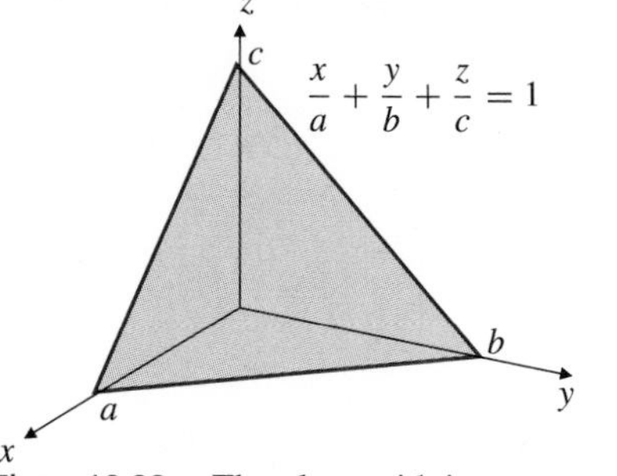

Figure 10.29 The plane with intercepts a, b, and c on the coordinate axes

EXAMPLE 2 Find an equation of the plane that passes through the three points $P = (1, 1, 0)$, $Q = (0, 2, 1)$, and $R = (3, 2, -1)$.

Solution We need to find a vector, $\mathbf{n}$, normal to the plane. Such a vector will be perpendicular to the vectors $\overrightarrow{PQ} = -\mathbf{i} + \mathbf{j} + \mathbf{k}$ and $\overrightarrow{PR} = 2\mathbf{i} + \mathbf{j} - \mathbf{k}$. Therefore, we can use

$$\mathbf{n} = \overrightarrow{PQ} \times \overrightarrow{PR} = \begin{vmatrix} \mathbf{i} & \mathbf{j} & \mathbf{k} \\ -1 & 1 & 1 \\ 2 & 1 & -1 \end{vmatrix} = -2\mathbf{i} + \mathbf{j} - 3\mathbf{k}.$$

We can use this normal vector together with the coordinates of any one of the three given points to write the equation of the plane. Using point P leads to the equation $-2(x-1) + 1(y-1) - 3(z-0) = 0$, or

$$2x - y + 3z = 1.$$

You can check that using either Q or R leads to the same equation. (If the cross product $\overrightarrow{PQ} \times \overrightarrow{PR}$ had been the zero vector, what would have been true about the three points P, Q, and R? Would they have determined a unique plane?)

EXAMPLE 3 Show that the two planes $x - y = 3$ and $x + y + z = 0$ intersect, and find a vector, $\mathbf{v}$, parallel to their line of intersection.

Solution The two planes have normal vectors

$$\mathbf{n}_1 = \mathbf{i} - \mathbf{j} \qquad \text{and} \qquad \mathbf{n}_2 = \mathbf{i} + \mathbf{j} + \mathbf{k},$$

respectively. Since these vectors are not parallel, the planes are not parallel, and they intersect in a straight line perpendicular to both $\mathbf{n}_1$ and $\mathbf{n}_2$. This line must therefore be parallel to

$$\mathbf{v} = \mathbf{n}_1 \times \mathbf{n}_2 = \begin{vmatrix} \mathbf{i} & \mathbf{j} & \mathbf{k} \\ 1 & -1 & 0 \\ 1 & 1 & 1 \end{vmatrix} = -\mathbf{i} - \mathbf{j} + 2\mathbf{k}.$$

A family of planes intersecting in a straight line is called a **pencil of planes**. (See Figure 10.30.) Such a pencil of planes is determined by any two nonparallel planes in it, since these have a unique line of intersection. If the two nonparallel planes have equations

$$A_1x + B_1y + C_1z = D_1 \qquad \text{and} \qquad A_2x + B_2y + C_2z = D_2,$$

then, for any value of the real number λ, the equation

$$A_1x + B_1y + C_1z - D_1 + \lambda(A_2x + B_2y + C_2z - D_2) = 0$$

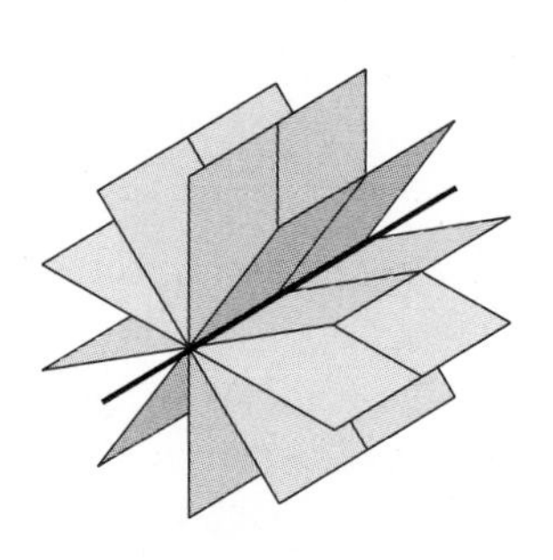

Figure 10.30 A pencil of planes

represents a plane in the pencil. To see this, observe that the equation is linear, and so represents a plane, and that any point (x, y, z) satisfying the equations of both given planes also satisfies this equation for any value of λ. Any plane in the pencil except the second defining plane, $A_2x + B_2y + C_2z = D_2$, can be obtained by suitably choosing the value of λ.

EXAMPLE 4 Find an equation of the plane passing through the line of intersection of the two planes

$$x + y - 2z = 6 \qquad \text{and} \qquad 2x - y + z = 2$$

and also passing through the point $(-2, 0, 1)$.

Solution For any constant λ, the equation

$$x + y - 2z - 6 + \lambda(2x - y + z - 2) = 0$$

represents a plane and is satisfied by the coordinates of all points on the line of intersection of the given planes. This plane passes through the point $(-2, 0, 1)$ if $-2 - 2 - 6 + \lambda(-4 + 1 - 2) = 0$, that is, if $\lambda = -2$. The equation of the required plane therefore simplifies to $3x - 3y + 4z + 2 = 0$. (This solution would not have worked if the given point had been on the second plane, $2x - y + z = 2$. Why?)

Lines in 3-Space

As we observed above, any two nonparallel planes in $\mathbb{R}^3$ determine a unique (straight) line of intersection, and a vector parallel to this line can be obtained by taking the cross product of normal vectors to the two planes.

Suppose that $\mathbf{r}_0 = x_0\mathbf{i} + y_0\mathbf{j} + z_0\mathbf{k}$ is the position vector of point P_0 and $\mathbf{v} = a\mathbf{i} + b\mathbf{j} + c\mathbf{k}$ is a nonzero vector. There is a unique line passing through P_0 parallel to $\mathbf{v}$. If $\mathbf{r} = x\mathbf{i} + y\mathbf{j} + z\mathbf{k}$ is the position vector of any other point P on the line, then $\mathbf{r} - \mathbf{r}_0$ lies along the line and so is parallel to $\mathbf{v}$. (See Figure 10.31.) Thus, $\mathbf{r} - \mathbf{r}_0 = t\mathbf{v}$ for some real number t. This equation, usually rewritten in the form

$$\mathbf{r} = \mathbf{r}_0 + t\mathbf{v},$$

is called the **vector parametric equation of the straight line**. All points on the line can be obtained as the parameter t ranges from $-\infty$ to ∞. The vector $\mathbf{v}$ is called a **direction vector** of the line.

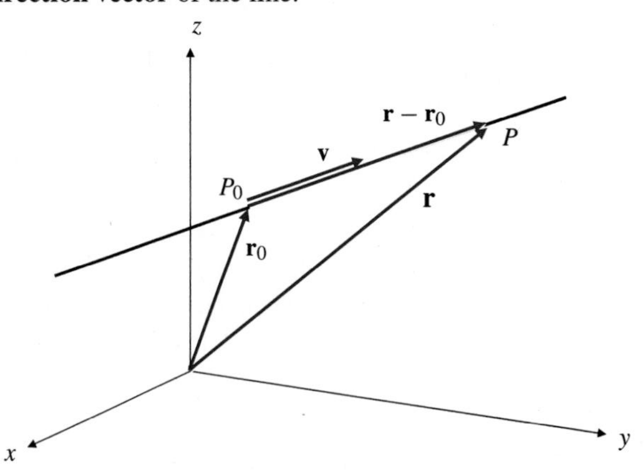

Figure 10.31 The line through P_0 parallel to $\mathbf{v}$

Breaking the vector parametric equation down into its components yields the **scalar parametric** equations of the line:

$$\begin{cases} x = x_0 + at \\ y = y_0 + bt \\ z = z_0 + ct. \end{cases} \qquad (-\infty < t < \infty)$$

These appear to be *three* linear equations, but the parameter t can be eliminated to give *two* linear equations in x, y, and z. If $a \neq 0$, $b \neq 0$, and $c \neq 0$, then we can solve each of the scalar equations for t and so obtain

$$\frac{x - x_0}{a} = \frac{y - y_0}{b} = \frac{z - z_0}{c},$$

which is called the **standard form** for the equations of the straight line through (x_0, y_0, z_0) parallel to $\mathbf{v}$. The standard form must be modified if any component of $\mathbf{v}$ vanishes. For example, if $c = 0$, the equations are

$$\frac{x - x_0}{a} = \frac{y - y_0}{b}, \qquad z = z_0.$$

Note that none of the above equations for straight lines is unique; each depends on the particular choice of the point (x_0, y_0, z_0) on the line. In general, you can always use the equations of two nonparallel planes to represent their line of intersection.

EXAMPLE 5 **(Equations of straight lines)**

(a) The equations

$$\begin{cases} x = 2 + t \\ y = 3 \\ z = -4t \end{cases}$$

represent the straight line through $(2, 3, 0)$ parallel to the vector $\mathbf{i} - 4\mathbf{k}$.

(b) The straight line through $(1, -2, 3)$ perpendicular to the plane $x - 2y + 4z = 5$ is parallel to the normal vector $\mathbf{i} - 2\mathbf{j} + 4\mathbf{k}$ of the plane. Therefore, the line has vector parametric equation

$$\mathbf{r} = \mathbf{i} - 2\mathbf{j} + 3\mathbf{k} + t(\mathbf{i} - 2\mathbf{j} + 4\mathbf{k}),$$

or scalar parametric equations

$$\begin{cases} x = 1 + t \\ y = -2 - 2t \\ z = 3 + 4t. \end{cases}$$

Its standard form equations are

$$\frac{x - 1}{1} = \frac{y + 2}{-2} = \frac{z - 3}{4}.$$

EXAMPLE 6 Find a direction vector for the line of intersection of the two planes $x + y - z = 0$ and $y + 2z = 6$, and find a set of equations for the line in standard form.

Solution The two planes have respective normals $\mathbf{n}_1 = \mathbf{i} + \mathbf{j} - \mathbf{k}$ and $\mathbf{n}_2 = \mathbf{j} + 2\mathbf{k}$. Thus, a direction vector of their line of intersection is

$$\mathbf{v} = \mathbf{n}_1 \times \mathbf{n}_2 = 3\mathbf{i} - 2\mathbf{j} + \mathbf{k}.$$

We need to know one point on the line in order to write equations in standard form. We can find a point by assigning a value to one coordinate and calculating the other two from the given equations. For instance, taking $z = 0$ in the two equations we are led to $y = 6$ and $x = -6$, so $(-6, 6, 0)$ is one point on the line. Thus, the line has standard form equations

$$\frac{x + 6}{3} = \frac{y - 6}{-2} = z.$$

This answer is not unique; the coordinates of any other point on the line could be used in place of $(-6, 6, 0)$. You could even find a direction vector $\mathbf{v}$ by subtracting the position vectors of two different points on the line.

Distances

The **distance** between two geometric objects always means the minimum distance between two points, one in each object. In the case of *flat* objects like lines and planes defined by linear equations, such minimum distances can usually be determined by geometric arguments without having to use calculus.

EXAMPLE 7 (Distance from a point to a plane)

(a) Find the distance from the point $P_0 = (x_0, y_0, z_0)$ to the plane $\mathcal{P}$ having equation $Ax + By + Cz = D$.

(b) What is the distance from $(2, -1, 3)$ to the plane $2x - 2y - z = 9$?

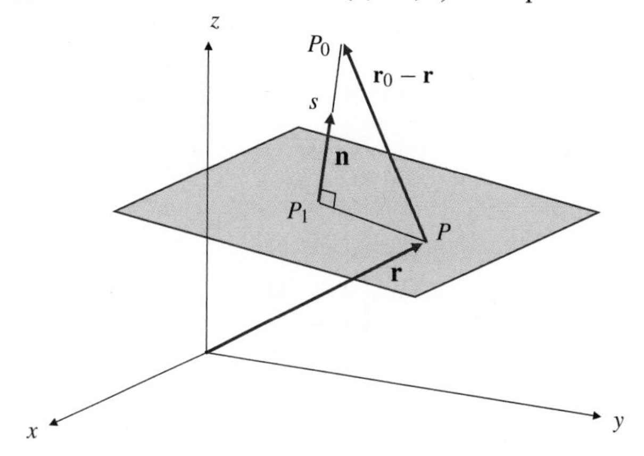

Figure 10.32 The distance from P_0 to the plane $\mathcal{P}$ is the length of the vector projection of PP_0 along the normal **n** to $\mathcal{P}$, where P is any point on $\mathcal{P}$

Solution

(a) Let $\mathbf{r}_0$ be the position vector of P_0, and let $\mathbf{n} = A\mathbf{i} + B\mathbf{j} + C\mathbf{k}$ be the normal to $\mathcal{P}$. Let P_1 be the point on $\mathcal{P}$ that is closest to P_0. Then $\overrightarrow{P_1P_0}$ is perpendicular to $\mathcal{P}$ and so is parallel to $\mathbf{n}$. The distance from P_0 to $\mathcal{P}$ is $s = |\overrightarrow{P_1P_0}|$. If P, having position vector $\mathbf{r}$, is any point on $\mathcal{P}$, then s is the length of the projection of $\overrightarrow{PP_0} = \mathbf{r}_0 - \mathbf{r}$ in the direction of $\mathbf{n}$. (See Figure 10.32.) Thus,

$$s = \left| \frac{\overrightarrow{PP_0} \bullet \mathbf{n}}{|\mathbf{n}|} \right| = \frac{|(\mathbf{r}_0 - \mathbf{r}) \bullet \mathbf{n}|}{|\mathbf{n}|} = \frac{|\mathbf{r}_0 \bullet \mathbf{n} - \mathbf{r} \bullet \mathbf{n}|}{|\mathbf{n}|}.$$

Since $P = (x, y, z)$ lies on $\mathcal{P}$, we have $\mathbf{r} \bullet \mathbf{n} = Ax + By + Cz = D$. In terms of the coordinates (x_0, y_0, z_0) of P_0, we can therefore represent the distance from P_0 to $\mathcal{P}$ as

$$s = \frac{|Ax_0 + By_0 + Cz_0 - D|}{\sqrt{A^2 + B^2 + C^2}}.$$

(b) The distance from $(2, -1, 3)$ to the plane $2x - 2y - z = 9$ is

$$s = \frac{|2(2) - 2(-1) - 1(3) - 9|}{\sqrt{2^2 + (-2)^2 + (-1)^2}} = \frac{|-6|}{3} = 2 \text{ units}.$$

EXAMPLE 8 (Distance from a point to a line)

(a) Find the distance from the point P_0 to the straight line $\mathcal{L}$ through P_1 parallel to the nonzero vector **v**.

(b) What is the distance from $(2, 0, -3)$ to the line $\mathbf{r} = \mathbf{i} + (1 + 3t)\mathbf{j} - (3 - 4t)\mathbf{k}$?

Solution

(a) Let $\mathbf{r}_0$ and $\mathbf{r}_1$ be the position vectors of P_0 and P_1, respectively. The point P_2 on $\mathcal{L}$ that is closest to P_0 is such that P_2P_0 is perpendicular to $\mathcal{L}$. The distance from P_0 to $\mathcal{L}$ is

$$s = |P_2P_0| = |P_1P_0|\sin\theta = |\mathbf{r}_0 - \mathbf{r}_1|\sin\theta,$$

where θ is the angle between $\mathbf{r}_0 - \mathbf{r}_1$ and $\mathbf{v}$. (See Figure 10.33(a).) Since

$$|(\mathbf{r}_0 - \mathbf{r}_1) \times \mathbf{v}| = |\mathbf{r}_0 - \mathbf{r}_1|\,|\mathbf{v}|\,\sin\theta,$$

we have

$$s = \frac{|(\mathbf{r}_0 - \mathbf{r}_1) \times \mathbf{v}|}{|\mathbf{v}|}.$$

(b) The line $\mathbf{r} = \mathbf{i} + (1 + 3t)\mathbf{j} - (3 - 4t)\mathbf{k}$ passes through $P_1 = (1, 1, -3)$ and is parallel to $\mathbf{v} = 3\mathbf{j} + 4\mathbf{k}$. The distance from $P_0 = (2, 0, -3)$ to this line is

$$\begin{aligned} s &= \frac{\big|\big((2-1)\mathbf{i} + (0-1)\mathbf{j} + (-3+3)\mathbf{k}\big) \times (3\mathbf{j} + 4\mathbf{k})\big|}{\sqrt{3^2 + 4^2}} \\ &= \frac{|(\mathbf{i} - \mathbf{j}) \times (3\mathbf{j} + 4\mathbf{k})|}{5} = \frac{|-4\mathbf{i} - 4\mathbf{j} + 3\mathbf{k}|}{5} = \frac{\sqrt{41}}{5} \text{ units.} \end{aligned}$$

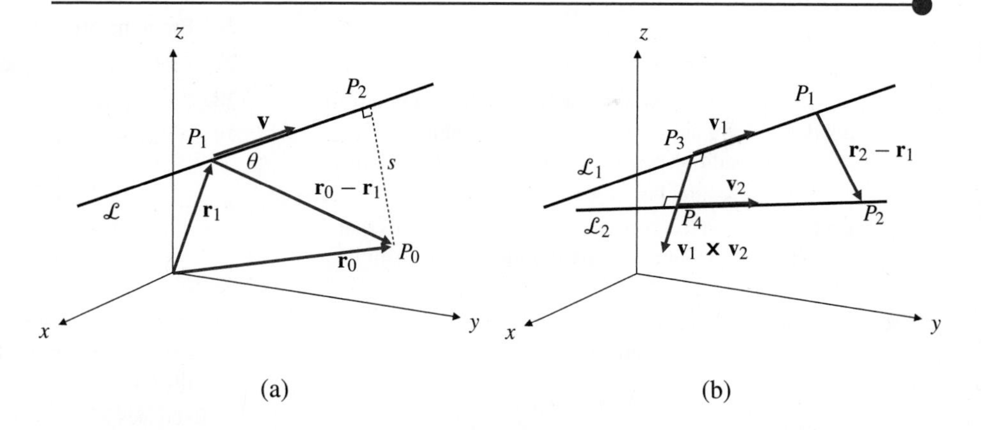

Figure 10.33

(a) The distance from P_0 to the line $\mathcal{L}$ is $s = |P_0P_1|\sin\theta$

(b) The distance between the lines $\mathcal{L}_1$ and $\mathcal{L}_2$ is the length of the projection of P_1P_2 along the vector $\mathbf{v}_1 \times \mathbf{v}_2$

EXAMPLE 9 **(The distance between two lines)** Find the distance between the two lines $\mathcal{L}_1$ through point P_1 parallel to vector $\mathbf{v}_1$ and $\mathcal{L}_2$ through point P_2 parallel to vector $\mathbf{v}_2$.

Solution Let $\mathbf{r}_1$ and $\mathbf{r}_2$ be the position vectors of points P_1 and P_2, respectively. If P_3 and P_4 (with position vectors $\mathbf{r}_3$ and $\mathbf{r}_4$) are the points on $\mathcal{L}_1$ and $\mathcal{L}_2$, respectively, that are closest to one another, then $\overrightarrow{P_3P_4}$ is perpendicular to both lines and is therefore parallel to $\mathbf{v}_1 \times \mathbf{v}_2$. (See Figure 10.33(b).) $\overrightarrow{P_3P_4}$ is the vector projection of $\overrightarrow{P_1P_2} = \mathbf{r}_2 - \mathbf{r}_1$ along $\mathbf{v}_1 \times \mathbf{v}_2$. Therefore, the distance $s = |\overrightarrow{P_3P_4}|$ between the lines is given by

$$s = |\mathbf{r}_4 - \mathbf{r}_3| = \frac{|(\mathbf{r}_2 - \mathbf{r}_1) \bullet (\mathbf{v}_1 \times \mathbf{v}_2)|}{|\mathbf{v}_1 \times \mathbf{v}_2|}.$$

EXERCISES 10.4

1. A single equation involving the coordinates (x, y, z) need not always represent a two-dimensional "surface" in $\mathbb{R}^3$. For example, $x^2 + y^2 + z^2 = 0$ represents the single point $(0, 0, 0)$, which has dimension zero. Give examples of single equations in x, y, and z that represent

(a) a (one-dimensional) straight line,

(b) the whole of $\mathbb{R}^3$,

(c) no points at all (i.e., the empty set).

In Exercises 2–9, find equations of the planes satisfying the given conditions.

2. Passing through $(0, 2, -3)$ and normal to the vector $4\mathbf{i} - \mathbf{j} - 2\mathbf{k}$

3. Passing through the origin and having normal $\mathbf{i} - \mathbf{j} + 2\mathbf{k}$

4. Passing through $(1, 2, 3)$ and parallel to the plane $3x + y - 2z = 15$

5. Passing through the three points $(1, 1, 0)$, $(2, 0, 2)$, and $(0, 3, 3)$

6. Passing through the three points $(-2, 0, 0)$, $(0, 3, 0)$, and $(0, 0, 4)$

7. Passing through $(1, 1, 1)$ and $(2, 0, 3)$ and perpendicular to the plane $x + 2y - 3z = 0$

8. Passing through the line of intersection of the planes $2x + 3y - z = 0$ and $x - 4y + 2z = -5$, and passing through the point $(-2, 0, -1)$

9. Passing through the line $x + y = 2$, $y - z = 3$, and perpendicular to the plane $2x + 3y + 4z = 5$

10. Under what geometric condition will three distinct points in $\mathbb{R}^3$ not determine a unique plane passing through them? How can this condition be expressed algebraically in terms of the position vectors, $\mathbf{r}_1$, $\mathbf{r}_2$, and $\mathbf{r}_3$, of the three points?

11. Give a condition on the position vectors of four points that guarantees that the four points are *coplanar*, that is, all lie on one plane.

Describe geometrically the one-parameter families of planes in Exercises 12–14. (λ is a real parameter.)

12. $x + y + z = \lambda$.

! 13. $x + \lambda y + \lambda z = \lambda$.

! 14. $\lambda x + \sqrt{1 - \lambda^2}\, y = 1$.

In Exercises 15–19, find equations of the line specified in vector and scalar parametric forms and in standard form.

15. Through the point $(1, 2, 3)$ and parallel to $2\mathbf{i} - 3\mathbf{j} - 4\mathbf{k}$

16. Through $(-1, 0, 1)$ and perpendicular to the plane $2x - y + 7z = 12$

17. Through the origin and parallel to the line of intersection of the planes $x + 2y - z = 2$ and $2x - y + 4z = 5$

18. Through $(2, -1, -1)$ and parallel to each of the two planes $x + y = 0$ and $x - y + 2z = 0$

19. Through $(1, 2, -1)$ and making equal angles with the positive directions of the coordinate axes

In Exercises 20–22, find the equations of the given line in standard form.

20. $\mathbf{r} = (1 - 2t)\mathbf{i} + (4 + 3t)\mathbf{j} + (9 - 4t)\mathbf{k}$.

21. $\begin{cases} x = 4 - 5t \\ y = 3t \\ z = 7 \end{cases}$

22. $\begin{cases} x - 2y + 3z = 0 \\ 2x + 3y - 4z = 4 \end{cases}$

23. If $P_1 = (x_1, y_1, z_1)$ and $P_2 = (x_2, y_2, z_2)$, show that the equations

$$\begin{cases} x = x_1 + t(x_2 - x_1) \\ y = y_1 + t(y_2 - y_1) \\ z = z_1 + t(z_2 - z_1) \end{cases}$$

represent a line through P_1 and P_2.

24. What points on the line in Exercise 23 correspond to the parameter values $t = -1$, $t = 1/2$, and $t = 2$? Describe their locations.

25. Under what conditions on the position vectors of four distinct points P_1, P_2, P_3, and P_4 will the straight line through P_1 and P_2 intersect the straight line through P_3 and P_4 at a unique point?

Find the required distances in Exercises 26–29.

26. From the origin to the plane $x + 2y + 3z = 4$

27. From $(1, 2, 0)$ to the plane $3x - 4y - 5z = 2$

28. From the origin to the line $x + y + z = 0$, $2x - y - 5z = 1$

29. Between the lines

$$\begin{cases} x + 2y = 3 \\ y + 2z = 3 \end{cases} \quad \text{and} \quad \begin{cases} x + y + z = 6 \\ x - 2z = -5 \end{cases}$$

30. Show that the line $x - 2 = \dfrac{y + 3}{2} = \dfrac{z - 1}{4}$ is parallel to the plane $2y - z = 1$. What is the distance between the line and the plane?

In Exercises 31–32, describe the one-parameter families of straight lines represented by the given equations. (λ is a real parameter.)

! 31. $(1 - \lambda)(x - x_0) = \lambda(y - y_0)$, $z = z_0$.

! 32. $\dfrac{x - x_0}{\sqrt{1 - \lambda^2}} = \dfrac{y - y_0}{\lambda} = z - z_0$.

33. Why does the factored second-degree equation

$$(A_1x + B_1y + C_1z - D_1)(A_2x + B_2y + C_2z - D_2) = 0$$

represent a pair of planes rather than a single straight line?

10.5 Quadric Surfaces

The most general second-degree equation in three variables is

$$Ax^2 + By^2 + Cz^2 + Dxy + Exz + Fyz + Gx + Hy + Iz = J.$$

We will not attempt the (rather difficult) task of classifying all the surfaces that can be represented by such an equation, but will examine some interesting special cases. Let us observe at the outset that if the above equation can be factored in the form

$$(A_1x + B_1y + C_1z - D_1)(A_2x + B_2y + C_2z - D_2) = 0,$$

then the graph is, in fact, a pair of planes,

$$A_1x + B_1y + C_1z = D_1 \quad \text{and} \quad A_2x + B_2y + C_2z = D_2,$$

or one plane if the two linear equations represent the same plane. This is considered a degenerate case. Where such factorization is not possible, the surface (called a **quadric surface**) will not be flat, although there may still be straight lines that lie on the surface. Nondegenerate quadric surfaces fall into the following six categories.

Spheres. The equation $x^2 + y^2 + z^2 = a^2$ represents a sphere of radius a centred at the origin. More generally,

$$(x - x_0)^2 + (y - y_0)^2 + (z - z_0)^2 = a^2$$

represents a sphere of radius a centred at the point (x_0, y_0, z_0). If a quadratic equation in x, y, and z has equal coefficients for the x^2, y^2, and z^2 terms and has no other second-degree terms, then it will represent, if any surface at all, a sphere. The centre can be found by completing the squares as for circles in the plane.

Cylinders. The equation $x^2 + y^2 = a^2$, being independent of z, represents a **right-circular cylinder** of radius a and axis along the z-axis. (See Figure 10.34(a).) The intersection of the cylinder with the horizontal plane $z = k$ is the circle with equations

$$\begin{cases} x^2 + y^2 = a^2 \\ z = k. \end{cases}$$

Quadric cylinders also come in other shapes: elliptic, parabolic, and hyperbolic. For instance, $z = x^2$ represents a parabolic cylinder with vertex line along the y-axis. (See Figure 10.34(b).) In general, an equation in two variables only will represent a cylinder in 3-space.

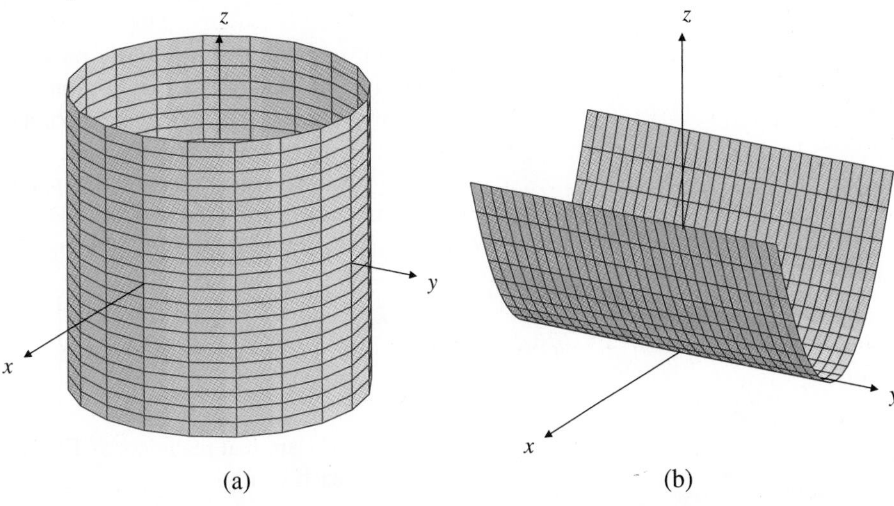

Figure 10.34
(a) The circular cylinder $x^2 + y^2 = a^2$
(b) The parabolic cylinder $z = x^2$

Cones. The equation $z^2 = x^2 + y^2$ represents a **right-circular cone** with axis along the z-axis. The surface is generated by rotating about the z-axis the line $z = y$ in the yz-plane. This *generator* makes an angle of 45° with the axis of the cone. Cross-sections of the cone in planes parallel to the xy-plane are circles. (See Figure 10.35(a).) The equation $x^2 + y^2 = a^2z^2$ also represents a right-circular cone with vertex at the origin and axis along the z-axis but having semi-vertical angle $\alpha = \tan^{-1} a$. A circular cone has plane cross-sections that are elliptical, parabolic, and hyperbolic. Conversely, any nondegenerate quadric cone has a direction perpendicular to which the cross-sections of the cone are circular. In that sense, every quadric cone is a circular cone, although it may be *oblique* rather than right-circular in that the line joining the centres of the circular cross-sections need not be perpendicular to those cross-sections. (See Exercise 24.)

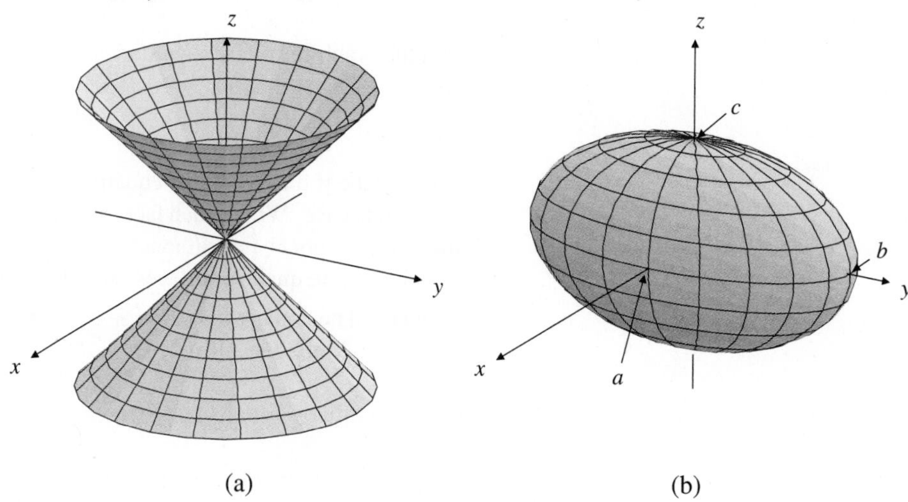

Figure 10.35

(a) The circular cone $a^2z^2 = x^2 + y^2$

(b) The ellipsoid $\dfrac{x^2}{a^2} + \dfrac{y^2}{b^2} + \dfrac{z^2}{c^2} = 1$

Ellipsoids. The equation

$$\frac{x^2}{a^2} + \frac{y^2}{b^2} + \frac{z^2}{c^2} = 1$$

represents an **ellipsoid** with *semi-axes* a, b, and c. (See Figure 10.35(b).) The surface is oval, and it is enclosed inside the rectangular parallelepiped $-a \le x \le a$, $-b \le y \le b$, $-c \le z \le c$. If $a = b = c$, the ellipsoid is a sphere. In general, all plane cross-sections of ellipsoids are ellipses. This is easy to see for cross-sections parallel to coordinate planes, but somewhat harder to see for other planes.

Paraboloids. The equations

$$z = \frac{x^2}{a^2} + \frac{y^2}{b^2} \qquad \text{and} \qquad z = \frac{x^2}{a^2} - \frac{y^2}{b^2}$$

represent, respectively, an **elliptic paraboloid** and a **hyperbolic paraboloid**. (See Figure 10.36(a) and (b).) Cross-sections in planes $z = k$ (k being a positive constant) are ellipses (circles if $a = b$) and hyperbolas, respectively. Parabolic reflective mirrors have the shape of circular paraboloids. The hyperbolic paraboloid is a **ruled surface**. (A ruled surface is one through every point of which there passes a straight line lying wholly on the surface. Cones and cylinders are also examples of ruled surfaces.) There are two one-parameter families of straight lines that lie on the hyperbolic paraboloid:

$$\begin{cases} \lambda z = \dfrac{x}{a} - \dfrac{y}{b} \\[2mm] \dfrac{1}{\lambda} = \dfrac{x}{a} + \dfrac{y}{b} \end{cases} \qquad \text{and} \qquad \begin{cases} \mu z = \dfrac{x}{a} + \dfrac{y}{b} \\[2mm] \dfrac{1}{\mu} = \dfrac{x}{a} - \dfrac{y}{b} \end{cases},$$

where λ and μ are real parameters. Every point on the hyperbolic paraboloid lies on one line of each family.

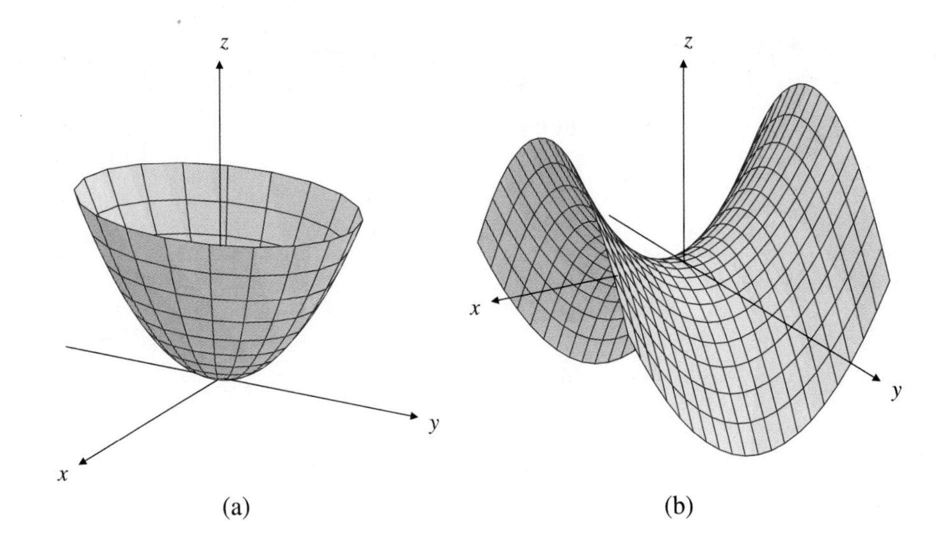

Figure 10.36

(a) The elliptic paraboloid $z = \dfrac{x^2}{a^2} + \dfrac{y^2}{b^2}$

(b) The hyperbolic paraboloid $z = \dfrac{x^2}{a^2} - \dfrac{y^2}{b^2}$

Hyperboloids. The equation

$$\frac{x^2}{a^2} + \frac{y^2}{b^2} - \frac{z^2}{c^2} = 1$$

represents a surface called a **hyperboloid of one sheet**. (See Figure 10.37(a).) The equation

$$\frac{x^2}{a^2} + \frac{y^2}{b^2} - \frac{z^2}{c^2} = -1$$

represents a **hyperboloid of two sheets**. (See Figure 10.37(b).) Both surfaces

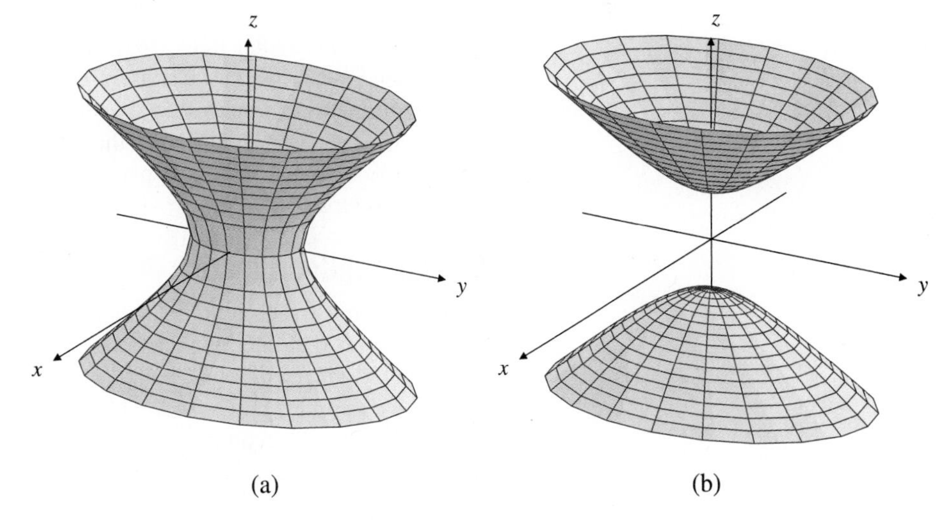

Figure 10.37

(a) The hyperboloid of one sheet $\dfrac{x^2}{a^2} + \dfrac{y^2}{b^2} - \dfrac{z^2}{c^2} = 1$

(b) The hyperboloid of two sheets $\dfrac{x^2}{a^2} + \dfrac{y^2}{b^2} - \dfrac{z^2}{c^2} = -1$

have elliptical cross-sections in horizontal planes and hyperbolic cross-sections in vertical planes. Both are *asymptotic* to the elliptic cone with equation

$$\frac{x^2}{a^2} + \frac{y^2}{b^2} = \frac{z^2}{c^2};$$

they approach arbitrarily close to the cone as they recede arbitrarily far away from the origin. Like the hyperbolic paraboloid, the hyperboloid of one sheet is a ruled surface.

EXERCISES 10.5

Identify the surfaces represented by the equations in Exercises 1–16 and sketch their graphs.

1. $x^2 + 4y^2 + 9z^2 = 36$

2. $x^2 + y^2 + 4z^2 = 4$

3. $2x^2 + 2y^2 + 2z^2 - 4x + 8y - 12z + 27 = 0$

4. $x^2 + 4y^2 + 9z^2 + 4x - 8y = 8$

5. $z = x^2 + 2y^2$

6. $z = x^2 - 2y^2$

7. $x^2 - y^2 - z^2 = 4$

8. $-x^2 + y^2 + z^2 = 4$

9. $z = xy$

10. $x^2 + 4z^2 = 4$

11. $x^2 - 4z^2 = 4$

12. $y = z^2$

13. $x = z^2 + z$

14. $x^2 = y^2 + 2z^2$

15. $(z - 1)^2 = (x - 2)^2 + (y - 3)^2$

16. $(z - 1)^2 = (x - 2)^2 + (y - 3)^2 + 4$

Describe and sketch the geometric objects represented by the systems of equations in Exercises 17–20.

17. $\begin{cases} x^2 + y^2 + z^2 = 4 \\ x + y + z = 1 \end{cases}$

18. $\begin{cases} x^2 + y^2 = 1 \\ z = x + y \end{cases}$

19. $\begin{cases} z^2 = x^2 + y^2 \\ z = 1 + x \end{cases}$

20. $\begin{cases} x^2 + 2y^2 + 3z^2 = 6 \\ y = 1 \end{cases}$

21. Find two one-parameter families of straight lines that lie on the hyperboloid of one sheet

$$\frac{x^2}{a^2} + \frac{y^2}{b^2} - \frac{z^2}{c^2} = 1.$$

22. Find two one-parameter families of straight lines that lie on the hyperbolic paraboloid $z = xy$.

23. The equation $2x^2 + y^2 = 1$ represents a cylinder with elliptical cross-sections in planes perpendicular to the z-axis. Find a vector **a** perpendicular to which the cylinder has circular cross-sections.

24. The equation $z^2 = 2x^2 + y^2$ represents a cone with elliptical cross-sections in planes perpendicular to the z-axis. Find a vector **a** perpendicular to which the cone has circular cross-sections. *Hint:* Do Exercise 23 first and use its result.

10.6 Cylindrical and Spherical Coordinates

Polar coordinates provide a useful alternative to plane Cartesian coordinates for describing plane regions with circular symmetry or bounded by arcs of circles centred at the origin and radial lines from the origin. Similarly, there are two commonly encountered alternatives to Cartesian coordinates in 3-space. They generalize plane polar coordinates to 3-space and are suitable for describing regions with cylindrical or spherical symmetry. We introduce these two coordinate systems here, but won't make much use of them until the latter part of Chapter 14 when we will learn how to integrate over such regions.

Cylindrical Coordinates

Among the most useful alternatives to Cartesian coordinates in 3-space is the coordinate systems that directly generalizes plane polar coordinates by replacing only the horizontal x and y coordinates with the polar coordinates r and θ, while leaving the vertical z coordinate untouched. This system is called **cylindrical coordinates.** Each point in 3-space has cylindrical coordinates $[r, \theta, z]$ related to its Cartesian coordinates (x, y, z) by the transformation

$$x = r\cos\theta, \quad y = r\sin\theta, \quad z = z.$$

Figure 10.38 shows how a point P is located by its cylindrical coordinates $[r, \theta, z]$ as well as by its Cartesian coordinates (x, y, z). Note that the distance from P to the z-axis is r, while the distance from P to the origin is

$$d = \sqrt{r^2 + z^2} = \sqrt{x^2 + y^2 + z^2}.$$

EXAMPLE 1 The point with Cartesian coordinates $(1, 1, 1)$ has cylindrical coordinates $[\sqrt{2}, \pi/4, 1]$. The point with Cartesian coordinates $(0, 2, -3)$ has cylindrical coordinates $[2, \pi/2, -3]$. The point with cylindrical coordinates $[4, -\pi/3, 5]$ has Cartesian coordinates $(2, -2\sqrt{3}, 5)$.

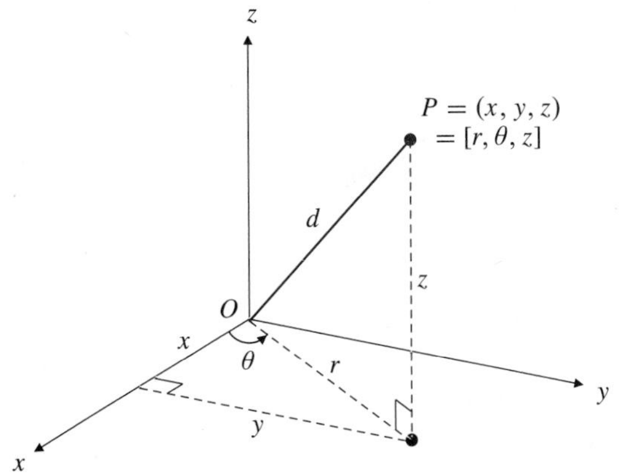

Figure 10.38 The cylindrical coordinates of a point

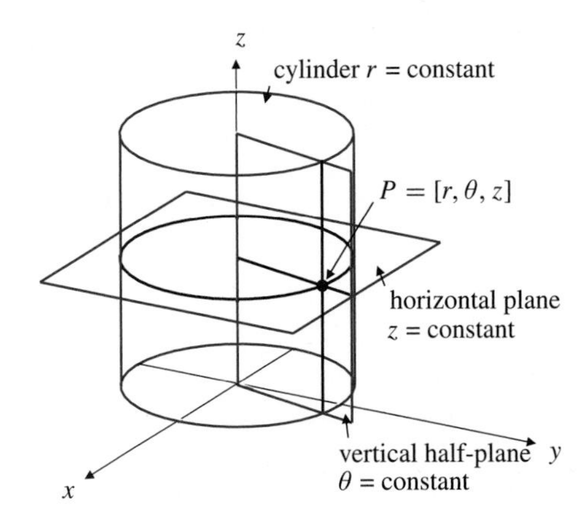

Figure 10.39 The coordinate surfaces for cylindrical coordinates

Just as planes with equations $x =$ constant, $y =$ constant, and $z =$ constant are the *coordinate surfaces* of the Cartesian coordinate system in 3-space, the coordinate surfaces in cylindrical coordinates are:

the r-surfaces with equations $r =$ constant (vertical circular cylinders centred on the z-axis),

the θ-surfaces $\theta =$ constant (vertical half-planes with edge along the z-axis), and

the z-surfaces $z =$ constant (horizontal planes).

See Figure 10.39. Cylindrical coordinates lend themselves to representing domains that are bounded by such surfaces and, in particular, to problems with axial symmetry (around the z-axis).

The coordinate curves in the cylindrical coordinate system are intersections of pairs of coordinate surfaces.

The r-curves are the intersections of the planes $\theta =$ constant and $z =$ constant, and so are horizontal radial lines emanating from the z-axis.

The θ-curves are intersections of the cylinders $r =$ constant and planes $z =$ constant, and so are horizontal circles centred on the z-axis.

The z-curves are intersections of the cylinders $r =$ constant and the half-planes $\theta =$ constant, and so are vertical straight lines.

EXAMPLE 2 Identify the surfaces whose equations in cylindrical coordinates are:

(a) $z = r$, (b) $z = r\cos\theta$, (c) $r = 2\cos\theta$.

Solution

(a) $z = r^2$ represents the circular paraboloid with Cartesian equation $z = x^2 + y^2$. It has vertex at the origin and axis of symmetry along the positive z-axis.

(b) $z = r\cos\theta$ represents the plane with Cartesian equation $z = x$. It contains the y-axis and the point with Cartesian coordinates $(1, 0, 1)$.

(c) $r = 2\cos\theta$ can be rewritten $r^2 = 2r\cos\theta$, so represents the vertical surface with Cartesian equation $x^2 + y^2 = 2x$. This is a circular cylinder of radius 1 with central axis along the vertical line through the point $(1, 0, 0)$ (in Cartesian coordinates).

EXAMPLE 3 Describe the curves whose equations in cylindrical coordinates are:

(a) $\begin{cases} r = z \\ z = 1 + r\cos\theta \end{cases}$, (b) $\begin{cases} \theta = \pi/2 \\ r^2 + z^2 = 4 \end{cases}$.

Solution

(a) The curve is the parabola in which the plane $z = 1 + x$ intersects the right-circular half-cone $z = \sqrt{x^2 + y^2}$. Since the plane is parallel to the line $z = x$, which is a generator of the cone, the intersection must be a parabola rather than an ellipse or a hyperbola. (See Section 8.1.)

(b) $\theta = \pi/2$ represents the half of the yz-plane where $y \ge 0$. $r^2 + z^2 = 4$ represents a sphere of radius 2 centred at the origin. Thus, this curve is the semicircle with cartesian equation $y = \sqrt{4 - z^2}$ in the plane $x = 0$.

Spherical Coordinates

In the system of **spherical coordinates** a point P in 3-space is represented by the ordered triple $[R, \phi, \theta]$, where R is the distance from P to the origin O, ϕ (Greek "phi") is the angle the radial line OP makes with the positive direction of the z-axis, and θ is the angle between the plane containing P and the z-axis and the xz-plane. (See Figure 10.40.) It is conventional to consider spherical coordinates restricted in such a way that $R \ge 0$, $0 \le \phi \le \pi$, and $0 \le \theta < 2\pi$ (or $-\pi < \theta \le \pi$). Every point not on the z-axis then has exactly one spherical coordinate representation, and the transformation from Cartesian coordinates (x, y, z) to spherical coordinates $[R, \phi, \theta]$ is one-to-one off the z-axis. Using the right-angled triangles in the figure, we can see that this transformation is given by:

$$\begin{aligned} x &= R\sin\phi\cos\theta \\ y &= R\sin\phi\sin\theta \\ z &= R\cos\phi. \end{aligned}$$

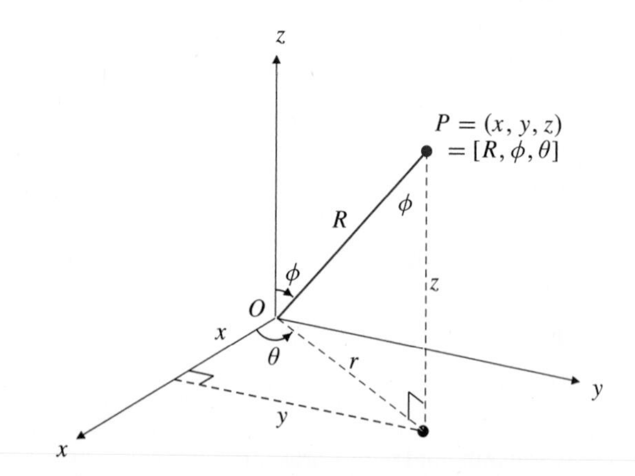

Figure 10.40 The spherical coordinates of a point

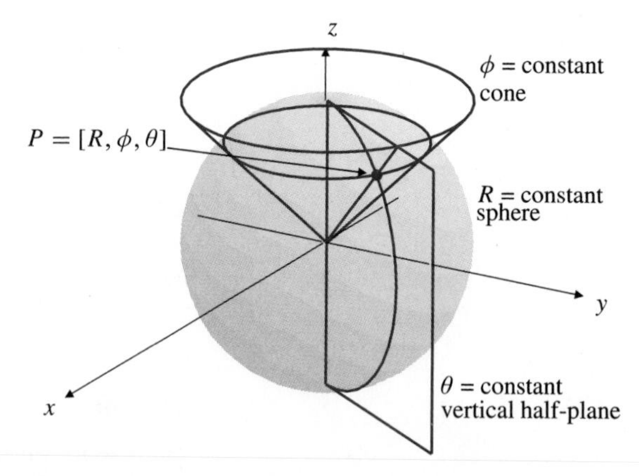

Figure 10.41 The coordinate surfaces for spherical coordinates

Observe that

$$R^2 = x^2 + y^2 + z^2 = r^2 + z^2$$

and that the r coordinate in cylindrical coordinates is related to R and ϕ by

$$r = \sqrt{x^2 + y^2} = R \sin \phi.$$

Thus, also

$$\tan \phi = \frac{r}{z} = \frac{\sqrt{x^2 + y^2}}{z} \qquad \text{and} \qquad \tan \theta = \frac{y}{x}.$$

If $\phi = 0$ or $\phi = \pi$, then $r = 0$, so the θ coordinate is irrelevant at points on the z-axis. Some coordinate surfaces for spherical coordinates are shown in Figure 10.41.

The R-surfaces ($R =$ constant) are spheres centred at the origin.

The ϕ-surfaces ($\phi =$ constant) are nappes of circular cones with the z-axis as axis.

The θ-surfaces ($\theta =$ constant) are vertical half-planes with edge along the z-axis.

Similarly, pairs of coordinate surfaces intersect in coordinate curves along which only one of the coordinates varies.

The R-curves (along which R varies) are the intersections of ϕ- and θ-surfaces, and so are radial lines emanating from the origin.

The ϕ-curves (along which ϕ varies) are the intersections of R- and θ-surfaces, and so are vertical semicircles centred at the origin and beginning and ending on the z-axis.

The θ-curves (along which θ varies) are the intersections of the R- and ϕ-surfaces, and thus are horizontal circles with centres on the z-axis.

If we take a coordinate system with origin at the centre of the earth, z-axis through the north pole, and x-axis through the intersection of the Greenwich meridian and the equator, then the earth's surface is (roughly speaking) a R-surface. It's intersections with the ϕ-surfaces are θ-curves on the earth's surface and are called *parallels of latitude*. The intersections of the surface of the earth with the θ-surfaces are ϕ-curves called *meridians of longitude*. Since latitude is measured from 90° at the north pole to $-90°$ at the south pole, while ϕ is measured from 0 at the north pole to π ($= 180°$) at the south pole, the coordinate ϕ is frequently referred to as the **colatitude** coordinate; θ is the **longitude** coordinate. Observe that θ has the same significance in spherical coordinates as it does in cylindrical coordinates.

Spherical coordinates are suited to problems involving spherical symmetry and, in particular, to regions bounded by spheres centred at the origin, circular cones with axes along the z-axis, and vertical planes containing the z-axis.

EXAMPLE 4 Find:

(a) the Cartesian coordinates of the point P with spherical coordinates $[2, \pi/3, \pi/2]$, and

(b) the spherical coordinates of the point Q with Cartesian coordinates $(1, 1, \sqrt{2})$.

Solution

(a) If $R = 2$, $\phi = \pi/3$, and $\theta = \pi/2$, then

$$\begin{aligned} x &= 2 \sin(\pi/3) \cos(\pi/2) = 0 \\ y &= 2 \sin(\pi/3) \sin(\pi/2) = \sqrt{3} \\ z &= 2 \cos(\pi/3) = 1. \end{aligned}$$

The Cartesian coordinates of P are $(0, \sqrt{3}, 1)$.

(b) Given that

$$\begin{aligned} R\sin\phi\cos\theta &= x = 1\\ R\sin\phi\sin\theta &= y = 1\\ R\cos\phi &= z = \sqrt{2}, \end{aligned}$$

we calculate that $R^2 = 1+1+2 = 4$, so $R = 2$. Also $r^2 = 1+1 = 2$, so $r = \sqrt{2}$. Thus, $\tan\phi = r/z = 1$, so $\phi = \pi/4$. Also, $\tan\theta = y/x = 1$, so $\theta = \pi/4$ or $5\pi/4$. Since $x > 0$, we must have $\theta = \pi/4$. The spherical coordinates of Q are $[2, \pi/4, \pi/4]$.

Remark You may wonder why we write spherical coordinates in the order R, ϕ, θ rather than R, θ, ϕ. The reason, which will not become apparent until Chapter 16, concerns the triad of unit vectors at any point P, taken in coordinate order and tangent to the corresponding coordinate curve in the direction of increase of that coordinate. The order R, ϕ, θ ensures that this triad is a *right-handed basis* rather than a left-handed one.

EXERCISES 10.6

1. Convert the Cartesian coordinates $(2, -2, 1)$ to cylindrical coordinates and to spherical coordinates.
2. Convert the cylindrical coordinates $[2, \pi/6, -2]$ to Cartesian coordinates and to spherical coordinates.
3. Convert the spherical coordinates $[4, \pi/3, 2\pi/3]$ to Cartesian coordinates and to cylindrical coordinates.
4. A point P has spherical coordinates $[1, \phi, \theta]$ and cylindrical coordinates $[r, \pi/4, r]$. Find the Cartesian coordinates of the point.

Describe the sets of points in 3-space that satisfy the equations in Exercises 5–14. Here, r, θ, R, and ϕ denote the appropriate cylindrical or spherical coordinates.

5. $\theta = \pi/2$
6. $\phi = 2\pi/3$
7. $\phi = \pi/2$
8. $R = 4$
9. $r = 4$
10. $R = z$
11. $R = r$
12. $R = 2x$
13. $R = 2\cos\phi$
14. $r = 2\cos\theta$

10.7 A Little Linear Algebra

Differential calculus is essentially the study of linear approximations to functions. The tangent line to the graph $y = f(x)$ at $x = x_0$ provides the "best linear approximation" to $f(x)$ near x_0. Differentiation of functions of several variables can also be viewed as a process of finding *best linear approximations.* Therefore, the language of linear algebra can be very useful for expressing certain concepts in the calculus of several variables.

Linear algebra is a vast subject and is usually studied independently of calculus. This is unfortunate because understanding the relationship between the two subjects can greatly enhance your understanding and appreciation of each of them. Knowledge of linear algebra, and therefore familiarity with the material covered in this section, is *not essential* for fruitful study of the rest of this book. However, we shall occasionally comment on the significance of the subject at hand from the point of view of linear algebra. To this end we need only a little of the terminology and content of this subject, especially that part pertaining to matrix manipulation and systems of linear equations. In the rest of this section we present an outline of this material. Some students will already be familiar with it; others will encounter it later. We make no attempt at completeness here and refer interested students to standard linear algebra texts for proofs of some assertions. Students proceeding beyond this book to further study of advanced calculus and differential equations will certainly need a much more extensive background in linear algebra.

Matrices

An $m \times n$ **matrix** $\mathcal{A}$ is a rectangular array of mn numbers arranged in m rows and n columns. If a_{ij} is the element in the ith row and the jth column, then

$$\mathcal{A} = \begin{pmatrix} a_{11} & a_{12} & \cdots & a_{1n} \\ a_{21} & a_{22} & \cdots & a_{2n} \\ \vdots & \vdots & & \vdots \\ a_{m1} & a_{m2} & \cdots & a_{mn} \end{pmatrix}.$$

Sometimes, as a shorthand notation, we write $\mathcal{A} = (a_{ij})$. In this case i is assumed to range from 1 to m and j from 1 to n. If $m = n$, we say that $\mathcal{A}$ is a square matrix. The elements a_{ij} of the matrices we use in this book will always be real numbers.

The **transpose** of an $m \times n$ matrix $\mathcal{A}$ is the $n \times m$ matrix $\mathcal{A}^T$ whose rows are the columns of $\mathcal{A}$:

$$\mathcal{A}^T = \begin{pmatrix} a_{11} & a_{21} & \cdots & a_{m1} \\ a_{12} & a_{22} & \cdots & a_{m2} \\ \vdots & \vdots & & \vdots \\ a_{1n} & a_{2n} & \cdots & a_{mn} \end{pmatrix}.$$

Matrix $\mathcal{A}$ is called **symmetric** if $\mathcal{A}^T = \mathcal{A}$. Symmetric matrices are necessarily square. Observe that $(\mathcal{A}^T)^T = \mathcal{A}$ for *every* matrix $\mathcal{A}$. Frequently we want to consider an n-vector $\mathbf{x}$ as an $n \times 1$ matrix having n rows and one column:

$$\mathbf{x} = \begin{pmatrix} x_1 \\ x_2 \\ \vdots \\ x_n \end{pmatrix}.$$

As such, $\mathbf{x}$ is called a **column vector**. $\mathbf{x}^T$ then has one row and n columns and is called a **row vector**:

$$\mathbf{x}^T = (x_1\ x_2\ \cdots\ x_n).$$

Note that $\mathbf{x}$ and $\mathbf{x}^T$ have the same components, so they are identical as vectors even though they appear differently as matrices.

Most of the usefulness of matrices depends on the following definition of matrix multiplication, which enables two arrays to be combined into a single one in a manner that preserves linear relationships.

DEFINITION 7

Multiplying matrices

If $\mathcal{A} = (a_{ij})$ is an $m \times n$ matrix and $\mathcal{B} = (b_{ij})$ is an $n \times p$ matrix, then the product $\mathcal{A}\mathcal{B}$ is the $m \times p$ matrix $\mathcal{C} = (c_{ij})$ with elements given by

$$c_{ij} = \sum_{k=1}^{n} a_{ik}b_{kj}, \qquad i = 1, \ldots, m, \quad j = 1, \ldots, p.$$

That is, c_{ij} is the *dot product* of the ith row of $\mathcal{A}$ and the jth column of $\mathcal{B}$ (both of which are n-vectors).

Note that only *some* pairs of matrices can be multiplied. The product $\mathcal{A}\mathcal{B}$ is only defined if the number of columns of $\mathcal{A}$ is equal to the number of rows of $\mathcal{B}$.

EXAMPLE 1

$$\begin{pmatrix} 1 & 0 & 3 \\ 2 & 1 & -1 \end{pmatrix}\begin{pmatrix} 2 & 1 & 1 & 0 \\ 0 & -1 & 3 & 1 \\ 1 & 0 & 4 & 5 \end{pmatrix} = \begin{pmatrix} 5 & 1 & 13 & 15 \\ 3 & 1 & 1 & -4 \end{pmatrix}$$

The left factor has 2 rows and 3 columns, and the right factor has 3 rows and 4 columns. Therefore, the product has 2 rows and 4 columns. The element in the first row and third column of the product, 13, is the dot product of the first row, $(1, 0, 3)$, of the left factor and the third column, $(1, 3, 4)$, of the second factor:

$$1 \times 1 + 0 \times 3 + 3 \times 4 = 13.$$

With a little practice you can easily calculate the elements of a matrix product by simultaneously running your left index finger across rows of the left factor and your right index finger down columns of the right factor while taking the dot products.

EXAMPLE 2

$$\begin{pmatrix} 1 & 2 & 3 \\ 0 & 1 & -1 \\ -2 & 3 & 0 \end{pmatrix}\begin{pmatrix} x \\ y \\ z \end{pmatrix} = \begin{pmatrix} x + 2y + 3z \\ y - z \\ -2x + 3y \end{pmatrix}$$

The product of a 3×3 matrix with a column 3-vector is a column 3-vector.

Matrix multiplication is *associative.* This means that

$$\mathcal{A}(\mathcal{B}\mathcal{C}) = (\mathcal{A}\mathcal{B})\mathcal{C}$$

(provided $\mathcal{A}$, $\mathcal{B}$, and $\mathcal{C}$ have dimensions compatible with the formation of the various products); therefore, it makes sense to write $\mathcal{A}\mathcal{B}\mathcal{C}$. However, matrix multiplication is *not commutative.* Indeed, if $\mathcal{A}$ is an $m \times n$ matrix and $\mathcal{B}$ is an $n \times p$ matrix, then the product $\mathcal{A}\mathcal{B}$ is defined, but the product $\mathcal{B}\mathcal{A}$ is not defined unless $m = p$. Even if $\mathcal{A}$ and $\mathcal{B}$ are square matrices of the same size, it is not necessarily true that $\mathcal{A}\mathcal{B} = \mathcal{B}\mathcal{A}$.

EXAMPLE 3

$$\begin{pmatrix} 1 & 2 \\ 3 & 0 \end{pmatrix}\begin{pmatrix} 1 & -1 \\ 1 & 1 \end{pmatrix} = \begin{pmatrix} 3 & 1 \\ 3 & -3 \end{pmatrix} \quad \text{but} \quad \begin{pmatrix} 1 & -1 \\ 1 & 1 \end{pmatrix}\begin{pmatrix} 1 & 2 \\ 3 & 0 \end{pmatrix} = \begin{pmatrix} -2 & 2 \\ 4 & 2 \end{pmatrix}$$

The reader should verify that if the product $\mathcal{A}\mathcal{B}$ is defined, then the transpose of the product is the product of the transposes *in the reverse order*:

$$(\mathcal{A}\mathcal{B})^T = \mathcal{B}^T\mathcal{A}^T.$$

Determinants and Matrix Inverses

In Section 10.3 we introduced 2×2 and 3×3 determinants as certain algebraic expressions associated with 2×2 and 3×3 square arrays of numbers. In general, it is possible to define the determinant $\det(\mathcal{A})$ for any square matrix. For an $n \times n$ matrix $\mathcal{A}$ we continue to denote

$$\det(\mathcal{A}) = \begin{vmatrix} a_{11} & a_{12} & \cdots & a_{1n} \\ a_{21} & a_{22} & \cdots & a_{2n} \\ \vdots & \vdots & \ddots & \vdots \\ a_{n1} & a_{n2} & \cdots & a_{nn} \end{vmatrix}.$$

We will not attempt to give a formal definition of the determinant here but will note that the properties of determinants stated for the 3×3 case in Section 10.3 continue to be true. In particular, an $n \times n$ determinant can be expanded in minors about any row or column and so expressed as a sum of multiples of $(n-1) \times (n-1)$ determinants. The expansion in minors of the $n \times n$ determinant $\det(\mathcal{A})$ about its ith row is a sum of n terms:

$$\det(\mathcal{A}) = \sum_{j=1}^{n} (-1)^{i+j} a_{ij} A_{ij},$$

where A_{ij} is the $(n-1) \times (n-1)$ determinant obtained by deleting the ith row and jth column from $\mathcal{A}$. Continuing this process, we can eventually reduce the evaluation of any $n \times n$ determinant to the evaluation of (perhaps many) 2×2 or 3×3 determinants. It is important to realize that the "diagonal" method for evaluating 2×2 or 3×3 determinants does not extend to 4×4 or higher-order determinants.

EXAMPLE 4 Here is the expansion of a certain 4×4 determinant about its third column:

$$\begin{vmatrix} 2 & 1 & 0 & 1 \\ 1 & 0 & 1 & 1 \\ 3 & 0 & 0 & 2 \\ -1 & 1 & 1 & 0 \end{vmatrix} = -1 \begin{vmatrix} 2 & 1 & 1 \\ 3 & 0 & 2 \\ -1 & 1 & 0 \end{vmatrix} - 1 \begin{vmatrix} 2 & 1 & 1 \\ 1 & 0 & 1 \\ 3 & 0 & 2 \end{vmatrix}$$

$$= -\left(-3 \begin{vmatrix} 1 & 1 \\ 1 & 0 \end{vmatrix} - 2 \begin{vmatrix} 2 & 1 \\ -1 & 1 \end{vmatrix}\right) - \left(-1 \begin{vmatrix} 1 & 1 \\ 3 & 2 \end{vmatrix}\right)$$

$$= 3(0-1) + 2(2+1) + 1(2-3) = 2.$$

Since the third column had only two nonzero elements, the expansion has only two nonzero terms involving 3×3 determinants. The first of these was then expanded about its second row, and the other about its second column.

In addition to the properties stated in Section 10.3, determinants have two other very important properties, which are stated in the following theorem.

THEOREM 3

If $\mathcal{A}$ and $\mathcal{B}$ are $n \times n$ matrices, then

(a) $\det(\mathcal{A}^T) = \det(\mathcal{A})$ and

(b) $\det(\mathcal{A}\mathcal{B}) = \det(\mathcal{A})\det(\mathcal{B})$.

We will not attempt any proof of this or other theorems in this section. The reader is referred to texts on linear algebra. Part (a) is not very difficult to prove, even in the case of general n. Part (b) cannot really be proved in general without a formal definition of determinant. However, the reader should verify (b) for 2×2 matrices by direct calculation.

We say that the square matrix $\mathcal{A}$ is **singular** if $\det(\mathcal{A}) = 0$. If $\det(\mathcal{A}) \neq 0$, we say that $\mathcal{A}$ is **nonsingular** or **invertible**.

Remark If $\mathcal{A}$ is a 3×3 matrix, then $\det(\mathcal{A})$ is the scalar triple product of the rows of $\mathcal{A}$, and its absolute value is the volume of the parallelepiped spanned by those rows. Therefore, $\mathcal{A}$ is nonsingular if and only if its rows span a parallelepiped of positive volume; the row vectors cannot all lie in the same plane. The same may be said of the columns of $\mathcal{A}$.

In general, an $n \times n$ matrix is singular if its rows (or columns), considered as vectors, satisfy one or more linear equations of the form

$$c_1\mathbf{x}_1 + c_2\mathbf{x}_2 + \cdots + c_n\mathbf{x}_n = \mathbf{0},$$

with at least one nonzero coefficient c_i. A set of vectors satisfying such a linear equation is called **linearly dependent** because one of the vectors can always be expressed as a linear combination of the others; if $c_1 \neq 0$, then

$$\mathbf{x}_1 = -\frac{c_2}{c_1}\mathbf{x}_2 - \frac{c_3}{c_1}\mathbf{x}_3 - \cdots - \frac{c_n}{c_1}\mathbf{x}_n.$$

All linear combinations of the vectors in a linearly dependent set of n vectors in $\mathbb{R}^n$ must lie in a **subspace** of dimension lower than n. Conversely, a set of m vectors in $\mathbb{R}^n$ (where $m \leq n$) is called **linearly independent** if the only linear combination of them that equals the zero vector is the one with all coefficients equal to zero; that is

$$c_1\mathbf{x}_1 + c_2\mathbf{x}_2 + \cdots + c_m\mathbf{x}_m = \mathbf{0} \quad \Longrightarrow \quad c_i = 0 \quad \text{for} \quad 1 \leq i \leq m.$$

Such a set of vectors spans (constitutes a basis of) a subspace space of dimension m in $\mathbb{R}^n$ unless $m = n$, in which case the set spans $\mathbb{R}^n$ itself.

The $n \times n$ **identity matrix** is the matrix

$$\boldsymbol{I} = \begin{pmatrix} 1 & 0 & \cdots & 0 \\ 0 & 1 & \cdots & 0 \\ \vdots & \vdots & \ddots & \vdots \\ 0 & 0 & \cdots & 1 \end{pmatrix}$$

with "1" in every position on the **main diagonal** and "0" in every other position. Evidently, $\boldsymbol{I}$ commutes with every $n \times n$ matrix: $\boldsymbol{I}\mathcal{A} = \mathcal{A}\boldsymbol{I} = \mathcal{A}$. Also $\det(\boldsymbol{I}) = 1$. The identity matrix plays the same role in matrix algebra that the number 1 plays in arithmetic.

Any nonzero number x has a reciprocal x^{-1} such that $xx^{-1} = x^{-1}x = 1$. A similar situation holds for square matrices. The **inverse** of a *nonsingular* square matrix $\mathcal{A}$ is a nonsingular square matrix $\mathcal{A}^{-1}$ satisfying

$$\mathcal{A}\mathcal{A}^{-1} = \mathcal{A}^{-1}\mathcal{A} = \boldsymbol{I}.$$

THEOREM 4

Every nonsingular square matrix $\mathcal{A}$ has a *unique* inverse $\mathcal{A}^{-1}$. Moreover, the inverse satisfies

(a) $\det(\mathcal{A}^{-1}) = \dfrac{1}{\det(\mathcal{A})}$,

(b) $(\mathcal{A}^{-1})^T = (\mathcal{A}^T)^{-1}$.

We will not have much cause to calculate inverses, but we note that it can be done by solving systems of linear equations, as the following simple example illustrates.

EXAMPLE 5 Show that the matrix $\mathcal{A} = \begin{pmatrix} 1 & -1 \\ 1 & 1 \end{pmatrix}$ is nonsingular and find its inverse.

Solution $\det(\mathcal{A}) = \begin{vmatrix} 1 & -1 \\ 1 & 1 \end{vmatrix} = 1 + 1 = 2$. Therefore, $\mathcal{A}$ is nonsingular and invertible. Let $\mathcal{A}^{-1} = \begin{pmatrix} a & b \\ c & d \end{pmatrix}$. Then $\mathcal{A}\mathcal{A}^{-1} = \boldsymbol{I}$, that is,

$$\begin{pmatrix} 1 & 0 \\ 0 & 1 \end{pmatrix} = \begin{pmatrix} 1 & -1 \\ 1 & 1 \end{pmatrix}\begin{pmatrix} a & b \\ c & d \end{pmatrix} = \begin{pmatrix} a-c & b-d \\ a+c & b+d \end{pmatrix},$$

so a, b, c, and d must satisfy the systems of equations

$$\begin{cases} a - c = 1 \\ a + c = 0 \end{cases} \qquad \begin{cases} b - d = 0 \\ b + d = 1. \end{cases}$$

Evidently, $a = b = d = 1/2$, $c = -1/2$, and

$$\mathcal{A}^{-1} = \begin{pmatrix} \frac{1}{2} & \frac{1}{2} \\ -\frac{1}{2} & \frac{1}{2} \end{pmatrix}.$$

Remark The same technique used in Example 5 can be used to show that the general 2×2 matrix $\mathcal{A} = \begin{pmatrix} a & b \\ c & d \end{pmatrix}$ is nonsingular (and therefore invertible) provided $D = ad - bc \neq 0$, and in this case

$$\mathcal{A}^{-1} = \begin{pmatrix} \dfrac{d}{D} & \dfrac{-b}{D} \\ \dfrac{-c}{D} & \dfrac{a}{D} \end{pmatrix}.$$

Generally, matrix inversion is not carried out by the method of Example 5 but rather by an orderly process of performing operations on the rows of the matrix to transform it into the identity. When the same operations are performed on the rows of the identity matrix, the inverse of the original matrix results. See a text on linear algebra for a description of the method. A singular matrix has no inverse.

Linear Transformations

A function $\mathbf{F}$ whose domain is the m-dimensional space $\mathbb{R}^m$ and whose range is contained in the n-dimensional space $\mathbb{R}^n$ is called a **linear transformation from $\mathbb{R}^m$ to $\mathbb{R}^n$** if it satisfies

$$\mathbf{F}(\lambda\mathbf{x} + \mu\mathbf{y}) = \lambda\mathbf{F}(\mathbf{x}) + \mu\mathbf{F}(\mathbf{y})$$

for all points $\mathbf{x}$ and $\mathbf{y}$ in $\mathbb{R}^m$ and all real numbers λ and μ. To such a linear transformation $\mathbf{F}$ there corresponds an $n \times m$ matrix $\mathcal{F}$ such that for all $\mathbf{x}$ in $\mathbb{R}^m$,

$$\mathbf{F}(\mathbf{x}) = \mathcal{F}\mathbf{x},$$

or, expressed in terms of the components of $\mathbf{x}$,

$$\mathbf{F}(x_1, x_2, \cdots, x_m) = \mathcal{F}\begin{pmatrix} x_1 \\ x_2 \\ \vdots \\ x_m \end{pmatrix}.$$

We say that $\mathcal{F}$ is a **matrix representation** of the linear transformation $\mathbf{F}$. If $m = n$ so that $\mathbf{F}$ maps $\mathbb{R}^m$ into itself, then $\mathcal{F}$ is a square matrix. In this case $\mathcal{F}$ is nonsingular if and only if $\mathbf{F}$ is one-to-one and has the whole of $\mathbb{R}^m$ as range.

A composition of linear transformations is still a linear transformation and will have a matrix representation. The real motivation lying behind the definition of matrix multiplication is that the matrix representation of a *composition* of linear transformations is the *product* of the individual matrix representations of the transformations being composed.

THEOREM 5

If $\mathbf{F}$ is a linear transformation from $\mathbb{R}^m$ to $\mathbb{R}^n$ represented by the $n \times m$ matrix $\mathcal{F}$, and if $\mathbf{G}$ is a linear transformation from $\mathbb{R}^n$ to $\mathbb{R}^p$ represented by the $p \times n$ matrix $\mathcal{G}$, then the composition $\mathbf{G} \circ \mathbf{F}$ defined by

$$\mathbf{G} \circ \mathbf{F}(x_1, x_2, \ldots, x_m) = \mathbf{G}\Big(\mathbf{F}(x_1, x_2, \ldots, x_m)\Big)$$

is itself a linear transformation from $\mathbb{R}^m$ to $\mathbb{R}^p$ represented by the $p \times m$ matrix $\mathcal{G}\mathcal{F}$. That is,

$$\mathbf{G}\Big(\mathbf{F}(\mathbf{x})\Big) = \mathcal{G}\mathcal{F}\mathbf{x}.$$

Linear Equations

A system of n linear equations in n unknowns:

$$\begin{aligned} a_{11}x_1 + a_{12}x_2 + \cdots + a_{1n}x_n &= b_1 \\ a_{21}x_1 + a_{22}x_2 + \cdots + a_{2n}x_n &= b_2 \\ &\vdots \\ a_{n1}x_1 + a_{n2}x_2 + \cdots + a_{nn}x_n &= b_n \end{aligned}$$

can be written compactly as a single matrix equation,

$$\mathcal{A}\mathbf{x} = \mathbf{b},$$

where

$$\mathcal{A} = \begin{pmatrix} a_{11} & a_{12} & \cdots & a_{1n} \\ a_{21} & a_{22} & \cdots & a_{2n} \\ \vdots & \vdots & \ddots & \vdots \\ a_{n1} & a_{n2} & \cdots & a_{nn} \end{pmatrix}, \qquad \mathbf{x} = \begin{pmatrix} x_1 \\ x_2 \\ \vdots \\ x_n \end{pmatrix}, \qquad \text{and} \quad \mathbf{b} = \begin{pmatrix} b_1 \\ b_2 \\ \vdots \\ b_n \end{pmatrix}.$$

Compare the equation $\mathcal{A}\mathbf{x} = \mathbf{b}$ with the equation $ax = b$ for a single unknown x. The equation $ax = b$ has the unique solution $x = a^{-1}b$ provided $a \neq 0$. By analogy, the linear system $\mathcal{A}\mathbf{x} = \mathbf{b}$ has a unique solution given by

$$\mathbf{x} = \mathcal{A}^{-1}\mathbf{b},$$

provided $\mathcal{A}$ is nonsingular. To see this, just multiply both sides of the equation $\mathcal{A}\mathbf{x} = \mathbf{b}$ on the left by $\mathcal{A}^{-1}$; $\mathbf{x} = I\mathbf{x} = \mathcal{A}^{-1}\mathcal{A}\mathbf{x} = \mathcal{A}^{-1}\mathbf{b}$.

If $\mathcal{A}$ is singular, then the system $\mathcal{A}\mathbf{x} = \mathbf{b}$ may or may not have a solution, and if a solution exists it will not be unique. Consider the case $\mathbf{b} = \mathbf{0}$ (the zero vector). Then any vector $\mathbf{x}$ perpendicular to all the rows of $\mathcal{A}$ will satisfy the system. Since the rows of $\mathcal{A}$ lie in a space of dimension less than n (because $\det(\mathcal{A}) = 0$), there will be at least a line of such vectors $\mathbf{x}$. Thus, solutions of $\mathcal{A}\mathbf{x} = \mathbf{0}$ are not unique if $\mathcal{A}$ is singular. The same must be true of the system $\mathcal{A}^T\mathbf{y} = \mathbf{0}$; there will be nonzero vectors $\mathbf{y}$ satisfying it if $\mathcal{A}$ is singular. But then, if the system $\mathcal{A}\mathbf{x} = \mathbf{b}$ has any solution $\mathbf{x}$, we must have

$$(\mathbf{y} \bullet \mathbf{b}) = \mathbf{y}^T\mathbf{b} = \mathbf{y}^T\mathcal{A}\mathbf{x} = (\mathbf{x}^T\mathcal{A}^T\mathbf{y})^T = (\mathbf{x}^T\mathbf{0})^T = (0).$$

Hence, $\mathcal{A}\mathbf{x} = \mathbf{b}$ can only have solutions for those vectors $\mathbf{b}$ that are perpendicular to *every* solution $\mathbf{y}$ of $\mathcal{A}^T\mathbf{y} = \mathbf{0}$.

A system of m linear equations in n unknowns may or may not have any solutions if $n < m$. It will have solutions if some $m - n$ of the equations are *linear combinations* (sums of multiples) of the other n equations. If $n > m$, then we can try to solve the m equations for m of the variables, allowing the solutions to depend on the other $n - m$ variables. Such a solution exists if the determinant of the coefficients of the m variables for which we want to solve is not zero. This is a special case of the **Implicit Function Theorem**, which we will examine in Section 12.8.

EXAMPLE 6 Solve $\begin{cases} 2x + y - 3z = 4 \\ x + 2y + 6z = 5 \end{cases}$ for x and y in terms of z.

Solution The system can be expressed in the form

$$\mathcal{A}\begin{pmatrix} x \\ y \end{pmatrix} = \begin{pmatrix} 4 + 3z \\ 5 - 6z \end{pmatrix}, \qquad \text{where} \quad \mathcal{A} = \begin{pmatrix} 2 & 1 \\ 1 & 2 \end{pmatrix}.$$

$\mathcal{A}$ has determinant 3 and inverse $\mathcal{A}^{-1} = \begin{pmatrix} 2/3 & -1/3 \\ -1/3 & 2/3 \end{pmatrix}$. Thus,

$$\begin{pmatrix} x \\ y \end{pmatrix} = \mathcal{A}^{-1}\begin{pmatrix} 4+3z \\ 5-6z \end{pmatrix} = \begin{pmatrix} 2/3 & -1/3 \\ -1/3 & 2/3 \end{pmatrix}\begin{pmatrix} 4+3z \\ 5-6z \end{pmatrix} = \begin{pmatrix} 1+4z \\ 2-5z \end{pmatrix}.$$

The solution is $x = 1+4z$, $y = 2-5z$. (Of course, this solution could have been found by elimination of x or y from the given equations without using matrix methods.)

The following theorem states a result of some theoretical importance expressing the solution of the system $\mathcal{A}\mathbf{x} = \mathbf{b}$ for nonsingular $\mathcal{A}$ in terms of determinants.

THEOREM 6

Cramer's Rule

Let $\mathcal{A}$ be a nonsingular $n \times n$ matrix. Then the solution $\mathbf{x}$ of the system

$$\mathcal{A}\mathbf{x} = \mathbf{b}$$

has components given by

$$x_1 = \frac{\det(\mathcal{A}_1)}{\det(\mathcal{A})}, \quad x_2 = \frac{\det(\mathcal{A}_2)}{\det(\mathcal{A})}, \quad \ldots, \quad x_n = \frac{\det(\mathcal{A}_n)}{\det(\mathcal{A})},$$

where $\mathcal{A}_j$ is the matrix $\mathcal{A}$ with its jth column replaced by the column vector $\mathbf{b}$. That is,

$$\det(\mathcal{A}_j) = \begin{vmatrix} a_{11} & \cdots & a_{1(j-1)} & b_1 & a_{1(j+1)} & \cdots & a_{1n} \\ a_{21} & \cdots & a_{2(j-1)} & b_2 & a_{2(j+1)} & \cdots & a_{2n} \\ \vdots & & \vdots & \vdots & \vdots & & \vdots \\ a_{n1} & \cdots & a_{n(j-1)} & b_n & a_{n(j+1)} & \cdots & a_{nn} \end{vmatrix}.$$

The following example provides a concrete illustration of the use of Cramer's Rule to solve a specific linear system. However, Cramer's Rule is primarily used in a more general (theoretical) context; it is not efficient to use determinants to calculate solutions of linear systems.

EXAMPLE 7 Find the point of intersection of the three planes

$$\begin{aligned} x + y + 2z &= 1 \\ 3x + 6y - z &= 0 \\ x - y - 4z &= 3. \end{aligned}$$

Solution The solution of the linear system above provides the coordinates of the intersection point. The determinant of the coefficient matrix of this system is

$$\det(\mathcal{A}) = \begin{vmatrix} 1 & 1 & 2 \\ 3 & 6 & -1 \\ 1 & -1 & -4 \end{vmatrix} = -32,$$

so the system does have a unique solution. We have

$$x = \frac{1}{-32}\begin{vmatrix} 1 & 1 & 2 \\ 0 & 6 & -1 \\ 3 & -1 & -4 \end{vmatrix} = \frac{-64}{-32} = 2,$$

$$y = \frac{1}{-32}\begin{vmatrix} 1 & 1 & 2 \\ 3 & 0 & -1 \\ 1 & 3 & -4 \end{vmatrix} = \frac{32}{-32} = -1,$$

$$z = \frac{1}{-32}\begin{vmatrix} 1 & 1 & 1 \\ 3 & 6 & 0 \\ 1 & -1 & 3 \end{vmatrix} = \frac{0}{-32} = 0.$$

The intersection point is $(2, -1, 0)$.

Quadratic Forms, Eigenvalues, and Eigenvectors

If $\mathbf{x}$ is a column vector in $\mathbb{R}^n$ and $\mathcal{A} = (a_{ij})$ is an $n \times n$, real, symmetric matrix (i.e., $a_{ij} = a_{ji}$ for $1 \le i, j \le n$), then the expression

$$Q(\mathbf{x}) = \mathbf{x}^T \mathcal{A} \mathbf{x} = \sum_{i,j=1}^{n} a_{ij} x_i x_j$$

is called a **quadratic form** on $\mathbb{R}^n$ corresponding to the matrix $\mathcal{A}$. Observe that $Q(\mathbf{x})$ is a real number for every n-vector $\mathbf{x}$.

We say that $\mathcal{A}$ is **positive definite** if $Q(\mathbf{x}) > 0$ for every nonzero vector $\mathbf{x}$. Similarly, $\mathcal{A}$ is **negative definite** if $Q(\mathbf{x}) < 0$ for every nonzero vector $\mathbf{x}$. We say that $\mathcal{A}$ is **positive semidefinite** (or **negative semidefinite**) if $Q(\mathbf{x}) \ge 0$ (or $Q(\mathbf{x}) \le 0$) for every nonzero vector $\mathbf{x}$.

If $Q(\mathbf{x}) > 0$ for some nonzero vectors $\mathbf{x}$ while $Q(\mathbf{x}) < 0$ for other such $\mathbf{x}$ (i.e., if $\mathcal{A}$ is neither positive semidefinite nor negative semidefinite), then we will say that $\mathcal{A}$ is **indefinite**.

EXAMPLE 8 The expression $Q(x, y, z) = 3x^2 + 2y^2 + 5z^2 - 2xy + 4xz + 2yz$ is a quadratic form on $\mathbb{R}^3$ corresponding to the symmetric matrix

$$\mathcal{A} = \begin{pmatrix} 3 & -1 & 2 \\ -1 & 2 & 1 \\ 2 & 1 & 5 \end{pmatrix}.$$

Observe how the elements of the matrix are obtained from the coefficients of Q; the coefficients of x^2, y^2, and z^2 form the main diagonal elements, while the coefficients of the product terms are cut in half and half is put in each of the two corresponding symmetric off-diagonal positions.

The matrix $\mathcal{A}$ is positive definite since $Q(x, y, z)$ can be rewritten in the form

$$Q(x, y, z) = x^2 + (x - y)^2 + (x + 2z)^2 + (y + z)^2,$$

from which it is apparent that $Q(x, y, z) \ge 0$ for all (x, y, z) and $Q(x, y, z) = 0$ only if $x = y = z = 0$.

In Section 13.1 we will use the positive or negative definiteness of certain matrices to classify critical points of functions of several variables as local maxima and minima. Useful criteria for definiteness can be expressed in terms of the *eigenvalues* of the matrix $\mathcal{A}$.

We say that λ is an **eigenvalue** of the $n \times n$ square matrix $\mathcal{A} = (a_{ij})$ if there exists a *nonzero* column vector $\mathbf{x}$ such that $\mathcal{A}\mathbf{x} = \lambda\mathbf{x}$, or, equivalently,

$$(\mathcal{A} - \lambda I)\mathbf{x} = \mathbf{0},$$

where I is the $n \times n$ identity matrix. The nonzero vector $\mathbf{x}$ is called an **eigenvector** of $\mathcal{A}$ corresponding to the eigenvalue λ and can exist only if $\mathcal{A} - \lambda I$ is a singular matrix, that is, if

$$\det(\mathcal{A} - \lambda I) = \begin{vmatrix} a_{11} - \lambda & a_{12} & \cdots & a_{1n} \\ a_{21} & a_{22} - \lambda & \cdots & a_{2n} \\ \vdots & \vdots & \ddots & \vdots \\ a_{n1} & a_{n2} & \cdots & a_{nn} - \lambda \end{vmatrix} = 0.$$

The eigenvalues of $\mathcal{A}$ must satisfy this nth-degree polynomial equation, so they can be either real or complex. The following theorems are proved in standard linear algebra texts.

THEOREM 7

If $\mathcal{A} = \left(a_{ij}\right)_{i,j=1}^{n}$ is a real, symmetric matrix, then

(a) all the eigenvalues of $\mathcal{A}$ are real,

(b) all the eigenvalues of $\mathcal{A}$ are nonzero if $\det(\mathcal{A}) \neq 0$,

(c) $\mathcal{A}$ is positive definite if all its eigenvalues are positive,

(d) $\mathcal{A}$ is negative definite if all its eigenvalues are negative,

(e) $\mathcal{A}$ is positive semidefinite if all its eigenvalues are nonnegative,

(f) $\mathcal{A}$ is negative semidefinite if all its eigenvalues are nonpositive,

(g) $\mathcal{A}$ is indefinite if it has at least one positive eigenvalue and at least one negative eigenvalue.

THEOREM 8

Let $\mathcal{A} = \left(a_{ij}\right)_{i,j=1}^{n}$ be a real symmetric matrix and consider the determinants

$$D_i = \begin{vmatrix} a_{11} & a_{12} & \cdots & a_{1i} \\ a_{21} & a_{22} & \cdots & a_{2i} \\ \vdots & \vdots & \ddots & \vdots \\ a_{i1} & a_{i2} & \cdots & a_{ii} \end{vmatrix} \qquad \text{for } 1 \le i \le n.$$

Thus, $D_1 = a_{11}$, $D_2 = \begin{vmatrix} a_{11} & a_{12} \\ a_{21} & a_{22} \end{vmatrix} = a_{11}a_{22} - a_{12}a_{21} = a_{11}a_{22} - a_{12}^2$, etc.

(a) If $D_i > 0$ for $1 \le i \le n$, then $\mathcal{A}$ is positive definite.

(b) If $D_i > 0$ for even numbers i in $\{1, 2, \dots, n\}$, and $D_i < 0$ for odd numbers i in $\{1, 2, \dots, n\}$, then $\mathcal{A}$ is negative definite.

(c) If $\det(\mathcal{A}) = D_n \neq 0$ but neither of the above conditions hold, then $Q(\mathbf{x})$ is indefinite.

(d) If $\det(\mathcal{A}) = 0$, then $\mathcal{A}$ is not positive or negative definite and may be semidefinite or indefinite.

EXAMPLE 9 For the matrix $\mathcal{A}$ of Example 8, we have

$$D_1 = 3 > 0, \quad D_2 = \begin{vmatrix} 3 & -1 \\ -1 & 2 \end{vmatrix} = 5 > 0, \quad D_3 = \begin{vmatrix} 3 & -1 & 2 \\ -1 & 2 & 1 \\ 2 & 1 & 5 \end{vmatrix} = 10 > 0,$$

which reconfirms that the quadratic form of that exercise is positive definite.

EXERCISES 10.7

Evaluate the matrix products in Exercises 1–4.

1. $\begin{pmatrix} 3 & 0 & -2 \\ 1 & 1 & 2 \\ -1 & 1 & -1 \end{pmatrix}\begin{pmatrix} 2 & 1 \\ 3 & 0 \\ 0 & -2 \end{pmatrix}$

2. $\begin{pmatrix} 1 & 1 & 1 \\ 0 & 1 & 1 \\ 0 & 0 & 1 \end{pmatrix}\begin{pmatrix} 1 & 1 & 1 \\ 0 & 1 & 1 \\ 0 & 0 & 1 \end{pmatrix}$

3. $\begin{pmatrix} a & b \\ c & d \end{pmatrix}\begin{pmatrix} w & x \\ y & z \end{pmatrix}$

4. $\begin{pmatrix} w & x \\ y & z \end{pmatrix}\begin{pmatrix} a & b \\ c & d \end{pmatrix}$

5. Evaluate $\mathcal{A}\mathcal{A}^T$ and $\mathcal{A}^2 = \mathcal{A}\mathcal{A}$, where

$$\mathcal{A} = \begin{pmatrix} 1 & 1 & 1 & 1 \\ 0 & 1 & 1 & 1 \\ 0 & 0 & 1 & 1 \\ 0 & 0 & 0 & 1 \end{pmatrix}.$$

6. Evaluate $\mathbf{x}\mathbf{x}^T$, $\mathbf{x}^T\mathbf{x}$, and $\mathbf{x}^T\mathcal{A}\mathbf{x}$, where

$$\mathbf{x} = \begin{pmatrix} x \\ y \\ z \end{pmatrix} \quad \text{and} \quad \mathcal{A} = \begin{pmatrix} a & p & q \\ p & b & r \\ q & r & c \end{pmatrix}.$$

Evaluate the determinants in Exercises 7–8.

7. $\begin{vmatrix} 2 & 3 & -1 & 0 \\ 4 & 0 & 2 & 1 \\ 1 & 0 & -1 & 1 \\ -2 & 0 & 0 & 1 \end{vmatrix}$ **8.** $\begin{vmatrix} 1 & 1 & 1 & 1 \\ 1 & 2 & 3 & 4 \\ -2 & 0 & 2 & 4 \\ 3 & -3 & 2 & -2 \end{vmatrix}$

9. Show that if $\mathcal{A} = (a_{ij})$ is an $n \times n$ matrix for which $a_{ij} = 0$ whenever $i > j$, then $\det(\mathcal{A}) = \prod_{k=1}^{n} a_{kk}$, the product of the elements on the main diagonal of $\mathcal{A}$.

10. Show that $\begin{vmatrix} 1 & 1 \\ x & y \end{vmatrix} = y - x$, and

$$\begin{vmatrix} 1 & 1 & 1 \\ x & y & z \\ x^2 & y^2 & z^2 \end{vmatrix} = (y - x)(z - x)(z - y).$$

Try to generalize this result to the $n \times n$ case.

11. Verify the associative law $(\mathcal{AB})\mathcal{C} = \mathcal{A}(\mathcal{BC})$ by direct calculation for three arbitrary 2×2 matrices.

12. Show that $\det(\mathcal{A}^T) = \det(\mathcal{A})$ for $n \times n$ matrices by induction on n. Start with the 2×2 case.

13. Verify by direct calculation that $\det(\mathcal{AB}) = \det(\mathcal{A})\det(\mathcal{B})$ holds for two arbitrary 2×2 matrices.

14. Let $\mathcal{A}_\theta = \begin{pmatrix} \cos\theta & \sin\theta \\ -\sin\theta & \cos\theta \end{pmatrix}$. Show that $(\mathcal{A}_\theta)^T = (\mathcal{A}_\theta)^{-1} = \mathcal{A}_{-\theta}$.

15. Verify by using matrix multiplication that the inverse of the matrix $\mathcal{A}$ in the remark following Example 5 is as specified there.

16. For what values of the variables x and y is the matrix $\mathcal{B} = \begin{pmatrix} x & y \\ x^2 & y^2 \end{pmatrix}$ invertible, and what is its inverse?

Find the inverses of the matrices in Exercises 17–18.

17. $\begin{pmatrix} 1 & 1 & 1 \\ 0 & 1 & 1 \\ 0 & 0 & 1 \end{pmatrix}$ **18.** $\begin{pmatrix} 1 & 0 & -1 \\ -1 & 1 & 0 \\ 2 & 1 & 3 \end{pmatrix}$

19. Use your result from Exercise18 to solve the linear system

$$\begin{cases} x - z = -2 \\ -x + y = 1 \\ 2x + y + 3z = 13. \end{cases}$$

20. Solve the system of Exercise 19 by using Cramer's Rule.

21. Solve the system $\begin{cases} x_1 + x_2 + x_3 + x_4 = 0 \\ x_1 + x_2 + x_3 - x_4 = 4 \\ x_1 + x_2 - x_3 - x_4 = 6 \\ x_1 - x_2 - x_3 - x_4 = 2. \end{cases}$

22. Verify Theorem 5 for the special case where **F** and **G** are linear transformations from $\mathbb{R}^2$ to $\mathbb{R}^2$.

In Exercises 23–28, classify the given symmetric matrices as positive or negative definite, positive or negative semidefinite, or indefinite.

23. $\begin{pmatrix} -1 & 1 \\ 1 & -2 \end{pmatrix}$ **24.** $\begin{pmatrix} 1 & 2 & 0 \\ 2 & 1 & 0 \\ 0 & 0 & 1 \end{pmatrix}$

25. $\begin{pmatrix} 2 & 1 & 1 \\ 1 & 2 & 1 \\ 1 & 1 & 2 \end{pmatrix}$ **26.** $\begin{pmatrix} 1 & 1 & 0 \\ 1 & 1 & 0 \\ 0 & 0 & 1 \end{pmatrix}$

27. $\begin{pmatrix} 1 & 0 & 1 \\ 0 & 1 & -1 \\ 1 & -1 & 1 \end{pmatrix}$ **28.** $\begin{pmatrix} 2 & 0 & 1 \\ 0 & 4 & -1 \\ 1 & -1 & 1 \end{pmatrix}$

10.8 Using Maple for Vector and Matrix Calculations

The use of a computer algebra system can free us from much of the tedious calculation needed to do calculus. This is especially true of calculations in multivariable and vector calculus, where the calculations can quickly become unmanageable as the number of variables increases. This author's colleague, Dr. Robert Israel, has written an excellent book, *Calculus, the Maple Way*, to show how Maple can be used effectively for doing calculus involving both single-variable and multivariable functions.

In this book we will occasionally call on the power of Maple to carry out calculations involving functions of several variables and vector-valued functions of one or more variables. This section illustrates some of the most basic techniques for calculating with vectors and matrices. The examples here were calculated using Maple 10, but Maple 6 or later should give similar output.

Most of Maple's capability to deal with vectors and matrices is not in its kernel but is written into a package of procedures called **LinearAlgebra**. Therefore, it is customary to load this package at the beginning of a session where it will be needed:

```
> with(LinearAlgebra):
```

One usually completes a Maple command with a semicolon rather than a colon. You can use a colon to suppress output. Had we used a semicolon to complete the command the result would have produced a list of all the procedures defined in the LinearAlgebra package.

Maple also includes a second linear algebra package called **linalg**, but it is inferior to LinearAlgebra, especially for heavy-duty numerical calculations using large matrices; it is also somewhat more difficult to use. However, the linalg package was present in releases of Maple earlier than release 6, and is still present in release 9. We will not make any use of linalg here, but it was used instead of LinearAlgebra in the fifth edition of this book.

Vectors

There are several ways to define vectors in Maple; the easiest are to use the `Vector([,])` or `<,>` constructions, where a comma-separated list of the components of the vector is placed in the square or angle brackets. Both of these constructions produce column vectors:

```
> Uc := Vector([1,2,3]); Vc := <a,b,c>;
```

$$Uc := \begin{bmatrix} 1 \\ 2 \\ 3 \end{bmatrix}$$

$$Vc := \begin{bmatrix} a \\ b \\ c \end{bmatrix}$$

You can use `Vector[row]([,])` to produce a row vector; alternatively, you can define a row vector using angle brackets with "|" to separate the components:

```
> Ur := Vector[row]([1,2,3]); Vr := <a|b|c>;
```

$$Ur := [1, 2, 3]$$

$$Vr := [a, b, c]$$

Vectors can be of any dimension; simply include the appropriate number of commas or | separated components. You can also use the `Vector()` construct with two arguments, the first a positive integer giving the dimension of the vector and the second either a square-bracket-enclosed list of components or an assignment rule giving the value of the ith component:

```
> <5|-2|3|x>; W := Vector[row](5, i -> i^2);
```

$$[5, -2, 3, x]$$

$$W := [1, 4, 9, 16, 25]$$

We can also construct a vector with arbitrary components like this:

```
> X := Vector(2, symbol=x);
  Y := Vector[row](4, symbol=y);
```

$$X := \begin{bmatrix} x_1 \\ x_2 \end{bmatrix}$$

$$Y := [y_1, y_2, y_3, y_4]$$

The components of a vector can be referenced by appending the index of the component, enclosed in square brackets, to the name or constructor of the vector. The fourth component of vector `W` above is `W[4]`:

```
> W[4]; Vector(16, i -> 3*i - 1)[10]; X[2]+Y[3];
```

$$16$$
$$29$$
$$x_2 + y_3$$

Vectors of the same dimension and type (row or column) can be added, subtracted, and multiplied by scalars using the ordinary operators $+$, $-$, and $*$:

```
> Uc + Vc; Vc - 3*Uc;
```

$$\begin{bmatrix} 1+a \\ 2+b \\ 3+c \end{bmatrix}$$

$$\begin{bmatrix} a-3 \\ b-6 \\ c-9 \end{bmatrix}$$

For most vector calculations it doesn't matter whether you think of vectors as row or column vectors, but it does make a difference for some LinearAlgebra operators; if you try to add a row vector to a column vector, or two vectors of different dimensions, you will get an error message.

The LinearAlgebra package also defines the product functions `DotProduct` and `CrossProduct`, each of which takes two vector arguments. For DotProduct, the arguments must be of the same but arbitrary dimension. For CrossProduct, both arguments must have dimension 3. However, neither requires both arguments to be of the same type (row or column). The cross product will be a column vector unless both its arguments are row vectors.

As defined in the LinearAlgebra package, DotProduct can produce some strange results. Consider the following:

```
> DotProduct(Uc,Vc); DotProduct(Vc,Uc);
  DotProduct(Ur,Vr);
```

$$a + 2b + 3c$$
$$\bar{a} + 2\bar{b} + 3\bar{c}$$
$$\bar{a} + 2\bar{b} + 3\bar{c}$$

What is going on here? The bars on the unknown quantities a, b, and c denote complex conjugates of these quantities; The LinearAlgebra package is designed to meet the needs of a great many users of linear algebra, not just calculus students for whom all vectors are assumed to have real components. In fact, `DotProduct(U,V)` sums the products of the complex congugates of the components of `U` and the unconjugated components of `V` if both vectors are column vectors, and vice versa if both are row vectors. In the first example above, the components of `Uc` are real numbers so no conjugates appeared over them; in the other two cases it is the components of `Vc` or `Vr` that require conjugation, and since Maple doesn't know that these are real, it puts on the bars. To avoid this difficulty when using real vectors, include "`conjugate=false`" as a third argument when using DotProduct from the LinearAlgebra package:

```
> DotProduct(Ur,Vr, conjugate=false);
```

$$a + 2b + 3c$$

It is also possible to use a dot "." as a binary operator to calculate a dot product. However, dot also represents matrix multiplication, so you must use a row vector to the left of the dot and a column vector to the right to be sure of getting a dot product.

```
> <1|2|3>.<a,b,c>; <1,2,3>.<a|b|c>;
```

$$a + 2b + 3c$$

$$\begin{bmatrix} a & b & c \\ 2a & 2b & 2c \\ 3a & 3b & 3c \end{bmatrix}$$

LinearAlgebra also has a CrossProduct function, which applies only to 3-vectors. It does not matter whether either of the arguments is a row or column vector. This function can be called using either `CrossProduct(U,V)` or `U &x V`.

```
> CrossProduct(Uc,Vc); Ur &x Vr;
```

$$\begin{bmatrix} 2c - 3b \\ 3a - c \\ b - 2a \end{bmatrix}$$

$$[2c{-}3b, 3a - c, b - 2a]$$

LinearAlgebra has a function `Norm()` for calculating the length of a vector. Unfortunately, Maple knows many different definitions for the length of a vector. The one we use is the Euclidean length. The Euclidean length of a vector `V` is calculated by `Norm(V,Euclidean)` or `Norm(V,2)`. (In the latter case the 2 stands for the fact that we use the *square root* of the sum of the *squares* of the components to find the length.)

```
> Norm(Ur,Euclidean); Norm(<1,-1,2,-3,1>,2);
```

$$\sqrt{14}$$

$$4$$

You can use `Normalize(U,Euclidean)` of `Normalize(U,2)` to find a unit vector in the same direction as `U`. Of course, you could always just multiply `U` by the scalar which is the reciprocal of its length:

```
> Normalize(<2|-2|1>,2); (1/Norm(Uc,2))*Uc;
```

$$\left[\frac{2}{3}, \frac{-2}{3}, \frac{1}{3}\right]$$

$$\begin{bmatrix} \frac{1}{14}\sqrt{14} \\ \frac{1}{7}\sqrt{14} \\ \frac{3}{14}\sqrt{14} \end{bmatrix}$$

LinearAlgebra has a function `VectorAngle` to give the angle between two vectors. It doesn't matter whether either vector is a row or column. The result will be in radian measure so you will have to multiply it by $180/\pi$ to get the angle in degrees.

```
> VectorAngle(<2,2,1>,<1,-2,2>);
```

$$\frac{1}{2}\pi$$

To further illustrate these ideas, let us get Maple to calculate an equation of the plane through $(2, 1, -1)$ perpendicular to the line of intersection of the two planes $2x + 3y + z = 5$ and $3x - 2y - 4z = 1$.

```
> (<2|3|1> &x <3|-2|-4>) . (<x,y,z>-<2,1,-1>) = 0;
```

$$-10x - 4 + 11y - 13z = 0$$

or, as we would write it, $10x - 11y + 13z = -4$. Note how we used the cross product of two row vectors (which is itself, therefore, a row vector) to the left of the "." and a difference of two column vectors (which is itself a column vector) to the right of the "." for calculating the dot product.

Finally, let us use Maple to verify the identity

$$(\mathbf{U} \times \mathbf{V}) \times \mathbf{W} = (\mathbf{W} \bullet \mathbf{U})\mathbf{V} - (\mathbf{W} \bullet \mathbf{V})\mathbf{U}.$$

First, we define **U**, **V**, and **W** to be vectors with arbitrary components. In view of the two dot products on the right-hand side of the identity, we make **W** a row vector and the other two column vectors:

```
> U := Vector(3,symbol=u);
  V := Vector(3, symbol=v);
  W := Vector[row](3, symbol=w);
```

$$U := \begin{bmatrix} u_1 \\ u_2 \\ u_3 \end{bmatrix}$$

$$V := \begin{bmatrix} v_1 \\ v_2 \\ v_3 \end{bmatrix}$$

$$W := [w_1, w_2, w_3]$$

Now we only need to subtract the right side of the identity from the left side and simplify the result:

```
> simplify((U &x V) &x W - (W . U)*V + (W . V)*U);
```

$$\begin{bmatrix} 0 \\ 0 \\ 0 \end{bmatrix}$$

The result is the zero vector, thus confirming the identity.

Remark Maple 8 and later releases have a new package called VectorCalculus, which provides greater functionality than the LinearAlgebra package for dealing with vector-valued functions and functions of vector variables. We will be illustrating the use of this package in later chapters, but note here that it also defines the vector operations considered above but not all of the matrix functions considered below. VectorCalculus reports vectors as linear combinations of basis vectors rather than as row or column matrices. The default bases it uses consist of vectors e_x, e_y, e_z (rather than **i**, **j**, **k**) for vectors of dimension up to 3, but $e_{x1}, e_{x2}, \ldots$ for dimensions higher than 3. Nevertheless, although it is not apparent from the way VectorCalculus displays vectors, it still maintains the distinction between row and column vectors and won't let you add a row vector to a column vector. A big advantage of the VectorCalculus package over LinearAlgebra is that VectorCalculus uses the usual definition of dot product (even when using the "." notation), so that the order of factors in a dot product is irrelevant and no complex conjugation is used. If you want to use the VectorCalculus package and still have access to all the matrix operations provided by LinearAlgebra, load the VectorCalculus package *after* the LinearAlgebra package, so that its new definitions of vector operations will replace those of the LinearAlgebra package.

```
> with(LinearAlgebra):
  with(VectorCalculus):
```

Even with output suppressed, the second `with` above produces a few lines of "warnings" mainly about the changed definitions of some vector operations.

```
> V1 := <2,-3,4>; V2 := <a|b|c>; V3 := <2,-3,4,-5,6>;
```

$$V1 := 2e_x - 3e_y + 4e_z$$

$$V2 := ae_x + be_y + ce_z$$

$$V3 := 2e_{x1} - 3e_{x2} + 4e_{x3} - 5e_{x4} + 6e_{x5}$$

```
> V1.V2; V2.V1;
```

$$2a - 3b + 4c$$
$$2a - 3b + 4c$$

Because `V1` is a column vector and `V2` is a row vector, any attempt to calculate a linear combination of these vectors will generate an error, as will attempts to calculate `M.V2` or `V1.M` if `M` is a 3×3 matrix. Of course, `M.V1` will work fine, as will `V2.M`, although the result will be a one-row matrix rather than a vector. We will examine VectorCalculus further in later chapters.

Matrices

The LinearAlgebra package also provides a variety of ways to define and manipulate matrices. We can define a matrix as a column vector whose elements are row vectors, or as a row vector whose elements are column vectors:

```
> <<1|1|1>,<2|1|3>>; <<1,2>|<1,1>|<1,3>>;
```

$$\begin{bmatrix} 1 & 1 & 1 \\ 2 & 1 & 3 \end{bmatrix}$$

$$\begin{bmatrix} 1 & 1 & 1 \\ 2 & 1 & 3 \end{bmatrix}$$

You can also use the `Matrix` function to define a matrix. This function can either be supplied with a list of lists specifying the rows of the matrix, or two positive integers (the number of rows and columns, respectively) and a rule for calculating the element in the ith row and jth column.

```
> L := Matrix([[1,1,1],[2,1,3]]);
  M := Matrix(3,3, (i,j) -> i-j);
```

$$L := \begin{bmatrix} 1 & 1 & 1 \\ 2 & 1 & 3 \end{bmatrix}$$

$$M := \begin{bmatrix} 0 & -1 & -2 \\ 1 & 0 & -1 \\ 2 & 1 & 0 \end{bmatrix}$$

A matrix P with 2 rows and four columns having arbitrary elements $p_{i,j}$ can be constructed as follows:

```
> P := Matrix(2,4,symbol=p);
```

$$P := \begin{bmatrix} p_{1,1} & p_{1,2} & p_{1,3} & p_{1,4} \\ p_{2,1} & p_{2,2} & p_{2,3} & p_{2,4} \end{bmatrix}$$

As with vectors, particular elements in a matrix can be accessed by including the row and column indices in square brackets following the name of the matrix.

```
> P[1,2] := Pi; P[1,4]+P[2,4]; P;
```

$$P_{1,2} := \pi$$
$$p_{1,4} + p_{2,4}$$

$$\begin{bmatrix} p_{1,1} & \pi & p_{1,3} & p_{1,4} \\ p_{2,1} & p_{2,2} & p_{2,3} & p_{2,4} \end{bmatrix}$$

There are also shorthand constructs for special kinds of matrices, such as ones with all zero entries, identity (square) matrices, and diagonal matrices:

```
> Matrix(2,3); IdentityMatrix(3);
  DiagonalMatrix([a,b,c]);
```

$$\begin{bmatrix} 0 & 0 & 0 \\ 0 & 0 & 0 \end{bmatrix}$$

$$\begin{bmatrix} 1 & 0 & 0 \\ 0 & 1 & 0 \\ 0 & 0 & 1 \end{bmatrix}$$

$$\begin{bmatrix} a & 0 & 0 \\ 0 & b & 0 \\ 0 & 0 & c \end{bmatrix}$$

The transpose T of the matrix L is obtained by using the `Transpose` function, or, more simply, `T := L^%T`.

```
> T := Transpose(L);
```

$$T := \begin{bmatrix} 1 & 2 \\ 1 & 1 \\ 1 & 3 \end{bmatrix}$$

The product, AB, of two matrices A and B is calculated using the binary operator `.`; that is, we calculate `A.B`. Of course, the number of columns of A must be equal to the number of rows of B.

```
> L.T; T.L;
```

$$\begin{bmatrix} 3 & 6 \\ 6 & 14 \end{bmatrix}$$

$$\begin{bmatrix} 5 & 3 & 7 \\ 3 & 2 & 4 \\ 7 & 4 & 10 \end{bmatrix}$$

The determinant and inverse of a square matrix are calculated with the `Determinant` and `MatrixInverse` functions.

```
> A := <<1|1|1>,<2|1|3>,<1|1|2>>;
  DetA := Determinant(A); Ainv := MatrixInverse(A);
```

$$A := \begin{bmatrix} 1 & 1 & 1 \\ 2 & 1 & 3 \\ 1 & 1 & 2 \end{bmatrix}$$

$$DetA := -1$$

$$Ainv := \begin{bmatrix} 1 & 1 & -2 \\ 1 & -1 & 1 \\ -1 & 0 & 1 \end{bmatrix}$$

```
> A.Ainv = Ainv.A;
```

$$\begin{bmatrix} 1 & 0 & 0 \\ 0 & 1 & 0 \\ 0 & 0 & 1 \end{bmatrix} = \begin{bmatrix} 1 & 0 & 0 \\ 0 & 1 & 0 \\ 0 & 0 & 1 \end{bmatrix}$$

Linear Equations

A set of n linear equations in n variables can be written in the form $A\mathbf{X} = \mathbf{B}$, where A is an $n \times n$ matrix and $\mathbf{X}$ and $\mathbf{B}$ are column n-vectors. Thus, the solution can be calculated as $\mathbf{X} = A^{-1}\mathbf{B}$. For example, the system

$$x + y + z = 2, \qquad 2x + y + 3z = 9, \qquad x + y + 2z = 1$$

has the matrix A defined above as its coefficient matrix, and $\mathbf{B}$ the column vector `<2,9,1>`. The solution of the system is:

```
> X : = Ainv.<2,9,1>;
```

$$X := \begin{bmatrix} 9 \\ -6 \\ -1 \end{bmatrix}$$

that is, $x = 9$, $y = -6$, $z = -1$. LinearAlgebra provides a simpler way of solving the system $A\mathbf{X} = \mathbf{B}$; we just need to use the function `LinearSolve(A,B)`:

```
> X := LinearSolve(A,<2,9,1>);
```

$$X := \begin{bmatrix} 9 \\ -6 \\ -1 \end{bmatrix}$$

`LinearSolve` is better at solving linear systems than is matrix inversion, since it can solve some systems for which the matrix is singular or not square. Consider the two systems

$$\begin{aligned} x + y &= 1 \\ 2x + 2y &= 2 \end{aligned} \qquad \text{and} \qquad \begin{aligned} x + y &= 1 \\ 2x + 2y &= 1 \end{aligned}$$

The first system has a one-parameter family of solutions $x = 1 - t$, $y = t$ for arbitrary t. The second system is inconsistent and has no solutions.

```
> L := Matrix([[1,1],[2,2]]); B1 := <1,2>; B2 := <1,1>;
```

$$L := \begin{bmatrix} 1 & 1 \\ 2 & 2 \end{bmatrix}$$

$$B1 := \begin{bmatrix} 1 \\ 2 \end{bmatrix}$$

$$B2 := \begin{bmatrix} 1 \\ 1 \end{bmatrix}$$

```
> X := LinearSolve(L,B1,free=t);
```

$$X := \begin{bmatrix} 1 - t_2 \\ t_2 \end{bmatrix}$$

The extra argument `free=t` was included to force LinearSolve to use subscripted t variables for any parameters. It is always safe to include an argument of this type; omitting it can cause output that looks somewhat strange. (Try it and see.) If the system has a unique solution, the `free=t` parameter will just be ignored.

```
> X := LinearSolve(L,B2,free=t);

Error, (in LinearSolve) inconsistent system
```

Eigenvalues and Eigenvectors

The LinearAlgebra package has procedures for finding the eigenvalues and eigenvectors of matrices. For a real symmetric matrix, the eigenvalues are always real.

```
> K := Matrix([[3,1,-1],[1,4,1],[-1,1,3]]);
```

$$K := \begin{bmatrix} 3 & 1 & -1 \\ 1 & 4 & 1 \\ -1 & 1 & 3 \end{bmatrix}$$

```
> Eigenvalues(K);
```

$$\begin{bmatrix} 4 \\ 3 + \sqrt{3} \\ 3 - \sqrt{3} \end{bmatrix}$$

The `Eigenvalues` function produces a column vector of the eigenvalues of the square matrix that is its argument. In this example all three eigenvalues are positive, so K is a positive definite matrix. Our main use for eigenvalues will be the classification of critical points of functions of several variables. This use does not require knowledge of the corresponding eigenvectors, but if we did need to know them, we could have used the function `Eigenvectors(K)` instead. The output would then have consisted of two items separated by a comma. The first item would be the column vector of eigenvalues of K; the second would be a square matrix whose columns are the eigenvectors corresponding to those eigenvalues. (Corresponding to an eigenvalue having multiplicity m there would be m linearly independent columns in the matrix.)

```
> Eigenvectors(K);
```

$$\begin{bmatrix} 4 \\ 3+\sqrt{3} \\ 3-\sqrt{3} \end{bmatrix}, \begin{bmatrix} -1 & -\frac{(-2+\sqrt{3})\sqrt{3}}{-3+2\sqrt{3}} & -\frac{(-2-\sqrt{3})\sqrt{3}}{-3-2\sqrt{3}} \\ 0 & -\frac{-3+\sqrt{3}}{-3+2\sqrt{3}} & -\frac{-3-\sqrt{3}}{-3-2\sqrt{3}} \\ 1 & 1 & 1 \end{bmatrix}$$

Maple isn't always good at spotting simplifications. If you follow the above Maple command with `simplify(%[2])`, you will see that the top row in the matrix of eigenvectors is, in fact, much simpler than it looks.

Remark All the matrices and vectors used in the examples of this section were of very small dimension. The LinearAlgebra package is capable of dealing with large matrices with hundreds of rows and columns, but for such matrices it is best to avoid simple expressions like `2*M-3*N` and `M.N` for linear combinations and products of matrices, and use instead `MatrixAdd(M,N,2,-3)` and `MatrixMatrixMultiply(M,N)`, which are much more efficient in their calculations. Similarly, use `MatrixVectorMultiply(M,X)` rather than `M.X` if `X` is a column vector and `ScalarMultiply(M,c)` rather than `c*M` if `c` is a number.

EXERCISES 10.8

Use Maple to calculate the quantities in Exercises 1–2.

1. The distance between the line through $(3, 0, 2)$ parallel to the vector $2\mathbf{i}+\mathbf{j}-2\mathbf{k}$ and the line through $(1, 2, 4)$ parallel to $\mathbf{i}+3\mathbf{j}+4\mathbf{k}$

2. The angle (in degrees) between the vector $\mathbf{i}-\mathbf{j}+2\mathbf{k}$ and the plane through the origin containing the vectors $\mathbf{i}-2\mathbf{j}-3\mathbf{k}$ and $2\mathbf{i}+3\mathbf{j}+4\mathbf{k}$

Use Maple to verify the identities in Exercises 3–4.

3. $\mathbf{U}\bullet(\mathbf{V}\times\mathbf{W})=\mathbf{V}\bullet(\mathbf{W}\times\mathbf{U})=\mathbf{W}\bullet(\mathbf{U}\times\mathbf{V})$

4. $(\mathbf{U}\times\mathbf{V})\times(\mathbf{U}\times\mathbf{W})=(\mathbf{U}\bullet(\mathbf{V}\times\mathbf{W}))\mathbf{U}$

In Exercises 5–10, define Maple functions to produce the indicated results. You may use functions already defined in **LinearAlgebra**.

5. A function `sp(U,V)` that gives the scalar projection of vector **U** along the nonzero vector **V**

6. A function `vp(U,V)` that gives the vector projection of vector **U** along the nonzero vector **V**

7. A function `ang(U,V)` that gives the angle between the nonzero vectors **U** and **V** in degrees as a decimal number

8. A function `unitn(U,V)` that gives a unit vector normal to the two nonparallel vectors **U** and **V** in 3-space

9. A function `VolT(U,V,W)` that gives the volume of the tetrahedron in 3-space that is spanned by the vectors **U**, **V**, and **W**

10. A function `dist(A,B)` giving the distance between two points having position vectors **A** and **B**. Use your function to find the distance between $[1, 1, 1, 1]$ and $[3, -1, 2, 5]$

In Exercises 11–12, use `LinearSolve` to solve the systems.

11. $$\begin{cases} u+2v+3x+4y+5z=20 \\ 6u-v+6x+2y-3z=0 \\ 2u+8v-8x-2y+z=6 \\ u+v+x+y+z=5 \\ 10u-3v+3x-2y+2z=5 \end{cases}$$

12. $$\begin{cases} u+v+x+y=10 \\ u+y+z=10 \\ u+x+y=8 \\ u+v+x+z=11 \\ v+y-z=1 \end{cases}$$

13. Evaluate the determinant of the coefficient matrix for the system in Exercise 11.

14. Find the eigenvalues of the coefficient matrix for the system in Exercise 12. Quote your answers as decimal numbers (use `evalf`) to 5 decimal places. Do you think any of them are really complex?

15. Find the inverse of the matrix

$$A=\begin{bmatrix} 1 & 1/2 & 1/3 \\ 1/2 & 1/3 & 1/4 \\ 1/3 & 1/4 & 1/5 \end{bmatrix}.$$

16. Find, in decimal form (using `evalf(Eigenvals(A))`, the eigenvalues of the matrix A of Exercise 15 and the Eigenvalues of its inverse. Use `Digits := 10`. How do you account for the fact that some of the eigenvalues appear to be complex? What relationship appears to exist between the eigenvalues of A and those of its inverse?

CHAPTER REVIEW

Key Ideas

- **What is each of the following?**
  - ◇ a neighbourhood ◇ an open set ◇ a closed set
  - ◇ the boundary of a set ◇ the interior of a set
  - ◇ a vector in 3-space ◇ the dot product of vectors
  - ◇ the cross product of two vectors in $\mathbb{R}^3$
  - ◇ a scalar triple product ◇ a vector triple product
  - ◇ a matrix ◇ a determinant
  - ◇ a plane ◇ a straight line ◇ a cone
  - ◇ a cylinder ◇ an ellipsoid ◇ a paraboloid
  - ◇ a hyperboloid of 1 sheet ◇ a hyperboloid of 2 sheets
  - ◇ the transpose of a matrix ◇ the inverse of a matrix
  - ◇ a linear transformation ◇ an eigenvalue of a matrix
- **What is the angle between the vectors u and v?**
- **How do you calculate u × v, given the components of u and v?**
- **What is an equation of the plane through P_0 having normal vector N?**
- **What is an equation of the straight line through P_0 parallel to a?**
- **Given two 3 × 3 matrices A and B, how do you calculate AB?**
- **What is the distance from P_0 to the plane $Ax + By + Cz + D = 0$?**
- **What is Cramer's Rule, and how is it used?**

Review Exercises

Describe the sets of points in 3-space that satisfy the given equations or inequalities in Exercises 1–18.

1. $x + 3z = 3$ **2.** $y - z \ge 1$

3. $x + y + z \ge 0$ **4.** $x - 2y - 4z = 8$

5. $y = 1 + x^2 + z^2$ **6.** $y = z^2$

7. $x = y^2 - z^2$ **8.** $z = xy$

9. $x^2 + y^2 + 4z^2 < 4$ **10.** $x^2 + y^2 - 4z^2 = 4$

11. $x^2 - y^2 - 4z^2 = 0$ **12.** $x^2 - y^2 - 4z^2 = 4$

13. $(x - z)^2 + y^2 = 1$ **14.** $(x - z)^2 + y^2 = z^2$

15. $\begin{cases} x + 2y = 0 \\ z = 3 \end{cases}$ **16.** $\begin{cases} x + y + 2z = 1 \\ x + y + z = 0 \end{cases}$

17. $\begin{cases} x^2 + y^2 + z^2 = 4 \\ x + y + z = 3 \end{cases}$ **18.** $\begin{cases} x^2 + z^2 \le 1 \\ x - y \ge 0 \end{cases}$

Find equations of the planes and lines specified in Exercises 19–28.

19. The plane through the origin perpendicular to the line

$$\frac{x-1}{2} = \frac{y+3}{-1} = \frac{z+2}{3}$$

20. The plane through $(2, -1, 1)$ and $(1, 0, -1)$ parallel to the line in Exercise 19

21. The plane through $(2, -1, 1)$ perpendicular to the planes $x - y + z = 0$ and $2x + y - 3z = 2$

22. The plane through $(-1, 1, 0)$, $(0, 4, -1)$, and $(2, 0, 0)$

23. The plane containing the line of intersection of the planes $x + y + z = 0$ and $2x + y - 3z = 2$, and passing through the point $(2, 0, 1)$

24. The plane containing the line of intersection of the planes $x + y + z = 0$ and $2x + y - 3z = 2$, and perpendicular to the plane $x - 2y - 5z = 17$

25. The vector parametric equation of the line through $(2, 1, -1)$ and $(-1, 0, 1)$

26. Standard form equations of the line through $(1, 0, -1)$ parallel to each of the planes $x - y = 3$ and $x + 2y + z = 1$

27. Scalar parametric equations of the line through the origin perpendicular to the plane $3x - 2y + 4z = 5$

28. The vector parametric equation of the line that joins points on the two lines

$$\mathbf{r} = (1 + t)\mathbf{i} - t\mathbf{j} - (2 + 2t)\mathbf{k}$$
$$\mathbf{r} = 2t\mathbf{i} + (t - 2)\mathbf{j} - (1 + 3t)\mathbf{k}$$

and is perpendicular to both those lines

Express the given conditions or quantities in Exercises 29–30 in terms of dot and cross products.

29. The three points with position vectors $\mathbf{r}_1$, $\mathbf{r}_2$, and $\mathbf{r}_3$ all lie on a straight line.

30. The four points with position vectors $\mathbf{r}_1$, $\mathbf{r}_2$, $\mathbf{r}_3$, and $\mathbf{r}_4$ do not all lie on a plane.

31. Find the area of the triangle with vertices $(1, 2, 1)$, $(4, -1, 1)$, and $(3, 4, -2)$.

32. Find the volume of the tetrahedron with vertices $(1, 2, 1)$, $(4, -1, 1)$, $(3, 4, -2)$, and $(2, 2, 2)$.

33. Show that the matrix

$$\mathcal{A} = \begin{pmatrix} 1 & 0 & 0 & 0 \\ 2 & 1 & 0 & 0 \\ 3 & 2 & 1 & 0 \\ 4 & 3 & 2 & 1 \end{pmatrix}$$

has an inverse, and find the inverse $\mathcal{A}^{-1}$.

34. Let $\mathcal{A} = \begin{pmatrix} 1 & 1 & 1 \\ 2 & 1 & 0 \\ 1 & 0 & -1 \end{pmatrix}$. What condition must the vector $\mathbf{b}$ satsify in order that the equation $\mathcal{A}\mathbf{x} = \mathbf{b}$ has solutions $\mathbf{x}$? What are the solutions $\mathbf{x}$ if $\mathbf{b}$ satisfies the condition?

35. Is the matrix $\begin{pmatrix} 3 & -1 & 1 \\ -1 & 1 & -1 \\ 1 & -1 & 2 \end{pmatrix}$ positive or negative definite or neither?

Challenging Problems

1. Show that the distance d from point P to the line AB can be expressed in terms of the position vectors of P, A, and B by

$$d = \frac{|(\mathbf{r}_A - \mathbf{r}_P) \times (\mathbf{r}_B - \mathbf{r}_P)|}{|\mathbf{r}_A - \mathbf{r}_B|}$$

❓ **2.** For any vectors $\mathbf{u}$, $\mathbf{v}$, $\mathbf{w}$, and $\mathbf{x}$, show that

$$(\mathbf{u}\times\mathbf{v})\times(\mathbf{w}\times\mathbf{x}) = \Big((\mathbf{u}\times\mathbf{v})\bullet\mathbf{x}\Big)\mathbf{w} - \Big((\mathbf{u}\times\mathbf{v})\bullet\mathbf{w}\Big)\mathbf{x}$$
$$= \Big((\mathbf{w}\times\mathbf{x})\bullet\mathbf{u}\Big)\mathbf{v} - \Big((\mathbf{w}\times\mathbf{x})\bullet\mathbf{v}\Big)\mathbf{u}.$$

In particular, show that

$$(\mathbf{u}\times\mathbf{v})\times(\mathbf{u}\times\mathbf{w}) = \Big((\mathbf{u}\times\mathbf{v})\bullet\mathbf{w}\Big)\mathbf{u}.$$

❓ **3.** Show that the area A of a triangle with vertices $(x_1, y_1, 0)$, $(x_2, y_2, 0)$, and $(x_3, y_3, 0)$ in the xy-plane is given by

$$A = \frac{1}{2}\left|\begin{vmatrix} x_1 & y_1 & 1 \\ x_2 & y_2 & 1 \\ x_3 & y_3 & 1 \end{vmatrix}\right|.$$

❓ **4.** (a) If L_1 and L_2 are two **skew** (i.e., nonparallel and nonintersecting) lines, show that there is a pair of parallel planes P_1 and P_2 such that L_1 lies in P_1 and L_2 lies in P_2.

(b) Find parallel planes containing the following two lines: L_1 through points $(1, 1, 0)$ and $(2, 0, 1)$ and L_2 through points $(0, 1, 1)$ and $(1, 2, 2)$.

❓ **5.** What condition must the vectors $\mathbf{a}$ and $\mathbf{b}$ satisfy to ensure that the equation $\mathbf{a}\times\mathbf{x} = \mathbf{b}$ has solutions? If this condition is satisfied, find all solutions of the equation. Describe the set of solutions.

CHAPTER 11

Vector Functions and Curves

"Philosophy is written in this grand book — I mean the universe — which stands continually open to our gaze, but it cannot be understood unless one first learns to comprehend the language and interpret the characters in which it is written. It is written in the language of mathematics, and its characters are triangles, circles, and other geometrical figures, without which it is humanly impossible to understand a single word of it; without these, one is wandering about in a dark labyrinth."

Galileo Galilei 1564–1642

Introduction

This chapter is concerned with functions of a single real variable that have *vector* values. Such functions can be thought of as parametric representations of curves, and we will examine them from both a *kinematic* point of view (involving position, velocity, and acceleration of a moving particle) and a *geometric* point of view (involving tangents, normals, curvature, and torsion). Finally, we will work through a simple derivation of Kepler's laws of planetary motion.

11.1 Vector Functions of One Variable

In this section we will examine several aspects of differential and integral calculus as applied to **vector-valued functions** of a single real variable. Such functions can be used to represent curves parametrically. It is natural to interpret a vector-valued function of the real variable t as giving the position, at time t, of a point or "particle" moving around in space. Derivatives of this *position vector* are then other vector-valued functions giving the velocity and acceleration of the particle. To motivate the study of vector functions, we will consider such a vectorial description of motion in 3-space. Some of our examples will involve motion in the plane; in this case the third components of the vectors will be 0 and will be omitted.

If a particle moves around in 3-space, its motion can be described by giving the three coordinates of its position as functions of time t:

$$x = x(t), \qquad y = y(t), \qquad \text{and} \qquad z = z(t).$$

It is more convenient, however, to replace these three equations by a single vector equation,

$$\mathbf{r} = \mathbf{r}(t),$$

giving the position vector of the moving particle as a function of t. (Recall that the position vector of a point is the vector from the origin to that point.) In terms of the standard basis vectors $\mathbf{i}$, $\mathbf{j}$, and $\mathbf{k}$, the position of the particle at time t is

$$\text{position:} \quad \mathbf{r} = \mathbf{r}(t) = x(t)\,\mathbf{i} + y(t)\,\mathbf{j} + z(t)\,\mathbf{k}.$$

As t increases, the particle moves along a *path*, a curve $\mathcal{C}$ in 3-space. If $z(t) = 0$, then $\mathcal{C}$ is a plane curve in the xy-plane. We assume that $\mathcal{C}$ is a *continuous curve*; the particle cannot instantaneously jump from one point to a distant point. This is equivalent to requiring that the component functions $x(t)$, $y(t)$, and $z(t)$ are continuous functions of t, and we therefore say that $\mathbf{r}(t)$ is a continuous vector function of t.

In the time interval from t to $t + \Delta t$, the particle moves from position $\mathbf{r}(t)$ to position $\mathbf{r}(t + \Delta t)$. Therefore, its **average velocity** is

$$\frac{\mathbf{r}(t + \Delta t) - \mathbf{r}(t)}{\Delta t},$$

which is a vector parallel to the secant vector from $\mathbf{r}(t)$ to $\mathbf{r}(t + \Delta t)$. If the average velocity has a limit as $\Delta t \to 0$, then we say that $\mathbf{r}$ is **differentiable** at t, and we call the limit the (instantaneous) **velocity** of the particle at time t. We denote the velocity vector by $\mathbf{v}(t)$:

$$\text{velocity:} \quad \mathbf{v}(t) = \lim_{\Delta t \to 0} \frac{\mathbf{r}(t + \Delta t) - \mathbf{r}(t)}{\Delta t} = \frac{d}{dt}\mathbf{r}(t).$$

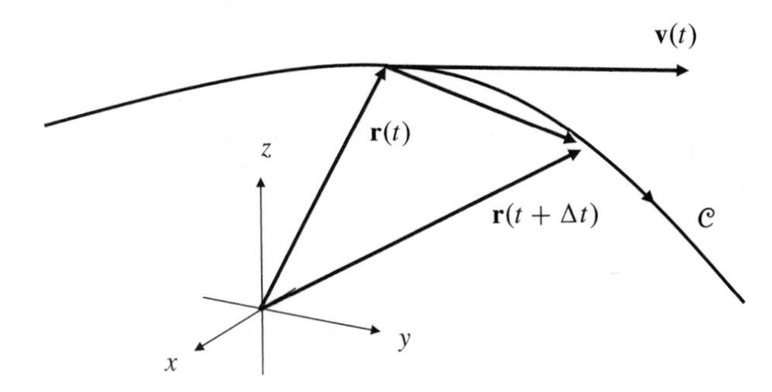

Figure 11.1 The velocity $\mathbf{v}(t)$ is the derivative of the position $\mathbf{r}(t)$ and is tangent to the path of motion at the point with position vector $\mathbf{r}(t)$

This velocity vector has direction tangent to the path $\mathcal{C}$ at the point $\mathbf{r}(t)$ (see Figure 11.1), and it points in the direction of motion. The length of the velocity vector, $v(t) = |\mathbf{v}(t)|$, is called the **speed** of the particle:

$$\text{speed:} \quad v(t) = |\mathbf{v}(t)|.$$

Wherever the velocity vector exists, is continuous, and does not vanish, the path $\mathcal{C}$ is a **smooth** curve; that is, it has a continuously turning tangent line. The path may not be smooth at points where the velocity is zero, even if the components of the velocity vector are smooth functions of t.

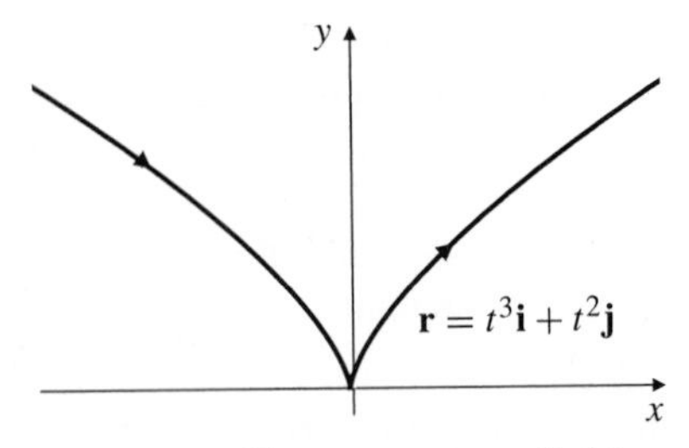

Figure 11.2 The components of $\mathbf{r}(t)$ are smooth functions of t, but the curve fails to be smooth at the origin, where $\mathbf{v} = \mathbf{0}$

EXAMPLE 1 Consider the plane curve $\mathbf{r} = t^3\mathbf{i} + t^2\mathbf{j}$. Its component functions t^3 and t^2 have continuous derivatives of all orders. However, the curve is not smooth at the origin ($t = 0$), where its velocity $\mathbf{v} = 3t^2\mathbf{i} + 2t\mathbf{j} = \mathbf{0}$. (See Figure 11.2.) The curve is smooth at all other points where $\mathbf{v}(t) \neq \mathbf{0}$.

The rules for addition and scalar multiplication of vectors imply that

$$\begin{aligned}\mathbf{v} &= \frac{d\mathbf{r}}{dt}\\ &= \lim_{\Delta t \to 0}\left(\frac{x(t + \Delta t) - x(t)}{\Delta t}\mathbf{i} + \frac{y(t + \Delta t) - y(t)}{\Delta t}\mathbf{j} + \frac{z(t + \Delta t) - z(t)}{\Delta t}\mathbf{k}\right)\\ &= \frac{dx}{dt}\mathbf{i} + \frac{dy}{dt}\mathbf{j} + \frac{dz}{dt}\mathbf{k}.\end{aligned}$$

Thus, the vector function $\mathbf{r}$ is differentiable at t if and only if its three scalar components, x, y, and z, are differentiable at t. In general, vector functions can be differentiated (or integrated) by differentiating (or integrating) their component functions, provided that the basis vectors with respect to which the components are taken are fixed in space and not changing with time.

Continuing our analysis of the moving particle, we define the **acceleration** of the particle to be the time derivative of the velocity:

$$\text{acceleration:}\quad \mathbf{a}(t) = \frac{d\mathbf{v}}{dt} = \frac{d^2\mathbf{r}}{dt^2}.$$

Newton's Second Law of Motion asserts that this acceleration is proportional to, and *in the same direction as*, the force $\mathbf{F}$ causing the motion: if the particle has mass m, then the law is expressed by the *vector equation* $\mathbf{F} = m\mathbf{a}$.

EXAMPLE 2 Describe the curve $\mathbf{r} = t\mathbf{i} + t^2\mathbf{j} + t^3\mathbf{k}$. Find the velocity and acceleration vectors for this curve at $(1, 1, 1)$.

Solution Since the scalar parametric equations for the curve are

$$x = t, \qquad y = t^2, \qquad \text{and} \qquad z = t^3,$$

which satisfy $y = x^2$ and $z = x^3$, the curve is the curve of intersection of the two cylinders $y = x^2$ and $z = x^3$. At any time t the velocity and acceleration vectors are given by

$$\mathbf{v} = \frac{d\mathbf{r}}{dt} = \mathbf{i} + 2t\mathbf{j} + 3t^2\mathbf{k},$$
$$\mathbf{a} = \frac{d\mathbf{v}}{dt} = 2\mathbf{j} + 6t\mathbf{k}.$$

The point $(1, 1, 1)$ on the curve corresponds to $t = 1$, so the velocity and acceleration at that point are $\mathbf{v} = \mathbf{i} + 2\mathbf{j} + 3\mathbf{k}$ and $\mathbf{a} = 2\mathbf{j} + 6\mathbf{k}$, respectively.

EXAMPLE 3 Find the velocity, speed, and acceleration, and describe the motion of a particle whose position at time t is

$$\mathbf{r} = 3\cos\omega t\,\mathbf{i} + 4\cos\omega t\,\mathbf{j} + 5\sin\omega t\,\mathbf{k}.$$

Solution The velocity, speed, and acceleration are readily calculated:

$$\mathbf{v} = \frac{d\mathbf{r}}{dt} = -3\omega\sin\omega t\,\mathbf{i} - 4\omega\sin\omega t\,\mathbf{j} + 5\omega\cos\omega t\,\mathbf{k}$$
$$v = |\mathbf{v}| = 5\omega$$
$$\mathbf{a} = \frac{d\mathbf{v}}{dt} = -3\omega^2\cos\omega t\,\mathbf{i} - 4\omega^2\cos\omega t\,\mathbf{j} - 5\omega^2\sin\omega t\,\mathbf{k} = -\omega^2\mathbf{r}.$$

Observe that $|\mathbf{r}| = 5$. Therefore, the path of the particle lies on the sphere with equation $x^2 + y^2 + z^2 = 25$. Since $x = 3\cos\omega t$ and $y = 4\cos\omega t$, the path also lies on the vertical plane $4x = 3y$. Hence, the particle moves around a circle of radius 5 centred at the origin and lying in the plane $4x = 3y$. Observe also that $\mathbf{r}$ is periodic with period $2\pi/\omega$. Therefore, the particle makes one revolution around the circle in time $2\pi/\omega$. The acceleration is always in the direction of $-\mathbf{r}$, that is, toward the origin. The term **centripetal acceleration** is used to describe such a "centre-seeking" acceleration.

EXAMPLE 4 **(The projectile problem)** Describe the path followed by a particle experiencing a constant downward acceleration, $-g\mathbf{k}$, caused by gravity. Assume that at time $t = 0$ the particle is at position $\mathbf{r}_0$ and its velocity is $\mathbf{v}_0$.

Solution If the position of the particle at time t is $\mathbf{r}(t)$, then its acceleration is $d^2\mathbf{r}/dt^2$. The position of the particle can be found by solving the *initial-value problem*

$$\frac{d^2\mathbf{r}}{dt^2} = -g\mathbf{k}, \qquad \left.\frac{d\mathbf{r}}{dt}\right|_{t=0} = \mathbf{v}_0, \qquad \mathbf{r}(0) = \mathbf{r}_0.$$

We integrate the differential equation twice. Each integration introduces a *vector* constant of integration that we can determine from the given data by evaluating at $t = 0$:

$$\frac{d\mathbf{r}}{dt} = -gt\mathbf{k} + \mathbf{v}_0$$

$$\mathbf{r} = -\frac{gt^2}{2}\mathbf{k} + \mathbf{v}_0 t + \mathbf{r}_0.$$

The latter equation represents a parabola in the vertical plane passing through the point with position vector $\mathbf{r}_0$ and containing the vector $\mathbf{v}_0$. (See Figure 11.3.) The parabola has scalar parametric equations

$$x = u_0 t + x_0,$$
$$y = v_0 t + y_0,$$
$$z = -\frac{gt^2}{2} + w_0 t + z_0,$$

where $\mathbf{r}_0 = x_0\mathbf{i} + y_0\mathbf{j} + z_0\mathbf{k}$ and $\mathbf{v}_0 = u_0\mathbf{i} + v_0\mathbf{j} + w_0\mathbf{k}$.

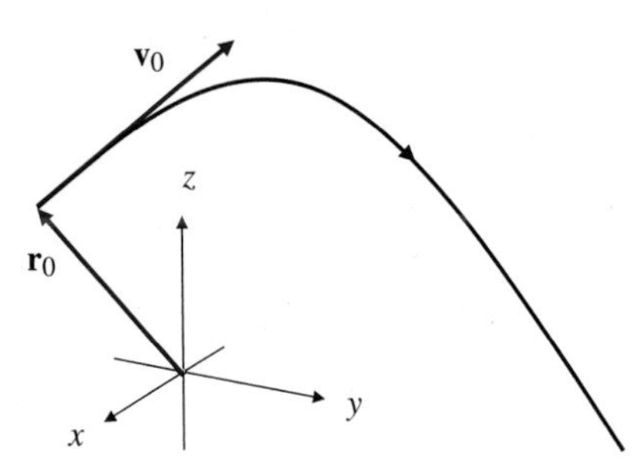

Figure 11.3 The path of a projectile fired from position $\mathbf{r}_0$ with velocity $\mathbf{v}_0$

EXAMPLE 5 An object moves to the right along the plane curve $y = x^2$ with constant speed $v = 5$. Find the velocity and acceleration of the object when it is at the point $(1, 1)$.

Solution The position of the object at time t is

$$\mathbf{r} = x\mathbf{i} + x^2\mathbf{j},$$

where x, the x-coordinate of the object's position, is a function of t. The object's velocity, speed, and acceleration at time t are given by

$$\mathbf{v} = \frac{d\mathbf{r}}{dt} = \frac{dx}{dt}\mathbf{i} + 2x\frac{dx}{dt}\mathbf{j} = \frac{dx}{dt}(\mathbf{i} + 2x\mathbf{j}),$$

$$v = |\mathbf{v}| = \left|\frac{dx}{dt}\right|\sqrt{1 + (2x)^2} = \frac{dx}{dt}\sqrt{1 + 4x^2},$$

$$\mathbf{a} = \frac{d\mathbf{v}}{dt} = \frac{d^2x}{dt^2}(\mathbf{i} + 2x\mathbf{j}) + 2\left(\frac{dx}{dt}\right)^2\mathbf{j}.$$

(In the speed calculation we used $|dx/dt| = dx/dt$ because the object is moving to the right.) We are given that the speed is constant; $v = 5$. Therefore,

$$\frac{dx}{dt} = \frac{5}{\sqrt{1+4x^2}}.$$

When $x = 1$, we have $dx/dt = 5/\sqrt{1+4} = \sqrt{5}$, so the velocity of the object at that point is $\mathbf{v} = \sqrt{5}\mathbf{i} + 2\sqrt{5}\mathbf{j}$. Now we can calculate

$$\begin{aligned}\frac{d^2x}{dt^2} &= \frac{d}{dt}\frac{5}{\sqrt{1+4x^2}} = \left(\frac{d}{dx}\frac{5}{\sqrt{1+4x^2}}\right)\frac{dx}{dt}\\ &= -\frac{5}{2(1+4x^2)^{3/2}}(8x)\frac{5}{\sqrt{1+4x^2}} = -\frac{100x}{(1+4x^2)^2}.\end{aligned}$$

At $x = 1$, we have $d^2x/dt^2 = -4$. Thus, the acceleration at that point is $\mathbf{a} = -4(\mathbf{i} + 2\mathbf{j}) + 10\mathbf{j} = -4\mathbf{i} + 2\mathbf{j}$.

Remark Note that we used x as the parameter for the curve in the above example, so we could use t for time. If you want to analyze motion along a curve $\mathbf{r} = \mathbf{r}(t)$, where t is just a parameter, not necessarily time, then you will have to use a different symbol, say τ (Greek "tau"), for time. The physical *velocity* and *acceleration* of a particle moving along the curve are then

$$\mathbf{v} = \frac{d\mathbf{r}}{d\tau} = \frac{dt}{d\tau}\frac{d\mathbf{r}}{dt} \quad \text{and} \quad \mathbf{a} = \frac{d\mathbf{v}}{d\tau} = \frac{d^2t}{d\tau^2}\frac{d\mathbf{r}}{dt} + \left(\frac{dt}{d\tau}\right)^2\frac{d^2\mathbf{r}}{dt^2}.$$

Be careful how you interpret t in a problem where time is meaningful.

Differentiating Combinations of Vectors

Vectors and scalars can be combined in a variety of ways to form other vectors or scalars. Vectors can be added and multiplied by scalars and can be factors in dot and cross products. Appropriate differentiation rules apply to all such combinations of vector and scalar functions; we summarize them in the following theorem.

THEOREM 1

Differentiation rules for vector functions

Let $\mathbf{u}(t)$ and $\mathbf{v}(t)$ be differentiable vector-valued functions, and let $\lambda(t)$ be a differentiable scalar-valued function. Then $\mathbf{u}(t) + \mathbf{v}(t)$, $\lambda(t)\mathbf{u}(t)$, $\mathbf{u}(t) \bullet \mathbf{v}(t)$, $\mathbf{u}(t) \times \mathbf{v}(t)$, and $\mathbf{u}\big(\lambda(t)\big)$ are differentiable, and

(a)	$\dfrac{d}{dt}\Big(\mathbf{u}(t)+\mathbf{v}(t)\Big)=\mathbf{u}'(t)+\mathbf{v}'(t)$
(b)	$\dfrac{d}{dt}\Big(\lambda(t)\mathbf{u}(t)\Big)=\lambda'(t)\mathbf{u}(t)+\lambda(t)\mathbf{u}'(t)$
(c)	$\dfrac{d}{dt}\Big(\mathbf{u}(t)\bullet\mathbf{v}(t)\Big)=\mathbf{u}'(t)\bullet\mathbf{v}(t)+\mathbf{u}(t)\bullet\mathbf{v}'(t)$
(d)	$\dfrac{d}{dt}\Big(\mathbf{u}(t)\times\mathbf{v}(t)\Big)=\mathbf{u}'(t)\times\mathbf{v}(t)+\mathbf{u}(t)\times\mathbf{v}'(t)$
(e)	$\dfrac{d}{dt}\Big(\mathbf{u}\big(\lambda(t)\big)\Big)=\lambda'(t)\mathbf{u}'\big(\lambda(t)\big).$

Also, at any point where $\mathbf{u}(t)\neq\mathbf{0}$,

(f) $$\frac{d}{dt}|\mathbf{u}(t)|=\frac{\mathbf{u}(t)\bullet\mathbf{u}'(t)}{|\mathbf{u}(t)|}.$$

Remark Formulas (b), (c), and (d) are versions of the Product Rule. Formula (e) is a version of the Chain Rule. Formula (f) is also a case of the Chain Rule applied to $|\mathbf{u}|=\sqrt{\mathbf{u}\bullet\mathbf{u}}$. All have the obvious form. Note that the order of the factors is the same in the terms on both sides of the cross product formula (d). It is essential that the order be preserved because, unlike the dot product or the product of a vector with a scalar, the cross product is *not commutative.*

Remark The formula for the derivative of a cross product is a special case of that for the derivative of a 3×3 determinant. (See Section 10.3.) Since every term in the expansion of a determinant of any order is a product involving one element from each row (or column), the general Product Rule implies that the derivative of an $n\times n$ determinant whose elements are functions will be the sum of n such $n\times n$ determinants, each with the elements of one of the rows (or columns) differentiated. For the 3×3 case we have

$$\frac{d}{dt}\begin{vmatrix} a_{11}(t) & a_{12}(t) & a_{13}(t)\\ a_{21}(t) & a_{22}(t) & a_{23}(t)\\ a_{31}(t) & a_{32}(t) & a_{33}(t)\end{vmatrix}=\begin{vmatrix} a'_{11}(t) & a'_{12}(t) & a'_{13}(t)\\ a_{21}(t) & a_{22}(t) & a_{23}(t)\\ a_{31}(t) & a_{32}(t) & a_{33}(t)\end{vmatrix}$$
$$+\begin{vmatrix} a_{11}(t) & a_{12}(t) & a_{13}(t)\\ a'_{21}(t) & a'_{22}(t) & a'_{23}(t)\\ a_{31}(t) & a_{32}(t) & a_{33}(t)\end{vmatrix}+\begin{vmatrix} a_{11}(t) & a_{12}(t) & a_{13}(t)\\ a_{21}(t) & a_{22}(t) & a_{23}(t)\\ a'_{31}(t) & a'_{32}(t) & a'_{33}(t)\end{vmatrix}.$$

EXAMPLE 6 Show that the speed of a moving particle remains constant over an interval of time if and only if the acceleration is perpendicular to the velocity throughout that interval.

Solution Since $\big(v(t)\big)^2=\mathbf{v}(t)\bullet\mathbf{v}(t)$, we have

$$\begin{aligned}2v(t)\frac{dv}{dt}&=\frac{d}{dt}\big(v(t)\big)^2=\frac{d}{dt}\Big(\mathbf{v}(t)\bullet\mathbf{v}(t)\Big)\\ &=\mathbf{a}(t)\bullet\mathbf{v}(t)+\mathbf{v}(t)\bullet\mathbf{a}(t)=2\mathbf{v}(t)\bullet\mathbf{a}(t).\end{aligned}$$

If we assume that $v(t)\neq0$, it follows that $dv/dt=0$ if and only if $\mathbf{v}\bullet\mathbf{a}=0$. The speed is constant if and only if the velocity is perpendicular to the acceleration.

EXAMPLE 7 If $\mathbf{u}$ is three times differentiable, calculate and simplify the triple product derivative

$$\frac{d}{dt}\left(\mathbf{u}\bullet\left(\frac{d\mathbf{u}}{dt}\times\frac{d^2\mathbf{u}}{dt^2}\right)\right).$$

Solution Using various versions of the Product Rule, we calculate

$$\begin{aligned}\frac{d}{dt}\left(\mathbf{u}\bullet\left(\frac{d\mathbf{u}}{dt}\times\frac{d^2\mathbf{u}}{dt^2}\right)\right)&=\frac{d\mathbf{u}}{dt}\bullet\left(\frac{d\mathbf{u}}{dt}\times\frac{d^2\mathbf{u}}{dt^2}\right)+\mathbf{u}\bullet\left(\frac{d^2\mathbf{u}}{dt^2}\times\frac{d^2\mathbf{u}}{dt^2}\right)+\mathbf{u}\bullet\left(\frac{d\mathbf{u}}{dt}\times\frac{d^3\mathbf{u}}{dt^3}\right)\\&=0+0+\mathbf{u}\bullet\left(\frac{d\mathbf{u}}{dt}\times\frac{d^3\mathbf{u}}{dt^3}\right)=\mathbf{u}\bullet\left(\frac{d\mathbf{u}}{dt}\times\frac{d^3\mathbf{u}}{dt^3}\right).\end{aligned}$$

The first term vanishes because $d\mathbf{u}/dt$ is perpendicular to its cross product with another vector; the second term vanishes because of the cross product of identical vectors.

EXERCISES 11.1

In Exercises 1–14, find the velocity, speed, and acceleration at time t of the particle whose position is $\mathbf{r}(t)$. Describe the path of the particle.

1. $\mathbf{r}=\mathbf{i}+t\mathbf{j}$

2. $\mathbf{r}=t^2\mathbf{i}+\mathbf{k}$

3. $\mathbf{r}=t^2\mathbf{j}+t\mathbf{k}$

4. $\mathbf{r}=\mathbf{i}+t\mathbf{j}+t\mathbf{k}$

5. $\mathbf{r}=t^2\mathbf{i}-t^2\mathbf{j}+\mathbf{k}$

6. $\mathbf{r}=t\mathbf{i}+t^2\mathbf{j}+t^2\mathbf{k}$

7. $\mathbf{r}=a\cos t\,\mathbf{i}+a\sin t\,\mathbf{j}+ct\mathbf{k}$

8. $\mathbf{r}=a\cos\omega t\,\mathbf{i}+b\mathbf{j}+a\sin\omega t\,\mathbf{k}$

9. $\mathbf{r}=3\cos t\,\mathbf{i}+4\cos t\,\mathbf{j}+5\sin t\,\mathbf{k}$

10. $\mathbf{r}=3\cos t\,\mathbf{i}+4\sin t\,\mathbf{j}+t\mathbf{k}$

11. $\mathbf{r}=ae^t\mathbf{i}+be^t\mathbf{j}+ce^t\mathbf{k}$

12. $\mathbf{r}=at\cos\omega t\,\mathbf{i}+at\sin\omega t\,\mathbf{j}+b\ln t\,\mathbf{k}$

13. $\mathbf{r}=e^{-t}\cos(e^t)\mathbf{i}+e^{-t}\sin(e^t)\mathbf{j}-e^t\mathbf{k}$

14. $\mathbf{r}=a\cos t\sin t\,\mathbf{i}+a\sin^2 t\,\mathbf{j}+a\cos t\,\mathbf{k}$

15. A particle moves around the circle $x^2+y^2=25$ at constant speed, making one revolution in 2 s. Find its acceleration when it is at $(3,4)$.

16. A particle moves to the right along the curve $y=3/x$. If its speed is 10 when it passes through the point $\left(2,\frac{3}{2}\right)$, what is its velocity at that time?

17. A point P moves along the curve of intersection of the cylinder $z=x^2$ and the plane $x+y=2$ in the direction of increasing y with constant speed $v=3$. Find the velocity of P when it is at $(1,1,1)$.

18. An object moves along the curve $y=x^2$, $z=x^3$, with constant vertical speed $dz/dt=3$. Find the velocity and acceleration of the object when it is at the point $(2,4,8)$.

19. A particle moves along the curve $\mathbf{r}=3u\mathbf{i}+3u^2\mathbf{j}+2u^3\mathbf{k}$ in the direction corresponding to increasing u and with a constant speed of 6. Find the velocity and acceleration of the particle when it is at the point $(3,3,2)$.

20. A particle moves along the curve of intersection of the cylinders $y=-x^2$ and $z=x^2$ in the direction in which x increases. (All distances are in centimetres.) At the instant when the particle is at the point $(1,-1,1)$, its speed is 9 cm/s, and that speed is increasing at a rate of 3 cm/s^2. Find the velocity and acceleration of the particle at that instant.

21. Show that if the dot product of the velocity and acceleration of a moving particle is positive (or negative), then the speed of the particle is increasing (or decreasing).

22. Verify the formula for the derivative of a dot product given in Theorem 1(c).

23. Verify the formula for the derivative of a 3×3 determinant in the second remark following Theorem 1. Use this formula to verify the formula for the derivative of the cross product in Theorem 1.

24. If the position and velocity vectors of a moving particle are always perpendicular, show that the path of the particle lies on a sphere.

25. Generalize Exercise 24 to the case where the velocity of the particle is always perpendicular to the line joining the particle to a fixed point P_0.

26. What can be said about the motion of a particle at a time when its position and velocity satisfy $\mathbf{r}\bullet\mathbf{v}>0$? What can be said when $\mathbf{r}\bullet\mathbf{v}<0$?

In Exercises 27–32, assume that the vector functions encountered have continuous derivatives of all required orders.

27. Show that $\dfrac{d}{dt}\left(\dfrac{d\mathbf{u}}{dt}\times\dfrac{d^2\mathbf{u}}{dt^2}\right)=\dfrac{d\mathbf{u}}{dt}\times\dfrac{d^3\mathbf{u}}{dt^3}$.

28. Write the Product Rule for $\dfrac{d}{dt}\Big(\mathbf{u}\bullet(\mathbf{v}\times\mathbf{w})\Big)$.

29. Write the Product Rule for $\dfrac{d}{dt}\Big(\mathbf{u}\times(\mathbf{v}\times\mathbf{w})\Big)$.

30. Expand and simplify: $\dfrac{d}{dt}\left(\mathbf{u}\times\left(\dfrac{d\mathbf{u}}{dt}\times\dfrac{d^2\mathbf{u}}{dt^2}\right)\right)$.

31. Expand and simplify: $\dfrac{d}{dt}\Big((\mathbf{u}+\mathbf{u}'')\bullet(\mathbf{u}\times\mathbf{u}')\Big)$.

32. Expand and simplify: $\dfrac{d}{dt}\Big((\mathbf{u}\times\mathbf{u}')\bullet(\mathbf{u}'\times\mathbf{u}'')\Big)$.

33. If at all times t the position and velocity vectors of a moving particle satisfy $\mathbf{v}(t)=2\mathbf{r}(t)$, and if $\mathbf{r}(0)=\mathbf{r}_0$, find $\mathbf{r}(t)$ and the acceleration $\mathbf{a}(t)$. What is the path of motion?

34. Verify that $\mathbf{r}=\mathbf{r}_0\cos(\omega t)+(\mathbf{v}_0/\omega)\sin(\omega t)$ satisfies the initial-value problem

$$\frac{d^2\mathbf{r}}{dt^2}=-\omega^2\mathbf{r},\qquad \mathbf{r}'(0)=\mathbf{v}_0,\qquad \mathbf{r}(0)=\mathbf{r}_0.$$

(It is the unique solution.) Describe the path $\mathbf{r}(t)$. What is the path if $\mathbf{r}_0$ is perpendicular to $\mathbf{v}_0$?

35. **(Free fall with air resistance)** A projectile falling under gravity and slowed by air resistance proportional to its speed has position satisfying

$$\frac{d^2\mathbf{r}}{dt^2} = -g\mathbf{k} - c\frac{d\mathbf{r}}{dt},$$

where c is a positive constant. If $\mathbf{r} = \mathbf{r}_0$ and $d\mathbf{r}/dt = \mathbf{v}_0$ at time $t = 0$, find $\mathbf{r}(t)$. (*Hint:* Let $\mathbf{w} = e^{ct}(d\mathbf{r}/dt)$.) Show that the solution approaches that of the projectile problem given in this section as $c \to 0$.

11.2 Some Applications of Vector Differentiation

Many interesting problems in mechanics involve the differentiation of vector functions. This section is devoted to a brief discussion of a few of these.

Motion Involving Varying Mass

The **momentum p** of a moving object is the product of its (scalar) mass m and its (vector) velocity $\mathbf{v}$; $\mathbf{p} = m\mathbf{v}$. Newton's Second Law of Motion states that the rate of change of *momentum* is equal to the external force acting on the object:

$$\mathbf{F} = \frac{d\mathbf{p}}{dt} = \frac{d}{dt}\Big(m\mathbf{v}\Big).$$

It is only when the mass of the object remains constant that this law reduces to the more familiar $\mathbf{F} = m\mathbf{a}$. When mass is changing you must deal with momentum rather than acceleration.

EXAMPLE 1 **(The changing velocity of a rocket)** A rocket accelerates by burning its onboard fuel. If the exhaust gases are ejected with constant velocity $\mathbf{v}_e$ *relative to the rocket*, and if the rocket ejects $p\%$ of its initial mass while its engines are firing, by what amount will the velocity of the rocket change? Assume the rocket is in deep space so that gravitational and other external forces acting on it can be neglected.

Solution Since the rocket is not acted on by any external forces (i.e., $\mathbf{F} = \mathbf{0}$), Newton's law implies that the total momentum of the rocket and its exhaust gases will remain constant. At time t the rocket has mass $m(t)$ and velocity $\mathbf{v}(t)$. At time $t + \Delta t$ the rocket's mass is $m + \Delta m$ (where $\Delta m < 0$), its velocity is $\mathbf{v} + \Delta\mathbf{v}$, and the mass $-\Delta m$ of exhaust gases has escaped with velocity $\mathbf{v} + \mathbf{v}_e$ (relative to a coordinate system fixed in space). Equating total momenta at t and $t + \Delta t$ we obtain

$$(m + \Delta m)(\mathbf{v} + \Delta\mathbf{v}) + (-\Delta m)(\mathbf{v} + \mathbf{v}_e) = m\mathbf{v}.$$

Simplifying this equation and dividing by Δt gives

$$(m + \Delta m)\frac{\Delta\mathbf{v}}{\Delta t} = \frac{\Delta m}{\Delta t}\mathbf{v}_e,$$

and, on taking the limit as $\Delta t \to 0$,

$$m\frac{d\mathbf{v}}{dt} = \frac{dm}{dt}\mathbf{v}_e.$$

Suppose that the engine fires from $t = 0$ to $t = T$. By the Fundamental Theorem of Calculus, the velocity of the rocket will change by

$$\begin{aligned}\mathbf{v}(T) - \mathbf{v}(0) &= \int_0^T \frac{d\mathbf{v}}{dt}\,dt = \left(\int_0^T \frac{1}{m}\frac{dm}{dt}\,dt\right)\mathbf{v}_e \\ &= \Big(\ln m(T) - \ln m(0)\Big)\mathbf{v}_e = -\ln\left(\frac{m(0)}{m(T)}\right)\mathbf{v}_e.\end{aligned}$$

Since $m(0) > m(T)$, we have $\ln\big(m(0)/m(T)\big) > 0$ and, as was to be expected, the change in velocity of the rocket is in the opposite direction to the exhaust velocity $\mathbf{v}_e$. If $p\%$ of the mass of the rocket is ejected during the burn, then the velocity of the rocket will change by the amount $-\mathbf{v}_e \ln(100/(100-p))$.

Remark It is interesting that this model places no restriction on how great a velocity the rocket can achieve, provided that a sufficiently large percentage of its initial mass is fuel. See Exercise 1 at the end of the section.

Circular Motion

The angular speed Ω of a rotating body is its rate of rotation measured in radians per unit time. For instance, a lighthouse lamp rotating at a rate of three revolutions per minute has an angular speed of $\Omega = 6\pi$ radians per minute. It is useful to represent the rate of rotation of a rigid body about an axis in terms of an **angular velocity** vector rather than just the scalar angular speed. The angular velocity vector, $\mathbf{\Omega}$, has magnitude equal to the angular speed, Ω, and direction along the axis of rotation such that if the extended right thumb points in the direction of $\mathbf{\Omega}$, then the fingers surround the axis in the direction of rotation.

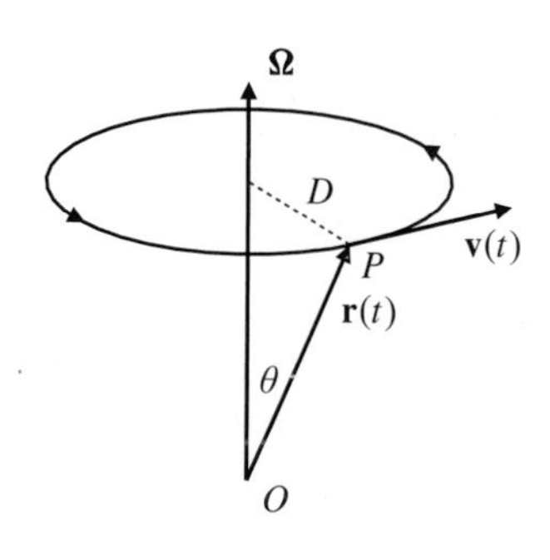

Figure 11.4 Rotation with angular velocity $\mathbf{\Omega}$: $\mathbf{v} = \mathbf{\Omega} \times \mathbf{r}$

If the origin of the coordinate system is on the axis of rotation, and $\mathbf{r} = \mathbf{r}(t)$ is the position vector at time t of a point P in the rotating body, then P moves around a circle of radius $D = |\mathbf{r}(t)| \sin\theta$, where θ is the (constant) angle between $\mathbf{\Omega}$ and $\mathbf{r}(t)$. (See Figure 11.4.) Thus, P travels a distance $2\pi D$ in time $2\pi/\Omega$, and its linear speed is

$$\frac{\text{distance}}{\text{time}} = \frac{2\pi D}{2\pi/\Omega} = \Omega D = |\mathbf{\Omega}||\mathbf{r}(t)|\sin\theta = |\mathbf{\Omega} \times \mathbf{r}(t)|.$$

Since the direction of $\mathbf{\Omega}$ was defined so that $\mathbf{\Omega} \times \mathbf{r}(t)$ would point in the direction of motion of P, the linear velocity of P at time t is given by

$$\frac{d\mathbf{r}}{dt} = \mathbf{v}(t) = \mathbf{\Omega} \times \mathbf{r}(t).$$

EXAMPLE 2 The position vector $\mathbf{r}(t)$ of a moving particle P satisfies the initial-value problem

$$\begin{cases} \dfrac{d\mathbf{r}}{dt} = 2\mathbf{i} \times \mathbf{r}(t) \\ \mathbf{r}(0) = \mathbf{i} + 3\mathbf{j}. \end{cases}$$

Find $\mathbf{r}(t)$ and describe the motion of P.

Solution There are two ways to solve this problem. We will do it both ways.

METHOD I. By the discussion above, the given differential equation is consistent with rotation about the x-axis with angular velocity $2\mathbf{i}$, so that the angular speed is 2, and the motion is counterclockwise as seen from far out on the positive x-axis. Therefore, the particle P moves on a circle in a plane x = constant and centred on the x-axis. Since P is at $(1, 3, 0)$ at time $t = 0$, the plane of motion is $x = 1$, and the radius of the circle is 3. Therefore, the circle has a parametric equation of the form

$$\mathbf{r} = \mathbf{i} + 3\cos(\lambda t)\mathbf{j} + 3\sin(\lambda t)\mathbf{k}.$$

P travels once around this circle (2π radians) in time $t = 2\pi/\lambda$, so the angular speed is λ. Therefore, $\lambda = 2$ and the motion of the particle is given by

$$\mathbf{r} = \mathbf{i} + 3\cos(2t)\mathbf{j} + 3\sin(2t)\mathbf{k}.$$

METHOD II. Break the given vector differential equation into components:

$$\frac{dx}{dt}\mathbf{i} + \frac{dy}{dt}\mathbf{j} + \frac{dz}{dt}\mathbf{k} = 2\mathbf{i} \times (x\mathbf{i} + y\mathbf{j} + z\mathbf{k}) = -2z\mathbf{j} + 2y\mathbf{k}$$

$$\frac{dx}{dt} = 0, \qquad \frac{dy}{dt} = -2z, \qquad \frac{dz}{dt} = 2y.$$

The first equation implies that x = constant. Since $x(0) = 1$, we have $x(t) = 1$ for all t. Differentiate the second equation with respect to t and substitute the third equation. This leads to the equation of simple harmonic motion for y,

$$\frac{d^2y}{dt^2} = -2\frac{dz}{dt} = -4y,$$

for which a general solution is

$$y = A\cos(2t) + B\sin(2t).$$

Thus, $z = -\frac{1}{2}(dy/dt) = A\sin(2t) - B\cos(2t)$. Since $y(0) = 3$ and $z(0) = 0$, we have $A = 3$ and $B = 0$. Thus, the particle P travels counterclockwise around the circular path

$$\mathbf{r} = \mathbf{i} + 3\cos(2t)\mathbf{j} + 3\sin(2t)\mathbf{k}$$

in the plane $x = 1$ with angular speed 2.

Remark Newton's Second Law states that $\mathbf{F} = (d/dt)(m\mathbf{v}) = d\mathbf{p}/dt$, where $\mathbf{p} = m\mathbf{v}$ is the (linear) momentum of a particle of mass m moving under the influence of a force $\mathbf{F}$. This law may be reformulated in a manner appropriate for describing rotational motion as follows. If $\mathbf{r}(t)$ is the position of the particle at time t, then, since $\mathbf{v} \times \mathbf{v} = \mathbf{0}$,

$$\frac{d}{dt}(\mathbf{r} \times \mathbf{p}) = \frac{d}{dt}\Big(\mathbf{r} \times (m\mathbf{v})\Big) = \mathbf{v} \times (m\mathbf{v}) + \mathbf{r} \times \frac{d}{dt}(m\mathbf{v}) = \mathbf{r} \times \mathbf{F}.$$

The quantities $\mathbf{H} = \mathbf{r} \times (m\mathbf{v})$ and $\mathbf{T} = \mathbf{r} \times \mathbf{F}$ are, respectively, the **angular momentum** of the particle about the origin and the **torque** of $\mathbf{F}$ about the origin. We have shown that

$$\mathbf{T} = \frac{d\mathbf{H}}{dt};$$

the torque of the external forces is equal to the rate of change of the angular momentum of the particle. This is the analogue for rotational motion of $\mathbf{F} = d\mathbf{p}/dt$.

Rotating Frames and the Coriolis Effect

The procedure of differentiating a vector function by differentiating its components is valid only if the basis vectors themselves do not depend on the variable of differentiation. In some situations in mechanics this is not the case. For instance, in modelling large-scale weather phenomena the analysis is affected by the fact that a coordinate system fixed with respect to the earth is, in fact, rotating (along with the earth) relative to directions fixed in space.

In order to understand the effect that the rotation of the coordinate system has on representations of velocity and acceleration, let us consider two Cartesian coordinate frames (i.e., systems of axes with corresponding unit basis vectors), a "fixed" frame with basis $\{\mathbf{I}, \mathbf{J}, \mathbf{K}\}$, not rotating with the earth, and a rotating frame with basis $\{\mathbf{i}, \mathbf{j}, \mathbf{k}\}$ attached to the earth and therefore rotating with the same angular speed as the earth, namely, $\pi/12$ radians/hour. Let us take the origin of the fixed frame to be at the centre of the earth, with $\mathbf{K}$ pointing north. Then the angular velocity of the earth is $\mathbf{\Omega} = (\pi/12)\mathbf{K}$. The fixed frame is being carried along with the earth in its orbit around the sun, but it is not rotating with the earth, and, since the earth's orbital rotation around the sun has angular speed only 1/365th of the angular speed of its rotation about its axis, we can ignore the much smaller effect of the motion of the earth along its orbit.

Let us take the origin of the rotating frame to be at the location of an observer on the surface of the earth, say, at point P_0 with position vector $\mathbf{R}_0$ with respect to the fixed frame.[1] Assume that P_0 has colatitude ϕ (the angle between $\mathbf{R}_0$ and $\mathbf{K}$) satisfying $0 < \phi < \pi$, so that P_0 is not at either the north pole or the south pole. Let us assume that $\mathbf{i}$ and $\mathbf{j}$ point, respectively, due east and north at P_0. Thus, $\mathbf{k}$ must point directly upward there. (See Figure 11.5.)

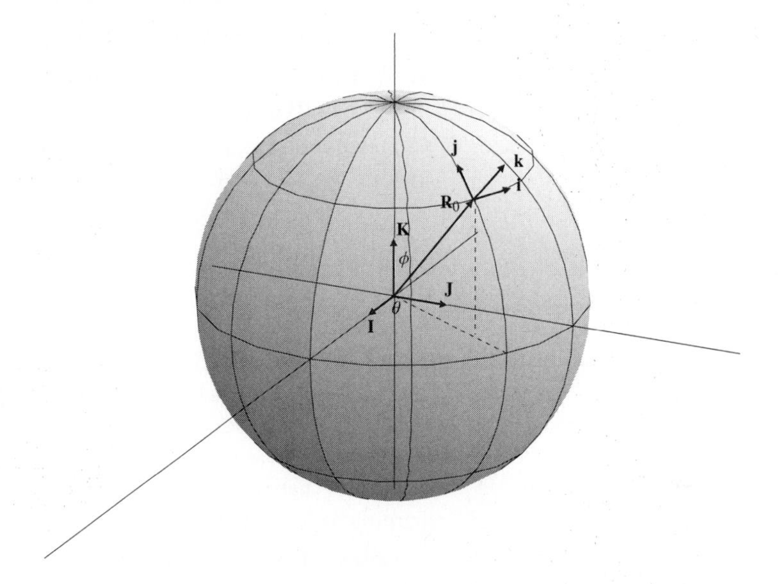

Figure 11.5 The fixed and local frames

Since each of the vectors $\mathbf{i}$, $\mathbf{j}$, $\mathbf{k}$, and $\mathbf{R}_0$ is rotating with the earth (with angular velocity $\mathbf{\Omega}$), we have, as shown earlier in this section,

$$\frac{d\mathbf{i}}{dt} = \mathbf{\Omega} \times \mathbf{i}, \quad \frac{d\mathbf{j}}{dt} = \mathbf{\Omega} \times \mathbf{j}, \quad \frac{d\mathbf{k}}{dt} = \mathbf{\Omega} \times \mathbf{k}, \quad \text{and} \quad \frac{d\mathbf{R}_0}{dt} = \mathbf{\Omega} \times \mathbf{R}_0.$$

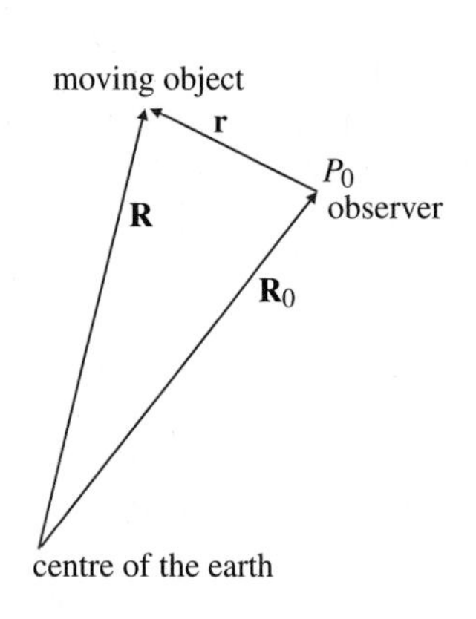

Figure 11.6 Position vectors relative to the fixed and rotating frames

Any vector function can be expressed in terms of either basis. Let us denote by $\mathbf{R}(t)$, $\mathbf{V}(t)$, and $\mathbf{A}(t)$ the position, velocity, and acceleration of a moving object with respect to the fixed frame, and by $\mathbf{r}(t)$, $\mathbf{v}(t)$, and $\mathbf{a}(t)$ the same quantities with respect to the rotating frame. Thus,

$$\begin{aligned}
\mathbf{R} &= X\mathbf{I} + Y\mathbf{J} + Z\mathbf{K}, & \mathbf{r} &= x\mathbf{i} + y\mathbf{j} + z\mathbf{k},\\
\mathbf{V} &= \frac{dX}{dt}\mathbf{I} + \frac{dY}{dt}\mathbf{J} + \frac{dZ}{dt}\mathbf{K}, & \mathbf{v} &= \frac{dx}{dt}\mathbf{i} + \frac{dy}{dt}\mathbf{j} + \frac{dz}{dt}\mathbf{k},\\
\mathbf{A} &= \frac{d^2X}{dt^2}\mathbf{I} + \frac{d^2Y}{dt^2}\mathbf{J} + \frac{d^2Z}{dt^2}\mathbf{K}, & \mathbf{a} &= \frac{d^2x}{dt^2}\mathbf{i} + \frac{d^2y}{dt^2}\mathbf{j} + \frac{d^2z}{dt^2}\mathbf{k}.
\end{aligned}$$

How are the rotating-frame values of these vectors related to the fixed-frame values? Since the origin of the rotating frame is at $\mathbf{R}_0$, we have (see Figure 11.6)

$$\mathbf{R} = \mathbf{R}_0 + \mathbf{r}.$$

When we differentiate with respect to time, we must remember that $\mathbf{R}_0$, $\mathbf{i}$, $\mathbf{j}$, and $\mathbf{k}$ all depend on time. Therefore,

$$\begin{aligned}
\mathbf{V} = \frac{d\mathbf{R}}{dt} &= \frac{d\mathbf{R}_0}{dt} + \frac{dx}{dt}\mathbf{i} + x\frac{d\mathbf{i}}{dt} + \frac{dy}{dt}\mathbf{j} + y\frac{d\mathbf{j}}{dt} + \frac{dz}{dt}\mathbf{k} + z\frac{d\mathbf{k}}{dt}\\
&= \mathbf{v} + \mathbf{\Omega} \times \mathbf{R}_0 + x\mathbf{\Omega} \times \mathbf{i} + y\mathbf{\Omega} \times \mathbf{j} + z\mathbf{\Omega} \times \mathbf{k}\\
&= \mathbf{v} + \mathbf{\Omega} \times \mathbf{R}_0 + \mathbf{\Omega} \times \mathbf{r}\\
&= \mathbf{v} + \mathbf{\Omega} \times \mathbf{R}.
\end{aligned}$$

[1] The authors are grateful to Professor Lon Rosen for suggesting this approach to the analysis of the rotating frame.

Similarly,

$$\begin{aligned}
\mathbf{A} &= \frac{d\mathbf{V}}{dt} = \frac{d}{dt}(\mathbf{v} + \boldsymbol{\Omega} \times \mathbf{R}) \\
&= \frac{d^2x}{dt^2}\mathbf{i} + \frac{dx}{dt}\frac{d\mathbf{i}}{dt} + \frac{d^2y}{dt^2}\mathbf{j} + \frac{dy}{dt}\frac{d\mathbf{j}}{dt} + \frac{d^2z}{dt^2}\mathbf{k} + \frac{dz}{dt}\frac{d\mathbf{k}}{dt} + \boldsymbol{\Omega} \times \frac{d\mathbf{R}}{dt} \\
&= \mathbf{a} + \boldsymbol{\Omega} \times \mathbf{v} + \boldsymbol{\Omega} \times (\mathbf{V}) \\
&= \mathbf{a} + 2\boldsymbol{\Omega} \times \mathbf{v} + \boldsymbol{\Omega} \times (\boldsymbol{\Omega} \times \mathbf{R}).
\end{aligned}$$

The term $2\boldsymbol{\Omega} \times \mathbf{v}$ is called the **Coriolis acceleration**, and the term $\boldsymbol{\Omega} \times (\boldsymbol{\Omega} \times \mathbf{R})$ is called the **centripetal acceleration**.

Suppose our moving object has mass m and is acted on by an external force $\mathbf{F}$. By Newton's Second Law,

$$\mathbf{F} = m\mathbf{A} = m\mathbf{a} + 2m\boldsymbol{\Omega} \times \mathbf{v} + m\boldsymbol{\Omega} \times (\boldsymbol{\Omega} \times \mathbf{R}),$$

or, equivalently,

$$\mathbf{a} = \frac{\mathbf{F}}{m} - 2\boldsymbol{\Omega} \times \mathbf{v} - \boldsymbol{\Omega} \times (\boldsymbol{\Omega} \times \mathbf{R}).$$

To the observer on the rotating earth, the object appears to be subject to $\mathbf{F}$ and to two other forces, a **Coriolis force**, whose value per unit mass is $-2\boldsymbol{\Omega} \times \mathbf{v}$, and a **centrifugal force**, whose value per unit mass is $-\boldsymbol{\Omega} \times (\boldsymbol{\Omega} \times \mathbf{R})$. The centrifugal and Coriolis forces are not "real" forces acting on the object. They are fictitious forces that compensate for the fact that we are measuring acceleration with respect to a frame that we are regarding as fixed, although it is really rotating and hence accelerating.

Observe that the centrifugal force points directly away from the polar axis of the earth. It represents the effect that the moving object wants to continue moving in a straight line and "fly off" from the earth rather than continuing to rotate along with the observer. This force is greatest at the equator (where $\boldsymbol{\Omega}$ is perpendicular to $\mathbf{R}$), but it is of very small magnitude: $|\boldsymbol{\Omega}|^2|\mathbf{R}_0| \approx 0.003g$.

The Coriolis force is quite different in nature from the centrifugal force. In particular, it is zero if the observer perceives the object to be at rest. It is perpendicular to both the velocity of the object and the polar axis of the earth, and its magnitude can be as large as $2|\boldsymbol{\Omega}||\mathbf{v}|$; and, in particular, it can be larger than that of the centrifugal force if $|\mathbf{v}|$ is sufficiently large.

EXAMPLE 3 **(Winds around the eye of a storm)** The circulation of winds around a storm centre is an example of the Coriolis effect. The eye of a storm is an area of low pressure sucking air toward it. The direction of rotation of the earth is such that the angular velocity $\boldsymbol{\Omega}$ points north and is parallel to the earth's axis of rotation. At any point P on the surface of the earth we can express $\boldsymbol{\Omega}$ as a sum of tangential (to the earth's surface) and normal components (see Figure 11.7(a)),

$$\boldsymbol{\Omega}(P) = \boldsymbol{\Omega}_T(P) + \boldsymbol{\Omega}_N(P).$$

If P is in the northern hemisphere, $\boldsymbol{\Omega}_N(P)$ points upward (away from the centre of the earth). At such a point the Coriolis "force" $\mathbf{C} = -2\boldsymbol{\Omega}(P) \times \mathbf{v}$ on a particle of air moving with horizontal velocity $\mathbf{v}$ would itself have horizontal and normal components

$$\mathbf{C} = -2\boldsymbol{\Omega}_T \times \mathbf{v} - 2\boldsymbol{\Omega}_N \times \mathbf{v} = \mathbf{C}_N + \mathbf{C}_T.$$

The normal component of the Coriolis force has negligible effect, since air is not free to travel great distances vertically. However, the tangential component of the Coriolis force, $\mathbf{C}_T = -2\boldsymbol{\Omega}_N \times \mathbf{v}$, is 90° to the right of $\mathbf{v}$ (i.e., clockwise from $\mathbf{v}$). Therefore, particles of air that are being sucked toward the eye of the storm experience Coriolis deflection to the right and so actually spiral into the eye in a counterclockwise direction. The opposite is true in the southern hemisphere, where the normal component $\boldsymbol{\Omega}_N$ is downward (into the earth). The suction force $\mathbf{F}$, the velocity $\mathbf{v}$, and the component of the Coriolis force tangential to the earth's surface, $\mathbf{C}_T$, are shown at two positions on the path of an air particle spiralling around a low-pressure area in the northern hemisphere in Figure 11.7(b).

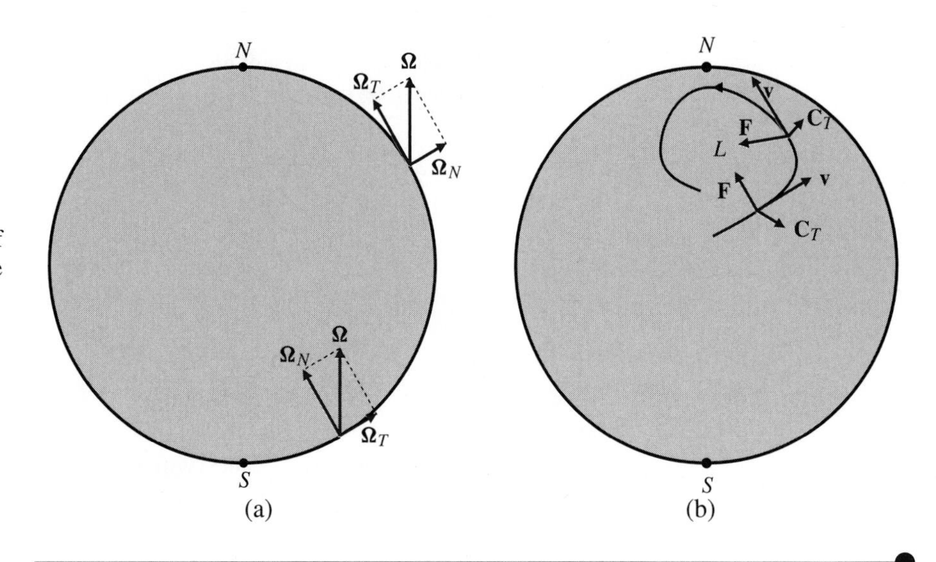

Figure 11.7

(a) Tangential and normal components of the angular velocity of the earth in the northern and southern hemispheres

(b) In the northern hemisphere the tangential Coriolis force deflects winds to the right of the path toward the low-pressure area L so the winds move counterclockwise around the centre of L

Remark Strong winds spiralling inward around low-pressure areas are called **cyclones**. Strong winds spiralling outward around high-pressure areas are called **anticyclones**. The latter spiral counterclockwise in the southern hemisphere and clockwise in the northern hemisphere. The Coriolis effect also accounts for the high-velocity eastward-flowing jet streams in the upper atmosphere at midlatitudes in both hemispheres, the energy being supplied by the rising of warm tropical air and its subsequent moving toward the poles.

The relationships between the basis vectors in the fixed and rotating frames can be used to analyze many phenomena. Recall that $\mathbf{R}_0$ makes angle ϕ with $\mathbf{K}$. Suppose the projection of $\mathbf{R}_0$ onto the equatorial plane (containing $\mathbf{I}$ and $\mathbf{J}$) makes angle θ with $\mathbf{I}$ as shown in Figure 11.5. Careful consideration of that figure should convince you that

$$\begin{aligned}
\mathbf{i} &= -\sin\theta\mathbf{I} + \cos\theta\mathbf{J} \\
\mathbf{j} &= -\cos\phi\cos\theta\mathbf{I} - \cos\phi\sin\theta\mathbf{J} + \sin\phi\mathbf{K} \\
\mathbf{k} &= \sin\phi\cos\theta\mathbf{I} + \sin\phi\sin\theta\mathbf{J} + \cos\phi\mathbf{K}.
\end{aligned}$$

Similarly, or by solving the above equations for **I**, **J**, and **K**,

$$\begin{aligned}
\mathbf{I} &= -\sin\theta\mathbf{i} - \cos\phi\cos\theta\mathbf{j} + \sin\phi\cos\theta\mathbf{k} \\
\mathbf{J} &= \cos\theta\mathbf{i} - \cos\phi\sin\theta\mathbf{j} + \sin\phi\sin\theta\mathbf{k} \\
\mathbf{K} &= \sin\phi\mathbf{j} + \cos\phi\mathbf{k}.
\end{aligned}$$

Note that as the earth rotates on its axis, ϕ remains constant while θ increases at $(\pi/12)$ radians/hour.

EXAMPLE 4 Suppose that the direction to the sun lies in the plane of **I** and **K**, and makes angle σ with **I**. Thus, the sun lies in the direction of the vector

$$\mathbf{S} = \cos\sigma\mathbf{I} + \sin\sigma\mathbf{K}.$$

($\sigma = 0$ at the March and September equinoxes, and $\sigma \approx 23.5°$ and $-23.5°$ at the June and December solstices.) Find the length of the day (the time between sunrise and sunset) for an observer at colatitude ϕ.

Solution The sun will be "up" for the observer if the angle between **S** and **k** does not exceed $\pi/2$, that is, if $\mathbf{S} \bullet \mathbf{k} \geq 0$. Thus, daytime corresponds to

$$\cos\sigma\,\sin\phi\cos\theta + \sin\sigma\,\cos\phi \geq 0,$$

or, equivalently, $\cos\theta \geq -\dfrac{\tan\sigma}{\tan\phi}$. Sunup and sundown occur where equality occurs, namely, when

$$\theta = \theta_0 = \pm\cos^{-1}\left(-\frac{\tan\sigma}{\tan\phi}\right)$$

if such values exist. (They will exist if $\phi \geq \sigma \geq 0$ or if $\pi - \phi \geq -\sigma \geq 0$.) In this case, daytime for the observer lasts

$$\frac{2\theta_0}{2\pi} \times 24 = \frac{24}{\pi}\cos^{-1}\left(-\frac{\tan\sigma}{\tan\phi}\right) \text{ hours.}$$

For instance, on June 21st at the Arctic Circle (so $\phi = \sigma$), daytime lasts $(24/\pi)\cos^{-1}(-1) = 24$ hours.

EXERCISES 11.2

1. What fraction of its total initial mass would the rocket considered in Example 1 have to burn as fuel in order to accelerate in a straight line from rest to the speed of its own exhaust gases? to twice that speed?

2. When run at maximum power output, the motor in a self-propelled tank car can accelerate the full car (mass M kg) along a horizontal track at a m/s^2. The tank is full at time zero, but the contents pour out of a hole in the bottom at rate k kg/s thereafter. If the car is at rest at time zero and full forward power is turned on at that time, how fast will it be moving at any time t before the tank is empty?

3. Solve the initial-value problem

$$\frac{d\mathbf{r}}{dt} = \mathbf{k} \times \mathbf{r}, \qquad \mathbf{r}(0) = \mathbf{i} + \mathbf{k}.$$

Describe the curve $\mathbf{r} = \mathbf{r}(t)$.

4. An object moves so that its position vector $\mathbf{r}(t)$ satisfies

$$\frac{d\mathbf{r}}{dt} = \mathbf{a} \times \big(\mathbf{r}(t) - \mathbf{b}\big)$$

and $\mathbf{r}(0) = \mathbf{r}_0$. Here, **a**, **b**, and $\mathbf{r}_0$ are given constant vectors with $\mathbf{a} \neq \mathbf{0}$. Describe the path along which the object moves.

The Coriolis effect

5. A satellite is in a low, circular, polar orbit around the earth (i.e., passing over the north and south poles). It makes one revolution every two hours. An observer standing on the earth at the equator sees the satellite pass directly overhead. In what direction does it seem to the observer to be moving? From the observer's point of view, what is the approximate value of the Coriolis force acting on the satellite?

6. Repeat Exercise 5 for an observer at a latitude of 45° in the northern hemisphere.

7. Describe the tangential and normal components of the Coriolis force on a particle moving with horizontal velocity **v** at (a) the north pole, (b) the south pole, and (c) the equator. In general, what is the effect of the normal component of the Coriolis force near the eye of a storm?

8. **(The location of sunrise and sunset)** Extend the argument in Example 4 to determine where on the horizon of the observer at P_0 the sun will rise and set. Specifically, if μ is the angle between **j** and **S** (the direction to the sun) at sunrise or sunset, show that

$$\cos\mu = \frac{\sin\sigma}{\sin\phi}.$$

For example, if $\sigma = 0$ (the equinoxes), then $\mu = \pi/2$ at all colatitudes ϕ; the sun rises due east and sets due west on those days.

9. Vancouver, Canada, has latitude 49.2° N, so its colatitude is 40.8°. How long is the sun visible in Vancouver on June 21st? Or rather, how long would it be visible if it weren't raining and if there were not so many mountains around? At what angle away from north would the sun rise and set?

10. Repeat Exercise 9 for Umeå, Sweden (latitude 63.5° N).

11.3 Curves and Parametrizations

In this section we will consider curves as geometric objects rather than as paths of moving particles. Everyone has an intuitive idea of what a curve is, but it is difficult to give a formal definition of a curve as a geometric object (i.e., as a certain kind of set of points) without involving the concept of parametric representation. We will avoid this difficulty by continuing to regard a curve in 3-space as the set of points whose positions are given by the position vector function

$$\mathbf{r} = \mathbf{r}(t) = x(t)\mathbf{i} + y(t)\mathbf{j} + z(t)\mathbf{k}, \qquad a \le t \le b.$$

However, the parameter t need no longer represent time or any other specific physical quantity.

EXAMPLE 1 Use $t = y$ to parametrize the part of the line of intersection of the two planes $y = 2x - 4$ and $z = 3x + 1$ from $(2, 0, 7)$ to $(3, 2, 10)$.

Solution We need to express all three coordinates of an arbitrary point on the line as functions of $t = y$. Since $y = t$, the equation $y = 2x - 4$ assures us that $x = \frac{1}{2}(y+4) = \frac{1}{2}(t+4)$. Then the equation $z = 3x+1$ gives $z = \frac{3}{2}(t+4)+1 = \frac{3}{2}t+7$. Since the line segment goes from $y = 0$ to $y = 2$, the required parametrization is

$$\mathbf{r} = \frac{t+4}{2}\mathbf{i} + t\mathbf{j} + \left(\frac{3}{2}t + 7\right)\mathbf{k}, \quad 0 \le t \le 2.$$

EXAMPLE 2 The plane $x + y = 1$ intersects the paraboloid $z = x^2 + y^2$ in a parabola. Parametrize the whole parabola using $t = x$ as parameter. Could $t = y$ have been used as parameter? What about $t = z$?

Solution From the equations of the two surfaces defining the parabola, we have $y = 1 - x = 1 - t$, and $z = x^2 + y^2 = 1 - 2t + 2t^2$. Thus, the required parametrization is

$$\mathbf{r} = t\mathbf{i} + (1 - t)\mathbf{j} + (1 - 2t + 2t^2)\mathbf{k}, \quad -\infty < t < \infty.$$

We could use $t = y$ instead of $t = x$ as the parameter; in this case the parametrization would be $\mathbf{r} = (1 - t)\mathbf{i} + t\mathbf{j} + (1 - 2t + 2t^2)\mathbf{k}$, $\ -\infty < t < \infty$. However, if we try to use $t = z$ as parameter, we would have to solve the system of equations $x + y = 1$, $x^2 + y^2 = t$ for x and y. This system has two possible solutions, each corresponding to a different half of the parabola starting at the lowest point $\left(\frac{1}{2}, \frac{1}{2}, \frac{1}{2}\right)$ because there are two points on the parabola at each height $z > \frac{1}{2}$. The whole parabola cannot be parametrized using z as the parameter.

Curves can be very pathological. For instance, there exist continuous curves that pass through every point in a cube. It is difficult to think of such a curve as a one-dimensional object. In order to avoid such strange objects we *assume* hereafter that the defining function $\mathbf{r}(t)$ has a *continuous* first derivative, $d\mathbf{r}/dt$, which we will continue to call "velocity" and denote by $\mathbf{v}(t)$ by analogy with the physical case where t is time. (We also continue to call $v(t) = |\mathbf{v}(t)|$ the "speed.") As we will see later, this implies that the curve has an *arc length* between any two points corresponding to parameter values t_1 and t_2; if $t_1 < t_2$, this arc length is

$$\int_{t_1}^{t_2} v(t)\,dt = \int_{t_1}^{t_2} |\mathbf{v}(t)|\,dt = \int_{t_1}^{t_2} \left|\frac{d\mathbf{r}}{dt}\right| dt.$$

Frequently we will want $\mathbf{r}(t)$ to have continuous derivatives of higher order. Whenever needed, we will assume that the "acceleration," $\mathbf{a}(t) = d^2\mathbf{r}/dt^2$, and even the third derivative, $d^3\mathbf{r}/dt^3$, are continuous. Of course, most of the curves we encounter in practice have parametrizations with continuous derivatives of all orders.

It must be recalled, however, that no assumptions on the continuity of derivatives of the function $\mathbf{r}(t)$ are sufficient to guarantee that the curve $\mathbf{r} = \mathbf{r}(t)$ is a "smooth" curve. It may fail to be smooth at a point where $\mathbf{v} = \mathbf{0}$. (See Example 1 in Section 11.1.) We will show in the next section that if, besides being continuous, the velocity vector $\mathbf{v}(t)$ is *never the zero vector*, then the curve $\mathbf{r} = \mathbf{r}(t)$ is **smooth** in the sense that it has a continuously turning tangent line.

Although we have said that a curve is a set of points given by a parametric equation $\mathbf{r} = \mathbf{r}(t)$, there is no *unique* way of representing a given curve parametrically. Just as two cars can travel the same highway at different speeds, stopping and starting at different places, so too can the same curve be defined by different parametrizations; a given curve can have infinitely many different parametrizations.

EXAMPLE 3 Show that each of the vector functions

$$\begin{aligned} \mathbf{r}_1(t) &= \sin t\,\mathbf{i} + \cos t\,\mathbf{j}, && (-\pi/2 \le t \le \pi/2), \\ \mathbf{r}_2(t) &= (t-1)\,\mathbf{i} + \sqrt{2t - t^2}\,\mathbf{j}, && (0 \le t \le 2), \quad \text{and} \\ \mathbf{r}_3(t) &= t\sqrt{2 - t^2}\,\mathbf{i} + (1 - t^2)\,\mathbf{j}, && (-1 \le t \le 1) \end{aligned}$$

all represent the same curve. Describe the curve.

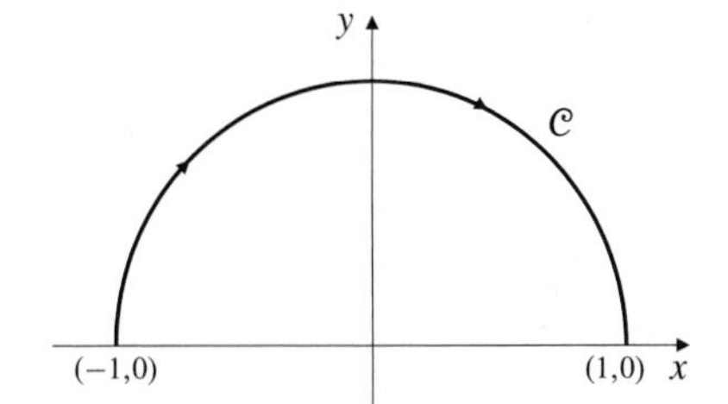

Figure 11.8 Three parametrizations of the semicircle $\mathcal{C}$ are given in Example 3

Solution All three functions represent points in the xy-plane. The function $\mathbf{r}_1(t)$ starts at the point $(-1, 0)$ with position vector $\mathbf{r}_1(-\pi/2) = -\mathbf{i}$ and ends at the point $(1, 0)$ with position vector $\mathbf{i}$. It lies in the half of the xy-plane where $y \ge 0$ (because $\cos t \ge 0$ for $(-\pi/2 \le t \le \pi/2)$). Finally, all points on the curve are at distance 1 from the origin:

$$|\mathbf{r}_1(t)| = \sqrt{(\sin t)^2 + (\cos t)^2} = 1.$$

Therefore, $\mathbf{r}_1(t)$ represents the semicircle $y = \sqrt{1 - x^2}$ in the xy-plane traversed from left to right.

The other two functions have the same properties: both graphs lie in $y \ge 0$,

$$\begin{aligned} \mathbf{r}_2(0) &= -\mathbf{i}, & \mathbf{r}_2(2) &= \mathbf{i}, & |\mathbf{r}_2(t)| &= \sqrt{(t-1)^2 + 2t - t^2} = 1, \\ \mathbf{r}_3(-1) &= -\mathbf{i}, & \mathbf{r}_3(1) &= \mathbf{i}, & |\mathbf{r}_3(t)| &= \sqrt{t^2(2 - t^2) + (1 - t^2)^2} = 1. \end{aligned}$$

Thus, all three functions represent the same semicircle (see Figure 11.8). Of course, the three parametrizations trace out the curve with different velocities.

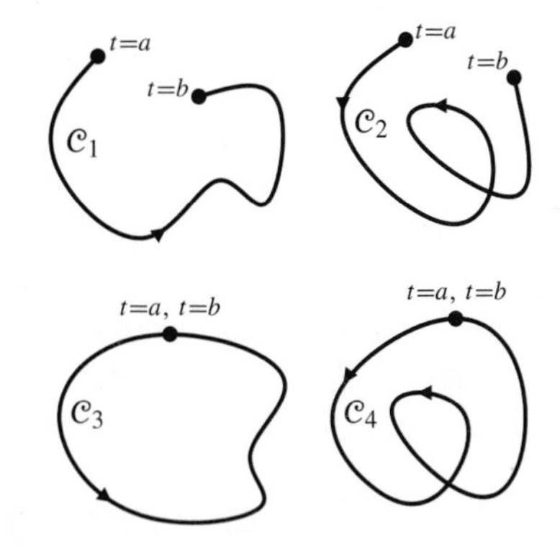

Figure 11.9 Curves $\mathcal{C}_1$ and $\mathcal{C}_3$ are non–self-intersecting
Curves $\mathcal{C}_2$ and $\mathcal{C}_4$ intersect themselves
Curves $\mathcal{C}_1$ and $\mathcal{C}_2$ are not closed
Curves $\mathcal{C}_3$ and $\mathcal{C}_4$ are closed
Curve $\mathcal{C}_3$ is a simple closed curve

The curve $\mathbf{r} = \mathbf{r}(t)$, $(a \le t \le b)$, is called a **closed curve** if $\mathbf{r}(a) = \mathbf{r}(b)$, that is, if the curve begins and ends at the same point. The curve $\mathcal{C}$ is **non–self-intersecting** if there exists some parametrization $\mathbf{r} = \mathbf{r}(t)$, $(a \le t \le b)$, of $\mathcal{C}$ that is one-to-one except that the endpoints could be the same:

$$\mathbf{r}(t_1) = \mathbf{r}(t_2) \quad a \le t_1 < t_2 \le b \quad \Longrightarrow t_1 = a \quad \text{and} \quad t_2 = b.$$

Such a curve can be closed, but otherwise does not intersect itself; it is then called a **simple closed curve**. Circles and ellipses are examples of simple closed curves. Every parametrization of a particular curve determines one of two possible **orientations** corresponding to the direction along the curve in which the parameter is increasing. Figure 11.9 illustrates these concepts. All three parametrizations of the semicircle in Example 3 orient the semicircle clockwise as viewed from a point above the xy-plane. This orientation is shown by the arrowheads on the curve in Figure 11.8. The same semicircle could be given the opposite orientation by, for example, the parametrization

$$\mathbf{r}(t) = \cos t\,\mathbf{i} + \sin t\,\mathbf{j}, \qquad 0 \le t \le \pi.$$

Parametrizing the Curve of Intersection of Two Surfaces

Frequently, a curve is specified as the intersection of two surfaces with given Cartesian equations. We may want to represent the curve by parametric equations. There is no unique way to do this, but if one of the given surfaces is a cylinder parallel to a coordinate axis (so its equation is independent of one of the variables), we can begin by parametrizing that surface. The following examples clarify the method.

EXAMPLE 4 Parametrize the curve of intersection of the plane $x + 2y + 4z = 4$ and the elliptic cylinder $x^2 + 4y^2 = 4$.

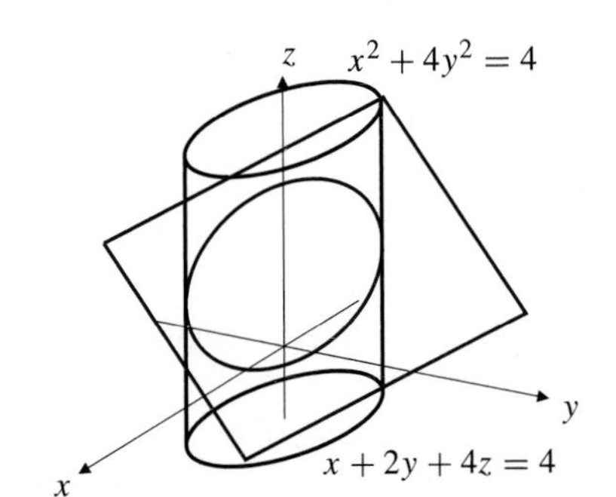

Figure 11.10 The curve of intersection of an oblique plane and an elliptic cylinder

Solution We begin with the equation $x^2 + 4y^2 = 4$, which is independent of z. It can be parametrized in many ways; one convenient way is

$$x = 2\cos t, \qquad y = \sin t, \qquad (0 \le t \le 2\pi).$$

The equation of the plane can then be solved for z, so that z can be expressed in terms of t:

$$z = \frac{1}{4}(4 - x - 2y) = 1 - \frac{1}{2}(\cos t + \sin t).$$

Thus, the given surfaces intersect in the curve (see Figure 11.10)

$$\mathbf{r} = 2\cos t\,\mathbf{i} + \sin t\,\mathbf{j} + \left(1 - \frac{\cos t + \sin t}{2}\right)\mathbf{k}, \qquad (0 \le t \le 2\pi).$$

EXAMPLE 5 Find a parametric representation of the curve of intersection of the two surfaces

$$x^2 + y + z = 2 \qquad \text{and} \qquad xy + z = 1.$$

Solution Here, neither given equation is independent of a variable, but we can obtain a third equation representing a surface containing the curve of intersection of the two given surfaces by subtracting the two given equations to eliminate z:

$$x^2 + y - xy = 1.$$

This equation is readily parametrized. If, for example, we let $x = t$, then

$$t^2 + y(1 - t) = 1, \qquad \text{so} \quad y = \frac{1 - t^2}{1 - t} = 1 + t.$$

Either of the given equations can then be used to express z in terms of t:

$$z = 1 - xy = 1 - t(1 + t) = 1 - t - t^2.$$

Thus, a possible parametrization of the curve is

$$\mathbf{r} = t\mathbf{i} + (1 + t)\mathbf{j} + (1 - t - t^2)\mathbf{k}.$$

Of course, this answer is not unique. Many other parametrizations can be found for the curve, providing orientations in either direction.

Arc Length

We now consider how to define and calculate the length of a curve. Let $\mathcal{C}$ be a bounded, continuous curve specified by

$$\mathbf{r} = \mathbf{r}(t), \qquad a \le t \le b.$$

Subdivide the closed interval $[a, b]$ into n subintervals by points

$$a = t_0 < t_1 < t_2 < \cdots < t_{n-1} < t_n = b.$$

The points $\mathbf{r}_i = \mathbf{r}(t_i)$, $(0 \le i \le n)$, subdivide $\mathcal{C}$ into n arcs. If we use the chord length $|\mathbf{r}_i - \mathbf{r}_{i-1}|$ as an approximation to the arc length between $\mathbf{r}_{i-1}$ and $\mathbf{r}_i$, then the sum

$$s_n = \sum_{i=1}^{n} |\mathbf{r}_i - \mathbf{r}_{i-1}|$$

approximates the length of $\mathcal{C}$ by the length of a polygonal line. (See Figure 11.11.) Evidently, any such approximation is less than or equal to the actual length of $\mathcal{C}$. We say that $\mathcal{C}$ is **rectifiable** if there exists a constant K such that $s_n \le K$ for every n and every choice of the points t_i. In this case, the completeness axiom of the real number system assures us that there will be a smallest such number K. We call this smallest K the **length** of $\mathcal{C}$ and denote it by s. Let $\Delta t_i = t_i - t_{i-1}$ and $\Delta \mathbf{r}_i = \mathbf{r}_i - \mathbf{r}_{i-1}$. Then s_n can be written in the form

$$s_n = \sum_{i=1}^{n} \left| \frac{\Delta \mathbf{r}_i}{\Delta t_i} \right| \Delta t_i.$$

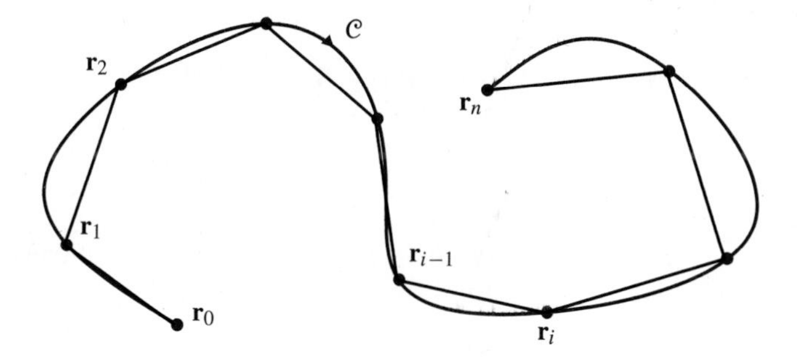

Figure 11.11 A polygonal approximation to a curve $\mathcal{C}$. The length of the polygonal line cannot exceed the length of the curve. In this figure the points on the curve are labelled with their position vectors, but the origin and these vectors are not themselves shown

If $\mathbf{r}(t)$ has a continuous derivative $\mathbf{v}(t)$, then

$$s = \lim_{\substack{n\to\infty \\ \max \Delta t_i \to 0}} s_n = \int_a^b \left| \frac{d\mathbf{r}}{dt} \right| dt = \int_a^b |\mathbf{v}(t)|\, dt = \int_a^b v(t)\, dt.$$

In kinematic terms, this formula states that the distance travelled by a moving particle is the integral of its speed.

Remark Although the above formula is expressed in terms of the parameter t, the arc length, as defined above, is a strictly geometric property of the curve $\mathcal{C}$. It is independent of the particular parametrization used to represent $\mathcal{C}$. See Exercise 27 at the end of this section.

If $s(t)$ denotes the arc length of that part of $\mathcal{C}$ corresponding to parameter values in $[a, t]$, then

$$\frac{ds}{dt} = \frac{d}{dt} \int_a^t v(\tau)\, d\tau = v(t),$$

so that the **arc length element** for $\mathcal{C}$ is given by

$$ds = v(t)\,dt = \left|\frac{d}{dt}\mathbf{r}(t)\right|\,dt.$$

The length of $\mathcal{C}$ is the integral of these arc length elements; we write

$$\int_{\mathcal{C}} ds = \text{length of } \mathcal{C} = \int_a^b v(t)\,dt.$$

Several familiar formulas for arc length follow from the above formula by using specific parametrizations of curves. For instance, the arc length element ds for the Cartesian plane curve $y = f(x)$ on $[a, b]$ is obtained by using x as parameter; here, $\mathbf{r} = x\mathbf{i} + f(x)\mathbf{j}$, so $\mathbf{v} = \mathbf{i} + f'(x)\mathbf{j}$ and

$$ds = \sqrt{1 + \big(f'(x)\big)^2}\,dx.$$

Similarly, the arc length element ds for a plane polar curve $r = g(\theta)$ can be calculated from the parametrization

$$\mathbf{r}(\theta) = g(\theta)\cos\theta\,\mathbf{i} + g(\theta)\sin\theta\,\mathbf{j}.$$

It is

$$ds = \sqrt{\big(g(\theta)\big)^2 + \big(g'(\theta)\big)^2}\,d\theta.$$

EXAMPLE 6 Find the length s of that part of the **circular helix**

$$\mathbf{r} = a\cos t\,\mathbf{i} + a\sin t\,\mathbf{j} + bt\,\mathbf{k}$$

between the points $(a, 0, 0)$ and $(a, 0, 2\pi b)$.

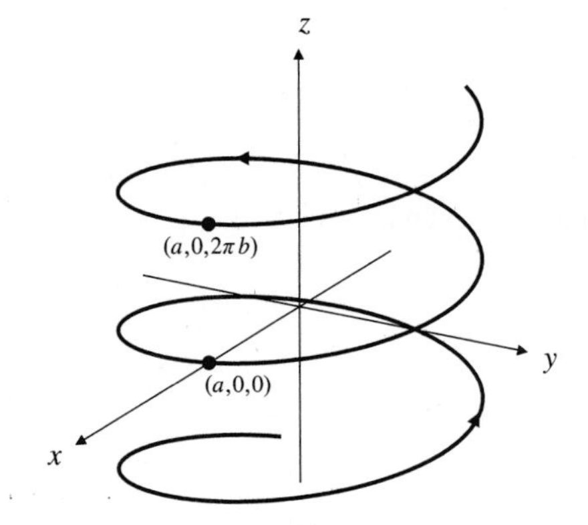

Figure 11.12 The helix
$x = a\cos t$
$y = a\sin t$
$z = bt$

Solution This curve spirals around the z-axis, rising as it turns. (See Figure 11.12.) It lies on the surface of the circular cylinder $x^2 + y^2 = a^2$. We have

$$\mathbf{v} = \frac{d\mathbf{r}}{dt} = -a\sin t\,\mathbf{i} + a\cos t\,\mathbf{j} + b\mathbf{k}$$
$$v = \sqrt{a^2 + b^2},$$

so that in terms of the parameter t the helix is traced out at constant speed. The required length s corresponds to parameter interval $[0, 2\pi]$. Thus,

$$s = \int_0^{2\pi} v(t)\,dt = \int_0^{2\pi} \sqrt{a^2 + b^2}\,dt = 2\pi\sqrt{a^2 + b^2}.$$

Piecewise Smooth Curves

As observed earlier, a parametric curve $\mathcal{C}$ given by $\mathbf{r} = \mathbf{r}(t)$ can fail to be smooth at points where $d\mathbf{r}/dt = \mathbf{0}$. If there are finitely many such points, we will say that the curve is piecewise smooth.

In general, a **piecewise smooth curve** $\mathcal{C}$ consists of a finite number of smooth arcs, $\mathcal{C}_1, \mathcal{C}_2, \ldots, \mathcal{C}_k$, as shown in Figure 11.13.

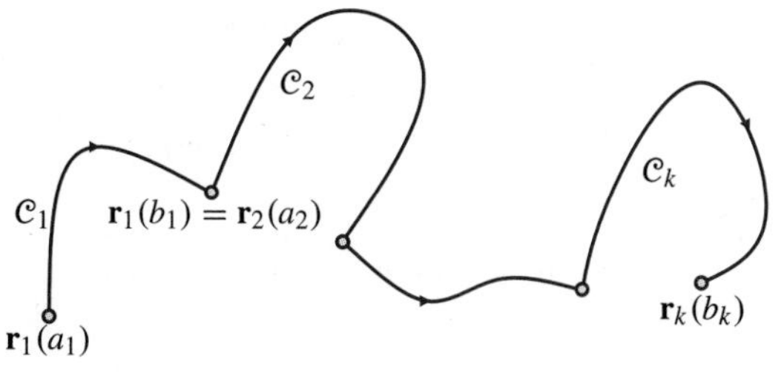

Figure 11.13 A piecewise smooth curve

In this case we express $\mathcal{C}$ as the sum of the individual arcs:

$$\mathcal{C} = \mathcal{C}_1 + \mathcal{C}_2 + \cdots + \mathcal{C}_k.$$

Each arc $\mathcal{C}_i$ can have its own parametrization

$$\mathbf{r} = \mathbf{r}_i(t), \qquad (a_i \le t \le b_i),$$

where $\mathbf{v}_i = d\mathbf{r}_i/dt \ne \mathbf{0}$ for $a_i < t < b_i$. The fact that $\mathcal{C}_{i+1}$ must begin at the point where $\mathcal{C}_i$ ends requires the conditions

$$\mathbf{r}_{i+1}(a_{i+1}) = \mathbf{r}_i(b_i) \qquad \text{for } 1 \le i \le k-1.$$

If also $\mathbf{r}_k(b_k) = \mathbf{r}_1(a_1)$, then $\mathcal{C}$ is a closed piecewise smooth curve.

The length of a piecewise smooth curve $\mathcal{C} = \mathcal{C}_1 + \mathcal{C}_2 + \cdots + \mathcal{C}_k$ is the sum of the lengths of its component arcs:

$$\text{length of } \mathcal{C} = \sum_{i=1}^{k} \int_{a_i}^{b_i} \left| \frac{d\mathbf{r}_i}{dt} \right| dt.$$

The Arc-Length Parametrization

The selection of a particular parameter in terms of which to specify a given curve will usually depend on the problem in which the curve arises; there is no one "right way" to parametrize a curve. However, there is one parameter that is "natural" in that it arises from the geometry (shape and size) of the curve itself and not from any particular coordinate system in which the equation of the curve is to be expressed. This parameter is the *arc length* measured from some particular point (the *initial point*) on the curve. The position vector of an arbitrary point P on the curve can be specified as a function of the arc length s along the curve from the initial point P_0 to P,

$$\mathbf{r} = \mathbf{r}(s).$$

This equation is called an **arc-length parametrization** or **intrinsic parametrization** of the curve. Since $ds = v(t)\,dt$ for any parametrization $\mathbf{r} = \mathbf{r}(t)$, for the arc-length parametrization we have $ds = v(s)\,ds$. Thus $v(s) = 1$, identically; *a curve parametrized in terms of arc length is traced at unit speed.* Although it is seldom easy (and usually not possible) to find $\mathbf{r}(s)$ explicitly when the curve is given in terms of some other parameter, smooth curves always have such parametrizations (see Exercise 28 at the end of this section), and they will prove useful when we develop the fundamentals of the *differential geometry* for 3-space curves in the next section.

Suppose that a curve is specified in terms of an arbitrary parameter t. If the arc length over a parameter interval $[t_0, t]$,

$$s = s(t) = \int_{t_0}^{t} \left| \frac{d}{d\tau}\mathbf{r}(\tau) \right| d\tau,$$

can be evaluated explicitly, and if the equation $s = s(t)$ can be explicitly solved for t as a function of s ($t = t(s)$), then the curve can be reparametrized in terms of arc length by substituting for t in the original parametrization:

$$\mathbf{r} = \mathbf{r}(t(s)).$$

EXAMPLE 7 Parametrize the circular helix

$$\mathbf{r} = a\cos t\mathbf{i} + a\sin t\mathbf{j} + bt\mathbf{k}$$

in terms of the arc length measured from the point $(a, 0, 0)$ in the direction of increasing t. (See Figure 11.12.)

Solution The initial point corresponds to $t = 0$. As shown in Example 6, we have $ds/dt = \sqrt{a^2 + b^2}$, so

$$s = s(t) = \int_0^t \sqrt{a^2 + b^2}\, d\tau = \sqrt{a^2 + b^2}\, t.$$

Therefore, $t = s/\sqrt{a^2 + b^2}$, and the arc-length parametrization is

$$\mathbf{r}(s) = a\cos\left(\frac{s}{\sqrt{a^2+b^2}}\right)\mathbf{i} + a\sin\left(\frac{s}{\sqrt{a^2+b^2}}\right)\mathbf{j} + \frac{bs}{\sqrt{a^2+b^2}}\mathbf{k}.$$

EXERCISES 11.3

In Exercises 1–4, find the required parametrization of the first quadrant part of the circular arc $x^2 + y^2 = a^2$.

1. In terms of the y-coordinate, oriented counterclockwise

2. In terms of the x-coordinate, oriented clockwise

3. In terms of the angle between the tangent line and the positive x-axis, oriented counterclockwise

4. In terms of arc length measured from $(0, a)$, oriented clockwise

5. The cylinders $z = x^2$ and $z = 4y^2$ intersect in two curves, one of which passes through the point $(2, -1, 4)$. Find a parametrization of that curve using $t = y$ as parameter.

6. The plane $x + y + z = 1$ intersects the cylinder $z = x^2$ in a parabola. Parametrize the parabola using $t = x$ as parameter.

In Exercises 7–10, parametrize the curve of intersection of the given surfaces. *Note*: the answers are not unique.

7. $x^2 + y^2 = 9$ and $z = x + y$

8. $z = \sqrt{1 - x^2 - y^2}$ and $x + y = 1$

9. $z = x^2 + y^2$ and $2x - 4y - z - 1 = 0$

10. $yz + x = 1$ and $xz - x = 1$

11. The plane $z = 1 + x$ intersects the cone $z^2 = x^2 + y^2$ in a parabola. Try to parametrize the parabola using as parameter: (a) $t = x$, (b) $t = y$, and (c) $t = z$. Which of these choices for t leads to a single parametrization that represents the whole parabola? What is that parametrization? What happens with the other two choices?

12. The plane $x + y + z = 1$ intersects the sphere $x^2 + y^2 + z^2 = 1$ in a circle $\mathcal{C}$. Find the centre $\mathbf{r}_0$ and radius r of $\mathcal{C}$. Also find two perpendicular unit vectors $\hat{\mathbf{v}}_1$ and $\hat{\mathbf{v}}_2$ parallel to the plane of $\mathcal{C}$. (*Hint:* To be specific, show that $\hat{\mathbf{v}}_1 = (\mathbf{i} - \mathbf{j})/\sqrt{2}$ is one such vector; then find a second that is perpendicular to $\hat{\mathbf{v}}_1$.) Use your results to construct a parametrization of $\mathcal{C}$.

13. Find the length of the curve $\mathbf{r} = t^2\mathbf{i} + t^2\mathbf{j} + t^3\mathbf{k}$ from $t = 0$ to $t = 1$.

14. For what values of the parameter λ is the length $s(T)$ of the curve $\mathbf{r} = t\mathbf{i} + \lambda t^2\mathbf{j} + t^3\mathbf{k}$, $(0 \le t \le T)$ given by $s(T) = T + T^3$?

15. Express the length of the curve $\mathbf{r} = at^2\,\mathbf{i} + bt\,\mathbf{j} + c\ln t\,\mathbf{k}$, $(1 \le t \le T)$, as a definite integral. Evaluate the integral if $b^2 = 4ac$.

16. Describe the parametric curve $\mathcal{C}$ given by

$$x = a\cos t\sin t, \qquad y = a\sin^2 t, \qquad z = bt.$$

What is the length of $\mathcal{C}$ between $t = 0$ and $t = T > 0$?

17. Find the length of the conical helix given by the parametrization $\mathbf{r} = t\cos t\,\mathbf{i} + t\sin t\,\mathbf{j} + t\,\mathbf{k}$, $(0 \le t \le 2\pi)$. Why is the curve called a conical helix?

18. Describe the intersection of the sphere $x^2 + y^2 + z^2 = 1$ and the elliptic cylinder $x^2 + 2z^2 = 1$. Find the total length of this intersection curve.

19. Let $\mathcal{C}$ be the curve $x = e^t\cos t$, $y = e^t\sin t$, $z = t$ between $t = 0$ and $t = 2\pi$. Find the length of $\mathcal{C}$.

20. Find the length of the piecewise smooth curve $\mathbf{r} = t^3\mathbf{i} + t^2\mathbf{j}$, $(-1 \le t \le 2)$.

21. Describe the piecewise smooth curve $\mathcal{C} = \mathcal{C}_1 + \mathcal{C}_2$, where $\mathbf{r}_1(t) = t\mathbf{i} + t\mathbf{j}$, $(0 \le t \le 1)$, and $\mathbf{r}_2(t) = (1 - t)\mathbf{i} + (1 + t)\mathbf{j}$, $(0 \le t \le 1)$.

22. A cable of length L and circular cross-section of radius a is wound around a cylindrical spool of radius b with no overlapping and with the adjacent windings touching one another. What length of the spool is covered by the cable?

In Exercises 23–26, reparametrize the given curve in the same orientation in terms of arc length measured from the point where $t = 0$.

23. $\mathbf{r} = At\mathbf{i} + Bt\mathbf{j} + Ct\mathbf{k}$, $\qquad (A^2 + B^2 + C^2 > 0)$

24. $\mathbf{r} = e^t\mathbf{i} + \sqrt{2}t\mathbf{j} - e^{-t}\mathbf{k}$

25. $\mathbf{r} = a\cos^3 t\,\mathbf{i} + a\sin^3 t\,\mathbf{j} + b\cos 2t\,\mathbf{k}, \qquad (0 \le t \le \frac{\pi}{2})$

26. $\mathbf{r} = 3t\cos t\,\mathbf{i} + 3t\sin t\,\mathbf{j} + 2\sqrt{2}t^{3/2}\mathbf{k}$

27. Let $\mathbf{r} = \mathbf{r}_1(t)$, $(a \le t \le b)$, and $\mathbf{r} = \mathbf{r}_2(u)$, $(c \le u \le d)$, be two parametrizations of the same curve $\mathcal{C}$, each one-to-one on its domain and each giving $\mathcal{C}$ the same orientation (so that $\mathbf{r}_1(a) = \mathbf{r}_2(c)$ and $\mathbf{r}_1(b) = \mathbf{r}_2(d)$). Then for each t in $[a, b]$ there is a unique $u = u(t)$ such that $\mathbf{r}_2(u(t)) = \mathbf{r}_1(t)$. Show that

$$\int_a^b \left|\frac{d}{dt}\mathbf{r}_1(t)\right| dt = \int_c^d \left|\frac{d}{du}\mathbf{r}_2(u)\right| du,$$

and thus that the length of $\mathcal{C}$ is independent of parametrization.

28. If the curve $\mathbf{r} = \mathbf{r}(t)$ has continuous, nonvanishing velocity $\mathbf{v}(t)$ on the interval $[a, b]$, and if t_0 is some point in $[a, b]$, show that the function

$$s = g(t) = \int_{t_0}^t |\mathbf{v}(u)|\, du$$

is an increasing function on $[a, b]$ and so has an inverse:

$$t = g^{-1}(s) \iff s = g(t).$$

Hence, show that the curve can be parametrized in terms of arc length measured from $\mathbf{r}(t_0)$.

11.4 Curvature, Torsion, and the Frenet Frame

In this section and the next we develop the fundamentals of differential geometry of curves in 3-space. We will introduce several new scalar and vector functions associated with a curve $\mathcal{C}$. The most important of these are the curvature and torsion of the curve and a right-handed triad of mutually perpendicular unit vectors forming a basis at any point on the curve, and called the Frenet frame. The curvature measures the rate at which a curve is turning (away from its tangent line) at any point. The torsion measures the rate at which the curve is twisting (out of the plane in which it is turning) at any point.

The Unit Tangent Vector

The velocity vector $\mathbf{v}(t) = d\mathbf{r}/dt$ is tangent to the parametric curve $\mathbf{r} = \mathbf{r}(t)$ at the point $\mathbf{r}(t)$ and points in the direction of the orientation of the curve there. Since we are assuming that $\mathbf{v}(t) \ne \mathbf{0}$, we can find a **unit tangent vector**, $\hat{\mathbf{T}}(t)$, at $\mathbf{r}(t)$ by dividing $\mathbf{v}(t)$ by its length:

$$\hat{\mathbf{T}}(t) = \frac{\mathbf{v}(t)}{v(t)} = \frac{d\mathbf{r}}{dt}\Big/\left|\frac{d\mathbf{r}}{dt}\right|.$$

Recall that a curve parametrized in terms of arc length, $\mathbf{r} = \mathbf{r}(s)$, is traced at unit speed; $v(s) = 1$. In terms of arc-length parametrization, the unit tangent vector is

$$\hat{\mathbf{T}}(s) = \frac{d\mathbf{r}}{ds}.$$

EXAMPLE 1 Find the unit tangent vector, $\hat{\mathbf{T}}$, for the circular helix of Example 6 of Section 11.3, in terms of both t and the arc-length parameter s.

Solution In terms of t we have

$$\begin{aligned}
\mathbf{r} &= a\cos t\,\mathbf{i} + a\sin t\,\mathbf{j} + bt\mathbf{k}\\
\mathbf{v}(t) &= -a\sin t\,\mathbf{i} + a\cos t\,\mathbf{j} + b\mathbf{k}\\
v(t) &= \sqrt{a^2\sin^2 t + a^2\cos^2 t + b^2} = \sqrt{a^2+b^2}\\
\hat{\mathbf{T}}(t) &= -\frac{a}{\sqrt{a^2+b^2}}\sin t\,\mathbf{i} + \frac{a}{\sqrt{a^2+b^2}}\cos t\,\mathbf{j} + \frac{b}{\sqrt{a^2+b^2}}\mathbf{k}.
\end{aligned}$$

In terms of the arc-length parameter (see Example 7 of Section 11.3)

$$\begin{aligned}\mathbf{r}(s) &= a\cos\left(\frac{s}{\sqrt{a^2+b^2}}\right)\mathbf{i} + a\sin\left(\frac{s}{\sqrt{a^2+b^2}}\right)\mathbf{j} + \frac{bs}{\sqrt{a^2+b^2}}\mathbf{k}\\ \hat{\mathbf{T}}(s) = \frac{d\mathbf{r}}{ds} &= -\frac{a}{\sqrt{a^2+b^2}}\sin\left(\frac{s}{\sqrt{a^2+b^2}}\right)\mathbf{i} + \frac{a}{\sqrt{a^2+b^2}}\cos\left(\frac{s}{\sqrt{a^2+b^2}}\right)\mathbf{j}\\ &\quad + \frac{b}{\sqrt{a^2+b^2}}\mathbf{k}.\end{aligned}$$

Remark If the curve $\mathbf{r} = \mathbf{r}(t)$ has a continuous, nonvanishing velocity $\mathbf{v}(t)$, then the unit tangent vector $\hat{\mathbf{T}}(t)$ is a continuous function of t. The angle $\theta(t)$ between $\hat{\mathbf{T}}(t)$ and any fixed unit vector $\hat{\mathbf{u}}$ is also continuous in t:

$$\theta(t) = \cos^{-1}(\hat{\mathbf{T}}(t) \bullet \hat{\mathbf{u}}).$$

Thus, as asserted previously, the curve is *smooth* in the sense that it has a continuously turning tangent line. The rate of this turning is quantified by the curvature, which we introduce now.

Curvature and the Unit Normal

In the rest of this section we will deal abstractly with a curve $\mathcal{C}$ parametrized in terms of arc length measured from some point on it:

$$\mathbf{r} = \mathbf{r}(s).$$

In the next section we return to curves with arbitrary parametrizations and apply the principles developed in this section to specific problems. Throughout we assume that the parametric equations of curves have continuous derivatives up to third order on the intervals where they are defined.

Having unit length, the tangent vector $\hat{\mathbf{T}}(s) = d\mathbf{r}/ds$ satisfies $\hat{\mathbf{T}}(s) \bullet \hat{\mathbf{T}}(s) = 1$. Differentiating this equation with respect to s we get

$$2\hat{\mathbf{T}}(s) \bullet \frac{d\hat{\mathbf{T}}}{ds} = 0,$$

so that $d\hat{\mathbf{T}}/ds$ is perpendicular to $\hat{\mathbf{T}}(s)$.

DEFINITION 1

Curvature and radius of curvature

The **curvature** of $\mathcal{C}$ at the point $\mathbf{r}(s)$ is the length of $d\hat{\mathbf{T}}/ds$ there. It is denoted by κ, the Greek letter "kappa":

$$\kappa(s) = \left|\frac{d\hat{\mathbf{T}}}{ds}\right|.$$

The **radius of curvature**, denoted ρ, the Greek letter "rho," is the reciprocal of the curvature:

$$\rho(s) = \frac{1}{\kappa(s)}.$$

As we will see below, the curvature of $\mathcal{C}$ at $\mathbf{r}(s)$ measures the rate of turning of the tangent line to the curve there. The radius of curvature is the radius of the circle through $\mathbf{r}(s)$ that most closely approximates the curve $\mathcal{C}$ near that point.

According to its definition, $\kappa(s) \geq 0$ everywhere on $\mathcal{C}$. If $\kappa(s) \neq 0$, we can divide $d\hat{\mathbf{T}}/ds$ by its length, $\kappa(s)$, and obtain a unit vector $\hat{\mathbf{N}}(s)$ in the same direction. This unit vector is called the **unit principal normal** to $\mathcal{C}$ at $\mathbf{r}(s)$, or, more commonly, just the **unit normal**:

$$\hat{\mathbf{N}}(s) = \frac{1}{\kappa(s)} \frac{d\hat{\mathbf{T}}}{ds} = \frac{d\hat{\mathbf{T}}}{ds} \Big/ \left|\frac{d\hat{\mathbf{T}}}{ds}\right|.$$

Note that $\hat{\mathbf{N}}(s)$ is perpendicular to $\mathcal{C}$ at $\mathbf{r}(s)$ and points in the direction that $\hat{\mathbf{T}}$, and therefore $\mathcal{C}$, is turning. The principal normal is not defined at points where the curvature $\kappa(s)$ is zero. For instance, a straight line has no principal normal. Figure 11.14(a) shows $\hat{\mathbf{T}}$ and $\hat{\mathbf{N}}$ at a point on a typical curve.

EXAMPLE 2 Let $a > 0$. Show that the curve $\mathcal{C}$ given by

$$\mathbf{r} = a\cos\left(\frac{s}{a}\right)\mathbf{i} + a\sin\left(\frac{s}{a}\right)\mathbf{j}$$

is a circle in the xy-plane having radius a and centre at the origin and that it is parametrized in terms of arc length. Find the curvature, the radius of curvature, and the unit tangent and principal normal vectors at any point on $\mathcal{C}$.

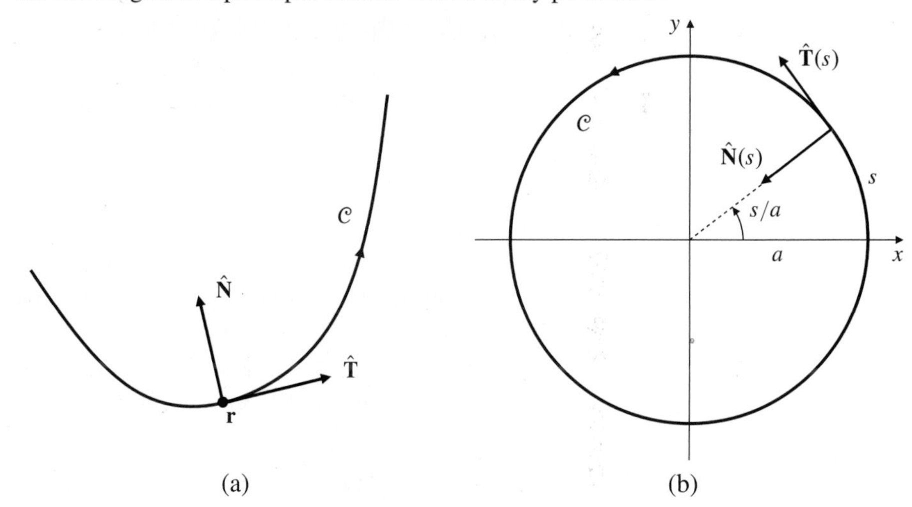

Figure 11.14

(a) The unit tangent and principal normal vectors for a curve

(b) The unit tangent and principal normal vectors for a circle

Solution Since

$$|\mathbf{r}(s)| = a\sqrt{\left(\cos\left(\frac{s}{a}\right)\right)^2 + \left(\sin\left(\frac{s}{a}\right)\right)^2} = a,$$

$\mathcal{C}$ is indeed a circle of radius a centred at the origin in the xy-plane. Since the speed

$$\left|\frac{d\mathbf{r}}{ds}\right| = \left|-\sin\left(\frac{s}{a}\right)\mathbf{i} + \cos\left(\frac{s}{a}\right)\mathbf{j}\right| = 1,$$

the parameter s must represent arc length; hence the unit tangent vector is

$$\hat{\mathbf{T}}(s) = -\sin\left(\frac{s}{a}\right)\mathbf{i} + \cos\left(\frac{s}{a}\right)\mathbf{j}.$$

Therefore,

$$\frac{d\hat{\mathbf{T}}}{ds} = -\frac{1}{a}\cos\left(\frac{s}{a}\right)\mathbf{i} - \frac{1}{a}\sin\left(\frac{s}{a}\right)\mathbf{j}$$

and the curvature and radius of curvature at $\mathbf{r}(s)$ are

$$\kappa(s) = \left|\frac{d\hat{\mathbf{T}}}{ds}\right| = \frac{1}{a}, \qquad \rho(s) = \frac{1}{\kappa(s)} = a.$$

Finally, the unit principal normal is

$$\hat{\mathbf{N}}(s) = -\cos\left(\frac{s}{a}\right)\mathbf{i} - \sin\left(\frac{s}{a}\right)\mathbf{j} = -\frac{1}{a}\mathbf{r}(s).$$

Note that the curvature and radius of curvature are constant; the latter is in fact the radius of the circle. The circle and its unit tangent and normal vectors at a typical point are sketched in Figure 11.14(b). Note that $\hat{\mathbf{N}}$ points toward the centre of the circle.

Remark Another observation can be made about the above example. The position vector $\mathbf{r}(s)$ makes angle $\theta = s/a$ with the positive x-axis; therefore, $\hat{\mathbf{T}}(s)$ makes the same angle with the positive y-axis. Therefore, the rate of rotation of $\hat{\mathbf{T}}$ with respect to s is

$$\frac{d\theta}{ds} = \frac{1}{a} = \kappa.$$

That is, κ is the rate at which $\hat{\mathbf{T}}$ is turning (measured with respect to arc length). This observation extends to a general smooth curve.

THEOREM 2

Curvature is the rate of turning of the unit tangent

Let $\kappa > 0$ on an interval containing s, and let $\Delta\theta$ be the angle between $\hat{\mathbf{T}}(s + \Delta s)$ and $\hat{\mathbf{T}}(s)$, the unit tangent vectors at neighbouring points on the curve. Then

$$\kappa(s) = \lim_{\Delta s \to 0} \left|\frac{\Delta\theta}{\Delta s}\right|.$$

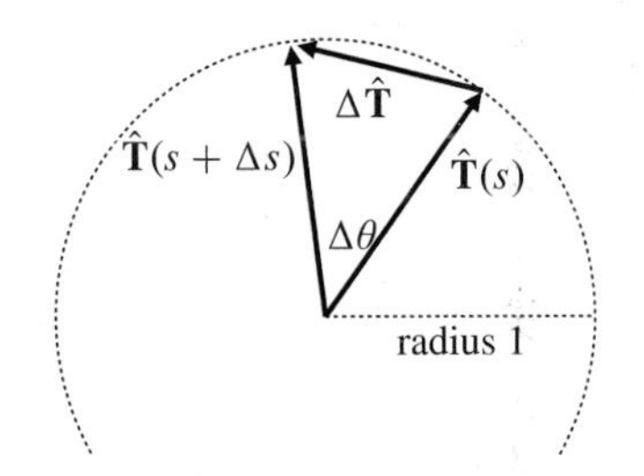

Figure 11.15 $|\Delta\hat{\mathbf{T}}| \approx |\Delta\theta|$ for small values of $|\Delta s|$

PROOF Let $\Delta\hat{\mathbf{T}} = \hat{\mathbf{T}}(s + \Delta s) - \hat{\mathbf{T}}(s)$. Because both $\hat{\mathbf{T}}(s)$ and $\hat{\mathbf{T}}(s + \Delta s)$ are unit vectors, $|\Delta\hat{\mathbf{T}}/\Delta\theta|$ is the ratio of the length of a chord to the length of the corresponding arc on a circle of radius 1. (See Figure 11.15.) Thus,

$$\lim_{\Delta s \to 0} \left|\frac{\Delta\hat{\mathbf{T}}}{\Delta\theta}\right| = 1 \qquad \text{and}$$

$$\kappa(s) = \lim_{\Delta s \to 0} \left|\frac{\Delta\hat{\mathbf{T}}}{\Delta s}\right| = \lim_{\Delta s \to 0} \left|\frac{\Delta\hat{\mathbf{T}}}{\Delta\theta}\right| \left|\frac{\Delta\theta}{\Delta s}\right| = \lim_{\Delta s \to 0} \left|\frac{\Delta\theta}{\Delta s}\right|.$$

The unit tangent $\hat{\mathbf{T}}$ and unit normal $\hat{\mathbf{N}}$ at a point $\mathbf{r}(s)$ on a curve $\mathcal{C}$ are regarded as having their tails at that point. They are perpendicular, and $\hat{\mathbf{N}}$ points in the direction toward which $\hat{\mathbf{T}}(s)$ turns as s increases. The plane passing through $\mathbf{r}(s)$ and containing the vectors $\hat{\mathbf{T}}(s)$ and $\hat{\mathbf{N}}(s)$ is called the **osculating plane** of $\mathcal{C}$ at $\mathbf{r}(s)$ (from the Latin *osculum*, meaning *kiss*). For a *plane curve*, such as the circle in Example 2, the osculating plane is just the plane containing the curve. For more general three-dimensional curves the osculating plane varies from point to point; at any point it is the plane that comes closest to containing the part of the curve near that point. The osculating plane is not properly defined at a point where $\kappa(s) = 0$, although if such points are isolated, it can sometimes be defined as a limit of osculating planes for neighbouring points.

Still assuming that $\kappa(s) \neq 0$, let

$$\mathbf{r}_c(s) = \mathbf{r}(s) + \rho(s)\hat{\mathbf{N}}(s).$$

For each s the point with position vector $\mathbf{r}_c(s)$ lies in the osculating plane of $\mathcal{C}$ at $\mathbf{r}(s)$, on the concave side of $\mathcal{C}$ and at distance $\rho(s)$ from $\mathbf{r}(s)$. It is called the **centre of curvature** of $\mathcal{C}$ for the point $\mathbf{r}(s)$. The circle in the osculating plane having centre at the centre of curvature and radius equal to the radius of curvature $\rho(s)$ is called the **osculating circle** for $\mathcal{C}$ at $\mathbf{r}(s)$. Among all circles that pass through the point $\mathbf{r}(s)$, the osculating circle is the one that best describes the behaviour of $\mathcal{C}$ near that point. Of course, the osculating circle of a circle at any point is the same circle. A typical example of an osculating circle is shown in Figure 11.16.

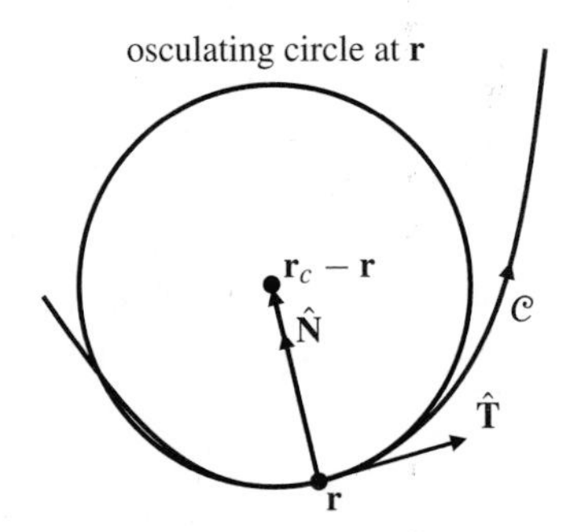

Figure 11.16 An osculating circle

Torsion and Binormal, the Frenet-Serret Formulas

At any point $\mathbf{r}(s)$ on the curve $\mathcal{C}$ where $\hat{\mathbf{T}}$ and $\hat{\mathbf{N}}$ are defined, a third unit vector, the **unit binormal** $\hat{\mathbf{B}}$, is defined by the formula

$$\hat{\mathbf{B}} = \hat{\mathbf{T}} \times \hat{\mathbf{N}}.$$

Note that $\hat{\mathbf{B}}(s)$ is normal to the osculating plane of $\mathcal{C}$ at $\mathbf{r}(s)$; if $\mathcal{C}$ is a plane curve, then $\hat{\mathbf{B}}$ is a constant vector, independent of s on any interval where $\kappa(s) \neq 0$. At each point $\mathbf{r}(s)$ on $\mathcal{C}$, the three vectors $\{\hat{\mathbf{T}}, \hat{\mathbf{N}}, \hat{\mathbf{B}}\}$ constitute a right-handed basis of mutually perpendicular unit vectors like the standard basis $\{\mathbf{i}, \mathbf{j}, \mathbf{k}\}$. (See Figure 11.17.) This basis is called the **Frenet frame** for $\mathcal{C}$ at the point $\mathbf{r}(s)$. Note that

$$\hat{\mathbf{B}} \times \hat{\mathbf{T}} = \hat{\mathbf{N}} \text{ and } \hat{\mathbf{N}} \times \hat{\mathbf{B}} = \hat{\mathbf{T}}.$$

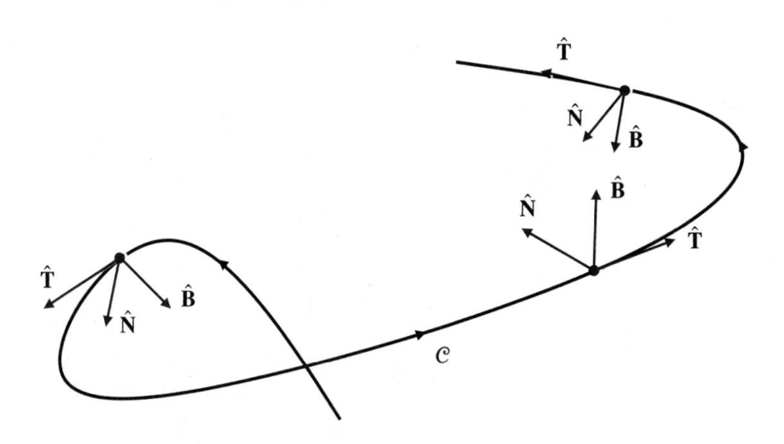

Figure 11.17 The Frenet frame $\{\hat{\mathbf{T}}, \hat{\mathbf{N}}, \hat{\mathbf{B}}\}$ at some points on $\mathcal{C}$

Since $1 = \hat{\mathbf{B}}(s) \bullet \hat{\mathbf{B}}(s)$, then $\hat{\mathbf{B}}(s) \bullet (d\hat{\mathbf{B}}/ds) = 0$, and $d\hat{\mathbf{B}}/ds$ is perpendicular to $\hat{\mathbf{B}}(s)$. Also, differentiating $\hat{\mathbf{B}} = \hat{\mathbf{T}} \times \hat{\mathbf{N}}$ we obtain

$$\frac{d\hat{\mathbf{B}}}{ds} = \frac{d\hat{\mathbf{T}}}{ds} \times \hat{\mathbf{N}} + \hat{\mathbf{T}} \times \frac{d\hat{\mathbf{N}}}{ds} = \kappa\hat{\mathbf{N}} \times \hat{\mathbf{N}} + \hat{\mathbf{T}} \times \frac{d\hat{\mathbf{N}}}{ds} = \hat{\mathbf{T}} \times \frac{d\hat{\mathbf{N}}}{ds}.$$

Therefore, $d\hat{\mathbf{B}}/ds$ is also perpendicular to $\hat{\mathbf{T}}$. Being perpendicular to both $\hat{\mathbf{T}}$ and $\hat{\mathbf{B}}$, $d\hat{\mathbf{B}}/ds$ must be parallel to $\hat{\mathbf{N}}$. This fact is the basis for our definition of torsion.

DEFINITION 2

Torsion

On any interval where $\kappa(s) \neq 0$ there exists a function $\tau(s)$ such that

$$\frac{d\hat{\mathbf{B}}}{ds} = -\tau(s)\hat{\mathbf{N}}(s).$$

The number $\tau(s)$ is called the **torsion** of $\mathcal{C}$ at $\mathbf{r}(s)$.

The torsion measures the degree of twisting that the curve exhibits near a point, that is, the extent to which the curve fails to be planar. It may be positive or negative, depending on the right-handedness or left-handedness of the twisting. We will present an example later in this section.

Theorem 2 has an analogue for torsion, for which the proof is similar. It states that the absolute value of the torsion, $|\tau(s)|$, at point $\mathbf{r}(s)$ on the curve $\mathcal{C}$ is the rate of turning of the unit binormal:

$$\lim_{\Delta s \to 0} \left| \frac{\Delta \psi}{\Delta s} \right| = |\tau(s)|,$$

where $\Delta\psi$ is the angle between $\hat{\mathbf{B}}(s + \Delta s)$ and $\hat{\mathbf{B}}(s)$.

EXAMPLE 3 **(The circular helix)** As observed in Example 7 of Section 11.3, the parametric equation

$$\mathbf{r}(s) = a\cos(cs)\mathbf{i} + a\sin(cs)\mathbf{j} + bcs\mathbf{k}, \qquad \text{where } c = \frac{1}{\sqrt{a^2+b^2}},$$

represents a circular helix wound on the cylinder $x^2 + y^2 = a^2$ and parametrized in terms of arc length. Assume $a > 0$. Find the curvature and torsion functions $\kappa(s)$ and $\tau(s)$ for this helix and also the unit vectors comprising the Frenet frame at any point $\mathbf{r}(s)$ on the helix.

Solution In Example 1 we calculated the unit tangent vector to be

$$\hat{\mathbf{T}}(s) = -ac\sin(cs)\mathbf{i} + ac\cos(cs)\mathbf{j} + bc\mathbf{k}.$$

Differentiating again leads to

$$\frac{d\hat{\mathbf{T}}}{ds} = -ac^2\cos(cs)\mathbf{i} - ac^2\sin(cs)\mathbf{j},$$

so that the curvature of the helix is

$$\kappa(s) = \left|\frac{d\hat{\mathbf{T}}}{ds}\right| = ac^2 = \frac{a}{a^2+b^2},$$

and the unit normal vector is

$$\hat{\mathbf{N}}(s) = \frac{1}{\kappa(s)}\frac{d\hat{\mathbf{T}}}{ds} = -\cos(cs)\mathbf{i} - \sin(cs)\mathbf{j}.$$

Now we have

$$\begin{aligned}\hat{\mathbf{B}}(s) = \hat{\mathbf{T}}(s) \times \hat{\mathbf{N}}(s) &= \begin{vmatrix} \mathbf{i} & \mathbf{j} & \mathbf{k} \\ -ac\sin(cs) & ac\cos(cs) & bc \\ -\cos(cs) & -\sin(cs) & 0 \end{vmatrix} \\ &= bc\sin(cs)\mathbf{i} - bc\cos(cs)\mathbf{j} + ac\mathbf{k}.\end{aligned}$$

Differentiating this formula leads to

$$\frac{d\hat{\mathbf{B}}}{ds} = bc^2\cos(cs)\mathbf{i} + bc^2\sin(cs)\mathbf{j} = -bc^2\hat{\mathbf{N}}(s).$$

Therefore, the torsion is given by

$$\tau(s) = -(-bc^2) = \frac{b}{a^2+b^2}.$$

Remark Observe that the curvature $\kappa(s)$ and the torsion $\tau(s)$ are both constant (i.e., independent of s) for a circular helix. In the above example, $\tau > 0$ (assuming that $b > 0$). This corresponds to the fact that the helix is *right-handed.* (See Figure 11.12 in the previous section.) If you grasp the helix with your right hand so your fingers surround it in the direction of increasing s (counterclockwise, looking down from the positive z-axis), then your thumb also points in the axial direction corresponding to increasing s (the upward direction). Had we started with a left-handed helix, such as

$$\mathbf{r} = a\sin t\,\mathbf{i} + a\cos t\,\mathbf{j} + bt\mathbf{k}, \qquad (a, b > 0),$$

we would have obtained $\tau = -b/(a^2+b^2)$.

Making use of the formulas $d\hat{\mathbf{T}}/ds = \kappa\hat{\mathbf{N}}$ and $d\hat{\mathbf{B}}/ds = -\tau\hat{\mathbf{N}}$, we can calculate $d\hat{\mathbf{N}}/ds$ as well:

$$\begin{aligned}\frac{d\hat{\mathbf{N}}}{ds} = \frac{d}{ds}(\hat{\mathbf{B}} \times \hat{\mathbf{T}}) &= \frac{d\hat{\mathbf{B}}}{ds} \times \hat{\mathbf{T}} + \hat{\mathbf{B}} \times \frac{d\hat{\mathbf{T}}}{ds} \\ &= -\tau\hat{\mathbf{N}} \times \hat{\mathbf{T}} + \kappa\hat{\mathbf{B}} \times \hat{\mathbf{N}} = -\kappa\hat{\mathbf{T}} + \tau\hat{\mathbf{B}}.\end{aligned}$$

Together, the three formulas

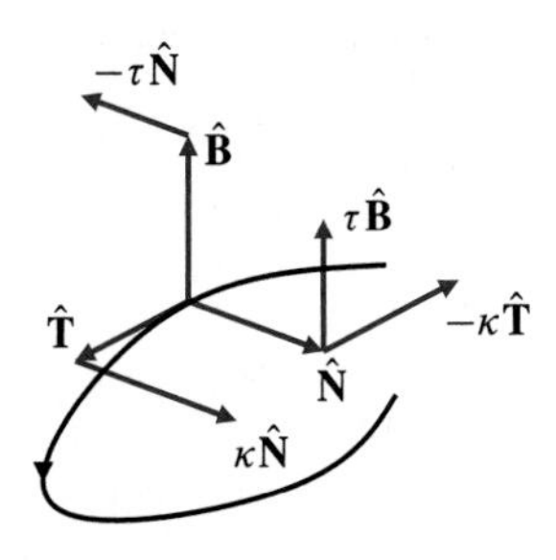

Figure 11.18 $\hat{\mathbf{T}}$, $\hat{\mathbf{N}}$, and $\hat{\mathbf{B}}$, and their directions of change

$$\frac{d\hat{\mathbf{T}}}{ds} = \kappa\hat{\mathbf{N}}$$
$$\frac{d\hat{\mathbf{N}}}{ds} = -\kappa\hat{\mathbf{T}} + \tau\hat{\mathbf{B}}$$
$$\frac{d\hat{\mathbf{B}}}{ds} = -\tau\hat{\mathbf{N}}$$

are known as the **Frenet–Serret formulas**. (See Figure 11.18.) They are of fundamental importance in the theory of curves in 3-space. The Frenet–Serret formulas can be written in matrix form as follows:

$$\frac{d}{ds}\begin{pmatrix}\hat{\mathbf{T}}\\ \hat{\mathbf{N}}\\ \hat{\mathbf{B}}\end{pmatrix} = \begin{pmatrix}0 & \kappa & 0\\ -\kappa & 0 & \tau\\ 0 & -\tau & 0\end{pmatrix}\begin{pmatrix}\hat{\mathbf{T}}\\ \hat{\mathbf{N}}\\ \hat{\mathbf{B}}\end{pmatrix}.$$

Using the Frenet–Serret formulas, we can show that the shape of a curve with nonvanishing curvature is completely determined by the curvature and torsion functions $\kappa(s)$ and $\tau(s)$.

THEOREM

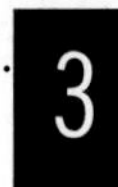

The Fundamental Theorem of Space Curves

Let $\mathcal{C}_1$ and $\mathcal{C}_2$ be two curves, both of which have the same nonvanishing curvature function $\kappa(s)$ and the same torsion function $\tau(s)$. Then the curves are congruent. That is, one can be moved rigidly (translated and rotated) so as to coincide exactly with the other.

PROOF We require $\kappa \neq 0$ because $\hat{\mathbf{N}}$ and $\hat{\mathbf{B}}$ are not defined where $\kappa = 0$. Move $\mathcal{C}_2$ rigidly so that its initial point coincides with the initial point of $\mathcal{C}_1$ and so that the Frenet frames of both curves coincide at that point. Let $\hat{\mathbf{T}}_1$, $\hat{\mathbf{T}}_2$, $\hat{\mathbf{N}}_1$, $\hat{\mathbf{N}}_2$, $\hat{\mathbf{B}}_1$, and $\hat{\mathbf{B}}_2$ be the unit tangents, normals, and binormals for the two curves. Let

$$f(s) = \hat{\mathbf{T}}_1(s) \bullet \hat{\mathbf{T}}_2(s) + \hat{\mathbf{N}}_1(s) \bullet \hat{\mathbf{N}}_2(s) + \hat{\mathbf{B}}_1(s) \bullet \hat{\mathbf{B}}_2(s).$$

We calculate the derivative of $f(s)$ using the Product Rule and the Frenet–Serret formulas:

$$\begin{aligned} f'(s) &= \hat{\mathbf{T}}_1' \bullet \hat{\mathbf{T}}_2 + \hat{\mathbf{T}}_1 \bullet \hat{\mathbf{T}}_2' + \hat{\mathbf{N}}_1' \bullet \hat{\mathbf{N}}_2 + \hat{\mathbf{N}}_1 \bullet \hat{\mathbf{N}}_2' + \hat{\mathbf{B}}_1' \bullet \hat{\mathbf{B}}_2 + \hat{\mathbf{B}}_1 \bullet \hat{\mathbf{B}}_2' \\ &= \kappa\hat{\mathbf{N}}_1 \bullet \hat{\mathbf{T}}_2 + \kappa\hat{\mathbf{T}}_1 \bullet \hat{\mathbf{N}}_2 - \kappa\hat{\mathbf{T}}_1 \bullet \hat{\mathbf{N}}_2 + \tau\hat{\mathbf{B}}_1 \bullet \hat{\mathbf{N}}_2 - \kappa\hat{\mathbf{N}}_1 \bullet \hat{\mathbf{T}}_2 \\ &\qquad + \tau\hat{\mathbf{N}}_1 \bullet \hat{\mathbf{B}}_2 - \tau\hat{\mathbf{N}}_1 \bullet \hat{\mathbf{B}}_2 - \tau\hat{\mathbf{B}}_1 \bullet \hat{\mathbf{N}}_2 \\ &= 0. \end{aligned}$$

Therefore, $f(s)$ is constant. Since the frames coincide at $s = 0$, the constant must, in fact, be 3:

$$\hat{\mathbf{T}}_1(s) \bullet \hat{\mathbf{T}}_2(s) + \hat{\mathbf{N}}_1(s) \bullet \hat{\mathbf{N}}_2(s) + \hat{\mathbf{B}}_1(s) \bullet \hat{\mathbf{B}}_2(s) = 3.$$

However, each dot product cannot exceed 1 since the factors are unit vectors. Therefore, each dot product must be equal to 1. In particular, $\hat{\mathbf{T}}_1(s) \bullet \hat{\mathbf{T}}_2(s) = 1$ for all s; hence,

$$\frac{d\mathbf{r}_1}{ds} = \hat{\mathbf{T}}_1(s) = \hat{\mathbf{T}}_2(s) = \frac{d\mathbf{r}_2}{ds}.$$

Integrating with respect to s and using the fact that both curves start from the same point when $s = 0$, we obtain $\mathbf{r}_1(s) = \mathbf{r}_2(s)$ for all s, which is what we wanted to show. ■

Remark It is a consequence of the above theorem that any curve having nonzero constant curvature and constant torsion must, in fact, be a circle (if the torsion is zero) or a circular helix (if the torsion is nonzero). See Exercises 7 and 8 below.

EXERCISES 11.4

Find the unit tangent vector $\hat{\mathbf{T}}(t)$ for the curves in Exercises 1–4.

1. $\mathbf{r} = t\mathbf{i} - 2t^2\mathbf{j} + 3t^3\mathbf{k}$
2. $\mathbf{r} = a \sin \omega t\, \mathbf{i} + a \cos \omega t\, \mathbf{k}$
3. $\mathbf{r} = \cos t \sin t\, \mathbf{i} + \sin^2 t\, \mathbf{j} + \cos t\, \mathbf{k}$
4. $\mathbf{r} = a \cos t\, \mathbf{i} + b \sin t\, \mathbf{j} + t\mathbf{k}$
5. Show that if $\kappa(s) = 0$ for all s, then the curve $\mathbf{r} = \mathbf{r}(s)$ is a straight line.
6. Show that if $\tau(s) = 0$ for all s, then the curve $\mathbf{r} = \mathbf{r}(s)$ is a plane curve. *Hint:* Show that $\mathbf{r}(s)$ lies in the plane through $\mathbf{r}(0)$ with normal $\hat{\mathbf{B}}(0)$.
7. Show that if $\kappa(s) = C$ is a positive constant and $\tau(s) = 0$ for all s, then the curve $\mathbf{r} = \mathbf{r}(s)$ is a circle. *Hint:* Find a circle having the given constant curvature. Then use Theorem 3.
8. Show that if the curvature $\kappa(s)$ and the torsion $\tau(s)$ are both nonzero constants, then the curve $\mathbf{r} = \mathbf{r}(s)$ is a circular helix. *Hint:* Find a helix having the given curvature and torsion.

11.5 Curvature and Torsion for General Parametrizations

The formulas developed above for curvature and torsion as well as for the unit normal and binormal vectors are not very useful if the curve we want to analyze is not expressed in terms of the arc-length parameter. We will now consider how to find these quantities in terms of a general parametrization $\mathbf{r} = \mathbf{r}(t)$. We will express them all in terms of the velocity, $\mathbf{v}(t)$, the speed, $v(t) = |\mathbf{v}(t)|$, and the acceleration, $\mathbf{a}(t)$. First, observe that

$$\mathbf{v} = \frac{d\mathbf{r}}{dt} = \frac{d\mathbf{r}}{ds}\frac{ds}{dt} = v\hat{\mathbf{T}}$$

$$\mathbf{a} = \frac{d\mathbf{v}}{dt} = \frac{dv}{dt}\hat{\mathbf{T}} + v\frac{d\hat{\mathbf{T}}}{dt}$$

$$= \frac{dv}{dt}\hat{\mathbf{T}} + v\frac{d\hat{\mathbf{T}}}{ds}\frac{ds}{dt} = \frac{dv}{dt}\hat{\mathbf{T}} + v^2\kappa\hat{\mathbf{N}}$$

$$\mathbf{v} \times \mathbf{a} = v\frac{dv}{dt}\hat{\mathbf{T}} \times \hat{\mathbf{T}} + v^3\kappa\hat{\mathbf{T}} \times \hat{\mathbf{N}} = v^3\kappa\hat{\mathbf{B}}.$$

Note that $\hat{\mathbf{B}}$ is in the direction of $\mathbf{v} \times \mathbf{a}$. From these formulas we obtain useful formulas for $\hat{\mathbf{T}}$, $\hat{\mathbf{B}}$, and κ:

$$\hat{\mathbf{T}} = \frac{\mathbf{v}}{v}, \qquad \hat{\mathbf{B}} = \frac{\mathbf{v} \times \mathbf{a}}{|\mathbf{v} \times \mathbf{a}|}, \qquad \kappa = \frac{|\mathbf{v} \times \mathbf{a}|}{v^3}.$$

There are several ways to calculate $\hat{\mathbf{N}}$. Perhaps the easiest is

$$\hat{\mathbf{N}} = \hat{\mathbf{B}} \times \hat{\mathbf{T}}.$$

Sometimes it may be easier to use $\dfrac{d\hat{\mathbf{T}}}{dt} = \dfrac{d\hat{\mathbf{T}}}{ds}\dfrac{ds}{dt} = v\dfrac{d\hat{\mathbf{T}}}{ds} = v\kappa\hat{\mathbf{N}}$ to calculate

$$\hat{\mathbf{N}} = \frac{1}{v\kappa}\frac{d\hat{\mathbf{T}}}{dt} = \frac{\rho}{v}\frac{d\hat{\mathbf{T}}}{dt} = \frac{d\hat{\mathbf{T}}}{dt}\bigg/\left|\frac{d\hat{\mathbf{T}}}{dt}\right|.$$

The torsion remains to be calculated. Observe that

$$\frac{d\mathbf{a}}{dt} = \frac{d}{dt}\left(\frac{dv}{dt}\hat{\mathbf{T}} + v^2\kappa\hat{\mathbf{N}}\right).$$

This differentiation will produce several terms. The only one that involves $\hat{\mathbf{B}}$ is the one that comes from evaluating $v^2\kappa(d\hat{\mathbf{N}}/dt) = v^3\kappa(d\hat{\mathbf{N}}/ds) = v^3\kappa(\tau\hat{\mathbf{B}} - \kappa\hat{\mathbf{T}})$. Therefore,

$$\frac{d\mathbf{a}}{dt} = \lambda\hat{\mathbf{T}} + \mu\hat{\mathbf{N}} + v^3\kappa\tau\hat{\mathbf{B}},$$

for certain scalars λ and μ. Since $\mathbf{v} \times \mathbf{a} = v^3 \kappa \hat{\mathbf{B}}$, it follows that

$$(\mathbf{v} \times \mathbf{a}) \bullet \frac{d\mathbf{a}}{dt} = (v^3\kappa)^2 \tau = |\mathbf{v} \times \mathbf{a}|^2 \tau.$$

Hence,

$$\tau = \frac{(\mathbf{v} \times \mathbf{a}) \bullet (d\mathbf{a}/dt)}{|\mathbf{v} \times \mathbf{a}|^2}.$$

EXAMPLE 1 Find the curvature, the torsion, and the Frenet frame at a general point on the curve

$$\mathbf{r} = (t + \cos t)\mathbf{i} + (t - \cos t)\mathbf{j} + \sqrt{2} \sin t\mathbf{k}.$$

Describe this curve.

Solution We calculate the various quantities using the recipe given above. First, the preliminaries:

$$\begin{aligned}
\mathbf{v} &= (1 - \sin t)\mathbf{i} + (1 + \sin t)\mathbf{j} + \sqrt{2}\cos t\mathbf{k} \\
\mathbf{a} &= -\cos t\mathbf{i} + \cos t\mathbf{j} - \sqrt{2}\sin t\mathbf{k} \\
\frac{d\mathbf{a}}{dt} &= \sin t\mathbf{i} - \sin t\mathbf{j} - \sqrt{2}\cos t\mathbf{k} \\
\mathbf{v} \times \mathbf{a} &= \begin{vmatrix} \mathbf{i} & \mathbf{j} & \mathbf{k} \\ 1 - \sin t & 1 + \sin t & \sqrt{2}\cos t \\ -\cos t & \cos t & -\sqrt{2}\sin t \end{vmatrix} \\
&= -\sqrt{2}(1 + \sin t)\mathbf{i} - \sqrt{2}(1 - \sin t)\mathbf{j} + 2\cos t\mathbf{k} \\
(\mathbf{v} \times \mathbf{a}) \bullet \frac{d\mathbf{a}}{dt} &= -\sqrt{2}\sin t(1 + \sin t) + \sqrt{2}\sin t(1 - \sin t) - 2\sqrt{2}\cos^2 t \\
&= -2\sqrt{2} \\
v = |\mathbf{v}| &= \sqrt{2 + 2\sin^2 t + 2\cos^2 t} = 2 \\
|\mathbf{v} \times \mathbf{a}| &= \sqrt{2(2 + 2\sin^2 t) + 4\cos^2 t} = \sqrt{8} = 2\sqrt{2}.
\end{aligned}$$

Thus, we have

$$\begin{aligned}
\kappa &= \frac{|\mathbf{v} \times \mathbf{a}|}{v^3} = \frac{2\sqrt{2}}{8} = \frac{1}{2\sqrt{2}} \\
\tau &= \frac{(\mathbf{v} \times \mathbf{a}) \bullet (d\mathbf{a}/dt)}{|\mathbf{v} \times \mathbf{a}|^2} = \frac{-2\sqrt{2}}{(2\sqrt{2})^2} = -\frac{1}{2\sqrt{2}} \\
\hat{\mathbf{T}} &= \frac{\mathbf{v}}{v} = \frac{1 - \sin t}{2}\mathbf{i} + \frac{1 + \sin t}{2}\mathbf{j} + \frac{1}{\sqrt{2}}\cos t\mathbf{k} \\
\hat{\mathbf{B}} &= \frac{\mathbf{v} \times \mathbf{a}}{|\mathbf{v} \times \mathbf{a}|} = -\frac{1 + \sin t}{2}\mathbf{i} - \frac{1 - \sin t}{2}\mathbf{j} + \frac{1}{\sqrt{2}}\cos t\mathbf{k} \\
\hat{\mathbf{N}} &= \hat{\mathbf{B}} \times \hat{\mathbf{T}} = -\frac{1}{\sqrt{2}}\cos t\mathbf{i} + \frac{1}{\sqrt{2}}\cos t\mathbf{j} - \sin t\mathbf{k}.
\end{aligned}$$

Since the curvature and torsion are both constant (they are therefore constant when expressed in terms of any parametrization), the curve must be a circular helix by Theorem 3. It is left-handed, since $\tau < 0$. By Example 3 in Section 11.4, it is congruent to the helix

$$\mathbf{r} = a\cos t\mathbf{i} + a\sin t\mathbf{j} + bt\mathbf{k},$$

provided that $a/(a^2 + b^2) = 1/(2\sqrt{2}) = -b/(a^2 + b^2)$. Solving these equations gives $a = \sqrt{2}$ and $b = -\sqrt{2}$, so the helix is wound on a cylinder of radius $\sqrt{2}$. The axis of this cylinder is the line $x = y$, $z = 0$, as can be seen by inspecting the components of $\mathbf{r}(t)$.

EXAMPLE 2 **(Curvature of the graph of a function of one variable)** Find the curvature of the plane curve with equation $y = f(x)$ at an arbitrary point $(x, f(x))$ on the curve.

Solution The graph can be parametrized: $\mathbf{r} = x\mathbf{i} + f(x)\mathbf{j}$. Thus,

$$\begin{aligned} \mathbf{v} &= \mathbf{i} + f'(x)\mathbf{j}, \\ \mathbf{a} &= f''(x)\mathbf{j}, \\ \mathbf{v} \times \mathbf{a} &= f''(x)\mathbf{k}. \end{aligned}$$

Therefore, the curvature is

$$\kappa(x) = \frac{|\mathbf{v} \times \mathbf{a}|}{v^3} = \frac{|f''(x)|}{\left(1 + (f'(x))^2\right)^{3/2}}.$$

Tangential and Normal Acceleration

In the formula obtained earlier for the acceleration in terms of the unit tangent and normal,

$$\mathbf{a} = \frac{dv}{dt}\hat{\mathbf{T}} + v^2\kappa\hat{\mathbf{N}},$$

the term $(dv/dt)\hat{\mathbf{T}}$ is called the **tangential** acceleration, and the term $v^2\kappa\hat{\mathbf{N}}$ is called the **normal** or **centripetal** acceleration. This latter component is directed toward the centre of curvature and its magnitude is $v^2\kappa = v^2/\rho$. Highway, railway, and roller-coaster designers attempt to bank curves in such a way that the resultant of the corresponding centrifugal force, $-m(v^2/\rho)\hat{\mathbf{N}}$, and the weight, $-mg\mathbf{k}$, of the vehicle will be normal to the surface at a desired speed.

EXAMPLE 3 **Banking a curve.** A level, curved road lies along the curve $y = x^2$ in the horizontal xy-plane. Find, as a function of x, the angle at which the road should be banked (i.e., the angle between the vertical and the normal to the surface of the road) so that the resultant of the centrifugal and gravitational ($-mg\mathbf{k}$) forces acting on the vehicle travelling at constant speed v_0 along the road is always normal to the surface of the road.

Solution By Example 2 the path of the road, $y = x^2$, has curvature

$$\kappa = \frac{\left|d^2y/dx^2\right|}{\left(1 + (dy/dx)^2\right)^{3/2}} = \frac{2}{(1 + 4x^2)^{3/2}}.$$

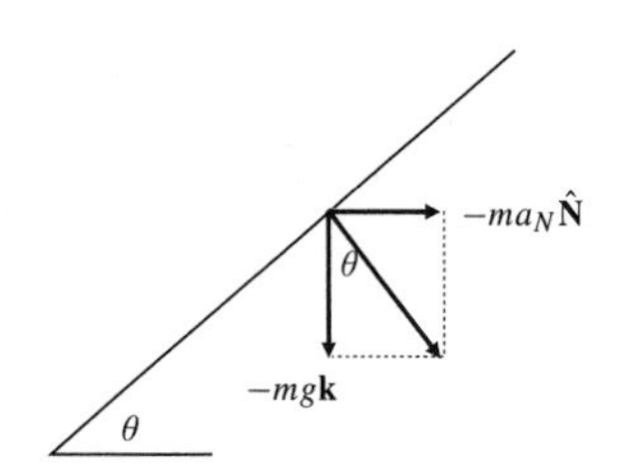

Figure 11.19 Banking a curve on a roadway

The normal component of the acceleration of a vehicle travelling at speed v_0 along the road is

$$a_N = v_0^2\kappa = \frac{2v_0^2}{(1 + 4x^2)^{3/2}}.$$

If the road is banked at angle θ (see Figure 11.19), then the resultant of the centrifugal force $-ma_N\hat{\mathbf{N}}$ and the gravitational force $-mg\mathbf{k}$ is normal to the roadway provided

$$\tan\theta = \frac{ma_N}{mg}, \qquad \text{that is,} \qquad \theta = \tan^{-1}\frac{2v_0^2}{g(1 + 4x^2)^{3/2}}.$$

Remark The definition of centripetal acceleration given above is consistent with the one that arose in the discussion of rotating frames in Section 11.2. If $\mathbf{r}(t)$ is the position of a moving particle at time t, we can regard the motion at any instant as being a rotation about the centre of curvature, so that the angular velocity must be $\boldsymbol{\Omega} = \Omega\hat{\mathbf{B}}$. The linear velocity is $\mathbf{v} = \boldsymbol{\Omega} \times (\mathbf{r} - \mathbf{r}_c) = v\hat{\mathbf{T}}$, so the speed is $v = \Omega\rho$, and $\boldsymbol{\Omega} = (v/\rho)\hat{\mathbf{B}}$. As developed in Section 11.2, the centripetal acceleration is

$$\boldsymbol{\Omega} \times (\boldsymbol{\Omega} \times (\mathbf{r} - \mathbf{r}_c)) = \boldsymbol{\Omega} \times \mathbf{v} = \frac{v^2}{\rho}\hat{\mathbf{B}} \times \hat{\mathbf{T}} = \frac{v^2}{\rho}\hat{\mathbf{N}}.$$

Evolutes

The centre of curvature $\mathbf{r}_c(t)$ of a given curve can itself trace out another curve as t varies. This curve is called the **evolute** of the given curve $\mathbf{r}(t)$.

EXAMPLE 4 Find the evolute of the exponential spiral

$$\mathbf{r} = ae^{-t}\cos t\,\mathbf{i} + ae^{-t}\sin t\,\mathbf{j}.$$

Solution The curve is a plane curve so $\tau = 0$. We will take a shortcut to the curvature and the unit normal without calculating $\mathbf{v} \times \mathbf{a}$. First, we calculate

$$\begin{aligned}
\mathbf{v} &= ae^{-t}\Big(-(\cos t + \sin t)\mathbf{i} - (\sin t - \cos t)\mathbf{j}\Big)\\
\frac{ds}{dt} &= v = \sqrt{2}ae^{-t}\\
\hat{\mathbf{T}}(t) &= \frac{1}{\sqrt{2}}\Big(-(\cos t + \sin t)\mathbf{i} - (\sin t - \cos t)\mathbf{j}\Big)\\
\frac{d\hat{\mathbf{T}}}{ds} &= \frac{1}{(ds/dt)}\frac{d\hat{\mathbf{T}}}{dt} = \frac{1}{2ae^{-t}}\Big((\sin t - \cos t)\mathbf{i} - (\cos t + \sin t)\mathbf{j}\Big)\\
\kappa(t) &= \left|\frac{d\hat{\mathbf{T}}}{ds}\right| = \frac{1}{\sqrt{2}ae^{-t}}.
\end{aligned}$$

It follows that the radius of curvature is $\rho(t) = \sqrt{2}ae^{-t}$. Since $d\hat{\mathbf{T}}/ds = \kappa\hat{\mathbf{N}}$, we have $\hat{\mathbf{N}} = \rho(d\hat{\mathbf{T}}/ds)$. The centre of curvature is

$$\begin{aligned}
\mathbf{r}_c(t) &= \mathbf{r}(t) + \rho(t)\hat{\mathbf{N}}(t)\\
&= \mathbf{r}(t) + \rho^2\frac{d\hat{\mathbf{T}}}{ds}\\
&= ae^{-t}\Big(\cos t\,\mathbf{i} + \sin t\,\mathbf{j}\Big)\\
&\quad + 2a^2e^{-2t}\frac{1}{2ae^{-t}}\Big((\sin t - \cos t)\mathbf{i} - (\cos t + \sin t)\mathbf{j}\Big)\\
&= ae^{-t}\Big(\sin t\,\mathbf{i} - \cos t\,\mathbf{j}\Big)\\
&= ae^{-t}\Big(\cos(t - \frac{\pi}{2})\mathbf{i} + \sin(t - \frac{\pi}{2})\mathbf{j}\Big).
\end{aligned}$$

Thus, interestingly, the evolute of the exponential spiral is the same exponential spiral rotated 90° clockwise in the plane. (See Figure 11.20(a).)

An Application to Track (or Road) Design

Model trains frequently come with two kinds of track sections: straight and curved. The curved sections are arcs of a circle of radius R, and the track is intended to be laid out in the shape shown in Figure 11.20(b); AB and CD are straight, and BC and DA are semicircles. The track looks smooth, but is it smooth enough?

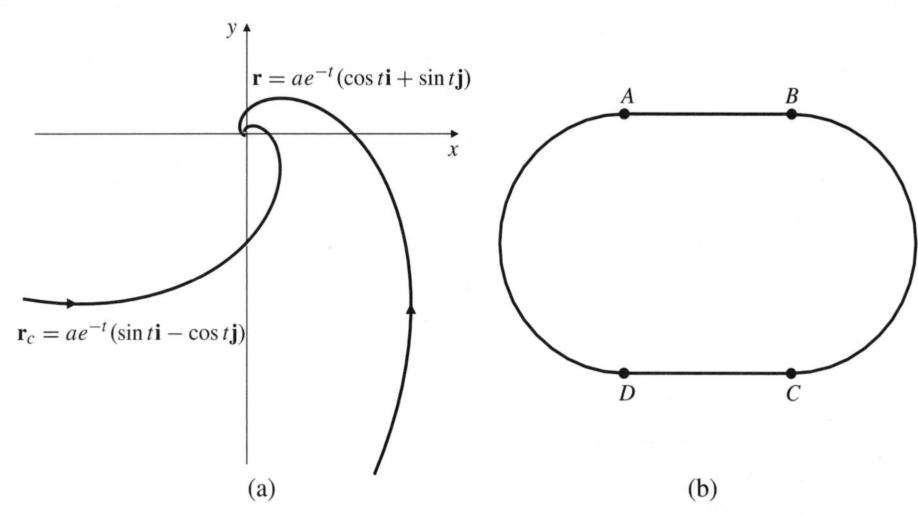

Figure 11.20

(a) The evolute of an exponential spiral is another exponential spiral

(b) The shape of a model train track

The track is held together by friction, and occasionally it can come apart as the train is racing around. It is especially likely to come apart at the points A, B, C, and D. To see why, assume that the train is travelling at constant speed v. Then the tangential acceleration, $(dv/dt)\hat{\mathbf{T}}$, is zero, and the total acceleration is just the centripetal acceleration, $\mathbf{a} = (v^2/\rho)\hat{\mathbf{N}}$. Therefore, $|\mathbf{a}| = 0$ along the straight sections, and $|\mathbf{a}| = v^2\kappa = v^2/R$ on the semicircular sections. The acceleration is *discontinuous* at the points A, B, C, and D, and the reactive force exerted by the train on the track is also discontinuous at these points. There is a "shock" or "jolt" as the train enters or leaves a curved part of the track. In order to avoid such stress points, tracks should be designed so that the curvature varies continuously from point to point.

EXAMPLE 5 Existing track along the negative x-axis and along the ray $y = x - 1$, $x \geq 2$, is to be joined smoothly by track along the transition curve $y = f(x)$, $0 \leq x \leq 2$, where $f(x)$ is a polynomial of degree as small as possible. Find $f(x)$ so that a train moving along the track will not experience discontinuous acceleration at the joins.

Solution The situation is shown in Figure 11.21. The polynomial $f(x)$ must be chosen so that the track is continuous, has continuous slope, and has continuous curvature at $x = 0$ and $x = 2$. Since the curvature of $y = f(x)$ is

$$\kappa = |f''(x)|\Big(1 + (f'(x))^2\Big)^{-3/2},$$

we need only arrange that f, f', and f'' take the same values at $x = 0$ and $x = 2$ that the straight sections do there:

$$\begin{aligned} &f(0) = 0, \qquad f'(0) = 0, \qquad f''(0) = 0, \\ &f(2) = 1, \qquad f'(2) = 1, \qquad f''(2) = 0. \end{aligned}$$

These six independent conditions suggest we should try a polynomial of degree 5 involving six arbitrary coefficients:

$$\begin{aligned} f(x) &= A + Bx + Cx^2 + Dx^3 + Ex^4 + Fx^5 \\ f'(x) &= B + 2Cx + 3Dx^2 + 4Ex^3 + 5Fx^4 \\ f''(x) &= 2C + 6Dx + 12Ex^2 + 20Fx^3. \end{aligned}$$

The three conditions at $x = 0$ imply that $A = B = C = 0$. Those at $x = 2$ imply that

$$\begin{aligned} 8D + 16E + \ \ 32F &= f(2) \ = 1 \\ 12D + 32E + \ \ 80F &= f'(2) = 1 \\ 12D + 48E + 160F &= f''(2) = 0. \end{aligned}$$

This system has solution $D = 1/4$, $E = -1/16$, and $F = 0$, so we should use $f(x) = (x^3/4) - (x^4/16)$.

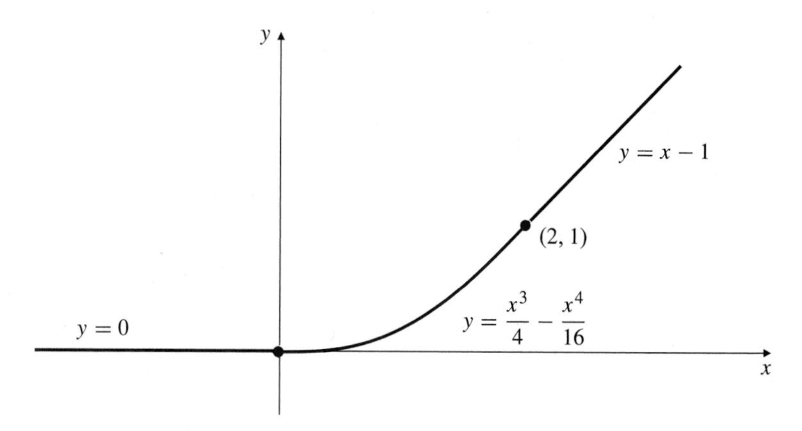

Figure 11.21 Joining two straight tracks with a curved track

Remark Road and railroad builders do not usually use polynomial graphs as transition curves. Other kinds of curves called **clothoids** and **lemniscates** are usually used. (See Exercise 7 in the Review Exercises at the end of this chapter.)

Maple Calculations

Calculations of the sort done in this section for fairly simple curves can become quite oppressive for more complicated curves. As usual, Maple can come to the rescue. Before the advent of the VectorCalculus package in Maple 8, which introduces a true vector data structure, defining a vector-valued function was a bit tricky: one had to use something like

```
> R := [t -> f(t), t -> g(t), t -> h(t)];
```

rather than the more obvious

```
> R := t -> [f(t), g(t), h(t)];
```

because in the latter the function `R` was not considered to be a vector, even though its values are vectors.

In the following we assume the LinearAlgebra and VectorCalculus packages have been loaded (in that order):

```
> with(LinearAlgebra): with(VectorCalculus):
```

Here is how we might define a vector-valued function representing a circular helix:

```
> R := t -> <a*cos(t), a*sin(t), b*t>;
```

The output from this definition (which we omit here) may appear a bit cryptic at first; it asserts that `R` is defined as a procedure "VectorCalculus-<,>" whose arguments are three "VectorCalculus-$*$" procedures for the products that represent the three components of `R`. Calling the function generates the expected results.

```
> R(t); R(Pi);
```

$$a\cos(t)\,e_x + a\sin(t)\,e_y + b\,t\,e_z$$
$$-a\,e_x + b\,\pi\,e_z$$

Velocity, acceleration, and speed functions can now be defined in the obvious way and the results used to find these quantities at any point:

```
> V := D(R): A := D(V):
> v := t -> Norm(V(t),2):
> V(t); A(t); v(t);
```

$$-a\sin(t)\,e_x + a\cos(t)\,e_y + b\,e_z$$
$$-a\cos(t)\,e_x - a\sin(t)\,e_y$$

$$\sqrt{|a\sin(t)|^2 + |a\cos(t)|^2 + |b|^2}$$

No attempt to simplify the last expression has much effect unless we tell Maple that a and b are real numbers. In fact, it is useful for purposes of simplification to tell Maple that a, b, and t are all real, and to suppress Maple's urge to beat us over the head with that fact by subsequently placing a tilde (˜) after each of these variables in all subsequent output. We can accomplish this with

```
> assume(a::real, b::real, t::real);
> interface(showassumed=0);
> simplify(v(t));
```

$$\sqrt{b^2+a^2}$$

The VectorCalculus package has a function called TNBFrame whose output is a list of functions generating the unit tangent, principal normal, and binormal vectors $\hat{\mathbf{T}}(t)$, $\hat{\mathbf{N}}(t)$, and $\hat{\mathbf{B}}(t)$. We can use the components of this list to define each vector:

```
> T := TNBFrame(R,t)[1]:  T(t);
```

$$-\frac{a\,\sin(t)}{\sqrt{b^2+a^2}}\,e_x+\frac{a\,\cos(t)}{\sqrt{b^2+a^2}}\,e_y+\frac{b}{\sqrt{b^2+a^2}}\,e_z$$

```
> N := TNBFrame(R,t)[2]:  N(t);
```

$$-\frac{a\,\cos(t)}{|a|}\,e_x-\frac{a\,\sin(t)}{|a|}\,e_y$$

```
> B := TNBFrame(R,t)[3]:  B(t);
```

$$\frac{b\,a\,\sin(t)}{\sqrt{b^2+a^2}\,|a|}\,e_x-\frac{b\,a\,\cos(t)}{\sqrt{b^2+a^2}\,|a|}\,e_y+\left(\frac{a^2\,\sin(t)^2}{\sqrt{b^2+a^2}\,|a|}+\frac{a^2\,\cos(t)^2}{\sqrt{b^2+a^2}\,|a|}\right)e_z$$

```
> simplify(%);
```

$$\frac{b\,a\,\sin(t)}{\sqrt{b^2+a^2}\,|a|}\,e_x-\frac{b\,a\,\cos(t)}{\sqrt{b^2+a^2}\,|a|}\,e_y+\frac{|a|}{\sqrt{b^2+a^2}}\,e_z$$

VectorCalculus also defines Curvature and Torsion functions that can be invoked as follows:

```
> simplify(Curvature(R,t)(t));
```

$$\frac{|a|}{\sqrt{b^2+a^2}}$$

```
> simplify(Torsion(R,t)(t));
```

$$\frac{b}{\sqrt{b^2+a^2}}$$

In Maple 8 and some releases of Maple 9 the expression generated for the torsion had an absolute value on the numerator (i.e., $|b|$ instead of b). This was in error if $b<0$; the torsion should be negative in this case. In fact, the result above would make the third Serret-Frenet formula false as we can see from

```
> simplify(diff(B(t),t) + tau*N(t));
```

$$\frac{a\,\cos(t)\left(b-\tau\sqrt{b^2+a^2}\right)}{\sqrt{b^2+a^2}\,|a|}\,e_x+\frac{a\,\sin(t)\left(b-\tau\sqrt{b^2+a^2}\right)}{\sqrt{b^2+a^2}\,|a|}\,e_y$$

This must be $\mathbf{0}$, but will be zero only if $\tau=b/\sqrt{b^2+a^2}$. This error has been corrected in Maple 10 and more recent releases.

EXERCISES 11.5

Find the radius of curvature of the curves in Exercises 1–4 at the points indicated.

1. $y=x^2$ at $x=0$ and at $x=\sqrt{2}$

2. $y=\cos x$ at $x=0$ and at $x=\pi/2$

3. $\mathbf{r}=2t\mathbf{i}+(1/t)\mathbf{j}-2t\mathbf{k}$ at $(2,1,-2)$

4. $\mathbf{r}=t^3\mathbf{i}+t^2\mathbf{j}+t\mathbf{k}$ at the point where $t=1$

Find the Frenet frames $\{\hat{\mathbf{T}}, \hat{\mathbf{N}}, \hat{\mathbf{B}}\}$ for the curves in Exercises 5–6 at the points indicated.

5. $\mathbf{r} = t\mathbf{i} + t^2\mathbf{j} + 2\mathbf{k}$ at $(1, 1, 2)$

6. $\mathbf{r} = t\mathbf{i} + t^2\mathbf{j} + t\mathbf{k}$ at $(1, 1, 1)$

In Exercises 7–8, find the unit tangent, normal, and binormal vectors and the curvature and torsion at a general point on the given curve.

7. $\mathbf{r} = t\mathbf{i} + \dfrac{t^2}{2}\mathbf{j} + \dfrac{t^3}{3}\mathbf{k}$ **8.** $\mathbf{r} = e^t(\cos t\,\mathbf{i} + \sin t\,\mathbf{j} + \mathbf{k})$

9. Find the curvature and torsion of the parametric curve

$$x = 2 + \sqrt{2}\cos t, \quad y = 1 - \sin t, \quad z = 3 + \sin t$$

at an arbitrary point t. What is the curve?

10. A particle moves along the plane curve $y = \sin x$ in the direction of increasing x with constant horizontal speed $dx/dt = k$. Find the tangential and normal components of the acceleration of the particle when it is at position x.

11. Find the unit tangent, normal and binormal, and the curvature and torsion for the curve

$$\mathbf{r} = \sin t \cos t\,\mathbf{i} + \sin^2 t\,\mathbf{j} + \cos t\,\mathbf{k}$$

at the points (a) $t = 0$ and (b) $t = \pi/4$.

12. A particle moves on an elliptical path in the xy-plane so that its position at time t is $\mathbf{r} = a\cos t\mathbf{i} + b\sin t\mathbf{j}$. Find the tangential and normal components of its acceleration at time t. At what points is the tangential acceleration zero?

13. Find the maximum and minimum values for the curvature of the ellipse $x = a\cos t$, $y = b\sin t$, where $a > b > 0$.

14. A bead of mass m slides without friction down a wire bent in the shape of the curve $y = x^2$, under the influence of the gravitational force $-mg\mathbf{j}$. The speed of the bead is v as it passes through the point $(1, 1)$. Find, at that instant, the magnitude of the normal acceleration of the bead and the rate of change of its speed.

15. Find the curvature of the plane curve $y = e^x$ at x. Find the equation of the evolute of this curve.

16. Show that the curvature of the plane polar graph $r = f(\theta)$ at a general point θ is

$$\kappa(\theta) = \frac{\left|2\big(f'(\theta)\big)^2 + \big(f(\theta)\big)^2 - f(\theta)f''(\theta)\right|}{\left[\big(f'(\theta)\big)^2 + \big(f(\theta)\big)^2\right]^{3/2}}.$$

17. Find the curvature of the cardioid $r = a(1 - \cos\theta)$.

18. Find the curve $\mathbf{r} = \mathbf{r}(t)$ for which $\kappa(t) = 1$ and $\tau(t) = 1$ for all t, and $\mathbf{r}(0) = \hat{\mathbf{T}}(0) = \mathbf{i}$, $\hat{\mathbf{N}}(0) = \mathbf{j}$, and $\hat{\mathbf{B}}(0) = \mathbf{k}$.

19. Suppose the curve $\mathbf{r} = \mathbf{r}(t)$ satisfies $\dfrac{d\mathbf{r}}{dt} = \mathbf{c} \times \mathbf{r}(t)$, where $\mathbf{c}$ is a constant vector. Show that the curve is the circle in which the plane through $\mathbf{r}(0)$ normal to $\mathbf{c}$ intersects a sphere with radius $|\mathbf{r}(0)|$ centred at the origin.

20. Find the evolute of the circular helix $\mathbf{r} = a\cos t\,\mathbf{i} + a\sin t\,\mathbf{j} + bt\mathbf{k}$.

21. Find the evolute of the parabola $y = x^2$.

22. Find the evolute of the ellipse $x = 2\cos t$, $y = \sin t$.

23. Find the polynomial $f(x)$ of lowest degree so that track along $y = f(x)$ from $x = -1$ to $x = 1$ joins with existing straight tracks $y = -1$, $x \le -1$ and $y = 1$, $x \ge 1$ sufficiently smoothly that a train moving at constant speed will not experience discontinuous acceleration at the joins.

24. Help out model train manufacturers. Design a track segment $y = f(x)$, $-1 \le x \le 0$, to provide a jolt-free link between a straight track section $y = 1$, $x \le -1$, and a semicircular arc section $x^2 + y^2 = 1$, $x \ge 0$.

25. If the position $\mathbf{r}$, velocity $\mathbf{v}$, and acceleration $\mathbf{a}$ of a moving particle satisfy $\mathbf{a}(t) = \lambda(t)\mathbf{r}(t) + \mu(t)\mathbf{v}(t)$, where $\lambda(t)$ and $\mu(t)$ are scalar functions of time t, and if $\mathbf{v} \times \mathbf{a} \ne \mathbf{0}$, show that the path of the particle lies in a plane.

Use Maple in Exercises 26–31. Make sure to load the LinearAlgebra and VectorCalculus packages.

In Exercises 26–29, determine the curvature and torsion functions for the given curves. Because of the problem with the `Torsion` function in some versions of the VectorCalculus package (as mentioned at the end of this section), you may want to use the formulas derived from the derivatives of position to determine it, and probably the curvature as well. Try to describe the curve.

26. $\mathbf{r}(t) = \cos(t)\mathbf{i} + 2\sin(t)\mathbf{j} + \cos(t)\mathbf{k}$. Why should you not be surprised at the value of the torsion? What are the maximum and minimum curvatures? Describe the curve.

27. $\mathbf{r}(t) = (t - \sin t)\mathbf{i} + (1 - \cos t)\mathbf{j} + t\mathbf{k}$. Are the curvature and torsion continuous for all t?

28. $\mathbf{r}(t) = \cos(t)\cos(2t)\mathbf{i} + \cos(t)\sin(2t)\mathbf{j} + \sin(t)\mathbf{k}$. Show that the curve lies on the sphere $x^2 + y^2 + z^2 = 1$. What is the minimum value of its curvature?

29. $\mathbf{r}(t) = (t + \cos t)\mathbf{i} + (t + \sin t)\mathbf{j} + (1 + t - \cos t)\mathbf{k}$.

In Exercises 30–31, define new Maple functions to calculate the requested items. Assume the LinearAlgebra and VectorCalculus packages are loaded.

30. The `evolute(R)(t)` whose value at `R` is the function whose value at `t` is the position vector of the centre of curvature of the curve `R` for the point `R(t)`.

31. A function `tanline(R)(t,u)` whose value at **R** is the function whose value at `(t,u)` is the position vector of the point on the tangent line to the curve `R` at `t` at distance `u` from `R(t)` in the direction of increasing `t`.

11.6 Kepler's Laws of Planetary Motion

The German mathematician and astronomer Johannes Kepler (1571–1630) was a student and colleague of Danish astronomer Tycho Brahe (1546–1601). Over a lifetime of observing the positions of planets without the aid of a telescope, Brahe compiled a vast amount of data, which Kepler analyzed. Although Polish astronomer Nicolaus Copernicus (1473–1543) had postulated that the earth and other planets moved around the sun, the religious and philosophical climate in Europe at the end of the sixteenth century still favoured explaining the motion of heavenly bodies in terms of circular orbits around the earth. It was known that planets such as Mars could not move on circular orbits centred at the earth, but models were proposed in which they moved on other circles (epicycles) whose centres moved on circles centred at the earth.

Brahe's observations of Mars were sufficiently detailed that Kepler realized that no simple model based on circles could be made to conform very closely with the actual orbit. He was, however, able to fit a more general quadratic curve, an ellipse with one focus at the sun. Based on this success and on Brahe's data on other planets, he formulated the following three laws of planetary motion:

Kepler's Laws

1. The planets move on elliptical orbits with the sun at one focus.
2. The radial line from the sun to a planet sweeps out equal areas in equal times.
3. The squares of the periods of revolution of the planets around the sun are proportional to the cubes of the major axes of their orbits.

Kepler's statement of the third law actually says that the squares of the periods of revolution of the planets are proportional to the cubes of their mean distances from the sun. The mean distance of points on an ellipse from a focus of the ellipse is equal to the semi-major axis. (See Exercise 17 at the end of this section.) Therefore, the two statements are equivalent.

The choice of ellipses was reasonable once it became clear that circles would not work. The properties of the conic sections were well understood, having been developed by the Greek mathematician Apollonius of Perga around 200 BC. Nevertheless, based, as it was, on observations rather than theory, Kepler's formulation of his laws without any causal explanation was a truly remarkable feat. The theoretical underpinnings came later when Newton, with the aid of his newly created calculus, showed that Kepler's laws implied an inverse square gravitational force. (See Review Exercises 14–16 at the end of this chapter.) Newton believed that his universal gravitational law also implied Kepler's laws, but his writings fail to provide a proof that is convincing by today's standards.[1]

Later in this section we will derive Kepler's laws from the gravitational law by an elegant method that exploits vector differentiation to the fullest. First, however, we need to attend to some preliminaries.

Ellipses in Polar Coordinates

The polar coordinates $[r, \theta]$ of a point in the plane whose distance r from the origin is ε times its distance $p - r\cos\theta$ from the line $x = p$ (see Figure 11.22) satisfy the equation $r = \varepsilon(p - r\cos\theta)$, or, solving for r,

[1] There are interesting articles debating the historical significance of Newton's work by Robert Weinstock, Curtis Wilson, and others in *The College Mathematics Journal,* vol. 25, No. 3, 1994.

$$r = \frac{\ell}{1 + \varepsilon \cos\theta},$$

where $\ell = \varepsilon p$. As observed in Sections 8.1 and 8.5, for $0 \le \varepsilon < 1$ this equation represents an ellipse having **eccentricity** ε. (It is a circle if $\varepsilon = 0$.) To see this, let us transform the equation to Cartesian coordinates:

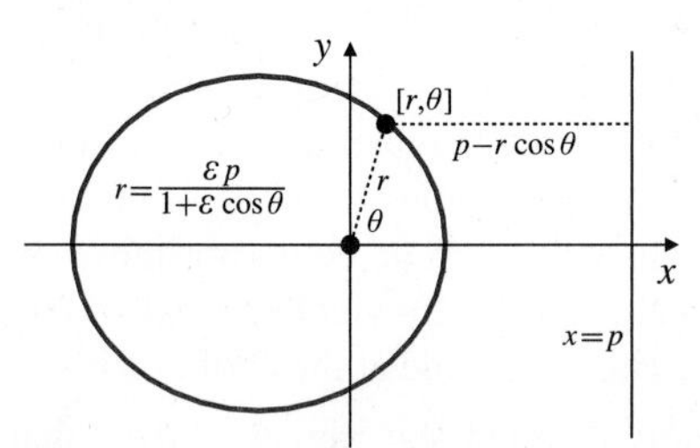

Figure 11.22 An ellipse with focus at the origin, directrix $x = p$, and eccentricity ε

$$x^2 + y^2 = r^2 = \varepsilon^2(p - r\cos\theta)^2 = \varepsilon^2(p - x)^2 = \varepsilon^2(p^2 - 2px + x^2).$$

With some algebraic manipulation, this equation can be juggled into the form

$$\frac{\left(x + \dfrac{\varepsilon\ell}{1-\varepsilon^2}\right)^2}{\left(\dfrac{\ell}{1-\varepsilon^2}\right)^2} + \frac{y^2}{\left(\dfrac{\ell}{\sqrt{1-\varepsilon^2}}\right)^2} = 1,$$

which can be recognized as an ellipse with **centre** at the point $C = (-c, 0)$, where $c = \varepsilon\ell/(1 - \varepsilon^2)$, and semi-axes a and b given by

$$a = \frac{\ell}{1-\varepsilon^2} \qquad \textbf{(semi-major axis)},$$
$$b = \frac{\ell}{\sqrt{1-\varepsilon^2}} \qquad \textbf{(semi-minor axis)}.$$

The Cartesian equation of the ellipse shows that the curve is symmetric about the lines $x = -c$ and $y = 0$ and so has a second focus at $F = (-2c, 0)$ and a second directrix with equation $x = -2c - p$. (See Figure 11.23.) The ends of the major axis are $A = (a - c, 0)$ and $A' = (-a - c, 0)$, and the ends of the minor axis are $B = (-c, b)$ and $B' = (-c, -b)$.

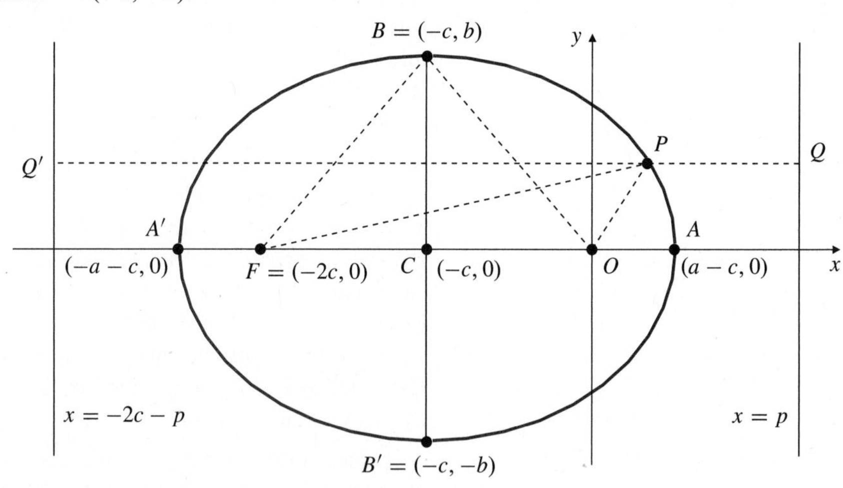

Figure 11.23 The sum of the distances from any point P on the ellipse to the two foci O and F is constant, ε times the distance between the directrices

If P is any point on the ellipse, then the distance OP is ε times the distance PQ from P to the right directrix. Similarly, the distance FP is ε times the distance $Q'P$ from P to the left directrix. Thus, the sum of the focal radii $OP + FP$ is the constant $\varepsilon Q'Q = \varepsilon(2c + 2p)$, regardless of where P is on the ellipse. Letting P be A or B we get for this sum

$$2a = (a - c) + (a + c) = OA + FA = OB + FB = 2\sqrt{b^2 + c^2}.$$

It follows that

$$a^2 = b^2 + c^2, \qquad c = \sqrt{a^2 - b^2} = \frac{\ell\varepsilon}{1-\varepsilon^2} = \varepsilon a.$$

The number ℓ is called the **semi-latus rectum** of the ellipse; the latus rectum is the width measured along the line through a focus, perpendicular to the major axis. (See Figure 11.24.)

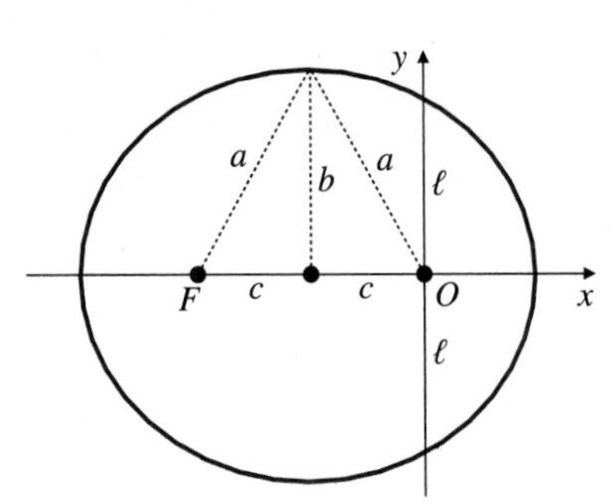

Figure 11.24 Some parameters of an ellipse

Remark The polar equation $r = \ell/(1 + \varepsilon\cos\theta)$ represents a *bounded* curve only if $\varepsilon < 1$; in this case we have $\ell/(1+\varepsilon) \le r \le \ell/(1-\varepsilon)$ for all directions θ. If $\varepsilon = 1$, the equation represents a parabola, and if $\varepsilon > 1$, a hyperbola. It is possible for objects to travel on parabolic or hyperbolic orbits, but they will approach the sun only once, rather than continue to loop around it. Some comets have hyperbolic orbits.

Polar Components of Velocity and Acceleration

Let $\mathbf{r}(t)$ be the position vector at time t of a particle P moving in the xy-plane. We construct two unit vectors at P, the vector $\hat{\mathbf{r}}$ points in the direction of the position vector $\mathbf{r}$, and the vector $\hat{\boldsymbol{\theta}}$ is rotated 90° counterclockwise from $\hat{\mathbf{r}}$. (See Figure 11.25.) If P has polar coordinates $[r, \theta]$, then $\hat{\mathbf{r}}$ points in the direction of increasing r at P, and $\hat{\boldsymbol{\theta}}$ points in the direction of increasing θ. Evidently,

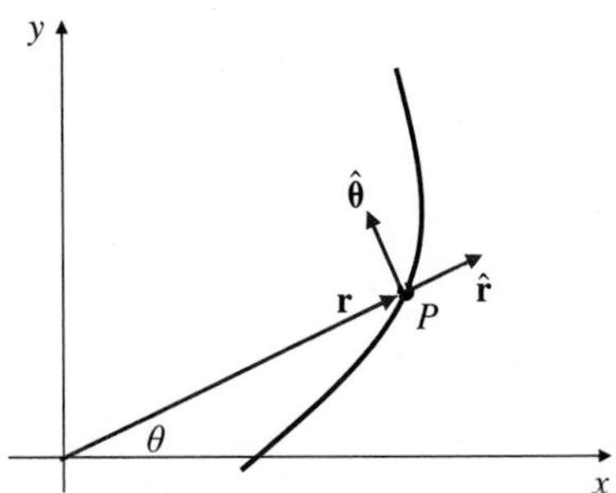

Figure 11.25 Basis vectors in the direction of increasing r and θ

$$\hat{\mathbf{r}} = \cos\theta\,\mathbf{i} + \sin\theta\,\mathbf{j}$$
$$\hat{\boldsymbol{\theta}} = -\sin\theta\,\mathbf{i} + \cos\theta\,\mathbf{j}.$$

Note that $\hat{\mathbf{r}}$ and $\hat{\boldsymbol{\theta}}$ do not depend on r but only on θ:

$$\frac{d\hat{\mathbf{r}}}{d\theta} = \hat{\boldsymbol{\theta}} \qquad \text{and} \qquad \frac{d\hat{\boldsymbol{\theta}}}{d\theta} = -\hat{\mathbf{r}}.$$

The pair $\{\hat{\mathbf{r}}, \hat{\boldsymbol{\theta}}\}$ forms a reference frame (a basis) at P so that vectors in the plane can be expressed in terms of these two unit vectors. The $\hat{\mathbf{r}}$ component of a vector is called the **radial component**, and the $\hat{\boldsymbol{\theta}}$ component is called the **transverse component**. The frame varies from point to point, so we must remember that $\hat{\mathbf{r}}$ and $\hat{\boldsymbol{\theta}}$ are both functions of t. In terms of this moving frame, the position $\mathbf{r}(t)$ of P can be expressed very simply:

$$\mathbf{r} = r\hat{\mathbf{r}},$$

where $r = r(t) = |\mathbf{r}(t)|$ is the distance from P to the origin at time t.

We are going to differentiate this equation with respect to t in order to express the velocity and acceleration of P in terms of the moving frame. Along the path of motion, $\mathbf{r}$ can be regarded as a function of either θ or t; θ is itself a function of t. To avoid confusion, let us adopt a notation that is used extensively in mechanics and that resembles the notation originally used by Newton in his calculus.

A dot over a quantity denotes the time derivative of that quantity. Two dots denote the second derivative with respect to time. Thus,

$$\dot{u} = du/dt \qquad \text{and} \qquad \ddot{u} = d^2u/dt^2.$$

First, let us record the time derivatives of the vectors $\hat{\mathbf{r}}$ and $\hat{\boldsymbol{\theta}}$. By the Chain Rule, we have

$$\dot{\hat{\mathbf{r}}} = \frac{d\hat{\mathbf{r}}}{d\theta}\frac{d\theta}{dt} = \dot{\theta}\hat{\boldsymbol{\theta}},$$
$$\dot{\hat{\boldsymbol{\theta}}} = \frac{d\hat{\boldsymbol{\theta}}}{d\theta}\frac{d\theta}{dt} = -\dot{\theta}\hat{\mathbf{r}}.$$

Now the velocity of P is

$$\mathbf{v} = \dot{\mathbf{r}} = \frac{d}{dt}(r\hat{\mathbf{r}}) = \dot{r}\hat{\mathbf{r}} + r\dot{\theta}\hat{\boldsymbol{\theta}}.$$

Polar components of velocity:

The **radial component of velocity** is $\dot{r}$.
The **transverse component of velocity** is $r\dot{\theta}$.

Since $\hat{\mathbf{r}}$ and $\hat{\boldsymbol{\theta}}$ are perpendicular unit vectors, the speed of P is given by

$$v = |\mathbf{v}| = \sqrt{\dot{r}^2 + r^2\dot{\theta}^2}.$$

Similarly, the acceleration of P can be expressed in terms of radial and transverse components:

$$\begin{aligned}\mathbf{a} = \dot{\mathbf{v}} = \ddot{\mathbf{r}} &= \frac{d}{dt}(\dot{r}\hat{\mathbf{r}} + r\dot{\theta}\hat{\boldsymbol{\theta}}) \\ &= \ddot{r}\hat{\mathbf{r}} + \dot{r}\dot{\theta}\hat{\boldsymbol{\theta}} + \dot{r}\dot{\theta}\hat{\boldsymbol{\theta}} + r\ddot{\theta}\hat{\boldsymbol{\theta}} - r\dot{\theta}^2\hat{\mathbf{r}} \\ &= (\ddot{r} - r\dot{\theta}^2)\hat{\mathbf{r}} + (r\ddot{\theta} + 2\dot{r}\dot{\theta})\hat{\boldsymbol{\theta}}.\end{aligned}$$

Polar components of acceleration:

The **radial component of acceleration** is $\ddot{r} - r\dot{\theta}^2$.
The **transverse component of acceleration** is $r\ddot{\theta} + 2\dot{r}\dot{\theta}$.

Central Forces and Kepler's Second Law

Polar coordinates are most appropriate for analyzing motion due to a **central force** that is always directed toward (or away from) a single point, the origin: $\mathbf{F} = \lambda(\mathbf{r})\mathbf{r}$, where the scalar $\lambda(\mathbf{r})$ depends on the position $\mathbf{r}$ of the object. If the velocity and acceleration of the object are $\mathbf{v} = \dot{\mathbf{r}}$ and $\mathbf{a} = \dot{\mathbf{v}}$, then Newton's Second Law of Motion ($\mathbf{F} = m\mathbf{a}$) says that $\mathbf{a}$ is parallel to $\mathbf{r}$. Therefore,

$$\frac{d}{dt}(\mathbf{r} \times \mathbf{v}) = \dot{\mathbf{r}} \times \mathbf{v} + \mathbf{r} \times \dot{\mathbf{v}} = \mathbf{v} \times \mathbf{v} + \mathbf{r} \times \mathbf{a} = \mathbf{0} + \mathbf{0} = \mathbf{0},$$

and $\mathbf{r} \times \mathbf{v} = \mathbf{h}$, a constant vector representing the object's angular momentum per unit mass about the origin. This says that $\mathbf{r}$ is always perpendicular to $\mathbf{h}$, so motion due to a central force always takes place in a *plane* through the origin having normal $\mathbf{h}$.

If we choose the z-axis to be in the direction of $\mathbf{h}$ and let $|\mathbf{h}| = h$, then $\mathbf{h} = h\mathbf{k}$, and the path of the object is in the xy-plane. In this case the position and velocity of the object satisfy

$$\mathbf{r} = r\hat{\mathbf{r}} \qquad \text{and} \qquad \mathbf{v} = \dot{r}\hat{\mathbf{r}} + r\dot{\theta}\hat{\boldsymbol{\theta}}.$$

Since $\hat{\mathbf{r}} \times \hat{\boldsymbol{\theta}} = \mathbf{k}$, we have

$$h\mathbf{k} = \mathbf{r} \times \mathbf{v} = r\dot{r}\hat{\mathbf{r}} \times \hat{\mathbf{r}} + r^2\dot{\theta}\hat{\mathbf{r}} \times \hat{\boldsymbol{\theta}} = r^2\dot{\theta}\mathbf{k}.$$

Hence, for any motion under a central force,

$$r^2\dot{\theta} = h \qquad \text{(a constant for the path of motion).}$$

This formula is equivalent to Kepler's Second Law; if $A(t)$ is the area in the plane of motion bounded by the orbit and radial lines $\theta = \theta_0$ and $\theta = \theta(t)$, then

$$A(t) = \frac{1}{2}\int_{\theta_0}^{\theta(t)} r^2\, d\theta,$$

so that

$$\frac{dA}{dt} = \frac{dA}{d\theta}\frac{d\theta}{dt} = \frac{1}{2}r^2\dot{\theta} = \frac{h}{2}.$$

Thus, area is being swept out at the constant rate $h/2$, and equal areas are swept out in equal times. Note that this law does not depend on the magnitude or direction of the force on the moving object other than the fact that it is *central*. You can also derive the equation $r^2\dot{\theta} = h$ (constant) directly from the fact that the transverse acceleration is zero:

$$\frac{d}{dt}(r^2\dot{\theta}) = 2r\dot{r}\dot{\theta} + r^2\ddot{\theta} = r(2\dot{r}\dot{\theta} + r\ddot{\theta}) = 0.$$

EXAMPLE 1 An object moves along the polar curve $r = 1/\theta$ under the influence of a force attracting it toward the origin. If the speed of the object is v_0 at the instant when $\theta = 1$, find the magnitude of the acceleration of the object at any point on its path as a function of its distance r from the origin.

Solution Since the force is central, we know that the transverse acceleration is zero and that $r^2\dot{\theta} = h$ is constant. Differentiating the equation of the path with respect to time and expressing the result in terms of r, we obtain

$$\dot{r} = -\frac{1}{\theta^2}\dot{\theta} = -r^2\frac{h}{r^2} = -h.$$

Hence, the radial component of acceleration is

$$a_r = \ddot{r} - r(\dot{\theta})^2 = 0 - r\frac{h^2}{r^4} = -\frac{h^2}{r^3}.$$

At $\theta = 1$ we have $r = 1$, so $\dot{\theta} = h$. At that instant the square of the speed is

$$v_0^2 = \dot{r}^2 + r^2\dot{\theta}^2 = h^2 + h^2 = 2h^2.$$

Hence, $h^2 = v_0^2/2$, and, at any point of its path, the magnitude of the acceleration of the object is

$$|a_r| = \frac{v_0^2}{2r^3}.$$

Derivation of Kepler's First and Third Laws

The planets and the sun move around their common centre of mass. Since the sun is vastly more massive than the planets, that centre of mass is quite close to the centre of the sun. For example, the joint centre of mass of the sun and the earth lies inside the sun. For the following derivation we will take the sun and a planet as *point masses* and consider the sun to be fixed at the origin. We will specify the directions of the coordinate axes later, when the need arises.

According to Newton's law of gravitation, the force that the sun exerts on a planet of mass m whose position vector is $\mathbf{r}$ is

$$\mathbf{F} = -\frac{km}{r^2}\hat{\mathbf{r}} = -\frac{km}{r^3}\mathbf{r},$$

where k is a positive constant depending on the mass of the sun, and $\hat{\mathbf{r}} = \mathbf{r}/r$.

As observed above, the fact that the force on the planet is always directed toward the origin implies that $\mathbf{r} \times \mathbf{v}$ is constant. We choose the direction of the z-axis so that $\mathbf{r} \times \mathbf{v} = h\mathbf{k}$, so the motion will be in the xy-plane and $r^2\dot{\theta} = h$. We have not yet specified the directions of the x- and y-axes but will do so shortly. Using polar coordinates in the xy-plane, we calculate

$$\frac{d\mathbf{v}}{d\theta} = \frac{\dot{\mathbf{v}}}{\dot{\theta}} = \frac{-\dfrac{k}{r^2}\hat{\mathbf{r}}}{\dfrac{h}{r^2}} = -\frac{k}{h}\hat{\mathbf{r}}.$$

Since $d\hat{\boldsymbol{\theta}}/d\theta = -\hat{\mathbf{r}}$, we can integrate the differential equation above to find $\mathbf{v}$:

$$\mathbf{v} = -\frac{k}{h}\int \hat{\mathbf{r}}\,d\theta = \frac{k}{h}\hat{\boldsymbol{\theta}} + \mathbf{C},$$

where $\mathbf{C}$ is a vector constant of integration. Therefore, we have shown that

$$|\mathbf{v} - \mathbf{C}| = \frac{k}{h}.$$

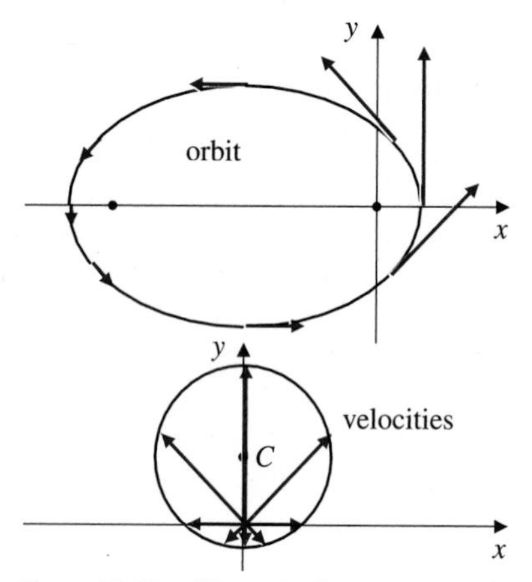

Figure 11.26 The velocity vectors define a circle

This result, known as **Hamilton's Theorem**, says that as a planet moves around its orbit, its velocity vector (when positioned with its tail at the origin) traces out a *circle* with centre at point C having position vector $\mathbf{C}$. It is perhaps surprising that there is a circle associated with the orbit of a planet after all. Only it is not the *position* vector that moves on a circle but the *velocity* vector. (See Figure 11.26.)

Recall that so far we have specified only the position of the origin and the direction of the z-axis. Therefore, the xy-plane is determined but not the directions of the x-axis or the y-axis. Let us choose these axes in the xy-plane so that $\mathbf{C}$ is in the direction of the y-axis; say $\mathbf{C} = (\varepsilon k/h)\mathbf{j}$, where ε is a positive constant. We therefore have

$$\mathbf{v} = \frac{k}{h}(\hat{\boldsymbol{\theta}} + \varepsilon\mathbf{j}).$$

The position of the x-axis is now determined by the fact that the three vectors $\mathbf{i}$, $\mathbf{j}$, and $\mathbf{k}$ are mutually perpendicular and form a right-handed basis. We calculate $\mathbf{r} \times \mathbf{v}$ again. Remember that $\mathbf{r} = r\cos\theta\mathbf{i} + r\sin\theta\mathbf{j}$, and also $\mathbf{r} = r\hat{\mathbf{r}}$:

$$\begin{aligned} h\mathbf{k} = \mathbf{r} \times \mathbf{v} &= \frac{k}{h}(r\hat{\mathbf{r}} \times \hat{\boldsymbol{\theta}} + r\varepsilon\cos\theta\mathbf{i} \times \mathbf{j} + r\varepsilon\sin\theta\mathbf{j} \times \mathbf{j}) \\ &= \frac{k}{h}r(1 + \varepsilon\cos\theta)\mathbf{k}. \end{aligned}$$

Thus, $h = \dfrac{kr}{h}(1 + \varepsilon\cos\theta)$, or, solving for r,

$$r = \frac{h^2/k}{1 + \varepsilon\cos\theta}.$$

This is the polar equation of the orbit. If $\varepsilon < 1$, it is an ellipse with one focus at the origin (the sun) and with parameters given by

Semi-latus rectum:	$\ell = \dfrac{h^2}{k}$
Semi-major axis:	$a = \dfrac{h^2}{k(1-\varepsilon^2)} = \dfrac{\ell}{1-\varepsilon^2}$
Semi-minor axis:	$b = \dfrac{h^2}{k\sqrt{1-\varepsilon^2}} = \dfrac{\ell}{\sqrt{1-\varepsilon^2}}$
Semi-focal separation:	$c = \sqrt{a^2 - b^2} = \dfrac{\varepsilon\ell}{1-\varepsilon^2}.$

We have deduced Kepler's First Law! The choices we made for the coordinate axes result in **perihelion** (the point on the orbit that is closest to the sun) being on the positive x-axis ($\theta = 0$).

EXAMPLE 2 A planet's orbit has eccentricity ε (where $0 < \varepsilon < 1$), and its speed at perihelion is v_P. Find its speed v_A at **aphelion** (the point on its orbit farthest from the sun).

Solution At perihelion and aphelion the planet's radial velocity $\dot{r}$ is zero (since r is minimum or maximum), so the velocity is entirely transverse. Thus, $v_P = r_P\dot{\theta}_P$ and $v_A = r_A\dot{\theta}_A$. Since $r^2\dot{\theta} = h$ has the same value at all points of the orbit, we have

$$r_P v_P = r_P^2\dot{\theta}_P = h = r_A^2\dot{\theta}_A = r_A v_A.$$

The planet's orbit has equation

$$r = \frac{\ell}{1 + \varepsilon\cos\theta},$$

so perihelion corresponds to $\theta = 0$ and aphelion to $\theta = \pi$:

$$r_P = \frac{\ell}{1+\varepsilon} \qquad \text{and} \qquad r_A = \frac{\ell}{1-\varepsilon}.$$

Therefore, $v_A = \dfrac{r_P}{r_A} v_P = \dfrac{1-\varepsilon}{1+\varepsilon} v_P.$

We can obtain Kepler's Third Law from the other two as follows. Since the radial line from the sun to a planet sweeps out area at a constant rate $h/2$, the total area A enclosed by the orbit is $A = (h/2)T$, where T is the period of revolution. The area of an ellipse with semi-axes a and b is $A = \pi ab$. Since $b^2 = \ell a = h^2 a/k$, we have

$$T^2 = \frac{4}{h^2}A^2 = \frac{4}{h^2}\pi^2a^2b^2 = \frac{4\pi^2}{k}a^3.$$

Note how the final expression for T^2 does not depend on h, which is a constant for the orbit of any one planet, but varies from planet to planet. The constant $4\pi^2/k$ does not depend on the particular planet. (k depends on the mass of the sun and a universal gravitational constant.) Thus,

$$T^2 = \frac{4\pi^2}{k}a^3$$

says that the square of the period of a planet is proportional to the cube of the length, $2a$, of the major axis of its orbit, the proportionality extending over all the planets. This is Kepler's Third Law. Modern astronomical data show that T^2/a^3 varies by only about three-tenths of one percent over the solar system's known planets.

Conservation of Energy

Solving the second-order differential equation of motion $\mathbf{F} = m\ddot{\mathbf{r}}$ to find the orbit of a planet requires two integrations. In the above derivation we exploited properties of the cross product to make these integrations easy. More traditional derivations of Kepler's laws usually begin with separating the radial and transverse components in the equation of motion:

$$\ddot{r} - r\dot{\theta}^2 = -\frac{k}{r^2}, \qquad r\ddot{\theta} + 2\dot{r}\dot{\theta} = 0.$$

As observed earlier, the second equation above implies that $r^2\dot{\theta} = h =$ constant, which is Kepler's Second Law. This can be used to eliminate θ from the first equation to give

$$\ddot{r} - \frac{h^2}{r^3} = -\frac{k}{r^2}.$$

Therefore,

$$\frac{d}{dt}\left(\frac{\dot{r}^2}{2} + \frac{h^2}{2r^2}\right) = \dot{r}\left(\ddot{r} - \frac{h^2}{r^3}\right) = -\frac{k}{r^2}\dot{r}.$$

If we integrate this equation, we obtain

$$\frac{1}{2}\left(\dot{r}^2 + \frac{h^2}{r^2}\right) - \frac{k}{r} = E.$$

This is a **conservation of energy** law. The first term on the left is $v^2/2$, the kinetic energy (per unit mass) of the planet. The term $-k/r$ is the potential energy per unit mass. It is difficult to integrate this equation and to find r as a function of t. In any event, we really want r as a function of θ so that we can recognize that we have an ellipse. Another way to obtain this is suggested in Exercise 18 below.

Remark The procedure used above to demonstrate Kepler's laws in fact shows that if any object moves under the influence of a force that attracts it toward the origin (or repels it away from the origin) and has magnitude proportional to the reciprocal of the square of distance from the origin, then the object must move in a plane orbit whose shape is a conic section. If the total energy E defined above is negative, then the orbit is *bounded* and must therefore be an ellipse. If $E = 0$, the orbit is a parabola. If $E > 0$, the orbit is a hyperbola. Hyperbolic orbits are typical for repulsive forces but may also occur for attractions if the object has high enough velocity (exceeding the *escape velocity*). See Exercise 22 for an example.

EXERCISES 11.6

1. **(Polar ellipses)** Fill in the details of the calculation suggested in the text to transform the polar equation of an ellipse, $r = \ell/(1 + \varepsilon \cos\theta)$, where $0 < \varepsilon < 1$, to Cartesian coordinates in a form showing the centre and semi-axes explicitly.

Polar components of velocity and acceleration

2. A particle moves on the circle with polar equation $r = k$, $(k > 0)$. What are the radial and transverse components of its velocity and acceleration? Show that the transverse component of the acceleration is equal to the rate of change of the speed of the particle.
3. Find the radial and transverse components of velocity and acceleration of a particle moving at unit speed along the exponential spiral $r = e^\theta$. Express your answers in terms of the angle θ.
4. If a particle moves along the polar curve $r = \theta$ under the influence of a central force attracting it to the origin, find the magnitude of the acceleration as a function of r and the speed of the particle.
5. An object moves along the polar curve $r = \theta^{-2}$ under the influence of a force attracting it toward the origin. If the speed of the object is v_0 at the instant when $\theta = 1$, find the magnitude of the acceleration of the object at any point on its path as a function of its distance r from the origin.

Deductions from Kepler's laws

6. The mean distance from the earth to the sun is approximately 150 million km. Comet Halley approaches perihelion (comes closest to the sun) in its elliptical orbit approximately every 76 years. Estimate the major axis of the orbit of Comet Halley.
7. The mean distance from the moon to the earth is about 385,000 km, and its period of revolution around the earth is about 27 days (the sidereal month). At approximately what distance from the centre of the earth, and in what plane, should a communications satellite be inserted into circular orbit if it must remain directly above the same position on the earth at all times?
8. An asteroid is in a circular orbit around the sun. If its period of revolution is T, find the radius of its orbit.
9. If the asteroid in Exercise 8 is instantaneously stopped in its orbit, it will fall toward the sun. How long will it take to get there? *Hint:* You can do this question easily if instead you regard the asteroid as *almost* stopped, so that it goes into a highly eccentric elliptical orbit whose major axis is a bit greater than the radius of the original circular orbit.
10. Find the eccentricity of an asteroid's orbit if the asteroid's speed at perihelion is twice its speed at aphelion.
11. Show that the orbital speed of a planet is constant if and only if the orbit is circular. *Hint:* Use the conservation of energy identity.
12. A planet's distance from the sun at perihelion is 80% of its distance at aphelion. Find the ratio of its speeds at perihelion and aphelion and the eccentricity of its orbit.
13. As a result of a collision, an asteroid originally in a circular orbit about the sun suddenly has its velocity cut in half, so that it falls into an elliptical orbit with maximum distance from the sun equal to the radius of the original circular orbit. Find the eccentricity of its new orbit.
14. If the speeds of a planet at perihelion and aphelion are v_P and v_A, respectively, what is its speed when it is at the ends of the minor axis of its orbit?
15. What fraction of its "year" (i.e., the period of its orbit) does a planet spend traversing the half of its orbit that is closest to the sun? Give your answer in terms of the eccentricity ε of the planet's orbit.

16. Suppose that a planet is travelling at speed v_0 at an instant when it is at distance r_0 from the sun. Show that the period of the planet's orbit is

$$T = \frac{2\pi}{\sqrt{k}}\left(\frac{2}{r_0} - \frac{v_0^2}{k}\right)^{-3/2}.$$

Hint: The quantity $\frac{k}{r} - \frac{1}{2}v^2$ is constant at all points of the orbit, as shown in the discussion of conservation of energy. Find the value of this expression at perihelion in terms of the semi-major axis, a.

17. The sum of the distances from a point P on an ellipse $\mathcal{E}$ to the foci of $\mathcal{E}$ is the constant $2a$, the length of the major axis of the ellipse. Use this fact in a *geometric* argument to show that the mean distance from points P to one focus of $\mathcal{E}$ is a. That is, show that

$$\frac{1}{c(\mathcal{E})}\int_{\mathcal{E}} r\,ds = a,$$

where $c(\mathcal{E})$ is the circumference of $\mathcal{E}$, and r is the distance from a point on $\mathcal{E}$ to one focus.

18. (A direct approach to Kepler's First Law) The result of eliminating θ between the equations for the radial and transverse components of acceleration for a planet is

$$\ddot{r} - \frac{h^2}{r^3} = -\frac{k}{r^2}.$$

Show that the change of dependent and independent variables:

$$r(t) = \frac{1}{u(\theta)}, \qquad \theta = \theta(t),$$

transforms this equation to the simpler equation

$$\frac{d^2u}{d\theta^2} + u = \frac{k}{h^2}.$$

Show that the solution of this equation is

$$u = \frac{k}{h^2}\Big(1 + \varepsilon\cos(\theta - \theta_0)\Big),$$

where ε and θ_0 are constants. Hence, show that the orbit is elliptical if $|\varepsilon| < 1$.

19. (What if gravitation were an inverse cube law?) Use the technique of Exercise 18 to find the trajectory of an object of unit mass attracted to the origin by a force of magnitude $f(r) = k/r^3$. Are there any orbits that do not approach infinity or the origin as $t \to \infty$?

20. Use the conservation of energy formula to show that if $E < 0$ the orbit must be bounded; that is, it cannot get arbitrarily far away from the origin.

21. (Polar hyperbolas) If $\varepsilon > 1$, then the equation

$$r = \frac{\ell}{1 + \varepsilon\cos\theta}$$

represents a hyperbola rather than an ellipse. Sketch the hyperbola, find its centre and the directions of its asymptotes, and determine its semi-transverse axis, its semi-conjugate axis, and semi-focal separation in terms of ℓ and ε.

22. (Hyperbolic orbits) A meteor travels from infinity on a hyperbolic orbit passing near the sun. At a very large distance from the sun it has speed v_∞. The asymptotes of its orbit pass at perpendicular distance D from the sun. (See Figure 11.27.) Show that the angle δ through which the meteor's path is deflected by the gravitational attraction of the sun is given by

$$\cot\left(\frac{\delta}{2}\right) = \frac{Dv_\infty^2}{k}.$$

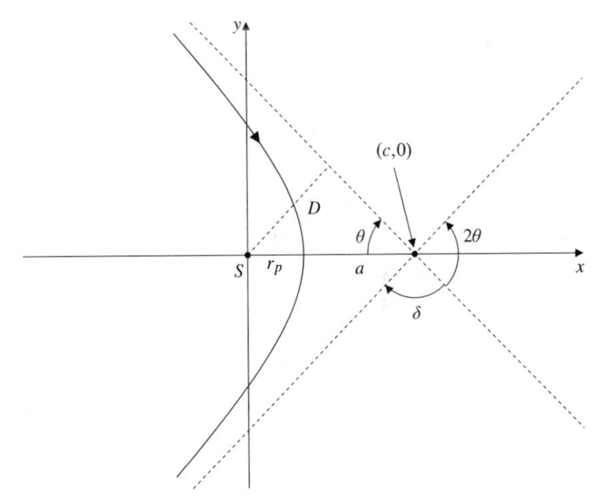

Figure 11.27 Path of a meteor

(*Hint:* You will need the result of Exercise 21.) The same analysis and results hold for electrostatic attraction or repulsion; $f(r) = \pm k/r^2$ in that case also. The constant k depends on the charges of two particles, and r is the distance between them.

CHAPTER REVIEW

Key Ideas

- **What is a vector function of a real variable, and why does it represent a curve?**
- **State the Product Rule for the derivative of $\mathbf{u}(t) \bullet \big(\mathbf{v}(t) \times \mathbf{w}(t)\big)$.**
- **What do the following terms mean?**
  - ◇ angular velocity
  - ◇ angular momentum

- ◇ centripetal acceleration ◇ Coriolis acceleration
- ◇ arc-length parametrization ◇ central force

- **Find the following quantities associated with a parametric curve $\mathcal{C}$ with parametrization $\mathbf{r} = \mathbf{r}(t)$, $(a \le t \le b)$.**
  - ◇ the velocity $\mathbf{v}(t)$ ◇ the speed $v(t)$
  - ◇ the arc length ◇ the acceleration $\mathbf{a}(t)$
  - ◇ the unit tangent $\hat{\mathbf{T}}(t)$ ◇ the unit normal $\hat{\mathbf{N}}(t)$
  - ◇ the curvature $\kappa(t)$ ◇ the radius of curvature $\rho(t)$
  - ◇ the osculating plane ◇ the osculating circle
  - ◇ the unit binormal $\hat{\mathbf{B}}(t)$ ◇ the torsion $\tau(t)$
  - ◇ the tangential acceleration ◇ the normal acceleration
  - ◇ the evolute
- **State the Frenet–Serret formulas.**
- **State Kepler's laws of planetary motion.**
- **What are the radial and transverse components of velocity and acceleration?**

Review Exercises

1. If $\mathbf{r}(t)$, $\mathbf{v}(t)$, and $\mathbf{a}(t)$ represent the position, velocity, and acceleration at time t of a particle moving in 3-space, and if, at every time t, the $\mathbf{a}$ is perpendicular to both $\mathbf{r}$ and $\mathbf{v}$, show that the vector $\mathbf{r}(t) - t\mathbf{v}(t)$ has constant length.

2. Describe the parametric curve
$$\mathbf{r} = t\cos t\mathbf{i} + t\sin t\mathbf{j} + (2\pi - t)\mathbf{k},$$
$(0 \le t \le 2\pi)$, and find its length.

3. A particle moves along the curve of intersection of the surfaces $y = x^2$ and $z = 2x^3/3$ with constant speed $v = 6$. It is moving in the direction of increasing x. Find its velocity and acceleration when it is at the point $(1, 1, 2/3)$.

4. A particle moves along the curve $y = x^2$ in the xy-plane so that at time t its speed is $v = t$. Find its acceleration at time $t = 3$ if it is at the point $(\sqrt{2}, 2)$ at that time.

5. Find the curvature and torsion at a general point of the curve $\mathbf{r} = e^t\mathbf{i} + \sqrt{2}t\mathbf{j} + e^{-t}\mathbf{k}$.

6. A particle moves on the curve of Exercise 5 so that it is at position $\mathbf{r}(t)$ at time t. Find its normal acceleration and tangential acceleration at any time t. What is its minimum speed?

7. **(A clothoid curve)** The plane curve $\mathcal{C}$ in Figure 11.28 has parametric equations
$$x(s) = \int_0^s \cos\frac{kt^2}{2}\,dt \quad \text{and} \quad y(s) = \int_0^s \sin\frac{kt^2}{2}\,dt.$$
Verify that s is, in fact, the arc length along $\mathcal{C}$ measured from $(0, 0)$ and that the curvature of $\mathcal{C}$ is given by $\kappa(s) = ks$. Because the curvature changes linearly with distance along the curve, such curves, called *clothoids*, are useful for joining track sections of different curvatures.

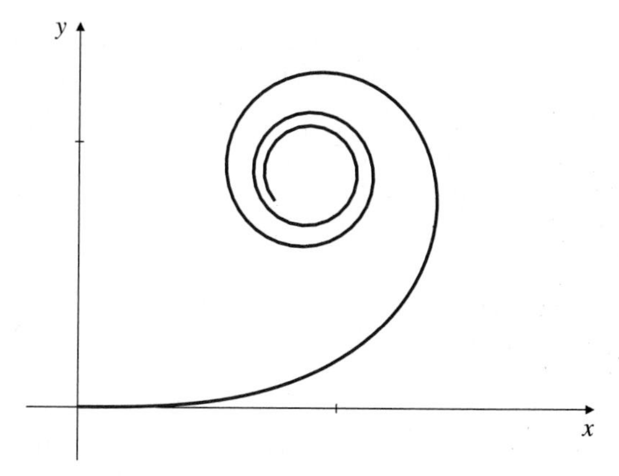

Figure 11.28 A clothoid curve

8. A particle moves along the polar curve $r = e^{-\theta}$ with constant angular speed $\dot{\theta} = k$. Express its velocity and acceleration in terms of radial and transverse components depending only on the distance r from the origin.

Some properties of cycloids

Exercises 9–12 all deal with the cycloid
$$\mathbf{r} = a(t - \sin t)\mathbf{i} + a(1 - \cos t)\mathbf{j}.$$
Recall that this curve is the path of a point on the circumference of a circle of radius a rolling along the x-axis.

9. Find the arc length $s = s(T)$ of the part of the cycloid from $t = 0$ to $t = T \le 2\pi$.

10. Find the arc-length parametrization $\mathbf{r} = \mathbf{r}(s)$ of the arch $0 \le t \le 2\pi$ of the cycloid, with s measured from the point $(0, 0)$.

11. Find the *evolute* of the cycloid; that is, find parametric equations of the centre of curvature $\mathbf{r} = \mathbf{r}_C(t)$ of the cycloid. Show that the evolute is the same cycloid translated πa units to the right and $2a$ units downward.

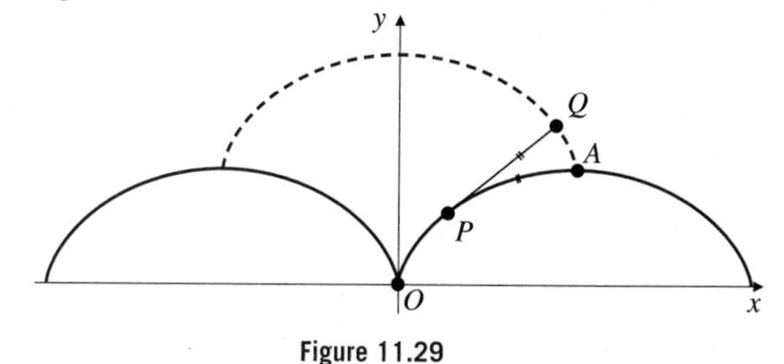

Figure 11.29

12. A string of length $4a$ has one end fixed at the origin and is wound along the arch of the cycloid to the right of the origin. Since that arch has total length $8a$, the free end of the string lies at the highest point A of the arch. Find the path followed by the free end Q of the string as it is unwound from the cycloid and is held taught during the unwinding. (See Figure 11.29.) If the string leaves the cycloid at P, then
$$(\text{arc } OP) + PQ = 4a.$$
The path of Q is called the *involute* of the cycloid. Show that, like the evolute, the involute is also a translate of the original cycloid. In fact, the cycloid is the evolute of its involute.

13. Let P be a point in 3-space with spherical coordinates (R, ϕ, θ). Suppose that P is not on the z-axis. Find a triad of mutually perpendicular unit vectors, $\{\hat{\mathbf{R}}, \hat{\boldsymbol{\phi}}, \hat{\boldsymbol{\theta}}\}$, at P in the directions of increasing R, ϕ, and θ, respectively. Is the triad right- or left-handed?

Kepler's laws imply Newton's law of gravitation

In Exercises 14–16, it is *assumed* that a planet of mass m moves in an elliptical orbit $r = \ell/(1 + \varepsilon\cos\theta)$, with focus at the origin (the sun), under the influence of a force $\mathbf{F} = \mathbf{F}(\mathbf{r})$ that depends only on the position of the planet.

14. Assuming Kepler's Second Law, show that $\mathbf{r} \times \mathbf{v} = \mathbf{h}$ is constant and, hence, that $r^2\dot{\theta} = h$ is constant.

15. Use Newton's Second Law of Motion ($\mathbf{F} = m\ddot{\mathbf{r}}$) to show that $\mathbf{r} \times \mathbf{F}(\mathbf{r}) = \mathbf{0}$. Therefore $\mathbf{F}(\mathbf{r})$ is parallel to $\mathbf{r}$: $\mathbf{F}(\mathbf{r}) = -f(\mathbf{r})\,\hat{\mathbf{r}}$, for some scalar-valued function $f(\mathbf{r})$, and the transverse component of $\mathbf{F}(\mathbf{r})$ is zero.

16. By direct calculation of the radial acceleration of the planet, show that $f(\mathbf{r}) = mh^2/(\ell r^2)$, where $r = |\mathbf{r}|$. Thus, $\mathbf{F}$ is an attraction to the origin, proportional to the mass of the planet, and inversely proportional to the square of its distance from the sun.

Challenging Problems

1. Let P be a point on the surface of the earth at 45° north latitude. Use a coordinate system with origin at P and basis vectors $\mathbf{i}$ and $\mathbf{j}$ pointing east and north, respectively, so that $\mathbf{k}$ points vertically upward.

(a) Express the angular velocity $\mathbf{\Omega}$ of the earth in terms of the basis vectors at P. What is the magnitude Ω of $\mathbf{\Omega}$ in radians per second?

(b) Find the Coriolis acceleration $\mathbf{a}_C = 2\mathbf{\Omega} \times \mathbf{v}$ of an object falling vertically with speed v above P.

(c) If the object in (b) drops from rest from a height of 100 m above P, approximately where will it strike the ground? Ignore air resistance but not the Coriolis acceleration. Since the Coriolis acceleration is much smaller than the gravitational acceleration in magnitude, you can use the vertical velocity as a good approximation to the actual velocity of the object at any time during its fall.

2. (The spin of a baseball) When a ball is thrown with spin about an axis that is not parallel to its velocity, it experiences a lateral acceleration due to differences in friction along its sides. This spin acceleration is given by $\mathbf{a}_s = k\mathbf{S} \times \mathbf{v}$, where $\mathbf{v}$ is the velocity of the ball, $\mathbf{S}$ is the angular velocity of its spin, and k is a positive constant depending on the surface of the ball. Suppose that a ball for which $k = 0.001$ is thrown horizontally along the x-axis with an initial speed of 70 ft/s and a spin of 1,000 radians/s about a vertical axis. Its velocity $\mathbf{v}$ must satisfy

$$\begin{cases} \dfrac{d\mathbf{v}}{dt} = (0.001)(1{,}000\mathbf{k}) \times \mathbf{v} - 32\mathbf{k} = \mathbf{k} \times \mathbf{v} - 32\mathbf{k} \\ \mathbf{v}(0) = 70\mathbf{i}, \end{cases}$$

since the acceleration of gravity is 32 ft/s^2.

(a) Show that the components of $\mathbf{v} = v_1\mathbf{i} + v_2\mathbf{j} + v_3\mathbf{k}$ satisfy

$$\begin{cases} \dfrac{dv_1}{dt} = -v_2 \\ v_1(0) = 70 \end{cases} \quad \begin{cases} \dfrac{dv_2}{dt} = v_1 \\ v_2(0) = 0 \end{cases} \quad \begin{cases} \dfrac{dv_3}{dt} = -32 \\ v_3(0) = 0. \end{cases}$$

(b) Solve these equations, and find the position of the ball t s after it is thrown. Assume that it is thrown from the origin at time $t = 0$.

(c) At $t = 1/5$ s, how far, and in what direction, has the ball deviated from the parabolic path it would have followed if it had been thrown without spin?

3. (Charged particles moving in magnetic fields) Magnetic fields exert forces on moving charged particles. If a particle of mass m and charge q is moving with velocity $\mathbf{v}$ in a magnetic field $\mathbf{B}$, then it experiences a force $\mathbf{F} = q\mathbf{v} \times \mathbf{B}$, and hence its velocity is governed by the equation

$$m\,\frac{d\mathbf{v}}{dt} = q\mathbf{v} \times \mathbf{B}.$$

For this exercise, suppose that the magnetic field is constant and vertical, say, $\mathbf{B} = B\mathbf{k}$ (as, e.g., in a cathode-ray tube). If the moving particle has initial velocity $\mathbf{v}_0$, then its velocity at time t is determined by

$$\begin{cases} \dfrac{d\mathbf{v}}{dt} = \omega\mathbf{v} \times \mathbf{k}, \qquad \text{where } \omega = \dfrac{qB}{m} \\ \mathbf{v}(0) = \mathbf{v}_0. \end{cases}$$

(a) Show that $\mathbf{v} \bullet \mathbf{k} = \mathbf{v}_0 \bullet \mathbf{k}$ and $|\mathbf{v}| = |\mathbf{v}_0|$ for all t.

(b) Let $\mathbf{w}(t) = \mathbf{v}(t) - (\mathbf{v}_0 \bullet \mathbf{k})\mathbf{k}$, so that $\mathbf{w}$ is perpendicular to $\mathbf{k}$ for all t. Show that $\mathbf{w}$ satisfies

$$\begin{cases} \dfrac{d^2\mathbf{w}}{dt^2} = -\omega^2\mathbf{w} \\ \mathbf{w}(0) = \mathbf{v}_0 - (\mathbf{v}_0 \bullet \mathbf{k})\mathbf{k} \\ \mathbf{w}'(0) = \omega\mathbf{v}_0 \times \mathbf{k}. \end{cases}$$

(c) Solve the initial-value problem in (b) for $\mathbf{w}(t)$, and hence find $\mathbf{v}(t)$.

(d) Find the position vector $\mathbf{r}(t)$ of the particle at time t if it is at the origin at time $t = 0$. Verify that the path of the particle is, in general, a circular helix. Under what circumstances is the path a straight line? a circle?

4. (The tautochrone) The parametric equations

$$x = a(\theta - \sin\theta) \quad \text{and} \quad y = a(\cos\theta - 1)$$

(for $0 \le \theta \le 2\pi$), describe an arch of the cycloid followed by a point on a circle of radius a rolling along the underside of the x-axis. Suppose the curve is made of wire along which a bead can slide without friction. (See Figure 11.30.) If the bead slides from rest under gravity, starting at a point having parameter value θ_0, show that the time it takes for the bead to fall to the lowest point on the arch (corresponding to $\theta = \pi$) is a *constant*, independent of the starting position θ_0. Thus, two such beads released simultaneously from different positions along the wire will always collide at the lowest point. For this reason, the cycloid is sometimes called the *tautochrone*, from the Greek for "constant time." *Hint:* When the bead has fallen from height $y(\theta_0)$ to height $y(\theta)$, its speed is $v = \sqrt{2g\big(y(\theta_0) - y(\theta)\big)}$. (Why?) The time for the bead to fall to the bottom is

$$T = \int_{\theta=\theta_0}^{\theta=\pi} \frac{1}{v}\,ds,$$

where ds is the arc length element along the cycloid.

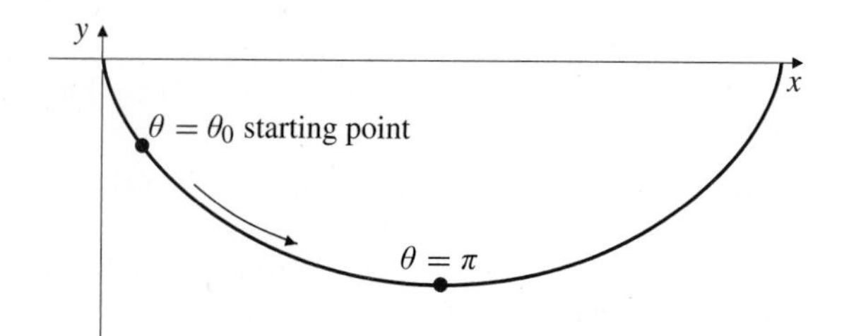

Figure 11.30

5. **(The Drop of Doom)** An amusement park ride that used to be located at the West Edmonton Mall in Alberta, Canada, gives thrill seekers a taste of free-fall. It consists of a car moving along a track consisting of straight vertical and horizontal sections joined by a smooth curve. The car drops from the top and falls vertically under gravity for $10 - 2\sqrt{2} \approx 7.2$ m before entering the curved section at B. (See Figure 11.31.) It falls another $2\sqrt{2} \approx 2.8$ m as it whips around the curve and into the horizontal section DE at ground level, where brakes are applied to stop it. (Thus, the total vertical drop from A to D or E is 10 m, a figure, like the others in this problem, chosen for mathematical convenience rather than engineering precision.) For purposes of this problem it is helpful to take the coordinate axes at a 45° angle to the vertical, so that the two straight sections of the track lie along the graph $y = |x|$. The curved section then goes from $(-2, 2)$ to $(2, 2)$ and can be taken to be symmetric about the y-axis. With this coordinate system, the gravitational acceleration is in the direction of $\mathbf{i} - \mathbf{j}$.

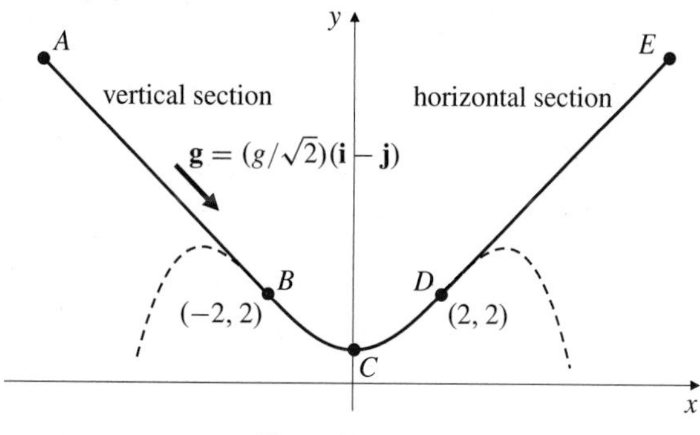

Figure 11.31

(a) Find a fourth-degree polynomial whose graph can be used to link the two straight sections of track without producing discontinuous accelerations for the falling car. (Why is fourth degree adequate?)

(b) Ignoring friction and air resistance, how fast is the car moving when it enters the curve at B? at the midpoint C of the curve? and when it leaves the curve at D?

(c) Find the magnitude of the normal acceleration and of the total acceleration of the car as it passes through C.

6. **(A chase problem)** A fox and a hare are running in the xy-plane. Both are running at the same speed v. The hare is running up the y-axis; at time $t = 0$ it is at the origin. The fox is always running straight toward the hare. At time $t = 0$ the fox is at the point $(a, 0)$, where $a > 0$. Let the fox's position at time t be $\big(x(t), y(t)\big)$.

(a) Verify that the tangent to the fox's path at time t has slope

$$\frac{dy}{dx} = \frac{y(t) - vt}{x(t)}.$$

(b) Show that the equation of the path of the fox satisfies the equation

$$x\frac{d^2y}{dx^2} = \sqrt{1 + \left(\frac{dy}{dx}\right)^2}.$$

Hint: Differentiate the equation in (a) with respect to t. On the left side note that $(d/dt) = (dx/dt)(d/dx)$.

(c) Solve the equation in (b) by substituting $u(x) = dy/dx$ and separating variables. Note that $y = 0$ and $u = 0$ when $x = a$.

7. Suppose the earth is a perfect sphere of radius a. You set out from the point on the equator whose spherical coordinates are $(R, \phi, \theta) = (a, \pi/2, 0)$ and travel on the surface of the earth at constant speed v, always moving toward the northeast (45° east of north).

(a) Will you ever get to the north pole? If so, how long will it take to get there?

(b) Find the functions $\phi(t)$ and $\theta(t)$ that are the angular spherical coordinates of your position at time $t > 0$.

(c) How many times does your path cross the meridian $\theta = 0$?

CHAPTER 12

Partial Differentiation

> I have a very wide command of matters mathematical,
> I understand equations both the simple and quadratical.
> About binomial theorem I'm teeming with a lot of news,
> And many cheerful facts about the square on the hypotenuse.

William Schwenck Gilbert 1836–1911
from *The Pirates of Penzance*

Introduction

This chapter is concerned with extending the idea of the derivative to real functions of a vector variable, that is, to functions depending on several real variables. Although differentiation is carried out one variable at a time, the relationship between derivatives with respect to different variables makes the analysis of such functions much more complicated and subtle than in the single-variable case.

12.1 Functions of Several Variables

The notation $y = f(x)$ is used to indicate that the variable y depends on the single real variable x, that is, that y is a function of x. The domain of such a function f is a set of real numbers. Many quantities can be regarded as depending on more than one real variable and thus to be functions of more than one variable. For example, the volume of a circular cylinder of radius r and height h is given by $V = \pi r^2 h$; we say that V is a function of the two variables r and h. If we choose to denote this function by f, then we would write $V = f(r, h)$ where

$$f(r, h) = \pi r^2 h, \qquad (r \geq 0, \quad h \geq 0).$$

Thus, f is a function of two variables having as *domain* the set of points in the rh-plane with coordinates (r, h) satisfying $r \geq 0$ and $h \geq 0$. Similarly, the relationship $w = f(x, y, z) = x + 2y - 3z$ defines w as a function of the three variables x, y, and z, with domain the whole of $\mathbb{R}^3$, or, if we state explicitly, some particular subset of $\mathbb{R}^3$.

By analogy with the corresponding definition for functions of one variable, we define a function of n variables as follows:

DEFINITION 1

A **function** f of n real variables is a rule that assigns a *unique* real number $f(x_1, x_2, \ldots, x_n)$ to each point $(x_1, x_2, \ldots, x_n)$ in some subset $\mathcal{D}(f)$ of $\mathbb{R}^n$. $\mathcal{D}(f)$ is called the **domain** of f. The set of real numbers $f(x_1, x_2, \ldots, x_n)$ obtained from points in the domain is called the **range** of f.

As for functions of one variable, the **domain convention** specifies that the domain of a function of n variables is the largest set of points $(x_1, x_2, \ldots, x_n)$ for which $f(x_1, x_2, \ldots, x_n)$ makes sense as a real number, unless that domain is explicitly stated to be a smaller set.

Most of the examples we consider hereafter will be functions of two or three independent variables. When a function f depends on two variables, we will usually call these independent variables x and y, and we will use z to denote the dependent variable that represents the value of the function; that is, $z = f(x, y)$. We will normally use x, y, and z as the independent variables of a function of three variables and w as the value of the function: $w = f(x, y, z)$. Some definitions will be given, and some theorems will be stated (and proved) only for the two-variable case, but extensions to three or more variables will usually be obvious.

Graphs

The graph of a *function* f of one variable (i.e., the graph of the *equation* $y = f(x)$) is the set of points in the xy-plane having coordinates $\big(x, f(x)\big)$, where x is in the domain of f. Similarly, the graph of a *function* f of two variables (i.e., the graph of the *equation* $z = f(x, y)$) is the set of points in 3-space having coordinates $\big(x, y, f(x, y)\big)$, where (x, y) belongs to the domain of f. This graph is a surface in $\mathbb{R}^3$ lying above (if $f(x, y) > 0$) or below (if $f(x, y) < 0$) the domain of f in the xy-plane. (See Figure 12.1.) The graph of a function of three variables is a three-dimensional *hypersurface* in 4-space, $\mathbb{R}^4$. In general, the graph of a function of n variables is an n-dimensional *surface* in $\mathbb{R}^{n+1}$. We will not attempt to *draw* graphs of functions of more than two variables!

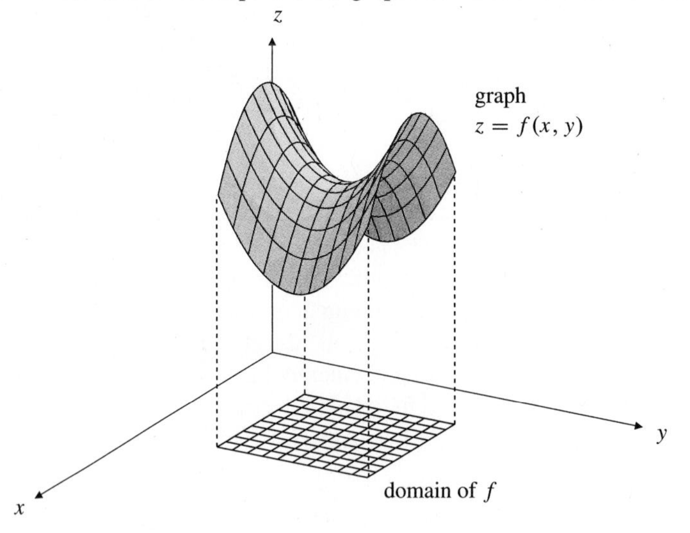

Figure 12.1 The graph of $f(x, y)$ is the surface with equation $z = f(x, y)$ defined for points (x, y) in the domain of f

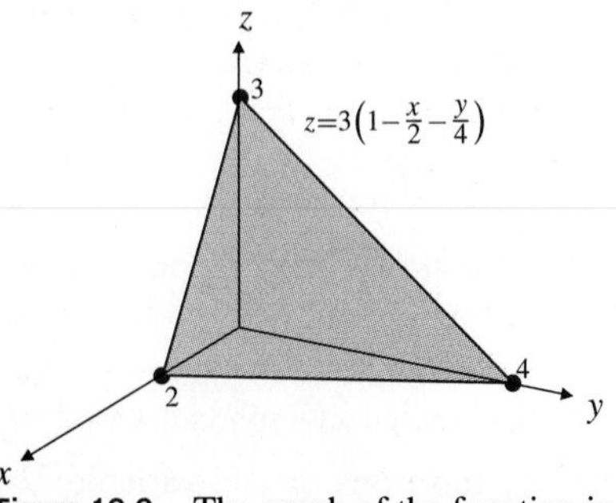

Figure 12.2 The graph of the function in Example 1

EXAMPLE 1 Consider the function

$$f(x, y) = 3\left(1 - \frac{x}{2} - \frac{y}{4}\right), \qquad (0 \le x \le 2, \quad 0 \le y \le 4 - 2x).$$

The graph of f is the plane triangular surface with vertices at $(2, 0, 0)$, $(0, 4, 0)$, and $(0, 0, 3)$. (See Figure 12.2.) If the domain of f had not been explicitly stated to be a particular set in the xy-plane, the graph would have been the whole plane through these three points.

EXAMPLE 2 Consider $f(x, y) = \sqrt{9 - x^2 - y^2}$. The expression under the square root cannot be negative, so the domain is the disk $x^2+y^2 \leq 9$ in the xy-plane.

If we square the equation $z = \sqrt{9 - x^2 - y^2}$, we can rewrite the result in the form $x^2 + y^2 + z^2 = 9$. This is a sphere of radius 3 centred at the origin. However, the graph of f is only the upper hemisphere where $z \geq 0$. (See Figure 12.3.)

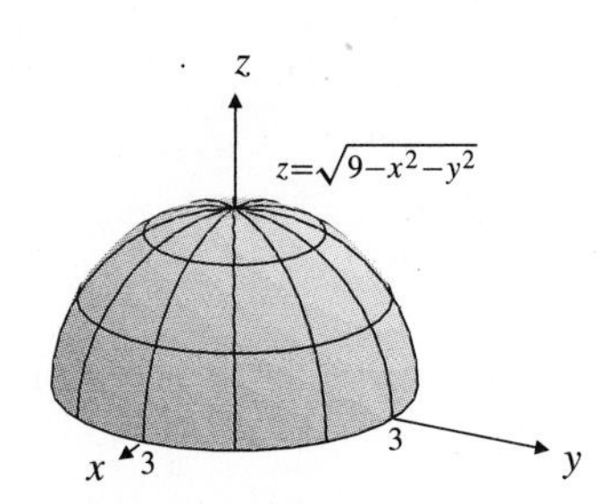

Figure 12.3 The graph of the funtion in Example 2 is a hemisphere

Since it is necessary to project the surface $z = f(x, y)$ onto a two-dimensional page, most such graphs are difficult to sketch without considerable artistic talent and training. Nevertheless, you should always try to visualize such a graph and sketch it as best you can. Sometimes it is convenient to sketch only part of a graph, for instance, the part lying in the first octant. It is also helpful to determine (and sketch) the intersections of the graph with various planes, especially the coordinate planes, and planes parallel to the coordinate planes. (See Figure 12.1.)

Some mathematical software packages will produce plots of three-dimensional graphs to help you get a feeling for how the corresponding functions behave. Figure 12.1 is an example of such a computer-drawn graph, as is Figure 12.4 below. Along with most of the other mathematical graphics in this book, both were produced using the mathematical graphics software package **MG**. Later in this section we discuss how to use Maple to produce such graphs.

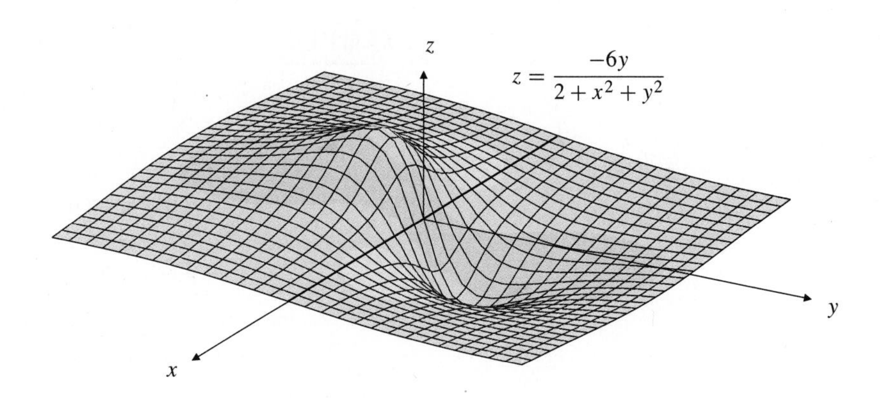

Figure 12.4 The graph of $z = \dfrac{-6y}{2 + x^2 + y^2}$

Level Curves

Another way to represent the function $f(x, y)$ graphically is to produce a two-dimensional *topographic map* of the surface $z = f(x, y)$. In the xy-plane we sketch the curves $f(x, y) = C$ for various values of the constant C. These curves are called **level curves** of f because they are the vertical projections onto the xy-plane of the curves in which the graph $z = f(x, y)$ intersects the horizontal (level) planes $z = C$. The graph and some level curves of the function $f(x, y) = x^2 + y^2$ are shown in Figure 12.5. The graph is a circular paraboloid in 3-space, which is a smooth surface. The level curves of f are circles centred at the origin in the xy-plane. Observe, however, that the function $g(x, y) = \sqrt{x^2 + y^2}$ has the same family of circles as its level curves (though for different values of C), but the graph of g is a circular cone with vertex at the origin and is therefore not smooth there. We can *not* infer from the smoothness of the level curves of a function that the graph of the function is smooth.

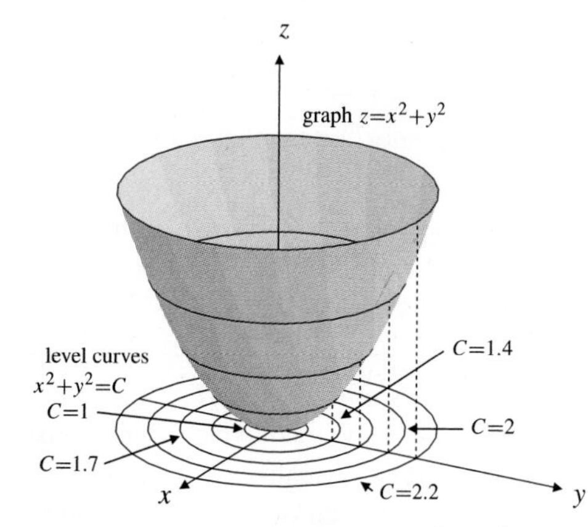

Figure 12.5 The graph of $f(x, y) = x^2 + y^2$ and some level curves of f

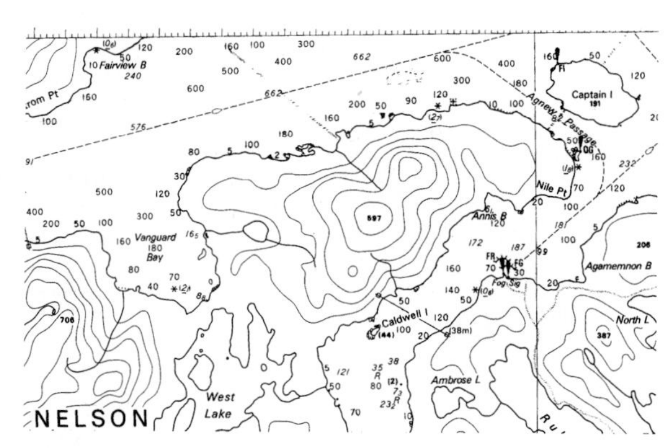

Figure 12.6 Level curves (contours) representing elevation in a topographic map

EXAMPLE 3 The contour curves in the topographic map in Figure 12.6 show the elevations, in 100 m increments above sea level, on part of Nelson Island on the British Columbia coast. Since these contours are drawn for equally spaced values of C, the spacing of the contours themselves conveys information about the relative steepness at various places on the mountains; the land is steepest where the contour lines are closest together. Observe also that the streams shown cross the contours at right angles. They take the route of steepest descent. Isotherms (curves of constant temperature) and isobars (curves of constant pressure) on weather maps are also examples of level curves.

EXAMPLE 4 The level curves of the function $f(x, y) = 3\left(1 - \frac{x}{2} - \frac{y}{4}\right)$ of Example 1 are the segments of the straight lines

$$3\left(1 - \frac{x}{2} - \frac{y}{4}\right) = C \qquad \text{or} \qquad \frac{x}{2} + \frac{y}{4} = 1 - \frac{C}{3}, \qquad (0 \le C \le 3),$$

which lie in the first quadrant. Several such level curves are shown in Figure 12.7(a). They correspond to equally spaced values of C, and their equal spacing indicates the uniform steepness of the graph of f in Figure 12.2.

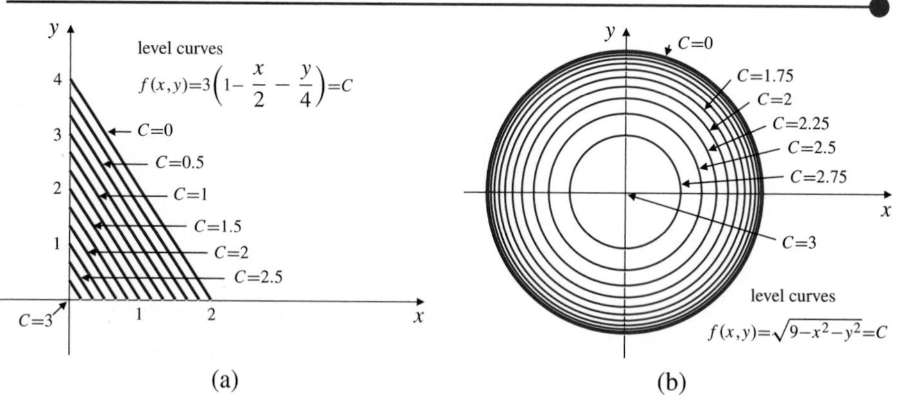

Figure 12.7

(a) Level curves of $3\left(1 - \frac{x}{2} - \frac{y}{4}\right)$

(b) Level curves of $\sqrt{9 - x^2 - y^2}$

EXAMPLE 5 The level curves of the function $f(x, y) = \sqrt{9 - x^2 - y^2}$ of Example 2 are the concentric circles

$$\sqrt{9 - x^2 - y^2} = C \qquad \text{or} \qquad x^2 + y^2 = 9 - C^2, \qquad (0 \le C \le 3).$$

Observe the spacing of these circles in Figure 12.7(b); they are plotted for several equally spaced values of C. The bunching of the circles as $C \to 0+$ indicates the steepness of the hemispherical surface that is the graph of f. (See Figure 12.3.)

A function determines its level curves with any given spacing between consecutive values of C. However, level curves only determine the function if *all of them* are known.

EXAMPLE 6 The level curves of the function $f(x, y) = x^2 - y^2$ are the curves $x^2 - y^2 = C$. For $C = 0$ the level "curve" is the pair of straight lines $x = y$ and $x = -y$. For other values of C the level curves are rectangular hyperbolas with these lines as asymptotes. (See Figure 12.8(a).) The graph of f is the saddle-like hyperbolic paraboloid in Figure 12.8(b).

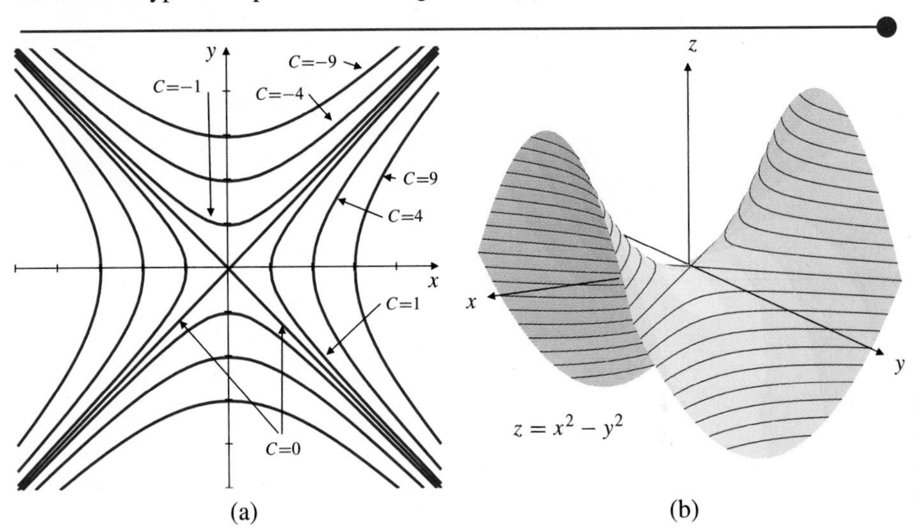

Figure 12.8
(a) Level curves of $x^2 - y^2$
(b) The graph of $x^2 - y^2$

EXAMPLE 7 Describe and sketch some level curves of the function $z = g(x, y)$ defined by $z \geq 0$, and $x^2 + (y - z)^2 = 2z^2$. Also sketch the graph of the function g.

Solution The level curve $z = g(x, y) = C$ (where C is a positive constant) has equation $x^2 + (y - C)^2 = 2C^2$ and is, therefore, a circle of radius $\sqrt{2}C$ centred at $(0, C)$. Level curves for C in increments of 0.1 from 0 to 1 are shown in Figure 12.9(a). These level curves intersect rays from the origin at equal spacing (the spacing is different for different rays) indicating that the surface $z = g(x, y)$ is an oblique cone. See Figure 12.9(b).

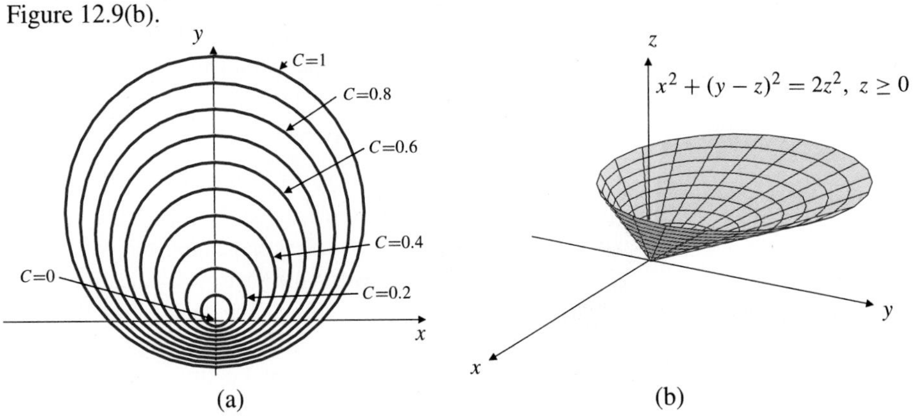

Figure 12.9
(a) Level curves of $z = g(x, y)$ for Example 7
(b) The graph of $z = g(x, y)$

Although the *graph* of a function $f(x, y, z)$ of three variables cannot easily be drawn (it is a three-dimensional *hypersurface* in 4-space), such a function has **level surfaces** in 3-space that can, perhaps, be drawn. These level surfaces have equations of the form $f(x, y, z) = C$ for various choices of the constant C. For instance, the level surfaces of the function $f(x, y, z) = x^2 + y^2 + z^2$ are concentric spheres centred at the origin. Figure 12.10 shows a few level surfaces of the function $f(x, y, z) = x^2 - z$. They are parabolic cylinders.

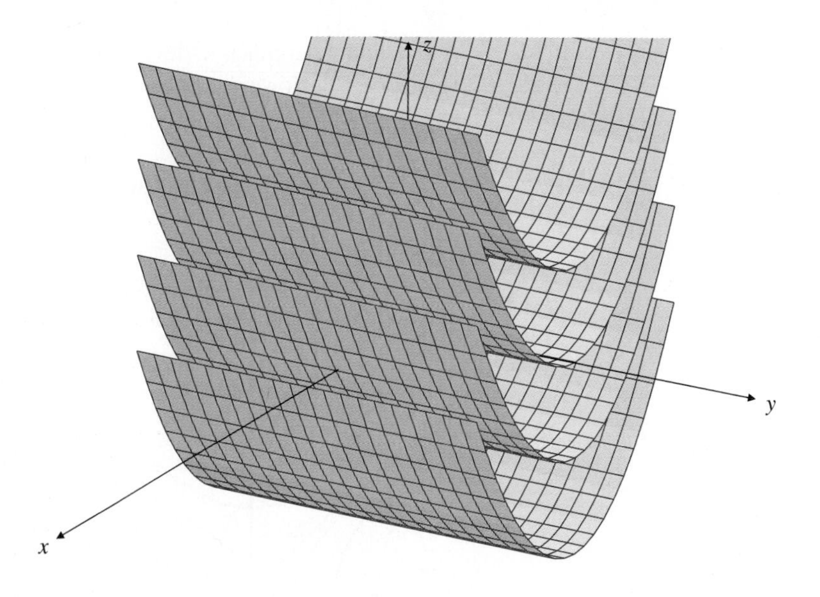

Figure 12.10 Level surfaces of $f(x, y, z) = x^2 - z$

Using Maple Graphics

Like many mathematical software packages, Maple has several plotting routines to help you visualize the behaviour of functions of two and three variables. We mention only a few of them here; there are many more. Most of the plotting routines are in the **plots** package, so you should begin any Maple session where you want to use them with the input

```
> with(plots);
```

To save space, we won't show any of the plot output here. You will need to play with modifications to the various plot commands to obtain the kind of output you desire.

The graph of a function $f(x, y)$ of two variables (or an expression in x and y) can be plotted over a rectangle in the xy-plane with a call to the **plot3d** routine. For example,

```
> f := -6*y/(2+x^2+y^2);
> plot3d(f, x=-6..6, y=-6..6);
```

will plot a surface similar to the one in Figure 12.4 but without axes and viewed from a steeper angle. You can add many kinds of options to the command to change the output. For instance,

```
> plot3d(f, x=-6..6, y=-6..6, axes=boxed,
    orientation=[30,70]);
```

will plot the same surface within a three-dimensional rectangular box with scales on three of its edges indicating the coordinate values. (If we had said `axes=normal` instead, we would have gotten the usual coordinate axes through the origin, but they tend to be more difficult to see against the background of the surface, so `axes=boxed` is usually preferable.) The option `orientation=[30,70]` results in the plot's being viewed from the direction making angle 70° with the z-axis and lying in a plane containing the z-axis making an angle 30° with the xz-plane. (The default value of the orientation is [45, 45] if the option is not specified.) By default, the surface plotted by plot3d is ruled by two families of curves, representing its intersection with vertical planes $x = a$ and $y = b$ for several equally spaced values of a and b, and it is coloured opaquely so that hidden parts do not show.

Instead of plot3d, you can use **contourplot3d** to get a plot of the surface ruled by contours on which the value of the function is constant. If you don't get enough contours by default, you can include a `contours=n` option to specify the number you want.

```
> contourplot3d(f, x=-6..6, y=-6..6, axes=boxed,
   contours=24);
```

The contours are the projections of the level curves onto the graph of the surface. Alternatively, you can get a two-dimensional plot of the level curves themselves using **contourplot**:

```
> contourplot(f, x=-6..6, y=-6..6, axes=normal,
   contours=24);
```

Other options you may want to include with plot3d or contourplot3d are

(a) `view=zmin..zmax` to specify the range of values of the function (i.e., z) to show in the plot.

(b) `grid=[m,n]` to specify the number of x and y values at which to evaluate the function. If your plot doesn't look smooth enough, try $m = n = 20$ or 30 or even higher values.

The graph of an equation, $f(x, y) = 0$, in the xy-plane can be generated without solving the equation for x or y first, by using **implicitplot**.

```
> implicitplot(x^3-y^2-5*x*y-x-5, x=-6..7, y=-5..6);
```

will produce the graph of $x^3 - y^2 - 5xy - x - 5 = 0$ on the rectangle $-6 \le x \le 7$, $-5 \le y \le 6$. There is also an **implicitplot3d** routine to plot the surface in 3-space having an equation of the form $f(x, y, z) = 0$. For this routine you must specify ranges for all three variables;

```
> implicitplot3d(x^2+y^2-z^2-1, x=-4..4, y=-4..4,
   z=-3..3, axes=boxed);
```

plots the hyperboloid $z^2 = x^2 + y^2 - 1$.

Finally, we observe that Maple is no more capable than we are of drawing graphs of functions of three or more variables, since it doesn't have four-dimensional plot capability. The best we can do is plot a set of level surfaces for such a function:

```
> implicitplot3d({z-x^2-2,z-x^2,z-x^2+2},x=-2..2,
   y=-2..2, z=-2..5, axes=boxed);
```

It is possible to construct a sequence of *plot structures* and assign them to, say, the elements of a list variable, without actually plotting them. Then all the plots can be plotted simultaneously using the **display** function.

```
> for c from -1 to 1 do
   p[c] := implicitplot3d(z^2-x^2-y^2-2*c, x=-3..3,
   y=-3..3, z=0..2, color=COLOR(RGB,(1+c)/2,(1-c)/2,1))
   od:
> display([seq(p[c],c=-1..1)], axes=boxed,
   orientation=[30,40]);
```

Note that the command creating the plots is terminated with a colon rather than the usual semicolon. If you don't suppress the output in this way, you will get vast amounts of meaningless numerical output as the plots are constructed. The `color=...` option is an attempt to give the three plots different colours so they can be distinguished from each other.

EXERCISES 12.1

Specify the domains of the functions in Exercises 1–10.

1. $f(x, y) = \dfrac{x + y}{x - y}$

2. $f(x, y) = \sqrt{xy}$

3. $f(x, y) = \dfrac{x}{x^2 + y^2}$

4. $f(x, y) = \dfrac{xy}{x^2 - y^2}$

5. $f(x, y) = \sqrt{4x^2 + 9y^2 - 36}$

6. $f(x, y) = \dfrac{1}{\sqrt{x^2 - y^2}}$

7. $f(x, y) = \ln(1 + xy)$

8. $f(x, y) = \sin^{-1}(x + y)$

9. $f(x, y, z) = \dfrac{xyz}{x^2 + y^2 + z^2}$

10. $f(x, y, z) = \dfrac{e^{xyz}}{\sqrt{xyz}}$

Sketch the graphs of the functions in Exercises 11–18.

11. $f(x, y) = x, \quad (0 \le x \le 2, \quad 0 \le y \le 3)$

12. $f(x, y) = \sin x, \quad (0 \le x \le 2\pi, \quad 0 \le y \le 1)$

13. $f(x, y) = y^2, \quad (-1 \le x \le 1, \quad -1 \le y \le 1)$

14. $f(x, y) = 4 - x^2 - y^2, \quad (x^2 + y^2 \le 4, \ x \ge 0, \ y \ge 0)$

15. $f(x, y) = \sqrt{x^2 + y^2}$

16. $f(x, y) = 4 - x^2$

17. $f(x, y) = |x| + |y|$

18. $f(x, y) = 6 - x - 2y$

Sketch some of the level curves of the functions in Exercises 19–26.

19. $f(x, y) = x - y$

20. $f(x, y) = x^2 + 2y^2$

21. $f(x, y) = xy$

22. $f(x, y) = \dfrac{x^2}{y}$

23. $f(x, y) = \dfrac{x - y}{x + y}$

24. $f(x, y) = \dfrac{y}{x^2 + y^2}$

25. $f(x, y) = xe^{-y}$

26. $f(x, y) = \sqrt{\dfrac{1}{y} - x^2}$

Exercises 27–28 refer to Figure 12.11, which shows contours of a hilly region with heights given in metres.

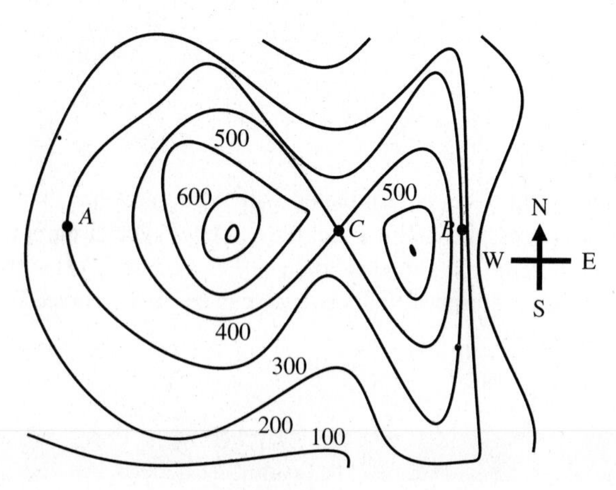

Figure 12.11

27. At which of the points A or B is the landscape steeper? How do you know?

28. Describe the topography of the region near point C.

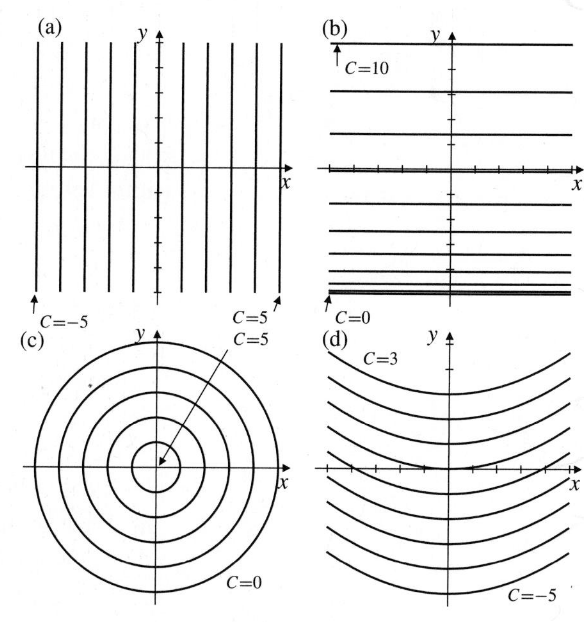

Figure 12.12

Describe the graphs of the functions $f(x, y)$ for which families of level curves $f(x, y) = C$ are shown in the figures referred to in Exercises 29–32. Assume that each family corresponds to equally spaced values of C and that the behaviour of the family is representative of all such families for the function.

29. See Figure 12.12(a).

30. See Figure 12.12(b).

31. See Figure 12.12(c).

32. See Figure 12.12(d).

33. Are the curves $y = (x - C)^2$ level curves of a function $f(x, y)$? What property must a family of curves in a region of the xy-plane have to be the family of level curves of a function defined in the region?

34. If we assume $z \ge 0$, the equation $4z^2 = (x - z)^2 + (y - z)^2$ defines z as a function of x and y. Sketch some level curves of this function. Describe its graph.

35. Find $f(x, y)$ if each level curve $f(x, y) = C$ is a circle centred at the origin and having radius
(a) C (b) C^2 (c) $\sqrt{C}$ (d) $\ln C$.

36. Find $f(x, y, z)$ if for each constant C the level surface $f(x, y, z) = C$ is a plane having intercepts C^3, $2C^3$, and $3C^3$ on the x-axis, the y-axis, and the z-axis, respectively.

Describe the level surfaces of the functions specified in Exercises 37–41.

37. $f(x, y, z) = x^2 + y^2 + z^2$

38. $f(x, y, z) = x + 2y + 3z$

39. $f(x, y, z) = x^2 + y^2$

40. $f(x, y, z) = \dfrac{x^2 + y^2}{z^2}$

41. $f(x, y, z) = |x| + |y| + |z|$

42. Describe the "level hypersurfaces" of the function

$$f(x, y, z, t) = x^2 + y^2 + z^2 + t^2.$$

Use Maple or other computer graphing software to plot the graphs and the level curves of the functions in Exercises 43–48.

43. $\dfrac{1}{1+x^2+y^2}$

44. $\dfrac{\cos x}{1+y^2}$

45. $\dfrac{y}{1+x^2+y^2}$

46. $\dfrac{x}{(x^2-1)^2+y^2}$

47. xy

48. $\dfrac{1}{xy}$

12.2 Limits and Continuity

Before reading this section you should review the concepts of neighbourhood, open and closed sets, and boundary and interior points introduced in Section 10.1.

The concept of the limit of a function of several variables is similar to that for functions of one variable. For clarity we present the definition for functions of two variables only; the general case is similar.

We might say that $f(x, y)$ approaches the limit L as the point (x, y) approaches the point (a, b), and write

$$\lim_{(x,y)\to(a,b)} f(x, y) = L,$$

if all points of a neighbourhood of (a, b), except possibly the point (a, b) itself, belong to the domain of f, and if $f(x, y)$ approaches L as (x, y) approaches (a, b). However, it is more convenient to define the limit in such a way that (a, b) can be a boundary point of the domain of f. Thus, our formal definition will generalize the one-dimensional notion of one-sided limit as well.

DEFINITION 2

Definition of Limit

We say that $\lim_{(x,y)\to(a,b)} f(x, y) = L$, provided that

(i) every neighbourhood of (a, b) contains points of the domain of f different from (a, b), and

(ii) for every positive number ϵ there exists a positive number $\delta = \delta(\epsilon)$ such that $|f(x, y) - L| < \epsilon$ holds whenever (x, y) is in the domain of f and satisfies $0 < \sqrt{(x-a)^2 + (y-b)^2} < \delta$.

Condition (i) is included in Definition 2 because it is not appropriate to consider limits at *isolated* points of the domain of f, that is, points with neighbourhoods that contain no other points of the domain.

If a limit exists it is unique. For a single-variable function f, the existence of $\lim_{x\to a} f(x)$ implies that $f(x)$ approaches the same finite number as x approaches a from either the right or the left. Similarly, for a function of two variables, we can have $\lim_{(x,y)\to(a,b)} f(x, y) = L$ only if $f(x, y)$ approaches the same number L no matter how (x, y) approaches (a, b) in the domain of f. In particular, (x, y) can approach (a, b) along any curve that lies in $\mathcal{D}(f)$. It is not necessary that $L = f(a, b)$ even if $f(a, b)$ is defined. The examples below illustrate these assertions.

All the usual laws of limits extend to functions of several variables in the obvious way. For example, if $\lim_{(x,y)\to(a,b)} f(x, y) = L$, $\lim_{(x,y)\to(a,b)} g(x, y) = M$, and every neighbourhood of (a, b) contains points in $\mathcal{D}(f) \cap \mathcal{D}(g)$ other than (a, b), then

$$\lim_{(x,y)\to(a,b)} \big(f(x,y) \pm g(x,y)\big) = L \pm M,$$
$$\lim_{(x,y)\to(a,b)} f(x,y)\,g(x,y) = LM,$$
$$\lim_{(x,y)\to(a,b)} \frac{f(x,y)}{g(x,y)} = \frac{L}{M}, \qquad \text{provided } M \neq 0.$$

Also, if $F(t)$ is continuous at $t = L$, then

$$\lim_{(x,y)\to(a,b)} F\big(f(x,y)\big) = F(L).$$

EXAMPLE 1

(a) $$\lim_{(x,y)\to(2,3)} \left(2x - y^2\right) = 4 - 9 = -5,$$

(b) $$\lim_{(x,y)\to(a,b)} x^2y = a^2b,$$

(c) $$\lim_{(x,y)\to(\pi/3,2)} y\sin\left(\frac{x}{y}\right) = 2\sin\left(\frac{\pi}{6}\right) = 1.$$

EXAMPLE 2 The function $f(x,y) = \sqrt{1 - x^2 - y^2}$ has limit $f(a,b)$ at all points (a,b) of its domain, the *closed* disk $x^2 + y^2 \le 1$, and is therefore considered to be *continuous* on its domain. Of course, (x,y) can approach points of the bounding circle $x^2 + y^2 = 1$ only from within the disk.

The following examples show that the requirement that $f(x,y)$ approach the same limit *no matter how* (x,y) approaches (a,b) can be very restrictive, and makes limits in two or more variables much more subtle than in the single-variable case.

EXAMPLE 3 Investigate the limiting behaviour of $f(x,y) = \dfrac{2xy}{x^2 + y^2}$ as (x,y) approaches $(0,0)$.

Solution Note that $f(x,y)$ is defined at all points of the xy-plane except the origin $(0,0)$. We can still ask whether $\lim_{(x,y)\to(0,0)} f(x,y)$ exists. If we let (x,y) approach $(0,0)$ along the x-axis ($y = 0$), then $f(x,y) = f(x,0) \to 0$ (because $f(x,0) = 0$ identically). Thus, $\lim_{(x,y)\to(0,0)} f(x,y)$ must be 0 if it exists at all. Similarly, at all points of the y-axis we have $f(x,y) = f(0,y) = 0$. However, at points of the line $x = y$, f has a different constant value; $f(x,x) = 1$. Since the limit of $f(x,y)$ is 1 as (x,y) approaches $(0,0)$ along this line, it follows that $f(x,y)$ cannot have a limit at the origin. That is,

$$\lim_{(x,y)\to(0,0)} \frac{2xy}{x^2 + y^2} \qquad \text{does not exist.}$$

Observe that $f(x,y)$ has a constant value on any ray from the origin (on the ray $y = kx$ the value is $2k/(1 + k^2)$), but these values differ on different rays. The level curves of f are the rays from the origin (with the origin itself removed). It is difficult to sketch the graph of f near the origin. The first octant part of the graph is the "hood-shaped" surface in Figure 12.13(a).

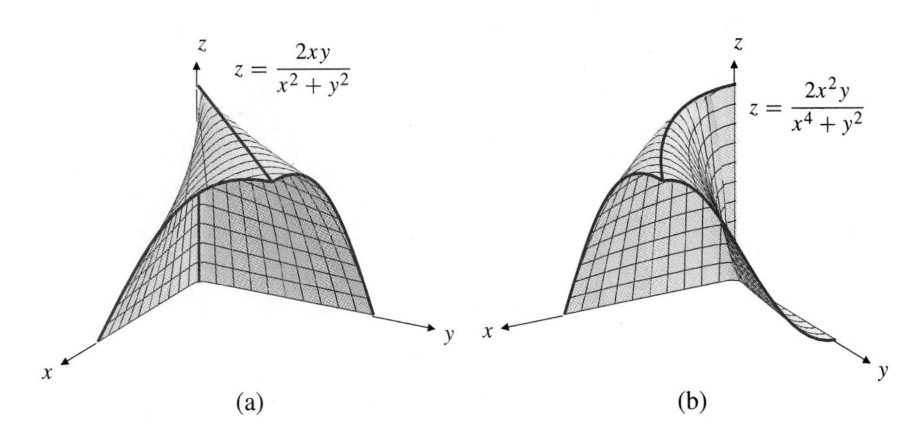

Figure 12.13

(a) $f(x, y)$ has different limits as $(x, y) \to (0, 0)$ along different straight lines

(b) $f(x, y)$ has the same limit 0 as $(x, y) \to (0, 0)$ along any straight line but has limit 1 as $(x, y) \to (0, 0)$ along $y = x^2$

EXAMPLE 4 Investigate the limiting behaviour of $f(x, y) = \dfrac{2x^2y}{x^4 + y^2}$ as (x, y) approaches $(0, 0)$.

Solution As in Example 3, $f(x, y)$ vanishes identically on the coordinate axes, so $\lim_{(x,y)\to(0,0)} f(x, y)$ must be 0 if it exists at all. If we examine $f(x, y)$ at points of the ray $y = kx$, we obtain

$$f(x, kx) = \frac{2kx^3}{x^4 + k^2x^2} = \frac{2kx}{x^2 + k^2} \to 0, \qquad \text{as} \quad x \to 0 \quad (k \neq 0).$$

Thus, $f(x, y) \to 0$ as $(x, y) \to (0, 0)$ along *any* straight line through the origin. We might be tempted to conclude, therefore, that $\lim_{(x,y)\to(0,0)} f(x, y) = 0$, but this is incorrect. Observe the behaviour of $f(x, y)$ along the curve $y = x^2$:

$$f(x, x^2) = \frac{2x^4}{x^4 + x^4} = 1.$$

Thus, $f(x, y)$ does not approach 0 as (x, y) approaches the origin along this curve, so $\lim_{(x,y)\to(0,0)} f(x, y)$ does not exist. The level curves of f are pairs of parabolas of the form $y = kx^2$, $y = x^2/k$ with the origin removed. See Figure 12.13(b) for the first octant part of the graph of f.

EXAMPLE 5 Show that the function $f(x, y) = \dfrac{x^2y}{x^2 + y^2}$ does have a limit at the origin; specifically,

$$\lim_{(x,y)\to(0,0)} \frac{x^2y}{x^2 + y^2} = 0.$$

Solution This function is also defined everywhere except at the origin. Observe that since $x^2 \leq x^2 + y^2$, we have

$$|f(x, y) - 0| = \left|\frac{x^2y}{x^2 + y^2}\right| \leq |y| \leq \sqrt{x^2 + y^2},$$

which approaches zero as $(x, y) \to (0, 0)$. (See Figure 12.14.) Formally, if $\epsilon > 0$ is given and we take $\delta = \epsilon$, then $|f(x, y) - 0| < \epsilon$ whenever $0 < \sqrt{x^2 + y^2} < \delta$, so $f(x, y)$ has limit 0 as $(x, y) \to (0, 0)$ by Definition 2.

As for functions of one variable, continuity of a function f at a point of its domain is defined directly in terms of the limit. (See, for instance, Example 2.)

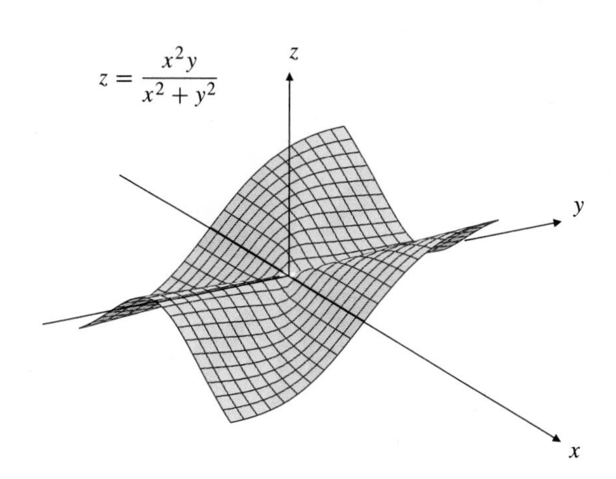

Figure 12.14 $\displaystyle\lim_{(x,y)\to(0,0)} \frac{x^2y}{x^2+y^2} = 0$

DEFINITION 3

> The function $f(x, y)$ is continuous at the point (a, b) if
>
> $$\lim_{(x,y)\to(a,b)} f(x, y) = f(a, b).$$

It remains true that sums, differences, products, quotients, and compositions of continuous functions are continuous. The functions of Examples 3 and 4 above are continuous wherever they are defined, that is, at all points except the origin. There is no way to define $f(0, 0)$ so that these functions become continuous at the origin. They show that the continuity of the single-variable functions $f(x, b)$ at $x = a$ and $f(a, y)$ at $y = b$ does *not* imply that $f(x, y)$ is continuous at (a, b). In fact, even if $f(x, y)$ is continuous along every straight line through (a, b), it still need not be continuous at (a, b). (See Exercises 16–17 below.) Note, however, that the function $f(x, y)$ of Example 5, although not defined at the origin, has a continuous extension to that point. If we extend the domain of f by defining $f(0, 0) = \lim_{(x,y)\to(0,0)} f(x, y) = 0$, then f is continuous on the whole xy-plane.

As for functions of one variable, the existence of a limit of a function at a point does not imply that the function is continuous at that point. The function

$$f(x, y) = \begin{cases} 0 & \text{if } (x, y) \neq (0, 0) \\ 1 & \text{if } (x, y) = (0, 0) \end{cases}$$

satisfies $\lim_{(x,y)\to(0,0)} f(x, y) = 0$, which is not equal to $f(0, 0)$, so f is not continuous at $(0, 0)$. Of course, we can *make* f continuous at $(0, 0)$ by redefining its value at that point to be 0.

EXERCISES 12.2

In Exercises 1–12, evaluate the indicated limit or explain why it does not exist.

1. $\displaystyle\lim_{(x,y)\to(2,-1)} xy + x^2$

2. $\displaystyle\lim_{(x,y)\to(0,0)} \sqrt{x^2+y^2}$

3. $\displaystyle\lim_{(x,y)\to(0,0)} \frac{x^2+y^2}{y}$

4. $\displaystyle\lim_{(x,y)\to(0,0)} \frac{x}{x^2+y^2}$

5. $\displaystyle\lim_{(x,y)\to(1,\pi)} \frac{\cos(xy)}{1-x-\cos y}$

6. $\displaystyle\lim_{(x,y)\to(0,1)} \frac{x^2(y-1)^2}{x^2+(y-1)^2}$

7. $\displaystyle\lim_{(x,y)\to(0,0)} \frac{y^3}{x^2+y^2}$

8. $\displaystyle\lim_{(x,y)\to(0,0)} \frac{\sin(x-y)}{\cos(x+y)}$

9. $\displaystyle\lim_{(x,y)\to(0,0)} \frac{\sin(xy)}{x^2+y^2}$

10. $\displaystyle\lim_{(x,y)\to(1,2)} \frac{2x^2-xy}{4x^2-y^2}$

11. $\displaystyle\lim_{(x,y)\to(0,0)} \frac{x^2y^2}{x^2+y^4}$

12. $\displaystyle\lim_{(x,y)\to(0,0)} \frac{x^2y^2}{2x^4+y^4}$

13. How can the function

$$f(x, y) = \frac{x^2+y^2-x^3y^3}{x^2+y^2}, \qquad (x, y) \neq (0, 0),$$

be defined at the origin so that it becomes continuous at all points of the xy-plane?

14. How can the function

$$f(x, y) = \frac{x^3-y^3}{x-y}, \qquad (x \neq y),$$

be defined along the line $x = y$ so that the resulting function is continuous on the whole xy-plane?

15. What is the domain of

$$f(x, y) = \frac{x - y}{x^2 - y^2}?$$

Does $f(x, y)$ have a limit as $(x, y) \to (1, 1)$? Can the domain of f be extended so that the resulting function is continuous at $(1, 1)$? Can the domain be extended so that the resulting function is continuous everywhere in the xy-plane?

16. Given a function $f(x, y)$ and a point (a, b) in its domain, define single-variable functions g and h as follows:

$$g(x) = f(x, b), \qquad h(y) = f(a, y).$$

If g is continuous at $x = a$ and h is continuous at $y = b$, does it follow that f is continuous at (a, b)? Conversely, does the continuity of f at (a, b) guarantee the continuity of g at a and the continuity of h at b? Justify your answers.

17. Let $\mathbf{u} = u\mathbf{i} + v\mathbf{j}$ be a unit vector, and let

$$f_{\mathbf{u}}(t) = f(a + tu, b + tv)$$

be the single-variable function obtained by restricting the domain of $f(x, y)$ to points of the straight line through (a, b) parallel to $\mathbf{u}$. If $f_{\mathbf{u}}(t)$ is continuous at $t = 0$ for every unit vector $\mathbf{u}$, does it follow that f is continuous at (a, b)? Conversely, does the continuity of f at (a, b) guarantee the continuity of $f_{\mathbf{u}}(t)$ at $t = 0$? Justify your answers.

18. What condition must the nonnegative integers m, n, and p satisfy to guarantee that $\lim_{(x,y)\to(0,0)} x^m y^n/(x^2 + y^2)^p$ exists? Prove your answer.

19. What condition must the constants a, b, and c satisfy to guarantee that $\lim_{(x,y)\to(0,0)} xy/(ax^2 + bxy + cy^2)$ exists? Prove your answer.

20. Can the function $f(x, y) = \dfrac{\sin x \sin^3 y}{1 - \cos(x^2 + y^2)}$ be defined at $(0, 0)$ in such a way that it becomes continuous there? If so, how?

21. Use two- and three-dimensional mathematical graphing software to examine the graph and level curves of the function $f(x, y)$ of Example 3 on the region $-1 \le x \le 1$, $-1 \le y \le 1$, $(x, y) \ne (0, 0)$. How would you describe the behaviour of the graph near $(x, y) = (0, 0)$?

22. Use two- and three-dimensional mathematical graphing software to examine the graph and level curves of the function $f(x, y)$ of Example 4 on the region $-1 \le x \le 1$, $-1 \le y \le 1$, $(x, y) \ne (0, 0)$. How would you describe the behaviour of the graph near $(x, y) = (0, 0)$?

23. The graph of a single-variable function $f(x)$ that is continuous on an interval is a curve that has no *breaks* in it there and that intersects any vertical line through a point in the interval exactly once. What analogous statement can you make about the graph of a bivariate function $f(x, y)$ that is continuous on a region of the xy-plane?

12.3 Partial Derivatives

In this section we begin the process of extending the concepts and techniques of single-variable calculus to functions of more than one variable. It is convenient to begin by considering the rate of change of such functions with respect to one variable at a time. Thus, a function of n variables has n *first-order partial derivatives,* one with respect to each of its independent variables. For a function of two variables, we make this precise in the following definition:

DEFINITION 4

The **first partial derivatives** of the function $f(x, y)$ **with respect to the variables** x and y are the functions $f_1(x, y)$ and $f_2(x, y)$ given by

$$f_1(x, y) = \lim_{h\to 0} \frac{f(x + h, y) - f(x, y)}{h},$$

$$f_2(x, y) = \lim_{k\to 0} \frac{f(x, y + k) - f(x, y)}{k},$$

provided these limits exist.

Each of the two partial derivatives is the limit of a Newton quotient in one of the variables. Observe that $f_1(x, y)$ is just the ordinary first derivative of $f(x, y)$ considered as a function of x only, regarding y as a constant parameter. Similarly, $f_2(x, y)$ is the first derivative of $f(x, y)$ considered as a function of y alone, with x held fixed.

EXAMPLE 1 If $f(x, y) = x^2 \sin y$, then

$$f_1(x, y) = 2x \sin y \qquad \text{and} \qquad f_2(x, y) = x^2 \cos y.$$

The subscripts 1 and 2 in the notations for the partial derivatives refer to the first and second variables of f. For functions of one variable we use the notation f' for the derivative; the *prime* ($'$) denotes differentiation with respect to the only variable on which f depends. For functions f of two variables, we use f_1 or f_2 to show the variable of differentiation. Do not confuse these subscripts with subscripts used for other purposes (e.g., to denote the components of vectors).

The partial derivative $f_1(a, b)$ measures the rate of change of $f(x, y)$ with respect to x at $x = a$ while y is held fixed at b. In graphical terms, the surface $z = f(x, y)$ intersects the vertical plane $y = b$ in a curve. If we take horizontal and vertical lines through the point $(0, b, 0)$ as coordinate axes in the plane $y = b$, then the curve has equation $z = f(x, b)$, and its slope at $x = a$ is $f_1(a, b)$. (See Figure 12.15.) Similarly, $f_2(a, b)$ represents the rate of change of f with respect to y at $y = b$ with x held fixed at a. The surface $z = f(x, y)$ intersects the vertical plane $x = a$ in a curve $z = f(a, y)$ whose slope at $y = b$ is $f_2(a, b)$. (See Figure 12.16.)

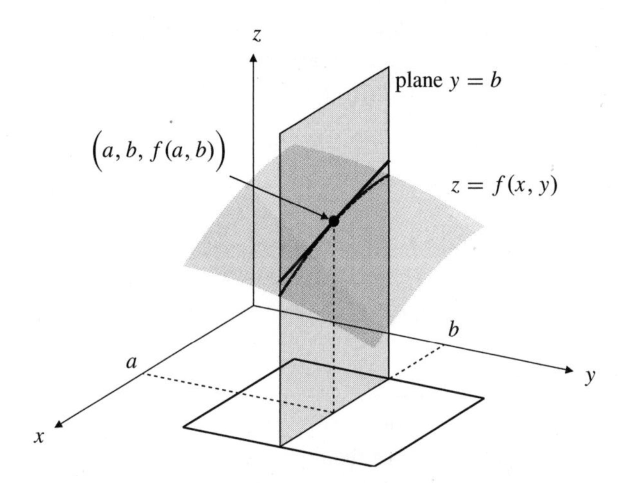

Figure 12.15 $f_1(a, b)$ is the slope of the curve of intersection of $z = f(x, y)$ and the vertical plane $y = b$ at $x = a$

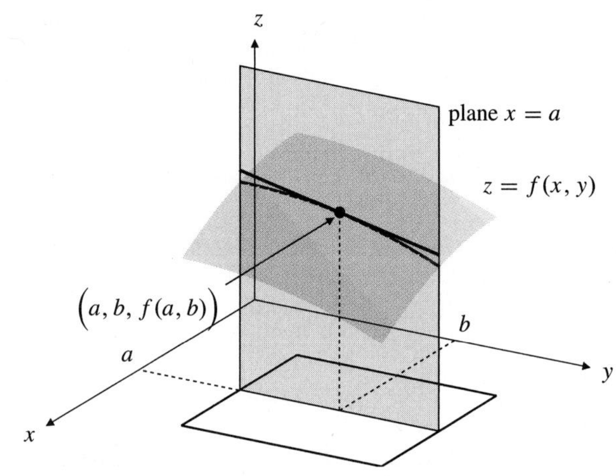

Figure 12.16 $f_2(a, b)$ is the slope of the curve of intersection of $z = f(x, y)$ and the vertical plane $x = a$ at $y = b$

Various notations can be used to denote the partial derivatives of $z = f(x, y)$ considered as functions of x and y:

Notations for first partial derivatives

$$\frac{\partial z}{\partial x} = \frac{\partial}{\partial x} f(x, y) = f_1(x, y) = D_1 f(x, y)$$

$$\frac{\partial z}{\partial y} = \frac{\partial}{\partial y} f(x, y) = f_2(x, y) = D_2 f(x, y)$$

The symbol $\partial/\partial x$ should be read as "partial with respect to x" so $\partial z/\partial x$ is "partial z with respect to x." The reason for distinguishing ∂ (pronounced "die") from the d of ordinary derivatives of single-variable functions will be made clear later. Similar notations can be used to denote the values of partial derivatives at a particular point (a, b):

Values of partial derivatives

$$\left.\frac{\partial z}{\partial x}\right|_{(a,b)} = \left.\left(\frac{\partial}{\partial x} f(x, y)\right)\right|_{(a,b)} = f_1(a, b) = D_1 f(a, b)$$

$$\left.\frac{\partial z}{\partial y}\right|_{(a,b)} = \left.\left(\frac{\partial}{\partial y} f(x, y)\right)\right|_{(a,b)} = f_2(a, b) = D_2 f(a, b)$$

BEWARE! Read the paragraph at the right carefully. It explains why, at least for the time being, we are using subscripts 1 and 2 instead of subscripts x and y for the partial derivatives of $f(x, y)$. Later on, and especially when we are discussing partial differential equations or dealing with vector-valued functions for which numerical subscripts normally represent components, we will prefer to use letter subscripts for partial derivatives.

Some authors prefer to use f_x, $D_x f$, or $\partial f/\partial x$, and f_y, $D_y f$, or $\partial f/\partial y$, instead of f_1 and f_2. However, this can lead to problems of ambiguity when compositions of functions arise. For instance, suppose $f(x, y) = x^2 y$. What should $f_x(x^2, xy)$ mean? By $f_1(x^2, xy)$ we clearly mean to evaluate the partial derivative of $f(u, v) = u^2 v$ with respect to its first variable u and evaluate the result at $u = x^2$ and $v = xy$:

$$f_1(x^2, xy) = \left.\left(\frac{\partial}{\partial u} f(u, v)\right)\right|_{u=x^2, v=xy} = 2uv\bigg|_{u=x^2, v=xy} = (2)(x^2)(xy) = 2x^3 y.$$

But does $f_x(x^2, xy)$ mean the same thing? One could argue that

$$f_x(x^2, xy) = \frac{\partial}{\partial x}\Big(f(x^2, xy)\Big) = \frac{\partial}{\partial x}\Big((x^2)^2(xy)\Big) = \frac{\partial}{\partial x}(x^5 y) = 5x^4 y.$$

In order to avoid such ambiguities we usually prefer to use f_1 and f_2 instead of f_x and f_y. (However, in some situations where no confusion is likely to occur we may still use the notations f_x and f_y, and also $D_x f$, $D_y f$, $\partial f/\partial x$ and $\partial f/\partial y$.)

All the standard differentiation rules for sums, products, reciprocals, and quotients continue to apply to partial derivatives.

EXAMPLE 2 Find $\partial z/\partial x$ and $\partial z/\partial y$ if $z = x^3 y^2 + x^4 y + y^4$.

Solution $\partial z/\partial x = 3x^2 y^2 + 4x^3 y$ and $\partial z/\partial y = 2x^3 y + x^4 + 4y^3$.

EXAMPLE 3 Find $f_1(0, \pi)$ if $f(x, y) = e^{xy} \cos(x + y)$.

Solution

$$f_1(x, y) = y\, e^{xy} \cos(x + y) - e^{xy} \sin(x + y),$$
$$f_1(0, \pi) = \pi\, e^0 \cos(\pi) - e^0 \sin(\pi) = -\pi.$$

The single-variable version of the Chain Rule also continues to apply to, say, $f\big(g(x, y)\big)$, where f is a function of only one variable having derivative f':

$$\frac{\partial}{\partial x} f\big(g(x, y)\big) = f'\big(g(x, y)\big)\, g_1(x, y), \qquad \frac{\partial}{\partial y} f\big(g(x, y)\big) = f'\big(g(x, y)\big)\, g_2(x, y).$$

We will develop versions of the Chain Rule for more complicated compositions of multivariate functions in Section 12.5.

EXAMPLE 4 If f is an everywhere differentiable function of one variable, show that $z = f(x/y)$ satisfies the *partial differential equation*

$$x\frac{\partial z}{\partial x} + y\frac{\partial z}{\partial y} = 0.$$

Solution By the (single-variable) Chain Rule,

$$\frac{\partial z}{\partial x} = f'\left(\frac{x}{y}\right)\left(\frac{1}{y}\right) \quad \text{and} \quad \frac{\partial z}{\partial y} = f'\left(\frac{x}{y}\right)\left(\frac{-x}{y^2}\right).$$

Hence,

$$x\frac{\partial z}{\partial x} + y\frac{\partial z}{\partial y} = f'\left(\frac{x}{y}\right)\left(x \times \frac{1}{y} + y \times \frac{-x}{y^2}\right) = 0.$$

Definition 4 can be extended in the obvious way to cover functions of more than two variables. If f is a function of n variables $x_1, x_2, \ldots, x_n$, then f has n first partial derivatives, $f_1(x_1, x_2, \ldots, x_n)$, $f_2(x_1, x_2, \ldots, x_n)$, $\ldots$, $f_n(x_1, x_2, \ldots, x_n)$, one with respect to each variable.

EXAMPLE 5

$$\frac{\partial}{\partial z}\left(\frac{2xy}{1 + xz + yz}\right) = -\frac{2xy}{(1 + xz + yz)^2}(x + y).$$

Again, all the standard differentiation rules are applied to calculate partial derivatives.

Remark If a single-variable function $f(x)$ has a derivative $f'(a)$ at $x = a$, then f is necessarily continuous at $x = a$. This property does *not* extend to partial derivatives. Even if all the first partial derivatives of a function of several variables exist at a point, the function may still fail to be continuous at that point. See Exercise 36 below.

Tangent Planes and Normal Lines

If the graph $z = f(x, y)$ is a "smooth" surface near the point P with coordinates $\big(a, b, f(a, b)\big)$, then that graph will have a **tangent plane** and a **normal line** at P. The normal line is the line through P that is perpendicular to the surface; for instance, a line joining a point on a sphere to the centre of the sphere is normal to the sphere. Any nonzero vector that is parallel to the normal line at P is called a normal vector to the surface at P. The tangent plane to the surface $z = f(x, y)$ at P is the plane through P that is perpendicular to the normal line at P.

Let us assume that the surface $z = f(x, y)$ has a *nonvertical* tangent plane (and therefore a *nonhorizontal* normal line) at point P. (Later in this chapter we will state precise conditions that guarantee that the graph of a function has a nonvertical tangent plane at a point.) The tangent plane intersects the vertical plane $y = b$ in a straight line that is tangent at P to the curve of intersection of the surface $z = f(x, y)$ and the plane $y = b$. (See Figures 12.15 and 12.17.) This line has slope $f_1(a, b)$, so it is parallel to the vector $\mathbf{T}_1 = \mathbf{i} + f_1(a, b)\mathbf{k}$. Similarly, the tangent plane intersects the vertical plane $x = a$ in a straight line having slope $f_2(a, b)$. This line is therefore parallel to the vector $\mathbf{T}_2 = \mathbf{j} + f_2(a, b)\mathbf{k}$. It follows that the tangent plane, and therefore the surface $z = f(x, y)$ itself, has normal vector

$$\mathbf{n} = \mathbf{T}_2 \times \mathbf{T}_1 = \begin{vmatrix} \mathbf{i} & \mathbf{j} & \mathbf{k} \\ 0 & 1 & f_2(a, b) \\ 1 & 0 & f_1(a, b) \end{vmatrix} = f_1(a, b)\mathbf{i} + f_2(a, b)\mathbf{j} - \mathbf{k}.$$

A normal vector to $z = f(x, y)$ at $\big(a, b, f(a, b)\big)$ is

$$\mathbf{n} = f_1(a, b)\mathbf{i} + f_2(a, b)\mathbf{j} - \mathbf{k}.$$

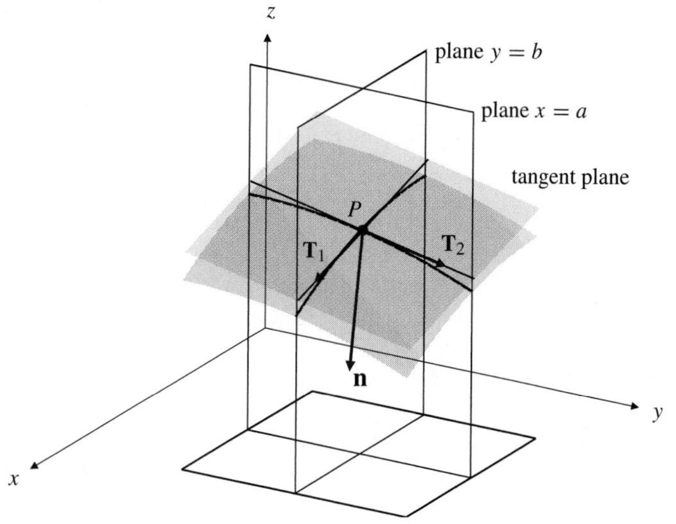

Figure 12.17 The tangent plane and a normal vector to $z = f(x, y)$ at $P = \big(a, b, f(a, b)\big)$

Since the tangent plane passes through $P = (a, b, f(a, b))$, it has equation

$$f_1(a, b)(x - a) + f_2(a, b)(y - b) - (z - f(a, b)) = 0,$$

or, equivalently,

An equation of the tangent plane to $z = f(x, y)$ at $\big(a, b, f(a, b)\big)$ is

$$z = f(a, b) + f_1(a, b)(x - a) + f_2(a, b)(y - b).$$

We shall obtain this result by a different method in Section 12.7.

The normal line to $z = f(x, y)$ at $\big(a, b, f(a, b)\big)$ has direction vector $f_1(a, b)\mathbf{i} + f_2(a, b)\mathbf{j} - \mathbf{k}$ and so has equations

$$\frac{x - a}{f_1(a, b)} = \frac{y - b}{f_2(a, b)} = \frac{z - f(a, b)}{-1}$$

with suitable modifications if either $f_1(a, b) = 0$ or $f_2(a, b) = 0$.

EXAMPLE 6 Find a normal vector and equations of the tangent plane and normal line to the graph $z = \sin(xy)$ at the point where $x = \pi/3$ and $y = -1$.

Solution The point on the graph has coordinates $(\pi/3, -1, -\sqrt{3}/2)$. Now

$$\frac{\partial z}{\partial x} = y\cos(xy) \qquad \text{and} \qquad \frac{\partial z}{\partial y} = x\cos(xy).$$

At $(\pi/3, -1)$ we have $\partial z/\partial x = -1/2$ and $\partial z/\partial y = \pi/6$. Therefore, the surface has normal vector $\mathbf{n} = -(1/2)\mathbf{i} + (\pi/6)\mathbf{j} - \mathbf{k}$ and tangent plane

$$z = \frac{-\sqrt{3}}{2} - \frac{1}{2}\left(x - \frac{\pi}{3}\right) + \frac{\pi}{6}(y + 1),$$

or, more simply, $3x - \pi y + 6z = 2\pi - 3\sqrt{3}$. The normal line has equation

$$\frac{x - \frac{\pi}{3}}{\frac{-1}{2}} = \frac{y + 1}{\frac{\pi}{6}} = \frac{z + \frac{\sqrt{3}}{2}}{-1} \quad \text{or} \quad \frac{6x - 2\pi}{-3} = \frac{6y + 6}{\pi} = \frac{6z + 3\sqrt{3}}{-6}.$$

EXAMPLE 7 What horizontal plane is tangent to the surface

$$z = x^2 - 4xy - 2y^2 + 12x - 12y - 1,$$

and what is the point of tangency?

Solution A plane is horizontal only if its equation is of the form $z = k$, that is, it is independent of x and y. Therefore, we must have $\partial z/\partial x = \partial z/\partial y = 0$ at the point of tangency. The equations

$$\frac{\partial z}{\partial x} = 2x - 4y + 12 = 0$$
$$\frac{\partial z}{\partial y} = -4x - 4y - 12 = 0$$

have solution $x = -4$, $y = 1$. For these values we have $z = -31$, so the required tangent plane has equation $z = -31$ and the point of tangency is $(-4, 1, -31)$.

Distance from a Point to a Surface: A Geometric Example

EXAMPLE 8 Find the distance from the point $(3, 0, 0)$ to the hyperbolic paraboloid with equation $z = x^2 - y^2$.

Solution This is an optimization problem of a sort we will deal with in a more systematic way in the next chapter. However, such problems involving minimizing distances from points to surfaces can frequently be solved using geometric methods.

If $Q = (X, Y, Z)$ is the point on the surface $z = x^2 - y^2$ that is closest to $P = (3, 0, 0)$, then the vector $\overrightarrow{PQ} = (X - 3)\mathbf{i} + Y\mathbf{j} + Z\mathbf{k}$ must be normal to the surface at Q. (See Figure 12.18(a).) Using the partial derivatives of $z = x^2 - y^2$, we know that the vector $\mathbf{n} = 2X\mathbf{i} - 2Y\mathbf{j} - \mathbf{k}$ is normal to the surface at Q. Thus, $\overrightarrow{PQ}$ must be parallel to $\mathbf{n}$, and $\overrightarrow{PQ} = t\mathbf{n}$ for some scalar t. Separated into components, this vector equation states that

$$X - 3 = 2Xt, \qquad Y = -2Yt, \qquad \text{and} \qquad Z = -t.$$

The middle equation implies that either $Y = 0$ or $t = -\frac{1}{2}$. We must consider both of these possibilities.

CASE I If $Y = 0$, then

$$X = \frac{3}{1 - 2t} \qquad \text{and} \qquad Z = -t.$$

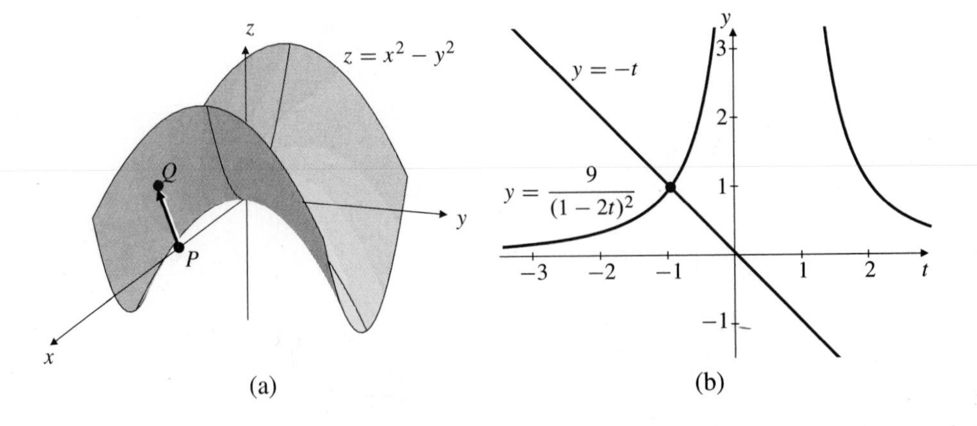

Figure 12.18

(a) If Q is the point on $z = x^2 - y^2$ closest to P, then $\overrightarrow{PQ}$ is normal to the surface

(b) Equation $-t = \dfrac{9}{(1 - 2t)^2}$ has only one real root, $t = -1$

But $Z = X^2 - Y^2$, so we must have

$$-t = \frac{9}{(1-2t)^2}.$$

This is a cubic equation in t, so we might expect to have to solve it numerically, for instance, by using Newton's Method. However, if we try small integer values of t, we will quickly discover that $t = -1$ is a solution. The graphs of both sides of the equation are shown in Figure 12.18(b). They show that $t = -1$ is the only real solution. Calculating the corresponding values of X and Z, we obtain $(1, 0, 1)$ as a candidate for Q. The distance from this point to P is $\sqrt{5}$.

CASE II If $t = -1/2$, then $X = 3/2$, $Z = 1/2$, and $Y = \pm\sqrt{X^2 - Z} = \pm\sqrt{7}/2$, and the distance from these points to P is $\sqrt{17}/2$.

Since $\frac{17}{4} < 5$, the points $(3/2, \pm\sqrt{7}/2, 1/2)$ are the points on $z = x^2 - y^2$ closest to $(3, 0, 0)$, and the distance from $(3, 0, 0)$ to the surface is $\sqrt{17}/2$ units.

EXERCISES 12.3

In Exercises 1–10, find all the first partial derivatives of the function specified, and evaluate them at the given point.

1. $f(x, y) = x - y + 2, \quad (3, 2)$

2. $f(x, y) = xy + x^2, \quad (2, 0)$

3. $f(x, y, z) = x^3 y^4 z^5, \quad (0, -1, -1)$

4. $g(x, y, z) = \dfrac{xz}{y+z}, \quad (1, 1, 1)$

5. $z = \tan^{-1}\left(\dfrac{y}{x}\right), \quad (-1, 1)$

6. $w = \ln(1 + e^{xyz}), \quad (2, 0, -1)$

7. $f(x, y) = \sin(x\sqrt{y}), \quad \left(\dfrac{\pi}{3}, 4\right)$

8. $f(x, y) = \dfrac{1}{\sqrt{x^2 + y^2}}, \quad (-3, 4)$

9. $w = x^{(y \ln z)}, \quad (e, 2, e)$

10. $g(x_1, x_2, x_3, x_4) = \dfrac{x_1 - x_2^2}{x_3 + x_4^2}, \quad (3, 1, -1, -2)$

In Exercises 11–12, calculate the first partial derivatives of the given functions at $(0, 0)$. You will have to use Definition 4.

11. $f(x, y) = \begin{cases} \dfrac{2x^3 - y^3}{x^2 + 3y^2}, & \text{if } (x, y) \neq (0, 0) \\ 0, & \text{if } (x, y) = (0, 0). \end{cases}$

12. $f(x, y) = \begin{cases} \dfrac{x^2 - 2y^2}{x - y}, & \text{if } x \neq y \\ 0, & \text{if } x = y. \end{cases}$

In Exercises 13–22, find equations of the tangent plane and normal line to the graph of the given function at the point with specified values of x and y.

13. $f(x, y) = x^2 - y^2$ at $(-2, 1)$

14. $f(x, y) = \dfrac{x - y}{x + y}$ at $(1, 1)$

15. $f(x, y) = \cos(x/y)$ at $(\pi, 4)$

16. $f(x, y) = e^{xy}$ at $(2, 0)$

17. $f(x, y) = \dfrac{x}{x^2 + y^2}$ at $(1, 2)$

18. $f(x, y) = y\, e^{-x^2}$ at $(0, 1)$

19. $f(x, y) = \ln(x^2 + y^2)$ at $(1, -2)$

20. $f(x, y) = \dfrac{2xy}{x^2 + y^2}$ at $(0, 2)$

21. $f(x, y) = \tan^{-1}(y/x)$ at $(1, -1)$

22. $f(x, y) = \sqrt{1 + x^3 y^2}$ at $(2, 1)$

23. Find the coordinates of all points on the surface with equation $z = x^4 - 4xy^3 + 6y^2 - 2$ where the surface has a horizontal tangent plane.

24. Find all horizontal planes that are tangent to the surface with equation $z = xye^{-(x^2+y^2)/2}$. At what points are they tangent?

In Exercises 25–31, show that the given function satisfies the given partial differential equation.

25. $z = x\, e^y, \quad x\dfrac{\partial z}{\partial x} = \dfrac{\partial z}{\partial y}$

26. $z = \dfrac{x + y}{x - y}, \quad x\dfrac{\partial z}{\partial x} + y\dfrac{\partial z}{\partial y} = 0$

27. $z = \sqrt{x^2 + y^2}, \quad x\dfrac{\partial z}{\partial x} + y\dfrac{\partial z}{\partial y} = z$

28. $w = x^2 + yz, \quad x\dfrac{\partial w}{\partial x} + y\dfrac{\partial w}{\partial y} + z\dfrac{\partial w}{\partial z} = 2w$

29. $w = \dfrac{1}{x^2 + y^2 + z^2}, \quad x\dfrac{\partial w}{\partial x} + y\dfrac{\partial w}{\partial y} + z\dfrac{\partial w}{\partial z} = -2w$

30. $z = f(x^2 + y^2)$, where f is any differentiable function of one variable,

$$y\frac{\partial z}{\partial x} - x\frac{\partial z}{\partial y} = 0.$$

31. $z = f(x^2 - y^2)$, where f is any differentiable function of one variable,

$$y\frac{\partial z}{\partial x} + x\frac{\partial z}{\partial y} = 0.$$

32. Give a formal definition of the three first partial derivatives of the function $f(x, y, z)$.

33. What is an equation of the "tangent hyperplane" to the graph $w = f(x, y, z)$ at $\big(a, b, c, f(a, b, c)\big)$?

34. Find the distance from the point $(1, 1, 0)$ to the circular paraboloid with equation $z = x^2 + y^2$.

35. Find the distance from the point $(0, 0, 1)$ to the elliptic paraboloid having equation $z = x^2 + 2y^2$.

36. Let $f(x, y) = \begin{cases} \dfrac{2xy}{x^2 + y^2}, & \text{if } (x, y) \neq (0, 0) \\ 0, & \text{if } (x, y) = (0, 0). \end{cases}$

Note that f is not continuous at $(0, 0)$. (See Example 3 of Section 12.2.) Therefore, its graph is not smooth there. Show, however, that $f_1(0, 0)$ and $f_2(0, 0)$ both exist. Hence, the existence of partial derivatives does not imply that a function of several variables is continuous. This is in contrast to the single-variable case.

37. Determine $f_1(0, 0)$ and $f_2(0, 0)$ if they exist, where

$$f(x, y) = \begin{cases} (x^3 + y)\sin\dfrac{1}{x^2 + y^2}, & \text{if } (x, y) \neq (0, 0) \\ 0, & \text{if } (x, y) = (0, 0). \end{cases}$$

38. Calculate $f_1(x, y)$ for the function in Exercise 37. Is $f_1(x, y)$ continuous at $(0, 0)$?

39. Let $f(x, y) = \begin{cases} \dfrac{x^3 - y^3}{x^2 + y^2}, & \text{if } (x, y) \neq (0, 0) \\ 0, & \text{if } (x, y) = (0, 0). \end{cases}$

Calculate $f_1(x, y)$ and $f_2(x, y)$ at all points (x, y) in the plane. Is f continuous at $(0, 0)$? Are f_1 and f_2 continuous at $(0, 0)$?

40. Let $f(x, y, z) = \begin{cases} \dfrac{xy^2z}{x^4 + y^4 + z^4}, & \text{if } (x, y, z) \neq (0, 0, 0) \\ 0, & \text{if } (x, y, z) = (0, 0, 0). \end{cases}$

Find $f_1(0, 0, 0)$, $f_2(0, 0, 0)$, and $f_3(0, 0, 0)$. Is f continuous at $(0, 0, 0)$? Are f_1, f_2, and f_3 continuous at $(0, 0, 0)$?

12.4 Higher-Order Derivatives

Partial derivatives of second and higher orders are calculated by taking partial derivatives of already calculated partial derivatives. The order in which the differentiations are performed is indicated in the notations used. If $z = f(x, y)$, we can calculate *four* partial derivatives of second order, namely, two **pure** second partial derivatives with respect to x or y,

$$\frac{\partial^2 z}{\partial x^2} = \frac{\partial}{\partial x}\frac{\partial z}{\partial x} = f_{11}(x, y) = f_{xx}(x, y),$$
$$\frac{\partial^2 z}{\partial y^2} = \frac{\partial}{\partial y}\frac{\partial z}{\partial y} = f_{22}(x, y) = f_{yy}(x, y),$$

and two **mixed** second partial derivatives with respect to x and y,

$$\frac{\partial^2 z}{\partial x \partial y} = \frac{\partial}{\partial x}\frac{\partial z}{\partial y} = f_{21}(x, y) = f_{yx}(x, y),$$
$$\frac{\partial^2 z}{\partial y \partial x} = \frac{\partial}{\partial y}\frac{\partial z}{\partial x} = f_{12}(x, y) = f_{xy}(x, y).$$

Again, we remark that the notations f_{11}, f_{12}, f_{21}, and f_{22} are usually preferable to f_{xx}, f_{xy}, f_{yx}, and f_{yy}, although the latter are often used in partial differential equations. Note that f_{12} indicates differentiation of f *first* with respect to its first variable and *then* with respect to its second variable; f_{21} indicates the opposite order of differentiation. The subscript closest to f indicates which differentiation occurs first.

Similarly, if $w = f(x, y, z)$, then

$$\frac{\partial^5 w}{\partial y \partial x \partial y^2 \partial z} = \frac{\partial}{\partial y}\frac{\partial}{\partial x}\frac{\partial}{\partial y}\frac{\partial}{\partial y}\frac{\partial w}{\partial z} = f_{32212}(x, y, z) = f_{zyyxy}(x, y, z).$$

EXAMPLE 1 Find the four second partial derivatives of $f(x, y) = x^3y^4$.

Solution

$$f_1(x, y) = 3x^2y^4, \qquad f_2(x, y) = 4x^3y^3,$$
$$f_{11}(x, y) = \frac{\partial}{\partial x}(3x^2y^4) = 6xy^4, \qquad f_{21}(x, y) = \frac{\partial}{\partial x}(4x^3y^3) = 12x^2y^3,$$
$$f_{12}(x, y) = \frac{\partial}{\partial y}(3x^2y^4) = 12x^2y^3, \qquad f_{22}(x, y) = \frac{\partial}{\partial y}(4x^3y^3) = 12x^3y^2.$$

EXAMPLE 2 Calculate $f_{223}(x, y, z)$, $f_{232}(x, y, z)$, and $f_{322}(x, y, z)$ for the function $f(x, y, z) = e^{x-2y+3z}$.

Solution

$$\begin{aligned}
f_{223}(x, y, z) &= \frac{\partial}{\partial z}\frac{\partial}{\partial y}\frac{\partial}{\partial y}e^{x-2y+3z} \\
&= \frac{\partial}{\partial z}\frac{\partial}{\partial y}\left(-2e^{x-2y+3z}\right) \\
&= \frac{\partial}{\partial z}\left(4e^{x-2y+3z}\right) = 12\,e^{x-2y+3z}, \\
f_{232}(x, y, z) &= \frac{\partial}{\partial y}\frac{\partial}{\partial z}\frac{\partial}{\partial y}e^{x-2y+3z} \\
&= \frac{\partial}{\partial y}\frac{\partial}{\partial z}\left(-2e^{x-2y+3z}\right) \\
&= \frac{\partial}{\partial y}\left(-6e^{x-2y+3z}\right) = 12\,e^{x-2y+3z}, \\
f_{322}(x, y, z) &= \frac{\partial}{\partial y}\frac{\partial}{\partial y}\frac{\partial}{\partial z}e^{x-2y+3z} \\
&= \frac{\partial}{\partial y}\frac{\partial}{\partial y}\left(3e^{x-2y+3z}\right) \\
&= \frac{\partial}{\partial y}\left(-6e^{x-2y+3z}\right) = 12\,e^{x-2y+3z}.
\end{aligned}$$

In both of the examples above observe that the mixed partial derivatives taken with respect to the same variables but in different orders turned out to be equal. This is not a coincidence. It will always occur for sufficiently smooth functions. In particular, the mixed partial derivatives involved are required to be *continuous*. The following theorem presents a more precise statement of this important phenomenon.

THEOREM 1

Equality of mixed partials

Suppose that two mixed nth-order partial derivatives of a function f involve the same differentiations but in different orders. If those partials are continuous at a point P, and if f and all partials of f of order less than n are continuous in a neighbourhood of P, then the two mixed partials are equal at the point P.

PROOF We shall prove only a representative special case, showing the equality of $f_{12}(a, b)$ and $f_{21}(a, b)$ for a function f of two variables, provided f_{12} and f_{21} are defined and f_1, f_2, and f are continuous throughout a disk of positive radius centred at (a, b), and f_{12} and f_{21} are continuous at (a, b). Let h and k have sufficiently small absolute values that the point $(a + h, b + k)$ lies in this disk. Then so do all points of the rectangle with sides parallel to the coordinate axes and diagonally opposite corners at (a, b) and $(a + h, b + k)$. (See Figure 12.19.)

Let $Q = f(a+h, b+k) - f(a+h, b) - f(a, b+k) + f(a, b)$ and define single-variable functions $u(x)$ and $v(y)$ by

$$u(x) = f(x, b+k) - f(x, b) \qquad \text{and} \qquad v(y) = f(a+h, y) - f(a, y).$$

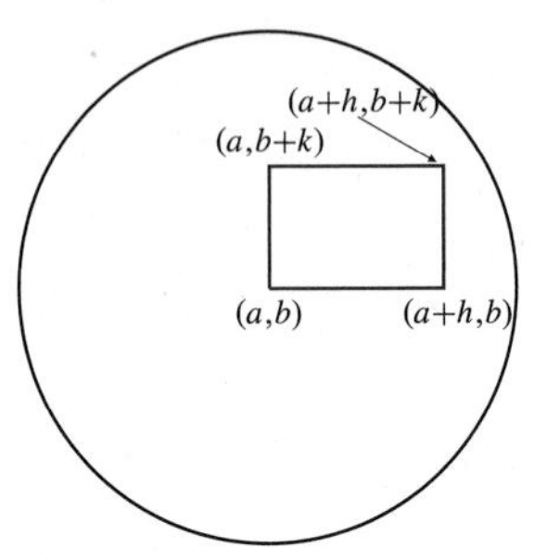

Figure 12.19 A rectangle contained in the disk where f and certain partials are continuous

Evidently, $Q = u(a+h) - u(a)$ and also $Q = v(b+k) - v(b)$. By the (single-variable) Mean-Value Theorem, there exists a number θ_1 satisfying $0 < \theta_1 < 1$ (so that $a + \theta_1 h$ lies between a and $a + h$) such that

$$Q = u(a+h) - u(a) = h\,u'(a+\theta_1 h) = h\big[f_1(a+\theta_1 h, b+k) - f_1(a+\theta_1 h, b)\big].$$

Now we apply the Mean-Value Theorem again, this time to f_1 considered as a function of its second variable, and obtain another number θ_2 satisfying $0 < \theta_2 < 1$ such that

$$f_1(a+\theta_1 h, b+k) - f_1(a+\theta_1 h, b) = k\, f_{12}(a+\theta_1 h, b+\theta_2 k).$$

BEWARE! The Mean-Value Theorem is used four times in this proof, each time to write a difference of the form $g(p+m) - g(p)$ in the form $g'(c)m$, where c is some number between p and $p+m$. It is convenient to write c in the form $p + \theta m$, where θ is some number between 0 and 1.

Thus, $Q = hk\, f_{12}(a+\theta_1 h, b+\theta_2 k)$. Two similar applications of the Mean-Value Theorem to $Q = v(b+k) - v(b)$ lead to $Q = hk\, f_{21}(a+\theta_3 h, b+\theta_4 k)$, where θ_3 and θ_4 are two numbers each between 0 and 1. Equating these two expressions for Q and cancelling the common factor hk, we obtain

$$f_{12}(a+\theta_1 h, b+\theta_2 k) = f_{21}(a+\theta_3 h, b+\theta_4 k).$$

Since f_{12} and f_{21} are continuous at (a, b), we can let h and k approach zero to obtain $f_{12}(a, b) = f_{21}(a, b)$, as required. ■

Exercise 16 below develops an example of a function for which f_{12} and f_{21} exist but are not continuous at (0, 0), and for which $f_{12}(0, 0) \neq f_{21}(0, 0)$.

Remark **Partial Derivatives in Maple** When you use the Maple function **diff** to calculate a derivative, you must include the name of the variable of differentiation. For example, `diff(x^2+y^3, x)` gives the result $2x$. It doesn't matter that the function being differentiated depends on more than one variable since you are telling Maple to differentiate with respect to x. If you wanted the derivative with respect to y, you would input `diff(x^2+y^3,y)` and the output would be $3y^2$. In this context, there is no distinction between ordinary and partial derivatives. There is, however, a difference when you want to apply a *differential operator* to a function f. If f is a function of one variable, you can denote its derivative f' in Maple by `D(f)`. For example,

```
> f := x -> sin(2*x); fprime := D(f);
```

$$f := x \to \sin(2\,x)$$

$$fprime := x \to 2\,\cos(2\,x)$$

The input `fprime(Pi/6)` will now give the output 1, as expected.

If f is a function of two (or more) variables, then `D(f)` no longer makes sense; do we mean f_1 or f_2? We distinguish the two (or more) first partials by using subscripts with D.

```
> f := (x,y) -> exp(3*y)*sin(2*x);
```

$$f := (x, y) \to e^{(3y)} * \sin(2\,x)$$

```
> fone := D[1](f); ftwo := D[2](f);
```

$$fone := (x, y) \to 2e^{(3y)} * \cos(2\,x)$$

$$ftwo := (x, y) \to 3e^{(3y)} * \sin(2\,x)$$

Higher-order partials are denoted with multiple subscripts (within one set of square brackets).

```
> D[1,1,2](f)(Pi/4, 0);
```

$$-12$$

You don't need to worry about the order of the subscripts in a mixed partial. Maple assumes the partials are continuous, even if it doesn't know what the function is. Even if `g` has not been assigned any meaning during the current Maple session, the input `D[1,2](g)(x,y)-D[2,1](g)(x,y);` produces the output 0.

The Laplace and Wave Equations

Many important and interesting phenomena are modelled by functions of several variables that satisfy certain *partial differential equations*. In the following examples we encounter two particular partial differential equations that arise frequently in mathematics and the physical sciences. Exercises 17–19 below introduce another such equation with important applications.

EXAMPLE 3 Show that for any real number k the functions

$$z = e^{kx}\cos(ky) \qquad \text{and} \qquad z = e^{kx}\sin(ky)$$

satisfy the partial differential equation

$$\frac{\partial^2 z}{\partial x^2} + \frac{\partial^2 z}{\partial y^2} = 0$$

at every point in the xy-plane.

Solution For $z = e^{kx}\cos(ky)$ we have

$$\frac{\partial z}{\partial x} = k\,e^{kx}\cos(ky), \qquad \frac{\partial z}{\partial y} = -k\,e^{kx}\sin(ky),$$

$$\frac{\partial^2 z}{\partial x^2} = k^2 e^{kx}\cos(ky), \qquad \frac{\partial^2 z}{\partial y^2} = -k^2 e^{kx}\cos(ky).$$

Thus,

$$\frac{\partial^2 z}{\partial x^2} + \frac{\partial^2 z}{\partial y^2} = k^2 e^{kx}\cos(ky) - k^2 e^{kx}\cos(ky) = 0.$$

The calculation for $z = e^{kx}\sin(ky)$ is similar.

Remark The partial differential equation in the above example is called the (two-dimensional) **Laplace equation**. A function of two variables having continuous second partial derivatives in a region of the plane is said to be **harmonic** there if it satisfies Laplace's equation. Such functions play a critical role in the theory of differentiable functions of a *complex variable* (see Appendix II) and are used to model various physical quantities such as steady-state temperature distributions, fluid flows, and electric and magnetic potential fields. Harmonic functions have many interesting properties. They have derivatives of all orders, and they are *analytic;* that is, they are the sums of their (multivariable) Taylor series. Moreover, a harmonic function can achieve maximum and minimum values only on the boundary of its domain. Laplace's equation, and therefore harmonic functions, can be considered in any number of dimensions. (See Exercises 13 and 14 below.)

EXAMPLE 4 If f and g are any twice-differentiable functions of one variable, show that

$$w = f(x - ct) + g(x + ct)$$

satisfies the partial differential equation

$$\frac{\partial^2 w}{\partial t^2} = c^2 \frac{\partial^2 w}{\partial x^2}.$$

Solution Using the Chain Rule for functions of one variable, we obtain

$$\frac{\partial w}{\partial t} = -c\, f'(x-ct) + c\, g'(x+ct), \qquad \frac{\partial w}{\partial x} = f'(x-ct) + g'(x+ct),$$
$$\frac{\partial^2 w}{\partial t^2} = c^2\, f''(x-ct) + c^2\, g''(x+ct), \qquad \frac{\partial^2 w}{\partial x^2} = f''(x-ct) + g''(x+ct).$$

Thus, w satisfies the given differential equation.

Remark The partial differential equation in the above example is called the (one-dimensional) **wave equation**. If t measures time, then $f(x-ct)$ represents a waveform travelling to the right along the x-axis with speed c. (See Figure 12.20.) Similarly, $g(x+ct)$ represents a waveform travelling to the left with speed c. Unlike the solutions of Laplace's equation that must be infinitely differentiable, solutions of the wave equation need only have enough derivatives to satisfy the differential equation. The functions f and g are otherwise arbitrary.

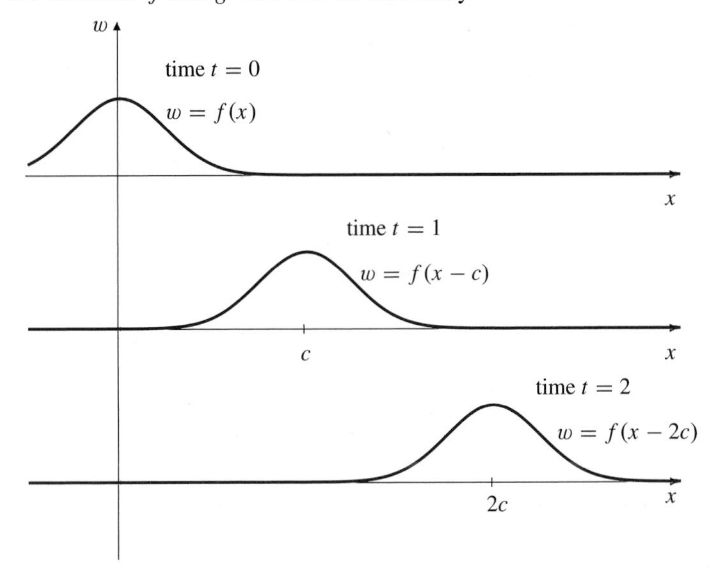

Figure 12.20 $w = f(x-ct)$ represents a waveform moving to the right with speed c

EXERCISES 12.4

In Exercises 1–6, find all the second partial derivatives of the given function.

1. $z = x^2(1+y^2)$

2. $f(x,y) = x^2 + y^2$

3. $w = x^3y^3z^3$

4. $z = \sqrt{3x^2+y^2}$

5. $z = x\,e^y - y\,e^x$

6. $f(x,y) = \ln\big(1+\sin(xy)\big)$

7. How many mixed partial derivatives of order 3 can a function of three variables have? If they are all continuous, how many different values can they have at one point? Find the mixed partials of order 3 for $f(x,y,z) = x\,e^{xy}\cos(xz)$ that involve two differentiations with respect to z and one with respect to x.

Show that the functions in Exercises 8–12 are harmonic in the plane regions indicated.

8. $f(x,y) = A(x^2-y^2) + Bxy$ in the whole plane (A and B are constants.)

9. $f(x,y) = 3x^2y - y^3$ in the whole plane (Can you think of another polynomial of degree 3 in x and y that is also harmonic?)

10. $f(x,y) = \dfrac{x}{x^2+y^2}$ everywhere except at the origin

11. $f(x,y) = \ln(x^2+y^2)$ everywhere except at the origin

12. $\tan^{-1}(y/x)$ except at points on the y-axis

13. Show that $w = e^{3x+4y}\sin(5z)$ is harmonic in all of $\mathbb{R}^3$, that is, it satisfies everywhere the 3-dimensional Laplace equation

$$\frac{\partial^2 w}{\partial x^2} + \frac{\partial^2 w}{\partial y^2} + \frac{\partial^2 w}{\partial z^2} = 0.$$

14. Assume that $f(x,y)$ is harmonic in the xy-plane. Show that each of the functions $z\,f(x,y)$, $x\,f(y,z)$, and $y\,f(z,x)$ is harmonic in the whole of $\mathbb{R}^3$. What condition should the constants a, b, and c satisfy to ensure that $f(ax+by, cz)$ is harmonic in $\mathbb{R}^3$?

15. Let the functions $u(x, y)$ and $v(x, y)$ have continuous second partial derivatives and satisfy the **Cauchy–Riemann equations**

$$\frac{\partial u}{\partial x} = \frac{\partial v}{\partial y} \quad \text{and} \quad \frac{\partial v}{\partial x} = -\frac{\partial u}{\partial y}.$$

Show that u and v are both harmonic.

16. Let $F(x, y) = \begin{cases} \dfrac{2xy(x^2 - y^2)}{x^2 + y^2}, & \text{if } (x, y) \neq (0, 0) \\ 0, & \text{if } (x, y) = (0, 0) \end{cases}$

Calculate $F_1(x, y)$, $F_2(x, y)$, $F_{12}(x, y)$, and $F_{21}(x, y)$ at points $(x, y) \neq (0, 0)$. Also calculate these derivatives at $(0, 0)$. Observe that $F_{21}(0, 0) = 2$ and $F_{12}(0, 0) = -2$. Does this result contradict Theorem 1? Explain why.

The heat (diffusion) equation

17. Show that the function $u(x, t) = t^{-1/2} e^{-x^2/4t}$ satisfies the partial differential equation

$$\frac{\partial u}{\partial t} = \frac{\partial^2 u}{\partial x^2}.$$

This equation is called the **one-dimensional heat equation** because it models heat diffusion in an insulated rod (with $u(x, t)$ representing the temperature at position x at time t) and other similar phenomena.

18. Show that the function $u(x, y, t) = t^{-1} e^{-(x^2+y^2)/4t}$ satisfies the **two-dimensional heat equation**

$$\frac{\partial u}{\partial t} = \frac{\partial^2 u}{\partial x^2} + \frac{\partial^2 u}{\partial y^2}.$$

19. By comparing the results of Exercises 17 and 18, guess a solution to the **three-dimensional heat equation**

$$\frac{\partial u}{\partial t} = \frac{\partial^2 u}{\partial x^2} + \frac{\partial^2 u}{\partial y^2} + \frac{\partial^2 u}{\partial z^2}.$$

Verify your guess. (If you're feeling lazy, use Maple.)

Biharmonic functions

A function $u(x, y)$ with continuous partials of fourth order is **biharmonic** if $\dfrac{\partial^2 u}{\partial x^2} + \dfrac{\partial^2 u}{\partial y^2}$ is a harmonic function.

20. Show that $u(x, y)$ is biharmonic if and only if it satisfies the biharmonic equation

$$\frac{\partial^4 u}{\partial x^4} + 2\frac{\partial^4 u}{\partial x^2 \partial y^2} + \frac{\partial^4 u}{\partial y^4} = 0$$

21. Verify that $u(x, y) = x^4 - 3x^2y^2$ is biharmonic.

22. Show that if $u(x, y)$ is harmonic, then $v(x, y) = xu(x, y)$ and $w(x, y) = yu(x, y)$ are biharmonic.

Use the result of Exercise 22 to show that the functions in Exercises 23–25 are biharmonic.

23. $x e^x \sin y$

24. $y \ln(x^2 + y^2)$

25. $\dfrac{xy}{x^2 + y^2}$

26. Propose a definition of a biharmonic function of three variables, and prove results analogous to those of Exercises 20 and 22 for biharmonic functions $u(x, y, z)$.

27. Use Maple to verify directly that the function of Exercise 25 is biharmonic.

12.5 The Chain Rule

The Chain Rule for functions of one variable is a formula that gives the derivative of a composition $f\big(g(x)\big)$ of two functions f and g:

$$\frac{d}{dx} f\big(g(x)\big) = f'\big(g(x)\big)g'(x).$$

The situation for several variables is more complicated. If f depends on more than one variable, and any of those variables can be functions of one or more other variables, we cannot expect a simple formula for partial derivatives of the composition to cover all possible cases. We must come to think of the Chain Rule as a *procedure for differentiating compositions* rather than as a formula for their derivatives. In order to motivate a formulation of the Chain Rule for functions of two variables, we begin with a concrete example.

EXAMPLE 1 Suppose you are hiking in a mountainous region for which you have a map. Let (x, y) be the coordinates of your position on the map (i.e., the horizontal coordinates of your actual position in the region). Let $z = f(x, y)$ denote the height of land (above sea level, say) at position (x, y). Suppose you are walking along a trail so that your position at time t is given by $x = u(t)$ and

$y = v(t)$. (These are parametric equations of the trail on the map.) At time t your altitude above sea level is given by the composite function

$$z = f\big(u(t), v(t)\big) = g(t),$$

a function of only one variable. How fast is your altitude changing with respect to time at time t?

Solution The answer is the derivative of $g(t)$:

$$\begin{aligned} g'(t) &= \lim_{h\to 0} \frac{g(t+h) - g(t)}{h} = \lim_{h\to 0} \frac{f\big(u(t+h), v(t+h)\big) - f\big(u(t), v(t)\big)}{h} \\ &= \lim_{h\to 0} \frac{f\big(u(t+h), v(t+h)\big) - f\big(u(t), v(t+h)\big)}{h} \\ &\qquad + \lim_{h\to 0} \frac{f\big(u(t), v(t+h)\big) - f\big(u(t), v(t)\big)}{h}. \end{aligned}$$

We added 0 to the numerator of the Newton quotient in a creative way so as to separate the quotient into the sum of two quotients, in the first of which the difference of values of f involves only the first variable of f, and in the second of which the difference involves only the second variable of f. The single-variable Chain Rule suggests that the sum of the two limits above is

$$g'(t) = f_1\big(u(t), v(t)\big)u'(t) + f_2\big(u(t), v(t)\big)v'(t).$$

The above formula is the Chain Rule for $\dfrac{d}{dt} f\big(u(t), v(t)\big)$. In terms of Leibniz notation we have

A version of the Chain Rule

If z is a function of x and y with continuous first partial derivatives, and if x and y are differentiable functions of t, then

$$\frac{dz}{dt} = \frac{\partial z}{\partial x}\frac{dx}{dt} + \frac{\partial z}{\partial y}\frac{dy}{dt}.$$

Note that there are two terms in the expression for dz/dt (or $g'(t)$), one arising from each variable of f that depends on t.

Now consider a function f of two variables, x and y, each of which is in turn a function of two other variables, s and t:

$$z = f(x, y), \qquad \text{where} \qquad x = u(s, t) \quad \text{and} \quad y = v(s, t).$$

We can form the composite function

$$z = f\big(u(s, t), v(s, t)\big) = g(s, t).$$

For instance, if $f(x, y) = x^2 + 3y$, where $u(s, t) = st^2$ and $v(s, t) = s - t$, then $g(s, t) = s^2t^4 + 3(s - t)$.

Let us assume that f, u, and v have first partial derivatives with respect to their respective variables and that those of f are continuous. Then g has first partial derivatives given by

$$\begin{aligned} g_1(s, t) &= f_1\big(u(s, t), v(s, t)\big)u_1(s, t) + f_2\big(u(s, t), v(s, t)\big)v_1(s, t), \\ g_2(s, t) &= f_1\big(u(s, t), v(s, t)\big)u_2(s, t) + f_2\big(u(s, t), v(s, t)\big)v_2(s, t). \end{aligned}$$

These formulas can be expressed more simply using Leibniz notation:

Another version of the Chain Rule

If z is a function of x and y with continuous first partial derivatives, and if x and y depend on s and t, then

$$\frac{\partial z}{\partial s} = \frac{\partial z}{\partial x}\frac{\partial x}{\partial s} + \frac{\partial z}{\partial y}\frac{\partial y}{\partial s},$$
$$\frac{\partial z}{\partial t} = \frac{\partial z}{\partial x}\frac{\partial x}{\partial t} + \frac{\partial z}{\partial y}\frac{\partial y}{\partial t}.$$

This can be deduced from the version obtained in Example 1 by allowing u and v there to depend on two variables, but holding one of them fixed while we differentiate with respect to the other. A more formal proof of this simple but representative case of the Chain Rule will be given in the next section.

The two equations in the box above can be combined into a single matrix equation:

$$\begin{pmatrix} \dfrac{\partial z}{\partial s} & \dfrac{\partial z}{\partial t} \end{pmatrix} = \begin{pmatrix} \dfrac{\partial z}{\partial x} & \dfrac{\partial z}{\partial y} \end{pmatrix} \begin{pmatrix} \dfrac{\partial x}{\partial s} & \dfrac{\partial x}{\partial t} \\ \dfrac{\partial y}{\partial s} & \dfrac{\partial y}{\partial t} \end{pmatrix}.$$

We will comment on the significance of this matrix form at the end of the next section.

In general, if z is a function of several "primary" variables, and each of these depends on some "secondary" variables, then the partial derivative of z with respect to one of the secondary variables will have several terms, one for the contribution to the derivative arising from each of the primary variables on which z depends.

Remark Note the significance of the various subscripts denoting partial derivatives in the functional form of the Chain Rule:

$$g_1(s,t) = f_1\big(u(s,t), v(s,t)\big)u_1(s,t) + f_2\big(u(s,t), v(s,t)\big)v_1(s,t).$$

The "1" in $g_1(s,t)$ refers to differentiation with respect to s, the first variable on which g depends. By contrast, the "1" in $f_1(u(s,t), v(s,t))$ refers to differentiation with respect to x, the first variable on which f depends. (This derivative is then evaluated at $x = u(s,t),\ y = v(s,t)$.)

EXAMPLE 2 If $z = \sin(x^2 y)$, where $x = st^2$ and $y = s^2 + \dfrac{1}{t}$, find $\partial z/\partial s$ and $\partial z/\partial t$

(a) by direct substitution and the single-variable form of the Chain Rule, and

(b) by using the (two-variable) Chain Rule.

Solution

(a) By direct substitution:

$$z = \sin\left((st^2)^2\left(s^2 + \frac{1}{t}\right)\right) = \sin(s^4t^4 + s^2t^3),$$
$$\frac{\partial z}{\partial s} = (4s^3t^4 + 2st^3)\cos(s^4t^4 + s^2t^3),$$
$$\frac{\partial z}{\partial t} = (4s^4t^3 + 3s^2t^2)\cos(s^4t^4 + s^2t^3).$$

(b) Using the Chain Rule:

$$\begin{aligned}
\frac{\partial z}{\partial s} &= \frac{\partial z}{\partial x}\frac{\partial x}{\partial s} + \frac{\partial z}{\partial y}\frac{\partial y}{\partial s} \\
&= \big(2xy\cos(x^2y)\big)t^2 + \big(x^2\cos(x^2y)\big)2s \\
&= \left(2st^2\left(s^2 + \frac{1}{t}\right)t^2 + 2s^3t^4\right)\cos(s^4t^4 + s^2t^3) \\
&= (4s^3t^4 + 2st^3)\cos(s^4t^4 + s^2t^3),
\end{aligned}$$

$$\begin{aligned}
\frac{\partial z}{\partial t} &= \frac{\partial z}{\partial x}\frac{\partial x}{\partial t} + \frac{\partial z}{\partial y}\frac{\partial y}{\partial t} \\
&= \big(2xy\cos(x^2y)\big)2st + \big(x^2\cos(x^2y)\big)\left(\frac{-1}{t^2}\right) \\
&= \left(2st^2(s^2 + \frac{1}{t})2st + s^2t^4\left(\frac{-1}{t^2}\right)\right)\cos(s^4t^4 + s^2t^3) \\
&= (4s^4t^3 + 3s^2t^2)\cos(s^4t^4 + s^2t^3).
\end{aligned}$$

Note that we still had to use direct substitution on the derivatives obtained in (b) in order to show that the values were the same as those obtained in (a).

EXAMPLE 3 Find $\frac{\partial}{\partial x} f(x^2y, x + 2y)$ and $\frac{\partial}{\partial y} f(x^2y, x + 2y)$ in terms of the partial derivatives of f, assuming that these partial derivatives are continuous.

Solution We have

$$\begin{aligned}
\frac{\partial}{\partial x} f(x^2y, x+2y) &= f_1(x^2y, x+2y)\frac{\partial}{\partial x}(x^2y) + f_2(x^2y, x+2y)\frac{\partial}{\partial x}(x+2y) \\
&= 2xyf_1(x^2y, x+2y) + f_2(x^2y, x+2y), \\
\frac{\partial}{\partial y} f(x^2y, x+2y) &= f_1(x^2y, x+2y)\frac{\partial}{\partial y}(x^2y) + f_2(x^2y, x+2y)\frac{\partial}{\partial y}(x+2y) \\
&= x^2f_1(x^2y, x+2y) + 2f_2(x^2y, x+2y).
\end{aligned}$$

EXAMPLE 4 Express the partial derivatives of $z = h(s,t) = f\big(g(s,t)\big)$ in terms of the derivative f' of f and the partial derivatives of g.

Solution The partial derivatives of h can be calculated using the single-variable version of the Chain Rule: if $x = g(s,t)$, then $z = f(x)$ and

$$\begin{aligned}
h_1(s,t) &= \frac{\partial z}{\partial s} = \frac{dz}{dx}\frac{\partial x}{\partial s} = f'\big(g(s,t)\big)g_1(s,t), \\
h_2(s,t) &= \frac{\partial z}{\partial t} = \frac{dz}{dx}\frac{\partial x}{\partial t} = f'\big(g(s,t)\big)g_2(s,t).
\end{aligned}$$

The following example involves a hybrid application of the Chain Rule to a function that depends both directly and indirectly on the variable of differentiation.

EXAMPLE 5 Find dz/dt, where $z = f(x,y,t)$, $x = g(t)$, and $y = h(t)$. (Assume that f, g, and h all have continuous derivatives.)

Solution Since z depends on t through each of the three variables of f, there will be three terms in the appropriate Chain Rule:

$$\begin{aligned}
\frac{dz}{dt} &= \frac{\partial z}{\partial x}\frac{dx}{dt} + \frac{\partial z}{\partial y}\frac{dy}{dt} + \frac{\partial z}{\partial t} \\
&= f_1(x,y,t)g'(t) + f_2(x,y,t)h'(t) + f_3(x,y,t).
\end{aligned}$$

Remark In the above example we can easily distinguish between the meanings of the symbols dz/dt and $\partial z/\partial t$. If, however, we had been dealing with the situation

$$z = f(x, y, s, t), \qquad \text{where} \quad x = g(s, t) \quad \text{and} \quad y = h(s, t),$$

then the meaning of the symbol $\partial z/\partial t$ would be unclear; it could refer to the simple partial derivative of f with respect to its fourth primary variable (i.e., $f_4(x, y, s, t)$), or it could refer to the derivative of the composite function $f(g(s, t), h(s, t), s, t)$. Three of the four primary variables of f depend on t and, therefore, contribute to the rate of change of z with respect to t. The partial derivative $f_4(x, y, s, t)$ denotes the contribution of only one of these three variables. It is conventional to use $\partial z/\partial t$ to denote the whole derivative of the *composite* function with respect to the secondary variable t:

$$\begin{aligned}\frac{\partial z}{\partial t} &= \frac{\partial}{\partial t} f(g(s, t), h(s, t), s, t) \\ &= f_1(x, y, s, t)g_2(s, t) + f_2(x, y, s, t)h_2(s, t) + f_4(x, y, s, t).\end{aligned}$$

When it is necessary, we can denote the contribution coming from the primary variable t by

$$\left(\frac{\partial z}{\partial t}\right)_{x,y,s} = \frac{\partial}{\partial t} f(x, y, s, t) = f_4(x, y, s, t).$$

Here, the subscripts denote those primary variables of f being *held fixed*, that is, whose contributions to the rate of change of z with respect to t are being *ignored.* Of course, in the situation described above, $(\partial z/\partial t)_s$ means the same as $\partial z/\partial t$.

In applications, the variables that contribute to a particular partial derivative will usually be clear from the context. The following example contains such an application. This is an example of a procedure called *differentiation following the motion.*

EXAMPLE 6 Atmospheric temperature depends on position and time. If we denote position by three spatial coordinates x, y, and z (measured in kilometres) and time by t (measured in hours), then the temperature T °C is a function of four variables, $T(x, y, z, t)$.

(a) If a thermometer is attached to a weather balloon that moves through the atmosphere on a path with parametric equations $x = f(t)$, $y = g(t)$, and $z = h(t)$, what is the rate of change at time t of the temperature T recorded by the thermometer?

(b) Find the rate of change of the recorded temperature at time $t = 1$ if

$$T(x, y, z, t) = \frac{xy}{1+z}(1+t),$$

and if the balloon moves along the curve

$$x = t, \qquad y = 2t, \qquad z = t - t^2.$$

Solution

(a) Here, the rate of change of the thermometer reading depends on the change in position of the thermometer as well as increasing time. Thus, none of the four variables of T can be ignored in the differentiation. The rate is given by

$$\frac{dT}{dt} = \frac{\partial T}{\partial x}\frac{dx}{dt} + \frac{\partial T}{\partial y}\frac{dy}{dt} + \frac{\partial T}{\partial z}\frac{dz}{dt} + \frac{\partial T}{\partial t}.$$

The term $\partial T/\partial t$ refers only to the rate of change of the temperature with respect to time at a fixed position in the atmosphere. The other three terms arise from the motion of the balloon.

(b) The values of the three coordinates and their derivatives at $t = 1$ are $x = 1$, $y = 2$, $z = 0$, $dx/dt = 1$, $dy/dt = 2$, and $dz/dt = -1$. Also, at $t = 1$,

$$\frac{\partial T}{\partial x} = \frac{y}{1+z}(1+t) = 4, \qquad \frac{\partial T}{\partial z} = \frac{-xy}{(1+z)^2}(1+t) = -4,$$
$$\frac{\partial T}{\partial y} = \frac{x}{1+z}(1+t) = 2, \qquad \frac{\partial T}{\partial t} = \frac{xy}{1+z} = 2.$$

Thus,

$$\left.\frac{dT}{dt}\right|_{t=1} = (4)(1) + (2)(2) + (-4)(-1) + 2 = 14.$$

The recorded temperature is increasing at a rate of 14 °C/h at time $t = 1$.

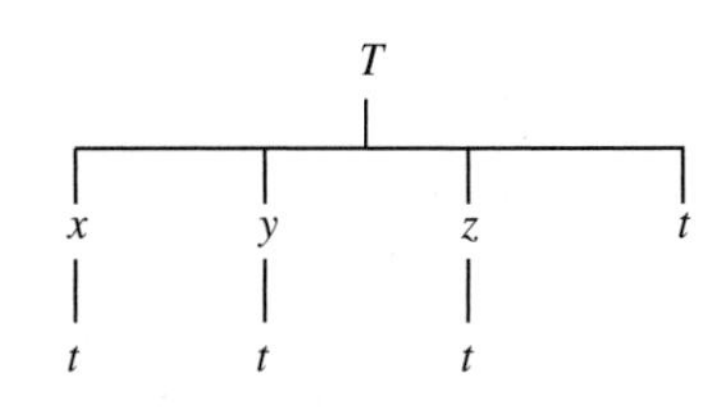

Figure 12.21 Chart showing the dependence of T on t in Example 6

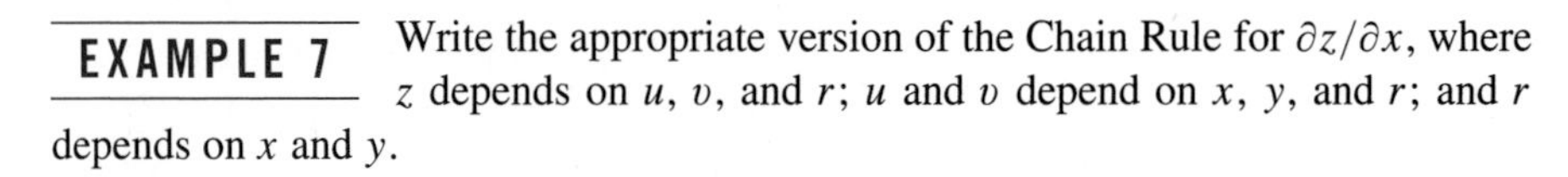

The discussion and examples above show that the Chain Rule for functions of several variables can take different forms depending on the numbers of variables of the various functions being composed. As an aid in determining the correct form of the Chain Rule in a given situation you can construct a chart showing which variables depend on which. Figure 12.21 shows such a chart for the temperature function of Example 6. The Chain Rule for dT/dt involves a term for every route from T to t in the chart. The route from T through x to t produces the term $\frac{\partial T}{\partial x}\frac{dx}{dt}$ and so on.

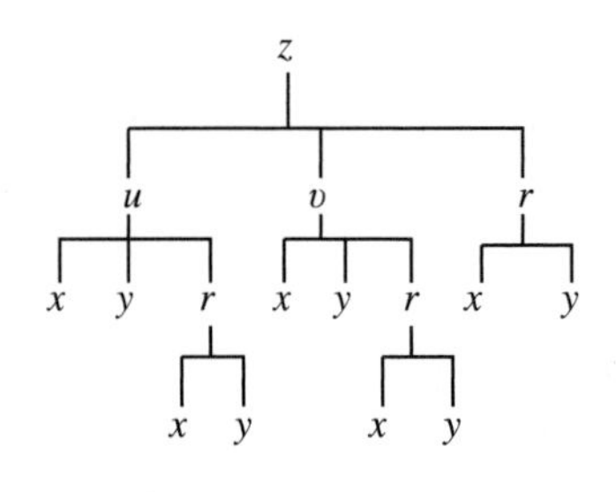

Figure 12.22 Dependence chart for Example 7

EXAMPLE 7 Write the appropriate version of the Chain Rule for $\partial z/\partial x$, where z depends on u, v, and r; u and v depend on x, y, and r; and r depends on x and y.

Solution The appropriate chart is shown in Figure 12.22. There are five routes from z to x:

$$\frac{\partial z}{\partial x} = \frac{\partial z}{\partial u}\frac{\partial u}{\partial x} + \frac{\partial z}{\partial u}\frac{\partial u}{\partial r}\frac{\partial r}{\partial x} + \frac{\partial z}{\partial v}\frac{\partial v}{\partial x} + \frac{\partial z}{\partial v}\frac{\partial v}{\partial r}\frac{\partial r}{\partial x} + \frac{\partial z}{\partial r}\frac{\partial r}{\partial x}.$$

Homogeneous Functions

A function $f(x_1, \ldots, x_n)$ is said to be **positively homogeneous of degree** k if, for every point $(x_1, x_2, \ldots, x_n)$ in its domain and every real number $t > 0$, we have

$$f(tx_1, tx_2, \ldots, tx_n) = t^k f(x_1, \ldots, x_n).$$

For example,

$$f(x, y) = x^2 + xy - y^2 \quad \text{is positively homogeneous of degree 2,}$$
$$f(x, y) = \sqrt{x^2 + y^2} \quad \text{is positively homogeneous of degree 1,}$$
$$f(x, y) = \frac{2xy}{x^2 + y^2} \quad \text{is positively homogeneous of degree 0,}$$
$$f(x, y, z) = \frac{x - y + 5z}{yz - z^2} \quad \text{is positively homogeneous of degree } -1,$$
$$f(x, y) = x^2 + y \quad \text{is not positively homogeneous.}$$

Observe that a positively homogeneous function of degree 0 remains constant along rays from the origin. More generally, along such rays a positively homogeneous function of degree k grows or decays proportionally to the kth power of distance from the origin.

THEOREM 2

Euler's Theorem

If $f(x_1, \ldots, x_n)$ has continuous first partial derivatives and is positively homogeneous of degree k, then

$$\sum_{i=1}^{n} x_i f_i(x_1, \ldots, x_n) = kf(x_1, \ldots, x_n).$$

PROOF Differentiate the equation $f(tx_1, tx_2, \ldots, tx_n) = t^k f(x_1, \ldots, x_n)$ with respect to t to get

$$x_1 f_1(tx_1, \ldots, tx_n) + x_2 f_2(tx_1, \ldots, tx_n) + \ldots + x_n f_n(tx_1, \ldots, tx_n) = k\,t^{k-1} f(x_1, \ldots, x_n).$$

Now substitute $t = 1$ to get the desired result.

Note that Exercises 26–29 in Section 12.3 illustrate this theorem.

Higher-Order Derivatives

Applications of the Chain Rule to higher-order derivatives can become quite complicated. It is important to keep in mind at each stage which variables are independent of one another.

EXAMPLE 8 Calculate $\dfrac{\partial^2}{\partial x \partial y} f(x^2 - y^2, xy)$ in terms of partial derivatives of the function f. Assume that the second-order partials of f are continuous.

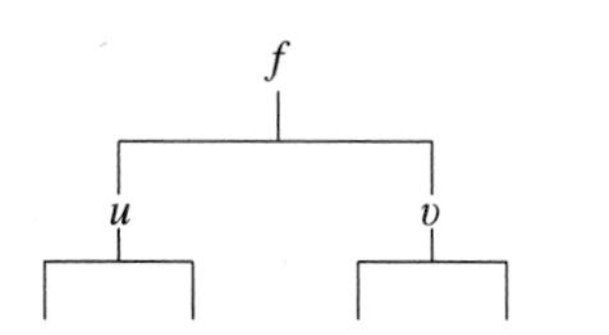

Figure 12.23 Chart showing the dependence of f on x and y through the primary variables u and v in Example 8

Solution In this problem symbols for the primary variables on which f depends are not stated explicitly. Let them be u and v. (See Figure 12.23.) The problem therefore asks us to find

$$\frac{\partial^2}{\partial x \partial y} f(u, v), \qquad \text{where} \quad u = x^2 - y^2 \quad \text{and} \quad v = xy.$$

First differentiate with respect to y:

$$\frac{\partial}{\partial y} f(u, v) = -2y f_1(u, v) + x f_2(u, v).$$

Now differentiate this result with respect to x. Note that the second term on the right is a product of two functions of x, so we need to use the Product Rule:

$$\begin{aligned}\frac{\partial^2}{\partial x \partial y} f(u, v) &= -\,2y\Big(2x f_{11}(u, v) + y f_{12}(u, v)\Big)\\ &\quad + f_2(u, v) + x\Big(2x f_{21}(u, v) + y f_{22}(u, v)\Big)\\ &= f_2(u, v) - 4xy f_{11}(u, v) + 2(x^2 - y^2) f_{12}(u, v) + xy f_{22}(u, v).\end{aligned}$$

In the last step we have used the fact that the mixed partials of f are continuous, so we could equate f_{12} and f_{21}.

Review the above calculation very carefully and make sure you understand what is being done at each step. Note that all the derivatives of f that appear are evaluated at $(u, v) = (x^2 - y^2, xy)$, not at (x, y), because x and y are not themselves the primary variables on which f depends.

Remark The kind of calculation done in the above example (and the following ones) is easily carried out by a computer algebra system. In Maple:

```
> g := (x,y) -> f(x^2 - y^2, x*y):
  simplify(D[1,2](g)(x,y));
```

$$\begin{aligned} -4yD_{1,1}(f)(x^2-y^2,xy)x - 2 * D_{1,2}(f)(x^2-y^2,xy)y^2 \\ + 2D_{1,2}(f)(x^2-y^2,xy)x^2 \\ + xD_{2,2}(f)(x^2-y^2,xy)y + D_2(f)(x^2-y^2,xy) \end{aligned}$$

which, on close inspection, is the same answer we calculated in the example.

EXAMPLE 9 If $f(x, y)$ is harmonic, show that $f(x^2-y^2, 2xy)$ is also harmonic.

Solution Let $u = x^2 - y^2$ and $v = 2xy$. If $z = f(u, v)$, then

$$\begin{aligned} \frac{\partial z}{\partial x} &= 2xf_1(u,v) + 2yf_2(u,v), \\ \frac{\partial z}{\partial y} &= -2yf_1(u,v) + 2xf_2(u,v), \\ \frac{\partial^2 z}{\partial x^2} &= 2f_1(u,v) + 2x\big(2xf_{11}(u,v) + 2yf_{12}(u,v)\big) \\ &\quad + 2y\big(2xf_{21}(u,v) + 2yf_{22}(u,v)\big) \\ &= 2f_1(u,v) + 4x^2f_{11}(u,v) + 8xyf_{12}(u,v) + 4y^2f_{22}(u,v), \end{aligned}$$

$$\begin{aligned} \frac{\partial^2 z}{\partial y^2} &= -2f_1(u,v) - 2y\big(-2yf_{11}(u,v) + 2xf_{12}(u,v)\big) \\ &\quad + 2x\big(-2yf_{21}(u,v) + 2xf_{22}(u,v)\big) \\ &= -2f_1(u,v) + 4y^2f_{11}(u,v) - 8xyf_{12}(u,v) + 4x^2f_{22}(u,v). \end{aligned}$$

Therefore,

$$\frac{\partial^2 z}{\partial x^2} + \frac{\partial^2 z}{\partial y^2} = 4(x^2+y^2)\big(f_{11}(u,v) + f_{22}(u,v)\big) = 0$$

because f is harmonic. Thus, $z = f(x^2-y^2, 2xy)$ is a harmonic function of x and y.

In the following example we show that the two-dimensional Laplace differential equation (see Example 3 in Section 12.4) takes the form

$$\frac{\partial^2 z}{\partial r^2} + \frac{1}{r}\frac{\partial z}{\partial r} + \frac{1}{r^2}\frac{\partial^2 z}{\partial \theta^2} = 0$$

when stated for a function z expressed in terms of polar coordinates r and θ.

EXAMPLE 10 **(Laplace's equation in polar coordinates)** If $z = f(x, y)$ has continuous partial derivatives of second order, and if $x = r\cos\theta$ and $y = r\sin\theta$, show that

$$\frac{\partial^2 z}{\partial r^2} + \frac{1}{r}\frac{\partial z}{\partial r} + \frac{1}{r^2}\frac{\partial^2 z}{\partial \theta^2} = \frac{\partial^2 z}{\partial x^2} + \frac{\partial^2 z}{\partial y^2}.$$

Solution It is possible to do this in two different ways; we can start with either side and use the Chain Rule to show that it is equal to the other side. Here, we will calculate the partial derivatives with respect to r and θ that appear on the left side and express them in terms of partial derivatives with respect to x and y. The other approach, involving expressing partial derivatives with respect to x and y in terms of partial derivatives with respect to r and θ, is a little more difficult. (See Exercise 24 at the end of this section.) However, we would have to do it that way if we were not given the form of the differential equation in polar coordinates and had to find it.

First, note that

$$\frac{\partial x}{\partial r} = \cos\theta, \qquad \frac{\partial x}{\partial \theta} = -r\sin\theta, \qquad \frac{\partial y}{\partial r} = \sin\theta, \qquad \frac{\partial y}{\partial \theta} = r\cos\theta.$$

Thus,

BEWARE! This is a difficult but important example. Examine each step carefully to make sure you understand what is being done.

$$\frac{\partial z}{\partial r} = \frac{\partial z}{\partial x}\frac{\partial x}{\partial r} + \frac{\partial z}{\partial y}\frac{\partial y}{\partial r} = \cos\theta\,\frac{\partial z}{\partial x} + \sin\theta\,\frac{\partial z}{\partial y}.$$

Now differentiate with respect to r again. Remember that r and θ are independent variables, so the factors $\cos\theta$ and $\sin\theta$ can be regarded as constants. However, $\partial z/\partial x$ and $\partial z/\partial y$ depend on x and y and, therefore, on r and θ.

$$\begin{aligned}\frac{\partial^2 z}{\partial r^2} &= \cos\theta\,\frac{\partial}{\partial r}\frac{\partial z}{\partial x} + \sin\theta\,\frac{\partial}{\partial r}\frac{\partial z}{\partial y}\\ &= \cos\theta\left(\cos\theta\,\frac{\partial^2 z}{\partial x^2} + \sin\theta\,\frac{\partial^2 z}{\partial y\partial x}\right) + \sin\theta\left(\cos\theta\,\frac{\partial^2 z}{\partial x\partial y} + \sin\theta\,\frac{\partial^2 z}{\partial y^2}\right)\\ &= \cos^2\theta\,\frac{\partial^2 z}{\partial x^2} + 2\cos\theta\sin\theta\,\frac{\partial^2 z}{\partial x\partial y} + \sin^2\theta\,\frac{\partial^2 z}{\partial y^2}.\end{aligned}$$

We have used the equality of mixed partials in the last line. Similarly,

$$\frac{\partial z}{\partial \theta} = -r\,\sin\theta\,\frac{\partial z}{\partial x} + r\,\cos\theta\,\frac{\partial z}{\partial y}.$$

When we differentiate a second time with respect to θ, we can regard r as constant, but each term above is still a product of two functions that depend on θ. Thus,

$$\begin{aligned}\frac{\partial^2 z}{\partial \theta^2} &= -r\left(\cos\theta\,\frac{\partial z}{\partial x} + \sin\theta\,\frac{\partial}{\partial\theta}\frac{\partial z}{\partial x}\right) + r\left(-\sin\theta\,\frac{\partial z}{\partial y} + \cos\theta\,\frac{\partial}{\partial\theta}\frac{\partial z}{\partial y}\right)\\ &= -r\,\frac{\partial z}{\partial r} - r\,\sin\theta\left(-r\,\sin\theta\,\frac{\partial^2 z}{\partial x^2} + r\,\cos\theta\,\frac{\partial^2 z}{\partial y\partial x}\right)\\ &\quad + r\,\cos\theta\left(-r\,\sin\theta\,\frac{\partial^2 z}{\partial x\partial y} + r\,\cos\theta\,\frac{\partial^2 z}{\partial y^2}\right)\\ &= -r\,\frac{\partial z}{\partial r} + r^2\left(\sin^2\theta\,\frac{\partial^2 z}{\partial x^2} - 2\sin\theta\cos\theta\,\frac{\partial^2 z}{\partial x\partial y} + \cos^2\theta\,\frac{\partial^2 z}{\partial y^2}\right).\end{aligned}$$

Combining these results, we obtain the desired formula:

$$\frac{\partial^2 z}{\partial r^2} + \frac{1}{r}\frac{\partial z}{\partial r} + \frac{1}{r^2}\frac{\partial^2 z}{\partial \theta^2} = \frac{\partial^2 z}{\partial x^2} + \frac{\partial^2 z}{\partial y^2}.$$

EXERCISES 12.5

In Exercises 1–4, write appropriate versions of the Chain Rule for the indicated derivatives.

1. $\partial w/\partial t$ if $w = f(x, y, z)$, where $x = g(s, t)$, $y = h(s, t)$, and $z = k(s, t)$

2. $\partial w/\partial t$ if $w = f(x, y, z)$, where $x = g(s)$, $y = h(s, t)$, and $z = k(t)$

3. $\partial z/\partial u$ if $z = g(x, y)$, where $y = f(x)$ and $x = h(u, v)$

4. dw/dt if $w = f(x, y)$, $x = g(r, s)$, $y = h(r, t)$, $r = k(s, t)$, and $s = m(t)$

5. If $w = f(x, y, z)$, where $x = g(y, z)$ and $y = h(z)$, state appropriate versions of the Chain Rule for $\dfrac{dw}{dz}$, $\left(\dfrac{\partial w}{\partial z}\right)_x$, and $\left(\dfrac{\partial w}{\partial z}\right)_{x,y}$.

6. Use two different methods to calculate $\partial u/\partial t$ if $u = \sqrt{x^2 + y^2}$, $x = e^{st}$, and $y = 1 + s^2 \cos t$.

7. Use two different methods to calculate $\partial z/\partial x$ if $z = \tan^{-1}(u/v)$, $u = 2x + y$, and $v = 3x - y$.

8. Use two methods to calculate dz/dt given that $z = txy^2$, $x = t + \ln(y + t^2)$, and $y = e^t$.

In Exercises 9–12, find the indicated derivatives, assuming that the function $f(x, y)$ has continuous first partial derivatives.

9. $\dfrac{\partial}{\partial x} f(2x, 3y)$

10. $\dfrac{\partial}{\partial x} f(2y, 3x)$

11. $\dfrac{\partial}{\partial x} f(y^2, x^2)$

12. $\dfrac{\partial}{\partial y} f\Big(yf(x, t), f(y, t)\Big)$

13. Suppose that the temperature T in a certain liquid varies with depth z and time t according to the formula $T = e^{-t}z$. Find the rate of change of temperature with respect to time at a point that is moving through the liquid so that at time t its depth is $f(t)$. What is this rate if $f(t) = e^t$? What is happening in this case?

14. Suppose the strength E of an electric field in space varies with position (x, y, z) and time t according to the formula $E = f(x, y, z, t)$. Find the rate of change with respect to time of the electric field strength measured by an instrument moving along the helix $x = \sin t$, $y = \cos t$, $z = t$.

In Exercises 15–20, assume that f has continuous partial derivatives of all orders.

15. If $z = f(x, y)$, where $x = 2s + 3t$ and $y = 3s - 2t$, find

(a) $\dfrac{\partial^2 z}{\partial s^2}$, (b) $\dfrac{\partial^2 z}{\partial s \partial t}$, and (c) $\dfrac{\partial^2 z}{\partial t^2}$.

16. If $f(x, y)$ is harmonic, show that $f\left(\dfrac{x}{x^2 + y^2}, \dfrac{-y}{x^2 + y^2}\right)$ is also harmonic.

17. If $x = t \sin s$ and $y = t \cos s$, find $\dfrac{\partial^2}{\partial s \partial t} f(x, y)$.

18. Find $\dfrac{\partial^3}{\partial x \partial y^2} f(2x + 3y, xy)$ in terms of partial derivatives of the function f.

19. Find $\dfrac{\partial^2}{\partial y \partial x} f(y^2, xy, -x^2)$ in terms of partial derivatives of the function f.

20. Find $\dfrac{\partial^3}{\partial t^2 \partial s} f(s^2 - t, s + t^2)$ in terms of partial derivatives of the function f.

21. Suppose that $u(x, y)$ and $v(x, y)$ have continuous second partial derivatives and satisfy the Cauchy–Riemann equations

$$\frac{\partial u}{\partial x} = \frac{\partial v}{\partial y} \quad \text{and} \quad \frac{\partial v}{\partial x} = -\frac{\partial u}{\partial y}.$$

Suppose also that $f(u, v)$ is a harmonic function of u and v. Show that $f\Big(u(x, y), v(x, y)\Big)$ is a harmonic function of x and y. *Hint:* u and v are harmonic functions by Exercise 15 in Section 12.4.

22. If $r^2 = x^2 + y^2 + z^2$, verify that $u(x, y, z) = 1/r$ is harmonic throughout $\mathbb{R}^3$ except at the origin.

23. If $x = e^s \cos t$, $y = e^s \sin t$, and $z = u(x, y) = v(s, t)$, show that

$$\frac{\partial^2 z}{\partial s^2} + \frac{\partial^2 z}{\partial t^2} = (x^2 + y^2)\left(\frac{\partial^2 z}{\partial x^2} + \frac{\partial^2 z}{\partial y^2}\right).$$

24. (Converting Laplace's equation to polar coordinates) The transformation to polar coordinates, $x = r \cos\theta$, $y = r \sin\theta$, implies that $r^2 = x^2 + y^2$ and $\tan\theta = y/x$. Use these equations to show that

$$\frac{\partial r}{\partial x} = \cos\theta \qquad \frac{\partial r}{\partial y} = \sin\theta$$
$$\frac{\partial \theta}{\partial x} = -\frac{\sin\theta}{r} \qquad \frac{\partial \theta}{\partial y} = \frac{\cos\theta}{r}.$$

Use these formulas to help you express $\dfrac{\partial^2 u}{\partial x^2} + \dfrac{\partial^2 u}{\partial y^2}$ in terms of partials of u with respect to r and θ, and hence reprove the formula for the Laplace differential equation in polar coordinates given in Example 10.

25. If $u(x, y) = r^2 \ln r$, where $r^2 = x^2 + y^2$, verify that u is a biharmonic function by showing that

$$\left(\frac{\partial^2}{\partial x^2} + \frac{\partial^2}{\partial y^2}\right)\left(\frac{\partial^2 u}{\partial x^2} + \frac{\partial^2 u}{\partial y^2}\right) = 0.$$

26. If $f(x, y)$ is positively homogeneous of degree k and has continuous partial derivatives of second order, show that

$$x^2 f_{11}(x, y) + 2xy f_{12}(x, y) + y^2 f_{22}(x, y) = k(k - 1) f(x, y).$$

27. Generalize the result of Exercise 26 to functions of n variables.

28. Generalize the results of Exercises 26 and 27 to expressions involving mth-order partial derivatives of the function f.

Exercises 29–30 revisit Exercise 16 of Section 12.4. Let

$$F(x, y) = \begin{cases} \dfrac{2xy(x^2 - y^2)}{x^2 + y^2}, & \text{if } (x, y) \neq (0, 0) \\ 0, & \text{if } (x, y) = (0, 0). \end{cases}$$

29. (a) Show that $F(x, y) = -F(y, x)$ for all (x, y).
(b) Show that $F_1(x, y) = -F_2(y, x)$ and $F_{12}(x, y) = -F_{21}(y, x)$ for $(x, y) \neq (0, 0)$.
(c) Show that $F_1(0, y) = -2y$ for all y and, hence, that $F_{12}(0, 0) = -2$.
(d) Deduce that $F_2(x, 0) = 2x$ and $F_{21}(0, 0) = 2$.

30. (a) Use Exercise 29(b) to find $F_{12}(x, x)$ for $x \neq 0$.
(b) Is $F_{12}(x, y)$ continuous at $(0, 0)$? Why?

31. Use the change of variables $\xi = x + ct$, $\eta = x$ to transform the partial differential equation

$$\frac{\partial u}{\partial t} = c\,\frac{\partial u}{\partial x}, \qquad (c = \text{constant}),$$

into the simpler equation $\partial v/\partial \eta = 0$, where $v(\xi, \eta) = v(x + ct, x) = u(x, t)$. This equation says that $v(\xi, \eta)$ does not depend on η, so $v = f(\xi)$ for some arbitrary differentiable function f. What is the corresponding "general solution" $u(x, t)$ of the original partial differential equation?

32. Having considered Exercise 31, guess a "general solution" $w(r, s)$ of the second-order partial differential equation

$$\frac{\partial^2}{\partial r \partial s}\, w(r, s) = 0.$$

Your answer should involve two arbitrary functions.

33. Use the change of variables $r = x + ct$, $s = x - ct$, $w(r, s) = u(x, t)$ to transform the one-dimensional wave equation

$$\frac{\partial^2 u}{\partial t^2} = c^2\,\frac{\partial^2 u}{\partial x^2}$$

to a simpler form. Now use the result of Exercise 32 to find the *general solution* of this wave equation in the form given in Example 4 in Section 12.4.

34. Show that the initial-value problem for the one-dimensional wave equation

$$\begin{cases} u_{tt}(x, t) = c^2 u_{xx}(x, t) \\ u(x, 0) = p(x) \\ u_t(x, 0) = q(x) \end{cases}$$

has the solution

$$u(x, t) = \frac{1}{2}\Big[p(x - ct) + p(x + ct)\Big] + \frac{1}{2c}\int_{x-ct}^{x+ct} q(s)\, ds.$$

(Note that we have used subscripts x and t instead of 1 and 2 to denote the partial derivatives here. This is common usage in dealing with partial differential equations.)

Remark The initial-value problem in Exercise 34 gives the small lateral displacement $u(x, t)$ at position x at time t of a vibrating string held under tension along the x-axis. The function $p(x)$ gives the *initial* displacement at position x, that is, the displacement at time $t = 0$. Similarly, $q(x)$ gives the initial velocity at position x. Observe that the position at time t depends only on values of these initial data at points no further than ct units away. This is consistent with the previous observation that the solutions of the wave equation represent waves travelling with speed c.

Redo the examples and exercises listed in Exercises 35–40 using Maple to do the calculations.

35. Example 10
36. Exercise 16
37. Exercise 19
38. Exercise 20
39. Exercise 23
40. Exercise 34

12.6 Linear Approximations, Differentiability, and Differentials

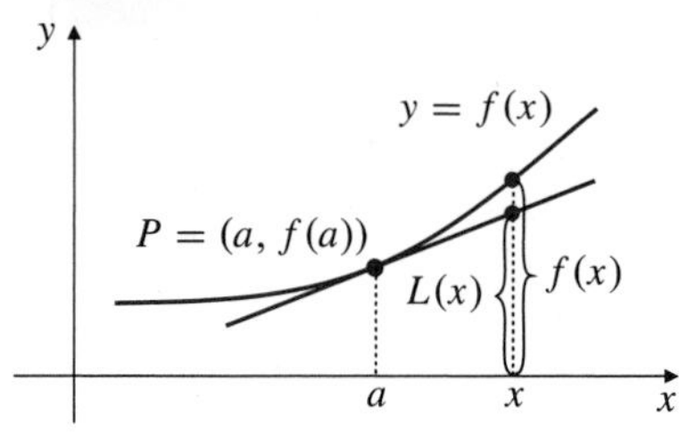

Figure 12.24 The linearization of f at $x = a$

As observed in Section 4.9, the tangent line to the graph $y = f(x)$ at $x = a$ provides a convenient approximation for values of $f(x)$ for x near a (see Figure 12.24):

$$f(x) \approx L(x) = f(a) + f'(a)(x - a).$$

Here, $L(x)$ is the **linearization** of f at a; its graph is the tangent line to $y = f(x)$ there. The mere existence of $f'(a)$ is sufficient to guarantee that the error in the approximation (the vertical distance between the curve and tangent at x) is small compared with the distance $h = x - a$ between a and x, that is,

$$\begin{aligned}\lim_{h\to 0} \frac{f(a+h) - L(a+h)}{h} &= \lim_{h\to 0} \frac{f(a+h) - f(a) - f'(a)h}{h} \\ &= \lim_{h\to 0} \frac{f(a+h) - f(a)}{h} - f'(a) \\ &= f'(a) - f'(a) = 0.\end{aligned}$$

Similarly, the tangent plane to the graph of $z = f(x, y)$ at (a, b) is $z = L(x, y)$, where

$$L(x, y) = f(a, b) + f_1(a, b)(x - a) + f_2(a, b)(y - b)$$

is the **linearization** of f at (a, b). We can use $L(x, y)$ to approximate values of $f(x, y)$ near (a, b):

$$f(x, y) \approx L(x, y) = f(a, b) + f_1(a, b)(x - a) + f_2(a, b)(y - b).$$

EXAMPLE 1 Find an approximate value for $f(x, y) = \sqrt{2x^2 + e^{2y}}$ at $(2.2, -0.2)$.

Solution It is convenient to use the linearization at $(2, 0)$, where the values of f and its partials are easily evaluated:

$$\begin{aligned} & & f(2,0) &= 3, \\ f_1(x, y) &= \frac{2x}{\sqrt{2x^2 + e^{2y}}}, & f_1(2, 0) &= \frac{4}{3}, \\ f_2(x, y) &= \frac{e^{2y}}{\sqrt{2x^2 + e^{2y}}}, & f_2(2, 0) &= \frac{1}{3}. \end{aligned}$$

Thus, $L(x, y) = 3 + \frac{4}{3}(x - 2) + \frac{1}{3}(y - 0)$, and

$$f(2.2, -0.2) \approx L(2.2, -0.2) = 3 + \frac{4}{3}(2.2 - 2) + \frac{1}{3}(-0.2 - 0) = 3.2.$$

(For the sake of comparison, $f(2.2, -0.2) \approx 3.2172$ to 4 decimal places.)

Unlike the single-variable case, the mere existence of the partial derivatives $f_1(a, b)$ and $f_2(a, b)$ does not even imply that f is continuous at (a, b), let alone that the error in the linearization is small compared with the distance $\sqrt{(x - a)^2 + (y - b)^2}$ between (a, b) and (x, y). We adopt this latter condition as our definition of what it means for a function of two variables to be *differentiable* at a point.

DEFINITION 5

We say that the function $f(x, y)$ is **differentiable** at the point (a, b) if

$$\lim_{(h,k)\to(0,0)} \frac{f(a + h, b + k) - f(a, b) - h\, f_1(a, b) - k f_2(a, b)}{\sqrt{h^2 + k^2}} = 0.$$

This definition and the following theorems can be generalized to functions of any number of variables in the obvious way. For the sake of simplicity, we state them for the two-variable case only.

The function $f(x, y)$ is differentiable at the point (a, b) if and only if the surface $z = f(x, y)$ has a *nonvertical tangent plane* at (a, b). This implies that $f_1(a, b)$ and $f_2(a, b)$ must exist and that f must be continuous at (a, b). (Recall, however, that the existence of the partial derivatives does *not* even imply that f is continuous, let alone differentiable.) In particular, the function is *continuous* wherever it is differentiable. We will prove a two-variable version of the Mean-Value Theorem and use it to show that functions are differentiable wherever they have *continuous* first partial derivatives.

THEOREM 3

A Mean-Value Theorem

If $f_1(x, y)$ and $f_2(x, y)$ are continuous in a neighbourhood of the point (a, b), and if the absolute values of h and k are sufficiently small, then there exist numbers θ_1 and θ_2, each between 0 and 1, such that

$$f(a + h, b + k) - f(a, b) = hf_1(a + \theta_1 h, b + k) + kf_2(a, b + \theta_2 k).$$

PROOF The proof of this theorem is very similar to that of Theorem 1 in Section 12.4, so we give only a sketch here. The reader can fill in the details. Write

$$f(a+h,b+k)-f(a,b)=\big(f(a+h,b+k)-f(a,b+k)\big)+\big(f(a,b+k)-f(a,b)\big),$$

and then apply the single-variable Mean-Value Theorem separately to $f(x,b+k)$ on the interval between a and $a+h$, and to $f(a,y)$ on the interval between b and $b+k$ to get the desired result.

THEOREM 4

If f_1 and f_2 are continuous in a neighbourhood of the point (a,b), then f is differentiable at (a,b).

PROOF Using Theorem 3 and the facts that

$$\left|\frac{h}{\sqrt{h^2+k^2}}\right|\le 1 \qquad \text{and} \qquad \left|\frac{k}{\sqrt{h^2+k^2}}\right|\le 1,$$

we estimate

$$\begin{aligned}
&\left|\frac{f(a+h,b+k)-f(a,b)-hf_1(a,b)-kf_2(a,b)}{\sqrt{h^2+k^2}}\right|\\
&\quad=\left|\frac{h}{\sqrt{h^2+k^2}}\Big(f_1(a+\theta_1 h,b+k)-f_1(a,b)\Big)\right.\\
&\qquad\left.+\frac{k}{\sqrt{h^2+k^2}}\Big(f_2(a,b+\theta_2 k)-f_2(a,b)\Big)\right|\\
&\quad\le\big|f_1(a+\theta_1 h,b+k)-f_1(a,b)\big|+\big|f_2(a,b+\theta_2 k)-f_2(a,b)\big|.
\end{aligned}$$

Since f_1 and f_2 are continuous at (a,b), each of these latter terms approaches 0 as h and k approach 0. This is what we needed to prove.

We illustrate differentiability with an example where we can calculate directly the error in the tangent plane approximation.

EXAMPLE 2 Calculate $f(x+h,y+k)-f(x,y)-f_1(x,y)h-f_2(x,y)k$ if $f(x,y)=x^3+xy^2$.

Solution Since $f_1(x,y)=3x^2+y^2$ and $f_2(x,y)=2xy$, we have

$$\begin{aligned}
f(x+h,y+k)&-f(x,y)-f_1(x,y)h-f_2(x,y)k\\
&=(x+h)^3+(x+h)(y+k)^2-x^3-xy^2-(3x^2+y^2)h-2xyk\\
&=3xh^2+h^3+2yhk+hk^2+xk^2.
\end{aligned}$$

Observe that the result above is a polynomial in h and k with no term of degree less than 2 in these variables. Therefore, this difference approaches zero like the *square* of the distance $\sqrt{h^2+k^2}$ from (x,y) to $(x+h,y+k)$ as $(h,k)\to(0,0)$, so the condition for differentiability is certainly satisfied:

$$\lim_{(h,k)\to(0,0)}\frac{3xh^2+h^3+2yhk+hk^2+xk^2}{\sqrt{h^2+k^2}}=0.$$

This quadratic behaviour is the case for any function f with continuous *second* partial derivatives. (See Exercise 23 below.)

Proof of the Chain Rule

We are now able to give a formal statement and proof of a simple but representative case of the Chain Rule for multivariate functions.

THEOREM 5

A Chain Rule

Let $z = f(x, y)$, where $x = u(s, t)$ and $y = v(s, t)$. Suppose that

(i) $u(a, b) = p$ and $v(a, b) = q$,

(ii) the first partial derivatives of u and v exist at the point (a, b), and

(iii) f is differentiable at the point (p, q).

Then $z = w(s, t) = f(u(s, t), v(s, t))$ has first partial derivatives with respect to s and t at (a, b), and

$$\begin{aligned} w_1(a, b) &= f_1(p, q)u_1(a, b) + f_2(p, q)v_1(a, b), \\ w_2(a, b) &= f_1(p, q)u_2(a, b) + f_2(p, q)v_2(a, b). \end{aligned}$$

That is,

$$\frac{\partial z}{\partial s} = \frac{\partial z}{\partial x}\frac{\partial x}{\partial s} + \frac{\partial z}{\partial y}\frac{\partial y}{\partial s} \qquad \text{and} \qquad \frac{\partial z}{\partial t} = \frac{\partial z}{\partial x}\frac{\partial x}{\partial t} + \frac{\partial z}{\partial y}\frac{\partial y}{\partial t}.$$

PROOF Define a function E of two variables as follows: $E(0, 0) = 0$, and if $(h, k) \neq (0, 0)$, then

$$E(h, k) = \frac{f(p + h, q + k) - f(p, q) - hf_1(p, q) - kf_2(p, q)}{\sqrt{h^2 + k^2}}.$$

Observe that $E(h, k)$ is continuous at $(0, 0)$ because f is differentiable at (p, q). Now,

$$f(p + h, q + k) - f(p, q) = hf_1(p, q) + kf_2(p, q) + \sqrt{h^2 + k^2}\, E(h, k).$$

In this formula put $h = u(a + \sigma, b) - u(a, b)$ and $k = v(a + \sigma, b) - v(a, b)$ and divide by σ to obtain

$$\begin{aligned} \frac{w(a + \sigma, b) - w(a, b)}{\sigma} &= \frac{f(u(a + \sigma, b), v(a + \sigma, b)) - f(u(a, b), v(a, b))}{\sigma} \\ &= \frac{f(p + h, q + k) - f(p, q)}{\sigma} \\ &= f_1(p, q)\frac{h}{\sigma} + f_2(p, q)\frac{k}{\sigma} + \sqrt{\left(\frac{h}{\sigma}\right)^2 + \left(\frac{k}{\sigma}\right)^2}\, E(h, k). \end{aligned}$$

We want to let σ approach 0 in this formula. Note that

$$\lim_{\sigma \to 0} \frac{h}{\sigma} = \lim_{\sigma \to 0} \frac{u(a + \sigma, b) - u(a, b)}{\sigma} = u_1(a, b),$$

and, similarly, $\lim_{\sigma \to 0}(k/\sigma) = v_1(a, b)$. Since $(h, k) \to (0, 0)$ if $\sigma \to 0$, we have

$$w_1(a, b) = f_1(p, q)u_1(a, b) + f_2(p, q)v_1(a, b).$$

The proof for w_2 is similar. ■

Differentials

If the first partial derivatives of a function $z = f(x_1, \ldots, x_n)$ exist at a point, we may construct a **differential** dz or df of the function at that point in a manner similar to that used for functions of one variable:

$$\begin{aligned} dz = df &= \frac{\partial z}{\partial x_1}\, dx_1 + \frac{\partial z}{\partial x_2}\, dx_2 + \cdots + \frac{\partial z}{\partial x_n}\, dx_n \\ &= f_1(x_1, \ldots, x_n)\, dx_1 + \cdots + f_n(x_1, \ldots, x_n)\, dx_n. \end{aligned}$$

Here, the differential dz is considered to be a function of the $2n$ independent variables $x_1, x_2, \ldots, x_n, dx_1, dx_2, \ldots, dx_n$.

For a *differentiable* function f, the differential df is an approximation to the change Δf in value of the function given by

$$\Delta f = f(x_1 + dx_1, \ldots, x_n + dx_n) - f(x_1, \ldots, x_n).$$

The error in this approximation is small compared with the distance between the two points in the domain of f; that is,

$$\frac{\Delta f - df}{\sqrt{(dx_1)^2 + \cdots + (dx_n)^2}} \to 0 \quad \text{if all } dx_i \to 0, \qquad (1 \le i \le n).$$

In this sense, differentials are just another way of looking at linearization.

EXAMPLE 3 Estimate the percentage change in the period $T = 2\pi\sqrt{\dfrac{L}{g}}$ of a simple pendulum if the length, L, of the pendulum increases by 2% and the acceleration of gravity, g, decreases by 0.6%.

Solution We calculate the differential of T:

$$\begin{aligned} dT &= \frac{\partial T}{\partial L}\, dL + \frac{\partial T}{\partial g}\, dg \\ &= \frac{2\pi}{2\sqrt{Lg}}\, dL - \frac{2\pi\sqrt{L}}{2g^{3/2}}\, dg. \end{aligned}$$

We are given that $dL = \dfrac{2}{100} L$ and $dg = -\dfrac{6}{1{,}000} g$. Thus,

$$dT = \frac{1}{100} 2\pi\sqrt{\frac{L}{g}} - \left(-\frac{6}{1{,}000}\right)\frac{2\pi}{2}\sqrt{\frac{L}{g}} = \frac{13}{1{,}000} T.$$

Therefore, the period T of the pendulum increases by 1.3%.

Functions from *n*-Space to *m*-Space

(*This is an optional topic.*) A vector $\mathbf{f} = (f_1, f_2, \ldots, f_m)$ of m functions, each depending on n variables $(x_1, x_2, \ldots, x_n)$, defines a *transformation* (i.e., a function) from $\mathbb{R}^n$ to $\mathbb{R}^m$; specifically, if $\mathbf{x} = (x_1, x_2, \ldots, x_n)$ is a point in $\mathbb{R}^n$, and

$$\begin{aligned} y_1 &= f_1(x_1, x_2, \ldots, x_n) \\ y_2 &= f_2(x_1, x_2, \ldots, x_n) \\ &\ \vdots \\ y_m &= f_m(x_1, x_2, \ldots, x_n), \end{aligned}$$

then $\mathbf{y} = (y_1, y_2, \ldots, y_m)$ is the point in $\mathbb{R}^m$ that corresponds to $\mathbf{x}$ under the transformation $\mathbf{f}$. We can write these equations more compactly as

$$\mathbf{y} = \mathbf{f}(\mathbf{x}).$$

Information about the rate of change of $\mathbf{y}$ with respect to $\mathbf{x}$ is contained in the various partial derivatives $\partial y_i/\partial x_j$, $(1 \le i \le m,\ 1 \le j \le n)$, and is conveniently organized into an $m \times n$ matrix, $D\mathbf{f}(\mathbf{x})$, called the **Jacobian matrix** of the transformation $\mathbf{f}$:

$$Df(\mathbf{x}) = \begin{pmatrix} \frac{\partial y_1}{\partial x_1} & \frac{\partial y_1}{\partial x_2} & \cdots & \frac{\partial y_1}{\partial x_n} \\ \frac{\partial y_2}{\partial x_1} & \frac{\partial y_2}{\partial x_2} & \cdots & \frac{\partial y_2}{\partial x_n} \\ \vdots & \vdots & & \vdots \\ \frac{\partial y_m}{\partial x_1} & \frac{\partial y_m}{\partial x_2} & \cdots & \frac{\partial y_m}{\partial x_n} \end{pmatrix}$$

If the partial derivatives in the Jacobian matrix are continuous, we say that $\mathbf{f}$ is **differentiable** at $\mathbf{x}$. In this case the linear transformation (see Section 10.7) represented by the Jacobian matrix is called **the derivative** of the transformation $\mathbf{f}$.

Remark We can regard the scalar-valued function of two variables, $f(x, y)$ say, as a transformation from $\mathbb{R}^2$ to $\mathbb{R}$. Its derivative is then the linear transformation with matrix

$$Df(x, y) = \big(f_1(x, y), f_2(x, y)\big).$$

It is not our purpose to enter into a study of such *vector-valued functions of a vector variable* at this point, but we can observe here that the Jacobian matrix of the composition of two such transformations is the matrix product of their Jacobian matrices.

To see this, let $\mathbf{y} = \mathbf{f}(\mathbf{x})$ be a transformation from $\mathbb{R}^n$ to $\mathbb{R}^m$ as described above, and let $\mathbf{z} = \mathbf{g}(\mathbf{y})$ be another such transformation from $\mathbb{R}^m$ to $\mathbb{R}^k$ given by

$$\begin{aligned} z_1 &= g_1(y_1, y_2, \ldots, y_m) \\ z_2 &= g_2(y_1, y_2, \ldots, y_m) \\ &\vdots \\ z_k &= g_k(y_1, y_2, \ldots, y_m), \end{aligned}$$

which has the $k \times m$ Jacobian matrix

$$Dg(\mathbf{y}) = \begin{pmatrix} \frac{\partial z_1}{\partial y_1} & \frac{\partial z_1}{\partial y_2} & \cdots & \frac{\partial z_1}{\partial y_m} \\ \frac{\partial z_2}{\partial y_1} & \frac{\partial z_2}{\partial y_2} & \cdots & \frac{\partial z_2}{\partial y_m} \\ \vdots & \vdots & & \vdots \\ \frac{\partial z_k}{\partial y_1} & \frac{\partial z_k}{\partial y_2} & \cdots & \frac{\partial z_k}{\partial y_m} \end{pmatrix}.$$

Then the composition $\mathbf{z} = \mathbf{g} \circ \mathbf{f}(\mathbf{x}) = \mathbf{g}\big(\mathbf{f}(\mathbf{x})\big)$ given by

$$\begin{aligned} z_1 &= g_1\big(f_1(x_1, \ldots, x_n), \ldots, f_m(x_1, \ldots, x_n)\big) \\ z_2 &= g_2\big(f_1(x_1, \ldots, x_n), \ldots, f_m(x_1, \ldots, x_n)\big) \\ &\vdots \\ z_k &= g_k\big(f_1(x_1, \ldots, x_n), \ldots, f_m(x_1, \ldots, x_n)\big) \end{aligned}$$

has, according to the Chain Rule, the $k \times n$ Jacobian matrix

$$\begin{pmatrix} \frac{\partial z_1}{\partial x_1} & \frac{\partial z_1}{\partial x_2} & \cdots & \frac{\partial z_1}{\partial x_n} \\ \frac{\partial z_2}{\partial x_1} & \frac{\partial z_2}{\partial x_2} & \cdots & \frac{\partial z_2}{\partial x_n} \\ \vdots & \vdots & & \vdots \\ \frac{\partial z_k}{\partial x_1} & \frac{\partial z_k}{\partial x_2} & \cdots & \frac{\partial z_k}{\partial x_n} \end{pmatrix} = \begin{pmatrix} \frac{\partial z_1}{\partial y_1} & \frac{\partial z_1}{\partial y_2} & \cdots & \frac{\partial z_1}{\partial y_m} \\ \frac{\partial z_2}{\partial y_1} & \frac{\partial z_2}{\partial y_2} & \cdots & \frac{\partial z_2}{\partial y_m} \\ \vdots & \vdots & & \vdots \\ \frac{\partial z_k}{\partial y_1} & \frac{\partial z_k}{\partial y_2} & \cdots & \frac{\partial z_k}{\partial y_m} \end{pmatrix} \begin{pmatrix} \frac{\partial y_1}{\partial x_1} & \frac{\partial y_1}{\partial x_2} & \cdots & \frac{\partial y_1}{\partial x_n} \\ \frac{\partial y_2}{\partial x_1} & \frac{\partial y_2}{\partial x_2} & \cdots & \frac{\partial y_2}{\partial x_n} \\ \vdots & \vdots & & \vdots \\ \frac{\partial y_m}{\partial x_1} & \frac{\partial y_m}{\partial x_2} & \cdots & \frac{\partial y_m}{\partial x_n} \end{pmatrix}$$

This is, in fact, the Chain Rule for compositions of transformations:

$$D(\mathbf{g} \circ \mathbf{f})(\mathbf{x}) = D\mathbf{g}(\mathbf{f}(\mathbf{x}))D\mathbf{f}(\mathbf{x}),$$

and exactly mimics the one-variable Chain Rule $D(g \circ f)(x) = Dg(f(x))Df(x)$.

The transformation $\mathbf{y} = \mathbf{f}(\mathbf{x})$ also defines a vector $d\mathbf{y}$ of differentials of the variables y_i in terms of the vector $d\mathbf{x}$ of differentials of the variables x_j. Writing $d\mathbf{y}$ and $d\mathbf{x}$ as column vectors we have

$$d\mathbf{y} = \begin{pmatrix} dy_1 \\ dy_2 \\ \vdots \\ dy_m \end{pmatrix} = \begin{pmatrix} \frac{\partial y_1}{\partial x_1} & \frac{\partial y_1}{\partial x_2} & \cdots & \frac{\partial y_1}{\partial x_n} \\ \frac{\partial y_2}{\partial x_1} & \frac{\partial y_2}{\partial x_2} & \cdots & \frac{\partial y_2}{\partial x_n} \\ \vdots & \vdots & & \vdots \\ \frac{\partial y_m}{\partial x_1} & \frac{\partial y_m}{\partial x_2} & \cdots & \frac{\partial y_m}{\partial x_n} \end{pmatrix} \begin{pmatrix} dx_1 \\ dx_2 \\ \vdots \\ dx_n \end{pmatrix} = D\mathbf{f}(\mathbf{x})d\mathbf{x}.$$

EXAMPLE 4 Find the Jacobian matrix $D\mathbf{f}(1, 0)$ for the transformation from $\mathbb{R}^2$ to $\mathbb{R}^3$ given by

$$\mathbf{f}(x, y) = \big(xe^y + \cos(\pi y), x^2, x - e^y\big)$$

and use it to find an approximate value for $\mathbf{f}(1.02, 0.01)$.

Solution $D\mathbf{f}(x, y)$ is the 3×2 matrix whose jth row consists of the partial derivatives of the jth component of $\mathbf{f}$ with respect to x and y. Thus,

$$D\mathbf{f}(1, 0) = \begin{pmatrix} e^y & xe^y - \pi \sin(\pi y) \\ 2x & 0 \\ 1 & -e^y \end{pmatrix}\Bigg|_{(1,0)} = \begin{pmatrix} 1 & 1 \\ 2 & 0 \\ 1 & -1 \end{pmatrix}.$$

Since $\mathbf{f}(1, 0) = (2, 1, 0)$ and $d\mathbf{x} = \begin{pmatrix} 0.02 \\ 0.01 \end{pmatrix}$, we have

$$d\mathbf{f} = D\mathbf{f}(1, 0)\, d\mathbf{x} = \begin{pmatrix} 1 & 1 \\ 2 & 0 \\ 1 & -1 \end{pmatrix} \begin{pmatrix} 0.02 \\ 0.01 \end{pmatrix} = \begin{pmatrix} 0.03 \\ 0.04 \\ 0.01 \end{pmatrix}.$$

Therefore, $\mathbf{f}(1.02, 0.01) \approx (2.03, 1.04, 0.01)$.

For transformations between spaces of the same dimension (say from $\mathbb{R}^n$ to $\mathbb{R}^n$), the corresponding Jacobian matrices are square and have determinants. These Jacobian determinants will play an important role in our consideration of implicit functions and inverse functions in Section 12.8 and in changes of variables in multiple integrals in Chapter 14.

Maple's **VectorCalculus** package has a function **Jacobian** that takes two inputs, a list (or vector) of expressions and a list of variables, and produces the Jacobian matrix of the partial derivatives of those expressions with respect to the variables. For example,

```
> with(VectorCalculus):
> Jacobian([x*y*exp(z), (x+2*y)*cos(z)],[x,y,z]);
```

$$\begin{bmatrix} ye^z & xe^z & xye^z \\ \cos(z) & 2\cos(z) & -(x + 2y)\sin(z) \end{bmatrix}$$

VectorCalculus has only been included since Maple 8. If you have an earlier release use `linalg` instead, and the function `jacobian`.

Differentials in Applications

Differentials are sometimes used as an alternative representation for differentiable functions. This is particularly so in the field of thermodynamics. In thermodynamics, physical states of thermodynamic equilibrium are expressed mathematically in terms of the existence of a function,

$$E = E(S, V, N_1, \ldots, N_n),$$

where E is internal energy, S is entropy, V is volume, and the N_i are numbers of atoms or molecules of type i.

These quantities are interpreted physically, but they are just independent variables in a function to which normal mathematical rules apply. Discussion of the physical meaning of a quantity like entropy, for example, is largely beyond the scope of this book. (One might remark that entropy is a logarithmic measure of the number of underlying physical states that appear indistinguishable on human scales, but such a description is completely unnecessary for this discussion.) $E(S, V, N_1, \ldots, N_n)$ is known as a *function of state*. Any explicit equation relating thermodynamic variables is also known as an *equation of state*.

Thermodynamics allows for any number of such variables to define the state. There can be others than those indicated for different physical systems. All such variables are additive in that, for example, the energy of two physical systems together is simply the sum of the energies of each system. The same is true for volume, entropy, and number. These additive variables are called *extensive* variables. In thermodynamics they are referred to as *state variables* or as *state functions*. That is because any one of the other variables can be expressed as a function of E and the remaining variables. For example, $S = S(E, V, N_1, \ldots, N_n)$.

Differentials appear in thermodynamics as the normal way to express the existence of a state function. In writing

$$dE = \frac{\partial E}{\partial S}\, dS + \frac{\partial E}{\partial V}\, dV + \frac{\partial E}{\partial N_1}\, dN_1 + \cdots + \frac{\partial E}{\partial N_n}\, dN_n\,,$$

we are saying that E depends on the variables whose differentials appear on the right side of the equation. In fact, everything is so effectively done with differentials that often no explicit function E is needed or even known.

Historically the differential was also meant to convey an intuitive sense of change in time, even though mathematically it is simply the differential of a function. In fact this historical interpretation can be quite confusing, because, paradoxically, the existence of the function of state, and its differential, means the physical system is in *thermodynamic equilibrium*, which can be described as a time-independent condition of a physical system. If it were not in (timeless) thermodynamic equilibrium, there would be no state function and no corresponding differentials. The resolution of the paradox is to stick to the mathematics, remembering that the differential only depicts a change in the values of variables and not any external process.

So, for example, the state equation has nothing to do with whether some process is slow or not. Differentials in this case, do not suggest a physical process any more than the differential of any other function does. The differential only expresses the content of the function, so it has nothing to do with the physical processes that cause changes, or with whether any change is carried out slowly (reversible processes) or not.

The partial derivatives that appear in the differential form of the state equation also have explicit physical interpretations: $\dfrac{\partial E}{\partial S}$ is *temperature* T, $-\dfrac{\partial E}{\partial V}$ is *pressure* P, and the quantities $\dfrac{\partial E}{\partial N_i}$ are known as *chemical potentials*, μ_i. These partial derivatives represent slopes on the graph of the function of state, and as such they are not additive. It makes no sense, for example, to add temperatures. Physically, these slopes define a condition rather than an amount. These non-additive quantities are called *intensive* variables.

With these definitions substituted, the differential form of the equation of state becomes,

$$dE = T\,dS - P\,dV + \mu_1\,dN_1 + \cdots + \mu_n\,dN_n\,,$$

which is known as the *Gibbs equation*. However, despite the special treatment, this expression remains simply the differential of $E(S, V, N_1, \dots, N_n)$. The Gibbs equation is a fundamental starting point in many thermodynamical problems.

Another related, and well-known, equation of differentials is the Gibbs-Duhem equation,

$$0 = S\,dT - V\,dP + N_1\,d\mu_1 + \cdots + N_n\,d\mu_n$$

This remarkable equation indicates that the intensive variables of thermodynamics are not independent of each other. It holds because the additivity of the extensive variables implies that the function of state, $E = E(S, V, N_1, \dots, N_n)$, is homogeneous of degree 1. (See Exercise 24.)

Differentials and Legendre Transformations

It is often useful to shift the dependence of a function on one or more of its independent variables to dependence on, instead, the derivatives of the function with respect to these variables. Consider, for example, the function $y = f(x)$, and denote its derivative by p; that is, $p = f'(x)$. If we let $u = px - f(x)$, and calculate the differential of u, treating x and p as independent variables, we obtain

$$du = p\,dx + x\,dp - f'(x)\,dx = p\,dx + x\,dp - p\,dx = x\,dp.$$

Since there is no dx term remaining in this differential, u does not depend explicitly on x, but only on p. Let us therefore define $f^*(p) = u = px - f(x)$. $f^*(p)$ is called the Legendre transformation of $f(x)$ with respect to x, and the two variables x and p are said to be **conjugate** to one another. Observe that

$$f(x) + f^*(p) = px,$$

and the symmetry of this equation indicates that f must also be the Legendre transformation of f^*; $f^{**} = f$. In fact, taking the partial derivatives of the equation with respect to x and p we obtain the symmetric relationships

$$f'(x) = p \quad \text{and} \quad (f^*)'(p) = x$$

from which it is apparent that f' and $(f^*)'$ are inverse functions;

$$f'\big((f^*)'(p)\big) = p, \qquad (f^*)'\big(f'(x)\big) = x.$$

Remark The above definition of f^* clearly shows the symmetry in its relationship with f. An alternative transformation, $-f^*(p)$ (i.e., the function $f(x) - px$) shifts dependence between a variable and the derivative of the function just as effectively, although it does not share this symmetry. In some fields, particularly thermodynamics, this alternative is known as the Legendre transformation instead.

EXAMPLE 5 Calculate the Legendre transformation $f^*(p)$ of the function $f(x) = e^x$.

Solution Here $p = f'(x) = e^x$, so $x = \ln p$. Therefore,

$$f^*(p) = px - f(x) = p\ln p - p.$$

For functions of several variables, Legendre transformations can be taken with respect to one or more of the independent variables. If $u = f(x, y)$, $p = f_1(x, y)$ and $q = f_2(x, y)$, and if $w = px + qy - u$, then

$$dw = p\,dx + x\,dp + q\,dy + y\,dq - f_1(x, y)\,dx - f_2(x, y)\,dy = x\,dp + y\,dq$$

and w does not depend explicitly on x or y, but only on p and q. We can call $w(p, q)$ (or $-w(p, q)$ if we are doing thermodynamics) the Legendre transformation of $f(x, y)$ with respect to x and y, and treat both $\{x, p\}$ and $\{y, q\}$ as conjugate pairs of variables. Observe that

$$\begin{aligned} f_1(x, y) &= p \\ f_2(x, y) &= q \end{aligned} \quad \text{and} \quad \begin{aligned} w_1(p, q) &= x \\ w_2(p, q) &= y. \end{aligned}$$

Returning to thermodynamics, the Gibbs equation tells us that E depends on S, V, and N_i. Since $T = \dfrac{\partial E}{\partial S}$, T and S are conjugate and we can express energy in terms of temperature rather than entropy by using an (alternative) Legendre transformation. Let $F = E - TS$. Then

$$dF = dE - S\,dT - T\,dS = -S\,dT - P\,dV + \mu_1\,dN_1 + \cdots + \mu_n\,dN_n.$$

Thus, $F = F(T, V, N_1, \ldots, N_n)$. F, is known as the *Helmholtz free energy*, which is called a *thermodynamic potential*. It can be more practical to use F, which depends explicitly on T, rather than E when an experiment is run at constant temperature.

Legendre transformations can be done in terms of any or all of the conjugate pairs. In the case of the Helmholtz free energy, only the conjugates T and S are used. Other specific Legendre transformations lead to other thermodynamic *potentials*. For example, the *Gibbs free energy*, $G = E - TS + PV$, is widely used in chemistry, where processes normally take place at constant temperature and pressure. (See Exercise 30 below.)

Legendre transformations are very important in other areas of classical and modern physics. Historically they appear in classical mechanics, where the functional expression of the energy, known as the Hamiltonian, is expressed in terms of Legendre transformations of a function known as the Lagrangian. (See Exercise 32 for a problem developing this relationship.) These notions extend to modern physics which is often cast in terms of Lagrangians.

EXERCISES 12.6

In Exercises 1–6, use suitable linearizations to find approximate values for the given functions at the points indicated.

1. $f(x, y) = x^2 y^3$ at $(3.1, 0.9)$

2. $f(x, y) = \tan^{-1}\left(\dfrac{y}{x}\right)$ at $(3.01, 2.99)$

3. $f(x, y) = \sin(\pi xy + \ln y)$ at $(0.01, 1.05)$

4. $f(x, y) = \dfrac{24}{x^2 + xy + y^2}$ at $(2.1, 1.8)$

5. $f(x, y, z) = \sqrt{x + 2y + 3z}$ at $(1.9, 1.8, 1.1)$

6. $f(x, y) = x\,e^{y+x^2}$ at $(2.05, -3.92)$

In Exercises 7–10 write the differential of the given function and use it to estimate the value of the function at the given point by starting with a known value at a nearby point.

7. $z = x^2 e^{3y}$, at $x = 3.05$, $y = -0.02$

8. $g(s, t) = s^2/t$, $g(2.1, 1.9)$

9. $F(x, y, z) = \sqrt{x^2 + y + 2 + z^2}$, $F(0.7, 2.6, 1.7)$

10. $u = x\sin(x + y)$, at $x = \dfrac{\pi}{2} + \dfrac{1}{20}$, $y = \dfrac{\pi}{2} - \dfrac{1}{30}$

11. The edges of a rectangular box are each measured to within an accuracy of 1% of their values. What is the approximate maximum percentage error in

(a) the calculated volume of the box,

(b) the calculated area of one of the faces of the box, and

(c) the calculated length of a diagonal of the box?

12. The radius and height of a right-circular conical tank are measured to be 25 ft and 21 ft, respectively. Each measurement is accurate to within 0.5 in. By about how much can the calculated volume of the tank be in error?

13. By approximately how much can the calculated area of the conical surface of the tank in Exercise 12 be in error?

14. Two sides and the contained angle of a triangular plot of land are measured to be 224 m, 158 m, and 64°, respectively. The length measurements were accurate to within 0.4 m and the angle measurement to within 2°. What is the approximate maximum percentage error if the area of the plot is calculated from these measurements?

15. The angle of elevation of the top of a tower is measured at two points A and B on the ground in the same direction from the base of the tower. The angles are 50° at A and 35° at B, each measured to within 1°. The distance AB is measured to be 100 m with error at most 0.1%. What is the calculated height of the building, and by about how much can it be in error? To which of the three measurements is the calculated height most sensitive?

16. By approximately what percentage will the value of $w = \dfrac{x^2y^3}{z^4}$ increase or decrease if x increases by 1%, y increases by 2%, and z increases by 3%?

17. Find the Jacobian matrix for the transformation $\mathbf{f}(r, \theta) = (x, y)$, where

$$x = r\cos\theta \quad \text{and} \quad y = r\sin\theta.$$

(Although (r, θ) can be regarded as *polar coordinates* in the xy-plane, they are Cartesian coordinates in their own $r\theta$-plane.)

18. Find the Jacobian matrix for the transformation $\mathbf{f}(R, \phi, \theta) = (x, y, z)$, where

$$x = R\sin\phi\cos\theta, \quad y = R\sin\phi\sin\theta, \quad z = R\cos\phi.$$

Here, (R, ϕ, θ) are *spherical coordinates* in xyz-space, as introduced in Section 10.6.

19. Find the Jacobian matrix $D\mathbf{f}(x, y, z)$ for the transformation of $\mathbb{R}^3$ to $\mathbb{R}^2$ given by

$$\mathbf{f}(x, y, z) = (x^2 + yz, y^2 - x\ln z).$$

Use $D\mathbf{f}(2, 2, 1)$ to help you find an approximate value for $\mathbf{f}(1.98, 2.01, 1.03)$.

20. Find the Jacobian matrix $D\mathbf{g}(1, 3, 3)$ for the transformation of $\mathbb{R}^3$ to $\mathbb{R}^3$ given by

$$\mathbf{g}(r, s, t) = (r^2 s, r^2 t, s^2 - t^2)$$

and use the result to find an approximate value for $\mathbf{g}(0.99, 3.02, 2.97)$.

21. Prove that if $f(x, y)$ is differentiable at (a, b), then $f(x, y)$ is continuous at (a, b).

22. Prove the following version of the Mean-Value Theorem: If $f(x, y)$ has first partial derivatives continuous near every point of the straight line segment joining the points (a, b) and $(a + h, b + k)$, then there exists a number θ satisfying $0 < \theta < 1$ such that

$$\begin{aligned} f(a + h, b + k) = &f(a, b) + hf_1(a + \theta h, b + \theta k) \\ &+ kf_2(a + \theta h, b + \theta k). \end{aligned}$$

(*Hint:* Apply the single-variable Mean-Value Theorem to $g(t) = f(a + th, b + tk)$.) Why could we not have used this result in place of Theorem 3 to prove Theorem 4 and hence the version of the Chain Rule given in this section?

23. Generalize Exercise 22 as follows: show that, if $f(x, y)$ has continuous partial derivatives of second order near the point (a, b), then there exists a number θ satisfying $0 < \theta < 1$ such that, for h and k sufficiently small in absolute value,

$$\begin{aligned} f(a + h, b + k) = &f(a, b) + hf_1(a, b) + kf_2(a, b) \\ &+ h^2 f_{11}(a + \theta h, b + \theta k) \\ &+ 2hkf_{12}(a + \theta h, b + \theta k) \\ &+ k^2 f_{22}(a + \theta h, b + \theta k). \end{aligned}$$

Hence, show that there is a constant K such that for all values of h and k that are sufficiently small in absolute value,

$$\begin{aligned} &\left|f(a + h, b + k) - f(a, b) - hf_1(a, b) - kf_2(a, b)\right| \\ &\quad \le K(h^2 + k^2). \end{aligned}$$

Thermodynamics and Legendre Transformations

24. Use the Gibbs equation

$$dE = T\,dS - P\,dV + \mu_1\,dN_1 + \cdots + \mu_n\,dN_n$$

and the fact that, being additive in its extensive variables, $E = E(S, V, N_1, \ldots, N_n)$ is necessarily homogeneous of degree 1, to establish the Gibbs-Duhem equation

$$0 = S\,dT - V\,dP + N_1\,d\mu_1 + \cdots + N_n\,d\mu_n.$$

(*Hint:*use Euler's Theorem, Theorem 2 of Section 12.5.)

25. The equation of state for an ideal gas in the form of $E = E(S, V, N)$, using extensive variables only, is rarely quoted. It is

$$E = \frac{3h^2N}{4\pi m}\left(\frac{N}{V}\right)^{2/3} e^{\left(\frac{2S}{3Nk} - \frac{5}{3}\right)}.$$

However, it is common to see $PV = NkT$, or $E = \frac{3}{2}NkT$ instead. Here k is the Boltzmann constant, h is Planck's constant, and m is the mass of one atom. Deduce these common forms from the explicit formula for E given as a function of S, V, and N.

26. If $f''(x) > 0$ for all x, show that the Legendre transformation $f^*(p)$ is the maximum value of the function $g(x) = px - f(x)$ considered as a function of x alone with p fixed.

In Exercises 27–29 give an explicit formula for the Legendre transformation $f^*(p)$ of the given function $f(x)$.

27. $f(x) = x^2$

28. $f(x) = x^4$

29. $f(x) = \ln(2 + 3x)$

30. Use differentials to show that the Gibbs free energy, $G = E - TS + PV$ depends on T and P alone when the numbers of molecules of each type are fixed. Determine the partial derivatives of G with respect to the new variables T and P.

31. Entropy can be written as a function, $S = S(E, V, N_1, \cdots, N_n)$. Legendre transformations can be performed on it too, although they are not so well-known. The resulting functions are called *Massieu-Planck functions*. Show that one of these, the Massieu's potential, $\Phi = S - \frac{1}{T}E$, depends on temperature instead of energy.

32. In classical mechanics, the energy of a system is expressed in terms of a function called the *Hamiltonian*. When the energy is independent of time, the Hamiltonian depends only on the positions, q_i, and the momenta, p_i of the particles in the system, that is, $H = H(q_1, \cdots, q_n, p_1, \cdots, p_n)$. There is also another function, called the *Lagrangian*, that depends on the positions q_i and the velocities $\dot{q}_i$, that is, $L = L(q_1, \cdots, q_n, \dot{q}_1, \cdots, \dot{q}_n)$, such that the Hamiltonian is a Legendre transformation of the Lagrangian with respect to the velocity variables:

$$H(q_1, \cdots, q_n, p_1, \cdots, p_n) = \sum_i p_i \dot{q}_i - L(q_1, \cdots, q_n, \dot{q}_1, \cdots, \dot{q}_n).$$

(a) What variables are conjugate in this Legendre transformation? What partial derivatives of L are implicitly determined by it?

(b) In the absence of external forces, the principle of least action requires that $\dfrac{\partial L}{\partial q_i} = \dot{p}_i$. By taking the differential of H and using the result of part (a), show that $\dfrac{\partial H}{\partial q_i} = -\dot{p}_i$ and $\dfrac{\partial H}{\partial p_i} = \dot{q}_i$. These are known as Hamilton's equations.

(c) Use Hamilton's equations to show that the Hamiltonian, $\frac{1}{2}(q^2 + p^2)$, represents a harmonic oscillator because it is equivalent to the differential equation $\ddot{q} + q = 0$.

Gradients and Directional Derivatives

A first partial derivative of a function of several variables gives the rate of change of that function with respect to distance measured in the direction of one of the coordinate axes. In this section we will develop a method for finding the rate of change of such a function with respect to distance measured in *any direction* in the domain of the function.

To begin, it is useful to combine the first partial derivatives of a function into a single *vector function* called a **gradient**. For simplicity, we will develop and interpret the gradient for functions of two variables. Extension to functions of three or more variables is straightforward and will be discussed later in this section.

DEFINITION

At any point (x, y) where the first partial derivatives of the function $f(x, y)$ exist, we define the **gradient vector** $\nabla f(x, y) = \mathbf{grad}\, f(x, y)$ by

$$\nabla f(x, y) = \mathbf{grad}\, f(x, y) = f_1(x, y)\mathbf{i} + f_2(x, y)\mathbf{j}.$$

Recall that $\mathbf{i}$ and $\mathbf{j}$ denote the unit basis vectors from the origin to the points $(1, 0)$ and $(0, 1)$, respectively. The symbol ∇, called *del* or *nabla*, is a *vector differential operator*:

$$\nabla = \mathbf{i}\frac{\partial}{\partial x} + \mathbf{j}\frac{\partial}{\partial y}.$$

We can *apply* this operator to a function $f(x, y)$ by writing the operator to the left of the function. The result is the gradient of the function

$$\nabla f(x, y) = \left(\mathbf{i}\frac{\partial}{\partial x} + \mathbf{j}\frac{\partial}{\partial y}\right) f(x, y) = f_1(x, y)\mathbf{i} + f_2(x, y)\mathbf{j}.$$

We will make extensive use of the del operator in Chapter 16.

EXAMPLE 1 If $f(x, y) = x^2 + y^2$, then $\nabla f(x, y) = 2x\mathbf{i} + 2y\mathbf{j}$. In particular, $\nabla f(1, 2) = 2\mathbf{i} + 4\mathbf{j}$. Observe that this vector is perpendicular to the tangent line $x + 2y = 5$ to the circle $x^2 + y^2 = 5$ at $(1, 2)$. This circle is the level curve of f that passes through the point $(1, 2)$. (See Figure 12.25.) As the following theorem shows, this perpendicularity is not a coincidence.

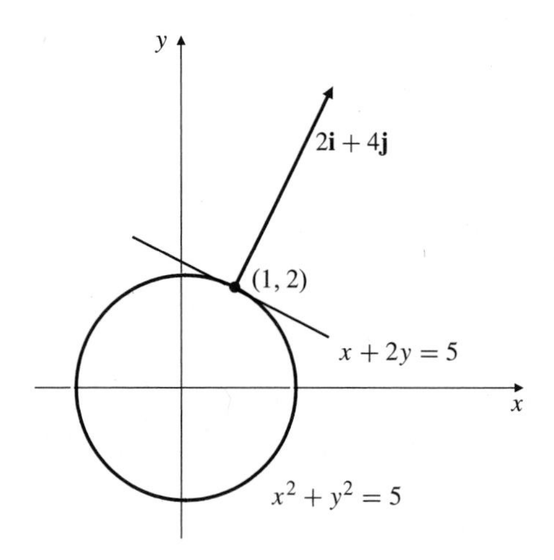

Figure 12.25 The gradient of $f(x, y) = x^2 + y^2$ at $(1, 2)$ is normal to the level curve of f through $(1, 2)$

THEOREM 6

If $f(x, y)$ is differentiable at the point (a, b) and $\nabla f(a, b) \neq \mathbf{0}$, then $\nabla f(a, b)$ is a normal vector to the level curve of f that passes through (a, b).

PROOF Let $\mathbf{r} = \mathbf{r}(t) = x(t)\mathbf{i} + y(t)\mathbf{j}$ be a parametrization of the level curve of f such that $x(0) = a$ and $y(0) = b$. Then for all t near 0, $f\big(x(t), y(t)\big) = f(a, b)$. Differentiating this equation with respect to t using the Chain Rule, we obtain

$$f_1\big(x(t), y(t)\big)\frac{dx}{dt} + f_2\big(x(t), y(t)\big)\frac{dy}{dt} = 0.$$

At $t = 0$ this says that $\nabla f(a, b) \bullet \left.\dfrac{d\mathbf{r}}{dt}\right|_{t=0} = 0$; that is, ∇f is perpendicular to the tangent vector $d\mathbf{r}/dt$ to the level curve at (a, b).

Directional Derivatives

The first partial derivatives $f_1(a, b)$ and $f_2(a, b)$ give the rates of change of $f(x, y)$ at (a, b) measured in the directions of the positive x- and y-axes, respectively. If we want to know how fast $f(x, y)$ changes value as we move through the domain of f at (a, b) in some other direction, we require a more general **directional derivative**. We can specify the direction by means of a nonzero vector. It is most convenient to use a *unit vector*.

DEFINITION 7

Let $\mathbf{u} = u\mathbf{i} + v\mathbf{j}$ be a unit vector, so that $u^2 + v^2 = 1$. The **directional derivative** of $f(x, y)$ at (a, b) in the direction of $\mathbf{u}$ is the rate of change of $f(x, y)$ with respect to distance measured at (a, b) along a ray in the direction of $\mathbf{u}$ in the xy-plane. (See Figure 12.26.) This directional derivative is given by

$$D_{\mathbf{u}} f(a, b) = \lim_{h \to 0+} \frac{f(a + hu, b + hv) - f(a, b)}{h}.$$

It is also given by

$$D_{\mathbf{u}} f(a, b) = \frac{d}{dt} f(a + tu, b + tv)\bigg|_{t=0}$$

if the derivative on the right side exists.

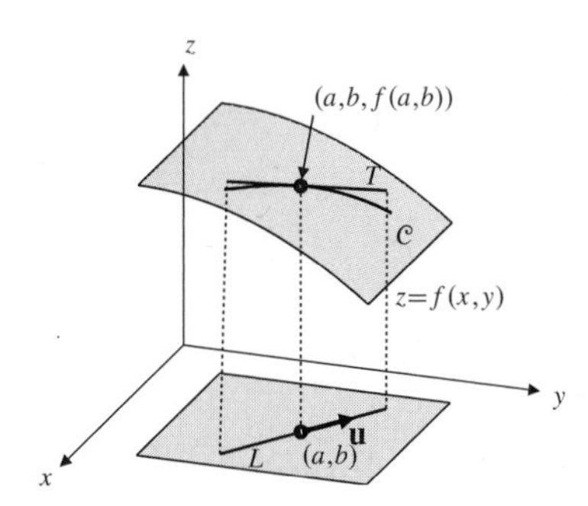

Figure 12.26 Unit vector $\mathbf{u}$ determines a line L through (a, b) in the domain of f. The vertical plane containing L intersects the graph of f in a curve $\mathcal{C}$ whose tangent T at $(a, b, f(a, b))$ has slope $D_{\mathbf{u}} f(a, b)$

Remark This is nothing more than the basic derivative in one variable disguised by the complications arising when $\mathbf{u}$ is not parallel to either coordinate axis. The line L through (a, b) parallel to $\mathbf{u}$ is given by the position vector $\mathbf{r}(t) = a\mathbf{i} + b\mathbf{j} + t\mathbf{u}$. If we regard L as a single coordinate axis with position on it given by the coordinate t, and ignore the rest of the 2-dimensional space, then $f(x(t), y(t)) = g(t)$ along L and

$$D_{\mathbf{u}} f(x(t), y(t)) = \frac{d}{dt} f\big(x(t), y(t)\big) = \frac{dg(t)}{dt},$$

for any t along L. Similarly, if we return to the original axes and choose a direction parallel to either of them, then the directional derivatives become the corresponding first partials: $D_{\mathbf{i}} f(a, b) = f_1(a, b)$, $D_{\mathbf{j}} f(a, b) = f_2(a, b)$, $D_{-\mathbf{i}} f(a, b) = -f_1(a, b)$, and $D_{-\mathbf{j}} f(a, b) = -f_2(a, b)$. The following theorem shows how the gradient can be used to calculate any directional derivative.

THEOREM

Using the gradient to find directional derivatives

If f is differentiable at (a, b) and $\mathbf{u} = u\mathbf{i} + v\mathbf{j}$ is a unit vector, then the directional derivative of f at (a, b) in the direction of $\mathbf{u}$ is given by

$$D_{\mathbf{u}} f(a, b) = \mathbf{u} \bullet \nabla f(a, b).$$

PROOF By the Chain Rule:

$$\begin{aligned} D_{\mathbf{u}} f(a, b) &= \frac{d}{dt} f(a + tu, b + tv)\bigg|_{t=0} \\ &= u f_1(a, b) + v f_2(a, b) = \mathbf{u} \bullet \nabla f(a, b). \end{aligned}$$

■

We already know that having partial derivatives at a point does not imply that a function is continuous there, let alone that it is differentiable. The same can be said about directional derivatives. It is possible for a function to have a directional derivative in every direction at a given point and still not be continuous at that point. See Exercise 37 for an example of such a function.

Given any nonzero vector $\mathbf{v}$, we can always obtain a unit vector in the same direction by dividing $\mathbf{v}$ by its length. The directional derivative of f at (a, b) in the direction of $\mathbf{v}$ is therefore given by

$$D_{\mathbf{v}/|\mathbf{v}|} f(a, b) = \frac{\mathbf{v}}{|\mathbf{v}|} \bullet \nabla f(a, b).$$

Remark When trying to understand why $\mathbf{u}$ must be a unit vector for calculating a directional derivative by the formula in Theorem 7, it helps to think of the directional derivative as a simple derivative with respect to a parameter t along the line L, described by a position vector $\mathbf{r}(t)$ as described in the remark preceeding the statement of the theorem. As in that remark, we have

$$\frac{dg(t)}{dt} = \frac{df(x(t), y(t))}{dt} = f_1(x, y)x'(t) + f_2(x, y)y'(t) = \nabla f \bullet \frac{d\mathbf{r}(t)}{dt}.$$

While this is true for any parameter t, a directional derivative along L is the rate of change with respect to *distance* or *arc length*, $s = t$. Given the formula for arc length in terms of a parameter from Section 8.4, it follows that $\mathbf{r}'(t)$ must be a unit vector:

$$|\mathbf{u}| = \left|\frac{d\mathbf{r}(t)}{dt}\right| = \sqrt{\left(x'(t)\right)^2 + \left(y'(t)\right)^2} = \frac{ds}{dt} = 1.$$

EXAMPLE 2 Find the rate of change of $f(x, y) = y^4 + 2xy^3 + x^2y^2$ at $(0, 1)$ measured in each of the following directions:

(a) $\mathbf{i} + 2\mathbf{j}$, (b) $\mathbf{j} - 2\mathbf{i}$, (c) $3\mathbf{i}$, (d) $\mathbf{i} + \mathbf{j}$.

Solution We calculate

$$\nabla f(x, y) = (2y^3 + 2xy^2)\mathbf{i} + (4y^3 + 6xy^2 + 2x^2y)\mathbf{j},$$
$$\nabla f(0, 1) = 2\mathbf{i} + 4\mathbf{j}.$$

(a) The unit vector in the direction of $\mathbf{i} + 2\mathbf{j}$ is $\dfrac{\mathbf{i} + 2\mathbf{j}}{\sqrt{5}}$. Thus, the directional derivative of f at $(0, 1)$ in that direction is

$$\frac{\mathbf{i} + 2\mathbf{j}}{\sqrt{5}} \bullet (2\mathbf{i} + 4\mathbf{j}) = \frac{2 + 8}{\sqrt{5}} = 2\sqrt{5}.$$

Observe that $\mathbf{i} + 2\mathbf{j}$ points in the same direction as $\nabla f(0, 1)$ so the directional derivative is positive and equal to the length of $\nabla f(0, 1)$.

(b) The unit vector in the direction of $\mathbf{j} - 2\mathbf{i}$ is $\dfrac{\mathbf{j} - 2\mathbf{i}}{\sqrt{5}}$. Thus, the directional derivative of f at $(0, 1)$ in that direction is

$$\frac{-2\mathbf{i} + \mathbf{j}}{\sqrt{5}} \bullet (2\mathbf{i} + 4\mathbf{j}) = \frac{-4 + 4}{\sqrt{5}} = 0.$$

Since $\mathbf{j} - 2\mathbf{i}$ is perpendicular to $\nabla f(0, 1)$, it is tangent to the level curve of f through $(0, 1)$, so the directional derivative in that direction is zero.

(c) The unit vector in the direction of $3\mathbf{i}$ is just $\mathbf{i}$, so the directional derivative of f at $(0, 1)$ in that direction is

$$\mathbf{i} \bullet (2\mathbf{i} + 4\mathbf{j}) = 2.$$

As noted previously, the directional derivative of f in the direction of the positive x-axis is just $f_1(0, 1)$.

(d) The unit vector in the direction of $\mathbf{i} + \mathbf{j}$ is $\dfrac{\mathbf{i} + \mathbf{j}}{\sqrt{2}}$, so the directional derivative of f at $(0, 1)$ in that direction is

$$\frac{\mathbf{i} + \mathbf{j}}{\sqrt{2}} \bullet (2\mathbf{i} + 4\mathbf{j}) = \frac{2 + 4}{\sqrt{2}} = 3\sqrt{2}.$$

If we move along the surface $z = f(x, y)$ through the point $(0, 1, 1)$ in a direction making horizontal angles of 45° with the positive directions of the x- and y-axes, we would be rising at a rate of $3\sqrt{2}$ vertical units per horizontal unit moved.

Remark A direction in the plane can be specified by a polar angle. The direction making angle ϕ with the positive direction of the x-axis corresponds to the unit vector (see Figure 12.27)

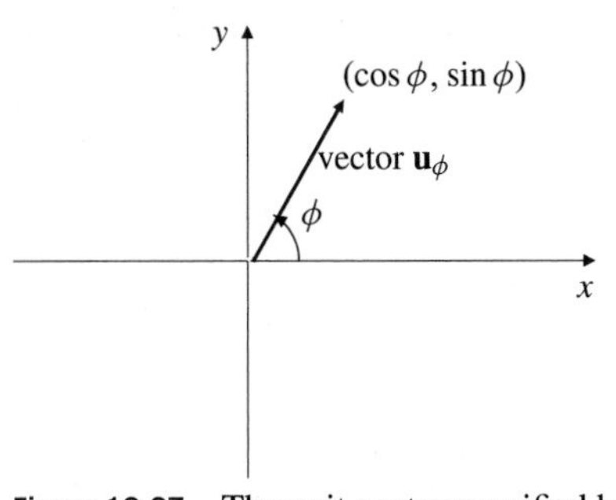

Figure 12.27 The unit vector specified by a polar angle ϕ

$$\mathbf{u}_\phi = \cos\phi\,\mathbf{i} + \sin\phi\,\mathbf{j},$$

so the directional derivative of f at (x, y) in that direction is

$$D_\phi f(x,y) = D_{\mathbf{u}_\phi} f(x,y) = \mathbf{u}_\phi \bullet \nabla f(x,y) = f_1(x,y)\cos\phi + f_2(x,y)\sin\phi.$$

Note the use of the symbol $D_\phi f(x, y)$ to denote a derivative of f with respect to *distance* measured in the direction ϕ.

As observed in the previous example, Theorem 7 provides a useful interpretation for the gradient vector. For any unit vector $\mathbf{u}$ we have

$$D_{\mathbf{u}} f(a,b) = \mathbf{u} \bullet \nabla f(a,b) = |\nabla f(a,b)|\,\cos\theta,$$

where θ is the angle between the vectors $\mathbf{u}$ and $\nabla f(a, b)$. Since $\cos\theta$ only takes on values between -1 and 1, $D_{\mathbf{u}} f(a, b)$ only takes on values between $-|\nabla f(a, b)|$ and $|\nabla f(a, b)|$. Moreover, $D_{\mathbf{u}} f(a, b) = -|\nabla f(a, b)|$ if and only if $\mathbf{u}$ points in the opposite direction to $\nabla f(a, b)$ (so that $\cos\theta = -1$), and $D_{\mathbf{u}} f(a, b) = |\nabla f(a, b)|$ if and only if $\mathbf{u}$ points in the same direction as $\nabla f(a, b)$ (so that $\cos\theta = 1$). The directional derivative is zero in the direction $\theta = \pi/2$; this is the direction of the (tangent line to the) level curve of f through (a, b).

We summarize these properties of the gradient as follows:

Geometric properties of the gradient vector

(i) At (a, b), $f(x, y)$ increases most rapidly in the direction of the gradient vector $\nabla f(a, b)$. The maximum rate of increase is $|\nabla f(a, b)|$.

(ii) At (a, b), $f(x, y)$ decreases most rapidly in the direction of $-\nabla f(a, b)$. The maximum rate of decrease is $|\nabla f(a, b)|$.

(iii) The rate of change of $f(x, y)$ at (a, b) is zero in directions tangent to the level curve of f that passes through (a, b).

Look again at the topographic map in Figure 12.6 in Section 12.1. The streams on the map flow in the direction of steepest descent, that is, in the direction of $-\nabla f$, where f measures the elevation of land. The streams therefore cross the contours (the level curves of f) at right angles. Like the stream, an experienced skier might choose a downhill path close to the direction of the negative gradient, while a novice skier would prefer to stay closer to the level curves.

EXAMPLE 3 The temperature at position (x, y) in a region of the xy-plane is T °C, where

$$T(x,y) = x^2 e^{-y}.$$

In what direction at the point (2, 1) does the temperature increase most rapidly? What is the rate of increase of f in that direction?

Solution We have

$$\nabla T(x,y) = 2x\,e^{-y}\mathbf{i} - x^2 e^{-y}\mathbf{j},$$

$$\nabla T(2,1) = \frac{4}{e}\mathbf{i} - \frac{4}{e}\mathbf{j} = \frac{4}{e}(\mathbf{i} - \mathbf{j}).$$

At (2, 1), $T(x, y)$ increases most rapidly in the direction of the vector $\mathbf{i} - \mathbf{j}$. The rate of increase in this direction is $|\nabla T(2, 1)| = 4\sqrt{2}/e$ °C/unit distance.

EXAMPLE 4 A hiker is standing beside a stream on the side of a mountain, examining her map of the region. The height of land (in metres) at any point (x, y) is given by the function

$$h(x, y) = \frac{20{,}000}{3 + x^2 + 2y^2},$$

where x and y (in kilometres) denote the coordinates of the point on the hiker's map. The hiker is at the point $(3, 2)$.

(a) What is the direction of flow of the stream at $(3, 2)$ on the hiker's map? How fast is the stream descending at her location?

(b) Find the equation of the path of the stream on the hiker's map.

(c) At what angle to the path of the stream (on the map) should the hiker set out if she wishes to climb at a 15° inclination to the horizontal?

(d) Make a sketch of the hiker's map, showing some curves of constant elevation, and showing the stream.

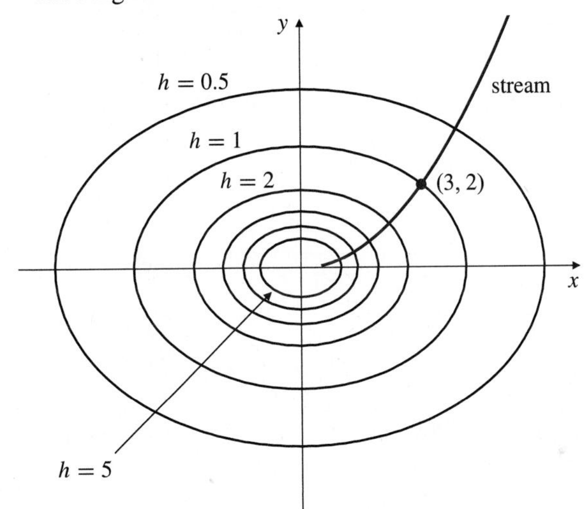

Figure 12.28 The hiker's map. Unlike most mountains, this one has perfectly elliptical contours.

Solution

(a) We begin by calculating the gradient of h and its length at $(3, 2)$:

$$\nabla h(x, y) = -\frac{20{,}000}{(3 + x^2 + 2y^2)^2}(2x\mathbf{i} + 4y\mathbf{j}),$$
$$\nabla h(3, 2) = -100(3\mathbf{i} + 4\mathbf{j}),$$
$$|\nabla h(3, 2)| = 500.$$

The stream is flowing in the direction whose horizontal projection at $(3, 2)$ is $-\nabla h(3, 2)$, that is, in the horizontal direction of the vector $3\mathbf{i} + 4\mathbf{j}$. The stream is descending at a rate of 500 m/km, that is, 0.5 m per horizontal metre travelled.

(b) Coordinates on the map are the coordinates (x, y) in the domain of the height function h. We can find an equation of the path of the stream on a map of the region by setting up a differential equation for a change of position along the path. If the vector $d\mathbf{r} = dx\,\mathbf{i} + dy\,\mathbf{j}$ is tangent to the path of the stream at point (x, y) on the map, then $d\mathbf{r}$ is parallel to $\nabla h(x, y)$. Hence, the components of these two vectors are proportional:

$$\frac{dx}{2x} = \frac{dy}{4y} \qquad \text{or} \qquad \frac{dy}{y} = \frac{2dx}{x}.$$

Integrating both sides of this equation, we get $\ln y = 2\ln x + \ln C$, or $y = Cx^2$. Since the path of the stream passes through $(3, 2)$, we have $C = 2/9$ and the equation is $9y = 2x^2$.

(c) Suppose the hiker moves away from $(3, 2)$ in the direction of the unit vector $\mathbf{u}$. She will be ascending at an inclination of $15°$ if the directional derivative of h in the direction of $\mathbf{u}$ is $1{,}000 \tan 15° \approx 268$. (The 1,000 compensates for the fact that the vertical units are metres while the horizontal units are kilometres.) If θ is the angle between $\mathbf{u}$ and the upstream direction, then

$$500 \cos\theta = |\nabla h(3,2)| \cos\theta = D_{\mathbf{u}} h(3,2) \approx 268.$$

Hence, $\cos\theta \approx 0.536$ and $\theta \approx 57.6°$. She should set out in a direction making a horizontal angle of about $58°$ with the upstream direction.

(d) A suitable sketch of the map is given in Figure 12.28.

EXAMPLE 5 Find the second directional derivative of $f(x, y)$ in the direction making angle ϕ with the positive x-axis.

Solution As observed earlier, the first directional derivative is

$$D_\phi f(x,y) = (\cos\phi\,\mathbf{i} + \sin\phi\,\mathbf{j}) \bullet \nabla f(x,y) = f_1(x,y)\cos\phi + f_2(x,y)\sin\phi.$$

The second directional derivative is therefore

$$\begin{aligned}
D_\phi^2 f(x,y) &= D_\phi\left(D_\phi f(x,y)\right) \\
&= (\cos\phi\,\mathbf{i} + \sin\phi\,\mathbf{j}) \bullet \nabla\Big(f_1(x,y)\cos\phi + f_2(x,y)\sin\phi\Big) \\
&= \Big(f_{11}(x,y)\cos\phi + f_{21}(x,y)\sin\phi\Big)\cos\phi \\
&\qquad + \Big(f_{12}(x,y)\cos\phi + f_{22}(x,y)\sin\phi\Big)\sin\phi \\
&= f_{11}(x,y)\cos^2\phi + 2f_{12}(x,y)\cos\phi\sin\phi + f_{22}(x,y)\sin^2\phi.
\end{aligned}$$

Note that if $\phi = 0$ or $\phi = \pi$ (so the directional derivative is in a direction parallel to the x-axis), then $D_\phi^2 f(x,y) = f_{11}(x,y)$. Similarly, $D_\phi^2 f(x,y) = f_{22}(x,y)$ if $\phi = \pi/2$ or $3\pi/2$.

Rates Perceived by a Moving Observer

Suppose that an observer is moving around in the xy-plane measuring the value of a function $f(x, y)$ defined in the plane as he passes through each point (x, y). (For instance, $f(x, y)$ might be the temperature at (x, y).) If the observer is moving with velocity $\mathbf{v}$ at the instant when he passes through the point (a, b), how fast would he observe $f(x, y)$ to be changing at that moment?

At the moment in question the observer is moving in the direction of the unit vector $\mathbf{v}/|\mathbf{v}|$. The rate of change of $f(x, y)$ at (a, b) in that direction is

$$D_{\mathbf{v}/|\mathbf{v}|} f(a,b) = \frac{\mathbf{v}}{|\mathbf{v}|} \bullet \nabla f(a,b)$$

measured in units of f per unit distance in the xy-plane. To convert this rate to units of f per unit time, we must multiply by the speed of the observer, $|\mathbf{v}|$ units of distance per unit time. Thus, the time rate of change of $f(x, y)$ as measured by the observer passing through (a, b) is

$$|\mathbf{v}|\,\frac{\mathbf{v}}{|\mathbf{v}|} \bullet \nabla f(a,b) = \mathbf{v} \bullet \nabla f(a,b).$$

It is natural to extend our use of the symbol $D_{\mathbf{v}} f(a, b)$ to represent this rate even though $\mathbf{v}$ is not (necessarily) a unit vector. Thus, we have established the following principle.

The rate of change of $f(x, y)$ at (a, b) as measured by an observer moving through (a, b) with velocity $\mathbf{v}$ is

$$D_{\mathbf{v}} f(a, b) = \mathbf{v} \bullet \nabla f(a, b)$$

units of f per unit time.

If the hiker in Example 4 moves away from $(3, 2)$ with horizontal velocity $\mathbf{v} = -\mathbf{i} - \mathbf{j}$ km/h, then she will be rising at a rate of

$$\mathbf{v} \bullet \nabla h(3, 2) = (-\mathbf{i} - \mathbf{j}) \bullet \left(-\frac{1}{10}(3\mathbf{i} + 4\mathbf{j})\right) = \frac{7}{10} \text{ km/h.}$$

As defined here, $D_{\mathbf{v}} f$ is the spatial component of the derivative of f following the motion. See Example 6 in Section 12.5. The rate of change of the reading on the moving thermometer in that example can be expressed as

$$\frac{dT}{dt} = D_{\mathbf{v}} T(x, y, z, t) + \frac{\partial T}{\partial t},$$

where $\mathbf{v}$ is the velocity of the moving thermometer and $D_{\mathbf{v}} T = \mathbf{v} \bullet \nabla T$. The gradient is being taken with respect to the *three spatial variables* only. (See below for the gradient in 3-space.)

The Gradient in Three and More Dimensions

By analogy with the two-dimensional case, a function $f(x_1, x_2, \ldots, x_n)$ of n variables possessing first partial derivatives has gradient given by

$$\nabla f(x_1, x_2, \ldots, x_n) = \frac{\partial f}{\partial x_1}\mathbf{e}_1 + \frac{\partial f}{\partial x_2}\mathbf{e}_2 + \cdots + \frac{\partial f}{\partial x_n}\mathbf{e}_n,$$

where $\mathbf{e}_j$ is the unit vector from the origin to the unit point on the jth coordinate axis. In particular, for a function of three variables,

$$\nabla f(x, y, z) = \frac{\partial f}{\partial x}\mathbf{i} + \frac{\partial f}{\partial y}\mathbf{j} + \frac{\partial f}{\partial z}\mathbf{k}.$$

The level surface of $f(x, y, z)$ passing through (a, b, c) has a tangent plane there if f is differentiable at (a, b, c) and $\nabla f(a, b, c) \neq \mathbf{0}$.

For functions of any number of variables, the vector $\nabla f(P_0)$ is normal to the "level surface" of f passing through the point P_0 (i.e., the (hyper)surface with equation $f(x_1, \ldots, x_n) = f(P_0)$), and, if f is differentiable at P_0, the rate of change of f at P_0 in the direction of the unit vector $\mathbf{u}$ is given by $\mathbf{u} \bullet \nabla f(P_0)$. Equations of tangent planes to surfaces in 3-space can be found easily with the aid of gradients.

EXAMPLE 6 Let $f(x, y, z) = x^2 + y^2 + z^2$.

(a) Find $\nabla f(x, y, z)$ and $\nabla f(1, -1, 2)$.

(b) Find an equation of the tangent plane to the sphere $x^2 + y^2 + z^2 = 6$ at the point $(1, -1, 2)$.

(c) What is the maximum rate of increase of f at $(1, -1, 2)$?

(d) What is the rate of change with respect to distance of f at $(1, -1, 2)$ measured in the direction from that point toward the point $(3, 1, 1)$?

Solution

(a) $\nabla f(x, y, z) = 2x\mathbf{i} + 2y\mathbf{j} + 2z\mathbf{k}$, so $\nabla f(1, -1, 2) = 2\mathbf{i} - 2\mathbf{j} + 4\mathbf{k}$.

(b) The required tangent plane has $\nabla f(1, -1, 2)$ as normal. (See Figure 12.29(a).) Therefore, its equation is given by $2(x - 1) - 2(y + 1) + 4(z - 2) = 0$ or, more simply, $x - y + 2z = 6$.

(c) The maximum rate of increase of f at $(1, -1, 2)$ is $|\nabla f(1, -1, 2)| = 2\sqrt{6}$, and it occurs in the direction of the vector $\mathbf{i} - \mathbf{j} + 2\mathbf{k}$.

(d) The direction from $(1, -1, 2)$ toward $(3, 1, 1)$ is specified by $2\mathbf{i} + 2\mathbf{j} - \mathbf{k}$. The rate of change of f with respect to distance in this direction is

$$\frac{2\mathbf{i} + 2\mathbf{j} - \mathbf{k}}{\sqrt{4 + 4 + 1}} \bullet (2\mathbf{i} - 2\mathbf{j} + 4\mathbf{k}) = \frac{4 - 4 - 4}{3} = -\frac{4}{3};$$

that is, f decreases at rate $4/3$ of a unit per horizontal unit moved.

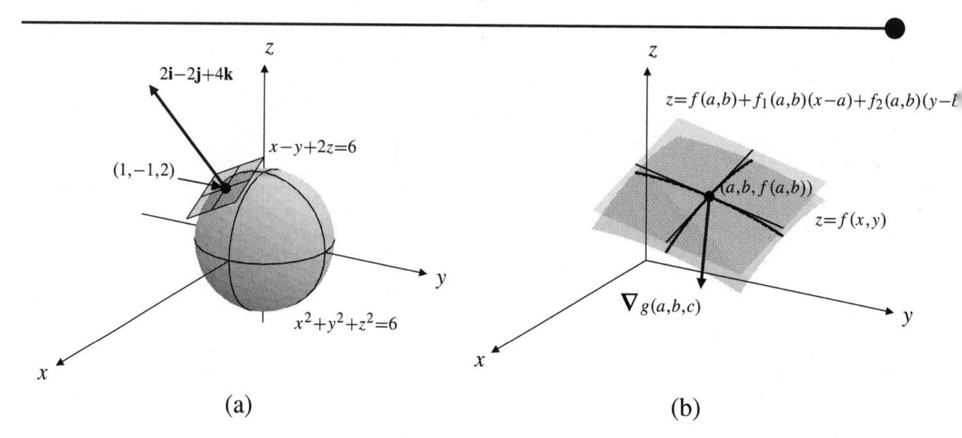

Figure 12.29

(a) The level surface $f(x, y, z) = 6$ for Example 6 and its tangent plane at $(1, -1, 2)$

(b) The gradient of $f(x, y) - z$ at $(a, b, f(a, b))$ is normal to the tangent plane to $z = f(x, y)$ at that point. See Example 7

EXAMPLE 7 The graph of a *function* $f(x, y)$ of two variables is the graph of the *equation* $z = f(x, y)$ in 3-space. This surface is also the level surface $g(x, y, z) = 0$ of the 3-variable function

$$g(x, y, z) = f(x, y) - z.$$

If f is differentiable at (a, b) and $c = f(a, b)$, then g is differentiable at (a, b, c), and

$$\nabla g(a, b, c) = f_1(a, b)\mathbf{i} + f_2(a, b)\mathbf{j} - \mathbf{k}$$

is normal to $g(x, y, z) = 0$ at (a, b, c). (Note that $\nabla g(a, b, c) \neq \mathbf{0}$, since its z component is -1.) It follows that the graph of f has nonvertical tangent plane at (a, b) given by

$$f_1(a, b)(x - a) + f_2(a, b)(y - b) - (z - c) = 0,$$

or

$$z = f(a, b) + f_1(a, b)(x - a) + f_2(a, b)(y - b).$$

(See Figure 12.29(b).) This result was obtained by a different argument in Section 12.3.

BEWARE! Make sure you understand the difference between the graph of a function and a level curve or level surface of that function. (See the discussion following this example.) Here, the surface $z = f(x, y)$ is the *graph* of the function f, but it is also a *level surface* of a *different* function g.

Students sometimes confuse graphs of functions with level curves or surfaces of those functions. In the above example, we are talking about a *level surface* of the function $g(x, y, z)$ that happens to coincide with the *graph* of a different function, $f(x, y)$. Do not confuse that surface with the graph of g, which is a three-dimensional *hypersurface* in 4-space having equation $w = g(x, y, z)$. Similarly, do not confuse the tangent *plane* to the graph of $f(x, y)$ (i.e., the plane obtained in the above example) with the tangent *line* to the level curve of $f(x, y)$ passing through (a, b) and lying in the xy-plane. This line has an equation involving only x and y: $f_1(a, b)(x - a) + f_2(a, b)(y - b) = 0$.

EXAMPLE 8 Find a vector tangent to the curve of intersection of the two surfaces $z = x^2 - y^2$ and $xyz + 30 = 0$ at the point $(-3, 2, 5)$.

Solution The coordinates of the given point satisfy the equations of both surfaces so the point lies on the curve of intersection of the two surfaces. A vector tangent to this curve at that point will be perpendicular to the normals to both surfaces, that is, to the vectors

$$\mathbf{n}_1 = \nabla(x^2 - y^2 - z)\Big|_{(-3,2,5)} = 2x\mathbf{i} - 2y\mathbf{j} - \mathbf{k}\Big|_{(-3,2,5)} = -6\mathbf{i} - 4\mathbf{j} - \mathbf{k},$$

$$\mathbf{n}_2 = \nabla(xyz + 30)\Big|_{(-3,2,5)} = (yz\mathbf{i} + xz\mathbf{j} + xy\mathbf{k})\Big|_{(-3,2,5)} = 10\mathbf{i} - 15\mathbf{j} - 6\mathbf{k}.$$

For the tangent vector $\mathbf{T}$ we can therefore use the cross product of these normals:

$$\mathbf{T} = \mathbf{n}_1 \times \mathbf{n}_2 = \begin{vmatrix} \mathbf{i} & \mathbf{j} & \mathbf{k} \\ -6 & -4 & -1 \\ 10 & -15 & -6 \end{vmatrix} = 9\mathbf{i} - 46\mathbf{j} + 130\mathbf{k}.$$

Remark Maple's **VectorCalculus** package defines a function **Gradient** that takes a pair of arguments—an expression and a list of variables—and produces the gradient of the expression with respect to those variables:

```
> with(VectorCalculus):
> f := x^2+y^3+z^4; G := Gradient(f, [x,y,z]);
```

$$f := x^2 + y^3 + z^4$$

$$G := 2\,x\,\bar{e}_x + 3\,y^2\,\bar{e}_y + 4\,z^3\,\bar{e}_z$$

Although the result for `G` looks like a vector, it is actually something different, namely a **vector field** which is a vector-valued function of a vector variable. This fact is conveyed by the bars that appear over the basis vectors in the output. We will deal extensively with vector fields in Chapters 15 and 16 and will say little about them here except to note that evaluating the Gradient at a particular point requires the `evalVF` function, which takes two arguments: a vector field and a vector at which to evaluate it.

```
> evalVF(G,<2,3,-1>);
```

$$4\,e_x + 27\,e_y - 4\,e_z$$

Observe that the output is a vector, not a vector field; there are no bars on the basis vectors.

If you want to define a gradient function (let us call it `grad`) such that you would get the above value by using the input `grad(f)(2,3,-1)`, you could use

```
> grad := g -> ((u,v,w) ->
>  evalVF(Gradient(g,[x,y,z]),<u,v,w>));
```

EXERCISES 12.7

In Exercises 1–6, find:

(a) the gradient of the given function at the point indicated,

(b) an equation of the plane tangent to the graph of the given function at the point whose x and y coordinates are given, and

(c) an equation of the straight line tangent, at the given point, to the level curve of the given function passing through that point.

1. $f(x, y) = x^2 - y^2$ at $(2, -1)$

2. $f(x, y) = \dfrac{x - y}{x + y}$ at $(1, 1)$

3. $f(x, y) = \dfrac{x}{x^2 + y^2}$ at $(1, 2)$

4. $f(x, y) = e^{xy}$ at $(2, 0)$

5. $f(x, y) = \ln(x^2 + y^2)$ at $(1, -2)$

6. $f(x, y) = \sqrt{1 + xy^2}$ at $(2, -2)$

In Exercises 7–9, find an equation of the tangent plane to the level surface of the given function that passes through the given point.

7. $f(x, y, z) = x^2y + y^2z + z^2x$ at $(1, -1, 1)$

8. $f(x, y, z) = \cos(x + 2y + 3z)$ at $\left(\dfrac{\pi}{2}, \pi, \pi\right)$

9. $f(x, y, z) = y\,e^{-x^2} \sin z$ at $(0, 1, \pi/3)$

In Exercises 10–13, find the rate of change of the given function at the given point in the specified direction.

10. $f(x, y) = 3x - 4y$ at $(0, 2)$ in the direction of the vector $-2\mathbf{i}$

11. $f(x, y) = x^2y$ at $(-1, -1)$ in the direction of the vector $\mathbf{i} + 2\mathbf{j}$

12. $f(x, y) = \dfrac{x}{1 + y}$ at $(0, 0)$ in the direction of the vector $\mathbf{i} - \mathbf{j}$

13. $f(x, y) = x^2 + y^2$ at $(1, -2)$ in the direction making a (positive) angle of $60°$ with the positive x-axis

14. Let $f(x, y) = \ln|\mathbf{r}|$, where $\mathbf{r} = x\mathbf{i} + y\mathbf{j}$. Show that $\nabla f = \dfrac{\mathbf{r}}{|\mathbf{r}|^2}$.

15. Let $f(x, y, z) = |\mathbf{r}|^{-n}$, where $\mathbf{r} = x\mathbf{i} + y\mathbf{j} + z\mathbf{k}$. Show that $\nabla f = \dfrac{-n\mathbf{r}}{|\mathbf{r}|^{n+2}}$.

❷16. Show that, in terms of polar coordinates (r, θ) (where $x = r\cos\theta$ and $y = r\sin\theta$), the gradient of a function $f(r, \theta)$ is given by

$$\nabla f = \frac{\partial f}{\partial r}\hat{\mathbf{r}} + \frac{1}{r}\frac{\partial f}{\partial \theta}\hat{\boldsymbol{\theta}},$$

where $\hat{\mathbf{r}}$ is a unit vector in the direction of the position vector $\mathbf{r} = x\,\mathbf{i} + y\,\mathbf{j}$, and $\hat{\boldsymbol{\theta}}$ is a unit vector at right angles to $\hat{\mathbf{r}}$ in the direction of increasing θ.

17. In what directions at the point $(2, 0)$ does the function $f(x, y) = xy$ have rate of change -1? Are there directions in which the rate is -3? How about -2?

18. In what directions at the point (a, b, c) does the function $f(x, y, z) = x^2 + y^2 - z^2$ increase at half of its maximal rate at that point?

19. Find $\nabla f(a, b)$ for the differentiable function $f(x, y)$ given the directional derivatives

$$D_{(\mathbf{i}+\mathbf{j})/\sqrt{2}}f(a, b) = 3\sqrt{2} \text{ and } D_{(3\mathbf{i}-4\mathbf{j})/5}f(a, b) = 5.$$

20. If $f(x, y)$ is differentiable at (a, b), what condition should angles ϕ_1 and ϕ_2 satisfy in order that the gradient $\nabla f(a, b)$ can be determined from the values of the directional derivatives $D_{\phi_1}f(a, b)$ and $D_{\phi_2}f(a, b)$?

21. The temperature $T(x, y)$ at points of the xy-plane is given by $T(x, y) = x^2 - 2y^2$.

(a) Draw a contour diagram for T showing some isotherms (curves of constant temperature).

(b) In what direction should an ant at position $(2, -1)$ move if it wishes to cool off as quickly as possible?

(c) If the ant moves in that direction at speed k (units distance per unit time), at what rate does it experience the decrease of temperature?

(d) At what rate would the ant experience the decrease of temperature if it moved from $(2, -1)$ at speed k in the direction of the vector $-\mathbf{i} - 2\mathbf{j}$?

(e) Along what curve through $(2, -1)$ should the ant move in order to continue to experience maximum rate of cooling?

22. Find an equation of the curve in the xy-plane that passes through the point $(1, 1)$ and intersects all level curves of the function $f(x, y) = x^4 + y^2$ at right angles.

23. Find an equation of the curve in the xy-plane that passes through the point $(2, -1)$ and that intersects every curve with equation of the form $x^2y^3 = K$ at right angles.

24. Find the second directional derivative of $e^{-x^2-y^2}$ at the point $(a, b) \neq (0, 0)$ in the direction directly away from the origin.

25. Find the second directional derivative of $f(x, y, z) = xyz$ at $(2, 3, 1)$ in the direction of the vector $\mathbf{i} - \mathbf{j} - \mathbf{k}$.

26. Find a vector tangent to the curve of intersection of the two cylinders $x^2 + y^2 = 2$ and $y^2 + z^2 = 2$ at the point $(1, -1, 1)$.

27. Repeat Exercise 26 for the surfaces $x + y + z = 6$ and $x^2 + y^2 + z^2 = 14$ and the point $(1, 2, 3)$.

28. The temperature in 3-space is given by

$$T(x, y, z) = x^2 - y^2 + z^2 + xz^2.$$

At time $t = 0$ a fly passes through the point $(1, 1, 2)$, flying along the curve of intersection of the surfaces $z = 3x^2 - y^2$ and $2x^2 + 2y^2 - z^2 = 0$. If the fly's speed is 7, what rate of temperature change does it experience at $t = 0$?

❷29. State and prove a version of Theorem 6 for a function of three variables.

30. What is the level surface of $f(x, y, z) = \cos(x + 2y + 3z)$ that passes through (π, π, π)? What is the tangent plane to that level surface at that point? (Compare this exercise with Exercise 8 above.)

❷31. If $\nabla f(x, y) = 0$ throughout the disk $x^2 + y^2 < r^2$, prove that $f(x, y)$ is constant throughout the disk.

❷32. Theorem 6 implies that the level curve of $f(x, y)$ passing through (a, b) is smooth (has a tangent line) at (a, b) provided f is differentiable at (a, b) and satisfies $\nabla f(a, b) \neq \mathbf{0}$. Show that the level curve need not be smooth at (a, b) if $\nabla f(a, b) = \mathbf{0}$. (*Hint:* Consider $f(x, y) = y^3 - x^2$ at $(0, 0)$.)

❷33. If $\mathbf{v}$ is a nonzero vector, express $D_{\mathbf{v}}(D_{\mathbf{v}}f)$ in terms of the components of $\mathbf{v}$ and the second partials of f. What is the interpretation of this quantity for a moving observer?

■34. An observer moves so that his position, velocity, and acceleration at time t are given by the formulas $\mathbf{r}(t) = x(t)\,\mathbf{i} + y(t)\,\mathbf{j} + z(t)\,\mathbf{k}$, $\mathbf{v}(t) = d\mathbf{r}/dt$, and $\mathbf{a}(t) = d\mathbf{v}/dt$. If the temperature in the vicinity of the observer depends only on position, $T = T(x, y, z)$, express the second time derivative of temperature as measured by the observer in terms of $D_{\mathbf{v}}$ and $D_{\mathbf{a}}$.

35. Repeat Exercise 34 but with T depending explicitly on time as well as position: $T = T(x, y, z, t)$.

36. Let $f(x, y) = \begin{cases} \dfrac{\sin(xy)}{\sqrt{x^2 + y^2}}, & \text{if } (x, y) \neq (0, 0) \\ 0, & \text{if } (x, y) = (0, 0). \end{cases}$

(a) Calculate $\nabla f(0, 0)$.

(b) Use the definition of directional derivative to calculate $D_{\mathbf{u}} f(0, 0)$, where $\mathbf{u} = (\mathbf{i} + \mathbf{j})/\sqrt{2}$.

(c) Is $f(x, y)$ differentiable at $(0, 0)$? Why?

37. Let $f(x, y) = \begin{cases} 2x^2 y/(x^4 + y^2), & \text{if } (x, y) \neq (0, 0) \\ 0, & \text{if } (x, y) = (0, 0). \end{cases}$

Use the definition of directional derivative as a limit (Definition 7) to show that $D_{\mathbf{u}} f(0, 0)$ exists for every unit vector $\mathbf{u} = u\mathbf{i} + v\mathbf{j}$ in the plane. Specifically, show that $D_{\mathbf{u}} f(0, 0) = 0$ if $v = 0$, and $D_{\mathbf{u}} f(0, 0) = 2u^2/v$ if $v \neq 0$. However, as was shown in Example 4 in Section 12.2, $f(x, y)$ has no limit as $(x, y) \to (0, 0)$, so it is not continuous there. Even if a function has directional derivatives in all directions at a point, it may not be continuous at that point.

12.8 Implicit Functions

When we study the calculus of functions of one variable, we encounter examples of functions that are defined implicitly as solutions of equations in two variables. Suppose, for example, that $F(x, y) = 0$ is such an equation. Suppose that the point (a, b) satisfies the equation and that F has continuous first partial derivatives (and so is differentiable) at all points near (a, b). Can the equation be solved for y as a function of x near (a, b)? That is, does there exist a function $y(x)$ defined in some interval $I = (a - h, a + h)$ (where $h > 0$) satisfying $y(a) = b$ and such that

$$F\big(x, y(x)\big) = 0$$

holds for all x in the interval I? If there is such a function $y(x)$, we can try to find its derivative at $x = a$ by differentiating the equation $F(x, y) = 0$ implicitly with respect to x, and evaluating the result at (a, b):

$$F_1(x, y) + F_2(x, y)\,\frac{dy}{dx} = 0,$$

so that

$$\left.\frac{dy}{dx}\right|_{x=a} = -\frac{F_1(a, b)}{F_2(a, b)}, \qquad \text{provided} \qquad F_2(a, b) \neq 0.$$

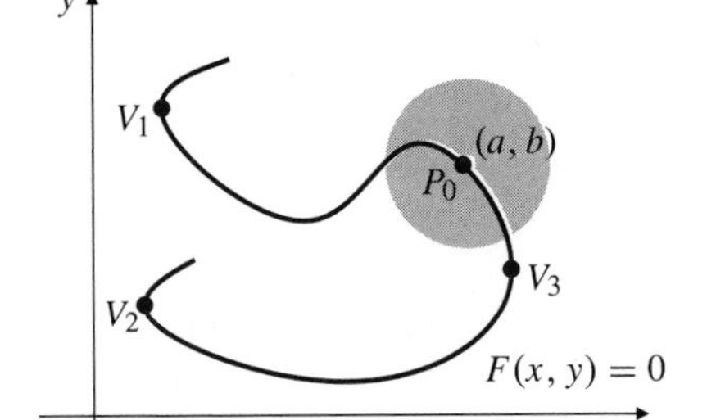

Figure 12.30 The equation $F(x, y) = 0$ can be solved for y as a function of x near P_0 or near any other point except the three points where the curve has a vertical tangent

Observe, however, that the condition $F_2(a, b) \neq 0$ required for the calculation of $y'(a)$ will itself guarantee that the solution $y(x)$ exists. This condition, together with the differentiability of $F(x, y)$ near (a, b), implies that the level curve $F(x, y) = F(a, b)$ has *nonvertical* tangent lines near (a, b), so some part of the level curve near (a, b) must be the graph of a function of x. (See Figure 12.30; the part of the curve $F(x, y) = 0$ in the shaded disk centred at $P_0 = (a, b)$ is the graph of a function $y(x)$ because vertical lines meet that part of the curve only once. The only points on the curve where a disk with that property cannot be drawn are the three points V_1, V_2, and V_3, where the curve has a vertical tangent, that is, where $F_2(x, y) = 0$.) This is a special case of the Implicit Function Theorem, which we will state more generally later in this section.

A similar situation holds for equations involving several variables. We can, for example, ask whether the equation

$$F(x, y, z) = 0$$

defines z as a function of x and y (say, $z = z(x, y)$) near some point P_0 with coordinates (x_0, y_0, z_0) satisfying the equation. If so, and if F has continuous first partials near P_0, then the partial derivatives of z can be found at (x_0, y_0) by implicit differentiation of the equation $F(x, y, z) = 0$ with respect to x and y:

$$F_1(x, y, z) + F_3(x, y, z)\,\frac{\partial z}{\partial x} = 0 \quad \text{and} \quad F_2(x, y, z) + F_3(x, y, z)\,\frac{\partial z}{\partial y} = 0,$$

so that

$$\left.\frac{\partial z}{\partial x}\right|_{(x_0,y_0)} = -\frac{F_1(x_0, y_0, z_0)}{F_3(x_0, y_0, z_0)} \quad\text{and}\quad \left.\frac{\partial z}{\partial y}\right|_{(x_0,y_0)} = -\frac{F_2(x_0, y_0, z_0)}{F_3(x_0, y_0, z_0)},$$

provided $F_3(x_0, y_0, z_0) \neq 0$. Since F_3 is the z component of the gradient of F, this condition implies that the level surface of F through P_0 does not have a horizontal normal vector, so it is not vertical (i.e., it is not parallel to the z-axis). Therefore, part of the surface near P_0 must indeed be the graph of a function $z = z(x, y)$. Similarly, $F(x, y, z) = 0$ can be solved for x as a function of y and z near points where $F_1 \neq 0$ and for $y = y(x, z)$ near points where $F_2 \neq 0$.

EXAMPLE 1 Near what points on the sphere $x^2 + y^2 + z^2 = 1$ can the equation of the sphere be solved for z as a function of x and y? Find $\partial z/\partial x$ and $\partial z/\partial y$ at such points.

Solution The sphere is the level surface $F(x, y, z) = 0$ of the function

$$F(x, y, z) = x^2 + y^2 + z^2 - 1.$$

The above equation can be solved for $z = z(x, y)$ near $P_0 = (x_0, y_0, z_0)$, provided that P_0 is not on the *equator* of the sphere, that is, the circle $x^2 + y^2 = 1$, $z = 0$. The equator consists of those points that satisfy $F_3(x, y, z) = 0$. If P_0 is not on the equator, then it is on either the upper or the lower hemisphere. The upper hemisphere has equation $z = z(x, y) = \sqrt{1 - x^2 - y^2}$, and the lower hemisphere has equation $z = z(x, y) = -\sqrt{1 - x^2 - y^2}$.

If $z \neq 0$, we can calculate the partial derivatives of the solution $z = z(x, y)$ by implicitly differentiating the equation of the sphere: $x^2 + y^2 + z^2 = 1$:

$$2x + 2z\frac{\partial z}{\partial x} = 0, \quad\text{so}\quad \frac{\partial z}{\partial x} = -\frac{x}{z},$$

$$2y + 2z\frac{\partial z}{\partial y} = 0, \quad\text{so}\quad \frac{\partial z}{\partial y} = -\frac{y}{z}.$$

Systems of Equations

Experience with linear equations shows us that systems of such equations can generally be solved for as many variables as there are equations in the system. We would expect, therefore, that a pair of equations in several variables might determine two of those variables as functions of the remaining ones. For instance, we might expect the two equations

$$\begin{cases} F(x, y, z, w) = 0 \\ G(x, y, z, w) = 0 \end{cases}$$

to possess, near some point that satisfies them, solutions of one or more of the forms

$$\begin{cases} x = x(z, w) \\ y = y(z, w), \end{cases} \quad \begin{cases} x = x(y, w) \\ z = z(y, w), \end{cases} \quad \begin{cases} x = x(y, z) \\ w = w(y, z), \end{cases}$$

$$\begin{cases} y = y(x, w) \\ z = z(x, w), \end{cases} \quad \begin{cases} y = y(x, z) \\ w = w(x, z), \end{cases} \quad \begin{cases} z = z(x, y) \\ w = w(x, y). \end{cases}$$

Where such solutions exist, we should be able to differentiate the given system of equations implicitly to find partial derivatives of the solutions.

If you are given a single equation $F(x, y, z) = 0$ and asked to find $\partial x/\partial z$, you would understand that x is intended to be a function of the remaining variables y and z, so there would be no chance of misinterpreting which variable is to be held constant in calculating the partial derivative. Suppose, however, that you are asked to calculate $\partial x/\partial z$ given the system $F(x, y, z, w) = 0$, $G(x, y, z, w) = 0$. The question implies that x is one of the dependent variables and z is one of the independent variables, but does not imply which of y and w is the other dependent variable and which is the other independent variable. In short, which of the situations

$$\begin{cases} x = x(z, w) \\ y = y(z, w) \end{cases} \quad \text{and} \quad \begin{cases} x = x(y, z) \\ w = w(y, z) \end{cases}$$

are we dealing with? As it stands, the question is ambiguous. To avoid this ambiguity, we can specify *in the notation for the partial derivative* which variable is to be regarded as the other independent variable and therefore *held fixed* during the differentiation. Thus,

$$\left(\frac{\partial x}{\partial z}\right)_w \quad \text{implies the interpretation} \quad \begin{cases} x = x(z, w) \\ y = y(z, w), \end{cases}$$

$$\left(\frac{\partial x}{\partial z}\right)_y \quad \text{implies the interpretation} \quad \begin{cases} x = x(y, z) \\ w = w(y, z). \end{cases}$$

EXAMPLE 2 Given the equations $F(x, y, z, w) = 0$ and $G(x, y, z, w) = 0$, where F and G have continuous first partial derivatives, calculate $(\partial x/\partial z)_w$.

Solution We differentiate the two equations with respect to z, regarding x and y as functions of z and w, and holding w fixed:

$$F_1\frac{\partial x}{\partial z} + F_2\frac{\partial y}{\partial z} + F_3 = 0$$
$$G_1\frac{\partial x}{\partial z} + G_2\frac{\partial y}{\partial z} + G_3 = 0$$

(Note that the terms $F_4(\partial w/\partial z)$ and $G_4(\partial w/\partial z)$ are not present because w and z are independent variables, and w is being held fixed during the differentiation.) The pair of equations above is linear in $\partial x/\partial z$ and $\partial y/\partial z$. Eliminating $\partial y/\partial z$ (or using Cramer's Rule, Theorem 6 of Section 10.7), we obtain

$$\left(\frac{\partial x}{\partial z}\right)_w = -\frac{F_3G_2 - F_2G_3}{F_1G_2 - F_2G_1}.$$

In the light of the examples considered above, you should not be too surprised to learn that the nonvanishing of the denominator $F_1G_2 - F_2G_1$ at some point $P_0 = (x_0, y_0, z_0, w_0)$ satisfying the system $F = 0$, $G = 0$ is sufficient to guarantee that the system does indeed have a solution of the form $x = x(z, w)$, $y = y(z, w)$ near P_0. We will not, however, attempt to prove this fact here.

EXAMPLE 3 Let x, y, u, and v be related by the equations

$$\begin{cases} u = x^2 + xy - y^2 \\ v = 2xy + y^2. \end{cases}$$

Find (a) $(\partial x/\partial u)_v$ and (b) $(\partial x/\partial u)_y$ at the point where $x = 2$ and $y = -1$.

Solution

(a) To calculate $(\partial x/\partial u)_v$ we regard x and y as functions of u and v and differentiate the given equations with respect to u, holding v constant:

$$1 = \frac{\partial u}{\partial u} = (2x+y)\frac{\partial x}{\partial u} + (x-2y)\frac{\partial y}{\partial u}$$
$$0 = \frac{\partial v}{\partial u} = 2y\frac{\partial x}{\partial u} + (2x+2y)\frac{\partial y}{\partial u}$$

At $x = 2$, $y = -1$ we have

$$1 = 3\frac{\partial x}{\partial u} + 4\frac{\partial y}{\partial u}$$
$$0 = -2\frac{\partial x}{\partial u} + 2\frac{\partial y}{\partial u}.$$

Eliminating $\partial y/\partial u$ leads to the result $(\partial x/\partial u)_v = 1/7$.

(b) To calculate $(\partial x/\partial u)_y$ we regard x and v as functions of y and u and differentiate the given equations with respect to u, holding y constant:

$$1 = \frac{\partial u}{\partial u} = (2x+y)\frac{\partial x}{\partial u}, \qquad \frac{\partial v}{\partial u} = 2y\frac{\partial x}{\partial u}.$$

At $x = 2$, $y = -1$ the first equation immediately gives $(\partial x/\partial u)_y = 1/3$.

Choosing Dependent and Independent Variables

Some applications involve several variables that must satisfy a smaller number of equations. The question naturally arises concerning which variables should be considered independent. Typically, if there are n equations to be satisfied by $n+m$ variables, we can choose any n of the variables to be considered as functions of the remaining m independent variables; in theory, at least, it is possible to solve the n equations for any n of the variables. However, applications often come with conventions that prefer one set of variables over others. For example, in Section 12.6, we introduced the extensive variables in thermodynamics, which are referred to as *proper variables* by some authors. In mechanics there are also preferred variables which are called *canonical.*

However, as alternative selections of independent and dependent variables are mathematically sound, and often useful, they cannot be excluded by such conventions. As discussed in Section 12.6, a single component gas involves seven variables, energy E, entropy S, volume V, temperature T, pressure P, number N of molecules, and chemical potential, μ. Among these seven variables there hold four equations. In the usual formulation E is a function of three independent variables S, V, and N, while P, T, and μ are partial derivatives of that function and so functions of the same three independent variables:

$$E = f(S,V,N), \quad T = \frac{\partial E}{\partial S} = f_1(S,V,N),$$
$$P = -\frac{\partial E}{\partial V} = -f_2(S,V,N), \quad \mu = \frac{\partial E}{\partial N} = f_3(S,V,N).$$

But it is just as reasonable (and sometimes preferable) to consider the independent variables to be T, V, and N with the other four variables being dependent on these three. In particular, we would have $E = g(T,V,N)$ and similar representations would hold for S, P, and μ. Confusion arises because f and g are sometimes casually written as if they are the same function (e.g. $E = E(S,V,N)$ or $E = E(T,V,N)$). They normally are not the same function. As shown in Exercise 25 of Section 12.6, the energy of an ideal gas can be written both as

$$E = f(S,V,N) = \frac{3h^2N}{4\pi m}\left(\frac{N}{V}\right)^{2/3} e^{\left(\frac{2S}{3Nk}-\frac{5}{3}\right)}. \qquad (*)$$

and

$$E = g(T, V, N) = \frac{3}{2}NkT.$$

Clearly, f and g are different functions, in different variables, that nevertheless produce the same energy. Observe, also, that while g is allowed to depend on V it is actually independent of V, while f is not. These functions may produce the same value, but the differences in their partial derivatives cannot be overlooked. We have

$$\left(\frac{\partial E}{\partial V}\right)_{S,N} = -P, \quad \text{and} \quad \left(\frac{\partial E}{\partial V}\right)_{T,N} = 0.$$

Thus, it is necessary to specify what variables are independent in the notation for the partial derivative, if it is not otherwise completely clear. The energy of an ideal gas will not change with volume, provided the temperature and number of molecules remain constant. On the other hand, energy will decrease as volume increases if the entropy and number of molecules remain constant.

EXAMPLE 4 Use the explicit formula $(*)$ for $E = f(S, V, N)$ and the definitions of T and P as partial derivatives of f to calculate $\left(\frac{\partial P}{\partial S}\right)_{V,N}$ and $\left(\frac{\partial T}{\partial V}\right)_{S,N}$, thus showing that these partial derivatives differ only in sign.

Solution Using formulas $(*)$ we obtain

$$\begin{aligned} P = -\frac{\partial E}{\partial V} = -f_2(S, V, N) &= -\frac{3h^2N^{5/3}}{4\pi m}\left(-\frac{2}{3}V^{-5/3}\right)e^{\left(\frac{2S}{3Nk}-\frac{5}{3}\right)} \\ &= \frac{h^2}{2\pi m}\left(\frac{N}{V}\right)^{5/3}e^{\left(\frac{2S}{3Nk}-\frac{5}{3}\right)}, \end{aligned}$$

so that

$$\left(\frac{\partial P}{\partial S}\right)_{V,N} = \frac{2}{3Nk}\frac{h^2}{2\pi m}\left(\frac{N}{V}\right)^{5/3}e^{\left(\frac{2S}{3Nk}-\frac{5}{3}\right)} = \frac{2}{3Nk}P.$$

Similarly,

$$T = \frac{\partial E}{\partial S} = f_1(S, V, N) = \frac{2}{3Nk}E = \frac{h^2}{2\pi mk}\left(\frac{N}{V}\right)^{2/3}e^{\left(\frac{2S}{3Nk}-\frac{5}{3}\right)},$$

so that

$$\left(\frac{\partial T}{\partial V}\right)_{S,N} = -\frac{2}{3}\frac{h^2}{2\pi mk}\left(\frac{N^{2/3}}{V^{5/3}}\right)e^{\left(\frac{2S}{3Nk}-\frac{5}{3}\right)} = -\frac{2}{3Nk}P.$$

Therefore, $\left(\frac{\partial P}{\partial S}\right)_{V,N} = -\left(\frac{\partial T}{\partial V}\right)_{S,N}$.

It is no accident that $\left(\frac{\partial T}{\partial V}\right)_{S,N} = -\left(\frac{\partial P}{\partial S}\right)_{V,N}$. Since $T = f_1(S, V, N)$ and $P = -f_2(S, V, N)$, equality of mixed partials (Theorem 1 of Section 12.4) assures us that

$$\left(\frac{\partial T}{\partial V}\right)_{S,N} = f_{12}(S, V, N) = f_{21}(S, V, N) = -\left(\frac{\partial P}{\partial S}\right)_{V,N}.$$

This is one of the general relationships between partial derivatives in thermodynamics known as the *Maxwell relations*. Note that the subscripts S, V, and N in the two partial derivatives involved tell us that we are regarding those three variables as the independent variables on which the remaining variables T, P, E, and μ depend. There are three more Maxwell relations that can be derived in terms of the Legendre transformations of E (with N held fixed) introduced in Section 12.6. These are presented in Exercises 32–34 at the end of this section.

Jacobian Determinants

Partial derivatives obtained by implicit differentiation of systems of equations are fractions, the numerators and denominators of which are conveniently expressed in terms of certain determinants called *Jacobians.*

DEFINITION 8

The **Jacobian determinant** (or simply the **Jacobian**) of the two functions, $u = u(x, y)$ and $v = v(x, y)$, with respect to two variables, x and y, is the determinant

$$\frac{\partial(u, v)}{\partial(x, y)} = \begin{vmatrix} \dfrac{\partial u}{\partial x} & \dfrac{\partial u}{\partial y} \\ \dfrac{\partial v}{\partial x} & \dfrac{\partial v}{\partial y} \end{vmatrix}.$$

Similarly, the Jacobian of two functions, $F(x, y, \ldots)$ and $G(x, y, \ldots)$, with respect to the variables, x and y, is the determinant

$$\frac{\partial(F, G)}{\partial(x, y)} = \begin{vmatrix} \dfrac{\partial F}{\partial x} & \dfrac{\partial F}{\partial y} \\ \dfrac{\partial G}{\partial x} & \dfrac{\partial G}{\partial y} \end{vmatrix} = \begin{vmatrix} F_1 & F_2 \\ G_1 & G_2 \end{vmatrix}.$$

The definition above can be extended in the obvious way to give the Jacobian of n functions (or variables) with respect to n variables. For example, the Jacobian of three functions, F, G, and H, with respect to three variables, x, y, and z, is the determinant

$$\frac{\partial(F, G, H)}{\partial(x, y, z)} = \begin{vmatrix} F_1 & F_2 & F_3 \\ G_1 & G_2 & G_3 \\ H_1 & H_2 & H_3 \end{vmatrix}.$$

Jacobians are the determinants of the square Jacobian matrices corresponding to transformations of $\mathbb{R}^n$ to $\mathbb{R}^n$ as discussed briefly in Section 12.6.

EXAMPLE 5 In terms of Jacobians, the value of $(\partial x/\partial z)_w$, obtained from the system of equations

$$F(x, y, z, w) = 0, \qquad G(x, y, z, w) = 0$$

in Example 2, can be expressed in the form

$$\left(\frac{\partial x}{\partial z}\right)_w = -\frac{\dfrac{\partial(F, G)}{\partial(z, y)}}{\dfrac{\partial(F, G)}{\partial(x, y)}}.$$

Observe the pattern here. The denominator is the Jacobian of F and G with respect to the two *dependent* variables, x and y. The numerator is the same Jacobian except that the dependent variable x is replaced by the independent variable z.

The pattern observed above is general. We state it formally in the Implicit Function Theorem below.

The Implicit Function Theorem

The Implicit Function Theorem guarantees that systems of equations can be solved for certain variables as functions of other variables under certain circumstances, and it provides formulas for the partial derivatives of the solution functions. Before stating it, we consider a simple illustrative example.

EXAMPLE 6 Consider the system of linear equations

$$F(x, y, s, t) = a_1x + b_1y + c_1s + d_1t + e_1 = 0$$
$$G(x, y, s, t) = a_2x + b_2y + c_2s + d_2t + e_2 = 0.$$

This system can be written in matrix form:

$$\mathcal{A}\begin{pmatrix} x \\ y \end{pmatrix} + \mathcal{C}\begin{pmatrix} s \\ t \end{pmatrix} + \mathcal{E} = \begin{pmatrix} 0 \\ 0 \end{pmatrix},$$

where

$$\mathcal{A} = \begin{pmatrix} a_1 & b_1 \\ a_2 & b_2 \end{pmatrix}, \qquad \mathcal{C} = \begin{pmatrix} c_1 & d_1 \\ c_2 & d_2 \end{pmatrix}, \quad \text{and} \quad \mathcal{E} = \begin{pmatrix} e_1 \\ e_2 \end{pmatrix}.$$

The equations can be solved for x and y as functions of s and t provided $\det(\mathcal{A}) \neq 0$; this implies the existence of the inverse matrix $\mathcal{A}^{-1}$ (Theorem 4 of Section 10.7), so

$$\begin{pmatrix} x \\ y \end{pmatrix} = -\mathcal{A}^{-1}\left(\mathcal{C}\begin{pmatrix} s \\ t \end{pmatrix} + \mathcal{E}\right).$$

Observe that $\det(\mathcal{A}) = \partial(F, G)/\partial(x, y)$, so the nonvanishing of this Jacobian guarantees that the equations can be solved for x and y.

THEOREM 8

The Implicit Function Theorem

Consider a system of n equations in $n + m$ variables,

$$\begin{cases} F_{(1)}(x_1, x_2, \ldots, x_m, y_1, y_2, \ldots, y_n) = 0 \\ F_{(2)}(x_1, x_2, \ldots, x_m, y_1, y_2, \ldots, y_n) = 0 \\ \quad\vdots \\ F_{(n)}(x_1, x_2, \ldots, x_m, y_1, y_2, \ldots, y_n) = 0, \end{cases}$$

and a point $P_0 = (a_1, a_2, \ldots, a_m, b_1, b_2, \ldots, b_n)$ that satisfies the system. Suppose each of the functions $F_{(i)}$ has continuous first partial derivatives with respect to each of the variables x_j and y_k, $(i = 1, \ldots, n,\ j = 1, \ldots, m,\ k = 1, \ldots, n)$, near P_0. Finally, suppose that

$$\left.\frac{\partial(F_{(1)}, F_{(2)}, \ldots, F_{(n)})}{\partial(y_1, y_2, \ldots, y_n)}\right|_{P_0} \neq 0.$$

Then the system can be solved for $y_1, y_2, \ldots, y_n$ as functions of $x_1, x_2, \ldots, x_m$ near P_0. That is, there exist functions

$$\phi_1(x_1, \ldots, x_m), \ldots, \phi_n(x_1, \ldots, x_m)$$

such that

$$\phi_j(a_1, \ldots, a_m) = b_j, \qquad (j = 1, \ldots, n),$$

and such that the equations

$$F_{(1)}\Big(x_1, \ldots, x_m, \phi_1(x_1, \ldots, x_m), \ldots, \phi_n(x_1, \ldots, x_m)\Big) = 0,$$
$$F_{(2)}\Big(x_1, \ldots, x_m, \phi_1(x_1, \ldots, x_m), \ldots, \phi_n(x_1, \ldots, x_m)\Big) = 0,$$
$$\vdots$$
$$F_{(n)}\Big(x_1, \ldots, x_m, \phi_1(x_1, \ldots, x_m), \ldots, \phi_n(x_1, \ldots, x_m)\Big) = 0,$$

hold for all $(x_1, \ldots, x_m)$ sufficiently near $(a_1, \ldots, a_m)$.

Moreover,

$$\frac{\partial \phi_i}{\partial x_j} = \left(\frac{\partial y_i}{\partial x_j}\right)_{x_1, \ldots, x_{j-1}, x_{j+1}, \ldots, x_m} = -\frac{\dfrac{\partial(F_{(1)}, F_{(2)}, \ldots, F_{(n)})}{\partial(y_1, \ldots, x_j, \ldots, y_n)}}{\dfrac{\partial(F_{(1)}, F_{(2)}, \ldots, F_{(n)})}{\partial(y_1, \ldots, y_i, \ldots, y_n)}}.$$

Remark The formula for the partial derivatives is a consequence of Cramer's Rule (Theorem 6 of Section 10.7) applied to the n linear equations in the n unknowns $\partial y_1/\partial x_j, \ldots, \partial y_n/\partial x_j$ obtained by differentiating each of the equations in the given system with respect to x_j.

EXAMPLE 7 Show that the system

$$\begin{cases} xy^2 + xzu + yv^2 = 3 \\ x^3yz + 2xv - u^2v^2 = 2 \end{cases}$$

can be solved for (u, v) as a (vector) function of (x, y, z) near the point P_0 where $(x, y, z, u, v) = (1, 1, 1, 1, 1)$, and find the value of $\partial v/\partial y$ for the solution at $(x, y, z) = (1, 1, 1)$.

Solution Let $\begin{cases} F(x, y, z, u, v) = xy^2 + xzu + yv^2 - 3 \\ G(x, y, z, u, v) = x^3yz + 2xv - u^2v^2 - 2 \end{cases}$. Then

$$\left.\frac{\partial(F, G)}{\partial(u, v)}\right|_{P_0} = \left|\begin{matrix} xz & 2yv \\ -2uv^2 & 2x - 2u^2v \end{matrix}\right|_{P_0} = \left|\begin{matrix} 1 & 2 \\ -2 & 0 \end{matrix}\right| = 4.$$

Since this Jacobian is not zero, the Implicit Function Theorem assures us that the given equations can be solved for u and v as functions of x, y, and z, that is, for $(u, v) = \mathbf{f}(x, y, z)$. Since

$$\left.\frac{\partial(F, G)}{\partial(u, y)}\right|_{P_0} = \left|\begin{matrix} xz & 2xy + v^2 \\ -2uv^2 & x^3z \end{matrix}\right|_{P_0} = \left|\begin{matrix} 1 & 3 \\ -2 & 1 \end{matrix}\right| = 7,$$

we have

$$\left(\frac{\partial v}{\partial y}\right)_{x,z} = -\left.\frac{\dfrac{\partial(F, G)}{\partial(u, y)}}{\dfrac{\partial(F, G)}{\partial(u, v)}}\right|_{P_0} = -\frac{7}{4}.$$

Remark If all we wanted in this example was to calculate $\partial v/\partial y$, it would have been easier to use the technique of Example 3 and differentiate the given equations directly with respect to y, holding x and z fixed.

EXAMPLE 8 If the equations $x = u^2 + v^2$ and $y = uv$ are solved for u and v in terms of x and y, find, where possible,

$$\frac{\partial u}{\partial x}, \quad \frac{\partial u}{\partial y}, \quad \frac{\partial v}{\partial x}, \quad \text{and} \quad \frac{\partial v}{\partial y}.$$

Hence, show that $\dfrac{\partial(u, v)}{\partial(x, y)} = 1 \Big/ \dfrac{\partial(x, y)}{\partial(u, v)}$, provided the denominator is not zero.

Solution The given equations can be rewritten in the form

$$F(u, v, x, y) = u^2 + v^2 - x = 0$$
$$G(u, v, x, y) = uv - y = 0.$$

Let

$$J = \frac{\partial(F, G)}{\partial(u, v)} = \left|\begin{matrix} 2u & 2v \\ v & u \end{matrix}\right| = 2(u^2 - v^2) = \frac{\partial(x, y)}{\partial(u, v)}.$$

If $u^2 \neq v^2$, then $J \neq 0$ and we can calculate the required partial derivatives:

$$\frac{\partial u}{\partial x} = -\frac{1}{J}\frac{\partial(F,G)}{\partial(x,v)} = -\frac{1}{J}\begin{vmatrix} -1 & 2v \\ 0 & u \end{vmatrix} = \frac{u}{2(u^2-v^2)}$$
$$\frac{\partial u}{\partial y} = -\frac{1}{J}\frac{\partial(F,G)}{\partial(y,v)} = -\frac{1}{J}\begin{vmatrix} 0 & 2v \\ -1 & u \end{vmatrix} = \frac{-2v}{2(u^2-v^2)}$$
$$\frac{\partial v}{\partial x} = -\frac{1}{J}\frac{\partial(F,G)}{\partial(u,x)} = -\frac{1}{J}\begin{vmatrix} 2u & -1 \\ v & 0 \end{vmatrix} = \frac{-v}{2(u^2-v^2)}$$
$$\frac{\partial v}{\partial y} = -\frac{1}{J}\frac{\partial(F,G)}{\partial(u,y)} = -\frac{1}{J}\begin{vmatrix} 2u & 0 \\ v & -1 \end{vmatrix} = \frac{2u}{2(u^2-v^2)}.$$

Thus,

$$\frac{\partial(u,v)}{\partial(x,y)} = \frac{1}{J^2}\begin{vmatrix} u & -2v \\ -v & 2u \end{vmatrix} = \frac{J}{J^2} = \frac{1}{J} = \frac{1}{\dfrac{\partial(x,y)}{\partial(u,v)}}.$$

Remark Note in the above example that $\partial u/\partial x \neq 1/(\partial x/\partial u)$. This should be contrasted with the single-variable situation where, if $y = f(x)$ and $dy/dx \neq 0$, then $x = f^{-1}(y)$ and $dx/dy = 1/(dy/dx)$. This is another reason for distinguishing between ∂ and d. It is the Jacobian rather than any single partial derivative that takes the place of the ordinary derivative in such situations.

Remark Let us look briefly at the general case of invertible transformations from $\mathbb{R}^n$ to $\mathbb{R}^n$. Suppose that $\mathbf{y} = \mathbf{f}(\mathbf{x})$ and $\mathbf{z} = \mathbf{g}(\mathbf{y})$ are both functions from $\mathbb{R}^n$ to $\mathbb{R}^n$ whose components have continuous first partial derivatives. As shown in Section 12.6, the Chain Rule implies that

$$\begin{pmatrix} \dfrac{\partial z_1}{\partial x_1} & \cdots & \dfrac{\partial z_1}{\partial x_n} \\ \vdots & \ddots & \vdots \\ \dfrac{\partial z_n}{\partial x_1} & \cdots & \dfrac{\partial z_n}{\partial x_n} \end{pmatrix} = \begin{pmatrix} \dfrac{\partial z_1}{\partial y_1} & \cdots & \dfrac{\partial z_1}{\partial y_n} \\ \vdots & \ddots & \vdots \\ \dfrac{\partial z_n}{\partial y_1} & \cdots & \dfrac{\partial z_n}{\partial y_n} \end{pmatrix} \begin{pmatrix} \dfrac{\partial y_1}{\partial x_1} & \cdots & \dfrac{\partial y_1}{\partial x_n} \\ \vdots & \ddots & \vdots \\ \dfrac{\partial y_n}{\partial x_1} & \cdots & \dfrac{\partial y_n}{\partial x_n} \end{pmatrix}.$$

This is just the Chain Rule for the composition $\mathbf{z} = \mathbf{g}\big(\mathbf{f}(\mathbf{x})\big)$. It follows from Theorem 3(b) of Section 10.7 that the determinants of these matrices satisfy a similar equation:

$$\frac{\partial(z_1\cdots z_n)}{\partial(x_1\cdots x_n)} = \frac{\partial(z_1\cdots z_n)}{\partial(y_1\cdots y_n)}\,\frac{\partial(y_1\cdots y_n)}{\partial(x_1\cdots x_n)}.$$

If $\mathbf{f}$ is one-to-one and $\mathbf{g}$ is the inverse of $\mathbf{f}$, then $\mathbf{z} = \mathbf{g}\big(\mathbf{f}(\mathbf{x})\big) = \mathbf{x}$, and $\partial(z_1\cdots z_n)/\partial(x_1\cdots x_n) = 1$, the determinant of the identity matrix. Thus,

$$\frac{\partial(x_1\cdots x_n)}{\partial(y_1\cdots y_n)} = \frac{1}{\dfrac{\partial(y_1\cdots y_n)}{\partial(x_1\cdots x_n)}}.$$

In fact, the nonvanishing of either of these determinants is sufficient to guarantee that $\mathbf{f}$ is one-to-one and has an inverse. This is a special case of the Implicit Function Theorem.

We will encounter Jacobians again when we study transformations of coordinates in multiple integrals in Chapter 14.

EXERCISES 12.8

In Exercises 1–12, calculate the indicated derivative from the given equation(s). What condition on the variables will guarantee the existence of a solution that has the indicated derivative? Assume that any general functions F, G, and H have continuous first partial derivatives.

1. $\dfrac{dx}{dy}$ if $xy^3 + x^4y = 2$

2. $\dfrac{\partial x}{\partial y}$ if $xy^3 = y - z$

3. $\dfrac{\partial z}{\partial y}$ if $z^2 + xy^3 = \dfrac{xz}{y}$

4. $\dfrac{\partial y}{\partial z}$ if $e^{yz} - x^2z\ln y = \pi$

5. $\dfrac{\partial x}{\partial w}$ if $x^2y^2 + y^2z^2 + z^2t^2 + t^2w^2 - xw = 0$

6. $\dfrac{dy}{dx}$ if $F(x, y, x^2 - y^2) = 0$

7. $\dfrac{\partial u}{\partial x}$ if $G(x, y, z, u, v) = 0$

8. $\dfrac{\partial z}{\partial x}$ if $F(x^2 - z^2, y^2 + xz) = 0$

9. $\dfrac{\partial w}{\partial t}$ if $H(u^2w, v^2t, wt) = 0$

10. $\left(\dfrac{\partial y}{\partial x}\right)_u$ if $xyuv = 1$ and $x + y + u + v = 0$

11. $\left(\dfrac{\partial x}{\partial y}\right)_z$ if $x^2 + y^2 + z^2 + w^2 = 1$, and $x + 2y + 3z + 4w = 2$

12. $\dfrac{du}{dx}$ if $x^2y + y^2u - u^3 = 0$ and $x^2 + yu = 1$

13. If $x = u^3 + v^3$ and $y = uv - v^2$ are solved for u and v in terms of x and y, evaluate

$$\frac{\partial u}{\partial x}, \quad \frac{\partial u}{\partial y}, \quad \frac{\partial v}{\partial x}, \quad \frac{\partial v}{\partial y}, \quad \text{and} \quad \frac{\partial(u, v)}{\partial(x, y)}$$

at the point where $u = 1$ and $v = 1$.

14. Near what points (r, s) can the transformation

$$x = r^2 + 2s, \qquad y = s^2 - 2r$$

be solved for r and s as functions of x and y? Calculate the values of the first partial derivatives of the solution at the origin.

15. Evaluate the Jacobian $\partial(x, y)/\partial(r, \theta)$ for the transformation to polar coordinates: $x = r\cos\theta$, $y = r\sin\theta$. Near what points (r, θ) is the transformation one-to-one and therefore invertible to give r and θ as functions of x and y?

16. Evaluate the Jacobian $\partial(x, y, z)/\partial(R, \phi, \theta)$, where

$$x = R\sin\phi\cos\theta, \quad y = R\sin\phi\sin\theta, \quad \text{and } z = R\cos\phi.$$

This is the transformation from Cartesian to spherical coordinates in 3-space that we discussed in Section 10.6. Near what points is the transformation one-to-one and hence invertible to give R, ϕ, and θ as functions of x, y, and z?

17. Show that the equations

$$\begin{cases} xy^2 + zu + v^2 = 3 \\ x^3z + 2y - uv = 2 \\ xu + yv - xyz = 1 \end{cases}$$

can be solved for x, y, and z as functions of u and v near the point P_0 where $(x, y, z, u, v) = (1, 1, 1, 1, 1)$, and find $(\partial y/\partial u)_v$ at $(u, v) = (1, 1)$.

18. Show that the equations $\begin{cases} xe^y + uz - \cos v = 2 \\ u\cos y + x^2v - yz^2 = 1 \end{cases}$ can be solved for u and v as functions of x, y, and z near the point P_0 where $(x, y, z) = (2, 0, 1)$ and $(u, v) = (1, 0)$, and find $(\partial u/\partial z)_{x,y}$ at $(x, y, z) = (2, 0, 1)$.

19. Find dx/dy from the system

$$F(x, y, z, w) = 0, \quad G(x, y, z, w) = 0, \quad H(x, y, z, w) = 0.$$

20. Given the system

$$\begin{aligned} F(x, y, z, u, v) &= 0 \\ G(x, y, z, u, v) &= 0 \\ H(x, y, z, u, v) &= 0, \end{aligned}$$

how many possible interpretations are there for $\partial x/\partial y$? Evaluate them.

21. Given the system

$$\begin{aligned} F(x_1, x_2, \ldots, x_8) &= 0 \\ G(x_1, x_2, \ldots, x_8) &= 0 \\ H(x_1, x_2, \ldots, x_8) &= 0, \end{aligned}$$

how many possible interpretations are there for the partial $\dfrac{\partial x_1}{\partial x_2}$? Evaluate $\left(\dfrac{\partial x_1}{\partial x_2}\right)_{x_4, x_6, x_7, x_8}$.

22. If $F(x, y, z) = 0$ determines z as a function of x and y, calculate $\partial^2 z/\partial x^2$, $\partial^2 z/\partial x\partial y$, and $\partial^2 z/\partial y^2$ in terms of the partial derivatives of F.

23. If $x = u + v$, $y = uv$, and $z = u^2 + v^2$ define z as a function of x and y, find $\partial z/\partial x$, $\partial z/\partial y$, and $\partial^2 z/\partial x\partial y$.

24. A certain gas satisfies the law $pV = T - \dfrac{4p}{T^2}$, where p = pressure, V = volume, and T = temperature.

(a) Calculate $\partial T/\partial p$ and $\partial T/\partial V$ at the point where $p = V = 1$ and $T = 2$.

(b) If measurements of p and V yield the values $p = 1 \pm 0.001$ and $V = 1 \pm 0.002$, find the approximate maximum error in the calculated value $T = 2$.

25. If $F(x, y, z) = 0$, show that $\left(\dfrac{\partial x}{\partial y}\right)_z \left(\dfrac{\partial y}{\partial z}\right)_x \left(\dfrac{\partial z}{\partial x}\right)_y = -1.$

Derive analogous results for $F(x, y, z, u) = 0$ and for $F(x, y, z, u, v) = 0$. What is the general case?

26. If the equations $F(x, y, u, v) = 0$ and $G(x, y, u, v) = 0$ are solved for x and y as functions of u and v, show that

$$\frac{\partial(x, y)}{\partial(u, v)} = \frac{\partial(F, G)}{\partial(u, v)} \Big/ \frac{\partial(F, G)}{\partial(x, y)}.$$

27. If the equations $x = f(u, v)$, $y = g(u, v)$ can be solved for u and v in terms of x and y, show that

$$\frac{\partial(u, v)}{\partial(x, y)} = 1 \Big/ \frac{\partial(x, y)}{\partial(u, v)}.$$

Hint: Use the result of Exercise 26.

28. If $x = f(u, v)$, $y = g(u, v)$, $u = h(r, s)$, and $v = k(r, s)$, then x and y can be expressed as functions of r and s. Verify by direct calculation that

$$\frac{\partial(x, y)}{\partial(r, s)} = \frac{\partial(x, y)}{\partial(u, v)} \frac{\partial(u, v)}{\partial(r, s)}.$$

This is a special case of the Chain Rule for Jacobians.

29. Two functions, $f(x, y)$ and $g(x, y)$, are said to be functionally dependent if one is a function of the other; that is, if there exists a single-variable function $k(t)$ such that $f(x, y) = k\Big(g(x, y)\Big)$ for all x and y. Show that in this case $\partial(f, g)/\partial(x, y)$ vanishes identically. Assume that all necessary derivatives exist.

30. Prove the converse of Exercise 29 as follows: Let $u = f(x, y)$ and $v = g(x, y)$, and suppose that $\partial(u, v)/\partial(x, y) = \partial(f, g)/\partial(x, y)$ is identically zero for all x and y. Show that $(\partial u/\partial x)_v$ is identically zero. Hence u, considered as a function of x and v, is independent of x; that is, $u = k(v)$ for some function k of one variable. Why does this imply that f and g are functionally dependent?

Thermodynamics Problems

31. Use the different versions of the equation of state, presented in this section, to determine explicit functions u and v such that $S = u(E, V, N)$ and $S = v(T, V, N)$.

In Exercises 32–34 verify the given Maxwell relation by using a suitable Legendre transformation (see the Thermodynamics subsection of Section 12.6) to involve the appropriate set of independent variables.

32. $\left(\frac{\partial P}{\partial T}\right)_{V,N} = \left(\frac{\partial S}{\partial V}\right)_{T,N}.$

33. $\left(\frac{\partial V}{\partial S}\right)_{P,N} = \left(\frac{\partial T}{\partial P}\right)_{S,N}.$

34. $\left(\frac{\partial S}{\partial P}\right)_{T,N} = -\left(\frac{\partial V}{\partial T}\right)_{P,N}.$

12.9 Taylor's Formula, Taylor Series, and Approximations

As is the case for functions of one variable, power series representations and their partial sums (Taylor polynomials) can provide an efficient method for determining the behaviour of a smooth function of several variables near a point in its domain. In this section we will look briefly at the extension of Taylor's Formula and Taylor series to such functions. We will do this for functions of n variables as it is no more difficult to do this than to treat the special case $n = 2$.

As a starting point, recall Taylor's Formula for a function $F(t)$ with continuous derivatives of order up to $m + 1$ on the interval $[0, 1]$. (See Theorem 12 in Section 4.10, and put $f = F$, $a = 0$, $x = h = 1$, and $s = \theta$ in the version of Taylor's Formula given there.)

$$F(1) = F(0) + F'(0) + \frac{F''(0)}{2!} + \cdots + \frac{F^{(m)}(0)}{m!} + \frac{F^{(m+1)}(\theta)}{(m+1)!},$$

where θ is some number between 0 and 1. (The last term in the formula is the *Lagrange* form of the remainder.)

To simplify the manipulation of many variables, irrespective of how many there are, it is convenient to introduce the idea of a function of a vector, which is an intuitively straightforward extension from functions of scalars. If $\mathbf{x}$ has components $(x_1, x_2, \ldots, x_n)$, then $f(\mathbf{x})$ just means $f(x_1, x_2, \ldots, x_n)$, a function of n variables

Now suppose that $\mathbf{a} = (a_1, a_2, \ldots, a_n)$ and $\mathbf{h} = (h_1, h_2, \ldots, h_n)$ belong to $\mathbb{R}^n$. If f is a function of $\mathbf{x} \in \mathbb{R}^n$ that has continuous partial derivatives of orders up to $m + 1$ in an open set containing the line segment joining $\mathbf{a}$ and $\mathbf{a} + \mathbf{h}$, we can apply the above formula to

$$F(t) = f(\mathbf{a} + t\mathbf{h}), \qquad (0 \le t \le 1).$$

By the Chain Rule we will have

$$\begin{aligned} F'(t) &= h_1 f_{h_1}(\mathbf{a} + t\mathbf{h}) + h_2 f_{h_2}(\mathbf{a} + t\mathbf{h}) + \cdots + h_n f_{h_n}(\mathbf{a} + t\mathbf{h}) \\ &= (\mathbf{h} \bullet \nabla) f(\mathbf{a} + t\mathbf{h}), \end{aligned}$$

where

$$(\mathbf{h} \bullet \nabla) f(\mathbf{a} + t\mathbf{h}) = \Big((h_1 D_1 + h_2 D_2 + \cdots + h_n D_n) f(\mathbf{x})\Big)\Big|_{\mathbf{x}=\mathbf{a}+t\mathbf{h}}$$

and $D_j = \partial/\partial x_j$, $(1 \le p \le n)$. Similarly,

$$F''(t) = h_1 h_1 f_{11}(\mathbf{a} + t\mathbf{h}) + h_1 h_2 f_{12}(\mathbf{a} + t\mathbf{h}) + \cdots + h_n h_n f_{nn}(\mathbf{a} + t\mathbf{h})$$
$$= \left(\mathbf{h} \bullet \nabla\right)^2 f(\mathbf{a} + t\mathbf{h})$$
$$\vdots$$
$$F^{(j)}(t) = \left(\mathbf{h} \bullet \nabla\right)^j f(\mathbf{a} + t\mathbf{h})$$

Thus $F(1) = f(\mathbf{a}+\mathbf{h})$, $\quad F(0) = f(\mathbf{a})$, $\quad$ and $\quad F^{(j)}(0) = (\mathbf{h}\bullet\nabla)^j f(\mathbf{a})$. The Taylor formula given above thus says that

$$f(\mathbf{a} + \mathbf{h}) = f(\mathbf{a}) + \mathbf{h} \bullet \nabla f(\mathbf{a}) + \frac{(\mathbf{h} \bullet \nabla)^2 f(\mathbf{a})}{2!} + \cdots + \frac{(\mathbf{h} \bullet \nabla)^m f(\mathbf{a})}{m!}$$
$$+ \frac{(\mathbf{h} \bullet \nabla)^{m+1} f(\mathbf{a} + \theta\mathbf{h})}{(m+1)!}$$
$$= \sum_{j=0}^{m} \frac{(\mathbf{h} \bullet \nabla)^j f(\mathbf{a})}{j!} + \frac{(\mathbf{h} \bullet \nabla)^{m+1} f(\mathbf{a} + \theta\mathbf{h})}{(m+1)!}$$
$$= P_m(\mathbf{h}) + R_m(\mathbf{h}, \theta).$$

This is Taylor's Formula for f about $\mathbf{x} = \mathbf{a}$. $P_m(\mathbf{h})$ is a polynomial of degree m in the components of $\mathbf{h}$ $P_m(\mathbf{h})$ is called the mth degree Taylor polynomial of f about $\mathbf{x} = \mathbf{a}$. The term corresponding to j in the summation defining P_m is, if not zero, a polynomial of degree exactly j in the components of $\mathbf{h}$, whose coefficients are jth order partial derivatives of f evaluated at $\mathbf{x} = \mathbf{a}$. The remainder term $R_m(\mathbf{h}, \theta)$ is also a polynomial in the components of $\mathbf{h}$, each of whose terms if not zero has degree exactly $m + 1$, but its coefficients are $(m+1)$st order partial derivatives of f evaluated at an indeterminate point $\mathbf{a} + \theta\mathbf{h}$ along the line segment between $\mathbf{a}$ and $\mathbf{a} + \mathbf{h}$.

Sometimes it is useful to replace the explicit remainder in Taylor's Formula with a Big-O term that is bounded by a multiple of $|\mathbf{h}|^{m+1}$ as $|\mathbf{h}| \to 0$. (See Section 4.10.)

$$f(\mathbf{a}+\mathbf{h}) = f(\mathbf{a})+\mathbf{h}\bullet\nabla f(\mathbf{a})+\frac{(\mathbf{h} \bullet \nabla)^2 f(\mathbf{a})}{2!}+\cdots+\frac{(\mathbf{h} \bullet \nabla)^m f(\mathbf{a})}{m!}+O(|\mathbf{h}|^{m+1}).$$

If all partial derivatives of f are continuous, and if there exists a positive number r such that whenever $|\mathbf{h}| < r$ we have for all $\theta \in [0, 1]$,

$$\lim_{m\to\infty} R_{m+1}(\mathbf{h}, \theta) = 0,$$

then we can represent $f(\mathbf{a} + \mathbf{h})$ as the sum of the Taylor series

$$f(\mathbf{a} + \mathbf{h}) = \sum_{j=0}^{\infty} \frac{(\mathbf{h} \bullet \nabla)^j f(\mathbf{a})}{j!}.$$

Remark An alternative approach is to develop Taylor's formula with directional derivatives. Following Section 12.7, a function $g(s)$ is introduced, where $s - s_0$ is distance, measured along a line L in direction $\mathbf{u}$, from the point on L corresponding to $s = s_0$. As in Section 4.10, a Taylor formula for $g(s)$ is

$$g(s) = g(s_0)+g'(s_0)(s-s_0)+\frac{1}{2}g''(s_0)(s-s_0)^2+\cdots+\frac{1}{2}g^{(m)}(s_0)(s-s_0)^2+O\big(|s-s_0|^{m+1}$$

Since $d/ds = \mathbf{u}\bullet\nabla$ is the directional derivative operation in direction $\mathbf{u}$, the directional derivative extends to all orders in the Taylor expansion in s. We may choose $g(s) = f(\mathbf{a} + (s - s_0)\mathbf{u})$, where $(s - s_0)\mathbf{u} = \mathbf{h}$. It follows that $|\mathbf{h}|^n = |s - s_0|^n$ and

$$g(s) = f(\mathbf{a}+\mathbf{h}) = f(\mathbf{a})+(\mathbf{h}\bullet\nabla)f(\mathbf{a})+\frac{(\mathbf{h} \bullet \nabla)^2 f(\mathbf{a})}{2!}\cdots+\frac{(\mathbf{h} \bullet \nabla)^m f(\mathbf{a})}{m!}+O(|\mathbf{h}|^{m+1})$$

as above.

EXAMPLE 1 Let us illustrate the above ideas with a simple special case. If f is a function of two variables, x and y having continuous partial derivatives of order up to 4 in the disk $(x-a)^2+(y-b)^2 \le r^2$, then for $\mathbf{h}=(h,k)$ in $\mathbb{R}^2$ satisfying $h^2+k^2<r$ we have

$$\begin{aligned} f(a+h,b+k) &= P_3(h,k)+R_3(h,k,\theta) \\ &= f(a,b)+(hD_1+kD_2)f(a,b)+\frac{1}{2!}(hD_1+kD_2)^2 f(a,b) \\ &\quad +\frac{1}{3!}(hD_1+kD_2)^3 f(a,b)+R_3(h,k,\theta) \\ &= f(a,b)+hf_1(a,b)+kf_2(a,b)+\frac{1}{2!}\Big(h^2 f_{11}(a,b)+2hkf_{12}(a,b)+k^2 f_{22}(a,b)\Big) \\ &\quad +\frac{1}{3!}\Big(h^3 f_{111}(a,b)+3h^2kf_{112}(a,b)+3hk^2 f_{122}(a,b)+k^3 f_{222}(a,b)\Big)+R_3(h,k,\theta), \end{aligned}$$

where

$$R_3(h,k,\theta)=\frac{1}{4!}(hD_1+kD_2)^4 f(a+\theta h, b+\theta k)=O\big((h^2+k^2)^2\big).$$

Note that since $0<\theta<1$, all the 4th order partial derivatives of f are bounded on the line segment from (a,b) to $(a+\theta h, b+\theta k)$. This is why the remainder term is $O\big((h^2+k^2)^2\big)$.

We stress that the expression $(hD_1+kD_2)^j f(a,b)$ means first calculate $(hD_1+kD_2)^j f(x,y)$ and then evaluate the result at $(x,y)=(a,b)$.

As for functions of one variable, the Taylor polynomial of degree m, provides the "best" nth-degree polynomial approximation to $f(x,y)$ near (a,b). For $n=1$ this approximation reduces to the tangent plane approximation

$$f(x,y)\approx f(a,b)+f_1(a,b)(x-a)+f_2(a,b)(y-b).$$

EXAMPLE 2 Find a second-degree polynomial approximation to the function $f(x,y)=\sqrt{x^2+y^3}$ near the point $(1,2)$, and use it to estimate the value of $\sqrt{(1.02)^2+(1.97)^3}$.

Solution For the second-degree approximation we need the values of the partial derivatives of f up to second order at $(1,2)$. We have

$$\begin{aligned} f(x,y) &= \sqrt{x^2+y^3} & f(1,2) &= 3 \\ f_1(x,y) &= \frac{x}{\sqrt{x^2+y^3}} & f_1(1,2) &= \frac{1}{3} \\ f_2(x,y) &= \frac{3y^2}{2\sqrt{x^2+y^3}} & f_2(1,2) &= 2 \\ f_{11}(x,y) &= \frac{y^3}{(x^2+y^3)^{3/2}} & f_{11}(1,2) &= \frac{8}{27} \\ f_{12}(x,y) &= \frac{-3xy^2}{2(x^2+y^3)^{3/2}} & f_{12}(1,2) &= -\frac{2}{9} \\ f_{22}(x,y) &= \frac{12x^2y+3y^4}{4(x^2+y^3)^{3/2}} & f_{22}(1,2) &= \frac{2}{3}. \end{aligned}$$

Thus,

$$f(1+h,2+k)\approx 3+\frac{1}{3}h+2k+\frac{1}{2!}\Big(\frac{8}{27}h^2+2\Big(-\frac{2}{9}\Big)hk+\frac{2}{3}k^2\Big)$$

or, setting $x = 1 + h$ and $y = 2 + k$,

$$f(x, y) = 3 + \frac{1}{3}(x-1) + 2(y-2) + \frac{4}{27}(x-1)^2 - \frac{2}{9}(x-1)(y-2) + \frac{1}{3}(y-2)^2.$$

This is the required second-degree Taylor polynomial for f near $(1, 2)$. Therefore,

$$\begin{aligned}\sqrt{(1.02)^2 + (1.97)^3} &= f(1 + 0.02, 2 - 0.03)\\ &\approx 3 + \frac{1}{3}(0.02) + 2(-0.03) + \frac{4}{27}(0.02)^2\\ &\quad - \frac{2}{9}(0.02)(-0.03) + \frac{1}{3}(-0.03)^2\\ &\approx 2.947\,159\,3.\end{aligned}$$

(For comparison purposes: the true value is 2.947 163 6... The approximation is accurate to 6 significant figures.)

As observed for functions of one variable, it is not usually necessary to calculate derivatives in order to determine the coefficients in a Taylor series or Taylor polynomial. It is often much easier to perform algebraic manipulations on known series. For instance, the above example could have been done by writing f in the form

$$\begin{aligned}f(1+h, 2+k) &= \sqrt{(1+h)^2 + (2+k)^3}\\ &= \sqrt{9 + 2h + h^2 + 12k + 6k^2 + k^3}\\ &= 3\sqrt{1 + \frac{2h + h^2 + 12k + 6k^2 + k^3}{9}}\end{aligned}$$

and then applying the binomial expansion

$$\sqrt{1+t} = 1 + \frac{1}{2}t + \frac{1}{2!}\left(\frac{1}{2}\right)\left(-\frac{1}{2}\right)t^2 + \cdots$$

with $t = \dfrac{2h + h^2 + 12k + 6k^2 + k^3}{9}$ to obtain the terms up to second degree in the variables h and k.

EXAMPLE 3 Find the Taylor polynomial of degree 3 in powers of x and y for the function $f(x, y) = e^{x-2y}$.

Solution The required Taylor polynomial will be the Taylor polynomial of degree 3 for e^t evaluated at $t = x - 2y$:

$$\begin{aligned}P_3(x, y) &= 1 + (x - 2y) + \frac{1}{2!}(x-2y)^2 + \frac{1}{3!}(x-2y)^3\\ &= 1 + x - 2y + \frac{1}{2}x^2 - 2xy + 2y^2 + \frac{1}{6}x^3 - x^2y + 2xy^2 - \frac{4}{3}y^3.\end{aligned}$$

Remark Maple can, of course, be used to compute multivariate Taylor polynomials with its function **mtaylor**, which, depending on the Maple version, may have to be read in from the Maple library before it can be used if it is not part of the Maple kernel.

```
> readlib(mtaylor):
```

Arguments fed to `mtaylor` are as follows:

(a) an expression involving the expansion variables

(b) a list whose elements are either variable names or equations of the form `variable=value` giving the coordinates of the point about which the expansion is calculated. (Just naming a variable is equivalent to using the equation `variable=0`.)

(c) (optionally) a positive integer m forcing the order of the computed Taylor polynomial to be less than m. If m is not specified, the value of Maple's global variable "Order" is used. The default value is 6.

A few examples should suffice.

```
> mtaylor(cos(x+y^2),[x,y]);
```

$$1-\frac{1}{2}x^2-y^2x+\frac{1}{24}x^4-\frac{1}{2}y^4+\frac{1}{6}y^2x^3$$

```
> mtaylor(cos(x+y^2),[x=Pi,y],5);
```

$$-1+\frac{1}{2}(x-\pi)^2+y^2(x-\pi)-\frac{1}{24}(x-\pi)^4+\frac{1}{2}y^4$$

```
> mtaylor(g(x,y),[x=a,y=b],3);
```

$$g(a,b)+D_1(g)(a,b)(x-a)+D_2(g)(a,b)(y-b)+\frac{1}{2}D_{1,1}(g)(a,b)(x-a)^2$$
$$+(x-a)D_{1,2}(g)(a,b)(y-b)+\frac{1}{2}D_{2,2}(g)(a,b)(y-b)^2$$

The function `mtaylor` can be a bit quirky. It has a tendency to expand linear terms; for example, in an expansion about $x=1$ and $y=-2$, it may rewrite terms $2+(x-1)+2(y+2)$ in the form $5+x+2y$.

Approximating Implicit Functions

In the previous section we saw how to determine whether an equation in several variables could be solved for one of those variables as a function of the others. Even when such a solution is known to exist, it is not usually possible to find an exact formula for it. However, if the equation involves only smooth functions, then the solution will have a Taylor series. We can determine at least the first several coefficients in that series and thus obtain a useful approximation to the solution. The following example shows the technique.

EXAMPLE 4 Show that the equation $\sin(x+y)=xy+2x$ has a solution of the form $y=f(x)$ near $x=0$ satisfying $f(0)=0$, and find the terms up to fourth degree for the Taylor series for $f(x)$ in powers of x.

Solution The given equation can be written in the form $F(x,y)=0$, where

$$F(x,y)=\sin(x+y)-xy-2x.$$

Since $F(0,0)=0$ and $F_2(0,0)=\cos(0)=1\neq 0$, the equation has a solution $y=f(x)$ near $x=0$ satisfying $f(0)=0$ by the Implicit Function Theorem. It is not possible to calculate $f(x)$ exactly, but it will have a Maclaurin series of the form

$$y=f(x)=a_1x+a_2x^2+a_3x^3+a_4x^4+\cdots.$$

(There is no constant term because $f(0)=0$.) We can substitute this series into the given equation and keep track of terms up to degree 4 in order to calculate the coefficients a_1, a_2, a_3, and a_4. For the left side we use the Maclaurin series for sin to obtain

$$\begin{aligned}
\sin(x+y) &= \sin\Big((1+a_1)x + a_2x^2 + a_3x^3 + a_4x^4 + \cdots\Big) \\
&= (1+a_1)x + a_2x^2 + a_3x^3 + a_4x^4 + \cdots \\
&\quad - \frac{1}{3!}\Big((1+a_1)x + a_2x^2 + \cdots\Big)^3 + \cdots \\
&= (1+a_1)x + a_2x^2 + \Big(a_3 - \frac{1}{6}(1+a_1)^3\Big)x^3 \\
&\quad + \Big(a_4 - \frac{3}{6}(1+a_1)^2 a_2\Big)x^4 + \cdots.
\end{aligned}$$

The right side is

$$xy + 2x = 2x + a_1x^2 + a_2x^3 + a_3x^4 + \cdots.$$

Equating coefficients of like powers of x, we obtain

$$\begin{aligned}
1 + a_1 &= 2 & a_1 &= 1 \\
a_2 &= a_1 & a_2 &= 1 \\
a_3 - \frac{1}{6}(1+a_1)^3 &= a_2 & a_3 &= \frac{7}{3} \\
a_4 - \frac{1}{2}(1+a_1)^2 a_2 &= a_3 & a_4 &= \frac{13}{3}.
\end{aligned}$$

Thus,

$$y = f(x) = x + x^2 + \frac{7}{3}x^3 + \frac{13}{3}x^4 + \cdots.$$

(We could have obtained more terms in the series by keeping track of higher powers of x in the substitution process.)

Remark From the series for $f(x)$ obtained above, we can determine the values of the first four derivatives of f at $x = 0$. Remember that

$$a_k = \frac{f^{(k)}(0)}{k!}.$$

We have, therefore,

$$\begin{aligned}
f'(0) &= a_1 = 1 & f''(0) &= 2!a_2 = 2 \\
f'''(0) &= 3!a_3 = 14 & f^{(4)}(0) &= 4!a_4 = 104.
\end{aligned}$$

We could have done the example by first calculating these derivatives by implicit differentiation of the given equation and then determining the series coefficients from them. This would have been a much more difficult way to do it. (Try it and see.)

EXERCISES 12.9

In Exercises 1–6, find the Taylor series for the given function about the indicated point.

1. $f(x, y) = \dfrac{1}{2 + xy^2}$, $(0, 0)$

2. $f(x, y) = \ln(1 + x + y + xy)$, $(0, 0)$

3. $f(x, y) = \tan^{-1}(x + xy)$, $(0, -1)$

4. $f(x, y) = x^2 + xy + y^3$, $(1, -1)$

5. $f(x, y) = e^{x^2+y^2}$, $(0, 0)$

6. $f(x, y) = \sin(2x + 3y), \quad (0, 0)$

In Exercises 7–12, find Taylor polynomials of the indicated degree for the given functions near the given point. After calculating them by hand, try to get the same results using Maple's `mtaylor` function.

7. $f(x, y) = \dfrac{1}{2 + x - 2y}$, degree 3, near $(2, 1)$

8. $f(x, y) = \ln(x^2 + y^2)$, degree 3, near $(1, 0)$

9. $f(x, y) = \displaystyle\int_0^{x+y^2} e^{-t^2}\, dt$, degree 3, near $(0, 0)$

10. $f(x, y) = \cos(x + \sin y)$, degree 4, near $(0, 0)$

11. $f(x, y) = \dfrac{\sin x}{y}$, degree 2, near $(\frac{\pi}{2}, 1)$

12. $f(x, y) = \dfrac{1 + x}{1 + x^2 + y^4}$, degree 2, near $(0, 0)$

In Exercises 13–14, show that, for x near the indicated point $x = a$, the given equation has a solution of the form $y = f(x)$ taking on the indicated value at that point. Find the first three nonzero terms of the Taylor series for $f(x)$ in powers of $x - a$.

13. $x \sin y = y + \sin x$, near $x = 0$, with $f(0) = 0$

14. $\sqrt{1 + xy} = 1 + x + \ln(1 + y)$, near $x = 0$, with $f(0) = 0$

15. Show that the equation $x + 2y + z + e^{2z} = 1$ has a solution of the form $z = f(x, y)$ near $x = 0$, $y = 0$, where $f(0, 0) = 0$. Find the Taylor polynomial of degree 2 for $f(x, y)$ in powers of x and y.

16. Use series methods to find the value of the partial derivative $f_{112}(0, 0)$ given that $f(x, y) = \arctan(x + y)$.

17. Use series methods to evaluate

$$\frac{\partial^{4n}}{\partial x^{2n} \partial y^{2n}} \frac{1}{1 + x^2 + y^2}\bigg|_{(0,0)}.$$

CHAPTER REVIEW

Key Ideas

- **What do the following sentences and phrases mean?**
  - ◇ $\mathcal{S}$ is the graph of $f(x, y)$.
  - ◇ $\mathcal{C}$ is a level curve of $f(x, y)$.
  - ◇ $\lim_{(x,y)\to(a,b)} f(x, y) = L$.
  - ◇ $f(x, y)$ is continuous at (a, b).
  - ◇ the partial derivative $(\partial/\partial x) f(x, y)$
  - ◇ the tangent plane to $z = f(x, y)$ at (a, b)
  - ◇ pure second partials ◇ mixed second partials
  - ◇ $f(x, y)$ is a harmonic function.
  - ◇ $L(x, y)$ is the linearization of $f(x, y)$ at (a, b).
  - ◇ the differential of $z = f(x, y)$
  - ◇ $f(x, y)$ is differentiable at (a, b).
  - ◇ the gradient of $f(x, y)$ at (a, b)
  - ◇ the directional derivative of $f(x, y)$ at (a, b) in direction $\mathbf{v}$
  - ◇ the Jacobian determinant $\partial(x, y)/\partial(u, v)$
- **Under what conditions are two mixed partial derivatives equal?**
- **State the Chain Rule for $z = f(x, y)$, where $x = g(u, v)$, and $y = h(u, v)$.**
- **Describe the process of calculating partial derivatives of implicitly defined functions.**
- **What is the Taylor series of $f(x, y)$ about (a, b)?**

Review Exercises

1. Sketch some level curves of the function $x + \dfrac{4y^2}{x}$.

2. Sketch some isotherms (curves of constant temperature) for the temperature function

$$T = \frac{140 + 30x^2 - 60x + 120y^2}{8 + x^2 - 2x + 4y^2} \quad (°\text{C}).$$

What is the coolest location?

3. Sketch some level curves of the polynomial function $f(x, y) = x^3 - 3xy^2$. Why do you think the graph of this function is called a *monkey saddle*?

4. Let $f(x, y) = \begin{cases} \dfrac{x^3}{x^2 + y^2}, & \text{if } (x, y) \neq (0, 0) \\ 0, & \text{if } (x, y) = (0, 0). \end{cases}$

Calculate each of the following partial derivatives or explain why it does not exist: $f_1(0, 0)$, $f_2(0, 0)$, $f_{21}(0, 0)$, $f_{12}(0, 0)$.

5. Let $f(x, y) = \dfrac{x^3 - y^3}{x^2 - y^2}$. Where is $f(x, y)$ continuous? To what additional set of points does $f(x, y)$ have a continuous extension? In particular, can f be extended to be continuous at the origin? Can f be defined at the origin in such a way that its first partial derivatives exist there?

6. The surface $\mathcal{S}$ is the graph of the function $z = f(x, y)$, where $f(x, y) = e^{x^2 - 2x - 4y^2 + 5}$.

   (a) Find an equation of the tangent plane to $\mathcal{S}$ at the point $(1, -1, 1)$.

   (b) Sketch a representative sample of the level curves of the function $f(x, y)$.

7. Consider the surface $\mathcal{S}$ with equation $x^2 + y^2 + 4z^2 = 16$.

   (a) Find an equation for the tangent plane to $\mathcal{S}$ at the point (a, b, c) on $\mathcal{S}$.

   (b) For which points (a, b, c) on $\mathcal{S}$ does the tangent plane to $\mathcal{S}$ at (a, b, c) pass through the point $(0, 0, 4)$? Describe this set of points geometrically.

   (c) For which points (a, b, c) on $\mathcal{S}$ is the tangent plane to $\mathcal{S}$ at (a, b, c) parallel to the plane $x + y + 2\sqrt{2}z = 97$?

8. Two variable resistors, R_1 and R_2, are connected in parallel so that their combined resistance, R, is given by

$$\frac{1}{R} = \frac{1}{R_1} + \frac{1}{R_2}.$$

If $R_1 = 100$ ohms $\pm 5\%$ and $R_2 = 25$ ohms $\pm 2\%$, by approximately what percentage can the calculated value of their combined resistance $R = 20$ ohms be in error?

9. You have measured two sides of a triangular field and the angle between them. The side measurements are 150 m and 200 m, each accurate to within ± 1 m. The angle measurement is $30°$, accurate to within $\pm 2°$. What area do you calculate for the field, and what is your estimate of the maximum percentage error in this area?

10. Suppose that $T(x, y, z) = x^3 y + y^3 z + z^3 x$ gives the temperature at the point (x, y, z) in 3-space.

(a) Calculate the directional derivative of T at $(2, -1, 0)$ in the direction toward the point $(1, 1, 2)$.

(b) A fly is moving through space with constant speed 5. At time $t = 0$ the fly crosses the surface $2x^2 + 3y^2 + z^2 = 11$ at right angles at the point $(2, -1, 0)$, moving in the direction of increasing temperature. Find dT/dt at $t = 0$ as experienced by the fly.

11. Consider the function $f(x, y, z) = x^2 y + yz + z^2$.

(a) Find the directional derivative of f at $(1, -1, 1)$ in the direction of the vector $\mathbf{i} + \mathbf{k}$.

(b) An ant is crawling on the plane $x + y + z = 1$ through $(1, -1, 1)$. Suppose it crawls so as to keep f constant. In what direction is it going as it passes through $(1, -1, 1)$?

(c) Another ant crawls on the plane $x + y + z = 1$, moving in the direction of the greatest rate of increase of f. Find its direction as it goes through $(1, -1, 1)$.

12. Let $f(x, y, z) = (x^2 + z^2) \sin \dfrac{\pi xy}{2} + yz^2$. Let P_0 be the point $(1, 1, -1)$.

(a) Find the gradient of f at P_0.

(b) Find the linearization $L(x, y, z)$ of f at P_0.

(c) Find an equation for the tangent plane at P_0 to the level surface of f through P_0.

(d) If a bird flies through P_0 with speed 5, heading directly toward the point $(2, -1, 1)$, what is the rate of change of f as seen by the bird as it passes through P_0?

(e) In what direction from P_0 should the bird fly at speed 5 to experience the greatest rate of increase of f?

13. Verify that for any constant, k, the function $u(x, y) = k\big(\ln \cos(x/k) - \ln \cos(y/k)\big)$ satisfies the *minimal surface equation*

$$(1 + u_x^2)u_{yy} - u u_x u_y u_{xy} + (1 + u_y^2)u_{xx} = 0.$$

14. The equations $F(x, y, z) = 0$ and $G(x, y, z) = 0$ can define any two of the variables x, y, and z as functions of the remaining variable. Show that

$$\frac{dx}{dy}\frac{dy}{dz}\frac{dz}{dx} = 1.$$

15. The equations $\begin{cases} x = u^3 - uv \\ y = 3uv + 2v^2 \end{cases}$ define u and v as functions of x and y near the point P where $(u, v, x, y) = (-1, 2, 1, 2)$.

(a) Find $\dfrac{\partial u}{\partial x}$ and $\dfrac{\partial u}{\partial y}$ at P.

(b) Find the approximate value of u when $x = 1.02$ and $y = 1.97$.

16. The equations $\begin{cases} u = x^2 + y^2 \\ v = x^2 - 2xy^2 \end{cases}$ define x and y implicitly as functions of u and v for values of (x, y) near $(1, 2)$ and values of (u, v) near $(5, -7)$.

(a) Find $\dfrac{\partial x}{\partial u}$ and $\dfrac{\partial y}{\partial u}$ at $(u, v) = (5, -7)$.

(b) If $z = \ln(y^2 - x^2)$, find $\dfrac{\partial z}{\partial u}$ at $(u, v) = (5, -7)$.

Challenging Problems

1. (a) If the graph of a function $f(x, y)$ that is differentiable at (a, b) contains part of a straight line through (a, b), show that the line lies in the tangent plane to $z = f(x, y)$ at (a, b).

(b) If $g(t)$ is a differentiable function of t, describe the surface $z = yg(x/y)$ and show that all its tangent planes pass through the origin.

2. A particle moves in 3-space in such a way that its direction of motion at any point is perpendicular to the level surface of

$$f(x, y, z) = 4 - x^2 - 2y^2 + 3z^2$$

through that point. If the path of the particle passes through the point $(1, 1, 8)$, show that it also passes through $(2, 4, 1)$. Does it pass through $(3, 7, 0)$?

3. (The Laplace operator in spherical coordinates) If $u(x, y, z)$ has continuous second partial derivatives and

$$v(R, \phi, \theta) = u(R \sin\phi \cos\theta, R \sin\phi \sin\theta, R \cos\phi),$$

show that

$$\frac{\partial^2 v}{\partial R^2} + \frac{2}{R}\frac{\partial v}{\partial R} + \frac{\cot\phi}{R^2}\frac{\partial v}{\partial \phi} + \frac{1}{R^2}\frac{\partial^2 v}{\partial \phi^2} + \frac{1}{R^2 \sin^2\phi}\frac{\partial^2 v}{\partial \theta^2}$$
$$= \frac{\partial^2 u}{\partial x^2} + \frac{\partial^2 u}{\partial y^2} + \frac{\partial^2 u}{\partial z^2}.$$

You can do this by hand, but it is a lot easier using computer algebra.

4. (Spherically expanding waves) If f is a twice differentiable function of one variable and $R = \sqrt{x^2 + y^2 + z^2}$, show that $u(x, y, z, t) = \dfrac{f(R - ct)}{R}$ satisfies the three-dimensional wave equation

$$\frac{\partial^2 u}{\partial t^2} = c^2\left(\frac{\partial^2 u}{\partial x^2} + \frac{\partial^2 u}{\partial y^2} + \frac{\partial^2 u}{\partial z^2}\right).$$

What is the geometric significance of this solution as a function of increasing time t? *Hint:* You may want to use the result of Exercise 3. In this case $v(R, \phi, \theta)$ is independent of ϕ and θ.

CHAPTER 13

Applications of Partial Derivatives

> I don't know what I may seem to the world, but as to myself, I seem to have been only like a boy playing on the sea-shore and diverting myself in now and then finding a smoother pebble or a prettier shell than ordinary, whilst the great ocean of truth lay all undiscovered before me.

Isaac Newton 1642–1727

Introduction

In this chapter we will discuss some of the ways partial derivatives contribute to the understanding and solution of problems in applied mathematics. Many such problems can be put in the context of determining maximum or minimum values for functions of several variables, and the first four sections of this chapter deal with that subject. The remaining sections discuss some miscellaneous problems involving the differentiation of functions with respect to parameters, and also Newton's Method for approximating solutions of systems of nonlinear equations. Much of the material in this chapter may be considered *optional*. Only Sections 13.1–13.3 contain *core material*, and even parts of those sections can be omitted (e.g., the discussion of linear programming in Section 13.2).

13.1 Extreme Values

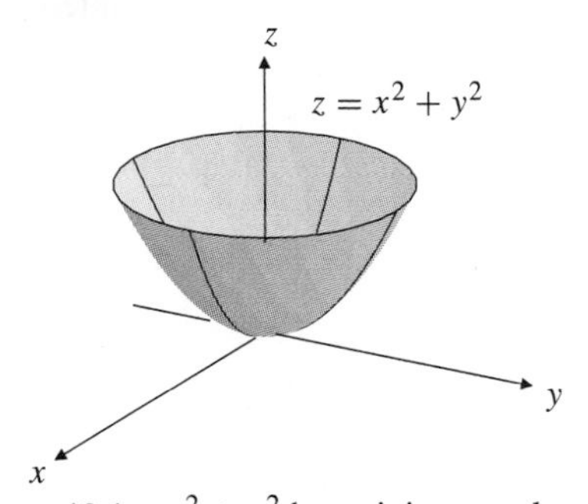

Figure 13.1 $x^2 + y^2$ has minimum value 0 at the origin

The function $f(x, y) = x^2 + y^2$, part of whose graph is shown in Figure 13.1, has a minimum value of 0; this value occurs at the origin $(0, 0)$ where the graph has a horizontal tangent plane. Similarly, the function $g(x, y) = 1 - x^2 - y^2$, part of whose graph appears in Figure 13.2, has a maximum value of 1 at $(0, 0)$. What techniques could be used to discover these facts if they were not evident from a diagram? Finding maximum and minimum values of functions of several variables is, like its single-variable counterpart, the crux of many applications of advanced calculus to problems that arise in other disciplines. Unfortunately, this problem is often much more complicated than in the single-variable case. Our discussion will begin by developing the techniques for functions of two variables. Some of the techniques extend to functions of more variables in obvious ways. The extension of those that do not will be discussed later in this section.

Let us begin by reviewing what we know about the single-variable case. Recall that a function $f(x)$ has a *local maximum value* (or a *local minimum value*) at a point a in its domain if $f(x) \le f(a)$ (or $f(x) \ge f(a)$) for all x in the domain of f that are *sufficiently close* to a. If the appropriate inequality holds *for all* x in the domain of f, then we say that f has an *absolute maximum* (or *absolute minimum*) value at a. Moreover, such local or absolute extreme values can occur only at points of one of the following three types:

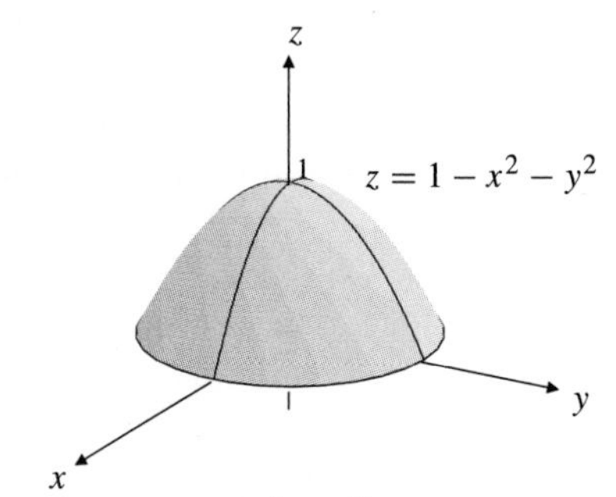

Figure 13.2 $1 - x^2 - y^2$ has maximum value 1 at the origin

(a) critical points, where $f'(x) = 0$,

(b) singular points, where $f'(x)$ does not exist, or

(c) endpoints of the domain of f.

A similar situation exists for functions of several variables. We say that a function of two variables has a **local maximum** or **relative maximum** value at the point (a, b) in its domain if $f(x, y) \leq f(a, b)$ for all points (x, y) in the domain of f that are *sufficiently close* to the point (a, b). If the inequality holds *for all* (x, y) in the domain of f, then we say that f has a **global maximum** or **absolute maximum** value at (a, b). Similar definitions hold for local (relative) and absolute (global) minimum values. In practice, the word *absolute* or *global* is usually omitted, and we refer simply to *the* maximum or *the* minimum value of f.

The following theorem shows that there are three possibilities for points where extreme values can occur, analogous to those for the single-variable case.

THEOREM 1 **Necessary conditions for extreme values**

A function $f(x, y)$ can have a local or absolute extreme value at a point (a, b) in its domain only if (a, b) is one of the following:

(a) a **critical point** of f, that is, a point satisfying $\nabla f(a, b) = \mathbf{0}$,

(b) a **singular point** of f, that is, a point where $\nabla f(a, b)$ does not exist, or

(c) a **boundary point** of the domain of f.

PROOF Suppose that (a, b) belongs to the domain of f. If (a, b) is not on the boundary of the domain of f, then it must belong to the interior of that domain, and if (a, b) is not a singular point of f, then $\nabla f(a, b)$ exists. Finally, if (a, b) is not a critical point of f, then $\nabla f(a, b) \neq \mathbf{0}$, so f has a positive directional derivative in the direction of $\nabla f(a, b)$ and a negative directional derivative in the direction of $-\nabla f(a, b)$; that is, f is increasing as we move from (a, b) in one direction and decreasing as we move in the opposite direction. Hence, f cannot have either a maximum or a minimum value at (a, b). Therefore, any point where an extreme value occurs must be either a critical point or a singular point of f, or a boundary point of the domain of f. ∎

Note that Theorem 1 remains valid with unchanged proof for functions of any number of variables. Of course, Theorem 1 does not guarantee that a given function will have any extreme values. It only tells us where to look to find any that may exist. Theorem 2, below, provides conditions that guarantee the existence of absolute maximum and minimum values for a continuous function. It is analogous to the Max-Min Theorem for functions of one variable. The proof is beyond the scope of this book; an interested student should consult an elementary text on mathematical analysis.

A set in $\mathbb{R}^n$ is **bounded** if it is contained inside some *ball* $x_1^2 + x_2^2 + \cdots + x_n^2 \leq R^2$ of finite radius R. A set on the real line is bounded if it is contained in an interval of finite length.

THEOREM 2 **Sufficient conditions for extreme values**

If f is a *continuous* function of n variables whose domain is a *closed* and *bounded* set in $\mathbb{R}^n$, then the range of f is a bounded set of real numbers, and there are points in its domain where f takes on absolute maximum and minimum values. ∎

EXAMPLE 1 The function $f(x, y) = x^2 + y^2$ (see Figure 13.1) has a critical point at $(0, 0)$, since $\nabla f = 2x\mathbf{i} + 2y\mathbf{j}$ and both components of ∇f vanish at $(0, 0)$. Since

$$f(x, y) > 0 = f(0, 0) \quad \text{if} \quad (x, y) \neq (0, 0),$$

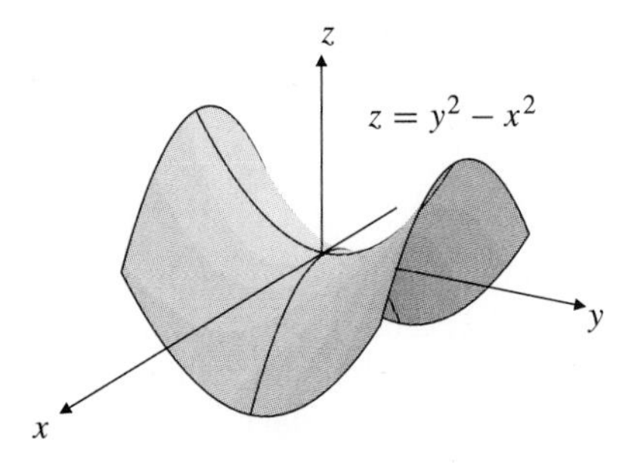

Figure 13.3 $y^2 - x^2$ has a saddle point at (0, 0)

f must have (absolute) minimum value 0 at that point. If the domain of f is not restricted, f has no maximum value. Similarly, $g(x, y) = 1 - x^2 - y^2$ has (absolute) maximum value 1 at its critical point (0, 0). (See Figure 13.2.)

EXAMPLE 2 The function $h(x, y) = y^2 - x^2$ also has a critical point at (0, 0) but has neither a local maximum nor a local minimum value at that point. Observe that $h(0, 0) = 0$ but $h(x, 0) < 0$ and $h(0, y) > 0$ for all nonzero values of x and y. (See Figure 13.3.) The graph of h is a hyperbolic paraboloid. In view of the shape of this surface, we call the critical point (0, 0) a *saddle point* of h.

In general, we will somewhat loosely call any *interior critical point* of the domain of a function f of several variables a **saddle point** if f does not have a local maximum or minimum value there. Even for functions of two variables, the graph will not always look like a saddle near a saddle point. For instance, the function $f(x, y) = -x^3$ has a whole line of saddle points along the y-axis (see Figure 13.4), although its graph does not resemble a saddle anywhere. These points resemble inflection points of a function of one variable. Saddle points are higher-dimensional analogues of such horizontal inflection points.

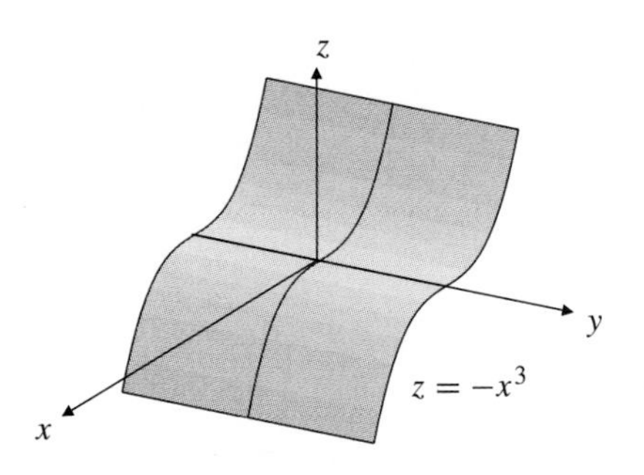

Figure 13.4 A line of saddle points

EXAMPLE 3 The function $f(x, y) = \sqrt{x^2 + y^2}$ has no critical points but does have a singular point at (0, 0) where it has a local (and absolute) minimum value, zero. The graph of f is (one nappe of) a circular cone. (See Figure 13.5(a).)

EXAMPLE 4 The function $f(x, y) = 1 - x$ is defined everywhere in the xy-plane and has no critical or singular points. ($\nabla f(x, y) = -\mathbf{i}$ at every point (x, y).) Therefore f has no extreme values. However, if we restrict the domain of f to the points in the disk $x^2 + y^2 \le 1$ (a closed bounded set in the xy-plane), then f does have absolute maximum and minimum values, as it must by Theorem 2. The maximum value is 2 at the boundary point $(-1, 0)$ and the minimum value is 0 at $(1, 0)$. (See Figure 13.5(b).)

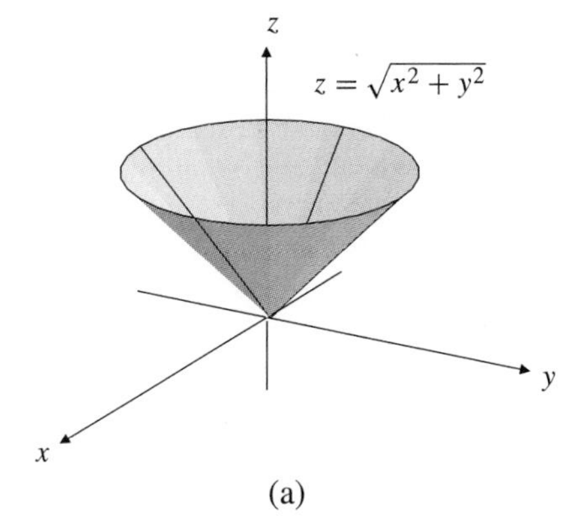

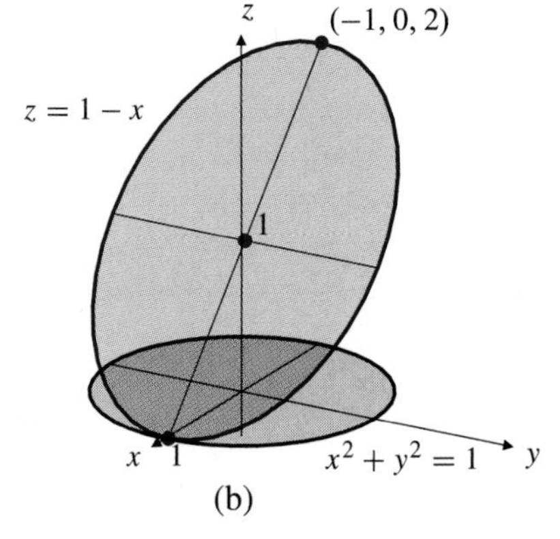

Figure 13.5
(a) $\sqrt{x^2 + y^2}$ has a minimum value at the singular point (0, 0)
(b) When restricted to the disk $x^2 + y^2 \le 1$, the function $1 - x$ has maximum and minimum values at boundary points

Classifying Critical Points

The above examples were very simple ones; it was immediately obvious in each case whether the function had a local maximum, local minimum, or a saddle point at the critical or singular point. For more complicated functions, it may be harder to classify the interior critical points. In theory, such a classification can always be made by considering the difference

$$\Delta f = f(a+h, b+k) - f(a,b)$$

for small values of h and k, where (a, b) is the critical point in question. If the difference is always nonnegative (or nonpositive) for small h and k, then f must have a local minimum (or maximum) at (a, b); if the difference is negative for some points (h, k) arbitrarily near $(0, 0)$ and positive for others, then f must have a saddle point at (a, b).

EXAMPLE 5 Find and classify the critical points of $f(x, y) = 2x^3 - 6xy + 3y^2$.

Solution The critical points must satisfy the system of equations:

$$\begin{aligned} 0 = f_1(x, y) &= 6x^2 - 6y \qquad \Longleftrightarrow \qquad x^2 = y \\ 0 = f_2(x, y) &= -6x + 6y \qquad \Longleftrightarrow \qquad x = y. \end{aligned}$$

Together, these equations imply that $x^2 = x$ so that $x = 0$ or $x = 1$. Therefore, the critical points are $(0, 0)$ and $(1, 1)$.

Consider $(0, 0)$. Here Δf is given by

$$\Delta f = f(h, k) - f(0, 0) = 2h^3 - 6hk + 3k^2.$$

Since $f(h, 0) - f(0, 0) = 2h^3$ is positive for small positive h and negative for small negative h, f cannot have a maximum or minimum value at $(0, 0)$. Therefore, $(0, 0)$ is a saddle point.

Now consider $(1, 1)$. Here Δf is given by

$$\begin{aligned} \Delta f &= f(1+h, 1+k) - f(1, 1) \\ &= 2(1+h)^3 - 6(1+h)(1+k) + 3(1+k)^2 - (-1) \\ &= 2 + 6h + 6h^2 + 2h^3 - 6 - 6h - 6k - 6hk + 3 + 6k + 3k^2 + 1 \\ &= 6h^2 - 6hk + 3k^2 + 2h^3 \\ &= 3(h-k)^2 + h^2(3+2h). \end{aligned}$$

Both terms in the latter expression are nonnegative if $|h| < 3/2$, and they are not both zero unless $h = k = 0$. Hence, $\Delta f > 0$ for small h and k, and f has a local minimum value -1 at $(1, 1)$.

The method used to classify critical points in the above example takes on a "brute force" aspect if the function involved is more complicated. However, there is a *second derivative test* similar to that for functions of one variable. The n-variable version is the subject of the following theorem, the proof of which is based on properties of quadratic forms presented in Section 10.7.

THEOREM 3 **A second derivative test**

Suppose that $\mathbf{a} = (a_1, a_2, \ldots, a_n)$ is a critical point of $f(\mathbf{x}) = f(x_1, x_2, \ldots, x_n)$ and is interior to the domain of f. Also, suppose that all the second partial derivatives of f are continuous throughout a neighbourhood of $\mathbf{a}$, so that the **Hessian matrix**

$$\mathcal{H}(\mathbf{x}) = \begin{pmatrix} f_{11}(\mathbf{x}) & f_{12}(\mathbf{x}) & \cdots & f_{1n}(\mathbf{x}) \\ f_{21}(\mathbf{x}) & f_{22}(\mathbf{x}) & \cdots & f_{2n}(\mathbf{x}) \\ \vdots & \vdots & \ddots & \vdots \\ f_{n1}(\mathbf{x}) & f_{n2}(\mathbf{x}) & \cdots & f_{nn}(\mathbf{x}) \end{pmatrix}$$

is also continuous in that neighbourhood. Note that the continuity of the partials guarantees that $\mathcal{H}$ is a symmetric matrix.

(a) If $\mathcal{H}(\mathbf{a})$ is positive definite, then f has a local minimum at $\mathbf{a}$.

(b) If $\mathcal{H}(\mathbf{a})$ is negative definite, then f has a local maximum at $\mathbf{a}$.

(c) If $\mathcal{H}(\mathbf{a})$ is indefinite, then f has a saddle point at $\mathbf{a}$.

(d) If $\mathcal{H}(\mathbf{a})$ is neither positive nor negative definite, nor indefinite, this test gives no information.

PROOF Let $g(t) = f(\mathbf{a} + t\mathbf{h})$ for $0 \le t \le 1$, where $\mathbf{h}$ is an n-vector. Then

$$g'(t) = \sum_{i=1}^{n} f_i(\mathbf{a} + t\mathbf{h})\, h_i$$

$$g''(t) = \sum_{i=1}^{n}\sum_{j=1}^{n} f_{ij}(\mathbf{a} + t\mathbf{h})\, h_i\, h_j = \mathbf{h}^T \mathcal{H}(\mathbf{a} + t\mathbf{h})\mathbf{h}.$$

(In the latter expression, $\mathbf{h}$ is being treated as a column vector.) We apply Taylor's Formula with Lagrange remainder to g to write

$$g(1) = g(0) + g'(0) + \frac{1}{2}g''(\theta)$$

for some θ between 0 and 1. Thus,

$$f(\mathbf{a} + \mathbf{h}) = f(\mathbf{a}) + \sum_{i=1}^{n} f_i(\mathbf{a})\, h_i + \frac{1}{2}\mathbf{h}^T \mathcal{H}(\mathbf{a} + \theta\mathbf{h})\mathbf{h}.$$

Since $\mathbf{a}$ is a critical point of f, $f_i(\mathbf{a}) = 0$ for $1 \le i \le n$, so

$$f(\mathbf{a} + \mathbf{h}) - f(\mathbf{a}) = \frac{1}{2}\mathbf{h}^T \mathcal{H}(\mathbf{a} + \theta\mathbf{h})\mathbf{h}.$$

If $\mathcal{H}(\mathbf{a})$ is positive definite, then, by the continuity of $\mathcal{H}$, so is $\mathcal{H}(\mathbf{a} + \theta\mathbf{h})$ for $|\mathbf{h}|$ sufficiently small. Therefore, $f(\mathbf{a} + \mathbf{h}) - f(\mathbf{a}) > 0$ for nonzero $\mathbf{h}$, proving (a).

Parts (b) and (c) are proved similarly. The functions $f(x, y) = x^4 + y^4$, $g(x, y) = -x^4 - y^4$, and $h(x, y) = x^4 - y^4$ all fall under part (d) and show that in this case a function can have a minimum, a maximum, or a saddle point.

■

Remark As mentioned in Section 12.9, the second derivative term $\mathbf{h}^T \mathcal{H}(\mathbf{a} + t\mathbf{h})\mathbf{h}$ is a second directional derivative. It can be thought of as a simple second derivative with respect to a single variable along a line L through $\mathbf{a}$ lying in the domain of f in the direction given by $\mathbf{h}$. This direction is not necessarily parallel to the given coordinate axes. Viewed as a simple second derivative, Theorem 9 from Section 4.5 tells us that the sign of this term determines the concavity of the curve in which the vertical plane containing L intersects the graph of f. This concavity makes sense even if $\mathbf{a}$ is not a critical point of f, and can vary as the direction of $\mathbf{h}$ changes. Therefore the Hessian can tell us about the concavity of the entire surface.

EXAMPLE 6 Find and classify the critical points of the function $f(x, y, z) = x^2y + y^2z + z^2 - 2x$.

Solution The equations that determine the critical points are

$$0 = f_1(x, y, z) = 2xy - 2,$$
$$0 = f_2(x, y, z) = x^2 + 2yz,$$
$$0 = f_3(x, y, z) = y^2 + 2z.$$

The third equation implies $z = -y^2/2$, and the second then implies $y^3 = x^2$. From the first equation we get $y^{5/2} = 1$. Thus, $y = 1$ and $z = -\frac{1}{2}$. Since $xy = 1$, we must have $x = 1$. The only critical point is $P = (1, 1, -\frac{1}{2})$. Evaluating the second partial derivatives of f at this point, we obtain the Hessian matrix

$$\mathcal{H} = \begin{pmatrix} 2 & 2 & 0 \\ 2 & -1 & 2 \\ 0 & 2 & 2 \end{pmatrix}.$$

Since

$$2 > 0, \quad \begin{vmatrix} 2 & 2 \\ 2 & -1 \end{vmatrix} = -6 < 0, \quad \begin{vmatrix} 2 & 2 & 0 \\ 2 & -1 & 2 \\ 0 & 2 & 2 \end{vmatrix} = -20 < 0,$$

$\mathcal{H}$ is indefinite by Theorem 8 of Section 10.7, so P is a saddle point of f.

Remark Applying the test (given in Theorem 8 of Section 10.7) for positive or negative definiteness or indefiniteness of a real symmetric matrix to the Hessian matrix for a function of two variables, we can paraphrase the second derivative test Theorem 3 for such a function:

Suppose that (a, b) is a critical point of the function $f(x, y)$ that is interior to the domain of f. Suppose also that the second partial derivatives of f are continuous in a neighbourhood of (a, b) and have at that point the values

$$A = f_{11}(a, b), \quad B = f_{12}(a, b) = f_{21}(a, b), \quad \text{and} \quad C = f_{22}(a, b).$$

(a) If $B^2 - AC < 0$ and $A > 0$, then f has a local minimum value at (a, b).
(b) If $B^2 - AC < 0$ and $A < 0$, then f has a local maximum value at (a, b).
(c) If $B^2 - AC > 0$, then f has a saddle point at (a, b).
(d) If $B^2 - AC = 0$, this test provides no information; f may have a local maximum or a local minimum value or a saddle point at (a, b).

EXAMPLE 7 Reconsider Example 5 and use the second derivative test to classify the two critical points $(0, 0)$ and $(1, 1)$ of $f(x, y) = 2x^3 - 6xy + 3y^2$.

Solution We have

$$f_{11}(x, y) = 12x, \quad f_{12}(x, y) = -6, \text{ and } \quad f_{22}(x, y) = 6.$$

At $(0, 0)$ we therefore have

$$A = 0, \quad B = -6, \quad C = 6, \text{ and } \quad B^2 - AC = 36 > 0,$$

so $(0, 0)$ is a saddle point. At $(1, 1)$ we have

$$A = 12 > 0, \quad B = -6, \quad C = 6, \text{ and } \quad B^2 - AC = -36 < 0,$$

so f must have a local minimum at $(1, 1)$.

EXAMPLE 8 Find and classify the critical points of

$$f(x, y) = xy\, e^{-(x^2+y^2)/2}.$$

Does f have absolute maximum and minimum values? Why?

Solution We begin by calculating the first- and second-order partial derivatives of the function f:

$$\begin{aligned}
f_1(x, y) &= y(1 - x^2)\, e^{-(x^2+y^2)/2}, \\
f_2(x, y) &= x(1 - y^2)\, e^{-(x^2+y^2)/2}, \\
f_{11}(x, y) &= xy(x^2 - 3)\, e^{-(x^2+y^2)/2}, \\
f_{12}(x, y) &= (1 - x^2)(1 - y^2)\, e^{-(x^2+y^2)/2}, \\
f_{22}(x, y) &= xy(y^2 - 3)\, e^{-(x^2+y^2)/2}.
\end{aligned}$$

At any critical point $f_1 = 0$ and $f_2 = 0$, so the critical points are the solutions of the system of equations

$$y(1-x^2) = 0 \qquad \text{(A)}$$
$$x(1-y^2) = 0. \qquad \text{(B)}$$

Equation (A) says that $y = 0$ or $x = \pm 1$. If $y = 0$, then equation (B) says that $x = 0$. If either $x = -1$ or $x = 1$, then equation (B) forces $y = \pm 1$. Thus, there are five points satisfying both equations: $(0,0)$, $(1,1)$, $(1,-1)$, $(-1,1)$, and $(-1,-1)$. We classify them using the second derivative test.

At $(0,0)$ we have $A = C = 0$, $B = 1$, so that $B^2 - AC = 1 > 0$. Thus, f has a saddle point at $(0,0)$.

At $(1,1)$ and $(-1,-1)$ we have $A = C = -2/e < 0$, $B = 0$. It follows that $B^2 - AC = -4/e^2 < 0$. Thus, f has local maximum values at these points. The value of f is $1/e$ at each point.

At $(1,-1)$ and $(-1,1)$ we have $A = C = 2/e > 0$, $B = 0$. If follows that $B^2 - AC = -4/e^2 < 0$. Thus, f has local minimum values at these points. The value of f at each of them is $-1/e$.

Indeed, f has absolute maximum and minimum values, namely, the values obtained above as local extrema. To see why, observe that $f(x,y)$ approaches 0 as the point (x,y) recedes to infinity in any direction because the negative exponential dominates the power factor xy for large $x^2 + y^2$. Pick a number between 0 and the local maximum value $1/e$ found above, say, the number $1/(2e)$. For some R, we must have $|f(x,y)| \le 1/(2e)$ whenever $x^2 + y^2 \ge R^2$. On the closed disk $x^2 + y^2 \le R^2$, f must have absolute maximum and minimum values by Theorem 2. These cannot occur on the boundary circle $x^2 + y^2 = R^2$ because $|f|$ is smaller there ($\le 1/(2e)$) than it is at the critical points considered above. Since f has no singular points, the absolute maximum and minimum values for the disk, and therefore for the whole plane, must occur at those critical points.

EXAMPLE 9 Find the shape of a rectangular box with no top having given volume V and the least possible total surface area of its five faces.

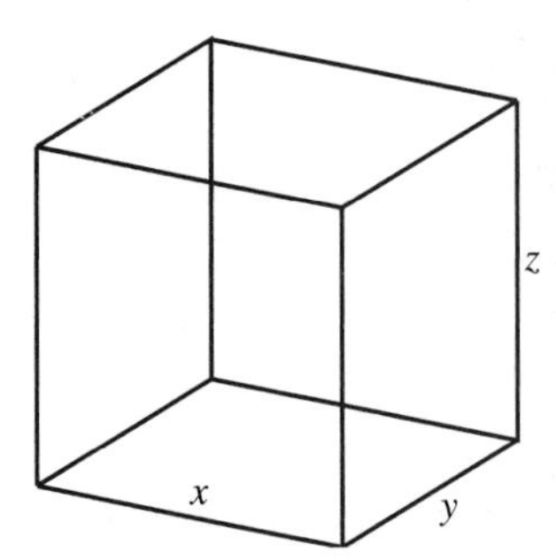

Figure 13.6 Dimensions of a box

Solution If the horizontal dimensions of the box are x, y, and its height is z (see Figure 13.6), then we want to minimize

$$S = xy + 2yz + 2xz$$

subject to the restriction that $xyz = V$, the required volume. We can use this restriction to reduce the number of variables on which S depends, for instance, by substituting

$$z = \frac{V}{xy}.$$

Then S becomes a function of the two variables x and y:

$$S = S(x,y) = xy + \frac{2V}{x} + \frac{2V}{y}.$$

A real box has positive dimensions, so the domain of S should consist of only those points (x,y) that satisfy $x > 0$ and $y > 0$. If either x or y approaches 0 or ∞, then $S \to \infty$, so the minimum value of S must occur at a critical point. (S has no singular points.) For critical points we solve the equations

$$0 = \frac{\partial S}{\partial x} = y - \frac{2V}{x^2} \qquad \Longleftrightarrow \qquad x^2 y = 2V,$$
$$0 = \frac{\partial S}{\partial y} = x - \frac{2V}{y^2} \qquad \Longleftrightarrow \qquad xy^2 = 2V.$$

Thus, $x^2y - xy^2 = 0$, or $xy(x-y) = 0$. Since $x > 0$ and $y > 0$, this implies that $x = y$. Therefore, $x^3 = 2V$, $x = y = (2V)^{1/3}$, and $z = V/(xy) = 2^{-2/3}V^{1/3} = x/2$. Since there is only one critical point, it must minimize S. (Why?) The box having minimal surface area has a square base but is only half as high as its horizontal dimensions.

Remark The preceding problem is a *constrained* extreme value problem in three variables; the equation $xyz = V$ is a *constraint* limiting the freedom of x, y, and z. We used the constraint to eliminate one variable, z, and so to reduce the problem to a *free* (i.e., *unconstrained*) problem in two variables. In Section 13.3 we will develop a more powerful method for solving constrained extreme value problems.

EXERCISES 13.1

In Exercises 1–17, find and classify the critical points of the given functions.

1. $f(x, y) = x^2 + 2y^2 - 4x + 4y$

2. $f(x, y) = xy - x + y$

3. $f(x, y) = x^3 + y^3 - 3xy$

4. $f(x, y) = x^4 + y^4 - 4xy$

5. $f(x, y) = \dfrac{x}{y} + \dfrac{8}{x} - y$

6. $f(x, y) = \cos(x + y)$

7. $f(x, y) = x \sin y$

8. $f(x, y) = \cos x + \cos y$

9. $f(x, y) = x^2 y e^{-(x^2+y^2)}$

10. $f(x, y) = \dfrac{xy}{2 + x^4 + y^4}$

11. $f(x, y) = x e^{-x^3+y^3}$

12. $f(x, y) = \dfrac{x^2}{x^2 + y^2}$

13. $f(x, y) = \dfrac{xy}{x^2 + y^2}$

14. $f(x, y) = \dfrac{1}{1 - x + y + x^2 + y^2}$

15. $f(x, y) = \left(1 + \dfrac{1}{x}\right)\left(1 + \dfrac{1}{y}\right)\left(\dfrac{1}{x} + \dfrac{1}{y}\right)$

16. $f(x, y, z) = xyz - x^2 - y^2 - z^2$

17. $f(x, y, z) = xy + x^2z - x^2 - y - z^2$

18. Show that $f(x, y, z) = 4xyz - x^4 - y^4 - z^4$ has a local maximum value at the point $(1, 1, 1)$.

19. Find the maximum and minimum values of

$f(x, y) = xy e^{-x^2-y^4}$.

20. Find the maximum and minimum values of

$f(x, y) = \dfrac{x}{(1 + x^2 + y^2)}$.

21. Find the maximum and minimum values of

$f(x, y, z) = xyz e^{-x^2-y^2-z^2}$. How do you know that such extreme values exist?

22. Find the minimum value of $f(x, y) = x + 8y + \dfrac{1}{xy}$ in the first quadrant $x > 0$, $y > 0$. How do you know that a minimum exists?

23. Postal regulations require that the sum of the height and girth (horizontal perimeter) of a package should not exceed L units. Find the largest volume of a rectangular box that can satisfy this requirement.

24. The material used to make the bottom of a rectangular box is twice as expensive per unit area as the material used to make the top or side walls. Find the dimensions of the box of given volume V for which the cost of materials is minimum.

25. Find the volume of the largest rectangular box (with faces parallel to the coordinate planes) that can be inscribed inside the ellipsoid

$$\frac{x^2}{a^2} + \frac{y^2}{b^2} + \frac{z^2}{c^2} = 1.$$

26. Find the three positive numbers a, b, and c, whose sum is 30 and for which the expression ab^2c^3 is maximum.

27. Find the critical points of the function $z = g(x, y)$ that satisfies the equation $e^{2zx-x^2} - 3e^{2zy+y^2} = 2$.

28. Classify the critical points of the function g in the previous exercise.

29. Let $f(x, y) = (y - x^2)(y - 3x^2)$. Show that the origin is a critical point of f and that the restriction of f to every straight line through the origin has a local minimum value at the origin. (That is, show that $f(x, kx)$ has a local minimum value at $x = 0$ for every k and that $f(0, y)$ has a local minimum value at $y = 0$.) Does $f(x, y)$ have a local minimum value at the origin? What happens to f on the curve $y = 2x^2$? What does the second derivative test say about this situation?

30. Verify by completing the square (i.e., without appealing to Theorem 8 of Section 10.7) that the quadratic form

$$Q(u, v) = (x, y)\begin{pmatrix} A & B \\ B & C \end{pmatrix}\begin{pmatrix} x \\ y \end{pmatrix} = Au^2 + 2Buv + Cv^2$$

is positive definite if $A > 0$ and $\begin{vmatrix} A & B \\ B & C \end{vmatrix} > 0$, negative definite if $A < 0$ and $\begin{vmatrix} A & B \\ B & C \end{vmatrix} > 0$, and indefinite if $\begin{vmatrix} A & B \\ B & C \end{vmatrix} < 0$. This gives independent confirmation of the assertion in the remark preceding Example 7.

31. State and prove (using square completion arguments rather than appealing to Theorem 8 of Section 10.7) a result analogous to that of Exercise 30 for a quadratic form $Q(u, v, w)$ involving three variables. What are the implications of this for a critical point (a, b, c) of a function $f(x, y, z)$ all of whose second partial derivatives are known at (a, b, c)?

13.2 Extreme Values of Functions Defined on Restricted Domains

Much of the previous section was concerned with techniques for determining whether a critical point of a function provides a local maximum or minimum value or is a saddle point. In this section we address the problem of determining absolute maximum and minimum values for functions that have them—usually functions whose domains are restricted to subsets of $\mathbb{R}^2$ (or $\mathbb{R}^n$) having nonempty interiors. In Example 8 of Section 13.1 we had to *prove* that the given function had absolute extreme values. If, however, we are dealing with a continuous function on a domain that is closed and bounded, then we can rely on Theorem 2 to guarantee the existence of such extreme values, but we will always have to check boundary points as well as any interior critical or singular points to find them. The following examples illustrate the technique.

How to find extreme values of a continuous function f on a closed, bounded domain D

1. Find any critical or singular points of f on the interior or D.
2. Find any points on the boundary of D where f might have extreme values. To do this you can parametrize the whole boundary, or parts of it, and express f as a function of the parameter(s). If you break the boundary into pieces, you must consider the endpoints of those pieces. Section 13.3 will present another alternative for analyzing f on the boundary of D.
3. Evaluate f at all the points found in steps 1 and 2.

EXAMPLE 1 Find the maximum and minimum values of $f(x, y) = 2xy$ on the closed disk $x^2 + y^2 \le 4$. (See Figure 13.7.)

Solution Since f is continuous and the disk is closed, f must have absolute maximum and minimum values at some points of the disk. The first partial derivatives of f are

$$f_1(x, y) = 2y \qquad \text{and} \qquad f_2(x, y) = 2x,$$

so there are no singular points, and the only critical point is $(0, 0)$, where f has the value 0.

We must still consider values of f on the boundary circle $x^2 + y^2 = 4$. We can express f as a function of a single variable on this circle by using a convenient parametrization of the circle, say,

$$x = 2\cos t, \qquad y = 2\sin t, \qquad (-\pi \le t \le \pi).$$

We have

$$f\Big(2\cos t, 2\sin t\Big) = 8\cos t \sin t = g(t).$$

We must find any extreme values of $g(t)$. We can do this in either of two ways. If we rewrite $g(t) = 4\sin 2t$, it is clear that $g(t)$ has maximum value 4 (at $t = \frac{\pi}{4}$ and $-\frac{3\pi}{4}$) and minimum value -4 (at $t = -\frac{\pi}{4}$ and $\frac{3\pi}{4}$). Alternatively, we can differentiate g to find its critical points:

$$0 = g'(t) = -8\sin^2 t + 8\cos^2 t \iff \tan^2 t = 1$$
$$\iff t = \pm\frac{\pi}{4} \text{ or } \pm\frac{3\pi}{4},$$

which again yield the maximum value 4 and the minimum value -4. (It is not necessary to check the endpoints $t = -\pi$ and $t = \pi$; since g is everywhere differentiable and is periodic with period π, any absolute maximum or minimum will occur at a critical point.)

In any event, f has maximum value 4 at the boundary points $(\sqrt{2}, \sqrt{2})$ and $(-\sqrt{2}, -\sqrt{2})$ and minimum value -4 at the boundary points $(\sqrt{2}, -\sqrt{2})$ and $(-\sqrt{2}, \sqrt{2})$. It is easily shown by the Second Derivative Test (or otherwise) that the interior critical point $(0, 0)$ is a saddle point. (See Figure 13.7.)

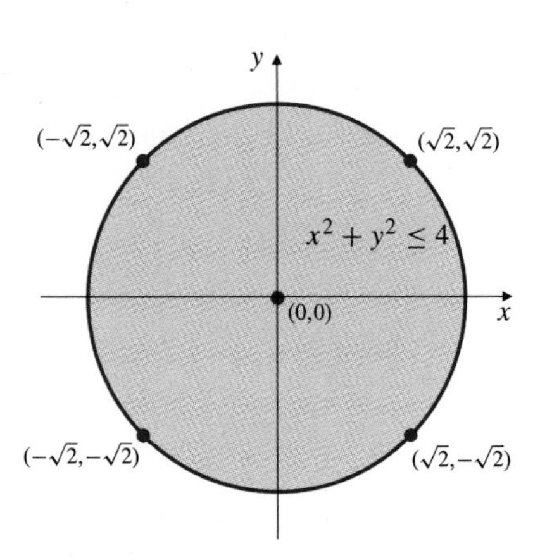

Figure 13.7 Points that are candidates for extreme values in Example 1

EXAMPLE 2 Find the extreme values of the function $f(x, y) = x^2 y e^{-(x+y)}$ on the triangular region T given by $x \geq 0$, $y \geq 0$, and $x + y \leq 4$.

Solution First, we look for critical points:

$$0 = f_1(x, y) = xy(2 - x)e^{-(x+y)} \quad \Longleftrightarrow \quad x = 0,\ y = 0, \text{ or } x = 2,$$
$$0 = f_2(x, y) = x^2(1 - y)e^{-(x+y)} \quad \Longleftrightarrow \quad x = 0 \text{ or } y = 1.$$

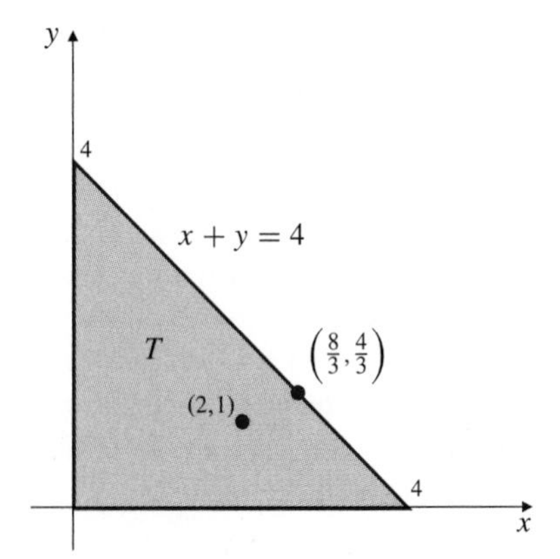

Figure 13.8 Points of interest in Example 2

The critical points are $(0, y)$ for any y and $(2, 1)$. Only $(2, 1)$ is an interior point of T. (See Figure 13.8.) $f(2, 1) = 4/e^3 \approx 0.199$. The boundary of T consists of three straight line segments. On two of these, the coordinate axes, f is identically zero. The third segment is given by

$$y = 4 - x, \qquad 0 \leq x \leq 4,$$

so the values of f on this segment can be expressed as a function of x alone:

$$g(x) = f(x, 4 - x) = x^2(4 - x)e^{-4}, \qquad 0 \leq x \leq 4.$$

Note that $g(0) = g(4) = 0$ and $g(x) > 0$ if $0 < x < 4$. The critical points of g are given by $0 = g'(x) = (8x - 3x^2)e^{-4}$, so they are $x = 0$ and $x = 8/3$. We have

$$g\left(\frac{8}{3}\right) = f\left(\frac{8}{3}, \frac{4}{3}\right) = \frac{256}{27}e^{-4} \approx 0.174 < f(2, 1).$$

We conclude that the maximum value of f over the region T is $4/e^3$ and that it occurs at the interior critical point $(2, 1)$. The minimum value of f is zero and occurs at all points of the two perpendicular boundary segments. Note that f has neither a local maximum nor a local minimum at the boundary point $(8/3, 4/3)$, although g has a local maximum there. Of course, that point is not a saddle point of f either; it is not a critical point of f.

EXAMPLE 3 Among all triangles with vertices on a given circle, find those that have the largest area.

Solution Intuition tells us that the equilateral triangles must have the largest area. However, proving this can be quite difficult unless a good choice of variables in which to set up the problem analytically is made. With a suitable choice of units and axes we can assume the circle is $x^2 + y^2 = 1$ and that one vertex of the triangle is the point P with coordinates $(1, 0)$. Let the other two vertices, Q and R, be as shown in Figure 13.9. There is no harm in assuming that Q lies on the upper semicircle and R on the lower, and that the origin O is inside triangle PQR. Let PQ and PR make angles θ and ϕ, respectively, with the negative direction of the x-axis. Clearly $0 \leq \theta \leq \pi/2$ and $0 \leq \phi \leq \pi/2$. The lines from O to Q and R make equal angles ψ with the line QR, where $2\theta + 2\phi + 2\psi = \pi$. Dropping perpendiculars from O to the three sides of the triangle PQR, we can write the area A of the triangle as the sum of the areas of six small, right-angled triangles:

$$\begin{aligned} A &= 2 \times \frac{1}{2}\sin\theta\cos\theta + 2 \times \frac{1}{2}\sin\phi\cos\phi + 2 \times \frac{1}{2}\sin\psi\cos\psi \\ &= \frac{1}{2}\big(\sin 2\theta + \sin 2\phi + \sin 2\psi\big). \end{aligned}$$

Since $2\psi = \pi - 2(\theta + \phi)$, we express A as a function of the two variables θ and ϕ:

$$A = A(\theta, \phi) = \frac{1}{2}\big(\sin 2\theta + \sin 2\phi + \sin 2(\theta + \phi)\big).$$

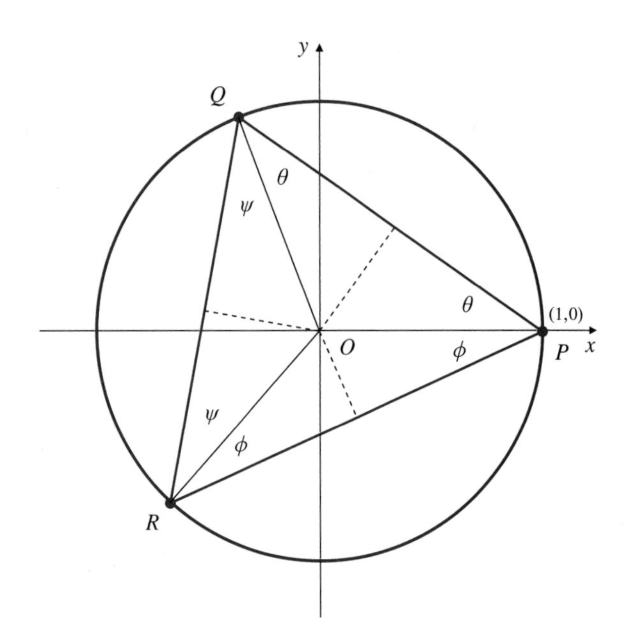

Figure 13.9 Where should Q and R be to ensure that triangle PQR has maximum area?

The domain of A is the triangle $\theta \ge 0$, $\phi \ge 0$, $\theta + \phi \le \pi/2$. $A = 0$ at the vertices of the triangle and is positive elsewhere. (See Figure 13.10.) We show that the maximum value of $A(\theta, \phi)$ on any edge of the triangle is 1 and occurs at the midpoint of that edge. On the edge $\theta = 0$ we have

$$A(0, \phi) = \frac{1}{2}\big(\sin 2\phi + \sin 2\phi\big) = \sin 2\phi \le 1 = A(0, \pi/4).$$

Similarly, on $\phi = 0$, $A(\theta, 0) \le 1 = A(\pi/4, 0)$. On the edge $\theta + \phi = \pi/2$ we have

$$\begin{aligned} A\Big(\theta, \frac{\pi}{2} - \theta\Big) &= \frac{1}{2}\big(\sin 2\theta + \sin(\pi - 2\theta)\big) \\ &= \sin 2\theta \le 1 = A\Big(\frac{\pi}{4}, \frac{\pi}{4}\Big). \end{aligned}$$

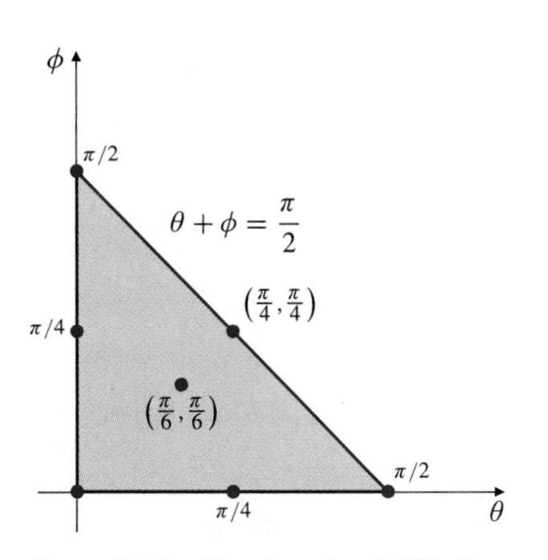

Figure 13.10 The domain of $A(\theta, \phi)$

We must now check for any interior critical points of $A(\theta, \phi)$. (There are no singular points.) For critical points we have

$$\begin{aligned} 0 &= \frac{\partial A}{\partial \theta} = \cos 2\theta + \cos(2\theta + 2\phi), \\ 0 &= \frac{\partial A}{\partial \phi} = \cos 2\phi + \cos(2\theta + 2\phi), \end{aligned}$$

so the critical points satisfy $\cos 2\theta = \cos 2\phi$ and, hence, $\theta = \phi$. We now substitute this equation into either of the above equations to determine θ:

$$\begin{aligned} \cos 2\theta + \cos 4\theta &= 0 \\ 2\cos^2 2\theta + \cos 2\theta - 1 &= 0 \\ (2\cos 2\theta - 1)(\cos 2\theta + 1) &= 0 \\ \cos 2\theta = \frac{1}{2} \quad &\text{or} \quad \cos 2\theta = -1. \end{aligned}$$

The only solution leading to an interior point of the domain of A is $\theta = \phi = \pi/6$. Note that

$$A\Big(\frac{\pi}{6}, \frac{\pi}{6}\Big) = \frac{1}{2}\left(\frac{\sqrt{3}}{2} + \frac{\sqrt{3}}{2} + \frac{\sqrt{3}}{2}\right) = \frac{3\sqrt{3}}{4} > 1;$$

this interior critical point maximizes the area of the inscribed triangle. Finally, observe that for $\theta = \phi = \pi/6$, we also have $\psi = \pi/6$, so the largest triangle is indeed equilateral.

Remark Since the area A of the inscribed triangle must have a maximum value (A is continuous and its domain is closed and bounded), a strictly geometric argument can be used to show that the largest triangle is equilateral. If an inscribed triangle has two unequal sides, its area can be made larger by moving the common vertex of these two sides along the circle to increase its perpendicular distance from the opposite side of the triangle.

Linear Programming

Linear programming is a branch of linear algebra that develops systematic techniques for finding maximum or minimum values of a *linear function* subject to several *linear inequality constraints.* Such problems arise frequently in management science and operations research. Because of their linear nature they do not usually involve calculus in their solution; linear programming is frequently presented in courses on *finite mathematics.* We will not attempt any formal study of linear programming here, but we will make a few observations for comparison with the more general nonlinear extreme-value problems considered above that involve calculus in their solution.

The inequality $ax + by \le c$ is an example of a linear inequality in two variables. The *solution set* of this inequality consists of a half-plane lying on one side of the straight line $ax + by = c$. The solution set of a system of several two-variable linear inequalities is an intersection of such half-planes, so it is a *convex* region of the plane bounded by a *polygonal line.* If it is a bounded set, then it is a convex polygon together with its interior. (A set is called **convex** if it contains the entire line segment between any two of its points. On the real line the convex sets are intervals.)

Let us examine a simple concrete example that involves only two variables and a few constraints.

EXAMPLE 4 Find the maximum value of $F(x, y) = 2x + 7y$ subject to the constraints $x + 2y \le 6$, $\quad 2x + y \le 6$, $\quad x \ge 0$, $\quad$ and $\quad y \ge 0$.

Solution The solution set $\mathcal{S}$ of the system of four constraint inequalities is shown in Figure 13.11. It is the quadrilateral region with vertices $(0, 0)$, $(3, 0)$, $(2, 2)$, and $(0, 3)$. Several level curves of the linear function F are also shown in the figure. They are parallel straight lines with slope $-\frac{2}{7}$. We want the line that gives F the greatest value and that still intersects $\mathcal{S}$. Evidently this is the line $F = 21$ that passes through the vertex $(0, 3)$ of $\mathcal{S}$. The maximum value of F subject to the constraints is 21.

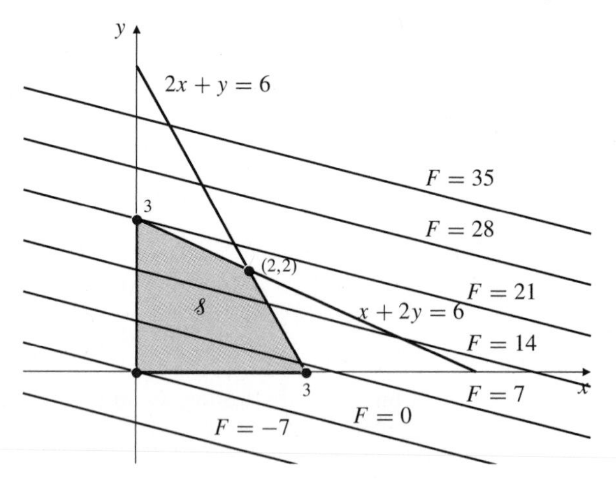

Figure 13.11 The shaded region is the solution set for the constraint inequalities in Example 4

As this simple example illustrates, a *linear* function with domain restricted by *linear inequalities* does not achieve maximum or minimum values at points in the interior of its domain (if that domain has an interior). Any such extreme value occurs at a boundary point of the domain or a set of such boundary points. Where an extreme value occurs at a set of boundary points, that set will *always* contain at least one vertex.

This phenomenon holds in general for extreme-value problems for linear functions in any number of variables with domains restricted by any number of linear inequalities. For problems involving three variables the domain will be a convex region of $\mathbb{R}^3$ bounded by planes. For a problem involving n variables the domain will be a convex region in $\mathbb{R}^n$ bounded by $(n-1)$-dimensional hyperplanes. Such *polyhedral* regions still have vertices (where n hyperplanes intersect), and maximum or minimum values of linear functions subject to the constraints will still occur at subsets of the boundary containing such vertices. These problems can therefore be solved by evaluating the linear function to be extremized (it is called the **objective function**) at all the vertices and selecting the greatest or least value.

In practice, linear programming problems can involve hundreds or even thousands of variables and even more constraints. Such problems need to be solved with computers, but even then it is extremely inefficient, if not impossible, to calculate all the vertices of the constraint solution set and the values of the objective function at them. Much of the study of linear programming therefore centres on devising techniques for getting to (or at least near) the optimizing vertex in as few steps as possible. Usually this involves criteria whereby large numbers of vertices can be rejected on geometric grounds. We will not delve into such techniques here but will content ourselves with one more example to illustrate, in a very simple case, how the underlying geometry of a problem can be used to reduce the number of vertices that must be considered.

EXAMPLE 5 A tailor has 230 m of a certain fabric and has orders for up to 20 suits, up to 30 jackets, and up to 40 pairs of slacks to be made from the fabric. Each suit requires 6 m, each jacket 3 m, and each pair of slacks 2 m of the fabric. If the tailor's profit is \$20 per suit, \$14 per jacket, and \$12 per pair of slacks, how many of each should he make to realize the maximum profit from his supply of the fabric?

Solution Suppose he makes x suits, y jackets, and z pairs of slacks. Then his profit will be

$$P = 20x + 14y + 12z.$$

The constraints posed in the problem are

$$\begin{aligned} x &\geq 0, & x &\leq 20, \\ y &\geq 0, & y &\leq 30, \\ z &\geq 0, & z &\leq 40, \end{aligned}$$

$$6x + 3y + 2z \leq 230.$$

The last inequality is due to the limited supply of fabric. The solution set is shown in Figure 13.12. It has 10 vertices, $A, B, \ldots, J$. Since P increases in the direction of the vector $\nabla P = 20\mathbf{i} + 14\mathbf{j} + 12\mathbf{k}$, which points into the first octant, its maximum value cannot occur at any of the vertices $A, B, \ldots, G$. (Think about why.) Thus, we need look only at the vertices H, I, and J.

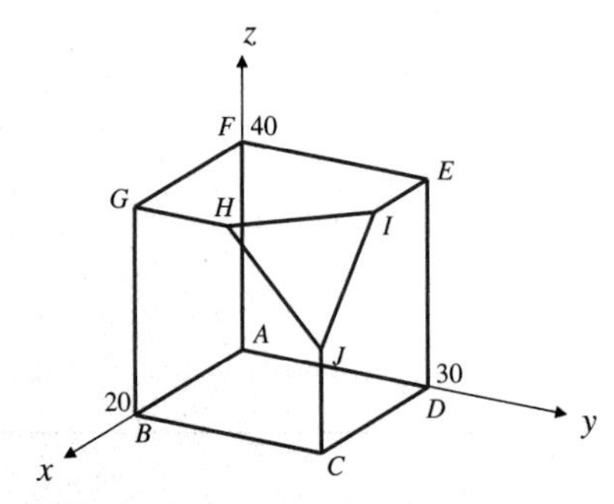

Figure 13.12 The convex set of points satisfying the constraints in Example 5

$$\begin{aligned} H &= (20, 10, 40), & P &= 1{,}020 \text{ at } H. \\ I &= (10, 30, 40), & P &= 1{,}100 \text{ at } I. \\ J &= (20, 30, 10), & P &= 940 \text{ at } J. \end{aligned}$$

Thus, the tailor should make 10 suits, 30 jackets, and 40 pairs of slacks to realize the maximum profit, \$1,100, from the fabric.

EXERCISES 13.2

1. Find the maximum and minimum values of $f(x, y) = x - x^2 + y^2$ on the rectangle $0 \le x \le 2$, $0 \le y \le 1$.

2. Find the maximum and minimum values of $f(x, y) = xy - 2x$ on the rectangle $-1 \le x \le 1, 0 \le y \le 1$.

3. Find the maximum and minimum values of $f(x, y) = xy - y^2$ on the disk $x^2 + y^2 \le 1$.

4. Find the maximum and minimum values of $f(x, y) = x + 2y$ on the disk $x^2 + y^2 \le 1$.

5. Find the maximum value of $f(x, y) = xy - x^3y^2$ over the square $0 \le x \le 1, 0 \le y \le 1$.

6. Find the maximum and minimum values of $f(x, y) = xy(1 - x - y)$ over the triangle with vertices $(0, 0)$, $(1, 0)$, and $(0, 1)$.

7. Find the maximum and minimum values of $f(x, y) = \sin x \cos y$ on the closed triangular region bounded by the coordinate axes and the line $x + y = 2\pi$.

8. Find the maximum value of

$$f(x, y) = \sin x \sin y \sin(x + y)$$

over the triangle bounded by the coordinate axes and the line $x + y = \pi$.

9. The temperature at all points in the disk $x^2 + y^2 \le 1$ is given by

$$T = (x + y)\, e^{-x^2-y^2}.$$

Find the maximum and minimum temperatures at points of the disk.

10. Find the maximum and minimum values of

$$f(x, y) = \frac{x - y}{1 + x^2 + y^2}$$

on the upper half-plane $y \ge 0$.

11. Find the maximum and minimum values of $xy^2 + yz^2$ over the ball $x^2 + y^2 + z^2 \le 1$.

12. Find the maximum and minimum values of $xz + yz$ over the ball $x^2 + y^2 + z^2 \le 1$.

13. Consider the function $f(x, y) = xy\, e^{-xy}$ with domain the first quadrant: $x \ge 0,\ y \ge 0$. Show that $\lim_{x\to\infty} f(x, kx) = 0$. Does f have a limit as (x, y) recedes arbitrarily far from the origin in the first quadrant? Does f have a maximum value in the first quadrant?

14. Repeat Exercise 13 for the function $f(x, y) = xy^2\, e^{-xy}$.

15. In a certain community there are two breweries in competition, so that sales of each negatively affect the profits of the other. If brewery A produces x litres of beer per month and brewery B produces y litres per month, then brewery A's monthly profit $\$P$ and brewery B's monthly profit $\$Q$ are assumed to be

$$P = 2x - \frac{2x^2 + y^2}{10^6},$$
$$Q = 2y - \frac{4y^2 + x^2}{2 \times 10^6}.$$

Find the sum of the profits of the two breweries if each brewery independently sets its own production level to maximize its own profit and assumes its competitor does likewise. Find the sum of the profits if the two breweries cooperate to determine their respective productions to maximize that sum.

16. Equal angle bends are made at equal distances from the two ends of a 100 m long straight length of fence so the resulting three-segment fence can be placed along an existing wall to make an enclosure of trapezoidal shape. What is the largest possible area for such an enclosure?

17. Maximize $Q(x, y) = 2x + 3y$ subject to the constraints $x \ge 0$, $y \ge 0$, $y \le 5$, $x + 2y \le 12$, and $4x + y \le 12$.

18. Minimize $F(x, y, z) = 2x + 3y + 4z$ subject to the constraints $x \ge 0$, $y \ge 0$, $z \ge 0$, $x + y \ge 2$, $y + z \ge 2$, and $x + z \ge 2$.

19. A textile manufacturer produces two grades of wool-cotton-polyester fabric. The deluxe grade has composition (by weight) 20% wool, 50% cotton, and 30% polyester, and it sells for \$3 per kilogram. The standard grade has composition 10% wool, 40% cotton, and 50% polyester, and sells for \$2 per kilogram. If he has in stock 2,000 kg of wool and 6,000 kg each of cotton and polyester, how many kilograms of fabric of each grade should he manufacture to maximize his revenue?

20. A 10-hectare parcel of land is zoned for building densities of 6 detached houses per hectare, 8 duplex units per hectare, or 12 apartments per hectare. The developer who owns the land can make a profit of \$40,000 per house, \$20,000 per duplex unit, and \$16,000 per apartment that he builds. Municipal bylaws require him to build at least as many apartments as the total of houses and duplex units. How many of each type of dwelling should he build to maximize his profit?

13.3 Lagrange Multipliers

A constrained extreme-value problem is one in which the variables of the function to be maximized or minimized are not completely independent of one another, but must satisfy one or more constraint equations or inequalities. For instance, the problems

$$\textbf{maximize} \quad f(x, y) \quad \textbf{subject to} \quad g(x, y) = C$$

and

$$\textbf{minimize} \quad f(x, y, z, w) \quad \textbf{subject to} \quad g(x, y, z, w) = C_1, \quad \textbf{and} \quad h(x, y, z, w) = C_2$$

have, respectively, one and two constraint equations, while the problem

$$\textbf{maximize} \quad f(x, y, z) \quad \textbf{subject to} \quad g(x, y, z) \le C$$

has a single constraint inequality.

Generally, inequality constraints can be regarded as restricting the domain of the function to be extremized to a smaller set that still has interior points. Section 13.2 was devoted to such problems. In each of the first three examples of that section we looked for *free* (i.e., *unconstrained*) extreme values in the interior of the domain, and we also examined the boundary of the domain, which was specified by one or more *constraint equations*. In Example 1 we parametrized the boundary and expressed the function to be extremized as a function of the parameter, thus reducing the boundary case to a free problem in one variable instead of a constrained problem in two variables. In Example 2 the boundary consisted of three line segments, on two of which the function was obviously zero. We solved the equation for the third boundary segment for y in terms of x, again in order to express the values of $f(x, y)$ on that segment as a function of one free variable. A similar approach was used in Example 3 to deal with the triangular boundary of the domain of the area function $A(\theta, \phi)$.

The reduction of extremization problems with equation constraints to free problems with fewer independent variables is only feasible when the constraint equations can be solved either explicitly for some variables in terms of others or parametrically for all variables in terms of some parameters. It is often very difficult or impossible to solve the constraint equations, so we need another technique.

The Method of Lagrange Multipliers

A technique for finding extreme values of $f(x, y)$ subject to the equality constraint $g(x, y) = 0$ is based on the following theorem:

THEOREM 4

Suppose that f and g have continuous first partial derivatives near the point $P_0 = (x_0, y_0)$ on the curve $\mathcal{C}$ with equation $g(x, y) = 0$. Suppose also that, when restricted to points on $\mathcal{C}$, the function $f(x, y)$ has a local maximum or minimum value at P_0. Finally, suppose that

(i) P_0 is not an endpoint of $\mathcal{C}$, and

(ii) $\nabla g(P_0) \ne \mathbf{0}$.

Then there exists a number λ_0 such that (x_0, y_0, λ_0) is a critical point of the **Lagrange function**

$$L(x, y, \lambda) = f(x, y) + \lambda g(x, y).$$

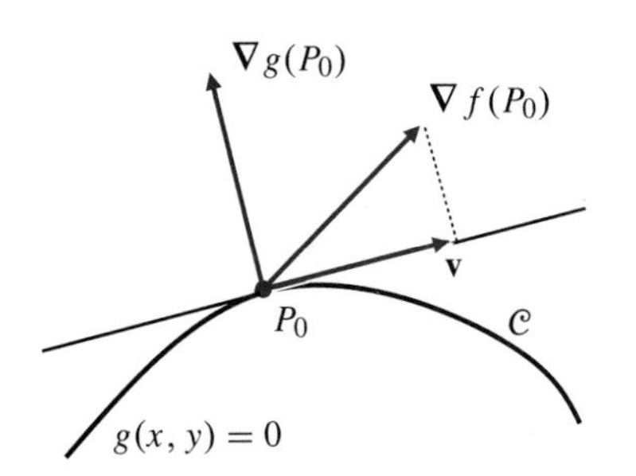

Figure 13.13 If $\nabla f(P_0)$ is not a multiple of $\nabla g(P_0)$, then $\nabla f(P_0)$ has a nonzero projection $\mathbf{v}$ tangent to the level curve of g through P_0

PROOF Together, (i) and (ii) imply that $\mathcal{C}$ is smooth enough to have a tangent line at P_0 and that $\nabla g(P_0)$ is normal to that tangent line. If $\nabla f(P_0)$ is not parallel to $\nabla g(P_0)$, then $\nabla f(P_0)$ has a nonzero vector projection $\mathbf{v}$ along the tangent line to $\mathcal{C}$ at P_0. (See Figure 13.13.) Therefore, f has a positive directional derivative at P_0 in the direction of $\mathbf{v}$ and a negative directional derivative in the opposite direction. Thus, $f(x, y)$ increases or decreases as we move away from P_0 along $\mathcal{C}$ in the direction of $\mathbf{v}$ or $-\mathbf{v}$, and f cannot have a maximum or minimum value at P_0. Since we are assuming that f *does* have an extreme value at P_0, it must be that $\nabla f(P_0)$ is parallel to $\nabla g(P_0)$. Since $\nabla g(P_0) \ne \mathbf{0}$, there must exist a real number λ_0 such that $\nabla f(P_0) = -\lambda_0 \nabla g(P_0)$, or

$$\nabla(f + \lambda_0 g)(P_0) = \mathbf{0}.$$

The two components of the above vector equation assert that $\partial L/\partial x = 0$ and $\partial L/\partial y = 0$ at (x_0, y_0, λ_0). The third equation that must be satisfied by a critical point of L is $\partial L/\partial \lambda = g(x, y) = 0$. This is satisfied at (x_0, y_0, λ_0) because P_0 lies on $\mathcal{C}$. Thus, (x_0, y_0, λ_0) is a critical point of $L(x, y, \lambda)$.

Theorem 4 suggests that to find candidates for points on the curve $g(x, y) = 0$ at which $f(x, y)$ is maximum or minimum, we should look for critical points of the Lagrange function

$$L(x, y, \lambda) = f(x, y) + \lambda g(x, y).$$

At any critical point of L we must have

$$\left.\begin{aligned} 0 &= \frac{\partial L}{\partial x} = f_1(x, y) + \lambda g_1(x, y), \\ 0 &= \frac{\partial L}{\partial y} = f_2(x, y) + \lambda g_2(x, y), \end{aligned}\right\} \quad \text{that is, } \nabla f \text{ is parallel to } \nabla g,$$

$$\text{and} \quad 0 = \frac{\partial L}{\partial \lambda} = g(x, y), \qquad \text{the constraint equation.}$$

Note, however, that it is *assumed* that the constrained problem *has a solution.* Theorem 4 does not guarantee that a solution exists; it only provides a means for finding a solution already known to exist. It is usually necessary to satisfy yourself that the problem you are trying to solve has a solution before using this method to find the solution.

Let us put the method to a concrete test:

EXAMPLE 1 Find the shortest distance from the origin to the curve $x^2 y = 16$.

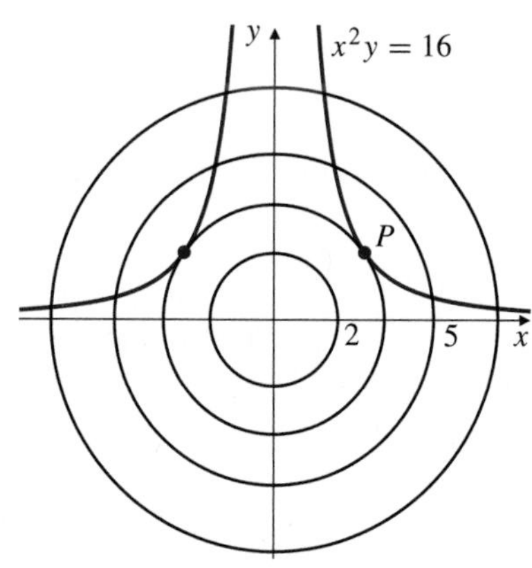

Figure 13.14 The level curve of the function representing the square of distance from the origin is tangent to the curve $x^2 y = 16$ at the two points on that curve that are closest to the origin

Solution The graph of $x^2 y = 16$ is shown in Figure 13.14. There appear to be two points on the curve that are closest to the origin and no points that are farthest from the origin. (The curve is unbounded.) To find the closest points it is sufficient to minimize the *square* of the distance from the point (x, y) on the curve to the origin. (It is easier to work with the square of the distance rather than the distance itself, which involves a square root and so is harder to differentiate.) Thus, we want to solve the problem

$$\textbf{minimize} \quad f(x, y) = x^2 + y^2 \quad \textbf{subject to} \quad g(x, y) = x^2 y - 16 = 0.$$

Let $L(x, y, \lambda) = x^2 + y^2 + \lambda(x^2 y - 16)$. For critical points of L we want

$$0 = \frac{\partial L}{\partial x} = 2x + 2\lambda xy = 2x(1 + \lambda y) \qquad \text{(A)}$$

$$0 = \frac{\partial L}{\partial y} = 2y + \lambda x^2 \qquad \text{(B)}$$

$$0 = \frac{\partial L}{\partial \lambda} = x^2 y - 16. \qquad \text{(C)}$$

Equation (A) requires that either $x = 0$ or $\lambda y = -1$. However, $x = 0$ is inconsistent with equation (C). Therefore $\lambda y = -1$. From equation (B) we now have

$$0 = 2y^2 + \lambda y x^2 = 2y^2 - x^2.$$

Thus, $x = \pm\sqrt{2}y$, and (C) now gives $2y^3 = 16$, so $y = 2$. There are, therefore, two candidates for points on $x^2 y = 16$ closest to the origin, $(\pm 2\sqrt{2}, 2)$. Both of these points are at distance $\sqrt{8 + 4} = 2\sqrt{3}$ units from the origin, so this must be the minimum distance from the origin to the curve. Some level curves of $x^2 + y^2$ are shown, along with the constraint curve $x^2 y = 16$, in Figure 13.14. Observe how the constraint curve is tangent to the level curve passing through the minimizing points $(\pm 2\sqrt{2}, 2)$, reflecting the fact that the two curves have parallel normals there.

Remark In the above example we could, of course, have solved the constraint equation for $y = 16/x^2$, substituted into f, and thus reduced the problem to one of finding the (unconstrained) minimum value of

$$F(x) = f\left(x, \frac{16}{x^2}\right) = x^2 + \frac{256}{x^4}.$$

The reader is invited to verify that this gives the same result.

The number λ that occurs in the Lagrange function is called a **Lagrange multiplier**. The technique for solving an extreme-value problem with equation constraints by looking for critical points of an unconstrained problem in more variables (the original variables plus a Lagrange multiplier corresponding to each constraint equation) is called the **method of Lagrange multipliers**. It can be expected to give results as long as the function to be maximized or minimized (called the **objective function** or **cost function**) and the constraint equations have *smooth* graphs in a neighbourhood of the points where the extreme values occur, and these points are not on *edges* of those graphs. See Example 3 and Exercise 26 below.

EXAMPLE 2 Find the points on the curve $17x^2 + 12xy + 8y^2 = 100$ that are closest to and farthest away from the origin.

Solution The quadratic form on the left side of the equation above is positive definite, as can be seen by completing a square. Hence, the curve is bounded and must have points closest to and farthest from the origin. (In fact, the curve is an ellipse with centre at the origin and oblique principal axes. The problem asks us to find the ends of the major and minor axes.)

Again, we want to extremize $x^2 + y^2$ subject to an equation constraint. The Lagrange function in this case is

$$L(x, y, \lambda) = x^2 + y^2 + \lambda(17x^2 + 12xy + 8y^2 - 100),$$

and its critical points are given by

$$0 = \frac{\partial L}{\partial x} = 2x + \lambda(34x + 12y) \qquad \text{(A)}$$

$$0 = \frac{\partial L}{\partial y} = 2y + \lambda(12x + 16y) \qquad \text{(B)}$$

$$0 = \frac{\partial L}{\partial \lambda} = 17x^2 + 12xy + 8y^2 - 100. \qquad \text{(C)}$$

Solving each of equations (A) and (B) for λ and equating the two expressions for λ obtained, we get

$$\frac{-2x}{34x + 12y} = \frac{-2y}{12x + 16y} \qquad \text{or} \qquad 12x^2 + 16xy = 34xy + 12y^2.$$

This equation simplifies to

$$2x^2 - 3xy - 2y^2 = 0. \qquad \text{(D)}$$

We multiply equation (D) by 4 and add the result to equation (C) to get $25x^2 = 100$, so that $x = \pm 2$. Finally, we substitute each of these values of x into (D) and obtain (for each) two values of y from the resulting quadratics:

$$\text{For } x = 2: \quad y^2 + 3y - 4 = 0, \quad (y-1)(y+4) = 0. \qquad \text{For } x = -2: \quad y^2 - 3y - 4 = 0, \quad (y+1)(y-4) = 0.$$

We therefore obtain four candidate points: $(2, 1)$, $(-2, -1)$, $(2, -4)$, and $(-2, 4)$. The first two points are closest to the origin (they are the ends of the minor axis of the ellipse); the other two are farthest from the origin (the ends of the major axis). (See Figure 13.15.)

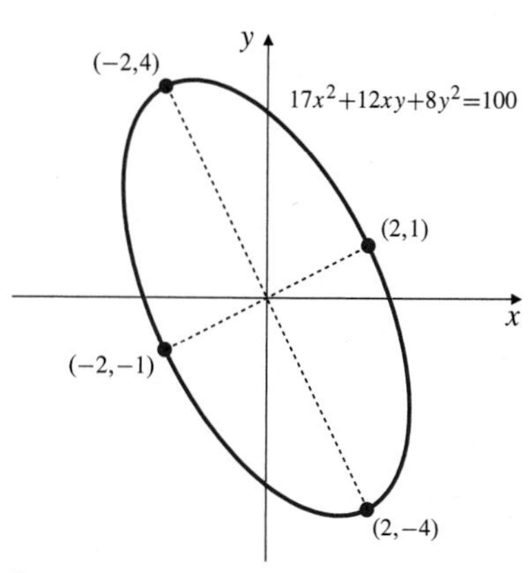

Figure 13.15 The points on the ellipse that are closest to and farthest from the origin

Considering the geometric underpinnings of the method of Lagrange multipliers, we would not expect the method to work if the level curves of the functions involved are not smooth or if the maximum or minimum occurs at an endpoint of the constraint curve. One of the pitfalls of the method is that the level curves of functions may not be smooth, even though the functions themselves have partial derivatives. Problems can occur where a gradient vanishes, as the following example shows.

EXAMPLE 3 Find the minimum value of $f(x, y) = y$ subject to the constraint equation $g(x, y) = y^3 - x^2 = 0$.

Solution The semicubical parabola $y^3 = x^2$ has a cusp at the origin. (See Figure 13.16.) Clearly, $f(x, y) = y$ has minimum value 0 at that point. Suppose, however, that we try to solve the problem using the method of Lagrange multipliers. The Lagrange function here is

$$L(x, y, \lambda) = y + \lambda(y^3 - x^2),$$

which has critical points given by

$$\begin{aligned} -2\lambda x &= 0, \\ 1 + 3\lambda y^2 &= 0, \\ y^3 - x^2 &= 0. \end{aligned}$$

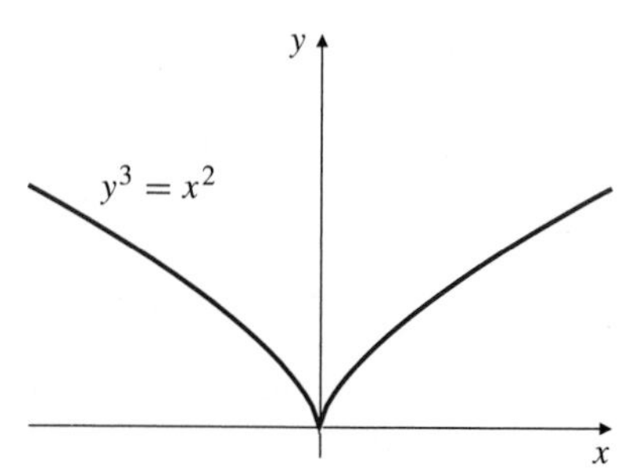

Figure 13.16 The minimum of y occurs at a point on the curve where the curve has no tangent line

Observe that $y = 0$ cannot satisfy the second equation, and, in fact, the three equations have *no solution* (x, y, λ). (The first equation implies either $\lambda = 0$ or $x = 0$, but neither of these is consistent with the other two equations.)

Remark The method of Lagrange multipliers breaks down in the above example because $\nabla g = \mathbf{0}$ at the solution point, and therefore the curve $g(x, y) = 0$ need not be smooth there. (In this case, it isn't smooth!) The geometric condition that ∇f should be parallel to ∇g at the solution point is meaningless in this case. When applying the method of Lagrange multipliers, be aware that an extreme value may occur at

(i) a critical point of the Lagrange function,

(ii) a point where $\nabla g = \mathbf{0}$,

(iii) a point where ∇f or ∇g does not exist, or

(iv) an "endpoint" of the constraint set.

This situation is similar to that for extreme values of a function f of one variable, which can occur at a critical point of f, a singular point of f, or an endpoint of the domain of f.

EXAMPLE 4 Find the maximum and minimum values of $f(x, y, z) = xy^2z^3$ on the ball $x^2 + y^2 + z^2 \le 1$.

Solution Since $f_1(x, y, z) = y^2z^3 = 0$ only if either $y = 0$ or $z = 0$, there can be no critical points of f where $f(x, y, z) \ne 0$. Evidently (x, y, z) is positive at some points in the ball, and negative at others, so no interior critical points can provide a maximum or minimum value for f on the ball. Therefore, these extreme values must occur on the boundary sphere $x^2 + y^2 + z^2 = 1$. To find them we look for critical points of the Lagrange function

$$L(x, y, z, \lambda) = xy^2z^3 + \lambda(x^2 + y^2 + z^2 - 1), \qquad x \ne 0, y \ne 0, z, \ne 0.$$

Thus we calculate:

$$\begin{aligned}
0 &= \frac{\partial L}{\partial x} = y^2z^3 + 2\lambda x &&\iff \quad \frac{y^2z^3}{x} = -2\lambda \\
0 &= \frac{\partial L}{\partial y} = 2xyz^3 + 2\lambda y &&\iff \quad 2xz^3 = -2\lambda \\
0 &= \frac{\partial L}{\partial z} = 3xy^2z^2 + 2\lambda z &&\iff \quad 3xy^2z = -2\lambda \\
0 &= \frac{\partial L}{\partial \lambda} = x^2 + y^2 + z^2 - 1.
\end{aligned}$$

Eliminating λ from pairs of the first three equations leads to

$$\frac{y^2z^3}{x} = 2xz^3 = 3xy^2z,$$

which, since none of x, y, and z can be zero, shows that at a critical point we must have $y^2 = 2x^2$ and $z^2 = (3/2)y^2 = 3x^2$. Substituting these into the final (constraint) equation above, we obtain $x^2 + 2x^2 + 3x^2 = 1$, so

$$x^2 = \frac{1}{6}, \qquad y^2 = \frac{1}{3}, \qquad z^2 = \frac{1}{2}.$$

Each of these squares has two square roots, leading to eight critical points (x, y, z) for L, one in each octant of $\mathbb{R}^3$. At the one in the first octant (and at three others) f has the value

$$f(x, y, z) = \left(\frac{1}{\sqrt{6}}\right)\left(\frac{1}{3}\right)\left(\frac{1}{2\sqrt{2}}\right) = \frac{1}{6\sqrt{3}}.$$

This is the maximum value of f on the ball. The minimum value is $-1/(6\sqrt{3})$ and it occurs at the remaining four critical points. of f.

Problems with More than One Constraint

Next consider a three-dimensional problem requiring us to find a maximum or minimum value of a function of three variables subject to two equation constraints:

extremize $f(x, y, z)$ **subject to** $g(x, y, z) = 0$ and $h(x, y, z) = 0$.

Again, we assume that the problem has a solution, say, at the point $P_0 = (x_0, y_0, z_0)$, and that the functions f, g, and h have continuous first partial derivatives near P_0. Also, we assume that $\mathbf{T} = \nabla g(P_0) \times \nabla h(P_0) \neq \mathbf{0}$. These conditions imply that the surfaces $g(x, y, z) = 0$ and $h(x, y, z) = 0$ are smooth near P_0 and are not tangent to each other there, so they must intersect in a curve $\mathcal{C}$ that is smooth near P_0. The curve $\mathcal{C}$ has tangent vector $\mathbf{T}$ at P_0. The same geometric argument used in the proof of Theorem 4 again shows that $\nabla f(P_0)$ must be perpendicular to $\mathbf{T}$. (If not, then it would have a nonzero vector projection along $\mathbf{T}$, and f would have nonzero directional derivatives in the directions $\pm\mathbf{T}$ and would therefore increase and decrease as we moved away from P_0 along $\mathcal{C}$ in opposite directions.)

Since $\nabla g(P_0)$ and $\nabla h(P_0)$ are nonzero and both are perpendicular to $\mathbf{T}$ (see Figure 13.17), $\nabla f(P_0)$ must lie in the plane spanned by these two vectors and hence must be a linear combination of them:

$$\nabla f(x_0, y_0, z_0) = -\lambda_0 \nabla g(x_0, y_0, z_0) - \mu_0 \nabla h(x_0, y_0, z_0)$$

for some constants λ_0 and μ_0. It follows that $(x_0, y_0, z_0, \lambda_0, \mu_0)$ is a critical point of the Lagrange function

$$L(x, y, z, \lambda, \mu) = f(x, y, z) + \lambda g(x, y, z) + \mu h(x, y, z).$$

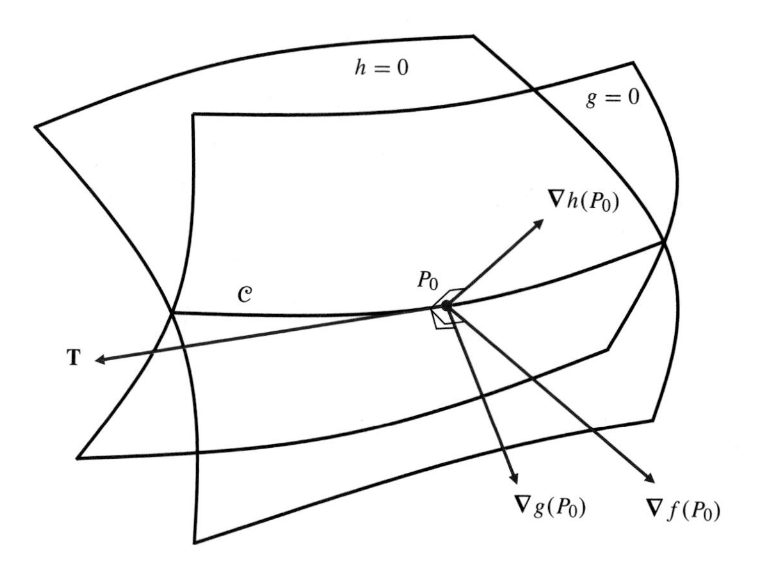

Figure 13.17 At P_0, ∇f, ∇g, and ∇h are all perpendicular to **T**. Thus, ∇f is in the plane spanned by ∇g and ∇h.

We look for triples (x, y, z) that extremize $f(x, y, z)$ subject to the two constraints $g(x, y, z) = 0$ and $h(x, y, z) = 0$ among the points (x, y, z, λ, μ) that are critical points of the above Lagrange function, and we therefore solve the system of equations

$$\begin{aligned} f_1(x, y, z) + \lambda g_1(x, y, z) + \mu h_1(x, y, z) &= 0, \\ f_2(x, y, z) + \lambda g_2(x, y, z) + \mu h_2(x, y, z) &= 0, \\ f_3(x, y, z) + \lambda g_3(x, y, z) + \mu h_3(x, y, z) &= 0, \\ g(x, y, z) &= 0, \\ h(x, y, z) &= 0. \end{aligned}$$

Solving such a system can be very difficult. It should be noted that, in using the method of Lagrange multipliers instead of solving the constraint equations, we have traded the problem of having to solve two equations for two variables as *functions* of a third one for a problem of having to solve five equations for *numerical* values of five unknowns.

EXAMPLE 5 Find the maximum and minimum values of $f(x, y, z) = xy + 2z$ on the circle that is the intersection of the plane $x + y + z = 0$ and the sphere $x^2 + y^2 + z^2 = 24$.

Solution The function f is continuous, and the circle is a closed bounded set in 3-space. Therefore, maximum and minimum values must exist. We look for critical points of the Lagrange function

$$L = xy + 2z + \lambda(x + y + z) + \mu(x^2 + y^2 + z^2 - 24).$$

Setting the first partial derivatives of L equal to zero, we obtain

$$\begin{aligned} y + \lambda + 2\mu x &= 0, &\text{(A)} \\ x + \lambda + 2\mu y &= 0, &\text{(B)} \\ 2 + \lambda + 2\mu z &= 0, &\text{(C)} \\ x + y + z &= 0, &\text{(D)} \\ x^2 + y^2 + z^2 - 24 &= 0. &\text{(E)} \end{aligned}$$

When none of the equations factors, try to combine two or more of them to produce an equation that does factor.

Subtracting (A) from (B) we get $(x - y)(1 - 2\mu) = 0$. Therefore, either $\mu = \frac{1}{2}$ or $x = y$. We analyze both possibilities.

CASE I If $\mu = \frac{1}{2}$, we obtain from (B) and (C)

$$x + \lambda + y = 0 \qquad \text{and} \qquad 2 + \lambda + z = 0.$$

Thus, $x + y = 2 + z$. Combining this with (D), we get $z = -1$ and $x + y = 1$. Now, by (E), $x^2 + y^2 = 24 - z^2 = 23$. Since $x^2 + y^2 + 2xy = (x + y)^2 = 1$, we have $2xy = 1 - 23 = -22$ and $xy = -11$. Now $(x - y)^2 = x^2 + y^2 - 2xy = 23 + 22 = 45$, so $x - y = \pm 3\sqrt{5}$. Combining this with $x + y = 1$, we obtain two critical points arising from $\mu = \frac{1}{2}$, namely, $\big((1 + 3\sqrt{5})/2, (1 - 3\sqrt{5})/2, -1\big)$ and $\big((1 - 3\sqrt{5})/2, (1 + 3\sqrt{5})/2, -1\big)$. At both of these points we find that $f(x, y, z) = xy + 2z = -11 - 2 = -13$.

CASE II If $x = y$, then (D) implies that $z = -2x$, and (E) then gives $6x^2 = 24$, so $x = \pm 2$. Therefore, points $(2, 2, -4)$ and $(-2, -2, 4)$ must be considered. We have $f(2, 2, -4) = 4 - 8 = -4$ and $f(-2, -2, 4) = 4 + 8 = 12$.

We conclude that the maximum value of f on the circle is 12, and the minimum value is -13.

EXERCISES 13.3

1. Use the method of Lagrange multipliers to maximize x^3y^5 subject to the constraint $x + y = 8$.

2. Find the shortest distance from the point $(3, 0)$ to the parabola $y = x^2$,

(a) by reducing to an unconstrained problem in one variable, and

(b) by using the method of Lagrange multipliers.

3. Find the distance from the origin to the plane $x + 2y + 2z = 3$,

(a) using a geometric argument (no calculus),

(b) by reducing the problem to an unconstrained problem in two variables, and

(c) using the method of Lagrange multipliers.

4. Find the maximum and minimum values of the function $f(x, y, z) = x + y - z$ over the sphere $x^2 + y^2 + z^2 = 1$.

5. Use the Lagrange multiplier method to find the greatest and least distances from the point $(2, 1, -2)$ to the sphere with equation $x^2 + y^2 + z^2 = 1$. (Of course, the answer could be obtained more easily using a simple geometric argument.)

6. Find the shortest distance from the origin to the surface $xyz^2 = 2$.

7. Find a, b, and c so that the volume $V = 4\pi abc/3$ of an ellipsoid $\dfrac{x^2}{a^2} + \dfrac{y^2}{b^2} + \dfrac{z^2}{c^2} = 1$ passing through the point $(1, 2, 1)$ is as small as possible.

8. Find the ends of the major and minor axes of the ellipse $3x^2 + 2xy + 3y^2 = 16$.

9. Find the maximum and minimum values of $f(x, y, z) = xyz$ on the sphere $x^2 + y^2 + z^2 = 12$.

10. Find the maximum and minimum values of $x + 2y - 3z$ over the ellipsoid $x^2 + 4y^2 + 9z^2 \le 108$.

11. Find the distance from the origin to the surface $xy^2z^4 = 32$.

12. Find the maximum value of $\sum_{i=1}^{n} x_i$ on the n-sphere $\sum_{i=1}^{n} x_i^2 = 1$ in $\mathbb{R}^n$.

13. Find the maximum and minimum values of the function $f(x, y, z) = x$ over the curve of intersection of the plane $z = x + y$ and the ellipsoid $x^2 + 2y^2 + 2z^2 = 8$.

14. Find the maximum and minimum values of $f(x, y, z) = x^2 + y^2 + z^2$ on the ellipse formed by the intersection of the cone $z^2 = x^2 + y^2$ and the plane $x - 2z = 3$.

15. Find the maximum and minimum values of $f(x, y, z) = 4 - z$ on the ellipse formed by the intersection of the cylinder $x^2 + y^2 = 8$ and the plane $x + y + z = 1$.

16. Find the maximum and minimum values of $f(x, y, z) = x + y^2z$ subject to the constraints $y^2 + z^2 = 2$ and $z = x$.

17. Use the method of Lagrange multipliers to find the shortest distance between the straight lines $x = y = z$ and $x = -y$, $z = 2$. (There are, of course, much easier ways to get the answer. This is an object lesson in the folly of shooting sparrows with cannons.)

18. Find the most economical shape of a rectangular box with no top.

19. Find the maximum volume of a rectangular box with faces parallel to the coordinate planes if one corner is at the origin and the diagonally opposite corner lies on the plane $4x + 2y + z = 2$.

20. Find the maximum volume of a rectangular box with faces parallel to the coordinate planes if one corner is at the origin and the diagonally opposite corner is on the first octant part of the surface $xy + 2yz + 3xz = 18$.

21. A rectangular box having no top and having a prescribed volume V m^3 is to be constructed using two different materials. The material used for the bottom and front of the box is five times as costly (per square metre) as the material used for the back and the other two sides. What should be the dimensions of the box to minimize the cost of materials?

22. Find the maximum and minimum values of $xy + z^2$ on the ball $x^2 + y^2 + z^2 \le 1$. Use Lagrange multipliers to treat the boundary case.

23. Repeat Exercise 22 but handle the boundary case by parametrizing the sphere $x^2 + y^2 + z^2 = 1$ using

$$x = \sin\phi\cos\theta, \quad y = \sin\phi\sin\theta, \quad z = \cos\phi,$$

where $0 \le \phi \le \pi$ and $0 \le \theta \le 2\pi$.

24. If α, β, and γ are the angles of a triangle, show that

$$\sin\frac{\alpha}{2}\sin\frac{\beta}{2}\sin\frac{\gamma}{2} \le \frac{1}{8}.$$

For what triangles does equality occur?

25. Suppose that f and g have continuous first partial derivatives throughout the xy-plane, and suppose that $g_2(a, b) \ne 0$. This implies that the equation $g(x, y) = g(a, b)$ defines y implicitly as a function of x near the point (a, b). Use the Chain Rule to show that if $f(x, y)$ has a local extreme value at (a, b) subject to the constraint $g(x, y) = g(a, b)$, then for some number λ the point (a, b, λ) is a critical point of the function

$$L(x, y, \lambda) = f(x, y) + \lambda g(x, y).$$

This constitutes a more formal justification of the method of Lagrange multipliers in this case.

26. What is the shortest distance from the point $(0, -1)$ to the curve $y = \sqrt{1 - x^2}$? Can this problem be solved by the Lagrange multiplier method? Why?

27. Example 3 showed that the method of Lagrange multipliers might fail to find a point that extremizes $f(x, y)$ subject to the constraint $g(x, y) = 0$ if $\nabla g = \mathbf{0}$ at the extremizing point. Can the method also fail if $\nabla f = \mathbf{0}$ at the extremizing point? Why?

13.4 Lagrange Multipliers in *n*-Space

In this section we will show how the method of Lagrange multipliers extends to the problem of finding local extreme values of a function f of n real variables, that is, of a vector variable $\mathbf{x} = (x_1, x_2, \ldots, x_n)$,

$$f(\mathbf{x}) = f(x_1, x_2, \ldots, x_n),$$

subject to $m \le n - 1$ constraints

$$g_1(\mathbf{x}) = 0, \quad g_2(\mathbf{x}) = 0, \quad \ldots, \quad g_m(\mathbf{x}) = 0,$$

where $1 \le m \le n - 1$.

Here the indices on g denote different functions, *not* partial derivatives.

In what follows we will assume that f and each of the functions g_i, $(1 \le i \le m)$, is smooth in the sense that its partial derivatives of orders up to 3 are all continuous. We also assume that for each i, the gradient $\nabla g_i \ne \mathbf{0}$ at any point $\mathbf{x}$ where $g_i(\mathbf{x}) = 0$. This means that for each i, the set of points $\mathbf{x}$ in $\mathbb{R}^n$ satisfying $g_i(\mathbf{x}) = 0$ is a smooth hypersurface of dimension $n - 1$ (called an $n - 1$-dimensional **manifold**).

In $\mathbb{R}^2$ a manifold of dimension 1 is just a curve. In $\mathbb{R}^3$ a manifold of dimension 1 is a curve; a manifold of dimension 2 is a surface. We are introducing the term manifold here to avoid having to use different terms to distinguish between curves, surfaces, and smooth subsets of dimension up to $n - 1$ in an n-dimensional space $\mathbb{R}^n$.

The intersection $\mathcal{M}$ of all m of these manifolds (i.e., the set of points satisfying all m constraint equations) will be a surface in $\mathbb{R}^n$ called the **constraint manifold** for the extremization problem. $\mathcal{M}$ will have dimension $n - m$ provided that the set of normal vectors $\nabla g_i(\mathbf{x})$, $(1 \le i \le m)$, is **linearly independent** at each point $\mathbf{x}$ on $\mathcal{M}$; that is, if an equation of the form

$$c_1\nabla g_1(\mathbf{a}) + c_2\nabla g_2(\mathbf{a}) + \cdots + c_m\nabla g_m(\mathbf{a}) = \mathbf{0}$$

holds, then every coefficient $c_i = 0$ for $1 \le i \le m$. The subspace of $\mathbb{R}^n$ spanned by the m gradient vectors $\nabla g_i(\mathbf{a})$, $(1 \le i \le m)$, is the m-dimensional space $\mathcal{N}$ normal to $\mathcal{M}$ at $\mathbf{a}$. In particular, if $m = 1$, then $\mathcal{M}$ has dimension $n - 1$ and the normal space $\mathcal{N}$ has dimension 1. If $m = n - 1$, then $\mathcal{M}$ has dimension 1 (and so is a curve in $\mathbb{R}^n$) and the normal space $\mathcal{N}$ is an $n - 1$-dimensional hyperplane perpendicular to that curve at the point $\mathbf{a}$. The tangent space $\mathcal{T}$ to $\mathcal{M}$ at $\mathbf{a}$ is the subspace of $\mathbb{R}^n$ consisting of all vectors perpendicular to the normal space $\mathcal{N}$. Equivalently, $\mathcal{T}$ consists of all points on lines through $\mathbf{a}$ that are tangent to $\mathcal{M}$ at $\mathbf{a}$. Like $\mathcal{M}$, its tangent space $\mathcal{T}$ has dimension $n - m$. For example, in $\mathbb{R}^3$, the normal space to a surface (2-dimensional manifold) at a point is just the normal line to the surface at that point. The tangent space is the plane perpendicular to the normal line at that point. Similarly, the normal space to a curve (1-dimensional manifold) at a point is the plane normal to the curve at that point and the tangent space is the tangent line to the curve there. (See Section 17.3 for more discussion of these ideas.)

The concept of a tangent space, $\mathcal{T}$, is simply the extension to higher dimensions of the tangent line in Section 2.1 and the tangent plane in section 12.3. Similarly, the normal space $\mathcal{N}$ extends the concept of normal line or normal plane.

Under the conditions described above, we will show that if f, when restricted to points on the constraint manifold $\mathcal{M}$, has a local extreme value at point $\mathbf{a}$, then $\mathbf{a}$ must be a critical point of the Lagrange function

$$L(\mathbf{x}) = f(\mathbf{x}) + \sum_{i=1}^{m} \lambda_i g_i(\mathbf{x})$$

for some values of the m Lagrange multipliers $\lambda_1, \lambda_2, \ldots, \lambda_m$. Then we will show that if $\mathbf{a}$ is any critical point of L on $\mathcal{M}$, the $n \times n$ Hessian matrix of second partial derivatives of L can be reduced to an $(n-m) \times (n-m)$ Hessian matrix on the $(n-m)$-dimensional space $\mathcal{T}$ tangent to $\mathcal{M}$ at $\mathbf{a}$ to provide a second derivative test for classifying the critical point $\mathbf{a}$. This test is presented in the following theorem. It is analagous to the test for unconstrained extrema given in Theorem 3 in Section 13.1.[1]

THEOREM 5

Suppose that the functions $f(\mathbf{x})$ and $g_i(\mathbf{x})$ for $1 \le i \le m$ have continuous partial derivatives of order up to 3 in a neighbourhood of point $\mathbf{a}$ on the constraint manifold $\mathcal{M}$ having equations $g_i(\mathbf{x}) = 0$, $(1 \le i \le m)$. Suppose also that the m vectors $\nabla g_i(\mathbf{a})$ are linearly independent in $\mathbb{R}^n$.

(a) **Necessary Conditions for a local extreme value:** If f, when restricted to points on $\mathcal{M}$, has a local maximum or a local minimum value at $\mathbf{a}$, then there exist numbers $\lambda_1, \lambda_2, \ldots, \lambda_m$ such that $\mathbf{a}$ is a critical point of the Lagrange function.

$$L(\mathbf{x}) = f(\mathbf{x}) + \sum_{i=1}^{n} \lambda_i g_i(\mathbf{x}). \qquad (*)$$

(b) **Second Derivative Test** Suppose a Lagrange function of the type (*) has a critical point at $\mathbf{a}$ on $\mathcal{M}$. Let $\mathcal{H}$ be the Hessian matrix of second partial derivatives of L with respect to the components of $\mathbf{x}$, evaluated at $\mathbf{x} = \mathbf{a}$:

$$\mathcal{H} = \begin{pmatrix} L_{11}(\mathbf{a}) & L_{12}(\mathbf{a}) & \cdots & L_{1n}(\mathbf{a}) \\ L_{21}(\mathbf{a}) & L_{22}(\mathbf{a}) & \cdots & L_{2n}(\mathbf{a}) \\ \vdots & \vdots & \ddots & \vdots \\ L_{n1}(\mathbf{a}) & L_{n2}(\mathbf{a}) & \cdots & L_{nn}(\mathbf{a}) \end{pmatrix}.$$

Let $\mathbf{u} = (u_1, u_2, \ldots, u_n)$ belong to the space $\mathcal{T}$ tangent to $\mathcal{M}$ at $\mathbf{a}$. For purposes of matrix multiplication we regard $\mathbf{u}$ as a column vector having transpose $\mathbf{u}^T$, a row vector. If the quadratic form

$$Q(\mathbf{u}) = \sum_{i=1}^{n}\sum_{j=1}^{n} L_{ij}(\mathbf{a})\, u_i\, u_j = \mathbf{u}^T \mathcal{H} \mathbf{u}$$

is positive (or negative) definite when restricted to vectors $\mathbf{u} \in \mathcal{T}$, then the restriction of f to $\mathcal{M}$ has a *local minimum* (or a *local maximum*) at $\mathbf{x} = \mathbf{a}$.

(c) **The restricted Hessian** If $\mathcal{H}$ is positive definite (or negative definite) on $\mathbb{R}^n$, then $Q(\mathbf{u})$ will be positive (or negative) definite on all of $\mathbb{R}^n$, and so on $\mathcal{T}$. If not, we can calculate a Hessian matrix restricted to $\mathcal{T}$ as follows. Since $\mathcal{M}$ has dimension $n-m$, so does $\mathcal{T}$. Let $\mathbf{u}_1, \mathbf{u}_2, \ldots, \mathbf{u}_{n-m}$ be an orthonormal basis for $\mathcal{T}$, that is, a basis consisting of mutually perpendicular unit vectors. Let $\mathcal{E}$ be the $n \times (n-m)$ matrix whose ith column consists of the components of the vector $\mathbf{u}_i$, $(1 \le i \le n-m)$. If $\mathcal{E}^T$ is the $(n-m) \times n$ transpose of $\mathcal{E}$, then the $(n-m) \times (n-m)$ matrix

$$\mathcal{H}_{\mathcal{T}} = \mathcal{E}^T\, \mathcal{H}\, \mathcal{E}$$

[1] This discussion is similar to the presentation of M. A. H. Nerenberg's paper: "The Second Derivative Test for Constrained Extremum Problems," *Int. J. Math. Educ. Sci. Technol.*, 1991, Vol. 22.

defines a quadratic form on $\mathcal{T}$ that restricts $\mathcal{H}$ to $\mathcal{T}$. Any vector $\mathbf{u} \in \mathcal{T}$ can be written $\mathbf{u} = \sum_{i=1}^{n-m} u_i \mathbf{u}_i$. Then

$$Q(\mathbf{u}) = \mathbf{u}^{\mathrm{T}} \mathcal{H} \mathbf{u} = \sum_{i=1}^{n-m} \sum_{j=1}^{n-m} \left(\mathcal{H}_{\mathcal{T}}\right)_{ij} u_i u_j,$$

where $\left(\mathcal{H}_{\mathcal{T}}\right)_{ij}$ is the element in the ith row and jth column of $\mathcal{H}_{\mathcal{T}}$. When restricted to $\mathcal{M}$, f will have a local minimum, a local maximum, or saddle behaviour at $\mathbf{a}$ if $\mathcal{H}_{\mathcal{T}}$ is positive definite, or negative definite or indefinite. (See, for example, Theorem 7 or Theorem 8 of Section 10.7.) If $\mathcal{H}_{\mathcal{T}}$ is neither definite nor indefinite, this test will give no information about the nature of the critical point $\mathbf{a}$.

PROOF (a) If $\nabla f(\mathbf{a})$ does not lie in the normal space $\mathcal{N}$ to $\mathcal{M}$ at $\mathbf{a}$, then it will have a nonzero projection $\mathbf{v}$ on $\mathcal{T}$, and f will have a positive directional derivative at $\mathbf{a}$ in the direction of $\mathbf{v}$ and a negative directional derivative in the direction of $-\mathbf{v}$, contradicting the assumption that when restricted to $\mathcal{M}$, f has a local extreme value at $\mathbf{a}$. Thus, $\nabla f(\mathbf{a}) \in \mathcal{N}$. Since the m vectors $\nabla g_i(\mathbf{a})$ span $\mathcal{N}$, there must exist numbers λ_i, $(1 \le i \le m)$ such that

$$\nabla f(\mathbf{a}) = -\sum_{i=1}^{m} \lambda_i \nabla g_i(\mathbf{a})$$

and so $\mathbf{a}$ is a critical point of the Lagrangian function $(*)$.

(b) Now let $\mathbf{a}$ and $\mathbf{a}+\mathbf{h}$ be two points on $\mathcal{M}$ and suppose that $\mathbf{a}$ is a critical point of the Lagrange function $(*)$ for some values of the multipliers λ_i, $(1 \le i \le m)$. Because of the smoothness assumptions made on f and the constraint functions g_i, Taylor's Formula (Section 12.9) gives

$$f(\mathbf{a}+\mathbf{h}) - f(\mathbf{a}) = \mathbf{h} \bullet \nabla f(\mathbf{a}) + \frac{1}{2}(\mathbf{h} \bullet \nabla)^2 f(\mathbf{a}) + O(|\mathbf{h}|^3),$$

$$g_i(\mathbf{a}+\mathbf{h}) - g_i(\mathbf{a}) = \mathbf{h} \bullet \nabla g_i(\mathbf{a}) + \frac{1}{2}(\mathbf{h} \bullet \nabla)^2 g_i(\mathbf{a}) + O(|\mathbf{h}|^3), \quad (1 \le i \le m).$$

Noting that $g_i(\mathbf{a}+\mathbf{h}) - g_i(\mathbf{a}) = 0$ and $\nabla L(\mathbf{a}) = 0$, multiplying the second formula above by λ_i, summing, and adding the result to the first formula, we get

$$\begin{aligned} f(\mathbf{a}+\mathbf{h}) - f(\mathbf{a}) &= \frac{1}{2}(\mathbf{h} \bullet \nabla)^2 L(\mathbf{a}) + O(|\mathbf{h}|^3) \\ &= \frac{1}{2} \sum_{i=1}^{n} \sum_{j=1}^{n} h_i h_j L_{ij}(\mathbf{a}) \\ &= \mathbf{h}^{\mathrm{T}} \mathcal{H} \mathbf{h} + O(|h|^3), \end{aligned}$$

where, in the final quadratic form expression, we are regarding $\mathbf{h}$ as a column vector with transpose $\mathbf{h}^{\mathrm{T}}$. Now let $\mathbf{h} = t\mathbf{e} = t\left(\mathbf{e}_{\mathcal{T}} + \mathbf{e}_{\mathcal{N}}\right)$, where $\mathbf{e}$ is a unit vector, and $\mathbf{e}_{\mathcal{T}}$ and $\mathbf{e}_{\mathcal{N}}$ are its projections onto $\mathcal{T}$ and $\mathcal{N}$, respectively. The smoothness of $\mathcal{M}$ shows that the angle θ between $\mathbf{e}$ and $\mathbf{e}_{\mathcal{N}}$ approaches $\pi/2$ as $t \to 0$. Accordingly, $\lim_{t\to 0} |\mathbf{e}_{\mathcal{T}}| = 1$ and $\lim_{t\to 0} |\mathbf{e}_{\mathcal{N}}| = 0$. For small enough positive t, therefore, $|\mathbf{h}_{\mathcal{N}}| < t|\mathbf{h}| = t^2$. Thus,

$$\begin{aligned} f(\mathbf{a}+\mathbf{h}) - f(\mathbf{a}) &= \frac{1}{2}\left(\mathbf{h}_{\mathcal{T}} + \mathbf{h}_{\mathcal{N}}\right)^{\mathrm{T}} \mathcal{H}(\mathbf{h}_{\mathcal{T}} + \mathbf{h}_{\mathcal{N}}) + O(|h|^3) \\ &= \frac{t^2}{2} \mathbf{u}_{\mathcal{T}}^{\mathrm{T}} \mathcal{H} \mathbf{u}_{\mathcal{T}} + O(t^3). \end{aligned}$$

For small t the t^2 term dominates the $O(t^3)$ term, which now also contains three terms from the previous line that involve at least one copy of $\mathbf{h}_{\mathcal{N}}$. Hence f, when restricted to $\mathcal{M}$, will have a minimum (or maximum) value at $\mathbf{a}$ if The Hessian matrix $\mathcal{H}$ is positive (or negative) definite on $\mathcal{T}$.

(c) Observe that the element in the ith row and jth column of $\mathcal{H}_\mathcal{T}$ is

$$\left(\mathcal{H}_\mathcal{T}\right)_{ij} = \mathbf{u}_i^T \mathcal{H} \mathbf{u}_j.$$

If $\mathbf{u} = u_1\mathbf{u}_1 + u_2\mathbf{u}_2 + \cdots + u_{n-m}\mathbf{u}_{n-m}$ is an arbitrary vector in $\mathcal{T}$, then

$$Q(\mathbf{u}) = \mathbf{u}^T \mathcal{H}\mathbf{u} = \sum_{i=1}^{n-m}\sum_{j=1}^{n-m} u_i\, u_j\, \mathbf{u}_i^T \mathcal{H}\mathbf{u}_j = \sum_{i=1}^{n-m}\sum_{j=1}^{n-m} \left(\mathcal{H}_\mathcal{T}\right)_{ij} u_i\, u_j.$$

Thus, Q is positive definite (or negative definite, or indefinite) on $\mathcal{T}$ provided the restricted Hessian matrix $\mathcal{H}_\mathcal{T}$ is positive definite (or negative definite, or indefinite). This completes the proof.

Remark Suppose $m = n - 1$, so that $\mathcal{M}$ is a one-dimensional curve in $\mathbb{R}^n$. Its tangent space $\mathcal{T}$ at $\mathbf{a}$ is a one-dimensional straight line, spanned by a single nonzero vector $\mathbf{u}$ that is normal to the $n-1$ gradients $\nabla g_i(\mathbf{a})$. In this case, the test boils down to looking at the sign of a single number, $\mathbf{u}^T\mathcal{H}\mathbf{u}$. For example, if $n = 2$ and $m = 1$, so that $\mathbf{x} = (x, y)$ and $\mathbf{a} = (a, b)$, then $\mathbf{u}$ must be normal to $\nabla g(\mathbf{a}) = g_x(a, b)\mathbf{i} + g_y(a, b)\mathbf{j}$. Evidently, $\mathbf{u} = g_y\mathbf{i} - g_x\mathbf{j}$ will do, and we examine the number

$$\begin{aligned} Q &= (g_y, -g_x)\begin{pmatrix} L_{xx} & L_{xy} \\ L_{yx} & L_{yy} \end{pmatrix}\begin{pmatrix} g_y \\ -g_x \end{pmatrix} \\ &= \Big(g_y L_{xx} - 2g_x g_y L_{xy} + g_x L_{yy}\Big)\Big|_{(a,b)}. \end{aligned}$$

If $Q > 0$ ($Q < 0$), then there will be a local minimum (maximum) at (a, b).

The following examples illustrate the use of Theorem 5 in classifying critical points for constrained extrema.

EXAMPLE 1 The entropy S of a system that can exist in n states is given by $S = -\sum_{i=1}^n p_i \ln p_i$, where each p_i satisfies $0 < p_i < 1$ and is the probability the system is in the ith state. S is subject to two constraints: $\sum_{i=1}^n p_i = 1$, and $\sum_{i=1}^n p_i E_i = E$, where the E_i and E are constants. (E_i is the energy of the ith state and E is the average energy.) Show that attempting to extremize S subject to these constraints leads to a maximum value for S.

Solution The Lagrange fuction for this problem is

$$L(p_1, \ldots, p_n) = -\left(\sum_{i=1}^n p_i \ln p_i\right) + \lambda\left[\left(\sum_{i=1}^n p_i\right) - 1\right] + \mu\left[\left(\sum_{i=1}^n p_i E_i\right) - E\right].$$

The critical points are given by

$$\frac{\partial L}{\partial p_i} = -\ln p_i - 1 + \lambda + \mu E_i, \qquad (1 \le i \le n)$$

and the two constraint equations. Solving the first equation for p_i, we obtain $p_i = C\exp(\mu E_i)$ for $1 \le i \le n$, where the constants C and μ can be found by substituting these values into the two constraint equations and solving. There is just the one critical point. Observe that

$$\frac{\partial^2 S}{\partial p_i \partial p_j} = \begin{cases} -\dfrac{1}{p_i} & \text{if } i = j \\ 0 & \text{if } i \ne j \end{cases}$$

and so the (unconstrained) Hessian matrix $\mathcal{H}$ has its only nonzero elements on the main diagonal, and these are all negative at the critical point. Accordingly, $\mathcal{H}$ is negative definite (by either of Theorems 7 and 8 of Section 10.7), and we don't need to worry about restricting $\mathcal{H}$ to the (tangent space to) the constraint manifold. The critical point gives S a local maximum value. Since, $\lim_{p_i \to 0+} p_i \ln p_i = 0$ and there are no other critical points, the local maximum must, in fact, be an absolute maximum.

EXAMPLE 2 Find the minimum distance between the circle $x^2 + y^2 = 2$ and the line $x + y = 4$.

Solution We really don't need to use such fancy theory to solve this problem. It is geometrically evident in Figure 13.18 that the two closest points are $B = (1, 1)$ on the circle and $A = (2, 2)$ on the line. We shall, however, treat it as a problem of minimizing (the square of) the distance between two arbitrary points, (x_1, y_1) on the circle and (x_2, y_2) on the line:

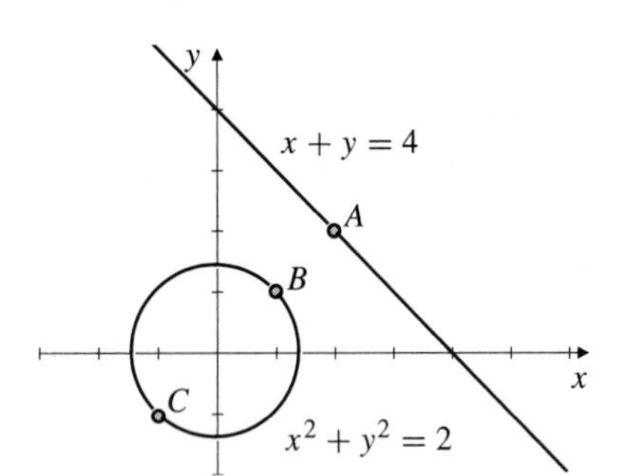

Figure 13.18 Clearly the distance between the circle and the line is $\sqrt{2}$ units, the distance between $B = (1, 1)$ and $A = (2, 2)$

$$\textbf{minimize} \quad S = (x_1 - x_2)^2 + (y_1 - y_2)^2$$
$$\textbf{subject to} \quad x_1^2 + y_1^2 - 2 = 0 \quad \text{and} \quad x_2 + y_2 - 4 = 0.$$

The Lagrange function is

$$L = (x_1 - x_2)^2 + (y_1 - y_2)^2 + \lambda(x_1^2 + y_1^2 - 2) + \mu(x_2 + y_2 - 4).$$

Since S and the constraint functions involve four variables, this is a problem in $\mathbb{R}^4$. Since there are two constraints, L depends on six variables, so its critical points satisfy

$$\begin{aligned}
0 &= \frac{\partial L}{\partial x_1} = 2(x_1 - x_2) + 2\lambda x_1 \\
0 &= \frac{\partial L}{\partial y_1} = 2(y_1 - y_2) + 2\lambda y_1 \\
0 &= \frac{\partial L}{\partial x_2} = -2(x_1 - x_2) + \mu \\
0 &= \frac{\partial L}{\partial y_2} = -2(y_1 - y_2) + \mu \\
0 &= \frac{\partial L}{\partial \lambda} = x_1^2 + y_1^2 - 2 \\
0 &= \frac{\partial L}{\partial \mu} = x_2 + y_2 - 4.
\end{aligned}$$

We leave it to the reader to show that $L(x_1, y_1, x_2, y_2, \lambda, \mu)$ has two critical points: $P = (1, 1, 2, 2, -1, -2)$, and $Q = (-1, -1, 2, 2, 3, -2)$. The Hessian matrices at P and Q are

$$\mathcal{H}(P) = \begin{pmatrix} 4 & 0 & -2 & 0 \\ 0 & 4 & 0 & -2 \\ -2 & 0 & 2 & 0 \\ 0 & -2 & 0 & 2 \end{pmatrix}, \qquad \mathcal{H}(Q) = \begin{pmatrix} -4 & 0 & -2 & 0 \\ 0 & -4 & 0 & -2 \\ -2 & 0 & 2 & 0 \\ 0 & -2 & 0 & 2 \end{pmatrix}.$$

In order to calculate the restrictions of these Hessians to the space tangent to the constraint manifold at each of P and Q, we need orthonormal bases for those tangent spaces. Let $\mathbf{e}_1$, $\mathbf{e}_2$, $\mathbf{e}_3$, and $\mathbf{e}_4$ be the standard basis vectors for the space $\mathbb{R}^4$ of coordinates (x_1, y_1, x_2, y_2) As luck would have it, the normal vectors at P and Q are $\nabla(x^2 + y^2 - 2) = 2x\mathbf{e}_1 + 2y\mathbf{e}_2 = \pm 2(\mathbf{e}_1 + \mathbf{e}_2)$ and $\nabla(x + y - 4) = \mathbf{e}_3 + \mathbf{e}_4$, so the normal spaces at both points are the same 2-dimensional subspace of $\mathbb{R}^4$, and the two perpendicular unit vectors

$$\mathbf{u}_1 = \frac{\mathbf{e}_1 - \mathbf{e}_2}{\sqrt{2}} \quad \text{and} \quad \mathbf{u}_2 = \frac{\mathbf{e}_3 - \mathbf{e}_4}{\sqrt{2}},$$

being perpendicular to both those normals, constitute an orthonormal basis for the 2-dimensional tangent space at each point. At both points we can use

$$\mathcal{E} = \frac{1}{\sqrt{2}} \begin{pmatrix} 1 & 0 \\ -1 & 0 \\ 0 & 1 \\ 0 & -1 \end{pmatrix}.$$

The restriction of $\mathcal{H}$ to the tangent space at P is

$$\mathcal{H}_{P\mathcal{T}} = \mathcal{E}^T \mathcal{H}(P)\mathcal{E} = \begin{pmatrix} 4 & -2 \\ -2 & 2 \end{pmatrix},$$

and that at Q is

$$\mathcal{H}_{Q\mathcal{T}} = \mathcal{E}^T \mathcal{H}(Q)\mathcal{E} = \begin{pmatrix} -4 & -2 \\ -2 & 2 \end{pmatrix}.$$

The eigenvalues of these 2×2 matrices are easily calculated. (See Section 10.7.) For $\mathcal{H}_{P\mathcal{T}}$ they are $3 \pm \sqrt{5}$, both positive. Therefore $\mathcal{H}_{PT}$ is positive definite by Theorem 7 of Section 10.7, and S has a local minimum value at P. It is also the absolute minimum, as observed in Figure 13.18. The minimum distance is the distance from $A = (2, 2)$ to $B = (1, 1)$, that is, $\sqrt{2}$ units.

For $\mathcal{H}_{Q\mathcal{T}}$ the eigenvalues are $-1 \pm \sqrt{13}$, which have opposite signs. Therefore, $\mathcal{H}_{QT}$ is indefinite and S has neither a local minimum nor a local maximum at Q. At first this may seem strange; it may appear that S should have a local maximum at Q; if point C in the figure moves along the circle away from $(-1, -1)$, its distance from $A = (2, 2)$ is decreasing. However, if A moves along the line away from $(2, 2)$, its distance from $C = (-1, -1)$ is increasing. Thus, Q really is a saddle point of the constrained problem. Of course, there is no absolute maximum distance since the line is unbounded.

Using Maple to Solve Constrained Extremal Problems

As the previous Example indicates, the classification of critical points for constrained problems can be quite computationally intensive. Our next Example will show how to make use of Maple to relieve some of the burden.

EXAMPLE 3 Find and classify the critical points of the Lagrange function for the problem

$$\textbf{Extremize} \quad F(x, y, z) = x^3 + y^3 + z^3 \quad \textbf{subject to} \quad \frac{1}{x} + \frac{1}{y} + \frac{1}{z} = 1.$$

Solution We begin by loading two Maple packages defining routines useful in what follows.

```
> with(LinearAlgebra): with(VectorCalculus):
```

The colons suppress output from these `with` commands. We will not reproduce here the results of the next few commands either, as their output just restates the input. First we define expressions for F and G and the Lagrange function L. We do not need these to be Maple functions, so we just set them up as expressions.

```
> F := x^3 + y^3 + z^3;
> G := (1/x) + (1/y) + (1/z);
> L := F + lambda*(G - 1);
```

Newer versions of Maple will use the symbol λ in place of `lambda` in the output. Some of the commands used below require us to list the variables to which the command should be applied. We require two sets of variables, the space variables x, y, z, and all variables, which includes the λ as well.

```
> spvars := [x, y, z]; allvars := [x, y, z, lambda];
```

Now we can get down to business. We calculate the gradient of L with respect to all four variables. The list `allvars` is required by the Gradient command in the VectorCalculus package.

```
> GrL := Gradient(L, allvars);
```

$$GrL := \left(3x^2 - \frac{\lambda}{x^2}\right)\bar{\mathbf{e}}_x + \left(3y^2 - \frac{\lambda}{y^2}\right)\bar{\mathbf{e}}_y + \left(3z^2 - \frac{\lambda}{z^2}\right)\bar{\mathbf{e}}_z + \left(\frac{1}{x} + \frac{1}{y} + \frac{1}{z} - 1\right)\bar{\mathbf{e}}_\lambda$$

In the output above the vectors $\bar{\mathbf{e}}_x$, $\bar{\mathbf{e}}_y$, $\bar{\mathbf{e}}_z$, and $\bar{\mathbf{e}}_\lambda$ denote the standard basis in the 4-space of variables x, y, z, and λ. To find the critical points of L we need to solve a set of four equations obtained by setting the four components of `GrL` equal to zero. We construct the list of these equations as follows.

```
> eqns := [seq(GrL[i]=0, i=1..4)];
```

$$eqns := \left[3x^2 - \frac{\lambda}{x^2} = 0,\ 3y^2 - \frac{\lambda}{y^2} = 0,\ 3z^2 - \frac{\lambda}{z^2} = 0,\ \frac{1}{x} + \frac{1}{y} + \frac{1}{z} - 1 = 0\right]$$

We now attempt to solve these equations for all four variables using Maple's `solve` command.

```
> solns := solve(eqns,allvars);
```

This command produces several lines of output, not all of which we reproduce here. The output consists of a list of eight lists, each of which provides one solution for the four variables. Only four of those solutions consist entirely of real numbers (in fact, integers). The other four involve expressions like `RootOf(_Z^2+1)` and `RootOf(5-2_Z+_Z^2)`, both of which represent complex numbers and are of no use to us. The four real critical points of L are

$$[x = 3, y = 3, z = 3, \lambda = 343], \quad [x = 1, y = 1, z = -1, \lambda = 3],$$
$$[x = 1, y = -1, z = 1, \lambda = 3], \quad [x = -1, y = 1, z = 1, \lambda = 3],$$

that is, the points $P = (3, 3, 3, 343)$, $Q = (1, 1, -1, 3)$, $R = (1, -1, 1, 3)$, and $S = (-1, 1, 1, 3)$. When we did this calculation, the four real solutions were the first, second, fourth, and fifth ones in the `solns` list. Thus, P was `solns[1]` and Q was `solns[2]`. Now we need to classify these four points. By the symmetry of F and G in the spatial variables x, y, and z, the points Q, R and S will be of the same type, so we need only look at P and Q. The VectorCalculus package has a function for calculating Hessian matrices.

```
> H := Hessian(L, spvars);
```

$$H := \begin{pmatrix} 6x + \frac{2\lambda}{x^3} & 0 & 0 \\ 0 & 6y + \frac{2\lambda}{y^3} & 0 \\ 0 & 0 & 6z + \frac{2\lambda}{z^3} \end{pmatrix}$$

At P and Q these Hessians are, respectively,

```
> HP := eval(H, solns[1]); HQ := eval(H, solns[2]):
```

$$HP := \begin{pmatrix} 36 & 0 & 0 \\ 0 & 36 & 0 \\ 0 & 0 & 36 \end{pmatrix}$$

$$HQ := \begin{pmatrix} 12 & 0 & 0 \\ 0 & 12 & 0 \\ 0 & 0 & -12 \end{pmatrix}$$

Both matrices are diagonal so the diagonal elements are the eigenvalues. HP is positive definite, so the constrained problem must have a local minimum at P. However, HQ is indefinite so we have to consider the restriction of HQ to the tangent plane $\mathcal{T}$ to the constraint manifold at Q to determine the nature of Q. A vector normal to $\mathcal{T}$ is given by

```
> NQ := subs([x=1,y=1,z=-1],Gradient(G,spvars));
```

$$NQ := -\bar{\mathbf{e}}_x - \bar{\mathbf{e}}_y - \bar{\mathbf{e}}_z$$

We need two linearly independent vectors each normal to NQ. Evidently, two such vectors are

```
> v1 := <1,-1,0>; v2 := <1,0,-1>;
```

$$v1 := \bar{\mathbf{e}}_x - \bar{\mathbf{e}}_y$$

$$v2 := \bar{\mathbf{e}}_x - \bar{\mathbf{e}}_z$$

We can now use the GramSchmidt function in the Linear Algebra package to generate an orthonormal basis for $\mathcal{T}$.

```
> B := GramSchmidt([v1, v2], normalized);
```

$$B := \left[\begin{bmatrix} \frac{1}{2}\sqrt{2} \\ -\frac{1}{2}\sqrt{2} \\ 0 \end{bmatrix}, \begin{bmatrix} \frac{1}{6}\sqrt{6} \\ \frac{1}{6}\sqrt{6} \\ -\frac{1}{3}\sqrt{6} \end{bmatrix}\right]$$

Now we convert B into the matrix E needed for calculating the restricted Hessian at Q.

```
> E := convert(B, Matrix);
```

$$E := \begin{bmatrix} \frac{1}{2}\sqrt{2} & \frac{1}{6}\sqrt{6} \\ -\frac{1}{2}\sqrt{2} & \frac{1}{6}\sqrt{6} \\ 0 & -\frac{1}{3}\sqrt{6} \end{bmatrix}$$

The transpose of E is `Transpose(E)`, so the restricted Hessian at Q is given by

```
> HQT := (Transpose(E)).HQ.E;
```

$$HQT := \begin{bmatrix} 12 & 0 \\ 0 & -4 \end{bmatrix}$$

This matrix is diagonal and clearly indefinite, so we conclude that F, when restricted to the constraint manifold, has saddle behaviour rather than a local maximum or minimum at Q, and by symmetry also at R and S.

Remark There are two places in the above use of Maple where difficulties can arise for other constrained problems. Firstly, depending on the functions involved, Maple's `solve` routine may not be able to solve the system of equations for the critical point of the Lagrange function. If so, you should try the floating point `fsolve` routine, but this may only give one solution even though there are many. Secondly, if F is a function of n variables, and is subject to $m \le n$ constraints, the tangent to the constraint manifold will have dimension $n - m$ and you will need to first find $m - n$ linearly independent vectors, each normal to the m gradients of the constraint functions, in order to apply the `GramSchmidt` routine to generate an orthonormal basis for T. This can usually be done by solving an underdetermined system of m linear equations in n unknowns.

Significance of Lagrange Multiplier Values

It would seem that the actual value of a Lagrange multiplier is of little significance for the process of solving constrained extreme value problems. However, it is significant if we want to determine the sensitivity of the extreme value to changes in the value of a parameter on which a constraint function depends.

Consider, for example, the problem of extremizing $f(x, y)$ subject to the constraint $g(x, y, p) = 0$. Here p is a parameter in the constraint equation which is beyond our control and so does not enter into the process of finding the extreme value of f. If

f has an extreme value at (a,b), then for some λ, (a,b,λ) is a critical point of the Lagrange function

$$L = f(x,y) + \lambda g(x,y,p)$$

and so a, b, and λ are determined by the three equations

$$\begin{aligned} f_1(a,b) &= -\lambda g_1(a,b,p) \\ f_2(a,b) &= -\lambda g_2(a,b,p) \\ g(a,b,p) &= 0. \end{aligned}$$

The solution of these equations for a, b, and λ results in all three being functions of p. How does the extreme value $f(a,b)$ change if p changes? Observe that

$$\begin{aligned} \frac{d}{dp} f(a,b) &= f_1(a,b)\frac{da}{dp} + f_2(a,b)\frac{db}{dp} \\ &= -\lambda\left(g_1(a,b)\frac{da}{dp} + g_2(a,b)\frac{db}{dp}\right). \end{aligned}$$

But, since $g(a,b,p) = 0$, we have

$$0 = \frac{d}{dp} g(a,b,p) = g_1(a,b)\frac{da}{dp} + g_2(a,b)\frac{db}{dp} + g_3(a,b,p).$$

Thus,

$$\frac{d}{dp} f(a,b) = \lambda\, g_3(a,b,p).$$

The extreme value of f changes at a rate λ times the rate of change of the function g with respect to the parameter p at the point where the extreme value occurs.

Nonlinear Programming

When we looked for extreme values of functions f on restricted domains R in Section 13.2, we had to look separately for critical points of f in the interior of R and then for critical points of the restriction of f to the boundary of R. The interior of R is typically specified by one or more inequality constraints of the form $g < 0$, while the boundary corresponds to equation constraints of the form $g = 0$ (for which Lagrange multipliers can be used).

It is possible to unify these approaches into a single method for finding extreme values of functions defined on regions specified by inequalities of the form $g \le 0$.

Consider, for example, the problem of finding extreme values of $f(x,y)$ over the region R specified by $g(x,y) \le 0$. We can proceed by trying to find critical points of the four-variable function

$$L(x,y,\lambda,u) = f(x,y) + \lambda\big(g(x,y) + u^2\big).$$

Such critical points must satisfy the four equations

$$0 = \frac{\partial L}{\partial x} = f_1(x,y) + \lambda g_1(x,y), \qquad \text{(A)}$$

$$0 = \frac{\partial L}{\partial y} = f_2(x,y) + \lambda g_2(x,y), \qquad \text{(B)}$$

$$0 = \frac{\partial L}{\partial \lambda} = g(x,y) + u^2, \qquad \text{(C)}$$

$$0 = \frac{\partial L}{\partial u} = 2\lambda u. \qquad \text{(D)}$$

Suppose that (x, y, λ, u) satisfies these equations. We consider two cases:

CASE I $u \neq 0$. Then (D) implies that $\lambda = 0$, (C) implies that $g(x, y) = -u^2 < 0$, and (A) and (B) imply that $f_1(x, y) = 0$ and $f_2(x, y) = 0$. Thus, (x, y) is an interior critical point of f.

CASE II $u = 0$. Then (C) implies that $g(x, y) = 0$, and (A) and (B) imply that $\nabla f(x, y) = -\lambda \nabla g(x, y)$, so that (x, y) is a boundary point candidate for the location of the extreme value.

This technique can be extended to the problem of finding extreme values of a function of n variables, $\mathbf{x} = (x_1, x_2, \ldots, x_n)$, over the intersection R of m regions R_j defined by inequality constraints of the form $g_j(\mathbf{x}) \le 0$.

$$\textbf{extremize} \quad f(\mathbf{x}) \quad \textbf{subject to} \quad g_1(\mathbf{x}) \le 0, \quad \ldots \quad g_m(\mathbf{x}) \le 0.$$

In this case we look for critical points of the $(n + 2m)$-variable Lagrange function

$$L(\mathbf{x}, \lambda_1, \ldots, \lambda_m, u_1, \ldots, u_m) = f(\mathbf{x}) + \sum_{j=1}^{m} \lambda_j \big(g_j(\mathbf{x}) + u_j^2\big).$$

The critical points will satisfy $n + 2m$ equations

$$\nabla f(\mathbf{x}) = -\sum_{j=1}^{m} \lambda_j \nabla g_j(\mathbf{x}), \qquad (n \text{ equations})$$

$$g_j(\mathbf{x}) = -u_j^2, \qquad (1 \le j \le m), \qquad (m \text{ equations})$$

$$2\lambda_j u_j = 0, \qquad (1 \le j \le m). \qquad (m \text{ equations})$$

The last m equations show that $\lambda_j = 0$ for any j for which $u_j \neq 0$. If all $u_j \neq 0$, then $\mathbf{x}$ is a critical point of f interior to R. Otherwise, some of the u_j will be zero, say, those corresponding to j in a subset J of $\{1, 2, \ldots, m\}$. In this case, $\mathbf{x}$ will lie on the part of the boundary of R consisting of points lying on the boundaries of each of the regions R_j for which $j \in J$, and ∇f will be a linear combination of the corresponding gradients ∇g_j:

$$\nabla f(\mathbf{x}) = -\sum_{j \in J} \lambda_j \nabla g_j(\mathbf{x}).$$

These are known as **Kuhn-Tucker conditions**, and this technique for solving extreme-value problems on restricted domains is called **nonlinear programming**.

EXERCISES 13.4

1. Find the maximum and minimum values of the n-variable function $x_1 + x_2 + \cdots + x_n$ subject to the constraint $x_1^2 + x_2^2 + \cdots + x_n^2 = 1$.

2. Repeat Exercise 1 for the function $x_1 + 2x_2 + 3x_3 + \cdots + nx_n$ with the same constraint.

3. Find a finite local extreme value of $S = \sum_{i=1}^{10} x_i^2$ subject to the two constraints $\sum_{i=1}^{10} x_i = 10$ and $\sum_{i=1}^{10} i x_i = 55$. Is the extreme value a local maximum or a local minimum? Is it absolute?

4. Repeat Exercise 3 except replace the second constraint with $\sum_{i=1}^{10} i x_i = 60$.

5. Find and classify the three critical points for the Lagrange function

$$L(x, y, u, v, \lambda, \mu) = S + \lambda(y - x^2) + \mu(v - 2u^2 - 1)$$

corresponding to the problem:

$$\textbf{extremize} \quad S = (x - u)^2 + (y - v)^2$$
$$\textbf{subject to} \quad y = x^2 \quad \textbf{and} \quad v = 2u^2 + 1.$$

What is the minimum distance between the curves $y = x^2$ and $y = 2x^2 + 1$?

13.5 The Method of Least Squares

Important optimization problems arise in the statistical analysis of experimental data. Frequently, experiments are designed to measure the values of one or more quantities supposed to be constant, or to demonstrate a supposed functional relationship between variable quantities. Experimental error is usually present in the measurements, and experiments need to be repeated several times in order to arrive at *mean* or *average* values of the quantities being measured.

Consider a very simple example. An experiment to measure a certain physical constant c is repeated n times, yielding the values $c_1, c_2, \ldots, c_n$. If none of the measurements is suspected of being faulty, intuition tells us that we should use the mean value $\bar{c} = (c_1 + c_2 + \cdots + c_n)/n$ as the value of c determined by the experiments. Let us see how this intuition can be justified.

Various methods for determining c from the data values are possible. We could, for instance, choose c to minimize the sum T of its distances from the data points:

$$T = |c - c_1| + |c - c_2| + \cdots + |c - c_n|.$$

This is unsatisfactory for a number of reasons. Since absolute values have singular points, it is difficult to determine the minimizing value of c. More importantly, c may not be determined uniquely. If $n = 2$, any point in the interval between c_1 and c_2 will give the same minimum value to T. (See Exercise 24 below for a generalization of this phenomenon.)

A more promising approach is to minimize the sum S of *squares* of the distances from c to the data points:

$$S = (c - c_1)^2 + (c - c_2)^2 + \cdots + (c - c_n)^2 = \sum_{i=1}^{n} (c - c_i)^2.$$

S is known as the **cost function** or **objective function**. It is well known in the theory of optimization that the objective function is not unique, and that the outcome depends on the choice of objective function. There is no reason why, for example, we could not choose to minimize the sum of the fourth powers of the distances from c to the data points instead. However, the second power is both convenient and traditional. In this type of analysis, we simply hope that other cost functions will produce results that are not too different.

S is convenient because second-degree polynomials have linear derivatives, meaning that the emerging expressions are linear equations, about which so much powerful and straightforward mathematical machinery is easily available. To see this, we note that $S(c)$ is smooth, and its (unconstrained) minimum value will occur at a critical point $\bar{c}$ given by

$$0 = \left.\frac{dS}{dc}\right|_{c=\bar{c}} = \sum_{i=1}^{n} 2(\bar{c} - c_i) = 2n\bar{c} - 2\sum_{i=1}^{n} c_i.$$

Thus $\bar{c}$ is the *mean* of the data values:

$$\bar{c} = \frac{1}{n}\sum_{i=1}^{n} c_i = \frac{c_1 + c_2 + \cdots + c_n}{n}.$$

The technique used to obtain $\bar{c}$ above is an example of what is called the **method of least squares**. It has the following geometric interpretation. If the data values $c_1, c_2, \ldots, c_n$ are regarded as components of a vector $\mathbf{c}$ in $\mathbb{R}^n$, and $\mathbf{w}$ is the vector with components $1, 1, \ldots, 1$, then the vector projection of $\mathbf{c}$ in the direction of $\mathbf{w}$,

$$\mathbf{c_w} = \frac{\mathbf{c} \bullet \mathbf{w}}{|\mathbf{w}|^2}\,\mathbf{w} = \frac{c_1 + c_2 + \cdots + c_n}{n}\,\mathbf{w},$$

has all its components equal to the average of the data values. Thus, determining c from the data by the method of least squares corresponds to finding the vector projection of the data vector onto the one-dimensional subspace of $\mathbb{R}^n$ spanned by $\mathbf{w}$. Had there been no error in the measurements c_i, then $\mathbf{c}$ would have been equal to $c\mathbf{w}$.

Linear Regression

In scientific investigations it is often believed that the response of a system is a certain kind of function of one or more input variables. An investigator can set up an experiment to measure the response of the system for various values of those variables in order to determine the parameters of the function.

For example, suppose that the response y of a system is suspected to depend on the input x according to the linear relationship

$$y = ax + b,$$

where the values of a and b are unknown. An experiment set up to measure values of y corresponding to several values of x yields n data points, (x_i, y_i), $i = 1, 2, \ldots, n$. If the supposed linear relationship is valid, these data points should lie *approximately* along a straight line, but not exactly on one because of experimental error. Suppose the points are as shown in Figure 13.19. The linear relationship seems reasonable in this case. We want to find values of a and b so that the straight line $y = ax + b$ "best" fits the data.

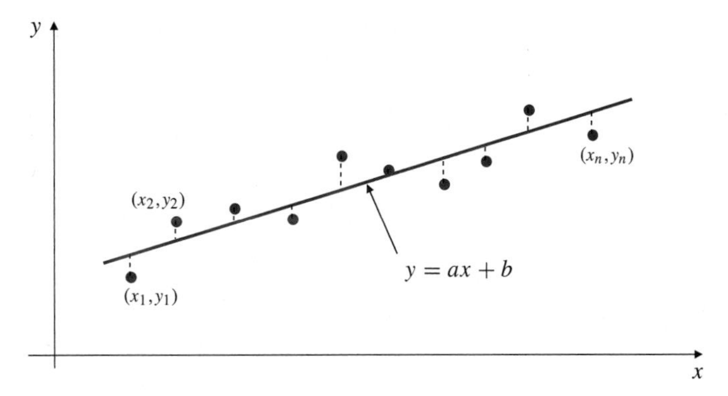

Figure 13.19 Fitting a straight line through experimental data

In this situation the method of least squares requires that a and b be chosen to minimize the sum S of the squares of the vertical displacements of the data points from the line:

$$S = \sum_{i=1}^{n} (y_i - ax_i - b)^2.$$

This is an unconstrained minimum problem in two variables, a and b. The minimum will occur at a critical point of S that satisfies

$$0 = \frac{\partial S}{\partial a} = -2 \sum_{i=1}^{n} x_i (y_i - ax_i - b),$$

$$0 = \frac{\partial S}{\partial b} = -2 \sum_{i=1}^{n} (y_i - ax_i - b).$$

These equations can be rewritten

$$\begin{aligned} \left(\sum_{i=1}^{n} x_i^2\right) a \quad &+ \quad \left(\sum_{i=1}^{n} x_i\right) b \quad = \quad \sum_{i=1}^{n} x_i y_i, \\ \left(\sum_{i=1}^{n} x_i\right) a \quad &+ \qquad\qquad n b \quad = \quad \sum_{i=1}^{n} y_i. \end{aligned}$$

Solving this pair of linear equations, we obtain the desired parameters:

$$a = \frac{n\left(\sum_{i=1}^{n} x_i y_i\right) - \left(\sum_{i=1}^{n} x_i\right)\left(\sum_{i=1}^{n} y_i\right)}{n\left(\sum_{i=1}^{n} x_i^2\right) - \left(\sum_{i=1}^{n} x_i\right)^2} = \frac{\overline{xy} - \bar{x}\bar{y}}{\overline{x^2} - (\bar{x})^2},$$

$$b = \frac{\left(\sum_{i=1}^{n} x_i^2\right)\left(\sum_{i=1}^{n} y_i\right) - \left(\sum_{i=1}^{n} x_i\right)\left(\sum_{i=1}^{n} x_i y_i\right)}{n\left(\sum_{i=1}^{n} x_i^2\right) - \left(\sum_{i=1}^{n} x_i\right)^2} = \frac{\overline{x^2}\bar{y} - \bar{x}\,\overline{xy}}{\overline{x^2} - (\bar{x})^2}.$$

In these formulas, we have used a bar to indicate the mean value of a quantity; thus, $\overline{xy} = (1/n)\sum_{i=1}^{n} x_i y_i$, and so on.

This procedure for fitting the "best" straight line through data points by the method of least squares is called **linear regression**, and the line $y = ax + b$ obtained in this way is called the **empirical regression line** corresponding to the data. Some scientific calculators with statistical features provide for linear regression by accumulating the sums of x_i, y_i, x_i^2, and $x_i y_i$ in various registers and keeping track of the number n of data points entered in another register. At any time it has available the information necessary to calculate a and b and the value of y corresponding to any given x.

EXAMPLE 1 Find the empirical regression line for the data $(x, y) = (0, 2.10)$, $(1, 1.92)$, $(2, 1.84)$, and $(3, 1.71)$, $(4, 1.64)$. What is the predicted value of y at $x = 5$?

Solution We have

$$\bar{x} = \frac{0 + 1 + 2 + 3 + 4}{5} = 2,$$

$$\bar{y} = \frac{2.10 + 1.92 + 1.84 + 1.71 + 1.64}{5} = 1.842,$$

$$\overline{xy} = \frac{(0)(2.10) + (1)(1.92) + (2)(1.84) + (3)(1.71) + (4)(1.64)}{5} = 3.458,$$

$$\overline{x^2} = \frac{0^2 + 1^2 + 2^2 + 3^2 + 4^2}{5} = 6.$$

Therefore,

$$a = \frac{3.458 - (2)(1.842)}{6 - 2^2} = -0.113,$$

$$b = \frac{(6)(1.842) - (2)(3.458)}{6 - 2^2} = 2.068,$$

and the empirical regression line is

$$y = 2.068 - 0.113x.$$

The predicted value of y at $x = 5$ is $2.068 - 0.113 \times 5 = 1.503$.

Remark Linear regression can also be interpreted in terms of vector projection. The data points define two vectors $\mathbf{x}$ and $\mathbf{y}$ in $\mathbb{R}^n$ with components $x_1, x_2, \ldots, x_n$ and $y_1, y_2, \ldots, y_n$, respectively. Let $\mathbf{w}$ be the vector with components $1, 1, \ldots, 1$. Finding the coefficients a and b for the regression line corresponds to finding the orthogonal projection of $\mathbf{y}$ onto the two-dimensional subspace (plane) in $\mathbb{R}^n$ spanned by $\mathbf{x}$ and $\mathbf{w}$. (See Figure 13.20.) This projection is $\mathbf{p} = a\mathbf{x} + b\mathbf{w}$. In fact, the two equations obtained above by setting the partial derivatives of S equal to zero are just the two conditions

$$(\mathbf{y} - \mathbf{p}) \bullet \mathbf{x} = 0,$$
$$(\mathbf{y} - \mathbf{p}) \bullet \mathbf{w} = 0,$$

Figure 13.20 $\mathbf{p} = a\mathbf{x} + b\mathbf{w}$ is the projection of $\mathbf{y}$ onto the plane spanned by $\mathbf{x}$ and $\mathbf{w}$

stating that $\mathbf{y}$ minus its projection onto the subspace is perpendicular to the subspace. The angle between $\mathbf{y}$ and this $\mathbf{p}$ provides a measure of how well the empirical regression line fits the data; the smaller the angle, the better the fit.

Linear regression can be used to find specific functional relationships of types other than linear if suitable transformations are applied to the data.

EXAMPLE 2 Find the values of constants K and s for which the curve $y = Kx^s$ best fits the experimental data points (x_i, y_i), $i = 1, 2, \ldots, n$. (Assume all data values are positive.)

Solution Observe that the required functional form corresponds to a linear relationship between $\ln y$ and $\ln x$:

$$\ln y = \ln K + s \ln x.$$

If we determine the parameters a and b of the empirical regression line $\eta = a\xi + b$ corresponding to the transformed data $(\xi_i, \eta_i) = (\ln x_i, \ln y_i)$, then $s = a$ and $K = e^b$ are the required values.

Remark It should be stressed that the constants K and s obtained by the method used in the solution above are not the same as those that would be obtained by direct application of the least squares method to the untransformed problem, that is, by minimizing $\sum_{i=1}^{n}(y_i - Kx_i^s)^2$. This latter problem cannot readily be solved. (Try it!)

Generally, the method of least squares is applied to fit an equation in which the response is expressed as a sum of constants times functions of one or more input variables. The constants are determined as critical points of the sum of squared deviations of the actual response values from the values predicted by the equation.

Applications of the Least Squares Method to Integrals

The method of least squares can be used to find approximations to reasonably well-behaved (say, piecewise continuous) functions as sums of constants times specified functions. The idea is to choose the constants to minimize the *integral* of the square of the difference.

For example, suppose we want to approximate the continuous function $f(x)$ over the interval $[0, 1]$ by a linear function $g(x) = px + q$. The method of least squares would require that p and q be chosen to minimize the integral

$$I(p, q) = \int_0^1 \Big(f(x) - px - q\Big)^2 \, dx.$$

Assuming that we can "differentiate through the integral" (we will investigate this issue in Section 13.6), the critical point of $I(p, q)$ can be found from

$$\begin{aligned} 0 &= \frac{\partial I}{\partial p} = -2\int_0^1 x\Big(f(x) - px - q\Big)\, dx, \\ 0 &= \frac{\partial I}{\partial q} = -2\int_0^1 \Big(f(x) - px - q\Big)\, dx. \end{aligned}$$

Thus,

$$\begin{aligned} \frac{p}{3} + \frac{q}{2} &= \int_0^1 xf(x)\, dx, \\ \frac{p}{2} + q &= \int_0^1 f(x)\, dx, \end{aligned}$$

and solving this linear system for p and q we get

$$p = \int_0^1 (12x - 6) f(x)\, dx,$$
$$q = \int_0^1 (4 - 6x) f(x)\, dx.$$

The following example concerns the approximation of a function by a **trigonometric polynomial**. Such approximations form the basis for the study of **Fourier series**, which are of fundamental importance in the solution of boundary-value problems for the Laplace, heat, and wave equations and other partial differential equations that arise in applied mathematics. (See Section 9.9.)

EXAMPLE 3 Use a least squares integral to approximate $f(x)$ by the sum

$$\sum_{k=1}^{n} b_k \sin kx$$

on the interval $0 \le x \le \pi$.

Solution We want to choose the constants to minimize

$$I = \int_0^{\pi} \left(f(x) - \sum_{k=1}^{n} b_k \sin kx \right)^2 dx.$$

For each $1 \le j \le n$, we have

$$0 = \frac{\partial I}{\partial b_j} = -2 \int_0^{\pi} \left(f(x) - \sum_{k=1}^{n} b_k \sin kx \right) \sin jx\, dx.$$

Thus,

$$\sum_{k=1}^{n} b_k \int_0^{\pi} \sin kx \sin jx\, dx = \int_0^{\pi} f(x) \sin jx\, dx.$$

However, if $j \ne k$, then $\sin kx \sin jx$ is an even function, so that

$$\begin{aligned}\int_0^{\pi} \sin kx \sin jx\, dx &= \frac{1}{2} \int_{-\pi}^{\pi} \sin kx \sin jx\, dx \\ &= \frac{1}{4} \int_{-\pi}^{\pi} \big(\cos(k - j)x - \cos(k + j)x\big)\, dx = 0.\end{aligned}$$

If $j = k$, then we have

$$\int_0^{\pi} \sin^2 jx\, dx = \frac{1}{2} \int_0^{\pi} (1 - \cos 2jx)\, dx = \frac{\pi}{2},$$

so that

$$b_j = \frac{2}{\pi} \int_0^{\pi} f(x) \sin jx\, dx.$$

Remark The series

$$\sum_{k=1}^{\infty} b_k \sin kx, \quad \text{where} \quad b_k = \frac{2}{\pi}\int_0^{\pi} f(x)\sin kx\,dx, \quad k = 1, 2, \ldots,$$

is called the **Fourier sine series** representation of $f(x)$ on the interval $(0, \pi)$. If f is continuous on $[0, \pi]$, it can be shown that

$$\lim_{n\to\infty} \int_0^{\pi} \left(f(x) - \sum_{k=1}^{n} b_k \sin kx \right)^2 dx = 0,$$

but more than just continuity is required of f to ensure that this Fourier sine series converges to $f(x)$ at each point of $(0, \pi)$. Such questions are studied in *harmonic analysis.* Similarly, the series

$$\frac{a_0}{2} + \sum_{k=1}^{\infty} a_k \cos kx, \quad \text{where} \quad a_k = \frac{2}{\pi}\int_0^{\pi} f(x)\cos kx\,dx, \quad k = 0, 1, 2, \ldots,$$

is called the **Fourier cosine series** representation of $f(x)$ on the interval $(0, \pi)$.

Remark Representing a function as the sum of a Fourier series is analogous to representing a vector as a linear combination of basis vectors. If we think of continuous functions on the interval $[0, \pi]$ as "vectors" with addition and scalar multiplication defined pointwise:

$$(f + g)(x) = f(x) + g(x), \qquad (cf)(x) = cf(x),$$

and with the "dot product" defined as

$$f \bullet g = \int_0^{\pi} f(x)g(x)\,dx,$$

then the functions $e_k(x) = \sqrt{2/\pi}\,\sin kx$ form a "basis." As shown in the example above, $e_j \bullet e_j = 1$, and if $k \neq j$, then $e_k \bullet e_j = 0$. Thus, these "basis vectors" are "mutually perpendicular unit vectors." The Fourier sine coefficients b_j of a function f are the components of f with respect to that basis.

EXERCISES 13.5

1. A generator is to be installed in a factory to supply power to n machines located at positions (x_i, y_i), $i = 1, 2, \ldots, n$. Where should the generator be located to minimize the sum of the squares of its distances from the machines?

2. The relationship $y = ax^2$ is known to hold between certain variables. Given the experimental data (x_i, y_i), $i = 1, 2, \ldots, n$, determine a value for a by the method of least squares.

3. Repeat Exercise 2 but with the relationship $y = ae^x$.

4. Use the method of least squares to find the plane $z = ax + by + c$ that best fits the data (x_i, y_i, z_i), $i = 1, 2, \ldots, n$.

5. Repeat Exercise 4 using a vector projection argument instead of the method of least squares.

In Exercises 6–11, show how to adapt linear regression to determine the two parameters p and q so that the given relationship fits the experimental data (x_i, y_i), $i = 1, 2, \ldots, n$. In which of these situations are the values of p and q obtained identical to those obtained by direct application of the method of least squares with no change of variable?

6. $y = p + qx^2$
7. $y = pe^{qx}$
8. $y = \ln(p + qx)$
9. $y = px + qx^2$
10. $y = \sqrt{px + q}$
11. $y = pe^x + qe^{-x}$

12. Find the parabola of the form $y = p + qx^2$ that best fits the data $(x, y) = (1, 0.11)$, $(2, 1.62)$, $(3, 4.07)$, $(4, 7.55)$, $(6, 17.63)$, and $(7, 24.20)$. No value of y was measured at $x = 5$. What value would you predict at this point?

13. Use the method of least squares to find constants a, b, and c so that the relationship $y = ax^2 + bx + c$ best describes the experimental data (x_i, y_i), $i = 1, 2, \ldots, n$, $(n \geq 3)$. How is

this situation interpreted in terms of vector projection?

14. How can the result of Exercise 13 be used to fit a curve of the form $y = pe^x + q + re^{-x}$ through the same data points?

15. Find the value of the constant a for which the function $f(x) = ax^2$ best approximates the function $g(x) = x^3$ on the interval $[0, 1]$, in the sense that the integral

$$I = \int_0^1 \left(f(x) - g(x)\right)^2 dx$$

is minimized. What is the minimum value of I?

16. Find a to minimize $I = \int_0^\pi \left(ax(\pi - x) - \sin x\right)^2 dx$. What is the minimum value of the integral?

17. Repeat Exercise 15 with the function $f(x) = ax^2 + b$ and the same g. Find a and b.

18. Find a, b, and c to minimize $\int_0^1 (x^3 - ax^2 - bx - c)^2\, dx$. What is the minimum value of the integral?

19. Find a and b to minimize $\int_0^\pi (\sin x - ax^2 - bx)^2\, dx$.

20. Find a, b, and c to minimize the integral

$$J = \int_{-1}^1 \left(x - a\sin\pi x - b\sin 2\pi x - c\sin 3\pi x\right)^2 dx.$$

21. Find constants a_j, $j = 0, 1, \ldots, n$, to minimize

$$\int_0^\pi \left(f(x) - \frac{a_0}{2} - \sum_{k=1}^n a_k \cos kx\right)^2 dx.$$

22. Find the Fourier sine series for the function $f(x) = x$ on $0 < x < \pi$. Assuming the series does converge to x on the interval $(0, \pi)$, to what function would you expect the series to converge on $(-\pi, 0)$?

23. Repeat Exercise 22 but obtaining instead a Fourier cosine series.

24. Suppose $x_1, x_2, \ldots, x_n$ satisfy $x_i \le x_j$ whenever $i < j$. Find x that minimizes $\sum_{i=1}^n |x - x_i|$. Treat the cases n odd and n even separately. For what values of n is x unique? *Hint:* Use no calculus in this problem.

13.6 Parametric Problems

In this section we will briefly examine three unrelated situations in which we want to differentiate a function with respect to a parameter rather than one of the basic variables of the function. Such situations arise frequently in mathematics and its applications.

Differentiating Integrals with Parameters

The Fundamental Theorem of Calculus shows how to differentiate a definite integral with respect to the upper limit of integration:

$$\frac{d}{dx}\int_a^x f(t)\, dt = f(x).$$

We are going to look at a different problem about differentiating integrals. If the integrand of a definite integral also depends on variables other than the variable of integration, then the integral will be a function of those other variables. How are we to find the derivative of such a function? For instance, consider the function $F(x)$ defined by

$$F(x) = \int_a^b f(x, t)\, dt.$$

We would like to be able to calculate $F'(x)$ by taking the derivative inside the integral:

$$F'(x) = \frac{d}{dx}\int_a^b f(x, t)\, dt = \int_a^b \frac{\partial}{\partial x} f(x, t)\, dt.$$

Observe that we use d/dx outside the integral and $\partial/\partial x$ inside; this is because the integral is a function of x only, but the integrand f is a function of both x and t. If the integrand depends on more than one parameter, then partial derivatives would be needed inside and outside the integral:

$$\frac{\partial}{\partial x}\int_a^b f(x,y,t)\,dt = \int_a^b \frac{\partial}{\partial x} f(x,y,t)\,dt.$$

The operation of taking a derivative with respect to a parameter inside the integral, or *differentiating through the integral*, as it is usually called, seems plausible. We differentiate sums term by term, and integrals are the limits of sums. However, both the differentiation and integration operations involve the taking of limits (limits of Newton quotients for derivatives, limits of Riemann sums for integrals). Differentiating through the integral requires changing the order in which the two limits are taken and, therefore, requires justification.

We have already seen another example of change of order of limits. When we assert that two mixed partial derivatives with respect to the same variables are equal,

$$\frac{\partial^2 f}{\partial x \partial y} = \frac{\partial^2 f}{\partial y \partial x},$$

we are, in fact, saying that limits corresponding to differentiation with respect to x and y can be taken in either order with the same result. This is not true in general; we proved it under the assumption that both of the mixed partials were *continuous*. (See Theorem 1 and Exercise 16 of Section 12.4.) In general, some assumptions are required to justify the interchange of limits. The following theorem gives one set of conditions that justify the interchange of limits involved in differentiating through the integral.

THEOREM 6

Differentiating through an integral

Suppose that for every x satisfying $c < x < d$, the following conditions hold:

(i) the integrals

$$\int_a^b f(x,t)\,dt \qquad \text{and} \qquad \int_a^b f_1(x,t)\,dt$$

both exist (either as proper or convergent improper integrals).

(ii) $f_{11}(x,t)$ exists and satisfies

$$|f_{11}(x,t)| \le g(t), \qquad a < t < b,$$

where

$$\int_a^b g(t)\,dt = K < \infty.$$

Then for each x satisfying $c < x < d$, we have

$$\frac{d}{dx}\int_a^b f(x,t)\,dt = \int_a^b \frac{\partial}{\partial x} f(x,t)\,dt.$$

PROOF Let

$$F(x) = \int_a^b f(x,t)\,dt.$$

If $c < x < d$, $h \ne 0$, and $|h|$ is sufficiently small that $c < x+h < d$, then, by Taylor's Formula,

$$f(x+h,t) = f(x,t) + hf_1(x,t) + \frac{h^2}{2} f_{11}(x+\theta h, t)$$

for some θ between 0 and 1. Therefore,

$$\begin{aligned}
&\left|\frac{F(x+h)-F(x)}{h} - \int_a^b f_1(x,t)\,dt\right| \\
&\qquad = \left|\int_a^b \frac{f(x+h,t)-f(x,t)}{h}\,dt - \int_a^b f_1(x,t)\,dt\right| \\
&\qquad \le \int_a^b \left|\frac{f(x+h,t)-f(x,t)}{h} - f_1(x,t)\right| dt \\
&\qquad = \int_a^b \left|\frac{h}{2} f_{11}(x+\theta h, t)\right| dt \\
&\qquad \le \frac{h}{2}\int_a^b g(t)\,dt = \frac{Kh}{2} \to 0 \text{ as } h \to 0.
\end{aligned}$$

Therefore,

$$F'(x) = \lim_{h\to 0} \frac{F(x+h)-F(x)}{h} = \int_a^b f_1(x,t)\,dt,$$

which is the desired result.

Remark It can be shown that the conclusion of Theorem 6 also holds under the sole assumption that $f_1(x,t)$ is continuous on the *closed, bounded* rectangle $c \le x \le d$, $a \le t \le b$. We cannot prove this here; the proof depends on a subtle property called *uniform continuity* possessed by continuous functions on closed bounded sets in $\mathbb{R}^n$. (See Appendix IV for the case $n = 1$.) In any event, Theorem 6 is more useful for our purposes because it allows for improper integrals.

EXAMPLE 1 Evaluate $\displaystyle\int_0^\infty t^n\, e^{-t}\, dt$.

Solution Starting with the convergent improper integral

$$\int_0^\infty e^{-s}\,ds = \lim_{R\to\infty} \left.\frac{e^{-s}}{-1}\right|_0^R = \lim_{R\to\infty}(1-e^{-R}) = 1,$$

we introduce a parameter by substituting $s = xt$, $ds = x\,dt$ (where $x > 0$) and get

$$\int_0^\infty e^{-xt}\,dt = \frac{1}{x}.$$

Now differentiate n times (each resulting integral converges):

$$\begin{aligned}
\int_0^\infty -t\, e^{-xt}\,dt &= -\frac{1}{x^2}, \\
\int_0^\infty (-t)^2\, e^{-xt}\,dt &= (-1)^2\frac{2}{x^3}, \\
&\ \ \vdots \\
\int_0^\infty (-t)^n\, e^{-xt}\,dt &= (-1)^n\frac{n!}{x^{n+1}}.
\end{aligned}$$

Putting $x = 1$, we get

$$\int_0^\infty t^n\, e^{-t}\,dt = n!.$$

Note that this result could be obtained by integration by parts (n times) or a reduction formula. This method is a little easier.

Remark The reader should check that the function $f(x,t) = t^k e^{-xt}$ satisfies the conditions of Theorem 6 for $x > 0$ and $k \ge 0$. We will normally not make a point of this.

EXAMPLE 2 Evaluate $F(x,y) = \displaystyle\int_0^\infty \frac{e^{-xt} - e^{-yt}}{t}\,dt$ for $x > 0,\ y > 0$.

Solution We have

$$\frac{\partial F}{\partial x} = -\int_0^\infty e^{-xt}\,dt = -\frac{1}{x} \quad\text{and}\quad \frac{\partial F}{\partial y} = \int_0^\infty e^{-yt}\,dt = \frac{1}{y}.$$

It follows that

$$F(x,y) = -\ln x + C_1(y) \quad\text{and}\quad F(x,y) = \ln y + C_2(x).$$

Comparing these two formulas for F, we are forced to conclude that $C_1(y) = \ln y + C$ for some constant C. Therefore,

$$F(x,y) = \ln y - \ln x + C = \ln\frac{y}{x} + C.$$

Since $F(1,1) = 0$, we must have $C = 0$ and $F(x,y) = \ln(y/x)$.

Remark We can combine Theorem 6 and the Fundamental Theorem of Calculus to differentiate an integral with respect to a parameter that appears in the limits of integration as well as in the integrand. If

$$F(x,b,a) = \int_a^b f(x,t)\,dt,$$

then, by the Chain Rule,

$$\frac{d}{dx}F\big(x, b(x), a(x)\big) = \frac{\partial F}{\partial x} + \frac{\partial F}{\partial b}\frac{db}{dx} + \frac{\partial F}{\partial a}\frac{da}{dx}.$$

Accordingly, we have

$$\begin{aligned}&\frac{d}{dx}\int_{a(x)}^{b(x)} f(x,t)\,dt \\ &= \int_{a(x)}^{b(x)} \frac{\partial}{\partial x} f(x,t)\,dt + f\big(x,b(x)\big)b'(x) - f\big(x,a(x)\big)a'(x).\end{aligned}$$

We require that $a(x)$ and $b(x)$ be differentiable at x, and for the application of Theorem 6, that $a \le a(x) \le b$ and $a \le b(x) \le b$ for all x satisfying $c < x < d$.

EXAMPLE 3 Solve the *integral equation*

$$f(x) = a - \int_b^x (x-t) f(t)\,dt.$$

Solution Assume, for the moment, that the equation has a sufficiently well-behaved solution to allow for differentiation through the integral. Differentiating twice, we get

$$\begin{aligned}f'(x) &= -(x-x)f(x) - \int_b^x f(t)\,dt = -\int_b^x f(t)\,dt, \\ f''(x) &= -f(x).\end{aligned}$$

The latter equation is the differential equation of simple harmonic motion. Observe that the given equation for f and that for f' imply the initial conditions

$$f(b) = a \qquad \text{and} \qquad f'(b) = 0.$$

Accordingly, we write the general solution of $f''(x) = -f(x)$ in the form

$$f(x) = A\cos(x - b) + B\sin(x - b).$$

The initial conditions then imply $A = a$ and $B = 0$, so the required solution is $f(x) = a\cos(x - b)$. Finally, we note that this function is indeed smooth enough to allow the differentiations through the integral and is, therefore, the solution of the given integral equation. (If you wish, verify it in the integral equation.)

Envelopes

An equation $f(x, y, c) = 0$ that involves a parameter c as well as the variables x and y represents a family of curves in the xy-plane. Consider, for instance, the family

$$f(x, y, c) = \frac{x}{c} + cy - 2 = 0.$$

This family consists of straight lines with intercepts $(2c, 2/c)$ on the coordinate axes. Several of these lines are sketched in Figure 13.21. It appears that there is a curve to which all these lines are tangent. This curve is called the *envelope* of the family of lines.

In general, a curve $\mathcal{C}$ is called the **envelope** of the family of curves with equations $f(x, y, c) = 0$ if, for each value of c, the curve $f(x, y, c) = 0$ is tangent to $\mathcal{C}$ at some point depending on c.

For the family of lines in Figure 13.21 it appears that the envelope may be the rectangular hyperbola $xy = 1$. We will verify this after developing a method for determining the equation of the envelope of a family of curves. We assume that the function $f(x, y, c)$ has continuous first partials and that the envelope is a smooth curve.

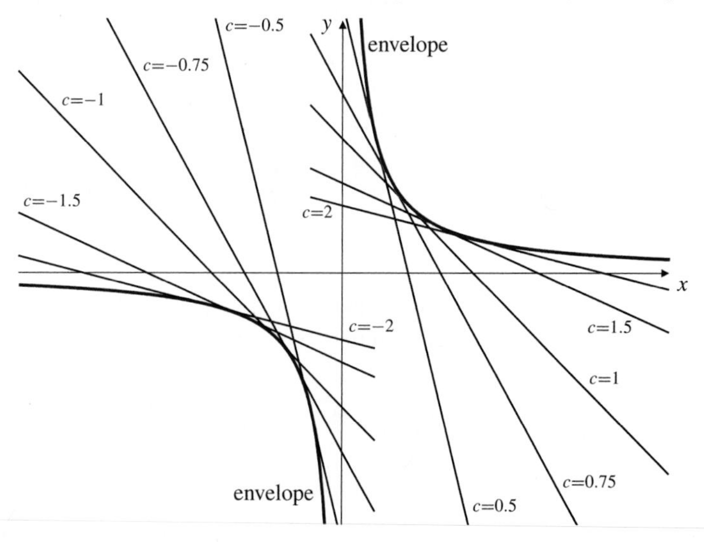

Figure 13.21 A family of straight lines and their envelope

For each c, the curve $f(x, y, c) = 0$ is tangent to the envelope at a point (x, y) that depends on c. Let us express this dependence in the explicit form $x = g(c)$, $y = h(c)$; these equations are parametric equations of the envelope. Since (x, y) lies on the curve $f(x, y, c) = 0$, we have

BEWARE! This is a subtle argument. Take your time and try to understand each step in the development.

$$f\big(g(c), h(c), c\big) = 0.$$

Differentiating this equation with respect to c, we obtain

$$f_1 g'(c) + f_2 h'(c) + f_3 = 0, \qquad (*)$$

where the partials of f are evaluated at $\big(g(c), h(c), c\big)$.

The slope of the curve $f(x, y, c) = 0$ at $\big(g(c), h(c), c\big)$ can be obtained by differentiating its equation implicitly with respect to x:

$$f_1 + f_2 \frac{dy}{dx} = 0.$$

On the other hand, the slope of the envelope $x = g(c)$, $y = h(c)$ at that point is $dy/dx = h'(c)/g'(c)$. Since the curve and the envelope are tangent at $f\big(g(c), h(c), c\big)$, these slopes must be equal. Therefore,

$$f_1 + f_2 \frac{h'(c)}{g'(c)} = 0, \qquad \text{so} \qquad f_1 g'(c) + f_2 h'(c) = 0.$$

Combining this with equation $(*)$ we get $f_3(x, y, c) = 0$ at all points of the envelope.

The equation of the envelope can be found by eliminating c between the two equations

$$f(x, y, c) = 0 \qquad \text{and} \qquad \frac{\partial}{\partial c} f(x, y, c) = 0.$$

EXAMPLE 4 Find the envelope of the family of straight lines

$$f(x, y, c) = \frac{x}{c} + cy - 2 = 0.$$

Solution We eliminate c between the equations

$$f(x, y, c) = \frac{x}{c} + cy - 2 = 0 \qquad \text{and} \qquad f_3(x, y, c) = -\frac{x}{c^2} + y = 0.$$

These equations can be easily solved and give $x = c$ and $y = 1/c$. Hence, they imply that the envelope is $xy = 1$, as we conjectured earlier.

EXAMPLE 5 Find the envelope of the family of circles

$$(x - c)^2 + y^2 = c.$$

Solution Here, $f(x, y, c) = (x - c)^2 + y^2 - c$. The equation of the envelope is obtained by eliminating c from the pair of equations

$$f(x, y, c) = (x - c)^2 + y^2 - c = 0,$$
$$\frac{\partial}{\partial c} f(x, y, c) = -2(x - c) - 1 = 0.$$

From the second equation, $x = c - \frac{1}{2}$, and then from the first, $y^2 = c - \frac{1}{4}$. Hence, the envelope is the parabola

$$x = y^2 - \frac{1}{4}.$$

This envelope and some of the circles in the family are sketched in Figure 13.22.

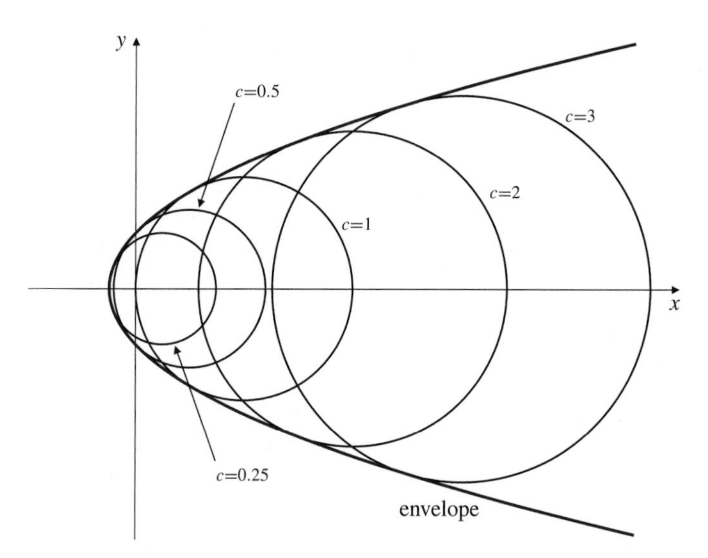

Figure 13.22 Circles $(x - c)^2 + y^2 = c$ and their envelope

A similar technique can be used to find the envelope of a family of surfaces. This will be a surface tangent to each member of the family.

EXAMPLE 6 **(The Mach cone)** Suppose that sound travels at speed c in still air and that a supersonic aircraft is travelling at speed $v > c$ along the x-axis, so that its position at time t is $(vt, 0, 0)$. Find the envelope at time t of the sound waves created by the aircraft at previous times. See Figure 13.23.

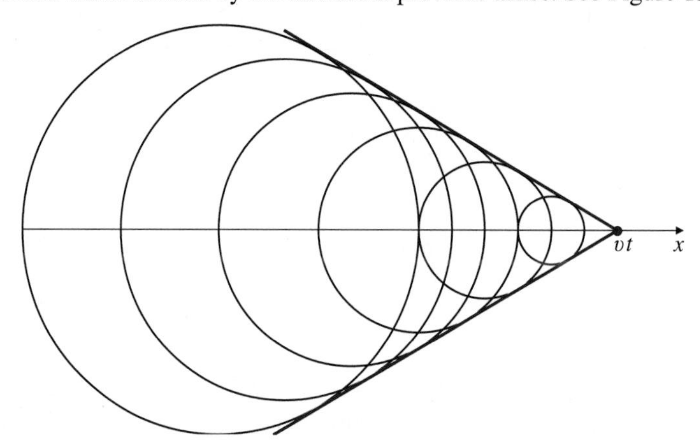

Figure 13.23 The Mach cone

Solution The sound created by the aircraft at time $\tau < t$ spreads out as a spherical wave front at speed c. The centre of this wave front is $(v\tau, 0, 0)$, the position of the aircraft at time τ. At time t the radius of this wave front is $c(t - \tau)$, so its equation is

$$f(x, y, z, \tau) = (x - v\tau)^2 + y^2 + z^2 - c^2(t - \tau)^2 = 0. \qquad (*)$$

At time t the envelope of all these wave fronts created at earlier times τ is obtained by eliminating the parameter τ from the above equation and the equation

$$\frac{\partial}{\partial \tau} f(x, y, z, \tau) = -2v(x - v\tau) + 2c^2(t - \tau) = 0.$$

Solving this latter equation for τ, we get $\tau = \dfrac{vx - c^2 t}{v^2 - c^2}$. Thus,

$$\begin{aligned} x - v\tau &= x - \frac{v^2 x - vc^2 t}{v^2 - c^2} = \frac{c^2}{v^2 - c^2}(vt - x) \\ t - \tau &= t - \frac{vx - c^2 t}{v^2 - c^2} = \frac{v}{v^2 - c^2}(vt - x). \end{aligned}$$

We substitute these two expressions into equation $(*)$ to eliminate τ:

$$\frac{c^4}{(v^2-c^2)^2}(vt-x)^2 + y^2 + z^2 - \frac{c^2v^2}{(v^2-c^2)^2}(vt-x)^2 = 0$$

$$y^2 + z^2 = \frac{c^2}{(v^2-c^2)^2}(v^2-c^2)(vt-x)^2 = \frac{c^2}{v^2-c^2}(vt-x)^2.$$

The envelope is the cone

$$x = vt - \frac{\sqrt{v^2-c^2}}{c}\sqrt{y^2+z^2},$$

which extends backward in the x direction from its vertex at $(vt, 0, 0)$, the position of the aircraft at time t. This is called the **Mach cone**. The sound of the aircraft cannot be heard at any point until the cone reaches that point.

Equations with Perturbations

In applied mathematics one frequently encounters intractable equations for which at least approximate solutions are desired. Sometimes such equations result from adding an extra term to what would otherwise be a simple and easily solved equation. This extra term is called a **perturbation** of the simpler equation. Often the perturbation has a coefficient smaller than the other terms in the equation; that is, it is a **small perturbation**. If this is the case, you can find approximate solutions to the perturbed equation by replacing the small coefficient by a parameter and calculating Maclaurin polynomials in that parameter. One example should serve to clarify the method.

EXAMPLE 7 Find an approximate solution of the equation

$$y + \frac{1}{50}\ln(1+y) = x^2.$$

Solution Without the logarithm term, the equation would clearly have the solution $y = x^2$. Let us replace the coefficient $1/50$ with the parameter ϵ and look for a solution $y = y(x, \epsilon)$ to the equation

$$y + \epsilon \ln(1+y) = x^2 \qquad (*)$$

in the form

$$y = y(x, \epsilon) = y(x, 0) + \epsilon y_\epsilon(x, 0) + \frac{\epsilon^2}{2!}y_{\epsilon\epsilon}(x, 0) + \cdots,$$

where the subscripts ϵ denote derivatives with respect to ϵ. We shall calculate the terms up to second order in ϵ. Evidently $y(x, 0) = x^2$. Differentiating equation (*) twice with respect to ϵ and evaluating the results at $\epsilon = 0$, we obtain

$$\frac{\partial y}{\partial \epsilon} + \ln(1+y) + \frac{\epsilon}{1+y}\frac{\partial y}{\partial \epsilon} = 0,$$

$$\frac{\partial^2 y}{\partial \epsilon^2} + \frac{2}{1+y}\frac{\partial y}{\partial \epsilon} + \epsilon\frac{\partial}{\partial \epsilon}\left(\frac{1}{1+y}\frac{\partial y}{\partial \epsilon}\right) = 0,$$

$$y_\epsilon(x, 0) = -\ln(1+x^2),$$

$$y_{\epsilon\epsilon}(x, 0) = \frac{2}{1+x^2}\ln(1+x^2).$$

Hence,

$$y(x, \epsilon) = x^2 - \epsilon\ln(1+x^2) + \frac{\epsilon^2}{1+x^2}\ln(1+x^2) + \cdots,$$

and the given equation has the approximate solution

$$y \approx x^2 - \frac{\ln(1+x^2)}{50} + \frac{\ln(1+x^2)}{2{,}500(1+x^2)}.$$

Similar perturbation techniques can be used for systems of equations and for differential equations.

EXERCISES 13.6

1. Let $F(x) = \int_0^1 t^x\,dt = \frac{1}{x+1}$ for $x > -1$. By repeated differentiation of F evaluate the integral

$$\int_0^1 t^x(\ln t)^n\,dt.$$

2. By replacing t with xt in the well-known integral

$$\int_{-\infty}^{\infty} e^{-t^2}\,dt = \sqrt{\pi},$$

and differentiating with respect to x, evaluate

$$\int_{-\infty}^{\infty} t^2 e^{-t^2}\,dt \quad \text{and} \quad \int_{-\infty}^{\infty} t^4 e^{-t^2}\,dt.$$

3. Evaluate $\int_{-\infty}^{\infty} \frac{e^{-xt^2} - e^{-yt^2}}{t^2}\,dt$ for $x > 0$, $y > 0$.

4. Evaluate $\int_0^1 \frac{t^x - t^y}{\ln t}\,dt$ for $x > -1$, $y > -1$.

5. Given that $\int_0^{\infty} e^{-xt}\sin t\,dt = \frac{1}{1+x^2}$ for $x > 0$ (which can be shown by integration by parts), evaluate

$$\int_0^{\infty} te^{-xt}\sin t\,dt \quad \text{and} \quad \int_0^{\infty} t^2 e^{-xt}\sin t\,dt.$$

6. Referring to Exercise 5, for $x > 0$ evaluate

$$F(x) = \int_0^{\infty} e^{-xt}\frac{\sin t}{t}\,dt.$$

Show that $\lim_{x\to\infty} F(x) = 0$ and hence evaluate the integral

$$\int_0^{\infty} \frac{\sin t}{t}\,dt = \lim_{x\to 0} F(x).$$

7. Evaluate $\int_0^{\infty} \frac{dt}{x^2+t^2}$ and use the result to help you evaluate

$$\int_0^{\infty} \frac{dt}{(x^2+t^2)^2} \quad \text{and} \quad \int_0^{\infty} \frac{dt}{(x^2+t^2)^3}.$$

8. Evaluate $\int_0^{x} \frac{dt}{x^2+t^2}$ and use the result to help you evaluate

$$\int_0^{x} \frac{dt}{(x^2+t^2)^2} \quad \text{and} \quad \int_0^{x} \frac{dt}{(x^2+t^2)^3}.$$

9. Find $f^{(n+1)}(a)$ if $f(x) = 1 + \int_a^x (x-t)^n f(t)\,dt$.

Solve the integral equations in Exercises 10–12.

10. $f(x) = Cx + D + \int_0^x (x-t)f(t)\,dt$

11. $f(x) = x + \int_0^x (x-2t)f(t)\,dt$

12. $f(x) = 1 + \int_0^1 (x+t)f(t)\,dt$

Find the envelopes of the families of curves in Exercises 13–18.

13. $y = 2cx - c^2$

14. $y - (x-c)\cos c = \sin c$

15. $x\cos c + y\sin c = 1$

16. $\frac{x}{\cos c} + \frac{y}{\sin c} = 1$

17. $y = c + (x-c)^2$

18. $(x-c)^2 + (y-c)^2 = 1$

19. Does every one-parameter family of curves in the plane have an envelope? Try to find the envelope of $y = x^2 + c$.

20. For what values of k does the family of curves $x^2 + (y-c)^2 = kc^2$ have an envelope?

21. Try to find the envelope of the family $y^3 = (x+c)^2$. Are the curves of the family tangent to the envelope? What have you actually found in this case? Compare with Example 3 of Section 13.3.

22. Show that if a two-parameter family of surfaces $f(x, y, z, \lambda, \mu) = 0$ has an envelope, then the equation of that envelope can be obtained by eliminating λ and μ from the three equations

$$f(x, y, z, \lambda, \mu) = 0,$$
$$\frac{\partial}{\partial\lambda} f(x, y, z, \lambda, \mu) = 0,$$
$$\frac{\partial}{\partial\mu} f(x, y, z, \lambda, \mu) = 0.$$

23. Find the envelope of the two-parameter family of planes

$$x\sin\lambda\cos\mu + y\sin\lambda\sin\mu + z\cos\lambda = 1.$$

24. Find the envelope of the two-parameter family of spheres

$$(x-\lambda)^2+(y-\mu)^2+z^2=\frac{\lambda^2+\mu^2}{2}.$$

In Exercises 25–27, find the terms up to second power in ϵ in the solution y of the given equation.

25. $y+\epsilon\sin\pi y=x$

26. $y^2+\epsilon e^{-y^2}=1+x^2$

27. $2y+\dfrac{\epsilon x}{1+y^2}=1$

28. Use perturbation methods to evaluate y with error less than 10^{-8} given that $y+(y^5/100)=1/2$.

29. Use perturbation methods to find approximate values for x and y from the system $x+2y+\dfrac{1}{100}e^{-x}=3$, $x-y+\dfrac{1}{100}e^{-y}=0$. Calculate all terms up to second order in $\epsilon=1/100$.

13.7 Newton's Method

A frequently encountered problem in applied mathematics is to determine, to some desired degree of accuracy, a root (i.e., a solution r) of an equation of the form

$$f(r)=0.$$

Such a root is called a **zero** of the function f. In Section 4.2 we introduced Newton's Method, a simple but powerful method for determining roots of functions that are sufficiently smooth. The method involves *guessing* an approximate value x_0 for a root r of the function f, and then calculating successive approximations $x_1, x_2, \ldots$, using the formula

$$x_{n+1}=x_n-\frac{f(x_n)}{f'(x_n)},\qquad n=0,1,2,\cdots.$$

If the initial guess x_0 is not too far from r, and if $|f'(x)|$ is *not too small* and $|f''(x)|$ is *not too large* near r, then the successive approximations $x_1, x_2, \ldots$ will converge very rapidly to r. Recall that each new approximation x_{n+1} is obtained as the x-intercept of the tangent line drawn to the graph of f at the previous approximation, x_n. The tangent line to the graph $y=f(x)$ at $x=x_n$ has equation

$$y-f(x_n)=f'(x_n)(x-x_n).$$

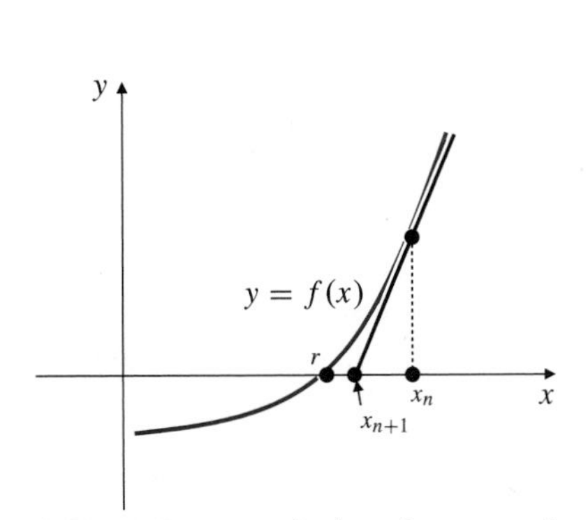

Figure 13.24 x_{n+1} is the x-intercept of the tangent at x_n

(See Figure 13.24.) The x-intercept, x_{n+1}, of this line is determined by setting $y=0$, $x=x_{n+1}$ in this equation, so is given by the formula in the shaded box above.

Newton's Method can be extended to finding solutions of systems of m equations in m variables. We will show here how to adapt the method to find approximations to a solution (x,y) of the pair of equations

$$\begin{cases} f(x,y)=0\\ g(x,y)=0,\end{cases}$$

starting from an initial guess (x_0,y_0). Under auspicious circumstances, we will observe the same rapid convergence of approximations to the root that typifies the single-variable case.

The idea is as follows. The two surfaces $z=f(x,y)$ and $z=g(x,y)$ intersect in a curve which itself intersects the xy-plane at the point whose coordinates are the desired solution. If (x_0,y_0) is near that point, then the tangent planes to the two surfaces at (x_0,y_0) will intersect in a straight line. This line meets the xy-plane at a point (x_1,y_1) that should be even closer to the solution point than was (x_0,y_0). We can easily determine (x_1,y_1). The tangent planes to $z=f(x,y)$ and $z=g(x,y)$ at (x_0,y_0) have equations

$$z=f(x_0,y_0)+f_1(x_0,y_0)(x-x_0)+f_2(x_0,y_0)(y-y_0),$$
$$z=g(x_0,y_0)+g_1(x_0,y_0)(x-x_0)+g_2(x_0,y_0)(y-y_0).$$

The line of intersection of these two planes meets the xy-plane at the point (x_1, y_1) satisfying

$$f_1(x_0, y_0)(x_1 - x_0) + f_2(x_0, y_0)(y_1 - y_0) + f(x_0, y_0) = 0,$$
$$g_1(x_0, y_0)(x_1 - x_0) + g_2(x_0, y_0)(y_1 - y_0) + g(x_0, y_0) = 0.$$

Solving these two equations for x_1 and y_1, we obtain

$$x_1 = x_0 - \left.\frac{fg_2 - f_2 g}{f_1 g_2 - f_2 g_1}\right|_{(x_0, y_0)} = x_0 - \left.\frac{\begin{vmatrix} f & f_2 \\ g & g_2 \end{vmatrix}}{\begin{vmatrix} f_1 & f_2 \\ g_1 & g_2 \end{vmatrix}}\right|_{(x_0, y_0)},$$

$$y_1 = y_0 - \left.\frac{f_1 g - f g_1}{f_1 g_2 - f_2 g_1}\right|_{(x_0, y_0)} = y_0 - \left.\frac{\begin{vmatrix} f_1 & f \\ g_1 & g \end{vmatrix}}{\begin{vmatrix} f_1 & f_2 \\ g_1 & g_2 \end{vmatrix}}\right|_{(x_0, y_0)}.$$

Observe that the denominator in each of these expressions is the Jacobian determinant $\partial(f, g)/\partial(x, y)\big|_{(x_0, y_0)}$. This is another instance where the Jacobian is the appropriate multivariable analogue of the derivative of a function of one variable.

Continuing in this way, we generate successive approximations (x_n, y_n) according to the formulas

$$x_{n+1} = x_n - \left.\frac{\begin{vmatrix} f & f_2 \\ g & g_2 \end{vmatrix}}{\begin{vmatrix} f_1 & f_2 \\ g_1 & g_2 \end{vmatrix}}\right|_{(x_n, y_n)},$$

$$y_{n+1} = y_n - \left.\frac{\begin{vmatrix} f_1 & f \\ g_1 & g \end{vmatrix}}{\begin{vmatrix} f_1 & f_2 \\ g_1 & g_2 \end{vmatrix}}\right|_{(x_n, y_n)}.$$

We stop when the desired accuracy has been achieved.

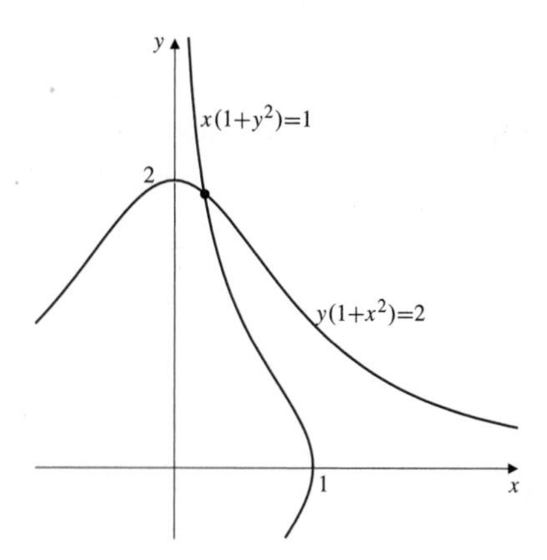

Figure 13.25 The two graphs intersect near (0.2, 1.8)

EXAMPLE 1 Find the root of the system $x(1 + y^2) - 1 = 0$, $y(1 + x^2) - 2 = 0$ with sufficient accuracy to ensure that the left sides of the equations vanish to the sixth decimal place.

Solution A sketch of the graphs of the two equations (see Figure 13.25) in the xy-plane indicates that the system has only one root near the point (0.2, 1.8). Application of Newton's Method requires successive computations of the quantities

$$f(x, y) = x(1 + y^2) - 1, \quad f_1(x, y) = 1 + y^2, \quad f_2(x, y) = 2xy,$$
$$g(x, y) = y(1 + x^2) - 2, \quad g_1(x, y) = 2xy, \quad g_2(x, y) = 1 + x^2.$$

Using a calculator or computer, we can calculate successive values of (x_n, y_n) starting from $x_0 = 0.2$, $y_0 = 1.8$:

Table 1. Root near (0.2, 1.8)

n	x_n	y_n	$f(x_n, y_n)$	$g(x_n, y_n)$
0	0.200 000	1.800 000	−0.152 000	−0.128 000
1	0.216 941	1.911 349	0.009 481	0.001 303
2	0.214 827	1.911 779	−0.000 003	0.000 008
3	0.214 829	1.911 769	0.000 000	0.000 000

The values in Table 1 were calculated sequentially in a spreadsheet by the method suggested below. They were rounded for inclusion in the table but the unrounded values were used in subsequent calculations. If you actually use the (rounded) values of x_n and y_n given in the table to calculate $f(x_n, y_n)$ and $g(x_n, y_n)$, your results may vary slightly.

The desired approximations to the root are the x_n and y_n values in the last line of the above table. Note the rapidity of convergence. However, many function evaluations are needed for each iteration of the method. For large systems Newton's Method is computationally too inefficient to be practical. Other methods requiring more iterations but many fewer calculations per iteration are used in practice.

Implementing Newton's Method Using a Spreadsheet

A computer spreadsheet is an ideal environment in which to calculate Newton's Method approximations. For a pair of equations in two unknowns such as the system in Example 1, you can proceed as follows:

(i) In the first nine cells of the first row (A1–I1) put the labels `n`, `x`, `y`, `f`, `g`, `f1`, `f2`, `g1`, and `g2`.

(ii) In cells A2–A9 put the numbers $0, 1, 2, \ldots, 7$.

(iii) In cells B2 and C2 put the starting values x_0 and y_0.

(iv) In cells D2–I2 put formulas for calculating $f(x, y), g(x, y), \ldots, g_2(x, y)$ in terms of values of x and y assumed to be in B2 and C2.

(v) In cells B3 and C3 store the Newton's Method formulas for calculating x_1 and y_1 in terms of the values x_0 and y_0, using values calculated in the second row. For instance, cell B3 should contain the formula

```
+B2-(D2*I2-G2*E2)/(F2*I2-G2*H2).
```

(vi) Replicate the formulas in cells D2–I2 to cells D3–I3.

(vii) Replicate the formulas in cells B3–I3 to the cells B4–I9.

You can now inspect the successive approximations x_n and y_n in columns B and C. To use different starting values, just replace the numbers in cells B2 and C2. To solve a different system of (two) equations, replace the contents of cells D2–I2. You may wish to save this spreadsheet for reuse with the exercises below or other systems you may want to solve later.

Remark While a detailed analysis of the convergence of Newton's Method approximations is beyond the scope of this book, a few observations can be made. At each step in the approximation process we must divide by J, the Jacobian determinant of f and g with respect to x and y evaluated at the most recently obtained approximation. Assuming that the functions and partial derivatives involved in the formulas are continuous, the larger the value of J at the actual solution, the more likely are the approximations to converge to the solution, and to do so rapidly. If J vanishes (or is very small) at the solution, the successive approximations may not converge, even if the initial guess is quite close to the solution. Even if the first partials of f and g are large at the solution, their Jacobian may be small if their gradients are nearly parallel there. Thus, we cannot expect convergence to be rapid when the curves $f(x, y) = 0$ and $g(x, y) = 0$ intersect at a very small angle.

Newton's Method can be applied to systems of m equations in m variables; the formulas are the obvious generalizations of those for two functions given above.

EXERCISES 13.7

Find the solutions of the systems in Exercises 1–6, so that the left-hand sides of the equations vanish up to 6 decimal places. These can be done with the aid of a scientific calculator, but that approach will be very time consuming. It is much easier to program the Newton's Method formulas on a computer to generate the required approximations. In each case try to determine reasonable *initial guesses* by sketching graphs of the equations.

1. $y - e^x = 0, \quad x - \sin y = 0$

2. $x^2 + y^2 - 1 = 0, \quad y - e^x = 0$ (two solutions)

3. $x^4 + y^2 - 16 = 0, \quad xy - 1 = 0$ (four solutions)

4. $x^2 - xy + 2y^2 = 10, \quad x^3y^2 = 2$ (four solutions)

5. $y - \sin x = 0, \quad x^2 + (y+1)^2 - 2 = 0$ (two solutions)

6. $\sin x + \sin y - 1 = 0, \quad y^2 - x^3 = 0$ (two solutions)

7. Write formulas for obtaining successive Newton's Method approximations to a solution of the system

$$f(x,y,z) = 0, \quad g(x,y,z) = 0, \quad h(x,y,z) = 0,$$

starting from an initial guess (x_0, y_0, z_0).

8. Use the formulas from Exercise 7 to find the first octant intersection point of the surfaces $y^2 + z^2 = 3$, $x^2 + z^2 = 2$, and $x^2 - z = 0$.

9. The equations $y - x^2 = 0$ and $y - x^3 = 0$ evidently have the solutions $x = y = 0$ and $x = y = 1$. Try to obtain these solutions using the two-variable form of Newton's Method with starting values:
(a) $x_0 = y_0 = 0.1$, and (b) $x_0 = y_0 = 0.9$.
How many iterations are required to obtain 6-decimal-place accuracy for the appropriate solution in each case?
How do you account for the difference in the behaviour of Newton's Method for these equations near $(0,0)$ and $(1,1)$?

13.8 Calculations with Maple

The calculations involved in solving systems of equations involving several variables can be very lengthy, even if the number of variables is small. In particular, locating critical points of a function of n variables involves solving a system of n (usually nonlinear) equations in n unknowns. In such situations the effective use of a computer algebra system like Maple can be very helpful. In this optional (and brief) section we present examples of how to use Maple's "fsolve" routine to solve systems of nonlinear equations and to find and classify critical points and thereby solve extreme-value problems.

Solving Systems of Equations

Maple has a procedure called **fsolve** built into its kernel (no package needs to be loaded to access it) that attempts to find floating-point real solutions to systems n equations in n variables. (For a single polynomial equation in one variable it will try to find all the real roots, but it may miss some.) For our purposes, an equation consists of either a single expression f in the variables (in which case the equation is taken to be $f = 0$) or else two expressions joined by an equal sign as in $f = g$. The procedure takes two or three arguments. The first is a set of n equations separated by commas. The set is enclosed in braces. The second argument is a set (also enclosed in braces) listing the n variables for which the equations are to be solved. (The number of variables in the equations must equal the number of equations.) The elements of the second set may consist of equations of the form "variable = initial guess," where the initial guess is a number we have reason to believe is *close* to the actual solution. It may not always be possible to make a good initial guess at the values of the variables, so, if we like, we can include a third argument specifing intervals of values of the variables in which to search for a solution. For example, to find a solution to the system $x^2 + y^3 = 3$, $x\sin(y) - y\cos(x) = 0$ near $(1, 2)$ we could try

```
> Digits := 6:
> fsolve({x^2+y^2=3, x*sin(y)-y*cos(x)}, {x=1, y=2});
```

$$\{x = 0.909510, \ y = 1.47404\}$$

If we had been unable to specify an initial guess, but instead had looked for a solution with x and y in $[0, 2]$, we would have got the same answer:

```
> fsolve({x^2+y^2=3, x*sin(y)-y*cos(x)},
  {x,y}, {x=0..2, y=0..2});
```

$$\{x = 0.909510,\ y = 1.47404\}$$

In fact, not specifying an initial guess or even search intervals would have led to the same outcome

```
> fsolve({x^2+y^2=3, x*sin(y)-y*cos(x)}, {x,y});
```

$$\{y = 1.47404,\ x = 0.909510\},$$

although, for its own private reasons, Maple chose to report the values of x and y in the opposite order this time. Had we specified a different search interval, we might have gotten a different result:

```
> fsolve({x^2+y^2=3, x*sin(y)-y*cos(x)},
  {x,y}, {x=0..2, y=0..1});
```

$$\{y = 0.,\ x = 1.73205\}$$

or even no solution at all, if there is in fact no solution in the given intervals.

```
> fsolve({x^2+y^2=3, x*sin(y)-y*cos(x)},
  {x,y}, {x=0..1, y=0..1});
```

$$fsolve(\{x^2 + y^2 = 3, s\sin(y) - y\sin(x)\}, \{x, y\}, \{x = (0..1), y = (0..1)\})$$

Using fsolve efficiently usually requires us to have some idea where solutions can be found. If the number of variables is 2 or 3, Maple's graphical routines can often be used to help us find approximate locations of solutions.

EXAMPLE 1 Solve the system

$$\begin{cases} x^2 + y^4 = 1 \\ z = x^3 y \\ e^x = 2y - z. \end{cases}$$

Solution We begin by defining the set of equations.

```
> eqns := {x^2+y^4=1, z=x^3*y, exp(x)=2*y-z};
```

$$eqns := \left\{x^2 + y^4 = 1, z = x^3\,y, \mathbf{e}^x = 2y - z\right\}$$

What are we to use for initial guesses? The first equation cannot be satisfied by any points outside the square $-1 \le x \le 1$, $-1 \le y \le 1$, so we need only consider starting values for x and y inside this square. The second equation then forces z to lie between -1 and 1 also. We could just try many initial guesses that satisfy these conditions and see what we get using fsolve. Alternatively, we can make several implicit plots of the three equations for fixed values of z between -1 and 1, looking for cases where the three curves come close to having a common intersection point:

```
> with(plots):
  for z from -1 by .2 to 1 do print("z =", z);
  implicitplot({x^2+y^4-1, z-x^3*y, exp(x)-2*y+z},
  x=-1.5 ..  1.5, y=-1.5 ..  1.5) od;
```

These commands produce 11 graphs of the three equations, considered as depending on x and y for z values ranging from -1 to 1 in steps of 0.2. Two of them are shown in Figure 13.26 and Figure 13.27. They correspond to $z = -0.2$ and $z = 0.2$ and indicate that the three equations likely have solutions near $(-1, 0.2, -0.2)$ and $(0.5, 0.9, 0.2)$. We run `fsolve` with these starting values and then substitute the resulting output into the three equations to check that the equations are satisfied. We limit Maple's output to 6 significant figures rather than the default 10:

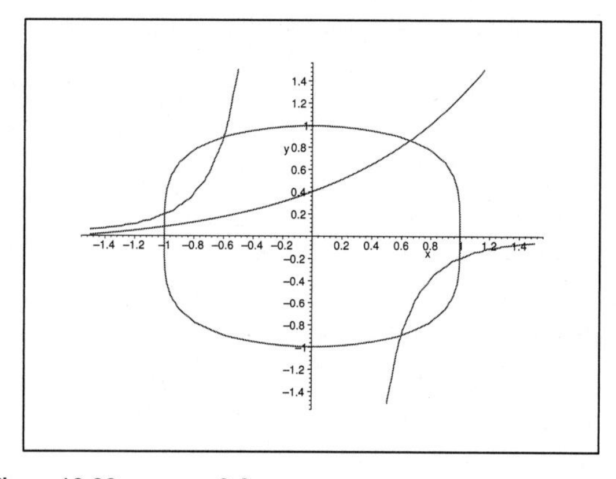

Figure 13.26 $z = -0.2$

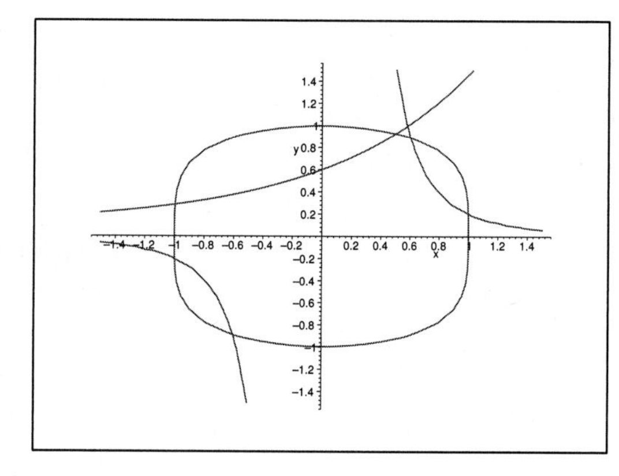

Figure 13.27 $z = 0.2$

```
> Digits := 6:
  vars := {x=-1, y=0.2, z=-0.2}:
  sols := fsolve(eqns,vars);
  evalf(subs(sols,eqns));
```

$$sols := \{x = -.999887, y = 0.122654, z = -.122613\}$$

$$\{-.122613 = -.122612, 1.00000 = 1., 0.367921 = 0.367921\}$$

```
> vars := {x=0.5, y=0.9, z=0.2}:
  sols := fsolve(eqns,vars);
  evalf(subs(sols,eqns));
```

$$sols := \{z = 0.138432, x = 0.531836, y = 0.920243\}$$

$$\{0.138432 = 0.138432, 1.00000 = 1., 1.70205 = 1.70206\}$$

We have found the two solutions to 6 significant digits.

Finding and Classifying Critical Points

Finding the critical points of a function of several variables amounts to solving the system of equations obtained by setting the first partial derivatives of the function to zero. The following example illustrates how this can be accomplished using Maple's `fsolve` routine. Since we also want to classify the critical points, we will find the eigenvalues of the Hessian matrix of the function at each critical point to determine whether that matrix is positive definite, negative definite, or indefinite.

Because the VectorCalculus package contains a procedure `Hessian` for calculating the Hessian matrix and the LinearAlgebra package contains a procedure `Eigenvalues` for determining the eigenvalues of a square matrix, we will either have to load both these packages or else call the procedures using `VectorCalculus[Hessian]` and `LinearAlgebra[Eigenvalues]`, respectively. As we need nothing else from these packages here, we will do it the latter way. If you have a version earlier than Maple 8, be aware that the older linalg package has procedures `hessian` and `eigenvals` that will do the same job.

EXAMPLE 2 Find and classify the critical points of

$$(x^2 + xy + 5y^2 + x - y)e^{-(x^2+y^2)}.$$

Solution We begin by defining f to be the expression above, which involves only the two variables x and y. We don't need f to be a function, so just define it as an expression.

```
> f := (x^2+x*y+5*y^2+x-y)*exp(-(x^2+y^2));
```

$$f := (x^2 + xy + 5y^2 + x - y)\mathbf{e}^{-(x^2+y^2)}$$

Next, we define H to be the Hessian matrix for f with respect to the variables x and y. Since this produces several lines of output, we will suppress the output.

`> H := VectorCalculus[Hessian](f,[x,y]):` The equations we want to solve to find the critical points of f are

```
> eqns := {diff(f,x)=0, diff(f,y)=0}:
```

Again we have surpressed output. We could have omitted the " = 0" from each equation; it would have been assumed.

Now comes the hard part: where do we look for solutions? Plotting some level curves of f can suggest likely locations for critical points.

```
> plots[contourplot](f,x=-3..3, y=-3..3, grid=[50,50],
  contours=16);
```

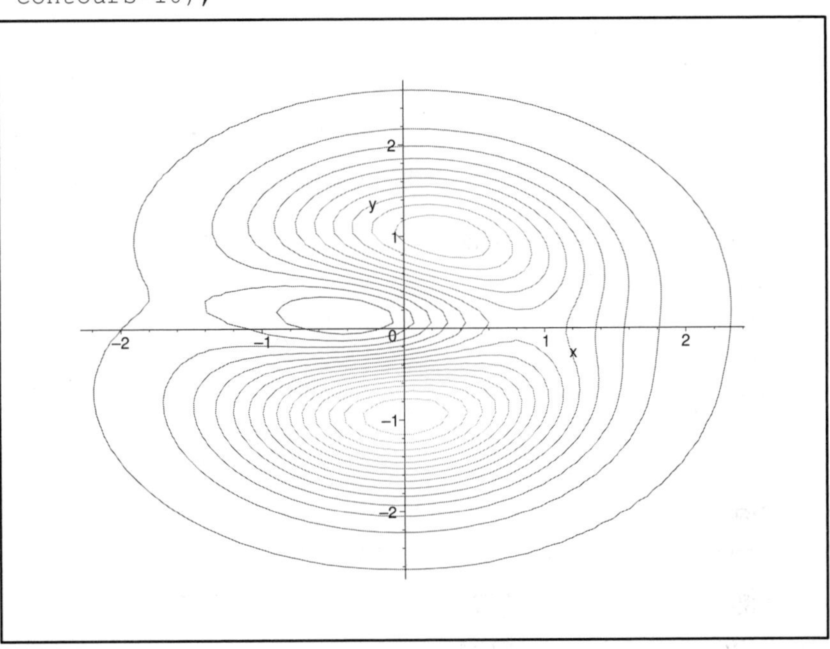

Figure 13.28 Contours of $f(x, y)$ in Example 2.

The contour plot (Figure 13.28) suggests that there are five critical points, three local extrema near $(0.3, 1)$, $(0, -1)$, and $(-0.6, 0.1)$ and two saddle points near $(1, 0)$ and $(-1.6, 0.2)$. We use each of these as initial guesses with `fsolve`. For each we first run `fsolve` to find the critical point. Then we find the value of f at that point. Finally, we calculate the eigenvalues of the Hessian of f to determine the nature of the critical point. We set Maple for 6 significant figures again.

```
> Digits := 6:
```

(a) Near the point $(0.3, 1)$:

```
> sols := fsolve(eqns,{x=0.3, y=1}); evalf(subs(sols,f));
```

$$sols := \{x = 0.275057, y = 1.00132\}$$
$$1.57773$$

```
> LinearAlgebra[Eigenvalues](subs(sols,H));
```

$$\begin{bmatrix} -2.41894 \\ -6.61497 \end{bmatrix}$$

Since both eigenvalues are negative, f has a local maximum value $1.577\,73$ at the critical point $(0.275\,057, 1.001\,32)$.

(b) Near the point $(0, -1)$:

```
> sols := fsolve(eqns,{x=0, y=-1}); evalf(subs(sols,f));
```

$$sols := \{y = -.955506, x = 0.00492113\}$$
$$2.21553$$

```
> LinearAlgebra[Eigenvalues](subs(sols,H));
```

$$\begin{bmatrix} -3.58875 \\ -8.54885 \end{bmatrix}$$

Since both eigenvalues are negative, f has a local maximum value 2.215 533 at the critical point (0.004 921 13, −0.955 506).

(c) Near the point (−0.6, 0.1):

```
> sols := fsolve(eqns,{x=-0.6, y=0.1});
  evalf(subs(sols,f));
```

$$sols := \{y = 0.132977, x = -.421365\}$$
$$-.283329$$

```
> LinearAlgebra[Eigenvalues](subs(sols,H));
```

$$\begin{bmatrix} 8.90194 \\ 2.32438 \end{bmatrix}$$

Since both eigenvalues are positive, f has a local minimum value −0.283 329 at the critical point (−0.421 365, 0.132 977).

(d) Near the point (1, 0):

```
> sols := fsolve(eqns,{x=1, y=0}); evalf(subs(sols,f));
```

$$sols := \{y = 0.0207852, x = 0.858435\}$$
$$0.762810$$

```
> LinearAlgebra[Eigenvalues](subs(sols,H));
```

$$\begin{bmatrix} 3.28636 \\ -2.84680 \end{bmatrix}$$

Since the Hessian has both positive and negative eigenvalues, f has a saddle point at (0.858 435, 0.020 785 2). Its value there is 0.762 810.

(e) Near the point (−1.6, 0.2):

```
> sols := fsolve(eqns,{x=-1.6, y=0.2});
  evalf(subs(sols,f));
```

$$sols := \{y = 0.292686, x = -1.58082\}$$
$$0.0445843$$

```
> LinearAlgebra[Eigenvalues](subs(sols,H));
```

$$\begin{bmatrix} 0.673365 \\ -.407579 \end{bmatrix}$$

Since the Hessian has both positive and negative eigenvalues, f has a saddle point at (−1.580 82, 0.292 686). Its value there is 0.044 584 3.

The negative exponential in the definition of f ensures that $f \to 0$ as $x^2 + y^2 \to \infty$. Assuming that we have found all the critical points of f, the value at the critical point in (b) must be an absolute maximum and that in (c) must be an absolute minimum.

Remark The most difficult part of using `fsolve` for large systems is determining suitable starting values for the roots or critical points. Graphical means are really only suitable for small systems (one, two, or three equations), and even then it is important to analyze the equations or functions involved for clues on where the roots or critical points may be. Here are some possibilities to consider:

1. Sometimes some of the equations will be simple enough that they can be solved for some variables and thus used to reduce the size of the system. We could have used the second equation in Example 1 to eliminate z from the first and third equations and, hence, reduced the system to two equations in two unknowns.

2. The system might result from adding a small extra term to a simpler system, the location of whose roots is known. In this case you can use those known roots as initial guesses.
3. Always be alert for equations limiting the possible values of some variables. For instance, in Example 1 the equation $x^2 + y^4 = 1$ limited x and y to the interval $[-1, 1]$.

EXERCISES 13.8

In Exercises 1–2, solve the given systems of equations by using Maple's `fsolve` routine. Quote the solutions to 5 significant figures. Be alert for simple substitutions that can reduce the number of equations that must be fed to `fsolve`.

1. $\begin{cases} x^2 + y^2 + z^2 = 1 \\ z = xy \\ 6xz = 1 \end{cases}$

2. $\begin{cases} x^4 + y^2 + z^2 = 1 \\ y = \sin z \\ z + z^3 + z^4 = x + y \end{cases}$

In Exercises 3–6, use `fsolve` to calculate the requested results. In each case quote the results to 5 significant digits.

3. Find the maximum and minimum values and their locations for $f(x, y) = (xy - x - 2y)/((1 + x^2 + y^2)^2)$. Use a contour plot to help you determine suitable starting points.

4. Evidently, $f = 1 - 10x^4 - 8y^4 - 7z^4$ has maximum value 1 at $(0, 0, 0)$. Find the absolute maximum value of $h = f + g$, where $g = yz - xyz - x - 2y + z$ by starting at various points near $(0, 0, 0)$.

5. Find the minimum value of

$$f = x^2 + y^2 + z^2 + 0.2xy - 0.3xz + 4x - y.$$

6. Find the maximum and minimum values of

$$f(x, y, z) = \frac{x + 1.1y - 0.9z + 1}{1 + x^2 + y^2 + z^2}.$$

13.9 Entropy in Statistical Mechanics and Information Theory

Entropy was introduced in Chapter 12 as an independent variable in a function that determines internal energy. Many feel compelled to ask what entropy means physically. It is curious that when carefully examined, thermodynamic energy is no less intuitively mysterious from a physical point of view, but few feel moved to subject it to the same level of scrutiny. Nonetheless physicists have delved extensively into the microscopic origins of both of these quantities through the subject of **statistical mechanics**.

This section presents gateway applications (marked) that pertain to entropy and represent entries into two distinct fields without attempting comprehensive treatments. First elementary multivariate calculus leads to a statistical mechanical view of **entropy**. This not only turns out to be surprisingly simple, but it also has an unanticipated broad scope, as often happens with mathematics. An important example of this is the distinct field of **information theory** where entropy becomes the central object. As an entry to the subject we illustrate with the elementary example of data compression.

Boltzmann Entropy

The main city graveyard of Vienna is a fascinating place. It is the final resting place of many important historical figures, including the famous scientist Ludwig Boltzmann. His tombstone has an equation carved into it, that relates a quantity S, known as **entropy**, to a single quantity W, known as **statistical weight**. W represents the number of ways that atomic and molecular positions and momenta can be rearranged without apparently changing how a physical system appears to us in our every day world.

Entropy is how we keep track of all of these invisible possibilities in thermodynamics. It has the key property that the overall entropy of two completely independent physical systems is just the sum of the entropies of each system evaluated separately. On the other hand, the number of ways one system can be arranged is independent of the other system, so the overall statistical weight of the independent pair of systems viewed as a whole is just the product of the statistical weights from each system. The size of the statistical weights are so large in reality that one can very effectively treat them as continuous variables and entropy as a differentiable function of them. We will use these properties to deduce the unique equation, valid for all systems, that you will find on Boltzmann's tombstone when you make your visit to the main city graveyard of Vienna.

We seek a unique function of the form $S = f(W)$ valid for all physical systems. Accordingly, two independent systems, labeled 1 and 2, will have entropies given by $S_1 = f(W_1)$ and $S_2 = f(W_2)$ in terms of their statistical weights W_1 and W_2.

Because of additivity $S = S_1 + S_2$. Because of independence, for every state in system 1 there are W_2 states in independent system 2 so the number of states for both systems combined is $W = W_1 W_2$. Thus, $S = f(W) = f(W_1 W_2)$. Since S_1 does not depend on W_2 and S_2 does not depend on W_1, it follows that

$$\frac{dS_1}{dW_1} = \frac{\partial S}{\partial W1} = f'(W)W_2 \quad \text{and} \quad \frac{dS_2}{dW_2} = \frac{\partial S}{\partial W_2} = f'(W)W_1,$$

and so

$$W_1 \frac{dS_1}{dW_1} = W_2 \frac{dS_2}{dW_2}.$$

Since the left side of this equation is independent of W_2 and the right side is independent of W_1, both sides sides must be independent of both variables and so must be a constant k. Hence, $S_1 = k \ln W_1 + C_1$ and $S_2 = k \ln W_2 + C_2$. The only way to make the function of statistical weight independent of the system is to require that entropy vanish when there is only one way to arrange the system; that is, when the statistical weight is 1, the entropy must be 0, and $C_1 = C_2 = 0$ (see Exercise 1 below). Thus, generally,

$$S = k \ln W,$$

which is Boltzmann's epitaph. A quibble is that the actual epitaph precedes the use of the ln notation for the natural logarithm. So what is actually carved in the stone is

$$S = k.\log_e W$$

where the unorthodox "period" so carved clearly denotes multiplication.

The positive constant k is known as the Boltzmann constant, which is regarded as one of the fundamental constants of nature. If entropy has units of energy per temperature, as we deduce from the definition of temperature in Chapter 12, then those are also the units of k. In modern physics, k is often written k_B to distinguish it from other uses for the symbol. But because temperature scales are discretionary to an extent, k can just as easily be set to 1 with suitable units.

Shannon Entropy

An equivalent form of entropy can be expressed in terms of probabilities. We introduced it in Example 1 of Section 13.4. Although this form originated in physics, also dating back to Boltzmann, it was made most famous by Claude Shannon in the 1950's in his creation, **Information Theory**. Thus, it became widely known as Shannon entropy. We adopt this slightly ahistorical usage for that reason.

To deduce this form, first consider an ensemble of M identical systems, each with entropy S. Each system has the same internal probability p_i of being in any particular state i. The number of systems in state i is $Mp_i = m_i$, where $\sum_i p_i = 1$. The number of ways that the collection of M systems can have m_1 systems in state 1, m_2 systems in state 2, etc., is

$$W_M = \frac{M!}{m_1! m_2! \cdots m_i! \cdots}.$$

By additivity, the combined entropy of the M systems is

$$S_M = MS = k \ln W_M = k(\ln M! - \sum_i \ln m_i!).$$

If we use the Modified Stirling Formula $\ln m! \approx m \ln m - m$ (see Exercise 45 of Section 9.6) to approximate the factorials, the above expression becomes

$$\begin{aligned} S_M = MS &\approx k\left(M \ln M - M - \sum_i (m_i \ln m_i - m_i)\right) \\ &= k\left(M \ln M - M - \sum_i (Mp_i \ln Mp_i - Mp_i)\right) \\ &= k\left(M \ln M - M - (M \ln M - M)\sum_i p_i - \sum_i (Mp_i \ln p_i)\right), \end{aligned}$$

so that, on division by M, we set

$$S = -k \sum_i p_i \ln p_i.$$

Example 1 in Section 13.4 illustrated this with two constraints. This representation of Entropy also retains a useful maximum property when constrained in terms of probability alone,

$$\textbf{extremize}: \quad S = -k \sum_{i=1}^{n} p_i \ln p_i \quad \text{where} \quad \sum_{i=1}^{n} p_i = 1$$

The maximum value occurs when all the probabilities p_i are equal to $1/n$. We expect a maximum principle to persist because in Example 1 of Section 13.4 it became apparent that any attempt to find a critical point of the entropy of the system led to a maximum value of the entropy.

Information Theory

The joining of probability with the maximum property of entropy led, amazingly, to an understanding of the general limits of transmission and encoding of signals, which was the origin of the subject of information theory. In information theory, entropy is expressed in a superficially different manner than it is in statistical mechanics. Instead of the natural logarithm, log to the base 2 is normally, but not necessarily, used and the constant is set to 1,

$$S = H(p_1, \ldots, p_N) = -\sum_{i=1}^{N} p_i \log_2 p_i, \quad \text{where} \quad \sum_{i=1}^{N} p_i = 1.$$

H is the customary notation for the *Shannon* or *information entropy* (or just *information*). It is a common problem that one must dress up universal concepts in different clothes as they pass between different fields. Although H has been widely adopted by users of information theory, its use to denote entropy in this probabilistic form actually dates back to Boltzmann, who articulated his early ideas in his historically famous "H-theorem." In this form, H can be viewed as the mean value of $\log_2 1/p_i$ which is called the *self information* or the *surprisal*.

Of course, these superficial changes do not alter the basic properties of entropy. H still has the maximum property, and it is still additive. That is, two independent systems with entropies, H_1 and H_2 can be regarded as a single system with entropy $H_1 + H_2$. The proof of this is left as an exercise. (See Exercise 3.)

If one has a sequence of n bits to represent a number in base 2 then there are 2^n possible numbers and the probability, when all probabilities are the same, of any one number is 2^{-n}. This is the maximum entropy scenario. In this case for n bits

$$H = -\sum_{i=1}^{2^n} 2^{-n} \log_2 2^{-n} = n,$$

Thus, the maximum information entropy is nothing more than the number of bits in the sequence.

We can use a string of bits to send a message, even though we send messages with an alphabet instead of bits. We can imagine a message as being a string of x characters. Each character is drawn from an alphabet of y letters. If all letters are equally likely, the number of possible messages is y^x. This means that the entropy of the message string is

$$H_m = -\sum_{i=1}^{y^x} y^{-x} \log_2 y^{-x} = -y^x\, y^{-x}\,(-x) \log_2 y = x \log_2 y$$

On the other hand, the entropy of each letter is given by

$$H_l = -\sum_{i=1}^{y} y^{-1} \log_2 y^{-1} = -y\, y^{-1}\,(-1) \log_2 y = \log_2 y.$$

By the additivity of entropy, since the message consists of x such letters, it's entropy must satisfy

$$H_m = x H_l = x \log_2 y,$$

agreeing with the earlier calculation.

We can use a bit string to assign a specific string of bits to represent one member of an alphabet. The now-classical example is ASCII (American Standard Code for Information Interchange), which in its original form had $2^7 = 128$ characters in its alphabet. This included the regular English alphabet in upper- and lower-case, numbers, punctuation marks, and other special characters. For ASCII, in the unlikely case where all characters were equally probable, the entropy of our x-character message would be $H_m = 7x$. More generally, for an m-bit alphabet (i.e., $y = 2^m$,)

$$H_m = mx.$$

Thus, what seemed to be the maximum entropy for the message string, turns out to be equivalent to the maximum entropy of the entire binary string, since $mx = n$.

The relationship between entropy and bit string length only holds in the case of equal probabilities with symbols represented by equal numbers of bits. We could, for example, change the probability structure by making some of the characters in the alphabet more likely than others, while using the same rules for sending the message in terms of bits. In that case, the number of bits would be unchanged but the entropy would not be given by the length of the bit string any longer. Instead of the length of the string, the entropy is given by

$$H = -x \sum_{i=1}^{y} p_i \log_2 p_i.$$

Here p_i is the probability of character i in an alphabet of y characters forming a message string of x characters.

This allows the possibility of compression. We might, on average, send a particular message with fewer than mx bits. This can be done quite simply by allowing the alphabet to be represented by unique bit strings of varying size. The improbable characters are assigned to longer bit strings, while the probable ones are assigned to shorter ones. A theorem from Information Theory says that the best compression possible is given by

$$\frac{-\sum_{i=1}^{y} p_i \log_2 p_i}{\log_2 y},$$

which is just the ratio of entropies.

EXAMPLE 1 Suppose our alphabet has only $y = 8$ characters, using the letters A through H for convenience. Then $\log_2 y = 3$, and we must use an *average* of 3 bits per character. Suppose, however, that the probabilities of the characters are as follows

char	A	B	C	D	E	F	G	H
prob	$\frac{1}{4}$	$\frac{1}{4}$	$\frac{1}{8}$	$\frac{1}{8}$	$\frac{1}{8}$	$\frac{1}{16}$	$\frac{1}{32}$	$\frac{1}{32}$

Note that the sum of the probabilities is indeed 1. Then the best compression is

$$\frac{1}{3}\left(\frac{2}{4}+\frac{2}{4}+\frac{3}{8}+\frac{3}{8}+\frac{3}{8}+\frac{4}{16}+\frac{5}{32}+\frac{5}{32}\right) \approx 0.896$$

using three figures of accuracy.

If we represent A by the string 10, B by 11, C by 001, D by 000, E by 010, F by 0111, G by 01101, and H by 01100, any string of bits can be uniquely decoded by a simple algorithm. One such algorithm for decoding the string character by character is:

Read the first two bits.
If 1 0 then A (done)
If 1 1 then B (done)
If 0 0 then read the 3rd bit
    If 1 then C (done)
    If 0 then D (done)
If 0 1 then read the 3rd bit
    If 0 then E (done)
    If 1 then read the 4th bit
        If 1 then F (done)
        If 0 then read the 5th bit
            If 1 then G (done)
            If 0 then H (done)

Repeat the above starting with the first unread bits for remaining characters. Here, in conformance to the entropy structure for average string length 3 bits, 2 bits correspond to the most probable letters while 5 bits represent the least. The optimal encoding scheme is not unique; all that is needed to be optimal is to assign the numbers of bits in such a way that the correct encoding can be deduced. If we use this encoding structure, we will on average have 89.6% of the message length of a scheme that assigns exactly 3 bits to every one of the characters in the 8-character alphabet.

Compression is just one simple application. There are also many other important results of information theory such as the transmission capacity on noisy channels, data analysis methods, and much more.

EXERCISES 13.9

1. Using properties of entropy, show that the only value for the universal constant C that satisfies the expression $S = k \ln W + C$ for all independent physical systems is $C = 0$.

2. Prove the form of the maximum property of entropy made in the text: If $k > 0$, $0 \le p_i \le 1$ for $1 \le i \le n$, and $\sum_{i=1}^{n} p_i = 1$, then $-k \sum_{i=1}^{n} p_i \ln p_i$ has a maximum value when $p_i = 1/n$ for each i.

3. Given two independent systems with information entropies

$$H_1 = -\sum_{i=1}^{I} p_i \log_2 p_i \ ; \quad \sum_{i=1}^{I} p_i = 1,$$

$$H_2 = -\sum_{j=1}^{J} q_j \log_2 q_j \ ; \quad \sum_{j=1}^{J} q_j = 1,$$

show that sum of the entropies is also the entropy for the system,

$$H = -\sum_{k=1}^{K} \pi_k \log_2 \pi_k \ ; \quad \sum_{k=1}^{K} \pi_k = 1,$$

formed by interpreting both independent systems as subsystems of a single larger system. *Hint:* for each (i, j) satisfying $1 \le i \le I$ and $1 \le j \le J$, there is a unique $k = i + I(j-1)$ satisfying $1 \le k \le K = IJ$. Show that $H_1 + H_2 = H$, provided $\pi_k = p_i q_j$.

4. Find an optimal binary compression for a 4-character alphabet a, b, c, d with probabilities $1/2, 1/4, 1/8, 1/8$, and state the average compression.

5. The statistical weight W for N distinct atoms distributed among N states is just $N!$. But suppose these states form M groupings, each grouping with distinct energy ϵ_i per atom, such that the atoms within each grouping may be exchanged without observable consequence. The physical condition of the system can then be specified by knowing only the number of atoms in each of these groupings, $n_1, n_2, \ldots, n_M$, where $\sum_{i=1}^{M} n_i = N$. The statistical weight then becomes

$$\frac{N!}{n_1! n_2! \cdots n_M!}.$$

Assuming all n_i are large, use the Modified Stirling approximation $\ln n! \approx n \ln n - n$ to show that maximizing the entropy $S = k \ln W$ subject to the constraints of having a fixed total number N of atoms, and a fixed total energy $\sum_{i=1}^{M} n_i \epsilon_i = E$, leads to the relationship

$$n_i = A e^{-B\epsilon_i},$$

where the constants A and B are determined by the values of the Lagrange multipliers for the constrained extremal problem for S, and hence by the two constraints.

6. The result of the previous problem holds for other classes of particles, for instance molecules of an ideal gas, provided the energies of the particles are mainly the kinetic energies of their translational motions. In that result, we can let N and M grow very large in such a way that the largest gap between adjacent values of ϵ_j approaches zero in length. In the limit, the kinetic energy of each atom is a function of its mass m and speed v: $\epsilon = \frac{1}{2} m v^2$.

Consider for the moment only the part of the kinetic energy of the particle due to its velocity $u\mathbf{i}$ in the x direction. The number of atoms for which the x-component of velocity is u will be given by a density function $n(u)$ satisfying, by the result of the previous exercise,

$$n(u) = A\, e^{-Bmu^2/2}.$$

(a) Show that $p(u) = \dfrac{n(u)}{N}$ is a normally distributed probability density function. What are the values of the mean and variance of u? (See Definition 7 in Section 7.8 and the following discussion.) Express the value of A in terms of B, m, and N.

(b) Find the expectation of u^2 for the random variable u, and hence the expected value of the part of the kinetic energy of a random particle in our system due to its motion in the x direction. What is the expected value of the total kinetic energy of a random particle in the system, and of all the particles?

(c) Use the formula $E = \dfrac{3}{2} NkT$ (from Exercise 25 of Section 12.6 or the discussion preceding Example 4 in Section 12.8), expressing the energy of an ideal gas at absolute temperature T and consisting of N molecules, to find the value of B. (Here k is the Boltzmann constant.) Hence, show that the probability density function for the number of particles having velocity $\mathbf{v}$ in an ideal gas is

$$p(\mathbf{v}) = \left(\frac{m}{2\pi kT}\right)^{\frac{3}{2}} e^{-\frac{m(|\mathbf{v}|^2)}{2kT}}.$$

This is known as the Maxwell-Boltzmann distribution.

CHAPTER REVIEW

Key Ideas

- **What is meant by the following terms?**
  - ◇ a critical point of $f(x, y)$
  - ◇ a singular point of $f(x, y)$
  - ◇ an absolute maximum value of $f(x, y)$
  - ◇ a local minimum value of $f(x, y)$
  - ◇ a saddle point of $f(x, y)$
  - ◇ a quadratic form

◇ a constraint

◇ linear programming

◇ an envelope of a family of curves

- **State the second derivative test for a critical point of $f(x, y)$.**
- **Describe the method of Lagrange multipliers.**
- **Describe the method of least squares.**
- **Describe Newton's Method for two equations.**

Review Exercises

In Exercises 1–4, find and classify all the critical points of the given functions.

1. $xy\,e^{-x+y}$

2. $x^2y - 2xy^2 + 2xy$

3. $\frac{1}{x} + \frac{4}{y} + \frac{9}{4 - x - y}$

4. $x^2y(2 - x - y)$

5. Let $f(x, y, z) = x^2 + y^2 + z^2 + \frac{1}{x^2 + y^2 + z^2}$. Does f have a minimum value? If so, what is it and where is it assumed?

6. Show that $x^2 + y^2 + z^2 - xy - xz - yz$ has a local minimum value at $(0, 0, 0)$. Is the minimum value assumed anywhere else?

7. Find the absolute maximum and minimum values of $f(x, y) = xye^{-x^2-4y^2}$. Justify your answer.

8. Let $f(x, y) = (4x^2 - y^2)e^{-x^2+y^2}$.

(a) Find the maximum and minimum values of $f(x, y)$ on the xy-plane.

(b) Find the maximum and minimum values of $f(x, y)$ on the wedge-shaped region $0 \le y \le 3x$.

9. A wire of length L cm is cut into at most three pieces, and each piece is bent into a square. What is the (a) minimum and (b) maximum of the sum of the areas of the squares?

10. A delivery service will accept parcels in the shape of rectangular boxes the sum of whose girth and height is at most 120 in. (The girth is the perimeter of a horizontal cross-section.) What is the largest possible volume of such a box?

11. Find the area of the smallest ellipse $\frac{x^2}{a^2} + \frac{y^2}{b^2} = 1$ that contains the rectangle $-1 \le x \le 1$, $-2 \le y \le 2$.

12. Find the volume of the smallest ellipsoid

$$\frac{x^2}{a^2} + \frac{y^2}{b^2} + \frac{z^2}{c^2} = 1$$

that contains the rectangular box $-1 \le x \le 1$, $-2 \le y \le 2$, $-3 \le z \le 3$.

13. Find the volume of the smallest region of the form

$$0 \le z \le a\left(1 - \frac{x^2}{b^2} - \frac{y^2}{c^2}\right)$$

that contains the box $-1 \le x \le 1, -2 \le y \le 2, 0 \le z \le 2$.

14. A window has the shape of a rectangle surmounted by an isosceles triangle. What are the dimensions x, y, and z of the window (see Figure 13.29) if its perimeter is L and its area is maximum?

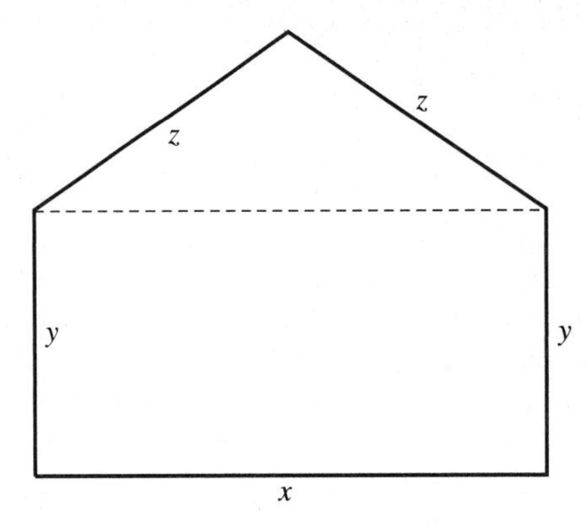

Figure 13.29

15. A widget manufacturer determines that if she manufactures x thousands of widgets per month and sells the widgets for y dollars each, then her monthly profit (in thousands of dollars) will be $P = xy - \frac{1}{27}x^2y^3 - x$. If her factory is capable of producing at most 3,000 widgets per month, and government regulations prevent her from charging more than \$2 per widget, how many should she manufacture, and how much should she charge for each, to maximize her monthly profit?

16. Find the envelope of the family of curves $y = (x-c)^3 + 3c$.

17. Find an approximate solution $y(x, \epsilon)$ of the equation $y + \epsilon x e^y = -2x$ having terms up to second degree in ϵ.

18. (a) Calculate $G'(y)$ if $G(y) = \int_0^\infty \frac{\tan^{-1}(xy)}{x}\,dx$.

(b) Evaluate $\int_0^\infty \frac{\tan^{-1}(\pi x) - \tan^{-1} x}{x}\,dx$. *Hint:* This integral is $G(\pi) - G(1)$.

Challenging Problems

1. (Fourier series)
Show that the constants a_k, $(k = 0, 1, 2, \ldots, n)$, and b_k, $(k = 1, 2, \ldots, n)$, which minimize the integral

$$I_n = \int_{-\pi}^{\pi} \left[f(x) - \frac{a_0}{2} - \sum_{k=0}^{n} \Big(a_k \cos kx + b_k \sin kx \Big) \right]^2 dx,$$

are given by

$$a_k = \frac{1}{\pi} \int_{\pi}^{\pi} f(x) \cos kx\,dx, \qquad b_k = \frac{1}{\pi} \int_{\pi}^{\pi} f(x) \sin kx\,dx.$$

Note that these numbers, called the **Fourier coefficients** of f on $[-\pi, \pi]$, do not depend on n. If they can be calculated for all positive integers k, then the series

$$\frac{a_0}{2} + \sum_{k=0}^{\infty} \Big(a_k \cos kx + b_k \sin kx \Big)$$

is called the **(full-range) Fourier series** of f on $[-\pi, \pi]$. (See Section 9.9.)

2. This is a continuation of Problem 1. Find the (full range) Fourier coefficients a_k and b_k of

$$f(x) = \begin{cases} 0 & \text{if } -\pi \le x < 0 \\ x & \text{if } 0 \le x \le \pi. \end{cases}$$

What is the minimum value of I_n in this case? How does it behave as $n \to \infty$?

3. Evaluate $\displaystyle\int_0^x \frac{\ln(tx+1)}{1+t^2}\,dt$.

4. **(Steiner's problem)** The problem of finding a point in the plane (or a higher-dimensional space) that minimizes the sum of its distances from n given points is very difficult. The case $n = 3$ is known as Steiner's problem. If $P_1P_2P_3$ is a triangle whose largest angle is less than 120°, there is a point Q inside the triangle so that the lines QP_1, QP_2, and QP_3 make equal 120° angles with one another. Show that the sum of the distances from the vertices of the triangle to a point P is minimum when $P = Q$. *Hint:* First show that if $P = (x, y)$ and $P_i = (x_i, y_i)$, then

$$\frac{d|PP_i|}{dx} = \cos\theta_i \quad \text{and} \quad \frac{d|PP_i|}{dy} = \sin\theta_i,$$

where θ_i is the angle between $\overrightarrow{P_iP}$ and the positive direction of the x-axis. Hence, show that the minimal point P satisfies two trigonometric equations involving θ_1, θ_2, and θ_3. Then try to show that any two of those angles differ by $\pm 2\pi/3$. Where should P be taken if the triangle has an angle of 120° or greater?

CHAPTER 14

Multiple Integration

> "Do you know what a mathematician is?" Lord Kelvin asked a class. He then stepped to the board and wrote
>
> $$\int_{-\infty}^{\infty} e^{-x^2}\,dx = \sqrt{\pi}.$$
>
> Putting his finger on what he had written, he turned to the class. "A mathematician is one to whom that is as obvious as that 'twice two makes four' is to you."

William Thomson Kelvin 1824–1907
anecdote from *Men of Mathematics* by E. Bell

Introduction

In this chapter we extend the concept of the definite integral to functions of several variables. Defined as limits of Riemann sums, like the one-dimensional definite integral, such multiple integrals can be evaluated using successive single definite integrals. They are used to represent and calculate quantities specified in terms of densities in regions of the plane or spaces of higher dimension. In the simplest instance, the volume of a three-dimensional region is given by a *double integral* of its height over the two-dimensional plane region that is its base.

14.1 Double Integrals

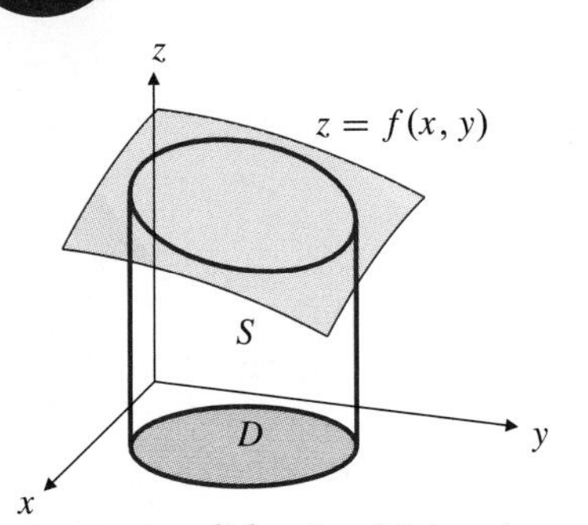

Figure 14.1 A *solid* region S lying above domain D in the xy-plane and below the surface $z = f(x, y)$

The definition of the definite integral, $\int_a^b f(x)\,dx$, is motivated by the *standard area problem*, namely, the problem of finding the area of the plane region bounded by the curve $y = f(x)$, the x-axis, and the lines $x = a$ and $x = b$. Similarly, we can motivate the double integral of a function of two variables over a domain D in the plane by means of the *standard volume problem* of finding the volume of the three-dimensional region S bounded by the surface $z = f(x, y)$, the xy-plane, and the cylinder parallel to the z-axis passing through the boundary of D. (See Figure 14.1. D is called the **domain of integration**.) We will call such a three-dimensional region S a "solid," although we are not implying that it is filled with any particular substance. We will define the double integral of $f(x, y)$ over the domain D,

$$\iint_D f(x, y)\,dA,$$

in such a way that its value will give the volume of the solid S whenever D is a "reasonable" domain and f is a "reasonable" function with positive values.

Let us start with the case where D is a closed rectangle with sides parallel to the coordinate axes in the xy-plane, and f is a bounded function on D. If D consists of the points (x, y) such that $a \le x \le b$ and $c \le y \le d$, we can form a **partition** P of D into small rectangles by partitioning each of the intervals $[a, b]$ and $[c, d]$, say by points

$$a = x_0 < x_1 < x_2 < \cdots < x_{m-1} < x_m = b,$$
$$c = y_0 < y_1 < y_2 < \cdots < y_{n-1} < y_n = d.$$

The partition P of D then consists of the mn rectangles R_{ij} $(1 \le i \le m,\ 1 \le j \le n)$, consisting of points (x, y) for which $x_{i-1} \le x \le x_i$ and $y_{j-1} \le y \le y_j$. (See Figure 14.2.)

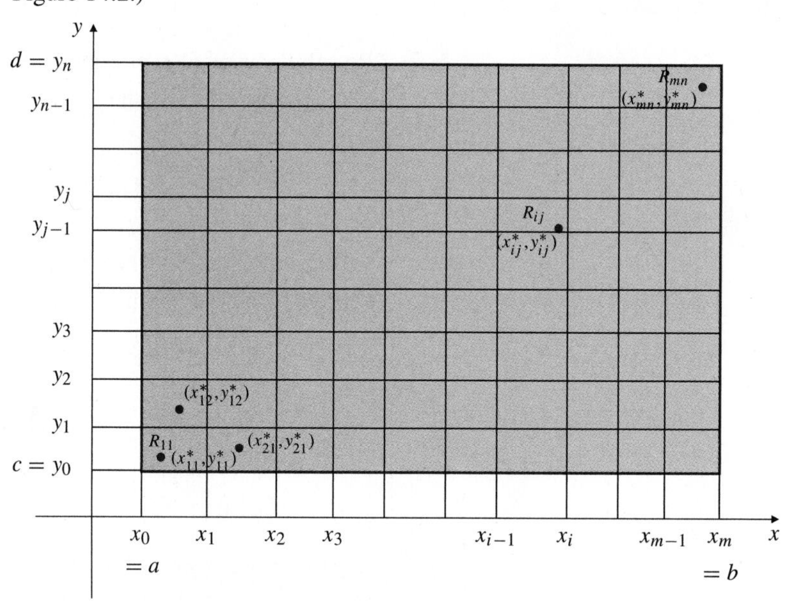

Figure 14.2 A partition of D (the large shaded rectangle) into smaller rectangles R_{ij} $(1 \le i \le m,\ 1 \le j \le n)$

The rectangle R_{ij} has area

$$\Delta A_{ij} = \Delta x_i \Delta y_j = (x_i - x_{i-1})(y_j - y_{j-1})$$

and *diameter* (i.e., diagonal length)

$$\operatorname{diam}(R_{ij}) = \sqrt{(\Delta x_i)^2 + (\Delta y_j)^2} = \sqrt{(x_i - x_{i-1})^2 + (y_j - y_{j-1})^2}.$$

The **norm** of the partition P is the largest of these subrectangle diameters:

$$\|P\| = \max_{\substack{1 \le i \le m \\ 1 \le j \le n}} \operatorname{diam}(R_{ij}).$$

Now we pick an arbitrary point (x_{ij}^*, y_{ij}^*) in each of the rectangles R_{ij} and form the **Riemann sum**

$$R(f, P) = \sum_{i=1}^{m} \sum_{j=1}^{n} f(x_{ij}^*, y_{ij}^*)\, \Delta A_{ij},$$

which is the sum of mn terms, one for each rectangle in the partition. (Here, the *double summation* indicates the sum as i goes from 1 to m of terms, each of which is itself a sum as j goes from 1 to n.) The term corresponding to rectangle R_{ij} is, if $f(x_{ij}^*, y_{ij}^*) \ge 0$, the volume of the rectangular box whose base is R_{ij} and whose height is the value of f at (x_{ij}^*, y_{ij}^*). (See Figure 14.3.) Therefore, for positive functions f, the Riemann sum $R(f, P)$ approximates the volume above D and under the graph of f. The double integral of f over D is defined to be the limit of such Riemann sums, provided the limit exists as $\|P\| \to 0$ independently of how the points (x_{ij}^*, y_{ij}^*) are chosen. We make this precise in the following definition.

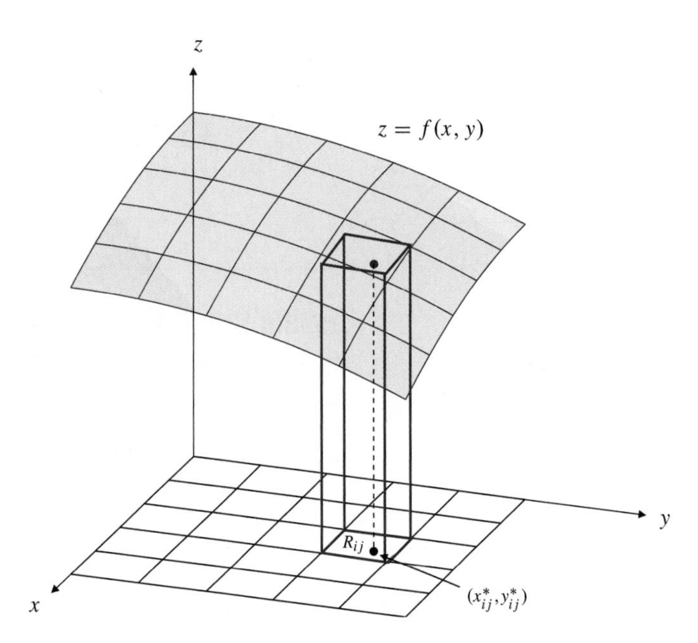

Figure 14.3 A rectangular box above rectangle R_{ij}. The Riemann sum is a sum of volumes of such boxes

DEFINITION 1

The double integral over a rectangle

We say that f is **integrable** over the rectangle D and has **double integral**

$$I = \iint_D f(x, y)\, dA,$$

if for every positive number ϵ there exists a number δ depending on ϵ, such that

$$|R(f, P) - I| < \epsilon$$

holds for every partition P of D satisfying $\|P\| < \delta$ and for all choices of the points (x_{ij}^*, y_{ij}^*) in the subrectangles of P.

The dA that appears in the expression for the double integral is an *area element*. It represents the limit of the $\Delta A = \Delta x\, \Delta y$ in the Riemann sum and can also be written $dx\, dy$ or $dy\, dx$, the order being unimportant. When we evaluate double integrals by *iteration* in the next section, dA will be replaced with a product of differentials dx and dy, and the order will be important.

As is true for functions of one variable, functions that are continuous on D are integrable on D. Of course, many bounded but discontinuous functions are also integrable, but an exact description of the class of integrable functions is beyond the scope of this text.

EXAMPLE 1 Let D be the square $0 \le x \le 1$, $0 \le y \le 1$. Use a Riemann sum corresponding to the partition of D into four smaller squares with points selected at the centre of each to find an approximate value for

$$\iint_D (x^2 + y)\, dA.$$

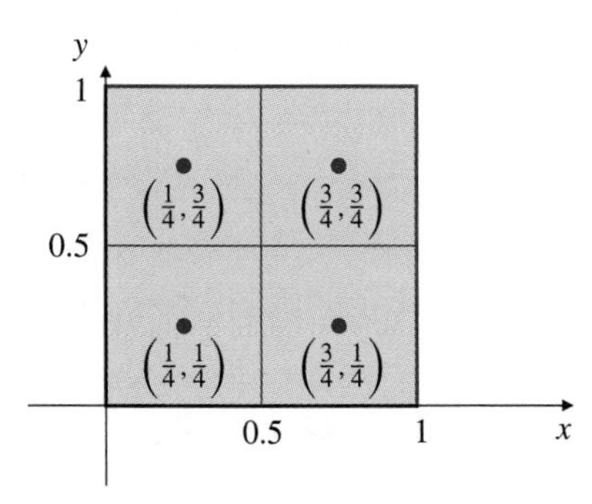

Figure 14.4 The partitioned square of Example 1

Solution The required partition P is formed by the lines $x = 1/2$ and $y = 1/2$, which divide D into four squares, each of area $\Delta A = 1/4$. The centres of these squares are the points $\left(\frac{1}{4}, \frac{1}{4}\right)$, $\left(\frac{1}{4}, \frac{3}{4}\right)$, $\left(\frac{3}{4}, \frac{1}{4}\right)$, and $\left(\frac{3}{4}, \frac{3}{4}\right)$. (See Figure 14.4.) Therefore, the required approximation is

$$\iint_D (x^2 + y)\,dA \approx R(x^2 + y, P) = \left(\frac{1}{16} + \frac{1}{4}\right)\frac{1}{4} + \left(\frac{1}{16} + \frac{3}{4}\right)\frac{1}{4} + \left(\frac{9}{16} + \frac{1}{4}\right)\frac{1}{4} + \left(\frac{9}{16} + \frac{3}{4}\right)\frac{1}{4} = \frac{13}{16} = 0.8125.$$

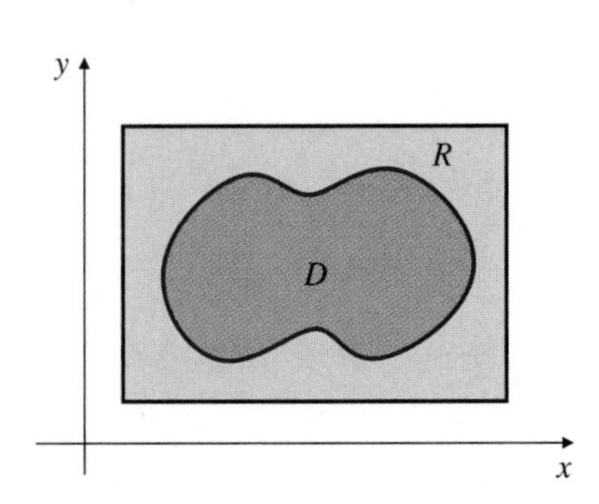

Figure 14.5 Bounded domain D is a subset of rectangle R

Double Integrals over More General Domains

It is often necessary to use double integrals of bounded functions $f(x, y)$ over domains that are not rectangles. If the domain D is *bounded*, we can choose a rectangle R with sides parallel to the coordinate axes such that D is contained inside R. (See Figure 14.5.) If $f(x, y)$ is defined on D, we can extend its domain to be R by defining $f(x, y) = 0$ for points in R that are outside of D. The integral of f over D can then be defined to be the integral of the extended function over the rectangle R.

DEFINITION 2

If $f(x, y)$ is defined and bounded on domain D, let $\hat{f}$ be the extension of f that is zero everywhere outside D:

$$\hat{f}(x, y) = \begin{cases} f(x, y), & \text{if } (x, y) \text{ belongs to } D \\ 0, & \text{if } (x, y) \text{ does not belong to } D. \end{cases}$$

If D is a *bounded* domain, then it is contained in some rectangle R with sides parallel to the coordinate axes. If $\hat{f}$ is integrable over R, we say that f is **integrable** over D and define the **double integral** of f over D to be

$$\iint_D f(x, y)\,dA = \iint_R \hat{f}(x, y)\,dA.$$

This definition makes sense because the values of $\hat{f}$ in the part of R outside of D are all zero, so do not contribute anything to the value of the integral. However, even if f is continuous on D, $\hat{f}$ will not be continuous on R unless $f(x, y) \to 0$ as (x, y) approaches the boundary of D. Nevertheless, if f and D are "well-behaved," the integral will exist. We cannot delve too deeply into what constitutes *well-behaved*, but assert, without proof, the following theorem that will assure us that most of the double integrals we encounter do, in fact, exist.

THEOREM 1

If f is continuous on a *closed, bounded* domain D whose boundary consists of finitely many curves of finite length, then f is integrable on D.

According to Theorem 2 of Section 13.1, a continuous function is bounded if its domain is closed and bounded. Generally, however, it is not necessary to restrict our domains to be closed. If D is a bounded domain and $\text{int}(D)$ is its interior (an open set), and if f is integrable on D, then

$$\iint_D f(x, y)\,dA = \iint_{\text{int}(D)} f(x, y)\,dA.$$

We will discuss *improper double integrals* of unbounded functions or over unbounded domains in Section 14.3.

Properties of the Double Integral

Some properties of double integrals are analogous to properties of the one-dimensional definite integral and require little comment: if f and g are integrable over D, and if L and M are constants, then

(a) $\displaystyle\iint_D f(x,y)\,dA = 0$ if D has zero area.

(b) **Area of a domain:** $\displaystyle\iint_D 1\,dA =$ area of D (because it is the volume of a cylinder with base D and height 1).

(c) **Integrals representing volumes:**
If $f(x,y) \geq 0$ on D, then $\displaystyle\iint_D f(x,y)\,dA = V \geq 0$, where V is the volume of the solid lying vertically above D and below the surface $z = f(x,y)$.

(d) If $f(x,y) \leq 0$ on D, then $\displaystyle\iint_D f(x,y)\,dA = -V \leq 0$, where V is the volume of the solid lying vertically below D and above the surface $z = f(x,y)$.

(e) **Linear dependence on the integrand:**

$$\iint_D \Big(Lf(x,y)+Mg(x,y)\Big)\,dA = L\iint_D f(x,y)\,dA + M\iint_D g(x,y)\,dA.$$

(f) **Inequalities are preserved:**
If $f(x,y) \leq g(x,y)$ on D, then $\displaystyle\iint_D f(x,y)\,dA \leq \iint_D g(x,y)\,dA$.

(g) **The triangle inequality:** $\displaystyle\left|\iint_D f(x,y)\,dA\right| \leq \iint_D |f(x,y)|\,dA$.

(h) **Additivity of domains:** If $D_1, D_2, \ldots, D_k$ are nonoverlapping domains on each of which f is integrable, then f is integrable over the union $D = D_1 \cup D_2 \cup \cdots \cup D_k$ and

$$\iint_D f(x,y)\,dA = \sum_{j=1}^{k} \iint_{D_j} f(x,y)\,dA.$$

Nonoverlapping domains can share boundary points but have no interior points in common.

Double Integrals by Inspection

As yet, we have not said anything about how to *evaluate* a double integral. The main technique for doing this, called *iteration*, will be developed in the next section, but it is worth pointing out that double integrals can sometimes be evaluated using symmetry arguments or by interpreting them as volumes that we already know.

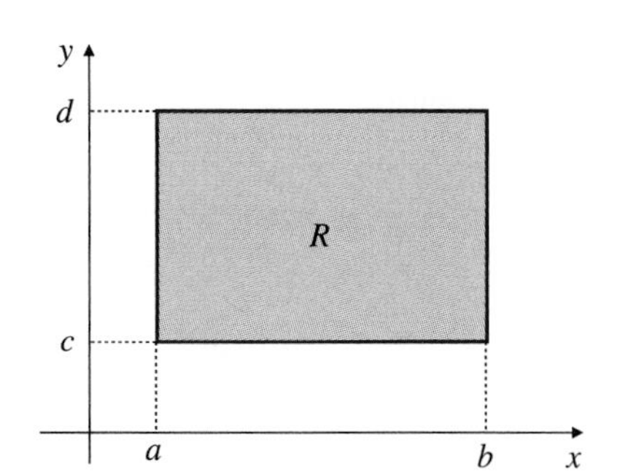

Figure 14.6 The base of a rectangular box

EXAMPLE 2 If R is the rectangle $a \leq x \leq b$, $c \leq y \leq d$, then

$$\iint_R 3\,dA = 3 \times \text{area of } R = 3(b-a)(d-c).$$

Here, the integrand is $f(x,y) = 3$, and the integral is equal to the volume of the solid box of height 3 whose base is the rectangle R. (See Figure 14.6.)

EXAMPLE 3 Evaluate $I = \displaystyle\iint_{x^2+y^2\leq 1} (\sin x + y^3 + 4)\,dA$.

Solution The integral can be expressed as the sum of three integrals by property (e) of double integrals:

$$I = \iint_{x^2+y^2 \le 1} \sin x \, dA + \iint_{x^2+y^2 \le 1} y^3 \, dA + \iint_{x^2+y^2 \le 1} 4 \, dA$$
$$= I_1 + I_2 + I_3.$$

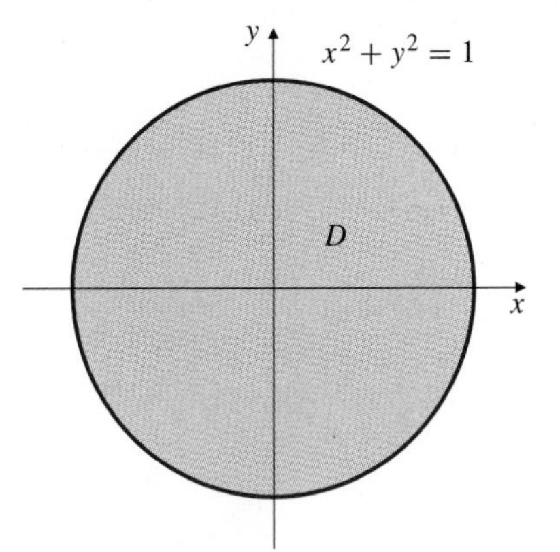

Figure 14.7 The disk is symmetric about both coordinate axes

The domain of integration (Figure 14.7) is a circular disk of radius 1 centred at the origin. Since $f(x, y) = \sin x$ is an *odd* function of x, its graph bounds as much volume below the xy-plane in the region $x < 0$ as it does above the xy-plane in the region $x > 0$. These two contributions to the double integral cancel, so $I_1 = 0$. Note that symmetry of *both* the domain *and* the integrand is necessary for this argument.

Similarly, $I_2 = 0$ because y^3 is an odd function and D is symmetric about the x-axis.

Finally,

$$I_3 = \iint_D 4 \, dA = 4 \times \text{area of } D = 4\pi.$$

Thus, $I = 0 + 0 + 4\pi = 4\pi$.

EXAMPLE 4 If D is the disk of Example 3, the integral

$$\iint_D \sqrt{1 - x^2 - y^2} \, dA$$

represents the volume of a hemisphere of radius 1 and so has the value $2\pi/3$.

When evaluating double integrals, always be alert for situations such as those in the above examples. You can save much time by not trying to calculate an integral whose value should be obvious without calculation.

EXERCISES 14.1

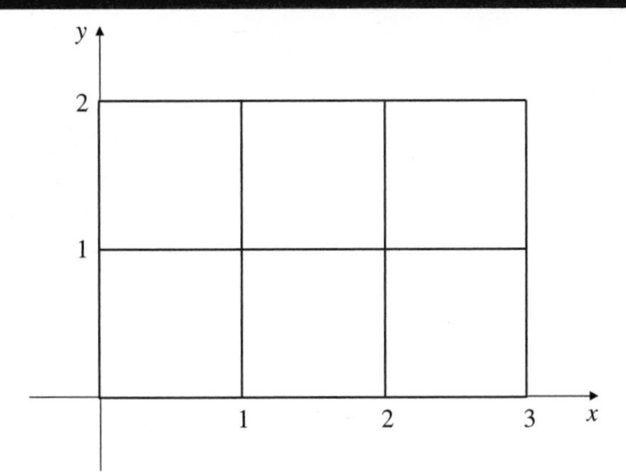

Figure 14.8

Exercises 1–6 refer to the double integral

$$I = \iint_D (5 - x - y) \, dA,$$

where D is the rectangle $0 \le x \le 3$, $0 \le y \le 2$. P is the partition of D into six squares of side 1 as shown in Figure 14.8. In Exercises 1–5, calculate the Riemann sums for I corresponding to the given choices of points (x^*_{ij}, y^*_{ij}).

1. (x^*_{ij}, y^*_{ij}) is the upper-left corner of each square.

2. (x^*_{ij}, y^*_{ij}) is the upper-right corner of each square.

3. (x^*_{ij}, y^*_{ij}) is the lower-left corner of each square.

4. (x^*_{ij}, y^*_{ij}) is the lower-right corner of each square.

5. (x^*_{ij}, y^*_{ij}) is the centre of each square.

6. Evaluate I by interpreting it as a volume.

In Exercises 7–10, D is the disk $x^2 + y^2 \le 25$, and P is the partition of the square $-5 \le x \le 5$, $-5 \le y \le 5$ into one hundred 1×1 squares, as shown in Figure 14.9. Approximate the double integral

$$J = \iint_D f(x, y) \, dA,$$

where $f(x, y) = 1$ by calculating the Riemann sums $R(f, P)$ corresponding to the indicated choice of points in the small squares. *Hint:* Using symmetry will make the job easier.

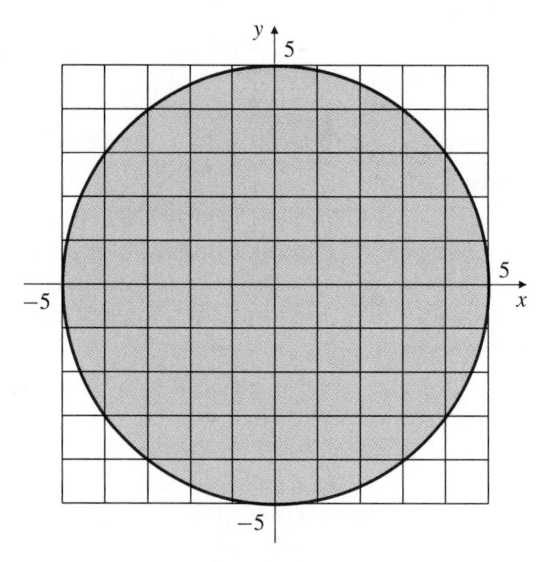

Figure 14.9

7. (x_{ij}^*, y_{ij}^*) is the corner of each square closest to the origin.
8. (x_{ij}^*, y_{ij}^*) is the corner of each square farthest from the origin.
9. (x_{ij}^*, y_{ij}^*) is the centre of each square.
10. Evaluate J.
11. Repeat Exercise 5 using the integrand e^x instead of $5 - x - y$.
12. Repeat Exercise 9 using $f(x, y) = x^2 + y^2$ instead of $f(x, y) = 1$.

In Exercises 13–22, evaluate the given double integral by inspection.

13. $\iint_R dA$, where R is the rectangle $-1 \le x \le 3$, $-4 \le y \le 1$
14. $\iint_D (x + 3)\, dA$, where D is the half-disk $0 \le y \le \sqrt{4 - x^2}$
15. $\iint_T (x + y)\, dA$, where T is the parallelogram having the points $(2, 2)$, $(1, -1)$, $(-2, -2)$, and $(-1, 1)$ as vertices
16. $\iint_{|x|+|y|\le 1} \left(x^3 \cos(y^2) + 3 \sin y - \pi\right) dA$
17. $\iint_{x^2+y^2\le 1} (4x^2y^3 - x + 5)\, dA$
18. $\iint_{x^2+y^2\le a^2} \sqrt{a^2 - x^2 - y^2}\, dA$
19. $\iint_{x^2+y^2\le a^2} (a - \sqrt{x^2 + y^2})\, dA$
20. $\iint_S (x + y)\, dA$, where S is the square $0 \le x \le a, 0 \le y \le a$
21. $\iint_T (1 - x - y)\, dA$, where T is the triangle with vertices $(0, 0)$, $(1, 0)$, and $(0, 1)$
22. $\iint_R \sqrt{b^2 - y^2}\, dA$, where R is the rectangle $0 \le x \le a, 0 \le y \le b$

14.2 Iteration of Double Integrals in Cartesian Coordinates

The existence of the double integral $\iint_D f(x, y)\, dA$ depends on f and the domain D. As we shall see, evaluation of double integrals is easiest when the domain of integration is of *simple* type.

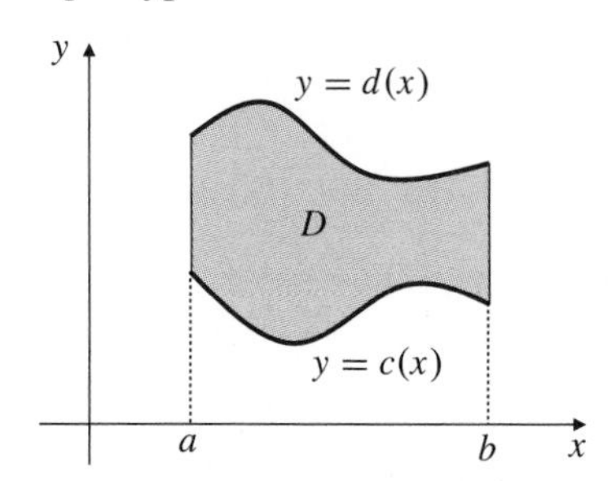

Figure 14.10 A y-simple domain

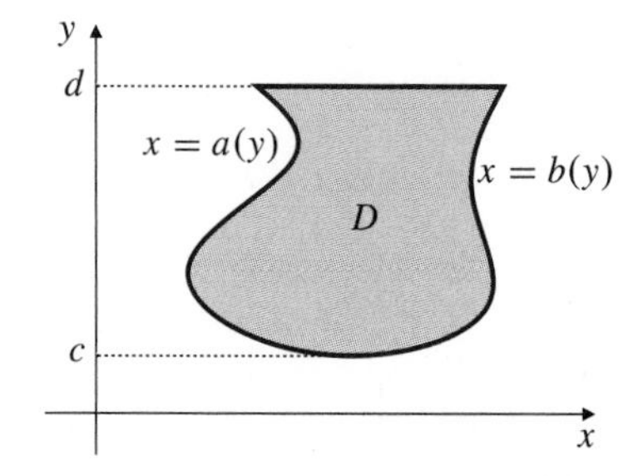

Figure 14.11 An x-simple domain

We say that the domain D in the xy-plane is y-**simple** if it is bounded by two vertical lines $x = a$ and $x = b$, and two continuous graphs $y = c(x)$ and $y = d(x)$ between these lines. (See Figure 14.10.) Lines parallel to the y-axis intersect a y-simple domain in an *interval* (possibly a single point) if at all. Similarly, D is x-**simple** if it is bounded by horizontal lines $y = c$ and $y = d$, and two continuous graphs $x = a(y)$ and $x = b(y)$ between these lines. (See Figure 14.11.) Many of the domains over which we will take integrals are y-simple, x-simple, or both. For example, rectangles, triangles, and disks are both x-simple and y-simple. Those domains that are neither one nor the other will usually be unions of finitely many nonoverlapping subdomains that are both x-simple and y-simple. We will call such domains **regular**. The shaded region in Figure 14.12 is divided into four subregions, each of which is both x-simple and y-simple.

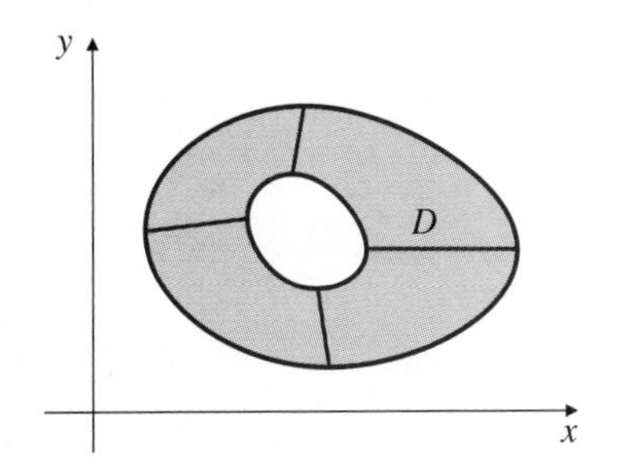

Figure 14.12 A regular domain

It can be shown that a bounded, continuous function $f(x, y)$ is integrable over a bounded x-simple or y-simple domain and, therefore, over any regular domain.

Unlike the examples in the previous section, most double integrals cannot be evaluated by inspection. We need a technique for evaluating double integrals similar to the technique for evaluating single definite integrals in terms of antiderivatives. Since the double integral represents a volume, we can evaluate it for simple domains by a slicing technique.

Suppose, for instance, that D is y-simple and is bounded by $x = a$, $x = b$, $y = c(x)$, and $y = d(x)$, as shown in Figure 14.13(a). Then $\iint_D f(x, y)\,dA$ represents (at least for positive f) the volume of the solid region inside the vertical cylinder through the boundary of D and between the xy-plane and the surface $z = f(x, y)$. Consider the cross-section of this solid in the vertical plane perpendicular to the x-axis at position x. Note that x is constant in that plane. If we use the projections of the y- and z-axes onto the plane as coordinate axes there, the cross-section is a plane region bounded by vertical lines $y = c(x)$ and $y = d(x)$, by the horizontal line $z = 0$, and by the curve $z = f(x, y)$. The area of the cross-section is therefore given by

$$A(x) = \int_{c(x)}^{d(x)} f(x, y)\,dy.$$

The double integral $\iint_D f(x, y)\,dA$ is obtained by summing the volumes of "thin" slices of area $A(x)$ and thickness dx between $x = a$ and $x = b$ and is therefore given by

$$\iint_D f(x, y)\,dA = \int_a^b A(x)\,dx = \int_a^b \left(\int_{c(x)}^{d(x)} f(x, y)\,dy \right) dx.$$

Notationally, it is common to omit the large parentheses and write

$$\iint_D f(x, y)\,dA = \int_a^b \int_{c(x)}^{d(x)} f(x, y)\,dy\,dx,$$

or

$$\iint_D f(x, y)\,dA = \int_a^b dx \int_{c(x)}^{d(x)} f(x, y)\,dy. \qquad (*)$$

The latter form $(*)$ shows more clearly which variable corresponds to which limits of integration.

RELAX! Do not be confused by the position of the dx in the formula $(*)$. Although up until now we have been in the habit of writing the integral of a function $A(x)$ from $x = a$ to $x = b$ in the form $\int_a^b A(x)\,dx$, there is no reason we can not write the dx before instead of after the $A(x)$: $\int_a^b A(x)\,dx = \int_a^b dx\,A(x)$. When $A(x)$ is itself an integral in a different variable, as it is in $(*)$, writing the dx closer to its own integral sign can be useful. It is still understood that the y integral must be done first as its integrand and limits can both depend on x so the result will be a function $A(x)$ of x.

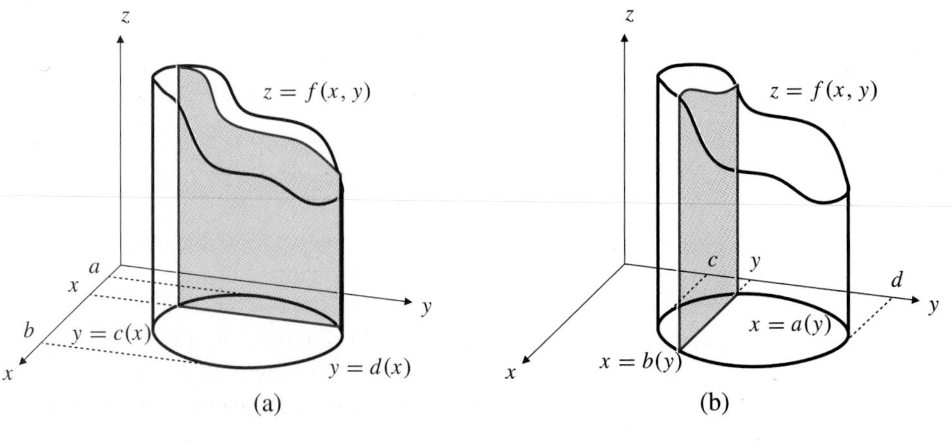

Figure 14.13

(a) In integrals over y-simple domains, slices should be perpendicular to the x-axis

(b) In integrals over x-simple domains, slices should be perpendicular to the y-axis

The expressions on the right-hand sides of the above formulas are called **iterated** integrals. **Iteration** is the process of reducing the problem of evaluating a double (or multiple) integral to one of evaluating two (or more) successive single definite integrals. In the above iteration, the integral

$$\int_{c(x)}^{d(x)} f(x, y)\, dy$$

is called the **inner** integral since it must be evaluated first. It is evaluated using standard techniques, treating x as a constant. The result of this evaluation is a function of x alone (note that both the integrand and the limits of the inner integral can depend on x) and is the integrand of the **outer** integral in which x is the variable of integration.

For double integrals over x-simple domains, we can slice perpendicularly to the y-axis and obtain an iterated integral with the outer integral in the y direction. (See Figure 14.13(b).) We summarize the above discussion in the following theorem whose formal proof we will, however, not give.

THEOREM 2 **Iteration of double integrals**

If $f(x, y)$ is continuous on the bounded y-simple domain D given by $a \le x \le b$ and $c(x) \le y \le d(x)$, then

$$\iint_D f(x, y)\, dA = \int_a^b dx \int_{c(x)}^{d(x)} f(x, y)\, dy.$$

Similarly, if f is continuous on the x-simple domain D given by $c \le y \le d$ and $a(y) \le x \le b(y)$, then

$$\iint_D f(x, y)\, dA = \int_c^d dy \int_{a(y)}^{b(y)} f(x, y)\, dx.$$

In scientific literature, double integrals and integrals in higher dimensional spaces are often represented with a single integral sign, for instance,

$$\int_D f(x, y)\, dx\, dy.$$

We will use multiple integral signs in Chapters 14–16, but will use single integral signs in Chapter 17, where integrals in $\mathbb{R}^n$ are considered.

Remark The symbol dA in the double integral is replaced in the iterated integrals by the dx and the dy. Accordingly, dA is frequently written $dx\, dy$ or $dy\, dx$ even in the double integral. The three expressions

$$\iint_D f(x, y)\, dx\, dy, \qquad \iint_D f(x, y)\, dy\, dx, \qquad \text{and} \qquad \iint_D f(x, y)\, dA$$

all stand for the double integral of f over D. Only when the double integral is iterated does the order of dx and dy become important. Later in this chapter we will iterate double integrals in polar coordinates, and dA will take the form $r\, dr\, d\theta$.

It is not always necessary to make a three-dimensional sketch of the solid volume represented by a double integral. In order to iterate the integral properly (in one direction or the other) it is usually sufficient to make a sketch of *the domain D* over which the integral is taken. The direction of iteration can be shown by a line along which the inner integral is taken. The following examples illustrate this.

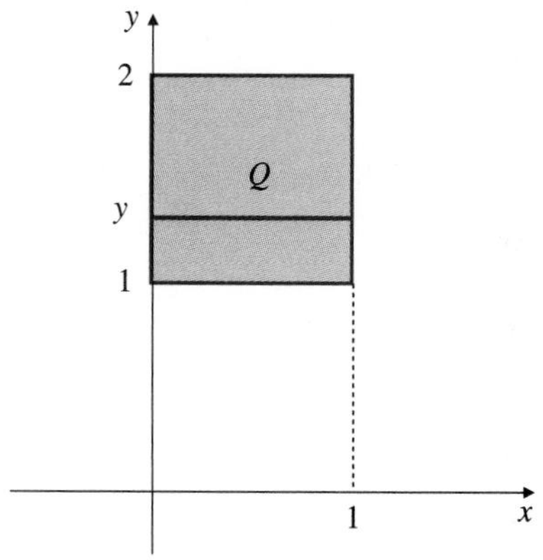

Figure 14.14 The horizontal line through Q indicates iteration with the inner integral in the x direction

EXAMPLE 1 Find the volume of the solid lying above the square Q defined by $0 \le x \le 1$ and $1 \le y \le 2$ and below the plane $z = 4 - x - y$.

Solution The square Q is both x-simple and y-simple, so the double integral giving the volume can be iterated in either direction. We will do it both ways just for practice. The horizontal line at height y in Figure 14.14 suggests that we first integrate with respect to x along this line (from 0 to 1) and then integrate the result with respect to y from 1 to 2. Iterating the double integral in this direction, we calculate

$$\begin{aligned}
\text{Volume above } Q &= \iint_Q (4 - x - y)\,dA \\
&= \int_1^2 dy \int_0^1 (4 - x - y)\,dx \\
&= \int_1^2 dy \left(4x - \frac{x^2}{2} - xy\right)\Bigg|_{x=0}^{x=1} \\
&= \int_1^2 \left(\frac{7}{2} - y\right) dy \\
&= \left(\frac{7y}{2} - \frac{y^2}{2}\right)\Bigg|_1^2 = 2 \text{ cubic units.}
\end{aligned}$$

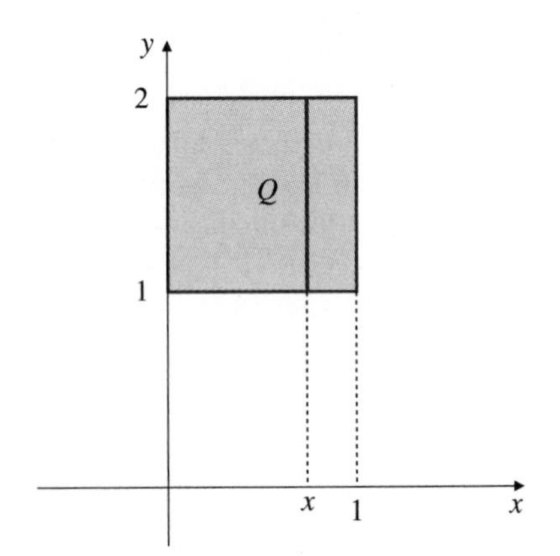

Figure 14.15 The vertical line through Q indicates iteration with the inner integral in the y direction

Using the opposite iteration, as illustrated in Figure 14.15, we calculate

$$\begin{aligned}
\text{Volume above } Q &= \iint_Q (4 - x - y)\,dA \\
&= \int_0^1 dx \int_1^2 (4 - x - y)\,dy \\
&= \int_0^1 dx \left(4y - xy - \frac{y^2}{2}\right)\Bigg|_{y=1}^{y=2} \\
&= \int_0^1 \left(\frac{5}{2} - x\right) dx \\
&= \left(\frac{5x}{2} - \frac{x^2}{2}\right)\Bigg|_0^1 = 2 \text{ cubic units.}
\end{aligned}$$

It is comforting to get the same answer both ways! Note that because Q is a rectangle with sides parallel to the coordinate axes, the limits of the inner integrals do not depend on the variables of the outer integrals in either iteration. This cannot be expected to happen with more general domains.

EXAMPLE 2 Evaluate $\iint_T xy\,dA$ over the triangle T with vertices $(0, 0)$, $(1, 0)$, and $(1, 1)$.

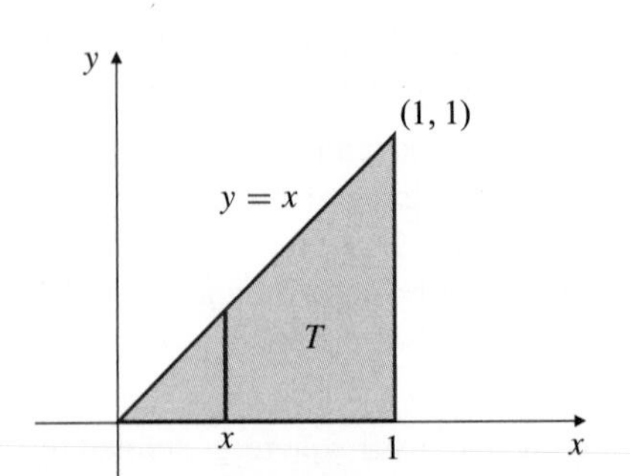

Figure 14.16 The triangular domain T with vertical line indicating iteration with inner integral in the y direction

Solution The triangle T is shown in Figure 14.16. It is both x-simple and y-simple. Using the iteration corresponding to slicing in the direction shown in the figure, we obtain:

$$\begin{aligned}
\iint_T xy\,dA &= \int_0^1 dx \int_0^x xy\,dy \\
&= \int_0^1 dx \left(\frac{xy^2}{2}\right)\Bigg|_{y=0}^{y=x} \\
&= \int_0^1 \frac{x^3}{2}\,dx = \frac{x^4}{8}\Bigg|_0^1 = \frac{1}{8}.
\end{aligned}$$

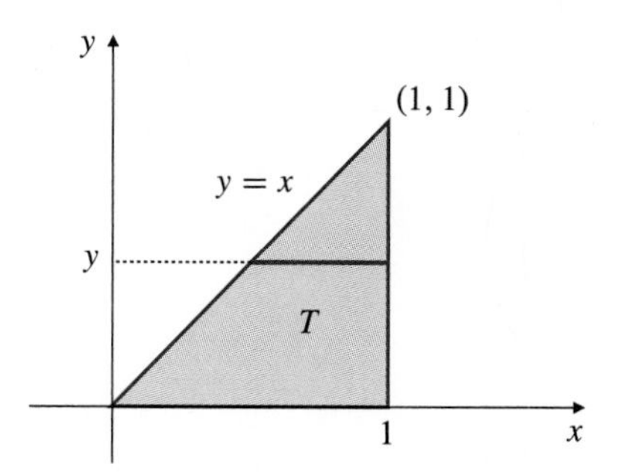

Figure 14.17 The triangular domain T with horizontal line indicating iteration with inner integral in the x direction

Iteration in the other direction (Figure 14.17) leads to the same value:

$$\begin{aligned}\iint_T xy\,dA &= \int_0^1 dy \int_y^1 xy\,dx \\ &= \int_0^1 dy \left(\frac{yx^2}{2}\right)\Bigg|_{x=y}^{x=1} \\ &= \int_0^1 \frac{y}{2}(1-y^2)\,dy \\ &= \left(\frac{y^2}{4} - \frac{y^4}{8}\right)\Bigg|_0^1 = \frac{1}{8}.\end{aligned}$$

In both of the examples above, the double integral could be evaluated easily using either possible iteration. (We did them both ways just to illustrate that fact.) It often occurs, however, that a double integral is easily evaluated if iterated in one direction and very difficult, or impossible, if iterated in the other direction. Sometimes you will even encounter iterated integrals whose evaluation requires that they be expressed as double integrals and then reiterated in the opposite direction.

EXAMPLE 3 Evaluate the iterated integral $I = \int_0^1 dx \int_{\sqrt{x}}^1 e^{y^3}\,dy$.

Solution We cannot antidifferentiate e^{y^3} to evaluate the inner integral in this iteration, so we express I as a double integral and identify the region over which it is taken:

$$I = \iint_D e^{y^3}\,dA,$$

where D is the region shown in Figure 14.18. Reiterating with the x integration on the inside we get

$$\begin{aligned}I &= \int_0^1 dy \int_0^{y^2} e^{y^3}\,dx \\ &= \int_0^1 e^{y^3}\,dy \int_0^{y^2} dx \\ &= \int_0^1 y^2 e^{y^3}\,dy = \frac{e^{y^3}}{3}\Bigg|_0^1 = \frac{e-1}{3}.\end{aligned}$$

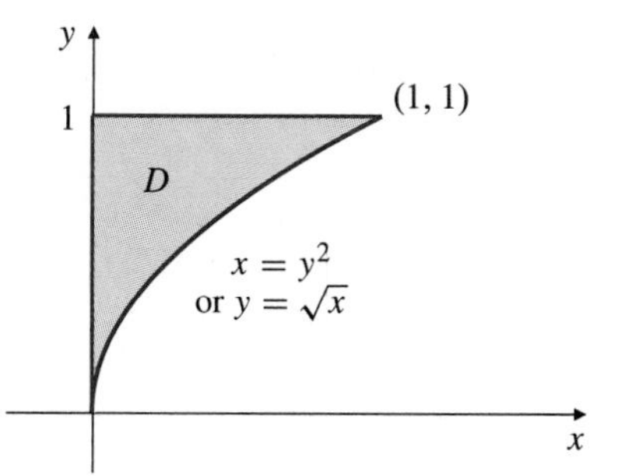

Figure 14.18 The region corresponding to the iterated integral in Example 3

The following is an example of the calculation of the volume of a somewhat awkward solid. Even though it is not always necessary to sketch solids to find their volumes, you are encouraged to sketch them whenever possible. When we encounter triple integrals over three-dimensional regions later in this chapter, it will usually be necessary to sketch the regions. Get as much practice as you can.

EXAMPLE 4 Sketch and find the volume of the solid bounded by the planes $y = 0$, $z = 0$, and $z = a - x + y$ and the parabolic cylinder $y = a - (x^2/a)$, where a is a positive constant.

Solution The solid is shown in Figure 14.19. Its base is the parabolic segment D in the xy-plane bounded by $y = 0$ and $y = a - (x^2/a)$, so the volume of the solid is given by

$$V = \iint_D (a - x + y)\,dA = \iint_D (a + y)\,dA.$$

(Note how we used symmetry to drop the x term from the integrand. This term is an odd function of x, and D is symmetric about the y-axis.) Iterating the double integral in the direction suggested by the slice shown in the figure, we obtain

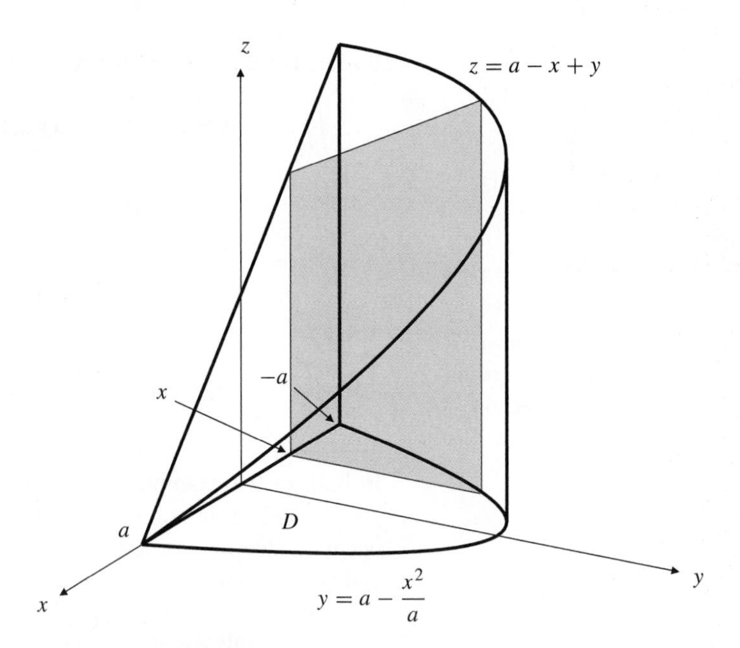

Figure 14.19 The solid in Example 4, sliced perpendicularly to the x-axis

$$\begin{aligned}
V &= \int_{-a}^{a} dx \int_{0}^{a-(x^2/a)} (a+y)\,dy \\
&= \int_{-a}^{a} \left(ay + \frac{y^2}{2} \right)\Bigg|_{y=0}^{y=a-(x^2/a)} dx \\
&= \int_{-a}^{a} \left[a^2 - x^2 + \frac{1}{2}\left(a^2 - 2x^2 + \frac{x^4}{a^2} \right) \right] dx \\
&= 2\int_{0}^{a} \left[\frac{3}{2}a^2 - 2x^2 + \frac{x^4}{2a^2} \right] dx \\
&= \left(3a^2x - \frac{4x^3}{3} + \frac{x^5}{5a^2} \right)\Bigg|_{0}^{a} \\
&= 3a^3 - \frac{4}{3}a^3 + \frac{1}{5}a^3 = \frac{28}{15}a^3 \text{ cubic units.}
\end{aligned}$$

Remark Maple's **int** routine can be nested to evaluate iterated double (or multiple) integrals symbolically. For instance, the iterated integral for the volume V calculated in Example 4 above can be calculated via the Maple command

```
> V = int(int(a+y, y=0..a - x^2/a), x=-a..a);
```

$$V = \frac{28}{15}a^3.$$

Recall that "int" has an *inert* form "Int," which prints the integral without attempting to evaluate it symbolically. For instance, we can print an equation for the reiterated integral in the solution of Example 3 using the command

```
> Int(Int(exp(y^3),x=0..y^2),y=0..1)
  =int(int(exp(y^3),x=0..y^2),y=0..1);
```

$$\int_0^1 \int_0^{y^2} e^{(y^3)}\,dx\,dy = \frac{1}{3}e - \frac{1}{3}$$

If you want Maple to approximate an iterated integral without first trying to evaluate it symbolically, just ask it to `evalf` the inert form.

```
> evalf(Int(Int(exp(y^3),x=0..y^2),y=0..1));
```

.5727606095

Of course, Maple can't evaluate all integrals in symbolic form. If we replace $\exp(y^3)$ in the iterated integral above with $\exp(x^3)$, recent versions of Maple will just return the inert form as the answer, being unable to calculate the inner integral.

```
> Int(Int(exp(x^3),x=0..y^2),y=0..1)
  =int(int(exp(x^3),x=0..y^2),y=0..1);
```

$$\int_0^1 \int_0^{y^2} e^{(x^3)}\,dx\,dy = \int_0^1 \int_0^{y^2} e^{(x^3)}\,dx\,dy$$

Again, we can force numerical approximation by using `evalf` on the inert form.

```
> Int(Int(exp(x^3),x=0..y^2),y=0..1)
  =evalf(Int(Int(exp(x^3),x=0..y^2),y=0..1));
```

$$\int_0^1 \int_0^{y^2} e^{(x^3)}\,dx\,dy = .3668032540$$

In recent versions of Maple it is not necessary to use the inert form of the integral with `evalf`, but some earlier versions could produce strange values (e.g., complex numbers for values of evidently real integrals) if you did not use the inert form. Software like Maple is constantly being revised and tweaked so that in unusual circumstances different versions of the software can lead to different results.

EXERCISES 14.2

In Exercises 1–4, calculate the given iterated integrals.

1. $\displaystyle\int_0^1 dx \int_0^x (xy + y^2)\,dy$ **2.** $\displaystyle\int_0^1 \int_0^y (xy + y^2)\,dx\,dy$

3. $\displaystyle\int_0^\pi \int_{-x}^x \cos y\,dy\,dx$ **4.** $\displaystyle\int_0^2 dy \int_0^y y^2 e^{xy}\,dx$

In Exercises 5–14, evaluate the double integrals by iteration.

5. $\displaystyle\iint_R (x^2 + y^2)\,dA$, where R is the rectangle $0 \le x \le a$, $0 \le y \le b$

6. $\displaystyle\iint_R x^2y^2\,dA$, where R is the rectangle of Exercise 5

7. $\displaystyle\iint_S (\sin x + \cos y)\,dA$, where S is the square $0 \le x \le \pi/2$, $0 \le y \le \pi/2$

8. $\displaystyle\iint_T (x - 3y)\,dA$, where T is the triangle with vertices $(0, 0)$, $(a, 0)$, and $(0, b)$

9. $\displaystyle\iint_R xy^2\,dA$, where R is the finite region in the first quadrant bounded by the curves $y = x^2$ and $x = y^2$

10. $\displaystyle\iint_D x\cos y\,dA$, where D is the finite region in the first quadrant bounded by the coordinate axes and the curve $y = 1 - x^2$

11. $\displaystyle\iint_D \ln x\,dA$, where D is the finite region in the first quadrant bounded by the line $2x + 2y = 5$ and the hyperbola $xy = 1$

12. $\displaystyle\iint_T \sqrt{a^2 - y^2}\,dA$, where T is the triangle with vertices $(0, 0)$, $(a, 0)$, and (a, a)

13. $\displaystyle\iint_R \frac{x}{y}e^y\,dA$, where R is the region $0 \le x \le 1$, $x^2 \le y \le x$

14. $\displaystyle\iint_T \frac{xy}{1 + x^4}\,dA$, where T is the triangle with vertices $(0, 0)$, $(1, 0)$, and $(1, 1)$

In Exercises 15–18, sketch the domain of integration and evaluate the given iterated integrals.

15. $\displaystyle\int_0^1 dy \int_y^1 e^{-x^2}\,dx$ **16.** $\displaystyle\int_0^{\pi/2} dy \int_y^{\pi/2} \frac{\sin x}{x}\,dx$

17. $\displaystyle\int_0^1 dx \int_x^1 \frac{y^\lambda}{x^2 + y^2}\,dy \quad (\lambda > 0)$

18. $\displaystyle\int_0^1 dx \int_x^{x^{1/3}} \sqrt{1 - y^4}\,dy$

In Exercises 19–28, find the volumes of the indicated solids.

19. Under $z = 1 - x^2$ and above the region $0 \le x \le 1$, $0 \le y \le x$

20. Under $z = 1 - x^2$ and above the region $0 \le y \le 1$, $0 \le x \le y$

21. Under $z = 1 - x^2 - y^2$ and above the region $x \ge 0$, $y \ge 0$, $x + y \le 1$

22. Under $z = 1 - y^2$ and above $z = x^2$

23. Under the surface $z = 1/(x + y)$ and above the region in the xy-plane bounded by $x = 1$, $x = 2$, $y = 0$, and $y = x$

24. Under the surface $z = x^2\sin(y^4)$ and above the triangle in the xy-plane with vertices $(0, 0)$, $(0, \pi^{1/4})$, and $(\pi^{1/4}, \pi^{1/4})$

25. Above the xy-plane and under the surface $z = 1 - x^2 - 2y^2$

26. Above the triangle with vertices $(0, 0)$, $(a, 0)$, and $(0, b)$, and under the plane $z = 2 - (x/a) - (y/b)$

27. Inside the two cylinders $x^2 + y^2 = a^2$ and $y^2 + z^2 = a^2$

28. Inside the cylinder $x^2 + 2y^2 = 8$, above the plane $z = y - 4$, and below the plane $z = 8 - x$

29. Suppose that $f(x, t)$ and $f_1(x, t)$ are continuous on the rectangle $a \le x \le b$ and $c \le t \le d$. Let

$$g(x) = \int_c^d f(x, t)\,dt \quad \text{and} \quad G(x) = \int_c^d f_1(x, t)\,dt.$$

Show that $g'(x) = G(x)$ for $a < x < b$. *Hint:* Evaluate $\int_a^x G(u)\,du$ by reversing the order of iteration. Then differentiate the result. This is a different version of Theorem 6 of Section 13.6.

30. Let $F'(x) = f(x)$ and $G'(x) = g(x)$ on the interval $a \le x \le b$. Let T be the triangle with vertices (a, a), (b, a), and (b, b). By iterating $\iint_T f(x)g(y)\,dA$ in both directions, show that

$$\int_a^b f(x)G(x)\,dx$$
$$= F(b)G(b) - F(a)G(a) - \int_a^b g(y)F(y)\,dy.$$

(This is an alternative derivation of the formula for integration by parts.)

31. Use Maple's **int** routine or similar routines in other computer algebra systems to evaluate the iterated integrals in Exercises 1–4 or the iterated integrals you constructed in the remaining exercises above.

14.3 Improper Integrals and a Mean-Value Theorem

To simplify matters, the definition of the double integral given in Section 14.1 required that the domain D be bounded and that the integrand f be bounded on D. As in the single-variable case, **improper double integrals** can arise if either the domain of integration is unbounded or the integrand is unbounded near any point of the domain or its boundary.

Improper Integrals of Positive Functions

An improper integral of a function f satisfying $f(x, y) \ge 0$ on the domain D must either exist (i.e., converge to a finite value) or be infinite (diverge to infinity). Convergence or divergence of improper double integrals of such *nonnegative* functions can be determined by iterating them and determining the convergence or divergence of any single improper integrals that result.

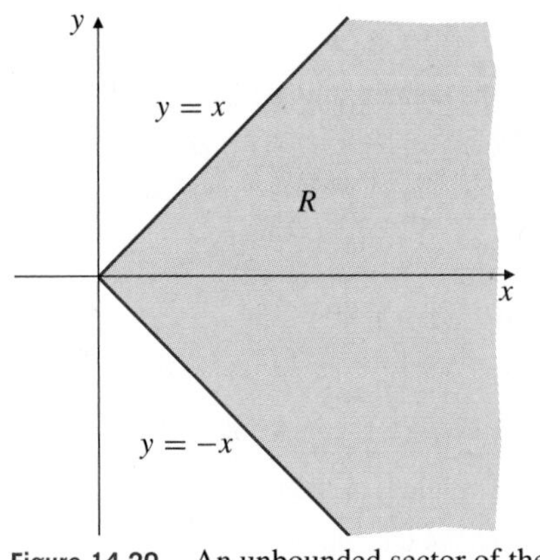

Figure 14.20 An unbounded sector of the plane

EXAMPLE 1 Evaluate $I = \iint_R e^{-x^2}\,dA$. Here, R is the region where $x \ge 0$ and $-x \le y \le x$. (See Figure 14.20.)

Solution We iterate with the outer integral in the x direction:

$$\begin{aligned} I &= \int_0^\infty dx \int_{-x}^x e^{-x^2}\,dy \\ &= \int_0^\infty e^{-x^2}\,dx \int_{-x}^x dy \\ &= 2\int_0^\infty xe^{-x^2}\,dx. \end{aligned}$$

This is an improper integral that can be expressed as a limit:

$$\begin{aligned} I &= 2 \lim_{r\to\infty} \int_0^r xe^{-x^2}\,dx \\ &= 2 \lim_{r\to\infty} \left(-\frac{1}{2}e^{-x^2}\right)\Bigg|_0^r \\ &= \lim_{r\to\infty} (1 - e^{-r^2}) = 1. \end{aligned}$$

The given integral converges; its value is 1.

EXAMPLE 2 If D is the region lying above the x-axis, under the curve $y = 1/x$, and to the right of the line $x = 1$, determine whether the double integral

$$\iint_D \frac{dA}{x+y}$$

converges or diverges.

Solution The region D is sketched in Figure 14.21. We have

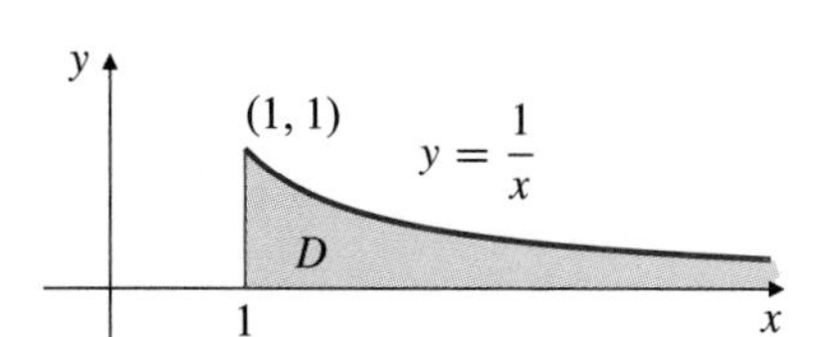

Figure 14.21 The domain of the integrand in Example 2

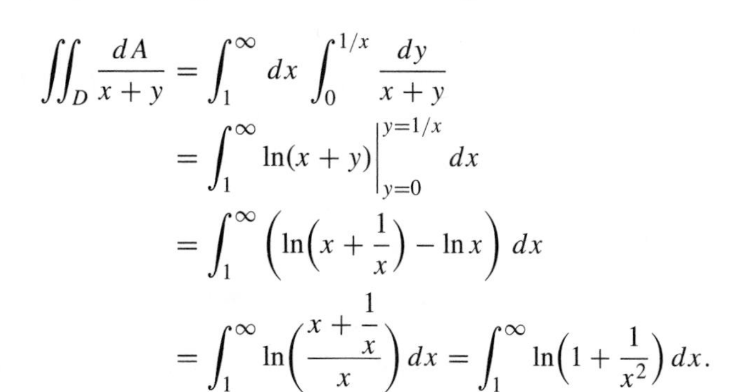

$$\begin{aligned}
\iint_D \frac{dA}{x+y} &= \int_1^\infty dx \int_0^{1/x} \frac{dy}{x+y} \\
&= \int_1^\infty \ln(x+y)\Big|_{y=0}^{y=1/x} dx \\
&= \int_1^\infty \left(\ln\left(x+\frac{1}{x}\right) - \ln x\right) dx \\
&= \int_1^\infty \ln\left(\frac{x+\frac{1}{x}}{x}\right) dx = \int_1^\infty \ln\left(1+\frac{1}{x^2}\right) dx.
\end{aligned}$$

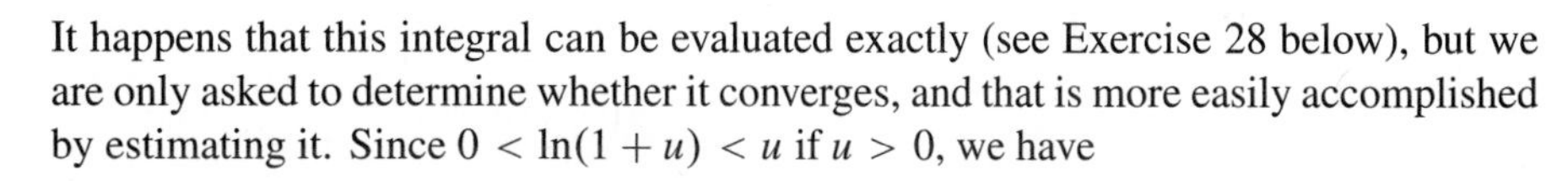

It happens that this integral can be evaluated exactly (see Exercise 28 below), but we are only asked to determine whether it converges, and that is more easily accomplished by estimating it. Since $0 < \ln(1+u) < u$ if $u > 0$, we have

$$0 < \iint_D \frac{dA}{x+y} < \int_1^\infty \frac{1}{x^2}\, dx = 1.$$

Therefore, the given integral converges, and its value lies between 0 and 1.

EXAMPLE 3 Evaluate $\displaystyle\iint_D \frac{1}{(x+y)^2}\, dA$, where D is the region $0 \le x \le 1$, $0 \le y \le x^2$.

Solution The integral is improper because the integrand is unbounded as (x, y) approaches $(0, 0)$, a boundary point of D. (See Figure 14.22.) Nevertheless, iteration leads to a proper integral:

$$\begin{aligned}
\iint_D \frac{1}{(x+y)^2}\, dA &= \lim_{c\to 0+} \int_c^1 dx \int_0^{x^2} \frac{1}{(x+y)^2}\, dy \\
&= \lim_{c\to 0+} \int_c^1 dx \left(-\frac{1}{x+y}\right)\Big|_{y=0}^{y=x^2} \\
&= \lim_{c\to 0+} \int_c^1 \left(\frac{1}{x} - \frac{1}{x^2+x}\right) dx \\
&= \int_0^1 \frac{1}{x+1}\, dx = \ln(x+1)\Big|_0^1 = \ln 2.
\end{aligned}$$

Figure 14.22 The function $\dfrac{1}{(x+y)^2}$ is unbounded on D

EXAMPLE 4 Determine the convergence or divergence of $I = \displaystyle\iint_D \frac{dA}{xy}$, where D is the bounded region in the first quadrant lying between the line $y = x$ and the parabola $y = x^2$.

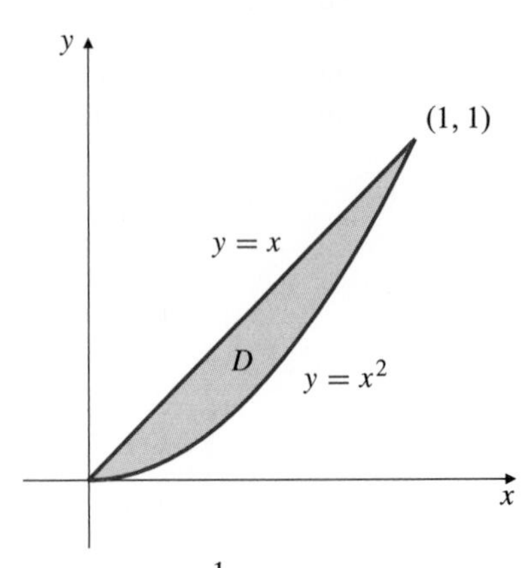

Figure 14.23 $\dfrac{1}{xy}$ is unbounded on the domain D

Solution The domain D is shown in Figure 14.23. Again, the integral is improper because the integrand $1/(xy)$ is unbounded as (x, y) approaches the boundary point $(0, 0)$. We have

$$\begin{aligned} I = \iint_D \frac{dA}{xy} &= \int_0^1 \frac{dx}{x} \int_{x^2}^{x} \frac{dy}{y} \\ &= \int_0^1 \frac{1}{x}(\ln x - \ln x^2)\, dx = -\int_0^1 \frac{\ln x}{x}\, dx. \end{aligned}$$

If we substitute $x = e^{-t}$ in this integral, we obtain

$$I = -\int_{\infty}^{0} \frac{-t}{e^{-t}}(-e^{-t})\, dt = \int_0^{\infty} t\, dt,$$

which diverges to infinity.

Remark In each of the examples above, the integrand was nonnegative on the domain of integration. Nonpositive integrands could have been handled similarly, but we cannot deal here with the convergence of general improper double integrals with integrands $f(x, y)$ that take both positive and negative values on the domain D of the integral. We remark, however, that such an integral cannot converge unless

$$\iint_E f(x, y)\, dA$$

is finite for every bounded, regular subdomain E of D. We cannot, in general, determine the convergence of the given integral by looking at the convergence of iterations. The double integral may diverge even if its iterations converge. (See Exercise 21 below.) In fact, opposite iterations may even give different values. This happens because of cancellation of infinite volumes of opposite sign. (Similar behaviour in one dimension is exemplified by the integral $\int_{-1}^{1} dx/x$, which does not exist, although it represents the difference between "equal" but infinite areas.) It can be shown (for a large class of functions containing, for example, continuous functions) that an improper double integral of $f(x, y)$ over D converges if the integral of $|f(x, y)|$ over D converges:

$$\iint_D |f(x, y)|\, dA \text{ converges} \quad \Rightarrow \iint_D f(x, y)\, dA \text{ converges.}$$

In this case any iterations will converge to the same value. Such double integrals are called **absolutely convergent** by analogy with absolutely convergent infinite series.

A Mean-Value Theorem for Double Integrals

Let D be a set in the xy-plane that is closed and bounded and has positive area $A = \iint_D dA$. Suppose that $f(x, y)$ is continuous on D. Then there exist points (x_1, y_1) and (x_2, y_2) in D where f assumes minimum and maximum values (see Theorem 2 of Section 13.1); that is,

$$f(x_1, y_1) \le f(x, y) \le f(x_2, y_2)$$

for all points (x, y) in D. If we integrate this inequality over D, we obtain

$$\begin{aligned} f(x_1, y_1)A &= \iint_D f(x_1, y_1)\, dA \\ &\le \iint_D f(x, y)\, dA \le \iint_D f(x_2, y_2)\, dA = f(x_2, y_2)A. \end{aligned}$$

Therefore, dividing by A, we find that the *number*

$$\bar{f} = \frac{1}{A} \iint_D f(x, y)\, dA$$

lies between the minimum and maximum values of f on D:

$$f(x_1, y_1) \le \bar{f} \le f(x_2, y_2).$$

A set D in the plane is said to be **connected** if any two points in it can be joined by a continuous parametric curve $x = x(t)$, $y = y(t)$, $(0 \le t \le 1)$, lying in D. Suppose this curve joins (x_1, y_1) (where $t = 0$) and (x_2, y_2) (where $t = 1$). Let $g(t)$ satisfy

$$g(t) = f\big(x(t), y(t)\big), \quad 0 \le t \le 1.$$

Then g is continuous and takes the values $f(x_1, y_1)$ at $t = 0$ and $f(x_2, y_2)$ at $t = 1$. By the Intermediate-Value Theorem there exists a number t_0 between 0 and 1 such that $\bar{f} = g(t_0) = f(x_0, y_0)$, where $x_0 = x(t_0)$ and $y_0 = y(t_0)$. Thus, we have found a point (x_0, y_0) in D such that

$$\frac{1}{\text{area of } D} \iint_D f(x, y)\, dA = f(x_0, y_0).$$

We have therefore proved the following version of the Mean-Value Theorem.

THEOREM 3

A Mean-Value Theorem for double integrals

If the function $f(x, y)$ is continuous on a closed, bounded, connected set D in the xy-plane, then there exists a point (x_0, y_0) in D such that

$$\iint_D f(x, y)\, dA = f(x_0, y_0) \times (\text{area of } D).$$

By analogy with the definition of average value for one-variable functions, we make the following definition:

DEFINITION 3

The **average value** or **mean value** of an integrable function $f(x, y)$ over the set D is the number

$$\bar{f} = \frac{1}{\text{area of } D} \iint_D f(x, y)\, dA.$$

If $f(x, y) \ge 0$ on D, then the cylinder with base D and constant height $\bar{f}$ has volume equal to that of the solid region lying above D and below the surface $z = f(x, y)$. It is often very useful to interpret a double integral in terms of the average value of the function which is its integrand.

EXAMPLE 5 The average value of x over a domain D having area A is

$$\bar{x} = \frac{1}{A} \iint_D x\, dA.$$

Of course, $\bar{x}$ is just the x-coordinate of the centroid of the region D.

EXAMPLE 6 A large number of points (x, y) are chosen at random in the triangle T with vertices $(0, 0)$, $(1, 0)$, and $(1, 1)$. What is the approximate average value of $x^2 + y^2$ for these points?

Solution The approximate average value of $x^2 + y^2$ for the randomly chosen points will be the average value of that function over the triangle, namely,

$$\frac{1}{1/2}\iint_T (x^2+y^2)\,dA = 2\int_0^1 dx \int_0^x (x^2+y^2)\,dy$$
$$= 2\int_0^1 \left(x^2y+\frac{1}{3}y^3\right)\Bigg|_{y=0}^{y=x} dx = \frac{8}{3}\int_0^1 x^3\,dx = \frac{2}{3}.$$

EXAMPLE 7 Let (a, b) be an interior point of a domain D on which $f(x, y)$ is continuous. For sufficiently small positive r, the closed circular disk D_r with centre at (a, b) and radius r is contained in D. Show that

$$\lim_{r\to 0}\frac{1}{\pi r^2}\iint_{D_r} f(x,y)\,dA = f(a,b).$$

Solution If D_r is contained in D, then by Theorem 3

$$\frac{1}{\pi r^2}\iint_{D_r} f(x,y)\,dA = f(x_0,y_0)$$

for some point (x_0, y_0) in D_r. As $r \to 0$, the point (x_0, y_0) approaches (a, b). Since f is continuous at (a, b), we have $f(x_0, y_0) \to f(a, b)$. Thus,

$$\lim_{r\to 0}\frac{1}{\pi r^2}\iint_{D_r} f(x,y)\,dA = f(a,b).$$

EXERCISES 14.3

In Exercises 1–12, determine whether the given integral converges or diverges. Try to evaluate those that converge.

1. $\displaystyle\iint_Q e^{-x-y}\,dA$, where Q is the first quadrant of the xy-plane

2. $\displaystyle\iint_Q \frac{dA}{(1+x^2)(1+y^2)}$, where Q is the first quadrant of the xy-plane

3. $\displaystyle\iint_S \frac{y}{1+x^2}\,dA$, where S is the strip $0 < y < 1$ in the xy-plane

4. $\displaystyle\iint_T \frac{1}{x\sqrt{y}}\,dA$ over the triangle T with vertices $(0, 0)$, $(1, 1)$, and $(1, 2)$

5. $\displaystyle\iint_Q \frac{x^2+y^2}{(1+x^2)(1+y^2)}\,dA$, where Q is the first quadrant of the xy-plane

6. $\displaystyle\iint_H \frac{1}{1+x+y}\,dA$, where H is the half-strip $0 \le x < \infty$, $0 < y < 1$

7. $\displaystyle\iint_{\mathbb{R}^2} e^{-(|x|+|y|)}\,dA$

8. $\displaystyle\iint_{\mathbb{R}^2} e^{-|x+y|}\,dA$

9. $\displaystyle\iint_T \frac{1}{x^3}e^{-y/x}\,dA$, where T is the region satisfying $x \ge 1$ and $0 \le y \le x$

10. $\displaystyle\iint_T \frac{dA}{x^2+y^2}$, where T is the region in Exercise 9

11. $\displaystyle\iint_Q e^{-xy}\,dA$, where Q is the first quadrant of the xy-plane

12. $\displaystyle\iint_R \frac{1}{x}\sin\frac{1}{x}\,dA$, where R is the region $2/\pi \le x < \infty$, $0 \le y \le 1/x$

13. Evaluate

$$I = \iint_S \frac{dA}{x+y},$$

where S is the square $0 \le x \le 1$, $0 \le y \le 1$,

(a) by direct iteration of the double integral,

(b) by using the symmetry of the integrand and the domain to write

$$I = 2\iint_T \frac{dA}{x+y},$$

where T is the triangle with vertices $(0, 0)$, $(1, 0)$, and $(1, 1)$.

14. Find the volume of the solid lying above the square S of Exercise 13 and under the surface $z = 2xy/(x^2 + y^2)$.

In Exercises 15–20, a and b are given real numbers, D_k is the region $0 \le x \le 1, 0 \le y \le x^k$, and R_k is the region $1 \le x < \infty$, $0 \le y \le x^k$. Find all real values of k for which the given integral converges.

15. $\displaystyle\iint_{D_k} \frac{dA}{x^a}$ **16.** $\displaystyle\iint_{D_k} y^b\, dA$

17. $\displaystyle\iint_{R_k} x^a\, dA$ **18.** $\displaystyle\iint_{R_k} \frac{dA}{y^b}$

19. $\displaystyle\iint_{D_k} x^a y^b\, dA$ **20.** $\displaystyle\iint_{R_k} x^a y^b\, dA$

21. Evaluate both iterations of the improper integral

$$\iint_S \frac{x-y}{(x+y)^3}\, dA,$$

where S is the square $0 < x < 1, 0 < y < 1$. Show that the above improper double integral does not exist, by considering

$$\iint_T \frac{x-y}{(x+y)^3}\, dA,$$

where T is that part of the square S lying under the line $x = y$.

In Exercises 22–24, find the average value of the given function over the given region.

22. x^2 over the rectangle $a \le x \le b, c \le y \le d$

23. $x^2 + y^2$ over the triangle $0 \le x \le a, 0 \le y \le a - x$

24. $1/x$ over the region $0 \le x \le 1, x^2 \le y \le \sqrt{x}$

25. Find the average distance from points in the quarter-disk $x^2 + y^2 \le a^2, x \ge 0, y \ge 0$, to the line $x + y = 0$.

26. Does $f(x, y) = x$ have an average value over the region $0 \le x < \infty, 0 \le y \le \dfrac{1}{1+x^2}$? If so, what is it?

27. Does $f(x, y) = xy$ have an average value over the region $0 \le x < \infty, 0 \le y \le \dfrac{1}{1+x^2}$? If so, what is it?

28. Find the exact value of the integral in Example 2. *Hint:* Integrate by parts in $\int_1^\infty \ln\left(1 + (1/x^2)\right) dx$.

29. Let (a, b) be an interior point of a domain D on which the function $f(x, y)$ is continuous. For small enough $h^2 + k^2$ the rectangle R_{hk} with vertices (a, b), $(a + h, b)$, $(a, b + k)$, and $(a + h, b + k)$ is contained in D. Show that

$$\lim_{(h,k)\to(0,0)} \frac{1}{hk} \iint_{R_{hk}} f(x, y) = f(a, b).$$

Hint: See Example 7.

30. (Another proof of equality of mixed partials) Suppose that $f_{12}(x, y)$ and $f_{21}(x, y)$ are continuous in a neighbourhood of the point (a, b). Without assuming the equality of these mixed partial derivatives, show that

$$\iint_R f_{12}(x, y)\, dA = \iint_R f_{21}(x, y)\, dA,$$

where R is the rectangle with vertices (a, b), $(a + h, b)$, $(a, b + k)$, and $(a + h, b + k)$ and $h^2 + k^2$ is sufficiently small. Now use the result of Exercise 29 to show that $f_{12}(a, b) = f_{21}(a, b)$. (This reproves Theorem 1 of Section 12.4. However, in that theorem we only assumed continuity of the mixed partials *at* (a, b). Here, we assume the continuity *at all points sufficiently near* (a, b).)

14.4 Double Integrals in Polar Coordinates

For many applications of double integrals, either the domain of integration, the integrand function, or both may be more easily expressed in terms of polar coordinates than in terms of Cartesian coordinates. Recall that a point P with Cartesian coordinates (x, y) can also be located by its polar coordinates $[r, \theta]$, where r is the distance from P to the origin O, and θ is the angle OP makes with the positive direction of the x-axis. (Positive angles θ are measured counterclockwise.) The polar and Cartesian coordinates of P are related by the transformations (see Figure 14.24)

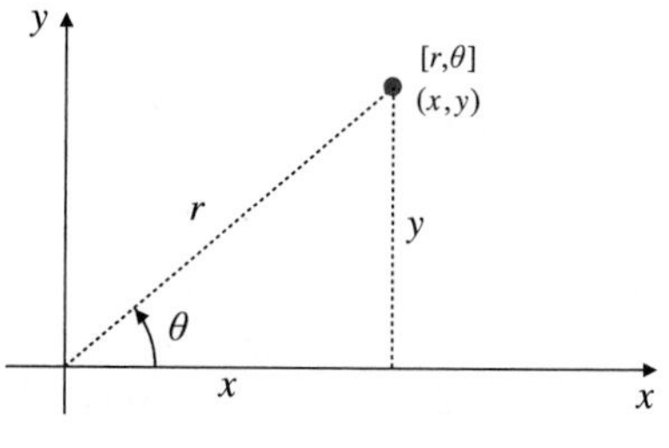

$$x = r\cos\theta, \qquad r^2 = x^2 + y^2,$$
$$y = r\sin\theta, \qquad \tan\theta = y/x.$$

Figure 14.24 Polar–Cartesian conversions

Consider the problem of finding the volume V of the solid region lying above the xy-plane and beneath the paraboloid $z = 1 - x^2 - y^2$. Since the paraboloid intersects the xy-plane in the circle $x^2 + y^2 = 1$, the volume is given in Cartesian coordinates by

$$V = \iint_{x^2+y^2\le 1} (1 - x^2 - y^2)\, dA = \int_{-1}^{1} dx \int_{-\sqrt{1-x^2}}^{\sqrt{1-x^2}} (1 - x^2 - y^2)\, dy.$$

Evaluating this iterated integral would require considerable effort. However, we can express the same volume in terms of polar coordinates as

$$V = \iint_{r \leq 1} (1 - r^2)\, dA.$$

In order to iterate this integral, we have to know the form that the *area element* dA takes in polar coordinates.

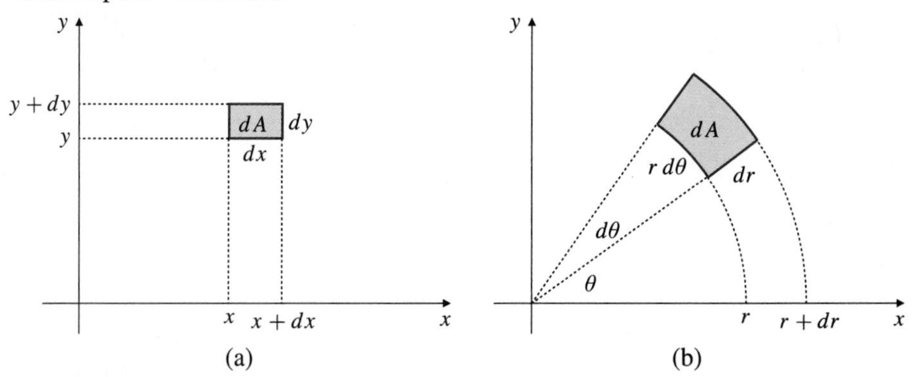

Figure 14.25
(a) $dA = dx\, dy$ in Cartesian coordinates
(b) $dA = r\, dr\, d\theta$ in polar coordinates

In the Cartesian formula for V, the area element $dA = dx\, dy$ represents the area of the "infinitesimal" region bounded by the coordinate lines at x, $x + dx$, y, and $y + dy$. (See Figure 14.25(a).) In the polar formula, the area element dA should represent the area of the "infinitesimal" region bounded by the coordinate circles with radii r and $r + dr$, and coordinate rays from the origin at angles θ and $\theta + d\theta$. (See Figure 14.25(b).) Observe that dA is approximately the area of a rectangle with dimensions dr and $r\, d\theta$. The error in this approximation becomes negligible *compared with the size of* dA as dr and $d\theta$ approach zero. Thus, in transforming a double integral between Cartesian and polar coordinates, the area element transforms according to the formula

$$dx\, dy = dA = r\, dr\, d\theta.$$

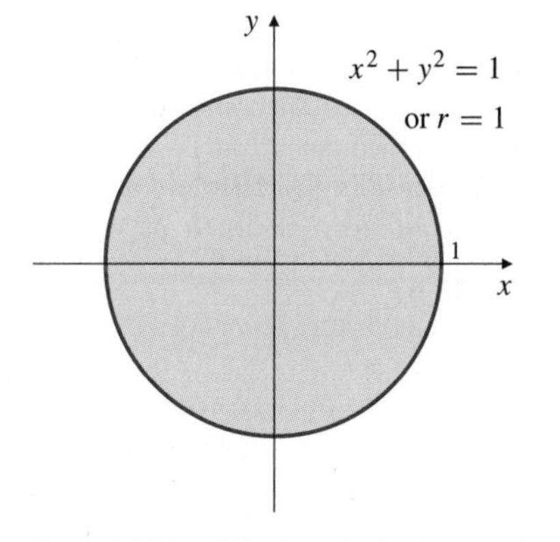

Figure 14.26 The domain in the xy-plane

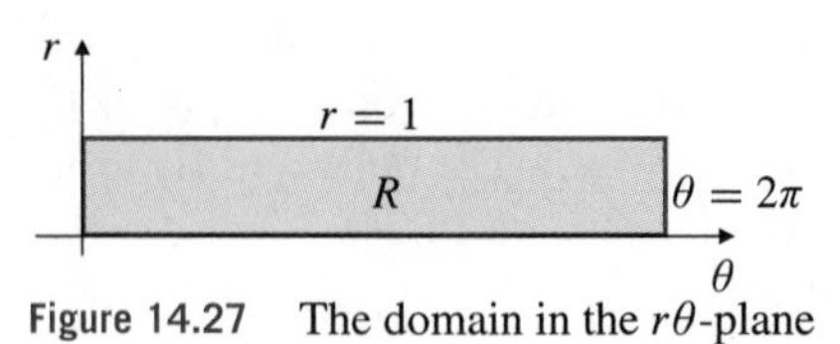

Figure 14.27 The domain in the $r\theta$-plane

In order to iterate the polar form of the double integral for V considered above, we can regard the domain of integration as a set in a plane having *Cartesian coordinates* r and θ. In the xy Cartesian plane the domain is a disk $r \leq 1$ (see Figure 14.26), but in the $r\theta$ Cartesian plane (with perpendicular r- and θ-axes) the domain is the rectangle R specified by $0 \leq r \leq 1$ and $0 \leq \theta \leq 2\pi$. (See Figure 14.27.) The area element in the $r\theta$-plane is $dA^* = dr\, d\theta$, so area is not preserved under the transformation to polar coordinates ($dA = r\, dA^*$). Thus, the polar integral for V is really a Cartesian integral in the $r\theta$-plane, with integrand modified by the inclusion of an extra factor r to compensate for the change of area. It can be evaluated by standard iteration methods:

$$V = \iint_R (1 - r^2) r\, dA^* = \int_0^{2\pi} d\theta \int_0^1 (1 - r^2) r\, dr$$
$$= \int_0^{2\pi} \left(\frac{r^2}{2} - \frac{r^4}{4} \right) \bigg|_0^1 d\theta = \frac{\pi}{2} \text{ units}^3.$$

Remark It is not necessary to sketch the region R in the $r\theta$-plane. We are used to thinking of polar coordinates in terms of distances and angles in the xy-plane and can easily understand from looking at the disk in Figure 14.26 that the iteration of the integral in polar coordinates corresponds to $0 \leq \theta \leq 2\pi$ and $0 \leq r \leq 1$. That is, we should be able to write the iteration

$$V = \int_0^{2\pi} d\theta \int_0^1 (1 - r^2) r\, dr$$

directly from consideration of the domain of integration in the xy-plane.

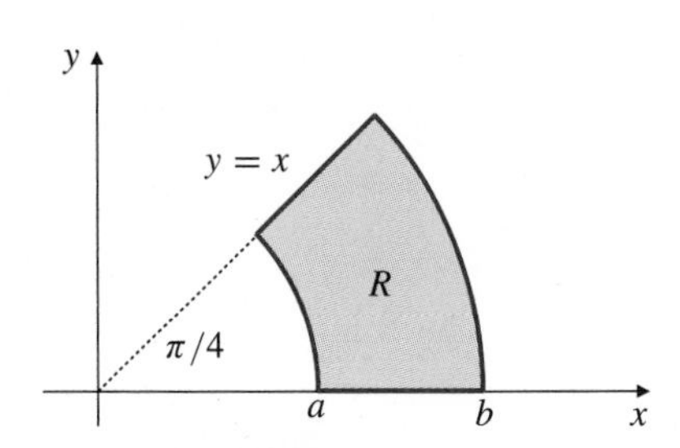

Figure 14.28 Region R corresponds to a rectangle in the $r\theta$-plane

EXAMPLE 1 If R is that part of the annulus $0 < a^2 \le x^2 + y^2 \le b^2$ lying in the first quadrant and below the line $y = x$, evaluate

$$I = \iint_R \frac{y^2}{x^2}\, dA.$$

Solution Figure 14.28 shows the region R. It is specified in polar coordinates by $0 \le \theta \le \pi/4$ and $a \le r \le b$. Since

$$\frac{y^2}{x^2} = \frac{r^2 \sin^2\theta}{r^2 \cos^2\theta} = \tan^2\theta,$$

we have

$$\begin{aligned} I &= \int_0^{\pi/4} \tan^2\theta\, d\theta \int_a^b r\, dr \\ &= \frac{1}{2}(b^2 - a^2) \int_0^{\pi/4} \left(\sec^2\theta - 1\right) d\theta \\ &= \frac{1}{2}(b^2 - a^2)(\tan\theta - \theta)\Big|_0^{\pi/4} \\ &= \frac{1}{2}(b^2 - a^2)\left(1 - \frac{\pi}{4}\right) = \frac{4 - \pi}{8}(b^2 - a^2). \end{aligned}$$

EXAMPLE 2 **(Area of a polar region)** Derive the formula for the area of the polar region R bounded by the curve $r = f(\theta)$ and the rays $\theta = \alpha$ and $\theta = \beta$. (See Figure 14.29.)

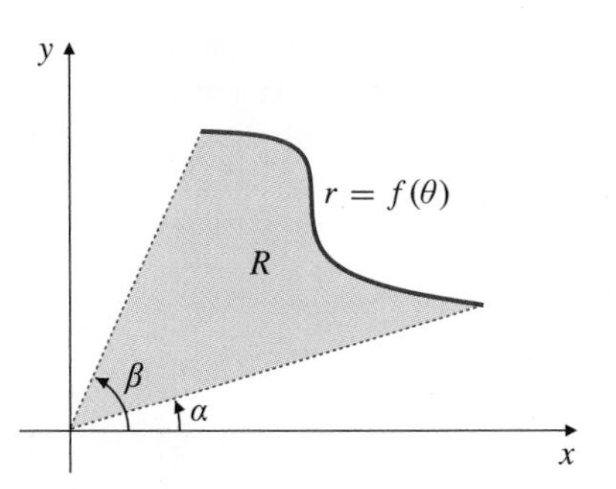

Figure 14.29 A standard area problem for polar coordinates

Solution The area A of R is numerically equal to the volume of a cylinder of height 1 above the region R:

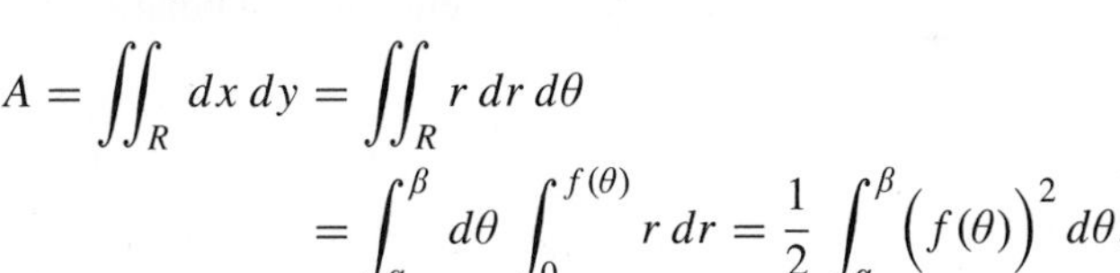

$$\begin{aligned} A &= \iint_R dx\, dy = \iint_R r\, dr\, d\theta \\ &= \int_\alpha^\beta d\theta \int_0^{f(\theta)} r\, dr = \frac{1}{2}\int_\alpha^\beta \left(f(\theta)\right)^2 d\theta. \end{aligned}$$

Observe that the inner integral in the iteration involves integrating r along the ray specified by θ from 0 to $f(\theta)$.

There is no firm rule as to whether one should or should not convert a double integral from Cartesian to polar coordinates. In Example 1 above, the conversion was strongly suggested by the shape of the domain but was also indicated by the fact that the integrand, y^2/x^2, becomes a function of θ alone when converted to polar coordinates. It is usually wise to switch to polar coordinates if the switch simplifies the iteration (i.e., if the *domain* is "simpler" when expressed in terms of polar coordinates), even if the form of the integrand is made more complicated.

EXAMPLE 3 Find the volume of the solid lying in the first octant, inside the cylinder $x^2 + y^2 = a^2$, and under the plane $z = y$.

Solution The solid is shown in Figure 14.30. The base is a quarter disk, which is expressed in polar coordinates by the inequalities $0 \le \theta \le \pi/2$ and $0 \le r \le a$. The height is given by $z = y = r\sin\theta$. The solid has volume

$$V = \int_0^{\pi/2} d\theta \int_0^a (r\sin\theta) r\, dr = \int_0^{\pi/2} \sin\theta\, d\theta \int_0^a r^2\, dr = \frac{1}{3}a^3 \text{ units}^3.$$

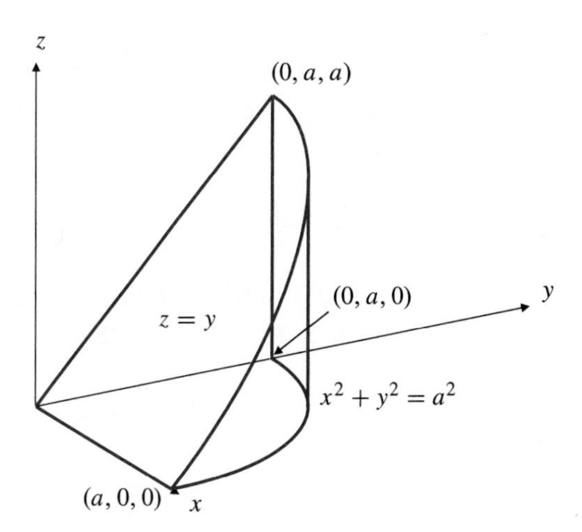

Figure 14.30 This volume is easily calculated using iteration in polar coordinates

The following example establishes the value of a definite integral that plays a very important role in probability theory and statistics. (See the quotation on the first page of this chapter.) It is interesting that this single-variable integral cannot be evaluated by the techniques of single-variable calculus.

EXAMPLE 4 **(A Very Important Integral)** Show that

$$\int_{-\infty}^{\infty} e^{-x^2}\,dx = \sqrt{\pi}.$$

Solution The improper integral (call it I) converges, and its value does not depend on what symbol we use for the variable of integration. Therefore, we can express the square of the integral as a product of two identical integrals but with their variables of integration named differently. We then interpret this product as an improper double integral and reiterate it in polar coordinates:

$$\begin{aligned} I^2 = \left(\int_{-\infty}^{\infty} e^{-x^2}\,dx\right)^2 &= \int_{-\infty}^{\infty} e^{-x^2}\,dx \int_{-\infty}^{\infty} e^{-y^2}\,dy \\ &= \iint_{\mathbb{R}^2} e^{-(x^2+y^2)}\,dA \\ &= \int_0^{2\pi} d\theta \int_0^{\infty} e^{-r^2}\, r\,dr \\ &= 2\pi \lim_{R\to\infty}\left(-\frac{1}{2}e^{-r^2}\right)\Bigg|_0^R = \pi. \end{aligned}$$

Thus, $I = \sqrt{\pi}$ as asserted.

Note that the r integral in the iteration above is a convergent improper integral; it was evaluated with the aid of the substitution $u = r^2$.

As our final example of iteration in polar coordinates, let us try something a little more demanding.

EXAMPLE 5 Find the volume of the solid region lying inside both the sphere $x^2 + y^2 + z^2 = 4a^2$ and the cylinder $x^2 + y^2 = 2ay$, where $a > 0$.

Solution The sphere is centred at the origin and has radius $2a$. The equation of the cylinder becomes

$$x^2 + (y - a)^2 = a^2$$

if we complete the square in the y terms. Thus, it is a vertical circular cylinder of radius a having its axis along the vertical line through $(0, a, 0)$. The z-axis lies on the cylinder. One-quarter of the required volume lies in the first octant. This part is shown in Figure 14.31.

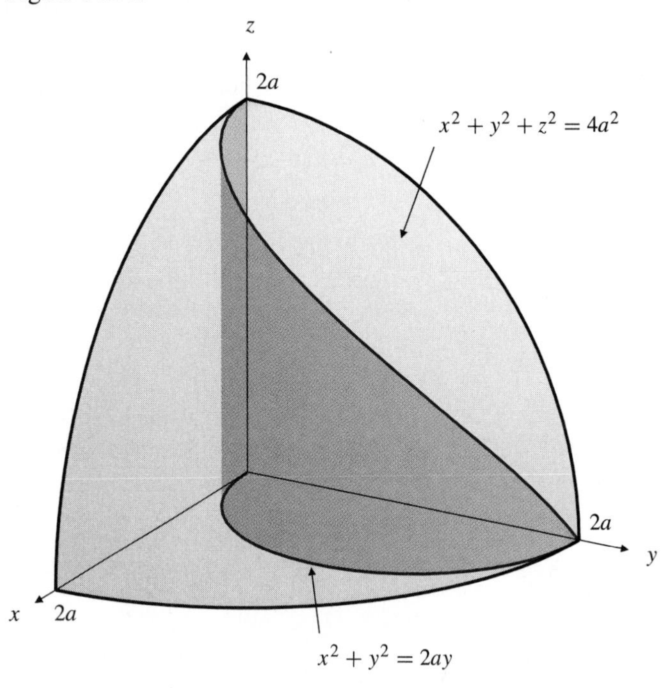

Figure 14.31 The first octant part of the intersection of the cylinder $x^2 + y^2 = 2ay$ and the sphere $x^2 + y^2 + z^2 = 4a^2$

If we use polar coordinates in the xy-plane, then the sphere has equation $r^2 + z^2 = 4a^2$ and the cylinder has equation $r^2 = 2ar\sin\theta$ or, more simply, $r = 2a\sin\theta$. The first octant portion of the volume lies above the region specified by the inequalities $0 \le \theta \le \pi/2$ and $0 \le r \le 2a\sin\theta$. Therefore, the total volume is

$$\begin{aligned}
V &= 4\int_0^{\pi/2} d\theta \int_0^{2a\sin\theta} \sqrt{4a^2 - r^2}\, r\, dr && \text{Let } u = 4a^2 - r^2 \\
&= 2\int_0^{\pi/2} d\theta \int_{4a^2\cos^2\theta}^{4a^2} \sqrt{u}\, du \\
&= \frac{4}{3}\int_0^{\pi/2} (8a^3 - 8a^3\cos^3\theta)\, d\theta && \text{Let } v = \sin\theta \\
&= \frac{16}{3}\pi a^3 - \frac{32}{3}a^3 \int_0^1 (1 - v^2)\, dv \\
&= \frac{16}{3}\pi a^3 - \frac{64}{9}a^3 = \frac{16}{9}(3\pi - 4)a^3 \text{ cubic units.}
\end{aligned}$$

Change of Variables in Double Integrals

The transformation of a double integral to polar coordinates is just a special case of a general change of variables formula for double integrals. Suppose that x and y are expressed as functions of two other variables u and v by the equations

$$x = x(u, v)$$
$$y = y(u, v).$$

We regard these equations as defining a **transformation** (or mapping) from points (u, v) in a uv-Cartesian plane to points (x, y) in the xy-plane. (See Figure 14.32.) We

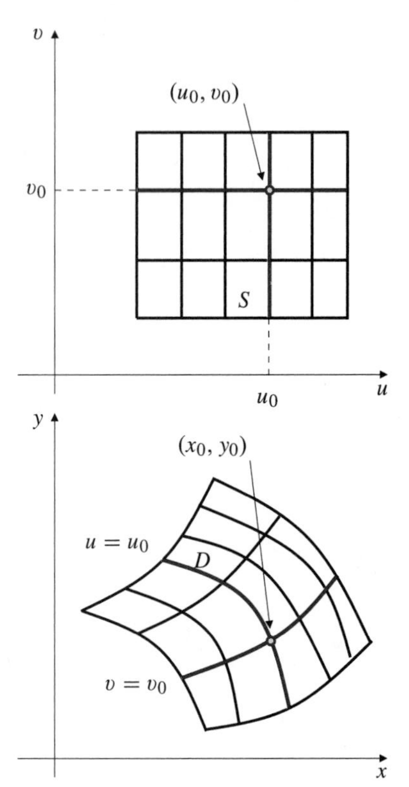

Figure 14.32 Under the transformation $\begin{cases} x = x(u, v) \\ y = y(u, v) \end{cases}$ the lines $u = u_0$ and $v = v_0$ in the uv-plane get mapped to the curves $\begin{cases} x = x(u_0, v) \\ y = y(u_0, v) \end{cases}$ and $\begin{cases} x = x(u, v_0) \\ y = y(u, v_0) \end{cases}$ in the xy-plane, which we still label as $u = u_0$ and $v = v_0$. The point (u_0, v_0) is mapped to the point (x_0, y_0).

say that the transformation is *one-to-one* from the set S in the uv-plane *onto* the set D in the xy-plane provided:

(i) every point in S gets mapped to a point in D,

(ii) every point in D is the image of a point in S, and

(iii) different points in S get mapped to different points in D.

If the transformation is one-to-one, the defining equations can be solved for u and v as functions of x and y, and the resulting **inverse transformation,**

$$u = u(x, y)$$
$$v = v(x, y),$$

is one-to-one from D onto S.

Let us assume that the functions $x(u, v)$ and $y(u, v)$ have continuous first partial derivatives and that the Jacobian determinant

$$\frac{\partial(x, y)}{\partial(u, v)} \neq 0 \quad \text{at} \quad (u, v).$$

As noted in Section 12.8, the Implicit Function Theorem implies that the transformation is one-to-one near (u, v) and the inverse transformation also has continuous first partial derivatives and nonzero Jacobian satisfying

$$\frac{\partial(u, v)}{\partial(x, y)} = \frac{1}{\dfrac{\partial(x, y)}{\partial(u, v)}} \quad \text{on } D.$$

EXAMPLE 6 The transformation $x = r\cos\theta$, $y = r\sin\theta$ to polar coordinates has Jacobian

$$\frac{\partial(x, y)}{\partial(r, \theta)} = \begin{vmatrix} \cos\theta & -r\sin\theta \\ \sin\theta & r\cos\theta \end{vmatrix} = r.$$

Near any point except the origin (where $r = 0$) the transformation is one-to-one. (In fact, it is one-to-one from any set in the $r\theta$-plane that does not contain more than one point where $r = 0$ and lies in, say, the strip $0 \le \theta < 2\pi$.)

A one-to-one transformation can be used to transform the double integral

$$\iint_D f(x, y)\, dA$$

to a double integral over the corresponding set S in the uv-plane. Under the transformation, the integrand $f(x, y)$ becomes $g(u, v) = f\big(x(u, v), y(u, v)\big)$. We must discover how to express the area element $dA = dx\, dy$ in terms of the area element $du\, dv$ in the uv-plane.

If the value of u is fixed, say $u = c$, the equations

$$x = x(c, v) \quad \text{and} \quad y = y(c, v)$$

define a parametric curve (with v as parameter) in the xy-plane. This curve is called a u-curve corresponding to the value $u = c$. Similarly, for fixed $v = c$ the equations

$$x = x(u, c) \quad \text{and} \quad y = y(u, c)$$

define a parametric curve (with parameter u) called a v-curve. Consider the differential area element bounded by the u-curves corresponding to nearby values u and $u + du$ and the v-curves corresponding to nearby values v and $v + dv$. Since these curves are smooth, for small values of du and dv the area element is approximately a parallelogram, and its area is approximately

$$dA = |\overrightarrow{PQ} \times \overrightarrow{PR}|,$$

where P, Q, and R are the points shown in Figure 14.33. The error in this approximation becomes negligible compared with dA as du and dv approach zero.

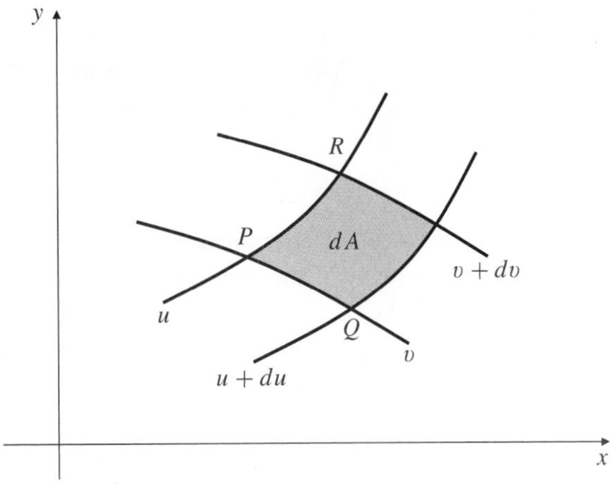

Figure 14.33 The image in the xy-plane of the area element $du\,dv$ in the uv-plane

Now $\overrightarrow{PQ} = dx\,\mathbf{i} + dy\,\mathbf{j}$, where

$$dx = \frac{\partial x}{\partial u}du + \frac{\partial x}{\partial v}dv \quad \text{and} \quad dy = \frac{\partial y}{\partial u}du + \frac{\partial y}{\partial v}dv.$$

However, $dv = 0$ along the v-curve PQ, so

$$\overrightarrow{PQ} = \frac{\partial x}{\partial u}du\,\mathbf{i} + \frac{\partial y}{\partial u}du\,\mathbf{j}.$$

Similarly,

$$\overrightarrow{PR} = \frac{\partial x}{\partial v}dv\,\mathbf{i} + \frac{\partial y}{\partial v}dv\,\mathbf{j}.$$

Hence,

$$dA = \left\| \begin{vmatrix} \mathbf{i} & \mathbf{j} & \mathbf{k} \\ \frac{\partial x}{\partial u}du & \frac{\partial y}{\partial u}du & 0 \\ \frac{\partial x}{\partial v}dv & \frac{\partial y}{\partial v}dv & 0 \end{vmatrix} \right\| = \left| \frac{\partial(x, y)}{\partial(u, v)} \right| du\,dv;$$

that is, the absolute value of the Jacobian $\partial(x, y)/\partial(u, v)$ is the ratio between corresponding area elements in the xy-plane and the uv-plane:

$$dA = dx\,dy = \left| \frac{\partial(x, y)}{\partial(u, v)} \right| du\,dv.$$

The following theorem summarizes the change of variables procedure for a double integral.

THEOREM 4

Change of variables formula for double integrals

Let $x = x(u, v)$, $y = y(u, v)$ be a one-to-one transformation from a domain S in the uv-plane onto a domain D in the xy-plane. Suppose that the functions x and y, and their first partial derivatives with respect to u and v, are continuous in S. If $f(x, y)$ is integrable on D, and if $g(u, v) = f(x(u, v), y(u, v))$, then g is integrable on S and

$$\iint_D f(x, y)\,dx\,dy = \iint_S g(u, v) \left| \frac{\partial(x, y)}{\partial(u, v)} \right| du\,dv.$$

Remark It is not necessary that S or D be closed or that the transformation be one-to-one on the boundary of S. The transformation to polar coordinates maps the rectangle $0 < r < 1$, $0 \le \theta < 2\pi$ one-to-one onto the punctured disk $0 < x^2 + y^2 < 1$ and, as in the first example in this section, we can transform an integral over the closed disk $x^2 + y^2 \le 1$ to one over the closed rectangle $0 \le r \le 1$, $0 \le \theta \le 2\pi$.

EXAMPLE 7 Use an appropriate change of variables to find the area of the elliptic disk E given by

$$\frac{x^2}{a^2}+\frac{y^2}{b^2}\le 1.$$

Solution Under the transformation $x = au$, $y = bv$, the elliptic disk E is the one-to-one image of the circular disk D given by $u^2+v^2 \le 1$. Assuming $a > 0$ and $b > 0$, we have

$$dx\,dy = \left|\frac{\partial(x,y)}{\partial(u,v)}\right| du\,dv = \left|\begin{vmatrix} a & 0 \\ 0 & b \end{vmatrix}\right| du\,dv = ab\,du\,dv.$$

Therefore, the area of E is given by

$$\iint_E 1\,dx\,dy = \iint_D ab\,du\,dv = ab \times (\text{area of } D) = \pi ab \text{ square units.}$$

It is often tempting to try to use the change of variable formula to transform the domain of a double integral into a rectangle so that iteration will be easy. As the following example shows, this usually involves defining the inverse transformation (u and v in terms of x and y). Remember that inverse transformations have reciprocal Jacobians.

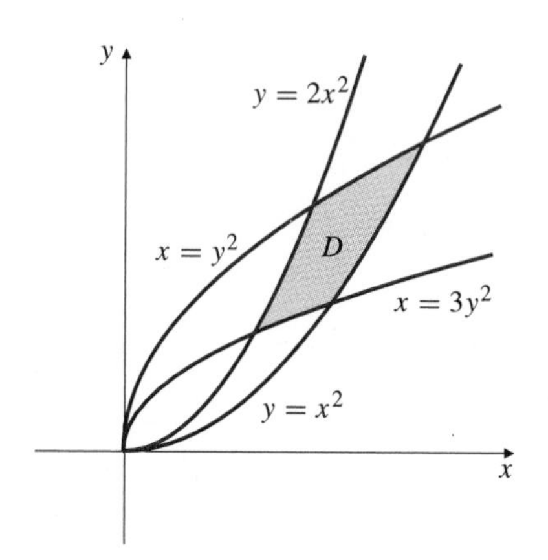

Figure 14.34 The region D of Example 8

EXAMPLE 8 Find the area of the finite plane region bounded by the four parabolas $y = x^2$, $y = 2x^2$, $x = y^2$, and $x = 3y^2$.

Solution The region, call it D, is sketched in Figure 14.34. Let

$$u = \frac{x^2}{y} \qquad \text{and} \qquad v = \frac{y^2}{x}.$$

Then the region D corresponds to the rectangle R in the uv-plane given by $\frac{1}{2} \le u \le 1$ and $\frac{1}{3} \le v \le 1$. (See Figure 14.35.) Since

$$\frac{\partial(u,v)}{\partial(x,y)} = \begin{vmatrix} 2x/y & -x^2/y^2 \\ -y^2/x^2 & 2y/x \end{vmatrix} = 4 - 1 = 3,$$

we have

$$\left|\frac{\partial(x,y)}{\partial(u,v)}\right| = \frac{1}{3}$$

and so the area of D is given by

$$\iint_D dx\,dy = \iint_R \frac{1}{3}\,du\,dv = \frac{1}{3}\times\frac{1}{2}\times\frac{2}{3} = \frac{1}{9} \text{ square units.}$$

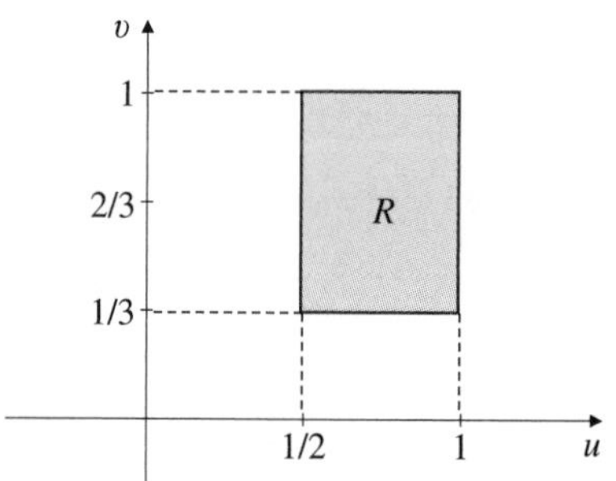

Figure 14.35 The transformed region R for Example 8

EXAMPLE 9 Evaluate $I = \displaystyle\iint_D \frac{y}{x}\,dx\,dy$, where D is the shaded region in Figure 14.36.

Figure 14.36 Domain D, Example 9

Solution We use the change of variables $u = x^2 + 4y^2$, $v = y/x$, so that the region R in the uv plane that corresponds to D is the rectangle $0 \le u \le 4$, $0 \le v \le 1$. (See Figure 14.37.) Since

$$\frac{\partial(u,v)}{\partial(x,y)} = \begin{vmatrix} 2x & 8y \\ -y/x^2 & 1/x \end{vmatrix} = 2 + 8\frac{y^2}{x^2} = 2 + 8v^2,$$

we have $\dfrac{\partial(x,y)}{\partial(u,v)} = \dfrac{1}{2+8v^2}$, and so

$$I = \iint_R \frac{v}{2+8v^2}\,du\,dv = \int_0^4 du \int_0^1 \frac{v}{2+8v^2}\,dv \qquad w = 2+8v^2, \quad dw = 16v\,dv$$

$$= \frac{4}{16}\int_2^1 0\frac{dw}{w} = \frac{1}{4}(\ln 10 - \ln 2) = \frac{1}{4}\ln 5.$$

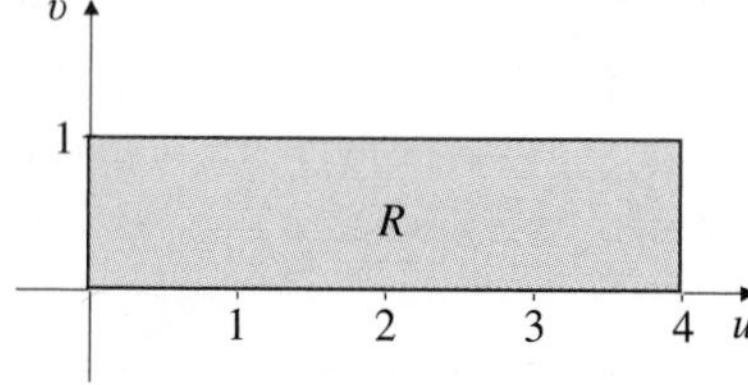

Figure 14.37 Region R, Example 9

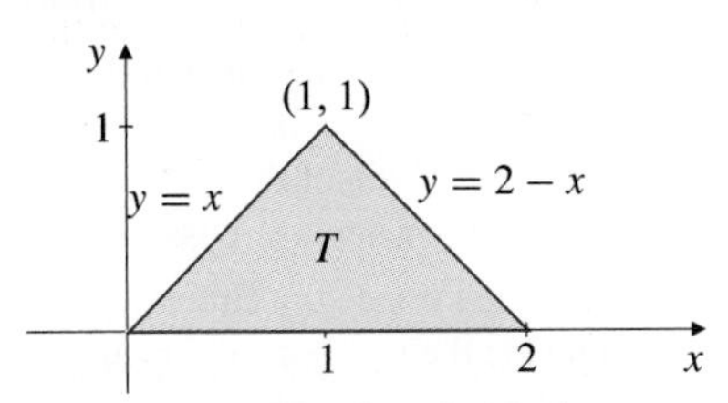

Figure 14.38 The domain T of Example 10

EXAMPLE 10 Evaluate

$$I = \iint_T (x+y)^3\, dx\, dy$$

over the triangle T with vertices $(0,0)$, $(1,1)$, and $(2,0)$.

Solution The triangle is shown in Figure 14.38. The transformation $u = y - x$, $v = y + x$ is linear so its image in the uv-plane is also a triangle R, this one with vertices $(0,0)$, $(0,2)$, and $(-2,2)$. (See Figure 14.39.) Since

$$\frac{\partial(u,v)}{\partial(x,y)} = \begin{vmatrix} -1 & 1 \\ 1 & 1 \end{vmatrix} = -2,$$

we have

$$dx\, dy = \left|\frac{\partial(x,y)}{\partial(u,v)}\right| du\, dv = \frac{1}{2}\, du\, dv$$

and we can calculate I as

$$I = \frac{1}{2}\iint_R v^3\, du\, dv = \frac{1}{2}\int_0^2 v^3\, dv \int_{-v}^0 du = \frac{1}{2}\int_0^2 v^4\, dv = \frac{16}{5}.$$

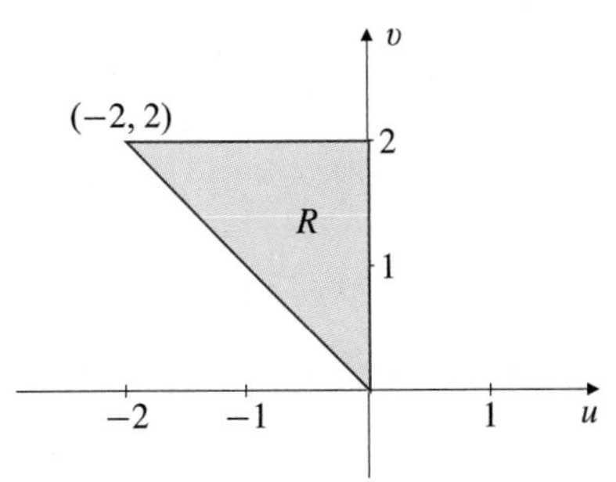

Figure 14.39 The transformed region R for Example 10

The following example shows what can happen if a transformation of the domain of a double integral is not one-to-one.

EXAMPLE 11 Let D be the square $0 \le x \le 1, 0 \le y \le 1$ in the xy-plane, and let S be the square $0 \le u \le 1, 0 \le v \le 1$ in the uv-plane. Show that the transformation

$$x = 4u - 4u^2, \qquad y = v$$

maps S onto D, and use it to transform the integral $I = \iint_D dx\, dy$. Compare the value of I with that of the transformed integral.

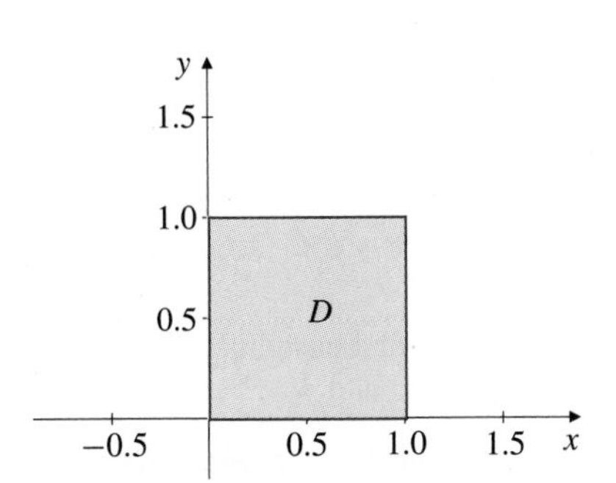

Figure 14.40 The square domain D of Example 11

Solution Since $x = 4u - 4u^2 = 1 - (1-2u)^2$, the minimum value of x on the interval $0 \le u \le 1$ is 0 (at $u = 0$ and $u = 1$), and the maximum value is 1 (at $u = \frac{1}{2}$). Therefore, $x = 4u - 4u^2$ maps the interval $0 \le u \le 1$ onto the interval $0 \le x \le 1$. Since $y = v$ clearly maps $0 \le v \le 1$ onto $0 \le y \le 1$, the given transformation maps S onto D. Since

$$dx\, dy = \left|\frac{\partial(x,y)}{\partial(u,v)}\right| du\, dv = \left|\begin{vmatrix} 4-8u & 0 \\ 0 & 1 \end{vmatrix}\right| du\, dv = |4-8u|\, du\, dv,$$

transforming I leads to the integral

$$J = \iint_S |4-8u|\, du\, dv = 4\int_0^1 dv \int_0^1 |1-2u|\, du = 8\int_0^{1/2} (1-2u)\, du = 2.$$

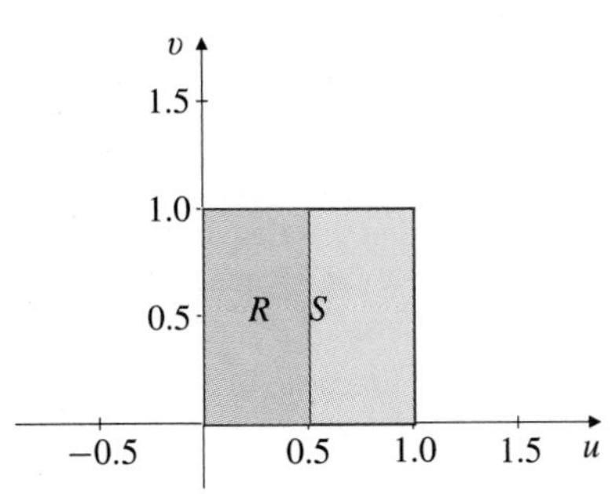

Figure 14.41 The square S and its left half, the rectangle R, for Example 11

However, $I = \iint_D dx\, dy$ = area of D = 1. The reason that $J \ne I$ is that the transformation is not one-to-one from S onto D; it actually maps S onto D twice. The rectangle R defined by $0 \le u \le \frac{1}{2}$ and $0 \le v \le 1$ (i.e., the left half of S) is mapped one-to-one onto D by the transformation, so the appropriate transformed integral is $\iint_R |4-8u|\, du\, dv$, which is equal to I.

EXERCISES 14.4

In Exercises 1–6, evaluate the given double integral over the disk D given by $x^2+y^2 \le a^2$, where $a>0$.

1. $\iint_D (x^2+y^2)\,dA$ **2.** $\iint_D \sqrt{x^2+y^2}\,dA$

3. $\iint_D \frac{1}{\sqrt{x^2+y^2}}\,dA$ **4.** $\iint_D |x|\,dA$

5. $\iint_D x^2\,dA$ **6.** $\iint_D x^2y^2\,dA$

In Exercises 7–10, evaluate the given double integral over the quarter-disk Q given by $x\ge 0$, $y\ge 0$, and $x^2+y^2\le a^2$, where $a>0$.

7. $\iint_Q y\,dA$ **8.** $\iint_Q (x+y)\,dA$

9. $\iint_Q e^{x^2+y^2}\,dA$ **10.** $\iint_Q \frac{2xy}{x^2+y^2}\,dA$

11. Evaluate $\iint_S (x+y)\,dA$, where S is the region in the first quadrant lying inside the disk $x^2+y^2\le a^2$ and under the line $y=\sqrt{3}x$.

12. Find $\iint_S x\,dA$, where S is the disk segment $x^2+y^2\le 2$, $x\ge 1$.

13. Evaluate $\iint_T (x^2+y^2)\,dA$, where T is the triangle with vertices $(0,0)$, $(1,0)$, and $(1,1)$.

14. Evaluate $\iint_{x^2+y^2\le 1} \ln(x^2+y^2)\,dA$.

15. Find the average distance from the origin to points in the disk $x^2+y^2\le a^2$.

16. Find the average value of $e^{-(x^2+y^2)}$ over the annular region $0<a\le\sqrt{x^2+y^2}\le b$.

17. For what values of k, and to what value, does the integral $\iint_{x^2+y^2\le 1}\frac{dA}{(x^2+y^2)^k}$ converge?

18. For what values of k, and to what value, does the integral $\iint_{\mathbb{R}^2}\frac{dA}{(1+x^2+y^2)^k}$ converge?

19. Evaluate $\iint_D xy\,dA$, where D is the plane region satisfying $x\ge 0$, $0\le y\le x$, and $x^2+y^2\le a^2$.

20. Evaluate $\iint_C y\,dA$, where C is the upper half of the cardioid disk $r\le 1+\cos\theta$.

21. Find the volume lying between the paraboloids $z=x^2+y^2$ and $3z=4-x^2-y^2$.

22. Find the volume lying inside both the sphere $x^2+y^2+z^2=a^2$ and the cylinder $x^2+y^2=ax$.

23. Find the volume lying inside both the sphere $x^2+y^2+z^2=2a^2$ and the cylinder $x^2+y^2=a^2$.

24. Find the volume of the region lying above the xy-plane, inside the cylinder $x^2+y^2=4$ and below the plane $z=x+y+4$.

25. Find the volume of the region lying inside all three of the circular cylinders $x^2+y^2=a^2$, $x^2+z^2=a^2$, and $y^2+z^2=a^2$. *Hint:* Make a good sketch of the first octant part of the region, and use symmetry whenever possible.

26. Find the volume of the region lying inside the circular cylinder $x^2+y^2=2y$ and inside the parabolic cylinder $z^2=y$.

27. Many points are chosen at random in the disk $x^2+y^2\le 1$. Find the approximate average value of the distance from these points to the nearest side of the smallest square that contains the disk.

28. Find the average value of x over the segment of the disk $x^2+y^2\le 4$ lying to the right of $x=1$. What is the centroid of the segment?

29. Find the volume enclosed by the ellipsoid

$$\frac{x^2}{a^2}+\frac{y^2}{b^2}+\frac{z^2}{c^2}=1.$$

30. Find the volume of the region in the first octant below the paraboloid

$$z=1-\frac{x^2}{a^2}-\frac{y^2}{b^2}.$$

Hint: Use the change of variables $x=au$, $y=bv$.

31. Evaluate $\iint_{|x|+|y|\le a} e^{x+y}\,dA$.

32. Find $\iint_P (x^2+y^2)\,dA$, where P is the parallelogram bounded by the lines $x+y=1$, $x+y=2$, $3x+4y=5$, and $3x+4y=6$.

33. Find the area of the region in the first quadrant bounded by the curves $xy=1$, $xy=4$, $y=x$, and $y=2x$.

34. Evaluate $\iint_R (x^2+y^2)\,dA$, where R is the region in the first quadrant bounded by $y=0$, $y=x$, $xy=1$, and $x^2-y^2=1$.

35. Let T be the triangle with vertices $(0,0)$, $(1,0)$, and $(0,1)$. Evaluate the integral $\iint_T e^{(y-x)/(y+x)}\,dA$,

(a) by transforming to polar coordinates, and

(b) by using the transformation $u=y-x$, $v=y+x$.

36. Use the method of Example 7 to find the area of the region inside the ellipse $4x^2+9y^2=36$ and above the line $2x+3y=6$.

37. (The error function) The error function, $\mathrm{Erf}(x)$, is defined for $x\ge 0$ by

$$\mathrm{Erf}(x)=\frac{2}{\sqrt{\pi}}\int_0^x e^{-t^2}\,dt.$$

Show that $\left(\mathrm{Erf}(x)\right)^2=\frac{4}{\pi}\int_0^{\pi/4}\left(1-e^{-x^2/\cos^2\theta}\right)d\theta.$

Hence deduce that $\mathrm{Erf}(x)\ge\sqrt{1-e^{-x^2}}$.

38. (The gamma and beta functions) The gamma function $\Gamma(x)$ and the beta function $B(x, y)$ are defined by

$$\Gamma(x) = \int_0^\infty t^{x-1}e^{-t}\,dt, \quad (x > 0),$$

$$B(x, y) = \int_0^1 t^{x-1}(1-t)^{y-1}\,dt, \quad (x > 0,\ y > 0).$$

The gamma function satisfies

$$\Gamma(x+1) = x\Gamma(x) \qquad \text{and}$$

$$\Gamma(n+1) = n!, \qquad (n = 0, 1, 2, \ldots).$$

Deduce the following further properties of these functions:

(a) $\Gamma(x) = 2\int_0^\infty s^{2x-1}e^{-s^2}\,ds, \qquad (x > 0),$

(b) $\Gamma\left(\frac{1}{2}\right) = \sqrt{\pi}, \qquad \Gamma\left(\frac{3}{2}\right) = \frac{1}{2}\sqrt{\pi},$

(c) If $x > 0$ and $y > 0$, then

$$B(x, y) = 2\int_0^{\pi/2} \cos^{2x-1}\theta \sin^{2y-1}\theta\,d\theta,$$

(d) $B(x, y) = \dfrac{\Gamma(x)\Gamma(y)}{\Gamma(x+y)}.$

14.5 Triple Integrals

Now that we have seen how to extend definite integration to two-dimensional domains, the extension to three (or more) dimensions is straightforward. For a bounded function $f(x, y, z)$ defined on a rectangular box B $(x_0 \le x \le x_1,\ y_0 \le y \le y_1,\ z_0 \le z \le z_1)$, the **triple integral** of f over B,

$$\iiint_B f(x, y, z)\,dV \qquad \text{or} \qquad \iiint_B f(x, y, z)\,dx\,dy\,dz,$$

can be defined as a suitable limit of Riemann sums corresponding to partitions of B into subboxes by planes parallel to each of the coordinate planes. We omit the details. Triple integrals over more general domains are defined by extending the function to be zero outside the domain and integrating over a rectangular box containing the domain.

Again, we remark that triple and other multiple integrals are often represented with a single integral sign, for example,

$$\int_R f(x, y, z)\,dx\,dy\,dz,$$

in scientific literature, and in Chapter 17 of this book.

All the properties of double integrals mentioned in Section 14.1 have analogues for triple integrals. In particular, a continuous function is integrable over a closed, bounded domain. If $f(x, y, z) = 1$ on the domain D, then the triple integral gives the volume of D:

$$\text{Volume of } D = \iiint_D dV.$$

The triple integral of a positive function $f(x, y, z)$ can be interpreted as the "hypervolume" (i.e., the four-dimensional volume) of a region in 4-space having the set D as its three-dimensional "base" and having its top on the hypersurface $w = f(x, y, z)$. This is not a particularly useful interpretation; many more useful ones arise in applications. For instance, if $\rho(x, y, z)$ represents the density (mass per unit volume) at position (x, y, z) in a substance occupying the domain D in 3-space, then the mass m of the solid is the "sum" of mass elements $dm = \rho(x, y, z)\,dV$ occupying volume elements dV:

$$\text{mass} = \iiint_D \rho(x, y, z)\,dV.$$

Some triple integrals can be evaluated by inspection, using symmetry and known volumes.

EXAMPLE 1 Evaluate

$$\iiint_{x^2+y^2+z^2 \le a^2} (2 + x - \sin z)\,dV.$$

Solution The domain of integration is the ball of radius a centred at the origin. The integral of 2 over this ball is twice the ball's volume, that is, $8\pi a^3/3$. The integrals of x and $\sin z$ over the ball are both zero, since both functions are odd in one of the variables and the domain is symmetric about each coordinate plane. (For instance, for every volume element dV in the half of the ball where $x > 0$, there is a corresponding element in the other half where x has the same size but the opposite sign. The contributions from these two elements cancel one another.) Thus,

$$\iiint_{x^2+y^2+z^2 \le a^2} (2 + x - \sin z)\, dV = \frac{8}{3}\pi a^3 + 0 + 0 = \frac{8}{3}\pi a^3.$$

Most triple integrals are evaluated by an iteration procedure similar to that used for double integrals. We slice the domain D with a plane parallel to one of the coordinate planes, double integrate the function with respect to two variables over that slice, and then integrate the result with respect to the remaining variable. Some examples should clarify the procedure.

EXAMPLE 2 Let B be the rectangular box $0 \le x \le a$, $0 \le y \le b$, $0 \le z \le c$. Evaluate

$$I = \iiint_B (xy^2 + z^3)\, dV.$$

Solution As indicated in Figure 14.42(a), we will slice with planes perpendicular to the z-axis, so the z integral will be outermost in the iteration. The slices are rectangles, so the double integrals over them can be immediately iterated also. We do it with the y integral outer and the x integral inner, as suggested by the line shown in the slice.

$$\begin{aligned}
I &= \int_0^c dz \int_0^b dy \int_0^a (xy^2 + z^3)\, dx \\
&= \int_0^c dz \int_0^b dy \left(\frac{x^2y^2}{2} + xz^3 \right)\bigg|_{x=0}^{x=a} \\
&= \int_0^c dz \int_0^b \left(\frac{a^2y^2}{2} + az^3 \right) dy \\
&= \int_0^c dz \left(\frac{a^2y^3}{6} + ayz^3 \right)\bigg|_{y=0}^{y=b} \\
&= \int_0^c \left(\frac{a^2b^3}{6} + abz^3 \right) dz \\
&= \left(\frac{a^2b^3z}{6} + \frac{abz^4}{4} \right)\bigg|_{z=0}^{z=c} = \frac{a^2b^3c}{6} + \frac{abc^4}{4}.
\end{aligned}$$

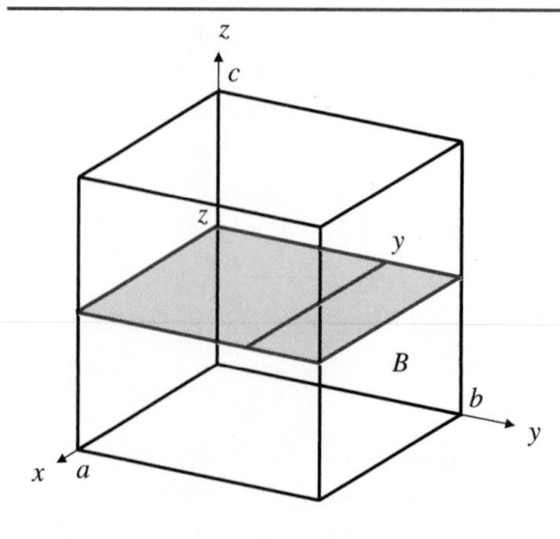

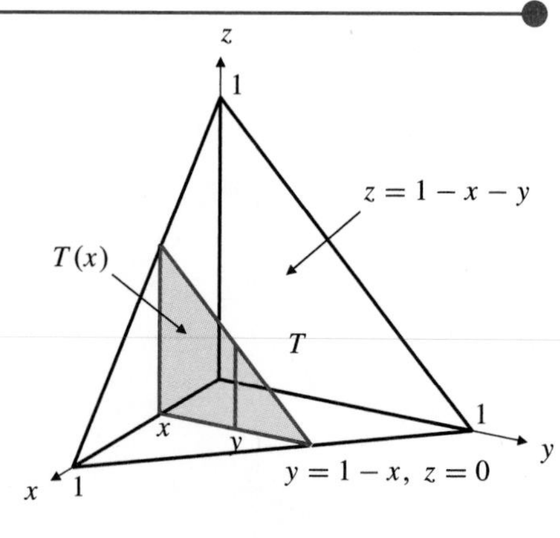

Figure 14.42
(a) The iteration in Example 2
(b) The iteration in Example 3

EXAMPLE 3 If T is the tetrahedron with vertices $(0, 0, 0)$, $(1, 0, 0)$, $(0, 1, 0)$, and $(0, 0, 1)$, evaluate $I = \iiint_T y\,dV$.

Solution The tetrahedron is shown in Figure 14.42(b). The plane slice in the plane normal to the x-axis at position x is the triangle $T(x)$ shown in that figure; x is constant and y and z are variables in the slice. The double integral of y over $T(x)$ is a function of x. We evaluate it by integrating first in the z direction and then in the y direction as suggested by the vertical line shown in the slice:

$$\begin{aligned}\iint_{T(x)} y\,dA &= \int_0^{1-x} dy \int_0^{1-x-y} y\,dz\\ &= \int_0^{1-x} y(1-x-y)\,dy\\ &= \left((1-x)\frac{y^2}{2} - \frac{y^3}{3}\right)\Bigg|_0^{1-x} = \frac{1}{6}(1-x)^3.\end{aligned}$$

The value of the triple integral I is the integral of this expression with respect to the remaining variable x, to sum the contributions from all such slices between $x = 0$ and $x = 1$:

$$I = \int_0^1 \frac{1}{6}(1-x)^3\,dx = -\frac{1}{24}(1-x)^4\Bigg|_0^1 = \frac{1}{24}.$$

In the above solution we carried out the iteration in two steps in order to show the procedure clearly. In practice, triple integrals are iterated in one step, with no explicit mention made of the double integral over the slice. Thus, using the iteration suggested by Figure 14.42(b), we would immediately write

$$I = \int_0^1 dx \int_0^{1-x} dy \int_0^{1-x-y} y\,dz.$$

The evaluation proceeds as above, starting with the right (i.e., inner) integral, followed by the middle integral and then the left (outer) integral. The triple integral represents the "sum" of elements $y\,dV$ over the three-dimensional region T. The above iteration corresponds to "summing" (i.e., integrating) first along a vertical line (the z integral), then summing these one-dimensional sums in the y direction to get the double sum of all elements in the plane slice, and finally summing these double sums in the x direction to add up the contributions from all the slices. The iteration can be carried out in other directions; there are six possible iterations corresponding to different orders of doing the x, y, and z integrals. The other five are

$$I = \int_0^1 dx \int_0^{1-x} dz \int_0^{1-x-z} y\,dy,$$

$$I = \int_0^1 dy \int_0^{1-y} dx \int_0^{1-x-y} y\,dz,$$

$$I = \int_0^1 dy \int_0^{1-y} dz \int_0^{1-y-z} y\,dx,$$

$$I = \int_0^1 dz \int_0^{1-z} dx \int_0^{1-x-z} y\,dy,$$

$$I = \int_0^1 dz \int_0^{1-z} dy \int_0^{1-y-z} y\,dx.$$

You should verify these by drawing diagrams analogous to Figure 14.42(b). Of course, all six iterations give the same result.

It is sometimes difficult to visualize the region of 3-space over which a given triple integral is taken. In such situations try to determine the *projection* of that region on one or other of the coordinate planes. For instance, if a region R is bounded by two surfaces with given equations, combining these equations to eliminate one variable will yield the equation of a cylinder (not necessarily circular) with axis parallel to the axis of the eliminated variable. This cylinder will then determine the projection of R onto the coordinate plane perpendicular to that axis. The following example illustrates the use of this technique to find a volume bounded by two surfaces. The volume is expressed as a triple integral with unit integrand.

EXAMPLE 4 Find the volume of the region R lying below the plane $z = 3 - 2y$ and above the paraboloid $z = x^2 + y^2$.

Solution The region R is shown in Figure 14.43. The two surfaces bounding R intersect on the vertical cylinder $x^2 + y^2 = 3 - 2y$, or $x^2 + (y+1)^2 = 4$. If D is the circular disk in which this cylinder intersects the xy-plane, then partial iteration gives

$$V = \iiint_R dV = \iint_D dx\,dy \int_{x^2+y^2}^{3-2y} dz.$$

Figure 14.43 shows a slice of R corresponding to a further iteration of the double integral over D:

$$V = \int_{-3}^{1} dy \int_{-\sqrt{3-2y-y^2}}^{\sqrt{3-2y-y^2}} dx \int_{x^2+y^2}^{3-2y} dz,$$

but there is an easier way to iterate the double integral. Since D is a circular disk of radius 2 and centre $(0, -1)$, we can use polar coordinates with centre at that point (i.e., $x = r\cos\theta$, $y = -1 + r\sin\theta$). Thus,

$$\begin{aligned} V &= \iint_D (3 - 2y - x^2 - y^2)\,dx\,dy \\ &= \iint_D \left(4 - x^2 - (y+1)^2\right) dx\,dy \\ &= \int_0^{2\pi} d\theta \int_0^2 (4 - r^2) r\,dr = 2\pi \left(2r^2 - \frac{r^4}{4}\right)\Bigg|_0^2 = 8\pi \text{ cubic units.} \end{aligned}$$

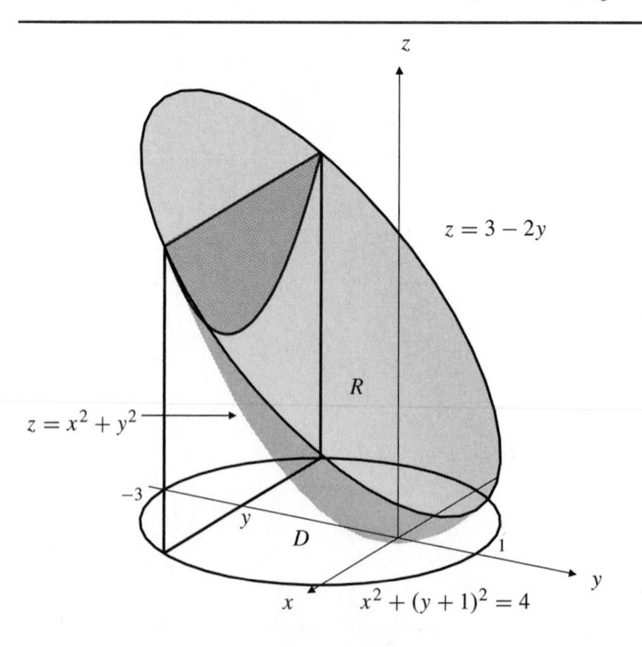

Figure 14.43 The volume above a paraboloid and under a slanting plane

As was the case for double integrals, it is sometimes necessary to reiterate a given iterated integral so that the integrations are performed in a different order. This task is most easily accomplished if we can translate the given iteration into a sketch of the region of integration. The ability to deduce the shape of the region from the limits in the iterated integral is a skill that you can acquire with a little practice. You should first determine the projection of the region on a coordinate plane, namely, the plane of the two variables in the outer integrals of the given iteration.

It is also possible to reiterate an iterated integral in a different order by manipulating the limits of integration algebraically. We will illustrate both approaches (graphical and algebraic) in the following examples.

EXAMPLE 5 Express the iterated integral

$$I = \int_0^1 dy \int_y^1 dz \int_0^z f(x, y, z)\, dx$$

as a triple integral, and sketch the region over which it is taken. Reiterate the integral in such a way that the integrations are performed in the order: first y, then z, then x (i.e., the opposite order to the given iteration).

Solution We express I as an uniterated triple integral:

$$I = \iiint_R f(x, y, z)\, dV.$$

The outer integral in the given iteration shows that the region R lies between the planes $y = 0$ and $y = 1$. For each such value of y, z must lie between y and 1. Therefore, R lies below the plane $z = 1$ and above the plane $z = y$, and the projection of R onto the yz-plane is the triangle with vertices $(0, 0, 0)$, $(0, 0, 1)$, and $(0, 1, 1)$. Through any point $(0, y, z)$ in this triangle, a line parallel to the x-axis intersects R between $x = 0$ and $x = z$. Thus, the solid is bounded by the five planes $x = 0$, $y = 0$, $z = 1$, $y = z$, and $z = x$. It is sketched in Figure 14.44(a), with slice and line corresponding to the given iteration.

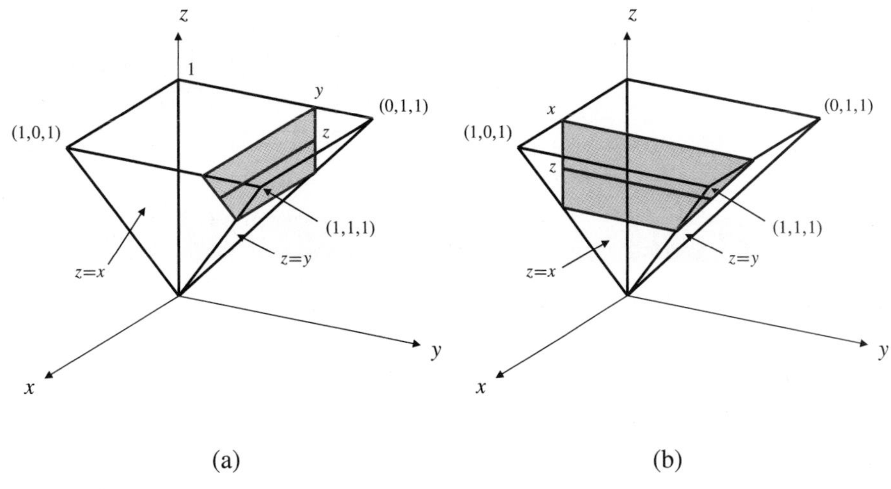

Figure 14.44
(a) The solid region for the triple integral in Example 5 sliced corresponding to the given iteration
(b) The same solid sliced to conform to the desired iteration

The required iteration corresponds to the slice and line shown in Figure 14.44(b). Therefore, it is

$$I = \int_0^1 dx \int_x^1 dz \int_0^z f(x, y, z)\, dy.$$

EXAMPLE 6 Use algebra to write an iteration of the integral

$$I = \int_0^1 dx \int_x^1 dy \int_x^y f(x, y, z)\, dz$$

with the order of integrations reversed.

Solution From the given iteration we can write three sets of inequalities satisfied by the outer variable x, the middle variable y, and the inner variable z. We write these in order as follows:

$0 \le x \le 1$	inequalities for x
$x \le y \le 1$	inequalities for y
$x \le z \le y$	inequalities for z.

Note that the limits for each variable can be constant or can depend only on variables whose inequalities are on lines above the line for that variable. (In this case, the limits for x must both be constant, those for y can depend on x, and those for z can depend on both x and y.) This is a requirement for iterated integrals; outer integrals cannot depend on the variables of integration of the inner integrals.

We want to construct an equivalent set of inequalities with those for z on the top line, then those for y, then those for x on the bottom line. The limits for z must be constants. From the inequalities above we determine that $0 \le x \le z$ and $z \le y \le 1$. Thus z must satisfy $0 \le z \le 1$. The inequalities for y can depend on z. Since $z \le y$ and $y \le 1$, we have $z \le y \le 1$. Finally, the limits for x can depend on both y and z. We have $0 \le x$, $x \le y$, and $x \le z$. Since we have already determined that $z \le y$, we must have $0 \le x \le z$. Thus, the revised inequalities are

$0 \le z \le 1$	inequalities for z
$z \le y \le 1$	inequalities for y
$0 \le x \le z$	inequalities for x

and the required iteration is

$$I = \int_0^1 dz \int_z^1 dy \int_0^z f(x, y, z)\, dx.$$

EXERCISES 14.5

In Exercises 1–12, evaluate the triple integrals over the indicated region. Be alert for simplifications and auspicious orders of iteration.

1. $\iiint_R (1 + 2x - 3y)\, dV$, over the box $-a \le x \le a$, $-b \le y \le b$, $-c \le z \le c$

2. $\iiint_B xyz\, dV$, over the box B given by $0 \le x \le 1$, $-2 \le y \le 0$, $1 \le z \le 4$

3. $\iiint_D (3 + 2xy)\, dV$, over the solid hemispherical dome D given by $x^2 + y^2 + z^2 \le 4$ and $z \ge 0$

4. $\iiint_R x\, dV$, over the tetrahedron bounded by the coordinate planes and the plane $\dfrac{x}{a} + \dfrac{y}{b} + \dfrac{z}{c} = 1$

5. $\iiint_R (x^2 + y^2)\, dV$, over the cube $0 \le x, y, z \le 1$

6. $\iiint_R (x^2 + y^2 + z^2)\, dV$, over the cube of Exercise 5

7. $\iiint_R (xy + z^2)\, dV$, over the set $0 \le z \le 1 - |x| - |y|$

8. $\iiint_R yz^2 e^{-xyz}\, dV$, over the cube $0 \le x, y, z \le 1$

9. $\iiint_R \sin(\pi y^3)\,dV$, over the pyramid with vertices (0, 0, 0), (0, 1, 0), (1, 1, 0), (1, 1, 1), and (0, 1, 1)

10. $\iiint_R y\,dV$, over that part of the cube $0 \le x, y, z \le 1$ lying above the plane $y + z = 1$ and below the plane $x + y + z = 2$

11. $\iiint_R \dfrac{1}{(x+y+z)^3}\,dV$, over the region bounded by the six planes $z = 1$, $z = 2$, $y = 0$, $y = z$, $x = 0$, and $x = y + z$

12. $\iiint_R \cos x \cos y \cos z\,dV$, over the tetrahedron defined by $x \ge 0$, $y \ge 0$, $z \ge 0$, and $x + y + z \le \pi$

13. Evaluate $\iiint_{\mathbb{R}^3} e^{-x^2-2y^2-3z^2}\,dV$. *Hint:* Use the result of Example 4 of Section 14.4.

14. Find the volume of the region lying inside the cylinder $x^2 + 4y^2 = 4$, above the xy-plane, and below the plane $z = 2 + x$.

15. Find $\iiint_T x\,dV$, where T is the tetrahedron bounded by the planes $x = 1$, $y = 1$, $z = 1$, and $x + y + z = 2$.

16. Sketch the region R in the first octant of 3-space that has finite volume and is bounded by the surfaces $x = 0$, $z = 0$, $x + y = 1$, and $z = y^2$. Write six different iterations of the triple integral of $f(x, y, z)$ over R.

In Exercises 17–20, express the given iterated integral as a triple integral and sketch the region over which it is taken. Reiterate the integral, so that the outermost integral is with respect to x and the innermost is with respect to z.

17. $\displaystyle\int_0^1 dz \int_0^{1-z} dy \int_0^1 f(x, y, z)\,dx$

18. $\displaystyle\int_0^1 dz \int_z^1 dy \int_0^y f(x, y, z)\,dx$

19. $\displaystyle\int_0^1 dz \int_z^1 dx \int_0^{x-z} f(x, y, z)\,dy$

20. $\displaystyle\int_0^1 dy \int_0^{\sqrt{1-y^2}} dz \int_{y^2+z^2}^1 f(x, y, z)\,dx$

21. Repeat Exercise 17 using the method of Example 6.

22. Repeat Exercise 18 using the method of Example 6.

23. Repeat Exercise 19 using the method of Example 6.

24. Repeat Exercise 20 using the method of Example 6.

25. Rework Example 5 using the method of Example 6.

26. Rework Example 6 using the method of Example 5.

In Exercises 27–28, evaluate the given iterated integral by reiterating it in a different order. (You will need to make a good sketch of the region.)

27. $\displaystyle\int_0^1 dz \int_z^1 dx \int_0^x e^{x^3}\,dy$

28. $\displaystyle\int_0^1 dx \int_0^{1-x} dy \int_y^1 \frac{\sin(\pi z)}{z(2-z)}\,dz$

29. Define the average value of an integrable function $f(x, y, z)$ over a region R of 3-space. Find the average value of $x^2 + y^2 + z^2$ over the cube $0 \le x \le 1$, $0 \le y \le 1$, $0 \le z \le 1$.

30. State a Mean-Value Theorem for triple integrals analogous to Theorem 3 of Section 14.3. Use it to prove that if $f(x, y, z)$ is continuous near the point (a, b, c) and if $B_\epsilon(a, b, c)$ is the ball of radius ϵ centred at (a, b, c), then

$$\lim_{\epsilon \to 0} \frac{3}{4\pi\epsilon^3} \iiint_{B_\epsilon(a,b,c)} f(x, y, z)\,dV = f(a, b, c).$$

14.6 Change of Variables in Triple Integrals

The change of variables formula for a double integral extends to triple (and higher-order) integrals. Consider the transformation

$$x = x(u, v, w),$$
$$y = y(u, v, w),$$
$$z = z(u, v, w),$$

where x, y, and z have continuous first partial derivatives with respect to u, v, and w. Near any point where the Jacobian $\partial(x, y, z)/\partial(u, v, w)$ is nonzero, the transformation scales volume elements according to the formula

$$dV = dx\,dy\,dz = \left|\frac{\partial(x, y, z)}{\partial(u, v, w)}\right| du\,dv\,dw.$$

Thus, if the transformation is one-to-one from a domain S in uvw-space onto a domain D in xyz-space, and if

$$g(u, v, w) = f\big(x(u, v, w), y(u, v, w), z(u, v, w)\big),$$

then

$$\iiint_D f(x, y, z)\,dx\,dy\,dz = \iiint_S g(u, v, w) \left|\frac{\partial(x, y, z)}{\partial(u, v, w)}\right| du\,dv\,dw.$$

The proof is similar to that of the two-dimensional case given in Section 14.4. See Exercise 21 at the end of this section.

EXAMPLE 1 Under the change of variables $x = au$, $y = bv$, $z = cw$, where $a, b, c > 0$, the solid ellipsoid E given by

$$\frac{x^2}{a^2} + \frac{y^2}{b^2} + \frac{z^2}{c^2} \le 1$$

becomes the ball B given by $u^2 + v^2 + w^2 \le 1$. The Jacobian of this transformation is

$$\frac{\partial(x, y, z)}{\partial(u, v, w)} = \begin{vmatrix} a & 0 & 0 \\ 0 & b & 0 \\ 0 & 0 & c \end{vmatrix} = abc,$$

so the volume of the ellipsoid is given by

$$\begin{aligned}\text{Volume of } E &= \iiint_E dx\,dy\,dz \\ &= \iiint_B abc\,du\,dv\,dw = abc \times (\text{Volume of } B) \\ &= \frac{4}{3}\pi abc \text{ cubic units.}\end{aligned}$$

Cylindrical Coordinates

In Section 10.6 we introduced the system of cylindrical coordinates r, θ, z in 3-space, related to Cartesian coordinates by the transformation

$$x = r\cos\theta, \quad y = r\sin\theta, \quad z = z.$$

The geometric significance of these coordinates are shown in Figure 14.45, and the coordinate surfaces are illustrated in Figure 14.46.

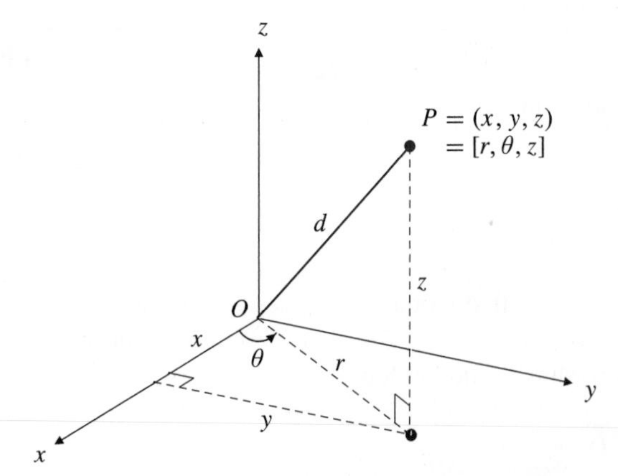

Figure 14.45 The cylindrical coordinates of a point

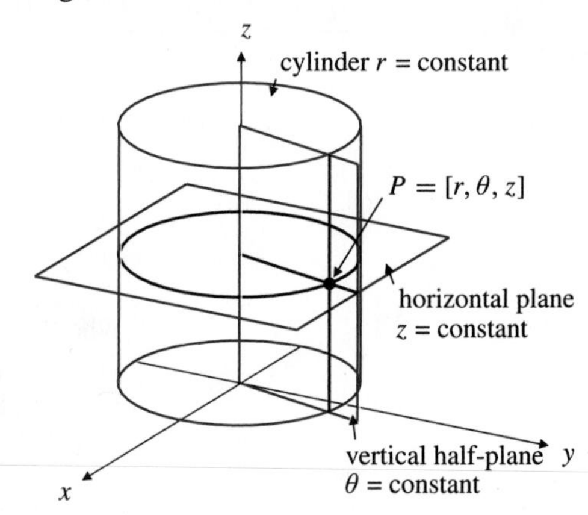

Figure 14.46 The coordinate surfaces for cylindrical coordinates

As noted previously, cylindrical coordinates lend themselves to representing domains that are bounded by such surfaces and, in general, to problems with axial symmetry (around the z-axis).

The **volume element in cylindrical coordinates** is

$$dV = r\,dr\,d\theta\,dz,$$

which is easily seen by examining the infinitesimal "box" bounded by the coordinate surfaces corresponding to values $r, r+dr, \theta, \theta+d\theta, z$, and $z+dz$ (see Figure 14.47) or by calculating the Jacobian

$$\frac{\partial(x,y,z)}{\partial(r,\theta,z)} = \begin{vmatrix} \cos\theta & -r\sin\theta & 0 \\ \sin\theta & r\cos\theta & 0 \\ 0 & 0 & 1 \end{vmatrix} = r.$$

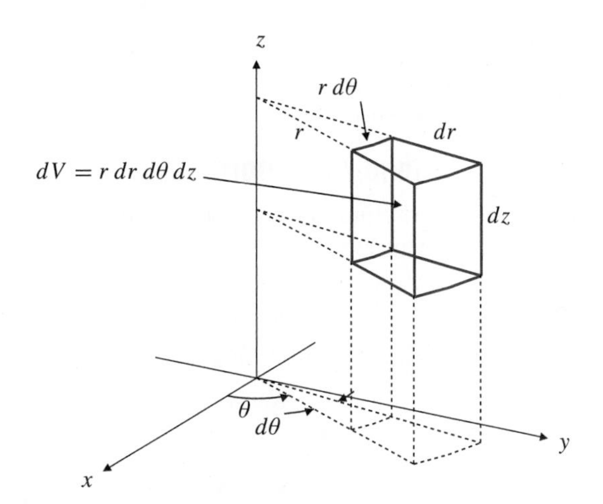

Figure 14.47 The volume element in cylindrical coordinates

EXAMPLE 2 Evaluate $\iiint_D (x^2+y^2)\,dV$ over the first octant region bounded by the cylinders $x^2+y^2=1$ and $x^2+y^2=4$ and the planes $z=0$, $z=1$, $x=0$, and $x=y$.

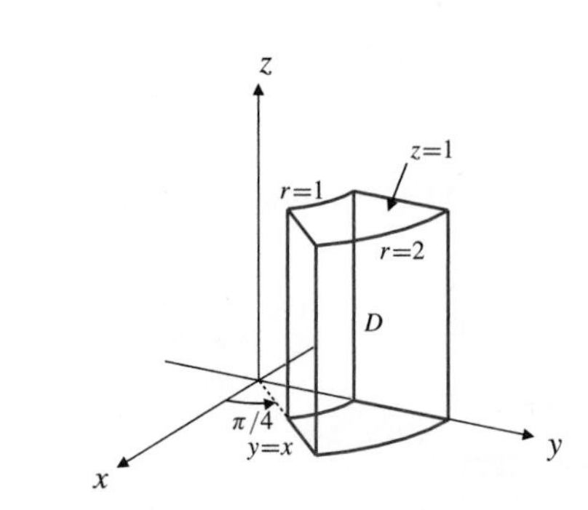

Figure 14.48

Solution In terms of cylindrical coordinates, the region is bounded by $r=1$, $r=2$, $\theta=\pi/4$, $\theta=\pi/2$, $z=0$, and $z=1$. (See Figure 14.48. It is a rectangular coordinate box in $r\theta z$-space.) Since the integrand is $x^2+y^2=r^2$, the integral is

$$\begin{aligned}\iiint_D (x^2+y^2)\,dV &= \int_0^1 dz \int_{\pi/4}^{\pi/2} d\theta \int_1^2 r^2\,r\,dr \\ &= (1-0)\left(\frac{\pi}{2}-\frac{\pi}{4}\right)\left(\frac{2^4}{4}-\frac{1^4}{4}\right) = \frac{15}{16}\pi.\end{aligned}$$

This integral would have been much more difficult to evaluate using Cartesian coordinates.

EXAMPLE 3 Use a triple integral to find the volume of the solid region inside the sphere $x^2+y^2+z^2=6$ and above the paraboloid $z=x^2+y^2$.

Solution One-quarter of the required volume lies in the first octant. (See region R in Figure 14.49.) The two surfaces intersect on the vertical cylinder

$$6-x^2-y^2=z^2=(x^2+y^2)^2,$$

or, in terms of cylindrical coordinates, $6-r^2=r^4$, that is,

$$\begin{aligned}&r^4+r^2-6=0\\&(r^2+3)(r^2-2)=0.\end{aligned}$$

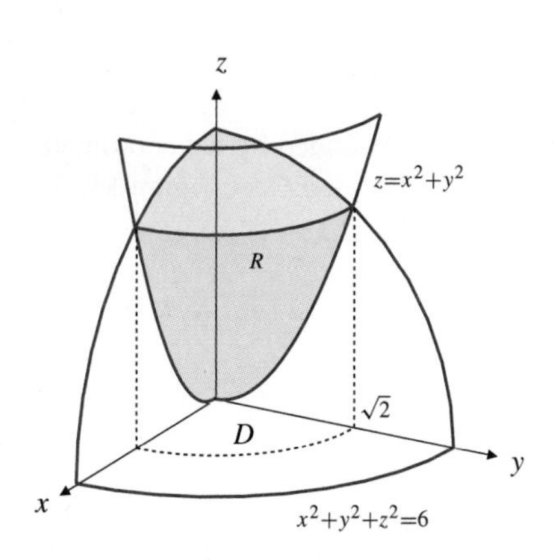

Figure 14.49 This figure shows one-quarter of the solid region R and its projection, one-quarter of the disk D in the xy-plane for Example 3

The only relevant solution to this equation is $r = \sqrt{2}$. Thus, the required volume lies above the disk D of radius $\sqrt{2}$ centred at the origin in the xy-plane. The total volume V of the region is

$$\begin{aligned} V = \iiint_R dV &= \int_0^{2\pi} d\theta \int_0^{\sqrt{2}} r\,dr \int_{r^2}^{\sqrt{6-r^2}} dz \\ &= 2\pi \int_0^{\sqrt{2}} \left(r\sqrt{6-r^2} - r^3\right) dr \\ &= 2\pi \left[-\frac{1}{3}(6-r^2)^{3/2} - \frac{r^4}{4}\right]\Bigg|_0^{\sqrt{2}} \\ &= 2\pi \left[\frac{6\sqrt{6}}{3} - \frac{8}{3} - 1\right] = \frac{2\pi}{3}(6\sqrt{6} - 11) \text{ cubic units.} \end{aligned}$$

Spherical Coordinates

Also introduced in Section 10.6 is the system of spherical coordinates related to Cartesian coordinates x, y, z, and cylindrical coordinates r, θ, z by the equations

$$\begin{aligned} x &= R\sin\phi\cos\theta \\ y &= R\sin\phi\sin\theta \\ z &= R\cos\phi, \end{aligned}$$

$$R^2 = x^2 + y^2 + z^2 = r^2 + z^2,$$

$$r = \sqrt{x^2 + y^2} = R\sin\phi,$$

$$\tan\phi = \frac{r}{z} = \frac{\sqrt{x^2+y^2}}{z} \quad \text{and} \quad \tan\theta = \frac{y}{x}.$$

These relationships are illustraded in Figure 14.50, and the coordinate surfaces in spherical coordinates are illustrated in Figure 14.51.

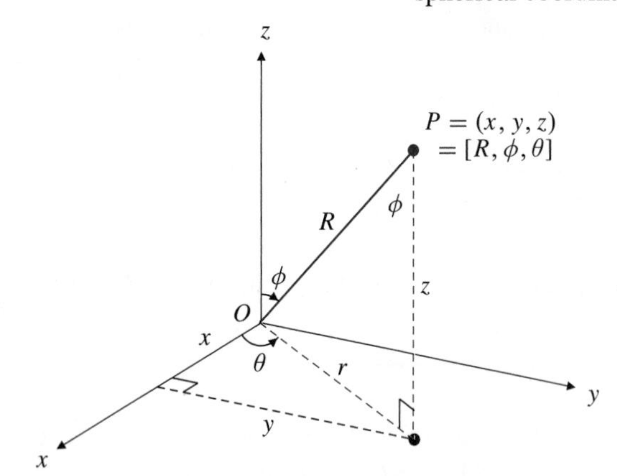

Figure 14.50 The spherical coordinates of a point

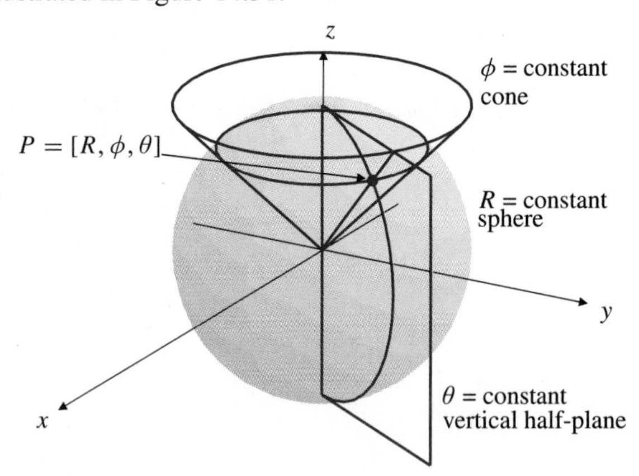

Figure 14.51 The coordinate surfaces for spherical coordinates

The **volume element in spherical coordinates** is

$$dV = R^2 \sin\phi \, dR \, d\phi \, d\theta.$$

To see this, observe that the infinitesimal coordinate box bounded by the coordinate surfaces corresponding to values R, $R + dR$, ϕ, $\phi + d\phi$, θ, and $\theta + d\theta$ has dimensions dR, $R\,d\phi$, and $R\sin\phi\,d\theta$. (See Figure 14.52.) Alternatively, the Jacobian of the transformation can be calculated:

$$\begin{aligned}\frac{\partial(x,y,z)}{\partial(R,\phi,\theta)} &= \begin{vmatrix} \sin\phi\cos\theta & R\cos\phi\cos\theta & -R\sin\phi\sin\theta \\ \sin\phi\sin\theta & R\cos\phi\sin\theta & R\sin\phi\cos\theta \\ \cos\phi & -R\sin\phi & 0 \end{vmatrix} \\ &= \cos\phi \begin{vmatrix} R\cos\phi\cos\theta & -R\sin\phi\sin\theta \\ R\cos\phi\sin\theta & R\sin\phi\cos\theta \end{vmatrix} \\ &\quad + R\sin\phi \begin{vmatrix} \sin\phi\cos\theta & -R\sin\phi\sin\theta \\ \sin\phi\sin\theta & R\sin\phi\cos\theta \end{vmatrix} \\ &= \cos\phi(R^2\sin\phi\cos\phi) + R\sin\phi(R\sin^2\phi) \\ &= R^2\sin\phi. \end{aligned}$$

Spherical coordinates are suited to problems involving spherical symmetry and, in particular, to regions bounded by spheres centred at the origin, circular cones with axes along the z-axis, and vertical planes containing the z-axis.

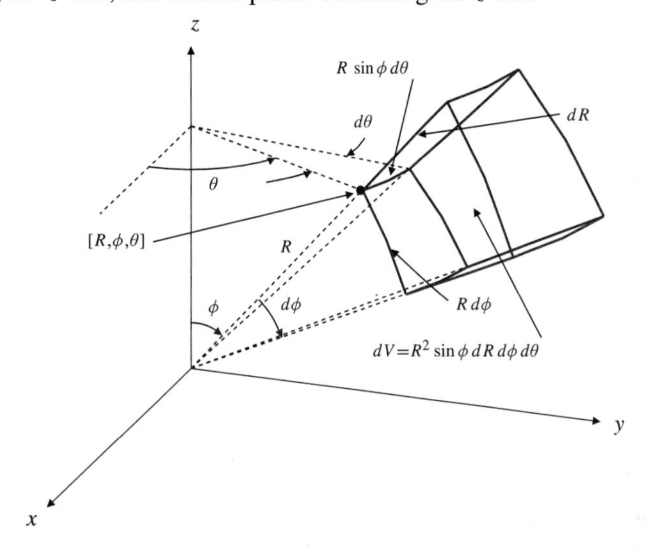

Figure 14.52 The volume element in spherical coordinates

EXAMPLE 4 A solid half-ball H of radius a has density ρ (mass per unit volume) depending on the distance R from the centre of the base disk. The density is given by $\rho = k(2a - R)$, where k is a constant. Find the mass of the half-ball.

Solution Choosing coordinates with origin at the centre of the base, and so that the half-ball lies above the xy-plane, we calculate the mass m as follows:

$$\begin{aligned} m &= \iiint_H k(2a - R)\,dV = \iiint_H k(2a - R)\,R^2\sin\phi\,dR\,d\phi\,d\theta \\ &= k\int_0^{2\pi} d\theta \int_0^{\pi/2} \sin\phi\,d\phi \int_0^a (2a - R)\,R^2\,dR \\ &= 2k\pi \times 1 \times \left(\frac{2a}{3}R^3 - \frac{1}{4}R^4\right)\Bigg|_0^a = \frac{5}{6}\pi k a^4 \text{ units.} \end{aligned}$$

Remark In the above example, both the integrand and the region of integration exhibited spherical symmetry, so the choice of spherical coordinates to carry out the integration was most appropriate. The mass could have been evaluated in cylindrical coordinates. The iteration in that system is

$$m = \int_0^{2\pi} d\theta \int_0^a r\,dr \int_0^{\sqrt{a^2-r^2}} k\left(2a - \sqrt{r^2+z^2}\right)dz$$

and is difficult to evaluate. It is even more difficult in Cartesian coordinates:

$$m = 4\int_0^a dx \int_0^{\sqrt{a^2-x^2}} dy \int_0^{\sqrt{a^2-x^2-y^2}} k\left(2a - \sqrt{x^2+y^2+z^2}\right) dz.$$

The choice of coordinate system can greatly affect the difficulty of computation of a multiple integral.

Many problems will have elements of spherical and axial symmetry. In such cases it may not be clear whether it would be better to use spherical or cylindrical coordinates. In such doubtful cases the integrand is usually the best guide. Use cylindrical or spherical coordinates according to whether the integrand involves x^2+y^2 or $x^2+y^2+z^2$.

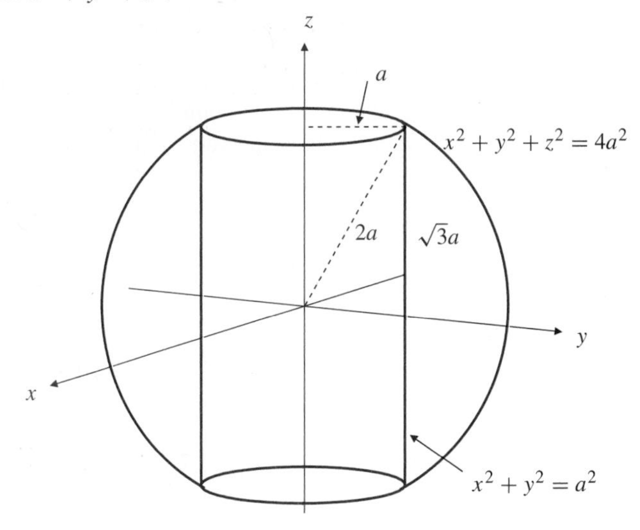

Figure 14.53 A solid ball with a cylindrical hole through it

EXAMPLE 5 The moment of inertia about the z-axis of a solid of density ρ occupying the region R is given by the integral

$$I = \iiint_R (x^2+y^2)\rho\, dV.$$

(See Section 14.7.) Calculate that moment of inertia for a solid of unit density occupying the region inside the sphere $x^2+y^2+z^2 = 4a^2$ and outside the cylinder $x^2+y^2 = a^2$.

Solution See Figure 14.53. In terms of spherical coordinates the required moment of inertia is

$$I = 2\int_0^{2\pi} d\theta \int_{\pi/6}^{\pi/2} \sin\phi\, d\phi \int_{a/\sin\phi}^{2a} R^2 \sin^2\phi\, R^2\, dR.$$

In terms of cylindrical coordinates it is

$$I = 2\int_0^{2\pi} d\theta \int_a^{2a} r\, dr \int_0^{\sqrt{4a^2-r^2}} r^2\, dz.$$

The latter formula looks somewhat easier to evaluate. We continue with it. Evaluating the θ and z integrals, we get

$$I = 4\pi \int_a^{2a} r^3\sqrt{4a^2-r^2}\, dr.$$

Making the substitution $u = 4a^2 - r^2$, $du = -2r\, dr$, we obtain

$$I = 2\pi \int_0^{3a^2} (4a^2-u)\sqrt{u}\, du = 2\pi\left(4a^2\frac{u^{3/2}}{3/2} - \frac{u^{5/2}}{5/2}\right)\Bigg|_0^{3a^2} = \frac{44}{5}\sqrt{3}\pi a^5.$$

EXERCISES 14.6

In Exercises 1–9, find the volumes of the indicated regions.

1. Inside the cone $z = \sqrt{x^2 + y^2}$ and inside the sphere $x^2 + y^2 + z^2 = a^2$
2. Above the surface $z = (x^2 + y^2)^{1/4}$ and inside the sphere $x^2 + y^2 + z^2 = 2$
3. Between the paraboloids $z = 10 - x^2 - y^2$ and $z = 2(x^2 + y^2 - 1)$
4. Inside the paraboloid $z = x^2 + y^2$ and inside the sphere $x^2 + y^2 + z^2 = 12$
5. Above the xy-plane, inside the cone $z = 2a - \sqrt{x^2 + y^2}$, and inside the cylinder $x^2 + y^2 = 2ay$
6. Above the xy-plane, under the paraboloid $z = 1 - x^2 - y^2$, and in the wedge $-x \le y \le \sqrt{3}x$
7. In the first octant, between the planes $y = 0$ and $y = x$, and inside the ellipsoid $\frac{x^2}{a^2} + \frac{y^2}{b^2} + \frac{z^2}{c^2} = 1$. *Hint:* Use the change of variables suggested in Example 1.
8. Bounded by the hyperboloid $\frac{x^2}{a^2} + \frac{y^2}{b^2} - \frac{z^2}{c^2} = 1$ and the planes $z = -c$ and $z = c$
9. Above the xy-plane and below the paraboloid
$$z = 1 - \frac{x^2}{a^2} - \frac{y^2}{b^2}$$

10. Evaluate $\iiint_R (x^2 + y^2 + z^2)\,dV$, where R is the cylinder $0 \le x^2 + y^2 \le a^2, 0 \le z \le h$.
11. Find $\iiint_B (x^2 + y^2)\,dV$, where B is the ball given by $x^2 + y^2 + z^2 \le a^2$.
12. Find $\iiint_B (x^2 + y^2 + z^2)\,dV$, where B is the ball of Exercise 11.
13. Find $\iiint_R (x^2 + y^2 + z^2)\,dV$, where R is the region that lies above the cone $z = c\sqrt{x^2 + y^2}$ and inside the sphere $x^2 + y^2 + z^2 = a^2$.
14. Evaluate $\iiint_R (x^2 + y^2)\,dV$ over the region R of Exercise 13.
15. Find $\iiint_R z\,dV$, over the region R satisfying $x^2 + y^2 \le z \le \sqrt{2 - x^2 - y^2}$.
16. Find $\iiint_R x\,dV$ and $\iiint_R z\,dV$, over that part of the hemisphere $0 \le z \le \sqrt{a^2 - x^2 - y^2}$ that lies in the first octant.
17. Find $\iiint_R x\,dV$ and $\iiint_R z\,dV$ over that part of the cone
$$0 \le z \le h\left(1 - \frac{\sqrt{x^2 + y^2}}{a}\right)$$
that lies in the first octant.
18. Find the volume of the region inside the ellipsoid $\frac{x^2}{a^2} + \frac{y^2}{b^2} + \frac{z^2}{c^2} = 1$ and above the plane $z = b - y$.
19. Show that for cylindrical coordinates the Laplace equation
$$\frac{\partial^2 u}{\partial x^2} + \frac{\partial^2 u}{\partial y^2} + \frac{\partial^2 u}{\partial z^2} = 0$$
is given by
$$\frac{\partial^2 u}{\partial r^2} + \frac{1}{r}\frac{\partial u}{\partial r} + \frac{1}{r^2}\frac{\partial^2 u}{\partial \theta^2} + \frac{\partial^2 u}{\partial z^2} = 0.$$
20. Show that in spherical coordinates the Laplace equation is given by
$$\frac{\partial^2 u}{\partial R^2} + \frac{2}{R}\frac{\partial u}{\partial R} + \frac{\cot\phi}{R^2}\frac{\partial u}{\partial \phi} + \frac{1}{R^2}\frac{\partial^2 u}{\partial \phi^2} + \frac{1}{R^2 \sin^2\phi}\frac{\partial^2 u}{\partial \theta^2} = 0.$$
21. If x, y, and z are functions of u, v, and w with continuous first partial derivatives and nonvanishing Jacobian at (u, v, w), show that they map an infinitesimal volume element in uvw-space bounded by the coordinate planes u, $u + du$, v, $v + dv$, w, and $w + dw$ into an infinitesimal "parallelepiped" in xyz-space having volume
$$dx\,dy\,dz = \left|\frac{\partial(x, y, z)}{\partial(u, v, w)}\right| du\,dv\,dw.$$
Hint: Adapt the two-dimensional argument given in Section 14.4. What three vectors from the point $P = (x(u, v, w), y(u, v, w), z(u, v, w))$ span the parallelepiped?

14.7 Applications of Multiple Integrals

When we express the volume V of a region R in 3-space as an integral,
$$V = \iiint_R dV,$$

we are regarding V as a "sum" of infinitely many *infinitesimal elements of volume,* that is, as the limit of the sum of volumes of smaller and smaller nonoverlapping subregions into which we subdivide R. This idea of representing sums of infinitesimal elements of quantities by integrals has many applications.

For example, if a rigid body of constant density ρ g/cm^3 occupies a volume V cm^3, then its mass is $m = \rho V$ g. If the density is not constant but varies continuously over the region R of 3-space occupied by the rigid body, say $\rho = \rho(x, y, z)$, we can still regard the density as being constant on an infinitesimal element of R having volume dV. The mass of this element is therefore $dm = \rho(x, y, z)\,dV$, and the mass of the whole body is calculated by integrating these mass elements over R:

$$m = \iiint_R \rho(x, y, z)\,dV.$$

Similar formulas apply when the rigid body is one- or two-dimensional, and its density is given in units of mass per unit length or per unit area. In such cases single or double integrals are needed to sum the individual elements of mass. All this works because mass is "additive"; that is, the mass of a composite object is the sum of the masses of the parts that comprise the object. The surface areas, gravitational forces, moments, and energies we consider in this section all have this additivity property.

The Surface Area of a Graph

We can use a double integral over a domain D in the xy-plane to add up surface area elements and thereby calculate the total area of the surface $\mathcal{S}$ with equation $z = f(x, y)$ defined for (x, y) in D. We assume that f has continuous first partial derivatives in D, so that $\mathcal{S}$ is smooth and has a nonvertical tangent plane at $P = \big(x, y, f(x, y)\big)$ for any (x, y) in D. The vector

$$\mathbf{n} = -f_1(x, y)\mathbf{i} - f_2(x, y)\mathbf{j} + \mathbf{k}$$

is an upward normal to $\mathcal{S}$ at P. An area element dA at position (x, y) in the xy-plane has a *vertical projection* onto $\mathcal{S}$ whose area dS is $\sec\gamma$ times the area dA, where γ is the angle between $\mathbf{n}$ and $\mathbf{k}$. (See Figure 14.54.)

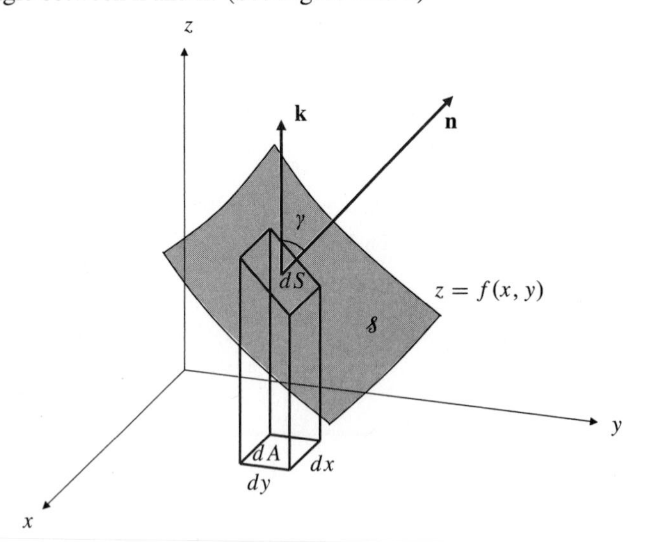

Figure 14.54 The surface area element dS on the surface $z = f(x, y)$ is $\sec\gamma$ times as large as its vertical projection dA onto the xy-plane

Since

$$\cos\gamma = \frac{\mathbf{n} \bullet \mathbf{k}}{|\mathbf{n}||\mathbf{k}|} = \frac{1}{\sqrt{1 + \big(f_1(x, y)\big)^2 + \big(f_2(x, y)\big)^2}},$$

we have

$$dS = \sqrt{1 + \left(\frac{\partial z}{\partial x}\right)^2 + \left(\frac{\partial z}{\partial y}\right)^2}\, dA.$$

Therefore, the area of $\mathcal{S}$ is

$$S = \iint_D \sqrt{1 + \left(\frac{\partial z}{\partial x}\right)^2 + \left(\frac{\partial z}{\partial y}\right)^2}\, dA.$$

EXAMPLE 1 Find the area of that part of the hyperbolic paraboloid $z = x^2 - y^2$ that lies inside the cylinder $x^2 + y^2 = a^2$.

Solution Since $\partial z/\partial x = 2x$ and $\partial z/\partial y = -2y$, the surface area element is

$$dS = \sqrt{1 + 4x^2 + 4y^2}\, dA = \sqrt{1 + 4r^2}\, r\, dr\, d\theta.$$

The required surface area is the integral of dS over the disk $r \leq a$:

$$\begin{aligned} S &= \int_0^{2\pi} d\theta \int_0^a \sqrt{1 + 4r^2}\, r\, dr \qquad \text{Let } u = 1 + 4r^2 \\ &= (2\pi)\frac{1}{8}\int_1^{1+4a^2} \sqrt{u}\, du \\ &= \frac{\pi}{4}\left(\frac{2}{3}\right) u^{3/2}\Big|_1^{1+4a^2} = \frac{\pi}{6}\left((1 + 4a^2)^{3/2} - 1\right) \text{ square units.} \end{aligned}$$

The Gravitational Attraction of a Disk

Newton's universal law of gravitation asserts that two point masses m_1 and m_2, separated by a distance s, attract one another with a force

$$F = \frac{km_1m_2}{s^2},$$

k being a universal constant. The force on each mass is directed toward the other, along the line joining the two masses. Suppose that a flat disk D of radius a, occupying the region $x^2 + y^2 \leq a^2$ of the xy-plane, has constant *areal density* σ (units of mass per unit area). Let us calculate the total force of attraction that this disk exerts upon a mass m located at the point $(0, 0, b)$ on the positive z-axis. The total force is a vector quantity. Although the various mass elements on the disk are in different directions from the mass m, symmetry indicates that the net force will be in the direction toward the centre of the disk, that is, toward the origin. Thus, the total force will be $-F\mathbf{k}$, where F is the magnitude of the force.

We will calculate F by integrating the vertical component dF of the force of attraction on m due to the mass $\sigma\, dA$ in an area element dA on the disk. If the area element is at the point with polar coordinates $[r, \theta]$, and if the line from this point to $(0, 0, b)$ makes angle ψ with the z-axis as shown in Figure 14.55, then the vertical component of the force of attraction of the mass element $\sigma\, dA$ on m is

$$dF = \frac{km\sigma\, dA}{r^2 + b^2}\cos\psi = km\sigma b\frac{dA}{(r^2 + b^2)^{3/2}}.$$

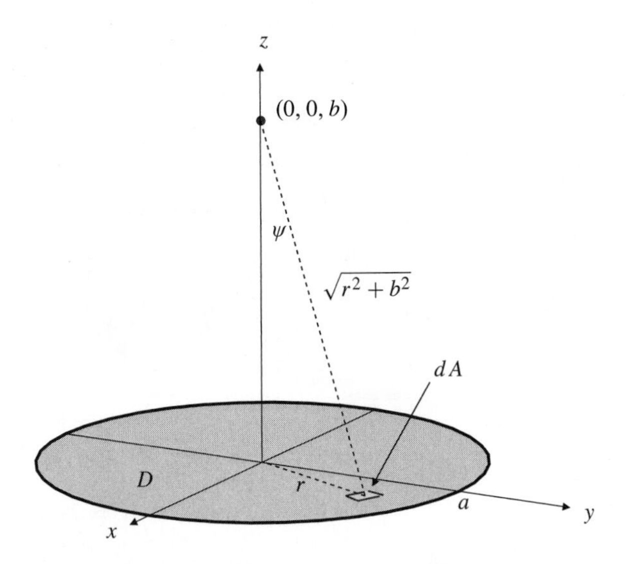

Figure 14.55 Each mass element $\sigma\, dA$ attracts m along a different line

Accordingly, the total vertical force of attraction of the disk on m is

$$\begin{aligned} F &= km\sigma b \iint_D \frac{dA}{(r^2+b^2)^{3/2}} \\ &= km\sigma b \int_0^{2\pi} d\theta \int_0^a \frac{r\,dr}{(r^2+b^2)^{3/2}} \qquad \text{Let } u = r^2+b^2 \\ &= \pi km\sigma b \int_{b^2}^{a^2+b^2} u^{-3/2}\,du \\ &= \pi km\sigma b \left(\frac{-2}{\sqrt{u}}\right)\Bigg|_{b^2}^{a^2+b^2} = 2\pi km\sigma \left(1 - \frac{b}{\sqrt{a^2+b^2}}\right). \end{aligned}$$

Remark If we let a approach infinity in the above formula, we obtain the formula $F = 2\pi km\sigma$ for the force of attraction of a plane of areal density σ on a mass m located at distance b from the plane. Observe that F does not depend on b. Try to reason on physical grounds why this should be so.

Remark The force of attraction on a point mass due to suitably symmetric solid objects (such as balls, cylinders, and cones) having constant density ρ (units of mass per unit volume) can be found by integrating elements of force contributed by thin, disk-shaped slices of the solid. See Exercises 14–17.

Moments and Centres of Mass

The centre of mass of a rigid body is that point (fixed in the body) at which the body can be supported so that in the presence of a constant gravitational field it will not experience any unbalanced torques that will cause it to rotate. The torques experienced by a mass element dm in the body can be expressed in terms of the **moments** of dm about the three coordinate planes. If the body occupies a region R in 3-space and has continuous volume density $\rho(x, y, z)$, then the mass element $dm = \rho(x, y, z)\, dV$ that occupies the volume element dV is said to have **moments** $(x - x_0)\, dm$, $(y - y_0)\, dm$, and $(z - z_0)\, dm$ about the planes $x = x_0$, $y = y_0$, and $z = z_0$, respectively. Thus, the total moments of the body about these three planes are

$$\begin{aligned} M_{x=x_0} &= \iiint_R (x - x_0)\rho(x, y, z)\, dV = M_{x=0} - x_0 m \\ M_{y=y_0} &= \iiint_R (y - y_0)\rho(x, y, z)\, dV = M_{y=0} - y_0 m \\ M_{z=z_0} &= \iiint_R (z - z_0)\rho(x, y, z)\, dV = M_{z=0} - z_0 m, \end{aligned}$$

where $m = \iiint_R \rho\, dV$ is the mass of the body and $M_{x=0}$, $M_{y=0}$, and $M_{z=0}$ are the moments about the coordinate planes $x = 0$, $y = 0$, and $z = 0$, respectively. The **centre of mass** $\bar{P} = (\bar{x}, \bar{y}, \bar{z})$ of the body is that point for which $M_{x=\bar{x}}$, $M_{y=\bar{y}}$, and $M_{z=\bar{z}}$ are all equal to zero. Thus,

Centre of mass

The centre of mass of a solid occupying region R of 3-space and having continuous density $\rho(x, y, z)$ (units of mass per unit volume) is the point $(\bar{x}, \bar{y}, \bar{z})$ with coordinates given by

$$\bar{x} = \frac{M_{x=0}}{m} = \frac{\displaystyle\iiint_R x\rho\, dV}{\displaystyle\iiint_R \rho\, dV}, \qquad \bar{y} = \frac{M_{y=0}}{m} = \frac{\displaystyle\iiint_R y\rho\, dV}{\displaystyle\iiint_R \rho\, dV},$$

$$\bar{z} = \frac{M_{z=0}}{m} = \frac{\displaystyle\iiint_R z\rho\, dV}{\displaystyle\iiint_R \rho\, dV}.$$

These formulas can be combined into a single vector formula for the position vector $\bar{\mathbf{r}} = \bar{x}\mathbf{i} + \bar{y}\mathbf{j} + \bar{z}\mathbf{k}$ of the centre of mass in terms of the position vector $\mathbf{r} = x\mathbf{i} + y\mathbf{j} + z\mathbf{k}$ of an arbitrary point in R,

$$\bar{\mathbf{r}} = \frac{M_{x=0}\mathbf{i} + M_{y=0}\mathbf{j} + M_{z=0}\mathbf{k}}{m} = \frac{\displaystyle\iiint_R \rho\,\mathbf{r}\, dV}{\displaystyle\iiint_R \rho\, dV},$$

where the integral of the vector function $\rho\,\mathbf{r}$ is understood to mean the vector whose components are the integrals of the components of $\rho\,\mathbf{r}$.

Remark Similar expressions hold for distributions of mass over regions in the plane or over intervals on a line. We use the appropriate areal or line densities and double or single definite integrals.

Remark If the density is constant, it cancels out of the expressions for the centre of mass. In this case the centre of mass is a *geometric* property of the region R and is called the **centroid** or **centre of gravity** of that region.

EXAMPLE 2 Find the centroid of the tetrahedron T bounded by the coordinate planes and the plane

$$\frac{x}{a} + \frac{y}{b} + \frac{z}{c} = 1.$$

Solution The density is assumed to be constant, so we may take it to be unity. The mass of T is thus equal to its volume: $m = V = abc/6$. The moment of T about the yz-plane is (see Figure 14.56):

$$\begin{aligned} M_{x=0} &= \iiint_T x\, dV \\ &= \int_0^a x\, dx \int_0^{b(1-\frac{x}{a})} dy \int_0^{c(1-\frac{x}{a}-\frac{y}{b})} dz \end{aligned}$$

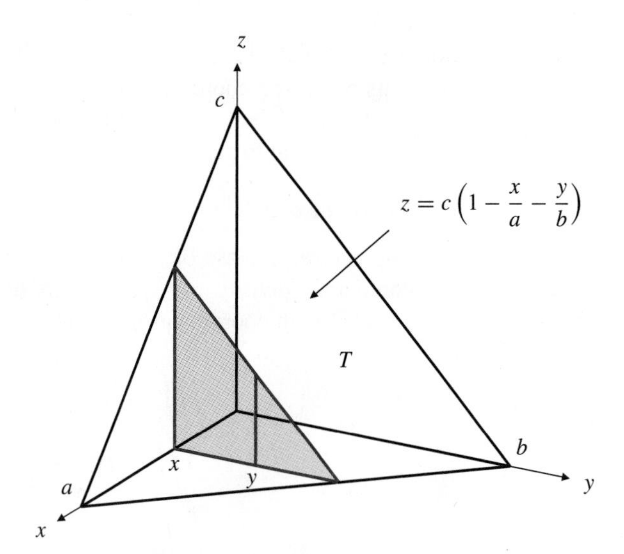

Figure 14.56 Iteration diagram for a triple integral over the tetrahedron of Example 2

$$
\begin{aligned}
&= c \int_0^a x\,dx \int_0^{b(1-\frac{x}{a})} \left(1 - \frac{x}{a} - \frac{y}{b}\right) dy \\
&= c \int_0^a x\left[\left(1 - \frac{x}{a}\right)y - \frac{y^2}{2b}\right]\Bigg|_{y=0}^{y=b(1-\frac{x}{a})} dx \\
&= \frac{bc}{2} \int_0^a x\left(1 - \frac{x}{a}\right)^2 dx \\
&= \frac{bc}{2}\left[\frac{x^2}{2} - \frac{2}{3}\frac{x^3}{a} + \frac{x^4}{4a^2}\right]\Bigg|_0^a = \frac{a^2bc}{24}.
\end{aligned}
$$

Thus, $\bar{x} = M_{x=0}/m = a/4$. By symmetry, the centroid of T is $\left(\frac{a}{4}, \frac{b}{4}, \frac{c}{4}\right)$.

EXAMPLE 3 Find the centre of mass of a solid occupying the region S that satisfies $x \geq 0$, $y \geq 0$, $z \geq 0$, and $x^2 + y^2 + z^2 \leq a^2$, if the density at distance R from the origin is kR.

Solution The mass of the solid is distributed symmetrically in the first octant part of the ball $R \leq a$ so that the centre of mass, $(\bar{x}, \bar{y}, \bar{z})$, must satisfy $\bar{x} = \bar{y} = \bar{z}$. The mass of the solid is

$$m = \iiint_S kR\,dV = \int_0^{\pi/2} d\theta \int_0^{\pi/2} \sin\phi\,d\phi \int_0^a (kR)R^2\,dR = \frac{\pi k a^4}{8}.$$

The moment about the xy-plane is

$$
\begin{aligned}
M_{z=0} &= \iiint_S zkR\,dV = \iiint_S (kR)R\cos\phi\,R^2\sin\phi\,dR\,d\phi\,d\theta \\
&= \frac{k}{2}\int_0^{\pi/2} d\theta \int_0^{\pi/2} \sin(2\phi)\,d\phi \int_0^a R^4\,dR = \frac{k\pi a^5}{20}.
\end{aligned}
$$

Hence, $\bar{z} = \frac{k\pi a^5}{20} \bigg/ \frac{k\pi a^4}{8} = \frac{2a}{5}$, and the centre of mass is $\left(\frac{2a}{5}, \frac{2a}{5}, \frac{2a}{5}\right)$.

Moment of Inertia

The **kinetic energy** of a particle of mass m moving with speed v is

$$\mathrm{KE} = \frac{1}{2}mv^2.$$

The mass of the particle measures its *inertia*, which is twice the energy it has when its speed is one unit.

If the particle is moving in a circle of radius D, its motion can be described in terms of its **angular speed**, Ω, measured in radians per unit time. In one revolution the particle travels a distance $2\pi D$ in time $2\pi/\Omega$. Thus, its (translational) speed v is related to its angular speed by

$$v = \Omega D.$$

Suppose that a rigid body is rotating with angular speed Ω about an axis L. If (at some instant) the body occupies a region R and has density $\rho = \rho(x, y, z)$, then each mass element $dm = \rho\, dV$ in the body has kinetic energy

$$d\mathrm{KE} = \frac{1}{2}v^2\, dm = \frac{1}{2}\rho\Omega^2 D^2\, dV,$$

where $D = D(x, y, z)$ is the perpendicular distance from the volume element dV to the axis of rotation L. The total kinetic energy of the rotating body is therefore

$$\mathrm{KE} = \frac{1}{2}\Omega^2 \iiint_R D^2\rho\, dV = \frac{1}{2}I\Omega^2,$$

where

$$I = \iiint_R D^2\rho\, dV.$$

I is called the **moment of inertia** of the rotating body about the axis L. The moment of inertia plays the same role in the expression for kinetic energy of rotation (in terms of angular speed) that the mass does in the expression for kinetic energy of translation (in terms of linear speed). The moment of inertia is twice the kinetic energy of the body when it is rotating with unit angular speed.

If the entire mass of the rotating body were concentrated at a distance D_0 from the axis of rotation, then its kinetic energy would be $\frac{1}{2}mD_0^2\Omega^2$. The **radius of gyration $\bar{D}$** is the value of D_0 for which this energy is equal to the actual kinetic energy $\frac{1}{2}I\Omega^2$ of the rotating body. Thus, $m\bar{D}^2 = I$, and the radius of gyration is

$$\bar{D} = \sqrt{I/m} = \left(\frac{\displaystyle\iiint_R D^2\rho\, dV}{\displaystyle\iiint_R \rho\, dV}\right)^{1/2}.$$

EXAMPLE 4 **(The acceleration of a rolling ball)**

(a) Find the moment of inertia and radius of gyration of a solid ball of radius a and constant density ρ about a diameter of that ball.

(b) With what linear acceleration will the ball roll (without slipping) down a plane inclined at angle α to the horizontal?

Solution

(a) We take the z-axis as the diameter and integrate in cylindrical coordinates over the ball B of radius a centred at the origin. Since the density ρ is constant, we have

$$\begin{aligned} I &= \rho \iiint_B r^2\, dV \\ &= \rho \int_0^{2\pi} d\theta \int_0^a r^3\, dr \int_{-\sqrt{a^2-r^2}}^{\sqrt{a^2-r^2}} dz \\ &= 4\pi\rho \int_0^a r^3 \sqrt{a^2-r^2}\, dr \qquad \text{Let } u = a^2 - r^2 \\ &= 2\pi\rho \int_0^{a^2} (a^2 - u)\sqrt{u}\, du \\ &= 2\pi\rho \left(\frac{2}{3} a^2 u^{3/2} - \frac{2}{5} u^{5/2} \right) \bigg|_0^{a^2} = \frac{8}{15}\pi\rho a^5. \end{aligned}$$

Since the mass of the ball is $m = \frac{4}{3}\pi\rho a^3$, the radius of gyration is

$$\bar{D} = \sqrt{\frac{I}{m}} = \sqrt{\frac{2}{5}}\, a.$$

(b) We can determine the acceleration of the ball by using conservation of total (kinetic plus potential) energy. When the ball is rolling down the plane with speed v, its centre is moving with speed v and losing height at a rate $v \sin\alpha$. (See Figure 14.57.) Since the ball is not slipping, it is rotating about a horizontal axis through its centre with angular speed $\Omega = v/a$. Hence, its kinetic energy (due to translation and rotation) is

$$\begin{aligned} \text{KE} &= \frac{1}{2} m v^2 + \frac{1}{2} I \Omega^2 \\ &= \frac{1}{2} m v^2 + \frac{1}{2}\frac{2}{5} m a^2 \frac{v^2}{a^2} = \frac{7}{10} m v^2. \end{aligned}$$

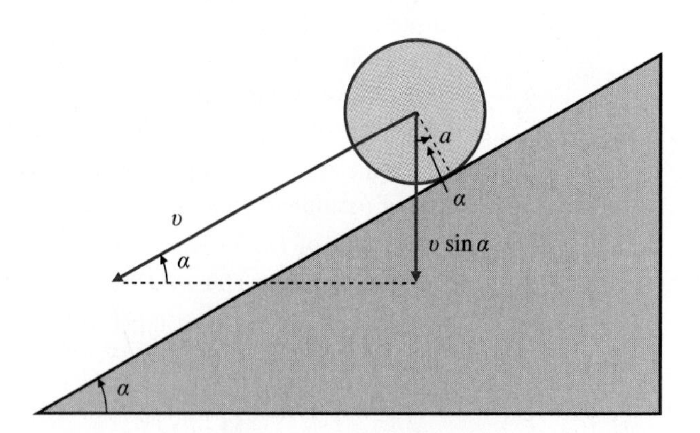

Figure 14.57 The actual velocity and the vertical velocity of a ball rolling down an incline as in Example 4

When the centre of the ball is at height h (above some reference height), the ball has (gravitational) potential energy

$$\text{PE} = mgh.$$

(This is the work that must be done against a constant gravitational force $F = mg$ to raise it to height h.) Since total energy is conserved,

$$\frac{7}{10} m v^2 + mgh = \text{constant}.$$

Differentiating with respect to time t, we obtain

$$0 = \frac{7}{10} m\, 2v \frac{dv}{dt} + mg \frac{dh}{dt} = \frac{7}{5} mv \frac{dv}{dt} - mgv \sin\alpha.$$

Thus, the ball rolls down the incline with acceleration $\dfrac{dv}{dt} = \dfrac{5}{7} g \sin\alpha$.

Remark **Integrals of higher multiplicity.** Just like higher order derivatives, it is easy to imagine the need for multiple integrations beyond just triple integrals. For instance, in physics we must consider both position and momentum of a particle to understand its behaviour. Each of these requires three coordinates, so a total of six coordinates are needed. Integrals may have to be taken over all six.

Suppose we know that the number of particles per unit interval in three space coordinates, x_1, x_2, x_3 and per unit momentum in three momentum coordinates, p_1, p_2, p_3 is $N(x_1, x_2, x_3, p_1, p_2, p_3)$. If the energy, $\epsilon(p_1, p_2, p_3)$, per particle is defined by its momentum then the total energy of the system of particles is given by the repeated integral,

$$\int\!\!\int\!\!\int\!\!\int\!\!\int\!\!\int N\epsilon \, dx_1\, dx_2\, dx_3\, dp_1\, dp_2\, dp_3$$

where the domain of integration is over the entire six dimensional space. Clearly, this notation is a bit clumsy.

Of course, we don't stop with six dimensions. The numbers of integrations can be arbitrarily large. In kinetic theory, for example, one may imagine spaces where there are six coordinates for every particle. If the number of particles is typically very large (e.g., 10^{23}), integrals over that space might involve 6×10^{23} integrations. Clearly, writing an integration sign for each coordinate is not just clumsy, it is impossible and pointless in such cases.

One alternative is to represent the integral of function $f(x) = f(x_1, x_2, \ldots, x_n)$ over a domain D in n dimensional space as

$$\int \cdots \int_D f(x)\, dx \qquad \text{or} \qquad \int \cdots \int_D f(x)\, dV,$$

where $dx = dV = dx_1\, dx_2 \cdots dx_n$. But the dots really don't convey anything new, so the integral is often written

$$\int_D f(x)\, dV, \qquad \text{or even} \qquad \int f(x)\, dV,$$

where the space and domain of the integration are simply described in the surrounding text. This is the common approach for all types of integrals in advanced texts. However, for introductory material with three or fewer iterations of integration, it remains helpful to denote numbers of integrations involved symbolically.

EXERCISES 14.7

Surface area problems

Use double integrals to calculate the areas of the surfaces in Exercises 1–9.

1. The part of the plane $z = 2x + 2y$ inside the cylinder $x^2 + y^2 = 1$
2. The part of the plane $5z = 3x - 4y$ inside the elliptic cylinder $x^2 + 4y^2 = 4$
3. The hemisphere $z = \sqrt{a^2 - x^2 - y^2}$
4. The half-ellipsoidal surface $z = 2\sqrt{1 - x^2 - y^2}$
5. The conical surface $3z^2 = x^2 + y^2$, $0 \le z \le 2$

6. The paraboloid $z = 1 - x^2 - y^2$ in the first octant

7. The part of the surface $z = y^2$ above the triangle with vertices $(0, 0)$, $(0, 1)$, and $(1, 1)$

8. The part of the surface $z = \sqrt{x}$ above the region $0 \le x \le 1$, $0 \le y \le \sqrt{x}$

9. The part of the cylindrical surface $x^2 + z^2 = 4$ that lies above the region $0 \le x \le 2$, $0 \le y \le x$

10. Show that the parts of the surfaces $z = 2xy$ and $z = x^2 + y^2$ that lie in the same vertical cylinder have the same area.

11. Show that the area S of the part of the paraboloid $z = \frac{1}{2}(x^2 + y^2)$ lying above the square $-1 \le x \le 1$, $-1 \le y \le 1$ is given by

$$S = \frac{8}{3}\int_0^{\pi/4} (1 + \sec^2\theta)^{3/2}\, d\theta - \frac{2\pi}{3},$$

and use numerical methods to evaluate the area to 3 decimal places.

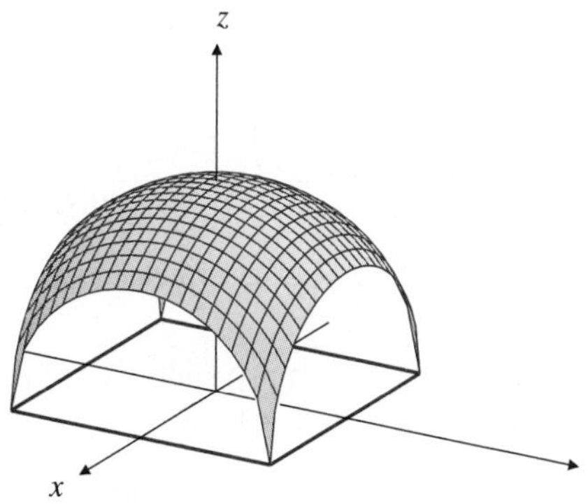

Figure 14.58

12. The canopy shown in Figure 14.58 is the part of the hemisphere of radius $\sqrt{2}$ centred at the origin that lies above the square $-1 \le x \le 1$, $-1 \le y \le 1$. Find its area. *Hint:* It is possible to get an exact solution by first finding the area of the part of the sphere $x^2 + y^2 + z^2 = 2$ that lies above the plane $z = 1$. If you do the problem directly by integrating the surface area element over the square, you may encounter an integral that you can't evaluate exactly, and you will have to use numerical methods.

Mass and gravitational attraction

13. Find the mass of a spherical planet of radius a whose density at distance R from the centre is $\rho = A/(B + R^2)$.

In Exercises 14–17, find the gravitational attraction that the given object exerts on a mass m located at $(0, 0, b)$. Assume the object has constant density ρ. In each case you can obtain the answer by integrating the contributions made by disks of thickness dz, making use of the formula for the attraction exerted by the disk obtained in the text.

14. The ball $x^2 + y^2 + z^2 \le a^2$, where $a < b$

15. The cylinder $x^2 + y^2 \le a^2$, $0 \le z \le h$, where $h < b$

16. The cone $0 \le z \le b - (\sqrt{x^2 + y^2})/a$

17. The half-ball $0 \le z \le \sqrt{a^2 - x^2 - y^2}$, where $a < b$

Centres of mass and centroids

18. Find the centre of mass of an object occupying the cube $0 \le x, y, z \le a$ with density given by $\rho = x^2 + y^2 + z^2$.

Find the centroids of the regions in Exercises 19–22.

19. The prism $x \ge 0$, $y \ge 0$, $x + y \le 1$, $0 \le z \le 1$

20. The unbounded region $0 \le z \le e^{-(x^2+y^2)}$

21. The first octant part of the ball $x^2 + y^2 + z^2 \le a^2$

22. The region inside the cube $0 \le x, y, z \le 1$ and under the plane $x + y + z = 2$

Moments of inertia

23. Explain *in physical terms* why the acceleration of the ball rolling down the incline in Example 4 does not approach g (the acceleration due to gravity) as the angle of incline, α, approaches 90°.

Find the moments of inertia and radii of gyration of the solid objects in Exercises 24–32. Assume constant density in all cases.

24. A circular cylinder of base radius a and height h about the axis of the cylinder

25. A circular cylinder of base radius a and height h about a diameter of the base of the cylinder

26. A right circular cone of base radius a and height h about the axis of the cone

27. A right circular cone of base radius a and height h about a diameter of the base of the cone

28. A cube of edge length a about an edge of the cube

29. A cube of edge length a about a diagonal of a face of the cube

30. A cube of edge length a about a diagonal of the cube

31. The rectangular box $-a \le x \le a$, $-b \le y \le b$, $-c \le z \le c$ about the z-axis

32. The region between the two concentric cylinders $x^2 + y^2 = a^2$ and $x^2 + y^2 = b^2$ (where $0 < a < b$) and between $z = 0$ and $z = c$ about the z-axis

33. A ball of radius a has constant density ρ. A cylindrical hole of radius $b < a$ is drilled through the centre of the ball. Find the mass of the remaining part of the ball and its moment of inertia about the axis of the hole.

34. With what acceleration will a solid cylinder of base radius a, height h, and constant density ρ roll (without slipping) down a plane inclined at angle α to the horizontal?

35. Repeat Exercise 34 for the ball with the cylindrical hole in Exercise 33. Assume that the axis of the hole remains horizontal while the ball rolls.

36. A rigid pendulum of mass m swings about point A on a horizontal axis. Its moment of inertia about that axis is I. The centre of mass C of the pendulum is at distance a from A. When the pendulum hangs at rest, C is directly under A. (Why?) Suppose the pendulum is swinging. Let $\theta = \theta(t)$ measure the angular displacement of the line AC from the vertical at time t. ($\theta = 0$ when the pendulum is in its rest position.) Use a conservation of energy argument similar to that in Example 4 to show that

$$\frac{1}{2}I\left(\frac{d\theta}{dt}\right)^2 - mga\cos\theta = \text{constant}$$

and, hence, differentiating with respect to t, that

$$\frac{d^2\theta}{dt^2} + \frac{mga}{I}\sin\theta = 0.$$

This is a nonlinear differential equation, and it is not easily solved. However, for small oscillations ($|\theta|$ small) we can use the approximation $\sin\theta \approx \theta$. In this case the differential equation is that of simple harmonic motion. What is the period?

37. Let L_0 be a straight line passing through the centre of mass of a rigid body B of mass m. Let L_k be a straight line parallel to and k units distant from L_0. If I_0 and I_k are the moments of inertia of B about L_0 and L_k, respectively, show that $I_k = I_0 + k^2 m$. Hence, a body always has smallest moment of inertia about an axis through its centre of mass. *Hint:* Assume that the z-axis coincides with L_0 and that L_k passes through the point $(k, 0, 0)$.

38. Reestablish the expression for the total kinetic energy of the rolling ball in Example 4 by regarding the ball at any instant as rotating about a horizontal line through its point of contact with the inclined plane. Use the result of Exercise 37.

39. **(Products of inertia)** A rigid body with density ρ is placed with its centre of mass at the origin and occupies a region R of 3-space. Suppose the six second moments P_{xx}, P_{yy}, P_{zz}, P_{xy}, P_{xz}, and P_{yz} are all known, where

$$P_{xx} = \iiint_R x^2 \rho \, dV, \quad P_{xy} = \iiint_R xy\rho \, dV, \quad \ldots.$$

(There exist tables giving these six moments for bodies of many standard shapes. They are called products of inertia.) Show how to express the moment of inertia of the body about any axis through the origin in terms of these six second moments. (If this result is combined with that of Exercise 37, the moment of inertia about *any* axis can be found.)

CHAPTER REVIEW

Key Ideas

- **What do the following terms and phrases mean?**
  - ◇ a Riemann sum for $f(x, y)$ on $a \le x \le b, c \le y \le d$
  - ◇ $f(x, y)$ is integrable on $a \le x \le b, c \le y \le d$
  - ◇ the double integral of $f(x, y)$ over $a \le x \le b, c \le y \le d$
  - ◇ iteration of a double integral
  - ◇ the average value of $f(x, y)$ over region R
  - ◇ the area element in polar coordinates
  - ◇ a triple integral
  - ◇ the volume element in cylindrical coordinates
  - ◇ the volume element in spherical coordinates
  - ◇ the surface area of the graph of $z = f(x, y)$
  - ◇ the moment of inertia of a solid about an axis
- **Describe how to change variables in a double integral.**
- **How do you calculate the centroid of a solid region?**
- **How do you calculate the moment of inertia of a solid about an axis?**

Review Exercises

1. Evaluate $\iint_R (x + y)\, dA$, over the first-quadrant region lying under $x = y^2$ and above $y = x^2$.

2. Evaluate $\iint_P (x^2+y^2)\, dA$, where P is the parallelogram with vertices $(0, 0)$, $(2, 0)$, $(3, 1)$, and $(1, 1)$.

3. Find $\iint_S (y/x)\, dA$, where S is the part of the disk $x^2+y^2 \le 4$ in the first quadrant and under the line $y = x$.

4. Consider the iterated integral

$$I = \int_0^{\sqrt{3}} dy \int_{y/\sqrt{3}}^{\sqrt{4-y^2}} e^{-x^2-y^2}\, dx.$$

(a) Write I as a double integral $\iint_R e^{-x^2-y^2}\, dA$, and sketch the region R over which the double integral is taken.

(b) Write I as an iterated integral with the order of integrations reversed from that of the given iteration.

(c) Write I as an iterated integral in polar coordinates.

(d) Evaluate I.

5. Find the constant $k > 0$ such that the volume of the region lying inside the sphere $x^2 + y^2 + z^2 = a^2$ and above the cone $z = k\sqrt{x^2 + y^2}$ is one-quarter of the volume contained by the whole sphere.

6. Reiterate the integral

$$I = \int_0^2 dy \int_0^y f(x, y)\, dx + \int_2^6 dy \int_0^{\sqrt{6-y}} f(x, y)\, dx$$

with the y integral on the inside.

7. Let $J = \int_0^1 dz \int_0^z dy \int_0^y f(x, y, z)\, dx$. Express J as an iterated integral where the integrations are to be performed in the following order: first z, then y, then x.

8. An object in the shape of a right-circular cone has height 10 m and base radius 5 m. Its density is proportional to the square of the distance from the base and equals 3,000 kg/m^3 at the vertex.

(a) Find the mass of the object.

(b) Express the moment of inertia of the object about its central axis as an iterated integral.

9. Find the average value of $f(t) = \int_t^a e^{-x^2}\, dx$ over the interval $0 \le t \le a$.

10. Find the average value of the function $f(x, y) = \lfloor x + y \rfloor$ over the quarter-disk $x \ge 0$, $y \ge 0$, $x^2 + y^2 \le 4$. (Recall that $\lfloor x \rfloor$ denotes the greatest integer less than or equal to x.)

11. Let D be the smaller of the two solid regions bounded by the surfaces

$$z = \frac{x^2 + y^2}{a} \quad \text{and} \quad x^2 + y^2 + z^2 = 6a^2,$$

where a is a positive constant. Find $\iiint_D (x^2 + y^2)\, dV$.

12. Find the moment of inertia about the z-axis of a solid V of density 1 if V is specified by the inequalities $0 \le z \le \sqrt{x^2 + y^2}$ and $x^2 + y^2 \le 2ay$, where $a > 0$.

13. The rectangular solid $0 \le x \le 1$, $0 \le y \le 2$, $0 \le z \le 1$ is cut into two pieces by the plane $2x + y + z = 2$. Let D be the piece that includes the origin. Find the volume of D and $\bar{z}$, the z-coordinate of the centroid of D.

14. A solid S consists of those points (x, y, z) that lie in the first octant and satisfy $x + y + 2z \le 2$ and $y + z \le 1$. Find the volume of S and the x-coordinate of its centroid.

15. Find $\iiint_S z\, dV$, where S is the portion of the first octant that is above the plane $x + y - z = 1$ and below the plane $z = 1$.

16. Find the area of that part of the plane $z = 2x$ that lies inside the paraboloid $z = x^2 + y^2$.

17. Find the area of that part of the paraboloid $z = x^2 + y^2$ that lies below the plane $z = 2x$. Express the answer as a single integral, and evaluate it to 3 decimal places.

18. Find the volume of the smaller of the two regions into which the plane $x + y + z = 1$ divides the interior of the ellipsoid $x^2 + 4y^2 + 9z^2 = 36$. *Hint:* First change variables so that the ellipsoid becomes a ball. Then replace the plane by a plane with a simpler equation passing the same distance from the origin.

Challenging Problems

1. The plane $(x/a) + (y/b) + (z/c) = 1$ (where $a > 0$, $b > 0$, and $c > 0$) divides the solid ellipsoid

$$\frac{x^2}{a^2} + \frac{y^2}{b^2} + \frac{z^2}{c^2} \le 1$$

into two unequal pieces. Find the volume of the smaller piece.

2. Find the area of the part of the plane $(x/a) + (y/b) + (z/c) = 1$ (where $a > 0$, $b > 0$, and $c > 0$) that lies inside the ellipsoid

$$\frac{x^2}{a^2} + \frac{y^2}{b^2} + \frac{z^2}{c^2} \le 1.$$

3. (a) Expand $1/(1 - xy)$ as a geometric series, and hence show that

$$\int_0^1 \int_0^1 \frac{1}{1 - xy}\, dx\, dy = \sum_{n=1}^{\infty} \frac{1}{n^2}.$$

(b) Similarly, express the following integrals as sums of series:

(i) $\displaystyle\int_0^1 \int_0^1 \frac{1}{1 + xy}\, dx\, dy,$

(ii) $\displaystyle\int_0^1 \int_0^1 \int_0^1 \frac{1}{1 - xyz}\, dx\, dy\, dz,$

(iii) $\displaystyle\int_0^1 \int_0^1 \int_0^1 \frac{1}{1 + xyz}\, dx\, dy\, dz.$

4. Let P be the parallelepiped bounded by the three pairs of parallel planes $\mathbf{a} \bullet \mathbf{r} = 0$, $\mathbf{a} \bullet \mathbf{r} = d_1 > 0$, $\mathbf{b} \bullet \mathbf{r} = 0$, $\mathbf{b} \bullet \mathbf{r} = d_2 > 0$, $\mathbf{c} \bullet \mathbf{r} = 0$, and $\mathbf{c} \bullet \mathbf{r} = d_3 > 0$, where $\mathbf{a}$, $\mathbf{b}$, and $\mathbf{c}$ are constant vectors, and $\mathbf{r} = x\mathbf{i} + y\mathbf{j} + z\mathbf{k}$. Show that

$$\iiint_P (\mathbf{a} \bullet \mathbf{r})(\mathbf{b} \bullet \mathbf{r})(\mathbf{c} \bullet \mathbf{r})\, dx\, dy\, dz = \frac{(d_1 d_2 d_3)^2}{8|\mathbf{a} \bullet (\mathbf{b} \times \mathbf{c})|}.$$

Hint: Make the change of variables $u = \mathbf{a} \bullet \mathbf{r}$, $v = \mathbf{b} \bullet \mathbf{r}$, $w = \mathbf{c} \bullet \mathbf{r}$.

5. A hole whose cross-section is a square of side 2 is punched through the middle of a ball of radius 2. Find the volume of the remaining part of the ball.

6. Find the volume bounded by the surface with equation $x^{2/3} + y^{2/3} + z^{2/3} = a^{2/3}$.

7. Find the volume bounded by the surface $|x|^{1/3} + |y|^{1/3} + |z|^{1/3} = |a|^{1/3}$.

CHAPTER 15

Vector Fields

"Take some more tea," the March Hare said to Alice, very earnestly.
"I've had nothing yet," Alice replied, in an offended tone, "so I can't take more."
"You mean you can't take less," said the Hatter: "it's very easy to take more than nothing."

Lewis Carroll (Charles Lutwidge Dodgson) 1832–1898
from *Alice's Adventures in Wonderland*

Introduction

This chapter and the next are concerned mainly with vector-valued functions of a vector variable, typically functions whose domains and ranges lie in the plane or in 3-space. Such functions are frequently called *vector fields.* Applications of vector fields often involve integrals taken, not along axes or over regions in the plane or 3-space, but rather over curves and surfaces. We will introduce such line and surface integrals in this chapter. The next chapter will be devoted to developing versions of the Fundamental Theorem of Calculus for integrals of vector fields.

15.1 Vector and Scalar Fields

A function whose domain and range are subsets of Euclidean 3-space, $\mathbb{R}^3$, is called a **vector field**. Thus, a vector field $\mathbf{F}$ associates a vector $\mathbf{F}(x, y, z)$ with each point (x, y, z) in its domain. The three components of $\mathbf{F}$ are scalar-valued (real-valued) functions $F_1(x, y, z)$, $F_2(x, y, z)$, and $F_3(x, y, z)$, and $\mathbf{F}(x, y, z)$ can be expressed in terms of the standard basis in $\mathbb{R}^3$ as

$$\mathbf{F}(x, y, z) = F_1(x, y, z)\mathbf{i} + F_2(x, y, z)\mathbf{j} + F_3(x, y, z)\mathbf{k}.$$

(Note that the subscripts here represent *components* of a vector, *not* partial derivatives.) If $F_3(x, y, z) = 0$ and F_1 and F_2 are independent of z, then $\mathbf{F}$ reduces to

$$\mathbf{F}(x, y) = F_1(x, y)\mathbf{i} + F_2(x, y)\mathbf{j}$$

and so is called a **plane vector field**, or a vector field in the xy-plane. We will frequently make use of position vectors in the arguments of vector fields. The position vector of (x, y, z) is $\mathbf{r} = x\mathbf{i} + y\mathbf{j} + z\mathbf{k}$, and we can write $\mathbf{F}(\mathbf{r})$ as a shorthand for $\mathbf{F}(x, y, z)$. In the context of discussion of vector fields, a scalar-valued function of a vector variable (i.e., a function of several real variables as considered in the context of Chapters 12–14) is frequently called a **scalar field**. Thus, the components of a vector field are scalar fields.

Many of the results we prove about vector fields require that the field be smooth in some sense. We will call a vector field **smooth** wherever its component scalar fields have continuous partial derivatives of all orders. (For most purposes, however, second order would be sufficient.)

Vector fields arise in many situations in applied mathematics. Let us list some:

(a) The gravitational field $\mathbf{F}(x, y, z)$ due to some object is the force of attraction that the object exerts on a unit mass located at position (x, y, z).

(b) The electrostatic force field $\mathbf{E}(x, y, z)$ due to an electrically charged object is the electrical force that the object exerts on a unit charge at position (x, y, z). (The force may be either an attraction or a repulsion.)

(c) The velocity field $\mathbf{v}(x, y, z)$ in a moving fluid (or solid) is the velocity of motion of the particle at position (x, y, z). If the motion is not "steady state," then the velocity field will also depend on time: $\mathbf{v} = \mathbf{v}(x, y, z, t)$.

(d) The gradient $\nabla f(x, y, z)$ of any scalar field f gives the direction and magnitude of the greatest rate of increase of f at (x, y, z). In particular, a *temperature gradient*, $\nabla T(x, y, z)$, is a vector field giving the direction and magnitude of the greatest rate of increase of temperature T at the point (x, y, z) in a heat-conducting medium. *Pressure gradients* provide similar information about the variation of pressure in a fluid such as an air mass or an ocean.

(e) The unit radial and unit transverse vectors $\hat{\mathbf{r}}$ and $\hat{\boldsymbol{\theta}}$ are examples of vector fields in the xy-plane. Both are defined at all points of the plane except the origin.

EXAMPLE 1 **(The gravitational field of a point mass)** The gravitational force field due to a point mass m located at point P_0 having position vector $\mathbf{r}_0$ is

$$\begin{aligned}\mathbf{F}(x, y, z) = \mathbf{F}(\mathbf{r}) &= \frac{-km}{|\mathbf{r} - \mathbf{r}_0|^3}(\mathbf{r} - \mathbf{r}_0) \\ &= -km\frac{(x - x_0)\mathbf{i} + (y - y_0)\mathbf{j} + (z - z_0)\mathbf{k}}{\left((x - x_0)^2 + (y - y_0)^2 + (z - z_0)^2\right)^{3/2}},\end{aligned}$$

where $k > 0$ is a constant. $\mathbf{F}$ points toward the point $\mathbf{r}_0$ and has magnitude

$$|\mathbf{F}| = km/|\mathbf{r} - \mathbf{r}_0|^2.$$

Some vectors in a plane section of the field are shown graphically in Figure 15.1. Each represents the value of the field at the position of its tail. The lengths of the vectors indicate that the strength of the force increases the closer you get to P_0. However, the vectors have a schematic meaning relative to each other; they do not imply actual distances in the plane.

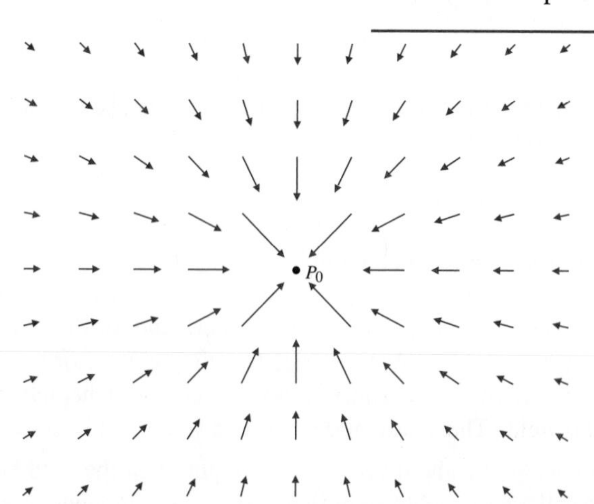

Figure 15.1 The gravitational field of a point mass located at P_0

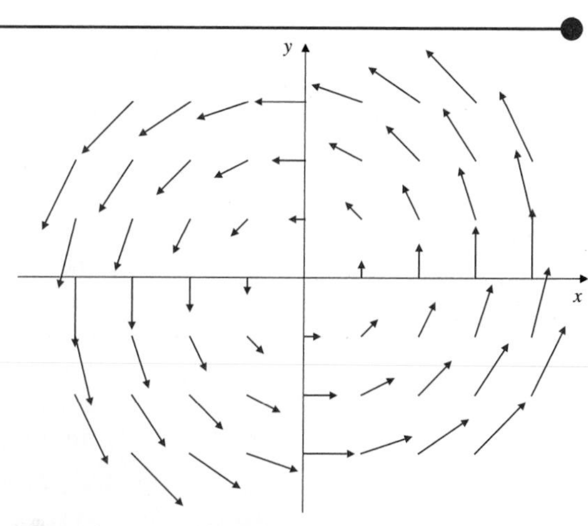

Figure 15.2 The velocity field of a rigid body rotating about the z-axis

Remark The electrostatic field **F** due to a point charge q at P_0 is given by the same formula as the gravitational field above, except with $-m$ replaced by q. The reason for the opposite sign is that like charges repel each other whereas masses attract each other.

EXAMPLE 2 The velocity field of a solid rotating about the z-axis with angular velocity $\mathbf{\Omega} = \Omega\mathbf{k}$ is

$$\mathbf{v}(x, y, z) = \mathbf{v}(\mathbf{r}) = \mathbf{\Omega} \times \mathbf{r} = -\Omega y\mathbf{i} + \Omega x\mathbf{j}.$$

Being the same in all planes normal to the z-axis, **v** can be regarded as a plane vector field. Some vectors of the field are shown in Figure 15.2.

Field Lines (Integral Curves, Trajectories, Streamlines)

The graphical representations of vector fields such as those shown in Figures 15.1 and 15.2 and the wind velocity field over a hill shown in Figure 15.3 suggest a pattern of motion through space or in the plane. Whether or not the field is a velocity field, we can interpret it as such and ask what path will be followed by a corresponding particle, initially at some point, whose velocity is given by the field. The path will be a curve to which the field is tangent at every point. Such curves are called **field lines, integral curves**, or **trajectories** for the given vector field. In the specific case where the vector field gives the velocity in a fluid flow, the field lines are also called **streamlines** or **flow lines** of the flow; some of these are shown for the air flow in Figure 15.3. For a force field, the field lines are called **lines of force**.

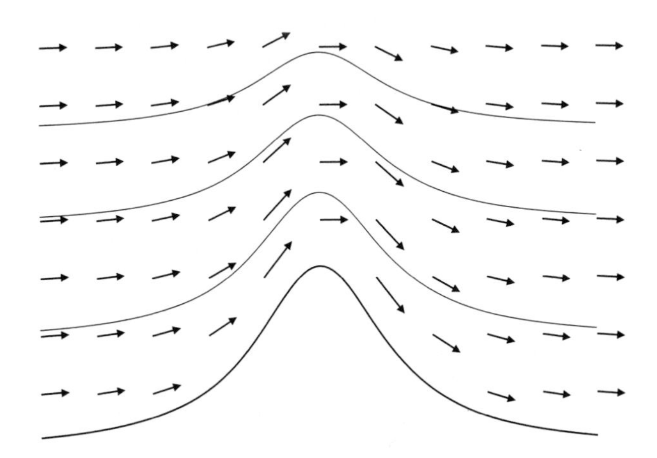

Figure 15.3 The velocity field and some streamlines of wind blowing over a hill

The field lines of **F** do not depend on the magnitude of **F** at any point but only on the direction of the field. If the field line through some point has parametric equation $\mathbf{r} = \mathbf{r}(t)$, then its tangent vector $d\mathbf{r}/dt$ must be parallel to $\mathbf{F}(\mathbf{r}(t))$ for all t. Thus,

$$\frac{d\mathbf{r}}{dt} = \lambda(t)\mathbf{F}\big(\mathbf{r}(t)\big).$$

For *some* vector fields, this differential equation can be integrated to find the field lines. If we break the equation into components,

$$\frac{dx}{dt} = \lambda(t)F_1(x, y, z), \quad \frac{dy}{dt} = \lambda(t)F_2(x, y, z), \quad \frac{dz}{dt} = \lambda(t)F_3(x, y, z),$$

we can obtain equivalent differential expressions for $\lambda(t)\,dt$ and hence write the differential equation for the field lines in the form

$$\frac{dx}{F_1(x, y, z)} = \frac{dy}{F_2(x, y, z)} = \frac{dz}{F_3(x, y, z)}.$$

If multiplication of these differential equations by some function puts them in the form

$$P(x)\,dx = Q(y)\,dy = R(z)\,dz,$$

then we can integrate all three expressions to find the field lines.

EXAMPLE 3 Find the field lines of the gravitational force field of Example 1:

$$\mathbf{F}(x,y,z) = -km\,\frac{(x-x_0)\mathbf{i} + (y-y_0)\mathbf{j} + (z-z_0)\mathbf{k}}{\big((x-x_0)^2 + (y-y_0)^2 + (z-z_0)^2\big)^{3/2}}.$$

Solution The vector in the numerator of the fraction gives the direction of **F**. Therefore, the field lines satisfy the system

$$\frac{dx}{x-x_0} = \frac{dy}{y-y_0} = \frac{dz}{z-z_0}.$$

Integrating all three expressions leads to

$$\ln|x-x_0| + \ln C_1 = \ln|y-y_0| + \ln C_2 = \ln|z-z_0| + \ln C_3,$$

or, on taking exponentials,

$$C_1(x-x_0) = C_2(y-y_0) = C_3(z-z_0).$$

This represents two families of planes all passing through $P_0 = (x_0, y_0, z_0)$. The field lines are the intersections of planes from each of the families, so they are straight lines through the point P_0. (This is a *two-parameter* family of lines; any one of the constants C_i that is nonzero can be divided out of the equations above.) The nature of the field lines should also be apparent from the plot of the vector field in Figure 15.1.

EXAMPLE 4 Find the field lines of the velocity field $\mathbf{v} = \Omega(-y\mathbf{i} + x\mathbf{j})$ of Example 2.

Solution The field lines satisfy the differential equation

$$\frac{dx}{-y} = \frac{dy}{x}.$$

We can separate variables in this equation to get $x\,dx = -y\,dy$. Integration then gives $x^2/2 = -y^2/2 + C/2$, or $x^2 + y^2 = C$. Thus, the field lines are circles centred at the origin in the xy-plane, as is also apparent from the vector field plot in Figure 15.2. If we regard $\mathbf{v}$ as a vector field in 3-space, we find that the field lines are horizontal circles centred on the z-axis:

$$x^2 + y^2 = C_1, \qquad z = C_2.$$

Our ability to find field lines depends on our ability to solve differential equations and, in 3-space, systems of differential equations.

EXAMPLE 5 Find the field lines of $\mathbf{F} = xz\mathbf{i} + 2x^2z\mathbf{j} + x^2\mathbf{k}$.

Solution The field lines satisfy $\dfrac{dx}{xz} = \dfrac{dy}{2x^2z} = \dfrac{dz}{x^2}$, or, equivalently

$$dy = 2x\,dx \qquad \text{and} \qquad dy = 2z\,dz.$$

The field lines are the curves of intersection of the two families $y = x^2 + C_1$ and $y = z^2 + C_2$ of parabolic cylinders.

Vector Fields in Polar Coordinates

A vector field in the plane can be expressed in terms of polar coordinates in the form

$$\mathbf{F} = \mathbf{F}(r, \theta) = F_r(r, \theta)\hat{\mathbf{r}} + F_\theta(r, \theta)\hat{\boldsymbol{\theta}},$$

where $\hat{\mathbf{r}}$ and $\hat{\boldsymbol{\theta}}$, defined everywhere except at the origin by

$$\hat{\mathbf{r}} = \cos\theta\mathbf{i} + \sin\theta\mathbf{j}$$
$$\hat{\boldsymbol{\theta}} = -\sin\theta\mathbf{i} + \cos\theta\mathbf{j},$$

are unit vectors in the direction of increasing r and θ at $[r, \theta]$. Note that $d\hat{\mathbf{r}}/d\theta = \hat{\boldsymbol{\theta}}$ and that $\hat{\boldsymbol{\theta}}$ is just $\hat{\mathbf{r}}$ rotated 90° counterclockwise. Also note that we are using F_r and F_θ to denote the components of $\mathbf{F}$ with respect to the basis $\{\hat{\mathbf{r}}, \hat{\boldsymbol{\theta}}\}$; the subscripts do not indicate partial derivatives. Here, $F_r(r, \theta)$ is called the *radial* component of $\mathbf{F}$, and $F_\theta(r, \theta)$ is called the *transverse* component.

A curve with polar equation $r = r(\theta)$ can be expressed in vector parametric form,

$$\mathbf{r} = r\hat{\mathbf{r}},$$

as we did in Section 11.6. This curve is a field line of $\mathbf{F}$ if its differential tangent vector

$$d\mathbf{r} = dr\,\hat{\mathbf{r}} + r\,\frac{d\hat{\mathbf{r}}}{d\theta}\,d\theta = dr\,\hat{\mathbf{r}} + r\,d\theta\,\hat{\boldsymbol{\theta}}$$

is parallel to the field vector $\mathbf{F}(r, \theta)$ at any point except the origin, that is, if $r = f(\theta)$ satisfies the differential equation

$$\frac{dr}{F_r(r, \theta)} = \frac{r\,d\theta}{F_\theta(r, \theta)}.$$

In specific cases we can find the field lines by solving this equation.

EXAMPLE 6 Sketch the vector field $\mathbf{F}(r, \theta) = \hat{\mathbf{r}} + \hat{\boldsymbol{\theta}}$, and find its field lines. Sketch several field lines.

Solution At each point $[r, \theta]$, the field vector bisects the angle between $\hat{\mathbf{r}}$ and $\hat{\boldsymbol{\theta}}$, making a counterclockwise angle of 45° with $\hat{\mathbf{r}}$. All of the vectors in the field have the same length, $\sqrt{2}$. Some of them are shown in Figure 15.4(a). They suggest that the field lines will spiral outward from the origin. Since $F_r(r, \theta) = F_\theta(r, \theta) = 1$ for this field, the field lines satisfy $dr = r\,d\theta$, or, dividing by $d\theta$, $dr/d\theta = r$. This is the differential equation of exponential growth and has solution $r = Ke^\theta$, or, equivalently, $r = e^{\theta+\alpha}$, where $\alpha = \ln K$ is a constant. Several such curves are shown in Figure 15.4(b).

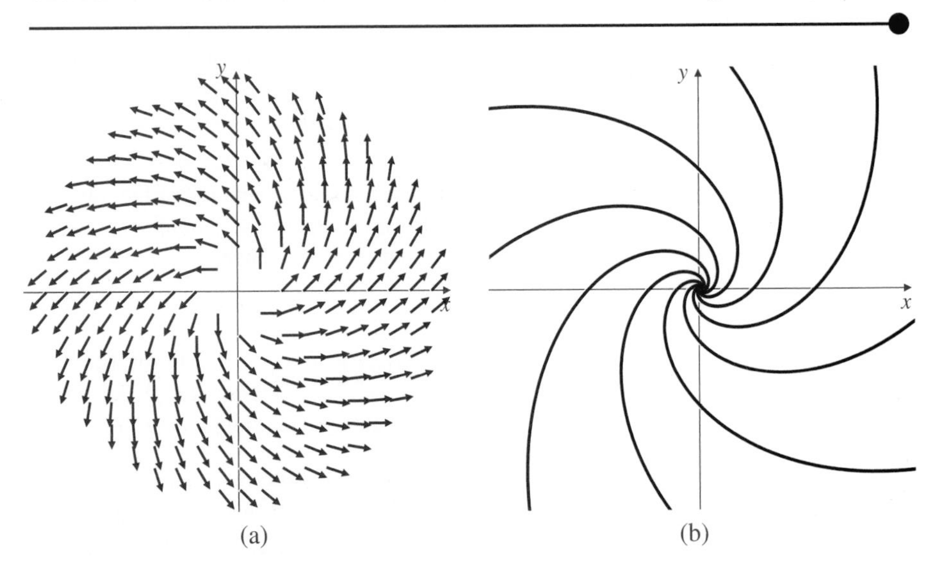

Figure 15.4
(a) The vector field $\mathbf{F} = \hat{\mathbf{r}} + \hat{\boldsymbol{\theta}}$
(b) Field lines of $\mathbf{F} = \hat{\mathbf{r}} + \hat{\boldsymbol{\theta}}$

Nonlinear Systems and Liapunov Functions

Many differential equations that arise in applications are nonlinear and cannot easily be solved. However, such an equation can sometimes be associated with a vector field in such a way that useful information about the behaviour of solutions of the differential equation can be obtained by examining the vector field.

An example is the Van der Pol equation $x'' - \mu(1-x^2)x' + x = 0$, which arises in connection with electrical circuits, where the independent variable on which x depends is time. If we take the constant $\mu = -1$, this equation can be rewritten as a first-order system of equations by substituting $x' = y$. The first-order system is

$$\begin{cases} x' = y \\ y' = -x + y(x^2 - 1), \end{cases}$$

and is associated with the vector field

$$\mathbf{F} = x'\mathbf{i} + y'\mathbf{j} = y\,\mathbf{i} + \big((-x + y(x^2 - 1)\big)\,\mathbf{j}.$$

We examine what the structure of this field implies about the solutions (x, y) of the linear system and hence about the Van der Pol equation.

One definitive property that fields and their associated trajectories have is the location and nature of "fixed points," which are the zeros of the vector field, or critical points of the first-order system. Since the "velocity" $\mathbf{F}(x, y) = \mathbf{0} = 0\mathbf{i} + 0\mathbf{j}$ there, movement along trajectories must stop at those points. Fixed points provide important insight into the solutions of differential equations and their visualization, helping us to have confidence in approximate solution methods. A key property of a fixed point is whether it is stable or not. Generally speaking, stability of a fixed point means that all trajectories near a fixed point trap any "particle" travelling on them so that it remains near the fixed point (**weak stability**), or, more stringently, so that it approaches the fixed point (**asymptotic stability**). For the case of the Van der Pol equation, $(x, y) = (0, 0)$ is clearly a fixed point.

Can we determine whether the fixed points of fields are stable or not, without solving the differential equations, and without resorting to approximate methods such as those used with computers? One powerful method for doing so is to use a Liapunov function in conjunction with the vector field. A **Liapunov function** is a positive function $V(x, y)$ that is decreasing toward the fixed point and that vanishes at the fixed point. One can always define many such functions for any point, but a Liapunov function must not only decrease, but must decrease along every trajectory of the vector field approaching the fixed point. Thus, for weak stability, we require $dV/dt = \nabla V \bullet \mathbf{F} \leq 0$ near the fixed point, and for asymptotic stability, we require $dV/dt < 0$ near the fixed point. Since ∇V is an outward normal to level curves of V that surround the fixed point, it follows for asymptotic stability that $\mathbf{F}$ points inward, across level curves of the Liapunov function, and thus "particles" moving along its trajectories get trapped in successively smaller domains surrounding the fixed point as t increases. Clearly, if the derivative is positive instead of negative, the fixed point is certainly unstable.

The mere *existence* of a Liapunov function with a negative derivative along trajectories of a vector field confirms stability of a fixed point. The entire test depends on a matter of existence. If a Liapunov function is not found, this does not prove or disprove stability.

EXAMPLE 7 **(A Liapunov function for a Van der Pol equation)** Show that the point $(0, 0)$ is an asymptotically stable fixed point of the Van der Pol vector field (case $\mu = -1$) given above.

Solution Substituting $(x, y) = (0, 0)$ into the vector field expressions of the Van der Pol equation, yields, $\mathbf{F} = x'\mathbf{i} + y'\mathbf{j} = y\mathbf{i} + (-x + y(x^2 - 1))\mathbf{j} = 0\mathbf{i} + 0\mathbf{j} = \mathbf{0}$, which confirms that $(0, 0)$ is a fixed point.

Note that a poor guess at a Liapunov function would be $V(x, y) = 2x^2 + y^2$. While it is positive and vanishes at $(0, 0)$, it fails to meet the requirement that its time derivative is always negative near $(0, 0)$: for instance, at points $(2y, y)$ arbitrarily close to $(0, 0)$, we have

$$\frac{dV}{dt} = 4xy + 2y\big(-x + y(x^2 - 1)\big) = 2y^2 + 8y^4 > 0.$$

We can do better with $V(x, y) = x^2 + y^2$. In this case,

$$\frac{dV}{dt} = 2xx' + 2yy' = 2xy + 2y\big(-x + y(x^2 - 1)\big) = 2y^2(x^2 - 1) \le 0$$

whenever $x^2 < 1$. This shows that the fixed point $(0, 0)$ is at least weakly stable, but it does not imply asymptotic stability because $dV/dt = 0$ if $y = 0$.

We could try something more general, like $V(x, y) = ax^2 + bxy + cy^2$ and attempt to choose the values of $a > 0$, $c > 0$, and b satisfying $b^2 < 4ac$ (why?) so that whenever (x, y) is sufficiently close (but not equal) to $(0, 0)$, we have $dV/dt < 0$. It turns out that we can make $V(x, y) = x^2 + xy + y^2$ work. For this V, we have

$$\begin{aligned} V(x, y) &= \left(x + \frac{y}{2}\right)^2 + \frac{3y^2}{4} > 0 \quad \text{if}(x, y) \ne (0, 0) \\ \frac{dV}{dt} &= 2x\frac{dx}{dt} + y\frac{dx}{dt} + x\frac{dy}{dt} + 2y\frac{dy}{dt} \\ &= 2xy + y^2 - x^2 + xy(x^2 - 1) - 2xy + 2y^2(x^2 - 1). \end{aligned}$$

If $x^2 < \frac{1}{4}$, then $-1 \le x^2 - 1 \le -\frac{3}{4}$, $xy(x^2 - 1) \le |x||y|$, and $2y^2(x^2 - 1) \le -\frac{3}{2}y^2$. Hence,

$$\frac{dV}{dt} \le -\frac{1}{2}y^2 - x^2 + |x||y| = -\left(|x| + \frac{|y|}{2}\right)^2 - \frac{y^2}{4} < 0,$$

unless $(x, y) = (0, 0)$. Thus, $(0, 0)$ is asymptotically stable.

Remark Sometimes the search for Liapunov functions can be very difficult, involving the use of computers to search for and then test candidate functions.

EXERCISES 15.1

In Exercises 1–8, sketch the given plane vector field and determine its field lines.

1. $\mathbf{F}(x, y) = x\mathbf{i} + x\mathbf{j}$

2. $\mathbf{F}(x, y) = x\mathbf{i} + y\mathbf{j}$

3. $\mathbf{F}(x, y) = y\mathbf{i} + x\mathbf{j}$

4. $\mathbf{F}(x, y) = \mathbf{i} + \sin x\,\mathbf{j}$

5. $\mathbf{F}(x, y) = e^x\mathbf{i} + e^{-x}\mathbf{j}$

6. $\mathbf{F}(x, y) = \nabla(x^2 - y)$

7. $\mathbf{F}(x, y) = \nabla \ln(x^2 + y^2)$

8. $\mathbf{F}(x, y) = \cos y\,\mathbf{i} - \cos x\,\mathbf{j}$

In Exercises 9–16, describe the streamlines of the given velocity fields.

9. $\mathbf{v}(x, y, z) = y\mathbf{i} - y\mathbf{j} - y\mathbf{k}$

10. $\mathbf{v}(x, y, z) = x\mathbf{i} + y\mathbf{j} - x\mathbf{k}$

11. $\mathbf{v}(x, y, z) = y\mathbf{i} - x\mathbf{j} + \mathbf{k}$

12. $\mathbf{v}(x, y, z) = \dfrac{x\mathbf{i} + y\mathbf{j}}{(1 + z^2)(x^2 + y^2)}$

13. $\mathbf{v}(x, y, z) = xz\mathbf{i} + yz\mathbf{j} + x\mathbf{k}$

14. $\mathbf{v}(x, y, z) = e^{xyz}(x\mathbf{i} + y^2\mathbf{j} + z\mathbf{k})$

15. $\mathbf{v}(x, y) = x^2\mathbf{i} - y\mathbf{j}$

16. $\mathbf{v}(x, y) = x\mathbf{i} + (x + y)\mathbf{j}$ *Hint:* Let $y = xv(x)$.

In Exercises 17–20, determine the field lines of the given polar vector fields.

17. $\mathbf{F} = \hat{\mathbf{r}} + r\hat{\boldsymbol{\theta}}$

18. $\mathbf{F} = \hat{\mathbf{r}} + \theta\hat{\boldsymbol{\theta}}$

19. $\mathbf{F} = 2\hat{\mathbf{r}} + \theta\hat{\boldsymbol{\theta}}$

20. $\mathbf{F} = r\hat{\mathbf{r}} - \hat{\boldsymbol{\theta}}$

21. Consider the Van der Pol equation with $\mu = 1$, so the corresponding vector field is $\mathbf{F} = y\mathbf{i} + \big(-x + y(1 - x^2)\big)\mathbf{j}$. Use $V(x, y) = x^2 - xy + y^2$ as in Example 7 to determine the stability of the the fixed point $(0, 0)$.

22. Consider the vector field of the Van der Pol equation when $\mu = 0$. Use the Liapunov function, $V(x, y) = x^2 + y^2$, to attempt to determine the stability of the fixed point (0,0). Explain the result.

23. In Example 7, using the simpler Liapunov function, $V(x, y) = x^2 + y^2$, we found $V' = 2y^2(x^2 - 1) \le 0$. This was not sufficient to establish asymptotic stability in itself because $V' = 0$ occurs when $y = 0$. Zeros of V' form a curve, in this case given by the entire x axis, which all occur when $x' = 0$. Curves defined by one component of the vector field vanishing are known as **nulclines**. The zeros of V' occur on one nulcline (i.e., $y = 0$). Write an expression for another nulcline of the Van der Pol vector field of Example 7.

24. Give an alternative solution to Example 7 by using the fact that the simpler Liapunov function in the previous exercise is given by $V = r^2$ in polar coordinates. Show explicitly that all trajectories of the Van der Pol field (for $\mu = -1$) crossing the x axis stop moving toward $(0, 0)$, by showing that $r(t)$ has a critical point. Then classify the associated critical point of $r(t)$ to demonstrate asymptotic stability.

15.2 Conservative Fields

Since the gradient of a scalar field is a vector field, it is natural to ask whether every vector field is the gradient of a scalar field. Given a vector field $\mathbf{F}(x, y, z)$, does there exist a scalar field $\phi(x, y, z)$ such that

$$\mathbf{F}(x, y, z) = \nabla\phi(x, y, z) = \frac{\partial \phi}{\partial x}\mathbf{i} + \frac{\partial \phi}{\partial y}\mathbf{j} + \frac{\partial \phi}{\partial z}\mathbf{k}\,?$$

The answer in general is "no." Only special vector fields can be written in this way.

DEFINITION

If $\mathbf{F}(x, y, z) = \nabla\phi(x, y, z)$ in a domain D, then we say that $\mathbf{F}$ is a **conservative** vector field in D, and we call the function ϕ a **(scalar) potential** for $\mathbf{F}$ on D. Similar definitions hold in the plane or in n-space.

Like antiderivatives, potentials are not determined uniquely; arbitrary constants can be added to them. Note that $\mathbf{F}$ is **conservative in a domain** D if and only if $\mathbf{F} = \nabla\phi$ at *every* point of D; the potential ϕ cannot have any singular points in D.

The equation $F_1(x, y, z)\,dx + F_2(x, y, z)\,dy + F_3(x, y, z)\,dz = 0$ is called an **exact** differential equation if the left side is the differential of a scalar function $\phi(x, y, z)$:

$$d\phi = F_1(x, y, z)\,dx + F_2(x, y, z)\,dy + F_3(x, y, z)\,dz.$$

In this case the differential equation has solutions given by $\phi(x, y, z) = C$ (constant). (See Section 17.2 for a discussion of exact equations in the plane.) Observe that the differential equation is exact if and only if the vector field $\mathbf{F} = F_1\mathbf{i} + F_2\mathbf{j} + F_3\mathbf{k}$ is conservative and that ϕ is the potential of $\mathbf{F}$.

Being scalar fields rather than vector fields, potentials for conservative vector fields are easier to manipulate algebraically than the vector fields themselves. For instance, a sum of potential functions is the potential function for the sum of the corresponding vector fields. A vector field can always be computed from its potential function by taking the gradient.

EXAMPLE 1 **(The gravitational field of a point mass is conservative)** Show that the gravitational field $\mathbf{F}(\mathbf{r}) = -km(\mathbf{r} - \mathbf{r}_0)/|\mathbf{r} - \mathbf{r}_0|^3$ of Example 1 in Section 15.1 is conservative wherever it is defined (i.e., everywhere in $\mathbb{R}^3$ except at $\mathbf{r}_0$), by showing that

$$\phi(x, y, z) = \frac{km}{|\mathbf{r} - \mathbf{r}_0|} = \frac{km}{\sqrt{(x - x_0)^2 + (y - y_0)^2 + (z - z_0)^2}}$$

is a potential function for $\mathbf{F}$.

Solution Observe that

$$\frac{\partial \phi}{\partial x} = \frac{-km(x-x_0)}{\left((x-x_0)^2+(y-y_0)^2+(z-z_0)^2\right)^{3/2}} = \frac{-km(x-x_0)}{|\mathbf{r}-\mathbf{r}_0|^3} = F_1(\mathbf{r}),$$

and similar formulas hold for the other partial derivatives of ϕ. It follows that $\nabla\phi(x, y, z) = \mathbf{F}(x, y, z)$ for $(x, y, z) \neq (x_0, y_0, z_0)$, and $\mathbf{F}$ is conservative except at $\mathbf{r}_0$.

Remark It is not necessary to write the expression $km/|\mathbf{r}-\mathbf{r}_0|$ in terms of the components of $\mathbf{r}-\mathbf{r}_0$ as we did in Example 1 in order to calculate its partial derivatives. Here is a useful formula for the derivative of the length of a vector function $\mathbf{F}$ with respect to a variable x:

$$\frac{\partial}{\partial x}|\mathbf{F}| = \frac{\mathbf{F} \bullet \left(\dfrac{\partial}{\partial x}\mathbf{F}\right)}{|\mathbf{F}|}.$$

To see why this is true, express $|\mathbf{F}| = \sqrt{\mathbf{F} \bullet \mathbf{F}}$, and calculate its derivative using the Chain Rule and the Product Rule:

$$\frac{\partial}{\partial x}|\mathbf{F}| = \frac{\partial}{\partial x}\sqrt{\mathbf{F} \bullet \mathbf{F}} = \frac{1}{2\sqrt{\mathbf{F} \bullet \mathbf{F}}}\, 2\mathbf{F} \bullet \left(\frac{\partial}{\partial x}\mathbf{F}\right) = \frac{\mathbf{F} \bullet \left(\frac{\partial}{\partial x}\mathbf{F}\right)}{|\mathbf{F}|}.$$

Compare this with the derivative of an absolute value of a function of one variable:

$$\frac{d}{dx}|f(x)| = \operatorname{sgn}(f(x))\, f'(x) = \frac{f(x)}{|f(x)|}\, f'(x).$$

In the context of Example 1, we have

$$\frac{\partial}{\partial x}\frac{km}{|\mathbf{r}-\mathbf{r}_0|} = \frac{-km}{|\mathbf{r}-\mathbf{r}_0|^2}\frac{\partial}{\partial x}|\mathbf{r}-\mathbf{r}_0| = \frac{-km}{|\mathbf{r}-\mathbf{r}_0|^2}\frac{(\mathbf{r}-\mathbf{r}_0) \bullet \mathbf{i}}{|\mathbf{r}-\mathbf{r}_0|} = \frac{-km(x-x_0)}{|\mathbf{r}-\mathbf{r}_0|^3},$$

with similar expressions for the other partials of $km/|\mathbf{r}-\mathbf{r}_0|$.

EXAMPLE 2 Show that the velocity field $\mathbf{v} = -\Omega y\mathbf{i} + \Omega x\mathbf{j}$ of rigid body rotation about the z-axis (see Example 2 of Section 15.1) is not conservative if $\Omega \neq 0$.

Solution There are two ways to show that no potential for $\mathbf{v}$ can exist. One way is to try to find a potential $\phi(x, y)$ for the vector field. We require

$$\frac{\partial \phi}{\partial x} = -\Omega y \qquad \text{and} \qquad \frac{\partial \phi}{\partial y} = \Omega x.$$

The first of these equations implies that $\phi(x, y) = -\Omega xy + C_1(y)$. (We have integrated with respect to x; the constant can still depend on y.) Similarly, the second equation implies that $\phi(x, y) = \Omega xy + C_2(x)$. Therefore, we must have $-\Omega xy + C_1(y) = \Omega xy + C_2(x)$, or $2\Omega xy = C_1(y) - C_2(x)$ for all (x, y). This is not possible for any choice of the single-variable functions $C_1(y)$ and $C_2(x)$ unless $\Omega = 0$.

Alternatively, if $\mathbf{v}$ has a potential ϕ, then we can form the mixed partial derivatives of ϕ from the two equations above and get

$$\frac{\partial^2 \phi}{\partial y \partial x} = -\Omega \qquad \text{and} \qquad \frac{\partial^2 \phi}{\partial x \partial y} = \Omega.$$

This is not possible if $\Omega \neq 0$ because the smoothness of **v** implies that its potential should be smooth, so the mixed partials should be equal. Thus, no such ϕ can exist; **v** is not conservative.

Example 2 suggests a condition that must be satisfied by any conservative plane vector field.

BEWARE! Do not confuse this **necessary condition** with a **sufficient condition** to guarantee that **F** is conservative. We will show later that more than just $\partial F_1/\partial y = \partial F_2/\partial x$ on D is necessary to guarantee that **F** is conservative on D.

Necessary condition for a conservative plane vector field

If $\mathbf{F}(x, y) = F_1(x, y)\mathbf{i} + F_2(x, y)\mathbf{j}$ is a conservative vector field in a domain D of the xy-plane, then the condition

$$\frac{\partial}{\partial y}F_1(x, y) = \frac{\partial}{\partial x}F_2(x, y)$$

must be satisfied at all points of D.

To see this, observe that

$$F_1\mathbf{i} + F_2\mathbf{j} = \mathbf{F} = \nabla\phi = \frac{\partial\phi}{\partial x}\mathbf{i} + \frac{\partial\phi}{\partial y}\mathbf{j}$$

implies the two scalar equations

$$F_1 = \frac{\partial\phi}{\partial x} \qquad \text{and} \qquad F_2 = \frac{\partial\phi}{\partial y},$$

and since the mixed partial derivatives of ϕ should be equal,

$$\frac{\partial F_1}{\partial y} = \frac{\partial^2\phi}{\partial y\partial x} = \frac{\partial^2\phi}{\partial x\partial y} = \frac{\partial F_2}{\partial x}.$$

A similar condition obtains for vector fields in 3-space.

Necessary conditions for a conservative vector field in 3-space

If $\mathbf{F}(x, y, z) = F_1(x, y, z)\mathbf{i} + F_2(x, y, z)\mathbf{j} + F_3(x, y, z)\mathbf{k}$ is a conservative vector field in a domain D in 3-space, then we must have, everywhere in D,

$$\frac{\partial F_1}{\partial y} = \frac{\partial F_2}{\partial x}, \qquad \frac{\partial F_1}{\partial z} = \frac{\partial F_3}{\partial x}, \qquad \frac{\partial F_2}{\partial z} = \frac{\partial F_3}{\partial y}.$$

Equipotential Surfaces and Curves

If $\phi(x, y, z)$ is a potential function for the conservative vector field **F**, then the *level surfaces* $\phi(x, y, z) = C$ of ϕ are called **equipotential surfaces** of **F**. Since $\mathbf{F} = \nabla\phi$ is normal to these surfaces (wherever it does not vanish), the field lines of **F** always intersect the equipotential surfaces at right angles. For instance, the equipotential surfaces of the gravitational force field of a point mass are spheres centred at the point; these spheres are normal to the field lines, which are straight lines passing through the point. Similarly, for a conservative plane vector field, the *level curves* of the potential function are called **equipotential curves** of the vector field. They are the **orthogonal trajectories** of the field lines; that is, they intersect the field lines at right angles.

EXAMPLE 3 Show that the vector field $\mathbf{F}(x, y) = x\mathbf{i} - y\mathbf{j}$ is conservative and find a potential function for it. Describe the field lines and the equipotential curves.

Solution Since $\partial F_1/\partial y = 0 = \partial F_2/\partial x$ everywhere in $\mathbb{R}^2$, we would expect $\mathbf{F}$ to be conservative. Any potential function ϕ must satisfy

$$\frac{\partial \phi}{\partial x} = F_1 = x \qquad \text{and} \qquad \frac{\partial \phi}{\partial y} = F_2 = -y.$$

The first of these equations gives

$$\phi(x, y) = \int x\, dx = \frac{1}{2}x^2 + C_1(y).$$

Observe that, since the integral is taken with respect to x, the "constant" of integration is allowed to depend on the other variable. Now we use the second equation to get

$$-y = \frac{\partial \phi}{\partial y} = C_1{}'(y) \quad \Rightarrow \quad C_1(y) = -\frac{1}{2}y^2 + C_2.$$

Thus, $\mathbf{F}$ is conservative and, for any constant C_2,

$$\phi(x, y) = \frac{x^2 - y^2}{2} + C_2$$

is a potential function for $\mathbf{F}$. The field lines of $\mathbf{F}$ satisfy

$$\frac{dx}{x} = -\frac{dy}{y} \quad \Rightarrow \quad \ln|x| = -\ln|y| + \ln C_3 \quad \Rightarrow \quad xy = C_3.$$

The field lines of $\mathbf{F}$ are thus rectangular hyperbolas with the coordinate axes as asymptotes. The equipotential curves constitute another family of rectangular hyperbolas, $x^2 - y^2 = C_4$, with the lines $x = \pm y$ as asymptotes. Curves of the two families intersect at right angles. (See Figure 15.5.) Note, however, that $\mathbf{F}$ does not specify a direction at the origin and the orthogonality breaks down there; in fact, neither family has a unique curve through that point.

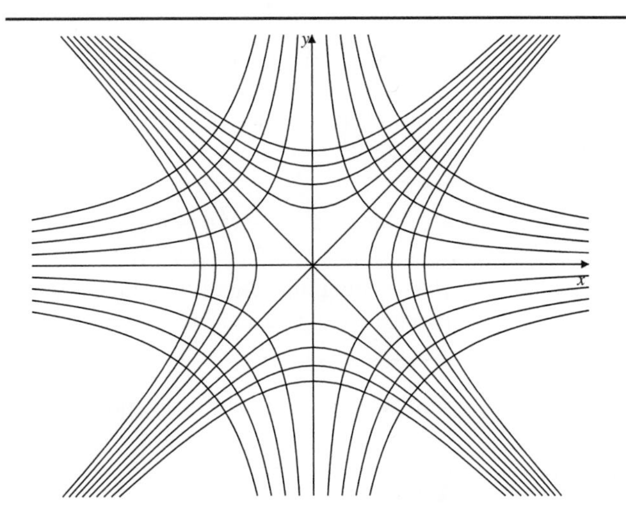

Figure 15.5 The field lines (black) and equipotential curves (colour) for the field $\mathbf{F} = x\mathbf{i} - y\mathbf{j}$

Remark In the above example we constructed the potential ϕ by first integrating $\partial\phi/\partial x = F_1$. We could equally well have started by integrating $\partial\phi/\partial y = F_2$, in which case the constant of integration would have depended on x. In the end, the same ϕ would have emerged.

EXAMPLE 4 Decide whether the vector field

$$\mathbf{F} = \left(xy - \sin z\right)\mathbf{i} + \left(\frac{1}{2}x^2 - \frac{e^y}{z}\right)\mathbf{j} + \left(\frac{e^y}{z^2} - x\cos z\right)\mathbf{k}$$

is conservative in $D = \{(x, y, z) : z \neq 0\}$, and find a potential if it is.

Solution Note that $\mathbf{F}$ is not defined when $z = 0$. However, since

$$\frac{\partial F_1}{\partial y} = x = \frac{\partial F_2}{\partial x}, \quad \frac{\partial F_1}{\partial z} = -\cos z = \frac{\partial F_3}{\partial x}, \quad \text{and} \quad \frac{\partial F_2}{\partial z} = \frac{e^y}{z^2} = \frac{\partial F_3}{\partial y},$$

$\mathbf{F}$ may still be conservative in domains not intersecting the xy-plane $z = 0$. If so, its potential ϕ should satisfy

$$\frac{\partial \phi}{\partial x} = xy - \sin z, \quad \frac{\partial \phi}{\partial y} = \frac{1}{2}x^2 - \frac{e^y}{z}, \quad \text{and} \quad \frac{\partial \phi}{\partial z} = \frac{e^y}{z^2} - x\cos z. \tag{$*$}$$

From the first equation of $(*)$,

$$\phi(x, y, z) = \int (xy - \sin z)\, dx = \frac{1}{2}x^2 y - x \sin z + C_1(y, z).$$

(Again, note that the constant of integration can be a function of any parameters of the integrand; it is constant only with respect to the variable of integration.) Using the second equation of $(*)$, we obtain

$$\frac{1}{2}x^2 - \frac{e^y}{z} = \frac{\partial \phi}{\partial y} = \frac{1}{2}x^2 + \frac{\partial C_1(y, z)}{\partial y}.$$

Thus,

$$C_1(y, z) = -\int \frac{e^y}{z}\, dy = -\frac{e^y}{z} + C_2(z)$$

and

$$\phi(x, y, z) = \frac{1}{2}x^2 y - x \sin z - \frac{e^y}{z} + C_2(z).$$

Finally, using the third equation of $(*)$,

$$\frac{e^y}{z^2} - x\cos z = \frac{\partial \phi}{\partial z} = -x\cos z + \frac{e^y}{z^2} + C_2{}'(z).$$

Thus, $C_2{}'(z) = 0$ and $C_2(z) = C$ (a constant). Indeed, $\mathbf{F}$ is conservative and, for any constant C,

$$\phi(x, y, z) = \frac{1}{2}x^2 y - x \sin z - \frac{e^y}{z} + C$$

is a potential function for $\mathbf{F}$ in the given domain D. C may have different values in the two regions $z > 0$ and $z < 0$ whose union constitutes D.

Remark If, in the above solution, the differential equation for $C_1(y, z)$ had involved x or if that for $C_2(z)$ had involved either x or y, we would not have been able to find ϕ. This did not happen because of the three conditions on the partials of F_1, F_2, and F_3 verified at the outset.

Remark The existence of a potential for a vector field depends on the *topology* of the domain of the field (i.e., whether the domain has *holes* in it and what kind of holes) as well as on the structure of the components of the field itself. (Even if the necessary conditions given above are satisfied, a vector field may not be conservative in a domain that has *holes*.) We will be probing further into the nature of conservative vector fields in Section 15.4 and in the next chapter; we will eventually show that the above *necessary conditions* are also *sufficient* to guarantee that $\mathbf{F}$ is conservative if the domain of $\mathbf{F}$ satisfies certain conditions. At this point, however, we give an example in which a plane vector field fails to be conservative on a domain where the necessary condition is, nevertheless, satisfied.

EXAMPLE 5 For $(x, y) \neq (0, 0)$, define a vector field $\mathbf{F}(x, y)$ and a scalar field $\theta(x, y)$ as follows:

$$\mathbf{F}(x, y) = \left(\frac{-y}{x^2 + y^2}\right)\mathbf{i} + \left(\frac{x}{x^2 + y^2}\right)\mathbf{j}$$

$$\theta(x, y) = \text{the polar angle } \theta \text{ of } (x, y) \text{ such that } 0 \leq \theta < 2\pi.$$

Thus, $x = r\cos\theta(x, y)$ and $y = r\sin\theta(x, y)$, where $r^2 = x^2 + y^2$. Verify the following:

(a) $\dfrac{\partial}{\partial y}F_1(x, y) = \dfrac{\partial}{\partial x}F_2(x, y)$ for $(x, y) \neq (0, 0)$.

(b) $\nabla\theta(x, y) = \mathbf{F}(x, y)$ for all $(x, y) \neq (0, 0)$ such that $0 < \theta < 2\pi$.

(c) $\mathbf{F}$ is not conservative on the whole xy-plane excluding the origin.

Solution

(a) We have $F_1 = \dfrac{-y}{x^2 + y^2}$ and $F_2 = \dfrac{x}{x^2 + y^2}$. Thus,

$$\begin{aligned}\frac{\partial}{\partial y}F_1(x, y) &= \frac{\partial}{\partial y}\left(-\frac{y}{x^2 + y^2}\right) = \frac{y^2 - x^2}{(x^2 + y^2)^2} = \frac{\partial}{\partial x}\left(\frac{x}{x^2 + y^2}\right) \\ &= \frac{\partial}{\partial x}F_2(x, y)\end{aligned}$$

for all $(x, y) \neq (0, 0)$.

(b) Differentiate the equations $x = r\cos\theta$ and $y = r\sin\theta$ implicitly with respect to x to obtain

$$\begin{aligned}1 &= \frac{\partial x}{\partial x} = \frac{\partial r}{\partial x}\cos\theta - r\sin\theta\frac{\partial\theta}{\partial x}, \\ 0 &= \frac{\partial y}{\partial x} = \frac{\partial r}{\partial x}\sin\theta + r\cos\theta\frac{\partial\theta}{\partial x}.\end{aligned}$$

Eliminating $\partial r/\partial x$ from this pair of equations and solving for $\partial\theta/\partial x$ leads to

$$\frac{\partial\theta}{\partial x} = -\frac{r\sin\theta}{r^2} = -\frac{y}{x^2 + y^2} = F_1.$$

Similarly, differentiation with respect to y produces

$$\frac{\partial\theta}{\partial y} = \frac{x}{x^2 + y^2} = F_2.$$

These formulas hold only if $0 < \theta < 2\pi$; θ is not even continuous on the positive x-axis; if $x > 0$, then

$$\lim_{y\to 0+}\theta(x, y) = 0 \qquad \text{but} \qquad \lim_{y\to 0-}\theta(x, y) = 2\pi.$$

Thus, $\nabla\theta = \mathbf{F}$ holds everywhere in the plane, except at points $(x, 0)$ where $x \geq 0$.

(c) Suppose that $\mathbf{F}$ is conservative on the whole plane excluding the origin. Then $\mathbf{F} = \nabla\phi$ there, for some scalar function $\phi(x, y)$. Then $\nabla(\theta - \phi) = \mathbf{0}$ for $0 < \theta < 2\pi$, and $\theta - \phi = C$ (constant), or $\theta = \phi + C$. The left side of this equation is discontinuous along the positive x-axis but the right side is not. Therefore, the two sides cannot be equal. This contradiction shows that $\mathbf{F}$ cannot be conservative on the whole plane, excluding the origin.

Remark Observe that the origin $(0, 0)$ is a *hole* in the domain of $\mathbf{F}$ in the above example. While $\mathbf{F}$ satisfies the necessary condition for being conservative everywhere except at this hole, you must remove from the domain of $\mathbf{F}$ a half-line (ray), or, more generally, a curve from the origin to infinity in order to get a potential function for $\mathbf{F}$. $\mathbf{F}$ is *not* conservative on any domain containing a curve that surrounds the origin. Exercises 22–24 of Section 15.4 will shed further light on this situation.

Sources, Sinks, and Dipoles

Imagine that 3-space is filled with an incompressible fluid emitted by a point source at the origin at a volume rate $dV/dt = 4\pi m$. (We say that the origin is a **source** of strength m.) By symmetry, the fluid flows outward on radial lines from the origin with equal speed at equal distances from the origin in all directions, and the fluid emitted at the origin at some instant $t = 0$ will at later time t be spread over a spherical surface of radius $r = r(t)$. All the fluid inside that sphere was emitted in the time interval $[0, t]$, so we have

$$\frac{4}{3}\pi r^3 = 4\pi mt.$$

Differentiating this equation with respect to t we obtain $r^2(dr/dt) = m$, and the outward speed of the fluid at distance r from the origin is $v(r) = m/r^2$. The velocity field of the moving fluid is therefore

$$\mathbf{v}(\mathbf{r}) = v(r)\frac{\mathbf{r}}{|\mathbf{r}|} = \frac{m}{r^3}\mathbf{r}.$$

This velocity field is conservative (except at the origin) and has potential

$$\phi(\mathbf{r}) = -\frac{m}{r}.$$

A **sink** is a negative source. A sink of strength m at the origin (which annihilates or sucks up fluid at a rate $dV/dt = 4\pi m$) has velocity field and potential given by

$$\mathbf{v}(\mathbf{r}) = -\frac{m}{r^3}\mathbf{r} \qquad \text{and} \qquad \phi(\mathbf{r}) = \frac{m}{r}.$$

The potentials or velocity fields of sources or sinks located at other points are obtained by translation of these formulas; for instance, the velocity field of a source of strength m at the point with position vector $\mathbf{r}_0$ is

$$\mathbf{v}(\mathbf{r}) = -\nabla\left(\frac{m}{|\mathbf{r} - \mathbf{r}_0|}\right) = \frac{m}{|\mathbf{r} - \mathbf{r}_0|^3}(\mathbf{r} - \mathbf{r}_0).$$

This should be compared with the gravitational force field due to a mass m at the origin. The two are the same except for sign and a constant related to units of measurement. For this reason we regard a point mass as a sink for its own gravitational field. Similarly, the electrostatic field due to a point charge q at $\mathbf{r}_0$ is the field of a source (or sink if $q < 0$) of strength proportional to q; if units of measurement are suitably chosen we have

$$\mathbf{E}(\mathbf{r}) = -\nabla\left(\frac{q}{|\mathbf{r} - \mathbf{r}_0|}\right) = \frac{q}{|\mathbf{r} - \mathbf{r}_0|^3}(\mathbf{r} - \mathbf{r}_0).$$

In general, the field lines of a vector field converge at a source or sink of that field.

A **dipole** is a system consisting of a source and a sink of equal strength m separated by a short distance ℓ. The product $\mu = m\ell$ is called the **dipole moment**, and the line containing the source and sink is called the **axis** of the dipole. Real physical dipoles, such as magnets, are frequently modelled by *ideal* dipoles that are the limits of such real dipoles as $m \to \infty$ and $\ell \to 0$ in such a way that the dipole moment μ remains constant.

EXAMPLE 6 Calculate the velocity field, $\mathbf{v}(x, y, z)$, associated with a dipole of moment μ located at the origin and having axis along the z-axis.

Solution We start with a source of strength m at position $(0, 0, \ell/2)$ and a sink of strength m at $(0, 0, -\ell/2)$. The potential of this system is

$$\phi(\mathbf{r}) = -m\left(\frac{1}{|\mathbf{r} - \frac{1}{2}\ell\mathbf{k}|} - \frac{1}{|\mathbf{r} + \frac{1}{2}\ell\mathbf{k}|}\right).$$

The potential of the ideal dipole is the limit of the potential of this system as $m \to \infty$ and $\ell \to 0$ in such a way that $m\ell = \mu$:

$$\begin{aligned}
\phi(\mathbf{r}) &= \lim_{\substack{\ell\to 0\\ ml=\mu}} -m\left(\frac{|\mathbf{r} + \frac{1}{2}\ell\mathbf{k}| - |\mathbf{r} - \frac{1}{2}\ell\mathbf{k}|}{|\mathbf{r} + \frac{1}{2}\ell\mathbf{k}|\,|\mathbf{r} - \frac{1}{2}\ell\mathbf{k}|}\right)\\
&= -\frac{\mu}{|\mathbf{r}|^2}\lim_{\ell\to 0}\frac{|\mathbf{r} + \frac{1}{2}\ell\mathbf{k}| - |\mathbf{r} - \frac{1}{2}\ell\mathbf{k}|}{\ell}
\end{aligned}$$

(now use l'Hôpital's Rule and the rule for differentiating lengths of vectors)

$$\begin{aligned}
&= -\frac{\mu}{|\mathbf{r}|^2}\lim_{\ell\to 0}\frac{\dfrac{(\mathbf{r} + \frac{1}{2}\ell\mathbf{k})\bullet\frac{1}{2}\mathbf{k}}{|\mathbf{r} + \frac{1}{2}\ell\mathbf{k}|} - \dfrac{(\mathbf{r} - \frac{1}{2}\ell\mathbf{k})\bullet(-\frac{1}{2}\mathbf{k})}{|\mathbf{r} - \frac{1}{2}\ell\mathbf{k}|}}{1}\\
&= -\frac{\mu}{|\mathbf{r}|^2}\lim_{\ell\to 0}\left(\frac{\frac{1}{2}z + \frac{1}{4}\ell}{|\mathbf{r} + \frac{1}{2}\ell\mathbf{k}|} + \frac{\frac{1}{2}z - \frac{1}{4}\ell}{|\mathbf{r} - \frac{1}{2}\ell\mathbf{k}|}\right)\\
&= -\frac{\mu z}{|\mathbf{r}|^3}.
\end{aligned}$$

The required velocity field is the gradient of this potential. We have

$$\begin{aligned}
\frac{\partial\phi}{\partial x} &= \frac{3\,\mu z}{|\mathbf{r}|^4}\frac{\mathbf{r}\bullet\mathbf{i}}{|\mathbf{r}|} = \frac{3\mu xz}{|\mathbf{r}|^5}\\
\frac{\partial\phi}{\partial y} &= \frac{3\mu yz}{|\mathbf{r}|^5}\\
\frac{\partial\phi}{\partial z} &= -\frac{\mu}{|\mathbf{r}|^3} + \frac{3\mu z^2}{|\mathbf{r}|^5} = \frac{\mu(2z^2 - x^2 - y^2)}{|\mathbf{r}|^5}\\
\mathbf{v}(\mathbf{r}) = \nabla\phi(\mathbf{r}) &= \frac{\mu}{|\mathbf{r}|^5}\big(3xz\mathbf{i} + 3yz\mathbf{j} + (2z^2 - x^2 - y^2)\mathbf{k}\big).
\end{aligned}$$

Some streamlines for a plane cross-section containing the z-axis are shown in Figure 15.6.

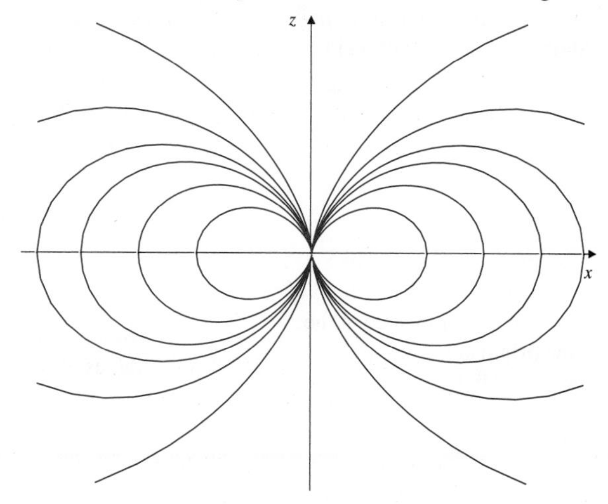

Figure 15.6 Streamlines of a dipole

EXERCISES 15.2

In Exercises 1–6, determine whether the given vector field is conservative, and find a potential if it is.

1. $\mathbf{F}(x, y, z) = x\mathbf{i} - 2y\mathbf{j} + 3z\mathbf{k}$
2. $\mathbf{F}(x, y, z) = y\mathbf{i} + x\mathbf{j} + z^2\mathbf{k}$
3. $\mathbf{F}(x, y) = \dfrac{x\mathbf{i} - y\mathbf{j}}{x^2 + y^2}$
4. $\mathbf{F}(x, y) = \dfrac{x\mathbf{i} + y\mathbf{j}}{x^2 + y^2}$
5. $\mathbf{F}(x, y, z) = (2xy - z^2)\mathbf{i} + (2yz + x^2)\mathbf{j} - (2zx - y^2)\mathbf{k}$
6. $\mathbf{F}(x, y, z) = e^{x^2+y^2+z^2}(xz\mathbf{i} + yz\mathbf{j} + xy\mathbf{k})$
7. Find the three-dimensional vector field with potential $\phi(\mathbf{r}) = \dfrac{1}{|\mathbf{r} - \mathbf{r}_0|^2}$.
8. Calculate $\nabla \ln |\mathbf{r}|$, where $\mathbf{r} = x\mathbf{i} + y\mathbf{j} + z\mathbf{k}$.
9. Show that the vector field

$$\mathbf{F}(x, y, z) = \frac{2x}{z}\mathbf{i} + \frac{2y}{z}\mathbf{j} - \frac{x^2 + y^2}{z^2}\mathbf{k}$$

is conservative, and find its potential. Describe the equipotential surfaces. Find the field lines of **F**.

10. Repeat Exercise 9 for the field

$$\mathbf{F}(x, y, z) = \frac{2x}{z}\mathbf{i} + \frac{2y}{z}\mathbf{j} + \left(1 - \frac{x^2 + y^2}{z^2}\right)\mathbf{k}.$$

11. Find the velocity field due to two sources of strength m, one located at $(0, 0, \ell)$ and the other at $(0, 0, -\ell)$. Where is the velocity zero? Find the velocity at any point $(x, y, 0)$ in the xy-plane. Where in the xy-plane is the speed greatest?
12. Find the velocity field for a system consisting of a source of strength 2 at the origin and a sink of strength 1 at $(0, 0, 1)$. Show that the velocity is vertical at all points of a certain sphere. Sketch the streamlines of the flow.

Exercises 13–18 provide an analysis of two-dimensional sources and dipoles similar to that developed for three dimensions in the text.

13. In 3-space filled with an incompressible fluid, we say that the z-axis is a **line source** of strength m if every interval Δz along that axis emits fluid at volume rate $dV/dt = 2\pi m \Delta z$. The fluid then spreads out symmetrically in all directions perpendicular to the z-axis. Show that the velocity field of the flow is

$$\mathbf{v} = \frac{m}{x^2 + y^2}(x\mathbf{i} + y\mathbf{j}).$$

14. The flow in Exercise 13 is two-dimensional because **v** depends only on x and y and has no component in the z direction. Regarded as a *plane* vector field, it is the field of a two-dimensional point source of strength m located at the origin (i.e., fluid is emitted at the origin at the *areal rate* $dA/dt = 2\pi m$). Show that the vector field is conservative, and find a potential function $\phi(x, y)$ for it.
15. Find the potential, ϕ, and the field, $\mathbf{F} = \nabla\phi$, for a two-dimensional dipole at the origin, with axis in the y direction and dipole moment μ. Such a dipole is the limit of a system consisting of a source of strength m at $(0, \ell/2)$ and a sink of strength m at $(0, -\ell/2)$, as $\ell \to 0$ and $m \to \infty$ such that $m\ell = \mu$.
16. Show that the equipotential curves of the two-dimensional dipole in Exercise 15 are circles tangent to the x-axis at the origin.
17. Show that the streamlines (field lines) of the two-dimensional dipole in Exercises 15 and 16 are circles tangent to the y-axis at the origin. *Hint:* It is possible to do this geometrically. If you choose to do it by setting up a differential equation, you may find the change of dependent variable

$$y = vx, \qquad \frac{dy}{dx} = v + x\frac{dv}{dx}$$

useful for integrating the equation.

18. Show that the velocity field of a line source of strength $2m$ can be found by integrating the (three-dimensional) velocity field of a point source of strength $m\,dz$ at $(0, 0, z)$ over the whole z-axis. Why does the integral correspond to a line source of strength $2m$ rather than strength m? Can the potential of the line source be obtained by integrating the potentials of the point sources?
19. Show that the gradient of a function expressed in terms of polar coordinates in the plane is

$$\nabla\phi(r, \theta) = \frac{\partial\phi}{\partial r}\hat{\mathbf{r}} + \frac{1}{r}\frac{\partial\phi}{\partial\theta}\hat{\boldsymbol{\theta}}.$$

(This is a repeat of Exercise 16 in Section 12.7.)

20. Use the result of Exercise 19 to show that a necessary condition for the vector field

$$\mathbf{F}(r, \theta) = F_r(r, \theta)\hat{\mathbf{r}} + F_\theta(r, \theta)\hat{\boldsymbol{\theta}}$$

(expressed in terms of polar coordinates) to be conservative is that

$$\frac{\partial F_r}{\partial\theta} - r\frac{\partial F_\theta}{\partial r} = F_\theta.$$

21. Show that $\mathbf{F} = r\sin 2\theta\hat{\mathbf{r}} + r\cos 2\theta\hat{\boldsymbol{\theta}}$ is conservative, and find a potential for it.
22. For what values of the constants α and β is the vector field

$$\mathbf{F} = r^2\cos\theta\hat{\mathbf{r}} + \alpha r^\beta \sin\theta\hat{\boldsymbol{\theta}}$$

conservative? Find a potential for **F** if α and β have these values.

15.3 Line Integrals

The definite integral, $\int_a^b f(x)\,dx$, represents the *total amount* of a quantity distributed along the x-axis between a and b in terms of the *line density*, $f(x)$, of that quantity at point x. The amount of the quantity in an *infinitesimal* interval of length dx at x is $f(x)\,dx$, and the integral adds up these infinitesimal contributions (or *elements*) to give the total amount of the quantity. Similarly, the integrals $\iint_D f(x,y)\,dA$ and $\iiint_R f(x,y,z)\,dV$ represent the total amounts of quantities distributed over regions D in the plane and R in 3-space in terms of the *areal* or *volume* densities of these quantities.

It may happen that a quantity is distributed with specified line density along a *curve* in the plane or in 3-space, or with specified areal density over a *surface* in 3-space. In such cases we require *line integrals* or *surface integrals* to add up the contributing elements and calculate the total quantity. We examine line integrals in this section and the next and surface integrals in Sections 15.5 and 15.6.

Let $\mathcal{C}$ be a bounded, continuous parametric curve in $\mathbb{R}^3$. Recall (from Section 11.1) that $\mathcal{C}$ is a *smooth curve* if it has a parametrization of the form

$$\mathbf{r} = \mathbf{r}(t) = x(t)\mathbf{i} + y(t)\mathbf{j} + z(t)\mathbf{k}, \qquad t \text{ in interval } I,$$

with "velocity" vector $\mathbf{v} = d\mathbf{r}/dt$ continuous and nonzero. We will call $\mathcal{C}$ a **smooth arc** if it is a smooth curve with *finite* parameter interval $I = [a,b]$.

In Section 11.3 we saw how to calculate the length of $\mathcal{C}$ by subdividing it into short arcs using points corresponding to parameter values

$$a = t_0 < t_1 < t_2 < \cdots < t_{n-1} < t_n = b,$$

adding up the lengths $|\Delta\mathbf{r}_i| = |\mathbf{r}_i - \mathbf{r}_{i-1}|$ of line segments joining these points, and taking the limit as the maximum distance between adjacent points approached zero. The length was denoted

$$\int_{\mathcal{C}} ds$$

and is a special example of a line integral along $\mathcal{C}$ having integrand 1.

The line integral of a general function $f(x,y,z)$ can be defined similarly. We choose a point (x_i^*, y_i^*, z_i^*) on the ith subarc and form the Riemann sum

$$S_n = \sum_{i=1}^{n} f(x_i^*, y_i^*, z_i^*)\,|\Delta\mathbf{r}_i|.$$

If this sum has a limit as $\max|\Delta\mathbf{r}_i| \to 0$, independent of the particular choices of the points (x_i^*, y_i^*, z_i^*), then we call this limit the **line integral** of f along $\mathcal{C}$ and denote it

$$\int_{\mathcal{C}} f(x,y,z)\,ds.$$

If $\mathcal{C}$ is a smooth arc and if f is continuous on $\mathcal{C}$, then the limit will certainly exist; its value is given by a definite integral of a continuous function, as shown in the next paragraph. It will also exist (for continuous f) if $\mathcal{C}$ is **piecewise smooth**, consisting of finitely many smooth arcs linked end to end; in this case the line integral of f along $\mathcal{C}$ is the sum of the line integrals of f along each of the smooth arcs. Improper line integrals can also be considered, where f has discontinuities or where the length of a curve is not finite.

Evaluating Line Integrals

The length of $\mathcal{C}$ was evaluated by expressing the arc length element $ds = |d\mathbf{r}/dt|\,dt$ in terms of a parametrization $\mathbf{r} = \mathbf{r}(t)$, $(a \le t \le b)$ of the curve, and integrating this from $t = a$ to $t = b$:

$$\text{length of } \mathcal{C} = \int_{\mathcal{C}} ds = \int_a^b \left|\frac{d\mathbf{r}}{dt}\right| dt.$$

More general line integrals are evaluated similarly:

$$\int_{\mathcal{C}} f(x, y, z)\, ds = \int_a^b f\big(\mathbf{r}(t)\big) \left|\frac{d\mathbf{r}}{dt}\right| dt.$$

Of course, all of the above discussion applies equally well to line integrals of functions $f(x, y)$ along curves $\mathcal{C}$ in the xy-plane.

Remark It should be noted that the value of the line integral of a function f along a curve $\mathcal{C}$ depends on f and $\mathcal{C}$ but not on the particular way $\mathcal{C}$ is parametrized. If $\mathbf{r} = \mathbf{r}^*(u)$, $\alpha \le u \le \beta$, is another parametrization of the same smooth curve $\mathcal{C}$, then any point $\mathbf{r}(t)$ on $\mathcal{C}$ can be expressed in terms of the new parametrization as $\mathbf{r}^*(u)$, where u depends on t: $u = u(t)$. If $\mathbf{r}^*(u)$ traces $\mathcal{C}$ in the same direction as $\mathbf{r}(t)$, then $u(a) = \alpha$, $u(b) = \beta$, and $du/dt \ge 0$; if $\mathbf{r}^*(u)$ traces $\mathcal{C}$ in the opposite direction, then $u(a) = \beta$, $u(b) = \alpha$, and $du/dt \le 0$. In either event,

$$\int_a^b f\big(\mathbf{r}(t)\big) \left|\frac{d\mathbf{r}}{dt}\right| dt = \int_a^b f\big(\mathbf{r}^*(u(t))\big) \left|\frac{d\mathbf{r}^*}{du}\frac{du}{dt}\right| dt = \int_\alpha^\beta f\big(\mathbf{r}^*(u)\big) \left|\frac{d\mathbf{r}^*}{du}\right| du.$$

Thus, the line integral is *independent of parametrization* of the curve $\mathcal{C}$. The following example illustrates this fact.

EXAMPLE 1 Evaluate $I = \displaystyle\int_{\mathcal{C}} (x^2 + y^2)\, ds$, where $\mathcal{C}$ is the straight line from the origin to the point $(2, 1)$.

Solution $\mathcal{C}$ can be parametrized $x = 2t$, $y = t$, for $0 \le t \le 1$, that is,

$$\mathbf{r} = 2t\mathbf{i} + t\mathbf{j},\ 0 \le t \le 1, \quad \text{so that} \quad ds = \left|\frac{d\mathbf{r}}{dt}\right| dt = |2\mathbf{i} + \mathbf{j}|\, dt = \sqrt{5}\, dt.$$

Thus, we have

$$I = \int_0^1 (4t^2 + t^2)\sqrt{5}\, dt = 5\sqrt{5} \int_0^1 t^2\, dt = \frac{5\sqrt{5}}{3}.$$

EXAMPLE 2 A circle of radius $a > 0$ has centre at the origin in the xy-plane. Let $\mathcal{C}$ be the half of this circle lying in the half-plane $y \ge 0$. Use two different parametrizations of $\mathcal{C}$ to find the moment of $\mathcal{C}$ about $y = 0$.

Solution We are asked to calculate $\displaystyle\int_{\mathcal{C}} y\, ds$.

$\mathcal{C}$ can be parametrized as $\mathbf{r} = a\cos t\mathbf{i} + a\sin t\mathbf{j}$, $(0 \le t \le \pi)$. Therefore,

$$\frac{d\mathbf{r}}{dt} = -a\sin t\mathbf{i} + a\cos t\mathbf{j} \qquad \text{and} \qquad \left|\frac{d\mathbf{r}}{dt}\right| = a,$$

and the moment of $\mathcal{C}$ about $y = 0$ is

$$\int_{\mathcal{C}} y\, ds = \int_0^\pi a\sin t\, a\, dt = -a^2\cos t\Big|_0^\pi = 2a^2.$$

$\mathcal{C}$ can also be parametrized $\mathbf{r} = x\mathbf{i} + \sqrt{a^2 - x^2}\mathbf{j}$, $(-a \le x \le a)$, for which we have

$$\frac{d\mathbf{r}}{dx} = \mathbf{i} - \frac{x}{\sqrt{a^2 - x^2}}\mathbf{j},$$

$$\left|\frac{d\mathbf{r}}{dx}\right| = \sqrt{1 + \frac{x^2}{a^2 - x^2}} = \frac{a}{\sqrt{a^2 - x^2}}.$$

Thus, the moment of $\mathcal{C}$ about $y = 0$ is

$$\int_{\mathcal{C}} y\, ds = \int_{-a}^{a} \sqrt{a^2 - x^2}\, \frac{a}{\sqrt{a^2 - x^2}}\, dx = a\int_{-a}^{a} dx = 2a^2.$$

It is comforting to get the same answer using different parametrizations. Unlike the line integrals of vector fields considered in the next section, the line integrals of scalar fields considered here do not depend on the direction (orientation) of $\mathcal{C}$. The two parametrizations of the semicircle were in opposite directions but still gave the same result.

Line integrals frequently lead to definite integrals that are very difficult or impossible to evaluate without using numerical techniques. Only very simple curves and ones that have been contrived to lead to simple expressions for ds are amenable to exact calculation of line integrals.

EXAMPLE 3 Find the centroid of the circular helix $\mathcal{C}$ given by

$$\mathbf{r} = a\cos t\,\mathbf{i} + a\sin t\,\mathbf{j} + bt\,\mathbf{k}, \qquad 0 \le t \le 2\pi.$$

Solution As we observed in Example 6 of Section 11.3, for this helix $ds = \sqrt{a^2 + b^2}\,dt$. On the helix we have $z = bt$, so its moment about $z = 0$ is

$$M_{z=0} = \int_{\mathcal{C}} z\, ds = b\sqrt{a^2 + b^2}\int_0^{2\pi} t\, dt = 2\pi^2 b\sqrt{a^2 + b^2}.$$

Since the helix has length $L = 2\pi\sqrt{a^2 + b^2}$, the z-component of its centroid is $M_{z=0}/L = \pi b$. The moment of the helix about $x = 0$ is

$$M_{x=0} = \int_{\mathcal{C}} x\, ds = a\sqrt{a^2 + b^2}\int_0^{2\pi} \cos t\, dt = 0,$$

$$M_{y=0} = \int_{\mathcal{C}} y\, ds = a\sqrt{a^2 + b^2}\int_0^{2\pi} \sin t\, dt = 0.$$

Thus, the centroid is $(0, 0, \pi b)$.

Sometimes a curve, along which a line integral is to be taken, is specified as the intersection of two surfaces with given equations. It is normally necessary to parametrize the curve in order to evaluate a line integral. Recall from Section 11.3 that if one of the surfaces is a cylinder parallel to one of the coordinate axes, it is usually easiest to begin by parametrizing that cylinder. (Otherwise, combine the equations to eliminate one variable and thus obtain such a cylinder on which the curve lies.)

EXAMPLE 4 Find the mass of a wire lying along the first octant part $\mathcal{C}$ of the curve of intersection of the elliptic paraboloid $z = 2 - x^2 - 2y^2$ and the parabolic cylinder $z = x^2$ between $(0, 1, 0)$ and $(1, 0, 1)$ (see Figure 15.7) if the density of the wire at position (x, y, z) is $\delta(x, y, z) = xy$.

Solution We need a convenient parametrization of $\mathcal{C}$. Since the curve $\mathcal{C}$ lies on the cylinder $z = x^2$ and x goes from 0 to 1, we can let $x = t$ and $z = t^2$. Thus, $2y^2 = 2 - x^2 - z = 2 - 2t^2$, so $y^2 = 1 - t^2$. Since $\mathcal{C}$ lies in the first octant, it can be parametrized by

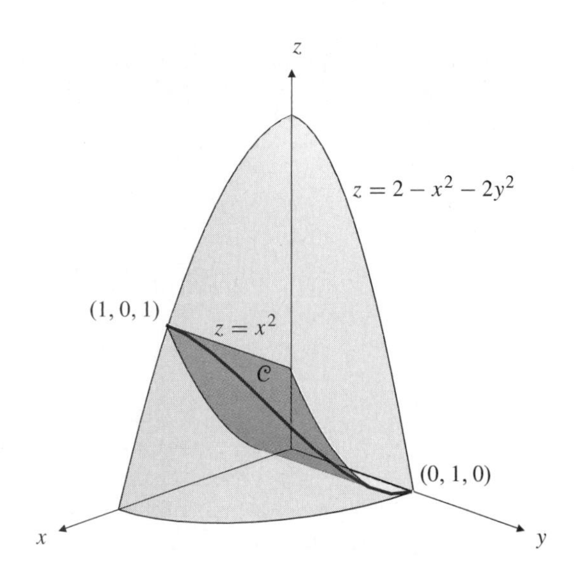

Figure 15.7 The curve of intersection of $z = x^2$ and $z = 2 - x^2 - 2y^2$

$$x = t, \qquad y = \sqrt{1 - t^2}, \qquad z = t^2, \qquad (0 \le t \le 1).$$

Then $dx/dt = 1$, $dy/dt = -t/\sqrt{1 - t^2}$, and $dz/dt = 2t$, so

$$ds = \sqrt{1 + \frac{t^2}{1 - t^2} + 4t^2}\, dt = \frac{\sqrt{1 + 4t^2 - 4t^4}}{\sqrt{1 - t^2}}\, dt.$$

Hence, the mass of the wire is

$$\begin{aligned}
m &= \int_{\mathcal{C}} xy\, ds = \int_0^1 t\sqrt{1 - t^2}\frac{\sqrt{1 + 4t^2 - 4t^4}}{\sqrt{1 - t^2}}\, dt \\
&= \int_0^1 t\sqrt{1 + 4t^2 - 4t^4}\, dt \qquad \text{Let } u = t^2 \\
&= \frac{1}{2}\int_0^1 \sqrt{1 + 4u - 4u^2}\, du \\
&= \frac{1}{2}\int_0^1 \sqrt{2 - (2u - 1)^2}\, du \qquad \text{Let } v = 2u - 1 \\
&= \frac{1}{4}\int_{-1}^1 \sqrt{2 - v^2}\, dv = \frac{1}{2}\int_0^1 \sqrt{2 - v^2}\, dv \\
&= \frac{1}{2}\left(\frac{\pi}{4} + \frac{1}{2}\right) = \frac{\pi + 2}{8}.
\end{aligned}$$

(The final integral above was evaluated by interpreting it as the area of part of a circle. You are invited to supply the details. It can also be done by the substitution $v = \sqrt{2}\sin w$.)

EXERCISES 15.3

In Exercises 1–2 evaluate the given line integral over the specified curve $\mathcal{C}$.

1. $\displaystyle\int_{\mathcal{C}} (x + y)\, ds, \quad \mathbf{r} = at\mathbf{i} + bt\mathbf{j} + ct\mathbf{k},\ 0 \le t \le m.$

2. $\displaystyle\int_{\mathcal{C}} y\, ds, \quad \mathbf{r} = t^2\mathbf{i} + t\mathbf{j} + t^2\mathbf{k},\ 0 \le t \le m.$

3. Show that the curve $\mathcal{C}$ given by

$$\mathbf{r} = a\cos t\sin t\,\mathbf{i} + a\sin^2 t\,\mathbf{j} + a\cos t\,\mathbf{k}, \quad (0 \le t \le \tfrac{\pi}{2}),$$

lies on a sphere centred at the origin. Find $\displaystyle\int_{\mathcal{C}} z\, ds$.

4. Let $\mathcal{C}$ be the conical helix with parametric equations $x = t\cos t$, $y = t\sin t$, $z = t$, $(0 \le t \le 2\pi)$. Find $\displaystyle\int_{\mathcal{C}} z\, ds$.

5. Find the mass of a wire along the curve

$$\mathbf{r} = 3t\mathbf{i} + 3t^2\mathbf{j} + 2t^3\mathbf{k}, \quad (0 \le t \le 1),$$

if the density at $\mathbf{r}(t)$ is $1 + t$ g/unit length.

6. Show that the curve $\mathcal{C}$ in Example 4 also has parametrization $x = \cos t$, $y = \sin t$, $z = \cos^2 t$, $(0 \le t \le \pi/2)$, and recalculate the mass of the wire in that example using this parametrization.

7. Find the moment of inertia about the z-axis (i.e., the value of $\delta \int_{\mathcal{C}} (x^2 + y^2)\, ds$), for a wire of constant density δ lying along the curve $\mathcal{C}$: $\mathbf{r} = e^t \cos t\mathbf{i} + e^t \sin t\mathbf{j} + t\mathbf{k}$, from $t = 0$ to $t = 2\pi$.

8. Evaluate $\int_{\mathcal{C}} e^z\, ds$, where $\mathcal{C}$ is the curve in Exercise 7.

9. Find $\int_{\mathcal{C}} x^2\, ds$ along the line of intersection of the two planes $x - y + z = 0$ and $x + y + 2z = 0$, from the origin to the point $(3, 1, -2)$.

10. Find $\int_{\mathcal{C}} \sqrt{1 + 4x^2z^2}\, ds$, where $\mathcal{C}$ is the curve of intersection of the surfaces $x^2 + z^2 = 1$ and $y = x^2$.

11. Find the mass and centre of mass of a wire bent in the shape of the circular helix $x = \cos t$, $y = \sin t$, $z = t$, $(0 \le t \le 2\pi)$, if the wire has line density given by $\delta(x, y, z) = z$.

12. Repeat Exercise 11 for the part of the wire corresponding to $0 \le t \le \pi$.

13. Find the moment of inertia about the y-axis of the curve $x = e^t$, $y = \sqrt{2}t$, $z = e^{-t}$, $(0 \le t \le 1)$, that is,

$$\int_{\mathcal{C}} (x^2 + z^2)\, ds.$$

14. Find the centroid of the curve in Exercise 13.

15. Find $\int_{\mathcal{C}} x\, ds$ along the first octant part of the curve of intersection of the cylinder $x^2 + y^2 = a^2$ and the plane $z = x$.

16. Find $\int_{\mathcal{C}} z\, ds$ along the part of the curve $x^2 + y^2 + z^2 = 1$, $x + y = 1$, where $z \ge 0$.

17. Find $\int_{\mathcal{C}} \frac{ds}{(2y^2 + 1)^{3/2}}$, where $\mathcal{C}$ is the parabola $z^2 = x^2 + y^2$, $x + z = 1$. *Hint:* Use $y = t$ as parameter.

18. Express as a definite integral, but do not try to evaluate, the value of $\int_{\mathcal{C}} xyz\, ds$, where $\mathcal{C}$ is the curve $y = x^2$, $z = y^2$ from $(0, 0, 0)$ to $(2, 4, 16)$.

19. The function

$$E(k, \phi) = \int_0^{\phi} \sqrt{1 - k^2 \sin^2 t}\, dt$$

is called the **elliptic integral function of the second kind**. The **complete elliptic integral** of the second kind is the function $E(k) = E(k, \pi/2)$. In terms of these functions, express the length of one complete revolution of the elliptic helix

$$x = a \cos t, \qquad y = b \sin t, \qquad z = ct,$$

where $0 < a < b$. What is the length of that part of the helix lying between $t = 0$ and $t = T$, where $0 < T < \pi/2$?

20. Evaluate $\int_L \frac{ds}{x^2 + y^2}$, where L is the entire straight line with equation $Ax + By = C$, $(C \ne 0)$. *Hint:* Use the symmetry of the integrand to replace the line with a line having a simpler equation but giving the same value to the integral.

15.4 Line Integrals of Vector Fields

In elementary physics the **work** done by a constant force of magnitude F in moving an object a distance d is defined to be the product of F and d: $W = Fd$. There is, however, a catch to this; it is understood that the force is exerted in the direction of motion of the object. If the object moves in a direction different from that of the force (because of some other forces acting on it), then the work done by the particular force is the product of the distance moved and the component of the force in the direction of motion. For instance, the work done by gravity in causing a 10 kg crate to slide 5 m down a ramp inclined at 45° to the horizontal is $W = 50g/\sqrt{2}$ N·m (where $g = 9.8$ m/s^2), since the scalar projection of the $10g$ N gravitational force on the crate in the direction of the ramp is $10g/\sqrt{2}$ N.

The work done by a *variable* force $\mathbf{F}(x, y, z) = \mathbf{F}(\mathbf{r})$, which depends continuously on position, in moving an object along a smooth curve $\mathcal{C}$ is the integral of *work elements* dW. The element dW corresponding to arc length element ds at position $\mathbf{r}$ on $\mathcal{C}$ is ds times the tangential component of the force $\mathbf{F}(\mathbf{r})$ along $\mathcal{C}$ in the direction of motion

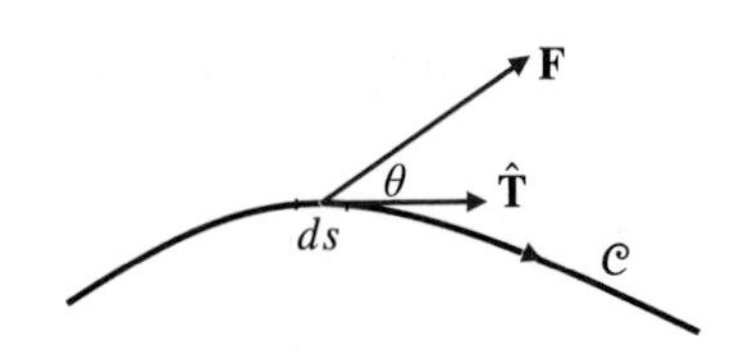

Figure 15.8 $dW = |\mathbf{F}|\cos\theta\,ds$
$= \mathbf{F} \bullet \hat{\mathbf{T}}\,ds$

(see Figure 15.8); since $\hat{\mathbf{T}} = d\mathbf{r}/ds$ is the unit tangent to $\mathcal{C}$,

$$dW = \mathbf{F}(\mathbf{r}) \bullet \hat{\mathbf{T}}\,ds = \mathbf{F} \bullet d\mathbf{r}.$$

Thus, the total work done by $\mathbf{F}$ in moving the object along $\mathcal{C}$ is

$$W = \int_{\mathcal{C}} \mathbf{F} \bullet \hat{\mathbf{T}}\,ds = \int_{\mathcal{C}} \mathbf{F} \bullet d\mathbf{r} = \int_{\mathcal{C}} F_1\,dx + F_2\,dy + F_3\,dz.$$

In general, if $\mathbf{F} = F_1\mathbf{i} + F_2\mathbf{j} + F_3\mathbf{k}$ is a continuous vector field, and $\mathcal{C}$ is an oriented smooth curve, then the **line integral of the tangential component of F** along $\mathcal{C}$ is

$$\begin{aligned}\int_{\mathcal{C}} \mathbf{F} \bullet d\mathbf{r} &= \int_{\mathcal{C}} \mathbf{F} \bullet \hat{\mathbf{T}}\,ds \\ &= \int_{\mathcal{C}} F_1(x, y, z)\,dx + F_2(x, y, z)\,dy + F_3(x, y, z)\,dz.\end{aligned}$$

Such a line integral is sometimes called, somewhat improperly, the line integral of $\mathbf{F}$ along $\mathcal{C}$. (It is not the line integral of $\mathbf{F}$, which should have a vector value, but rather the line integral of the *tangential component* of $\mathbf{F}$, which has a scalar value.) Unlike the line integral considered in the previous section, this line integral depends on the direction of the orientation of $\mathcal{C}$; reversing the direction of $\mathcal{C}$ causes this line integral to change sign.

If $\mathcal{C}$ is a closed curve, the line integral of the tangential component of $\mathbf{F}$ around $\mathcal{C}$ is also called the **circulation** of $\mathbf{F}$ around $\mathcal{C}$. The fact that the curve is closed is often indicated by a small circle drawn on the integral sign;

$$\oint_{\mathcal{C}} \mathbf{F} \bullet d\mathbf{r} \quad \text{denotes the circulation of } \mathbf{F} \text{ around the closed curve } \mathcal{C}.$$

Like the line integrals studied in the previous section, a line integral of a continuous vector field is converted into an ordinary definite integral by using a parametrization of the path of integration. For a smooth arc $\mathbf{r} = \mathbf{r}(t) = x(t)\mathbf{i} + y(t)\mathbf{j} + z(t)\mathbf{k}$, $(a \le t \le b)$, we have

$$\begin{aligned}\int_{\mathcal{C}} \mathbf{F} \bullet d\mathbf{r} &= \int_a^b \mathbf{F} \bullet \frac{d\mathbf{r}}{dt}\,dt \\ &= \int_a^b \bigg[F_1\Big(x(t), y(t), z(t)\Big)\frac{dx}{dt} + F_2\Big(x(t), y(t), z(t)\Big)\frac{dy}{dt} \\ &\qquad + F_3\Big(x(t), y(t), z(t)\Big)\frac{dz}{dt}\bigg]\,dt.\end{aligned}$$

Although this type of line integral changes sign if the orientation of $\mathcal{C}$ is reversed, it is otherwise independent of the particular parametrization used for $\mathcal{C}$. Again, a line integral over a piecewise smooth path is the sum of the line integrals over the individual smooth arcs constituting that path.

EXAMPLE 1 Let $\mathbf{F}(x, y) = y^2\mathbf{i} + 2xy\mathbf{j}$. Evaluate the line integral

$$\int_{\mathcal{C}} \mathbf{F} \bullet d\mathbf{r}$$

from $(0, 0)$ to $(1, 1)$ along

(a) the straight line $y = x$,

(b) the curve $y = x^2$, and

(c) the piecewise smooth path consisting of the straight line segments from $(0, 0)$ to $(0, 1)$ and from $(0, 1)$ to $(1, 1)$.

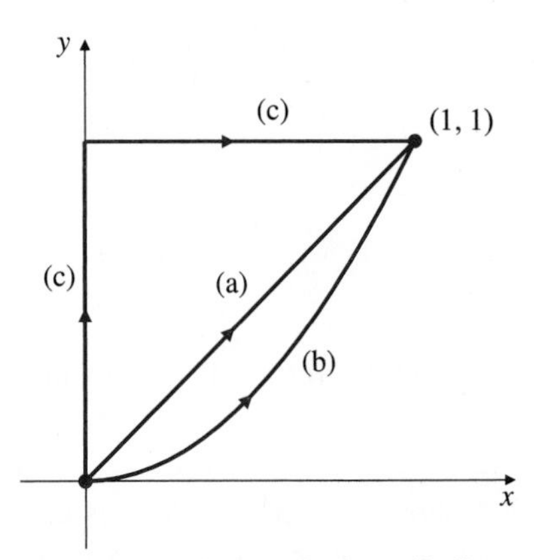

Figure 15.9 Three paths from $(0, 0)$ to $(1, 1)$

Solution The three paths are shown in Figure 15.9. The straight path (a) can be parametrized $\mathbf{r} = t\mathbf{i} + t\mathbf{j}$, $0 \le t \le 1$. Thus, $d\mathbf{r} = dt\mathbf{i} + dt\mathbf{j}$ and

$$\mathbf{F} \bullet d\mathbf{r} = (t^2\mathbf{i} + 2t^2\mathbf{j}) \bullet (\mathbf{i} + \mathbf{j})dt = 3t^2\, dt.$$

Therefore,

$$\int_{\mathcal{C}} \mathbf{F} \bullet d\mathbf{r} = \int_0^1 3t^2\, dt = t^3\Big|_0^1 = 1.$$

The parabolic path (b) can be parametrized $\mathbf{r} = t\mathbf{i} + t^2\mathbf{j}$, $0 \le t \le 1$, so that $d\mathbf{r} = dt\mathbf{i} + 2t\,dt\mathbf{j}$. Thus,

$$\mathbf{F} \bullet d\mathbf{r} = (t^4\mathbf{i} + 2t^3\mathbf{j}) \bullet (\mathbf{i} + 2t\mathbf{j})\, dt = 5t^4\, dt,$$

and

$$\int_{\mathcal{C}} \mathbf{F} \bullet d\mathbf{r} = \int_0^1 5t^4\, dt = t^5\Big|_0^1 = 1.$$

The third path (c) is made up of two segments, and we parametrize each separately. Let us use y as the parameter on the vertical segment (where $x = 0$ and $dx = 0$) and x as the parameter on the horizontal segment (where $y = 1$ and $dy = 0$):

$$\begin{aligned}\int_{\mathcal{C}} \mathbf{F} \bullet d\mathbf{r} &= \int_{\mathcal{C}} y^2\, dx + 2xy\, dy \\ &= \int_0^1 (0)\, dy + \int_0^1 (1)\, dx = 1.\end{aligned}$$

In view of these results, we might ask whether $\int_{\mathcal{C}} \mathbf{F} \bullet d\mathbf{r}$ is the same along *every* path from $(0, 0)$ to $(1, 1)$.

EXAMPLE 2 Let $\mathbf{F} = y\mathbf{i} - x\mathbf{j}$. Find $\int_{\mathcal{C}} \mathbf{F} \bullet d\mathbf{r}$ from $(1, 0)$ to $(0, -1)$ along

(a) the straight line segment joining these points and
(b) three-quarters of the circle of unit radius centred at the origin and traversed counterclockwise.

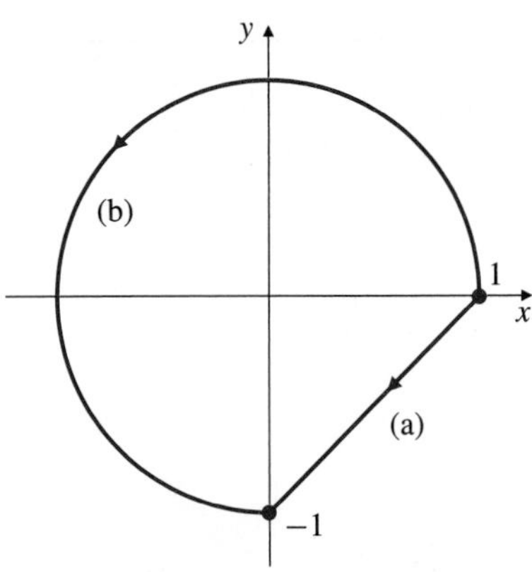

Figure 15.10 Two paths from $(1, 0)$ to $(0, -1)$

Solution Both paths are shown in Figure 15.10. The straight path (a) can be parametrized:

$$\mathbf{r} = (1 - t)\mathbf{i} - t\mathbf{j}, \qquad 0 \le t \le 1.$$

Thus, $d\mathbf{r} = -dt\mathbf{i} - dt\mathbf{j}$, and

$$\int_{\mathcal{C}} \mathbf{F} \bullet d\mathbf{r} = \int_0^1 \Big((-t)(-dt) - (1 - t)(-dt)\Big) = \int_0^1 dt = 1.$$

The circular path (b) can be parametrized:

$$\mathbf{r} = \cos t\,\mathbf{i} + \sin t\,\mathbf{j}, \qquad 0 \le t \le \frac{3\pi}{2},$$

so that $d\mathbf{r} = -\sin t\, dt\mathbf{i} + \cos t\, dt\mathbf{j}$. Therefore,

$$\mathbf{F} \bullet d\mathbf{r} = -\sin^2 t\,dt - \cos^2 t\,dt = -dt,$$

and we have

$$\int_{\mathcal{C}} \mathbf{F} \bullet d\mathbf{r} = -\int_0^{3\pi/2} dt = -\frac{3\pi}{2}.$$

In this case the line integral depends on the path from $(1, 0)$ to $(0, -1)$ along which the integral is taken.

Some readers may have noticed that in Example 1 above the vector field **F** is conservative, while in Example 2 it is not. Theorem 1 below confirms the link between *independence of path* for a line integral of the tangential component of a vector field and the existence of a scalar potential function for that field. This and subsequent theorems require specific assumptions on the nature of the domain of the vector field **F**, so we need to formulate some topological definitions.

Connected and Simply Connected Domains

Recall that a set S in the plane (or in 3-space) is open if every point in S is the centre of a disk (or a ball) having positive radius and contained in S. If S is open and B is a set (possibly empty) of boundary points of S, then the set $D = S \cup B$ is called a **domain**. A domain cannot contain isolated points. It may be closed, but it must have interior points near any of its boundary points. (See Section 10.1 for a discussion of open and closed sets and interior and boundary points.)

DEFINITION 2

A domain D is said to be **connected**, if every pair of points P and Q in D can be joined by a piecewise smooth curve lying in D.

For instance, the set of points (x, y) in the plane satisfying $x > 0$, $y > 0$, and $x^2 + y^2 \leq 4$ is a connected domain, but the set of points satisfying $|x| > 1$ is not connected. (There is no path from $(-2, 0)$ to $(2, 0)$ lying entirely in $|x| > 1$.) The set of points (x, y, z) in 3-space satisfying $0 < z < 1$ is a connected domain, but the set satisfying $z \neq 0$ is not.

A closed curve is **simple** if it has no self-intersections other than beginning and ending at the same point. (For example, a circle is a simple closed curve.) Imagine an elastic band stretched in the shape of such a curve. If the elastic is infinitely shrinkable, it can contract down to a single point.

DEFINITION 3

A **simply connected domain** D is a connected domain in which every *simple closed curve* can be continuously shrunk to a point in D without any part ever passing out of D.

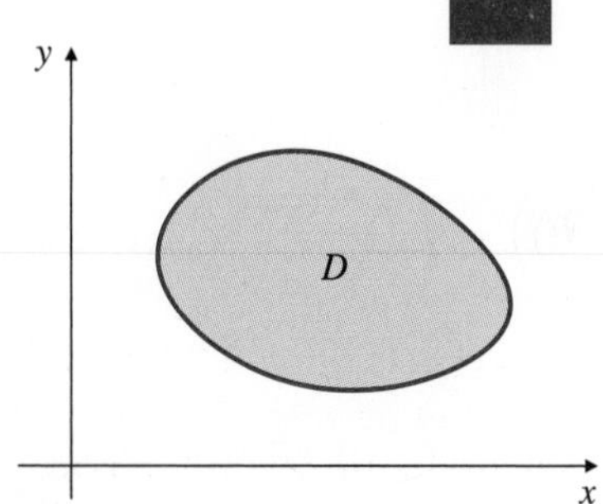

Figure 15.11 A simply connected domain

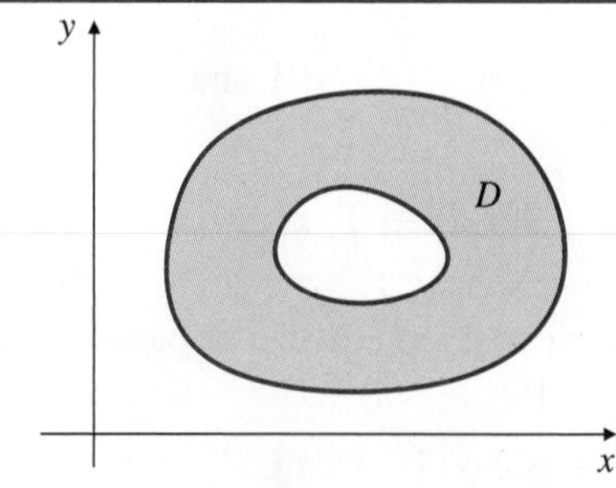

Figure 15.12 A connected domain that is not simply connected

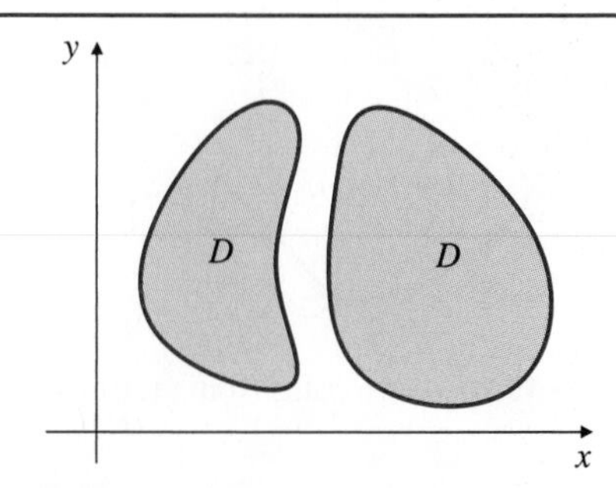

Figure 15.13 A domain that is not connected

Figure 15.11 shows a simply connected domain in the plane. Figure 15.12 shows a connected but not simply connected domain. (A closed curve surrounding the hole cannot be shrunk to a point without passing out of D.) The domain in Figure 15.13 is not even connected. It has two *components*; points in different components cannot be joined by a curve that lies in D.

In the plane, a simply connected domain D can have no holes, not even a hole consisting of a single point. The interior of every non–self-intersecting closed curve in such a domain D lies in D. For instance, the domain of the function $1/(x^2 + y^2)$ is not simply connected because the origin does not belong to it. (The origin is a "hole" in that domain.) In 3-space, a simply connected domain can have holes. The set of all points in $\mathbb{R}^3$ different from the origin is simply connected, as is the exterior of a ball. But the set of all points in $\mathbb{R}^3$ satisfying $x^2 + y^2 > 0$ is not simply connected. Neither is the interior of a doughnut (a *torus*). In general, each of the following conditions characterizes simply connected domains D:

(i) Any simple closed curve in D is the boundary of a "surface" lying in D.

(ii) If $\mathcal{C}_1$ and $\mathcal{C}_2$ are two curves in D having the same endpoints, then $\mathcal{C}_1$

can be continuously deformed into $\mathcal{C}_2$, while remaining in D throughout the deformation process.

Independence of Path

THEOREM 1 **Independence of path**

Let D be an open, connected domain, and let $\mathbf{F}$ be a smooth vector field defined on D. Then the following three statements are *equivalent* in the sense that, if any one of them is true, so are the other two:

(a) $\mathbf{F}$ is conservative in D.

(b) $\oint_{\mathcal{C}} \mathbf{F} \bullet d\mathbf{r} = 0$ for every piecewise smooth, closed curve $\mathcal{C}$ in D.

(c) Given any two points P_0 and P_1 in D, $\int_{\mathcal{C}} \mathbf{F} \bullet d\mathbf{r}$ has the same value for all piecewise smooth curves in D starting at P_0 and ending at P_1.

PROOF We will show that (a) implies (b), that (b) implies (c), and that (c) implies (a). It then follows that any one implies the other two.

Suppose (a) is true. Then $\mathbf{F} = \nabla\phi$ for some scalar potential function ϕ defined in D. Therefore,

$$\begin{aligned}\mathbf{F} \bullet d\mathbf{r} &= \left(\frac{\partial\phi}{\partial x}\mathbf{i} + \frac{\partial\phi}{\partial y}\mathbf{j} + \frac{\partial\phi}{\partial z}\mathbf{k}\right) \bullet \left(dx\,\mathbf{i} + dy\,\mathbf{j} + dz\mathbf{k}\right)\\ &= \frac{\partial\phi}{\partial x}\,dx + \frac{\partial\phi}{\partial y}\,dy + \frac{\partial\phi}{\partial z}\,dz = d\phi.\end{aligned}$$

If $\mathcal{C}$ is any piecewise smooth, closed curve, parametrized, say, by $\mathbf{r} = \mathbf{r}(t)$, $(a \le t \le b)$, then $\mathbf{r}(a) = \mathbf{r}(b)$, and

$$\int_{\mathcal{C}} \mathbf{F} \bullet d\mathbf{r} = \int_a^b \frac{d\phi\big(\mathbf{r}(t)\big)}{dt}\,dt = \phi\big(\mathbf{r}(b)\big) - \phi\big(\mathbf{r}(a)\big) = 0.$$

Thus, (a) implies (b).

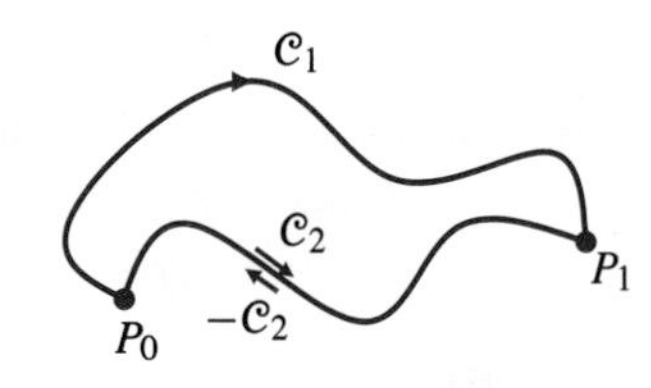

Figure 15.14 $\mathcal{C}_1 - \mathcal{C}_2 = \mathcal{C}_1 + (-\mathcal{C}_2)$ is a closed curve

Now suppose (b) is true. Let P_0 and P_1 be two points in D, and let $\mathcal{C}_1$ and $\mathcal{C}_2$ be two piecewise smooth curves in D from P_0 to P_1. Let $\mathcal{C} = \mathcal{C}_1 - \mathcal{C}_2$ denote the closed curve going from P_0 to P_1 along $\mathcal{C}_1$ and then back to P_0 along $\mathcal{C}_2$ in the opposite direction. (See Figure 15.14.) Since we are assuming that (b) is true, we have

$$0 = \oint_{\mathcal{C}} \mathbf{F} \bullet d\mathbf{r} = \int_{\mathcal{C}_1} \mathbf{F} \bullet d\mathbf{r} - \int_{\mathcal{C}_2} \mathbf{F} \bullet d\mathbf{r}.$$

Therefore,

$$\int_{\mathcal{C}_1} \mathbf{F} \bullet d\mathbf{r} = \int_{\mathcal{C}_2} \mathbf{F} \bullet d\mathbf{r},$$

and we have proved that (b) implies (c).

Finally, suppose that (c) is true. Let $P_0 = (x_0, y_0, z_0)$ be a fixed point in the domain D, and let $P = (x, y, z)$ be an arbitrary point in that domain. Define a function ϕ by

$$\phi(x, y, z) = \int_{\mathcal{C}} \mathbf{F} \bullet d\mathbf{r},$$

where $\mathcal{C}$ is some piecewise smooth curve in D from P_0 to P. (Under the hypotheses of the theorem such a curve exists, and, since we are assuming (c), the integral has the same value for all such curves. Therefore, ϕ is well defined in D.) We will show that $\nabla\phi = \mathbf{F}$ and thus establish that $\mathbf{F}$ is conservative and has potential ϕ.

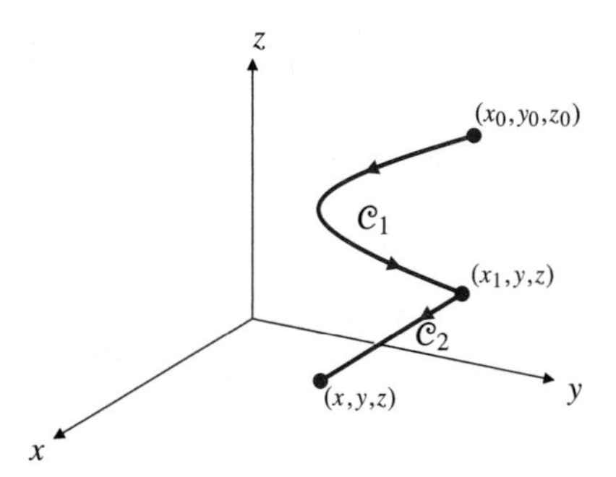

Figure 15.15 A special path from P_0 to P_1

It is sufficient to show that $\partial\phi/\partial x = F_1(x, y, z)$; the other two components are treated similarly. Since D is open, there is a ball of positive radius centred at P and contained in D. Pick a point (x_1, y, z) in this ball having $x_1 < x$. Note that the line from this point to P is parallel to the x-axis. Since we are free to choose the curve $\mathcal{C}$ in the integral defining ϕ, let us choose it to consist of two segments: $\mathcal{C}_1$, which is piecewise smooth and goes from (x_0, y_0, z_0) to (x_1, y, z), and $\mathcal{C}_2$, a straight line segment from (x_1, y, z) to (x, y, z). (See Figure 15.15.) Then

$$\phi(x, y, z) = \int_{\mathcal{C}_1} \mathbf{F} \bullet d\mathbf{r} + \int_{\mathcal{C}_2} \mathbf{F} \bullet d\mathbf{r}.$$

The first integral does not depend on x, so its derivative with respect to x is zero. The straight line path for the second integral is parametrized by $\mathbf{r} = t\mathbf{i} + y\mathbf{j} + z\mathbf{k}$, where $x_1 \le t \le x$ so $d\mathbf{r} = dt\mathbf{i}$. By the Fundamental Theorem of Calculus,

$$\frac{\partial\phi}{\partial x} = \frac{\partial}{\partial x}\int_{\mathcal{C}_2} \mathbf{F} \bullet d\mathbf{r} = \frac{\partial}{\partial x}\int_{x_1}^{x} F_1(t, y, z)\, dt = F_1(x, y, z),$$

which is what we wanted. Thus, $\mathbf{F} = \nabla\phi$ is conservative, and (c) implies (a). ∎

Remark It is very easy to evaluate the line integral of the tangential component of a *conservative* vector field along a curve $\mathcal{C}$, when you know a potential for $\mathbf{F}$. If $\mathbf{F} = \nabla\phi$, and $\mathcal{C}$ goes from P_0 to P_1, then

$$\int_{\mathcal{C}} \mathbf{F} \bullet d\mathbf{r} = \int_{\mathcal{C}} d\phi = \phi(P_1) - \phi(P_0).$$

As noted above, the value of the integral depends only on the endpoints of $\mathcal{C}$.

Remark In the next chapter we will add another item to the list of three conditions shown to be equivalent in Theorem 1, provided that the domain D is *simply connected*. For such a domain, each of the above three conditions in the theorem is equivalent to

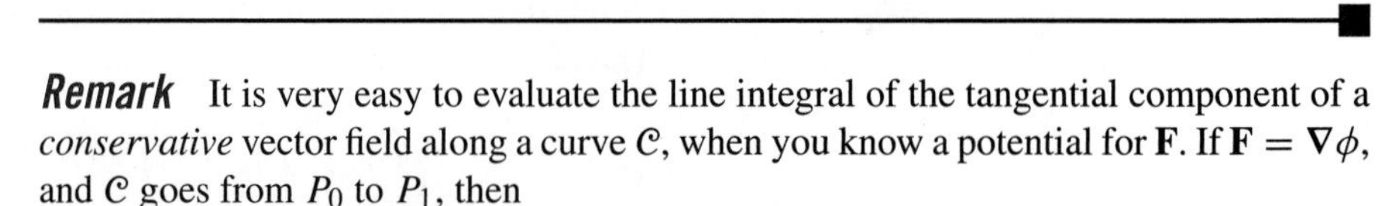

$$\frac{\partial F_1}{\partial y} = \frac{\partial F_2}{\partial x}, \qquad \frac{\partial F_1}{\partial z} = \frac{\partial F_3}{\partial x}, \qquad \text{and} \qquad \frac{\partial F_2}{\partial z} = \frac{\partial F_3}{\partial y}.$$

We already know that these equations are satisfied on a domain where $\mathbf{F}$ is conservative. Theorem 4 of Section 16.2 states that if these three equations hold on a simply connected domain, then $\mathbf{F}$ is conservative on that domain.

EXAMPLE 3 For what values of the constants A and B is the vector field

$$\mathbf{F} = Ax\sin(\pi y)\mathbf{i} + \big(x^2\cos(\pi y) + Bye^{-z}\big)\mathbf{j} + y^2e^{-z}\mathbf{k}$$

conservative? For this choice of A and B, evaluate $\int_{\mathcal{C}} \mathbf{F} \bullet d\mathbf{r}$, where $\mathcal{C}$ is

(a) the curve $\mathbf{r} = \cos t\mathbf{i} + \sin 2t\mathbf{j} + \sin^2 t\mathbf{k}$, $(0 \le t \le 2\pi)$, and

(b) the curve of intersection of the paraboloid $z = x^2 + 4y^2$ and the plane $z = 3x - 2y$ from $(0, 0, 0)$ to $(1, 1/2, 2)$.

Solution $\mathbf{F}$ cannot be conservative unless

$$\frac{\partial F_1}{\partial y} = \frac{\partial F_2}{\partial x}, \qquad \frac{\partial F_1}{\partial z} = \frac{\partial F_3}{\partial x}, \qquad \text{and} \qquad \frac{\partial F_2}{\partial z} = \frac{\partial F_3}{\partial y},$$

that is, unless

$$A\pi x\cos(\pi y) = 2x\cos(\pi y), \qquad 0 = 0, \qquad \text{and} \qquad -Bye^{-z} = 2ye^{-z}.$$

Thus, we require that $A = 2/\pi$ and $B = -2$. In this case, it is easily checked that

$$\mathbf{F} = \nabla\phi, \quad \text{where} \quad \phi(x, y, z) = \frac{x^2\sin(\pi y)}{\pi} - y^2e^{-z}.$$

For the curve (a) we have $\mathbf{r}(0) = \mathbf{i} = \mathbf{r}(2\pi)$, so this curve is a closed curve, and

$$\int_{\mathcal{C}} \mathbf{F} \bullet d\mathbf{r} = \oint_{\mathcal{C}} \nabla\phi \bullet d\mathbf{r} = 0.$$

Since the curve (b) starts at $(0, 0, 0)$ and ends at $(1, 1/2, 2)$, we have

$$\int_{\mathcal{C}} \mathbf{F} \bullet d\mathbf{r} = \left(\frac{x^2\sin(\pi y)}{\pi} - y^2e^{-z}\right)\Bigg|_{(0,0,0)}^{(1,1/2,2)} = \frac{1}{\pi} - \frac{1}{4e^2}.$$

The following example shows how to exploit the fact that

$$\int_{\mathcal{C}} \mathbf{F} \bullet d\mathbf{r}$$

is easily evaluated for conservative $\mathbf{F}$, even if the $\mathbf{F}$ we want to integrate isn't quite conservative.

EXAMPLE 4 Evaluate $I = \oint_{\mathcal{C}} (e^x\sin y + 3y)dx + (e^x\cos y + 2x - 2y)dy$ counterclockwise around the ellipse $4x^2 + y^2 = 4$.

Solution $I = \oint_{\mathcal{C}} \mathbf{F} \bullet d\mathbf{r}$, where $\mathbf{F}$ is the vector field

$$\mathbf{F} = \big(e^x\sin y + 3y\big)\mathbf{i} + \big(e^x\cos y + 2x - 2y\big)\mathbf{j}.$$

This vector field is not conservative, but it would be if the $3y$ term in F_1 were $2y$ instead; specifically, if

$$\phi(x, y) = e^x\sin y + 2xy - y^2,$$

then $\mathbf{F} = \nabla\phi + y\mathbf{i}$, the sum of a conservative part and a nonconservative part. Therefore, we have

$$I = \oint_{\mathcal{C}} \nabla\phi \bullet d\mathbf{r} + \oint_{\mathcal{C}} y\,dx.$$

The first integral is zero since $\nabla\phi$ is conservative and $\mathcal{C}$ is closed. For the second integral we parametrize $\mathcal{C}$ by $x = \cos t$, $y = 2\sin t$, $(0 \le t \le 2\pi)$, and obtain

$$I = \oint_{\mathcal{C}} y\,dx = -2\int_0^{2\pi} \sin^2 t\,dt = -2\int_0^{2\pi} \frac{1 - \cos(2t)}{2}\,dt = -2\pi.$$

EXERCISES 15.4

In Exercises 1–6, evaluate the line integral of the tangential component of the given vector field along the given curve.

1. $\mathbf{F}(x, y) = xy\mathbf{i} - x^2\mathbf{j}$ along $y = x^2$ from $(0, 0)$ to $(1, 1)$

2. $\mathbf{F}(x, y) = \cos x\,\mathbf{i} - y\mathbf{j}$ along $y = \sin x$ from $(0, 0)$ to $(\pi, 0)$

3. $\mathbf{F}(x, y, z) = y\mathbf{i} + z\mathbf{j} - x\mathbf{k}$ along the straight line from $(0, 0, 0)$ to $(1, 1, 1)$

4. $\mathbf{F}(x, y, z) = z\mathbf{i} - y\mathbf{j} + 2x\mathbf{k}$ along the curve $x = t$, $y = t^2$, $z = t^3$ from $(0, 0, 0)$ to $(1, 1, 1)$

5. $\mathbf{F}(x, y, z) = yz\mathbf{i} + xz\mathbf{j} + xy\mathbf{k}$ from $(-1, 0, 0)$ to $(1, 0, 0)$ along either direction of the curve of intersection of the cylinder $x^2 + y^2 = 1$ and the plane $z = y$

6. $\mathbf{F}(x, y, z) = (x - z)\mathbf{i} + (y - z)\mathbf{j} - (x + y)\mathbf{k}$ along the polygonal path from $(0, 0, 0)$ to $(1, 0, 0)$ to $(1, 1, 0)$ to $(1, 1, 1)$

7. Find the work done by the force field

$$\mathbf{F} = (x + y)\mathbf{i} + (x - z)\mathbf{j} + (z - y)\mathbf{k}$$

in moving an object from $(1, 0, -1)$ to $(0, -2, 3)$ along any smooth curve.

8. Evaluate $\oint_{\mathcal{C}} x^2y^2\,dx + x^3y\,dy$ counterclockwise around the square with vertices $(0, 0)$, $(1, 0)$, $(1, 1)$, and $(0, 1)$.

9. Evaluate

$$\int_{\mathcal{C}} e^{x+y}\sin(y + z)\,dx + e^{x+y}\Big(\sin(y + z) + \cos(y + z)\Big)\,dy + e^{x+y}\cos(y + z)\,dz$$

along the straight line segment from (0,0,0) to $(1, \frac{\pi}{4}, \frac{\pi}{4})$.

10. The field $\mathbf{F} = (axy + z)\mathbf{i} + x^2\mathbf{j} + (bx + 2z)\mathbf{k}$ is conservative. Find a and b, and find a potential for $\mathbf{F}$. Also, evaluate $\int_{\mathcal{C}} \mathbf{F} \bullet d\mathbf{r}$, where $\mathcal{C}$ is the curve from $(1, 1, 0)$ to $(0, 0, 3)$ that lies on the intersection of the surfaces $2x + y + z = 3$ and $9x^2 + 9y^2 + 2z^2 = 18$ in the octant $x \geq 0$, $y \geq 0$, $z \geq 0$.

11. Determine the values of A and B for which the vector field

$$\mathbf{F} = Ax\ln z\,\mathbf{i} + By^2z\,\mathbf{j} + \left(\frac{x^2}{z} + y^3\right)\mathbf{k}$$

is conservative. If $\mathcal{C}$ is the straight line from $(1, 1, 1)$ to $(2, 1, 2)$, find

$$\int_{\mathcal{C}} 2x\ln z\,dx + 2y^2z\,dy + y^3\,dz.$$

12. Find the work done by the force field

$$\mathbf{F} = (y^2\cos x + z^3)\mathbf{i} + (2y\sin x - 4)\mathbf{j} + (3xz^2 + 2)\mathbf{k}$$

in moving a particle along the curve $x = \sin^{-1}t$, $y = 1 - 2t$, $z = 3t - 1$, $(0 \leq t \leq 1)$.

13. If $\mathcal{C}$ is the intersection of $z = \ln(1 + x)$ and $y = x$ from $(0, 0, 0)$ to $(1, 1, \ln 2)$, evaluate

$$\int_{\mathcal{C}} \Big(2x\sin(\pi y) - e^z\Big)\,dx + \Big(\pi x^2\cos(\pi y) - 3e^z\Big)\,dy - xe^z\,dz.$$

14. Is each of the following sets a domain? a connected domain? a simply connected domain?

(a) the set of points (x, y) in the plane such that $x > 0$ and $y \geq 0$

(b) the set of points (x, y) in the plane such that $x = 0$ and $y \geq 0$

(c) the set of points (x, y) in the plane such that $x \neq 0$ and $y > 0$

(d) the set of points (x, y, z) in 3-space such that $x^2 > 1$

(e) the set of points (x, y, z) in 3-space such that $x^2 + y^2 > 1$

(f) the set of points (x, y, z) in 3-space such that $x^2 + y^2 + z^2 > 1$

In Exercises 15–19, evaluate the closed line integrals

$$\text{(a)} \quad \oint_{\mathcal{C}} x\,dy, \qquad \text{(b)} \quad \oint_{\mathcal{C}} y\,dx$$

around the given curves, all oriented counterclockwise.

15. The circle $x^2 + y^2 = a^2$

16. The ellipse $\dfrac{x^2}{a^2} + \dfrac{y^2}{b^2} = 1$

17. The boundary of the half-disk $x^2 + y^2 \leq a^2$, $y \geq 0$

18. The boundary of the square with vertices $(0, 0)$, $(1, 0)$, $(1, 1)$, and $(0, 1)$

19. The triangle with vertices $(0, 0)$, $(a, 0)$, and $(0, b)$

20. On the basis of your results for Exercises 15–19, guess the values of the closed line integrals

$$\text{(a)} \quad \oint_{\mathcal{C}} x\,dy, \qquad \text{(b)} \quad \oint_{\mathcal{C}} y\,dx$$

for any non–self-intersecting closed curve in the xy-plane. Prove your guess in the case that $\mathcal{C}$ bounds a region of the plane that is both x-simple and y-simple. (See Section 14.2.)

21. If f and g are scalar fields with continuous first partial derivatives in a connected domain D, show that

$$\int_{\mathcal{C}} f\nabla g \bullet d\mathbf{r} + \int_{\mathcal{C}} g\nabla f \bullet d\mathbf{r} = f(Q)g(Q) - f(P)g(P)$$

for any piecewise smooth curve in D from P to Q.

22. Evaluate

$$\frac{1}{2\pi}\oint_{\mathcal{C}} \frac{-y\,dx + x\,dy}{x^2 + y^2}$$

(a) counterclockwise around the circle $x^2 + y^2 = a^2$,

(b) clockwise around the square with vertices $(-1, -1)$, $(-1, 1)$, $(1, 1)$, and $(1, -1)$,

(c) counterclockwise around the boundary of the region $1 \leq x^2 + y^2 \leq 4$, $y \geq 0$.

23. Review Example 5 in Section 15.2 in which it was shown that

$$\frac{\partial}{\partial y}\left(\frac{-y}{x^2+y^2}\right) = \frac{\partial}{\partial x}\left(\frac{x}{x^2+y^2}\right),$$

for all $(x, y) \neq (0, 0)$. Why does this result, together with that of Exercise 22, not contradict the final assertion in the remark following Theorem 1?

24. (Winding number) Let $\mathcal{C}$ be a piecewise smooth curve in the xy-plane that does not pass through the origin. Let $\theta = \theta(x, y)$ be the polar angle coordinate of the point $P = (x, y)$ on $\mathcal{C}$, not restricted to an interval of length 2π, but varying continuously as P moves from one end of $\mathcal{C}$ to the other. As in Example 5 of Section 15.2, it happens that

$$\nabla\theta = -\frac{y}{x^2+y^2}\mathbf{i} + \frac{x}{x^2+y^2}\mathbf{j}.$$

If, in addition, $\mathcal{C}$ is a closed curve, show that

$$w(\mathcal{C}) = \frac{1}{2\pi}\oint_{\mathcal{C}} \frac{x\,dy - y\,dx}{x^2+y^2}$$

has an integer value. w is called the **winding number** of $\mathcal{C}$ about the origin.

15.5 Surfaces and Surface Integrals

This section and the next are devoted to integrals of functions defined over surfaces in 3-space. Before we can begin, it is necessary to make more precise just what is meant by the term "surface." Until now we have been treating surfaces in an intuitive way, either as the graphs of functions $f(x, y)$ or as the graphs of equations $f(x, y, z) = 0$.

A smooth curve is a *one-dimensional* object because points on it can be located by giving *one coordinate* (for instance, the distance from an endpoint). Therefore, the curve can be defined as the range of a vector-valued function of one real variable. A surface is a *two-dimensional* object; points on it can be located by using *two coordinates*, and it can be defined as the range of a vector-valued function of two real variables. We will call certain such functions parametric surfaces.

Parametric Surfaces

DEFINITION 4

A **parametric surface** in 3-space is a continuous function $\mathbf{r}$ defined on some rectangle R given by $a \le u \le b$, $c \le v \le d$ in the uv-plane and having values in 3-space:

$$\mathbf{r}(u, v) = x(u, v)\mathbf{i} + y(u, v)\mathbf{j} + z(u, v)\mathbf{k}, \qquad (u, v) \text{ in } R.$$

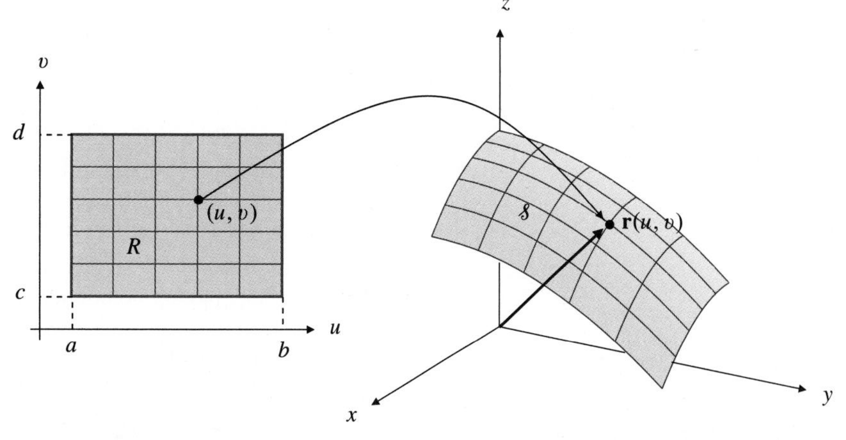

Figure 15.16 A parametric surface $\mathcal{S}$ defined on parameter region R. The *contour curves* on $\mathcal{S}$ correspond to the rulings of R

Actually, we think of the *range* of the function $\mathbf{r}(u, v)$ as being the parametric surface. It is a set $\mathcal{S}$ of points (x, y, z) in 3-space whose position vectors are the vectors $\mathbf{r}(u, v)$ for (u, v) in R. (See Figure 15.16.) If $\mathbf{r}$ is one-to-one, then the surface does not intersect itself. In this case $\mathbf{r}$ maps the boundary of the rectangle R (the four edges) onto a curve in 3-space, which we call the **boundary of the parametric surface**. The requirement that R be a rectangle is made only to simplify the discussion. Any connected, closed, bounded set in the uv-plane, having well-defined area and consisting of an open set together with its boundary points, would do as well. Thus, we will from time to time consider parametric surfaces over closed disks, triangles, or other such domains in the uv-plane. Being the range of a continuous function defined on a closed, bounded set, a parametric surface is always bounded in 3-space.

EXAMPLE 1 The graph of $z = f(x, y)$, where f has the rectangle R as its domain, can be represented as the parametric surface

$$\mathbf{r} = \mathbf{r}(u, v) = u\mathbf{i} + v\mathbf{j} + f(u, v)\mathbf{k}$$

for (u, v) in R. Its scalar parametric equations are

$$x = u, \qquad y = v, \qquad z = f(u, v), \qquad (u, v) \text{ in } R.$$

For such graphs it is sometimes convenient to identify the uv-plane with the xy-plane and write the equation of the surface in the form

$$\mathbf{r} = x\mathbf{i} + y\mathbf{j} + f(x, y)\mathbf{k}, \qquad (x, y) \text{ in } R.$$

EXAMPLE 2 Describe the surface

$$\mathbf{r} = a\cos u \sin v\,\mathbf{i} + a\sin u \sin v\,\mathbf{j} + a\cos v\,\mathbf{k}, \qquad (0 \le u \le 2\pi,\ 0 \le v \le \pi/2),$$

where $a > 0$. What is its boundary?

Solution Observe that if $x = a\cos u \sin v$, $y = a\sin u \sin v$, and $z = a\cos v$, then $x^2 + y^2 + z^2 = a^2$. Thus, the given parametric surface lies on the sphere of radius a centred at the origin. (Observe that u and v are the spherical coordinates θ and ϕ on the sphere.) The restrictions on u and v allow (x, y) to be any point in the disk $x^2 + y^2 \le a^2$ but force $z \ge 0$. Thus, the surface is the *upper half* of the sphere. The given parametrization is one-to-one on the open rectangle $0 < u < 2\pi$, $0 < v < \pi/2$, but not on the closed rectangle, since the edges $u = 0$ and $u = 2\pi$ get mapped onto the same points, and the entire edge $v = 0$ collapses to a single point. The boundary of the surface is still the circle $x^2 + y^2 = a^2$, $z = 0$, and corresponds to the edge $v = \pi/2$ of the rectangle.

Remark Surface parametrizations that are one-to-one only in the interior of the parameter domain R are still reasonable representations of the surface. However, as in Example 2, the boundary of the surface may be obtained from only part of the boundary of R, or there may be no boundary at all, in which case the surface is called a **closed surface**. For example, if the domain of $\mathbf{r}$ in Example 2 is extended to allow $0 \le v \le \pi$, then the surface becomes the entire sphere of radius a centred at the origin. The sphere is a closed surface, having no boundary curves.

Remark Like parametrizations of curves, parametrizations of surfaces are not unique. The hemisphere in Example 2 can also be parametrized:

$$\mathbf{r}(u, v) = u\mathbf{i} + v\mathbf{j} + \sqrt{a^2 - u^2 - v^2}\,\mathbf{k} \qquad \text{for } u^2 + v^2 \le a^2.$$

Here, the domain of $\mathbf{r}$ is a closed disk of radius a.

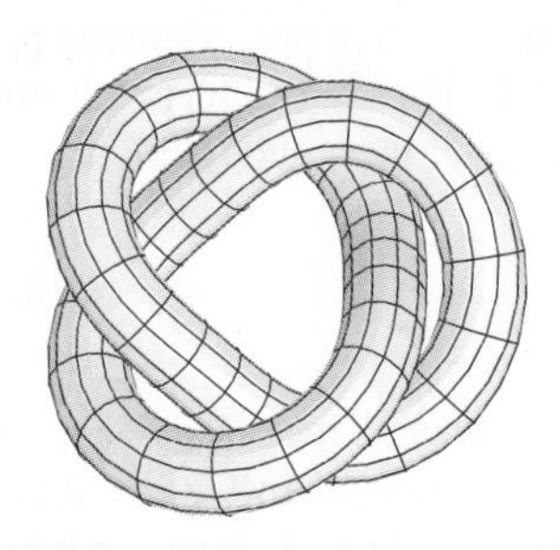

Figure 15.17 A tube in the shape of a trefoil knot

EXAMPLE 3 **(A tube around a curve)** If $\mathbf{r} = \mathbf{F}(t)$, $a \le t \le b$, is a parametric curve $\mathcal{C}$ in 3-space having unit normal $\hat{\mathbf{N}}(t)$ and binormal $\hat{\mathbf{B}}(t)$, then the parametric surface

$$\mathbf{r} = \mathbf{F}(u) + s\cos v\,\hat{\mathbf{N}}(u) + s\sin v\,\hat{\mathbf{B}}(u), \quad a \le u \le b, \quad 0 \le v \le 2\pi,$$

is a tube-shaped surface of radius s centred along the curve $\mathcal{C}$. (Why?) Figure 15.17 shows such a tube, having radius $s = 0.25$, around the curve

$$\mathbf{r} = \big(1 + 0.3\cos(3t)\big)\big(\cos(2t)\mathbf{i} + \sin(2t)\mathbf{j}\big) + 0.35\sin(3t)\mathbf{k}, \qquad 0 \le t \le 2\pi.$$

This closed curve is called a **trefoil knot**.

Composite Surfaces

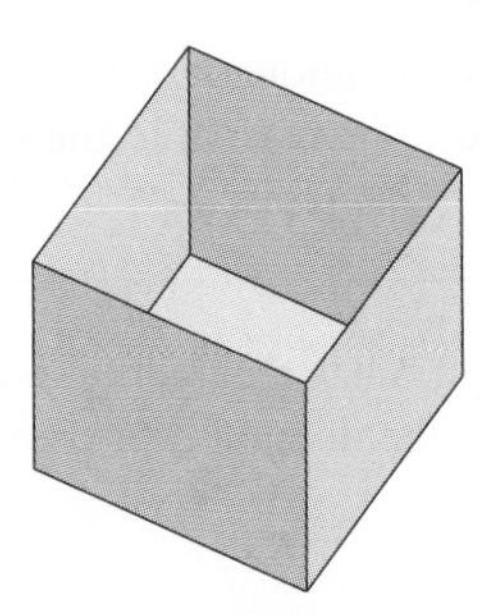

Figure 15.18 A composite surface obtained by joining five smooth parametric surfaces (squares) in pairs along edges. The four unpaired edges at the tops of the side faces make up the boundary of the composite surface

If two parametric surfaces are joined together along part or all of their boundary curves, the result is called a **composite surface**, or, thinking geometrically, just a **surface**. For example, a sphere can be obtained by joining two hemispheres along their boundary circles. In general, composite surfaces can be obtained by joining a finite number of parametric surfaces pairwise along edges. The surface of a cube consists of the six square faces joined in pairs along the edges of the cube. This surface is closed since there are no unjoined edges to comprise the boundary. If the top square face is removed, the remaining five form the surface of a cubical box with no top. The top edges of the four side faces now constitute the boundary of this composite surface. (See Figure 15.18.)

Surface Integrals

In order to define integrals of functions defined on a surface as limits of Riemann sums, we need to refer to the *areas* of regions on the surface. It is more difficult to define the area of a curved surface than it is to define the length of a curve. However, you will likely have a good idea of what area means for a region lying in a plane, and we examined briefly the problem of finding the area of the graph of a function $f(x, y)$ in Section 14.7. We will avoid difficulties by assuming that all the surfaces we will encounter are "smooth enough" that they can be subdivided into small pieces each of which is approximately planar. We can then approximate the surface area of each piece by a plane area and add up the approximations to get a Riemann sum approximation to the area of the whole surface. We will make more precise definitions of "smooth surface" and "surface area" later in this section. For the moment, we assume the reader has an intuitive feel for what they mean.

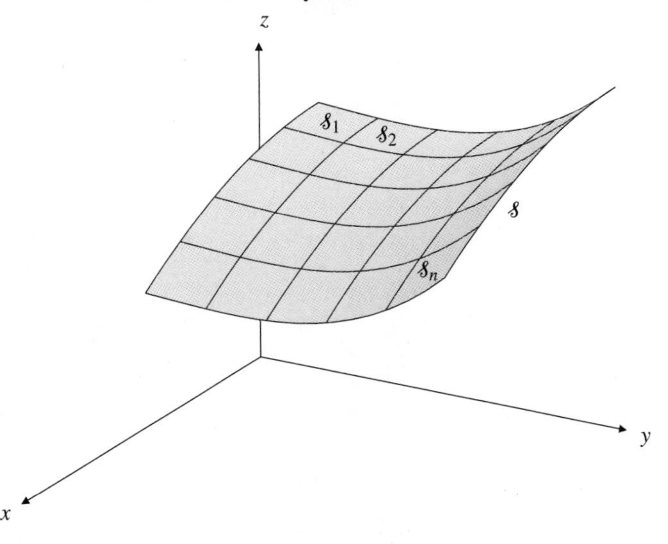

Figure 15.19 A partition of a parametric surface into many nonoverlapping pieces

Let $\mathcal{S}$ be a smooth surface of finite area in $\mathbb{R}^3$, and let $f(x, y, z)$ be a bounded function defined at all points of $\mathcal{S}$. If we subdivide $\mathcal{S}$ into small, nonoverlapping pieces, say $\mathcal{S}_1, \mathcal{S}_2, \ldots, \mathcal{S}_n$, where $\mathcal{S}_i$ has area ΔS_i (see Figure 15.19), we can form a **Riemann sum** R_n for f on $\mathcal{S}$ by choosing arbitrary points (x_i, y_i, z_i) in $\mathcal{S}_i$ and letting

$$R_n = \sum_{i=1}^{n} f(x_i, y_i, z_i)\, \Delta S_i.$$

If such Riemann sums have a unique limit as the diameters of all the pieces $\mathcal{S}_i$ approach zero, independently of how the points (x_i, y_i, z_i) are chosen, then we say that f is **integrable** on $\mathcal{S}$ and call the limit the **surface integral** of f over $\mathcal{S}$, denoting it by

$$\iint_{\mathcal{S}} f(x, y, z)\, dS.$$

Smooth Surfaces, Normals, and Area Elements

A surface is smooth if it has a unique tangent plane at any nonboundary point P. A nonzero vector $\mathbf{n}$ normal to that tangent plane at P is said to be normal to the surface at P. The following somewhat technical definition makes this precise.

DEFINITION 5

A set $\mathcal{S}$ in 3-space is a **smooth surface** if any point P in $\mathcal{S}$ has a neighbourhood N (an open ball of positive radius centred at P) that is the domain of a smooth function $g(x, y, z)$ satisfying:

(i) $N \cap S = \{Q \in N : g(Q) = 0\}$ and

(ii) $\nabla g(Q) \neq \mathbf{0}$, if Q is in $N \cap S$.

For example, the cone $x^2 + y^2 = z^2$, with the origin removed, is a smooth surface. Note that $\nabla(x^2 + y^2 - z^2) = \mathbf{0}$ at the origin, and the cone is not smooth there, since it does not have a unique tangent plane.

A parametric surface cannot satisfy the condition of the smoothness definition at its boundary points but will be called **smooth** if that condition is satisfied at all nonboundary points.

We can find the normal to a smooth parametric surface defined on parameter domain R as follows. If (u_0, v_0) is a point in the interior of R, then $\mathbf{r} = \mathbf{r}(u, v_0)$ and $\mathbf{r} = \mathbf{r}(u_0, v)$ are two curves on $\mathcal{S}$, intersecting at $\mathbf{r}_0 = \mathbf{r}(u_0, v_0)$ and having, at that point, tangent vectors (see Figure 15.20)

$$\left.\frac{\partial \mathbf{r}}{\partial u}\right|_{(u_0,v_0)} \quad \text{and} \quad \left.\frac{\partial \mathbf{r}}{\partial v}\right|_{(u_0,v_0)},$$

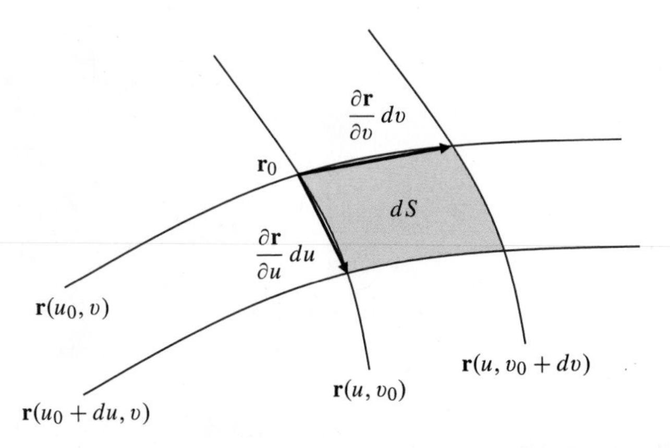

Figure 15.20 An area element dS on a parametric surface

respectively. Assuming these two tangent vectors are not parallel, their cross product $\mathbf{n}$, which is not zero, is *normal* to $\mathcal{S}$ at $\mathbf{r}_0$. Furthermore, the *area element* on $\mathcal{S}$ bounded by the four curves $\mathbf{r} = \mathbf{r}(u_0, v)$, $\mathbf{r} = \mathbf{r}(u_0 + du, v)$, $\mathbf{r} = \mathbf{r}(u, v_0)$, and $\mathbf{r} = \mathbf{r}(u, v_0 + dv)$ is an infinitesimal parallelogram spanned by the vectors $(\partial\mathbf{r}/\partial u)\, du$ and $(\partial\mathbf{r}/\partial v)\, dv$ (at (u_0, v_0)), and hence has area

$$dS = \left| \frac{\partial \mathbf{r}}{\partial u} \times \frac{\partial \mathbf{r}}{\partial v} \right| du\, dv.$$

Let us express the normal vector $\mathbf{n}$ and the area element dS in terms of the components of $\mathbf{r}$. Since

$$\frac{\partial \mathbf{r}}{\partial u} = \frac{\partial x}{\partial u}\mathbf{i} + \frac{\partial y}{\partial u}\mathbf{j} + \frac{\partial z}{\partial u}\mathbf{k} \qquad \text{and} \qquad \frac{\partial \mathbf{r}}{\partial v} = \frac{\partial x}{\partial v}\mathbf{i} + \frac{\partial y}{\partial v}\mathbf{j} + \frac{\partial z}{\partial v}\mathbf{k},$$

the **normal vector** to $\mathcal{S}$ at $\mathbf{r}(u, v)$ is

$$\mathbf{n} = \frac{\partial \mathbf{r}}{\partial u} \times \frac{\partial \mathbf{r}}{\partial v} = \begin{vmatrix} \mathbf{i} & \mathbf{j} & \mathbf{k} \\ \dfrac{\partial x}{\partial u} & \dfrac{\partial y}{\partial u} & \dfrac{\partial z}{\partial u} \\ \dfrac{\partial x}{\partial v} & \dfrac{\partial y}{\partial v} & \dfrac{\partial z}{\partial v} \end{vmatrix}$$

$$= \frac{\partial(y, z)}{\partial(u, v)}\mathbf{i} + \frac{\partial(z, x)}{\partial(u, v)}\mathbf{j} + \frac{\partial(x, y)}{\partial(u, v)}\mathbf{k}.$$

Also, the **area element** at a point $\mathbf{r}(u, v)$ on the surface is given by

$$dS = \left| \frac{\partial \mathbf{r}}{\partial u} \times \frac{\partial \mathbf{r}}{\partial v} \right| du\, dv$$

$$= \sqrt{\left(\frac{\partial(y, z)}{\partial(u, v)}\right)^2 + \left(\frac{\partial(z, x)}{\partial(u, v)}\right)^2 + \left(\frac{\partial(x, y)}{\partial(u, v)}\right)^2}\, du\, dv.$$

The area of the surface itself is the "sum" of these area elements:

$$\text{Area of } \mathcal{S} = \iint_{\mathcal{S}} dS.$$

In general, the surface integral of a function $f(\mathbf{r}) = f(x, y, z)$ over the surface $\mathcal{S}$ defined by the parametric equations $\mathbf{r} = \mathbf{r}(u, v)$ for (u, v) in the domain D of the uv-plane is given by

$$\iint_{\mathcal{S}} f\, dS = \iint_D f(\mathbf{r}(u, v)) \left| \frac{\partial \mathbf{r}}{\partial u} \times \frac{\partial \mathbf{r}}{\partial v} \right| du\, dv$$

$$= \iint_D f\big(x(u, v), y(u, v), z(u, v)\big)$$

$$\times \sqrt{\left(\frac{\partial(y, z)}{\partial(u, v)}\right)^2 + \left(\frac{\partial(z, x)}{\partial(u, v)}\right)^2 + \left(\frac{\partial(x, y)}{\partial(u, v)}\right)^2}\, du\, dv.$$

EXAMPLE 4 The graph $z = g(x, y)$ of a function g with continuous first partial derivatives in a domain D of the xy-plane can be regarded as a parametric surface $\mathcal{S}$ with parametrization

$$x = u, \qquad y = v, \qquad z = g(u, v), \qquad (u, v) \text{ in } D.$$

In this case,

$$\frac{\partial(y, z)}{\partial(u, v)} = -g_1(u, v), \qquad \frac{\partial(z, x)}{\partial(u, v)} = -g_2(u, v), \quad \text{and} \quad \frac{\partial(x, y)}{\partial(u, v)} = 1,$$

and, since the parameter region coincides with the domain D of g, the surface integral of $f(x, y, z)$ over $\mathcal{S}$ can be expressed as a double integral over D:

$$\iint_{\mathcal{S}} f(x, y, z)\, dS = \iint_D f(x, y, g(x, y)) \sqrt{1 + \big(g_1(x, y)\big)^2 + \big(g_2(x, y)\big)^2}\, dx\, dy.$$

As observed in Section 14.7, this formula can also be justified geometrically. The vector $\mathbf{n} = -g_1(x, y)\mathbf{i} - g_2(x, y)\mathbf{j} + \mathbf{k}$ is normal to $\mathcal{S}$ and makes angle γ with the positive z-axis, where

$$\cos\gamma = \frac{\mathbf{n} \bullet \mathbf{k}}{|\mathbf{n}|} = \frac{1}{\sqrt{1 + \big(g_1(x, y)\big)^2 + \big(g_2(x, y)\big)^2}}.$$

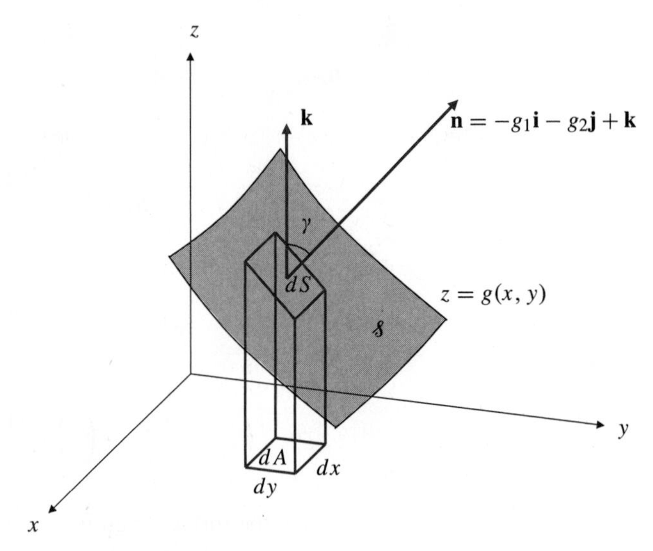

Figure 15.21 The surface area element dS and its projection onto the xy-plane

The surface area element dS must have area $1/\cos\gamma$ times the area $dx\, dy$ of its perpendicular projection onto the xy-plane. (See Figure 15.21.)

Evaluating Surface Integrals

We illustrate the use of the formulas given above for dS in calculating surface integrals.

EXAMPLE 5 Evaluate $\iint_{\mathcal{S}} z\, dS$ over the conical surface $z = \sqrt{x^2 + y^2}$ between $z = 0$ and $z = 1$.

Solution Since $z^2 = x^2 + y^2$ on the surface $\mathcal{S}$, we have $\partial z/\partial x = x/z$ and $\partial z/\partial y = y/z$. Therefore,

$$dS = \sqrt{1 + \frac{x^2}{z^2} + \frac{y^2}{z^2}}\, dx\, dy = \sqrt{\frac{z^2 + z^2}{z^2}}\, dx\, dy = \sqrt{2}\, dx\, dy.$$

(Note that we could have anticipated this result, since the normal to the cone always makes an angle of $\gamma = 45°$ with the positive z-axis; see Figure 15.22. Therefore, $dS = dx\, dy/\cos 45° = \sqrt{2}\, dx\, dy$.) Since $z = \sqrt{x^2 + y^2} = r$ on the conical surface, it is easiest to carry out the integration in polar coordinates:

$$\iint_{\mathcal{S}} z\, dS = \sqrt{2} \iint_{x^2+y^2 \le 1} z\, dx\, dy = \sqrt{2} \int_0^{2\pi} d\theta \int_0^1 r^2\, dr = \frac{2\sqrt{2}\pi}{3}.$$

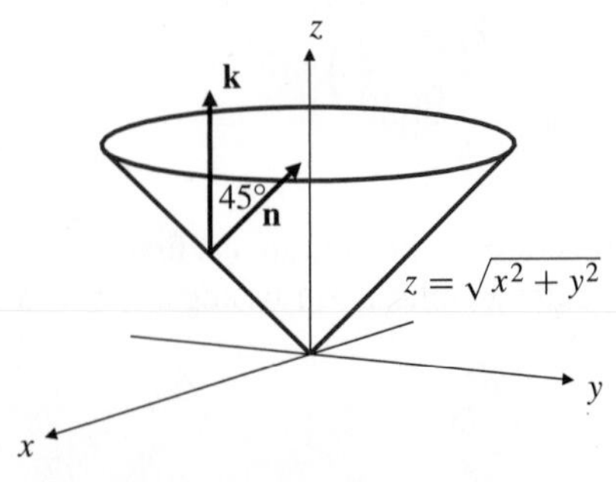

Figure 15.22 $dS = \sqrt{2}\, dx\, dy$ on this cone

EXAMPLE 6 Find the moment of inertia about the z-axis of the parametric surface $x = 2uv$, $y = u^2 - v^2$, $z = u^2 + v^2$, where $u^2 + v^2 \le 1$.

Solution We are asked to find $\iint_{\mathcal{S}} (x^2 + y^2)\, dS$. We have

$$\begin{aligned}
\frac{\partial(x, y)}{\partial(u, v)} &= \begin{vmatrix} 2v & 2u \\ 2u & -2v \end{vmatrix} = -4(u^2 + v^2), \\
\frac{\partial(z, x)}{\partial(u, v)} &= \begin{vmatrix} 2u & 2v \\ 2v & 2u \end{vmatrix} = 4(u^2 - v^2), \\
\frac{\partial(y, z)}{\partial(u, v)} &= \begin{vmatrix} 2u & -2v \\ 2u & 2v \end{vmatrix} = 8uv.
\end{aligned}$$

Therefore, the surface area element on $\mathcal{S}$ is given by

$$\begin{aligned}
dS &= 4\sqrt{(u^2 + v^2)^2 + (u^2 - v^2)^2 + 4u^2v^2}\, du\, dv \\
&= 4\sqrt{2(u^4 + v^4 + 2u^2v^2)}\, du\, dv = 4\sqrt{2}(u^2 + v^2)\, du\, dv.
\end{aligned}$$

Now $x^2 + y^2 = 4u^2v^2 + (u^2 - v^2)^2 = (u^2 + v^2)^2$. Thus,

$$\begin{aligned}
\iint_{\mathcal{S}} (x^2 + y^2)\, dS &= \iint_{u^2+v^2 \le 1} (u^2 + v^2)^2\, 4\sqrt{2}(u^2 + v^2)\, du\, dv \\
&= 4\sqrt{2} \int_0^{2\pi} d\theta \int_0^1 r^6\, r\, dr \qquad \text{(using polar coordinates)} \\
&= \sqrt{2}\pi.
\end{aligned}$$

This is the required moment of inertia.

Even though most surfaces we encounter can be easily parametrized, it is usually possible to obtain the surface area element dS geometrically rather than relying on the parametric formula. As we have seen above, if a surface has a one-to-one projection onto a region in the xy-plane, then the area element dS on the surface can be expressed as

$$dS = \left| \frac{1}{\cos\gamma} \right| dx\, dy = \frac{|\mathbf{n}|}{|\mathbf{n} \bullet \mathbf{k}|}\, dx\, dy,$$

where γ is the angle between the normal vector $\mathbf{n}$ to $\mathcal{S}$ and the positive z-axis. This formula is useful no matter how we obtain $\mathbf{n}$.

Consider a surface $\mathcal{S}$ with equation of the form $G(x, y, z) = 0$. As we discovered in Section 12.7, if G has continuous first partial derivatives that do not all vanish at a point (x, y, z) on $\mathcal{S}$, then the nonzero vector

$$\mathbf{n} = \nabla G(x, y, z)$$

is normal to $\mathcal{S}$ at that point. Since $\mathbf{n} \bullet \mathbf{k} = G_3(x, y, z)$, if $\mathcal{S}$ has a one-to-one projection onto the domain D in the xy-plane, then

$$dS = \left| \frac{\nabla G(x, y, z)}{G_3(x, y, z)} \right| dx\, dy,$$

and the surface integral of $f(x, y, z)$ over $\mathcal{S}$ can be expressed as a double integral over the domain D:

$$\iint_{\mathcal{S}} f(x, y, z)\, dS = \iint_D f(x, y, g(x, y)) \left| \frac{\nabla G(x, y, z)}{G_3(x, y, z)} \right| dx\, dy.$$

Of course, there are analogous formulas for area elements of surfaces (and integrals over surfaces) with one-to-one projections onto the xz-plane or the yz-plane. (G_3 is replaced by G_2 and G_1, respectively.)

EXAMPLE 7 Find the moment about $z = 0$, that is, $\iint_{\mathcal{S}} z\, dS$, where $\mathcal{S}$ is the hyperbolic bowl $z^2 = 1 + x^2 + y^2$ between the planes $z = 1$ and $z = \sqrt{5}$.

Solution $\mathcal{S}$ is given by $G(x, y, z) = 0$, where $G(x, y, z) = x^2 + y^2 - z^2 + 1$. It lies above the disk $x^2 + y^2 \le 4$ in the xy-plane. We have $\nabla G = 2x\mathbf{i} + 2y\mathbf{j} - 2z\mathbf{k}$, and $G_3 = -2z$. Hence, on $\mathcal{S}$, we have

$$z\, dS = z\, \frac{\sqrt{4x^2 + 4y^2 + 4z^2}}{2z}\, dx\, dy = \sqrt{1 + 2(x^2 + y^2)}\, dx\, dy,$$

and the required moment is

$$\begin{aligned}\iint_{\mathcal{S}} z\, dS &= \iint_{x^2+y^2\le 4} \sqrt{1 + 2(x^2 + y^2)}\, dx\, dy \\ &= \int_0^{2\pi} d\theta \int_0^2 \sqrt{1 + 2r^2}\, r\, dr = \frac{\pi}{3}(1 + 2r^2)^{3/2}\Big|_0^2 = \frac{26\pi}{3}.\end{aligned}$$

The next example illustrates a technique that can often reduce the effort needed to integrate over a cylindrical surface.

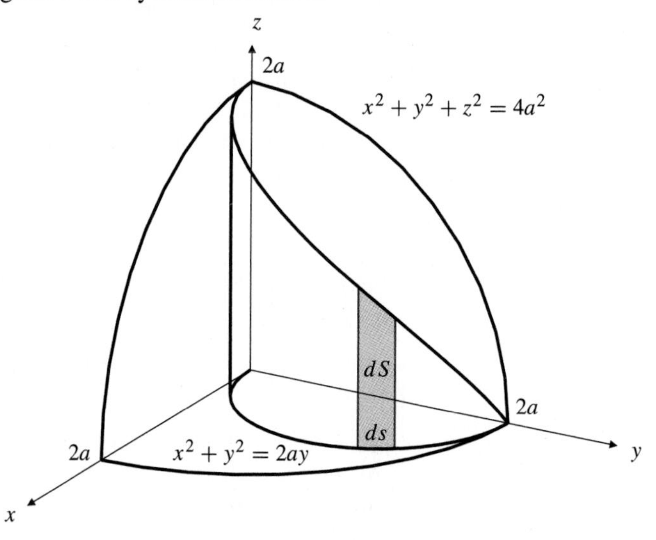

Figure 15.23 An area element on a cylinder. The z-coordinate has already been integrated

EXAMPLE 8 Find the area of that part of the cylinder $x^2 + y^2 = 2ay$ that lies inside the sphere $x^2 + y^2 + z^2 = 4a^2$.

Solution One quarter of the required area lies in the first octant. (See Figure 15.23.) Since the cylinder is generated by vertical lines, we can express an area element dS on it in terms of the length element ds along the curve $\mathcal{C}$ in the xy-plane having equation $x^2 + y^2 = 2ay$:

$$dS = z\, ds = \sqrt{4a^2 - x^2 - y^2}\, ds.$$

In expressing dS this way, we have already integrated dz, so only a single integral is needed to sum these area elements. Again, it is convenient to use polar coordinates in the xy-plane. In terms of polar coordinates, the curve $\mathcal{C}$ has equation $r = 2a\sin\theta$. Thus $dr/d\theta = 2a\cos\theta$ and $ds = \sqrt{r^2 + (dr/d\theta)^2}\,d\theta = 2a\,d\theta$. Therefore, the total surface area of that part of the cylinder that lies inside the sphere is given by

$$\begin{aligned} A &= 4\int_0^{\pi/2} \sqrt{4a^2 - r^2}\, 2a\,d\theta \\ &= 8a\int_0^{\pi/2} \sqrt{4a^2 - 4a^2\sin^2\theta}\,d\theta \\ &= 16a^2\int_0^{\pi/2} \cos\theta\,d\theta = 16a^2 \text{ square units.} \end{aligned}$$

Remark The area calculated in Example 8 can also be calculated by projecting the cylindrical surface in Figure 15.23 into the yz-plane. (This is the only coordinate plane you can use. Why?) See Exercise 6 below.

In spherical coordinates, ϕ and θ can be used as parameters on the spherical surface $R = a$. The area element on that surface can therefore be expressed in terms of these coordinates:

Area element on the sphere $R = a$: $\quad dS = a^2 \sin\phi\,d\phi\,d\theta.$

(See Figure 14.52 in Section 14.6 and Exercise 2 below.)

EXAMPLE 9 Find $\iint_{\mathcal{S}} z^2\,dS$ over the hemisphere $z = \sqrt{a^2 - x^2 - y^2}$.

Solution Since $z = a\cos\phi$ and the hemisphere corresponds to $0 \le \theta \le 2\pi$, and $0 \le \phi \le \dfrac{\pi}{2}$, we have

$$\begin{aligned} \iint_{\mathcal{S}} z^2\,dS &= \int_0^{2\pi} d\theta \int_0^{\pi/2} a^2\cos^2\phi\, a^2\sin\phi\,d\phi \\ &= 2\pi a^4\left(-\frac{1}{3}\cos^3\phi\right)\bigg|_0^{\pi/2} = \frac{2\pi a^4}{3}. \end{aligned}$$

Finally, if a composite surface $\mathcal{S}$ is composed of *smooth parametric surfaces* joined pairwise along their edges, then we call $\mathcal{S}$ a **piecewise smooth surface**. The surface integral of a function f over a piecewise smooth surface $\mathcal{S}$ is the sum of the surface integrals of f over the individual smooth surfaces comprising $\mathcal{S}$. We will encounter an example of this in the next section.

The Attraction of a Spherical Shell

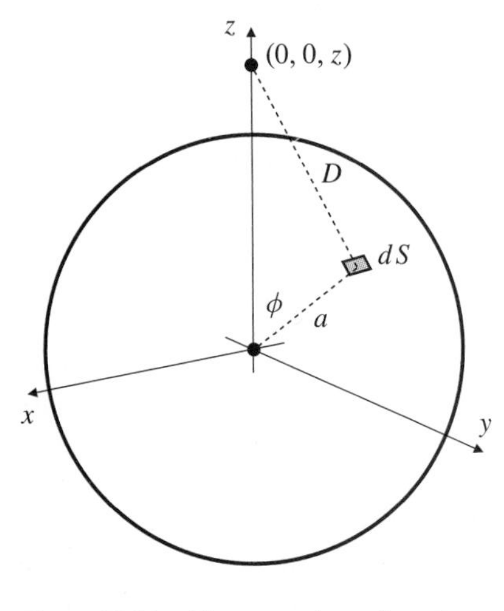

Figure 15.24 The attraction of a sphere

In Section 14.7 we calculated the gravitational attraction of a disk in the xy-plane on a mass m located at position $(0, 0, b)$ on the z-axis. Here, we undertake a similar calculation of the attractive force exerted on m by a spherical shell of radius a and areal density σ (units of mass per unit area) centred at the origin. This calculation would be more difficult if we tried to do it by integrating the vertical component of the force on m as we did in Section 14.7. It is greatly simplified if, instead, we use an integral to find the total *gravitational potential* $\Phi(0, 0, z)$ due to the sphere at position $(0, 0, z)$ and then calculate the force on m as $\mathbf{F} = m\nabla\Phi(0, 0, b)$.

By the Cosine Law, the distance from the point with spherical coordinates $[a, \phi, \theta]$ to the point $(0, 0, z)$ on the positive z-axis (see Figure 15.24) is

$$D = \sqrt{a^2 + z^2 - 2az\cos\phi}.$$

The area element $dS = a^2 \sin\phi\, d\phi\, d\theta$ at $[a, \phi, \theta]$ has mass $dm = \sigma\, dS$, and its gravitational potential at $(0, 0, z)$ (see Example 1 in Section 15.2) is

$$d\Phi(0,0,z) = \frac{k\,dm}{D} = \frac{k\sigma a^2 \sin\phi\, d\phi\, d\theta}{\sqrt{a^2 + z^2 - 2az\cos\phi}}.$$

For the total potential at $(0, 0, z)$ due to the sphere, we integrate $d\Phi$ over the surface of the sphere. Making the change of variables $u = a^2+z^2-2az\cos\phi$, $du = 2az\sin\phi\, d\phi$, we obtain

$$\begin{aligned}
\Phi(0,0,z) &= k\sigma a^2 \int_0^{2\pi} d\theta \int_0^{\pi} \frac{\sin\phi\, d\phi}{\sqrt{a^2+z^2-2az\cos\phi}} \\
&= 2\pi k\sigma a^2 \int_{(z-a)^2}^{(z+a)^2} \frac{1}{\sqrt{u}} \frac{du}{2az} \\
&= \frac{2\pi k\sigma a}{z} \sqrt{u}\,\Big|_{(z-a)^2}^{(z+a)^2} \\
&= \frac{2\pi k\sigma a}{z}\Big(z + a - |z-a|\Big) = \begin{cases} 4\pi k\sigma a^2/z & \text{if } z > a \\ 4\pi k\sigma a & \text{if } z < a. \end{cases}
\end{aligned}$$

The potential is constant inside the sphere and decreases proportionally to $1/z$ outside. The force on a mass m located at $(0, 0, b)$ is, therefore,

$$\mathbf{F} = m\nabla\Phi(0,0,b) = \begin{cases} -(4\pi k m\sigma a^2/b^2)\mathbf{k} & \text{if } b > a \\ \mathbf{0} & \text{if } b < a. \end{cases}$$

We are led to the somewhat surprising result that, if the mass m is anywhere inside the sphere, the net force of attraction of the sphere on it is zero. This is to be expected at the centre of the sphere, but away from the centre it appears that the larger forces due to parts of the sphere close to m are exactly cancelled by smaller forces due to parts farther away; these farther parts have larger area and therefore larger total mass. If m is outside the sphere, the sphere attracts it with a force of magnitude

$$F = \frac{kmM}{b^2},$$

where $M = 4\pi\sigma a^2$ is the total mass of the sphere. This is the same force that would be exerted by a point mass with the same mass as the sphere and located at the centre of the sphere.

Remark A solid ball of constant density, or density depending only on the distance from the centre (for instance, a planet), can be regarded as being made up of mass elements that are concentric spheres of constant density. Therefore, the attraction of such a ball on a mass m located outside the ball will also be the same as if the whole mass of the ball were concentrated at its centre. However, the attraction on a mass m located somewhere inside the ball will be that produced by only the part of the ball that is closer to the centre than m is. The maximum force of attraction will occur when m is right at the surface of the ball. If the density is constant, the magnitude of the force increases linearly with the distance from the centre (why?) up to the surface and then decreases with the square of the distance as m recedes from the ball. (See Figure 15.25.)

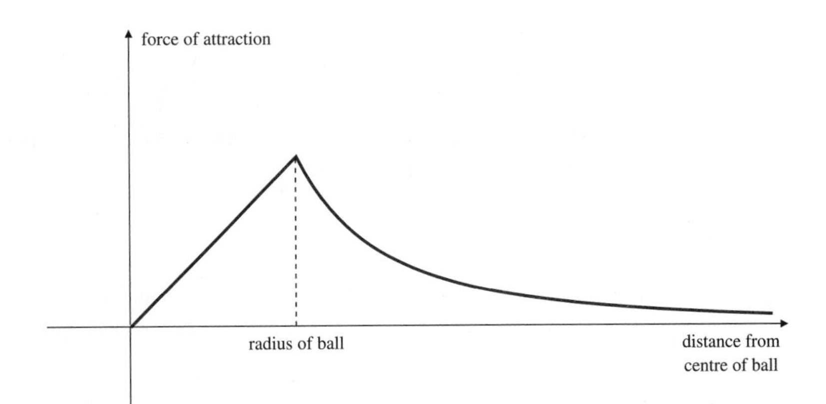

Figure 15.25 The force of attraction of a homogeneous solid ball on a particle located at varying distances from the centre of the ball

Remark All of the above discussion also holds for the electrostatic attraction or repulsion of a point charge by a uniform charge density over a spherical shell, which is also governed by an inverse square law. In particular, there is no net electrostatic force on a charge located inside the shell.

EXERCISES 15.5

1. Verify that on the curve with polar equation $r = g(\theta)$ the arc length element is given by

$$ds = \sqrt{(g(\theta))^2 + (g'(\theta))^2}\, d\theta.$$

What is the area element on the vertical cylinder given in terms of cylindrical coordinates by $r = g(\theta)$?

2. Verify that on the spherical surface $x^2 + y^2 + z^2 = a^2$ the area element is given in terms of spherical coordinates by $dS = a^2 \sin\phi\, d\phi\, d\theta$.

3. Find the area of the part of the plane $Ax + By + Cz = D$ lying inside the elliptic cylinder

$$\frac{x^2}{a^2} + \frac{y^2}{b^2} = 1.$$

4. Find the area of the part of the sphere $x^2 + y^2 + z^2 = 4a^2$ that lies inside the cylinder $x^2 + y^2 = 2ay$.

5. State formulas for the surface area element dS for the surface with equation $F(x, y, z) = 0$ valid for the case where the surface has a one-to-one projection on (a) the xz-plane and (b) the yz-plane.

6. Repeat the area calculation of Example 8 by projecting the part of the surface shown in Figure 15.23 onto the yz-plane and using the formula in Exercise 5(b).

7. Find $\iint_{\mathcal{S}} x\, dS$ over the part of the parabolic cylinder $z = x^2/2$ that lies inside the first octant part of the cylinder $x^2 + y^2 = 1$.

8. Find the area of the part of the cone $z^2 = x^2 + y^2$ that lies inside the cylinder $x^2 + y^2 = 2ay$.

9. Find the area of the part of the cylinder $x^2 + y^2 = 2ay$ that lies outside the cone $z^2 = x^2 + y^2$.

10. Find the area of the part of the cylinder $x^2 + z^2 = a^2$ that lies inside the cylinder $y^2 + z^2 = a^2$.

11. A circular cylinder of radius a is circumscribed about a sphere of radius a so that the cylinder is tangent to the sphere along the equator. Two planes, each perpendicular to the axis of the cylinder, intersect the sphere and the cylinder in circles. Show that the area of that part of the sphere between the two planes is equal to the area of the part of the cylinder between the two planes. Thus, the area of the part of a sphere between two parallel planes that intersect it depends only on the radius of the sphere and the distance between the planes, and not on the particular position of the planes.

12. Let $0 < a < b$. In terms of the elliptic integral functions defined in Exercise 19 of Section 15.3, find the area of that part of each of the cylinders $x^2 + z^2 = a^2$ and $y^2 + z^2 = b^2$ that lies inside the other cylinder.

13. Find $\iint_{\mathcal{S}} y\, dS$, where $\mathcal{S}$ is the part of the plane $z = 1 + y$ that lies inside the cone $z = \sqrt{2(x^2 + y^2)}$.

14. Find $\iint_{\mathcal{S}} y\, dS$, where $\mathcal{S}$ is the part of the cone $z = \sqrt{2(x^2 + y^2)}$ that lies below the plane $z = 1 + y$.

15. Find $\iint_{\mathcal{S}} xz\, dS$, where $\mathcal{S}$ is the part of the surface $z = x^2$ that lies in the first octant of 3-space and inside the paraboloid $z = 1 - 3x^2 - y^2$.

16. Find the mass of the part of the surface $z = \sqrt{2xy}$ that lies above the region $0 \le x \le 5$, $0 \le y \le 2$, if the areal density of the surface is $\sigma(x, y, z) = kz$.

17. Find the total charge on the surface

$$\mathbf{r} = e^u \cos v\mathbf{i} + e^u \sin v\mathbf{j} + u\mathbf{k}, \quad (0 \le u \le 1,\ 0 \le v \le \pi),$$

if the charge density on the surface is $\delta = \sqrt{1 + e^{2u}}$.

Exercises 18–19 concern **spheroids**, which are ellipsoids with two of their three semi-axes equal, say $a = b$:

$$\frac{x^2}{a^2} + \frac{y^2}{a^2} + \frac{z^2}{c^2} = 1.$$

18. Find the surface area of a **prolate spheroid**, where $0 < a < c$. A prolate spheroid has its two shorter semi-axes equal, like an American "pro football."

19. Find the surface area of an **oblate spheroid**, where $0 < c < a$. An oblate spheroid has its two longer semi-axes equal, like the earth.

20. Describe the parametric surface

$$x = au \cos v, \qquad y = au \sin v, \qquad z = bv,$$

($0 \le u \le 1$, $0 \le v \le 2\pi$), and find its area.

21. Evaluate $\iint_{\mathcal{P}} \frac{dS}{(x^2 + y^2 + z^2)^{3/2}}$, where $\mathcal{P}$ is the plane with equation $Ax + By + Cz = D$, $(D \ne 0)$.

22. A spherical shell of radius a is centred at the origin. Find the centroid of that part of the sphere that lies in the first octant.

23. Find the centre of mass of a right-circular conical shell of base radius a, height h, and constant areal density σ.

24. Find the gravitational attraction of a hemispherical shell of radius a and constant areal density σ on a mass m located at the centre of the base of the hemisphere.

25. Find the gravitational attraction of a circular cylindrical shell of radius a, height h, and constant areal density σ on a mass m located on the axis of the cylinder b units above the base.

In Exercises 26–28, find the moment of inertia and radius of gyration of the given object about the given axis. Assume constant areal density σ in each case.

26. A cylindrical shell of radius a and height h about the axis of the cylinder

27. A spherical shell of radius a about a diameter

28. A right-circular conical shell of base radius a and height h about the axis of the cone

29. With what acceleration will the spherical shell of Exercise 27 roll down a plane inclined at angle α to the horizontal? (Compare your result with that of Example 4(b) of Section 14.7.)

15.6 Oriented Surfaces and Flux Integrals

Surface integrals of normal components of vector fields play a very important role in vector calculus, similar to the role played by line integrals of tangential components of vector fields. Before we consider such surface integrals we need to define the *orientation* of a surface.

Oriented Surfaces

A smooth surface $\mathcal{S}$ in 3-space is said to be **orientable** if there exists a *unit vector field* $\hat{\mathbf{N}}(P)$ defined on $\mathcal{S}$ that varies continuously as P ranges over $\mathcal{S}$ and that is everywhere normal to $\mathcal{S}$. Any such vector field $\hat{\mathbf{N}}(P)$ determines an **orientation** of $\mathcal{S}$. The surface must have two sides since $\hat{\mathbf{N}}(P)$ can have only one value at each point P. The side out of which $\hat{\mathbf{N}}$ points is called the **positive side**; the other side is the **negative side**. An **oriented surface** is a smooth surface together with a particular choice of orienting unit normal vector field $\hat{\mathbf{N}}(P)$.

For example, if we define $\hat{\mathbf{N}}$ on the smooth surface $z = f(x, y)$ by

$$\hat{\mathbf{N}} = \frac{-f_1(x, y)\mathbf{i} - f_2(x, y)\mathbf{j} + \mathbf{k}}{\sqrt{1 + (f_1(x, y))^2 + (f_2(x, y))^2}},$$

then the top of the surface is the positive side. (See Figure 15.26.)

A smooth or piecewise smooth surface may be **closed** (i.e., it may have no boundary), or it may have one or more boundary curves. (The unit normal vector field $\hat{\mathbf{N}}(P)$ need not be defined at points of the boundary curves.)

An oriented surface $\mathcal{S}$ **induces an orientation** on any of its boundary curves $\mathcal{C}$; if we stand on the positive side of the surface $\mathcal{S}$ and walk around $\mathcal{C}$ in the direction of its orientation, then $\mathcal{S}$ will be on our left side. (See Figure 15.26(a) and (b).)

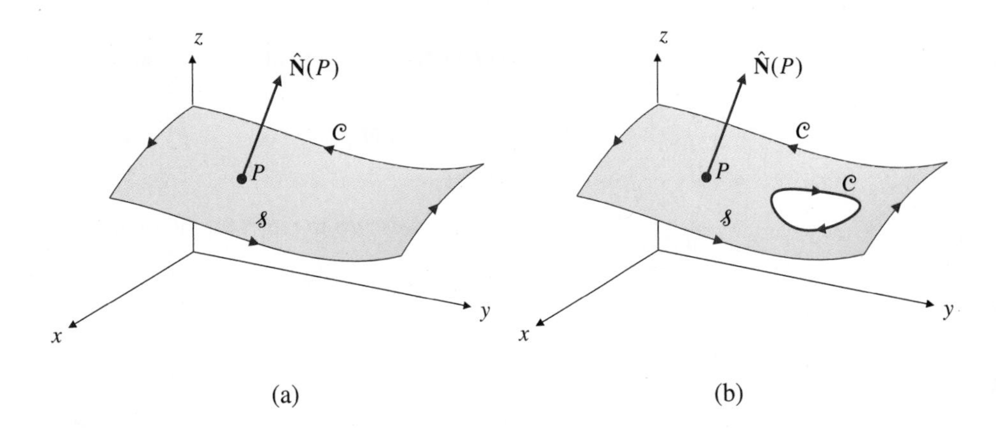

Figure 15.26 The boundary curves of an oriented surface are themselves oriented with the surface on the left

A *piecewise smooth* surface is **orientable** if, whenever two smooth component surfaces join along a common boundary curve $\mathcal{C}$, they induce *opposite* orientations along $\mathcal{C}$. This forces the normals $\hat{\mathbf{N}}$ to be on the same side of adjacent components. For instance, the surface of a cube is a piecewise smooth, closed surface, consisting of six smooth surfaces (the square faces) joined along edges. (See Figure 15.27.) If all of the faces are oriented so that their normals $\hat{\mathbf{N}}$ point out of the cube (or if they all point into the cube), then the surface of the cube itself is oriented.

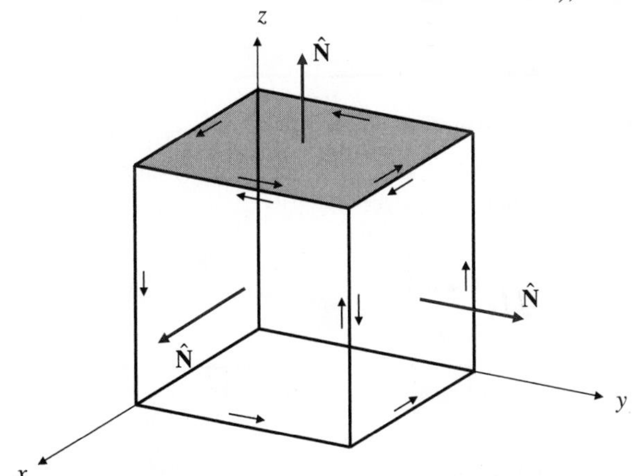

Figure 15.27 The surface of the cube is orientable; adjacent faces induce opposite orientations on their common edge

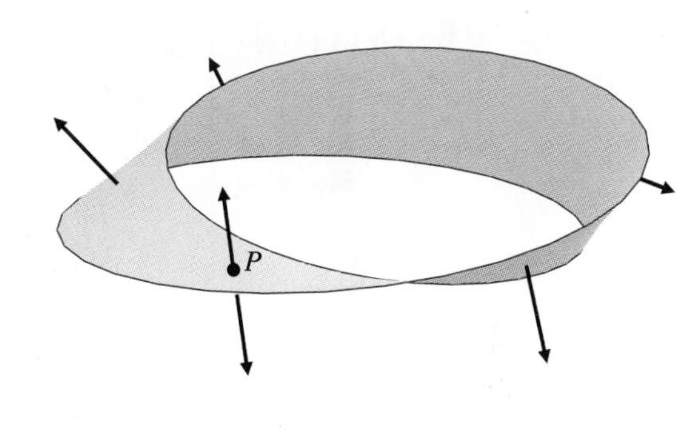

Figure 15.28 The Möbius band is not orientable; it has only one "side"

Not every surface can be oriented, even if it appears smooth. An orientable surface must have two sides. For example, a **Möbius band**, consisting of a strip of paper with ends joined together to form a loop, but with one end given a half twist before the ends are joined, has only one side (make one and see), so it cannot be oriented. (See Figure 15.28.) If a nonzero vector is moved around the band, starting at point P, so that it is always normal to the surface, then it can return to its starting position pointing in the opposite direction.

The Flux of a Vector Field Across a Surface

Suppose 3-space is filled with an incompressible fluid that flows with velocity field $\mathbf{v}$. Let $\mathcal{S}$ be an imaginary, smooth, oriented surface in 3-space. (We say $\mathcal{S}$ is *imaginary* because it does not impede the motion of the fluid; it is fixed in space and the fluid can move freely through it.) We calculate the rate at which fluid flows across $\mathcal{S}$. Let dS be a small area element at point P on the surface. The fluid crossing that element between time t and time $t + dt$ occupies a cylinder of base area dS and height $|\mathbf{v}(P)|\, dt\, \cos\theta$, where θ is the angle between $\mathbf{v}(P)$ and the normal $\hat{\mathbf{N}}(P)$. (See Figure 15.29.) This cylinder has (signed) volume $\mathbf{v}(P) \bullet \hat{\mathbf{N}}(P)\, dS\, dt$. The rate at which fluid crosses dS is

$\mathbf{v}(P) \bullet \hat{\mathbf{N}}(P)\, dS$, and the total rate at which it crosses $\mathcal{S}$ is given by the surface integral

$$\iint_{\mathcal{S}} \mathbf{v} \bullet \hat{\mathbf{N}}\, dS \qquad \text{or} \qquad \iint_{\mathcal{S}} \mathbf{v} \bullet d\mathbf{S},$$

where we use $d\mathbf{S}$ to represent the vector surface area element $\hat{\mathbf{N}}\, dS$.

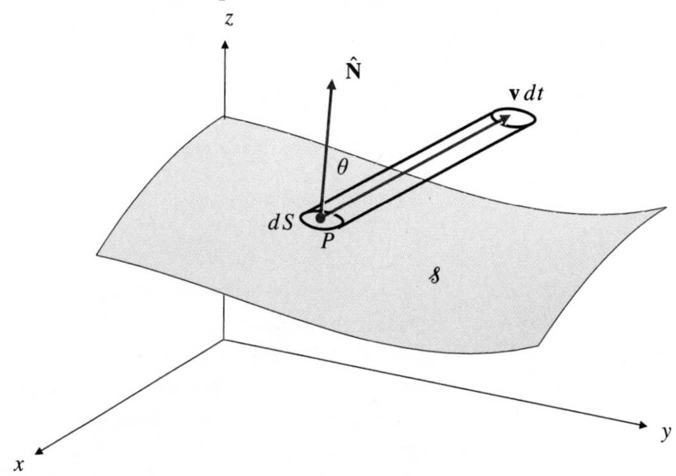

Figure 15.29 The fluid crossing dS in time dt fills the tube

DEFINITION 6

Flux of a vector field across an oriented surface

Given any continuous vector field $\mathbf{F}$, the **flux** of $\mathbf{F}$ across the orientable surface $\mathcal{S}$ is integral of the normal component of $\mathbf{F}$ over $\mathcal{S}$,

$$\iint_{\mathcal{S}} \mathbf{F} \bullet \hat{\mathbf{N}}\, dS \qquad \text{or} \qquad \iint_{\mathcal{S}} \mathbf{F} \bullet d\mathbf{S}.$$

When the surface is closed, the flux integral can be denoted by

$$\oiint_{\mathcal{S}} \mathbf{F} \bullet \hat{\mathbf{N}}\, dS \qquad \text{or} \qquad \oiint_{\mathcal{S}} \mathbf{F} \bullet d\mathbf{S}.$$

In this case we refer to the flux of $\mathbf{F}$ *out of* $\mathcal{S}$ if $\hat{\mathbf{N}}$ is the unit *exterior* normal, and the flux *into* $\mathcal{S}$ if $\hat{\mathbf{N}}$ is the unit *interior* normal.

EXAMPLE 1 Find the flux of the vector field $\mathbf{F} = m\mathbf{r}/|\mathbf{r}|^3$ out of a sphere $\mathcal{S}$ of radius a centred at the origin. (Here $\mathbf{r} = x\mathbf{i} + y\mathbf{j} + z\mathbf{k}$.)

Solution Since $\mathbf{F}$ is the field associated with a source of strength m at the origin (which produces $4\pi m$ units of fluid per unit time at the origin), the answer must be $4\pi m$. Let us calculate it anyway. We use spherical coordinates. At any point $\mathbf{r}$ on the sphere, with spherical coordinates $[a, \phi, \theta]$, the unit outward normal is $\hat{\mathbf{r}} = \mathbf{r}/|\mathbf{r}|$. Since the vector field is $\mathbf{F} = m\hat{\mathbf{r}}/a^2$ on the sphere, and since an area element is $dS = a^2 \sin\phi\, d\phi\, d\theta$, the flux of $\mathbf{F}$ out of the sphere is

$$\oiint_{\mathcal{S}} \left(\frac{m}{a^2}\hat{\mathbf{r}}\right) \bullet \hat{\mathbf{r}}\, a^2 \sin\phi\, d\phi\, d\theta = m \int_0^{2\pi} d\theta \int_0^{\pi} \sin\phi\, d\phi = 4\pi m.$$

EXAMPLE 2 Calculate the total flux of $\mathbf{F} = x\mathbf{i} + y\mathbf{j} + z\mathbf{k}$ outward through the surface of the solid cylinder $x^2 + y^2 \le a^2$, $-h \le z \le h$.

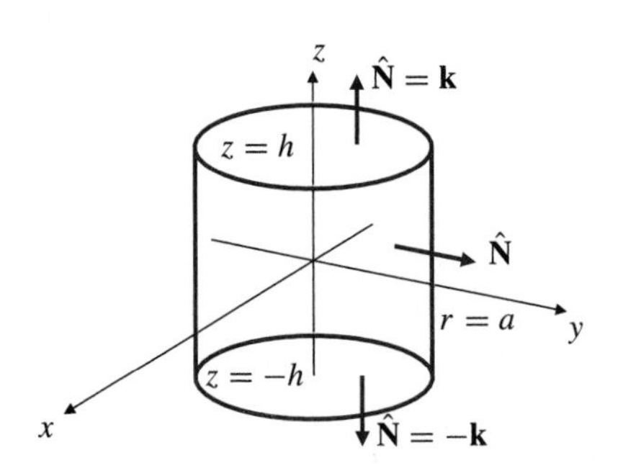

Figure 15.30 The three components of the surface of a solid cylinder with their outward normals

Solution The cylinder is shown in Figure 15.30. Its surface consists of top and bottom disks and the cylindrical side wall. We calculate the flux of $\mathbf{F}$ out of each. Naturally, we use cylindrical coordinates. On the top disk we have $z = h$, $\hat{\mathbf{N}} = \mathbf{k}$, and $dS = r\,dr\,d\theta$. Therefore, $\mathbf{F} \bullet \hat{\mathbf{N}}\,dS = hr\,dr\,d\theta$ and

$$\iint_{\text{top}} \mathbf{F} \bullet \hat{\mathbf{N}}\,dS = h \int_0^{2\pi} d\theta \int_0^a r\,dr = \pi a^2 h.$$

On the bottom disk we have $z = -h$, $\hat{\mathbf{N}} = -\mathbf{k}$, and $dS = r\,dr\,d\theta$. Therefore, $\mathbf{F} \bullet \hat{\mathbf{N}}\,dS = hr\,dr\,d\theta$ and

$$\iint_{\text{bottom}} \mathbf{F} \bullet \hat{\mathbf{N}}\,dS = \iint_{\text{top}} \mathbf{F} \bullet \hat{\mathbf{N}}\,dS = \pi a^2 h.$$

On the cylindrical wall $\mathbf{F} = a\cos\theta\,\mathbf{i} + a\sin\theta\,\mathbf{j} + z\mathbf{k}$, $\hat{\mathbf{N}} = \cos\theta\,\mathbf{i} + \sin\theta\,\mathbf{j}$, and $dS = a\,d\theta\,dz$. Thus, $\mathbf{F} \bullet \hat{\mathbf{N}}\,dS = a^2\,d\theta\,dz$ and

$$\iint_{\text{cylwall}} \mathbf{F} \bullet \hat{\mathbf{N}}\,dS = a^2 \int_0^{2\pi} d\theta \int_{-h}^{h} dz = 4\pi a^2 h.$$

The total flux of $\mathbf{F}$ out of the surface $\mathcal{S}$ of the cylinder is the sum of these three contributions:

$$\oiint_{\mathcal{S}} \mathbf{F} \bullet \hat{\mathbf{N}}\,dS = 6\pi a^2 h.$$

Calculating Flux Integrals

If $\mathcal{S}$ is a parametric surface given by $\mathbf{r} = \mathbf{r}(u, v)$ for (u, v) in domain D in the uv-plane, then, as shown in the previous section, the vector

$$\mathbf{n} = \frac{\partial \mathbf{r}}{\partial u} \times \frac{\partial \mathbf{r}}{\partial v} = \frac{\partial(y, z)}{\partial(u, v)}\mathbf{i} + \frac{\partial(z, x)}{\partial(u, v)}\mathbf{j} + \frac{\partial(x, y)}{\partial(u, v)}\mathbf{k}$$

is normal to $\mathcal{S}$, and $dS = |\mathbf{n}|\,du\,dv$ is an area element on $\mathcal{S}$. Accordingly, the vector area element for $\mathcal{S}$ is

$$d\mathbf{S} = \hat{\mathbf{N}}\,dS = \pm\frac{\mathbf{n}}{|\mathbf{n}|}\,|\mathbf{n}|\,du\,dv = \pm\mathbf{n}\,du\,dv,$$

where the sign must be chosen to reflect the desired orientation of $\mathcal{S}$. The flux of $\mathbf{F} = F_1(x, y, z)\,\mathbf{i} + F_2(x, y, z)\,\mathbf{j} + F_3(x, y, z)\,\mathbf{k}$ through $\mathcal{S}$ is given by

$$\begin{aligned}\iint_{\mathcal{S}} \mathbf{F} \bullet d\mathbf{S} &= \pm \iint_D \mathbf{F} \bullet \left(\frac{\partial \mathbf{r}}{\partial u} \times \frac{\partial \mathbf{r}}{\partial v}\right) du\,dv \\ &= \pm \iint_D \left(F_1 \frac{\partial(y, z)}{\partial(u, v)} + F_2 \frac{\partial(z, x)}{\partial(u, v)} + F_3 \frac{\partial(x, y)}{\partial(u, v)}\right) du\,dv.\end{aligned}$$

There are, of course, simpler versions of these formulas for surfaces of special types. For instance, let $\mathcal{S}$ be a smooth, oriented surface with a one-to-one projection onto a domain D in the xy-plane, and with equation of the form $G(x, y, z) = 0$. In Section 15.5 we showed that the surface area element on $\mathcal{S}$ could be written in the form

$$dS = \left|\frac{\nabla G}{G_3}\right| dx\,dy,$$

and hence surface integrals over $\mathcal{S}$ could be reduced to double integrals over the domain D. Flux integrals can be treated likewise. Depending on the orientation of $\mathcal{S}$, the unit normal $\hat{\mathbf{N}}$ can be written as

$$\hat{\mathbf{N}} = \pm\frac{\nabla G}{|\nabla G|}.$$

Thus, the vector area element $d\mathbf{S}$ can be written

$$d\mathbf{S} = \hat{\mathbf{N}}\,dS = \pm\frac{\nabla G(x,y,z)}{G_3(x,y,z)}\,dx\,dy.$$

The sign must be chosen to give $\mathcal{S}$ the desired orientation. If $G_3 > 0$ and we want the positive side of $\mathcal{S}$ to face upward, we should use the $+$ sign. Of course, similar formulas apply for surfaces with one-to-one projections onto the other coordinate planes.

EXAMPLE 3 Find the flux of $z\mathbf{i} + x^2\mathbf{k}$ upward through that part of the surface $z = x^2 + y^2$ lying above the square R defined by $-1 \le x \le 1$ and $-1 \le y \le 1$.

Solution For $F(x,y,z) = z - x^2 - y^2$ we have $\nabla F = -2x\mathbf{i} - 2y\mathbf{j} + \mathbf{k}$ and $F_3 = 1$. Thus,

$$d\mathbf{S} = (-2x\mathbf{i} - 2y\mathbf{j} + \mathbf{k})\,dx\,dy,$$

and the required flux is

$$\begin{aligned}\iint_{\mathcal{S}} (z\mathbf{i} + x^2\mathbf{k}) \bullet d\mathbf{S} &= \iint_R \left(-2x(x^2+y^2) + x^2\right) dx\,dy \\ &= \int_{-1}^{1} dx \int_{-1}^{1} (x^2 - 2x^3 - 2xy^2)\,dy \\ &= \int_{-1}^{1} 2x^2\,dx = \frac{4}{3}.\end{aligned}$$

(Two of the three terms in the double integral had zero integrals because of symmetry.)

For a surface $\mathcal{S}$ with equation $z = f(x,y)$ we have

$$\hat{\mathbf{N}} = \pm\frac{-\dfrac{\partial f}{\partial x}\mathbf{i} - \dfrac{\partial f}{\partial y}\mathbf{j} + \mathbf{k}}{\sqrt{1 + \left(\dfrac{\partial f}{\partial x}\right)^2 + \left(\dfrac{\partial f}{\partial y}\right)^2}} \quad \text{and}$$

$$dS = \sqrt{1 + \left(\frac{\partial f}{\partial x}\right)^2 + \left(\frac{\partial f}{\partial y}\right)^2}\,dx\,dy,$$

so that the vector area element on $\mathcal{S}$ is given by

$$d\mathbf{S} = \hat{\mathbf{N}}\,dS = \pm\left(-\frac{\partial f}{\partial x}\mathbf{i} - \frac{\partial f}{\partial y}\mathbf{j} + \mathbf{k}\right) dx\,dy.$$

Again, the $+$ sign corresponds to an upward normal.

EXAMPLE 4 Find the flux of $\mathbf{F} = y\mathbf{i} - x\mathbf{j} + 4\mathbf{k}$ upward through $\mathcal{S}$, where $\mathcal{S}$ is the part of the surface $z = 1 - x^2 - y^2$ lying in the first octant of 3-space.

Solution The vector area element corresponding to the upward normal on $\mathcal{S}$ is

$$d\mathbf{S} = \left(-\frac{\partial z}{\partial x}\mathbf{i} - \frac{\partial z}{\partial y}\mathbf{j} + \mathbf{k}\right) dx\,dy = (2x\mathbf{i} + 2y\mathbf{j} + \mathbf{k})\,dx\,dy.$$

The projection of $\mathcal{S}$ onto the xy-plane is the quarter-circular disk Q given by $x^2+y^2 \le 1$, $x \ge 0$, and $y \ge 0$. Thus, the flux of $\mathbf{F}$ upward through $\mathcal{S}$ is

$$\begin{aligned}\iint_{\mathcal{S}} \mathbf{F} \bullet d\mathbf{S} &= \iint_Q (2xy - 2xy + 4)\,dx\,dy \\ &= 4 \times (\text{area of } Q) = \pi.\end{aligned}$$

EXAMPLE 5 Find the flux of $\mathbf{F} = \dfrac{2x\mathbf{i} + 2y\mathbf{j}}{x^2 + y^2} + \mathbf{k}$ downward through the surface $\mathcal{S}$ defined parametrically by

$$\mathbf{r} = u\cos v\mathbf{i} + u\sin v\mathbf{j} + u^2\mathbf{k}, \qquad (0 \le u \le 1,\ 0 \le v \le 2\pi).$$

Solution First we calculate $d\mathbf{S}$:

$$\begin{aligned}
\frac{\partial \mathbf{r}}{\partial u} &= \cos v\mathbf{i} + \sin v\mathbf{j} + 2u\mathbf{k} \\
\frac{\partial \mathbf{r}}{\partial v} &= -u\sin v\mathbf{i} + u\cos v\mathbf{j} \\
\frac{\partial \mathbf{r}}{\partial u} \times \frac{\partial \mathbf{r}}{\partial v} &= -2u^2\cos v\mathbf{i} - 2u^2\sin v\mathbf{j} + u\mathbf{k}.
\end{aligned}$$

Since $u \ge 0$ on $\mathcal{S}$, the latter expression is an upward normal. We want a downward normal, so we use

$$d\mathbf{S} = (2u^2\cos v\mathbf{i} + 2u^2\sin v\mathbf{j} - u\mathbf{k})\,du\,dv.$$

On $\mathcal{S}$ we have

$$\mathbf{F} = \frac{2x\mathbf{i} + 2y\mathbf{j}}{x^2 + y^2} + \mathbf{k} = \frac{2u\cos v\mathbf{i} + 2u\sin v\mathbf{j}}{u^2} + \mathbf{k},$$

so the downward flux of $\mathbf{F}$ through $\mathcal{S}$ is

$$\iint_{\mathcal{S}} \mathbf{F} \bullet d\mathbf{S} = \int_0^{2\pi} dv \int_0^1 (4u - u)\,du = 3\pi.$$

EXERCISES 15.6

1. Find the flux of $\mathbf{F} = x\mathbf{i} + z\mathbf{j}$ out of the tetrahedron bounded by the coordinate planes and the plane $x + 2y + 3z = 6$.
2. Find the flux of $\mathbf{F} = x\mathbf{i} + y\mathbf{j} + z\mathbf{k}$ outward across the sphere $x^2 + y^2 + z^2 = a^2$.
3. Find the flux of the vector field of Exercise 2 out of the surface of the box $0 \le x \le a, 0 \le y \le b, 0 \le z \le c$.
4. Find the flux of the vector field $\mathbf{F} = y\mathbf{i} + z\mathbf{k}$ out across the boundary of the solid cone $0 \le z \le 1 - \sqrt{x^2 + y^2}$.
5. Find the flux of $\mathbf{F} = x\mathbf{i} + y\mathbf{j} + z\mathbf{k}$ upward through the part of the surface $z = a - x^2 - y^2$ lying above plane $z = b < a$.
6. Find the flux of $\mathbf{F} = x\mathbf{i} + x\mathbf{j} + \mathbf{k}$ upward through the part of the surface $z = x^2 - y^2$ inside the cylinder $x^2 + y^2 = a^2$.
7. Find the flux of $\mathbf{F} = y^3\mathbf{i} + z^2\mathbf{j} + x\mathbf{k}$ downward through the part of the surface $z = 4 - x^2 - y^2$ that lies above the plane $z = 2x + 1$.
8. Find the flux of $\mathbf{F} = z^2\mathbf{k}$ upward through the part of the sphere $x^2 + y^2 + z^2 = a^2$ in the first octant of 3-space.
9. Find the flux of $\mathbf{F} = x\mathbf{i} + y\mathbf{j}$ upward through the part of the surface $z = 2 - x^2 - 2y^2$ that lies above the xy-plane.
10. Find the flux of $\mathbf{F} = 2x\mathbf{i} + y\mathbf{j} + z\mathbf{k}$ upward through the surface $\mathbf{r} = u^2v\mathbf{i} + uv^2\mathbf{j} + v^3\mathbf{k}$, $(0 \le u \le 1, 0 \le v \le 1)$.
11. Find the flux of $\mathbf{F} = x\mathbf{i} + y\mathbf{j} + z^2\mathbf{k}$ upward through the surface $u\cos v\,\mathbf{i} + u\sin v\,\mathbf{j} + u\,\mathbf{k}$, $(0 \le u \le 2, 0 \le v \le \pi)$.
12. Find the flux of $\mathbf{F} = yz\mathbf{i} - xz\mathbf{j} + (x^2 + y^2)\mathbf{k}$ upward through the surface $\mathbf{r} = e^u\cos v\,\mathbf{i} + e^u\sin v\,\mathbf{j} + u\,\mathbf{k}$, where $0 \le u \le 1$ and $0 \le v \le \pi$.
13. Find the flux of $\mathbf{F} = m\mathbf{r}/|\mathbf{r}|^3$ out of the surface of the cube $-a \le x, y, z \le a$.
14. Find the flux of the vector field of Exercise 13 out of the box $1 \le x, y, z \le 2$. *Note:* This problem can be solved very easily using the Divergence Theorem of Section 16.4; the required flux is, in fact, zero. However, the object here is to do it by direct calculation of the surface integrals involved, and as such it is quite difficult. By symmetry, it is sufficient to evaluate the net flux out of the cube through any one of the three pairs of opposite faces; that is, you must calculate the flux through only two faces, say $z = 1$ and $z = 2$. Be prepared to work very hard to evaluate these integrals! When they are done, you may find the identities $2\arctan a = \arctan\left(2a/(1 - a^2)\right)$ and $\arctan a + \arctan(1/a)\pi/2$ useful for showing that the net

flux is zero.

15. Define the flux of a *plane* vector field across a piecewise smooth *curve*. Find the flux of $\mathbf{F} = x\mathbf{i} + y\mathbf{j}$ outward across

(a) the circle $x^2 + y^2 = a^2$, and

(b) the boundary of the square $-1 \le x, y \le 1$.

16. Find the flux of $\mathbf{F} = -(x\mathbf{i} + y\mathbf{j})/(x^2 + y^2)$ *inward* across each of the two curves in the previous exercise.

17. If $\mathcal{S}$ is a smooth, oriented surface in 3-space and $\hat{\mathbf{N}}$ is the unit vector field determining the orientation of $\mathcal{S}$, show that the flux of $\hat{\mathbf{N}}$ across $\mathcal{S}$ is the area of $\mathcal{S}$.

! 18. The Divergence Theorem presented in Section 16.4 implies that the flux of a constant vector field across any oriented, piecewise smooth, closed surface is zero. Prove this now for (a) a rectangular box, and (b) a sphere.

CHAPTER REVIEW

Key Ideas

- **What do the following terms and phrases mean?**
  - ◇ vector field
  - ◇ scalar field
  - ◇ field line
  - ◇ conservative field
  - ◇ scalar potential
  - ◇ equipotential
  - ◇ a source
  - ◇ a dipole
  - ◇ connected domain
  - ◇ simply connected
  - ◇ parametric surface
  - ◇ orientable surface
  - ◇ the line integral of f along curve $\mathcal{C}$
  - ◇ the line integral of the tangential component of $\mathbf{F}$ along $\mathcal{C}$
  - ◇ the flux of a vector field through a surface
- **How are the field lines of a conservative field related to its equipotential curves or surfaces?**
- **How is a line integral of a scalar field calculated?**
- **How is a line integral of the tangential component of a vector field calculated?**
- **When is a line integral between two points independent of the path joining those points?**
- **How is a surface integral of a scalar field calculated?**
- **How do you calculate the flux of a vector field through a surface?**

Review Exercises

1. Find $\displaystyle\int_{\mathcal{C}} \frac{1}{y}\, ds$, where $\mathcal{C}$ is the curve

$$x = t, \quad y = 2e^t, \quad z = e^{2t}, \quad (-1 \le t \le 1).$$

2. Let $\mathcal{C}$ be the part of the curve of intersection of the surfaces $z = x + y^2$ and $y = 2x$ from the origin to the point $(2, 4, 18)$. Evaluate $\displaystyle\int_{\mathcal{C}} 2y\, dx + x\, dy + 2\, dz$.

3. Find $\displaystyle\iint_{\mathcal{S}} x\, dS$, where $\mathcal{S}$ is that part of the cone $z = \sqrt{x^2 + y^2}$ in the region $0 \le x \le 1 - y^2$.

4. Find $\displaystyle\iint_{\mathcal{S}} xyz\, dS$ over the part of the plane $x + y + z = 1$ lying in the first octant.

5. Find the flux of $x^2y\mathbf{i} - 10xy^2\mathbf{j}$ upward through the surface $z = xy$, $0 \le x \le 1$, $0 \le y \le 1$.

6. Find the flux of $x\mathbf{i} + y\mathbf{j} + z\mathbf{k}$ downward through the part of the plane $x + 2y + 3z = 6$ lying in the first octant.

7. A bead of mass m slides down a wire in the shape of the curve $x = a \sin t$, $\quad y = a \cos t$, $\quad z = bt$, where $0 \le t \le 6\pi$.

(a) What is the work done by the gravitational force $\mathbf{F} = -mg\mathbf{k}$ on the bead during its descent?

(b) What is the work done against a resistance of constant magnitude R which directly opposes the motion of the bead during its descent?

8. For what values of the constants a, b, and c can you determine the value of the integral I of the tangential component of $\mathbf{F} = (axy + 3yz)\mathbf{i} + (x^2 + 3xz + by^2z)\mathbf{j} + (bxy + cy^3)\mathbf{k}$ along a curve from $(0, 1, -1)$ to $(2, 1, 1)$ without knowing exactly which curve? What is the value of the integral?

9. Let $\mathbf{F} = (x^2/y)\mathbf{i} + y\mathbf{j} + \mathbf{k}$.

(a) Find the field line of $\mathbf{F}$ that passes through $(1, 1, 0)$ and show that it also passes through $(e, e, 1)$.

(b) Find $\displaystyle\int_{\mathcal{C}} \mathbf{F} \bullet d\mathbf{r}$, where $\mathcal{C}$ is the part of the field line in (a) from $(1, 1, 0)$ to $(e, e, 1)$.

10. Consider the vector fields

$$\mathbf{F} = (1 + x)e^{x+y}\mathbf{i} + (xe^{x+y} + 2y)\mathbf{j} - 2z\mathbf{k},$$
$$\mathbf{G} = (1 + x)e^{x+y}\mathbf{i} + (xe^{x+y} + 2z)\mathbf{j} - 2y\mathbf{k}.$$

(a) Show that $\mathbf{F}$ is conservative by finding a potential for it.

(b) Evaluate $\displaystyle\int_{\mathcal{C}} \mathbf{G} \bullet d\mathbf{r}$, where $\mathcal{C}$ is given by

$$\mathbf{r} = (1 - t)e^t\mathbf{i} + t\mathbf{j} + 2t\mathbf{k}, \quad (0 \le t \le 1),$$

by taking advantage of the similarity between $\mathbf{F}$ and $\mathbf{G}$.

11. Find a plane vector field $\mathbf{F}(x, y)$ that satisfies the following conditions:

(i) The field lines of $\mathbf{F}$ are the curves $xy = C$.

(ii) $|\mathbf{F}(x, y)| = 1$ if $(x, y) \ne (0, 0)$.

(iii) $\mathbf{F}(1, 1) = (\mathbf{i} - \mathbf{j})/\sqrt{2}$.

(iv) $\mathbf{F}$ is continuous except at $(0, 0)$.

12. Let $\mathcal{S}$ be the part of the surface of the cylinder $y^2 + z^2 = 16$ that lies in the first octant and between the planes $x = 0$ and $x = 5$. Find the flux of $3z^2x\mathbf{i} - x\mathbf{j} - y\mathbf{k}$ away from the x-axis through $\mathcal{S}$.

Challenging Problems

1. Find the centroid of the surface

$$\mathbf{r} = (2 + \cos v)(\cos u\mathbf{i} + \sin u\mathbf{j}) + \sin v\mathbf{k},$$

where $0 \le u \le 2\pi$ and $0 \le v \le \pi$. Describe this surface.

2. A smooth surface $\mathcal{S}$ is given parametrically by

$$\begin{aligned}\mathbf{r} &= (\cos 2u)(2 + v\cos u)\mathbf{i} \\ &\quad + (\sin 2u)(2 + v\cos u)\mathbf{j} + v\sin u\mathbf{k},\end{aligned}$$

where $0 \le u \le 2\pi$ and $-1 \le v \le 1$. Show that for *every* smooth vector field $\mathbf{F}$ on $\mathcal{S}$,

$$\iint_{\mathcal{S}} \mathbf{F} \bullet \hat{\mathbf{N}}\,dS = 0,$$

where $\hat{\mathbf{N}} = \hat{\mathbf{N}}(u, v)$ is a unit normal vector field on $\mathcal{S}$ that depends continuously on (u, v). How do you explain this? *Hint:* Try to describe what the surface $\mathcal{S}$ looks like.

3. Recalculate the gravitational force exerted by a sphere of radius a and areal density σ centred at the origin on a point mass located at $(0, 0, b)$ by directly integrating the vertical component of the force due to an area element dS, rather than by integrating the potential as we did in the last part of Section 15.5. You will have to be quite creative in dealing with the resulting integral.

CHAPTER 16

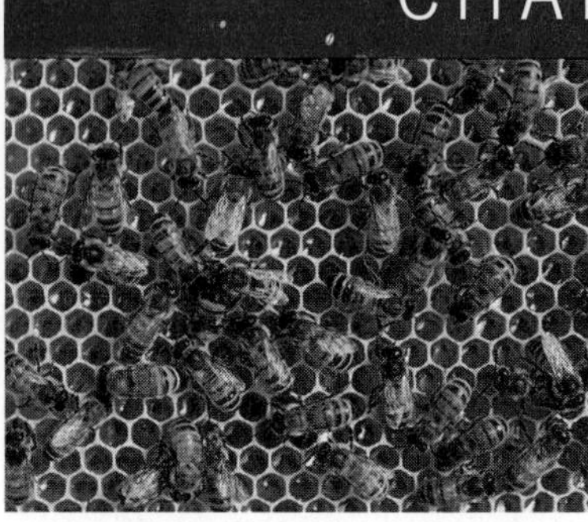

Vector Calculus

> Mathematicians are like Frenchmen: whenever you say something to them, they translate it into their own language, and at once it is something entirely different.
>
> Johann Wolfgang von Goethe 1749–1832
> from *Maxims and Reflections, 1829*

Introduction

In this chapter we develop two- and three-dimensional analogues of the one-dimensional Fundamental Theorem of Calculus. These analogues—Green's Theorem, Gauss's Divergence Theorem, and Stokes's Theorem—are of great importance both theoretically and in applications. They are phrased in terms of certain differential operators, divergence and curl, which are related to the gradient operator encountered in Section 12.7. The operators are introduced and their properties are derived in Sections 16.1 and 16.2. The rest of the chapter deals with the generalizations of the Fundamental Theorem of Calculus and their applications.

16.1 Gradient, Divergence, and Curl

First-order information about the rate of change of a three-dimensional scalar field, $f(x, y, z)$, is contained in the three first partial derivatives $\partial f/\partial x$, $\partial f/\partial y$, and $\partial f/\partial z$. The gradient,

$$\mathbf{grad}\, f(x, y, z) = \nabla f(x, y, z) = \frac{\partial f}{\partial x}\mathbf{i} + \frac{\partial f}{\partial y}\mathbf{j} + \frac{\partial f}{\partial z}\mathbf{k},$$

collects this information into a single vector-valued "derivative" of f. We will develop similar ways of conveying information about the rate of change of vector fields.

First-order information about the rate of change of the vector field

$$\mathbf{F}(x, y, z) = F_1(x, y, z)\mathbf{i} + F_2(x, y, z)\mathbf{j} + F_3(x, y, z)\mathbf{k}$$

is contained in nine first partial derivatives, three for each of the three components of the vector field $\mathbf{F}$:

$$\begin{array}{ccc} \dfrac{\partial F_1}{\partial x} & \dfrac{\partial F_1}{\partial y} & \dfrac{\partial F_1}{\partial z} \\[2mm] \dfrac{\partial F_2}{\partial x} & \dfrac{\partial F_2}{\partial y} & \dfrac{\partial F_2}{\partial z} \\[2mm] \dfrac{\partial F_3}{\partial x} & \dfrac{\partial F_3}{\partial y} & \dfrac{\partial F_3}{\partial z}. \end{array}$$

(Again, we stress that F_1, F_2, and F_3 denote the components of $\mathbf{F}$, not partial derivatives.) Two special combinations of these derivatives organize this information in particularly useful ways, as the gradient does for scalar fields. These are the **divergence** of $\mathbf{F}$ (**div F**) and the **curl** of $\mathbf{F}$ (**curl F**), defined as follows:

Divergence and curl

$$\mathbf{div}\,\mathbf{F} = \nabla \bullet \mathbf{F} = \frac{\partial F_1}{\partial x} + \frac{\partial F_2}{\partial y} + \frac{\partial F_3}{\partial z},$$

$$\begin{aligned} \mathbf{curl}\,\mathbf{F} &= \nabla \times \mathbf{F} \\ &= \left(\frac{\partial F_3}{\partial y} - \frac{\partial F_2}{\partial z}\right)\mathbf{i} + \left(\frac{\partial F_1}{\partial z} - \frac{\partial F_3}{\partial x}\right)\mathbf{j} + \left(\frac{\partial F_2}{\partial x} - \frac{\partial F_1}{\partial y}\right)\mathbf{k} \\ &= \begin{vmatrix} \mathbf{i} & \mathbf{j} & \mathbf{k} \\ \dfrac{\partial}{\partial x} & \dfrac{\partial}{\partial y} & \dfrac{\partial}{\partial z} \\ F_1 & F_2 & F_3 \end{vmatrix}. \end{aligned}$$

Note that the divergence of a vector field is a scalar field, while the curl is another vector field. Also observe the notation $\nabla \bullet \mathbf{F}$ and $\nabla \times \mathbf{F}$, which we will sometimes use instead of **div F** and **curl F**. This makes use of the *vector differential operator*

$$\nabla = \mathbf{i}\frac{\partial}{\partial x} + \mathbf{j}\frac{\partial}{\partial y} + \mathbf{k}\frac{\partial}{\partial z},$$

frequently called *del* or *nabla*. Just as the gradient of the scalar field f can be regarded as *formal scalar multiplication* of ∇ and f, so also can the divergence and curl of $\mathbf{F}$ be regarded as *formal dot* and *cross products* of ∇ with $\mathbf{F}$. When using ∇ the order of "factors" is important; the quantities on which ∇ acts must appear to the right of ∇. For instance, $\nabla \bullet \mathbf{F}$ and $\mathbf{F} \bullet \nabla$ do not mean the same thing; the former is a scalar field and the latter is a scalar differential operator:

BEWARE! Do not confuse the scalar field $\nabla \bullet \mathbf{F}$ with the scalar differential operator $\mathbf{F} \bullet \nabla$. They are quite different objects.

$$\mathbf{F} \bullet \nabla = F_1\frac{\partial}{\partial x} + F_2\frac{\partial}{\partial y} + F_3\frac{\partial}{\partial z}.$$

EXAMPLE 1 Find the divergence and curl of the vector field

$$\mathbf{F} = xy\mathbf{i} + (y^2 - z^2)\mathbf{j} + yz\mathbf{k}.$$

Solution We have

$$\mathbf{div}\,\mathbf{F} = \nabla \bullet \mathbf{F} = \frac{\partial}{\partial x}(xy) + \frac{\partial}{\partial y}(y^2 - z^2) + \frac{\partial}{\partial z}(yz) = y + 2y + y = 4y,$$

$$\begin{aligned} \mathbf{curl}\,\mathbf{F} = \nabla \times \mathbf{F} &= \begin{vmatrix} \mathbf{i} & \mathbf{j} & \mathbf{k} \\ \dfrac{\partial}{\partial x} & \dfrac{\partial}{\partial y} & \dfrac{\partial}{\partial z} \\ xy & y^2 - z^2 & yz \end{vmatrix} \\ &= \left[\frac{\partial}{\partial y}(yz) - \frac{\partial}{\partial z}(y^2 - z^2)\right]\mathbf{i} + \left[\frac{\partial}{\partial z}(xy) - \frac{\partial}{\partial x}(yz)\right]\mathbf{j} \\ &\quad + \left[\frac{\partial}{\partial x}(y^2 - z^2) - \frac{\partial}{\partial y}(xy)\right]\mathbf{k} = 3z\mathbf{i} - x\mathbf{k}. \end{aligned}$$

The divergence and curl of a two-dimensional vector field can also be defined: if $\mathbf{F}(x, y) = F_1(x, y)\mathbf{i} + F_2(x, y)\mathbf{j}$, then

$$\mathbf{div}\,\mathbf{F} = \frac{\partial F_1}{\partial x} + \frac{\partial F_2}{\partial y},$$
$$\mathbf{curl}\,\mathbf{F} = \left(\frac{\partial F_2}{\partial x} - \frac{\partial F_1}{\partial y}\right)\mathbf{k}.$$

Note that the curl of a two-dimensional vector field is still a 3-vector and is perpendicular to the plane of the field. Although **div** and **grad** are defined in all dimensions, **curl** is defined only in three dimensions and in the plane (provided we allow values in three dimensions).

EXAMPLE 2 Find the divergence and curl of $\mathbf{F} = xe^y\mathbf{i} - ye^x\mathbf{j}$.

Solution We have

$$\begin{aligned}\mathbf{div}\,\mathbf{F} = \nabla \bullet \mathbf{F} &= \frac{\partial}{\partial x}(xe^y) + \frac{\partial}{\partial y}(-ye^x) = e^y - e^x,\\ \mathbf{curl}\,\mathbf{F} = \nabla \times \mathbf{F} &= \left(\frac{\partial}{\partial x}(-ye^x) - \frac{\partial}{\partial y}(xe^y)\right)\mathbf{k}\\ &= -(ye^x + xe^y)\mathbf{k}.\end{aligned}$$

Interpretation of the Divergence

The value of the divergence of a vector field $\mathbf{F}$ at point P is, loosely speaking, a measure of the rate at which the field "diverges" or "spreads away" from P. This spreading away can be measured by the flux out of a small closed surface surrounding P. For instance, $\mathbf{div}\,\mathbf{F}(P)$ is the limit of the *flux per unit volume* out of smaller and smaller spheres centred at P.

THEOREM 1

The divergence as flux density

If $\hat{\mathbf{N}}$ is the unit outward normal on the sphere $\mathcal{S}_\epsilon$ of radius ϵ centred at point P, and if $\mathbf{F}$ is a smooth three-dimensional vector field, then

$$\mathbf{div}\,\mathbf{F}(P) = \lim_{\epsilon \to 0+} \frac{3}{4\pi\epsilon^3} \oiint_{\mathcal{S}_\epsilon} \mathbf{F} \bullet \hat{\mathbf{N}}\,dS.$$

PROOF Without loss of generality we assume that P is at the origin. We want to expand $\mathbf{F} = F_1\mathbf{i} + F_2\mathbf{j} + F_3\mathbf{k}$ in a Taylor series about the origin (a Maclaurin series). As shown in Section 12.9 for a function of two variables, the Maclaurin series for a scalar-valued function of three variables takes the form

$$f(x, y, z) = f(0, 0, 0) + \left.\frac{\partial f}{\partial x}\right|_{(0,0,0)} x + \left.\frac{\partial f}{\partial y}\right|_{(0,0,0)} y + \left.\frac{\partial f}{\partial z}\right|_{(0,0,0)} z + \cdots,$$

where "$\cdots$" represents terms of second and higher degree in x, y, and z. If we apply this formula to the components of $\mathbf{F}$, we obtain

$$\mathbf{F}(x, y, z) = \mathbf{F}_0 + \mathbf{F}_{x0}\,x + \mathbf{F}_{y0}\,y + \mathbf{F}_{z0}\,z + \cdots,$$

So as not to confuse partial derivatives with components of vectors, we are using subscripts $x0$, $y0$, and $z0$ here to denote the values of the first partial derivatives of $\mathbf{F}$ at $(0, 0, 0)$.

where

$$\begin{aligned}
\mathbf{F}_0 &= \mathbf{F}(0,0,0) \\
\mathbf{F}_{x0} &= \left.\frac{\partial \mathbf{F}}{\partial x}\right|_{(0,0,0)} = \left.\left(\frac{\partial F_1}{\partial x}\mathbf{i} + \frac{\partial F_2}{\partial x}\mathbf{j} + \frac{\partial F_3}{\partial x}\mathbf{k}\right)\right|_{(0,0,0)} \\
\mathbf{F}_{y0} &= \left.\frac{\partial \mathbf{F}}{\partial y}\right|_{(0,0,0)} = \left.\left(\frac{\partial F_1}{\partial y}\mathbf{i} + \frac{\partial F_2}{\partial y}\mathbf{j} + \frac{\partial F_3}{\partial y}\mathbf{k}\right)\right|_{(0,0,0)} \\
\mathbf{F}_{z0} &= \left.\frac{\partial \mathbf{F}}{\partial z}\right|_{(0,0,0)} = \left.\left(\frac{\partial F_1}{\partial z}\mathbf{i} + \frac{\partial F_2}{\partial z}\mathbf{j} + \frac{\partial F_3}{\partial z}\mathbf{k}\right)\right|_{(0,0,0)};
\end{aligned}$$

again, the "$\cdots$" represents the second- and higher-degree terms in x, y, and z. The unit normal on $\mathcal{S}_\epsilon$ is $\hat{\mathbf{N}} = (x\mathbf{i} + y\mathbf{j} + z\mathbf{k})/\epsilon$, so we have

$$\begin{aligned}
\mathbf{F} \bullet \hat{\mathbf{N}} = \frac{1}{\epsilon}\Big(&\mathbf{F}_0 \bullet \mathbf{i}\,x + \mathbf{F}_0 \bullet \mathbf{j}\,y + \mathbf{F}_0 \bullet \mathbf{k}\,z \\
&+ \mathbf{F}_{x0} \bullet \mathbf{i}\,x^2 + \mathbf{F}_{x0} \bullet \mathbf{j}\,xy + \mathbf{F}_{x0} \bullet \mathbf{k}\,xz \\
&+ \mathbf{F}_{y0} \bullet \mathbf{i}\,xy + \mathbf{F}_{y0} \bullet \mathbf{j}\,y^2 + \mathbf{F}_{y0} \bullet \mathbf{k}\,yz \\
&+ \mathbf{F}_{z0} \bullet \mathbf{i}\,xz + \mathbf{F}_{z0} \bullet \mathbf{j}\,yz + \mathbf{F}_{z0} \bullet \mathbf{k}\,z^2 + \cdots\Big).
\end{aligned}$$

We integrate each term within the parentheses over $\mathcal{S}_\epsilon$. By symmetry,

$$\begin{aligned}
&\iint_{\mathcal{S}_\epsilon} x\,dS = \iint_{\mathcal{S}_\epsilon} y\,dS = \iint_{\mathcal{S}_\epsilon} z\,dS = 0, \\
&\iint_{\mathcal{S}_\epsilon} xy\,dS = \iint_{\mathcal{S}_\epsilon} xz\,dS = \iint_{\mathcal{S}_\epsilon} yz\,dS = 0.
\end{aligned}$$

Also, by symmetry,

$$\begin{aligned}
\iint_{\mathcal{S}_\epsilon} x^2\,dS &= \iint_{\mathcal{S}_\epsilon} y^2\,dS = \iint_{\mathcal{S}_\epsilon} z^2\,dS \\
&= \frac{1}{3}\iint_{\mathcal{S}_\epsilon} (x^2 + y^2 + z^2)\,dS = \frac{1}{3}(\epsilon^2)(4\pi\epsilon^2) = \frac{4}{3}\pi\epsilon^4,
\end{aligned}$$

and the higher-degree terms have surface integrals involving ϵ^5 and higher powers. Thus,

$$\begin{aligned}
\frac{3}{4\pi\epsilon^3}\iint_{\mathcal{S}_\epsilon} \mathbf{F} \bullet \hat{\mathbf{N}}\,dS &= \mathbf{F}_{x0} \bullet \mathbf{i} + \mathbf{F}_{y0} \bullet \mathbf{j} + \mathbf{F}_{z0} \bullet \mathbf{k} + \epsilon\,(\cdots) \\
&= \nabla \bullet \mathbf{F}(0,0,0) + \epsilon\,(\cdots) \\
&\to \nabla \bullet \mathbf{F}(0,0,0)
\end{aligned}$$

as $\epsilon \to 0^+$. This is what we wanted to show.

Remark The spheres $\mathcal{S}_\epsilon$ in the above theorem can be replaced by other contracting families of piecewise smooth surfaces. For instance, if B is the surface of a rectangular box with dimensions Δx, Δy, and Δz containing P, then

$$\mathbf{div}\,\mathbf{F}(P) = \lim_{\Delta x, \Delta y, \Delta z \to 0} \frac{1}{\Delta x\,\Delta y\,\Delta z}\iint_B \mathbf{F} \bullet \hat{\mathbf{N}}\,dS.$$

See Exercise 12 below.

Remark In two dimensions, the value **div F**(P) represents the limiting *flux per unit area* outward across small, non–self-intersecting closed curves that enclose P. See Exercise 13 at the end of this section.

Let us return again to the interpretation of a vector field as a velocity field of a moving incompressible fluid. If the total flux of the velocity field outward across the boundary surface of a domain is positive (or negative), then the fluid must be produced (or annihilated) within that domain.

The vector field $\mathbf{F} = x\mathbf{i} + y\mathbf{j} + z\mathbf{k}$ of Example 2 in Section 15.6 has constant divergence, $\nabla \bullet \mathbf{F} = 3$. In that example we showed that the flux of $\mathbf{F}$ out of a certain cylinder of base radius a and height $2h$ is $6\pi a^2 h$, which is three times the volume of the cylinder. Exercises 2 and 3 of Section 15.6 confirm similar results for the flux of $\mathbf{F}$ out of other domains. This leads to another interpretation for the divergence: **div F**(P) is the *source strength per unit volume* of $\mathbf{F}$ at P. With this interpretation, we would expect, even for a vector field $\mathbf{F}$ with nonconstant divergence, that the total flux of $\mathbf{F}$ out of the surface $\mathcal{S}$ of a domain D would be equal to the total source strength of $\mathbf{F}$ within D; that is,

$$\oiint_{\mathcal{S}} \mathbf{F} \bullet \hat{\mathbf{N}}\, dS = \iiint_D \nabla \bullet \mathbf{F}\, dV.$$

This is the **Divergence Theorem**, which we will prove in Section 16.4.

EXAMPLE 3 Verify that the vector field $\mathbf{F} = m\mathbf{r}/|\mathbf{r}|^3$, due to a source of strength m at $(0, 0, 0)$, has zero divergence at all points in $\mathbb{R}^3$ except the origin. What would you expect to be the total flux of $\mathbf{F}$ outward across the boundary surface of a domain D if the origin lies outside D? if the origin is inside D?

Solution Since

$$\mathbf{F}(x, y, z) = \frac{m}{r^3}\left(x\mathbf{i} + y\mathbf{j} + z\mathbf{k}\right), \quad \text{where} \quad r^2 = x^2 + y^2 + z^2,$$

and since $\partial r/\partial x = x/r$, we have

$$\frac{\partial F_1}{\partial x} = m\frac{\partial}{\partial x}\left(\frac{x}{r^3}\right) = m\frac{r^3 - 3xr^2\left(\frac{x}{r}\right)}{r^6} = m\frac{r^2 - 3x^2}{r^5}.$$

Similarly,

$$\frac{\partial F_2}{\partial y} = m\frac{r^2 - 3y^2}{r^5} \quad \text{and} \quad \frac{\partial F_3}{\partial z} = m\frac{r^2 - 3z^2}{r^5}.$$

Adding these up, we get $\nabla \bullet \mathbf{F}(x, y, z) = 0$ if $r > 0$.

If the origin lies outside the domain D, then the source density of $\mathbf{F}$ in D is zero, so we would expect the total flux of $\mathbf{F}$ out of D to be zero. If the origin lies inside D, then D contains a source of strength m (producing $4\pi m$ cubic units of fluid per unit time), so we would expect the flux out of D to be $4\pi m$. See Example 1 of Section 15.6 and also Exercises 13 and 14 of that section for specific examples.

Distributions and Delta Functions

If $\ell(x)$ represents the line density (mass per unit length) of mass distributed on the x-axis, then the total mass so distributed is

$$m = \int_{-\infty}^{\infty} \ell(x)\, dx.$$

Now suppose that the only mass on the axis is a "point mass" $m = 1$ located at the origin. Then at all other points $x \neq 0$, the density is $\ell(x) = 0$, but we must still have

$$\int_{-\infty}^{\infty} \ell(x)\, dx = m = 1,$$

so $\ell(0)$ must be infinite. This is an ideal situation, a mathematical model. No real *function* $\ell(x)$ can have such properties; if a function is zero everywhere except at a single point, then any integral of that function will be zero. (Why?) (Also, no real mass can occupy just a single point.) Nevertheless, it is very useful to model real, isolated masses as point masses and to model their densities using **generalized functions** (also called **distributions**).

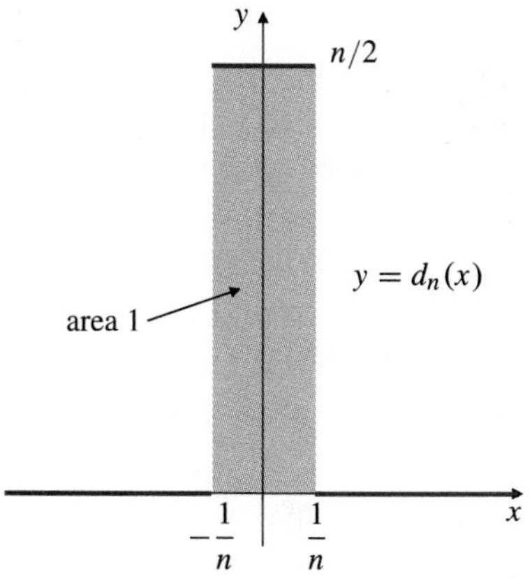

We can think of the density of a point mass 1 at $x = 0$ as the limit of large densities concentrated on small intervals. For instance, if

$$d_n(x) = \begin{cases} n/2 & \text{if } |x| \le 1/n \\ 0 & \text{if } |x| > 1/n \end{cases}$$

y

$n/2$

$y = d_n(x)$

area 1

$-\frac{1}{n}$ $\frac{1}{n}$ x

Figure 16.1 The functions $d_n(x)$ converge to $\delta(x)$ as $n \to \infty$

(see Figure 16.1), then for any smooth function $f(x)$ defined on $\mathbb{R}$ we have

$$\int_{-\infty}^{\infty} d_n(x)\, f(x)\, dx = \frac{n}{2} \int_{-1/n}^{1/n} f(x)\, dx.$$

Replace $f(x)$ in the integral on the right with its Maclaurin series:

$$f(x) = f(0) + \frac{f'(0)}{1!}x + \frac{f''(0)}{2!}x^2 + \cdots.$$

Since

$$\int_{-1/n}^{1/n} x^k\, dx = \begin{cases} 2/((k+1)n^{k+1}) & \text{if } k \text{ is even} \\ 0 & \text{if } k \text{ is odd,} \end{cases}$$

we can take the limit as $n \to \infty$ and obtain

$$\lim_{n\to\infty} \int_{-\infty}^{\infty} d_n(x)\, f(x)\, dx = f(0).$$

DEFINITION 1

The **Dirac distribution** $\delta(x)$ (also called the **Dirac delta function**, although it is really not a function) is the "limit" of the sequence $d_n(x)$ as $n \to \infty$. It is defined by the requirement that

$$\int_{-\infty}^{\infty} \delta(x) f(x)\, dx = f(0)$$

for every smooth function $f(x)$.

A formal change of variables shows that the delta function also satisfies

$$\int_{-\infty}^{\infty} \delta(x - t) f(t)\, dt = f(x).$$

EXAMPLE 4 In view of the fact that $\mathbf{F}(\mathbf{r}) = m\mathbf{r}/|\mathbf{r}|^3$ satisfies $\mathbf{div}\,\mathbf{F}(x, y, z) = 0$ for $(x, y, z) \neq (0, 0, 0)$ but produces a flux of $4\pi m$ out of any sphere centred at the origin, we can regard $\mathbf{div}\,\mathbf{F}(x, y, z)$ as a distribution

$$\mathbf{div}\,\mathbf{F}(x, y, z) = 4\pi m \delta(x)\delta(y)\delta(z).$$

In particular, integrating this distribution against $f(x, y, z) = 1$ over $\mathbb{R}^3$, we have

$$\begin{aligned}\iiint_{\mathbb{R}^3} \mathbf{div}\,\mathbf{F}(x, y, z)\,dV &= 4\pi m \int_{-\infty}^{\infty} \delta(x)\,dx \int_{-\infty}^{\infty} \delta(y)\,dy \int_{-\infty}^{\infty} \delta(z)\,dz \\ &= 4\pi m.\end{aligned}$$

The integral can equally well be taken over *any domain* in $\mathbb{R}^3$ that contains the origin in its interior, and the result will be the same. If the origin is outside the domain, the result will be zero. We will reexamine this situation after establishing the Divergence Theorem in Section 16.4.

A formal study of distributions is beyond the scope of this book; refer to more advanced textbooks on differential equations and engineering mathematics.

Interpretation of the Curl

Roughly speaking, **curl** $\mathbf{F}(P)$ measures the extent to which the vector field $\mathbf{F}$ "swirls" around P.

EXAMPLE 5 Consider the velocity field

$$\mathbf{v} = -\Omega y\mathbf{i} + \Omega x\mathbf{j}$$

of a solid rotating with angular speed Ω about the z-axis, that is, with angular velocity $\boldsymbol{\Omega} = \Omega\mathbf{k}$. (See Figure 15.2 in Section 15.1.) Calculate the circulation of this field around a circle $\mathcal{C}_\epsilon$ in the xy-plane centred at any point (x_0, y_0), having radius ϵ, and oriented counterclockwise. What is the relationship between this circulation and the curl of $\mathbf{v}$?

Solution The indicated circle has parametrization

$$\mathbf{r} = (x_0 + \epsilon\cos t)\mathbf{i} + (y_0 + \epsilon\sin t)\mathbf{j}, \qquad (0 \le t \le 2\pi),$$

and the circulation of $\mathbf{v}$ around it is given by

$$\begin{aligned}\oint_{\mathcal{C}_\epsilon} \mathbf{v} \bullet d\mathbf{r} &= \int_0^{2\pi} \Big(-\Omega(y_0 + \epsilon\sin t)(-\epsilon\sin t) + \Omega(x_0 + \epsilon\cos t)(\epsilon\cos t)\Big)\,dt \\ &= \int_0^{2\pi} \Big(\Omega\epsilon(y_0\sin t + x_0\cos t) + \Omega\epsilon^2\Big)\,dt \\ &= 2\Omega\pi\epsilon^2.\end{aligned}$$

Since

$$\mathbf{curl}\,\mathbf{v} = \nabla \times \mathbf{v} = \Big(\frac{\partial}{\partial x}(\Omega x) - \frac{\partial}{\partial y}(-\Omega y)\Big)\mathbf{k} = 2\Omega\mathbf{k} = 2\boldsymbol{\Omega},$$

the circulation is the product of $(\mathbf{curl}\,\mathbf{v}) \bullet \mathbf{k}$ and the area bounded by $\mathcal{C}_\epsilon$. Note that this circulation is constant for circles of any fixed radius; it does not depend on the position of the centre.

The calculations in the example above suggest that the curl of a vector field is a measure of the *circulation per unit area* in planes normal to the curl. A more precise version of this conjecture is stated in Theorem 2 below. We will not prove this theorem now because a proof at this stage would be quite complicated. (However, see Exercise 14 below for a special case.) A simple proof can be based on Stokes's Theorem; see Exercise 13 in Section 16.5.

THEOREM 2

The curl as circulation density

If $\mathbf{F}$ is a smooth vector field and $\mathcal{C}_\epsilon$ is a circle of radius ϵ centred at point P and bounding a disk $\mathcal{S}_\epsilon$ with unit normal $\hat{\mathbf{N}}$ (and orientation inherited from $\mathcal{C}_\epsilon$; see Figure 16.2), then

$$\lim_{\epsilon\to 0+} \frac{1}{\pi\epsilon^2} \oint_{\mathcal{C}_\epsilon} \mathbf{F} \bullet d\mathbf{r} = \hat{\mathbf{N}} \bullet \mathbf{curl}\, \mathbf{F}(P).$$

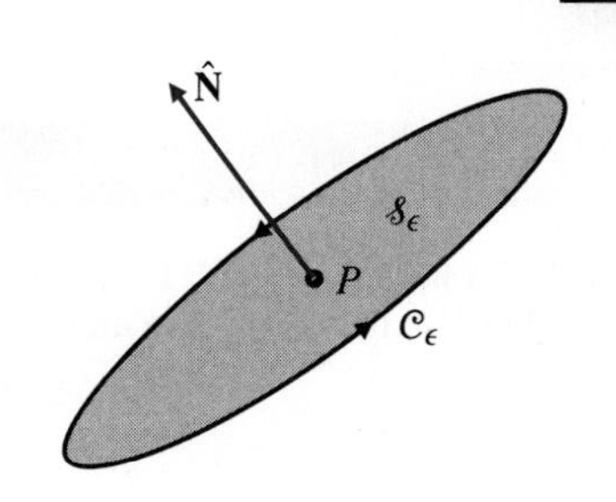

Figure 16.2 Illustrating Theorem 2

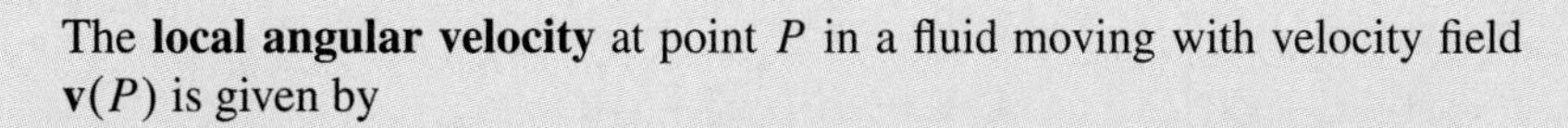

Example 5 also suggests the following definition for the *local* angular velocity of a moving fluid:

> The **local angular velocity** at point P in a fluid moving with velocity field $\mathbf{v}(P)$ is given by
>
> $$\mathbf{\Omega}(P) = \frac{1}{2}\mathbf{curl}\, \mathbf{v}(P).$$

Theorem 2 states that the local angular velocity $\mathbf{\Omega}(P)$ is that vector whose component in the direction of any unit vector $\hat{\mathbf{N}}$ is one-half of the limiting circulation per unit area around the (oriented) boundary circles of small circular disks centred at P and having normal $\hat{\mathbf{N}}$.

Not all vector fields with nonzero curl *appear* to circulate. The velocity field for the rigid body rotation considered in Example 5 appears to circulate around the axis of rotation, but the circulation around a circle in a plane perpendicular to that axis turned out to be independent of the position of the circle; it depended only on its area. The circle need not even surround the axis of rotation. The following example investigates a fluid velocity field whose streamlines are *straight lines* but that still has nonzero, constant curl and, therefore, constant local angular velocity.

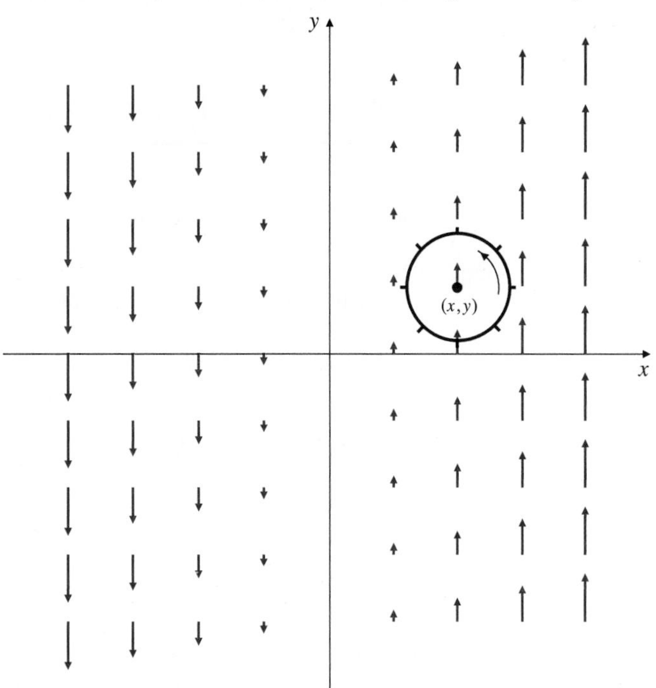

Figure 16.3 The paddle wheel is not only carried along but is set rotating by the flow

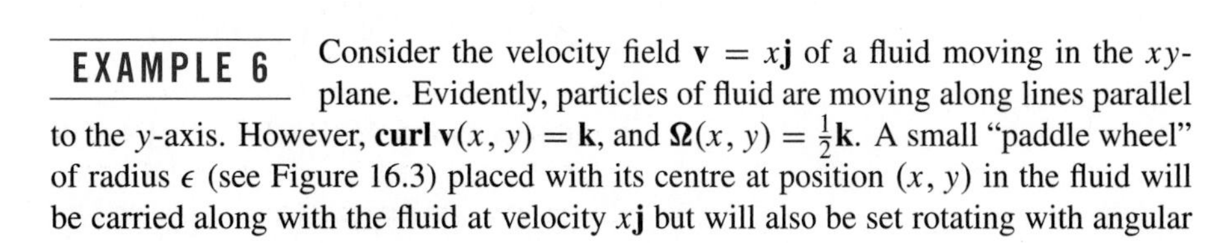

EXAMPLE 6 Consider the velocity field $\mathbf{v} = x\mathbf{j}$ of a fluid moving in the xy-plane. Evidently, particles of fluid are moving along lines parallel to the y-axis. However, $\mathbf{curl}\, \mathbf{v}(x, y) = \mathbf{k}$, and $\mathbf{\Omega}(x, y) = \frac{1}{2}\mathbf{k}$. A small "paddle wheel" of radius ϵ (see Figure 16.3) placed with its centre at position (x, y) in the fluid will be carried along with the fluid at velocity $x\mathbf{j}$ but will also be set rotating with angular

velocity $\mathbf{\Omega}(x, y) = \frac{1}{2}\mathbf{k}$, which is independent of its position. This angular velocity is due to the fact that the velocity of the fluid along the right side of the wheel exceeds that along the left side.

EXERCISES 16.1

In Exercises 1–11, calculate **div F** and **curl F** for the given vector fields.

1. $\mathbf{F} = x\mathbf{i} + y\mathbf{j}$
2. $\mathbf{F} = y\mathbf{i} + x\mathbf{j}$
3. $\mathbf{F} = y\mathbf{i} + z\mathbf{j} + x\mathbf{k}$
4. $\mathbf{F} = yz\mathbf{i} + xz\mathbf{j} + xy\mathbf{k}$
5. $\mathbf{F} = x\mathbf{i} + x\mathbf{k}$
6. $\mathbf{F} = xy^2\mathbf{i} - yz^2\mathbf{j} + zx^2\mathbf{k}$
7. $\mathbf{F} = f(x)\mathbf{i} + g(y)\mathbf{j} + h(z)\mathbf{k}$
8. $\mathbf{F} = f(z)\mathbf{i} - f(z)\mathbf{j}$
9. $\mathbf{F}(r, \theta) = r\mathbf{i} + \sin\theta\mathbf{j}$, where (r, θ) are polar coordinates in the plane
10. $\mathbf{F} = \hat{\mathbf{r}} = \cos\theta\mathbf{i} + \sin\theta\mathbf{j}$
11. $\mathbf{F} = \hat{\boldsymbol{\theta}} = -\sin\theta\mathbf{i} + \cos\theta\mathbf{j}$
12. Let **F** be a smooth, three-dimensional vector field. If $B_{a,b,c}$ is the surface of the box $-a \le x \le a$, $-b \le y \le b$, $-c \le z \le c$, with outward normal $\hat{\mathbf{N}}$, show that

$$\lim_{a,b,c\to 0+} \frac{1}{8abc} \oiint_{B_{a,b,c}} \mathbf{F} \bullet \hat{\mathbf{N}}\, dS = \nabla \bullet \mathbf{F}(0, 0, 0).$$

13. Let **F** be a smooth two-dimensional vector field. If $\mathcal{C}_\epsilon$ is the circle of radius ϵ centred at the origin, and $\hat{\mathbf{N}}$ is the unit outward normal to $\mathcal{C}_\epsilon$, show that

$$\lim_{\epsilon\to 0+} \frac{1}{\pi\epsilon^2} \oint_{\mathcal{C}_\epsilon} \mathbf{F} \bullet \hat{\mathbf{N}}\, ds = \mathbf{div}\, \mathbf{F}(0, 0).$$

14. Prove Theorem 2 in the special case that $\mathcal{C}_\epsilon$ is the circle in the xy-plane with parametrization $x = \epsilon\cos\theta$, $y = \epsilon\sin\theta$, $(0 \le \theta \le 2\pi)$. In this case $\hat{\mathbf{N}} = \mathbf{k}$. *Hint:* Expand $\mathbf{F}(x, y, z)$ in a vector Taylor series about the origin as in the proof of Theorem 1, and calculate the circulation of individual terms around $\mathcal{C}_\epsilon$.

16.2 Some Identities Involving Grad, Div, and Curl

There are numerous identities involving the functions

$$\mathbf{grad}\, f(x, y, z) = \nabla f(x, y, z) = \frac{\partial f}{\partial x}\mathbf{i} + \frac{\partial f}{\partial y}\mathbf{j} + \frac{\partial f}{\partial z}\mathbf{k},$$

$$\mathbf{div}\, \mathbf{F}(x, y, z) = \nabla \bullet \mathbf{F}(x, y, z) = \frac{\partial F_1}{\partial x} + \frac{\partial F_2}{\partial y} + \frac{\partial F_3}{\partial z},$$

$$\mathbf{curl}\, \mathbf{F}(x, y, z) = \nabla \times \mathbf{F}(x, y, z) = \begin{vmatrix} \mathbf{i} & \mathbf{j} & \mathbf{k} \\ \frac{\partial}{\partial x} & \frac{\partial}{\partial y} & \frac{\partial}{\partial z} \\ F_1 & F_2 & F_3 \end{vmatrix},$$

and the **Laplacian operator**, $\nabla^2 = \nabla \bullet \nabla$, defined for a scalar field ϕ by

$$\nabla^2\phi = \nabla \bullet \nabla\phi = \mathbf{div}\, \mathbf{grad}\, \phi = \frac{\partial^2\phi}{\partial x^2} + \frac{\partial^2\phi}{\partial y^2} + \frac{\partial^2\phi}{\partial z^2},$$

and for a vector field $\mathbf{F} = F_1\mathbf{i} + F_2\mathbf{j} + F_3\mathbf{k}$ by

$$\nabla^2\mathbf{F} = (\nabla^2 F_1)\mathbf{i} + (\nabla^2 F_2)\mathbf{j} + (\nabla^2 F_3)\mathbf{k}.$$

(The Laplacian operator, $\nabla^2 = (\partial^2/\partial x^2) + (\partial^2/\partial y^2) + (\partial^2/\partial z^2)$, is denoted by Δ in some books.) Recall that a function ϕ is called **harmonic** in a domain D if $\nabla^2\phi = 0$ throughout D. (See Section 12.4.)

We collect the most important identities together in the following theorem. Most of them are forms of the Product Rule. We will prove a few of the identities to illustrate the techniques involved (mostly brute-force calculation) and leave the rest as exercises. Note that two of the identities involve quantities like $(\mathbf{G} \bullet \nabla)\mathbf{F}$; this represents the vector obtained by applying the scalar differential operator $\mathbf{G} \bullet \nabla$ to the vector field $\mathbf{F}$:

$$(\mathbf{G} \bullet \nabla)\mathbf{F} = G_1 \frac{\partial \mathbf{F}}{\partial x} + G_2 \frac{\partial \mathbf{F}}{\partial y} + G_3 \frac{\partial \mathbf{F}}{\partial z}.$$

THEOREM 3 **Vector differential identities**

Let ϕ and ψ be scalar fields and $\mathbf{F}$ and $\mathbf{G}$ be vector fields, all assumed to be sufficiently smooth that all the partial derivatives in the identities are continuous. Then the following identities hold:

(a) $\nabla(\phi\psi) = \phi\nabla\psi + \psi\nabla\phi$

(b) $\nabla \bullet (\phi\mathbf{F}) = (\nabla\phi) \bullet \mathbf{F} + \phi(\nabla \bullet \mathbf{F})$

(c) $\nabla \times (\phi\mathbf{F}) = (\nabla\phi) \times \mathbf{F} + \phi(\nabla \times \mathbf{F})$

(d) $\nabla \bullet (\mathbf{F} \times \mathbf{G}) = (\nabla \times \mathbf{F}) \bullet \mathbf{G} - \mathbf{F} \bullet (\nabla \times \mathbf{G})$

(e) $\nabla \times (\mathbf{F} \times \mathbf{G}) = (\nabla \bullet \mathbf{G})\mathbf{F} + (\mathbf{G} \bullet \nabla)\mathbf{F} - (\nabla \bullet \mathbf{F})\mathbf{G} - (\mathbf{F} \bullet \nabla)\mathbf{G}$

(f) $\nabla(\mathbf{F} \bullet \mathbf{G}) = \mathbf{F} \times (\nabla \times \mathbf{G}) + \mathbf{G} \times (\nabla \times \mathbf{F}) + (\mathbf{F} \bullet \nabla)\mathbf{G} + (\mathbf{G} \bullet \nabla)\mathbf{F}$

(g) $\nabla \bullet (\nabla \times \mathbf{F}) = 0$ (**div curl** $= 0$)

(h) $\nabla \times (\nabla\phi) = \mathbf{0}$ (**curl grad** $= \mathbf{0}$)

(i) $\nabla \times (\nabla \times \mathbf{F}) = \nabla(\nabla \bullet \mathbf{F}) - \nabla^2\mathbf{F}$
(**curl curl** $=$ **grad div** $-$ Laplacian)

Identities (a)–(f) are versions of the Product Rule and are first-order identities involving only one application of ∇. Identities (g)–(i) are second-order identities. Identities (g) and (h) are equivalent to the equality of mixed partial derivatives and are especially important for the understanding of **div** and **curl**.

PROOF We will prove only identities (c), (e), and (g). The remaining proofs are similar to these.

(c) The first component ($\mathbf{i}$ component) of $\nabla \times (\phi\mathbf{F})$ is

$$\frac{\partial}{\partial y}(\phi F_3) - \frac{\partial}{\partial z}(\phi F_2) = \frac{\partial \phi}{\partial y} F_3 - \frac{\partial \phi}{\partial z} F_2 + \phi \frac{\partial F_3}{\partial y} - \phi \frac{\partial F_2}{\partial z}.$$

The first two terms on the right constitute the first component of $(\nabla\phi) \times \mathbf{F}$, and the last two terms constitute the first component of $\phi(\nabla \times \mathbf{F})$. Therefore, the first components of both sides of identity (c) are equal. The equality of the other components follows similarly.

(e) Again, it is sufficient to show that the first components of the vectors on both sides of the identity are equal. To calculate the first component of $\nabla \times (\mathbf{F} \times \mathbf{G})$ we need the second and third components of $\mathbf{F} \times \mathbf{G}$, which are

$$(\mathbf{F} \times \mathbf{G})_2 = F_3G_1 - F_1G_3 \quad \text{and} \quad (\mathbf{F} \times \mathbf{G})_3 = F_1G_2 - F_2G_1.$$

The first component of $\nabla \times (\mathbf{F} \times \mathbf{G})$ is therefore

$$\begin{aligned}
&\frac{\partial}{\partial y}(F_1G_2 - F_2G_1) - \frac{\partial}{\partial z}(F_3G_1 - F_1G_3) \\
&\qquad = \frac{\partial F_1}{\partial y}G_2 + F_1\frac{\partial G_2}{\partial y} - \frac{\partial F_2}{\partial y}G_1 - F_2\frac{\partial G_1}{\partial y} - \frac{\partial F_3}{\partial z}G_1 \\
&\qquad\quad - F_3\frac{\partial G_1}{\partial z} + \frac{\partial F_1}{\partial z}G_3 + F_1\frac{\partial G_3}{\partial z}.
\end{aligned}$$

The first components of the four terms on the right side of identity (e) are

$$\begin{aligned}
((\nabla \bullet \mathbf{G})\mathbf{F})_1 &= F_1\frac{\partial G_1}{\partial x} + F_1\frac{\partial G_2}{\partial y} + F_1\frac{\partial G_3}{\partial z} \\
((\mathbf{G} \bullet \nabla)\mathbf{F})_1 &= \frac{\partial F_1}{\partial x}G_1 + \frac{\partial F_1}{\partial y}G_2 + \frac{\partial F_1}{\partial z}G_3 \\
-((\nabla \bullet \mathbf{F})\mathbf{G})_1 &= -\frac{\partial F_1}{\partial x}G_1 - \frac{\partial F_2}{\partial y}G_1 - \frac{\partial F_3}{\partial z}G_1 \\
-((\mathbf{F} \bullet \nabla)\mathbf{G})_1 &= -F_1\frac{\partial G_1}{\partial x} - F_2\frac{\partial G_1}{\partial y} - F_3\frac{\partial G_1}{\partial z}.
\end{aligned}$$

When we add up all the terms in these four expressions, some cancel out and we are left with the same terms as in the first component of $\nabla \times (\mathbf{F} \times \mathbf{G})$.

(g) This is a straightforward calculation involving the equality of mixed partial derivatives:

$$\begin{aligned}
\nabla \bullet (\nabla \times \mathbf{F}) &= \frac{\partial}{\partial x}\left(\frac{\partial F_3}{\partial y} - \frac{\partial F_2}{\partial z}\right) + \frac{\partial}{\partial y}\left(\frac{\partial F_1}{\partial z} - \frac{\partial F_3}{\partial x}\right) \\
&\quad + \frac{\partial}{\partial z}\left(\frac{\partial F_2}{\partial x} - \frac{\partial F_1}{\partial y}\right) \\
&= \frac{\partial^2 F_3}{\partial x \partial y} - \frac{\partial^2 F_2}{\partial x \partial z} + \frac{\partial^2 F_1}{\partial y \partial z} - \frac{\partial^2 F_3}{\partial y \partial x} + \frac{\partial^2 F_2}{\partial z \partial x} - \frac{\partial^2 F_1}{\partial z \partial y} \\
&= 0.
\end{aligned}$$

Remark Two *triple product* identities for vectors were previously presented in Exercises 18 and 23 of Section 10.3:

$$\begin{aligned}
&\mathbf{a} \bullet (\mathbf{b} \times \mathbf{c}) = \mathbf{b} \bullet (\mathbf{c} \times \mathbf{a}) = \mathbf{c} \bullet (\mathbf{a} \times \mathbf{b}), \\
&\mathbf{a} \times (\mathbf{b} \times \mathbf{c}) = (\mathbf{a} \bullet \mathbf{c})\mathbf{b} - (\mathbf{a} \bullet \mathbf{b})\mathbf{c}.
\end{aligned}$$

While these are useful identities, they *cannot* be used to give simpler proofs of the identities in Theorem 3 by replacing one or other of the vectors with ∇. (Why?)

Scalar and Vector Potentials

Two special terms are used to describe vector fields for which either the divergence or the curl is identically zero.

DEFINITION 2

Solenoidal and irrotational vector fields

A vector field $\mathbf{F}$ is called **solenoidal** in a domain D if $\mathbf{div}\,\mathbf{F} = 0$ in D.
A vector field $\mathbf{F}$ is called **irrotational** in a domain D if $\mathbf{curl}\,\mathbf{F} = \mathbf{0}$ in D.

Part (h) of Theorem 3 says that $\mathbf{F} = \mathbf{grad}\,\phi \implies \mathbf{curl}\,\mathbf{F} = \mathbf{0}$. Thus,

Every conservative vector field is irrotational.

Part (g) of Theorem 3 says that $\mathbf{F} = \mathbf{curl}\,\mathbf{G} \implies \mathbf{div}\,\mathbf{F} = 0$. Thus,

The curl of any vector field is solenoidal.

The *converses* of these assertions hold if the domain of **F** satisfies certain conditions.

THEOREM 4

If **F** is a smooth, irrotational vector field on a simply connected domain D, then $\mathbf{F} = \nabla\phi$ for some scalar potential function defined on D, so **F** is conservative.

THEOREM 5

If **F** is a smooth, solenoidal vector field on a domain D with the property that every closed surface in D bounds a domain contained in D, then $\mathbf{F} = \mathbf{curl}\,\mathbf{G}$ for some vector field **G** defined on D. Such a vector field **G** is called a **vector potential** of the vector field **F**.

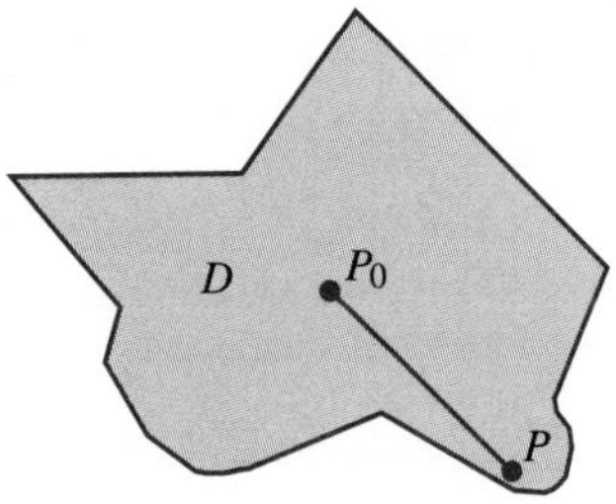

We cannot prove these results in their full generality at this point. However, both theorems have simple proofs in the special case where the domain D is **star-like**. A star-like domain is one for which there exists a point P_0 such that the line segment from P_0 to any point P in D lies wholly in D. (See Figure 16.4.) Both proofs are *constructive* in that they tell you how to find a potential.

Figure 16.4 The line segment from P_0 to any point in D lies in D

Proof of Theorem 4 for star-like domains. Without loss of generality, we can assume that P_0 is the origin. If $P = (x, y, z)$ is any point in D, then the straight line segment

$$\mathbf{r}(t) = tx\mathbf{i} + ty\mathbf{j} + tz\mathbf{k}, \qquad (0 \le t \le 1),$$

from P_0 to P lies in D. Define the function ϕ on D by

$$\begin{aligned}\phi(x, y, z) &= \int_0^1 \mathbf{F}\big(\mathbf{r}(t)\big) \bullet \frac{d\mathbf{r}}{dt}\,dt \\ &= \int_0^1 \Big(xF_1(\xi, \eta, \zeta) + yF_2(\xi, \eta, \zeta) + zF_3(\xi, \eta, \zeta)\Big)\,dt,\end{aligned}$$

where $\xi = tx$, $\eta = ty$, and $\zeta = tz$. We calculate $\partial\phi/\partial x$, making use of the fact that $\mathbf{curl}\,\mathbf{F} = \mathbf{0}$ to replace $(\partial/\partial\xi)F_2(\xi, \eta, \zeta)$ with $(\partial/\partial\eta)F_1(\xi, \eta, \zeta)$ and $(\partial/\partial\xi)F_3(\xi, \eta, \zeta)$ with $(\partial/\partial\zeta)F_1(\xi, \eta, \zeta)$:

$$\begin{aligned}\frac{\partial\phi}{\partial x} &= \int_0^1 \left(F_1(\xi, \eta, \zeta) + tx\frac{\partial F_1}{\partial\xi} + ty\frac{\partial F_2}{\partial\xi} + tz\frac{\partial F_3}{\partial\xi}\right) dt \\ &= \int_0^1 \left(F_1(\xi, \eta, \zeta) + tx\frac{\partial F_1}{\partial\xi} + ty\frac{\partial F_1}{\partial\eta} + tz\frac{\partial F_1}{\partial\zeta}\right) dt \\ &= \int_0^1 \frac{d}{dt}\Big(t\,F_1(\xi, \eta, \zeta)\Big)\,dt \\ &= \Big(t\,F_1(tx, ty, tz)\Big)\Big|_0^1 = F_1(x, y, z).\end{aligned}$$

Similarly, $\partial\phi/\partial y = F_2$ and $\partial\phi/\partial z = F_3$. Thus $\nabla\phi = \mathbf{F}$.

The details of the proof of Theorem 5 for star-like domains are similar to those of Theorem 4, and we relegate the proof to Exercise 18 below.

Note that vector potentials, when they exist, are *very* nonunique. Since **curl grad** ϕ is identically zero (Theorem 3(h)), an arbitrary conservative field can be added to **G** without changing the value of **curl G**. The following example illustrates just how much freedom you have in making simplifying assumptions when trying to find a vector potential.

EXAMPLE 1 Show that the vector field $\mathbf{F} = (x^2+yz)\mathbf{i} - 2y(x+z)\mathbf{j} + (xy+z^2)\mathbf{k}$ is solenoidal in $\mathbb{R}^3$ and find a vector potential for it.

Solution Since $\mathbf{div}\,\mathbf{F} = 2x - 2(x+z) + 2z = 0$ in $\mathbb{R}^3$, **F** is solenoidal. A vector potential **G** for **F** must satisfy $\mathbf{curl}\,\mathbf{G} = \mathbf{F}$; that is,

$$\frac{\partial G_3}{\partial y} - \frac{\partial G_2}{\partial z} = x^2 + yz,$$
$$\frac{\partial G_1}{\partial z} - \frac{\partial G_3}{\partial x} = -2xy - 2yz,$$
$$\frac{\partial G_2}{\partial x} - \frac{\partial G_1}{\partial y} = xy + z^2.$$

The three components of **G** have nine independent first partial derivatives, so there are nine "degrees of freedom" involved in their determination. The three equations above use up three of these nine degrees of freedom. That leaves six. Let us try to find a solution **G** with $G_2 = 0$ identically. This means that all three first partials of G_2 are zero, so we have used up three degrees of freedom in making this assumption. We have three left. The first equation now implies that

$$G_3 = \int (x^2 + yz)\,dy = x^2y + \frac{1}{2}y^2z + M(x, z).$$

(Since we were integrating with respect to y, the constant of integration can still depend on x and z.) We make a second simplifying assumption, that $M(x, z) = 0$. This uses up two more degrees of freedom, leaving one. From the second equation we have

$$\frac{\partial G_1}{\partial z} = \frac{\partial G_3}{\partial x} - 2xy - 2yz = 2xy - 2xy - 2yz = -2yz,$$

so

$$G_1 = -2\int yz\,dz = -yz^2 + N(x, y).$$

We cannot assume that $N(x, y) = 0$ identically because that would require two degrees of freedom and we have only one. However, the third equation implies

$$xy + z^2 = -\frac{\partial G_1}{\partial y} = z^2 - \frac{\partial N}{\partial y}.$$

Thus, $(\partial/\partial y)N(x, y) = -xy$; observe that the terms involving z have cancelled out. This happened because $\mathbf{div}\,\mathbf{F} = 0$. Had **F** not been solenoidal, we could not have determined N as a function of x and y only from the above equation. As it is, however, we have

$$N(x, y) = -\int xy\,dy = -\frac{1}{2}xy^2 + P(x).$$

We can use our last degree of freedom to choose $P(x)$ to be identically zero and hence obtain

$$\mathbf{G} = -\left(yz^2 + \frac{xy^2}{2}\right)\mathbf{i} + \left(x^2y + \frac{y^2z}{2}\right)\mathbf{k}$$

as the required vector potential for **F**. You can check that $\mathbf{curl}\,\mathbf{G} = \mathbf{F}$. Of course, other choices of simplifying assumptions would have led to very different functions **G**, which would have been equally correct.

In theoretical physics any particular choice of curl free term added to **G** is called a "gauge," and there is an elaborate theory known as "gauge theory," which explores the relative merits of such gauges and their relationships to each other.

Maple Calculations

The Maple **VectorCalculus** package defines routines for creating a vector field as well as calculating the gradient of a scalar field and the divergences and curl of a vector field. It will also calculate the Laplacian of a scalar or vector field and allow the use of the "del" operator in dot and cross products. Some of these capabilities are restricted to 3-dimensional vector fields. Let us begin by loading the package and declaring the type of coordinate system we will use and the names of the coordinates:

```
> with(VectorCalculus):
> SetCoordinates('cartesian'[x,y,z]);
```

$$cartesian_{x,y,z}$$

Setting the coordinates at the outset means we don't have to do it every time we call one of the procedures for handling vector fields, such as the `Gradient` procedure, which we illustrated at the end of Section 12.7. To calculate the gradient of a scalar expression in the variables x, y, and z, we could simply enter

```
> f := x^2 + x*y - z^3; G := Gradient(f);
```

$$f := x^2 + xy - z^3$$

$$G := (2x + y)\,\bar{e}_x + x\,\bar{e}_y - 3z^2\,\bar{e}_z$$

Maple shows that the result G is a *vector field* rather than just a *vector* by placing bars over the basis vectors. Maple treats vector fields and vectors as different kinds of objects; you can, for example, add two vector fields or two vectors, but you can't add a vector to a vector field. A vector field is a vector-valued function of a vector variable. To evaluate a vector field at a particular vector, you use the `evalVF` procedure:

```
> evalVF(G,<1,1,1>);
```

$$3\,e_x + e_y - 3\,e_z$$

You can define a vector field **F** with the `VectorField` procedure:

```
> F := VectorField(<x*y, 2*y*z, 3*x*z>);
```

$$F := x\,y\,\bar{e}_x + 2\,y\,z\,\bar{e}_y + 3\,x\,z\,\bar{e}_z$$

Then we can calculate the divergence or curl of **F** by using the `Divergence` or `Curl` procedures, or by dot or cross products with the `Del` operator:

```
> Divergence(F); Del.F;
```

$$y + 2\,z + 3\,x$$

$$y + 2\,z + 3\,x$$

```
> Curl(F); Del &x F;
```

$$-2\,y\,\bar{e}_x - 3\,z\,\bar{e}_y - x\,\bar{e}_z$$

$$-2\,y\,\bar{e}_x - 3\,z\,\bar{e}_y - x\,\bar{e}_z$$

We can verify the identities in Theorem 3 by using arbitrary scalar and vector fields:

```
> H := VectorField(<u(x,y,z),v(x,y,z),w(x,y,z)>);
```

$$H := u(x, y, z)\,\bar{e}_x + v(x, y, z)\,\bar{e}_y + w(x, y, z)\,\bar{e}_z$$

```
> Divergence(Curl(H)); Curl(Gradient(u(x,y,z)));
```

$$0$$

$$0\,\bar{e}_x$$

$0\,\bar{e}_x$ is VectorCalculus's way of denoting the zero vector field.

```
> Curl(Curl(H)) - Gradient(Divergence(H)) + Laplacian(H);
```

$$0\,\bar{e}_x$$

VectorCalculus also has procedures for finding the scalar potential of an irrotational vector field and the vector potential of a solenoidal vector field:

```
> ScalarPotential(VectorField(<x,y,z>));
```

$$\frac{1}{2}x^2 + \frac{1}{2}y^2 + \frac{1}{2}z^2$$

```
> VectorPotential(VectorField(<x^2, -x*y, -x*z>));
```

$$-x\,y\,z\,\bar{e}_x - x^2\,z\,\bar{e}_y$$

Neither procedure gives any output if you fail to feed it a vector field satisfying the appropriate condition (irrotational or solenoidal).

Finally, let us note that VectorCalculus is quite happy to deal with coordinate systems other than `'cartesian'[x,y,z]`. For instance,

```
> SetCoordinates('cylindrical'[r,theta,z]);
```

$$cylindrical_{r,\theta,z}$$

```
> Laplacian(u(r,theta,z));
```

$$\frac{\left(\frac{\partial}{\partial r}u(r,\theta,z)\right) + r\left(\frac{\partial^2}{\partial r^2}u(r,\theta,z)\right) + \frac{\frac{\partial^2}{\partial \theta^2}u(r,\theta,z)}{r} + r\left(\frac{\partial^2}{\partial z^2}u(r,\theta,z)\right)}{r}$$

which is not written as neatly as we would like, but is correct. Similarly, we can use coordinate systems `'spherical'[rho,phi,theta]` in 3-space and also `'polar'[r,theta]` in the plane.

EXERCISES 16.2

1. Prove part (a) of Theorem 3.
2. Prove part (b) of Theorem 3.
3. Prove part (d) of Theorem 3.
4. Prove part (f) of Theorem 3.
5. Prove part (h) of Theorem 3.
6. Prove part (i) of Theorem 3.
7. Given that the field lines of the vector field $\mathbf{F}(x, y, z)$ are parallel straight lines, can you conclude anything about **div F**? about **curl F**?
8. Let $\mathbf{r} = x\mathbf{i} + y\mathbf{j} + z\mathbf{k}$ and let $\mathbf{c}$ be a constant vector. Show that $\nabla \bullet (\mathbf{c} \times \mathbf{r}) = 0$, $\nabla \times (\mathbf{c} \times \mathbf{r}) = 2\mathbf{c}$, and $\nabla(\mathbf{c} \bullet \mathbf{r}) = \mathbf{c}$.
9. Let $\mathbf{r} = x\mathbf{i} + y\mathbf{j} + z\mathbf{k}$ and let $r = |\mathbf{r}|$. If f is a differentiable function of one variable, show that
$$\nabla \bullet (f(r)\mathbf{r}) = rf'(r) + 3f(r).$$
Find $f(r)$ if $f(r)\mathbf{r}$ is solenoidal for $r \neq 0$.
10. If the smooth vector field $\mathbf{F}$ is both irrotational and solenoidal on $\mathbb{R}^3$, show that the three components of $\mathbf{F}$ and the scalar potential for $\mathbf{F}$ are all harmonic functions in $\mathbb{R}^3$.
11. If $\mathbf{r} = x\mathbf{i} + y\mathbf{j} + z\mathbf{k}$ and $\mathbf{F}$ is smooth, show that
$$\nabla \times (\mathbf{F} \times \mathbf{r}) = \mathbf{F} - (\nabla \bullet \mathbf{F})\mathbf{r} + \nabla(\mathbf{F} \bullet \mathbf{r}) - \mathbf{r} \times (\nabla \times \mathbf{F}).$$
In particular, if $\nabla \bullet \mathbf{F} = 0$ and $\nabla \times \mathbf{F} = \mathbf{0}$, then
$$\nabla \times (\mathbf{F} \times \mathbf{r}) = \mathbf{F} + \nabla(\mathbf{F} \bullet \mathbf{r}).$$
12. If ϕ and ψ are harmonic functions, show that $\phi\nabla\psi - \psi\nabla\phi$ is solenoidal.
13. If ϕ and ψ are smooth scalar fields, show that
$$\nabla \times (\phi\nabla\psi) = -\nabla \times (\psi\nabla\phi) = \nabla\phi \times \nabla\psi.$$
14. Verify the identity
$$\nabla \bullet \Big(f(\nabla g \times \nabla h)\Big) = \nabla f \bullet (\nabla g \times \nabla h)$$
for smooth scalar fields f, g, and h.
15. If the vector fields $\mathbf{F}$ and $\mathbf{G}$ are smooth and conservative, show that $\mathbf{F} \times \mathbf{G}$ is solenoidal. Find a vector potential for $\mathbf{F} \times \mathbf{G}$.
16. Find a vector potential for $\mathbf{F} = -y\mathbf{i} + x\mathbf{j}$.

17. Show that $\mathbf{F} = xe^{2z}\mathbf{i} + ye^{2z}\mathbf{j} - e^{2z}\mathbf{k}$ is a solenoidal vector field, and find a vector potential for it.

18. Suppose $\mathbf{div}\,\mathbf{F} = 0$ in a domain D any point P of which can by joined to the origin by a straight line segment in D. Let $\mathbf{r} = tx\mathbf{i} + ty\mathbf{j} + tz\mathbf{k}$, $(0 \le t \le 1)$, be a parametrization of the line segment from the origin to (x, y, z) in D. If

$$\mathbf{G}(x, y, z) = \int_0^1 t\mathbf{F}(\mathbf{r}(t)) \times \frac{d\mathbf{r}}{dt}\, dt,$$

show that $\mathbf{curl}\,\mathbf{G} = \mathbf{F}$ throughout D. *Hint:* It is enough to check the first components of $\mathbf{curl}\,\mathbf{G}$ and $\mathbf{F}$. Proceed in a manner similar to the proof of Theorem 4.

19. Use the Maple VectorCalculus package to verify the identities (a)–(f) of Theorem 3. *Hint:* For expressions of the form $(\mathbf{F} \bullet \nabla)\mathbf{G}$ you will have to use

```
> F[1]*diff(G,x)+F[2]*diff(G,y)
>  +F[3]*diff(G,z)
```

because `Del` cannot be applied to a vector field except via a dot or cross product.

16.3 Green's Theorem in the Plane

The Fundamental Theorem of Calculus,

$$\int_a^b \frac{d}{dx} f(x)\, dx = f(b) - f(a),$$

expresses the integral, taken over the interval $[a, b]$, of the derivative of a single-variable function, f, as a *sum* of values of that function at the *oriented boundary* of the interval $[a, b]$, that is, at the two endpoints a and b, the former providing a *negative* contribution and the latter a *positive* one. The line integral of a conservative vector field over a curve $\mathcal{C}$ from A to B,

$$\int_{\mathcal{C}} \nabla\phi \bullet d\mathbf{r} = \phi(B) - \phi(A),$$

has a similar interpretation; $\nabla\phi$ is a derivative, and the curve $\mathcal{C}$, although lying in a two- or three-dimensional space, is intrinsically a one-dimensional object, and the points A and B constitute its boundary.

Green's Theorem is a two-dimensional version of the Fundamental Theorem of Calculus that expresses the *double integral* of a certain kind of derivative of a two-dimensional vector field $\mathbf{F}(x, y)$, namely, the $\mathbf{k}$-component of $\mathbf{curl}\,\mathbf{F}$, over a region R in the xy-plane as a line integral (i.e., a "sum") of the tangential component of $\mathbf{F}$ around the curve $\mathcal{C}$ which is the oriented boundary of R:

$$\iint_R \mathbf{curl}\,\mathbf{F} \bullet \mathbf{k}\, dA = \oint_{\mathcal{C}} \mathbf{F} \bullet d\mathbf{r},$$

or, more explicitly,

$$\iint_R \left(\frac{\partial F_2}{\partial x} - \frac{\partial F_1}{\partial y}\right) dx\, dy = \oint_{\mathcal{C}} F_1(x, y)\, dx + F_2(x, y)\, dy.$$

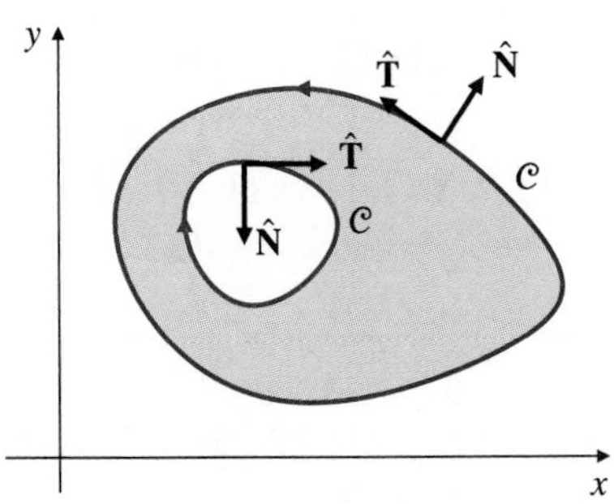

Figure 16.5 A plane domain with positively oriented boundary

For this formula to hold, $\mathcal{C}$ must be the oriented boundary of R considered as a surface with orientation provided by $\hat{\mathbf{N}} = \mathbf{k}$. Thus, $\mathcal{C}$ is oriented with R on the left as we move around $\mathcal{C}$ in the direction of its orientation. We will call such a curve positively oriented with respect to R. In particular, if $\mathcal{C}$ is a simple closed curve bounding R, then $\mathcal{C}$ is oriented counterclockwise. Of course, R may have holes, and the boundaries of the holes will be oriented clockwise. In any case, the unit tangent $\hat{\mathbf{T}}$ and unit exterior (pointing out of R) normal $\hat{\mathbf{N}}$ on $\mathcal{C}$ satisfy $\hat{\mathbf{N}} = \hat{\mathbf{T}} \times \mathbf{k}$. See Figure 16.5.

THEOREM 6 **Green's Theorem**

Let R be a regular, closed region in the xy-plane whose boundary, $\mathcal{C}$, consists of one or more piecewise smooth, simple closed curves that are positively oriented with respect to R. If $\mathbf{F} = F_1(x, y)\mathbf{i} + F_2(x, y)\mathbf{j}$ is a smooth vector field on R, then

$$\oint_{\mathcal{C}} F_1(x, y)\, dx + F_2(x, y)\, dy = \iint_R \left(\frac{\partial F_2}{\partial x} - \frac{\partial F_1}{\partial y} \right) dA.$$

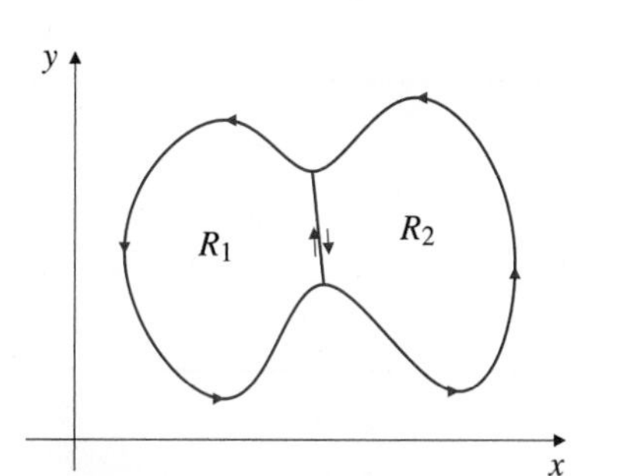

Figure 16.6 Green's Theorem holds for the union of R_1 and R_2 if it holds for each of those regions

PROOF Recall that a regular region can be divided into nonoverlapping subregions that are both x-simple and y-simple. (See Section 14.2.) When two such regions share a common boundary curve, they induce opposite orientations on that curve, so the sum of the line integrals over the boundaries of the subregions is just the line integral over the boundary of the whole region. (See Figure 16.6.) The double integrals over the subregions also add to give the double integral over the whole region. It therefore suffices to show that the formula holds for a region R that is both x-simple and y-simple.

Since R is y-simple, it is specified by inequalities of the form $a \le x \le b$, $f(x) \le y \le g(x)$, with the bottom boundary $y = f(x)$ oriented left to right and the upper boundary $y = g(x)$ oriented right to left. (See Figure 16.7.) Thus,

$$\begin{aligned} -\iint_R \frac{\partial F_1}{\partial y}\, dx\, dy &= -\int_a^b dx \int_{f(x)}^{g(x)} \frac{\partial F_1}{\partial y}\, dy \\ &= \int_a^b \Big(-F_1\big(x, g(x)\big) + F_1\big(x, f(x)\big) \Big)\, dx. \end{aligned}$$

On the other hand, since $dx = 0$ on the vertical sides of R, and the top boundary is traversed from b to a, we have

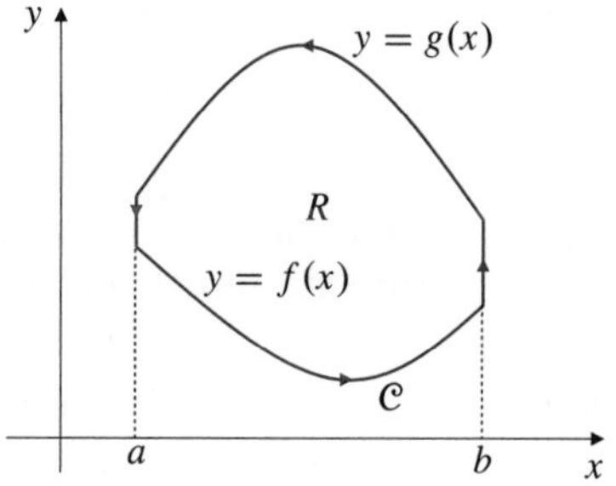

$$\oint_{\mathcal{C}} F_1(x, y)\, dx = \int_a^b \Big(F_1\big(x, f(x)\big) - F_1\big(x, g(x)\big) \Big)\, dx = \iint_R -\frac{\partial F_1}{\partial y}\, dx\, dy.$$

Similarly, since R is x-simple, $\displaystyle\oint_{\mathcal{C}} F_2\, dy = \iint_R \frac{\partial F_2}{\partial x}\, dx\, dy$, so

$$\oint_{\mathcal{C}} F_1(x, y)\, dx + F_2(x, y)\, dy = \iint_R \left(\frac{\partial F_2}{\partial x} - \frac{\partial F_1}{\partial y} \right) dA.$$

Figure 16.7 $\oint_{\mathcal{C}} F_1\, dx = -\iint_R \dfrac{\partial F_1}{\partial y}\, dA$ for this y-simple region R

EXAMPLE 1 **(Area bounded by a simple closed curve)** For any of the three vector fields

$$\mathbf{F} = x\mathbf{j}, \qquad \mathbf{F} = -y\mathbf{i}, \qquad \text{and} \qquad \mathbf{F} = \frac{1}{2}(-y\mathbf{i} + x\mathbf{j}),$$

we have $(\partial F_2/\partial x) - (\partial F_1/\partial y) = 1$. If $\mathcal{C}$ is a positively oriented, piecewise smooth, simple closed curve bounding a region R in the plane, then, by Green's Theorem,

$$\oint_{\mathcal{C}} x\, dy = -\oint_{\mathcal{C}} y\, dx = \frac{1}{2} \oint_{\mathcal{C}} x\, dy - y\, dx = \iint_R 1\, dA = \text{area of } R.$$

EXAMPLE 2 Use the result of the previous example to calculate the area of the elliptic disk bounded by the curve $\mathcal{C}$ given by

$$\mathbf{r} = 3(\cos t + \sin t)\,\mathbf{i} + 2(\sin t - \cos t)\,\mathbf{j}, \quad 0 \le t \le 2\pi.$$

Solution The parametrization of $\mathcal{C}$ gives

$$x = 3(\cos t + \sin t), \qquad y = 2(\sin t - \cos t),$$
$$dx = 3(-\sin t + \cos t)\,dt, \qquad dy = 2(\cos t + \sin t)\,dt,$$

so that $x\,dy - y\,dx = 6\big((\cos t + \sin t)^2 + (\sin t - \cos t)^2\big)\,dt = 12\,dt$. Thus, by the third formula for the area given in the previous example, the disk has

$$\text{area} = \frac{1}{2}\int_{\mathcal{C}} x\,dy - y\,dx$$
$$= \frac{1}{2}\int_0^{2\pi} 12\,dt = 12\pi \quad \text{square units.}$$

EXAMPLE 3 Evaluate $I = \oint_{\mathcal{C}} (x - y^3)\,dx + (y^3 + x^3)\,dy$, where $\mathcal{C}$ is the positively oriented boundary of the quarter-disk Q: $0 \le x^2 + y^2 \le a^2$, $x \ge 0$, $y \ge 0$.

Solution We use Green's Theorem to calculate I:

$$I = \iint_Q \left(\frac{\partial}{\partial x}(y^3 + x^3) - \frac{\partial}{\partial y}(x - y^3)\right) dA$$
$$= 3\iint_Q (x^2 + y^2)\,dA = 3\int_0^{\pi/2} d\theta \int_0^a r^3\,dr = \frac{3}{8}\pi a^4.$$

EXAMPLE 4 Let $\mathcal{C}$ be a positively oriented, simple closed curve in the xy-plane, bounding a region R and not passing through the origin. Show that

$$\oint_{\mathcal{C}} \frac{-y\,dx + x\,dy}{x^2 + y^2} = \begin{cases} 0 & \text{if the origin is outside } R \\ 2\pi & \text{if the origin is inside } R. \end{cases}$$

Solution First, if $(x, y) \ne (0, 0)$, then, by direct calculation,

$$\frac{\partial}{\partial x}\left(\frac{x}{x^2 + y^2}\right) - \frac{\partial}{\partial y}\left(\frac{-y}{x^2 + y^2}\right) = 0.$$

If the origin is not in R, then Green's Theorem implies that

$$\oint_{\mathcal{C}} \frac{-y\,dx + x\,dy}{x^2 + y^2} = \iint_R \left[\frac{\partial}{\partial x}\left(\frac{x}{x^2 + y^2}\right) - \frac{\partial}{\partial y}\left(\frac{-y}{x^2 + y^2}\right)\right] dx\,dy = 0.$$

Now suppose the origin is in R. Since it is assumed that the origin is not on $\mathcal{C}$, it must be an interior point of R. The interior of R is open, so there exists $\epsilon > 0$ such that the circle $\mathcal{C}_\epsilon$ of radius ϵ centred at the origin is in the interior of R. Let $\mathcal{C}_\epsilon$ be oriented negatively (clockwise). By direct calculation (see Exercise 22(a) of Section 15.4) it is easily shown that

$$\oint_{\mathcal{C}_\epsilon} \frac{-y\,dx + x\,dy}{x^2 + y^2} = -2\pi.$$

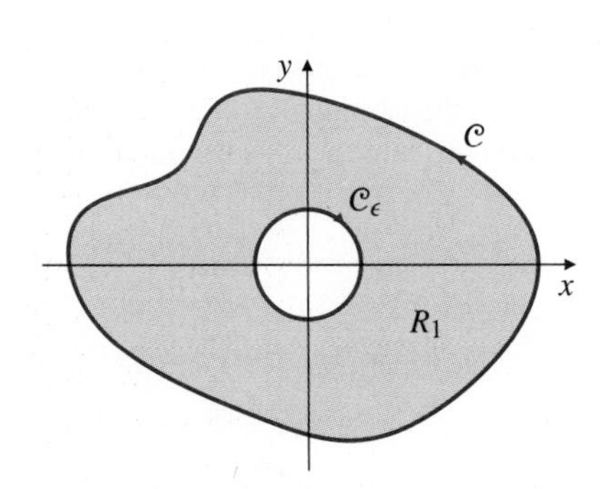

Figure 16.8 The origin does not lie in R_1

Together $\mathcal{C}$ and $\mathcal{C}_\epsilon$ form the positively oriented boundary of a region R_1 that excludes the origin. (See Figure 16.8.) So, by Green's Theorem,

$$\oint_{\mathcal{C}} \frac{-y\,dx + x\,dy}{x^2 + y^2} + \oint_{\mathcal{C}_\epsilon} \frac{-y\,dx + x\,dy}{x^2 + y^2} = 0.$$

The desired result now follows:

$$\oint_{\mathcal{C}} \frac{-y\,dx + x\,dy}{x^2 + y^2} = -\oint_{\mathcal{C}_\epsilon} \frac{-y\,dx + x\,dy}{x^2 + y^2} = -(-2\pi) = 2\pi.$$

The Two-Dimensional Divergence Theorem

The following theorem is an alternative formulation of the two-dimensional Fundamental Theorem of Calculus. In this case we express the double integral of $\mathbf{div}\,\mathbf{F}$ (a derivative of $\mathbf{F}$) over R as a single integral of the outward normal component of $\mathbf{F}$ on the boundary $\mathcal{C}$ of R.

THEOREM

The Divergence Theorem in the Plane

Let R be a regular, closed region in the xy-plane whose boundary, $\mathcal{C}$, consists of one or more piecewise smooth, simple closed curves. Let $\hat{\mathbf{N}}$ denote the unit outward (from R) normal field on $\mathcal{C}$. If $\mathbf{F} = F_1(x, y)\mathbf{i} + F_2(x, y)\mathbf{j}$ is a smooth vector field on R, then

$$\iint_R \mathbf{div}\,\mathbf{F}\,dA = \oint_{\mathcal{C}} \mathbf{F} \bullet \hat{\mathbf{N}}\,ds.$$

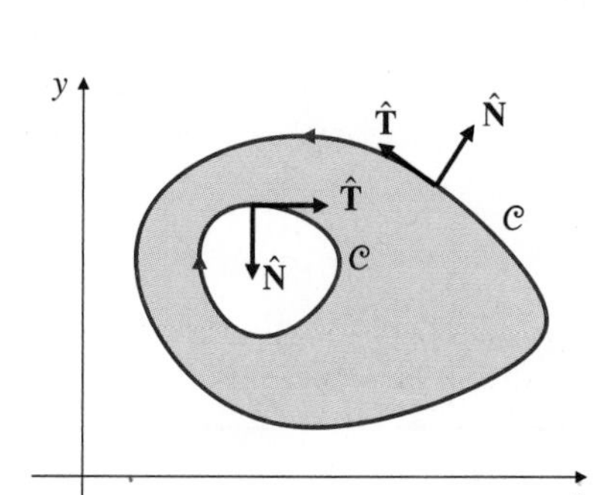

Figure 16.9 $\hat{\mathbf{N}} = \hat{\mathbf{T}} \times \mathbf{k}$

PROOF As observed in the second paragraph of this section, $\hat{\mathbf{N}} = \hat{\mathbf{T}} \times \mathbf{k}$, where $\hat{\mathbf{T}}$ is the unit tangent field in the positive direction on $\mathcal{C}$. If $\hat{\mathbf{T}} = T_1\mathbf{i} + T_2\mathbf{j}$, then $\hat{\mathbf{N}} = T_2\mathbf{i} - T_1\mathbf{j}$. (See Figure 16.9.) Now let $\mathbf{G}$ be the vector field with components $G_1 = -F_2$ and $G_2 = F_1$. Then $\mathbf{G} \bullet \hat{\mathbf{T}} = \mathbf{F} \bullet \hat{\mathbf{N}}$ and, by Green's Theorem,

$$\begin{aligned}\iint_R \mathbf{div}\,\mathbf{F}\,dA &= \iint_R \left(\frac{\partial F_1}{\partial x} + \frac{\partial F_2}{\partial y}\right) dA \\ &= \iint_R \left(\frac{\partial G_2}{\partial x} - \frac{\partial G_1}{\partial y}\right) dA \\ &= \oint_{\mathcal{C}} \mathbf{G} \bullet d\mathbf{r} = \oint_{\mathcal{C}} \mathbf{G} \bullet \hat{\mathbf{T}}\,ds = \oint_{\mathcal{C}} \mathbf{F} \bullet \hat{\mathbf{N}}\,ds.\end{aligned}$$

■

EXERCISES 16.3

1. Evaluate $\oint_{\mathcal{C}} (\sin x + 3y^2)\,dx + (2x - e^{-y^2})\,dy$, where $\mathcal{C}$ is the boundary of the half-disk $x^2 + y^2 \le a^2$, $y \ge 0$, oriented counterclockwise.

2. Evaluate $\oint_{\mathcal{C}} (x^2 - xy)\,dx + (xy - y^2)\,dy$ clockwise around the triangle with vertices (0, 0), (1, 1), and (2, 0).

3. Evaluate $\oint_{\mathcal{C}} \left(x\sin(y^2) - y^2\right) dx + \left(x^2 y\cos(y^2) + 3x\right) dy$, where $\mathcal{C}$ is the counterclockwise boundary of the trapezoid with vertices $(0, -2)$, $(1, -1)$, $(1, 1)$, and $(0, 2)$.

4. Evaluate $\oint_{\mathcal{C}} x^2 y\,dx - xy^2\,dy$, where $\mathcal{C}$ is the clockwise boundary of the region $0 \le y \le \sqrt{9 - x^2}$.

5. Use a line integral to find the plane area enclosed by the curve $\mathbf{r} = a\cos^3 t\,\mathbf{i} + b\sin^3 t\,\mathbf{j}$, $(0 \le t \le 2\pi)$.

6. We deduced the two-dimensional Divergence Theorem from Green's Theorem. Reverse the argument and use the two-dimensional Divergence Theorem to prove Green's Theorem.

7. Sketch the plane curve $\mathcal{C}$: $\mathbf{r} = \sin t\,\mathbf{i} + \sin 2t\,\mathbf{j}$, $(0 \le t \le 2\pi)$. Evaluate $\oint_{\mathcal{C}} \mathbf{F} \bullet d\mathbf{r}$, where $\mathbf{F} = ye^{x^2}\mathbf{i} + x^3 e^y\mathbf{j}$.

8. If $\mathcal{C}$ is the positively oriented boundary of a plane region R having area A and centroid $(\bar{x}, \bar{y})$, interpret geometrically the line integral $\oint_{\mathcal{C}} \mathbf{F} \bullet d\mathbf{r}$, where (a) $\mathbf{F} = x^2\mathbf{j}$, (b) $\mathbf{F} = xy\mathbf{i}$, and (c) $\mathbf{F} = y^2\mathbf{i} + 3xy\mathbf{j}$.

9. **(Average values of harmonic functions)** If $u(x, y)$ is harmonic in a domain containing a disk of radius r with boundary C_r, then the average value of u around the circle is the value of u at the centre. Prove this by showing that the derivative of the average value with respect to r is zero using the Divergence Theorem and the harmonicity of u, and the fact that the limit of the average value as $r \to 0$ is the value of u at the centre.

16.4 The Divergence Theorem in 3-Space

The **Divergence Theorem** (also called **Gauss's Theorem**) is one of two important versions of the Fundamental Theorem of Calculus in $\mathbb{R}^3$. (The other is Stokes's Theorem, presented in the next section.)

In the Divergence Theorem, the integral of the *derivative* $\mathbf{div}\,\mathbf{F} = \nabla \bullet \mathbf{F}$ over a domain in 3-space is expressed as the flux of $\mathbf{F}$ out of the surface of that domain. It therefore closely resembles the two-dimensional version, Theorem 7, given in the previous section. The theorem holds for a general class of domains in $\mathbb{R}^3$ that are bounded by piecewise smooth closed surfaces. However, we will restrict our statement and proof of the theorem to domains of a special type. Extending the concept of an x-simple plane domain defined in Section 14.2, we say the three-dimensional domain D is x-**simple** if it is bounded by a piecewise smooth surface $\mathcal{S}$ and if every straight line parallel to the x-axis and passing through an interior point of D meets $\mathcal{S}$ at exactly two points. Similar definitions hold for y-simple and z-simple, and we call the domain D **regular** if it is a union of finitely many, nonoverlapping subdomains, each of which is x-simple, y-simple, and z-simple.

THEOREM 8 **The Divergence Theorem (Gauss's Theorem)**

Let D be a regular, three-dimensional domain whose boundary $\mathcal{S}$ is an oriented, closed surface with unit normal field $\hat{\mathbf{N}}$ pointing out of D. If $\mathbf{F}$ is a smooth vector field defined on D, then

$$\iiint_D \mathbf{div}\,\mathbf{F}\,dV = \oiint_{\mathcal{S}} \mathbf{F} \bullet \hat{\mathbf{N}}\,dS.$$

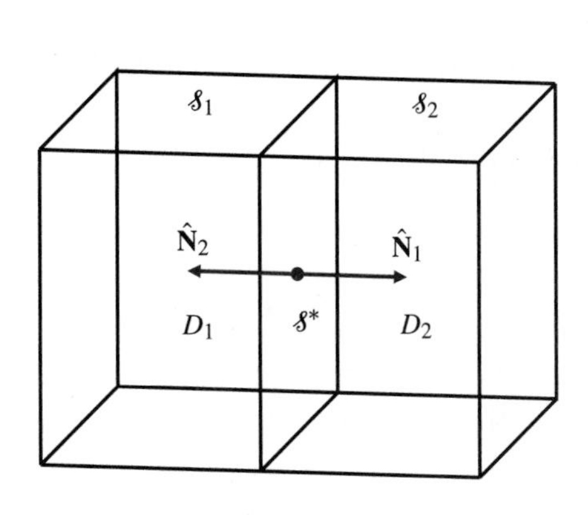

Figure 16.10 A union of abutting domains

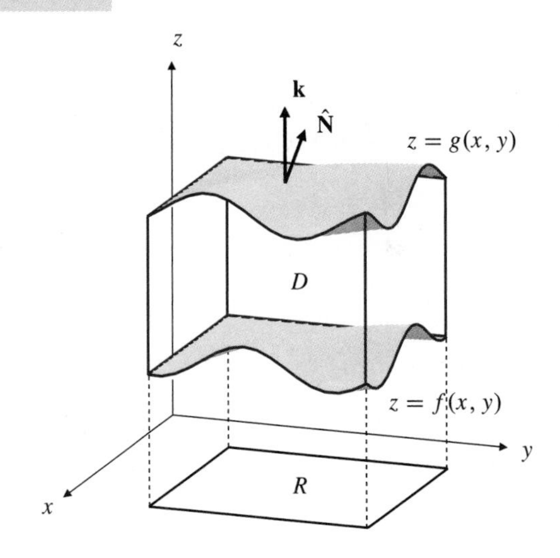

Figure 16.11 A z-simple domain

PROOF Since the domain D is a union of finitely many nonoverlapping domains that are x-simple, y-simple, and z-simple, it is sufficient to prove the theorem for a subdomain of D with this property. To see this, suppose, for instance, that D and $\mathcal{S}$ are each divided into two parts, D_1 and D_2, and $\mathcal{S}_1$ and $\mathcal{S}_2$, by a surface $\mathcal{S}^*$ slicing through D. (See Figure 16.10.) $\mathcal{S}^*$ is part of the boundary of both D_1 and D_2, but the exterior normals, $\hat{\mathbf{N}}_1$ and $\hat{\mathbf{N}}_2$, of the two subdomains point in opposite directions on either side of $\mathcal{S}^*$. If the formula in the theorem holds for both subdomains,

$$\iiint_{D_1} \mathbf{div}\,\mathbf{F}\,dV = \oiint_{\mathcal{S}_1 \cup \mathcal{S}^*} \mathbf{F} \bullet \hat{\mathbf{N}}_1\,dS$$

$$\iiint_{D_2} \mathbf{div}\,\mathbf{F}\,dV = \oiint_{\mathcal{S}_2 \cup \mathcal{S}^*} \mathbf{F} \bullet \hat{\mathbf{N}}_2\,dS,$$

then, adding these equations, we get

$$\iiint_D \operatorname{div} \mathbf{F}\, dV = \oiint_{\mathcal{S}_1 \cup \mathcal{S}_2} \mathbf{F} \bullet \hat{\mathbf{N}}\, dS = \oiint_{\mathcal{S}} \mathbf{F} \bullet \hat{\mathbf{N}}\, dS;$$

the contributions from $\mathcal{S}^*$ cancel out because on that surface $\hat{\mathbf{N}}_2 = -\hat{\mathbf{N}}_1$.

For the rest of this proof we assume, therefore, that D is x-, y-, and z-simple. Since D is z-simple, it lies between the graphs of two functions defined on a region R in the xy-plane; if (x, y, z) is in D, then (x, y) is in R and $f(x, y) \le z \le g(x, y)$. (See Figure 16.11.) We have

$$\begin{aligned}\iiint_D \frac{\partial F_3}{\partial z}\, dV &= \iint_R dx\, dy \int_{f(x,y)}^{g(x,y)} \frac{\partial F_3}{\partial z}\, dz \\ &= \iint_R \Big(F_3\big(x, y, g(x, y)\big) - F_3\big(x, y, f(x, y)\big)\Big)\, dx\, dy.\end{aligned}$$

Now

$$\oiint_{\mathcal{S}} \mathbf{F} \bullet \hat{\mathbf{N}}\, dS = \oiint_{\mathcal{S}} \Big(F_1\, \mathbf{i} \bullet \hat{\mathbf{N}} + F_2\, \mathbf{j} \bullet \hat{\mathbf{N}} + F_3\, \mathbf{k} \bullet \hat{\mathbf{N}}\Big)\, dS.$$

Only the last term involves F_3, and it can be split into three integrals, over the top surface $z = g(x, y)$, the bottom surface $z = f(x, y)$, and vertical side wall lying above the boundary of R:

$$\oiint_{\mathcal{S}} F_3(x, y, z)\, \mathbf{k} \bullet \hat{\mathbf{N}}\, dS = \left(\iint_{\text{top}} + \iint_{\text{bottom}} + \iint_{\text{side}}\right) F_3(x, y, z)\, \mathbf{k} \bullet \hat{\mathbf{N}}\, dS.$$

On the side wall, $\mathbf{k} \bullet \hat{\mathbf{N}} = 0$, so that integral is zero. On the top surface, $z = g(x, y)$, and the vector area element is

$$\hat{\mathbf{N}}\, dS = \left(-\frac{\partial g}{\partial x}\mathbf{i} - \frac{\partial g}{\partial y}\mathbf{j} + \mathbf{k}\right) dx\, dy.$$

Accordingly,

$$\iint_{\text{top}} F_3(x, y, z)\, \mathbf{k} \bullet \hat{\mathbf{N}}\, dS = \iint_R F_3\big(x, y, g(x, y)\big)\, dx\, dy.$$

Similarly, we have

$$\iint_{\text{bottom}} F_3(x, y, z)\, \mathbf{k} \bullet \hat{\mathbf{N}}\, dS = -\iint_R F_3\big(x, y, f(x, y)\big)\, dx\, dy;$$

the negative sign occurs because $\hat{\mathbf{N}}$ points down rather than up on the bottom. Thus, we have shown that

$$\iiint_D \frac{\partial F_3}{\partial z}\, dV = \oiint_{\mathcal{S}} F_3\, \mathbf{k} \bullet \hat{\mathbf{N}}\, dS.$$

Similarly, because D is also x-simple and y-simple,

$$\begin{aligned}\iiint_D \frac{\partial F_1}{\partial x}\, dV &= \oiint_{\mathcal{S}} F_1\, \mathbf{i} \bullet \hat{\mathbf{N}}\, dS \\ \iiint_D \frac{\partial F_2}{\partial y}\, dV &= \oiint_{\mathcal{S}} F_2\, \mathbf{j} \bullet \hat{\mathbf{N}}\, dS.\end{aligned}$$

Adding these three results, we get

$$\iiint_D \operatorname{div} \mathbf{F}\, dV = \oiint_{\mathcal{S}} \mathbf{F} \bullet \hat{\mathbf{N}}\, dS.$$

The Divergence Theorem can be used in both directions to simplify explicit calculations of surface integrals or volumes. We give examples of each.

EXAMPLE 1 Let $\mathbf{F} = bxy^2\mathbf{i} + bx^2y\mathbf{j} + (x^2 + y^2)z^2\mathbf{k}$, and let $\mathcal{S}$ be the closed surface bounding the solid cylinder R defined by $x^2 + y^2 \le a^2$ and $0 \le z \le b$. Find $\oiint_{\mathcal{S}} \mathbf{F} \bullet d\mathbf{S}$.

Solution By the Divergence Theorem,

$$\begin{aligned}\oiint_{\mathcal{S}} \mathbf{F} \bullet d\mathbf{S} &= \iiint_R \mathbf{div}\,\mathbf{F}\,dV = \iiint_R (x^2 + y^2)(b + 2z)\,dV \\ &= \int_0^b (b + 2z)\,dz \int_0^{2\pi} d\theta \int_0^a r^2\,r\,dr \\ &= (b^2 + b^2)2\pi(a^4/4) = \pi a^4 b^2.\end{aligned}$$

EXAMPLE 2 Evaluate $\oiint_{\mathcal{S}} (x^2+y^2)\,dS$, where $\mathcal{S}$ is the sphere $x^2+y^2+z^2 = a^2$. Use the Divergence Theorem.

Solution On $\mathcal{S}$ we have

$$\hat{\mathbf{N}} = \frac{\mathbf{r}}{a} = \frac{x\mathbf{i} + y\mathbf{j} + z\mathbf{k}}{a}.$$

We would like to choose $\mathbf{F}$ so that $\mathbf{F} \bullet \hat{\mathbf{N}} = x^2 + y^2$. Observe that $\mathbf{F} = a(x\mathbf{i} + y\mathbf{j})$ will do. If B is the ball bounded by $\mathcal{S}$, then

$$\begin{aligned}\oiint_{\mathcal{S}} (x^2 + y^2)\,dS &= \oiint_{\mathcal{S}} \mathbf{F} \bullet \hat{\mathbf{N}}\,dS = \iiint_B \mathbf{div}\,\mathbf{F}\,dV \\ &= \iiint_B 2a\,dV = (2a)\frac{4}{3}\pi a^3 = \frac{8}{3}\pi a^4.\end{aligned}$$

EXAMPLE 3 By using the Divergence Theorem with $\mathbf{F} = x\mathbf{i}+y\mathbf{j}+z\mathbf{k}$, calculate the volume of a cone having base area A and height h. The base can be any smoothly bounded plane region.

Solution Let the vertex of the cone be at the origin and the base in the plane $z = h$ as shown in Figure 16.12. The solid cone C has surface consisting of two parts: the conical wall $\mathcal{S}$ and the base region D that has area A. Since $\mathbf{F}(x, y, z)$ points directly away from the origin at any point $(x, y, z) \ne (0, 0, 0)$, we have $\mathbf{F} \bullet \hat{\mathbf{N}} = 0$ on $\mathcal{S}$. On D, we have $\hat{\mathbf{N}} = \mathbf{k}$ and $z = h$, so $\mathbf{F} \bullet \hat{\mathbf{N}} = z = h$ on the base of the cone. Since $\mathbf{div}\,\mathbf{F}(x, y, z) = 1 + 1 + 1 = 3$, we have, by the Divergence Theorem,

$$\begin{aligned}3V = \iiint_C \mathbf{div}\,\mathbf{F}\,dV &= \iint_{\mathcal{S}} \mathbf{F} \bullet \hat{\mathbf{N}}\,dS + \iint_D \mathbf{F} \bullet \hat{\mathbf{N}}\,dS \\ &= 0 + h \iint_D dS = Ah.\end{aligned}$$

Thus, $V = \frac{1}{3}Ah$, the well-known formula for the volume of a cone.

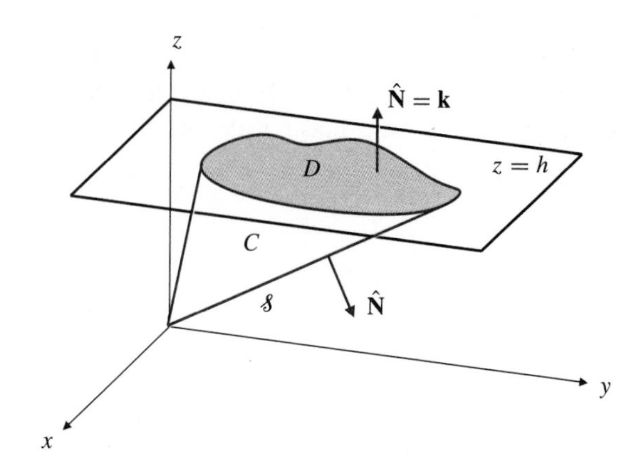

Figure 16.12 A cone with an arbitrarily shaped base

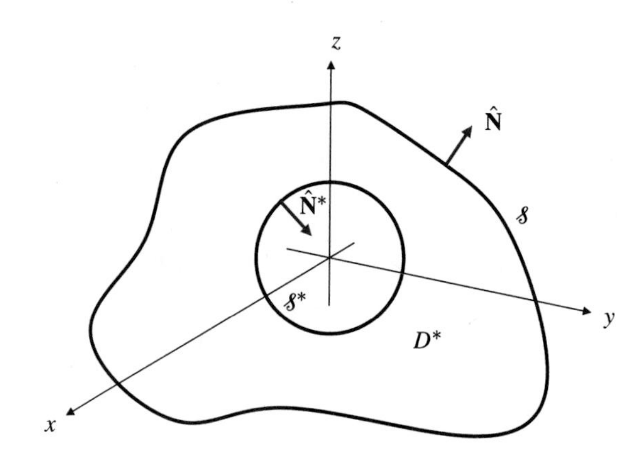

Figure 16.13 A solid domain with a spherical cavity

EXAMPLE 4 Let $\mathcal{S}$ be the surface of an arbitrary regular domain D in 3-space that contains the origin in its interior. Find

$$\oiint_{\mathcal{S}} \mathbf{F} \bullet \hat{\mathbf{N}}\, dS,$$

where $\mathbf{F}(\mathbf{r}) = m\mathbf{r}/|\mathbf{r}|^3$ and $\hat{\mathbf{N}}$ is the unit outward normal on $\mathcal{S}$. (See Figure 16.13.)

Solution Since $\mathbf{F}$ and, therefore, $\mathbf{div}\,\mathbf{F}$ are undefined at the origin, we cannot apply the Divergence Theorem directly. To overcome this problem we use a little trick. Let $\mathcal{S}^*$ be a small sphere centred at the origin bounding a ball contained wholly in D. (See Figure 16.13.) Let $\hat{\mathbf{N}}^*$ be the unit normal on $\mathcal{S}^*$ pointing *into* the sphere, and let D^* be that part of D that lies outside $\mathcal{S}^*$. As shown in Example 3 of Section 16.1, $\mathbf{div}\,\mathbf{F} = 0$ on D^*. Also,

$$\oiint_{\mathcal{S}^*} \mathbf{F} \bullet \hat{\mathbf{N}}^*\, dS = -4\pi m$$

is the flux of $\mathbf{F}$ *inward* through the sphere $\mathcal{S}^*$. (See Example 1 of Section 15.6.) Therefore,

$$\begin{aligned} 0 = \iiint_{D^*} \mathbf{div}\,\mathbf{F}\, dV &= \oiint_{\mathcal{S}} \mathbf{F} \bullet \hat{\mathbf{N}}\, dS + \oiint_{\mathcal{S}^*} \mathbf{F} \bullet \hat{\mathbf{N}}^*\, dS \\ &= \oiint_{\mathcal{S}} \mathbf{F} \bullet \hat{\mathbf{N}}\, dS - 4\pi m, \end{aligned}$$

so $\oiint_{\mathcal{S}} \mathbf{F} \bullet \hat{\mathbf{N}}\, dS = 4\pi m.$

EXAMPLE 5 Find the flux of $\mathbf{F} = x\mathbf{i} + y^2\mathbf{j} + z\mathbf{k}$ upward through the first octant part $\mathcal{S}$ of the cylindrical surface $x^2 + z^2 = a^2$, $0 \le y \le b$.

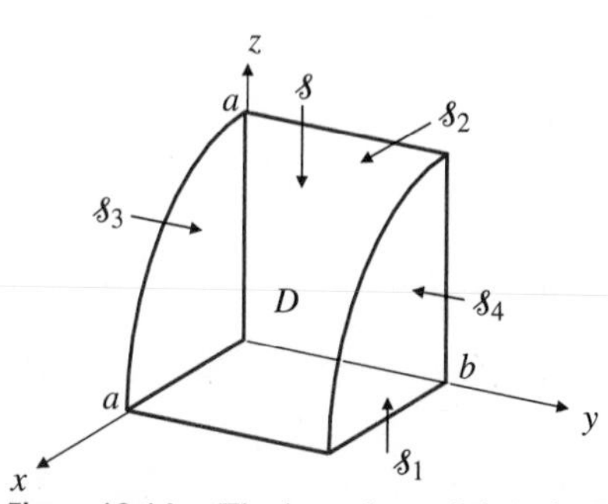

Figure 16.14 The boundary of domain D has five faces, one curved and four planar

Solution $\mathcal{S}$ is one of five surfaces that form the boundary of the solid region D shown in Figure 16.14. The other four surfaces are planar: $\mathcal{S}_1$ lies in the plane $z = 0$, $\mathcal{S}_2$ lies in the plane $x = 0$, $\mathcal{S}_3$ lies in the plane $y = 0$, and $\mathcal{S}_4$ lies in the plane $y = b$. Orient all these surfaces with normal $\hat{\mathbf{N}}$ pointing out of D. On $\mathcal{S}_1$ we have $\hat{\mathbf{N}} = -\mathbf{k}$, so $\mathbf{F} \bullet \hat{\mathbf{N}} = -z = 0$ on $\mathcal{S}_1$. Similarly, $\mathbf{F} \bullet \hat{\mathbf{N}} = 0$ on $\mathcal{S}_2$ and $\mathcal{S}_3$. On $\mathcal{S}_4$, $y = b$ and $\hat{\mathbf{N}} = \mathbf{j}$, so $\mathbf{F} \bullet \hat{\mathbf{N}} = y^2 = b^2$ there. If $\mathcal{S}_{\text{tot}}$ denotes the whole boundary of D, then

$$\begin{aligned} \oiint_{\mathcal{S}_{\text{tot}}} \mathbf{F} \bullet \hat{\mathbf{N}}\, dS &= \iint_{\mathcal{S}} \mathbf{F} \bullet \hat{\mathbf{N}}\, dS + 0 + 0 + 0 + \iint_{\mathcal{S}_4} \mathbf{F} \bullet \hat{\mathbf{N}}\, dS \\ &= \iint_{\mathcal{S}} \mathbf{F} \bullet \hat{\mathbf{N}}\, dS + \frac{\pi a^2 b^2}{4}. \end{aligned}$$

On the other hand, by the Divergence Theorem,

$$\oiint_{\mathcal{S}_{\text{tot}}} \mathbf{F} \bullet \hat{\mathbf{N}}\, dS = \iiint_D \operatorname{div} \mathbf{F}\, dV = \iiint_D (2 + 2y)\, dV = 2V + 2V\bar{y},$$

where $V = \pi a^2 b/4$ is the volume of D, and $\bar{y} = b/2$ is the y-coordinate of the centroid of D. Combining these results, the flux of $\mathbf{F}$ upward through $\mathcal{S}$ is

$$\oiint_{\mathcal{S}} \mathbf{F} \bullet \hat{\mathbf{N}}\, dS = \frac{2\pi a^2 b}{4}\left(1 + \frac{b}{2}\right) - \frac{\pi a^2 b^2}{4} = \frac{\pi a^2 b}{2}.$$

Among the examples above, Example 4 is the most significant and the one that best represents the way that the Divergence Theorem is used in practice. It is predominantly a theoretical tool, rather than a tool for calculation. We will look at some applications in Section 16.6.

Variants of the Divergence Theorem

Other versions of the Fundamental Theorem of Calculus can be derived from the Divergence Theorem. Two are given in the following theorem.

THEOREM 9

If D satisfies the conditions of the Divergence Theorem and has surface $\mathcal{S}$, and if $\mathbf{F}$ is a smooth vector field and ϕ is a smooth scalar field, then

(a) $$\iiint_D \operatorname{curl} \mathbf{F}\, dV = -\oiint_{\mathcal{S}} \mathbf{F} \times \hat{\mathbf{N}}\, dS,$$

(b) $$\iiint_D \operatorname{grad} \phi\, dV = \oiint_{\mathcal{S}} \phi \hat{\mathbf{N}}\, dS.$$

PROOF Observe that both of these formulas are equations of *vectors*. They are derived by applying the Divergence Theorem to $\mathbf{F} \times \mathbf{c}$ and $\phi\mathbf{c}$, respectively, where $\mathbf{c}$ is an arbitrary constant vector. We give the details for formula (a) and leave (b) as an exercise.

Using Theorem 3(d), we calculate

$$\nabla \bullet (\mathbf{F} \times \mathbf{c}) = (\nabla \times \mathbf{F}) \bullet \mathbf{c} - \mathbf{F} \bullet (\nabla \times \mathbf{c}) = (\nabla \times \mathbf{F}) \bullet \mathbf{c}.$$

Also, by the scalar triple product identity (see Exercise 18 of Section 10.3),

$$(\mathbf{F} \times \mathbf{c}) \bullet \hat{\mathbf{N}} = (\hat{\mathbf{N}} \times \mathbf{F}) \bullet \mathbf{c} = -(\mathbf{F} \times \hat{\mathbf{N}}) \bullet \mathbf{c}.$$

Therefore,

$$\begin{aligned}
\left(\iiint_D \operatorname{curl} \mathbf{F}\, dV + \oiint_{\mathcal{S}} \mathbf{F} \times \hat{\mathbf{N}}\, dS\right) \bullet \mathbf{c}& \\
&= \iiint_D (\nabla \times \mathbf{F}) \bullet \mathbf{c}\, dV - \oiint_{\mathcal{S}} (\mathbf{F} \times \mathbf{c}) \bullet \hat{\mathbf{N}}\, dS \\
&= \iiint_D \operatorname{div} (\mathbf{F} \times \mathbf{c})\, dV - \oiint_{\mathcal{S}} (\mathbf{F} \times \mathbf{c}) \bullet \hat{\mathbf{N}}\, dS = 0.
\end{aligned}$$

Since $\mathbf{c}$ is arbitrary, the vector in the large parentheses must be the zero vector. (If $\mathbf{c} \bullet \mathbf{a} = 0$ for every vector $\mathbf{c}$, then $\mathbf{a} = \mathbf{0}$.) This establishes formula (a).

EXERCISES 16.4

In Exercises 1–4, use the Divergence Theorem to calculate the flux of the given vector field out of the sphere $\mathcal{S}$ with equation $x^2 + y^2 + z^2 = a^2$, where $a > 0$.

1. $\mathbf{F} = x\mathbf{i} - 2y\mathbf{j} + 4z\mathbf{k}$

2. $\mathbf{F} = ye^z\mathbf{i} + x^2e^z\mathbf{j} + xy\mathbf{k}$

3. $\mathbf{F} = (x^2 + y^2)\mathbf{i} + (y^2 - z^2)\mathbf{j} + z\mathbf{k}$

4. $\mathbf{F} = x^3\mathbf{i} + 3yz^2\mathbf{j} + (3y^2z + x^2)\mathbf{k}$

In Exercises 5–8, evaluate the flux of $\mathbf{F} = x^2\mathbf{i} + y^2\mathbf{j} + z^2\mathbf{k}$ outward across the boundary of the given solid region.

5. The ball $(x - 2)^2 + y^2 + (z - 3)^2 \le 9$

6. The solid ellipsoid $x^2 + y^2 + 4(z - 1)^2 \le 4$

7. The tetrahedron $x + y + z \le 3$, $x \ge 0$, $y \ge 0$, $z \ge 0$

8. The cylinder $x^2 + y^2 \le 2y$, $0 \le z \le 4$

9. Let A be the area of a region D forming part of the surface of a sphere of radius R centred at the origin, and let V be the volume of the solid cone C consisting of all points on line segments joining the centre of the sphere to points in D. Show that $V = \dfrac{1}{3}AR$ by applying the Divergence Theorem to $\mathbf{F} = x\mathbf{i} + y\mathbf{j} + z\mathbf{k}$.

10. Let $\phi(x, y, z) = xy + z^2$. Find the flux of $\nabla\phi$ upward through the triangular planar surface $\mathcal{S}$ with vertices at $(a, 0, 0)$, $(0, b, 0)$, and $(0, 0, c)$.

11. A conical domain with vertex $(0, 0, b)$ and axis along the z-axis has as base a disk of radius a in the xy-plane. Find the flux of

$$\mathbf{F} = (x + y^2)\mathbf{i} + (3x^2y + y^3 - x^3)\mathbf{j} + (z + 1)\mathbf{k}$$

upward through the conical part of the surface of the domain.

12. Find the flux of $\mathbf{F} = (y + xz)\mathbf{i} + (y + yz)\mathbf{j} - (2x + z^2)\mathbf{k}$ upward through the first octant part of the sphere $x^2 + y^2 + z^2 = a^2$.

13. Let D be the region $x^2 + y^2 + z^2 \le 4a^2$, $x^2 + y^2 \ge a^2$. The surface $\mathcal{S}$ of D consists of a cylindrical part, $\mathcal{S}_1$, and a spherical part, $\mathcal{S}_2$. Evaluate the flux of

$$\mathbf{F} = (x + yz)\mathbf{i} + (y - xz)\mathbf{j} + (z - e^x \sin y)\mathbf{k}$$

out of D through (a) the whole surface $\mathcal{S}$, (b) the surface $\mathcal{S}_1$, and (c) the surface $\mathcal{S}_2$.

14. Evaluate $\displaystyle\iint_{\mathcal{S}} (3xz^2\mathbf{i} - x\mathbf{j} - y\mathbf{k}) \bullet \hat{\mathbf{N}}\,dS$, where $\mathcal{S}$ is that part of the cylinder $y^2 + z^2 = 1$ that lies in the first octant and between the planes $x = 0$ and $x = 1$.

15. A solid region R has volume V and centroid at the point $(\bar{x}, \bar{y}, \bar{z})$. Find the flux of

$$\mathbf{F} = (x^2 - x - 2y)\mathbf{i} + (2y^2 + 3y - z)\mathbf{j} - (z^2 - 4z + xy)\mathbf{k}$$

out of R through its surface.

16. The plane $x + y + z = 0$ divides the cube $-1 \le x \le 1$, $-1 \le y \le 1$, $-1 \le z \le 1$ into two parts. Let the lower part (with one vertex at $(-1, -1, -1)$) be D. Sketch D. Note that it has seven faces, one of which is hexagonal. Find the flux of $\mathbf{F} = x\mathbf{i} + y\mathbf{j} + z\mathbf{k}$ out of D through each of its faces.

17. Let $\mathbf{F} = (x^2 + y + 2 + z^2)\mathbf{i} + (e^{x^2} + y^2)\mathbf{j} + (3 + x)\mathbf{k}$. Let $a > 0$, and let $\mathcal{S}$ be the part of the spherical surface $x^2 + y^2 + z^2 = 2az + 3a^2$ that is above the xy-plane. Find the flux of $\mathbf{F}$ outward across $\mathcal{S}$.

18. A pile of wet sand having total volume 5π covers the disk $x^2 + y^2 \le 1$, $z = 0$. The momentum of water vapour is given by $\mathbf{F} = \mathbf{grad}\,\phi + \mu\,\mathbf{curl}\,\mathbf{G}$, where $\phi = x^2 - y^2 + z^2$ is the water concentration, $\mathbf{G} = \frac{1}{3}(-y^3\mathbf{i} + x^3\mathbf{j} + z^3\mathbf{k})$, and μ is a constant. Find the flux of $\mathbf{F}$ upward through the top surface of the sand pile.

In Exercises 19–29, D is a three-dimensional domain satisfying the conditions of the Divergence Theorem, and $\mathcal{S}$ is its surface. $\hat{\mathbf{N}}$ is the unit outward (from D) normal field on $\mathcal{S}$. The functions ϕ and ψ are smooth scalar fields on D. Also, $\partial\phi/\partial n$ denotes the first directional derivative of ϕ in the direction of $\hat{\mathbf{N}}$ at any point on $\mathcal{S}$:

$$\frac{\partial\phi}{\partial n} = \nabla\phi \bullet \hat{\mathbf{N}}.$$

❷ 19. Show that $\displaystyle\oiint_{\mathcal{S}} \mathbf{curl}\,\mathbf{F} \bullet \hat{\mathbf{N}}\,dS = 0$, where $\mathbf{F}$ is an arbitrary smooth vector field.

❷ 20. Show that the volume V of D is given by

$$V = \frac{1}{3}\oiint_{\mathcal{S}} (x\mathbf{i} + y\mathbf{j} + z\mathbf{k}) \bullet \hat{\mathbf{N}}\,dS.$$

❷ 21. If D has volume V, show that

$$\bar{\mathbf{r}} = \frac{1}{2V}\oiint_{\mathcal{S}} (x^2 + y^2 + z^2)\hat{\mathbf{N}}\,dS$$

is the position vector of the centre of gravity of D.

❷ 22. Show that $\displaystyle\oiint_{\mathcal{S}} \nabla\phi \times \hat{\mathbf{N}}\,dS = 0$.

❷ 23. If $\mathbf{F}$ is a smooth vector field on D, show that

$$\iiint_D \phi\,\mathbf{div}\,\mathbf{F}\,dV + \iiint_D \nabla\phi \bullet \mathbf{F}\,dV = \oiint_{\mathcal{S}} \phi\mathbf{F} \bullet \hat{\mathbf{N}}\,dS.$$

Hint: Use Theorem 3(b) from Section 16.2.

Properties of the Laplacian operator

24. If $\nabla^2\phi = 0$ in D and $\phi(x, y, z) = 0$ on $\mathcal{S}$, show that $\phi(x, y, z) = 0$ in D. *Hint:* Let $\mathbf{F} = \nabla\phi$ in Exercise 23.

❷ 25. (Uniqueness for the Dirichlet problem) The Dirichlet problem for the Laplacian operator is the boundary-value problem

$$\begin{cases} \nabla^2 u(x, y, z) = f(x, y, z) & \text{on } D \\ u(x, y, z) = g(x, y, z) & \text{on } \mathcal{S}, \end{cases}$$

where f and g are given functions defined on D and $\mathcal{S}$, respectively. Show that this problem can have at most one solution $u(x, y, z)$. *Hint:* Suppose there are two solutions, u and v, and apply Exercise 24 to their difference $\phi = u - v$.

26. (The Neumann problem) If $\nabla^2\phi = 0$ in D and $\partial\phi/\partial n = 0$ on $\mathcal{S}$, show that $\nabla\phi(x, y, z) = 0$ on D. The Neumann problem for the Laplacian operator is the boundary-value problem

$$\begin{cases} \nabla^2 u(x, y, z) = f(x, y, z) & \text{on } D \\ \dfrac{\partial}{\partial n} u(x, y, z) = g(x, y, z) & \text{on } \mathcal{S}, \end{cases}$$

where f and g are given functions defined on D and $\mathcal{S}$, respectively. Show that, if D is connected, then any two solutions of the Neumann problem must differ by a constant on D.

27. Verify that $\displaystyle\iiint_D \nabla^2\phi\,dV = \oiint_{\mathcal{S}} \frac{\partial\phi}{\partial n}\,dS$.

28. Verify that

$$\iiint_D \left(\phi\nabla^2\psi - \psi\nabla^2\phi\right) dV = \oiint_{\mathcal{S}} \left(\phi\frac{\partial\psi}{\partial n} - \psi\frac{\partial\phi}{\partial n}\right) dS.$$

29. By applying the Divergence Theorem to $\mathbf{F} = \phi\mathbf{c}$, where $\mathbf{c}$ is an arbitrary constant vector, show that

$$\iiint_D \nabla\phi\,dV = \oiint_{\mathcal{S}} \phi\hat{\mathbf{N}}\,dS.$$

30. Let P_0 be a fixed point, and for each $\epsilon > 0$ let D_ϵ be a domain with boundary $\mathcal{S}_\epsilon$ satisfying the conditions of the Divergence Theorem. Suppose that the maximum distance from P_0 to points P in D_ϵ approaches zero as $\epsilon \to 0+$. If D_ϵ has volume $\text{vol}(D_\epsilon)$, show that

$$\lim_{\epsilon\to 0+} \frac{1}{\text{vol}(D_\epsilon)} \oiint_{\mathcal{S}_\epsilon} \mathbf{F} \bullet \hat{\mathbf{N}}\,dS = \mathbf{div}\,\mathbf{F}(P_0).$$

This generalizes Theorem 1 of Section 16.1.

16.5 Stokes's Theorem

If we regard a region R in the xy-plane as a surface in 3-space with normal field $\hat{\mathbf{N}} = \mathbf{k}$, the Green's Theorem formula can be written in the form

$$\oint_{\mathcal{C}} \mathbf{F} \bullet d\mathbf{r} = \iint_R \mathbf{curl}\,\mathbf{F} \bullet \hat{\mathbf{N}}\,dS.$$

Stokes's Theorem given below generalizes this to nonplanar surfaces.

THEOREM

Stokes's Theorem

Let $\mathcal{S}$ be a piecewise smooth, oriented surface in 3-space, having unit normal field $\hat{\mathbf{N}}$ and boundary $\mathcal{C}$ consisting of one or more piecewise smooth, closed curves with orientation inherited from $\mathcal{S}$. If $\mathbf{F}$ is a smooth vector field defined on an open set containing $\mathcal{S}$, then

$$\oint_{\mathcal{C}} \mathbf{F} \bullet d\mathbf{r} = \iint_{\mathcal{S}} \mathbf{curl}\,\mathbf{F} \bullet \hat{\mathbf{N}}\,dS.$$

PROOF An argument similar to those given in the proofs of Green's Theorem and the Divergence Theorem shows that if $\mathcal{S}$ is decomposed into finitely many nonoverlapping subsurfaces, then it is sufficient to prove that the formula above holds for each of them. (If subsurfaces $\mathcal{S}_1$ and $\mathcal{S}_2$ meet along the curve $\mathcal{C}^*$, then $\mathcal{C}^*$ inherits opposite orientations as part of the boundaries of $\mathcal{S}_1$ and $\mathcal{S}_2$, so the line integrals along $\mathcal{C}^*$ cancel out. See Figure 16.15(a).) We can subdivide $\mathcal{S}$ into enough smooth subsurfaces that each one has a one-to-one normal projection onto a coordinate plane. We will establish the formula for one such subsurface, which we will now call $\mathcal{S}$.

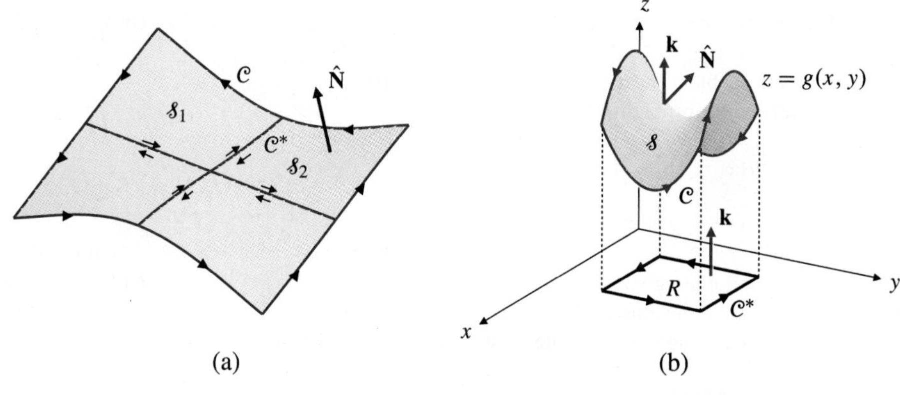

Figure 16.15

(a) Stokes's Theorem holds for a composite surface comprised of nonoverlapping subsurfaces for which it is true

(b) A surface with a one-to-one projection on the xy-plane

Without loss of generality, assume that $\mathcal{S}$ has a one-to-one normal projection onto the xy-plane and that its normal field $\hat{\mathbf{N}}$ points upward. Therefore, on $\mathcal{S}$, z is a smooth function of x and y, say $z = g(x, y)$, defined for (x, y) in a region R of the xy-plane. The boundaries $\mathcal{C}$ of $\mathcal{S}$ and $\mathcal{C}^*$ of R are both oriented counterclockwise as seen from a point high on the z-axis. (See Figure 16.15(b).) The normal field on $\mathcal{S}$ is

$$\hat{\mathbf{N}} = \frac{-\dfrac{\partial g}{\partial x}\mathbf{i} - \dfrac{\partial g}{\partial y}\mathbf{j} + \mathbf{k}}{\sqrt{1 + \left(\dfrac{\partial g}{\partial x}\right)^2 + \left(\dfrac{\partial g}{\partial y}\right)^2}},$$

and the surface area element on $\mathcal{S}$ is expressed in terms of the area element $dA = dx\,dy$ in the xy-plane as

$$dS = \sqrt{1 + \left(\frac{\partial g}{\partial x}\right)^2 + \left(\frac{\partial g}{\partial y}\right)^2}\, dA.$$

Therefore,

$$\iint_{\mathcal{S}} \mathbf{curl\,F} \bullet \hat{\mathbf{N}}\, dS = \iint_R \left[\left(\frac{\partial F_3}{\partial y} - \frac{\partial F_2}{\partial z}\right)\left(-\frac{\partial g}{\partial x}\right) + \left(\frac{\partial F_1}{\partial z} - \frac{\partial F_3}{\partial x}\right)\left(-\frac{\partial g}{\partial y}\right) + \left(\frac{\partial F_2}{\partial x} - \frac{\partial F_1}{\partial y}\right)\right] dA.$$

Since $z = g(x, y)$ on $\mathcal{C}$, we have $dz = \dfrac{\partial g}{\partial x}\, dx + \dfrac{\partial g}{\partial y}\, dy$. Thus,

$$\begin{aligned}
\oint_{\mathcal{C}} \mathbf{F} \bullet d\mathbf{r} &= \oint_{\mathcal{C}^*} \left[F_1(x, y, z)\, dx + F_2(x, y, z)\, dy + F_3(x, y, z)\left(\frac{\partial g}{\partial x}\, dx + \frac{\partial g}{\partial y}\, dy\right)\right] \\
&= \oint_{\mathcal{C}^*} \left(\left[F_1(x, y, z) + F_3(x, y, z)\frac{\partial g}{\partial x}\right] dx + \left[F_2(x, y, z) + F_3(x, y, z)\frac{\partial g}{\partial y}\right] dy \right).
\end{aligned}$$

We now apply Green's Theorem in the xy-plane to obtain

$$\begin{aligned}
\oint_{\mathcal{C}} \mathbf{F} \bullet d\mathbf{r} &= \iint_R \left(\frac{\partial}{\partial x}\left[F_2(x, y, z) + F_3(x, y, z)\frac{\partial g}{\partial y}\right] - \frac{\partial}{\partial y}\left[F_1(x, y, z) + F_3(x, y, z)\frac{\partial g}{\partial x}\right]\right) dA \\
&= \iint_R \left(\frac{\partial F_2}{\partial x} + \frac{\partial F_2}{\partial z}\frac{\partial g}{\partial x} + \frac{\partial F_3}{\partial x}\frac{\partial g}{\partial y} + \frac{\partial F_3}{\partial z}\frac{\partial g}{\partial x}\frac{\partial g}{\partial y} + F_3\frac{\partial^2 g}{\partial x \partial y} \right. \\
&\qquad \left. - \frac{\partial F_1}{\partial y} - \frac{\partial F_1}{\partial z}\frac{\partial g}{\partial y} - \frac{\partial F_3}{\partial y}\frac{\partial g}{\partial x} - \frac{\partial F_3}{\partial z}\frac{\partial g}{\partial y}\frac{\partial g}{\partial x} - F_3\frac{\partial^2 g}{\partial y \partial x}\right) dA.
\end{aligned}$$

Observe that four terms in the final integrand cancel out, and the remaining terms are equal to the terms in the expression for $\iint_{\mathcal{S}} \mathbf{curl}\,\mathbf{F} \bullet \hat{\mathbf{N}}\, dS$ calculated above. Therefore, the proof is complete.

Remark If $\mathbf{curl}\,\mathbf{F} = \mathbf{0}$ on a domain D with the property that every piecewise smooth, non–self-intersecting, closed curve in D is the boundary of a piecewise smooth surface in D, then Stokes's Theorem assures us that $\oint_{\mathcal{C}} \mathbf{F} \bullet d\mathbf{r} = 0$ for *every* such curve $\mathcal{C}$; therefore $\mathbf{F}$ must be conservative. A simply connected domain D does have the property specified above. We will not attempt a formal proof of this topological fact here, but it should seem plausible if you recall the definition of simple connectedness. A closed curve $\mathcal{C}$ in a simply connected domain D must be able to shrink to a point in D without ever passing out of D. In so shrinking, it traces out a surface in D. This is why Theorem 4 of Section 16.2 is valid for simply connected domains.

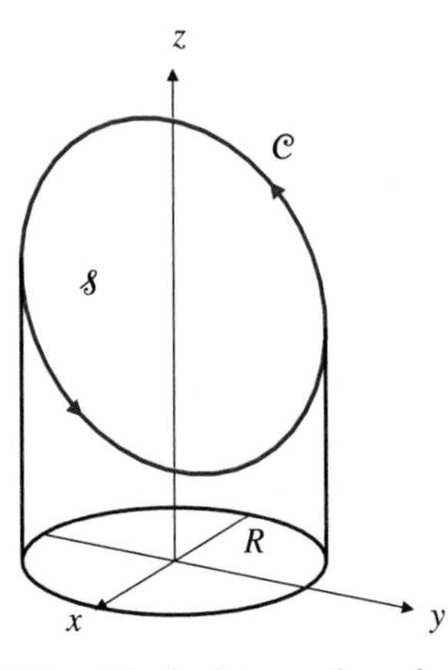

Figure 16.16 $\mathcal{C}$ is the intersection of a vertical cylinder and an oblique plane

EXAMPLE 1 Evaluate $\oint_{\mathcal{C}} \mathbf{F} \bullet d\mathbf{r}$, where $\mathbf{F} = -y^3\mathbf{i} + x^3\mathbf{j} - z^3\mathbf{k}$, and $\mathcal{C}$ is the curve of intersection of the cylinder $x^2 + y^2 = 1$ and the plane $2x+2y+z = 3$ oriented so as to have a counterclockwise projection onto the xy-plane.

Solution $\mathcal{C}$ is the oriented boundary of an elliptic disk $\mathcal{S}$ that lies in the plane $2x + 2y + z = 3$ and has the circular disk R: $x^2 + y^2 \le 1$ as projection onto the xy-plane. (See Figure 16.16.) On $\mathcal{S}$ we have

$$\hat{\mathbf{N}}\, dS = (2\mathbf{i} + 2\mathbf{j} + \mathbf{k})\, dx\, dy.$$

Also,

$$\mathbf{curl}\,\mathbf{F} = \begin{vmatrix} \mathbf{i} & \mathbf{j} & \mathbf{k} \\ \dfrac{\partial}{\partial x} & \dfrac{\partial}{\partial y} & \dfrac{\partial}{\partial z} \\ -y^3 & x^3 & -z^3 \end{vmatrix} = 3(x^2 + y^2)\mathbf{k}.$$

Thus, by Stokes's Theorem,

$$\begin{aligned} \oint_{\mathcal{C}} \mathbf{F} \bullet d\mathbf{r} &= \iint_{\mathcal{S}} \mathbf{curl}\,\mathbf{F} \bullet \hat{\mathbf{N}}\, dS \\ &= \iint_R 3(x^2 + y^2)\, dx\, dy = 2\pi \int_0^1 3r^2\, r\, dr = \frac{3\pi}{2}. \end{aligned}$$

As with the Divergence Theorem, the principal importance of Stokes's Theorem is as a theoretical tool. However, it can also simplify the calculation of circulation integrals such as the one in the previous example. It is not difficult to imagine integrals whose evaluation would be impossibly difficult without the use of Stokes's Theorem or the Divergence Theorem. In the following example we use Stokes's Theorem twice, but the result could be obtained just as easily by using the Divergence Theorem.

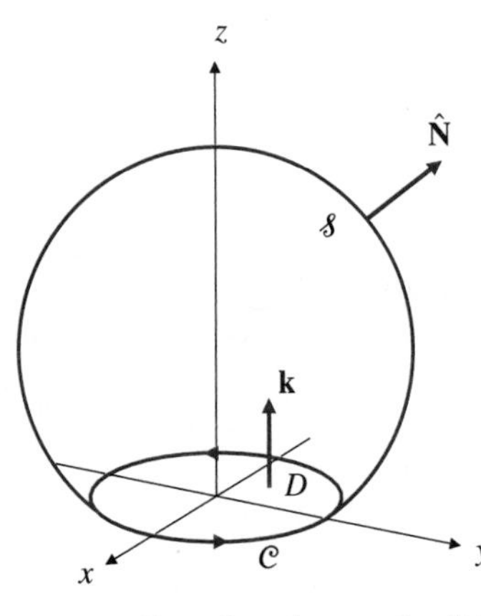

Figure 16.17 Part of a sphere and a disk with the same boundary

EXAMPLE 2 Find $I = \displaystyle\iint_{\mathcal{S}} \mathbf{curl}\,\mathbf{F} \bullet \hat{\mathbf{N}}\, dS$, where $\mathcal{S}$ is that part of the sphere $x^2 + y^2 + (z-2)^2 = 8$ that lies above the xy-plane, $\hat{\mathbf{N}}$ is the unit outward normal field on $\mathcal{S}$, and

$$\mathbf{F} = y^2 \cos xz\, \mathbf{i} + x^3 e^{yz}\mathbf{j} - e^{-xyz}\mathbf{k}.$$

Solution The boundary, $\mathcal{C}$, of $\mathcal{S}$ is the circle $x^2 + y^2 = 4$ in the xy-plane, oriented counterclockwise as seen from the positive z-axis. (See Figure 16.17.) This curve is also the oriented boundary of the plane disk D: $x^2 + y^2 \le 4$, $z = 0$, with normal field $\hat{\mathbf{N}} = \mathbf{k}$. Thus, two applications of Stokes's Theorem give

$$I = \iint_{\mathcal{S}} \mathbf{curl}\,\mathbf{F} \bullet \hat{\mathbf{N}}\, dS = \oint_{\mathcal{C}} \mathbf{F} \bullet d\mathbf{r} = \iint_D \mathbf{curl}\,\mathbf{F} \bullet \mathbf{k}\, dA.$$

On D we have

$$\begin{aligned}\mathbf{curl\,F} \bullet \mathbf{k} &= \left(\frac{\partial}{\partial x}\left(x^3 e^{yz}\right) - \frac{\partial}{\partial y}\left(y^2 \cos xz\right)\right)\bigg|_{z=0} \\ &= 3x^2 - 2y.\end{aligned}$$

By symmetry, $\iint_D y\,dA = 0$, so

$$I = 3\iint_D x^2\,dA = 3\int_0^{2\pi} \cos^2\theta\,d\theta \int_0^2 r^3\,dr = 12\pi.$$

Remark A surface $\mathcal{S}$ satisfying the conditions of Stokes's Theorem may no longer do so if a single point is removed from it. An isolated boundary point of a surface is not an orientable curve, and Stokes's Theorem may therefore break down for such a surface. Consider, for example, the vector field

$$\mathbf{F} = \frac{\hat{\boldsymbol{\theta}}}{r} = -\frac{y}{x^2+y^2}\mathbf{i} + \frac{x}{x^2+y^2}\mathbf{j},$$

which is defined on the *punctured disk* D satisfying $0 < x^2+y^2 \leq a^2$. (See Example 4 in Section 16.3.) If D is oriented with upward normal $\mathbf{k}$, then its boundary consists of the oriented, smooth, closed curve, $\mathcal{C}$, given by $x = a\cos\theta$, $y = a\sin\theta$, $(0 \leq \theta \leq 2\pi)$, and the isolated point $(0, 0)$. We have

$$\begin{aligned}\oint_{\mathcal{C}} \mathbf{F} \bullet d\mathbf{r} &= \int_0^{2\pi} \left(\frac{-\sin\theta}{a}\mathbf{i} + \frac{\cos\theta}{a}\mathbf{j}\right) \bullet (-a\sin\theta\mathbf{i} + a\cos\theta\mathbf{j})\,d\theta \\ &= \int_0^{2\pi} (\sin^2\theta + \cos^2\theta)\,d\theta = 2\pi.\end{aligned}$$

However,

$$\mathbf{curl\,F} = \left[\frac{\partial}{\partial x}\left(\frac{x}{x^2+y^2}\right) - \frac{\partial}{\partial y}\left(-\frac{y}{x^2+y^2}\right)\right]\mathbf{k} = \mathbf{0}$$

identically on D. Thus,

$$\iint_D \mathbf{curl\,F} \bullet \hat{\mathbf{N}}\,dS = 0,$$

and the conclusion of Stokes's Theorem fails in this case.

EXERCISES 16.5

1. Evaluate $\oint_{\mathcal{C}} xy\,dx + yz\,dy + zx\,dz$ around the triangle with vertices $(1, 0, 0)$, $(0, 1, 0)$, and $(0, 0, 1)$, oriented clockwise as seen from the point $(1, 1, 1)$.

2. Evaluate $\oint_{\mathcal{C}} y\,dx - x\,dy + z^2\,dz$ around the curve $\mathcal{C}$ of intersection of the cylinders $z = y^2$ and $x^2 + y^2 = 4$, oriented counterclockwise as seen from a point high on the z-axis.

3. Evaluate $\iint_{\mathcal{S}} \mathbf{curl\,F} \bullet \hat{\mathbf{N}}\,dS$, where $\mathcal{S}$ is the hemisphere $x^2 + y^2 + z^2 = a^2$, $z \geq 0$ with outward normal, and $\mathbf{F} = 3y\mathbf{i} - 2xz\mathbf{j} + (x^2 - y^2)\mathbf{k}$.

4. Evaluate $\iint_{\mathcal{S}} \mathbf{curl\,F} \bullet \hat{\mathbf{N}}\,dS$, where $\mathcal{S}$ is the surface $x^2 + y^2 + 2(z-1)^2 = 6$, $z \geq 0$, $\hat{\mathbf{N}}$ is the unit outward (away from the origin) normal on $\mathcal{S}$, and

$$\mathbf{F} = (xz - y^3\cos z)\mathbf{i} + x^3 e^z\mathbf{j} + xyz\,e^{x^2+y^2+z^2}\mathbf{k}.$$

5. Use Stokes's Theorem to show that

$$\oint_{\mathcal{C}} y\,dx + z\,dy + x\,dz = \sqrt{3}\,\pi a^2,$$

where $\mathcal{C}$ is the suitably oriented intersection of the surfaces $x^2 + y^2 + z^2 = a^2$ and $x + y + z = 0$.

6. Evaluate $\oint_{\mathcal{C}} \mathbf{F} \bullet d\mathbf{r}$ around the curve

$$\mathbf{r} = \cos t\,\mathbf{i} + \sin t\,\mathbf{j} + \sin 2t\,\mathbf{k}, \quad (0 \le t \le 2\pi),$$

where

$$\mathbf{F} = (e^x - y^3)\mathbf{i} + (e^y + x^3)\mathbf{j} + e^z\mathbf{k}.$$

Hint: Show that $\mathcal{C}$ lies on the surface $z = 2xy$.

7. Find the circulation of $\mathbf{F} = -y\mathbf{i} + x^2\mathbf{j} + z\mathbf{k}$ around the oriented boundary of the part of the paraboloid $z = 9 - x^2 - y^2$ lying above the xy-plane and having normal field pointing upward.

8. Evaluate $\oint_{\mathcal{C}} \mathbf{F} \bullet d\mathbf{r}$, where

$$\mathbf{F} = ye^x\mathbf{i} + (x^2 + e^x)\mathbf{j} + z^2e^z\mathbf{k},$$

and $\mathcal{C}$ is the curve

$$\mathbf{r}(t) = (1 + \cos t)\mathbf{i} + (1 + \sin t)\mathbf{j} + (1 - \cos t - \sin t)\mathbf{k}$$

for $0 \le t \le 2\pi$. *Hint:* Use Stokes's Theorem, observing that $\mathcal{C}$ lies in a certain plane and has a circle as its projection onto the xy-plane. The integral can also be evaluated by using the techniques of Section 15.4.

9. Let $\mathcal{C}_1$ be the straight line joining $(-1, 0, 0)$ to $(1, 0, 0)$, and let $\mathcal{C}_2$ be the semicircle $x^2 + y^2 = 1$, $z = 0$, $y \ge 0$. Let $\mathcal{S}$ be a smooth surface joining $\mathcal{C}_1$ to $\mathcal{C}_2$ having upward normal, and let

$$\mathbf{F} = (\alpha x^2 - z)\mathbf{i} + (xy + y^3 + z)\mathbf{j} + \beta y^2(z + 1)\mathbf{k}.$$

Find the values of α and β for which $I = \iint_{\mathcal{S}} \mathbf{F} \bullet d\mathbf{S}$ is independent of the choice of $\mathcal{S}$, and find the value of I for these values of α and β.

10. Let $\mathcal{C}$ be the curve $(x - 1)^2 + 4y^2 = 16$, $2x + y + z = 3$, oriented counterclockwise when viewed from high on the z-axis. Let

$$\mathbf{F} = (z^2 + y^2 + \sin x^2)\mathbf{i} + (2xy + z)\mathbf{j}) + (xz + 2yz)\mathbf{k}.$$

Evaluate $\oint_{\mathcal{C}} \mathbf{F} \bullet d\mathbf{r}$.

11. If $\mathcal{C}$ is the oriented boundary of surface $\mathcal{S}$, and ϕ and ψ are arbitrary smooth scalar fields, show that

$$\oint_{\mathcal{C}} \phi \nabla \psi \bullet d\mathbf{r} = -\oint_{\mathcal{C}} \psi \nabla \phi \bullet d\mathbf{r} = \iint_{\mathcal{S}} (\nabla \phi \times \nabla \psi) \bullet \hat{\mathbf{N}}\,dS.$$

Is $\nabla \phi \times \nabla \psi$ solenoidal? Find a vector potential for it.

12. Let $\mathcal{C}$ be a piecewise smooth, simple closed plane curve in $\mathbb{R}^3$, which lies in a plane with unit normal $\hat{\mathbf{N}} = a\mathbf{i} + b\mathbf{j} + c\mathbf{k}$ and has orientation inherited from that of the plane. Show that the plane area enclosed by $\mathcal{C}$ is

$$\frac{1}{2}\oint_{\mathcal{C}} (bz - cy)\,dx + (cx - az)\,dy + (ay - bx)\,dz.$$

13. Use Stokes's Theorem to prove Theorem 2 of Section 16.1.

16.6 Some Physical Applications of Vector Calculus

In this section we will show how the theory developed in this chapter can be used to model concrete applied mathematical problems. We will look at two areas of application—fluid dynamics and electromagnetism—and will develop a few of the fundamental vector equations underlying these disciplines. Our purpose is to illustrate the techniques of vector calculus in applied contexts, rather than to provide any complete or even coherent introductions to the disciplines themselves.

Fluid Dynamics

Suppose that a region of 3-space is filled with a fluid (liquid or gas) in motion. Two approaches can be taken to describe the motion. We could attempt to determine the position, $\mathbf{r} = \mathbf{r}(a, b, c, t)$ at any time t, of a "particle" of fluid that was located at the point (a, b, c) at time $t = 0$. This is the Lagrange approach. Alternatively, we could attempt to determine the velocity, $\mathbf{v}(x, y, z, t)$, the density, $\rho(x, y, z, t)$, and other physical variables such as the pressure, $p(x, y, z, t)$, at any time t at any point (x, y, z) in the region occupied by the fluid. This is the Euler approach.

We will examine the latter method and describe how the Divergence Theorem can be used to translate some fundamental physical laws into equivalent mathematical equations. We assume throughout that the velocity, density, and pressure vary smoothly in all their variables and that the fluid is an *ideal fluid*, that is, nonviscous (it doesn't stick to itself), homogeneous, and isotropic (it has the same properties at all points and in all directions). Such properties are not always shared by real fluids, so we are dealing with a simplified mathematical model that does not always correspond exactly to the behaviour of real fluids.

Consider an imaginary closed surface $\mathcal{S}$ in the fluid, bounding a domain D. We call $\mathcal{S}$ "imaginary" because it is not a barrier that impedes the flow of the fluid in any way; it is just a means to concentrate our attention on a particular part of the fluid. It is fixed in space and does not move with the fluid. Let us assume that the fluid is being neither created nor destroyed anywhere (in particular, there are no sources or sinks), so the law of **conservation of mass** tells us that the rate of change of the mass of fluid in D equals the rate at which fluid enters D across $\mathcal{S}$.

The mass of fluid in volume element dV located at position (x, y, z) at time t is $\rho(x, y, z, t)\, dV$, so the mass in D at time t is $\iiint_D \rho\, dV$. This mass changes at rate

$$\frac{\partial}{\partial t} \iiint_D \rho\, dV = \iiint_D \frac{\partial \rho}{\partial t}\, dV.$$

As we noted in Section 15.6, the volume of fluid passing *out* of D through area element dS at position (x, y, z) in the interval from time t to $t + dt$ is given by $\mathbf{v}(x, y, z, t) \bullet \hat{\mathbf{N}}\, dS\, dt$, where $\hat{\mathbf{N}}$ is the unit normal at (x, y, z) on $\mathcal{S}$ pointing out of D. Hence, the mass crossing dS outward in that time interval is $\rho\mathbf{v} \bullet \hat{\mathbf{N}}\, dS\, dt$, and the *rate* at which mass is flowing out of D across $\mathcal{S}$ at time t is

$$\oiint_{\mathcal{S}} \rho\mathbf{v} \bullet \hat{\mathbf{N}}\, dS.$$

The rate at which mass is flowing *into* D is the negative of the above rate. Since mass is conserved, we must have

$$\iiint_D \frac{\partial \rho}{\partial t}\, dV = -\oiint_{\mathcal{S}} \rho\mathbf{v} \bullet \hat{\mathbf{N}}\, dS = -\iiint_D \mathbf{div}\,(\rho\mathbf{v})\, dV,$$

where we have used the Divergence Theorem to replace the surface integral with a volume integral. Thus,

$$\iiint_D \left(\frac{\partial \rho}{\partial t} + \mathbf{div}\,(\rho\mathbf{v}) \right) dV = 0.$$

This equation must hold for *any* domain D in the fluid.

If a continuous function f satisfies $\iiint_D f(P)\, dV = 0$ for every domain D, then $f(P) = 0$ at all points P, for if there were a point P_0 such that $f(P_0) \neq 0$ (say $f(P_0) > 0$), then, by continuity, f would be positive at all points in some sufficiently small ball B centred at P_0, and $\iiint_B f(P)\, dV$ would be greater than 0. Applying this principle, we must have

$$\frac{\partial \rho}{\partial t} + \mathbf{div}\,(\rho\mathbf{v}) = 0$$

throughout the fluid. This is called the **equation of continuity** for the fluid. It is equivalent to conservation of mass. Observe that if the fluid is **incompressible** then ρ is a constant, independent of both time and spatial position. In this case $\partial\rho/\partial t = 0$, and $\mathbf{div}\,(\rho\mathbf{v}) = \rho\, \mathbf{div}\, \mathbf{v}$. Therefore, the equation of continuity for an incompressible fluid is simply

$$\mathbf{div}\,\mathbf{v} = 0.$$

The motion of the fluid is governed by Newton's Second Law, which asserts that the rate of change of momentum of any part of the fluid is equal to the sum of the forces applied to that part. Again, let us consider the part of the fluid in a domain D. At any time t its momentum is $\iiint_D \rho\mathbf{v}\,dV$ and is changing at the rate

$$\iiint_D \frac{\partial}{\partial t}(\rho\mathbf{v})\,dV.$$

This change is due partly to momentum crossing $\mathcal{S}$ into or out of D (the momentum of the fluid crossing $\mathcal{S}$), partly to the pressure exerted on the fluid in D by the fluid outside, and partly to any external *body forces* (such as gravity or electromagnetic forces) acting on the fluid. Let us examine each of these causes in turn.

Momentum is transferred across $\mathcal{S}$ into D at the rate

$$-\oiint_{\mathcal{S}} \mathbf{v}(\rho\mathbf{v} \bullet \hat{\mathbf{N}})\,dS.$$

The pressure on the fluid in D is exerted across $\mathcal{S}$ in the direction of the inward normal $-\hat{\mathbf{N}}$. Thus, this part of the force on the fluid in D is

$$-\oiint_{\mathcal{S}} p\hat{\mathbf{N}}\,dS.$$

The body forces are best expressed in terms of the *force density* (force per unit mass), $\mathbf{F}$. The total body force on the fluid in D is therefore

$$\iiint_D \rho\mathbf{F}\,dV.$$

Newton's Second Law now implies that

$$\iiint_D \frac{\partial}{\partial t}(\rho\mathbf{v})\,dV = -\oiint_{\mathcal{S}} \mathbf{v}(\rho\mathbf{v} \bullet \hat{\mathbf{N}})\,dS - \oiint_{\mathcal{S}} p\hat{\mathbf{N}}\,dS + \iiint_D \rho\mathbf{F}\,dV.$$

Again, we would like to convert the surface integrals to triple integrals over D. If we use the results of Exercise 29 of Section 16.4 and Exercise 2 below, we get

$$\oiint_{\mathcal{S}} p\hat{\mathbf{N}}\,dS = \iiint_D \nabla p\,dV,$$

$$\oiint_{\mathcal{S}} \mathbf{v}(\rho\mathbf{v} \bullet \hat{\mathbf{N}})\,dS = \iiint_D \Big(\rho(\mathbf{v} \bullet \nabla)\mathbf{v} + \mathbf{v}\,\mathbf{div}\,(\rho\mathbf{v})\Big)\,dV.$$

Accordingly, we have

$$\iiint_D \left(\rho\frac{\partial\mathbf{v}}{\partial t} + \mathbf{v}\frac{\partial\rho}{\partial t} + \mathbf{v}\,\mathbf{div}\,(\rho\mathbf{v}) + \rho(\mathbf{v} \bullet \nabla)\mathbf{v} + \nabla p - \rho\mathbf{F}\right)dV = \mathbf{0}.$$

The second and third terms in the integrand cancel out by virtue of the continuity equation. Since D is arbitrary, we must therefore have

$$\rho\frac{\partial\mathbf{v}}{\partial t} + \rho(\mathbf{v} \bullet \nabla)\mathbf{v} = -\nabla p + \rho\mathbf{F}.$$

This is the **equation of motion** of the fluid. Observe that it is not a *linear* partial differential equation; the second term on the left is not linear in $\mathbf{v}$.

Electromagnetism

In 3-space there are defined two vector fields that determine the electric and magnetic forces that would be experienced by a unit charge at a particular point if it is moving with unit speed. (These vector fields are determined by electric charges and currents present in the space.) A charge q_0 at position $\mathbf{r} = x\mathbf{i} + y\mathbf{j} + z\mathbf{k}$ moving with velocity $\mathbf{v}_0$ experiences an electric force $q_0\mathbf{E}(\mathbf{r})$, where $\mathbf{E}$ is the **electric field**, and a magnetic force $q_0\mathbf{v}_0 \times \mathbf{B}(\mathbf{r})$, where $\mathbf{B}$ is the **magnetic field**. We will look briefly at each of these fields but will initially restrict ourselves to considering *static* situations. Electric fields produced by static charge distributions and magnetic fields produced by static electric currents do not depend on time. Later we will consider the interaction between the two fields when they are time-dependent.

Electrostatics

Experimental evidence shows that the value of the electric field at any point $\mathbf{r}$ is the vector sum of the fields caused by any elements of charge located in 3-space. A "point charge" q at position $\mathbf{s} = \xi\mathbf{i} + \eta\mathbf{j} + \zeta\mathbf{k}$ generates the electric field

$$\mathbf{E}(\mathbf{r}) = \frac{q}{4\pi\epsilon_0}\frac{\mathbf{r}-\mathbf{s}}{|\mathbf{r}-\mathbf{s}|^3} \qquad \text{(Coulomb's Law)},$$

where $\epsilon_0 \approx 8.85\times10^{-12}$ coulombs^2/N·m^2 is a physical constant called the **permittivity of free space**. This is just the field due to a point source of strength $q/4\pi\epsilon_0$ at $\mathbf{s}$. Except at $\mathbf{r} = \mathbf{s}$ the field is conservative, with potential

$$\phi(\mathbf{r}) = -\frac{q}{4\pi\epsilon_0}\frac{1}{|\mathbf{r}-\mathbf{s}|},$$

so for $\mathbf{r} \neq \mathbf{s}$ we have $\mathbf{curl}\,\mathbf{E} = \mathbf{0}$. Also $\mathbf{div}\,\mathbf{E} = 0$, except at $\mathbf{r} = \mathbf{s}$ where it is infinite; in terms of the Dirac distribution, $\mathbf{div}\,\mathbf{E} = (q/\epsilon_0)\rho(x-\xi)\rho(y-\eta)\rho(z-\zeta)$. (See Section 16.1.) The flux of $\mathbf{E}$ outward across the surface $\mathcal{S}$ of any region R containing q is

$$\oiint_{\mathcal{S}} \mathbf{E}\bullet\hat{\mathbf{N}}\,dS = \frac{q}{\epsilon_0},$$

by analogy with Example 4 of Section 16.4.

Given a *charge distribution* of density $\rho(\xi,\eta,\zeta)$ in 3-space (so that the charge in volume element $dV = d\xi\,d\eta\,d\zeta$ at $\mathbf{s}$ is $dq = \rho\,dV$), the flux of $\mathbf{E}$ out of $\mathcal{S}$ due to the charge in R is

$$\oiint_{\mathcal{S}} \mathbf{E}\bullet\hat{\mathbf{N}}\,dS = \frac{1}{\epsilon_0}\iiint_R dq = \frac{1}{\epsilon_0}\iiint_R \rho\,dV.$$

If we apply the Divergence Theorem to the surface integral, we obtain

$$\iiint_R \left(\mathbf{div}\,\mathbf{E} - \frac{\rho}{\epsilon_0}\right) dV = 0,$$

and since R is an arbitrary region,

$$\mathbf{div}\,\mathbf{E} = \frac{\rho}{\epsilon_0}.$$

This is the differential form of Gauss's Law. See Exercise 3 below.

The potential due to a charge distribution of density $\rho(\mathbf{s})$ in the region R is

$$\begin{aligned}\phi(\mathbf{r}) &= -\frac{1}{4\pi\epsilon_0}\iiint_R \frac{\rho(\mathbf{s})}{|\mathbf{r}-\mathbf{s}|}\,dV \\ &= -\frac{1}{4\pi\epsilon_0}\iiint_R \frac{\rho(\xi,\eta,\zeta)\,d\xi\,d\eta\,d\zeta}{\sqrt{(x-\xi)^2+(y-\eta)^2+(z-\zeta)^2}}.\end{aligned}$$

If ρ is continuous and vanishes outside a bounded region, the triple integral is convergent everywhere (see Exercise 4 below), so $\mathbf{E} = \nabla\phi$ is conservative throughout 3-space. Thus, at all points,

$$\mathbf{curl}\,\mathbf{E} = \mathbf{0}.$$

Since $\mathbf{div}\,\mathbf{E} = \mathbf{div}\,\nabla\phi = \nabla^2\phi$, the potential ϕ satisfies **Poisson's equation**

$$\nabla^2\phi = \frac{\rho}{\epsilon_0}.$$

In particular, ϕ is harmonic in regions of space where no charge is distributed.

Magnetostatics

Magnetic fields are produced by moving charges, that is, by currents. Suppose that a constant electric current, I, is flowing in a filament along the curve $\mathcal{F}$. It has been determined experimentally that the magnetic fields produced at position $\mathbf{r} = x\mathbf{i}+y\mathbf{j}+z\mathbf{k}$ by the elements of current $dI = I\,ds$ along the filament add vectorially and that the element at position $\mathbf{s} = \xi\mathbf{i} + \eta\mathbf{j} + \zeta\mathbf{k}$ produces the field

$$d\mathbf{B}(\mathbf{r}) = \frac{\mu_0 I}{4\pi}\,\frac{d\mathbf{s} \times (\mathbf{r} - \mathbf{s})}{|\mathbf{r} - \mathbf{s}|^3} \qquad \text{(the Biot–Savart Law),}$$

where $\mu_0 \approx 1.26 \times 10^{-6}$ N/ampere^2 is a physical constant called the **permeability of free space**, and $d\mathbf{s} = \hat{\mathbf{T}}\,ds$, $\hat{\mathbf{T}}$ being the unit tangent to $\mathcal{F}$ in the direction of the current. Under the reasonable assumption that charge is not created or destroyed anywhere, the filament $\mathcal{F}$ must form a closed circuit, and the total magnetic field at $\mathbf{r}$ due to the current flowing in the circuit is

$$\mathbf{B} = \frac{\mu_0 I}{4\pi} \oint_{\mathcal{F}} \frac{d\mathbf{s} \times (\mathbf{r} - \mathbf{s})}{|\mathbf{r} - \mathbf{s}|^3}.$$

Let $\mathbf{A}$ be the vector field defined by

$$\mathbf{A}(\mathbf{r}) = \frac{\mu_0 I}{4\pi} \oint_{\mathcal{F}} \frac{d\mathbf{s}}{|\mathbf{r} - \mathbf{s}|},$$

for all $\mathbf{r}$ not on the filament $\mathcal{F}$. If we make use of the fact that

$$\nabla\left(\frac{1}{|\mathbf{r} - \mathbf{s}|}\right) = -\frac{\mathbf{r} - \mathbf{s}}{|\mathbf{r} - \mathbf{s}|^3},$$

and the vector identity $\nabla \times (\phi\mathbf{F}) = (\nabla\phi) \times \mathbf{F} + \phi(\nabla \times \mathbf{F})$ (with $\mathbf{F}$ the vector $d\mathbf{s}$, which does not depend on $\mathbf{r}$), we can calculate the curl of $\mathbf{A}$:

$$\nabla \times \mathbf{A} = \frac{\mu_0 I}{4\pi} \oint_{\mathcal{F}} \nabla\left(\frac{1}{|\mathbf{r} - \mathbf{s}|}\right) \times d\mathbf{s} = \frac{\mu_0 I}{4\partial} \oint_{\mathcal{F}} -\frac{\mathbf{r} - \mathbf{s}}{|\mathbf{r} - \mathbf{s}|^3} \times d\mathbf{s} = \mathbf{B}(\mathbf{r}).$$

Thus, $\mathbf{A}$ is a vector potential for $\mathbf{B}$, and $\mathbf{div}\,\mathbf{B} = 0$ at points off the filament. We can also verify by calculation that $\mathbf{curl}\,\mathbf{B} = \mathbf{0}$ off the filament. (See Exercises 9–11 below.)

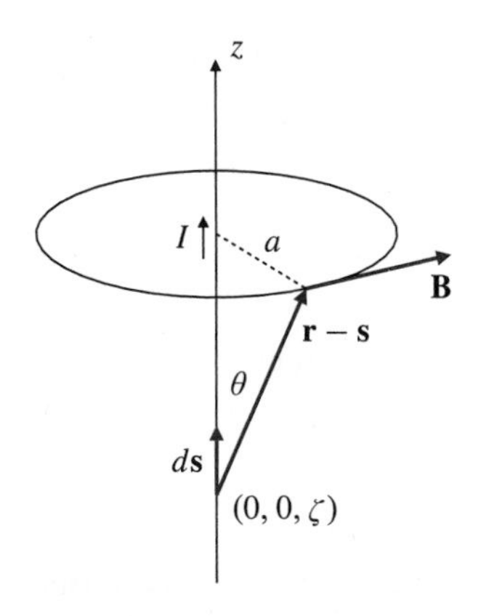

Figure 16.18 The magnetic field due to current in a vertical filament

Imagine a circuit consisting of a straight filament along the z-axis with return at infinite distance. The field $\mathbf{B}$ at a finite point will then just be due to the current along the z-axis, where the current I is flowing in the direction of $\mathbf{k}$, say. The currents in all elements $d\mathbf{s}$ produce, at $\mathbf{r}$, fields in the same direction, normal to the plane containing $\mathbf{r}$ and the z-axis. (See Figure 16.18.) Therefore, the field strength $B = |\mathbf{B}|$ at a distance a from the z-axis is obtained by integrating the elements

$$dB = \frac{\mu_0 I}{4\pi}\,\frac{\sin\theta\,d\zeta}{a^2 + (\zeta - z)^2} = \frac{\mu_0 I}{4\pi}\,\frac{a\,d\zeta}{\left(a^2 + (\zeta - z)^2\right)^{3/2}}.$$

We have

$$B = \frac{\mu_0 I a}{4\pi} \int_{-\infty}^{\infty} \frac{d\zeta}{\left(a^2 + (\zeta - z)^2\right)^{3/2}} \qquad \text{(Let } \zeta - z = a \tan\phi.\text{)}$$

$$= \frac{\mu_0 I}{4\pi a} \int_{-\pi/2}^{\pi/2} \cos\phi \, d\phi = \frac{\mu_0 I}{2\pi a}.$$

The field lines of $\mathbf{B}$ are evidently horizontal circles centred on the z-axis. If $\mathcal{C}_a$ is such a circle, having radius a, then the circulation of $\mathbf{B}$ around $\mathcal{C}_a$ is

$$\oint_{\mathcal{C}_a} \mathbf{B} \bullet d\mathbf{r} = \frac{\mu_0 I}{2\pi a} 2\pi a = \mu_0 I.$$

Observe that the circulation calculated above is independent of a. In fact, if $\mathcal{C}$ is any closed curve that encircles the z-axis once counterclockwise (as seen from above), then $\mathcal{C}$ and $-\mathcal{C}_a$ comprise the oriented boundary of a washer-like surface $\mathcal{S}$ with a hole in it through which the filament passes. Since $\mathbf{curl}\,\mathbf{B} = \mathbf{0}$ on $\mathcal{S}$, Stokes's Theorem guarantees that

$$\oint_{\mathcal{C}} \mathbf{B} \bullet d\mathbf{r} = \oint_{\mathcal{C}_a} \mathbf{B} \bullet d\mathbf{r} = \mu_0 I.$$

Furthermore, when $\mathcal{C}$ is very small (and therefore very close to the filament), most of the contribution to the circulation of $\mathbf{B}$ around it comes from the part of the filament that is very close to $\mathcal{C}$. It therefore does not matter whether the filament is straight or infinitely long. For any closed-loop filament carrying a current, the circulation of the magnetic field around the oriented boundary of a surface through which the filament passes is equal to μ_0 times the current flowing in the loop. This is **Ampère's Circuital Law**. The surface is oriented with normal on the side out of which the current is flowing.

Now let us replace the filament with a more general current specified by a vector density, $\mathbf{J}$. This means that at any point $\mathbf{s}$ the current is flowing in the direction $\mathbf{J}(\mathbf{s})$ and that the current crossing an area element dS with unit normal $\hat{\mathbf{N}}$ is $\mathbf{J} \bullet \hat{\mathbf{N}}\, dS$. The circulation of $\mathbf{B}$ around the boundary $\mathcal{C}$ of surface $\mathcal{S}$ is equal to the total current flowing across $\mathcal{S}$, so

$$\oint_{\mathcal{C}} \mathbf{B} \bullet d\mathbf{r} = \mu_0 \iint_{\mathcal{S}} \mathbf{J} \bullet \hat{\mathbf{N}}\, dS.$$

By using Stokes's Theorem, we can replace the line integral with another surface integral and so obtain

$$\iint_{\mathcal{S}} (\mathbf{curl}\,\mathbf{B} - \mu_0 \mathbf{J}) \bullet \hat{\mathbf{N}}\, dS = 0.$$

Since $\mathcal{S}$ is arbitrary, we must have, at all points,

$$\mathbf{curl}\,\mathbf{B} = \mu_0 \mathbf{J},$$

which is the pointwise version of Ampère's Circuital Law. It can be readily checked that, if

$$\mathbf{A}(\mathbf{r}) = \frac{\mu_0}{4\pi} \iiint_R \frac{\mathbf{J}(\mathbf{s})}{|\mathbf{r} - \mathbf{s}|}\, dV,$$

then $\mathbf{B} = \mathbf{curl}\,\mathbf{A}$ (so that $\mathbf{A}$ is a vector potential for the magnetic field $\mathbf{B}$). Here, R is the region of 3-space where $\mathbf{J}$ is nonzero. If $\mathbf{J}$ is continuous and vanishes outside a bounded set, then the triple integral converges for all $\mathbf{r}$ (see Exercise 4 below), and $\mathbf{B}$ is everywhere solenoidal:

$$\mathbf{div}\,\mathbf{B} = 0.$$

Maxwell's Equations

The four equations obtained above for static electric and magnetic fields,

$$\mathbf{div}\,\mathbf{E} = \rho/\epsilon_0 \qquad \mathbf{div}\,\mathbf{B} = 0$$
$$\mathbf{curl}\,\mathbf{E} = \mathbf{0} \qquad \mathbf{curl}\,\mathbf{B} = \mu_0\,\mathbf{J},$$

require some modification if the fields $\mathbf{E}$ and $\mathbf{B}$ depend on time. Gauss's Law $\mathbf{div}\,\mathbf{E} = \rho/\epsilon_0$ remains valid, as does $\mathbf{div}\,\mathbf{B} = 0$, which expresses the fact that there are no *known* magnetic *sources* or *sinks* (i.e., magnetic *monopoles*). The field lines of $\mathbf{B}$ must be closed curves.

It was observed by Michael Faraday that the circulation of an electric field around a simple closed curve $\mathcal{C}$ corresponds to a change in the magnetic flux

$$\Phi = \iint_{\mathcal{S}} \mathbf{B} \bullet \hat{\mathbf{N}}\,dS$$

through any oriented surface $\mathcal{S}$ having boundary $\mathcal{C}$, according to the formula

$$\frac{d\Phi}{dt} = -\oint_{\mathcal{C}} \mathbf{E} \bullet d\mathbf{r}.$$

Applying Stokes's Theorem to the line integral, we obtain

$$\iint_{\mathcal{S}} \mathbf{curl}\,\mathbf{E} \bullet \hat{\mathbf{N}}\,dS = \oint_{\mathcal{C}} \mathbf{E} \bullet d\mathbf{r} = -\frac{d}{dt}\iint_{\mathcal{S}} \mathbf{B} \bullet \hat{\mathbf{N}}\,dS = -\iint_{\mathcal{S}} \frac{\partial \mathbf{B}}{\partial t} \bullet \hat{\mathbf{N}}\,dS.$$

Since $\mathcal{S}$ is arbitrary, we obtain the differential form of Faraday's Law:

$$\mathbf{curl}\,\mathbf{E} = -\frac{\partial \mathbf{B}}{\partial t}.$$

The electric field is irrotational only if the magnetic field is constant in time.

The differential form of Ampère's Law, $\mathbf{curl}\,\mathbf{B} = \mu_0\,\mathbf{J}$, also requires modification. If the electric field depends on time, then so will the current density $\mathbf{J}$. Assuming conservation of charge (charges are not produced or destroyed), we can show, by an argument identical to that used to obtain the continuity equation for fluid motion earlier in this section, that the rate of change of charge density satisfies

$$\frac{\partial \rho}{\partial t} = -\mathbf{div}\,\mathbf{J}.$$

(See Exercise 5 below.) This is inconsistent with Ampère's Law because $\mathbf{div}\,\mathbf{curl}\,\mathbf{B} = 0$, while $\mathbf{div}\,\mathbf{J} \neq 0$ when ρ depends on time. Note, however, that $\rho = \epsilon_0 \mathbf{div}\,\mathbf{E}$ implies that

$$-\mathbf{div}\,\mathbf{J} = \frac{\partial \rho}{\partial t} = \epsilon_0 \mathbf{div}\,\frac{\partial \mathbf{E}}{\partial t},$$

so $\mathbf{div}\,\big(\mathbf{J} + \epsilon_0 \partial\mathbf{E}/\partial t\big) = 0$. This suggests that, for the nonstatic case, Ampère's Law becomes

$$\mathbf{curl}\,\mathbf{B} = \mu_0\,\mathbf{J} + \mu_0\epsilon_0 \frac{\partial \mathbf{E}}{\partial t},$$

which indicates (as was discovered by Maxwell) that magnetic fields are not just produced by currents, but also by changing electric fields.

Together, the four equations

$$\mathbf{div}\,\mathbf{E} = \rho/\epsilon_0 \qquad \mathbf{div}\,\mathbf{B} = 0$$
$$\mathbf{curl}\,\mathbf{E} = -\frac{\partial \mathbf{B}}{\partial t} \qquad \mathbf{curl}\,\mathbf{B} = \mu_0\,\mathbf{J} + \mu_0\epsilon_0 \frac{\partial \mathbf{E}}{\partial t}$$

are known as **Maxwell's equations**. They govern the way electric and magnetic fields are produced in 3-space by the presence of charges and currents. Observe that $\sqrt{\mu_0\epsilon_0} = 1/c^2$, where $c \approx 2.99 \times 10^8$ m/s, which is the speed of light in a vacuum. (See Exercise 15.)

EXERCISES 16.6

1. **(Archimedes' principle)** A solid occupying region R with surface $\mathcal{S}$ is immersed in a liquid of constant density ρ. The pressure at depth h in the liquid is $\rho g h$, so the pressure satisfies $\nabla p = \rho\mathbf{g}$, where $\mathbf{g}$ is the (vector) constant acceleration of gravity. Over each surface element dS on $\mathcal{S}$ the pressure of the fluid exerts a force $-p\hat{\mathbf{N}}\,dS$ on the solid.

(a) Show that the resultant "buoyancy force" on the solid is

$$\mathbf{B} = -\iiint_R \rho\mathbf{g}\,dV.$$

Thus, the buoyancy force has the same magnitude as, and opposite direction to, the weight of the liquid displaced by the solid. This is Archimedes' principle.

(b) Extend the above result to the case where the solid is only partly submerged in the fluid.

2. By breaking the vector $\mathbf{F}(\mathbf{G} \bullet \hat{\mathbf{N}})$ into its separate components and applying the Divergence Theorem to each separately, show that

$$\oiint_{\mathcal{S}} \mathbf{F}(\mathbf{G} \bullet \hat{\mathbf{N}})\,dS = \iiint_D \Big(\mathbf{F}\,\mathbf{div}\,\mathbf{G} + (\mathbf{G} \bullet \nabla)\mathbf{F}\Big)\,dV,$$

where $\hat{\mathbf{N}}$ is the unit outward normal on the surface $\mathcal{S}$ of the domain D.

3. **(Gauss's Law)** Show that the flux of the electric field $\mathbf{E}$ outward through a closed surface $\mathcal{S}$ in 3-space is $1/\epsilon_0$ times the total charge enclosed by $\mathcal{S}$.

4. If $\mathbf{s} = \xi\mathbf{i} + \eta\mathbf{j} + \zeta\mathbf{k}$ and $f(\xi, \eta, \zeta)$ is continuous on $\mathbb{R}^3$ and vanishes outside a bounded region, show that, for any fixed $\mathbf{r}$,

$$\iiint_{\mathbb{R}^3} \frac{|f(\xi, \eta, \zeta)|}{|\mathbf{r} - \mathbf{s}|}\,d\xi\,d\eta\,d\zeta \le \text{constant}.$$

This shows that the potentials for the electric and magnetic fields corresponding to continuous charge and current densities that vanish outside bounded regions exist everywhere in $\mathbb{R}^3$. *Hint:* Without loss of generality you can assume $\mathbf{r} = \mathbf{0}$ and use spherical coordinates.

5. The electric charge density, ρ, in 3-space depends on time as well as position if charge is moving around. The motion is described by the current density, $\mathbf{J}$. Derive the **continuity equation**

$$\frac{\partial\rho}{\partial t} = -\mathbf{div}\,\mathbf{J}$$

from the fact that charge is conserved.

6. If $\mathbf{b}$ is a constant vector, show that

$$\nabla\left(\frac{1}{|\mathbf{r} - \mathbf{b}|}\right) = -\frac{\mathbf{r} - \mathbf{b}}{|\mathbf{r} - \mathbf{b}|^3}.$$

7. If $\mathbf{a}$ and $\mathbf{b}$ are constant vectors, show that for $\mathbf{r} \ne \mathbf{b}$,

$$\mathbf{div}\left(\mathbf{a} \times \frac{\mathbf{r} - \mathbf{b}}{|\mathbf{r} - \mathbf{b}|^3}\right) = 0.$$

Hint: Use identities (d) and (h) from Theorem 3 of Section 16.2.

8. Use the result of Exercise 7 to give an alternative proof that

$$\mathbf{div} \oint_{\mathcal{F}} \frac{d\mathbf{s} \times (\mathbf{r} - \mathbf{s})}{|\mathbf{r} - \mathbf{s}|^3} = 0.$$

Note that **div** refers to the $\mathbf{r}$ variable.

9. If $\mathbf{a}$ and $\mathbf{b}$ are constant vectors, show that for $\mathbf{r} \ne \mathbf{b}$,

$$\mathbf{curl}\left(\mathbf{a} \times \frac{\mathbf{r} - \mathbf{b}}{|\mathbf{r} - \mathbf{b}|^3}\right) = -(\mathbf{a} \bullet \nabla)\frac{\mathbf{r} - \mathbf{b}}{|\mathbf{r} - \mathbf{b}|^3}.$$

Hint: Use identity (e) from Theorem 3 of Section 16.2.

10. If $\mathbf{F}$ is any smooth vector field, show that

$$\oint_{\mathcal{F}} (d\mathbf{s} \bullet \nabla)\mathbf{F}(\mathbf{s}) = \mathbf{0}$$

around any closed loop $\mathcal{F}$. *Hint:* The gradients of the components of $\mathbf{F}$ are conservative.

11. Verify that if $\mathbf{r}$ does not lie on $\mathcal{F}$, then

$$\mathbf{curl} \oint_{\mathcal{F}} \frac{d\mathbf{s} \times (\mathbf{r} - \mathbf{s})}{|\mathbf{r} - \mathbf{s}|^3} = \mathbf{0}.$$

Here, **curl** is taken with respect to the $\mathbf{r}$ variable.

12. Verify the formula $\mathbf{curl}\,\mathbf{A} = \mathbf{B}$, where $\mathbf{A}$ is the magnetic vector potential defined in terms of the steady-state current density $\mathbf{J}$.

13. If $\mathbf{A}$ is the vector potential for the magnetic field produced by a steady current in a closed-loop filament, show that $\mathbf{div}\,\mathbf{A} = 0$ off the filament.

14. If $\mathbf{A}$ is the vector potential for the magnetic field produced by a steady, continuous current density, show that $\mathbf{div}\,\mathbf{A} = 0$ everywhere. Hence, show that $\mathbf{A}$ satisfies the vector Poisson equation $\nabla^2\mathbf{A} = -\mathbf{J}$.

15. Show that in a region of space containing no charges ($\rho = 0$) and no currents ($\mathbf{J} = \mathbf{0}$), both $\mathbf{U} = \mathbf{E}$ and $\mathbf{U} = \mathbf{B}$ satisfy the wave equation

$$\frac{\partial^2\mathbf{U}}{\partial t^2} = c^2\nabla^2\mathbf{U},$$

where $c = \sqrt{1/(\epsilon_0\mu_0)} \approx 3 \times 10^8$ m/s.

16. As shown in this section, the static versions of Maxwell's equations needed revision when the fields **E** and **B** were allowed to depend on time. Show that the expression $\mathbf{E} = -\nabla\phi$ is no longer consistent with Maxwell's equations because the **E** field is no longer irrotational. Why does **curl A** = **B** continue to hold?

17. While the nonstatic Maxwell equations are not compatible with $\mathbf{E} = -\nabla\phi$, show that they are compatible with the equation

$$\mathbf{E} = -\nabla\phi - \frac{\partial \mathbf{A}}{\partial t}.$$

18. **(Heat flow in 3-space)** The internal energy, E, of a volume element dV within a homogeneous solid is $\rho c T\, dV$, where ρ and c are constants (the density and specific heat of the solid material), and $T = T(x, y, z, t)$ is the temperature at time t at position (x, y, z) in the solid. Heat always flows in the direction of the negative temperature gradient and at a rate proportional to the size of that gradient. Thus, the rate of flow of heat energy across a surface element dS with normal $\hat{\mathbf{N}}$ is $-k\nabla T \bullet \hat{\mathbf{N}}\, dS$, where k is also a constant depending on the material of the solid (the coefficient of thermal conductivity). Use "conservation of heat energy" to show that for any region R with surface $\mathcal{S}$ within the solid

$$\rho c \iiint_R \frac{\partial T}{\partial t}\, dV = k \oiint_{\mathcal{S}} \nabla T \bullet \hat{\mathbf{N}}\, dS,$$

where $\hat{\mathbf{N}}$ is the unit outward normal on $\mathcal{S}$. Hence, show that heat flow within the solid is governed by the partial differential equation

$$\frac{\partial T}{\partial t} = \frac{k}{\rho c}\nabla^2 T = \frac{k}{\rho c}\left(\frac{\partial^2 T}{\partial x^2} + \frac{\partial^2 T}{\partial y^2} + \frac{\partial^2 T}{\partial z^2}\right).$$

16.7 Orthogonal Curvilinear Coordinates

In this optional section we will derive formulas for the gradient of a scalar field and the divergence and curl of a vector field in terms of coordinate systems more general than the Cartesian coordinate system used in the earlier sections of this chapter. In particular, we will express these quantities in terms of the cylindrical and spherical coordinate systems introduced in Section 14.6.

We denote by xyz-space the usual system of Cartesian coordinates (x, y, z) in $\mathbb{R}^3$. A different system of coordinates $[u, v, w]$ in xyz-space can be defined by a continuous transformation of the form

$$x = x(u, v, w), \qquad y = y(u, v, w), \qquad z = z(u, v, w).$$

If the transformation is one-to-one from a region D in uvw-space onto a region R in xyz-space, then a point P in R can be represented by a triple $[u, v, w]$, the (Cartiesian) coordinates of the unique point Q in uvw-space that the transformation maps to P. In this case we say that the transformation defines a **curvilinear coordinate system** in R and call $[u, v, w]$ the **curvilinear coordinates** of P with respect to that system. Note that $[u, v, w]$ are Cartesian coordinates in their own space (uvw-space); they are curvilinear coordinates in xyz-space.

Typically, we relax the requirement that the transformation defining a curvilinear coordinate system be one-to-one, that is, that every point P in R should have a unique set of curvilinear coordinates. It is reasonable to require the transformation to be only *locally one-to-one*. Thus, there may be more than one point Q that gets mapped to a point P by the transformation, but only one in any suitably small subregion of D. For example, in the plane polar coordinate system

$$x = r\cos\theta, \qquad y = r\sin\theta,$$

the transformation is locally one-to-one from D, the half of the $r\theta$-plane where $0 < r < \infty$, to the region R consisting of all points in the xy-plane except the origin. Although, say, $[1, 0]$ and $[1, 2\pi]$ are polar coordinates of the same point in the xy-plane, they are not close together in D. Observe, however, that there is still a problem with the origin, which can be represented by $[0, \theta]$ for *any* θ. Since the transformation is not even locally one-to-one at $r = 0$, we regard the origin of the xy-plane as a **singular point** for the polar coordinate system in the plane.

EXAMPLE 1 The **cylindrical coordinate system** $[r, \theta, z]$ in $\mathbb{R}^3$ is defined by the transformation

$$x = r\cos\theta, \qquad y = r\sin\theta, \qquad z = z,$$

where $r \ge 0$. (See Section 10.6.) This transformation maps the half-space D given by $r > 0$ onto all of xyz-space excluding the z-axis, and it is locally one-to-one. We regard $[r, \theta, z]$ as cylindrical polar coordinates in all of xyz-space but call points on the z-axis singular points of the system since the points $[0, \theta, z]$ are identical for any θ.

EXAMPLE 2 The **spherical coordinate system** $[R, \phi, \theta]$ is defined by the transformation

$$x = R\sin\phi\cos\theta, \qquad y = R\sin\phi\sin\theta, \qquad z = R\cos\phi,$$

where $R \ge 0$ and $0 \le \phi \le \pi$. (See Section 10.6.) The transformation maps the region D in $R\phi\theta$-space given by $R > 0$, $0 < \phi < \pi$ in a locally one-to-one way onto xyz-space excluding the z-axis. The point with Cartesian coordinates $(0, 0, z)$ can be represented by the spherical coordinates $[0, \phi, \theta]$ for arbitrary ϕ and θ if $z = 0$, by $[z, 0, \theta]$ for arbitrary θ if $z > 0$, and by $[|z|, \pi, \theta]$ for arbitrary θ if $z < 0$. Thus, all points of the z-axis are singular for the spherical coordinate system.

Coordinate Surfaces and Coordinate Curves

Let $[u, v, w]$ be a curvilinear coordinate system in xyz-space, and let P_0 be a nonsingular point for the system. Thus, the transformation

$$x = x(u, v, w), \qquad y = y(u, v, w), \qquad z = z(u, v, w)$$

is locally one-to-one near P_0. Let P_0 have curvilinear coordinates $[u_0, v_0, w_0]$. The plane with equation $u = u_0$ in uvw-space gets mapped by the transformation to a surface in xyz-space passing through P_0. We call this surface a u-surface and still refer to it by the equation $u = u_0$; it has parametric equations

$$x = x(u_0, v, w), \qquad y = y(u_0, v, w), \qquad z = z(u_0, v, w)$$

with parameters v and w. Similarly, the v-surface $v = v_0$ and the w-surface $w = w_0$ pass through P_0; they are the images of the planes $v = v_0$ and $w = w_0$ in uvw-space.

Orthogonal curvilinear coordinates

We say that $[u, v, w]$ is an **orthogonal curvilinear coordinate system** in xyz-space if, for every nonsingular point P_0 in xyz-space, each of the three **coordinate surfaces** $u = u_0$, $v = v_0$, and $w = w_0$ intersects the other two at P_0 at right angles.

It is tacitly assumed that the coordinate surfaces are smooth at all nonsingular points, so we are really assuming that their normal vectors are mutually perpendicular. Figure 16.19 shows the coordinate surfaces through P_0 for a typical orthogonal curvilinear coordinate system.

Pairs of coordinate surfaces through a point intersect along a **coordinate curve** through that point. For example, the coordinate surfaces $v = v_0$ and $w = w_0$ intersect along the u-**curve** with parametric equations

$$x = x(u, v_0, w_0), \quad y = y(u, v_0, w_0), \quad \text{and} \quad z = z(u, v_0, w_0),$$

where the parameter is u. A unit vector $\hat{\mathbf{u}}$ tangent to the u-curve through P_0 is normal to the coordinate surface $u = u_0$ there. Similar statements hold for unit vectors $\hat{\mathbf{v}}$ and $\hat{\mathbf{w}}$. For an orthogonal curvilinear coordinate system, the three vectors $\hat{\mathbf{u}}$, $\hat{\mathbf{v}}$, and $\hat{\mathbf{w}}$ form a basis of mutually perpendicular unit vectors at any nonsingular point P_0. (See Figure 16.19.) We call this basis the **local basis** at P_0.

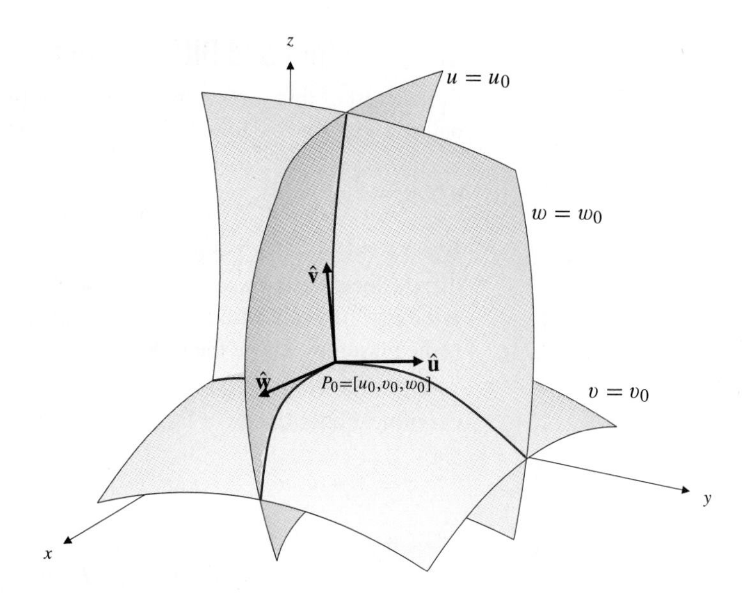

Figure 16.19 u-, v-, and w-coordinate surfaces

EXAMPLE 3 For the cylindrical coordinate system (see Figure 16.20), the coordinate surfaces are:

circular cylinders with axis along the z-axis	(r-surfaces),
vertical half-planes radiating from the z-axis	(θ-surfaces),
horizontal planes	(z-surfaces).

The coordinate curves are:

horizontal straight half-lines radiating from the z-axis	(r-curves),
horizontal circles with centres on the z-axis	(θ-curves),
vertical straight lines	(z-curves).

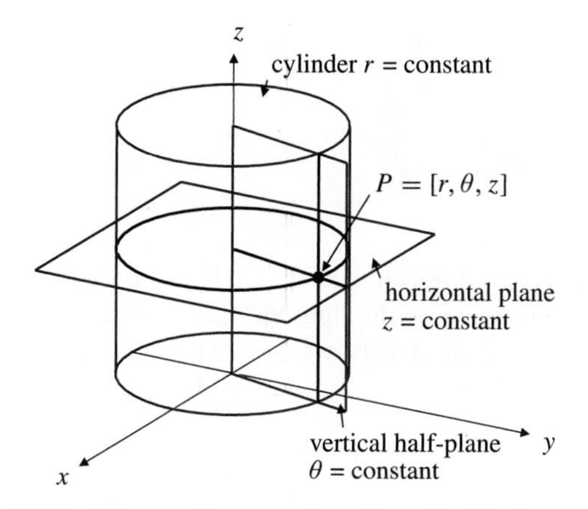

Figure 16.20 The coordinate surfaces for cylindrical coordinates

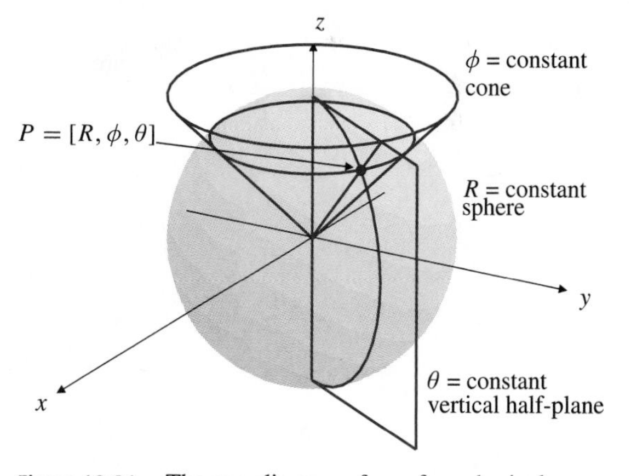

Figure 16.21 The coordinate surfaces for spherical coordinates

EXAMPLE 4 For the spherical coordinate system (see Figure 16.21), the coordinate surfaces are:

spheres centred at the origin	(R-surfaces),
vertical circular cones with vertices at the origin	(ϕ-surfaces),
vertical half-planes radiating from the z-axis	(θ-surfaces).

The coordinate curves are:

half-lines radiating from the origin	(R-curves),
vertical semicircles with centres at the origin	(ϕ-curves),
horizontal circles with centres on the z-axis	(θ-curves).

Scale Factors and Differential Elements

For the rest of this section we assume that $[u, v, w]$ are **orthogonal** curvilinear coordinates in xyz-space defined via the transformation

$$x = x(u, v, w), \qquad y = y(u, v, w), \qquad z = z(u, v, w).$$

We also assume that the coordinate surfaces are smooth at any nonsingular point and that the local basis vectors $\hat{\mathbf{u}}$, $\hat{\mathbf{v}}$, and $\hat{\mathbf{w}}$ at any such point form a right-handed triad. This is the case for both cylindrical and spherical coordinates. For spherical coordinates, this is the reason we chose the order of the coordinates as $[R, \phi, \theta]$, rather than $[R, \theta, \phi]$.

The **position vector** of a point P in xyz-space can be expressed in terms of the curvilinear coordinates:

$$\mathbf{r} = x(u, v, w)\mathbf{i} + y(u, v, w)\mathbf{j} + z(u, v, w)\mathbf{k}.$$

If we hold $v = v_0$ and $w = w_0$ fixed and let u vary, then $\mathbf{r} = \mathbf{r}(u, v_0, w_0)$ defines a u-curve in xyz-space. At any point P on this curve, the vector

$$\frac{\partial \mathbf{r}}{\partial u} = \frac{\partial x}{\partial u}\mathbf{i} + \frac{\partial y}{\partial u}\mathbf{j} + \frac{\partial z}{\partial u}\mathbf{k}$$

is tangent to the u-curve at P. In general, the three vectors

$$\frac{\partial \mathbf{r}}{\partial u}, \qquad \frac{\partial \mathbf{r}}{\partial v}, \qquad \text{and} \qquad \frac{\partial \mathbf{r}}{\partial w}$$

are tangent, respectively, to the u-curve, the v-curve, and the w-curve through P. They are also normal, respectively, to the u-surface, the v-surface, and the w-surface through P, so they are mutually perpendicular. (See Figure 16.19.) The lengths of these tangent vectors are called the *scale factors* of the coordinate system.

The **scale factors** of the orthogonal curvilinear coordinate system $[u, v, w]$ are the three functions

$$h_u = \left|\frac{\partial \mathbf{r}}{\partial u}\right|, \qquad h_v = \left|\frac{\partial \mathbf{r}}{\partial v}\right|, \qquad h_w = \left|\frac{\partial \mathbf{r}}{\partial w}\right|.$$

The scale factors are nonzero at a nonsingular point P of the coordinate system, so the local basis at P can be obtained by dividing the tangent vectors to the coordinate curves by their lengths. As noted previously, we denote the local basis vectors by $\hat{\mathbf{u}}$, $\hat{\mathbf{v}}$, and $\hat{\mathbf{w}}$. Thus,

$$\frac{\partial \mathbf{r}}{\partial u} = h_u\hat{\mathbf{u}}, \qquad \frac{\partial \mathbf{r}}{\partial v} = h_v\hat{\mathbf{v}}, \qquad \text{and} \qquad \frac{\partial \mathbf{r}}{\partial w} = h_w\hat{\mathbf{w}}.$$

The basis vectors $\hat{\mathbf{u}}$, $\hat{\mathbf{v}}$, and $\hat{\mathbf{w}}$ will form a right-handed triad provided we have chosen a suitable order for the coordinates u, v, and w.

EXAMPLE 5 For cylindrical coordinates we have $\mathbf{r} = r\cos\theta\mathbf{i} + r\sin\theta\mathbf{j} + z\mathbf{k}$, so

$$\frac{\partial \mathbf{r}}{\partial r} = \cos\theta\,\mathbf{i} + \sin\theta\,\mathbf{j}, \qquad \frac{\partial \mathbf{r}}{\partial \theta} = -r\sin\theta\,\mathbf{i} + r\cos\theta\,\mathbf{j}, \qquad \text{and} \qquad \frac{\partial \mathbf{r}}{\partial z} = \mathbf{k}.$$

Thus, the scale factors for the cylindrical coordinate system are given by

$$h_r = \left|\frac{\partial \mathbf{r}}{\partial r}\right| = 1, \quad h_\theta = \left|\frac{\partial \mathbf{r}}{\partial \theta}\right| = r, \quad \text{and} \quad h_z = \left|\frac{\partial \mathbf{r}}{\partial z}\right| = 1,$$

and the local basis consists of the vectors

$$\hat{\mathbf{r}} = \cos\theta\,\mathbf{i} + \sin\theta\,\mathbf{j}, \qquad \hat{\boldsymbol{\theta}} = -\sin\theta\,\mathbf{i} + \cos\theta\,\mathbf{j}, \qquad \hat{\mathbf{z}} = \mathbf{k}.$$

See Figure 16.22. The local basis is right-handed.

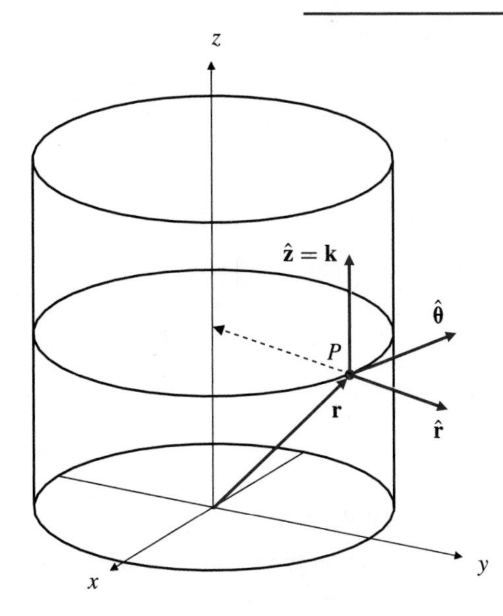

Figure 16.22 The local basis for cylindrical coordinates

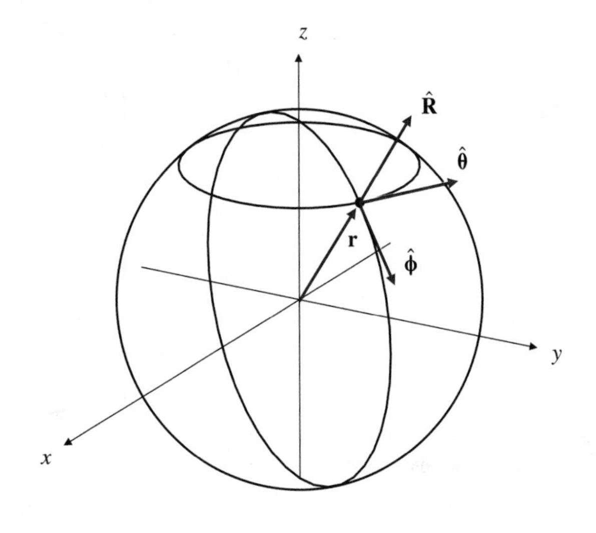

Figure 16.23 The local basis for spherical coordinates

EXAMPLE 6 For spherical coordinates we have

$$\mathbf{r} = R\sin\phi\cos\theta\mathbf{i} + R\sin\phi\sin\theta\mathbf{j} + R\cos\phi\mathbf{k}.$$

Thus, the tangent vectors to the coordinate curves are

$$\frac{\partial \mathbf{r}}{\partial R} = \sin\phi\cos\theta\,\mathbf{i} + \sin\phi\sin\theta\,\mathbf{j} + \cos\phi\,\mathbf{k},$$
$$\frac{\partial \mathbf{r}}{\partial \phi} = R\cos\phi\cos\theta\,\mathbf{i} + R\cos\phi\sin\theta\,\mathbf{j} - R\sin\phi\,\mathbf{k},$$
$$\frac{\partial \mathbf{r}}{\partial \theta} = -R\sin\phi\sin\theta\,\mathbf{i} + R\sin\phi\cos\theta\,\mathbf{j},$$

and the scale factors are given by

$$h_R = \left|\frac{\partial \mathbf{r}}{\partial R}\right| = 1, \quad h_\phi = \left|\frac{\partial \mathbf{r}}{\partial \phi}\right| = R, \quad \text{and} \quad h_\theta = \left|\frac{\partial \mathbf{r}}{\partial \theta}\right| = R\sin\phi.$$

The local basis consists of the vectors

$$\hat{\mathbf{R}} = \sin\phi\cos\theta\,\mathbf{i} + \sin\phi\sin\theta\,\mathbf{j} + \cos\phi\,\mathbf{k}$$
$$\hat{\boldsymbol{\phi}} = \cos\phi\cos\theta\,\mathbf{i} + \cos\phi\sin\theta\,\mathbf{j} - \sin\phi\,\mathbf{k}$$
$$\hat{\boldsymbol{\theta}} = -\sin\theta\,\mathbf{i} + \cos\theta\,\mathbf{j}.$$

See Figure 16.23. The local basis is right-handed.

The volume element in an orthogonal curvilinear coordinate system is the volume of an infinitesimal *coordinate box* bounded by pairs of u-, v-, and w-surfaces corresponding to values u and $u+du$, v and $v+dv$, and w and $w+dw$, respectively. See Figure 16.24. Since these coordinate surfaces are assumed smooth, and since they intersect at right angles, the coordinate box is rectangular, and is spanned by the vectors

$$\frac{\partial \mathbf{r}}{\partial u}\,du = h_u\,du\,\hat{\mathbf{u}}, \quad \frac{\partial \mathbf{r}}{\partial v}\,dv = h_v\,dv\,\hat{\mathbf{v}}, \quad \text{and} \quad \frac{\partial \mathbf{r}}{\partial w}\,dw = h_w\,dw\,\hat{\mathbf{w}}.$$

Therefore, the volume element is given by

$$dV = h_u h_v h_w \, du \, dv \, dw.$$

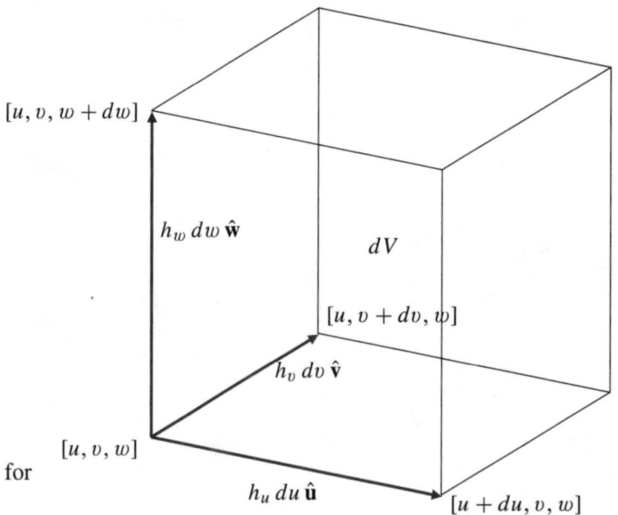

Figure 16.24 The volume element for orthogonal curvilinear coordinates

Furthermore, the surface area elements on the u-, v-, and w-surfaces are the areas of the appropriate faces of the coordinate box:

Area elements on coordinate surfaces

$$dS_u = h_v h_w \, dv \, dw, \qquad dS_v = h_u h_w \, du \, dw, \qquad dS_w = h_u h_v \, du \, dv.$$

The arc length elements along the u-, v-, and w-coordinate curves are the edges of the coordinate box:

Arc length elements on coordinate curves

$$ds_u = h_u \, du, \qquad ds_v = h_v \, dv, \qquad ds_w = h_w \, dw.$$

EXAMPLE 7 For cylindrical coordinates, the volume element, as shown in Section 14.6, is

$$dV = h_r h_\theta h_z \, dr \, d\theta \, dz = r \, dr \, d\theta \, dz.$$

The surface area elements on the cylinder r = constant, the half-plane θ = constant, and the plane z = constant are, respectively,

$$dS_r = r \, d\theta \, dz, \qquad dS_\theta = dr \, dz, \qquad \text{and} \qquad dS_z = r \, dr \, d\theta.$$

EXAMPLE 8 For spherical coordinates, the volume element, as developed in Section 14.6, is

$$dV = h_R h_\phi h_\theta \, dR \, d\phi \, d\theta = R^2 \sin\phi \, dR \, d\phi \, d\theta.$$

The area element on the sphere R = constant is

$$dS_R = h_\phi h_\theta \, d\phi \, d\theta = R^2 \sin\phi \, d\phi \, d\theta.$$

The area element on the cone ϕ = constant is

$$dS_\phi = h_R h_\theta \, dR \, d\theta = R \sin\phi \, dR \, d\theta.$$

The area element on the half-plane θ = constant is

$$dS_\theta = h_R h_\phi \, dR \, d\phi = R \, dR \, d\phi.$$

Grad, Div, and Curl in Orthogonal Curvilinear Coordinates

The gradient ∇f of a scalar field f can be expressed in terms of the local basis at any point P with curvilinear coordinates $[u, v, w]$ in the form

$$\nabla f = f_u \hat{\mathbf{u}} + f_v \hat{\mathbf{v}} + f_w \hat{\mathbf{w}}.$$

In order to determine the coefficients f_u, f_v, and f_w in this formula, we will compare two expressions for the directional derivative of f along an arbitrary curve in xyz-space.

If the curve $\mathcal{C}$ has parametrization $\mathbf{r} = \mathbf{r}(s)$ in terms of arc length, then the directional derivative of f along $\mathcal{C}$ is given by

$$\frac{df}{ds} = \frac{\partial f}{\partial u}\frac{du}{ds} + \frac{\partial f}{\partial v}\frac{dv}{ds} + \frac{\partial f}{\partial w}\frac{dw}{ds}.$$

On the other hand, this directional derivative is also given by $df/ds = \nabla f \bullet \hat{\mathbf{T}}$, where $\hat{\mathbf{T}}$ is the unit tangent vector to $\mathcal{C}$. We have

$$\begin{aligned}\hat{\mathbf{T}} = \frac{d\mathbf{r}}{ds} &= \frac{\partial \mathbf{r}}{\partial u}\frac{du}{ds} + \frac{\partial \mathbf{r}}{\partial v}\frac{dv}{ds} + \frac{\partial \mathbf{r}}{\partial w}\frac{dw}{ds} \\ &= h_u \frac{du}{ds}\hat{\mathbf{u}} + h_v \frac{dv}{ds}\hat{\mathbf{v}} + h_w \frac{dw}{ds}\hat{\mathbf{w}}.\end{aligned}$$

Thus,

$$\frac{df}{ds} = \nabla f \bullet \hat{\mathbf{T}} = f_u h_u \frac{du}{ds} + f_v h_v \frac{dv}{ds} + f_w h_w \frac{dw}{ds}.$$

Comparing these two expressions for df/ds along $\mathcal{C}$, we see that

$$f_u h_u = \frac{\partial f}{\partial u}, \qquad f_v h_v = \frac{\partial f}{\partial v}, \qquad f_w h_w = \frac{\partial f}{\partial w}.$$

Therefore, we have shown that

The gradient in orthogonal curvilinear coordinates

$$\nabla f = \frac{1}{h_u}\frac{\partial f}{\partial u}\hat{\mathbf{u}} + \frac{1}{h_v}\frac{\partial f}{\partial v}\hat{\mathbf{v}} + \frac{1}{h_w}\frac{\partial f}{\partial w}\hat{\mathbf{w}}.$$

EXAMPLE 9 In terms of cylindrical coordinates, the gradient of the scalar field $f(r, \theta, z)$ is

$$\nabla f(r, \theta, z) = \frac{\partial f}{\partial r}\hat{\mathbf{r}} + \frac{1}{r}\frac{\partial f}{\partial \theta}\hat{\boldsymbol{\theta}} + \frac{\partial f}{\partial z}\mathbf{k}.$$

EXAMPLE 10 In terms of spherical coordinates, the gradient of the scalar field $f(R, \phi, \theta)$ is

$$\nabla f(R, \phi, \theta) = \frac{\partial f}{\partial R}\hat{\mathbf{R}} + \frac{1}{R}\frac{\partial f}{\partial \phi}\hat{\boldsymbol{\phi}} + \frac{1}{R\sin\phi}\frac{\partial f}{\partial \theta}\hat{\boldsymbol{\theta}}.$$

Now consider a vector field $\mathbf{F}$ expressed in terms of the curvilinear coordinates:

$$\mathbf{F}(u, v, w) = F_u(u, v, w)\hat{\mathbf{u}} + F_v(u, v, w)\hat{\mathbf{v}} + F_w(u, v, w)\hat{\mathbf{w}}.$$

The flux of **F** out of the infinitesimal coordinate box of Figure 16.24 is the sum of the fluxes of **F** out of the three pairs of opposite surfaces of the box. The flux out of the u-surfaces corresponding to u and $u + du$ is

$$\begin{aligned}
&\mathbf{F}(u+du, v, w) \bullet \hat{\mathbf{u}}\, dS_u - \mathbf{F}(u, v, w) \bullet \hat{\mathbf{u}}\, dS_u \\
&\quad = \big(F_u(u+du, v, w) h_v(u+du, v, w) h_w(u+du, v, w) \\
&\qquad - F_u(u, v, w) h_v(u, v, w) h_w(u, v, w)\big)\, dv\, dw \\
&\quad = \frac{\partial}{\partial u}\big(h_v h_w F_u\big)\, du\, dv\, dw.
\end{aligned}$$

Similar expressions hold for the fluxes out of the other pairs of coordinate surfaces.

The divergence at P of **F** is the flux *per unit volume* out of the infinitesimal coordinate box at P. Thus, it is given by

The divergence in orthogonal curvilinear coordinates

$$\begin{aligned}
\operatorname{div} \mathbf{F}(u, v, w) = \frac{1}{h_u h_v h_w} \bigg[&\frac{\partial}{\partial u}\big(h_v h_w F_u(u, v, w)\big) \\
&+ \frac{\partial}{\partial v}\big(h_u h_w F_v(u, v, w)\big) + \frac{\partial}{\partial w}\big(h_u h_v F_w(u, v, w)\big) \bigg].
\end{aligned}$$

EXAMPLE 11 For cylindrical coordinates, $h_r = h_z = 1$, and $h_\theta = r$. Thus, the divergence of $\mathbf{F} = F_r \hat{\mathbf{r}} + F_\theta \hat{\boldsymbol{\theta}} + F_z \mathbf{k}$ is

$$\begin{aligned}
\operatorname{div} \mathbf{F} &= \frac{1}{r}\left[\frac{\partial}{\partial r}(r F_r) + \frac{\partial}{\partial \theta} F_\theta + \frac{\partial}{\partial z}(r F_z)\right] \\
&= \frac{\partial F_r}{\partial r} + \frac{1}{r} F_r + \frac{1}{r}\frac{\partial F_\theta}{\partial \theta} + \frac{\partial F_z}{\partial z}.
\end{aligned}$$

EXAMPLE 12 For spherical coordinates, $h_R = 1$, $h_\phi = R$, and $h_\theta = R \sin\phi$. The divergence of the vector field $\mathbf{F} = F_R \hat{\mathbf{R}} + F_\phi \hat{\boldsymbol{\phi}} + F_\theta \hat{\boldsymbol{\theta}}$ is

$$\begin{aligned}
\operatorname{div} \mathbf{F} &= \frac{1}{R^2 \sin\phi}\left[\frac{\partial}{\partial R}(R^2 \sin\phi\, F_R) + \frac{\partial}{\partial \phi}(R \sin\phi\, F_\phi) + \frac{\partial}{\partial \theta}(R\, F_\theta)\right] \\
&= \frac{1}{R^2}\frac{\partial}{\partial R}(R^2 F_R) + \frac{1}{R \sin\phi}\frac{\partial}{\partial \phi}(\sin\phi\, F_\phi) + \frac{1}{R \sin\phi}\frac{\partial F_\theta}{\partial \theta} \\
&= \frac{\partial F_R}{\partial R} + \frac{2}{R} F_R + \frac{1}{R}\frac{\partial F_\phi}{\partial \phi} + \frac{\cot\phi}{R} F_\phi + \frac{1}{R \sin\phi}\frac{\partial F_\theta}{\partial \theta}.
\end{aligned}$$

To calculate the curl of a vector field expressed in terms of orthogonal curvilinear coordinates we can make use of some previously obtained vector identities. First, observe that the gradient of the scalar field $f(u, v, w) = u$ is $\hat{\mathbf{u}}/h_u$, so that $\hat{\mathbf{u}} = h_u \nabla u$. Similarly, $\hat{\mathbf{v}} = h_v \nabla v$ and $\hat{\mathbf{w}} = h_w \nabla w$. Therefore, the vector field

$$\mathbf{F} = F_u \hat{\mathbf{u}} + F_v \hat{\mathbf{v}} + F_w \hat{\mathbf{w}}$$

can be written in the form

$$\mathbf{F} = F_u h_u \nabla u + F_v h_v \nabla v + F_w h_w \nabla w.$$

Using the identity **curl** $(f\nabla g) = \nabla f \times \nabla g$ (see Exercise 13 of Section 16.2), we can calculate the curl of each term in the expression above. We have

$$\begin{aligned}\mathbf{curl}\,(F_u h_u \nabla u) &= \nabla(F_u h_u) \times \nabla u \\ &= \left[\frac{1}{h_u}\frac{\partial}{\partial u}(F_u h_u)\hat{\mathbf{u}} + \frac{1}{h_v}\frac{\partial}{\partial v}(F_u h_u)\hat{\mathbf{v}} + \frac{1}{h_w}\frac{\partial}{\partial w}(F_u h_u)\hat{\mathbf{w}}\right] \times \frac{\hat{\mathbf{u}}}{h_u} \\ &= \frac{1}{h_u h_w}\frac{\partial}{\partial w}(F_u h_u)\hat{\mathbf{v}} - \frac{1}{h_u h_v}\frac{\partial}{\partial v}(F_u h_u)\hat{\mathbf{w}} \\ &= \frac{1}{h_u h_v h_w}\left[\frac{\partial}{\partial w}(F_u h_u)(h_v\hat{\mathbf{v}}) - \frac{\partial}{\partial v}(F_u h_u)(h_w\hat{\mathbf{w}})\right].\end{aligned}$$

We have used the facts that $\hat{\mathbf{u}} \times \hat{\mathbf{u}} = \mathbf{0}$, $\hat{\mathbf{v}} \times \hat{\mathbf{u}} = -\hat{\mathbf{w}}$, and $\hat{\mathbf{w}} \times \hat{\mathbf{u}} = \hat{\mathbf{v}}$ to obtain the result above. This is why we assumed that the curvilinear coordinate system was right-handed.

Corresponding expressions can be calculated for the other two terms in the formula for **curl F**. Combining the three terms, we conclude that the curl of

$$\mathbf{F} = F_u\hat{\mathbf{u}} + F_v\hat{\mathbf{v}} + F_w\hat{\mathbf{w}}$$

is given by

The curl in orthogonal curvilinear coordinates

$$\mathbf{curl}\,\mathbf{F}(u, v, w) = \frac{1}{h_u h_v h_w}\begin{vmatrix} h_u\hat{\mathbf{u}} & h_v\hat{\mathbf{v}} & h_w\hat{\mathbf{w}} \\ \dfrac{\partial}{\partial u} & \dfrac{\partial}{\partial v} & \dfrac{\partial}{\partial w} \\ F_u h_u & F_v h_v & F_w h_w \end{vmatrix}.$$

EXAMPLE 13 For cylindrical coordinates, the curl of $\mathbf{F} = F_r\hat{\mathbf{r}} + F_\theta\hat{\boldsymbol{\theta}} + F_z\mathbf{k}$ is given by

$$\begin{aligned}\mathbf{curl}\,\mathbf{F} &= \frac{1}{r}\begin{vmatrix} \hat{\mathbf{r}} & r\hat{\boldsymbol{\theta}} & \mathbf{k} \\ \dfrac{\partial}{\partial r} & \dfrac{\partial}{\partial \theta} & \dfrac{\partial}{\partial z} \\ F_r & rF_\theta & F_z \end{vmatrix} \\ &= \left(\frac{1}{r}\frac{\partial F_z}{\partial \theta} - \frac{\partial F_\theta}{\partial z}\right)\hat{\mathbf{r}} + \left(\frac{\partial F_r}{\partial z} - \frac{\partial F_z}{\partial r}\right)\hat{\boldsymbol{\theta}} + \left(\frac{\partial F_\theta}{\partial r} + \frac{F_\theta}{r} - \frac{1}{r}\frac{\partial F_r}{\partial \theta}\right)\mathbf{k}.\end{aligned}$$

EXAMPLE 14 For spherical coordinates, the curl of $\mathbf{F} = F_R\hat{\mathbf{R}} + F_\phi\hat{\boldsymbol{\phi}} + F_\theta\hat{\boldsymbol{\theta}}$ is given by

$$\begin{aligned}\mathbf{curl}\,\mathbf{F} &= \frac{1}{R^2\sin\phi}\begin{vmatrix} \hat{\mathbf{R}} & R\hat{\boldsymbol{\phi}} & R\sin\phi\hat{\boldsymbol{\theta}} \\ \dfrac{\partial}{\partial R} & \dfrac{\partial}{\partial \phi} & \dfrac{\partial}{\partial \theta} \\ F_R & RF_\phi & R\sin\phi F_\theta \end{vmatrix} \\ &= \frac{1}{R\sin\phi}\left[\frac{\partial}{\partial \phi}(\sin\phi F_\theta) - \frac{\partial F_\phi}{\partial \theta}\right]\hat{\mathbf{R}} \\ &\quad + \frac{1}{R\sin\phi}\left[\frac{\partial F_R}{\partial \theta} - \sin\phi\frac{\partial}{\partial R}(RF_\theta)\right]\hat{\boldsymbol{\phi}} \\ &\quad + \frac{1}{R}\left[\frac{\partial}{\partial R}(RF_\phi) - \frac{\partial F_R}{\partial \phi}\right]\hat{\boldsymbol{\theta}}\end{aligned}$$

$$= \frac{1}{R\sin\phi}\left[(\cos\phi)F_\theta + (\sin\phi)\frac{\partial F_\theta}{\partial \phi} - \frac{\partial F_\phi}{\partial \theta}\right]\hat{\mathbf{R}}$$
$$+ \frac{1}{R\sin\phi}\left[\frac{\partial F_R}{\partial \theta} - (\sin\phi)F_\theta - (R\sin\phi)\frac{\partial F_\theta}{\partial R}\right]\hat{\boldsymbol{\phi}}$$
$$+ \frac{1}{R}\left[F_\phi + R\frac{\partial F_\phi}{\partial R} - \frac{\partial F_R}{\partial \phi}\right]\hat{\boldsymbol{\theta}}.$$

EXERCISES 16.7

In Exercises 1–2, calculate the gradients of the given scalar fields expressed in terms of cylindrical or spherical coordinates.

1. $f(r,\theta,z) = r\theta z$ **2.** $f(R,\phi,\theta) = R\phi\theta$

In Exercises 3–8, calculate **div F** and **curl F** for the given vector fields expressed in terms of cylindrical coordinates or spherical coordinates.

3. $\mathbf{F}(r,\theta,z) = r\hat{\mathbf{r}}$ **4.** $\mathbf{F}(r,\theta,z) = r\hat{\boldsymbol{\theta}}$

5. $\mathbf{F}(R,\phi,\theta) = \sin\phi\hat{\mathbf{R}}$ **6.** $\mathbf{F}(R,\phi,\theta) = R\hat{\boldsymbol{\phi}}$

7. $\mathbf{F}(R,\phi,\theta) = R\hat{\boldsymbol{\theta}}$ **8.** $\mathbf{F}(R,\phi,\theta) = R^2\hat{\mathbf{R}}$

9. Let $x = x(u,v)$, $y = y(u,v)$ define orthogonal curvilinear coordinates (u,v) in the xy-plane. Find the scale factors, local basis vectors, and area element for the system of coordinates (u,v).

10. Continuing Exercise 9, express the gradient of a scalar field $f(u,v)$ and the divergence and curl of a vector field $\mathbf{F}(u,v)$ in terms of the curvilinear coordinates.

11. Express the gradient of the scalar field $f(r,\theta)$ and the divergence and curl of a vector field $\mathbf{F}(r,\theta)$ in terms of plane polar coordinates (r,θ).

12. The transformation $x = a\cosh u\cos v$, $y = a\sinh u\sin v$ defines **elliptical coordinates** in the xy-plane. This coordinate system has singular points at $x = \pm a$, $y = 0$.

(a) Show that the v-curves, u = constant, are ellipses with foci at the singular points.

(b) Show that the u-curves, v = constant, are hyperbolas with foci at the singular points.

(c) Show that the u-curve and the v-curve through a nonsingular point intersect at right angles.

(d) Find the scale factors h_u and h_v and the area element dA for the elliptical coordinate system.

13. Describe the coordinate surfaces and coordinate curves of the system of elliptical cylindrical coordinates in xyz-space defined by

$$x = a\cosh u\cos v, \qquad y = a\sinh u\sin v, \qquad z = z.$$

14. The Laplacian $\nabla^2 f$ of a scalar field f can be calculated as $\mathbf{div}\,\nabla f$. Use this method to calculate the Laplacian of the function $f(r,\theta,z)$ expressed in terms of cylindrical coordinates. (This repeats Exercise 19 of Section 14.6.)

15. Calculate the Laplacian $\nabla^2 f = \mathbf{div}\,\nabla f$ for the function $f(R,\phi,\theta)$, expressed in terms of spherical coordinates. (This repeats Exercise 20 of Section 14.6 but is now much easier.)

16. Calculate the Laplacian $\nabla^2 f = \mathbf{div}\,\nabla f$ for a function $f(u,v,w)$ expressed in terms of arbitrary orthogonal curvilinear coordinates (u,v,w).

CHAPTER REVIEW

Key Ideas

- **What do the following terms mean?**
  - ◇ the divergence of a vector field **F**
  - ◇ the curl of a vector field **F**
  - ◇ **F** is solenoidal
  - ◇ **F** is irrotational
  - ◇ a scalar potential
  - ◇ a vector potential
  - ◇ orthogonal curvilinear coordinates
- **State the following theorems:**
  - ◇ the Divergence Theorem
  - ◇ Green's Theorem
  - ◇ Stokes's Theorem

Review Exercises

1. If $\mathbf{F} = x^2z\mathbf{i} + (y^2z + 3y)\mathbf{j} + x^2\mathbf{k}$, find the flux of **F** across the part of the ellipsoid $x^2 + y^2 + 4z^2 = 16$, where $z \ge 0$, oriented with upward normal.

2. Let $\mathcal{S}$ be the part of the cylinder $x^2 + y^2 = 2ax$ between the horizontal planes $z = 0$ and $z = b$, where $b > 0$. Find the flux of $\mathbf{F} = x\mathbf{i} + \cos(z^2)\mathbf{j} + e^z\mathbf{k}$ outward through $\mathcal{S}$.

3. Find $\oint_{\mathcal{C}} (3y^2 + 2xe^{y^2})\,dx + (2x^2ye^{y^2})\,dy$ counterclockwise around the boundary of the parallelogram with vertices (0, 0), (2, 0), (3, 1), and (1, 1).

4. If $\mathbf{F} = -z\mathbf{i}+x\mathbf{j}+y\mathbf{k}$, what are the possible values of $\oint_{\mathcal{C}} \mathbf{F}\bullet d\mathbf{r}$ around circles of radius a in the plane $2x + y + 2z = 7$?

5. Let $\mathbf{F}$ be a smooth vector field in 3-space and suppose that, for every $a > 0$, the flux of $\mathbf{F}$ out of the sphere of radius a centred at the origin is $\pi(a^3 + 2a^4)$. Find the divergence of $\mathbf{F}$ at the origin.

6. Let $\mathbf{F} = -y\mathbf{i} + x\cos(1 - x^2 - y^2)\mathbf{j} + yz\mathbf{k}$. Find the flux of **curl F** upward through a surface whose boundary is the curve $x^2 + y^2 = 1$, $z = 2$.

7. Let $\mathbf{F}(\mathbf{r}) = r^{\lambda}\mathbf{r}$, where $\mathbf{r} = x\mathbf{i}+y\mathbf{j}+z\mathbf{k}$ and $r = |\mathbf{r}|$. For what value(s) of λ is $\mathbf{F}$ solenoidal on an open subset of 3-space? Is $\mathbf{F}$ solenoidal on all of 3-space for any value of λ?

8. Given that $\mathbf{F}$ satisfies $\mathbf{curl}\,\mathbf{F} = \mu\mathbf{F}$ on 3-space, where μ is a nonzero constant, show that $\nabla^2\mathbf{F} + \mu^2\mathbf{F} = \mathbf{0}$.

9. Let P be a polyhedron in 3-space having n planar faces, F_1, $F_2, \ldots, F_n$. Let $\mathbf{N}_i$ be normal to F_i in the direction outward from P, and let $\mathbf{N}_i$ have length equal to the area of face F_i. Show that

$$\sum_{i=1}^{n} \mathbf{N}_i = \mathbf{0}.$$

Also, state a version of this result for a plane polygon P.

10. Around what simple, closed curve C in the xy-plane does the vector field

$$\mathbf{F} = (2y^3 - 3y + xy^2)\mathbf{i} + (x - x^3 + x^2y)\mathbf{j}$$

have the greatest circulation?

11. Through what closed, oriented surface in $\mathbb{R}^3$ does the vector field

$$\mathbf{F} = (4x + 2x^3z)\mathbf{i} - y(x^2 + z^2)\mathbf{j} - (3x^2z^2 + 4y^2z)\mathbf{k}$$

have the greatest flux?

12. Find the maximum value of

$$\oint_C \mathbf{F}\bullet d\mathbf{r},$$

where $\mathbf{F} = xy^2\mathbf{i} + (3z - xy^2)\mathbf{j} + (4y - x^2y)\mathbf{k}$, and C is a simple closed curve in the plane $x + y + z = 1$ oriented counterclockwise as seen from high on the z-axis. What curve C gives this maximum?

Challenging Problems

1. **(The expanding universe)** Let $\mathbf{v}$ be the large-scale velocity field of matter in the universe. (*Large-scale* means on the scale of intergalactic distances; *small-scale* motion such as that of planetary systems about their suns, and even stars about galactic centres, has been averaged out.) Assume that $\mathbf{v}$ is a smooth vector field. According to present astronomical theory, the distance between any two points is increasing, and the rate of increase is proportional to the distance between the points. The constant of proportionality, C, is called *Hubble's constant.* In terms of $\mathbf{v}$, if $\mathbf{r}_1$ and $\mathbf{r}_2$ are two points, then

$$\big(\mathbf{v}(\mathbf{r}_2) - \mathbf{v}(\mathbf{r}_1)\big)\bullet(\mathbf{r}_2 - \mathbf{r}_1) = C|\mathbf{r}_2 - \mathbf{r}_1|^2.$$

Show that **div v** is constant, and find the value of the constant in terms of Hubble's constant. *Hint:* Find the flux of $\mathbf{v}(\mathbf{r})$ out of a sphere of radius ϵ centred at $\mathbf{r}_1$ and take the limit as ϵ approaches zero.

2. **(Solid angle)** Two rays from a point P determine an angle at P whose measure in radians is equal to the length of the arc of the circle of radius 1 with centre at P lying between the two rays. Similarly, an arbitrarily shaped half-cone K with vertex at P determines a **solid angle** at P whose measure in **steradians** (*stereo* + *radians*) is the area of that part of the sphere of radius 1 with centre at P lying within K. For example, the first octant of $\mathbb{R}^3$ is a half-cone with vertex at the origin. It determines a solid angle at the origin measuring

$$4\pi \times \frac{1}{8} = \frac{\pi}{2} \text{ steradians},$$

since the area of the unit sphere is 4π. (See Figure 16.25.)

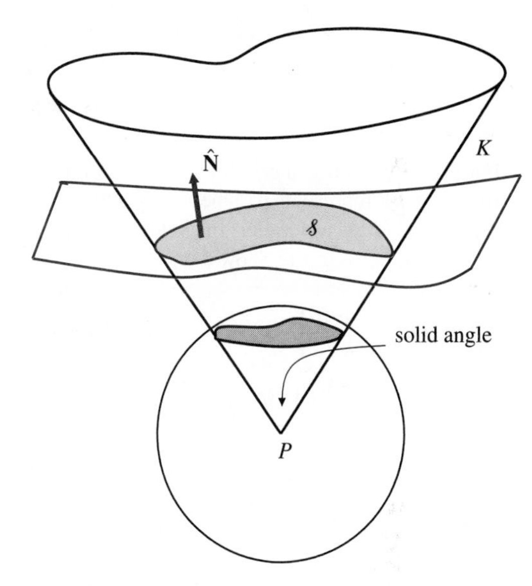

Figure 16.25

(a) Find the steradian measure of the solid angle at the vertex of a right-circular half-cone whose generators make angle α with its central axis.

(b) If a smooth, oriented surface intersects the general half-cone K but not at its vertex P, let $\mathcal{S}$ be the part of the surface lying within K. Orient $\mathcal{S}$ with normal pointing away from P. Show that the steradian measure of the solid angle at P determined by K is the flux of $\mathbf{r}/|\mathbf{r}|^3$ through $\mathcal{S}$, where $\mathbf{r}$ is the vector from P to the point (x, y, z).

Integrals over moving domains

By the Fundamental Theorem of Calculus, the derivative with respect to time t of an integral of $f(x, t)$ over a "moving interval" $[a(t), b(t)]$ is given by

$$\frac{d}{dt}\int_{a(t)}^{b(t)} f(x,t)\,dx = \int_{a(t)}^{b(t)} \frac{\partial}{\partial t} f(x,t)\,dx + f(b(t),t)\frac{db}{dt} - f(a(t),t)\frac{da}{dt}.$$

The next three problems, suggested by Luigi Quartapelle of the Politecnico di Milano, provide various extensions of this one-dimensional result to higher dimensions. The calculations are somewhat lengthy, so you may want to get some help from Maple or another computer algebra system.

3. (Rate of change of circulation along a moving curve)

(a) Let $\mathbf{F}(\mathbf{r}, t)$ be a smooth vector field in $\mathbb{R}^3$ depending on a parameter t, and let

$$\mathbf{G}(s,t) = \mathbf{F}\Big(\mathbf{r}(s,t), t\Big) = \mathbf{F}\Big(x(s,t), y(s,t), z(x,t), t\Big),$$

where $\mathbf{r}(s,t) = x(s,t)\mathbf{i} + y(s,t)\mathbf{j} + z(s,t)\mathbf{k}$ has continuous partial derivatives of second order. Show that

$$\frac{\partial}{\partial t}\left(\mathbf{G} \bullet \frac{\partial \mathbf{r}}{\partial s}\right) - \frac{\partial}{\partial s}\left(\mathbf{G} \bullet \frac{\partial \mathbf{r}}{\partial t}\right) = \frac{\partial \mathbf{F}}{\partial t} \bullet \frac{\partial \mathbf{r}}{\partial s} + \left((\nabla \times \mathbf{F}) \times \frac{\partial \mathbf{r}}{\partial t}\right) \bullet \frac{\partial \mathbf{r}}{\partial s}.$$

Here, the curl $\nabla \times \mathbf{F}$ is taken with respect to the position vector $\mathbf{r}$.

(b) For fixed t (which you can think of as time), $\mathbf{r} = \mathbf{r}(s,t)$, $(a \le s \le b)$, represents parametrically a curve C_t in $\mathbb{R}^3$. The curve moves as t varies; the velocity of any point on C_t is $\mathbf{v}_C(s,t) = \partial\mathbf{r}/\partial t$. Show that

$$\frac{d}{dt}\int_{C_t} \mathbf{F} \bullet d\mathbf{r} = \int_{C_t} \frac{\partial \mathbf{F}}{\partial t} \bullet d\mathbf{r} + \int_{C_t} \Big((\nabla \times \mathbf{F}) \times \mathbf{v}_C\Big) \bullet d\mathbf{r} + \mathbf{F}\Big(\mathbf{r}(b,t), t\Big) \bullet \mathbf{v}_C(b,t) - \mathbf{F}\Big(\mathbf{r}(a,t), t\Big) \bullet \mathbf{v}_C(a,t).$$

Hint: Write

$$\begin{aligned}\frac{d}{dt}\int_{C_t} \mathbf{F} \bullet d\mathbf{r} &= \int_a^b \frac{\partial}{\partial t}\left(\mathbf{G} \bullet \frac{\partial \mathbf{r}}{\partial s}\right) ds \\ &= \int_a^b \left[\frac{\partial}{\partial s}\left(\mathbf{G} \bullet \frac{\partial \mathbf{r}}{\partial t}\right) + \left(\frac{\partial}{\partial t}\left(\mathbf{G} \bullet \frac{\partial \mathbf{r}}{\partial s}\right) - \frac{\partial}{\partial s}\left(\mathbf{G} \bullet \frac{\partial \mathbf{r}}{\partial t}\right)\right)\right] ds.\end{aligned}$$

Now use the result of (a).

4. (Rate of change of flux through a moving surface) Let S_t be a moving surface in $\mathbb{R}^3$ smoothly parametrized (for each t) by

$$\mathbf{r} = \mathbf{r}(u, v, t) = x(u,v,t)\mathbf{i} + y(u,v,t)\mathbf{j} + z(u,v,t)\mathbf{k},$$

where (u, v) belongs to a parameter region R in the uv-plane. Let $\mathbf{F}(\mathbf{r}, t) = F_1\mathbf{i} + F_2\mathbf{j} + F_3\mathbf{k}$ be a smooth 3-vector function, and let $\mathbf{G}(u, v, t) = \mathbf{F}(\mathbf{r}(u, v, t), t)$.

(a) Show that

$$\begin{aligned}&\frac{\partial}{\partial t}\left(\mathbf{G} \bullet \left[\frac{\partial \mathbf{r}}{\partial u} \times \frac{\partial \mathbf{r}}{\partial v}\right]\right) - \frac{\partial}{\partial u}\left(\mathbf{G} \bullet \left[\frac{\partial \mathbf{r}}{\partial t} \times \frac{\partial \mathbf{r}}{\partial v}\right]\right) - \frac{\partial}{\partial v}\left(\mathbf{G} \bullet \left[\frac{\partial \mathbf{r}}{\partial u} \times \frac{\partial \mathbf{r}}{\partial t}\right]\right) \\ &= \frac{\partial \mathbf{F}}{\partial t} \bullet \left[\frac{\partial \mathbf{r}}{\partial u} \times \frac{\partial \mathbf{r}}{\partial v}\right] + (\nabla \bullet \mathbf{F})\frac{\partial \mathbf{r}}{\partial t} \bullet \left[\frac{\partial \mathbf{r}}{\partial u} \times \frac{\partial \mathbf{r}}{\partial v}\right].\end{aligned}$$

(b) If C_t is the boundary of S_t with orientation corresponding to that of S_t, use Green's Theorem to show that

$$\begin{aligned}&\iint_R \left[\frac{\partial}{\partial u}\left(\mathbf{G} \bullet \left[\frac{\partial \mathbf{r}}{\partial t} \times \frac{\partial \mathbf{r}}{\partial v}\right]\right) + \frac{\partial}{\partial v}\left(\mathbf{G} \bullet \left[\frac{\partial \mathbf{r}}{\partial u} \times \frac{\partial \mathbf{r}}{\partial t}\right]\right)\right] du\, dv \\ &= \oint_{C_t} \left(\mathbf{F} \times \frac{\partial \mathbf{r}}{\partial t}\right) \bullet d\mathbf{r}.\end{aligned}$$

(c) Combine the results of (a) and (b) to show that

$$\begin{aligned}&\frac{d}{dt}\iint_{S_t} \mathbf{F} \bullet \hat{\mathbf{N}}\, dS \\ &= \iint_{S_t} \frac{\partial \mathbf{F}}{\partial t} \bullet \hat{\mathbf{N}}\, dS + \iint_{S_t} (\nabla \bullet \mathbf{F})\mathbf{v}_S \bullet \hat{\mathbf{N}}\, dS \\ &\quad + \oint_{C_t} (\mathbf{F} \times \mathbf{v}_C) \bullet d\mathbf{r},\end{aligned}$$

where $\mathbf{v}_S = \partial\mathbf{r}/\partial t$ on S_t is the velocity of S_t, $\mathbf{v}_C = \partial\mathbf{r}/\partial t$ on C_t is the velocity of C_t, and $\hat{\mathbf{N}}$ is the unit normal field on S_t corresponding to its orientation.

5. (Rate of change of integrals over moving volumes) Let S_t be the position at time t of a smooth, closed surface in $\mathbb{R}^3$ that varies smoothly with t and bounds at any time t a region D_t. If $\hat{\mathbf{N}}(\mathbf{r}, t)$ denotes the unit outward (from D_t) normal field on S_t, and $\mathbf{v}_S(\mathbf{r}, t)$ is the velocity of the point $\mathbf{r}$ on S_t at time t, show that

$$\frac{d}{dt}\iiint_{D_t} f\, dV = \iiint_{D_t} \frac{\partial f}{\partial t}\, dV + \oiint_{S_t} f\mathbf{v}_S \bullet \hat{\mathbf{N}}\, dS$$

holds for smooth functions $f(\mathbf{r}, t)$. *Hint:* Let ΔD_t consist of the points through which S_t passes as t increases to $t + \Delta t$. The volume element dV in ΔD_t can be expressed in terms of the area element dS on S_t by

$$dV = \mathbf{v} \bullet \hat{\mathbf{N}}\, dS\, \Delta t.$$

Show that

$$\begin{aligned}&\frac{1}{\Delta t}\left[\iiint_{D_{t+\Delta t}} f(\mathbf{r}, t + \Delta t)\, dV - \iiint_{D_t} f(\mathbf{r}, t)\, dV\right] \\ &= \iiint_{D_t} \frac{f(\mathbf{r}, t + \Delta t) - f(\mathbf{r}, t)}{\Delta t}\, dV \\ &\quad + \frac{1}{\Delta t}\iiint_{\Delta D_t} f(\mathbf{r}, t)\, dV \\ &\quad + \iiint_{\Delta D_t} \frac{f(\mathbf{r}, t + \Delta t) - f(\mathbf{r}, t)}{\Delta t}\, dV,\end{aligned}$$

and show that the last integral $\to 0$ as $\Delta t \to 0$.

CHAPTER 17

Differential Forms and Exterior Calculus

> "The miracle of the appropriateness of the language of mathematics for the formulation of the laws of physics is a wonderful gift which we neither understand nor deserve."

Eugene P. Wigner 1902–1995
from *The Unreasonable Effectiveness of Mathematics in the Natural Sciences, (1960)*

Introduction

In S. P. Thompson's classic 1914 text, *Calculus Made Easy* (2nd ed.), he playfully described the "d" in a differential as a "dreadful" symbol. He concluded it was best to think of "d" as an operation that takes "a little bit of." Thus, the ubiquitous intuition about differentials being vaguely "small" has a long history that belies the historical, but ultimately successful, struggle of mathematicians to escape from "infinitesimals." Our definitions of differentials in Sections 2.2 and 12.6 made it quite clear that differentials are just new independent and dependent variables that can have any values, not just small ones. It is only when we have used differentials to approximate the changes in values of functions that we have thought of differentials as small in order that the errors in the approximations be small. We have also seen differentials used in contexts where smallness is neither implied nor desirable, for example, in the applications in Sections 12.6 and 13.9.

This chapter focuses on differentials and develops a new kind of "calculus" called **exterior calculus** that enables differentials to play a much greater role in applications in the physical and other sciences. It amounts to a rethink of how calculus is traditionally done. Sections 17.1 and 17.2 set up the mechanics of "k-forms" and "differential forms," (which are fields of k-forms analogous to vector fields) and the operators "wedge product" and "exterior derivative" that act on them. These are analogous to differential calculus, while the remaining three Sections 17.3–17.5 constitute a rethink of integration. Section 17.3 defines manifolds and bridges the classical multiple integral to integrals of differential forms. A central issue in integration is orientation, which differential forms naturally take into account in any dimension. This is the subject of Section 17.4. Section 17.5 revisits the classical integration theorems of advanced calculus, showing them in a unified light in the Generalized Stokes Theorem.

Differentials and Vectors

Differentials have properties similar to vectors. Consider the differential of the function $f(x, y, z)$ and its gradient ∇f:

$$df = \frac{\partial f}{\partial x}\,dx + \frac{\partial f}{\partial y}\,dy + \frac{\partial f}{\partial z}\,dz$$

$$\nabla f = \frac{\partial f}{\partial x}\,\mathbf{i} + \frac{\partial f}{\partial y}\,\mathbf{j} + \frac{\partial f}{\partial z}\,\mathbf{k}.$$

The expression for df appears to expand df as a linear combination of "basis vectors" dx, dy, and dz, which play the same role as **i**, **j** and **k** do in the expression for ∇f; they both imply direction as well as magnitude. We will come to regard differentials as elements of vector spaces in this chapter.

The idea of a differential having direction (orientation) is implicit in the definition of the definite integral in Chapter 5. $\int_a^b f(x)\,dx$ is the integral of the differential form $f(x)\,dx$ over the interval $[a, b]$ oriented from a to b. Reversing this orientation results in the integral changing sign: $\int_b^a f(x)\,dx = -\int_a^b f(x)\,dx$. Our definitions of double and triple integrals in Chapter 14 involved no such "built-in" concept of orientation; for instance, we treated the area element in $\mathbb{R}^2$ as $dA = dx\,dy = dy\,dx$. This meant that the orientation concept had to be artificially built in to the statements of two- and three-dimensional versions of the Fundamental Theorem of Calculus in Chapter 16. This deficiency will be remedied in this chapter by the introduction of a new kind of product (the wedge product), where we will replace the inadequate product $dx\,dy$ with $dx \wedge dy$, which is antisymmetric in the sense that $dy \wedge dx = -dx \wedge dy$. This will, in turn, make it possible to define integrals over "manifolds" of any dimension and obtain a single version of the Fundamental Theorem of Calculus that applies in any dimension.

Other than representing the dreaded "little bit of area," dA has no meaning; it is not the differential of anything, and neither is $dx\,dy$.

Derivatives versus Differentials

It is a peculiarity of the conventional language that, except in special cases, when we speak of differential equations we are actually speaking of equations between derivatives and not equations between differentials. Exterior calculus inverts this. The exterior derivative defined in Section 17.2 is properly a kind of differential and not a derivative as the term is conventionally used in calculus. The exterior derivative (i.e., "d"), together with the notion of products of forms, allows for a new kind of object. One can, loosely speaking, take the differential of a differential in a meaningful way. This is something completely new. By forming independent bases in their own vector spaces, k-forms retain the ability to "separate" (into components) that vectors have. Thus, differential equations can be replaced by equivalent equations in differentials of k-forms.

17.1 k-Forms

In this section, we develop the notion of forms and their products, known as wedge products. Let the n vectors $\mathbf{e}_1 = (1, 0, 0, \ldots, 0)$, $\mathbf{e}_2 = (0, 1, 0, \ldots, 0)$, $\ldots$, and $\mathbf{e}_n = (0, 0, 0, \ldots, 1)$ be the standard basis for the n-dimensional real vector space $\mathbb{R}^n$. A function that maps a real vector space into $\mathbb{R}$ is called a "functional." In physical examples, such as integrals for energy, functionals are commonly encountered on vector spaces of functions (infinite dimensional function spaces), but in the following definition we introduce a functional on the finite dimensional vector space $\mathbb{R}^n$.

DEFINITION

A real-valued function ϕ defined on $\mathbb{R}^n$ is called a **1-form** (or a **linear functional**) on $\mathbb{R}^n$ if, whenever **x** and **y** belong to $\mathbb{R}^n$ and a and b are real numbers, then

$$\phi(a\mathbf{x} + b\mathbf{y}) = a\phi(\mathbf{x}) + b\phi(\mathbf{y}).$$

For example, if $\mathbf{x} = x_1\mathbf{e}_1 + x_2\mathbf{e}_2 + \cdots + x_n\mathbf{e}_n$, then the function ϕ defined by

$$\phi(\mathbf{x}) = a_1x_1 + a_2x_2 + \cdots + a_nx_n = \mathbf{a} \bullet \mathbf{x}$$

is a 1-form on $\mathbb{R}^n$ for any $\mathbf{a} \in \mathbb{R}^n$. In fact, every 1-form on $\mathbb{R}^n$ is of this type, because, if ϕ is an arbitrary 1-form on $\mathbb{R}^n$ and we let $a_i = \phi(\mathbf{e}_i)$, then by linearity,

$$\phi(\mathbf{x}) = \phi\left(\sum_{i=1}^{n} x_i\,\mathbf{e}_i\right) = \sum_{i=1}^{n} x_i\phi(\mathbf{e}_i) = \mathbf{x} \bullet \mathbf{a} = \mathbf{a} \bullet \mathbf{x}.$$

The set of all 1-forms on $\mathbb{R}^n$ is denoted $\Lambda_1(\mathbb{R}^n)$ and is a real vector space called the **dual space** of $\mathbb{R}^n$. If ϕ and ψ are 1-forms and $\theta = u\phi + v\psi$, where u and v are real numbers, then, as noted above, $\phi(\mathbf{x}) = \mathbf{a} \bullet \mathbf{x}$, and $\psi(\mathbf{x}) = \mathbf{b} \bullet \mathbf{x}$ for certain n-vectors $\mathbf{a}$ and $\mathbf{b}$, so

$$\theta(\mathbf{x}) = u\phi(\mathbf{x}) + v\psi(\mathbf{x}) = u\mathbf{a} \bullet \mathbf{x} + v\mathbf{b} \bullet \mathbf{x} = (u\mathbf{a} + v\mathbf{b}) \bullet \mathbf{x},$$

and θ is a 1-form corresponding to the vector $u\mathbf{a} + v\mathbf{b}$.

Now we make an important definition that appears to give "differentials" a new role to play, rather than just being new independent and dependent variables in a differentiation process. Being a vector space, $\Lambda_1(\mathbb{R}^n)$ must itself have a basis.

DEFINITION 2

Differentials as basis vectors for 1-forms

For $1 \le i \le n$, let dx_i be the 1-form that assigns to $\mathbf{v} \in \mathbb{R}^n$ its ith component v_i:

$$dx_i(\mathbf{v}) = v_i \qquad \text{for all } \mathbf{v} \in \mathbb{R}^n.$$

Since any 1-form ϕ on $\mathbb{R}^n$ can be written in the form

$$\phi(\mathbf{v}) = \sum_{i=1}^{n} \phi(\mathbf{e}_i)\, v_i = \sum_{i=1}^{n} \phi(\mathbf{e}_i)\, dx_i(\mathbf{v}),$$

we can therefore write $\phi = \sum_{i=1}^{n} \phi(\mathbf{e}_i)\, dx_i$.

Thus, the differentials dx_i for $1 \le i \le n$ constitute a basis for $\Lambda_1(\mathbb{R}^n)$, which we will call the standard basis. $\Lambda_1(\mathbb{R}^n)$ must therefore also be an n-dimensional vector space.

We have now departed from the convention up to this point of depicting differentials on both sides of any equality. It is not necessary that a 1-form be the differential of some function.

Bilinear Forms and 2-Forms

The **Cartesian Product** $\mathbb{R}^n \times \mathbb{R}^n = \{(x, y) : x, y \in \mathbb{R}^n\}$ is a vector space of dimension $2n$. A **bilinear form** ϕ on $\mathbb{R}^n$ is a map from $\mathbb{R}^n \times \mathbb{R}^n$ into $\mathbb{R}$ such that $\phi(\mathbf{x}, \mathbf{y})$ is linear in $\mathbf{x}$ for each fixed $\mathbf{y}$ and linear in $\mathbf{y}$ for each fixed $\mathbf{x}$; that is,

$$\phi(a\mathbf{x} + b\mathbf{y}, \mathbf{z}) = a\phi(\mathbf{x}, \mathbf{z}) + b\phi(\mathbf{y}, \mathbf{z})$$
$$\phi(\mathbf{x}, a\mathbf{y} + b\mathbf{z}) = a\phi(\mathbf{x}, \mathbf{y}) + b\phi(\mathbf{x}, \mathbf{z})$$

holds for all $a, b \in \mathbb{R}$ and all $\mathbf{x}, \mathbf{y}, \mathbf{z} \in \mathbb{R}^n$.

EXAMPLE 1 If $\mathbf{x} \in \mathbb{R}^n$ and $\mathbf{y} \in \mathbb{R}^n$ are row vectors (so that the transpose $\mathbf{y}^T$ is a column vector), and if $\mathcal{A} = (a_{ij})$ is a real $n \times n$ matrix, then

$$\phi(\mathbf{x}, \mathbf{y}) = \mathbf{x}\,\mathcal{A}\,\mathbf{y}^T = \sum_{i=1}^{n}\sum_{j=1}^{n} x_1 a_{ij} y_j$$

is a bilinear form on $\mathbb{R}^n$. In fact, every bilinear ϕ form on $\mathbb{R}^n$ can be expressed in this way, where $a_{ij} = \phi(\mathbf{e}_i, \mathbf{e}_j)$.

DEFINITION 3

A **2-form** on $\mathbb{R}^n$ is a bilinear form on $\mathbb{R}^n$ that is also **antisymmetric** (or **skew-symmetric**) in the sense that for every $(\mathbf{x}, \mathbf{y}) \in \mathbb{R}^n \times \mathbb{R}^n$,

$$\phi(\mathbf{y}, \mathbf{x}) = -\phi(\mathbf{x}, \mathbf{y}).$$

The set of all such 2-forms on $\mathbb{R}^n$ is denoted $\Lambda_2(\mathbb{R}^n)$ and is a vector space.

EXAMPLE 2 Let ϕ and ψ be two 1-forms on $\mathbb{R}^n$, (i.e., ϕ and ψ belong to $\Lambda_1(\mathbb{R}^n)$). Then the expression $\xi = \phi \wedge \psi$ defined by

$$\xi(\mathbf{x}, \mathbf{y}) = (\phi \wedge \psi)(\mathbf{x}, \mathbf{y}) = \begin{vmatrix} \phi(\mathbf{x}) & \phi(\mathbf{y}) \\ \psi(\mathbf{x}) & \psi(\mathbf{y}) \end{vmatrix} = \phi(\mathbf{x})\psi(\mathbf{y}) - \phi(\mathbf{y})\psi(\mathbf{x})$$

is bilinear and antisymmetric, and so is a 2-form on $\mathbb{R}^n$. (This follows at once from properties of the determinant.) The symbol $\wedge$ is called a **wedge product**. This wedge product is a function of two vectors. (Later we will encounter wedge products that have more than two arguments.) For example, in terms of elementary 1-forms,

$$(dx_i \wedge dx_j)(\mathbf{x}, \mathbf{y}) = \begin{vmatrix} x_i & y_i \\ x_j & y_j \end{vmatrix} = x_i y_j - x_j y_i.$$

Note that the antisymmetric property of the wedge product implies that $\phi \wedge \phi = 0$ (the zero 2-form) for any $\phi \in \Lambda_1(\mathbb{R}^n)$.

Let $\phi \in \Lambda_2(\mathbb{R}^n)$, and let $\mathbf{x} = \sum_{i=1}^n x_i \mathbf{e}_i$ and $\mathbf{y} = \sum_{j=1}^n y_j \mathbf{e}_j$ belong to $\mathbb{R}^n$. If $a_{ij} = \phi(\mathbf{e}_i, \mathbf{e}_j)$, then the numbers a_{ij} satisfy $a_{ji} = -a_{ij}$ and $a_{ii} = 0$; that is, the matrix (a_{ij}) is antisymmetric. Therefore, we have, using the bilinearity of ϕ,

$$\begin{aligned} \phi(\mathbf{x}, \mathbf{y}) &= \sum_{i=1}^n \sum_{j=1}^n x_i\, y_j\, \phi(\mathbf{e}_i, \mathbf{e}_j) = \sum_{i=1}^n \sum_{j=1}^n a_{ij}\, x_i\, y_j \\ &= \sum_{1 \le i < j \le n} \big(a_{ij}\, x_i\, y_j - a_{ij}\, x_j\, y_i\big) = \sum_{1 \le i < j \le n} a_{ij} \big(dx_i \wedge dx_j\big)(\mathbf{x}, \mathbf{y}). \end{aligned}$$

Moreover, if $\sum_{1 \le i < j \le n} a_{ij}\, dx_i \wedge dx_j(\mathbf{x}, \mathbf{y}) = 0$ for all choices of n-vectors $\mathbf{x}$ and $\mathbf{y}$, then, taking $\mathbf{x} = \mathbf{e}_i$ and $\mathbf{y} = \mathbf{e}_j$, we obtain $a_{ij} = 0$ for all choices of i and j satisfying $1 \le i < j \le n$. Thus, we have proved the following:

THEOREM 1

The **elementary 2-forms** $dx_i \wedge dx_j$, where $1 \le i < j \le n$, constitute a basis for $\Lambda_2(\mathbb{R}^n)$, which is therefore a real vector space having dimension $\binom{n}{2} = \dfrac{n(n-1)}{2}$.

While we have been explicitly stating the functional dependence on two vectors $\mathbf{x}$ and $\mathbf{y}$ to this point, we will take this as understood unless explicitly needed.

EXAMPLE 3 Let ϕ and ψ be two 1-forms on $\mathbb{R}^3$, say,

$$\begin{aligned} \phi &= a_1\, dx_1 + a_2\, dx_2 + a_3\, dx_3 \\ \psi &= b_1\, dx_1 + b_2\, dx_2 + b_3\, dx_3. \end{aligned}$$

Expand the 2-form $\phi \wedge \psi$ in terms of the three basis vectors $dx_2 \wedge dx_3$, $dx_3 \wedge dx_1$ (which is just $-\, dx_1 \wedge dx_3$), and $dx_1 \wedge dx_2$ of $\Lambda_2(\mathbb{R}^3)$. What vector in $\mathbb{R}^3$ does the result correspond to if we regard the three basis vectors above as corresponding to $\mathbf{i} = \mathbf{e}_1$, $\mathbf{j} = \mathbf{e}_2$, and $\mathbf{k} = \mathbf{e}_3$ in $\mathbb{R}^3$?

Solution We have

$$\begin{aligned}\phi \wedge \psi &= a_1b_1\,dx_1 \wedge dx_1 + a_1b_2\,dx_1 \wedge dx_2 + a_1b_3\,dx_1 \wedge dx_3\\ &\quad + a_2b_1\,dx_2 \wedge dx_1 + a_2b_2\,dx_2 \wedge dx_2 + a_2b_3\,dx_2 \wedge dx_3\\ &\quad + a_3b_1\,dx_3 \wedge dx_1 + a_3b_2\,dx_3 \wedge dx_2 + a_3b_3\,dx_3 \wedge dx_3\\ &= (a_2b_3 - a_3b_2)\,dx_2 \wedge dx_3 + (a_3b_1 - a_1b_3)\,dx_3 \wedge dx_1\\ &\quad + (a_1b_2 - a_2b_1)\,dx_1 \wedge dx_2.\end{aligned}$$

The coefficients here are those of the *cross product* of $\mathbf{a} = a_1\mathbf{i} + a_2\mathbf{j} + a_3\mathbf{k}$ and $\mathbf{b} = b_1\mathbf{i} + b_2\mathbf{j} + b_3\mathbf{k}$ in $\mathbb{R}^3$. Thus, the wedge product mapping of $\Lambda_1(\mathbb{R}^3) \times \Lambda_1(\mathbb{R}^3)$ into $\Lambda_2(\mathbb{R}^3)$ corresponds to the cross product mapping of $\mathbb{R}^3 \times \mathbb{R}^3$ into $\mathbb{R}^3$.

Remark Note that $n = 3$ is a unique case in that it is the only one with the bases for $\Lambda_2(\mathbb{R}^n)$ and $\Lambda_1(\mathbb{R}^n)$ having the same dimension. In a sense, this is what makes cross products possible in $\mathbb{R}^3$.

k-Forms

DEFINITION 4

A ***k*-form** on $\mathbb{R}^n$ is a multilinear antisymmetric functional ϕ defined on the Cartesian product $(\mathbb{R}^n)^k = \mathbb{R}^n \times \mathbb{R}^n \times \cdots \times \mathbb{R}^n$ (k factors $\mathbb{R}^n$). That is, ϕ maps $(\mathbb{R}^n)^k$ into $\mathbb{R}$ and satisfies the two conditions:

(a) **multilinearity:** $\phi(\mathbf{v}_1, \ldots, \mathbf{v}_k)$ is linear in each of the vectors $\mathbf{v}_i$ with the others held fixed.

$$\begin{aligned}&\phi(\mathbf{v}_1, \ldots, \mathbf{v}_{i-1}, (a\mathbf{u} + b\mathbf{w}), \mathbf{v}_{i+1}, \ldots, \mathbf{v}_k)\\ &\quad = a\phi(\mathbf{v}_1, \ldots, \mathbf{v}_{i-1}, \mathbf{u}, \mathbf{v}_{i+1}, \ldots, \mathbf{v}_k)\\ &\qquad + b\phi(\mathbf{v}_1, \ldots, \mathbf{v}_{i-1}, \mathbf{w}, \mathbf{v}_{i+1}, \ldots, \mathbf{v}_k)\end{aligned}$$

for all real numbers a and b and vectors $\mathbf{u}$ and $\mathbf{w}$ in $\mathbb{R}^n$, and

(b) **antisymmetry:** if any two arguments of ϕ have their positions switched, the value of ϕ changes sign.

$$\phi(\mathbf{v}_1, \ldots, \mathbf{v}_i, \ldots, \mathbf{v}_j, \ldots \mathbf{v}_k) = -\phi(\mathbf{v}_1, \ldots, \mathbf{v}_j, \ldots, \mathbf{v}_i, \ldots \mathbf{v}_k).$$

The vector space of all k-forms on $\mathbb{R}^n$ is denoted $\Lambda_k(\mathbb{R}^n)$.

If $k > 2$, we need to extend the notion of antisymmetry to allow for exchanges involving more than two arguments. We call a rearrangement of the numbers $\{1, 2, 3, \ldots, k\}$ a **permutation**. Such permutations can always be constructed by successive reversals of pairs of the numbers. The reversal (ij) exchanges the numbers i and j (where $j \neq i$). For example, the rearrangement π that maps $\{1, 2, 3\}$ to $\{2, 3, 1\}$ can be regarded as first switching 1 and 2 (producing $\{2, 1, 3\}$) and then switching 1 and 3 to get $\{2, 3, 1\}$. π can be regarded as representing a sequence of reversals to achieve the rearrangement. We describe π as the "product" of these reversals: $\pi = (12)(13)$, where (12) and (13) depict the specific reversals for a particular rearrangement. Of course, such a representation is not unique; it is also true that $\pi = (13)(23)$. However, if a permutation π can be expressed as a product of an even (or odd) number of reversals, then all ways of expressing it as a product of reversals will involve an even (or odd) number, and we say that the permutation itself is *even* (or *odd*). Accordingly, we define the **sign** of the permutation π as

This use of sgn to denote the sign for an even or odd permutation should not be confused with the signum function of Section P.5, neither should π, the permutation, be confused with the number π.

$$\operatorname{sgn}(\pi) = \begin{cases} 1 & \text{if } \pi \text{ is an even permutation}\\ -1 & \text{if } \pi \text{ is an odd permutation.}\end{cases}$$

It follows that the antisymmetry property of a k-form ϕ can be generalized as follows: if π is any permutation of the numbers $\{1, 2, \dots, k\}$, then

$$\phi(\mathbf{v}_{\pi(1)}, \mathbf{v}_{\pi(2)} \dots, \mathbf{v}_{\pi(k)}) = \operatorname{sgn}(\pi)\, \phi(\mathbf{v}_1, \mathbf{v}_2, \dots, \mathbf{v}_k).$$

We can now extend the definition of the wedge product to allow for k factors. Let dx_i be the 1-form introduced earlier in this section: $dx_i(\mathbf{x}) = x_i$ for all $\mathbf{x} \in \mathbb{R}^n$. We define the **elementary k-forms**

$$(dx_{i_1} \wedge dx_{i_2} \wedge \cdots \wedge dx_{i_k})(\mathbf{v}_1, \mathbf{v}_2, \dots, \mathbf{v}_k) = \begin{vmatrix} dx_{i_1}(\mathbf{v}_1) & dx_{i_1}(\mathbf{v}_2) & \cdots & dx_{i_1}(\mathbf{v}_k) \\ dx_{i_2}(\mathbf{v}_1) & dx_{i_2}(\mathbf{v}_2) & \cdots & dx_{i_2}(\mathbf{v}_k) \\ \vdots & \vdots & \ddots & \vdots \\ dx_{i_k}(\mathbf{v}_1) & dx_{i_k}(\mathbf{v}_2) & \cdots & dx_{i_k}(\mathbf{v}_k) \end{vmatrix}$$

$$= \begin{vmatrix} v_{1i_1} & v_{2i_1} & \cdots & v_{ki_1} \\ v_{1i_2} & v_{2i_2} & \cdots & v_{ki_2} \\ \vdots & \vdots & \ddots & \vdots \\ v_{1i_k} & v_{2i_k} & \cdots & v_{ki_k} \end{vmatrix}.$$

Remark While the above formula makes sense for all positive integers k, the resulting determinant will be zero if $k > n$. Since there are only n distinct 1-forms dx_i, if $k > n$ at least two of the subscripts $i_1, i_2, \dots, i_k$ will be equal and so the determinant will have at least two identical rows. The same applies for $k \le n$; if any two of the factors in $dx_{i_1} \wedge dx_{i_2} \wedge \cdots \wedge dx_{i_k}$ are identical, then the wedge product is the zero k-form.

Remark For $k \le n$, let the numbers $i_1, i_2, \dots, i_k$ satisfy $1 \le i_1 < i_2 < \cdots < i_k \le n$. If π is a permutation of those numbers, then

$$dx_{\pi(i_1)} \wedge dx_{\pi(i_2)} \wedge \cdots \wedge dx_{\pi(i_k)} = \operatorname{sgn}(\pi)\, dx_{i_1} \wedge dx_{i_2} \wedge \cdots \wedge dx_{i_k}.$$

As observed earlier for 2-forms, the collection of all wedge products of the form $dx_{i_1} \wedge dx_{i_2} \wedge \cdots \wedge dx_{i_k}$, where $i_1, i_2, \dots, i_k$, satisfy $1 \le i_1 < i_2 < \cdots < i_k \le n$, constitute a basis for $\Lambda_k(\mathbb{R}^n)$, which therefore has dimension $\binom{n}{k} = \dfrac{n!}{(n-k)!\,k!}$. In particular, $\Lambda_n(\mathbb{R}^n)$ has dimension 1; it is spanned by the single form $dx_1 \wedge dx_2 \wedge \cdots \wedge dx_n$.

The wedge product of an arbitrary k-form ϕ and ℓ-form ψ can now be calculated using the bases of $\Lambda_k(\mathbb{R}^n)$ and $\Lambda_\ell(\mathbb{R}^n)$. If

$$\phi = \sum_{1 \le i_1 < i_2 \cdots < i_k \le n} a_{i_1 i_2 \cdots i_k} dx_{i_1} \wedge dx_{i_2} \wedge \cdots \wedge dx_{i_k}$$

$$\psi = \sum_{1 \le j_1 < j_2 \cdots < j_\ell \le n} b_{j_1 j_2 \cdots j_\ell} dx_{j_1} \wedge dx_{j_2} \wedge \cdots \wedge dx_{j_\ell},$$

then

$$\phi \wedge \psi = \sum_{\substack{1 \le i_1 < \cdots < i_k \le n \\ 1 \le j_1 < \cdots < j_\ell \le n}} a_{i_1 i_2 \cdots i_k}\, b_{j_1 j_2 \cdots j_\ell}\, dx_{i_1} \wedge dx_{i_2} \wedge \cdots \wedge dx_{i_k} \wedge dx_{j_1} \wedge dx_{j_2} \wedge \cdots \wedge dx_{j_\ell}.$$

The result is a $(k+\ell)$-form. Any terms on the right side for which one of the dx_is is identical to one of the dx_js will be zero. This will happen to all terms if $k + \ell > n$.

Assuming that ϕ, ϕ_1, and ϕ_2 are k-forms, that ψ is an ℓ-form, that χ is an m-form, and that a and b are real numbers, then the wedge product has the following properties:

(a) It is *linear* in each of its arguments:

$$(a\phi_1 + b\phi_2) \wedge \psi = a\,\phi_1 \wedge \psi + b\,\phi_2 \wedge \psi.$$

(b) It is *associative*:

$$(\phi \wedge \psi) \wedge \chi = \phi \wedge (\psi \wedge \chi).$$

so this triple product can be written unambiguously as $\phi \wedge \psi \wedge \chi$,

(c) It is *skew-commutative*:

$$\phi \wedge \psi = (-1)^{k\ell}\, \psi \wedge \phi.$$

EXAMPLE 4 We summarize the description of all k-forms on $\mathbb{R}^3$ as follows:

(a) **1-forms** $\Lambda_1(\mathbb{R}^3)$ has dimension 3. It consists of forms of the type

$$\phi = a_1\, dx_1 + a_2\, dx_2 + a_3\, dx_3, \quad \text{where each } a_i \in \mathbb{R}.$$

(b) **2-forms** $\Lambda_2(\mathbb{R}^3)$ also has dimension 3. It consists of forms of the type

$$\psi = b_1\, dx_2 \wedge dx_3 + b_2\, dx_3 \wedge dx_1 + b_3\, dx_1 \wedge dx_2\, quad \text{where each } b_i \in \mathbb{R}.$$

(c) **3-forms** $\Lambda_3(\mathbb{R}^3)$ has dimension 1. It consists of forms of the type

$$\chi = c\, dx_1 \wedge dx_2 \wedge dx_3, \quad \text{where } c \in \mathbb{R}.$$

(d) **higher-order forms** If $k \geq 4$, then $\Lambda_k(\mathbb{R}^3) = \{0\}$, the zero k-form that maps a k-tuple of vectors in $\mathbb{R}^3$ to the number 0.

EXAMPLE 5 Calculate and simplify $\phi \wedge \psi$, where

$$\phi = a_1\, dx_1 + a_2\, dx_2 + a_3\, dx_3 \in \Lambda_1(\mathbb{R}^3),$$
$$\psi = b_1\, dx_2 \wedge dx_3 + b_2\, dx_3 \wedge dx_1 + b_3\, dx_1 \wedge dx_2 \in \Lambda_2(\mathbb{R}^3).$$

Interpret the result in terms of the vectors $\mathbf{a} = a_1\mathbf{i} + a_2\mathbf{j} + a_3\mathbf{k}$ and $\mathbf{b} = b_1\mathbf{i} + b_2\mathbf{j} + b_3\mathbf{k}$ in $\mathbb{R}^3$.

Solution By linearity and skew-symmetry, we have

$$\begin{aligned}
\phi \wedge \psi &= a_1b_1\, dx_1 \wedge dx_2 \wedge dx_3 + a_1b_2\, dx_1 \wedge dx_3 \wedge dx_1 + a_1b_3\, dx_1 \wedge dx_1 \wedge dx_2 \\
&\quad + a_2b_1\, dx_2 \wedge dx_2 \wedge dx_3 + a_2b_2\, dx_2 \wedge dx_3 \wedge dx_1 + a_2b_3\, dx_2 \wedge dx_1 \wedge dx_2 \\
&\quad + a_3b_1\, dx_3 \wedge dx_2 \wedge dx_3 + a_3b_2\, dx_3 \wedge dx_3 \wedge dx_1 + a_3b_3\, dx_3 \wedge dx_1 \wedge dx_2 \\
&= (a_1b_1 + a_2b_2 + a_3b_3)\, dx_1 \wedge dx_2 \wedge dx_3 \\
&= (\mathbf{a} \bullet \mathbf{b})\, dx_1 \wedge dx_2 \wedge dx_3.
\end{aligned}$$

As a map from $\Lambda_1(\mathbb{R}^3) \times \Lambda_2(\mathbb{R}^3)$ into $\Lambda_3(\mathbb{R}^3)$, the wedge product corresponds to the dot product in $\mathbb{R}^3$.

Forms on a Vector Space

Everything said above about forms on $\mathbb{R}^n$ can be applied to any n-dimensional real vector space provided we redefine the elementary forms so that $dx_i(\mathbf{v})$ selects the ith component of the vector $\mathbf{v} \in V$ with respect to some particular basis of V. For our purposes, we will be mainly interested in the case where V is an m-dimensional subspace of $\mathbb{R}^n$, where $m < n$. In this case, it is possible to restrict the elementary forms $dx_{i_1} \wedge \cdots \wedge dx_{i_k}$ in $\Lambda_k(\mathbb{R}^n)$ so that they apply only to vectors in V. In this restriction, there will generally be fewer independent forms, because $\Lambda_k(V)$ will have smaller dimension $\binom{m}{k}$ than $\Lambda_k(\mathbb{R}^n)$.

For example, the set of points (x, y, z) in $\mathbb{R}^3$ satisfying $x - y + z = 0$ constitutes a 2-dimensional subspace V of $\mathbb{R}^3$. Clearly, $\mathbf{v}_1 = (a, b, b - a)$ and $\mathbf{v}_2 = (p, q, q - p)$ belong to V. However, observe that the restrictions to V of three elementary 2-forms in $\Lambda_2(\mathbb{R}^3)$ evaluate at these vectors to give $dx \wedge dy(\mathbf{v}_1, \mathbf{v}_2) = aq - bp$, $dx \wedge dz(\mathbf{v}_1, \mathbf{v}_2) = aq - bp$, $dy \wedge dz(\mathbf{v}_1, \mathbf{v}_2) = aq - bp$. There is only one independent elementary form in $\Lambda_2(V)$, which has dimension $\binom{2}{2} = 1$.

EXERCISES 17.1

In Exercises 1–4 calculate and simplify the wedge product of the given forms ϕ and ψ. Write the dx_is in the answers in increasing subscript order.

1. $\phi = a_1\,dx_2 \wedge dx_3 + a_2\,dx_3 \wedge dx_4 + a_3\,dx_4 \wedge dx_1$
$\psi = b_1\,dx_1 \wedge dx_2 + b_2\,dx_3 \wedge dx_4$

2. $\phi = dx_2 \wedge dx_3 \wedge dx_4$
$\psi = dx_1 + dx_3 + dx_4$

3. $\phi = dx_1 + 2\,dx_2 + 3\,dx_3 + 4\,dx_4 + 5\,dx_5$
$\psi = dx_1 \wedge dx_2 \wedge dx_3 \wedge dx_4 + 2\,dx_2 \wedge dx_3 \wedge dx_4 \wedge dx_5$

4. $\phi = a_1\,dx_1 + a_2\,dx_2 + a_3 dx_3 + a_4\,dx_4$
$\psi = b_1\,dx_2 \wedge dx_3 + b_2\,dx_3 \wedge dx_4 + b_3\,dx_4 \wedge dx_1 + b_4\,dx_1 \wedge dx_2$

5. For what values of k is the permutation π that maps $\{1, 2, \ldots, k\}$ to $\{2, 3, \ldots, k, 1\}$ even? odd? Express π as a product of reversals.

6. Verify that if ϕ is a k-form and ψ is an ℓ-form, then $\phi \wedge \psi = (-1)^{k\ell}\psi \wedge \phi$.

7. Let $\mathbf{u} = (1, 1, 0, 0)$, $\mathbf{v} = (1, 0, 1, 0)$, and $\mathbf{w} = (1, 0, 0, 1)$ in $\mathbb{R}^4$. Evaluate (a) $dx_1 \wedge dx_2(\mathbf{u}, \mathbf{v})$
(b) $dx_1 \wedge dx_2 \wedge dx_3(\mathbf{u}, \mathbf{v}, \mathbf{w})$
(c) $dx_3 \wedge dx_4 \wedge dx_1(\mathbf{u}, \mathbf{v}, \mathbf{w})$
(d) $dx_3 \wedge dx_2 \wedge dx_4(\mathbf{u}, \mathbf{v}, \mathbf{w})$

8. Let $\mathbf{e}_i$, $(1 \le i \le 4)$ be the standard basis vectors for $\mathbb{R}^4$. Let

$$\mathbf{v}_1 = \mathbf{e}_1 + 2\mathbf{e}_2 + 3\mathbf{e}_3 - 4\mathbf{e}_4 \qquad \mathbf{v}_2 = 2\mathbf{e}_1 + 3\mathbf{e}_2 - 4\mathbf{e}_3$$
$$\mathbf{v}_3 = 3\mathbf{e}_1 - 4\mathbf{e}_2 \qquad \mathbf{v}_4 = 4\mathbf{e}_1$$

Evaluate $\phi(\mathbf{v}_1, \ldots, \mathbf{v}_4)$ if $\phi = dx_1 \wedge dx_2 \wedge dx_3 \wedge dx_4$.

17.2 Differential Forms and the Exterior Derivative

Just as we extended the notion of vector to define vector fields as vector-valued functions of position in a domain in $\mathbb{R}^2$ or $\mathbb{R}^3$, so we can also extend the notion of k-form to define k-form fields as k-form-valued functions of position in a domain in $\mathbb{R}^n$. These fields will be called **differential forms**.

DEFINITION

For $k \ge 1$, a **differential k-form** on a domain D (an open set) in $\mathbb{R}^n$ is a smooth function Φ from D into $\Lambda_k(\mathbb{R}^n)$. Thus, for each $\mathbf{x} \in D$, $\Phi(\mathbf{x})$ is a k-form on $\mathbb{R}^n$ that can be expressed as a linear combination of the the $\binom{n}{k}$ standard basis vectors of $\Lambda_k(\mathbb{R}^n)$. The coefficients of this linear combination (i.e., the coefficients of $\Phi(\mathbf{x})$) will be smooth real-valued functions of $\mathbf{x}$:

$$\Phi(\mathbf{x}) = \sum_{1 \le i_1 < i_2 < \cdots < i_k \le n} a_{i_1 i_2 \cdots i_k}(\mathbf{x})\,dx_{i_1} \wedge dx_{i_2} \wedge \cdots \wedge dx_{i_k}.$$

A **differential 0-form** on D is a smooth real-valued function f on D.

For simplicity, we take "smooth" to mean that the coefficients have continuous partial derivatives of all orders (or at least all orders we need to calculate in a given situation).

For $k \ge 0$, we denote the set of all differential $\boldsymbol{k}$-forms on D by $\mathcal{F}_k(D)$.

EXAMPLE 1 A differential 1-form on $D \subset \mathbb{R}^n$ can be expressed as

$$\Phi(\mathbf{x}) = \sum_{i=1}^{n} a_i(x_1, x_2, \ldots, x_n)\,dx_i,$$

where the coefficients a_i are functions of $\mathbf{x} \in D$. Of course, for any $\mathbf{x} \in D$, $\Phi(\mathbf{x})$ is a 1-form on $\mathbb{R}^n$, whose value at $\mathbf{v} \in \mathbb{R}^n$ is given by

$$\Phi(\mathbf{x})(\mathbf{v}) = \sum_{i=1}^{n} a_i(\mathbf{x})\, dx_i(\mathbf{v}) = \sum_{i=1}^{n} a_i(\mathbf{x})\, v_i = \mathbf{a}(\mathbf{x}) \bullet \mathbf{v},$$

where $\mathbf{a}(\mathbf{x})$ is the n-vector field with components $a_i(\mathbf{x})$.

More generally, if Ψ is the differential k-form on D given by

$$\Psi = \Psi(\mathbf{x}) = \sum_{1 \le i_1 < i_2 < \cdots < i_k \le n} a_{i_1 i_2 \cdots i_k}(\mathbf{x})\, dx_{i_1} \wedge dx_{i_2} \wedge \cdots \wedge dx_{i_k},$$

where the coefficients $a_{i_1 i_2 \cdots i_k}$ are functions of $\mathbf{x} \in D$, then the value of $\Psi(\mathbf{x})$ at the sequence of vectors $\{\mathbf{v}_1, \mathbf{v}_2, \ldots, \mathbf{v}_k\}$ in $\mathbb{R}^n$ is

$$\begin{aligned}\Psi(\mathbf{x})(\mathbf{v}_1, \ldots, \mathbf{v}_k) &= \sum_{1 \le i_1 < i_2 < \cdots < i_k \le n} a_{i_1 i_2 \cdots i_k}(\mathbf{x})\, (dx_{i_1} \wedge dx_{i_2} \wedge \cdots \wedge dx_{i_k})(\mathbf{v}_1, \ldots, \mathbf{v}_k) \\ &= \sum_{1 \le i_1 < i_2 < \cdots < i_k \le n} a_{i_1 i_2 \cdots i_k}(\mathbf{x}) \begin{vmatrix} v_{1i_1} & v_{2i_1} & \cdots & v_{ki_1} \\ v_{1i_2} & v_{2i_2} & \cdots & v_{ki_2} \\ \vdots & \vdots & \ddots & \vdots \\ v_{1i_k} & v_{2i_k} & \cdots & v_{ki_k} \end{vmatrix}.\end{aligned}$$

The wedge product of k-forms extends in the obvious way (pointwise on D) to differential k-forms with the added requirement that if f is a differential 0-form on D, then $f \wedge \Phi = f\Phi$ for any differential k-form Φ; the coefficients of $f \wedge \Phi$ are just the coefficients of Φ multiplied by f. Observe that for any two differential forms Φ and Ψ, and any differential 0-form f, we have

$$(f\Phi) \wedge \Psi = f(\Phi \wedge \Psi) = \Phi \wedge (f\Psi).$$

The Exterior Derivative

The following definition is central to the study of differential forms; in a sense it justifies our use of the symbols dx_i in the bases of the spaces of k-forms, and the use of the term "differential k-form" to describe a form-field.

DEFINITION 6

The **exterior derivative** of a differential 0-form (that is, a function) f on domain $D \subset \mathbb{R}^n$ is the differential 1-form df given by

$$df(\mathbf{x}) = \sum_{i=1}^{n} \frac{\partial f}{\partial x_i}\, dx_i.$$

If Φ is an arbitrary differential k-form on D:

$$\Phi(\mathbf{x}) = \sum_{1 \le i_1 < i_2 < \cdots < i_k \le n} a_{i_1 i_2 \cdots i_k}(\mathbf{x})\, dx_{i_1} \wedge dx_{i_2} \wedge \cdots \wedge dx_{i_k},$$

then its exterior derivative, $d\Phi$, is the differential $(k+1)$-form given by

$$d\Phi(\mathbf{x}) = \sum_{1 \le i_1 < i_2 < \cdots < i_k \le n} \big(da_{i_1 i_2 \cdots i_k}(\mathbf{x})\big) \wedge dx_{i_1} \wedge dx_{i_2} \wedge \cdots \wedge dx_{i_k}.$$

The exterior differential operator d maps $\mathcal{F}_k(D)$ into $\mathcal{F}_{k+1}(D)$.

In ordinary calculus "d" denotes the differential of a function (i.e., of a 0-form). Here we have extended "d" to apply to any differential form to give a new form of one higher order. Do not confuse "d" with "D," which can represent a derivative in ordinary calculus.

It is worth stressing that the exterior derivative of a differential 0-form f coincides with the ordinary differential of f. The coefficients of df are just those of the gradient **grad** (f). If $\mathbf{v} \in \mathbb{R}^n$, then

$$df(\mathbf{x})(\mathbf{v}) = \sum_{i=1}^{n} \frac{\partial f}{\partial x_i}\, dx_i(\mathbf{v}) = \sum_{i=1}^{n} \frac{\partial f}{\partial x_i}\, v_i = \mathbf{grad}\,(f) \bullet \mathbf{v},$$

illustrating again that a 1-form on $\mathbb{R}^n$ is just a dot product with a fixed vector. But it is more remarkable that it yields a clear meaning to the differential of a differential. This is something completely new.

THEOREM 2

Properties of the exterior derivative If Φ and Ψ are differential k-forms and Ω is a differential ℓ form on domain $D \subset \mathbb{R}^n$, and if a and b are real numbers, then

(a) $d(a\Phi + b\Psi) = a\, d\Phi + b\, d\Psi$; that is, the operator d is linear from $\mathcal{F}_k(D)$ into $\mathcal{F}_{k+1}(D)$.

(b) $d(\Phi \wedge \Omega) = (d\Phi) \wedge \Omega + (-1)^k\, \Phi \wedge (d\Omega)$; (a Product Rule).

(c) $d^2\Phi = d(d\Phi) = \{0\}$, the zero differential form. That is, $d^2 = 0$.

$d^2 f = ddf$ makes no sense in terms of classical differentials. It is only in the context of wedge products and differential forms that d^2 makes sense. d^2 maps every differential k-form to the zero $(k+2)$-form.

PROOF The proofs of parts (a) and (b) are elementary, and left as exercises for the reader. For (c) we proceed as follows. If Φ is a differential k-form, then $\Phi(\mathbf{x})$ is a sum of terms of the form $a(\mathbf{x})\, dx_{i_1} \wedge dx_{i_2} \wedge \ldots \wedge dx_{i_k}$. By part (a), it is sufficient to prove that $d^2\Phi$ is zero for any one such term. We have

$$\begin{aligned} d^2\, a(\mathbf{x})\, dx_{i_1} \wedge dx_{i_2} \wedge \ldots \wedge dx_{i_k} &= d\left(\sum_{j=1}^{n} \frac{\partial a}{\partial x_j}\, dx_j\right) \wedge dx_{i_1} \wedge dx_{i_2} \wedge \ldots \wedge dx_{i_k} \\ &= \left(\sum_{\ell=1}^{n} \sum_{j=1}^{n} \frac{\partial^2 a}{\partial x_\ell \partial x_j}\, dx_\ell \wedge dx_j\right) \wedge dx_{i_1} \wedge dx_{i_2} \wedge \ldots \wedge dx_{i_k}. \end{aligned}$$

The expression in the large parentheses is zero because the smoothness assumption on partial derivatives of a implies that

$$\frac{\partial^2 a}{\partial x_\ell \partial x_j} = \frac{\partial^2 a}{\partial x_j \partial x_\ell} \qquad \text{and} \qquad dx_\ell \wedge dx_j = -dx_j \wedge dx_\ell.$$

■

Remark The power of wedge products and exterior derivatives begins to become evident in the above proof, which holds only if the d-operator has been applied twice to a differential form. It is clear that the exterior derivative of a general differential form is not necessarily zero. We will, of course, have $d\Phi = 0$ if Φ is a differential n-form on $\mathbb{R}^n$. (Why?)

EXAMPLE 2 Let $\Phi = F_1\, dx + F_2\, dy + F_3\, dz$ belong to $\mathcal{F}_1(\mathbb{R}^3)$. Calculate and simplify $d\Phi$. What does the result correspond to if we identify Φ with the vector field $\mathbf{F} = F_1\mathbf{i} + F_2\mathbf{j} + F_3\mathbf{k}$ and the differential 2-form $d\Phi$ with the vector field having components that are the coefficients of $dy \wedge dz$, $dz \wedge dx$, and $dx \wedge dy$?

Solution Here we are using (x, y, z) instead of (x_1, x_2, x_3) as coordinates in $\mathbb{R}^3$. We have

$$\begin{aligned} d\Phi &= dF_1 \wedge dx + dF_2 \wedge dy + dF_3 \wedge dz \\ &= \frac{\partial F_1}{\partial x} dx \wedge dx + \frac{\partial F_1}{\partial y} dy \wedge dx + \frac{\partial F_1}{\partial z} dz \wedge dx \\ &\quad + \frac{\partial F_2}{\partial x} dx \wedge dy + \frac{\partial F_2}{\partial y} dy \wedge dy + \frac{\partial F_2}{\partial z} dz \wedge dy \\ &\quad + \frac{\partial F_3}{\partial x} dx \wedge dz + \frac{\partial F_3}{\partial y} dy \wedge dz + \frac{\partial F_3}{\partial z} dz \wedge dz \\ &= \left(\frac{\partial F_3}{\partial y} - \frac{\partial F_2}{\partial z}\right) dy \wedge dz + \left(\frac{\partial F_1}{\partial z} - \frac{\partial F_3}{\partial x}\right) dz \wedge dx \\ &\quad + \left(\frac{\partial F_2}{\partial x} - \frac{\partial F_1}{\partial y}\right) dx \wedge dy. \end{aligned}$$

Thus, d maps $\mathcal{F}_1(\mathbb{R}^3)$ into $\mathcal{F}_2(\mathbb{R}^3)$ by taking the differential 1-form $F_1 dx + F_2 dy + F_3 dz$ into the differential 2-form whose coefficients are the components of the vector field **curl F**, where $\mathbf{F} = F_1\mathbf{i} + F_2\mathbf{j} + F_3\mathbf{k}$.

EXAMPLE 3 Let $\Psi = F_1\, dy \wedge dz + F_2\, dz \wedge dx + F_3\, dx \wedge dy$ belong to $\mathcal{F}_2(\mathbb{R}^3)$. Calculate and simplify $d\Psi$. What is the coefficient of $dx \wedge dy \wedge dz$ in terms of the vector field $\mathbf{F} = F_1\mathbf{i} + F_2\mathbf{j} + F_3\mathbf{k}$?

Solution We have

$$\begin{aligned} d\Psi &= dF_1 \wedge dy \wedge dz + dF_2 \wedge dz \wedge dx + dF_3 \wedge dx \wedge dy \\ &= \frac{\partial F_1}{\partial x} dx \wedge dy \wedge dz + \frac{\partial F_1}{\partial y} dy \wedge dy \wedge dz + \frac{\partial F_1}{\partial z} dz \wedge dy \wedge dz \\ &\quad + \frac{\partial F_2}{\partial x} dx \wedge dz \wedge dx + \frac{\partial F_2}{\partial y} dy \wedge dz \wedge dx + \frac{\partial F_2}{\partial z} dz \wedge dz \wedge dx \\ &\quad + \frac{\partial F_3}{\partial x} dx \wedge dx \wedge dy + \frac{\partial F_3}{\partial y} dy \wedge dx \wedge dy + \frac{\partial F_3}{\partial z} dz \wedge dx \wedge dy \\ &= \left(\frac{\partial F_1}{\partial x} + \frac{\partial F_2}{\partial y} + \frac{\partial F_3}{\partial z}\right) dx \wedge dy \wedge dz \\ &= (\mathbf{div\, F})\, dx \wedge dy \wedge dz. \end{aligned}$$

Here d maps $\mathcal{F}_2(\mathbb{R}^3)$ into $\mathcal{F}_3(\mathbb{R}^3)$ by taking the differential 2-form with coefficients F_1, F_2, and F_3 to the 3-form with coefficient the **divergence** of the vector field $F_1\mathbf{i} + F_2\mathbf{j} + F_3\mathbf{k}$.

EXAMPLE 4 In Section 12.6 we encountered the Gibbs form of the equation of state for a thermodynamical system. For a system involving only one type of molecule it is

$$dE = T\, dS - P\, dV + \mu\, dN,$$

where E is energy; $\dfrac{\partial E}{\partial S}$ is temperature, T; $-\dfrac{\partial E}{\partial V}$ is pressure, P; and $\dfrac{\partial E}{\partial N}$ is the chemical potential, μ. Here V is volume, S is entropy, and N is the number of molecules. Take the exterior derivative of this 1-form to find the Maxwell relation $\dfrac{\partial P}{\partial S} = -\dfrac{\partial T}{\partial V}$ of Example 4 in Section 12.8.

Solution Here we are using (S, V, N) instead of (x_1, x_2, x_3) as independent variables. Following Example 2,

$$\begin{aligned}
0 = d^2E &= dT \wedge dS - dP \wedge dV + d\mu \wedge dN \\
&= \frac{\partial T}{\partial S} dS \wedge dS + \frac{\partial T}{\partial V} dV \wedge dS + \frac{\partial T}{\partial N} dN \wedge dS \\
&\quad - \frac{\partial P}{\partial S} dS \wedge dV - \frac{\partial P}{\partial V} dV \wedge dV - \frac{\partial P}{\partial N} dN \wedge dV \\
&\quad + \frac{\partial \mu}{\partial S} dS \wedge dN + \frac{\partial \mu}{\partial V} dV \wedge dN + \frac{\partial \mu}{\partial N} dN \wedge dN \\
&= \left(\frac{\partial \mu}{\partial V} + \frac{\partial P}{\partial N}\right) dV \wedge dN + \left(\frac{\partial T}{\partial N} - \frac{\partial \mu}{\partial S}\right) dN \wedge dS \\
&\quad - \left(\frac{\partial P}{\partial S} + \frac{\partial T}{\partial V}\right) dS \wedge dV.
\end{aligned}$$

Since the wedge products in the final line above are linearly independent, we conclude that

$$\frac{\partial P}{\partial S} = -\frac{\partial T}{\partial V}; \qquad \frac{\partial T}{\partial N} = \frac{\partial \mu}{\partial S}; \qquad \frac{\partial \mu}{\partial V} = -\frac{\partial P}{\partial N}.$$

The first of these is the Maxwell relation from Example 4 of Section 12.8. The other two are additional relations not previously mentioned because Maxwell relations are traditionally used for fixed N, but they are no less valid.

1-Forms and Legendre Transformations

As we saw in Chapter 12, thermodynamic variables come in conjugate pairs. For S, V, and N in the energy 1-form in Example 4, T, $-P$, and μ are the respective conjugate variables. A conjugate variable is defined here as the function in front of the differential of that variable that ensures the product has a positive sign. In Section 12.6 we found that Legendre transformations, such as $F = E - TS$, led to 1-forms in a new quantity, which was the exterior derivative of a zero-form in a new set of variables. Clearly,

$$dF = dE - T\,dS - S\,dT = -S\,dT - P\,dV + \mu\,dN,$$

so that F is a function of T, V, and N, instead of S, V, and N. The conjugate variables are now $-S$, $-P$, and μ. Note that the Legendre transformation introduces a sign change for the new conjugate variable. See Section 12.6 for details.

Once we realize that, we can use Legendre transformations of E to construct a 0-form that depends on any three of the six variables S, V, N, T, P, and μ that we may choose, provided the three chosen variables do not include a variable and its conjugate (and we account for the sign change due to Legendre transformations). It is easy to generate any of the many Maxwell relations using the properties of the wedge product and the exterior derivative. Note that it is only necessary to know that the 0-form exists and how many variables have been swapped between independent variables and conjugate variables, not what the new function is specifically, because d^2 will eliminate it. $d^2 = 0$ helps explain why thermodynamics potentials, as these Legendre transformations of energy are known, are better understood for their properties under differential operations than for their actual values.

Deducing more Maxwell relations employing wedge products and exterior derivatives is a topic for the exercises.

Maxwell's Equations Revisited

James Clerk Maxwell is most famous for his four differential equations governing electromagnetism, which are known as Maxwell's equations and are described in Section

16.6. These are not to be confused with the Maxwell relations of thermodynamics. Maxwell's equations are four partial differential equations in the magnetic field vector **B** and the electric field vector **E**:

$$\begin{aligned}
\nabla \bullet \mathbf{E} &= \frac{\rho}{\epsilon_0} \\
\nabla \times \mathbf{B} &= \mu_0 \mathbf{J} + \frac{1}{c^2}\frac{\partial \mathbf{E}}{\partial t} \\
\nabla \bullet \mathbf{B} &= 0 \\
\nabla \times \mathbf{E} &= -\frac{\partial \mathbf{B}}{\partial t},
\end{aligned}$$

where ρ and $\mathbf{J}$ are charge density and charge current density (i.e., current per unit area) respectively, ϵ_0 and μ_0 are constants, and $c = 1/\sqrt{\epsilon_0 \mu_0}$ is the speed of light.

EXAMPLE 5 Consider the two 2-forms constructed from the six components of **B** and **E** within $\mathbb{R}^4$ (known in physics as "space-time") as follows:

$$\begin{aligned}
F =& B_x\, dy \wedge dz + B_y\, dz \wedge dx + B_z\, dx \wedge dy + E_x\, dx \wedge dt + E_y\, dy \wedge dt \\
&+ E_z\, dz \wedge dt \\
G =& \frac{E_x}{c^2}\, dy \wedge dz + \frac{E_y}{c^2}\, dz \wedge dx + \frac{E_z}{c^2}\, dx \wedge dy - B_x\, dx \wedge dt - B_y\, dy \wedge dt \\
&- B_z\, dz \wedge dt
\end{aligned}$$

Show that the equation $dF = 0$ is equivalent to the last two Maxwell equations above. The first two Maxwell equations are related to G in a somewhat more complicated way. (See Exercise 17 for the details, and Exercise 18 for further implications of this approach.)

Solution

$$\begin{aligned}
dF =& d\big(B_x\, dy \wedge dz + B_y\, dz \wedge dx + B_z\, dx \wedge dy + E_x\, dx \wedge dt + E_y\, dy \wedge dt \\
&+ E_z\, dz \wedge dt\big) \\
=& \left(\frac{\partial B_x}{\partial t} dt + \frac{\partial B_x}{\partial x} dx + \frac{\partial B_x}{\partial y} dy + \frac{\partial B_x}{\partial z} dz\right) \wedge dy \wedge dz + \dots
\end{aligned}$$

There are six such terms in total. In each case only two terms in the brackets will survive when wedged with the associated 2-from because no wedge factors are repeated. Grouping the surviving terms, we obtain only four distinct 3-forms in the expansion of dF, namely,

$$\begin{aligned}
&\left(\frac{\partial B_x}{\partial x} + \frac{\partial B_y}{\partial y} + \frac{\partial B_z}{\partial z}\right) dx \wedge dy \wedge dz + \left(\frac{\partial B_x}{\partial t} + \frac{\partial E_z}{\partial y} - \frac{\partial E_y}{\partial z}\right) dt \wedge dy \wedge dz \\
&+ \left(\frac{\partial B_y}{\partial t} + \frac{\partial E_x}{\partial z} - \frac{\partial E_z}{\partial x}\right) dt \wedge dz \wedge dx + \left(\frac{\partial B_z}{\partial t} + \frac{\partial E_y}{\partial x} - \frac{\partial E_x}{\partial y}\right) dt \wedge dx \wedge dy
\end{aligned}$$

The coefficients for the respective 3-forms vanish if the latter two of Maxwell's equations hold. The first coefficient vanishes because $\nabla \bullet \mathbf{B} = 0$ while the remaining three represent the components of $\nabla \times \mathbf{E} + \partial \mathbf{B}/\partial t$. Thus, the latter two Maxwell equations are equivalent to $dF = 0$. The remaining two Maxwell equations can be expressed in the form $dG = H$, where H will be determined in Exercise 17.

Closed and Exact Forms

A differential k-form Φ is said to be **closed** if $d\Phi = 0$ (the zero $(k+1)$-form). Depending on the context, closed forms are analogous to irrotational or solenoidal vector fields. Since $d^2 = 0$, every exterior derivative is a closed form.

A differential k-form Φ is **exact** if $\Phi = d\Psi$ for some $(k-1)$-form Ψ. For $k = 1$, exact forms are analogous to conservative vector fields.

Every exact differential form is closed. Depending on the domain of the form, the converse of this statement may or may not be true. It is true for a smooth differential k-form (where $k \geq 1$) defined on a domain in $\mathbb{R}^k$ that can be shrunk to a point. We will not attempt to prove this here. A slightly weaker version is stated below for star-like domains. (See the discussion following Theorems 4 and 5 in Section 16.2.)

THEOREM 3

Poincaré's Lemma Let Φ be a smooth closed differential k-form defined on a star-like domain D in $\mathbb{R}^k$. Then Φ is exact on D.

We will not attempt a full proof of this theorem either, but suggest a proof for the special case $k = 1$ in Exercise 14 below.

EXERCISES 17.2

In Exercises 1–4 calculate the exterior derivatives of the given differential forms.

1. $\Phi = x^2\,dx + y^2\,dz$ in $\mathbb{R}^3$.

2. $f = x\,e^{2y}\sin(3z)$ in $\mathbb{R}^3$.

3. $\Psi = x_1\,dx_2 \wedge dx_3 + x_2\,dx_1 \wedge dx_4 + (x_3 + x_4)\,dx_1 \wedge dx_2$ in $\mathbb{R}^4$.

4. $\Theta = x_1x_2x_3\,dx_1 \wedge dx_3 \wedge dx_5 + x_3x_4x_5\,dx_2 \wedge dx_4 \wedge dx_5$.

5. Let Φ be the following differential 1-form: $\Phi = e^{2y}\sin(3z)\,dx + 2x\,e^{2y}\sin(3z)\,dy + 3x\,e^{2y}\cos(3z)\,dz$. Directly calculate $d\Phi$. Why are you not surprised at the result? (See Exercise 2.)

6. Repeat the previous exercise for the differential 3-form $\Phi = x_1x_3\,dx_1 \wedge dx_2 \wedge dx_3 \wedge dx_5 + x_4x_5\,dx_2 \wedge dx_3 \wedge dx_4 \wedge dx_5$. (See Exercise 4.)

7. Verify Theorem 2 (a). **8.** Verify Theorem 2 (b).

9. Generalize part (b) of Theorem 2 to a wedge product $\Phi \wedge \Psi \wedge \Theta$ of a differential k-form Φ, ℓ-form Ψ, and m-form Θ.

10. (A Leibniz Rule) Generalize the previous exercise to the wedge product $\Phi_1 \wedge \Phi_2 \wedge \cdots \wedge \Phi_m$, where Φ_i is a differential k_i-form for $1 \leq i \leq m$.

11. What vector differential identity (see Theorem 3 of Section 16.2) follows immediately from applying Theorem 2(c) and Example 2 to the differential 0-form f on $\mathbb{R}^3$?

12. What vector differential identity (see Theorem 3 of Section 16.2) follows immediately from applying Theorem 2(c) and Example 3 to the differential 1-form $F_1dx + F_2dy + F_3dz$ on $\mathbb{R}^3$?

Exercises 13–14 set up the proof of Poincaré's Lemma for differential 1-forms on star-like domains in $\mathbb{R}^k$.

13. Let $\Phi = \sum_{i=1}^{k} a_i(\mathbf{x})\,dx_i$ be a differential 1-form in $\mathbb{R}^k$. If $d\Phi = 0$, the zero differential 2-form on $\mathbb{R}^k$, show that

$$\frac{\partial a_i(\mathbf{x})}{\partial x_j} = \frac{\partial a_j(\mathbf{x})}{\partial x_i} \quad \text{for} \quad 1 \leq i, j \leq k.$$

14. Let D be a domain in $\mathbb{R}^k$ which is star-like with respect to a point x_0. (See the discussion following Theorems 4 and 5 in Section 16.2.) If $\Phi = \sum_{i=1}^{k} a_i(\mathbf{x})\,dx_i$ is a differential 1-form in D that satisfies $d\Phi = 0$, show that $\Phi = df$ for some differential 0-form f. *Hint:* Specifically, show that the function f defined for $x \in D$ by

$$f(\mathbf{x}) = \int_0^1 \sum_{i=1}^{k} x_i\,a_i\Big(\mathbf{x}_0 + t(\mathbf{x} - \mathbf{x}_0)\Big)\,dt$$

satisfies $df = \Phi$.

15. The thermodynamic variables (S, V, N) and their respective conjugates $(T, -P, \mu)$ were presented following Example 4. Use the wedge product structure and the fact that Legendre transformations (Section 12.6) ensure that an exact 1-form exists for any three variables selected from either set, excluding conjugate pairs, to determine how many equations between partial derivatives (i.e., Maxwell relations) are possible in the sense of Example 4.

16. Use exterior calculus and Legendre transformation considerations to generate Maxwell relations corresponding to the following wedge products:

(a) $dT \wedge dN$ (b) $dS \wedge d\mu$

(c) $dT \wedge -dP$ (d) $dT \wedge dV$

(e) $-dP \wedge dS$

17. (a) Find the exterior derivative of G from Example 5.

(b) Find a 3-form, H, such that the equation $dG = H$ implies the first two Maxwell equations listed above Example 5. Under what physical conditions is G a closed 2-form? What does this imply about $dF = 0$?

18. (Conservation of charge) Take the exterior derivative of $dG = H$ to find a differential equation in charge density ρ and charge current density $\mathbf{J}$ only, which expresses conservation of charge. Use the fact that $\mu_0\epsilon_0c^2 = 1$.

19. According to Exercise 17 in Section 16.6, the vector potential $\mathbf{A}$ and the scalar potential, ϕ satisfied $\mathbf{E} = -\nabla\phi - \frac{\partial \mathbf{A}}{\partial t}$ and $\mathbf{B} = \nabla \times \mathbf{A}$ in the fully time-varying case. The components of $\mathbf{A}$ and ϕ may be combined to form a "four vector" in space-time known as an " electromagnetic four-potential": $(A_x, A_y, A_z, -\phi)$. A 1-form is naturally created from the components of the four-potential, and the physical units in the potential equation for $\mathbf{E}$ suggest the following configuration:

$$\psi = -\phi\,dt + A_x\,dx + A_y\,dy + A_z\,dz.$$

Show that $d\psi = F$ and thus that $dF = 0$.

20. (The connection between F and G) Instead of using (x, y, z, t) as coordinates in space-time, it is considered more physically natural to use coordinates like (x, y, z, ct) all four of which have the same units (length). (Note: in theoretical physics, it is sometimes convenient to choose physical units so that $c = 1$ to avoid this issue.) Express the 2-forms F and G using coordinate ct instead of t. Using the fact that the elementary 2-forms from which F and G are constructed form a basis in the six-dimensional vector space of 2-forms in 4 variables, show that the vectors $\mathbf{F}$ and $\mathbf{G}$, having the same components as the coefficients of F and G respectively, satisfy $\mathbf{F} \bullet \mathbf{G} = 0$. Thus $\mathbf{F}$ and $\mathbf{G}$ are orthogonal, and in this sense the first two of Maxwell's equations listed above Example 5 may be regarded as orthogonal to, and hence independent of, the remaining two.

17.3 Integration on Manifolds

This section introduces the language of manifolds. It also introduces parametrizations to link integrals of differential forms to specific iterated integrals. While the concepts of vector calculus were adequate for extending the Fundamental Theorem of Calculus to functions in $\mathbb{R}^2$ and $\mathbb{R}^3$, they do not lend themselves to higher-dimensional problems. The natural setting for integration in $\mathbb{R}^n$ (which we will not encounter until Section 17.4) is the integral of a differential k-form Φ over a k-dimensional manifold $\mathcal{M}$:

This is a departure from notation in classical integral calculus because the "d" is hidden in Φ.

$$\int_{\mathcal{M}} \Phi.$$

A brief discussion of manifolds in $\mathbb{R}^n$ and their tangent and normal spaces was given in Section 13.4. We amplify this further here.

Smooth Manifolds

The graph of a function $\mathbf{f}$ from $\mathbb{R}^m$ into $\mathbb{R}^n$ is the set of all points $(\mathbf{x}, \mathbf{y}) \in \mathbb{R}^m \times \mathbb{R}^n = \mathbb{R}^{m+n}$ satisfying $\mathbf{y} = \mathbf{f}(\mathbf{x})$. The graph is smooth if all first-order partial derivatives of all n components of $\mathbf{f}$ exist and are continuous. (See Section 12.6 for a brief discussion of such functions.)

We need to introduce a general term like "manifold" because terms like "curve" and "surface", which worked in three or fewer dimensions, do not encompass all the smooth objects in higher dimensions. Roughly speaking, a smooth manifold of dimension k in $\mathbb{R}^n$ (where $m \le n$) is a subset $\mathcal{M}$ of $\mathbb{R}^n$ that is *locally* the graph of a smooth $(n-k)$-vector-valued function of k variables. To be more precise,

DEFINITION 7

A subset $\mathcal{M}$ of $\mathbb{R}^n$ is a **smooth manifold of dimension $k \le n$**, or, more simply, **a k-manifold in $\mathbb{R}^n$**, if for every point $\mathbf{x} \in \mathcal{M}$ there exists an open set U in $\mathbb{R}^n$ containing $\mathbf{x}$, and a smooth function $\mathbf{f}$ from U into $\mathbb{R}^{n-k}$ such that the following two conditions hold:

i) the part of $\mathcal{M}$ inside U is specified by the equation $\mathbf{f}(\mathbf{x}) = \mathbf{0}$, and

ii) the linear transformation $D\mathbf{f}(\mathbf{x})$ from $\mathbb{R}^n$ into $\mathbb{R}^{n-k}$ given by the Jacobian matrix

$$\frac{\partial(f_1, \ldots, f_{n-k})}{\partial(x_1, \ldots, x_n)} = \begin{pmatrix} \dfrac{\partial f_1}{\partial x_1} & \cdots & \dfrac{\partial f_1}{\partial x_n} \\ \vdots & \ddots & \vdots \\ \dfrac{\partial f_{n-k}}{\partial x_1} & \cdots & \dfrac{\partial f_{n-k}}{\partial x_n} \end{pmatrix}$$

is onto $\mathbb{R}^{n-k}$. (This is equivalent to asserting that the rows of the Jacobian matrix are linearly independent.)

EXAMPLE 1 (a) The graph $y = f(x_1, x_2, \dots, x_n)$ of a smooth real-valued function f is a smooth n-manifold in $\mathbb{R}^{n+1}$, that is, a smooth hypersurface in $\mathbb{R}^{n+1}$.

(b) An open set $\mathcal{M}$ in $\mathbb{R}^n$ is a smooth n-manifold in $\mathbb{R}^n$. Since $\mathcal{M}$ is open, we can take $U = \mathcal{M}$ and use the trivial function $f(\mathbf{x}) = 0$ from $\mathcal{M} \cap U$ into $\{0\}$, the zero-dimensional subspace of $\mathbb{R}^n$.

(c) Although $x^{1/3}$ is not a smooth function on $\mathbb{R}$, the curve $y = x^{1/3}$ is a smooth 1-manifold in $\mathbb{R}^2$ because it coincides with the curve $x = y^3$, and y^3 *is* a smooth function on $\mathbb{R}$.

(d) The sphere S with equation $x^2 + y^2 + z^2 = 1$ is a smooth 2-manifold in $\mathbb{R}^3$. Any point on the sphere is the centre of an open ball U whose radius is sufficiently small that the projection of U onto at least one of the coordinate planes, say the plane $x = 0$, lies inside S. The intersection $U \cap S$ will then be given by one of the two equations $x = \pm\sqrt{1 - y^2 - z^2}$ and will be smooth.

Strictly speaking, the two ways of describing smooth manifolds given at the right need only apply locally to pieces of the manifold rather than to the manifold as a whole. More about this in the next section.

There are two ways a smooth k-manifold $\mathcal{M}$ in $\mathbb{R}^n$ can be described:

(a) By requiring that its points $\mathbf{x} = (x_1, \dots, x_n)$ satisfy a set of $n - k$ independent equations in $(x_1, \dots, x_n)$:

$$\mathbf{f}(x_1, x_2, \dots, x_n) = \mathbf{0}, \qquad \text{where} \quad \mathbf{f} = (f_1, f_2, \dots f_{n-k}).$$

This was the method used to describe the constraint manifold in Section 13.4. Each equation represents an $(n - 1)$-dimensional surface in $\mathbb{R}^n$ and so reduces the dimension by 1. The equations are independent if the gradients $\nabla(f_i)$, $(1 \le i \le n - k)$ are linearly independent at every point $\mathbf{x} \in \mathcal{M}$. In this case, the dimension will be reduced by $n - k$ and so it will be k.

(b) By using a parametrization, that is, a mapping $\mathbf{x} = \mathbf{p}(\mathbf{u})$ from an open set $U \subset \mathbb{R}^k$ into $\mathbb{R}^n$ that satisfies

i) $\mathbf{p}(\mathbf{u})$ is one-to-one from U onto $\mathcal{M}$, and

ii) the linear transformation $D\mathbf{p}(\mathbf{u})$ from $\mathbb{R}^k$ into $\mathbb{R}^n$ with Jacobian matrix

$$J(\mathbf{u}) = \frac{\partial(x_1, \dots, x_n)}{\partial(u_1, \dots, u_k)} = \begin{pmatrix} \dfrac{\partial x_1}{\partial u_1} & \cdots & \dfrac{\partial x_1}{\partial u_k} \\ \vdots & \ddots & \vdots \\ \dfrac{\partial x_n}{\partial u_1} & \cdots & \dfrac{\partial x_n}{\partial u_k} \end{pmatrix}$$

is one-to-one. (See Section 12.6. This condition requires that the $n \times k$ matrix $J(\mathbf{u})$ have k linearly independent columns for each $\mathbf{u} \in U$.) Later in this section we will relax these conditions to allow slightly less restrictive parametrizations to be used for integration purposes.

Both descriptions have their good and bad features. For the equations description, it is easy to check whether a given point lies on the manifold, but hard to find a point on it. For the parametric description, it is easy to find points on the manifold but hard to check whether a given point lies on it.

EXAMPLE 2 Consider the two equations $f(x, y, z) = x^2 + z^2 - 1 = 0$ and $g(x, y, z) = x + y + z - 1 = 0$ in $\mathbb{R}^3$. Since $\nabla(f) = 2x\mathbf{i} + 2z\mathbf{k}$ and $\nabla(g) = \mathbf{i} + \mathbf{j} + \mathbf{k}$ are never linearly dependent, the two equations define a smooth manifold of dimension $3 - 2 = 1$, that is, a smooth curve in $\mathbb{R}^3$. (If you think about it for a moment, you will realize that this curve is an ellipse.)

EXAMPLE 3 Show that the following parametric equations define a smooth 2-manifold in $\mathbb{R}^4$.

$$(x_1, x_2, x_3, x_4) = \mathbf{x} = \mathbf{p}(\mathbf{u}) = (u_1^2 + u_2, 2u_1 + u_2^2, 3u_1 + u_2, u_1), \quad (0 < u_1, u_2 < 1)$$

Solution The Jacobian matrix of the transformation $\mathbf{x} = \mathbf{p}(\mathbf{u})$ is

$$J = \begin{pmatrix} \frac{\partial x_1}{\partial u_1} & \frac{\partial x_1}{\partial u_2} \\ \frac{\partial x_2}{\partial u_1} & \frac{\partial x_2}{\partial u_2} \\ \frac{\partial x_3}{\partial u_1} & \frac{\partial x_3}{\partial u_2} \\ \frac{\partial x_4}{\partial u_1} & \frac{\partial x_4}{\partial u_2} \end{pmatrix} = \begin{pmatrix} 2u_1 & 1 \\ 2 & 2u_2 \\ 3 & 1 \\ 1 & 0 \end{pmatrix}.$$

Since all the partials in J are continuous, the transformation is smooth. Since all but one of them is positive on the square $(0, 1) \times (0, 1)$, it is easily seen that the transformation is one-to-one; different points (u_1, u_2) in the square give different points in $\mathbb{R}^4$. Finally, since the last two rows of J are linearly independent for every (u_1, u_2), the range of the linear transformation $D\mathbf{p}(\mathbf{u})$ having matrix J is a two-dimensional subspace of $\mathbb{R}^4$. Thus, the range of the transformation is a 2-manifold in $\mathbb{R}^4$.

At every point $\mathbf{x}$ on a smooth k-manifold $\mathcal{M}$ in $\mathbb{R}^n$ there will exist a k-dimensional **tangent space** $T_{\mathbf{x}}(\mathcal{M})$ consisting of all vectors in $\mathbb{R}^n$ that are tangent to $\mathcal{M}$ at $\mathbf{x}$, and also an $(n-k)$-dimensional **normal space** $N_{\mathbf{x}}(\mathcal{M})$ consisting of all vectors in $\mathbb{R}^n$ that are normal to the tangent space, and therefore to $\mathcal{M}$ at $\mathbf{x}$.

For a manifold specified by $n-k$ equations $f_i(\mathbf{x}) = 0$, $1 \le i \le n-k$, the normal space will be spanned by the $n-k$ gradient vectors of the functions f_i evaluated at $\mathbf{x}$.

For manifolds specified by a parametrization $\mathbf{x}(\mathbf{u})$, the tangent space at $\mathbf{x}$ is spanned by the k-vectors $\partial\mathbf{x}/\partial u_i$, $(1 \le i \le k)$.

BEWARE! In the rest of this section, we are going to revert to the classic approach (from Chapter 14) to multiple integrals of functions as limits of Riemann sums, and extend them to $\mathbb{R}^n$. In so doing we will be writing volume elements dV_n as though they had meaning as differentials, which they do not. In Section 17.4, we will climb back on the wagon and properly define the integral of a differential form over a manifold.

Integration in *n* Dimensions

The definition of a double integral given in Section 14.1 (or a triple integral in Section 14.5) can be extended to integrals of real-valued functions $f(\mathbf{x}) = f(x_1, x_2, \ldots, x_n)$ over suitable domains in $\mathbb{R}^n$. First, we consider the rectangular domain $R = \{\mathbf{x} \in \mathbb{R}^n : a_i \le x_i \le b_i,\ 1 \le i \le n\}$, which we consider to have n-volume $\Pi_{i=1}^n (b_i - a_i)$. If f is continuous on R, we define the integral of f over R to be the limit of a suitable Riemann sum:

$$\int_R f(\mathbf{x})\, dV_n = \lim \sum_{i=1}^{N} f(\mathbf{x}_i)\, \mathrm{vol}_n(R_i),$$

where the sum is taken over a partition of R into N subrectangles R_i of volume $\mathrm{vol}_n(R_i)$ and $\mathbf{x}_i$ is a point in R_i. The limit is taken as $N \to \infty$ in such a way that the maximum dimension of the hyperrectangles R_i approaches zero. Note that we use a single integral sign rather than an n-fold one, which is rather too awkward.

It is easier to write

$$\int_R f(\mathbf{x})\, dV_n$$

than it is to write

$$\underbrace{\int\!\!\int \cdots \int_R}_{n} f(\mathbf{x})\, dV_n.$$

If f is defined in a domain $D \subset \mathbb{R}^n$ that is "sufficiently nice," we can find a hyperrectangle R containing D and define

$$\int_D f(\mathbf{x})\, dV_n = \int_R \hat{f}(\mathbf{x})\, dV_n,$$

where $\hat{f}$ is defined to be $f(x)$ if $x \in D$ and 0 otherwise. Even if f is continuous on D, it will likely be discontinuous on the boundary ∂D of D, so "How nice is sufficiently nice?" is a question that will have to be answered.

Some simple integrals over domains in $\mathbb{R}^n$ can be evaluated by the technique of *iteration* used to evaluate double and triple integrals in Chapter 14.

The "n-volume element" dV_n that we warned about in this chapter's introduction, is also written as $d\mathbf{x}$ or $dx_1\,dx_2\,\ldots\,dx_n$ sometimes. The latter notation is perhaps even more unsatisfactory than dV_n, as it suggests the differential of a vector, which it certainly is not. However, it does indicate what variables are being integrated, which is useful in this context. Another popular alternative, (never used in this book, including here) is to write $d^n x$. This also has its problems, as we are *not* speaking of an n-fold exterior derivative of x.

EXAMPLE 4 Evaluate $\int_Q x_1 x_2 \cdots x_n\, d\mathbf{x}$ over the hypercube $Q = \{\mathbf{x} \in \mathbb{R}^n : 0 \le x_i \le 1,\ 0 \le i \le n\}$.

Solution This integral iterates into n identical single integrals:

$$\int_Q x_1 x_2 \cdots x_n\, d\mathbf{x} = \int_0^1 x_1\, dx_1 \int_0^1 x_2\, dx_2 \cdots \int_0^1 x_n\, dx_n = \left(\frac{1}{2}\right)^n = \frac{1}{2^n}.$$

Sets of k-Volume Zero

We are used to manifolds having zero area in $\mathbb{R}^2$ (e.g., curves), or zero volume in $\mathbb{R}^3$, (e.g., curves and surfaces). In higher dimensions we have no such classical terminology for describing the "volume" of manifolds or their subsets, that may be zero in higher dimensional spaces. Accordingly, we make the following definition.

DEFINITION 8

Sets of k-volume zero in $\mathbb{R}^n$ Let $1 \le k \le n$. For each positive integer m, let $\mathcal{Q}_m$ be a partition of $\mathbb{R}^n$ into n-dimensional cubes each having edge length $1/2^m$. If S is a bounded subset of $\mathbb{R}^n$ we say that S has k-volume 0 if

$$\lim_{m\to\infty} \sum_{\substack{Q\in\mathcal{Q}_m \\ Q\cap S\ne\emptyset}} \frac{1}{2^{km}} = 0.$$

The sum is taken over only those cubes $Q \in \mathcal{Q}_m$ that contain points of S.

If S is unbounded, let $S_r = \{\mathbf{x} \in S : |\mathbf{x}| \le r\}$. We say that S has k-volume zero if S_r has k-volume zero for every positive r.

It can be shown that a smooth m-manifold in $\mathbb{R}^n$ has k-volume 0 provided $m < k \le n$.

If the boundary ∂D of a bounded open set $D \subset \mathbb{R}^n$ is a smooth $(n-1)$-manifold in $\mathbb{R}^n$, then ∂D has n-volume zero and will contribute nothing to the integral of a function f continuous on the closed, bounded set $D \cup \partial D$. Thus $\int_D f(\mathbf{x})\, dV_n$ will exist in this case.

Parametrizing and Integrating over a Smooth Manifold

In order to define integrals over a smooth k-manifold $\mathcal{M}$ in $\mathbb{R}^n$, where $k < n$ (such as, for example, curves in $\mathbb{R}^2$ and $\mathbb{R}^3$, and surfaces in $\mathbb{R}^3$), we need to parametrize the manifold using a smooth, one-to-one mapping from an open set in $\mathbb{R}^k$ onto $\mathcal{M}$. This approach generalizes the technique used to evaluate line and surface integrals in Chapter 15.

Unfortunately, the definition of parametrization given earlier in this section is a bit too restrictive; it rules out, for example, the parametrization $x = \cos u \cos v$, $y = \cos u \sin v$, $z = \sin u$, $0 \le u \le \pi$, $-\pi < v \le \pi$ of the sphere $x^2 + y^2 + z^2 = 1$ in $\mathbb{R}^3$. We can fix this by slightly easing the restrictions on parametrizations as follows.

DEFINITION 9

Smooth parametrization of a manifold
Let $\mathcal{M} \subset \mathbb{R}^n$ be a smooth, k-manifold in $\mathbb{R}^n$. Let U be a subset of $\mathbb{R}^k$ having boundary ∂U with k-volume 0. Let S be a subset of U with k-volume 0 such that $U - S = \{x \in U : x \notin S\}$ is open in $\mathbb{R}^k$. Suppose $\mathbf{p}$ is a mapping from U into $\mathbb{R}^n$ satisfying the following conditions:

i) $\mathbf{p}(S)$ has k-volume 0,

ii) $\mathcal{M} \subset \mathbf{p}(U)$,

iii) $\mathbf{p}(U - S) \subset \mathcal{M}$,

iv) $\mathbf{p}$ is one-to-one and differentiable on $U - S$, and

v) the derivative $D\mathbf{p}(\mathbf{u})$ is one-to-one from $\mathbb{R}^k$ onto the tangent space $T_{\mathbf{p}(\mathbf{u})}(\mathcal{M})$.

Then we say that $\mathbf{p}$ is a **smooth parametrization of $\mathcal{M}$ over U**, and that it is a **strict parametrization over $U - S$.**

These conditions are satisfied for the parametrization of the unit sphere in the paragraph preceding the definition if we take $U = \{(u, v) : 0 \le u \le \pi, -\pi < v \le \pi\}$ and $S = \{(u, v) : u = 0 \text{ or } u = \pi \text{ or } v = \pi\}$.

As was done for surface integrals in Section 15.5, we can evaluate an integral of a function $f(\mathbf{x})$ of n variables defined on a k-manifold $\mathcal{M}$ in $\mathbb{R}^n$ by transforming it into an integral of a function of k variables over a domain in $\mathbb{R}^k$. We do this by using a smooth parametrization $\mathbf{x} = \mathbf{p}(\mathbf{u})$. A differential volume element $dV_k(\mathbf{u}) = du_1\, du_2 \ldots du_k$ at point $\mathbf{u} \in \mathbb{R}^k$ is a k-dimensional rectangular box with corner at $\mathbf{u}$ spanned by the vectors $du_1\,\mathbf{e}_1,\ du_2\,\mathbf{e}_2,\ \ldots,\ du_k\,\mathbf{e}_k$. The derivative $D\mathbf{p}(\mathbf{u})$ transforms this volume element to a k-dimensional parallelogram in the tangent space $T_{\mathbf{p}(\mathbf{u})}(\mathcal{M})$, the k-volume of which provides the volume element $dV_k\big(\mathbf{p}(\mathbf{u})\big)$ on $\mathcal{M}$ at $\mathbf{p}(\mathbf{u})$.

DEFINITION 10

***k*-Parallelograms** A k-parallelogram at $\mathbf{y} \in \mathbb{R}^n$ spanned by the k vectors $\mathbf{v}_1, \ldots, \mathbf{v}_k$ is the set $P^k_{\mathbf{y}}(\mathbf{v}_1, \ldots, \mathbf{v}_k)$ of points $\mathbf{x} \in \mathbb{R}^n$ such that

$$\mathbf{x} = \mathbf{y} + \sum_{i=1}^{k} t_i \mathbf{v}_i, \quad \text{where } 0 < t_i < 1,\ 1 \le i \le k.$$

$P^k_{\mathbf{y}}(\mathbf{v}_1, \ldots, \mathbf{v}_k)$ is a k-manifold in $\mathbb{R}^n$.

Remark The k-volume of $P^k_{\mathbf{y}}(\mathbf{v}_1, \ldots, \mathbf{v}_k)$ is given by $\sqrt{G_k(\mathbf{v}_1, \ldots, \mathbf{v}_k)}$, where

$$G_k(\mathbf{v}_1, \ldots, \mathbf{v}_k) = \begin{vmatrix} \mathbf{v}_1 \bullet \mathbf{v}_1 & \mathbf{v}_2 \bullet \mathbf{v}_1 & \cdots & \mathbf{v}_k \bullet \mathbf{v}_1 \\ \mathbf{v}_1 \bullet \mathbf{v}_2 & \mathbf{v}_2 \bullet \mathbf{v}_2 & \cdots & \mathbf{v}_k \bullet \mathbf{v}_2 \\ \vdots & \vdots & \ddots & \vdots \\ \mathbf{v}_1 \bullet \mathbf{v}_k & \mathbf{v}_2 \bullet \mathbf{v}_k & \cdots & \mathbf{v}_k \bullet \mathbf{v}_k \end{vmatrix}.$$

See Exercises 7–9 for a suggestion on how to prove this fact. In particular, if $n = k$, so that the vectors $\mathbf{v}_i$ are all in $\mathbb{R}^k$, then the k-volume of $P^k_{\mathbf{y}}(\mathbf{v}_1, \ldots, \mathbf{v}_k)$ is given by $|\det(A)|$, where A is the $k \times k$ square matrix whose columns are the components of the vectors $\mathbf{v}_i$,

$$A = \begin{pmatrix} \vdots & \vdots & \cdots & \vdots \\ \mathbf{v}_1 & \mathbf{v}_2 & \cdots & \mathbf{v}_k \\ \vdots & \vdots & \cdots & \vdots \end{pmatrix}$$

because $G_k(\mathbf{v}_1, \ldots, \mathbf{v}_k) = \det(A^T A) = \big(\det(A)\big)^2$.

The derivative of the transformation $\mathbf{x} = \mathbf{p}(\mathbf{u})$ is the linear transformation of $\mathbb{R}^k$ to the tangent space $T_{\mathbf{p}(\mathbf{u})}(\mathcal{M})$ given by the $n \times k$ Jacobian matrix

$$J(\mathbf{u}) = \frac{\partial(x_1, \ldots, x_n)}{\partial(u_1, \ldots, u_k)} = \begin{pmatrix} \dfrac{\partial x_1}{\partial u_1} & \cdots & \dfrac{\partial x_1}{\partial u_k} \\ \vdots & \ddots & \vdots \\ \dfrac{\partial x_n}{\partial u_1} & \cdots & \dfrac{\partial x_n}{\partial u_k} \end{pmatrix}.$$

The columns of this matrix are the k vectors that span the k-parallelogram, which is the image of the k-cube spanned by the standard basis vectors in $\mathbb{R}^k$ under the parametrization $\mathbf{p}$. It follows that the k volume element at $\mathbf{p}(\mathbf{u})$ on $\mathcal{M}$ is given by

$$dV_k\big(\mathbf{p}(\mathbf{u})\big) = \sqrt{G_k\left(\frac{\partial \mathbf{p}}{\partial u_1}, \frac{\partial \mathbf{p}}{\partial u_2}, \ldots, \frac{\partial \mathbf{p}}{\partial u_k}\right)}\, dV_k(\mathbf{u}),$$

or, since the matrix $J(\mathbf{u})^T J(\mathbf{u})$ is a square $k \times k$ matrix that has the same elements as the determinant G_k,

$$dV_k\big(p(\mathbf{u})\big) = \sqrt{\det(J(\mathbf{u})^T J(\mathbf{u}))}\, du_1\, du_2 \cdots du_k.$$

Now suppose we want to integrate a function $f(\mathbf{x}) = f(x_1, x_2, \ldots, x_n)$ over a smooth k-manifold $\mathcal{M}$ parametrized by the mapping $\mathbf{p}$ as described in Definition 9. We want to transform the integral of f over $\mathcal{M}$ to an equivalent integral of $f\big(\mathbf{p}(\mathbf{u})\big)$ over U in $\mathbb{R}^k$. Since $p(S)$ has k-volume 0, it is sufficient to integrate g over $U - S$, where $\mathbf{p}$ is one-to-one and differentiable. Thus,

$$\int_{\mathcal{M}} f(\mathbf{x})\, dV_k(\mathbf{x}) = \int_{U-S} f\big(\mathbf{p}(\mathbf{u})\big) \sqrt{\det(J(\mathbf{u})^T J(\mathbf{u}))}\, d\mathbf{u}.$$

In particular, the k-volume of the k-manifold $\mathcal{M}$ is given by

$$\int_{U-S} \sqrt{\det(J(\mathbf{u})^T J(\mathbf{u}))}\, d\mathbf{u}.$$

The same simplification observed above (when $n = k$) for the volume of a k-parallelogram in $\mathbb{R}^k$ occurs if the k-manifold is "flat," that is, if it is an open subset of $\mathbb{R}^k$. In this case, the parametrization $\mathbf{p}(\mathbf{u})$ is just a transformation of coordinates in $\mathbb{R}^k$, and the Jacobian matrix of the derivative $D\mathbf{p}(\mathbf{u})$ is just a $k \times k$ square matrix $J(\mathbf{u})$, whose determinant is equal to that of its transpose. It follows that

$$\det(J(\mathbf{u})^T J(\mathbf{u})) = \big(\det(J(\mathbf{u}))\big)^2$$

and so the transformed volume element is

$$|\det(J(\mathbf{u})|\, d\mathbf{u} = \left|\frac{\partial(x_1, x_2, \ldots, x_k)}{\partial(u_1, u_2, \ldots, u_k)}\right| du_1\, du_2 \cdots du_k.$$

This is just the k-dimensional analogue of the general change-of-variables area and volume elements for double and triple integrals given in Sections 14.4 and 14.6.

EXERCISES 17.3

In Exercises 1–4 find the k-volumes of the k-parallelograms in $\mathbb{R}^4$ spanned by the vectors with the given components.

1. $k = 2$, $\mathbf{v}_1 = (1, 2, 1, 0)$, $\mathbf{v}_2 = (2, -1, 0, -1)$
2. $k = 2$, $\mathbf{v}_1 = (1, 1, 1, 1)$, $\mathbf{v}_2 = (1, -1, -1, 0)$
3. $k = 3$, $\mathbf{v}_1 = (1, 1, 0, 0)$, $\mathbf{v}_2 = (0, 1, 1, 0)$, $\mathbf{v}_3 = (0, 0, 1, 1)$
4. $k = 4$, $\mathbf{v}_1 = (1, 0, 0, 0)$, $\mathbf{v}_2 = (1, 1, 0, 0)$, $\mathbf{v}_3 = (0, 0, 1, 1)$, $\mathbf{v}_4 = (1, 0, 1, 0)$
5. Find $\int_{\mathcal{M}} (x_1 + x_2)\, dV_2(\mathbf{x})$, where $\mathcal{M}$ is the 2-manifold in $\mathbb{R}^4$ given parametrically by $\mathbf{x} = (u_1 + u_2, u_1 - u_2, u_1^2, 1 + u_2)$ for $0 < u_1 < 1$, $0 < u_2 < 1$.
6. Find $\int_{\mathcal{M}} \sqrt{1 + x_1^2 + x_3^2}\, dV_2(\mathbf{x})$, where $\mathcal{M}$ is the 2-manifold in $\mathbb{R}^4$ given parametrically by $\mathbf{x} = \left(u_1 \cos(u_2), \sqrt{3}u_1, u_1 \sin(u_2), u_2\right)$ for $0 < u_1 < 1$, $0 < u_2 < \pi/2$.

Exercises 7–9 provide a proof of the claim made concerning the k-volume of a k-parallelogram in $\mathbb{R}^n$ following Definition 10. They concern the determinant function $G_k(\mathbf{v}_1, \dots, \mathbf{v}_k)$ defined there.

7. If $\mathbf{v}_1, \mathbf{v}_2, \dots, \mathbf{v}_k$ are k vectors in $\mathbb{R}^n$, where $n \geq k$, and M is the $n \times k$ matrix whose jth column consists of the components of $\mathbf{v}_j$, show that $\det(M^T M) = G_k(\mathbf{v}_1, \dots, \mathbf{v}_k)$.
8. Show that the 2-volume (i.e., area A) of the parallelogram spanned by the vectors $\mathbf{v}_1$ and $\mathbf{v}_2$ is given by $A = \sqrt{G_2(\mathbf{v}_1, \mathbf{v}_2)}$.
9. Complete the proof of the formula for the k-volume of a k-parallelogram spanned by the vectors $\mathbf{v}_1, \dots, \mathbf{v}_k$ by induction on k. The case $k = 1$ is trivial, and the case $k = 2$ is done in the previous exercise. A $(k + 1)$-parallelogram has $2(k + 1)$ faces, each of which is a k-parallelogram. The $(k + 1)$-volume of the $(k + 1)$-parallelogram is the k-volume of one of its faces (say, spanned by $\mathbf{v}_1, \dots, \mathbf{v}_k$) multiplied by the length h of the perpendicular projection of the remaining edge $\mathbf{v}_{k+1}$ onto the k-dimensional subspace containing that face. You will find Cramer's Rule (Theorem 6 of Section 10.7) useful in finding h^2.
10. Let Φ be the k-form $\Phi = dx_{i_1} \wedge dx_{i_2} \wedge \cdots \wedge dx_{i_k}$, $1 \leq i_1 < i_2 < \cdots < i_k \leq n$. Show that the k-volume of the projection P of the k-parallelogram in $\mathbb{R}^n$ spanned by the vectors $\mathbf{v}_1, \mathbf{v}_2, \dots, \mathbf{v}_k$ in $\mathbb{R}^n$ onto the k-dimensional coordinate plane in $\mathbb{R}^n$ spanned by $\mathbf{e}_{i_1}, \mathbf{e}_{i_2}, \dots, \mathbf{e}_{i_k}$ is given by $|\Phi(\mathbf{v}_1, \mathbf{v}_2, \dots, \mathbf{v}_k)|$.

17.4 Orientations, Boundaries, and Integration of Forms

Oriented Manifolds

As noted at the beginning of this chapter, it was the fact that an interval on the real line has a natural orientation (left to right) that enabled us to formulate the Fundamental Theorem of Calculus for functions of one variable. We now examine how to specify the orientation of a manifold.

DEFINITION

If V is a k-dimensional vector space, then any nonzero k-form ω on V defines an **orientation** for V. Any k-dimensional vector space has only two orientations. Since $\Lambda_k(V)$ is one dimensional, any nonzero k-form on V will be a multiple of ω by a nonzero real number, either positive or negative. The positive multiples of ω provide the same orientation for V as ω; the negative multiples provide the **opposite** orientation for V.

For example, $\omega = dx_1 \wedge dx_2 \wedge \cdots \wedge dx_n$ provides an orientation for $\mathbb{R}^n$ that gives the value $+1$ to the standard basis $\mathbf{e}_1, \mathbf{e}_2, \dots, \mathbf{e}_n$. If $n = 3$, this orientation is shared by any basis satisfying the "right-hand-rule." (See Section 10.1.) We usually refer to the orientation of $\mathbb{R}^n$ given by ω as the "positive" orientation.

The tangent space $T_{\mathbf{x}}(\mathcal{M})$ at any point x on a smooth, k-manifold in $\mathbb{R}^n$ is itself a k-dimensional vector subspace of $\mathbb{R}^n$ and so has one of two possible orientations. If we can select a non-zero k-form on each such subspace in a way that varies smoothly with $\mathbf{x}$, they will constitute a smooth k-form field on $\mathcal{M}$, which then orients the manifold.

DEFINITION 12

Supose that at each $\mathbf{x}$ on the k-manifold $\mathcal{M}$, there exists a nonzero k-form $\omega_{\mathbf{x}}$ that varies smoothly with $\mathbf{x}$ and orients the tangent space $T_{\mathbf{x}}(\mathcal{M})$, then we say that $\mathcal{M}$ is **orientable** and that the differential k-form field $\omega(\mathbf{x}) = \omega_{\mathbf{x}}$ **orients** $\mathcal{M}$.

EXAMPLE 1 **(Orienting a curve in $\mathbb{R}^n$)** A smooth curve (1-manifold) $\mathcal{C}$ in $\mathbb{R}^n$ is orientable if there exists a nonvanishing, smoothly varying tangent vector field $\mathbf{t}(\mathbf{x})$ on $\mathcal{C}$. Since the tangent space $T_{\mathbf{x}}(\mathcal{C})$ is one-dimensional, the differential 1-form whose value at any $\mathbf{x}$ on $\mathcal{C}$ and any $\mathbf{v} \in \mathbb{R}^n$ is given by $\omega(\mathbf{x})(\mathbf{v}) = \mathbf{t}(\mathbf{x}) \bullet \mathbf{v}$ orients $\mathcal{C}$. It specifies the positive direction of $\mathcal{C}$ at $\mathbf{x}$ as the direction of $\mathbf{t}(\mathbf{x})$. The opposite (negative) direction would be specified by the nonvanishing tangent field $-\mathbf{t}(\mathbf{x})$.

EXAMPLE 2 **(Orienting a hypersurface in $\mathbb{R}^n$)** A smooth $(n-1)$-manifold $\mathcal{M}$ in $\mathbb{R}^n$ (also called a hypersurface) is orientable if there exists a nonvanishing, smoothly varying normal vector field $\mathbf{n}(\mathbf{x})$ on $\mathcal{M}$. For instance, if $\mathcal{M}$ is specified by the single equation $g(\mathbf{x}) = 0$, and $\mathbf{grad}\, g(\mathbf{x}) \neq \mathbf{0}$ anywhere on $\mathcal{M}$, then $\mathbf{n}(\mathbf{x}) = \mathbf{grad}\, g(\mathbf{x})$ can provide an orientation for $\mathcal{M}$. Since the tangent space $T_{\mathbf{x}}(\mathcal{M})$ to $\mathcal{M}$ at $\mathbf{x}$ has dimension $n-1$, any $n-1$ linearly independent vectors $\mathbf{v}_1, \mathbf{v}_2, \ldots, \mathbf{v}_{n-1}$ in $T_{\mathbf{x}}(\mathcal{M})$ will be perpendicular to $\mathbf{n}$, and the differential $(n-1)$-form

$$\omega(\mathbf{x})(\mathbf{v}_1, \mathbf{v}_2, \ldots, \mathbf{v}_{n-1}) = \begin{vmatrix} \vdots & \vdots & \cdots & \vdots \\ \mathbf{n}(\mathbf{x}) & \mathbf{v}_1 & \cdots & \mathbf{v}_{n-1} \\ \vdots & \vdots & \cdots & \vdots \end{vmatrix}$$

orients $\mathcal{M}$. It specifies a "positive side" of $\mathcal{M}$, out of which $\mathbf{n}(\mathbf{x})$ points, and a negative side, out of which $-\mathbf{n}(\mathbf{x})$ points.

EXAMPLE 3 **(Orienting an open set in $\mathbb{R}^n$)** An open set $\mathcal{M}$ in $\mathbb{R}^n$ is an n-manifold. At any point $\mathbf{x} \in \mathcal{M}$, the tangent space $T_{\mathbf{x}}(\mathcal{M}) = \mathbb{R}^n$ and the normal space is zero-dimensional subspace $\{0\}$. For any vectors $\mathbf{v}_1, \mathbf{v}_2, \ldots, \mathbf{v}_n$ in $\mathbb{R}^n$, the differential n-form

$$dx_1 \wedge dx_2 \wedge \cdots \wedge dx_n(\mathbf{v}_1, \mathbf{v}_2, \ldots, \mathbf{v}_n) = \begin{vmatrix} \vdots & \vdots & \cdots & \vdots \\ \mathbf{v}_1 & \mathbf{v}_2 & \cdots & \mathbf{v}_n \\ \vdots & \vdots & \cdots & \vdots \end{vmatrix}$$

orients $\mathcal{M}$.

It is useful to regard a single point $\mathbf{x}$ in $\mathbb{R}^n$ as a 0-dimensional manifold. In this case choosing an orientation comes down to choosing a sign to attach to the point. One orientation of $\mathbf{x}$ is $+\mathbf{x}$; the opposite orientation is $-\mathbf{x}$.

When we say that an oriented point is $+\mathbf{x}$ or $-\mathbf{x}$, we don't intend the orientation signs "+" or "−" to mean scalar multiplication of the vector $\mathbf{x}$ by 1 or -1. Rather, we mean that the value of a function f at $+\mathbf{x}$ is $f(\mathbf{x})$, while the value of f at $-\mathbf{x}$ is $-f(\mathbf{x})$. This will be important when we evaluate the "integral" (i.e., sum) of a 0-form over the oriented boundary of an interval $[a, b]$ later in this chapter.

Not every smooth manifold is orientable. The Möbius band illustrated in Section 15.6 has only one side and is not orientable.

The following example illustrates how you can orient a k-manifold $\mathcal{M}$ in $\mathbb{R}^n$, where $1 < k < n-1$. The idea is to find a basis of $n-k$ vectors for the normal space at an arbitrary point $\mathbf{x}$ on $\mathcal{M}$ and use them to define a differential k-form that orients $\mathcal{M}$.

EXAMPLE 4 Find an orientation for the 2-manifold $\mathcal{M}$ in $\mathbb{R}^4$ specified by the equations $f(\mathbf{x}) = x_1 + x_3 = 0$ and $g(\mathbf{x}) = x_2 - x_4^2 = 0$.

Solution At any point $\mathbf{x}$ satisfying the two equations, the vectors $\mathbf{grad}\, f = \mathbf{e}_1 + \mathbf{e}_3$ and $\mathbf{grad}\, g = \mathbf{e}_2 - 2x_4\mathbf{e}_4$ are normal to $\mathcal{M}$ and are clearly linearly independent. Thus, they span the 2-dimensional normal space to $\mathcal{M}$ at $\mathbf{x}$. If $\mathbf{u}$ and $\mathbf{v}$ are linearly independent vectors in the tangent space $T_{\mathbf{x}}(\mathcal{M})$, then the differential 2-form

$$\omega(\mathbf{x})(\mathbf{u},\mathbf{v}) = \begin{vmatrix} 1 & 0 & u_1 & v_1 \\ 0 & 1 & u_2 & v_2 \\ 1 & 0 & u_3 & v_3 \\ 0 & -2x_4 & u_4 & v_4 \end{vmatrix}$$

defines an orientation for $\mathcal{M}$. Observe the first two columns are the components of the two independent normals, and the determinant depends smoothly on $\mathbf{x}$. To verify that it is never zero, observe that the vectors $\mathbf{t} = \mathbf{e}_1 - \mathbf{e}_3$ and $\mathbf{w} = 2x_4\mathbf{e}_2 + \mathbf{e}_4$ are perpendicular to each other and to each of the two normals. They must therefore span the 2-dimensional tangent space to $\mathcal{M}$ at $\mathbf{x}$. Direct calculation shows that

$$\omega(\mathbf{x})(\mathbf{t},\mathbf{w}) = -2(1+4x_4^2) < 0$$

for all $\mathbf{x}$. If $\mathbf{u} = \alpha\mathbf{t} + \beta\mathbf{w}$ and $\mathbf{v} = \gamma\mathbf{t} + \delta\mathbf{w}$, where $\alpha\delta - \beta\gamma \neq 0$ so that $\mathbf{u}$ and $\mathbf{v}$ are linearly independent, then

$$\begin{aligned} \omega(\mathbf{x})(\mathbf{u},\mathbf{v}) &= \begin{vmatrix} 1 & 0 & \alpha t_1 + \beta w_1 & \gamma t_1 + \delta w_1 \\ 0 & 1 & \alpha t_2 + \beta w_2 & \gamma t_2 + \delta w_2 \\ 1 & 0 & \alpha t_3 + \beta w_3 & \gamma t_3 + \delta w_3 \\ 0 & -2x_4 & \alpha t_4 + \beta w_4 & \gamma t_4 + \delta w_4 \end{vmatrix} \\ &= \begin{vmatrix} 1 & 0 & t_1 & w_1 \\ 0 & 1 & t_2 & w_2 \\ 1 & 0 & t_3 & w_3 \\ 0 & -2x_4 & t_4 & w_4 \end{vmatrix} \begin{vmatrix} 1 & 0 & 0 & 0 \\ 0 & 1 & 0 & 0 \\ 0 & 0 & \alpha & \gamma \\ 0 & 0 & \beta & \delta \end{vmatrix} = -(2+4x_4^2)(\alpha\delta - \beta\gamma), \end{aligned}$$

since the determinant of a product is the product of the determinants. The result is nonzero and has constant sign for all $\mathbf{x} \in \mathcal{M}$, so ω orients $\mathcal{M}$. If $\alpha\delta - \beta\gamma < 0$, the positive orientation will be given by using the ordered pair $(\mathbf{t},\mathbf{w})$ as a basis for the tangent space $T_{\mathbf{x}}(\mathcal{M})$. Otherwise, use $(\mathbf{w},\mathbf{t})$.

Pieces-with-Boundary of a Manifold

The extension of the Fundamental Theorem of Calculus (the Generalized Stokes Theorem) that we will develop in the next section relates the integral of the exterior derivative $d\Phi$ of a differential $(k-1)$-form Φ over a subset M of an oriented k-manifold in $\mathbb{R}^n$ to the integral of Φ over the suitably oriented boundary ∂M of S: $\int_M d\Phi = \int_{\partial M} \Phi$. We must now clarify some of these terms, in particular, the kind of set M must be to enable the evaluation of integrals over its boundary. Boundaries of open sets can be very pathological, and we will have to restrict them somehow.

A manifold $\mathcal{M}$ in $\mathbb{R}^n$ does not itself contain any boundary points, but a subset M of $\mathcal{M}$ can have a boundary contained in $\mathcal{M}$. Specifically, **the boundary** ∂M of M in $\mathcal{M}$ consists of all points $\mathbf{x} \in \mathcal{M}$ such that every open set $U \subset \mathbb{R}^n$ containing $\mathbf{x}$ also contains points $\mathbf{y} \neq \mathbf{x}$ in M and points $\mathbf{y} \neq \mathbf{x}$ that are in $\mathcal{M}$ but not in M. The boundary may or may not be a subset of M.

EXAMPLE 5 (a) The sphere $x^2 + y^2 + z^2 = 1$ in $\mathbb{R}^3$ (a smooth 2-manifold) has no boundary, but its upper hemisphere (the subset H of the sphere where $z \geq 0$) has a boundary ∂H consisting of all points on the circle $x^2 + y^2 = 1$, $z = 0$. For this example, the boundary is a smooth manifold of dimension 1, and H contains its boundary.

(b) All of $\mathbb{R}^2$ is a 2-manifold $\mathcal{M}$ in $\mathbb{R}^2$. It has no boundary, but the square subset $Q = \{(x, y) \in \mathbb{R}^2 : 0 \leq x \leq 1,\ 0 \leq y \leq 1\}$ does have a boundary. ∂Q consists of all points on the four edges of square. This boundary is not a smooth manifold, but if we omit the four corners of the square, each of the remaining straight line segments is a 1-dimensional manifold in $\mathbb{R}^2$. Again, $\partial Q \subset Q$.

It would be nice if we only had to deal with boundaries that are smooth, but such an assumption is too restrictive for our purposes.

DEFINITION 13

Smooth and nonsmooth boundary points Let M be a subset of a k-manifold $\mathcal{M}$ in $\mathbb{R}^n$. A point $\mathbf{x}$ on the boundary ∂M is a **smooth boundary point** of M if there exists an open set $U \subset \mathbb{R}^n$ containing $\mathbf{x}$, a smooth function $\mathbf{f}$ mapping U into $\mathbb{R}^{n-k}$, and a smooth function g mapping U into $\mathbb{R}$, such that:

i) $\mathcal{M} \cap U = \{\mathbf{y} \in U : \mathbf{f}(\mathbf{y}) = 0\}$,

ii) $M \cap U = \{\mathbf{y} \in U : \mathbf{f}(\mathbf{y}) = 0 \quad \text{and} \quad g(\mathbf{y}) \geq 0\}$, and

iii) If $\mathbf{h}(\mathbf{x}) = \big(\mathbf{f}(\mathbf{x}), g(\mathbf{x})\big)$, then $D\mathbf{h}(\mathbf{x})$ maps $\mathbb{R}^n$ onto $\mathbb{R}^{n-k+1}$.

The set of all smooth boundary points of M constitutes the **smooth boundary** of M. The points of ∂M that are not smooth boundary points constitute the **nonsmooth boundary** of M.

Remark The smooth boundary of a subset M of a smooth k-manifold in $\mathbb{R}^n$ consists of one or more smooth $(k-1)$-manifolds in $\mathbb{R}^n$.

DEFINITION

A **piece-with-boundary** of a k-manifold $\mathcal{M}$ in $\mathbb{R}^n$ is a closed (in $\mathbb{R}^n$) subset M of $\mathcal{M}$ satisfying

i) the nonsmooth part of the boundary of M has $(k-1)$-volume zero, and

ii) for every point $\mathbf{x} \in \partial M$, there is an open set $U \subset \mathbb{R}^n$ such that $\{\mathbf{y} \in \partial M \cap U\}$ has finite $(k-1)$-volume.

Evidently, both of the subsets in Example 5 above are *pieces-with-boundary* of their respective manifolds. The smooth boundary of the hemisphere in part (a) is the whole circle.[1] The smooth boundary of the square region in part (b) consists of the four sides of the square excluding the corner points which are nonsmooth boundary points.

The tangent space to the smooth boundary ∂M of a piece-with-boundary M of k-manifold $\mathcal{M}$ at $\mathbf{x}$ is a $(k-1)$-dimensional subspace of the k-dimensional tangent space to $\mathcal{M}$ at $\mathbf{x}$. Therefore, there exists a 1-dimensional space which is tangent to $\mathcal{M}$ at $\mathbf{x}$ but normal to ∂M at $\mathbf{x}$. This normal space is spanned by $\mathbf{grad}\, g(\mathbf{x})$ where g is the function in the definition of smooth boundary. Since $g(\mathbf{x}) = 0$ and $g(\mathbf{y}) \geq 0$ for $\mathbf{y} \in M$, $\mathbf{grad}\, g(\mathbf{x})$ is a normal pointing into M and $-\mathbf{grad}\, g(\mathbf{x})$ is an outward pointing normal. (Note that condition (iii) of Definition 13 guarantees that $\mathbf{grad}\, g(\mathbf{x}) \neq \mathbf{0}$.) We always use an **outer normal** to orient the smooth boundary of M, describing the result as the **orientation inherited from the orientation of M.**

DEFINITION

Inherited orientation of the smooth boundary If M is a piece-with-boundary of an oriented (by ω) k-manifold in $\mathbb{R}^n$, the $(k-1)$-dimensional smooth boundary of M inherits the orientation $\partial\omega$ given by the differential $(k-1)$-form

$$\partial\omega(\mathbf{x})(\mathbf{v}_1, \ldots, \mathbf{v}_{k-1}) = \omega(\mathbf{x})(\mathbf{n}(\mathbf{x}), \mathbf{v}_1, \ldots, \mathbf{v}_{k-1}),$$

where $\mathbf{n}(\mathbf{x})$ is a normal field on the smooth part of ∂M that points out of M.

We have already seen this situation when considering Green's Theorem in $\mathbb{R}^2$, Stokes' Theorem in $\mathbb{R}^3$, and the Divergence Theorem in both $\mathbb{R}^2$ and $\mathbb{R}^3$ in Chapter 16. The following examples confirm that the definition above gives the same result as the orientations used there.

[1] Like some other terms used here, "piece-with-boundary" was introduced by John and Barbara Hubbard in their text *Vector Calculus, Linear Algebra, and Differential Forms*, 2nd ed., Prentice Hall, 2002.

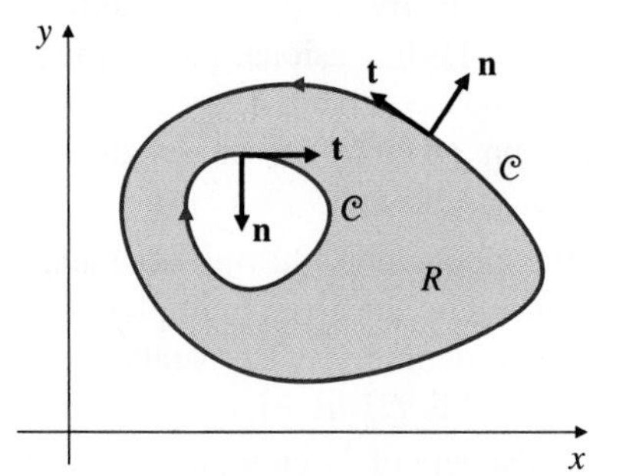

Figure 17.1 The orientation of $\mathcal{C}$ inherited from the standard orientation of R

EXAMPLE 6 A region R in $\mathbb{R}^2$ bounded by one or more piecewise smooth closed curves $\mathcal{C}$ is a piece-with-boundary of the 2-manifold $\mathbb{R}^2$. If we assume that $\mathbb{R}^2$ is oriented by the 2-form $\omega = dx \wedge dy$ so that $\omega(\mathbf{i}, \mathbf{j}) = 1$, then $\omega(\mathbf{n}, \mathbf{t})$ will be positive whenever $\mathbf{n}$ is an outward (from R) normal to $\mathcal{C}$ and $\mathbf{t}$ is a tangent to $\mathcal{C}$ in the direction of the orientation of $\mathcal{C}$. See Figure 17.1. The positive direction of $\mathcal{C}$ given by $\mathbf{t}$ is 90° counterclockwise from the outward normal.

EXAMPLE 7 Let $\mathcal{S}$ be a smooth surface (2-manifold) in $\mathbb{R}^3$. Let $\mathbf{n} = n_1\mathbf{i} + n_2\mathbf{j} + n_3\mathbf{k}$ be a nonvanishing, smoothly varying normal vector field on $\mathcal{S}$. As suggested in Example 2 above, $\mathcal{S}$ can be oriented with the differential 2-form

$$\omega(\mathbf{x})(\mathbf{v}_1, \mathbf{v}_2) = \begin{vmatrix} n_1(\mathbf{x}) & v_{11} & v_{21} \\ n_2(\mathbf{x}) & v_{12} & v_{22} \\ n_3(\mathbf{x}) & v_{13} & v_{23} \end{vmatrix}$$

for $\mathbf{v}_1$ and $\mathbf{v}_2$ in the tangent space to $\mathcal{S}$ at $\mathbf{x}$.

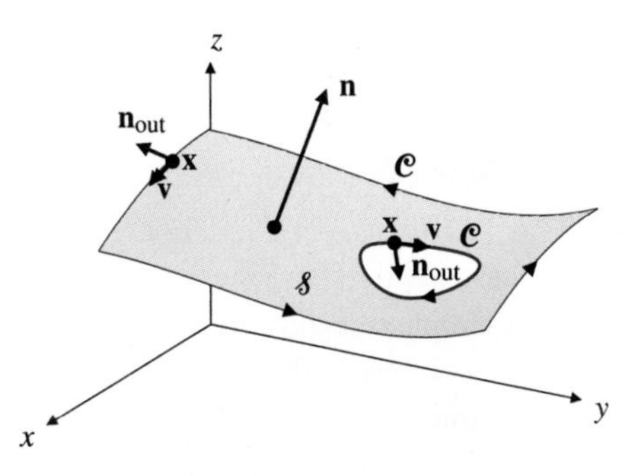

Figure 17.2 Boundary orientation inherited from the orientation of a smooth surface in $\mathbb{R}^3$.

If S is a piece-with-boundary of $\mathcal{S}$ having smooth boundary consisting of pieces of curves (1-manifolds), the boundary ∂S inherits the orientation given by the differential 1-form $\partial\omega(\mathbf{x})(\mathbf{v}) = \omega(\mathbf{x})(\mathbf{n}_{\text{out}}, \mathbf{v})$, where n_{out} is an outward (from S) normal to the smooth boundary at $\mathbf{x}$ and $\mathbf{v}$ is tangent to that boundary. It follows that $\partial\omega(\mathbf{x})(\mathbf{v})$ is the value of the 3×3 determinant whose columns, in order, are the components of the normal field $\mathbf{n}$ orienting $\mathcal{S}$, the components of $\mathbf{n}_{\text{out}}$ (which is tangent to $\mathcal{S}$ but normal to the boundary of S), and the components of $\mathbf{v}$. Assuming that $\mathbb{R}^3$ has the standard basis, the vectors $\mathbf{n}$, $\mathbf{n}_{\text{out}}$, and $\mathbf{v}$ form a right-handed-triad, so the boundary orientation is such that if we stand erect on the smooth boundary of S (head upward in the direction of $\mathbf{n}$), facing out of S (i.e., in the direction of $\mathbf{n}_{\text{out}}$), then the positive direction of the boundary of S will be to our left. See Figure 17.2.

EXAMPLE 8 Consider a cube Q in $\mathbb{R}^3$ with the standard orientation given by $\omega = dx \wedge dy \wedge dz$. ∂Q consists of 6 square faces (smooth boundary) and 12 edges together with their endpoints (nonsmooth boundary). Each square face is oriented with an outward normal $\mathbf{n}$ inherited from ω. In turn, each square face induces an orientation on its four edges. That orientation is counterclockwise as seen from a point outside the cube in the direction of the normal for that face. Note that every edge of the cube is part of the smooth boundary of two of the square faces and those faces induce opposite orientations on that edge. See Figure 17.3. This "cancellation" suggests the observation that the boundary of a boundary of a piece-with-boundary is empty.

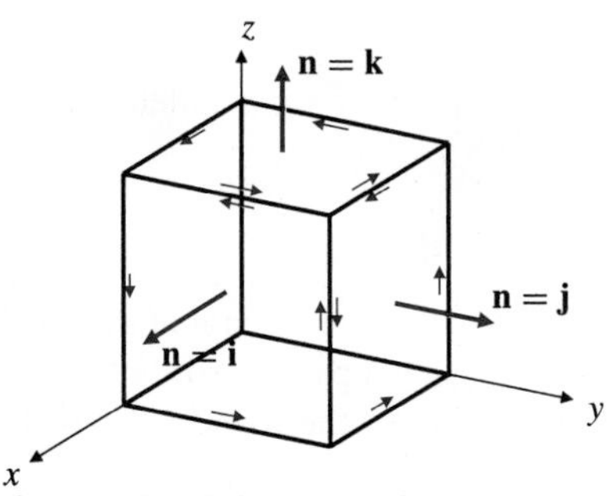

Figure 17.3 Orientation of 3 faces of a cube in $\mathbb{R}^3$, and of their edges

We can calculate the orientation of the six faces of the cube. The front face of the cube in Figure 17.3 has normal $\mathbf{n} = \mathbf{i}$. Accordingly, its orientation is given by

$$\begin{aligned} \omega_{\text{front}}(\mathbf{v}_1, \mathbf{v}_2) = \omega(\mathbf{i}, \mathbf{v}_1, \mathbf{v}_2) &= \begin{vmatrix} 1 & v_{11} & v_{21} \\ 0 & v_{12} & v_{22} \\ 0 & v_{13} & v_{23} \end{vmatrix} \\ &= v_{12}v_{23} - v_{13}v_{22} = dy \wedge dz(\mathbf{v}_1, \mathbf{v}_2). \end{aligned}$$

On the other hand, the normal for the back face is $-\mathbf{i}$ so the orientation for that face is

$$\omega_{\text{back}}(\mathbf{v}_1, \mathbf{v}_2) = \omega(-\mathbf{i}, \mathbf{v}_1, \mathbf{v}_2) = -dy \wedge dz(\mathbf{v}_1, \mathbf{v}_2).$$

Observe that the sum of the front and back orientations is 0. A similar situation holds for the sum of the orientations of the left and right side square faces, and the top and bottom square faces. See Exercise 1 and Exercise 2.

When the smooth boundary of a piece-with-boundary of a smooth, oriented k-manifold consists of several disjoint pieces-with-boundary of $(k-1)$-manifolds, together with some nonsmooth sets where these $(k-1)$-dimensional pieces join in pairs, as is the case in Example 8, it is useful to write the smooth boundary as a "sum" of these disjoint pieces, each with its proper orientation, given by a $+$ or $-$ sign.

EXAMPLE 9 Denote by $Q^k_{\mathbf{y}}(h\mathbf{e}_1, \ldots, h\mathbf{e}_k)$ the k-cube in $\mathbb{R}^k$ having edge length $h > 0$, one corner at $\mathbf{y}$, and spanned by the given multiples of the standard basis vectors in $\mathbb{R}^k$. (This is by analogy with the definition of a k-parallelogram given in Definition 10.) The cube shown in Figure 17.3 is $Q^3_{\mathbf{0}}(h\mathbf{i}, h\mathbf{j}, h\mathbf{k})$. The cube $Q^k_{\mathbf{y}}(h\mathbf{e}_1, \ldots, h\mathbf{e}_k)$ has smooth oriented boundary consisting of $2k$ cubes of dimension $(k-1)$ oriented by the direction of their outward normals. The boundary cubes come in pairs of opposite ones; for example, the pair with normals $\pm\mathbf{e}_j$ are (using a hat $\widehat{\ }$ to indicate a missing component)

$$Q^{k-1}_{\mathbf{y}+a\mathbf{e}_j}(h\mathbf{e}_1, \ldots, \widehat{h\mathbf{e}_j}, \ldots, h\mathbf{e}_k) \quad \text{and} \quad Q^{k-1}_{\mathbf{y}}(h\mathbf{e}_1, \ldots, \widehat{h\mathbf{e}_j}, \ldots, h\mathbf{e}_k),$$

and these have orientations given by

$$\begin{aligned} \omega(\mathbf{e}_j, h\mathbf{e}_1, \ldots, \widehat{h\mathbf{e}_j}, \ldots, h\mathbf{e}_k) &= (-1)^{j-1}h^{k-1}\omega(\mathbf{e}_1, \ldots, \mathbf{e}_j, \ldots, \mathbf{e}_k), \quad \text{and} \\ \omega(-\mathbf{e}_j, h\mathbf{e}_1, \ldots, \widehat{h\mathbf{e}_j}, \ldots, h\mathbf{e}_k) &= -(-1)^{j-1}h^{k-1}\omega(\mathbf{e}_1, \ldots, \mathbf{e}_j, \ldots, \mathbf{e}_k). \end{aligned}$$

The factors $(-1)^{j-1}$ account for the fact that $j-1$ simple transpositions and needed to move the normals $+\mathbf{e}_j$ and $-\mathbf{e}_j$ into the missing positions in these orientations so they are consistent with the standard positive orientation ω of $\mathbb{R}^k$ and therefore of the given k-cube. Accordingly, the smooth oriented boundary of $Q^k_{\mathbf{y}}(h\mathbf{e}_1, \ldots, h\mathbf{e}_k)$ can be expressed as the sum

$$\begin{aligned} &\partial Q^k_{\mathbf{y}}(h\mathbf{e}_1, \ldots, h\mathbf{e}_k) \\ &= \sum_{j=1}^{k}(-1)^{j-1}\Big(Q^{k-1}_{\mathbf{y}+h\mathbf{e}_j}(h\mathbf{e}_1, \ldots, \widehat{h\mathbf{e}_j}, \ldots, h\mathbf{e}_k) - Q^{k-1}_{\mathbf{y}}(h\mathbf{e}_1, \ldots, \widehat{h\mathbf{e}_j}, \ldots, h\mathbf{e}_k)\Big). \end{aligned}$$

Remark A similar formula holds for the oriented boundary of a k-parallelogram in $\mathbb{R}^k$; just replace the Qs with Ps and the vectors $a\mathbf{e}_j$ with $\mathbf{v}_j$, $1 \le j \le k$.

Integrating a Differential Form over a Manifold

As we did for functions in Section 17.3, we are going to define the integral of a smooth differential k-form over a smooth k-manifold in $\mathbb{R}^n$ by using a parametrization of the manifold over a set in $\mathbb{R}^k$. Now, however, the orientation of the manifold must be preserved by the parametrization.

DEFINITION 16

Orientation preserving parametrizations Let $\mathcal{M} \subset \mathbb{R}^n$ be a smooth k-manifold in $\mathbb{R}^n$ oriented by the differential k-form $\omega(\mathbf{x})$. Suppose $\mathbf{p}$ is a smooth parametrization of $\mathcal{M}$ over a subset $U \subset \mathbb{R}^k$, and that it is strict on $U - S$ where $S \subset U$ has k-volume zero (as specified in Definition 9 in Section 17.3). We say that $\mathbf{p}$ is **orientation preserving** if for all $u \in U - S$,

$$\omega(\mathbf{p}(\mathbf{u}))\left(\frac{\partial \mathbf{p}(\mathbf{u})}{\partial u_1}, \frac{\partial \mathbf{p}(\mathbf{u})}{\partial u_2}, \ldots, \frac{\partial \mathbf{p}(\mathbf{u})}{\partial u_k}\right) > 0.$$

If the inequality above is reversed, we say $\mathbf{p}$ is **orientation reversing**.

The definition of the integral of a differential k-form over a k-manifold is similar to that of a function over a manifold given in the previous section except that the k-form now plays the role of both the integrand and the volume element.

DEFINITION 17

Integration of a differential k-form over a k-manifold
Let $\mathbf{p}$, mapping $U \subset \mathbb{R}^k$ into $\mathbb{R}^n$, be an orientation preserving, smooth parametrization of the k-manifold $\mathcal{M} \subset \mathbb{R}^n$ oriented by the differential k-form ω. If Φ is a smooth differential k-form defined in an open set in $\mathbb{R}^n$ containing $\mathcal{M}$, we define the integral of Φ over $\mathcal{M}$ as

$$\int_{\mathcal{M}} \Phi = \int_U \Phi\left(\frac{\partial \mathbf{p}(\mathbf{u})}{\partial u_1}, \frac{\partial \mathbf{p}(\mathbf{u})}{\partial u_2}, \ldots, \frac{\partial \mathbf{p}(\mathbf{u})}{\partial u_k}\right) du_1\, du_2 \cdots du_k.$$

The following is a trivial, but important, example.

EXAMPLE 10 If $\Phi = f(x_1, \ldots, x_k)\, dx_1 \wedge \cdots \wedge dx_k$ and $\mathcal{M}$ is a k-manifold (an open set) in $\mathbb{R}^k$, show that

$$\int_{\mathcal{M}} \Phi = \int_{\mathcal{M}} f(x_1, x_2, \ldots, x_k)\, dx_1\, dx_2 \cdots dx_k.$$

Solution We use the identity parametrization $\mathbf{p}(\mathbf{u})$ given by $x_i = p_i(\mathbf{u}) = u_i$, so that $\partial \mathbf{p}(\mathbf{u})/\partial u_i = \mathbf{e}_i$, the ith standard basis vector in $\mathbb{R}^k$. Observe that $dx_1 \wedge \cdots \wedge dx_n(\mathbf{e}_1, \ldots, \mathbf{e}_k) = 1$ (the determinant of the $k \times k$ identity matrix), so we have

$$\int_{\mathcal{M}} \Phi = \int_{\mathcal{M}} f(u_1, u_2, \ldots, u_n)\, du_1\, du_2 \cdots du_n$$

which is the desired result if we replace the u_is with x_i's.

EXAMPLE 11 Let $\mathbf{p}(\mathbf{u})$ be a parametrization of an $(n-1)$-manifold (hypersurface) $\mathcal{S}$ in $\mathbb{R}^n$ over a domain $U \subset \mathbb{R}^{n-1}$. Show that

$$\mathbf{n} = \sum_{i=1}^{n} (-1)^{i-1} \frac{\partial(x_1, \ldots, \widehat{x_i}, \ldots, x_n)}{\partial(u_1, \ldots, u_{n-1})}\, \mathbf{e}_i \qquad (*)$$

is normal to $\mathcal{S}$ at $\mathbf{p}(\mathbf{u})$, and that

$$dS = dV_{n-1} = |\mathbf{n}|\, du_1\, du_2 \ldots du_{n-1}$$

is the "area element" (actually $(n-1)$ volume element) on $\mathcal{S}$ expressed in terms of the parameters $\mathbf{u}$. The case $n = 3$ of this result was proved in Section 15.5.

Solution The vector $\mathbf{n}$ given by $(*)$ is just the expansion in minors about the first row of the determinant

$$\begin{vmatrix} \mathbf{e}_1 & \mathbf{e}_2 & \cdots & \mathbf{e}_n \\ \dfrac{\partial x_1}{\partial u_1} & \dfrac{\partial x_2}{\partial u_1} & \cdots & \dfrac{\partial x_n}{\partial u_1} \\ \vdots & \vdots & \ddots & \vdots \\ \dfrac{\partial x_1}{\partial u_{n-1}} & \dfrac{\partial x_2}{\partial u_{n-1}} & \cdots & \dfrac{\partial x_n}{\partial u_{n-1}} \end{vmatrix}.$$

The vectors $\mathbf{v}_i = (\partial\mathbf{x}/\partial u_i)$ are the last $n-1$ rows of the above determinant and are linearly independent and tangent to $\mathcal{S}$ at $\mathbf{p}(\mathbf{u})$. Hence, $\mathbf{n}$ is normal to each of those vectors and so to $\mathcal{S}$. Also, the $n-1$ tangent vectors $\mathbf{v}_i\,du_i$ span an $(n-1)$-dimensional parallelogram that is the area element on $\mathcal{S}$ at $\mathbf{p}(\mathbf{u})$ corresponding to the element $du_1\,du_2\ldots du_{n-1}$ in $\mathbb{R}^{n-1}$. This parallelogram has $(n-1)$-volume $dS = |\mathbf{n}|\,du_1\,du_2\ldots du_{n-1}$ in $\mathbb{R}^{n-1}$.

Remark If $\mathcal{S} = \partial\mathcal{M}$ where $\mathcal{M}$ is an n-dimensional oriented manifold (open set) in $\mathbb{R}^n$ with the standard orientation, then $\mathbf{n}$ is the normal on $\mathcal{S}$ pointing outward from $\mathcal{M}$. In this case, if $\mathbf{F}$ is a vector field in $\mathbb{R}^n$ and $\hat{\mathbf{N}} = \mathbf{n}/|\mathbf{n}|$ is the unit outward normal field on $\mathcal{S}$, then

$$\begin{aligned}\mathbf{F}\bullet\hat{\mathbf{N}}\,dS &= \mathbf{F}\bullet\mathbf{n}\,du_1\,du_2\cdots du_{n-1}\\ &= \sum_{i=1}^{n}(-1)^{i-1}F_i(\mathbf{p}(\mathbf{u}))\,\frac{\partial(x_1,\ldots,\widehat{x_i},\ldots,x_n)}{\partial(u_1,\ldots,u_{i-1})}\,du_1\,du_2\cdots du_{n-1}\end{aligned}$$

is the **flux** of $\mathbf{F}$ out of $\mathcal{M}$ through the $(n-1)$ volume element dS.

EXERCISES 17.4

Exercises 1–4 refer to faces of the cube Q in $\mathbb{R}^3$ considered in Example 8.

1. Show that the orientation of the top face of Q is given by $dx\wedge dy$. What is the orientation of the bottom face?
2. Show that the orientation of the right face of Q is given by $dx\wedge dz = -dz\wedge dx$. What is the orientation of the left face?
3. Review the calculation of the orientations of the front and back faces of the cube Q in Example 8. Show that
$$\omega_{\text{front}}(\mathbf{v}_1,\mathbf{v}_2) = dx(\mathbf{v}_1\times\mathbf{v}_2) = -\omega_{\text{back}}(\mathbf{v}_1,\mathbf{v}_2).$$
4. As in the previous exercise, reexpress the orientations of the top and bottom faces of Q from Exercise 1 and the right and left faces of Q from Exercise 2 as differential 1-forms evaluated at the cross product of $\mathbf{v}_1$ and $\mathbf{v}_2$.
5. The 2-manifold $\mathcal{M}$ in $\mathbb{R}^4$ given by the equations $x_1+x_2=0$ and $x_3+x_4=0$, where $0<x_1<1$ and $0<x_3<1$, has normals $\mathbf{e}_1+\mathbf{e}_2$ and $\mathbf{e}_3+\mathbf{e}_4$. It is oriented by the 2-form
$$\omega(\mathbf{v}_1,\mathbf{v}_2) = \begin{vmatrix}1 & 0 & v_{11} & v_{21}\\ 1 & 0 & v_{12} & v_{22}\\ 0 & 1 & v_{13} & v_{23}\\ 0 & 1 & v_{14} & v_{24}\end{vmatrix}.$$
Is the parametrization $(x_1,x_2,x_3,x_4) = \mathbf{x} = \mathbf{p}(\mathbf{u}) = (u_1,-u_1,u_2,-u_2)$ orientation preserving for $\mathcal{M}$? If not, give an example of a parametrization that would be.
6. Using the orientation preserving parametrization for the manifold $\mathcal{M}$ of the previous exercise, evaluate $\displaystyle\int_{\mathcal{M}}\Phi$ where $\Phi = x_2x_4\,dx_1\wedge dx_3$.
7. Let S be a piece-with-boundary of a smooth hypersurface ($(k-1)$-manifold) in $\mathbb{R}^k$ given by equation $x_i = g_i(x_1,\ldots,x_{i-1},x_{i+1},\ldots,x_k)$. Let $\Phi = dx_1\wedge\cdots\wedge dx_{i-1}\wedge dx_{i+1}\wedge\cdots\wedge dx_k$. Show that $\int_S\Phi$ is (apart from sign due to the orientation of S) the $(k-1)$-volume of the projection of S on the coordinate hyperplane $x_i=0$.
8. Let M be a convex open set in $\mathbb{R}^k$ with boundary ∂M, and let Φ be a constant $(k-1)$-form on $\mathbb{R}^k$ (i.e., all its coefficients are constant). Show that $\int_{\partial M}\Phi = 0$. *Hint:* For each i, M can be described as lying between two surfaces of the form considered in the previous exercise, both of which have the same projection M_i on $x_i=0$.

17.5 The Generalized Stokes Theorem

The previous four sections have developed much new machinery, forms and differential forms, the exterior derivative, manifolds and their boundaries and orientations, and integrals of functions and differential forms over manifolds and their boundaries. This has all been done with one ultimate goal in mind—namely, the provision of a generalized version of the Fundamental Theorem of Calculus that holds and appears the same in any number of dimensions. Without further ado, here it is.

THEOREM 4

The Generalized Stokes Theorem (GST) If M is a closed, bounded, piece-with-boundary of an oriented k-manifold $\mathcal{M}$ in $\mathbb{R}^n$, and Φ is a smooth differential $(k-1)$-form defined in an open set containing M, then

$$\int_M d\Phi = \int_{\partial M} \Phi,$$

where ∂M has the orientation inherited from $\mathcal{M}$. It is understood that the boundary integral is really taken over the smooth part of the boundary ∂M.

Remark While we will not prove this theorem in its full generality here, we will prove it for a significant special case from which the general case can, with some effort, be deduced. We will also give a somewhat handwaving argument that should convince you of the validity of the general case. Then we will show how the major theorems of vector calculus are all special cases of the Generalized Stokes Theorem.

Remark The requirement that M be bounded is not necessarily restrictive. If M is the union of non-overlapping, bounded pieces-with-boundary of $\mathcal{M}$ we can add the results of the theorem applied to the individual pieces to get the integral of $d\Phi$ over the whole piece. Where two pieces abut along parts of their boundaries, those parts will have opposite orientations inherited from $\mathcal{M}$, so their contributions to the sum of the boundary integrals will cancel, leaving only the contributions from the parts that are part of the boundary of the union. If M is unbounded but the sums taken over those bounded pieces contained in the ball of radius r in $\mathbb{R}^n$ approach limits as $r \to \infty$, the GST will still hold for M.

Remark Let us confirm that the Fundamental Theorem of Calculus really is a special case of the Generalized Stokes Theorem. Let $M = [a, b]$ be a subset of the 1-manifold $\mathbb{R}$ oriented from a to b. The boundary ∂M of M consists of the two points a and b, each of which is a 0-dimensional manifold; in this case the "outward" (from M) direction is $+$ at b and $-$ at a. Thus $\partial M = \{-a, +b\}$. If f is a smooth function (0-form) on M, then its exterior derivative is $df = f'(x)\,dx$. We have

$$\int_a^b f'(x)\,dx = \int_M df = \int_{\partial M} f = \big(+f(b)\big) + \big(-f(a)\big) = f(b) - f(a).$$

Again, we stress that if the oriented boundary of ∂M consists of the oriented points $-a$ and $+b$, then

$$\int_{\partial M} f = +f(b) - f(a),$$

not $f(b) + f(-a)$.

Proof of Theorem 4 for a *k*-Cube

Let Φ be a differential $(k-1)$-form on $\mathbb{R}^k$. Then

$$\Phi = \sum_{i=1}^{k} a_i(x_1, \ldots, x_k)\,dx_1 \wedge \cdots \wedge \widehat{dx_i} \wedge \cdots \wedge dx_k,$$

where, again, the hat $\widehat{\ }$ indicates a missing factor. The exterior derivative of Φ is the differential k-form given by

$$d\Phi = \sum_{i=1}^{k} \left(\sum_{j=1}^{k} \frac{\partial a_i}{\partial x_j}\,dx_j \right) \wedge dx_1 \wedge \cdots \wedge \widehat{dx_i} \wedge \cdots \wedge dx_k.$$

The only nonzero terms in this double sum are those for which $j = i$, so we have

$$\begin{aligned} d\Phi &= \sum_{i=1}^{k} \frac{\partial a_i}{\partial x_i}\,dx_i \wedge dx_1 \wedge \cdots \wedge \widehat{dx_i} \wedge \cdots \wedge dx_k \\ &= \sum_{i=1}^{k} (-1)^{i-1} \frac{\partial a_i}{\partial x_i}\,dx_1 \wedge \cdots \wedge dx_i \wedge \cdots \wedge dx_k, \end{aligned}$$

since $i-1$ reversals are required to move dx_i from the front of the list to fill in the missing ith position.

Now let $Q = Q^k_{\mathbf{y}}(h\mathbf{e}_1, \ldots, h\mathbf{e}_k)$ be the cube of edge length h in $\mathbb{R}^k$ described in Example 9 in Section 17.4. Then

$$\begin{aligned}\int_Q d\Phi &= \sum_{i=1}^{k}(-1)^{i-1}\int_Q \frac{\partial a_i}{\partial x_i}\, dx_1 \wedge \cdots \wedge dx_k \\ &= \sum_{i=1}^{k}(-1)^{i-1}\int_Q \frac{\partial a_i}{\partial x_i}\, dx_1\, dx_2 \cdots dx_k\end{aligned}$$

by the result of Example 10 in Section 17.4. Let Q_i be the projection of Q on the coordinate plane with normal $\mathbf{e}_i$, that is, $Q_i = \{\mathbf{x} \in \mathbb{R}^k : x_i = 0\; y_j \le x_j \le y_j + h,\ j \ne i\}$. We can iterate the above integral to obtain

$$\begin{aligned}\int_Q d\Phi &= \sum_{i=1}^{k}(-1)^{i-1}\int_{Q_i} dx_1 \ldots \widehat{dx_i} \ldots dx_n \int_{y_i}^{y_i+h} \frac{\partial a_i}{\partial x_i}\, dx_i \\ &= \sum_{i=1}^{k}(-1)^{i-1}\int_{Q_i}\Big(a_i(x_1, \ldots, y_i + h, \ldots, x_k) \\ &\qquad\qquad - a_i(x_1, \ldots, y_i, \ldots, x_k)\Big)\, dx_1 \ldots, \widehat{dx_i} \ldots dx_k \\ &= \int_{\partial Q} \Phi\end{aligned}$$

because the k pairs of $(k-1)$-cube faces of the oriented boundary ∂Q of Q are given by

$$\sum_{i=1}^{k}(-1)^{i-1}\Big(Q^{k-1}_{\mathbf{y}+h\mathbf{e}_i}(h\mathbf{e}_1, \ldots, \widehat{h\mathbf{e}_i}, \ldots, h\mathbf{e}_k) - Q^{k-1}_{\mathbf{y}}(h\mathbf{e}_1, \ldots, \widehat{h\mathbf{e}_i}, \ldots, h\mathbf{e}_k)\Big).$$

Remark Although the proof above was carried out for a k-cube in $\mathbb{R}^k$, the result extends to a k-cube in $\mathbb{R}^n$. If the cube is spanned by k mutually perpendicular unit vectors, an invertible linear transformation of coordinates in $\mathbb{R}^n$ can be found that maps those vectors to the first k basis vectors $\mathbf{e}_1, \ldots, \mathbf{e}_k$ so that the coordinates of $\mathbf{x} \in Q$ satisfy x_i = constant in Q for $i > k$. If Φ is a $(k-1)$-form on Q, its coefficients will not vary with those coordinates, and those of its exterior derivative $d\Phi$ will be a multiple of $dx_1 \wedge \cdots \wedge dx_k$.

Remark If the coefficients of the differential $(k-1)$ form Φ are smooth, and if, for all small $h > 0$, the k-cube $\{Q_h\}$ has edge length h and contains the point $\mathbf{y}$, then

$$\lim_{h\to 0}\frac{1}{h^k}\int_{\partial Q_h} \Phi = \lim_{h\to 0}\frac{1}{h^k}\int_{Q_h} d\Phi = d\Phi(\mathbf{y}). \qquad (\dagger)$$

Some writers use $(\dagger)$ as the definition of the exterior derivative $d\Phi$. Since Q_h has k-volume h^k, the second equality is not surprising. But the $(k-1)$-cubes ($2k$ of them) that form ∂Q_k have total $(k-1)$-volume $2kh^{k-1}$, so it is more surprising that $\lim_{h\to 0}(1/h^k)\int_{\partial Q_h} \Phi$ should be finite.

Completing the Proof

While we will not give a detailed proof of the GST here, we will make several observations about extending the proof to wider classes of domains.

(a) The proof above extends with minimal change to k-dimensional rectangles.

(b) An invertible linear transformation can map a k-parallelogram in $\mathbb{R}^n$ to a k-cube, so the GST holds for k-parallelograms.

(c) Let M be a piece-with-boundary of a k-manifold. If M is a union of nonoverlapping k-cubes (or k-rectangles), then $\int_M d\Phi$ will be the sum of the integrals over those cubes. However, where two such cubes abut along parts of their smooth boundaries, they induce opposite orientations, so that those contributions to the boundary integral of Φ will cancel and the sum of the integrals over the boundaries of the cubes will reduce to the boundary integral on ∂M.

This suggests that we can approximate M by nonoverlapping cubes of small edge length h and obtain

$$\int_M d\Phi \approx \sum_i \int_{Q_i} d\Phi = \sum_i \int_{\partial Q} \Phi \approx \int_{\partial M} \Phi. \qquad (*)$$

The error in the first approximation in $(*)$ approaches 0 as $h \to 0$ because the number of cubes near the boundary grows of order $h^{-(n-1)}$, while the volume of each decreases of order h^n. The error in the second approximation in $(*)$ cannot be similarly argued to decrease with h because the $(k-1)$-volumes of the uncanceled parts of the boundaries of the cubes may remain relatively larger than the $(k-1)$-volume of ∂M. However, we can exploit the assumed smoothness of Φ to compensate for this. Expanding (the coefficients of) Φ in Taylor series about a point $\mathbf{y}$ near the boundary ∂M we obtain $\Phi(\mathbf{x}) = \Phi_0(\mathbf{y}) + O(|\mathbf{x}-\mathbf{y}|)$ for $\mathbf{x}$ near $\mathbf{y}$. If we can fill the region between ∂M and the set of cubes used above to approximate $\int_M d\Phi$ with convex sets of diameter of order h abutting the cubes and numbering of order $h^{-(n-1)}$, and use the fact that the integral of Φ_0 over the boundary of such convex sets is zero (see Exercise 8 in the previous section), we can still have the error in the second approximation decreasing of order h.

(d) **Strict parametrization** Let U be an open set in $\mathbb{R}^k$ and let $\mathbf{x} = \mathbf{p}(\mathbf{u})$ be a one-to-one, orientation-preserving parametrization over U of a subset of $\mathcal{M}$ containing the closed piece-with-boundary M. Suppose that $M = \mathbf{p}(Q)$, where Q is a closed k-cube in U with edges parallel to the standard basis vectors in $\mathbb{R}^k$, and that $\partial M = \mathbf{p}(\partial Q)$. If Φ is a smooth differential $(k-1)$-form on $\mathcal{M}$, then

$$\int_M d\Phi = \int_Q d\Phi\left(\frac{\partial\mathbf{p}(\mathbf{u})}{\partial u_1}, \frac{\partial\mathbf{p}(\mathbf{u})}{\partial u_2}, \ldots, \frac{\partial\mathbf{p}(\mathbf{u})}{\partial u_k}\right) du_1\, du_2 \cdots du_k.$$

The $2k$ faces of Q consist of k pairs B_i of $(k-1)$-dimensional cubes, with orientation inherited from Q, and such that the coordinate x_i is constant in each cube of the pair B_i. It follows that

$$\begin{aligned}
&\int_{\partial M} \Phi \\
&= \sum_{i=1}^{k} (-1)^{i-1} \int_{B_i} \Phi\left(\frac{\partial\mathbf{p}(\mathbf{u})}{\partial u_1}, \ldots, \widehat{\frac{\partial\mathbf{p}(\mathbf{u})}{\partial u_i}}, \ldots, \frac{\partial\mathbf{p}(\mathbf{u})}{\partial u_k}\right) du_1 \cdots \widehat{du_i} \cdots du_k \\
&= \int_{\partial M} \Phi.
\end{aligned}$$

EXAMPLE 1 Evaluate the integral of the differential form

$$\Phi = (x_1^2 + x_4^2)\, dx_1 \wedge dx_2 \wedge dx_3 + (x_1^2 + x_3^2)\, dx_2 \wedge dx_3 \wedge dx_4$$

over the oriented boundary of the spherical cylinder C in $\mathbb{R}^4$ consisting of those points $\mathbf{x}$ satisfying $(x_1+1)^2 + x_2^2 + x_3^2 \le 9$ and $0 \le x_4 \le 1$.

Solution Direct evaluation of the integral of Φ by parametrizing the "cylindrical wall" $(x_1+1)^2 + x_2^2 + x_3^2 = 9$, $0 \le x_4 \le 1$, and then doing the same with the ends of the cylinder, $(x_1+1)^2 + x_2^2 + x_3^2 \le 9$, $x_4 = 0$ or $x_4 = 1$, while not impossible, would be somewhat time consuming. It is much easier to use the GST. Observe that

$$\begin{aligned}
d\Phi &= 2x_4\, dx_4 \wedge dx_1 \wedge dx_2 \wedge dx_3 + 2x_1\, dx_1 \wedge dx_2 \wedge dx_3 \wedge dx_4 \\
&= 2(x_1 - x_4)\, dx_1 \wedge dx_2 \wedge dx_3 \wedge dx_4.
\end{aligned}$$

When integrating a function times an orienting differential k-form in $\mathbb{R}^k$, the wedge products can be dropped. (See Example 10 in Section 17.4.) The integral is a normal k-fold integral that can be iterated by the usual techniques.

Since $dx_1 \wedge \cdots \wedge dx_4$ provides the standard orientation for $\mathbb{R}^4$, we have

$$\int_{\partial C} \Phi = \int_C d\Phi = \int_C 2(x_1 - x_4)\, dx_1\, dx_2\, dx_3\, dx_4$$
$$= 2\int_B x_1\, dx_1\, dx_2\, dx_3 \int_0^1 dx_4 - 2\int_B dx_1\, dx_2\, dx_3 \int_0^1 x_4\, dx_4,$$

where B is the ball in $\mathbb{R}^3$ with with centre at $(-1, 0, 0)$ and radius 3, having volume

$$V_B = \frac{4}{3}\pi 3^3 = 36\pi.$$

The integral of x_1 over B is V_B times the x_1-coordinate of the centroid (i.e., the centre) of B. Accordingly,

$$\int_{\partial C} = (2)(36\pi)(-1) \int_0^1 dx_4 - (2)(36\pi) \int_0^1 x_4\, dx_4 = -108\pi.$$

Sometimes it is helpful to use the GST to evaluate the integral of a form over only part of the surface of a region.

EXAMPLE 2 Let $\Phi = yz\, dx \wedge dy + zx\, dy \wedge dz + xy\, dz \wedge dx$. Evaluate $\int_P \Phi$, where P is the part of the plane $x + y + z = 1$ lying in the first octant of $\mathbb{R}^3$. Assume that P is oriented with upward normal.

Solution S is part of the boundary of the tetrahedron T with vertices at $(0, 0, 0)$, $(1, 0, 0)$, $(0, 1, 0)$, and $(0, 0, 1)$. The other three parts of the boundary of T are triangles in the three coordinate planes. Observe $\Phi = 0$ on each of those three triangles. (For instance, on $z = 0$ we have $dz = 0$, so all three terms of Φ are zero.) Since the assumed normal on P is outward from T, we have

$$\int_P \Phi = \int_{\partial T} \Phi = \int_T d\Phi = \int_T (y + z + x)\, dx \wedge dy \wedge dz.$$

By symmetry,

$$\int_P \Phi = 3\int_T z\, dx\, dy\, dz = 3\int_0^1 z\, dz \int_0^{1-z} dy \int_0^{1-y-z} dx = \frac{1}{8}.$$

(We have omitted the details of evaluating the iterated integral.)

The Classical Theorems of Vector Calculus

EXAMPLE 3 **Line integrals of conservative fields** Let C be a piece-with-boundary of a smooth curve (1-manifold) in $\mathbb{R}^n$ oriented so that C runs from $\mathbf{a}$ to $\mathbf{b}$. Let f be continuously differentiable on an open set containing C. Let $\hat{\mathbf{T}}$ be the unit tangent vector field on C in the direction of its orientation, and let ds be the arc length element on C. Then

$$\int_C \mathbf{grad}\, f \bullet \hat{\mathbf{T}}\, ds = f(\mathbf{b}) - f(\mathbf{a}).$$

Solution Let Φ be the 0-form $f(\mathbf{x})$ so that

$$d\Phi = \sum_{i=1}^{n} \frac{\partial f}{\partial x_i} dx_i.$$

If C is parametrized by $\mathbf{x} = \mathbf{p}(u)$ for $u \in [a, b]$ with $\mathbf{p}(a) = \mathbf{a}$ and $\mathbf{p}(b) = \mathbf{b}$, then $\hat{\mathbf{T}}(\mathbf{x}) = (d\mathbf{x}/du)/|d\mathbf{x}/du|$ and $ds = |d\mathbf{x}/du|\, du$, so that

$$\begin{aligned}\int_C \mathbf{grad}\, f \bullet \hat{\mathbf{T}}\, ds &= \int_{[a,b]} \mathbf{grad}\, f(\mathbf{p}(u)) \bullet \frac{d\mathbf{x}}{du}\, du \\ &= \sum_{i=1}^{n} \int_{[a,b]} \frac{\partial f(\mathbf{p}(u))}{\partial x_i} \frac{dx_i}{du}\, du \\ &= \int_C d\Phi = \int_{\partial C} \Phi = +f(\mathbf{b}) + \big(-f(\mathbf{a})\big) = f(\mathbf{b}) - f(\mathbf{a}),\end{aligned}$$

since ∂C is the 0-manifold consisting of the two oriented points $+\mathbf{b}$ and $-\mathbf{a}$.

EXAMPLE 4 **Stokes's Theorem and Green's Theorem** Let S be a piece-with-boundary of a smooth surface (2-manifold) in $\mathbb{R}^3$, oriented with unit normal field $\hat{\mathbf{N}}$, and let C be the piecewise-smooth, closed bounding curve of S with inherited orientation given by a unit tangent field $\hat{\mathbf{T}}$. Let $\mathbf{F} = F_1(\mathbf{x})\,\mathbf{i} + F_2(\mathbf{x})\,\mathbf{j} + F_3(\mathbf{x})\,\mathbf{k}$ have components that are continuously differentiable in an open set in $\mathbb{R}^3$ containing S. If dS and ds denote the area element on S and the arc length element on C, then

$$\int_S \mathbf{curl}\,\mathbf{F} \bullet \hat{\mathbf{N}}\, dS = \int_C \mathbf{F} \bullet \hat{\mathbf{T}}\, ds.$$

Solution Let $\Phi = F_1\, dx + F_2\, dy + F_3\, dz$. As shown in Example 2 in Section 17.2,

$$\begin{aligned}d\Phi = &\left(\frac{\partial F_3}{\partial y} - \frac{\partial F_2}{\partial z}\right) dy \wedge dz + \left(\frac{\partial F_1}{\partial z} - \frac{\partial F_3}{\partial x}\right) dz \wedge dx \\ &+ \left(\frac{\partial F_2}{\partial x} - \frac{\partial F_1}{\partial y}\right) dx \wedge dy,\end{aligned}$$

while $\mathbf{curl}\,\mathbf{F}$ has the same components as $d\Phi$ has coefficients;

$$\mathbf{curl}\,\mathbf{F} = \left(\frac{\partial F_3}{\partial y} - \frac{\partial F_2}{\partial z}\right)\mathbf{i} + \left(\frac{\partial F_1}{\partial z} - \frac{\partial F_3}{\partial x}\right)\mathbf{j} + \left(\frac{\partial F_2}{\partial x} - \frac{\partial F_1}{\partial y}\right)\mathbf{k}.$$

Now suppose $x = p_1(u, v)$, $y = p_2(u, v)$, $z = p_3(u, v)$ is a smooth, orientation-preserving parametrization of S over a set U in $\mathbb{R}^2$ (the uv-plane). Then

$$\begin{aligned}dy \wedge dz &= \left(\frac{\partial y}{\partial u} du + \frac{\partial y}{\partial v} dv\right) \wedge \left(\frac{\partial z}{\partial u} du + \frac{\partial z}{\partial v} dv\right) \\ &= \left(\frac{\partial y}{\partial u}\frac{\partial z}{\partial v} - \frac{\partial z}{\partial u}\frac{\partial y}{\partial v}\right) du \wedge dv = \frac{\partial(y, z)}{\partial(u, v)}\, du \wedge dv.\end{aligned}$$

Similarly,

$$dz \wedge dx = \frac{\partial(z, x)}{\partial(u, v)}\, du \wedge dv \quad \text{and} \quad dx \wedge dy = \frac{\partial(x, y)}{\partial(u, v)}\, du \wedge dv.$$

By Example 11 in Section 17.4, a normal vector and surface area element on S are given by

$$\begin{aligned}\mathbf{n} &= \frac{\partial(y, z)}{\partial(u, v)}\mathbf{i} + \frac{\partial(z, x)}{\partial(u, v)}\mathbf{j} + \frac{\partial(x, y)}{\partial(u, v)}\mathbf{k} \\ dS &= |\mathbf{n}|\, du\, dv.\end{aligned}$$

Thus, $d\Phi = \mathbf{curl\,F} \bullet \mathbf{n}\,du \wedge dv = \mathbf{curl\,F} \bullet \hat{\mathbf{N}}dS$ and

$$\int_S \mathbf{F} \bullet \hat{\mathbf{N}}\,dS = \int_S d\Phi = \int_C \Phi$$

by the GST.

Now let $\mathbf{x}(t) = x(t)\,\mathbf{i} + y(t)\,\mathbf{j} + z(t)\,\mathbf{k}$, $a \le t \le b$ be an orientation-preserving parametrization of C. Since C is a closed curve, $\mathbf{x}(a) = \mathbf{x}(b)$. The unit tangent vector in the direction of C is $\hat{\mathbf{T}}(t) = (d\mathbf{x}/dt)/|d\mathbf{x}/dt|$ and the arc length element on C is $|d\mathbf{x}/dt|$. Accordingly,

$$\Phi = F_1\,\frac{dx}{dt}\,dt + F_1\,\frac{dy}{dt}\,dt + F_1\,\frac{dz}{dt}\,dt = \mathbf{F} \bullet \frac{d\mathbf{x}}{dt}\,dt = \mathbf{F} \bullet \hat{\mathbf{T}}ds$$

and

$$\int_C \Phi = \int_a^b \mathbf{F}(\mathbf{x}(t)) \bullet \frac{d\mathbf{x}}{dt}\,dt = \int_C \mathbf{F} \bullet \hat{\mathbf{T}}\,ds.$$

Remark Green's Theorem in $\mathbb{R}^2$ is just a special case of Stokes's Theorem where S and C lie in the xy-plane, $\mathbf{F}$ is independent of z, and $F_3 = 0$.

EXAMPLE 5 **The Divergence Theorem** Let M be an open set in $\mathbb{R}^n$, equipped with the standard orientation $dx_1 \wedge \cdots \wedge dx_n$, and having a piecewise smooth $(n-1)$-dimensional boundary manifold ∂M equipped with an outward unit normal field $\hat{\mathbf{N}}$. If $\mathbf{F} = \sum_{i=1}^n F_i(\mathbf{x})\,\mathbf{e}_i$ is a smooth vector field defined on M, show that the GST implies

$$\int_M \mathbf{div\,F}(\mathbf{x})\,d\mathbf{x} = \int_{\partial M} \mathbf{F} \bullet \hat{\mathbf{N}}\,dS,$$

where $\mathbf{div\,F}(\mathbf{x}) = \displaystyle\sum_{j=1}^n \frac{\partial F_j(\mathbf{x})}{\partial x_j}$ and dS is the "area" ($(n-1)$-volume) element on ∂M.

Solution Let $\Phi = \sum_{i=1}^n (-1)^{i-1}\,F_i(\mathbf{x})\,dx_1 \cdots \widehat{dx_i} \cdots dx_n$ be a differential $(n-1)$-form on $\mathbb{R}^n$. Then we have

$$\begin{aligned}
d\Phi &= \sum_{i=1}^n (-1)^{i-1} \left(\sum_{j=1}^n \frac{\partial F_i}{\partial x_j}\,dx_j \right) \wedge dx_1 \cdots \widehat{dx_i} \cdots dx_n \\
&= \sum_{i=1}^n (-1)^{i-1}\,\frac{\partial F_i}{\partial x_i}\,dx_j \wedge dx_1 \cdots \widehat{dx_i} \cdots dx_n \\
&= \left(\sum_{i=1}^n \frac{\partial F_i}{\partial x_i} \right) dx_1\,dx_2 \cdots dx_n \\
&= (\mathbf{div\,F})\,dx_1\,dx_2 \cdots dx_n.
\end{aligned}$$

Thus,

$$\int_M \mathbf{div\,F}\,dx_1\,dx_2 \cdots dx_n = \int_M \Phi.$$

On the other hand, if ∂M has a smooth parametrization $\mathbf{x} = \mathbf{p}(\mathbf{u})$ over a domain $U \subset \mathbb{R}^{n-1}$, then using the formulas for the normal $\mathbf{n}$ and surface area element dS given in Example 11 in Section 17.4, we have

$$\begin{aligned}
\mathbf{F} \bullet \hat{\mathbf{N}}\,dS &= \mathbf{F} \bullet \mathbf{n}\,du_1 \cdots du_{n-1} \\
&= \sum_{i=1}^n (-1)^{i-1} F_i\big(\mathbf{p}(\mathbf{u})\big)\,\frac{\partial(x_1, \ldots, \widehat{x_i}, \ldots, x_n)}{\partial(u_1, \ldots, u_{n-1})}\,du_1 \cdots du_{n-1}.
\end{aligned}$$

But this latter expression is just the parametrized version of Φ, since

$$dx_1 \wedge \cdots \wedge \widehat{dx_i} \wedge \cdots \wedge dx_n = \frac{\partial(x_1, \ldots, \widehat{x_i}, \ldots, x_n)}{\partial(u_1, \ldots, u_{n-1})} \, du_1 \cdots du_{n-1}.$$

Thus, $\displaystyle\int_{\partial M} \mathbf{F} \bullet \hat{\mathbf{N}} \, dS = \int_{\partial M} \Phi$ and the Divergence Theorem holds in $\mathbb{R}^n$.

EXERCISES 17.5

1. If Φ is a constant differential $(k-1)$-form defined in a neighbourhood of a smooth k-manifold $\mathcal{M}$ in $\mathbb{R}^n$, show that $\displaystyle\int_{\partial M} \Phi = 0$ for any piece-with-boundary M of $\mathcal{M}$. The GST gives a simple proof of this assertion, first made under restrictive conditions on M in Exercise 8 in Section 17.4.

2. Let $\Phi = \sum_{i=1}^{k} (-1)^{i-1} x_i \, dx_1 \wedge \cdots \wedge \widehat{dx_i} \wedge \cdots \wedge dx_k$ and let M be a piece-with-boundary of a k-manifold in $\mathbb{R}^n$ (where $n \geq k$). Show that the k-volume $V_k(M)$ of M is given by

$$V_k(M) = \frac{1}{k} \int_{\partial M} \Phi.$$

In Exercises 3–6, find the integral of the given differential form Φ over the oriented boundary of the given domain D.

3. $\Phi = x \, dy \wedge dz + yz \, dx \wedge dz$,
$D = \{(x, y, z) \in \mathbb{R}^3 : 1 \leq x, y, z, \leq 1\}$

4. $\Phi = (x_1 + x_4^2) dx_2 \wedge dx_3 \wedge dx_4 + (x_2^2 + x_3 x_4) dx_1 \wedge dx_3 \wedge dx_4$,
$D = \{\mathbf{x} \in \mathbb{R}^4 : 0 \leq x_i \leq i, \ 1 \leq i \leq 4\}$.

5. $\Phi = x_1^3 \, dx_2 \wedge dx_3 \wedge dx_4 - x_2^3 \, dx_3 \wedge dx_4 \wedge dx_1 + x_3 \, dx_4 \wedge dx_1 \wedge dx_2$,
$D = \{\mathbf{x} \in \mathbb{R}^4 : x_1^2 + x_2^2 \leq 4, \ x_3^2 + x_4^2 \leq 9\}$.

6. $\Phi = (x_1^2 + \cdots + x_6^2) \, dx_1 \wedge dx_3 \wedge dx_4 \wedge dx_5 \wedge dx_6$, $D = \{\mathbf{x} \in \mathbb{R}^6 : x_1, x_2 \geq 0, \ x_1 + x_2 \leq 1, \ 0 \leq x_3, x_4, x_5, x_6 \leq 1\}$.

CHAPTER 18

Ordinary Differential Equations

> In order to solve this differential equation you look at it until the solution occurs to you.

George Polyá 1887–1985
from *How to Solve It* Princeton, 1945

> Science is a differential equation. Religion is a boundary condition.

Alan Turing 1912–1954
quoted in *Theories of Everything* by J. D. Barrow

Introduction

A **differential equation** (or **DE**) is an equation that involves one or more derivatives of an unknown function. Solving the differential equation means finding a function (or every such function) that satisfies the differential equation.

Many physical laws and relationships between quantities studied in various scientific disciplines are expressed mathematically as differential equations. For example, Newton's Second Law of Motion ($F = ma$) states that the position $x(t)$ at time t of an object of constant mass m subjected to a force $F(t)$ must satisfy the differential equation (equation of motion):

$$m\frac{d^2x}{dt^2} = F(t).$$

Similarly, the biomass $m(t)$ at time t of a bacterial culture growing in a uniformly supporting medium changes at a rate proportional to the biomass:

$$\frac{dm}{dt} = km(t),$$

which is the differential equation of exponential growth (or, if $k < 0$, exponential decay). Because differential equations arise so extensively in the abstract modelling of concrete phenomena, such equations and techniques for solving them are at the heart of applied mathematics. Indeed, most of the existing mathematical literature is either directly involved with differential equations or is motivated by problems arising in the study of such equations. Because of this, we have introduced various differential equations, terms for their description, and techniques for their solution at several places in the development of calculus throughout this book. This final chapter provides a more unified framework for a brief introduction to the study of ordinary differential equations. Some material from earlier sections (notably Sections 7.9 and 3.7) forms a natural part of this chapter; you will be referred back to these sections at the appropriate time. This chapter is, of necessity, relatively short. Students of mathematics and its applications usually take one or more full courses on differential equations, and even then hardly scratch the surface of the subject.

18.1 Classifying Differential Equations

Differential equations are classified in several ways. The most significant classification is based on the number of variables with respect to which derivatives appear in the equation. An **ordinary differential equation (ODE)** is one that involves derivatives with respect to only one variable. Both of the examples given above are ordinary differential equations. A **partial differential equation (PDE)** is one that involves partial derivatives of the unknown function with respect to more than one variable. For example, the **one-dimensional wave equation**

$$\frac{\partial^2 u}{\partial t^2} = c^2 \frac{\partial^2 u}{\partial x^2}$$

models the lateral displacement $u(x, t)$ at position x at time t of a stretched vibrating string. (See Section 12.4.) We will not discuss partial differential equations in this chapter.

Differential equations are also classified with respect to **order**. The order of a differential equation is the order of the highest-order derivative present in the equation. The one-dimensional wave equation is a second-order PDE. The following example records the order of two ODEs.

EXAMPLE 1

$$\frac{d^2y}{dx^2} + x^3 y = \sin x \qquad \text{has order 2,}$$

$$\frac{d^3y}{dx^3} + 4x\left(\frac{dy}{dx}\right)^2 = y\frac{d^2y}{dx^2} + e^y \qquad \text{has order 3.}$$

Like any equation, a differential equation can be written in the form $F = 0$, where F is a function. For an ODE, the function F can depend on the independent variable (usually called x or t), the unknown function (usually y), and any derivatives of the unknown function up to the order of the equation. For instance, an nth-order ODE can be written in the form

$$F(x, y, y', y'', \ldots, y^{(n)}) = 0.$$

An important special class of differential equations consists of those that are **linear**. An nth-order linear ODE has the form

$$\begin{aligned} a_n(x)y^{(n)}(x) &+ a_{n-1}(x)y^{(n-1)}(x) + \cdots \\ &+ a_2(x)y''(x) + a_1(x)y'(x) + a_0(x)y(x) = f(x). \end{aligned}$$

Each term in the expression on the left side is the product of a *coefficient* that is a function of x, and a second factor that is either y or one of the derivatives of y. The term on the right does not depend on y; it is called the **nonhomogeneous term**. Observe that no term on the left side involves any power of y or its derivatives other than the first power, and y and its derivatives are never multiplied together.

A linear ODE is said to be **homogeneous** if all of its terms involve the unknown function y, that is, if $f(x) = 0$. If $f(x)$ is not identically zero, the equation is **nonhomogeneous**.

EXAMPLE 2 In Example 1 the first DE, $\dfrac{d^2y}{dx^2} + x^3y = \sin x$, is linear. Here, the coefficients are $a_2(x) = 1$, $a_1(x) = 0$, $a_0(x) = x^3$, and the nonhomogeneous term is $f(x) = \sin x$. Although it can be written in the form

$$\frac{d^3y}{dx^3} + 4x\left(\frac{dy}{dx}\right)^2 - y\frac{d^2y}{dx^2} - e^y = 0,$$

the second equation is *not linear* (we say it is **nonlinear**) because the second term involves the square of a derivative of y, the third term involves the product of y and one of its derivatives, and the fourth term is not y times a function of x. The equation

$$(1 + x^2)\frac{d^3y}{dx^3} + \sin x\,\frac{d^2y}{dx^2} - 4\frac{dy}{dx} + y = 0$$

is a linear equation of order 3. The coefficients are $a_3(x) = 1 + x^2$, $a_2(x) = \sin x$, $a_1(x) = -4$, and $a_0(x) = 1$. Since $f(x) = 0$, this equation is *homogeneous*.

The following theorem states that any *linear combination* of solutions of a linear, homogeneous DE is also a solution. This is an extremely important fact about linear, homogeneous DEs.

THEOREM 1

If $y = y_1(x)$ and $y = y_2(x)$ are two solutions of the linear, homogeneous DE

$$a_n y^{(n)} + a_{n-1}y^{(n-1)} + \cdots + a_2y'' + a_1y' + a_0y = 0,$$

then so is the linear combination

$$y = Ay_1(x) + By_2(x)$$

for any values of the constants A and B.

PROOF We are given that

$$\begin{aligned} &a_n y_1^{(n)} + a_{n-1}y_1^{(n-1)} + \cdots + a_2y_1'' + a_1y_1' + a_0y_1 = 0 \qquad \text{and} \\ &a_n y_2^{(n)} + a_{n-1}y_2^{(n-1)} + \cdots + a_2y_2'' + a_1y_2' + a_0y_2 = 0. \end{aligned}$$

Multiplying the first equation by A and the second by B and adding the two gives

$$\begin{aligned} &a_n(Ay_1^{(n)} + By_2^{(n)}) + a_{n-1}(Ay_1^{(n-1)} + By_2^{(n-1)}) \\ &\quad + \cdots + a_2(Ay_1'' + By_2'') + a_1(Ay_1' + By_2') + a_0(Ay_1 + By_2) = 0. \end{aligned}$$

Thus, $y = Ay_1(x) + By_2(x)$ is also a solution of the equation.

The same kind of proof can be used to verify the following theorem.

THEOREM 2

If $y = y_1(x)$ is a solution of the linear, homogeneous equation

$$a_n y^{(n)} + a_{n-1}y^{(n-1)} + \cdots + a_2y'' + a_1y' + a_0y = 0$$

and $y = y_2(x)$ is a solution of the linear, nonhomogeneous equation

$$a_n y^{(n)} + a_{n-1}y^{(n-1)} + \cdots + a_2y'' + a_1y' + a_0y = f(x),$$

then $y = y_1(x) + y_2(x)$ is also a solution of the same linear, nonhomogeneous equation.

We will make extensive use of the two theorems above when we discuss second-order linear equations in Sections 18.4–18.6.

EXAMPLE 3 Verify that $y = \sin 2x$ and $y = \cos 2x$ satisfy the DE $y'' + 4y = 0$. Find a solution $y(x)$ of that DE that satisfies the *initial conditions* $y(0) = 2$ and $y'(0) = -4$.

Solution If $y = \sin 2x$, then $y'' = \dfrac{d}{dx}(2\cos 2x) = -4\sin 2x = -4y$. Thus, $y'' + 4y = 0$. A similar calculation shows that $y = \cos 2x$ also satisfies the DE. Since the DE is linear and homogeneous, the function

$$y = A\sin 2x + B\cos 2x$$

is a solution for any values of the constants A and B. We want $y(0) = 2$, so we need $2 = A\sin 0 + B\cos 0 = B$. Thus $B = 2$. Also,

$$y' = 2A\cos 2x - 2B\sin 2x.$$

We want $y'(0) = -4$, so $-4 = 2A\cos 0 - 2B\sin 0 = 2A$. Thus, $A = -2$ and the required solution is $y = -2\sin 2x + 2\cos 2x$.

Remark Let $P_n(r)$ be the nth-degree polynomial in the variable r given by

$$P_n(r) = a_n(x)r^n + a_{n-1}(x)r^{n-1} + \cdots + a_2(x)r^2 + a_1(x)r + a_0(x),$$

with coefficients depending on the variable x. We can write the nth-order linear ODE with coefficients $a_k(x)$, $(0 \le k \le n)$, and nonhomogeneous term $f(x)$ in the form

$$P_n(D)y(x) = f(x),$$

where D stands for the *differential operator* d/dx. The left side of the equation above denotes the application of the nth-order differential operator

$$P_n(D) = a_n(x)D^n + a_{n-1}(x)D^{n-1} + \cdots + a_2(x)D^2 + a_1(x)D + a_0(x)$$

to the function $y(x)$. For example,

$$a_k(x)D^k y(x) = a_k(x)\frac{d^k y}{dx^k}.$$

It is often useful to write linear DEs in terms of differential operators in this way.

Remark Unfortunately, the term *homogeneous* is used in more than one way in the study of differential equations. Certain ODEs that are not necessarily linear are called homogeneous for a different reason than the one applying for linear equations above. We will encounter equations of this type in Section 18.2.

EXERCISES 18.1

In Exercises 1–10, state the order of the given DE, and whether it is linear or nonlinear. If it is linear, is it homogeneous or nonhomogeneous?

1. $\dfrac{dy}{dx} = 5y$

2. $\dfrac{d^2y}{dx^2} + x = y$

3. $y\dfrac{dy}{dx} = x$

4. $y''' + xy' = x\sin x$

5. $y'' + x\sin x\, y' = y$

6. $y'' + 4y' - 3y = 2y^2$

7. $\dfrac{d^3y}{dt^3} + t\dfrac{dy}{dt} + t^2y = t^3$

8. $\cos x\dfrac{dx}{dt} + x\sin t = 0$

9. $y^{(4)} + e^x y'' = x^3 y'$

10. $x^2y'' + e^x y' = \dfrac{1}{y}$

11. Verify that $y = \cos x$ and $y = \sin x$ are solutions of the DE $y'' + y = 0$. Are any of the following functions solutions? (a) $\sin x - \cos x$, (b) $\sin(x + 3)$, (c) $\sin 2x$. Justify your answers.

12. Verify that $y = e^x$ and $y = e^{-x}$ are solutions of the DE $y'' - y = 0$. Are any of the following functions solutions? (a) $\cosh x = \frac{1}{2}(e^x + e^{-x})$, (b) $\cos x$, (c) x^e. Justify your answers.

13. $y_1 = \cos(kx)$ is a solution of $y'' + k^2 y = 0$. Guess and verify another solution y_2 that is not a multiple of y_1. Then find a solution that satisfies $y(\pi/k) = 3$ and $y'(\pi/k) = 3$.

14. $y_1 = e^{kx}$ is a solution of $y'' - k^2 y = 0$. Guess and verify another solution y_2 that is not a multiple of y_1. Then find a solution that satisfies $y(1) = 0$ and $y'(1) = 2$.

15. Find a solution of $y'' + y = 0$ that satisfies $y(\pi/2) = 2y(0)$ and $y(\pi/4) = 3$. *Hint:* See Exercise 11.

16. Find two values of r such that $y = e^{rx}$ is a solution of $y'' - y' - 2y = 0$. Then find a solution of the equation that satisfies $y(0) = 1$, $y'(0) = 2$.

17. Verify that $y = x$ is a solution of $y'' + y = x$, and find a solution y of this DE that satisfies $y(\pi) = 1$ and $y'(\pi) = 0$. *Hint:* Use Exercise 11 and Theorem 2.

18. Verify that $y = -e$ is a solution of $y'' - y = e$, and find a solution y of this DE that satisfies $y(1) = 0$ and $y'(1) = 1$. *Hint:* Use Exercise 12 and Theorem 2.

18.2 Solving First-Order Equations

In this section we will develop techniques for solving several types of first-order ODEs, specifically,

1. separable equations,
2. linear equations,
3. homogeneous equations, and
4. exact equations.

Most first-order equations are of the form

$$\frac{dy}{dx} = f(x, y).$$

Solving such differential equations typically involves integration; indeed, the process of solving a DE is called *integrating* the DE. Nevertheless, solving DEs is usually more complicated than just writing down an integral and evaluating it. The only kind of DE that can be solved that way is the simplest kind of first-order, linear DE that can be written in the form

$$\frac{dy}{dx} = f(x).$$

The solution is then just the antiderivative of f:

$$y = \int f(x)\,dx.$$

Separable Equations

The next simplest kind of equation to solve is a so-called **separable equation**. A separable equation is one of the form

$$\frac{dy}{dx} = f(x)g(y),$$

where the derivative dy/dx is a product of a function of x alone times a function of y alone, rather than a more general function of the two variables x and y.

A thorough discussion of separable equations with examples and exercises can be found in Section 7.9; we will not repeat it here. If you have not studied that material, please do so now.

First-Order Linear Equations

A first-order **linear** differential equation is one of the type

$$\frac{dy}{dx} + p(x)y = q(x),$$

where $p(x)$ and $q(x)$ are given functions, which we assume to be continuous. The equation is *homogeneous* (in the sense described in Section 18.1) provided that $q(x)$ is 0 for all x. In that case, the given linear equation is separable:

$$\frac{dy}{y} = -p(x)\,dx,$$

which can be solved by integrating both sides. Nonhomogeneous first-order linear equations can be solved by a procedure involving the calculation of an integrating factor.

The technique for solving first-order linear differential equations, along with several examples and exercises, can be found in Section 7.9. If you have not studied that material, please do so now.

First-Order Homogeneous Equations

A first-order DE of the form

$$\frac{dy}{dx} = f\left(\frac{y}{x}\right)$$

is said to be **homogeneous**. This is a *different* use of the term homogeneous from that in the previous section, which applied only to linear equations. Here, homogeneous refers to the fact that y/x, and therefore $g(x, y) = f(y/x)$ is *homogeneous of degree 0* in the sense described after Example 7 in Section 12.5. Such a homogeneous equation can be transformed into a separable equation (and therefore solved) by means of a change of dependent variable. If we set

$$v = \frac{y}{x}, \qquad \text{or equivalently} \qquad y = xv(x),$$

then we have

$$\frac{dy}{dx} = v + x\,\frac{dv}{dx},$$

and the original differential equation transforms into

$$\frac{dv}{dx} = \frac{f(v) - v}{x},$$

which is separable.

EXAMPLE 1 Solve the equation

$$\frac{dy}{dx} = \frac{x^2 + xy}{xy + y^2}.$$

Solution The equation is homogeneous. (Divide the numerator and denominator of the right-hand side by x^2 to see this.) If $y = vx$ the equation becomes

$$v + x\,\frac{dv}{dx} = \frac{1 + v}{v + v^2} = \frac{1}{v},$$

or

$$x\frac{dv}{dx} = \frac{1-v^2}{v}.$$

Separating variables and integrating, we calculate

$$\begin{aligned}
\int \frac{v\,dv}{1-v^2} &= \int \frac{dx}{x} && \text{Let } u = 1-v^2 \\
-\frac{1}{2}\int \frac{du}{u} &= \int \frac{dx}{x} \\
-\ln|u| &= 2\ln|x| + C_1 = \ln C_2 x^2 && (C_1 = \ln C_2) \\
\frac{1}{|u|} &= C_2 x^2 \\
|1-v^2| &= \frac{C_3}{x^2} && (C_3 = 1/C_2) \\
\left|1-\frac{y^2}{x^2}\right| &= \frac{C_3}{x^2}.
\end{aligned}$$

The solution is best expressed in the form $x^2 - y^2 = C_4$. However, near points where $y \neq 0$, the equation can be solved for y as a function of x.

Exact Equations

A first-order differential equation expressed in differential form as

$$M(x, y)\,dx + N(x, y)\,dy = 0,$$

which is equivalent to $\dfrac{dy}{dx} = -\dfrac{M(x, y)}{N(x, y)}$, is said to be **exact** if the left-hand side is the differential of a function $\phi(x, y)$:

$$d\phi(x, y) = M(x, y)\,dx + N(x, y)\,dy.$$

The function ϕ is called an **integral function** of the differential equation. The level curves $\phi(x, y) = C$ of ϕ are the **solution curves** of the differential equation. For example, the differential equation

$$x\,dx + y\,dy = 0$$

has solution curves given by

$$x^2 + y^2 = C$$

since $d(x^2 + y^2) = 2(x\,dx + y\,dy) = 0$.

Remark The condition that the differential equation $M\,dx + N\,dy = 0$ should be exact is just the condition that the vector field

$$\mathbf{F} = M(x, y)\,\mathbf{i} + N(x, y)\,\mathbf{j}$$

should be *conservative*; the integral function of the differential equation is then the potential function of the vector field. (See Section 15.2.)

A **necessary condition** for the exactness of the DE $M\,dx + N\,dy = 0$ is that

$$\frac{\partial M}{\partial y} = \frac{\partial N}{\partial x};$$

this just says that the mixed partial derivatives $\frac{\partial^2 \phi}{\partial x \partial y}$ and $\frac{\partial^2 \phi}{\partial y \partial x}$ of the integral function ϕ must be equal.

Once you know that an equation is exact, you can often guess the integral function. In any event, ϕ can always be found by the same method used to find the potential of a conservative vector field in Section 15.2.

EXAMPLE 2 Verify that the DE

$$(2x + \sin y - ye^{-x})\,dx + (x \cos y + \cos y + e^{-x})\,dy = 0$$

is exact and find its solution curves.

Solution Here, $M = 2x + \sin y - ye^{-x}$ and $N = x \cos y + \cos y + e^{-x}$. Since

$$\frac{\partial M}{\partial y} = \cos y - e^{-x} = \frac{\partial N}{\partial x},$$

the DE is exact. We want to find ϕ so that

$$\frac{\partial \phi}{\partial x} = M = 2x + \sin y - ye^{-x} \quad \text{and} \quad \frac{\partial \phi}{\partial y} = N = x \cos y + \cos y + e^{-x}.$$

Integrate the first equation with respect to x, being careful to allow the constant of integration to depend on y:

$$\phi(x, y) = \int (2x + \sin y - ye^{-x})\,dx = x^2 + x \sin y + ye^{-x} + C_1(y).$$

Now substitute this expression into the second equation:

$$x \cos y + \cos y + e^{-x} = \frac{\partial \phi}{\partial y} = x \cos y + e^{-x} + C_1'(y).$$

Thus, $C_1'(y) = \cos y$, and $C_1(y) = \sin y + C_2$. (It is because the original DE was exact that the equation for $C_1'(y)$ turned out to be independent of x; this had to happen or we could not have found C_1 as a function of y only.) Choosing $C_2 = 0$, we find that $\phi(x, y) = x^2 + x \sin y + ye^{-x} + \sin y$ is an integral function for the given DE. The solution curves for the DE are the level curves

$$x^2 + x \sin y + ye^{-x} + \sin y = C.$$

Integrating Factors

Any ordinary differential equation of order 1 and degree 1 can be expressed in differential form: $M\,dx + N\,dy = 0$. However, this latter equation will usually not be exact. It *may* be possible to multiply the equation by an **integrating factor** $\mu(x, y)$ so that the resulting equation

$$\mu(x, y)\,M(x, y)\,dx + \mu(x, y)\,N(x, y)\,dy = 0$$

is exact. In general, such integrating factors are difficult to find; they must satisfy the partial differential equation

$$M(x, y)\frac{\partial \mu}{\partial y} - N(x, y)\frac{\partial \mu}{\partial x} = \mu(x, y)\left(\frac{\partial N}{\partial x} - \frac{\partial M}{\partial y}\right),$$

which follows from the necessary condition for exactness stated above. We will not try to solve this equation here.

Sometimes it happens that a differential equation has an integrating factor depending on only one of the two variables. Suppose, for instance, that $\mu(x)$ is an integrating factor for $M\,dx + N\,dy = 0$. Then $\mu(x)$ must satisfy the ordinary differential equation

$$N(x,y)\frac{d\mu}{dx} = \mu(x)\left(\frac{\partial M}{\partial y} - \frac{\partial N}{\partial x}\right),$$

or

$$\frac{1}{\mu(x)}\frac{d\mu}{dx} = \frac{\dfrac{\partial M}{\partial y} - \dfrac{\partial N}{\partial x}}{N(x,y)}.$$

This equation can be solved (by integration) for μ as a function of x alone *provided that the right-hand side is independent of* y.

EXAMPLE 3 Show that $(x + y^2)\,dx + xy\,dy = 0$ has an integrating factor depending only on x, find it, and solve the equation.

Solution Here $M = x + y^2$ and $N = xy$. Since

$$\frac{\dfrac{\partial M}{\partial y} - \dfrac{\partial N}{\partial x}}{N(x,y)} = \frac{2y - y}{xy} = \frac{1}{x}$$

does not depend on y, the equation has an integrating factor depending only on x. This factor is given by $d\mu/\mu = dx/x$. Evidently, $\mu = x$ is a suitable integrating factor; if we multiply the given differential equation by x, we obtain

$$0 = (x^2 + xy^2)\,dx + x^2y\,dy = d\left(\frac{x^3}{3} + \frac{x^2y^2}{2}\right).$$

The solution is therefore $2x^3 + 3x^2y^2 = C$.

Remark Of course, it may be possible to find an integrating factor depending on y instead of x. See Exercises 17–19 below. It is also possible to look for integrating factors that depend on specific combinations of x and y, for instance, xy. See Exercise 20.

EXERCISES 18.2

See Section 7.9 for exercises on separable equations and linear equations.

Solve the homogeneous differential equations in Exercises 1–6.

1. $\dfrac{dy}{dx} = \dfrac{x+y}{x-y}$

2. $\dfrac{dy}{dx} = \dfrac{xy}{x^2+2y^2}$

3. $\dfrac{dy}{dx} = \dfrac{x^2+xy+y^2}{x^2}$

4. $\dfrac{dy}{dx} = \dfrac{x^3+3xy^2}{3x^2y+y^3}$

5. $x\dfrac{dy}{dx} = y + x\cos^2\left(\dfrac{y}{x}\right)$

6. $\dfrac{dy}{dx} = \dfrac{y}{x} - e^{-y/x}$

7. Find an equation of the curve in the xy-plane that passes through the point $(2, 3)$ and has, at every point (x, y) on it, slope $2x/(1+y^2)$.

8. Repeat Problem 7 for the point $(1, 3)$ and slope $1 + (2y/x)$.

9. Show that the change of variables $\xi = x - x_0$, $\eta = y - y_0$ transforms the equation

$$\frac{dy}{dx} = \frac{ax+by+c}{ex+fy+g}$$

into the homogeneous equation

$$\frac{d\eta}{d\xi} = \frac{a\xi+b\eta}{e\xi+f\eta},$$

provided (x_0, y_0) is the solution of the system

$$ax + by + c = 0$$
$$ex + fy + g = 0.$$

10. Use the technique of Exercise 9 to solve the equation $\dfrac{dy}{dx} = \dfrac{x + 2y - 4}{2x - y - 3}$.

Show that the DEs in Exercises 11–14 are exact, and solve them.

11. $(xy^2 + y)\,dx + (x^2y + x)\,dy = 0$

12. $(e^x \sin y + 2x)\,dx + (e^x \cos y + 2y)\,dy = 0$

13. $e^{xy}(1 + xy)\,dx + x^2e^{xy}\,dy = 0$

14. $\left(2x + 1 - \dfrac{y^2}{x^2}\right) dx + \dfrac{2y}{x}\,dy = 0$

Show that the DEs in Exercises 15–16 admit integrating factors that are functions of x alone. Then solve the equations.

15. $(x^2 + 2y)\,dx - x\,dy = 0$

16. $(xe^x + x \ln y + y)\,dx + \left(\dfrac{x^2}{y} + x \ln x + x \sin y\right) dy = 0$

17. What condition must the coefficients $M(x, y)$ and $N(x, y)$ satisfy if the equation $M\,dx + N\,dy = 0$ is to have an integrating factor of the form $\mu(y)$, and what DE must the integrating factor satisfy?

18. Find an integrating factor of the form $\mu(y)$ for the equation

$$2y^2(x + y^2)\,dx + xy(x + 6y^2)\,dy = 0,$$

and hence solve the equation. *Hint:* See Exercise 17.

19. Find an integrating factor of the form $\mu(y)$ for the equation $y\,dx - (2x + y^3e^y)\,dy = 0$, and hence solve the equation. *Hint:* See Exercise 17.

20. What condition must the coefficients $M(x, y)$ and $N(x, y)$ satisfy if the equation $M\,dx + N\,dy = 0$ is to have an integrating factor of the form $\mu(xy)$, and what DE must the integrating factor satisfy?

21. Find an integrating factor of the form $\mu(xy)$ for the equation

$$\left(x \cos x + \frac{y^2}{x}\right) dx - \left(\frac{x \sin x}{y} + y\right) dy = 0,$$

and hence solve the equation. *Hint:* See Exercise 20.

18.3 Existence, Uniqueness, and Numerical Methods

A general first-order differential equation of the form

$$\frac{dy}{dx} = f(x, y)$$

specifies a slope $f(x, y)$ at every point (x, y) in the domain of f, and therefore represents a **slope field**. Such a slope field can be represented graphically by drawing short line segments of the indicated slope at many points in the xy-plane. Slope fields resemble vector fields, but the segments are usually drawn having the same length and without arrowheads. Figure 18.1 portrays the slope field for the differential equation

$$\frac{dy}{dx} = x - y.$$

Solving a typical initial-value problem

$$\begin{cases} \dfrac{dy}{dx} = f(x, y) \\ y(x_0) = y_0 \end{cases}$$

involves finding a function $y = \phi(x)$ such that

$$\phi'(x) = f\big(x, \phi(x)\big) \qquad \text{and} \qquad \phi(x_0) = y_0.$$

The graph of the equation $y = \phi(x)$ is a curve passing through (x_0, y_0) that is tangent to the slope-field at each point. Such curves are called **solution curves** of the differential equation. Figure 18.1 shows four solution curves for $y' = x - y$ corresponding to the initial conditions $y(0) = C$, where $C = -2, -1, 0$, and 1.

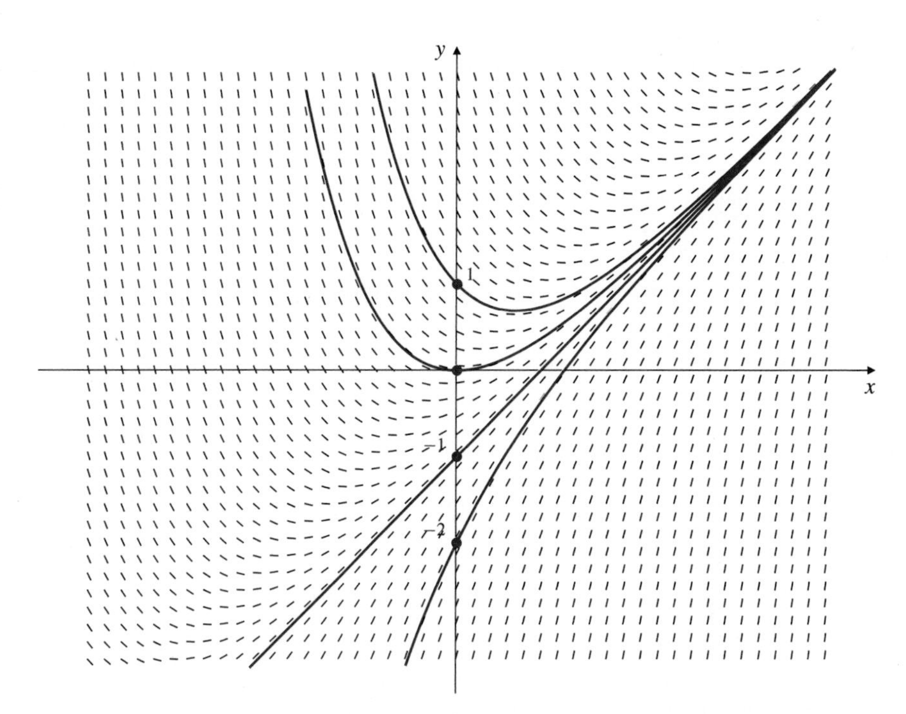

Figure 18.1 The slope field for the DE $y' = x - y$ and four solution curves for this DE

The DE $y' = x - y$ is linear and can be solved explicitly by the method of Section 18.2. Indeed, the solution satisfying $y(0) = C$ is $y = x - 1 + (C + 1)e^{-x}$. Most differential equations of the form $y' = f(x, y)$ cannot be solved for y as an explicit function of x, so we must use numerical approximation methods to find the value of a solution function $\phi(x)$ at particular points.

Existence and Uniqueness of Solutions

Even if we cannot calculate an explicit solution of an initial-value problem, it is important to know when the problem has a solution and whether that solution is unique.

THEOREM 3

An existence and uniqueness theorem for first-order initial-value problems

Suppose that $f(x, y)$ and $f_2(x, y) = (\partial/\partial y) f(x, y)$ are continuous on a rectangle R of the form $a \le x \le b$, $c \le y \le d$, containing the point (x_0, y_0) in its interior. Then there exists a number $\delta > 0$ and a *unique* function $\phi(x)$ defined and having a continuous derivative on the interval $(x_0 - \delta, x_0 + \delta)$ such that $\phi(x_0) = y_0$ and $\phi'(x) = f\big(x, \phi(x)\big)$ for $x_0 - \delta < x < x_0 + \delta$. In other words, the initial-value problem

$$\begin{cases} \dfrac{dy}{dx} = f(x, y) \\ y(x_0) = y_0 \end{cases} \qquad (*)$$

has a unique solution on $(x_0 - \delta, x_0 + \delta)$.

We give only an outline of the proof here. Any solution $y = \phi(x)$ of the initial-value problem $(*)$ must also satisfy the **integral equation**

$$\phi(x) = y_0 + \int_{x_0}^{x} f\big(t, \phi(t)\big)\, dt, \qquad (**)$$

and, conversely, any solution of the integral equation $(**)$ must also satisfy the initial-value problem $(*)$. A sequence of approximations $\phi_n(x)$ to a solution of $(**)$ can be constructed as follows:

$$\phi_0(x) = y_0$$

$$\phi_{n+1}(x) = y_0 + \int_{x_0}^{x} f\big(t, \phi_n(t)\big)\, dt \qquad \text{for} \quad n = 0, 1, 2, \ldots$$

(These are called **Picard iterations.**) The proof of Theorem 3 involves showing that

$$\lim_{n\to\infty} \phi_n(x) = \phi(x)$$

exists on an interval $(x_0 - \delta, x_0 + \delta)$ and that the resulting limit $\phi(x)$ satisfies the integral equation $(**)$. The details can be found in more advanced texts on differential equations and analysis.

Remark Some initial-value problems can have nonunique solutions. For example, the functions $y_1(x) = x^3$ and $y_2(x) = 0$ both satisfy the initial-value problem

$$\begin{cases} \dfrac{dy}{dx} = 3y^{2/3} \\ y(0) = 0. \end{cases}$$

In this case $f(x, y) = 3y^{2/3}$ is continuous on the whole xy-plane. However, $\partial f/\partial y = 2y^{-1/3}$ is not continuous on the x-axis and is therefore not continuous on any rectangle containing $(0, 0)$ in its interior. The conditions of Theorem 3 are not satisfied, and the initial-value problem has a solution but not a unique one.

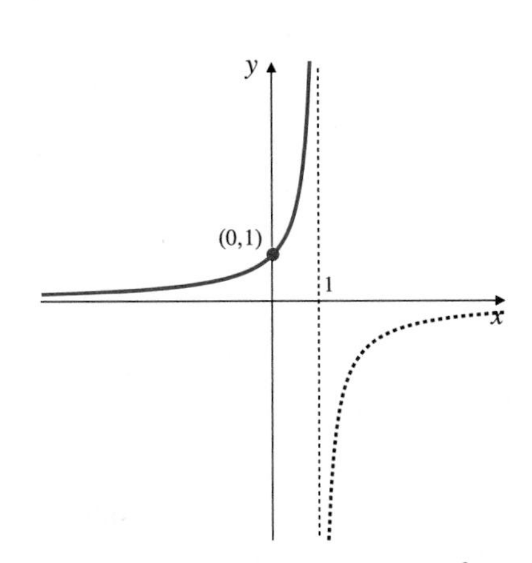

Figure 18.2 The solution to $y' = y^2$, $y(0) = 1$ is the part of the curve $y = 1/(1 - x)$ to the left of the vertical asymptote at $x = 1$

Remark The unique solution $y = \phi(x)$ to the initial-value problem $(*)$ guaranteed by Theorem 3 may not be defined on the whole interval $[a, b]$ because it can "escape" from the rectangle R through the top or bottom edges. Even if $f(x, y)$ and $(\partial/\partial y)f(x, y)$ are continuous on the whole xy-plane, the solution may not be defined on the whole real line. For example,

$$y = \frac{1}{1-x} \quad \text{satisfies the initial-value problem} \quad \begin{cases} \dfrac{dy}{dx} = y^2 \\ y(0) = 1 \end{cases}$$

but only for $x < 1$. Starting from $(0, 1)$, we can follow the solution curve as far as we want to the left of $x = 0$, but to the right of $x = 0$ the curve recedes to ∞ as $x \to 1-$. (See Figure 18.2.) It makes no sense to regard the part of the curve to the right of $x = 1$ as part of the solution curve to the initial-value problem.

Numerical Methods

Suppose that the conditions of Theorem 3 are satisfied, so we know that the initial-value problem

$$\begin{cases} \dfrac{dy}{dx} = f(x, y) \\ y(x_0) = y_0 \end{cases}$$

has a unique solution $y = \phi(x)$ on some interval containing x_0. Even if we cannot solve the differential equation and find $\phi(x)$ explicitly, we can still try to find approximate values y_n for $\phi(x_n)$ at a sequence of points

$$x_0, \quad x_1 = x_0 + h, \quad x_2 = x_0 + 2h, \quad x_3 = x_0 + 3h, \quad \ldots$$

starting at x_0. Here, $h > 0$ (or $h < 0$) is called the **step size** of the approximation scheme. In the remainder of this section we will describe three methods for constructing the approximations $\{y_n\}$:

1. the Euler method,
2. the improved Euler method, and
3. the fourth-order Runge–Kutta method.

Each of these methods starts with the given value of y_0 and provides a formula for constructing y_{n+1} when you know y_n. The three methods are listed above in increasing order of the complexity of their formulas, but the more complicated formulas produce much better approximations for any given step size h.

The **Euler method** involves approximating the solution curve $y = \phi(x)$ by a polygonal line (a sequence of straight line segments joined end to end), where each segment has horizontal length h and slope determined by the value of $f(x, y)$ at the end of the previous segment. Thus, if $x_n = x_0 + nh$, then

$$\begin{aligned} y_1 &= y_0 + f(x_0, y_0)h \\ y_2 &= y_1 + f(x_1, y_1)h \\ y_3 &= y_2 + f(x_2, y_2)h \end{aligned}$$

and, in general,

Iteration formulas for Euler's method

$$x_{n+1} = x_n + h, \qquad y_{n+1} = y_n + hf(x_n, y_n).$$

EXAMPLE 1 Use Euler's method to find approximate values for the solution of the initial-value problem

$$\begin{cases} \dfrac{dy}{dx} = x - y \\ y(0) = 1 \end{cases}$$

on the interval [0, 1] using

(a) 5 steps of size $h = 0.2$, and

(b) 10 steps of size $h = 0.1$.

Calculate the error at each step, given that the problem (which involves a linear equation and so can be solved explicitly) has solution $y = \phi(x) = x - 1 + 2e^{-x}$.

Solution

(a) Here we have $f(x, y) = x - y$, $x_0 = 0$, $y_0 = 1$, and $h = 0.2$, so that

$$x_n = \frac{n}{5}, \qquad y_{n+1} = y_n + 0.2(x_n - y_n),$$

and the error is $e_n = \phi(x_n) - y_n$ for $n = 0, 1, 2, 3, 4$, and 5. The results of the calculation, which was done easily using a computer spreadsheet program, are presented in Table 1.

Table 1. Euler approximations with $h = 0.2$

n	x_n	y_n	$f(x_n, y_n)$	y_{n+1}	$e_n = \phi(x_n) - y_n$
0	0.0	1.000 000	−1.000 000	0.800 000	0.000 000
1	0.2	0.800 000	−0.600 000	0.680 000	0.037 462
2	0.4	0.680 000	−0.280 000	0.624 000	0.060 640
3	0.6	0.624 000	−0.024 000	0.619 200	0.073 623
4	0.8	0.619 200	0.180 800	0.655 360	0.079 458
5	1.0	0.655 360	0.344 640		0.080 399

The exact solution $y = \phi(x)$ and the polygonal line representing the Euler approximation are shown in Figure 18.3. The approximation lies below the solution curve, as is reflected in the positive values in the last column of Table 1, representing the error at each step.

(b) Here we have $h = 0.1$, so that

$$x_n = \frac{n}{10}, \qquad y_{n+1} = y_n + 0.1(x_n - y_n)$$

for $n = 0, 1, \ldots, 10$. Again we present the results in tabular form:

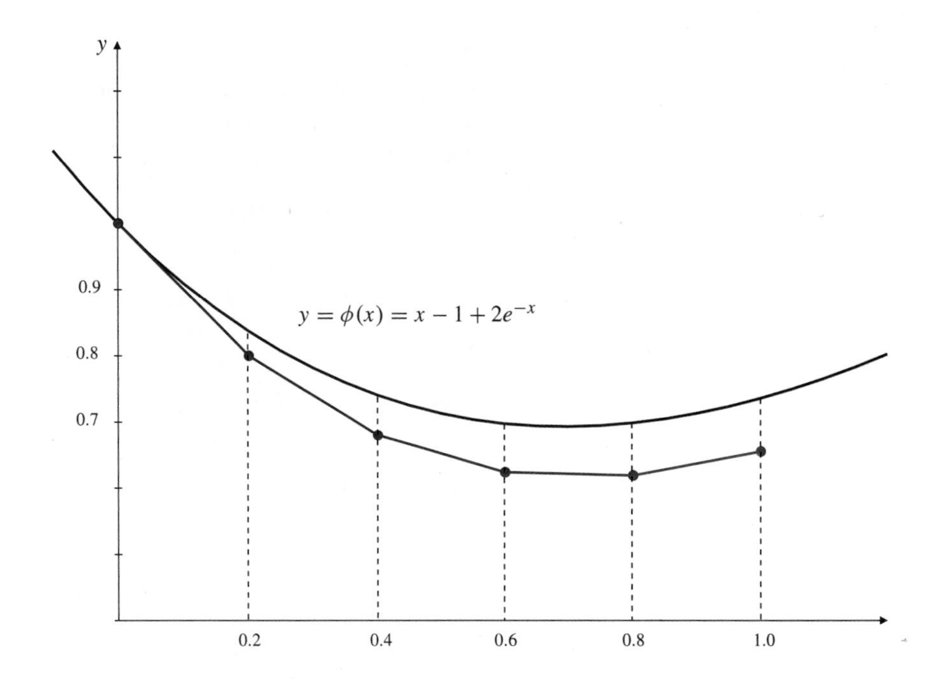

Figure 18.3 The solution $y = \phi(x)$ to $y' = x - y$, $y(0) = 1$ and an Euler approximation to it on $[0, 1]$ with step size $h = 0.2$

Table 2. Euler approximations with $h = 0.1$

n	x_n	y_n	$f(x_n, y_n)$	y_{n+1}	$e_n = \phi(x_n) - y_n$
0	0.0	1.000 000	−1.000 000	0.900 000	0.000 000
1	0.1	0.900 000	−0.800 000	0.820 000	0.009 675
2	0.2	0.820 000	−0.620 000	0.758 000	0.017 462
3	0.3	0.758 000	−0.458 000	0.712 200	0.023 636
4	0.4	0.712 200	−0.312 200	0.680 980	0.028 440
5	0.5	0.680 980	−0.180 980	0.662 882	0.032 081
6	0.6	0.662 882	−0.062 882	0.656 594	0.034 741
7	0.7	0.656 594	0.043 406	0.660 934	0.036 577
8	0.8	0.660 934	0.139 066	0.674 841	0.037 724
9	0.9	0.674 841	0.225 159	0.697 357	0.038 298
10	1.0	0.697 357	0.302 643		0.038 402

Observe that the error at the end of the first step is about one-quarter of the error at the end of the first step in part (a), but the final error at $x = 1$ is only about half as large as in part (a). This behaviour is characteristic of Euler's method.

If we decrease the step size h, it takes more steps ($n = |x - x_0|/h$) to get from the starting point x_0 to a particular value x where we want to know the value of the solution. For Euler's method it can be shown that the error at each step decreases on average proportionally to h^2, but the errors can accumulate from step to step, so the error at x can be expected to decrease proportionally to $nh^2 = |x - x_0|h$. This is consistent with the results of Example 1. Decreasing h and so increasing n is costly in terms of computing resources, so we would like to find ways of reducing the error without decreasing the step size. This is similar to developing better techniques than the Trapezoid Rule for evaluating definite integrals numerically.

The **improved Euler method** is a step in this direction. The accuracy of the Euler method is hampered by the fact that the slope of each segment in the approximating polygonal line is determined by the value of $f(x, y)$ at one endpoint of the segment. Since f varies along the segment, we would expect to do better by using, say, the average value of $f(x, y)$ at the two ends of the segment, that is, by calculating y_{n+1} from the formula

$$y_{n+1} = y_n + h\frac{f(x_n, y_n) + f(x_{n+1}, y_{n+1})}{2}.$$

Unfortunately, y_{n+1} appears on both sides of this equation, and we can't usually solve the equation for y_{n+1}. We can get around this difficulty by replacing y_{n+1} on the right side by its Euler approximation $y_n + hf(x_n, y_n)$. The resulting formula is the basis for the improved Euler method.

Iteration formulas for the improved Euler method

$$x_{n+1} = x_n + h$$
$$u_{n+1} = y_n + h\, f(x_n, y_n)$$
$$y_{n+1} = y_n + h\,\frac{f(x_n, y_n) + f(x_{n+1}, u_{n+1})}{2}.$$

EXAMPLE 2 Use the improved Euler method with $h = 0.2$ to find approximate values for the solution to the initial-value problem of Example 1 on $[0, 1]$. Compare the errors with those obtained by the Euler method.

Solution Table 3 summarizes the calculation of five steps of the improved Euler method for $f(x, y) = x - y$, $x_0 = 0$, and $y_0 = 1$.

Table 3. Improved Euler approximations with $h = 0.2$

n	x_n	y_n	u_{n+1}	y_{n+1}	$e_n = \phi(x_n) - y_n$
0	0.0	1.000 000	0.800 000	0.840 000	0.000 000
1	0.2	0.840 000	0.712 000	0.744 800	−0.002 538
2	0.4	0.744 800	0.675 840	0.702 736	−0.004 160
3	0.6	0.702 736	0.682 189	0.704 244	−0.005 113
4	0.8	0.704 244	0.723 395	0.741 480	−0.005 586
5	1.0	0.741 480	0.793 184		−0.005 721

Observe that the errors are considerably less than one-tenth those obtained in Example 1(a). Of course, more calculations are necessary at each step, but the number of evaluations of $f(x, y)$ required is only twice the number required for Example 1(a). As for numerical integration, if f is complicated, it is these function evaluations that constitute most of the computational "cost" of computing numerical solutions.

Remark It can be shown for well-behaved functions f that the error at each step in the improved Euler method is bounded by a multiple of h^3 rather than h^2 as for the (unimproved) Euler method. Thus the cumulative error at x can be bounded by a constant times $|x - x_0|h^2$. If Example 2 is repeated with 10 steps of size $h = 0.1$, the error at $n = 10$ (i.e., at $x = 1$) is $-0.001\,323$, which is about one-fourth the size of the error at $x = 1$ with $h = 0.2$.

The fourth-order Runge–Kutta method further improves upon the improved Euler method, but at the expense of requiring more complicated calculations at each step. It requires four evaluations of $f(x, y)$ at each step, but the error at each step is less than a constant times h^5, so the cumulative error decreases like h^4 as h decreases. Like the improved Euler method, this method involves calculating a certain kind of average slope for each segment in the polygonal approximation to the solution to the initial-value problem. We present the appropriate formulas below but cannot derive them here.

Iteration formulas for the Runge–Kutta method

$$\begin{aligned}
x_{n+1} &= x_n + h \\
p_n &= f(x_n, y_n) \\
q_n &= f\left(x_n + \frac{h}{2}, y_n + \frac{h}{2}p_n\right) \\
r_n &= f\left(x_n + \frac{h}{2}, y_n + \frac{h}{2}q_n\right) \\
s_n &= f(x_n + h, y_n + hr_n) \\
y_{n+1} &= y_n + h\,\frac{p_n + 2q_n + 2r_n + s_n}{6}.
\end{aligned}$$

EXAMPLE 3 Use the fourth-order Runge-Kutta method with $h = 0.2$ to find approximate values for the solution to the initial-value problem of Example 1 on $[0, 1]$. Compare the errors with those obtained by the Euler and improved Euler methods.

Solution Table 4 summarizes the calculation of five steps of the Runge–Kutta method for $f(x, y) = x - y$, $x_0 = 0$, and $y_0 = 1$ according to the formulas above. The table does not show the values of the intermediate quantities p_n, q_n, r_n, and s_n, but columns for these quantities were included in the spreadsheet in which the calculations were made.

Table 4. Fourth-order Runge–Kutta approximations with $h = 0.2$

n	x_n	y_n	$e_n = \phi(x_n) - y_n$
0	0.0	1.000 000	0.000 000 0
1	0.2	0.837 467	−0.000 005 2
2	0.4	0.740 649	−0.000 008 5
3	0.6	0.697 634	−0.000 010 4
4	0.8	0.698 669	−0.000 011 3
5	1.0	0.735 770	−0.000 011 6

The errors here are about 1/500 of the size of the errors obtained with the improved Euler method and about 1/7,000 of the size of the errors obtained with the Euler method. This great improvement was achieved at the expense of doubling the number of function evaluations required in the improved Euler method and quadrupling the number required in the Euler method. If we use 10 steps of size $h = 0.1$ in the Runge–Kutta method, the error at $x = 1$ is reduced to $-6.664\,82 \times 10^{-7}$, which is less than 1/16 of its value when $h = 0.2$.

Our final example shows what can happen with numerical approximations to a solution that is unbounded.

EXAMPLE 4 Obtain approximations at $x = 0.4$, $x = 0.8$, and $x = 1.0$ for solutions to the initial-value problem

$$\begin{cases} y' = y^2 \\ y(0) = 1 \end{cases}$$

using all three methods described above, and using step sizes $h = 0.2$, $h = 0.1$, and $h = 0.05$ for each method. What do the results suggest about the values of the solution at these points? Compare the results with the actual solution $y = 1/(1 - x)$.

Solution The various approximations are calculated using the various formulas described above for $f(x, y) = y^2$, $x_0 = 0$, and $y_0 = 1$. The results are presented in Table 5.

Table 5. Comparing methods and step sizes for $y' = y^2$, $y(0) = 1$

	$h = 0.2$	$h = 0.1$	$h = 0.05$
Euler			
$x = 0.4$	1.488 000	1.557 797	1.605 224
$x = 0.8$	2.676 449	3.239 652	3.793 197
$x = 1.0$	4.109 124	6.128 898	9.552 668
Improved Euler			
$x = 0.4$	1.640 092	1.658 736	1.664 515
$x = 0.8$	4.190 396	4.677 726	4.897 519
$x = 1.0$	11.878 846	22.290 765	43.114 668
Runge–Kutta			
$x = 0.4$	1.666 473	1.666 653	1.666 666
$x = 0.8$	4.965 008	4.996 663	4.999 751
$x = 1.0$	41.016 258	81.996 399	163.983 395

Little useful information can be read from the Euler results. The improved Euler results suggest that the solution exists at $x = 0.4$ and $x = 0.8$, but likely not at $x = 1$. The Runge–Kutta results confirm this and suggest that $y(0.4) = 5/3$ and $y(0.8) = 5$, which are the correct values provided by the actual solution $y = 1/(1 - x)$. They also suggest very strongly that the solution "blows up" at (or near) $x = 1$.

EXERCISES 18.3

A computer is almost essential for doing most of these exercises. The calculations are easily done with a spreadsheet program in which formulas for calculating the various quantities involved can be replicated down columns to automate the iteration process.

1. Use the Euler method with step sizes (a) $h = 0.2$, (b) $h = 0.1$, and (c) $h = 0.05$ to approximate $y(2)$ given that $y' = x + y$ and $y(1) = 0$.

2. Repeat Exercise 1 using the improved Euler method.

3. Repeat Exercise 1 using the Runge–Kutta method.

4. Use the Euler method with step sizes (a) $h = 0.2$ and (b) $h = 0.1$ to approximate $y(2)$ given that $y' = xe^{-y}$ and $y(0) = 0$.

5. Repeat Exercise 4 using the improved Euler method.

6. Repeat Exercise 4 using the Runge–Kutta method.

7. Use the Euler method with (a) $h = 0.2$, (b) $h = 0.1$, and (c) $h = 0.05$ to approximate $y(1)$ given that $y' = \cos y$ and $y(0) = 0$.

8. Repeat Exercise 7 using the improved Euler method.

9. Repeat Exercise 7 using the Runge–Kutta method.

10. Use the Euler method with (a) $h = 0.2$, (b) $h = 0.1$, and (c) $h = 0.05$ to approximate $y(1)$ given that $y' = \cos(x^2)$ and $y(0) = 0$.

11. Repeat Exercise 10 using the improved Euler method.

12. Repeat Exercise 10 using the Runge–Kutta method.

Solve the integral equations in Exercises 13–14 by rephrasing them as initial-value problems.

13. $y(x) = 2 + \int_1^x \big(y(t)\big)^2\,dt$. *Hint:* Find $\frac{dy}{dx}$ and $y(1)$.

14. $u(x) = 1 + 3\int_2^x t^2 u(t)\,dt$. *Hint:* Find $\frac{du}{dx}$ and $u(2)$.

15. The methods of this section can be used to approximate definite integrals numerically. For example,

$$I = \int_a^b f(x)\,dx$$

is given by $I = y(b)$, where

$$y' = f(x), \qquad \text{and} \qquad y(a) = 0.$$

Show that one step of the Runge–Kutta method with $h = b - a$ gives the same result for I as Simpson's Rule (Section 6.7) with two subintervals of length $h/2$.

16. If $\phi(0) = A \ge 0$ and $\phi'(x) \ge k\phi(x)$ on $[0, X]$, where $k > 0$ and $X > 0$ are constants, show that $\phi(x) \ge Ae^{kx}$ on $[0, X]$. *Hint:* Calculate $(d/dx)(\phi(x)/e^{kx})$.

17. Consider the three initial-value problems

(A)	$u' = u^2$	$u(0) = 1$
(B)	$y' = x + y^2$	$y(0) = 1$
(C)	$v' = 1 + v^2$	$v(0) = 1$

(a) Show that the solution of (B) remains between the solutions of (A) and (C) on any interval $[0, X]$ where solutions of all three problems exist. *Hint:* We must have $u(x) \ge 1$, $y(x) \ge 1$, and $v(x) \ge 1$ on $[0, X]$. (Why?) Apply the result of Exercise 16 to $\phi = y - u$ and to $\phi = v - y$.

(b) Find explicit solutions for problems (A) and (C). What can you conclude about the solution to problem (B)?

(c) Use the Runge–Kutta method with $h = 0.05$, $h = 0.02$, and $h = 0.01$ to approximate the solution to (B) on $[0, 1]$. What can you conclude now?

18.4 Differential Equations of Second Order

The general second-order ordinary differential equation is of the form

$$F\left(\frac{d^2y}{dx^2}, \frac{dy}{dx}, y, x\right) = 0$$

for some function F of four variables. When such an equation can be solved explicitly for y as a function of x, the solution typically involves two integrations and therefore two arbitrary constants. A unique solution usually results from prescribing the values of the solution y and its first derivative $y' = dy/dx$ at a particular point. Such a prescription constitutes an **initial-value problem** for the second-order equation.

Equations Reducible to First Order

A second-order equation of the form

$$F\left(\frac{d^2y}{dx^2}, \frac{dy}{dx}, x\right) = 0$$

that does not involve the unknown function y explicitly (except through its derivatives) can be reduced to a first-order equation by a change of dependent variable; if $v = dy/dx$, then the equation can be written

$$F\left(\frac{dv}{dx}, v, x\right) = 0.$$

This first-order equation in v may be amenable to the techniques described in earlier sections. If an explicit solution $v = v(x)$ can be found and integrated, then the function

$$y = \int v(x)\,dx$$

is an explicit solution of the given equation.

EXAMPLE 1 Solve the initial-value problem

$$\frac{d^2y}{dx^2} = x\left(\frac{dy}{dx}\right)^2, \qquad y(0) = 1, \qquad y'(0) = -2.$$

Solution If we let $v = dy/dx$, the given differential equation becomes

$$\frac{dv}{dx} = xv^2,$$

which is a separable first-order equation. Thus,

$$\begin{aligned} \frac{dv}{v^2} &= x\,dx \\ -\frac{1}{v} &= \frac{x^2}{2} + \frac{C_1}{2} \\ v &= -\frac{2}{x^2 + C_1}. \end{aligned}$$

The initial condition $y'(0) = -2$ implies that $v(0) = -2$ and so $C_1 = 1$. Therefore,

$$y = -2\int \frac{dx}{x^2+1} = -2\tan^{-1}x + C_2.$$

The initial condition $y(0) = 1$ implies that $C_2 = 1$, so the solution of the given initial-value problem is $y = 1 - 2\tan^{-1}x$.

A second-order equation of the form

$$F\left(\frac{d^2y}{dx^2}, \frac{dy}{dx}, y\right) = 0$$

that does not explicitly involve the independent variable x can be reduced to a first-order equation by a change of both dependent and independent variables. Again let $v = dy/dx$, but regard v as a function of y rather than x; $v = v(y)$. Then

$$\frac{d^2y}{dx^2} = \frac{dv}{dx} = \frac{dv}{dy}\frac{dy}{dx} = v\frac{dv}{dy}$$

by the Chain Rule. Hence, the given differential equation becomes

$$F\left(v\frac{dv}{dy}, v, y\right) = 0,$$

which is a first-order equation for v as a function of y. If this equation can be solved for $v = v(y)$, there still remains the problem of solving the separable equation $(dy/dx) = v(y)$ for y as a function of x.

EXAMPLE 2 Solve the equation $y\dfrac{d^2y}{dx^2} = \left(\dfrac{dy}{dx}\right)^2$.

Solution The change of variable $dy/dx = v(y)$ leads to the equation

$$yv\frac{dv}{dy} = v^2,$$

which is separable, $dv/v = dy/y$, and has solution $v = C_1 y$. The equation

$$\frac{dy}{dx} = C_1 y$$

is again separable and leads to

$$\begin{aligned}\frac{dy}{y} &= C_1\,dx\\ \ln|y| &= C_1x + C_2\\ y &= \pm e^{C_1x+C_2} = C_3e^{C_1x}.\end{aligned}$$

Second-Order Linear Equations

The most frequently encountered ordinary differential equations arising in applications are second-order linear equations. The general second-order linear equation is of the form

$$a_2(x)\frac{d^2y}{dx^2} + a_1(x)\frac{dy}{dx} + a_0(x)y = f(x).$$

As remarked in Section 18.1, if $f(x) = 0$ identically, then we say that the equation is **homogeneous**. If the coefficients $a_2(x)$, $a_1(x)$, and $a_0(x)$ are continuous on an interval and $a_2(x) \neq 0$ there, then the homogeneous equation

$$a_2(x)\frac{d^2y}{dx^2} + a_1(x)\frac{dy}{dx} + a_0(x)y = 0$$

has a general solution of the form

$$y_h = C_1y_1(x) + C_2y_2(x),$$

where $y_1(x)$ and $y_2(x)$ are two **independent** solutions, that is, two solutions with the property that $C_1y_1(x) + C_2y_2(x) = 0$ for all x in the interval only if $C_1 = C_2 = 0$. (We will not prove this here.)

Whenever one solution, $y_1(x)$, of a homogeneous linear second-order equation is known, another independent solution (and therefore the general solution) can be found by substituting $y = v(x)y_1(x)$ into the differential equation. This leads to a first-order, linear, separable equation for v'.

EXAMPLE 3 Show that $y_1 = e^{-2x}$ is a solution of $y'' + 4y' + 4y = 0$, and find the general solution of this equation.

Solution Since $y_1' = -2e^{-2x}$ and $y_1'' = 4e^{-2x}$, we have

$$y_1'' + 4y_1' + 4y_1 = e^{-2x}(4 - 8 + 4) = 0,$$

so y_1 is indeed a solution of the given differential equation. To find the general solution, try $y = y_1v = e^{-2x}v(x)$. We have

$$\begin{aligned} y' &= -2e^{-2x}v + e^{-2x}v' \\ y'' &= 4e^{-2x}v - 4e^{-2x}v' + e^{-2x}v''. \end{aligned}$$

Substituting these expressions into the given DE, we obtain

$$\begin{aligned} 0 &= y'' + 4y' + 4y \\ &= e^{-2x}(4v - 4v' + v'' - 8v + 4v' + 4v) = e^{-2x}v''. \end{aligned}$$

Thus, $y = y_1v$ is a solution provided $v''(x) = 0$. This equation for v has the general solution $v = C_1 + C_2x$, so the given equation has the general solution

$$y = C_1e^{-2x} + C_2xe^{-2x} = C_1y_1(x) + C_2y_2(x),$$

where $y_2 = xe^{-2x}$ is a second solution of the DE, independent of y_1.

By Theorem 2 of Section 18.1, the general solution of the second-order, linear, nonhomogeneous equation (with $f(x) \neq 0$) is of the form

$$y = y_p(x) + y_h(x),$$

where $y_p(x)$ is any particular solution of the nonhomogeneous equation, and $y_h(x)$ is the general solution (as described above) of the corresponding homogeneous equation. In Section 18.6 we will discuss the solution of nonhomogeneous linear equations. First, however, in Section 18.5 we concentrate on some special classes of homogeneous, linear equations.

EXERCISES 18.4

1. Show that $y = e^x$ is a solution of $y'' - 3y' + 2y = 0$, and find the general solution of this DE.

2. Show that $y = e^{-2x}$ is a solution of $y'' - y' - 6y = 0$, and find the general solution of this DE.

3. Show that $y = x$ is a solution of $x^2y'' + 2xy' - 2y = 0$ on the interval $(0, \infty)$, and find the general solution on this interval.

4. Show that $y = x^2$ is a solution of $x^2y'' - 3xy' + 4y = 0$ on the interval $(0, \infty)$, and find the general solution on this interval.

5. Show that $y = x$ is a solution of the differential equation $x^2y'' - (2x + x^2)y' + (2 + x)y = 0$, and find the general solution of this equation.

6. Show that $y = x^{-1/2} \cos x$ is a solution of the Bessel equation with $\nu = 1/2$:

$$x^2y'' + xy' + \left(x^2 - \frac{1}{4}\right)y = 0.$$

Find the general solution of this equation.

First-order systems

7. A system of n first-order, linear, differential equations in n unknown functions $y_1, y_2, \cdots, y_n$ is written

$$\begin{aligned} y_1' &= a_{11}(x)y_1 + a_{12}(x)y_2 + \cdots + a_{1n}(x)y_n + f_1(x) \\ y_2' &= a_{21}(x)y_1 + a_{22}(x)y_2 + \cdots + a_{2n}(x)y_n + f_2(x) \\ &\vdots \\ y_n' &= a_{n1}(x)y_1 + a_{n2}(x)y_2 + \cdots + a_{nn}(x)y_n + f_n(x). \end{aligned}$$

Such a system is called an $n \times n$ **first-order linear system** and can be rewritten in vector-matrix form as $\mathbf{y}' = \mathcal{A}(x)\mathbf{y} + \mathbf{f}(x)$, where

$$\mathbf{y}(x) = \begin{pmatrix} y_1(x) \\ \vdots \\ y_n(x) \end{pmatrix}, \quad \mathbf{f}(x) = \begin{pmatrix} f_1(x) \\ \vdots \\ f_n(x) \end{pmatrix},$$

$$\mathcal{A}(x) = \begin{pmatrix} a_{11}(x) & \cdots & a_{1n}(x) \\ \vdots & \ddots & \vdots \\ a_{n1}(x) & \cdots & a_{nn}(x) \end{pmatrix}.$$

Show that the second-order, linear equation $y'' + a_1(x)y' + a_0(x)y = f(x)$ can be transformed into a 2×2 first-order system with $y_1 = y$ and $y_2 = y'$ having

$$\mathcal{A}(x) = \begin{pmatrix} 0 & 1 \\ -a_0(x) & -a_1(x) \end{pmatrix}, \quad \mathbf{f}(x) = \begin{pmatrix} 0 \\ f(x) \end{pmatrix}.$$

8. Generalize Exercise 7 to transform an nth-order linear equation

$$y^{(n)} + a_{n-1}(x)y^{(n-1)} + a_{n-2}(x)y^{(n-2)} + \cdots + a_0(x)y = f(x)$$

into an $n \times n$ first-order system.

9. If $\mathcal{A}$ is an $n \times n$ constant matrix, and if there exists a scalar λ and a nonzero constant vector $\mathbf{v}$ for which $\mathcal{A}\mathbf{v} = \lambda\mathbf{v}$, show that $\mathbf{y} = C_1e^{\lambda x}\mathbf{v}$ is a solution of the homogeneous system $\mathbf{y}' = \mathcal{A}\mathbf{y}$.

10. Show that the determinant $\begin{vmatrix} 2-\lambda & 1 \\ 2 & 3-\lambda \end{vmatrix}$ is zero for two distinct values of λ. For each of these values find a nonzero vector $\mathbf{v}$ that satisfies the condition $\begin{pmatrix} 2 & 1 \\ 2 & 3 \end{pmatrix}\mathbf{v} = \lambda\mathbf{v}$. Hence, solve the system

$$y_1' = 2y_1 + y_2, \qquad y_2' = 2y_1 + 3y_2.$$

18.5 Linear Differential Equations with Constant Coefficients

A differential equation of the form

$$a\,y'' + b\,y' + cy = 0, \qquad (*)$$

where a, b, and c are constants and $a \neq 0$, is said to be a **linear, homogeneous, second-order equation with constant coefficients**.

A thorough discussion of techniques for solving such equations, together with examples, exercises, and applications to the study of simple and damped harmonic motion, can be found in Section 3.7; we will not repeat that discussion here. If you have not studied it, please do so now.

We will, however, extend the treatment to cover linear, constant coefficient differential equations of higher order.

Constant-Coefficient Equations of Higher Order

Because in most applications of equation $(*)$ the dependent variable represents time, we will, as we did in Section 3.7, regard y as a function of t rather than x, so that the prime symbol $(')$ denotes the derivative d/dt. The basic result of Section 3.7 was that the function $y = e^{rt}$ was a solution of $(*)$ provided that r satisfies the **auxiliary equation**

$$ar^2 + br + c = 0. \qquad (**)$$

The auxiliary equation is quadratic and can have either

(a) two distinct real roots, r_1 and r_2 (if $b^2 > 4ac$), in which case $(*)$ has general solution $y = C_1 e^{r_1 t} + C_2 e^{r_2 t}$,

(b) a single repeated real root r (if $b^2 = 4ac$), in which case $(*)$ has general solution $y = (C_1 + C_2 t)e^{rt}$, or

(c) a pair of complex conjugate roots, $r = k \pm i\omega$ with k and ω real (if $b^2 < 4ac$), in which case $(*)$ has general solution $y = e^{kt}\big(C_1 \cos(\omega t) + C_2 \sin(\omega t)\big)$.

The situation is analogous for higher-order linear, homogeneous DEs with constant coefficients. We describe the procedure without offering any proofs. If

$$P_n(r) = a_n r^n + a_{n-1} r^{n-1} + \cdots + a_2 r^2 + a_1 r + a_0$$

is a polynomial of degree n with constant coefficients a_j, $(0 \le j \le n)$, and $a_n \ne 0$, then the DE

$$P_n(D)y = 0, \qquad (\dagger)$$

where $D = d/dt$ can be solved by substituting $y = e^{rt}$ and obtaining the *auxiliary equation* $P_n(r) = 0$. This polynomial equation has n roots (see Appendix II) some of which may be equal and some or all of which can be complex. If the coefficients of the polynomial $P_n(r)$ are all real then any complex roots must occur in complex conjugate pairs $k \pm i\omega$ (with the same multiplicity), where k and ω are real.

The general solution of $(\dagger)$ can be expressed as a *linear combination* of n independent particular solutions

$$y = C_1 y_1(t) + C_2 y_2(t) + \cdots + C_n y_n(t),$$

where the C_j are arbitrary constants. The independent solutions $y_1, y_2, \ldots, y_n$ are constructed as follows:

1. If r_1 is a k-fold real root of the auxiliary equation (i.e., if $(r - r_1)^k$ is a factor of $P_n(r)$), then

$$e^{r_1 t}, \quad te^{r_1 t}, \quad t^2 e^{r_1 t}, \quad \ldots, \quad t^{k-1} e^{r_1 t}$$

are k independent solutions of $(\dagger)$.

2. If $r = a + ib$ and $r = a - ib$ (where a and b are real) constitute a k-fold pair of complex conjugate roots of the auxiliary equation (i.e., if $[(r - a)^2 + b^2]^k$ is a factor of $P_n(r)$), then

$$\begin{array}{llll} e^{at} \cos bt, & te^{at} \cos bt, & \ldots, & t^{k-1} e^{at} \cos bt, \\ e^{at} \sin bt, & te^{at} \sin bt, & \ldots, & t^{k-1} e^{at} \sin bt \end{array}$$

are $2k$ independent solutions of $(\dagger)$.

EXAMPLE 1 Solve (a) $y^{(4)} - 16y = 0$ and (b) $y^{(5)} - 2y^{(4)} + y^{(3)} = 0$.

Solution The auxiliary equation for (a) is $r^4 - 16 = 0$, which factors down to $(r-2)(r+2)(r^2+4) = 0$ and, hence, has roots $r = 2, -2, 2i$, and $-2i$. Thus, the DE (a) has general solution

$$y = C_1 e^{2t} + C_2 e^{-2t} + C_3 \cos(2t) + C_4 \sin(2t)$$

for arbitrary constants C_1, C_2, C_3, and C_4.

The auxiliary equation for (b) is $r^5 - 2r^4 + r^3$, which factors to $r^3(r-1)^2 = 0$, and so has roots $r = 0,\ 0,\ 0,\ 1,\ 1$. The general solution of the DE (b) is

$$y = C_1 + C_2 t + C_3 t^2 + C_4 e^t + C_5 t e^t,$$

where $C_1, \ldots, C_5$ are arbitrary constants.

EXAMPLE 2 What are the order and the general solution of the constant-coefficient, linear, homogeneous DE whose auxiliary equation is

$$(r+4)^3(r^2+4r+13)^2 = 0?$$

Solution The auxiliary equation has degree 7 so the DE is of seventh order. Since $r^2 + 4r + 13 = (r+2)^2 + 9$, which has roots $-2 \pm 3i$, the DE must have the general solution

$$\begin{aligned} y =\,& C_1 e^{-4t} + C_2 t e^{-4t} + C_3 t^2 e^{-4t} \\ &+ C_4 e^{-2t}\cos(3t) + C_5 e^{-2t}\sin(3t) + C_6 t e^{-2t}\cos(3t) + C_7 t e^{-2t}\sin(3t). \end{aligned}$$

Euler (Equidimensional) Equations

A homogeneous, linear equation of the form

$$ax^2\frac{d^2y}{dx^2} + bx\frac{dy}{dx} + cy = 0$$

is called an **Euler equation** or an **equidimensional equation**, the latter term being appropriate since all the terms in the equation have the same dimension (i.e., they are measured in the same units), provided that the constants a, b, and c all have the same dimension. The coefficients of an Euler equation are *not constant,* but there is a technique for solving these equations that is similar to that for solving equations with constant coefficients, so we include a brief discussion of these equations in this section. As in the case of constant coefficient equations, we assume that the constants a, b, and c are real numbers and that $a \neq 0$. Even so, the leading coefficient, ax^2, does vanish at $x = 0$ (which is called a **singular point** of the equation), and this can cause solutions to fail to be defined at $x = 0$. We will solve the equation in the interval $x > 0$; the same solution will also hold for $x < 0$ provided we replace x by $|x|$ in the solution.

Let us search for solutions in $x > 0$ given by powers of x; if

$$y = x^r, \qquad \frac{dy}{dx} = rx^{r-1}, \qquad \frac{d^2y}{dx^2} = r(r-1)x^{r-2},$$

then the Euler equation becomes

$$\big(ar(r-1) + br + c\big)x^r = 0.$$

This will be satisfied for all $x > 0$, provided that r satisfies the **auxiliary equation**

$$ar(r-1)+br+c=0 \quad \text{or, equivalently,} \quad ar^2+(b-a)r+c=0.$$

As for constant coefficient equations, there are three possibilities.

CASE I. If $(b-a)^2 \ge 4ac$, then the auxiliary equation has two real roots:

$$r_1 = \frac{a-b+\sqrt{(b-a)^2-4ac}}{2a},$$
$$r_2 = \frac{a-b-\sqrt{(b-a)^2-4ac}}{2a}.$$

In this case, the Euler equation has the general solution

$$y = C_1 x^{r_1} + C_2 x^{r_2}, \qquad (x>0).$$

The general solution is usually quoted in the form

$$y = C_1|x|^{r_1} + C_2|x|^{r_2},$$

which is valid in any interval not containing $x=0$ and may even be valid on intervals containing the origin if, for example, r_1 and r_2 are nonnegative integers.

EXAMPLE 3 Solve the initial-value problem

$$2x^2y'' - xy' - 2y = 0, \qquad y(1)=5, \qquad y'(1)=0.$$

Solution The auxiliary equation is $2r(r-1)-r-2=0$, that is, $2r^2-3r-2=0$, or $(r-2)(2r+1)=0$, and has roots $r=2$ and $r=-(1/2)$. Thus, the general solution of the differential equation (valid for $x>0$) is

$$y = C_1x^2 + C_2x^{-1/2}.$$

The initial conditions imply that

$$5 = y(1) = C_1 + C_2 \quad \text{and} \quad 0 = y'(1) = 2C_1 - \frac{1}{2}C_2.$$

Therefore, $C_1 = 1$ and $C_2 = 4$, and the initial-value problem has solution

$$y = x^2 + \frac{4}{\sqrt{x}}, \qquad (x>0).$$

CASE II. If $(b-a)^2 = 4ac$, then the auxiliary equation has one double root, namely, the root $r=(a-b)/2a$. It is left to the reader to verify that in this case the transformation $y=x^r v(x)$ leads to the general solution

$$y = C_1x^r + C_2x^r \ln x, \qquad (x>0),$$

or, more generally,

$$y = C_1|x|^r + C_2|x|^r \ln|x|, \qquad (x \ne 0).$$

CASE III. If $(b-a)^2 < 4ac$, then the auxiliary equation has complex conjugate roots:

$$r = \alpha \pm i\beta, \qquad \text{where } \alpha = \frac{a-b}{2a}, \quad \beta = \frac{\sqrt{4ac-(b-a)^2}}{2a}.$$

The corresponding powers x^r can be expressed in real form in a manner similar to that used for constant coefficient equations; we have

$$\begin{aligned} x^{\alpha \pm i\beta} = e^{(\alpha \pm i\beta)\ln x} &= e^{\alpha \ln x}\big[\cos(\beta \ln x) \pm i \sin(\beta \ln x)\big] \\ &= x^{\alpha}\cos(\beta \ln x) \pm i x^{\alpha}\sin(\beta \ln x). \end{aligned}$$

Accordingly, the Euler equation has the general solution

$$y = C_1|x|^{\alpha}\cos(\beta \ln|x|) + C_2|x|^{\alpha}\sin(\beta \ln|x|).$$

EXAMPLE 4 Solve the DE $x^2y'' - 3xy' + 13y = 0$.

Solution The DE has the auxiliary equation $r(r-1) - 3r + 13 = 0$, that is, $r^2 - 4r + 13 = 0$, which has roots $r = 2 \pm 3i$. The DE, therefore, has the general solution

$$y = C_1x^2\cos(3\ln|x|) + C_2x^2\sin(3\ln|x|).$$

Remark Euler equations can be transformed into constant coefficient equations by using a simple change of variable. See Exercise 14 for the details.

EXERCISES 18.5

Exercises involving the solution of second-order, linear, homogeneous equations with constant coefficients can be found at the end of Section 3.7.

Find general solutions of the DEs in Exercises 1–4.

1. $y''' - 4y'' + 3y' = 0$

2. $y^{(4)} - 2y'' + y = 0$ **3.** $y^{(4)} + 2y'' + y = 0$

4. $y^{(4)} + 4y^{(3)} + 6y'' + 4y' + y = 0$

5. Show that $y = e^{2t}$ is a solution of

$$y''' - 2y' - 4y = 0$$

(where $'$ denotes d/dt), and find the general solution of this DE.

6. Write the general solution of the linear, constant-coefficient DE having auxiliary equation $(r^2 - r - 2)^2(r^2 - 4)^2 = 0$.

Find general solutions to the Euler equations in Exercises 7–12.

7. $x^2y'' - xy' + y = 0$ **8.** $x^2y'' - xy' - 3y = 0$

9. $x^2y'' + xy' - y = 0$ **10.** $x^2y'' - xy' + 5y = 0$

11. $x^2y'' + xy' = 0$ **12.** $x^2y'' + xy' + y = 0$

13. Solve the DE $x^3y''' + xy' - y = 0$ in the interval $x > 0$.

14. Show that the change of variables $x = e^t$, $z(t) = y(e^t)$, transforms the Euler equation

$$ax^2\frac{d^2y}{dx^2} + bx\frac{dy}{dx} + cy = 0$$

into the constant coefficient equation

$$a\frac{d^2z}{dt^2} + (b-a)\frac{dz}{dt} + cz = 0.$$

15. Use the transformation $x = e^t$ of the previous exercise to solve the Euler equation

$$x^2\frac{d^2y}{dx^2} - x\frac{dy}{dx} + 2y = 0, \qquad (x > 0).$$

18.6 Nonhomogeneous Linear Equations

We now consider the problem of solving the nonhomogeneous second-order differential equation

$$a_2(x)\frac{d^2y}{dx^2} + a_1(x)\frac{dy}{dx} + a_0(x)y = f(x). \qquad (*)$$

We assume that two independent solutions, $y_1(x)$ and $y_2(x)$, of the corresponding homogeneous equation

$$a_2(x)\frac{d^2y}{dx^2} + a_1(x)\frac{dy}{dx} + a_0(x)y = 0$$

are known. The function $y_h(x) = C_1y_1(x) + C_2y_2(x)$, which is the general solution of the homogeneous equation, is called the **complementary function** for the nonhomogeneous equation. Theorem 2 of Section 18.1 suggests that the general solution of the nonhomogeneous equation is of the form

$$y = y_p(x) + y_h(x) = y_p(x) + C_1y_1(x) + C_2y_2(x),$$

where $y_p(x)$ is any **particular solution** of the nonhomogeneous equation. All we need to do is find *one solution* of the nonhomogeneous equation, and we can write the general solution.

There are two common methods for finding a particular solution y_p of the nonhomogeneous equation $(*)$:

1. the method of undetermined coefficients, and
2. the method of variation of parameters.

The first of these hardly warrants being called a *method*; it just involves making an educated guess about the form of the solution as a sum of terms with unknown coefficients and substituting this guess into the equation to determine the coefficients. This method works well for simple DEs, especially ones with constant coefficients. The nature of the *guess* depends on the nonhomogeneous term $f(x)$, but can also be affected by the solution of the corresponding homogeneous equation. A few examples will illustrate the ideas involved.

EXAMPLE 1 Find the general solution of $y'' + y' - 2y = 4x$.

Solution Because the nonhomogeneous term $f(x) = 4x$ is a first-degree polynomial, we "guess" that a particular solution can be found that is also such a polynomial. Thus, we try

$$y = Ax + B, \qquad y' = A, \qquad y'' = 0.$$

Substituting these expressions into the given DE, we obtain

$$\begin{aligned} 0 + A - 2(Ax + B) &= 4x \qquad \text{or} \\ -(2A + 4)x + (A - 2B) &= 0. \end{aligned}$$

This latter equation will be satisfied for all x provided $2A + 4 = 0$ and $A - 2B = 0$. Thus, we require $A = -2$ and $B = -1$; a particular solution of the given DE is

$$y_p(x) = -2x - 1.$$

Since the corresponding homogeneous equation $y''+y'-2y = 0$ has auxiliary equation $r^2 + r - 2 = 0$ with roots $r = 1$ and $r = -2$, the given DE has the general solution

$$y = y_p(x) + C_1e^x + C_2e^{-2x} = -2x - 1 + C_1e^x + C_2e^{-2x}.$$

EXAMPLE 2 Find general solutions of the equations (where $'$ denotes d/dt)

(a) $y'' + 4y = \sin t$,

(b) $y'' + 4y = \sin(2t)$,

(c) $y'' + 4y = \sin t + \sin(2t)$.

Solution

(a) Let us look for a particular solution of the form

$$\begin{aligned} y &= A\sin t + B\cos t \qquad \text{so that} \\ y' &= A\cos t - B\sin t \\ y'' &= -A\sin t - B\cos t. \end{aligned}$$

Substituting these expressions into the DE $y'' + 4y = \sin t$, we get

$$-A\sin t - B\cos t + 4A\sin t + 4B\cos t = \sin t,$$

which is satisfied for all x if $3A = 1$ and $3B = 0$. Thus $A = 1/3$ and $B = 0$. Since the homogeneous equation $y'' + 4y = 0$ has general solution $y = C_1\cos(2t) + C_2\sin(2t)$, the given nonhomogeneous equation has the general solution

$$y = \frac{1}{3}\sin t + C_1\cos(2t) + C_2\sin(2t).$$

(b) Motivated by our success in part (a), we might be tempted to try for a particular solution of the form $y = A\sin(2t) + B\cos(2t)$, but that won't work, because this function is a solution of the homogeneous equation, so we would get $y'' + 4y = 0$ for any choice of A and B. In this case it is useful to try

$$y = At\sin(2t) + Bt\cos(2t).$$

We have

$$\begin{aligned} y' &= A\sin(2t) + 2At\cos(2t) + B\cos(2t) - 2Bt\sin(2t) \\ &= (A - 2Bt)\sin(2t) + (B + 2At)\cos(2t) \\ y'' &= -2B\sin(2t) + 2(A - 2Bt)\cos(2t) + 2A\cos(2t) \\ &\quad - 2(B + 2At)\sin(2t) \\ &= -4(B + At)\sin(2t) + 4(A - Bt)\cos(2t). \end{aligned}$$

Substituting into $y'' + 4y = \sin(2t)$ leads to

$$\begin{aligned} -4(B + At)\sin(2t) + 4(A - Bt)\cos(2t) + 4At\sin(2t) + 4Bt\cos(2t) \\ = \sin(2t). \end{aligned}$$

Observe that the terms involving $t\sin(2t)$ and $t\cos(2t)$ cancel out, and we are left with

$$-4B\sin(2t) + 4A\cos(2t) = \sin(2t),$$

which is satisfied for all x if $A = 0$ and $B = -1/4$. Hence, the general solution for part (b) is

$$y = -\frac{1}{4}t\cos(2t) + C_1\cos(2t) + C_2\sin(2t).$$

(c) Since the homogeneous equation is the same for (a), (b), and (c), and the nonhomogeneous term in equation (c) is the sum of the nonhomogeneous terms in equations (a) and (b), the sum of particular solutions of (a) and (b) is a particular solution of (c). (This is because the equation is *linear.*) Thus, the general solution of equation (c) is

$$y = \frac{1}{3}\sin t - \frac{1}{4}t\cos(2t) + C_1\cos(2t) + C_2\sin(2t).$$

We summarize the appropriate forms to try for particular solutions of constant-coefficient equations as follows:

Trial solutions for constant-coefficient equations

Let $A_n(x)$, $B_n(x)$, and $P_n(x)$ denote the nth-degree polynomials

$$\begin{aligned} A_n(x) &= a_0 + a_1x + a_2x^2 + \cdots + a_nx^n \\ B_n(x) &= b_0 + b_1x + b_2x^2 + \cdots + b_nx^n \\ P_n(x) &= p_0 + p_1x + p_2x^2 + \cdots + p_nx^n. \end{aligned}$$

To find a particular solution $y_p(x)$ of the second-order linear, constant-coefficient, nonhomogeneous DE

$$a_2\frac{d^2y}{dx^2} + a_1\frac{dy}{dx} + a_0y = f(x),$$

use the following forms:

If $f(x) = P_n(x)$, try $y_p = x^m A_n(x)$.
If $f(x) = P_n(x)e^{rx}$, try $y_p = x^m A_n(x)e^{rx}$.
If $f(x) = P_n(x)e^{rx}\cos(kx)$, try $y_p = x^m e^{rx}[A_n(x)\cos(kx) + B_n(x)\sin(kx)]$.
If $f(x) = P_n(x)e^{rx}\sin(kx)$, try $y_p = x^m e^{rx}[A_n(x)\cos(kx) + B_n(x)\sin(kx)]$,
where m is the smallest of the integers 0, 1, and 2, that ensures that no term of y_p is a solution of the corresponding homogeneous equation

$$a_2\frac{d^2y}{dx^2} + a_1\frac{dy}{dx} + a_0y = 0.$$

Resonance

For $\lambda > 0$, $\lambda \neq 1$, the solution $y_\lambda(t)$ of the initial-value problem

$$\begin{cases} y'' + y = \sin(\lambda t) \\ y(0) = 0 \\ y'(0) = 1 \end{cases}$$

can be determined by first looking for a particular solution of the DE having the form $y = A\sin(\lambda t)$, and then adding the complementary function $y = B\cos t + C\sin t$. The calculations give $A = 1/(1-\lambda^2)$, $B = 0$, $C = (1-\lambda-\lambda^2)/(1-\lambda^2)$, so

$$y_\lambda(t) = \frac{\sin(\lambda t) + (1-\lambda-\lambda^2)\sin t}{1-\lambda^2}.$$

For $\lambda = 1$ the nonhomogeneous term in the DE is a solution of the homogeneous equation $y'' + y = 0$, so we must try for a particular solution of the form $y = At\cos t + Bt\sin t$. In this case, the solution of the initial-value problem is

$$y_1(t) = \frac{3\sin t - t\cos t}{2}.$$

(This solution can also be found by calculating $\lim_{\lambda\to 1} y_\lambda(t)$ using l'Hôpital's Rule.) Observe that this solution is unbounded; the amplitude of the oscillations becomes larger and larger as t increases. In contrast, the solutions $y_\lambda(t)$ for $\lambda \neq 1$ are bounded for all t, although they can become quite large for some values of t if λ is close to 1. The graphs of the solutions $y_{0.9}(t)$, $y_{0.95}(t)$, and $y_1(t)$ on the interval $-10 \le t \le 100$ are shown in Figure 18.4.

The phenomenon illustrated here is called **resonance**. Vibrating mechanical systems have natural frequencies at which they will vibrate. If you try to force them to vibrate at a different frequency, the amplitude of the vibrations will themselves vary sinusoidally over time, producing an effect known as **beats**. The amplitudes of the beats can grow quite large, and the period of the beats lengthens as the forcing frequency approaches the natural frequency of the system. If the system has no resistive damping (the one illustrated here has no damping), then forcing vibrations at the natural frequency will cause the system to vibrate at ever increasing amplitudes.

As a concrete example, if you push a child on a swing, the swing will rise highest if your pushes are timed to have the same frequency as the natural frequency of the swing. Resonance is used in the design of tuning circuits of radios; the circuit is tuned (ususally by a variable capacitor) so that its natural frequency of oscillation is the frequency of the station being tuned in. The circuit then responds much more strongly to the signal received from that station than to others on different frequencies.

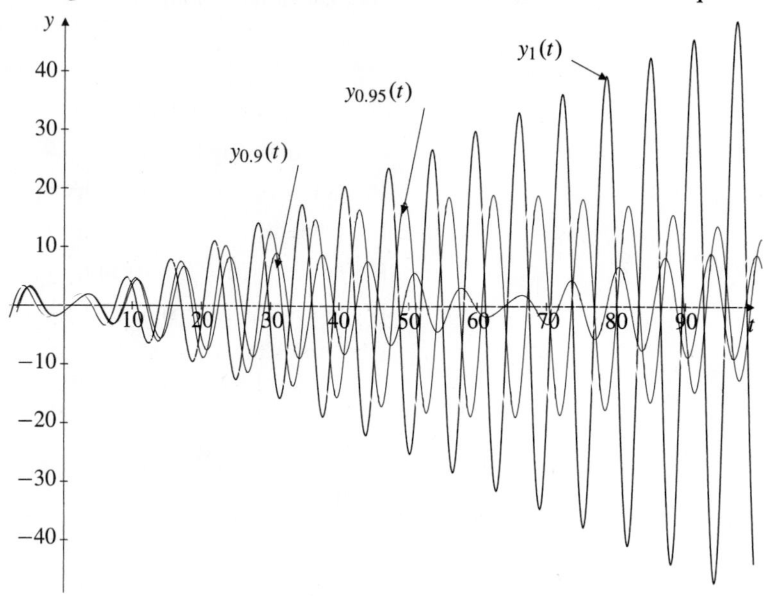

Figure 18.4 Resonance

Variation of Parameters

A more formal method for finding a particular solution $y_p(x)$ of the nonhomogeneous equation

$$a_2(x)\frac{d^2y}{dx^2} + a_1(x)\frac{dy}{dx} + a_0(x)y = f(x), \qquad (*)$$

when we know two independent solutions, $y_1(x)$ and $y_2(x)$, of the corresponding homogeneous equation is to replace the constants in the complementary function by functions, that is, search for y_p in the form

$$y_p = u_1(x)y_1(x) + u_2(x)y_2(x).$$

Requiring y_p to satisfy the given nonhomogeneous DE $(*)$ provides one equation that must be satisfied by the two unknown functions u_1 and u_2. We are free to require them to satisfy a second equation also. To simplify the calculations below, we choose this second equation to be

$$u_1'(x)y_1(x) + u_2'(x)y_2(x) = 0.$$

Now we have

$$y_p' = u_1'y_1 + u_1y_1' + u_2'y_2 + u_2y_2' = u_1y_1' + u_2y_2'$$
$$y_p'' = u_1'y_1' + u_1y_1'' + u_2'y_2' + u_2y_2''.$$

Substituting these expressions into the given DE, we obtain

$$a_2(u_1'y_1' + u_2'y_2') + u_1(a_2y_1'' + a_1y_1' + a_0y_1) + u_2(a_2y_2'' + a_1y_2' + a_0y_2)$$
$$= a_2(u_1'y_1' + u_2'y_2') = f(x),$$

because y_1 and y_2 satisfy the homogeneous equation. Therefore, u_1' and u_2' satisfy the pair of equations

$$u_1'(x)y_1(x) + u_2'(x)y_2(x) = 0,$$
$$u_1'(x)y_1'(x) + u_2'(x)y_2'(x) = \frac{f(x)}{a_2(x)}.$$

We can solve these two equations for the unknown functions u_1' and u_2' by Cramer's Rule (Theorem 6 of Section 10.7), or otherwise, and obtain

$$u_1' = -\frac{y_2(x)}{W(x)}\frac{f(x)}{a_2(x)}, \qquad u_2' = \frac{y_1(x)}{W(x)}\frac{f(x)}{a_2(x)},$$

where $W(x)$, called the **Wronskian** of y_1 and y_2, is the determinant

$$W(x) = \begin{vmatrix} y_1(x) & y_2(x) \\ y_1'(x) & y_2'(x) \end{vmatrix}.$$

Then u_1 and u_2 can be found by integration.

EXAMPLE 3 Find the general solution of $y'' + y = \tan x$.

Solution The homogeneous equation $y'' + y = 0$ has general solution

$$y_h = C_1 \cos x + C_2 \sin x.$$

A particular solution $y_p(x)$ of the nonhomogeneous equation can be found in the form

$$y_p = u_1(x)\cos x + u_2(x)\sin x,$$

where u_1 and u_2 satisfy

$$u_1'(x)\cos x + u_2'(x)\sin x = 0$$
$$-u_1'(x)\sin x + u_2'(x)\cos x = \tan x.$$

Solving these equations for $u_1'(x)$ and $u_2'(x)$, we obtain

$$u_1'(x) = -\frac{\sin^2 x}{\cos x}, \qquad u_2'(x) = \sin x.$$

Therefore,

$$u_1(x) = -\int \frac{\sin^2 x}{\cos x}\,dx = \int (\cos x - \sec x)\,dx = \sin x - \ln(\sec x + \tan x)$$
$$u_2(x) = -\cos x.$$

Hence, $y_p = \sin x \cos x - \cos x \ln(\sec x + \tan x) - \cos x \sin x = -\cos x \ln(\sec x + \tan x)$ is a particular solution of the nonhomogeneous equation, and the general solution is

$$y = C_1 \cos x + C_2 \sin x - \cos x \ln(\sec x + \tan x).$$

Note that no arbitrary constants were included when we integrated u_1' and u_2' to produce u_1 and u_2 as they would have produced terms in the general solution that are already included in y_h.

Remark This method for solving the nonhomogeneous equation is called the **method of variation of parameters**. It is completely general and extends to higher-order equations in a reasonable way, but it can be computationally somewhat difficult. We would not likely have been able to "guess" the form of the particular solution in the above example, so we could not have used the method discussed earlier in this section to solve this equation.

Maple Calculations

Maple has a `dsolve` routine for solving (some) differential equations and initial-value problems. This routine takes as input a DE and, if desired, initial conditions for it. We illustrate for the equation $y'' + 2y' + 5y = 25t + 20$ (assuming that the independent variable is t):

```
> DE := (D@@2)(y)(t)+2*D(y)(t)+5*y(t)=25*t+20;
```

$$DE := D^{(2)}(y)(t) + 2D(y)(t) + 5y(t) = 25t + 20$$

```
> dsolve(DE, y(t));
```

$$y(t) = e^{(-t)}\sin(2t)\_C2 + e^{(-t)}\cos(2t)\_C1 + 2 + 5t$$

Note Maple's use of $\_C1$ and $\_C2$ for arbitrary constants. For an initial-value problem we supply the DE and its initial conditions to `dsolve` as a single list or set argument enclosed in square brackets or braces:

```
> dsolve([DE, y(0)=3, D(y)(0)=-2], y(t));
```

$$y(t) = -3e^{(-t)}\sin(2t) + e^{(-t)}\cos(2t) + 2 + 5t$$

You might think that this output indicates that y has been defined as a function of t and you can find a decimal value for, say, $y(1)$ by giving the input `evalf(y(1))`. But this won't work. In fact, the output of the `dsolve` is just an equation with left side the symbol $y(t)$. We can, however, use this output to define y as a function of t as follows:

```
> y := unapply(op(2,%),t);
```

$$y := t \to -3e^{(-t)}\sin(2t) + e^{(-t)}\cos(2t) + 2 + 5t$$

The `op(2,%)` in the `unapply` command refers to the second operand of the previous result (i.e., the right side of equation output from the `dsolve`). `unapply(f,t)` converts an expression `f` to a function of t. To confirm:

```
> evalf(y(1));
```

$$5.843372646$$

EXERCISES 18.6

Find general solutions for the nonhomogeneous equations in Exercises 1–12 by the method of undetermined coefficients.

1. $y'' + y' - 2y = 1$
2. $y'' + y' - 2y = x$
3. $y'' + y' - 2y = e^{-x}$
4. $y'' + y' - 2y = e^{x}$
5. $y'' + 2y' + 5y = x^2$
6. $y'' + 4y = x^2$
7. $y'' - y' - 6y = e^{-2x}$
8. $y'' + 4y' + 4y = e^{-2x}$
9. $y'' + 2y' + 2y = e^{x}\sin x$
10. $y'' + 2y' + 2y = e^{-x}\sin x$
11. $y'' + y' = 4 + 2x + e^{-x}$
12. $y'' + 2y' + y = xe^{-x}$
13. Repeat Exercise 3 using the method of variation of parameters.
14. Repeat Exercise 4 using the method of variation of parameters.
15. Find a particular solution of the form $y = Ax^2$ for the Euler equation $x^2y'' + xy' - y = x^2$, and hence obtain the general solution of this equation on the interval $(0, \infty)$.
16. For what values of r can the Euler equation $x^2y'' + xy' - y = x^r$ be solved by the method of Exercise 15? Find a particular solution for each such r.
17. Try to guess the form of a particular solution for $x^2y'' + xy' - y = x$, and hence obtain the general solution for this equation on the interval $(0, \infty)$.

In Exercises 18–20 find the general solution on the interval $(0, \infty)$ of the given DE using variation of parameters.

18. $x^2y'' + xy' - y = x$ **19.** $y'' - 2y' + y = \dfrac{e^x}{x}$

20. $y'' + 4y' + 4y = \dfrac{e^{-2x}}{x^2}$

21. Consider the nonhomogeneous, linear equation

$$x^2y'' - (2x + x^2)y' + (2 + x)y = x^3.$$

Use the fact that $y_1(x) = x$ and $y_2(x) = xe^x$ are independent solutions of the corresponding homogeneous equation (see Exercise 5 of Section 18.4) to find the general solution of this nonhomogeneous equation.

22. Consider the nonhomogeneous, Bessel equation

$$x^2y'' + xy' + \left(x^2 - \frac{1}{4}\right)y = x^{3/2}.$$

Use the fact that $y_1(x) = x^{-1/2}\cos x$ and $y_2(x) = x^{-1/2}\sin x$ are independent solutions of the corresponding homogeneous equation (see Exercise 6 of Section 18.4) to find the general solution of this nonhomogeneous equation.

18.7 Series Solutions of Differential Equations

In Section 18.5 we developed a recipe for solving second-order, linear, homogeneous differential equations with constant coefficients:

$$ay'' + by' + cy = 0$$

and Euler equations of the form

$$ax^2y'' + bxy' + cy = 0.$$

Many of the second-order, linear, homogeneous differential equations that arise in applications do not have constant coefficients and are not of Euler type. If the coefficient functions of such an equation are sufficiently well-behaved, we can often find solutions in the form of power series (Taylor series). Such series solutions are frequently used to define new functions, whose properties are deduced partly from the fact that they solve particular differential equations. For example, Bessel functions of order ν (Greek "nu") are defined to be certain series solutions of Bessel's differential equation

$$x^2y'' + xy' + (x^2 - \nu^2)y = 0.$$

Series solutions for second-order homogeneous linear differential equations are most easily found near an **ordinary point** of the equation. This is a point $x = a$ such that the equation can be expressed in the form

$$y'' + p(x)y' + q(x)y = 0,$$

where the functions $p(x)$ and $q(x)$ are **analytic** at $x = a$. (Recall that a function f is analytic at $x = a$ if $f(x)$ can be expressed as the sum of its Taylor series in powers of $x - a$ in an interval of positive radius centred at $x = a$.) Thus, we assume

$$p(x) = \sum_{n=0}^{\infty} p_n(x-a)^n,$$

$$q(x) = \sum_{n=0}^{\infty} q_n(x-a)^n,$$

with both series converging in some interval of the form $a - R < x < a + R$. Frequently $p(x)$ and $q(x)$ are polynomials, so they are analytic everywhere. A change of independent variable $\xi = x - a$ will put the point $x = a$ at the origin $\xi = 0$, so we can assume that $a = 0$.

The following example illustrates the technique of series solution around an ordinary point.

EXAMPLE 1 Find two independent series solutions in powers of x for the Hermite equation

$$y'' - 2xy' + \nu y = 0.$$

For what values of ν does the equation have a polynomial solution?

Solution We try for a power series solution of the form

$$\begin{aligned} y &= \sum_{n=0}^{\infty} a_n x^n = a_0 + a_1 x + a_2 x^2 + a_3 x^3 + \cdots, \qquad \text{so that} \\ y' &= \sum_{n=1}^{\infty} n a_n x^{n-1} \\ y'' &= \sum_{n=2}^{\infty} n(n-1) a_n x^{n-2} = \sum_{n=0}^{\infty} (n+2)(n+1) a_{n+2} x^n. \end{aligned}$$

(We have replaced n by $n+2$ in order to get x^n in the sum for y''.) We substitute these expressions into the differential equation to get

$$\sum_{n=0}^{\infty} (n+2)(n+1) a_{n+2} x^n - 2 \sum_{n=1}^{\infty} n a_n x^n + \nu \sum_{n=0}^{\infty} a_n x^n = 0$$

$$\text{or} \quad 2a_2 + \nu a_0 + \sum_{n=1}^{\infty} \Big[(n+2)(n+1) a_{n+2} - (2n - \nu) a_n \Big] x^n = 0.$$

This identity holds for all x provided that the coefficient of every power of x vanishes; that is,

$$a_2 = -\frac{\nu a_0}{2}, \qquad a_{n+2} = \frac{(2n - \nu) a_n}{(n+2)(n+1)}, \qquad (n = 1, 2, \cdots).$$

The latter of these formulas is called a **recurrence relation**.

We can choose a_0 and a_1 to have any values; then the above conditions determine all the remaining coefficients a_n, $(n \geq 2)$. We can get one solution by choosing, for instance, $a_0 = 1$ and $a_1 = 0$. Then, by the recurrence relation,

$$a_3 = 0, \quad a_5 = 0, \quad a_7 = 0, \quad \cdots, \qquad \text{and}$$

$$\begin{aligned} a_2 &= -\frac{\nu}{2} \\ a_4 &= \frac{(4-\nu)a_2}{4 \times 3} = -\frac{\nu(4-\nu)}{2 \times 3 \times 4} = -\frac{\nu(4-\nu)}{4!} \\ a_6 &= \frac{(8-\nu)a_4}{6 \times 5} = -\frac{\nu(4-\nu)(8-\nu)}{6!} \\ &\cdots \end{aligned}$$

The pattern is obvious here:

$$a_{2n} = -\frac{\nu(4-\nu)(8-\nu)\cdots(4n-4-\nu)}{(2n)!}, \qquad (n = 1, 2, \cdots).$$

One solution to the Hermite equation is

$$y_1 = 1 + \sum_{n=1}^{\infty} -\frac{\nu(4-\nu)(8-\nu)\cdots(4n-4-\nu)}{(2n)!} x^{2n}.$$

We observe that if $\nu = 4n$ for some non-negative integer n, then y_1 is an even polynomial of degree $2n$, because $a_{2n+2} = 0$ and all subsequent even coefficients therefore also vanish.

The second solution, y_2, can be found in the same way, by choosing $a_0 = 0$ and $a_1 = 1$. It is

$$y_2 = x + \sum_{n=1}^{\infty} \frac{(2-\nu)(6-\nu)\cdots(4n-2-\nu)}{(2n+1)!} x^{2n+1},$$

and it is an odd polynomial of degree $2n+1$ if $\nu = 4n+2$.

Both of these series solutions converge for all x. The ratio test can be applied directly to the recurrence relation. Since consecutive nonzero terms of each series are of the form $a_n x^n$ and $a_{n+2}x^{n+2}$, we calculate

$$\rho = \lim_{n\to\infty} \left| \frac{a_{n+2}x^{n+2}}{a_n x^n} \right| = |x|^2 \lim_{n\to\infty} \left| \frac{a_{n+2}}{a_n} \right| = |x|^2 \lim_{n\to\infty} \left| \frac{2n-\nu}{(n+2)(n+1)} \right| = 0$$

for every x, so the series converges by the ratio test.

If $x = a$ is not an ordinary point of the equation

$$y'' + p(x)y' + q(x)y = 0,$$

then it is called a **singular point** of that equation. This means that at least one of the functions $p(x)$ and $q(x)$ is not analytic at $x = a$. If, however, $(x-a)p(x)$ and $(x-a)^2q(x)$ are analytic at $x = a$, then the singular point is said to be a **regular singular point**. For example, the origin $x = 0$ is a regular singular point of Bessel's equation,

$$x^2y'' + xy' + (x^2 - \nu^2)y = 0,$$

since $p(x) = 1/x$ and $q(x) = (x^2 - \nu^2)/x^2$ satisfy $xp(x) = 1$ and $x^2q(x) = x^2 - \nu^2$, which are both polynomials and therefore analytic.

The solutions of differential equations are usually not analytic at singular points. However, it is still possible to find at least one series solution about such a point. The method involves searching for a series solution of the form x^μ times a power series; that is,

$$y = (x-a)^\mu \sum_{n=0}^{\infty} a_n(x-a)^n = \sum_{n=0}^{\infty} a_n(x-a)^{n+\mu}, \qquad \text{where } a_0 \neq 0.$$

Substitution into the differential equation produces a quadratic **indicial equation**, which determines one or two values of μ for which such solutions can be found, and a **recurrence relation** enabling the coefficients a_n to be calculated for $n \geq 1$. If the indicial roots are not equal and do not differ by an integer, two independent solutions can be calculated. If the indicial roots are equal or differ by an integer, one such solution can be calculated (corresponding to the larger indicial root), but finding a second independent solution (and so the general solution) requires techniques beyond the scope of this book. The reader is referred to standard texts on differential equations for more discussion and examples. We will content ourselves here with one final example.

EXAMPLE 2 Find one solution, in powers of x, of Bessel's equation of order $\nu = 1$, namely,

$$x^2y'' + xy' + (x^2 - 1)y = 0.$$

Solution We try

$$\begin{aligned} y &= \sum_{n=0}^{\infty} a_n x^{\mu+n} \\ y' &= \sum_{n=0}^{\infty} (\mu+n) a_n x^{\mu+n-1} \\ y'' &= \sum_{n=0}^{\infty} (\mu+n)(\mu+n-1) a_n x^{\mu+n-2}. \end{aligned}$$

Substituting these expressions into the Bessel equation, we get

$$\sum_{n=0}^{\infty} \Big[\big((\mu+n)(\mu+n-1) + (\mu+n) - 1\big) a_n x^n + a_n x^{n+2} \Big] = 0$$

$$\sum_{n=0}^{\infty} \Big[(\mu+n)^2 - 1 \Big] a_n x^n + \sum_{n=2}^{\infty} a_{n-2} x^n = 0$$

$$(\mu^2 - 1) a_0 + \big((\mu+1)^2 - 1\big) a_1 x + \sum_{n=2}^{\infty} \Big[\big((\mu+n)^2 - 1\big) a_n + a_{n-2} \Big] x^n = 0.$$

All of the terms must vanish. Since $a_0 \neq 0$ (we may take $a_0 = 1$), we obtain

$$\begin{aligned} &\mu^2 - 1 = 0, && \text{the indicial equation} \\ &[(\mu+1)^2 - 1] a_1 = 0, && \\ &a_n = -\frac{a_{n-2}}{(\mu+n)^2 - 1}, \quad (n \geq 2). && \text{the recurrence relation} \end{aligned}$$

Evidently $\mu = \pm 1$; therefore $a_1 = 0$. If we take $\mu = 1$, then the recurrence relation is $a_n = -a_{n-2}/(n)(n+2)$. Thus,

$$a_3 = 0, \quad a_5 = 0, \quad a_7 = 0, \quad \cdots$$

$$a_2 = \frac{-1}{2 \times 4}, \quad a_4 = \frac{1}{2 \times 4 \times 4 \times 6}, \quad a_6 = \frac{-1}{2 \times 4 \times 4 \times 6 \times 6 \times 8}, \quad \ldots.$$

Again the pattern is obvious:

$$a_{2n} = \frac{(-1)^n}{2^{2n} n!(n+1)!},$$

and one solution of the Bessel equation of order 1 is

$$y = \sum_{n=0}^{\infty} \frac{(-1)^n}{2^{2n} n!(n+1)!} x^{2n+1}.$$

By the ratio test, this series converges for all x.

Remark Observe that if we tried to calculate a second solution using $\mu = -1$, we would get the recurrence relation

$$a_n = -\frac{a_{n-2}}{n(n-2)},$$

and we would be unable to calculate a_2. This shows what can happen if the indicial roots differ by an integer.

EXERCISES 18.7

1. Find the general solution of $y'' = (x-1)^2 y$ in the form of a power series $y = \sum_{n=0}^{\infty} a_n (x-1)^n$.
2. Find the general solution of $y'' = xy$ in the form of a power series $y = \sum_{n=0}^{\infty} a_n x^n$ with a_0 and a_1 arbitrary.
3. Solve the initial-value problem

$$\begin{cases} y'' + xy' + 2y = 0 \\ y(0) = 1 \\ y'(0) = 2. \end{cases}$$

4. Find the solution of $y'' + xy' + y = 0$ that satisfies $y(0) = 1$ and $y'(0) = 0$.
5. Find the first three nonzero terms in a power series solution in powers of x for the initial-value problem $y'' + (\sin x)y = 0$, $y(0) = 1$, $y'(0) = 0$.
6. Find the solution, in powers of x, for the initial-value problem

$$(1-x^2)y'' - xy' + 9y = 0, \quad y(0) = 0, \quad y'(0) = 1.$$

7. Find two power series solutions in powers of x for $3xy'' + 2y' + y = 0$.
8. Find one power series solution for the Bessel equation of order $\nu = 0$, that is, the equation $xy'' + y' + xy = 0$.

CHAPTER REVIEW

Key Ideas

- **What do the following phrases mean?**
  - ◇ an ordinary DE
  - ◇ a partial DE
  - ◇ the general solution of a DE
  - ◇ a linear combination of solutions of a DE
  - ◇ the order of a DE
  - ◇ a linear DE
  - ◇ a separable DE
  - ◇ an exact DE
  - ◇ an integrating factor
  - ◇ a constant coefficient DE
  - ◇ an Euler equation
  - ◇ an auxiliary equation
- **Describe how to solve:**
  - ◇ a separable DE
  - ◇ a first-order, linear DE
  - ◇ a homogeneous, first-order DE
  - ◇ a constant coefficient DE
  - ◇ an Euler equation
- **What conditions imply that an initial-value problem for a first-order DE has a unique solution near the initial point?**
- **Describe the following methods for solving first-order DEs numerically:**
  - ◇ the Euler method
  - ◇ the improved Euler method
  - ◇ the fourth-order Runge–Kutta method
- **Describe the following methods for solving a nonhomogeneous, linear DE:**
  - ◇ undetermined coefficients
  - ◇ variation of parameters
- **What are an ordinary point and a regular singular point of a linear, second-order DE? Describe how series can be used to solve such an equation near such a point.**

Review Exercises

Find the general solutions of the differential equations in Exercises 1–16.

1. $\dfrac{dy}{dx} = 2xy$
2. $\dfrac{dy}{dx} = e^{-y}\sin x$
3. $\dfrac{dy}{dx} = x + 2y$
4. $\dfrac{dy}{dx} = \dfrac{x^2+y^2}{2xy}$
5. $\dfrac{dy}{dx} = \dfrac{x+y}{y-x}$
6. $\dfrac{dy}{dx} = -\dfrac{y+e^x}{x+e^y}$
7. $\dfrac{d^2y}{dt^2} = \left(\dfrac{dy}{dt}\right)^2$
8. $2\dfrac{d^2y}{dt^2} + 5\dfrac{dy}{dt} + 2y = 0$
9. $4y'' - 4y' + 5y = 0$
10. $2x^2y'' + y = 0$
11. $t^2\dfrac{d^2y}{dt^2} - t\dfrac{dy}{dt} + 5y = 0$
12. $\dfrac{d^3y}{dt^3} + 8\dfrac{d^2y}{dt^2} + 16\dfrac{dy}{dt} = 0$
13. $\dfrac{d^2y}{dx^2} - 5\dfrac{dy}{dx} + 6y = e^x + e^{3x}$
14. $\dfrac{d^2y}{dx^2} - 5\dfrac{dy}{dx} + 6y = xe^{2x}$
15. $\dfrac{d^2y}{dx^2} + 2\dfrac{dy}{dx} + y = x^2$
16. $x\dfrac{d^2y}{dx^2} - 2y = x^3$

Solve the initial-value problems in Exercises 17–26.

17. $\begin{cases} \dfrac{dy}{dx} = \dfrac{x^2}{y^2} \\ y(2) = 1 \end{cases}$

18. $\begin{cases} \dfrac{dy}{dx} = \dfrac{y^2}{x^2} \\ y(2) = 1 \end{cases}$

19. $\begin{cases} \dfrac{dy}{dx} = \dfrac{xy}{x^2+y^2} \\ y(0) = 1 \end{cases}$

20. $\begin{cases} \dfrac{dy}{dx} + (\cos x)y = 2\cos x \\ y(\pi) = 1 \end{cases}$

21. $\begin{cases} y'' + 3y' + 2y = 0 \\ y(0) = 1 \\ y'(0) = 2 \end{cases}$

22. $\begin{cases} y'' + 2y' + (1+\pi^2)y = 0 \\ y(1) = 0 \\ y'(1) = \pi \end{cases}$

23. $\begin{cases} y'' + 10y' + 25y = 0 \\ y(1) = e^{-5} \\ y'(1) = 0 \end{cases}$

24. $\begin{cases} x^2y'' - 3xy' + 4y = 0 \\ y(e) = e^2 \\ y'(e) = 0 \end{cases}$

25. $\begin{cases} \dfrac{d^2y}{dt^2} + 4y = 8e^{2t} \\ y(0) = 1 \\ y'(0) = -2 \end{cases}$

26. $\begin{cases} 2\dfrac{d^2y}{dx^2} + 5\dfrac{dy}{dx} - 3y = 6 + 7e^{x/2} \\ y(0) = 0 \\ y'(0) = 1 \end{cases}$

27. For what values of the constants A and B is the equation

$$[(x + A)e^x \sin y + \cos y]\,dx + x[e^x \cos y + B \sin y]\,dy = 0$$

exact? What is the general solution of the equation if A and B have these values?

28. Find a value of n for which x^n is an integrating factor for

$$(x^2 + 3y^2)\,dx + xy\,dy = 0,$$

and solve the equation.

29. Show that $y = x$ is a solution of

$$x^2y'' - x(2 + x\cot x)y' + (2 + x\cot x)y = 0,$$

and find the general solution of this equation.

30. Use the method of variation of parameters and the result of Exercise 29 to find the general solution of the nonhomogeneous equation

$$x^2y'' - x(2 + x\cot x)y' + (2 + x\cot x)y = x^3 \sin x.$$

31. Suppose that $f(x, y)$ and $\dfrac{\partial}{\partial y} f(x, y)$ are continuous on the whole xy-plane and that $f(x, y)$ is bounded there, say $|f(x, y)| \le K$. Show that no solution of $y' = f(x, y)$ can have a vertical asymptote. Describe the region in the plane in which the solution to the initial-value problem

$$\begin{cases} y' = f(x, y) \\ y(x_0) = y_0 \end{cases}$$

must remain.

APPENDIX I

Complex Numbers

"Old Macdonald had a farm,
Minus E-squared 0."

a mathematically simplified children's song

Many of the problems to which mathematics is applied involve the solution of equations. Over the centuries the number system had to be expanded many times to provide solutions for more and more kinds of equations. The natural numbers

$$\mathbb{N} = \{1, 2, 3, 4, \ldots\}$$

are inadequate for the solutions of equations of the form

$$x + n = m, \qquad (m, n \in \mathbb{N}).$$

Zero and negative numbers can be added to create the integers

$$\mathbb{Z} = \{\ldots, -3, -2, -1, 0, 1, 2, 3, \ldots\}$$

in which that equation has the solution $x = m - n$ even if $m < n$. (Historically, this extension of the number system came much later than some of those mentioned below.) Some equations of the form

$$nx = m, \qquad (m, n \in \mathbb{Z}, \quad n \neq 0),$$

cannot be solved in the integers. Another extension is made to include numbers of the form m/n, thus producing the set of rational numbers

$$\mathbb{Q} = \left\{\frac{m}{n} : m, n \in \mathbb{Z}, \quad n \neq 0\right\}.$$

Every linear equation

$$ax = b, \qquad (a, b \in \mathbb{Q}, \quad a \neq 0),$$

has a solution $x = b/a$ in $\mathbb{Q}$, but the quadratic equation

$$x^2 = 2$$

has no solution in $\mathbb{Q}$, as was shown in Section P.1. Another extension enriches the rational numbers to the real numbers $\mathbb{R}$ in which some equations like $x^2 = 2$ have solutions. However, other quadratic equations, for instance,

$$x^2 = -1$$

do not have solutions, even in the real numbers, so the extension process is not complete. In order to be able to solve any quadratic equation, we need to extend the real number system to a larger set, which we call the **complex number system**. In this appendix we will define complex numbers and develop some of their basic properties.

Definition of Complex Numbers

We begin by defining the symbol i, called **the imaginary unit,**[1] to have the property

$$i^2 = -1.$$

Thus, we could also call i the **square root of** -1 and denote it $\sqrt{-1}$. Of course, i is not a real number; no real number has a negative square.

DEFINITION 1

> A **complex number** is an expression of the form
>
> $$a + bi \qquad \text{or} \qquad a + ib,$$
>
> where a and b are *real numbers*, and i is the imaginary unit.

For example, $3+2i$, $\frac{7}{2}-\frac{2}{3}i$, $i\pi = 0+i\pi$, and $-3 = -3+0i$ are all complex numbers. The last of these examples shows that every real number can be regarded as a complex number. (We will normally use $a + bi$ unless b is a complicated expression, in which case we will write $a + ib$ instead. Either form is acceptable.)

It is often convenient to represent a complex number by a single letter; w and z are frequently used for this purpose. If a, b, x, and y are real numbers, and

$$w = a + bi \qquad \text{and} \qquad z = x + yi,$$

then we can refer to the complex numbers w and z. Note that $w = z$ if and only if $a = x$ and $b = y$. Of special importance are the complex numbers

$$0 = 0 + 0i, \qquad 1 = 1 + 0i, \qquad \text{and} \qquad i = 0 + 1i.$$

DEFINITION

> If $z = x + yi$ is a complex number (where x and y are real), we call x the **real part** of z and denote it $\mathrm{Re}\,(z)$. We call y the **imaginary part** of z and denote it $\mathrm{Im}\,(z)$:
>
> $$\mathrm{Re}\,(z) = \mathrm{Re}\,(x + yi) = x, \qquad \mathrm{Im}\,(z) = \mathrm{Im}\,(x + yi) = y.$$

Note that both the real and imaginary parts of a complex number are real numbers:

$$\begin{aligned} &\mathrm{Re}\,(3 - 5i) = 3 && \mathrm{Im}\,(3 - 5i) = -5 \\ &\mathrm{Re}\,(2i) = \mathrm{Re}\,(0 + 2i) = 0 && \mathrm{Im}\,(2i) = \mathrm{Im}\,(0 + 2i) = 2 \\ &\mathrm{Re}\,(-7) = \mathrm{Re}\,(-7 + 0i) = -7 && \mathrm{Im}\,(-7) = \mathrm{Im}\,(-7 + 0i) = 0. \end{aligned}$$

Graphical Representation of Complex Numbers

Since complex numbers are constructed from pairs of real numbers (their real and imaginary parts), it is natural to represent complex numbers graphically as points in a Cartesian plane. We use the point with coordinates (a, b) to represent the complex number $w = a + ib$. In particular, the origin $(0, 0)$ represents the complex number 0, the point $(1, 0)$ represents the complex number $1 = 1 + 0i$, and the point $(0, 1)$ represents the point $i = 0 + 1i$. (See Figure I.1.)

---

[1] In some fields, for example, electrical engineering, the imaginary unit is denoted j instead of i. Like "negative," "surd," and "irrational," the term "imaginary" suggests the distrust that greeted the new kinds of numbers when they were first introduced.

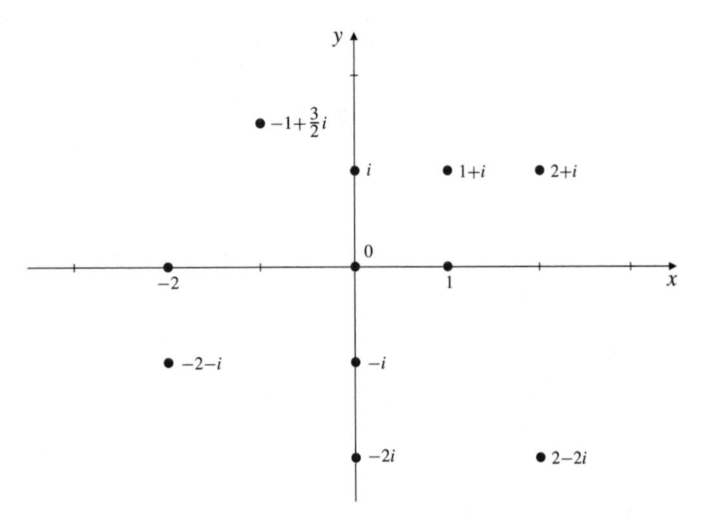

Figure I.1 An Argand diagram representing the complex plane

Such a representation of complex numbers as points in a plane is called an **Argand diagram**. Since each complex number is represented by a unique point in the plane, the set of all complex numbers is often referred to as the **complex plane**. The symbol $\mathbb{C}$ is used to represent the set of all complex numbers and, equivalently, the complex plane:

$$\mathbb{C} = \{x + yi \ : \ x, \ y, \in \mathbb{R}\}.$$

The points on the x-axis of the complex plane correspond to real numbers ($x = x + 0i$), so the x-axis is called the **real axis**. The points on the y-axis correspond to **pure imaginary** numbers ($yi = 0 + yi$), so the y-axis is called the **imaginary axis**.

It can be helpful to use the *polar coordinates* of a point in the complex plane.

DEFINITION 3

The distance from the origin to the point (a, b) corresponding to the complex number $w = a + bi$ is called the **modulus** of w and is denoted by $|w|$ or $|a + bi|$:

$$|w| = |a + bi| = \sqrt{a^2 + b^2}.$$

DEFINITION 4

If the line from the origin to (a, b) makes angle θ with the positive direction of the real axis (with positive angles measured counterclockwise), then we call θ an **argument** of the complex number $w = a + bi$ and denote it by $\arg(w)$ or $\arg(a + bi)$. (See Figure I.2.)

The modulus of a complex number is always real and nonnegative. It is positive unless the complex number is 0. Modulus plays a similar role for complex numbers that absolute value does for real numbers. Indeed, sometimes modulus is called absolute value.

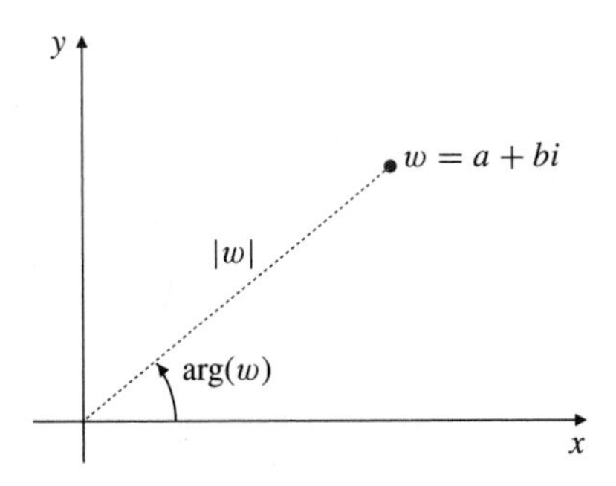

Figure I.2 The modulus and argument of a complex number

Arguments of complex numbers are not unique. If $w = a + bi \neq 0$, then any two possible values for $\arg(w)$ differ by an integer multiple of 2π. The symbol $\arg(w)$ actually represents not a single number, but a set of numbers. When we write $\arg(w) = \theta$, we are saying that the set $\arg(w)$ contains all numbers of the form $\theta + 2k\pi$, where k is an integer. Similarly, the statement $\arg(z) = \arg(w)$ says that two sets are identical.

If $w = a + bi$, where $a = \operatorname{Re}(w) \neq 0$, then

$$\tan \arg(w) = \tan \arg(a + bi) = \frac{b}{a}.$$

This means that $\tan\theta = b/a$ for every θ in the set $\arg(w)$.

It is sometimes convenient to restrict $\theta = \arg(w)$ to an interval of length 2π, say, the interval $0 \le \theta < 2\pi$, or $-\pi < \theta \le \pi$, so that nonzero complex numbers will have unique arguments. We will call the value of $\arg(w)$ in the interval $-\pi < \theta \le \pi$ the **principal argument** of w and denote it $\operatorname{Arg}(w)$. Every complex number w except 0 has a unique principal argument $\operatorname{Arg}(w)$.

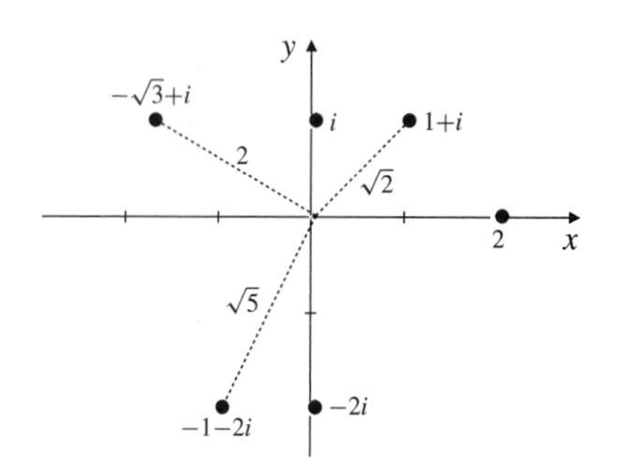

Figure I.3 Some complex numbers with their moduli

EXAMPLE 1 **(Some moduli and principal arguments)** See Figure I.3.

$$|2| = 2 \qquad \operatorname{Arg}(2) = 0$$
$$|1+i| = \sqrt{2} \qquad \operatorname{Arg}(1+i) = \pi/4$$
$$|i| = 1 \qquad \operatorname{Arg}(i) = \pi/2$$
$$|-2i| = 2 \qquad \operatorname{Arg}(-2i) = -\pi/2$$
$$|-\sqrt{3}+i| = 2 \qquad \operatorname{Arg}(-\sqrt{3}+i) = 5\pi/6$$
$$|-1-2i| = \sqrt{5} \qquad \operatorname{Arg}(-1-2i) = -\pi + \tan^{-1}(2).$$

BEWARE! Review the cautionary remark at the end of the discussion of the arctangent function in Section 3.5; different programs implement the two-variable arctangent using different notations and/or order of variables.

Remark If $z = x + yi$ and $\operatorname{Re}(z) = x > 0$, then $\operatorname{Arg}(z) = \tan^{-1}(y/x)$. Many computer spreadsheets and mathematical software packages implement a two-variable arctan function denoted $\operatorname{atan2}(x, y)$ which gives the polar angle of (x, y) in the interval $(-\pi, \pi]$. Thus,

$$\operatorname{Arg}(x + yi) = \operatorname{atan2}(x, y).$$

Given the modulus $r = |w|$ and any value of the argument $\theta = \arg(w)$ of a complex number $w = a + bi$, we have $a = r\cos\theta$ and $b = r\sin\theta$, so w can be expressed in terms of its modulus and argument as

$$w = r\cos\theta + i\,r\sin\theta.$$

The expression on the right side is called the **polar representation** of w.

DEFINITION 5

The **conjugate** or **complex conjugate** of a complex number $w = a+bi$ is another complex number, denoted $\overline{w}$, given by

$$\overline{w} = a - bi.$$

EXAMPLE 2 $\overline{2-3i} = 2+3i, \qquad \overline{3} = 3, \qquad \overline{2i} = -2i.$

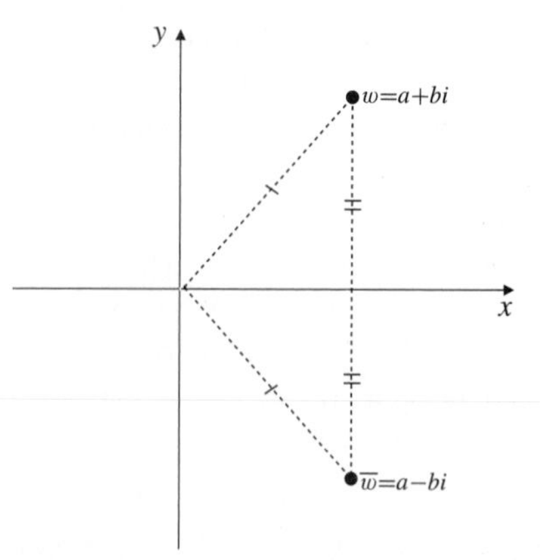

Figure I.4 A complex number and its conjugate are mirror images of each other in the real axis

Observe that

$$\operatorname{Re}(\overline{w}) = \operatorname{Re}(w) \qquad |\overline{w}| = |w|$$
$$\operatorname{Im}(\overline{w}) = -\operatorname{Im}(w) \qquad \arg(\overline{w}) = -\arg(w).$$

In an Argand diagram the point $\overline{w}$ is the reflection of the point w in the real axis. (See Figure I.4.)

Note that w is real ($\operatorname{Im}(w) = 0$) if and only if $\overline{w} = w$. Also, w is pure imaginary ($\operatorname{Re}(w) = 0$) if and only if $\overline{w} = -w$. (Here, $-w = -a - bi$ if $w = a + bi$.)

Complex Arithmetic

Like real numbers, complex numbers can be added, subtracted, multiplied, and divided. Two complex numbers are added or subtracted as though they are two-dimensional vectors whose components are their real and imaginary parts.

The sum and difference of complex numbers

If $w = a + bi$ and $z = x + yi$, where a, b, x, and y are real numbers, then

$$w + z = (a + x) + (b + y)i$$
$$w - z = (a - x) + (b - y)i.$$

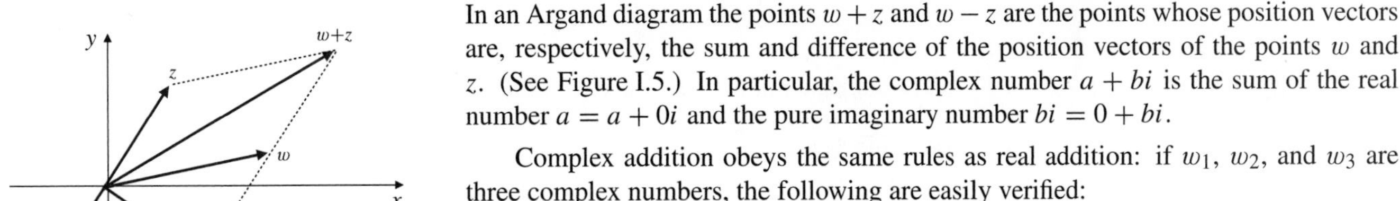

Figure I.5 Complex numbers are added and subtracted vectorially. Observe the parallelograms

In an Argand diagram the points $w + z$ and $w - z$ are the points whose position vectors are, respectively, the sum and difference of the position vectors of the points w and z. (See Figure I.5.) In particular, the complex number $a + bi$ is the sum of the real number $a = a + 0i$ and the pure imaginary number $bi = 0 + bi$.

Complex addition obeys the same rules as real addition: if w_1, w_2, and w_3 are three complex numbers, the following are easily verified:

$w_1 + w_2 = w_2 + w_1$	Addition is commutative.
$(w_1 + w_2) + w_3 = w_1 + (w_2 + w_3)$	Addition is associative.
$\|w_1 \pm w_2\| \le \|w_1\| + \|w_2\|$	the triangle inequality

Note that $|w_1 - w_2|$ is the distance between the two points w_1 and w_2 in the complex plane. Thus, the triangle inequality says that in the triangle with vertices w_1, $\mp w_2$ and 0, the length of one side is less than the sum of the other two.

It is also easily verified that the conjugate of a sum (or difference) is the sum (or difference) of the conjugates:

$$\overline{w + z} = \overline{w} + \overline{z}.$$

EXAMPLE 3 (a) If $w = 2 + 3i$ and $z = 4 - 5i$, then

$$w + z = (2 + 4) + (3 - 5)i = 6 - 2i$$
$$w - z = (2 - 4) + (3 - (-5))i = -2 + 8i.$$

(b) $3i + (1 - 2i) - (2 + 3i) + 5 = 4 - 2i.$

Multiplication of the complex numbers $w = a + bi$ and $z = x + yi$ is carried out by formally multiplying the binomial expressions and replacing i^2 by -1:

$$\begin{aligned} wz &= (a + bi)(x + yi) = ax + ayi + bxi + byi^2 \\ &= (ax - by) + (ay + bx)i. \end{aligned}$$

The product of complex numbers

If $w = a + bi$ and $z = x + yi$, where a, b, x, and y are real numbers, then

$$wz = (ax - by) + (ay + bx)i.$$

EXAMPLE 4 (a) $(2 + 3i)(1 - 2i) = 2 - 4i + 3i - 6i^2 = 8 - i.$

(b) $i(5 - 4i) = 5i - 4i^2 = 4 + 5i.$

(c) $(a + bi)(a - bi) = a^2 - abi + abi - b^2i^2 = a^2 + b^2.$

Part (c) of the example above shows that the square of the modulus of a complex number is the product of that number with its complex conjugate:

$$w\,\overline{w} = |w|^2.$$

Complex multiplication shares many properties with real multiplication. In particular, if w_1, w_2, and w_3 are complex numbers, then

$$w_1 w_2 = w_2 w_1 \qquad \text{Multiplication is commutative.}$$
$$(w_1 w_2) w_3 = w_1 (w_2 w_3) \qquad \text{Multiplication is associative.}$$
$$w_1 (w_2 + w_3) = w_1 w_2 + w_1 w_3 \qquad \text{Multiplication distributes over addition.}$$

The conjugate of a product is the product of the conjugates:

$$\overline{wz} = \overline{w}\,\overline{z}.$$

To see this, let $w = a + bi$ and $z = x + yi$. Then

$$\begin{aligned}\overline{wz} &= \overline{(ax - by) + (ay + bx)i}\\ &= (ax - by) - (ay + bx)i\\ &= (a - bi)(x - yi) = \overline{w}\,\overline{z}.\end{aligned}$$

It is particularly easy to determine the product of complex numbers expressed in polar form. If

$$w = r(\cos\theta + i\sin\theta) \qquad \text{and} \qquad z = s(\cos\phi + i\sin\phi),$$

where $r = |w|$, $\theta = \arg(w)$, $s = |z|$, and $\phi = \arg(z)$, then

$$\begin{aligned}wz &= rs(\cos\theta + i\sin\theta)(\cos\phi + i\sin\phi)\\ &= rs\big((\cos\theta\cos\phi - \sin\theta\sin\phi) + i(\sin\theta\cos\phi + \cos\theta\sin\phi)\big)\\ &= rs\big(\cos(\theta + \phi) + i\sin(\theta + \phi)\big).\end{aligned}$$

(See Figure I.6.) Since arguments are only determined up to integer multiples of 2π, we have proved that

The modulus and argument of a product

$$|wz| = |w||z| \qquad \text{and} \qquad \arg(wz) = \arg(w) + \arg(z).$$

The second of these equations says that the set $\arg(wz)$ consists of all numbers $\theta + \phi$, where θ belongs to the set $\arg(w)$ and ϕ to the set $\arg(z)$.

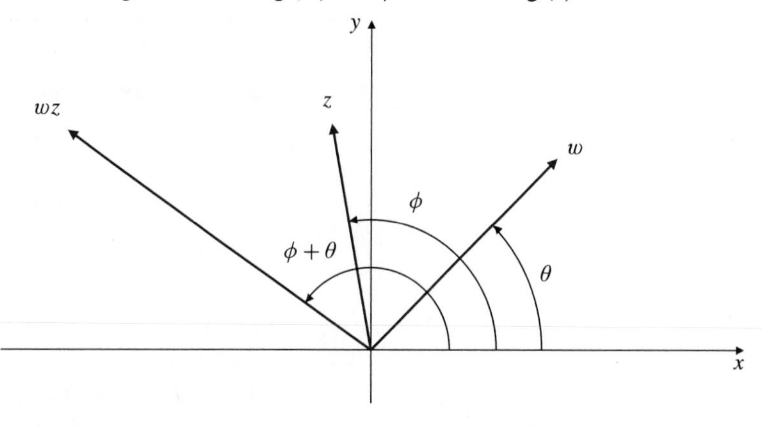

Figure I.6 The argument of a product is the sum of the arguments of the factors

More generally, if $w_1, w_2, \ldots, w_n$ are complex numbers, then

$$|w_1 w_2 \cdots w_n| = |w_1||w_2| \cdots |w_n|$$
$$\arg(w_1 w_2 \cdots w_n) = \arg(w_1) + \arg(w_2) + \cdots + \arg(w_n).$$

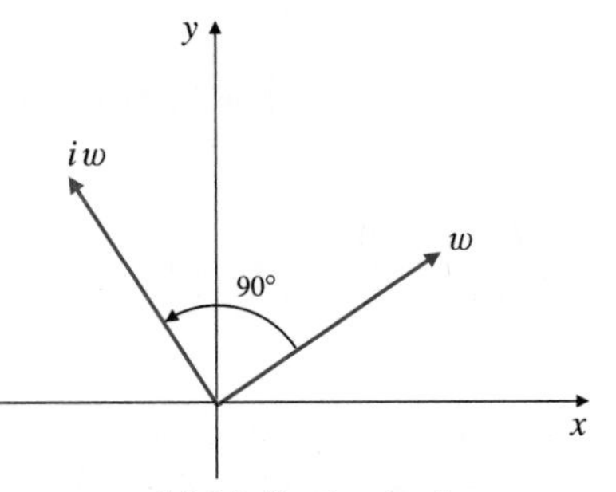

Figure I.7 Multiplication by i corresponds to counterclockwise rotation by 90°

Multiplication of a complex number by i has a particularly simple geometric interpretation in an Argand diagram. Since $|i| = 1$ and $\arg(i) = \pi/2$, multiplication of $w = a + bi$ by i leaves the modulus of w unchanged but increases its argument by $\pi/2$. (See Figure I.7.) Thus, multiplication by i rotates the position vector of w counterclockwise by 90° about the origin.

Let $z = \cos\theta + i\sin\theta$. Then $|z| = 1$ and $\arg(z) = \theta$. Since the modulus of a product is the product of the moduli of the factors and the argument of a product is the sum of the arguments of the factors, we have $|z^n| = |z|^n = 1$ and $\arg(z^n) = n\arg(z) = n\theta$. Thus,

$$z^n = \cos n\theta + i\sin n\theta,$$

and we have proved de Moivre's Theorem.

THEOREM 1

de Moivre's Theorem

$$(\cos\theta + i\sin\theta)^n = \cos n\theta + i\sin n\theta.$$

Remark Much of the study of complex-valued functions of a complex variable is beyond the scope of this book. However, in Appendix II we will introduce a complex version of the exponential function having the following property: if $z = x + iy$ (where x and y are real), then

$$e^z = e^{x+iy} = e^x e^{iy} = e^x(\cos y + i\sin y).$$

Thus, the modulus of e^z is $e^{\mathrm{Re}(z)}$, and $\mathrm{Im}(z)$ is a value of $\arg(e^z)$. In this context, de Moivre's Theorem just says

$$(e^{i\theta})^n = e^{in\theta}.$$

EXAMPLE 5 Express $(1+i)^5$ in the form $a + bi$.

Solution Since $|(1+i)^5| = |1+i|^5 = \left(\sqrt{2}\right)^5 = 4\sqrt{2}$, and $\arg\left((1+i)^5\right) = 5\arg(1+i) = \dfrac{5\pi}{4}$, we have

$$(1+i)^5 = 4\sqrt{2}\left(\cos\frac{5\pi}{4} + i\sin\frac{5\pi}{4}\right) = 4\sqrt{2}\left(-\frac{1}{\sqrt{2}} - \frac{1}{\sqrt{2}}i\right) = -4 - 4i.$$

de Moivre's Theorem can be used to generate trigonometric identities for multiples of an angle. For example, for $n = 2$ we have

$$\cos 2\theta + i\sin 2\theta = \left(\cos\theta + i\sin\theta\right)^2 = \cos^2\theta - \sin^2\theta + 2i\cos\theta\sin\theta.$$

Thus, $\cos 2\theta = \cos^2\theta - \sin^2\theta$, and $\sin 2\theta = 2\sin\theta\cos\theta$.

The **reciprocal** of the nonzero complex number $w = a + bi$ can be calculated by multiplying the numerator and denominator of the reciprocal expression by the conjugate of w:

$$w^{-1} = \frac{1}{w} = \frac{1}{a+bi} = \frac{a-bi}{(a+bi)(a-bi)} = \frac{a-bi}{a^2+b^2} = \frac{\overline{w}}{|w|^2}.$$

Since $|\overline{w}| = |w|$, and $\arg(\overline{w}) = -\arg(w)$, we have

$$\left|\frac{1}{w}\right| = \frac{|\overline{w}|}{|w|^2} = \frac{1}{|w|} \qquad \text{and} \qquad \arg\left(\frac{1}{w}\right) = -\arg(w).$$

The **quotient** z/w of two complex numbers $z = x + yi$ and $w = a + bi$ is the product of z and $1/w$, so

$$\frac{z}{w} = \frac{z\overline{w}}{|w|^2} = \frac{(x + yi)(a - bi)}{a^2 + b^2} = \frac{xa + yb + i(ya - xb)}{a^2 + b^2}.$$

We have

The modulus and argument of a quotient

$$\left|\frac{z}{w}\right| = \frac{|z|}{|w|} \quad \text{and} \quad \arg\left(\frac{z}{w}\right) = \arg(z) - \arg(w).$$

The set $\arg(z/w)$ consists of all numbers $\theta - \phi$ where θ belongs to the set $\arg(z)$ and ϕ to the set $\arg(w)$.

EXAMPLE 6 Simplify (a) $\dfrac{2 + 3i}{4 - i}$ and (b) $\dfrac{i}{1 + i\sqrt{3}}$.

Solution

(a) $$\frac{2 + 3i}{4 - i} = \frac{(2 + 3i)(4 + i)}{(4 - i)(4 + i)} = \frac{8 - 3 + (2 + 12)i}{4^2 + 1^2} = \frac{5}{17} + \frac{14}{17}i.$$

(b) $$\frac{i}{1 + i\sqrt{3}} = \frac{i(1 - i\sqrt{3})}{(1 + i\sqrt{3})(1 - i\sqrt{3})} = \frac{\sqrt{3} + i}{1^2 + 3} = \frac{\sqrt{3}}{4} + \frac{1}{4}i.$$

Alternatively, since $|1 + i\sqrt{3}| = 2$ and $\arg(1 + i\sqrt{3}) = \tan^{-1}\sqrt{3} = \dfrac{\pi}{3}$, the quotient in (b) has modulus $\dfrac{1}{2}$ and argument $\dfrac{\pi}{2} - \dfrac{\pi}{3} = \dfrac{\pi}{6}$. Thus,

$$\frac{i}{1 + i\sqrt{3}} = \frac{1}{2}\left(\cos\frac{\pi}{6} + i\sin\frac{\pi}{6}\right) = \frac{\sqrt{3}}{4} + \frac{1}{4}i.$$

Roots of Complex Numbers

If a is a positive real number, there are two distinct real numbers whose square is a. These are usually denoted

$\sqrt{a}$ (the positive square root of a) and

$-\sqrt{a}$ (the negative square root of a).

Every nonzero complex number $z = x + yi$ (where $x^2 + y^2 > 0$) also has two square roots; if w_1 is a complex number such that $w_1^2 = z$, then $w_2 = -w_1$ also satisfies $w_2^2 = z$. Again, we would like to single out one of these roots and call it $\sqrt{z}$.

Let $r = |z|$, so that $r > 0$. Let $\theta = \operatorname{Arg}(z)$. Thus, $-\pi < \theta \le \pi$. Since

$$z = r\big(\cos\theta + i\sin\theta\big),$$

the complex number

$$w = \sqrt{r}\left(\cos\frac{\theta}{2} + i\sin\frac{\theta}{2}\right)$$

clearly satisfies $w^2 = z$. We call this w the **principal square root** of z and denote it $\sqrt{z}$. The two solutions of the equation $w^2 = z$ are, thus, $w = \sqrt{z}$ and $w = -\sqrt{z}$.

Observe that the real part of $\sqrt{z}$ is always nonnegative, since $\cos(\theta/2) \ge 0$ for $-\pi/2 < \theta \le \pi/2$. In this interval $\sin(\theta/2) = 0$ only if $\theta = 0$, in which case $\sqrt{z}$ is real and positive.

EXAMPLE 7

(a) $\sqrt{4} = \sqrt{4(\cos 0 + i \sin 0)} = 2.$

(b) $\sqrt{i} = \sqrt{1\left(\cos\frac{\pi}{2} + i\sin\frac{\pi}{2}\right)} = \cos\frac{\pi}{4} + i\sin\frac{\pi}{4} = \frac{1}{\sqrt{2}} + \frac{1}{\sqrt{2}}i.$

(c) $\sqrt{-4i} = \sqrt{4\left[\cos\left(-\frac{\pi}{2}\right) + i\sin\left(-\frac{\pi}{2}\right)\right]} = 2\left[\cos\left(-\frac{\pi}{4}\right) + i\sin\left(-\frac{\pi}{4}\right)\right]$
$= \sqrt{2} - i\sqrt{2}.$

(d) $\sqrt{-\frac{1}{2} + i\frac{\sqrt{3}}{2}} = \sqrt{\cos\frac{2\pi}{3} + i\sin\frac{2\pi}{3}} = \cos\frac{\pi}{3} + i\sin\frac{\pi}{3} = \frac{1}{2} + \frac{\sqrt{3}}{2}i.$

Given a nonzero complex number z, we can find n distinct complex numbers w that satisfy $w^n = z$. These n numbers are called nth roots of z. For example, if $z = 1 = \cos 0 + i \sin 0$, then each of the numbers

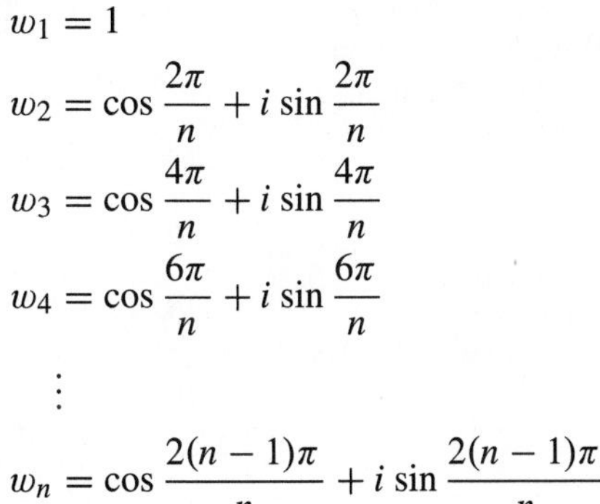

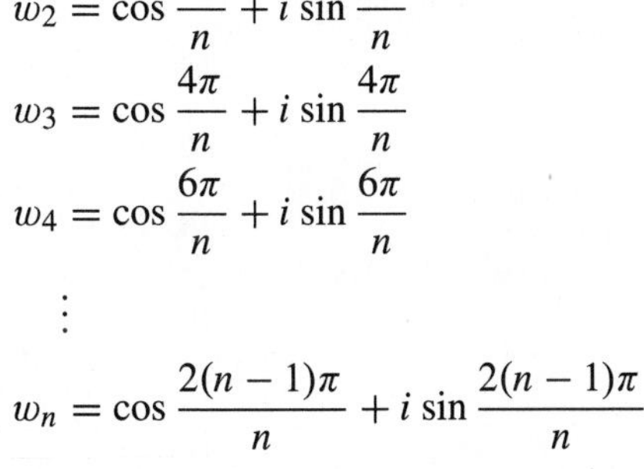

$$
\begin{aligned}
w_1 &= 1 \\
w_2 &= \cos\frac{2\pi}{n} + i\sin\frac{2\pi}{n} \\
w_3 &= \cos\frac{4\pi}{n} + i\sin\frac{4\pi}{n} \\
w_4 &= \cos\frac{6\pi}{n} + i\sin\frac{6\pi}{n} \\
&\vdots \\
w_n &= \cos\frac{2(n-1)\pi}{n} + i\sin\frac{2(n-1)\pi}{n}
\end{aligned}
$$

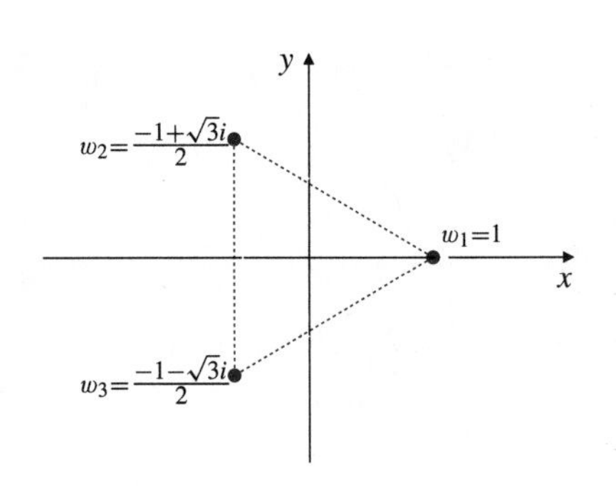

Figure I.8 The cube roots of unity

satisfies $w^n = 1$ so is an nth root of 1. (These numbers are usually called the nth roots of unity.) Figure I.8 shows the three cube roots of 1. Observe that they are at the three vertices of an equilateral triangle with centre at the origin and one vertex at 1. In general, the n nth roots of unity lie on a circle of radius 1 centred at the origin, and at the vertices of a regular n-sided polygon with one vertex at 1.

If z is any nonzero complex number, and θ is the principal argument of z $(-\pi < \theta \le \pi)$, then the number

$$w_1 = |z|^{1/n}\left(\cos\frac{\theta}{n} + i\sin\frac{\theta}{n}\right)$$

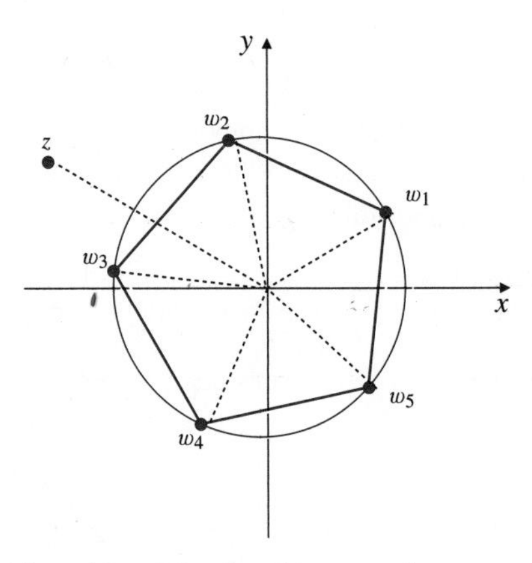

Figure I.9 The five 5th roots of z

is called the **principal** nth root of z. All the nth roots of z are on the circle of radius $|z|^{1/n}$ centred at the origin and are at the vertices of a regular n-sided polygon with one vertex at w_1. (See Figure I.9.) The other nth roots are

$$
\begin{aligned}
w_2 &= |z|^{1/n}\left(\cos\frac{\theta + 2\pi}{n} + i\sin\frac{\theta + 2\pi}{n}\right) \\
w_3 &= |z|^{1/n}\left(\cos\frac{\theta + 4\pi}{n} + i\sin\frac{\theta + 4\pi}{n}\right) \\
&\vdots \\
w_n &= |z|^{1/n}\left(\cos\frac{\theta + 2(n-1)\pi}{n} + i\sin\frac{\theta + 2(n-1)\pi}{n}\right).
\end{aligned}
$$

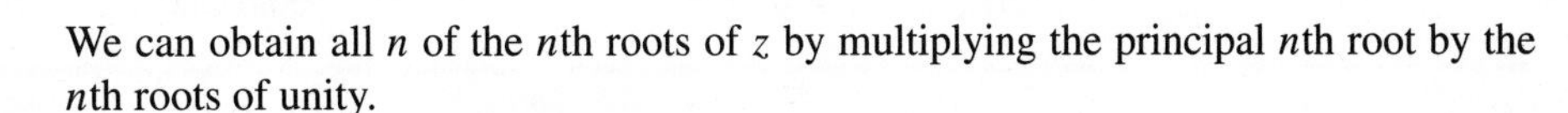

We can obtain all n of the nth roots of z by multiplying the principal nth root by the nth roots of unity.

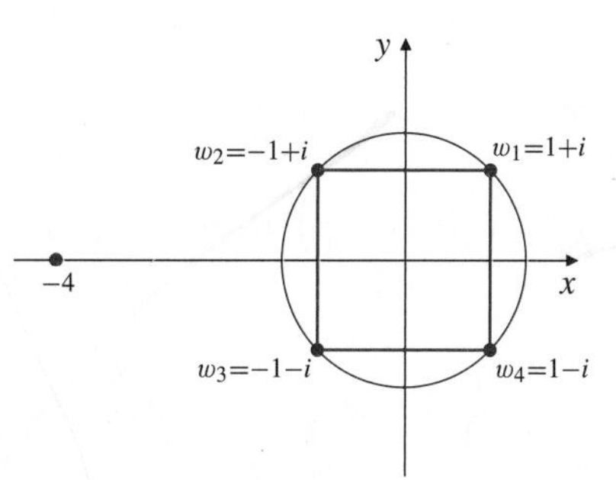

Figure I.10 The four 4th roots of -4

EXAMPLE 8 Find the 4th roots of -4. Sketch them in an Argand diagram.

Solution Since $|-4|^{1/4} = \sqrt{2}$ and $\arg(-4) = \pi$, the principal 4th root of -4 is

$$w_1 = \sqrt{2}\left(\cos\frac{\pi}{4} + i\sin\frac{\pi}{4}\right) = 1 + i.$$

The other three 4th roots are at the vertices of a square with centre at the origin and one vertex at $1 + i$. (See Figure I.10.) Thus the other roots are

$$w_2 = -1 + i, \qquad w_3 = -1 - i, \qquad w_4 = 1 - i.$$

EXERCISES: APPENDIX I

In Exercises 1–4, find the real and imaginary parts (Re (z) and Im (z)) of the given complex numbers z, and sketch the position of each number in the complex plane (i.e., in an Argand diagram).

1. $z = -5 + 2i$

2. $z = 4 - i$

3. $z = -\pi i$

4. $z = -6$

In Exercises 5–15, find the modulus $r = |z|$ and the principal argument $\theta = \mathrm{Arg}\,(z)$ of each given complex number z, and express z in terms of r and θ.

5. $z = -1 + i$

6. $z = -2$

7. $z = 3i$

8. $z = -5i$

9. $z = 1 + 2i$

10. $z = -2 + i$

11. $z = -3 - 4i$

12. $z = 3 - 4i$

13. $z = \sqrt{3} - i$

14. $z = -\sqrt{3} - 3i$

15. $z = 3\cos\dfrac{4\pi}{5} + 3i\sin\dfrac{4\pi}{5}$

16. If $\mathrm{Arg}\,(z) = 3\pi/4$ and $\mathrm{Arg}\,(w) = \pi/2$, find $\mathrm{Arg}\,(zw)$.

17. If $\mathrm{Arg}\,(z) = -5\pi/6$ and $\mathrm{Arg}\,(w) = \pi/4$, find $\mathrm{Arg}\,(z/w)$.

In Exercises 18–23, express in the form $z = x + yi$ the complex number z whose modulus and argument are given.

18. $|z| = 2, \quad \arg(z) = \pi$

19. $|z| = 5, \quad \arg(z) = \tan^{-1}\dfrac{3}{4}$

20. $|z| = 1, \quad \arg(z) = \dfrac{3\pi}{4}$

21. $|z| = \pi, \quad \arg(z) = \dfrac{\pi}{6}$

22. $|z| = 0, \quad \arg(z) = 1$

23. $|z| = \dfrac{1}{2}, \quad \arg(z) = -\dfrac{\pi}{3}$

In Exercises 24–27, find the complex conjugates of the given complex numbers.

24. $5 + 3i$

25. $-3 - 5i$

26. $4i$

27. $2 - i$

Describe geometrically (or make a sketch of) the set of points z in the complex plane satisfying the given equations or inequalities in Exercises 28–33.

28. $|z| = 2$

29. $|z| \le 2$

30. $|z - 2i| \le 3$

31. $|z - 3 + 4i| \le 5$

32. $\arg z = \dfrac{\pi}{3}$

33. $\pi \le \arg(z) \le \dfrac{7\pi}{4}$

Simplify the expressions in Exercises 34–43.

34. $(2 + 5i) + (3 - i)$

35. $i - (3 - 2i) + (7 - 3i)$

36. $(4 + i)(4 - i)$

37. $(1 + i)(2 - 3i)$

38. $(a + bi)(\overline{2a - bi})$

39. $(2 + i)^3$

40. $\dfrac{2 - i}{2 + i}$

41. $\dfrac{1 + 3i}{2 - i}$

42. $\dfrac{1 + i}{i(2 + 3i)}$

43. $\dfrac{(1 + 2i)(2 - 3i)}{(2 - i)(3 + 2i)}$

44. Prove that $\overline{z + w} = \overline{z} + \overline{w}$.

45. Prove that $\overline{\left(\dfrac{z}{w}\right)} = \dfrac{\overline{z}}{\overline{w}}$.

46. Express each of the complex numbers $z = 3 + i\sqrt{3}$ and $w = -1 + i\sqrt{3}$ in polar form (i.e., in terms of its modulus and argument). Use these expressions to calculate zw and z/w.

47. Repeat Exercise 46 for $z = -1 + i$ and $w = 3i$.

48. Use de Moivre's Theorem to find a trigonometric identity for $\cos 3\theta$ in terms of $\cos\theta$ and one for $\sin 3\theta$ in terms of $\sin\theta$.

49. Describe the solutions, if any, of the equations (a) $\overline{z} = 2/z$ and (b) $\overline{z} = -2/z$.

50. For positive real numbers a and b it is always true that $\sqrt{ab} = \sqrt{a}\sqrt{b}$. Does a similar identity hold for $\sqrt{zw}$, where z and w are complex numbers? *Hint:* Consider $z = w = -1$.

51. Find the three cube roots of -1.

52. Find the three cube roots of $-8i$.

53. Find the three cube roots of $-1 + i$.

54. Find all the fourth roots of 4.

55. Find all complex solutions of the equation $z^4 + 1 - i\sqrt{3} = 0$.

56. Find all solutions of $z^5 + a^5 = 0$, where a is a positive real number.

57. Show that the sum of the n nth roots of unity is zero. *Hint:* Show that these roots are all powers of the principal root.

APPENDIX II

Complex Functions

> The shortest path between two truths in the real domain passes through the complex domain.

Jacques Hadamard 1865–1963
quoted in *The Mathematical Intelligencer, v 13, 1991*

Most of this book is concerned with developing the properties of **real functions**, that is, functions of one or more real variables, having values that are themselves real numbers or vectors with real components. The definition of *function* given in Section P.4 can be paraphrased to allow for complex-valued functions of a complex variable.

DEFINITION 1

A **complex function** f is a rule that assigns a unique complex number $f(z)$ to each number z in some set of complex numbers (called the **domain** of the function).

Typically, we will use $z = x + yi$ to denote a general point in the domain of a complex function and $w = u + vi$ to denote the value of the function at z; if $w = f(z)$, then the real and imaginary parts of w ($u = \mathrm{Re}\,(w)$ and $v = \mathrm{Im}\,(w)$) are real-valued functions of z, and hence real-valued functions of the two real variables x and y:

$$u = u(x, y), \qquad v = v(x, y).$$

For example, the complex function $f(z) = z^2$, whose domain is the whole complex plane $\mathbb{C}$, assigns the value z^2 to the complex number z. If $w = z^2$ (where $w = u + vi$ and $z = x + yi$), then

$$u + vi = (x + yi)^2 = x^2 - y^2 + 2xyi,$$

so that

$$u = \mathrm{Re}\,(z^2) = x^2 - y^2 \qquad \text{and} \qquad v = \mathrm{Im}\,(z^2) = 2xy.$$

It is not convenient to draw the *graph* of a complex function. The graph of $w = f(z)$ would have to be drawn in a four-dimensional (real) space, since two dimensions (a z-plane) are required for the independent variable, and two more dimensions (a w-plane) are required for the dependent variable. Instead, we can graphically represent the behaviour of a complex function $w = f(z)$ by drawing the z-plane and the w-plane separately, and showing the image in the w-plane of certain, appropriately chosen sets of points in the z-plane. For example, Figure II.1 illustrates the fact that for the function $w = z^2$ the image of the quarter-disk $|z| \le a$, $0 \le \arg(z) \le \frac{\pi}{2}$ is the half-disk $|w| \le a^2$, $0 \le \arg(w) \le \pi$. To see why this is so, observe that if $z = r(\cos\theta + i\sin\theta)$, then $w = r^2(\cos 2\theta + i\sin 2\theta)$. Thus, the function maps the circle $|z| = r$ onto the circle $|w| = r^2$ and the radial line $\arg(z) = \theta$ onto the radial line $\arg(w) = 2\theta$.

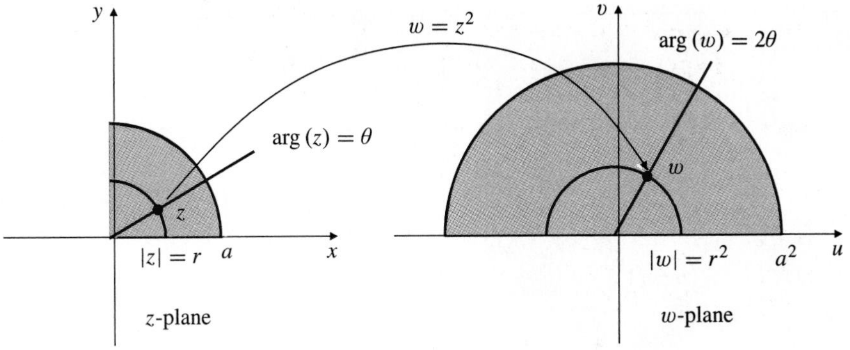

Figure II.1 The function $w = z^2$ maps a quarter-disk of radius a to a half-disk of radius a^2 by squaring the modulus and doubling the argument of each point z

Limits and Continuity

The concepts of limit and continuity carry over from real functions to complex functions in an obvious way providing we use $|z_1 - z_2|$ as the distance between the complex numbers z_1 and z_2. We say that

$$\lim_{z \to z_0} f(z) = \lambda$$

provided we can ensure that $|f(z) - \lambda|$ is as small as we wish by taking z sufficiently close to z_0. Formally,

DEFINITION 2

We say that $f(z)$ tends to the **limit** λ as z approaches z_0, and we write

$$\lim_{z \to z_0} f(z) = \lambda,$$

if for every positive real number ϵ there exists a positive real number δ (depending on ϵ), such that

$$0 < |z - z_0| < \delta \qquad \Longrightarrow \qquad |f(z) - \lambda| < \epsilon.$$

DEFINITION 3

The complex function $f(z)$ is **continuous** at $z = z_0$ if $\lim_{z \to z_0} f(z)$ exists and equals $f(z_0)$.

All the laws of limits and continuity apply as for real functions. Polynomials, that is, functions of the form

$$P(z) = a_0 + a_1 z + a_2 z^2 + \cdots + a_n z^n,$$

are continuous at every point of the complex plane. Rational functions, that is, functions of the form

$$R(z) = \frac{P(z)}{Q(z)},$$

where $P(z)$ and $Q(z)$ are polynomials, are continuous everywhere except at points where $Q(z) = 0$. Integer powers z^n are continuous except at the origin if $n < 0$. The situation for fractional powers is more complicated. For example, $\sqrt{z}$ (the principal square root) is continuous except at points $z = x < 0$. The function $f(z) = \bar{z}$ is continuous everywhere, because

$$|\bar{z} - \overline{z_0}| = |\overline{z - z_0}| = |z - z_0|.$$

The Complex Derivative

The definition of derivative is the same as for real functions:

DEFINITION 4

The complex function f is **differentiable** at z and has **derivative** $f'(z)$ there, provided

$$\lim_{h \to 0} \frac{f(z+h) - f(z)}{h} = f'(z)$$

exists.

Note, however, that in this definition h is a complex number. The limit must exist no matter how h approaches 0 in the complex plane. This fact has profound implications. The existence of a derivative in this sense forces the function f to be much better behaved than is necessary for a differentiable real function. For example, it can be shown that if $f'(z)$ exists for all z in an open region D in $\mathbb{C}$, then f has derivatives of *all* orders throughout D. Moreover, such a function is the sum of its Taylor series

$$f(z) = f(z_0) + f'(z_0)(z - z_0) + \frac{f''(z_0)}{2!}(z - z_0)^2 + \cdots$$

about any point z_0 in D; the series has positive radius of convergence R and converges in the disk $|z - z_0| < R$. For this reason, complex functions that are differentiable on open sets in $\mathbb{C}$ are usually called **analytic functions**. It is beyond the scope of this introductory appendix to prove these assertions. They are proved in courses and texts on complex analysis.

The usual differentiation rules apply:

$$\begin{aligned}
&\frac{d}{dz}\big(Af(z) + Bg(z)\big) = Af'(z) + Bg'(z) \\
&\frac{d}{dz}\big(f(z)g(z)\big) = f'(z)g(z) + f(z)g'(z) \\
&\frac{d}{dz}\left(\frac{f(z)}{g(z)}\right) = \frac{g(z)f'(z) - f(z)g'(z)}{\big(g(z)\big)^2} \\
&\frac{d}{dz}f\big(g(z)\big) = f'\big(g(z)\big)g'(z).
\end{aligned}$$

As one would expect, the derivative of $f(z) = z^n$ is $f'(z) = nz^{n-1}$.

EXAMPLE 1 Show that the function $f(z) = \overline{z}$ is not differentiable at any point.

Solution We have

$$\begin{aligned}
f'(z) &= \lim_{h \to 0} \frac{\overline{z + h} - \overline{z}}{h} \\
&= \lim_{h \to 0} \frac{\overline{z + h - z}}{h} = \lim_{h \to 0} \frac{\overline{h}}{h}.
\end{aligned}$$

But $\overline{h}/h = 1$ if h is real, and $\overline{h}/h = -1$ if h is pure imaginary. Since there are real and pure imaginary numbers arbitrarily close to 0, the limit above does not exist, so $f'(z)$ does not exist.

The following theorem links the existence of the derivative of a complex function $f(z)$ with certain properties of its real and imaginary parts $u(x, y)$ and $v(x, y)$.

THEOREM 1 **The Cauchy–Riemann equations**

If $f(z) = u(x, y) + iv(x, y)$ is differentiable at $z = x + yi$, then u and v satisfy the Cauchy–Riemann equations

$$\frac{\partial u}{\partial x} = \frac{\partial v}{\partial y}, \qquad \frac{\partial v}{\partial x} = -\frac{\partial u}{\partial y}.$$

Conversely, if u and v are sufficiently smooth (say, if they have continuous second partial derivatives near (x, y)), and if u and v satisfy the Cauchy–Riemann equations at (x, y), then f is differentiable at $z = x + yi$ and

$$f'(z) = \frac{\partial u}{\partial x} + i\frac{\partial v}{\partial x}.$$

PROOF First, assume that f is differentiable at z. Letting $h = s + ti$, we have

$$\begin{aligned} f'(z) &= \lim_{h\to 0} \frac{f(z+h) - f(z)}{h} \\ &= \lim_{(s,t)\to(0,0)} \left[\frac{u(x+s, y+t) - u(x, y)}{s+it} + i\,\frac{v(x+s, y+t) - v(x, y)}{s+it} \right]. \end{aligned}$$

The limit must be independent of the path along which h approaches 0. Letting $t = 0$, so that $h = s$ approaches 0 along the real axis, we obtain

$$f'(z) = \lim_{s\to 0} \left[\frac{u(x+s, y) - u(x, y)}{s} + i\,\frac{v(x+s, y) - v(x, y)}{s} \right] = \frac{\partial u}{\partial x} + i\,\frac{\partial v}{\partial x}.$$

Similarly, letting $s = 0$, so that $h = ti$ approaches 0 along the imaginary axis, we obtain

$$\begin{aligned} f'(z) &= \lim_{t\to 0} \left[\frac{u(x, y+t) - u(x, y)}{it} + i\,\frac{v(x, y+t) - v(x, y)}{it} \right] \\ &= \lim_{t\to 0} \left[\frac{v(x, y+t) - v(x, y)}{t} - i\,\frac{u(x, y+t) - u(x, y)}{t} \right] \\ &= \frac{\partial v}{\partial y} - i\,\frac{\partial u}{\partial y}. \end{aligned}$$

Equating these two expressions for $f'(z)$, we see that

$$\frac{\partial u}{\partial x} = \frac{\partial v}{\partial y}, \qquad \frac{\partial v}{\partial x} = -\frac{\partial u}{\partial y}.$$

To prove the converse, we use the result of Exercise 22 of Section 12.6. Since u and v are assumed to have continuous second partial derivatives, we must have

$$\begin{aligned} u(x+s, y+t) - u(x, y) &= s\frac{\partial u}{\partial x} + t\frac{\partial u}{\partial y} + O(s^2 + t^2) \\ v(x+s, y+t) - v(x, y) &= s\frac{\partial v}{\partial x} + t\frac{\partial v}{\partial y} + O(s^2 + t^2), \end{aligned}$$

where we have used the Big-O notation (see Definition 9 of Section 4.10); the expression $O(\lambda)$ denotes a term satisfying $|O(\lambda)| \le K|\lambda|$ for some constant K. Thus, if u and v satisfy the Cauchy–Riemann equations, then

$$\begin{aligned} \frac{f(z+h) - f(z)}{h} &= \frac{s\dfrac{\partial u}{\partial x} + t\dfrac{\partial u}{\partial y} + i\left(s\dfrac{\partial v}{\partial x} + t\dfrac{\partial v}{\partial y}\right) + O(s^2 + t^2)}{s+it} \\ &= \frac{(s+it)\dfrac{\partial u}{\partial x} + i(s+it)\dfrac{\partial v}{\partial x}}{s+it} + O(\sqrt{s^2 + t^2}) \\ &= \frac{\partial u}{\partial x} + i\,\frac{\partial v}{\partial x} + O(\sqrt{s^2 + t^2}). \end{aligned}$$

Thus, we may let $h = s + ti$ approach 0 and obtain

$$f'(z) = \frac{\partial u}{\partial x} + i\,\frac{\partial v}{\partial x}.$$

It follows immediately from the Cauchy–Riemann equations that the real and imaginary parts of a differentiable complex function are real harmonic functions:

$$\frac{\partial^2 u}{\partial x^2} + \frac{\partial^2 u}{\partial y^2} = 0, \qquad \frac{\partial^2 v}{\partial x^2} + \frac{\partial^2 v}{\partial y^2} = 0.$$

(See Exercise 15 of Section 12.4.)

The Exponential Function

Consider the function

$$f(z) = e^x \cos y + ie^x \sin y,$$

where $z = x + yi$. The real and imaginary parts of $f(z)$,

$$u(x, y) = \mathrm{Re}\,(f(z)) = e^x \cos y \qquad \text{and} \qquad v(x, y) = \mathrm{Im}\,(f(z)) = e^x \sin y,$$

satisfy the Cauchy–Riemann equations

$$\frac{\partial u}{\partial x} = e^x \cos y = \frac{\partial v}{\partial y} \qquad \text{and} \qquad \frac{\partial v}{\partial x} = e^x \sin y = -\frac{\partial u}{\partial y}$$

everywhere in the z-plane. Therefore, $f(z)$ is differentiable (analytic) everywhere and satisfies

$$f'(z) = \frac{\partial u}{\partial x} + i\frac{\partial v}{\partial x} = e^x \cos y + ie^x \sin y = f(z).$$

Evidently $f(0) = 1$, and $f(z) = e^x$ if $z = x$ is a real number. It is therefore natural to denote the function $f(z)$ as the exponential function e^z.

The complex exponential function

$$e^z = e^x(\cos y + i \sin y) \qquad \text{for} \quad z = x + yi.$$

In particular, if $z = yi$ is pure imaginary, then

$$e^{yi} = \cos y + i \sin y,$$

a fact that can also be obtained by separating the real and imaginary parts of the Maclaurin series for e^{yi}:

$$\begin{aligned} e^{yi} &= 1 + (yi) + \frac{(yi)^2}{2!} + \frac{(yi)^3}{3!} + \frac{(yi)^4}{4!} + \frac{(yi)^5}{5!} + \cdots \\ &= \left(1 - \frac{y^2}{2!} + \frac{y^4}{4!} - \cdots\right) + i\left(y - \frac{y^3}{3!} + \frac{y^5}{5!} - \cdots\right) \\ &= \cos y + i \sin y. \end{aligned}$$

Observe that

$$\begin{aligned} |e^z| &= \sqrt{e^{2x}\left(\cos^2 y + \sin^2 y\right)} = e^x, \\ \arg(e^z) &= \arg(e^{yi}) = \arg(\cos y + i \sin y) = y, \\ \overline{e^z} &= e^x \cos y - ie^x \sin y = e^x \cos(-y) + ie^x \sin(-y) = e^{\bar{z}}. \end{aligned}$$

In summary:

Properties of the exponential function

If $z = x + yi$ then $\overline{e^z} = e^{\bar{z}}$. Also,

$$\operatorname{Re}(e^z) = e^x \cos y, \qquad |e^z| = e^x,$$
$$\operatorname{Im}(e^z) = e^x \sin y, \qquad \arg(e^z) = y.$$

EXAMPLE 2 Sketch the image in the w-plane of the rectangle $R : a \le x \le b, \quad c \le y \le d$ in the z-plane under the transformation $w = e^z$.

Solution The vertical lines $x = a$ and $x = b$ get mapped to the concentric circles $|w| = e^a$ and $|w| = e^b$. The horizontal lines $y = c$ and $y = d$ get mapped to the radial lines $\arg(w) = c$ and $\arg(w) = d$. Thus, the rectangle R gets mapped to the polar region P shown in Figure II.2.

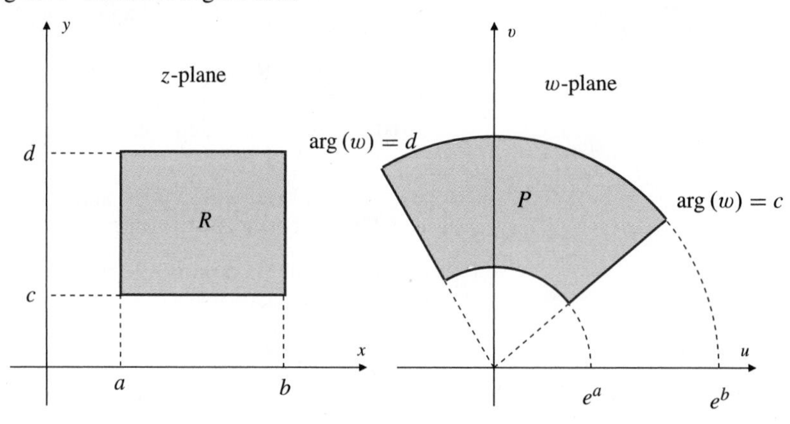

Figure II.2 Under the exponential function $w = e^z$, vertical lines get mapped to circles centred at the origin, and horizontal lines get mapped to half-lines radiating from the origin

Note that if $d - c \ge 2\pi$, then the image of R will be the entire annular region $e^a \le |w| \le e^b$, which may be covered more than once. The exponential function e^z is periodic with period $2\pi i$:

$$e^{z+2\pi i} = e^z \qquad \text{for all} \quad z,$$

and is therefore not one-to-one on the whole complex plane. However, $w = e^z$ is one-to-one from any horizontal strip of the form

$$-\infty < x < \infty, \qquad c < y \le c + 2\pi$$

onto the whole w-plane excluding the origin.

The Fundamental Theorem of Algebra

As observed at the beginning of Appendix I, extending the number system to include complex numbers allows larger classes of equations to have solutions. We conclude this appendix by verifying that polynomial equations always have solutions in the complex numbers.

A **complex polynomial** of degree n is a function of the form

$$P_n(z) = a_n z^n + a_{n-1} z^{n-1} + \cdots + a_2 z^2 + a_1 z + a_0,$$

where $a_0, a_1, \ldots, a_n$ are complex numbers and $a_n \ne 0$. The numbers a_i $(0 \le i \le n)$ are called the **coefficients** of the polynomial. If they are all real numbers, then $P_n(x)$ is called a real polynomial.

A complex number z_0 that satisfies the equation $P(z_0) = 0$ is called a **zero** or **root** of the polynomial. Every polynomial of degree 1 has a zero: if $a_1 \neq 0$, then $a_1 z + a_0$ has zero $z = -a_0/a_1$. This zero is real if a_1 and a_0 are both real.

Similarly, every complex polynomial of degree 2 has two zeros. If the polynomial is given by

$$P_2(z) = a_2 z^2 + a_1 z + a_0$$

(where $a_2 \neq 0$), then the zeros are given by the *quadratic formula*

$$z = z_1 = \frac{-a_1 - \sqrt{a_1^2 - 4a_2a_0}}{2a_2} \quad \text{and} \quad z = z_2 = \frac{-a_1 + \sqrt{a_1^2 - 4a_2a_0}}{2a_2}.$$

In this case, $P_2(z)$ has two linear factors:

$$P_2(z) = a_2(z - z_1)(z - z_2).$$

Even if $a_1^2 - 4a_2a_0 = 0$, so that $z_1 = z_2$, we still regard the polynomial as having two (equal) zeros, one corresponding to each factor. If the coefficients a_0, a_1, and a_2 are all real numbers, the zeros will be real provided $a_1^2 \geq 4a_2a_0$. When real coefficients satisfy $a_1^2 < 4a_2a_0$ then the zeros are complex, in fact, complex conjugates: $z_2 = \overline{z_1}$.

EXAMPLE 3 Solve the equation $z^2 + 2iz - (1 + i) = 0$.

Solution The zeros of this equation are

$$\begin{aligned} z &= \frac{-2i \pm \sqrt{-4 + 4(1+i)}}{2} \\ &= -i \pm \sqrt{i} \\ &= -i \pm \frac{1+i}{\sqrt{2}} = \frac{1}{\sqrt{2}}\big(1 + (1 - \sqrt{2})i\big) \quad \text{or} \quad -\frac{1}{\sqrt{2}}\big(1 + (1 + \sqrt{2})i\big). \end{aligned}$$

The Fundamental Theorem of Algebra asserts that every complex polynomial of positive degree has a complex zero.

THEOREM 2 **The Fundamental Theorem of Algebra**

If $P(z) = a_n z^n + a_{n-1} z^{n-1} + \cdots + a_1 z + a_0$ is a complex polynomial of degree $n \geq 1$, then there exists a complex number z_1 such that $P(z_1) = 0$.

PROOF (We will only give an informal sketch of the proof.) We can assume that the coefficient of z^n in $P(z)$ is $a_n = 1$, since we can divide the equation $P(z) = 0$ by a_n without changing its solutions. We can also assume that $a_0 \neq 0$; if $a_0 = 0$, then $z = 0$ is certainly a zero of $P(z)$. Thus, we deal with the polynomial

$$P(z) = z^n + Q(z),$$

where $Q(z)$ is a polynomial of degree less than n having a nonzero constant term. If R is sufficiently large, then $|Q(z)|$ will be less than R^n for all numbers z satisfying $|z| = R$. As z moves around the circle $|z| = R$ in the z-plane, $w = z^n$ moves around the circle $|w| = R^n$ in the w-plane (n times). Since the distance from z^n to $P(z)$ is equal to $|P(z) - z^n| = |Q(z)| < R^n$, it follows that the image of the circle $|z| = R$ under the transformation $w = P(z)$ is a curve that winds around the origin n times. (If you walk around a circle of radius r n times, with your dog on a leash of length less than r, and your dog returns to his starting point, then he must also go around the centre of the circle n times.) This situation is illustrated for the particular case

$$P(z) = z^3 + z^2 - iz + 1, \qquad |z| = 2$$

in Figure II.3. The image of $|z| = 2$ is the large curve in the w-plane that winds around the origin three times. As R decreases, the curve traced out by $w = P(z)$ for $|z| = R$ changes continuously. For R close to 0, it is a small curve staying close to the constant term a_0 of $P(z)$. For small enough R the curve will not enclose the origin. (In Figure II.3 the image of $|z| = 0.3$ is the small curve staying close to the point 1 in the w-plane.) Thus, for some value of R, say $R = R_1$, the curve must pass through the origin. That is, there must be a complex number z_1, with $|z_1| = R_1$, such that $P(z_1) = 0$.

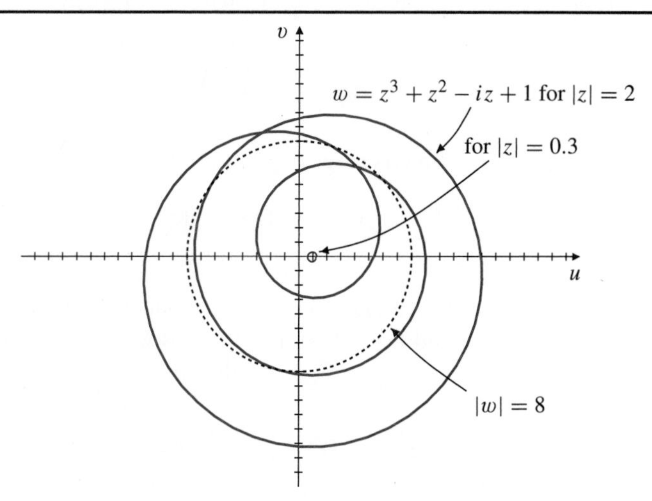

Figure II.3 The image of the circle $|z| = 2$ winds around the origin in the w-plane three times, but the image of $|z| = 0.3$ does not wind around the origin at all

Remark The above proof suggests that there should be n such solutions of the equation $P(z) = 0$; the curve has to go from winding around the origin n times to winding around the origin 0 times as R decreases toward 0. We can establish this as follows. $P(z_1) = 0$ implies that $z - z_1$ is a factor of $P(z)$:

$$P(z) = (z - z_1)P_{n-1}(z),$$

where P_{n-1} is a polynomial of degree $n - 1$. If $n > 1$, then P_{n-1} must also have a zero, z_2, by the Fundamental Theorem. We can continue this argument inductively to obtain n zeros and factor $P(z)$ into a product of the constant a_n and n linear factors:

$$P(z) = a_n(z - z_1)(z - z_2)\cdots(z - z_n).$$

Of course, some of the zeros can be equal.

Remark If P is a real polynomial, that is, one whose coefficients are all real numbers, then $\overline{P(z)} = P(\bar{z})$. Therefore, if z_1 is a nonreal zero of $P(z)$, then so is $z_2 = \overline{z_1}$:

$$P(z_2) = P(\overline{z_1}) = \overline{P(z_1)} = \bar{0} = 0.$$

Real polynomials can have complex zeros, but they must always occur in complex conjugate pairs. Every real polynomial of odd degree must have at least one real zero.

EXAMPLE 4 Show that $z_1 = -i$ is a zero of the polynomial $P(z) = z^4 + 5z^3 + 7z^2 + 5z + 6$, and find all the other zeros of this polynomial.

Solution First observe that $P(z_1) = P(-i) = 1 + 5i - 7 - 5i + 6 = 0$, so $z_1 = -i$ is indeed a zero. Since the coefficients of $P(z)$ are real, $z_2 = i$ must also be a zero. Thus, $z + i$ and $z - i$ are factors of $P(z)$, and so is

$$(z + i)(z - i) = z^2 + 1.$$

Dividing $P(z)$ by $z^2 + 1$, we obtain

$$\frac{P(z)}{z^2 + 1} = z^2 + 5z + 6 = (z + 2)(z + 3).$$

Thus, the four zeros of $P(z)$ are $z_1 = -i$, $z_2 = i$, $z_3 = -2$, and $z_4 = -3$.

EXERCISES: APPENDIX II

In Exercises 1–12, the z-plane region D consists of the complex numbers $z = x + yi$ that satisfy the given conditions. Describe (or sketch) the image R of D in the w-plane under the given function $w = f(z)$.

1. $0 \le x \le 1,\ 0 \le y \le 2;\quad w = \bar{z}$.

2. $x + y = 1;\quad w = \bar{z}$.

3. $1 \le |z| \le 2,\ \dfrac{\pi}{2} \le \arg z \le \dfrac{3\pi}{4};\quad w = z^2$.

4. $0 \le |z| \le 2,\ 0 \le \arg(z) \le \dfrac{\pi}{2};\quad w = z^3$.

5. $0 < |z| \le 2,\ 0 \le \arg(z) \le \dfrac{\pi}{2};\quad w = \dfrac{1}{z}$.

6. $\dfrac{\pi}{4} \le \arg(z) \le \dfrac{\pi}{3};\quad w = -iz$.

7. $\arg(z) = -\dfrac{\pi}{3};\quad w = \sqrt{z}$.

8. $x = 1;\quad w = z^2$.

9. $y = 1;\quad w = z^2$.

10. $x = 1;\quad w = \dfrac{1}{z}$.

11. $-\infty < x < \infty,\ \dfrac{\pi}{4} \le y \le \dfrac{\pi}{2};\quad w = e^z$.

12. $0 < x < \dfrac{\pi}{2},\ 0 < y < \infty;\quad w = e^{iz}$.

In Exercises 13–16, verify that the real and imaginary parts of each function $f(z)$ satisfy the Cauchy–Riemann equations, and thus find $f'(z)$.

13. $f(z) = z^2$.

14. $f(z) = z^3$.

15. $f(z) = \dfrac{1}{z}$.

16. $f(z) = e^{z^2}$.

17. Use the fact that $e^{yi} = \cos y + i \sin y$ (for real y) to show that

$$\cos y = \frac{e^{yi} + e^{-yi}}{2} \quad \text{and} \quad \sin y = \frac{e^{yi} - e^{-yi}}{2i}.$$

Exercise 16 suggests that we define complex functions

$$\cos z = \frac{e^{zi} + e^{-zi}}{2} \quad \text{and} \quad \sin z = \frac{e^{zi} - e^{-zi}}{2i},$$

as well as extend the definitions of the hyperbolic functions to

$$\cosh z = \frac{e^{z} + e^{-z}}{2} \quad \text{and} \quad \sinh z = \frac{e^{z} - e^{-z}}{2}.$$

Exercises 18–26 develop properties of these functions and relationships between them.

18. Show that $\cos z$ and $\sin z$ are periodic with period 2π, and that $\cosh z$ and $\sinh z$ are periodic with period $2\pi i$.

19. Show that $(d/dz)\sin z = \cos z$ and $(d/dz)\cos z = -\sin z$. What are the derivatives of $\sinh z$ and $\cosh z$?

20. Verify the identities $\cos z = \cosh(iz)$ and $\sin z = -i\sinh(iz)$. What are the corresponding identities for $\cosh z$ and $\sinh(z)$ in terms of cos and sin?

21. Find all complex zeros of $\cos z$ (i.e., all solutions of $\cos z = 0$).

22. Find all complex zeros of $\sin z$.

23. Find all complex zeros of $\cosh z$ and $\sinh z$.

24. Show that $\mathrm{Re}(\cosh z) = \cosh x \cos y$ and $\mathrm{Im}(\cosh z) = \sinh x \sin y$.

25. Find the real and imaginary parts of $\sinh z$.

26. Find the real and imaginary parts of $\cos z$ and $\sin z$.

Find the zeros of the polynomials in Exercises 27–32.

27. $P(z) = z^2 + 2iz$

28. $P(z) = z^2 - 2z + i$

29. $P(z) = z^2 + 2z + 5$

30. $P(z) = z^2 - 2iz - 1$

31. $P(z) = z^3 - 3iz^2 - 2z$

32. $P(z) = z^4 - 2z^2 + 4$

33. The polynomial $P(z) = z^4 + 1$ has two pairs of complex conjugate zeros. Find them, and hence express $P(z)$ as a product of two quadratic factors with real coefficients.

In Exercises 34–36, check that the given number z_1 is a zero of the given polynomial, and find all the zeros of the polynomial.

34. $P(z) = z^4 - 4z^3 + 12z^2 - 16z + 16;\quad z_1 = 1 - \sqrt{3}i$.

35. $P(z) = z^5 + 3z^4 + 4z^3 + 4z^2 + 3z + 1;\quad z_1 = i$.

36. $P(z) = z^5 - 2z^4 - 8z^3 + 8z^2 + 31z - 30;$ $z_1 = -2 + i$.

37. Show that the image of the circle $|z| = 2$ under the mapping $w = z^4 + z^3 - 2iz - 3$ winds around the origin in the w-plane four times.

APPENDIX III

Continuous Functions

> Geometry may sometimes appear to take the lead over analysis, but in fact precedes it only as a servant goes before his master to clear the path and light him on the way. The interval between the two is as wide as between empiricism and science, as between the understanding and the reason, or as between the finite and the infinite.

J. J. Sylvester 1814–1897
from *Philosophic Magazine, 1866*

The development of calculus depends in an essential way on the concept of the limit of a function and thereby on properties of the real number system. In Chapter 1 we presented these notions in an intuitive way and did not attempt to prove them except in Section 1.5, where the *formal* definition of limit was given and used to verify some elementary limits and prove some simple properties of limits.

Many of the results on limits and continuity of functions stated in Chapter 1 may seem quite obvious; most students and users of calculus are not bothered by applying them without proof. Nevertheless, mathematics is a highly logical and rigorous discipline, and any statement, however obvious, that cannot be proved by strictly logical arguments from acceptable assumptions must be considered suspect. In this appendix we build upon the formal definition of limit given in Section 1.5 and combine it with the notion of *completeness* of the real number system first encountered in Section P.1 to give formal proofs of the very important results about continuous functions stated in Theorems 8 and 9 of Section 1.4, the Max-Min Theorem and the Intermediate-Value Theorem. Most of our development of calculus in this book depends essentially on these two theorems.

The branch of mathematics that deals with proofs such as these is called mathematical analysis. This subject is usually not pursued by students in introductory calculus courses but is postponed to higher years and studied by students in majors or honours programs in mathematics. It is hoped that some of this material will be of value to honours-level calculus courses and individual students with a deeper interest in understanding calculus.

Limits of Functions

At the heart of mathematical analysis is the formal definition of limit, Definition 8 in Section 1.5, which we restate as follows:

The formal definition of limit

We say that $\lim_{x\to a} f(x) = L$ if for every positive number ϵ there exists a positive number δ, depending on ϵ (i.e., $\delta = \delta(\epsilon)$), such that

$$0 < |x - a| < \delta \quad \Longrightarrow \quad |f(x) - L| < \epsilon.$$

Section 1.5 was marked "optional" because understanding the material presented there was not essential for learning calculus. However, that material is an *essential* prerequisite for this appendix. It is highly recommended that you go back to Section 1.5 and read it carefully, paying special attention to Examples 2 and 4, and attempt at least Exercises 31–36. These exercises provide proofs for the standard laws of limits stated in Section 1.2.

Continuous Functions

Consider the following definitions of continuity, which are equivalent to those given in Section 1.4.

DEFINITION 1

Continuity of a function at a point

A function f, defined on an open interval containing the point a, is said to be continuous at the point a if

$$\lim_{x\to a} f(x) = f(a);$$

that is, if for every $\epsilon > 0$ there exists $\delta > 0$ such that if $|x - a| < \delta$, then $|f(x) - f(a)| < \epsilon$.

DEFINITION 2

Continuity of a function on an interval

A function f is continuous on an interval if it is continuous at every point of that interval. In the case of an endpoint of a closed interval, f need only be continuous on one side. Thus, f is continuous on the interval $[a, b]$ if

$$\lim_{t\to x} f(t) = f(x)$$

for each x satisfying $a < x < b$, and

$$\lim_{t\to a+} f(t) = f(a) \qquad \text{and} \qquad \lim_{t\to b-} f(t) = f(b).$$

These concepts are illustrated in Figure III.1.

Some important results about continuous functions are collected in Theorems 6 and 7 of Section 1.4, which we restate here:

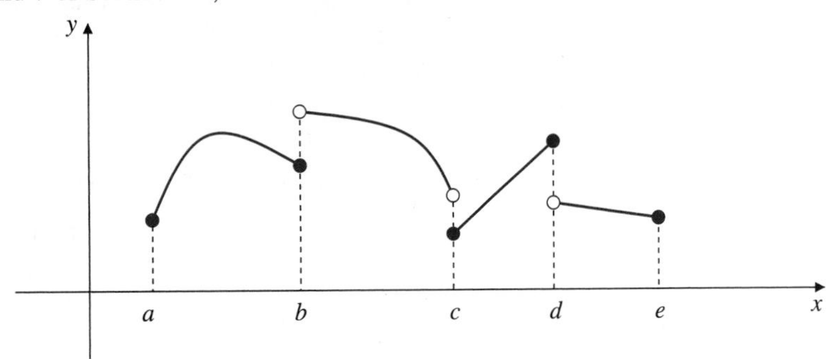

Figure III.1 f is continuous on the intervals $[a, b]$, (b, c), $[c, d]$, and $(d, e]$

THEOREM 

Combining continuous functions

(a) If f and g are continuous at the point a, then so are $f + g$, $f - g$, fg, and, if $g(a) \neq 0$, f/g.

(b) If f is continuous at the point L and if $\lim_{x\to a} g(x) = L$, then we have

$$\lim_{x\to a} f\big(g(x)\big) = f(L) = f\big(\lim_{x\to a} g(x)\big).$$

In particular, if g is continuous at the point a (so that $L = g(a)$), then $\lim_{x\to a} f\big(g(x)\big) = f\big(g(a)\big)$, that is, $f \circ g(x) = f\big(g(x)\big)$ is continuous at $x = a$.

(c) The functions $f(x) = C$ (constant) and $g(x) = x$ are continuous on the whole real line.

(d) For any rational number r the function $f(x) = x^r$ is continuous at every real number where it is defined.

PROOF Part (a) is just a restatement of various rules for combining limits; for example,

$$\lim_{x \to a} f(x)g(x) = \left(\lim_{x \to a} f(x)\right)\left(\lim_{x \to a} g(x)\right) = f(a)g(a).$$

Part (b) can be proved as follows. Let $\epsilon > 0$ be given. Since f is continuous at L, there exists $k > 0$ such that $|f(g(x)) - f(L)| < \epsilon$ whenever $|g(x) - L| < k$. Since $\lim_{x \to a} g(x) = L$, there exists $\delta > 0$ such that if $0 < |x - a| < \delta$, then $|g(x) - L| < k$. Hence, if $0 < |x - a| < \delta$, then $|f(g(x)) - f(L)| < \epsilon$, and $\lim_{x \to a} f(g(x)) = f(L)$.

The proofs of (c) and (d) are left to the student in Exercises 3–9 at the end of this appendix.

Completeness and Sequential Limits

DEFINITION 3

A real number u is said to be an **upper bound** for a nonempty set S of real numbers if $x \leq u$ for every x in S.
The number u^* is called the **least upper bound** or **supremum** of S if u^* is an upper bound for S and $u^* \leq u$ for every upper bound u of S. The supremum of S is usually denoted $\sup(S)$.
Similarly, ℓ is a **lower bound** for S if $\ell \leq x$ for every x in S. The number ℓ^* is the **greatest lower bound** or **infimum** of S if ℓ^* is a lower bound for S and $\ell \leq \ell^*$ for every lower bound ℓ of S. The infimum of S is denoted $\inf(S)$.

EXAMPLE 1 Set $S_1 = [2, 3]$ and $S_2 = (2, \infty)$. Any number $u \geq 3$ is an upper bound for S_1. S_2 has no upper bound; we say that it is not bounded above. The least upper bound of S_1 is $\sup(S_1) = 3$. Any real number $\ell \leq 2$ is a lower bound for both S_1 and S_2. The greatest lower bound of each set is 2: $\inf(S_1) = \inf(S_2) = 2$. Note that the least upper bound and greatest lower bound of a set may or may not belong to that set.

We now recall the completeness axiom for the real number system, which we discussed briefly in Section P.1.

The completeness axiom for the real numbers

A nonempty set of real numbers that has an upper bound must have a least upper bound.
Equivalently, a nonempty set of real numbers having a lower bound must have a greatest lower bound.

We stress that this is an *axiom* to be assumed without proof. It cannot be deduced from the more elementary algebraic and order properties of the real numbers. These other properties are shared by the rational numbers, a set that is not complete. The completeness axiom is essential for the proof of the most important results about continuous functions, in particular, for the Max-Min Theorem and the Intermediate-Value Theorem. Before attempting these proofs, however, we must develop a little more machinery.

In Section 9.1 we stated a version of the completeness axiom that pertains to *sequences* of real numbers; specifically, that an increasing sequence that is bounded above converges to a limit. We begin by verifying that this follows from the version stated above. (Both statements are, in fact, equivalent.) As noted in Section 9.1, the sequence

$$\{x_n\} = \{x_1, x_2, x_3, \ldots\}$$

is a function on the positive integers, that is, $x_n = x(n)$. We say that the sequence converges to the limit L, and we write $\lim x_n = L$, if the corresponding function $x(t)$ satisfies $\lim_{t\to\infty} x(t) = L$ as defined above. More formally,

DEFINITION 4

Limit of a sequence

We say that $\lim x_n = L$ if for every positive number ϵ there exists a positive number $N = N(\epsilon)$ such that $|x_n - L| < \epsilon$ holds whenever $n \ge N$.

THEOREM 2

If $\{x_n\}$ is an increasing sequence that is bounded above, that is,

$$x_{n+1} \ge x_n \qquad \text{and} \qquad x_n \le K \qquad \text{for } n = 1, 2, 3, \ldots,$$

then $\lim x_n = L$ exists. (Equivalently, if $\{x_n\}$ is decreasing and bounded below, then $\lim x_n$ exists.)

PROOF Let $\{x_n\}$ be increasing and bounded above. The set S of real numbers x_n has an upper bound, K, and so has a least upper bound, say $L = \sup(S)$. Thus,$x_n \le L$ for every n, and if $\epsilon > 0$, then there exists a positive integer N such that $x_N > L - \epsilon$. (Otherwise, $L - \epsilon$ would be an upper bound for S that is lower than the least upper bound.) If $n \ge N$, then we have $L - \epsilon < x_N \le x_n \le L$, so $|x_n - L| < \epsilon$. Thus, $\lim x_n = L$. The proof for a decreasing sequence that is bounded below is similar.

THEOREM

If $a \le x_n \le b$ for each n, and if $\lim x_n = L$, then $a \le L \le b$.

PROOF Suppose that $L > b$. Let $\epsilon = L - b$. Since $\lim x_n = L$, there exists n such that $|x_n - L| < \epsilon$. Thus, $x_n > L - \epsilon = L - (L - b) = b$, which is a contradiction, since we are given that $x_n \le b$. Thus, $L \le b$. A similar argument shows that $L \ge a$.

THEOREM 4

If f is continuous on $[a, b]$, if $a \le x_n \le b$ for each n, and if $\lim x_n = L$, then $\lim f(x_n) = f(L)$.

The proof is similar to that of Theorem 1(b), and is left as Exercise 15 at the end of this appendix.

Continuous Functions on a Closed, Finite Interval

We are now in a position to prove the main results about continuous functions on closed, finite intervals.

THEOREM

The Boundedness Theorem

If f is continuous on $[a, b]$, then f is bounded there; that is, there exists a constant K such that $|f(x)| \le K$ if $a \le x \le b$.

PROOF We show that f is bounded above; a similar proof shows that f is bounded below. For each positive integer n let S_n be the set of points x in $[a, b]$ such that $f(x) > n$:

$$S_n = \{x : a \le x \le b \quad \text{and} \quad f(x) > n\}.$$

We would like to show that S_n is empty for some n. It would then follow that $f(x) \le n$ for all x in $[a, b]$; that is, n would be an upper bound for f on $[a, b]$.

Suppose, to the contrary, that S_n is nonempty for every n. We will show that this leads to a contradiction. Since S_n is bounded below (a is a lower bound), by completeness S_n has a greatest lower bound; call it x_n. (See Figure III.2.) Evidently, $a \le x_n$. Since $f(x) > n$ at some point of $[a, b]$ and f is continuous at that point, $f(x) > n$ on some interval contained in $[a, b]$. Hence, $x_n < b$. It follows that $f(x_n) \ge n$. (If $f(x_n) < n$, then by continuity $f(x) < n$ for some distance to the right of x_n, and x_n could not be the greatest lower bound of S_n.)

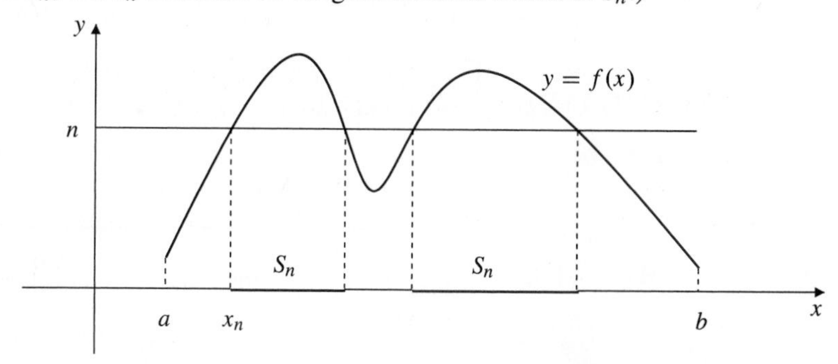

Figure III.2 The set S_n

For each n we have $S_{n+1} \subset S_n$. Therefore, $x_{n+1} \ge x_n$ and $\{x_n\}$ is an increasing sequence. Being bounded above (b is an upper bound) this sequence converges, by Theorem 2. Let $\lim x_n = L$. By Theorem 3, $a \le L \le b$. Since f is continuous at L, $\lim f(x_n) = f(L)$ exists by Theorem 4. But since $f(x_n) \ge n$, $\lim f(x_n)$ cannot exist. This contradiction completes the proof.

THEOREM

The Max-Min Theorem

If f is continuous on $[a, b]$, then there are points v and u in $[a, b]$ such that for any x in $[a, b]$ we have

$$f(v) \le f(x) \le f(u);$$

that is, f assumes maximum and minimum values on $[a, b]$.

PROOF By Theorem 5 we know that the set $S = \{f(x) : a \le x \le b\}$ has an upper bound and, therefore, by the completeness axiom, a least upper bound. Call this least upper bound M. Suppose that there exists no point u in $[a, b]$ such that $f(u) = M$. Then by Theorem 1(a), $1/(M - f(x))$ is continuous on $[a, b]$. By Theorem 5, there exists a constant K such that $1/(M - f(x)) \le K$ for all x in $[a, b]$. Thus $f(x) \le M - 1/K$, which contradicts the fact that M is the *least* upper bound for the values of f. Hence, there must exist some point u in $[a, b]$ such that $f(u) = M$. Since M is an upper bound for the values of f on $[a, b]$, we have $f(x) \le f(u) = M$ for all x in $[a, b]$.

The proof that there must exist a point v in $[a, b]$ such that $f(x) \ge f(v)$ for all x in $[a, b]$ is similar.

THEOREM

The Intermediate-Value Theorem

If f is continuous on $[a, b]$ and s is a real number lying between the numbers $f(a)$ and $f(b)$, then there exists a point c in $[a, b]$ such that $f(c) = s$.

PROOF To be specific, we assume that $f(a) < s < f(b)$. (The proof for the case $f(a) > s > f(b)$ is similar.) Let $S = \{x : a \le x \le b \text{ and } f(x) \le s\}$. S is nonempty (a belongs to S) and bounded above (b is an upper bound), so by completeness S has a least upper bound; call it c.

Suppose that $f(c) > s$. Then $c \ne a$ and, by continuity, $f(x) > s$ on some interval $(c - \delta, c]$ where $\delta > 0$. But this says $c - \delta$ is an upper bound for S lower than the least upper bound, which is impossible. Thus $f(c) \le s$.

Suppose $f(c) < s$. Then $c \neq b$ and, by continuity, $f(x) < s$ on some interval of the form $[c, c+\delta)$ for some $\delta > 0$. But this says that $[c, c+\delta) \subset S$, which contradicts the fact that c is an upper bound for S. Hence we cannot have $f(c) < s$. Therefore, $f(c) = s$. ■

For more discussion of these theorems and some applications, see Section 1.4.

EXERCISES: APPENDIX III

1. Let $a < b < c$ and suppose that $f(x) \leq g(x)$ for $a \leq x \leq c$. If $\lim_{x \to b} f(x) = L$ and $\lim_{x \to b} g(x) = M$, prove that $L \leq M$. *Hint:* Assume that $L > M$ and deduce that $f(x) > g(x)$ for all x sufficiently near b. This contradicts the condition that $f(x) \leq g(x)$ for $a \leq x \leq b$.

2. If $f(x) \leq K$ on the intervals $[a, b)$ and $(b, c]$, and if $\lim_{x \to b} f(x) = L$, prove that $L \leq K$.

3. Use the formal definition of limit to prove that $\lim_{x \to 0+} x^r = 0$ for any positive, rational number r.

Prove the assertions in Exercises 4–9.

4. $f(x) = C$ (constant) and $g(x) = x$ are both continuous on the whole real line.

5. Every polynomial is continuous on the whole real line.

6. A rational function (quotient of polynomials) is continuous everywhere except where the denominator is 0.

7. If n is a positive integer and $a > 0$, then $f(x) = x^{1/n}$ is continuous at $x = a$.

8. If $r = m/n$ is a rational number, then $g(x) = x^r$ is continuous at every point $a > 0$.

9. If $r = m/n$, where m and n are integers and n is odd, show that $g(x) = x^r$ is continuous at every point $a < 0$. If $r \geq 0$, show that g is continuous at 0 also.

10. Prove that $f(x) = |x|$ is continuous on the real line.

Use the definitions from Chapter 3 for the functions in Exercises 11–14 to show that these functions are continuous on their respective domains.

11. $\sin x$

12. $\cos x$

13. $\ln x$

14. e^x

15. Prove Theorem 4.

16. Suppose that every function that is continuous and bounded on $[a, b]$ must assume a maximum value and a minimum value on that interval. Without using Theorem 5, prove that every function f that is continuous on $[a, b]$ must be bounded on that interval. *Hint:* Show that $g(t) = t/(1+|t|)$ is continuous and increasing on the real line. Then consider $g\big(f(x)\big)$.

APPENDIX IV

The Riemann Integral

> It seems to be expected of every pilgrim up the slopes of the mathematical Parnassus, that he will at some point or other of his journey sit down and invent a definite integral or two towards the increase of the common stock.
>
> J. J. Sylvester 1814–1897

In Section 5.3 we defined the definite integral $\int_a^b f(x)\,dx$ of a function f that is continuous on the finite, closed interval $[a, b]$. The integral was defined as a kind of "limit" of Riemann sums formed by partitioning the interval $[a, b]$ into small subintervals. In this appendix we will reformulate the definition of the integral so that it can be used for functions that are not necessarily continuous; in the following discussion we assume only that f is **bounded** on $[a, b]$. Later we will prove Theorem 2 of Section 5.3, which asserts that any continuous function is integrable.

Recall that a **partition** P of $[a, b]$ is a finite, ordered set of points $P = \{x_0, x_1, x_2, \ldots, x_n\}$, where $a = x_0 < x_1 < x_2 < \cdots < x_{n-1} < x_n = b$. Such a partition subdivides $[a, b]$ into n subintervals $[x_0, x_1]$, $[x_1, x_2]$, $\ldots$, $[x_{n-1}, x_n]$, where $n = n(P)$ depends on the partition. The length of the jth subinterval $[x_{j-1}, x_j]$ is $\Delta x_j = x_j - x_{j-1}$.

Suppose that the function f is bounded on $[a, b]$. Given any partition P, the n sets $S_j = \{f(x) : x_{j-1} \le x \le x_j\}$ have least upper bounds M_j and greatest lower bounds m_j, $(1 \le j \le n)$, so that

$$m_j \le f(x) \le M_j \qquad \text{on} \qquad [x_{j-1}, x_j].$$

We define upper and lower Riemann sums for f corresponding to the partition P to be

$$U(f, P) = \sum_{j=1}^{n(P)} M_j \Delta x_j \qquad \text{and}$$

$$L(f, P) = \sum_{j=1}^{n(P)} m_j \Delta x_j.$$

(See Figure IV.1.) Note that if f is continuous on $[a, b]$, then m_j and M_j are, in fact, the minimum and maximum values of f over $[x_{j-1}, x_j]$ (by Theorem 6 of Appendix III); that is, $m_j = f(l_j)$ and $M_j = f(u_j)$, where $f(l_j) \le f(x) \le f(u_j)$ for $x_{j-1} \le x \le x_j$.

If P is any partition of $[a, b]$ and we create a new partition P^* by adding new subdivision points to those of P, thus subdividing the subintervals of P into smaller ones, then we call P^* a **refinement** of P.

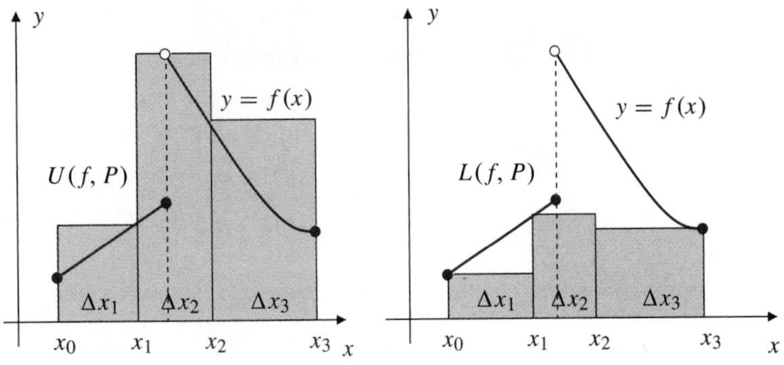

Figure IV.1 Upper and lower sums corresponding to the partition $P = \{x_0, x_1, x_2, x_3\}$

THEOREM 1

If P^* is a refinement of P, then $L(f, P^*) \geq L(f, P)$ and $U(f, P^*) \leq U(f, P)$.

PROOF If S and T are sets of real numbers, and $S \subset T$, then any lower bound (or upper bound) of T is also a lower bound (or upper bound) of S. Hence, the greatest lower bound of S is at least as large as that of T, and the least upper bound of S is no greater than that of T.

Let P be a given partition of $[a, b]$ and form a new partition P' by adding one subdivision point to those of P, say, the point k dividing the jth subinterval $[x_{j-1}, x_j]$ of P into two subintervals $[x_{j-1}, k]$ and $[k, x_j]$. (See Figure IV.2.) Let m_j, m_j', and m_j'' be the greatest lower bounds of the sets of values of $f(x)$ on the intervals $[x_{j-1}, x_j]$, $[x_{j-1}, k]$, and $[k, x_j]$, respectively. Then $m_j \leq m_j'$ and $m_j \leq m_j''$. Thus, $m_j(x_j - x_{j-1}) \leq m_j'(k - x_{j-1}) + m_j''(x_j - k)$, so $L(f, P) \leq L(f, P')$.

If P^* is a refinement of P, it can be obtained by adding one point at a time to those of P and thus $L(f, P) \leq L(f, P^*)$. We can prove that $U(f, P) \geq U(f, P^*)$ in a similar manner.

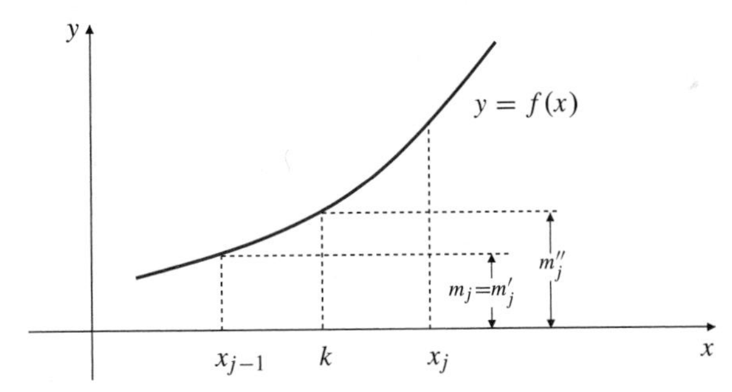

Figure IV.2 Adding one point to a partition

THEOREM 2

If P and P' are any two partitions of $[a, b]$, then $L(f, P) \leq U(f, P')$.

PROOF Combine the subdivision points of P and P' to form a new partition P^*, which is a refinement of both P and P'. Then by Theorem 1,

$$L(f, P) \leq L(f, P^*) \leq U(f, P^*) \leq U(f, P').$$

No lower sum can exceed any upper sum.

Theorem 2 shows that the set of values of $L(f, P)$ for fixed f and various partitions P of $[a, b]$ is a bounded set; any upper sum is an upper bound for this set. By completeness, the set has a least upper bound, which we shall denote I_*. Thus, $L(f, P) \leq I_*$ for any partition P. Similarly, there exists a greatest lower bound I^* for the set of values of $U(f, P)$ corresponding to different partitions P. It follows that $I_* \leq I^*$. (See Exercise 4 at the end of this appendix.)

DEFINITION 1

The Riemann integral

If f is bounded on $[a, b]$ and $I_* = I^*$, then we say that f is **Riemann integrable**, or simply **integrable** on $[a, b]$, and denote by

$$\int_a^b f(x)\,dx = I_* = I^*$$

the **(Riemann) integral** of f on $[a, b]$.

The following theorem provides a convenient test for determining whether a given bounded function is integrable:

THEOREM 3

The bounded function f is integrable on $[a, b]$ if and only if for every positive number ϵ there exists a partition P of $[a, b]$ such that $U(f, P) - L(f, P) < \epsilon$.

PROOF Suppose that for every $\epsilon > 0$ there exists a partition P of $[a, b]$ such that $U(f, P) - L(f, P) < \epsilon$, then

$$I^* \le U(f, P) < L(f, P) + \epsilon \le I_* + \epsilon.$$

Since $I^* < I_* + \epsilon$ must hold for every $\epsilon > 0$, it follows that $I^* \le I_*$. Since we already know that $I^* \ge I_*$, we have $I^* = I_*$ and f is integrable on $[a, b]$.

Conversely, if $I^* = I_*$ and $\epsilon > 0$ are given, we can find a partition P' such that $L(f, P') > I_* - \epsilon/2$, and another partition P'' such that $U(f, P'') < I^* + \epsilon/2$. If P is a common refinement of P' and P'', then by Theorem 1 we have that $U(f, P) - L(f, P) \le U(f, P'') - L(f, P') < (\epsilon/2) + (\epsilon/2) = \epsilon$, as required.

EXAMPLE 1 Let $f(x) = \begin{cases} 0 & \text{if } 0 \le x < 1 \text{ or } 1 < x \le 2 \\ 1 & \text{if } x = 1. \end{cases}$

Show that f is integrable on $[0, 2]$ and find $\int_0^2 f(x)\,dx$.

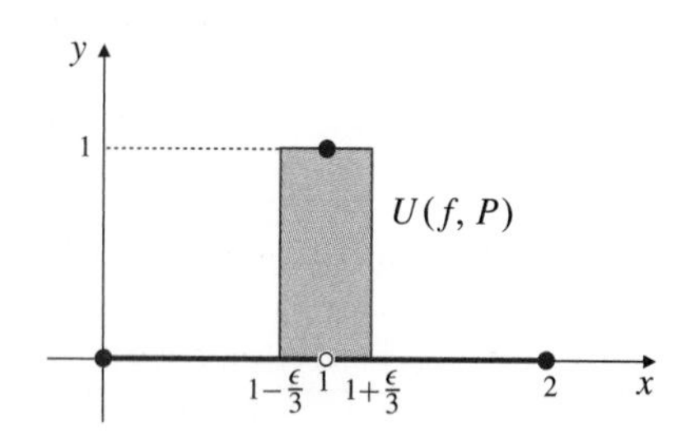

Figure IV.3 Constructing a small upper sum for a nonnegative function that is positive at only one point

Solution Let $\epsilon > 0$ be given. Let $P = \{0,\ 1 - \epsilon/3,\ 1 + \epsilon/3,\ 2\}$. Then $L(f, P) = 0$ since $f(x) = 0$ at points of each of these subintervals into which P subdivides $[0, 2]$. (See Figure IV.3.) Since $f(1) = 1$, we have

$$U(f, P) = 0\left(1 - \frac{\epsilon}{3}\right) + 1\left(\frac{2\epsilon}{3}\right) + 0\left(2 - \left(1 + \frac{\epsilon}{3}\right)\right) = \frac{2\epsilon}{3}.$$

Hence, $U(f, P) - L(f, P) < \epsilon$ and f is integrable on $[0, 2]$. Since $L(f, P) = 0$ for every partition, $\int_0^2 f(x)\,dx = I_* = 0$.

EXAMPLE 2 Let $f(x)$ be defined on $[0, 1]$ by

$$f(x) = \begin{cases} 1 & \text{if } x \text{ is rational} \\ 0 & \text{if } x \text{ is irrational.} \end{cases}$$

Show that f is not integrable on $[0, 1]$.

PROOF Every subinterval of $[0, 1]$ having positive length contains both rational and irrational numbers. Hence, for any partition P of $[0, 1]$ we have $L(f, P) = 0$ and $U(f, P) = 1$. Thus, $I_* = 0$ and $I^* = 1$, so f is not integrable on $[0, 1]$.

Uniform Continuity

When we assert that a function f is continuous on the interval I, we imply that for every x in that interval and every $\epsilon > 0$, we can find a positive number δ (depending on *both* x and ϵ) such that $|f(y) - f(x)| < \epsilon$ whenever $|y - x| < \delta$ and y lies in I. If it is possible to find a such a number δ *independent of* x *and so depending only on* ϵ such that $|f(y) - f(x)| < \epsilon$ holds whenever x and y belong to I and satisfy $|y - x| < \delta$, we say that that f is **uniformly continuous** on the interval I. Such is the case for a closed finite interval.

THEOREM 4

If f is continuous on the closed, finite interval $[a, b]$, then f is uniformly continuous on that interval.

PROOF Let $\epsilon > 0$ be given. Define numbers x_n in $[a, b]$ and subsets S_n of $[a, b]$ as follows:

$$x_1 = a$$
$$S_1 = \left\{x : x_1 < x \le b \text{ and } |f(x) - f(x_1)| \ge \frac{\epsilon}{3}\right\}.$$

If S_1 is empty, stop; otherwise, let

$$x_2 = \text{ the greatest lower bound of } S_1$$
$$S_2 = \left\{x : x_2 < x \le b \text{ and } |f(x) - f(x_2)| \ge \frac{\epsilon}{3}\right\}.$$

If S_2 is empty, stop; otherwise, proceed to define x_3 and S_3 analogously. We proceed in this way as long as we can; if x_n and S_n have been defined and S_n is not empty, we define

$$x_{n+1} = \text{ the greatest lower bound of } S_n$$
$$S_{n+1} = \left\{x : x_{n+1} < x \le b \text{ and } |f(x) - f(x_{n+1})| \ge \frac{\epsilon}{3}\right\}.$$

At any stage where S_n is not empty, the continuity of f at x_n assures us that $x_{n+1} > x_n$ and $|f(x_{n+1}) - f(x_n)| = \epsilon/3$.

We must consider two possibilities for the above procedure: either S_n is empty for some n, or S_n is nonempty for every n.

Suppose S_n is nonempty for every n. Then we have constructed an infinite, increasing sequence $\{x_n\}$ in $[a, b]$ that, being bounded above (by b), must have a limit by completeness (Theorem 2 of Appendix II). Let $\lim x_n = x^*$. We have $a \le x^* \le b$. Since f is continuous at x^*, there exists $\delta > 0$ such that $|f(x) - f(x^*)| < \epsilon/8$ whenever $|x - x^*| < \delta$ and x lies in $[a, b]$. Since $\lim x_n = x^*$, there exists a positive integer N such that $|x_n - x^*| < \delta$ whenever $n \ge N$. For such n we have

$$\begin{aligned}\frac{\epsilon}{3} = |f(x_{n+1}) - f(x_n)| &= |f(x_{n+1}) - f(x^*) + f(x^*) - f(x_n)| \\ &\le |f(x_{n+1}) - f(x^*)| + |f(x_n) - f(x^*)| \\ &< \frac{\epsilon}{8} + \frac{\epsilon}{8} = \frac{\epsilon}{4},\end{aligned}$$

which is clearly impossible. Thus, S_n must, in fact, be empty for some n.

Suppose that S_N is empty. Thus, S_n is nonempty for $n < N$, and the procedure for defining x_n stops with x_N. Since S_{N-1} is not empty, $x_N < b$. In this case define $x_{N+1} = b$ and let

$$\delta = \min\{x_2 - x_1, x_3 - x_2, \ldots, x_{N+1} - x_N\}.$$

The minimum of a finite set of positive numbers is a positive number, so $\delta > 0$. If x lies in $[a, b]$, then x lies in one of the intervals $[x_1, x_2]$, $[x_2, x_3]$, ..., $[x_N, x_{N+1}]$. Suppose x lies in $[x_k, x_{k+1}]$. If y is in $[a, b]$ and $|y - x| < \delta$, then y lies in either the same subinterval as x or in an adjacent one; that is, y lies in $[x_j, x_{j+1}]$, where $j = k - 1, k$, or $k + 1$. Thus,

$$\begin{aligned} |f(y) - f(x)| &= |f(y) - f(x_j) + f(x_j) - f(x_k) + f(x_k) - f(x)| \\ &\le |f(y) - f(x_j)| + |f(x_j) - f(x_k)| + |f(x_k) - f(x)| \\ &< \frac{\epsilon}{3} + \frac{\epsilon}{3} + \frac{\epsilon}{3} = \epsilon, \end{aligned}$$

which was to be proved.

We are now in a position to prove that a continuous function is integrable.

THEOREM 5

If f is continuous on $[a, b]$, then f is integrable on $[a, b]$.

PROOF By Theorem 4, f is uniformly continuous on $[a, b]$. Let $\epsilon > 0$ be given. Let $\delta > 0$ be such that $|f(x) - f(y)| < \epsilon/(b-a)$ whenever $|x - y| < \delta$ and x and y belong to $[a, b]$. Choose a partition $P = \{x_0, x_1, \ldots, x_n\}$ of $[a, b]$ for which each subinterval $[x_{j-1}, x_j]$ has length $\Delta x_j < \delta$. Then the greatest lower bound, m_j, and the least upper bound, M_j, of the set of values of $f(x)$ on $[x_{j-1}, x_j]$ satisfy $M_j - m_j < \epsilon/(b - a)$. Accordingly,

$$U(f, P) - L(f, P) < \frac{\epsilon}{b - a} \sum_{j=1}^{n(P)} \Delta x_j = \frac{\epsilon}{b - a}(b - a) = \epsilon.$$

Thus f is integrable on $[a, b]$, as asserted.

EXERCISES: APPENDIX IV

1. Let $f(x) = \begin{cases} 1 & \text{if } 0 \le x \le 1 \\ 0 & \text{if } 1 < x \le 2 \end{cases}$. Prove that f is integrable on $[0, 2]$ and find the value of $\int_0^2 f(x)\,dx$.

2. Let $f(x) = \begin{cases} 1 & \text{if } x = 1/n, \quad n = 1, 2, 3, \ldots \\ 0 & \text{for all other values of } x. \end{cases}$
   Show that f is integrable over $[0, 1]$ and find the value of the integral $\int_0^1 f(x)\,dx$.

3. Let $f(x) = 1/n$ if $x = m/n$, where m, n are integers having no common factors, and let $f(x) = 0$ if x is an irrational number. Thus, $f(1/2) = 1/2$, $f(1/3) = f(2/3) = 1/3$, $f(1/4) = f(3/4) = 1/4$, etc. Show that f is integrable on $[0, 1]$ and find $\int_0^1 f(x)\,dx$. *Hint:* Show that for any $\epsilon > 0$, only finitely many points of the graph of f over $[0, 1]$ lie above the line $y = \epsilon$.

4. Prove that I_* and I^* defined in the paragraph following Theorem 2 satisfy $I_* \le I^*$ as claimed there.

Properties of the Riemann Integral

In Exercises 5–8, you are asked to provide proofs of properties of the Riemann integral that were stated for the definite integral of a continuous function in Theorem 3 of Section 5.4.

5. Prove that if f and g are bounded and integrable on $[a, b]$, and A and B are constants, then $Af + Bg$ is integrable on $[a, b]$ and
$$\int_a^b \big(Af(x)+Bg(x)\big)\,dx = A\int_a^b f(x)\,dx + B\int_a^b g(x)\,dx.$$

6. Prove that if f is bounded and integrable on an interval containing a, b, and c, then
$$\int_a^b f(x)\,dx + \int_b^c f(x)\,dx = \int_a^c f(x)\,dx.$$

7. Prove that if f and g are bounded and integrable on the interval $[a, b]$ (where $a < b$) and $f(x) \le g(x)$ for $a \le x \le b$, then
$$\int_a^b f(x)\,dx \le \int_a^b g(x)\,dx.$$
   Also, if $|f|$ is bounded and integrable on $[a, b]$,
$$\left|\int_a^b f(x)\,dx\right| \le \int_a^b |f(x)|\,dx$$

8. If f is bounded and integrable on $[-a, a]$, where $a > 0$, then

(a) if f is an odd function, then $\int_{-a}^{a} f(x)\,dx = 0$, or

(b) if f is an even function, then

$$\int_{-a}^{a} f(x)\,dx = 2\int_{0}^{a} f(x)\,dx.$$

9. Use the definition of uniform continuity given in the paragraph preceding Theorem 4 to prove that $f(x) = \sqrt{x}$ is uniformly continuous on $[0, 1]$. Do not use Theorem 4 itself.

10. Show directly from the definition of uniform continuity (without using Theorem 5 of Appendix III) that a function f uniformly continuous on a closed, finite interval is necessarily bounded there.

11. If f is bounded and integrable on $[a, b]$, prove that $F(x) = \int_a^x f(t)\,dt$ is uniformly continuous on $[a, b]$. (If f were continuous, we would have a stronger result; F would be differentiable on (a, b) and $F'(x) = f(x)$ (which is the Fundamental Theorem of Calculus).)

APPENDIX V

Doing Calculus with Maple

> "I think, therefore I am."
>
> René Descartes 1596–1650
> *Discourse on Method*

> "AI [Artificial Intelligences] think, therefore I am."
>
> David Braue
> *APC Magazine, November 2003*

Computer algebra systems like Maple and Mathematica are capable of doing most of the tedious calculations involved in doing calculus, especially the very intensive calculations required by many applied problems. (They cannot, of course, do the thinking for you; you must still fully understand what you are doing and what are the limitations of such programs.) Throughout this text we have inserted material illustrating how to use **Maple** to do common calculus-oriented calculations. These insertions range in length from single paragraphs and remarks to entire sections. To help you locate the Maple material appropriate for specific topics, we include below a list pointing to the text sections containing Maple examples and the pages where they occur.

Note, however, that this material assumes you are familiar with the basics of starting a Maple session, preferably with a graphical user interface which typically displays the prompt > when it is waiting for your input. In this book the input is shown in colour. It normally concludes with a semicolon (;) followed by pressing the `<enter>` key, which we omit from our examples. The output is typically printed by Maple centred in the window; we show it in black. For instance,

```
> factor(x^2-x-2);
```

$$(x + 1)(x - 2)$$

Output can be supressed by using a colon (:) instead of a semicolon at the end of the input.

The authors used Maple 10 for preparing the Maple examples in this edition. They should work equally well in later editions. These examples are by no means complete or exhaustive. For a more complete treatment of Maple as a tool for doing calculus, the authors highly recommend the excellent Maple lab manual *Calculus: The Maple Way*, written by Professor Robert Israel of the University of British Columbia. Like this book, it is published by Pearson Canada under the Addison-Wesley logo.

List of Maple Examples and Discussion

Answers to Odd-Numbered Exercises

Chapter P

Preliminaries

Section P.1 (page 10)

1. $0.\overline{2}$

3. $4/33$

5. $1/7 = 0.\overline{142857}$, $2/7 = 0.\overline{285714}$, $3/7 = 0.\overline{428571}$, $4/7 = 0.\overline{571428}$, $5/7 = 0.\overline{714285}$, $6/7 = 0.\overline{857142}$

7. $[0, 5]$

9. $(-\infty, -6) \cup (-5, \infty)$

11. $(-2, \infty)$

13. $(-\infty, -2)$

15. $(-\infty, 5/4]$

17. $(0, \infty)$

19. $(-\infty, 5/3) \cup (2, \infty)$

21. $[0, 2]$

23. $(-2, 0) \cup (2, \infty)$

25. $[-2, 0) \cup [4, \infty)$

27. $x = -3,\ 3$

29. $t = -1/2,\ -9/2$

31. $s = -1/3,\ 17/3$

33. $(-2, 2)$

35. $[-1, 3]$

37. $\left(\dfrac{5}{3}, 3\right)$

39. $[0, 4]$

41. $x > 1$

43. true if $a \geq 0$, false if $a < 0$

Section P.2 (page 16)

1. $\Delta x = 4$, $\Delta y = -3$, dist $= 5$

3. $\Delta x = -4$, $\Delta y = -4$, dist $= 4\sqrt{2}$

5. $(2, -4)$

7. circle, centre $(0, 0)$, radius 1

9. points inside and on circle, centre $(0, 0)$, radius 1

11. points on and above the parabola $y = x^2$

13. (a) $x = -2$, (b) $y = 5/3$

15. $y = x + 2$

17. $y = 2x + b$

19. above

21. $y = 3x/2$

23. $y = (7 - x)/3$

25. $y = \sqrt{2} - 2x$

27. 4, 3,

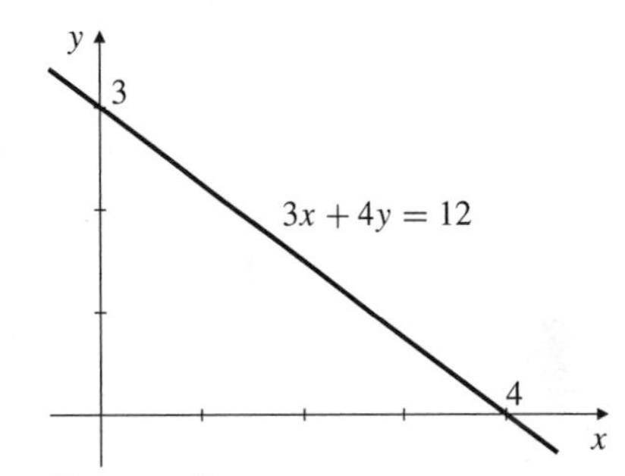

29. $\sqrt{2}$, $-2/\sqrt{3}$

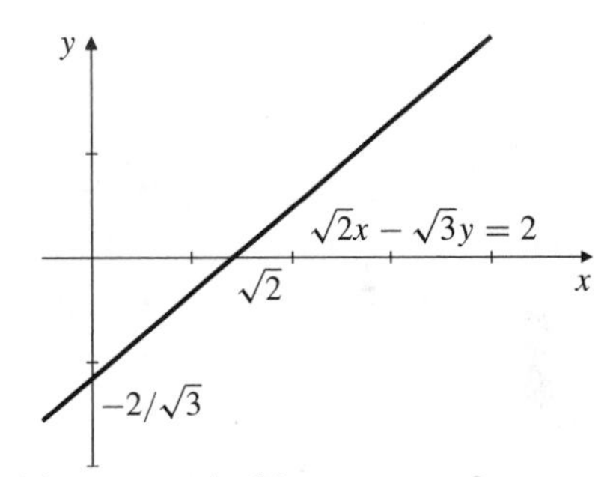

31. (a) $y = x - 1$, (b) $y = -x + 3$

33. $(2, -3)$

37. 5

39. \$23,000

43. $(-2, -2)$

45. $\left(\frac{1}{3}(x_1 + 2x_2), \frac{1}{3}(y_1 + 2y_2)\right)$

47. circle, centre $(2, 0)$, radius 4

49. perp. if $k = -8$, parallel if $k = 1/2$

Section P.3 (page 22)

1. $x^2 + y^2 = 16$

3. $x^2 + y^2 + 4x = 5$

5. $(1, 0)$, 2

7. $(1, -2)$, 3

9. exterior of circle, centre $(0, 0)$, radius 1

11. closed disk, centre $(-1, 0)$, radius 2

13. washer shaped region between the circles of radius 1 and 2 centred at $(0, 0)$

15. first quadrant region lying inside the two circles of radius 1 having centres at $(1, 0)$ and $(0, 1)$

17. $x^2 + y^2 + 2x - 4y < 1$

19. $x^2 + y^2 < 2$, $x \geq 1$

21. $x^2 = 16y$

23. $y^2 = 8x$

25. $(0, 1/2)$, $y = -1/2$

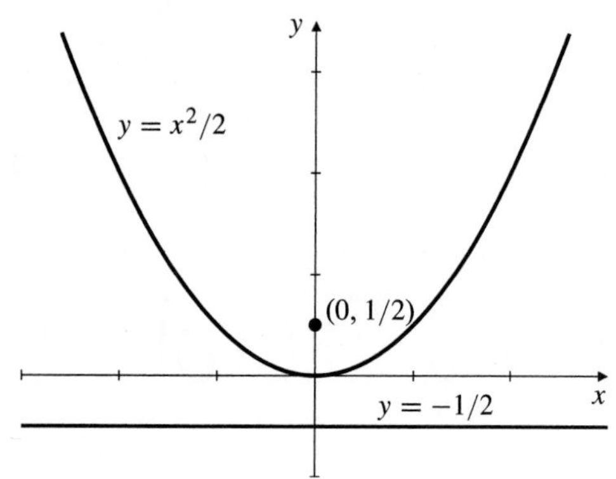

27. $(-1, 0)$, $x = 1$

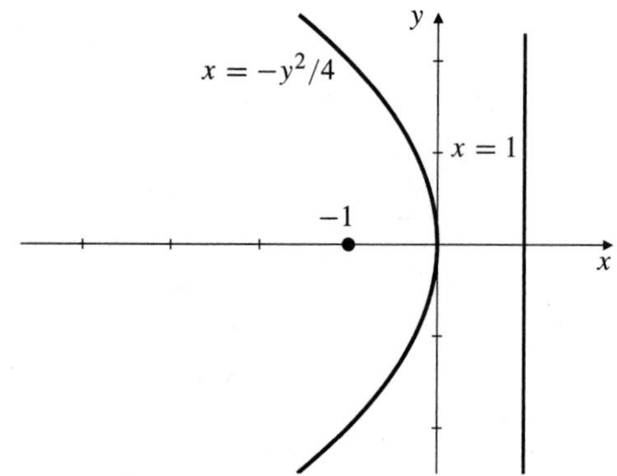

29. (a) $y = x^2 - 3$, (b) $y = (x-4)^2$, (c) $y = (x-3)^2 + 3$, (d) $y = (x-4)^2 - 2$

31. $y = \sqrt{(x/3) + 1}$

33. $y = \sqrt{(3x/2) + 1}$

35. $y = -(x+1)^2$

37. $y = (x-2)^2 - 2$

39. $(2, 7)$, $(1, 4)$

41. $(4, -3)$, $(-4, 3)$

43. ellipse, centre $(0, 0)$, semiaxes $2, 1$

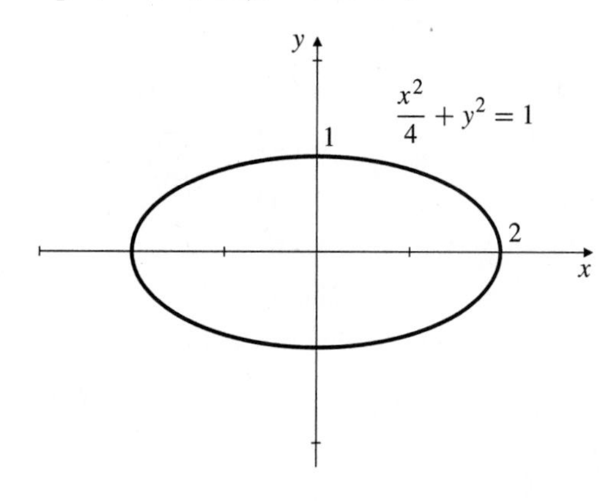

45. ellipse, centre $(3, -2)$, semiaxes $3, 2$

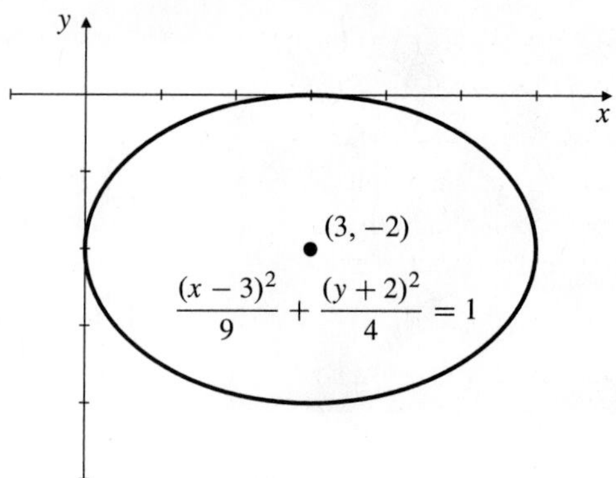

47. hyperbola, centre $(0, 0)$, asymptotes $x = \pm 2y$, vertices $(\pm 2, 0)$

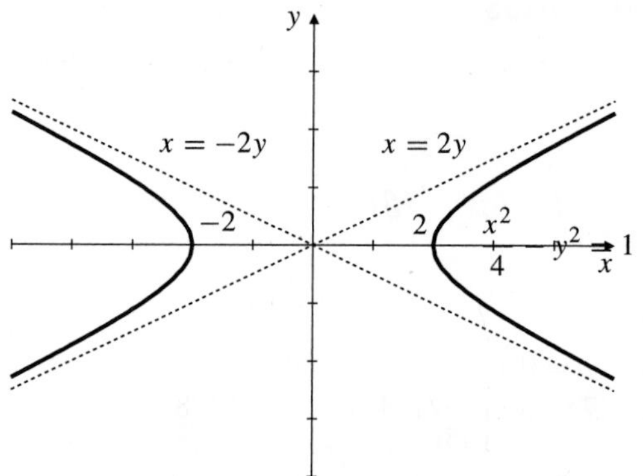

49. rectangular hyperbola, asymptotes $x = 0$ and $y = 0$, vertices $(2, -2)$ and $(-2, 2)$

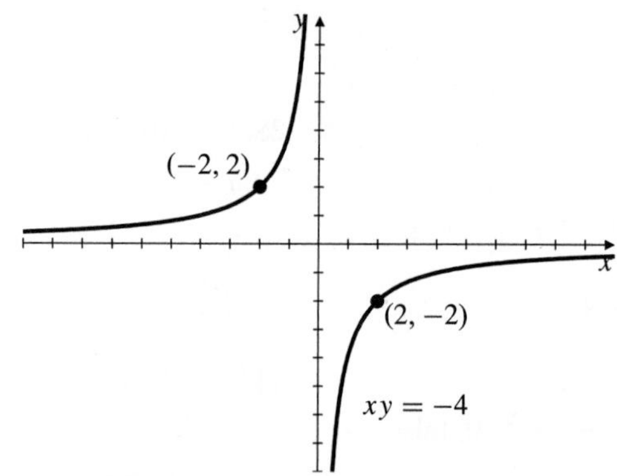

51. (a) reflecting the graph in the y-axis, (b) reflecting the graph in the x-axis

53.

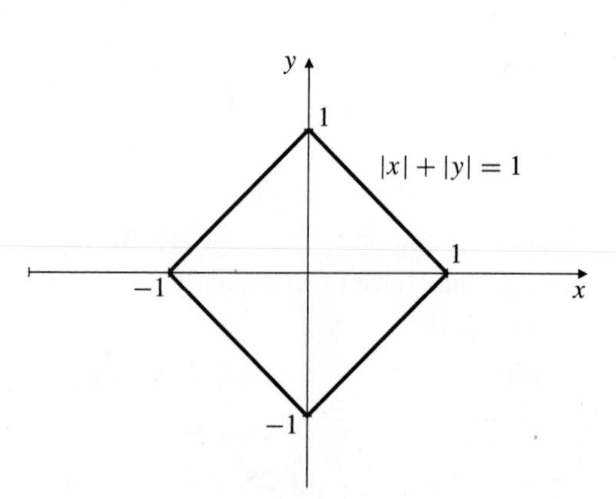

Section P.4 (page 32)

1. $\mathcal{D}(f) = \mathbb{R}$, $\mathcal{R}(f) = [1, \infty)$

3. $\mathcal{D}(G) = (-\infty, 4]$, $\mathcal{R}(g) = [0, \infty)$

5. $\mathcal{D}(h) = (-\infty, 2)$, $\mathcal{R}(h) = (-\infty, \infty)$

7. Only (ii) is the graph of a function. Vertical lines can meet the others more than once.

11. even, sym. about y-axis **13.** odd, sym. about $(0, 0)$

15. sym. about $(2, 0)$ **17.** sym. about $x = 3$

19. even, sym. about y-axis

21. no symmetry

23.

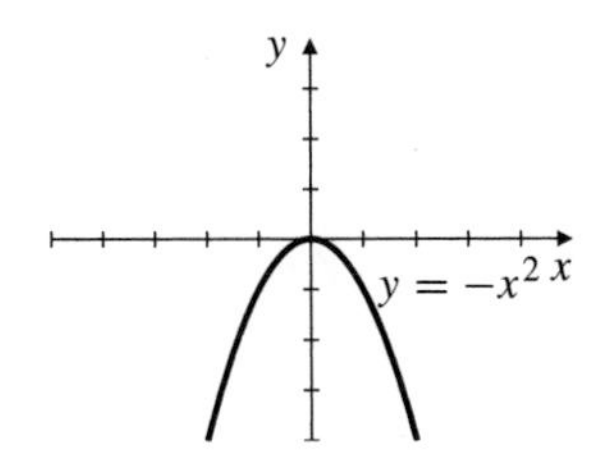

25.

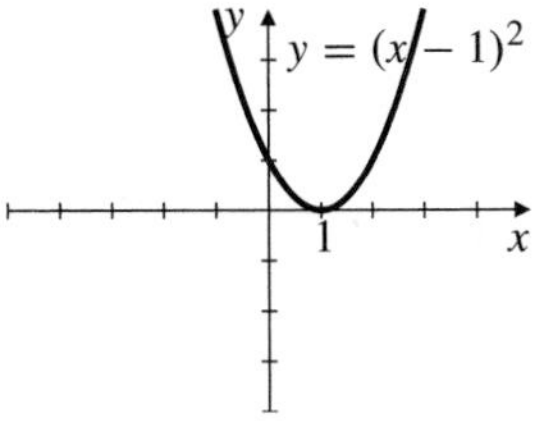

27.

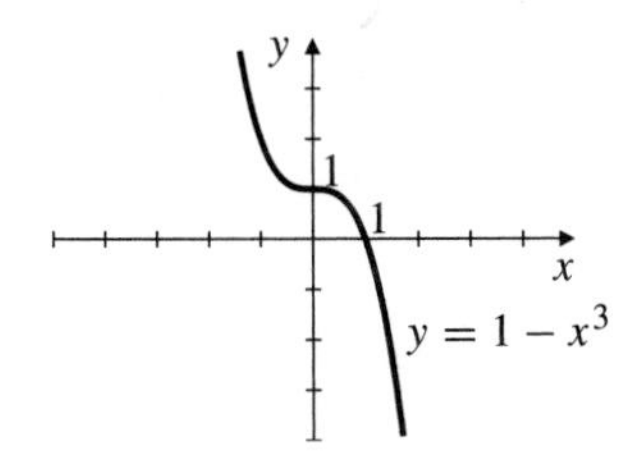

29.

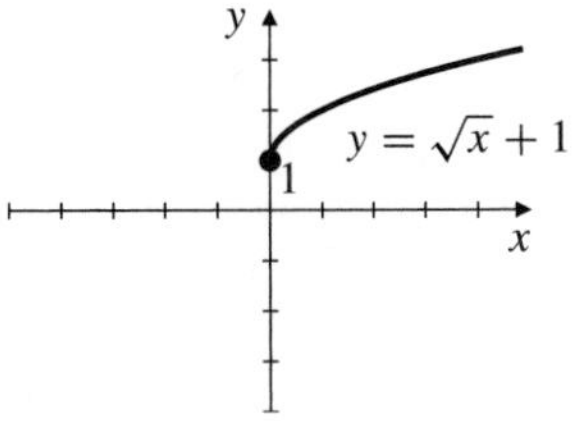

31.

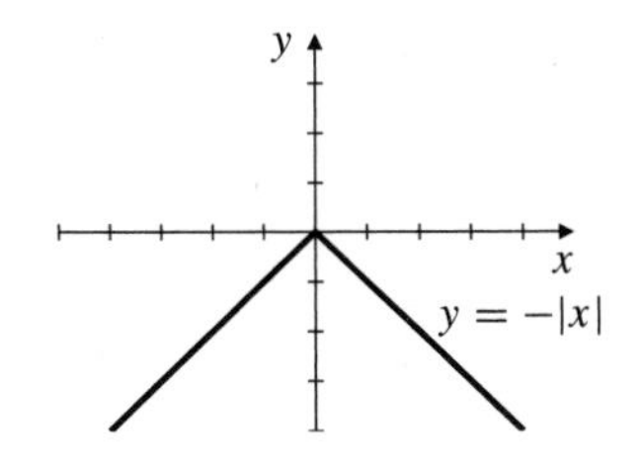

33.

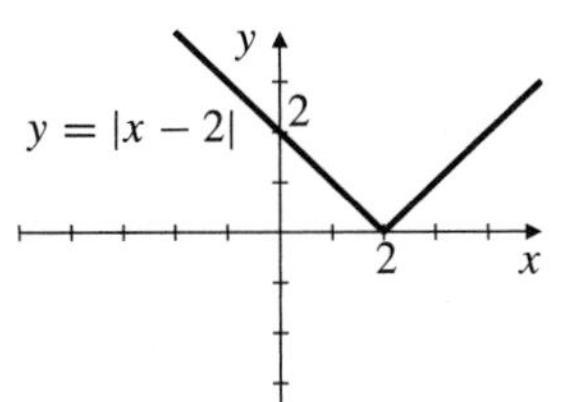

35.

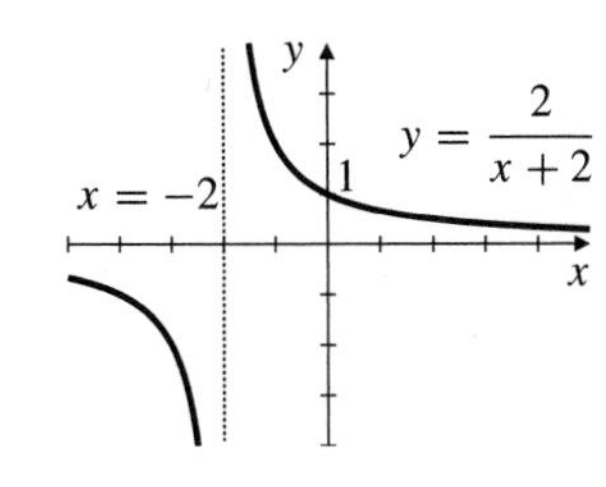

37.

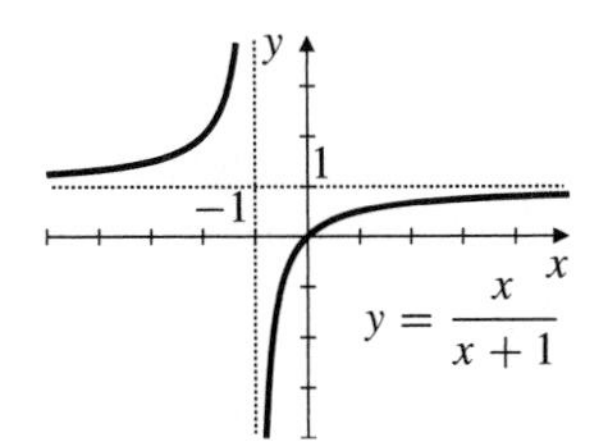

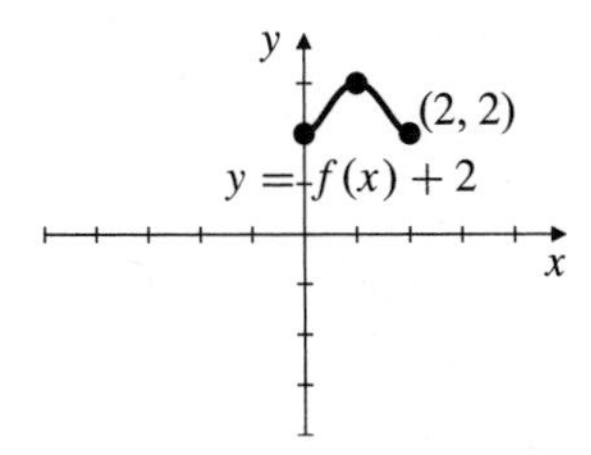

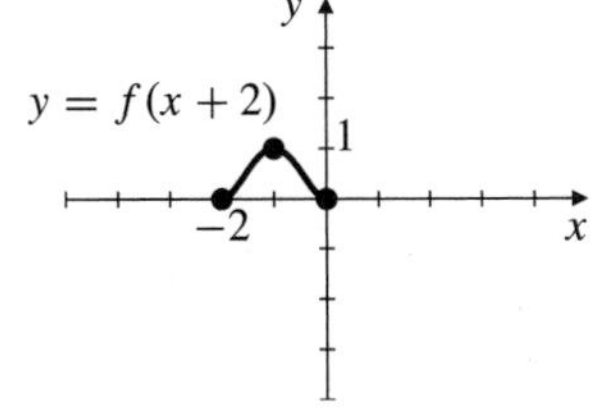

39. $\mathcal{D} = [0, 2]$, $\mathcal{R} = [2, 3]$

41. $\mathcal{D} = [-2, 0]$, $\mathcal{R} = [0, 1]$

43. $\mathcal{D} = [0, 2]$, $\mathcal{R} = [-1, 0]$

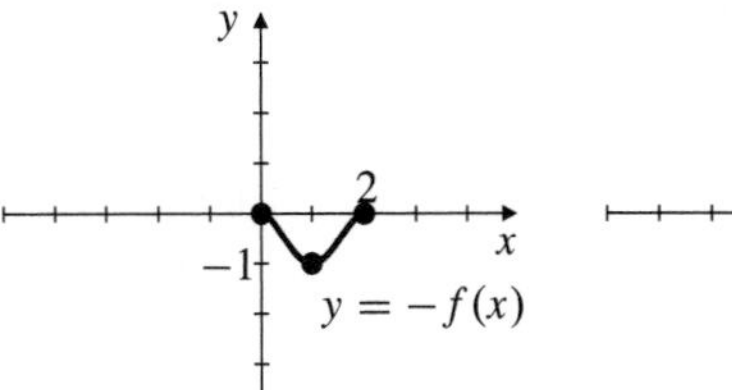

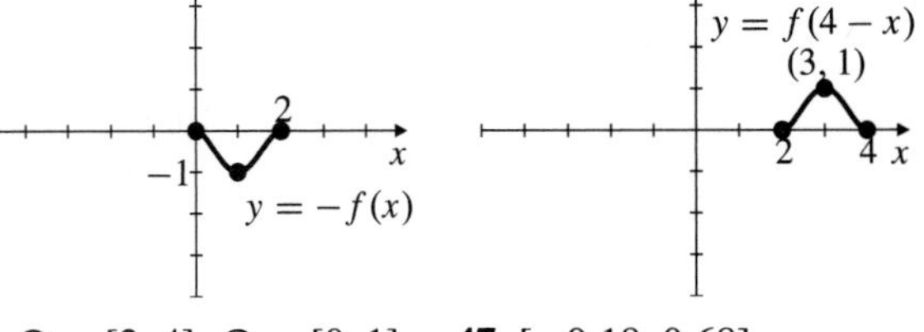

45. $\mathcal{D} = [2, 4]$, $\mathcal{R} = [0, 1]$ **47.** $[-0.18, 0.68]$

49. $y = 3/2$

51. $(2, 1)$, $y = x - 1$, $y = 3 - x$

53. $f(x) = 0$

Section P.5 (page 38)

1. The domains of $f + g$, $f - g$, fg, and g/f are $[1, \infty)$. The domain of f/g is $(1, \infty)$.
$(f + g)(x) = x + \sqrt{x - 1}$
$(f - g)(x) = x - \sqrt{x - 1}$
$(fg)(x) = x\sqrt{x - 1}$
$(f/g)(x) = x/\sqrt{x - 1}$
$(g/f)(x) = \sqrt{x - 1}/x$

3.

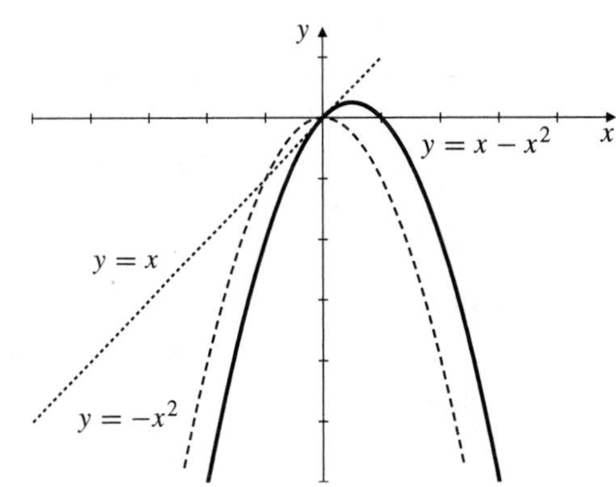

5.

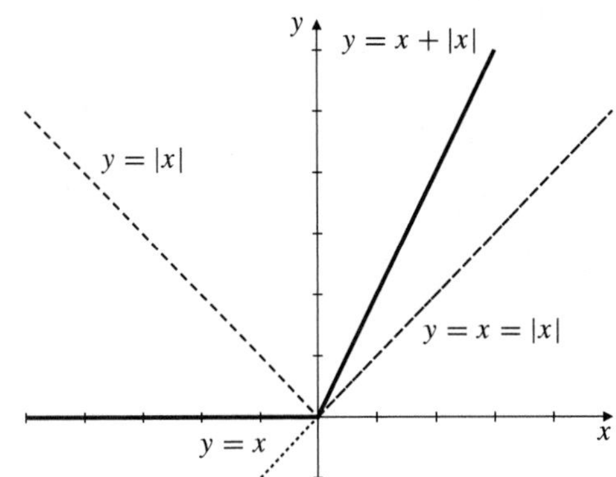

7. (a) 2, (b) 22, (c) $x^2 + 2$, (d) $x^2 + 10x + 22$, (e) 5, (f) -2, (g) $x + 10$, (h) $x^4 - 6x^2 + 6$

9. (a) $(x - 1)/x$, $x \neq 0, 1$,
(b) $1/(1 - \sqrt{x - 1})$ on $[1, 2) \cup (2, \infty)$,
(c) $\sqrt{x/(1 - x)}$, on $[0, 1)$
(d) $\sqrt{\sqrt{x - 1} - 1}$, on $[2, \infty)$

11. $(x + 1)^2$ **13.** x^2

15. $1/(x - 1)$ **19.** $\mathcal{D} = [0, 2]$, $\mathcal{R} = [0, 2]$

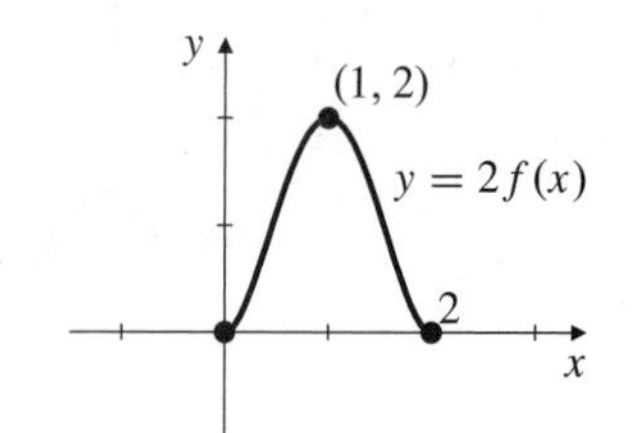

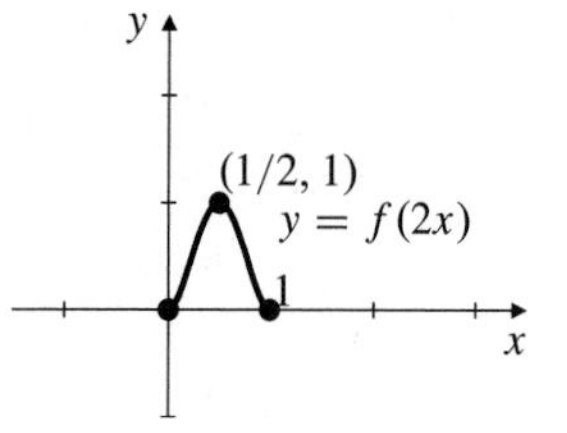

21. $\mathcal{D} = [0, 1]$, $\mathcal{R} = [0, 1]$

23. $\mathcal{D} = [-4, 0]$, $\mathcal{R} = [1, 2]$

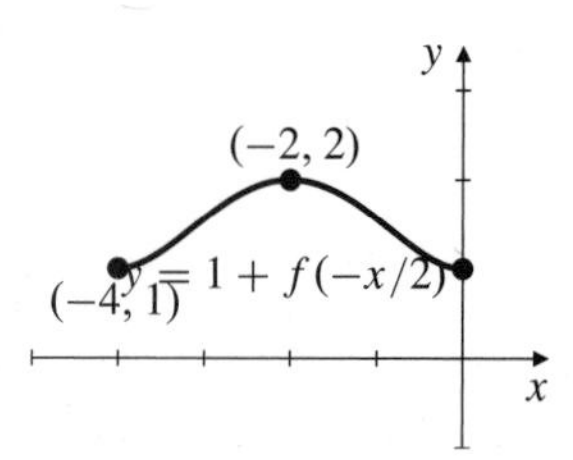

25.

27. (a) $A = 0$, B arbitrary, or $A = 1$, $B = 0$
(b) $A = -1$, B arbitrary, or $A = 1$, $B = 0$

29. all integers

31.

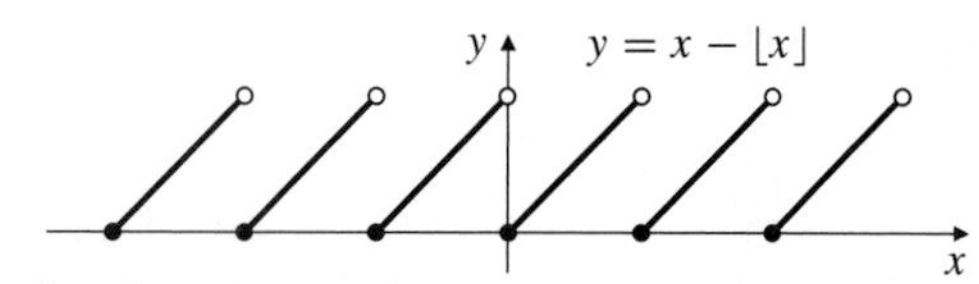

33. f^2, g^2, $f \circ f$, $f \circ g$, $g \circ f$ are even
fg, f/g, g/f, $g \circ g$ are odd
$f + g$ is neither, unless either $f(x) = 0$ or $g(x) = 0$.

Section P.6 (page 45)

1. roots -5 and -2; $(x + 5)(x + 2)$

3. roots $-1 \pm i$; $(x + 1 - i)(x + 1 + i)$

5. roots $1/2$ (double) and $-1/2$ (double); $(2x - 1)^2(2x + 1)^2$

7. roots $-1, \frac{1}{2} \pm \frac{\sqrt{3}}{2} i$; $(x+1)\left(x - \frac{1}{2} + \frac{\sqrt{3}}{2} i\right)\left(x - \frac{1}{2} - \frac{\sqrt{3}}{2} i\right)$

9. roots 1 (triple) and -1 triple; $(x - 1)^3(x + 1)^3$

11. roots $-2, i, -i, 1 + \sqrt{3}i, 1 - \sqrt{3}i$; $(x + 2)(x - i)(x + i)(x - 1 - \sqrt{3}i)(x - 1 + \sqrt{3}i)$

13. all real numbers

15. all real numbers except 0 and -1

17. $x + \dfrac{2x - 1}{x^2 - 2}$

19. $x - 2 + \dfrac{x + 6}{x^2 + 2x + 3}$

21. $P(x) = (x^2 - 2x + 2)(x^2 + 2x + 2)$

Section P.7 (page 57)

1. $-1/\sqrt{2}$

3. $\sqrt{3}/2$

5. $(\sqrt{3} - 1)/(2\sqrt{2})$

7. $-\cos x$

9. $-\cos x$

11. $1/(\sin x \cos x)$

17. $3 \sin x - 4 \sin^3 x$

19. period π

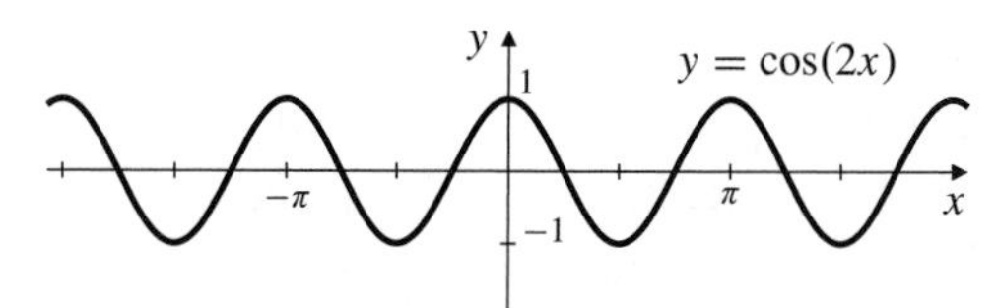

21. period 2

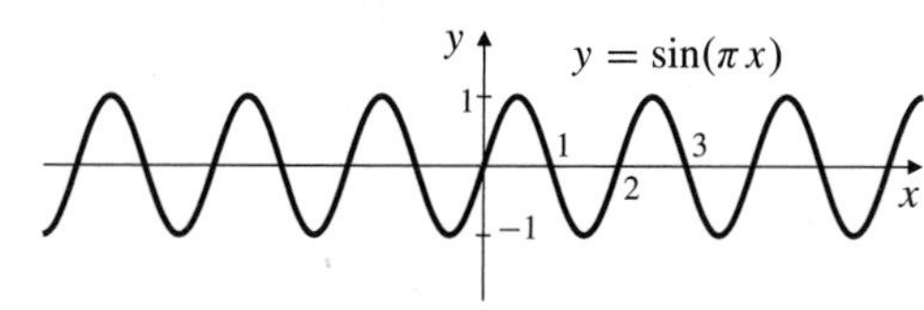

23.

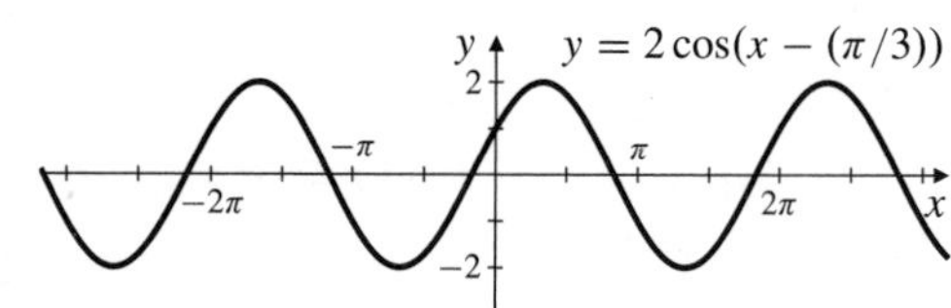

25. $\cos\theta = -4/5$, $\tan\theta = -3/4$

27. $\sin\theta = -2\sqrt{2}/3$, $\tan\theta = -2\sqrt{2}$

29. $\cos\theta = -\sqrt{3}/2$, $\tan\theta = 1/\sqrt{3}$

31. $a = 1$, $b = \sqrt{3}$

33. $b = 5/\sqrt{3}$, $c = 10/\sqrt{3}$

35. $a = b \tan A$

37. $a = b \cot B$

39. $c = b \sec A$

41. $\sin A = \sqrt{c^2 - b^2}/c$

43. $\sin B = 3/(4\sqrt{2})$

45. $\sin B = \sqrt{135}/16$

47. $6/(1 + \sqrt{3})$

49. $b = 4 \sin 40°/\sin 70° \approx 2.736$

51. approx. 16.98 m

Chapter 1
Limits and Continuity

Section 1.1 (page 63)

1. $((t + h)^2 - t^2)/h$ m/s

3. 4 m/s

5. -3 m/s, 3 m/s, 0 m/s

7. to the left, stopped, to the right

9. height 2, moving down

11. -1 ft/s, weight moving downward

13. day 45

Section 1.2 (page 71)

1. (a) 1, (b) 0, (c) 1

3. 1

5. 0

7. 1

9. $2/3$

11. 0

13. 0

15. does not exist

17. $1/6$

19. 0

21. -1

23. does not exist

25. 2

27. $3/8$

29. $-1/2$

31. $8/3$

33. $1/4$ **35.** $1/\sqrt{2}$

37. $2x$ **39.** $-1/x^2$

41. $1/(2\sqrt{x})$ **43.** 1

45. $1/2$ **47.** 1

49. 0 **51.** 2

53. does not exist **55.** does not exist

57. $-1/(2a)$ **59.** 0

61. -2 **63.** π^2

65. (a) 0, (b) 8, (c) 9, (d) -3

67. 5 **69.** 1

71. 0.7071 **73.** $\lim_{x\to 0} f(x) = 0$

75. 2

77. $x^{1/3} < x^3$ on $(-1, 0)$ and $(1, \infty)$, $x^{1/3} > x^3$ on $(-\infty, -1)$ and $(0, 1)$, $\lim_{x\to a} h(x) = a$ for $a = -1, 0,$ and 1

Section 1.3 (page 78)

1. $1/2$ **3.** $-3/5$

5. 0 **7.** -3

9. $-2/\sqrt{3}$ **11.** does not exist

13. $+\infty$ **15.** 0

17. $-\infty$ **19.** $-\infty$

21. ∞ **23.** $-\infty$

25. ∞ **27.** $-\sqrt{2}/4$

29. -2 **31.** -1

33. horiz: $y = 0$, $y = -1$, vert: $x = 0$

35. 1 **37.** 1

39. $-\infty$ **41.** 2

43. -1 **45.** 1

47. 3 **49.** does not exist

51. 1

53. $C(t)$ has a limit at every real t except at the integers. $\lim_{t\to t_0-} C(t) = C(t_0)$ everywhere, but

$$\lim_{t\to t_0+} C(t) = \begin{cases} C(t_0) & \text{if } t_0 \text{ not integral} \\ C(t_0) + 1.5 & \text{if } t_0 \text{ an integer} \end{cases}$$

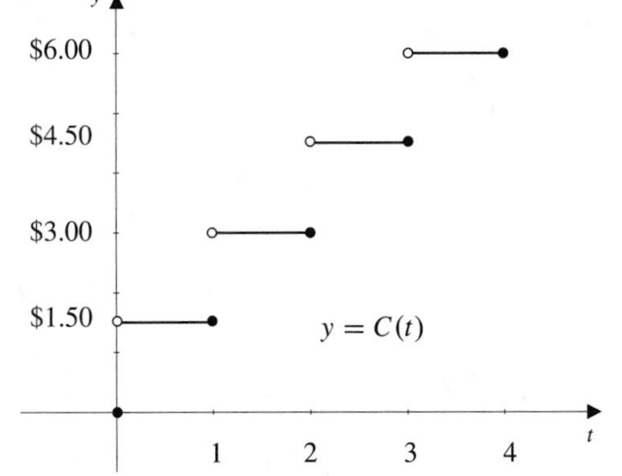

55. (a) B, (b) A, (c) A, (d) A

Section 1.4 (page 87)

1. at -2, right cont. and cont., at -1 disc., at 0 disc. but left cont., at 1 disc. and right cont., at 2 disc.

3. no abs. max, abs. min 0 **5.** no

7. cont. everywhere

9. cont. everywhere except at $x = 0$, disc. at $x = 0$

11. cont. everywhere except at the integers, discontinuous but left-continuous at the integers

13. 4, $x + 2$ **15.** $1/5$, $(t-2)/(t+2)$

17. $k = 8$ **19.** no max, min = 0

21. 16 **23.** 5

25. f positive on $(-1, 0)$ and $(1, \infty)$; f negative on $(-\infty, -1)$ and $(0, 1)$

27. f positive on $(-\infty, -2)$, $(-1, 1)$ and $(2, \infty)$; f negative on $(-2, -1)$ and $(1, 2)$

35. max 1.593 at -0.831, min -0.756 at 0.629

37. max $31/3 \approx 10.333$ at $x = 3$, min 4.762 at $x = 1.260$

39. 0.682

41. -0.6367326508, 1.409624004

Section 1.5 (page 92)

1. between 12 °C and 20 °C

3. (1.99, 2.01) **5.** (0.81, 1.21)

7. $\delta = 0.01$ **9.** $\delta \approx 0.0165$

Review Exercises (page 93)

1. 13 **3.** 12

5. 4 **7.** does not exist

9. does not exist **11.** $-\infty$

13. $12\sqrt{3}$ **15.** 0

17. does not exist **19.** $-1/3$

21. $-\infty$ **23.** ∞

25. does not exist **27.** 0

29. 2 **31.** no disc.

33. disc. and left cont. at 2

35. disc. and right cont. at $x = 1$

37. no disc.

Challenging Problems (page 94)

1. to the right **3.** $-1/4$

5. 3 **7.** T, F, T, F, F

Chapter 2 Differentiation

Section 2.1 (page 100)

1. $y = 3x - 1$ **3.** $y = 8x - 13$

5. $y = 12x + 24$ **7.** $x - 4y = -5$

9. $x - 4y = -2$ **11.** $y = 2x_0x - x_0^2$

13. no **15.** yes, $x = -2$

17. yes, $x = 0$

19. (a) $3a^2$; (b) $y = 3x - 2$ and $y = 3x + 2$

21. $(1, 1)$, $(-1, 1)$ **23.** $k = 3/4$

25. horiz. tangent at $(0, 0)$, $(3, 108)$, $(5, 0)$

27. horiz. tangent at $(-0.5, 1.25)$, no tangents at $(-1, 1)$ and $(1, -1)$

29. horiz. tangent at $(0, -1)$

31. no, consider $y = x^{2/3}$ at $(0, 0)$

Section 2.2 (page 107)

1. **3.**

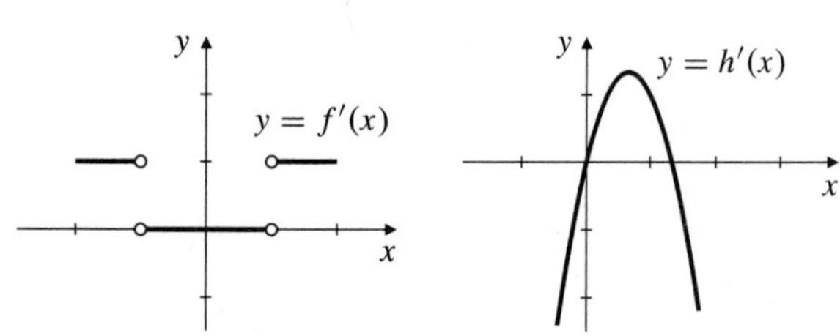

5. on $[-2, 2]$ except at $x = -1$ and $x = 1$

7. slope positive for $x < 1.5$, negative for $x > 1.5$; horizontal tangent at $x = 1.5$

9. singular points at $x = -1,\ 0,\ 1$, horizontal tangents at about $x = \pm 0.57$

11. (a) $y' = 2x - 3$, (b) $dy = (2x - 3)\,dx$

13. (a) $f'(x) = 3x^2$, (b) $df(x) = 3x^2\,dx$

15. (a) $g'(x) = -\dfrac{4}{(2+x)^2}$, (b) $dg(x) = -\dfrac{4}{(2+x)^2}\,dx$

17. (a) $F'(t) = \dfrac{1}{\sqrt{2t+1}}$, (b) $dF(t) = \dfrac{1}{\sqrt{2t+1}}\,dt$

19. (a) $y' = 1 - \dfrac{1}{x^2}$, (b) $dy = \left(1 - \dfrac{1}{x^2}\right)dx$

21. (a) $F'(x) = -\dfrac{x}{(1+x^2)^{3/2}}$, (b) $dF(x) = -\dfrac{x}{(1+x^2)^{3/2}}\,dx$

23. (a) $y' = -\dfrac{1}{2(1+x)^{3/2}}$, (b) $dy = -\dfrac{1}{2(1+x)^{3/2}}\,dx$

25. Define $f(0) = 0$, f is not differentiable at 0

27. at $x = -1$ and $x = -2$

29.

x	$\dfrac{f(x)-f(2)}{x-2}$
1.9	−0.26316
1.99	−0.25126
1.999	−0.25013
1.9999	−0.25001

x	$\dfrac{f(x)-f(2)}{x-2}$
2.1	−0.23810
2.01	−0.24876
2.001	−0.24988
2.0001	−0.24999

$$\frac{d}{dx}\left(\frac{1}{x}\right)\bigg|_{x=2} = -\frac{1}{4}$$

31. $x - 6y = -15$

33. $y = \dfrac{2}{a^2+a} - \dfrac{2(2a+1)}{(a^2+a)^2}(t-a)$

35. $22t^{21}$, all t **37.** $-(1/3)x^{-4/3}$, $x \neq 0$

39. $(119/4)s^{115/4}$, $s \geq 0$ **41.** -16

43. $1/(8\sqrt{2})$ **45.** $y = a^2x - a^3 + \dfrac{1}{a}$

47. $y = 6x - 9$ and $y = -2x - 1$

49. $\dfrac{1}{2\sqrt{2}}$ **53.** $f'(x) = \frac{1}{3}x^{-2/3}$

Section 2.3 (page 115)

1. $6x - 5$ **3.** $2Ax + B$

5. $\frac{1}{3}s^4 - \frac{1}{5}s^2$

7. $\frac{1}{3}t^{-2/3} + \frac{1}{2}t^{-3/4} + \frac{3}{5}t^{-4/5}$

9. $x^{2/3} + x^{-8/5}$ **11.** $\dfrac{5}{2\sqrt{x}} - \frac{3}{2}\sqrt{x} - \frac{5}{6}x^{3/2}$

13. $-\dfrac{2x+5}{(x^2+5x)^2}$ **15.** $\dfrac{\pi^2}{(2-\pi t)^2}$

17. $(4x^2 - 3)/x^4$

19. $-t^{-3/2} + (1/2)t^{-1/2} + (3/2)\sqrt{t}$

21. $-\dfrac{24}{(3+4x)^2}$ **23.** $\dfrac{1}{\sqrt{t}(1-\sqrt{t})^2}$

25. $\dfrac{ad - bc}{(cx+d)^2}$

27. $10 + 70x + 150x^2 + 96x^3$

29. $2x(\sqrt{x}+1)(5x^{2/3}-2) + \dfrac{1}{2\sqrt{x}}(x^2+4)(5x^{2/3}-2)$ $+\dfrac{10}{3}x^{-1/3}(x^2+4)(\sqrt{x}+1)$

31. $\dfrac{6x+1}{(6x^2+2x+1)^2}$ **33.** -1

35. 20 **37.** $-\dfrac{1}{2}$

39. $-\dfrac{1}{18\sqrt{2}}$ **41.** $y = 4x - 6$

43. $(1, 2)$ and $(-1, -2)$ **45.** $\left(-\frac{1}{2}, \frac{4}{3}\right)$

47. $y = b - \dfrac{b^2x}{4}$

49. $y = 12x - 16$, $y = 3x + 2$

51. $x/\sqrt{x^2+1}$

Section 2.4 (page 120)

1. $12(2x+3)^5$ **3.** $-20x(4-x^2)^9$

5. $\dfrac{30}{t^2}\left(2 + \dfrac{3}{t}\right)^{-11}$ **7.** $\dfrac{12}{(5-4x)^2}$

9. $-2x\,\mathrm{sgn}\,(1-x^2)$ **11.** $\begin{cases} 8 & \text{if } x > 1/4 \\ 0 & \text{if } x < 1/4 \end{cases}$

13. $\dfrac{-3}{2\sqrt{3x+4}(2+\sqrt{3x+4})^2}$

15. $-\dfrac{5}{3}\left(1 - \dfrac{1}{(u-1)^2}\right)\left(u + \dfrac{1}{u-1}\right)^{-8/3}$

17.

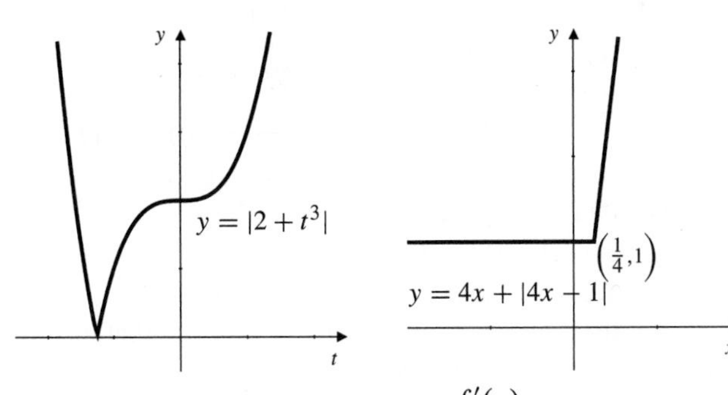

23. $(5-2x)f'(5x-x^2)$ **25.** $\dfrac{f'(x)}{\sqrt{3+2f(x)}}$

27. $\dfrac{1}{\sqrt{x}}f'(3+2\sqrt{x})$

29. $15f'(4-5t)f'(2-3f(4-5t))$

31. $\dfrac{3}{2\sqrt{2}}$ **33.** 102

35. $-6\left(1-\frac{15}{2}(3x)^4\left((3x)^5-2\right)^{-3/2}\right) \times \left(x+\left((3x)^5-2\right)^{-1/2}\right)^{-7}$

37. $y=2^{3/2}-\sqrt{2}(x+1)$ **39.** $y=\frac{1}{27}+\frac{5}{162}(x+2)$

41. $\dfrac{x(x^4+2x^2-2)}{(x^2+1)^{5/2}}$ **43.** 857,592

45. no; yes; both functions are equal to x^2.

Section 2.5 (page 125)

3. $-3\sin 3x$ **5.** $\pi\sec^2\pi x$

7. $3\csc^2(4-3x)$ **9.** $r\sin(s-rx)$

11. $2\pi x\cos(\pi x^2)$ **13.** $\dfrac{-\sin x}{2\sqrt{1+\cos x}}$

15. $-(1+\cos x)\sin(x+\sin x)$

17. $(3\pi/2)\sin^2(\pi x/2)\cos(\pi x/2)$

19. $a\cos 2at$ **21.** $2\cos(2x)+2\sin(2x)$

23. $\sec^2 x-\csc^2 x$ **25.** $\tan^2 x$

27. $-t\sin t$ **29.** $1/(1+\cos x)$

31. $2x\cos(3x)-3x^2\sin(3x)$

33. $2x[\sec(x^2)\tan^2(x^2)+\sec^3(x^2)]$

35. $-\sec^2 t\sin(\tan t)\cos(\cos(\tan t))$

39. $y=\pi-x,\ y=x-\pi$

41. $y=1-(x-\pi)/4,\ y=1+4(x-\pi)$

43. $y=\dfrac{1}{\sqrt{2}}+\dfrac{\pi}{180\sqrt{2}}(x-45)$

45. $\pm(\pi/4,1)$ **49.** yes, (π,π)

51. yes, $(2\pi/3,(2\pi/3)+\sqrt{3})$, $(4\pi/3,(4\pi/3)-\sqrt{3})$

53. 2 **55.** 1

57. 1/2

59. infinitely many, 0.336508, 0.161228

Section 2.6 (page 130)

1. $\begin{cases} y'=-14(3-2x)^6, \\ y''=168(3-2x)^5, \\ y'''=-1680(3-2x)^4 \end{cases}$

3. $\begin{cases} y'=-12(x-1)^{-3}, \\ y''=36(x-1)^{-4}, \\ y'''=-144(x-1)^{-5} \end{cases}$

5. $\begin{cases} y'=\frac{1}{3}x^{-2/3}+\frac{1}{3}x^{-4/3}, \\ y''=-\frac{2}{9}x^{-5/3}-\frac{4}{9}x^{-7/3} \\ y'''=\frac{10}{27}x^{-8/3}+\frac{28}{27}x^{-10/3} \end{cases}$

7. $\begin{cases} y'=\frac{5}{2}x^{3/2}+\frac{3}{2}x^{-1/2} \\ y''=\frac{15}{4}x^{1/2}-\frac{3}{4}x^{-3/2} \\ y'''=\frac{15}{8}x^{-1/2}+\frac{9}{8}x^{-5/2} \end{cases}$

9. $y'=\sec^2 x$, $y''=2\sec^2 x\tan x$, $y'''=4\sec^2 x\tan^2 x+2\sec^4 x$

11. $y'=-2x\sin(x^2)$, $y''=-2\sin(x^2)-4x^2\cos(x^2)$, $y'''=-12x\cos(x^2)+8x^3\sin(x^2)$

13. $(-1)^n n!x^{-(n+1)}$ **15.** $n!(2-x)^{-(n+1)}$

17. $(-1)^n n!b^n(a+bx)^{-(n+1)}$

19. $f^{(n)}=\begin{cases}(-1)^k a^n\cos(ax) & \text{if } n=2k \\ (-1)^{k+1}a^n\sin(ax) & \text{if } n=2k+1\end{cases}$ where $k=0,1,2,\ldots$

21. $f^{(n)}=(-1)^k[a^n x\sin(ax)-na^{n-1}\cos(ax)]$ if $n=2k$, or $(-1)^k[a^n x\cos(ax)+na^{n-1}\sin(ax)]$ if $n=2k+1$, where $k=0,1,2,\ldots$

23. $-\dfrac{1\times 3\times 5\times\cdots\times(2n-3)}{2^n}3^n(1-3x)^{-(2n-1)/2}$, $(n=2,3,\ldots)$

Section 2.7 (page 136)

1. $-0.0025,\ 0.4975$ **3.** $-1/40,\ -1/40$

5. 4% **7.** −4%

9. 1% **11.** 6%

13. 8 ft²/ft

15. $1/\sqrt{\pi A}$ units/square unit

17. 16π m³/m

19. $\dfrac{dC}{dA}=\sqrt{\dfrac{\pi}{A}}$ length units/area unit

21. (a) 10,500 L/min, 3,500 L/min, (b) 7,000 L/min

23. decreases at 1/8 pound/mi

25. (a) \$300, (b) $C(101)-C(100)=\$299.50$

27. (a) −\$2.00, (b) \$9.11

Section 2.8 (page 143)

1. $c=\dfrac{a+b}{2}$ **3.** $c=\pm\dfrac{2}{\sqrt{3}}$

9. Incr. $x>0$, decr. $x<0$

11. Incr. on $(-\infty,-4)$ and $(0,\infty)$, decr. on $(-4,0)$

13. inc. on $\left(-\infty, -\frac{2}{\sqrt{3}}\right)$ and $\left(\frac{2}{\sqrt{3}}, \infty\right)$, dec. on $\left(-\frac{2}{\sqrt{3}}, \frac{2}{\sqrt{3}}\right)$

15. inc. on $(-2, 0)$ and $(2, \infty)$; dec. on $(-\infty, -2)$ and $(0, 2)$

17. inc. on $(-\infty, 3)$ and $(5, \infty)$; dec. on $(3, 5)$

19. inc. on $(-\infty, \infty)$ **23.** 0.535898, 7.464102

25. 0, -0.518784

Section 2.9 (page 148)

1. $\dfrac{1-y}{2+x}$ **3.** $\dfrac{2x+y}{3y^2-x}$

5. $\dfrac{2-2xy^3}{3x^2y^2+1}$ **7.** $-\dfrac{3x^2+2xy}{x^2+4y}$

9. $2x+3y=5$ **11.** $y=x$

13. $y = 1 - \dfrac{4}{4-\pi}\left(x - \dfrac{\pi}{4}\right)$

15. $y = 2 - x$ **17.** $\dfrac{2(y-1)}{(1-x)^2}$

19. $\dfrac{(2-6y)(1-3x^2)^2}{(3y^2-2y)^3} - \dfrac{6x}{3y^2-2y}$

21. $-a^2/y^3$ **23.** 0

25. -26

Section 2.10 (page 154)

1. $5x + C$ **3.** $\frac{2}{3}x^{3/2} + C$

5. $\frac{1}{4}x^4 + C$ **7.** $-\cos x + C$

9. $a^2x - \frac{1}{3}x^3 + C$ **11.** $\frac{4}{3}x^{3/2} + \frac{9}{4}x^{4/3} + C$

13. $\frac{1}{12}x^4 - \frac{1}{6}x^3 + \frac{1}{2}x^2 - x + C$

15. $\frac{1}{2}\sin(2x) + C$ **17.** $\dfrac{-1}{1+x} + C$

19. $\frac{1}{3}(2x+3)^{3/2} + C$ **21.** $-\cos(x^2) + C$

23. $\tan x - x + C$ **25.** $(x + \sin x \cos x)/2 + C$

27. $y = \frac{1}{2}x^2 - 2x + 3$, all x

29. $y = 2x^{3/2} - 15$, $(x > 0)$

31. $y = \dfrac{A}{3}(x^3 - 1) + \dfrac{B}{2}(x^2 - 1) + C(x-1) + 1$, (all x)

33. $y = \sin x + (3/2)$, (all x)

35. $y = 1 + \tan x$, $-\pi/2 < x < \pi/2$

37. $y = x^2 + 5x - 3$, (all x)

39. $y = \dfrac{x^5}{20} - \dfrac{x^2}{2} + 8$, (all x)

41. $y = 1 + x - \cos x$, (all x)

43. $y = 3x - \dfrac{1}{x}$, $(x > 0)$

45. $y = -\dfrac{7\sqrt{x}}{2} + \dfrac{18}{\sqrt{x}}$, $(x > 0)$

Section 2.11 (page 160)

1. (a) $t > 2$, (b) $t < 2$, (c) all t, (d) no t, (e) $t > 2$, (f) $t < 2$, (g) 2, (h) 0

3. (a) $t < -2/\sqrt{3}$ or $t > 2/\sqrt{3}$, (b) $-2/\sqrt{3} < t < 2/\sqrt{3}$, (c) $t > 0$, (d) $t < 0$, (e) $t > 2/\sqrt{3}$ or $-2/\sqrt{3} < t < 0$, (f) $t < -2/\sqrt{3}$ or $0 < t < 2/\sqrt{3}$, (g) $\pm 12/\sqrt{3}$ at $t = \pm 2/\sqrt{3}$, (h) 12

5. acc = 9.8 m/s^2 downward at all times; max height = 4.9 m; ball strikes ground at 9.8 m/s

7. time 27.8 s; distance 771.6 m

9. $4h$ m, $\sqrt{2}v_0$ m/s **11.** 400 ft

13. 0.833 km

15. $v = \begin{cases} 2t & \text{if } 0 < t \le 2 \\ 4 & \text{if } 2 < t < 8 \\ 20 - 2t & \text{if } 8 \le t < 10 \end{cases}$

v is continuous for $0 < t < 10$.

$a = \begin{cases} 2 & \text{if } 0 < t < 2 \\ 0 & \text{if } 2 < t < 8 \\ -2 & \text{if } 8 < t < 10 \end{cases}$

a is continuous except at $t = 2$ and $t = 8$. Maximum velocity 4 is attained for $2 \le t \le 8$.

17. 7 s **19.** 448 ft

Review Exercises (page 161)

1. $18x + 6$ **3.** -1

5. $6\pi x + 12y = 6\sqrt{3} + \pi$

7. $\dfrac{\cos x - 1}{(x - \sin x)^2}$ **9.** $x^{-3/5}(4 - x^{2/5})^{-7/2}$

11. $-2\theta \sec^2\theta \tan\theta$ **13.** $20x^{19}$

15. $-\sqrt{3}$ **17.** $-2xf'(3 - x^2)$

19. $2f'(2x)\sqrt{g(x/2)} + \dfrac{f(2x)\,g'(x/2)}{4\sqrt{g(x/2)}}$

21. $f'(x + (g(x))^2)(1 + 2g(x)g'(x))$

23. $\cos x\, f'(\sin x)\, g(\cos x) - \sin x\, f(\sin x)\, g'(\cos x)$

25. $7x + 10y = 24$ **27.** $\dfrac{x^3}{3} - \dfrac{1}{x} + C$

29. $2\tan x + 3\sec x + C$ **31.** $4x^3 + 3x^4 - 7$

33. $I_1 = x\sin x + \cos x + C$, $I_2 = \sin x - x\cos x + C$

35. $y = 3x$

37. points $k\pi$ and $k\pi/(n+1)$ where k is any integer

39. $(0, 0)$, $(\pm 1/\sqrt{2}, 1/2)$, dist. $= \sqrt{3}/2$ units

41. (a) $k = g/R$ **43.** 15.3 m

45. 80 ft/s or about 55 mph

Challenging Problems (page 162)

3. (a) 0, (b) 3/8, (c) 12, (d) -48, (e) 3/7, (f) 21

13. $f(m) = C - (m - B)^2/(4A)$

17. (a) $3b^2 > 8ac$

19. (a) 3 s, (b) $t = 7$ s, (c) $t = 12$ s, (d) about 13.07 m/s^2, (e) 197.5 m, (f) 60.3 m.

Chapter 3 Transcendental Functions

Section 3.1 (page 169)

1. $f^{-1}(x) = x + 1$
$\mathcal{D}(f^{-1}) = \mathcal{R}(f) = \mathcal{R}(f^{-1}) = \mathcal{D}(f) = \mathbb{R}$

3. $f^{-1}(x) = x^2 + 1$, $\mathcal{D}(f^{-1}) = \mathcal{R}(f) = [0, \infty)$, $\mathcal{R}(f^{-1}) = \mathcal{D}(f) = [1, \infty)$

5. $f^{-1}(x) = x^{1/3}$
$\mathcal{D}(f^{-1}) = \mathcal{R}(f) = \mathcal{R}(f^{-1}) = \mathcal{D}(f) = \mathbb{R}$

7. $f^{-1}(x) = -\sqrt{x}$, $\mathcal{D}(f^{-1}) = \mathcal{R}(f) = [0, \infty)$, $\mathcal{R}(f^{-1}) = \mathcal{D}(f) = (-\infty, 0]$

9. $f^{-1}(x) = \frac{1}{x} - 1$, $\mathcal{D}(f^{-1}) = \mathcal{R}(f) = \{x : x \neq 0\}$, $\mathcal{R}(f^{-1}) = \mathcal{D}(f) = \{x : x \neq -1\}$

11. $f^{-1}(x) = \frac{1-x}{2+x}$,
$\mathcal{D}(f^{-1}) = \mathcal{R}(f) = \{x : x \neq -2\}$,
$\mathcal{R}(f^{-1}) = \mathcal{D}(f) = \{x : x \neq -1\}$

13. $g^{-1}(x) = f^{-1}(x + 2)$ **15.** $k^{-1}(x) = f^{-1}\left(-\frac{x}{3}\right)$

17. $p^{-1}(x) = f^{-1}\left(\frac{1}{x} - 1\right)$

19. $r^{-1}(x) = \frac{1}{4}\left(3 - f^{-1}\left(\frac{1-x}{2}\right)\right)$

21. $f^{-1}(x) = \begin{cases} \sqrt{x-1} & \text{if } x >= 1 \\ x - 1 & \text{if } x < 1 \end{cases}$

23. $h^{-1}(x) = \begin{cases} \sqrt{x-1} & \text{if } x \geq 1 \\ \sqrt{1-x} & \text{if } x < 1 \end{cases}$

25. $g^{-1}(1) = 2$ **29.** $\left(f^{-1}\right)'(2) = 1/4$

31. 2.23362 **33.** $\mathbb{R}$, 1

35. $c = 1$, a, b arbitrary, or $a = b = 0$, $c = -1$.

37. no

Section 3.2 (page 174)

1. $\sqrt{3}$ **3.** x^6

5. 3 **7.** $-2x$

9. x **11.** 1

13. 1 **15.** 2

17. $\log_a(x^4 + 4x^2 + 3)$ **19.** 4.728804...

21. $x = (\log_{10} 5)/(\log_{10}(4/5)) \approx -7.212567$

23. $x = 3^{1/5} = 10^{(\log_{10} 3)/5} \approx 1.24573$

29. 1/2 **31.** 0

33. ∞

Section 3.3 (page 182)

1. $\sqrt{e}$ **3.** x^5

5. $-3x$ **7.** $\ln \frac{64}{81}$

9. $\ln\left(x^2(x-2)^5\right)$ **11.** $x = \frac{\ln 2}{\ln(3/2)}$

13. $x = \frac{\ln 5 - 9 \ln 2}{2 \ln 2}$ **15.** $0 < x < 2$

17. $3 < x < 7/2$ **19.** $5e^{5x}$

21. $(1 - 2x)e^{-2x}$ **23.** $\frac{3}{3x - 2}$

25. $\frac{e^x}{1 + e^x}$ **27.** $\frac{e^x - e^{-x}}{2}$

29. e^{x+e^x} **31.** $e^x(\sin x + \cos x)$

33. $\frac{1}{x \ln x}$ **35.** $2x \ln x$

37. $(2 \ln 5)5^{2x+1}$ **39.** $t^x x^t \ln t + t^{x+1} x^{t-1}$

41. $\frac{b}{(bs + c) \ln a}$

43. $x^{\sqrt{x}}\left(\frac{1}{\sqrt{x}}\left(\frac{1}{2} \ln x + 1\right)\right)$

45. $\sec x$ **47.** $-\frac{1}{\sqrt{x^2 + a^2}}$

49. $f^{(n)}(x) = e^{ax}(na^{n-1} + a^n x)$, $n = 1, 2, 3, \ldots$

51. $y' = 2xe^{x^2}$, $y'' = 2(1 + 2x^2)e^{x^2}$,
$y''' = 4(3x + 2x^3)e^{x^2}$, $y^{(4)} = 4(3 + 12x^2 + 4x^4)e^{x^2}$

53. $f'(x) = x^{x^2+1}(2 \ln x + 1)$,
$g'(x) = x^{x^x} x^x \left(\ln x + (\ln x)^2 + \frac{1}{x}\right)$;
g grows more rapidly than does f.

55. $f'(x) = f(x)\left(\frac{1}{x-1} + \frac{1}{x-2} + \frac{1}{x-3} + \frac{1}{x-4}\right)$

57. $f'(2) = \frac{556}{3675}$, $f'(1) = \frac{1}{6}$

59. f inc. for $x < 1$, dec. for $x > 1$

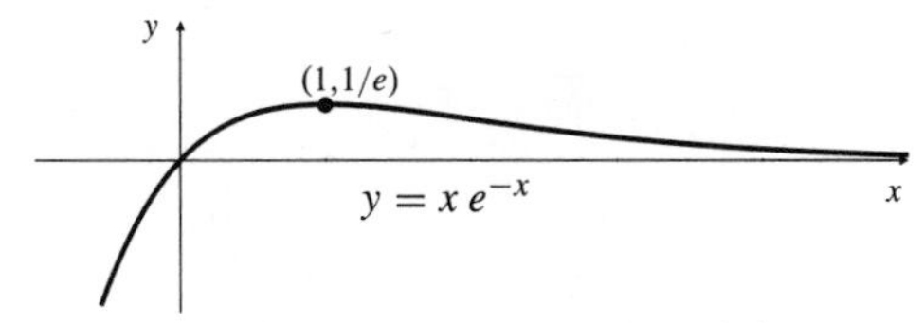

61. $y = ex$ **63.** $y = 2e \ln 2(x - 1)$

65. $-1/e^2$

67. $f'(x) = (A + B) \cos \ln x + (B - A) \sin \ln x$,
$\int \cos \ln x \, dx = \frac{x}{2}(\cos \ln x + \sin \ln x)$,
$\int \sin \ln x \, dx = \frac{x}{2}(\sin \ln x - \cos \ln x)$

69. (a) $F_{2B,-2A}(x)$; (b) $-2e^x(\cos x + \sin x)$

Section 3.4 (page 189)

1. 0 **3.** 2

5. 0 **7.** 0

9. 566 **11.** 29.15 years

13. 160.85 years **15.** 4,139 g

17. $7,557.84 **19.** about 14.7 years

21. about 142

23. (a) $f(x) = Ce^{bx} - (a/b)$,
(b) $y = (y_0 + (a/b))e^{bx} - (a/b)$

25. 22.35 °C **27.** 6.84 min

31. $(0, -(1/k)\ln(y_0/(y_0 - L)))$, solution $\to -\infty$

33. about 7,671 cases, growing at about 3,028 cases/week

Section 3.5 (page 198)

1. $\pi/3$ **3.** $-\pi/4$

5. 0.7 **7.** $-\pi/3$

9. $\dfrac{\pi}{2} + 0.2$ **11.** $2/\sqrt{5}$

13. $\sqrt{1 - x^2}$ **15.** $\dfrac{1}{\sqrt{1 + x^2}}$

17. $\dfrac{\sqrt{1 - x^2}}{x}$ **19.** $\dfrac{1}{\sqrt{2 + x - x^2}}$

21. $\dfrac{-\operatorname{sgn} a}{\sqrt{a^2 - (x - b)^2}}$ **23.** $\tan^{-1} t + \dfrac{t}{1 + t^2}$

25. $2x \tan^{-1} x + 1$

27. $\dfrac{\sqrt{1 - 4x^2}\sin^{-1} 2x - 2\sqrt{1 - x^2}\sin^{-1} x}{\sqrt{1 - x^2}\sqrt{1 - 4x^2}\left(\sin^{-1} 2x\right)^2}$

29. $\dfrac{x}{\sqrt{(1 - x^4)\sin^{-1} x^2}}$ **31.** $\sqrt{\dfrac{a - x}{a + x}}$

33. $\dfrac{\pi - 2}{\pi - 1}$

37. $\dfrac{d}{dx}\csc^{-1} x = -\dfrac{1}{|x|\sqrt{x^2 - 1}}$

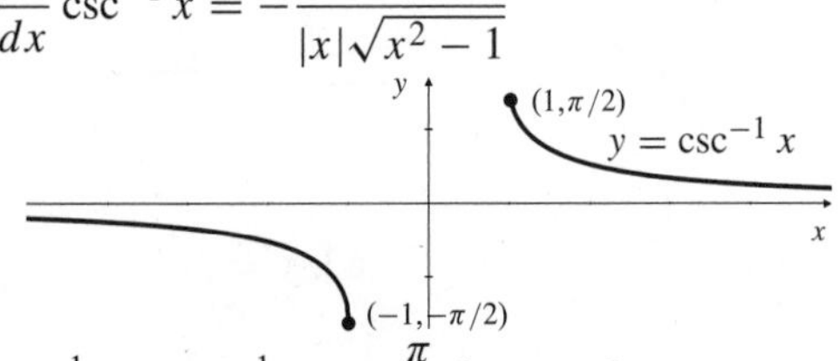

39. $\tan^{-1} x + \cot^{-1} x = -\dfrac{\pi}{2}$ for $x < 0$

41. cont. everywhere, differentiable except at $n\pi$ for integers n

43. continuous and differentiable everywhere except at odd multiples of $\pi/2$

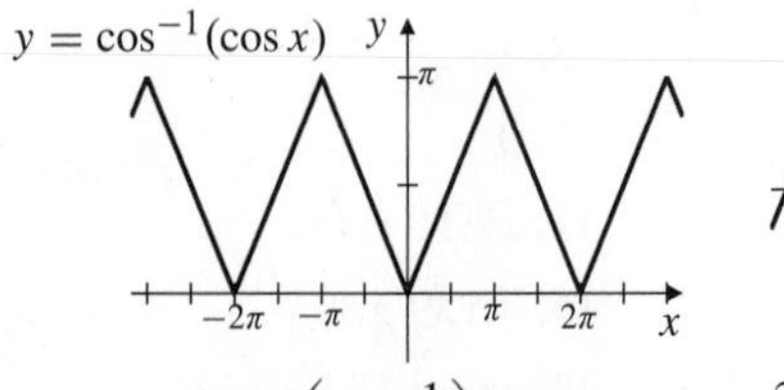

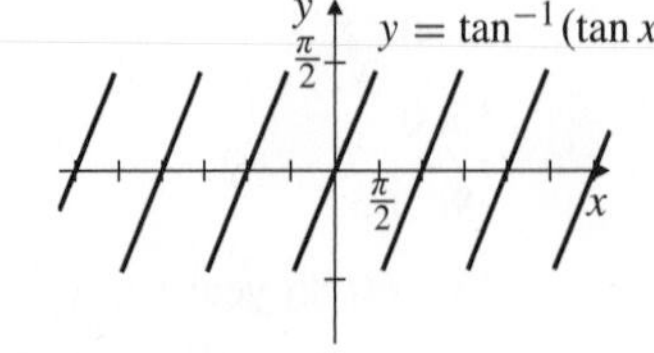

49. $\tan^{-1}\left(\dfrac{x - 1}{x + 1}\right) - \tan^{-1} x = \dfrac{3\pi}{4}$ on $(-\infty, -1)$

51. $f'(x) = 1 - \operatorname{sgn}(\cos x)$

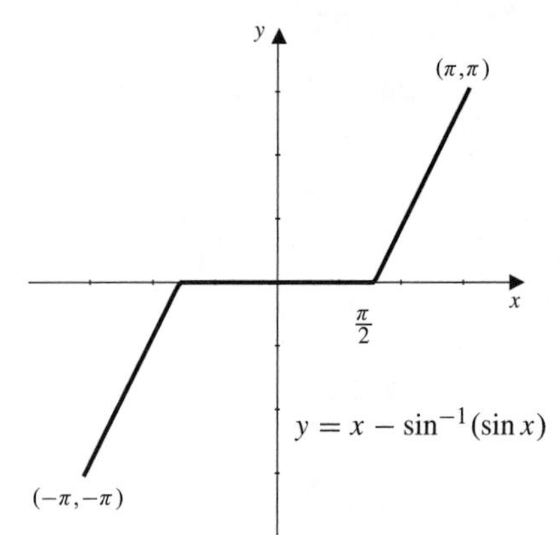

53. $y = \dfrac{1}{3}\tan^{-1}\dfrac{x}{3} + 2 - \dfrac{\pi}{12}$

55. $y = 4\sin^{-1}\dfrac{x}{5}$

Section 3.6 (page 203)

3. $\tanh(x + y) = \dfrac{\tanh x + \tanh y}{1 + \tanh x \tanh y}$

$\tanh(x - y) = \dfrac{\tanh x - \tanh y}{1 - \tanh x \tanh y}$

5. $\dfrac{d}{dx}\sinh^{-1}(x) = \dfrac{1}{\sqrt{x^2 + 1}}$,

$\dfrac{d}{dx}\cosh^{-1}(x) = \dfrac{1}{\sqrt{x^2 - 1}}$,

$\dfrac{d}{dx}\tanh^{-1}(x) = \dfrac{1}{1 - x^2}$,

$\displaystyle\int \frac{dx}{\sqrt{x^2 + 1}} = \sinh^{-1}(x) + C$,

$\displaystyle\int \frac{dx}{\sqrt{x^2 - 1}} = \cosh^{-1}(x) + C \quad (x > 1)$,

$\displaystyle\int \frac{dx}{1 - x^2} = \tanh^{-1}(x) + C \quad (-1 < x < 1)$

7. (a) $\dfrac{x^2 - 1}{2x}$; (b) $\dfrac{x^2 + 1}{2x}$; (c) $\dfrac{x^2 - 1}{x^2 + 1}$; (d) x^2

9. domain $(0, 1]$, range $[0, \infty)$, derivative $-1/(x\sqrt{1 - x^2})$

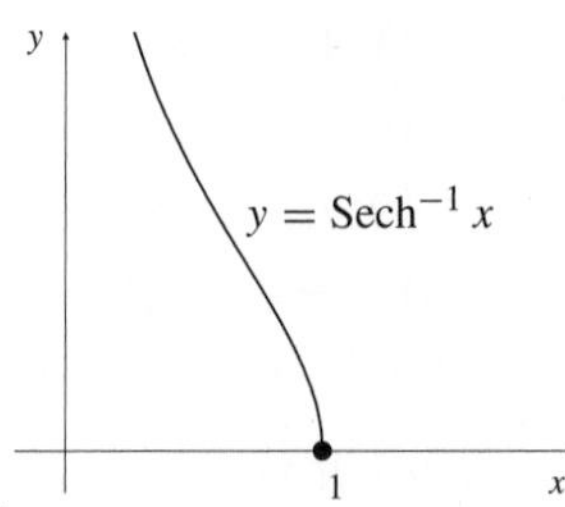

11. $f_{A,B} = g_{A+B,A-B}$; $g_{C,D} = f_{(C+D)/2,(C-D)/2}$

13. $y = y_0 \cosh k(x - a) + \dfrac{v_0}{k}\sinh k(x - a)$

Section 3.7 (page 210)

1. $y = Ae^{-5t} + Be^{-2t}$ **3.** $y = A + Be^{-2t}$

5. $y = (A + Bt)e^{-4t}$

7. $y = (A\cos t + B\sin t)e^{3t}$

9. $y = (A\cos 2t + B\sin 2t)e^{-t}$
11. $y = (A\cos\sqrt{2}t + B\sin\sqrt{2}t)e^{-t}$
13. $y = \frac{6}{7}e^{t/2} + \frac{1}{7}e^{-3t}$
15. $y = e^{-2t}(2\cos t + 6\sin t)$
25. $y = \frac{3}{10}\sin(10t)$, circ freq 10, freq $\frac{10}{2\pi}$, per $\frac{2\pi}{10}$, amp $\frac{3}{10}$

33. $y = e^{3-t}[2\cos(2(t-3)) + sin(2(t-3))]$

35. $y = \dfrac{c}{k^2}(1 - \cos(kx) + a\cos(kx) + \dfrac{b}{k}\sin(kx)$

Review Exercises (page 211)

1. $1/3$ **3.** both limits are 0
5. max $1/\sqrt{2e}$, min $-1/\sqrt{2e}$
7. $f(x) = 3e^{(x^2/2)-2}$
9. (a) about 13.863%, (b) about 68 days
11. e^{2x} **13.** y=x
15. 13.8165% approx.
17. $\cos^{-1}x = \frac{\pi}{2} - \sin^{-1}x$, $\cot^{-1}x = \operatorname{sgn} x \sin^{-1}(1/\sqrt{x^2+1})$, $\csc^{-1}x = \sin^{-1}(1/x)$
19. 15 °C

Challenging Problems (page 212)

Chapter 4 More Applications of Differentiation

Section 4.1 (page 218)

1. 32 cm^2/min
3. increasing at 160π cm^2/s
5. (a) $1/(6\pi r)$ km/hr, (b) $1/(6\sqrt{\pi A})$ km/hr
7. $1/(180\pi)$ cm/s **9.** 2 cm^2/s
11. increasing at 2 cm^3/s **13.** increasing at rate 12
15. increasing at rate $2/\sqrt{5}$
17. $45\sqrt{3}$ km/h **19.** 1/3 m/s, 5/6 m/s
21. 100 tons/day **23.** $16\frac{4}{11}$ min after 3:00
25. $1/(18\pi)$ m/min
27. $9/(6250\pi)$ m/min, 4.64 m
29. 8 m/min **31.** dec. at 126.9 km/h
33. 1/8 units/s **35.** $\sqrt{3}/16$ m/min
37. (a) down at 24/125 m/s, (b) right at 7/125 m/s
39. dec. at 0.0197 rad/s **41.** 0.047 rad/s

Section 4.2 (page 227)

1. 0.351734 **3.** 0.95025
5. 0.45340 **7.** 1.41421356237
9. 0.453397651516
11. 1.64809536561, 2.352392647658
13. 0.510973429389
15. infinitely many, 4.49340945791
19. max 1, min −0.11063967219 ...
21. $x_1 = -a$, $x_2 = a = x_0$. Look for a root half way between x_0 and x_1
23. $x_n = (-1/2)^n \to 0$ (root) as $n \to \infty$.

Section 4.3 (page 233)

1. $3/4$ **3.** a/b
5. 1 **7.** 1
9. 0 **11.** $-3/2$
13. 1 **15.** $-1/2$
17. ∞ **19.** $2/\pi$
21. -2 **23.** a
25. 1 **27.** $-1/2$
29. e^{-2} **31.** 0
33. $f''(x)$

Section 4.4 (page 239)

1. abs min 1 at $x = -1$; abs max 3 at $x = 1$
3. abs min 1 at $x = -1$; no max
5. abs min -1 at $x = 0$; abs max 8 at $x = 3$; loc max 3 at $x = -2$
7. abs min $a^3 + a - 4$ at $x = a$; abs max $b^3 + b - 4$ at $x = b$
9. abs max $b^5 + b^3 + 2b$ at $x = b$; no min value
11. no max or min values
13. max 3 at $x = -2$, min 0 at $x = 1$
15. abs max 1 at $x = 0$; no min value
17. no max or min value
19. loc max at $x = -1$; loc min at $x = 1$

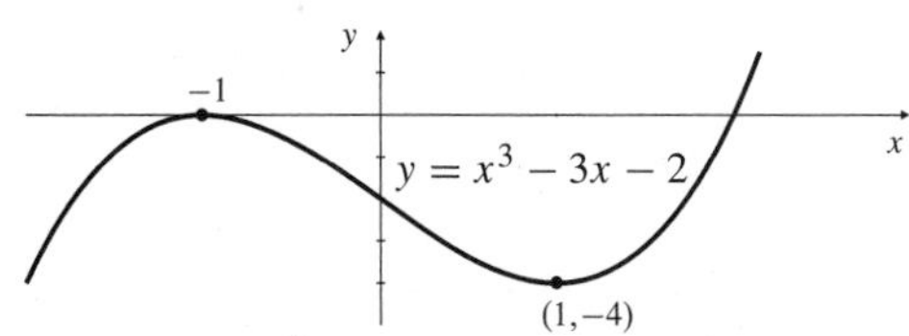

21. loc max at $x = \frac{3}{5}$; loc min at $x = 1$; critical point $x = 0$ is neither max nor min

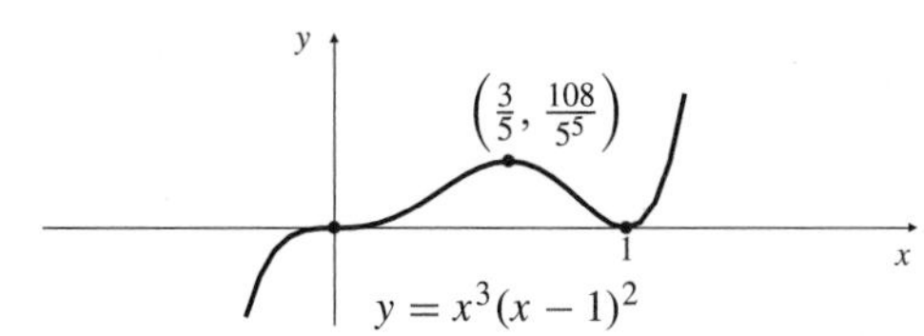

23. loc max at $x = -1$ and $x = 1/\sqrt{5}$; loc min at $x = 1$ and $x = -1/\sqrt{5}$

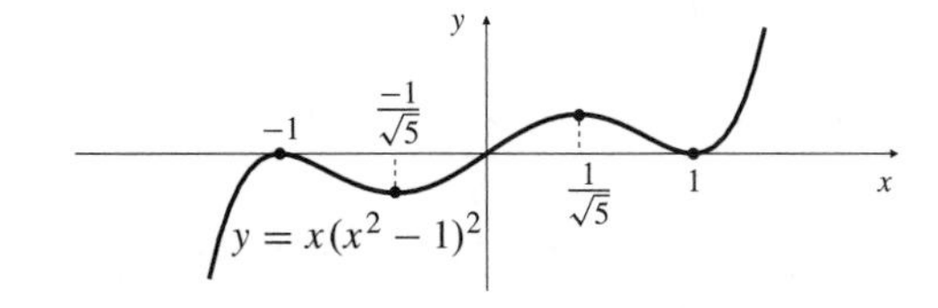

25. abs min at $x = 0$

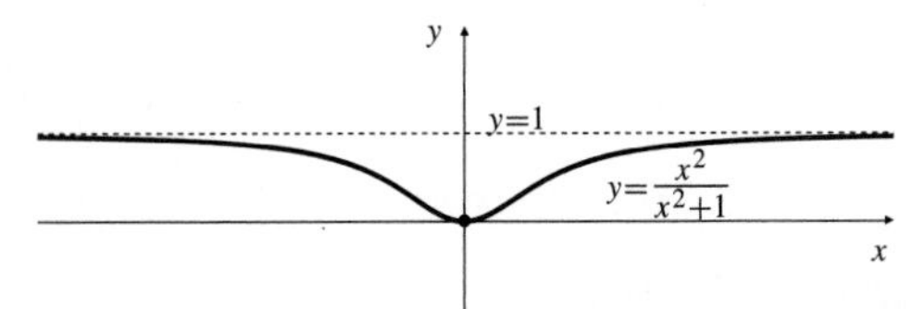

27. loc min at CP $x = -1$ and endpoint SP $x = \sqrt{2}$; loc max at CP $x = 1$ and endpoint SP $x = -\sqrt{2}$

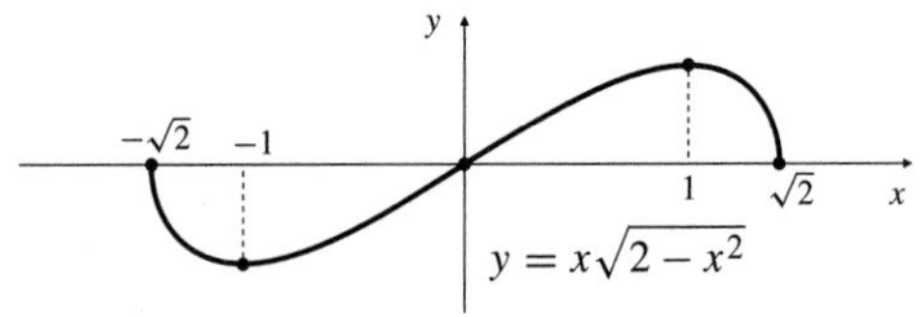

29. loc max at $x = 2n\pi - \dfrac{\pi}{3}$; loc min at $x = 2n\pi + \dfrac{\pi}{3}$ $(n = 0, \pm1, \pm2, \ldots)$

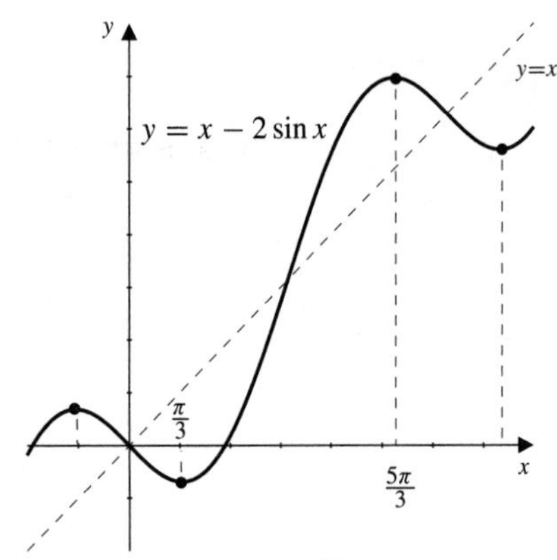

31. loc max at CP $x = \sqrt{3}/2$ and endpoint SP $x = -1$; loc min at CP $x = -\sqrt{3}/2$ and endpoint SP $x = 1$

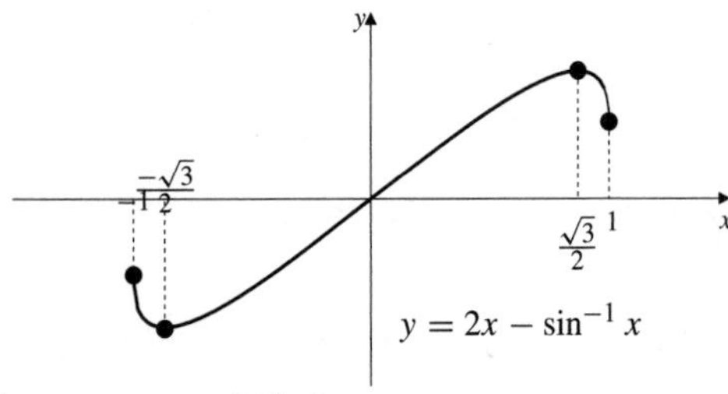

33. abs max at $x = 1/\ln 2$

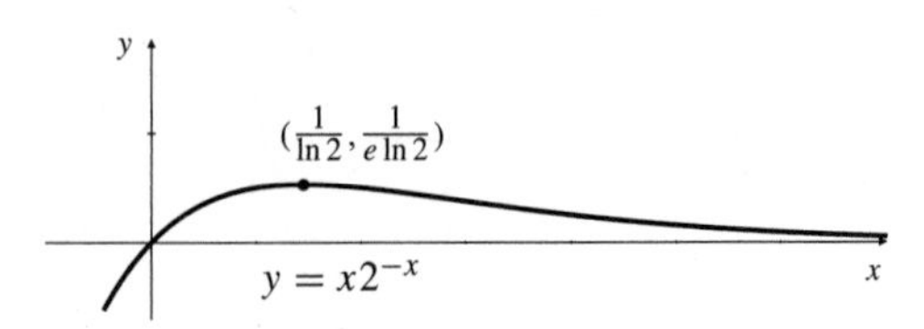

35. abs max at $x = e$

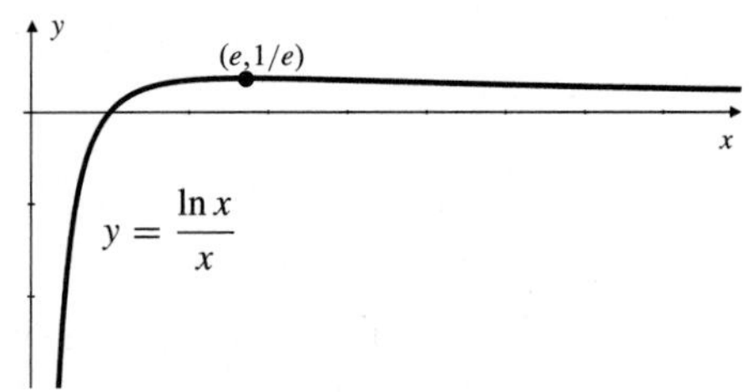

37. loc max at CP $x = 0$; abs min at SPs $x = \pm1$

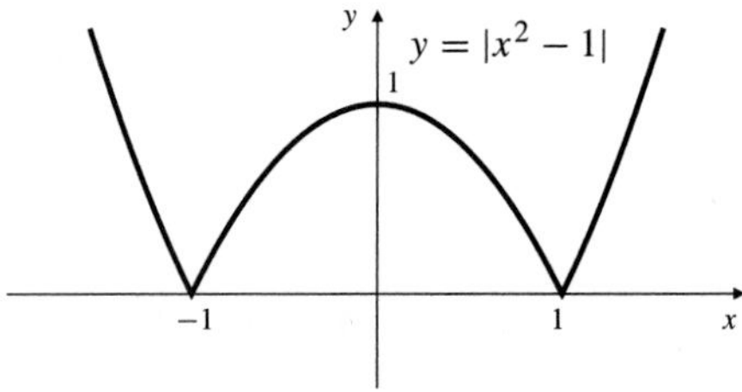

39. abs max at CPs $x = (2n+1)\pi/2$; abs min at SPs $x = n\pi$ $(n = 0, \pm1, \pm2, \ldots)$

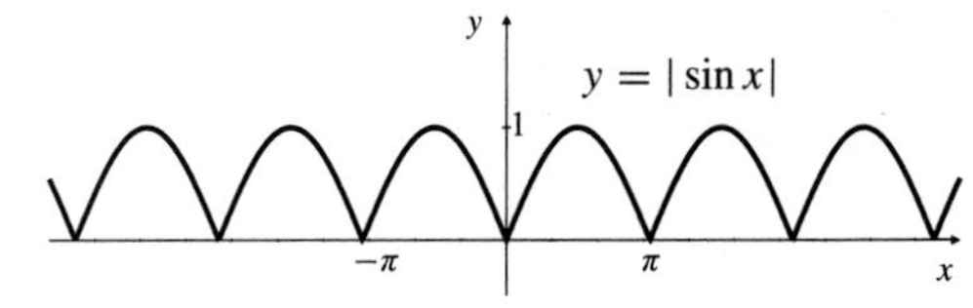

41. no max or min

43. max 2, min -2

45. has min, no max

47. yes, no

Section 4.5 (page 244)

1. concave down on $(0, \infty)$

3. concave up on $\mathbb{R}$

5. concave down on $(-1, 0)$ and $(1, \infty)$; concave up on $(-\infty, -1)$ and $(0, 1)$; inflection $x = -1, 0, 1$

7. concave down on $(-1, 1)$; concave up on $(-\infty, -1)$ and $(1, \infty)$; inflection $x = \pm1$

9. concave down on $(-2, -2/\sqrt{5})$ and $(2/\sqrt{5}, 2)$; concave up on $(-\infty, -2)$, $(-2/\sqrt{5}, 2/\sqrt{5})$ and $(2, \infty)$; inflection $x = \pm2,\ \pm2/\sqrt{5}$

11. concave down on $(2n\pi, (2n+1)\pi)$; concave up on $((2n-1)\pi, 2n\pi)$, $(n = 0, \pm1, \pm2, \ldots)$; inflection $x = n\pi$

13. concave down on $\left(n\pi, (n+\frac{1}{2})\pi\right)$; concave up on $\left((n-\frac{1}{2})\pi, n\pi\right)$; inflection $x = n\pi/2$, $(n = 0, \pm1, \pm2, \ldots)$

15. concave down on $(0, \infty)$, up on $(-\infty, 0)$; inflection $x = 0$

17. concave down on $(-1/\sqrt{2}, 1/\sqrt{2})$, up on $(-\infty, -1/\sqrt{2})$ and $(1/\sqrt{2}, \infty)$; inflection $x = \pm1/\sqrt{2}$

19. concave down on $(-\infty, -1)$ and $(1, \infty)$; conc up on $(-1, 1)$; inflection $x = \pm1$

21. concave down on $(-\infty, 4)$, up on $(4, \infty)$; inflection $x = 4$

23. no concavity, no inflections

25. loc min at $x = 2$; loc max at $x = \frac{2}{3}$

27. loc min at $x = 1/\sqrt[4]{3}$; loc max at $-1/\sqrt[4]{3}$

29. loc max at $x = 1$; loc min at $x = -1$ (both abs)

31. loc (and abs) min at $x = 1/e$

33. loc min at $x = 0$; inflections at $x = \pm 2$ (not discernible by Second Derivative Test)

35. abs min at $x = 0$; abs max at $x = \pm 1/\sqrt{2}$

39. If n is even, f_n has a min and g_n has a max at $x = 0$. If n is odd both have inflections at $x = 0$.

Section 4.6 (page 252)

1. (a) g, (b) f'', (c) f, (d) f'

3. (a) $k(x)$, (b) $g(x)$, (c) $f(x)$, (d) $h(x)$

5.

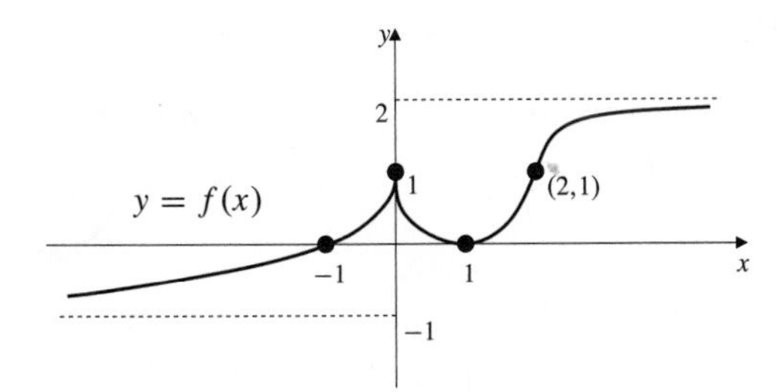

7.

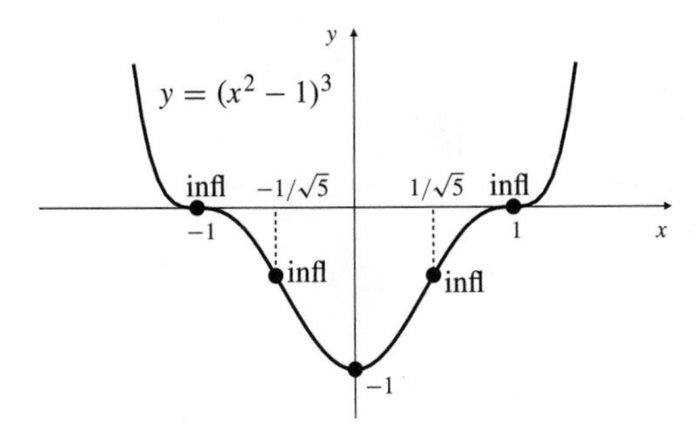

9.

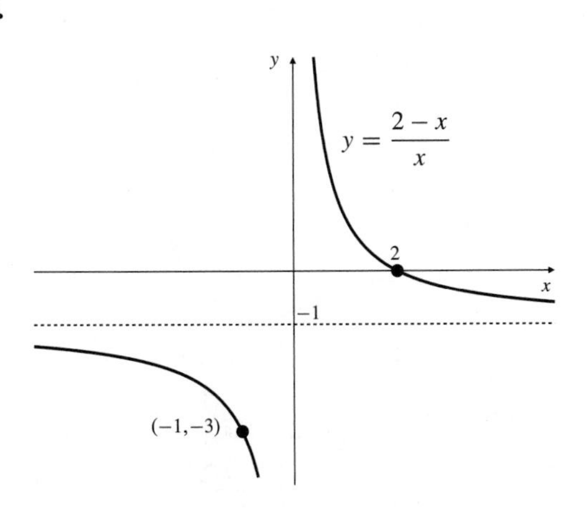

11.

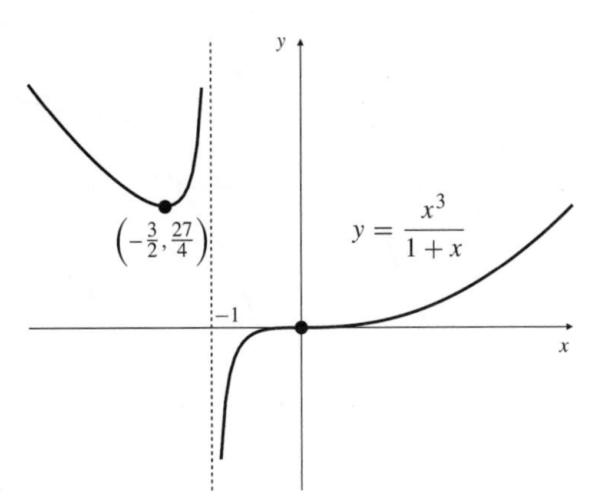

13.

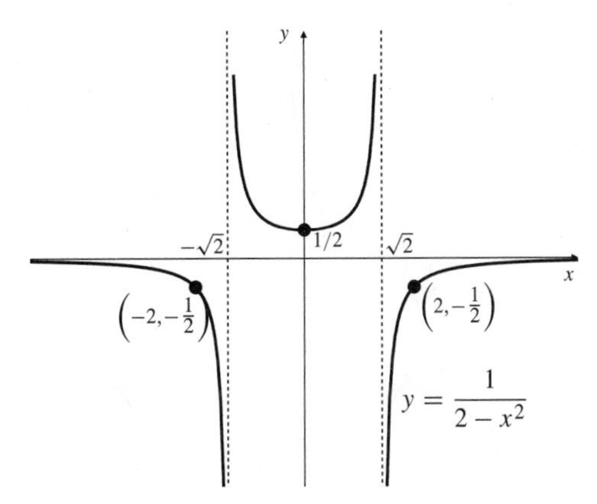

15.

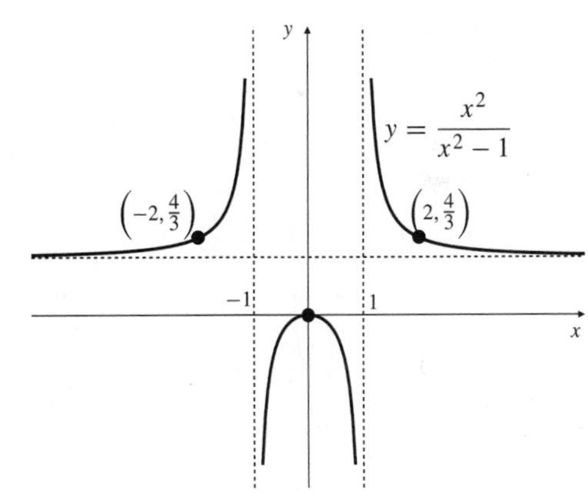

17.

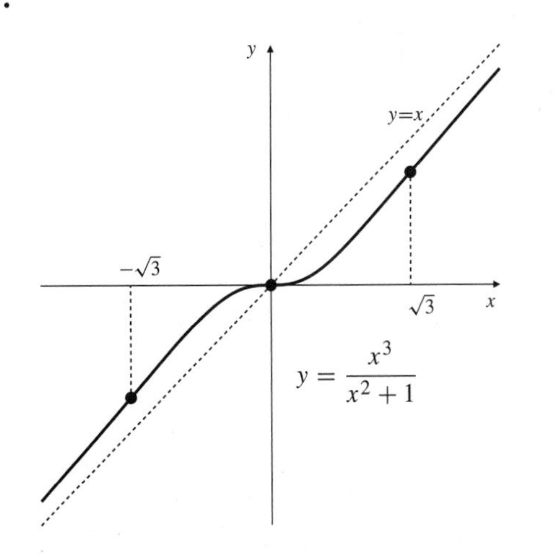

19.

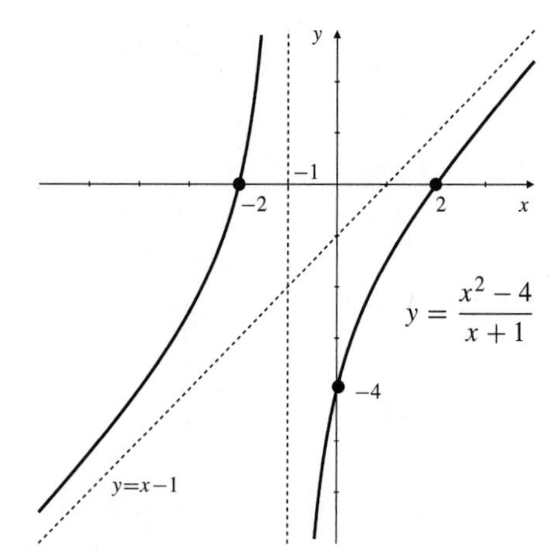

21.

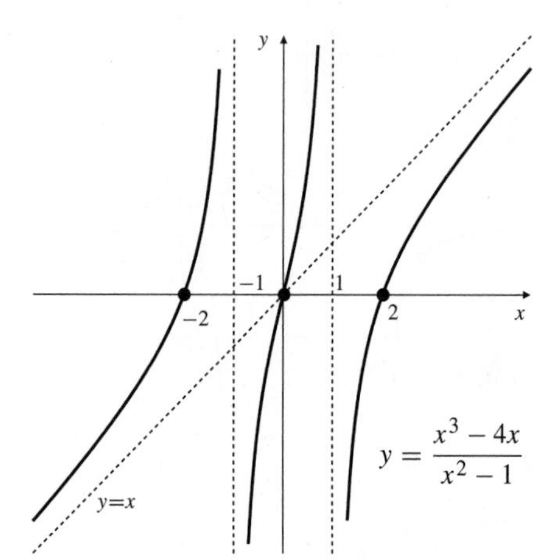

23.

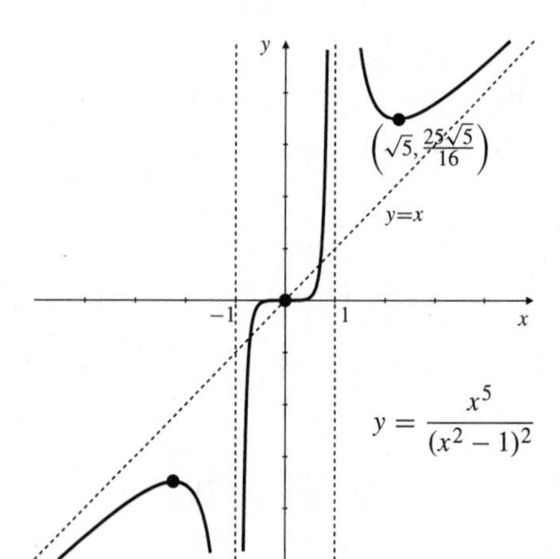

25.

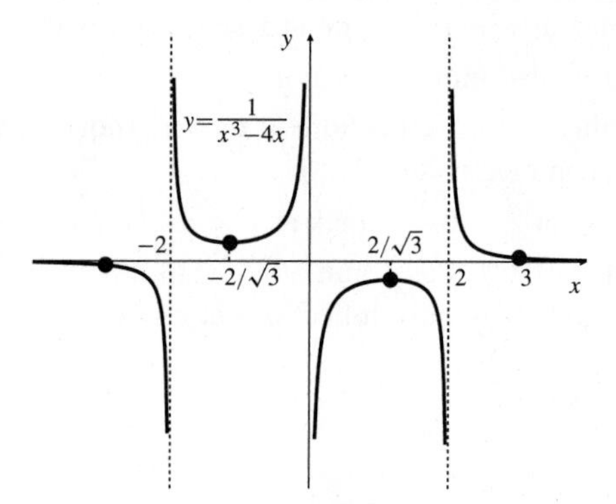

27.

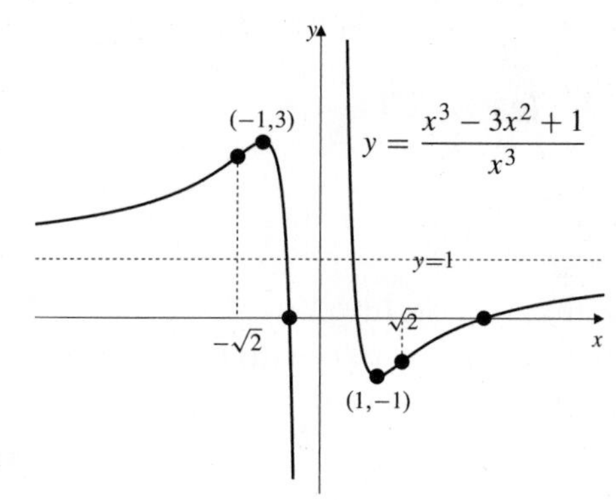

29.

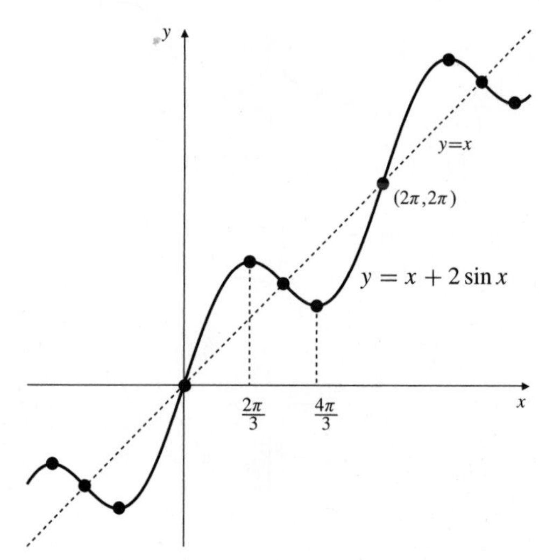

31.

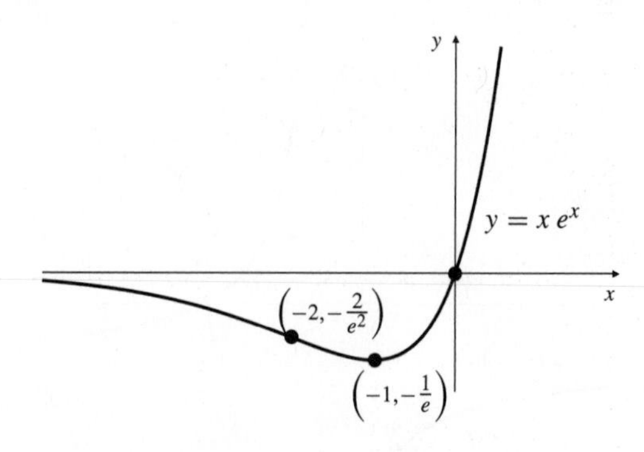

33.

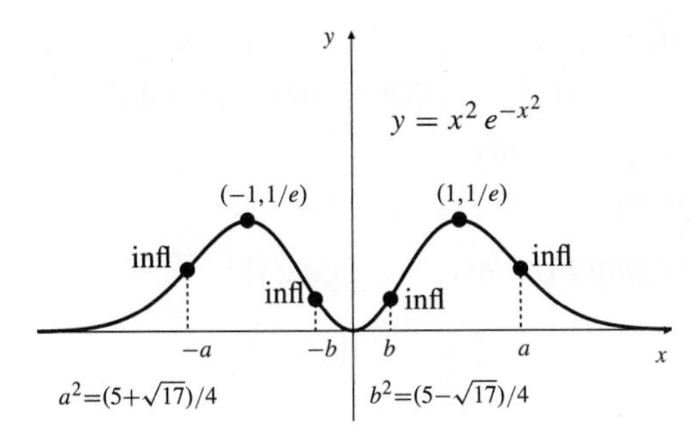

35.

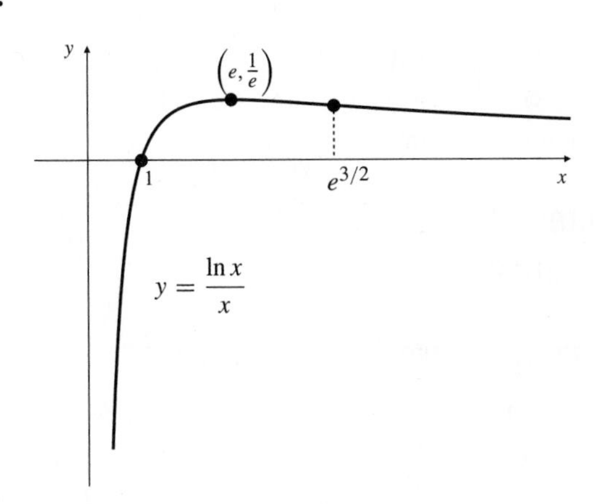

37.

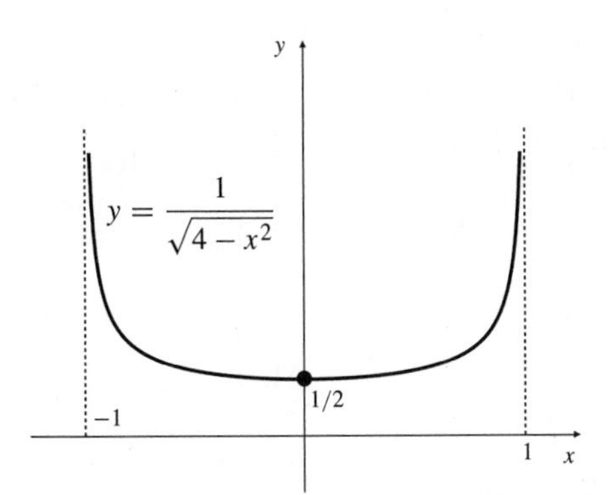

39.

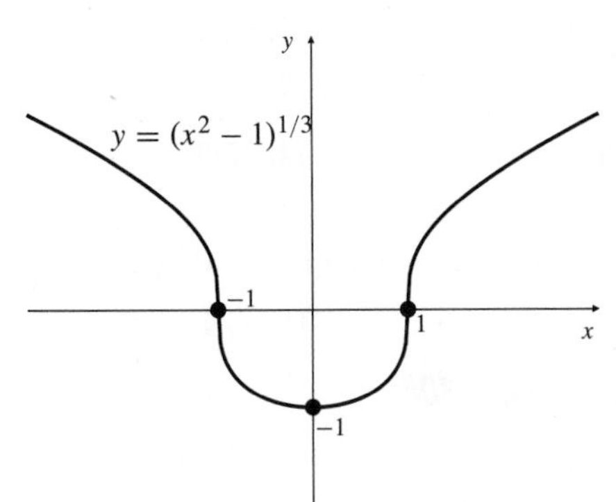

41. $y = 0$. Curve crosses asymptote at $x = n\pi$ for every integer n.

Section 4.7 (page 258)

5. 10^{-324}

Section 4.8 (page 264)

1. $49/4$ **3.** 20 and 40

5. 71.45 **11.** R^2 sq. units

13. $2ab$ units^2 **15.** 50 cm^2

17. width $8 + 10\sqrt{2}$ m, height $4 + 5\sqrt{2}$ m

19. rebate \$250 **21.** point 5 km east of A

25. (a) 0 m, (b) $\pi/(4+\pi)$ m

27. $8\sqrt{3}$ units

29. $\left[(a^{2/3}+b^{2/3})^3+c^2\right]^{1/2}$ units

31. $3^{1/2}/2^{1/3}$ units

33. height $\dfrac{2R}{\sqrt{3}}$, radius $\sqrt{\dfrac{2}{3}}\,R$ units

35. base 2m × 2m, height 1 m

37. width $\dfrac{20}{4+\pi}$ m, height $\dfrac{10}{4+\pi}$ m

41. width R, depth $\sqrt{3}R$ **43.** $Q = 3L/8$

45. 750 cars **47.** $\dfrac{5000}{\pi}$ m^2; semicircle

49. $\dfrac{3\sqrt{3}a}{4}$ cm

Section 4.9 (page 272)

1. $6x - 9$ **3.** $2 - (x/4)$

5. $(7-2x)/27$ **7.** $\pi - x$

9. $(1/4) + (\sqrt{3}/2)(x - (\pi/6))$

11. about 8 cm^2 **13.** about 62.8 mi

15. $\sqrt{50} \approx \frac{99}{14} \approx 7.071429$, error < 0,
$|\text{error}| < \frac{1}{2744} \approx 0.0003644$, $(7.07106, 7.071429)$

17. $\sqrt[4]{85} \approx \frac{82}{27}$, error < 0, $|\text{error}| < \frac{1}{2\times 3^6}$, $(3.03635, 3.03704)$

19. $\cos 46° \approx \dfrac{1}{\sqrt{2}}\left(1 - \dfrac{\pi}{180}\right) \approx 0.694765$, error < 0,
$|\text{error}| < \dfrac{1}{2\sqrt{2}}\left(\dfrac{\pi}{180}\right)^2$, $(0.694658, 0.694765)$

21. $\sin(3.14) \approx \pi - 3.14$, error < 0,
$|\text{error}| < (\pi - 3.14)^3/2 < 2.02 \times 10^{-9}$,
$(\pi - 3.14 - (\pi - 3.14)^3/2, \pi - 3.14)$

23. $(7.07106, 7.07108)$, $\sqrt{50} \approx 7.07107$

25. $(0.80891, 0.80921)$, $\sqrt[4]{85} \approx 0.80906$

27. $3 \le f(3) \le 13/4$

29. $g(1.8) \approx 0.6$, $|\text{error}| < 0.0208$

31. about 1005 cm^3

Section 4.10 (page 280)

1. $1 - x + \frac{1}{2}x^2 - \dfrac{1}{6}x^3 + \frac{1}{24}x^4$

3. $\ln 2 + \dfrac{x-2}{2} - \dfrac{(x-2)^2}{8} + \dfrac{(x-3)^3}{24} - \dfrac{(x-2)^4}{64}$

5. $2 + \dfrac{x-4}{4} - \dfrac{(x-4)^2}{64} + \dfrac{3(x-4)^3}{1536}$

7. $P_n(x) = \frac{1}{3} - \frac{1}{9}(x-1) + \frac{1}{27}(x-1)^2 - \cdots + \frac{(-1)^n}{3^{n+1}}(x-1)^n$

9. $x^{1/3} \approx 2 + \frac{1}{12}(x-8) - \frac{1}{288}(x-8)^2$, $9^{1/3} \approx 2.07986$, $0 < \text{error} \le 5/(81 \times 256)$, $2.07986 < 9^{1/3} < 2.08010$

11. $\frac{1}{x} \approx 1-(x-1)+(x-1)^2$, $\frac{1}{1.02} \approx 0.9804$, $-(0.02)^3 \le \text{error} < 0$, $0.980392 \le \frac{1}{1.02} < 0.9804$

13. $e^x \approx 1 + x + \frac{1}{2}x^2$, $e^{-0.5} \approx 0.625$, $-\frac{1}{6}(0.5)^3 \le \text{error} < 0$, $0.604 \le e^{-0.5} < 0.625$

15. $\sin x = x - \frac{x^3}{3!} + \frac{x^5}{5!} - \frac{x^7}{7!} + R_7$; $R_7 = \frac{\sin c}{8!}x^8$ for some c between 0 and x

17. $\sin x = \frac{1}{\sqrt{2}}\left[1 + \left(x - \frac{\pi}{4}\right) - \frac{1}{2!}\left(x - \frac{\pi}{4}\right)^2 - \frac{1}{3!}\left(x - \frac{\pi}{4}\right)^3 + \frac{1}{4!}\left(x - \frac{\pi}{4}\right)^4\right] + R_4$; where $R_4 = \frac{\cos c}{5!}\left(x - \frac{\pi}{4}\right)^5$ for some c between x and $\pi/4$

19. $\ln x = (x-1) - \frac{(x-1)^2}{2} + \frac{(x-1)^3}{3} - \frac{(x-11)^4}{4} + \frac{(x-1)^5}{5} - \frac{(x-1)^6}{6} + R_6$; where $R_6 = \frac{(x-1)^7}{7c^7}$ for some c between 1 and x

21. $\frac{1}{e^3} + \frac{3}{e^3}(x+1) + \frac{9}{2e^3}(x+1)^2 + \frac{9}{2e^3}(x+1)^3$

23. $x^2 - \frac{1}{3}x^4$

25. $1 - 2x^2 + 4x^4 - 8x^6$

27. $P_n(x) = 0$ if $0 \le n \le 2$; $P_n(x) = x^3$ if $n \ge 3$

29. $x + \frac{x^3}{3!} + \frac{x^5}{5!} + \cdots + \frac{x^{2n+1}}{(2n+1)!}$

31. $e^{-x} = 1 - x + \frac{x^2}{2!} - \frac{x^3}{3!} + \cdots + (-1)^n\frac{x^n}{n!} + R_n$; where $R_n = (-1)^{n+1}\frac{e^{-X}x^{n+1}}{(n+1)!}$ for some X between 0 and x; $\frac{1}{e} \approx \frac{1}{2!} - \frac{1}{3!} + \cdots + \frac{1}{8!} \approx 0.36788$

33. $1 - 2x + x^2$ (f is its own best quadratic approximation; (error = 0). $g(x) \approx 4 + 3x + 2x^2$; error $= x^3$; since $g'''(x) = 6 = 3!$, therefore error $= \frac{g'''(c)}{3!}x^3$; no improvement is possible.

35. $P_n(x) = 1 + 2x + 3x^2 + \cdots + (n+1)x^n$

Section 4.11 (page 284)

1. No. No.

Review Exercises (page 285)

1. $6\%/\text{min}$

3. (a) $-1{,}600$ ohms/min, (b) $-1{,}350$ ohms/min

5. $2,000$

7. $32\pi R^3/81$ un^3

9. 9000 cm^3

11. approx 0.057 rad/s

13. about 9.69465 cm

15. 2.06%

17. $\frac{\pi}{4} + 0.0475 \approx 0.83290$, $|\text{error}| < 0.00011$

19. 0, 1.4055636328

21. approx. $(-1.1462, 0.3178)$

Challenging Problems (page 286)

1. (a) $\frac{dx}{dt} = \frac{k}{3}(x_0^3 - x^3)$, (b) $V_0/2$

3. (b) 11

5. (c) $y_0(1 - (t/T))^2$, (d) $(1 - (1/\sqrt{2}))T$

7. $P^2(3 - 2\sqrt{2})/4$

9. (a) $\cos^{-1}(r_2/r_1)^2$, (b) $\cos^{-1}(r_2/r_1)^4$.

11. approx 921 cm^3

Chapter 5 Integration

Section 5.1 (page 293)

1. $1^3 + 2^3 + 3^3 + 4^3$

3. $3 + 3^2 + 3^3 + \cdots + 3^n$

5. $\frac{(-2)^3}{1^2} + \frac{(-2)^4}{2^2} + \frac{(-2)^5}{3^2} + \cdots + \frac{(-2)^n}{(n-2)^2}$

7. $\sum_{i=5}^{9} i$

9. $\sum_{i=2}^{99}(-1)^i i^2$

11. $\sum_{i=0}^{n} x^i$

13. $\sum_{i=1}^{n}(-1)^{i-1}/i^2$

15. $\sum_{i=1}^{100}\sin(i-1)$

17. $n(n+1)(2n+7)/6$

19. $\frac{\pi(\pi^n - 1)}{\pi - 1} - 3n$

21. $\ln(n!)$

23. 400

25. $(x^{2n+1} + 1)/(x+1)$

27. $-4{,}949$

31. $2^m - 1$

33. $n/(n+1)$

Section 5.2 (page 299)

1. 3/2 sq. units

3. 6 sq. units

5. 26/3 sq. units

7. 15 sq. units

9. 4 sq. units

11. 32/3 sq. units

13. $3/(2\ln 2)$ sq. units

15. $\ln(b/a)$, follows from definition of ln

17. 0

19. $\pi/4$

Section 5.3 (page 305)

1. $L(f, P_8) = 7/4$, $U(f, P_8) = 9/4$

3. $L(f, P_4) = \frac{e^4 - 1}{e^2(e-1)} \approx 4.22$, $U(f, P_4) = \frac{e^4 - 1}{e(e-1)} \approx 11.48$

5. $L(f, P_6) = \frac{\pi}{6}(1 + \sqrt{3}) \approx 1.43$, $U(f, P_6) = \frac{\pi}{6}(3 + \sqrt{3}) \approx 2.48$

7. $L(f, P_n) = \dfrac{n-1}{2n}$, $U(f, P_n) = \dfrac{n+1}{2n}$, $\int_0^1 x\,dx = \dfrac{1}{2}$

9. $L(f, P_n) = \dfrac{(n-1)^2}{4n^2}$, $U(f, P_n) = \dfrac{(n+1)^2}{4n^2}$, $\int_0^1 x^3\,dx = \dfrac{1}{4}$

11. $\int_0^1 \sqrt{x}\,dx$ **13.** $\int_0^\pi \sin x\,dx$

15. $\int_0^1 \tan^{-1} x\,dx$

Section 5.4 (page 310)

1. 0 **3.** 8

5. $(b^2 - a^2)/2$ **7.** π

9. 0 **11.** 2π

13. 0 **15.** $(2\pi + 3\sqrt{3})/6$

17. 16 **19.** $32/3$

21. $(4 + 3\pi)/12$ **23.** $\ln 2$

25. $\ln 3$ **27.** 4

29. 1 **31.** $\pi/2$

33. 1 **35.** $11/6$

37. $\dfrac{\pi}{3} - \sqrt{3}$ **39.** $41/2$

41. $3/4$ **43.** $k = \bar{f}$

Section 5.5 (page 316)

1. 4 **3.** 1

5. 9 **7.** $80\frac{4}{5}$

9. $\dfrac{2-\sqrt{2}}{2\sqrt{2}}$ **11.** $(1/\sqrt{2}) - (1/2)$

13. $e^\pi - e^{-\pi}$ **15.** $(a^e - 1)/\ln a$

17. $\pi/2$ **19.** $\dfrac{\pi}{3}$

21. $\frac{1}{5}$ sq. units **23.** $\frac{32}{3}$ sq. units

25. $\frac{1}{6}$ sq. units **27.** $\frac{1}{3}$ sq. units

29. $\frac{1}{12}$ sq. units **31.** 2π sq. units

33. 3 **35.** $\frac{16}{3}$

37. $e - 1$ **39.** $\dfrac{\sin x}{x}$

41. $-2\dfrac{\sin x^2}{x}$ **43.** $\dfrac{\cos t}{1+t^2}$

45. $(\cos x)/(2\sqrt{x})$ **47.** $f(x) = \pi e^{\pi(x-1)}$

49. $1/x^2$ is not continuous (or even defined) at $x = 0$ so the Fundamental Theorem cannot be applied over $[-1, 1]$. Since $1/x^2 > 0$ on its domain, we would expect the integral to be positive if it exists at all. (It doesn't.)

51. $F(x)$ has a maximum value at $x = 1$ but no minimum value.

53. 2

Section 5.6 (page 324)

1. $-\frac{1}{2}e^{5-2x} + C$ **3.** $\frac{2}{9}(3x+4)^{3/2} + C$

5. $-\frac{1}{32}(4x^2+1)^{-4} + C$ **7.** $\frac{1}{2}e^{x^2} + C$

9. $\frac{1}{2}\tan^{-1}\left(\frac{1}{2}\sin x\right) + C$

11. $2\ln\left|e^{x/2} - e^{-x/2}\right| + C = \ln\left|e^x - 2 + e^{-x}\right| + C$

13. $-\frac{2}{5}\sqrt{4-5s} + C$ **15.** $\frac{1}{2}\sin^{-1}\left(\dfrac{t^2}{2}\right) + C$

17. $-\ln\left(1 + e^{-x}\right) + C$ **19.** $-\frac{1}{2}(\ln\cos x)^2 + C$

21. $\frac{1}{2}\tan^{-1}\dfrac{x+3}{2} + C$

23. $\frac{1}{8}\cos^8 x - \frac{1}{6}\cos^6 x + C$

25. $-\dfrac{1}{3a}\cos^3 ax + C$

27. $\frac{5}{16}x - \frac{1}{4}\sin 2x + \frac{3}{64}\sin 4x + \frac{1}{48}\sin^3 2x + C$

29. $\frac{1}{5}\sec^5 x + C$

31. $\frac{2}{3}(\tan x)^{3/2} + \frac{2}{7}(\tan x)^{7/2} + C$

33. $\frac{3}{8}\sin x - \frac{1}{4}\sin(2\sin x) + \frac{1}{32}\sin(4\sin x) + C$

35. $\frac{1}{3}\tan^3 x + C$

37. $-\frac{1}{9}\csc^9 x + \frac{2}{7}\csc^7 x - \frac{1}{5}\csc^5 x + C$

39. $\frac{14}{3}\sqrt{17} + \frac{2}{3}$ **41.** $3\pi/16$

43. $\ln 2$ **45.** $2,\ 2(\sqrt{2} - 1)$

47. $\pi/32$ sq. units

Section 5.7 (page 328)

1. $\dfrac{1}{6}$ sq. units **3.** $\dfrac{64}{3}$ sq. units

5. $\dfrac{125}{12}$ sq. units **7.** $\dfrac{1}{2}$ sq. units

9. $\dfrac{5}{12}$ sq. units **11.** $\dfrac{15}{8} - 2\ln 2$ sq. units

13. $\dfrac{\pi}{2} - \dfrac{1}{3}$ sq. units **15.** $\dfrac{4}{3}$ sq. units

17. $2\sqrt{2}$ sq. units **19.** $1 - \pi/4$ sq. units

21. $(\pi/8) - \ln\sqrt{2}$ sq. units

23. $(4\pi/3) - 2\ln(2 + \sqrt{3})$ sq. units

25. $(4/\pi) - 1$ sq. units **27.** $\dfrac{4}{3}$ sq. units

29. $\dfrac{e}{2} - 1$ sq. units

Review Exercises (page 329)

1. sum is $n(n+2)/(n+1)^2$

3. $20/3$ **5.** 4π

7. 0 **9.** 2

11. $\sin(t^2)$ **13.** $-4e^{\sin(4s)}$

15. $f(x) = -\frac{1}{2}e^{(3/2)(1-x)}$ **17.** $9/2$ sq. units

19. $3/10$ sq. units **21.** $(3\sqrt{3}/4) - 1$ sq. units

23. $(\frac{1}{6}\sin(2x^3+1)+C$ **25.** $98/3$

27. $(\pi/8)-(1/2)\tan^{-1}(1/2)$

29. $-\cos\sqrt{2s+1}+C$ **31.** min $-\pi/4$, no max

35. $x_1=\dfrac{\sqrt{3}-1}{2\sqrt{3}},\quad x_2=\dfrac{\sqrt{3}+1}{2\sqrt{3}}$

Challenging Problems (page 330)

Chapter 6
Techniques of Integration

Section 6.1 (page 337)

1. $x\sin x+\cos x+C$

3. $\dfrac{1}{\pi}x^2\sin\pi x+\dfrac{2}{\pi^2}x\cos\pi x-\dfrac{2}{\pi^3}\sin\pi x+C$

5. $\frac{1}{4}x^4\ln x-\frac{1}{16}x^4+C$

7. $x\tan^{-1}x-\frac{1}{2}\ln(1+x^2)+C$

9. $\left(\frac{1}{2}x^2-\frac{1}{4}\right)\sin^{-1}x+\frac{1}{4}x\sqrt{1-x^2}+C$

11. $\frac{7}{8}\sqrt{2}+\frac{3}{8}\ln(1+\sqrt{2})$

13. $\frac{1}{13}e^{2x}(2\sin 3x-3\cos 3x)+C$

15. $\ln(2+\sqrt{3})-\dfrac{\pi}{6}$ **17.** $x\tan x-\ln|\sec x|+C$

19. $\dfrac{x}{2}\big[\cos(\ln x)+\sin(\ln x)\big]+C$

21. $\ln x\big(\ln(\ln x)-1\big)+C$

23. $x\cos^{-1}x-\sqrt{1-x^2}+C$

25. $\dfrac{2\pi}{3}-\ln(2+\sqrt{3})$

27. $\frac{1}{2}(x^2+1)\left(\tan^{-1}x\right)^2-x\tan^{-1}x+\frac{1}{2}\ln(1+x^2)+C$

29. $\dfrac{1+e^{-\pi}}{2}$ square units

31. $I_n=x(\ln x)^n-nI_{n-1}$,
$I_4=x\left[(\ln x)^4-4(\ln x)^3+12(\ln x)^2-24(\ln x)+24\right]+C$

33. $I_n=-\dfrac{1}{n}\sin^{n-1}x\cos x+\dfrac{n-1}{n}I_{n-2}$,
$I_6=\dfrac{5x}{16}-\cos x\left[\frac{1}{6}\sin^5x+\frac{5}{24}\sin^3x+\frac{5}{16}\sin x\right]+C$,
$I_7=-\cos x\left[\frac{1}{7}\sin^6x+\frac{6}{35}\sin^4x+\frac{8}{35}\sin^2x+\frac{16}{35}\right]+C$

35. $I_n=\dfrac{x}{2a^2(n-1)(x^2+a^2)^{n-1}}+\dfrac{2n-3}{2a^2(n-1)}I_{n-1}$,
$I_3=\dfrac{x}{4a^2(x^2+a^2)^2}+\dfrac{3x}{8a^4(x^2+a^2)}+\dfrac{3}{8a^5}\tan^{-1}\dfrac{x}{a}+C$

37. Any conditions which guarantee that $f(b)g'(b)-f'(b)g(b)=f(a)g'(a)-f'(a)g(a)$ will suffice.

Section 6.2 (page 346)

1. $\ln|2x-3|+C$

3. $\dfrac{x}{\pi}-\dfrac{2}{\pi^2}\ln|\pi x+2|+C$

5. $\dfrac{1}{6}\ln\left|\dfrac{x-3}{x+3}\right|+C$ **7.** $\dfrac{1}{2a}\ln\left|\dfrac{a+x}{a-x}\right|+C$

9. $x-\frac{4}{3}\ln|x+2|+\frac{1}{3}\ln|x-1|+C$

11. $3\ln|x+1|-2\ln|x|+C$

13. $\dfrac{1}{3(1-3x)}+C$

15. $-\frac{1}{9}x-\dfrac{13}{54}\ln|2-3x|+\frac{1}{6}\ln|x|+C$

17. $\dfrac{1}{2a^2}\ln\dfrac{|x^2-a^2|}{x^2}+C$

19. $x+\dfrac{a}{3}\ln|x-a|-\dfrac{a}{6}\ln(x^2+ax+a^2)$
$-\dfrac{a}{\sqrt{3}}\tan^{-1}\dfrac{2x+a}{\sqrt{3}a}+C$

21. $\frac{1}{3}\ln|x|-\frac{1}{2}\ln|x-1|+\frac{1}{6}\ln|x-3|+C$

23. $\dfrac{1}{4}\ln\left|\dfrac{x+1}{x-1}\right|-\dfrac{x}{2(x^2-1)}+C$

25. $\dfrac{1}{27}\ln\left|\dfrac{x-3}{x}\right|+\dfrac{1}{9x}+\dfrac{1}{6x^2}+C$

27. $\dfrac{x}{4}-\dfrac{1}{4}\ln|e^x-2|-\dfrac{1}{2(e^x-2)}+K$

29. $\dfrac{A}{x-1}+\dfrac{B}{(x-1)^2}+\dfrac{3}{(x-1)^3}+\dfrac{D}{x+1}+\dfrac{Ex+F}{x^2+x+1}$

31. $x-4+\dfrac{A}{x+2}+\dfrac{B}{(x+2)^2}+\dfrac{C}{(x+2)^3}+\dfrac{D}{x-2}$

Section 6.3 (page 353)

1. $\frac{1}{2}\sin^{-1}(2x)+C$

3. $\frac{9}{2}\sin^{-1}\dfrac{x}{3}-\frac{1}{2}x\sqrt{9-x^2}+C$

5. $-\dfrac{\sqrt{9-x^2}}{9x}+C$

7. $-\sqrt{9-x^2}+\sin^{-1}\dfrac{x}{3}+C$

9. $\frac{1}{3}(9+x^2)^{3/2}-9\sqrt{9+x^2}+C$

11. $\dfrac{1}{a^2}\dfrac{x}{\sqrt{a^2-x^2}}+C$

13. $\dfrac{x}{\sqrt{a^2-x^2}}-\sin^{-1}\dfrac{x}{a}+C$

15. $\dfrac{1}{2}\sec^{-1}\dfrac{x}{2}+C$ **17.** $\frac{1}{3}\tan^{-1}\dfrac{x+1}{3}+C$

19. $\frac{1}{32}\tan^{-1}\dfrac{2x+1}{2}+\dfrac{1}{16}\dfrac{2x+1}{4x^2+4x+5}+C$

21. $a\sin^{-1}\dfrac{x-a}{a}-\sqrt{2ax-x^2}+C$

23. $\dfrac{3-x}{4\sqrt{3-2x-x^2}}+C$

25. $\frac{3}{8}\tan^{-1}x+\dfrac{3x^3+5x}{8(1+x^2)^2}+C$

27. $\frac{1}{2}\ln\left(1+\sqrt{1-x^2}\right)-\frac{1}{2}\ln|x|-\dfrac{\sqrt{1-x^2}}{2x^2}+C$

29. $2\sqrt{x}-4\ln(2+\sqrt{x})+C$

31. $\frac{6}{7}x^{7/6}-\frac{6}{5}x^{5/6}+\frac{3}{2}x^{2/3}+2x^{1/2}$
$-3x^{1/3}-6x^{1/6}+3\ln(1+x^{1/3})+6\tan^{-1}x^{1/6}+C$

33. $\dfrac{\pi}{6}-\dfrac{\sqrt{3}}{8}$ **35.** $\pi/3$

37. $\dfrac{t-1}{4(t^2+1)} - \frac{1}{4}\ln|t+1| + \frac{1}{8}\ln(t^2+1) + C$

39. $\dfrac{1}{3}\ln\left|\dfrac{1-\sqrt{1-x^2}}{x}\right| + \dfrac{1}{12}\ln\left(\dfrac{\left(2+\sqrt{1-x^2}\right)^2}{3+x^2}\right) + C$

41. $\dfrac{1}{\sqrt{1+x^2}} + \dfrac{1}{2}\ln\left|\dfrac{1-\sqrt{1+x^2}}{1+\sqrt{1+x^2}}\right| + C$

43. $\dfrac{2}{\sqrt{3}}\tan^{-1}\left(\dfrac{2\tan(\theta/2)+1}{\sqrt{3}}\right) + C$

45. $\dfrac{2}{\sqrt{5}}\tan^{-1}\left(\dfrac{\tan(\theta/2)}{\sqrt{5}}\right) + C$

47. $\dfrac{9}{2\sqrt{2}}\tan^{-1}\dfrac{1}{\sqrt{2}} - \dfrac{1}{2}$ square units

49. $a^2\cos^{-1}\left(\dfrac{b}{a}\right) - b\sqrt{a^2-b^2}$ square units

51. $\dfrac{25}{2}\left(\sin^{-1}\dfrac{4}{5} - \sin^{-1}\dfrac{3}{5}\right) - 12\ln\dfrac{4}{3}$ square units

53. $\dfrac{\ln(Y+\sqrt{1+Y^2})}{2}$ sq. units

Section 6.4 (page 359)

1. $\dfrac{3}{25}e^{3x}\sin(4x) + C$

3. $-\left(\dfrac{x^4}{2} + x^2 + 1\right)e^{-x^2} + C$

9. $\dfrac{x\sqrt{x^2-2}}{2} + \ln|x+\sqrt{x^2-2}| + C$

11. $-\sqrt{3t^2+5}/(5t) + C$

13. $(x^5/3125)(625(\ln x)^4 - 500(\ln x)^3 + 300(\ln x)^2 - 120\ln x + 24) + C$

15. $(1/6)(2x^2-x-3)\sqrt{2x-x^2} - (1/2)\sin^{-1}(1-x) + C$

17. $(x-2)/(4\sqrt{4x-x^2}) + C$

Section 6.5 (page 367)

1. $1/2$ **3.** $1/2$

5. $3\times 2^{1/3}$ **7.** $3/2$

9. 3 **11.** π

13. $1/2$ **15.** diverges to ∞

17. 2 **19.** diverges

21. 0 **23.** 1 sq. unit

25. $2\ln 2$ square units **29.** 2

31. diverges to ∞ **33.** converges

35. diverges to ∞ **37.** diverges to ∞

39. diverges **41.** diverges to ∞

Section 6.6 (page 375)

1. $T_4 = 4.75$,
$M_4 = 4.625$,
$T_8 = 4.6875$,
$M_8 = 4.65625$,
$T_{16} = 4.671875$,
Actual errors:
$I - T_4 \approx -0.0833333$,
$I - M_4 \approx 0.0416667$,
$I - T_8 \approx -0.0208333$,
$I - M_8 \approx 0.0104167$,
$I - T_{16} \approx -0.0052083$
Error estimates:
$|I - T_4| \le 0.0833334$,
$|I - M_4| \le 0.0416667$,
$|I - T_8| \le 0.0208334$,
$|I - M_8| \le 0.0104167$,
$|I - T_{16}| \le 0.0052084$

3. $T_4 = 0.9871158$,
$M_4 = 1.0064545$,
$T_8 = 0.9967852$,
$M_8 = 1.0016082$,
$T_{16} = 0.9991967$,
Actual errors:
$I - T_4 \approx 0.0128842$,
$I - M_4 \approx -0.0064545$,
$I - T_8 \approx 0.0032148$,
$I - M_8 \approx -0.0016082$,
$I - T_{16} \approx 0.0008033$
Error estimates:
$|I - T_4| \le 0.020186$,
$|I - M_4| \le 0.010093$,
$|I - T_8| \le 0.005047$,
$|I - M_8| \le 0.002523$,
$|I - T_{16}| \le 0.001262$

5. $T_4 = 46$, $T_8 = 46.7$

7. $T_4 = 3{,}000\ \text{km}^2$, $T_8 = 3{,}400\ \text{km}^2$

9. $T_4 \approx 2.02622$, $M_4 \approx 2.03236$,
$T_8 \approx 2.02929$, $M_8 \approx 2.02982$,
$T_{16} \approx 2.029555$

11. $M_8 \approx 1.3714136$, $T_{16} \approx 1.3704366$, $I \approx 1.371$

Section 6.7 (page 380)

1. $S_4 = S_8 = I$, Errors $= 0$

3. $S_4 \approx 1.0001346$, $S_8 \approx 1.0000083$,
$I - S_4 \approx -0.0001346$, $I - S_8 \approx -0.0000083$

5. 46.93

7. For $f(x) = e^{-x}$:
$|I - S_4| \le 0.000022$, $|I - S_8| \le 0.0000014$;
for $f(x) = \sin x$,
$|I - S_4| \le 0.00021$,
$|I - S_8| \le 0.000013$

9. $S_4 \approx 2.0343333$, $S_8 \approx 2.0303133$,
$S_{16} \approx 2.0296433$

Section 6.8 (page 386)

1. $3\displaystyle\int_0^1 \frac{u\,du}{1+u^3}$

3. $\displaystyle\int_{-\pi/2}^{\pi/2} e^{\sin\theta}\,d\theta$, or $2\displaystyle\int_0^1 \frac{e^{1-u^2}+e^{u^2-1}}{\sqrt{2-u^2}}\,du$

5. $4\displaystyle\int_0^1 \frac{dv}{\sqrt{(2-v^2)(2-2v^2+v^4)}}$

7. $T_2 \approx 0.603553$ $T_4 \approx 0.643283$,
$T_8 \approx 0.658130$, $T_{16} \approx 0.663581$;
Errors: $I - T_2 \approx 0.0631$, $I - T_4 \approx 0.0234$,
$I - T_8 \approx 0.0085$, $I - T_{16} \approx 0.0031$.
Errors do not decrease like $1/n^2$ because the second derivative of $f(x) = \sqrt{x}$ is not bounded on $[0, 1]$.

9. $I \approx 0.74684$ with error less than 10^{-4}; seven terms of the series are needed.

11. $A = 1$, $u = 1/\sqrt{3}$

13. $A = 5/9$, $B = 8/9$, $u = \sqrt{3/5}$

15. $R_1 \approx 0.7471805$, $R_2 \approx 0.7468337$,
$R_3 \approx 0.7468241$, $I \approx 0.746824$

17. $R_2 = \dfrac{2h}{45}\big(7y_0 + 32y_1 + 12y_2 + 32y_3 + 7y_4\big)$

Review Exercises on Techniques of Integration (page 388)

1. $\frac{2}{3}\ln|x+2| - \frac{1}{6}\ln|2x+1| + C$

3. $\frac{1}{4}\sin^4 x - \frac{1}{6}\sin^6 x + C$

5. $\dfrac{3}{4}\ln\left|\dfrac{2x-1}{2x+1}\right| + C$

7. $-\dfrac{1}{3}\left(\dfrac{\sqrt{1-x^2}}{x}\right)^3 + C$

9. $\frac{1}{5}\left(5x^3-2\right)^{1/3} + C$

11. $\frac{1}{16}\tan^{-1}\dfrac{x}{2} + \dfrac{x}{8(4+x^2)} + C$

13. $\dfrac{1}{2\ln 2}\big(2^x\sqrt{1+4^x} + \ln(2^x + \sqrt{1+4^x})\big) + C$

15. $\frac{1}{4}\tan^4 x + \frac{1}{6}\tan^6 x + C$

17. $-e^{-x}\left(\frac{2}{5}\cos 2x + \frac{1}{5}\sin 2x\right) + C$

19. $\dfrac{x}{10}\big(\cos(3\ln x) + 3\sin(3\ln x)\big) + C$

21. $\frac{1}{4}\big(\ln(1+x^2)\big)^2 + C$

23. $\sin^{-1}\dfrac{x}{\sqrt{2}} - \dfrac{x\sqrt{2-x^2}}{2} + C$

25. $\dfrac{1}{64}\left(-\dfrac{1}{7(4x+1)^7} + \dfrac{1}{4(4x+1)^8} - \dfrac{1}{9(4x+1)^9}\right) + C$

27. $-\frac{1}{4}\cos 4x + \frac{1}{6}\cos^3 4x - \frac{1}{20}\cos^5 4x + C$

29. $-\frac{1}{2}\ln(2e^{-x}+1) + C$

31. $-\frac{1}{2}\sin^2 x - 2\sin x - 4\ln(2-\sin x) + C$

33. $-\dfrac{\sqrt{1-x^2}}{x} + C$

35. $\frac{1}{48}(1-4x^2)^{3/2} - \frac{1}{16}\sqrt{1-4x^2} + C$

37. $\sqrt{x^2+1} + \ln(x + \sqrt{x^2+1}) + C$

39. $x + \frac{1}{3}\ln|x| + \frac{4}{3}\ln|x-3| - \frac{5}{3}\ln|x+3| + C$

41. $-\frac{1}{10}\cos^{10} x + \frac{1}{6}\cos^{12} x - \frac{1}{14}\cos^{14} x + C$

43. $\dfrac{1}{2}\ln|x^2+2x-1| - \dfrac{1}{2\sqrt{2}}\ln\left|\dfrac{x+1-\sqrt{2}}{x+1+\sqrt{2}}\right| + C$

45. $\frac{1}{3}x^3\sin^{-1}2x + \frac{1}{24}\sqrt{1-4x^2} - \frac{1}{72}(1-4x^2)^{3/2} + C$

47. $\frac{1}{128}\left(3x - \sin(4x) + \frac{1}{8}\sin(8x)\right)$

49. $\tan^{-1}\dfrac{\sqrt{x}}{2} + C$

51. $\dfrac{x^2}{2} - 2x + \dfrac{1}{4}\ln|x| + \dfrac{1}{2x} + \dfrac{15}{4}\ln|x+2| + C$

53. $-\frac{1}{2}\cos(2\ln x) + C$

55. $\frac{1}{2}\exp(2\tan^{-1}x) + C$

57. $\frac{1}{4}\big(\ln(3+x^2)\big)^2 + C$

59. $\frac{1}{2}\big(\sin^{-1}(x/2)\big)^2 + C$

61. $\sqrt{x^2+6x+10} - 2\ln(x+3+\sqrt{x^2+6x+10}) + C$

63. $\dfrac{2}{5(2+x^2)^{5/2}} - \dfrac{1}{3(2+x^2)^{3/2}} + C$

65. $\frac{6}{7}x^{7/6} - \frac{6}{5}x^{5/6} + 2\sqrt{x} - 6x^{1/6} + 6\tan^{-1}x^{1/6} + C$

67. $\frac{2}{3}x^{3/2} - x + 4\sqrt{x} - 4\ln(1+\sqrt{x}) + C$

69. $\dfrac{1}{2(4-x^2)} + C$

71. $\frac{1}{3}x^3\tan^{-1}x - \frac{1}{6}x^2 + \frac{1}{6}\ln(1+x^2) + C$

73. $\dfrac{1}{5}\ln\left|\dfrac{3\tan(x/2)-1}{\tan(x/2)+3}\right| + C$

75. $\frac{1}{2}\ln|\tan(x/2)| - \frac{1}{4}\big(\tan^{-1}(x/2)\big)^2 + C$
$= \dfrac{1}{4}\left(\ln\left|\dfrac{1-\cos x}{1+\cos x}\right| - \dfrac{1-\cos x}{1+\cos x}\right) + C$

77. $2\sqrt{x} - 2\tan^{-1}\sqrt{x} + C$

79. $\dfrac{1}{2}x^2 + \dfrac{4}{3}\ln|x-2| - \frac{2}{3}\ln(x^2+2x+4)$
$+ \dfrac{4}{\sqrt{3}}\tan^{-1}\dfrac{x+1}{\sqrt{3}} + C$

Review Exercises (Other) (page 389)

1. $I = \frac{1}{2}\big(xe^x\cos x + (x-1)e^x\sin x\big)$,
$J = \frac{1}{2}\big((1-x)e^x\cos x + xe^x\sin x\big)$

3. diverges to ∞

5. $-4/9$

9. 367,000 m^3

11. $T_8 = 1.61800$, $S_8 = 1.62092$, $I \approx 1.62$

13. (a) $T_4 = 5.526$, $S_4 = 5.504$; (b) $S_8 = 5.504$; (c) Yes, because $S_4 = S_8$, and Simpson's Rule is exact for cubics.

Challenging Problems (page 389)

1. (c) $I = \dfrac{1}{630}$, $\dfrac{22}{7} - \dfrac{1}{630} < \pi < \dfrac{22}{7} - \dfrac{1}{1260}$.

3. (a) $\dfrac{1}{\sqrt{3}}\tan^{-1}\left(\dfrac{2x+1}{\sqrt{3}}\right) + \dfrac{1}{\sqrt{3}}\tan^{-1}\left(\dfrac{2x-1}{\sqrt{3}}\right)$,
(b) $\dfrac{1}{\sqrt{2}}\tan^{-1}(\sqrt{2}x+1) + \dfrac{1}{\sqrt{2}}\tan^{-1}(\sqrt{2}x-1)$

7. (a) $a = 7/90$, $b = 16/45$, $c = 2/15$.
(b) one interval: approx 0.6321208750, two intervals: approx 0.6321205638, true val: 0.6321205588

Chapter 7 Applications of Integration

Section 7.1 (page 399)

1. $\frac{\pi}{5}$ cu. units **3.** $\frac{3\pi}{10}$ cu. units

5. (a) $\frac{16\pi}{15}$ cu. units, (b) $\frac{8\pi}{3}$ cu. units

7. (a) $\frac{27\pi}{2}$ cu. units, (b) $\frac{108\pi}{5}$ cu. units

9. (a) $\frac{15\pi}{4} - \frac{\pi^2}{8}$ cu. units, (b) $\pi(2 - \ln 2)$ cu. units

11. $\frac{10\pi}{3}$ cu. units **13.** about 35%

15. $\frac{\pi h}{3}\left(b^2 - 3a^2 + \frac{2a^3}{b}\right)$ cu. units

17. $\frac{\pi}{3}(a-b)^2(2a+b)$ cu. units

19. $\frac{4\pi ab^2}{3}$ cu. units

21. (a) $\pi/2$ cu. units, (b) 2π cu. units

23. $k > 2$ **25.** yes; no; $a^2b/2$ cm^3

27. about 1,537 cu. units **29.** $8192\pi/105$ cu. units

31. $R = \frac{h \sin\alpha}{\sin\alpha + \cos 2\alpha}$

Section 7.2 (page 403)

1. 6 m^3 **3.** $\pi/3$ units^3

5. 132 ft^3 **7.** $\pi a^2 h/2$ cm^3

9. $3z^2$ sq. units **11.** $\frac{16r^3}{3}$ cu. units

13. 72π cm^3 **15.** $\pi r^2(a+b)/2$ cu. units

17. $\frac{16,000}{3}$ cu. units **19.** $12\pi\sqrt{2}$ in^3

21. approx 97.28 cm^3

Section 7.3 (page 410)

1. $2\sqrt{5}$ units **3.** $52/3$ units

5. $(2/27)(13^{3/2} - 8)$ units **7.** 6 units

9. $(e^2 + 1)/4$ units **11.** $\sinh a$ units.

13. $\sqrt{17} + \frac{1}{4}\ln(4 + \sqrt{17})$ units

15. $6a$ units **17.** 1.0338 units

19. 1.0581

21. $(10^{3/2} - 1)\pi/27$ sq. units

23. $\frac{64\pi}{81}\left[\frac{(13/4)^{5/2} - 1}{5} - \frac{(13/4)^{3/2} - 1}{3}\right]$ sq. units

25. $2\pi\left(\sqrt{2} + \ln(1 + \sqrt{2})\right)$ sq. units

27. $2\pi\left(\frac{255}{16} + \ln 4\right)$ sq. units

29. $4\pi^2 ab$ sq. units

31. $8\pi\left(1 + \frac{\ln(2+\sqrt{3})}{2\sqrt{3}}\right)$ sq. units

33. $s = \frac{5}{\pi}\sqrt{4+\pi^2}\,E\left(\frac{\pi}{\sqrt{4+\pi^2}}\right)$

35. $k > -1$

37. (a) π cu. units; (c) "Covering" a surface with paint requires putting on a layer of constant thickness. Far enough to the right, the horn is thinner than any prescribed constant, so it can contain less paint than would be necessary to paint its surface.

Section 7.4 (page 417)

1. mass $\frac{2L}{\pi}$; centre of mass at $\bar{s} = \frac{L}{2}$

3. $m = \frac{1}{4}\pi\sigma_0 a^2$; $\bar{x} = \bar{y} = \frac{4a}{3\pi}$

5. $m = \frac{256k}{15}$; $\bar{x} = 0$, $\bar{y} = \frac{16}{7}$

7. $m = \frac{ka^3}{2}$; $\bar{x} = \frac{2a}{3}$, $\bar{y} = \frac{a}{2}$

9. $m = \int_a^b \sigma(x)\big(g(x) - f(x)\big)\,dx$;
$M_{x=0} = \int_a^b x\sigma(x)\big(g(x) - f(x)\big)\,dx$, $\bar{x} = M_{x=0}/m$,
$M_{y=0} = \frac{1}{2}\int_a^b \sigma(x)\big((g(x))^2 - (f(x))^2\big)\,dx$,
$\bar{y} = M_{y=0}/m$

11. Mass is $\frac{8}{3}\pi R^4$ kg.The centre of mass is along the line through the centre of the ball perpendicular to the plane, at a distance $R/10$ m from the centre of the ball on the side opposite the plane.

13. $m = \frac{1}{8}\pi\rho_0 a^4$; $\bar{x} = 16a/(15\pi)$, $\bar{y} = 0$, $\bar{z} = 8a/15$

15. $m = \frac{1}{3}k\pi a^3$; $\bar{x} = 0$, $y = \frac{3a}{2\pi}$

17. about $5.57C/k^{3/2}$

Section 7.5 (page 422)

1. $\left(\frac{4r}{3\pi}, \frac{4r}{3\pi}\right)$

3. $\left(\frac{\sqrt{2} - 1}{\ln(1+\sqrt{2})}, \frac{\pi}{8\ln(1+\sqrt{2})}\right)$

5. $\left(0, \frac{9\sqrt{3} - 4\pi}{4\pi - 3\sqrt{3}}\right)$ **7.** $\left(\frac{19}{9}, -\frac{1}{3}\right)$

9. The centroid is on the axis of symmetry of the hemisphere half way between the base plane and the vertex.

11. The centroid is on the axis of the cone, one quarter of the cone's height above the base plane.

13. $\left(\frac{\pi}{2}, \frac{\pi}{8}\right)$ **15.** $\left(\frac{2r}{\pi}, \frac{2r}{\pi}\right)$

17. $(8/9, 11/9)$ **19.** $(0, 2/(3(\pi + 2)))$

21. $(1, -2)$ **23.** $\frac{5\pi}{3}$ cu. units

25. $(0.71377, 0.26053)$ **27.** $\left(1, \frac{1}{5}\right)$

29. $\bar{x} = \dfrac{M_{x=0}}{A}$, $\bar{y} = \dfrac{M_{y=0}}{A}$,
where $A = \int_c^d (g(y) - f(y))\,dy$,
$M_{x=0} = \frac{1}{2}\int_c^d ((g(y))^2 - (f(y))^2)\,dy$,
$M_{y=0} = \int_c^d y(g(y) - f(y))\,dy$

31. diamond orientation, edge upward

Section 7.6 (page 429)

1. (a) 235,200 N, (b) 352,800 N

3. 6.12×10^8 N

5. 8.92×10^6 N

7. 7.056×10^5 N·m

9. $2450\pi a^3\left(a + \dfrac{8h}{3}\right)$ N·m

11. $\dfrac{19{,}600}{3} XR^3$ N·m

Section 7.7 (page 433)

1. \$11, 000

3. $\$8(\sqrt{x} - \ln(1 + \sqrt{x}))$

5. \$9,063.46

7. \$5,865.64

9. \$50,000

11. \$11,477.55

13. \$64,872.10

15. $\int_0^T e^{-\lambda(t)}P(t)\,dt$

17. about 23,300, \$11,890

Section 7.8 (page 445)

1. no more than \$2.47

3. \$6.81

5. $\mu \approx 3.5833$, $\sigma = 1.7059$, $\Pr(X \le 3) = 0.4833$

7. (a) eight triples (x, y, z) where $x, y, z \in \{H, T\}$
(b) $\Pr(H, H, H) = 0.166375$, $\Pr(H, H, T) = \Pr(H, T, H) = \Pr(T, H, H) = 0.136125$, $\Pr(H, T, T) = \Pr(T, H, T) = \Pr(T, T, H) = 0.111375$, $\Pr(T, T, T) = 0.091125$
(c) $f(0) = 0.911125$, $f(1) = 0.334125$, $f(2) = 0.408375$, $f(3) = 0.166375$
(d) 0.908875, (e) 1.650000

9. (a) $\dfrac{2}{9}$, (b) $\mu = 2$, $\sigma^2 = \dfrac{1}{2}$, $\sigma = \dfrac{1}{\sqrt{2}}$,
(c) $\dfrac{8}{9\sqrt{2}} \approx 0.63$

11. (a) 3, (b) $\mu = \dfrac{3}{4}$, $\sigma^2 = \dfrac{3}{80}$, $\sigma = \sqrt{\dfrac{3}{80}}$,
(c) $\dfrac{69}{20}\sqrt{\dfrac{3}{80}} \approx 0.668$

13. (a) 6 (b) $\mu = \dfrac{1}{2}$, $\sigma^2 = \dfrac{1}{20}$, $\sigma = \sqrt{\dfrac{1}{20}}$,
(c) $\dfrac{7}{5\sqrt{5}} \approx 0.626$

15. (a) $\dfrac{2}{\sqrt{\pi}}$, (b) $\mu = \dfrac{1}{\sqrt{\pi}} \approx 0.0.564$, $\sigma^2 = \dfrac{\pi - 2}{2\pi}$,
$\sigma = \sqrt{\dfrac{\pi - 2}{2\pi}} \approx 0.426$, (c) Pr$\approx 0.68$

19. (a) 0, (b) $e^{-3} \approx 0.05$, (c) ≈ 0.046

21. approximately 0.006

Section 7.9 (page 453)

1. $y^2 = Cx$

3. $x^3 - y^3 = C$

5. $Y = Ce^{t^2/2}$

7. $y = \pm 1$, $y = \dfrac{Ce^{2x} - 1}{Ce^{2x} + 1}$

9. $y = -\ln\left(Ce^{-2t} - \frac{1}{2}\right)$

11. $y = x^3 + Cx^2$

13. $y = \frac{3}{2} + Ce^{-2x}$

15. $y = x - 1 + Ce^{-x}$

17. $y = (1 + e^{1-10t})/10$

19. $y = (x + 2)e^{1/x}$

21. $y = \sqrt{4 + x^2}$

23. $y = \dfrac{2x}{1 + x}$, $(x > 0)$

25. b

27. If $a = b$ the given solution is indeterminate $0/0$; in this case the solution is $x = a^2kt/(1 + akt)$.

29. $v = \sqrt{\dfrac{mg}{k}}$, $v = \sqrt{\dfrac{mg}{k}}\,\dfrac{e^{2\sqrt{kg/mt}} - 1}{e^{2\sqrt{kg/mt}} + 1}$, $v \to \sqrt{\dfrac{mg}{k}}$

31. the hyperbolas $x^2 - y^2 = C$

Review Exercises (page 454)

1. about 833

3. $a \approx 1.1904$, $b \approx 0.0476$

5. $a = 2.1773$

7. $\left(\frac{8}{3\pi}, \frac{4}{3\pi}\right)$

9. about 27,726 N·cm

11. $y = 4(x - 1)^3$

13. \$8, 798.85

Challenging Problems (page 455)

1. (b) $\ln 2/(2\pi)$, (c) $\pi/(4k(k^2 + 1))$

3. $y = (r/h^3)x^3 - 3(r/h^2)x^2 + 3(r/h)x$

5. $b = -a = 27/2$

7. $1/\pi$

9. (a) $S(a, a, c) = 2\pi a^2 + \dfrac{2\pi ac^2}{\sqrt{a^2 - c^2}}\ln\left(\dfrac{a + \sqrt{a^2 - c^2}}{c}\right)$.
(b) $S(a, c, c) = 2\pi c^2 + \dfrac{2\pi a^2 c}{\sqrt{a^2 - c^2}}\cos^{-1}\left(\dfrac{c}{a}\right)$.
(c) $S(a, b, c) \approx \dfrac{b - c}{a - c}S(a, a, c) + \dfrac{a - b}{a - c}S(a, c, c)$.
(d) $S(3, 2, 1) \approx 49.595$.

Chapter 8
Conics, Parametric Curves, and Polar Curves

Section 8.1 (page 468)

1. $(x^2/5) + (y^2/9) = 1$

3. $(x - 2)^2 = 16 - 4y$

5. $3y^2 - x^2 = 3$

7. single point $(-1, 0)$

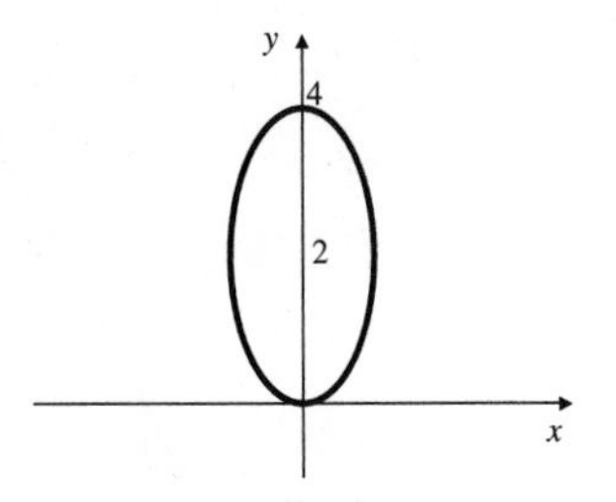

9. ellipse, centre $(0, 2)$

11. parabola, vertex $(-1, -4)$

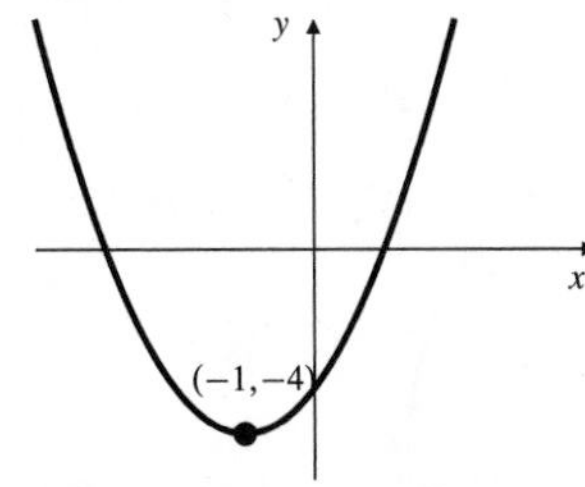

13. hyperbola, centre $\left(-\frac{3}{2}, 1\right)$
asymptotes
$2x+3 = \pm 2^{3/2}(y-1)$

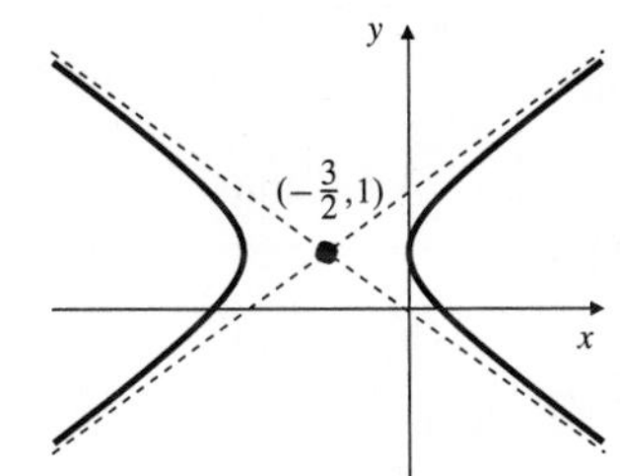

15. ellipse, centre $(1, -1)$

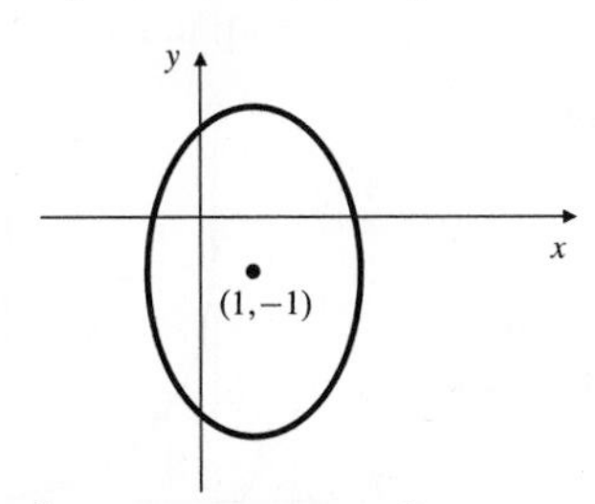

17. $y^2 - 8y = 16x$ or $y^2 - 8y = -4x$

19. rectangular hyperbola, centre $(1, -1)$,
semiaxes $a = b = \sqrt{2}$,
eccentricity $\sqrt{2}$,
foci $(\sqrt{2}+1, \sqrt{2}-1)$,
$(-\sqrt{2}+1, -\sqrt{2}-1)$,
asymptotes $x = 1, y = -1$

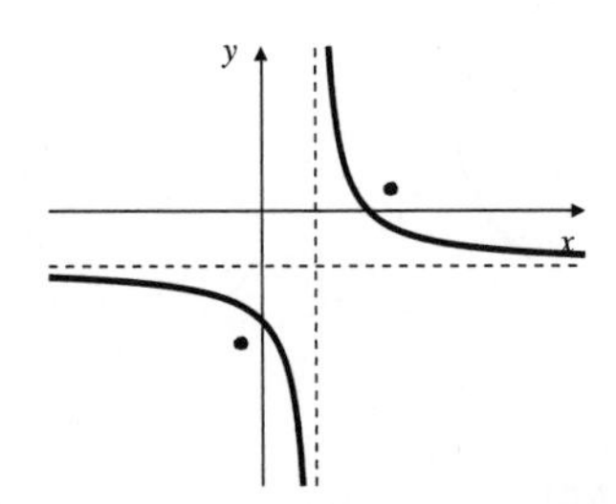

21. ellipse, centre (0,0),
semi-axes $a = 2, b = 1$,
foci $\pm\left(2\sqrt{\frac{3}{5}}, -\sqrt{\frac{3}{5}}\right)$

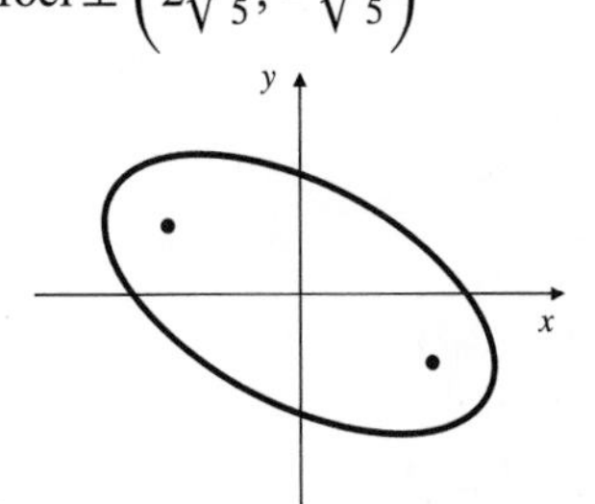

23. $(1-\varepsilon^2)x^2 + y^2 - 2p\varepsilon^2 x = \varepsilon^2 p^2$

Section 8.2 (page 474)

1. $y = (x-1)^2/4$

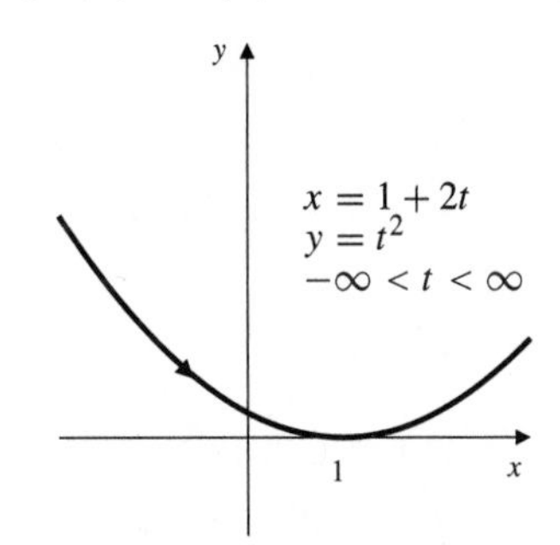

3. $y = (1/x) - 1$

5. $x^2 + y^2 = 9$

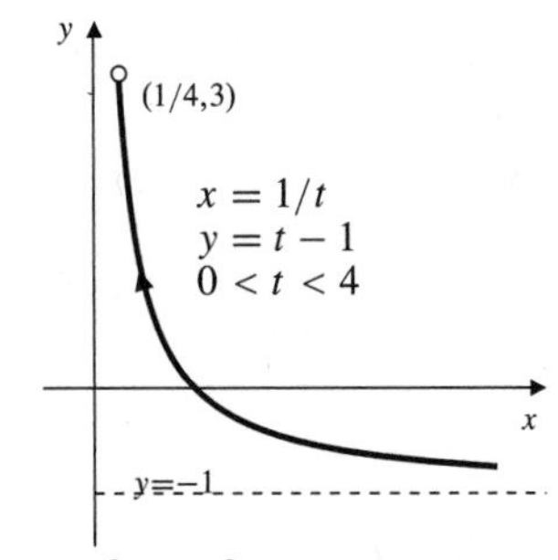

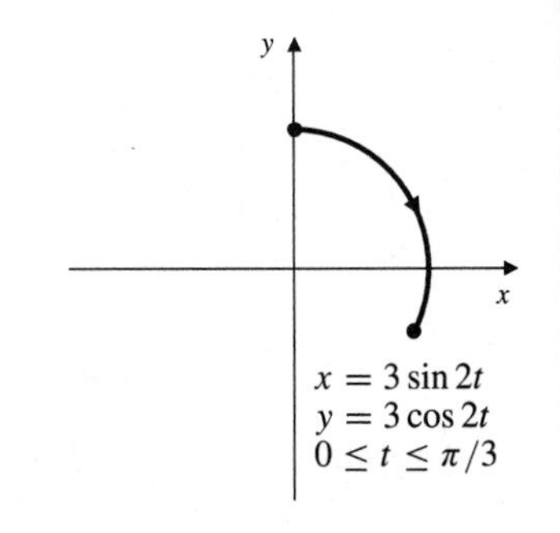

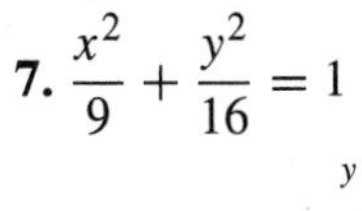

7. $\dfrac{x^2}{9} + \dfrac{y^2}{16} = 1$

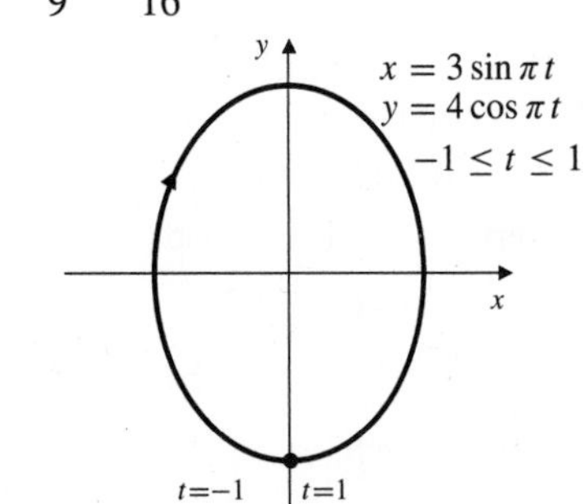

9. $x^{2/3} + y^{2/3} = 1$

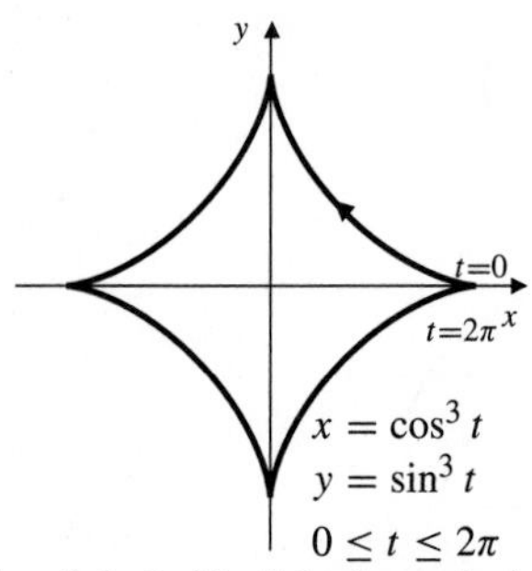

11. the right half of the hyperbola $x^2 - y^2 = 1$

13. the curve starts at the origin and spirals twice counterclockwise around the origin to end at $(4\pi, 0)$

15. $x = m/2, \quad y = m^2/4, \quad (-\infty < m < \infty)$

17. $x = a \sec t, \ y = a \sin t;$
$y^2 = a^2(x^2 - a^2)/x^2$

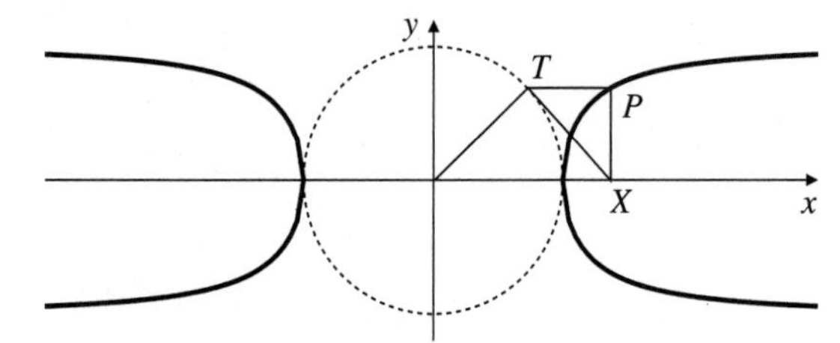

19. $x^3 + y^3 = 3xy$

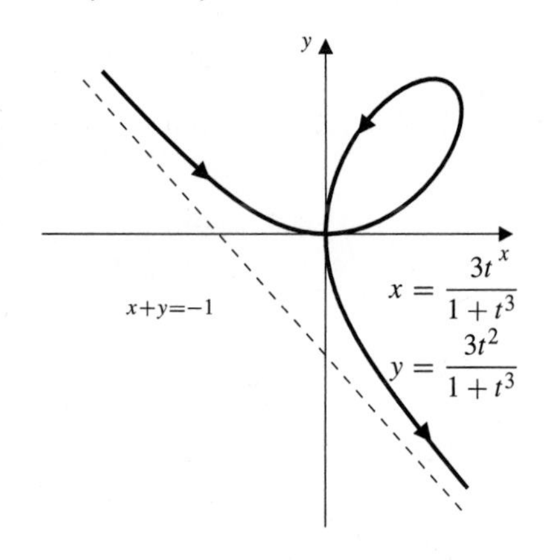

21.

23.

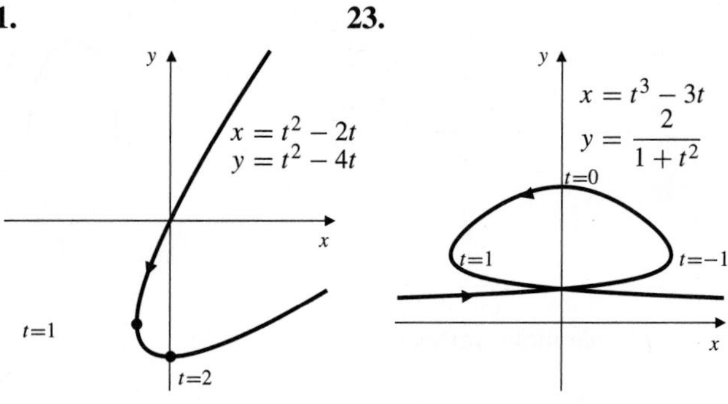

25.

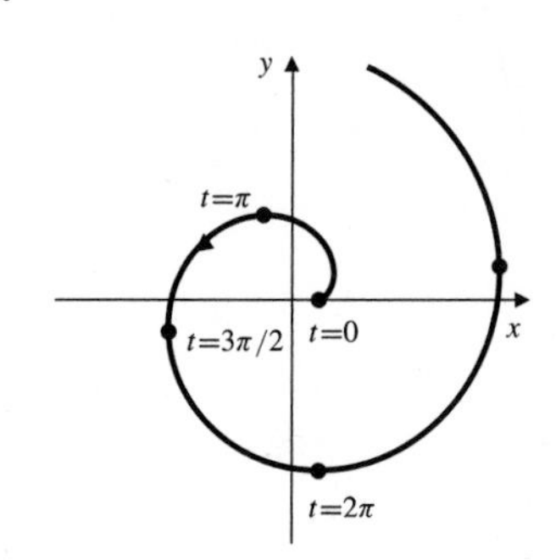

Section 8.3 (page 479)

1. vertical at $(1, -4)$

3. horizontal at $(0, -16)$ and $(8, 16)$; vertical at $(-1, -11)$

5. horizontal at $(0, 1)$, vertical at $(\pm 1/\sqrt{e}, 1/e)$

7. horiz. at $(0, \pm 1)$, vert. at $(\pm 1, 1/\sqrt{2})$ and $(\pm 1, -1/\sqrt{2})$

9. $-3/4$ **11.** $-1/2$

13. $x = t - 2, \ y = 4t - 2$ **15.** slopes ± 1

17. not smooth at $t = 0$

19. not smooth at $t = 0$

Section 8.4 (page 483)

1. $4\sqrt{2} - 2$ units **3.** $6a$ units

5. $\frac{8}{3}\left((1 + \pi^2)^{3/2} - 1\right)$ units

7. 4 units **9.** $8a$ units

11. $2\sqrt{2}\pi(1 + 2e^{\pi})/5$ sq. units

13. $72\pi(1 + \sqrt{2})/15$ sq. units

15. $256/15$ sq. units **17.** $1/6$ sq. units

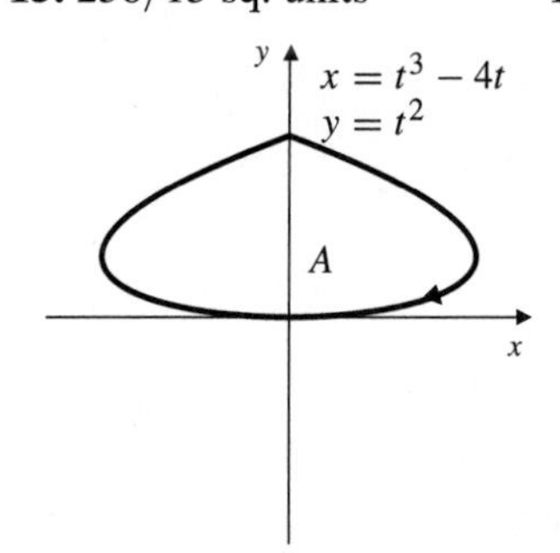

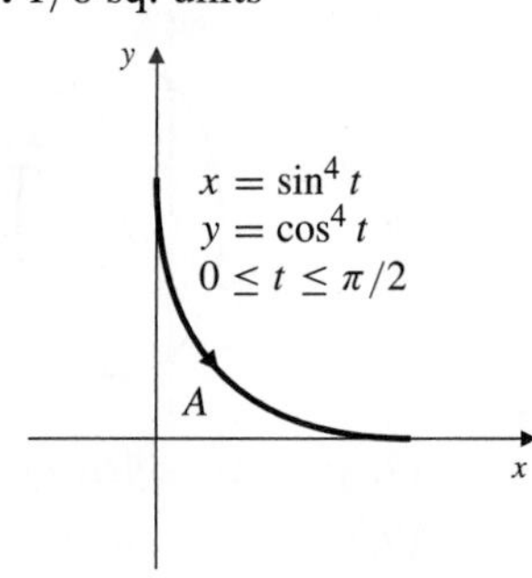

19. $9\pi/2$ sq. units

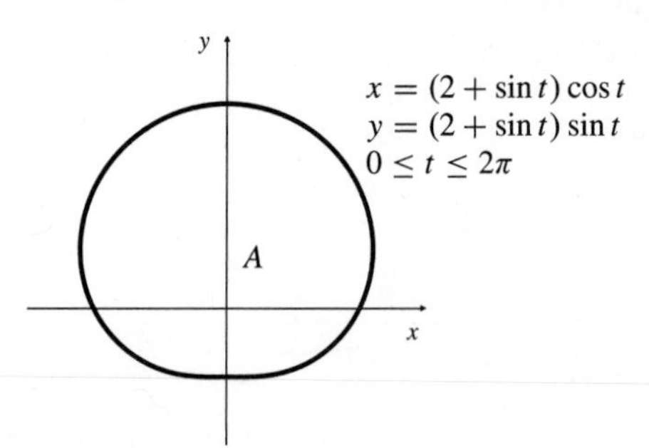

23. $32\pi a^3/105$ cu. units

Section 8.5 (page 489)

1. $x = 3$, vertical straight line

3. $3y - 4x = 5$, straight line

5. $2xy = 1$, rectangular hyperbola

7. $y = x^2 - x$, a parabola

9. $y^2 = 1 + 2x$, a parabola

11. $x^2 - 3y^2 - 8y = 4$, a hyperbola

13. **15.**

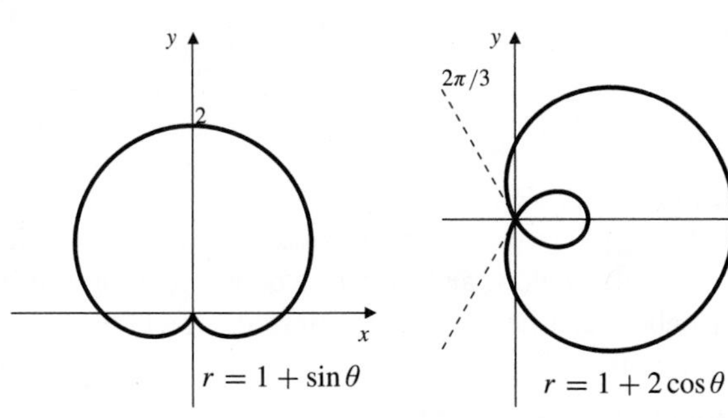

17. **19.**

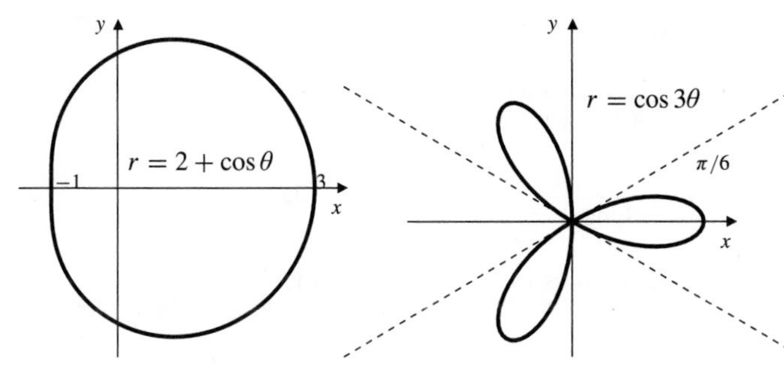

21. **23.** $r = \pm\sqrt{\sin 3\theta}$

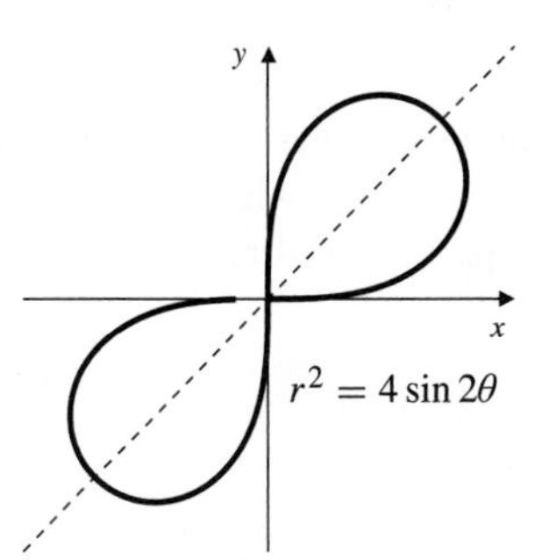

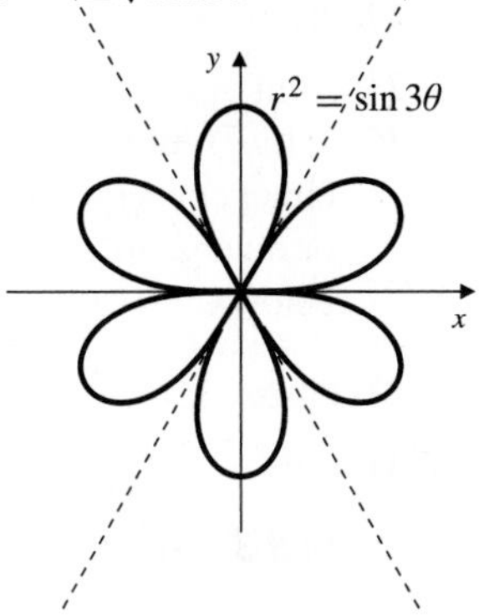

25. the origin and $[\sqrt{3}/2, \pi/3]$

27. the origin and $[3/2, \pm\pi/3]$

29. asymptote $y = 1$,
$r = 1/(\theta - \alpha)$ has
asymptote$(\cos\alpha)y - (\sin\alpha)x = 1$

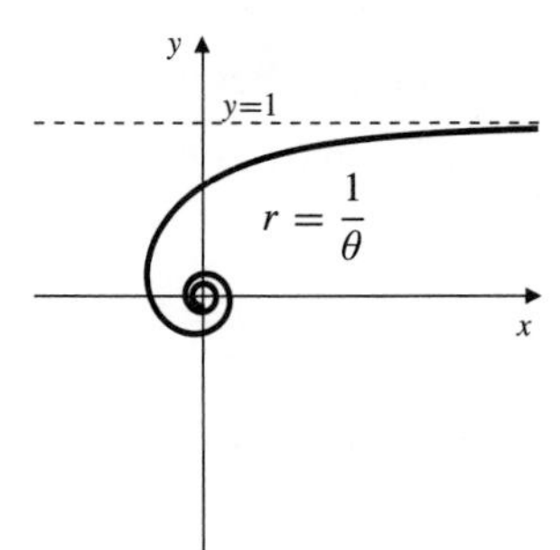

31. $x = f(\theta)\cos\theta, \quad y = f(\theta)\sin\theta$

39. $\ln\theta_1 = 1/\theta_1$, point $(-0.108461, 0.556676)$; $\ln\theta_2 = -1/(\theta_2 + \pi)$, point $(-0.182488, -0.178606)$

Section 8.6 (page 493)

1. π^2 sq. units

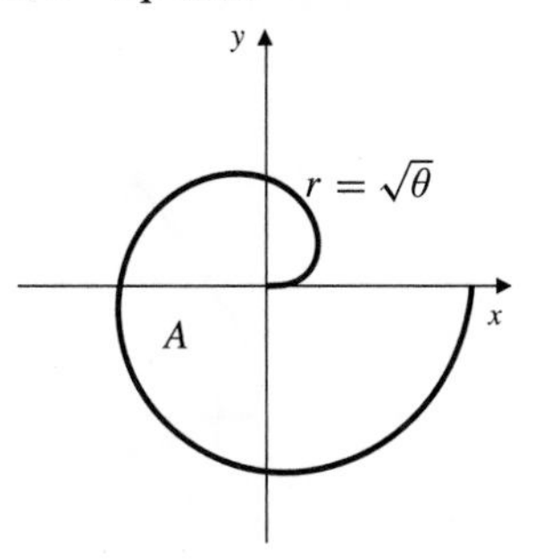

3. a^2 sq. units

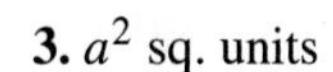

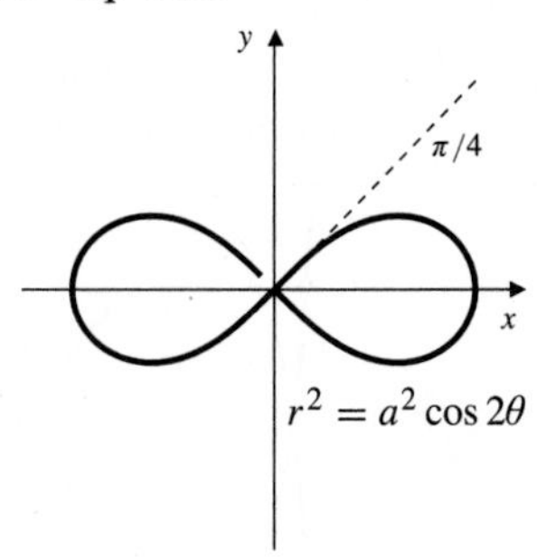

5. $\pi/2$ sq. units

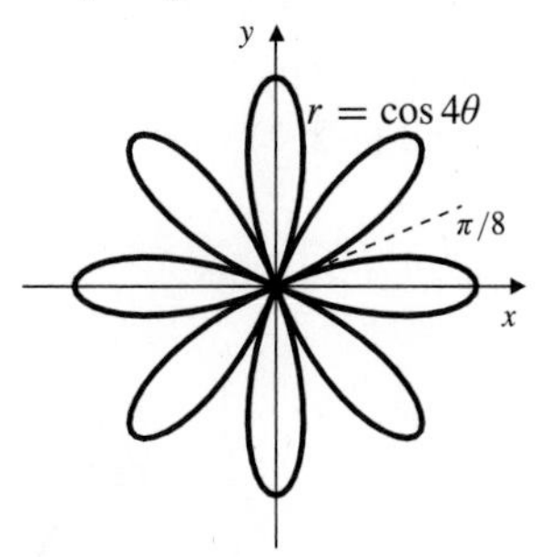

7. $2 + (\pi/4)$ sq. units

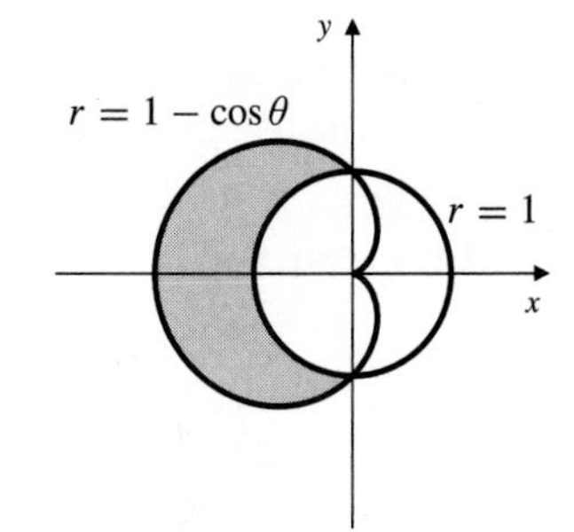

9. $\pi/4$ sq. units

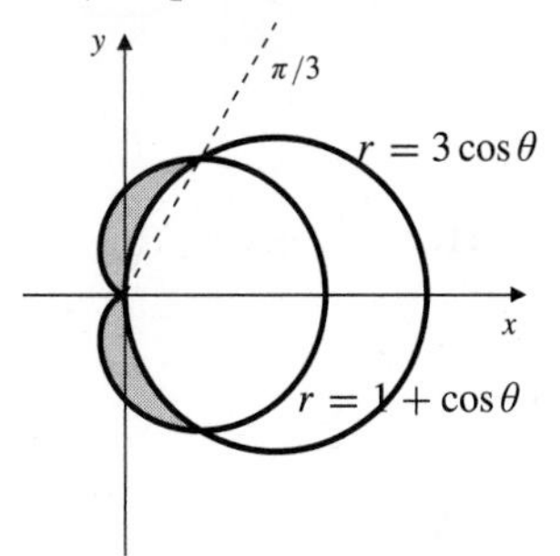

11. $\pi - \frac{3}{2}\sqrt{3}$ sq. units

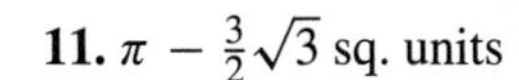

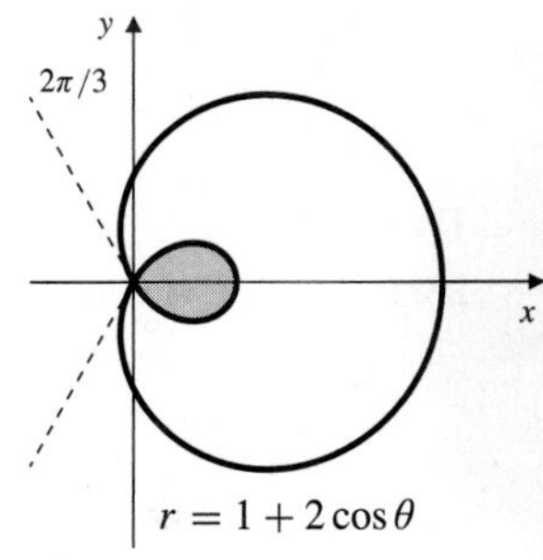

13. $\dfrac{\sqrt{1+a^2}}{a}\left(e^{a\pi} - e^{-a\pi}\right)$ units

17. $67.5°, -22.5°$

19. $90°$ at $(0,0)$,
$\pm 45°$ at $\left(1 - \dfrac{1}{\sqrt{2}}, \dfrac{\pi}{4}\right)$,
$\pm 135°$ at $\left(1 + \dfrac{1}{\sqrt{2}}, \dfrac{5\pi}{4}\right)$

21. horizontal at $\left(\pm\frac{\pi}{4}, \sqrt{2}\right)$, vertical at $(2, 0)$ and the origin

23. horizontal at $(0, 0)$, $\left(\frac{2}{3}\sqrt{2}, \pm\tan^{-1}\sqrt{2}\right)$,
$\left(\frac{2}{3}\sqrt{2}, \pi \pm \tan^{-1}\sqrt{2}\right)$,
vertical at $\left(0, \frac{\pi}{2}\right)$, $\left(\frac{2}{3}\sqrt{2}, \pm\tan^{-1}(1/\sqrt{2})\right)$,
$\left(\frac{2}{3}\sqrt{2}, \pi \pm \tan^{-1}(1/\sqrt{2})\right)$

25. horizontal at $\left(4, -\frac{\pi}{2}\right)$, $\left(1, \frac{\pi}{6}\right)$, $\left(1, \frac{5\pi}{6}\right)$,
vertical at $\left(3, -\frac{\pi}{6}\right)$, $\left(3, -\frac{5\pi}{6}\right)$, no tangent at $\left(0, \frac{\pi}{2}\right)$

Review Exercises (page 494)

1. ellipse, foci $(\pm 1, 0)$, semi-major axis $\sqrt{2}$, semi-minor axis 1

3. parabola, vertex $(4, 1)$, focus $(15/4, 1)$

5. straight line from $(0, 2)$ to $(2, 0)$

7. the parabola $y = x^2 - 1$ left to right

9. first quadrant part of ellipse $16x^2 + y^2 = 16$ from $(1, 0)$ to $(0, 4)$

11. horizontal tangents at $(2, \pm 2)$ (i.e. $t = \pm 1$)
vertical tangent at $(4, 0)$ (i.e. $t = 0$)

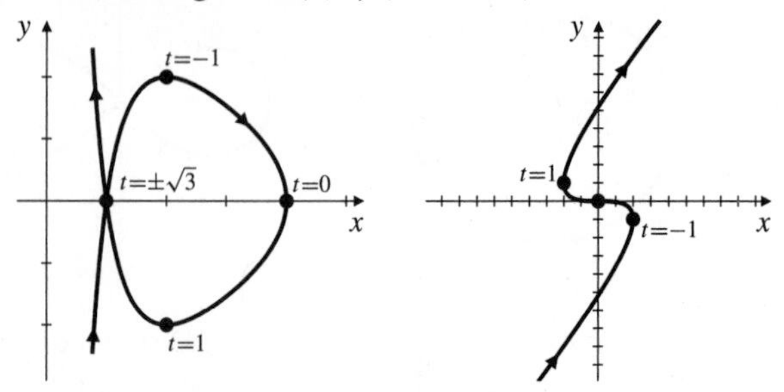

13. horizontal tangent at $(0, 0)$ (i.e. $t = 0$)
vertical tangents at $(2, -1)$ and $(-2, 1)$
(i.e. $t = \pm 1$)

15. 1/2 sq. units

17. $1 + e^2$ units

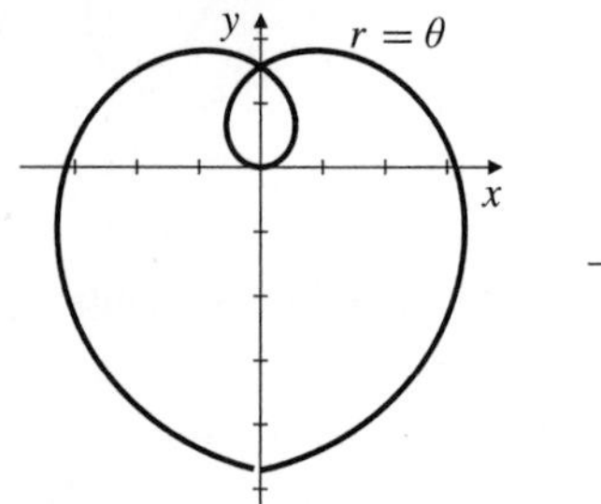

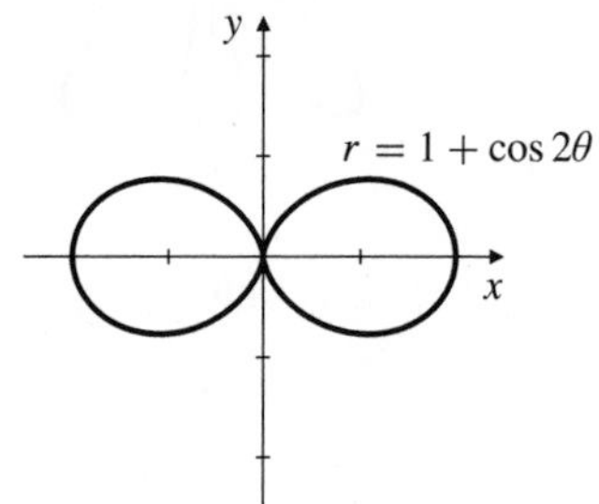

19. $r = \theta$

21. $r = 1 + \cos 2\theta$

23. $r = 1 + 2\cos 2\theta$

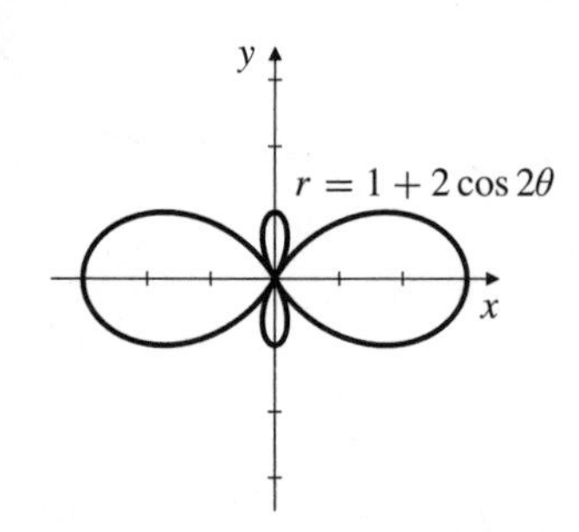

25. $\pi + (3\sqrt{3}/4)$ sq. units

27. $(\pi - 3)/2$ sq. units

Challenging Problems (page 494)

1. $16\pi \sec\theta$ cm^2

5. $40\pi/3$ ft^3

7. about 84.65 minutes

9. $r^2 = \cos(2\theta)$ is the inner curve; area between curves is 1/3 sq. units

Chapter 9
Sequences, Series, and Power Series

Section 9.1 (page 503)

1. bounded, positive, increasing, convergent to 2

3. bounded, positive, convergent to 4

5. bounded below, positive, increasing, divergent to infinity

7. bounded below, positive, increasing, divergent to infinity

9. bounded, positive, decreasing, convergent to 0

11. divergent

13. divergent

15. ∞

17. 0

19. 1

21. e^{-3}

23. 0

25. 1/2

27. 0

29. 0

31. $\lim_{n\to\infty} a_n = 5$

33. If $\{a_n\}$ is (ultimately) decreasing, then either it is bounded below, and therefore convergent, or else it is unbounded below and therefore divergent to negative infinity.

Section 9.2 (page 510)

1. $\dfrac{1}{2}$

3. $\dfrac{1}{(2+\pi)^8\big((2+\pi)^2 - 1\big)}$

5. $\dfrac{25}{4{,}416}$

7. $\dfrac{8e^4}{e-2}$

9. diverges to ∞

11. $\dfrac{3}{4}$

13. $\dfrac{1}{3}$

15. div. to ∞

17. div. to ∞

19. diverges

21. 14 m

25. If $\{a_n\}$ is ultimately negative, then the series $\sum a_n$ must either converge (if its partial sums are bounded below), or diverge to $-\infty$ (if its partial sums are not bounded below).

27. false, e.g. $\sum \dfrac{(-1)^n}{2^n}$

29. true

31. true

Section 9.3 (page 520)

1. converges

3. diverges to ∞

5. converges

7. diverges to ∞

9. converges

11. diverges to ∞

13. diverges to ∞

15. converges

17. converges

19. diverges to ∞

21. converges

23. converges

25. converges

27. $s_n + \dfrac{1}{3(n+1)^3} \le s \le s_n + \dfrac{1}{3n^3}$; $n = 6$

29. $s_n + \dfrac{2}{\sqrt{n+1}} \le s \le s_n + \dfrac{2}{\sqrt{n}}$; $n = 63$

31. $0 < s - s_n \le \dfrac{n+2}{2^n(n+1)!(2n+3)}$; $n = 4$

33. $0 < s - s_n \le \dfrac{2^n(4n^2+6n+2)}{(2n)!(4n^2+6n)}$; $n = 4$

39. converges, $a_n^{1/n} \to (1/e) < 1$

41. no info from ratio test, but series diverges to infinity since all terms exceed 1.

43. (b) $s \le \dfrac{2}{k(1-k)}$, $k = \frac{1}{2}$,

(c) $0 < s - s_n < \dfrac{(1+k)^{n+1}}{2^n k(1-k)}$, $k = \dfrac{n+2-\sqrt{n^2+8}}{2(n-1)}$ for $n \ge 2$

45. (a) 10, (b) 5, (c) 0.765

Section 9.4 (page 526)

1. conv. conditionally **3.** conv. conditionally

5. diverges **7.** conv. absolutely

9. conv. conditionally **11.** diverges

13. 999 **15.** 13

17. converges absolutely if $-1 < x < 1$, conditionally if $x = -1$, diverges elsewhere

19. converges absolutely if $0 < x < 2$, conditionally if $x = 2$, diverges elsewhere

21. converges absolutely if $-2 < x < 2$, conditionally if $x = -2$, diverges elsewhere

23. converges absolutely if $-\frac{7}{2} < x < \frac{1}{2}$, conditionally if $x = -\frac{7}{2}$, diverges elsewhere

25. AST does not apply directly, but does if we remove all the 0 terms; series converges conditionally

27. (a) false, e.g. $a_n = \dfrac{(-1)^n}{n}$,

(b) false, e.g. $a_n = \dfrac{\sin(n\pi/2)}{n}$, (see Exercise 25),

(c) true

29. converges absolutely for $-1 < x < 1$, conditionally if $x = -1$, diverges elsewhere

Section 9.5 (page 536)

1. centre 0, radius 1, interval $(-1, 1)$

3. centre -2, radius 2, interval $[-4, 0)$

5. centre $\frac{3}{2}$, radius $\frac{1}{2}$, interval $(1, 2)$

7. centre 0, radius ∞, interval $(-\infty, \infty)$

9. $\dfrac{1}{(1-x)^3} = \displaystyle\sum_{n=0}^{\infty} \frac{(n+1)(n+2)}{2} x^n$, $(-1 < x < 1)$

11. $\dfrac{1}{(1-x)^2} = \displaystyle\sum_{n=0}^{\infty} (n+1)x^n$, $(-1 < x < 1)$

13. $\dfrac{1}{(2-x)^2} = \displaystyle\sum_{n=0}^{\infty} \frac{n+1}{2^{n+2}} x^n$, $(-2 < x < 2)$

15. $\ln(2-x) = \ln 2 - \sum_{n=1}^{\infty} \dfrac{x^n}{2^n n}$, $(-2 \le x < 2)$

17. $\dfrac{1}{x^2} = \displaystyle\sum_{n=0}^{\infty} \frac{n+1}{2^{n+2}} (x+2)^n$, $(-4 < x < 0)$

19. $\dfrac{x^3}{1-2x^2} = \displaystyle\sum_{n=0}^{\infty} 2^n x^{2n+3}$, $\left(-\dfrac{1}{\sqrt{2}} < x < \dfrac{1}{\sqrt{2}}\right)$

21. $\left(-\frac{1}{4}, \frac{1}{4}\right)$; $\dfrac{1}{1+4x}$

23. $[-1, 1)$; $\frac{1}{3}$ if $x = 0$,

$-\dfrac{1}{x^3}\ln(1-x) - \dfrac{1}{x^2} - \dfrac{1}{2x}$ otherwise

25. $(-1, 1)$; $\dfrac{2}{(1-x^2)^2}$ **27.** $3/4$

29. $\pi^2(\pi+1)/(\pi-1)^3$ **31.** $\ln(3/2)$

Section 9.6 (page 545)

1. $e^{3x+1} = \sum_{n=0}^{\infty} \dfrac{3^n e}{n!} x^n$, (all x)

3. $\sin\left(x - \dfrac{\pi}{4}\right)$

$= \dfrac{1}{\sqrt{2}} \displaystyle\sum_{n=0}^{\infty} (-1)^n \left[-\frac{x^{2n}}{(2n)!} + \frac{x^{2n+1}}{(2n+1)!}\right]$, (all x)

5. $x^2 \sin\left(\dfrac{x}{3}\right) = \sum_{n=0}^{\infty} \dfrac{(-1)^n}{3^{2n+1}(2n+1)!} x^{2n+3}$, (all x)

7. $\sin x \cos x = \sum_{n=0}^{\infty} \dfrac{(-1)^n 2^{2n}}{(2n+1)!} x^{2n+1}$, (all x)

9. $\dfrac{1+x^3}{1+x^2} = 1 - x^2 + \displaystyle\sum_{n=2}^{\infty} (-1)^n \left(x^{2n-1} + x^{2n}\right)$, $(-1 < x < 1)$

11. $\ln \dfrac{1+x}{1-x} = 2 \displaystyle\sum_{n=1}^{\infty} \frac{x^{2n-1}}{2n-1}$, $(-1 < x < 1)$

13. $\cosh x - \cos x = 2\sum_{n=0}^{\infty} \dfrac{x^{4n+2}}{(4n+2)!}$, (all x)

15. $e^{-2x} = e^2 \sum_{n=0}^{\infty} \dfrac{(-1)^n 2^n}{n!} (x+1)^n$, (all x)

17. $\cos x = \sum_{n=0}^{\infty} \dfrac{(-1)^{n+1}}{(2n)!} (x-\pi)^{2n}$, (all x)

19. $\ln 4 + \sum_{n=1}^{\infty} \dfrac{(-1)^{n-1}}{4^n n} (x-2)^n$, $(-2 < x \le 6)$

21. $\sin x - \cos x =$

$\sqrt{2} \sum_{n=0}^{\infty} \dfrac{(-1)^n}{(2n+1)!} \left(x - \dfrac{\pi}{4}\right)^{2n+1}$, (all x)

23. $\dfrac{1}{x^2} = \dfrac{1}{4} \displaystyle\sum_{n=0}^{\infty} \frac{n+1}{2^n} (x+2)^n$, $(-4 < x < 0)$

25. $(x-1) + \sum_{n=2}^{\infty} \dfrac{(-1)^n}{n(n-1)} (x-1)^n$, $(0 \le x \le 2)$

27. $1 + \dfrac{x^2}{2} + \dfrac{5x^4}{24}$ **29.** $x + \dfrac{x^2}{2} - \dfrac{x^3}{6}$

31. $1 + \dfrac{x}{2} - \dfrac{x^2}{8}$ **33.** e^{x^2} (all x)

35. $\dfrac{e^x - e^{-x}}{2x} = \dfrac{\sinh x}{x}$ if $x \ne 0$, 1 if $x = 0$

37. (a) $1 + x + x^2$, (b) $3 + 3(x-1) + (x-1)^2$

Section 9.7 (page 549)

1. $\dfrac{1}{720}(0.2)^7$ **3.** 1.22140

5. 3.32011 **7.** 0.99619

9. -0.10533 **11.** 0.42262

13. 1.54306

15. $I(x) = \sum_{n=0}^{\infty} \dfrac{(-1)^n}{(2n+1)(2n+1)!} x^{2n+1}$, (all x)

17. $K(x) = \sum_{n=0}^{\infty} \dfrac{(-1)^n}{(n+1)^2} x^{n+1}$, $(-1 \le x \le 1)$

19. $M(x) = \sum_{n=0}^{\infty} \dfrac{(-1)^n}{(2n+1)(4n+1)} x^{4n+1}$, $(-1 \le x \le 1)$

21. 0.946 **23.** 2

25. $-3/25$ **27.** 0

Section 9.8 (page 554)

1. $\sqrt{1+x} = 1 + \sum_{n=1}^{\infty} \dfrac{(-1)^{n-1} 1 \times 3 \times 5 \times \cdots \times (2n-3)}{2^n n!} x^n$ $|x| < 1$

3. $\sqrt{4+x} = 2 + \dfrac{x}{4} + 2\sum_{n=2}^{\infty} (-1)^{n-1} \dfrac{1 \times 3 \times 5 \times \cdots \times (2n-3)}{2^{3n} n!} x^n$, $(-4 < x \le 4)$

5. $\sum_{n=0}^{\infty} (n+1)x^n$, $|x| < 1$

7. $\dfrac{\pi}{2} - x - \sum_{n=1}^{\infty} \dfrac{1 \times 3 \times 5 \times \cdots \times (2n-1)}{2^n n!(2n+1)} x^{2n+1}$, $(-1 < x < 1)$

Section 9.9 (page 560)

1. $2\pi/3$ **3.** π

5. $2\sum_{n=1}^{\infty} (-1)^{n-1}(\sin(nt))/n$

7. $\dfrac{1}{4} - \sum_{n=1}^{\infty} \left(\dfrac{2\cos((2n-1)\pi t)}{(2n-1)^2\pi^2} + \dfrac{(-1)^n \sin(n\pi t)}{n\pi} \right)$

9. 1

11. $2\sum_{n=1}^{\infty} \dfrac{(-1)^n}{n\pi} \sin(n\pi t)$

13. $\pi^2/8$

Review Exercises (page 561)

1. conv. to 0 **3.** div. to ∞

5. $\lim_{n\to\infty} a_n = \sqrt{2}$ **7.** $4\sqrt{2}/(\sqrt{2}-1)$

9. 2 **11.** converges

13. converges **15.** converges

17. conv. abs. **19.** conv. cond.

21. conv. abs. for x in $(-1, 5)$, cond. for $x = -1$, div. elsewhere

23. 1.202

25. $\sum_{n=0}^{\infty} x^n/3^{n+1}$, $|x| < 3$

27. $1 + \sum_{n=1}^{\infty} (-1)^{n-1} x^{2n}/(ne^n)$, $-\sqrt{e} < x \le \sqrt{e}$

29. $x + \sum_{n=1}^{\infty} (-1)^n 2^{2n-1} x^{2n+1}/(2n)!$, all x

31. $(1/2) + \sum_{n=1}^{\infty} \dfrac{(-1)^n\, 1 \times 4 \times 7 \times \cdots \times (3n-2)x^n}{2 \times 24^n\, n!}$, $-8 < x \le 8$

33. $\sum_{n=0}^{\infty} (-1)^n (x-\pi)^n/\pi^{n+1}$, $0 < x < 2\pi$

35. $1 + 2x + 3x^2 + \frac{10}{3}x^3$ **37.** $1 - \frac{1}{2}x^2 + \frac{5}{24}x^4$

39. $\begin{cases} \cos\sqrt{x} & \text{if } x \ge 0 \\ \cosh\sqrt{|x|} & \text{if } x < 0 \end{cases}$ **41.** $\pi^2/(\pi-1)^2$

43. $\ln(e/(e-1))$ **45.** $1/14$

47. 3, 0.49386 **49.** $\sum_{n=1}^{\infty} \dfrac{2}{n} \sin(nt)$

Challenging Problems (page 562)

5. (c) 1.645

7. (a) ∞, (c) e^{-x^2}, (d) $f(x) = e^{x^2} \int_0^x e^{-t^2}\,dt$

Chapter 10
Vectors and Coordinate Geometry in 3-Space

Section 10.1 (page 569)

1. 3 units **3.** $\sqrt{6}$ units

5. $|z|$ units; $\sqrt{y^2+z^2}$ units

7. $\cos^{-1}(-4/9) \approx 116.39°$

9. $\sqrt{3}/2$ sq. units **11.** $\sqrt{n-1}$ units

13. the half-space containing the origin and bounded by the plane passing through $(0, -1, 0)$ perpendicular to the y-axis.

15. the vertical plane (parallel to the z-axis) passing through $(1, 0, 0)$ and $(0, 1, 0)$.

17. the sphere of radius 2 centred at $(1, -2, 3)$.

19. the solid circular cylinder of radius 2 with axis along the x-axis.

21. the parabolic cylinder generated by translating the parabola $z = y^2$ in the yz-plane in the direction of the x-axis .

23. the plane through the points $(6, 0, 0)$, $(0, 3, 0)$ and $(0, 0, 2)$.

25. the straight line through $(1, 0, 0)$ and $(1, 1, 1)$.

27. the circle in which the sphere of radius 2 centred at the origin intersects the sphere of radius 2 with centre $(2, 0, 0)$.

29. the ellipse in which the plane $z = x$ intersects the circular cylinder of radius 1 and axis along the z-axis.

31. the part of the solid circular cylinder of radius 1 and axis along the z-axis lying above or on the plane $z = y$.

33. bdry $(0, 0)$ and $x^2 + y^2 = 1$; interior$= S$; S open

35. bdry of S is S; interior empty; S is closed

37. bdry — the spheres $x^2 + y^2 + z^2 = 1$ and $x^2 + y^2 + z^2 = 4$; interior — points between these spheres; S is closed

39. bdry of S is S, namely the line $x = y = z$; interior is empty; S closed

Section 10.2 (page 578)

1. (a) $3\mathbf{i} - 2\mathbf{j}$, (b) $-3\mathbf{i} + 2\mathbf{j}$, (c) $2\mathbf{i} - 5\mathbf{j}$, (d) $-2\mathbf{i} + 4\mathbf{j}$,(e) $-\mathbf{i} - 2\mathbf{j}$, (f) $4\mathbf{i} + \mathbf{j}$, (g) $-7\mathbf{i} + 20\mathbf{j}$, (h) $2\mathbf{i} - (5/3)\mathbf{j}$

3. a) $6\mathbf{i} - 10\mathbf{k}$, $8\mathbf{j}$, $-3\mathbf{i} + 20\mathbf{j} + 5\mathbf{k}$
b) $5\sqrt{2}$, $5\sqrt{2}$
c) $\frac{3}{5\sqrt{2}}\mathbf{i} \pm \frac{4}{5\sqrt{2}}\mathbf{j} - \frac{1}{\sqrt{2}}\mathbf{k}$ d) 18
e) $\cos^{-1}(9/25) \approx 68.9°$ f) $18/5\sqrt{2}$
g) $(27/25)\mathbf{i} + (36/25)\mathbf{j} - (9/5)\mathbf{k}$

9. from southwest at $50\sqrt{2}$ km/h.

11. head at angle θ to the east of AC, where
$$\theta = \sin^{-1}\frac{3}{2\sqrt{1+4k^2}}.$$
The trip is not possible if $k < \frac{1}{4}\sqrt{5}$. If $k > \frac{1}{4}\sqrt{5}$ there is a second possible heading, $\pi - \theta$, but the trip will take longer.

13. $t = 2$

15. $\cos^{-1}(2/\sqrt{6}) \approx 35.26°$, $90°$

17. $(\mathbf{i} + \mathbf{j} + \mathbf{k})/\sqrt{3}$

19. $\lambda = 1/2$, midpoint, $\lambda = 2/3$, 2/3 of way from P_1 to P_2, $\lambda = -1$, P_1 is midway between this point and P_2.

21. plane through point with position vector $(b/|\mathbf{a}|^2)\mathbf{a}$ perpendicular to $\mathbf{a}$.

23. $\mathbf{x} = 2\mathbf{i} - 3\mathbf{j} - 4\mathbf{k}$

25. $(|\mathbf{u}|\mathbf{v} + |\mathbf{v}|\mathbf{u})/\big||\mathbf{u}|\mathbf{v} + |\mathbf{v}|\mathbf{u}\big|$

31. $\mathbf{u} = (\mathbf{w} \bullet \mathbf{a}/|\mathbf{a}|^2)\mathbf{a}$, $\mathbf{v} = \mathbf{w} - \mathbf{u}$

33. $\mathbf{x} = (\mathbf{a} + K\hat{\mathbf{u}})/(2r)$, $\mathbf{y} = (\mathbf{a} - K\hat{\mathbf{u}})/(2s)$, where $K = \sqrt{|\mathbf{a}|^2 - 4rst}$ and $\hat{\mathbf{u}}$ is any unit vector

35. about 12.373 m **37.** about 19 m

Section 10.3 (page 586)

1. $5\mathbf{i} + 13\mathbf{j} + 7\mathbf{k}$ **3.** $\sqrt{6}$ sq. units

5. $\pm\frac{1}{3}(2\mathbf{i} - 2\mathbf{j} + \mathbf{k})$ **15.** 4/3 cubic units

17. $k = -6$

19. $\lambda = \dfrac{\mathbf{x} \bullet (\mathbf{v} \times \mathbf{w})}{\mathbf{u} \bullet (\mathbf{v} \times \mathbf{w})}$, $\mu = \dfrac{\mathbf{x} \bullet (\mathbf{w} \times \mathbf{u})}{\mathbf{u} \bullet (\mathbf{v} \times \mathbf{w})}$, $\nu = \dfrac{\mathbf{x} \bullet (\mathbf{u} \times \mathbf{v})}{\mathbf{u} \bullet (\mathbf{v} \times \mathbf{w})}$

21. $\mathbf{u} \times (\mathbf{v} \times \mathbf{w}) = -2\mathbf{i} + 7\mathbf{j} - 4\mathbf{k}$, $(\mathbf{u} \times \mathbf{v}) \times \mathbf{w} = \mathbf{i} + 9\mathbf{j} + 9\mathbf{k}$; the first is in the plane of $\mathbf{v}$ and $\mathbf{w}$, the second is in the plane of $\mathbf{u}$ and $\mathbf{v}$.

Section 10.4 (page 594)

1. a) $x^2 + y^2 + z^2 = z^2$; b) $x + y + z = x + y + z$;
c) $x^2 + y^2 + z^2 = -1$

3. $x - y + 2z = 0$ **5.** $7x + 5y - z = 12$

7. $x - 5y - 3z = -7$ **9.** $x + 6y - 5z = 17$

11. $(\mathbf{r}_1 - \mathbf{r}_2) \bullet [(\mathbf{r}_1 - \mathbf{r}_3) \times (\mathbf{r}_1 - \mathbf{r}_4)] = 0$

13. planes passing through the line $x = 0$, $y + z = 1$ (except the plane $y + z = 1$ itself)

15. $\mathbf{r} = (1 + 2t)\mathbf{i} + (2 - 3t)\mathbf{j} + (3 - 4t)\mathbf{k}$, $(-\infty < t < \infty)$
$x = 1 + 2t$, $y = 2 - 3t$, $z = 3 - 4t$, $(-\infty < t < \infty)$
$$\frac{x-1}{2} = \frac{y-2}{-3} = \frac{z-3}{-4}$$

17. $\mathbf{r} = t(7\mathbf{i} - 6\mathbf{j} - 5\mathbf{k})$; $x = 7t$, $y = -6t$, $z = -5t$; $x/7 = -y/6 = -z/5$

19. $\mathbf{r} = \mathbf{i} + 2\mathbf{j} - \mathbf{k} + t(\mathbf{i} + \mathbf{j} + \mathbf{k})$;
$x = 1 + t$, $y = 2 + t$, $z = -1 + t$;
$x - 1 = y - 2 = z + 1$

21. $\frac{x-4}{-5} = \frac{y}{3}$, $z = 7$

25. $\mathbf{r}_i \neq \mathbf{r}_j$, $(i, j = 1, \cdots, 4,\ i \neq j)$,
$\mathbf{v} = (\mathbf{r}_1 - \mathbf{r}_2) \times (\mathbf{r}_3 - \mathbf{r}_4) \neq 0$, $(\mathbf{r}_1 - \mathbf{r}_3) \bullet \mathbf{v} = 0$.

27. $7\sqrt{2}/10$ units **29.** $18/\sqrt{69}$ units

31. all lines parallel to the xy-plane and passing through (x_0, y_0, z_0).

33. (x, y, z) satisfies the quadratic if either $A_1x + B_1y + C_1z = D_1$ or $A_2x + B_2y + C_2z = D_2$.

Section 10.5 (page 598)

1. ellipsoid centred at the origin with semiaxes 6, 3 and 2 along the x-, y- and z-axes respectively.

3. sphere with centre $(1, -2, 3)$ and radius $1/\sqrt{2}$.

5. elliptic paraboloid with vertex at the origin, axis along the z-axis, and cross-section $x^2 + 2y^2 = 1$ in the plane $z = 1$.

7. hyperboloid of two sheets with vertices $(\pm 2, 0, 0)$ and circular cross-sections in planes $x = c$, $(c^2 > 4)$.

9. hyperbolic paraboloid — same as $z = x^2 - y^2$ but rotated 45° about the z-axis (counterclockwise as seen from above).

11. hyperbolic cylinder parallel to the y-axis, intersecting the xz-plane in the hyperbola $(x^2/4) - z^2 = 1$.

13. parabolic cylinder parallel to the y-axis.

15. circular cone with vertex $(2, 3, 1)$, vertical axis, and semi-vertical angle 45°.

17. circle in the plane $x + y + z = 1$ having centre $(1/3, 1/3, 1/3)$ and radius $\sqrt{11/3}$.

19. a parabola in the plane $z = 1 + x$ having vertex at $(-1/2, 0, 1/2)$ and axis along the line $z = 1 + x$, $y = 0$.

21. $\dfrac{y}{b} - \dfrac{z}{c} = \lambda\left(1 - \dfrac{x}{a}\right)$, $\dfrac{y}{b} + \dfrac{z}{c} = \dfrac{1}{\lambda}\left(1 + \dfrac{x}{a}\right)$;
$\dfrac{y}{b} - \dfrac{z}{c} = \mu\left(1 + \dfrac{x}{a}\right)$, $\dfrac{y}{b} + \dfrac{z}{c} = \dfrac{1}{\mu}\left(1 - \dfrac{x}{a}\right)$

23. $\mathbf{a} = \mathbf{i} \pm \mathbf{k}$ (or any multiple)

Section 10.6 (page 602)

1. cylindrical: $[2\sqrt{2}, -\pi/4, 1]$; spherical $[3, \cos^{-1}(1/3), -\pi/4]$

3. Cartesian: $(-\sqrt{3}, 3, 2)$; cylindrical: $[2\sqrt{3}, 2\pi/3, 2]$

5. the half-plane $x = 0$, $y > 0$

7. the xy-plane

9. the circular cylinder of radius 4 with axis along the z-axis

11. the xy-plane

13. sphere of radius 1 with centre $(0, 0, 1)$

Section 10.7 (page 611)

1. $\begin{pmatrix} 6 & 7 \\ 5 & -3 \\ 1 & 1 \end{pmatrix}$ **3.** $\begin{pmatrix} aw + by & ax + bz \\ cw + dy & cx + dz \end{pmatrix}$

5. $\mathcal{A}\mathcal{A}^T = \begin{pmatrix} 4 & 3 & 2 & 1 \\ 3 & 3 & 2 & 1 \\ 2 & 2 & 2 & 1 \\ 1 & 1 & 1 & 1 \end{pmatrix}$ $\mathcal{A}^2 = \begin{pmatrix} 1 & 2 & 3 & 4 \\ 0 & 1 & 2 & 3 \\ 0 & 0 & 1 & 2 \\ 0 & 0 & 0 & 1 \end{pmatrix}$

7. 36

17. $\begin{pmatrix} 1 & -1 & 0 \\ 0 & 1 & -1 \\ 0 & 0 & 1 \end{pmatrix}$

19. $x = 1,\ y = 2,\ z = 3$

21. $x_1 = 1,\ x_2 = 2,\ x_3 = -1,\ x_4 = -2$

23. neg. def.

25. pos. def.

27. indefinite

Section 10.8 (page 620)

1. 2 units

5.
```
sp:=(U,V)->DotProduct(
U,Normalize(V,2),conjugate=false)
```

7.
```
ang := (u,v) -> evalf(
(180/Pi)*VectorAngle(U,V))
```

9.
```
VolT:=(U,V,W)->(1/6)*abs(
DotProduct(U,(V &x W),conjugate=false))
```

11. $(u, v, x, y, z) = (1, 0, -1, 3, 2)$

13. -935

15. $\begin{bmatrix} 9 & -36 & 30 \\ -36 & 192 & -180 \\ 30 & -180 & 180 \end{bmatrix}$

Review Exercises (page 621)

1. plane parallel to y-axis through $(3, 0, 0)$ and $(0, 0, 1)$

3. all points on or above the plane through the origin with normal $\mathbf{i} + \mathbf{j} + \mathbf{k}$

5. circular paraboloid with vertex at $(0, 1, 0)$ and axis along the y-axis, opening in the direction of increasing y

7. hyperbolic paraboloid

9. points inside the ellipsoid with vertices at $(\pm 2, 0, 0)$, $(0, \pm 2, 0$, and $(0, 0, \pm 1)$

11. cone with axis along the x-axis, vertex at the origin, and elliptical cross-sections perpendicular to its axis

13. oblique circular cone (elliptic cone). Cross-sections in horizontal planes $z = k$ are circles of radius 1 with centres at $(k, 0, k)$

15. horizontal line through $(0, 0, 3)$ and $(2, -1, 3)$

17. circle of radius 1 centred at $(1, 1, 1)$ in plane normal to $\mathbf{i} + \mathbf{j} + \mathbf{k}$

19. $2x - y + 3z = 0$

21. $2x + 5y + 3z = 2$

23. $7x + 4y - 8z = 6$

25. $\mathbf{r} = (2 + 3t)\mathbf{i} + (1 + t)\mathbf{j} - (1 + 2t)\mathbf{k}$

27. $x = 3t,\ y = -2t,\ z = 4t$

29. $(\mathbf{r}_2 - \mathbf{r}_1) \times (\mathbf{r}_3 - \mathbf{r}_1) = \mathbf{0}$

31. $(3/2)\sqrt{34}$ sq. units

33. $\mathcal{A}^{-1} = \begin{pmatrix} 1 & 0 & 0 & 0 \\ -2 & 1 & 0 & 0 \\ 1 & -2 & 1 & 0 \\ 0 & 1 & -2 & 1 \end{pmatrix}$

35. pos. def.

Challenging Problems (page 621)

5. condition: $\mathbf{a} \bullet \mathbf{b} = 0$,
$\mathbf{x} = \dfrac{\mathbf{b} \times \mathbf{a}}{|\mathbf{a}|^2} + t\mathbf{a}$ (for any scalar t)

Chapter 11
Vector Functions and Curves

Section 11.1 (page 629)

1. $\mathbf{v} = \mathbf{j}$, $v = 1$, $\mathbf{a} = \mathbf{0}$, path is the line $x = 1, z = 0$

3. $\mathbf{v} = 2t\mathbf{j}+\mathbf{k}$, $v = \sqrt{4t^2 + 1}$, $\mathbf{a} = 2\mathbf{j}$, path is the parabola $y = z^2$, in the yz-plane

5. $\mathbf{v} = 2t\mathbf{i} - 2t\mathbf{j}$, $v = 2\sqrt{2}t$, $\mathbf{a} = 2\mathbf{i} - 2\mathbf{j}$, path is the straight half-line $x + y = 0,\ z = 1,\ (x \geq 0)$

7. $\mathbf{v} = -a \sin t\mathbf{i} + a \cos t\mathbf{j} + c\mathbf{k}$, $v = \sqrt{a^2 + c^2}$, $\mathbf{a} = -a \cos t\mathbf{i} - a \sin t\mathbf{j}$, path is a circular helix

9. $\mathbf{v} = -3 \sin t\,\mathbf{i} - 4 \sin t\,\mathbf{j} + 5 \cos t\,\mathbf{k}$, $v = 5$, $\mathbf{a} = -\mathbf{r}$, path is the circle of intersection of the plane $4x = 3y$ with the sphere $x^2 + y^2 + z^2 = 25$

11. $\mathbf{a} = \mathbf{v} = \mathbf{r}$, $v = \sqrt{a^2 + b^2 + c^2}\, e^t$, path is the straight line $\dfrac{x}{a} = \dfrac{y}{b} = \dfrac{z}{c}$

13. $\mathbf{v} = -(e^{-t} \cos e^t + \sin e^t)\mathbf{i}$
$\quad +(-e^{-t} \sin e^t + \cos e^t)\mathbf{j} - e^t\mathbf{k}$
$v = \sqrt{1 + e^{-2t} + e^{2t}}$
$\mathbf{a} = [(e^{-t} - e^t) \cos e^t + \sin e^t]\mathbf{i}$
$\quad +[(e^{-t} - e^t) \sin e^t - \cos e^t]\mathbf{j} - e^t\mathbf{k}$
The path is a spiral lying on the surface $z = -1/\sqrt{x^2 + y^2}$

15. $\mathbf{a} = -3\pi^2\mathbf{i} - 4\pi^2\mathbf{j}$

17. $\sqrt{3/2}(-\mathbf{i} + \mathbf{j} - 2\mathbf{k})$

19. $\mathbf{v} = 2\mathbf{i} + 4\mathbf{j} + 4\mathbf{k}$, $\mathbf{a} = -\frac{8}{9}(2\mathbf{i} + \mathbf{j} - 2\mathbf{k})$

29. $\dfrac{d}{dt}\big(\mathbf{u} \times (\mathbf{v} \times \mathbf{w})\big) = \dfrac{d\mathbf{u}}{dt} \times (\mathbf{v} \times \mathbf{w})$
$\quad + \mathbf{u} \times \big(\dfrac{d\mathbf{v}}{dt} \times \mathbf{w}\big) + \mathbf{u} \times \big(\mathbf{v} \times \dfrac{d\mathbf{w}}{dt}\big)$

31. $\mathbf{u}''' \bullet (\mathbf{u} \times \mathbf{u}')$

33. $\mathbf{r} = \mathbf{r}_0 e^{2t}$, $\mathbf{a} = 4\mathbf{r}_0 e^{2t}$; the path is a straight line through the origin in the direction of $\mathbf{r}_0$

35. $\mathbf{r} = \mathbf{r}_0 + \dfrac{1 - e^{-ct}}{c}\mathbf{v}_0 - \dfrac{g}{c^2}(ct + e^{-ct} - 1)\mathbf{k}$

Section 11.2 (page 636)

1. $\dfrac{e - 1}{e},\ \dfrac{e^2 - 1}{e^2}$

3. $\mathbf{r} = \cos t\mathbf{i} + \sin t\mathbf{j} + \mathbf{k}$; the curve is a circle of radius 1 in the plane $z = 1$

5. 4.76° west of south; $\dfrac{\pi^2 R}{72}$ towards the ground, where R is the radius of the earth

7. (a) tangential only, 90° counterclockwise from **v**.
(b) tangential only, 90° clockwise from **v**.
(c) normal only

9. 16.0 hours, 52.7°

Section 11.3 (page 643)

1. $x = \sqrt{a^2 - t^2},\ y = t,\ 0 \le t \le a$

3. $x = a\sin\theta,\ y = -a\cos\theta,\ \frac{\pi}{2} \le \theta \le \pi$

5. $\mathbf{r} = -2t\mathbf{i} + t\mathbf{j} + 4t^2\mathbf{k}$

7. $\mathbf{r} = 3\cos t\mathbf{i} + 3\sin t\mathbf{j} + 3(\cos t + \sin t)\mathbf{k}$

9. $\mathbf{r} = (1 + 2\cos t)\mathbf{i} - 2(1 - \sin t)\mathbf{j} + (9 + 4\cos t - 8\sin t)\mathbf{k}$

11. Choice (b) leads to $\mathbf{r} = \dfrac{t^2-1}{2}\mathbf{i} + t\mathbf{j} + \dfrac{t^2+1}{2}\mathbf{k}$, which represents the whole parabola. Choices (a) and (c) lead to separate parametrizations for the halves $y \ge 0$ and $y \le 0$ of the parabola. For (a) these are $\mathbf{r} = t\mathbf{i} \pm \sqrt{1+2t}\mathbf{j} + (1+t)\mathbf{k}$, $(t \ge -1/2)$

13. $(17\sqrt{17} - 16\sqrt{2})/27$ units

15. $\displaystyle\int_1^T \frac{\sqrt{4a^2t^4 + b^2t^2 + c^2}}{t}\,dt$ units;
$a(T^2 - 1) + c\ln T$ units

17. $\pi\sqrt{2 + 4\pi^2} + \ln(\sqrt{2}\pi + \sqrt{1 + 2\pi^2})$ units

19. $\sqrt{2e^{4\pi} + 1} - \sqrt{3} + \frac{1}{2}\ln\frac{e^{4\pi}+1-\sqrt{2e^{4\pi}+1}}{e^{4\pi}}$
$-\frac{1}{2}\ln(2 - \sqrt{3})$ units

21. straight line segments from $(0, 0)$ to $(1, 1)$, then to $(0, 2)$

23. $\mathbf{r} = \frac{1}{\sqrt{A^2+B^2+C^2}}(As\mathbf{i} + Bs\mathbf{j} + Cs\mathbf{k})$

25. $\mathbf{r} = a\left(1 - \dfrac{s}{K}\right)^{3/2}\mathbf{i} + a\left(\dfrac{s}{K}\right)^{3/2}\mathbf{j} + b\left(1 - \dfrac{2s}{K}\right)\mathbf{k}$,
$0 \le s \le K,\ K = (\sqrt{9a^2 + 16b^2})/2$

Section 11.4 (page 651)

1. $\hat{\mathbf{T}} = \frac{1}{\sqrt{1+16t^2+81t^4}}(\mathbf{i} - 4t\mathbf{j} + 9t^2\mathbf{k})$

3. $\hat{\mathbf{T}} = \frac{1}{\sqrt{1+\sin^2 t}}(\cos 2t\mathbf{i} + \sin 2t\mathbf{j} - \sin t\mathbf{k})$

Section 11.5 (page 657)

1. $1/2,\ 27/2$

3. $27/(4\sqrt{2})$

5. $\hat{\mathbf{T}} = (\mathbf{i} + 2\mathbf{j})/\sqrt{5},\ \hat{\mathbf{N}} = (-2\mathbf{i} + \mathbf{j})/\sqrt{5},\ \hat{\mathbf{B}} = \mathbf{k}$

7. $\hat{\mathbf{T}} = \dfrac{1}{\sqrt{1 + t^2 + t^4}}(\mathbf{i} + t\mathbf{j} + t^2\mathbf{k})$,
$\hat{\mathbf{B}} = \dfrac{1}{\sqrt{t^4 + 4t^2 + 1}}(t^2\mathbf{i} - 2t\mathbf{j} + \mathbf{k})$,
$\hat{\mathbf{N}} = \dfrac{-(t + 2t^3)\mathbf{i} + (1 - t^4)\mathbf{j} + (t^3 + 2t)\mathbf{k}}{\sqrt{t^4 + 4t^2 + 1}\sqrt{1 + t^2 + t^4}}$,
$\kappa = \dfrac{\sqrt{t^4 + 4t^2 + 1}}{(t^4 + t^2 + 1)^{3/2}},\quad \tau = \dfrac{2}{t^4 + 4t^2 + 1}$

9. $\kappa(t) = 1/\sqrt{2}$, $\tau(t) = 0$, curve is a circle in the plane $y + z = 4$, having centre $(2, 1, 3)$ and radius $\sqrt{2}$

11. (a) $\hat{\mathbf{T}} = \mathbf{i},\quad \hat{\mathbf{N}} = \dfrac{2\mathbf{j} - \mathbf{k}}{\sqrt{5}}$,
$\hat{\mathbf{B}} = \dfrac{\mathbf{j} + 2\mathbf{k}}{\sqrt{5}},\quad \kappa = \sqrt{5},\quad \tau = 0$
(b) $\hat{\mathbf{T}} = \sqrt{\frac{2}{3}}(\mathbf{j} - \frac{1}{\sqrt{2}}\mathbf{k}),\quad \hat{\mathbf{B}} = \frac{1}{\sqrt{13}}(-\mathbf{i} + 2\mathbf{j} + 2\sqrt{2}\mathbf{k})$,
$\hat{\mathbf{N}} = -\frac{1}{\sqrt{39}}(6\mathbf{i} + \mathbf{j} + \sqrt{2}\mathbf{k}),\quad \kappa = \frac{2\sqrt{39}}{9},\quad \tau = -\frac{6\sqrt{2}}{13}$

13. max a/b^2, min b/a^2

15. $\kappa = \dfrac{e^x}{(1 + e^{2x})^{3/2}}$, $\mathbf{r} = (x - 1 - e^{2x})\mathbf{i} + (2e^x + e^{-x})\mathbf{j}$

17. $\frac{3}{2\sqrt{2ar}}$

21. $\mathbf{r} = -4x^3\mathbf{i} + (3x^2 + \frac{1}{2})\mathbf{j}$

23. $f(x) = \frac{1}{8}(15x - 10x^3 + 3x^5)$

Section 11.6 (page 666)

3. velocity: $1/\sqrt{2}$, $1/\sqrt{2}$; acceleration: $-e^{-\theta}/2$, $e^{-\theta}/2$.

5. $|a_r| = \dfrac{v_0^2}{5}\left(\dfrac{2}{r^2} + \dfrac{1}{r^3}\right)$

7. 42,777 km, the equatorial plane

9. $\frac{T}{4\sqrt{2}}$

13. $3/4$

15. $(1/2) - (\epsilon/\pi)$

19. $r = A\sec\omega(\theta - \theta_0),\quad \omega^2 = 1 - (k/h^2)$ if $k < h^2$,
$r = 1/(A + B\theta)$ if $k = h^2$,
$r = A\,e^{\omega\theta} + B\,e^{-\omega\theta},\quad \omega^2 = (k/h^2) - 1,\quad$ if $k > h^2$;
there are no bounded orbits that do not approach the origin except in the case $k = h^2$ if $B = 0$ when there are circular orbits. (Now aren't you glad gravitation is an inverse square rather than an inverse cube attraction?)

21. centre $\left(\dfrac{\ell\epsilon}{\epsilon^2 - 1}, 0\right)$;
asymptotes in directions $\theta = \pm\cos^{-1}\left(-\dfrac{1}{\epsilon}\right)$;
semi-transverse axis $a = \dfrac{\ell}{\epsilon^2 - 1}$;
semi-conjugate axis $b = \dfrac{\ell}{\sqrt{\epsilon^2 - 1}}$;
semi-focal separation $c = \dfrac{\ell\epsilon}{\epsilon^2 - 1}$.

Review Exercises (page 668)

3. $\mathbf{v} = 2(\mathbf{i} + 2\mathbf{j} + 2\mathbf{k}),\ \mathbf{a} = (8/3)(-2\mathbf{i} - \mathbf{j} + 2\mathbf{k})$

5. $\kappa = \tau = \sqrt{2}/(e^t + e^{-t})^2$

9. $4a(1 - \cos(T/2))$ units

11. $\mathbf{r}_C(t) = a(t - \sin t)\mathbf{i} + a(1 - \cos t)\mathbf{j}$

13. $\hat{\boldsymbol{\rho}} = \sin\phi\cos\theta\mathbf{i} + \sin\phi\sin\theta\mathbf{j} + \cos\phi\mathbf{k}$ right-handed
$\hat{\boldsymbol{\phi}} = \cos\phi\cos\theta\mathbf{i} + \cos\phi\sin\theta\mathbf{j} - \sin\phi\mathbf{k}$
$\hat{\boldsymbol{\theta}} = -\sin\theta\mathbf{i} + \cos\theta\mathbf{j}$

Challenging Problems (page 669)

1. (a) $\mathbf{\Omega} = \Omega\dfrac{\mathbf{j}+\mathbf{k}}{\sqrt{2}}$, $\Omega \approx 7.272 \times 10^{-5}$.
(b) $\mathbf{a}_C = -\sqrt{2}\Omega v\mathbf{i}$.
(c) about 15.5 cm west of P.

3.
(c) $\mathbf{v}(t) = (\mathbf{v}_0 - (\mathbf{v}_0 \bullet \mathbf{k})\mathbf{k})\cos(\omega t) + (\mathbf{v}_0 \times \mathbf{k})\sin(\omega t) + (\mathbf{v}_0 \bullet \mathbf{k})\mathbf{k}$.
(d) Straight line if $\mathbf{v}_0$ is parallel to $\mathbf{k}$, circle if $\mathbf{v}_0$ is perpendicular to $\mathbf{k}$.

5. (a) $y = (48 + 24x^2 - x^4)/64$

7. (a) Yes, time $\pi a/(v\sqrt{2})$, (b) $\phi = \dfrac{\pi}{2} - \dfrac{vt}{a\sqrt{2}}$,
$\theta = \ln\left[\sec\left(\dfrac{vt}{a\sqrt{2}}\right) + \tan\left(\dfrac{vt}{a\sqrt{2}}\right)\right]$.
(c) infinitely often

Chapter 12
Partial Differentiation

Section 12.1 (page 677)

1. all (x, y) with $x \neq y$ **3.** all (x, y) except $(0, 0)$

5. all (x, y) satisfying $4x^2 + 9y^2 \geq 36$

7. all (x, y) with $xy > -1$

9. all (x, y, z) except $(0, 0, 0)$

11. $z = f(x, y) = x$

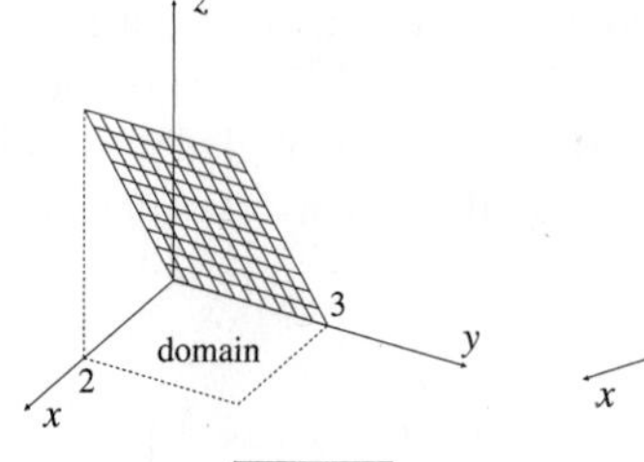

13. $z = f(x, y) = y^2$

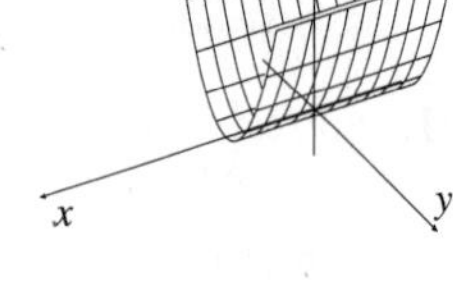

15. $f(x, y) = \sqrt{x^2 + y^2}$

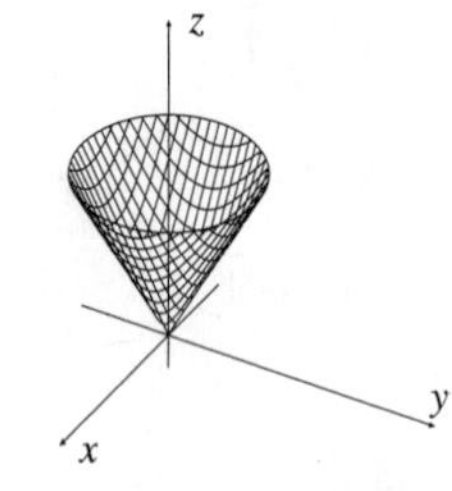

17. $f(x, y) = |x| + |y|$

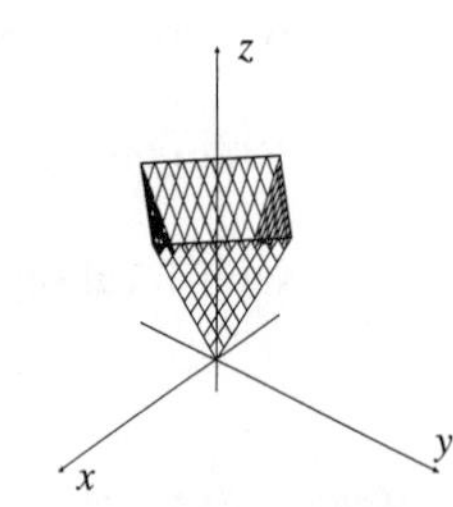

19. $f(x, y) = x - y = C$

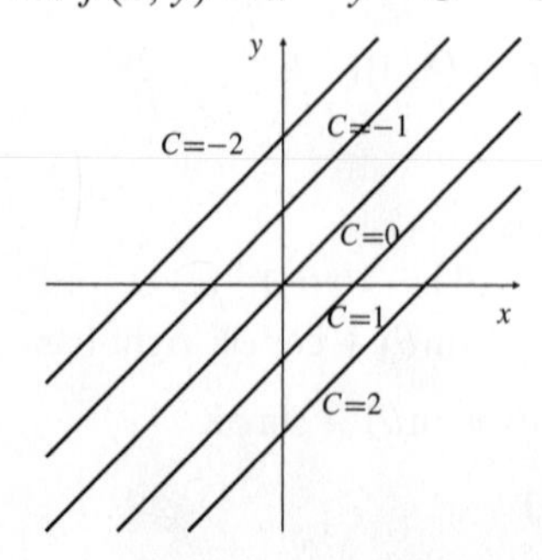

21. $f(x, y) = xy = C$

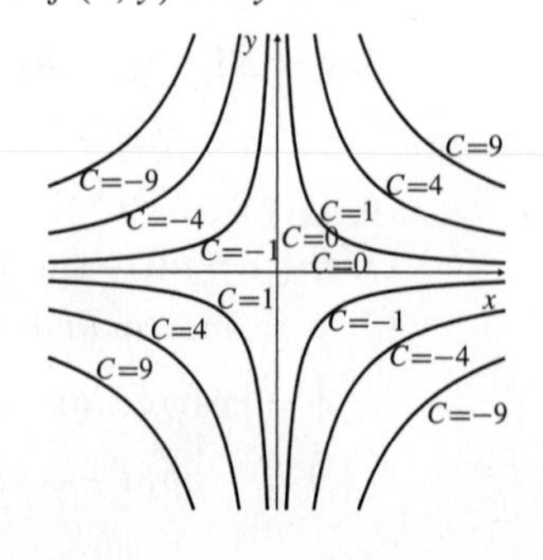

23. $f(x, y) = \dfrac{x-y}{x+y} = C$ **25.** $f(x, y) = xe^{-y} = C$

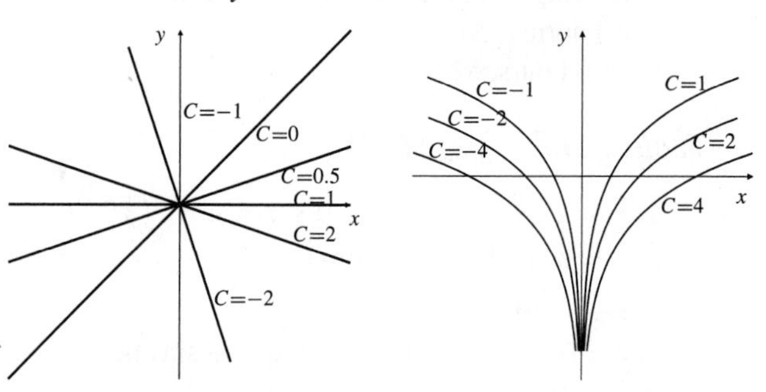

27. At B, because the contours are closer together there

29. a plane containing the y-axis, sloping uphill in the x direction

31. a right-circular cone with base in the xy-plane and vertex at height 5 on the z-axis

33. No, different curves of the family must not intersect in the region.

35. (a) $\sqrt{x^2 + y^2}$, (b) $(x^2 + y^2)^{1/4}$,
(c) $x^2 + y^2$, (d) $e^{\sqrt{x^2+y^2}}$

37. spheres centred at the origin

39. circular cylinders with axis along the z-axis

41. regular octahedra with vertices on the coordinate axes

Section 12.2 (page 682)

1. 2 **3.** does not exist

5. -1 **7.** 0

9. does not exist **11.** 0

13. $f(0, 0) = 1$

15. all (x, y) such that $x \neq \pm y$; yes; yes $f(x, x) = \frac{1}{2x}$ makes f continuous at (x, x) for $x \neq 0$; no, f has no continuous extension to the line $x + y = 0$.

17. no, yes **19.** $a = c = 0$, $b \neq 0$

23. a surface having no tears in it, meeting vertical lines through points of the region exactly once

Section 12.3 (page 689)

1. $f_1(x, y) = f_1(3, 2) = 1$, $f_2(x, y) = f_2(3, 2) = -1$

3. $f_1 = 3x^2y^4z^5$, $f_2 = 4x^3y^3z^5$, $f_3 = 5x^3y^4z^4$
All three vanish at $(0, -1, -1)$.

5. $\dfrac{\partial z}{\partial x} = \dfrac{-y}{x^2+y^2}$, $\dfrac{\partial z}{\partial y} = \dfrac{x}{x^2+y^2}$,
At $(-1, 1)$: $\dfrac{\partial z}{\partial x} = -\dfrac{1}{2}$, $\dfrac{\partial z}{\partial y} = -\dfrac{1}{2}$

7. $f_1 = \sqrt{y}\cos(x\sqrt{y})$, $f_2 = \dfrac{x\cos(x\sqrt{y})}{2\sqrt{y}}$,
At $(\pi/3, 4)$: $f_1 = -1$, $f_2 = -\pi/24$

9. $\dfrac{\partial w}{\partial x} = y\ \ln z\, x^{(y \ln z - 1)}$, $\dfrac{\partial w}{\partial y} = \ln x\ \ln z\, x^{y \ln z}$,
$\dfrac{\partial w}{\partial z} = \dfrac{y \ln x}{z} x^{y \ln z}$
At $(e, 2, e)$: $\dfrac{\partial w}{\partial x} = \dfrac{\partial w}{\partial z} = 2e$, $\dfrac{\partial w}{\partial y} = e^2$.

11. $f_1(0, 0 = 2,\ f_2(0, 0) = -1/3$

13. $z = -4x - 2y - 3;\qquad \frac{x+2}{-4} = \frac{y-1}{-2} = \frac{z-3}{-1}$

15. $z = \dfrac{1}{\sqrt{2}}\left(1 - \dfrac{x-\pi}{4} + \dfrac{\pi}{16}(y-4)\right);$

$$\frac{x-\pi}{-1/4\sqrt{2}} = \frac{y-4}{\pi/16\sqrt{2}} = \frac{z-1/\sqrt{2}}{-1}$$

17. $z = \frac{2}{5} + \frac{3x}{25} - \frac{4y}{25};\quad \frac{x-1}{3} = \frac{y-2}{-4} = \frac{z-1/5}{-25}$

19. $z = \ln 5 + \frac{2}{5}(x-1) - \frac{4}{5}(y+2);$

$$\frac{x-1}{2/5} = \frac{y+2}{-4/5} = \frac{z - \ln 5}{-1}$$

21. $z = \dfrac{x+y}{2} - \dfrac{\pi}{4};\quad 2(x-1) = 2(y+1) = -z - \dfrac{\pi}{4}$

23. $(0, 0),\quad (1, 1),\quad (-1, -1)$

33. $w = f(a, b, c) + f_1(a, b, c)(x-a) + f_2(a, b, c)(y - b) + f_3(a, b, c)(z - c)$

35. $\sqrt{7}/4$ units

37. $f_1(0, 0) = 1$, $f_2(0, 0)$ does not exist.

39. f is continuous at $(0, 0)$; f_1 and f_2 are not.

Section 12.4 (page 694)

1. $\dfrac{\partial^2 z}{\partial x^2} = 2(1+y^2),\quad \dfrac{\partial^2 z}{\partial x \partial y} = 4xy,\quad \dfrac{\partial^2 z}{\partial y^2} = 2x^2$

3. $\dfrac{\partial^2 w}{\partial x^2} = 6xy^3z^3,\quad \dfrac{\partial^2 w}{\partial y^2} = 6x^3yz^3,$

$\dfrac{\partial^2 w}{\partial z^2} = 6x^3y^3z,\quad \dfrac{\partial^2 w}{\partial x \partial y} = 9x^2y^2z^3,$

$\dfrac{\partial^2 w}{\partial x \partial z} = 9x^2y^3z^2,\quad \dfrac{\partial^2 w}{\partial y \partial z} = 9x^3y^2z^2$

5. $\dfrac{\partial^2 z}{\partial x^2} = -y\,e^x,\quad \dfrac{\partial^2 z}{\partial x \partial y} = e^y - e^x,\quad \dfrac{\partial^2 z}{\partial y^2} = x\,e^y$

7. $27,\ 10,\ x^2e^{xy}\big(xz\sin xz - (3+xy)\cos xz\big)$

19. $u(x, y, z, t) = t^{-3/2}e^{-(x^2+y^2+z^2)/4t}$

Section 12.5 (page 704)

1. $\dfrac{\partial w}{\partial t} = f_1g_2 + f_2h_2 + f_3k_2$

3. $\dfrac{\partial z}{\partial u} = g_1h_1 + g_2f'h_1$

5. $\dfrac{dw}{dz} = f_1g_1h' + f_1g_2 + f_2h' + f_3,$

$\left.\dfrac{\partial w}{\partial z}\right|_x = f_2h' + f_3,$

$\left.\dfrac{\partial w}{\partial z}\right|_{x,y} = f_3$

7. $\dfrac{\partial z}{\partial x} = \dfrac{-5y}{13x^2 - 2xy + 2y^2}$

9. $2f_1(2x, 3y)$

11. $2x\ f_2(y^2, x^2)$

13. $dT/dt = e^{-t}\big(f'(t) - f(t)\big)$; $dT/dt = 0$ if $f(t) = e^t$: in this case the decrease in T with time (at fixed depth) is exactly balanced by the increase in T with depth.

15. $4f_{11} + 12f_{12} + 9f_{22},\quad 6f_{11} + 5f_{12} - 6f_{22},$
$9f_{11} - 12f_{12} + 4f_{22}$

17. $f_1\ \cos s - f_2\ \sin s + f_{11}\,t\ \cos s\ \sin s$
$+ f_{12}\,t(\cos^2 s - \sin^2 s) - f_{22}\,t\ \sin s\ \cos s$

19. $f_2 + 2y^2f_{12} + xyf_{22} - 4xyf_{31} - 2x^2f_{32};$
all derivatives at $(y^2, xy, -x^2)$

27. $\sum_{i,j=1}^n x_i x_j f_{ij}(x_1, \cdots, x_n) = k(k-1)\,f(x_1, \cdots, x_n)$

31. $u(x, y) = f(x + ct)$

Section 12.6 (page 714)

1. 6.9

3. 0.0814

5. 2.967

7. $dz = 2xe^{3y}\,dx + 3x^2e^{3y}\,dy,\ 8.76$

9. $dF = \dfrac{x\,dx + y\,dy + z\,dz}{\sqrt{x^2 + y + 2 + z^2}},\ 3.1$

11. (a) 3%, (b) 2%, (c) 1%

13. 8.88 ft^2

15. 169 m, 24 m, most sensitive to angle at B

17. $\begin{pmatrix} \cos\theta & -r\sin\theta \\ \sin\theta & r\cos\theta \end{pmatrix}$

19. $\begin{pmatrix} 2x & z & y \\ -\ln z & 2y & -x/z \end{pmatrix},\quad (5.99, 3.98)$

27. $f^*(p) = p^2/4$

29. $f^*(p) = 1 - \dfrac{2p}{3} - \ln\left(\dfrac{3}{p}\right)$

Section 12.7 (page 725)

1. $4\mathbf{i} + 2\mathbf{j};\quad z = 4x + 2y - 3;\quad 2x + y = 3$

3. $(3\mathbf{i} - 4\mathbf{j})/25;\quad 3x - 4y - 25z + 10 = 0;$
$3x - 4y + 5 = 0$

5. $(2\mathbf{i} - 4\mathbf{j})/5;\ 2x - 4y - 5z = 10 - 5\ln 5;\ x - 2y = 5$

7. $x + y - 3z = -3$

9. $\sqrt{3}y + z = \sqrt{3} + \pi/3$

11. $\dfrac{4}{\sqrt{5}}$

13. $1 - 2\sqrt{3}$

17. in directions making angles $-30°$ or $-150°$ with positive x-axis; no; $-\mathbf{j}$.

19. $7\mathbf{i} - \mathbf{j}$

21. a)

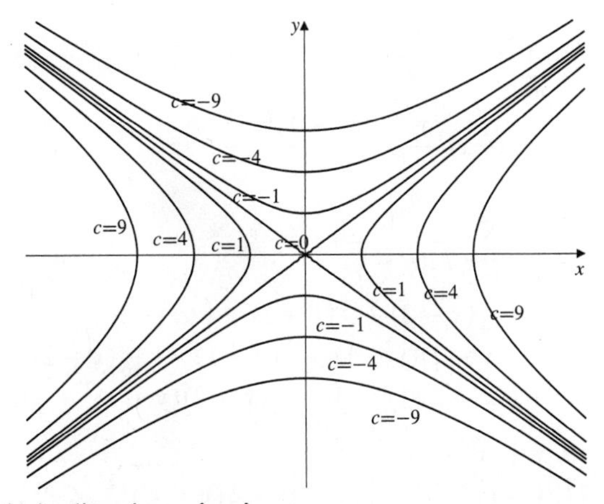

b) in direction $-\mathbf{i} - \mathbf{j}$
c) $4\sqrt{2k}$ deg/unit time
d) $12k/\sqrt{5}$ deg/unit time
e) $x^2y = -4$

23. $3x^2 - 2y^2 = 10$ **25.** $-4/3$

27. $\mathbf{i} - 2\mathbf{j} + \mathbf{k}$

33. $D_{\mathbf{v}}(D_{\mathbf{v}}f) = v_1^2 f_{11} + v_2^2 f_{22} + v_3^2 f_{33} + 2v_1v_2 f_{12} + 2v_1v_3 f_{13} + 2v_2v_3 f_{23}$.
This is the second derivative of f as measured by an observer moving with velocity $\mathbf{v}$.

35. $\dfrac{\partial^2 T}{\partial t^2} + 2D_{\mathbf{v}(t)}\left(\dfrac{\partial T}{\partial t}\right) + D_{\mathbf{a}(t)}T + D_{\mathbf{v}(t)}(D_{\mathbf{v}(t)}T)$

Section 12.8 (page 736)

1. $-\dfrac{x^4 + 3xy^2}{y^3 + 4x^3y}$, $y \neq 0$, $y^2 \neq -4x^3$

3. $\dfrac{3xy^4 + xz}{xy - 2y^2z}$, $y \neq 0$, $x \neq 2yz$

5. $\dfrac{x - 2t^2w}{2xy^2 - w}$, $w \neq 2xy^2$ **7.** $-\dfrac{\partial G/\partial x}{\partial G/\partial u}$, $\dfrac{\partial G}{\partial u} \neq 0$

9. $-\dfrac{v^2H_2 + wH_3}{u^2H_1 + tH_3}$, $u^2H_1 + tH_3 \neq 0$,
all derivatives at (u^2w, v^2t, wt)

11. $\dfrac{2w - 4y}{4x - w}$, $4x \neq w$ **13.** $\dfrac{1}{6}, \dfrac{1}{2}, \dfrac{1}{6}, -\dfrac{1}{2}, -\dfrac{1}{6}$

15. r; all points except the origin

17. $-3/2$

19. $-\dfrac{\partial(F,G,H)}{\partial(y,z,w)} \Big/ \dfrac{\partial(F,G,H)}{\partial(x,z,w)}$

21. 15; $-\dfrac{\partial(F,G,H)}{\partial(x_2,x_3,x_5)} \Big/ \dfrac{\partial(F,G,H)}{\partial(x_1,x_3,x_5)}$

23. $2(u+v)$, -2, 0

31. $S = \dfrac{3Nk}{2}\left(\ln\left[\dfrac{4\pi mE}{3h^2N}\left(\dfrac{V}{N}\right)^{2/3}\right] + \dfrac{5}{3}\right)$

$S = \dfrac{3Nk}{2}\left(\ln\left[\dfrac{2\pi mkT}{h^2}\left(\dfrac{V}{N}\right)^{2/3}\right] + \dfrac{5}{3}\right)$

Section 12.9 (page 742)

1. $\displaystyle\sum_{n=0}^{\infty}(-1)^n \frac{x^n y^{2n}}{2^{n+1}}$

3. $\displaystyle\sum_{n=0}^{\infty}(-1)^n \frac{x^{2n+1}(y+1)^{2n+1}}{2n+1}$

5. $\displaystyle\sum_{n=0}^{\infty}\sum_{k=0}^{n} \frac{1}{k!(n-k)!} x^{2k} y^{2n-2k}$

7. $\frac{1}{2} - \frac{1}{4}(x-2) + \frac{1}{2}(y-1) + \frac{1}{8}(x-2)^2 - \frac{1}{2}(x-2)(y-1) + \frac{1}{2}(y-1)^2 - \frac{1}{16}(x-2)^3 + \frac{3}{8}(x-2)^2(y-1) - \frac{3}{4}(x-2)(y-1)^2 + \frac{1}{2}(y-1)^3$

9. $x + y^2 - \frac{x^3}{3}$

11. $1 - (y-1) + (y-1)^2 - \frac{1}{2}(x - \frac{\pi}{2})^2$

13. $-x - x^2 - (5/6)x^3$

15. $-\dfrac{x}{3} - \dfrac{2y}{3} - \dfrac{2x^2}{27} - \dfrac{8xy}{27} - \dfrac{8y^2}{27}$

17. $\dfrac{[(2n)!]^3}{(n!)^2}$

Review Exercises (page 743)

1.

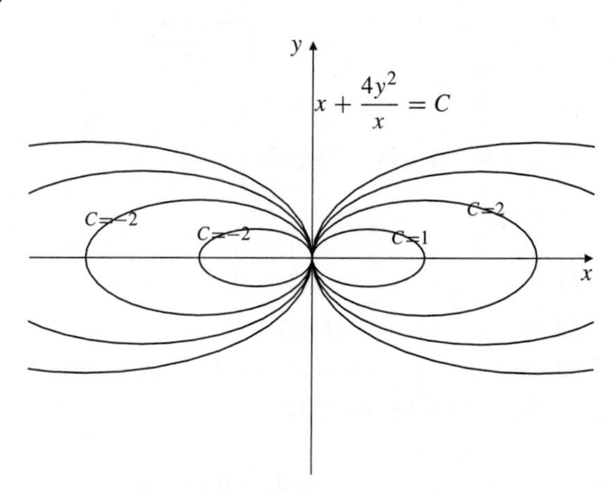

3.

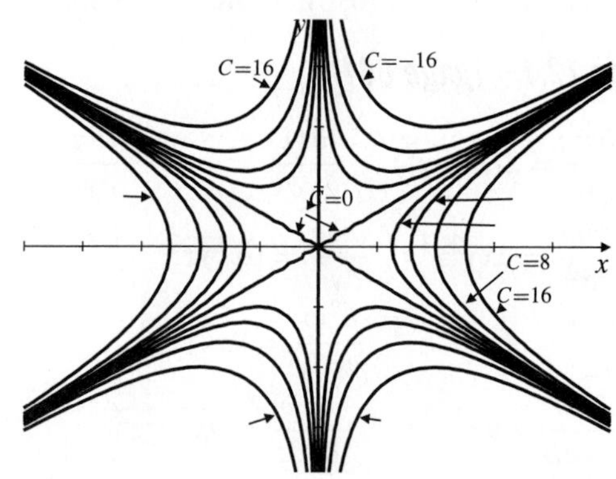

5. cont. except on lines $x = \pm y$; can be extended to $x = y$ except at the origin; if $f(0,0) = 0$ then $f_1(0,0) = f_2(0,0) = 1$

7. (a) $ax + by + 4cz = 16$,
(b) the circle $z = 1$, $x^2 + y^2 = 12$, (c) $\pm(2, 2, \sqrt{2})$

9. 7,500 m^2, 7.213%

11. (a) $-1/\sqrt{2}$, (b) dir. of $\pm(\mathbf{i} + 3\mathbf{j} - 4\mathbf{k}$, (c) dir. of $-7\mathbf{i} + 5\mathbf{j} + 2\mathbf{k}$

15. (a) $\partial u/\partial x = -5$, $\partial u/\partial y = 1$, (b) -1.13

Challenging Problems (page 744)

Chapter 13
Applications of Partial Derivatives

Section 13.1 (page 752)

1. $(2, -1)$, loc. (abs) min.

3. $(0, 0)$, saddle pt; $(1, 1)$, loc. min.

5. $(-4, 2)$, loc. max.

7. $(0, n\pi)$, $n = 0, \pm 1, \pm 2, \cdots$, all saddle points

9. $(0, a)$, $(a > 0)$, loc min; $(0, a)$, $(a < 0)$, loc max; $(0, 0)$ saddle point; $(\pm 1, 1/\sqrt{2})$, loc. (abs) max; $(\pm 1, -1/\sqrt{2})$, loc. (abs) min.

11. $(3^{-1/3}, 0)$, saddle pt.

13. max at (x, x), min at $(x, -x)$, $x \neq 0$

15. $(-1, -1)$, $(1, -1)$, $(-1, 1)$, saddle pts; $(-3, -3)$, loc. min.

17. $(1, 1, \frac{1}{2})$, saddle pt.

19. $(0, 0)$, saddle pt; $(\frac{1}{\sqrt{2}}, \frac{1}{\sqrt{2}})$, $(-\frac{1}{\sqrt{2}}, -\frac{1}{\sqrt{2}})$, loc. (abs) max; $(\frac{1}{\sqrt{2}}, -\frac{1}{\sqrt{2}})$, $(-\frac{1}{\sqrt{2}}, \frac{1}{\sqrt{2}})$, loc. (abs) min.

21. max $e^{-3/2}/2\sqrt{2}$, min $-e^{-3/2}/2\sqrt{2}$; f is continuous everywhere, and $f(x, y, z) \to 0$ as $x^2 + y^2 + z^2 \to \infty$.

23. $L^3/108$ cu. units

25. $8abc/(3\sqrt{3})$ cu. units

27. CPs are $(\sqrt{\ln 3}, -\sqrt{\ln 3})$ and $(-\sqrt{\ln 3}, \sqrt{\ln 3})$.

29. f does not have a local minimum at $(0, 0)$; the second derivative test is inconclusive ($B^2 = AC$).

Section 13.2 (page 758)

1. max 5/4, min -2

3. max $(\sqrt{2} - 1)/2$, min $-(\sqrt{2} + 1)/2$.

5. max $2/3\sqrt{3}$, min 0

7. max 1, min -1

9. max $1/\sqrt{e}$, min $-1/\sqrt{e}$

11. max 4/9, min $-4/9$

13. no limit; yes, max $f = e^{-1}$ (at all points of the curve $xy = 1$)

15. \$625,000, \$733,333

17. max 37/2 at (7/4,5)

19. 6667 kg deluxe, 6667 kg standard

Section 13.3 (page 765)

1. 84, 375

3. 1 unit

5. max 4 units, min 2 units

7. $a = \pm\sqrt{3}$, $b = \pm 2\sqrt{3}$, $c = \pm\sqrt{3}$

9. max 8, min -8

11. $\sqrt{7}$ units

13. max 2, min -2

15. max 7, min -1

17. $\dfrac{2\sqrt{6}}{3}$ units

19. $\frac{1}{6} \times \frac{1}{3} \times \frac{2}{3}$

21. width = $\left(\dfrac{2V}{15}\right)^{1/3}$, depth = 3×width, height = $\dfrac{5}{2}$×width

23. max 1, min $-\frac{1}{2}$

27. Method will not fail if $\nabla f = \mathbf{0}$ at extreme point; but we will have $\lambda = 0$.

Section 13.4 (page 775)

3. local and absolute minimum 10

5. $P = (0, 0, 0, 1, 2, -2)$ has saddle behaviour. Local minima at Q, $R = (\pm\sqrt{6}/2, 3/2, \pm\sqrt{6}/4, 7/4, 1/2, -1/2)$. Distance $\sqrt{7}/4$.

Section 13.5 (page 781)

1. at $(\bar{x}, \bar{y})$ where $\bar{x} = \left(\sum_{i=1}^{n} x_i\right)/n$, $\bar{y} = \left(\sum_{i=1}^{n} y_i\right)/n$

3. $a = \left(\sum_{i=1}^{n} y_i e^{x_i}\right) \Big/ \left(\sum_{i=1}^{n} e^{2x_i}\right)$

5. If $A = \sum x_i^2$, $B = \sum x_i y_i$, $C = \sum x_i$, $D = \sum y_i^2$, $E = \sum y_i$, $F = \sum x_i z_i$, $G = \sum y_i z_i$, and $H = \sum z_i$, then

$$\Delta = \begin{vmatrix} A & B & C \\ B & D & E \\ C & E & n \end{vmatrix}, \qquad a = \frac{1}{\Delta}\begin{vmatrix} F & B & C \\ G & D & E \\ H & E & n \end{vmatrix},$$

$$b = \frac{1}{\Delta}\begin{vmatrix} A & F & C \\ B & G & E \\ C & H & n \end{vmatrix}, \qquad c = \frac{1}{\Delta}\begin{vmatrix} A & B & F \\ B & D & G \\ C & E & H \end{vmatrix}$$

7. Use linear regression to fit $\eta = a + bx$ to the data $(x_i, \ln y_i)$. Then $p = e^a$, $q = b$. These are not the same values as would be obtained by minimizing the expression $\sum(y_i - pe^{qx_i})^2$.

9. Use linear regression to fit $\eta = a + b\xi$ to the data $\left(x_i, \dfrac{y_i}{x_i}\right)$. Then $p = a, q = b$. Not the same as minimizing $\sum(y_i - px_i - qx_i^2)^2$.

11. Use linear regression to fit $\eta = a + b\xi$ to the data $\left(e^{-2x_i}, \dfrac{y_i}{e^{x_i}}\right)$. Then $p = a, q = b$. Not the same as minimizing $\sum(y_i - pe^{x_i} - qe^{-x_i})^2$. Other answers are possible.

13. If $A = \sum x_i^4$, $B = \sum x_i^3$, $C = \sum x_i^2$, $D = \sum x_i$, $H = \sum x_i^2 y_i$, $I = \sum x_i y_i$, and $J = \sum y_i$, then

$$\Delta = \begin{vmatrix} A & B & C \\ B & C & D \\ C & D & n \end{vmatrix}, \qquad a = \frac{1}{\Delta}\begin{vmatrix} H & B & C \\ I & C & D \\ J & D & n \end{vmatrix},$$

$$b = \frac{1}{\Delta}\begin{vmatrix} A & H & C \\ B & I & D \\ C & J & n \end{vmatrix}, \qquad c = \frac{1}{\Delta}\begin{vmatrix} A & B & H \\ B & C & I \\ C & D & J \end{vmatrix}$$

15. $a = 5/6$, $I = 1/252$

17. $a = 15/16$, $b = -1/16$, $I = 1/448$

19. $a = \frac{20}{\pi^5}(\pi^2 - 16)$, $b = \frac{12}{\pi^4}(20 - \pi^2)$

21. $a_k = \frac{2}{\pi}\int_0^{\pi} f(x)\cos kx\,dx$, $(k = 0, 1, 2, \cdots)$

23. $\pi - \frac{4}{\pi}\sum_{k=0}^{\infty}\frac{\cos((2k+1)x)}{(2k+1)^2}$; $-x$

Section 13.6 (page 790)

1. $\frac{(-1)^n n!}{(x+1)^{n+1}}$

3. $2\sqrt{\pi}(\sqrt{y} - \sqrt{x})$

5. $\frac{2x}{(1+x^2)^2}$; $\frac{(6x^2-2)}{(1+x^2)^3}$

7. $\frac{\pi}{2x}$, assume $x > 0$; $\frac{\pi}{4x^3}$; $\frac{3\pi}{16x^5}$

9. $n!$

11. $f(x) = \int_0^x e^{-t^2/2}\,dt$

13. $y = x^2$

15. $x^2 + y^2 = 1$

17. $y = x - \frac{1}{4}$

19. no

21. no; a line of singular points

23. $x^2 + y^2 + z^2 = 1$

25. $y = x - \epsilon\sin(\pi x) + \frac{\pi\epsilon^2}{2}\sin(2\pi x) + \cdots$

27. $y = \frac{1}{2} - \frac{2}{5}\epsilon x - \frac{16}{125}\epsilon^2 x^2 + \cdots$

29. $x \approx 1 - \frac{1}{100e} - \frac{1}{30000e^2}$, $y \approx 1 - \frac{1}{30000e^2}$

Section 13.7 (page 794)

1. (0.797105, 2.219107)

3. $(\pm 0.2500305, \pm 3.9995115)$, $(\pm 1.9920783, \pm 0.5019883)$

5. $(0.3727730, 0.3641994)$, $(-1.4141606, -0.9877577)$

7. $x = x_0 - \frac{\Delta_1}{\Delta}$, $y = y_0 - \frac{\Delta_2}{\Delta}$, $z = z_0 - \frac{\Delta_3}{\Delta}$, where $\Delta = \left.\frac{\partial(f, g, h)}{\partial(x, y, z)}\right|_{(x_0, y_0, z_0)}$ and Δ_i is Δ with the ith column replaced with $\begin{matrix} f \\ g \\ h \end{matrix}$

9. 18 iterations near $(0, 0)$, 4 iterations near $(1, 1)$; the two curves are tangent at (0,0), but not at (1,1).

Section 13.8 (page 799)

1. $(\pm .45304, .81204, \pm .36789)$, $(\pm .96897, .17751, \pm .17200)$

3. local and absolute max 0.81042 at $(-0.33853, -0.52062)$; local and absolute min -0.66572 at $(0.13319, 0.53682)$

5. -4.5937

Review Exercises (page 805)

1. $(0, 0)$ saddle pt., $(1, -1)$ loc. min.

3. $(2/3, 4/3)$ loc. min; $(2, -4)$ and $(-1, 2)$ saddle points

5. yes, 2, on the sphere $x^2 + y^2 + z^2 = 1$

7. max $1/(4e)$, min $-1/(4e)$

9. (a) $L^2/48$ cm^2, (b) $L^2/16$ cm^2

11. 4π sq. units

13. 16π cu. units

15. 1,688 widgets, \$2.00 each

17. $y \approx -2x - \epsilon x e^{-2x} + \epsilon^2 x^2 e^{-4x}$

Challenging Problems (page 805)

3. $\frac{1}{2}\ln(1 + x^2)\tan^{-1} x$

Chapter 14 Multiple Integration

Section 14.1 (page 812)

1. 15

3. 21

5. 15

7. 96

9. 80

11. 36.6258

13. 20

15. 0

17. 5π

19. $\frac{\pi a^3}{3}$

21. $\frac{1}{6}$

Section 14.2 (page 819)

1. $5/24$

3. 4

5. $\frac{ab(a^2+b^2)}{3}$

7. π

9. $\frac{3}{56}$

11. $\frac{33}{8}\ln 2 - \frac{45}{16}$

13. $\frac{e-2}{2}$

15. $\frac{1}{2}\left(1 - \frac{1}{e}\right)$; region is a triangle with vertices $(0, 0)$, $(1, 0)$ and $(1, 1)$

17. $\frac{\pi}{4\lambda}$; region is a triangle with vertices $(0, 0)$, $(0, 1)$ and $(1, 1)$

19. $1/4$ cu. units

21. $1/3$ cu. units

23. $\ln 2$ cu. units

25. $\frac{\pi}{2\sqrt{2}}$ cu. units

27. $\frac{16a^3}{3}$ cu. units

Section 14.3 (page 824)

1. converges to 1

3. converges to $\pi/2$

5. diverges to ∞

7. converges to 4

9. converges to $1 - \frac{1}{e}$

11. diverges to ∞

13. converges to $2\ln 2$

15. $k > a - 1$

17. $k < -1 - a$

19. $k > -\frac{1 + a}{1 + b}$ (provided $b > -1$)

21. $\frac{1}{2}, -\frac{1}{2}$ (different answers are possible because the *double integral* does not exist.)

23. $\frac{a^2}{3}$

25. $\frac{4\sqrt{2}a}{3\pi}$

27. yes, $1/(2\pi)$

Section 14.4 (page 834)

1. $\pi a^4/2$

3. $2\pi a$

5. $\pi a^4/4$

7. $a^3/3$

9. $\pi(e^{a^2} - 1)/4$

11. $\frac{(\sqrt{3}+1)a^3}{6}$

13. $\frac{1}{3}$

15. $\frac{2a}{3}$

17. $k < 1$; $\frac{\pi}{1-k}$

19. $\frac{a^4}{16}$

21. $\frac{2\pi}{3}$ cu. units

23. $\frac{4\pi(2\sqrt{2}-1)a^3}{3}$ cu. units

25. $16[1 - (1/\sqrt{2})]a^3$ cu. units

27. $1 - \frac{4\sqrt{2}}{3\pi}$ units

29. $\frac{4}{3}\pi abc$ cu. units

31. $2a \sinh a$

33. $\frac{3\ln 2}{2}$ sq. units

35. $\frac{1}{4}(e - e^{-1})$

Section 14.5 (page 840)

1. $8abc$

3. 16π

5. $2/3$

7. $1/15$

9. $2/(3\pi)$

11. $\frac{3}{16}\ln 2$

13. $\pi\sqrt{\frac{\pi}{6}}$

15. $1/8$

17. $\int_0^1 dx \int_0^1 dy \int_0^{1-y} f(x,y,z)\,dz$

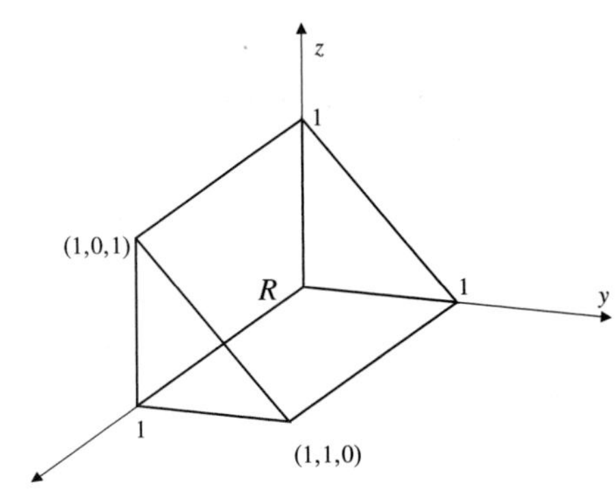

19. $\int_0^1 dx \int_0^x dy \int_0^{x-y} f(x,y,z)\,dz$

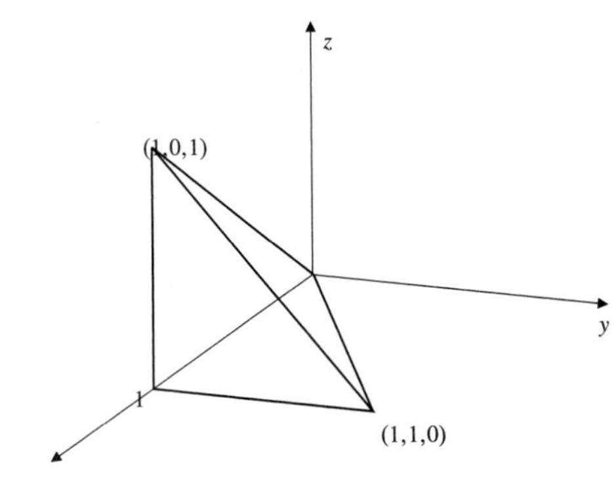

27. $(e-1)/3$

29. $\overline{f} = \dfrac{1}{\text{vol}(R)} \iiint_R f\,dV;\ 1$

Section 14.6 (page 847)

1. $\frac{2}{3}\pi a^3\left(1-\frac{1}{\sqrt{2}}\right)$ cu. units

3. 24π cu. units

5. $(2\pi - \dfrac{32}{9})a^3$ cu. units.

7. $\dfrac{abc}{3}\tan^{-1}\dfrac{a}{b}$ cu. units

9. $\frac{\pi ab}{2}$ cu. units

11. $\dfrac{8\pi a^5}{15}$

13. $\dfrac{2\pi a^5}{5}\left(1-\dfrac{c}{\sqrt{c^2+1}}\right)$

15. $\dfrac{7\pi}{12}$

17. $\frac{ha^3}{12};\quad \frac{\pi a^2 h^2}{48}$

Section 14.7 (page 855)

1. 3π sq. units

3. $2\pi a^2$ sq. units

5. $24\pi/\sqrt{3}$ sq. units

7. $(5\sqrt{5}-1)/12$ sq. units

9. 4 sq. units

11. 5.123

13. $4\pi A\left[a - \sqrt{B}\tan^{-1}\left(\frac{a}{\sqrt{B}}\right)\right]$ units

15. $2\pi km\rho(h + \sqrt{a^2+(b-h)^2} - \sqrt{a^2+b^2})$

17. $2\pi km\rho(h + \sqrt{a^2+(b-h)^2} - \sqrt{a^2+b^2})$

19. $\left(\frac{1}{3}, \frac{1}{3}, \frac{1}{2}\right)$

21. $\left(\frac{3a}{8}, \frac{3a}{8}, \frac{3a}{8}\right)$

23. The model still involves angular acceleration to spin the ball—it doesn't just fall. Part of the gravitational energy goes to producing this spin even in the limiting case.

25. $I = \pi\rho a^2 h\left(\frac{h^2}{3} + \frac{a^2}{4}\right), \quad \bar{D} = \left(\frac{h^2}{3} + \frac{a^2}{4}\right)^{1/2}$

27. $I = \frac{\pi\rho a^2 h}{3}\left(\frac{2h^2+3a^2}{20}\right), \quad \bar{D} = \left(\frac{2h^2+3a^2}{20}\right)^{1/2}$

29. $I = \frac{5a^5\rho}{12}, \quad \bar{D} = \sqrt{\frac{5}{12}}\,a$

31. $I = \frac{8}{3}\rho abc(a^2+b^2),\ \bar{D} = \sqrt{\frac{a^2+b^2}{3}}$

33. $m = \frac{4\pi}{3}\rho(a^2-b^2)^{3/2},\ I = \frac{1}{5}m(2a^2+3b^2)$

35. $\frac{5a^2 g \sin\alpha}{7a^2+3b^2}$

39. The moment of inertia about the line $\mathbf{r}(t) = At\mathbf{i} + Bt\mathbf{j} + Ct\mathbf{k}$ is

$$\frac{1}{A^2+B^2+C^2}\big((B^2+C^2)P_{xx} + (A^2+C^2)P_{yy} + (A^2+B^2)P_{zz} - 2ABP_{xy} - 2ACP_{xz} - 2BCP_{yz}\big).$$

Review Exercises (page 857)

1. $3/10$

3. $\ln 2$

5. $k = 1/\sqrt{3}$

7. $\displaystyle\int_0^1 dx \int_x^1 dy \int_y^1 f(x,y,z)\,dz$

9. $(1-e^{-a^2})/(2a)$

11. $\dfrac{8\pi}{15}(18\sqrt{6}-41)a^5$

13. vol = $7/12$, $\bar{z} = 11/28$

15. $17/24$

17. $\displaystyle\frac{1}{6}\int_0^{\pi/2}\left[(1+16\cos^2\theta)^{3/2} - 1\right]d\theta \approx 7.904$ sq. units

Challenging Problems (page 858)

1. $\pi abc\left(\dfrac{2}{3} - \dfrac{8}{9\sqrt{3}}\right)$ cu. units

3. (b) (i) $\sum_{n=1}^{\infty}(-1)^{n-1}1/n^2$, (ii) $\sum_{n=1}^{\infty}1/n^3$, (iii) $\sum_{n=1}^{\infty}(-1)^{n-1}1/n^3$

5. $4 - \tan^{-1}(\sqrt{2}) + \dfrac{32}{3}\tan^{-1}\left(\dfrac{5}{\sqrt{2}}\right) - \dfrac{4}{3}(7\pi + 2\sqrt{2})$
≈ 18.9348 cu. units

7. $a^3/210$ cu. units

Chapter 15

Vector Fields

Section 15.1 (page 865)

1. field lines: $y = x + C$

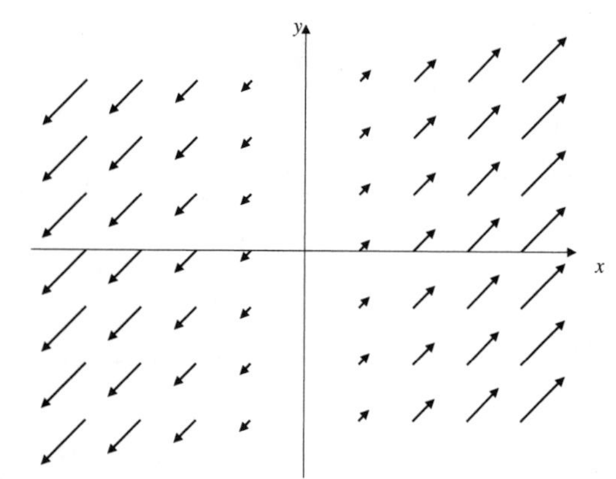

3. field lines: $y^2 = x^2 + C$

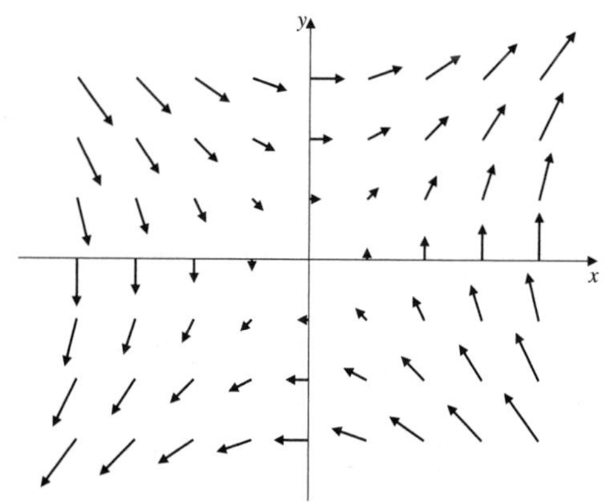

5. field lines: $y = -\frac{1}{2}e^{-2x} + C$

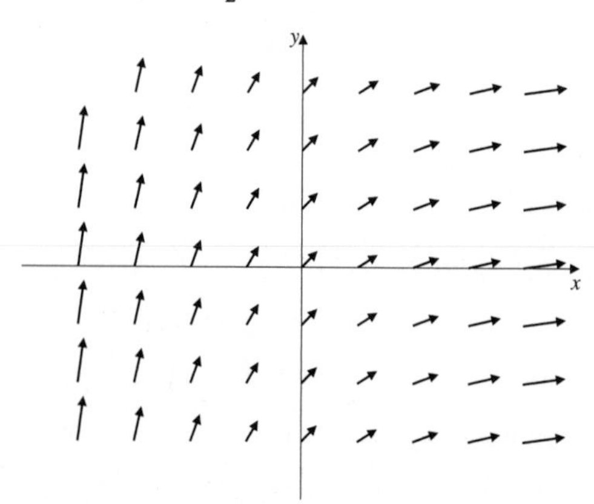

7. field lines: $y = Cx$

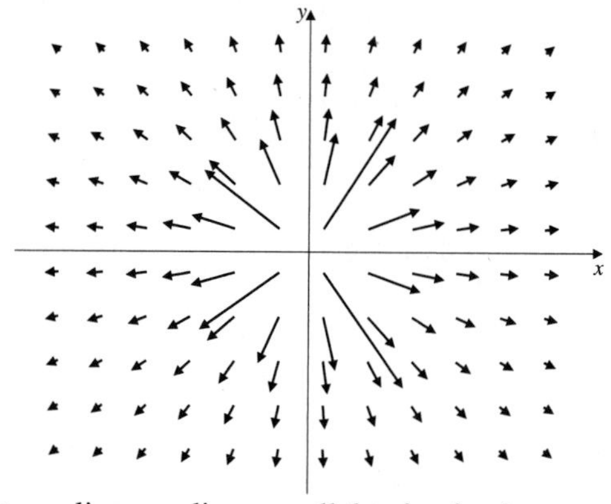

9. streamlines are lines parallel to $\mathbf{i} - \mathbf{j} - \mathbf{k}$

11. streamlines: $x^2 + y^2 = a^2,\ x = a \sin(z - b)$ (spirals)

13. $y = C_1 x,\ 2x = z^2 + C_2$

15. $y = Ce^{1/x}$

17. $r = \theta + C$

19. $r = C\theta^2$

21. unstable

23. $y' = 0$ or $y = \dfrac{x}{x^2 - 1}$

Section 15.2 (page 874)

1. conservative; $\dfrac{x^2}{2} - y^2 + \dfrac{3z^2}{2}$

3. not conservative

5. conservative; $x^2 y + y^2 z - z^2 x$

7. $-2\frac{\mathbf{r}-\mathbf{r}_0}{|\mathbf{r}-\mathbf{r}_0|^4}$

9. $(x^2 + y^2)/z$; equipotential surfaces are paraboloids $z = C(x^2+y^2)$; field lines are ellipses $x^2+y^2+2z^2 = A$, $y = Bx$ in vertical planes through the origin

11. $\mathbf{v} = \dfrac{m(x\mathbf{i} + y\mathbf{j} + (z - \ell)\mathbf{k})}{[x^2 + y^2 + (z - \ell)^2]^{3/2}} + \dfrac{m(x\mathbf{i} + y\mathbf{j} + (z + \ell)\mathbf{k})}{[x^2 + y^2 + (z + \ell)^2]^{3/2}}$, $\mathbf{v} = \mathbf{0}$ only at the origin; $\mathbf{v}(x, y, 0) = \dfrac{2m(x\mathbf{i} + y\mathbf{j})}{(x^2 + y^2 + \ell^2)^{3/2}}$; speed maximum on the circle $x^2 + y^2 = \ell^2/2,\ z = 0$

15. $\phi = -\frac{\mu y}{r^2}$, $\mathbf{F} = \frac{\mu(2xy\mathbf{i}+(y^2-x^2)\mathbf{j})}{r^4}$, $(r^2 = x^2 + y^2)$

21. $\phi = \frac{1}{2}r^2 \sin 2\theta$

Section 15.3 (page 878)

1. $\frac{(a+b)\sqrt{a^2+b^2+c^2}}{2}\, m^2$

3. $\frac{a^2}{2}\left(\sqrt{2} + \ln(1 + \sqrt{2})\right)$

5. 8 gm

7. $\frac{\delta}{6}\left((2e^{4\pi} + 1)^{3/2} - 3^{3/2}\right)$

9. $3\sqrt{14}$

11. $m = 2\sqrt{2}\pi^2,\ (0, -1/\pi, 4\pi/3)$

13. $(e^6 + 3e^4 - 3e^2 - 1)/(3e^3)$

15. $\left(\sqrt{2} + \ln(\sqrt{2} + 1)\right)a^2/2$

17. $\pi/\sqrt{2}$

19. $4\sqrt{b^2+c^2}\,E\left(\sqrt{\frac{b^2-a^2}{b^2+c^2}}\right)$;

$\sqrt{b^2+c^2}\,E\left(\sqrt{\frac{b^2-a^2}{b^2+c^2}}, T\right)$

Section 15.4 (page 886)

1. $-1/4$ **3.** $1/2$

5. 0 **7.** $19/2$

9. $e^{1+(\pi/4)}$

11. $A=2,\ B=3;\ 4\ln 2-\frac{1}{2}$

13. $-13/2$ **15.** a) πa^2, b) $-\pi a^2$

17. a) $\frac{\pi a^2}{2}$, b) $-\frac{\pi a^2}{2}$ **19.** a) $ab/2$, b) $-ab/2$

23. The plane with origin removed is not simply connected.

Section 15.5 (page 897)

1. $dS = ds\,dz = \sqrt{(g(\theta))^2+(g'(\theta))^2}\,d\theta\,dz$

3. $\frac{\pi ab\sqrt{A^2+B^2+C^2}}{|C|}$ sq. units $(C\neq 0)$

5. (a) $dS = |\nabla F/F_2|\,dx\,dz$, (b) $dS=|\nabla F/F_1|\,dy\,dz$

7. $\frac{\pi}{8}$ **9.** $16a^2$ sq. units

13. 2π **15.** $1/96$

17. $\pi(3e+e^3-4)/3$

19. $2\pi a^2+\frac{2\pi ac^2}{\sqrt{a^2-c^2}}\ln\left(\frac{a+\sqrt{a^2-c^2}}{c}\right)$ sq. units

21. $2\pi\sqrt{A^2+B^2+C^2}/|D|$

23. one-third of the way from the base to the vertex on the axis

25. $2\pi k\sigma ma\left(\frac{1}{\sqrt{a^2+(b-h)^2}}-\frac{1}{\sqrt{a^2+b^2}}\right)$

27. $I=\frac{8}{3}\pi\sigma a^4;\ \bar{D}=\sqrt{\frac{2}{3}}\,a$

29. $\frac{3}{5}\,g\sin\alpha$

Section 15.6 (page 903)

1. 6 **3.** $3abc$

5. $\pi(3a^2-4ab+b^2)/2$ **7.** 4π

9. $2\sqrt{2}\pi$ **11.** $4\pi/3$

13. $4\pi m$ **15.** a) $2\pi a^2$, b) 8

Review Exercises (page 904)

1. $(3e/2)-(3/(2e))$ **3.** $8\sqrt{2}/15$

5. 1

7. (a) $6\pi mgb$, (b) $6\pi R\sqrt{a^2+b^2}$

9. (b) e^2 **11.** $(x\mathbf{i}-y\mathbf{j})/\sqrt{x^2+y^2}$

Challenging Problems (page 904)

1. centroid $(0,0,2/\pi)$; upper half of the surface of the torus obtained by rotating the circle $(x-2)^2+z^2=1$, $y=0$, about the z-axis

Chapter 16
Vector Calculus

Section 16.1 (page 914)

1. $\mathbf{div}\,\mathbf{F}=2,\ \mathbf{curl}\,\mathbf{F}=\mathbf{0}$

3. $\mathbf{div}\,\mathbf{F}=0,\ \mathbf{curl}\,\mathbf{F}=-\mathbf{i}-\mathbf{j}-\mathbf{k}$

5. $\mathbf{div}\,\mathbf{F}=1,\ \mathbf{curl}\,\mathbf{F}=-\mathbf{j}$

7. $\mathbf{div}\,\mathbf{F}=f'(x)+g'(y)+h'(z),\quad \mathbf{curl}\,\mathbf{F}=\mathbf{0}$

9. $\mathbf{div}\,\mathbf{F}=\cos\theta\left(1+\frac{1}{r}\cos\theta\right)$;

$\mathbf{curl}\,\mathbf{F}=-\sin\theta\left(1+\frac{1}{r}\cos\theta\right)\mathbf{k}$

11. $\mathbf{div}\,\mathbf{F}=0;\quad \mathbf{curl}\,\mathbf{F}=(1/r)\mathbf{k}$

Section 16.2 (page 920)

7. **div F** can have any value, **curl F** must be normal to **F**

9. $f(r)=Cr^{-3}$

15. If $\mathbf{F}=\nabla\phi$ and $\mathbf{G}=\nabla\psi$ then $\nabla\times(\phi\nabla\psi)=\mathbf{F}\times\mathbf{G}$.

17. $\mathbf{G}=ye^{2z}\mathbf{i}+xye^{2z}\mathbf{k}$ is one possible vector potential.

Section 16.3 (page 924)

1. πa^2-4a^3 **3.** 9

5. $\frac{3\pi ab}{8}$ sq. units

7. 0

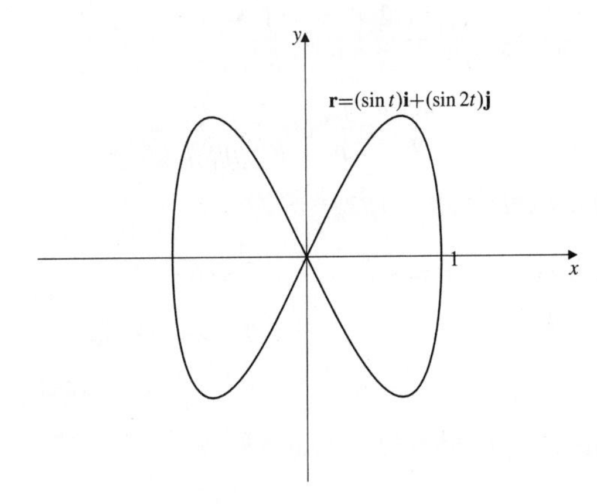

Section 16.4 (page 930)

1. $4\pi a^3$ **3.** $(4/3)\pi a^3$

5. 360π **7.** $81/4$

11. $\frac{2}{3}\pi a^2b+\frac{3}{10}\pi a^4b+\pi a^2$

13. (a) $12\sqrt{3}\pi a^3$, (b) $-4\sqrt{3}\pi a^3$, (c) $16\sqrt{3}\pi a^3$

15. $(6+2\bar{x}+4\bar{y}-2\bar{z})V$ **17.** $9\pi a^2$

Section 16.5 (page 934)

1. 1/2 **3.** $-3\pi a^2$

7. 9π

9. $\alpha=-\frac{1}{2},\ \beta=-3,\ I=-\frac{3}{8}\pi$

11. yes, $\phi\nabla\psi$

Section 16.7 (page 952)

1. $\nabla f=\theta z\hat{\mathbf{r}}+z\hat{\boldsymbol{\theta}}+r\theta\mathbf{k}$ **3.** $\mathbf{div\,F}=2,\ \mathbf{curl\,F}=\mathbf{0}$

5. $\mathbf{div\,F}=\dfrac{2\sin\phi}{R},\ \mathbf{curl\,F}=-\dfrac{\cos\phi}{R}\hat{\boldsymbol{\theta}}$

7. $\mathbf{div\,F}=0,\ \mathbf{curl\,F}=\cot\phi\hat{\mathbf{R}}-2\hat{\boldsymbol{\phi}}$

9. scale factors: $h_u=\left|\dfrac{\partial\mathbf{r}}{\partial u}\right|,\ h_v=\left|\dfrac{\partial\mathbf{r}}{\partial v}\right|$

local basis: $\hat{\mathbf{u}}=\dfrac{1}{h_u}\dfrac{\partial\mathbf{r}}{\partial u},\ \hat{\mathbf{v}}=\dfrac{1}{h_v}\dfrac{\partial\mathbf{r}}{\partial v}$

area element: $dA=h_uh_v\,du\,dv$

11.
$$\nabla f(r,\theta)=\frac{\partial f}{\partial r}\hat{\mathbf{r}}+\frac{1}{r}\frac{\partial r}{\partial\theta}\hat{\boldsymbol{\theta}}$$
$$\nabla\bullet\mathbf{F}(r,\theta)=\frac{\partial F_r}{\partial r}+\frac{1}{r}F_r+\frac{1}{r}\frac{\partial F_\theta}{\partial\theta}$$
$$\nabla\times\mathbf{F}(r,\theta)=\left(\frac{\partial F_\theta}{\partial r}+\frac{1}{r}F_\theta-\frac{1}{r}\frac{\partial F_r}{\partial\theta}\right)\mathbf{k}$$

13. u-surfaces: vertical elliptic cylinders with focal axes at $x=\pm a,\ y=0$
v-surfaces: vertical hyperbolic cylinders with focal axes at $x=\pm a,\ y=0$
z-surfaces: horizontal planes
u-curves: horizontal hyperbolas with foci $x=\pm a,\ y=0$
v-curves: horizontal ellipses with foci $x=\pm a,\ y=0$
z-curves: vertical straight lines

15.
$$\nabla f=\frac{\partial^2 f}{\partial R^2}+\frac{2}{R}\frac{\partial f}{\partial R}+\frac{1}{R^2}\frac{\partial^2 f}{\partial\phi^2}+\frac{\cot\phi}{R^2}\frac{\partial f}{\partial\phi}+\frac{1}{R^2\sin^2\phi}\frac{\partial^2 f}{\partial\theta^2}$$

Review Exercises (page 952)

1. 128π **3.** -6

5. 3/4 **7.** $\lambda=-3$, no

11. the ellipsoid $x^2+4y^2+z^2=4$ with outward normal

Challenging Problems (page 953)

1. $\mathbf{div\,v}=3C$

Chapter 17 Differential Forms and Exterior Calculus

Section 17.1 (page 962)

1. $\phi\wedge\psi=a_2b_1\,dx_1\wedge dx_2\wedge dx_3\wedge dx_4$

3. $\phi\wedge\psi=7\,dx_1\wedge dx_2\wedge dx_3\wedge dx_4\wedge dx_5$

5. $\pi=(12)(13)(14)\cdots(1k)$ is odd (even) if k is even (odd)

7. (a) -1, (b) 1, (c) 1, (d) -1

Section 17.2 (page 968)

1. $d\Phi=2y\,dy\wedge dz$

3. $d\Psi=2\,dx_1\wedge dx_2\wedge dx_3$

5. 0, the zero differential 2-form

9. $d(\Phi\wedge\Psi\wedge\Theta)=(d\Phi)\wedge\Psi\wedge\Theta+(-1)^k\Phi\wedge(d\Psi)\wedge\Theta+(-1)^{k+\ell}\Phi\wedge\Psi\wedge(d\Theta)$

11. $\mathbf{curl\,grad}\,f=\mathbf{0}$

Section 17.3 (page 975)

1. 6 square units **3.** 2 cubic units

5. $\dfrac{1}{18}(18\sqrt{18}-6\sqrt{6})$

Section 17.4 (page 982)

1. $\omega_{\text{bottom}}=-dx\wedge dy$

5. no; $\mathbf{x}=(u_2,-u_2,u_1,-u_1)$ is orientation preserving (non-unique answer)

Section 17.5 (page 989)

3. 1/2 **5.** $252\pi^2$

Chapter 18 Ordinary Differential Equations

Section 18.1 (page 993)

1. 1, linear, homogeneous **3.** 1, nonlinear

5. 2, linear, homogeneous

7. 3, linear, nonhomogeneous

9. 4, linear, homogeneous

11. (a) and (b) are solutions, (c) is not

13. $y_2=\sin(kx),\ y=-3(\cos(kx)+(3/k)\sin(kx))$

15. $y=\sqrt{2}(\cos x+2\sin x)$

17. $y=x+\sin x+(\pi-1)\cos x$

Section 18.2 (page 998)

1. $2\tan^{-1}(y/x)=\ln(x^2+y^2)+C$

3. $y=x\tan\big(\ln|x|+C\big)$ **5.** $y=x\tan^{-1}(\ln|Cx|)$

7. $y^3+3y-3x^2=24$ **11.** $2xy+x^2y^2=C$

13. $xe^{xy}=C$ **15.** $\ln|x|-\frac{y}{x^2}=C$

17. $\dfrac{1}{M}\left(\dfrac{\partial N}{\partial x}-\dfrac{\partial M}{\partial y}\right)$ must depend only on y

19. $\dfrac{1}{M}\left(\dfrac{\partial N}{\partial x}-\dfrac{\partial M}{\partial y}\right)$ must depend only on y.
$x-y^2e^y=Cy^2$

21. $\frac{1}{\mu}\frac{d\mu}{dx} = \frac{\frac{\partial N}{\partial x} - \frac{\partial M}{\partial y}}{xM - yN}$ must depend only on xy;
$\frac{\sin x}{y} - \frac{y}{x} = C$

Section 18.3 (page 1006)

1. (a) 1.97664, (b) 2.187485, (c) 2.306595
3. (a) 2.436502, (b) 2.436559, (c) 2.436563
5. (a) 1.097897, (b) 1.098401
7. (a) 0.89441, (b) 0.87996, (c) 0.872831
9. (a) 0.865766, (b) 0.865769, (c) 0.865769
11. (a) 0.898914, (b) 0.903122, (c) 0.904174
13. $y = 2/(3 - 2x)$
17. (b) $u = 1/(1 - x)$, $v = \tan(x + \frac{\pi}{4})$. $y(x)$ is defined at least on $[0, \pi/4)$ and satisfies $1/(1-x) \le y(x) \le \tan(x+\frac{\pi}{4})$ there.

Section 18.4 (page 1010)

1. $y = C_1e^x + C_2e^{2x}$ **3.** $y = C_1x + \frac{C_2}{x^2}$
5. $y = C_1x + C_2xe^x$

Section 18.5 (page 1014)

1. $y = C_1 + C_2e^t + C_3e^{3t}$
3. $y = C_1\cos t + C_2\sin t + C_3t\cos t + C_4t\sin t$
5. $y = C_1e^{2t} + C_2e^{-t}\cos t + C_3e^{-t}\sin t$
7. $y = Ax + Bx\ln x$ **9.** $y = Ax + \frac{B}{x}$
11. $y = A + B\ln x$
13. $y = C_1x + C_2x\ln x + C_3x(\ln x)^2$
15. $y = C_1x\cos(\ln x) + C_2x\sin(\ln x)$

Section 18.6 (page 1020)

1. $y = -\frac{1}{2} + C_1e^x + C_2e^{-2x}$
3. $y = -\frac{1}{2}e^{-x} + C_1e^x + C_2e^{-2x}$
5. $y = -\frac{2}{125} - \frac{4x}{25} + \frac{x^2}{5} + C_1e^{-x}\cos(2x) + C_2e^{-x}\sin(2x)$
7. $y = -\frac{1}{5}xe^{-2x} + C_1e^{-2x} + C_2e^{3x}$
9. $y = \frac{1}{8}e^x(\sin x - \cos x) + e^{-x}(C_1\cos x + C_2\sin x)$
11. $y = 2x + x^2 - xe^{-x} + C_1 + C_2e^{-x}$
15. $y_p = \frac{x^2}{3}$, $y = \frac{x^2}{3} + C_1x + \frac{C_2}{x}$
17. $y = \frac{1}{2}x\ln x + C_1x + \frac{C_2}{x}$
19. $y = C_2e^x + C_2xe^x + xe^x\ln x$
21. $y = -x^2 + C_1x + C_2xe^x$

Section 18.7 (page 1025)

1. $y = a_0\left(1 + \sum_{k=1}^{\infty}\frac{(x-1)^{4k}}{4(k!)(3)(7)\cdots(4k-1)}\right) + a_1\left(x - 1 + \sum_{k=1}^{\infty}\frac{(x-1)^{4k+1}}{4(k!)(5)(9)\cdots(4k+1)}\right)$
3. $y = \sum_{n=0}^{\infty}(-1)^n\left[\frac{2^n n!}{(2n)!}x^{2n} + \frac{1}{2^{n-1}n!}x^{2n+1}\right]$
5. $y = 1 - \frac{1}{6}x^3 + \frac{1}{120}x^5 + \cdots$
7. $y_1 = 1 + \sum_{k=1}^{\infty}\frac{(-1)^k x^k}{(k!)(2)(5)(8)\cdots(3k-1)}$,
$y_2 = x^{1/3}\left(1 + \sum_{k=0}^{\infty}\frac{(-1)^k x^k}{(k!)(4)(7)\cdots(3k+1)}\right)$

Review Exercises (page 1025)

1. $y = Ce^{x^2}$ **3.** $y = Ce^{2x} - \frac{x}{2} - \frac{1}{4}$
5. $x^2 + 2xy - y^2 = C$ **7.** $y = C_1 - \ln|t + C_2|$
9. $y = e^{x/2}(C_2\cos x + C_2\sin x)$
11. $y = C_1t\cos(2\ln|t|) + C_2t\sin(2\ln|t|)$
13. $y = \frac{1}{2}e^x + xe^{3x} + C_1e^{2x} + C_2e^{3x}$
15. $y = x^2 - 4x + 6 + C_1e^{-x} + C_2xe^{-x}$
17. $y = (x^3 - 7)^{1/3}$ **19.** $y = e^{x^2/2y^2}$
21. $y = 4e^{-t} - 3e^{-2t}$ **23.** $y = (5t - 4)e^{-5t}$
25. $y = e^{2t} - 2\sin(2t)$
27. $A = 1$, $B = -1$, $x(e^x\sin y + \cos y) = C$
29. $y = C_1x + C_2x\cos x$

Appendix I Complex Numbers (page A-10)

1. $\Re(z) = -5$, $\Im(z) = 2$ **3.** $\Re(z) = 0$, $\Im(z) = -\pi$
5. $|z| = \sqrt{2}$, $\theta = 3\pi/4$ **7.** $|z| = 3$, $\theta = \pi/2$
9. $|z| = \sqrt{5}$, $\theta = \tan^{-1}2$
11. $|z| = 5$, $\theta = -\pi + \tan^{-1}(4/3)$
13. $|z| = 2$, $\theta = -\pi/6$ **15.** $|z| = 3$, $\theta = 4\pi/5$
17. $11\pi/12$ **19.** $4 + 3i$
21. $\frac{\pi\sqrt{3}}{2} + \frac{\pi}{2}i$ **23.** $\frac{1}{4} - \frac{\sqrt{3}}{4}i$
25. $-3 + 5i$ **27.** $2 + i$
29. closed disk, radius 2, centre 0
31. closed disk, radius 5, centre $3 - 4i$
33. closed plane sector lying under $y = 0$ and to the left of $y = -x$
35. 4 **37.** $5 - i$
39. $2 + 11i$ **41.** $-\frac{1}{5} + \frac{7}{5}i$
43. 1
47. $zw = -3 - 3i$, $\frac{z}{w} = \frac{1+i}{3}$
49. (a) circle $|z| = \sqrt{2}$, (b) no solutions

51. $-1,\ \frac{1}{2} \pm \frac{\sqrt{3}}{2}\mathbf{i}$

53. $2^{1/6}(\cos\theta + i\sin\theta)$ where $\theta = \pi/4,\ 11\pi/12,\ 19\pi/12$

55. $\pm 2^{1/4}\left(\frac{\sqrt{3}}{2} + \frac{1}{2}i\right),\ \pm 2^{1/4}\left(\frac{1}{2} - \frac{\sqrt{3}}{2}i\right)$

Appendix II Complex Functions *(page A-19)*

1. $0 \le \Re(w) \le 1,\ -2 \le \Im(w) \le 0$

3. $1 \le |w| \le 4,\ \pi \le \arg w \le \dfrac{3\pi}{2}$

5. $\dfrac{1}{2} \le |w| < \infty,\ -\dfrac{\pi}{2} \le \arg w \le 0$

7. $\arg(w) = 5\pi/6$

9. parabola $v^2 = 4u + 4$

11. $u \ge 0,\ v \ge u$

13. $f'(z) = 2z$

15. $f'(z) = -1/z^2$

19. $\dfrac{d}{dz}\sinh z = \cosh z,\ \dfrac{d}{dz}\cosh z = \sinh z$

21. $z = \dfrac{\pi}{2} + k\pi,\ (k \in \mathbb{Z})$

23. zeros of $\cosh z$: $z = i\left(\dfrac{\pi}{2} + k\pi\right)\ (k \in \mathbb{Z})$
zeros of $\sinh z$: $z = k\pi i\ (k \in \mathbb{Z})$

25. $\Re(\sinh z) = \sinh x \cos y,\ \Im(\sinh z) = \cosh x \sin y$

27. $z = 0,\ -2i$

29. $z = -1 \pm 2i$

31. $z = 0,\ i,\ 2i$

33. $z = \dfrac{1 \pm i}{\sqrt{2}},\ z = \dfrac{-1 \pm i}{\sqrt{2}}$
$z^4 + 1 = (z^2 + \sqrt{2}z + 1)(z^2 - \sqrt{2}z + 1)$

35. $z = -1,\ -1,\ -1,\ i,\ -i$

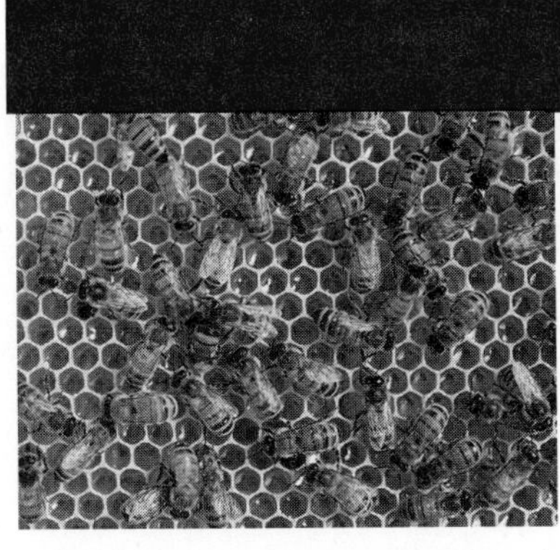

Index

INTEGRATION RULES

$$\int (Af(x) + Bg(x))\,dx = A\int f(x)\,dx + B\int g(x)\,dx$$

$$\int f'(g(x))\,g'(x)\,dx = f(g(x)) + C$$

$$\int U(x)\,dV(x) = U(x)\,V(x) - \int V(x)\,dU(x)$$

$$\int_a^b f'(x)\,dx = f(b) - f(a)$$

$$\frac{d}{dx}\int_a^x f(t)\,dt = f(x)$$

ELEMENTARY INTEGRALS

$$\int x^r\,dx = \frac{1}{r+1}x^{r+1} + C \text{ if } r \neq -1$$

$$\int \frac{dx}{x} = \ln|x| + C$$

$$\int e^x\,dx = e^x + C$$

$$\int a^x\,dx = \frac{a^x}{\ln a} + C$$

$$\int \sin x\,dx = -\cos x + C$$

$$\int \cos x\,dx = \sin x + C$$

$$\int \sec^2 x\,dx = \tan x + C$$

$$\int \csc^2 x\,dx = -\cot x + C$$

$$\int \sec x \tan x\,dx = \sec x + C$$

$$\int \csc x \cot x\,dx = -\csc x + C$$

$$\int \tan x\,dx = \ln|\sec x| + C$$

$$\int \cot x\,dx = \ln|\sin x| + C$$

$$\int \sec x\,dx = \ln|\sec x + \tan x| + C$$

$$\int \csc x\,dx = \ln|\csc x - \cot x| + C$$

$$\int \frac{dx}{\sqrt{a^2 - x^2}} = \sin^{-1}\frac{x}{a} + C \quad (a > 0,\ |x| < a)$$

$$\int \frac{dx}{a^2 + x^2} = \frac{1}{a}\tan^{-1}\frac{x}{a} + C \quad (a > 0)$$

$$\int \frac{dx}{a^2 - x^2} = \frac{1}{2a}\ln\left|\frac{x+a}{x-a}\right| + C \quad (a > 0)$$

$$\int \frac{dx}{x\sqrt{x^2 - a^2}} = \frac{1}{a}\sec^{-1}\left|\frac{x}{a}\right| + C \quad (a > 0,\ |x| > a)$$

TRIGONOMETRIC INTEGRALS

$$\int \sin^2 x\,dx = \frac{x}{2} - \frac{1}{4}\sin 2x + C$$

$$\int \cos^2 x\,dx = \frac{x}{2} + \frac{1}{4}\sin 2x + C$$

$$\int \tan^2 x\,dx = \tan x - x + C$$

$$\int \cot^2 x\,dx = -\cot x - x + C$$

$$\int \sec^3 x\,dx = \frac{1}{2}\sec x \tan x + \frac{1}{2}\ln|\sec x + \tan x| + C$$

$$\int \csc^3 x\,dx = -\frac{1}{2}\csc x \cot x + \frac{1}{2}\ln|\csc x - \cot x| + C$$

$$\int \sin ax \sin bx\,dx = \frac{\sin(a-b)x}{2(a-b)} - \frac{\sin(a+b)x}{2(a+b)} + C \text{ if } a^2 \neq b^2$$

$$\int \cos ax \cos bx\,dx = \frac{\sin(a-b)x}{2(a-b)} + \frac{\sin(a+b)x}{2(a+b)} + C \text{ if } a^2 \neq b^2$$

$$\int \sin ax \cos bx\,dx = -\frac{\cos(a-b)x}{2(a-b)} - \frac{\cos(a+b)x}{2(a+b)} + C \text{ if } a^2 \neq b^2$$

$$\int \sin^n x\,dx = -\frac{1}{n}\sin^{n-1} x \cos x + \frac{n-1}{n}\int \sin^{n-2} x\,dx$$

$$\int \cos^n x\,dx = \frac{1}{n}\cos^{n-1} x \sin x + \frac{n-1}{n}\int \cos^{n-2} x\,dx$$

$$\int \tan^n x\,dx = \frac{1}{n-1}\tan^{n-1} x - \int \tan^{n-2} x\,dx \text{ if } n \neq 1$$

$$\int \cot^n x\,dx = \frac{-1}{n-1}\cot^{n-1} x - \int \cot^{n-2} x\,dx \text{ if } n \neq 1$$

$$\int \sec^n x\,dx = \frac{1}{n-1}\sec^{n-2} x \tan x + \frac{n-2}{n-1}\int \sec^{n-2} x\,dx \text{ if } n \neq 1$$

$$\int \csc^n x\,dx = \frac{-1}{n-1}\csc^{n-2} x \cot x + \frac{n-2}{n-1}\int \csc^{n-2} x\,dx \text{ if } n \neq 1$$

$$\int \sin^n x \cos^m x\,dx = -\frac{\sin^{n-1} x \cos^{m+1} x}{n+m} + \frac{n-1}{n+m}\int \sin^{n-2} x \cos^m x\,dx \text{ if } n \neq -m$$

$$\int \sin^n x \cos^m x\,dx = \frac{\sin^{n+1} x \cos^{m-1} x}{n+m} + \frac{m-1}{n+m}\int \sin^n x \cos^{m-2} x\,dx \text{ if } m \neq -n$$

$$\int x \sin x\,dx = \sin x - x\cos x + C$$

$$\int x \cos x\,dx = \cos x + x\sin x + C$$

$$\int x^n \sin x\,dx = -x^n \cos x + n\int x^{n-1}\cos x\,dx$$

$$\int x^n \cos x\,dx = x^n \sin x - n\int x^{n-1}\sin x\,dx$$

INTEGRALS INVOLVING $\sqrt{x^2 \pm a^2}$ $(a > 0)$

(If $\sqrt{x^2 - a^2}$, assume $x > a > 0$.)

$$\int \sqrt{x^2 \pm a^2}\, dx = \frac{x}{2}\sqrt{x^2 \pm a^2} \pm \frac{a^2}{2} \ln|x + \sqrt{x^2 \pm a^2}| + C$$

$$\int \frac{dx}{\sqrt{x^2 \pm a^2}} = \ln|x + \sqrt{x^2 \pm a^2}| + C$$

$$\int \frac{\sqrt{x^2 + a^2}}{x}\, dx = \sqrt{x^2 + a^2} - a \ln\left|\frac{a + \sqrt{x^2 + a^2}}{x}\right| + C$$

$$\int \frac{\sqrt{x^2 - a^2}}{x}\, dx = \sqrt{x^2 - a^2} - a \tan^{-1}\frac{\sqrt{x^2 - a^2}}{a} + C$$

$$\int x^2\sqrt{x^2 \pm a^2}\, dx = \frac{x}{8}(2x^2 \pm a^2)\sqrt{x^2 \pm a^2} - \frac{a^4}{8} \ln|x + \sqrt{x^2 \pm a^2}| + C$$

$$\int \frac{x^2}{\sqrt{x^2 \pm a^2}}\, dx = \frac{x}{2}\sqrt{x^2 \pm a^2} \mp \frac{a^2}{2} \ln|x + \sqrt{x^2 \pm a^2}| + C$$

$$\int \frac{\sqrt{x^2 \pm a^2}}{x^2}\, dx = -\frac{\sqrt{x^2 \pm a^2}}{x} + \ln|x + \sqrt{x^2 \pm a^2}| + C$$

$$\int \frac{dx}{x^2\sqrt{x^2 \pm a^2}} = \mp\frac{\sqrt{x^2 \pm a^2}}{a^2 x} + C$$

$$\int \frac{dx}{(x^2 \pm a^2)^{3/2}} = \frac{\pm x}{a^2\sqrt{x^2 \pm a^2}} + C$$

$$\int (x^2 \pm a^2)^{3/2}\, dx = \frac{x}{8}(2x^2 \pm 5a^2)\sqrt{x^2 \pm a^2} + \frac{3a^4}{8} \ln|x + \sqrt{x^2 \pm a^2}| + C$$

INTEGRALS INVOLVING $\sqrt{a^2 - x^2}$ $(a > 0, |x| < a)$

$$\int \sqrt{a^2 - x^2}\, dx = \frac{x}{2}\sqrt{a^2 - x^2} + \frac{a^2}{2} \sin^{-1}\frac{x}{a} + C$$

$$\int \frac{\sqrt{a^2 - x^2}}{x}\, dx = \sqrt{a^2 - x^2} - a \ln\left|\frac{a + \sqrt{a^2 - x^2}}{x}\right| + C$$

$$\int \frac{x^2}{\sqrt{a^2 - x^2}}\, dx = -\frac{x}{2}\sqrt{a^2 - x^2} + \frac{a^2}{2} \sin^{-1}\frac{x}{a} + C$$

$$\int x^2\sqrt{a^2 - x^2}\, dx = \frac{x}{8}(2x^2 - a^2)\sqrt{a^2 - x^2} + \frac{a^4}{8} \sin^{-1}\frac{x}{a} + C$$

$$\int \frac{dx}{x^2\sqrt{a^2 - x^2}} = -\frac{\sqrt{a^2 - x^2}}{a^2 x} + C$$

$$\int \frac{\sqrt{a^2 - x^2}}{x^2}\, dx = -\frac{\sqrt{a^2 - x^2}}{x} - \sin^{-1}\frac{x}{a} + C$$

$$\int \frac{dx}{x\sqrt{a^2 - x^2}} = -\frac{1}{a} \ln\left|\frac{a + \sqrt{a^2 - x^2}}{x}\right| + C$$

$$\int \frac{dx}{(a^2 - x^2)^{3/2}} = \frac{x}{a^2\sqrt{a^2 - x^2}} + C$$

$$\int (a^2 - x^2)^{3/2}\, dx = \frac{x}{8}(5a^2 - 2x^2)\sqrt{a^2 - x^2} + \frac{3a^4}{8} \sin^{-1}\frac{x}{a} + C$$

INTEGRALS OF INVERSE TRIGONOMETRIC FUNCTIONS

$$\int \sin^{-1}x\, dx = x \sin^{-1}x + \sqrt{1 - x^2} + C$$

$$\int \tan^{-1}x\, dx = x \tan^{-1}x - \frac{1}{2} \ln(1 + x^2) + C$$

$$\int \sec^{-1}x\, dx = x \sec^{-1}x - \ln|x + \sqrt{x^2 - 1}| + C \quad (x > 1)$$

$$\int x \sin^{-1}x\, dx = \frac{1}{4}(2x^2 - 1)\sin^{-1}x + \frac{x}{4}\sqrt{1 - x^2} + C$$

$$\int x \tan^{-1}x\, dx = \frac{1}{2}(x^2 + 1)\tan^{-1}x - \frac{x}{2} + C$$

$$\int x \sec^{-1}x\, dx = \frac{x^2}{2} \sec^{-1}x - \frac{1}{2}\sqrt{x^2 - 1} + C \quad (x > 1)$$

$$\int x^n \sin^{-1}x\, dx = \frac{x^{n+1}}{n+1}\sin^{-1}x - \frac{1}{n+1}\int \frac{x^{n+1}}{\sqrt{1 - x^2}}\, dx + C \text{ if } n \neq -1$$

$$\int x^n \tan^{-1}x\, dx = \frac{x^{n+1}}{n+1}\tan^{-1}x - \frac{1}{n+1}\int \frac{x^{n+1}}{1 + x^2}\, dx + C \text{ if } n \neq -1$$

$$\int x^n \sec^{-1}x\, dx = \frac{x^{n+1}}{n+1}\sec^{-1}x - \frac{1}{n+1}\int \frac{x^n}{\sqrt{x^2 - 1}}\, dx + C \quad (n \neq -1,\ x >$$

EXPONENTIAL AND LOGARITHMIC INTEGRALS

$$\int x e^x\, dx = (x - 1)e^x + C$$

$$\int x^n e^x\, dx = x^n e^x - n\int x^{n-1} e^x\, dx$$

$$\int \ln x\, dx = x \ln x - x + C$$

$$\int x^n \ln x\, dx = \frac{x^{n+1}}{n+1} \ln x - \frac{x^{n+1}}{(n+1)^2} + C, \quad (n \neq -1)$$

$$\int x^n (\ln x)^m\, dx = \frac{x^{n+1}}{n+1}(\ln x)^m - \frac{m}{n+1}\int x^n (\ln x)^{m-1}\, dx \quad (n \neq -1)$$

$$\int e^{ax} \sin bx\, dx = \frac{e^{ax}}{a^2 + b^2}(a \sin bx - b \cos bx) + C$$

$$\int e^{ax} \cos bx\, dx = \frac{e^{ax}}{a^2 + b^2}(a \cos bx + b \sin bx) + C$$

INTEGRALS OF HYPERBOLIC FUNCTIONS

$$\int \sinh x\, dx = \cosh x + C$$

$$\int \cosh x\, dx = \sinh x + C$$

$$\int \tanh x\, dx = \ln(\cosh x) + C$$

$$\int \coth x\, dx = \ln|\sinh x| + C$$

$$\int \operatorname{sech} x\, dx = 2\tan^{-1}(e^x) + C$$

$$\int \operatorname{csch} x\, dx = \ln\left|\tanh\frac{x}{2}\right| + C$$

$$\int \sinh^2 x\, dx = \frac{1}{4} \sinh 2x - \frac{x}{2} + C$$

$$\int \cosh^2 x\, dx = \frac{1}{4} \sinh 2x + \frac{x}{2} + C$$

$$\int \tanh^2 x\, dx = x - \tanh x + C$$

$$\int \coth^2 x\, dx = x - \coth x + C$$

$$\int \operatorname{sech}^2 x\, dx = \tanh x + C$$

$$\int \operatorname{csch}^2 x\, dx = -\coth x + C$$

$$\int \operatorname{sech} x \tanh x\, dx = -\operatorname{sech} x + C$$

$$\int \operatorname{csch} x \coth x\, dx = -\operatorname{csch} x + C$$